CHILTON®

ASIAN

DIAGNOSTIC SERVICE
2005 Edition

THOMSON

DELMAR LEARNING™

Australia • Canada • Mexico • Singapore • Spain • United Kingdom • United States

THOMSON

DELMAR LEARNING

Chilton®

Asian Diagnostic Service

2005 Edition

Vice President, Technology and Trades SBU:
Alar Elken

Executive Director, Professional Business Unit:
Gregory L. Clayton

Publisher, Professional Business Unit:
David Koontz

Marketing Director:
Beth A. Lutz

Production Director:
Mary Ellen Black

Marketing Specialist
Brian McGrath

Marketing Coordinator
Marissa Mariella

Production Editor
Elizabeth Hough

Editorial Assistant
Christine Wade

Editor:
Timothy A. Crain

Publishing Assistant:
Paula Baillie

Cover Design:
Melinda Possinger

NOTICE TO THE READER

Publisher does not warrant or guarantee any of the products described herein or perform any independent analysis in connection with any of the product information contained herein. Publisher does not assume, and expressly disclaims, any obligation to obtain and include information other than that provided to it by the manufacturer.

The reader is expressly warned to consider and adopt all safety precautions that might be indicated by the activities herein and to avoid all potential hazards. By following the instructions contained herein, the reader willingly assumes all risks in connection with such instructions.

The publisher makes no representation or warranties of any kind, including but not limited to, the warranties of fitness for particular purpose or merchantability, nor are any such representations implied with respect to the material set forth herein, and the publisher takes no responsibility with respect to such material. The publisher shall not be liable for any special, consequential, or exemplary damages resulting, in whole or part, from the readers' use of, or reliance upon, this material.

TABLE OF CONTENTS

USING THIS MANUAL

Manufacturer and Model Coverage

This manual covers 1990-2003 Acura, Honda, Toyota and Lexus models equipped with first and second generation On-Board Diagnostics (OBD-I and OBD-II). Earlier OBD-I Acura and Honda models are also included.

Model Year Information

Every effort is made to gather current data from the Original Vehicle Manufacturers (OEMs) when they publish it. Different OEMs choose to release their new model information at different times of the year. Indeed, the same OEM can publish information early one season and late the next season. As a result, not all models are equally current when each edition of this manual is published.

Although information in this manual is based on industry sources and is as complete as possible at the time of publication, some vehicle manufacturers may make changes which cannot be included here. Information on late models may not be available in some circumstances. While striving for total accuracy, the publisher cannot assume responsibility for any errors, changes, or omissions that may occur in the compilation of this data.

Safety Notice

Proper service and repair procedures are vital to the safe, reliable operation of all motor vehicles, as well as the personal safety of those performing the repairs. This manual outlines procedures for diagnosing and serving vehicles using safe, effective methods. The procedures may contain many NOTES and CAUTIONS which should be followed along with standard safety procedures to reduce the possibility of personal injury or improper service which could damage the vehicle or compromise its safety.

Diagnostic procedures, tools, parts, and technician skill and experience vary widely. It is not possible to anticipate all conceivable ways or conditions under which vehicles may be serviced, or to provide cautions for all possible hazards that may result. Standard and accepted safety precautions and equipment should be used when handling toxic or flammable substances, and safety goggles or other protection should be used during any process that may cause sparking, material removal or projectiles.

Some procedures require the use of tools specially designed for a specific purpose. Before substituting another tool or procedure, you must be completely satisfied that neither your personal safety, nor the performance of the vehicle will be endangered.

Special Tools

Special tools are recommended by the vehicle manufacturer to perform specific jobs. When necessary, special tools may be referred to in the text by part number. These tools may be purchased, under the appropriate part number, from your local dealer or regional distributor, or an equivalent tool can be purchased locally from a tool supplier or parts outlet. Before substituting any tool for the one recommended, read the previous Safety Notice.

ACKNOWLEDGEMENT

The publisher would like to express appreciation to the following vehicle manufacturers for their assistance in producing this publication. No further reproduction or distribution of the material in this manual is allowed without the expressed written permission of the vehicle manufacturers and the publisher.

> American Honda Motor Co., including Acura and Honda Division
> Toyota Motor Sales USA, including Lexus and Toyota Division

Understanding On-Board Diagnostics

Introduction

OBD II OVERVIEW

The OBD II system was developed as a step toward compliance with California and Federal regulations that set standards for vehicle emission control monitoring for all automotive manufacturers. The primary goal of this system is to detect when the degradation or failure of a component or system will cause emissions to rise by 50%. Every manufacturer must meet OBD II standards by the 1996 model year. Some manufacturers began programs that were OBD II mandated as early as 1992, but most manufacturers began an OBD II phase-in period starting in 1994.

The changes to On-Board Diagnostics influenced by this new program include:

- Common Diagnostic Connector
- Expanded Malfunction Indicator Light Operation
- Common Trouble Code and Diagnostic Language
- Common Diagnostic Procedures
- New Emissions-Related Procedures, Logic and Sensors
- Expanded Emissions-Related Monitoring

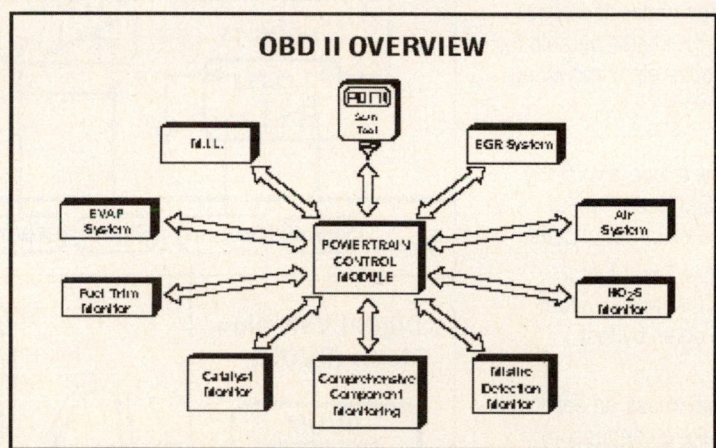

COMMON TERMINOLOGY

OBD II introduces common terms, connectors, diagnostic language and new emissions-related monitoring procedures. The most important benefit of OBD II is that all vehicles will have a common data output system with a common connector. This allows equipment Scan Tool manufacturers to read data from every vehicle and pull codes with common names and similar descriptions of fault conditions. In the future, emissions testing will require the use of an OBD II certifiable Scan Tool.

TECHNICIAN REQUIREMENTS

As an automotive repair technician, you should have a basic understanding of how to use the hand tools and meters necessary to effectively use the information in this OBD II manual.

■ **NOTE:** *Lack of basic knowledge of the Powertrain when performing test procedures could cause incorrect diagnosis or damage to Powertrain components. Do not attempt to diagnose a Powertrain problem without having this basic knowledge.*

ELECTRICITY AND ELECTRICAL CIRCUITS

You should understand basic electricity and know the meaning of voltage (volts), current (amps), and resistance (ohms). You should be able to identify a *Series* circuit as well as a *Parallel* circuit in an automotive wiring diagram. Refer to the examples in the Graphic to the right.

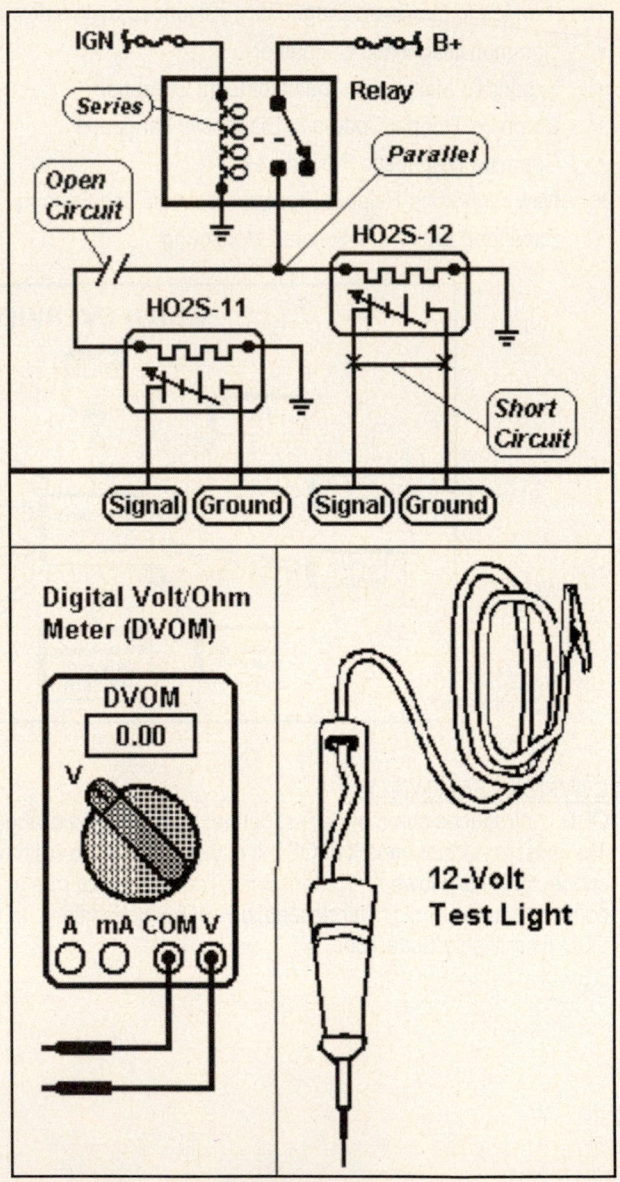

You should understand what happens in an electrical circuit with an open circuit or a shorted wire, and you should be able to identify an open or shorted circuit condition using a DVOM. You should also be able to read and understand an automotive electrical wiring diagram.

CIRCUIT TESTING TOOLS

You should have (and know how to operate) a 12v Test Light, DVOM, Lab Scope and Scan Tool to diagnose vehicle computers and electrical circuits.

You should know not to use a 12v Test Light to diagnose the Engine Controller Electrical system unless specifically instructed to do so by test procedures.

You should have and know when to use an applicable aftermarket connector kit (to make a connection) whenever test procedures call for a connector to be probed in order to make a measurement.

ELECTRICAL CIRCUITS

When you encounter a wiring problem during testing, and need to refer to electrical circuit information, you should be comfortable with this type of information:

- Wiring schematics (including circuit numbers and colors)
- Electrical component connector, splice and ground locations
- Wiring repair procedures and wiring repair parts information

OBD II System

History of OBD Systems

INTRODUCTION

Starting in 1978, several vehicle manufacturers introduced a new type of control for several vehicle systems and computer control of engine management systems. These computer-controlled systems included programs to test for problems in the engine mechanical area, electrical fault identification and tests to help diagnose the computer control system. Early attempts at diagnosis involved expensive and specialized diagnostic testers that hooked up externally to the computer in series with the wiring connector and monitored the input/output operations of the computer.

By early 1980, vehicle manufacturers had designed systems in which the onboard computer incorporated programs to monitor selected components, and to store a trouble code in its memory that could be retrieved at a later time. These trouble codes identified failure conditions that could be used to refer a technician to diagnostic repair charts or test procedures to help pinpoint the problem area.

EVOLUTION OF DAIMLER CHRYSLER COMPUTERIZED ENGINE CONTROLS

The evolution of Computerized Engine Controls on Chrysler vehicles equipped with fuel injection is highlighted in the Graphic below.

Computerized Engine Controls Evolution Graphic - Chrysler & Jeep

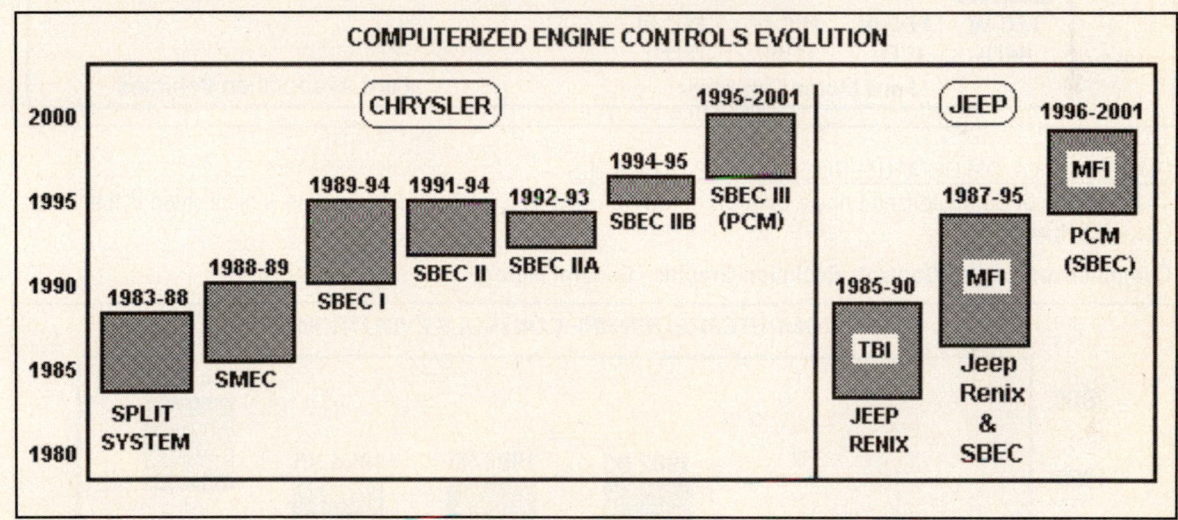

EVOLUTION OF FORD MOTOR COMPANY COMPUTERIZED ENGINE CONTROLS
The evolution of Computerized Engine Controls on Ford vehicles equipped with fuel injection is highlighted in the Graphic below.

Computerized Engine Controls Evolution Graphic - Ford Motor Company

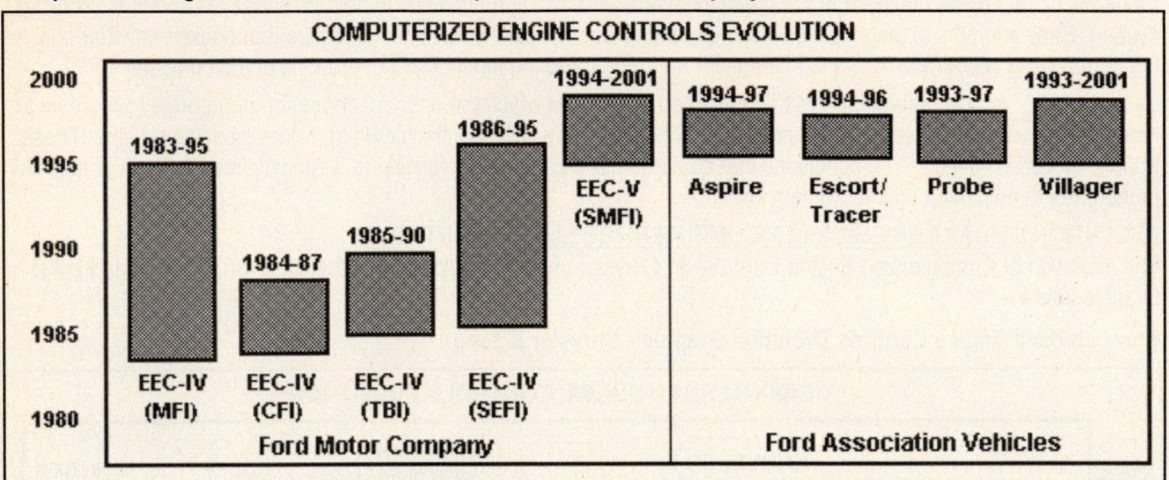

EVOLUTION OF GM COMPUTERIZED ENGINE CONTROLS
The evolution of Computerized Engine Controls on GM vehicles equipped with fuel injection is highlighted in the Graphic below.

Computerized Engine Controls Evolution Graphic - General Motors

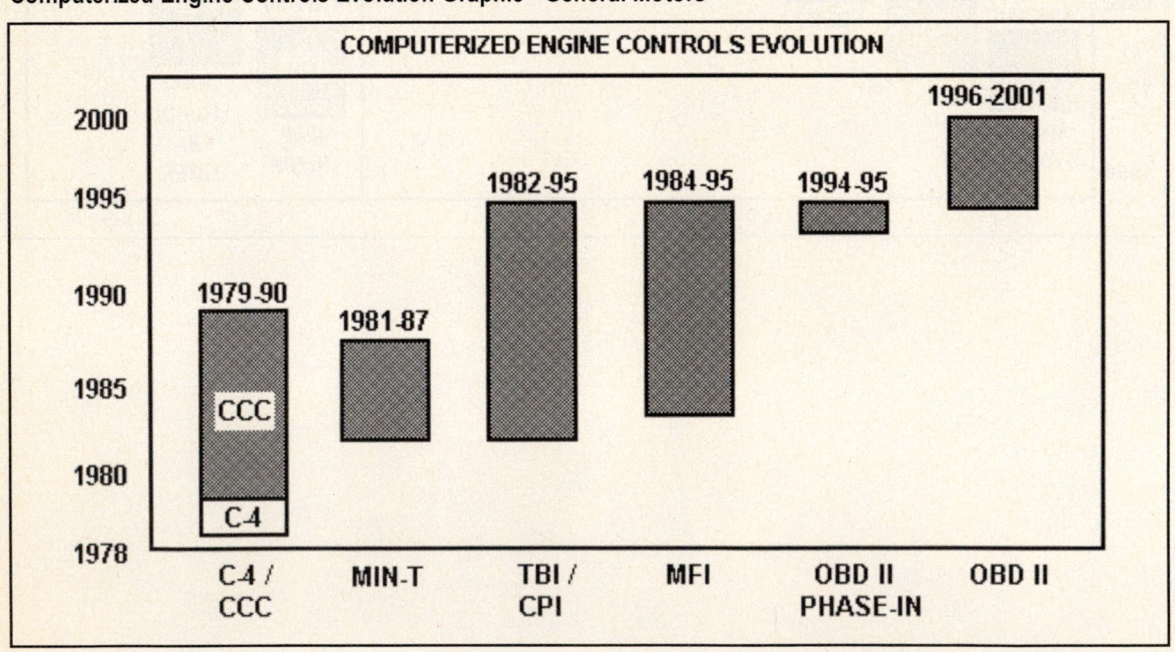

Computer Diagnostics

INTRODUCTION

General diagnostics of computers fall into these two categories:

• External Onboard Diagnostics
• Internal Onboard Diagnostics

The first level of diagnostics uses an external tool that taps into the computer and runs a series of diagnostic tests. This method of diagnostics was popular in the 1970's and was used in the 1980's on many European vehicles. The second level incorporates diagnostics into the circuit board of the computer and is known in the industry as "On-Board" diagnostics because the diagnostics are on the computer circuit board.

In 1980 General Motors incorporated an On-Board Computer Program where the "check engine" light came on to inform the vehicle owner that there was a fault in the computer system. The light was turned on when a diagnostic code was set to alert the driver that service was needed on the vehicle.

California formed a government agency, the California Air Resources Board (CARB) to monitor the air quality and establish regulations to reduce air pollution. The California Health and Safety Code authorized the Air Resources Board to adopt motor vehicle emissions standards and in-use performance standards that it finds necessary, cost effective and technologically feasible. In 1988, CARB required that all vehicles sold in California incorporate a system with an On-Board Diagnostic program where a "check engine" light would come on to notify the vehicle owner of a potential failure of computer sensors and/or their systems. This system is known as On-Board Diagnostics First Generation and is now referred to as OBD I.

PROBLEMS WITH OBD I SYSTEMS

One of the problems with OBD I was that the code retrieval methods varied from manufacturer to manufacturer and there was no consistency between systems. Most manufacturers looked at similar computer sensors and circuits, but codes were inconsistent and difficult to identify and define. Some manufacturers require special tools to retrieve trouble codes or required special test procedures for these tools which self-tested circuits and systems or energized the components for testing in the field.

SCAN TOOL INTRODUCTION

Domestic vehicle manufacturers (Chrysler, Ford and GM) designed their computers to have an accessible data line where a diagnostic tester could retrieve data on sensors and the status of operation for components.

These testers became known in the automotive repair industry as "Scan Tools" because they scanned the data on the computers and provided information for the technician.

Ford Motor Company developed a tester that would access codes, activate sensors and perform limited tests and adjustments, however they did not incorporate data stream features until 1988.

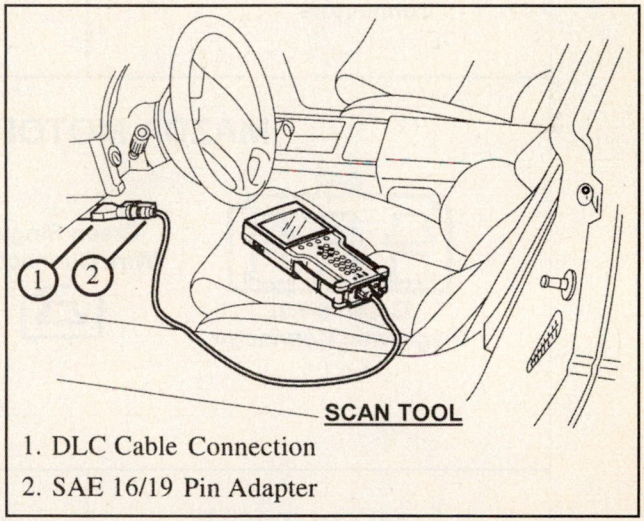

SCAN TOOL

1. DLC Cable Connection
2. SAE 16/19 Pin Adapter

OBD I SYSTEM CONNECTORS

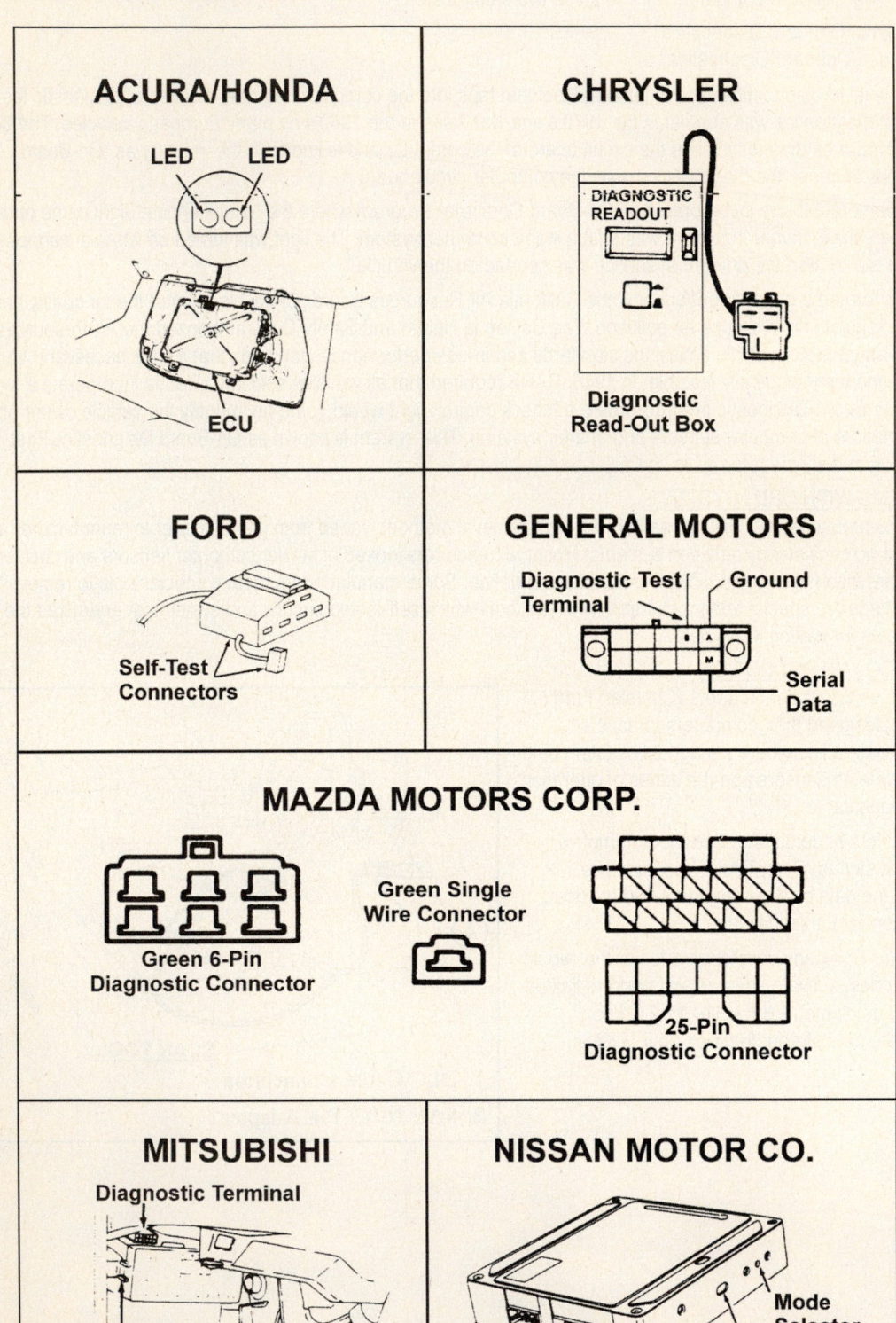

ACURA/HONDA

LED   LED

ECU

CHRYSLER

DIAGNOSTIC READOUT

Diagnostic Read-Out Box

FORD

Self-Test Connectors

GENERAL MOTORS

Diagnostic Test Terminal   **Ground**

Serial Data

MAZDA MOTORS CORP.

Green 6-Pin Diagnostic Connector

Green Single Wire Connector

25-Pin Diagnostic Connector

MITSUBISHI

Diagnostic Terminal

Left Kick Panel

NISSAN MOTOR CO.

Mode Selector

LED Monitor Lamps

Government Regulations

INTRODUCTION

The California Air Resources Board (CARB) conducted research on OBD I vehicle emissions and the study resulted in the following conclusions:

- The research found a significant number of pre-1988 vehicles with degraded emissions components. These components were not failing outright, but deterioration increased emissions levels. This problem did not usually set codes alerting the vehicle owner or technician that there was a problem, therefore the condition was not perceived as a problem in the field. However, CARB viewed this as a problem due to the increased emission levels.

- Vehicle testing programs found failures in Canister Purge systems and Secondary Air Management systems. Many of these failures occurred under road load conditions and were not quickly or easily detectable in the service bay. These failures resulted in increased emissions.

- Catalytic Converters were failing and vehicles were being driven with deteriorated catalysts. A leading cause of this failure was engine misfire.

- The On-Board Monitoring Systems did not detect fuel system faults that were responsible for increasing emissions even though fuel systems were deteriorated enough to have excessive emissions.

- The monitoring systems did not detect oxygen sensors that were "lazy" or slow in response. This condition was found to result in an increase in emissions levels.

- EGR monitoring did not verify if the system was operating within a range that could result in an increase of emissions. There was a need to monitor the flow of EGR gases through the system in order to verify the EGR passages were not clogged.

- Codes were different for each manufacturer and this was confusing for a technician working on different vehicles.

DEVELOPMENT OF OBD I STANDARDS

CARB reviewed the system of monitoring Engine Control sensors and systems developed by Chrysler Motors and General Motors. They incorporated this concept into their regulations.

The result was that the California regulations required that all vehicle manufacturers develop a set of diagnostics that would incorporate a system where codes and data are made available through a Scan Tool accessible to every technician.

These California standards, originally published in October of 1988, generally apply to 1994 and later passenger cars, light duty trucks and medium duty vehicles. Similar diesel and alternative fuel vehicle regulations took affect in 1996. After 1988, California made the decision to accept the Federal (EPA) OBD II regulations.

<u>FEDERAL TEST PROCEDURE</u>

OBD II requires that the on-board computer monitors vehicle emissions and in some cases perform an "Active" diagnostic test of those systems. These tests were developed by the EPA and are a reflection of the Federal Test Procedure (FTP).

The FTP is a series of programmed tests where a vehicle is driven through specific drive cycles while emissions are being monitored. These tests are conducted at various mileage levels and test emissions under very specific conditions. The amount of fuel in the gas tank is monitored, the type of fuel and octane level are all controlled. These tests are conducted on a dynamometer and are performed under hot and cold vehicle conditions. They are conducted under EPA supervision and are required to certify a vehicle for sale in the USA.

The OBD II system was designed to monitor these same systems and a Malfunction Indicator Lamp (MIL) must illuminate if a system or component either fails or deteriorates to a point where the vehicle emissions could rise beyond 1.5 times the FTP standard.

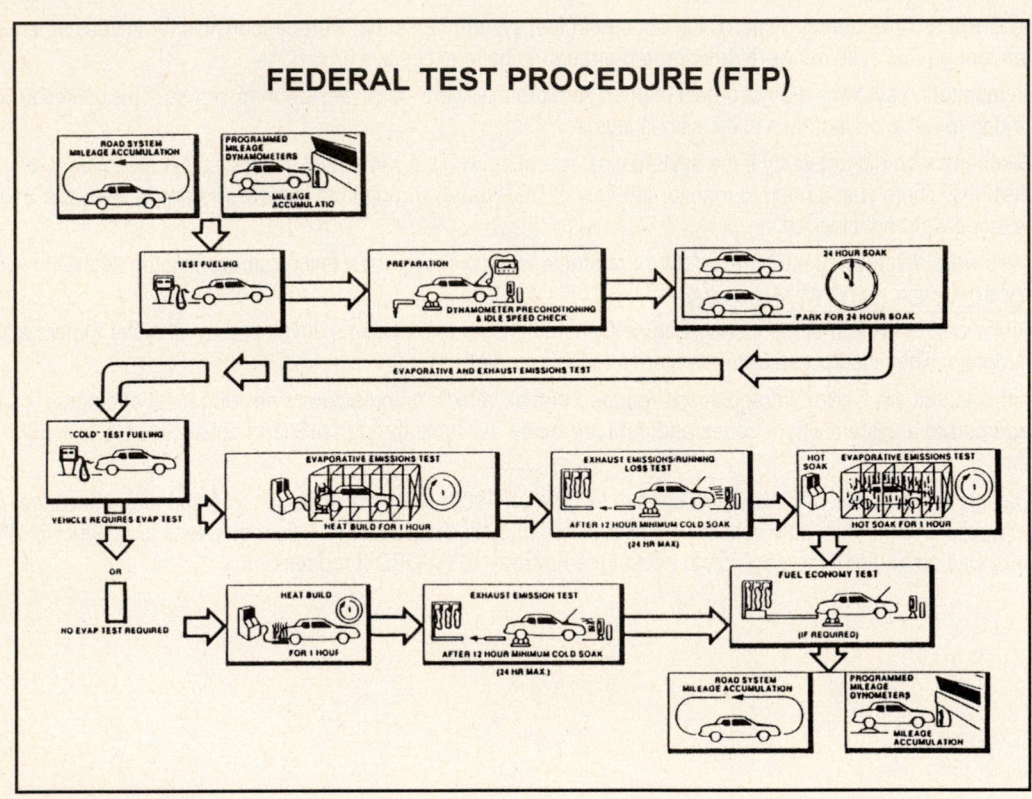

CARB Regulations

SUMMARY OF CARB REGULATIONS

- All vehicles are equipped with a Malfunction Indicator Light (MIL) that will remain "on" when certain faults are detected by the controller. If the MIL is "on", the program should be able to turn it "off" if the fault does not reappear on two consecutive trips. Codes can be erased from memory if the fault does not occur for 40 warmup periods.

- The catalytic converter closest to the engine (if more than one converter is on the vehicle) must be monitored to see if it has deteriorated.

- The oxygen sensor must be monitored for the output voltage and the response rate. The monitor should include the rich/lean transitions (cross counts or oxygen sensor switch rate) and should check the oxygen sensor transition from rich to lean and lean to rich. The oxygen sensor heater should be monitored for proper operation.

- The diagnostic system must monitor the engine for misfire and identify the specific cylinders misfiring. This system must include identification informing the technician that the catalyst is deteriorated, damaged, or has failed a Federal Test Procedure type drive cycle. If the engine is misfiring at a level that could cause damage to the converter catalyst, the MIL should flash when these conditions are present and the controller should switch to a backup program designed to reduce the level of misfire.

- The fuel delivery system should be monitored for its ability to control fuel. It should determine if the system has a condition that is over the range for optimal fuel control. This system is referred to as fuel trim monitor.

- The EGR system should be monitored for both low and high flow rates as well as verify that solenoids controlling the EGR are working properly.

- The EVAP System should be monitored for HC flow changes based on fuel tank fill level and the monitor should calculate for these differences.

- The Secondary Air Management system should be monitored for the control valve operation and for the airflow throughout the system.

- Any emissions-related component or system not otherwise described that provides input directly or indirectly to the processor should be monitored. This includes all input and output components.

EPA Regulations

The Federal Environmental Protection Act (EPA) also established diagnostic regulations, but in an effort to simplify the development process for vehicle manufacturers, the EPA decided to comply with the California OBD II regulations from 1994 to 1998.

SAE Forms Committees to Develop Standards

The Society of Automotive Engineers formed committees to help its member engineers coordinate efforts and develop a second generation of diagnostics. The new diagnostics, known as OBD II, would standardize the diagnostic connector, diagnostic data retrieval and code identification.

EPA Expands the SAE Standards

The EPA reviewed the SAE standards and added regulations requiring all manufacturers to meet Federal OBD II standards by 1996. To assist in development of expanded diagnostics, the EPA decided to use the California regulations as federal standards through 1998. Federal and California regulations established a two-year phase-in to take effect from 1994-1996 to allow for design and phase-in of expanded diagnostics.

Government (CARB & EPA) and SAE Regulations

OVERVIEW

The Society of Automobile Engineers (SAE), CARB and the EPA set the standards that relate to changes in the industry (terminology, common Scan Tool interface, etc.) while others set the diagnostic standard for how information is handled by vehicle controllers.

Industry Regulations

Government Mandated Regulations	SAE "Recommended" Compliance
J1930 - Industry Terminology Standardization	J2201 - Additional Guidelines for Generic Scan Tool Interface
J1978 - Standards for Generic OBD II Scan Tool Interface Protocol	J2190 - Enhanced Test Mode Standards
J2205 - Standards for Expanded Scan Tool Interface Protocol	J2008 - Guidelines for Repair Service Information (CARB Standards)
J2008 - Standards for Repair and Service Information (EPA Guidelines)	

On-Board Computer Regulations

Government Mandated Regulations	SAE "Recommended" Compliance
J2012 - Standards for a Diagnostic Trouble Code (DTC)	J2186 - CARB approved standards for anti-tamper procedures
J1962 - Standard 16-Pin diagnostic connector	J2178 - Scan Tool message strategy guidelines
J1979 - Standards for Diagnostic Test Modes	J2190 - Enhanced Test Mode Standards
J1850 - Scan Tool Communication guidelines for Class 'B' Data Interface	J1724 - Vehicle Electronic Identification Standards
J2186 - EPA mandated Anti-Tamper procedures	

CARB AND EPA REGULATIONS

The government agencies that set OBD regulations are CARB and the EPA. The tables below compare differences between the EPA and CARB requirements for OBD II.

Part One - Industry Regulations

CARB	EPA	Government Requirements
	X	1994 model year - all service information must conform to J1930.
X	X	Service manuals must publish a normal range for calculated load values and Mass Air Flow Rate at idle and at 2,500 RPM.
	X	The vehicle manufacturer is responsible for ensuring information is available even if the information is provided by an intermediary.
	X	The cost of repair information to the independent technician shall not exceed the lowest price that is available to a dealership.
	X	All other information available at a fair and reasonable price, otherwise it is considered not available (a fine of $25K per day could be applied).
	X	Electronic service information must be available by the 1998 model year.
X		Repair procedures must be available which allow effective diagnosis and repairs using a J1978 Generic Scan Tool and readily available repair tools.
X	X	J1978 Scan Tool compatibility - communication protocol. All serial data and enhanced tests must be available to a Scan Tool. Scan Tool must inform user which emissions systems are monitored. EPA added requirements that the VIN be accessible off the DLC.
	X	Requires Bi-directional diagnostic control of the computer be available on the Scan Tool meeting J2205 and J1979 standards.
	X	1996 model year - all service information must be in the J2008 format.
	X	Labeling requirements must meet J1877 and J1892 standards.

Part Two - On-Board Computer Regulations

CARB	EPA	Government Requirements
X	X	J2012 Diagnostic Trouble Codes - If the PCM detects a fault, it must set a code to identify a fault (uniform identification), include conditions that describe how the PCM reverts to default mode and erase the code after 40 warm up cycles with the MIL off.
X	X	J1962 Diagnostic Connector mounted on Instrument Panel driver's side of vehicle with standard pins for serial data, power and ground.
X	X	J1979 Diagnostic Test Mode Messages - defines standard messages for access to trouble codes, vehicle data stream and Freeze Frame data.
X	X	Scan Tool Interface for DLC - SAE J1850 serial data link required to access all emission-related data. Requires 3 byte headers (does not allow IBS or checksum).
X	X	Vehicle manufacturers must provide tampering deterrence for a PCM that is programmable (where the PROM is rewritten to change operating parameters). This must include write-protect standards for programmable memory and references J2186 - Data Link Security for write protect.
	X	Access to vehicle calibration data, odometer and keyless entry codes can be limited, but OEMs must provide "the best means available for providing non-dealer technicians with calibration data necessary to perform repairs".
X	X	Freeze Frame Data Stored with the first fault of any component or system and replace data if there is a subsequent fuel system or misfire fault. EPA added "airflow rate" to required Freeze Frame data.
X	X	Signal access to the required diagnostic data must be made available through the diagnostic connector using standard messages. Actual values should be identified separately from default or limp home values.
X	X	The vehicle must have only one Malfunction Indicator Light (MIL) for emission related problems. The MIL must remain "on" if a malfunction is detected and stays on until three trips indicate the fault is gone. The MIL must blink if a catalyst-threatening misfire condition is present. Note: A few manufacturers received exemptions from this standard for specific 1994-96 models.

Part Three - On-Board Computer Monitor Regulations

CARB	EPA	Government Requirements
X	X	Oxygen Sensor Monitor - It must check the output voltage and response rate for all oxygen sensors once per trip. The results of most recent oxygen sensor evaluation test must be available over the data link connector as serial data. The EPA added a requirement that the results of the most recent on-board monitoring data and test limits for all systems with specific Monitor evaluation tests must be made available.
X	X	Catalyst Monitor - it must verify the catalyst is functioning at steady state efficiency and that it does not deteriorate over 1.5X the standard. This test is done once per trip.
X	X	Misfire Monitor - It must run continuously under all conditions in order to identify a misfiring cylinder. On vehicles with SFI, it must cutoff fuel to a misfiring cylinder.
X		EGR System - It must monitor for both low and high EGR flow rate once per trip.
X		EVAP Purge Monitor - It must check the system function and for leaks once per trip.
X		Secondary AIR Management Monitor - It must perform a functional test of the AIR system and switching valves (a test for proper function and airflow once per trip).
X		Fuel System Monitor - It must check the ability of the controller to control fuel delivery.
X	X	Comprehensive Component Monitor - It must monitor all components or systems that send data to or receives data from the PCM. CARB requires a check for out-of-range signals and a functional response test of related outputs. CARB requires continuous monitoring while the EPA requires evaluation periodically once per "Drive Cycle".
	X	The system must monitor any deterioration or malfunction which occurs which can cause exhaust or evaporative emissions to increase 1.5X the Federal Test Standard.
X		Air Conditioning system must be monitored for loss of reactive refrigerant once per trip. Non-reactive refrigerant does not have to be monitored.

Explanation of SAE Standards

J1930 - Common Names for Components

J1930 established common nomenclature for emissions and computer-related components and systems. This includes common definitions, abbreviations and acronyms.

This standard is designed to provide the technician with a recognizable name for components that apply to all vehicles. This nomenclature has been determined to be beneficial for technicians who work on multiple lines of vehicles as well as vehicles from different manufacturers.

J2008 - Service Information Availability

J2008 requires that "all information" must be made available to "any person engaged" in the repair of the vehicle. The legislation is very specific and requires that "no such information on vehicle repair" may be withheld from any technician. J2008 is still being finalized and will continue to be interpreted by the vehicle manufacturer. It also sets standards for the organization of vehicle service information. This includes the data model, data type definition, graphics standards and electronic transmission of data.

EPA Guidelines on Repair Information

The EPA published guidelines that state that information availability requirements include 1994 model year vehicles, and that vehicle manufacturer must furnish to "any person" engaged in the repair or service of a motor vehicle with "all information" required to make emission related diagnosis and repairs. Includes, but is not limited to service manuals, technical service bulletins, vehicle recalls, engine control emissions system information, bi-directional control and training information.

None of this information may be withheld if provided directly or indirectly to the dealers. Information cost to independent technicians shall not exceed the lowest price of the same information to the dealerships. Other repair information must be made available at a "fair and reasonable" cost, otherwise it is considered unavailable.

J2205 - Expanded Diagnostic Protocol for Scan Tools

Some Scan Tools incorporate a protocol that allows the technician to access information not specifically required by OBD II standards. The information in the messages on this tool will be specified in factory service information and provided to the technician. Refer to the examples under Diagnostic Function in the Graphic to the right.

Source: OTC Scan Tool with a 1999 Pathfinder cartridge.

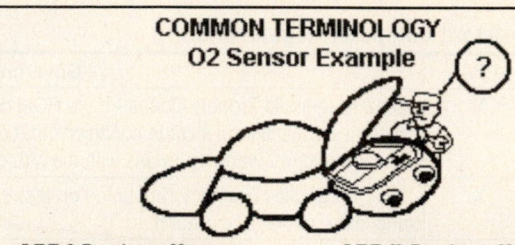

COMMON TERMINOLOGY
O2 Sensor Example

OBD I System Names	OBD II System Names
• O2 Sensor	• O2S-11 Bank 1 Sensor 1
• HEGO Sensor	• HO2S-11 Bank 1 Sensor 1
• EGO Sensor	• HO2S-12 Bank 1 Sensor 2
• LAMBDA Sensor	• HO2S-13 Bank 1 Sensor 3
• Feedback Sensor	• HO2S-21 Bank 2 Sensor 1
	• HO2S-22 Bank 2 Sensor 2

SCAN TOOL MENUS

Press:
> 1-OEM Tests
   2-OBD II

DIAGNOSTIC FUNCTION

Press:
> 1- Datastream
   2-Diagnostic Codes
   4-Record/Playback
   5-Special Test
   7-Monitor Setup

1-DATASTREAM

BARO	29.4" HG
BATT TEMP	49°F
ENGINE RPM	750
IAC DESIRED	37
IAC MOTOR	37

2-DIAGNOSTIC CODES

Press:
> 1-Read Codes
   2-Clear Codes
   3-Code History

EXPLANATION OF ONBOARD COMPUTER REGULATIONS

J1978 - Generic Scan Tool Usage

J1978 requires that all vehicle manufacturers make readily available to the automotive repair industry all data, codes and emissions-related information that can be accessed by a generic Scan Tool. The values for all trouble codes, sensors and components along with Freeze Frame data stored in the computer must be accessible for download to a Generic Scan Tool.

Once a Generic Scan Tool is connected to the 16-pin OBD II connector, it can retrieve certain data from the computer data stream, retrieve Freeze Frame data, read any 5-digit codes and clear these codes from memory.

The EPA expanded this regulation to include the ability to perform bi-directional diagnostic control. The EPA did not define "bi-directional", and the vehicle manufacturers requested that more specific standards be written and incorporated into J2205 after review by SAE committees.

J2178 sets the standards for how vehicle interface messages are displayed on the Generic Scan Tool.

J2201 - Generic Scan Tool Terminal Designation

J2201 sets additional guidelines for Generic Scan Tool interface and assigns the designation of terminals for voltage feed, ground and data transmission.

Generic Scan Tool Menu Example

The Scan Tool Parameter Identification (PID) Mode allows access to certain data values, analog and digital input and output signals, calculated values and system status information.

Generic Scan Tool Navigation

An example of how to navigate through the Scan Tool menus (Snap On example) to locate the Generic PID information is shown in the Graphic to the right.

There are 16 engine related parameters for this vehicle on a Generic OBD II Scan Tool. The parameters in the last frame of this example represent known good values.

Parameter ID (PID) Information

The proper sequence to follow to obtain a complete Generic PID list for this vehicle is shown in the Graphic.

1) Scroll through the main menu and line up the tilde (~) with the desired choice (in this case, GENERIC).
2) Scroll to CODES & DATA. Select it with the tilde (~).
3) Connect an OBD II K2 Adapter to the test connector to allow the tool to read OBD II Generic information.
4) Scroll through the menu and then line up the tilde (~) with the desired choice (DATA - NO CODES).

Source: Snap On Scan Tool with a 1999 cartridge.

SCAN TOOL MENUS

GM/SATURN (1980-1999)
CHRYSLER (1983-1999)
JEEP (1984-1999)
FORD (1981-1999)
~GENERIC OBD II

OBD II GENERIC SCREENS

MAIN MENU-EMISSIONS
[PRESS N FOR HELP]
~CODES & DATA MENU
CUSTOM SETUP

CONNECT OBD-II
K2 ADAPTER TO
16-PIN OBD II TEST
CONNECTOR.
NO REPAIR TIPS
AVAILABLE IN
GENERIC MODE.
PRESS Y TO CONTINUE

CODES & DATA MENU
CODES ONLY
~DATA (NO CODES)
O2 MONITORS
FREEZE FRAME
PENDING CODES

OBD II DATA
(CODES NOT AVAILABLE)

ENGINE RPM	720
THROTTLE(%)	16.4
FUEL SYS1	OL
FUEL SYS2	N/A
COOLANT(°F)	117
MAP("Hg)	29.00
IGN ADVANCE(°)	0.0
ST TRIM B1(%)	0.00
LT TRIM B1(%)	00.0
O2 B1-S1(V)	0.470
TRIM B1-S1(%)	0.00
O2 B1-S2(V)	0.510
TRIM B1-S2(%)	N/A
VEH SPEED(MPH)	0
ENG LOAD(%)	17.0
MIL STATUS	OFF

J1724 - Vehicle Electronic Identification

SAE has developed a recommended practice to provide electronic access to vehicle content information necessary to diagnose, service, test and repair passenger cars and light duty trucks. The SAE committee in charge of this area continues to look at a wide range of interpretations for this standard.

J1850 - Scan Tool Access to Emission Related Data

Access to emission related data must be made available on a standard Diagnostic Data Link (DDL) defined in the J1850 standard (Class 'A' data). There are also other systems on the vehicle that use PCM data. For example, the Climate Control Automatic Air Conditioning System uses signals from the ECT sensor to help determine when to operate the Air Conditioning and Electric Cooling Fan. SAE also developed standards for Vehicle Network and Multiplexing Data Communications (referred to as Class 'B' data). Class 'B' data communications use a system where data is transferred between one or more controllers (or Modules) to eliminate redundant sensors and other system duplication. The modules in this type of system form a multiplex of interactive systems.

Class 'B' Data Communication

Class 'B' data communications have to be able to perform all Class 'A' data functions. However, these two types of communication protocols differ from each other and usually do not communicate in the same format. Scan Tools that communicate with both formats will be available as this standard is defined further and may be made available from the vehicle manufacturer. This means that the Generic Scan Tool (GST) may not be able to access information from computers that control ABS, Air Conditioning, Steering and Suspension, Electronic Transmissions and other related systems.

J1962 - Common Diagnostic Connector

J1962 establishes a set of standards for the OBD II 16-Pin diagnostic connector. The 8 pins assigned by SAE include two pins for a Serial Data Link, two pins for an ISO 9141 Serial Data Link (European) and pins for battery power and ground.

J1979 - Diagnostic Test Mode Messages

Defines standard messages for access to trouble codes, vehicle data stream information and Freeze Frame data.

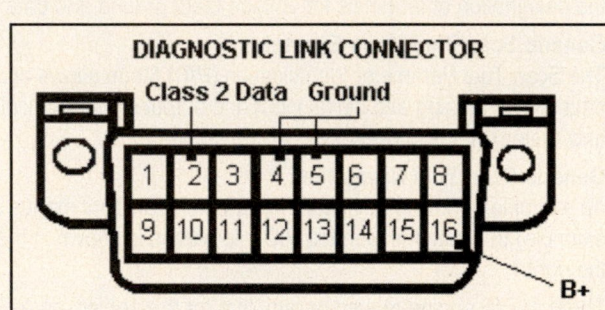

J2012 - Diagnostic Trouble Codes (DTC) Standardization

Diagnostic Trouble Code (DTC) is a term used to describe the method used when a vehicle computer detects a problem in a component or system that it is monitoring.

OBD I system trouble codes were one (1), two (2) or three (3) digit numbers. SAE J2012 set standards for trouble codes and definitions for emission-related systems.

OBD II codes use a five-digit code. OBD II codes begin with a letter followed by four numbers. Refer to the example in the Graphic to the right.

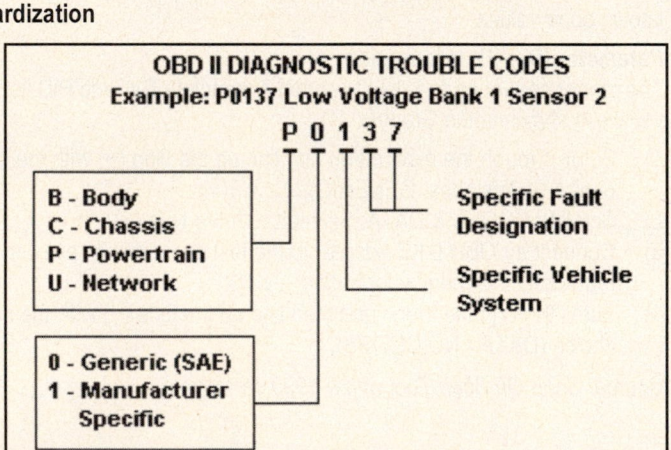

J2186 - Diagnostic Data Link Security Standards

Procedures used in tamper protection must discourage tampering yet allow for the service industry to reprogram or service the PCM as deemed necessary by a vehicle manufacturer. Legitimate service of the PROM will be referenced in later EPA regulations. This service can only be performed if the proper security codes are transmitted from the Scan Tool. However, normal Generic Scan Tool (GST) communications are not affected by this standard.

SAE J2186 defines EE Data Link Security standards for computers with electronically erasable and re-programmable PROMS. There are established standards intended to eliminate "hot rod" PROMS that could disable/defeat emissions-related control systems.

Engine operating parameters should not be changed without the use of manufacturer specialized tools, codes and procedures. CARB specified that any re-programmable computer coded system should include proven write-protection procedures and hardware. The Federal EPA requires that any re-programmable computer codes or operating parameters must be tamper resistant and must conform to SAE J2186 EE Data Link Security standards.

Tamper Protection Explanation

The possibility that On-Board programs may be tampered with from manufacturer specifications introduces the possibility of additional vehicle problems that would need to be diagnosed. Even before vehicles used computers, it was difficult to diagnose a driveability problem in a modified or performance vehicle. The potential of these changes could make it difficult for committees to balance between allowing individuals to install performance changes and protecting the repair industry from the changes.

There is a need for standards that would allow a PCM to identify a vehicle ID number and current calibration, whether the change is a factory update, or an aftermarket modification. This information could also be made accessible to a repair technician.

This would allow the technician to know at the start of a repair procedure that the diagnostic situation might not follow published diagnostic information. To avoid this situation, EPA proposed regulations that would force vehicle manufacturers to utilize complicated methods to deter unauthorized reprogramming. They would include executable routines that could have copyright protection with encrypted data and mandated electronic access by service facilities to an off-site computer maintained by the vehicle manufacturer. Access to an executable routine would be controlled by the vehicle manufacturer and made available only at one of their authorized dealers.

OBD II Warmup Cycle Definition

It is important to understand the meaning of the expression *warmup cycle*. Once the MIL is turned off and the fault that caused the code does not reappear, the OBD II code that was stored in PCM memory will be erased after 40 warmup cycles. The exceptions to this rule are codes related to a Fuel system or Misfire problem. These trouble codes require that 80 warmup cycles occur without the reoccurrence of the fault before the code will be erased from the PCM memory.

A warmup cycle is defined as engine operation (after a key off and engine cool-down period) in which engine temperature increases at least 40ºF and reaches at least 160ºF.

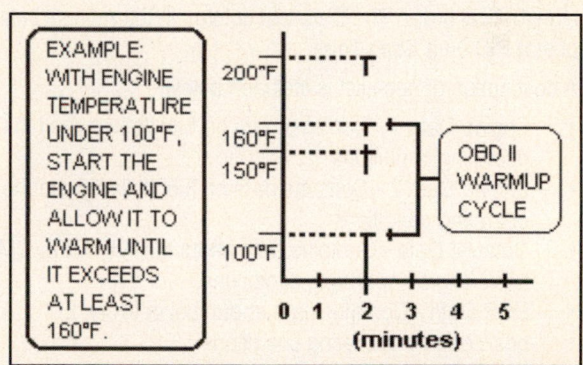

EXPLANATION OF SAE STANDARDS

J2190 - Enhanced Diagnostic Test Modes

This standard identifies test modes and diagnostics for issues not covered by the EPA and CARB regulations. They include Enhanced Test Modes (including an expanded diagnostic routine), and include the protocols required to establish the screens on the Scan Tool for these items:

- Request a diagnostic session
- Request trouble code related Freeze Frame data
- Request all diagnostic trouble codes
- Request status of Main Monitors, clear all test data
- Request diagnostic data, security access
- Disable or enable normal message transmission
- Request diagnostic data packets, test routine results
- Enter or exit diagnostic routines
- Substitute sensor values, substitute output controls
- Read or write to the PROM
- Messages from Enhanced Diagnostic Test Modes are available through an Enhanced Scan Tool, but may not be available through a Generic Scan Tool.

Scan Tool Enhanced Menu Example

An example of how to navigate through the Vetronix Scan Tool menus to locate the OEM PID information is shown in the Graphic. An example of first seven steps to follow is shown. Step (8) contains examples of PID data.

Parameter ID (PID) Information

The PID information for this vehicle is organized into various Data Lists (Engine Data 1, etc). Each PID is categorized into a particular list.

The parameters in the PCM PID Tables in this article are listed in alphabetical order. The Data List column in the manual indicates the location of that PID on a Scan Tool.

A description of each list is included below:

- Engine Data 1 - Contains data on fuel delivery and the basic engine operating conditions.
- Engine Data 2 - Contains data on fuel delivery and the basic engine operating conditions.
- Catalyst Data - Contains data about the A/C, CKP, CMP, KS, and the basic engine operating conditions.
- EGR Data - Contains data about fuel delivery, ECT, IAT, VTD, and the basic engine operating conditions.
- EVAP Data - Contains data that allows it to display parameters needed to verify EVAP system operation.
- HO2S Data - Contains data on the Oxygen sensor.
- Misfire Data - Contains data for Misfire diagnostics.
- Output Device Driver Data - Contains data specific to the ODD operation.

Source: Vetronix Scan Tool & 2000 Mass Storage Unit.

SCAN TOOL MENUS

(1)
```
SELECT APPLICATION

GLOBAL OBDII (MT)
GLOBAL OBDII (T1)
GM P/T
GM CHASSIS
GM BODY SYSTEMS
FORD P/T
FORD CHASSIS
CHRYSLER P/T
ACURA
CHRYSLER IMPORTS
```

(2)
```
  2000 SELECT:
F0:VIN   F4:CAD
F1:MFI   F5:CSFI
F2:TBI   F6:DIESL
```

(3)
```
F0: C-CAR
F1: F-CAR
F2: H-CAR
F3: W-CAR
```

(4)
```
SELECT MFI ENG.
3800 SFI (VIN=K)
  C,F,H,W-CAR?
2000 (YES/NO)
```

(5)
```
SELECT TRANS.
F0: 3 SPD AUTO
F1: 4 SPD AUTO
F2: 5 SPD MAN
```

(6)
```
SELECT MODE:↑↓
F0:Data List
F1:Capture Info.
F2:DTC
F3:Snapshot
F4:OBD Controls
F8:Information
```

(7)
```
SELECT DATA:↑↓
F0:Engine 1
F1:Engine 2
F4:Specific Eng.
F5:A/T
F9:Specific A/T
```

(8)
```
Inj. Pulse Width
    2.9 ms
Air/Fuel Ratio
   14.7:1
```

Malfunction Indicator Lamp

INTRODUCTION

The CARB and Federal EPA regulations require that a Malfunction Indicator Lamp (MIL) be illuminated when an emissions related fault is detected and that a Diagnostic Trouble Code be stored in the vehicle controller (PCM) memory.

Most vehicle manufacturers provided the "Check Engine" light diagnostics required by the 1988 California regulations in time to meet this deadline.

OBD II regulations established changes in the "Check Engine" light operation. A new universal term identified this "light" as a Malfunction Indicator Light (MIL). However, the light on the dash may still be identified with the term "Check Engine" or "Service Engine Soon" for ease of customer understanding.

OBD II guidelines set tight conditions for activating and de-activating the MIL (lamp). This strict set of guidelines has resulted in multiple "levels" of diagnostics with different criteria and conditions for *when* an emissions-related fault will cause the MIL to activate and set a code. Also, there are other codes available that will not cause the PCM to activate the MIL. The guidelines established how quickly the onboard diagnostics must be able to identify a fault, set the trouble code in memory and activate the MIL (lamp).

REGULATIONS FOR CLEARING CODES AND CONTROLLING THE MIL

There are strict regulations for conditions to turn off the light and to clear trouble codes. In the past, some vehicle manufacturers had the technician remove battery voltage from the computer to clear the codes. These new regulations contain significant changes in how and when the controller turns off the MIL. The vehicle must be driven under specific conditions while the emission systems are monitored. Once a fault is detected, the system or component that failed must pass three consecutive tests (three trips) without failing before the MIL will be turned off. OBD II regulations include:

- A standard to regulate how quickly a computer must identify a fault, activate the MIL and set a trouble code.
- A standard to regulate criteria that can turn off the MIL when a fault is not present.
- A standard that establishes how long a trouble code remains in the computer memory once the problem has been repaired and the code is cleared.
- A standard to regulate what information must be available from the vehicle manufacturer that the repair technician can use to assist them in identifying the cause of a fault (i.e., Scan Tool and Freeze Frame data).

FAULTS NOT RELATED TO EMISSION CONTROL SYSTEMS

On an OBD II system, the MIL is not activated unless the computer determines that a failure in a component or system will affect the emissions levels of the vehicle. In effect, this means that **only emissions-related faults (codes)** will cause the PCM to activate the MIL. Be aware that some driveability-related problems not related to emission control components or systems can cause a code to set without the PCM activating the MIL. However, any trouble codes associated with the fault will still be set in memory for a technician to access with an OBD II certified Scan Tool.

KEY POINTS

Just like with OBD I systems, there can be trouble codes without activating the MIL, and there can be failure conditions on some systems not related to emission controls that do not set a trouble code. However, on OBD II systems, when diagnosing any driveability or emissions-related problems, all codes are considered "hard" codes. You should first read and record the codes and related data, then make the repairs. Once these steps are done, you can clear the trouble codes and related Freeze Frame data.

UNDERSTANDING MIL CONDITIONS

The three (3) possible MIL conditions are explained next.

Condition 1: MIL Off

This condition indicates that the PCM has not detected any faults in an emission-related component or system, or that the MIL power or control circuit is not working properly.

Condition 2: MIL On Steady

This condition indicates a fault in an emission-related component or system that could increase tailpipe emissions.

Condition 3: MIL Flashing

This condition indicates either a misfire or fuel system related fault that could cause damage to a catalytic converter.

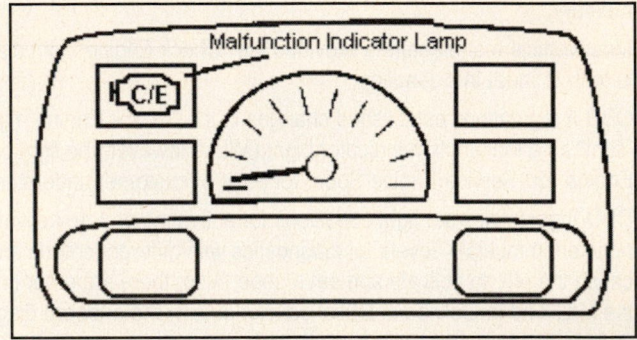

Note: *If a misfire condition exists with the MIL "on" steady, and the driver reaches a vehicle speed and load condition where the engine misfires at a level that could cause catalyst damage, the MIL will begin to flash. It will continue to flash until the engine speed and load conditions that caused that level of misfire subside. Then the MIL will return to the MIL "on" steady condition. This situation may result in a customer complaint as described next: "The MIL in my instrument cluster comes on and then flashes intermittently".*

ACTIONS OR CONDITIONS TO TURN OFF THE MIL

The PCM will turn off the MIL if any of the following actions or conditions occurs:

- The codes are cleared with a Generic or Proprietary Scan Tool
- Power to the PCM is removed (at the battery or with the PCM power fuse)
- A vehicle is driven on three consecutive trips **(including three warmup cycles)** and meets all of the particular code set conditions without the PCM detecting any faults

The PCM will set a code if a fault is detected that could cause tailpipe emissions to exceed 1.5 times the FTP Standard. However, the PCM will not de-activate the MIL until the vehicle has been driven on three consecutive trips with vehicle conditions similar to actual conditions present when the fault was detected. *This is not just three (3) vehicle startups and trips. It means three trips where certain engine operating conditions are met so that the OBD II Monitor that found the fault can "rerun" and pass that diagnostic test.*

Once the MIL is de-activated, the original code will remain in memory until forty warmup cycles are completed without the fault reappearing. A warmup cycle is defined as a trip where with an engine temperature change of at least 40°F, and where the engine temperature reaches at least 160°F.

SIMILAR CONDITIONS (FUEL TRIM AND MISFIRE CODES)

If a Fuel Control system (fuel trim) or misfire-related code is set, the vehicle must be driven under conditions similar to conditions present when the fault was detected before the PCM will de-activate the MIL (lamp). These "similar conditions: are described next:

- The vehicle must be driven with engine speed within 375 RPM of the engine speed stored in the Freeze Frame data when the code set
- The vehicle must be driven within engine load ± 10% of the engine load value stored in the Freeze Frame data when the code set
- The vehicle must be driven with engine temperature conditions similar to the temperature value stored in Freeze Frame data when the code set

Diagnostic Trouble Codes

INTRODUCTION

One of the key features in the OBD II system was an attempt to standardize the wording that describes a diagnostic trouble code or DTC (a term used to describe the method applied when the onboard controller recognizes and identifies a problem in one or more of the circuits or components that it monitors). As a point of review, keep in mind that the trouble codes used with OBD I systems consisted of codes identified with one (1), two (2), or three (3) digits. In effect, trouble codes were only identified with numbers.

DIAGNOSTIC TROUBLE CODES (5-DIGIT)

As previously discussed, SAE J2012 set standards for trouble codes and definitions for emission-related systems. OBD II trouble codes use a five-digit code, and these codes begin with a letter and are followed by four numbers. Since a letter is involved in the sequence, the correct way to read this type of code is with an OBD II certified Scan Tool.

The range of the code designations was designed to allow for future expansion and to allow for manufacturer specific usage on some systems. The illustration in the Graphic includes an explanation of OBD II Code Standardization for a DTC P0137.

UNIVERSAL CODE DESIGNATION EXPLANATION

The number in the thousandths position indicates that the trouble code is common to all manufacturers (a "P0" code).

Most vehicle manufacturers use this designation and then assign a common number and fault message to the problem. The code repair chart is not universal and service procedures will vary between the different vehicle manufacturers. However, the fault described in the code title is common to all systems on the vehicles (e.g., it was assigned a universal code designation). The first letter in the code identifies the system that controlled the device that failed (refer to the table below).

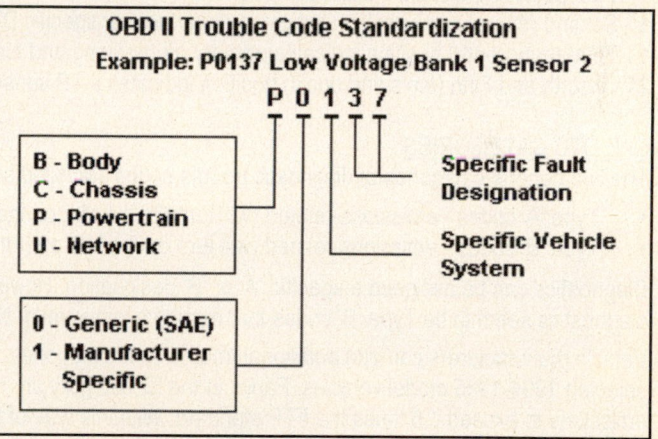

Code Description Table

System ID	System Description
B	Body Control System
C	Chassis Control System
P	Powertrain System
U	UART Data Link, Network Code

MANUFACTURER SPECIFIC DESIGNATION EXPLANATION

Vehicle manufacturers had some code conditions that are specific to the design of their individual system. Not all vehicle manufacturers have chosen to use P1xxx series codes due to differences in their basic systems, diagnostic strategy and their implementation.

These codes are designated as manufacturer specific codes (e.g., a "P1xxx" code), and each manufacturer can define the code and fault description for this designation. Although it was expected that each vehicle manufacturer would remain consistent across their product line, there has been considerable variation on P1xxx designations.

DTC NUMBERING EXPLANATION

The Number in the hundredth position indicates the specific vehicle system or subgroup that failed. This position should be consistent for P0xxx and P1xxx type trouble codes. An SAE committee established the numbers and systems listed below:

- **P0100** - Air Metering and Fuel System fault
- **P0200** - Fuel System (fuel injector only) fault
- **P0300** - Ignition System or Misfire fault
- **P0400** - Emissions Control System fault
- **P0500** - Idle Speed Control, Vehicle Speed Sensor fault
- **P0600** - Computer Output Circuit (relay, solenoid, etc.) fault
- **P0700** - Transaxle, Transmission faults

Note: *The "ten's" and "one's" in the numbers indicate the part of the system at fault.*

DTC NUMBERING EXAMPLE

DTC P1121 - GM Throttle Position Sensor Circuit Intermittent High Voltage:

P - First position indicates Powertrain DTC

1 - Second (thousandth) position indicates manufacturer specific DTC

1 - Third (hundredth) position indicates primary air metering and fuel system

21 - Fourth and Fifth (ten's and one's) position indicates a TP sensor fault

SAE DTC CATEGORIES

The two general categories of diagnostic trouble codes are listed below:

- Type 'A' codes - emissions-related (will turn On the MIL on the first failure)
- Type 'B' codes - emissions-related (will turn on the MIL after the second consecutive trip with a failure)

Diagnostics can be assigned a specific 'A' or 'B' designation. However, most vehicle manufacturer expanded diagnostics seem to be Type 'B' codes that require a minimum of two consecutive trips with a fault to activate the MIL.

Vehicle manufacturers can add additional trouble code categories. For example, GM has a 'D' category for a few selected 1994-1995 model vehicles. Faults in the 'D' category are non-emissions faults that will not cause tailpipe emissions to exceed 1.5 times the FTP standard. With this type of fault, the PCM does not activate any lamps or store any fault data in the Freeze Frame buffer (used for Type 'A' and Type 'B' faults).

COMMON CODE NAMES AND DESCRIPTIONS

OBD II guidelines set standards to universalize the Code Name and Description. These standards only apply to P0xxx codes. You need to be careful because there are several OBD II trouble codes where the same code number can have a different code title.

CODE VARIATION EXAMPLE

Note the use of the same code number for 2 different electrical faults in the table below.

DTC Number	Code Description & Conditions
P1641 (N/MIL) 1996-98 A, L, N & W Body: VIN M, X	**MIL Control Circuit Conditions** Key on, then the PCM received an improper voltage level on the MIL driver circuit (ODM 'A' output 1), condition met for 30 seconds. • Refer to the correct code repair chart.
P1641 (N/MIL) 1996 B Body: VIN P & W 1996 Y Body: VIN 5 & P	**Fan Control Relay 1 Control Circuit Conditions** No A/C or ECT codes set, engine speed over 600 rpm, then the PCM detected that the commanded state of the FC Relay 1 driver and Actual state did not match for 5 seconds. • Refer to the correct code repair chart.

DTC DESCRIPTOR DEFINITIONS

The SAE J2012 document further defines most circuit, component or system codes into the four basic categories explained next.

Circuit Malfunction - Indicates a fixed value or no response from the system. This descriptor can be used instead of a dual High/Low Voltage Code or used to indicate another failure mode.

Range/Performance - Indicates that the circuit is functional, but not operating normally. This descriptor may also indicate a stuck, erratic, intermittent or skewed value that could cause poor performance of an emission control circuit, component or system.

Low Input - Indicates that a signal circuit voltage, frequency or other measurement at a PCM input terminal is at or near zero. The test is made with the external circuit, component or system connected. The signal type is used in place of the word "input."

High Input - Indicates that a signal circuit voltage, frequency or other test measurement at a PCM input terminal is at or near full scale. This test is made with the external circuit, component, or system connected. Signal type is used in place of the word "input."

CONDITIONS TO CLEAR TROUBLE CODES

Diagnostic trouble codes are cleared from the PCM memory using several different methods (the actual method varies between vehicle manufacturers). An example of the Scan Tool navigation screens that appear on a 1999 Ford Windstar during a code clearing procedure is shown in the Graphic (Source: Snap On).

The list below contains a summary of a few of the methods that can be used to clear OBD II trouble codes. The actual conditions for each vehicle manufacturer must be determined and followed exactly.

- Regulations adopted with OBD II allow codes to be cleared by the PCM once 40 warmup cycles occur after the "last test failed" message clears and after 40 "last test passed" messages occur. Refer to Page 1-17 for the definition of a "warmup cycle".
- The Scan Tool can be used to clear any stored codes (and Freeze Frame data).
- On some vehicles, if battery voltage to the PCM is removed, the trouble codes, Freeze Frame data, "trip" or "drive cycle" status and I/M Readiness status will be lost. The battery voltage must be removed for 5 minutes or longer for this action to occur.

ν **NOTE:** *Do not clear the trouble codes unless the code repair chart diagnostic procedure instructs you to do so. Most manufacturers will clear Freeze Frame data (that can be used to diagnose the cause of the fault) at the same time a code is cleared. In effect, this step will result in the loss of the Freeze Frame data on most systems.*

Source: Snap On Scan Tool with a 1999 cartridge.

SCAN TOOL MENUS

SELECT 8th VIN CHAR.
VIN: --T--U--4----------
VEH: 1999 FORD VAN
ENG: 3.8L V6 EEC-V SEFI

SCROLL TO SELECT
THE SYSTEM:
~ENGINE & PCM
ABS
AIRBAG
GEM

SERVICE CODE MENU
KOEO SELF-TEST
~CLEAR CODES
MEMORY CODES

CLEAR CODES

THIS STEP WILL CLEAR
ALL TROUBLE CODES,
FREEZE FRAME DATA &
READINESS INFORMATION

ARE YOU SURE?
~ YES
NO

Diagnostic Routines

OBD I SYSTEM DIAGNOSTICS

One of the most important things to understand about the automotive repair industry is the fact that you have to continually learn new systems and new diagnostic routines (the test procedures designed to isolate a problem on a vehicle system). For OBD I and II systems, a diagnostic routine can be defined as a procedure (a series of steps) that you follow to find the cause of a problem, make a repair and then verify the problem is fixed.

CHANGES IN DIAGNOSTIC ROUTINES

In some cases, a new Engine Control system may be similar to an earlier system, but it can have more indepth control of vehicle emissions, input and output devices and it may include a diagnostic "monitor" embedded in the engine controller designed to run a thorough set of emission control system tests.

OBD I Diagnostic Flowchart

The OBD I Diagnostic Flowchart on this page can be used to find the cause of problems related to Engine Control system trouble codes or driveability symptoms detected on OBD I systems. It includes a step-by-step procedure to use to repair these systems. To compare this flowchart with the one used on OBD II systems, refer to the next page.

The steps in this flow chart should be followed as described below (from top to bottom).

- Do the Pre-Computer Checks.
- Check for any trouble codes stored in memory.
- Read the trouble codes - If trouble codes are set, record them and then clear the codes.
- Start the vehicle and see if the trouble code(s) reset. If they do, then use the correct trouble code repair chart to make the repair.
- If the codes do not reset, than the problem may be intermittent in nature. In this case, refer to the test steps used to find the cause of an intermittent fault (wiggle test).
- In no trouble codes are found at the initial check, then determine if a driveability symptom is present. If so, then refer to the approriate driveability symptom repair chart to make the repair. If the first symptom chart does not isolate the cause of the condition, then go on to another driveability symptom and follow that procedure to conclusion.
- If the problem is intermittent in nature, then refer to the special intermittent tests. Follow all available intermittent tests to determine the cause of this type of fault (usually an electrical connection problem).

OBD I DIAGNOSTIC FLOWCHART

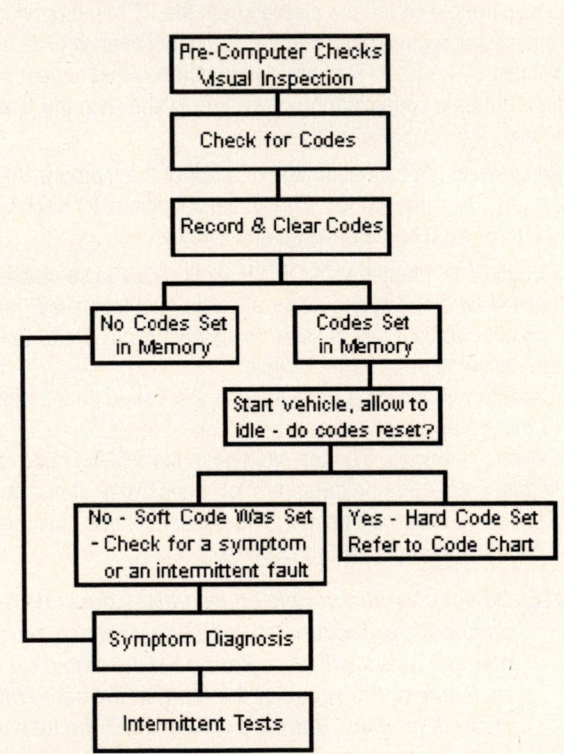

OBD II SYSTEM DIAGNOSTICS

The diagnostic approach used in OBD II systems is more complex than that of the one for OBD I systems. This complexity will effect how you approach diagnosing the vehicle. On an OBD II system, the onboard diagnostics will identify sensor faults (i.e., open, shorted or grounded circuits) as well as those that lose calibration. Another new test that arrived with OBD II is the rationality test (a test that checks whether the value for one input makes rational sense when compared against other sensor input values). The changes plus the use of OBD II Monitors have dramatically changed OBD II diagnostics.

The use of a repeatable test routine can help you quickly get to the root cause of a customer complaint, save diagnostic time and result in a higher percentage of properly repaired vehicles. You can use this Diagnostic Flow Chart to keep on track as you diagnose an Engine Control problem or a base engine fault on vehicles with OBD II.

OBD II Diagnostic Flowchart

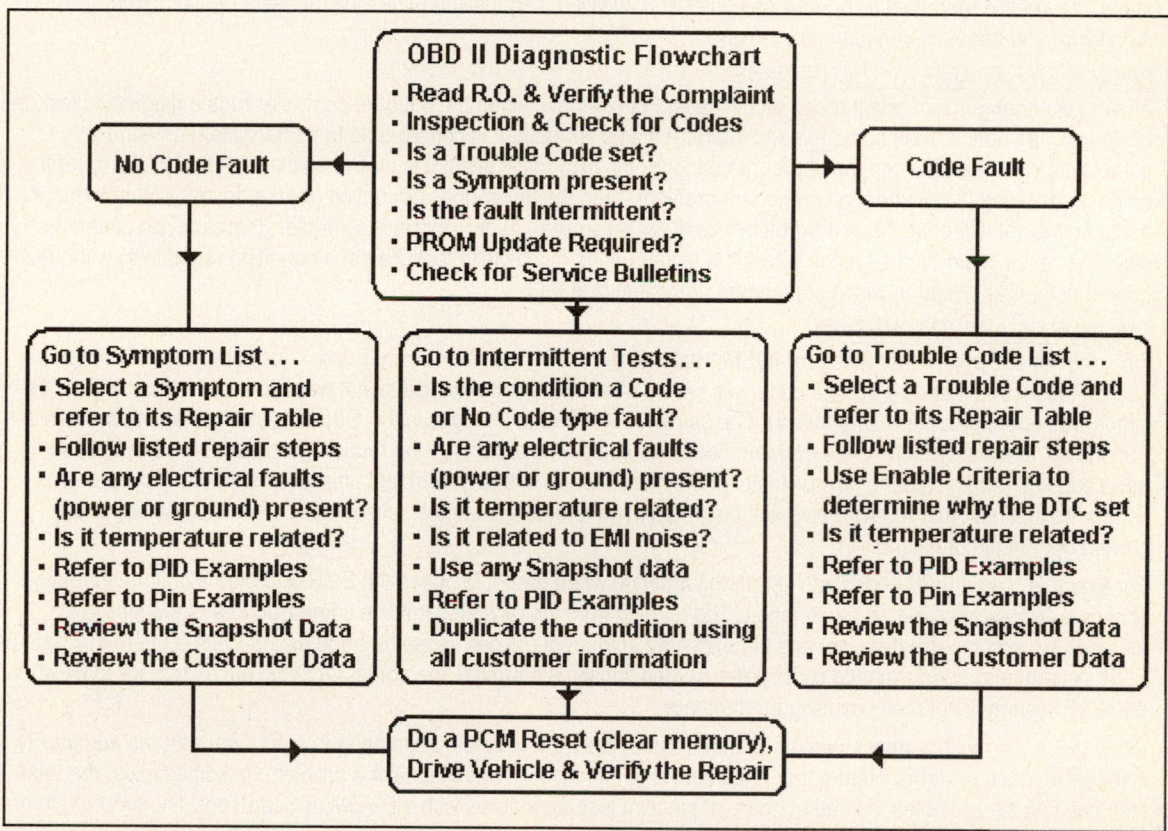

FLOW CHART STEPS

Here are some of the steps included in the Diagnostic Routine:

• Review the repair order and verify the customer complaint as described
• Perform a Visual Inspection of underhood or engine related items
• If the engine will not start, refer to No Start Tests
• If codes are set, refer to the trouble code list, select a code and use the repair chart
• If no codes are set, and a symptom is present, refer to the Symptom List
• Check for any related technical service bulletins (for both Code and No Code Faults)
• If the problem is intermittent in nature, refer to the special Intermittent Tests

Expanded Diagnostics

INTRODUCTION

The primary focus of OBD Expanded Diagnostics is to verify that all Emission Control systems continue to operate efficiently and that they do not deteriorate to a point where tailpipe emissions would increase to a point more than 1.5 times the FTP standard.

On a vehicle equipped with an OBD II system, all emissions faults must set diagnostic codes. However, instead of just identifying a circuit or component failure, this system will identify some problems where the component is deteriorating. These types of problems often occur before the vehicle driver notices a problem. If the fault is driveability related, the driver may not notice it while driving the vehicle. Some faults do not impact vehicle operation (and may not be noticed by the driver - they are not symptom-related faults).

All of these faults can be identified by an OBD II system when it monitors an emissions-related system and determines that it has deteriorated to a point where it would increase emissions beyond the FTP standard. Evaluation of these systems requires expanded programs that monitor all emissions-related components and systems for deterioration. In effect, the vehicle manufacturers have designed and installed diagnostic routines that activate the components and then look for changes in various input values.

CHANGES THAT AFFECT TECHNICIANS

All of these changes can affect repair technicians. For example, when the onboard controller runs a diagnostic test, the customer may notice (feel) a change and interpret it as a problem. It is conceivable that intermittent driveability complaints could result when these programs trigger the diagnostic tests. It is also possible that the MIL (lamp) will come on notifying the driver of a component problem when the driver does not notice (feel) a driveability problem. In addition, the vehicle could have a problem that is not recognized by the onboard controller (therefore, no codes are set). The result could be a situation where the owner brings a vehicle in for repair of a condition (symptom) without a trouble code that indicates which component or system is at fault.

DIAGNOSTIC APPROACHES

One original equipment manufacturer (OEM) refers to the OBD I system diagnostics as "normality" checks. In this interpretation, this means a trouble code was set when the "normal" electrical signal was too high or too low (i.e., the circuit was open, grounded, or shorted). The same OEM refers to changes in the OBD II system as a strategy with a "rationality" test. This refers to the onboard controller being able to monitor the range of a sensor in relationship to other sensors (along with the normal tests for electrical faults). In this type of test, this means that an input signal is compared against other inputs along with the information to determine if the sensor input makes sense under the current operating conditions.

For example, some EGR codes can set during a period when the PCM opens the EGR and then monitors the oxygen sensor response at cruise or deceleration. The reading from the oxygen sensor is compared to a range stored in memory. If the computer does not see the expected amount of oxygen sensor change as the EGR is opened, it sets an EGR system range/performance code. If the oxygen sensor is marginal, the computer could set a code for EGR when the EGR system is not really causing the problem.

While these diagnostics may seem more complicated, the OEM is in fact attempting to assist you with the complexity involved in these systems. Remember, the intent is to identify systems that have a problem. In some cases, the total test time can take minutes instead of hours of pinpoint test procedures with a DVOM or Scan Tool. The proper use of a DVOM, Scan Tool and Lab Scope must be understood to work on OBD II systems.

Expanded Diagnostics

SCAN TOOL INFORMATION

CARB regulations require that vehicle manufacturers make available to repair technicians procedures which allow effective emission related diagnostic and repair using a Generic Scan Tool. This regulation was developed into SAE J1978 that sets guidelines for a common Scan Tool to access On-Board information.

The actual information shown on a Scan Tool can vary between different vehicle systems. Each manufacturer emphasized certain programs and then displayed the information in their own format. SAE J2205 sets the standard for how the Scan Tool will interface with the computer and access the computer information. This standard was necessary because computers are interfacing in bi-directional formats on these vehicles.

There is a difference in the amount of information available on each brand of Scan Tool. Review how this information is accessed on your OBD II certified Scan Tool. The Scan Tool gets its power from the vehicle being tested and talks or "interfaces" with the vehicle diagnostic system or the diagnostic executive program.

TROUBLE CODE INFORMATION

The trouble code information on a Scan Tool includes:

- Current and History trouble codes
- MIL Requested Information ("MIL ON" data)
- Diagnostic test status (test run/test pass or fail)
- Last test pass or fail message
- Freeze Frame data for the 1st emission fault
- Some Scan Tools can display Failure Records

FREEZE FRAME INFORMATION

CARB and EPA regulations require that the controller store specific Freeze Frame (engine related) data when the first emission related fault is detected. The data stored in Freeze Frame can only be replaced by data from a trouble code with a higher priority (i.e., a trouble related to a Fuel system or Misfire Monitor fault).

The Freeze Frame has to contain data values that occurred at the time the code was set (these values are provided in standard units of measurements). As a result, OBD II systems record the data present at the time an emission related code is recorded and the MIL activated. This data can be accessed and displayed on a Scan Tool. Freeze Frame data is one frame or one instant in time. It records the data that set the code.

REQUIRED FREEZE FRAME DATA ITEMS
- Calculated load value, Engine Speed (rpm), Short and Long Term fuel trim values
- Fuel system pressure value (where applicable)
- Vehicle speed (MPH) & Closed / Open Loop status
- Engine coolant temperature and Intake manifold pressure
- Trouble Code that triggered the Freeze Frame
- If misfire code set - identify which cylinder is misfiring

```
┌─────────────────────────────┐
│  MAIN MENU (CHRYSLER)       │
│ ┌─────────────────────────┐ │
│ │ Press:                  │ │
│ │  OTHER SYSTEMS          │ │
│ │  FUNCTIONAL TESTS       │ │
│ │  CODES & DATA MENU      │ │
│ │  CUSTOM SETUP           │ │
│ │ ~SYSTEM TESTS           │ │
│ └─────────────────────────┘ │
│                             │
│     SYSTEM TESTS            │
│ ┌─────────────────────────┐ │
│ │ Press:                  │ │
│ │ SYSTEM TESTS:           │ │
│ │  EGR SYSTEMS TEST       │ │
│ │  GENERATOR FIELD TEST   │ │
│ │  INJ. KILL TEST         │ │
│ │  MISFIRE COUNTERS       │ │
│ │ ~PURGE VAPORS TEST      │ │
│ │  READ VIN               │ │
│ └─────────────────────────┘ │
│                             │
│   PURGE VAPORS TEST         │
│ ┌─────────────────────────┐ │
│ │ Press:                  │ │
│ │ * Y TO SWITCH BETWEEN   │ │
│ │ NORM, FLOW & BLOCK      │ │
│ │ PURGE STATUS __NORM     │ │
│ │ ENGINE RPM_____736     │ │
│ │ NO CODES IN THIS MODE   │ │
│ │ UPSTRM O2S(V)___0.63    │ │
│ │ DWNSTRM O2S(V)__0.14    │ │
│ │ PURGE(mA)_____120     │ │
│ │ ST ADAP(%)_____1.4     │ │
│ └─────────────────────────┘ │
└─────────────────────────────┘
```

Expanded Diagnostics

TROUBLE CODE "TEST CONDITIONS"

Some vehicle emission control components and systems are "continuously" monitored by the Comprehensive Component, Fuel System and Misfire Monitors while some of the OBD II Main Monitors only run their diagnostic tests after certain test conditions or enable criteria have been met (e.g., the EGR system and EVAP system Monitors).

Key Point - Certain code "test conditions" must be met to "run" certain Monitors, and the conditions vary by vehicle and engine configuration. Also, the information related to each trouble code contains the actual conditions present when that particular code set.

PGM-FI SYSTEM CONTENTS

PGM-FI SYSTEMS

Introduction
<u>ABOUT THIS MANUAL</u>
This manual was developed to provide technicians with information on the technical features and diagnostics of the Acura/Honda PGM-FI (MFI & TBI) system for model years 1985-03.

<u>MAIN FEATURES</u>
The main features of this manual are divided into these four sections:

- PGM-FI Systems & Special Tests (Contents Page 1-1)
- Acura PGM-FI Engine Control Systems (Contents Page 2-1)
- Honda PGM-FI Engine Control Systems (Contents Page 3-1)
- OBD II System Theory and Diagnostics (Contents Page 4-1)

Overview of Sections
Section 1 - The PGM-FI System Section contains an introduction to Acura/Honda PGM-FI systems and a suggested Diagnostic Routine for OBD I systems. It also contains articles on these subjects: the Control Box, Catalytic Converters, Exhaust & Fuel Systems, MAP Sensor Tests, Pin Voltage Tests and the VTEC System. The Reference subsection includes an Acura/Honda Glossary and code lists for OBD II System trouble codes.

Section 2 - The Acura section contains these Acura PGM-FI subsections: Air Intake, Catalyst & Exhaust, EGR, EVAP, Fuel, Idle Speed, Ignition, PCV and Secondary Air Injection systems. It also contains a subsection on Acura PGM-FI Diagnostics for OBD I systems. Examples of Pin Charts and Wiring Diagrams are in the Reference Section.

Section 3 - The Honda section contains these Honda PGM-FI subsections: Air Intake, Catalyst & Exhaust, EGR, EVAP, Fuel, Idle Speed, Ignition, PCV and Secondary Air Injection systems. It also contains a subsection on Honda PGM-FI Diagnostics for OBD I systems. Examples of Pin Charts and Wiring Diagrams are in the Reference Section.

Section 4 - The OBD II Section contains information on the OBD II system theory and diagnostics for Acura and Honda vehicle applications. This section includes explanations of the Catalyst, EGR, EVAP, Fuel, HO2S, Misfire and Secondary Air Monitors that include the major software and hardware pieces that make up the OBD II system. Some Scan Tool tips are included in this section.

<u>DIAGNOSTIC HELP</u>
Many of the subsections in this manual contain separate component test articles. These articles include "real world" test examples and results for use with a DVOM, Scan Tool, and Lab Scope.

The vehicle specific test information in this manual can be of great help during actual repair work and vehicle testing. The suggested tests in this manual combine theory with a practical hands-on testing approach. In addition, there are numerous articles in this manual that contain diagnostic help embedded in the PID Chart and Pin Chart examples.

<u>PIN VOLTAGE CHARTS, WIRING & VACUUM DIAGRAMS</u>
Pin Voltage Charts with examples of "known-good" values for several vehicles are included at the end of Sections 2, 3 and 4. There are also Wiring and Vacuum Diagram examples for the same vehicles so that the example PCM circuits can be researched.

Acura & Honda PGM-FI Systems

ACURA PGM-FI SYSTEMS

The information on this page provides an overview of the evolution of the Acura PGM-FI system. A quick review of this information reveals the Acura engine control systems from 1986-95 were equipped with OBD I system diagnostics. Vehicle applications for these years included the Integra, Legend, NSX and Vigor models.

ACURA PGM-FI SYSTEMS (OBD II)

The Acura 2.5TL was introduced in 1995 with OBD II system diagnostics. All of the 1996-03 vehicle applications, the Integra (1996), 2.2CL (1997), 2.3CL (1998), 2.5TL (1996-03), 3.0CL (1997-03), 3.2TL (1996-03), 3.5TL and the NSX (1995-03), were produced with OBD II system diagnostics. For information on Acura SLX models, refer to the article below.

HONDA PGM-FI SYSTEMS (OBD I)

The information on this page provides an overview of the evolution of the Honda PGM-FI system (Figure 1-01). A quick review of this information reveals that Honda engine control systems from 1985-95 were equipped with OBD I system diagnostics. Vehicle applications during these years were the Accord, Civic, Odyssey and Prelude models.

HONDA PGM-FI SYSTEMS (OBD II)

All of the 1996-03 vehicle applications, the Accord, Civic, CR-V, Odyssey and Prelude, were produced with the OBD II system diagnostics. For the introduction on Honda Passport models, see the article below.

ACURA SLX

In 1996 Acura produced the SLX model as a "badge" vehicle. The SLX uses a powertrain with OBD II diagnostics built by Isuzu Motor Co. This manual includes a small amount of information on 1996-99 SLX vehicles.

HONDA PASSPORT

In 1994, Honda introduced the Passport model as a "badge" vehicle. The Passport uses a powertrain built by Isuzu Motor Co with OBD II diagnostics. This manual includes a small amount of information on 1994-02 Passport vehicles.

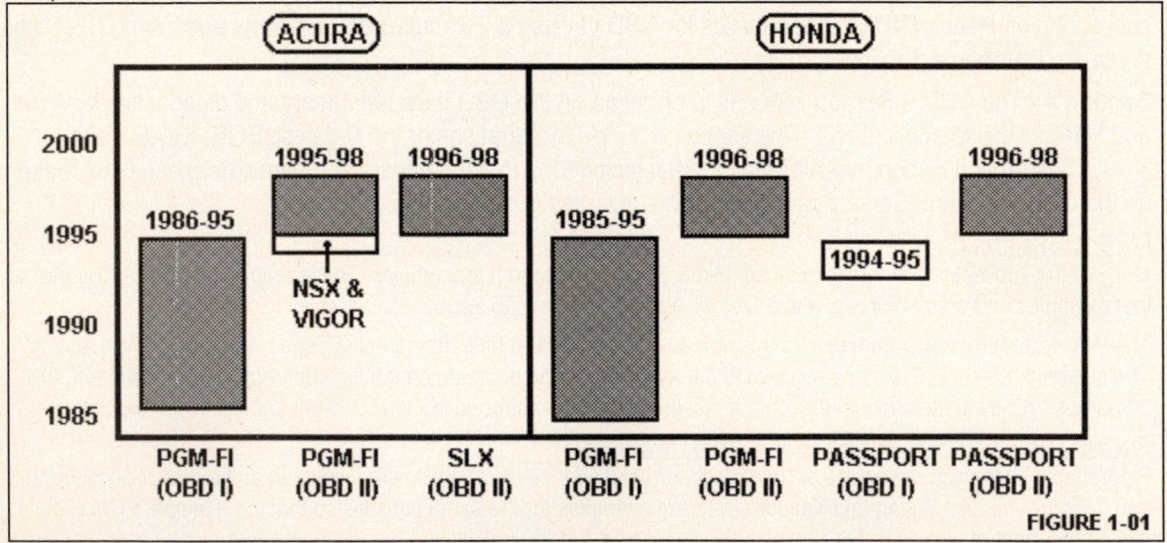

FIGURE 1-01

VIN Code Explanation

The vehicle identification number (VIN) code below is from a 1996 Accord Sedan (EX).

1 H G C D 5 6 3 1 V A 0 2 2 0 7 4
(1) (2) (3) (4) (5) (6) (7) (8) (9) (10) (11) (12) (13) (14) (15) (16) (17)

VIN CODE LEGEND

1 - Nation of origin (J-Japan, 1- USA)

2 - Manufacturer (H-HM LTD, HG- America)

3 - Vehicle type

4, 5, 6 - Line, body, engine

7 - Body type and transmission type

8 - Vehicle grade (equipment, etc.)

9 - Check digit

10 - Model year

11 - Factory code

12, 13, 14, 15, 16, 17 - Serial #

Vehicle Coverage - Acura Cars & Utility Vehicles

1986 Acura Applications	
Integra Hatchback & Sedan	
Engine: 1.6L I4 MFI (D16A1)	VIN DA1
Engine: 1.6L I4 MFI (D16A1)	VIN DA3
Legend Sedan	
Engine: 2.5L V6 MFI (C25A1)	VIN KA2
1987 Acura Applications	
Integra Hatchback & Sedan	
Engine: 1.6L I4 MFI (D16A1)	VIN DA1
Engine: 1.6L I4 MFI (D16A1)	VIN DA3
Legend Coupe & Sedan	
Engine: 2.7L V6 MFI (C27A1)	VIN KA3
Engine: 2.5L V6 MFI (C25A1)	VIN KA2
1988 Acura Applications	
Integra Hatchback & Sedan	
Engine: 1.6L I4 MFI (D16A1)	VIN DA1
Engine: 1.6L I4 MFI (D16A1)	VIN DA3
Legend Coupe & Sedan	
Engine: 2.7L V6 MFI (C27A1)	VIN KA3
Engine: 2.7L V6 MFI (C27A1)	VIN KA4
1989 Acura Applications	
Integra Hatchback & Sedan	
Engine: 1.6L I4 MFI (D16A1)	VIN DA1
Engine: 1.6L I4 MFI (D16A1)	VIN DA3
Legend Coupe & Sedan	
Engine: 2.7L V6 MFI (C27A1)	VIN KA3
Engine: 2.7L V6 MFI (C27A1)	VIN KA4
1990 Acura Applications	
Integra Hatchback & Sedan	
Engine: 1.8L I4 MFI (B18A1)	VIN DA9
Engine: 1.8L I4 MFI (B18A1)	VIN DB1
Legend Coupe & Sedan	
Engine: 2.7L V6 MFI (C27A1)	VIN KA3
Engine: 2.7L V6 MFI (C27A1)	VIN KA4
1991 Acura Applications	
Integra Hatchback & Sedan	
Engine: 1.8L I4 MFI (B18A1)	VIN DA9
Engine: 1.8L I4 MFI (B18A1)	VIN DB1
Legend Coupe & Sedan	
Engine: 3.2L V6 MFI (C32A1)	VIN KA8
Engine: 3.2L V6 MFI (C32A1)	VIN KA7
NSX (Coupe, Open Top)	
Engine: 3.0L V6 MFI (C30A1)	VIN NA1

Vehicle Coverage - Acura Cars & Utility Vehicles

1992 Acura Applications

Integra Hatchback & Sedan
Engine: 1.7L I4 DOHC MFI VTEC (B17A1) .... VIN DB2
Engine: 1.8L I4 DOHC MFI (B18A1) .... VIN DA9
Engine: 1.8L I4 MFI (B18A1) .... VIN DB1

Legend Coupe & Sedan (L, LS)
Engine: 3.2L V6 MFI (C32A1) .... VIN KA8
Engine: 3.2L V6 MFI (C32A1) .... VIN KA7

NSX (Coupe, Open Top)
Engine: 3.0L V6 MFI (C30A1) .... VIN NA1

Vigor (LS, GS)
Engine: 2.5L I5 MFI (G25A1) .... VIN CC2

1993 Acura Applications

Integra Hatchback & Sedan (RS, LS, GS, GSR, LSS)
Engine: 1.7L I4 DOHC SMFI VTEC (B17A1) .... VIN DB2
Engine: 1.8L I4 DOHC SMFI (B18A1) .... VIN DB1
Engine: 1.8L I4 DOHC SMFI (B18A1) .... VIN DA9

Legend Coupe & Sedan (L, LS)
Engine: 3.2L V6 MFI (C32A1) .... VIN KA8
Engine: 3.2L V6 MFI (C32A1) .... VIN KA7

NSX (Coupe, Open Top)
Engine: 3.0L V6 MFI (C30A1) .... VIN NA1

Vigor (LS, GS)
Engine: 2.5L I5 MFI (G25A1) .... VIN CC2

1994 Acura Applications

Integra Hatchback & Sedan (RS, LS, GSR)
Engine: 1.8L I4 DOHC SMFI (B18B1) .... VIN DC2
Engine: 1.8L I4 DOHC SMFI VTEC (B18C1) .... VIN DB4
Engine: 1.8L I4 DOHC SMFI (B18B1) .... VIN DC7
Engine: 1.8L I4 DOHC SMFI VTEC (B18C1) .... VIN DB8

Legend Coupe & Sedan (L, LS)
Engine: 3.2L V6 MFI (C32A1) .... VIN KA8
Engine: 3.2L V6 MFI (C32A1) .... VIN KA7

NSX (Coupe, Open Top)
Engine: 3.0L V6 MFI (C30A1) .... VIN NA1

Vigor (LS, GS)
Engine: 2.5L I5 MFI (G25A1) .... VIN CC2

1995 Acura Applications

Integra Hatchback & Sedan (RS, LS, GSR)
Engine: 1.8L I4 DOHC SMFI (B18B1) .... VIN DC2
Engine: 1.8L I4 DOHC SMFI VTEC (B18C1) .... VIN DB4
Engine: 1.8L I4 DOHC SMFI (B18B1) .... VIN DC7
Engine: 1.8L I4 DOHC SMFI VTEC (B18C1) .... VIN DB8

Legend Coupe & Sedan (L, LS)
Engine: 3.2L V6 MFI (C32A1) .... VIN KA8
Engine: 3.2L V6 MFI (C32A1) .... VIN KA7

2.5TL (Moon Roof, w/o Moon Roof)
Engine: 2.5L I5 MFI (G25A4) 2.5TL .... VIN UA2

NSX (Coupe, Open Top)
Engine: 3.0L V6 MFI (C30A1) NSX .... VIN NA1

Vehicle Coverage - Acura Cars & Utility Vehicles

1996 Acura Applications

Integra Hatchback (GS, GS-R, RS, LS, LS Special)
Engine: 1.8L I4 DOHC SMFI VTEC (B18C1) .... VIN DC2
Engine: 1.8L I4 DOHC SMFI (B18B1) .... VIN DC4

Integra Sedan (GS, GS-R, Type R, RS, LS)
Engine: 1.8L I4 DOHC SMFI (B18B1) .... VIN DB7
Engine: 1.8L I4 DOHC SMFI VTEC (B18C1) .... VIN DB8

2.5TL (Moon Roof, w/o Moon Roof)
Engine: 2.5L I5 MFI (G25A4) .... VIN UA2

3.2TL (Moon Roof, w/o Moon Roof)
Engine: 3.2L V6 MFI (C32A6) .... VIN UA3

3.5TL (L, L2, LS)
Engine: 3.5L V6 MFI (C35A1) .... VIN KA9

RL Sedan
Engine: 3.5L V6 MFI (C35A1) .... VIN KA9

NSX (Coupe, Open Top) Engine: 3.0L V6 MFI (C30A1) .... VIN NA1

SLX (J-4WD, M-4WD)
Engine: 3.2L V6 SOHC MFI (6VD1) .... VIN V

1997 Acura Applications

Integra Hatchback (GS, GS-R, Type R, RS, LS)
Engine: 1.8L I4 DOHC SMFI VTEC (B18C1) .... VIN DC2
Engine: 1.8L I4 DOHC SMFI (B18B1) .... VIN DC4

Integra Sedan (GS, GS-R, Type R, RS, LS)
Engine: 1.8L I4 DOHC SMFI (B18B1) .... VIN DB7
Engine: 1.8L I4 DOHC SMFI VTEC (B18C1) .... VIN DB8

2.2CL (Base Model, Premium Package)
Engine: 2.2L I4 SOHC SMFI VTEC (F22B1) .... VIN YA1

2.5TL (Moon Roof, w/o Moon Roof)
Engine: 2.5L I5 SMFI (G25A4) .... VIN UA2

3.0CL (Premium Package, w/o Premium Package)
Engine: 3.0L V6 SOHC SMFI VTEC (J30A1) .... VIN YA2

3.2TL (Moon Roof, w/o Moon Roof)
Engine: 3.2L V6 MFI (C32A6) .... VIN UA3

3.5TL (L, L2, LS)
Engine: 3.5L V6 MFI (C35A1) .... VIN KA9

RL Sedan
Engine: 3.5L V6 MFI (C35A1) .... VIN KA9

NSX (Coupe, Open Top) Engine: 3.0L V6 MFI (C30A1) .... VIN NA1

SLX (J-4WD, M-4WD)
Engine: 3.2L V6 SOHC MFI (6VD1) .... VIN V

1998 Acura Applications

Integra Hatchback (GS, GS-R, Type R, RS, LS)
Engine: 1.8L I4 DOHC SMFI VTEC (B18C1, B18C5) .... VIN DC2
Engine: 1.8L I4 DOHC SMFI 16v (B18B1) .... VIN DC4

Integra Sedan (GS, GS-R, Type R, RS, LS)
Engine: 1.8L I4 DOHC SMFI (B18B1) .... VIN DB7
Engine: 1.8L I4 DOHC SMFI VTEC (B18C1) .... VIN DB8

2.3CL (Base Model, Premium Package)
Engine: 2.3L I4 SOHC MFI VTEC (F23A1) .... VIN YA3

2.5TL (Moon Roof, w/o Moon Roof)
Engine: 2.5L I5 SMFI 20v (G25A4) .... VIN UA2

3.0CL (Premium Package, w/o Premium Package)
Engine: 3.0L V6 SOHC SMFI VTEC (J30A1) .... VIN YA2

3.2TL (Moon Roof, w/o Moon Roof)
Engine: 3.2L V6 SMFI 24v (C32A6) .... VIN UA3

3.5TL (L, L2, LS)
Engine: 3.5L V6 SMFI 24v (C35A1) .... VIN KA9

RL Sedan
Engine: 3.5L V6 MFI (C35A1) .... VIN KA9

NSX (Coupe, Open Top) Engine: 3.0L V6 MFI (C30A1) .... VIN NA1

SLX (J-4WD, M-4WD)
Engine: 3.5L V6 SOHC SMFI (6VE1) .... VIN X

1999 Acura Applications

Integra Hatchback (GS, GS-R, Type R, RS, LS)
Engine: 1.8L I4 DOHC SMFI VTEC (B18C1, B18C5) ....VIN DC2
Engine: 1.8L I4 DOHC SMFI 16v (B18B1) ....VIN DC4
Integra Sedan (GS, GS-R, Type R, RS, LS)
Engine: 1.8L I4 DOHC SMFI (B18B1) ....VIN DB7
Engine: 1.8L I4 DOHC SMFI VTEC (B18C1) ....VIN DB8
2.3CL (Base Model, Premium Package)
Engine: 2.3L I4 SOHC MFI VTEC (F23A1) ....VIN YA3
3.0CL (Premium Package, w/o Premium Package)
Engine: 3.0L V6 SOHC SMFI VTEC (J30A1) ....VIN YA2
3.2TL (Moon Roof, w/o Moon Roof)
Engine: 3.2L V6 SMFI 24v (J32A1) ....VIN UA5
3.5TL (L, L2, LS)
Engine: 3.5L V6 SMFI 24v (C35A1) ....VIN KA9
RL Sedan
Engine: 3.5L V6 MFI (C35A1) ....VIN KA9
NSX (Coupe, Open Top)
Engine: 3.0L V6 MFI (C30A1) ....VIN NA1
Engine: 3.2L V6 MFI (C32B1) ....VIN NA2
SLX (J-4WD, M-4WD)
Engine: 3.5L V6 SOHC SMFI (6VE1) ....VIN X

2000 Acura Applications

Integra Hatchback (GS, GS-R, Type R, RS, LS)
Engine: 1.8L I4 DOHC SMFI VTEC (B18C1, B18C5) ....VIN DC2
Engine: 1.8L I4 DOHC SMFI 16v (B18B1) ....VIN DC4
Integra Sedan (GS, GS-R, Type R, RS, LS)
Engine: 1.8L I4 DOHC SMFI (B18B1) ....VIN DB7
Engine: 1.8L I4 DOHC SMFI VTEC (B18C1) ....VIN DB8
2.3CL (Base Model, Premium Package)
Engine: 2.3L I4 SOHC MFI VTEC (F23A1) ....VIN YA3
3.0CL (Premium Package, w/o Premium Package)
Engine: 3.0L V6 SOHC SMFI VTEC (J30A1) ....VIN YA2
3.2TL (Moon Roof, w/o Moon Roof)
Engine: 3.2L V6 SMFI 24v (J32A1) ....VIN UA5
3.5TL (L, L2, LS)
Engine: 3.5L V6 SMFI 24v (C35A1) ....VIN KA9
RL Sedan
Engine: 3.5L V6 MFI (C35A1) ....VIN KA9
NSX (Coupe, Open Top)
Engine: 3.0L V6 MFI (C30A1) ....VIN NA1
Engine: 3.2L V6 MFI (C32B1) ....VIN NA2

2001 Acura Applications

Integra Hatchback (GS, GS-R, Type R, RS, LS)
Engine: 1.8L I4 DOHC SMFI VTEC (B18C1, B18C5) ....VIN DC2
Engine: 1.8L I4 DOHC SMFI 16v (B18B1) ....VIN DC4
Integra Sedan (GS, GS-R, Type R, RS, LS)
Engine: 1.8L I4 DOHC SMFI (B18B1) ....VIN DB7
Engine: 1.8L I4 DOHC SMFI VTEC (B18C1) ....VIN DB8
3.2CL (Premium Package, w/o Premium Package)
Engine: 3.2L V6 SOHC SMFI VTEC (J32A1) ....VIN YA4
3.2TL (Moon Roof, w/o Moon Roof)
Engine: 3.2L V6 SMFI 24v (J32A1) ....VIN UA5
3.5TL (L, L2, LS)
Engine: 3.5L V6 SMFI 24v (C35A1) ....VIN KA9
RL Sedan
Engine: 3.5L V6 MFI (C35A1) ....VIN KA9
NSX (Coupe, Open Top)
Engine: 3.0L V6 MFI (C30A1) ....VIN NA1
Engine: 3.2L V6 MFI (C32B1) ....VIN NA2
MDX (SUV)
Engine: 3.5L V6 MFI (J35A3) ....VIN YD1

2002 Acura Applications

Integra Hatchback (GS, GS-R, Type R, RS, LS)
Engine: 1.8L I4 DOHC SMFI VTEC (B18C1, B18C5) .... VIN DC2
Engine: 1.8L I4 DOHC SMFI 16v (B18B1) .... VIN DC4

Integra Sedan (GS, GS-R, Type R, RS, LS)
Engine: 1.8L I4 DOHC SMFI (B18B1) .... VIN DB7
Engine: 1.8L I4 DOHC SMFI VTEC (B18C1) .... VIN DB8

3.2CL (Premium Package, w/o Premium Package)
Engine: 3.2L V6 SOHC SMFI VTEC (J32A1) .... VIN YA4

3.2TL (Moon Roof, w/o Moon Roof)
Engine: 3.2L V6 SMFI 24v (J32A1) .... VIN UA5

3.5TL (L, L2, LS)
Engine: 3.5L V6 SMFI 24v (C35A1) .... VIN KA9

RL Sedan
Engine: 3.5L V6 MFI (C35A1) .... VIN KA9

NSX (Coupe, Open Top)
Engine: 3.0L V6 MFI (C30A1) .... VIN NA1
Engine: 3.2L V6 MFI (C32B1) .... VIN NA2

MDX (SUV)
Engine: 3.5L V6 MFI (J35A3) .... VIN YD1

RSX (Coup, Type S)
Engine: 2.0L I4 MFI (K20A3, K20A2) .... VIN DC5

2003 Acura Applications

Integra Hatchback (GS, GS-R, Type R, RS, LS)
Engine: 1.8L I4 DOHC SMFI VTEC (B18C1, B18C5) .... VIN DC2
Engine: 1.8L I4 DOHC SMFI 16v (B18B1) .... VIN DC4

Integra Sedan (GS, GS-R, Type R, RS, LS)
Engine: 1.8L I4 DOHC SMFI (B18B1) .... VIN DB7
Engine: 1.8L I4 DOHC SMFI VTEC (B18C1) .... VIN DB8

3.2CL (Premium Package, w/o Premium Package)
Engine: 3.2L V6 SOHC SMFI VTEC (J32A1) .... VIN YA4

3.2TL (Moon Roof, w/o Moon Roof)
Engine: 3.2L V6 SMFI 24v (J32A1) .... VIN UA5

3.5TL (L, L2, LS)
Engine: 3.5L V6 SMFI 24v (C35A1) .... VIN KA9

RL Sedan
Engine: 3.5L V6 MFI (C35A1) .... VIN KA9

NSX (Coupe, Open Top)
Engine: 3.0L V6 MFI (C30A1) .... VIN NA1
Engine: 3.2L V6 MFI (C32B1) .... VIN NA2

MDX (SUV)
Engine: 3.5L V6 MFI (J35A3) .... VIN YD1

RSX (Coup, Type S)
Engine: 2.0L I4 MFI (K20A3, K20A2) .... VIN DC5

Vehicle Coverage - Honda Cars

1985 Honda Applications

Accord
Engine: 1.8L I4 MFI (ES3) .... VIN AD

Civic CRX
Engine: 1.5L I4 MFI (EW3) .... VIN AF

Prelude
Engine: 2.0L I4 MFI (BT) .... VIN BB

1986 Honda Applications

Accord
Engine: 2.0L I4 MFI (BT) .... VIN BA

Civic CRX & Sedan
CRX Engine: 1.5L I4 MFI (EW3) .... VIN AF
Sedan Engine: 1.5L I4 MFI (EW1) .... VIN AK

Prelude
Engine: 2.0L I4 MFI (BT) .... VIN BB

1987 Honda Applications

Accord
Engine: 2.0L I4 MFI (A20A3) .... VIN CA5

Civic CRX, Hatchback & Sedan
CRX Engine: 1.5L I4 MFI (D15A3) .... VIN EC1
Hatchback Engine: 1.5L I4 MFI (D15A3) .... VIN EC3
Sedan Engine: 1.5L I4 MFI (D15A3) .... VIN EC4

Prelude
Engine: 2.0L I4 MFI (A20A3) .... VIN BA3

1988 Honda Applications

Accord
Engine: 2.0L I4 MFI (A20A3) .... VIN CA5

Civic CRX, Hatchback & Sedan
CRX Engine: 1.5L I4 TBI (D15B2) .... VIN ED8
CRX Engine: 1.5L I4 MFI (D15B6) .... VIN ED8
CRX Engine: 1.6L I4 MFI (D16A6) .... VIN ED9
Hatchback Engine: 1.5L I4 TBI (D15B1) .... VIN ED6
Hatchback Engine: 1.5L I4 TBI (D15B2) .... VIN ED6
Sedan Engine: 1.5L I4 TBI (D15B1) .... VIN ED3
Sedan Engine: 1.5L I4 TBI HP (D15B2) .... VIN ED3

Civic Wagon & Wagovan
Wagon Engine: 1.5L I4 TBI (D15B2) .... VIN EE2
Wagon Engine: 1.6L I4 MFI (D16A6) .... VIN EE4
Wagovan Engine: 1.5L I4 TBI (D15B2) .... VIN EY3

Prelude
Engine: 2.0L I4 MFI (B20A5) .... VIN BA4

1989 Honda Applications

Accord & Accord Coupe
Accord Engine: 2.0L I4 MFI (A20A3) .... VIN CA5
Coupe Engine: 2.0L I4 MFI (A20A3) .... VIN CA6

Civic CRX, Hatchback, Sedan, Wagon, Wagovan
CRX Engine: 1.5L I4 TBI (D15B2) .... VIN ED8
CRX Engine: 1.5L I4 MFI (D15B6) .... VIN ED8
CRX Engine: 1.6L I4 MFI (D16A6) .... VIN ED9
Hatchback Engine: 1.5L I4 TBI (D15B1) .... VIN ED6
Hatchback Engine: 1.5L I4 TBI (D15B2) .... VIN ED6
Hatchback Engine: 1.6L I4 MFI (D16A6) .... VIN ED7
Sedan Engine: 1.5L I4 TBI (D15B1) .... VIN ED3
Sedan Engine: 1.5L I4 TBI (D15B2) .... VIN ED3
Wagon Engine: 1.5L I4 TBI (D15B2) .... VIN EE2
Wagon Engine: 1.6L I4 MFI (D16A6) .... VIN EE4
Wagovan Engine: 1.5L I4 TBI (D15B2) .... VIN EY3

Prelude
Engine: 2.0L I4 MFI (B20A5) .... VIN BA4

Vehicle Coverage - Honda Cars

1990 Honda Applications

Accord
Engine: 2.2L I4 MFI Single Intake & Exhaust (F22A1) .... VIN CB7
Engine: 2.2L I4 MFI Dual Intake & Exhaust (F22A4) .... VIN CB7

Civic CRX, Hatchback & Sedan
CRX Engine: 1.5L I4 TBI SOHC (D15B2) .... VIN ED8
CRX Engine: 1.5L I4 SOHC MFI (D15B6) .... VIN ED8
CRX Engine: 1.6L I4 SOHC MFI (D16A6) .... VIN ED9
Hatchback Engine: 1.5L I4 TBI SOHC (D15B1) .... VIN ED6
Hatchback Engine: 1.5L I4 TBI SOHC (D15B2) .... VIN ED6
Hatchback Engine: 1.6L I4 MFI (D16A6) .... VIN ED7
Sedan Engine: 1.5L I4 TBI SOHC (D15B1) .... VIN ED3
Sedan Engine: 1.5L I4 TBI SOHC (D15B2) .... VIN ED3
Sedan Engine: 1.6L I4 SOHC MFI (D16A6) .... VIN ED4

Civic Wagon & 4WD Wagon
Wagon Engine: 1.5L I4 TBI 93 HP (D15B2) .... VIN EE2
4WD Wagon Engine: 1.6L I4 MFI (D16A6) .... VIN EE4

Prelude
Engine: 2.0L I4 DOHC MFI (B20A5) .... VIN BA4
Engine: 2.1L I4 DOHC MFI (B21A1) .... VIN BA4

1991 Honda Applications

Accord Coupe, Sedan & Accord Wagon (DX, LX, EX, SE)
Engine: 2.2L I4 MFI (F22A1 Single Intake & Exhaust) .... VIN CB7
Engine: 2.2L I4 MFI (F22A6 Dual Intake & Exhaust) .... VIN CB7
Engine: 2.2L I4 MFI (F22A1 Single Intake & Exhaust) .... VIN CB9
Engine: 2.2L I4 MFI (F22A6 Dual Intake & Exhaust) .... VIN CB9

Civic CRX, Hatchback, Sedan & Wagon
CRX Engine: 1.5L I4 TBI SOHC (D15B2) .... VIN ED8
CRX Engine: 1.5L I4 SOHC MFI (D15B6) .... VIN ED8
CRX Engine: 1.6L I4 SOHC MFI (D16A6) .... VIN ED9
Hatchback Engine: 1.5L I4 SOHC TBI (D15B1) .... VIN ED6
Hatchback Engine: 1.5L I4 SOHC TBI (D15B2) .... VIN ED6
Hatchback Engine: 1.6L I4 SOHC MFI (D16A6) .... VIN ED7
Sedan Engine: 1.5L I4 SOHC TBI (D15B1) .... VIN ED3
Sedan Engine: 1.5L I4 SOHC TBI (D15B2) .... VIN ED3
Sedan Engine: 1.6L I4 SOHC MFI (D16A6) .... VIN ED4
Wagon Engine: 1.5L I4 SOHC TBI (D15B2) .... VIN EE2
4WD Wagon Engine: 1.6L I4 MFI (D16A6) .... VIN EE4

Prelude (S, Si)
Engine: 2.0L I4 DOHC MFI (B20A5) .... VIN BA8
Engine: 2.1L I4 DOHC MFI (B21A1) .... VIN BB2

1992 Honda Applications

Accord Coupe, Sedan & Accord Wagon (DX, LX, EX, SE)
Engine: 2.2L I4 MFI (F22A1 Single Intake & Exhaust) .... VIN CB7
Engine: 2.2L I4 MFI (F22A6 Dual Intake & Exhaust) .... VIN CB7
Engine: 2.2L I4 MFI (F22A1 Single Intake & Exhaust) .... VIN CB9
Engine: 2.2L I4 MFI (F22A6 Dual Intake & Exhaust) .... VIN CB9

Civic Hatchback (CX, DX, Si, VX)
Engine: 1.5L I4 SOHC MFI 16v (D15B7) .... VIN EH2
Engine: 1.5L I4 SOHC MFI 8v (D15B8) .... VIN EH2
Engine: 1.5L I4 SOHC MFI 16v VTEC-E (D15Z1) .... VIN EH2
Engine: 1.6L I4 SOHC MFI 16v VTEC (D16Z6) .... VIN EH3

Civic Sedan (DX, EX, LX)
Engine: 1.6L I4 SOHC MFI 16v (D15B7) .... VIN EG8
Engine: 1.6L I4 SOHC MFI 16v VTEC (D16Z6) .... VIN EH9

Prelude (S, Si)
Engine: 2.2L I4 SOHC MFI (F22A1) .... VIN BA8
Engine: 2.3L I4 DOHC MFI (H23A1) .... VIN BB2

Vehicle Coverage - Honda Cars & Utility Vehicles

1993 Honda Applications

Accord Coupe, Sedan & Accord Wagon (DX, LX, EX, SE)
Engine: 2.2L I4 MFI (F22A1 Single Intake & Exhaust) ....VIN CB7
Engine: 2.2L I4 MFI (F22A6 Dual Intake & Exhaust) ....VIN CB7
Engine: 2.2L I4 MFI (F22A1 Single Intake & Exhaust) ....VIN CB9
Engine: 2.2L I4 MFI (F22A6 Dual Intake & Exhaust) ....VIN CB9

Civic Coupe (DX, EX) & Hatchback (CX, DX, VX, Si)
Engine: 1.5L I4 SOHC MFI 16v (D15B7) ....VIN EH2
Engine: 1.5L I4 SOHC MFI 8v (D15B8) ....VIN EH2
Engine: 1.5L I4 SOHC MFI 16v VTEC-E (D15Z1) ....VIN EH2
Engine: 1.6L I4 SOHC MFI 16v VTEC (D15Z6) ....VIN EH3

Civic Del Sol (S, Si)
Engine: 1.5L I4 SOHC SMFI 16v VTEC (D15B7) ....VIN EG1
Engine: 1.6L I4 SOHC SMFI 16 VTEC (D16Z6) ....VIN EH6

Civic Sedan (DX, LX, EX)
Engine: 1.5L I4 SOHC MFI 16v (D15B7) ....VIN EG8
Engine: 1.6L I4 SOHC MFI VTEC (D16Z6) ....VIN EH9

Prelude (S, Si, Si 4WS, Si VTEC)
Engine: 2.2L I4 SOHC SMFI (F22A1) ....VIN BA8
Engine: 2.2L I4 DOHC SMFI VTEC (H22A1) ....VIN BB1
Engine: 2.3L I4 DOHC SMFI (H23A1) ....VIN BB2

1994 Honda Applications

Accord Coupe, Sedan & Wagon (DX, EX, LX, LX w/ABS)
Engine: 2.2L I4 SOHC MFI VTEC (F22B1) ....VIN CD5
Engine: 2.2L I4 SOHC MFI (F22B2) ....VIN CD5
Engine: 2.2L I4 SOHC MFI VTEC (F22B1) ....VIN CD7
Engine: 2.2L I4 SOHC MFI (F22B2) ....VIN CD7
Engine: 2.2L I4 SOHC MFI VTEC (F22B1) ....VIN CE1
Engine: 2.2L I4 SOHC MFI (F22B2) ....VIN CE1

Civic Coupe (DX, EX, EX w/ABS)
Engine: 1.5L I4 SOHC SMFI 16v (D15B7) ....VIN EJ2
Engine: 1.6L I4 SOHC SMFI 16v VTEC (D15Z6) ....VIN EJ1

Civic Del Sol (S, Si, VTEC)
Engine (S): 1.5L I4 SOHC SMFI 16v VTEC (D15B7) ....VIN EG1
Engine (Si): 1.6L I4 DOHC SMFI 16v VTEC (B16A3) ....VIN EG2
Engine (VTEC): 1.6L I4 SOHC SMFI 16v VTEC (D16Z6) ....VIN EH6

Civic Hatchback (CX, DX, VX, Si, Si w/ABS)
Engine: 1.5L I4 SOHC SMFI 16v (D15B7) ....VIN EH2
Engine: 1.5L I4 SOHC SMFI 8v (D15B8) ....VIN EH2
Engine: 1.5L I4 SOHC SMFI 16v VTEC-E (D15Z1) ....VIN EH2
Engine: 1.6L I4 SOHC SMFI 16v VTEC (D15Z6) ....VIN EH3

Civic Sedan (DX, EX, LX, LX w/ABS)
Engine: 1.5L I4 SOHC SMFI 16v (D15B7) ....VIN EG8
Engine: 1.6L I4 SOHC SMFI 16v VTEC (D16Z6) ....VIN EG9

Passport G-Body 2WD, Y-Body 4WD
G-Body 2WD Engine: 2.6L I4 MFI (4ZE1) ....VIN E
Y-Body 2WD Engine: 3.2L V6 MFI (6VD1) ....VIN V
Y-Body 4WD Engine: 3.2L V6 MFI (6VD1) ....VIN V

Prelude (S, Si, Si VTEC)
Engine: 2.2L I4 SOHC SMFI (F22A1) ....VIN BA8
Engine: 2.2L I4 DOHC SMFI VTEC (H22A1) ....VIN BB1
Engine: 2.3L I4 DOHC SMFI (H23A1) ....VIN BB2

Vehicle Coverage - Honda Cars & Utility Vehicles

1995 Honda Applications

Accord Coupe, Sedan & Wagon (DX, EX, LX, LX w/ABS)
Engine: 2.2L I4 SOHC MFI (F22B1 VTEC, F22B2) ....VIN CD5
Engine: 2.2L I4 SOHC MFI (F22B1 VTEC, F22B2) ....VIN CD7
Engine: 2.7L V6 SOHC SMFI (C27A4) ....VIN CE6
Engine: 2.2L I4 SOHC MFI (F22B1 VTEC, F22B2) ....VIN CE1
Civic Coupe (DX, EX, EX w/ABS)
Engine: 1.5L I4 SOHC SMFI 16v (D15B7) ....VIN EJ2
Engine: 1.6L I4 SOHC SMFI 16v VTEC (D16Z6) ....VIN EJ1
Civic Del Sol (S, Si, VTEC)
Engine: 1.5L I4 SOHC SMFI 16v (D15B7) ....VIN EG1
Engine: 1.6L I4 DOHC SMFI 16v VTEC (B16A3) ....VIN EG2
Engine: 1.6L I4 SOHC SMFI 16v VTEC (D16Z6) ....VIN EH6
Civic Hatchback (CX, DX, Si, Si w/ABS, VX)
Engine: 1.5L I4 SOHC SMFI 16v (D15B7) ....VIN EH2
Engine: 1.5L I4 SOHC SMFI 8v (D15B87) ....VIN EH2
Engine: 1.5L I4 SOHC SMFI 16v VTEC-E (D15Z1) ....VIN EH2
Engine: 1.6L I4 SOHC SMFI 16v VTEC (D16Z6) ....VIN EH3
Civic Sedan (DX, EX, LX, LX w/ABS)
Engine: 1.5L I4 SOHC SMFI 16v (D15B7) ....VIN EG8
Engine: 1.6L I4 SOHC SMFI 16v VTEC (D16Z6) ....VIN EH9
Odyssey (EX, LX)
Engine: 2.2L I4 MFI (F22B6) ....VIN RA1
Passport G-Body 2WD, Y-Body 4WD
G-Body 2WD Engine: 2.6L I4 MFI (4ZE1) ....VIN E
G-Body 2WD Engine: 3.2L V6 MFI (6VD1) ....VIN V
Y-Body Engine: 3.2L V6 MFI (6VD1) ....VIN V
Prelude (S, Si, Si VTEC)
Engine: 2.2L I4 SOHC SMFI (F22A1) ....VIN BA8
Engine: 2.2L I4 DOHC SMFI VTEC (H22A1) ....VIN BB1
Engine: 2.3L I4 DOHC SMFI (H23A1) ....VIN BB2

1996 Honda Applications

Accord Coupe, Sedan & Wagon (DX, EX, LX, LX w/ABS)
Engine: 2.2L I4 SOHC MFI (F22B1 VTEC, F22B2) ....VIN CD5
Engine: 2.2L I4 SOHC MFI (F22B1 VTEC, F22B2) ....VIN CD7
Engine: 2.7L V6 SOHC MFI (C27A4) ....VIN CE6
Engine: 2.2L I4 SOHC SMFI (F22B1 VTEC, F22B2) ....VIN CE1
Civic Coupe (DX, DX w/ABS)
Engine: 1.6L I4 SOHC SMFI 16v (D16Y7) ....VIN EJ6
Engine: 1.6L I4 SOHC SMFI 16v VTEC-E (D16Y5) ....VIN EJ7
Engine: 1.6L I4 SOHC SMFI 16v VTEC (D16Y8) ....VIN EJ8
Civic Del Sol (S, Si, VTEC)
Engine: 1.5L I4 DOHC SMFI 16v (VTEC B16A2) ....VIN EG2
Engine: 1.6L I4 SOHC SMFI 16v (D16Y7) ....VIN EH6
Engine: 1.6L I4 SOHC SMFI 16v (VTEC D16&8) ....VIN EH6
Civic Hatchback (CX, DX)
Engine: 1.6L I4 SOHC SMFI 16v (D16Y7) ....VIN EJ6
Civic Sedan (DX, DX w/ABS, LX, LX w/ ABS)
Engine: 1.6L I4 SOHC SMFI 16v (D16Y7) ....VIN EJ6
Engine: 1.6L I4 SOHC SMFI 16v VTEC (D16Y8) ....VIN EJ8
Odyssey (EX, LX)
Engine: 2.2L I4 SOHC SMFI (F22B6) ....VIN RA1
Passport G-Body 2WD, Y-Body 4WD
G-Body 2WD Engine: 2.6L I4 MFI (4ZE1) ....VIN E
G-Body 2WD Engine: 3.2L V6 MFI (6VD1) ....VIN V
Y-Body 4WD Engine: 3.2L V6 MFI (6VD1) ....VIN V
Prelude (S, Si, Si VTEC)
Engine: 2.2L I4 SOHC SMFI (F22A1) ....VIN BA8
Engine: 2.2L I4 DOHC SMFI VTEC (H22A1) ....VIN BB1
Engine: 2.3L I4 DOHC SMFI (H23A1) ....VIN BB2

Vehicle Coverage - Honda Cars & Utility Vehicles

1997 Honda Applications

Accord Coupe, Sedan & Wagon (DX, EX, EX-L)
Coupe Engine: 2.2L I4 SOHC MFI (F22B1 VTEC, F22B2) .... VIN CD5
Sedan Engine: 2.2L I4 SOHC MFI (F22B1 VTEC, F22B2) .... VIN CD7
Sedan Engine: 2.7L V6 SOHC MFI (C27A4) .... VIN CE6
Wagon Engine: 2.2L I4 SOHC SMFI (F22B1 VTEC, F22B2) .... VIN CE1

Civic Coupe (DX, DX w/ABS, HX, HX w/ABS)
Engine: 1.6L I4 SOHC SMFI 16v (D16Y7) .... VIN EJ6
Engine: 1.6L I4 SOHC SMFI 16v VTEC-E (D16Y5) .... VIN EJ7
Engine: 1.6L I4 SOHC SMFI 16v VTEC (D16Y8) .... VIN EJ8

Civic Del Sol (S, Si, Si w/ABS)
Engine: 1.5L I4 DOHC SMFI 16v (VTEC B16A2) .... VIN EG2
Engine: 1.6L I4 SOHC SMFI 16v (D16Y7) .... VIN EH6
Engine: 1.6L I4 SOHC SMFI 16v (VTEC D16Y8) .... VIN EH6

Civic Hatchback (CX, DX, DX w/ABS)
Engine: 1.6L I4 SOHC SMFI 16v (D16Y7) .... VIN EJ6

Civic Sedan (DX, DX w/ABS, LX, LX w/ABS)
Engine: 1.6L I4 SOHC SMFI 16v (D16Y7) .... VIN EJ6
Engine: 1.6L I4 SOHC SMFI 16v VTEC (D16Y8) .... VIN EJ8

CR-V (LX, LX w/ABS)
Engine: 2.0L I4 DOHC SMFI (B20B4) .... VIN RD1

Odyssey (EX, LX)
Engine: 2.2L I4 SOHC SMFI (F22B6) .... VIN RA1

Passport G-Body 2WD, Y-Body 4WD
G-Body 2WD Engine: 2.6L I4 MFI (4ZE1) .... VIN E
G-Body 2WD Engine: 3.2L V6 MFI (6VD1) .... VIN V
Y-Body 4WD Engine: 3.2L V6 MFI (6VD1) .... VIN V

Prelude, Prelude SH
Engine: 2.2L I4 DOHC SMFI VTEC (H22A4) .... VIN BB6

1998 Honda Applications

Accord Coupe & Sedan (DX, EX, EX-L, EX-ULEV)
Engine: 2.3L I4 SOHC MFI VTEC (F23A1) .... VIN CG5
Engine: 2.3L I4 SOHC MFI VTEC (F23A4) .... VIN CG6
Engine: 2.3L I4 SOHC MFI (F23A5) .... VIN CF8
Engine: 3.0L V6 SOHC MFI VTEC (J30A1) .... VIN CG1

Civic Coupe (DX, DX w/ABS, HX, HX w/ABS)
Engine: 1.6L I4 SOHC SMFI 16v (D16Y7) .... VIN EJ6
Engine: 1.6L I4 SOHC SMFI 16v VTEC-E (D16Y5) .... VIN EJ7
Engine: 1.6L I4 SOHC SMFI 16v VTEC (D16Y8) .... VIN EJ8

Civic Hatchback (CX, DX, DX w/ABS)
Engine: 1.6L I4 SOHC SMFI 16v (D16Y7) .... VIN EJ6

Civic Sedan (DX, DX w/ABS, LX, LX w/ABS)
Engine: 1.6L I4 SOHC SMFI 16v (D16Y7) .... VIN EJ6
Engine: 1.6L I4 SOHC SMFI 16v VTEC (D16Y8) .... VIN EJ8

CR-V (LX, LX w/ABS)
Engine: 2.0L I4 DOHC SMFI (B20B4) .... VIN RD1

Odyssey (EX, LX)
Engine: 2.3L I4 SOHC SMFI (F23A7) .... VIN RA3

Passport G-Body 2WD, Y-Body 4WD
G-Body 2WD Engine: 3.2L V6 DOHC SMFI (6VD1) .... VIN W
Y-Body 4WD Engine: 3.2L V6 DOHC SMFI (6VD1) .... VIN W

Prelude Base, Prelude SH
Engine: 2.2L I4 DOHC SMFI 16v VTEC (H22A4) .... VIN BB6

Vehicle Coverage - Honda Cars & Utility Vehicles

1999 Honda Applications

Accord Coupe & Sedan (DX, EX, EX-L, EX-ULEV)
Engine: 2.3L I4 SOHC MFI VTEC (F23A1) .... VIN CG5
Engine: 2.3L I4 SOHC MFI VTEC (F23A4) .... VIN CG6
Engine: 2.3L I4 SOHC MFI (F23A5) .... VIN CF8
Engine: 3.0L V6 SOHC MFI VTEC (J30A1) .... VIN CG1

Civic Coupe (DX, DX w/ABS, HX, HX w/ABS)
Engine: 1.6L I4 SOHC SMFI 16v (D16Y7) .... VIN EJ6
Engine: 1.6L I4 SOHC SMFI 16v VTEC-E (D16Y5) .... VIN EJ7
Engine: 1.6L I4 SOHC SMFI 16v VTEC (D16Y8) .... VIN EJ8

Civic Hatchback (CX, DX, DX w/ABS)
Engine: 1.6L I4 SOHC SMFI 16v (D16Y7) .... VIN EJ6

Civic Sedan (DX, DX w/ABS, LX, LX w/ABS)
Engine: 1.6L I4 SOHC SMFI 16v (D16Y7) .... VIN EJ6
Engine: 1.6L I4 SOHC SMFI 16v VTEC (D16Y8) .... VIN EJ8

CR-V (LX, LX w/ABS)
Engine: 2.0L I4 DOHC SMFI (B20B4) .... VIN RD1

Odyssey (EX, LX)
Engine: 3.5L V6 SOHC SMFI (J35A1) .... VIN RL1

Passport G-Body 2WD, Y-Body 4WD
G-Body 2WD Engine: 3.2L V6 DOHC SMFI (6VD1) .... VIN W
Y-Body 4WD Engine: 3.2L V6 DOHC SMFI (6VD1) .... VIN W

Prelude Base, Prelude SH
Engine: 2.2L I4 DOHC SMFI 16v VTEC (H22A4) .... VIN BB6

2000 Honda Applications

Accord Coupe & Sedan (DX, EX, EX-L, EX-ULEV)
Engine: 2.3L I4 SOHC MFI VTEC (F23A1) .... VIN CG5
Engine: 2.3L I4 SOHC MFI VTEC (F23A4) .... VIN CG6
Engine: 2.3L I4 SOHC MFI (F23A5) .... VIN CF8
Engine: 3.0L V6 SOHC MFI VTEC (J30A1) .... VIN CG1

Civic Coupe (DX, DX w/ABS, HX, HX w/ABS)
Engine: 1.6L I4 SOHC SMFI 16v (D16Y7) .... VIN EJ6
Engine: 1.6L I4 SOHC SMFI 16v VTEC-E (D16Y5) .... VIN EJ7
Engine: 1.6L I4 SOHC SMFI 16v VTEC (D16Y8) .... VIN EJ8

Civic Hatchback (CX, DX, DX w/ABS)
Engine: 1.6L I4 SOHC SMFI 16v (D16Y7) .... VIN EJ6

Civic Sedan (DX, DX w/ABS, LX, LX w/ABS)
Engine: 1.6L I4 SOHC SMFI 16v (D16Y7) .... VIN EJ6
Engine: 1.6L I4 SOHC SMFI 16v VTEC (D16Y8) .... VIN EJ8

CR-V (LX, LX w/ABS)
Engine: 2.0L I4 DOHC SMFI (B20B4) .... VIN RD1

Odyssey (EX, LX)
Engine: 3.5L V6 SOHC SMFI (J35A1) .... VIN RL1

Passport G-Body 2WD, Y-Body 4WD
G-Body 2WD Engine: 3.2L V6 DOHC SMFI (6VD1) .... VIN W
Y-Body 4WD Engine: 3.2L V6 DOHC SMFI (6VD1) .... VIN W

Prelude Base, Prelude SH
Engine: 2.2L I4 DOHC SMFI 16v VTEC (H22A4) .... VIN BB6

Insight
Engine: 1.0L I3 SOHC MFI (ECA1) .... VIN ZE1

S2000
Engine: 2.0L I4 DOHC MFI (F20C1) .... VIN AP1

Vehicle Coverage - Honda Cars & Utility Vehicles

2001 Honda Applications	
Accord Coupe & Sedan (DX, EX, EX-L, EX-ULEV)	
Engine: 2.3L I4 SOHC MFI VTEC (F23A1)	VIN CG5
Engine: 2.3L I4 SOHC MFI VTEC (F23A4)	VIN CG6
Engine: 2.3L I4 SOHC MFI (F23A5)	VIN CF8
Engine: 3.0L V6 SOHC MFI VTEC (J30A1)	VIN CG1
Civic Coupe (DX, DX w/ABS, HX, HX w/ABS)	
Engine: 1.7L I4 SOHC SMFI 16v (D17A1)	VIN ES1
Engine: 1.7L I4 SOHC SMFI 16v VTEC (D17A2)	VIN ES2
Engine: 1.7L I4 SOHC SMFI 16v (D17A7)	VIN EN2
Civic Hatchback (CX, DX, DX w/ABS)	
Engine: 1.7L I4 SOHC SMFI 16v (D17A1)	VIN ES1
Civic Sedan (DX, DX w/ABS, LX, LX w/ABS)	
Engine: 1.7L I4 SOHC SMFI 16v (D17A1)	VIN ES1
Engine: 1.7L I4 SOHC SMFI 16v VTEC (D17A2)	VIN ES2
CR-V (LX, LX w/ABS)	
Engine: 2.0L I4 DOHC SMFI (B20B4)	VIN RD1
Odyssey (EX, LX)	
Engine: 3.5L V6 SOHC SMFI (J35A1)	VIN RL1
Passport G-Body 2WD, Y-Body 4WD	
G-Body 2WD Engine: 3.2L V6 DOHC SMFI (6VD1)	VIN W
Y-Body 4WD Engine: 3.2L V6 DOHC SMFI (6VD1)	VIN W
Prelude Base, Prelude SH	
Engine: 2.2L I4 DOHC SMFI 16v VTEC (H22A4)	VIN BB6
Prelude Base, Prelude SH	
Engine: 2.2L I4 DOHC SMFI 16v VTEC (H22A4)	VIN BB6
Insight	
Engine: 1.0L I3 SOHC MFI (ECA1)	VIN ZE1
S2000	
Engine: 2.0L I4 DOHC MFI (F20C1)	VIN AP1
2002 Honda Applications	
Accord Coupe & Sedan (DX, EX, EX-L, EX-ULEV)	
Engine: 2.3L I4 SOHC MFI VTEC (F23A1)	VIN CG5
Engine: 2.3L I4 SOHC MFI VTEC (F23A4)	VIN CG6
Engine: 2.3L I4 SOHC MFI (F23A5)	VIN CF8
Engine: 3.0L V6 SOHC MFI VTEC (J30A1)	VIN CG1
Civic Coupe (DX, DX w/ABS, HX, HX w/ABS)	
Engine: 1.7L I4 SOHC SMFI 16v (D17A1)	VIN ES1
Engine: 1.7L I4 SOHC SMFI 16v VTEC (D17A2)	VIN ES2
Engine: 1.7L I4 SOHC SMFI 16v (D17A7)	VIN EN2
Civic Hatchback (CX, DX, DX w/ABS)	
Engine: 1.7L I4 SOHC SMFI 16v (D17A1)	VIN ES1
Civic Sedan (DX, DX w/ABS, LX, LX w/ABS)	
Engine: 1.7L I4 SOHC SMFI 16v (D17A1)	VIN ES1
Engine: 1.7L I4 SOHC SMFI 16v VTEC (D17A2)	VIN ES2
CR-V (LX, LX w/ABS)	
Engine: 2.0L I4 DOHC SMFI (B20B4)	VIN RD1
Odyssey (EX, LX)	
Engine: 3.5L V6 SOHC SMFI (J35A1)	VIN RL1
Passport G-Body 2WD, Y-Body 4WD	
G-Body 2WD Engine: 3.2L V6 DOHC SMFI (6VD1)	VIN W
Y-Body 4WD Engine: 3.2L V6 DOHC SMFI (6VD1)	VIN W
Insight	
Engine: 1.0L I3 SOHC MFI (ECA1)	VIN ZE1
S2000	
Engine: 2.0L I4 DOHC MFI (F20C1)	VIN AP1

Vehicle Coverage - Honda Cars & Utility Vehicles

2003 Honda Applications	
Accord Coupe & Sedan (DX, EX, EX-L, EX-ULEV)	
Engine: 2.3L I4 SOHC MFI VTEC (F23A1)	VIN CG5
Engine: 2.3L I4 SOHC MFI VTEC (F23A4)	VIN CG6
Engine: 2.3L I4 SOHC MFI (F23A5)	VIN CF8
Engine: 3.0L V6 SOHC MFI VTEC (J30A1)	VIN CG1
Civic Coupe (DX, DX w/ABS, HX, HX w/ABS)	
Engine: 1.7L I4 SOHC SMFI 16v (D17A1)	VIN ES1
Engine: 1.7L I4 SOHC SMFI 16v VTEC (D17A2)	VIN ES2
Engine: 1.7L I4 SOHC SMFI 16v (D17A7)	VIN EN2
Civic Hatchback (CX, DX, DX w/ABS)	
Engine: 1.7L I4 SOHC SMFI 16v (D17A1)	VIN ES1
Civic Sedan (DX, DX w/ABS, LX, LX w/ABS)	
Engine: 1.7L I4 SOHC SMFI 16v (D17A1)	VIN ES1
Engine: 1.7L I4 SOHC SMFI 16v VTEC (D17A2)	VIN ES2
CR-V (LX, LX w/ABS)	
Engine: 2.0L I4 DOHC SMFI (B20B4)	VIN RD1
Odyssey (EX, LX)	
Engine: 3.5L V6 SOHC SMFI (J35A1)	VIN RL1
Passport G-Body 2WD, Y-Body 4WD	
G-Body 2WD Engine: 3.2L V6 DOHC SMFI (6VD1)	VIN W
Y-Body 4WD Engine: 3.2L V6 DOHC SMFI (6VD1)	VIN W
Insight	
Engine: 1.0L I3 SOHC MFI (ECA1)	VIN ZE1
S2000	
Engine: 2.0L I4 DOHC MFI (F20C1)	VIN AP1
Element (DX, EX)	
Engine: FWD 2.4L I4 DOHC MFI (K24A4)	VIN YH1
Engine: 4WD 2.4L I4 DOHC MFI (K24A4)	VIN YH2
Pilot (EX, EX-L, LX)	
Engine: 3.5L V6 SOHC MFI (J35A4)	VIN YF1

PGM-FI DIAGNOSTICS

Diagnostic Routine (OBD I)

INTRODUCTION

Proper diagnosis of an engine and/or computer related problem requires the use of a comprehensive plan of inspection, diagnosis, testing and repair, and a Road Test.

While there are many different time-tested repair strategies used by successful automotive repair technicians for diagnosis and repair of late model vehicles, all of these plans share some common steps. However, the sequence used to perform these steps seems to vary between repair technicians. A suggested procedure for Honda vehicles is provided below.

VISUAL INSPECTION

*   Inspect for disconnected wires or vacuum lines, check fluid levels

PRELIMINARY STEPS

*   Warm engine until cooling fan cycles, check idle speed and base timing

RETRIEVE DIAGNOSTIC DATA

*   Read, record and clear any trouble codes
*   If codes are set, start the vehicle in order to determine the code resets

FUEL SYSTEM CONTROL

*   Determine if the PCM has "control" of the fuel feedback system
*   Monitor the PCM and O_2 sensor at both idle and off idle engine speeds
*   Zirconia O_2 (18mm) sensor: should switch rich to lean in less than 100ms
*   Titania O_2 (14mm) sensor: should switch rich to lean in less than 35ms

DRIVEABILITY SYMPTOMS

*   If no codes are set, determine if a driveability symptom exists
*   If a symptom exists, verify that it is repeatable (not intermittent in nature)
*   Decide which engine subsystem is most likely the cause of the problem (i.e., the Base Engine, Emission Controls, Fuel or Ignition system)

INTERMITTENT FAULTS

*   If a code fault or driveability symptom cannot be easily verified, it may be an intermittent fault (these types of faults require special test procedures)

DRIVEABILITY PROBLEMS - CHECK THE EASY THINGS FIRST!

When attempting to solve a driveability problem, remember to check the easy things first. In other words, look for obvious causes of the complaint. Check for loose connections, computer or output device ground problems and lack of proper fuel control by the PCM and oxygen sensor.

TROUBLE CODE & SYMPTOM REPAIR CHARTS

It is important to understand that code repair charts and symptom-based repair charts do not include specific checks for faults *that should be located and repaired during an inspection of the Base Engine components*.

BUILD A DIAGNOSTIC FOUNDATION

The use of a repeatable routine can help to quickly get to the cause of a customer complaint, save diagnostic time and result in a higher percentage of properly repaired vehicles. Begin by reviewing the repair order and verifying the actual customer complaint. If a customer is complaining of a driveability symptom, the symptom must be present on the vehicle at the time of testing.

NOTE: ***The customer needs to describe or demonstrate the problem to the technician. No one can fix a problem they cannot duplicate or verify.***

DIAGNOSTIC FLOW CHART (OBD I)

This flow chart can be used as a quick reference to help keep on track during diagnosis of engine or engine computer faults on Honda vehicles with OBD I.

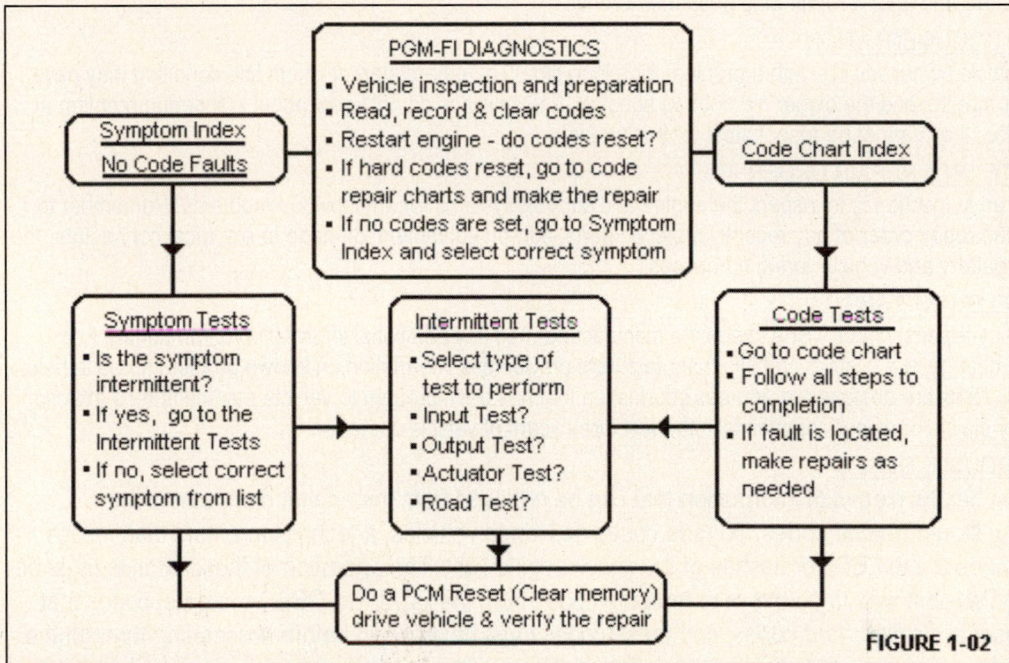

FIGURE 1-02

FLOW CHART STEPS

Here are the steps included in the Acura/Honda PGM-FI Diagnostics Routine:

- Review the repair order
- Perform a visual Inspection
- Verify the customer complaint (perform a Drive Cycle or Road Test)
- If the engine will not start (No Start symptom), follow the correct No Start diagnostic routine found in this manual
- Repair any stored codes before repairing a driveability symptom
- Test base engine areas (engine mechanical, fuel and ignition test areas)
- Check all related technical service bulletins (TSBs)
- If the problem is intermittent in nature, use the special Intermittent Tests in this manual

CHECK BASE ENGINE AREAS

Depending on the symptom, check or test the Battery and Charging system, Engine Mechanical, Emission Controls, Fuel and Ignition systems for proper operation. These systems are considered to be non-computer checks. Problems in these systems can result in a situation where the PCM sets a false code or can cause problems related to incorrect readings from sensors or output devices. Verify that all of the engine base settings such as base timing, minimum idle speed and TP sensor voltage are set to specification.

ROAD TESTS - WHY ARE THEY IMPORTANT?

If possible, perform a Road Test to verify the problem before any repairs. This step also helps a technician identify any other problems that could cause confusion between the customer and a technician after a repair is done due to a new symptom. Also, if a customer feels the vehicle is not performing correctly, a technician will have a sense of how the vehicle drove before and after a repair as a point of reference.

GOOD REPAIR DOCUMENTATION

Sometimes a vehicle owner will live with a problem for a long time (i.e., a hard start or rough idle condition may have been present for months and the owner did nothing about it). If technicians do not know about a lingering problem at the time of service, they cannot be responsible for this problem.

WHAT TO WRITE ON A REPAIR ORDER

The first step of an inspection is to inspect the engine and support systems for any obvious problems. Remember to make notes on the repair order of any recently replaced parts, add-on equipment, damage to electrical connectors, the condition of the battery and vehicle wiring harnesses.

CHECK FOR RELEVANT TSBS

Before starting any repairs, check TSBs to see if a manufacturer has any additional information on an update to a repair chart or a change to a system. Vehicle manufacturers provide this information on known problem areas to help fix their vehicles. TSBs are designed to address information that has been brought to vehicle manufacturer's attention through warranty claims or owner complaints in the first three years of vehicle operation.

CHECK FOR TROUBLE CODES

One of the most helpful pieces of information that can be retrieved from the engine computer is the presence of any stored trouble codes. To read codes on Honda vehicles, a technician enters diagnostics and reads changes to the LEDs or flashes of the check engine light. The operation of these circuits must be verified, and the easiest way to do this is to read the codes from the PCM. On OBD I systems, codes that immediately reset are called *hard codes*, and these codes must be repaired before proceeding through the Diagnostic Routine. The opposite of a hard code is a *soft code*. This is a code that does not easily reset, and these codes are treated as intermittent faults.

Driveability Symptom List

This page contains a list of symptom descriptions. To use the list, locate the symptom that matches a problem and refer to the suggested test areas. In some cases, the items listed under each symptom may not apply to all models, engines or vehicle systems. These procedures usually recommend testing a component or system on a vehicle.

Acura/Honda Symptom List

Symptom Description	Suggested Test Areas
Test 1 - Starting Concerns • No Crank • Hard Start, Long Crank, Erratic Crank • Stall After Start • No Start, Normal Crank • No Start, MIL is off due to VREF short to ground	• Check battery, battery circuits to starter • Check for a damaged flywheel, engine compression, base timing and idle speed • Check for a failed PGM-FI Main Relay • Check for distributor rotor 'punch-through' • Check for a faulty igniter unit • Check for VREF circuit shorted to ground
Test 2 Idle Concerns • Slow Return to Idle • Rolling Idle • Fast Idle • Low or Slow Idle	• Check for engine vacuum leaks • Check PCV valve, for excessive carbon • Check for a restricted exhaust • Check base idle speed, fuel pressure • Check throttle linkage for binding
Test 3 - Stalls, Quits Running • At idle • During acceleration • At cruise • At deceleration	• Check for engine vacuum leaks • Check PCV valve, for excessive carbon • Check for a restricted exhaust • Check base idle speed, fuel pressure • Check the throttle linkage for binding
Test 4 - Runs Rough • At idle • During acceleration • At cruise	• Check for vacuum leaks (intake manifold) • Check Ignition secondary components • Check base timing and idle speed setting • Check fuel pressure and fuel injectors • Check for excessive carbon buildup
Test 5 - Cuts-out, Misses • At idle • During acceleration • At cruise	• Check for vacuum leaks (intake manifold) • Check Ignition secondary components • Check fuel pressure and fuel injectors • Check for restricted exhaust
Test 6 - Bucks, Jerks • During acceleration • At cruise • During deceleration	• Check for vacuum leaks (intake manifold) • Check Ignition secondary components • Check fuel pressure and fuel injectors • Check for restricted exhaust
Test 7 - Backfires • At idle • During acceleration • At deceleration	• Check for vacuum leaks (intake manifold) • Check Ignition secondary components • Check the PCV valve and system • Check for engine mechanical parts
Test 8 - Lack Of Power, Loss Of Power • During acceleration • At cruise	• Check for vacuum leaks (intake manifold) • Check Ignition secondary components • Check fuel pressure and fuel injectors • Check for restricted exhaust
Test 9 - Spark Knock • During acceleration • At cruise	• Check base timing, ignition advance • Check Ignition secondary components • Check the PCV valve and system • Check for engine mechanical parts
Test 10 - Other Concerns • Diesels or Runs On • Poor Fuel Economy • Emissions Compliance • DTC Received During Emissions Test	• Check base timing, for restricted exhaust • Check fuel pressure, fuel injectors • Check ignition components, PCV system • Check cooling system, exhaust system and catalytic converter, AIR system
Test 11 - Engine • Base Engine	• Check engine mechanical components • Check camshaft and valve timing

Symptom Test 1 (Starting Concerns)

NOTE: *If the vehicle cranks and will not start, check the rotor for "punch-through" as this is a common failure area. Also test for a failed PGM-FI main relay (especially if there is no spark or fuel pressure). Check for a shorted ECT sensor (intermittent).*

PRELIMINARY CHECKS

Prior to starting this symptom test routine, inspect these underhood items:

* Battery charge and condition, starter current draw
* Verify engine cranks (turns over), and verify starter relay operation
* Check Air Intake system for restrictions (air inlet tubes, dirty air filter, etc.)

Symptom Test 1 - Step 1

STEP	ACTION	YES	NO
1	**Step description:** *No Start Condition only* • Check battery cables, state of charge. • **Check PGM-FI main relay operation.** • Inspect for a blown fuse (or a fusible link). • Does the engine crank normally?	If yes, go to Step 2.	If no, make needed repairs to the battery, starter, related circuits or engine mechanical condition.
2	**Step description:** *Test ignition secondary* • Inspect ignition secondary components for damage (**rotor punch-through**). • Check spark output with spark tester, use an engine analyzer to test ignition circuits. • Are ignition system faults suspected?	If yes, make any needed repairs to the ignition secondary. Then do a PCM reset and retest the system.	If no, go to Step 3.
3	**Step description:** *Test fuel System* • Inspect fuel delivery system for leaks. • Are any fuel system faults suspected?	If yes, make repairs to fuel system. Then do a PCM reset and retest.	If no, go to Step 4.
4	**Step description:** *Check exhaust system* • Check for leaking/damaged components. • Test exhaust system for leaks, or restriction (refer to Section One). • Is an exhaust restriction suspected?	If yes, locate the exhaust restriction, make any needed repairs. Then do a PCM reset and retest.	If no, go to Step 5.
5	**Step description:** *Check for hot engine* • Check for signs of an engine overheating condition (one that is hard to start). • Is the engine overheating?	If yes, make needed repairs to repair the overheating problem, retest for the symptom.	If no, go to Step 6.
6	**Step description:** *Check PCV System* • Inspect PCV system components for broken parts or loose connections. • Test PCV valve operation. • Are any PCV system faults suspected?	If yes, go to the Acura or Honda PCV system tests. Locate and repair the fault in the PCV system.	If no, go to Step 7.
7	**Step description:** *Check EVAP system* • Inspect for disconnected or damaged EVAP system components. • Are any EVAP system faults suspected?	If yes, go to Acura or Honda EVAP System tests, make repairs, retest the symptom.	If no, go to Step 8
8	**Step description:** *Inspect and test engine mechanical condition* • Test engine compression. • Test valve timing and timing chain timing. • Check for worn camshaft or valve train. • Check for manifold gasket leaks. • Are any Base Engine faults present?	If yes, make the needed repairs. Then do a PCM reset and retest the system.	If no, repeat the test steps from the beginning to locate and repair the starting concern.

Symptom Test 2 (Idle Concerns)

NOTE: *If the vehicle has a rough idle and the base idle speed and EACV or IAC valve are okay, check the engine for excessive carbon buildup.*

PRELIMINARY CHECKS

Prior to starting this symptom test routine, inspect these underhood items:

- All related vacuum lines for proper routing and integrity
- All related electrical connectors and wiring harnesses for faults
- Air Intake system for restrictions (air inlet tubes, dirty air filter, etc.)
- Intake manifold and components for leaks (EGR valve, EACV assembly)

Symptom Test 2 - Step 1

STEP	ACTION	YES	NO
1	**Step description:** *Verify the low idle speed condition* • Does the engine have a warm engine low idle speed condition in Park or Neutral	If yes (meaning the engine does have a low speed or rough idle problem), got to step 2.	If no, problem is not present at this time. Refer to Symptom List and select a symptom.
2	**Step description:** *Check for stall or near stall with the EACV motor disconnected* • Start engine at part throttle in P/N. • Does the engine start and run smoothly at part throttle in Park or Neutral?	If yes, repair the Idle Speed system.	If no, go to Step 3.
3	**Step description:** *Check/compare actual pin voltages to the values in pin voltage charts.* • Connect a Honda Test Harness or BOB. • Turn off all accessories. • Start engine and allow it to warmup. • Use a DVOM to read pin voltages that could affect idle speed (i.e., EACV, power and logic grounds). • Are all pin voltages within their normal range?	If yes (meaning all pin voltages are okay), go to Step 4.	If no, (meaning that one or more of the pin voltages are out of normal range), go to the Idle Speed system to make the needed repairs.
4	**Step description:** *Check ignition secondary* • Inspect/test ignition secondary components. • Check spark output with spark tester, use an engine analyzer to check ignition secondary. • Are ignition system faults suspected?	If yes, make any needed repairs to the ignition secondary. Then do a PCM reset and retest the system.	If no, go to Step 5.
5	**Step description:** *Check fuel delivery system* • Inspect fuel delivery system for leaks. • Are any fuel system faults suspected?	If yes, repair the fuel system. Then do a PCM reset and retest.	If no, go to Step 6.
6	**Step description:** *Check exhaust system* • Inspect for leaking or damaged components. • Test exhaust system for leaks, or restriction (refer to Catalyst & Exhaust system tests). • Is an exhaust restriction suspected?	If yes, locate the exhaust restriction, make any needed repairs. Then do a PCM reset and retest.	If no, go to Step 7.
7	**Step description:** *Check PCV valve & system* • Inspect PCV system components for broken parts, loose connections or disconnects. • Are any PCV system faults suspected?	If yes, go to the Acura or Honda PCV system tests. Make repairs and retest the system.	If no, go to Step 8.
8	**Step description:** *Check the EVAP system* • Inspect for disconnected or damaged EVAP system components. • Are any EVAP system faults suspected?	If yes, go to Acura or Honda EVAP System tests. Make repairs and retest the system.	If no, test Base Engine (compression, valve timing) for cause of the symptom.

Symptom Test 3 (Stalls, Quits Running)

NOTE: *If the vehicle stalls or quits running and the base idle speed and EACV or IAC valve are okay, check the engine for excessive carbon buildup.*

<u>PRELIMINARY CHECKS</u>

Prior to starting this symptom test routine, inspect these underhood items:

- All related vacuum lines for proper routing and integrity
- All related electrical connectors and wiring harnesses for faults
- Air Intake system for restrictions (air inlet tubes, dirty air filter, etc.)
- Intake manifold and components for leaks (EGR valve, EACV assembly)

Symptom Test 3 - Step 1

STEP	ACTION	YES	NO
1	**Step description:** *Verify that the engine stalls* • Start the engine and allow it to idle in P/N. • Does the engine stall or almost stall at idle speed in Park or Neutral?	If yes, check for any trouble codes. Are any codes set? If yes, go to Trouble Code Lists. If no codes are set, go to Step 4.	If no (meaning the engine does not stall or almost stall), go to Step 2.
2	**Step description:** *Check for a warm engine rough idle condition* • Does the engine have a warm engine rough idle condition in Park or Neutral?	If yes (meaning the engine does have a rough idle condition), got to step 4.	If no, recheck for any codes. If codes are set, go to Code Index. If no, go to Step 3.
3	**Step description:** *Check/compare actual pin voltages to the values in pin voltage charts.* • Connect a Honda Test Harness or BOB. • Turn off all accessories. • Start the engine and allow it to warmup until the electric cooling fan cycles. • Use a DVOM to read pin voltages that could affect idle speed (i.e., EACV, power and logic grounds). • Are all pin voltages within their normal range?	If yes (meaning all pin voltages are okay), go to Step 6.	If no, (meaning that one or more of the pin voltages are out of normal range), go to the Idle Speed system.
4	**Step description:** *Check for stall or near stall with the EACV motor disconnected* • Start engine at part throttle in P/N. • Does the engine start and run smoothly at part throttle in Park or Neutral?	If yes, go to the Idle Speed system to make the needed repairs.	If no, go to Step 6.
5	**Step description:** *Check EACV operation* • Start the engine and allow it to idle. • Disconnect the EACV connector. • Check for rpm drop or engine stall. • After testing, turn key off and reconnect the EACV. • Did the rpm drop or engine stall with the EACV connector removed?	If yes, go to Step 3 to read and record the appropriate pin chart values. Refer to the Pin Charts in the appropriate section for example Pin Charts.	If no, go to the Idle Speed system to make the needed repairs.
6	**Step description:** *Check ignition secondary* • Inspect the ignition secondary for damaged or defective components. • Check spark output with spark tester, use an engine analyzer to check ignition secondary. • Are ignition system faults suspected?	If yes, make any needed repairs to the ignition secondary. Then do a PCM reset and retest the system.	If no, go to Step 7.

Symptom Test 3 (Stalls, Quits Running)

Symptom Test 3 - Step 7

STEP	ACTION	YES	NO
7	**Step description:** *Check Fuel delivery system* • Inspect fuel delivery system for leaks. • Are any fuel system faults suspected?	If yes, make needed repairs to the fuel system. Then do a PCM reset and retest.	If no, go to Step 8.
8	**Step description:** *Check Exhaust system* • Inspect for leaking or damaged components. • Test the exhaust system for leaks, damage or an exhaust restriction (refer to Catalyst & Exhaust system tests in this section). • Is an exhaust restriction suspected?	If yes, locate the exhaust restriction, make any needed repairs. Then do a PCM reset and retest.	If no, go to Step 9.
9	**Step description:** *Check PCV valve & system* • Inspect PCV system components for broken parts, loose connections or disconnects. • Test PCV valve operation. • Are any PCV system faults suspected?	If yes, go to the Acura or Honda PCV system tests. Locate and repair the fault in the PCV system. The do a PCM reset and retest.	If no, go to Step 10.
10	**Step description:** *Check Secondary AIR system* • Inspect the Secondary AIR components for any broken parts, loose connections or disconnects. • Test the Secondary AIR system. • Are Secondary Air faults suspected?	If yes, go to Acura Secondary Air system tests. Locate the fault and make any needed repairs. Then do a PCM reset and retest.	If no, go to Step 11.
11	**Step description:** *Check the EVAP system* • Inspect for disconnected or damaged EVAP system components. • Are any EVAP system faults suspected?	If yes, go to the Acura or Honda EVAP System tests. Locate the fault and make any needed repairs. Then do a PCM reset and retest the system.	If no, test Base Engine items and Air Intake system in Acura or Honda articles. Test the transaxle or transmission operation and controls for faults.

Symptom Test 4 (Runs Rough)

NOTE: *If the vehicle has a runs rough condition, and all engine subsystems are okay after testing, check the engine for excessive carbon buildup.*

<u>PRELIMINARY CHECKS</u>

Prior to starting this symptom test routine, inspect these underhood items:

- All related vacuum lines for proper routing and integrity
- All related electrical connectors and wiring harnesses for faults
- Air Intake system for restrictions (air inlet tubes, dirty air filter, etc.)
- Intake manifold and components for leaks (EGR valve, EACV assembly)

Symptom Test 4 - Step 1

STEP	ACTION	YES	NO
1	**Step description:** *Verify the engine runs rough* • Start the engine and allow it to idle in P/N. • Does the engine run rough in Park or Neutral?	If yes, check for any trouble codes. Are any codes set? If yes, go to Trouble Code Lists. If there are no codes set, go to Step 3.	If no, go to Step 2.
2	**Step description:** *The rough running condition is not present at this time.* • Inspect various underhood items that could cause a rough running condition (i.e., EACV or IAC valve connections, ignition wiring, etc.). • Where any problems located?	If yes, correct the problems. Do a PCM reset and verify the rough idle condition is repaired.	If no (meaning that no obvious connection or other underhood problem were located, go to Step 3.
3	**Step description:** *Check for a runs rough condition with the EACV motor disconnected* • Start the engine and allow it to idle. • Disconnect the EACV connector. • Check for a rough running condition. • After testing, turn engine off and reconnect the EACV connector. • Did the engine run rough with the EACV or IAC connector removed?	If yes, go to Step 4 to read and record the appropriate pin chart values. Refer to the pin values in the Pin Charts.	If no, go to the Idle Speed system to make the needed repairs.
4	**Step description:** *Check/compare actual pin voltages to the values in pin voltage charts.* • Connect a Honda Test Harness or BOB. • Turn off all accessories. • Start the engine and allow it to warmup until the electric cooling fan cycles. • Use a DVOM to read pin voltages that could affect idle speed control (i.e., EACV, power grounds, engine coolant temperature, etc.). • Are all pin voltages within their normal range?	If yes (meaning all pin voltages are okay), go to Step 5.	If no, (meaning that one or more of the pin voltages are out of normal range), go to the engine subsystem or component related to the circuit that is out of range and repair the component or circuit.
5	**Step description:** *Check Ignition secondary* • Inspect the ignition secondary for damaged or defective components (is the rotor leaking?). • Check spark output with spark tester, use an engine analyzer to check ignition secondary. • Are any ignition system faults suspected?	If yes, make any needed repairs to the ignition secondary. Then do a PCM reset and retest the system.	If no, go to Step 6.
6	**Step description:** *Check Fuel delivery system* • Inspect fuel delivery system for leaks or low fuel pressure (check the regulator). • Are any fuel system faults suspected?	If yes, make needed repairs to the fuel system. Then do a PCM reset and retest.	If no, go to Step 7.

Symptom Test 4 (Runs Rough)

Symptom Test 4 - Step 6

STEP	ACTION	YES	NO
7	**Step description:** *Check Exhaust system* • Inspect for leaking or damaged components. • Test the exhaust system for leaks, damage or an exhaust restriction (refer to Catalyst & Exhaust system tests in this section). • Is an exhaust restriction suspected?	If yes, locate the exhaust restriction, make any needed repairs. Then do a PCM reset and retest.	If no, go to Step 8.
8	**Step description:** *Check PCV valve & system* • Inspect PCV system components for broken parts, loose connections or disconnects. • Test PCV valve operation. • Are any PCV system faults suspected?	If yes, go to the Acura or Honda PCV system tests. Locate and repair the fault in the PCV system. The do a PCM reset and retest.	If no, go to Step 9.
9	**Step description:** *Check Secondary AIR system* • Inspect the Secondary AIR components for any broken parts, loose connections or disconnects. • Test the Secondary AIR system. • Are Secondary Air faults suspected?	If yes, go to Acura Secondary Air system tests. Locate the fault and make any needed repairs. Then do a PCM reset and retest.	If no, go to Step 10.
10	**Step description:** *Check the EVAP system* • Inspect for disconnected or damaged EVAP system components. • Are any EVAP system faults suspected?	If yes, go to the Acura or Honda EVAP System tests. Locate the fault and make any needed repairs. Then do a PCM reset and retest the system.	If no, test Base Engine items and Air Intake system in Acura or Honda articles. Test the transaxle or transmission operation and controls for faults.

Symptom Test 5 (Cuts-out, Misses)

NOTE: *If the vehicle has a cuts-out or misses condition, and all engine subsystems are okay after testing, check the engine for excessive carbon buildup.*

PRELIMINARY CHECKS

Prior to starting this symptom test routine, inspect these underhood items:

- All related vacuum lines for proper routing and integrity
- All related electrical connectors and wiring harnesses for faults
- Air Intake system for restrictions (air inlet tubes, dirty air filter, etc.)
- Intake manifold and components for leaks (EGR valve, EACV assembly)
- Ignition secondary components (coil, coil wire, spark plugs and wires)
- Check Fuel Quality (contaminated fuel, wrong octane level or blend)

Symptom Test 5 - Step 1

STEP	ACTION	YES	NO
1	**Step description:** *Verify the cuts-out or misses condition* • Start the engine and allow it to idle in P/N. • Does the engine have a cuts-out or misses condition?	If yes, check for any trouble codes. Are any codes set? If yes, go to Trouble Code Lists. If there are no codes set, go to Step 3.	If no, go to Step 2.
2	**Step description:** *The cuts-out or missing condition is not present at this time.* • Inspect various underhood items that could cause a cuts-out or missing condition (i.e., the igniter, the fuel system or ignition secondary). • Where any problems located?	If yes, correct the problems. Do a PCM reset and verify that the cuts-out or misses condition is repaired.	If no (meaning that no obvious problems were found in the fuel or ignition system), go to Step 3.
3	**Step description:** *Check/compare actual pin voltages to the values in pin voltage charts.* • Connect a Honda Test Harness or BOB. • Turn off all accessories. • Start the engine and allow it to warmup until the electric cooling fan cycles. • Use a DVOM to read pin voltages that could affect idle speed control (i.e., EACV, power grounds, engine coolant temperature, etc.). • Are all pin voltages within their normal range?	If yes (meaning all pin voltages are okay), go to Step 4	If no, (meaning that one or more of the pin voltages are out of normal range), go to the engine subsystem or component related to the circuit that is out of range and repair the component or circuit.
4	**Step description:** *Check Ignition secondary* • Inspect the ignition secondary for damaged or defective components (is the rotor leaking?). • Check spark output with spark tester, use an engine analyzer to check ignition secondary. • Are any ignition system faults suspected?	If yes, make any needed repairs to the ignition secondary. Then do a PCM reset and retest the system.	If no, go to Step 5
5	**Step description:** *Check Fuel delivery system* • Inspect fuel delivery system for leaks or low fuel pressure (check the regulator). • Are any fuel system faults suspected?	If yes, make needed repairs to the fuel system. Then do a PCM reset and retest.	If no, go to Step 6
6	**Step description:** *Check Exhaust system* • Inspect for leaking or damaged components. • Test the exhaust system for leaks, damage or an exhaust restriction (refer to Catalyst & Exhaust system tests in this section). • Is an exhaust restriction suspected?	If yes, locate the exhaust restriction, make any needed repairs. Then do a PCM reset and retest.	If no, go to Step 7

Symptom Test 5 (Cuts-out, Misses)

Symptom Test 5 - Step 7

STEP	ACTION	YES	NO
7	**Step description:** *Check PCV valve & system* • Inspect PCV system components for broken parts, loose connections or disconnects. • Test PCV valve operation. • Are any PCV system faults suspected?	If yes, go to the Acura or Honda PCV system tests. Locate and repair the fault in the PCV system. The do a PCM reset and retest.	If no, go to Step 8
8	**Step description:** *Check Secondary AIR system* • Inspect the Secondary AIR components for any broken parts, loose connections or disconnects. • Test the Secondary AIR system. • Are Secondary Air faults suspected?	If yes, go to Acura Secondary Air system tests. Locate the fault and make any needed repairs. Then do a PCM reset and retest.	If no, go to Step 9
9	**Step description:** *Check the EVAP system* • Inspect for disconnected or damaged EVAP system components. • Are any EVAP system faults suspected?	If yes, go to the Acura or Honda EVAP System tests. Locate the fault and make any needed repairs. Then do a PCM reset and retest the system.	If no, test Base Engine items and Air Intake system in Acura or Honda articles. Test the transaxle or transmission operation and controls for faults.

Symptom Test 6 (Bucks or Jerks)

NOTE: *If the vehicle has a bucks or jerks condition, and all engine subsystems are okay after testing, check the engine for excessive carbon buildup.*

PRELIMINARY CHECKS

Prior to starting this symptom test routine, inspect these underhood items:

- All related vacuum lines for proper routing and integrity
- All related electrical connectors and wiring harnesses for faults
- Air Intake system for restrictions (air inlet tubes, dirty air filter, etc.)
- Intake manifold and components for leaks (EGR valve, EACV assembly)
- Ignition secondary components (coil, coil wire, spark plugs and wires)
- Check Fuel Quality (contaminated fuel, wrong octane level or blend)

Symptom Test 6 - Step 1

STEP	ACTION	YES	NO
1	**Step description:** *Verify the bucks or jerks condition* • Start the engine and allow it to idle in P/N. • Does the engine have a bucks or jerks condition?	If yes, check for any trouble codes. Are any codes set? If yes, go to Trouble Code Lists. If there are no codes set, go to Step 3.	If no, go to Step 2.
2	**Step description:** *The bucks or jerks condition is not present at this time.* • Inspect various underhood items that could cause a bucks or jerks condition (i.e., the igniter, the fuel system or ignition secondary). • Where any problems located?	If yes, correct the problems. Do a PCM reset and verify that the cuts-out or misses condition is repaired.	If no (meaning that no obvious problems were found in the fuel or ignition system), go to Step 3.
3	**Step description:** *Check/compare actual pin voltages to the values in pin voltage charts.* • Connect a Honda Test Harness or BOB. • Turn off all accessories. • Start the engine and allow it to warmup until the electric cooling fan cycles. • Use a DVOM to read pin voltages that could affect idle speed control (i.e., EACV, power grounds, engine coolant temperature, etc.). • Are all pin voltages within their normal range?	If yes (meaning all pin voltages are okay), go to Step 4	If no, (meaning that one or more of the pin voltages are out of normal range), go to the engine subsystem or component related to the circuit that is out of range and repair the component or circuit.
4	**Step description:** *Check Ignition secondary* • Inspect the ignition secondary for damaged or defective components (is the rotor leaking?). • Check spark output with spark tester, use an engine analyzer to check ignition secondary. • Are any ignition system faults suspected?	If yes, make any needed repairs to the ignition secondary. Then do a PCM reset and retest the system.	If no, go to Step 5
5	**Step description:** *Check Fuel delivery system* • Inspect fuel delivery system for leaks or low fuel pressure (check the regulator). • Are any fuel system faults suspected?	If yes, make needed repairs to the fuel system. Then do a PCM reset and retest.	If no, go to Step 6
6	**Step description:** *Check Exhaust system* • Inspect for leaking or damaged components. • Test the exhaust system for leaks, damage or an exhaust restriction (refer to Catalyst & Exhaust system tests in this section). • Is an exhaust restriction suspected?	If yes, locate the exhaust restriction, make any needed repairs. Then do a PCM reset and retest.	If no, go to Step 7

Symptom Test 6 (Bucks or Jerks)

Symptom Test 6 - Step 7

STEP	ACTION	YES	NO
7	**Step description:** *Check PCV valve & system* • Inspect PCV system components for broken parts, loose connections or disconnects. • Test PCV valve operation. • Are any PCV system faults suspected?	If yes, go to the Acura or Honda PCV system tests. Locate and repair the fault in the PCV system. The do a PCM reset and retest.	If no, go to Step 8
8	**Step description:** *Check Secondary AIR system* • Inspect the Secondary AIR components for any broken parts, loose connections or disconnects. • Test the Secondary AIR system. • Are Secondary Air faults suspected?	If yes, go to Acura Secondary Air system tests. Locate the fault and make any needed repairs. Then do a PCM reset and retest.	If no, go to Step 9
9	**Step description:** *Check the EVAP system* • Inspect for disconnected or damaged EVAP system components. • Are any EVAP system faults suspected?	If yes, go to the Acura or Honda EVAP System tests. Locate the fault and make any needed repairs. Then do a PCM reset and retest the system.	If no, test Base Engine items and Air Intake system in Acura or Honda articles. Test the transaxle or transmission operation and controls for faults.

Symptom Test 7 (Backfires)

NOTE: *If the vehicle has a backfire condition, and all engine subsystems are okay after testing, check the engine for excessive carbon buildup.*

PRELIMINARY CHECKS

Prior to starting this symptom test routine, inspect these underhood items:

*   All related vacuum lines for proper routing and integrity
*   All related electrical connectors and wiring harnesses for faults
*   Ignition secondary components (coil, coil wire, spark plugs and wires)
*   Spark plug wires for proper routing and for correct firing order
*   Verify that engine backfires at idle or during acceleration or deceleration

Symptom Test 7 - Step 1

STEP	ACTION	YES	NO
1	**Step description:** *Read & Record Codes* • Perform complete visual inspection, make any needed repairs. • Read and record all trouble codes. • Were any codes recorded?	If yes, go to Code List, select the code and make the repair.	If no, go to Step 2.
2	**Step description:** *Check Ignition secondary* • Inspect the ignition secondary for damaged or defective components. • Check spark output with spark tester, use an engine analyzer to check ignition secondary. • Are ignition system faults suspected?	If yes, make any needed repairs to the ignition secondary. Then do a PCM reset and retest the system.	If no, go to Step 3.
3	**Step description:** *Check Fuel delivery system* • Inspect fuel delivery system for leaks. • Are any fuel system faults suspected?	If yes, make needed repairs to the fuel system. Then do a PCM reset and retest.	If no, go to Step 4.
4	**Step description:** *Inspect and test engine mechanical condition* • Test engine compression. • Test valve timing and timing chain condition. • Check for worn camshaft or valve train. • Check for manifold gasket leaks. • Check for intake vacuum leaks. • Are any Base Engine faults present?	If yes, make the needed repairs. Then do a PCM reset and retest the system.	If no, go to Step 5.
5	**Step description:** *Check Exhaust system* • Inspect for leaking or damaged components. • Test the exhaust system for leaks, damage or an exhaust restriction (refer to Catalyst & Exhaust system tests in this section). • Is an exhaust restriction suspected?	If yes, locate the exhaust restriction, make any needed repairs. Then do a PCM reset and retest.	If no, confirm the test results from the steps in this chart. If okay, go to the Symptom List and select another set of tests to perform.

Symptom Test 8 (Lack of power)

NOTE: *If the vehicle has a lack of power condition, and all engine subsystems are okay after testing, check the engine for excessive carbon buildup.*

PRELIMINARY CHECKS

Prior to starting this symptom test routine, inspect these underhood items:

- All related vacuum lines for proper routing and integrity
- All related electrical connectors and wiring harnesses for faults
- Verify that engine has a lack of power or loss of power condition during acceleration or at cruise speeds

Symptom Test 8 - Step 1

STEP	ACTION	YES	NO
1	**Step description:** *Visual Inspection* • Inspect for a severely restricted air filter. • Inspect Air Intake system for restrictions. • Inspect for binding throttle linkage. • Inspect for plugged/restricted radiator. • Check transmission fluid level. • Were any problems found?	If yes, make the needed repairs. Then do a PCM reset and retest the system.	If no, go to Step 2.
2	**Step description:** *Perform Quick Check* • Perform complete visual inspection, make any needed repairs. • Read and record all trouble codes. • Were any trouble codes recorded?	If yes, go to Code List, select the code and make the repair.	If no, go to Step 3
3	**Step description:** *Check/compare actual PCM pin voltages to pin chart values in the manual.* • Connect a Honda Test Harness or BOB. • Turn off all accessories. • Start engine and warm it until the electric cooling fan cycles. • Use a DVOM to read pin voltages that could affect idle speed control (i.e., EACV, power grounds, engine coolant temperature, etc.). • Are all pin voltages within their normal range?	If yes (meaning all pin voltages are okay), go to Step 5.	If no, (meaning one or more of the pin charts were out of normal range), go to the repair information that relates to that device or to another Symptom Test to check/repair the device or component.
4	**Step description:** *Check Pressure Regulator Control solenoid (for Acura/Honda models with this device only)* • Inspect and check the pressure regulator control solenoid for damage. • Is the solenoid faulty or damaged?	If yes, replace or repair the damaged pressure regulator control solenoid. Then do a PCM reset and retest the system.	If no, Go to STEP 5
5	**Step description:** *Check fuel delivery system* • Inspect fuel delivery system for leaks. • Are any fuel system faults suspected?	If yes, repair the fuel system. Then do a PCM reset and retest.	If no, go to Step 6.
6	**Step description:** *Check ignition secondary* • Inspect the ignition secondary for damaged or defective components. • Check spark output with spark tester, use an engine analyzer to check ignition secondary. • Are ignition system faults suspected?	If yes, make any needed repairs to the ignition secondary. Then do a PCM reset and retest the system.	If no, go to Step 7.

Symptom Test 8 - Step 7

STEP	ACTION	YES	NO
7	**Step description:** *Check Exhaust system* • Inspect for leaking or damaged components. • Test the exhaust system for leaks, damage or an exhaust restriction (refer to Catalyst & Exhaust system tests in this section). • Is an exhaust restriction suspected?	If yes, locate the exhaust restriction, make any needed repairs. Then do a PCM reset and retest.	If no, confirm the test results from the steps in this chart. If okay, go to the Symptom List and select another set of tests to perform.
8	**Step description:** *Inspect and test engine mechanical condition* • Test engine compression. • Test valve timing and timing chain condition. • Check for worn camshaft or valve train. • Check for manifold gasket leaks. • Check for intake vacuum leaks. • Are any Base Engine faults present?	If yes, make the needed repairs. Then do a PCM reset and retest the system.	If no, go to Step 9.
9	**Step description:** *Check the operation of the Automatic Transmission* • Inspect for low fluid levels or burnt fluid. • Refer to Transmission Repair Manual. • Are any transmission faults present?	If yes, make the needed repairs. Then do a PCM reset and retest the system.	If no, go to Step 10.
10	**Step description:** *Check the Brake System* • Inspect for binding or dragging brakes. • Refer to the Brake Repair Manual. • Are any brake system faults present?	If yes, make the needed repairs. Then do a PCM reset and retest the system.	If no, go to Step 11
11	**Step description:** *Perform these additional inspections and tests* • Verify that ignition base timing is okay. • Check customer driving habits for excessive loads or towing conditions. • Check for faulty or slipping clutch (MT). • Check for low Charging system output. • Were any faults located in these steps?	If yes, make the needed repairs. Then do a PCM reset and retest the system.	If no, confirm the test results from the steps in this chart. If okay, go to the Symptom List and select another set of tests to perform.

INTERMITTENT TESTS

Intermittent problems can be difficult to diagnose because they may or may not be present while the vehicle is available to the repair technician. An intermittent fault can appear and leave so quickly that the vehicle computer does not detect a fault and store a trouble code in memory.

A computer will have to see the fault for a specific period of time before a fault is detected and a code set. While intermittent problems may appear to be occasional by nature, they usually occur under specific conditions. Therefore, the technician must identify and duplicate these conditions. Since an intermittent fault is difficult to duplicate, it is important to follow a logical, systematic routine (or checklist) when attempting to find the faulty system, circuit or component.

JP-29491 Probe
Outside Diameter: 3/32"
Inside Diameter: 5/64"
JP-29591 Probe
Outside Diameter: 1/8"
Inside Diameter: 3/32"
TO ORDER BACKPROBE TOOLS,
CONTACT J. S. POPPER, INC AT:
201-641-3252
FIGURE 1-03

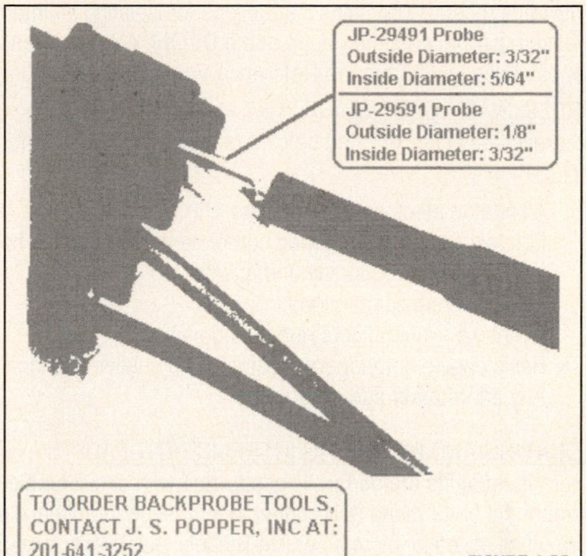

WHAT CAUSED THE FAULT?

Some intermittent faults can be due to a loose connection, a wiring problem or a warped circuit board. Some of these problems are caused by poor test procedures that can damage the male or female ends of a connector. To test for loose or damaged connection, take the male end of a used connector from a wiring harness and carefully push it into the female terminal to verify the opening is tight. There should be some resistance felt when inserting the male into the female part of the terminal.

THE WIGGLE TEST

A wiggle test can be used to locate some intermittent faults. To connect to the circuits, backprobe the actuator, sensor or switch and the PCM (Figure 1-03).

To perform this test, wiggle the suspect wiring, connector, and component while watching for a change on the DVOM. If it is available, use the Min/Max record mode on the DVOM to capture faults.

This is an excellent method to use to attempt to identify a condition that could cause an intermittent fault to occur. Refer to figure 1-04 for an example of how to perform this test on a brake switch.

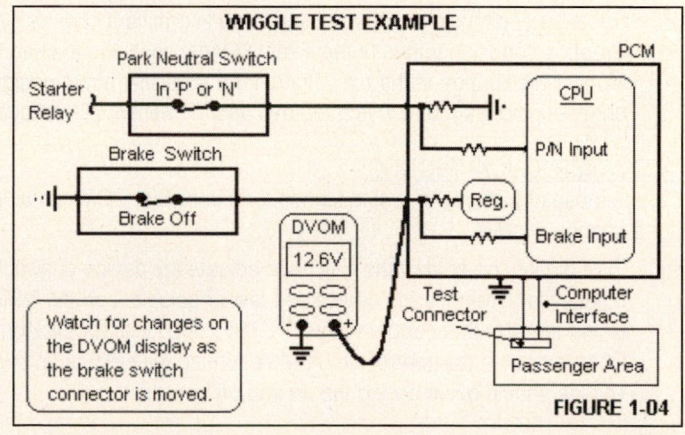

WIGGLE TEST EXAMPLE
Park Neutral Switch
PCM
Starter Relay
In 'P' or 'N'
CPU
P/N Input
Brake Switch
Brake Off
Reg.
DVOM
12.6V
Brake Input
Test Connector
Computer Interface
Passenger Area
Watch for changes on the DVOM display as the brake switch connector is moved.
FIGURE 1-04

INTERMITTENT TESTS

This test procedure contains instructions for isolating an intermittent fault while using a Scan Tool, Breakout Box, fuel pressure gauge, vacuum gauge and a DVOM. Actual values from the vehicle under test can be compared to a typical set of values taken from the Reference Values in this manual.

PRELIMINARY CHECKS

All preliminary checks listed below are required prior to performing this test for intermittent faults. Preliminary checks include:

- All related electrical connections
- For vacuum leaks in related components or mounting hardware
- Fuel level - both quantity and quality
- Ignition wiring connections
- Air intake system filters, tubes and gaskets
- Basic engine components (compression, valve and ignition timing, etc.)
- Any aftermarket add-on devices

LOCATING AND REPAIRING INTERMITTENT FAULTS

Intermittent faults are generally associated with circuit problems. In order to pinpoint the fault area, a particular component or its wiring and connectors must be thoroughly inspected and tested. Prior to starting your test sequence, turn off all accessories and vehicle lighting. Also, verify that the battery and vehicle charging system are free of problems as these areas can disguise or mask a problem. Several of these tests are discussed next.

CHANGE INPUT AND VERIFY OUTPUT RESPONSE

The purpose of this type of test is to monitor how the PCM and its output devices respond to changes in sensor or switch inputs.

- Connect the Honda Test Harness or BOB.
- Record any pin voltages that related to an intermittent code or symptom
- Create a condition to cause the selected input condition to change
- Monitor the change in the pin voltage for a particular actuator signal on the DVOM (i.e., increase the throttle angle under engine load, and watch the IAC and TP Sensor pin voltage changes)

ACTUATOR "CLICK" TESTING

The purpose of this type of test is to monitor a particular PCM controlled relay or solenoid while watching and listening for a change of state

- Turn the key on or start the engine to actuate the device or switch
- If necessary, remove the device and test or actuate it on the bench
- Listen to verify that certain relays (PGM-FI, A/C) actually click on and off. If a Breakout Box is connected to the PCM, measure the control circuit while turning the outputs on and then off. A voltage change to close to battery voltage should occur during the on and off transition.

TEST FOR OPEN CONDITIONS IN A HARNESS

The purpose of this type of test is to check a suspect wiring harness for an open circuit condition.

- Turn the key off and install the Honda Test Harness or BOB
- Remove the "suspect" component from the vehicle wiring harness
- Use a DVOM (ohmmeter function on the low range) to measure from one end of the suspect circuit at the test harness or BOB to the other end of the circuit at the particular component connector pin.
- The continuity test result should read under 5 ohms (If the test leads are okay and the ohmmeter is zeroed, the actual test results will read near 0.1 ohms).

TEST FOR SHORTED CONDITIONS IN A HARNESS

The purpose of this type of test is to check a suspect wiring harness for a short-to-power or short-to-ground condition.

- Turn the key off and disconnect the wiring harness from the PCM
- Remove the "suspect" component from its wiring harness connector.
- Use a DVOM (ohmmeter function on the high range) at the sensor harness connector to measure between the "suspect" circuit and signal return (ground) and to vehicle power or the voltage reference circuit.
- The continuity test result read more than 10,000 ohms (Infinity, >> or OL on some meters). If the actual reading is less than this amount, the two circuits are making contact somewhere in the wiring harness.

ROAD TEST PROCEDURE WORKSHEET

If it is available, transfer all appropriate "typical values" from reference information in the Pin Voltage Charts onto the Road Test Worksheet. Once this step is done, compare the "actual values" to the typical values that were recorded during or after the Road Test procedures.

Pin Chart Values	KOEO Actual Values	KOEO Typical Values	Hot Idle Actual Values	Hot Idle Typical Values	30 mph Actual Values	30 mph Typical Values	55 mph Actual Values	55 mph Typical Values
Measured Values								
Other Values								
Weather Conditions				Driving Route				

Catalytic Converter Tests

CONVERTER TEMPERATURE TEST

On Acura and Honda vehicles equipped with fuel injection, the efficiency of the converter can be determined by measuring the converter temperature at the inlet and outlet pipe connections. This test is based on the concept that when the converter is oxidizing hydrocarbons and carbon monoxide, it will create heat. In effect, if the converter is working properly, it will create a heat differential between its inlet and outlet connections.

CAUTION: **Block the drive wheels and connect the vehicle exhaust pipe(s) to a suitable vent system to purge exhaust fumes from the work area.**

TEST PROCEDURE

Another converter test is to measure the outside operating temperature of the converter exhaust pipes. This step requires a digital pyrometer (that can measure exhaust temperatures that exceed 500°F) or a Raytec digital pyrometer (Raynger # ST2, 3 or 6) that is placed *close to* the exhaust pipes.

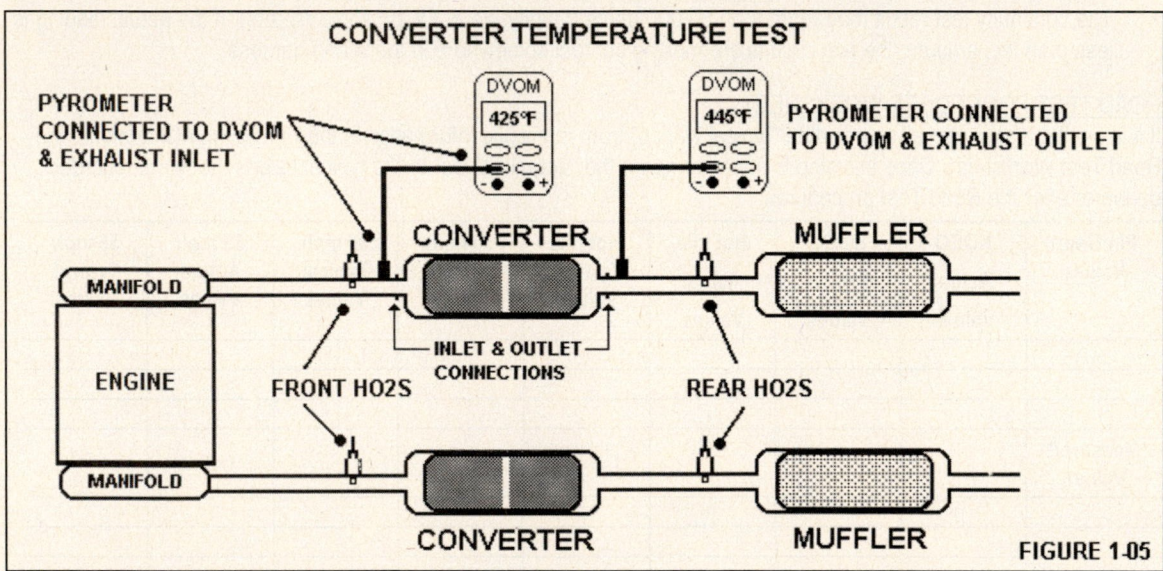

FIGURE 1-05

First, drive the vehicle to allow the engine and emission systems to reach normal temperature. Return to the service bay and run the engine at high speed (2000 rpm) in Neutral for one minute. Then measure the inlet pipe temperature at a point as close as possible to the converter. The inlet temperature should be at least 300°F to achieve converter light off. Next, check the outlet temperature by measuring at the outlet pipe as close as possible to the converter. Compare the readings to evaluate the converter.

SUMMARY

- Exhaust outlet pipe temperature is 5-10% hotter than the inlet - converter is working normally (Figure 1-05)
- Exhaust outlet pipe temperature is 200 degrees hotter than the inlet - the converter is overheating (find the cause of rich mixture or engine misfire)
- If the outlet temperature is the same as the inlet temperature, the converter may not have reached light-off temperature yet or the engine may be running so clean that it is not oxidizing. In these cases the converter MAY NOT BE FAULTY!
- To verify it is working, momentarily ground one spark plug lead (for no more than 30 seconds). If the outlet temperature does not exceed the inlet temperature under these conditions, the converter is not working.

Catalytic Converter Tests

CONVERTER OXYGEN LEVEL TEST

On Acura and Honda vehicles equipped with fuel injection and an O_2 feedback system, the converter efficiency can be determined by testing how much "free" Oxygen is in the exhaust with a 4 or 5-gas analyzer. The amount of oxygen in the exhaust indicates if the converter is using all of the free Oxygen. It is important to verify that all engine systems that affect engine performance (fuel, ignition and O_2 feedback) are operating correctly prior to testing. If there are any faults in these systems, this test will not work correctly.

This test requires the use of a calibrated gas analyzer that meets CA BAR-84 or BAR-90 specifications. These analyzers have the required response time needed to perform the Oxygen Test.

NOTE: A calibrated O_2 sensor will read from 20.8% to 21% oxygen at sea level.

PRELIMINARY STEPS

Inspect for leaks in the exhaust. Leaks will skew the test results to a point where they cannot be used. If the vehicle has air injection, disable the system prior to testing as additional air added to the exhaust will skew the results.

TEST PROCEDURE

- Warm the engine until it enters closed loop fuel control and the fan runs. Connect a known-good (calibrated) 4 or 5-gas analyzer to the exhaust.
- Raise engine speed to 2000 rpm and observe the exhaust readings. The readings should drop as the converter reaches its light-off temperature. As the numbers start to drop, watch the O_2 reading (it should go to 0%). Normally, the converter will use all of the free Oxygen. However, if there is not CO present, there may be a little Oxygen in the exhaust (very little).
- If there is free Oxygen left in the exhaust, stop testing and verify that the PCM is in control of the A/F mixture. If it is not, fix that problem first. If fuel control is okay, disconnect the O_2 sensor. Use a propane enrichment tool (safety valve working) to add propane to bring the CO up to around 0.5%.
- Once the O_2 reading on the gas analyzer stabilizes (with the gear selector in Neutral or Park), snap the throttle open and then let it drop back to idle. Monitor the rise in Oxygen level - - - it should not rise past 1.2% (Figure 1-06). Be sure to check the O2 reading on the gas analyzer with the CO climbing as the propane is added.

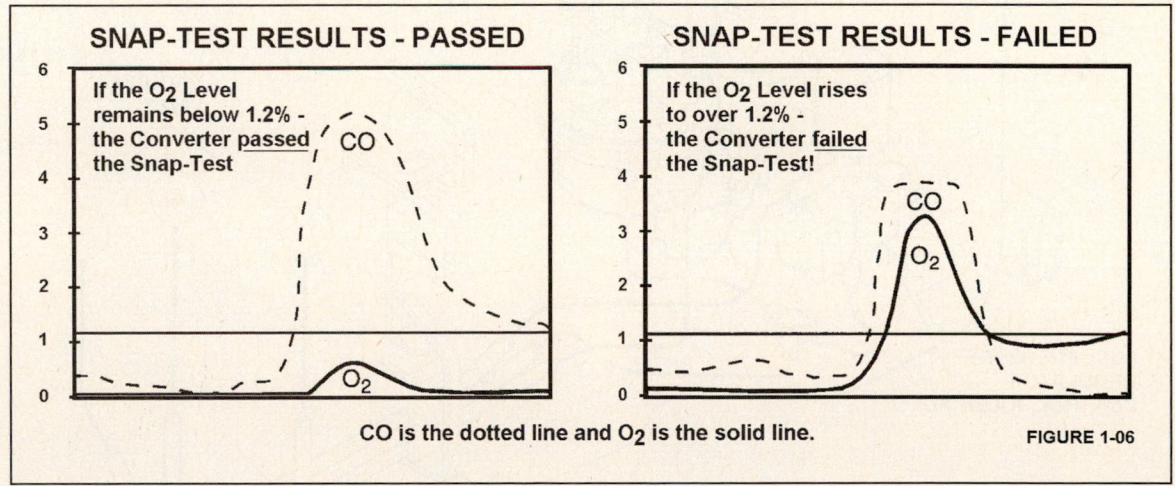

SNAP-TEST RESULTS - PASSED

If the O_2 Level remains below 1.2% - the Converter passed the Snap-Test

SNAP-TEST RESULTS - FAILED

If the O_2 Level rises to over 1.2% - the Converter failed the Snap-Test!

CO is the dotted line and O_2 is the solid line.

FIGURE 1-06

Control Box

INTRODUCTION

Some Acura and Honda applications use a Control Box to house various engine control system devices. Examples of devices located inside this box include:

- Constant Vacuum Control Valve
- Exhaust Gas Recirculation Control Solenoid
- Evaporative Emissions Control Solenoid
- Fuel Pressure Regulator Control Solenoid
- Manifold Absolute Pressure Sensor

COMPONENT EXAMPLES

On some applications, the electrical connections to various devices can be removed and checked from outside of the box (Figure 1-07). Also, vacuum lines that enter and exit the Control Box are numbered so that they can be easily identified in the repair steps that are part of trouble code and symptom repair charts (Figure 1-08).

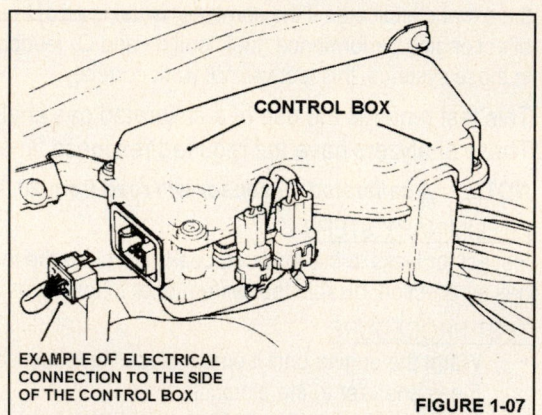

EXAMPLE OF ELECTRICAL CONNECTION TO THE SIDE OF THE CONTROL BOX

FIGURE 1-07

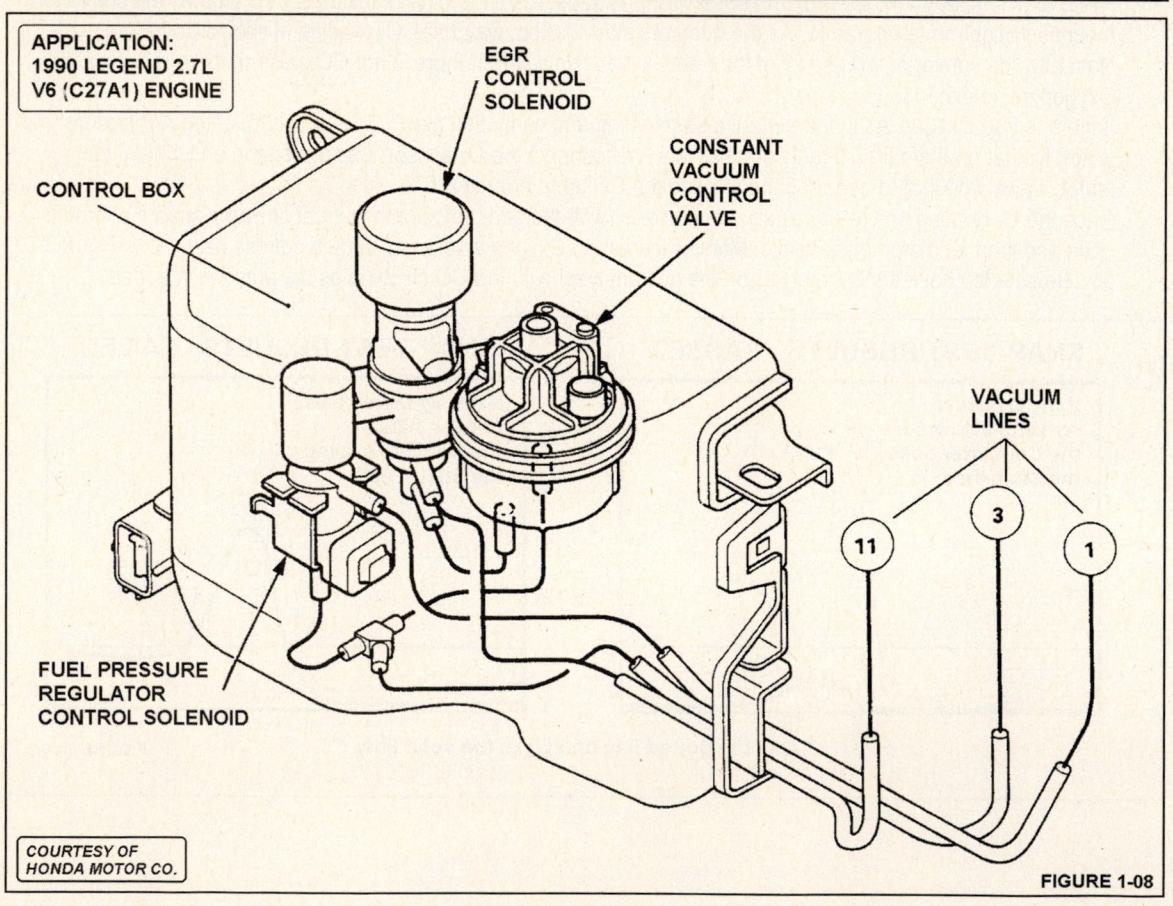

APPLICATION: 1990 LEGEND 2.7L V6 (C27A1) ENGINE

CONTROL BOX

EGR CONTROL SOLENOID

CONSTANT VACUUM CONTROL VALVE

VACUUM LINES

FUEL PRESSURE REGULATOR CONTROL SOLENOID

COURTESY OF HONDA MOTOR CO.

FIGURE 1-08

Drive By Wire System

INTRODUCTION

The Drive by Wire system used on NSX models is an electronic throttle control system.

SYSTEM COMPONENTS

This system consists of the following components:

A throttle valve control motor

A throttle position (TP) sensor (in the throttle body unit)\

An accelerator position (AP) sensor

The powertrain control module (PCM)

COMPONENT EXAMPLES

The components that make up this system in a 1995-97 NSX are shown in Figure 1-09.

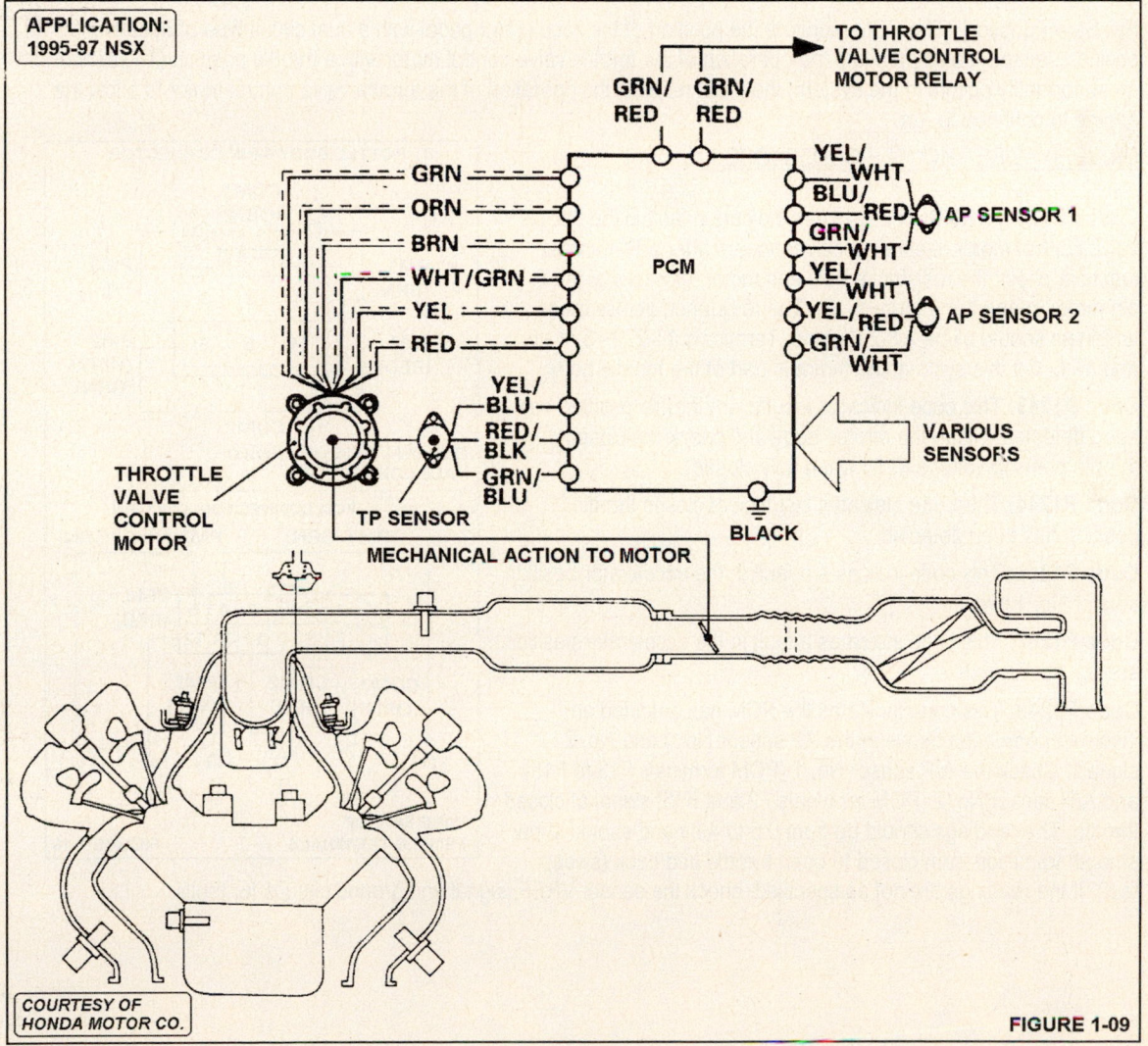

FIGURE 1-09

Drive By Wire System

DRIVE BY WIRE FUNCTIONS

Idle Control Function - With the engine at idle speed, the PCM controls the throttle valve control motor to maintain the proper idle speed according to engine loads

Acceleration Control Function - When the pedal is depressed, the PCM opens the throttle valve depending on the accelerator position (AP) sensor signals.

Cruise Control Function - The PCM controls the throttle valve control motor to maintain the "set speed" when the Cruise Control system is operating. The throttle valve control motor takes the place of the cruise control actuator.

Traction Control System (TCS) Function - If wheel spin occurs during acceleration or cornering, the TCS control unit requests the PCM to reduce engine power by retarding the ignition spark timing and closing the throttle valve. If the wheels lock during deceleration, the TCS control unit signals the PCM to open the throttle valve.

Engine Protection Function - When engine speeds exceed 8000 on a M/T vehicle or 7500 rpm on an A/T vehicle, the PCM controls the throttle valve (regardless of the accelerator position) in order to protect the engine from over-revving.

Fail Safe Function - The PCM monitors the position of the accelerator pedal with a dual circuit type of accelerator position sensor. It also monitors the operation of the throttle valve control motor with a throttle position (TP) sensor. If an abnormality occurs in the system, the PCM restricts the operation of the throttle valve control motor to allow the engine to continue to run.

DRIVE BY WIRE TROUBLE CODES (OBD II)

Codes P1241, P1242: These codes indicate a fault in the throttle valve control motor. If either code reappears after a PCM reset function, check the resistance between motor terminals at the 6-pin motor connector (Figure 1-10). The resistance across these terminals should all be 1.8-2.0 ohms: Terminals 1 - 2, 2 - 3, 4 - 5 and 5 - 6. On this system, the motor is part of the throttle body.

Code P1243: This code indicates insufficient throttle position has been detected. Clean the throttle body and check the closed throttle position voltage (0.3v when fully closed).

Code P1244: This code indicates insufficient closed throttle position has been detected.

Code P1246: This code indicates a fault in the accelerator position sensor No. 1 circuit.

Code P1247: This code indicates a fault in the accelerator position sensor No. 2 circuit.

Code P1248: This code indicates the PCM has detected an incorrect correlation between the AP sensor No. 1 and No. 2 signals. Check the A/P sensor No. 1 (PCM terminals F13 to F15) and A/P sensor No. 2 (PCM terminals F2 and F15) signal at closed throttle. The readings should be from 0.5 to 4.6v and should show smooth transition from closed to open throttle and back (sweep test). If the readings are not as specified, check the sensor VREF, signal and ground circuits for faults.

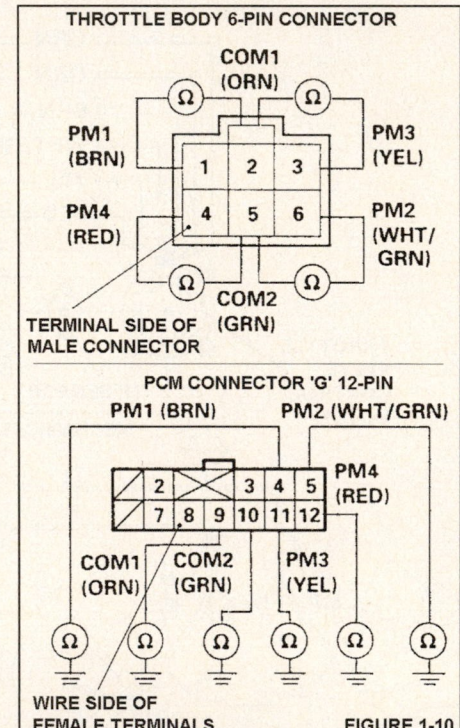

FIGURE 1-10

Exhaust System Tests

INTRODUCTION

The inspections and tests on this page should be performed in sequence to isolate the problem area in order to make the necessary repairs. Always start the test procedure by following the preliminary steps.

PRELIMINARY STEPS

Connect a known-good vacuum gauge to an intake manifold vacuum source. An alternate method is to connect a known-good pressure gauge to the exhaust system somewhere in front of the catalytic converter. Disable the EGR valve by removing and plugging the vacuum line to the valve.

CAUTION: Block the drive wheels and connect the vehicle exhaust pipe(s) to a suitable vent system to purge exhaust fumes from the work area.

EXHAUST BACKPRESSURE TEST - **Vacuum Gauge**
- Start the engine and allow it to warm until the electric cooling fan cycles.
- With the gearshift selector in Neutral and the drive wheels blocked, read and record the vacuum gauge reading at idle speed and again at high speed (around 2000 rpm) after one minute.
- The vacuum gauge reading at idle speed should not be more than one inch higher than the vacuum reading at high speed (2000 rpm). If the vacuum reading at high speed is more than one inch lower than the idle reading, an Exhaust system restriction is suspected. Refer to the Converter Inspection Procedure article.

EXHAUST BACKPRESSURE TEST - **Pressure Gauge**
- Connect a pressure gauge (i.e., exhaust backpressure gauge, fuel pressure gauge) into the exhaust stream at the exhaust pipe, EGR valve, the oxygen sensor (Figure 1-11) or at the air injection pump check valve (if equipped). Read the pressure gauge at idle and again at cruise speed for one minute, then record the gauge readings.
- Start the engine and allow it to warm until the electric cooling fan cycles.
- With the gearshift selector in Neutral and the drive wheels blocked, read and record the exhaust pressure reading at 2500 rpm after one minute. If the final reading is less than 1.5 psi at 2500 rpm, the exhaust system is not restricted. If the final pressure reading is from 1.5-2.75 psi, the system may be restricted. If the final reading is over 2.75 psi, the exhaust system is restricted and must be inspected (the restriction may not be the converter). Refer to the Converter Inspection Procedure article.

NOTE: If it is possible, test exhaust pressure in front of and after the converter. If the pressure remains too high, the restriction is after the converter.

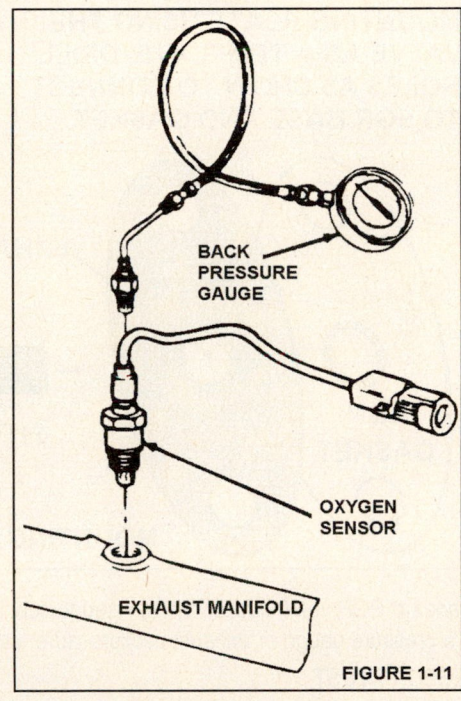

BACK PRESSURE GAUGE

OXYGEN SENSOR

EXHAUST MANIFOLD

FIGURE 1-11

Exhaust System Tests

RATTLE TEST PROCEDURE

Use a rubber mallet and tap lightly on the converter shell while listening for rattles. If it rattles when tapped, the converter substrate may be broken.

CONVERTER INSPECTION PROCEDURE

To inspect the inside of the catalytic converter, first allow the outside shell to cool, then remove it form the vehicle. Visually inspect the inside of the converter shell for plugging, melting or cracking of the catalyst. If more than 50% of the visible area is damaged or plugged, replace the converter. *When replacing a converter that has sustained heat damage, find the cause of the heat damage before replacing the unit. Retest the system after replacement.*

CAUTION: **Block the drive wheels and connect the vehicle exhaust pipe(s) to a suitable vent system to purge exhaust fumes from the work area.**

EGR VALVE CONNECTION

To measure the exhaust system pressure at the EGR valve requires that a special EGR adapter plate be constructed. Use the EGR valve as a template and fabricate a 1/8 to 1/4" steel plate that matches the mounting surface of the valve base. Drill two mounting holes and a hole in the plate for the exhaust port connection. Thread the hole to accept the appropriate fitting for the pressure gauge hose (Figure 1-12).

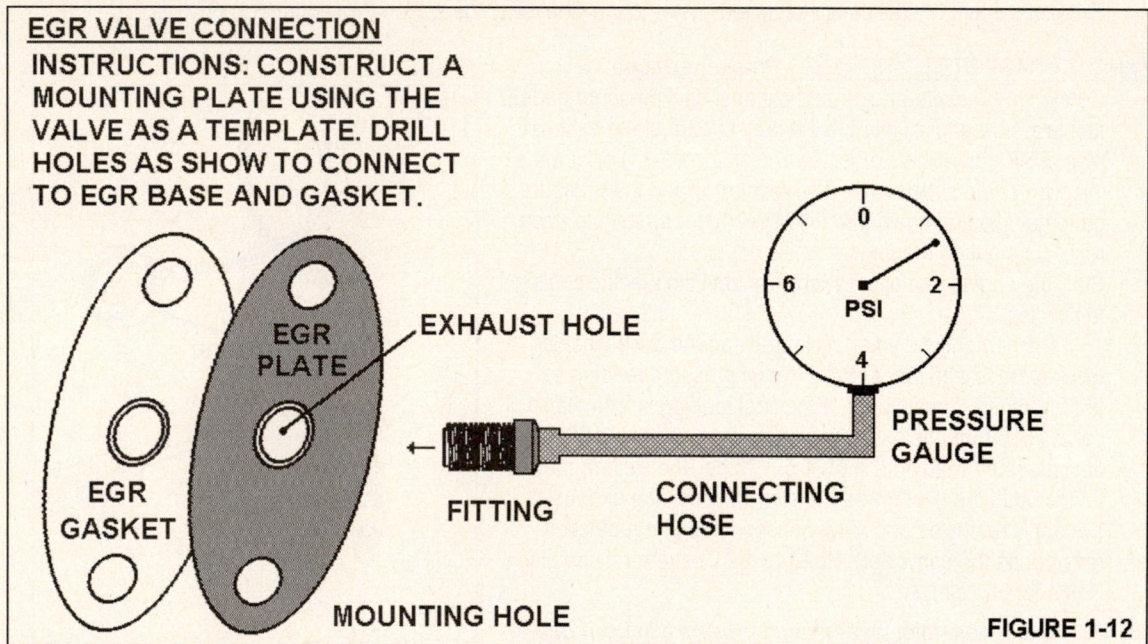

A standard EGR valve gasket can be used to seal the plate to the engine. Install the adapter plate, a new EGR gasket, and a pressure gauge or exhaust backpressure tester with a suitable adapter to the EGR valve opening.

PRESSURE TEST

Start the engine and read the exhaust pressure. If the final pressure reading is from 1.5-2.75 psi, the system may be restricted. If the final reading is over 2.75 psi, the system is restricted. To determine the location of the restriction, take pressure measurements after the converter, the muffler and/or resonator as needed to determine the exact location of the exhaust restriction. Remember that the exhaust restriction may not be in the catalytic converter!

Exhaust System Tests

<u>COMMON CONNECTION POINTS</u>

The four common connection points into the exhaust stream are outlined on this page. They are described in the articles on the next two pages.

CAUTION: *Block the drive wheels and connect the vehicle exhaust pipe(s) to a suitable vent system to purge exhaust fumes from the work area.*

<u>EXHAUST MANIFOLD CONNECTION</u>

Obtain an aftermarket kit that has adapters to safely tap into the exhaust system. First, make a small puncture into the exhaust pipe (not the exhaust manifold) in front of the converter. Next, install a vacuum nipple (and gauge) by threading the nipple into the hole in the exhaust pipe. If the pressure reads too high, make additional measurements between the converter and the muffler or between the muffler and resonator. When finished making pressure measurements, install a small bolt to seal any holes.

<u>AIR INJECTION PUMP CHECK VALVE CONNECTION</u>

In this case, the air pump check valve pipe must connect to the exhaust before the converter. Some check valve pipes connect directly into the converter. Do not connect and test into these connections. Remove the check valve and install a pressure gauge into the air injection rail. Be careful not to break or damage the Air Injection pipes (they could be rusty or brittle). If exhaust pressure is too high, make more measurements after the converter.

<u>OXYGEN SENSOR CONNECTION</u>

To test the exhaust pressure at the oxygen sensor, first carefully remove the sensor from the exhaust. Then install an exhaust backpressure tester (readily available from aftermarket suppliers) or construct an adapter using an 18-mm adapter from a fuel pressure gauge with a connecting fitting and a suitable pressure gauge (Figure 1-13).

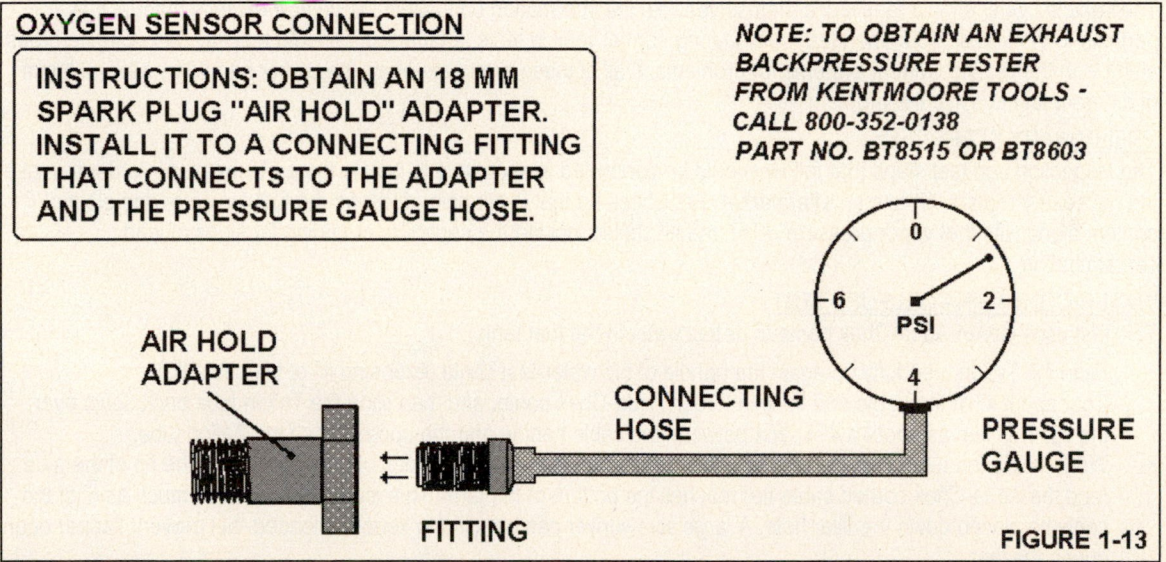

OXYGEN SENSOR CONNECTION

INSTRUCTIONS: OBTAIN AN 18 MM SPARK PLUG "AIR HOLD" ADAPTER. INSTALL IT TO A CONNECTING FITTING THAT CONNECTS TO THE ADAPTER AND THE PRESSURE GAUGE HOSE.

NOTE: TO OBTAIN AN EXHAUST BACKPRESSURE TESTER FROM KENTMOORE TOOLS - CALL 800-352-0138 PART NO. BT8515 OR BT8603

AIR HOLD ADAPTER

CONNECTING HOSE

PRESSURE GAUGE

FITTING

FIGURE 1-13

<u>PRESSURE TEST</u>

Start the engine and read the exhaust pressure at 2500 rpm in P/N for one minute. If the final pressure reading is from 1.5-2.75 psi, the system may be restricted. If the final reading is over 2.75 psi, the system is restricted. To determine the location of the restriction, take pressure measurements after the converter, the muffler and/or resonator as needed to determine the exact location of the exhaust restriction.

Fuel System Tests

INTRODUCTION

If any dirt, fuel with water or fuel with a high percentage of alcohol is detected, the fuel in the tank is contaminated and should not be used. Some driveability conditions are caused by contaminated gasoline or because low volatility fuel is used. *Therefore, the fuel quality should be checked along with the condition of the vehicle fuel as part of an overall Base Engine Test to locate potential causes of driveability symptoms.*

When working around automotive fuel systems that operate under high fuel pressure, pay particular attention to the CAUTION information that follows.

CAUTION: *Do not smoke while working on the fuel system. Keep all open flames and sparks away from the work area. Do not attempt to relieve the fuel pressure unless the engine is off.*

DRIVEABILITY SYMPTOMS

Problems with Fuel Quality could be the cause of any of these symptoms:

* Cold Engine - Hard Start
* Difficult to start when cold
* Hot Engine - Hard Start
* Misfire or Poor performance
* No Start Condition
* Rough idle or rough running conditions

NOTE: *The purpose of the Fuel Quality Test is to determine if a problem with the fuel could cause any of the driveability symptoms listed above.*

RECOMMENDED FUELS

The vehicle owner should familiarize themselves with the information on fuels in the vehicle owner's manual that pertains to the use of recommended fuels. During normal vehicle tests, a technician should test the Base Engine areas along with the Engine Control systems for problems. If all of these areas are okay, the owner should try several tanks of different fuel to solve the problem.

PRELIMINARY STEPS

The inspection and test steps that follow should be performed in sequence to isolate the problem area in order to make the necessary repairs. Obtain an aftermarket Gas Check Kit that is capable of testing the fuel for water and alcohol content along with fuel vapor pressure. Also, the kit should include the capability of testing for signs of lead contamination

WATER CONTAMINATED FUEL TEST

This test uses Green Aqua-Chek paste to detect water in the fuel tank.

* Slide the Teflon tube fully up again the handle of the water-test cable (exposing 4" of the cable tip).
* Coat about 4" of the cable and tip with Green Aqua-Chek paste, and then slide the Teflon tube back down over the tip. This leaves about a 4" space between the cable handle and the upper end of the Teflon tube.
* Work the Teflon tube, with the coated cable down in the gasoline filler neck of the vehicle until the tip of the tube (and the Aqua-Chek coated cable tip) reaches the bottom of the tank. This may require that as much as 6' of the cable be placed down the filler hole. A large screwdriver can be used to keep the leaded-fuel prevent flapper open during the test.

Fuel System Tests

<u>WATER CONTAMINATED FUEL TEST (CONTINUED)</u>

- Push the cable handle down fully against the top of the Teflon tube. This action pushes the coated tip out onto the tank bottom at the other end. Move the cable-and-tube assembly back and forth to move the tip around the bottom of the tank to find any water located there.
- Withdraw the handle about 4" from the Teflon tube. This action sheaths the Aqua-Chek paste for the trip out of the fuel tank.
- Pull out the cable and Teflon tube. Then, carefully push the coated tip out of the end of the tube and look for any bright Violet coloration of the Green Aqua-Chek paste.

Summary of Test Results - If the paste turns to a Violet color, then there is enough water on the gas tank bottom to cause a fuel system problem. If so, drain, remove and clean the fuel tank.

<u>LEADED GASOLINE TEST</u>

During this test, a piece of Plumbtesmo coated paper is used to detect the presence of lead deposit in the motor vehicle's tail pipe(s).

- Warm the engine to normal temperature and then turn off the engine. Next, verify that the tail pipe is not too hot to perform this test.
- If necessary, clean a section inside of tail pipe to remove any loose soot by gently wiping it with a damp cloth. Do not use a wet cloth!
- Remove the test paper from the test kit package and moisten it with distilled water. Do not saturate the paper as this may cause the paper to lose the re-active material impregnated into the paper.
- Quickly press the moistened paper firmly onto the cleared surface of the exhaust pipe. Hold the paper in place with a clothespin for one minute.

Summary of Test Result - If the paper changes to a bright Red color, this action indicates that leaded fuel has been used.

CAUTION: *Use care to avoid contamination of the Plumbtesmo test paper. Also, the technician should wash their hands prior to and between each test. Failing to wash their hands between tests could contaminate the cloth and clothespin used to hold the paper inside the tail pipe.*

<u>FUEL VAPOR PRESSURE TEST</u>

Correcting the Gasoline Sample - The vapor pressure test must be made using chilled gasoline. If the gasoline under test is from a hot fuel tank or any other warm container, the sample must placed into a chilled sample container, securely capped to prevent loss of light ends, then chilled using ice or chilled water. If ice is readily available, it should be used to chill the sample bottle before collecting the sample and then to chill the fuel sample container. Chilled drinking water is acceptable for this test procedure.

NOTE: *The procedure outlined next may give erroneous results, depending on type and quantity of alcohol and co-solvents that are in the gasoline.*

Fuel System Tests

FUEL VAPOR PRESSURE TEST

The test gauge assembly should be at room temperature when performing this test. Do not attempt to use the gauge outdoors in cold weather.

- Rinse the inside of the fuel cup and the air chamber with clean water and purge it with air by any convenient means (i.e., blowing it with low air pressure, using a siphon bulb or by blowing on it with your breath). Do not direct air under high pressure directly into the cavity of the gauge or chamber as possible damage to the gauge could occur.
- Place the fuel cup in chilled water.
- Fill the large cup with water warmed to about 110°F to within 3/4" of full. Fit the gauge/chamber assembly over the cup with the open-ended air chamber immersed in the hot water and let it stand for 3-5 minutes. This will condition the gauge/chamber assembly and allow it to be close to equilibrium when the temperature drops to 105°F (this is the temperature point at which the test should be run). Do not allow the temperature to drop below 105°F before starting the test. Add more hot water as needed.
- Do these steps as quickly as possible. Remove fuel cup from the chilled water, shake loose water from it and place in the fuel cup holder. Using a chilled syringe or dropper, withdraw the chilled gasoline from the sample bottle and fill the fuel cup to within 1/8" of top. If overflow occurs, wipe the liquid from the top of the fuel cup. Remove the gauge/chamber assembly from large cup holding it vertically at all times. Gently shake loose any water from the chamber, then move it over the fuel cup and quickly attach the cup and screw it on finger tight. Return the gauge/chamber assembly with the attached fuel cup to the water bath and let stand for two minutes.
- After two minutes (or longer if convenient), while holding the large cup and cover assembly together, tilt the unit with the gauge pointed down (about 45°) and shake vigorously for a few seconds - shaking along the direction that the unit is held. Turn upright, or as convenient to read the gauge, and read the pressure. Tap the gauge lightly as the pressure is read. Repeat the above until a stable reading is obtained (two or three times should be enough). Immediately after the last pressure reading, read the water temperature as accurately as possible.
- Optional Step - If the large cup and gauge/chamber cannot be held together during Step 4, do this step instead. After two minutes, remove the gauge/chamber assembly from the large cup. Tilt the assembly with the gauge pointed down (about 45°) and shake it vigorously 8 times - shaking along the direction that the assembly is held. After shaking, place the assembly back into the large cup, wait three minutes and read the pressure. Tap the gauge lightly as the pressure is read. Right after reading the pressure, read the temperature as accurately as possible.
- Record the pressure and temperature readings. Refer to the correction table in the test kit to obtain gasoline vapor pressure corrected to 100°F.

Fuel System Tests

FUEL VAPOR PRESSURE TEST (CONTINUED)

After each sample, the fuel cup should be emptied and both the fuel cup and air charnber purged with air several times to ensure removing the gasoline and associated vapors. Once all of the fuel samples are tested, the test apparatus should be dried to prevent corroding the aluminum parts.

Storing the Tester - Rinse the fuel cup and air chamber with hot water after each use. Shake loose water from each and blow dry for storage. Store the apparatus with the fuel cap not attached to air chamber.

REED VAPOR PRESSURE LIMITS

ASTM Standard D439 for Motor Gasoline lists five volatility classes for different temperatures. They are shown in the RVP limits table below.

REID		DAILY TEMPERATURE RANGE	
Class	Pressure	Low	High
A	9.0 psi	Above 60°F	Above 110°F
B	10.0 psi	Above 50°F	Above 110°F
C	11.5 psi	Above 40°F	Above 97°F
D	13.5 psi	Above 20°F	Above 85°F
E	15.0 psi	Below 20°F	Above 69°F

NOTE: This table is designed for sea level. For altitudes above sea level, add 2.4°F per 1000 feet to the temperature of the test location. Refer to ASTM document D439 for details on locality and yearly changes.

ETHANOL AND/OR METHANOL DETECTION TEST

Fuel blends that contain high amounts of Ethanol or Methanol can cause damage to Fuel system components. While this test can be used to detect if fuel contains enough alcohol to cause a symptom or a damaged component complaint, it cannot differentiate between the various alcohol types in fuels.

DETECTION TEST PROCEDURE

- Pour the suspect fuel into a clean glass container.
- Fill a graduated cylinder to the 10ml mark.
- Before starting the test, calibrate an eyedropper to determine how many eye drops equals 2.0ml.
- Next add 2ml of water to the graduated cylinder (count the number of drops from the eyedropper).
- Put the stopper into the cylinder and shake it vigorously for one minute. Relieve any pressure buildup by occasionally removing the stopper as alcohol dissolves in water (it drops to the bottom).
- Place the cylinder on flat surface for one minute.
- Read the number near the bottom of the cylinder at the boundary between the two liquids.

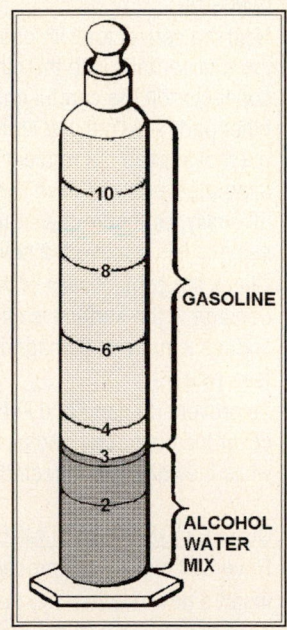

To determine the actual alcohol percentage, subtract 2 from the reading and multiply by 10. For instance, if the reading is 3.1ml, then 3.1 - 2 x 10 = 11% alcohol. If the increase in fuel volume is 0.2% or less, it can be assumed that the test fuel contains no alcohol.

Fuel System Tests

CAUTION: Keep open flames away from the work area. DO NOT SMOKE!

FUEL TUBE QUICK-CONNECT FITTINGS

Some Acura and Honda vehicles are equipped with a fuel tube quick disconnect fittings assembly that connects the in-tank fuel pump with the fuel feed pipe. During removal or installation procedures of the fuel pump or fuel tank, it is necessary to disconnect and reconnect the quick-connect fittings. During this procedure, look for the following:

- As the fuel tube quick-connect fittings assembly is not heat resistant, be careful not to damage it during welding or other heat generating repair procedures
- The fuel tube quick-connect fittings retainers are not acid-proof. Do not touch them with a shop towel that has come in contact with battery electrolyte. Replace the assembly if it comes in contact with electrolyte or similar acids.
- While disconnecting or connecting the unit, do not bend or twist the fuel tube quick disconnect fittings assembly excessively. Replace it if it is damaged in any way.

A disconnected quick-connect fitting can be reconnected, but the retainer on the mating pipe cannot be reused if it has been removed from the pipe. Replace the retainer when replacing the fuel pump or pipes, if it is damaged or if it has been removed from the pipe (Figure 1-14).

DISCONNECTION PROCEDURE

1) Remove the negative battery cable.
2) Relieve fuel pressure in the system. Remove the fuel cap to relieve tank pressure.
3) Check the fuel quick-connect fittings for dirt (clean as needed).
4) Hold the connector with one hand and press down the retainer tabs with the other hand, then pull the connector off (be careful not to damage the pipe or other parts - do not use tools). If the connector does not move, keep the retainer tabs pressed down and alternately pull and push the connector until it comes off easily. Do not remove the retainer from fuel pipe, as once it is removed, a new one must be installed.
5) Check the contact area of the pipe for dirt and damage. If the surface is dirty, clean it carefully. If the surface is rusty or damaged, replace the pump or fuel feed pipe.
6) To prevent damage and keep out any foreign matter, cover the connector and pipe end with plastic bags while they are disconnected.

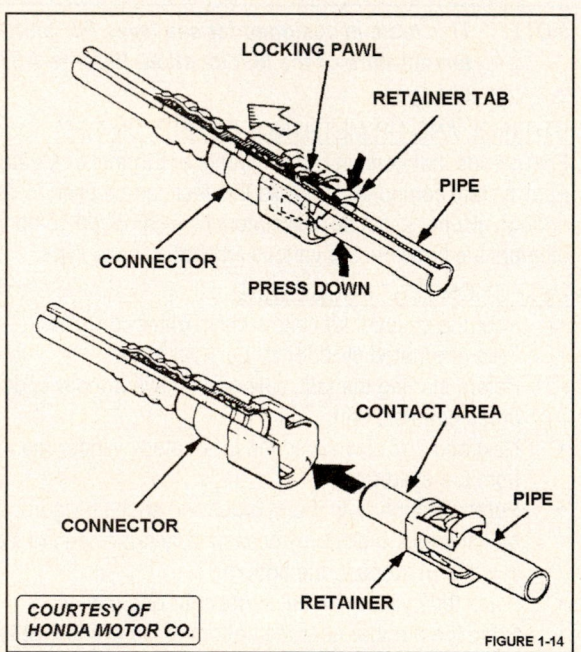

FIGURE 1-14

RECONNECTION PROCEDURE

1) Reverse the previous procedure
2) Insert a new retainer into the connector if the retainer is damaged at any time
3) Align the fittings with the pipe and align the retainer locking pawls with the connector groves during assembly and press the fittings onto the pipe until they click.
4) Reconnect the battery cable. Cycle the key to on and then check for any fuel leaks.

Fuel System Tests

<u>LOW FUEL INDICATOR SYSTEM</u>

Some Acura and Honda vehicles are equipped with a Low Fuel Indicator system. If the system is not working correctly, this article can be used to quickly check of the system.

<u>TEST PROCEDURE</u>

Check the Meter No. 25 fuse (7.5 amp) in the underdash fuse/relay box for a fault.

Park the vehicle on level ground

Drain fuel into an approved container. Then install the drain bolt with a new washer.

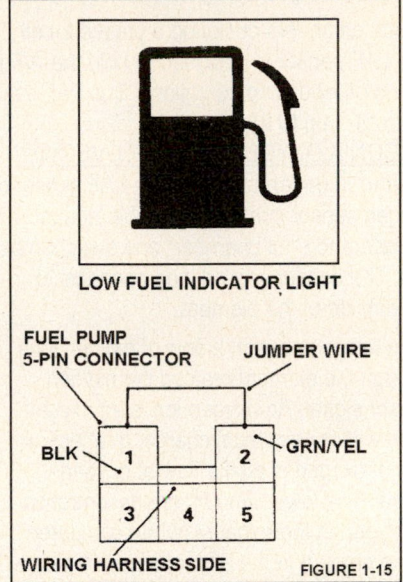

LOW FUEL INDICATOR LIGHT

FIGURE 1-15

Add less than 2.1 U.S. gallons (8 liters) of fuel and then turn the key on. The low fuel indicator should activate within four minutes. If the low fuel indicator light does not activate within four minutes, go to Step 9. If the light activates within four minutes, go to Step 5.

Gain access to the top of the fuel tank. For example, on CR-V models, remove the rear seat cushion or whatever it takes to gain access to the top of the tank.

Remove the fuel tank access panel from the floor and disconnect the 5-pin connector from the fuel pump.

Connect the No. 1 and No. 2 terminals with a jumper wire and observe the light. If the light activates (okay), test or replace the sending unit assembly (Figure 1-15).

If the light does not activate, check the GRN/YEL wire for an open circuit between the fuel unit and the gauge assembly. Also check the low fuel indicator for a blown light bulb. If the feed circuits and bulb are okay, check the ground circuit connection for an open condition and for high resistance. Make repairs as needed.

Add 1.1 U.S. gallons (4 liters) of fuel and turn the key on. The light should go off within four minutes.

*CAUTION: Keep open flames away from the work area. **DO NOT SMOKE!** Drain fuel only into an approved fuel container.*

Lean Air Fuel Sensor

INTRODUCTION

Starting in 1992, certain Honda Civic models were equipped with a new type of oxygen sensor called a Lean Air Fuel (LAF) sensor. A LAF sensor is used to allow a vehicle cruising at a steady speed to achieve improved fuel efficiency. This is accomplished by maintaining an A/F mixture that is leaner than 14.7:1.

Honda engineers designed a unique 2-cell Zirconia oxygen sensor to take advantage of the increased fuel efficiency. The LAF sensor is designed to hold the A/F mixture at 14.7:1 during certain driving conditions and leaner than 14.7:1 during other driving conditions. The LAF sensor can monitor exhaust oxygen conditions over a wider operating range (from 14:1 up to 23:1).

ZIRCONIA OXYGEN SENSOR OVERVIEW

In order to understand how the LAF sensor operates, a review of how a normal oxygen sensor works is in order. The oxygen sensor contains a thimble-shaped Zirconia element with two sides or electrodes. The inside of the element is the reference "air chamber" and this chamber is vented to atmosphere. The outside element is exposed to the exhaust gas. The sensor ground lead connects to the outside of the element.

In an oxygen sensor, a flow of oxygen ions through the element creates the oxygen sensor signal. An oxygen ion is an oxygen atom with an electrical charge. In effect, a flow of oxygen is also a flow of oxygen atoms. The lower amount or concentration of oxygen in the exhaust (when compared to the amount in the atmosphere) creates a flow of oxygen ions in the sensor. This action in turn produces the signal to the PCM.

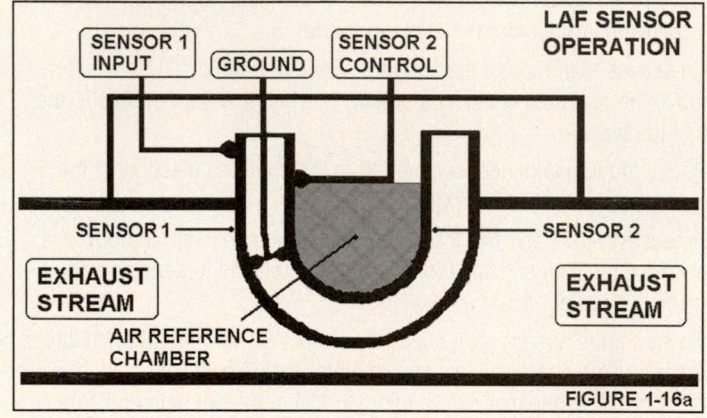

LAF SENSOR COMPONENTS

The LAF sensor has two (not one as in conventional sensors) Zirconia elements that share a common diffusion chamber. In effect, the LAF sensor contains two oxygen sensors mounted inside one unit. The heater control (HTRC) circuit is connected to ground at the sensor shell. The remaining circuits are used to control and monitor the two air chambers inside the unit. There are three chambers in the LAF sensor as follows:

- The first chamber (used to contact the exhaust flow)
- The diffusion chamber (between the elements)
- The air reference chamber

The first chamber is really the outside of the sensor (it contacts the exhaust). The diffusion chamber is the area between the two Zirconia elements. The air reference chamber is the other end. The basic operating principle behind the LAF sensor is that by controlling the amount of oxygen in the diffusion chamber, the operating range of the sensor can be controlled. One of the Zirconia elements acts as an oxygen pump. As previously discussed, the flow of oxygen ions inside an oxygen sensor creates the oxygen sensor signal and this explains how it works like a pump. An inverse flow of oxygen ions in an oxygen sensor also causes a similar action. In effect, a flow of electrons applied to the LAF sensor causes a flow of oxygen ions.

Lean Air Fuel Sensor

ZIRCONIA OXYGEN SENSOR OPERATION

The PCM monitors the voltage between sensor No. 1 input and the ground lead. The PCM is constantly attempting to hold the voltage difference between sensor No. 1 and the ground lead to a level of 450mv.

When the LAF sensor shifts rich, oxygen ions flow from the diffusion chamber to the exhaust. When this occurs, the voltage at the sensor No. 1 input increases. The PCM detects the voltage increase and reduces the voltage on the sensor No. 2 input. This causes the voltage on sensor No. 2 to shift to a more negative value than the ground voltage. This action causes sensor No. 2 to pump oxygen out of the diffusion chamber into the air reference chamber. When the oxygen content in the diffusion chamber drops, the voltage signal from sensor No. 1 drops. At the same time, the PCM reduces the voltage on sensor No. 2 input and this also results in a reduced fuel delivery command.

The change in fuel delivery command results in a leaner A/F mixture, and this causes the opposite affect on the LAF sensor. As the mixture goes leaner, oxygen ions in the exhaust flow into the diffusion chamber and the voltage on sensor No. 1 input decreases. The PCM detects this voltage decrease and increases the voltage on the sensor No. 2 input. The voltage on sensor No. 2 then goes to a more positive value than the ground reference. This action causes sensor No. 2 to pump oxygen into the diffusion chamber from the air reference section. The voltage between sensor No. 1 input and ground is consistently held at 450mv.

The PCM uses these signal changes to determine how rich or lean the exhaust stream is at any given moment by detecting how much amperage it takes the sensor No. 2 input to hold the sensor No. 1 input voltage to 450mv. This information allows the PCM to control the operation of the LAF sensor.

LAF SENSOR DIAGNOSTICS

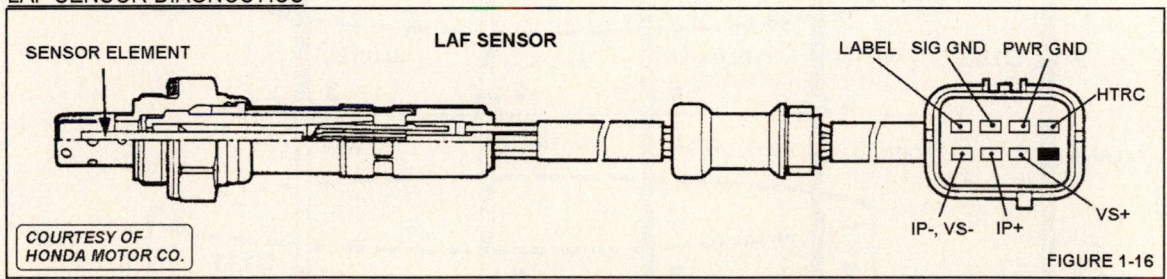

FIGURE 1-16

Problems in the LAF sensor are diagnosed in the same manner as any other oxygen sensor - through the use of the PGM-FI Diagnostics. A fault in a LAF sensor circuit or component will cause the PCM to set a trouble code in memory. A list of these available codes is shown next.

LAF SENSOR CODES- OBD I

Code 48 -This code indicates one of the LAF sensor circuits has an open or short to ground condition, or that the LAF sensor has failed.

LAF SENSOR CODES- OBD II

Codes P1162, 1163, 1168, 1169 -These codes indicates one of the LAF sensor circuits has an open or short to ground condition, or that the LAF sensor has failed.

PGM-FI DIAGNOSTICS

Lean Air Fuel Sensor

ZIRCONIA OXYGEN SENSOR SCHEMATIC

In Figure 1-17, note how the LAF sensor is connected to the PCM. The Zirconia elements are connected in parallel and there is a common ground lead. This ground is a reference point for the PCM (it should not be confused with body ground).

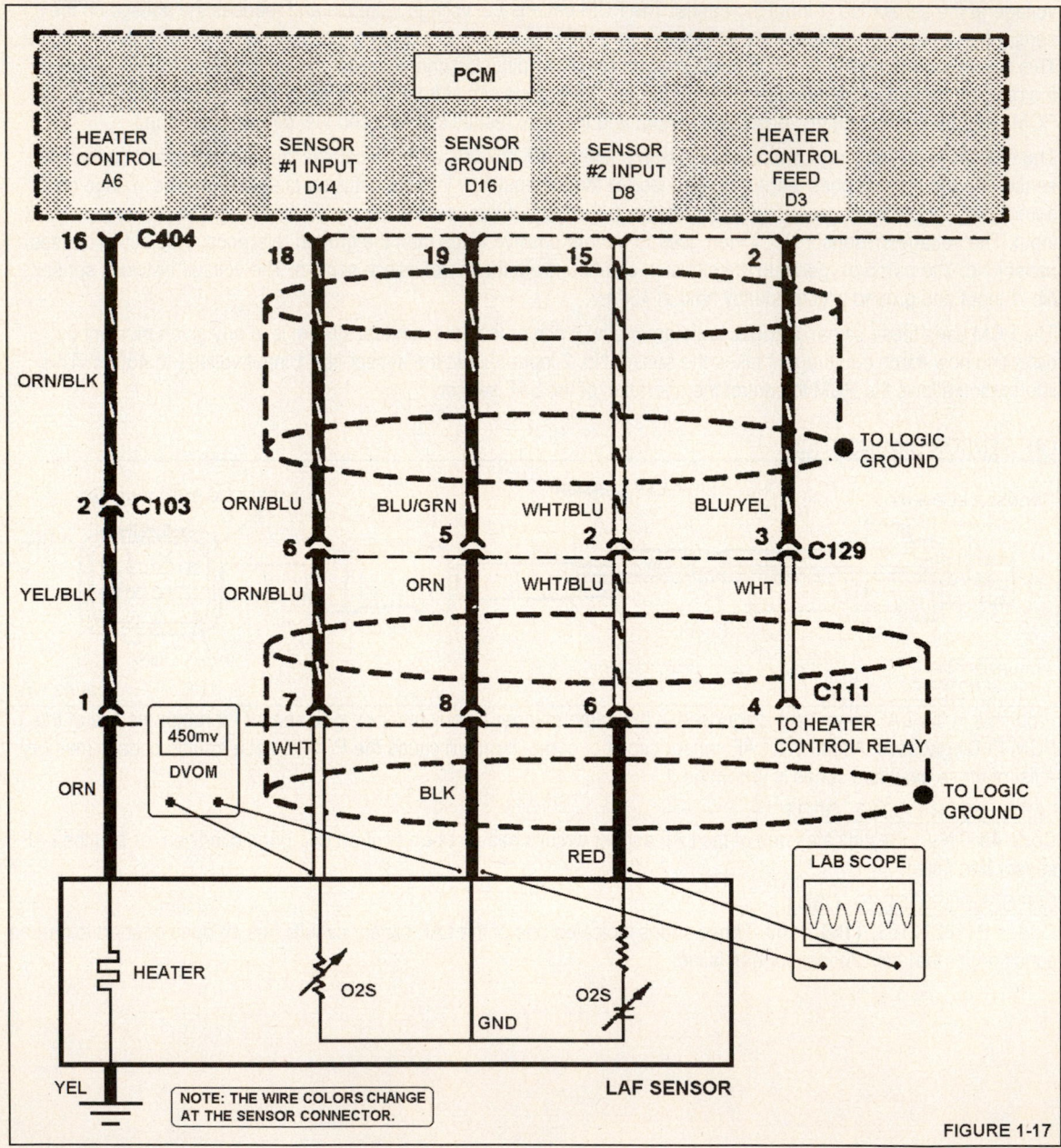

FIGURE 1-17

LAF SENSOR TESTS

Measure the voltage between sensor ground and body ground with the key on. It should read near 2.7v (Figure 1-17). During normal operation, a voltage measurement reading taken between Sensor No. 1 (D14) and Sensor Ground (D16) should read near 450 mv.

If a Lab Scope is connected between Sensor No. 2 (D8) and Sensor Ground (D16), it should show the normal oxygen sensor waveform. These two quick checks can help to establish that the LAF sensor is working properly.

MAP Sensor Tests

INTRODUCTION

This article includes information on how to test the MAP sensor for No Code or driveability faults. If a MAP sensor *hard code* is present, use the code repair chart to correct the condition. Problems related to the MAP sensor can appear in these areas:

- The Map sensor or any of its circuits (VREF (5v), signal (SIG) and sensor ground
- The vacuum source to the Map sensor (engine mechanical faults, leaks, moisture)

MAP SENSOR CIRCUIT TESTS

MAP sensor electrical operation consists of these three circuits: MAP sensor signal, VREF and the MAP sensor ground circuit.

MAP VREF circuit - Remove the MAP sensor connector, turn the key on and backprobe the VREF circuit (YEL/RED wire). A reading of 4.9-5.1v on the VREF circuit is normal. If the VREF reading reads 0v, check the wire for a short to ground or an open circuit condition. Make repairs as needed. If the circuit is okay, substitute a known-good PCM and retest the circuit. Replace the original PCM if the VREF reading is now within range.

MAP ground circuit - Turn the key on and connect the DVOM as follows. Connect one lead to the MAP sensor ground circuit in 3-P connector and the other lead to battery negative. The DVOM should read less than 50 millivolts if the ground circuit is okay.

MAP signal circuit - Check circuit continuity and for a short-to-ground condition. Voltage readings at the sensor should closely match any readings taken at the PCM or BOB.

COMPONENT TESTS

Dynamic Test - Connect a DVOM or Lab Scope to the signal wire. Read the voltage in KOEO and KOER modes. Compare the readings to the values in the Sensor Range Charts. Note the close relationship between the MAP signal and engine rpm signals at idle rpm (Figure 1-18a).

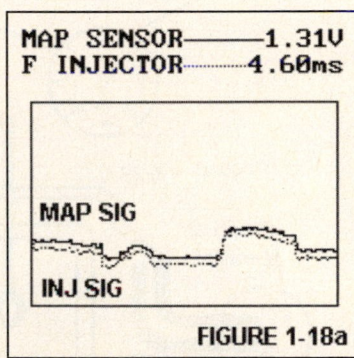

```
MAP SENSOR————1.31V
F INJECTOR————4.60ms




MAP SIG


INJ SIG
```
FIGURE 1-18a

Static Test - Connect a vacuum pump to the sensor and slowly apply vacuum to the sensor (sweep the sensor through its complete range). If the reading drops off or changes too quickly, replace the MAP sensor.

MAP SENSOR CALIBRATION

Connect the DVOM leads to the MAP signal and battery negative. Turn the key on, connect a vacuum pump to the sensor and slowly apply 10" Hg to the sensor. Record the reading at 10" Hg. At sea level, the reading should be 1.8-2.0v. At over 4000 feet, the reading should be under 1.8-2.0v (for each 1000 feet, the signal should drop 0.25v). If the voltage change is not as specified, replace the MAP sensor and repeat the test.

MAP SENSOR BACKUP STRATEGY

If the MAP sensor or a circuit fails, the PCM will shift to a backup strategy program that allows the vehicle to limp-in for repair. To test the PCM, remove the MAP sensor 3-P connector and attempt to start the engine. The PCM should detect that the MAP signal is missing, set a code and switch to its backup strategy. The engine should start and run.

TIPS: *Problems with the MAP sensor or its vacuum source can cause the engine to run extremely rich. If engine vacuum is low due to a mechanical problem, the MAP signal will be over 1.1v at idle (normal is 0.9-0.95v) and the engine will run rich. Examples of mechanical faults that can cause a low vacuum reading include a timing belt one tooth off, valves that are too tight, a restricted exhaust and the use of a performance camshaft. Any of these conditions will cause the engine to run rich!*

Map Sensor Tests

DUAL GAUGE VACUUM TEST

Problems related to the source of the vacuum to the MAP sensor are difficult to diagnose. The vacuum connection at either the manifold or throttle body is vulnerable to restrictions or plugging due to contamination from the A/F mixture or PCV valve flow that can buildup and partially cover the vacuum port opening to the MAP sensor vacuum line. The dual gauge vacuum test was designed to determine if the MAP sensor source vacuum is correct.

To verify that the source vacuum is not restricted or delayed to the MAP sensor requires a pair of vacuum gauges (analog type), two 'T' type vacuum fittings, and two short pieces of vacuum hose. First, check the engine vacuum source line for any damaged areas or sharp bends that could cause leakage or a restriction resulting in a low vacuum reading.

Next, verify that the vacuum line is not restricted or full of moisture. If the driveability problem happens only when the engine is cold or in sub-zero temperatures, suspect that moisture may be in the vacuum line and collecting in a low area where it freezes. After the engine warms up, the ice melts and the problem goes away. Check for the presence of moisture. Remove both ends of the vacuum line and apply shop air to the vacuum line. If plastic vacuum lines are used, inspect them for melted or collapsed housings.

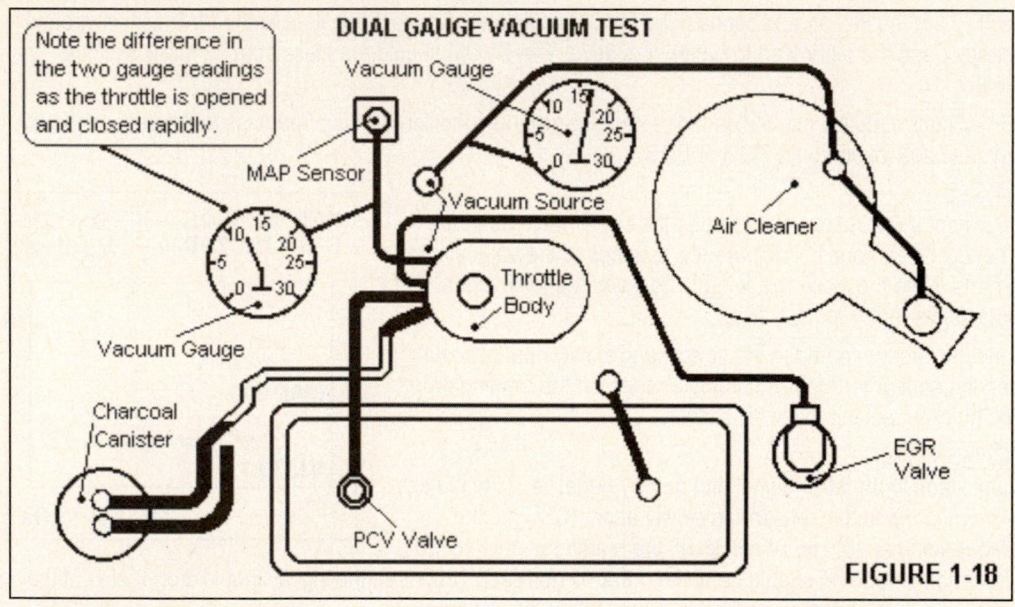

FIGURE 1-18

ENGINE RUNNING TEST

To test the source vacuum with the engine running, connect two vacuum gauges as shown in Figure 1-18. The gauges are connected at two separate vacuum sources. Place the transaxle in Park or Neutral and start the engine. Watch the operation of both gauges as the throttle is snapped open and closed rapidly (repeat this step several times). The gauges should track each other very closely. If the gauge on the MAP signal line lags behind the other gauge, clear the obstruction at the vacuum source. The problem may require throttle body removal to clean the MAP sensor vacuum source opening.

NOTE: *If a MAP vacuum signal is delayed, changes to fuel enrichment and timing will lag behind engine demands and result in tip-in hesitation or trailer-hitching on decel.*

Pin Voltage Tests

MPSI IBOB & OTC UB-80 BREAKOUT BOX

If a repair step requires a frequency, voltage or resistance measurement at the PCM wiring harness, there are several methods to use to accomplish this task. A Honda Test Harness can be used to allow for these measurements. An OTC UB-80 Breakout Box (Figure 1-19) or MPSI Intelligent Breakout Box (Figure 1-20) with the Honda adapters can also be used. Make sure to verify the ground circuit connections on the adapters.

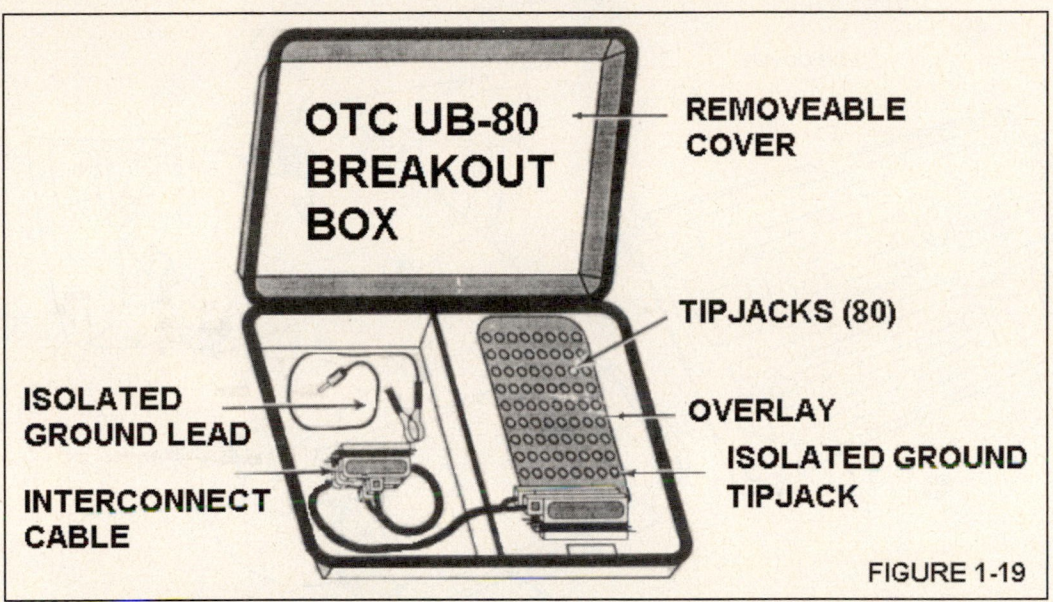

FIGURE 1-19

OTC UB-80 Breakout Box

MPSI Intelligent Breakout Box (IBOB)

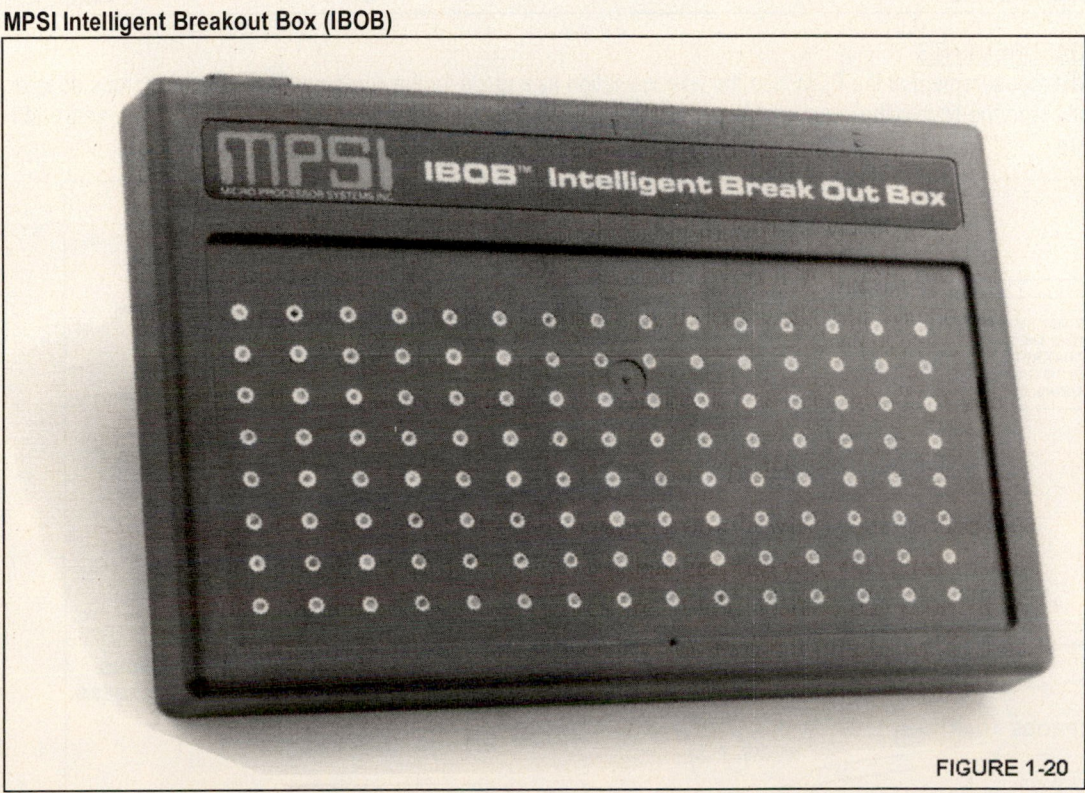

FIGURE 1-20

PGM-FI DIAGNOSTICS

Pin Voltage Tests
BACKPROBE TEST PROCEDURE
If a repair step requires a frequency, voltage or resistance measurement at the PCM wiring harness, carefully remove the PCM from its location at the right door sill molding on the right side kick panel (some models). Pull the carpet back to expose the PCM and remove the PCM cover (Figure 1-21).

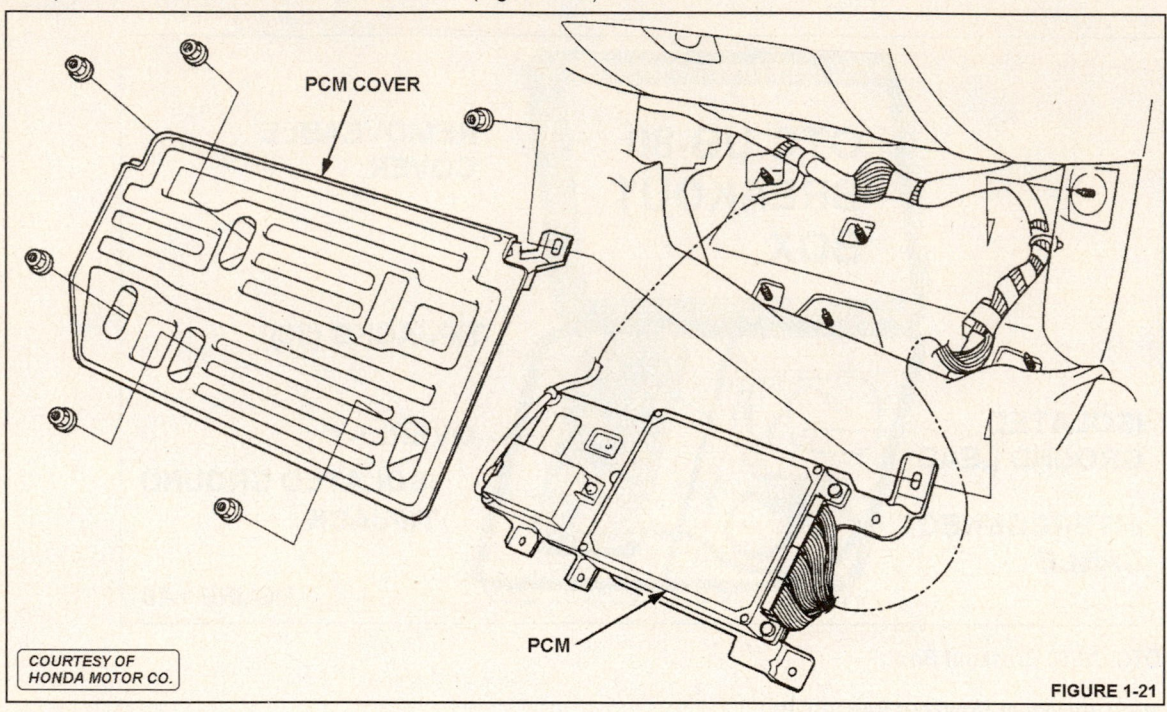

FIGURE 1-21

PCM MEASUREMENTS
To make a measurement at the PCM, use the wire insulation as a guide for the contoured tip of the backprobe adapter and gently slide the tip into the connector from the wire side until the probe tip comes in contact with the terminal end of the wire.

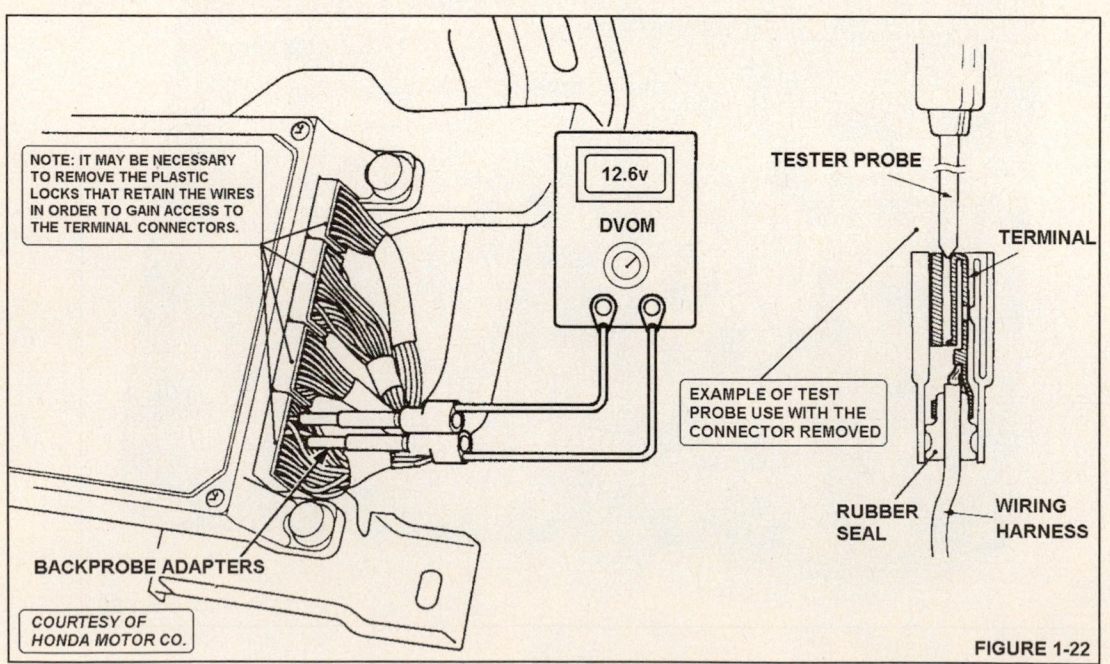

FIGURE 1-22

VTEC System Tests

<u>VTEC SOLENOID & SWITCH TESTS</u>

The Variable Valve Timing with Electronic Control (VTEC) system was introduced on the 1992 Honda Civic 1.5L and 1.6L MFI engine (D16Z6).

<u>ROCKER ARM INSPECTION</u>

To inspect the rocker arms on this engine, perform these steps (Figure 1-23).

*   Rotate the engine to place the No. 1 piston to top dead center (TDC).
*   Remove the valve cover to expose the intake mid-rocker arm and components.
*   Push the No. 1 cylinder intake mid-rocker arm manually and check that the intake mid-rocker arm moves independently of the primary and secondary intake rocker arms.
*   Check the intake mid-rocker arm of each cylinder at TDC. If the intake mid-rocker arm does not move, remove the primary, mid and secondary intake rocker arms as an assembly and check that the pistons in the primary and mid-rocker arms move smoothly. If any rocker arm needs replacing, replace the primary, secondary and mid-rocker arm as an assembly.
*   During re-assembly, apply oil to the spark plug tube oil seals with your finger when installing the valve cover to the engine.

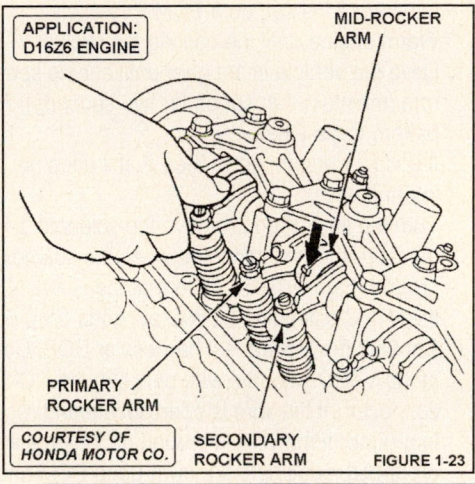

APPLICATION: D16Z6 ENGINE — MID-ROCKER ARM — PRIMARY ROCKER ARM — SECONDARY ROCKER ARM

COURTESY OF HONDA MOTOR CO.   FIGURE 1-23

<u>VTEC SOLENOID INSPECTION</u>

To inspect the VTEC solenoid valve, perform these inspection steps.

*   Remove the 1P connection from the VTEC solenoid valve and measure the resistance between the terminal and body ground.
*   If the reading is 14-30 ohms, remove the solenoid from the cylinder head and check the solenoid valve filter for clogging. If the reading is out of range, replace the solenoid.
*   If the filter is not clogged, push the VTEC solenoid valve with your finger to check its movement (Figure 1-24). If the valve movement is normal, check that the engine oil pressure is within specifications.

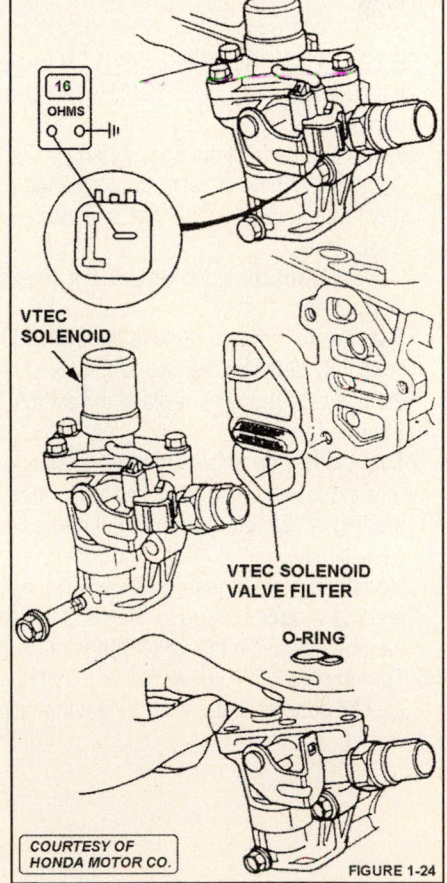

16 OHMS — VTEC SOLENOID — VTEC SOLENOID VALVE FILTER — O-RING

COURTESY OF HONDA MOTOR CO.   FIGURE 1-24

VTEC System Tests

CODE 21 - VTEC SOLENOID VALVE FAULT

- If a code 21 is set, do a PCM Reset (clear codes) and restart the engine.
- Warm engine until the cooling fan cycles.
- Drive the vehicle in 1st gear until engine speed exceeds 6000 rpm (maintain that speed for two seconds). Repeat the drive pattern three times.
- If code 21 resets, go to Step 5. If it does not reset, the fault is intermittent.
- Turn the key off and remove the solenoid 1-P connector and measure the resistance between the solenoid terminal and ground (Figure 1-25). If the reading is not 14-30 ohms, replace the VTEC solenoid. If it is okay, go to Step 6.
- Connect the Honda test harness or BOB. Check the continuity of the VTEC solenoid wire between the PCM and the

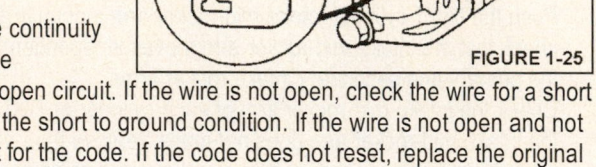

FIGURE 1-25

connector. If the wire is open, locate and repair the open circuit. If the wire is not open, check the wire for a short to ground. If the wire is grounded, locate and repair the short to ground condition. If the wire is not open and not grounded, substitute a known-good PCM and retest for the code. If the code does not reset, replace the original PCM.

CODE 22 - VTEC PRESSURE SWITCH FAULT

- If a code 22 is set, do a PCM Reset (clear codes) and restart the engine.
- Warm engine until the cooling fan cycles.
- Drive the vehicle in 1st gear to an engine speed over 6000 rpm (hold that speed for over two seconds). Repeat this drive pattern three times.
- If code 22 resets, go to Step 5. If it does not reset, the fault is intermittent.
- Turn the key off and remove the switch 2P connector. Test for continuity between the two terminals of the switch (Figure 1-26). If there is no continuity, replace the switch. If there is continuity, go to Step 6.
- Turn the key on and measure the voltage between the BLU/BLK wire in the switch harness connector and body ground. If the reading is 12v, go to Step 7. If it reads 0v, locate and repair the open circuit.
- Measure the voltage across the two pins in the connector. If the reading is near 12v, go to Step 8. If it reads 0v, locate and repair the open circuit in the BRN/BLK wire (to body ground).
- Turn the key off, remove the VTEC pressure switch and test the special adapter and switch to test the oil pressure.

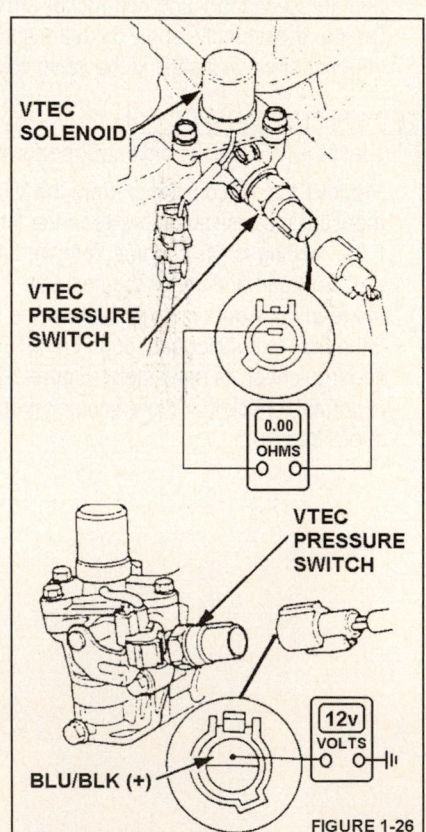

FIGURE 1-26

Reference Information

ECT and IAT Sensor Charts

NOTE: *These example charts were designed for use with the Honda Test Harness or a Breakout Box and a DVOM. Readings were collected from a "known good" vehicle at Key On, Engine Off unless otherwise noted.*

ECT Sensor Conversion Chart

Degrees F	Resistance	Voltage
0°F	14,000-20,000 ohms	4.0v
100°F	900-1200 ohms	2-3v
180°F	250-350 ohms	0.5-0.6v

IAT Sensor Conversion Chart

Degrees F	Resistance	Voltage
0°F	14,000-20,000 ohms	4.0v
100°F	900-1200 ohms	2-3v
192°F	150-350 ohms	0.49-0.55v

Atmospheric Pressure Sensor Chart

BARO Pressure (inches Hg)	Voltage
0" Hg	2.8-3.0v
5" Hg	2.3-2.5v
10" Hg	1.8-2.0v
15" Hg	1.3-1.5v
20"Hg	0.8-1.0v
25" Hg	0.3-0.5v

MAP Sensor Chart   Fuel Tank Pressure Chart

MAP Pressure (inches Hg)	Voltage		Fuel Tank Pressure (in. Hg)	Voltage
0" Hg	2.8-3.0v		0" Hg	2.8-3.0v
5" Hg	2.3-2.5v		5" Hg	2.3-2.5v
10" Hg	1.8-2.0v		10" Hg	1.8-2.0v
15" Hg	1.3-1.5v		15" Hg	1.3-1.5v
20"Hg	0.8-1.0v		20"Hg	0.8-1.0v
25" Hg	0.3-0.5v		25" Hg	0.3-0.5v

EGR Lift Sensor Chart

EGR Valve Position	Voltage
Valve Fully Closed (idle)	1.0-1.2v
Valve Fully Open (off-idle)	4.5v

Acura & Honda Trouble Code List (OBD I)

Instructions: Read and record all codes in memory. Perform a PCM reset function and do a Road Test. If the code resets, refer to the codes and diagnostics tips below as needed.

1986-95 Acura & Honda Models

CODE ID	CODE DESCRIPTION & DIAGNOSTIC TIPS
0	**PCM internal Fault** - Replace the PCM and recheck the code.
1	**Left Side (Front) O$_2$ Sensor Fault** - Monitor the O$_2$ sensor signal with a Lab Scope during acceleration (0.5-1.0v) and deceleration (0.0-0.4v). Inspect fuel filter for restriction and check fuel pressure and fuel pressure regulator.
2	**Right Side (Rear) O$_2$ Sensor Fault** - Monitor the O$_2$ sensor signal with a Lab Scope during acceleration (0.5-1.0v) and deceleration (0.0-0.4v). Inspect fuel filter for restriction and check fuel pressure and fuel pressure regulator.
3	**MAP Sensor Electrical Fault** - Test the MAP sensor circuits. Refer to the MAP Sensor Tests in Section 1 to test the circuits.
4	**Crank Sensor (in combination Crank/Cylinder Sensor) Fault** - Test sensor resistance and compare to specifications (500-1000 ohms). Compare sensor output to cranking specifications (250mv AC minimum).
5	**MAP Sensor Mechanical Fault** - Remove and inspect MAP sensor source vacuum hose. Check for restrictions, moisture and leaks. Turn the key on and apply vacuum directly to the MAP sensor connector. If the sensor holds vacuum, check the calibration at 10" of vacuum (0.8-2.0v).
6 (D4 lamp may be on also)	**Engine Coolant Temperature Sensor Fault** - Inspect sensor connections for damage or corrosion. Compare sensor reading at KOEO and compare to specifications (0.5-0.6v at a temperature of 180°F).
7 (D4 lamp may be on also)	**Throttle Angle Sensor Fault** - Inspect sensor connections for damage or corrosion. Compare sensor reading at KOEO to specifications (0.5v at closed throttle). Monitor sensor signals through its complete range with a Lab Scope while slowly depressing the throttle (KOEO sweep test).
8	**TDC Sensor (in combination Crank/CYL/TDC Sensor) Fault** - Test sensor resistance and compare to specifications (500-1200 ohms). Compare sensor output to cranking specifications (250mv AC minimum).
9	**Cylinder Sensor (in combination Crank/Cylinder Sensor) Fault** - Test sensor resistance and compare to specifications (500-1200 ohms). Compare sensor output to cranking specifications (75mv AC minimum).
10	**Intake Air Temperature Sensor Fault** - Inspect sensor connections for damage or corrosion. Compare sensor reading at KOEO and compare to specifications (2-3v at a temperature of 100°F)
11	**Idle Mixture Adjuster Sensor Fault** - Inspect IMA sensor connections for damage and corrosion. Check sensor wires for open or short to ground conditions. Sensor KOEO reading should be 0.5-4.5v.
12	**EGR Lift Sensor Fault** - Inspect EGR Lift sensor connections for damage and corrosion. With engine at idle speed, apply vacuum to EGR valve (the engine should stall). Sensor KOEO reading should be 1.0-1.2v.
13	**BARO or PA Sensor Electrical Fault** - Test the BARO sensor circuits. The BARO VREF should be 4.9-5.1v. The BARO GND circuit should read less than 100mv between the sensor and body ground. The BARO signal circuit should read about 2.8-3.0v at KOEO (sea level).
14	**Electronic Air Control Valve Fault** - Inspect the EACV (valve) connections. Check the circuits to the ignition feed and control wires for short to ground or open conditions. Compare EACV resistance to specification (6-20 ohms).
15	**Ignition Output Signal Fault** - Check the ignition output circuit for a short to ground or open condition. The KOEO voltage reading should be 12v. If the circuit is okay, the igniter or the PCM (switching transistor) may be at fault.

Acura & Honda Trouble Code List (OBD I)

Instructions: Read and record all codes in memory. Perform a PCM reset function and do a Road Test. If the code resets, refer to the codes and diagnostics tips below as needed.

1986-95 Acura & Honda Models

CODE ID	CODE DESCRIPTION & REPAIR TIPS
17 (D4 lamp may be on also)	**Vehicle Speed Sensor Fault** - Inspect VSS connections for damage. Monitor sensor signal at over 10 mph (it should vary 0-5 volts). Note: If the S4 lamp also blinks on A/T models, check the A/T control unit.
18	**Ignition Timing Adjuster Sensor Fault** - Inspect ITA connections for damage. Check sensor wires for open or short to ground conditions.
20 Non-MIL Code	**Electric Load Detector Fault** - Inspect the ELD connections for damage. Check the sensor wires between the ELD and the PCM for open or short to ground conditions.
23	**Left Side Knock Sensor Fault** - Remove the L/S Knock sensor connector and check the circuit for a short to ground or open circuit condition. If the wire is okay, replace the sensor and/or PCM and retest for the code.
30 Non-MIL Code	**A/T FI Signal 'A' Circuit Fault** - Remove the A/T controller 22-pin connector and check the A/T FI Signal 'A' circuit for a grounded or open condition. If it is okay, substitute a known-good PCM and retest. If fault is gone, replace original PCM.
31 Non-MIL Code	**A/T FI Signal 'B' Circuit Fault** - Remove the A/T controller 22-pin connector and check the A/T FI Signal 'B' circuit for a grounded or open condition. If it is okay, substitute a known-good PCM and retest. If fault is gone, replace original PCM.
41	**Left Side (Front) HO2S Heater Control Fault** - Check sensor connector for damage or corrosion. Measure HO2S heater resistance across the 'C' and 'D' terminals. Compare the reading to specifications (10-20 ohms).
42	**Right Side (Rear) HO2S Heater Control Fault** - Check sensor connector for damage or corrosion. Measure HO2S heater resistance across the 'C' and 'D' terminals (the reading should be from 10-20 ohms).
43	**Left Side (Front) Fuel System Fault** - Monitor the HO2S signal with a Lab Scope during acceleration (0.5-1.0v) and deceleration (0.0-0.4v). Inspect fuel filter for restriction and check fuel pressure and fuel pressure regulator. Check the sensor connector for damage and corrosion. Check the wire to PCM terminal D14 (WHT wire) for an open or grounded condition.
44	**Right Side (Rear) Fuel System Fault** - Monitor the HO2S signal with a Lab Scope during acceleration (0.5-1.0v) and deceleration (0.0-0.4v). Check sensor connector for damage and corrosion. Check HO2S circuit (RED/BLU wire) for circuit faults.
45	**Left Side (Front) A/F Ratio Fault** - Test the fuel pressure and check for a plugged fuel filter. If okay, replace the MAP sensor and retest. If now okay replace original MAP sensor. If code remains, reinstall MAP sensor. Test left side injectors for leaks.
46	**Right Side (Front) A/F Ratio Fault** - Test the fuel pressure and check for a plugged fuel filter. If okay, replace the MAP sensor and retest. If now okay replace original MAP sensor. If code remains, reinstall MAP sensor. Test right side injectors for leak.
48	**Linear A/F Sensor Fault** - Check the LAF sensor circuit for an open, short to ground or disconnected condition. Monitor the LAF sensor signal with a Lab Scope.
53	**Right Side Knock Sensor Fault** - Remove R/S Knock sensor connector and check the circuit for a short to ground or open circuit condition. If the wire is okay, replace the sensor and/or PCM and retest for the code.
54	**Crank Angle Sensor No. 2 Fault** - Test sensor resistance and compare to specifications (650-900 ohms). Check sensor connector for damage or dirt. Compare sensor output to cranking specifications (250mv AC minimum).
59	**Cylinder Position Sensor No. 2 Fault** - Test sensor resistance and compare to specifications (500-1000 ohms). Check sensor cranking output (250mv AC min.).

Acura & Honda Trouble Code List (OBD II)

Instructions: Read and record all codes in memory. Also, record the Freeze Frame data as this information can be used during trouble code diagnosis. If a PCM reset function is done, all codes and freeze frame data will be lost! Code descriptions and test conditions in this Code List are not meant to replace the use of a trouble code repair chart.

1995-03 Acura & Honda Models

CODE ID	CODE DESCRIPTION & TEST CONDITIONS
P0106 (2-T)	**MAP Sensor Circuit Range/Performance Fault** *Test Conditions & results:* Engine running for less than one second, then test started and the PCM detected a MAP sensor value of more than 11.8" Hg.
P0107 (1-T)	**MAP Sensor Circuit Low Voltage** *Test Conditions & results:* Key on, then test started and the PCM detected a MAP sensor value of near 0" Hg.
P0108 (1-T)	**MAP Sensor Circuit High Voltage** *Test Conditions & results:* Engine running, then test started and the PCM detected a MAP sensor value of near 29.9" Hg.
P0111 (2-T)	**IAT Sensor Circuit Range/Performance Fault** *Test Conditions & results:* Key on, then test started and the PCM detected an IAT sensor value that changed too much in too short a time period.
P0112 (1-T)	**IAT Sensor Circuit Low Voltage** *Test Conditions & results:* Key on, then test started and the PCM detected an IAT sensor value of over 302°F.
P0113 (1-T)	**IAT Sensor Circuit High Voltage** *Test Conditions & results:* Key on, then test started and the PCM detected an IAT sensor value of under -4°F.
P0116 (2-T)	**ECT Sensor Circuit Range/Performance Fault** *Test Conditions & results:* Key on, then test started and the PCM detected an ECT sensor value that changed too much in too short a time period.
P0117 (1-T)	**ECT Sensor Circuit Low Voltage** *Test Conditions & results:* Key on, then test started and the PCM detected an ECT sensor value of over 302°F.
P0118 (1-T)	**ECT Sensor Circuit High Voltage** *Test Conditions & results:* Key on, then test started and the PCM detected an ECT sensor value of under -4°F.
P0122 (1-T)	**TP Sensor Circuit Low Voltage** *Test Conditions & results:* Key On or engine running, engine warmed-up, then test started and PCM detected a closed throttle TP sensor value of under 10% (0.16v).
P0123 (1-T)	**TP Sensor Circuit High Voltage** *Test Conditions & Results:* Key On or engine running, engine warmup finished, then test started and PCM detected a WOT TP sensor value of over 90% (4.6v).
P0131 - Front (1-T)	**HO2S-1 Circuit Low Voltage (Bank 1 Sensor 1)** *Test Conditions & Results:* Engine warmup finished, vehicle running in closed loop with AT in (2) position (4th gear for MT), then test started and PCM detected a HO2S-1 voltage fixed below a stored limit.
P0132 - Front (1-T)	**HO2S1 Circuit High Voltage (Bank 1 Sensor 1)** *Test Conditions & Results:* Engine warmup finished, vehicle running in closed loop with AT in (2) position (4th gear for MT), then test started and PCM detected a HO2S-1 voltage fixed over a stored limit.
P0133 - Front (2-T)	**HO2S-1 Circuit Slow Response (Bank 1 Sensor 1)** *Test Conditions & Results:* Engine warmup finished, vehicle running in closed loop with AT in (2) position (4th gear for MT), then test started and PCM detected the HO2S-1 response time from 300-600 mv was too slow, or a too slow rich to lean or lean to rich response time.
P0135 - Front (1-T)	**HO2S-1 Heater Circuit Fault (Bank 1 Sensor 1)** *Test Conditions & Results:* Engine running, then test started and PCM detected an open or short in the HO2S-1 heater circuit.

Acura & Honda Trouble Code List (OBD II)

1995-03 Acura & Honda Models

CODE ID	CODE DESCRIPTION & TEST CONDITIONS
P0137 - Rear (1-T)	**HO2S-2 Circuit Low Voltage (Bank 1 Sensor 2)** *Test Conditions & Results:* Engine warmup finished, vehicle running in second gear in closed loop, then test started and PCM detected HO2S-2 voltage fixed at less than 100 mv.
P0138 - Rear (1-T)	**HO2S-2 Circuit High Voltage (Bank 1 Sensor 2)** • *Test Conditions & Results:* Engine warmup finished, vehicle running in second gear in closed loop, then test started and PCM detected a HO2S-2 voltage fixed at more than 600 mv.
P0139 - Rear (2-T)	**HO2S-2 Circuit Slow Response (Bank 1 Sensor 2)** *Test Conditions & Results:* Engine warmup finished, vehicle running in 2nd gear in closed loop, then test started and PCM detected the HO2S-2 voltage did not vary between 300-600 mv for 2 minutes.
P0141 - Rear (1-T)	**HO2S-2 Heater Circuit Fault (Bank 1 Sensor 2)** *Test Conditions & Results:* Engine running, then test started and PCM detected an open or short in the HO2S-2 heater circuit.
P0151 - Left Front (1-T)	**HO2S-1 Circuit Low Voltage (Bank 2 Sensor 1)** *Test Conditions & Results:* Engine warmup finished, vehicle running in second gear in closed loop, then test started and PCM detected a HO2S-1 voltage fixed at less than 100 mv.
P0152 - Left Front (1-T)	**HO2S-1 Circuit High Voltage (Bank 2 Sensor 1)** *Test Conditions & Results:* Engine warmup finished, vehicle running in second gear in closed loop, then test started and PCM detected a HO2S-1 voltage fixed at more than 900 mv.
P0153 -Left Front (2-T)	**HO2S-1 Circuit Slow Response (Bank 2 Sensor 1)** *Test Conditions & Results:* Engine warmup finished, vehicle running in second gear in closed loop, then test started and PCM detected the HO2S-1 voltage did not vary between 300-600 mv for 2 minutes.
P0155 - Left Front (1-T)	**HO2S-1 Heater Circuit Fault (Bank 2 Sensor 1)** *Test Conditions & Results:* Engine running, then test started and PCM detected an open or short in the HO2S-1 heater circuit.
P0157 - Left Rear (1-T)	**HO2S-2 Circuit Low Voltage (Bank 2 Sensor 2)** *Test Conditions & Results:* Engine warmup finished, vehicle running in second gear in closed loop, then test started and PCM detected a HO2S-2 voltage fixed at less than 300 mv.
P0158 - Left Rear (1-T)	**HO2S-2 Circuit High Voltage (Bank 2 Sensor 2)** *Test Conditions & Results:* Engine warmup finished, vehicle running in second gear in closed loop, then test started and PCM detected a HO2S-2 voltage fixed at more than 600 mv.
P0159 - Left Rear (2-T)	**HO2S-2 Circuit Slow Response (Bank 2 Sensor 2)** *Test Conditions & Results:* Engine warmup finished, vehicle running in second gear in closed loop, then test started and PCM detected the HO2S-2 voltage did not vary between 300-600 mv for 2 minutes.
P0161 - Left Rear (1-T)	**HO2S-2 Heater Circuit Fault (Bank 2 Sensor 2)** *Test Conditions & Results:* Engine running, then test started and PCM detected an open or short in the HO2S-2 heater circuit.
P0171, P0174 - (1-T)	**Fuel System Too Lean (P0171-Bank 1, P0174-Bank 2)** *Test Conditions & Results:* No engine system codes set, engine warmup finished, running in closed loop, then test started and PCM detected the Long Term Fuel Trim (LONGFT) exceeded the lean limit amount.
P0172, P0175 - (1-T)	**Fuel System Too Rich (P0171-Bank 1, P0175-Bank 2)** *Test Conditions & Results:* No engine codes set, engine running in closed loop, then test started and PCM detected the LONGFT number exceeded the rich limit amount.
P0301-P0306 - (1-T or 2-T)	**Cylinder 1, 2, 3, 4, 5 or 6 Misfire Detected** *Test Conditions & Results:* No engine system codes set, engine running, then test started and PCM detected a misfire in Cylinders 1, 2, 3, 4, 5 or 6.

Acura & Honda Trouble Code List (OBD II)

1995-03 Acura & Honda Models

CODE ID	CODE DESCRIPTION & TEST CONDITIONS
P0325 (2-T)	**Knock Sensor Circuit Fault (Rear)** *Test Conditions & Results:* Engine running, then test started and PCM detected an open or short circuit fault in the Rear Knock Sensor or circuits.
P0330 (2-T)	**Knock Sensor Circuit Fault (Front)** *Test Conditions & Results:* Engine running, then test started and PCM detected an open or short circuit fault in the Front Knock Sensor or circuits.
P0335 (2-T)	**CKP Sensor 'A' Circuit Fault (No Signal)** *Test Conditions & Results:* Engine running, then test started and PCM detected no signal output from CKP Sensor 'A'.
P0336 (2-T)	**CKP Sensor 'A' Circuit Fault (Intermittent Signal)** *Test Conditions & Results:* Engine running, then test started and PCM detected an intermittent signal from CKP Sensor 'A'.
P0401 - (2-T)	**EGR System, Insufficient Flow Detected** *Test Conditions & results:* Engine warmup finished, engine running in closed loop, VSS at 40-55 mph for 2 minutes in D$_4$, then decelerate to 35 mph with throttle closed, then test started and PCM detected insufficient EGR Flow during the test.
P0420 - (3-T)	**Catalyst Efficiency Below Threshold (Bank 1)** *Test Conditions & results:* P0137, P0138, P0141 not set, running in closed loop, VSS at 40-55 mph for 2 minutes, then decelerate to 35 mph with throttle closed, then test started and PCM detected excessive Bank 1 HO2S-12 activity for set time period.
P0430 - (3-T)	**Catalyst Efficiency Below Threshold (Bank 2)** *Test Conditions & results:* P0137, P0138, P0141 not set, engine running in closed loop, VSS at 40-55 mph for 2 minutes, then decelerate to 35 mph with throttle closed, then test started and PCM detected excessive Bank 2 HO2S-22 activity for set time period. Test fuel quality, for exhaust leaks, mechanical faults.
P0441 - (2-T)	**EVAP System Incorrect Purge Flow** *Test Conditions & results:* Engine temperature over 154°F, vehicle driven for 10 minutes, then vehicle gradually accelerated to over 50 mph, then test started and the PCM detected incorrect purge flow during the test (determined by EVAP switch).
P0452 (1-T)	**Fuel Tank Pressure Sensor Circuit Low Voltage** *Test Conditions & results:* Key On, then test started and PCM detected a Fuel Tank Pressure (FTP) sensor signal below the allowable range.
P0453 (1-T)	**Fuel Tank Pressure Sensor Circuit High Voltage** *Test Conditions & results:* Key On, then test started and PCM detected a Fuel Tank Pressure (FTP) sensor signal over the allowable range.
P0500 (1-T)	**Vehicle Speed Sensor Circuit Low Voltage** *Test Conditions & Results:* Engine running at road load, then test started and PCM detected a low voltage signal from the Vehicle Speed Sensor (VSS).
P0501 (2-T)	**Vehicle Speed Sensor Circuit Performance** *Test Conditions & Results:* Engine running at road load, then test started and PCM detected a low voltage signal from the Vehicle Speed Sensor (VSS).
P0505 (2-T)	**Idle Speed Control System Fault** *Test Conditions & Results:* Engine running at road load, then test started and PCM determined that the difference between Actual idle speed and Target idle speed exceeded a stored value in the PCM.
P0700 - D$_4$ Lamp Blinks (1-T)	**Automatic Transaxle Fault** *Test Conditions & Results:* Engine running, then test started and the PCM detected an A/T fault (P0700 sets when other TCM fault codes are set).
P0715 - Non MIL Code (1-T)	**TCM A/T Mainshaft Speed Sensor Circuit Fault** *Test Conditions & Results:* Key On or engine running, then test started and PCM detected an open or short in the Mainshaft Speed Sensor or its related circuits.
P0720 - D$_4$ Lamp Blinks (1-T)	**TCM A/T Countershaft Speed Sensor Circuit Fault** *Test Conditions & Results:* Key On or engine running, then test started and PCM detected a signal from TCM that indicated an open or short condition in the Countershaft Speed Sensor or its related circuits.

Acura & Honda Trouble Code List (OBD II)

1995-03 Acura & Honda Models

CODE ID	CODE DESCRIPTION & TEST CONDITIONS
P0725 - Non MIL Code (1-T)	**Automatic Transaxle Fault** *Test Conditions & Results:* Key On or engine running; then test started and PCM detected an A/T fault (PO725 sets when other Automatic Transaxle codes are set).
P0730 - Non MIL Code (1-T)	**TCM A/T Shift Control System Fault** *Test Conditions & Results:* No other A/T codes set, vehicle running at cruise speed, then test started and PCM detected the lockup clutch did not engage or disengage.
P0740 - Non MIL Code (1-T)	**TCM A/T Lockup Clutch System Fault** *Test Conditions & Results:* No other A/T codes set, vehicle running at cruise speed, then test started and PCM detected the lockup clutch did not engage or disengage.
P0753 - D4 Lamp Blinks (1-T)	**TCM A/T Lockup Solenoid 'A' Circuit Fault** *Test Conditions & Results:* Key On or engine running, then test started and PCM detected an open or short in the Solenoid Valve 'A' or its related circuits.
P0758 - D4 Lamp Blinks (1-T)	**TCM A/T Lockup Solenoid 'B' Circuit Fault** *Test Conditions & Results:* Key On or engine running, then test started and PCM detected an open or short in the Solenoid Valve 'B' or its related circuits.
P0763 - D4 Lamp & MIL Blink (1-T)	**TCM A/T Control Unit or Related Circuit Fault** *Test Conditions & Results:* Key On or engine running, then test started and PCM detected a fault in the TCM A/T Control Unit or one of its related circuits.
P1106 (2-T)	**Barometric Pressure Sensor Range/Performance Fault** *Test Conditions & Results:* Engine running; then test started and PCM detected a BARO sensor signal that did not change within a certain time period.
P1107 (1-T)	**Barometric Pressure Sensor Circuit Low Voltage** *Test Conditions & Results:* Key On, then test started and PCM detected a BARO sensor signal below the allowable range.
P1108 (1-T)	**Barometric Pressure Sensor Circuit High Voltage** *Test Conditions & Results:* Key On, then test started and PCM detected a BARO sensor signal above the allowable range.
P1121 (1-T)	**Throttle Position Sensor Signal Lower Than Expected** *Test Conditions & Results:* Engine running, then test started and PCM detected a TP sensor signal lower than the expected value.
P1122 (1-T)	**Throttle Position Sensor Signal Higher Than Expected** *Test Conditions & Results:* Engine running, then test started and PCM detected a TP sensor signal higher than the expected value.
P1128 (1-T)	**Manifold Absolute Pressure Sensor Lower Than Expected** *Test Conditions & Results:* Engine running, then test started and PCM detected a MAP sensor signal lower than the expected value.
P1129 (1-T)	**Manifold Absolute Pressure Sensor Higher Than Expected** *Test Conditions & Results:* Engine running, then test started and PCM detected a MAP sensor signal higher than the expected value.
P1149 (2-T) 1996-03 Honda	**HO2S-1 Performance (Bank 1 Sensor 1)** *Test Conditions & Results:* Engine warmup finished, engine running at steady throttle at 55 mph in 5th gear (MT), then test started and PCM detected the HO2S-1 response time from 300-600 mv was slow, or a slow rich to lean or lean to rich time.
P1162 (1-T) 1996-03 Honda	**HO2S-1 Circuit Fault (Bank 1 Sensor 1)** *Test Conditions & Results:* Engine warmup finished, engine running at 1500 rpm in 3rd gear (MT), then test started and PCM detected a HO2S-1 circuit fault due to an open or short circuit condition.
P1163 (2-T) 1996-03 Honda Civic: D16Y5 VTEC-E MT	**HO2S-1 Slow Response (Bank 1 Sensor 1)** *Test Conditions & Results:* Engine warmup finished, engine running at steady throttle at 55 mph in 5th gear (MT), then test started and PCM detected the HO2S-1 response time from 300-600 mv was slow, or a slow rich to lean or lean to rich time.

Acura & Honda Trouble Code List (OBD II)

1995-03 Acura & Honda Models

CODE ID	CODE DESCRIPTION & TEST CONDITIONS
P1164 (2-T) 1996-03 Honda	**HO2S-1 Circuit Fault (Bank 1 Sensor 1)** *Test Conditions & Results:* Engine warmup finished, engine running at 1500 rpm in 4th gear (MT) during normal acceleration and deceleration conditions, then test started and PCM detected a HO2S-1 voltage too high or too low circuit condition.
P1165 (2-T) 1996-03 Honda	**HO2S-1 Performance (Bank 1 Sensor 1)** *Test Conditions & Results:* Engine warmup finished, engine running at steady throttle at 55 mph in 5th gear (MT), then test started and PCM detected the HO2S-1 response time from 300-600 mv was slow, or a slow rich to lean or lean to rich time.
P1166 (1-T) 1996-03 Honda	**HO2S-1 Heater Circuit Fault (Bank 1 Sensor 1)** *Test Conditions & Results:* Engine running, then test started and PCM detected an open or short in the HO2S-1 heater circuit.
P1167 (1-T) 1996-03 Honda	**HO2S-1 Heater Circuit (VS+) Fault (Bank 1 Sensor 1)** *Test Conditions & Results:* Engine running, then test started and PCM detected an open in the HO2S-1 VS+ heater circuit.
P1168 (2-T) 1996-03 Honda	**HO2S-1 Label Circuit Low Voltage Fault (Bank 1 Sensor 1)** *Test Conditions & Results:* Engine warmup finished, engine running for over 2 minutes, then test started and PCM detected a HO2S-1 voltage that remained below a low threshold value for too long.
P1169 (2-T) 1996-03 Honda Civic: VTEC-E	**HO2S-1 Label Circuit High Voltage Fault (Bank 1 Sensor 1)** *Test Conditions & Results:* Engine running in closed loop for 2 minutes, then test started and PCM detected a HO2S-1 voltage over the high threshold too long.
P1201-P1206 (1-T or 2-T)	**Cylinder 1, 2, 3, 4, 5 Or 6 Misfire Detected** *Test Conditions & Results:* No engine system codes set, engine running, then test started and PCM detected a misfire in engine cylinder 1, 2, 3, 4, 5 or 6.
P1241 (1-T) 1995-97 NSX	**Throttle Valve Control Motor Circuit No. 1 Fault** *Test Conditions & Results:* Engine running, then test started and PCM detected an open or short circuit in the Throttle Valve Control Motor Circuit No. 1.
P1242 (2-T) 1995-97 NSX	**Throttle Valve Control Motor Circuit No. 2 Fault** *Test Conditions & Results:* Engine running, then test started and PCM detected an open or short circuit in the Throttle Valve Control Motor Circuit No. 2.
P1243 (2-T) 1995-97 NSX	**Insufficient Throttle Position Detected** *Test Conditions & Results:* Engine running at road load, with at least one wide open throttle event, then test started and PCM detected insufficient throttle position.
P1244 (2-T) 1995-97 NSX	**Insufficient Closed Throttle Position Detected** *Test Conditions & Results:* Engine running at road load, with at least one wide open throttle event, then test started and PCM detected insufficient closed throttle position.
P1246 (1-T) 1995-97 NSX	**Accelerator Position Sensor No. 1 Circuit Fault** *Test Conditions & Results:* Key On or engine running, then test started and PCM detected an open or short in the Accelerator Pedal Position sensor No. 1 circuit.
P1247 (1-T) 1995-97 NSX	**Accelerator Position Sensor No. 2 Circuit Fault** *Test Conditions & Results:* Key On or engine running, then test started and PCM detected an open or short in the Accelerator Pedal Position sensor '2' circuit.
P1248 (2-T) 1995-97 NSX	**Accelerator Pedal Position Sensor Correlation Fault** *Test Conditions & Results:* Key On or engine running, then test started and PCM detected an incorrect correlation between Accelerator Pedal Position Sensor No. 1 and Sensor No. 2 signals.
P1253 (1-T)	**VTEC System Circuit Fault** *Test Conditions & Results:* Engine running, then test started and PCM detected an open or short in the VTEC solenoid or its circuit.
P1259 (2-T) Rear Bank	**VTEC System Circuit Fault (Bank 1)** • *Test Conditions & Results:* Engine running in closed loop, accelerate in first gear to a speed over 6000 rpm for two seconds, then test started and PCM detected either an open or short in the right side Bank 1 VTEC pressure solenoid or switch.

Acura & Honda Trouble Code List (OBD II)

1995-03 Acura & Honda Models

CODE ID	CODE DESCRIPTION & TEST CONDITIONS
P1279 (2-T) Front Bank	**VTEC System Circuit Fault (Bank 2)** *Test Conditions & Results:* Engine running in closed loop, accelerate in first gear to a speed over 6000 rpm for two seconds, then test started and PCM detected either an open or short in the left side Bank 2 VTEC pressure solenoid or switch.
P1297 (1-T)	**Electrical Load Detector Circuit Low Voltage** *Test Conditions & Results:* Engine running, headlights On, then test started and PCM detected an Electrical Load Detector signal less than the allowable range.
P1298 (1-T)	**Electrical Load Detector Circuit High Voltage** *Test Conditions & Results:* Engine running, headlights On, then test started and PCM detected an Electrical Load Detector signal more than the allowable range.
P1300 (1-T or 2-T)	**Random Misfire Detected** *Test Conditions & Results:* No engine system codes set, engine running, then test started and PCM detected a random misfire fault in one or more cylinders.
P1301-P1306 (1-T or 2-T)	**Misfire Detected** *Test Conditions & Results:* No engine system codes set, engine running, then test started and PCM detected a misfire fault in Cylinder No. 1 to Cyl No. 6.
P1316 (1-T)	**Spark Plug Detection Module Circuit Fault (Bank 2)** *Test Conditions & Results:* No engine codes set, running in closed loop, then test started and PCM detected an open or short in the Bank 2 SPD Module or its circuit.
P1317 (1-T)	**Spark Plug Detection Module Circuit Fault (Bank 1)** *Test Conditions & Results:* No other codes set, running in closed loop, then test started and PCM detected an open or short in the Bank 1 SPD module or its circuit.
P1318 (1-T)	**Spark Plug Detection Module Reset Fault (Bank 2)** *Test Conditions & Results:* No other codes set, running in closed loop, then test started and PCM detected an open or short in the Bank 2 SPD Module Reset circuit.
P1319 (1-T)	**Spark Plug Detection Module Reset Fault (Bank 1)** *Test Conditions & Results:* No other codes set, running in closed loop, then test started and PCM detected an open or short in the Bank 1 SPD Module Reset circuit.
P1336 (1-T)	**Crankshaft Speed Fluctuation Sensor Signal Fault** *Test Conditions & Results:* Engine running, then test started and PCM detected an intermittent signal from the Crankshaft Speed Fluctuation sensor.
P1336 (1-T)	**Crankshaft Speed Fluctuation Sensor 'B' Signal Fault** *Test Conditions & Results:* Engine running, then test started and PCM detected an intermittent signal from Crankshaft Speed Fluctuation Sensor 'B'.
P1337 (1-T)	**Crankshaft Speed Fluctuation Sensor Circuit Fault (No Signal)** *Test Conditions & Results:* Engine running, then test started and PCM received no signal output from the Crankshaft Speed Fluctuation (CKF) sensor.
P1337 (1-T)	**Crankshaft Position Sensor 'B' Fault (No Signals)** *Test Conditions & Results:* Engine running then test started and PCM determined that no signal output was received from Crankshaft Position Sensor 'B'.
P1359 (1-T)	**Crankshaft Position Top Dead Center Sensor Fault** *Test Conditions & Results:* Engine running then test started and PCM detected a fault in the Crankshaft Position Top Dead Center sensor or its related circuits.
P1361 (1-T)	**Top Dead Center Sensor No. 1 Fault (Intermittent Signal)** *Test Conditions & Results:* Engine running, then test started and PCM detected an intermittent signal from the Top Dead Center (TDC) No. 1 sensor.
P1362 (1-T)	**Top Dead Center Sensor No. 1 Fault (No Signal)** *Test Conditions & Results:* Engine running, then test started and PCM detected no signal from the Top Dead Center (TDC) No. 1 sensor.
P1366 (1-T)	**Top Dead Center Sensor No. 2 Fault (Intermittent Signal)** *Test Conditions & Results:* Engine running, then test started and PCM detected an intermittent signal from the Top Dead Center (TDC) No. 2 sensor.
P1367 (1-T)	**Top Dead Center Sensor No. 2 Fault (No Signal)** *Test Conditions & Results:* Engine running, then test started and PCM detected no signal from the Top Dead Center (TDC) No. 2 sensor.

Acura & Honda Trouble Code List (OBD II)

1995-03 Acura & Honda Models

CODE ID	CODE DESCRIPTION & TEST CONDITIONS
P1381 (1-T)	**Camshaft Position Sensor Fault (Intermittent Signal)** *Test Conditions & Results:* Engine running, then test started and PCM detected an intermittent signal from the CMP sensor (CMP sensor 'A' for Acura).
P1382 (1-T)	**Camshaft Position Sensor Fault (No Signal)** *Test Conditions & Results:* Engine running, then test started and PCM detected no signals from the CMP sensor (CMP sensor 'A' for Acura).
P1386 (1-T)	**Camshaft Position Sensor 'B' Intermittent Signals** *Test Conditions & Results:* Engine running, then test started and PCM detected an intermittent signal from camshaft position sensor 'B'.
P1387 (1-T)	**Camshaft Position Sensor 'B' No Signals** *Test Conditions & Results:* Engine running, then test started and PCM detected no signals from Camshaft Position sensor 'B'.
P1456 (2-T)	**EVAP Control System Leak Detected (Fuel Tank Area)** *Test Conditions & Results:* Engine temperature over 154°F, IAT over 32°F and VSS over 5 mph for 2 minutes. Then the EVAP Vent and Control solenoids are controlled to run the test. Then the PCM detected an incorrect fuel tank pressure reading.
P1457 (2-T)	**EVAP Control System Leak Detected (Canister Area)** *Test Conditions & Results:* Engine temperature over 154°F, IAT over 32°F and VSS over 5 mph for 2 minutes. Then the EVAP Vent and Control solenoids are controlled to run the test. Then the PCM detected an incorrect fuel tank pressure reading.
P1459 (1-T)	**EVAP Purge Flow Switch Fault** *Test Conditions & Results:* Engine warmup finished, engine idling, then test started and PCM detected a fault in the EVAP Purge Flow Switch.
P1491 (2-T)	**EGR Valve Lift Sensor Insufficient Flow Detected** *Test Conditions & Results:* Vehicle at cruise speed in closed loop, engine speed from 1700-2500 rpm for 10 minutes, then test started and PCM detected an EGR Valve Lift sensor signal that indicated that insufficient EGR flow was present.
P1498 (1-T)	**EGR Valve Lift Sensor High Voltage** *Test Conditions & Results:* Key on or engine running, then test started and PCM detected an EGR Valve Lift sensor signal voltage above the allowable range.
P1508 (1-T)	**Idle Air Control Valve Circuit Fault** *Test Conditions & Results:* Engine running, then test started and PCM detected a fault in the Idle Air Control (IAC) valve or its related circuits.
P1509 (1-T)	**Idle Air Control Valve Circuit Fault** *Test Conditions & Results:* Engine running, then test started and PCM detected an open or short condition in the Idle Air Control (IAC) valve or its related circuits.
P1519 (1-T)	**Idle Air Control Valve Circuit Fault** *Test Conditions & Results:* Engine running, then test started and PCM detected an open or short condition in the Idle Air Control (IAC) valve or its related circuits.
P1607 (1-T)	**PCM Internal Fault 'A'** *Test Conditions & Results:* Key On, then test started and Internal Fault 'A' detected.
P1608 (1-T)	**PCM Internal Fault 'B'** *Test Conditions & Results:* Key On, then test started and Internal Fault 'B' detected.
P1655 (1-T)	**TMA or TMB Signal Line Fault** *Test Conditions & Results:* Engine running, then test started and PCM detected an open or short condition in the TMA or TMB circuit.
P1655 - D4 Lamp Blinks (1-T)	**SEFA or SEAF Signal Line Fault** *Test Conditions & Results:* Engine running, then test started and PCM detected an open or short condition in the SEFA or SEAF signal circuit.
P1671 (1-T)	**TCM A/T FI Data Line, No Signal** *Test Conditions & Results:* Engine idling for one minute, then test started and PCM detected no signal present at TCM FI Data Line.
P1672 (1-T)	**TCM A/T FI Data Line, Failure Detected** *Test Conditions & Results:* Engine idling for one minute, then test started and PCM detected a failure in the TCM FI Data Line.

Acura & Honda Trouble Code List (OBD II)

1995-03 Acura & Honda Models

CODE ID	CODE DESCRIPTION & TEST CONDITIONS
P1676 (1-T)	**TCM A/T FI Data Line, Failure Detected** *Test Conditions & Results:* Engine idling for one minute, then test started and PCM detected a failure in the TCM FI Data Line.
P1677 (1-T)	**TCM A/T FI Data Line, Failure Detected** *Test Conditions & Results:* Engine idling for one minute, then test started and PCM detected a failure in the TCM FI Data Line.
P1690 (1-T)	**TCSTB Data Line, Failure Detected** *Test Conditions & Results:* Engine idling for one minute, then test started and PCM detected a failure in the TCSTB Data Line.
P1696 (1-T)	**TCFC Line Low Input Detected** *Test Conditions & Results:* Engine idling for one minute, then test started and PCM detected a low input on the TCFC Data Line.
P1697 (1-T)	**TCFC Line High Input Detected** *Test Conditions & Results:* Engine idling for one minute, then test started and PCM detected a high input on the TCFC Data Line.
P1705 - D$_4$ Lamp Blinks (1-T)	**TCM A/T Gear Position Switch Circuit Shorted** *Test Conditions & Results:* Engine running, then test started and PCM detected a short in the A/T Gear Position Switch with no lockup clutch operation.
P1706 - Non MIL Code (1-T)	**TCM A/T Gear Position Switch Circuit Open** *Test Conditions & Results:* Engine running, then test started and PCM detected an open condition in the A/T Gear Position Switch with no lockup clutch operation.
P1738 - D$_4$ Lamp Blinks & MIL is on (1-T)	**TCM A/T Controller or Related Circuit Fault** *Test Conditions & Results:* Key On or engine running, then test started and PCM detected a fault in the A/T Controller or one of its related circuits.
P1739 - D$_4$ Lamp Blinks & MIL is on (1-T)	**TCM A/T Controller or Related Circuit Fault** *Test Conditions & Results:* Key On or engine running, then test started and PCM detected a fault in the A/T Controller or one of its related circuits.
P1753 - D$_4$ Lamp Blinks (1-T)	**TCM A/T Lockup Solenoid Valve 'A' Fault** *Test Conditions & Results:* Key On or engine running, then test started and PCM detected that the lockup clutch did not operate properly (may be stuck in 4th gear).
P1758 - D$_4$ Lamp Blinks (1-T)	**TCM A/T Lockup Solenoid Valve 'B' Fault** *Test Conditions & Results:* Key On or engine running, then test started and PCM detected that the lockup clutch operate properly (may be stuck in 1st or 4th gear).
P1768 - D$_4$ Lamp Blinks & MIL is on (1-T)	**TCM A/T Controller or Related Circuit Fault** *Test Conditions & Results:* Key On or engine running, then test started and PCM detected a fault in the A/T Controller or one of its related circuits.
P1773 - D$_4$ Lamp Blinks & MIL is on (1-T)	**TCM A/T Controller or Related Circuit Fault** *Test Conditions & Results:* Key On or engine running, then test started and PCM detected a fault in the A/T Controller or one of its related circuits.
P1790 - D$_4$ Lamp Blinks or Off (1-T)	**TCM A/T TP Sensor Circuit Fault** *Test Conditions & Results:* Engine running for 15 seconds, then test started and PCM detected a signal from TCM that a faulty TPS signal was received.
P1791 - D$_4$ Lamp Blinks (1-T)	**TCM A/T Vehicle Speed Sensor Circuit Fault** *Test Conditions & Results:* Engine running for 15 seconds, then test started and PCM detected a signal from TCM that a faulty VSS signal was received.
P1792 - D$_4$ Lamp Blinks (1-T)	**TCM A/T ECT Sensor Circuit Fault** *Test Conditions & Results:* Engine running for 5 seconds, then test started and PCM detected a signal from TCM that a faulty ECT signal was received.

NOTE: *In the Code ID column, (1-T) indicates this is a one-trip code, (2-T) indicates this is a two-trip code and (3-T) indicates this a three-trip code.*

INTRODUCTION

The following terms and acronyms are used throughout the articles in this manual. This glossary provides a brief description of these terms.

Accelerated Warmup System - When activated by the PCM, this system supplies bypass air into a dynamic chamber.

Airflow Meter - A device that detects the amount of intake airflow and sends a signal to the PCM.

Air Valve - A device that supplies bypass air for the dynamic chamber (a chamber that connects the intake runners) in order to increase engine idle speed during warm-up.

Atmospheric Pressure Sensor (PA) - A device that detects the amount of atmospheric pressure and sends a signal to the PCM.

Battery Voltage Signal - The PCM uses the battery voltage signal from the alternator 'FR' terminal to compensate for low voltage conditions that could result in poor fuel injector response. When the PCM detects a certain battery voltage value, it compensates by increasing injector opening time.

BARO or PA Sensor - The BARO sensor, located inside the PCM on late-model vehicles, is used to convert atmospheric pressure into a voltage signal. The PCM uses this signal to modify the basic fuel injector discharge duration.

Baud Rate - The rate at which a PCM is able to transfer and receive data. Baud rate is measured in bits per second.

Bypass Control Solenoid - Some Acura/Honda engine applications provide for two air intake paths to the intake manifold. The use of two intake paths allows the PCM to select the path length that will provide the most power or performance for any given engine speed. The PCM selects the air intake path to use based upon sensor input signals. The PCM controls bypass valves located in the intake manifold through the use of two bypass control solenoids and a bypass control diaphragm.

Catalyst - A material that promotes a chemical reaction and is not changed by the chemical reaction. The noble metals Platinum, Palladium and Rhodium are used in catalysts in automotive catalytic converters.

Circuit Opening Relay - A relay used on some engines in order to provide voltage to the fuel pump when the engine is running.

Closed Loop - An engine operating state in which the PCM controls the A/F mixture based upon signals received from the upstream or front oxygen sensor.

Combination Meter - This term is used to describe the instrument or gauge portion of the instrument panel.

Crank Angle Sensor (CAS) - The crank angle sensor, mounted in the distributor housing, consists of a combination TDC/CYL or a combination TDC/CYL/CRANK sensor. Depending upon design, it provides two or three separate signals to the PCM.

Crankshaft Position (CKP) Sensor - The CKP sensor, located in front of the engine along as part of the combination CKP/TDC sensor (on some models) or inside the distributor, is used to determine fuel injection and ignition timing for each cylinder. It is also used to detect engine speed.

Cylinder Position (CYL or CYP) Sensor - The CYL or CYP sensor, located inside the distributor on engines equipped with DI, is used to detect the position of engine cylinder No. 1. This signal is used as a reference for sequential fuel injection to each cylinder.

Dashpot Control Vacuum Solenoid - On some engine applications, the PCM (based upon ECT signals) controls a dashpot control vacuum solenoid in order to regulate vacuum to the outer side of the idle speed dashpot. When the solenoid is energized, the vacuum signal to the dashpot is blocked. This action causes spring pressure to slightly open the throttle valve, allowing for increased cold engine fast idle. The idle speed dashpot (inner side) controls throttle closure speed based upon spring pressure and a restricted manifold vacuum signal through an orifice. The PCM controls the outer side.

DI - This term is used to describe engines equipped with Distributor Ignition systems.

EI - This term is used to describe engines equipped with Distributorless Ignition systems.

Engine Coolant Temperature Sensor - The engine coolant temperature sensor uses a thermistor to measure differences in coolant temperature.

Electronic Air Control Valve (EACV) - The electronic air control valve (also called an electronic idle control valve on some engines) is controlled by the PCM in order to bypass additional air into the intake manifold. This use of the additional air allows the idle speed to increase and maintain normal idle.

EGR Control Solenoid - On some engine applications, the PCM controls the ground circuit to the EGR control solenoid in order to regulate the flow of recirculated exhaust gases back into the intake system.

EGR Lift Sensor - The EGR lift sensor senses the position of the EGR valve and relays that position information to the PCM.

Fast Idle Valve - During periods when the engine is cold, it requires higher idle speed to prevent erratic engine operation during warmup. A thermowax design fast idle valve allows addition engine rpm (not controlled by the PCM).

Heated Oxygen Sensor or Oxygen Sensor - The heated oxygen sensor detects the oxygen content of exhaust gas and produces a voltage signal proportional to that content. This signal is sent to the PCM.

Idle Mixture Adjuster (IMA) Sensor - The IMA sensor is mounted on the PCM housing. When the sensor adjusting screw is turned, the voltage signal to the PCM is changed. This action alters the fuel injector discharge duration.

Ignition Timing Adjuster (ITA) Connector - The ignition timing adjuster signal circuit is connected directly to the PCM. When the connector is jumped, the PCM defaults to base timing to allow for timing adjustments.

Immobilizer System - The Immobilizer (Security) System used on the 1998 Accord and Odyssey includes a set of encoded keys programmed to each vehicle. If a PCM is replaced on these vehicles without reprogramming the keys and PCM with a Honda Scan Tool, it will only start 12 times. This Immobilizer system is wired to the PCM.

Intake Air Temperature (IAT) Sensor - The intake air temperature sensor is a thermistor located in the intake manifold. It is used to measure the temperature of the incoming air stream and to provide a signal to the PCM.

LONGFT - The long term fuel trim adaptive memory stored by the PCM. This memory contains the injector pulsewidth compensation used to maintain minimum emissions output. LONGFT adaptive drives the SHRTFT adaptive to maximum operating efficiency.

Manifold Absolute Pressure (MAP) Sensor - The MAP sensor converts manifold air pressure readings into electrical voltage signals and sends a signal to the PCM.

Misfire - This term is defined as the lack of complete combustion in an engine cylinder.

Open Loop - An engine operating state in which the PCM controls the A/F mixture based upon a standard program in memory, and not in response to HO2S signals.

Oxygen Sensor Heater Control (HTRC) - Some oxygen sensors contain a heater that is controlled by the PCM in order to allow it to heat up rapidly and reduce emissions. The HTRC is connected to ignition power on most applications. On the 1998 2.3L I4 Accord engine, power to the HTRC circuit is controlled by a separate relay.

PGM-FI Main Relay - This relay contains two separate relays. One is energized whenever the key is on, and this action supplies power to the PCM, to the injectors, and to the second internal relay. The second relay is energized to supply power to the fuel pump for two seconds after the key is first turned on, and when the engine is running.

Powertrain Control Module (PCM) - The powertrain control module (old name engine control unit or ECU) is a solid state micro-computer that monitors engine conditions (sensors and switches) and controls the A/F ratio, fuel injection and ignition timing.

Primary or Front Heated Oxygen Sensor (HO2S-1) - This heated oxygen sensor, a Zirconia design sensor, is used to detect exhaust gas oxygen content, and to send signals to the PCM. Based upon these signals, the PCM varies fuel injection duration. To allow for quick warmup, and to stabilize its output, the HO2S-1 has an internal heater. This sensor is installed in the exhaust manifold or exhaust pipe. These

Purge Control Solenoid - The PCM uses the purge control solenoid (valve) to purge fuel vapors from the charcoal canister.

Secondary or Rear Heated Oxygen Sensor (HO2S-2) - This heated oxygen sensor, a Zirconia design, is used to detect exhaust gas oxygen content, and to send signals to the PCM. Based upon these signals, the PCM determines the relative condition of the catalytic converter (its ability to store free oxygen). To allow for quick warmup, and to stabilize its output, the HO2S-2 has an internal heater. It is installed after the converter.

SHRTFT - The short term fuel trim adaptive memory stored by the PCM. It contains the latest changes to injector pulsewidth compensation. The PCM uses SHRTFT memory along with HO2S signals to vary injector pulsewidth to obtain minimal emissions output.

Tandem Valve Control Solenoid - On Honda Civic engines equipped with throttle body injection, a tandem valve control solenoid is used to improve fuel atomization. Better fuel atomization is accomplished by regulating venturi vacuum to the tandem valve-actuating diaphragm through the PCM controlled tandem valve control solenoid.

Throttle Angle (TA) Sensor - This device is a mechanical variable resistor connected to the throttle shaft. On later models, this sensor is called a throttle position (TP) sensor.

Top Dead Center (TDC) Position Sensor - The TDC sensor, located in front of the engine as part of a combination TDC/CKP sensor (on some models) or inside the distributor, is used to determine ignition timing at initial startup. It is also used to determine ignition timing if the CKP sensor signal is abnormal or missing.

Throttle Position (TP) Sensor - This device is a mechanical variable resistor connected to the throttle shaft. On later models, this sensor is called a throttle position (TP) sensor.

Vacuum Control Solenoid - On some engine applications, the flow of vacuum to the outside distributor advance diaphragm is controlled by the PCM through the use of a vacuum control solenoid (valve).

Variable Valve Timing with Electronic Control (VTEC)- On some engine applications, the PCM controls the electronic operation of the engine valve timing through the use of a VTEC solenoid and VTEC vacuum switch.

ACURA CONTENTS

ACURA CONTENTS

ACURA PGM-FI System

Introduction

<u>EARLY SYSTEMS</u>

The Acura Programmed Fuel Injection (PGM-FI) system, introduced in 1986 on Integra and Legend models, is a computerized emission and fuel control system that provides optimum control of the engine and transmission (models with an Automatic Transaxle).

At the heart of the PGM-FI system is a Powertrain Control Module or PCM. The PCM was referred to as the Electronic Control Unit or ECU in earlier publications. The PCM includes a central processor unit (CPU) and digital control system in order to control various engine control systems. It is capable of accepting multiple inputs from various sensors and switches (see the lists that follow), and then performing fast calculations so that it can quickly decide how to control the various engine output devices.

<u>PGM-FI DIAGNOSTICS (1986-91 SYSTEMS)</u>

The first version of Acura PGM-FI systems (Integra from 1986-91, Legend from 1986-90) included a diagnostic system to control the operation of a light emitting diode(s) or LEDs on the PCM.

<u>PGM-FI DIAGNOSTICS (1991-95 SYSTEMS)</u>

The second version of Acura PGM-FI systems (Integra from 1992-95, Legend from 1991-95) included a diagnostic system to control the operation of a PGM-FI or Check Engine (C/E) light on the dash.

<u>OBD I SYSTEM DIAGNOSTICS</u>

In 1988, in order to meet government regulations, the OBD I system was added to the PGM-FI system. This first version of emission control diagnostics incorporates a set of strategies controlled by software inside the PCM. These strategies include control of the Check Engine light, code setting and clearing.

On 1992-95 Integra models, there is bi-directional communication with a Scan Tool that allows the tool to read any stored codes and a series of serial data items (i.e., ECT, IAT, HO2S, MAP, TP, RPM and VSS signals).

<u>Backup Function</u>

The PGM-FI system includes a backup function that is used to provide minimal engine operation should the CPU portion of the PCM fail.

<u>FAILSAFE FUNCTION</u>

The PGM-FI system also includes a failsafe function that is used if one of the main sensor inputs to the PCM is lost (i.e. ECT, MAP or TPS).

<u>SYSTEM HARDWARE & SOFTWARE</u>

The PGM-FI system is divided into two main parts, the system hardware and software.

The hardware components include:

All related actuators, relays, and solenoids
All related sensors, switches, interconnecting wires, connectors and terminals
The PGM-FI Main Relay and Powertrain Control Module or PCM

The software components include the programs that contain the strategies used by the PCM to control system outputs based on related inputs. These include these systems:

A/T Lockup Converter
Idle Speed Control
Ignition Spark Timing
Fuel Delivery Control
Powertrain Diagnostics

Introduction (continued)

COMPONENT LOCATION EXAMPLE

Component locations for a 1998 Acura 3.5L V6 (C35A1) engine with an Intake Air Bypass system (OBD II diagnostics) are shown in Figure 2-01.

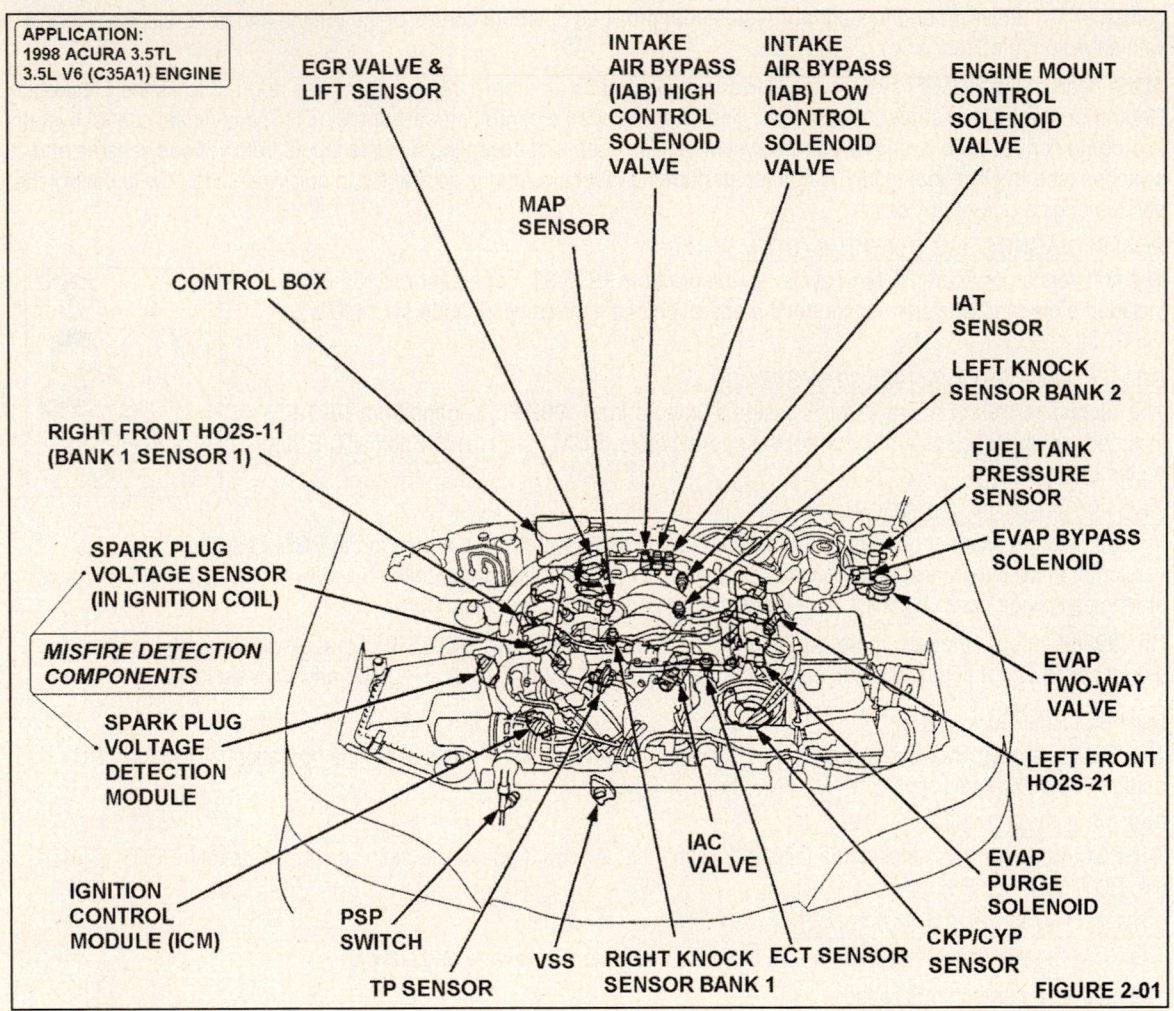

FIGURE 2-01

SENSOR INPUTS

The PCM in the PGM-FI system receives signals from these devices:

Barometric (PA) sensor (indicates atmospheric pressure)
ECT & IAT sensors (indicates engine coolant or air inlet temperature)
Crank, CYL and TDC sensors (located inside the distributor)
EGR Valve Lift sensor (indicates the amount of sensor opening)
MAP sensor (indicates the amount of engine load)
Oxygen sensors (indicate the amount of free oxygen in the exhaust)
Throttle Position sensor (indicates the throttle valve position)
Vehicle Speed sensor signal (indicates the vehicle speed)
Power Steering Pressure Switch (indicates load from the power steering)
Park Neutral Position Switch (Indicates the gear shift lever position)
Fuel Tank Pressure Sensor (Indicates the pressure in the fuel tank)

Introduction (continued)

<u>SWITCH INPUTS</u>

The PCM in the PGM-FI system receives signals from these switches:

Air Conditioning on signal (indicates the A/C clutch is engaged)
Automatic Transaxle Shift Position signal (indicated gearshift positions)
Alternator FR signal (indicates the alternator is charging)
Starter signal (from a circuit connected to the starter solenoid)
Battery Voltage signal (indicates the battery voltage at IGN 1 terminal)
Brake Switch signal (indicates if the brake switch is on or off)
Electronic Load Detector signal (indicates the amount of electronic load)
Heater Fan Switch signal (indicates if the heater blower is on or off)

<u>OUTPUT DEVICES</u>

The PCM in the PGM-FI system controls the operation of these devices

A/C Compressor Clutch operation (to turn the clutch on or off)
Air Induction Tandem Valve solenoid (to turn the solenoid on or off)
A/T Lockup (clutch) solenoid operation (to turn the solenoid on or off)
EGR Control solenoid (to turn the solenoid on or off)
Fast Idle Control solenoid operation (to turn the solenoid on or off)
Fuel injector control (to turn the fuel injectors on and off)
Ignition Igniter unit (to switch the coil primary on and off)
MFI Main Relay control (to turn the fuel pump on or off)
Purge Cutoff solenoid control (to turn the solenoid on or off)

<u>MISFIRE DETECTION</u>

The 1998 3.5L V6 engine is equipped with a Spark Plug Voltage Sensor and Spark Plug Voltage Detection Module. These devices are part of the Misfire Detection system.

<u>HEATED OXYGEN SENSOR IDENTIFICATION</u>

Acura I4, I5 and V6 engines from 1996-98 equipped with OBD II system diagnostics use both front and rear heated oxygen sensors (HO2S). The front (pre-catalyst) sensor is identified as HO2S-11.This designator indicates the pre-catalyst sensor located in cylinder bank 1 (identified as CYL 1 in the firing order). The rear sensor post-catalyst) is identified as HO2S-12 (the sensor after the catalyst). Refer to Figure 2-02 as needed.

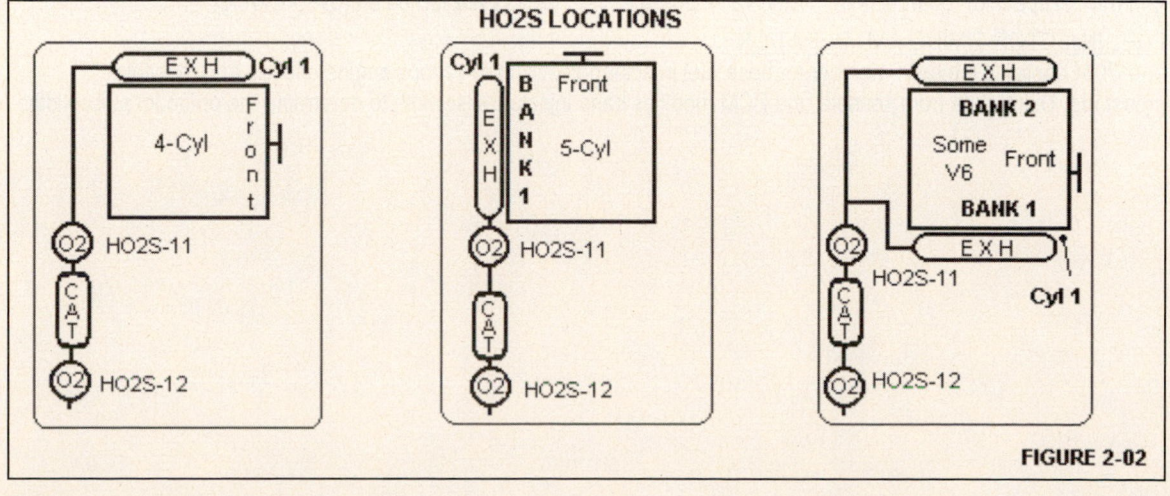

FIGURE 2-02

PGM-FI Controlled Functions

INTRODUCTION

The purpose of the PGM-FI system is to provide optimum control of the engine and transmission while meeting the objectives of regulations related to both the OBD I and OBD II systems. At the center of this system is a Powertrain Control Module (PCM) connected to various input and output devices through a wiring harness and several connectors. The PCM receives input information from various sensors and switches, and performs calculations based on information stored in Keep Alive Memory (KAM) tables in order to control various output devices such as actuators, relays, and solenoids.

On the PGM-FI system, the PCM controls the operation of the devices and functions of the items listed on the next two pages.

ALTERNATOR OUTPUT

The PCM controls the voltage generated at the alternator in accordance with the electrical load and driving mode. This action reduces engine load in order to improve the fuel economy of the vehicle.

A/C COMPRESSOR CLUTCH RELAY

When the PCM receives a demand for cooling from the air conditioning (A/C) system, it momentarily delays the command to energize the compressor clutch and enriches the A/F mixture to ensure smooth transition to A/C mode.

COLD FAST IDLE

With the engine is cold, the PCM supplies ground to the Fast Idle solenoid (valve). This action allows for additional bypass air (air that passes around the throttle plates) for cold engine idle stability.

EGR VALVE

When EGR is required for control of NOx emissions in the exhaust stream, the PCM grounds the EGR solenoid in order to allow a regulated amount of vacuum to flow to the EGR valve.

EVAP PURGE SOLENOID VALVE

When the PCM detects that engine temperature is over 122°F, it controls the EVAP purge solenoid valve that controls vacuum to the EVAP purge canister.

IDLE AIR CONTROL

When the PCM detects any of the following conditions, it controls current to the IAC valve in order to maintain the correct idle speed: A/C compressor on, a cold engine, transmission in gear, brake pedal depressed, high power steering load or high alternator charging rate.

FUEL CUT-OFF

During deceleration from engine speeds of over 950-1300 rpm (model specific) with the throttle valve closed, the PCM cuts off the fuel injectors to clean up tailpipe emissions and improve fuel economy. If engine speed exceeds 5500-6800 rpm (model specific) regardless of throttle valve position, the injectors are turned off (over-revving).

FUEL INJECTOR TIMING AND DURATION

The PCM contains tables in memory for base fuel injector pulsewidth at various engine speeds and manifold pressures. Using data from sensors, the PCM modifies base injector pulsewidth to determine the optimum pulsewidth.

FUEL PUMP CONTROL

When the ignition is turned on, the PCM supplies ground to the PGM-FI main relay that supplies current to the fuel pump for two seconds to pressurize the fuel system. Once the engine is running, the PCM continues to supply ground to the relay to supply current to the pump. With the engine off and key on, the PCM opens the ground circuit to the relay to cut current flow to the fuel pump.

IGNITION SPARK TIMING

The PCM contains data in long term memory that includes base ignition spark timing for various engine speeds and manifold pressures. The PCM changes ignition spark timing based on current engine temperature, engine load and throttle position inputs.

INTAKE AIR BYPASS SYSTEM

The Intake Air Bypass (IAB) system is used to achieve satisfactory power performance at both high and low engine speeds.

The additional performance is achieved by closing and opening two IAB system valves. High power at low engine speeds is achieved with the valves closed and high power at high engine speeds is achieved with the valve opened (at over 3200-3800 rpm).

LOCKUP SOLENOID CONTROL

Depending upon engine speed and throttle angle, the PCM grounds the A/T Lockup solenoid to control operation of the torque converter lockup clutch.

PURGE CUTOFF SOLENOID

STD Models - With engine temperature below 176°F, the PCM supplies ground to the EVAP Purge Cutoff solenoid (valve). This action cuts off the vacuum supply to the Purge Control valve.

VTEC SOLENOID CONTROL

On engines equipped with the VTEC system, the PCM controls the operation of the system through inputs from a VTEC pressure switch and by control of a VTEC solenoid valve (Figure 2-03).

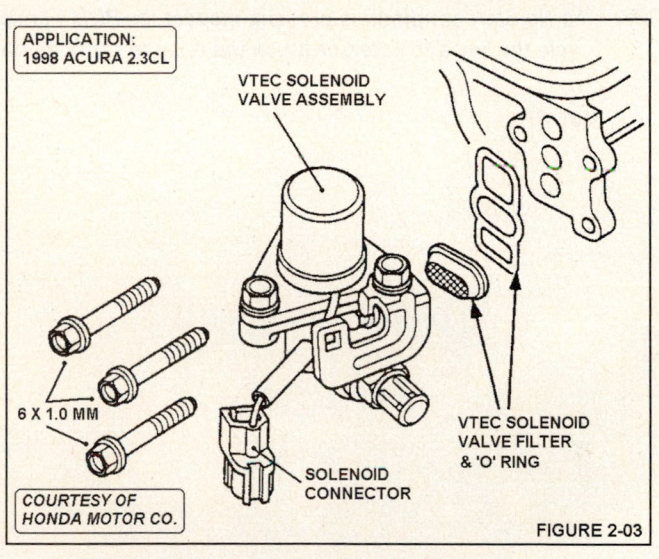

APPLICATION:
1998 ACURA 2.3CL

VTEC SOLENOID VALVE ASSEMBLY

6 X 1.0 MM

SOLENOID CONNECTOR

VTEC SOLENOID VALVE FILTER & 'O' RING

COURTESY OF HONDA MOTOR CO.

FIGURE 2-03

PGM-FI MAIN RELAY (1986-95)

INTRODUCTION

On 1986-95 Acura vehicles with fuel injection, the main relay consists of two relays inside one assembly located under the left side of the dash near the fuse/relay box or on a bracket. The main relay for the Acura SLX (Isuzu Trooper) is located on the right side engine compartment in the relay box.

MAIN RELAY OPERATION (EXCEPT FOR SLX MODELS)

With the key on, one relay is energized and this action supplies battery voltage to the PCM and to second relay. With the key on, the second relay is energized by the PCM for two seconds. The second relay is also energized whenever the engine is running. The second relay provides power to the fuel pump and fuel injectors. Refer to Figure 2-04 on the next page for a schematic of the main relay.

DRIVEABILITY SYMPTOMS

Problems with the main relay or its circuits could cause these symptoms:

No Start condition
Engine quits on the road and then restarts

TIPS: *If a No Start condition is present, inspect the PGM-FI relay for damage or corrosion. Have an assistant cycle the key and listen or touch the relay to verify its operation.*

PGM-FI MAIN RELAY (1986-95)

PRELIMINARY STEPS

All of the test steps in this procedure should be done in sequence in order to isolate the problem area. First, remove the main relay from the vehicle. Obtain a pair of jumper wires with connectors suitable to attach to the main relay and battery cables or posts.

PGM-FI MAIN RELAY TEST

With the relay removed, connect one jumper wire to battery positive and to main relay terminal No. 4. Connect another jumper wire to the relay terminal No. 8 and to body ground (Figure 2-04).

With the relay energized, use a DVOM to check for continuity between relay terminals No. 5 and No. 7. If there is continuity, go to Step 2. If there is no continuity, replace the relay.

Move the jumper wires to connect battery positive to main relay terminal No. 5 and then relay terminal No. 2 to ground. Check for continuity between main relay terminals No. 1 and No. 3. If there is no continuity, replace the main relay and retest.

Move the jumper wires to connect battery positive to main relay terminal No. 3 and then relay terminal No. 8 to ground. The jumper wire polarity is important due to an internal diode. Check for continuity between relay terminals No. 5 and No. 7. If there is no continuity, replace the main relay and retest. If there is continuity, test for battery power to the fuel pump or other devices. If these devices have power, go to Wiring Harness Tests.

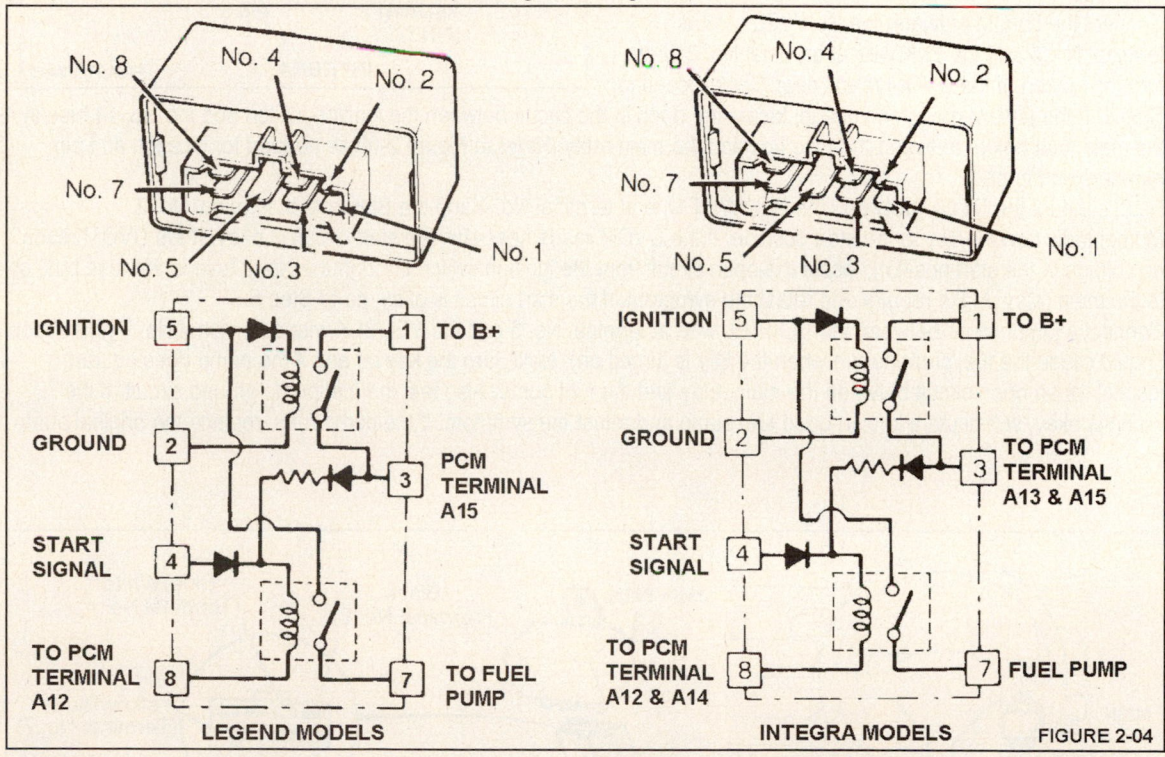

■ NOTE: *For applications other than the 1990 Integra and Legend models in this article, refer to other pin voltage charts to identify the pins and wire colors for all of the references to PCM terminals and wire colors in this article.*

PGM-FI MAIN RELAY (1986-95)

WIRING HARNESS TEST

The key should be off with the main relay removed to expose the relay wiring harness for testing during this test procedure.

Check for continuity between the BLK wire at main relay terminal No. 2 connector and body ground (Figure 2-05 for Integra). The DVOM should show continuity (less than 5 ohms of resistance). Repair any problems before continuing with this test).

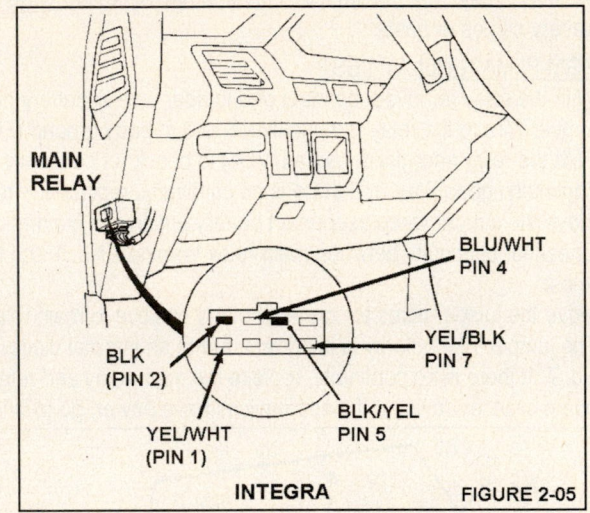

Connect a DVOM between the YEL/WHT wire at the harness terminal No. 1 and the BLK wire at terminal No. 2. If the DVOM reads battery voltage (okay), proceed to Step 3. If the DVOM reads no voltage, locate the open condition in the feed circuit from the battery through the PCM fuse (in the main fuse box) that connects to the main relay. Make any needed repairs and retest for the symptom.

Connect the DVOM between the BLK/YEL wire at terminal No. 5 and the BLK wire at terminal No. 2; then turn the key on. If the DVOM reads near 12v, proceed to Step 4. If the DVOM reads no voltage, locate the open in the circuit between the ignition switch and the No. 14 fuse in the main fuse box or between the fuse box and the main relay. Refer to Figure 2-06 as needed for location and pin numbers on Legend.

Next, connect the DVOM between the BLU/WHT wire at terminal No. 4 and the BLK wire at terminal No. 2. Momentarily turn the key to the start position. If the DVOM reads near 10v, the start circuit is okay. If the DVOM reads no voltage in the start position, locate the open circuit from the ignition switch through the No. 2 fuse in the fuse box, or to the main relay. Make repairs and retest the symptom. If the start circuit is okay, go to Step 5.

Connect a jumper wire between the BLK/YEL wire at terminal No. 5 and the YEL/BLK wire at terminal No. 7 (this action should cause the fuel pump to run when the key is turned on). Next, turn the key on and if the pump does not run, inspect for an open circuit between the main relay and the fuel pump. Also test the fuel pump ground circuit. If the wiring is okay, substitute a known-good fuel pump and retest the symptom. If the pump runs, replace the original pump.

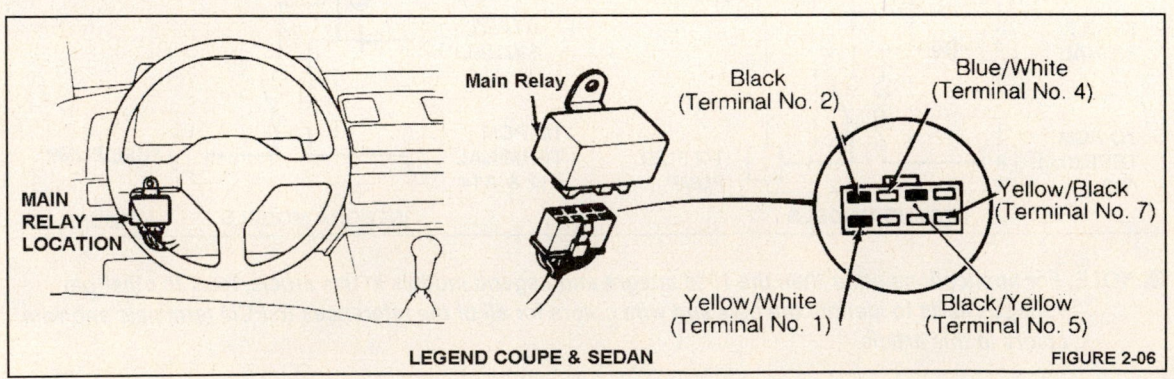

PGM-FI MAIN RELAY (1996-98)

INTRODUCTION

On 1996-98 Acura vehicles with fuel injection, the PGM-FI main relay consists of two different relays inside of one assembly. The main relay is located in these passenger compartment locations:

2.2CL - behind left side of instrument panel
2.3CL - behind left side of instrument panel (Figure 2-07)
2.5TL - behind left side of instrument panel
3.0CL - behind left side of instrument panel
3.2TL - behind left side of instrument panel
3.5RL - behind left side of instrument panel
Integra - behind left side of instrument panel
SLX (4-terminal) - In fuse/relay box at right side of the engine area

DRIVEABILITY SYMPTOMS

If the vehicle starts and runs, the PGM-FI main relay is operating. However, the relay can fail during vehicle use and cause it to stop on the road or fail to start.

A faulty main relay can cause any of the symptoms shown below.

No Start condition
Engine quits on the road and then restarts

PRELIMINARY STEPS

All of the test steps in this procedure should be done in sequence to quickly isolate the problem area.

First, carefully remove the main relay from the vehicle to allow for testing the wiring harness. Obtain a pair of jumper wires with connectors suitable to attach to the main relay and to the battery cables.

CAUTION: ***Do not smoke while working on the fuel system. Keep all open flames or sparks away from the work area. Do not attempt to relieve the fuel pressure unless the engine is off.***

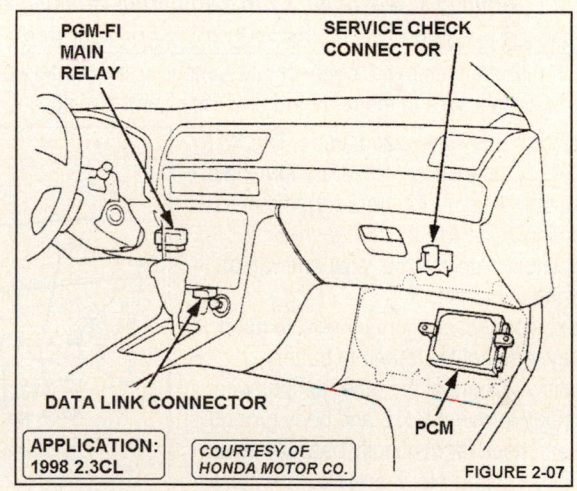

PGM-FI MAIN RELAY

SERVICE CHECK CONNECTOR

DATA LINK CONNECTOR

PCM

APPLICATION: 1998 2.3CL

COURTESY OF HONDA MOTOR CO.

FIGURE 2-07

<u>PGM-FI MAIN RELAY (1996-98)</u>

<u>MAIN RELAY OPERATION (ALL MODELS)</u>
With the key on, the first relay is energized. This action supplies battery voltage to the PCM, the fuel injectors and second relay. The second relay is energized by the PCM for two seconds when the key is in the on position. It is also energized with the engine running. The second relay provides power to the fuel pump and injectors. S1 and S2 identify switches and L1 and L2 identify load devices (Figure 2-08).

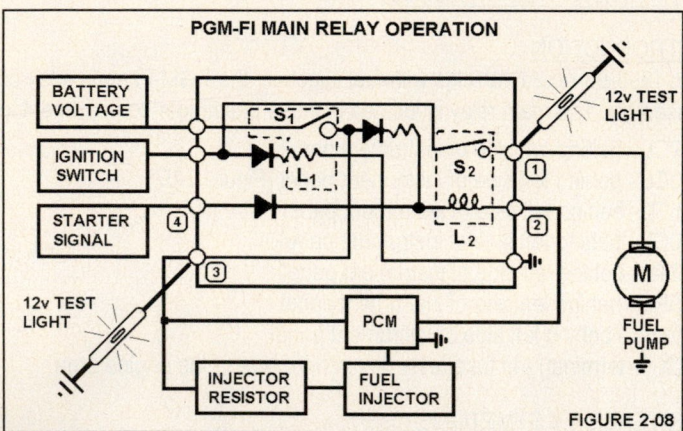

FIGURE 2-08

TIPS: *A quick test of the relay can be made with a 12v-test light (Figure 2-08). The light should glow for two seconds during cranking and with the key on at test point 1. It should glow dim during cranking and running at test point 2. It should glow at key on or engine running at test point 3 and also during cranking at test point 4.*

<u>MAIN RELAY TEST (ALL EXCEPT 3.2TL, 3.5RL MODELS)</u>
With the relay removed, connect one jumper wire to battery positive and to main relay terminal No. 2. Connect the other jumper wire to the relay terminal No. 1 and ground.

With the relay energized, use a DVOM to check for continuity from relay terminals No. 5 and No. 4 (Figure 2-09), If there is continuity, go to Step 2. If there is no continuity, replace the relay and retest for the symptom.

Next, connect one jumper wire to main relay terminal No. 5 and to battery positive. Connect the other jumper wire to relay terminal No. 3 and body ground. Then check for continuity between main relay terminals No. 7 and No. 6. If there is continuity, go to Step 3. If there is no continuity, replace the main relay and retest for the symptom.

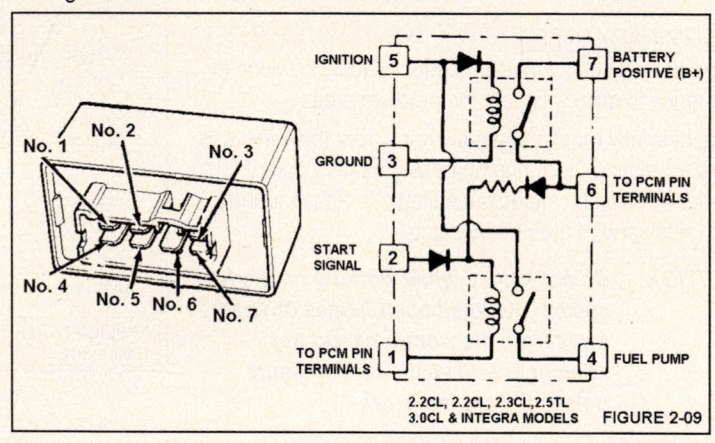

FIGURE 2-09

Next, connect one jumper wire to relay terminal No. 6 and to battery positive. Connect the other jumper wire to relay terminal No. 1 and body ground. Jumper wire polarity is important in this step due to an internal diode (Figure 2-09). Check for continuity between relay terminals No. 5 and No. 4. If there is no continuity, replace the main relay and retest for the symptom. If there is continuity, the relay is okay. To test the main relay wire harness and circuits (start, power and ground), go to the Wiring Harness Test on page 2-14.

■ **NOTE:** *For applications other than the 1996 2.5L and Integra models, refer to other pin voltage charts to identify the pins and wire colors for all references to PCM terminals and wire colors in this article.*

PGM-FI MAIN RELAY (1996-98)

MAIN RELAY TEST (3.2TL, 3.5RL MODELS)
To perform these tests, first remove the PGM-FI relay from under the dash.

Connect one jumper wire to relay terminal No. 3* and battery positive. Connect another jumper wire to the relay terminal No. 2* and battery negative or body ground. With the relay energized, use a DVOM to check for continuity from relay terminals No. 5* and No. 2 (Figure 2-10). If there is continuity, go to Step 2. If there is no continuity, replace the relay and retest for the symptom.
Next, connect one jumper wire to main relay terminal No. 4 and to battery positive. Connect the other jumper wire to relay terminal No. 7* and body ground. Then check for continuity between main relay terminals No. 3 and No. 1. If there is continuity, go to Step 3. If there is no continuity, replace the main relay and retest for the symptom.
Next, connect one jumper wire to relay terminal No. 1 and to battery positive. Connect the other jumper wire to relay terminal No. 2* and body ground. The jumper wire polarity is important due to an internal diode (Figure 2-10). Check for continuity between relay terminals No. 5* and No. 2. If there is no continuity, replace the main relay and retest for the symptom. If there is continuity, the main relay is okay. To test the main relay wire harness and circuits (start, power and ground), go to the Wiring Harness Test on the next page.

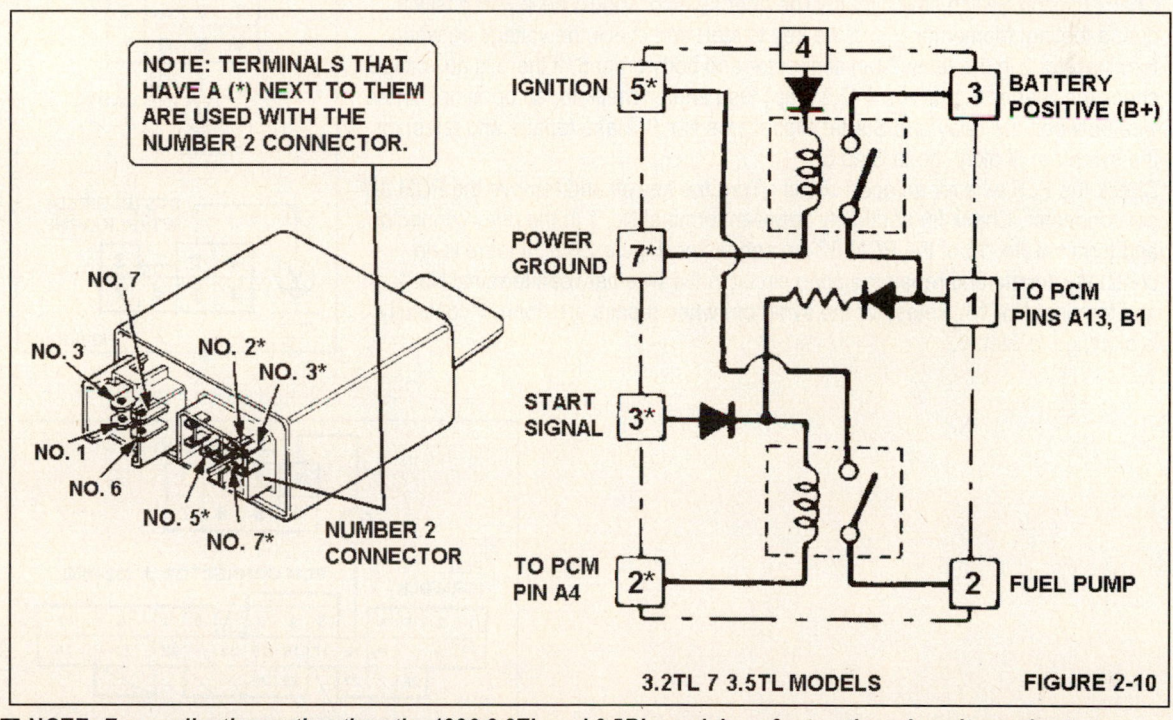

FIGURE 2-10

■ NOTE: *For applications other than the 1996 3.2TL and 3.5RL models, refer to other pin voltage charts to identify the pins and wire colors for all references to PCM terminals and wire colors in this article.*

PGM-FI MAIN RELAY (1996-98)

WIRING HARNESS TEST (1996 INTEGRA)

This test example is for a 1996 Integra. The key should be off with the main relay removed during this test procedure.

Check the ground circuit - Check for continuity between terminal No. 3 of the relay 7-Pconnector and ground (Figure 2-11). If continuity is not okay, repair the ground circuit and recheck the symptom. If okay, go to Step 2.

Check the battery feed circuit - Check the voltage reading between terminal No. 7 of the main relay 7-P connector and body ground. If there is no voltage, check the PCM (15-amp) fuse and/or locate the open feed circuit between the fuse and the relay. Make repairs and then retest for the symptom. If okay, go to Step 3.

Check ignition feed for an open circuit - Turn the key on and test voltage between terminal No. 5 in the relay 7-pin connector and body ground. If there is no voltage available, check the No. 2 Fuel Pump 15-amp fuse and/or locate the open or shorted wire between the relay and the fuse. Make repairs and retest for the symptom. If okay, go to Step 4.

Check Ignition Switch start circuit - The gear selector should be in P/N position during testing. Momentarily turn the key to start and check the voltage between terminal No. 2 in the relay 7-pin connector and body ground. If there is no voltage, check the Starter Signal No. 9 (7.5 amp) fuse and/or check for an open or shorted wire between the relay and Starter Signal fuse No. 9. Make repairs and retest for the symptom. If okay, go to Step 5.

Check the FLR wire for an open circuit - Turn the key off and remove the PCM 32-pin connector. Check for continuity between terminal No. 1 in the relay connector and terminal No. 16 of the PCM 32-pin connector (Figure 2-12). If there is no continuity, locate and repair the open circuit in the wire between terminal No. 1 and terminal No. 16. Retest for the symptom when repairs are done. If continuity is okay, go to Step 6.

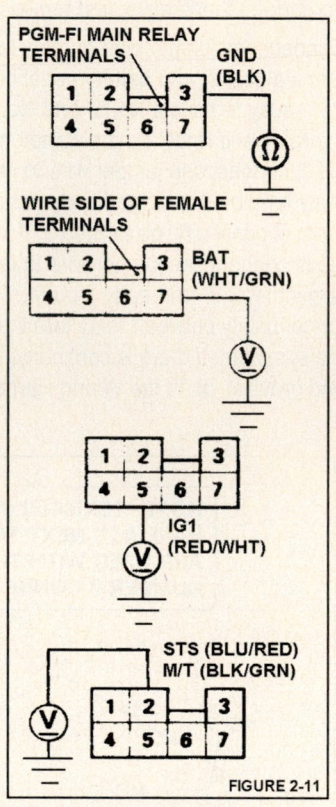

FIGURE 2-11

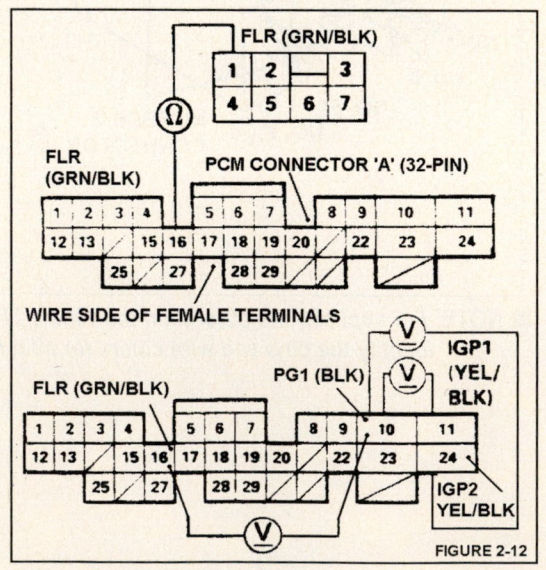

FIGURE 2-12

PGM-FI MAIN RELAY (1996-98)

WIRING HARNESS TEST (1996 INTEGRA - CONTINUED)

Check IGP1 and IGP2 circuits for an open condition -
Reconnect PCM 32-pin connector and the relay connector.
Turn the key on and measure the voltage between terminals
A11 and A10, and between terminals A24 and A10 at the PCM
(Figure 2-12). If no voltage is present, locate and repair the
open circuit between terminal A11 at the relay and/or replace
the main relay. Then retest for the symptom. If okay, go to
Step 7.

Check for open circuit in the PCM - Read the voltage between
terminals A16 and A10 at the PCM for two seconds just after
the key is turned on. If the reading less than 1.0v, remove the
main relay for testing. If the reading over 1.0v, substitute a
known-good PCM and retest. If the reading is now less than
1.0v, replace the original PCM.

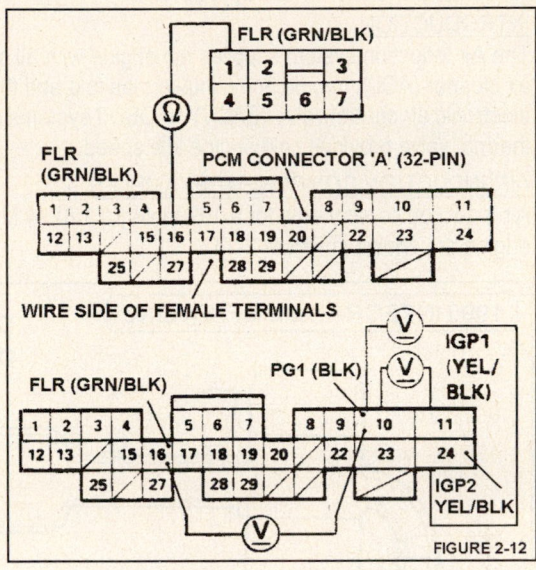

FIGURE 2-12

FUEL PUMP RELAY TEST - SLX MODELS (1996-98)

The PCM supplies power to the fuel pump relay for five seconds after the key is on. It also provides power to the relay
when it receives cranking or ignition reference signals.

First, remove the relay from the right-
side underhood relay box.
Connect one jumper wire to terminal No.
3 and battery positive. Connect another
jumper wire to the relay terminal No. 4
and battery negative.
Check the continuity from relay
terminals No. 1 and No. 2 with the relay
energized. If there is no continuity,
replace the relay and retest for the
symptom.
If continuity is okay, test the PCM main
relay and the fuel pump relay circuits.

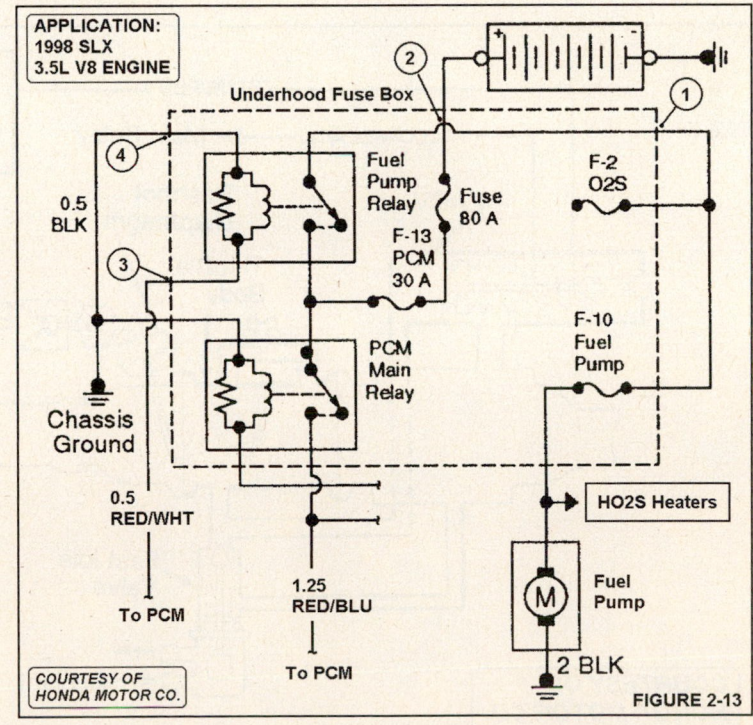

FIGURE 2-13

PGM-FI ENGINE CONTROLS

Air Intake & Throttle Body Systems

INTRODUCTION

The Air Induction system supplies the engine with all of its clean air needs. This system consists of an air intake pipe, air cleaner (ACL), throttle body, intake manifold and throttle control system. The PCM controls idle speed with an electronic air control valve (EACV) on OBD I systems or an idle speed control valve on OBD II systems. A fast idle thermal valve provides cold engine idle speed.

AIR INDUCTION COMPONENT EXAMPLES

Air Induction components for a 1990 Integra 1.8L I4 MFI engine and a 1990 Legend Coupe or Legend Sedan 2.7L MFI engine are shown in Figure 2-14.

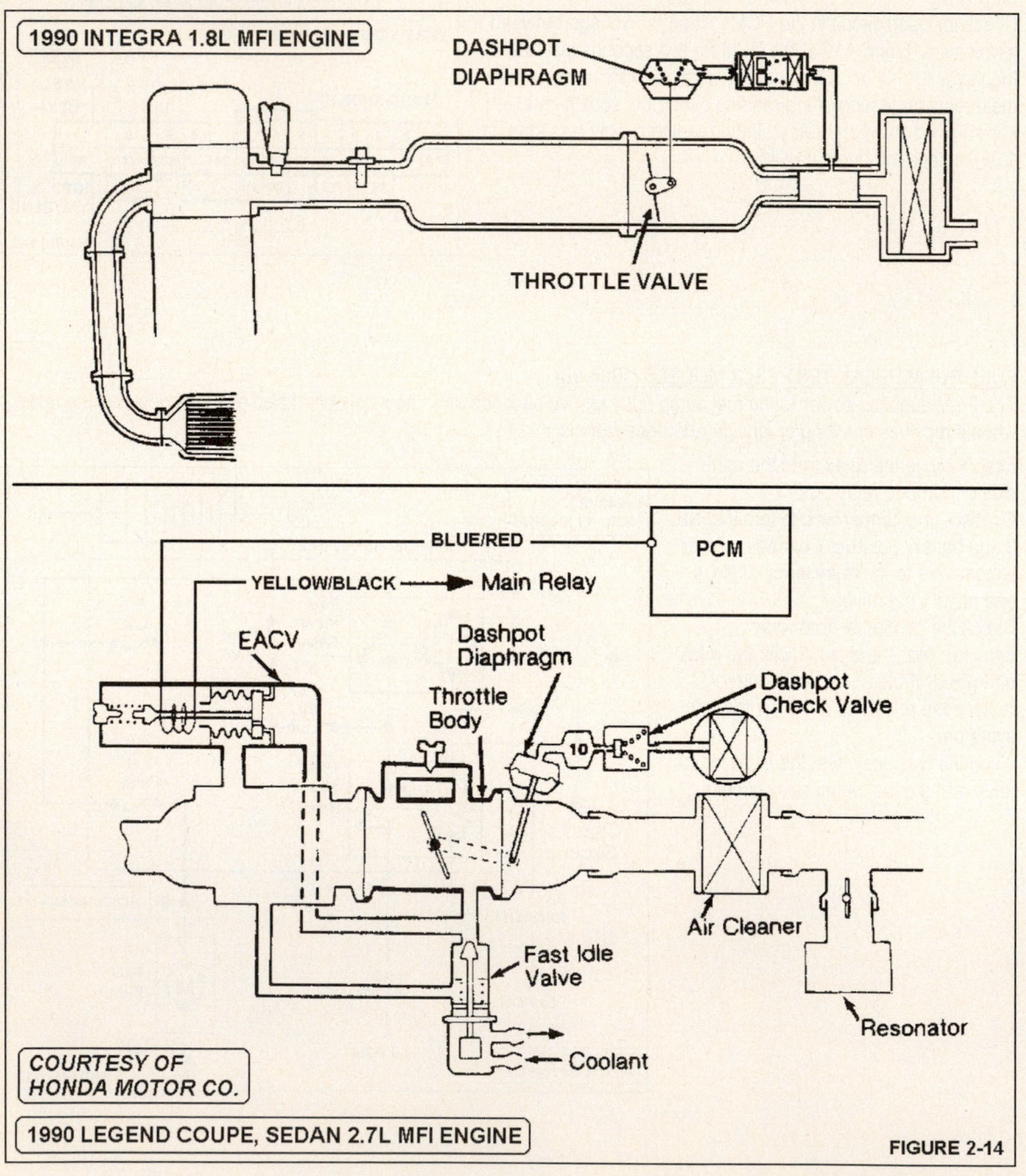

FIGURE 2-14

AIR INDUCTION COMPONENT EXAMPLES

Air Induction components for a 1998 Acura 3.0CL with a 3.0L V6 (J30A1) MFI engine are shown in Figure 2-15.

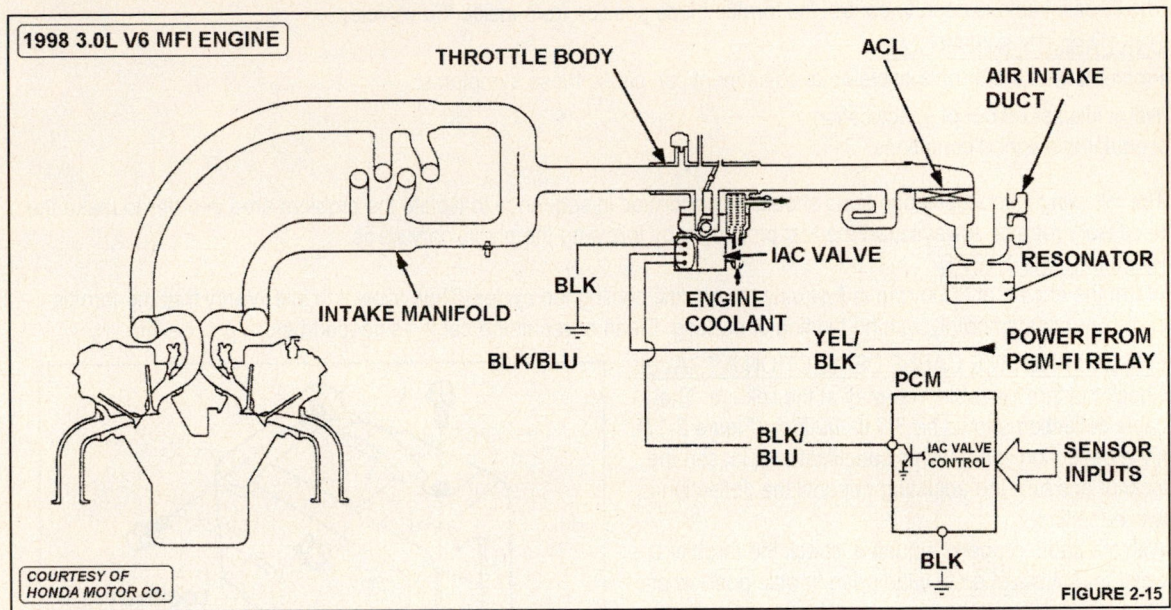

FIGURE 2-15

Air Induction components for a 1998 Acura 3.5CL with a 3.5L V6 (C35A1) MFI engine are shown in Figure 2-16.

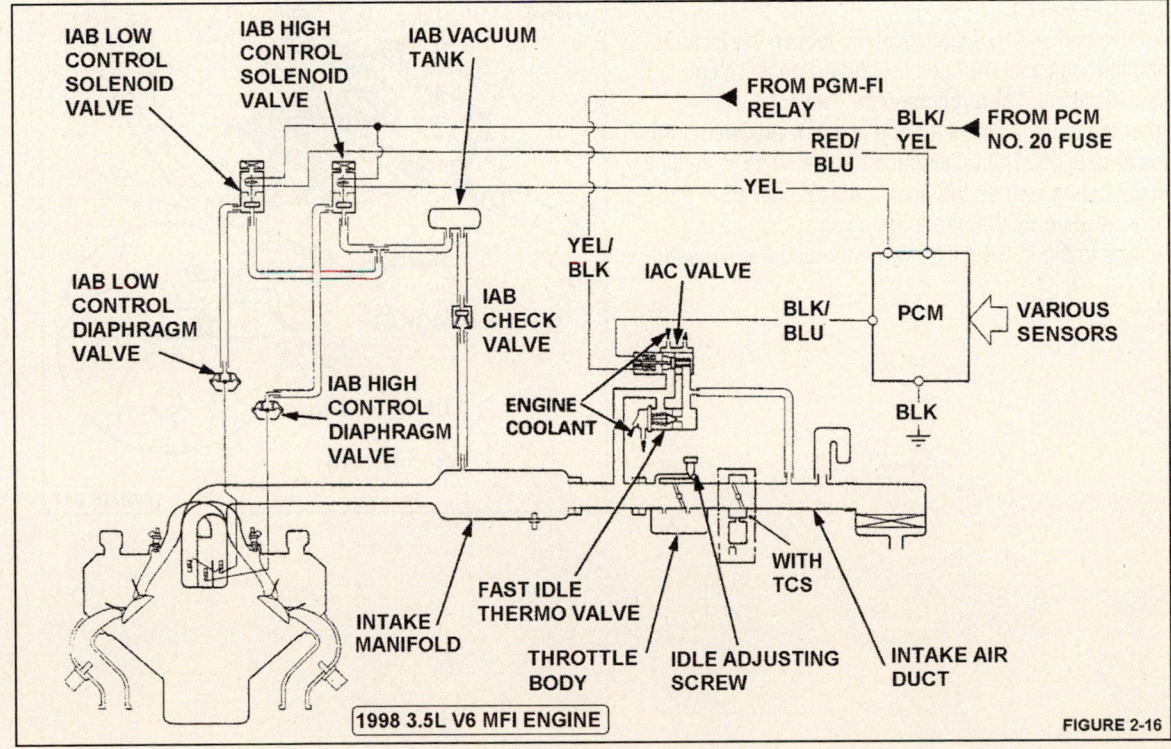

FIGURE 2-16

THROTTLE CABLE INSPECTION & ADJUSTMENT

The throttle cable is used to control the throttle blade position from inside the vehicle.

DRIVEABILITY SYMPTOMS

Problems with the throttle cable or its adjustment can cause these symptoms:

Warm idle speed out of specification

Irregular idle speed conditions

The following inspections and tests should be performed in sequence to isolate the problem area in order to make the necessary repairs. Always start the test procedure by following the preliminary steps.

PRELIMINARY STEPS

Warm the engine at 3000 rpm in P/N until the electric cooling fan cycles. Then allow it to idle. Verify that the throttle cable operates smoothly with no binding or sticking. Clean or repair the cable as needed before proceeding.

CHECK THROTTLE CABLE DEFLECTION & TRAVEL

Check the throttle cable free-play at the linkage. The cable deflection should be 3/8 to 1/2 inch (Figure 2-17). If the deflection is not within specifications, loosen the locknut and turn the adjusting nut until the deflection is within the limits.

With the cable properly adjusted, check the throttle valve for full travel up to wide open throttle position as the accelerator pedal is pushed to the floor. Verify the throttle valve returns to idle position. The throttle lever should rest against the throttle stop screw with the throttle released.

If deflection is out of specification, loosen the locknut and turn adjusting nut unit the deflection is within specifications. Then retighten the locknut.

After the cable is checked and properly adjusted, turn the engine off. Then check the throttle valve to verify that it opens fully as the accelerator is pushed to the floor. Also verify that the throttle valve to be sure it returns to the idle position when the pedal is released.

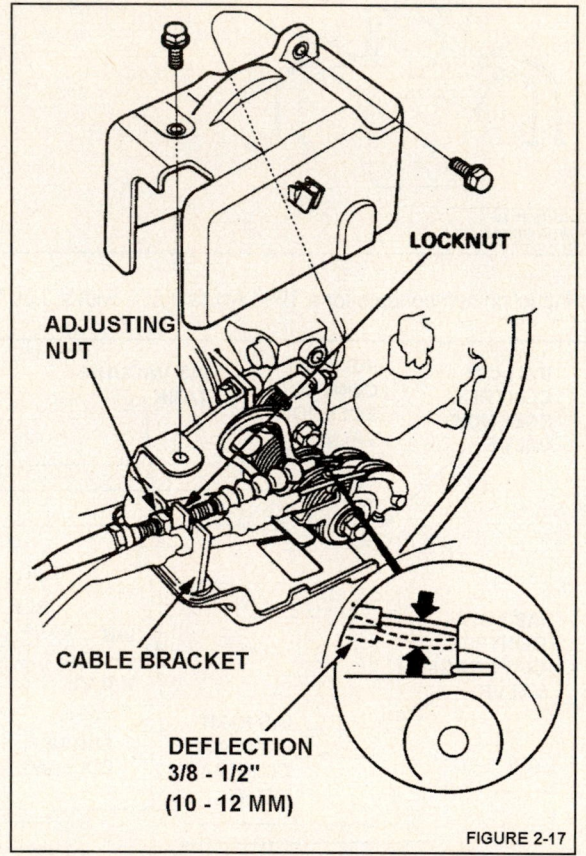

LOCKNUT

ADJUSTING NUT

CABLE BRACKET

DEFLECTION
3/8 - 1/2"
(10 - 12 MM)

FIGURE 2-17

THROTTLE CABLE INSPECTION & ADJUSTMENT

CAUTION: *Block the drive wheels and connect the vehicle exhaust pipe(s) to a suitable vent system to purge exhaust fumes from the work area.*

CABLE INSTALLATION

Open the throttle valve fully, then install the throttle cable to the throttle linkage and install the cable housing in the cable bracket (Figure 2-18). Start the engine in P/N and hold the engine speed at 3000 rpm with no load until the electric cooling fan cycles. Then allow the engine to return to a normal idle speed setting.

Hold the cable sheath while removing all slack from the cable.

Turn the adjusting nut until it is 3/8 inch away from the cable bracket (Figure 2-19).

Tighten the throttle cable locknut securely.

Recheck the cable deflection. It should now be 3/8 - 1/2 inch as shown in Figure 2-19. If either the cable adjustment or cable deflection is still not correct, refer to the previous article on Throttle Cable Inspection & Adjustment Procedures.

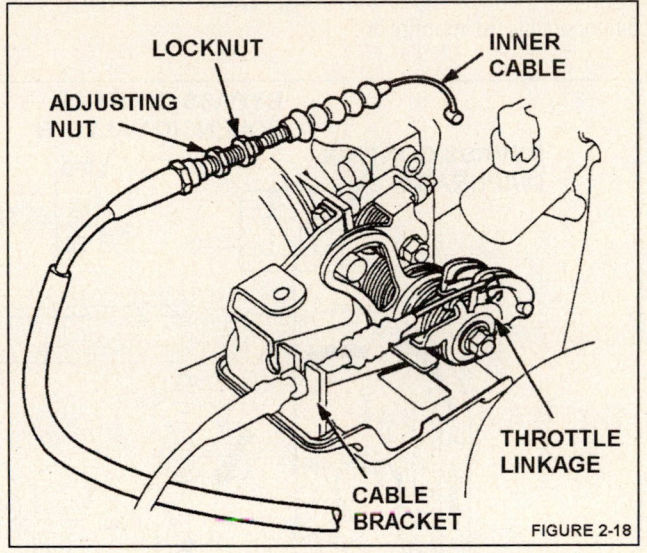

FIGURE 2-18

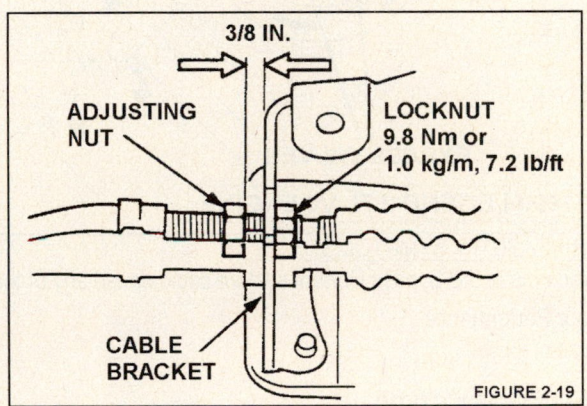

FIGURE 2-19

BYPASS CONTROL SYSTEM TESTS

INTRODUCTION

The Bypass Control system consists of two air intake paths provided in the intake manifold to allow the selection of the intake path length most favorable for a given engine speed. High torque at low engine speeds is achieved by using the long intake path and high power at high speed is achieved by using the short intake path. Refer to Figure 2-20 for component layout information.

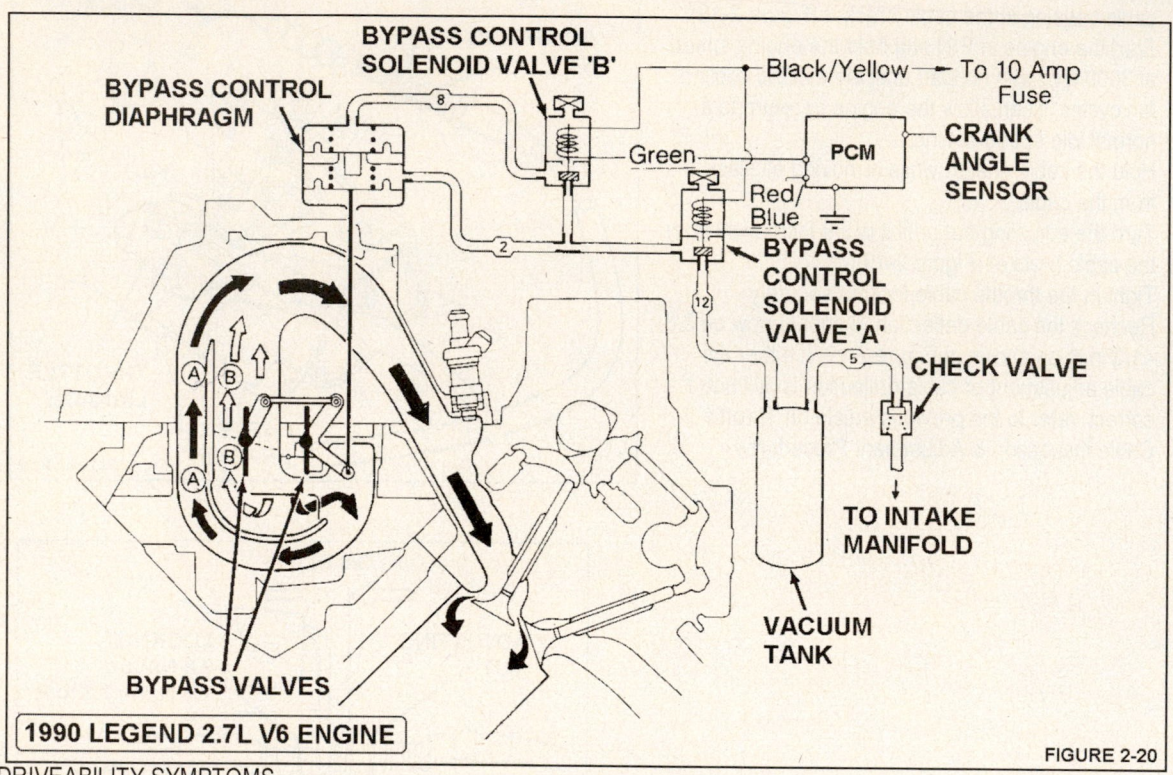

FIGURE 2-20

DRIVEABILITY SYMPTOMS

Problems in the Bypass Control system could cause any of these symptoms:

Poor Performance

Loss of Power

PRELIMINARY STEPS

The following inspections and tests should be performed in sequence to isolate the problem area in order to make the necessary repairs. Always start the test procedure by following the preliminary steps.

Carefully check the by-pass valve shaft for binding or sticking. Also, check the by-pass valve for smooth movement. Clean the shaft as required.

CAUTION: **Block the drive wheels and connect the vehicle exhaust pipe(s) to a suitable vent system to purge exhaust fumes from the work area.**

BYPASS CONTROL SYSTEM TESTS

■ **NOTE:** *The explanation and pictures in this article are from the information in the 1990 Legend Coupe and Sedan Repair Manual. To identify wire colors and PCM terminals on other models, refer to the appropriate pin voltage charts.*

CHECK THE BYPASS VALVE SHAFT FOR STICKING

Check that the stop cam of bypass valve 'A' is in close contact with the stopper with the No. 2 and No. 8 hoses are disconnected from the diaphragm. (Figure 2-21). This is the high-speed position.

Check that the stop cam of by-pass valve 'B' is in close contact with the full-closed screw with 8-10 inches of vacuum applied simultaneously to both diaphragms. This is the closed throttle or low engine speed position when both solenoids are normally energized.

If problems in full travel or binding are found, clean the linkage and throttle shafts with an approved cleaner. If the problem remains after cleaning the throttle shafts, the intake manifold will have to be removed and disassembled to locate and repair the problem.

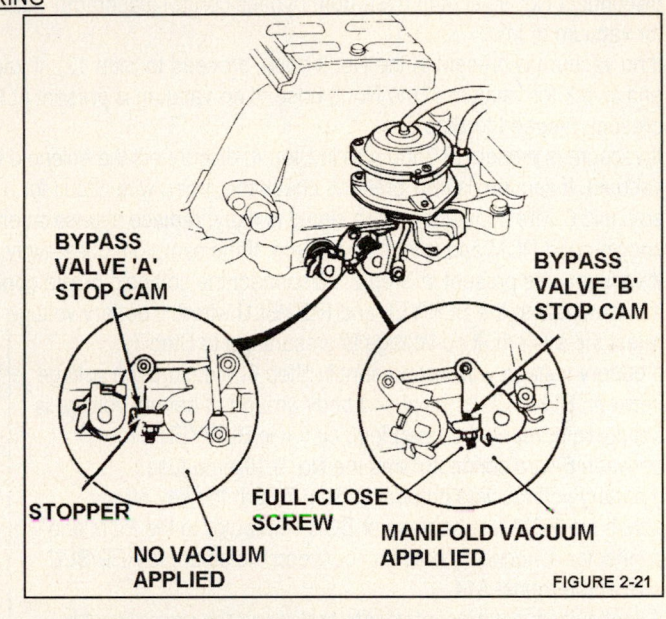

FIGURE 2-21

TEST VACUUM CIRCUITS, BYPASS VALVES & RELATED CIRCUIT OPERATION

Start the engine and allow it to idle. Remove the No. 2 vacuum hose from the bypass control diaphragm and attach a vacuum gauge to the hose (Figure 2-22). Check for vacuum. If vacuum is okay, go to Step 3.

If no vacuum is present, remove the No. 8 vacuum supply hose that connects to solenoid 'A' and check for vacuum. If vacuum is present, proceed to Step 6. If no vacuum is present, check vacuum line between the solenoid and intake manifold. Make repairs as needed and retest the system.

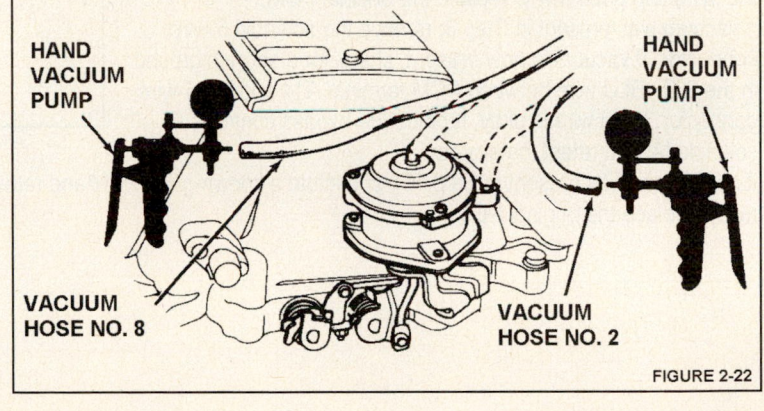

FIGURE 2-22

BYPASS CONTROL SYSTEM TESTS

TEST VACUUM CIRCUITS, BYPASS VALVES & RELATED CIRCUIT OPERATION

If vacuum was present in Step 1, raise the engine speed to 4000 rpm and check the vacuum at the No. 2 hose. If vacuum is present, proceed to Step 10). If no vacuum is present (okay), reconnect the No. 2 vacuum hose and disconnect No. 8 vacuum hose from bypass control diaphragm. Attach a vacuum gauge to the No. 8 hose and check for vacuum at idle.

If no vacuum is present at the No. 8 hose, proceed to Step 12). If vacuum is present, raise engine speed to 3500 rpm and check for vacuum at the No. 8 hose. If no vacuum is present at the No. 8 hose, the system is okay. If vacuum is present, proceed to Step 5.

If vacuum is present at 3500 rpm in Step 4, disconnect the solenoid 6-wire connector (Figure 2-23) and recheck for vacuum. If vacuum is now present, check the green wire circuit for a short to ground between PCM terminal B3 and the solenoid 6-wire connector. If the circuit is okay, replace bypass solenoid 'B'. If no vacuum is present, substitute a known-good PCM and retest the system. If the symptom goes away, replace the original PCM and retest.

If vacuum was present in Step 2, disconnect the solenoid 6-wire connector (Figure 2-23). Measure the voltage between the terminals to the BLK/YEL and RED/BLU wires. If battery voltage is present, replace bypass control solenoid 'A' and retest the system. If no voltage is present, go to Step 7.

If battery voltage was not present in Step 6, measure the voltage between BLK/YEL terminal and body ground. If battery voltage is not present, repair the open feed circuit in BLK/YEL wire between 6-wire connector and the No. 9 10-amp fuse.

If battery voltage was present in Step 7, turn the key off and attach the PCM test harness or BOB connector to the PCM and connector. Check for continuity between the solenoid RED/BLU wire and terminal A14.

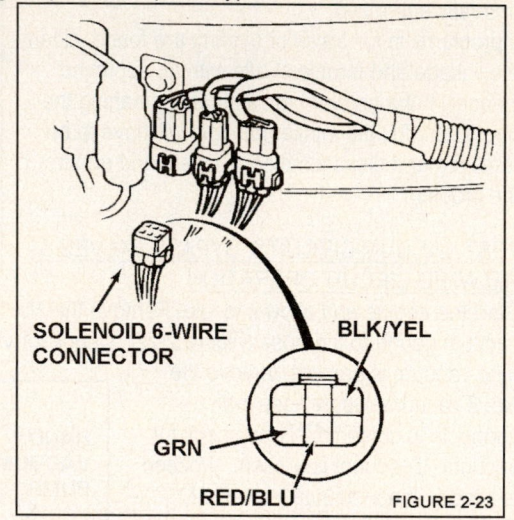

SOLENOID 6-WIRE CONNECTOR

BLK/YEL

GRN

RED/BLU

FIGURE 2-23

If continuity is not present, locate and repair the open circuit in the RED/BLU wire between PCM and connector. If continuity is present, substitute a known-good PCM and retest the system. If the symptom goes away, replace the original PCM.

If vacuum was present in Step 3, remove the solenoid 6-wire connector. If vacuum is now present, check for a short to ground in the RED/BLU wire between PCM terminal A14 and the 6-wire connector. If the wire is okay, replace the bypass control solenoid 'A' and retest the system.

If no vacuum was present in Step 10), substitute a known-good PCM and retest the system. If the symptom goes away, replace the original PCM.

BYPASS CONTROL SYSTEM TESTS

TEST VACUUM CIRCUITS, BYPASS VALVES & RELATED CIRCUIT OPERATION

If the gauge showed no vacuum was present in Step 4, remove the solenoid 6-P connector (Figure 2-23). Check for battery voltage between BLK/YEL and GRN wire terminals. If battery voltage is present, replace bypass control solenoid 'B' and retest system. If no voltage is present, proceed to step 13).

Measure the voltage between the BLK/YEL wire terminal and body ground. If battery voltage is not present, locate and repair the open circuit in the BLK/YEL wire between 6-P connector and No. 9 10-amp fuse.

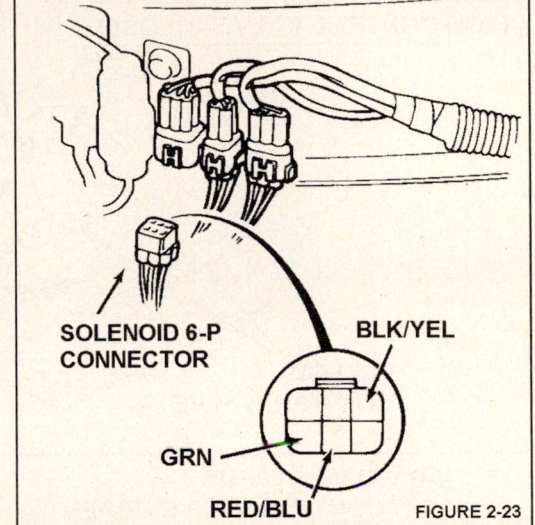

If battery voltage is present, turn the key off and connect the PCM test harness or BOB connector between the PCM and its connector. Check for continuity between the GRN wire in 6-P connector and PCM terminal B3 (Figure 2-24). If continuity is not present, locate and repair the open circuit in the GRN wire between the PCM and the connector, then retest the system.

If continuity is present, substitute a known-good PCM and retest the system. If the symptom goes away, replace the original PCM.

SOLENOID 6-P CONNECTOR

BLK/YEL

GRN

RED/BLU

FIGURE 2-23

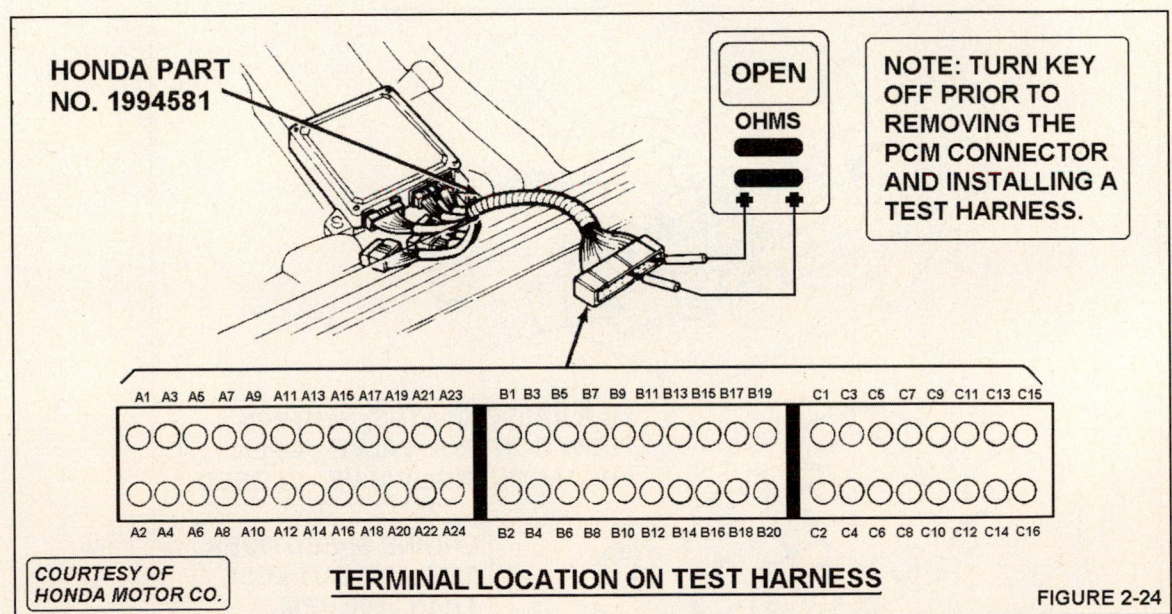

HONDA PART NO. 1994581

OPEN

OHMS

NOTE: TURN KEY OFF PRIOR TO REMOVING THE PCM CONNECTOR AND INSTALLING A TEST HARNESS.

A1 A3 A5 A7 A9 A11 A13 A15 A17 A19 A21 A23   B1 B3 B5 B7 B9 B11 B13 B15 B17 B19   C1 C3 C5 C7 C9 C11 C13 C15

A2 A4 A6 A8 A10 A12 A14 A16 A18 A20 A22 A24   B2 B4 B6 B8 B10 B12 B14 B16 B18 B20   C2 C4 C6 C8 C10 C12 C14 C16

TERMINAL LOCATION ON TEST HARNESS

FIGURE 2-24

INTAKE AIR BYPASS SYSTEM TESTS

The Intake Air Bypass system is used on some engines to improve power performance. This is achieved by opening and closing the intake air bypass (IAB) control valves. High torque at low engine speeds is achieved when the valves are closed. High power at high engine speeds is achieved when the valves are opened (Figure 2-25).

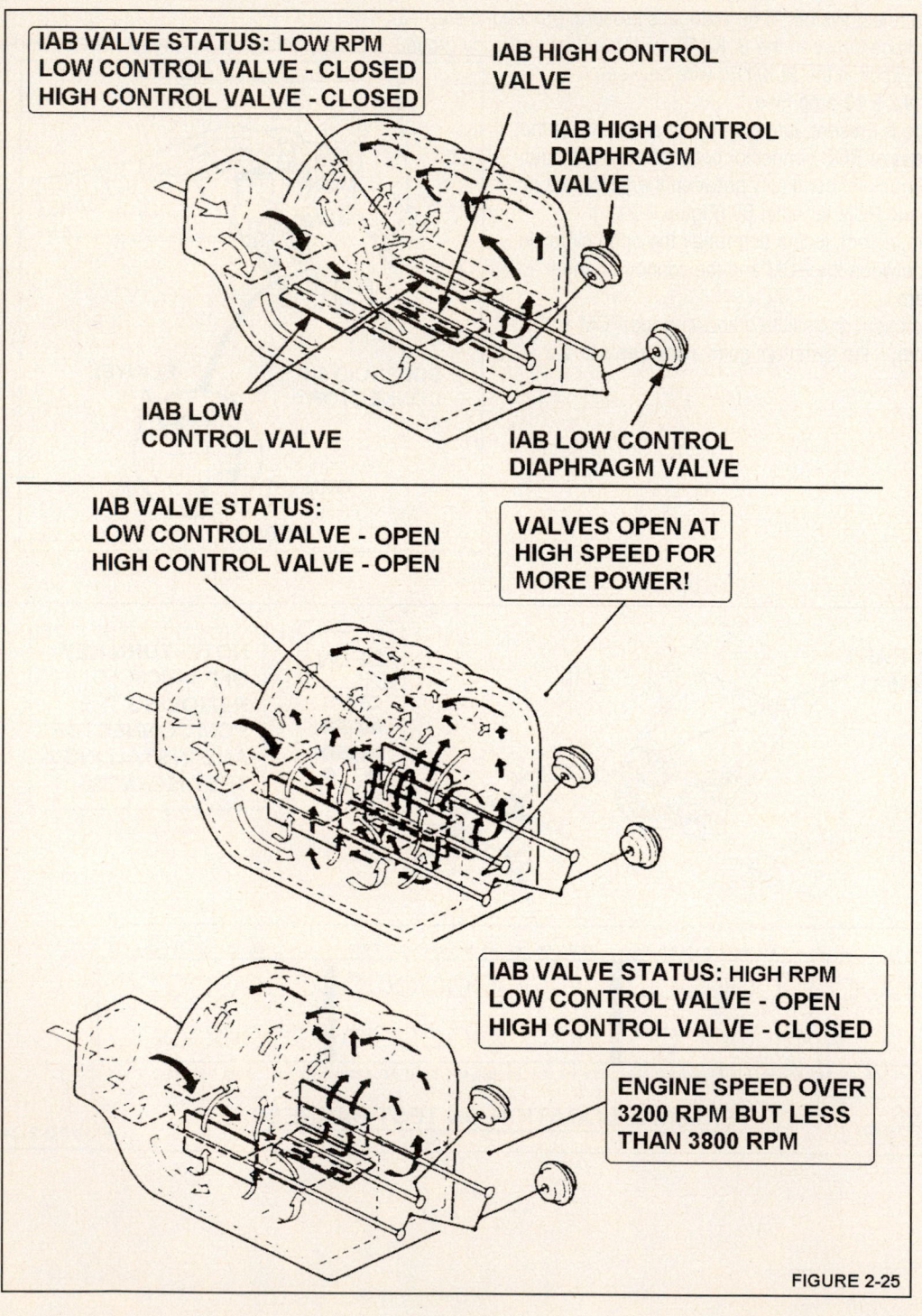

IAB VALVE STATUS: LOW RPM
LOW CONTROL VALVE - CLOSED
HIGH CONTROL VALVE - CLOSED

IAB HIGH CONTROL VALVE

IAB HIGH CONTROL DIAPHRAGM VALVE

IAB LOW CONTROL VALVE

IAB LOW CONTROL DIAPHRAGM VALVE

IAB VALVE STATUS:
LOW CONTROL VALVE - OPEN
HIGH CONTROL VALVE - OPEN

VALVES OPEN AT HIGH SPEED FOR MORE POWER!

IAB VALVE STATUS: HIGH RPM
LOW CONTROL VALVE - OPEN
HIGH CONTROL VALVE - CLOSED

ENGINE SPEED OVER 3200 RPM BUT LESS THAN 3800 RPM

FIGURE 2-25

INTAKE AIR BYPASS SYSTEM TESTS

INTAKE AIR BYPASS LOW CONTROL VALVE TESTS

Remove the No. 2 hose from the vacuum hose manifold (Figure 2-26). Attach a hand vacuum pump to the No. 2 hose that connects to the manifold and valve.

Apply vacuum and verify that the IAB low control diaphragm valve holds vacuum, and that as vacuum is applied and released, the IAB control diaphragm valve does not move in and out (it should move when the vacuum is released).

If the IAB control diaphragm valve does not hold vacuum or the IAB control diaphragm valve rod does not move in and out, replace the IAB control diaphragm and retest the system.

INTAKE AIR BYPASS HIGH CONTROL VALVE TEST

Remove the No. 8 hose from the vacuum hose manifold (Figure 2-26). Attach a hand vacuum pump to the No. 8 hose that connects to the manifold and valve.

Apply vacuum and verify that the IAB high control diaphragm valve holds vacuum, and that as vacuum is applied and released, the IAB control diaphragm valve does not move in and out (it should move when the vacuum is released).

If the IAB control diaphragm valve does not hold vacuum or the IAB control diaphragm valve rod moves in and out due to a leak, replace the IAB control diaphragm and retest the system.

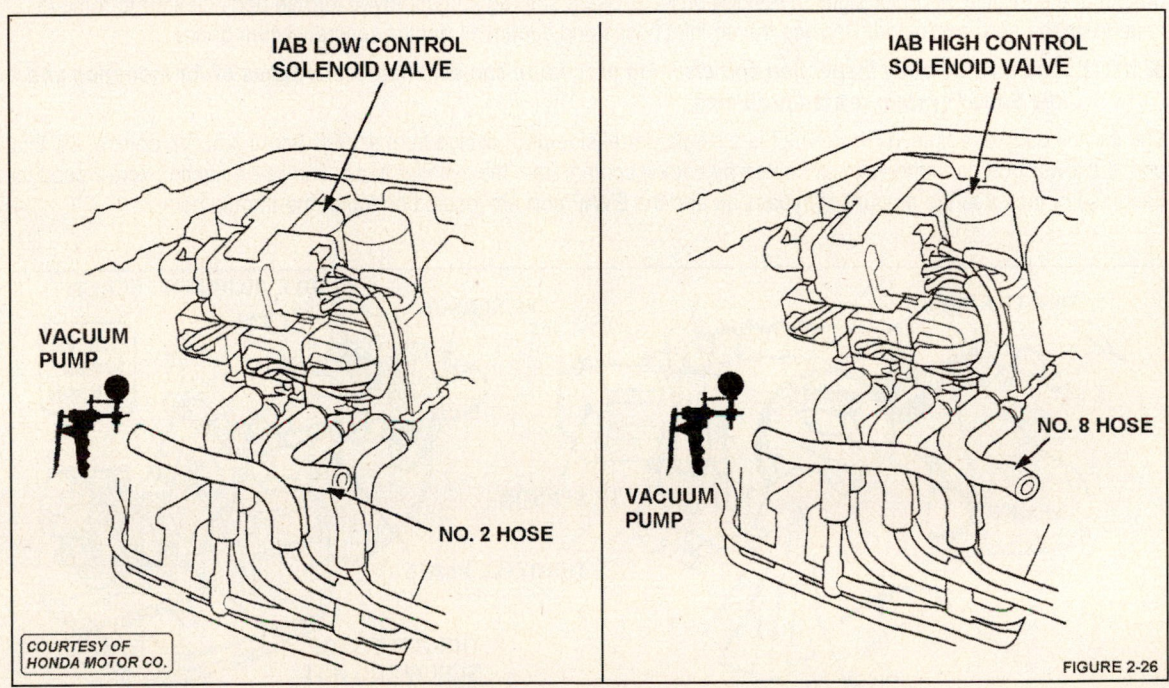

FIGURE 2-26

THROTTLE BODY SYSTEM TESTS

Newer Acura vehicles use a throttle body that is a single barrel, single-throttle plate design (Figure 2-27). The lower portion of the throttle body is heated by engine coolant that circulates from the engine cylinder head. The idle adjustment screw is preset at the factory and is used to increase or decrease the amount of bypass air. The EVAP canister purge port is located on top of the throttle body assembly.

DRIVEABILITY SYMPTOMS

Problems with the throttle body or its related vacuum supply lines could cause any of the these symptoms:

Frequent stalling at engine startup
Warm engine idle speed out of specification
Rough idle, stalling during gear shifting or deceleration
Loss of power or poor performance

PRELIMINARY STEPS

The following inspections and tests should be performed in sequence to isolate problems in the throttle body unit or related hoses. Always start the test procedure by following the preliminary steps.

Carefully remove the air inlet hoses or ducting between the throttle body and the air cleaner unit. With the engine off, use a flashlight to look inside the throttle body bore. Slowly open the throttle blade and inspect the throttle bore and bottom of the throttle blade for signs of residue buildup. Use a soft rag and approved throttle body cleaner to remove residue on the bore and blade. Replace the air inlet hoses and tighten all clamps securely when done.

■ NOTE: *The throttle body inspection and cleaning procedure can eliminate many types of Air Induction and Idle Speed system related problems.*

The throttle body example in Figure 2-27 is a single-barrel side-draft design from a 1998 Acura 3.5L V6 engine. On this unit, the lower portion of the valve is heated by engine coolant from the cylinder head. The idle adjusting screw used to decrease or increase the amount of bypass air and the EVAP port are located on top of the throttle body.

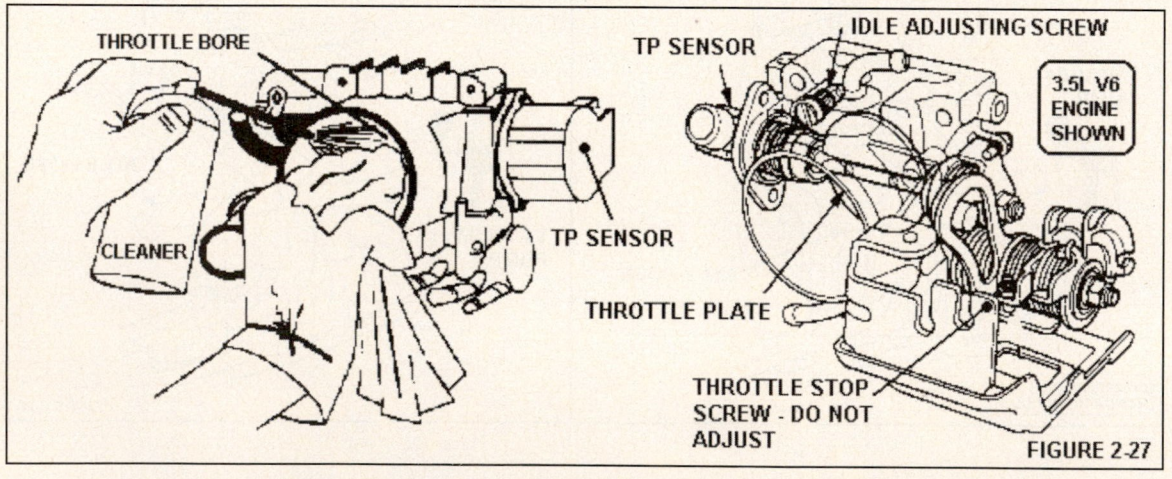

FIGURE 2-27

THROTTLE BODY SYSTEM TESTS

CAUTION: ***Block the drive wheels and connect the vehicle exhaust pipe(s) to a suitable vent system to purge exhaust fumes from the work area.***

CHECK THROTTLE VALVE, INSPECT VACUUM PORTS

Start the engine and bring it to normal operating temperature or until the electric cooling fan cycles.

Disconnect the vacuum hose at the top of the throttle body that runs to the EVAP canister. Connect a vacuum gauge to the exposed port.

Check the vacuum gauge reading at idle speed. The gauge should read no vacuum at idle. If vacuum is present, check throttle cable operation.

Open the throttle slightly and check that the gauge reads some vacuum. If the gauge registers no vacuum with the throttle slightly open, clean the throttle body bore and ports with an approved cleaner. Depending upon the severity of the residue inside the throttle body, it may need to be removed in order to clean it properly.

With the engine off, verify that the throttle cable operates smoothly without any binding or sticking. If okay, perform these inspections.

Inspect for excessive wear or play in the throttle valve shaft

With the throttle valve fully closed, check for a sticking or binding throttle lever (make repairs as needed)

Check for clearance between the throttle stop screw and throttle lever at fully closed position (there should be no clearance between them)

Replace the throttle body unit if there is excessive play or wear in the throttle valve shaft, the shaft is binding or sticking or it cannot be cleaned

THROTTLE STOP SCREW ADJUSTMENTS

Prior to making any adjustments to engine idle speed with the throttle stop screw (Figure 2-28), read the information in this article.

If the both the throttle body and blade are clean with no signs of worn or sticking parts, the normal PCM control of idle speed along with the minimum-air-rate adjustment should provide correct idle speed.

In this case, do not change the setting of the factory-set throttle stop screw since it cannot be reset properly in the field.

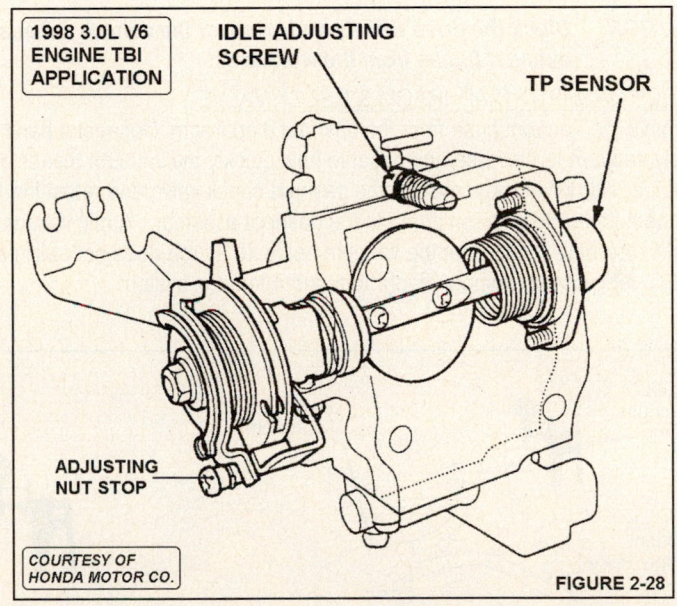

1998 3.0L V6 ENGINE TBI APPLICATION

IDLE ADJUSTING SCREW

TP SENSOR

ADJUSTING NUT STOP

COURTESY OF HONDA MOTOR CO.

FIGURE 2-28

DASHPOT SYSTEM TESTS

INTRODUCTION
The Dashpot assembly on the throttle body is used to allow the throttle valve to close slowly during shifting or periods of deceleration.

DRIVEABILITY SYMPTOMS
Problems in the dashpot diaphragm, check valve, related linkage or vacuum lines could cause any of the symptoms shown below.

High Idle Speed
Stall at Decel
Stall during shifting

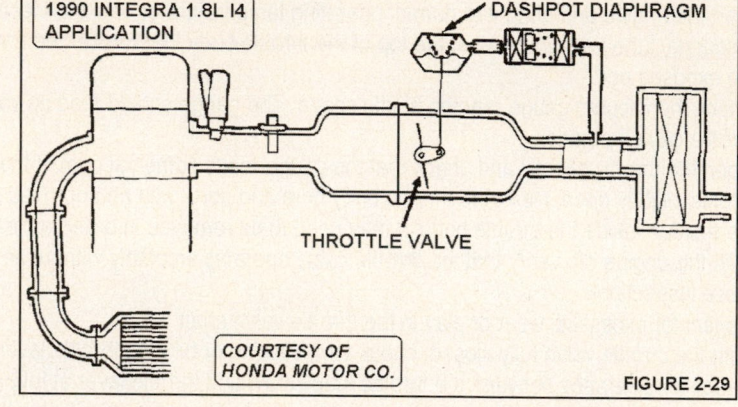

1990 INTEGRA 1.8L I4 APPLICATION

DASHPOT DIAPHRAGM

THROTTLE VALVE

COURTESY OF HONDA MOTOR CO.

FIGURE 2-29

PRELIMINARY STEPS
The following inspections and tests should be performed in sequence to isolate a problem in the Dashpot system in order to make the necessary repairs. Always start the test procedure by following the preliminary steps.

Carefully remove the vacuum hose connecting the air filter assembly to the dashpot diaphragm. Inspect the vacuum hose for leaks, restriction or a disconnection. Make repairs as needed and reconnect the hose when done.

CAUTION: *Block the drive wheels and connect the vehicle exhaust pipe(s) to a suitable vent system to purge exhaust fumes from the work area.*

CHECK DASHPOT DIAPHRAGM & CHECK VALVE
Remove the vacuum hose from the dashpot diaphragm. Connect a hand vacuum pump to the hose (Figure 2-30). Apply vacuum to the hose and observe how quickly the vacuum bleeds down. The vacuum should bleed down slowly. If it bleeds down quickly, replace the dashpot check valve and retest the system.
Connect the hand vacuum pump to the dashpot assembly. Apply vacuum to the diaphragm and observe that the control rod pulls in and that the vacuum holds. If the rod does not pull in or the diaphragm does not hold vacuum (it leaks), replace the dashpot diaphragm and retest the system.

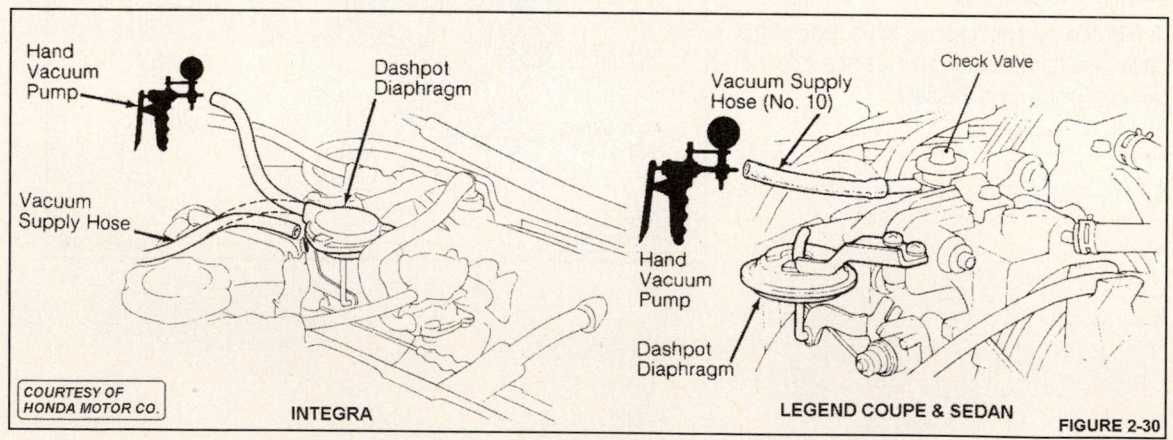

Hand Vacuum Pump

Dashpot Diaphragm

Vacuum Supply Hose

Vacuum Supply Hose (No. 10)

Check Valve

Hand Vacuum Pump

Dashpot Diaphragm

COURTESY OF HONDA MOTOR CO.

INTEGRA

LEGEND COUPE & SEDAN

FIGURE 2-30

RESONATOR CONTROL SYSTEM TESTS

INTRODUCTION

The Resonator Control system is designed to decrease air intake noise.

When the engine speed is below 3800 rpm on Legend models (3000 rpm on Integra models), the PCM supplies current to the Resonator Control solenoid valve. This action opens the normally closed solenoid valve to allow intake manifold vacuum to the Resonator Control diaphragm to help control air intake noise.

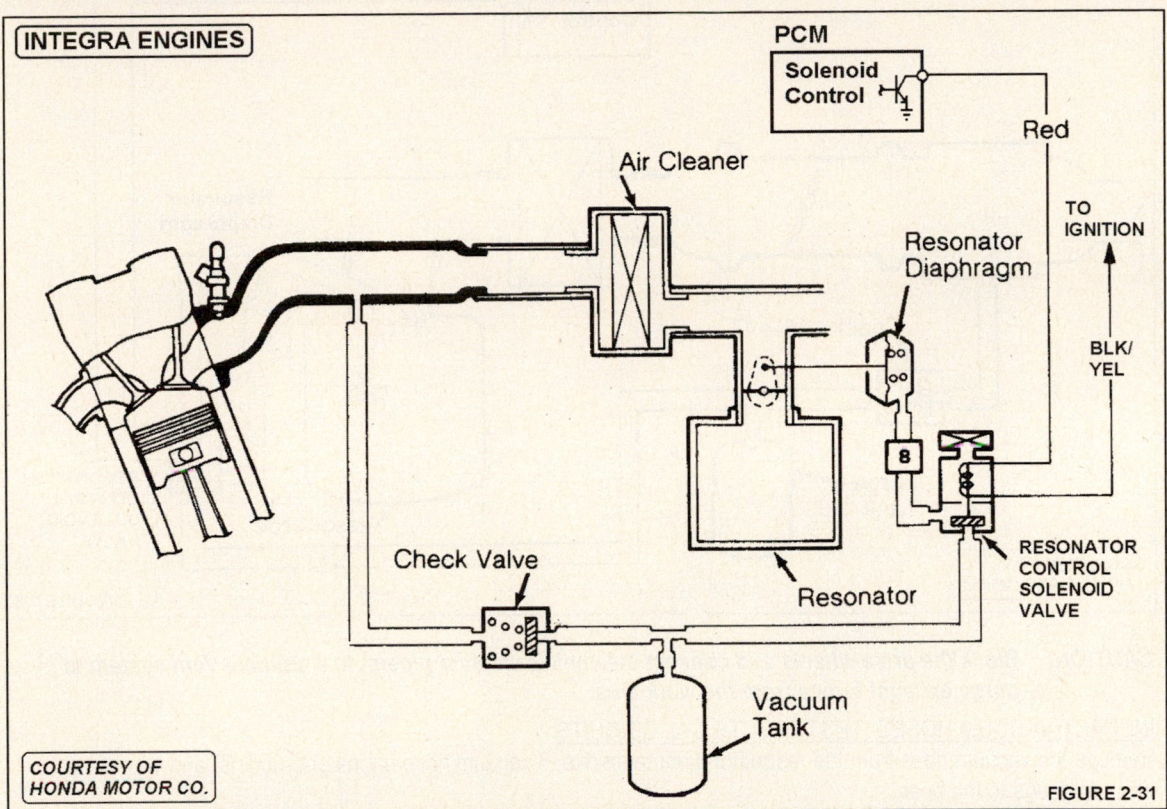

FIGURE 2-31

DRIVEABILITY SYMPTOMS

Problems with the Resonator Control system could cause these symptoms:

Excessive air induction (air cleaner) noise at idle and low speeds
Loss of power

■ NOTE:*For models other than 1990 Integra, Legend Coupe and Sedan discussed in this article, refer to other pin voltage charts for pin and wire color identification for all references to PCM terminals and wire colors.*

PRELIMINARY STEPS

The following inspections and tests should be performed in sequence to isolate the problem area in order to make the necessary repairs. Always start the test procedure by following the preliminary steps.

Inspect the resonator diaphragm linkage for binding and check vacuum hose routing and connections.

RESONATOR CONTROL SYSTEM TESTS

COMPONENT EXAMPLES

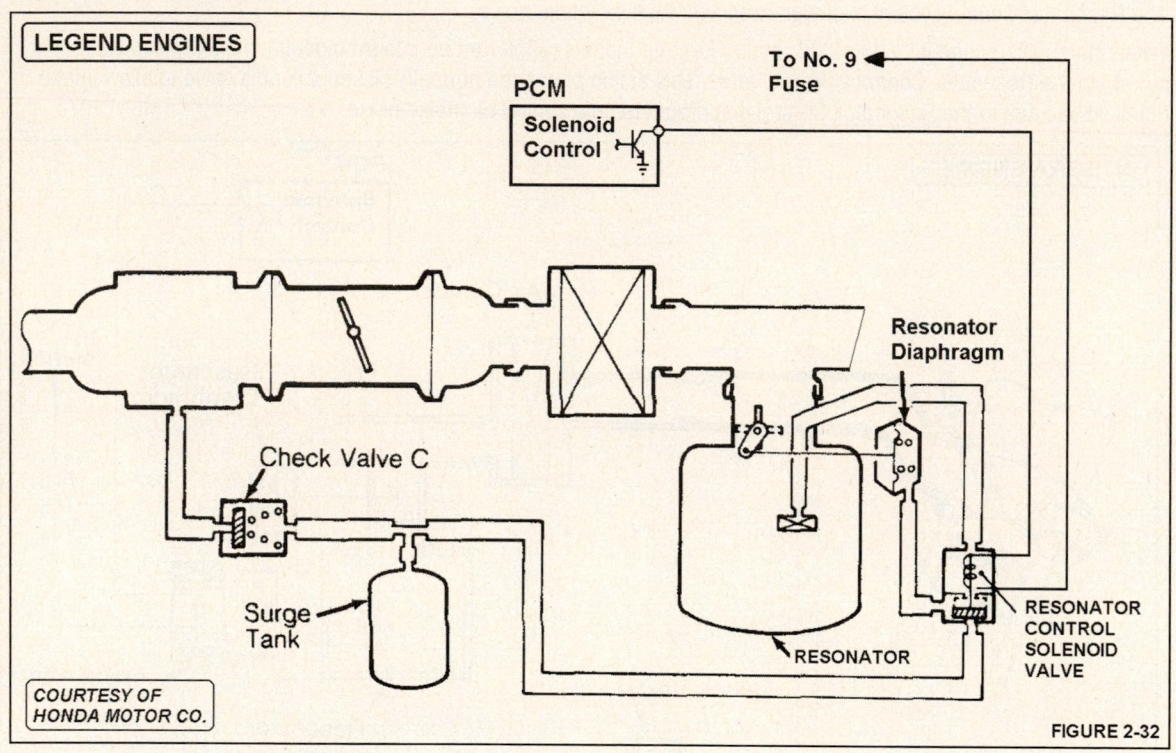

LEGEND ENGINES

PCM

Solenoid Control

To No. 9 Fuse

Resonator Diaphragm

Check Valve C

Surge Tank

RESONATOR

RESONATOR CONTROL SOLENOID VALVE

COURTESY OF HONDA MOTOR CO.

FIGURE 2-32

CAUTION: *Block the drive wheels and connect the vehicle exhaust pipe(s) to a suitable Vent system to purge exhaust fumes from the work area.*

INSPECT VACUUM HOSES, TEST ELECTRICAL CIRCUITS

Remove the vacuum hose from the resonator diaphragm (No. 8 vacuum hose on Integra models) and connect a vacuum gauge to the hose.

Start the engine and allow it to idle.

Observe the vacuum gauge reading. If the gauge reads vacuum (okay), raise engine speed to 3800 rpm (3000 rpm on Integra models) and again read the gauge connected to the resonator diaphragm vacuum hose. If the gauge reads no vacuum at high speed, the Resonator Control system is okay and no further testing is required. If the gauge reads vacuum at 3800 rpm (3000 rpm on Integra models), go to Step 8.

If the gauge reads no vacuum at idle speed in Step 3, remove the Resonator Control vacuum hose at the surge tank and use a vacuum gauge to check for vacuum to the tank. If there is no vacuum reading at the tank, locate and repair the blockage (or vacuum leak) between the surge tank and intake manifold. Return to Step 3 and continue testing.

RESONATOR CONTROL SYSTEM TESTS

INSPECT VACUUM HOSES, TEST ELECTRICAL CIRCUITS

If the gauge reads vacuum at the surge tank hose, disconnect the 2-wire plug (6-P connector on Integra models) from the Resonator Control solenoid (Figure 2-33). Use a DVOM to measure the voltage between the black/yellow terminal and red/blue terminal (red wire on Integra models). If the DVOM reads battery voltage, replace the faulty Resonator Control solenoid valve and retest the system.

If the DVOM reads zero volts, move the negative lead to a good body ground point and again read the DVOM. If the DVOM still reads zero volts, locate and repair the open circuit between the 2-P connector (6-P connector on Integra models) and the 10-amp fuse (No. 9 fuse on Legend Coupe and Legend Sedan, No. 4 fuse on Integra models) and retest the system.

If the DVOM reads battery voltage is Step 6, check the continuity of the red/blue wire (red on Integra models) between the 2-P connector (6-P on Integra models) and the PCM. If the wire is open, make needed repairs and retest the system. If the wire has continuity, substitute a known-good PCM and retest. If the symptom is gone, replace the original PCM.

If the gauge read vacuum at 3800 rpm (3000 rpm on Integra models) in step 3, then remove the 2-P connector (6-P on Integra models) and read the gauge. If the gauge reads vacuum, replace the Resonator Control solenoid valve (as it is leaking) and retest the system.

If the gauge reads no vacuum with the 2-P connector off (6-P connector on Integra models), turn the key off and disconnect the 'A' connector from the PCM. If the Honda test harness or BOB connector is installed at the PCM, leave it connected for the next step.

Use the DVOM to check for continuity to ground on the red/black wire (red on Integra models). This wire is the control circuit between PCM terminal A14 (terminal A10 on Integra) and the 2-P connector (6-P on Integra models). If continuity exists, locate and repair the short-to-ground on the red/black wire (red wire on Integra models) and retest the system.

If continuity is not present in Step 10, substitute a known-good PCM and retest. If the symptom goes away, replace the original PCM.

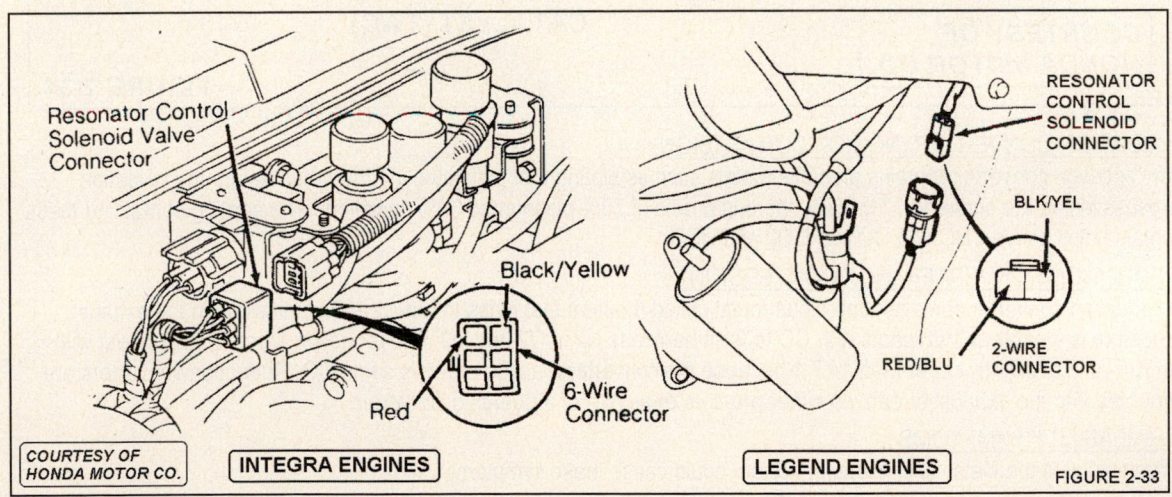

Resonator Control Solenoid Valve Connector

RESONATOR CONTROL SOLENOID CONNECTOR

BLK/YEL

Black/Yellow

Red

6-Wire Connector

RED/BLU

2-WIRE CONNECTOR

COURTESY OF HONDA MOTOR CO.

INTEGRA ENGINES

LEGEND ENGINES

FIGURE 2-33

Catalyst & Exhaust Systems

CATALYTIC CONVERTER & EXHAUST SYSTEM TESTS

INTRODUCTION

Acura vehicles with fuel injection are equipped with a three-way catalytic converter (TWC) to convert hydrocarbons (HC), carbon monoxide (CO) and oxides of nitrogen (NOx) in the exhaust into harmless gases composed of carbon dioxide (CO_2), dinitrogen (N_2) and water vapor. Some applications use two converters to reduce and oxidize the exhaust. The front converter, mounted close to the exhaust manifold, is called a pre-cat converter. The pre-cat converter comes up to light-off temperature very quickly.

The converter outside shell is made of stainless steel in order to allow it to sustain the high operating temperatures sustained during normal engine operation. Once the engine warms up, the converter outside shell can reach temperatures near 700°F and can reach 1400°F inside the shell. If misfire conditions occur, these temperatures rise and can damage the converter (Figure 2-34).

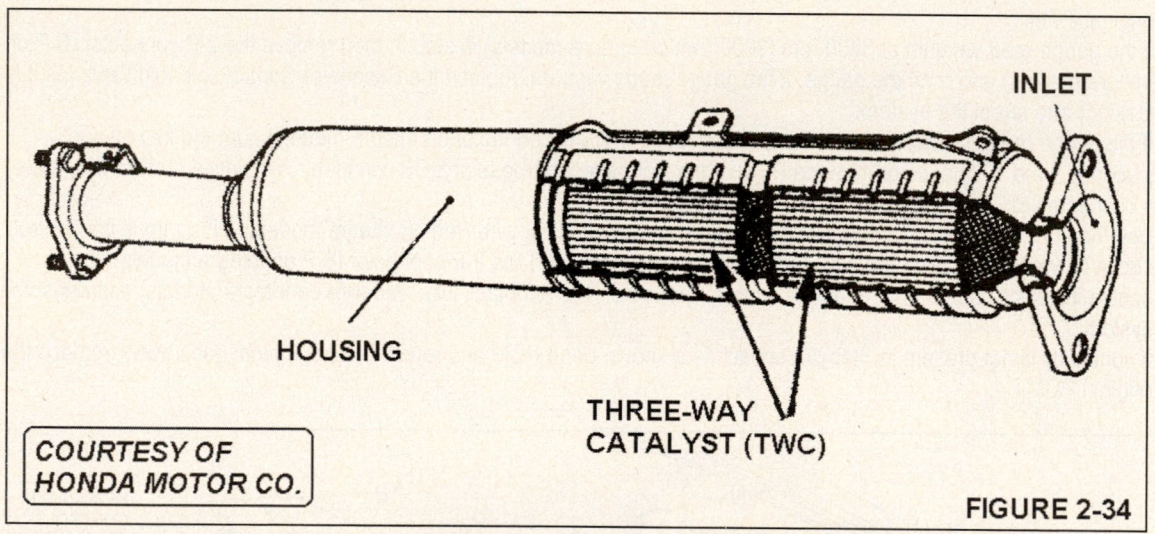

INLET

HOUSING

THREE-WAY
CATALYST (TWC)

COURTESY OF
HONDA MOTOR CO.

FIGURE 2-34

OXIDATION CONVERTER (NO NOX CONVERSION)

An Oxidation converter contains precious metals such as platinum or palladium in order to speed up the oxidation process inside the converter. Typically, there is a ratio of 70% platinum to 30% palladium. The coating containing these metals helps convert HC and CO into CO_2 and H_2O.

REDUCTION CONVERTER (NOX CONVERSION)

A reduction converter contains a precious metal called rhodium that helps to reduce NOx emissions in the exhaust. Rhodium is a catalyst that reacts with CO to form harmless N_2, CO_2 and O_2. The reduction process works best with an A/F ratio of slightly richer than 14.7:1 because the converter reaction requires some CO. Reduction converters are mounted into the exhaust stream as either pre-cats or as part of a dual-bed converter.

DRIVEABILITY SYMPTOMS

Problems with the Catalyst or Exhaust system could cause these symptoms:

Fails I/M or State Emissions Test
Loss of Power
Poor Engine Performance

Refer to Catalytic Converter or Exhaust System Tests in Section One for special tests.

EGR System
EGR SYSTEM TESTS

INTRODUCTION
The Exhaust Gas Recirculation (EGR) system is designed to reduce oxides of nitrogen emissions by recirculating exhaust gas through the EGR valve and the intake manifold into the combustion chambers (Figure 2-35).

SYSTEM COMPONENTS
On 1988-95 Acura vehicles (OBD I systems), a typical EGR system consists of the components listed below. For 1996-98 Acura applications (OBD II systems), refer to EGR system information in Section 4.

EGR valve with EGR valve lift sensor
Constant vacuum control (CVC) valve
EGR control solenoid valve
EGR related wiring and vacuum lines
The Powertrain Control Module (PCM)

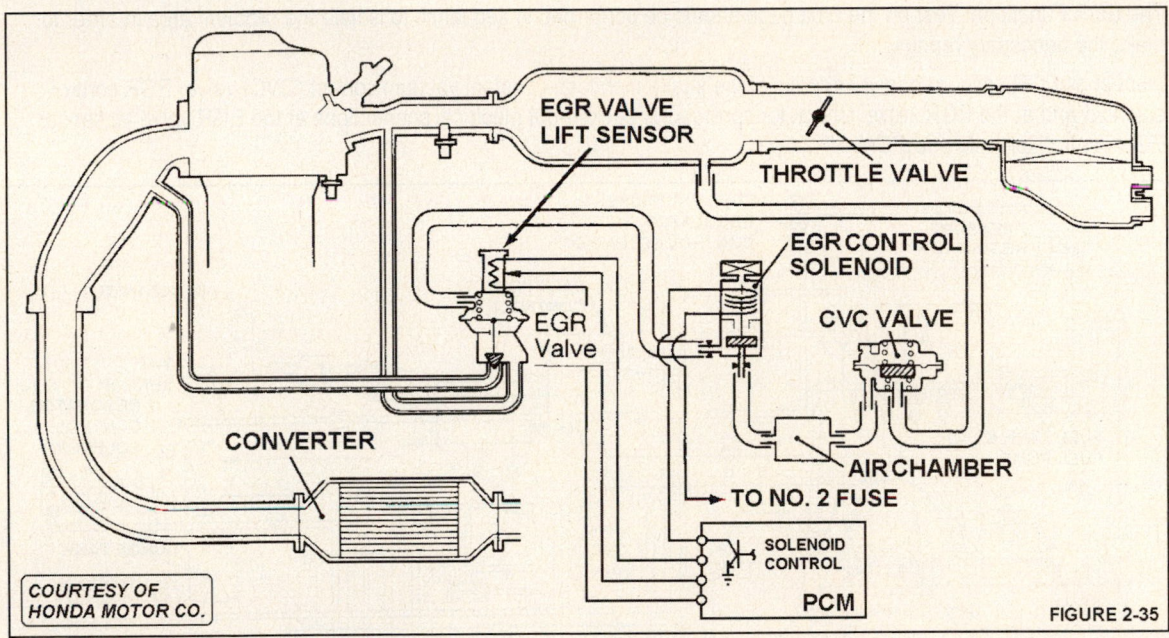

FIGURE 2-35

ACTUAL & IDEAL VALVE LIFT POSITION
The PCM contains memory tables that include the ideal EGR valve position (amount of EGR valve lift or opening) for various operating conditions.

The PCM determines the amount of EGR valve lift (or opening) through a signal from the EGR valve lift sensor. The actual valve lift position is compared to an ideal sensor valve lift value stored in memory.

The PCM determines the ideal EGR valve lift value through signals it receives from other sensors. If there is a difference between the actual and ideal valve lift values, the PCM turns off the EGR control solenoid valve to reduce the amount of vacuum applied to the EGR valve.

EGR System Tests

EGR SYSTEM DIAGNOSTICS

As discussed, the PCM monitors the operation of the EGR system through the amount of change in the EGR valve lift sensor. If the PCM detects that the amount of change in the sensor is too much or too little, or if it detects a shorted or open sensor circuit, it will set a hard Code 12 in memory and turn on the Check Engine light (or MIL).

The EGR Functional Test in this article can be used to locate symptom-related faults in one of the EGR components or circuits. In effect, the functional test provides a "manual" method of cycling the EGR components.

DRIVEABILITY SYMPTOMS

Problems in the EGR system can cause any of the symptoms listed below:

Loss of power or Detonation
Rough Idle or Frequent Stalling
Surges at Steady Cruise or Trailer-Hitching on Decel

PRELIMINARY STEPS

The EGR Functional Test on the next page should be performed in sequence to isolate the problem area in order to make the necessary repairs.

Inspect all EGR vacuum hoses for leaks at the intake manifold, constant vacuum control (CVC) valve, EGR control solenoid and at the EGR valve. Check for corroded or backed-out electrical connections at the EGR valve lift sensor, EGR control solenoid and PCM.

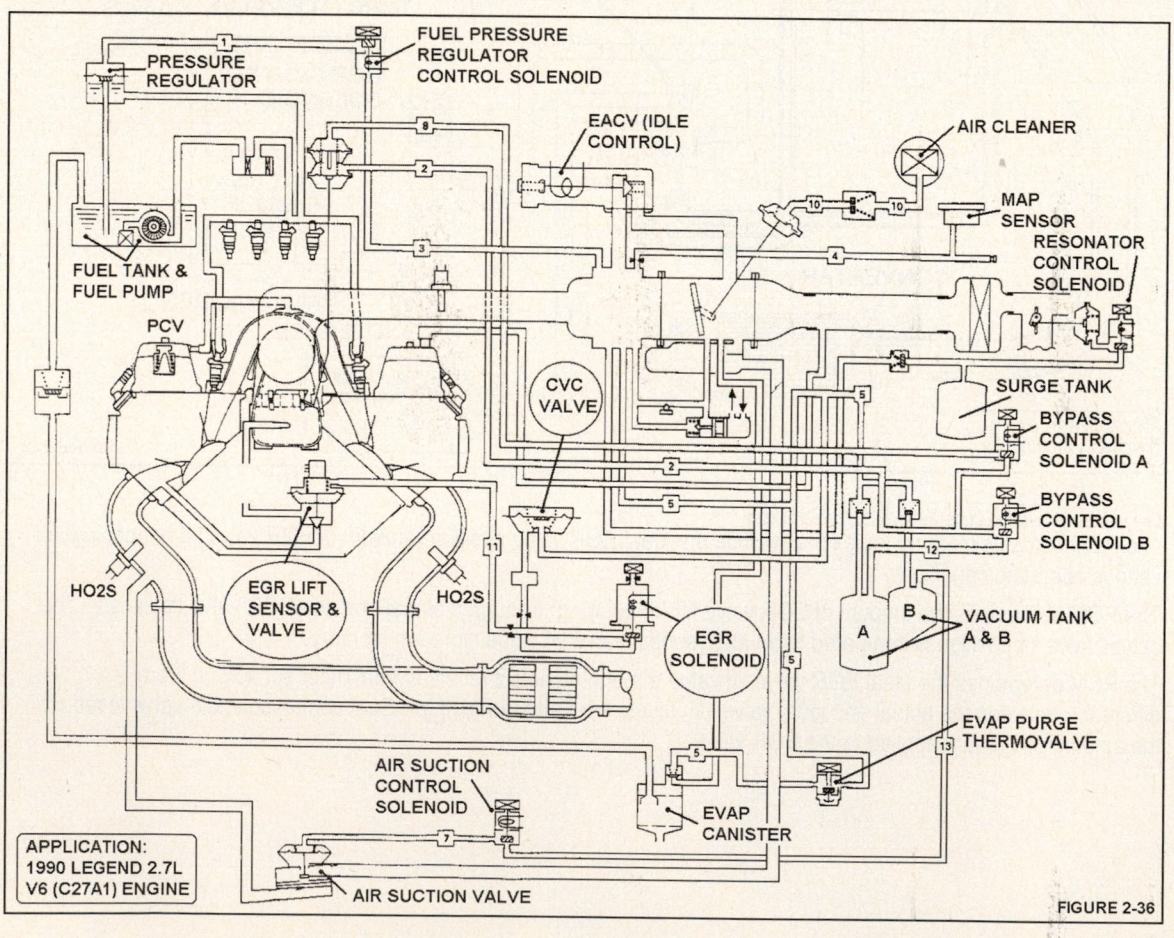

FIGURE 2-36

EGR SYSTEM TESTS

CAUTION: ***Block the drive wheels and connect the vehicle exhaust pipe(s) to a suitable vent system to purge exhaust fumes from the work area.***

EGR FUNCTIONAL TEST

Start the engine and run it at 3000 rpm in P/N until the cooling fan cycles.

Allow the engine to return to idle and remove the No. 11 vacuum hose (Figure 2-37) at the EGR valve or at the control box. Connect a hand vacuum pump to the valve nipple and apply 8-10" of vacuum to the valve. The EGR valve should hold vacuum and the engine should stall as vacuum is applied. If it does not, determine if the EGR valve is defective or if the intake manifold passages are plugged. Replace the valve or clean the passages as required and retest the system.

If the EGR valve operates normally in Step 2, move the vacuum gauge to the No. 11 hose (Figure 2-37) and check for vacuum. The gauge should read 0" of vacuum with the engine at idle speed (indicating that the PCM has the EGR control solenoid valve off). If the gauge shows any vacuum at this time, turn the engine off and perform the following steps:

Inspect EGR vacuum hoses to verify correct routing to the solenoid

Test EGR control solenoid valve for vacuum leaks (replace if faulty)

Remove the EGR valve and inspect for carbon around or under the valve seat. Clean the valve as needed and install it with a new gasket.

If an intermittent Code 12 is set, refer to the correct code repair chart

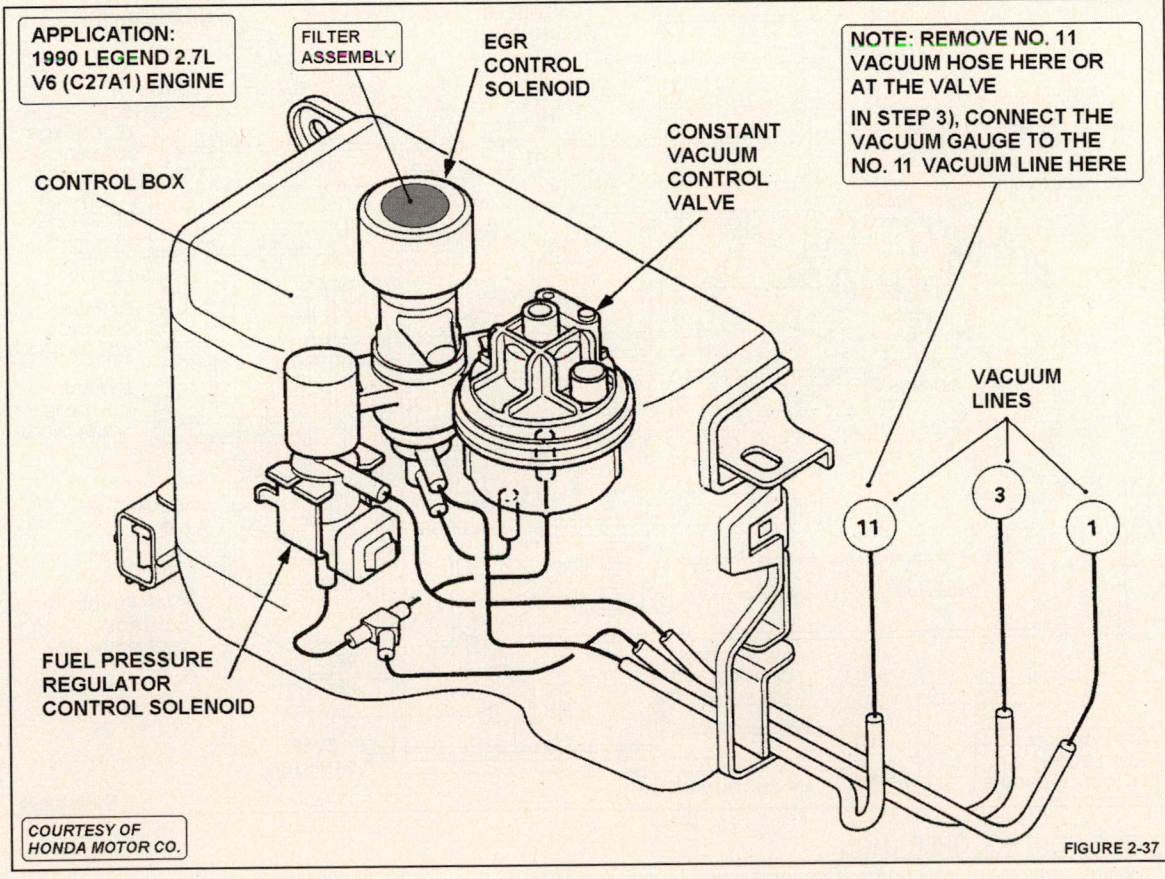

APPLICATION:
1990 LEGEND 2.7L
V6 (C27A1) ENGINE

FILTER ASSEMBLY

EGR CONTROL SOLENOID

CONSTANT VACUUM CONTROL VALVE

NOTE: REMOVE NO. 11 VACUUM HOSE HERE OR AT THE VALVE

IN STEP 3), CONNECT THE VACUUM GAUGE TO THE NO. 11 VACUUM LINE HERE

CONTROL BOX

VACUUM LINES

FUEL PRESSURE REGULATOR CONTROL SOLENOID

COURTESY OF HONDA MOTOR CO.

FIGURE 2-37

EVAP System
<u>EVAP SYSTEM TESTS</u>

<u>INTRODUCTION</u>
The Evaporative Emission Control (EVAP) system is designed to minimize the amount of Fuel system vapors released into the atmosphere.

<u>SYSTEM COMPONENTS</u>
On 1988-95 Acura vehicles (OBD I systems), a typical EVAP system consists of the components listed below. For 1996-98 Acura applications with OBD II diagnostics, refer to the EVAP system information in Section 4.

EVAP charcoal canister
Vapor purge control system
EVAP purge cut-off solenoid valve (Integra) or EVAP thermovalve (Legend)
Fuel tank vapor control system
Powertrain Control Module (PCM)

Component Examples for a 1990 Legend 2.7L V6 engine are shown in Figure 2-38.

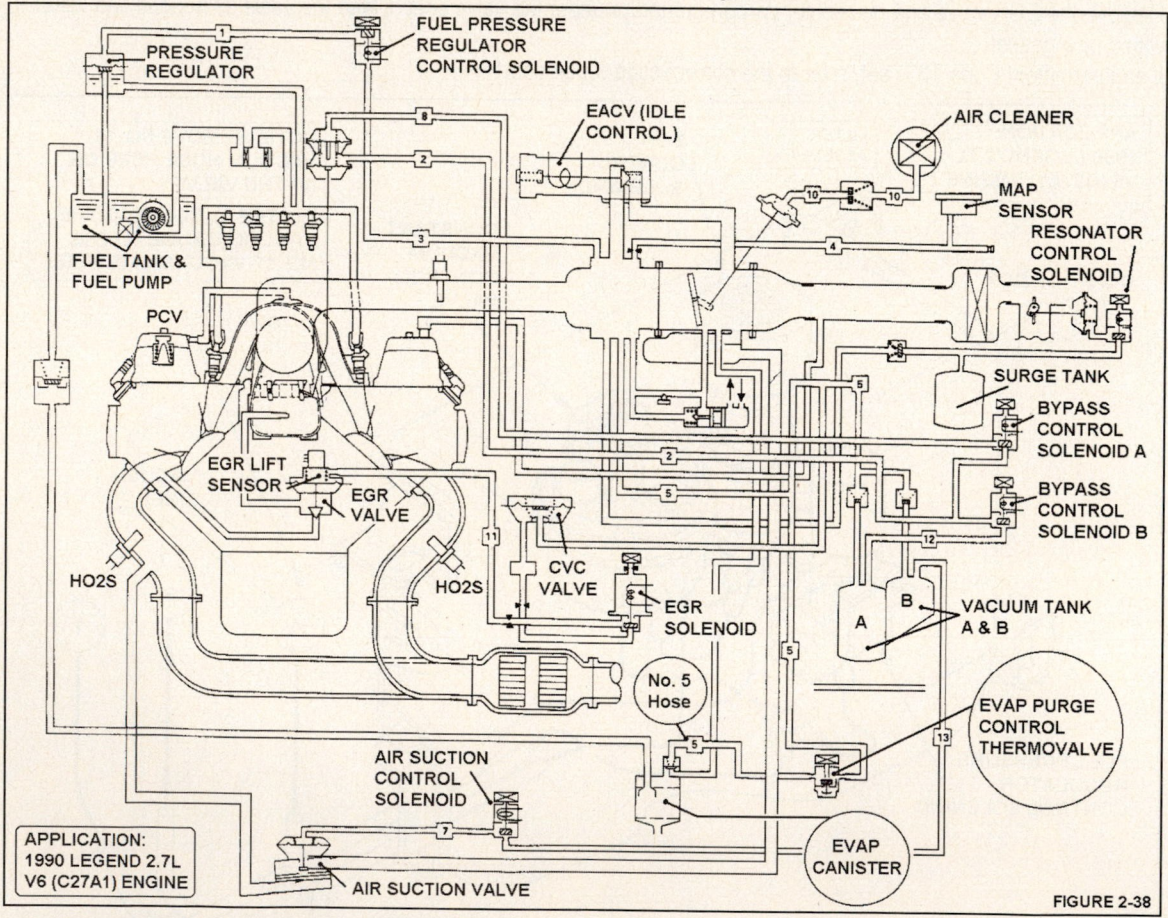

FIGURE 2-38

<u>PURGE CONTROL OPERATION</u>
On Acura vehicles, EVAP purge is accomplished by drawing fresh air through the canister using ported vacuum from the purge port on the throttle body. A purge control thermovalve controls the vacuum signal to the canister.

EVAP SYSTEM TESTS

PRELIMINARY STEPS

The inspection and test steps that follow should be performed in sequence to isolate a problem in the EVAP system in order to make the necessary repairs.

Inspect all EVAP vacuum hoses for cracks, leaks or disconnects at the intake manifold, EVAP canister and purge controls solenoid valve. Connect a suitable tachometer to the engine ignition system to read the engine speed. Check for corroded, loose or backed-out electrical connections at the EVAP purge control solenoid valve and the PCM before proceeding.

DRIVEABILITY SYMPTOMS

An inoperative EVAP system or fault could cause the following symptom:

Poor Performance with a Cold Engine
Fails I/M or State Emissions Test

EVAP SYSTEM TESTS

CAUTION: ***Block the drive wheels and connect the vehicle exhaust pipe(s) to an Exhaust Vent system in order to purge exhaust fumes from the work area.***

COLD ENGINE TESTS - INTEGRA & LEGEND MODELS

Remove the No. 5 vacuum hose (No. 7 on Integra models) at the EVAP purge control diaphragm valve on the canister. Connect a vacuum gauge to the hose (Figure 2-39). With engine temperature cold (below 158°F on Legend or below 140°F on Integra models), start the engine and allow it to idle. The gauge should not read vacuum at this time. On Legend models, if the gauge reads vacuum, replace the EVAP purge thermovalve and repeat the test. Then retest the system and if it passes on Legend models, go to the Warm Engine Tests. On Integra models, if the gauge reads vacuum, go to Step 3.

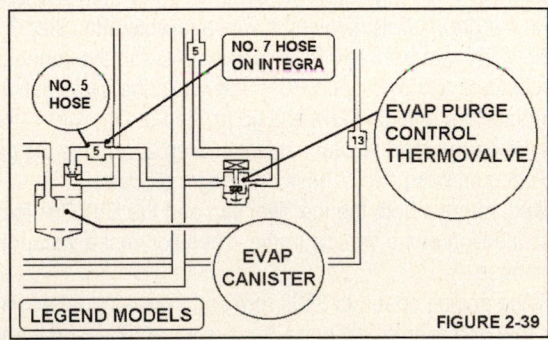

FIGURE 2-39

On Integra models, if the gauge reads vacuum in Step 2, disconnect the 4-P connector. Use a DVOM to measure the voltage between the BLK/YEL and GRN wire terminals. If battery voltage is present, verify the vacuum hose routing and for proper connections. If the vacuum hose is properly routed and connections are okay, replace the EVAP purge cutoff solenoid and retest the system. Then go to the Warm Engine Tests.

On Integra models, if battery voltage is not present in Step 3, use a DVOM to measure the voltage between the BLK/YEL wire in the 4-P connector and body ground. If the DVOM reads battery voltage, check for an open circuit between the PCM terminal A6 and the connector. If the wire is not open, substitute a known-good PCM and retest. If the symptom goes away, replace the original PCM and go to the Warm Engine Tests. If voltage is still not available at the connector, locate and repair the open circuit condition in the BLK/YEL wire between the 4-P connector and the No. 24 fuse. Then go to the Warm Engine Test.

EVAP SYSTEM TESTS

WARM ENGINE TESTS - INTEGRA & LEGEND MODELS

The next few test steps require that the engine be at normal operating temperature with the cooling fan cycling.

Disconnect the No. 5 vacuum hose (No. 7 on Integra models) at the purge control diaphragm valve at the canister. Connect a suitable vacuum gauge to the hose (Figure 2-40).

Start the engine and allow it to idle. The vacuum gauge should read vacuum with the engine at idle. On Integra models, if the gauge reads vacuum at idle within 10 seconds, go to Step 5. If the gauge reads no vacuum, go to Step 3. On Legend models, if the gauge reads no vacuum, replace the thermovalve and retest the system. Go to Step 5 when repairs are done.

On Integra 4-P wire connector (Figure 2-41). If the gauge reads vacuum now, verify the vacuum hose routing and check the connections. If the vacuum hose is properly routed and connections are okay, replace the purge cutoff solenoid valve and retest the system. Go to Step 4 when repairs are done.

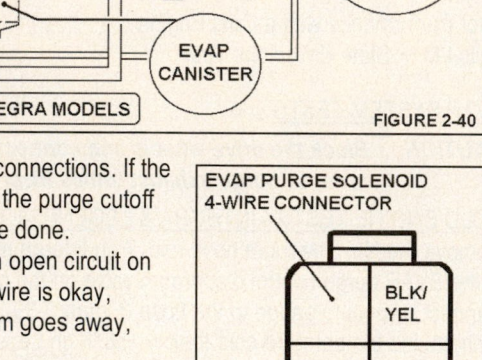

On Integra models, if vacuum was available after Step 3, test for an open circuit on the GRN wire between PCM terminal A6 and the connector. If the wire is okay, substitute a known-good PCM and retest the system. If the symptom goes away, replace the original PCM and go to Step 5.

If the gauge read vacuum at idle in Step 2, remove the gauge and reconnect the No. 5 vacuum hose (No. 7 hose on Integra models).

Next, remove both the fuel filler cap and the EVAP purge air hose from its connection at the vehicle frame. Then connect a vacuum gauge to the EVAP purge air hose.

Raise engine speed to 3500 rpm and maintain that engine speed for one minute. The gauge should read vacuum within one minute. If it does, the system is okay and the test is finished. Go to the Two-way Valve Test. If no vacuum is shown, replace the EVAP canister assembly and the fuel filler cap. Go to the Two-way Valve Test when repairs are done.

TWO WAY VALVE TEST

To test the two-way valve, remove the fuel filler cap and the vapor line from the fuel tank. Connect a 'T'-fitting to the vacuum hose. Then connect a hand held vacuum pump into the fitting (Figure 2-42).

Use the hand vacuum pump to slowly apply vacuum to the 'T'-fitting while observing the gauge. The vacuum reading should stabilize at 0.2-0.6" of vacuum. If the vacuum reading indicates that the valve opens below 0.2" of vacuum or above 0.6" of vacuum specification, replace the two-way valve and retest system. If the vacuum reading stabilizes as specified, go to Step 3.

Next, move the vacuum gauge hose from the vacuum side to the pressure side of the hand vacuum pump (Figure 2-42).

Slowly pressurize the vapor line while watching the pressure gauge. The pressure reading should stabilize at 1.0-2.0" of pressure (0.4-1.4" of pressure on Integra models).

If the pressure reading stabilizes for a moment within the specified range, the valve opened at the correct pressure value (this action indicates the valve is okay). If the valve opens below or above the specification, the two-way valve is faulty and must be replaced.

Replace the two-way valve and fuel filler cap. Retest the EVAP system when repairs are done.

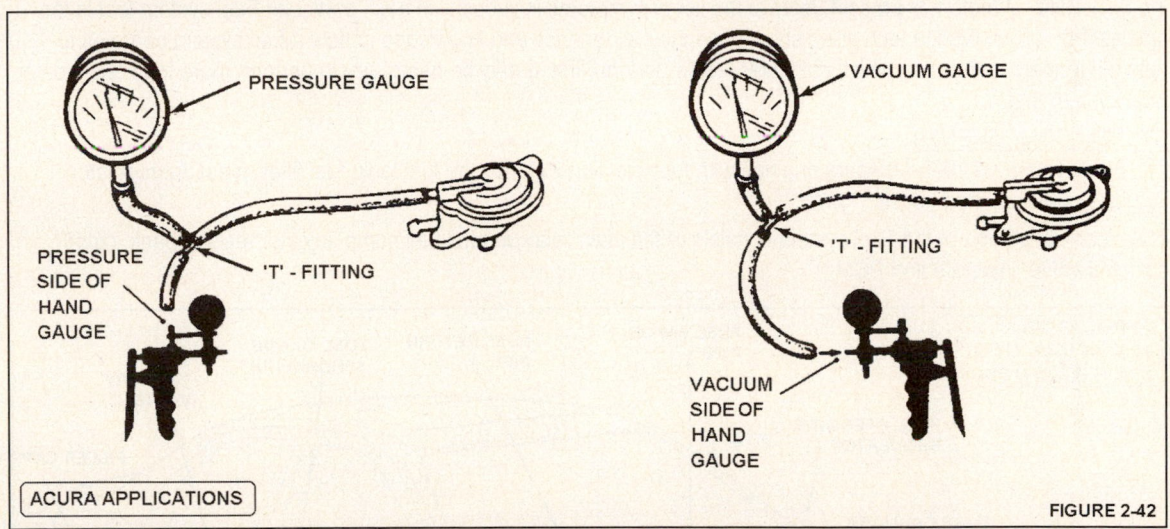

FIGURE 2-42

■ NOTE: *For applications other than 1990 Integra, Legend Coupe and Sedan models discussed in this article, refer to other pin voltage charts for pin and wire color identification for all references to PCM terminals and wire colors.*

Fuel System

INTRODUCTION

The Fuel system on Acura vehicles is designed to deliver fuel at a regulated high pressure to the fuel rail and injectors. Additionally, the Fuel system is also designed to cut off fuel pressure if the engine stops running.

FUEL SYSTEM TESTS

The Fuel system tests in these articles should be used only after these steps are done:

The complaint has been verified and a visual inspection is done
PGM-FI Diagnostics are completed when they show Pass (See NOTE)
Base Engine tests are done
An actual driveability symptom is present during vehicle testing
Related technical service bulletins have been searched in the TSB index

■ **NOTE:** *Turn the key on and watch the Check Engine or PGM-FI light. It should come on for two seconds and then go off. If the Check Engine light remains on, refer to the PGM-FI Diagnostics article in this section (Section 2).*

DRIVEABILITY SYMPTOM DIAGNOSIS

If a driveability symptom is present, refer to the list of symptoms included with each particular Fuel system test in the articles that follow. Also, a technician should use their experience and knowledge of how a fuel system operates to help guide them in deciding which particular test to perform first. It may be necessary to perform more than one type of Fuel system test.

SYSTEM COMPONENTS

The Fuel system on 1985-98 Acura engines with fuel injection includes the fuel tank, fuel filter, pressure regulator, injectors, fuel lines and hoses.

The electrical portion of the Fuel system consists of the main relay, in-tank fuel pump, injector resistor bank, cutoff solenoid valve, injectors and PCM.

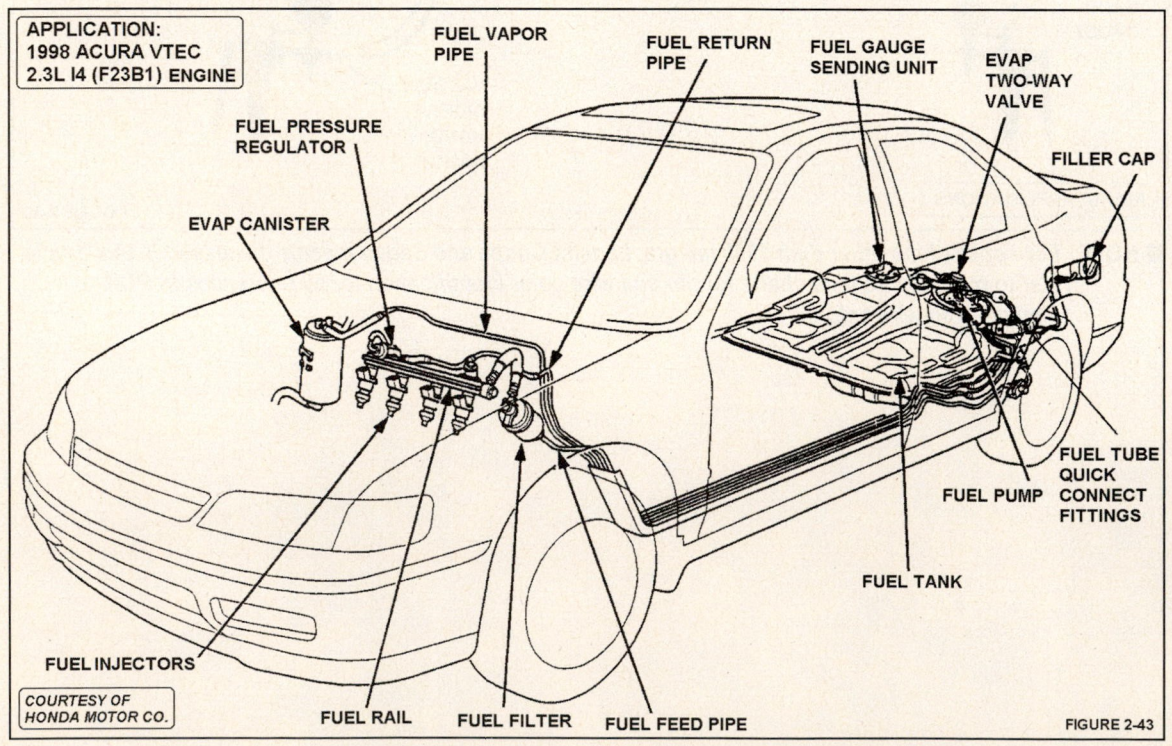

APPLICATION:
1998 ACURA VTEC
2.3L I4 (F23B1) ENGINE

FUEL VAPOR PIPE

FUEL RETURN PIPE

FUEL GAUGE SENDING UNIT

EVAP TWO-WAY VALVE

FUEL PRESSURE REGULATOR

FILLER CAP

EVAP CANISTER

FUEL TUBE QUICK CONNECT FITTINGS

FUEL PUMP

FUEL TANK

FUEL INJECTORS

COURTESY OF HONDA MOTOR CO.

FUEL RAIL   FUEL FILTER   FUEL FEED PIPE

FIGURE 2-43

FUEL PRESSURE TESTS

These repair procedures should be used whenever connecting or disconnecting a fuel pressure gauge to Acura vehicles with fuel injection. Automotive Fuel systems operate under high fuel pressure - always follow the precautions in the CAUTION note below.

CAUTION: *Keep all open flames or sparks away from the work area. Do not smoke during repairs and do not attempt to relieve the fuel pressure unless the engine is off.*

RELIEVING FUEL SYSTEM PRESSURE

Before any fuel lines or hoses or hoses are disconnected, any fuel pressure in the Fuel system must be relieved. This can be accomplished by loosening the service bolt (6 mm) located at the top of the fuel filter assembly.

Remove negative battery cable

Remove the fuel cap to relieve any pressure in the fuel tank

Use a box end wrench on the 6 mm service bolt at the top of the fuel filter while holding the banjo bolt with another end wrench

Place a rag or shop towel over the 6 mm service bolt and slowly loosen the service bolt one complete turn

FUEL PRESSURE GAUGE CONNECTIONS

A fuel pressure gauge (Honda Part No. 07406-0040001 or equivalent) can be attached at the hole where the service bolt connects (Figure 2-44). If the service bolt is loosened to relieve pressure or to connect a fuel pressure gauge, the washer installed between the service bolt and banjo bolt must be replaced. Also, whenever the service bolt is removed to disassemble the component, replace all washers.

Once fuel pressure is relieved, remove the service bolt on top of the fuel filter while holding the banjo bolt with a wrench. Attach a fuel pressure gauge and tighten it securely.

Start the engine and then check for signs of a fuel leak.

Next, watch the fuel pressure with the engine at idle speed and pressure regulator hose removed and plugged. The fuel pressure should read 35-41 psi with the vacuum hose to the regulator removed.

If fuel pressure is higher than specified, check for a pinched (or clogged) fuel return hose or fault in the fuel pressure regulator.

If fuel pressure is lower than specified, check for a clogged fuel filter, fault in the fuel pressure regulator, faulty fuel pump or leakage in the system.

■ NOTE: *Refer to the appropriate Fuel System Tests in this section to diagnose other parts of the system (i.e., Fuel Injector Test, Fuel Quality Tests, etc.).*

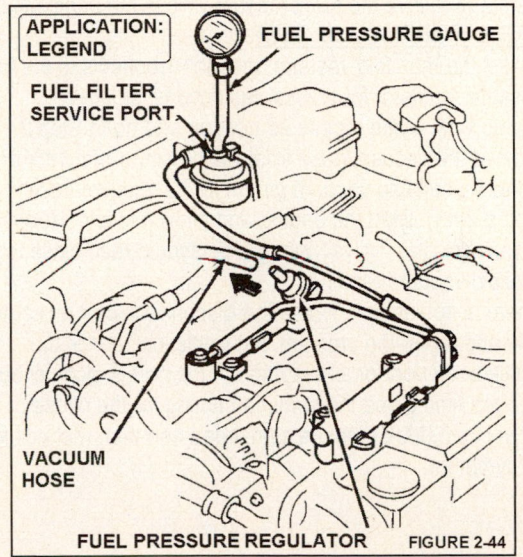

FIGURE 2-44

<u>FUEL PRESSURE REGULATOR TESTS</u>

<u>INTRODUCTION</u>
The fuel pressure regulator is used to maintain a constant fuel pressure to the fuel rail and fuel injectors. When the difference between fuel pressure and manifold vacuum exceeds 36 psi, the diaphragm in the pressure regulator is pushed upward to allow the excess fuel to be fed through the fuel return line back to the fuel tank.

<u>DRIVEABILITY SYMPTOMS</u>
If a driveability symptom is present, refer to the list of symptoms below and choose the symptom that best fits the particular problem for the vehicle under test. It may be necessary to perform more than one type of Fuel system test to locate the fault and complete the repair. Problems in the fuel pressure regulator could cause any of the symptoms listed below.

Frequent Stalling
Hard Start - Cold or Hot Engine
Loss of Power or Poor Performance
Fails I/M or State Emissions Test
Poor Fuel Economy

<u>PRELIMINARY STEPS</u>
The following inspection and test steps should be performed in sequence to isolate the problem area in order to make the necessary repairs. Always start the test procedure by performing the preliminary steps.

Attach a suitable fuel pressure gauge to the service port of the fuel filter.
CAUTION: Keep all open flames or sparks away from the work area. Do not smoke during repairs and do not attempt to relieve the fuel pressure unless the engine is off.

<u>REGULATOR VACUUM TEST</u>
Inspect the manifold vacuum hose that connects to the fuel pressure regulator for cracks, restrictions, or loose connections. Make repairs as needed and go to Step 2. Remove the vacuum hose from the pressure regulator and connect a suitable vacuum gauge to the vacuum hose (Figure 2-45). Start the engine and allow it to idle. Read the vacuum gauge. The vacuum gauge should read close to the manifold vacuum reading.

If there is no vacuum, locate the disconnected or misrouted hose and repair the problem. If it reads low, check the throttle body port for a restriction. If the port is okay, check the base timing and the camshaft timing for the cause of the low vacuum. Make repairs as needed and then recheck for the symptom.

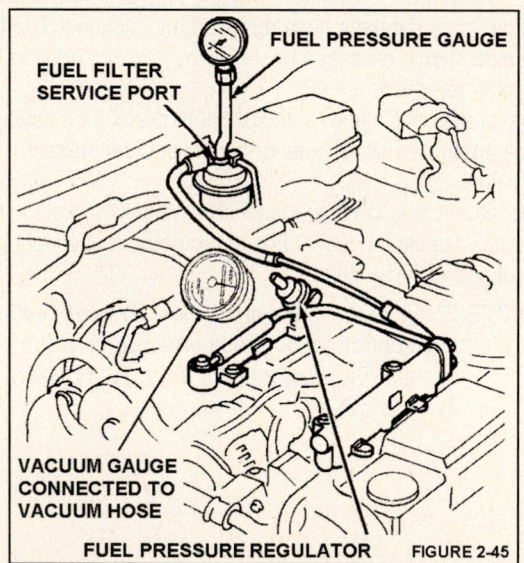

FUEL PRESSURE GAUGE
FUEL FILTER SERVICE PORT
VACUUM GAUGE CONNECTED TO VACUUM HOSE
FUEL PRESSURE REGULATOR   FIGURE 2-45

CAUTION: *Keep all open flames or sparks away from the work area. Do not smoke during repairs and do not attempt to relieve the fuel pressure unless the engine is off.*

REGULATOR PRESSURE TEST

Turn the engine off and attach a fuel pressure gauge to the service port of the fuel filter (if not already connected).

Restart the engine with the gauge installed and checks for leaks. Next, verify that the fuel pressure reading on the gauge rises each time the vacuum hose is connected and removed from the pressure regulator. A typical amount of change is between 8-9 psi. The fuel pressure reading should be as specified in the table below with the vacuum hose removed.

If the pressure is within specification and rises with the hose removed, the regulator is operating normally. If the pressure does not rise with the hose removed, wrap the fuel return hose with a shop towel and very carefully lightly pinch the fuel return hose (Figure 2-46). If the pressure remains the same, check the fuel filter (for restriction) and the fuel pump for low output. Make repairs as needed and retest for the symptom. If the pressure rises with the return lightly pinched, replace the pressure regulator. Retest the symptom when repairs are completed.

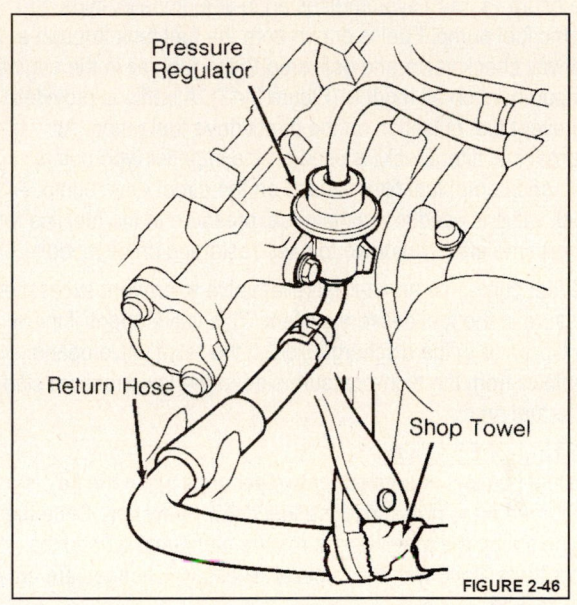

FIGURE 2-46

FUEL PRESSURE CHART - VACUUM HOSE REMOVED

Integra 1.7L I4 engine	48-56 psi
Integra 1.8L I4 engine	41-48 psi
Integra GS 1.8L I4 engine	44-51 psi
Integra GS-R 1.8L I4 engine	40-47 psi
Legend 2.7L I4 engine	37-46 psi
Legend 3.2L I4 engine	37-46 psi
Vigor 2.5L I5 engine	43-50 psi
SLX 3.5L V6 engine	41-46 psi
2.3CL (2.3L) I4 engine	33-41 psi
2.5TL (2.5L) I5 engine	32-40 psi
3.0CL (3.0L) V6 engine	38-45 psi
3.2TL (3.2L) V6 engine	35-41 psi
3.5RL (3.5L) V6 engine	35-41 psi

FUEL PUMP TESTS

INTRODUCTION

The Acura PGM-FI system uses an in-line, impeller-type electric fuel pump. Fuel is drawn from the fuel tank through a one-way check valve and delivered to the fuel rail in the engine compartment on all models (Figure 2-47). A baffle is provided to prevent fuel pulsation on the direct drive fuel pump. An external fuel filter (sock) is used on the impeller type pump, while an internal fuel filter is used on the direct drive pump. A check valve is used to maintain fuel pressure in the fuel line for a short time after shutdown to ease restarting on all models.

The fuel pump has an internal relief valve to prevent excessive pressure in the fuel delivery system. This valve opens if there is a blockage in the discharge side. If the relief valve opens, fuel flows from the high-pressure side to the low-pressure side of the fuel pump.

SYSTEM OPERATION

The fuel pump is energized for two seconds when the key is first turned on to pressurize the fuel delivery system. Once the engine starts, the main relay turns the fuel pump on and the motor turns along with the impeller. Pressure changes are created by the numerous grooves around the impeller (Figure 2-48).

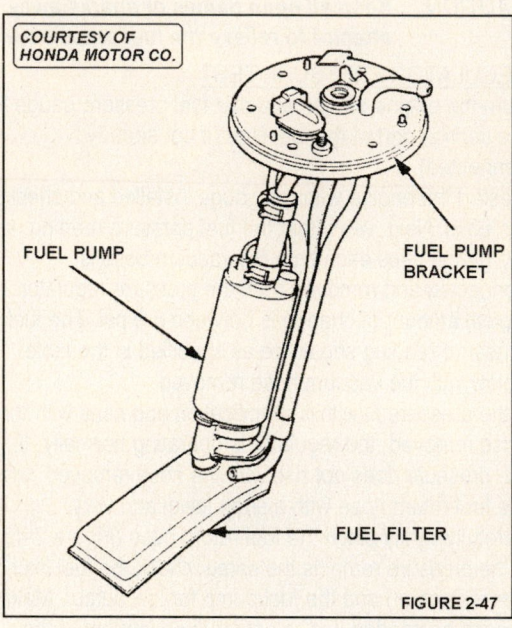

COURTESY OF HONDA MOTOR CO.

FUEL PUMP

FUEL PUMP BRACKET

FUEL FILTER

FIGURE 2-47

Fuel enters the inlet port and flows inside the motor form the pumping chamber and is forced through the discharge port via the check valve. If fuel flow is obstructed at the discharge side of the fuel line, the relief valve will open to bypass fuel to the inlet port to prevent high fuel pressure. Once the engine stops, the PCM turns off the fuel pump through the main relay due to the loss of ignition reference (rpm) signals. An internal check valve closes by spring action to retain residual pressure for quick restarts.

COMPONENT EXAMPLES

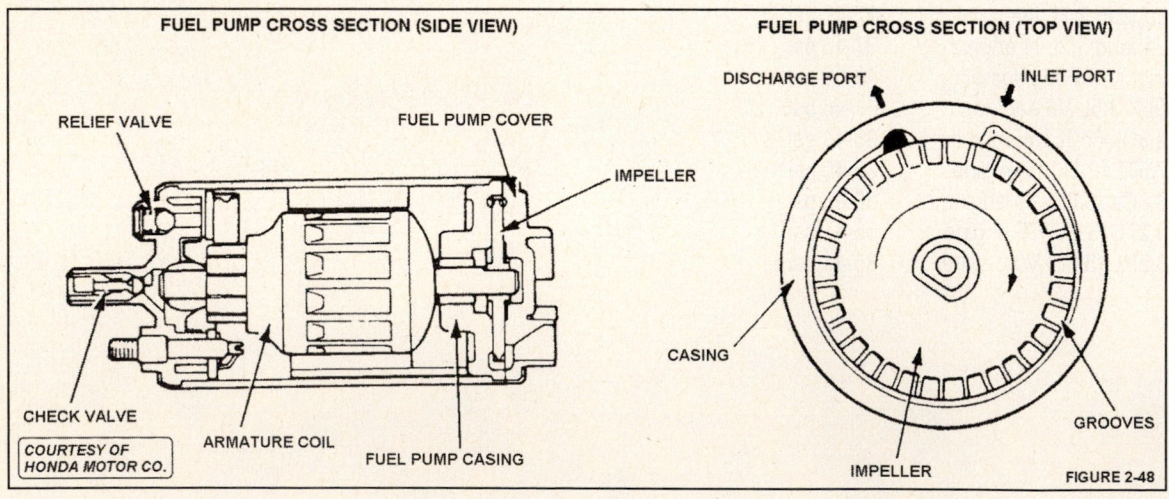

FUEL PUMP CROSS SECTION (SIDE VIEW)

RELIEF VALVE

FUEL PUMP COVER

IMPELLER

CHECK VALVE

ARMATURE COIL

FUEL PUMP CASING

COURTESY OF HONDA MOTOR CO.

FUEL PUMP CROSS SECTION (TOP VIEW)

DISCHARGE PORT

INLET PORT

CASING

GROOVES

IMPELLER

FIGURE 2-48

CAUTION: *Keep all open flames or sparks away from the work area. Do not smoke during repairs and do not attempt to relieve the fuel pressure unless the engine is off.*

PRELIMINARY STEPS

The inspection and test steps in the Fuel system tests on the next few pages should be performed in sequence in order to isolate the problem area and make the necessary repairs.

Start the test procedure by performing the preliminary steps.

On all models, attach a suitable fuel pressure gauge to the service port at the fuel filter or test point outlined in the service repair information (Figure 2-49).

Crank or start the engine and check the fuel pressure gauge connection for leaks before starting the test procedure.

FUEL PUMP TESTS WITH AN AMP PROBE

If a defective fuel pump is suspected, an amp probe with an inductive pickup can be used to monitor the operation of the fuel pump and its related circuits.

Examples of typical fuel pump motor waveforms that can be captured with an amp probe and Lab Scope are shown in Figure 2-50.

With the engine running, the fuel pumps on these systems should have a current draw near 4 amps (± 0.5 amps).

LAB SCOPE SETTINGS

Set the Lab Scope to these initial settings:

Volts per division: 1 volt
Time per division: 10 or 20ms
Trigger setting: 50% with a positive slope

Unplug the PGM-FI main relay and install a jumper to bypass the relay. Next, zero the amp probe for testing and then connect the amp probe around the jumper wire.

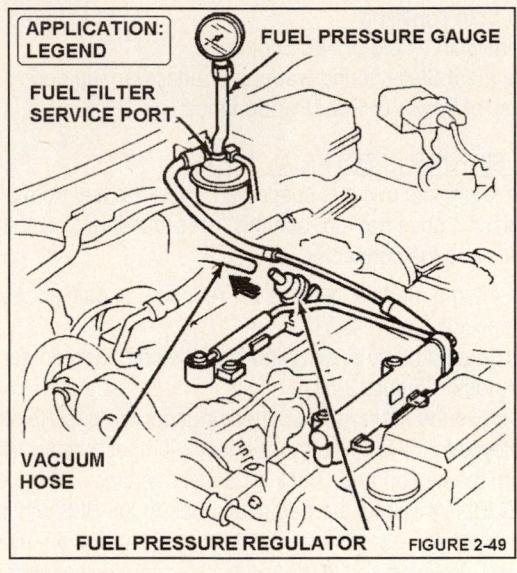

APPLICATION: LEGEND — FUEL PRESSURE GAUGE — FUEL FILTER SERVICE PORT — VACUUM HOSE — FUEL PRESSURE REGULATOR — FIGURE 2-49

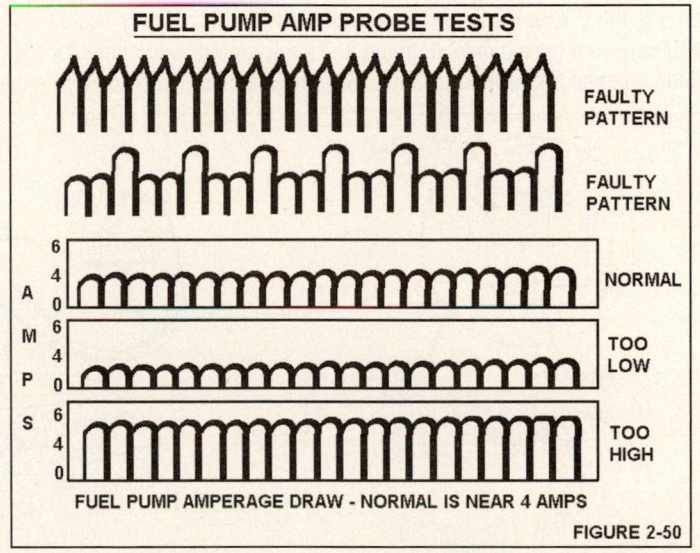

FUEL PUMP AMP PROBE TESTS — FAULTY PATTERN — FAULTY PATTERN — NORMAL — TOO LOW — TOO HIGH — AMPS — FUEL PUMP AMPERAGE DRAW - NORMAL IS NEAR 4 AMPS — FIGURE 2-50

DRIVEABILITY SYMPTOMS

Choose the symptom that matches the particular problem for the vehicle under test.

No Start condition
Hard Start - Cold or Hot Engine
Frequent Stalls during warmup or after warmup period
Loss of Power or Poor Performance

FUEL PUMP ELECTRICAL TESTS

If a faulty fuel pump is suspected, cycle the key from off to on while an assistant listens at the pump to determine if it runs. If it does not run, use this test procedure to test the fuel pump and main relay circuits. Turn the key off before removing the connector.

On Integra models, remove the rear seat to gain access to the fuel pump connector. Remove the pump 4-wire connector (Figure 2-51).

On Legend models, clear the luggage compartment for access to the fuel pump connector. Remove the fuel pump connector (Figure 2-51).

Remove the main relay connector under the dash to allow for testing.

To bypass the main relay, connect a jumper wire to the BLK/YEL and YEL/BLK wires at the relay connector.

Turn the key on and check for battery voltage at the fuel pump connector. Connect the DVOM positive probe to the YEL/BLK wire and the negative probe to the BLK wire at the wiring harness side connector. If there is no voltage for the first two seconds after the key is first turned on, move the negative lead to battery negative or body ground and retest. Turn the key to off and back on for two seconds while checking for voltage at the pump. If there is no voltage, refer to the Wiring Harness Tests in the PGM-FI Main Relay Tests.

If battery voltage is available to the fuel pump on the feed circuit for the first two seconds, the pump should run. If it does not, and the ground circuit is okay, replace the fuel pump and retest for the symptom.

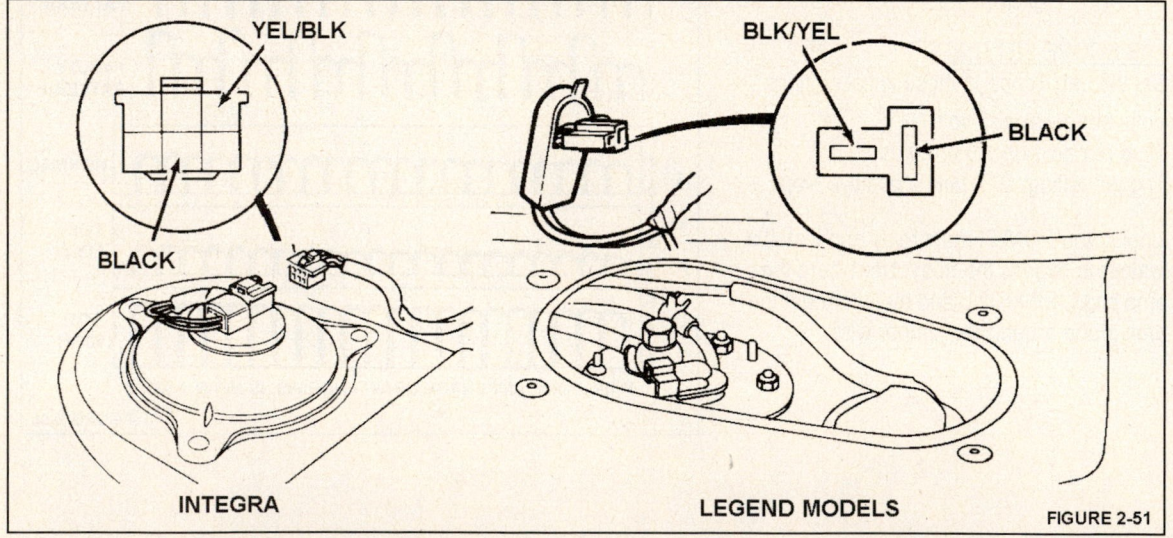

FIGURE 2-51

■ NOTE: *For vehicles other than 1990 Integra and Legend models referenced in this article, refer to other charts for correct wire colors and procedures.*

FUEL INJECTOR RESISTOR TESTS

Some Acura engines are equipped with a fuel injector resistor bank that is designed to lower the amount of current supplied to the injectors (Figure 2-52).

The resistor bank helps prevent damage to the fuel injector coils. The use of the resistor bank allows for a faster response time at the fuel injectors.

DRIVEABILITY SYMPTOMS

If a driveability symptom is present, refer to the list of symptoms below and choose the symptom that best fits the particular problem for the vehicle under test.

No Start condition
Misfire or Rough Running
Rough Idle
Loss of Power
Fails State Emission Test

PRELIMINARY STEPS

Turn the key off and remove the injector resistor bank. Disconnect the electrical connector.

Resistor Bank Test Steps
Use a DVOM to measure the resistance between terminal 'A' and terminals B, C, D and E (plus F and G on V6 engines). The DVOM should read from 5-7 ohms on each circuit at room temperature. If any of the readings indicate an open or shorted condition exists, replace the resistor bank and retest.
If the resistor bank is okay, turn the key on and measure the voltage between the resistor bank feed BLK/YEL wire and body ground. If the reading is 0v, check for an open circuit condition between the resistor bank and the PGM-FI main relay.
If the reading is near 12v, the resistor bank and its feed circuits are okay. Turn the key off and reconnect the resistor bank. Select another type of Fuel system test.

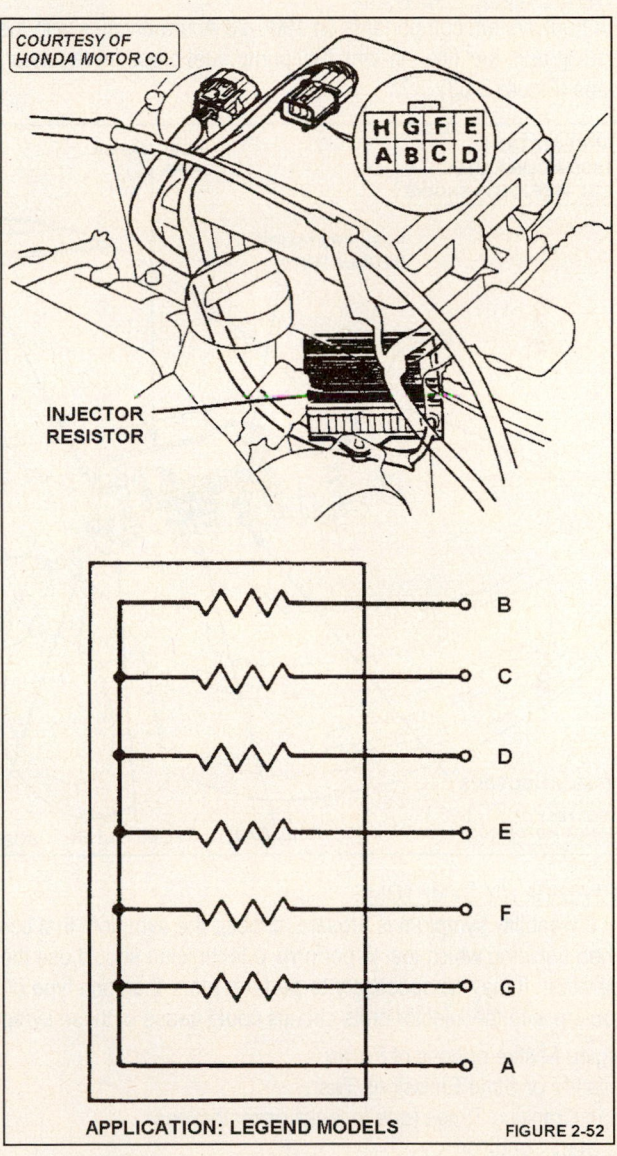

COURTESY OF HONDA MOTOR CO.

INJECTOR RESISTOR

APPLICATION: LEGEND MODELS

FIGURE 2-52

FUEL INJECTOR TESTS

INTRODUCTION

The Acura PGM-FI fuel injection system is equipped with fuel injectors that are solenoid-activated constant-stroke pintle design. The injector consists of a solenoid, plunger needle valve, and housing. When current is applied to the coil in the solenoid, the valve lifts up and pressurized fuel is injected close to the intake valve. The solenoid valve list is fixed, so with a constant fuel pressure, the amount of fuel delivered is determined by the length of time the injector is held open (referred to in this manual as injector pulsewidth).

SYSTEM COMPONENTS

The Fuel system components on 1986-98 Acura engines with fuel injection include the fuel tank, fuel cap, fuel gauge sending unit, fuel filter, in-tank fuel pump, fuel pressure regulator, fuel injectors, fuel pulsation damper, fuel pipes and hoses (Figure 2-53).

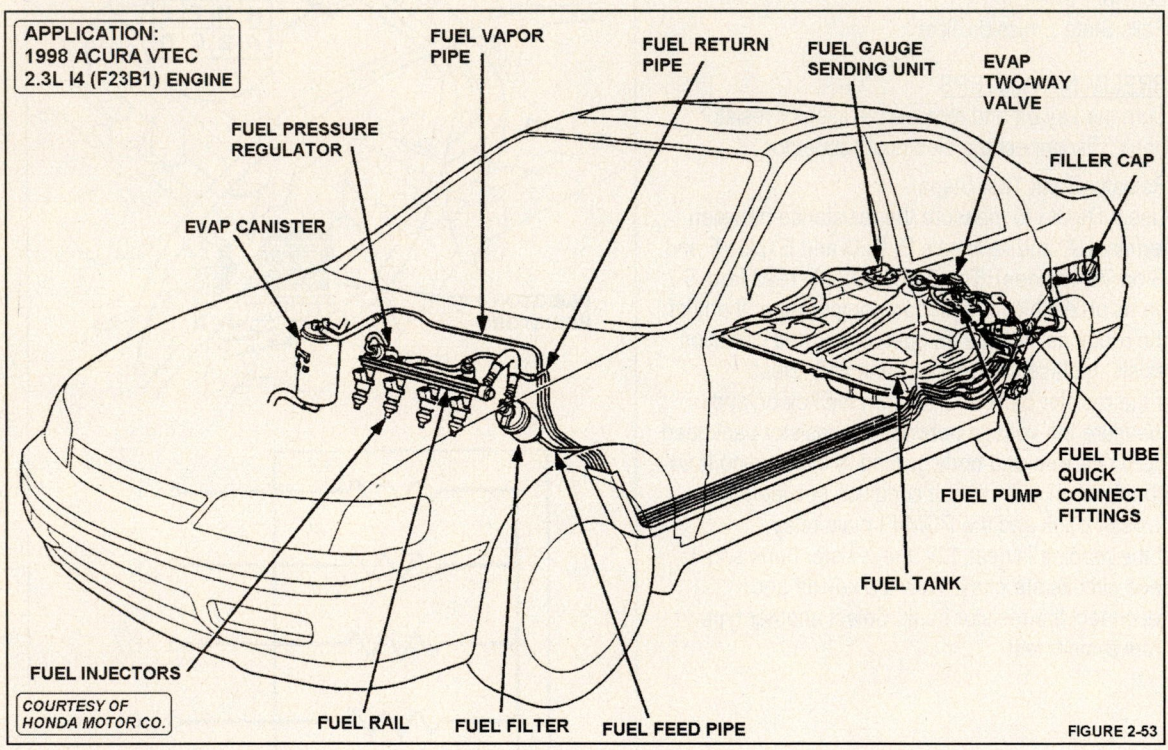

APPLICATION:
1998 ACURA VTEC
2.3L I4 (F23B1) ENGINE

COURTESY OF HONDA MOTOR CO.

FIGURE 2-53

DRIVEABILITY SYMPTOMS

If a driveability symptom is present, choose the symptom that best fits the particular problem for the vehicle under test. When deciding which test to perform, a technician should use their experience and understanding of Fuel system operation. It may be necessary to perform more than one type of test to locate the fault and complete the repair. A problem with the injector or its circuits could cause of these symptoms:

Engine Misfire or Loss of Power
Fails I/M or State Emissions Test
Long Cranking Times (due to leaks or restrictions)
No Start Condition
Rough idle with a cold or warm engine

PRELIMINARY STEPS

Perform all the preliminary steps in sequence to isolate the problem area. Check and adjust base timing, idle speed and idle CO% to specifications.

CAUTION: ***Block the drive wheels and connect the vehicle exhaust pipe(s) to a suitable vent system to purge exhaust fumes from the work area.***

FUEL INJECTOR & INJECTOR RESISTOR TESTS (FOR NO START CONDITION)

Turn the key off and remove the suspect fuel injector connector. Use a DVOM to measure the injector resistance. The injector resistance should be 1.5-2.5 ohms except for injectors without injector resistors that have a resistance of 10-13 ohms. If the injector resistance is not within specifications, replace the injector. Repeat this test procedure for all of the injectors. If okay, go to Step 2.

If all injectors pass the resistance test, go to the Fuel system tests to check the fuel pressure. If fuel pressure is okay, go to Step 3.

On models with an injector resistor bank, if injector resistance and the fuel pressure are within specifications, check for 12V to the resistor bank wire from the main relay. Also inspect for corroded or loose connections. Test the injector resistor bank for open circuit condition. If the resistor bank is okay, check the continuity of the wires between the injectors and the resistor bank. Make repairs and retest the symptom.

All models: Check connections and continuity of the injector control wires between the injectors and PCM. Make repairs and retest the symptom.

If the injector feed and control circuits to the injectors (Figure 2-54) are okay, test the main power and ground circuits to the PGM-FI main relay.

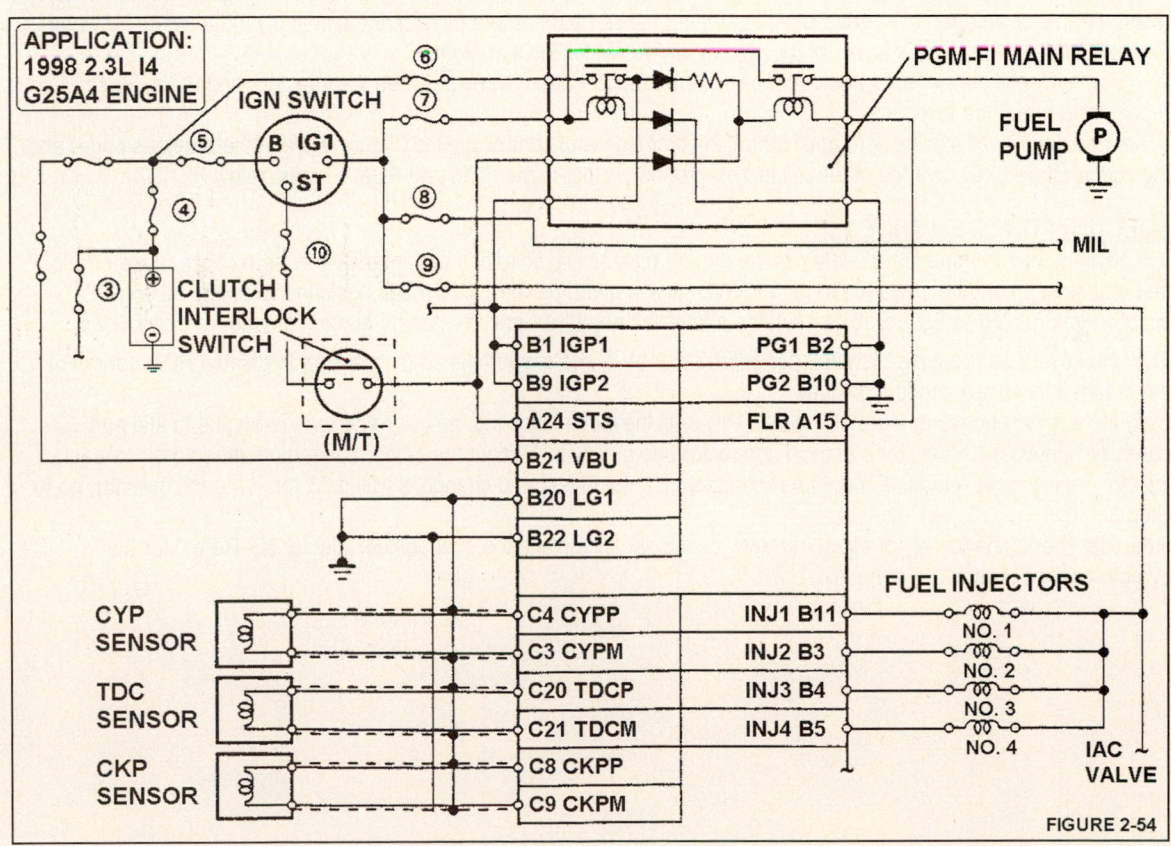

FIGURE 2-54

FUEL INJECTOR TESTS (MISFIRE CONDITION)

Start the engine and allow it to idle. Locate a stethoscope that can be used to monitor the operation of each fuel injector by listening to its clicking sound as the engine runs at idle (Figure 2-55). If any injector fails to make a normal clicking sound, replace the injector and listen to the new injector. If the new injector does not sound normal, go to Step 2 to continue the test sequence.

To isolate a fault that causes abnormal injector sounds, first check the injector battery feed voltage (YEL/BLK wire) at the injector resistor bank or at the injector. Inspect for short circuits, corrosion, or poor connections.

On models with an injector resistor bank, if injector resistance and the fuel pressure are within specifications, check for battery

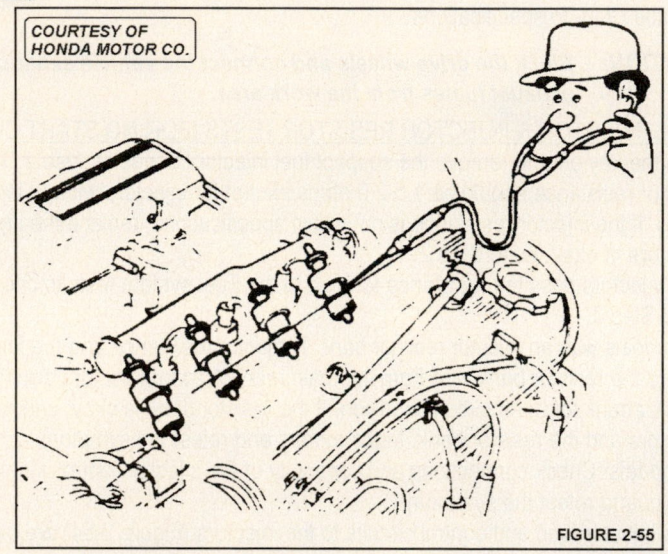

COURTESY OF HONDA MOTOR CO.

FIGURE 2-55

voltage on the resistor bank wire from the main relay. Also inspect for corroded or loose connections. Test the injector resistor bank for an open or shorted condition. If the resistor bank is okay, check the continuity of the wires between the injectors and the resistor bank. Make repairs and retest for the symptom.

All models: Check connections and continuity of the injector control wires between the injectors and PCM. Make repairs and retest the symptom.

If the feed circuit, resistor bank (if applicable), injector feed and control circuits (Figure 2-54 on the previous page) and the injectors are okay, test the main power and ground circuits to the PCM and PGM-FI main relay.

INJECTOR POWER BALANCE TEST

It is obvious that any problems related to the engine mechanical condition or the Ignition system could cause erratic test results in an engine power balance test. With this in mind, perform a thorough basic inspection of all engine components related to Base Engine systems and repair any faults prior to starting the test procedure that follows.

Turn the key off and remove the idle speed (EACV or IAC) and oxygen sensor connectors to disable PCM control of the engine idle speed and fuel mixture.

Start the engine and warm it at 2000 rpm in P/N until the electric cooling fan cycles. Allow the engine to idle and carefully remove each injector electrical connector one at a time. Record the amount of engine speed drop as each injector connector is removed and then reconnected. If the idle speed change is within 75 rpm for each cylinder, go to Step 3.

If the idle speed change is not equal, remove, check and/or replace the suspect fuel injector (s). Retest for the symptom when repairs are done.

FUEL INJECTOR WAVEFORMS

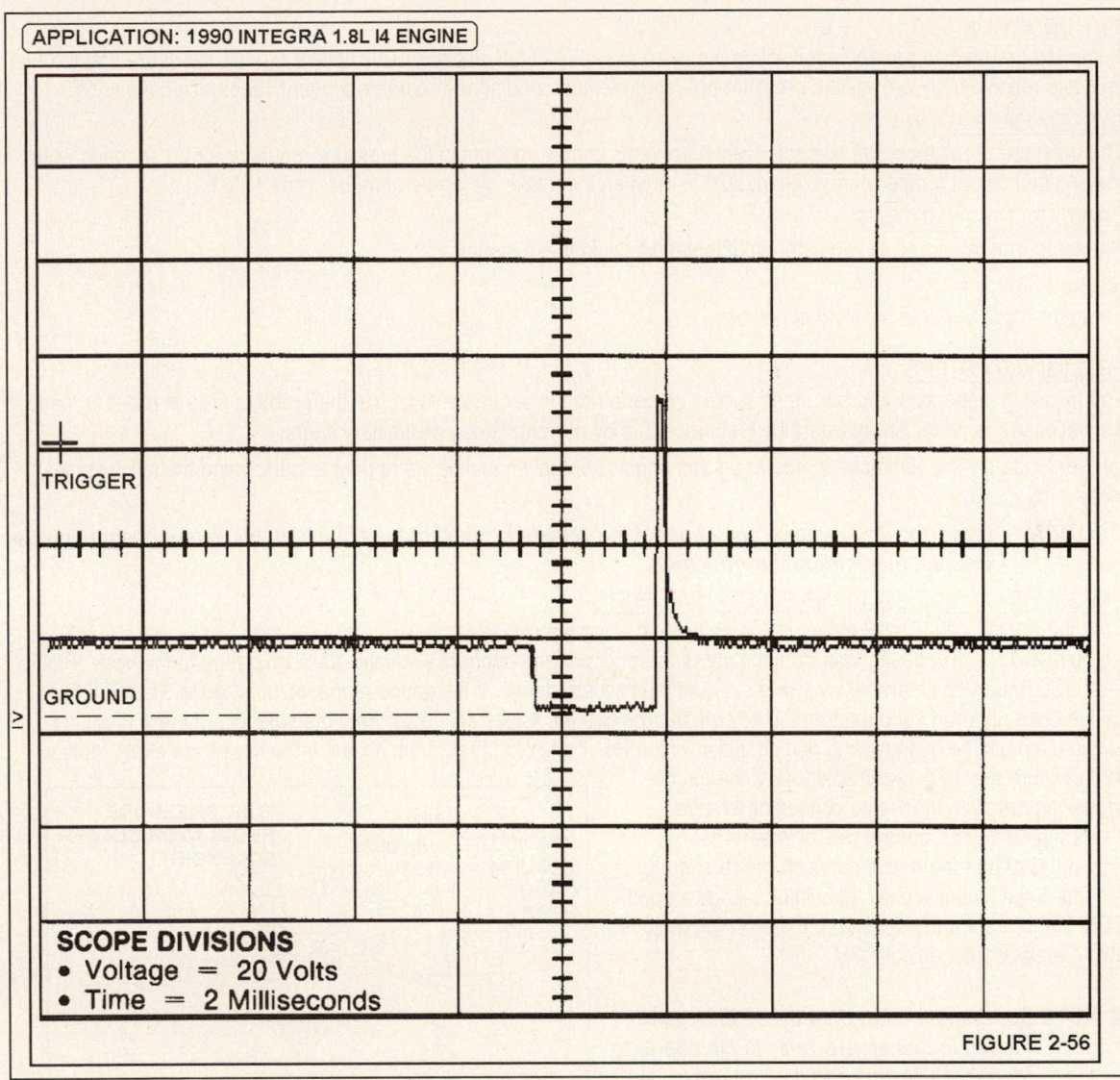

APPLICATION: 1990 INTEGRA 1.8L I4 ENGINE

TRIGGER

GROUND

SCOPE DIVISIONS
• Voltage = 20 Volts
• Time = 2 Milliseconds

FIGURE 2-56

PRESSURE REGULATOR CUTOFF SOLENOID - INTEGRA MODELS

INTRODUCTION
Some Integra models are equipped with a pressure regulator cutoff solenoid. This device is used to ensure that the fuel injectors receive high fuel system pressure hot (soak) restart conditions in order to prevent vapor lock conditions.

SYSTEM OPERATION
The pressure regulator cutoff solenoid (valve) prevents engine vacuum to the pressure regulator for 60 seconds when the engine coolant temperature exceeds 220°F or when the intake air temperature exceeds 175°F.

DRIVEABILITY SYMPTOMS
Problems with the pressure regulator cutoff solenoid could these symptoms:

No Start Condition
Frequent Stalls during hot restart conditions

PRELIMINARY STEPS
The following inspection and test steps should be performed in sequence to isolate the problem area in order to make the necessary repairs. Always start the test procedure by performing the preliminary steps.

Integra Models - The air intake temperature and engine coolant should be warm prior to performing the following test steps.

CAUTION: **Block the drive wheels and connect the vehicle exhaust pipe(s) to a suitable vent system to purge exhaust fumes from the work area.**

PRESSURE CUTOFF SOLENOID ELECTRICAL TESTS
Start the engine and allow it warm until the electric cooling fan cycles.
Remove the No. 1 vacuum hose from the pressure regulator and connect a suitable vacuum gauge to the hose (Figure 2-57). Start the engine and allow it to idle. Read the vacuum gauge. If the gauge reads vacuum, go to Step 3. If the gauge does not read vacuum, turn the key off and remove the 4-wire connector from the solenoid. If there is still no vacuum, check the hose routing and condition of the No. 1 and No. 20 vacuum hoses. If the hoses are okay, replace the pressure regulator cutoff solenoid. If the gauge reads vacuum with the 4-wire connector removed, check the GRN/YEL control wire between PCM terminal C10 and the 4-wire connector for a short to ground. If the circuit is okay, substitute a known-good PCM and retest for the symptom. If the symptom goes away, replace the original PCM.

■ NOTE:*For models other than the 1990 Legend modes in this article, refer to pin charts to identify PCM pins and wire colors.*

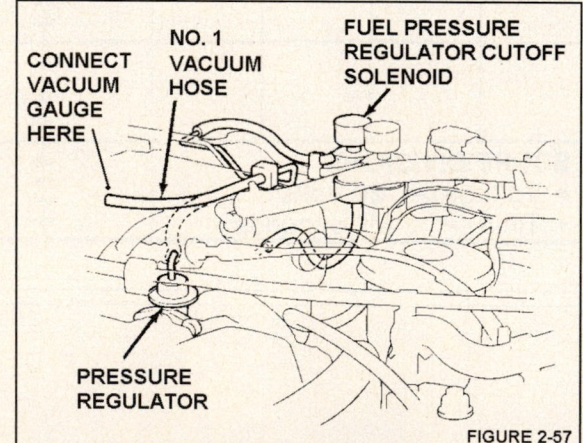

FIGURE 2-57

PRESSURE REGULATOR CUTOFF SOLENOID - INTEGRA MODELS

PRESSURE CUTOFF SOLENOID ELECTRICAL TESTS

If the gauge reads vacuum in Step 2, turn the key off and then remove the 4-P connector from the solenoid. Connect a jumper wire between terminal 'C' of the solenoid and battery positive (Figure 2-58). Use a second jumper wire to connect terminal 'D' of the solenoid to battery negative. Start the engine and allow it to idle. If the gauge reads vacuum, replace the solenoid. If the gauge does not read vacuum, go to Step 4.

If the gauge did not read vacuum in Step 3, measure the voltage between the BLK/YEL wire and body ground. If the DVOM reads 0-volts, locate and repair the open circuit in the BLK/YEL wire between fuse No. 24 and the 4-P connector. If the DVOM reads battery voltage, turn the key off and go to Step 5.

Reconnect the 4-P connector and connect the PCM test harness or the BOB to the PCM and its wiring harness. Start the engine and allow it to idle. Connect a jumper wire between terminal A10 and body ground. Read the vacuum gauge again. If the gauge reads manifold vacuum, substitute a known-good PCM and retest the symptom. If the fault is gone, replace the original PCM. If the gauge still reads no vacuum, locate and repair the open circuit in the GRN/YEL wire between PCM terminal A10 and the 4-P connector. Retest for the symptom when repairs are completed.

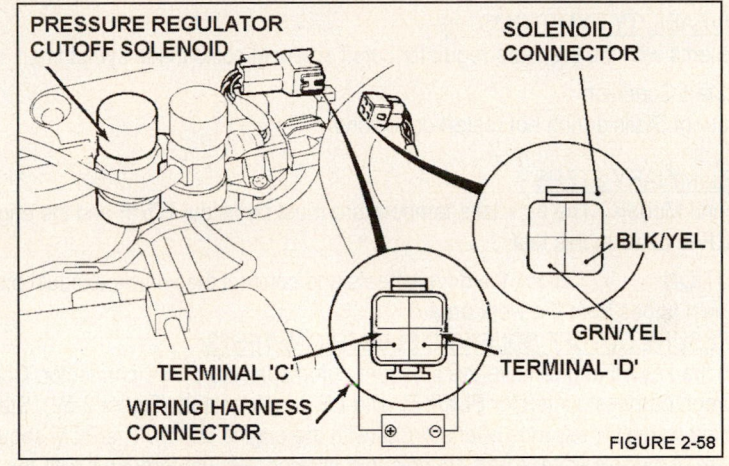

FIGURE 2-58

■ NOTE: *For applications other than the 1990 Integra models in this article, refer to the pin charts to identify PCM pins and wire colors.*

PRESSURE REGULATOR CUTOFF SOLENOID - LEGEND MODELS

INTRODUCTION
Some Legend models are equipped with a pressure regulator cutoff solenoid. This device is used to ensure that the fuel injectors receive high fuel system pressure during hot (soak) restart conditions in order to prevent vapor lock conditions.

SYSTEM OPERATION
The pressure regulator cutoff solenoid (valve) prevents engine vacuum to the pressure regulator for 60 seconds when the engine coolant temperature exceeds 220°F or when intake air temperature exceeds 175°F.

DRIVEABILITY SYMPTOMS
Problems with the pressure regulator cutoff solenoid could these symptoms:

No Start Condition
Frequent Stalls during hot restart conditions

PRELIMINARY STEPS
Legend Models - The air intake temperature must be below 175°F and the engine coolant temperature must be below 220°F to perform this test.

CAUTION: Block the drive wheels and connect the vehicle exhaust pipe(s) to a suitable vent system to purge exhaust fumes from the work area.

PRESSURE CUTOFF SOLENOID ELECTRICAL TESTS
Turn the key off and remove the pressure cutoff solenoid 6-wire connector. Connect the DVOM voltmeter leads between harness connector BLK/YEL and LG wire terminals (Figure 2-59). Start the engine and record the voltage reading within 60 seconds after startup (with the engine warm). The PCM should energize the solenoid for about one minute. If there is voltage and the solenoid allowed vacuum through it with the engine warm, replace the solenoid and retest for the symptom.

If there is no voltage in Step 1, leave the positive lead on the BLK/YEL wire and move the negative lead to body ground. If there still is no voltage, locate and repair the open circuit in the BLK/YEL wire between the 10 amp (No. 9) fuse and the solenoid. Retest the symptom when repairs are completed.

If there is voltage in Step 2 with the negative lead connected to body ground, check the LG wire for an open circuit between the solenoid and PCM terminal A13. If the wire is open, make the repair and retest. If the wire is okay, substitute a known-good PCM and retest. Retest for the symptom, and if it is gone, replace the original PCM.

■ NOTE: *For applications other than the 1990 Legend models in this article, refer to pin charts to identify PCM pins and wire colors.*

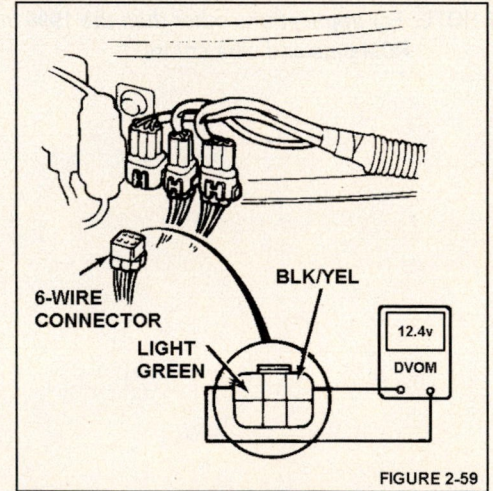

FIGURE 2-59

Idle Speed System
IDLE SPEED SYSTEM

INTRODUCTION
The PCM uses one of these devices listed below to control engine idle speed:

Electronic idle control valve (EICV)
Electronic air control valve (EACV)
Idle air control (IAC) valve

The PCM and idle speed valve control the engine idle speed at all times. Early fuel injection systems used an EICV (valve). It was replaced by the EACV (valve). In 1993, the EACV (valve) was renamed the idle air control (IAC) valve on fuel injected engines.

The PCM sends an electric current pulse to control the position of the valve in order to control the amount of bypass air that enters the intake manifold. When the solenoid is energized, the valve opens to maintain the correct idle speed for all conditions.

IDLE SPEED SYSTEM COMPONENT EXAMPLES - 1990 LEGEND 2.7L V6 ENGINE

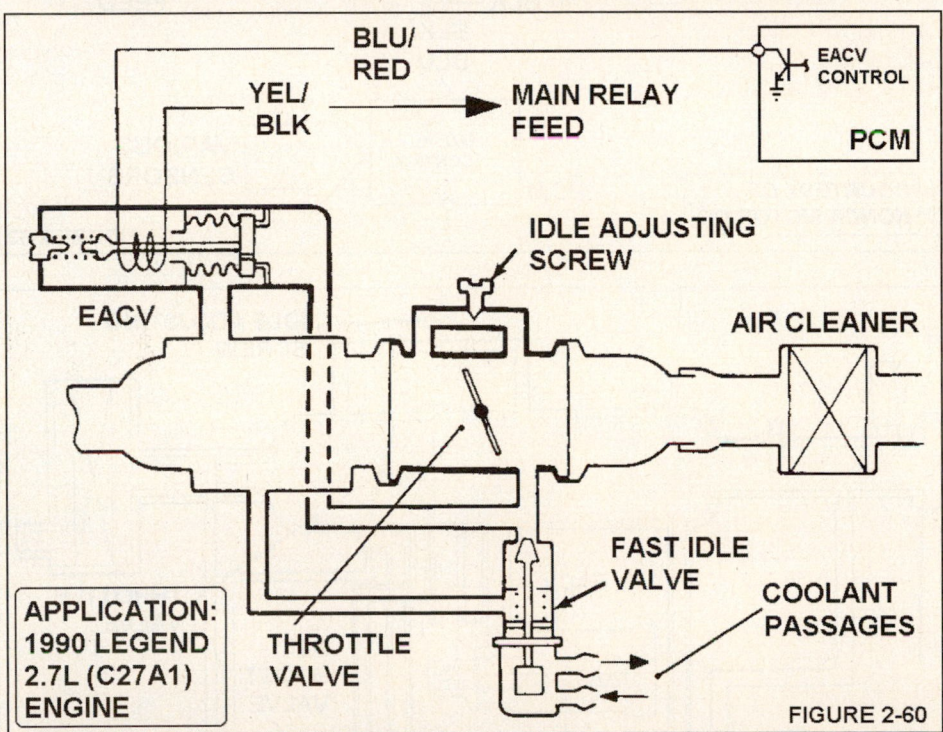

FIGURE 2-60

■ NOTE: *Idle Speed System test procedures cannot be used properly unless the condition is present at the time of testing. If the condition does not appear during testing, refer to Intermittent Tests in the PGM-FI Diagnostics articles in Section 1.*

IDLE SPEED SYSTEM BASICS
Idle Speed system tests should be used after these test steps are done:

The complaint is verified and a visual inspection is done
PGM-FI Diagnostics are completed and no trouble codes are stored
Base Engine tests are done
An actual driveability symptom is present during vehicle testing
Related technical service bulletins have been searched in the TSB index

IDLE SPEED SYSTEM

IDLE SPEED SYSTEM COMPONENT EXAMPLES

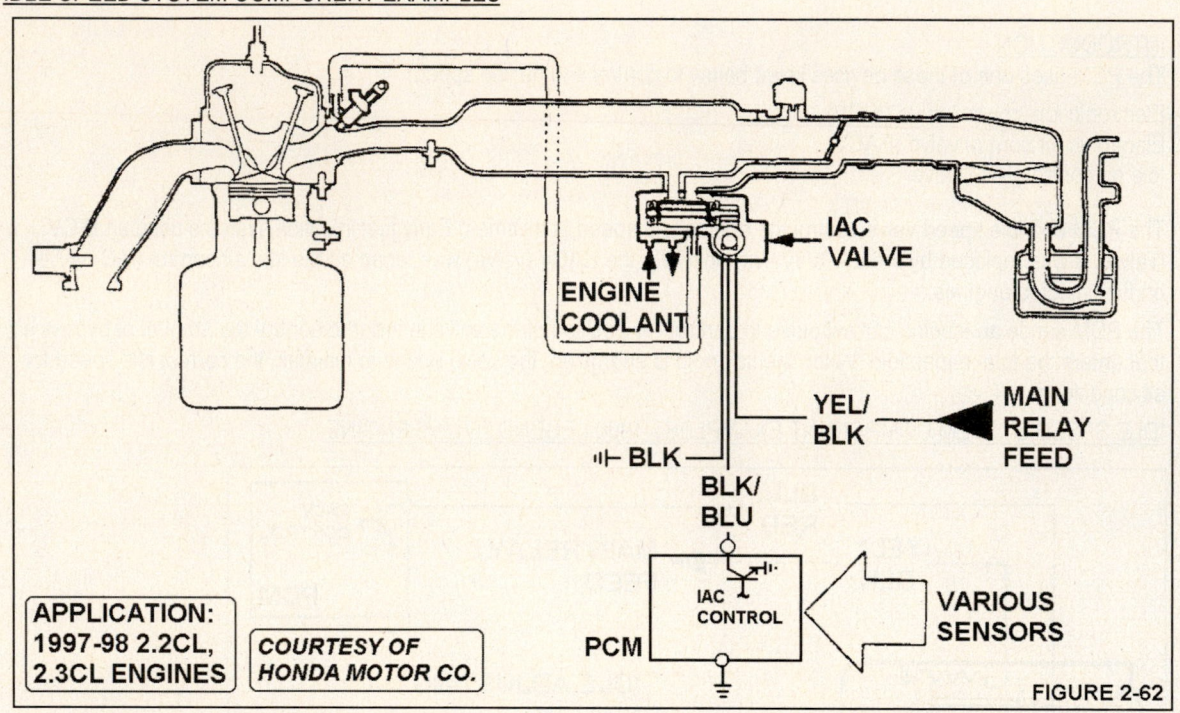

IAC VALVE

ENGINE COOLANT

YEL/ BLK → MAIN RELAY FEED

BLK

BLK/ BLU

IAC CONTROL

PCM

VARIOUS SENSORS

APPLICATION: 1997-98 2.2CL, 2.3CL ENGINES

COURTESY OF HONDA MOTOR CO.

FIGURE 2-62

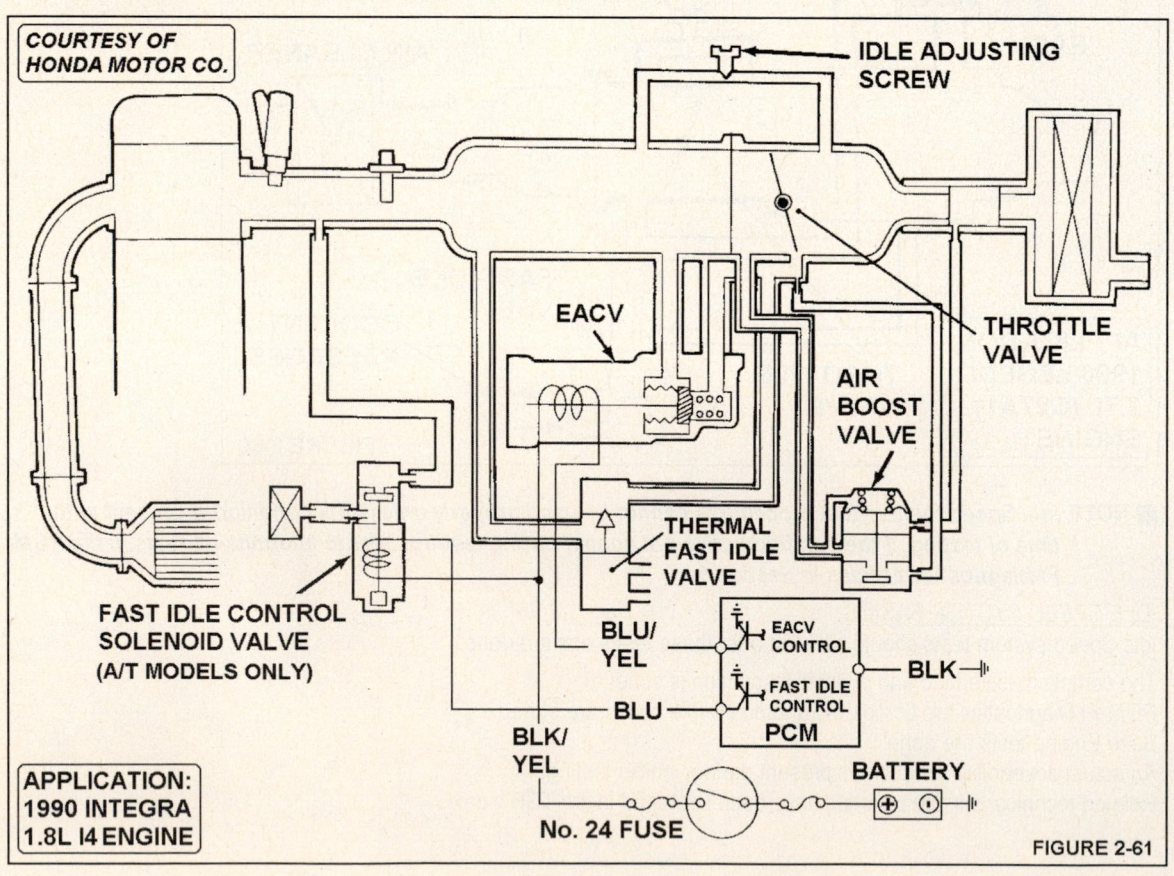

COURTESY OF HONDA MOTOR CO.

IDLE ADJUSTING SCREW

THROTTLE VALVE

EACV

AIR BOOST VALVE

FAST IDLE CONTROL SOLENOID VALVE (A/T MODELS ONLY)

THERMAL FAST IDLE VALVE

BLU/ YEL — EACV CONTROL

BLK

BLU — FAST IDLE CONTROL

PCM

BLK/ YEL

BATTERY

No. 24 FUSE

APPLICATION: 1990 INTEGRA 1.8L I4 ENGINE

FIGURE 2-61

IDLE SPEED SYSTEM TESTS

EACV / EICV / IAC VALVE OPERATION

As previously discussed, these valves are used to change the amount of air that passes the throttle plate in response to an electric current pulse from the PCM. When the valve motor is energized, the valve opens to maintain the proper idle speed.

After initial start-up, the valve is opened for a set time limit. The amount of air allowed to by-pass is increased to raise idle speed 150-300 rpm. Also, when the engine is cold, the valve is opened to obtain the proper fast idle speed. In effect, the PCM controls the amount of extra air allowed to bypass the throttle blade in relation to engine coolant temperature. When coolant temperature is below 86°F, a fast idle valve opens during warmup to prevent engine stalling.

IDLE SPEED SYSTEM TROUBLE CODES

If the PCM cannot control the operation of the idle speed valve due to an electrical fault, it will activate the Check Engine Light and set code 14 on OBD I systems or codes P0505 and P1519 on OBD II systems. If a code is present, use the code repair chart to repair the fault before using this test.

SWITCH TESTS

The PCM uses signals from several switches to determine how much engine speed is needed under additional engine load conditions. These switches include the A/C switch, A/T shift position and P/S switches. On 1985-95 applications (OBD I systems), there are no trouble codes assigned to these switches. These switches are diagnosed as a driveability symptom. Refer to the individual Switch Tests that start on Page 2-101 in this section for more information.

DRIVEABILITY SYMPTOM DIAGNOSIS

If a driveability symptom is present, refer to the list of symptoms included with each Idle Speed Control system test in the articles that follow. It may be necessary to do more than one Idle Speed system test to repair the fault.

A problem with the EACV, EICV or IAC valve or their related circuits could cause any of these symptoms:

Hard Start with a cold engine
Idle speed out of specification on a cold engine
Idle speed to high on a warm engine
Idle speed to low on a warm engine
Idle speed to low right after startup
Idle speed drops when vehicle is placed in gear
Idle speed drops as steering wheel is turned
Idle speed drops as air conditioning is enabled
Fails I/M or State Emissions Test

PRELIMINARY STEPS

Check and adjust ignition timing, base idle speed and idle CO percentage.

CAUTION: **Block the drive wheels and connect the vehicle exhaust pipe(s) to a suitable vent system to purge exhaust fumes from the work area.**

IDLE SPEED SYSTEM TESTS

EACV, EICV & IAC VALVE TESTS

If an idle speed driveability symptom is present, first check for any trouble codes related to the Idle Speed system.

If no codes are displayed during the PGM-FI Diagnostics check, a thorough check must be made of all mechanical and electrical devices that could cause idle speed symptoms (Figure 2-63).

NO-CODE FAULTS

The list below contains various inspection and test steps that should be used during diagnosis of No-Code faults on Idle Speed systems.

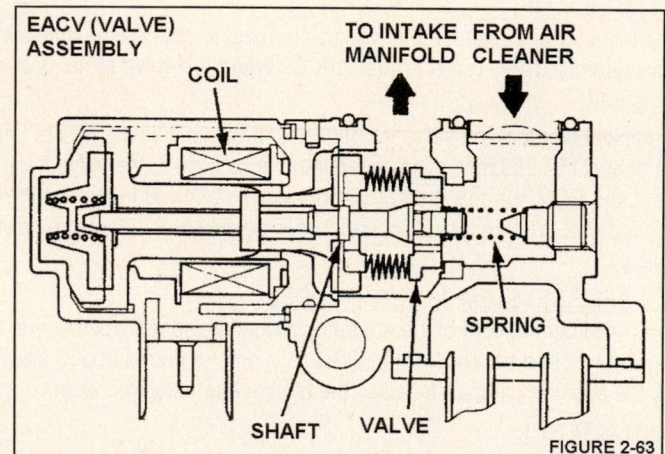

FIGURE 2-63

Verify base idle speed is set to underhood or other vehicle specifications

Inspect all hoses and connections to the EACV, EICV or IAC valve

Inspect the O-rings that seal the EACV, EICV or IAC valve

Test the starter signal input (in cranking position)

Test the alternator FR signal input

Test the automatic transaxle shift position signal input

Test the power steering pressure signal input

Test the air conditioning signal input

Test the brake switch signal input

Test the heater fan switch signal input

Test the manual transmission neutral switch signal input

Test the manual transmission clutch switch signal input

Test the operation of the fast idle valve (with the engine cold)

Test the fast idle control solenoid (automatic transaxle models)

Test the operation of the air boost valve

IDLE SPEED VALVE & PCM SUBSTITUTION

If no problems are found in the EACV, EICV or IAC valve, hoses, connections and O-rings, and all of the PCM input signals are correct, the next step is to substitute a known-good replacement valve to determine if the valve is faulty.

Replace the valve and adjust base idle speed as needed. Then retest for the original symptom. If the symptom goes away, leave the replacement valve on the vehicle.

If the symptom reappears with the replacement valve, substitute a known-good PCM and retest for the symptom. If the symptom goes away, replace the original PCM. If the symptom is still present with a different valve and PCM, return to the list of inspections and tests in the previous article and repeat the tests.

FAST IDLE VALVES

INTRODUCTION

Acura vehicles include fast idle control to prevent erratic running conditions with the engine is cold. A fast idle valve is used to raise idle speed during engine warmup. The thermowax plunger controls the air bypass portion of the fast idle valve. If it is working correctly, the fast idle valve should have no effect on a warm engine.

When the engine is extremely cold, engine coolant that surrounds the thermowax element in the valve makes contract with the internal plunger. This action allows extra air to enter the intake manifold and results in an increase in idle speed. When the engine warms up to normal operating temperature, the valve closes. This action cuts off the flow of additional air entering the intake manifold and results in a reduction in idle speed.

COMPONENT EXAMPLES

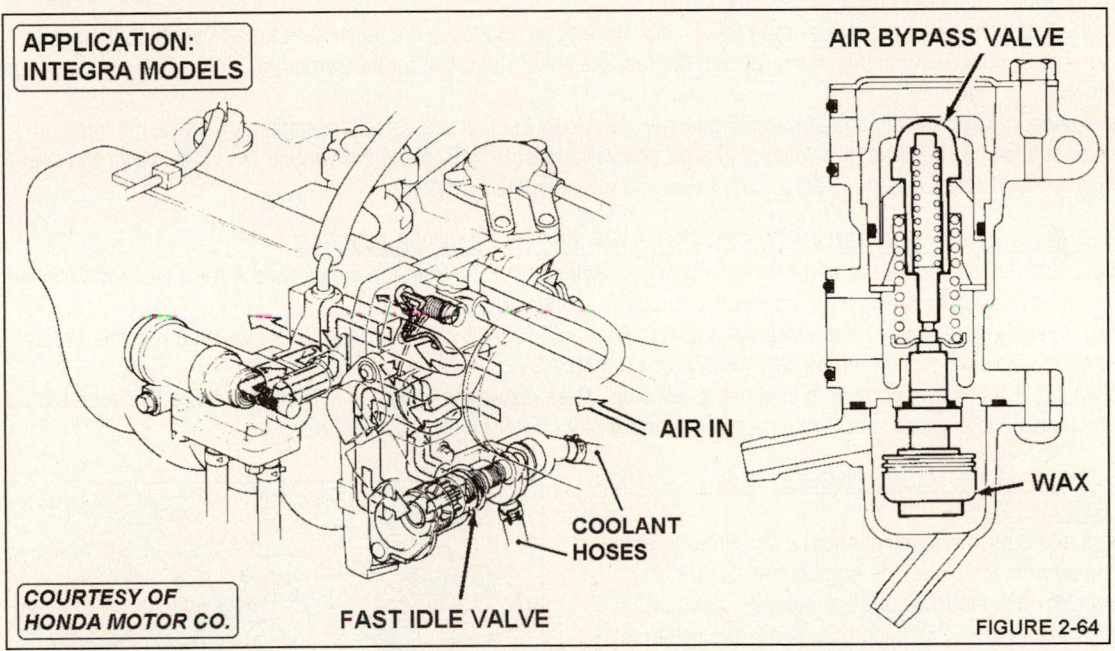

APPLICATION: INTEGRA MODELS

AIR BYPASS VALVE

AIR IN

COOLANT HOSES

FAST IDLE VALVE

WAX

COURTESY OF HONDA MOTOR CO.

FIGURE 2-64

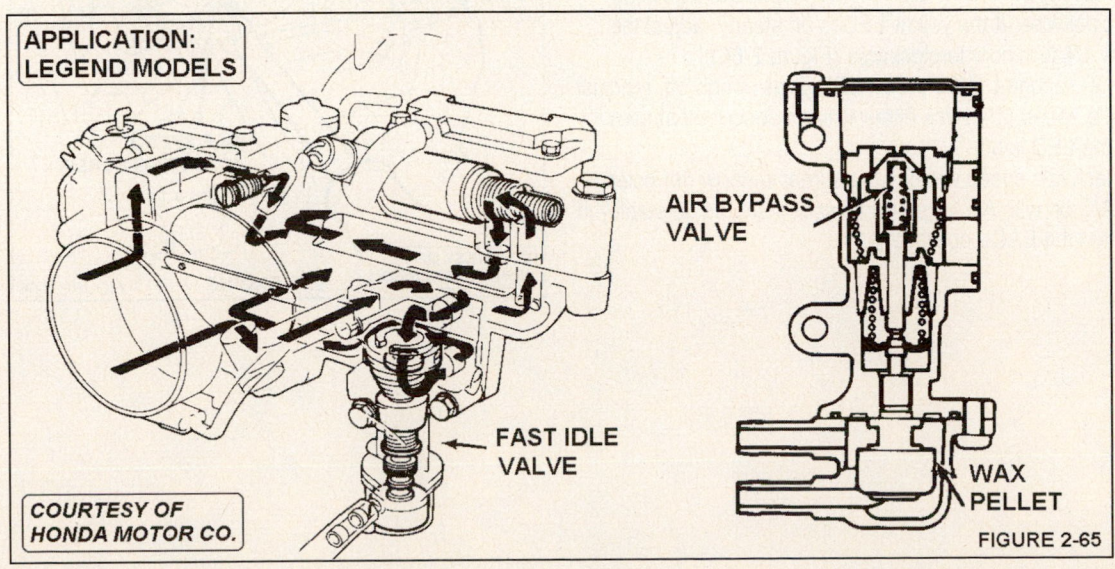

APPLICATION: LEGEND MODELS

AIR BYPASS VALVE

FAST IDLE VALVE

WAX PELLET

COURTESY OF HONDA MOTOR CO.

FIGURE 2-65

FAST IDLE VALVE TESTS

DRIVEABILITY SYMPTOM DIAGNOSIS
A problem with the fast idle valve could cause any of these symptoms:

Cold engine: hard Start or fast idle speed out of specification (too low)
Hot engine: idle speed remains too high after engine warmup period

PRELIMINARY STEPS
Block the drive wheels and purge the vehicle exhaust from the work area. Verify that the throttle blade is not held open due to sticking throttle linkage.

FAST IDLE VALVE TEST - INTEGRA & LEGEND MODELS
Carefully remove the cover from the fast idle valve.

Begin the test with the engine colder than 86°F. Start the engine and touch the valve seat area to verify that there is airflow. If air is not flowing at this point, replace the fast idle valve and retest for the symptom. If air is flowing (okay), go to Step 4.

Allow engine to warm until the electric cooling fan cycles. Verify the fast idle valve is completely closed. If it remains open, air suction can be felt at the valve seat area, the valve is leaking. Replace the valve to bring the warm idle speed within idle speed specifications. Replace the valve and retest for the symptom.

IDLE SPEED ADJUSTMENT PROCEDURES - INTEGRA, LEGEND & VIGOR MODELS
Turn accessories off. Connect a tachometer. Place gear selector in P/N. Start the engine and allow it to warm until the cooling fan cycles. Turn the key off and remove the EACV connector.

Restart engine (depress the throttle slightly) and stabilize engine speed at 1000 rpm, then slowly return to idle, Adjust idle speed to set speed. Turn off the key and reconnect the EACV.

Remove the PCM hazard or backup fuse for 10 seconds. Then restart the engine in P/N and compare idle speed to specifications. If idle speed is not within specifications, test the EACV (valve) operation.

IDLE SPEED ADJUSTMENT PROCEDURES - 1986-90 LEGEND MODELS
Turn off all accessories and connect a tachometer. Place gear selector in P/N. Start the engine and allow it to warmup. Access the PCM under the passenger seat. Observe the yellow LED. If the LED is off, idle speed is within specifications. If it is blinking, adjust the screw 1/4 turn clockwise. If the yellow LED is on steady, adjust the screw 1/4 turn counterclockwise (Figure 2-66).

Wait 30 seconds. If the yellow LED does not go off, readjust the idle speed (1/4 turns either way as described above) until the LED is off.

Recheck idle speed with lights on, rear window defroster and A/C on with A/T in gear (idle speed should be stable). If not, test the EACV operation.

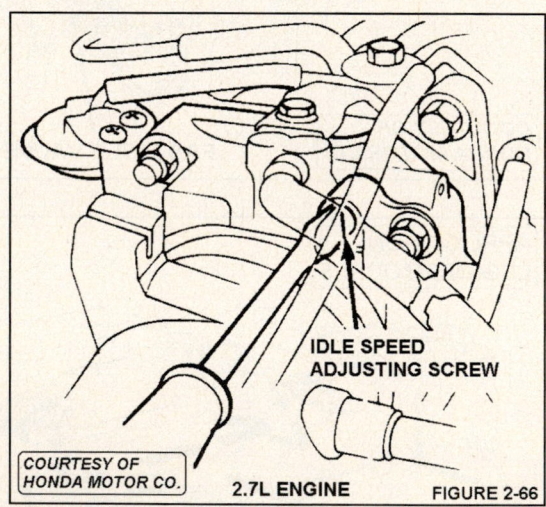

IDLE SPEED ADJUSTING SCREW

COURTESY OF HONDA MOTOR CO.   2.7L ENGINE   FIGURE 2-66

Ignition Systems

IGNITION SYSTEM DESCRIPTIONS

The Ignition system articles in this manual describe both distributor ignition and distributorless ignition systems. The acronyms include in these articles are:

DI - Distributor Ignition System (no previous acronym)

EI - Electronic Distributorless Ignition System (previous acronym was DIS)

IGNITION TIMING CONTROLS - TYPE ONE SYSTEM

Acura vehicles are equipped with two types of ignition timing controls:

1988-98 Acura vehicles are equipped with a Type One ignition timing control system. This type of Ignition system, found on both DI and EI systems, uses electronic spark advance with spark timing controlled by the PCM and/or the igniter. On models with a distributor, there is no centrifugal or vacuum spark advance mechanism. On this system, a 12-volt test light (and spark tester) can be used to toggle the igniter control circuit to test the operation of the ignition coil, igniter and primary circuits (Figure 2-67).

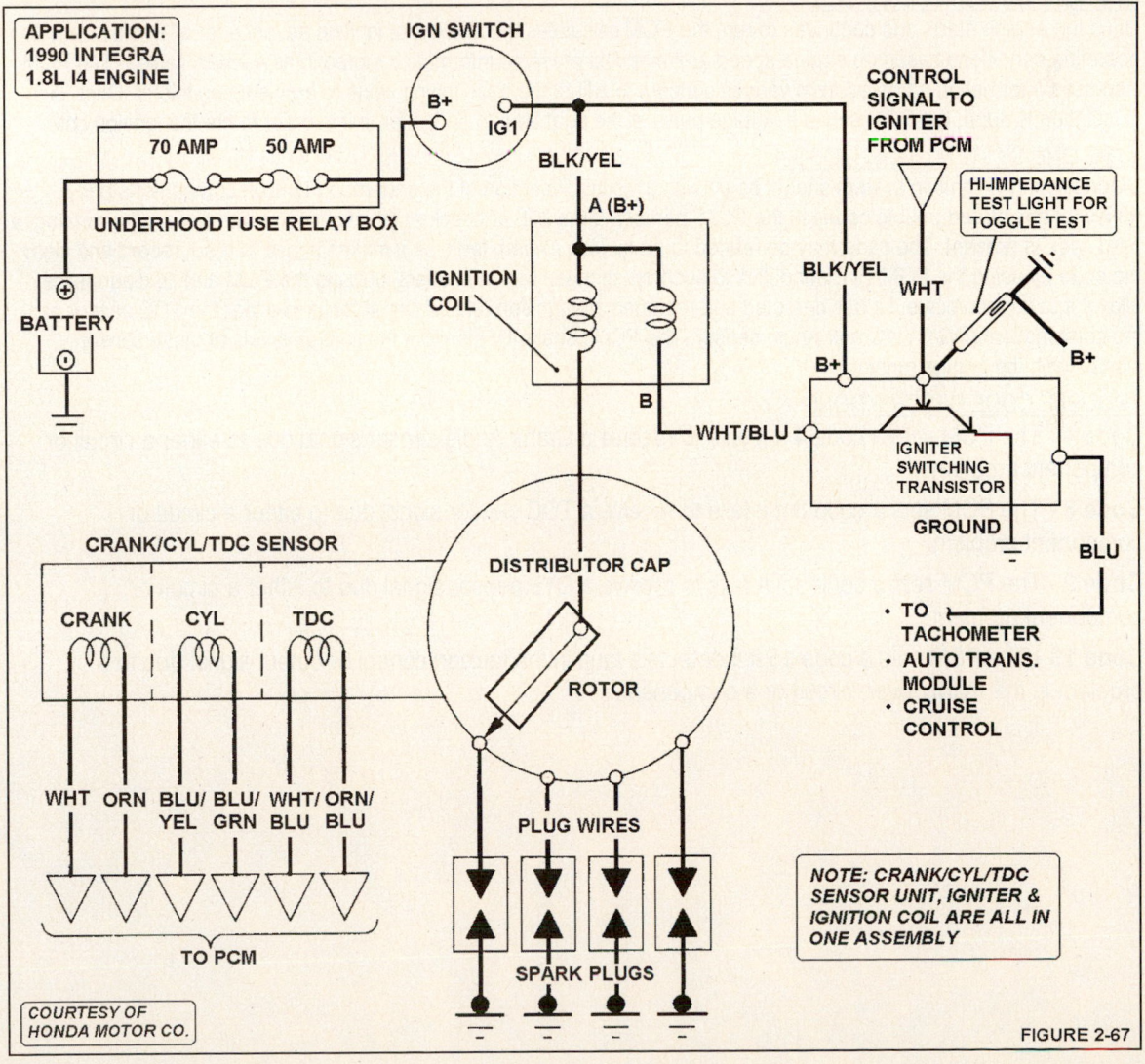

FIGURE 2-67

<u>TYPE ONE IGNITION SYSTEM</u>

<u>SYSTEM OPERATION</u>
On Type One Ignition systems, the PCM controls ignition timing advance and retard through all engine conditions. The PCM responds to changes detected from several engine sensors. Engine speed is calculated through inputs received from the one or more sensors (e.g., the TDC and Crank/Cylinder sensors). Throttle angle is determined from a throttle angle (TA) sensor on early systems and from a throttle position (TP) sensor on later systems. The PCM determines engine temperature from signals from the ECT sensor. Intake manifold pressure is determined from MAP sensor signals.

The PCM contains tables in memory that it uses to calculate when to fire the ignition coil(s). This type of ignition timing control uses complicated characteristics and schedules that cannot be achieved through the use of conventional centrifugal advance and a vacuum diaphragm.

<u>IGNITION TIMING IN CRANK MODE</u>
During engine cranking and immediately after start-up, ignition timing is fixed at 10° BTDC. The PCM determines that the engine is in crank mode through the starter signal input. While in crank mode, the PCM uses the TDC signal from the distributor to calculate engine speed and cylinder top dead center.

<u>IGNITION TIMING IN RUN MODE</u>
Once the engine starts and continues to run, the PCM calculates the amount of ignition advance for all engine-operating conditions based on engine speed and manifold pressure information stored in its memory tables. The PCM also uses information gathered from various sensors to adjust the base timing value to ambient conditions. Once this calculation is done, the PCM sends a voltage pulse at the right time to an igniter unit in order to fire the ignition coil.

<u>TYPE ONE SYSTEM CODES (OBD I)</u>
Diagnosis of the Ignition system should begin with a visual inspection of the underhood ignition components and a review of any stored trouble codes in the PCM memory. If the MIL or check engine light is on with the engine running, a hard code is present. The code may be related to an Ignition system fault, so it makes sense to read, record and clear the code by using the PGM-FI Diagnostics to accomplish these tasks. In effect, utilizing the PCM and its diagnostics allows it to communicate if it has detected and recorded any ignition related circuit faults in either the TDC sensor or the combination TDC/CYL/Crank Angle sensor. The PCM constantly monitors the voltage levels of these sensor signals with the engine running.

<u>TROUBLE CODE DESCRIPTIONS</u>
Code 4 - The PCM sets a code 4 if it fails to receive a Crank Angle sensor signal due to either a circuit or component problem.

Code 8 - The PCM sets a code 8 if it fails to receive a TDC sensor signal due to either a circuit or component problem.

Code 9 - The PCM sets a code 9 if it fails to receive a CYL sensor signal due to either a circuit or component problem.

Code 15 - The PCM sets a code 15 if it detects a fault in the ignition control or output signal due to a problem in the PCM driver, circuit or a component.

TYPE ONE SYSTEM TESTS - PRELIMINARY STEPS

Prior to testing the Ignition system for faults, verify that the battery open circuit is at least 12.4v and a cranking voltage of at least 9.6v. Test, replace or recharge the battery as needed before continuing the test sequence. The battery cables should not be removed at any time with the engine running.

Any high resistance in the igniter ground connection can cause Ignition system problems. With this in mind, verify that a clean and tight ground connection is present to the distributor body and that the distributor is properly grounded to the engine. Check the ground connection at the thermostat housing. Use a high impedance type DVOM to test the ground circuit resistance by measuring the ground circuit voltage drop with the key on and engine off. It should be less than 100 millivolts (100mv).

Connect a tachometer to the coil negative terminal to measure engine speed (not to the tachometer terminal). An inductive type tachometer that connects to any secondary ignition wire is another easy method of reading engine speed during testing.

■ **NOTE:** *Do not leave the key on for more than 3 minutes in KOEO mode. If it is left on too long, the ignition coil or igniter could overheat and fail.*

SPARK OUTPUT TEST

Remove the ignition coil secondary wire and inspect it for corrosion at both ends. If the wire is okay, connect the coil

wire end to the coil and connect the distributor cap end to a high-output spark tester. Connect the spark test clamp firmly to engine ground. If a coil wire is not available, remove a spark plug cable and connect a spark tester to the spark plug cable (plug end) and engine ground (Figure 2-68).

Crank the engine for several short periods of time. A remote starter switch can be used to crank the engine for short periods of time. Verify the distributor turns at this time.

If a consistent white or blue-white spark is not present at the spark tester connected to the coil wire, the coil wire may be faulty. To test the coil wire, measure the coil wire resistance. It should not exceed 10,000 ohms. If a spark tester is used at one or more spark plug cables and spark is not present or is inconsistent, measure the spark plug wire resistance. The spark plug wire resistance should not exceed 25,000 ohms. Also, inspect for signs of secondary problems in the distributor cap and rotor. If any problems are located during this step, replace the component and

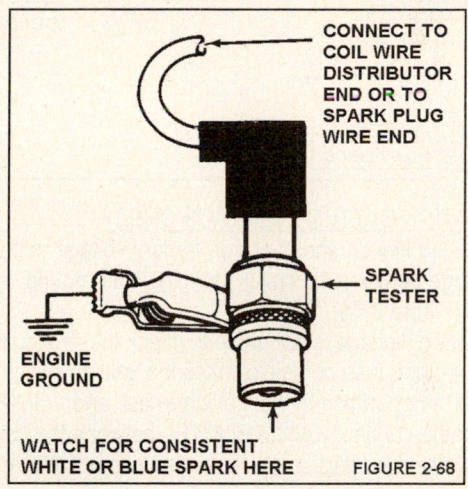

CONNECT TO COIL WIRE DISTRIBUTOR END OR TO SPARK PLUG WIRE END

SPARK TESTER

ENGINE GROUND

WATCH FOR CONSISTENT WHITE OR BLUE SPARK HERE   FIGURE 2-68

retest for the symptom. If the Ignition system passes the Spark Output Test, go to Inspect Ignition System Connections.

CAUTION: *Do not crank the engine for more than two seconds without stopping and starting over as fuel is injected into the engine during the spark output test. Cranking for long periods of time could flood the engine and cause hydrostatic lock-up.*

IGNITION SYSTEM CONNECTIONS

There are several connectors on Type One systems that need to be checked for clean and tight connections. Locate the ignition coil positive and negative terminals. Also locate the PCM ignition control (signal) wire that connects at the igniter. For example, the control signal is the WHT wire to terminals B15 and 17 on 1988-90 Integra, and the RED/BLU wire to terminals B15 and 16 on 1988-90 Legend models).

Carefully inspect all three wires at the distributor connections. Look for signs of a backed-out or loose connector, terminal corrosion or other problems. Clean, tighten and repair all connections as needed and repeat the Spark Output Test. If the spark output is still weak or there is no output, go to the Ignition Battery Voltage Test.

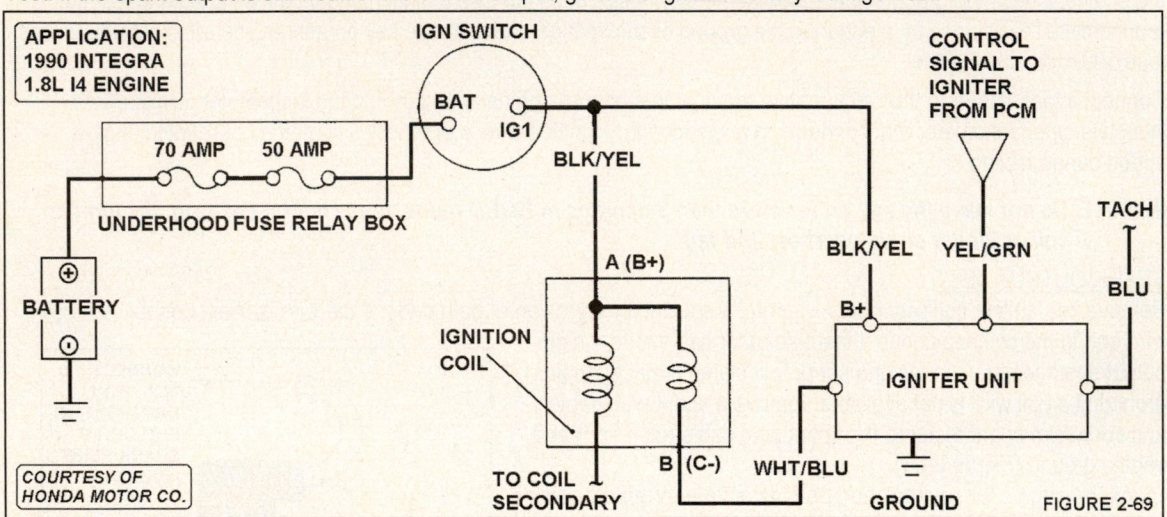

FIGURE 2-69

IGNITION BATTERY VOLTAGE CHECK

Turn the key on and check for battery voltage at the BLK/YEL wire to the igniter and ignition coil. Also check the voltage between the coil (+) terminal and ground. Inspect for a blown fuse or open circuit on the BLK/YEL wire to the coil (Figure 2-69).

If battery voltage is not present, check these circuits and components:

Inspect the fuse or fusible link to the ignition switch

Test the operation of the ignition switch and PGM-FI main relay

Test the continuity of the BLK/YEL between the coil and the igniter

If a fault is located, make repairs as needed and retest for the symptom. If no problems are located, go to Step 4.

Test the igniter ground circuit. Turn the key on and use a DVOM to measure the voltage drop between the igniter ground connection and engine ground at the distributor. The reading should be less than 100mv (this is the ground circuit for the coil primary). If the voltage drop is too high, check the following areas for the cause of high ground resistance:

The distributor ground and engine ground circuit to the battery

Repair any problems located and retest the system. If no problems are located in this step, go to the Ignition Coil Resistance Test.

IGNITION COIL RESISTANCE TEST

Select an appropriate low range on the DVOM to test the ignition coil primary circuit. Primary circuit specifications: 0.6-0.8 ohms at 70°F on a standard coil and 0.3-0.4 ohms at 70°F on an integral coil.

Select an appropriate ohmmeter high range on the DVOM to test the coil secondary circuit resistance: Integra models: 9,760-14,640 ohms at 70°F, Legend models: 9040-13,560 ohms at 70°F.

If either coil reading is out of the specification, replace the ignition coil and retest for the symptom. If the coil resistance readings are okay, go to the Igniter Voltage Tests.

TIPS: *If the coil output is low, inspect the coil for signs of internal or external arcing to the primary or ground circuit. Look for burnt spots that could indicate a problem. It is especially helpful to use the spark tester to lightly stress the ignition coil while observing the coil tower for signs of secondary or primary breakdown. Also, an ignition analyzer is an excellent tool to use to observe the coil primary and secondary operation with the spark tester installed. Some models have the tachometer circuit incorporated in the coil assembly. The resistance of the tachometer circuit on Legend models (measured between terminals 'B' and 'D' of the coil connector) is from 2090-2310 ohms.*

Test Review: If spark output is weak, the coil and coil wire should have been tested and replaced as needed. If the engine will not start and there is no spark output, the previous tests should have validated the feed, ground and coil circuits. These problems should have been repaired. If the vehicle still will not start at this point, the next step is to determine if the PCM is switching the igniter (and coil primary) on and off during cranking. Also, there should not be any ignition related codes set and the tachometer circuit should be tested to verify that it is not grounded. The next step is the Igniter Voltage Checks.

IGNITER VOLTAGE CHECKS - INTEGRA MODELS

To test the operation of the PCM and igniter, turn the key off and remove the five-wire connector.

Then turn the key on and measure the voltage between the BLK/YEL wire at the igniter harness connector and body ground (Figure 2-69 on the previous page). If the reading is near 12v, go to Step 3. If the reading is 0v, locate and repair the open circuit between the underhood main fuse box (both the 50 and 70 amp fuses are in series) and terminal 'A'. Then retest for the symptom.

If the reading was near 12v in Step 2, measure the voltage between the WHT/BLU wire at the igniter harness connector and body ground. If there is battery voltage, go to Step 4. If the reading is 0v, check for an open primary circuit in the coil or an open circuit in the WHT/BLU wire between the coil and igniter. Make needed repairs and retest for the symptom.

If the reading was near 12v in Step 3, test the continuity of the control signal WHT wire to the PCM, If it is open, make repairs and retest for the symptom. If the WHT wire is not open, test the igniter ground circuit resistance. If the igniter ground circuit is less than 5 ohms, and the WHT wire has continuity, replace the igniter and retest for the symptom.

IGNITER VOLTAGE CHECKS (LEGEND MODELS)

To test the operation of the PCM and igniter, turn the key off and remove the four-wire connector.

Then turn the key on and measure the voltage between the BLK/YEL wire at the igniter harness connector and body ground (Figure 2-70). If the reading is near 12v, go to Step 3. If the reading is 0v, locate and repair the open circuit between the fuse box and coil. Also check for continuity between terminals 'A' and 'C' on the coil. Then retest the symptom.

If the reading was near 12v in Step 2, measure the voltage between the BLU wire at the igniter harness connector and body ground. If there is battery voltage, go to Step 4. If the reading is 0v, check for an open primary circuit in the coil or an open circuit in the BLU wire between the coil and igniter. Make needed repairs and retest for the symptom.

If the reading was near 12v in Step 3, test the continuity of the control signal RED/BLU wire to the PCM. If it is open, make repairs and retest for the symptom. If the RED/BLU wire is not open, test the igniter ground circuit resistance. If the igniter ground circuit is okay and the WHT wire has continuity, replace the igniter and retest for the symptom.

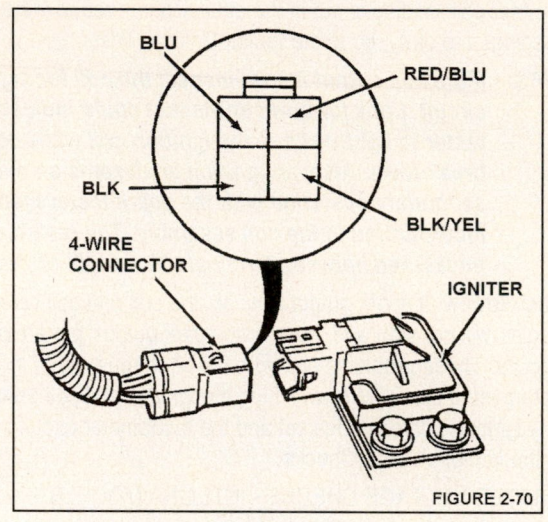

FIGURE 2-70

■ NOTE:*For models other than the 1988-90 Legend and Integra applications discussed in this article, refer to other pin charts for pin and wire color identification for all references to PCM terminals and wire colors.*

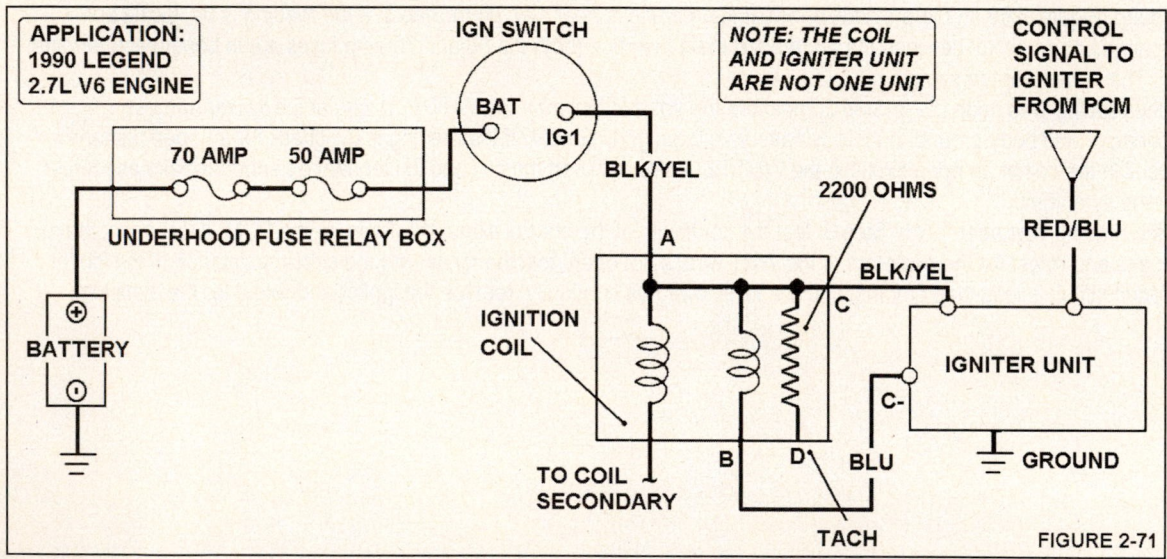

FIGURE 2-71

TYPE TWO IGNITION SYSTEM

1986-87 Integra & Legend models are equipped with a Type Two ignition timing system. This type of Ignition system is found on distributor ignition (DI) models only. In this system, the PCM controls spark timing in response to changes in engine coolant temperature and manifold vacuum in order to provide optimum ignition timing with the engine at both cold and normal engine-operating conditions (Figure 2-72).

The distributor uses a centrifugal and vacuum advance. The igniter controls spark advance in response to pickup coil signals. On this system, the PCM controls the vacuum advance diaphragm through one or more vacuum-operated solenoid valves.

TYPE TWO IGNITION SYSTEM COMPONENTS

The Type Two Ignition system consists of distributor vacuum advance diaphragms 'A' and 'B', ignition control solenoid valves 'A' and 'B', a vacuum reservoir and check valve, an igniter unit, an ignition coil, secondary components and the PCM. The magnetic distributor contains a pick-up coil with centrifugal advance, radio noise suppressor, distributor rotor and cap.

VACUUM ADVANCE OPERATION

The distributor in this Ignition system has two separate vacuum advance diaphragms that are operated with manifold vacuum (Figure 2-72). An ignition control solenoid valve is connected into the vacuum line that connects to advance diaphragm 'B' that connects to the distributor.

The PCM controls the on/off operation of the ignition control solenoid depending upon the information it receives from the engine temperature and manifold vacuum sensors.

Once the ignition control solenoid is opened, it allows the flow of vacuum to diaphragm 'B' in the distributor to provide additional spark advance to the amount of advance provided by diaphragm 'A'. The additional advance is provided to improve cold driveability. This type of ignition system does not include the use of a detonation sensor.

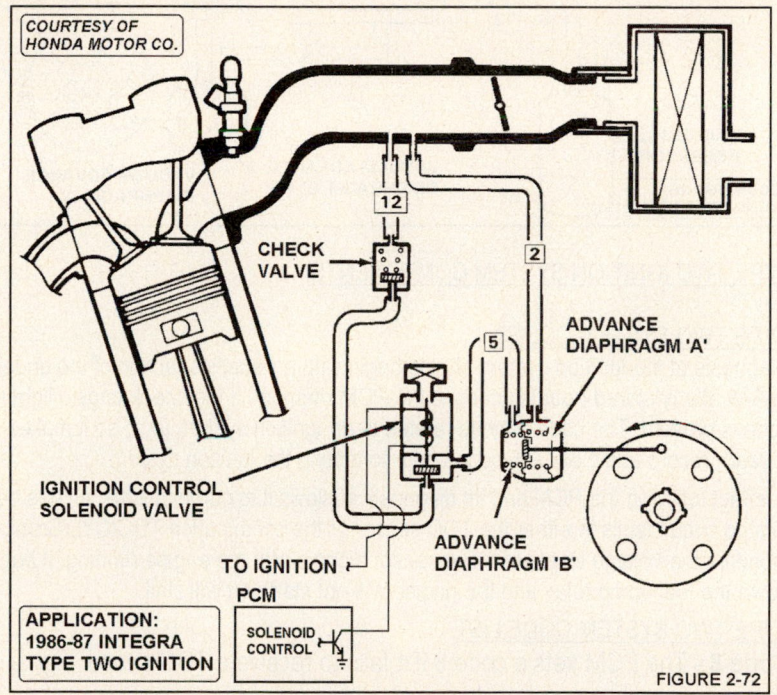

COURTESY OF HONDA MOTOR CO.

CHECK VALVE

IGNITION CONTROL SOLENOID VALVE

TO IGNITION PCM

ADVANCE DIAPHRAGM 'A'

ADVANCE DIAPHRAGM 'B'

APPLICATION: 1986-87 INTEGRA TYPE TWO IGNITION

SOLENOID CONTROL

FIGURE 2-72

Ignition System

TYPE TWO IGNITION SYSTEMS

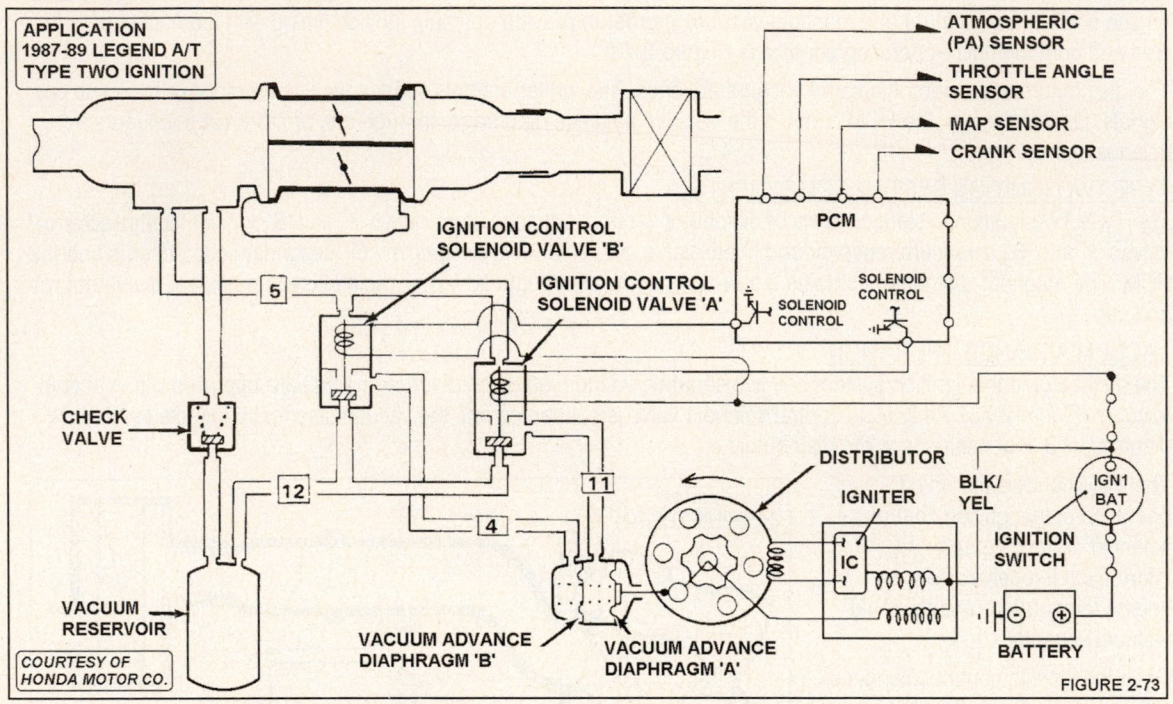

APPLICATION
1987-89 LEGEND A/T
TYPE TWO IGNITION

ATMOSPHERIC
(PA) SENSOR

THROTTLE ANGLE
SENSOR

MAP SENSOR

CRANK SENSOR

PCM

IGNITION CONTROL
SOLENOID VALVE 'B'

IGNITION CONTROL
SOLENOID VALVE 'A'

SOLENOID
CONTROL

SOLENOID
CONTROL

CHECK
VALVE

DISTRIBUTOR

IGNITER

BLK/
YEL

IGN1
BAT

IGNITION
SWITCH

VACUUM
RESERVOIR

VACUUM ADVANCE
DIAPHRAGM 'B'

VACUUM ADVANCE
DIAPHRAGM 'A'

BATTERY

COURTESY OF
HONDA MOTOR CO.

FIGURE 2-73

TYPE TWO IGNITION SYSTEM COMPONENTS

TYPE TWO SYSTEM CODES

Diagnosis of the Ignition system should begin with a visual inspection of the underhood ignition components and a review of any stored trouble codes in the PCM memory. If the check engine light is on with the engine running, a hard code is present. The code may be related to an Ignition system fault, so it makes sense to use the PGM-FI Diagnostics to read, record and clear any codes prior to testing the Ignition system.

In effect, utilizing the PCM and its diagnostics allows it to communicate if it has detected and recorded any ignition related circuit faults in either the TDC sensor or the combination TDC/CYL/Crank Angle sensor. The PCM constantly monitors the voltage levels of these sensor signals with the engine running. If both signals are lost, the PCM will shut down the fuel pump relay and the engine will not start or it will stall.

TYPE TWO SYSTEM CODE LIST

Code 8 - The PCM sets a code 8 if it fails to receive a TDC sensor signal due to either a circuit or component problem.

Code 9 - The PCM sets a code 9 if it fails to receive a CYL sensor signal due to either a circuit or component problem.

TYPE TWO SYSTEM TESTS

SPARK OUTPUT TEST

Remove the ignition coil secondary wire and inspect it for corrosion at both ends. If the wire is okay, connect the coil wire end to the coil and connect the distributor cap end to a low-output spark tester. Connect the spark test clamp firmly to engine ground. If a coil wire is not available, remove a spark plug cable and connect a spark tester to the spark plug cable (plug end) and engine ground (Figure 2-74).

Crank the engine for several short periods of time. A remote starter switch may be used for this step. Verify the distributor turns at this time.

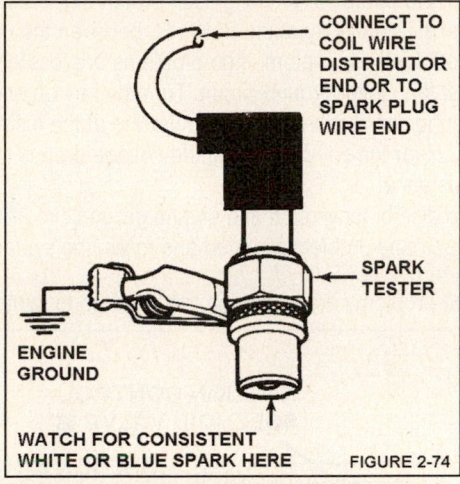

If a consistent white or blue-white spark is not present at the spark tester that is connected to the coil wire, the coil wire may be faulty. To test the coil wire, measure the coil wire resistance. It should not exceed 10,000 ohms. If a spark tester is used at one or more spark plug cables and spark is not present or is inconsistent, measure the spark plug wire resistance. The spark plug wire resistance should not exceed 25,000 ohms. Also, inspect for signs of secondary problems in the distributor cap and rotor. If any problems are located during this step, replace the component and retest for the symptom. If there is no spark output or weak output during this test, go to the Ignition System Connections step.

CAUTION: *Do not crank the engine for more than two seconds without stopping and starting over as fuel is injected into the engine during the spark output test. If the engine is cranked for long periods of time, engine hydrostatic lock-up could occur or the engine could become flooded.*

IGNITION SYSTEM CONNECTIONS

There are several connectors on Type Two Ignition systems that need to be checked for clean and tight connections. Locate and inspect the ignition coil positive and negative terminals. Also inspect the ignition control wire at the igniter assembly. Look for signs of any backed-out or loose connectors, terminal corrosion or other damage. Clean, tighten and repair connections as needed. Then repeat the Spark Output Test. If no connection problems were located during this test step, go to the Ignition Battery Voltage Test.

TYPE TWO SYSTEM TESTS

IGNITION BATTERY VOLTAGE CHECK

Turn the key on and check for battery voltage at the BLK/YEL wire to the igniter and ignition coil (Figure 2-75). Also check the voltage between the coil positive terminal and ground. Inspect for a blown fuse or open circuit on the BLK/YEL wire to the coil. If battery voltage is not present, check these circuits and components:

Inspect the fuse or fusible link to the ignition switch

Test the operation of the ignition switch and PGM-FI main relay

Test the continuity of the BLK/YEL between the coil and the igniter. If a fault is located, make repairs as needed and retest for the symptom. If no problems are located, go to Step 2.

Test the igniter ground circuit. Turn the key on and use a DVOM to measure the voltage drop between the igniter ground connection and engine ground at the distributor. The reading should be less than 100mv (this is the ground circuit for the coil primary). If the voltage drop is too high, check the following areas for the cause of high ground resistance:

The distributor ground and engine ground circuit to the battery

Repair any problems located and retest the system.

If no problems are located in this step, go to Ignition Coil Resistance Test.

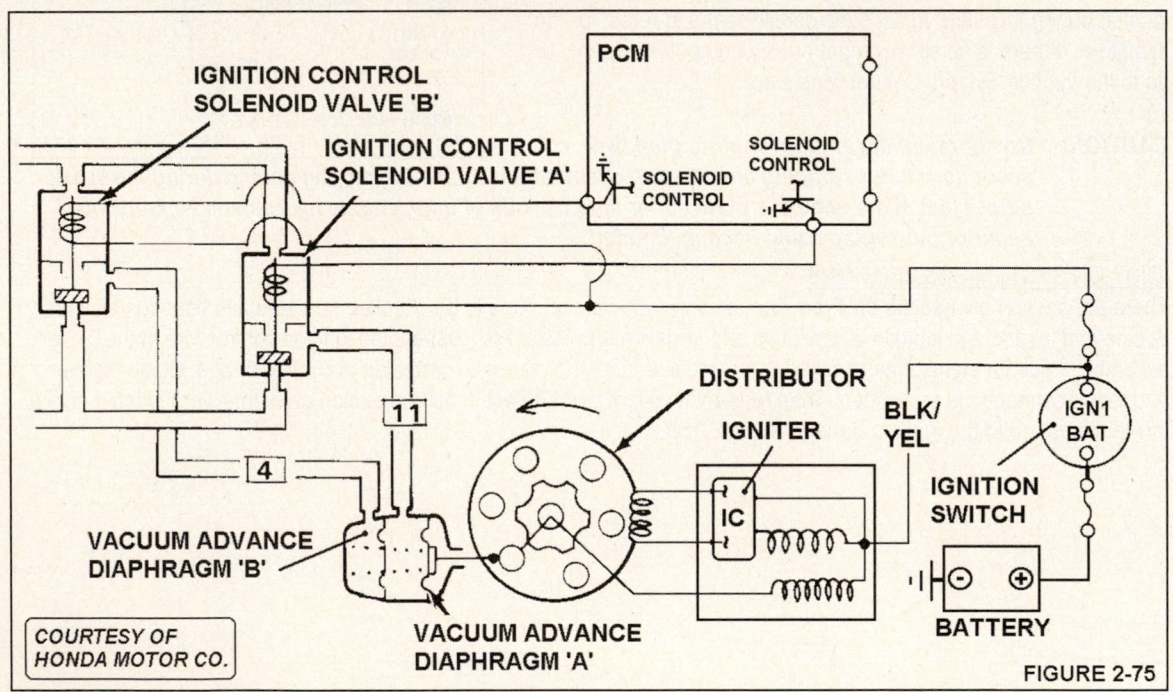

COURTESY OF HONDA MOTOR CO.

FIGURE 2-75

IGNITION COIL RESISTANCE TEST

Select an appropriate low range on the DVOM to test the ignition coil primary circuit. Coil primary circuit specifications: Integra - 1.2-1.5 ohms at 70°F, Legend - 0.32-0.39 ohms at 70°F.

Select the high range on the ohmmeter on the DVOM and test the resistance of the coil secondary circuit. On Integra models: 9,760-14,640 ohms at 70°F, Legend models: 9000-13,500 ohms at 70°F (Figure 2-76).

If either coil reading is out of range, replace the ignition coil and retest for the symptom. If the coil readings are okay, go to the Igniter Voltage Tests.

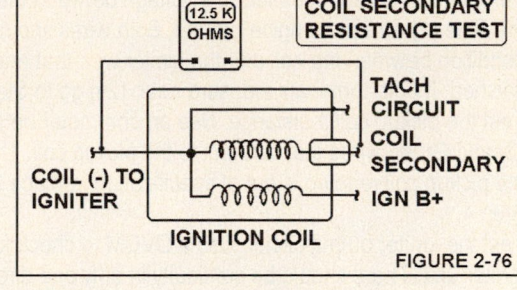

FIGURE 2-76

TIPS: ***If the coil output is low, inspect the coil for signs of internal or external arcing to the primary or ground circuit. Look for burnt spots that could indicate a problem. It is especially helpful to use the spark tester to lightly stress the ignition coil while observing the coil tower for signs of secondary or primary breakdown. Also, an ignition analyzer is an excellent tool to use to observe the operation of the ignition coil primary and secondary circuits with the spark tester installed.***

Test Review: If spark output is weak, the coil and coil wire should have been tested and replaced as needed. If the engine will not start and there is no spark output, the previous tests should have validated the feed, ground and coil circuits. These problems should have been repaired. If the vehicle still will not start at this point, the next step is to determine if the PCM is switching the igniter (and coil primary) on and off during cranking. There should not be any ignition related codes set. Also, the tachometer circuit should be tested to verify that it is not grounded. If okay, go to Igniter Tests.

IGNITER TESTS

To test the igniter, first remove the distributor cap to allow for testing. Then remove the igniter and its cover to expose the four terminals for testing (Figure 2-77).

Turn the key on and measure the voltage between the BLK/YEL wire and ground. Also measure the voltage between the Blue (1) wire and engine ground. Both wires should read near 12v. If they do not, locate and repair the open circuit condition between the coil and the igniter wire that reads 0v during this test. Retest for spark output when repairs are finished. If both terminals measure near 12v, go to Step 3.

Test the pickup coil resistance. Use an ohmmeter on the proper scale and measure the resistance between the Blue (2) and Green wires that connect to the pickup coil. The pickup coil resistance at 70°F should be from 650-750 ohms. If the pickup coil reading is out of specification, replace the coil and retest for the symptom. If the reading is okay, go to Step 4.

Test the igniter output circuit. Use a DVOM to check for continuity in both directions between terminals 'A' and 'B' of the igniter unit. There should be continuity in only one direction. Replace the igniter if continuity is present in both directions or if there is no continuity in either direction. Then retest for spark output. If continuity is okay, go to Step 5.

Test the igniter 'D' input circuit. Connect the DVOM positive probe to the 'D' terminal and the negative probe to the inside of the igniter case or ground. The resistance reading should be 50 Kohms or higher at 70°F. If the reading is out of range, replace the igniter (with its silicone grease). Then retest the spark output. If the igniter passes all of these checks and pickup coil resistance is okay, replace the igniter and retest the symptom.

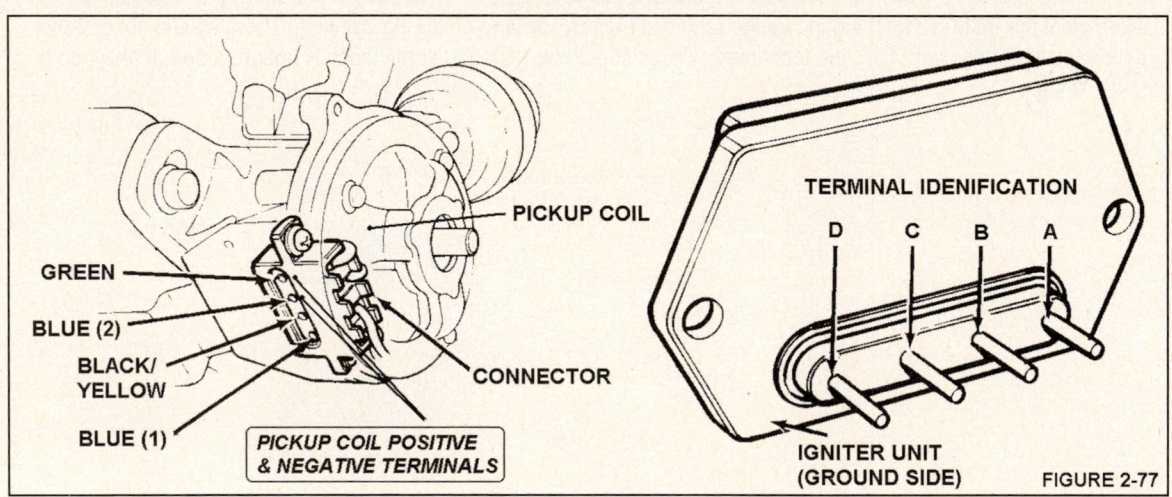

FIGURE 2-77

TYPE TWO SYSTEM SOLENOID TESTS

In a Type Two system, the distributor has two separate vacuum advance diaphragms that operate by manifold vacuum (Figure 2-78).

Diaphragm 'B' is equipped with an ignition control solenoid valve in the vacuum line. The PCM controls this solenoid once it receives information from the ECT sensor and MAP sensor.

Once the solenoid is opened, vacuum flows to diaphragm 'B' and this action allows additional spark advance in order to improve cold driveability.

DRIVEABILITY SYMPTOMS

This test procedure can be used to determine why distributor diaphragm 'B' did not receive any vacuum during the Ignition Timing Test (Legend 2.5L V6).

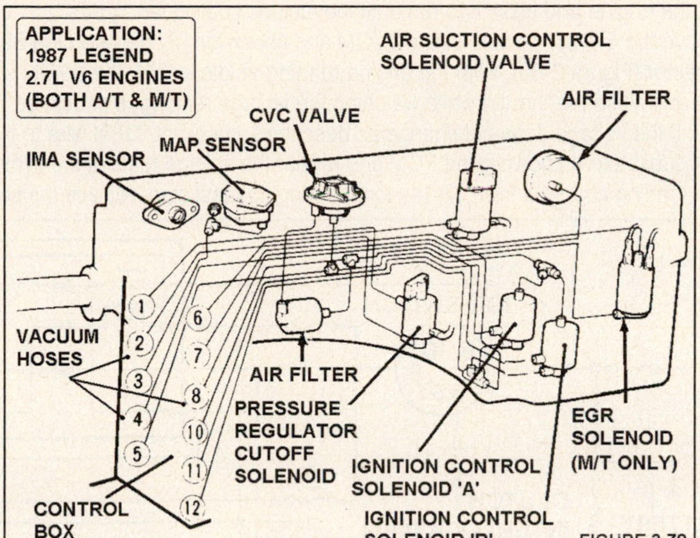

SOLENOID 'A' VACUUM TEST

Start the engine and allow it to run until the fan cycles.

Remove the No. 5 vacuum hose from the 3-way joint. Next, check for available vacuum at the hose end. If it is present, go to Step 3. If vacuum is not present, inspect vacuum connections and hoses for leaks. Inspect vacuum check valve for clogging. Make needed repairs and go to Step 3.

Remove the 6-wire solenoid connector and check for 12v between the BLK/YEL and ORN wire of the harness connector (Figure 2-79). With the engine at idle, if the DVOM reading is near 12v, inspect the vacuum hose to the control box. If the hose is okay, replace solenoid 'A' and retest for the symptom. If the reading is 0v, go to Step 4.

If the voltage read 0v in Step 3, test the voltage between the BLK/YEL wire at the connector and body ground with the key on. If the reading is still 0v, locate and repair the open circuit in the BLK/YEL wire between the solenoid valve harness connector and the No. 11 (10 amp) fuse. If the reading is 12v with the DVOM negative lead connected to body ground, check the ORN wire to the PCM for an open circuit. If the wire is okay, substitute a known-good PCM and repeat the voltage tests. If the reading is now correct, replace the original PCM.

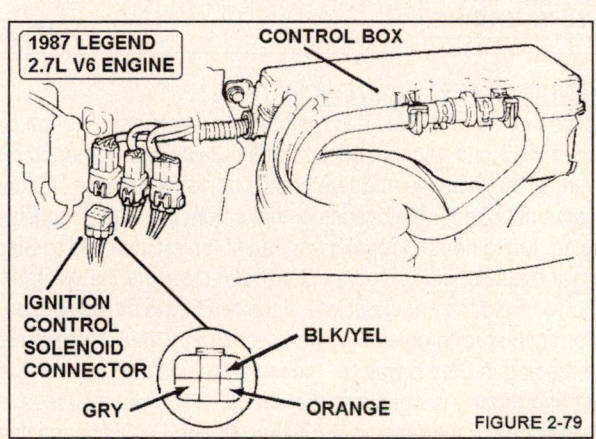

APPLICATION:
1987 LEGEND
2.7L V6 ENGINES
(BOTH A/T & M/T)

AIR SUCTION CONTROL SOLENOID VALVE

AIR FILTER

CVC VALVE

IMA SENSOR

MAP SENSOR

VACUUM HOSES

1 6
2 7
3 8
4 10
5 11
12

CONTROL BOX

AIR FILTER

PRESSURE REGULATOR CUTOFF SOLENOID

IGNITION CONTROL SOLENOID 'A'

IGNITION CONTROL SOLENOID 'B'

EGR SOLENOID (M/T ONLY)

FIGURE 3-78

1987 LEGEND 2.7L V6 ENGINE

CONTROL BOX

IGNITION CONTROL SOLENOID CONNECTOR

BLK/YEL

GRY

ORANGE

FIGURE 2-79

PCM CONTROL SOLENOID TEST

Start the engine and allow it to run until the electric cooling fan cycles.

Remove the 6-P connector. Use a DVOM and check for 12v between the BLK/YEL and ORN wire of the main harness connector (Figure 2-80). With the engine running at idle in P/N, the DVOM should read battery voltage. Next, rapidly open and close the throttle while watching the voltage reading. It should go from 12-14v to 0v as the throttle is snapped open. If the voltage does not change as described, check the ORN wire to the PCM for an open circuit. If the wire is okay, substitute a known-good PCM and repeat the voltage tests. If the symptom goes away, replace the original PCM. If the reading changed from 12-14v to 0v during this test step, replace the solenoid 'A' and retest for the symptom.

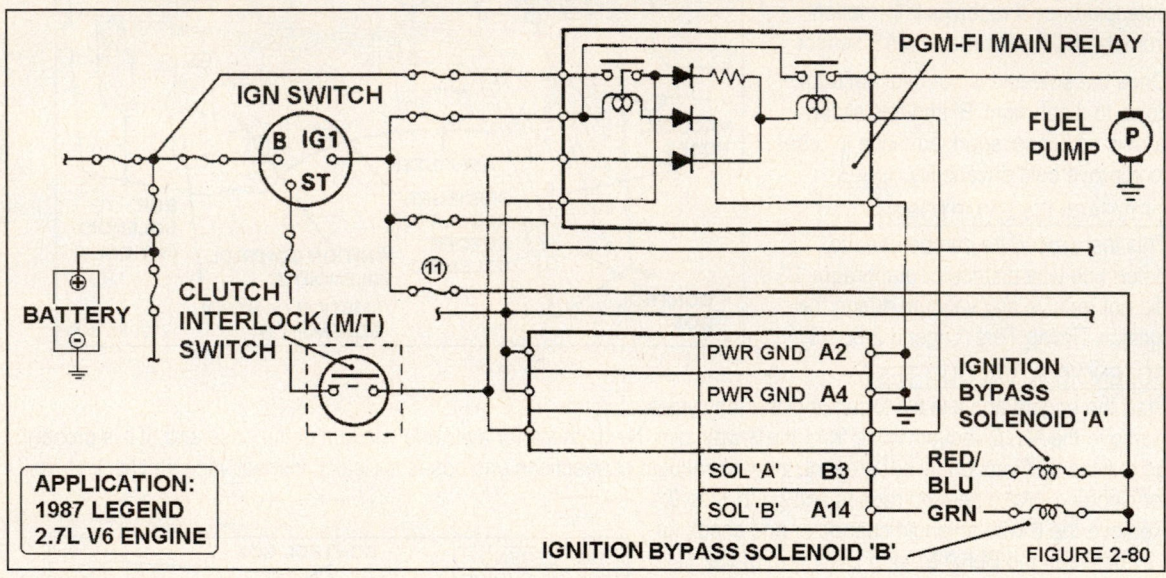

FIGURE 2-80

SOLENOID 'B' TEST (A/T VEHICLES)

Start the engine and allow it to run until the electric cooling fan cycles.

Remove the No. 5 vacuum hose from the 3-way joint connector. With the engine idling in Park or Neutral, check for vacuum to the 3-way joint (there should be vacuum to the 'T' connector). If there is no vacuum to the connector, check the vacuum hose for restrictions or leaks from the intake manifold to the connector. Also inspect the check valve for clogging. Make needed repairs and retest for vacuum. Go to Step 3 when repairs are completed.

Remove the solenoid 6-P connector. Next, measure the voltage between the BLK/YEL and GRN wires of the connector (with the engine idling). If the reading is 0v, go to Step 4. If the reading is 12-14v, inspect vacuum hose to the control box for problems. If the hose is okay, replace solenoid 'B' and retest for the symptom.

If the reading in Step 3 was 0v, measure the voltage between the BLK/YEL wire and body ground. If the reading is 0v, locate and repair the open circuit in the BLK/YEL wire between the solenoid valve and the No. 11 (10-amp) fuse in the dash fuse box. If the reading is 12-14v with the DVOM connected to body ground, locate and repair the open circuit in the GRN control circuit wire between the connector and the PCM. If the GRN wire is okay, substitute a known-good PCM and repeat the test. If the reading is now okay, replace the original PCM and retest the symptom.

SOLENOID 'B' TEST (M/T VEHICLES)

Start the engine and allow it to run until the electric cooling fan cycles.

Remove the No. 5 vacuum hose from the 'T' connector. With the engine idling in Park or Neutral, check for vacuum to the 3-way joint (there should be vacuum to the connector). If there is no vacuum to the 'T' connector, check the vacuum hose for restrictions or leaks between the intake manifold and the 'T' connector. Also inspect the check valve for clogging. Make needed repairs and retest for vacuum. Go to Step 3 when repairs are completed.

Remove the 6-wire solenoid connector. With the engine idling, use a DVOM to measure the voltage between the BLK/YEL and GRN wires of the connector. If the reading is 0v, go to Step 4. If the reading is near 12v, inspect the vacuum hose to the control box for leaks. If the hose is okay, replace solenoid 'B' and retest the system.

If the reading in Step 3 was 0v, measure the voltage between the BLK/YEL wire and body ground. If the reading remains at 0v, the feed circuit is open. Locate and repair the open circuit in the BLK/YEL wire between the solenoid valve and the No. 11 (10-amp) fuse in the dash fuse box. If the reading is 12-14v with the DVOM connected to body ground, locate and repair the open circuit in the GRN control circuit wire between the connector and the PCM. If the GRN wire is okay, substitute a known-good PCM and repeat the test. If the reading is now okay, replace the original PCM and retest the symptom.

PCM CONTROL SOLENOID TEST

Start the engine and allow it to run until the electric cooling fan cycles.

Remove the 6-wire solenoid connector. Connect a DVOM between the BLK/YEL and GRN wires of the harness connector to the solenoid. With the engine idling in Park or Neutral, slowly raise the engine speed to 3500 rpm. The DVOM should not read voltage during this part of the test. If it reads 12-14v, the PCM is okay and the solenoid is faulty. If this is the case, replace solenoid 'B' and repeat the voltage test. If the DVOM reads 0v, substitute a known-good PCM and repeat the test. If the DVOM reads 0v now, replace the original PCM and retest for the symptom.

■ **NOTE:** *For models other than the 1987 Legend applications discussed in this article, refer to other pin voltage charts for pin and wire color identification for all references to PCM terminals and wire colors.*

DIRECT IGNITION SYSTEM

SYSTEM OPERATION

The DIS System is designed to provide the engine optimum control of ignition timing by using the PCM to respond to various input signals as shown below:

Engine Speed - combination Crank and Cylinder sensor
Throttle Angle - throttle angle (TA) or throttle position (TP) sensors
Engine Temperature - Engine coolant temperature (ECT) sensor
Intake Manifold Pressure - MAP Sensor Signal (MAP) sensor

The PCM is capable of calculating ignition firing times with complicated characteristics and schedules. The PCM has tables in memory that contain optimum basic timing settings for various operating conditions based on engine speed and intake manifold pressure calculations. This distributorless ignition system does not use the waste-spark design of spark management.

The PCM uses information gathered from the other sensors to adjust this base timing value to ambient conditions. Once a calculation is accomplished, it sends a voltage pulse to the igniter unit to fire the ignition coils. Ignition systems that use the PCM to control the amount of spark advance are called Type One Systems.

The following Acura vehicles are equipped with a Direct Ignition System:

1991-98 NSX 3.0L (V6)
1991-95 Legend 3.2L (V6)
1996-97 SLX 3.0L (V6), 1998 3.5L SLX (V6)
1996-98 3.2TL (V6) & 3.5RL(V6)

SYSTEM COMPONENTS

The Direct Ignition System (DIS) includes a combination crank/cylinder sensor, 6 high-output ignition coils, dual knock sensors, an ignition timing adjuster, a 14-pin igniter unit and the PCM. In this system, each cylinder has its own high output ignition coil. This is an electronic spark advance (ESA) system where the PCM controls the spark timing for each individual cylinder (Figure 2-81).

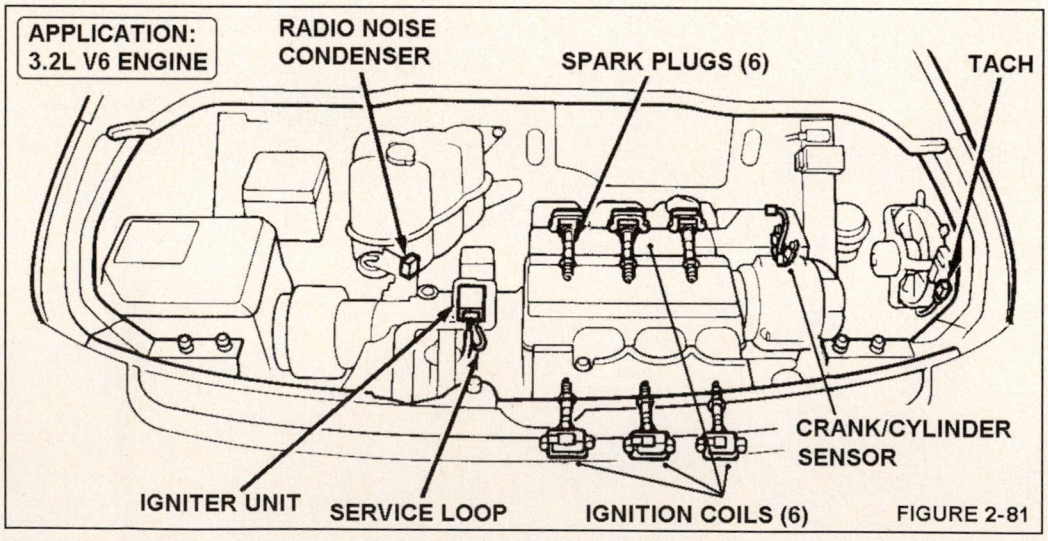

FIGURE 2-81

CRANK/CYLINDER (COMBINATION) SENSOR

The DIS (system) incorporates a combination crank angle sensor that contains a dual magnetic design crank and cylinder pickup unit (Figure 2-82). This pickup unit provides TDC and CYL position of companion cylinders in the V6 engine.

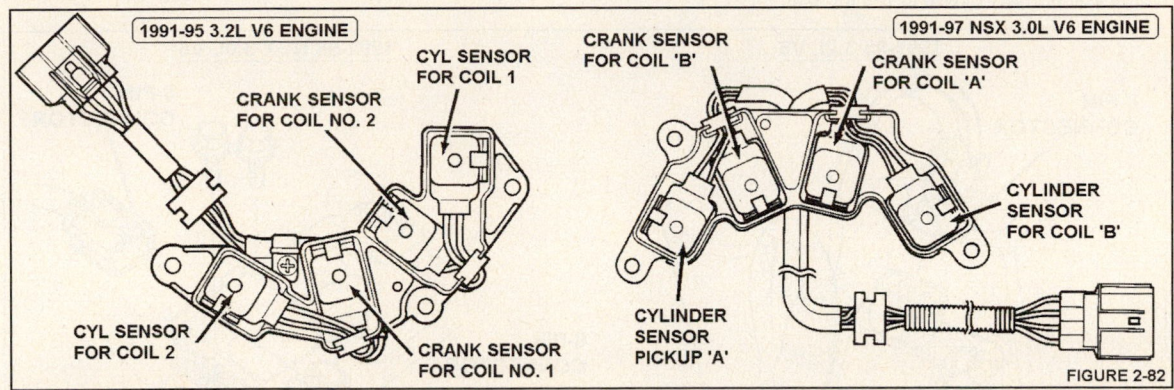

FIGURE 2-82

CRANK SENSORS

DIS crank sensors are identified as Sensor 'A' and 'B' or Sensor No. 1 and No. 2. The PCM uses crank sensor signals to obtain TDC information to determine fuel injection and spark timing for each cylinder and the engine speed (rpm).

CYLINDER SENSORS

DIS cylinder sensors are identified as Sensor 'A' and 'B' or Sensor No. 1 and No. 2. The PCM uses cylinder sensor signals to obtain cylinder position information in order to fire the fuel injectors in sequential order and to compute ignition spark timing.

IGNITION COILS

The DIS (system) incorporates the use of six individual high-output coils (one for each cylinder). The secondary circuits in this system are wired directly to the coil primary circuits. These coils are not waste-spark design coils.

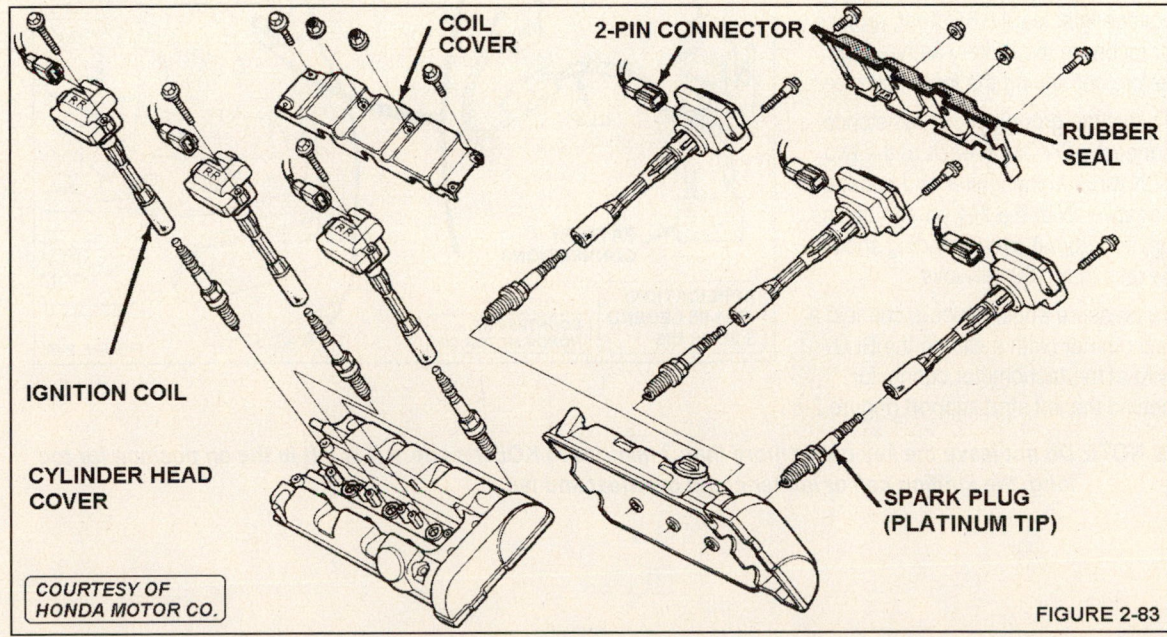

FIGURE 2-83

IGNITER UNIT

The igniter unit in this system is controlled by the PCM. The igniter is connected to six coil primary wires, six ignition control wires at the PCM and to two main ground wires (Figure 2-84). Once the PCM determines which cylinder to fire from the crank and cylinder sensor inputs, it triggers the correct igniter unit transistor and this action turns on the related coil primary circuit to fire the coil.

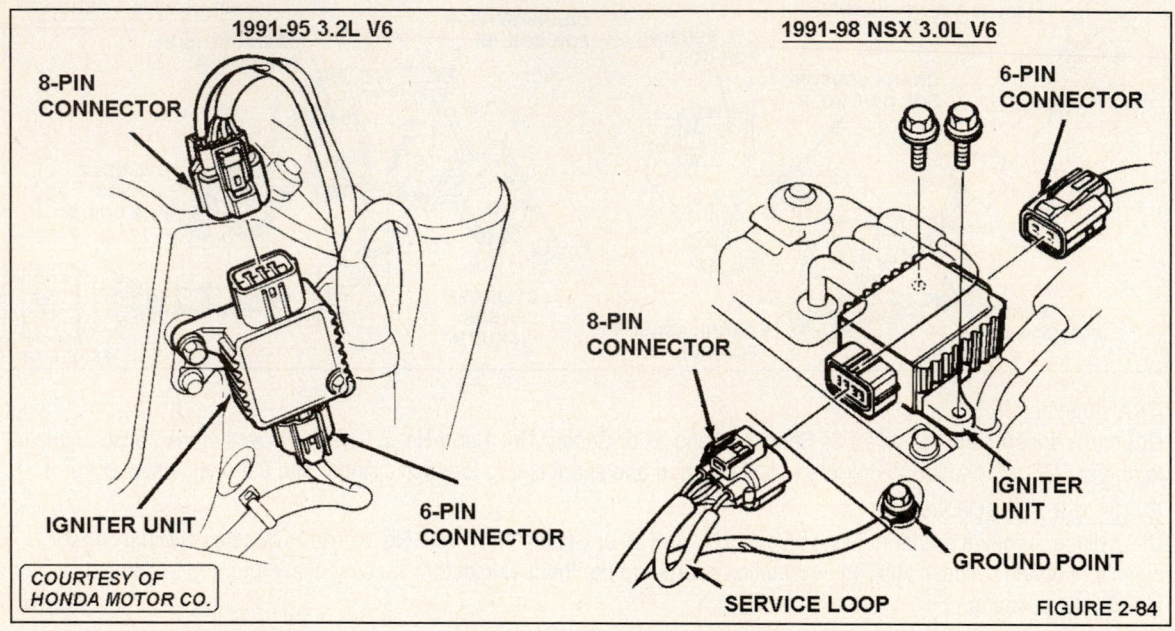

FIGURE 2-84

PRELIMINARY TEST STEPS

Prior to testing the Ignition system, verify that battery open circuit voltage is at least 12.4v and that cranking voltage is at least 9.6v. Test, replace or recharge the battery as needed before continuing the test sequence.

To test the ground circuit resistance, connect a DVOM to each of the two BLK wires at the igniter and to battery negative. With the key on and engine off, the ground circuit reading should be less than 100 millivolts.

To measure engine speed, connect a tachometer (with a clip) to the BLU wire at the tachometer connector behind the left strut support (Figure 2-85).

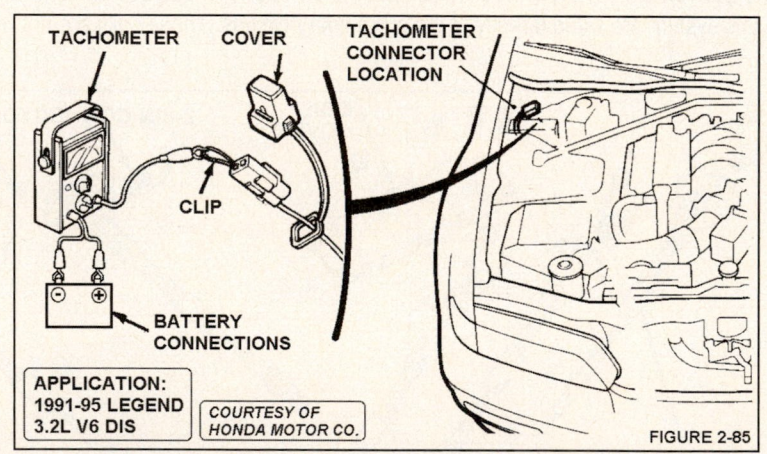

FIGURE 2-85

■ NOTE: *Do not leave the key on for more than 3 minutes in KOEO mode. If it is left in the on position for too long, the ignition coil or igniter could overheat and fail.*

DIS (SYSTEM) TROUBLE CODES

Diagnosis of the DIS (system) should begin with a visual inspection of the underhood ignition components and a review of any stored trouble codes in the PCM memory. If the check engine light is on with the engine running, a hard code is present. The code may be related to a DIS (system) fault, so it makes sense to use the PGM-FI Diagnostics to read, record and clear codes prior to testing. In effect, utilizing the PCM and its diagnostics allows it to communicate if it has detected and recorded any ignition related circuit faults in the combination Crank Angle sensor. The PCM constantly monitors the voltage levels of the signals from the crank and cylinder sensor with the engine running. If both signals are lost, the PCM will shut down the fuel pump relay and the engine will not start or it will stall.

DIS (SYSTEM) TROUBLE CODE DESCRIPTIONS

Code 4 - The PCM sets code 4 if it fails to receive a crank angle sensor No. 1 signal due to either a circuit or component problem.

Code 9 - The PCM sets a code 9 if it fails to receive a cylinder No. 1 signal due to either a circuit or component problem.

Code 15 - The PCM sets a code 15 if it detects a fault in the ignition control signal due to either a circuit or PCM driver problem.

Code 18 - The PCM sets a code 18 if it detects a fault in the ignition timing adjuster circuit or an internal PCM problem.

Code 23 - The PCM sets a code 23 if it detects a fault in the left side knock sensor due to either a circuit or component problem.

Code 53 - The PCM sets a code 53 if it detects a fault in the right side knock sensor due to either a circuit or component problem.

Code 54 - The PCM sets code 54 if it fails to receive a crank angle sensor No. 2 signal due to either a circuit or component problem.

Code 59 - The PCM sets a code 59 if it fails to receive a cylinder No. 2 signal due to either a circuit or component problem.

Refer to the DIS trouble codes examples on the next few pages.

Code 18 - IGNITION TIMING ADJUSTER

If a hard code 18 is set in memory, test the (ITA) circuits as follows.

Remove the 3-pin connector at the control box. Test the resistance across terminals 'A' and 'C' of the ITA unit. If the reading is not in the 3-6 Kohm range, replace the unit and retest for the hard code. If the reading is okay, go to Step 2.

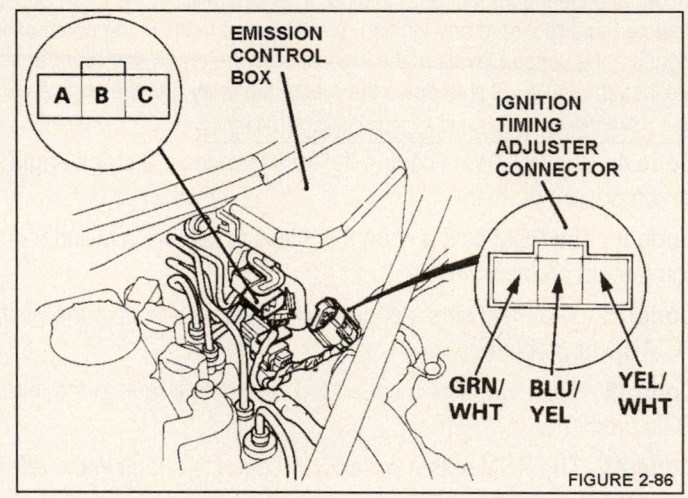

FIGURE 2-86

Measure the resistance across terminals 'B' and 'C' of the ITA. If the reading is in the 3.5-6.5 Kohms range, go to Step 3. If the sum is not within this range, replace the ITA and retest the code.

Turn the key on and measure the voltage between the YEL/WHT and GRN/WHT wires of the ITA harness (Figure 2-86). If the reading is around 5v, go to Step 4. If the reading is not okay, measure the voltage between the YEL/WHT wire and body ground. If it reads near 5v now, locate and repair the open circuit in the GRN/WHT wire. If the reading is still not near 5v, install the Honda test harness or BOB and test the same circuits on the PCM. If the DVOM reads 5v across the YEL/WHT and GRN/WHT wires, the circuit is okay. If the reading is 0v, substitute a known-good PCM and retest the voltage. If it reads 5v now, replace the original PCM and retest the code.

Turn the key off, reconnect the 3-pin connector to the ITA unit and then turn the key back on. Measure the voltage across PCM terminals D8 and D22. If the reading is not from 0.4-4.5v, locate and repair the open circuit or short to ground fault between PCM pin D8 and the ITA signal wire at the harness. If the reading is okay, substitute a known-good PCM and retest for the code. If the code is gone, replace the original PCM.

Code 23 or 53 - LEFT AND/OR RIGHT SIDE KNOCK SENSOR

If a hard code 23 or 53 is set, connect the test harness or BOB to the PCM.

Remove the knock sensor connector and check the D3 WHT (right side) and/or D4 RED/BLU (left side) sensor wire for continuity to ground. If the wire is shorted, locate and repair the short. If it is okay, go to Step 2.

Check the D3 and/or D4 wire for an open circuit from the 3-pin connector to the test harness or BOB. If the wire is open, locate and repair the open circuit and retest for the code. If the wire has continuity, go to Step 3.

Substitute another knock sensor. Run the engine at 3000-4000 rpm in P/N for 10 seconds to retest for the hard code. If the code does not reset, replace the original knock sensor and retest. If it resets, replace the PCM.

Code 15 - IGNITER CONTROL SIGNAL

If a hard code 15 is set, test the ignition feed circuits to the coils and the igniter control circuits as discussed next.

Turn the key on and measure the voltage at the BLK/YEL wires (the six connection points at the coils) and ground. If the reading is 0v, locate and repair the open circuit in the BLK/YEL wire between the battery, ignition switch, the No. 25 underhood fuse and the coil connector(s). If the reading is near 12v, go to Step 2.

To test the igniter control signal circuit, turn the key off and remove the igniter 8-pin connector (Figure 2-87). Use a DVOM and check for continuity between terminals 'C' and then 'F' to body ground (to test for an open ground circuit). If the circuit is open, locate and repair the open ground circuit. If the circuit has continuity, go to Step 3.

Remove the igniter 6-pin connector (Figure 2-88). Check for continuity between the six connectors in 6-pin connector and terminals A21, A22, B3, B4, B6 and B8 at the PCM (with a jumper wire long enough to check back to the PCM). If none of the control wires are open, check all six wires for a short to ground condition. Repair any open or ground control signal wires and retest for the hard code condition.

If no problems were detected with these test steps, substitute a known-good igniter and retest for the code. If the engine starts and the code does not reset, replace the original igniter.

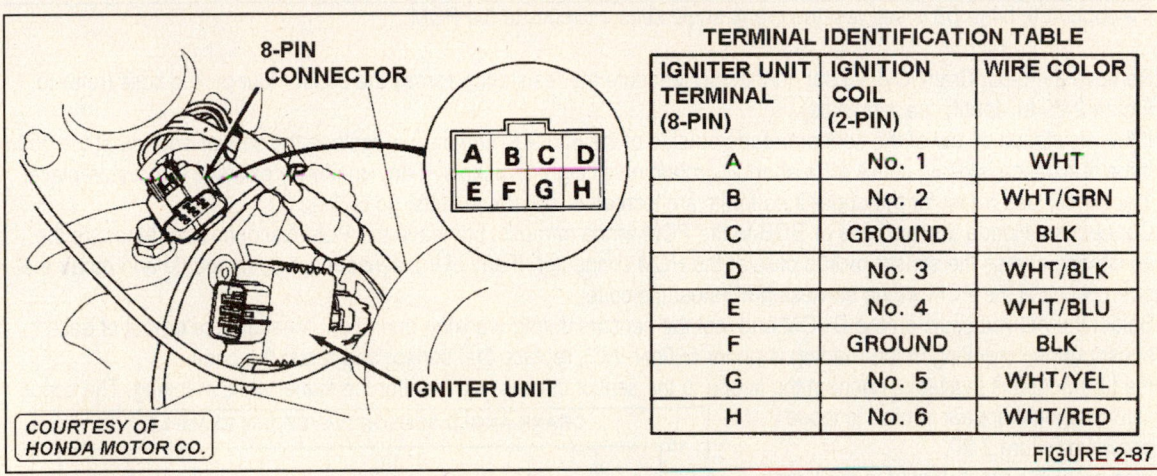

8-PIN CONNECTOR

A	B	C	D
E	F	G	H

IGNITER UNIT

COURTESY OF HONDA MOTOR CO.

TERMINAL IDENTIFICATION TABLE

IGNITER UNIT TERMINAL (8-PIN)	IGNITION COIL (2-PIN)	WIRE COLOR
A	No. 1	WHT
B	No. 2	WHT/GRN
C	GROUND	BLK
D	No. 3	WHT/BLK
E	No. 4	WHT/BLU
F	GROUND	BLK
G	No. 5	WHT/YEL
H	No. 6	WHT/RED

FIGURE 2-87

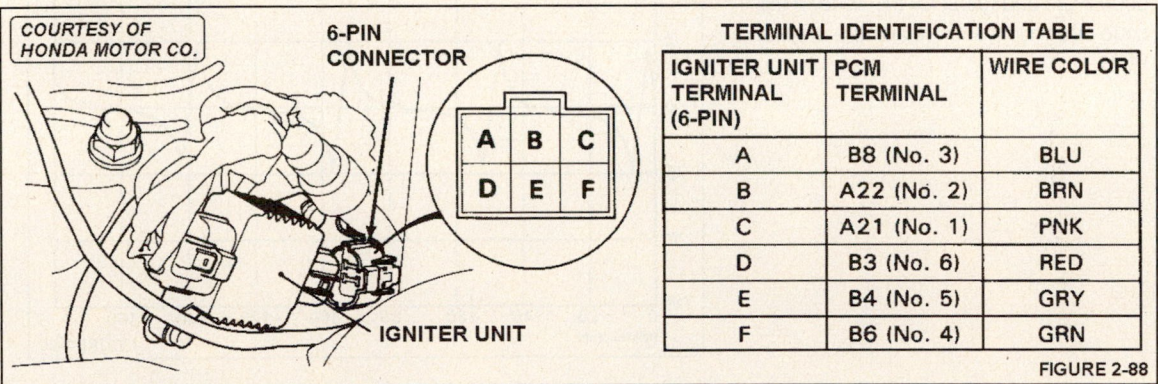

COURTESY OF HONDA MOTOR CO.

6-PIN CONNECTOR

A	B	C
D	E	F

IGNITER UNIT

TERMINAL IDENTIFICATION TABLE

IGNITER UNIT TERMINAL (6-PIN)	PCM TERMINAL	WIRE COLOR
A	B8 (No. 3)	BLU
B	A22 (No. 2)	BRN
C	A21 (No. 1)	PNK
D	B3 (No. 6)	RED
E	B4 (No. 5)	GRY
F	B6 (No. 4)	GRN

FIGURE 2-88

Codes 4, 9, 54 & 59 - CRANK ANGLE SENSOR SIGNALS

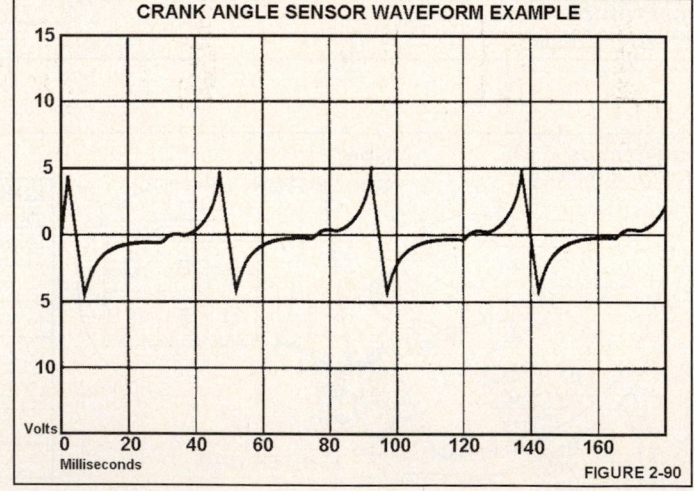

TERMINAL IDENTIFICATION TABLE

SENSOR NAME	CODE ID	SENSOR TERMINAL	PCM TERMINAL	WIRE COLOR
CRANK 1	4	A	B15	PNK
		B	B16	GRY
CRANK 2	54	C	B14	GRN/WHT
		D	B13	BLK/WHT
CYL 1	9	G	B12	BLU/YEL
		H	B11	YEL/GRN
CYL 2	59	E	B9	ORN
		F	B10	WHT

TERMINAL SIDE

COURTESY OF HONDA MOTOR CO.

FIGURE 2-89

If a code 4, 9, 54 or 59 is set, test the crank angle sensor circuits to the PCM.

Remove the 8-pin Crank/Cyl sensor connector. Measure the resistance across the sensor four pickup coils (refer to Figure 2-89 to identify the terminals).

If the resistance of any of the sensors reads outside of the 650-900 ohm range, replace the sensor and retest the code. If the resistance is okay, check for a short to ground on all eight terminals. If any continuity exists in this test, replace that sensor and retest for the code. If no faults are located during these tests, go to Step 3.

Connect the Honda test harness or BOB to the PCM wiring harness, but leave the PCM disconnected. Measure the resistance across the sensor pickup coils at the PCM connector. If any of the sensors are open, locate and repair the open circuit in the PCM wiring harness and retest the code.

Select the AC volt scale on the DVOM and test the sensors during cranking operation. Measure the output of each sensor during cranking. If the reading is under 500mv (AC), replace that sensor and retest the code.

If a Lab Scope is available, connect the scope to the sensor circuits and monitor the signal while cranking. The scope signals should appear similar to those shown in Figure 2-90.

If no problems are detected in the Crank, CYL or TDC sensor, substitute a known-good PCM and retest for the code. If the code does not reset, replace the original PCM.

CRANK ANGLE SENSOR WAVEFORM EXAMPLE

FIGURE 2-90

DIRECT IGNITION SYSTEM SPARK OUTPUT TESTS

Remove the left and right side covers and the coils. Install high-output spark testers to all six coils and clamp the testers firmly to engine ground.

Crank the engine for a few seconds while monitoring the coils for proper coil output. Also, verify that the crank angle sensor rotates at this time.

If the spark output is weak at any of the spark testers, replace it with a known-good coil. If the spark output is now okay, replace that coil assembly. Inspect for signs of arching at the coil units. If there is no spark output, go to Check DIS Connections.

CHECK DIS CONNECTIONS

Inspect the connections to the BLK/YEL wires to the DIS coils and the igniter control wires at the igniter assembly. Look for signs of any backed-out or loose connectors, terminal corrosion or other damage. Clean, tighten and repair connections as needed, then repeat the Spark Output Test. If all are okay, go to the Ignition Battery Voltage Test.

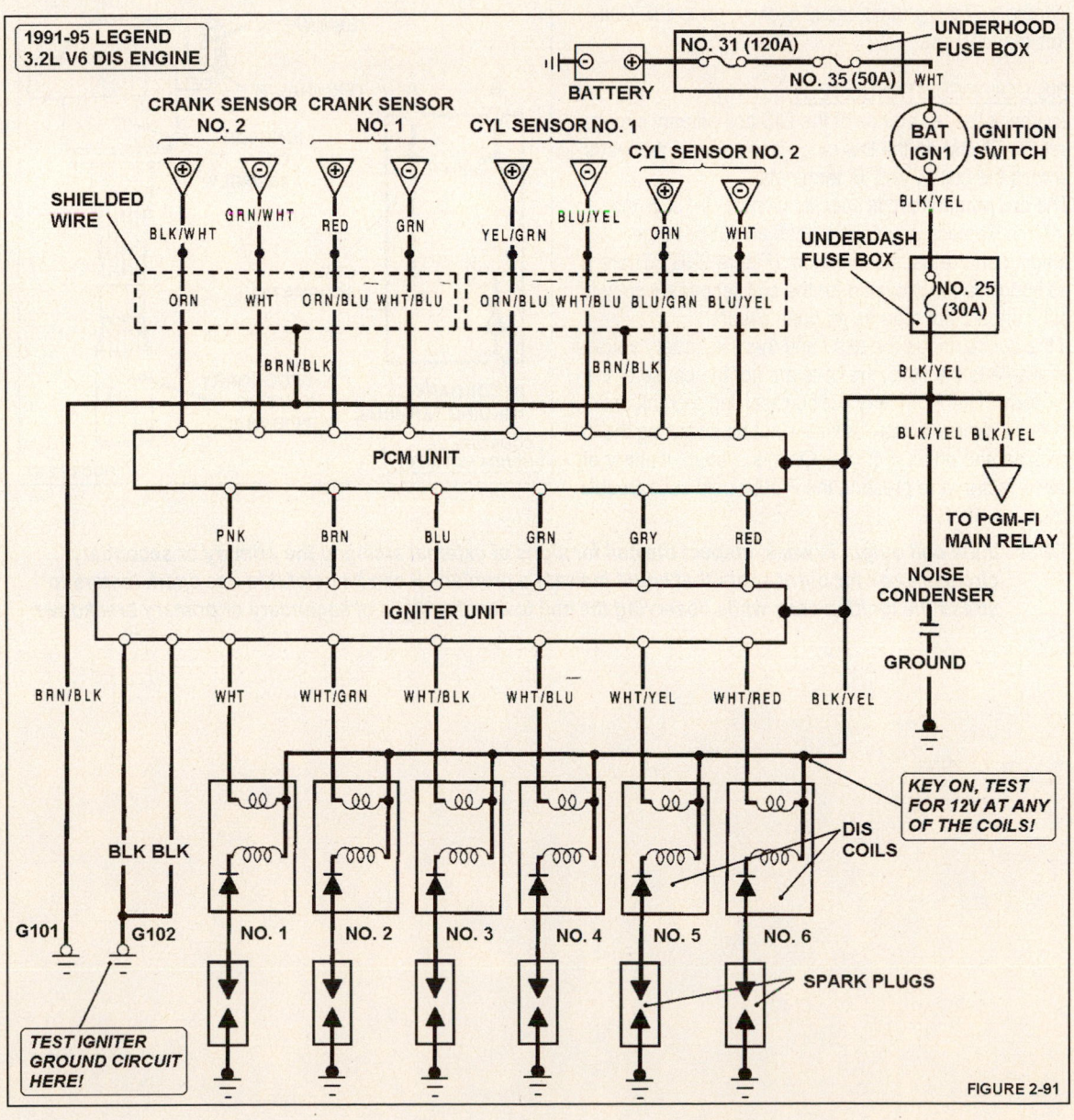

FIGURE 2-91

DIRECT IGNITION SYSTEM TESTS

DIS BATTERY VOLTAGE CHECK

This test steps checks for battery voltage to the DIS coils. The DIS coils receive battery voltage at the BLK/YEL wire from a connection between the ignition switch and PGM-FI main relay. Turn the key on and test for 12v between the any of the coil (+) terminals and ground (refer to Figure 2-91 on the previous page).

If the reading is 0v, inspect the 120 and 50 amp fusible links in the underhood relay fuse box and the No. 25 (30-amp) fuse in the underdash fuse box. If all of the fuses are okay, check the BLK/YEL wire to the coils (terminal 'A') for an open circuit condition. (Figure 2-92). If a fault is located, make repairs and retest for the symptom. If no faults are located, go to Step 3.

Test the igniter ground circuit. Turn the key on and use a DVOM to measure the voltage drop between the igniter ground connection and engine ground at the distributor. The reading should be less than 100mv (this is the ground circuit for the coil primary). If the voltage drop is too high, check the engine ground connections to locate the high resistance. Make repairs as needed and retest for the symptom. If the ground circuit is okay, go to DIS Coil Resistance Test.

DIS COIL RESISTANCE TEST

To check the resistance of the DIS coil primary circuit, set the DVOM on the low range and connect the leads across the coil 'A' and 'B' terminals.

The coil primary circuit specifications: 0.9-1.1 ohms (77°F). Replace any coil that has an out-of-range reading and retest the symptom (Figure 2-92). There is no specification provided for the coil secondary circuit, but it should not read as an open circuit.

If the coil readings are okay and there is battery voltage available to the coils, the coils are not the cause of the problem. The PCM may not be receiving a crank or cylinder sensor signal, or it may not be switching the coils on and off as required. On this system, if either of these faults were present, the PCM would set a trouble code.

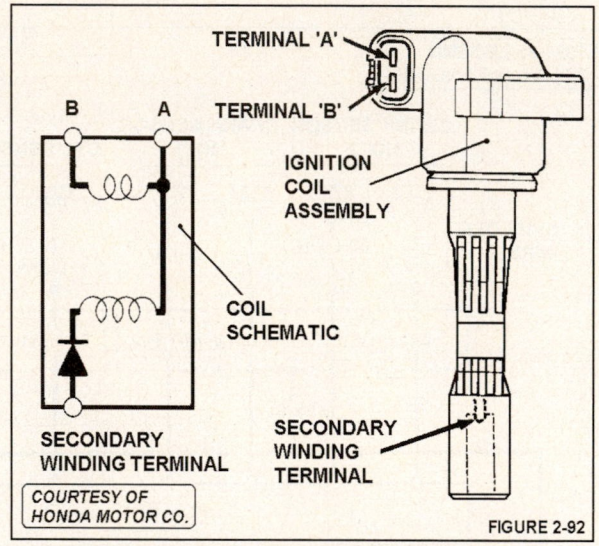

FIGURE 2-92

TIPS: *If the coil output is weak, inspect the coil for signs of external arcing to the primary or secondary circuits. Look for burnt spots that could indicate a problem. It can be helpful to use spark testers to stress the ignition coils while observing the coil towers for signs of secondary or primary breakdown*

PCV System

INTRODUCTION

The Positive Crankcase Ventilation (PCV) system is designed to prevent engine blow-by gases from escaping into the atmosphere (Figure 2-93).

The PCV valve contains a spring-loaded plunger. Once the engine starts to run, the plunger in the PCV valve is lifted in proportion to intake manifold vacuum. This action causes the blow-by gas to be drawn into the intake manifold.

PCV SYSTEM INSPECTION

Inspect the PCV hoses and connections for leaks and clogging.

Make any repairs as needed and continue testing the system.

PCV SYSTEM TEST

Start the engine and allow it to idle. Obtain a pair of pliers and a shop towel.

Next, lightly squeeze the hose between the PCV valve and the intake manifold with the pliers (or your fingers) and listen for a clicking sound from the valve (Figure 2-94).

If there is no clicking sound during this test, inspect the PCV valve grommet and hose for cracks or leaks.

If the grommet and hose are okay, replace the PCV valve and retest for the symptom.

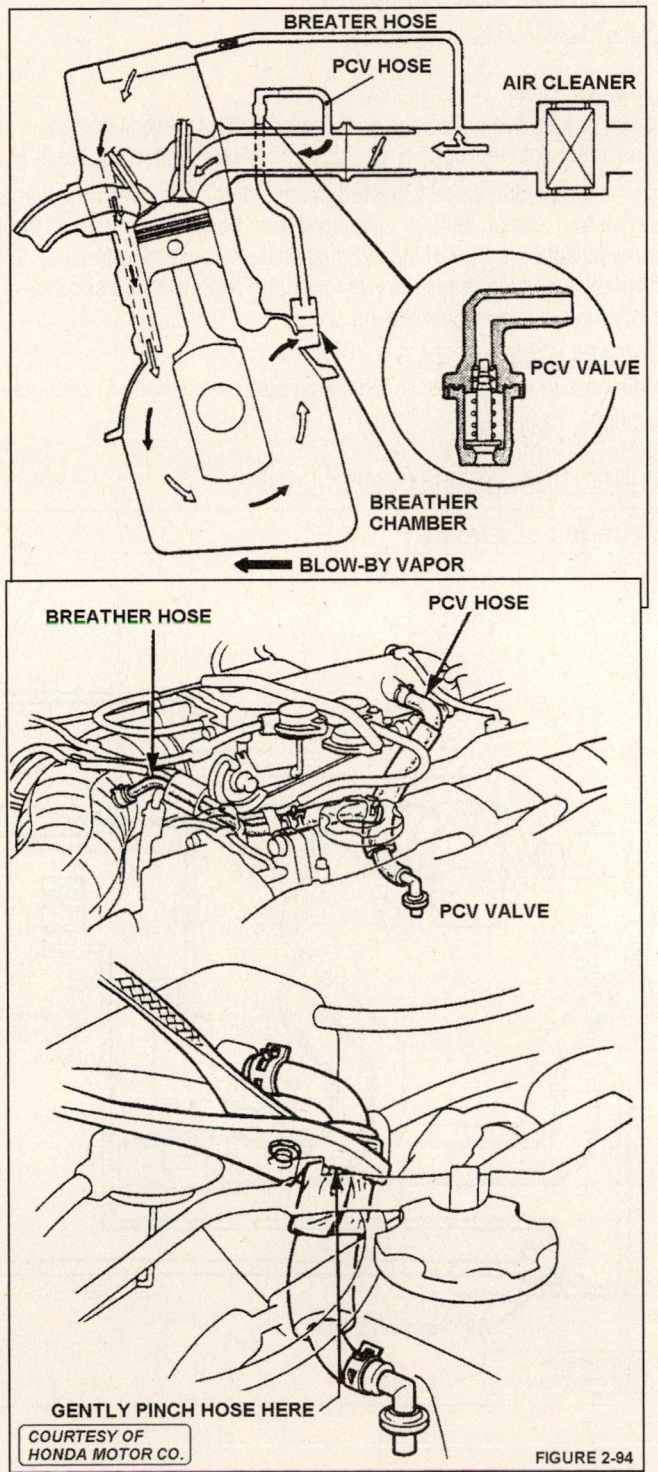

FIGURE 2-94

Secondary Air Injection System
AIR INJECTION SYSTEM TESTS

INTRODUCTION

In order to improve emissions performance, the Air Injection system supplies fresh air from the air cleaner into the exhaust manifold through an air suction (solenoid) valve (Figure 2-95).

With the air suction valve activated, manifold vacuum raises the diaphragm valve portion of the air suction valve so that fresh air from the air cleaner is induced into the exhaust manifold through a reed valve in the air suction valve. The normal pulsation action of the exhaust system causes the fresh air to enter the system. Air suction control solenoid activation is delayed after start-up for approximately 10-60 seconds. The actual amount of time delay depends upon the startup coolant temperature.

DRIVEABILITY SYMPTOMS

Problems with the Air Injection system could cause of the following symptom:

Fails I/M or State Emissions Test

AIR INJECTION SYSTEM COMPONENTS

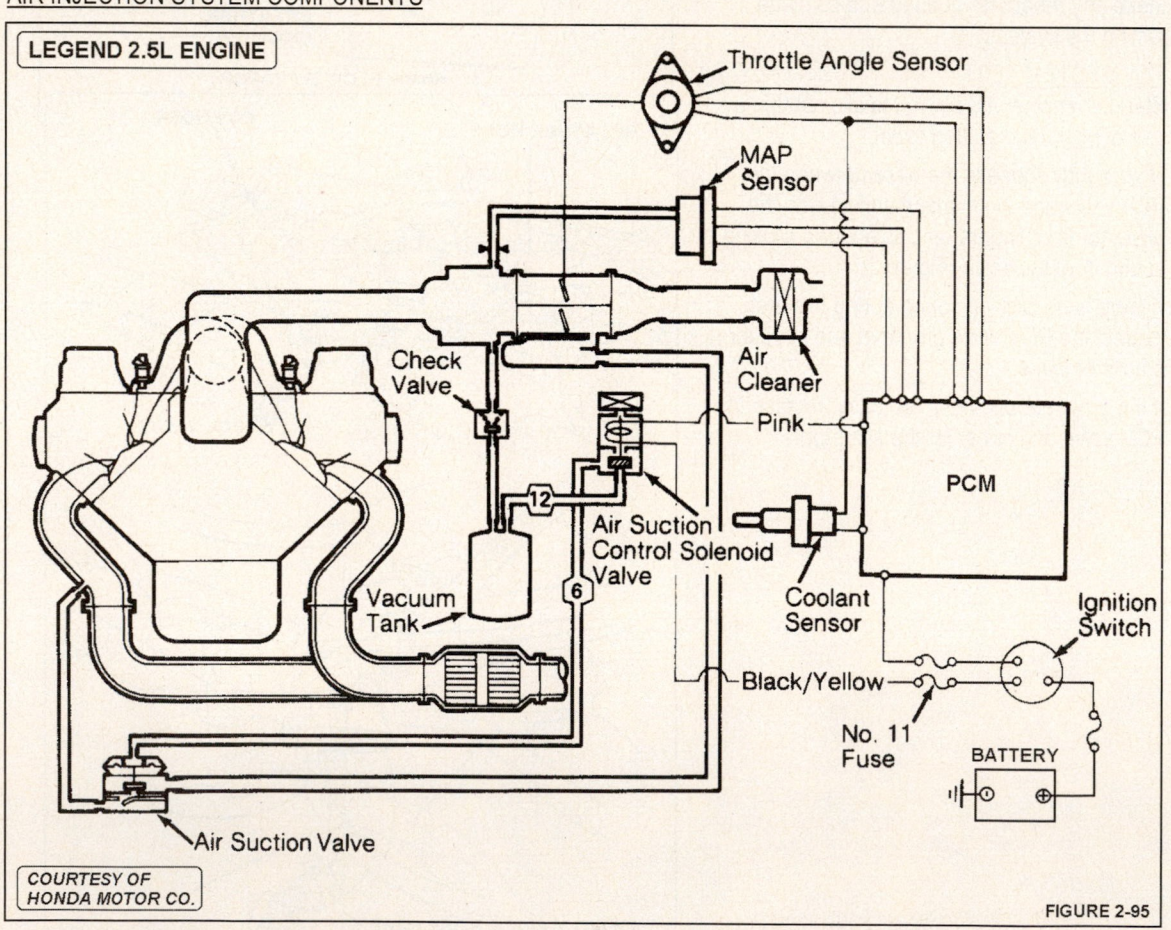

LEGEND 2.5L ENGINE

COURTESY OF HONDA MOTOR CO.

FIGURE 2-95

Secondary Air Injection System
AIR INJECTION SYSTEM TESTS

AIR INJECTION SYSTEM COMPONENTS

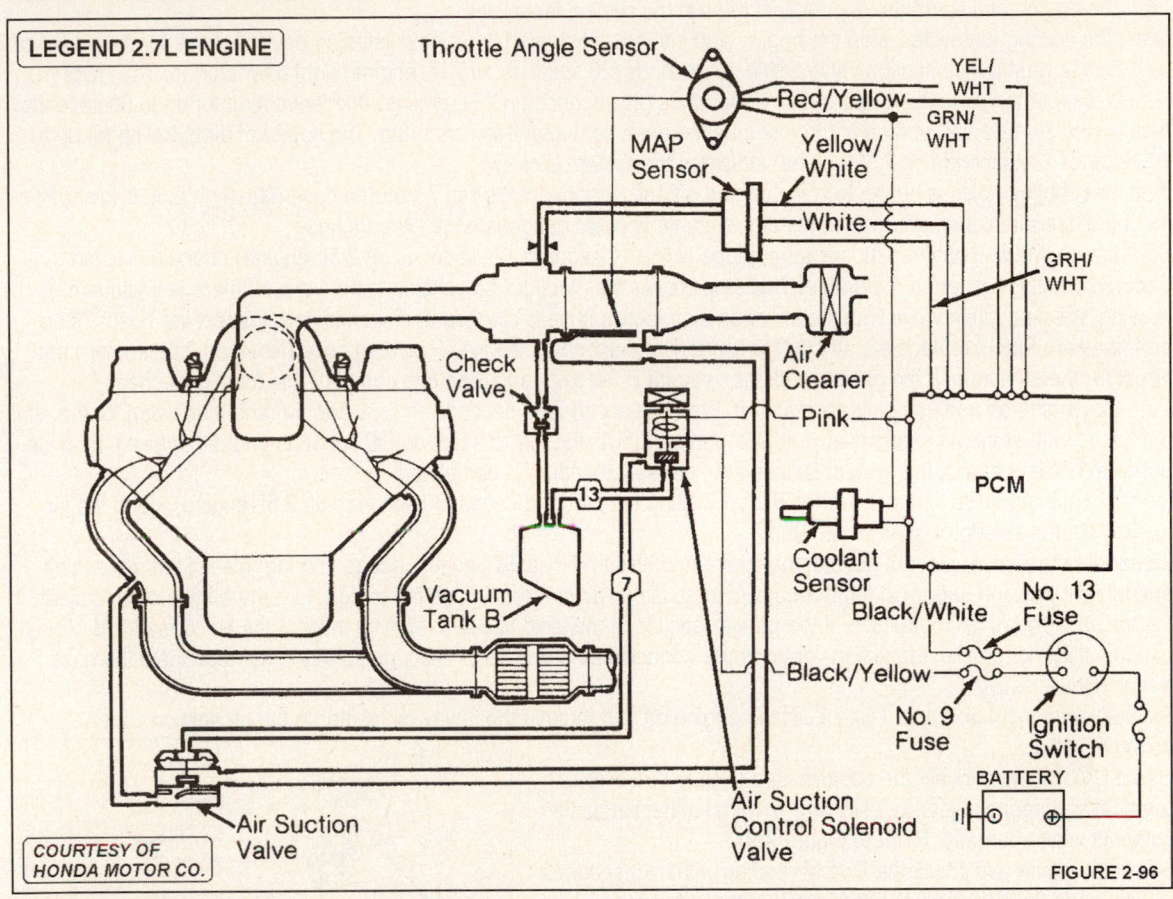

FIGURE 2-96

PRELIMINARY TESTS

The inspections and tests on the next few pages should be performed in sequence to isolate the problem area in order to make the necessary repairs. Always start the test procedure by following the preliminary steps.

Obtain a watch that is capable of a second reading and a listening device such as an automotive stethoscope.

CAUTION: **Block the drive wheels and connect the vehicle exhaust pipe(s) to a suitable vent system to purge exhaust fumes from the work area.**

Secondary Air Injection System
AIR INJECTION SYSTEM TESTS

AIR SUCTION VALVE TEST

Start the engine and warm the engine until the electric cooling fan cycles.

Once the cooling fan cycles, stop the engine and immediately restart it. Using a listening device, check for the sound of air bubbling past the air suction valve within 10 seconds (20 seconds on 2.5L engine) right after startup. The PCM should disable the air suction valve after 10 seconds (20 seconds on 2.5L engine) and leave it off for up to 60 seconds. This is why the listening portion of the test must be done right after engine startup. The sound of air bubbling for up to 10 seconds (20 seconds on 2.5L engine) indicates the system is okay.

If no air bubbling noise is heard, turn the engine off and disconnect the No. 7 vacuum hose (No. 6 on 2.5L engine) from the top of the air suction reed valve. Install a suitable vacuum gauge into the vacuum hose.

Start the engine and observe the vacuum gauge within 10 seconds (20 seconds on 2.5L engine) of engine restart. Vacuum should register on the gauge. This step proves that vacuum is available to the valve. If there is a vacuum reading, the air suction valve is defective or an air injection hose is plugged or disconnected. Inspect the hoses. If no problems are found, replace the air suction valve. Then reconnect the No. 7 vacuum hose (No. 6 on 2.5L engine) and retest for the symptom. If the gauge reads no vacuum in Step 4, go to Step 8 to complete the functional check.

Turn the engine off and immediately restart it. Wait 15 seconds (30 seconds on 2.5L engine) and then listen for the air bubbling noise at the air suction valve. It is important to wait the full 15 seconds (30 seconds on 2.5L engine). If no air bubbling noise is heard, the system is okay and the test procedure is completed.

If an air bubbling noise is heard in Step 6 after waiting the full 15 seconds (30 seconds on 2.5L engine), go to the air suction control solenoid valve test in Step 8.

Disconnect the No. 5 vacuum hose at the intake manifold and install a vacuum gauge into the hose. Start the engine and allow it to warm to normal temperature. If there is no vacuum reading on the gauge, turn the engine off and clean the vacuum port of its obstruction. If the gauge registers a manifold vacuum reading, inspect the No. 5 manifold vacuum hose for cracks, restrictions or improper connections. If the No. 5 vacuum hose is okay, reconnect the hose and continue testing.

Air Suction Control Solenoid Test - Turn the engine off and remove the 6-way connector to the air suction control solenoid valve.

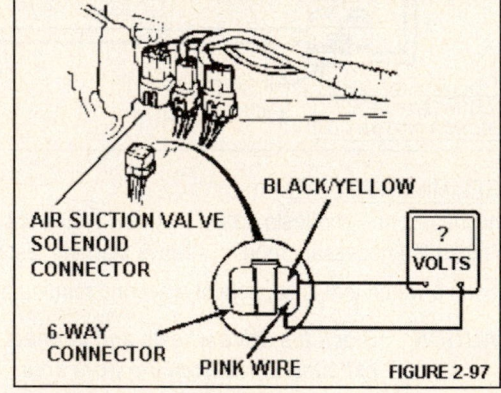

Use a DVOM and connect the positive lead to BLK/YEL wire terminal and negative lead to PNK wire terminal of the air suction solenoid wiring harness. Refer to Figure 2-97.

Start the engine and check the DVOM reading within 10 seconds (20 seconds on 2.5L engine) after engine restart. If voltage is present within 10 seconds (20 seconds on 2.5L engine) after the engine is restarted, (indicating that the PCM completed the circuit), replace the air suction control solenoid valve. If no voltage was present, go to the step 12).

AIR SUCTION VALVE SOLENOID CONNECTOR

BLACK/YELLOW

6-WAY CONNECTOR

PINK WIRE

? VOLTS

FIGURE 2-97

Secondary Air Injection System
AIR INJECTION SYSTEM TESTS

AIR SUCTION CONTROL SOLENOID TEST - CONTINUED

Turn the engine off and move the DVOM negative lead to a good body ground (Figure 2-98). Turn the key on and recheck the voltage reading. If voltage is not available, repair the open circuit in the black/yellow feed wire between the solenoid valve and the No. 9 10-amp fuse (No. 11 on 2.5L engine). When repairs are done, go to step 14 to continue testing.

If voltage was present in step 13), check for an open circuit in the pink wire between solenoid valve and PCM. If the wire is okay, substitute a known-good PCM and retest the system. If the symptom goes away, replace the original PCM and go on to Step 14).

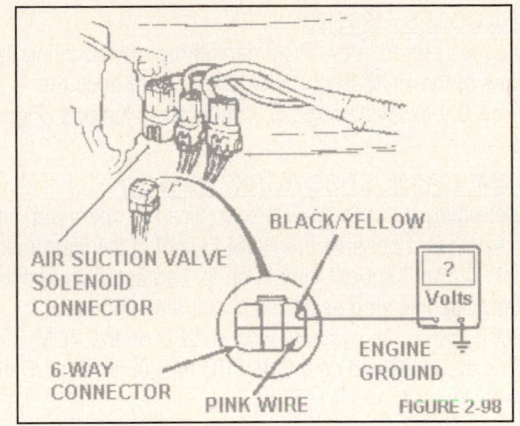

FIGURE 2-98

Disconnect the 6-wire air suction solenoid connector. Start the engine and allow it to warm up until the cooling fan cycles. Use a DVOM and connect the positive lead to BLK/YEL wire terminal and the negative lead to the PNK wire terminal. With the engine idling, observe the DVOM reading 15 seconds (30 seconds on 2.5L engine) after the engine is restarted. If voltage is not present, replace the solenoid valve and retest the system. If voltage is present, check for a short to ground in the pink wire between the solenoid valve and PCM terminal B1 (B15 on 2.5L engine). If the wire is okay, substitute a known-good PCM and retest the system. If the symptom goes away, replace the original PCM and retest the system.

SUMMARY OF TEST STEPS

Use steps 1 to 7 to perform a functional check of the system.
Use steps 8 to13 to verify that the PCM actually turns the solenoid on.
Use steps 14 to15 to verify that the PCM actually turns the solenoid off.

PGM – FI DIAGNOSTICS

Read, Record & Clear Codes

INTRODUCTION

The Acura PGM-FI system uses the PCM to perform its diagnostic functions. The PCM in this system is designed to detect failures in many of its systems or circuits and to set a trouble code in memory when a fault is detected. The articles that follow explain how to use the PGM-FI diagnostics.

READ CODES - 1985-90

Any codes in the PCM memory can be read by counting the flashes of the Red LED on the side of the PCM

READ CODES - 1991-95

Any codes stored in the PCM can be read by counting the flashes of the PGM-FI or check engine light once the Service Check System (SCS) connector is jumped (Figure 2-99).

DO THE DIAGNOSTICS WORK? (1985-95)

To determine if the PGM-FI diagnostics are operating, turn the key on and observe the PGM-FI light in the instrument panel. The light should remain on for two seconds and then go out. Use this step as a bulb check and diagnostic function check. On systems with an LED on the PCM, access the PCM and count the LED flashes to identify the codes.

On later systems, jumper the underdash SCS 2-pin connector and count the check engine light flashes to identify the codes.

PCM CHART NO. 1

If the PGM-FI or check engine light does not activate when the key is first turned on, refer to Chart No. 1 in this section.

PCM CHART NO. 2

If the PGM-FI or check engine light remains on for more than two seconds, but the light does not flash any codes, refer to Chart No. 2 in this section.

BACKUP MODE

If the PGM-FI or check engine light remains on, connect a jumper wire across the SCS 2-pin connector (Figure 2-100). If the MIL does not flash any codes, the PCM is operating in backup mode. In this case, substitute a known-good PCM and recheck the PGM-FI or C/E light operation. If the system now operates normally, replace the original PCM.

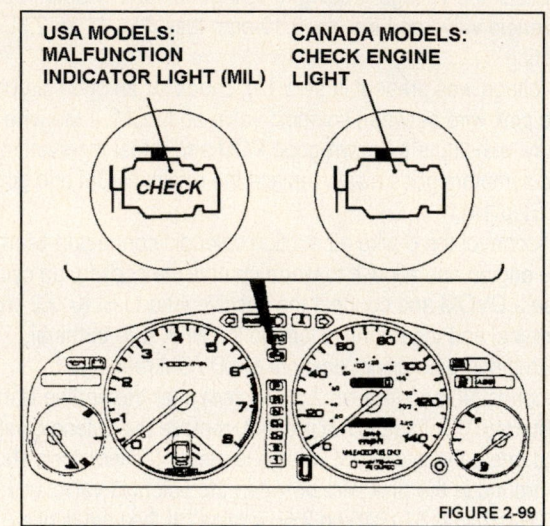

USA MODELS: MALFUNCTION INDICATOR LIGHT (MIL)

CANADA MODELS: CHECK ENGINE LIGHT

CHECK

FIGURE 2-99

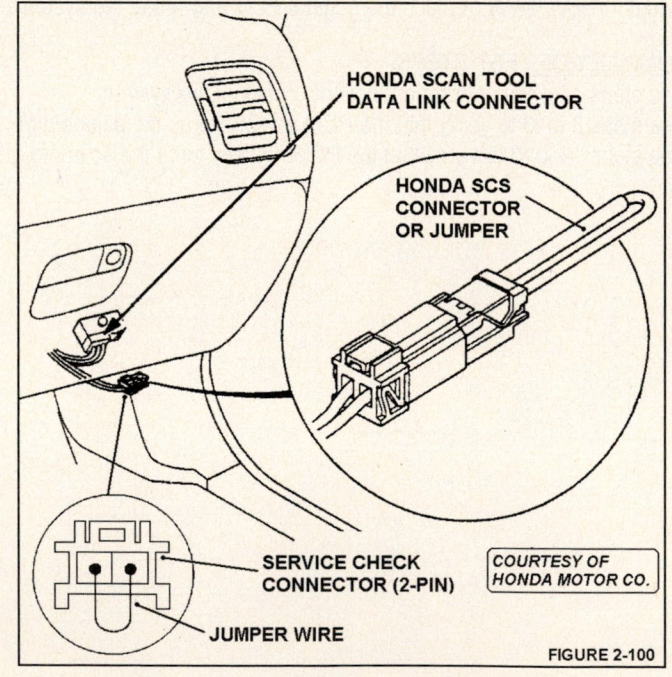

HONDA SCAN TOOL DATA LINK CONNECTOR

HONDA SCS CONNECTOR OR JUMPER

SERVICE CHECK CONNECTOR (2-PIN)

JUMPER WIRE

COURTESY OF HONDA MOTOR CO.

FIGURE 2-100

PCM RESET
To perform a PCM reset procedure (clear all codes from memory), do these test steps. Turn the key off and then remove the PCM power fuse from the underhood fuse/relay box for 10 seconds. This step resets the PCM. If the engine is started after doing this step, any codes that reset are hard codes.

HARD CODES
The term hard code refers to a trouble code that reappears immediately after a PCM reset is completed. On the PGM-FI system, if a hard code is present when the key is turned on, the PGM-FI or check engine light will remain on AFTER the bulb check time period (two seconds). In this case, the PCM should output the code(s) through the Red LED on the side of its case.

SOFT CODES
On the PGM-FI systems, the term soft code describes a code that appears before a PCM reset is done, but does not reappear after a PCM reset is done. A soft code is output in the same manner as a hard code is indicated - through the Red LED. The difference is that the light will remain off after the initial bulb check period.

A soft code can be recorded during the initial step of read, record and clear codes. In effect, a soft code is a code stored in memory from a previously detected fault that does not set after a PCM reset.

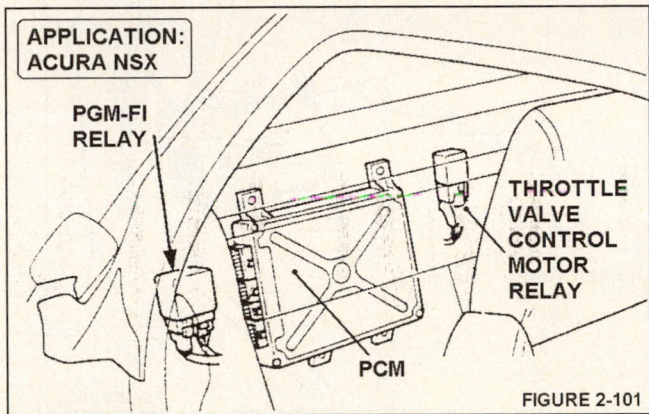

APPLICATION: ACURA NSX

PGM-FI RELAY

THROTTLE VALVE CONTROL MOTOR RELAY

PCM

FIGURE 2-101

PCM LOCATIONS
1986-89 Integra - under the passenger seat
1990-97 Integra - at right front footwell under the carpet
1987-98 Legend Coupe - at bottom front of passenger floor pan
1986-90 Legend Sedan - under the passenger seat
1991-95 Legend Sedan - at right front footwell under the carpet
1991-97 NSX - behind the rear seat back panels (Figure 2-101)
1992-94 Vigor - at bottom front of passenger floor pan
1997-98 2.2CL, 2.3CL, 3.0CL - below passenger side footrest
1995-97 2.5TL - below passenger side footrest
1996-98 3.2TL - below passenger side footrest
1996-98 SLX - behind center of dash below the radio

READ CODES - LED DISPLAY (1986-90)

On these early models, to read any stored codes locate the PCM and pull down the inspection window or remove it to allow the LED to be viewed. Turn the key on and observe the Red LED on the PCM. If codes are stored in PCM memory, they will be displayed as long and short pulses of the Red LED.

The PCM can initiate any number of long and short pulses through the flashing Red LED. These pulses correspond to component or circuit problems that have been detected and stored in memory.

READ CODES - CHECK ENGINE LIGHT OR MIL DISPLAY (1993-95)

On these models, to read any stored codes, locate the SCS connector and connect a jumper wire across the connector (Figure 2-102). Turn the key on and read the PGM-FI or check engine light flashes. This light is also referred to as the malfunction indicator lamp (MIL) in some literature.

If codes are stored in PCM memory, they will be displayed as long and short pulses of the PGM-FI or check engine light. The PCM can initiate any number of long and short pulses by altering the light on/off sequence. These pulses correspond to component or circuit problems that have been detected and stored in memory.

COUNTING CODES

Codes 1 - 9 are indicated by a series of short pulses while codes 10 - 43 are indicated by a series of long and short pulses of the LED. In this system, one long pulse of the LED is equal to ten short pulses. To determine the actual code number, add the long and short pulses together. An index of these codes is provided at the end of this article.

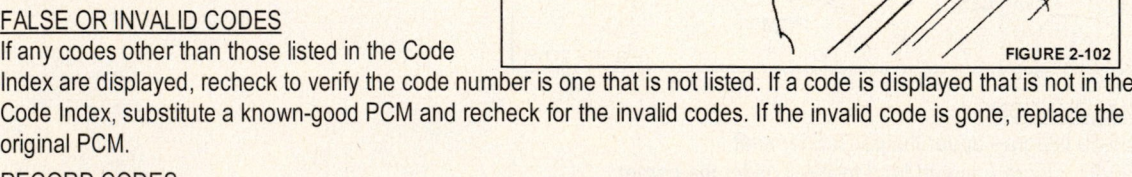

APPLICATION: 1993 LEGEND

SERVICE CHECK CONNECTOR

JUMPER WIRE

PCM

FIGURE 2-102

FALSE OR INVALID CODES

If any codes other than those listed in the Code Index are displayed, recheck to verify the code number is one that is not listed. If a code is displayed that is not in the Code Index, substitute a known-good PCM and recheck for the invalid codes. If the invalid code is gone, replace the original PCM.

RECORD CODES

Record any trouble codes that are displayed by the PCM. This is an important point as the next step is to do a PCM reset (clear the memory). Any codes that are not recorded will be lost. This information could be important if any of the codes were intermittent in nature. Refer to the previous information in this article to distinguish between a hard or soft type of trouble code.

PCM CHART NO. 1 (NO PGM-FI LIGHT)

Use this chart if the PGM-FI or C/E light does not activate for the first two seconds after the key is first turned on (the first step in the PGM-FI Diagnostics function).

CHART NO. 1 TEST PROCEDURE

Turn the key on and observe the oil pressure light. If it is on, go to Step 2. If the oil pressure light is off, check the No. 5 and No. 9 fuses (Figure 2-103). If it is blown, locate and repair the cause of the blown fuse and replace the fuse. If the fuses are okay, locate and repair the open circuit in the wire between the fuse and combination meter.

If the oil pressure light is on in Step 1, turn the key off and connect the Honda Test Harness or BOB to the PCM and wiring harness. Then connect a jumper wire between PCM terminal B6 and body ground. Turn the key on and observe the C/E light. If the light is on now, go to Step 3. If the light is still off, replace the bulb. If the bulb is okay, locate and repair the open circuit in the check engine light wire between the PCM and the combination meter.

If the C/E light was on in Step 3, test the PCM ground connections as follows. Turn the key on and use a DVOM to measure the voltage drop between PCM ground connections A4, A16 and A18 (A23, A24 and A26 on some later models) and body ground. If the voltage drop reading is less than 100mvolt on all terminals (okay), substitute a known-good PCM and retest the C/E light at KOEO for two seconds. If the C/E light works okay, replace the original PCM. If any of the readings are over 100mv, locate and repair the cause of the open or high resistance connection.

■ NOTE:*Inspect the connection at the thermostat housing carefully (it may need to be cleaned and reassembled to fix a ground circuit problem).*

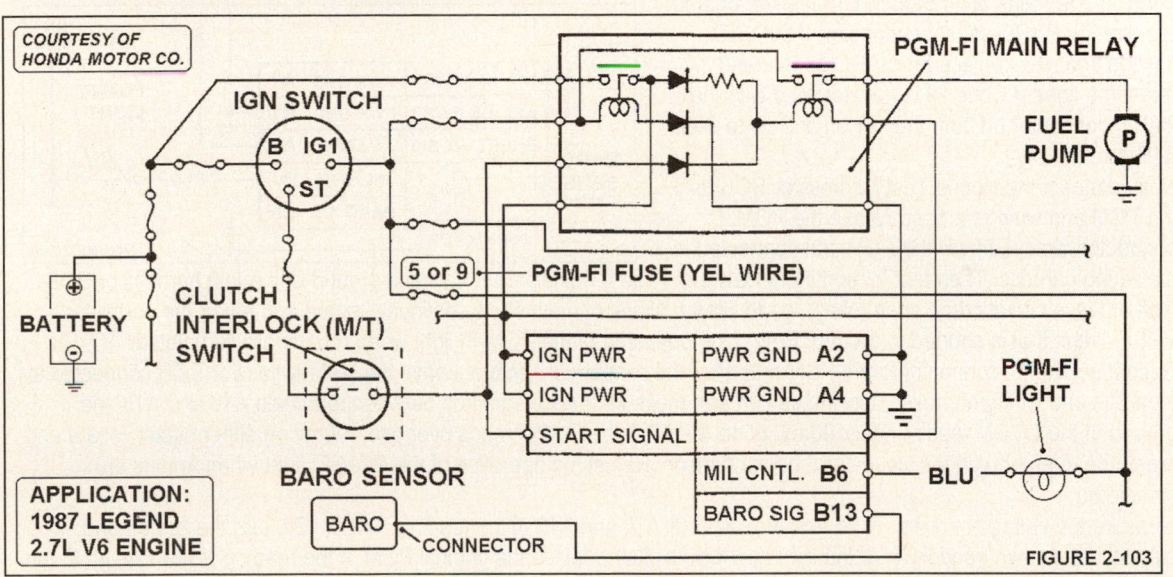

FIGURE 2-103

PCM CHART NO. 2 (PGM-FI LIGHT ON, NO CODES)

Use this chart if the PGM-FI or check engine light remains on for more than two seconds with the key on and no codes are output on the PCM red LED.

CHART NO. 2 TEST PROCEDURE

Attempt to start the engine. If it will not start, go to Step 3. If engine starts and the PGM-FI goes out after two seconds, the fault was intermittent. In this case, inspect for loose wires or connectors at the combination meter and PCM. Retest the system after the inspection. If the light remains on for over two seconds, go to Step 2.

If the light remains on for over two seconds in Step 1, check the PCM for any code output on the LED. Is the PCM flashing any codes? If so, refer to the code repair chart. If no codes are output, turn the key off. Next, remove the PCM 'B' connector. Turn the key on and observe the PGM-FI light. If the light is on, locate and repair the short to ground in the PGM-FI light wire between PCM terminal B6 and the combination meter. If light is off with the PCM 'B' connector removed, substitute a known-good PCM. If the PGM-FI light now works okay, replace the original PCM.

If the engine did not start in Step 1, turn the key off. Remove and inspect the fuse that feeds the PGM-FI light in the main fuse box. If the fuse is blown, locate and repair the short, then replace the fuse. If the fuse is okay, remove and inspect the correct fuse from the dash fuse box. If that fuse is blown, locate and repair the short and replace the fuse. If this fuse is okay, turn the key on. Next, remove the connectors to the MAP, TP and EGR Lift sensors (remove one at a time to observe any change). Leave the sensors disconnected as each one is removed. If the C/E light remains on, go to Step 4. If the light goes out after any of these sensors are disconnected, replace the faulty sensor and retest the light.

If the light remains on in Step 3, turn the key on and remove the BARO sensor connector. If the LED flashes code 13, replace the BARO sensor and retest the light. If Code 13 is not indicated with the BARO connector off, turn the key off and go to Step 5 (Figure 2-104).

Next, connect the Honda Test Harness or BOB to the PCM and wiring harness. Leave the PCM 'C' connector removed from the PCM, but connected to

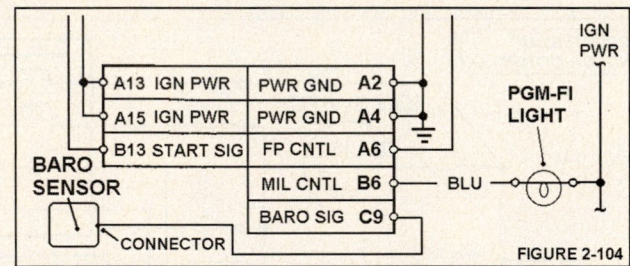

FIGURE 2-104

the wiring harness. Then test for continuity between PCM terminals C13, C15 and ground at the test harness or BOB. If all of these circuits read open (okay), go to Step 6. If any of them show continuity, locate and repair the particular VREF circuit that is shorted to ground. Retest the operation of the PGM-FI light when repairs are completed.

Reconnect the 'C' connector to the PCM and all of the sensor connectors. Leave the test harness or BOB connected to the PCM and wiring harness. Turn the key on and measure the voltage drop between terminals A16 and A18 and ground. If the DVOM reads under 100mv, go to Step 7. If the DVOM reads over one 100mv on either circuit, locate and repair the open or high resistance ground connection. Retest the operation of the PGM-FI light when repairs are completed.

Measure the voltage between PCM terminals A13 or A15 and A18. If the reading is near 12v, turn the key off and substitute a known-good PCM. If the light now works okay, replace the original PCM. If the reading is not 12v, go to the PGM-FI Main Relay Tests in this section.

CLEAR CODES (PCM RESET)

Turn the key off and remove the correct PCM fuse for 10 seconds (Figure 2-105), then replace the PCM fuse. The codes are cleared once the PCM reset function is completed.

■ **NOTE:** *On models with an anti-theft radio, do not remove the battery cable to do a PCM reset. This action will require that a radio code be entered to reset radio functions.*

PCM FAILURE

If the PGM-FI or check engine light remains on steady for over two seconds without flashing any codes, either the light circuit is grounded or the PCM is faulty. Remove the PCM and turn the key on. If the light remains on, the light circuit is grounded. If the light goes out, replace the PCM and retest the system. If the light does not activate at all, check the PGM-FI light circuit from the ignition switch through the fuse to the C/E light.

If the PGM-FI or check engine light remains on for over two seconds without flashing any codes, either the light circuit is grounded or the PCM is faulty. Remove the PCM and turn the key on. If the light remains on, the light circuit is grounded. If the light goes out, replace the PCM and retest the system. If the light does not activate at all, check the PGM-FI light circuit from the ignition switch through the fuse to the light.

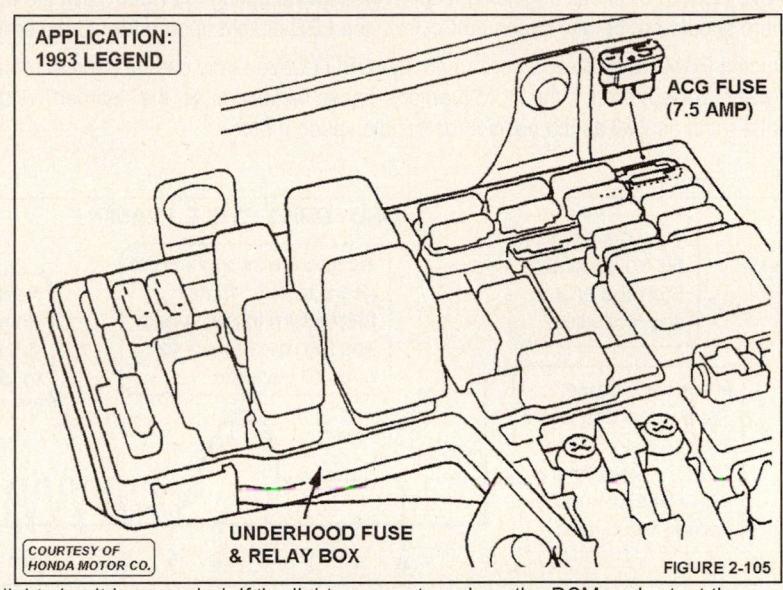

APPLICATION: 1993 LEGEND

ACG FUSE (7.5 AMP)

COURTESY OF HONDA MOTOR CO.

UNDERHOOD FUSE & RELAY BOX

FIGURE 2-105

PCM RESET FUSE IDENTIFICATION

Integra Models

1985-91: Hazard fuse in the underhood fuse/relay box

1992-98: Backup or Backup/ACC (7.5-amp) fuse in the underhood fuse/relay box

Legend Models

1986-90: Alternator Sensor fuse in the underdash fuse box

1989-95: ACG (7.5-amp) fuse in the underdash fuse box

NSX Models

1995-98: Clock (7.5-amp) fuse in underhood fuse/relay box

Vigor Models

1992-94: Backup (10-amp) fuse in underhood fuse/relay box

2.2CL Models

1997: Backup fuse in underhood fuse/relay box

2.3CL Models

1998: Backup fuse in underhood fuse/relay box

2.5TL Models

1995-98: Backup/Radio (10-amp No. 39) fuse in underhood fuse/relay box

3.2TL & 3.5RL Models

1995-98: Backup/Radio (10-amp No. 56) fuse in underhood fuse/relay box

DRIVE CYCLE

Once it is established that the PGM-FI diagnostics are working and that there are no codes stored in memory, perform a Road Test and attempt to recreate any symptoms or problems listed on the work order.

If the PGM-FI or check engine light is activated during a Road Test or drive cycle, a hard code should be output by the PCM on the LED or by the PGM-FI or check engine light during diagnostics.

If the PGM-FI or check engine light does not activate during the Road Test, the diagnostic has passed and therefore, there should not be any codes output from the LED or light upon return to the service bay.

Once a PCM Reset step is performed, the OBD I Drive Cycle can be used to rebuild the fuel control and idle speed control tables stored in the PCM memory. These tables contain the "learned" values used to control the fuel injector pulsewidth as well as the position of the idle speed motor.

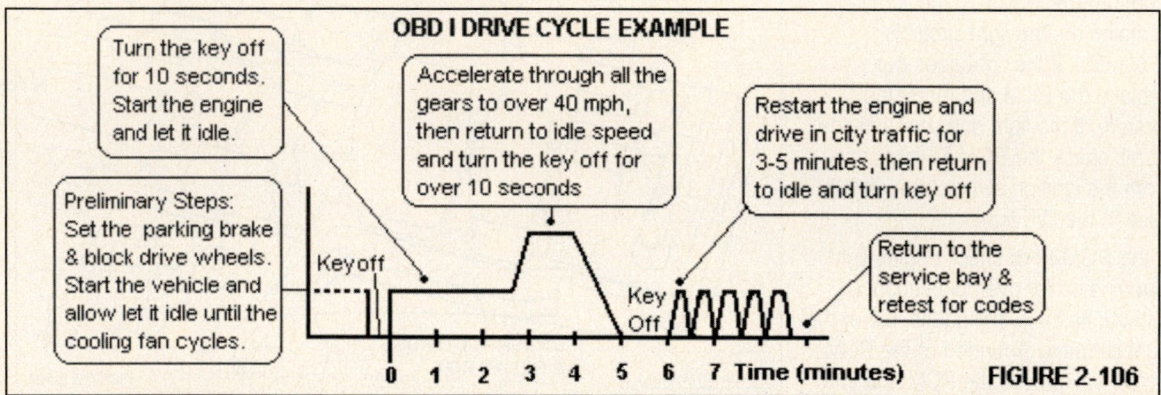

FIGURE 2-106

■ **NOTE:** *Go to the next page to review how to diagnose No Code Faults (driveability symptoms or intermittent faults).*

No Code Fault Diagnosis

DRIVEABILITY SYMPTOMS

If the PGM-FI diagnostics show Pass, attempt to determine if a driveability symptom is present during a Road Test. If a symptom is present, refer to the Driveability Symptom List in Section One to locate the correct symptom for a particular vehicle and problem.

Once the correct symptom is located, work down through the list of system and component checks. At this point, a technician can use their experience and knowledge of testing to determine which symptom test to run first. Choose a particular symptom description (e.g., Idle Concerns) and then refer to that index in the same section to find the driveability symptom. The index is in alphabetical order and the test sequence may require that more than one type of system of component be tested to repair the problem.

INTERMITTENT FAULTS

If no driveability symptoms appear during testing, determine from the customer invoice or previous experience during the Road Test if an intermittent fault may exist on the vehicle. If so, go to Intermittent Tests in Section One.

FINAL CHECKOUT PROCEDURE

Once all repairs are completed, do a PCM reset (clear the memory) and Road Test the vehicle one last time. Observe the PGM-FI or check engine light for any signs of a fault being detected or for signs of a driveability symptom that could be present. If a code sets or a symptom is present, repeat the appropriate procedure as outlined in the block chart below.

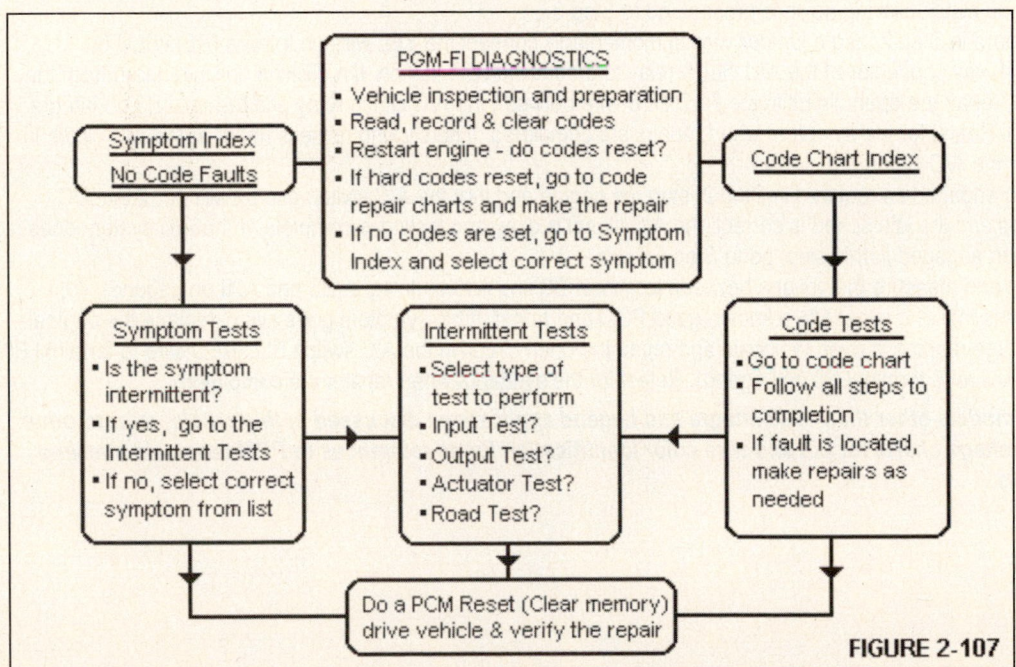

FIGURE 2-107

AIR CONDITIONER SIGNAL (INTEGRA & LEGEND MODELS)

INTRODUCTION
The PCM uses the air conditioning (A/C) signal to determine if the A/C clutch has been engaged. This signal is diagnosed as a symptom, not as a code.

DRIVEABILITY SYMPTOM DIAGNOSIS
A problem with the air conditioning signal can cause the engine idle speed to drop when the A/C is turned on with a warm engine.

AIR CONDITIONER SIGNAL TEST
Turn the key off and connect the Honda test harness or BOB to the PCM and wiring harness to allow for circuit testing of components and the wire harness.

The 'A' and 'C' connectors should be connected to the PCM and wiring harness (through the test harness or BOB). Leave the 'B' connector removed from the wiring harness for this step. Turn the key on and measure the voltage between terminals B3 on Integra (B8 on Legend) and A18. The DVOM should read 5v. If it reads 5v, go to Step 2. If the DVOM does not read 5v, substitute a known-good PCM and retest the voltage. If the DVOM now reads 5v, replace the original PCM.

Turn the key off and reconnect the 'B' connector at the test harness or BOB connector. Turn the key on and momentarily jumper the A/C clutch relay terminals B3 on Integra (B8 on Legend) to A18. An audible clicking noise should be heard from the A/C compressor clutch each time the connection is made. If the clutch clicking noise is heard, go to Step 4. If no clutch clicking noise is present, go to Step 3.

If no noise is heard in Step 2, use a jumper wire to momentarily connect the YEL wire on Integra (RED/BLU on Legend) at the 4-way connector of the A/C clutch relay to ground with the key on. If a clicking noise is heard from the A/C clutch now, repair the open circuit in the RED/BLU wire between the A/C clutch relay and terminal B3 on Integra (B8 on Legend). Retest for the symptom when repairs are completed. If no clicking noise is heard at the A/C clutch in this step, repair the A/C clutch.

If the A/C clutch engaged as required in Step 2, start the engine and turn the A/C switch and blower on. If the A/C system operates and the idle speed is correct, the A/C signal is okay and testing is complete. If the A/C system does not operate when engaged in this step, go to Step 5.

Turn the key on and measure the voltage between terminals B8 and A18 on Integra (B5 and A26 on Legend). If the voltage reads less than 1v, substitute a known-good PCM and retest. If the symptom goes away, replace the original PCM. If the voltage reading is over 1v, locate and repair the open circuit in the A/C switch BLU/RED wire to terminal B8 on Integra (PNK wire to terminal B8 on Legend). Retest for the symptom when repairs are completed.

■ **NOTE:** *For models other than 1990 Integra and Legend applications discussed in this article, refer to other pin voltage charts for pin and wire color identification for all references to PCM terminals and wire colors.*

ALTERNATOR 'FR' SIGNAL (INTEGRA & LEGEND MODELS)

INTRODUCTION
The PCM uses the alternator 'FR' signal to determine when the alternator is charging. This signal is diagnosed as a symptom (there are no OBD I codes).

DRIVEABILITY SYMPTOM DIAGNOSIS
A problem with the alternator 'FR' signal could cause the engine idle speed to drop when high electrical loads are enabled on the vehicle.

ALTERNATOR FR SIGNAL TEST
Turn the key off and connect the Honda test harness or BOB to the PCM and wiring harness to allow for circuit testing of components and the wire harness.

Leave the 'B' connector on the PCM but disconnected from the wiring harness. This allows for testing of the 'FR' circuit from the PCM without it connected to the alternator. Leave all other connectors installed so that the PCM is connected to power and ground.

Turn the key on and measure the voltage between terminals B14 and A18 at the PCM. If the voltage is near 5v, go to Step 3. If the voltage is below 4.75v, turn the key off and substitute a known-good PCM. Retest the voltage. If the reading is 5v, replace the original PCM.

If the reading was near 5v in Step 2, turn the key off and reconnect the 'B' connector to the PCM wiring harness. Restart the engine and allow it to warm until the cooling fan cycles. Connect the DVOM to terminals B14 and A18 and monitor the voltage with the headlight switch and rear defogger engaged. If voltage does not decrease as the load devices are engaged, go to Step 4. If the voltage decreases as the load devices are engaged, the alternator 'FR' signal is okay. Turn the key off and remove the Alternator Sense fuse for 10 seconds to reset the PCM memory.

With the key off, remove the 'B' connector at the PCM (do not remove it from the wiring harness). Next, disconnect the negative battery cable. Check for continuity (this would indicate a short circuit) between terminal B14 and body ground. If continuity is present, remove the GRN connector at the alternator and recheck for continuity. If continuity exists with the GRN connector removed, locate and repair the short in the BLU wire between the alternator harness and terminal B14 on Integra (WHT/RED wire between the alternator harness and terminal B14 on Legend). If the DVOM reads no continuity with the GRN connector off, repair the alternator. Make repairs and retest for the symptom. If the DVOM reads no continuity, go to Step 5.

If there is no continuity in Step 4, remove the GRN connector at the alternator. Jumper the BLU wire on Integra (WHT/RED wire on Legend) to body ground. If there is no continuity, locate and repair the open circuit in the BLU wire on Integra (BLU/RED wire on Legend) between the alternator connector and terminal B14. If continuity exists with the BLU wire on Integra (BLU/RED wire on Legend) jumped to ground (okay), repair the alternator. Retest for the symptom when repairs are completed.

■ NOTE: *For models other than 1990 Integra and Legend applications discussed in this article, refer to other pin voltage charts for pin and wire color identification for all references to PCM terminals and wire colors.*

START SIGNAL (INTEGRA & LEGEND MODELS)

INTRODUCTION

The PCM uses the ignition start signal to determine when the engine is in the crank mode. Once this signal is detected, the PCM controls the current signal to the EACV (valve) to provide for easy startup conditions. The start signal is diagnosed as a symptom (there are no OBD I codes).

DRIVEABILITY SYMPTOM DIAGNOSIS

A problem with the start signal could cause a low engine idle speed condition on initial startup with a warm engine.

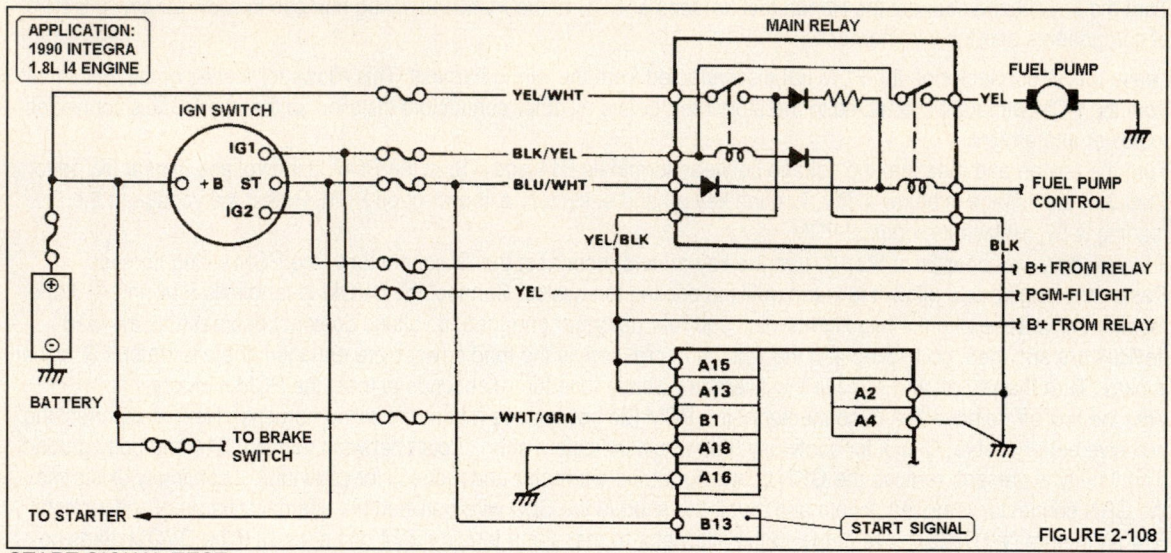

FIGURE 2-108

START SIGNAL TEST

Turn the key off and connect the Honda test harness or BOB to the PCM and wiring harness to allow for circuit testing of components and the wire harness.

Connect a DVOM between PCM terminals B13 and A18 in order to test the ignition start signal at startup (Figure 2-108). Turn the ignition switch quickly to start and watch the DVOM reading. If the DVOM reads from 9-11v until the engine starts, the starter signal is okay and the test is completed.

If the voltage reading in Step 1 in not from 9-11v during startup (cranking) conditions, inspect for a blown fuse in the start signal circuit. If the fuse is blown, locate and repair the cause of the blown fuse. Replace the fuse and retest for the symptom.

If the fuse is okay in Step 2, locate and repair the open circuit in the BLU/WHT wire between the PCM terminal B13 and the fuse or switch. Retest for the symptom when repairs are completed.

■ **NOTE:** *For models other than 1990 Integra and Legend applications discussed in this article, refer to other pin voltage charts for pin and wire color identification for all references to PCM terminals and wire colors.*

A/T SHIFT POSITION SWITCH (ALL MODELS)

The PCM uses the automatic transaxle (A/T) shift position signal to determine when the transaxle is shifted into park or neutral position. This signal is diagnosed as a symptom.

DRIVEABILITY SYMPTOM DIAGNOSIS

A problem with the A/T shift position could cause the engine idle speed to drop to low when the vehicle is shifted into a drive gear.

A/T SHIFT INDICATOR LAMP TEST

Turn the key off and connect the Honda test harness or BOB to the PCM and wiring harness. Turn the key on and leave the engine off. Watch the A/T shift indicator lights as the gearshift selector in moved between gears. If the lights do not work properly in each position, repair or replace the shift indicator switch. Then go to the A/T Shift Signal Test.

COMPONENT EXAMPLE

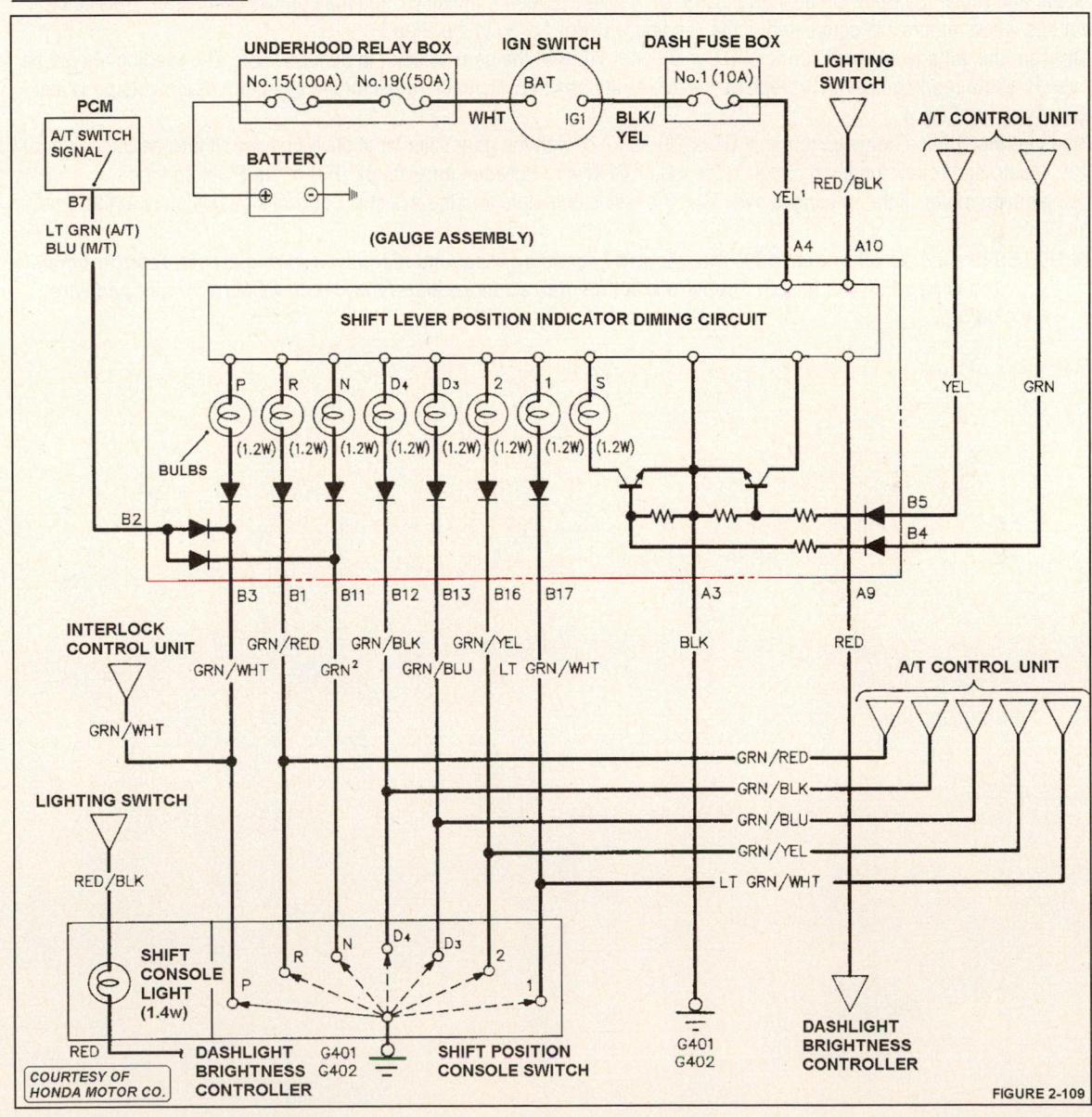

FIGURE 2-109

A/T SHIFT POSITION SIGNAL TEST

Turn the key off and connect the Honda test harness or BOB to the PCM and wiring harness to allow for circuit testing of components and the wire harness.

Leave the 'B' connector removed from the test harness, but connect all of the other connectors so that the PCM is connected to power and ground. This allows the A/T shift position signal to be isolated for testing. Measure the voltage between terminals B7 (B9 on some early Integra and Legend models) and A18. If the reading is over 10v, go to the Step 2. If the reading is under 10v, turn the key off and inspect the terminal B7 or B9 for a faulty connection at the PCM. If the connection is okay, substitute a known-good PCM and retest the voltage. If the reading is now over 10v, replace the original PCM and retest for the symptom.

Place the transmission gear selector in neutral position. Reconnect the 'B' connector to the test harness or BOB. Measure the voltage between terminal B7 or B9 and A18. The reading should be under 1v. If the reading is over 1v, locate and repair the open circuit in the LG or GRN wire between terminal B7 and the combination meter. Retest the voltage when repairs are completed. If the reading is under 1v, go to the Step 3.

Measure the voltage between terminal B7 or B9 and A18 with the gear selector in park position. The reading should be over 1v. If the reading is over 1v, replace the A/T shift position switch and retest for the symptom. If the voltage is less than 1v, go to Step 4.

Measure the voltage between terminal B7 or B9 and A18 with the gear selector in drive position. If the reading is under 10v, locate and repair the short circuit in the LG or GRN wire between terminal B7 (B9) at the PCM and the combination meter. If the reading is over 10v, the test is complete and the A/T shift position signal is okay at this time.

■ NOTE:*For models other than 1990 Integra and Legend applications discussed in this article, refer to other pin voltage charts for pin and wire color identification for all references to PCM terminals and wire colors.*

BRAKE SWITCH (INTEGRA & LEGEND MODELS)

The PCM uses the brake switch signal to determine when the brake pedal is depressed (i.e., a switch signal change from off to on).

DRIVEABILITY SYMPTOM DIAGNOSIS

A problem with the brake switch signal could cause the engine to stall or almost stall during deceleration conditions with the brake pedal depressed.

PRELIMINARY STEPS

Turn the key off and connect the Honda test harness or BOB between to the PCM and wiring harness to allow for proper circuit testing. Next, check the brake lights at the rear of the vehicle to determine if the lights are on all the time. If they are on all the time, repair the switch or circuit. When repairs are completed (or if the lights are off when they should be off), go to Step 1.

BRAKE SWITCH SIGNAL TEST (INTEGRA)

Depress the brake pedal and verify that the brake lights are on. If none of the brake lights are on, inspect the correct stop or horn fuse in the main fuse box. If the fuse is blown, locate and repair the short. Then go to Step 2.
If the fuse is okay, remove the brake switch 2-wire connector. Use a DVOM to check for continuity across the two switch terminals with the brake pedal depressed. If the meter reads no continuity, replace the faulty brake switch and then go to Step 3.
Measure the voltage between PCM terminals C10 and A18 (Figure 2-110). If the brake switch is working and properly adjusted, the DVOM will read battery voltage only with the brake pedal depressed. If the brake switch is okay, the test is finished and the test is completed. If the DVOM does not read battery voltage in this step with the pedal depressed, locate and repair the open signal circuit in the GRN/WHT wire between the brake switch and PCM terminal C10. Make repairs as needed and retest for the symptom.

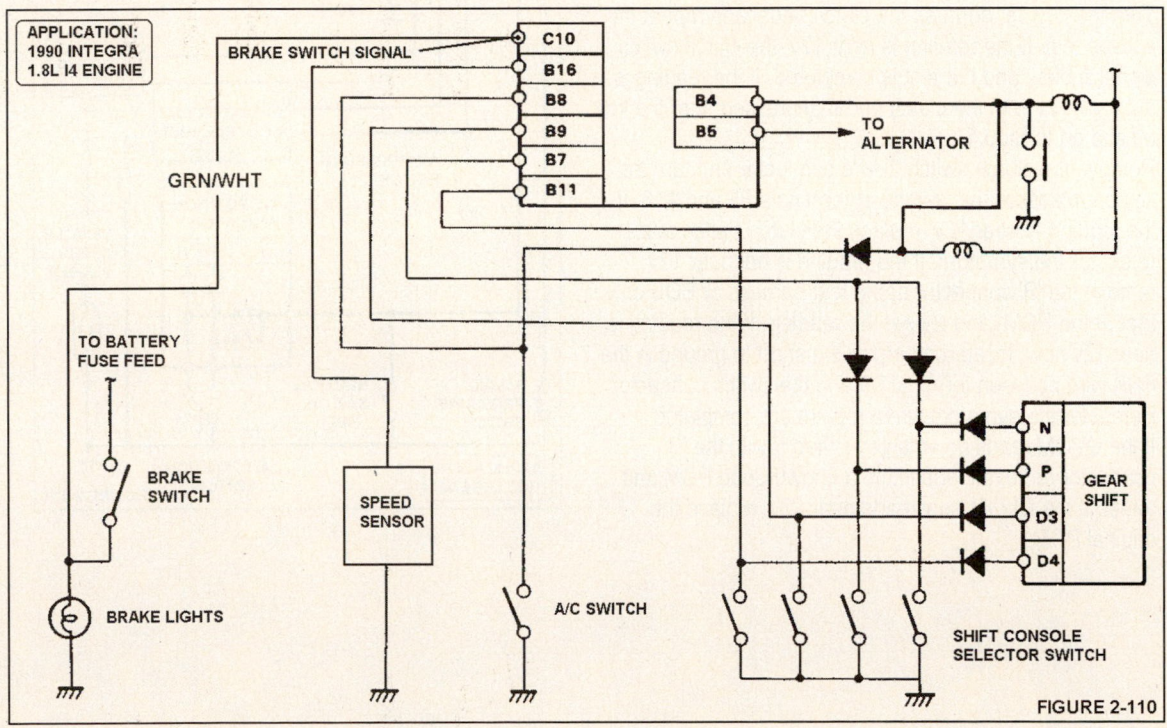

FIGURE 2-110

M/T CLUTCH SWITCH (LEGEND MODELS WITH M/T)

The PCM uses the clutch switch on Legend models with a M/T to determine when the clutch is depressed (i.e., a switch signal change from off to on).

DRIVEABILITY SYMPTOM DIAGNOSIS

A problem with the manual transaxle switch signal could cause the engine to stall or almost stall during conditions when the clutch is engaged in 1st gear.

PRELIMINARY STEPS

Turn the key off and connect the Honda test harness or BOB to the PCM and wiring harness to allow for circuit testing of components and the wire harness.

M/T CLUTCH SWITCH SIGNAL TEST

Turn the key on and engine off.

Measure the voltage between PCM terminals B9 and A18. If the reading is less than 1v, go to Step 4. If the reading is near 12v, turn the key off and remove the clutch switch 2-wire connector. Check for continuity across the switch terminals with the clutch pedal released. If the circuit is open, replace the clutch switch and retest for the symptom. If continuity exists, go to Step 3.

Turn the key on and measure the voltage between terminal B9 (PNK wire) and body ground (Figure 2-111). If the reading is near 12v, locate and repair the open ground circuit in the BLK wire between the switch and body ground. If the reading is not near 12v, repair the open circuit in the PNK wire between PCM terminal B9 and the switch connector. Retest for the symptom when repairs are completed.

Turn the key on, depress the clutch pedal and repeat the voltage test. If the reading is near 12v, the clutch switch signal is okay and the test is completed. If the reading is not near 12v with the clutch pedal depressed, turn the key off and go to Step 5.

Remove the clutch switch 2-wire connector and turn the key on. Measure the voltage at terminals B9 and A18. If the reading is near 12v, replace the clutch switch and retest for the symptom. If the reading is not near 12v, remove the 'B' connector at the test harness or BOB only (not at the PCM) and repeat the reading. If the reading is near 12v now, locate and repair the short to ground in the PNK wire between terminal B9 and the switch connector. Retest for the symptom when repairs are completed.

If the DVOM reads no voltage in Step 5 with the 'B' connector removed, substitute a known-good PCM and retest. If the DVOM now reads near 12v, replace the original PCM.

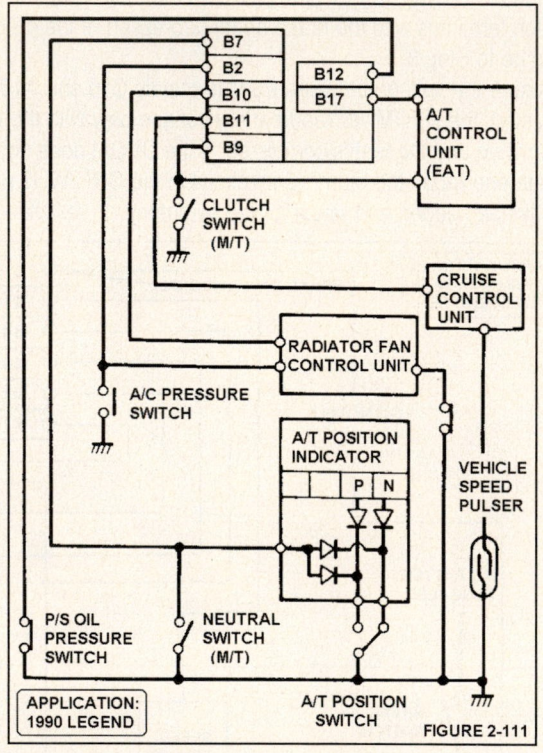

FIGURE 2-111

M/T NEUTRAL SWITCH (LEGEND MODELS WITH M/T)

The PCM uses the neutral switch on Legend models with a M/T to determine when the transaxle is in neutral (i.e., a switch signal change from off to on).

DRIVEABILITY SYMPTOM DIAGNOSIS

A problem with the manual transaxle neutral switch could cause the warm engine idle speed to be too high or too low when the transaxle is in neutral.

PRELIMINARY STEPS

Turn the key off and connect the Honda test harness or BOB to the PCM and wiring harness to allow for circuit testing of components and the wire harness.

M/T NEUTRAL SWITCH SIGNAL TEST

Turn the key on and engine off with the gear selector in neutral.

Measure the voltage between PCM terminals B7 and A18 (Figure 2-111). If the reading is not near 0v, go to Step 3. If the reading is near 0v, shift the transaxle into gear (engine off) and repeat the test. If the reading is not near 12v in gear, go to Step 4. If the reading is near 12v in gear, the M/T neutral switch is okay and the test is completed.

Remove the M/T neutral switch 2-wire connector. Use a jumper wire to connect the connector terminals (BLU wire to BLK wire). Repeat the voltage test from B7 to A18. If the reading is near 0v, locate and repair the open circuit in the BLU/LG wire between PCM terminal B7 and the switch connector or the open ground circuit in the BLK wire between the switch and body ground. Retest for the symptom when repairs are completed.

The transaxle should be in gear. Remove the 'B' connector from the test harness or BOB (not from the PCM). Repeat the voltage test from B7 to A18. If the reading remains near 0v, substitute a known-good PCM and repeat the test. If the reading is now near 12v, replace the original PCM.

If the reading is near 12v with the 'B' connector removed in Step 4, reconnect the 'B' connector to the test harness or BOB. Next, remove the M/T neutral switch 2-wire connector and repeat the voltage test. If the reading is near 12v, replace the M/T neutral switch and retest for the symptom. If the reading remains near 0v with the switch disconnected, locate and repair the short to ground in the BLU/LG wire between the PCM terminal B7 and the M/T neutral switch connector. Retest for the symptom when repairs are completed.

■ **NOTE:** *For models other than 1990 Legend applications discussed in this article, refer to other pin voltage charts for pin and wire color identification for all references to PCM terminals and wire colors.*

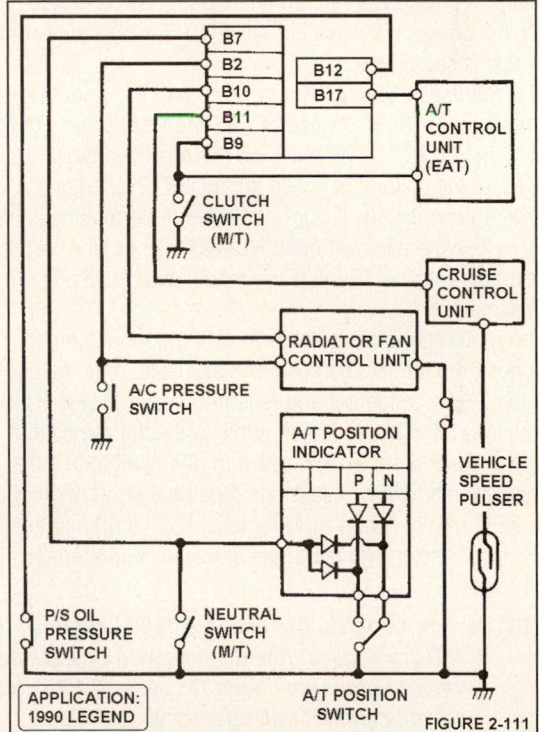

FIGURE 2-111

POWER STEERING SWITCH (ACURA & LEGEND MODELS)

The PCM uses the power steering (oil pressure) switch to determine when the power steering load is high (i.e., a switch signal change from off to on).

DRIVEABILITY SYMPTOM DIAGNOSIS

A problem with the power steering pressure switch could cause the engine idle speed to drop too low when the steering wheel is turned.

PRELIMINARY STEPS

Turn the key off and connect the Honda test harness or BOB to the PCM and wiring harness to allow for circuit testing of components and the wire harness.

POWER STEERING SWITCH SIGNAL TEST

Turn the key on and engine off.

Measure the voltage between PCM terminals B12 on Legend (B19 on Integra) and A18 (Figure 2-111). The reading should be less than 1v. If it is, go to Step 4. If the reading is near 12v, (the circuit is open), remove the P/S switch 2-wire connector.

Connect a jumper wire across the two switch terminals (RED wire to BLK wire) and repeat the voltage test. If the reading drops to less than 12v, replace the P/S pressure switch and retest for the symptom.

If the reading was less than 1v in Step 2, start the engine and slowly turn the steering wheel while monitoring the voltage reading between terminals B12 (B19) and A18. If the reading is near 12v while turning, the P/S with is okay and the test is completed.

If the reading in Step 4 was under 1v with the wheel turned, remove the PCM 'B' connector from the test harness or BOB (not from the PCM). To verify the PCM feed voltage, measure the voltage between terminal B12 (B19) and A18. If the reading is near 12v, go to Step 6. If the reading is not 12v, substitute a known-good PCM and repeat the test. If the reading is now 12v, replace the original PCM and retest the symptom.

If the reading was near 12v in Step 5, reconnect the 'B' connector to the test harness or BOB. Next, remove the P/S switch 2-wire connector and repeat the voltage test. If the reading is near 12v with the switch connector removed, replace the P/S switch and retest for the symptom. If the reading remains low, locate and repair the short to ground in the RED wire between PCM terminal B12 (B19) and the P/S switch connector. Retest for the symptom when repairs are completed.

■ **NOTE:** *For models other than the 1990 Legend and Integra applications discussed in this article, refer to other pin charts for pin and wire color identification for all references to PCM terminals and wire colors.*

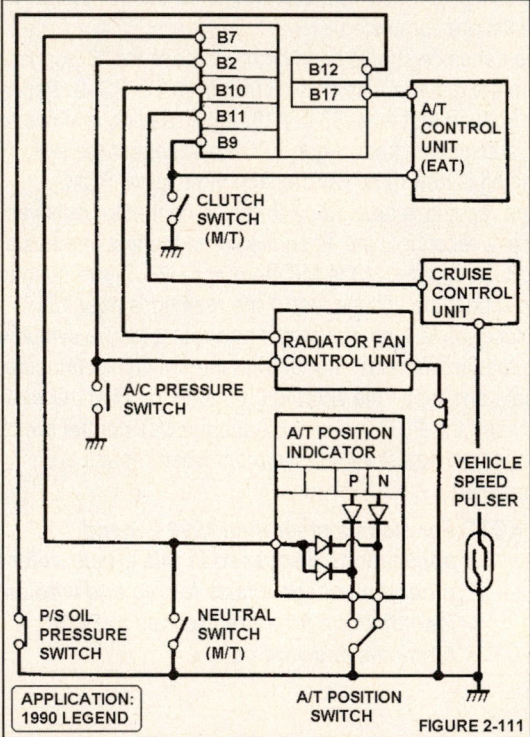

FIGURE 2-111

REFERENCE INFORMATION
INTEGRA PIN CHARTS
1993 Integra 1.7L I4 MFI VTEC VIN DB2 (All) 26-P 'A' Connector

PIN # / BOB #	COLOR	APPLICATION & ACRONYM	VALUE AT IDLE
A1	BRN	Injector No. 1	At hot idle: 2.0-3.3ms
A2	YEL	Injector No. 4	At hot idle: 2.0-3.3ms
A3	RED	Injector No. 2	At hot idle: 2.0-3.3ms
A4	GRN/YEL	VTEC Solenoid	At low rpm: <1v, at high rpm: 12v
A5	LT BLU	Injector No. 3	At hot idle: 2.0-3.3ms
A6	ORN/BLK	O2 Sensor Heater Control	HTRC On: <1v, Off: 12-14v
A7	GRN/BLK	Fuel Pump Relay Control	KOEO: 12v, KOER: <1v
A8	GRN/BLK	Fuel Pump Relay Control	KOEO: 12v, KOER: <1v
A9	BLU/YEL	Idle Air Control (IAC) Valve	At hot idle: 8-10v
A10-A12	---	---	---
A13	GRN/ORN	Check Engine Light	C/E Light On: <1v, Off: 12-14v
A14	---	---	---
A15	YEL	A/C Clutch Relay	A/C Relay On: <1v, Off: 12-14v
A16-A19	---	---	---
A20	GRN	Purge Cut-Off Solenoid	EVAP solenoid On: <1v, Off: 12-1v
A21	YEL/GRN	Igniter Control Signal	Idle: 10%, 2500 rpm: 60% d/cycle
A22	YEL/GRN	Igniter Control Signal	Idle: 10%, 2500 rpm: 60% d/cycle
A23, A24	BLK	Power Ground	<0.1v
A25	YEL/BLK	Ignition Power from Relay	12-14v
A26	BRN/BLK	Logic Ground	0.1v

1993 Integra 1.7L I4 MFI VTEC VIN DB2 (All) 16-P 'B' Connector

PIN # / BOB #	COLOR	APPLICATION & ACRONYM	VALUE AT IDLE
B1	YEL/BLK	Ignition Power from Relay	12-14v
B2	BRN/BLK	Logic Ground	<0.1v
B3, B4	---	---	---
B5	BLU/RED	A/C Pressure Switch	A/C & Blower On: 0.1v, Off: 12-14v
B6	---	---	---
B7	GRN	A/T Gear Position Indicator	In P/N: 0v, all others: 12-14v
B8	RED	Power Steering Pressure Switch	Wheel straight: 0v, turned: 12-14v
B9	BLU/WHT	Start Signal	KOEC (cranking): 9-11v
B10	YEL/RED	Vehicle Speed Sensor Signal	Moving at 50 mph: 60 Hz
B11	ORN	CYL Sensor (P)	KOER: 0.250mv (AC)
B12	WHT	CYL Sensor (M)	KOER: 0.250mv (AC)
B13	ORN/BLU	TDC Sensor (P)	KOER: 1.00v (AC)
B14	WHT/BLU	TDC Sensor (M)	KOER: 1.00v (AC)
B15	BLU/GRN	CKP Signal (P)	KOER: 0.900mv (AC)
B16	BLU/YEL	CKP Signal (M)	KOER: 0.900mv (AC)

■ NOTE: *When <1v is shown, this indicates the reading should be less than 1.0v DC.*

1993 Integra 1.7L I4 MFI VTEC VIN DB2 (All) 22-P 'C' Connector

PIN # / BOB #	COLOR	APPLICATION & ACRONYM	VALUE AT IDLE
D1	WHT/YEL	Keep Alive Power	12-14v
D2	GRN/WHT	Brake Switch Signal	Brake On: 12v, Off: 0v
D3	RED/BLU	Knock Sensor	No knock present: 2.5v
D4	BRN	Service Check Connector	Open: 4.80v, jumped: 0.1v
D5	---	---	---
D6	BLU/BLK	VTEC Pressure Switch	Low rpm: 0.1v, at high rpm: 12-14v
D7	LT BLU	Data Link Connector	---
D8	RED/WHT	TCM Signal	---
D9	BLU	Alternator 'FR' Signal	KOER: 2.5-3.5v (decreases w/load)
D10	---	---	---
D11	RED/BLU	Throttle Position (TP) Sensor	At hot idle: 0.49v
D12	---	---	---
D13	RED/WHT	Engine Coolant Temp. Sensor	At 180°F: 0.51v
D14	WHT	HO2S Signal	Hot idle: 0-1v, Accel: 0.5-1v
D15	RED/YEL	Intake Air Temperature Sensor	At 90-100°F: 1.3-1.5v
D16	BLU/WHT	TCM Signal	---
D17	WHT	MAP Sensor Signal	In P/N at idle: 0.9v (sea level)
D18	---	---	---
D19	YEL/RED	MAP Sensor VREF	4.9-5.1v
D20	YEL/WHT	Sensor VREF	4.9-5.1v
D21	GRN/WHT	MAP Sensor Ground	<0.1v
D22	GRN/WHT	Sensor Ground	<0.1v

■ **NOTE:** *When <1v is shown, this indicates the reading should be less than 1.0v DC.*

Standard Honda Colors and Abbreviations

Abbreviation	Color	Abbreviation	Color	Abbreviation	Color
BLK	Black	GRN	Green	PNK	Pink
BLU	Blue	LT BLU	Lt. Blue	TAN	Tan
BRN	Brown	LT GRN	Lt. Green	WHT	White
GRY	Gray	ORN	Orange	YEL	Yellow

26-PIN		16-PIN		22-PIN	
1 3 5 7 9 11 13 15 17 19 21 23 25		1 3 5 7 9 11 13 15		1 3 5 7 9 11 13 15 17 19 21	
2 4 6 8 10 12 14 16 18 20 22 24 26		2 4 6 8 10 12 14 16		2 4 6 8 10 12 14 16 18 20 22	
CONNECTOR 'A'		**CONNECTOR 'B'**		**CONNECTOR 'D'**	

WIRE SIDE OF HARNESS TERMINALS

1993 Integra 1.8L I4 MFI VIN DB1, DA9 (A/T) 26-P 'A' Connector

PIN # / BOB #	COLOR	APPLICATION & ACRONYM	VALUE AT IDLE
A1	BRN	Injector No. 1	At hot idle: 2.0-3.3ms
A2	YEL	Injector No. 4	At hot idle: 2.0-3.3ms
A3	RED	Injector No. 2	At hot idle: 2.0-3.3ms
A4	---	---	---
A5	LT BLU	Injector No. 3	At hot idle: 2.0-3.3ms
A6	ORN/BLK	O2 Sensor Heater Control	HTRC On: <1v, Off: 12-14v
A7	GRN/BLK	Fuel Pump Relay Control	KOEO: 12v, KOER: <1v
A8	GRN/BLK	Fuel Pump Relay Control	KOEO: 12v, KOER: <1v
A9	BLU/YEL	Idle Air Control (IAC) Motor	At hot idle: 8-10v
A10	GRN/YEL	Fuel Pressure Regulator Solenoid	FP solenoid On: <1v, Off: 12-14v
A11	RED	A/T: EGR Valve Lift Sensor	At hot idle: 1.2v
A12, A14	---	---	---
A13	GRN/ORN	Check Engine Light	C/E Light On: <1v, Off: 12-14v
A14	---	---	---
A15	YEL	A/C Clutch Relay	A/C Relay On: <1v, Off: 12-14v
A16-A19	---	---	---
A20	GRN	EVAP Purge Cutoff Solenoid	EVAP solenoid On: <1v, Off: 12-14v
A21	YEL/GRN	Igniter Control Signal	KOER: 12-14v
A22	YEL/GRN	Igniter Control Signal	KOER: 12-14v
A23, A24	BLK	Power Ground	<0.1v
A25	YEL/BLK	Ignition Power from Relay	12-14v
A26	BRN/BLK	Logic Ground	0.1v

1993 Integra 1.8L I4 MFI VIN DB1, DA9 (A/T) 16-P 'B' Connector

PIN # / BOB #	COLOR	APPLICATION & ACRONYM	VALUE AT IDLE
B1	YEL/BLK	Ignition Power from Relay	12-14v
B2	BRN/BLK	Power Ground	<0.1v
B3, B4, B6	---	---	---
B5	BLU/RED	A/C Pressure Switch	A/C & Blower On: 0.1v, Off: 12-14v
B7	GRN	A/T Park/Neutral Signal	In Neutral: 0v, all others: 12-14v
B8	RED	Power Steering Press. Switch	Wheel straight: 0v, turned: 12-14v
B9	BLU/WHT	Start Signal	KOEC (cranking): 9-11v
B10	YEL/RED	Vehicle Speed Sensor Signal	Moving at 50 mph: 60 Hz
B11	ORN	CYP Sensor (P)	KOER: 0.250mv (AC)
B12	WHT	CYP Sensor (M)	KOER: 0.250mv (AC)
B13	ORN/BLU	TDC Sensor (P)	KOER: 1.00v (AC)
B14	WHT/BLU	TDC Sensor (M)	KOER: 1.00v (AC)
B15	BLU/GRN	CKP Sensor (P)	KOER: 0.900mv (AC)
B16	BLU/YEL	CKP Sensor (M)	KOER: 0.900mv (AC)

■ **NOTE:** *When <1v is shown, this indicates the reading should be less than 1.0v DC.*

1993 Integra 1.8L I4 MFI VIN DB1, DA9 (A/T) 22-P 'D' Connector

PIN # / BOB #	COLOR	APPLICATION & ACRONYM	VALUE AT IDLE
D1	WHT/YEL	Keep Alive Power	12-14v
D2	GRN/WHT	Brake Switch Signal	Brake On: 12v, Off: 0v
D3, D5, D6	---	---	---
D4	BRN	Service Check Connector	Open: 4.80v, jumped: 0.1v
D7	LT BLU	Data Link Connector	---
D8	BLU	TCM Signal	---
D9	BLU	Alternator Control Signal	At idle: varies 7-9v
D10, D18	---	---	---
D11	RED/BLU	Throttle Position (TP) Sensor	At hot idle: 0.51v
D12	YEL	A/T: EGR Valve Lift Sensor	At hot idle: 1.2v
D13	RED/WHT	Engine Coolant Temp. Sensor	At 180°F: 0.53v
D14	WHT	O2 Sensor Heater Control	HTRC On: <1v, Off: 12-14v
D15	RED/YEL	Intake Air Temperature Sensor	At 90-100°F: 1.3v
D16	BLU/WHT	A/T Control Unit	Open: 12v, Closed: 0v
D17	WHT	MAP Sensor Signal	At idle in P/N: 0.9v (sea level)
D19	YEL/RED	MAP Sensor VREF	4.9-5.1v
D20	YEL/WHT	Sensor VREF	4.9-5.1v
D21	GRN/WHT	MAP Sensor Ground	<0.1v
D22	GRN/WHT	Sensor Ground	<0.1v

■ **NOTE:** *When <1v is shown, this indicates the reading should be less than 1.0v DC.*

Standard Honda Colors and Abbreviations

Abbreviation	Color	Abbreviation	Color	Abbreviation	Color
BLK	Black	GRN	Green	PNK	Pink
BLU	Blue	LT BLU	Lt. Blue	TAN	Tan
BRN	Brown	LT GRN	Lt. Green	WHT	White
GRY	Gray	ORN	Orange	YEL	Yellow

26-PIN

1	3	5	7	9	11	13	15	17	19	21	23	25
2	4	6	8	10	12	14	16	18	20	22	24	26

CONNECTOR 'A'

16-PIN

1	3	5	7	9	11	13	15
2	4	6	8	10	12	14	16

CONNECTOR 'B'

22-PIN

1	3	5	7	9	11	13	15	17	19	21
2	4	6	8	10	12	14	16	18	20	22

CONNECTOR 'D'

WIRE SIDE OF HARNESS TERMINALS

REFERENCE INFORMATION

Integra Wiring Diagrams - 1993 Integra (Part One)

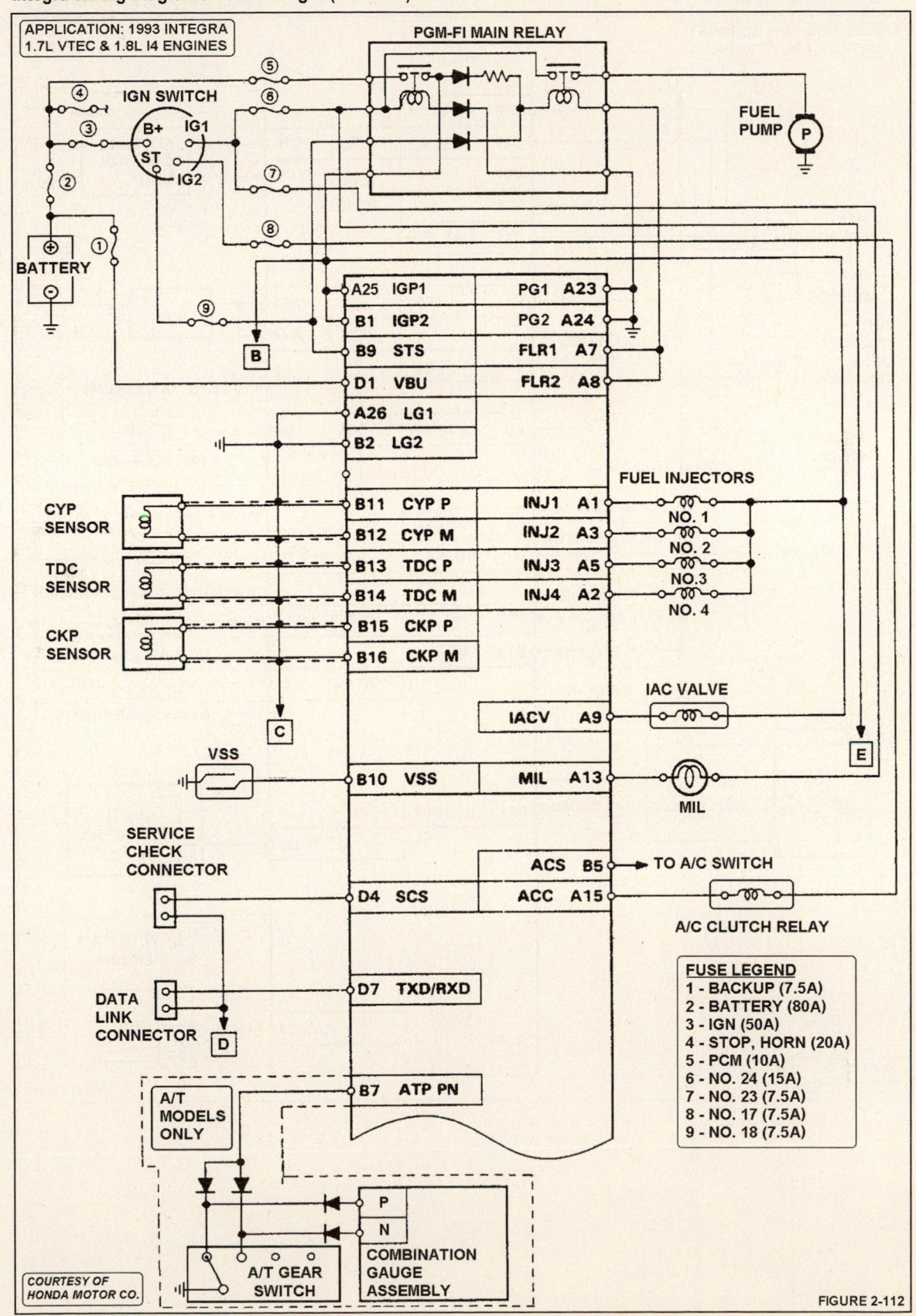

FIGURE 2-112

Integra Wiring Diagrams - 1993 Integra (Part Two)

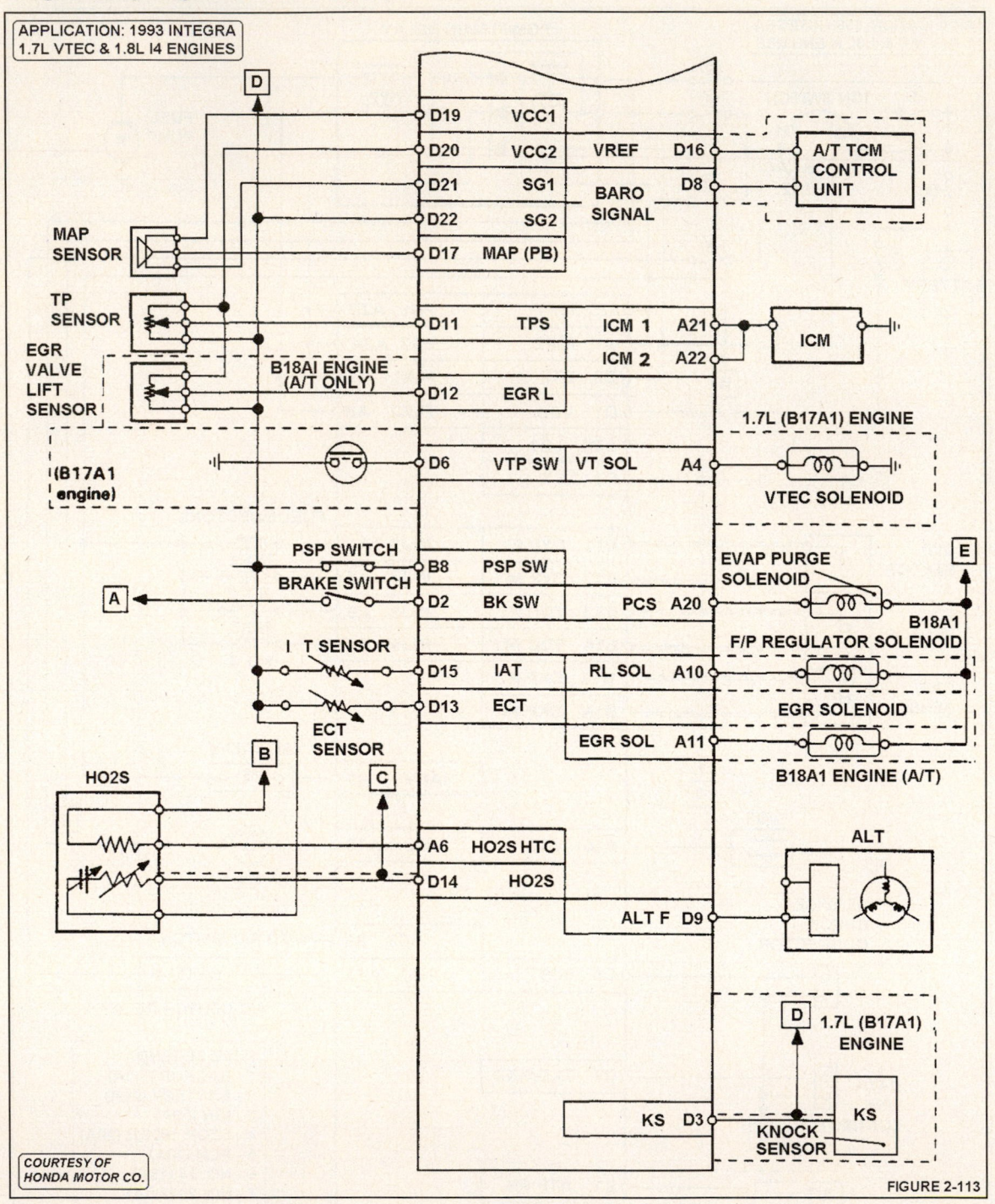

APPLICATION: 1993 INTEGRA
1.7L VTEC & 1.8L I4 ENGINES

COURTESY OF HONDA MOTOR CO.

FIGURE 2-113

REFERENCE INFORMATION

Integra Vacuum Diagram - 1993 Integra

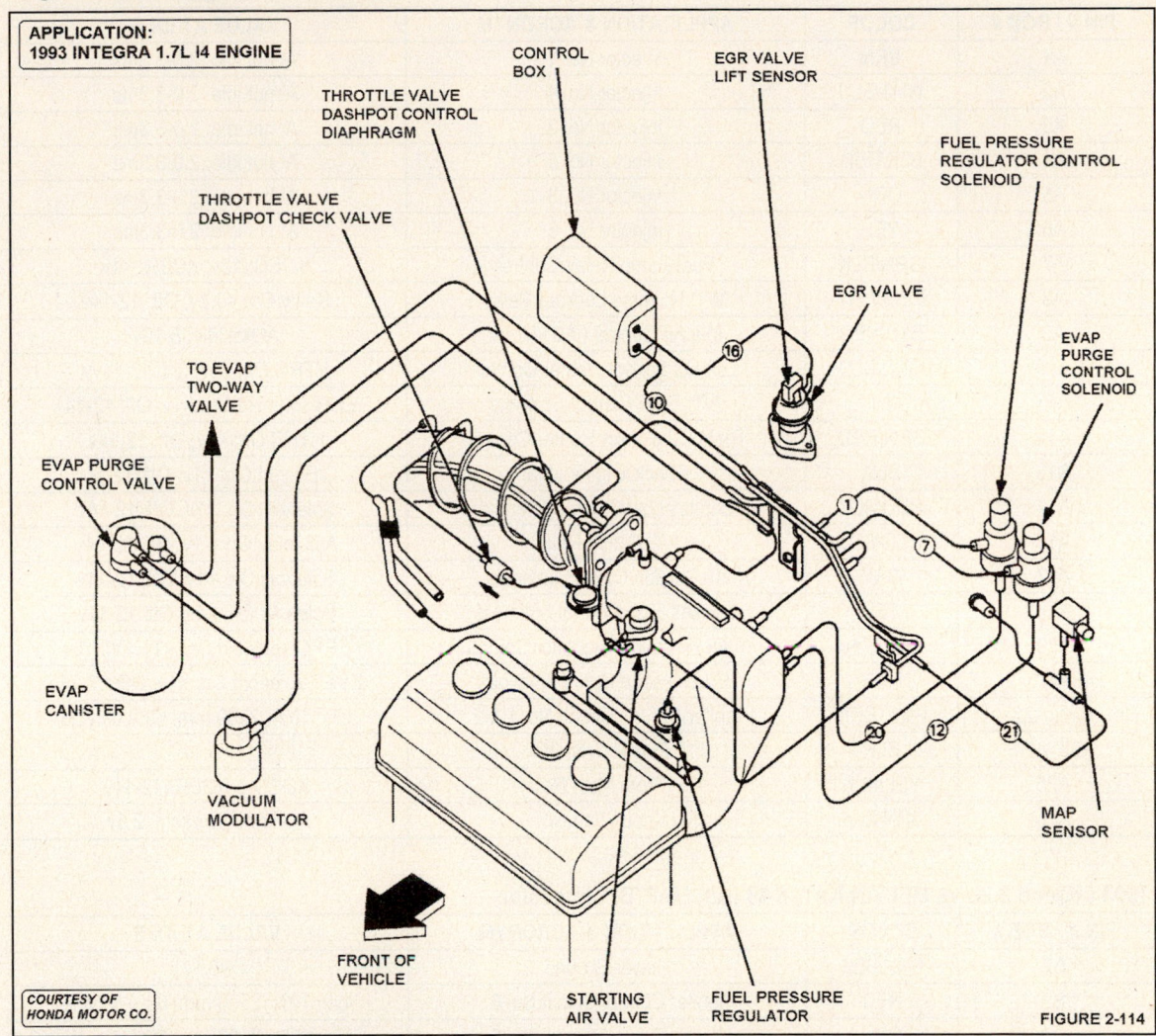

APPLICATION:
1993 INTEGRA 1.7L I4 ENGINE

CONTROL
BOX

EGR VALVE
LIFT SENSOR

THROTTLE VALVE
DASHPOT CONTROL
DIAPHRAGM

FUEL PRESSURE
REGULATOR CONTROL
SOLENOID

THROTTLE VALVE
DASHPOT CHECK VALVE

EGR VALVE

EVAP
PURGE
CONTROL
SOLENOID

TO EVAP
TWO-WAY
VALVE

EVAP PURGE
CONTROL VALVE

EVAP
CANISTER

VACUUM
MODULATOR

MAP
SENSOR

FRONT OF
VEHICLE

STARTING
AIR VALVE

FUEL PRESSURE
REGULATOR

COURTESY OF
HONDA MOTOR CO.

FIGURE 2-114

Legend Pin Charts

1993 Legend 3.2L V6 MFI VIN KA7, KA8 (All) 26-P 'A' Connector

PIN # / BOB #	COLOR	APPLICATION & ACRONYM	VALUE AT IDLE
A1	BRN	Injector No. 1	At hot idle: 2.0-3.3ms
A2	WHT/BLU	Injector No. 4	At hot idle: 2.0-3.3ms
A3	RED	Injector No. 2	At hot idle: 2.0-3.3ms
A4	BLK/RED	Injector No. 5	At hot idle: 2.0-3.3ms
A5	ORN	Injector No. 3	At hot idle: 2.0-3.3ms
A6	YEL	Injector No. 6	At hot idle: 2.0-3.3ms
A7	GRN/BLK	Fuel Pump Relay Control	KOEO: 12v, KOER: <1v
A8	RED/WHT	M/T: Reverse Lockout Relay	Relay On: <0.1v, Off: 12-14v
A9	BLU/RED	Idle Air Control (IAC) Motor	At hot idle: 8-10v
A10	GRN/BLU	Left O2 Sensor Heater Control	HTRC On: <1v, Off: 12-14vv
A11	WHT	A/T: EGR Control Solenoid	EGR solenoid On: <1v, Off: 12-14v
A12	GRN/RED	Right O2 Sensor Heater Control	HTRC On: <1v, Off: 12-14v
A13	BLU	Check Engine Light	C/E Light On: <1v, Off: 12-14v
A14	RED/BLU	Bypass Low Control Solenoid	Solenoid On: <1v, Off: 12-14v
A15	RED/BLU	A/C Clutch Relay	A/C relay On: <1v, Off: 12-14v
A17	GRY	Air Suction Control Solenoid	Solenoid On: <1v, Off: 12-14v
A18	YEL	Bypass High Control Solenoid	Solenoid On: <1v, Off: 12-14v
A19	LT GRN	Fuel Pressure Regulator Control	FPRC solenoid On: <1v, Off: 12v
A20	LT GRN	EVAP Purge Cutoff Solenoid	EVAP solenoid On: <1v, Off: 12-14v
A21, A22	PNK, BRN	Igniter Control Circuit No. 1 & 2	Idle: 10%, 2500 rpm: 60% d/cycle
A23, A24	BLK	Power Ground	<0.1v
A25	YEL/BLK	B+ From Main Relay	KOEO & KOER: 12-14v
A26	BRN/BLK	Knock Sensor	No knock present: 2.5v

1993 Legend 3.2L V6 MFI VIN KA7, KA8 (All) 16-P 'B' Connector

PIN # / BOB #	COLOR	APPLICATION & ACRONYM	VALUE AT IDLE
B2	BRN/BLK	Power Ground	<0.1v
B3	RED	Igniter Control Circuit No. 6	Idle: 10%, 2500 rpm: 60% d/cycle
B4	GRN	Igniter Control Circuit No. 5	Idle: 10%, 2500 rpm: 60% d/cycle
B5	GRN	Power Steering Press. Switch	Wheel straight: 0v, turned: 12-14v
B6	GRN	Igniter Control Circuit No. 4	Idle: 10%, 2500 rpm: 60% d/cycle
B7	LT GRN	M/T Neutral Switch	In Neutral: 0v, all others: 12-14v
B8	BLU	Igniter Control Circuit No. 3	Idle: 10%, 2500 rpm: 60% d/cycle
B9	BLU/GRN	CYL Sensor No. 2	KOER: 0.250mv (AC)
B10	BLU/YEL	CYL Sensor No. 2	KOER: 0.250mv (AC)
B11	ORN/BLU	CYL Sensor No. 1	KOER: 0.250mv (AC)
B12	WHT/BLU	CYL Sensor No. 1	KOER: 0.250mv (AC)
B13	ORN	CKP Signal No. 2	KOER: 0.900mv (AC)
B14	WHT	CKP Signal No. 2	KOER: 0.900mv (AC)
B15	ORN/BLU	CKP Signal No. 1	KOER: 0.900mv (AC)
B16	WHT/BLU	CKP Signal No. 1	KOER: 0.900mv (AC)

Legend Pin Charts

1993 Legend 3.2L V6 MFI VIN KA7, KA8 (All) 12-P 'C' Connector

Pin # / BOB #	Color	Application & Acronym	Value at Idle
C1	YEL/BLK	B+ From Main Relay	KOEO & KOER: 12-14v
C2	YEL/RED	Vehicle Speed Sensor	Moving at 50 mph: 60 Hz
C3	BLU/BLK	Cooling Fan Switch	Fan switch closed: 0v, open: 12-14v
C4	BLU	Tachometer Signal	KOER: 12-14v
C5	RED/BLU	A/C Pressure Switch	A/C & Blower On: 0.1v, Off: 12-14v
C6, C8, C10	---	---	---
C7	PNK	M/T Clutch Switch	In Neutral: 0v, all others: 12-14v
C9	WHT	Service Check Connector	Open: 4.80v, jumped: 0.1v
C11	BLK/WHT	Start Signal	KOEC (cranking): 9-11v
C12	---	---	---

1993 Legend 3.2L V6 MFI VIN KA7, KA8 (All) 22-P 'D' Connector

Pin # / BOB #	Color	Application & Acronym	Value at Idle
D1	YEL/BLU	Keep Alive Power	12-14v
D2	GRN/WHT	Brake Switch Signal	Brake On: 12v, Off: 0v
D3	WHT	Right Knock Sensor	No knock present: 2.5v
D4	RED/BLU	Left Knock Sensor	No knock present: 2.5v
D5, D6, D7	---	---	---
D8	BLU/YEL	Ignition Timing Adjustment	KOER: 0.4-4.5v
D9	WHT/RED	Alternator 'FR' Signal	KOER: 2.5-3.5v (decreases w/load)
D10	---	---	---
D11	RED/BLU	Throttle Position (TP) Sensor	At hot idle: 0.51v
D12	BLK/WHT	EGR Valve Lift Sensor (A/T)	At hot idle: 1.2v
D13	RED/WHT	Engine Coolant Temp. Sensor	At 180°F: 0.52v
D14	WHT	Left (Front) Oxygen Sensor	Hot idle: 0-1v, Accel: 0.5-1.0v
D15	RED/YEL	Intake Air Temperature Sensor	At 90-100°F: 1.3-1.5v
D16	RED/BLU	Right (Rear) Oxygen Sensor	Hot idle: 0-1v, Accel: 0.5-1.0v
D17	RED	Manifold Absolute Pressure Sensor	At idle in P/N: 0.9v (sea level)
D18	---	---	---
D19	YEL/WHT	MAP Sensor VREF	4.9-5.1v
D20	YEL/WHT	Sensor VREF	4.9-5.1v
D21	GRN/WHT	MAP Sensor Ground	<0.1v
D22	GRN/WHT	Sensor Ground	<0.1v

■ NOTE: *When <1v is shown, this indicates the reading should be less than 1.0v DC.*

26-PIN

| 1 | 3 | 5 | 7 | 9 | 11 | 13 | 15 | 17 | 19 | 21 | 23 | 25 |
| 2 | 4 | 6 | 8 | 10 | 12 | 14 | 16 | 18 | 20 | 22 | 24 | 26 |

CONNECTOR 'A'

16-PIN

| 1 | 3 | 5 | 7 | 9 | 11 | 13 | 15 |
| 2 | 4 | 6 | 8 | 10 | 12 | 14 | 16 |

CONNECTOR 'B'

12-PIN

| 1 | 3 | 5 | 7 | 9 | 11 |
| 2 | 4 | 6 | 8 | 10 | 12 |

CONNECTOR 'C'

22-PIN

| 1 | 3 | 5 | 7 | 9 | 11 | 13 | 15 | 17 | 19 | 21 |
| 2 | 4 | 6 | 8 | 10 | 12 | 14 | 16 | 18 | 20 | 22 |

CONNECTOR 'D'

WIRE SIDE OF HARNESS TERMINALS

Legend Wiring Diagrams - 1993 Legend (Part One)

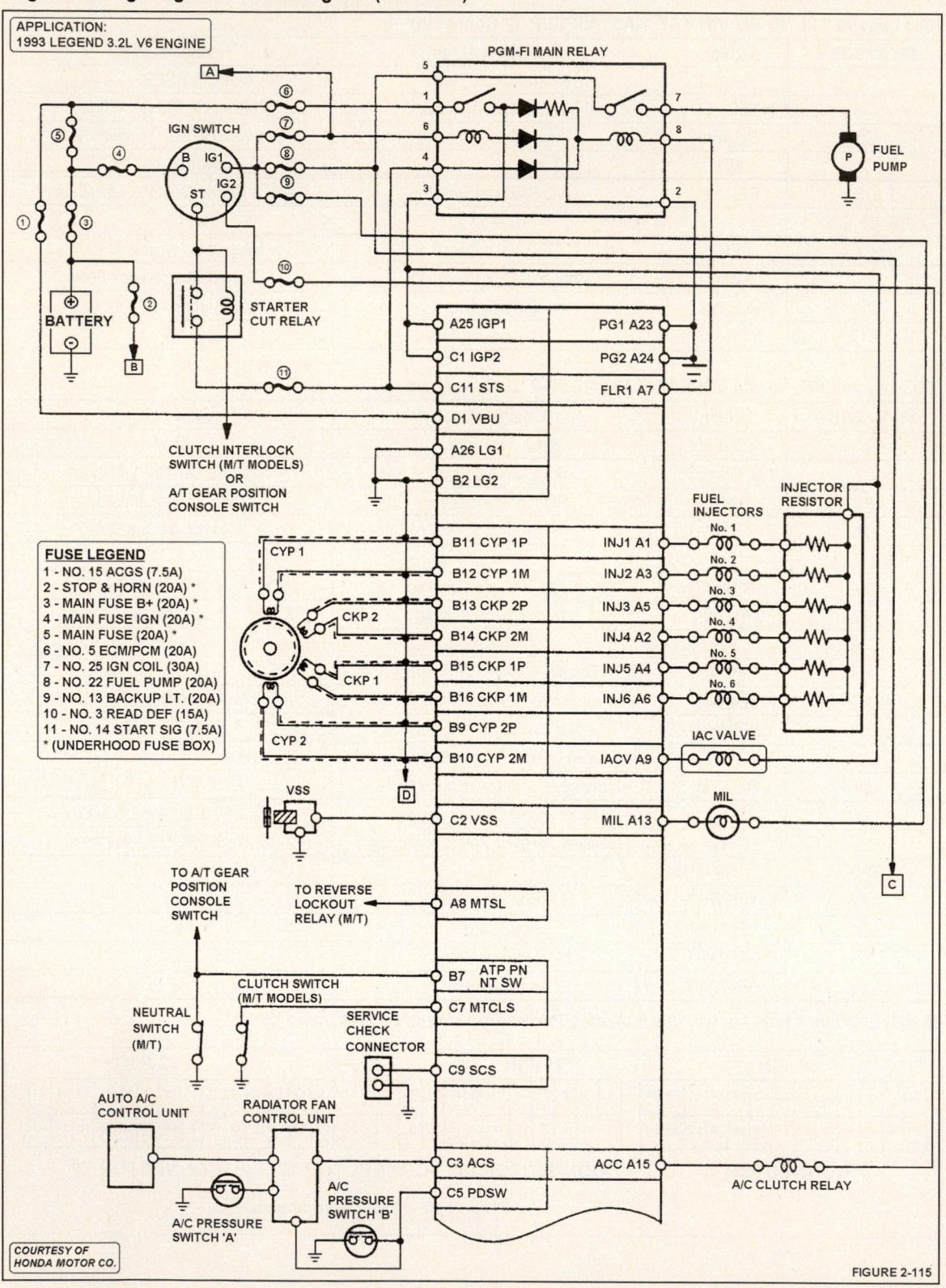

APPLICATION:
1993 LEGEND 3.2L V6 ENGINE

PGM-FI MAIN RELAY

IGN SWITCH

FUEL PUMP

BATTERY

STARTER CUT RELAY

CLUTCH INTERLOCK SWITCH (M/T MODELS) OR A/T GEAR POSITION CONSOLE SWITCH

FUSE LEGEND
1 - NO. 15 ACGS (7.5A)
2 - STOP & HORN (20A) *
3 - MAIN FUSE B+ (20A) *
4 - MAIN FUSE IGN (20A) *
5 - MAIN FUSE (20A) *
6 - NO. 5 ECM/PCM (20A)
7 - NO. 25 IGN COIL (30A)
8 - NO. 22 FUEL PUMP (20A)
9 - NO. 13 BACKUP LT. (20A)
10 - NO. 3 READ DEF (15A)
11 - NO. 14 START SIG (7.5A)
* (UNDERHOOD FUSE BOX)

A25 IGP1 — PG1 A23
C1 IGP2 — PG2 A24
C11 STS — FLR1 A7
D1 VBU
A26 LG1
B2 LG2

B11 CYP 1P — INJ1 A1
B12 CYP 1M — INJ2 A3
B13 CKP 2P — INJ3 A5
B14 CKP 2M — INJ4 A2
B15 CKP 1P — INJ5 A4
B16 CKP 1M — INJ6 A6
B9 CYP 2P
B10 CYP 2M — IACV A9

CYP 1
CKP 2
CKP 1
CYP 2

FUEL INJECTORS
No. 1
No. 2
No. 3
No. 4
No. 5
No. 6

INJECTOR RESISTOR

IAC VALVE

VSS
C2 VSS — MIL A13

MIL

TO A/T GEAR POSITION CONSOLE SWITCH

TO REVERSE LOCKOUT RELAY (M/T)
A8 MTSL

NEUTRAL SWITCH (M/T)

CLUTCH SWITCH (M/T MODELS)

B7 ATP PN NT SW
C7 MTCLS

SERVICE CHECK CONNECTOR
C9 SCS

AUTO A/C CONTROL UNIT

RADIATOR FAN CONTROL UNIT

C3 ACS — ACC A15
C5 PDSW

A/C PRESSURE SWITCH 'A'
A/C PRESSURE SWITCH 'B'

A/C CLUTCH RELAY

FIGURE 2-115

Legend Wiring Diagrams - 1993 Legend (Part Two)

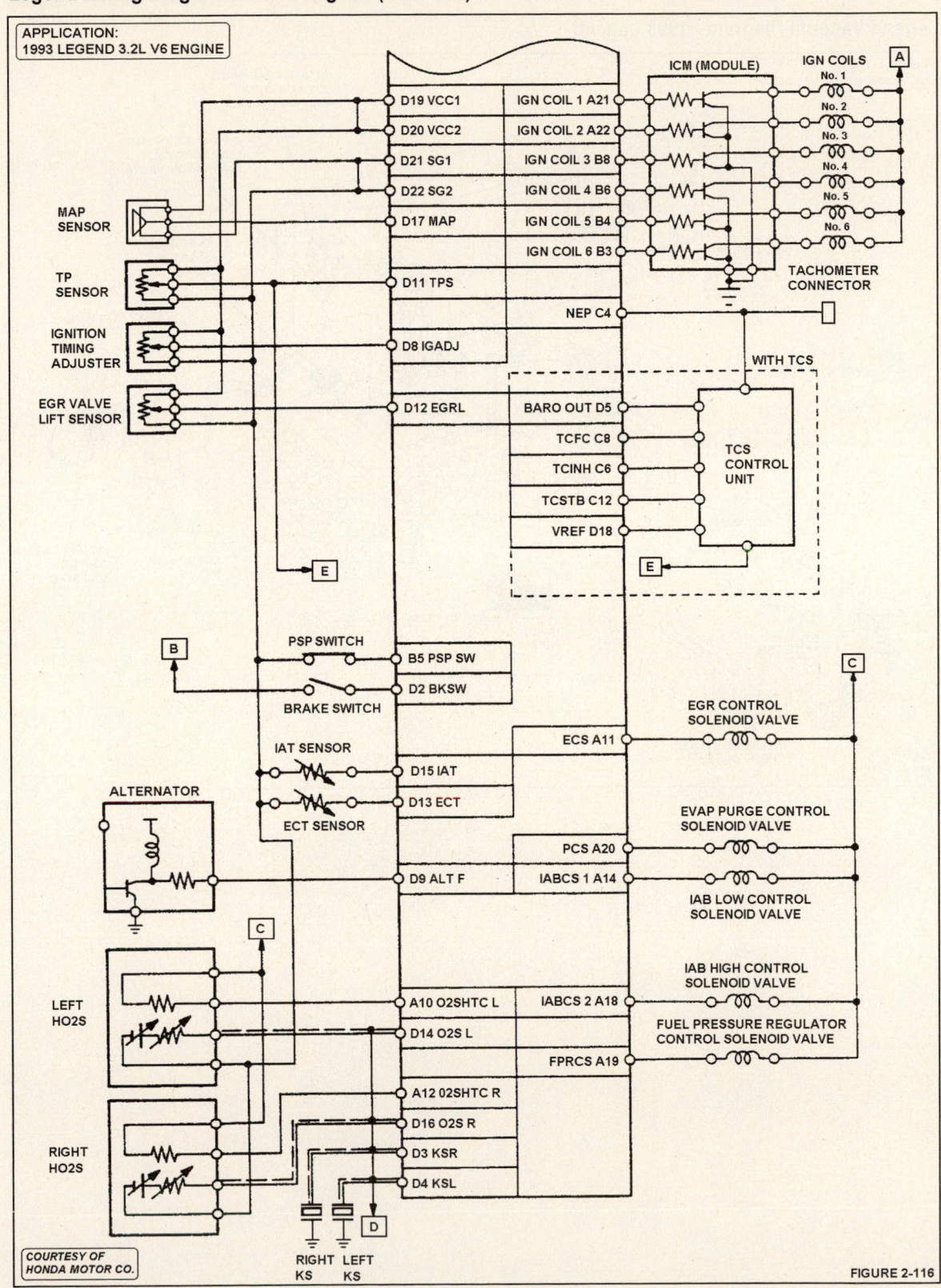

APPLICATION:
1993 LEGEND 3.2L V6 ENGINE

COURTESY OF
HONDA MOTOR CO.

FIGURE 2-116

Legend Vacuum Diagram - 1993 Legend

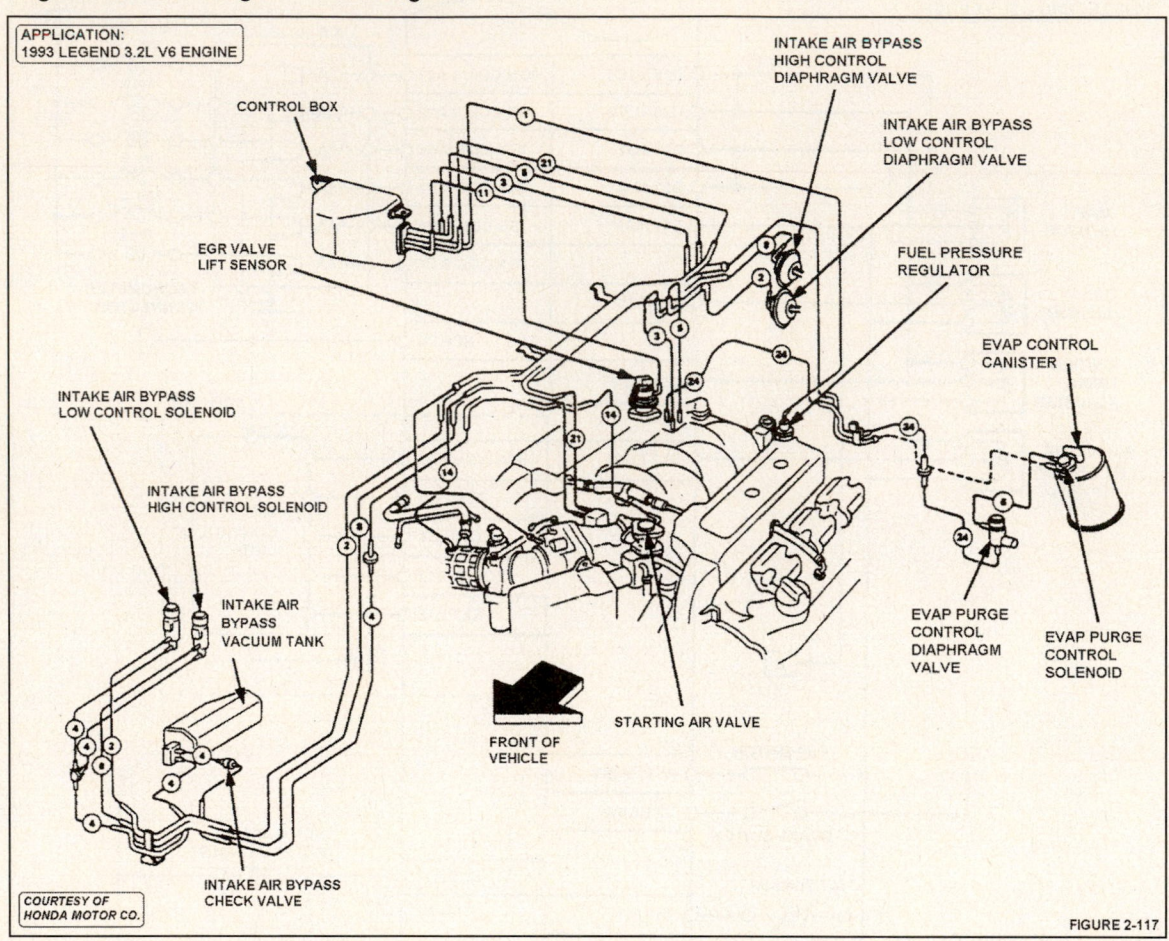

APPLICATION:
1993 LEGEND 3.2L V6 ENGINE

CONTROL BOX

INTAKE AIR BYPASS
HIGH CONTROL
DIAPHRAGM VALVE

INTAKE AIR BYPASS
LOW CONTROL
DIAPHRAGM VALVE

EGR VALVE
LIFT SENSOR

FUEL PRESSURE
REGULATOR

EVAP CONTROL
CANISTER

INTAKE AIR BYPASS
LOW CONTROL SOLENOID

INTAKE AIR BYPASS
HIGH CONTROL SOLENOID

INTAKE AIR
BYPASS
VACUUM TANK

EVAP PURGE
CONTROL
DIAPHRAGM
VALVE

EVAP PURGE
CONTROL
SOLENOID

STARTING AIR VALVE

FRONT OF
VEHICLE

INTAKE AIR BYPASS
CHECK VALVE

COURTESY OF
HONDA MOTOR CO.

FIGURE 2-117

Vigor Pin Charts

1993 Vigor 2.5L I5 MFI VIN CC2 (A/T) 26-P 'A' Connector

PIN # / BOB #	COLOR	APPLICATION & ACRONYM	VALUE AT IDLE
A1	BRN	Injector No. 1	At hot idle: 2.0-3.3ms
A2	YEL	Injector No. 4	At hot idle: 2.0-3.3ms
A3	RED	Injector No. 2	At hot idle: 2.0-3.3ms
A4	GRN	Injector No. 5	At hot idle: 2.0-3.3ms
A5	BLU	Injector No. 3	At hot idle: 2.0-3.3ms
A6	PNK/WHT	O2 Sensor Heater Control	HTRC On: <1v, Off: 12-14v
A7	GRN/BLK	Fuel Pump Relay Control	KOEO: 12v, KOER: <1v
A8	GRN/BLK	Fuel Pump Relay Control	KOEO: 12v, KOER: <1v
A9	BLK/BLU	Idle Air Control (IAC) Motor	At hot idle: 8-10v
A10	ORN	EVAP Purge Cutoff Solenoid	EVAP solenoid On: <1v, Off: 12-14v
A11	RED	A/T: EGR Control Solenoid	EGR Solenoid On: <1v, Off: 12-14v
A12	GRN/YEL	Fan Timing Unit Signal	---
A13	GRN/RED	Check Engine Light	C/E Light On: <1v, Off: 12-14v
A14	---	---	---
A15	RED/BLU	A/C Clutch Relay	A/C relay On: <1v, Off: 12-14v
A16	WHT/GRN	Alternator Control Signal	At idle: varies 7-9v
A17	BLU/RED	Bypass Solenoid Signal	Solenoid On: <1v, Off: 12-14v
A18	BLK/RED	Power Ground	<0.1v
A19, A20	---	---	---
A21	YEL/GRN	Igniter Control Circuit No. 1	Idle: 10%, 2500 rpm: 60% d/cycle
A22	YEL/GRN	Igniter Control Circuit No. 2	Idle: 10%, 2500 rpm: 60% d/cycle
A23, A24	BLK	Power Ground	<0.1v
A25	YEL/BLK	Ignition Power	12-14v
A26	BLK/RED	Shield Ground	<0.1v

1993 Vigor 2.5L I5 MFI VIN CC2 (A/T) 16-P 'B' Connector

PIN # / BOB #	COLOR	APPLICATION & ACRONYM	VALUE AT IDLE
B1	YEL/BLK	Ignition Power	12-14v
B2	BRN/BLK	Power Ground	<0.1v
B3	WHT/GRN	A/T Control Unit Signal	---
B4	WHT/RED	A/T Control Unit Signal	---
B5	BLU/BLK	A/C Pressure Switch	A/C & Blower On: 0.1v, Off: 12-14v
B6	---	---	---
B7	YEL/GRN	A/T Neutral Indicator	In Neutral: 0v, all others: 12-14v
B8	RED	Power Steering Press. Switch	Wheel straight: 0v, turned: 12-14v
B9	BLU/RED	Start Signal	KOEC (cranking): 9-11v
B10	ORN	Vehicle Speed Sensor Signal	Moving at 50 mph: 60 Hz
B11	ORN	CYL Sensor (P)	KOER: 0.250mv (AC)
B12	WHT	CYL Sensor (M)	KOER: 0.250mv (AC)
B13	ORN/BLU	TDC Sensor (P)	KOER: 1.00v (AC)
B14	WHT/BLU	TDC Sensor (M)	KOER: 1.00v (AC)
B15	YEL/GRN	CKP Signal (P)	KOER: 0.900mv (AC)
B16	BLU/YEL	CKP Signal (M)	KOER: 0.900mv (AC)

■ NOTE: *When <1v is shown, this indicates the reading should be less than 1.0v DC.*

Vigor Pin Charts

1993 Vigor 2.5L I5 MFI VIN CC2 (A/T) 12-P Connector

PIN # / BOB #	COLOR	APPLICATION & ACRONYM	VALUE AT IDLE
C1	ORN/WHT	Rear Knock Sensor	No knock present: 2.5v
C2	---	---	---
C3	RED/BLU	Front Knock Sensor	No knock present: 2.5v
C4-C12	---	---	---

1993 Vigor 2.5L I5 MFI VIN CC2 (A/T) 22-P Connector

PIN # / BOB #	COLOR	APPLICATION & ACRONYM	VALUE AT IDLE
D1	WHT/GRN	Keep Alive Power	12-14v
D2	GRN/WHT	Brake Switch Signal	Brake On: 12v, Off: 0v
D3	---	---	---
D4	BRN	Service Check Connector	Open: 4.80v, jumped: 0.1v
D5, D6, D7	---	---	---
D8	BRN	Ignition Timing Adjustment	KOER: 0.4-4.5v
D9	WHT/RED	Alternator 'FR' Signal	KOER: varies 1.5-3.5 w/load
D10	GRN/RED	Electronic Load Detector	KOER: varies 0.5-4.5v
D11	RED/YEL	Throttle Position (TP) Sensor	At hot idle: 0.49v
D12	WHT/BLK	A/T: EGR Valve Lift Sensor	At hot idle: 1.2v
D13	YEL/GRN	Engine Coolant Temp. Sensor	At 180°F: 0.52v
D14	WHT	Heated Oxygen Sensor	Hot idle: 0-1v, Accel: 0.5-1.0v
D15	WHT/YEL	Intake Air Temperature Sensor	At 90-100°F: 1.3-1.5v
D16	---	---	---
D17	WHT/BLU	Manifold Absolute Pressure Sensor	At idle in P/N: 0.9v (sea level)
D18	BLU/WHT	A/T Control Unit VREF	---
D19	YEL/WHT	MAP Sensor VREF	4.9-5.1v
D20	YEL/WHT	Sensor VREF	4.9-5.1v
D21	GRN/WHT	MAP Sensor Ground	<0.1v
D22	GRN/WHT	Sensor Ground	<0.1v

■ NOTE: *When <1v is shown, this indicates the reading should be less than 1.0v DC.*

Standard Honda Colors and Abbreviations

Abbreviation	Color	Abbreviation	Color	Abbreviation	Color
BLK	Black	GRN	Green	PNK	Pink
BLU	Blue	LT BLU	Lt. Blue	TAN	Tan
BRN	Brown	LT GRN	Lt. Green	WHT	White
GRY	Gray	ORN	Orange	YEL	Yellow

26-PIN

| 1 | 3 | 5 | 7 | 9 | 11 | 13 | 15 | 17 | 19 | 21 | 23 | 25 |
| 2 | 4 | 6 | 8 | 10 | 12 | 14 | 16 | 18 | 20 | 22 | 24 | 26 |

CONNECTOR 'A'

16-PIN

| 1 | 3 | 5 | 7 | 9 | 11 | 13 | 15 |
| 2 | 4 | 6 | 8 | 10 | 12 | 14 | 16 |

CONNECTOR 'B'

12-PIN

| 1 | 3 | 5 | 7 | 9 | 11 |
| 2 | 4 | 6 | 8 | 10 | 12 |

CONNECTOR 'C'

22-PIN

| 1 | 3 | 5 | 7 | 9 | 11 | 13 | 15 | 17 | 19 | 21 |
| 2 | 4 | 6 | 8 | 10 | 12 | 14 | 16 | 18 | 20 | 22 |

CONNECTOR 'D'

WIRE SIDE OF HARNESS TERMINALS

REFERENCE INFORMATION

Vigor Wiring Diagrams - 1993 Vigor (Part One)

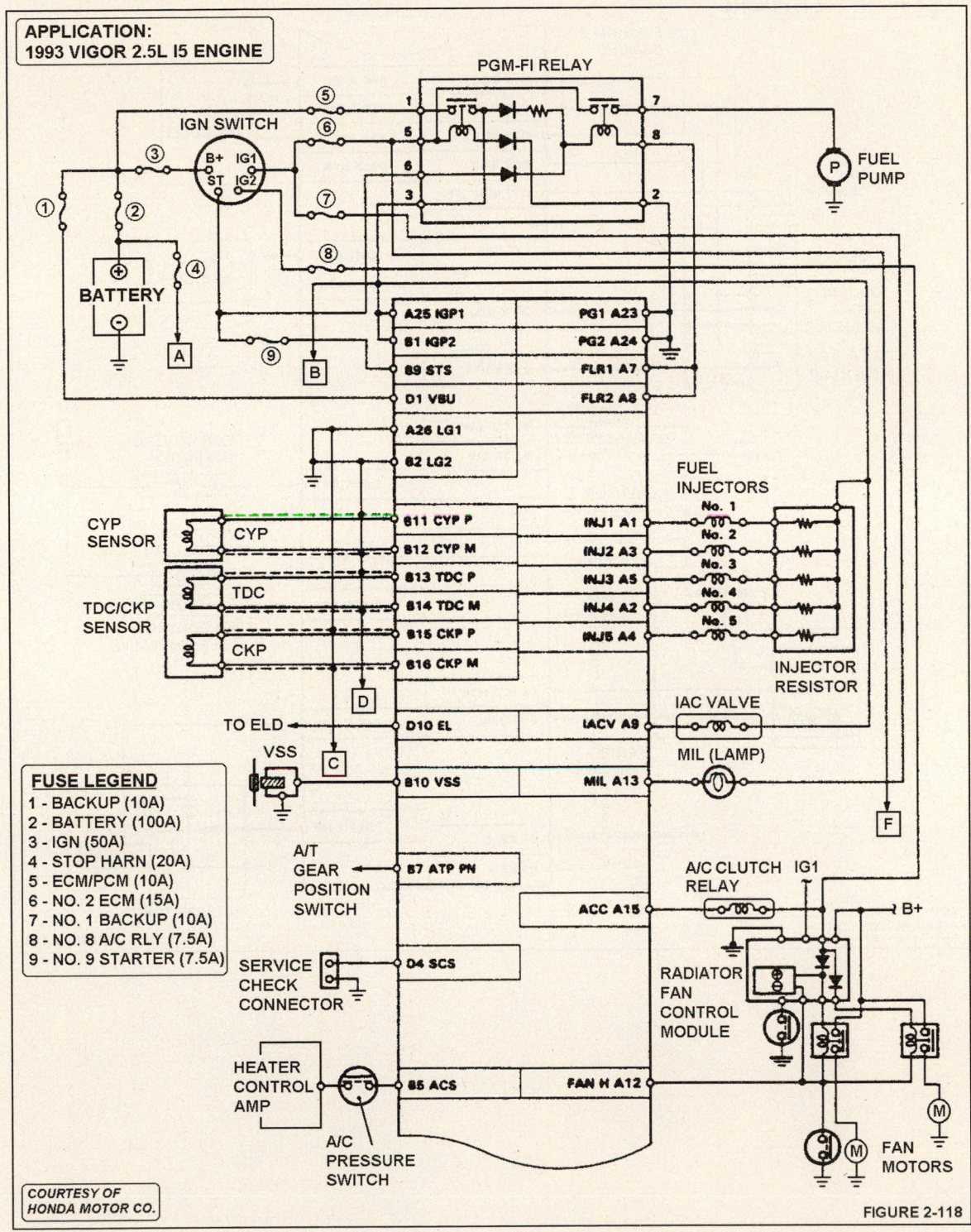

APPLICATION:
1993 VIGOR 2.5L I5 ENGINE

FUSE LEGEND
1 - BACKUP (10A)
2 - BATTERY (100A)
3 - IGN (50A)
4 - STOP HARN (20A)
5 - ECM/PCM (10A)
6 - NO. 2 ECM (15A)
7 - NO. 1 BACKUP (10A)
8 - NO. 8 A/C RLY (7.5A)
9 - NO. 9 STARTER (7.5A)

COURTESY OF
HONDA MOTOR CO.

FIGURE 2-118

Vigor Wiring Diagrams - 1993 Vigor (Part Two)

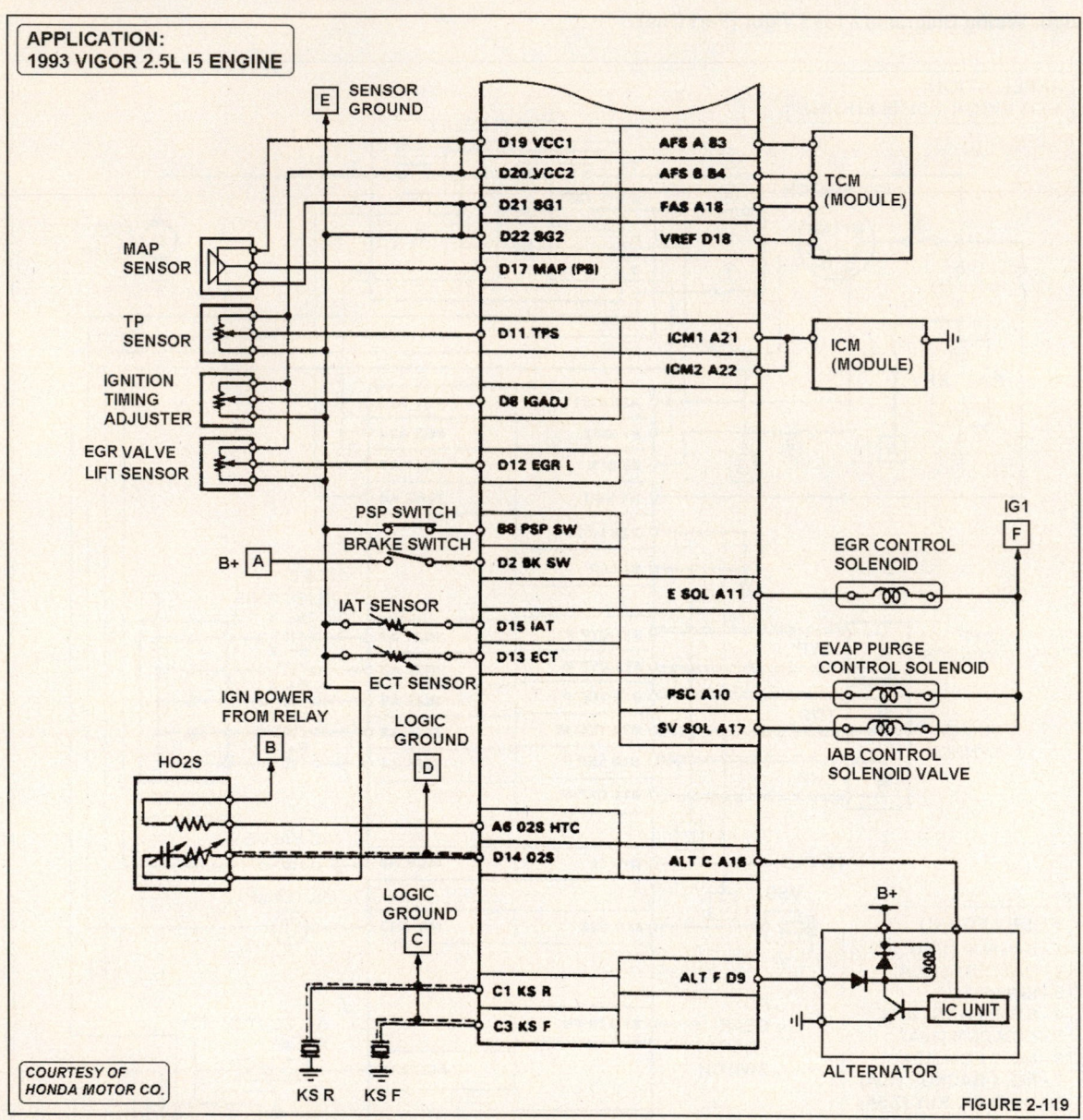

FIGURE 2-119

REFERENCE INFORMATION

Vigor Vacuum Diagram - 1993 Vigor

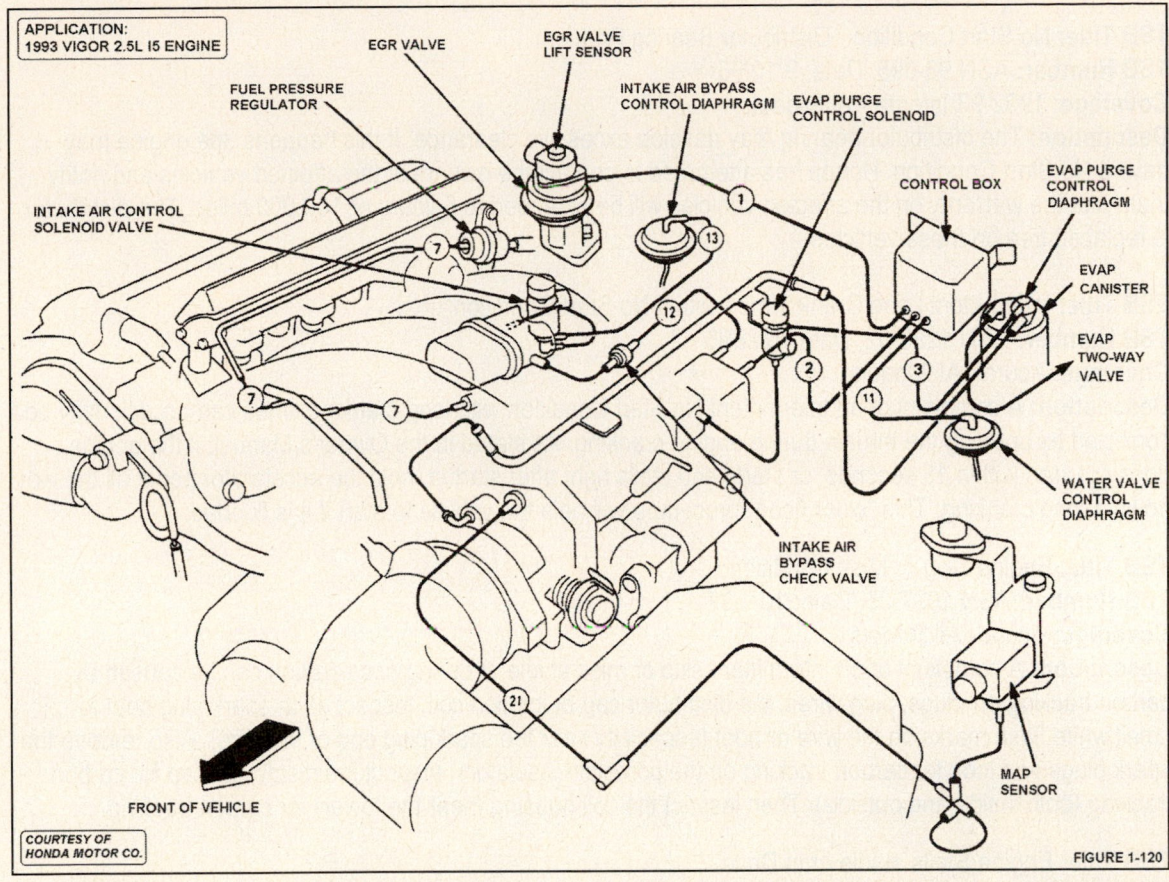

APPLICATION:
1993 VIGOR 2.5L I5 ENGINE

EGR VALVE

EGR VALVE
LIFT SENSOR

FUEL PRESSURE
REGULATOR

INTAKE AIR BYPASS
CONTROL DIAPHRAGM

EVAP PURGE
CONTROL SOLENOID

CONTROL BOX

EVAP PURGE
CONTROL
DIAPHRAGM

INTAKE AIR CONTROL
SOLENOID VALVE

EVAP
CANISTER

EVAP
TWO-WAY
VALVE

WATER VALVE
CONTROL
DIAPHRAGM

INTAKE AIR
BYPASS
CHECK VALVE

MAP
SENSOR

FRONT OF VEHICLE

COURTESY OF
HONDA MOTOR CO.

FIGURE 1-120

TECHNICAL SERVICE BULLETINS

Engine Performance TSB List

TSB Title: No Start Condition - Distributor Bearing Problem
TSB Number: ASN 96-095, Date: 9/10/96
Coverage: 1992-93 Integra - All Models
Description: The distributor bearing may develop excessive clearance. If this happens, the engine may have a No Start Condition. Honda has attempted to contact the owners of the affected vehicles and notify them that the warranty on the affected vehicles will be extended to 6 years or 100,000 miles. The distributor is replaced free on these vehicles.

TSB Title: Intermittent Long Crank Times with a No Start Condition
TSB Number: ASN 1295-02, Date: 12/1/95
Coverage: Acura - All Models
Description: A complaint of an intermittent No Start Condition with long cranking times can occasionally be corrected by opening the throttle during engine cranking. As stated in the Owner's Manual, if the engine does not start within 15 seconds, or starts and stalls right after startup, hold the accelerator pedal all the way down during cranking. This "clear flood" procedure will help the vehicle to start if it is flooded.

TSB Title: Engine Skip or Miss Condition
TSB Number: ASN 1097-05, Date: 10/1/97
Coverage: Acura - All Models
Description: A complaint of an intermittent skip or miss at idle or during acceleration can be caused by carbon tracking on plugs, plug wires, the distributor cap or ignition coil. Inspect each spark plug boot for small white flash marks on the wire or boot (especially near the spark plug end of the wire). Also remove the spark plugs and look for carbon tracking on the porcelain insulators. Inspect the distributor cap for carbon tracking (both inside and outside). Then inspect the coil housing (near the tower) for carbon tracking.

TSB Title: Engine Stalls at Idle or in Drive
TSB Number: ASN 0992-01, Date: 9/1/92
Coverage: Acura - 1986-92 Models
Description: Whenever an Acura remanufactured transmission is installed, an ATF filter is installed in-line with the transmission cooler line. The filter prevents any debris left in the transmission cooler or lines (after flushing) from contaminating the remanufactured transmission. However, debris left in the ATF can restrict or plug the ATF filter and result in a low engine idle speed or cause a stall when placed in gear. This symptom shows up more when the engine is completely warmed up. To confirm that the ATF filter is the cause of the symptom, remove the cooler hoses from the inlet and outlet pipes. Then connect a spare hose to bypass the entire transmission cooler circuit. If the symptom goes away with the spare hose installed, flush the system and replace the ATF filter.

TSB Title: New Terminal Kit
TSB Number: ASN 1093-07, Date: 10/1/93
Coverage: Acura - All Models
Description: A terminal Pin Kit 'B' (T/N 07QAZ-003020A) is available at all dealerships. The kit includes three sizes of shrink tube splices (shrink tubing with a butt connector inside), a special crimping tool, an electric heat gun and storage case. Additional terminals are available to fill the existing Terminal Pin Kit (T/N 07JAZ-003000A). The new part numbers are available as individual terminals, terminals with 8-inch (200mm) pigtails, and certain connector housings.

HONDA CONTENTS

Honda PGM-FI System

Introduction

EARLY SYSTEMS

The Honda Programmed Fuel Injection (PGM-FI) system, introduced in 1985 on Accord and CRX models, is a computerized emission and fuel control system that provides optimum control of the engine and transmission (models with an Automatic Transaxle).

At the heart of the PGM-FI system is a Powertrain Control Module or PCM. The PCM was referred to as the Electronic Control Unit or ECU in earlier publications. The PCM includes a central processor unit (CPU) and digital control system in order to control various engine control systems. It is capable of accepting multiple inputs from various sensors and switches (see the lists that follow), and then performing fast calculations so that it can quickly decide how to control the various engine output devices.

PGM-FI DIAGNOSTICS (1985-91 SYSTEMS)

The first version of Honda PGM-FI systems (Accord and Prelude from 1985-90, Civic from 1986-91) included a diagnostic system to control the operation of a light emitting diode or LED on the PCM.

PGM-FI DIAGNOSTICS (1991-95 SYSTEMS)

The second version of Honda PGM-FI systems (Accord and Prelude from 1991-95, Civic from 1992-95) included a diagnostic system to control the operation of a PGM-FI or Check Engine light on the dash.

OBD I SYSTEM DIAGNOSTICS

In 1988, in order to meet government regulations, the OBD I system was added to the PGM-FI system. This first version of emission control diagnostics incorporates a set of strategies controlled by software inside the PCM. These strategies include control of the Check Engine light, code setting and clearing.

Additionally, on 1992-95 Civic, 1994-95 Accord I4 and 1995 Accord V6 models, there is bi-directional communication with a Scan Tool that allows the tool to read any codes and a series of serial data items (i.e., ECT, IAT, HO2S, MAP, TP, RPM and VSS signals).

Backup Function

The PGM-FI system includes a backup function that is used to provide minimal engine operation should the CPU portion of the PCM fail.

FAILSAFE FUNCTION

The PGM-FI system also includes a failsafe function that is used if one of the main sensor inputs to the PCM is lost (i.e. ECT, MAP or TP sensor signals).

SYSTEM HARDWARE & SOFTWARE

The PGM-FI system is divided into two main parts, the system hardware and software.

The hardware components include:

- All related actuators, relays, and solenoids
- All related sensors, switches, interconnecting wires, connectors and terminals
- The PGM-FI Main Relay and Powertrain Control Module or PCM

The software components include the programs that contain the strategies used by the PCM to control system outputs based on related inputs. These include these systems:

- A/T Lockup Converter
- Idle Speed Control
- Ignition Spark Timing
- Fuel Delivery Control
- Powertrain Diagnostics

SENSOR INPUTS

The PCM in the PGM-FI system receives signals from these sensors:

- Air Temperature (TA) sensor (indicates inlet air temperature)
- Atmospheric or PA (Pressure) sensor (indicates atmospheric pressure)
- Engine Coolant Temperature sensor (indicates engine temperature)
- Crank, CYL and TDC sensors (located inside the distributor)
- EGR Valve Lift sensor (indicates the amount of EGR sensor opening)
- Electronic Load Detector sensor (indicates the amount of electronic load)
- MAP sensor (indicates the amount of engine load)
- Oxygen sensor (the output voltage of the Oxygen sensor)
- Throttle Angle or Position sensor (indicates the amount of throttle opening)
- Vehicle Speed sensor (indicates the vehicle speed)

SWITCH INPUTS

The PCM in the PGM-FI system receives signals from these switches:

- Air Conditioning switch (indicates the A/C clutch is engaged)
- Automatic Transaxle Shift Position switch (indicated gearshift positions)
- Alternator FR signal (indicates the alternator is charging)
- Starter signal (from a circuit connected to the starter solenoid)
- Battery Voltage signal (indicates the battery voltage at IGN 1 terminal)
- Brake Switch (indicates if the brake switch is On or Off)
- Heater Fan Switch (indicates if the heater blower is On or Off)

OUTPUT DEVICES

The PCM in the PGM-FI system controls the operation of these devices:

- A/C Compressor Clutch operation (turn the clutch On/Off)
- Air Induction Tandem Valve (turn the valve On/Off)
- A/T Lockup (clutch) solenoid operation (turn the solenoid On/Off)
- EGR Control solenoid (turn the solenoid On or Off)
- Fast Idle Control solenoid operation (turn the solenoid On/Off)
- Fuel injector On/Off operation
- Ignition Igniter Unit (switch the coil primary On/Off)
- MFI Main Relay On/Off operation (turn the fuel pump On/Off)
- Purge Cut-Off solenoid operation (turn the solenoid On/Off)

COMPONENT LOCATION EXAMPLES

Examples of component locations from a 1990 Civic CRX equipped with an automatic transaxle and 1.5L I4 TBI D15B2 engine are shown in Figure 3-01.

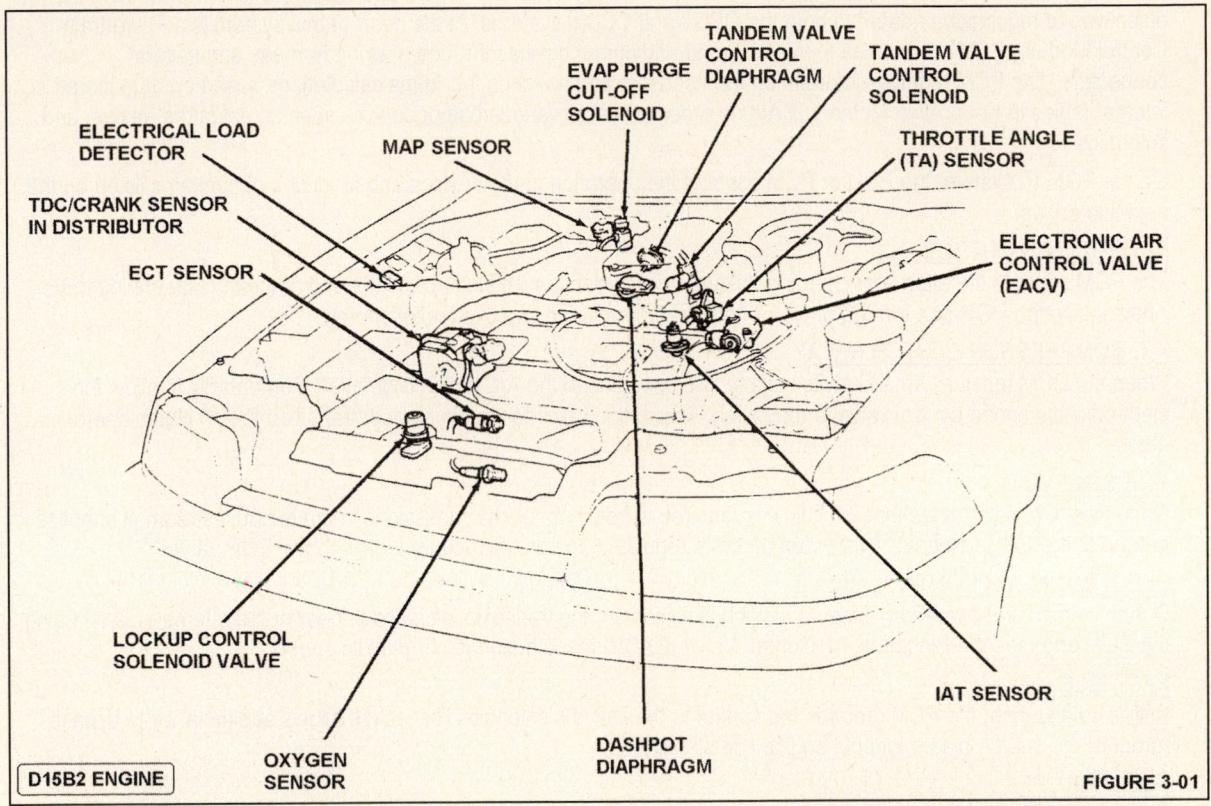

FIGURE 3-01

Examples of component locations from a 1994 Accord equipped with an automatic transaxle and 2.2L I4 MFI F22B1 engine are shown in Figure 3-02.

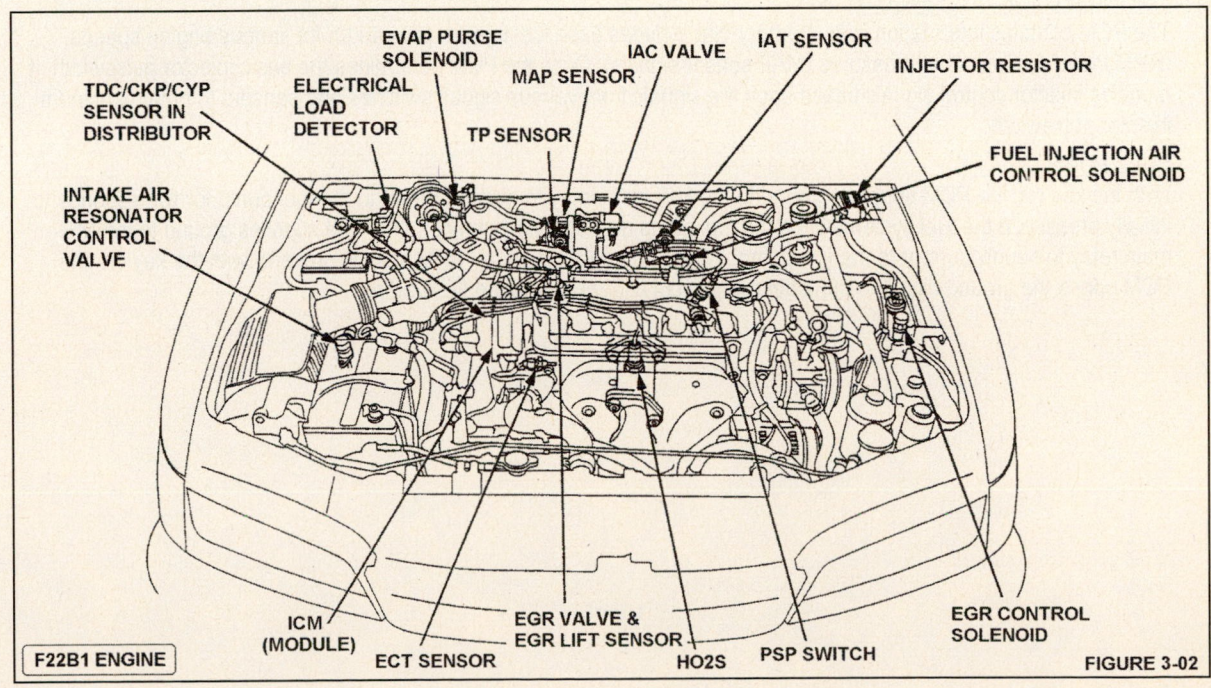

FIGURE 3-02

PGM-FI Controlled Functions

INTRODUCTION

The purpose of the PGM-FI system is to provide optimum control of the engine and transmission while meeting the objectives of regulations related to both the OBD I and OBD II systems. At the heart of this system is a Powertrain Control Module (PCM) connected to various input and output devices through a wiring harness and several connectors. The PCM receives information from sensors and switches, performs calculations based on data stored in internal tables in Keep Alive Memory (KAM) in order to control various output devices such as actuators, relays, and solenoids.

On the PGM-FI system, the ECU or PCM controls the operation of the devices and functions of the items listed on the next two pages.

ALTERNATOR OUTPUT

The PCM controls the alternator charging voltage in accordance with electrical loads and any particular driving mode. Additionally, the PCM can turn off or slow alternator operation to improve fuel economy.

A/C COMPRESSOR CLUTCH RELAY

When the PCM receives an A/C switch signal, the operation of the A/C clutch is delayed momentarily to allow the electronic idle speed control valve to reposition. This action helps to assure smooth transition to A/C clutch operation mode.

EGR SOLENOID

Once the PCM determines that EGR flow is required to help control NOx emissions in the exhaust stream, it supplies ground to the EGR solenoid. This action allows a regulated amount of vacuum to flow to the EGR valve.

ELECTRONIC AIR CONTROL VALVE

During cold engine operation, if the A/C clutch is turned on, the transaxle is placed in gear or the alternator is charging, the PCM engages the Electronic Air Control Valve (EACV) to maintain the correct idle speed.

FAST IDLE

With a cold engine, the PCM grounds the control to the fast idle solenoid. This action allows additional air to bypass (around) the throttle plates for cold engine idle stability.

FUEL CUTOFF

During deceleration from engine speeds of over 950-1300 rpm (model specific) with the throttle valve closed, the PCM cuts off the fuel injectors to clean up tailpipe emissions and improve fuel economy. If engine speed exceeds 5500-6800 rpm (model specific) regardless of throttle valve position, the injectors are turned off (over-revving).

FUEL INJECTOR PULSEWIDTH

The PCM contains information in its memory that includes base fuel injector pulsewidth for various engine speeds (RPM input) and manifold pressures (MAP sensor signals). Once the PCM determines the basic injector pulsewidth, it modifies injector control signals based upon the signals from various signal, switches and sensors to obtain optimum injector pulsewidth.

FUEL PUMP ON/OFF

With the key on, the PCM supplies ground to the MFI main relay to supply current to the fuel pump for two seconds to initially pressurize the fuel system. Once the fuel pump is energized (running), the PCM supplies ground to the MFI main relay to supply current to the fuel pump. When the engine is off (no RPM signals present) with the key on, the PCM opens the ground circuit to the main relay to cut current flow to the fuel pump.

IGNITION SPARK TIMING

The PCM contains data in long term memory that includes base ignition spark timing for various engine speeds and manifold pressures (MAP signals). The PCM will also change spark timing based on current engine temperature.

LOCKUP SOLENOID CONTROL

Depending upon engine speed and throttle angle, the PCM grounds the A/T Lockup solenoid to control operation of the torque converter lockup clutch.

PURGE CUTOFF SOLENOID

HF Models - With engine temperature below 104°F, the PCM supplies ground to the EVAP Purge Cutoff solenoid (valve). This action cuts off the vacuum supply to the Purge Control valve.

Si Models - With engine temperature below 135°F, the PCM supplies ground to the EVAP Purge Cutoff solenoid (valve). This action cuts off the vacuum supply to the Purge Control valve.

STD Models - With engine temperature below 176°F, the PCM supplies ground to the EVAP Purge Cutoff solenoid (valve). This action cuts off the vacuum supply to the Purge Control valve.

STARTUP FUEL ENRICHMENT

The PCM provides a richer A/F mixture at startup to stabilize the engine.

TANDEM SOLENOID VALVE

Depending upon engine temperature and speed, the PCM grounds the Tandem solenoid to open the Tandem valve. The valve is opened and closed to maintain good A/F mixture atomization by the main fuel injector.

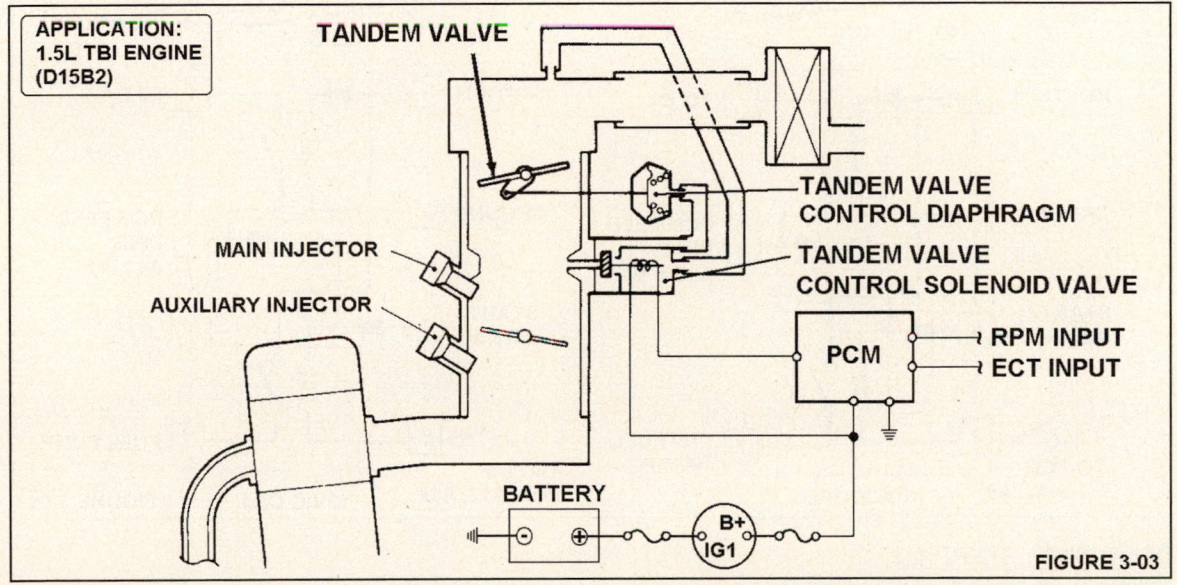

FIGURE 3-03

PGM-FI MAIN RELAY TESTS (1985-95)

INTRODUCTION

On 1985-95 Honda Accord and CRX STD models with fuel injection, the PGM-FI main relay is two relays inside one assembly mounted to the left side of the cowl. With the key on, a relay is energized and this action supplies battery voltage to the PCM and the second relay. The second relay is energized by the PCM for two seconds whenever the ignition is turned on. The second relay is also energized by the PCM when the engine is running. The second relay feeds power to the fuel pump and fuel injectors.

On 1985-95 Honda CRX HF & Si models with fuel injection, the PGM-FI main relay is two relays inside one assembly mounted to the left side of the cowl. With the key on, a relay is energized and this action supplies battery voltage to the PCM, the fuel injectors and the second relay. The second relay is energized by the PCM for two seconds whenever the ignition is turned on. The second, which feeds power only to the fuel pump on these applications, is also energized by the PCM when the engine is running.

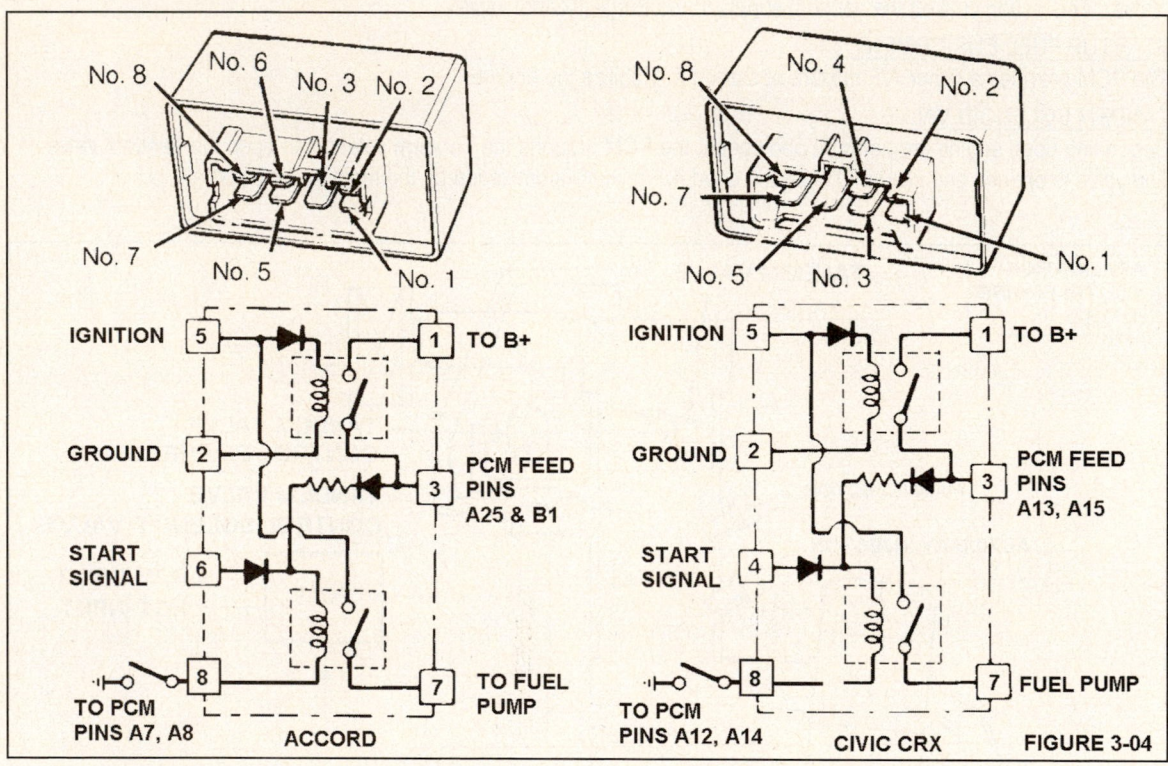

FIGURE 3-04

DRIVEABILITY SYMPTOMS

Problems with the PGM-FI main relay or its circuits could cause these symptoms:

*   No Start Condition or the engine quits on the road and then restarts

TIPS: *The PGM-FI relay was mounted with the terminals up instead of down on early Civic models. This position allowed moisture to enter the relay and caused No Start faults.*

PRELIMINARY STEPS

All of the test steps in this procedure should be done in sequence in order to isolate the problem area. First, carefully remove the main relay and its electrical connector from the vehicle. Obtain a pair of jumper wires with connectors suitable to attach to the main relay and battery cables or posts.

PGM-FI MAIN RELAY TESTS (1985-95)

MAIN RELAY TEST

With the relay removed, connect one jumper wire to battery positive (+) and to main relay terminal No. 4 (No. 6 on Accord models). Connect another jumper wire to the relay terminal No. 8 and to battery ground.

1) With the relay energized, use a DVOM to check for continuity between relay terminals No. 5 and No. 7. If there is continuity, go to Step 2. If there is no continuity, replace the main relay and retest the symptom.

2) Move the jumper wires to connect battery positive to main relay terminal No. 5 and then relay terminal No. 2 to ground. Check for continuity between main relay terminals No. 1 and No. 3. If there is no continuity, replace the main relay and retest (Figure 3-04 on the previous page).

3) Move the jumper wires to connect battery positive to relay terminal No. 3 and then relay terminal No. 8 to ground. Jumper wire polarity is important in this step due to an internal diode. Check for continuity between relay terminals No. 5 and No. 7. If there is no continuity, replace the main relay and retest. If there is continuity, test for battery power to the fuel pump or other devices. If these devices have power, go to Wiring Harness Tests on the next page.

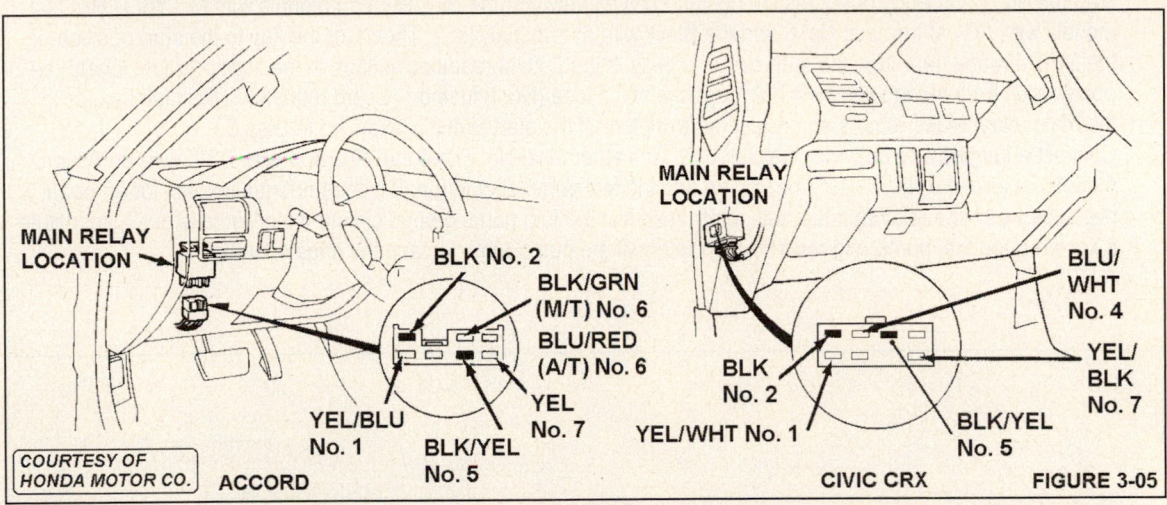

■ **NOTE:** *For applications other than the 1990 Accord, Civic CRX STD, HF and Si models, refer to other pin voltage charts to identify the pins and wire colors for all references to PCM terminals and wire colors in this article.*

PGM-FI MAIN RELAY TESTS (1985-95)

WIRING HARNESS TESTS

The key should be off with the main relay removed during this test procedure.

1) Use a DVOM to check for continuity between the BLK wire at main relay terminal No. 2 (relay connector) and body ground (Figure 3-05). The DVOM should show continuity (less than 5 ohms of resistance). Repair any problems before continuing.

2) Connect a DVOM between the YEL/WHT (YEL/BLU on Accord models) wire at the harness terminal No. 1 and the BLK wire at terminal No. 2 (Figure 3-05). If the DVOM reads battery voltage (okay), proceed to Step 3. If the DVOM reads no voltage, locate the open condition in the feed circuit from the battery through the PCM fuse (in the main fuse box) that connects to the main relay. Make repairs and retest for the symptom.

3) Connect the DVOM between the BLK/YEL wire at terminal No. 5 and the BLK wire at terminal No. 2 (Figure 3-05).

4) Turn the key on and if the DVOM reads battery voltage, proceed to Step 4. If the DVOM reads no voltage, locate the open in the circuit between the ignition switch and the No. 14 fuse (No. 2 on Accord models) fuse in the main fuse box or between the fuse box and the main relay.

5) Next, on Civic CRX models connect the DVOM between the BLU/WHT wire at terminal No. 4 and the BLK wire at terminal No. 2. On Accord models connect the DVOM between the BLK/GRN on models with M/T (BLU/RED on models with A/T) at terminal No. 6 and the Black wire at terminal No. 2. Then turn the key to the start position. If the DVOM reads near 10v, the start circuit is okay. If the DVOM reads no voltage in the start position, locate the open circuit from the ignition switch through the No. 2 fuse (No. 9 fuse on Accord models) in the fuse box, or to the main relay. Make repairs and retest the symptom. If the start circuit is okay, go to Step 5.

6) Connect a jumper wire between the BLK/YEL wire at terminal No. 5 and the YEL/BLK wire (YEL wire on Accord models) at terminal No. 7. Turn the key on and the fuel pump should run. If it does not run, inspect for an open circuit between the relay and the fuel pump. Also test the fuel pump ground circuit. If the wiring is okay, substitute a known-good fuel pump and retest the symptom. If the pump runs, replace the original pump.

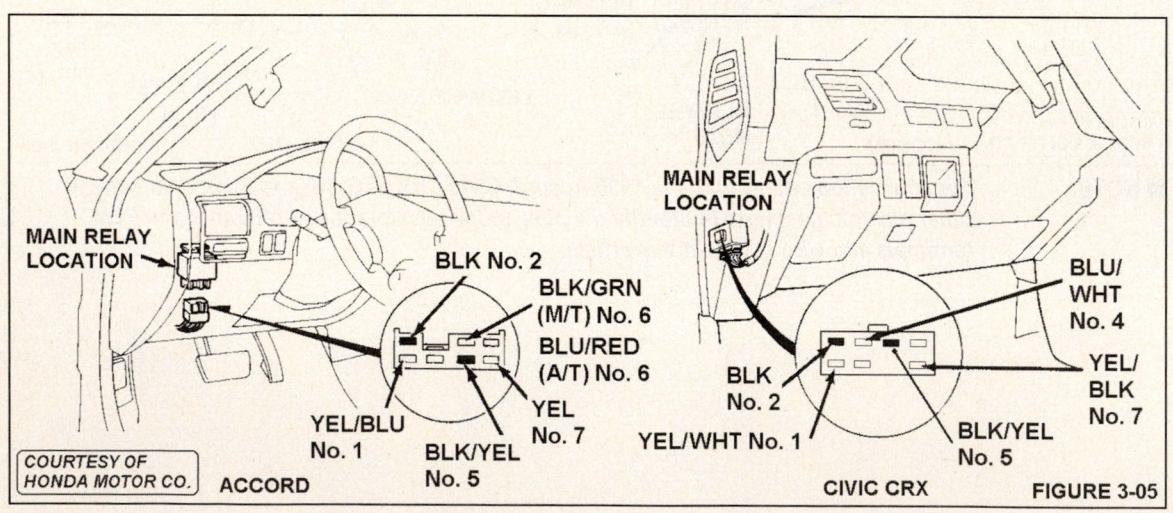

MAIN RELAY LOCATION

BLK No. 2
BLK/GRN (M/T) No. 6
BLU/RED (A/T) No. 6
YEL/BLU No. 1
YEL No. 7
BLK/YEL No. 5

COURTESY OF HONDA MOTOR CO.　ACCORD

MAIN RELAY LOCATION

BLU/ WHT No. 4
YEL/ BLK No. 7
BLK No. 2
BLK/YEL No. 5
YEL/WHT No. 1

CIVIC CRX　FIGURE 3-05

PGM-FI MAIN RELAY TESTS (1996-98)

INTRODUCTION

On 1996-98 Honda vehicles with fuel injection, the PGM-FI main relay consists of two relays inside one assembly. The main relay is located in these passenger compartment locations:

- Accord - behind the dash to the right of steering column (Figure 3-06)
- Civic & Del Sol - behind the right side of glove box
- Odyssey - behind the dash at the left side near the door
- Passport (Isuzu Rodeo) - In fuse/relay box (right side of engine)
- Prelude - behind the left side of dash near clutch bracket

MAIN RELAY OPERATION

With the key on, the first relay is energized and this action supplies battery voltage to the PCM and the second relay.

The second relay is energized by the PCM for two seconds whenever the key is in the on position. It is also energized with engine running. The second relay provides power to the fuel pump and injectors.

In Figure 3-07, S1 and S2 identify switch actions and L1 and L2 identify load operations.

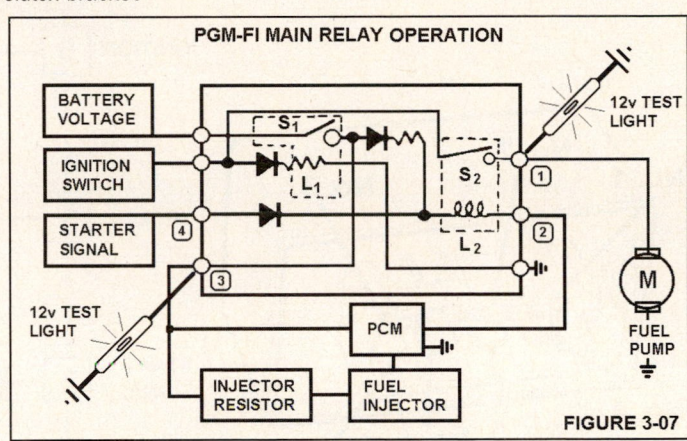

PGM-FI MAIN RELAY OPERATION

FIGURE 3-07

TIPS: *A quick test of the relay can be made with a 12v-test light (Figure 3-07). The light should glow for two seconds during cranking and with the key on at test point 1. It should glow dim during cranking and running at test point 2. It should glow at key on or engine running at test point 3 and also during cranking at test point 4.*

DRIVEABILITY SYMPTOMS

If the vehicle starts and runs, the PGM-FI main relay is okay. Problems with the main relay or its circuits could cause these symptoms:

- No Start Condition
- Engine quits on the road and then restarts

■ **NOTE:** *The purpose of this inspection and test sequence on the next page is to determine if a fault in the main relay is the cause of one of the symptoms listed above.*

PGM-FI Main Relay Tests (1996-98)

PRELIMINARY STEPS

Perform all of the test steps in sequence to isolate the problem area. First, remove the main relay and its electrical connector from the vehicle. Obtain a pair of jumper wires with connectors suitable to attach to the main relay and to the battery cables.

CAUTION: Keep all open flames or sparks away from the work area. Do not attempt to relieve the fuel pressure unless the engine is off.

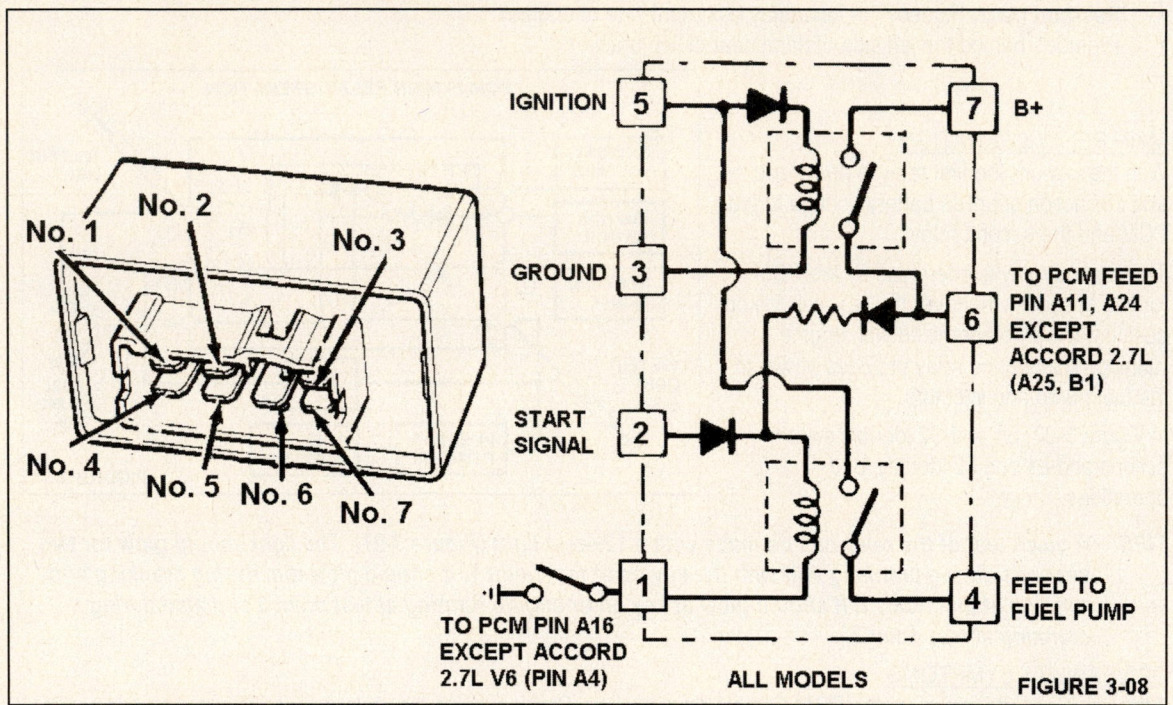

FIGURE 3-08

■ **NOTE:**   *For applications other than the 1996 Accord, Civic, Del Sol and Odyssey models, refer to other pin voltage charts to identify the pins and wire colors for all references to PCM terminals and wire colors in this article.*

MAIN RELAY TESTS

With the relay removed, connect one jumper wire to battery (+) and to relay terminal No. 2. Connect the other jumper wire to the relay terminal No. 1 and to battery ground.

1) With the relay energized, use a DVOM to check for continuity from relay terminals No. 5 and No. 4 (Figure 3-08), If there is continuity, go to Step 2. If there is no continuity, replace the relay and retest for the symptom.

2) Next, connect one jumper wire to relay terminal No. 5 and to battery (+). Connect the other jumper wire to relay terminal No. 3 and a good ground. Then check for continuity between main relay terminals No. 7 and No. 6. If there is continuity, go to Step 3. If there is no continuity, replace the main relay and retest for the symptom.

3) Next, connect one jumper wire to relay terminal No. 6 and to battery positive. Connect the other jumper wire to relay terminal No. 1 and a good ground. Jumper wire polarity is important in this step due to an internal diode (Figure 3-08). Check for continuity between relay terminals No. 5 and No. 4. If there is no continuity, replace the main relay and retest for the symptom. If there is continuity, the main relay is okay. To check the relay start, power and ground circuits, go to the Wiring Harness Tests article on the next page.

PGM-FI MAIN RELAY TESTS (1996-98)

WIRING HARNESS TESTS

The main relay is removed for this test procedure.

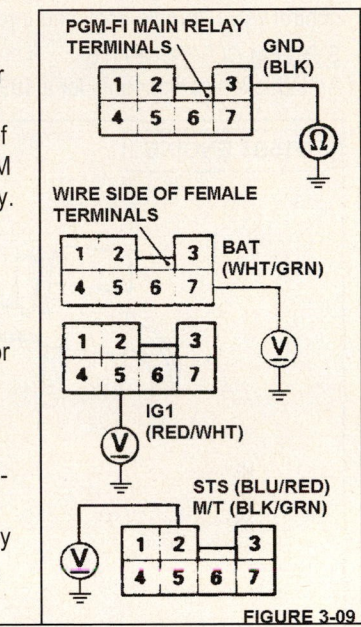

1) Check the ground circuit - Check for continuity between terminal No. 3 of the relay connector and battery ground (Figure 3-09). If continuity is not okay, repair the ground circuit and recheck the symptom. If okay, go to Step 2.

2) Check battery feed circuit - Check the voltage reading between terminal No. 7 of the main relay connector and body ground. If there is no voltage, check the PCM (15-amp) fuse and/or locate the open feed circuit between the fuse and the relay. Make repairs and then retest for the symptom. If okay, go to Step 3.

3) Check ignition feed for an open circuit - Turn the key on and test voltage between terminal No. 3 in the relay 7-pin connector and body ground. If there is no voltage available, check the No. 2 Fuel Pump 15-amp fuse and/or locate the open or shorted wire between the relay and the fuse. Make repairs and retest for the symptom. If okay, go to Step 4.

4) Check ignition switch start circuit - On vehicles with a M/T, depress the clutch during testing. On vehicles with an A/T, place gear selector in P/N. Momentarily turn the key to start and check the voltage between terminal No. 2 in the relay 7-pin connector and body ground. If there is no voltage, check the Starter Signal No. 9 (7.5-amp) fuse and/or check for an open or shorted wire between the relay and Starter Signal fuse No. 9. Make repairs and retest for the symptom. If okay, go to Step 5.

FIGURE 3-09

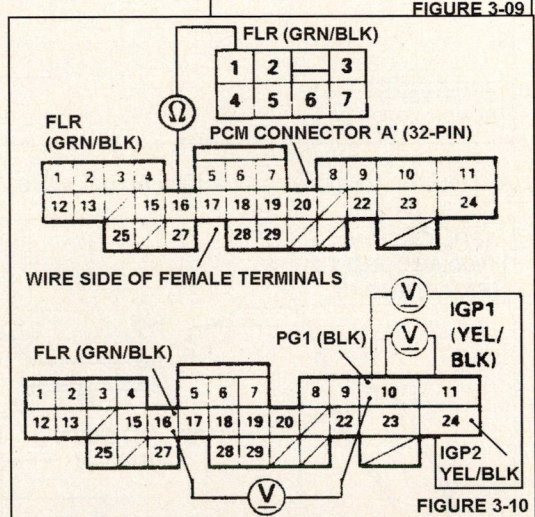

5) Check the FLR wire for an open circuit - Turn the key off and remove the PCM 32-pin connector. Check for continuity between terminal No. 1 in the relay connector and terminal No. 16 of the PCM 32-pin connector (Figure 3-10). If there is no continuity, locate and repair the open circuit in the wire between terminal No. 1 and terminal No. 16. Retest for the symptom when repairs are done. If continuity is okay, go to Step 6.

6) Check IGP1 and IGP2 circuits for an open condition - Reconnect PCM 32-pin connector and the main relay connector. Turn the key on and measure the voltage between terminals A11 and A10, and between terminals A24 and A10 at the PCM (Figure 3-10). If battery voltage is not available, locate and repair the open circuit between terminal A11 at the relay and/or replace the main relay. Then retest for the symptom. If okay, go to Step 6.

FIGURE 3-10

7) Check for an open circuit in the PCM - Read the voltage between terminals A16 and A10 at the PCM for two seconds just after the key is turned on. If the reading is 1.0v or less, substitute a known-good PCM and retest. If now okay, replace the original PCM and retest the symptom. If the voltage is okay, test the relay with it removed.

PGM-FI ENGINE CONTROLS

Air Intake & Throttle Body Systems

INTRODUCTION

The Air Induction system supplies clean air to the engine. It consists of an air cleaner, air intake pipe, electronic air control valve, throttle body and throttle control system, intake manifold and Tandem Control system (Civic CRX).

SYSTEM COMPONENTS

Air Induction components for a 1990 Civic CRX with a 1.5L I4 TBI engine (Figure 3-11).

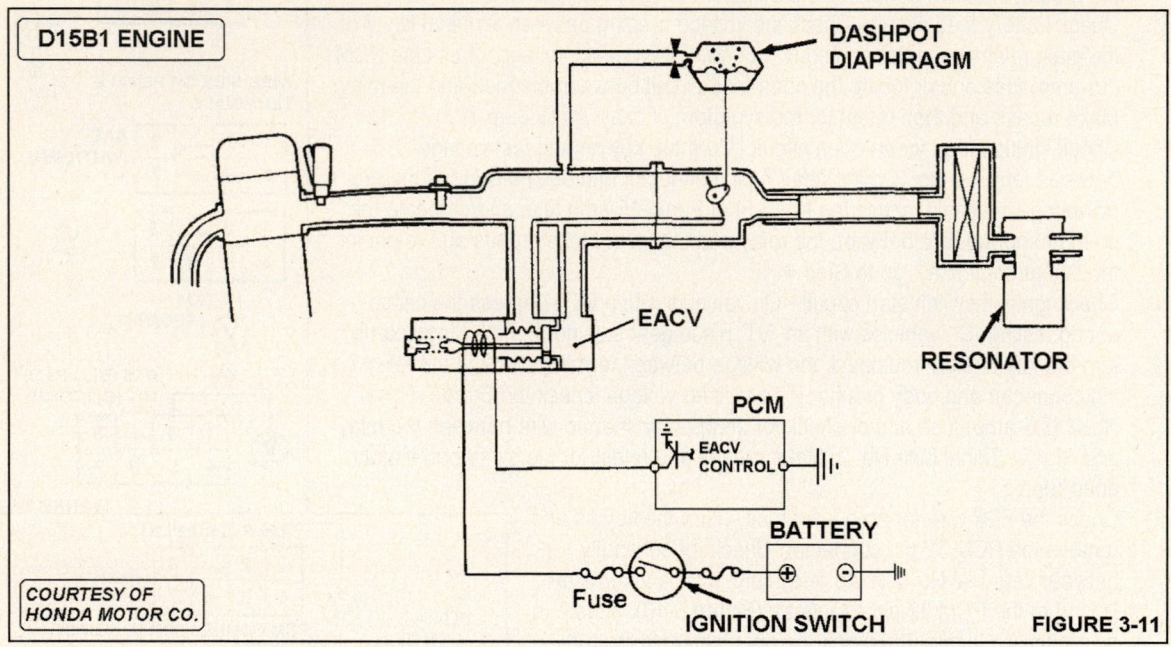

FIGURE 3-11

Air Induction components for a 1990 Accord with a 2.2L I4 MFI engine.

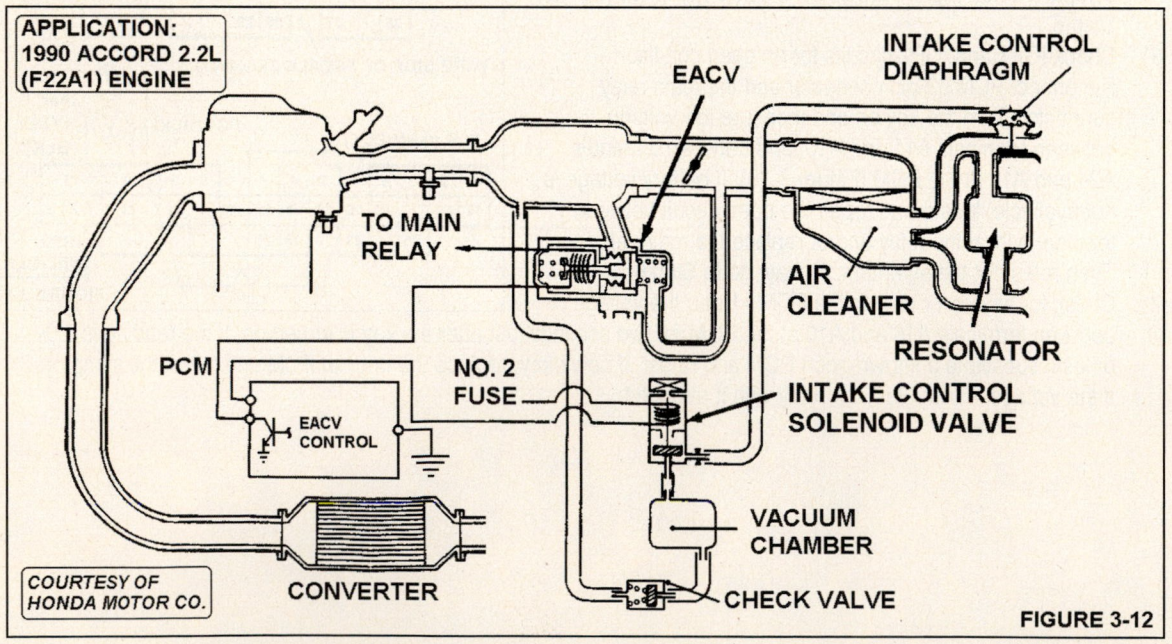

FIGURE 3-12

Air Intake & Throttle Body Systems
THROTTLE CABLE INSPECTION & TESTS

DRIVEABILITY SYMPTOMS
Faults in the throttle cable or its adjustment could cause these symptoms:

- Warm engine idle speed out of specification (too high or too low)
- Irregular engine idle speed condition

PRELIMINARY STEPS
The following inspections and tests should be performed in sequence to isolate the problem area in order to make the necessary repairs. Always start the test procedure by following the preliminary steps.

Warm the engine to normal operating temperature or until the electric cooling fan cycles. Verify that the throttle cable operates smoothly with no binding or sticking. Clean or repair the cable as needed before proceeding.

CHECK THROTTLE CABLE DEFLECTION & TRAVEL
1. Check the throttle cable free-play at the throttle linkage. The cable deflection should be 0.39-0.47 inch (Figure 3-13).
2. If the deflection is not within specifications, loosen the lock nut and turn the adjusting nut until cable deflection is within specified limits in step 1.
3. With the cable properly adjusted, check the throttle valve for full travel up to wide open throttle (WOT) position as the accelerator pedal is pushed to the floor. Also inspect that the throttle valve returns to the idle position (it should allow the throttle lever to rest against the throttle stop screw) as the throttle is released.

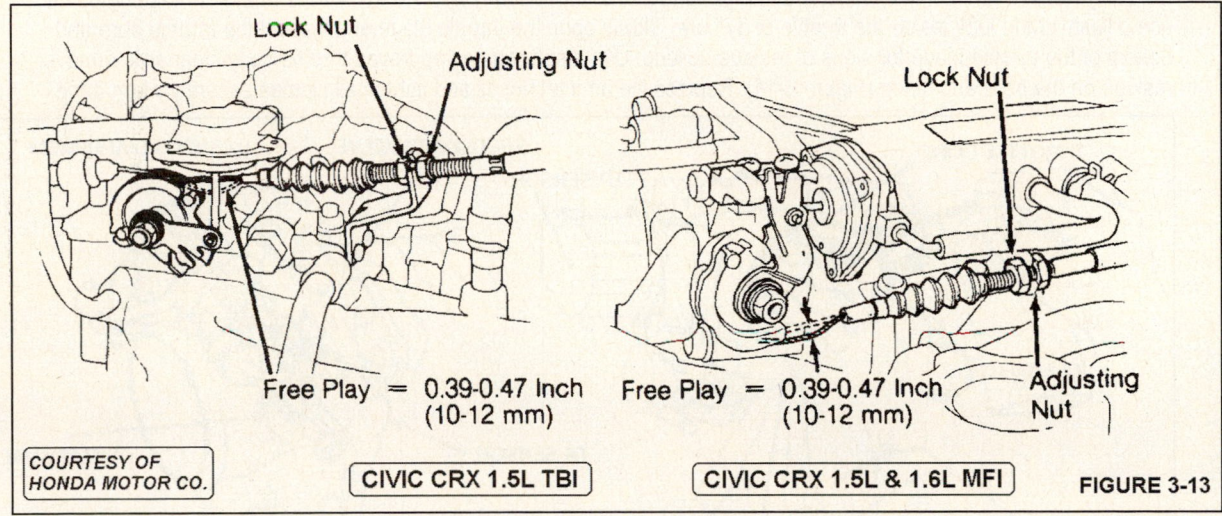

FIGURE 3-13

THROTTLE BODY INSPECTION & TESTS

INTRODUCTION

Early models with fuel injection used a two-barrel side-draft design with the primary air horn at the top (Figure 3-14). On later Accord and Prelude models, the throttle body is a single barrel, single-throttle plate design. On Civic and Civic CRX models, the throttle body is a single barrel, two-plate design. The lower portion of the throttle body is heated by engine coolant that circulates from the engine cylinder head. The idle adjustment screw is preset at the factory and is used to increase or decrease the amount of bypass air. The EVAP canister purge port is located on top of the throttle body assembly.

DRIVEABILITY SYMPTOMS

Problems with the throttle body or its related vacuum supply lines could be the cause of the these symptoms:

* Frequent stalling at engine startup
* Warm engine idle speed out of specification
* Rough idle
* Loss of power or poor performance
* Stalling during gear shifting or deceleration

PRELIMINARY STEPS

The following inspections and tests should be performed in sequence to isolate the problem area in order to make the necessary repairs. Always start the test procedure by following the preliminary steps.

Carefully remove the air inlet hoses or ducting between the throttle body and the air cleaner unit. With the engine off, use a flashlight to look inside the throttle body bore. Slowly open the throttle blade and inspect the throttle bore and bottom of the throttle blade for signs of residue buildup. Use a soft rag and approved throttle body cleaner to remove residue on the bore and blade (Figure 3-14). Replace the air inlet hoses and tighten all clamps securely when done.

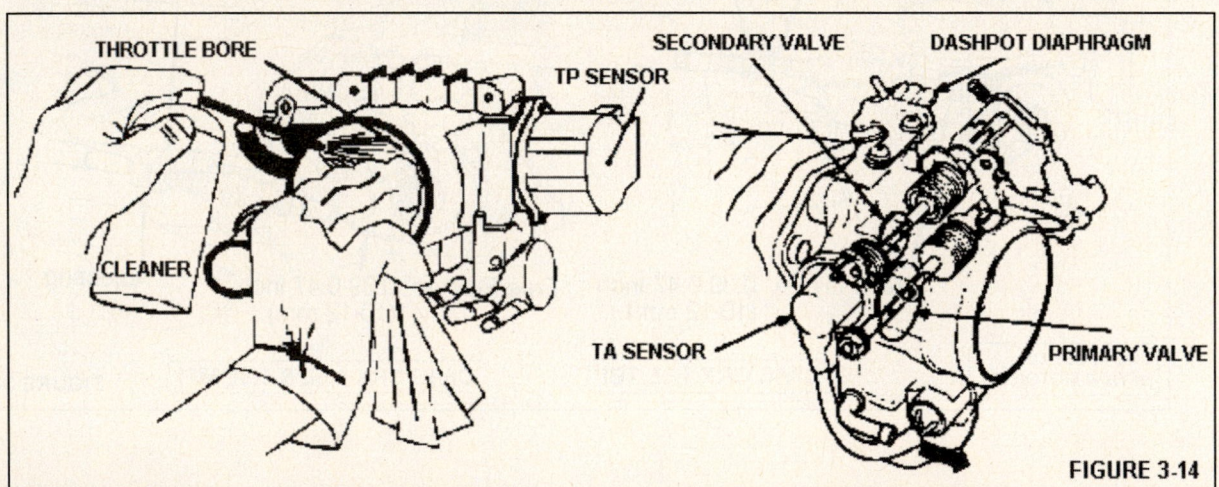

FIGURE 3-14

■ **NOTE:** *This simple inspection and cleaning procedure can eliminate many types of Air Induction system and idle speed related problems.*

CAUTION: *The following test steps require that the drive wheels are blocked and the vehicle exhaust system is purged from the work area for safety purposes.*

CHECK THROTTLE VALVE, INSPECT VACUUM PORTS

1) Start the engine and allow it to warm until the electric cooling fan cycles.
2) Disconnect the vacuum hose at the top of the throttle body that runs to the EVAP canister. Connect a vacuum gauge to the exposed port.
3) Check the vacuum gauge reading at idle speed. The gauge should read no vacuum at idle. If vacuum is present, check throttle cable operation.
4) Open the throttle slightly and check that the vacuum gauge reads some vacuum. If the gauge does not read some vacuum with the throttle slightly open, turn the engine off and clean the throttle body bore and ports with an approved cleaner. Depending upon the severity of the residue inside the throttle body, it may need to be removed so that it can be cleaned properly.
5) With the engine off, verify that the throttle cable operates smoothly without any binding or sticking. If okay, perform these inspections.
6) Inspect for excessive wear or play in the throttle valve shaft.
7) With the throttle valve fully closed, check for a sticking or binding throttle lever (make repairs as needed).
8) Check for clearance between the throttle stop screw and throttle lever at fully closed position (there should be no clearance between them).
9) Replace the throttle body unit if there is excessive play or wear in the throttle valve shaft, the shaft is binding or sticking or it cannot be cleaned.

IDLE OR THROTTLE STOP SCREW ADJUSTMENTS

Prior to making any adjustments to idle speed with the idle or throttle stop screw, read the information that follows.

If the throttle body and blade are clean with no signs of worn or sticking parts, the low idle speed (minimum-air-rate) and the normal PCM control of idle speed should provide correct idle speed. In this case, do not adjust the throttle stop screw since it cannot be reset properly in the field.

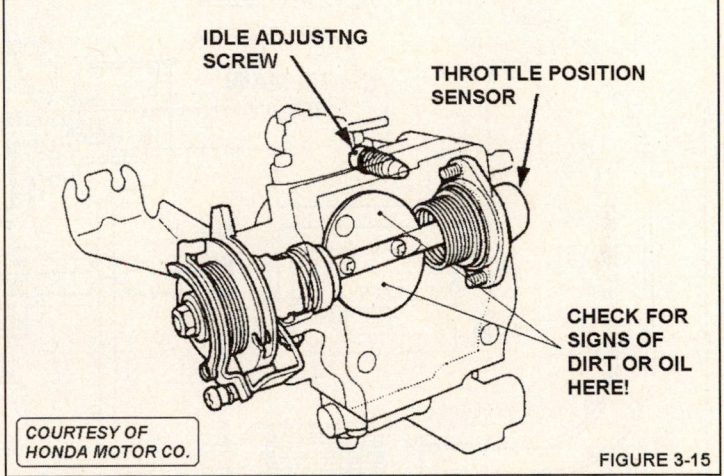

IDLE ADJUSTNG SCREW

THROTTLE POSITION SENSOR

CHECK FOR SIGNS OF DIRT OR OIL HERE!

COURTESY OF HONDA MOTOR CO.

FIGURE 3-15

INTAKE CONTROL SYSTEM INSPECTION & TESTS

INTRODUCTION

The Intake Control system on Accord models is designed to decrease intake air noise. With engine speed less than 3500 rpm, the PCM supplies current to the Intake Control solenoid valve. This actions causes the solenoid valve to open which allows intake manifold vacuum to apply to the intake control diaphragm to help control air intake noise.

DRIVEABILITY SYMPTOMS

Problems with the Intake Air Control system could cause these symptoms.

* Excessive air induction (air cleaner area) noise at idle and low speeds
* Loss of power

PRELIMINARY STEPS

The following inspections and tests should be performed in sequence to isolate the problem area in order to make the necessary repairs. Always start the test procedure by following the preliminary steps.

Carefully inspect the Intake Control diaphragm linkage for binding or sticking. Also check for proper vacuum hose routing and connections. Make repairs as needed and continue the test sequence.

COMPONENT EXAMPLES

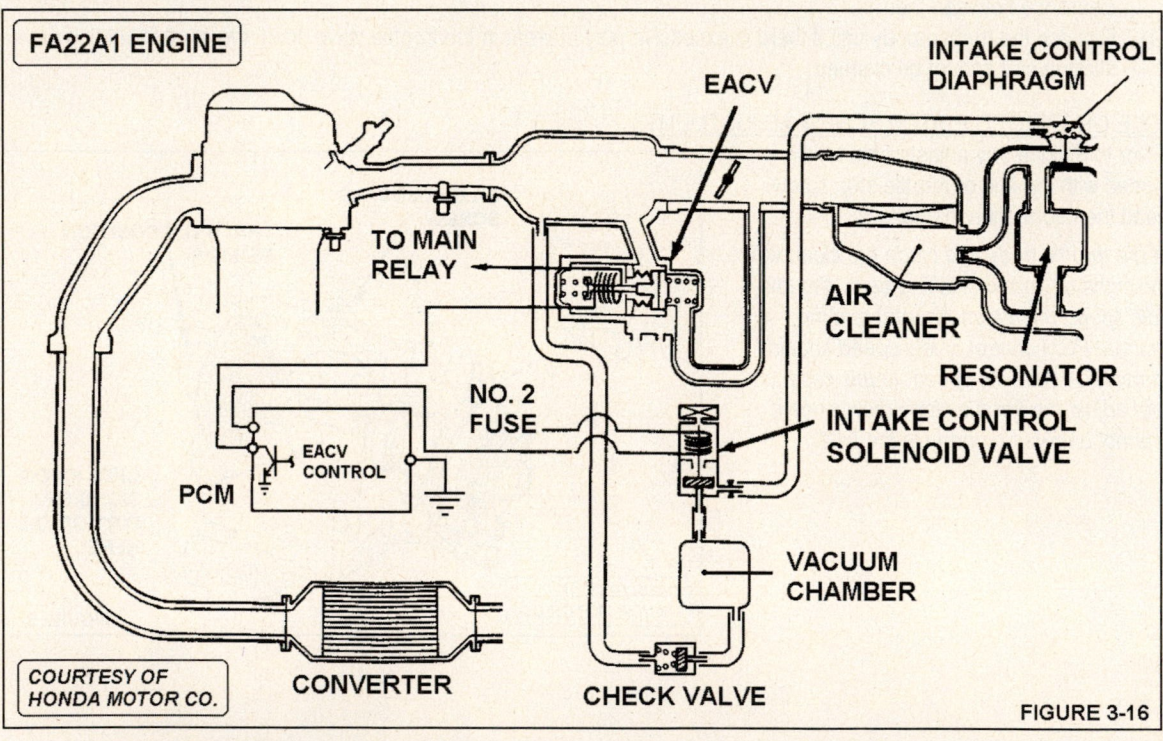

FIGURE 3-16

CAUTION: *The test steps that begin on the next page require that the drive wheels are blocked and the vehicle exhaust system is purged from the work area for safety purposes.*

INSPECT VACUUM HOSE & ELECTRICAL CIRCUITS

1) Start the engine and allow it to idle.
2) Carefully remove the upper vacuum hose from the Intake Control solenoid valve. Connect a vacuum gauge to the hose at the valve.
3) Observe the gauge reading. If the gauge connected to the valve reads vacuum (Figure 3-17), raise engine speed to 3700 rpm and again read the gauge. If the gauge shows no vacuum at 3700 rpm, connect a hand vacuum pump to the No. 8 vacuum hose that connects to the diaphragm and apply vacuum. If the gauge reads and holds vacuum, the system is okay. If the gauge loses vacuum and the hose is not leaking, replace the diaphragm unit and retest the system. If the gauge reads vacuum at 3700 rpm, go to Step 9.
4) If the vacuum gauge reads no vacuum at idle speed in Step 3, remove the lower vacuum hose from the Intake Control solenoid and connect a vacuum gauge (Figure 3-18). If there is no reading on the vacuum gauge, repair the blockage or vacuum leak between the valve and the intake manifold. If the hose is okay, clean the intake manifold vacuum port. Then return to Step 3 to continue testing.
5) If vacuum was recorded in Step 4, disconnect the 2-P connector from the Intake Control solenoid. Use a DVOM to measure the voltage between the BLK/YEL and white wire terminals in the connector. Refer to Figure 3-19 on the next page as needed. If battery voltage is present, replace the solenoid and retest the system.

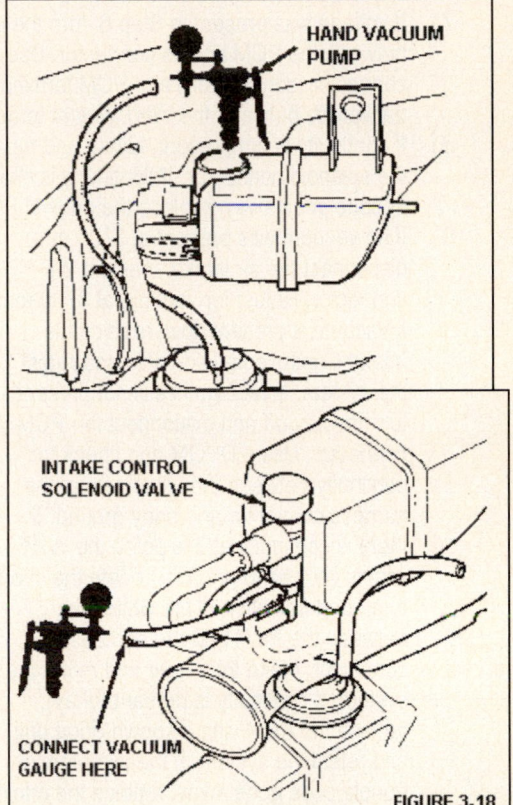

HAND VACUUM PUMP

INTAKE CONTROL SOLENOID VALVE

CONNECT VACUUM GAUGE HERE

FIGURE 3-18

6) If battery voltage was not present in Step 5, use a DVOM to measure the voltage between the solenoid BLK/YEL wire and body ground (Figure 3-19). If the DVOM reads no voltage, locate and repair the open circuit in the BLK/YEL wire between the harness connector and the No. 2 fuse.

7) If voltage was present in Step 6, turn the ignition off. Connect the Honda PCM test harness or BOB connector between the PCM and its connector. Use a DVOM to check the continuity between the white wire of the 2-P connector (control side) and PCM terminal A-19.If neither adapter is available, use a DVOM and carefully backprobe between the WHT wire at terminal A-19 and solenoid harness connector.

8) If continuity is not present, locate and repair the open circuit condition in the white wire between terminal A-19 and the solenoid connector. If continuity is okay, substitute a known-good PCM and retest the system. If the symptom or code goes away, replace the original PCM.

9) If no vacuum was present at 3700 rpm, disconnect the air intake solenoid 2-P connector. Retest for vacuum at 3700 rpm. If vacuum is present now, replace the Intake Control solenoid valve and retest the system. If vacuum is still not present, turn the key off and disconnect the PCM 'A' connector. Use a DVOM and check for continuity between the white wire of the harness connector and body ground. If there is continuity at this point, the WHT wire is shorted to ground. Locate the short to ground condition in the white wire between the PCM and the harness connector. Make the repair and retest the system. If continuity is present (okay), replace the PCM with a known-good unit and retest the system. If the symptom or trouble code goes away, replace the original PCM.

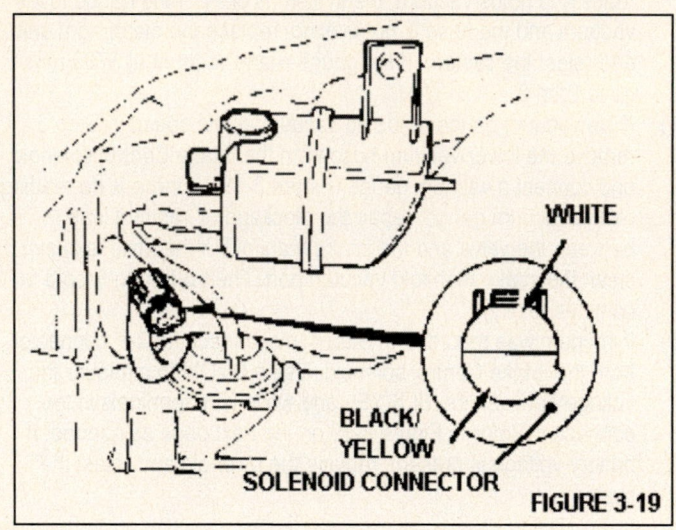

FIGURE 3-19

■ NOTE: *Refer to the pin voltage charts in this manual for models other than a 1990 Accord 2.2L MFI engine for proper pin and wire color identification for all references to PCM terminals in this article.*

TANDEM CONTROL SYSTEM INSPECTION & TESTS

INTRODUCTION (CIVIC CRX 1.5L TBI)

The Tandem Control system is designed to improve atomization of fuel from the main fuel injector in response to various operating conditions. When the tandem control solenoid is turned off, venturi vacuum is not applied to the vacuum chamber in the tandem valve control diaphragm. This allows the tandem valve to move to a nearly closed position. The narrow clearance between the tandem valve and inner wall of the throttle body generates rapid airflow that promotes improved atomization of fuel from the main injector.

When the tandem valve control solenoid is turned on, venturi vacuum is applied on the tandem diaphragm and the tandem valve opens in response to the venturi vacuum. Venturi vacuum represents the airflow rate through the venturi. This action causes improved atomization of A/F mixture, regardless of the airflow rate.

The solenoid is enabled if any of these conditions exist:

*   Engine coolant temperature is over 160°F
*   Engine coolant temperature is below 160°F and engine speed is over 1500 rpm on automatic transaxle applications
*   On manual transaxle applications, with engine coolant temperature less than 160°F, engine speed over 2000 rpm and atmospheric pressure below 21.7" Hg
*   On manual transaxle applications, with engine speed over 2800 rpm with a BARO pressure above 21.7" Hg

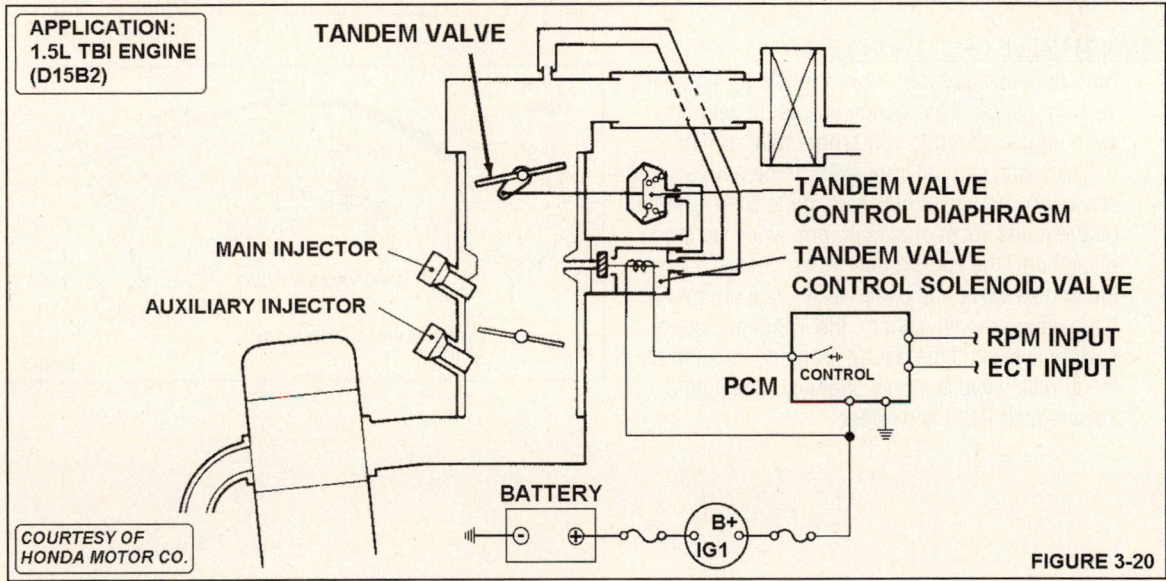

FIGURE 3-20

DRIVEABILITY SYMPTOMS

Problems with the Tandem Control system could cause these symptoms.

*   Frequent Stalling (during engine warmup)
*   Hard Start Cold
*   Loss of Power

PRELIMINARY STEPS

The following inspections and tests should be performed in sequence to isolate the problem area in order to make the necessary repairs. Always start the test procedure by following the preliminary steps.

Carefully disconnect the vacuum hose from the tandem valve control diaphragm. Then connect a vacuum gauge to the hose. The engine coolant temperature must be below 160°F in order to perform this test.

CAUTION: *The following test steps require that the drive wheels are blocked and the vehicle exhaust system is purged from the work area for safety purposes.*

TANDEM VALVE DIAPHRAGM INSPECTION

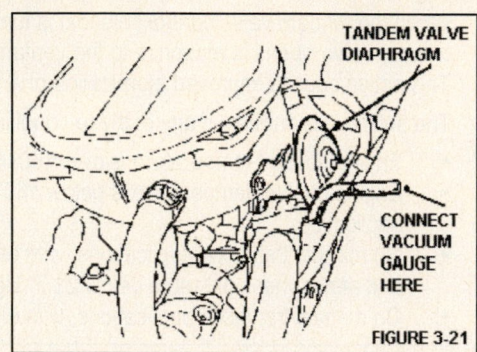

FIGURE 3-21

1)   Inspect the tandem valve shaft for binding or sticking conditions. Inspect the tandem valve for smooth movement. If problems are found, clean the linkage and shafts with an approved cleaner and then retest the system.

2)   Disconnect the vacuum hose from the control diaphragm and connect a hand held vacuum pump to the hose. Apply vacuum and check that the lever tang of the tandem valve is in close contact with the stopper when the valve is fully closed. If any faults are found that cannot be repaired, replace the tandem control diaphragm and retest.

TANDEM VALVE OPERATION TESTS

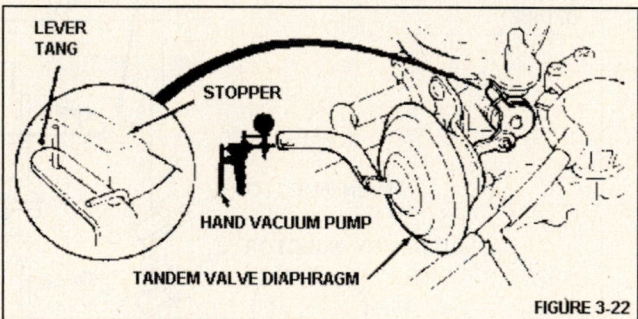

FIGURE 3-22

1)   Start the engine (cool), allow it to idle and read the vacuum gauge. If the gauge shows no vacuum, raise engine speed to 2000 rpm (3000 rpm on vehicles with manual transaxle). If the gauge shows no vacuum at high rpm, go to Step 4. If the gauge reads vacuum at high rpm, warm up the engine until the cooling fan cycles. Slowly open the throttle valve and observe the vacuum gauge. If the gauge reads vacuum, the Tandem Control system is okay. If the gauge reads no vacuum as the throttle valve is slowly opened, substitute a known-good PCM and retest.

TANDEM VALVE OPERATION TESTS (CONTINUED)

2) If the gauge reads vacuum in Step 2, remove the 2-P connector from the tandem valve solenoid. If the gauge reads vacuum with the solenoid unplugged, the valve is leaking. Replace the valve and retest the system.

3) If the gauge reads no vacuum in Step 2, turn the key off and disconnect the PCM 'B' connector. Use a DVOM to check for continuity to ground on the orange wire. If continuity is present, locate and repair the short to ground condition on the orange wire between terminal B2 in the PCM and the 2-P connector. Retest the system when repairs are done. If the orange wire is not grounded, substitute a known-good PCM and retest the system. If the symptom goes away, replace the original PCM.

4) If the gauge reads no vacuum at high speed in Step 1, remove the 2-P connector from the tandem valve. Raise engine speed to 2000 rpm (3000 rpm on M/T models) and use a DVOM to test for voltage between the ORN and BLK/YEL wires of the connector. If the DVOM reads battery voltage at high speed, turn the engine off and remove the solenoid valve from the throttle body. Inspect and clean the internal port as needed. If the port is clean, replace the solenoid valve and retest the system.

5) If the DVOM reads no voltage at high speed in Step 4, measure the voltage between the BLK/YEL wire and body ground (Figure 3-23). If it still reads no voltage, locate and repair the open feed circuit in the BLK/YEL wire between the No. 14 fuse and the 2-P connector. Retest the system when done.

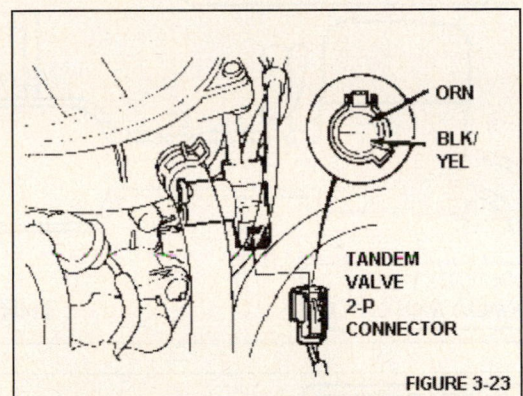

ORN

BLK/ YEL

TANDEM VALVE 2-P CONNECTOR

FIGURE 3-23

6) If the DVOM reads voltage between the BLK/YEL wire and body ground, connect the Honda PCM (ECU) test harness or BOB connector between the PCM and its connector. Use a DVOM to test the continuity of the ORN wire between the B2 terminal and the 2-P connector. If neither of the adapters are available, backprobe between PCM terminal B2 and the 2-P connector to check the continuity of the Orange wire. If the circuit is open, locate and repair the open circuit in the ORN wire between the PCM terminal B2 and the 2-P connector. If continuity is okay on the ORN wire, substitute a known-good PCM and retest. If the symptom or code is gone, replace the original PCM.

■ NOTE: *Refer to the pin voltage charts in this manual for models other than a 1990 Civic CRX 1.5L TBI engine for proper pin and wire color identification for all references to PCM terminals in this article.*

THROTTLE CONTROL SYSTEM (DASHPOT) INSPECTION & TESTS

INTRODUCTION

During engine cranking, the Throttle Control system uses the dashpot diaphragm to function as a throttle valve opener. During engine cranking, the spring in the dashpot diaphragm pushes the throttle valve open a certain amount to assist engine startup.

During idle mode, intake manifold vacuum is applied to the diaphragm to pull up the diaphragm rod to hold the throttle valve in the idle position.

On manual transaxle applications, the Throttle Control system also assists in preventing engine speed from dropping too low during shifting.

COMPONENT EXAMPLES

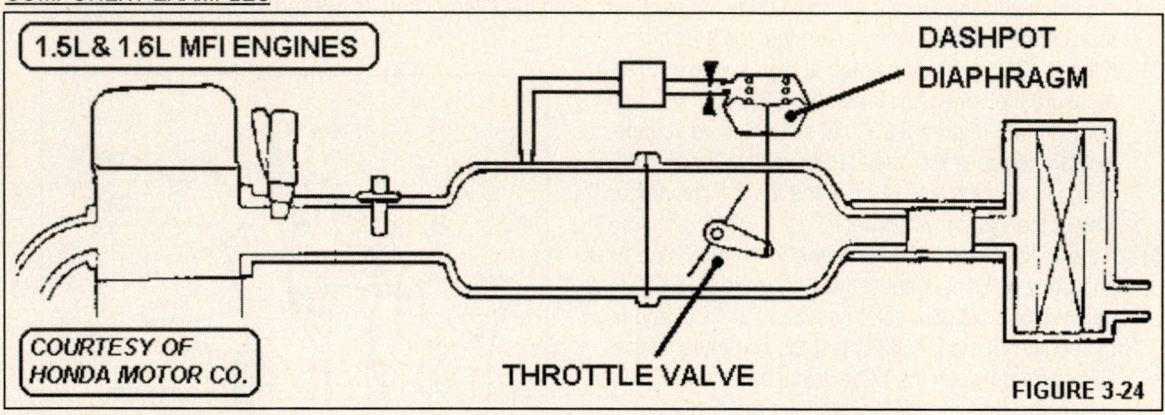

1.5L & 1.6L MFI ENGINES

DASHPOT DIAPHRAGM

THROTTLE VALVE

COURTESY OF HONDA MOTOR CO.

FIGURE 3-24

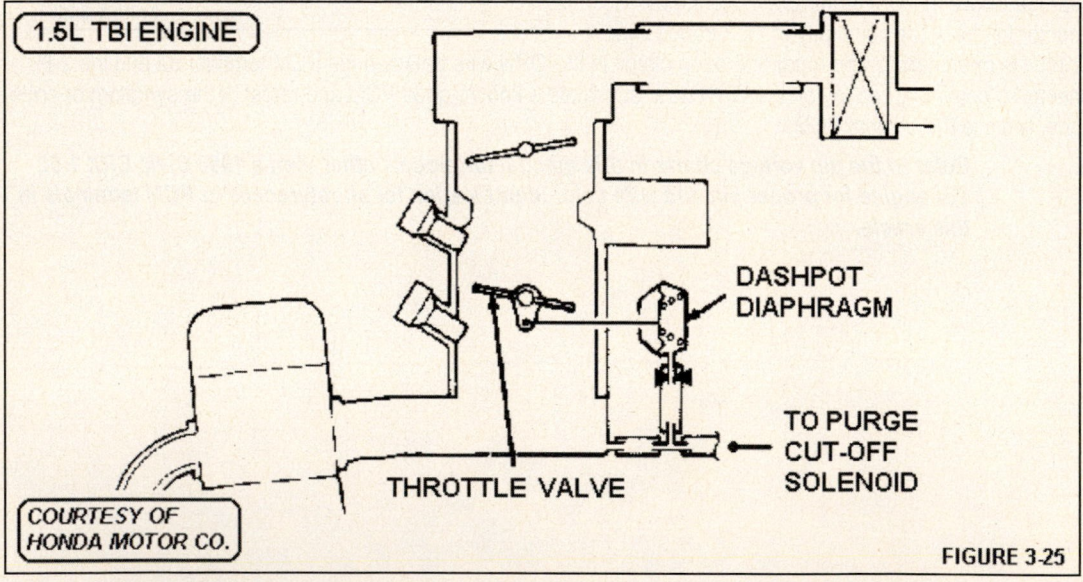

1.5L TBI ENGINE

DASHPOT DIAPHRAGM

TO PURGE CUT-OFF SOLENOID

THROTTLE VALVE

COURTESY OF HONDA MOTOR CO.

FIGURE 3-25

DRIVEABILITY SYMPTOMS

Problems with the Throttle Control system could cause these symptoms:

- Engine speed too high with a warm engine
- Engine speed drops to low during shifts on M/T applications
- Fast idle speed out of specifications

PRELIMINARY STEPS

The following inspections and tests should be performed in sequence to isolate the problem area in order to make the necessary repairs. Always start the test procedure by following the preliminary steps.

Carefully disconnect the No. 6 vacuum hose from the dashpot diaphragm. Then connect a tachometer to the engine.

CAUTION: *The following test steps require that the drive wheels are blocked and the vehicle exhaust system is purged from the work area for safety purposes.*

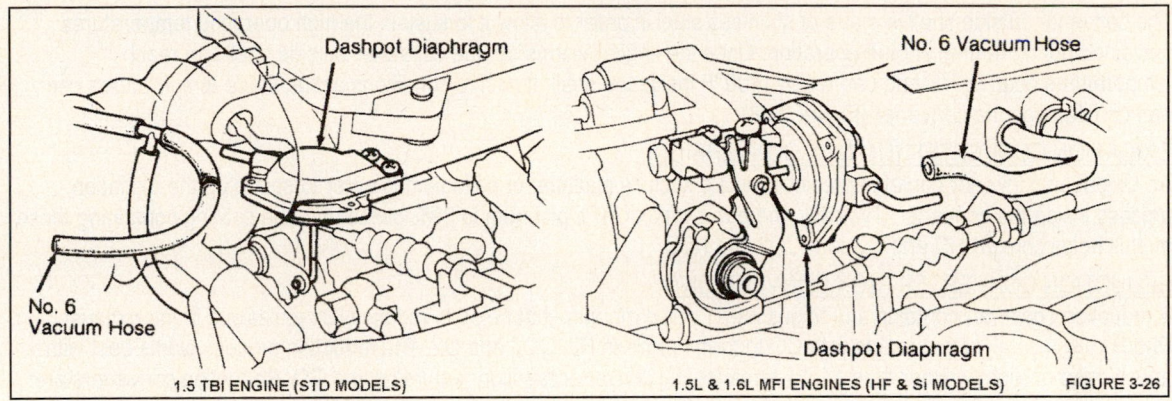

| 1.5 TBI ENGINE (STD MODELS) | 1.5L & 1.6L MFI ENGINES (HF & Si MODELS) | FIGURE 3-26 |

DASHPOT INSPECTION & TESTS

1) Start and warmup the engine until the cooling fan cycles.
2) Observe the engine speed at idle. It should be 2000-3000 rpm on both A/T and M/T models (vacuum hose to dashpot disconnected). If the engine speed is too high, adjust the rpm by bending the dashpot diaphragm linkage tab.
3) Reconnect the vacuum hose momentarily and observe the engine speed change. If the engine speed does not change, connect a vacuum gauge to the No. 6 vacuum hose and check for vacuum. The gauge should read vacuum. If the gauge reads no vacuum, check the No. 6 vacuum hose for proper connection, cracks or blockage. On 1.5L engines, inspect the vacuum T-fitting and replace it if faults are found. If there is vacuum present (okay), reconnect the No. 6 vacuum hose and check the base idle speed. Refer to the idle speed adjustment specifications as needed.

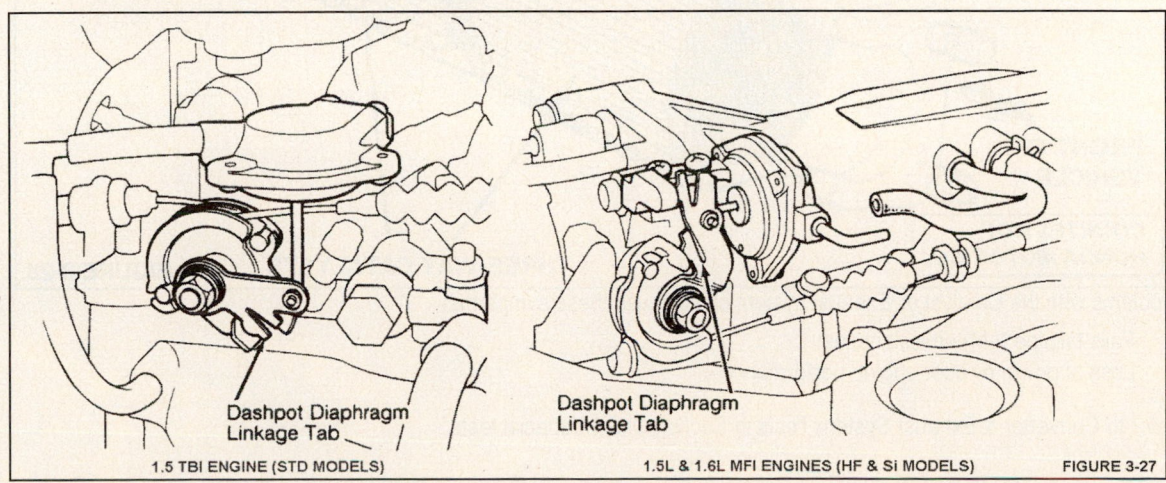

| 1.5 TBI ENGINE (STD MODELS) | 1.5L & 1.6L MFI ENGINES (HF & Si MODELS) | FIGURE 3-27 |

Catalyst & Exhaust Systems

INTRODUCTION

Honda vehicles with fuel injection are equipped with a three-way catalytic converter (TWC) to convert hydrocarbons (HC), carbon monoxide (CO) and oxides of nitrogen (NOx) in the exhaust into harmless gases composed of carbon dioxide (CO_2), dinitrogen (N_2) and water vapor. Some applications use two converters to reduce and oxidize the exhaust. The front converter, mounted close to the exhaust manifold, is called a pre-cat converter. The pre-cat converter comes up to light-off temperature very quickly.

The converter outside shell is made of stainless steel in order to allow it to sustain the high operating temperatures sustained during normal engine operation. Once the engine warms up, the converter outside shell can reach temperatures near 700°F and can reach 1400°F inside the shell. If misfire conditions occur, these temperatures rise and can damage the converter (Figure 3-28).

OXIDATION CONVERTER (NO NOX CONVERSION)

An Oxidation converter contains precious metals such as platinum or palladium in order to speed up the oxidation process inside the converter. Typically, there is a ratio of 70% platinum to 30% palladium. The coating containing these metals helps convert HC and CO into CO_2 and H_2O.

REDUCTION CONVERTER (NOX CONVERSION)

A reduction converter contains a precious metal called rhodium that helps to reduce NOx emissions in the exhaust. Rhodium is a catalyst that reacts with CO to form harmless N_2, CO_2 and O_2. The reduction process works best with an A/F ratio of slightly richer than 14.7:1 because the converter reaction requires some CO. Reduction converters are mounted into the exhaust stream as either pre-catalyst converters or as part of a dual-bed converter.

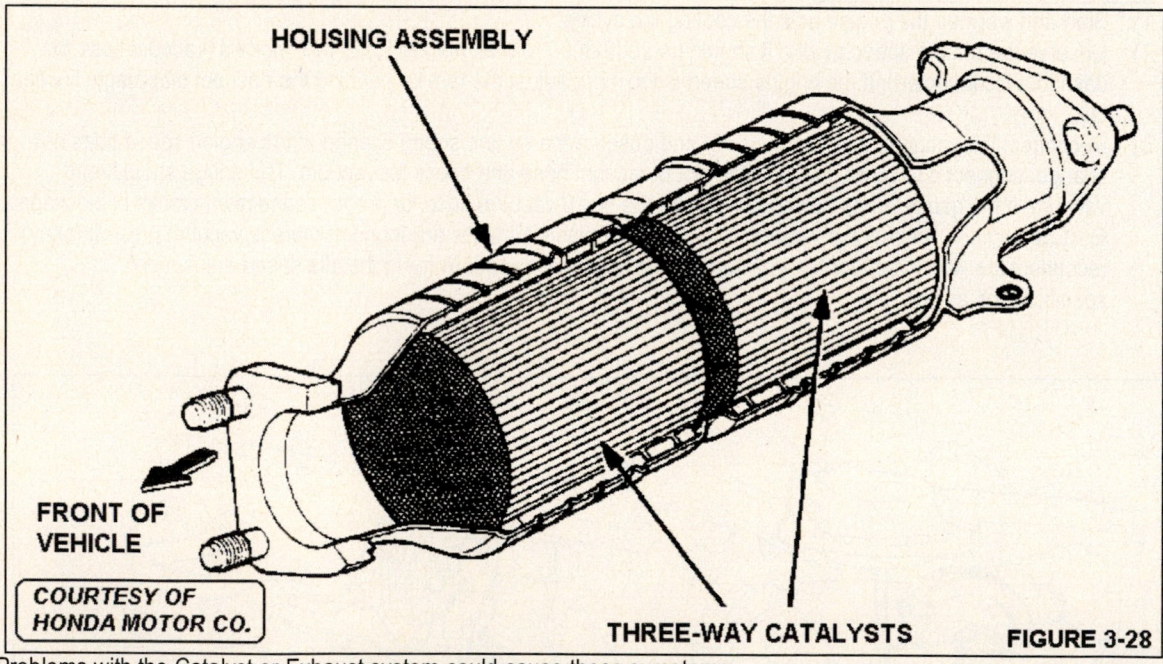

HOUSING ASSEMBLY

FRONT OF VEHICLE

COURTESY OF HONDA MOTOR CO.

THREE-WAY CATALYSTS

FIGURE 3-28

Problems with the Catalyst or Exhaust system could cause these symptoms:

- Fails tailpipe (I/M) emissions test
- Loss of power or poor engine performance

Refer to Converter & Exhaust System Tests in Section One for special tests.

EGR System

EGR SYSTEM INSPECTION & TESTS

INTRODUCTION

The Exhaust Gas Recirculation (EGR) system is designed to reduce oxides of nitrogen emissions by recirculating exhaust gas through the EGR valve and the intake manifold into the combustion chambers.

SYSTEM COMPONENTS

On 1988-95 Honda vehicles (OBD I systems), a typical EGR system consists of the components listed below. For 1996-98 Honda applications (OBD II systems), refer to EGR system information in Section 4.

- EGR valve with EGR valve lift sensor
- Constant vacuum control (CVC) valve
- EGR control solenoid valve
- EGR related wiring and vacuum lines
- The Powertrain Control Module (PCM)

An example of the EGR system components for a 1990 Accord with a 2.2L I4 (F22A1) engine is shown in Figure 3-29.

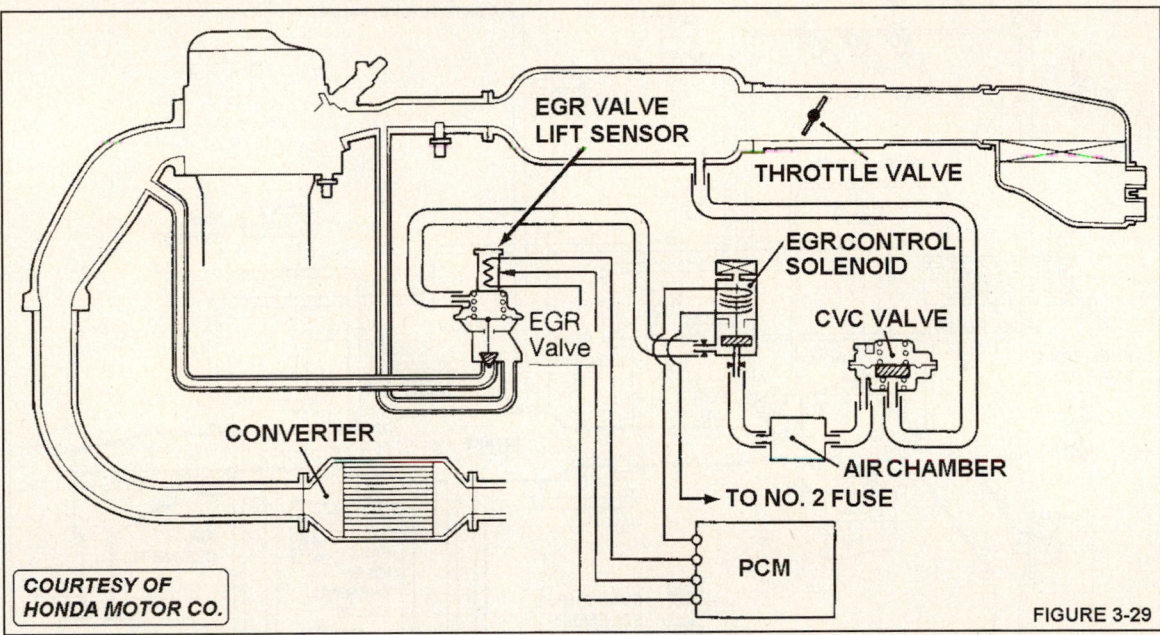

FIGURE 3-29

ACTUAL & IDEAL EGR VALVE LIFT POSITION

The PCM contains memory tables that include the ideal EGR valve position (amount of EGR valve lift or opening) for various operating conditions.

The PCM determines the amount of EGR valve lift (or opening) through a signal from the EGR valve lift sensor. The actual valve lift position is compared to an ideal sensor valve lift value stored in memory.

The PCM determines the ideal EGR valve lift value through signals it receives from other sensors. If there is a difference between the actual and ideal valve lift values, the PCM turns off the EGR control solenoid valve to reduce the amount of vacuum applied to the EGR valve.

EGR SYSTEM INSPECTION & TESTS

EGR SYSTEM DIAGNOSTICS

As discussed, the PCM monitors the operation of the EGR system through the amount of change in the EGR valve lift sensor. If the PCM detects that the amount of change in the sensor is too much or too little, or if it detects a circuit fault, it will set a Code 12 and turn on the Check Engine light (or MIL).

However, if an EGR related symptom is suspected, the EGR Functional Test in this article can be used to locate a fault in one of the EGR components or circuits. In effect, the functional test provides a "manual" method of cycling the EGR components.

DRIVEABILITY SYMPTOMS

Problems in the EGR system can cause any of the following symptoms:

*   Rough Idle or Frequent Stalling
*   Loss of power or Detonation

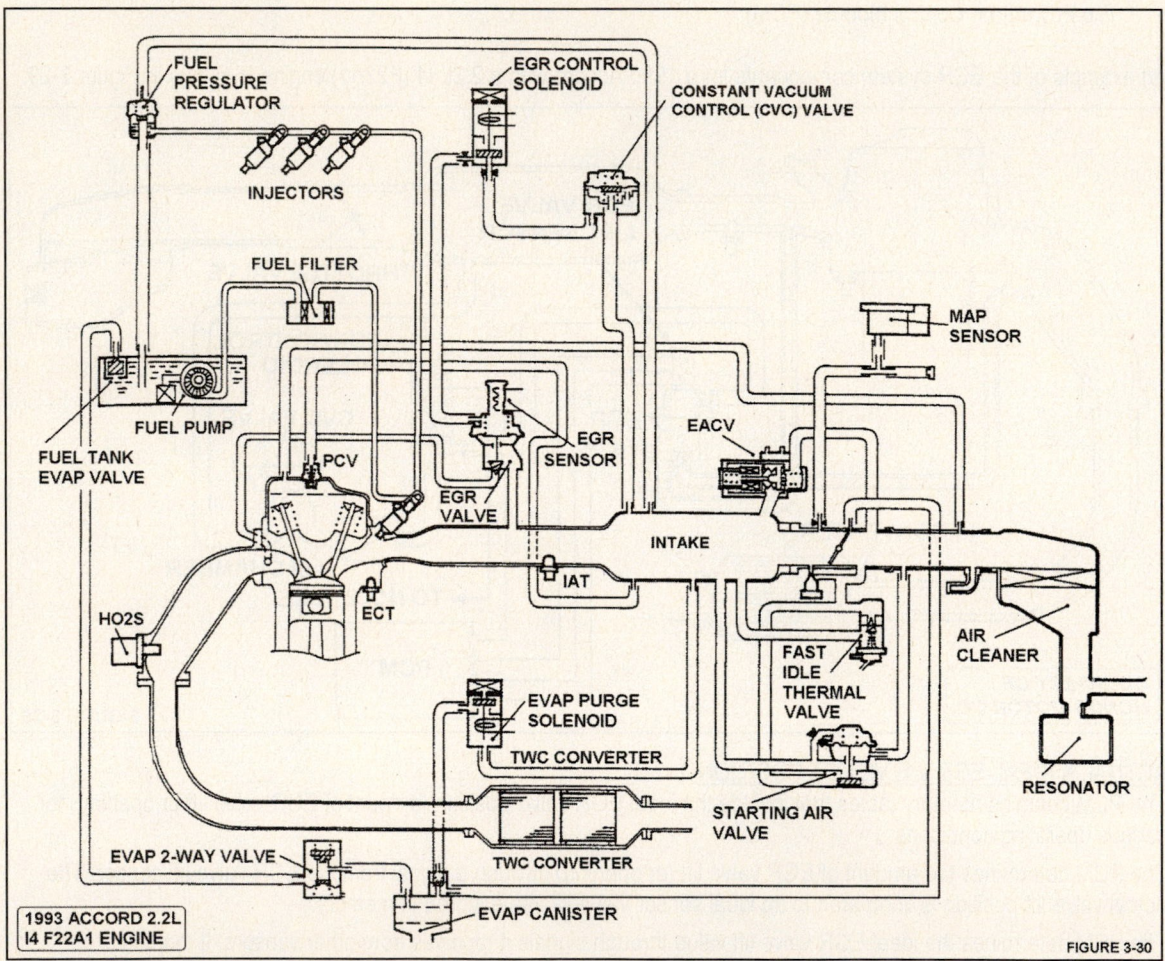

FIGURE 3-30

PRELIMINARY STEPS

The EGR Functional Test on the next page should be performed in sequence to isolate the problem area in order to make the necessary repairs. Inspect all EGR vacuum hoses for leaks at the intake manifold, constant vacuum control (CVC) valve, EGR control solenoid and at the EGR valve.

CAUTION: *The following test steps require that the drive wheels are blocked and the vehicle exhaust system is purged from the work area for safety purposes.*

<u>EGR FUNCTIONAL TEST</u>

1) Start the engine and run it at 2000 rpm in P/N until the electric cooling fan cycles.

2) Allow the engine to return to idle and remove the No. 16 vacuum hose (Figure 3-31) from the EGR valve. Connect a hand vacuum pump to the valve nipple and apply 8 to 10" of vacuum to the valve. The EGR valve should hold vacuum and the engine should stall as vacuum is applied. If it does not, determine if the EGR valve is defective or if the intake manifold passages are plugged. Replace the valve or clean the passages as required and retest.

3) If the EGR valve operates normally in Step 2, move the vacuum gauge to the No. 16 hose (Figure 3-31) and check for vacuum. The gauge should read 0" vacuum with the engine at idle speed (the PCM should have the EGR control solenoid valve turned off). If the gauge shows any vacuum at this time, turn the engine off and perform the following steps:

   • Inspect the EGR vacuum hoses to verify correct routing
   • If the EGR control solenoid valve is leaking vacuum, replace it
   • Remove the EGR valve and inspect for carbon around or under the valve seat. Clean the valve as needed and install it with a new gasket.

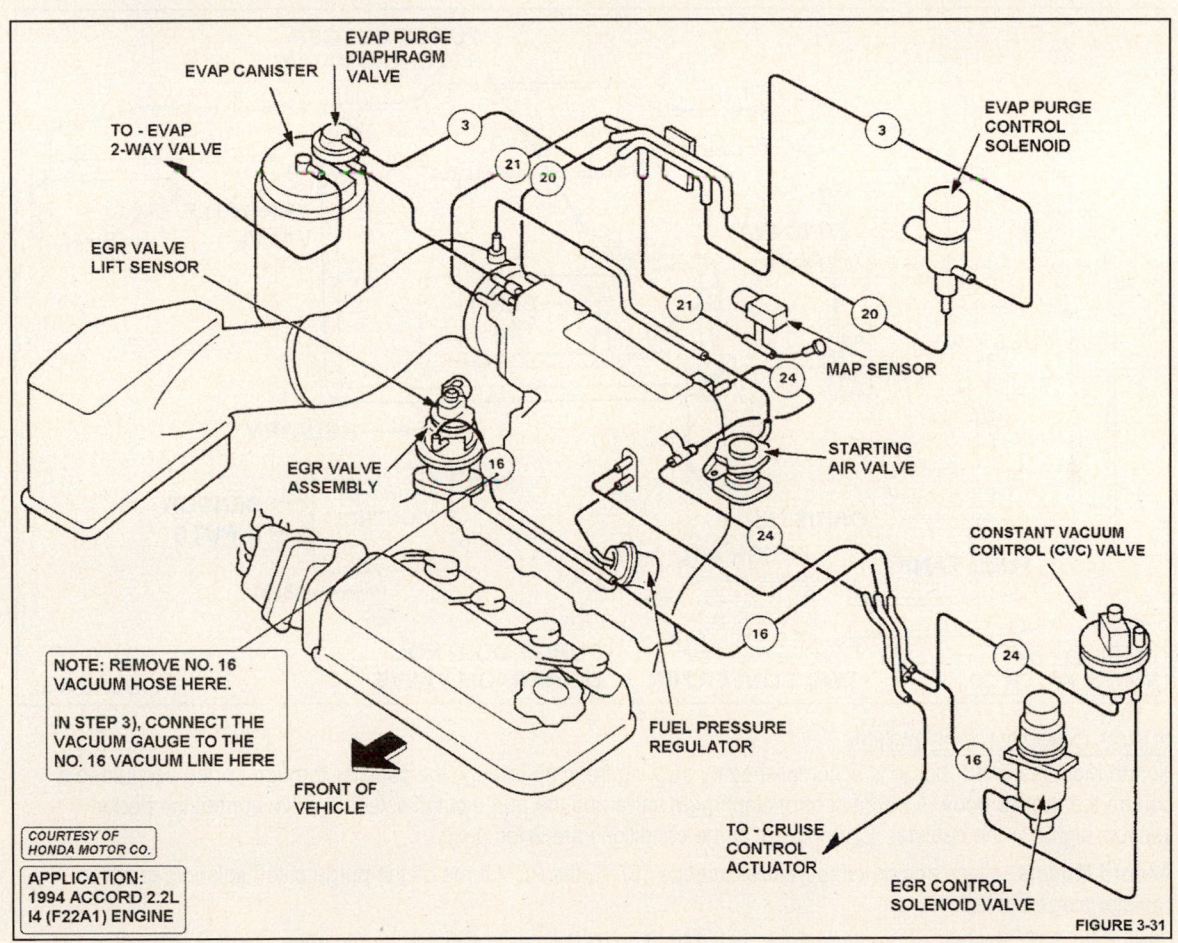

FIGURE 3-31

EVAP System
EVAP SYSTEM INSPECTION & TESTS

INTRODUCTION
The Evaporative Emission Control (EVAP) system is designed to minimize the amount of hydrocarbons (fuel vapors) released into the atmosphere.

SYSTEM COMPONENTS
On 1988-95 Honda vehicles (OBD I systems), a typical EVAP system consists of the components listed below. For 1996-98 Honda applications with OBD II diagnostics, refer to the EVAP system information in Section 4.

- EVAP charcoal canister
- Fuel tank vapor control system
- EVAP purge cutoff solenoid valve
- A Powertrain Control Module (PCM)

1990 ACCORD EVAP SYSTEM COMPONENT EXAMPLES

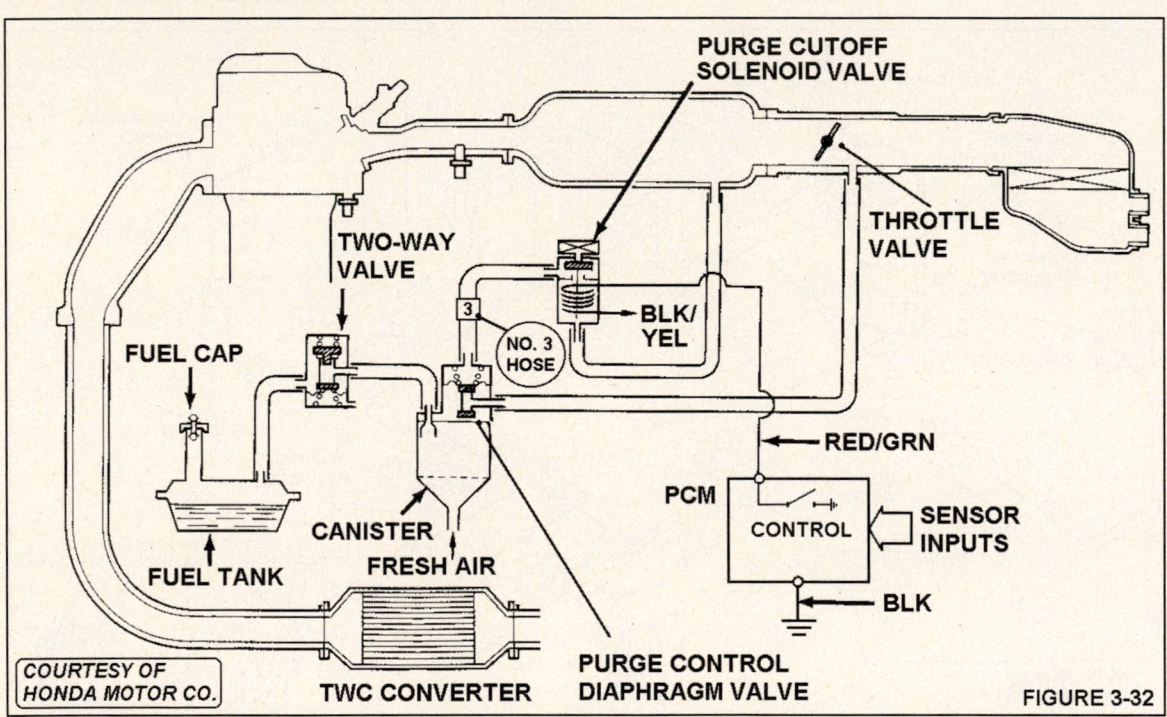

PURGE CUTOFF SOLENOID VALVE

THROTTLE VALVE

TWO-WAY VALVE

BLK/YEL

NO. 3 HOSE

FUEL CAP

RED/GRN

PCM

SENSOR INPUTS

CONTROL

BLK

CANISTER

FRESH AIR

FUEL TANK

PURGE CONTROL DIAPHRAGM VALVE

COURTESY OF HONDA MOTOR CO.

TWC CONVERTER

FIGURE 3-32

PURGE CONTROL OPERATION
Honda Models - EVAP purge is accomplished by drawing fresh air through the canister through ported vacuum to a port on the throttle body. A purge control diaphragm valve and the purge cutoff solenoid valve control the ported vacuum signal to the canister. Examples of purge conditions are listed next:

Accord Models - Once engine temperature exceeds 167°F, the PCM turns off the purge cutoff solenoid to allow canister purge operation.

Civic CRX STD Models - At temperatures over 176°F on STD models with the throttle open, the purge cutoff solenoid is turned Off 5 seconds after startup to allow purge.

Civic CRX HF & Si - At temperatures over 104°F on HF (35°F on Si) with the throttle open, the purge cutoff solenoid is turned Off 10 seconds after startup to allow purge.

EVAP SYSTEM INSPECTION & TESTS

DRIVEABILITY SYMPTOMS

Problems in the EVAP system or its components could cause these symptoms:

- Fails I/M or State Emission Test
- Poor Fuel Economy
- Poor Cold Engine Driveability

SYSTEM COMPONENTS

Examples of the EVAP system components for 1990 Civic CRX STD, HF & Si engines are shown in Figure 3-33.

PRELIMINARY STEPS

The inspection and test steps that follow should be performed in sequence to isolate the problem area in order to make repairs.

Inspect all EVAP vacuum hoses for cracks, leaks or disconnects at the intake manifold, canister and purge control solenoid valve.

Check for corroded, loose or backed-out electrical connections at the EVAP purge control solenoid valve and the PCM before proceeding.

Next, connect a suitable tachometer to the engine ignition system in order to make engine speed measurements.

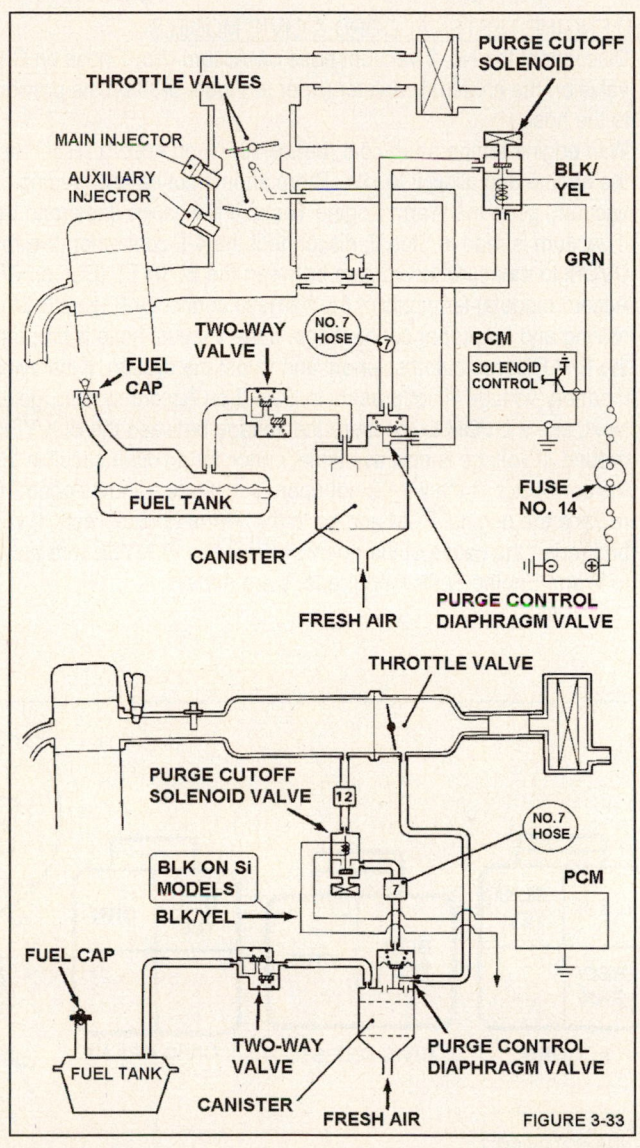

FIGURE 3-33

CAUTION: *The following test steps require that the drive wheels are blocked and the vehicle exhaust system is purged from the work area for safety purposes.*

COLD ENGINE TESTS - ACCORD & CIVIC MODELS

1)   Disconnect the No. 3 vacuum hose on Accord (No. 7 hose on Civic models) at the EVAP purge control diaphragm valve on the charcoal canister (refer to Figure 3-33 on the previous page) and connect a suitable vacuum gauge to the hose.

2)   With engine temperature cold (below 167°F on Accord, 176°F on Civic STD or 104°F on CRX HF models), start the engine and allow it to idle. There should not be any reading on the vacuum gauge. If the gauge does not read vacuum, go to the Warm Engine Tests. If the gauge does read vacuum, go to Step 3.

3)   If vacuum is read in Step 2, disconnect the 4-P connector (2-wire connector on CRX and STD models). Use a DVOM to measure the voltage between the BLK/YEL (BLK on CRX Si models) and the GRN wire (RED/GRN on Accord models) terminals of the harness connector (Figure 3-34). If battery voltage is present, verify vacuum hose routing and for proper connections. If the vacuum hose is properly routed and the connections are tight, replace the EVAP purge cutoff solenoid and retest the system. If the symptom goes away, go to the Warm Engine Tests. If battery voltage is not present in Step 3 on Accord models, go to Step 4.

4)   Next, use a DVOM to measure the voltage between the BLK/YEL wire in the 4-wire connector and a good body ground. If voltage is now available, check for an open circuit in the RED/GRN wire between PCM terminal A6 and the connector. If the wire is not open, substitute a known-good PCM and retest. If the symptom goes away, replace the original PCM and go to the Warm Engine Tests. If voltage is still not available at the connector, locate and repair the cause of the open circuit in the BLK/YEL wire between the 4-P connector and the No. 24 fuse. Go to Warm Engine Tests when repairs are done.

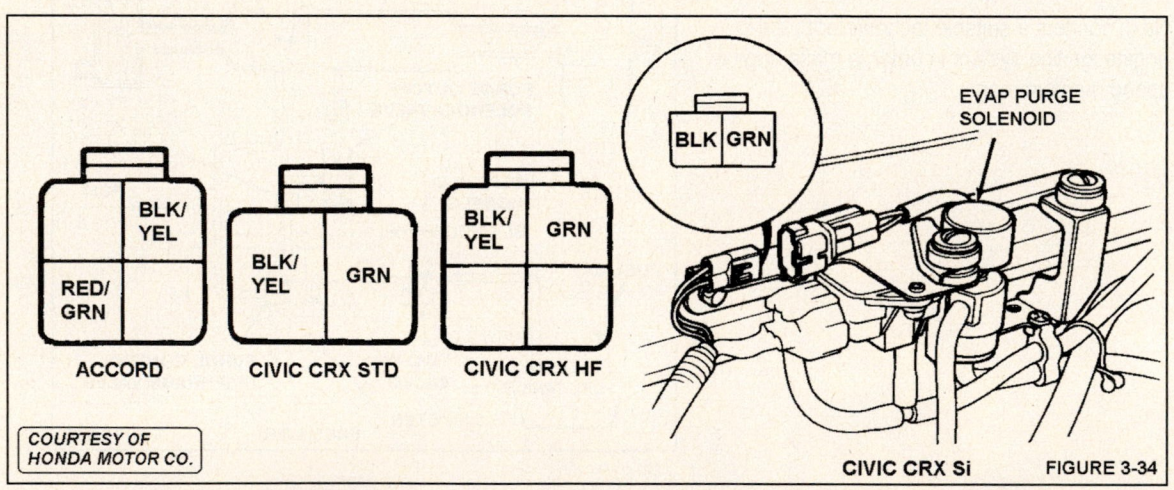

BLK GRN

EVAP PURGE
SOLENOID

BLK/
YEL

RED/
GRN

BLK/
YEL   GRN

BLK/
YEL   GRN

ACCORD   CIVIC CRX STD   CIVIC CRX HF

COURTESY OF
HONDA MOTOR CO.

CIVIC CRX Si   FIGURE 3-34

WARM ENGINE TESTS - ACCORD & CIVIC MODELS

The next few test steps require that the engine be at normal operating temperature with the cooling fan cycling.

1) Remove the No. 3 vacuum hose on Accord (No. 7 hose on Civics) at the canister purge control diaphragm (Figure 3-35). Connect a vacuum gauge to the hose.

2) Start the engine and allow it to idle. The gauge should read vacuum at idle speed. If it reads vacuum (within 10 seconds on CRX models), go to Step 5.

3) If the gauge reads no vacuum in Step 2, inspect the vacuum hose routing and connections. If they are okay, remove the solenoid 4-P connector (2-P connector on CRX models). If the gauge now reads vacuum, replace the purge cutoff solenoid and retest for the symptom. Go to Step 4 when repairs are completed. On CRX models, if vacuum is present in this step, go to Step 9.

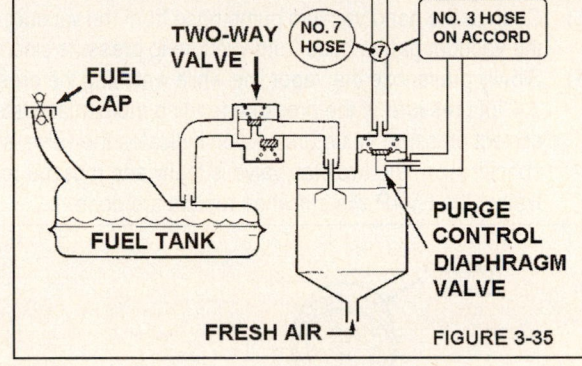

FIGURE 3-35

4) On Accord models, if the gauge reads vacuum after repairs are completed in Step 3, test the RED/GRN wire for an open circuit between PCM terminal A6 and the connector. If the wire is okay, substitute a known-good PCM and retest the system. If the symptom is gone, replace the original PCM and go to Step 5.

5) If the gauge reads vacuum at idle speed in Step 2, remove the gauge and reconnect the vacuum hose.

6) Next, remove both the fuel filler cap and the EVAP purge line hose from its connection at the vehicle frame and connect a vacuum gauge to the EVAP purge air hose.

7) Raise engine speed to 3500 rpm and maintain that engine speed for one minute. The gauge should read vacuum within one minute. If it does, the system is okay. Go to the Two-way Valve Test. If no vacuum is shown, replace the EVAP canister assembly on CRX models. Retest the EVAP system when the repairs are finished. On Accord models, if the gauge reads no vacuum after one minute, go to Step 8.

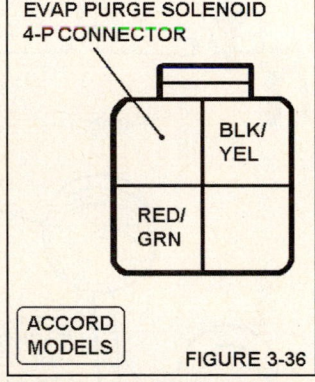

FIGURE 3-36

8) On Accord models, if the gauge reads no vacuum in Step 7, move the gauge to the fresh air fitting at the bottom of the canister and again raise engine speed to 3500 rpm. If the gauge still reads no vacuum after one minute, inspect the purge hose for restrictions. If the hose is okay, clean the purge port on the throttle body. If the gauge read vacuum after one minute, replace the canister and retest the system.

9) On Civic CRX models, if the gauge reads vacuum in Step 3 with the connector removed, turn the key off and remove the PCM 'A' connector. Use a DVOM to check for continuity between the GRN wire and ground. If continuity is present, locate and repair the short to ground condition in the GRN wire. If the wire is okay, substitute a known-good PCM and retest. If the symptom goes away, replace the original PCM.

TWO-WAY VALVE TEST

On all models, to test the two-way valve, remove the fuel filler cap and the vapor line at the fuel tank. Connect a T-fitting from the vacuum gauge with a suitable hand vacuum pump into the fitting (Figure 3-37).

1) Use the hand vacuum pump to slowly apply vacuum to the T-fitting while reading the vacuum gauge. The reading should stabilize at 0.2-0.6 inch of vacuum (H_2O). If the vacuum reading indicates that the valve opens at below 0.2 inches of vacuum or above 0.6 inches of vacuum, replace the two-way valve and retest system. If the vacuum reading stabilizes as specified, go to Step 2.

2) Change the hand vacuum pump hose from the vacuum fitting to the pressure fitting on the hand pump and move the vacuum gauge hose from vacuum to pressure side of the two-way valve (Figure 3-37).

3) Slowly pressurize the vapor line while watching the pressure gauge. The pressure reading should stabilize at 0.4-1.4" of pressure. If the pressure reading momentarily stabilizes at the specified range, the valve opened at the correct pressure value (this action indicates the valve is okay). If the valve opened above or below the specification, the two-way valve is faulty and must be replaced. Replace the two-way valve and the fuel filler cap. Retest the EVAP system when repairs are done.

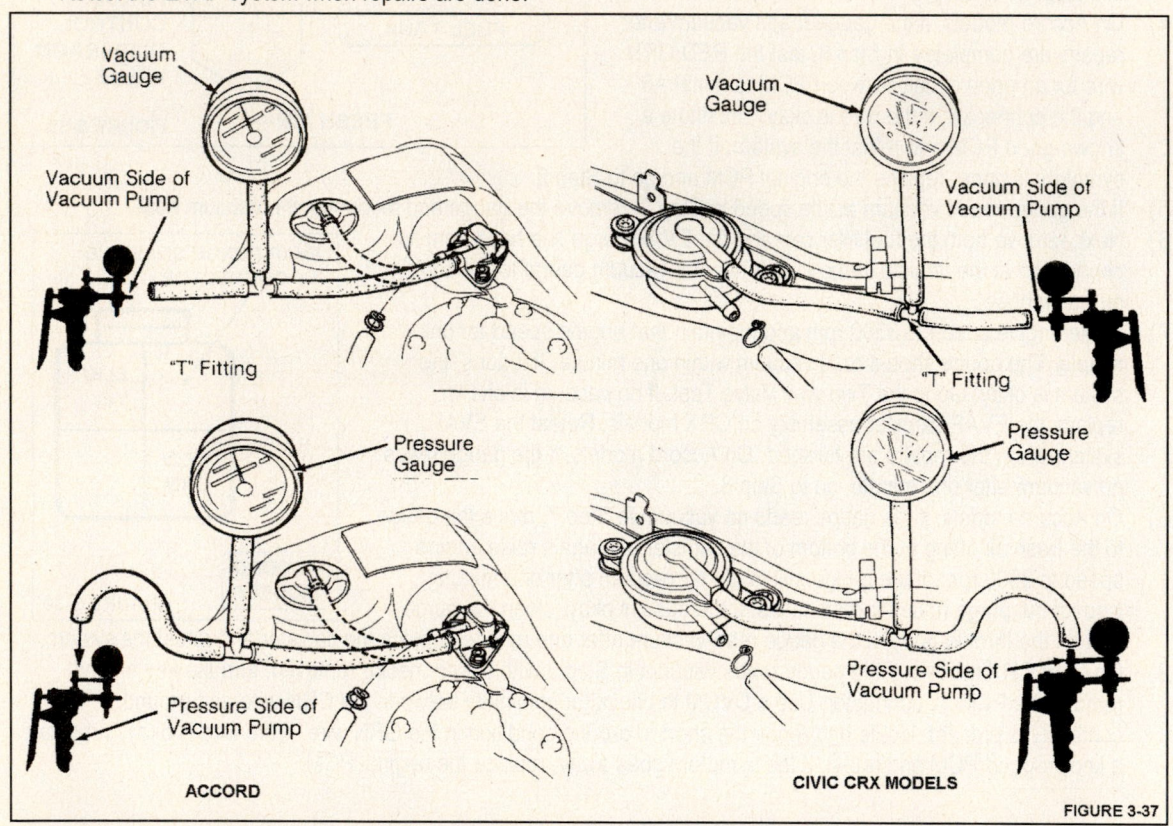

FIGURE 3-37

■ **NOTE:**   *For models other than 1990 Accord, Civic CRX STD, HF and Si applications discussed in this article, refer to other pin voltage charts for pin and wire color identification for all references to PCM terminals and wire colors.*

Fuel System
FUEL SYSTEM TESTS

INTRODUCTION

The Fuel system on Honda vehicles is designed to deliver fuel at a regulated high pressure to the fuel rail and injectors. Additionally, the Fuel system is also designed to cutoff fuel pressure if the engine stops running.

Fuel system tests are after these test steps have been performed:

- The complaint is verified and a visual inspection is done
- PGM-FI Diagnostics are completed when they show Pass (see NOTE)
- Base engine tests are done
- An actual driveability symptom is present during vehicle testing
- Related technical service bulletins have been searched in the TSB index

■ **NOTE:** *Turn the key on and watch the Check Engine or PGM-FI light. It should come on for two seconds and then go off. If the Check Engine light remains on, refer to the PGM-FI Diagnostics article in this section (Section 3).*

DRIVEABILITY SYMPTOM DIAGNOSIS

If a driveability symptom is present, refer to the list of symptoms included with each particular Fuel system test in the articles that follow. Also, a technician should use their experience and knowledge of how a Fuel system operates when deciding which test to perform first. It may be necessary to perform more than one type of Fuel system test to repair the fault.

SYSTEM COMPONENTS

The Fuel system on 1985-98 Honda engines with fuel injection includes the fuel tank, fuel filter, pressure regulator, injectors, fuel lines and hoses (Figure 3-38).

The electrical portion of the Fuel system consists of the main relay, in-tank fuel pump, injector resistor bank (some models), fuel injectors and the PCM.

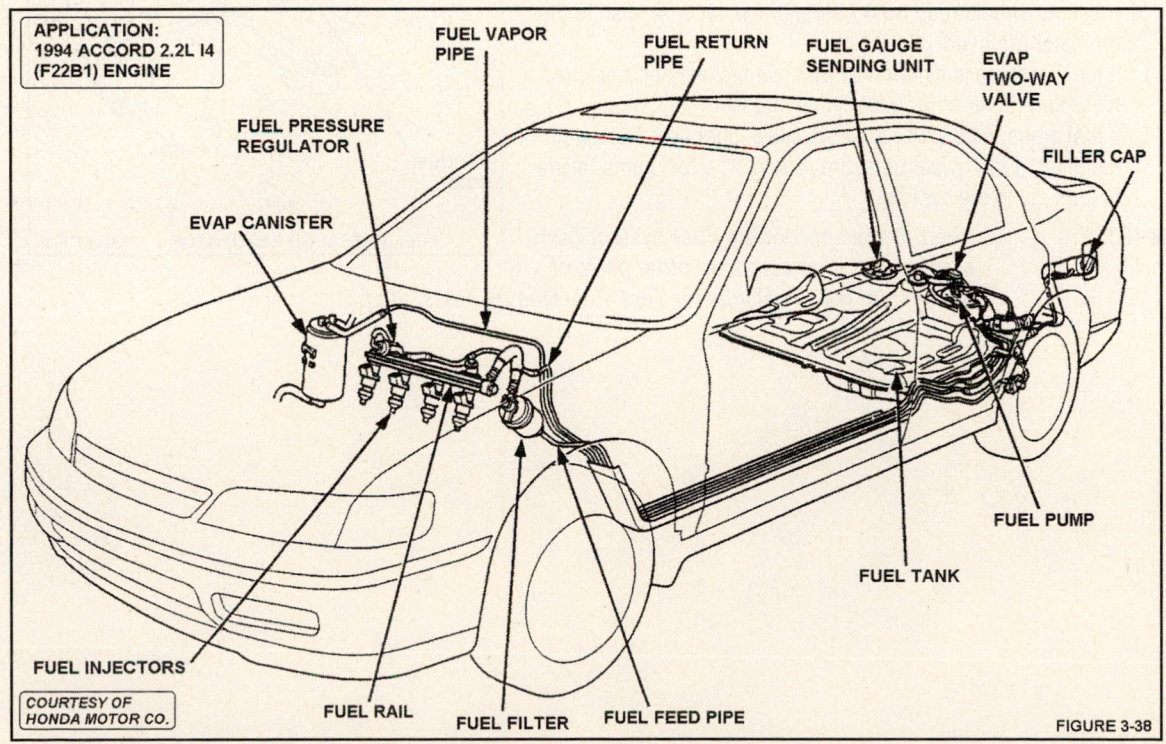

APPLICATION:
1994 ACCORD 2.2L I4
(F22B1) ENGINE

FUEL VAPOR PIPE
FUEL RETURN PIPE
FUEL GAUGE SENDING UNIT
EVAP TWO-WAY VALVE
FILLER CAP
FUEL PRESSURE REGULATOR
EVAP CANISTER
FUEL INJECTORS
FUEL RAIL
FUEL FILTER
FUEL FEED PIPE
FUEL TANK
FUEL PUMP

COURTESY OF HONDA MOTOR CO.

FIGURE 3-38

FUEL PRESSURE TESTS

These repair procedures should be used whenever connecting or disconnecting a fuel pressure gauge to Honda vehicles with fuel injection. When working around automotive fuel systems that operate under high fuel pressure, pay particular attention to the CAUTION information that follows.

CAUTION: *Keep all open flames or sparks away from the work area and do not smoke. Do not attempt to relieve the fuel pressure unless the engine is off.*

RELIEVING FUEL SYSTEM PRESSURE

Before any fuel lines or hoses are disconnected, carefully relieve any pressure in the Fuel system. This can be accomplished by loosening the service bolt (6 mm) located at the top of the fuel filter assembly.

1) Remove the negative battery cable
2) Remove the fuel cap to relieve any pressure in the fuel tank
3) Use a box end wrench on the 6 mm service bolt at the top of the fuel filter while holding the banjo bolt with another end wrench
4) Place a rag or shop towel over the 6 mm service bolt and slowly loosen the service bolt one complete turn

FUEL PRESSURE GAUGE CONNECTIONS

A fuel pressure gauge (Honda Part No. 07406-0040001 or equivalent) can be attached at the hole where the service bolt connects. If the service bolt is loosened to relieve pressure or to connect a fuel pressure gauge, the washer installed between the service bolt and banjo bolt must be replaced. Also, whenever the service bolt is removed to disassemble the component, replace all washers.

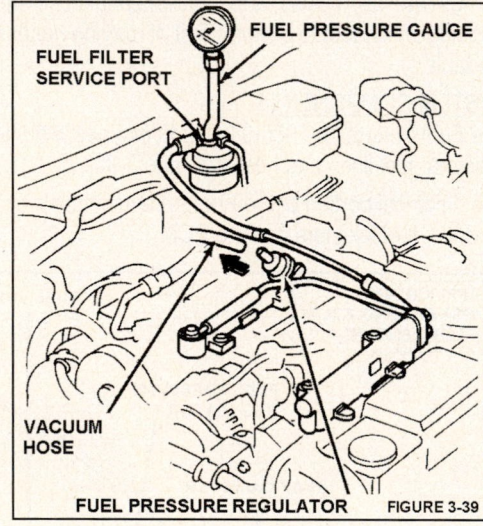

1) Once fuel pressure is relieved, remove the service bolt on top of the fuel filter while holding the banjo bolt with a wrench. Attach a fuel pressure gauge and tighten securely.
2) Start the engine and then check for signs of a fuel leak. Next, watch the fuel pressure with the engine at idle speed and pressure regulator hose removed and plugged. The fuel pressure should read 35-41 psi with the vacuum hose to the regulator removed.
3) If fuel pressure is higher than specified, check for a pinched fuel return hose or for a faulty fuel pressure regulator.
4) If fuel pressure is lower than specified, check for a clogged fuel filter, a faulty fuel pressure regulator, a fuel pump failure or leakage in the system.

■ **NOTE:** *Refer to the appropriate Fuel System Tests in this section to diagnose other parts of the system (i.e., Fuel Injector Test, Fuel Quality Tests, etc.).*

FUEL PRESSURE REGULATOR TESTS

INTRODUCTION

The fuel pressure regulator is used to maintain a constant fuel pressure to the fuel rail and fuel injectors. When the difference between fuel pressure and manifold vacuum exceeds 36 psi, the diaphragm in the pressure regulator is pushed upward to allow the excess fuel to be fed through the fuel return line back to the fuel tank.

DRIVEABILITY SYMPTOMS

If a driveability symptom is present, refer to the list of symptoms below and choose the symptom that best fits the particular problem for the vehicle under test. It may be necessary to perform more than one type of Fuel system test to locate the fault and complete the repair. Problems in the fuel pressure regulator can cause these symptoms:

- Fails I/M or State Emissions Test
- Frequent Stalling
- Hard Start - Cold Engine
- Loss of Power or Poor Performance
- Poor Fuel Economy

PRELIMINARY STEPS

The following inspection and test steps should be performed in sequence to isolate the problem area in order to make the necessary repairs. Always start the test procedure by performing the preliminary steps.

Attach a suitable fuel pressure gauge to the service port of the fuel filter on Civic CRX models. On all other models, attach a suitable fuel pressure gauge to the service port on the fuel rail at the pressure regulator.

CAUTION: ***Keep all open flames or sparks away from the work area and do not smoke. Do not attempt to relieve the fuel pressure unless the engine is off.***

REGULATOR VACUUM TEST

1) Inspect the manifold vacuum hose that connects to the fuel pressure regulator for cracks, restrictions, or proper connections. Make repairs as needed and go to Step 2.
2) Remove the vacuum hose from the pressure regulator and connect a suitable vacuum gauge to the hose (Figure 3-40). Start the engine and allow it to idle. Read the vacuum gauge. The gauge should read near manifold vacuum.
3) If there is no vacuum, locate the disconnected or misrouted hose and repair the problem. If it reads low, check the throttle body port for a restriction. Also recheck the base timing. If timing is okay, check the camshaft timing.
4) Make repairs as needed and then recheck for the symptom.

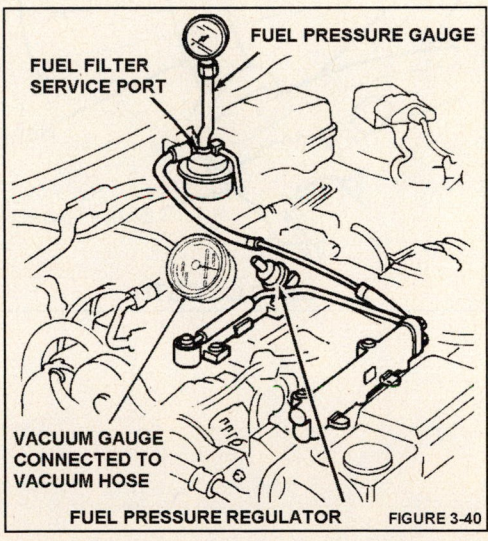

FIGURE 3-40

CAUTION: *Keep all open flames or sparks away from the work area and o not smoke. Do not attempt to relieve the fuel pressure unless the engine is off.*

1)   Turn the engine off and attach a fuel pressure gauge to the service port of the fuel filter or fuel rail on Accord model (if not already connected).

2)   Restart the engine with the gauge installed and checks for leaks. Next, verify that the fuel pressure reading on the gauge rises each time the vacuum hose is connected and removed from the pressure regulator. A typical amount of pressure change is from 8-9 psi. The pressure reading should be 36-41 psi with the vacuum hose off.

3)   If the pressure is within specification and rises with the hose removed, the regulator is operating normally. If the pressure does not rise with the hose removed, wrap the fuel return hose with a shop towel and very carefully lightly pinch the fuel return hose (Figure 3-41).

4)   If the pressure remains the same, check the fuel filter (for restriction) and the fuel pump for low output. Make repairs as needed and retest for the symptom. If the pressure rises with the return lightly pinched, replace the pressure regulator. Retest the symptom when repairs are completed.

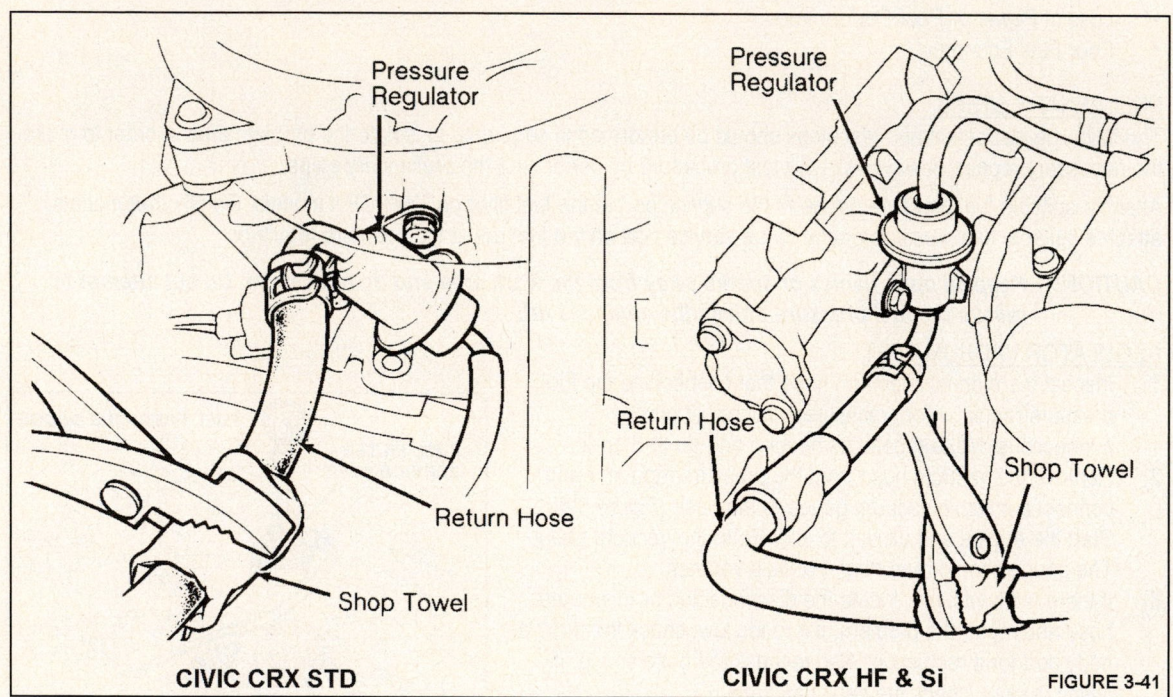

CIVIC CRX STD   **CIVIC CRX HF & Si**   FIGURE 3-41

FUEL PUMP TESTS

INTRODUCTION

The Honda PGM-FI system uses an in-line, impeller-type, electric fuel pump. Fuel is drawn from the fuel tank through a one-way check valve and delivered to the fuel rail in the engine compartment on all models (Figure 3-42a). A baffle is provided to prevent fuel pulsation on the direct drive fuel pump. An external fuel filter (sock) is used on the impeller type pump, while an internal fuel filter is used on the direct drive pump. A check valve is used to maintain fuel pressure in the fuel line for a short time after shutdown to ease restarting on all models.

The fuel pump has an internal relief valve to prevent excessive pressure in the fuel delivery system. This valve opens if there is a blockage in the discharge side. If the relief valve opens, fuel flows from the high-pressure side to the low-pressure side of the fuel pump.

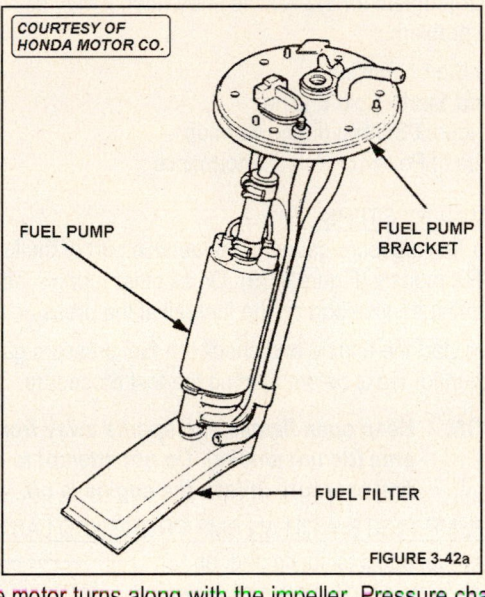

FIGURE 3-42a

SYSTEM OPERATION

The fuel pump is energized for two seconds when the key is first turned on to pressurize the fuel delivery system. Once the engine starts, the main relay turns the fuel pump on and the motor turns along with the impeller. Pressure changes are created by the numerous grooves around the impeller (Figure 3-42b).

Fuel enters the inlet port and flows inside the motor form the pumping chamber and is forced through the discharge port via the check valve. If fuel flow is obstructed at the discharge side of the fuel line, the relief valve will open to bypass fuel to the inlet port to prevent high fuel pressure. Once the engine stops, the PCM turns off the fuel pump through the main relay due to the loss of ignition reference (rpm) signals. An internal check valve closes by spring action to retain residual pressure for quick restarts.

COMPONENT EXAMPLES

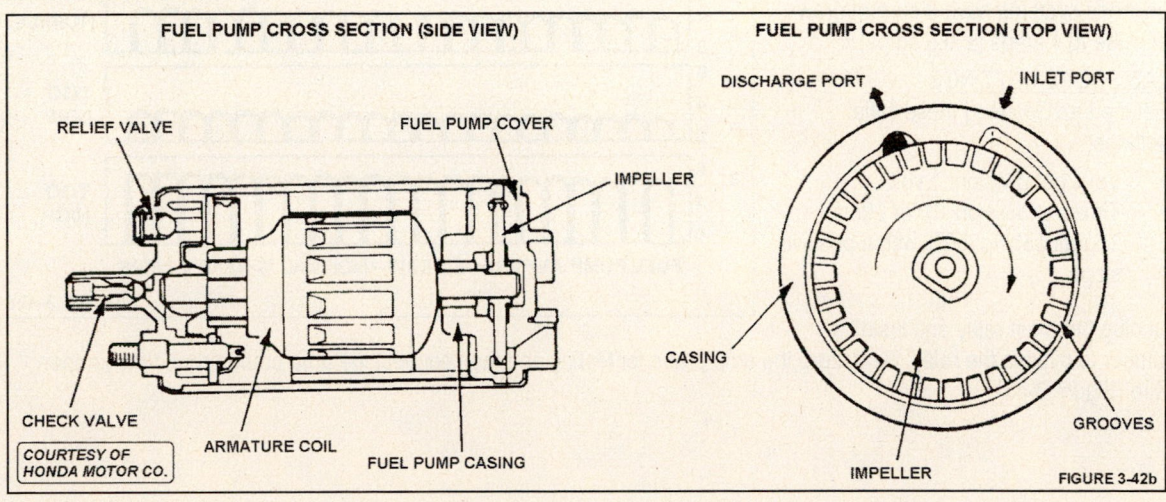

FIGURE 3-42b

DRIVEABILITY SYMPTOMS

If a driveability symptom is present, refer to the list of symptoms below to determine if one of the symptoms fits a particular problem for the vehicle under test. It may be necessary to perform more than one type of Fuel system test to repair a problem.

- No Start condition
- Hard Start - Cold Engine
- Frequent Stalling during warmup
- Loss of Power or Poor Performance

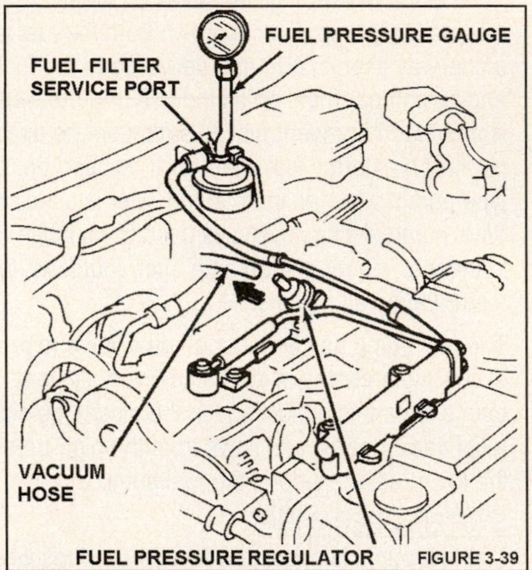

PRELIMINARY STEPS

Attach a fuel pressure gauge to the service port of the fuel filter on Civic CRX models (Figure 3-39). On all other models, attach the gauge to the service port on the fuel rail at the pressure regulator.

Crank or start the engine and check the fuel pressure gauge connection for leaks before starting the test procedure.

CAUTION: ***Keep open flames and sparks away from the work area (do not smoke). Do not attempt to relieve the fuel pressure unless the engine is off.***

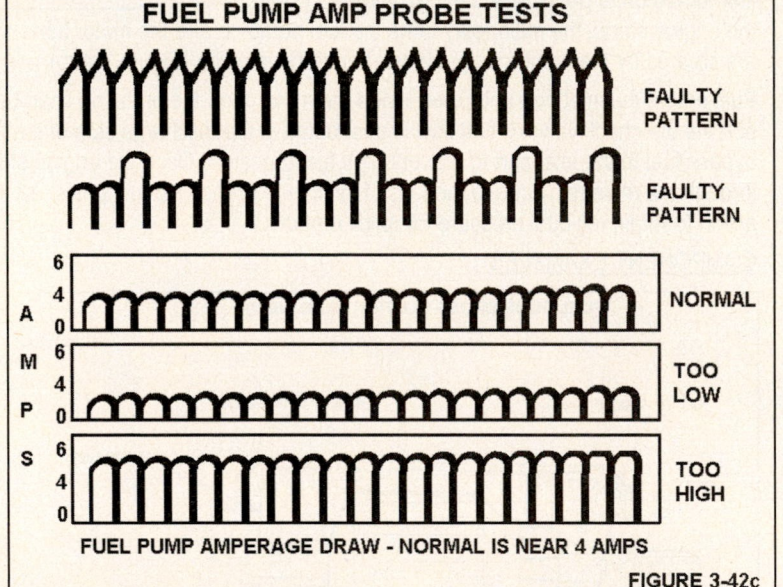

FIGURE 3-39

FUEL PUMP TESTS WITH AN AMP PROBE AND LAB SCOPE

If a faulty fuel pump is suspected, an amp probe with an inductive pickup can be used to monitor the operation of the fuel pump and its related circuits.

Examples of typical fuel pump motor waveforms that can be captured with an amp probe and Lab Scope are shown in Figure 3-42c.

With the engine running, the fuel pumps on these systems have a current draw of close to 4 amps (± 0.5 amps).

LAB SCOPE SETTINGS

Set the Lab Scope to these initial settings:

- Volts per division: 1 volt
- Time per division: 10 or 20ms
- Trigger setting: 50% with a positive slope

FUEL PUMP AMP PROBE TESTS

FAULTY PATTERN

FAULTY PATTERN

NORMAL

TOO LOW

TOO HIGH

FUEL PUMP AMPERAGE DRAW - NORMAL IS NEAR 4 AMPS

FIGURE 3-42c

Unplug the main relay and install a jumper to bypass the relay. Next, zero the amp probe for testing and then connect the amp probe around the jumper wire (Figure 3-43).

FUEL PUMP ELECTRICAL TESTS

If an inoperative fuel pump is suspected, cycle the key from off to on while an assistant listens at the pump to determine if it runs. If it does not run, use the following tests to test the fuel pump electrical circuits and relay. To gain access to the fuel pump on Civic and CR-V models, remove the rear seat. Clear the luggage compartment to gain access to the fuel pump on Accord models. Remove the fuel pump connector (Figure 3-42).

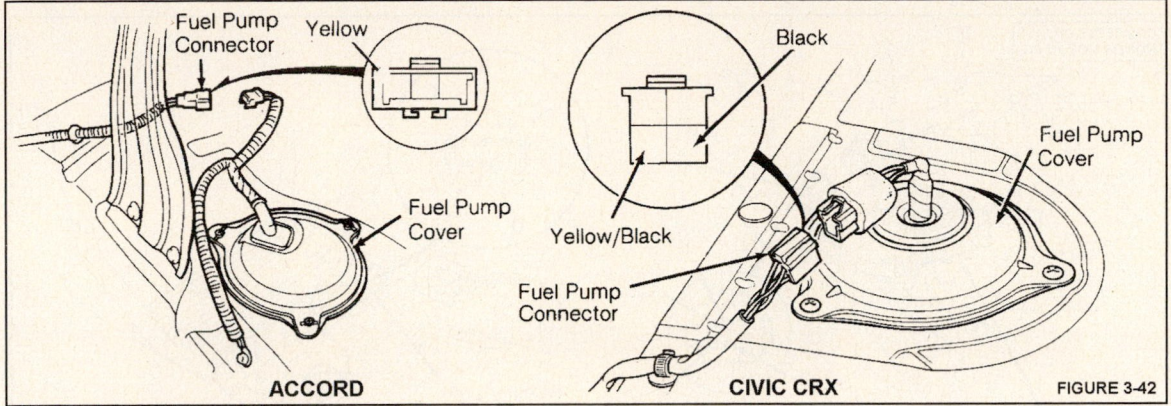

FIGURE 3-42

MAIN RELAY BYPASS TEST

1) Remove the PGM-FI main relay connector from under the dash to allow for testing.
2) To bypass the main relay, connect a jumper wire to the ignition power BLK/YEL and fuel pump YEL/BLK (or YEL on Accord) wires at the relay connector (Figure 3-43).
3) Turn the key on and check for 12v at the fuel pump connector. Connect the DVOM positive probe to the YEL/BLK wire and the negative probe to the BLK wire at the wiring harness side connector (to YEL and ground on Accord). If the reading is 0v for the first two seconds after the key is turned on, move the negative lead to body ground and retest. Cycle the key from off to on for two seconds during this test. If there is no voltage, refer to the Wiring Harness Tests in the Main Relay Tests.
4) If the DVOM reads near 12v at the pump for the first two seconds after the key is turned on and the pump does not run, replace the pump and retest for the symptom.
5) To watch the fuel pump waveform with a Lab Scope, reconnect the pump connector.

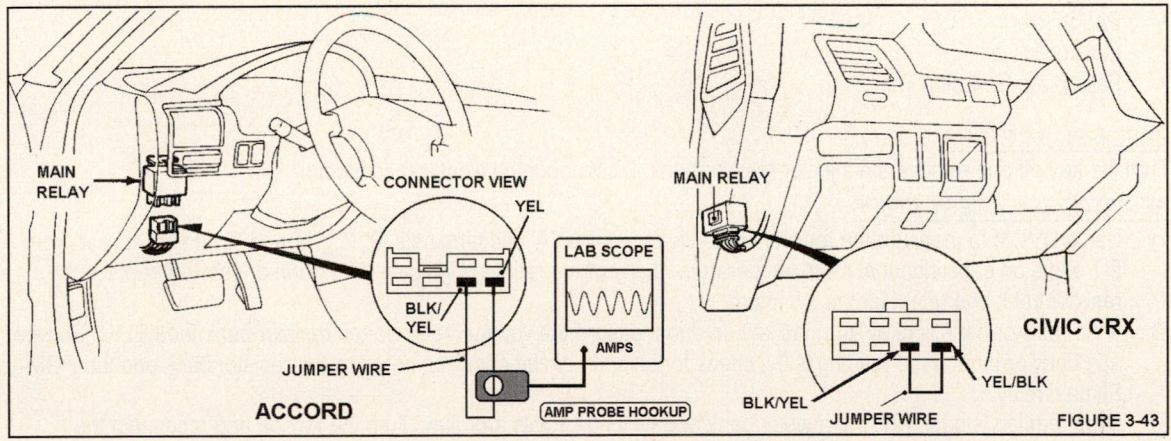

FIGURE 3-43

■ **NOTE:** *For vehicles other than the1990 Accord or Civic models referenced in this article, refer to other pin charts for correct wire colors and procedures.*

FUEL INJECTOR RESISTOR TESTS

Some Honda engines are equipped with a fuel injector resistor bank that is designed to lower the current supplied to the injectors (Figure 3-44a).

The resistor bank helps prevent damage to the fuel injector coils. The use of the resistor bank allows for a faster response time at the fuel injectors.

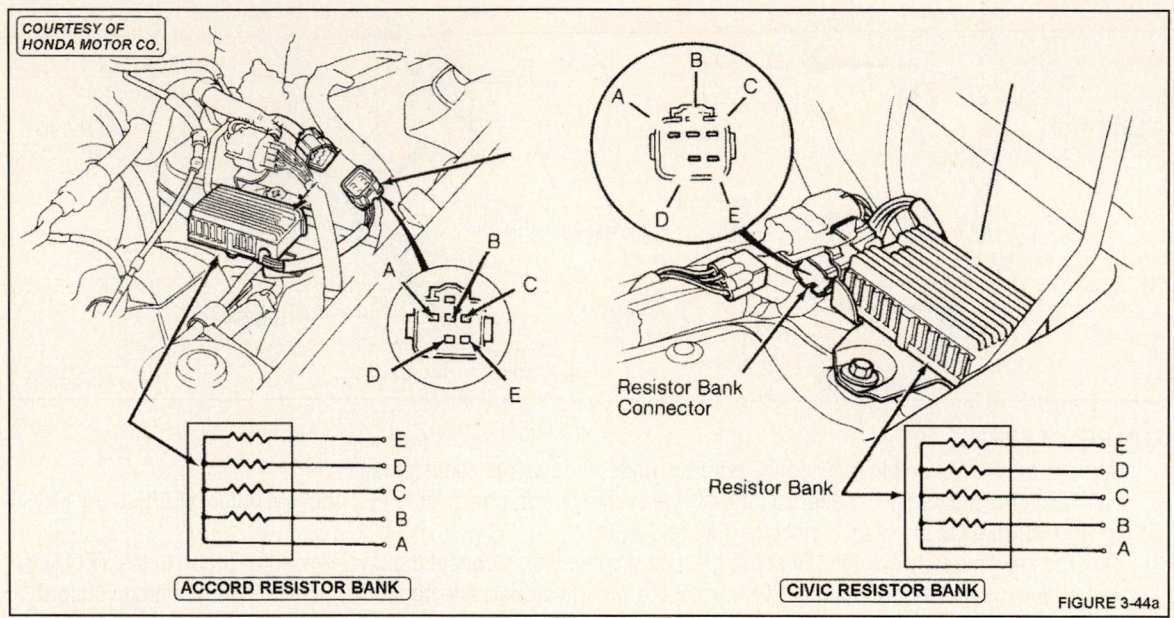

FIGURE 3-44a

DRIVEABILITY SYMPTOMS

If a driveability symptom is present, refer to the list of symptoms below and choose the symptom that best fits the particular problem for the vehicle under test.

- No Start condition
- Misfire or Rough Running
- Rough Idle
- Loss of Power
- Fails State Emission Test

PRELIMINARY STEPS

Turn the key off and remove the injector resistor bank. Disconnect the electrical connector.

RESISTOR BANK TEST STEPS

1) Use a DVOM to measure the resistance between terminal 'A' and terminals 'B', 'C', 'D' and 'E'. It should read from 5-7 ohms on each circuit at room temperature. If any of the readings indicate read open or shorted, replace the resistor bank and retest for the symptom.
2) If the resistor bank is okay, turn the key on and measure the voltage between the resistor bank feed BLK/YEL wire and body ground. If the reading is 0v, check for an open circuit condition between the resistor bank and the PGM-FI main relay.
3) If the reading is near 12v, the resistor bank and its feed circuits are okay. Turn the key off and reconnect the resistor bank. Select another type of Fuel system test.

FUEL INJECTOR TESTS

INTRODUCTION

The Honda PGM-FI fuel injection system is equipped with solenoid-activated constant-stroke pintle type fuel injectors. The injector consists of a solenoid, plunger needle valve, and housing. When current is applied to the coil in the solenoid, the valve lifts up and pressurized fuel is injected close to the intake valve. The solenoid valve list is fixed, so with a constant fuel pressure, the amount of fuel delivered is determined by the length of time the injector is held open (referred to in this manual as injector pulsewidth).

SYSTEM COMPONENTS

The Fuel system components on 1996-98 Civic engines with fuel injection includes the fuel tank, fuel cap, fuel gauge sending unit, fuel filter, in-tank fuel pump, fuel pressure regulator, fuel injectors, fuel pulsation damper, fuel pipes and hoses (Figure 3-44).

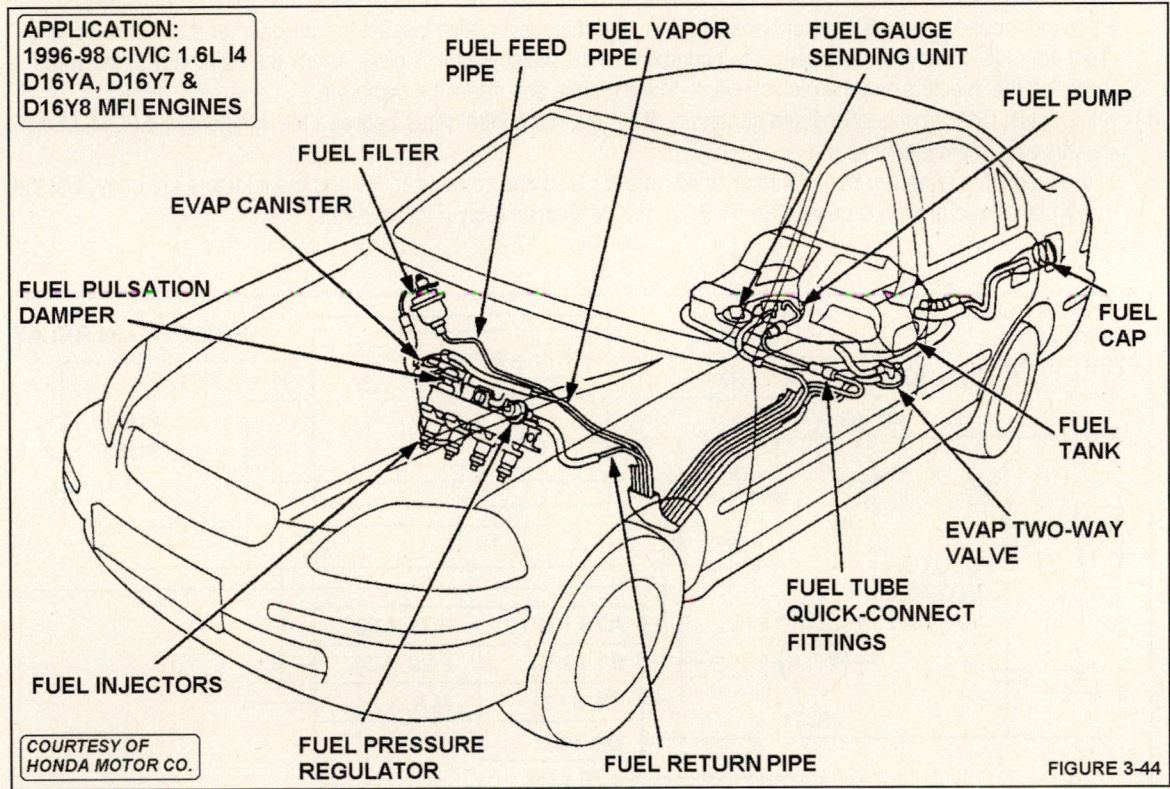

APPLICATION:
1996-98 CIVIC 1.6L I4
D16YA, D16Y7 &
D16Y8 MFI ENGINES

FUEL FEED PIPE • FUEL VAPOR PIPE • FUEL GAUGE SENDING UNIT • FUEL PUMP • FUEL FILTER • EVAP CANISTER • FUEL PULSATION DAMPER • FUEL CAP • FUEL TANK • EVAP TWO-WAY VALVE • FUEL TUBE QUICK-CONNECT FITTINGS • FUEL INJECTORS • FUEL PRESSURE REGULATOR • FUEL RETURN PIPE

COURTESY OF HONDA MOTOR CO.

FIGURE 3-44

DRIVEABILITY SYMPTOMS

If a driveability symptom is present, a technician should choose the symptom that best fits the particular problem for the vehicle under test. When deciding which test to perform, a technician should use their experience and understanding of Fuel system operation. It may be necessary to perform more than one type of test to locate the fault and complete the repair.

A problem with an injector or its circuits could cause any of these symptoms:

- No Start Condition
- Engine Misfire or Loss of Power
- Rough idle with a cold or warm engine
- Fails I/M or Emission Test

PRELIMINARY STEPS

These test procedures apply to the Accord and Civic models with multiport fuel injection. For Civic CRX STD models with a 1.5L TBI engine, refer to other repair charts. Check and adjust ignition timing, base idle speed and set idle CO% to specifications.

CAUTION: *The following test steps require that the drive wheels are blocked and the vehicle exhaust system is purged from the work area for safety purposes.*

FUEL INJECTOR & INJECTOR RESISTOR TESTS (FOR NO START CONDITION)

1) Turn the key off and remove the suspect fuel injector connector. Use a DVOM to measure the injector resistance (specification: 1.5-2.5 ohms). If the resistance is not okay, replace the injector. Repeat this test for all injectors and then go to Step 2.

2) If all injectors pass the resistance test, go to the Fuel system tests to check the fuel pressure. If fuel pressure is okay, go to Step 3.

3) On models with an injector resistor bank, if injector resistance and the fuel pressure are within specifications, check for near 12v on the resistor bank wire from the main relay. Also inspect for corroded or loose connections. Test the resistor bank for an open circuit condition. If the resistor bank is okay, check the continuity of the wires between the injectors and the resistor bank. Make repairs and retest the symptom.

4) All models: Check connections and continuity of the injector control wires between the injectors and PCM. Make repairs and retest the symptom.

5) If the feed circuit, resistor bank (if applicable), injector feed and control circuits and the injectors are okay, test the main power and ground circuits to the PCM and PGM-FI main relay (Figure 3-45).

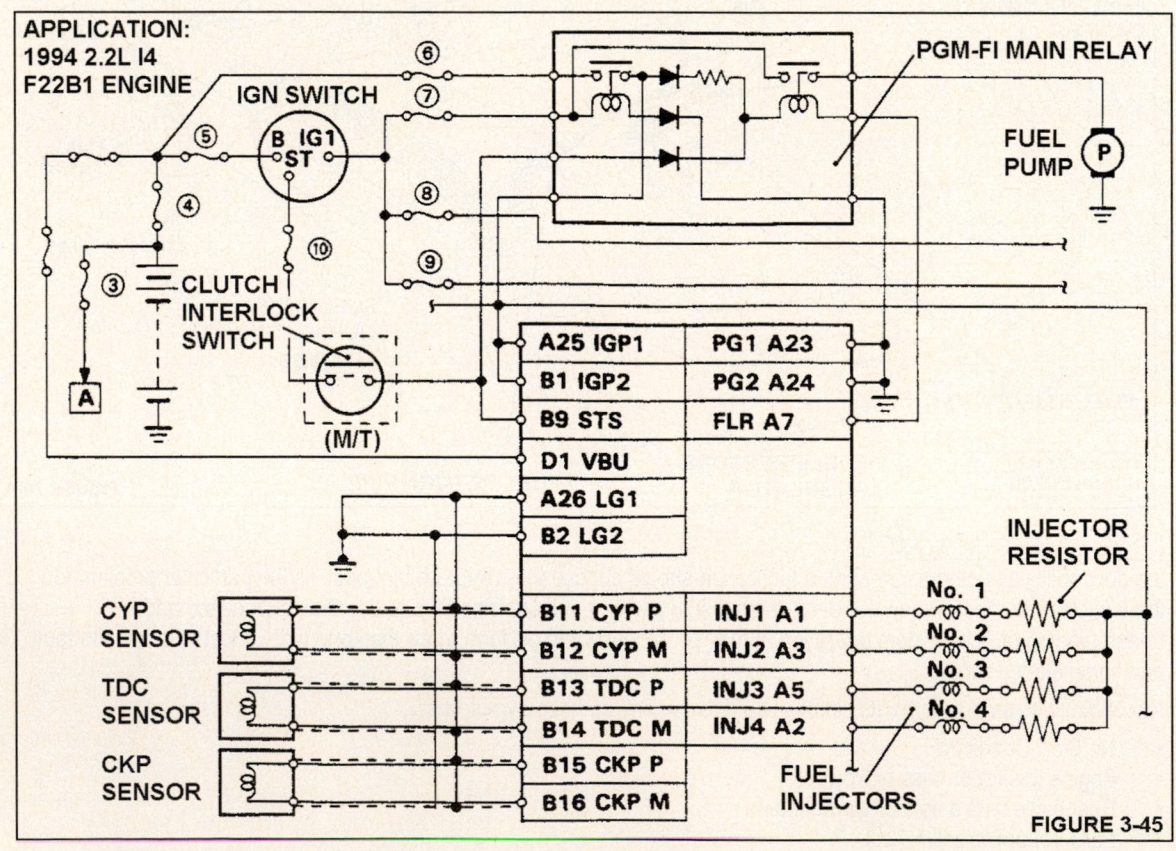

FIGURE 3-45

FUEL INJECTOR TESTS (MISFIRE CONDITION)

1) Start the engine and allow it to idle. Locate a stethoscope that can be used to monitor the operation of each fuel injector by listening to its clicking sound as the engine runs at idle. Listen to each injector for normal clicking sounds (Figure 3-46). Replace any injectors that are suspect, and listen to the new injector. If the new injector does not sound normal, go to Step 2 and continue the test sequence.

2) To isolate a fault that causes abnormal injector sounds, first check the injector battery feed voltage (YEL/BLK wire) at the injector resistor bank or at the injector. Inspect for short circuits, corrosion, or poor connections.

3) On models with an injector resistor bank, if injector resistance and the fuel pressure are within specifications, check for battery voltage on the resistor bank wire from the main relay. Also inspect for corroded or loose connections. Test the injector resistor bank for open circuit condition. If the resistor bank is okay, check the continuity of the wires between the injectors and the resistor bank. Make repairs and retest for the symptom.

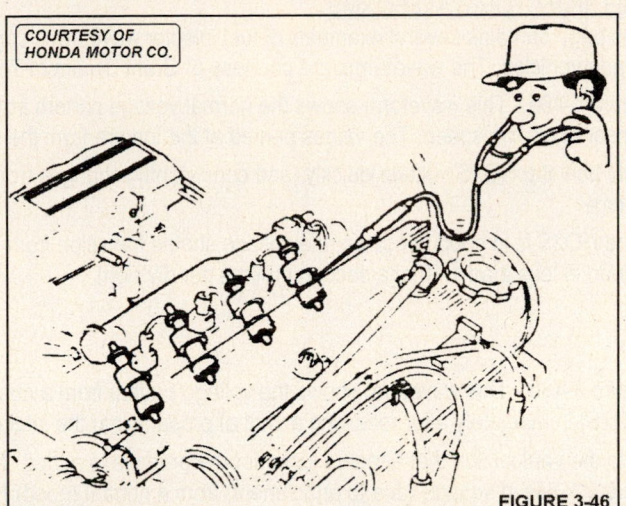

COURTESY OF
HONDA MOTOR CO.

FIGURE 3-46

4) All models: Check connections and continuity of the injector control wires between the injectors and PCM. Make repairs and retest the symptom.

5) If the feed circuit, resistor bank (if applicable), injector feed and control circuits and the injectors are okay, test the main power and ground circuits to the PCM and PGM-FI main relay.

INJECTOR POWER BALANCE TEST

It is obvious that problems related to the engine mechanical condition or the Ignition system could cause erratic test results in an engine power balance test. With this in mind, perform a thorough basic inspection of all engine components related to Base Engine systems and repair any faults prior to starting the test procedure that follows.

1) Turn the key off and remove the idle speed (EACV or IAC) and oxygen sensor connectors to disable the PCM control of engine idle speed.

2) Start the engine and warm it at 2000 rpm in P/N until the electric cooling fan cycles. Allow the engine to idle and carefully remove each injector electrical connector one at a time. Record the amount of engine speed drop as each injector connector is removed and then reconnected. If the idle speed change is within 75 rpm for each cylinder, go to Step 3.

3) If the idle speed change is not equal, remove, check and/or replace the suspect fuel injector(s). Retest for the symptom when repairs are done.

FUEL INJECTOR WAVEFORMS

This page contains several examples of fuel injector waveforms captured from various Honda vehicles. The waveforms are courtesy of Grant Swaim of Tech2Tech™.

Figure 2-46a - This waveform shows the normal voltage pattern from a heated oxygen sensor at off-idle speed. The values printed at the top are from the right of the screen.

Note how the HO2S voltage quickly (and consistently) changes from low to high to low voltage.

If the HO2S is in good condition, the voltage should transition from lean to rich or from rich to lean in less than 1/10 of a second (100ms per division).

Figure 3-46b - This waveform shows the voltage pattern from a heated oxygen sensor (HO2S) immediately after receiving a shot of propane into the intake manifold.

Note the vertical line that appears just after the propane is added. This waveform shows the quick rise in voltage (due to enrichment) from a normal (good) HO2S.

The same pattern should occur if the throttle is opened quickly with the engine hot and the gearshift in Park or Neutral position.

Figure 3-46c - This waveform shows the voltage pattern from a heated oxygen sensor (HO2S) immediately after a lean condition was created by momentarily removing the connector from a fuel injector. This waveform shows the quick decrease in sensor voltage (due to enleanment) from a normal (good) HO2S.

If a vacuum leak is created on a Speed Density system, it will only cause the engine speed to increase.

In this graphic, the fuel injector "plot" is the dotted line. The values printed at the top of the figure were captured at the far right of the screen.

Figure 3-46d - This waveform shows changes to the fuel injector pulsewidth and spark advance right after the air conditioning was turned on with the engine fully warmed up.

Note in this pattern how the fuel injector pulsewidth widened at the same time that the spark advance increased. In this figure, the solid line is plotting the actual fuel injector pulsewidth over time (it is not a fuel injector waveform).

■ **NOTE:**   *A Vetronix Mastertech™ Scan Tool with the Honda aftermarket cartridge was used in "graphing mode" to capture all of the figures on this page.*

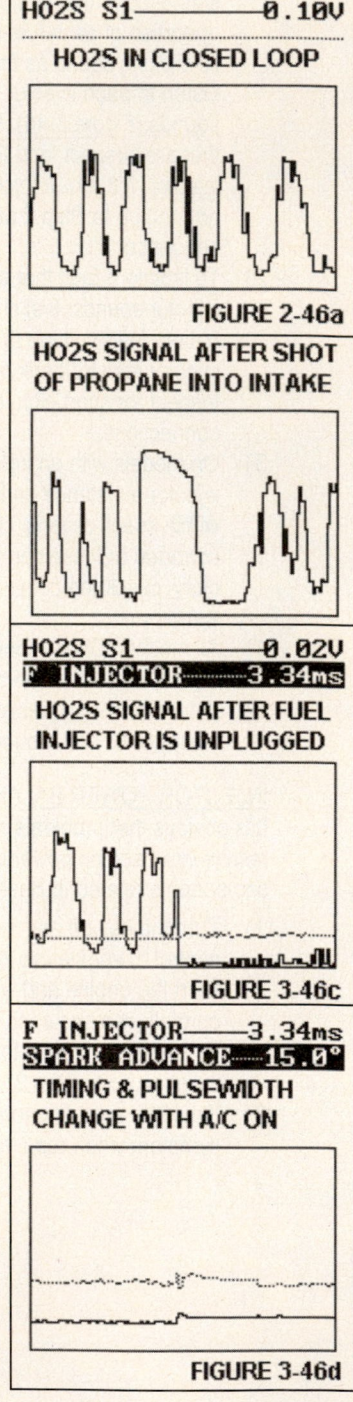

HO2S S1 ————— 0.10V
HO2S IN CLOSED LOOP

FIGURE 2-46a

HO2S SIGNAL AFTER SHOT OF PROPANE INTO INTAKE

HO2S S1 ————— 0.02V
F INJECTOR ———— 3.34ms
HO2S SIGNAL AFTER FUEL INJECTOR IS UNPLUGGED

FIGURE 3-46c

F INJECTOR ———— 3.34ms
SPARK ADVANCE ——— 15.0°
TIMING & PULSEWIDTH CHANGE WITH A/C ON

FIGURE 3-46d

Idle Speed System
IDLE SPEED SYSTEM TESTS

INTRODUCTION

The PCM uses on of the following devices to control engine idle speed:

- Electronic idle control valve (EICV
- Electronic air control valve (EACV)
- Idle air control (IAC) valve

Engine idle speed is controlled by the PCM and the EACV, EICV or IAC valve at all times. Early fuel injection systems used an EICV (valve), and it was replaced by the EACV (valve). In 1993, the idle air control (IAC) valve was introduced on Honda fuel injected engines.

All three devices (EACV, EICV and ISC) are controlled by the PCM to change the amount of bypass air that enters the intake manifold in response to electric current pulses sent by the PCM. When the solenoid is energized, the valve opens to maintain the correct idle speed for all conditions.

■ **NOTE:** *Idle Speed system test procedures cannot be used properly unless the symptom (condition) is present at the time of testing. If the condition does not appear during testing, refer to the Intermittent Tests in PGM-FI Diagnostics in this section.*

IDLE SPEED SYSTEM BASICS

Idle Speed system tests should be used after these test steps are done:

- The complaint is verified and a visual inspection is done
- PGM-FI Diagnostics are completed when they show Pass (see NOTE)
- Base engine tests are done
- An actual driveability symptom is present during vehicle testing
- Related technical service bulletins have been searched in the TSB index

■ **NOTE:** *Turn the key on and watch the Check Engine or PGM-FI light. It should come on for two seconds and then go off. If the Check Engine light remains on, refer to the PGM-FI Diagnostics article in this section (Section 3).*

COMPONENT EXAMPLES

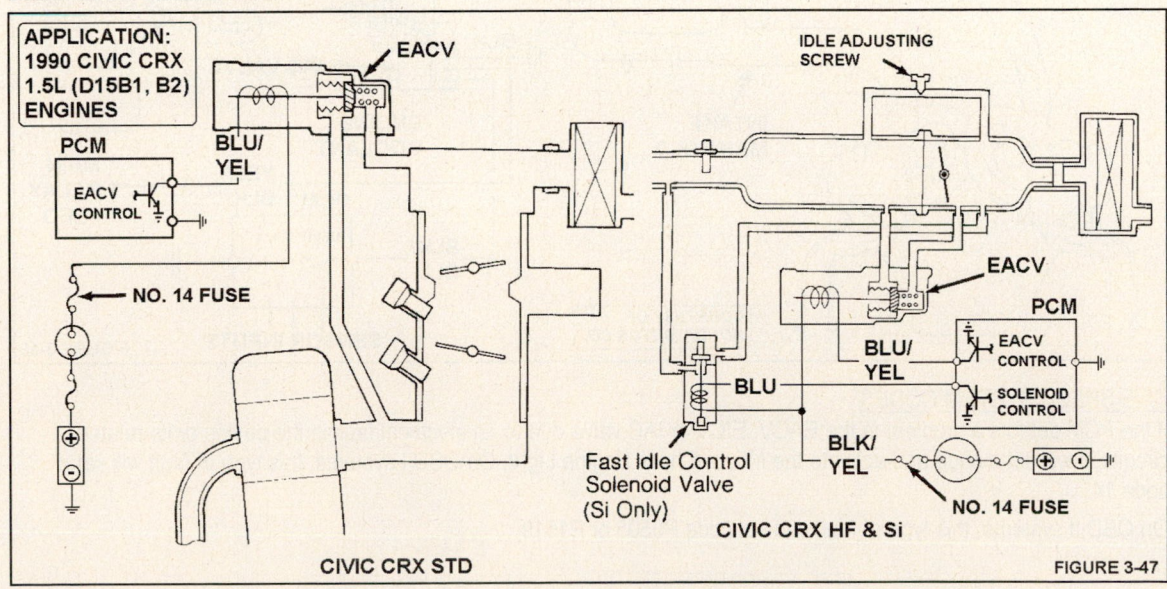

FIGURE 3-47

1990 ACCORD 2.2L I4 COMPONENT EXAMPLE

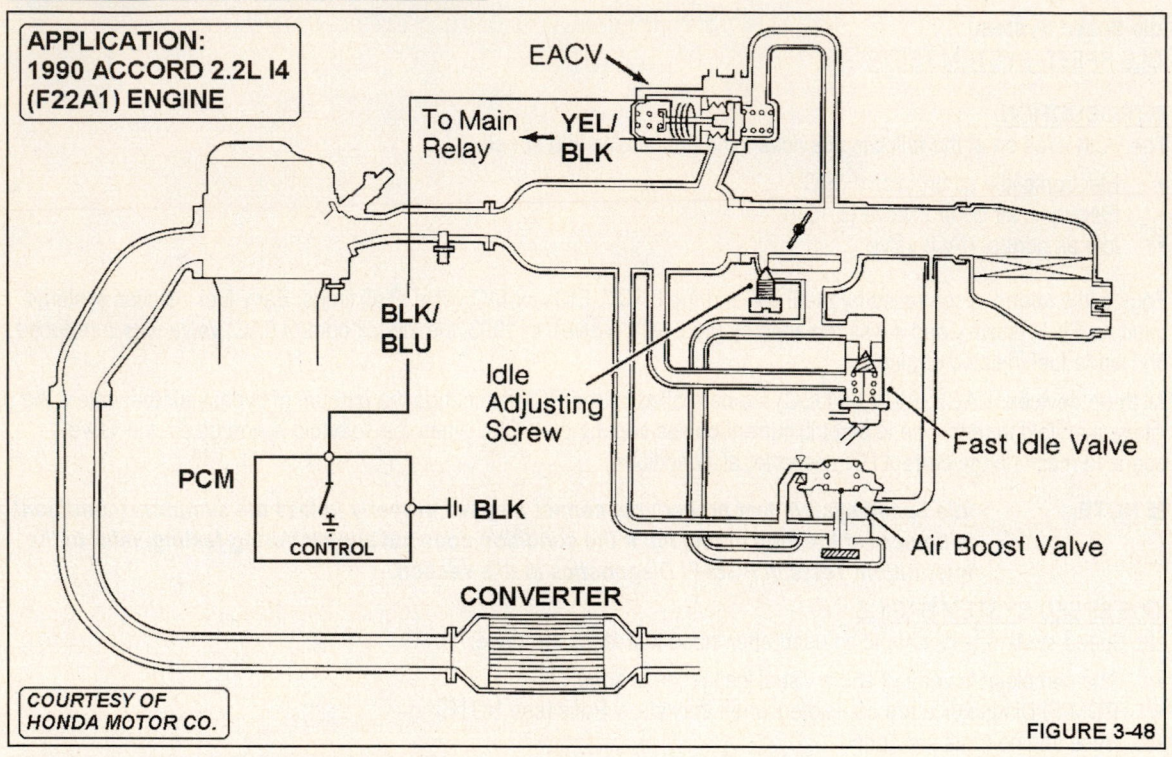

APPLICATION:
1990 ACCORD 2.2L I4
(F22A1) ENGINE

EACV

To Main Relay — YEL/BLK

BLK/BLU

PCM

CONTROL

BLK

Idle Adjusting Screw

Fast Idle Valve

Air Boost Valve

CONVERTER

COURTESY OF HONDA MOTOR CO.

FIGURE 3-48

1998 3.0L V6 ACCORD COMPONENT EXAMPLE

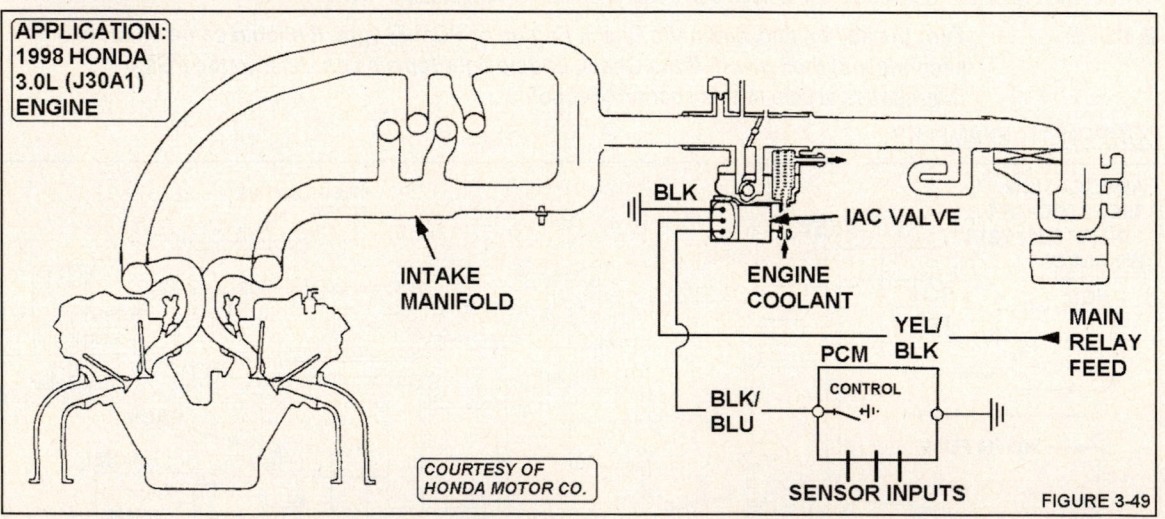

APPLICATION:
1998 HONDA
3.0L (J30A1)
ENGINE

INTAKE MANIFOLD

BLK

IAC VALVE

ENGINE COOLANT

YEL/BLK

MAIN RELAY FEED

PCM

CONTROL

BLK/BLU

SENSOR INPUTS

COURTESY OF HONDA MOTOR CO.

FIGURE 3-49

IDLE SPEED SYSTEM CODES

If the PCM detects a problem in the EACV, EICV or IAC valve due to an electrical fault in the device or its related circuits, it will set a code and activate the MIL or Check Engine Light. On OBD I systems, this type of fault will set a code 14.

On OBD II systems, this type of fault can set code P0505 or P1519.

EACV/EICV/IAC VALVE OPERATION

As previously discussed, these valves are used to change the amount of air that bypasses the intake manifold in response to an electric current pulse from the PCM. When the valve motor is energized, the valve opens to maintain the proper idle speed.

After initial start-up, the valve is opened for a set time limit. The amount of air allowed to by-pass is increased to raise idle speed 150-300 rpm. Also, when the engine is cold, the valve is opened to obtain the proper fast idle speed. In effect, the PCM controls the amount of extra air allowed to bypass the throttle blade in relation to engine coolant temperature. On some engines, when coolant temperature is less than 86°F, a fast idle valve also opens to prevent the idle speed from dropping to low.

SWITCH TESTS

The PCM uses signals from several switches to determine how much engine speed is needed under additional engine load conditions. These switches include the A/C switch, A/T shift position and P/S switches. On 1985-95 applications (OBD I systems), there are no trouble codes assigned to these switches. These switches are diagnosed as a driveability symptom. Refer to the Switch Tests in PGM-FI Diagnostics in this section.

DRIVEABILITY SYMPTOM DIAGNOSIS

If a driveability symptom is present, refer to the list of symptoms included with each Idle Speed Control system test in the articles that follow. Also, a technician should use their experience and knowledge of how an Idle Speed Control system operates when deciding which test to perform first. It may be necessary to do more than one Idle Speed system test to repair the fault.

A problem with the EACV, EICV or IAC valve or their related circuits could cause any of these symptoms:

- Hard Start with a cold engine
- Idle speed out of specification on a cold engine
- Idle speed to high on a warm engine
- Idle speed to low on a warm engine
- Idle speed to low right after startup
- Idle speed drops when vehicle is placed in gear
- Idle speed drops as steering wheel is turned
- Idle speed drops as air conditioning is enabled
- Fails I/M or State Emissions Test

PRELIMINARY STEPS

These test procedures apply to Honda Accord and Civic models equipped with fuel injection. Always start the test procedure by performing the preliminary steps and perform the test steps in sequence to isolate the problem area in order to make the necessary repairs. Check and adjust ignition timing, base idle speed and idle CO % to specifications (if applicable).

CAUTION: *The following test steps require that the drive wheels are blocked and the vehicle exhaust system is purged from the work area for safety purposes.*

EACV, EICV & IAC VALVE TESTS

If an idle speed driveability symptom is present and no idle speed trouble codes appear during PGM-FI Diagnostics, a thorough check must be made of all mechanical and electrical devices that could cause idle speed symptoms.

NO CODE FAULTS

The list below contains inspection and test steps that should be used to diagnose No Code faults related to the Idle Speed system.

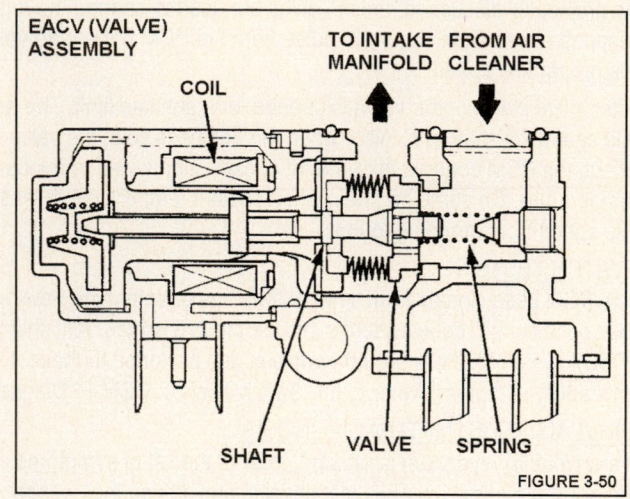

FIGURE 3-50

- Verify base idle speed is set to underhood or other specification
- Inspect all hoses and connections to the EACV, EICV or IAC valve
- Inspect the O-rings that seal the EACV, EICV or IAC valve
- Test the starter signal input (in cranking position)
- Test the alternator FR signal input
- Test the automatic transaxle shift position signal input
- Test the power steering pressure signal input
- Test the air conditioning signal input
- Test the brake switch signal input
- Test the heater fan switch signal input (Civic CRX HF models)
- Test the manual transmission neutral switch signal input
- Test the manual transmission clutch switch signal input
- Test the operation of the fast idle valve (with the engine cold)
- Test the fast idle control solenoid valve (automatic transaxle models)
- Test the operation of the air boost valve

IDLE SPEED VALVE & PCM SUBSTITUTION

If no problems are found in the EACV, EICV or IAC valve, hoses, connections and O-rings, and all of the PCM input signals are correct, replace the valve with a known -good replacement valve to determine if the valve is faulty.

Replace the valve and adjust base idle speed as needed. Then retest for the original symptom. If the symptom goes away, leave the replacement valve on the vehicle.

If the symptom reappears with the replacement valve, substitute a known-good PCM and retest for the symptom. If the symptom goes away, replace the original PCM. If the symptom is still present with a different valve and PCM, return to the inspection and test list in the previous article and repeat the tests.

FAST IDLE VALVE TESTS

INTRODUCTION

Honda vehicles include fast idle control to prevent erratic running conditions with a cold engine. A fast idle valve is used to raise idle speed during warmup. The thermowax plunger controls the air bypass portion of the valve. If it is working correctly, the valve should have no effect on a warm engine. Honda CRX Si vehicles use a solenoid valve.

When the engine is extremely cold, engine coolant that surrounds the thermowax element in the valve makes contact with the valve internal plunger. This action allows extra air to enter the intake manifold and results in an increase in idle speed. When the engine warms to normal operating temperature, the valve closes. This action cuts off the flow of additional air entering the intake manifold and results in a reduction in idle speed.

ACCORD ENGINE COMPONENT EXAMPLE

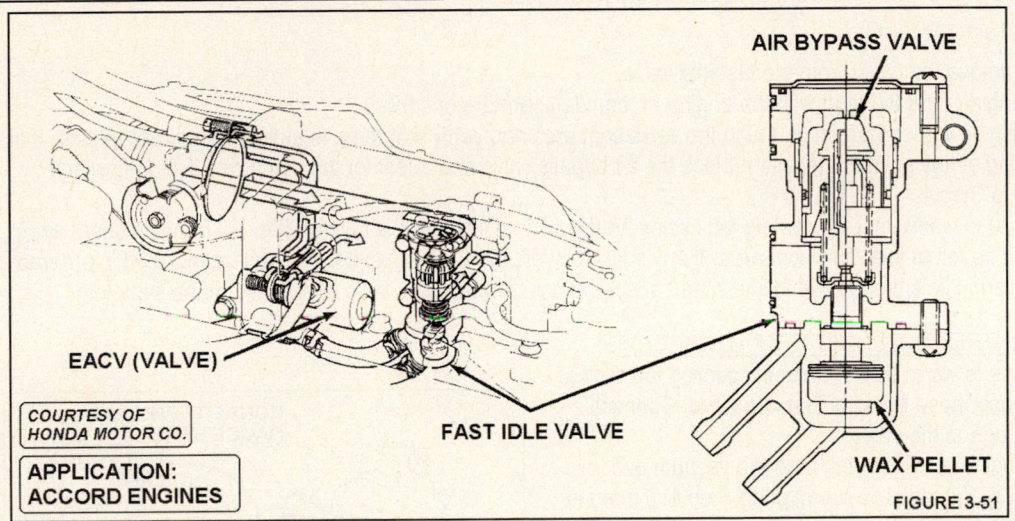

AIR BYPASS VALVE

EACV (VALVE)

FAST IDLE VALVE

WAX PELLET

COURTESY OF
HONDA MOTOR CO.

APPLICATION:
ACCORD ENGINES

FIGURE 3-51

CIVIC ENGINE COMPONENT EXAMPLE

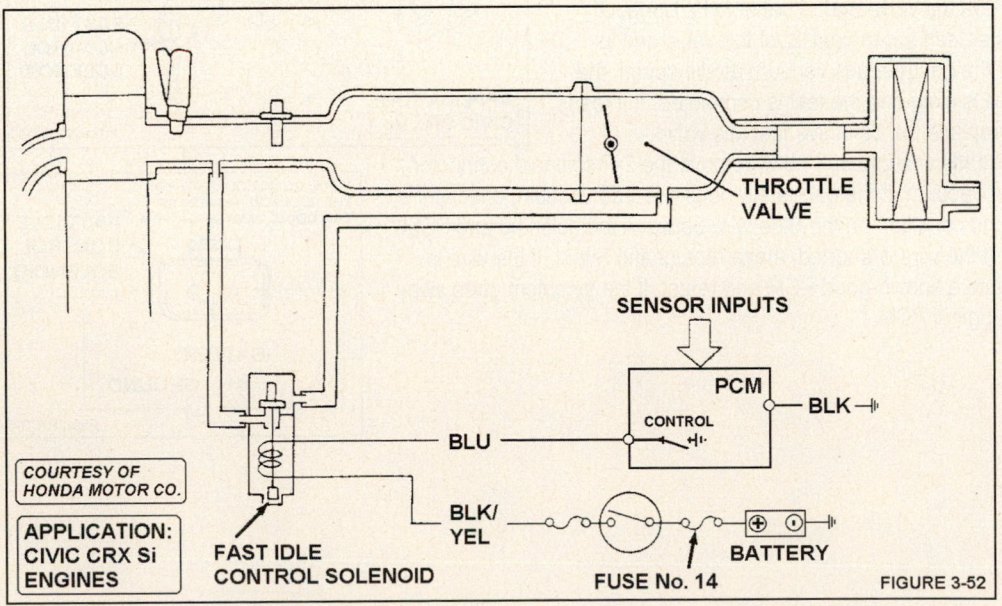

THROTTLE
VALVE

SENSOR INPUTS

PCM

CONTROL

BLK

BLU

COURTESY OF
HONDA MOTOR CO.

APPLICATION:
CIVIC CRX Si
ENGINES

FAST IDLE
CONTROL SOLENOID

BLK/
YEL

FUSE No. 14

BATTERY

FIGURE 3-52

DRIVEABILITY SYMPTOM DIAGNOSIS
A problem with the fast idle valve could cause any of these symptoms:

- Code engine: Hard Start or Fast idle speed out of specification
- Warm engine: Base idle speed too high

PRELIMINARY STEPS
The following inspection and test steps should be performed in sequence. Always start the test procedure by performing the preliminary steps.

Turn the key off. Next, verify that the throttle blade is not held open due to a sticking or binding (throttle linkage) condition.

CAUTION: *The following test steps require that the drive wheels are blocked and the vehicle exhaust system is purged from the work area for safety purposes.*

Fast Idle Valve Test - Accord

1) Carefully remove the cover from the fast idle valve.
2) Start the engine (and the test) with the engine at room temperature or colder.
3) With the engine colder than 86°F, touch the valve seat area and verify that there is airflow with a cold engine. If air is not flowing at this point in the test, replace the air bypass valve and retest for the symptom. If air is flowing (okay), go to Step 4.
4) Allow engine to warm until the cooling fan cycles. Verify the fast idle valve is fully closed. If it remains open, air suction can be felt at the valve seat area. If any suction is felt, the valve is leaking. It must be replaced in order to bring the warm idle speed to within idle speed specifications. Replace the valve and retest for the symptom.

FAST IDLE CONTROL SOLENOID TEST - CIVIC CRX SI

1) Allow engine to warm until the electric cooling fan cycles.
2) Remove upper hose from the fast idle valve. Connect vacuum gauge to the valve.
3) With the engine at idle speed, read the vacuum gauge (Figure 3-53). If it reads vacuum, go to Step 4. If there is no vacuum, turn the key off and remove the 2-P connector from the fast idle valve. Connect a jumper lead to terminal 'A' of the valve and to battery (+). Connect another jumper lead to terminal 'B' of the valve and to battery (-). If the gauge reads vacuum at idle speed, the fast idle valve is okay and the test is completed. If it reads no vacuum, replace the defective fast idle valve.

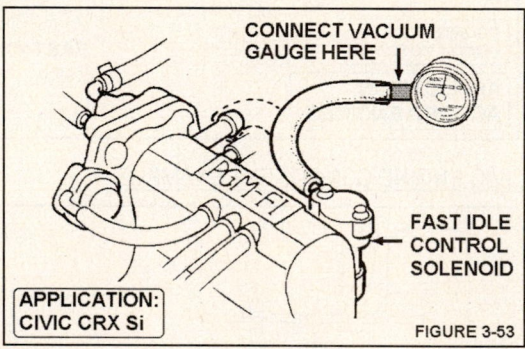

CONNECT VACUUM GAUGE HERE

FAST IDLE CONTROL SOLENOID

APPLICATION: CIVIC CRX Si

FIGURE 3-53

4) If the gauge reads vacuum in Step 3, remove the 2-P solenoid connector and retest for vacuum. If the gauge still does not read vacuum, check for a short to ground condition on the Blue wire between the solenoid and PCM terminal B2. If the wire is shorted, make repairs and retest. If the wire is okay, substitute a known-good PCM and retest. If the symptom goes away, replace the original PCM.

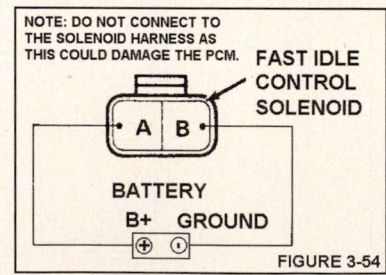

NOTE: DO NOT CONNECT TO THE SOLENOID HARNESS AS THIS COULD DAMAGE THE PCM.

FAST IDLE CONTROL SOLENOID

A B

BATTERY
B+ GROUND

FIGURE 3-54

IDLE SPEED ADJUSTMENT PROCEDURES - 1985-87 ACCORD 1.8L, CIVIC 1.5L

1) Turn off all accessories and connect a tachometer. Place the gear selector in P/N. Remove and plug vacuum hose from intake air diaphragm.

2) Start the engine and raise speed to 1000 rpm until the cooling fan cycles. On the 1.5L engine, the fast idle valve may need to be removed. Adjust the idle speed to 700-800 rpm by turning the idle adjusting screw. If the idle speed is too high and cannot be adjusted, check fast idle valve operation. Reconnect vacuum hose.

1988-95 ACCORD ADJUSTMENTS

1) Start engine and raise engine speed to 2000 rpm until cooling fan cycles. Turn the key off, turn off all lights and accessories and remove the EACV connector. Restart the engine with throttle slightly open, then slowly return to idle speed in P/N. Then adjust the idle speed by turning the idle adjusting screw to 550 rpm. Recheck the ignition timing and adjust as needed. Repeat the test starting at Step 1.

2) Turn the key off and reconnect the EACV or IAC valve connector. Then remove the Clock or Backup fuse for 10 seconds (Canada models: pull parking brake lever up).

3) Restart engine and bring it to 1000 rpm for one minute. Allow engine to idle in P/N. If idle speed is not within specifications (A/T & M/T: 700 rpm ± 50 rpm), test the operation of the EACV or other items that could affect engine idle speed.

1988-95 CIVIC & DEL SOL ADJUSTMENTS

1) Start engine and raise engine speed to 2000 rpm until cooling fan cycles. Turn the key off, turn off all lights and accessories and remove the EACV connector. Restart the engine with throttle slightly open, then slowly return to idle speed in P/N. Then adjust the idle speed by turning the idle adjusting screw to 420 rpm ± 50 rpm. Recheck the ignition timing and adjust as needed. Repeat the test starting at Step 1.

2) Turn the key off and reconnect the EACV or IAC valve connector. Then remove the Clock or Backup fuse for 10 seconds (Canada models: pull parking brake lever up).

3) Restart engine (with no load devices on) and check idle speed at idle in P/N to specifications: A/T: 700 rpm ± 50 or M/T: 600 rpm ± 50.

4) Then turn headlights on low and wait one minute, then recheck engine speed to specifications: A/T: 750 rpm ± 50 or M/T: 700 rpm ± 50. Then turn the headlights off and turn the A/C on with fan switch on high speed, and recheck engine speed to specifications: A/T & M/T: 810 rpm ± 50. If idle speed is not within the ranges shown, test the operation of the EACV or other items that could affect engine idle speed.

1988-95 PRELUDE ADJUSTMENTS

1) Start engine and raise engine speed to 2000 rpm until cooling fan cycles. Turn the key off, turn off all lights and accessories and remove the EACV connector. Restart the engine with throttle slightly open, then slowly return to idle speed in P/N. Then adjust the idle speed by turning the idle adjusting screw to 550 rpm ± 50. Check the ignition timing and adjust as needed. Repeat the test starting at Step 1.

2) Turn the key off and reconnect the EACV or IAC valve connector. Then remove the Clock or Backup fuse for 10 seconds (Canada models: pull parking brake lever up).

3) Restart engine and bring it to 1000 rpm for one minute. Allow engine to idle in P/N with no load. If idle speed is not within specifications (A/T & M/T: 700 rpm ± 50 rpm), test the operation of the EACV or other items that could affect engine idle speed.

IGNITION SYSTEM

IGNITION SYSTEM DESCRIPTIONS

The Ignition system articles in this manual describe both distributor ignition and distributorless ignition systems. The acronyms include in these articles are:

- DI - Distributor Ignition System (no previous acronym)
- EI - Electronic Distributorless Ignition System (previous acronym was DIS)

IGNITION TIMING CONTROLS - TYPE ONE AND TYPE TWO

Honda vehicles are equipped with two types of ignition timing controls:

Type One - 1988-98 models with the DI system. This system uses electronic spark advance (ESA) with timing controlled by the PCM and/or the igniter. On models with a distributor, there is no centrifugal or vacuum advance mechanism. On this system, a 12v test light (and spark tester) can be used to toggle the igniter control circuit to test the operation of the ignition coil, igniter and primary ignition circuits (Figure 3-55).

Type Two - 1985-87 models with distributor ignition (DI). This system uses a distributor with centrifugal and vacuum advance. The igniter controls the base spark advance and the PCM controls the vacuum advance diaphragm through vacuum solenoid valves.

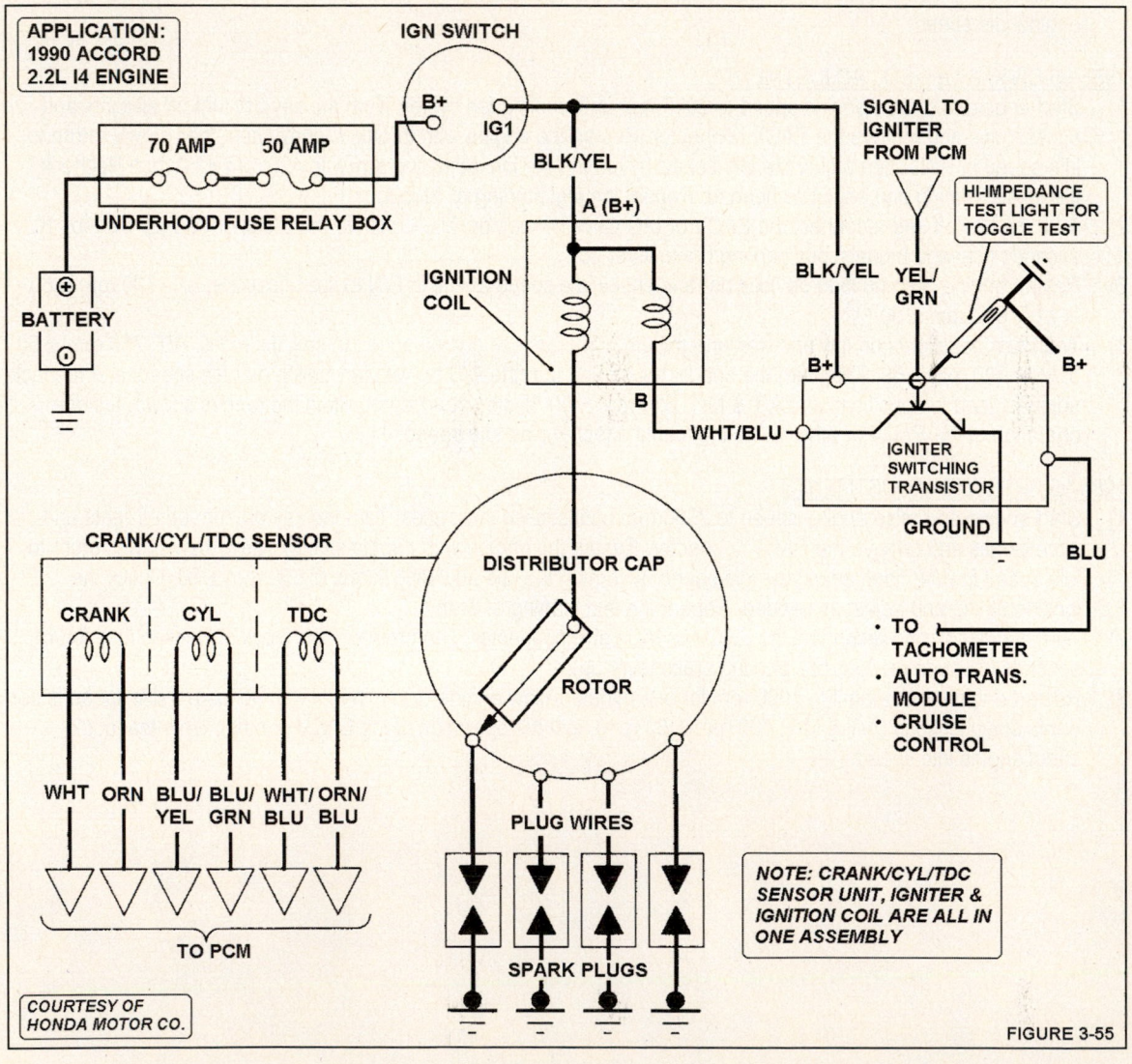

APPLICATION:
1990 ACCORD
2.2L I4 ENGINE

FIGURE 3-55

TYPE ONE SYSTEM OPERATION

Honda Ignition systems are designed to provide the engine optimum control of ignition timing by using a microcomputer (PCM) that responds to various input signals. In most cases, engine speed is calculated through inputs received from the one or more sensors (e.g., the TDC and Crank/Cylinder sensors). Throttle angle is determined from a throttle angle (TA) sensor on early systems and from a throttle position (TP) sensor on later systems. The PCM determines engine temperature from the coolant temperature sensor and the intake manifold pressure is determined from the MAP sensor.

The PCM contains tables in memory that it uses to calculate when to fire the ignition coil(s). This type of ignition timing control uses complicated characteristics and schedules that cannot be achieved through the use of conventional centrifugal advance and a vacuum diaphragm.

1990 ACCORD 2.2L I4 COMPONENT EXAMPLES

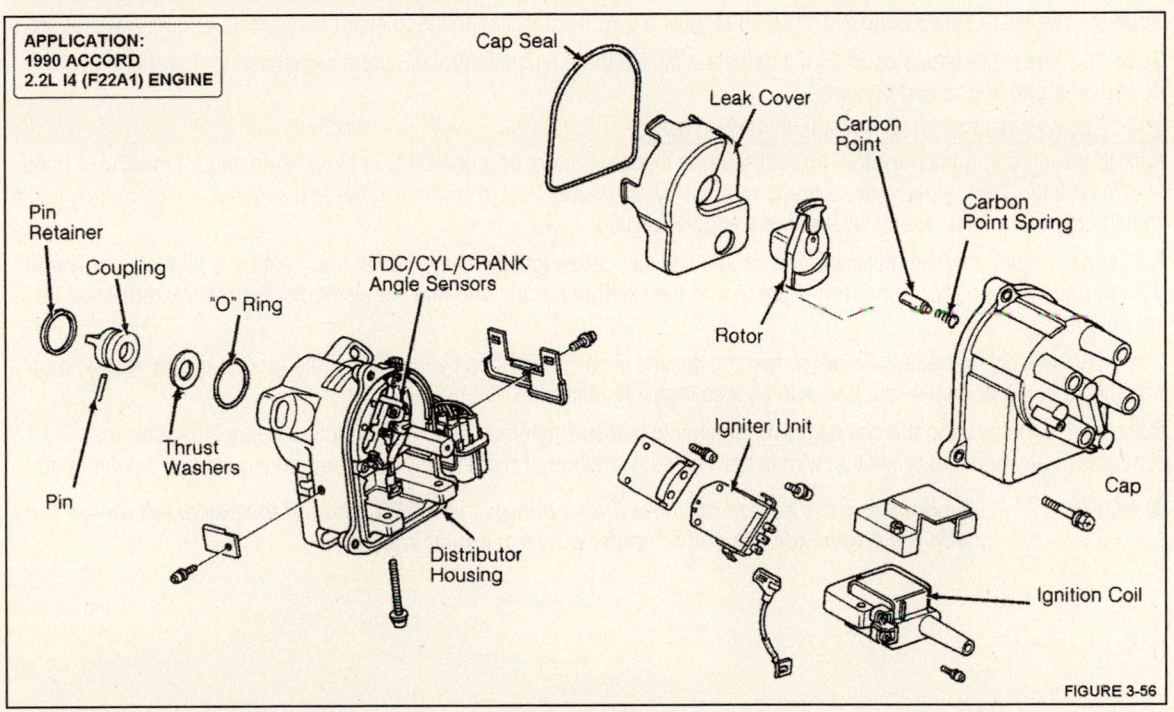

FIGURE 3-56

IGNITION TIMING IN CRANK MODE

During engine cranking and immediately after start-up, ignition timing is fixed at 10°BTDC. The PCM determines that the engine is in crank mode through the starter signal input. While in crank mode, the PCM uses the TDC signal from the distributor to calculate engine speed and cylinder top dead center.

IGNITION TIMING IN RUN MODE

Once the engine starts and continues to run, the PCM calculates the amount of ignition advance for all engine-operating conditions based on engine speed and manifold pressure information stored in its memory tables. The PCM also uses information gathered from various sensors to adjust the base timing value to ambient conditions. Once this calculation is done, the PCM sends a voltage pulse at the right time to an igniter unit in order to fire the ignition coil.

TYPE ONE SYSTEM CODES

Any diagnosis of the Ignition system should begin with a visual inspection of the underhood ignition components and a review of any stored trouble codes in the PCM memory. If the MIL or check engine light is on with the engine running, a hard code is present. The code may be related to an Ignition system fault, so it makes sense to read, record and clear the code by using the PGM-FI Diagnostics to accomplish these tasks. In effect, utilizing the PCM and its diagnostics allows it to communicate if it has detected and recorded any ignition related circuit faults in either the TDC sensor or the combination TDC/CYL/Crank Angle sensor. The PCM constantly monitors the voltage levels of these sensor signals with the engine running.

TYPE ONE SYSTEM CODE LIST

Code 4 - The PCM sets a code 4 if it fails to receive a Crank Angle sensor signal due to either a circuit or component problem.

Code 8 - The PCM sets a code 8 if it fails to receive a TDC sensor signal due to either a circuit or component problem.

Code 9 - The PCM sets a code 9 if it fails to receive a CYL sensor signal due to either a circuit or component problem.

Code 15 - The PCM sets a code 15 if it detects a fault in the ignition control or output signal due to a problem in the PCM driver, circuit or a component.

IGNITION SYSTEM TESTS - PRELIMINARY STEPS

Prior to testing the Ignition system for faults, verify that the battery open circuit is at least 12.4v and a cranking voltage of at least 9.6v. Test, replace or recharge the battery as needed before continuing the test sequence. The battery cable should not be removed at any time with the engine running.

Any high resistance in the igniter ground connection can cause Ignition system problems. With this in mind, verify that a clean and tight ground connection is present to the distributor body and that the distributor is properly grounded to the engine.

Use a DVOM (high impedance type) to test the ground circuit resistance by measuring the ground circuit voltage drop with the key on and engine off. It should be less than 100 millivolts (100mv).

Connect a tachometer to the coil negative terminal to measure engine speed. An inductive type tachometer that connects to any secondary ignition wire is another easy method of reading engine speed during Ignition system tests.

■ **NOTE:** *Do not leave the key on for more than 3 minutes in KOEO mode. If the key is left on too long with the engine off, the coil or igniter could overheat and fail.*

TYPE ONE SYSTEM - SPARK OUTPUT TEST

1) Remove the ignition coil secondary wire and inspect it for corrosion at both ends. If the wire is okay, connect the coil wire end to the coil and connect the distributor cap end to a high-output spark tester. Connect the spark test clamp firmly to engine ground. If a coil wire is not available, remove a spark plug cable and connect a spark tester to the spark plug end of the cable and to a secure engine ground (Figure 3-57).

2) Crank the engine for several short periods of time. A remote starter switch can be used to crank the engine for short periods of time. Verify the distributor turns at this time.

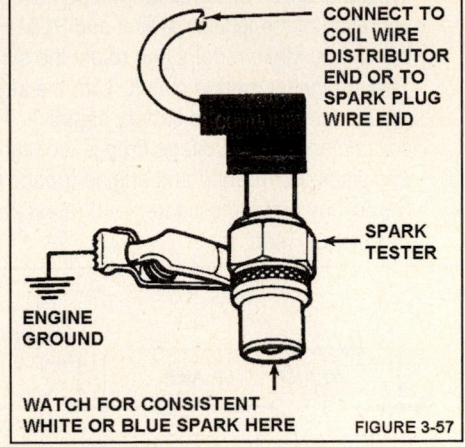

CONNECT TO COIL WIRE DISTRIBUTOR END OR TO SPARK PLUG WIRE END

SPARK TESTER

ENGINE GROUND

WATCH FOR CONSISTENT WHITE OR BLUE SPARK HERE

FIGURE 3-57

3) If a consistent white or blue-white spark is not present at the spark tester connected to the coil wire, the coil wire may be faulty. To test the coil wire, measure the coil wire resistance. It should not exceed 10,000 ohms. If a spark tester is used at one or more spark plug cables and spark is not present or is inconsistent, measure the spark plug wire resistance. The spark plug wire resistance should not exceed 25,000 ohms. Also, inspect for signs of secondary problems in the distributor cap and rotor. If any problems are located during this step, replace the component and retest for the symptom. If the Ignition system passes the Spark Output Test, go to Inspect Ignition System Connections.

CAUTION: *Do not crank the engine for over two seconds without stopping and starting over. It is important not to crank the engine for too long a period of time as fuel is injected into the engine during the spark output test and this action could flood the engine and cause hydrostatic lock-up.*

IGNITION SYSTEM CONNECTIONS

There are several connectors on Type One systems that need to be checked for clean and tight connections. Locate and check the ignition coil positive and negative terminals and the igniter control wires for problems. Examples of the terminal number and wire color changes for the Igniter Control Signals for Prelude models are shown below.

- 1988-90 2.0L Prelude igniter control signal terminals B15, B17 (WHT wires).
- 1990-91 2.1L Prelude igniter control signal terminals A21, A22 (WHT wires).
- 1992 2.1L Prelude igniter control signal terminals A21, A22 (YEL/GRN wires).
- 1992-95 2.2L Prelude igniter control signal terminals A21, A22 (YEL/GRN wires).
- 1992-95 2.3L Prelude igniter control signal terminals A21, A22 (YEL/GRN wires).
- 1996-98 2.2L Prelude igniter control signal terminals A20 (YEL/GRN wire).

Inspect all the Ignition system primary circuit wires at the distributor connections or at the PCM. Look for signs of a backed-out or loose connector, terminal corrosion or other problems that could be the cause of a No Spark Output condition.

Clean, tighten and repair all connections as needed and repeat the Spark Output Test. If the coil spark output is weak or missing, go to the Ignition Voltage Test.

TYPE ONE SYSTEM - IGNITION VOLTAGE TEST

1) Turn the key on and use a DVOM to test for battery voltage to the ignition circuit (BLK/YEL wire) to the igniter and the ignition coil (Figure 3-58). With the key on, use a DVOM to check for voltage between the coil positive terminal and ground. Inspect for a blown fuse or open circuit on the BLK/YEL wire to the coil. Use the correct Ignition system diagram to check for an open circuit in the fuse or fusible link to the ignition switch. Also check the operation of the ignition switch and PGM-FI main relay. Test the continuity of the BLK/YEL between the coil and the igniter. Make repairs and retest the symptom. If no faults were located, go to Step 2.

2) Test the igniter ground circuit. Turn the key on and use a DVOM to measure the voltage drop between the igniter ground connection and battery negative. The reading should be less than 100mv (this is the ground circuit for the coil primary). If the voltage drop is too high, check the following areas for the cause of high ground resistance:

3) The distributor ground and engine ground circuit to the battery

4) Repair any problems located and retest the system. If no problems are located in this step, go to the Coil Resistance Test.

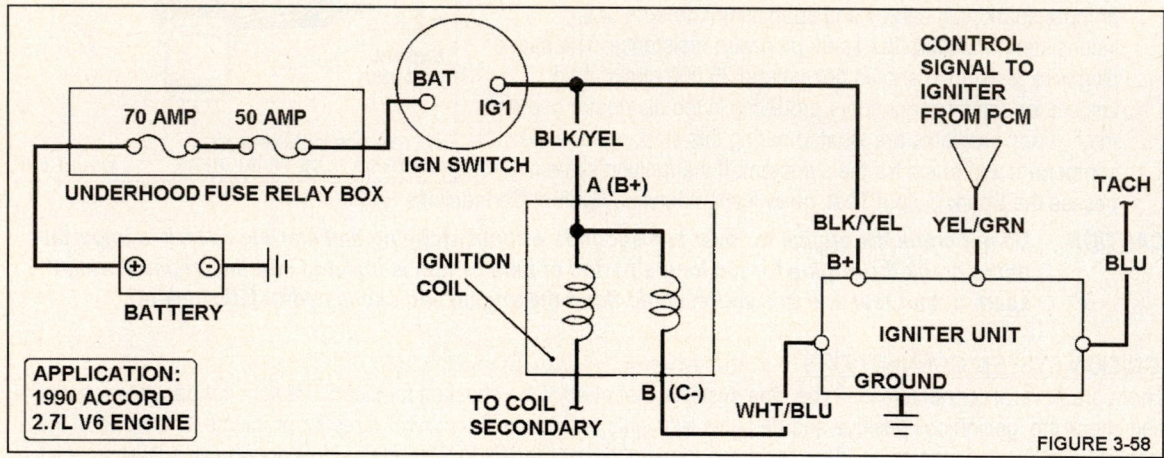

FIGURE 3-58

COIL RESISTANCE TEST

1) Select the appropriate low range on the DVOM to test the ignition coil primary circuit. Primary circuit specifications: 0.6-0.8 ohms on a standard coil and 0.3-0.4 ohms on an integral coil (both readings at 70°F).

2) Select the appropriate high range on the DVOM to test the coil secondary circuit. Secondary circuit specifications: 13.2-19.8 Kohms at 70°F.

3) If either coil reading is out of range, replace the coil and retest for the symptom. If the readings are in-range, go to the Igniter Voltage Tests.

TIPS: If the coil output is low, inspect the coil for signs of internal or external arcing to the primary or ground circuit. Look for burnt spots that could indicate a problem. It is especially helpful to use the spark tester to lightly stress the ignition coil while observing the coil tower for signs of secondary or primary breakdown. Also, an ignition analyzer is an excellent tool to use to observe the coil primary and secondary operation with the spark tester installed. Some models have the tachometer circuit incorporated in the coil assembly. The resistance of the tachometer circuit on some Honda coils (measured between terminals 'B' and 'D' of the connector) is from 2090-2310 ohms.

TYPE ONE SYSTEM - TEST REVIEW

If spark output was weak, the coil and coil wire will have been tested and replaced as needed. If the engine is a no start and there was no spark output, the previous tests would have validated the feed, ground and coil circuits. Any problems will have been repaired. If the vehicle still will not start at this point, the next step is to determine if the PCM is toggling or switching the igniter (and coil primary) on and off during cranking. Also, there should not be any ignition related codes set and the tachometer circuit should be tested to verify that it is not grounded prior to the Igniter Voltage Checks.

TYPE ONE SYSTEM - IGNITER VOLTAGE CHECKS

1) To test the PCM and igniter, turn the key off and remove the distributor cap, rotor and lead cover inside the distributor.
2) Remove the BLK/YEL, WHT/BLU, YEL/GRN (WHT on Civic & Accord) and BLU wires from the igniter to allow for circuit voltage testing. (Figure 3-59). Do not allow any of the disconnected wires to touch engine ground during testing with the key on.
3) Turn the key on and check for voltage between the BLK/YEL wire and body ground. If the DVOM reads battery voltage, go to Step 4. If the reading is 0v, locate and repair the open circuit in the BLK/YEL wire between the ignition switch and the igniter terminal. Retest for the symptom when repairs are completed.
4) With battery voltage available in Step 3, turn the key on and check for battery voltage between the WHT/BLU wire and ground. If there is battery voltage, go to Step 5. If the reading is 0v, check for an open primary circuit in the coil or for no feed voltage to the coil on the BLK/YEL wire. If these circuits are okay, check the continuity of the WHT/BLU wire between the coil and igniter terminal. Make repairs and retest for the symptom.
5) Turn the key off. Check for an open circuit in the YEL/GRN (WHT on Civic & Accord) wire between the PCM and igniter terminal. If this wire is okay, check for an open circuit in the BLU wire to the tachometer. If any faults are located, make

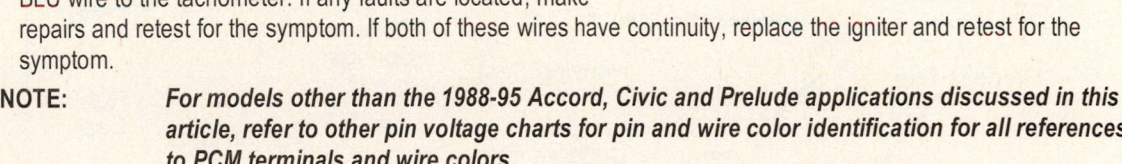

FIGURE 3-59

repairs and retest for the symptom. If both of these wires have continuity, replace the igniter and retest for the symptom.

■ NOTE: *For models other than the 1988-95 Accord, Civic and Prelude applications discussed in this article, refer to other pin voltage charts for pin and wire color identification for all references to PCM terminals and wire colors.*

<u>TYPE TWO SYSTEM - INTRODUCTION</u>

In a Type Two Ignition system the PCM controls spark timing in response to changes in engine coolant temperature and manifold vacuum. System components include a magnetic design distributor and pickup coil, distributor vacuum advance diaphragm, ignition coil, igniter, rotor, cap and the PCM (Figure 3-60).

<u>1987 ACCORD 2.0L I4 COMPONENT EXAMPLES</u>

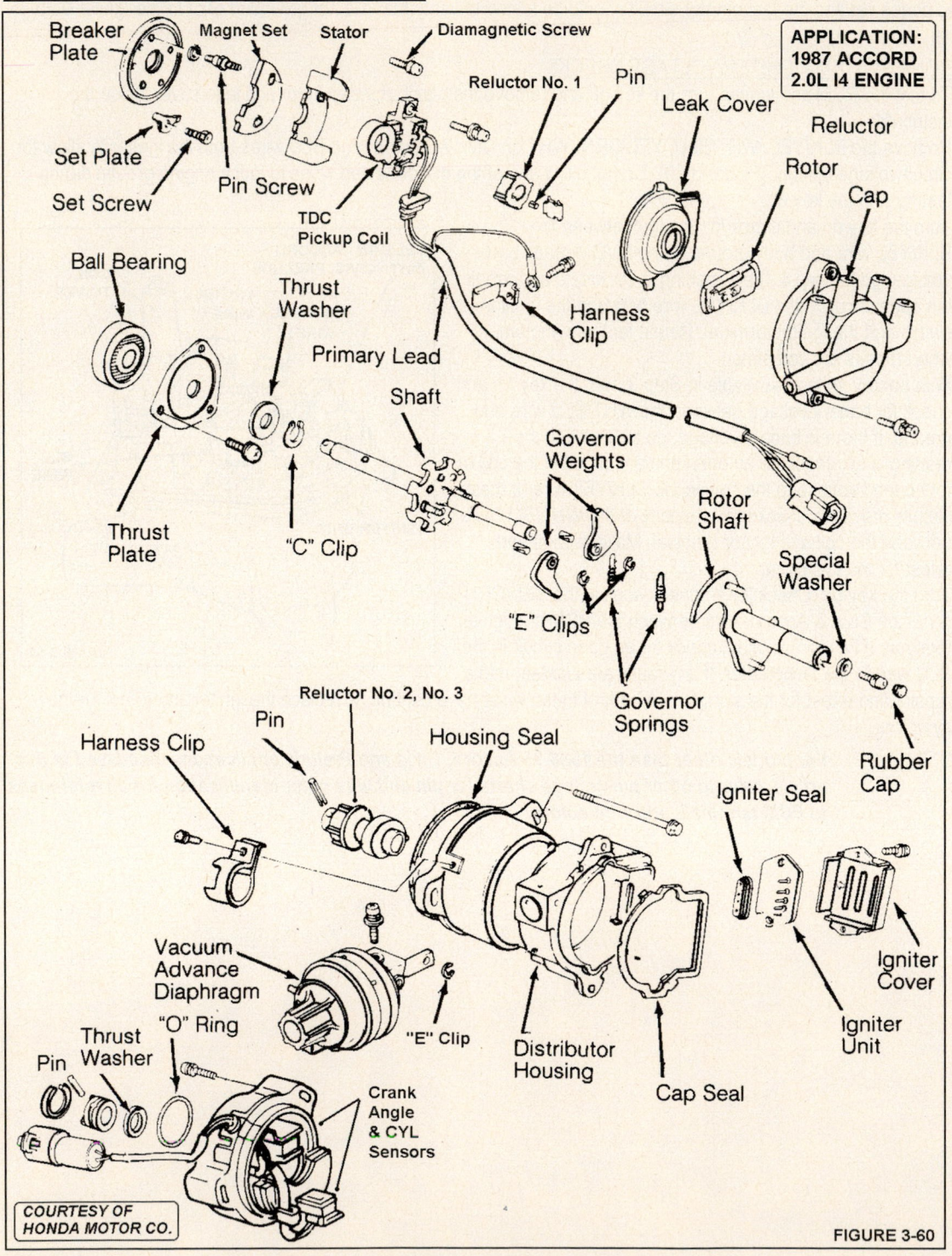

APPLICATION:
1987 ACCORD
2.0L I4 ENGINE

COURTESY OF
HONDA MOTOR CO.

FIGURE 3-60

TYPE TWO SYSTEM - VACUUM ADVANCE OPERATION

Distributor advance diaphragm 'B' is equipped with an ignition control solenoid valve in the vacuum line (Figure 3-61). The PCM controls the solenoid based on signals it receives from the various engine sensors. When the solenoid valve is opened (no control signal to the solenoid), it allows vacuum to diaphragm 'B' to improve cold driveability by providing additional ignition spark advance.

TYPE TWO SYSTEM - PRELIMINARY STEPS

Prior to testing the Ignition system for faults, verify that battery open circuit voltage is at least 12.4v and cranking voltage is at least 9.6v. Test, replace or recharge the battery as needed before continuing the test sequence. The battery cables should not be removed at any time with the engine running.

Also, high resistance in the igniter ground connection can cause Ignition system problems. With this in mind, verify that a clean and tight ground connection is present to the distributor body and that the distributor is properly grounded to the engine. Test the ground resistance with a DVOM. Measure the voltage drop from the igniter ground circuit to battery negative with the key on and engine off. The reading should be less than 100mv.

Connect a tachometer to the coil negative terminal to measure engine speed. An inductive type tachometer that connects to any secondary ignition wire is another easy method of reading engine speed during testing.

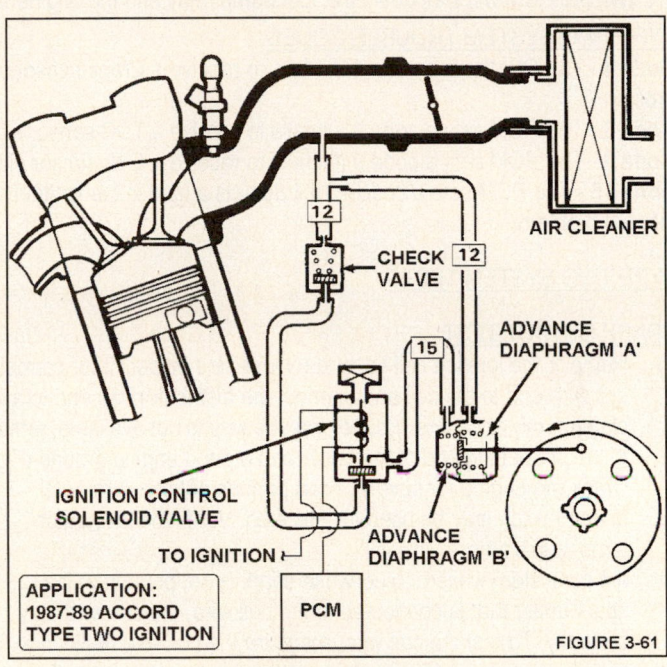

AIR CLEANER

CHECK VALVE

ADVANCE DIAPHRAGM 'A'

IGNITION CONTROL SOLENOID VALVE

TO IGNITION

ADVANCE DIAPHRAGM 'B'

APPLICATION:
1987-89 ACCORD
TYPE TWO IGNITION

PCM

FIGURE 3-61

■ **NOTE:** *Do not leave the key on for more than 3 minutes in KOEO mode. If the key is left on too long with the engine off, the coil or igniter could overheat and fail.*

TYPE TWO SYSTEM CODES (OBD I)

Diagnosis of the Ignition system should begin with a visual inspection of the underhood ignition components and a review of any stored trouble codes in the PCM memory. If the check engine light is on with the engine running, a hard code is present. The code may be related to an Ignition system fault, so it makes sense to use the PGM-FI Diagnostics to read, record and clear any codes prior. The diagnostics allow the PCM to record certain ignition-related faults in the TDC, CYL or Crank sensor. The PCM monitors the signals from these sensors with the engine running. If two signals are lost, the PCM will shut down the fuel pump relay and the engine will not start or it will stall.

TYPE TWO SYSTEM TROUBLE CODES

Code 4 - The PCM sets a code 4 if it fails to receive a Crank sensor signal due to either a circuit or component problem.

Code 8 - The PCM sets a code 8 if it fails to receive a TDC sensor signal due to either a circuit or component problem.

Code 9 - The PCM sets a code 9 if it fails to receive a CYL sensor signal due to either a circuit or component problem.

Code 15 - The PCM sets a code 15 if it detects a fault in the ignition control signal due to either a circuit or PCM internal problem.

TYPE TWO SYSTEM TESTS

SPARK OUTPUT TEST

1) Remove the ignition coil secondary wire and inspect it for corrosion at both ends. If the wire is okay, connect the coil wire end to the coil and connect the distributor cap end to a low-output spark tester. Connect the spark test clamp firmly to engine ground. If a coil wire is not available, remove a spark plug cable and connect a spark tester to the spark plug end of the cable and a good engine ground (Figure 3-57).

2) Crank the engine for several short periods of time. A remote starter switch may be used for this step. Verify the distributor turns at this time.

3) If a consistent white or blue-white spark is not present at the spark tester that is connected to the coil wire, the coil wire may be faulty. To test the coil wire, measure the coil wire resistance. It should not exceed 10,000 ohms. If a spark tester is used at one or more spark plug cables and spark is not present or is inconsistent, measure the spark plug wire resistance. The spark plug wire resistance should not exceed 25,000 ohms. Also, inspect for signs of secondary problems in the distributor cap and rotor. If any problems are located during this step, replace the component and retest for the symptom. If there is no spark output or weak output during this test, go to the Ignition System Connections step.

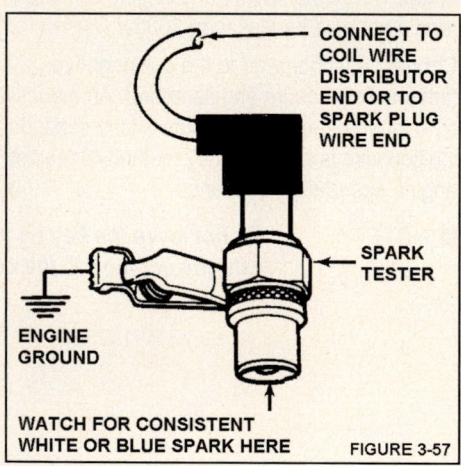

CONNECT TO COIL WIRE DISTRIBUTOR END OR TO SPARK PLUG WIRE END

SPARK TESTER

ENGINE GROUND

WATCH FOR CONSISTENT WHITE OR BLUE SPARK HERE

FIGURE 3-57

CAUTION: *Do not crank the engine for over two seconds without stopping and starting over as fuel is injected into the engine during the spark output test and this action could flood the engine and cause hydrostatic lock-up.*

IGNITION SYSTEM CONNECTIONS

There are several connectors on Type Two systems that need to be checked for clean and tight connections. Inspect all of the Ignition system primary circuit wires at the distributor and ignition coil connections. Look for signs of a backed-out or loose connector, terminal corrosion or other problems that could be the cause of a No Spark Output condition.

Clean, tighten and repair all connections as needed and repeat the Spark Output Test. If the coil spark output is weak or missing, go to the Ignition Voltage Test.

IGNITION VOLTAGE TEST

1) Turn the key on and check for battery voltage at the BLK/YEL wire to the igniter and ignition coil. Also check the voltage between the coil positive terminal and ground. Inspect for a blown fuse or open circuit on the BLK/YEL wire to the coil.
2) If battery voltage is not present, check the fuse or fusible link to the ignition switch for an open or blown circuit condition. Also inspect the operation of the ignition switch and PGM-FI main relay. Test the continuity of the BLK/YEL between the coil and the igniter.
3) If a fault is located, make repairs as needed and retest for the symptom. If no problems are located, go to Step 4.
4) Test the igniter ground circuit. Turn the key on and use a DVOM to measure the voltage drop between the igniter ground connection and battery negative. The reading should be less than 100mv (this is the ground circuit for the coil primary). If the voltage drop is too high, check the distributor ground and engine ground circuit (at the thermostat housing on some models) to the battery. Repair any problems located and retest for the symptom.

If no problems are located in this step, go to Coil Resistance Test on the next page.

COIL RESISTANCE TEST

1) Select an appropriate low range on the DVOM to test the ignition coil primary circuit. Coil primary circuit specifications: 1.2-1.5 ohms at 70°F.

2) Select an appropriate ohmmeter high range on the DVOM to test the coil secondary circuit resistance: 9,000-13,500 ohms at 70°F.

3) If either coil reading is out of range, replace the ignition coil and retest for the symptom. If the coil readings are okay, go to the Igniter Voltage Tests.

TIPS: *If the coil output is low, inspect the coil for signs of internal or external arcing to the primary or ground circuit. Look for burnt spots that could indicate a problem. It is especially helpful to use the spark tester to lightly stress the ignition coil while observing the coil tower for signs of secondary or primary breakdown. Also, an ignition analyzer is an excellent tool to use to observe the coil primary and secondary operation with the spark tester installed. The resistance of the tachometer circuit used on some models (measured between terminals 'B' and 'D' of the connector) is from 2000-2300 ohms.*

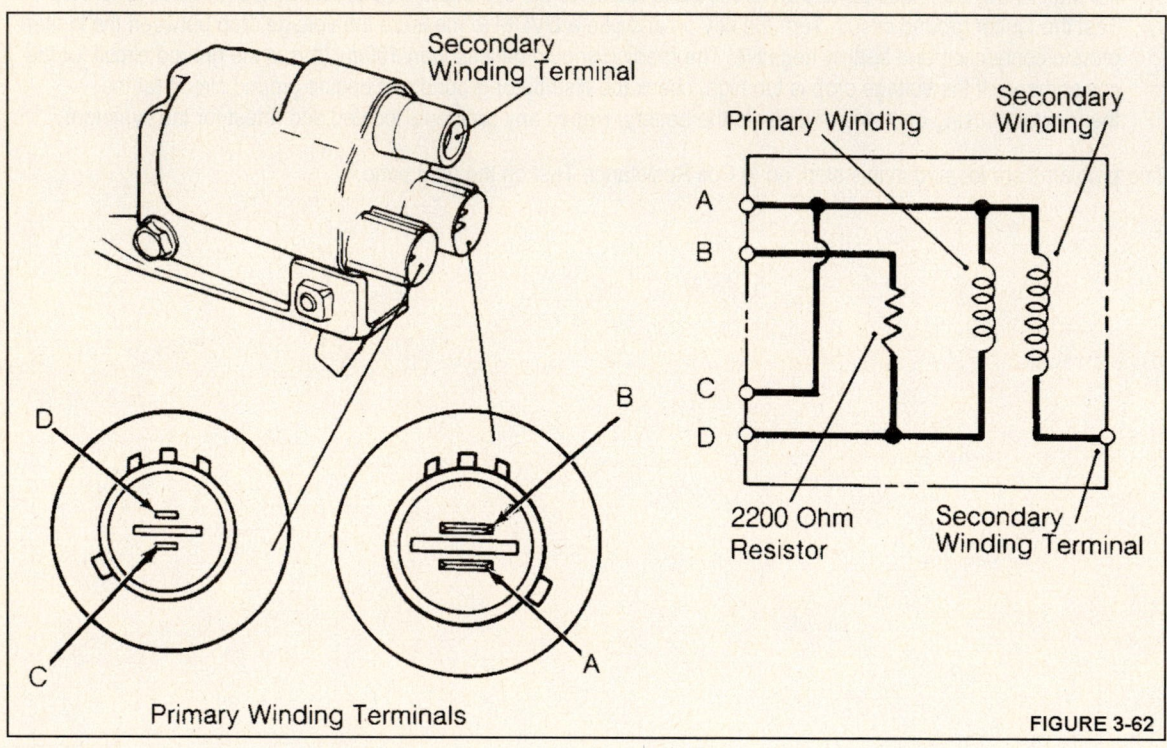

Primary Winding Terminals

FIGURE 3-62

TYPE TWO SYSTEM - TEST REVIEW

If spark output is weak, the coil and coil wire should have been tested and replaced. If the engine will not start and there is no spark output, the previous tests should have validated the feed, ground and coil circuits. These problems should have been repaired.

If the vehicle still will not start at this point, the next step is to determine if the PCM is switching the igniter (and coil primary) on and off during cranking. Also, there should not be any ignition related codes set and the tachometer circuit should be tested to verify that it is not grounded. There are two different types of distributors used on this system: Toyodenso and Hitachi. The Toyodenso unit is referred to as a TEC unit when ordering any parts. The Igniter Voltage Tests for these two distributors are discussed next.

TOYODENSO TYPE DISTRIBUTORS - IGNITER VOLTAGE TEST

1) To test the igniter voltage, take off the distributor cap and remove the igniter inside cover. Then pull out the igniter to expose the four terminals.

2) Turn the key on and measure the voltage from both the BLK/YEL and Blue (1) wires to battery negative (Figure 3-63). The DVOM should read 12v on both wires. If the reading is 0v, locate and repair the open circuit the faulty wire(s) and then retest for the symptom. If the reading is okay, go to the Pickup Coil Test.

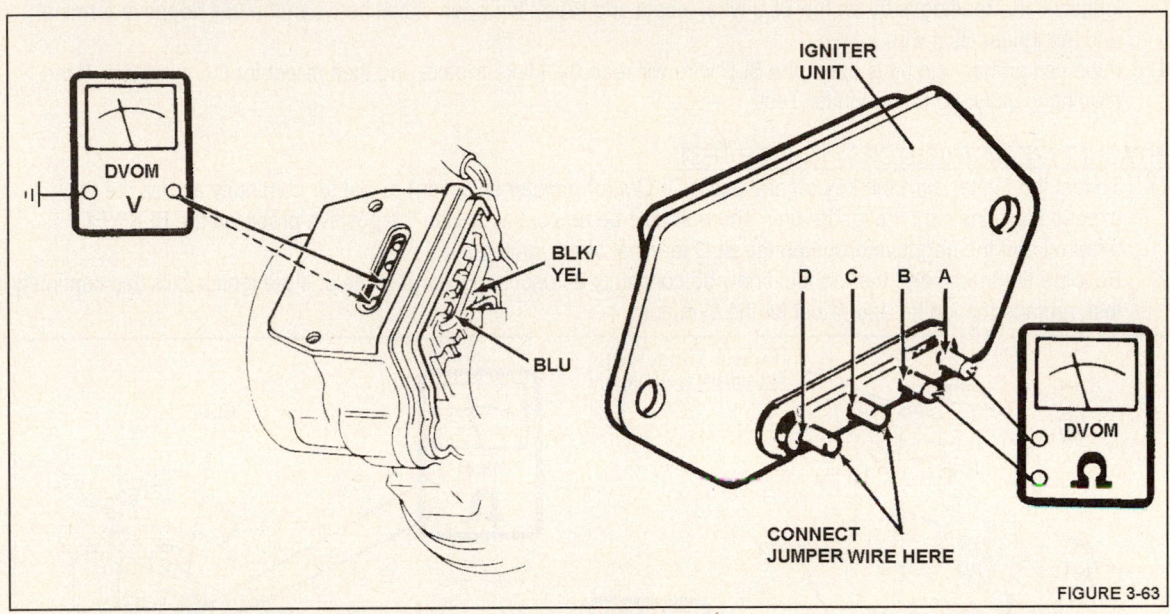

FIGURE 3-63

TOYODENSO TYPE DISTRIBUTORS - PICKUP COIL TEST

1) To test the pickup coil, select ohmmeter on the DVOM and connect between the BLU (1) and GRN wires to the pickup coil. The resistance should be 650-850 ohm (70°F).

2) If the reading is out of range, replace the pickup coil and retest for the symptom. If it is okay, go to Igniter Output Test.

TOYODENSO TYPE DISTRIBUTORS - IGNITER OUTPUT TEST

1) To test the igniter output circuit, first connect a jumper wire between the 'C' and 'D' terminals of the igniter. Then check for continuity in both directions between the igniter 'A' and 'B' terminals with a DVOM (Figure 3-63).

2) There should be continuity in only one direction. Replace the igniter if continuity is present in both directions. If it is okay, go to the Igniter Input Test.

TOYODENSO TYPE DISTRIBUTORS - IGNITER INPUT TEST

1) To test the igniter ground circuit, connect the DVOM positive lead to the 'D' terminal and the negative lead to the igniter inside case (this is ground).

2) The reading should be 50,000 ohms or higher (70°F). If the resistance is lower than the specified range, replace the igniter and retest for the symptom.

Summary: If the igniter passes all of the tests and the pickup coil passed its test, replace the igniter and retest for the symptom.

HITACHI TYPE DISTRIBUTORS - IGNITER VOLTAGE TEST

1) To test the igniter voltage, first turn the key off. Then remove the distributor cap, igniter inside cover and the lead wires from the igniter unit.

2) Turn the key on and measure the voltage from both the BLK/YEL and the BLU wires to battery negative. The reading should be 12v on both wires.

3) If the reading is 0v on the BLK/YEL wire, locate and repair the open circuit between the ignition switch and the igniter. If the reading is 0v on the BLU wire, locate and repair the open circuit between the coil negative terminal and the igniter BLU wire.

4) If the coil primary circuit is open, the BLU wire will read 0v. Make repairs and then retest for the symptom. If the reading is okay, go to the Igniter Test.

HITACHI TYPE DISTRIBUTORS - IGNITER TEST

1) To test the igniter, turn the key off and use a DVOM (ohmmeter selection) to test for continuity across the two exposed terminals on the igniter unit. There should be no continuity with the positive probe on the BLK/YEL terminal and the negative probe on the BLU terminal of the igniter.

2) Reverse the leads, and there should now be continuity across the igniter terminals. If the igniter fails this continuity test, replace the igniter and retest for the symptom.

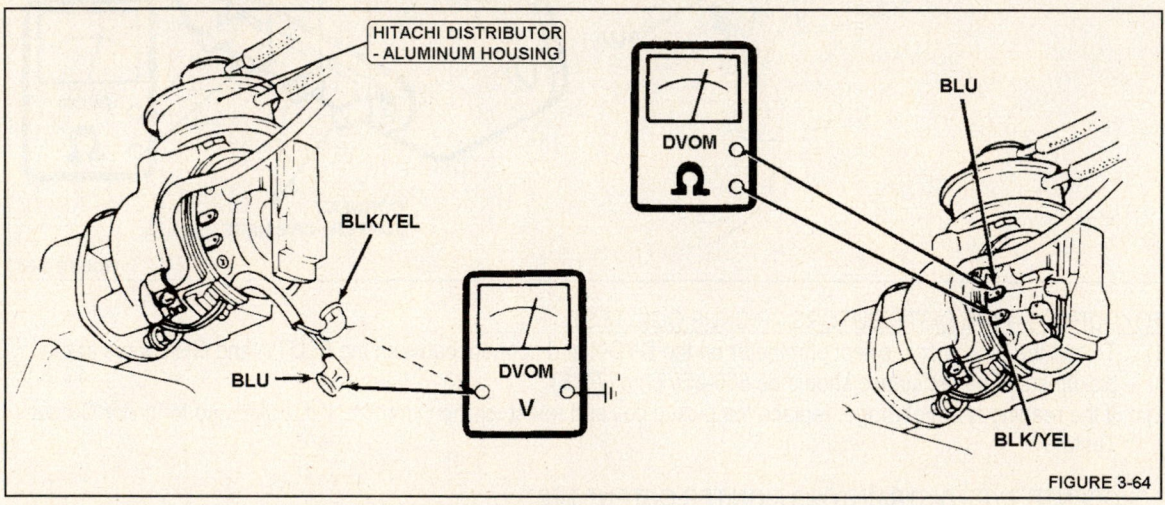

FIGURE 3-64

HITACHI TYPE DISTRIBUTORS - PICKUP COIL TEST

On this design of distributor, the pickup coil is part of the igniter assembly and cannot be tested separately.

Summary: If there is battery voltage to the igniter BLK/YEL and BLU wires, and the igniter passes the continuity test with the DVOM, the pickup coil may have failed. If there is no spark output, the igniter and pickup coil may have to be changed as a unit to resolve the no start condition.

TYPE TWO IGNITION CONTROL SOLENOID TESTS

The distributor in a Type Two system has two separate vacuum advance diaphragms that operate with manifold vacuum. An ignition control solenoid valve is connected into the Diaphragm 'B' vacuum line. The PCM controls the on/off operation solenoid once it receives information from various engine sensors. Once the solenoid is opened, it allows the flow of vacuum to diaphragm 'B' in the distributor to provide additional spark advance in to improve cold driveability.

DRIVEABILITY SYMPTOMS

Use this test if diaphragm 'B' fails the vacuum part of the Ignition Timing Test.

DIAPHRAGM 'A' VACUUM TEST

1) Start the engine and allow it to run until the fan cycles.

2) Remove the No. 12 vacuum hose from the diaphragm (Figure 3-65). Allow the engine to idle and check for available vacuum at the hose end. If vacuum is present, go to Step 3. If there is no vacuum, inspect for a disconnected or restricted hose. Check the vacuum check valve for clogging. Make repairs as needed and go to Step 3.

3) Reconnect the No. 12 vacuum hose to the diaphragm and then recheck the ignition timing. The spark timing should read 15°BTDC. If the timing does not advance to the specified amount, replace the diaphragm 'A' and retest the symptom. If the timing advanced, go to Step 4.

4) Next, remove the No. 15 vacuum hose and check for the presence of vacuum at idle speed (Figure 3-66). There should be vacuum present at the hose at idle speed. If there is no vacuum, go to Ignition Control Solenoid Test Two. If vacuum is present, with the gear selector in P/N, raise engine speed to 1500 rpm and recheck for vacuum. There should still be vacuum present at the hose. If there is no vacuum at 1500 rpm in P/N, go to Ignition Control Solenoid Test One. If vacuum is present, reconnect the No. 15 vacuum hose and the test is completed.

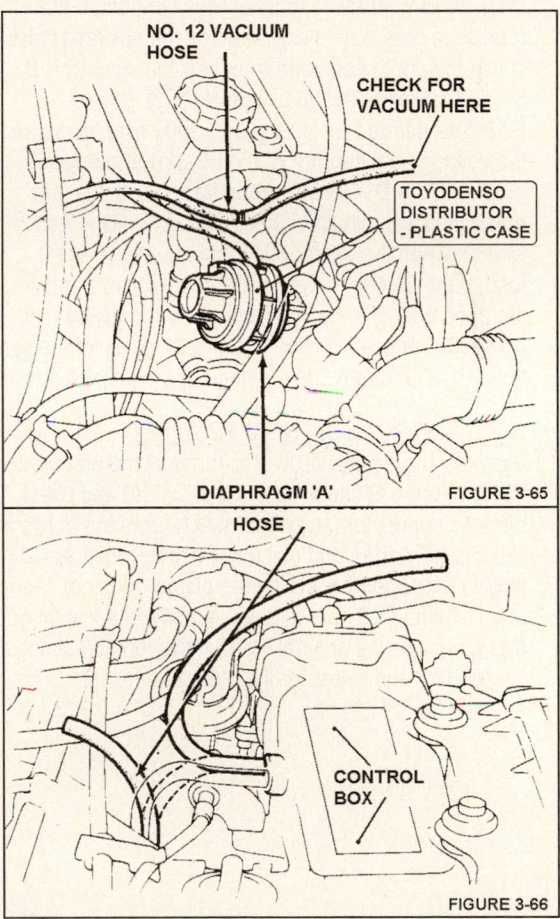

NO. 12 VACUUM HOSE

CHECK FOR VACUUM HERE

TOYODENSO DISTRIBUTOR - PLASTIC CASE

DIAPHRAGM 'A' HOSE

FIGURE 3-65

CONTROL BOX

FIGURE 3-66

TYPE TWO IGNITION CONTROL SOLENOID TEST ONE

1) Start the engine and allow it to run until the electric cooling fan cycles.
2) Remove the No. 12 vacuum hose to the intake manifold. Allow engine to idle and check for available vacuum at the intake manifold connection. If there is no vacuum present, clean the vacuum port connection and retest for vacuum. If vacuum is present, check the No. 12 vacuum hose for restriction, plugging, a cracked or disconnected hose. Make repairs as needed and go to Step 3.

3) Remove the 6-P solenoid connector at the control box (Figure 3-67). Raise engine speed to 1500 rpm, and check for battery voltage between the BLK/YEL and WHT wires at the solenoid harness connector. If the reading is near 12v (okay), replace the solenoid valve inside the control box and retest for the symptom. If the DVOM reads 0v, go to Step 4.

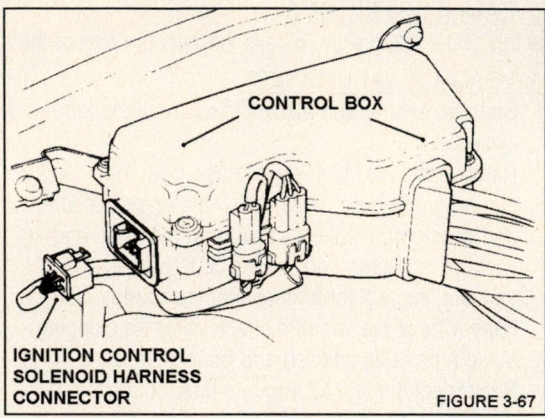

CONTROL BOX

IGNITION CONTROL
SOLENOID HARNESS
CONNECTOR

FIGURE 3-67

4) If the DVOM read 0v in Step 3 at 1500 rpm, measure the voltage between the BLK/YEL wire and engine ground. If the DVOM still reads near 0v, locate and repair the open circuit in the BLK/YEL ignition feed wire between the solenoid and the No. 1 (15-amp) fuse. If the DVOM reads near 12v, locate and repair the open circuit in the WHT control wire between the harness connector and the PCM and retest for the symptom. If the WHT wire is not open, substitute a known-good PCM and retest the voltage. If the voltage now reads near 12v at 1500 rpm, replace the original PCM.

IGNITION CONTROL SOLENOID TEST TWO

1) Start the engine and allow it to run until the electric cooling fan cycles.
2) Remove the 6-P connector. Use a DVOM and check for 12v between the BLK/YEL and ORN wire of the main harness connector (engine should be running at idle in Park or Neutral). The DVOM should read battery voltage. Next, rapidly open and close the throttle while watching the voltage reading. It should go from 12-14v to 0v as the throttle is snapped open. If the voltage does not change as described, check the ORN wire to the PCM for an open circuit. If the wire is okay, substitute a known-good PCM and repeat the voltage tests. If the symptom goes away, replace the original PCM. If the reading changed from 12-14v to 0v during this test step, replace the solenoid 'A' and retest for the symptom.

PCV System

INTRODUCTION

The Honda Positive Crankcase Ventilation (PCV) system is designed to prevent blow-by gas from escaping into the atmosphere.

The PCV valve contains a spring-loaded plunger. Once the engine starts to run, the plunger in the PCV valve is lifted in proportion to intake manifold vacuum.

This action causes the blow-by gas to be drawn into the intake manifold.

PCV SYSTEM INSPECTION

Inspect the PCV hoses and connections for leaks and clogging. Make any repairs as needed and continue testing the system.

PCV SYSTEM TEST

Start the engine and allow it to idle. Obtain a pair of pliers and a shop towel.

Next, lightly squeeze the hose between the PCV valve and the intake manifold with the pliers (or your fingers) and listen for a clicking sound from the valve.

If there is no clicking sound during this test, inspect the PCV valve grommet and hose for cracks or leaks.

If the grommet and hose are okay, replace the PCV valve and retest for the symptom.

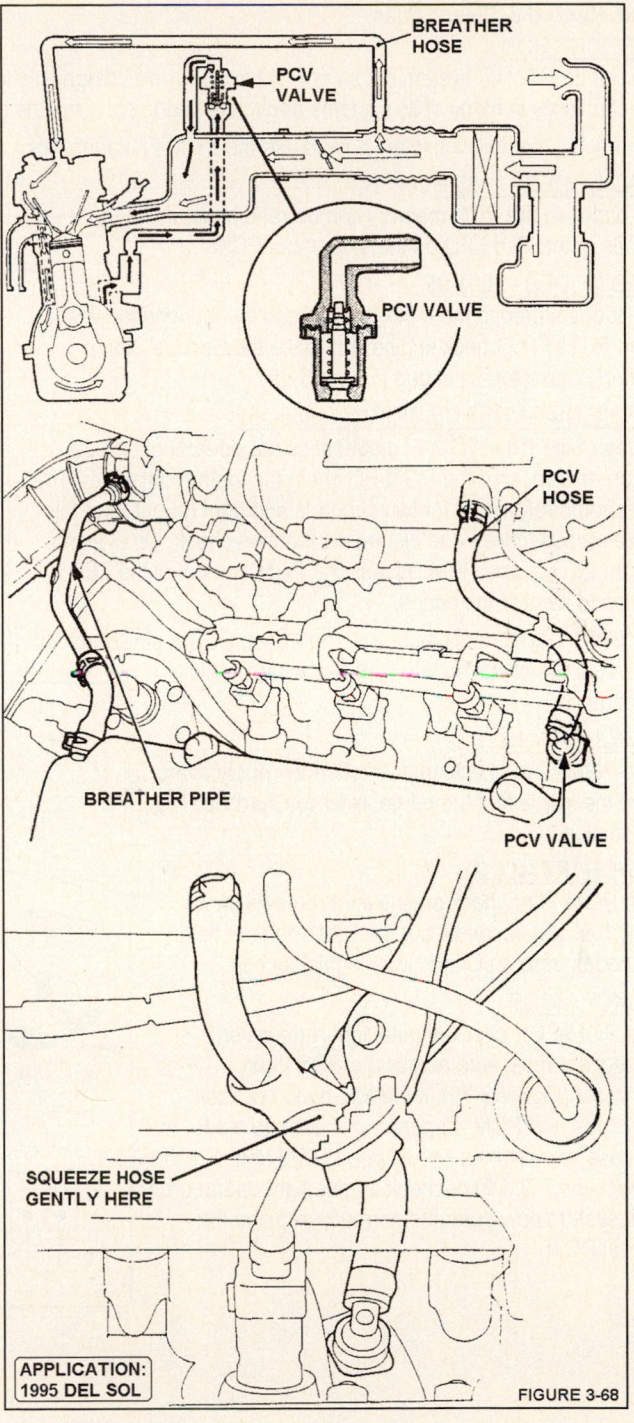

FIGURE 3-68

PGM – FI Diagnostics

Read, Record & Clear Codes

INTRODUCTION

The Honda PGM-FI system uses the PCM to perform its diagnostic functions. The PCM in this system is designed to detect failures in many of its systems or circuits and to set a trouble code in memory when a fault is detected.

The articles that follow explain how to use the PGM-FI diagnostics.

READ CODES - 1985-90

Any codes in the PCM memory can be read by counting the flashes of the Red LED on the side of the PCM

READ CODES - 1991-95

Any codes stored in the PCM can be read by counting the flashes of the PGM-FI or check engine light once the Service Check System connector is jumped (Figure 3-69).

DO THE DIAGNOSTICS WORK? (1985-95)

To determine if the PGM-FI diagnostics are operating, turn the key on and observe the PGM-FI light in the instrument panel. The light should remain on for two seconds and then go out. Use this step as a bulb check and diagnostic function check. On systems with an LED on the PCM, access the PCM and count the LED flashes to identify the codes.

On 1990-91 systems, jumper the underdash SCS 2-pin connector and count the check engine light flashes to identify the codes.

PCM CHART NO. 1

If the PGM-FI or check engine light does not activate when the key is first turned on, refer to Chart No. 1 in this section.

PCM CHART NO. 2

If the PGM-FI or check engine light remains on for more than two seconds, but the light does not flash any codes, refer to Chart No. 2 in this section.

BACKUP MODE

If the PGM-FI or check engine light remains on, connect a jumper wire across the SCS 2-pin connector (Figure 3-70). If the MIL does not flash any codes, the PCM is operating in backup mode. In this case, substitute a known-good PCM and recheck the PGM-FI or check engine light operation. If the system now operates normally, replace the original PCM.

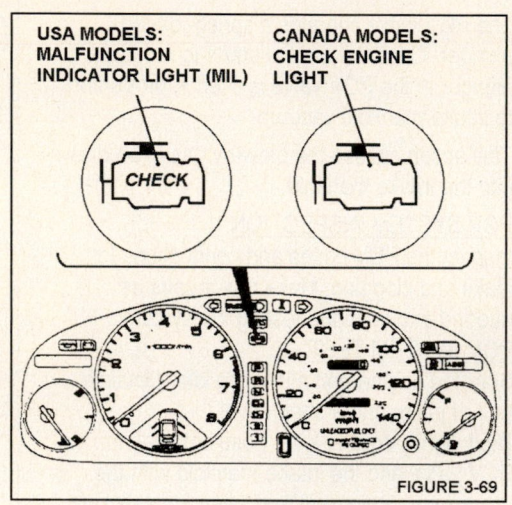

USA MODELS:
MALFUNCTION INDICATOR LIGHT (MIL)

CANADA MODELS:
CHECK ENGINE LIGHT

FIGURE 3-69

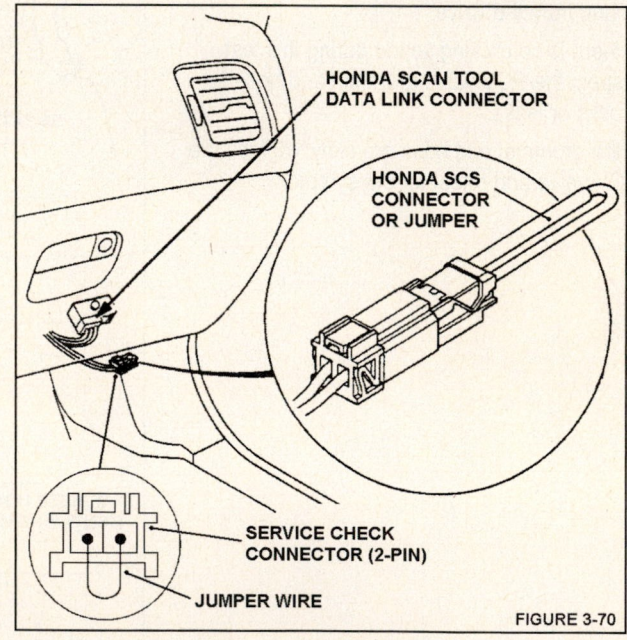

HONDA SCAN TOOL
DATA LINK CONNECTOR

HONDA SCS
CONNECTOR
OR JUMPER

SERVICE CHECK
CONNECTOR (2-PIN)

JUMPER WIRE

FIGURE 3-70

HARD CODES

The term hard code refers to a trouble code that reappears immediately after a PCM reset is completed. On the PGM-FI system, if a hard code is present when the key is turned on, the PGM-FI or check engine light will remain on AFTER the bulb check time period (two seconds). In this case, the PCM should output the code(s) through the Red LED on the side of its case.

SOFT CODES

On PGM-FI systems, the term *soft code* describes a code that appears before a PCM reset is done, but does not reappear after a PCM reset. A *soft code* is output in the same manner as a *hard code* is indicated - through the MIL or LED. The difference is that the light will remain off after the initial bulb check period. A *soft code* can be recorded during the initial step of read, record and clear codes. In effect, a *soft code* is a code stored in memory from an earlier fault that does not set after a PCM reset.

PCM RESET

To perform a PCM reset procedure (clear all codes from memory), do these test steps. Turn the key off and then remove the PCM power fuse from the underhood fuse/relay box for 10 seconds. This step resets the PCM. If the engine is started after doing this step, any codes that reset are hard codes.

PCM LOCATIONS

1985-89 Accord - under driver's seat

1990-98 Accord - at bottom front of passenger floor pan

1985-87 Civic - under passenger seat

1988-91 Civic - at bottom front of passenger floor pan

1992-98 Civic - at right side kick panel

1992-98 CR-V - behind the right center of dash (Figure 3-71)

1995-98 Odyssey - at bottom front of passenger floor pan

1994-98 Passport - at center of dash

1986-87 Prelude - behind trim panel, above left side of passenger seat

1988-98 Prelude - at bottom front of passenger floor pan

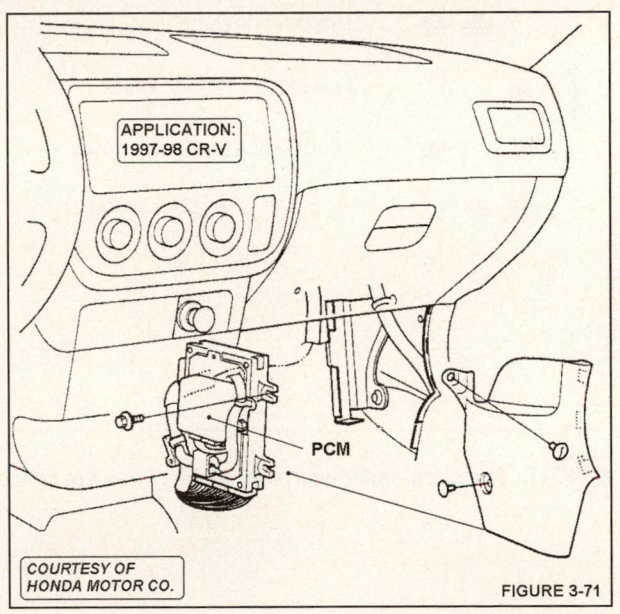

APPLICATION: 1997-98 CR-V

PCM

COURTESY OF HONDA MOTOR CO.

FIGURE 3-71

READ CODES - 4 LED DISPLAY (1985-87)

Some early Honda vehicles use a PCM with 4 LEDs on the side instead of one found on the later models. The models that use this system of code readout are 1985 Accord, 1986-87 Civic and 1985-87 Civic CRX.

To identify trouble code on this system, add up the number of flashes on the four LEDs (Figure 3-72).

If a hard code is set, the PGM-FI light on the dash will be activated and the PCM should flash the code number using the four LEDs on the side of the PCM. If a soft code is set, the PCM will retain the code number and continue to flash the LEDs to output the code. However, the PGM-FI light will not be activated if a soft or history code is in the system.

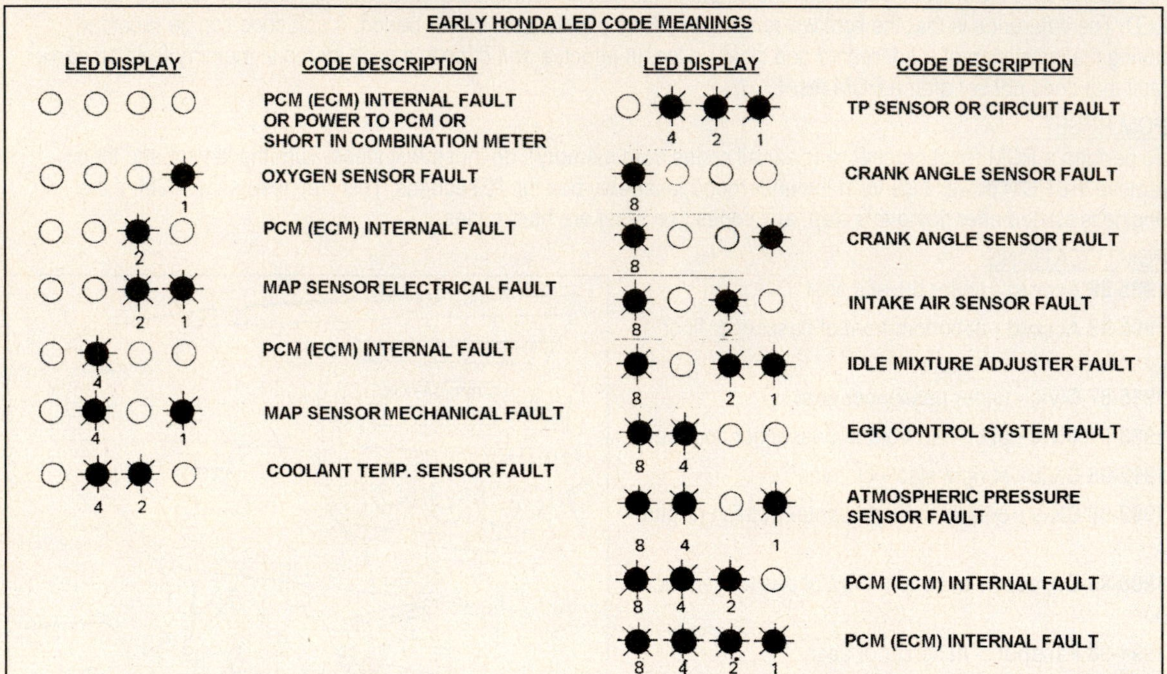

EARLY HONDA LED CODE MEANINGS

LED DISPLAY	CODE DESCRIPTION	LED DISPLAY	CODE DESCRIPTION
○ ○ ○ ○	PCM (ECM) INTERNAL FAULT OR POWER TO PCM OR SHORT IN COMBINATION METER	○ ● ● ● (4 2 1)	TP SENSOR OR CIRCUIT FAULT
○ ○ ○ ● (1)	OXYGEN SENSOR FAULT	● ○ ○ ○ (8)	CRANK ANGLE SENSOR FAULT
○ ○ ● ○ (2)	PCM (ECM) INTERNAL FAULT	● ○ ○ ● (8)	CRANK ANGLE SENSOR FAULT
○ ○ ● ● (2 1)	MAP SENSOR ELECTRICAL FAULT	● ○ ● ○ (8 2)	INTAKE AIR SENSOR FAULT
○ ● ○ ○ (4)	PCM (ECM) INTERNAL FAULT	● ○ ● ● (8 2 1)	IDLE MIXTURE ADJUSTER FAULT
○ ● ○ ● (4 1)	MAP SENSOR MECHANICAL FAULT	● ● ○ ○ (8 4)	EGR CONTROL SYSTEM FAULT
○ ● ● ○ (4 2)	COOLANT TEMP. SENSOR FAULT	● ● ○ ● (8 4 1)	ATMOSPHERIC PRESSURE SENSOR FAULT
		● ● ● ○ (8 4 2)	PCM (ECM) INTERNAL FAULT
		● ● ● ● (8 4 2 1)	PCM (ECM) INTERNAL FAULT

FIGURE 3-72

■ NOTE: *The code repair charts for these codes are similar to the charts for later models.*

READ CODES - LED DISPLAY (1986-87)

To read any stored codes, locate the PCM and pull down the inspection window or remove it to allow the Red LED to be viewed. Turn the key on and observe the LED on the PCM. If codes are stored in PCM memory, they will be displayed as long and short pulses of the LED. The PCM can initiate any number of long and short pulses through the flashing LED. These pulses correspond to component or circuit problems that have been detected and stored in memory.

READ CODES - CHECK ENGINE LIGHT (1993-95)

To read any stored codes with the check engine light, locate the SCS connector and connect a jumper wire across the connector (Figure 3-73). Turn the key on and read the PGM-FI or C/E light flashes. This light is also referred to as the malfunction indicator lamp (MIL) in some literature.

If codes are stored in PCM memory, they will be displayed as long and short pulses of the PGM-FI or check engine light. The PCM can initiate any number of long and short pulses by altering the light on/off sequence. These pulses correspond to component or circuit problems that have been detected and stored in memory.

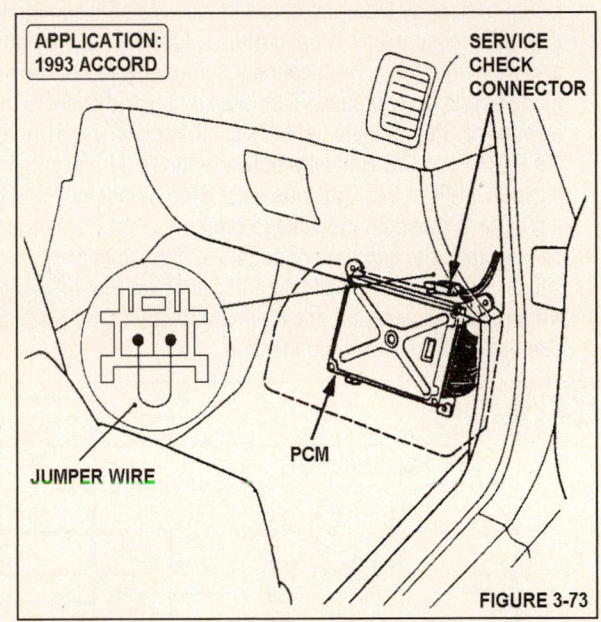

FIGURE 3-73

COUNTING CODES

Codes 1 - 9 are indicated by a series of short pulses while codes 10 - 43 are indicated by a series of long and short pulses of the LED. In this system, one long pulse of the LED is equal to ten short pulses. To determine the actual code number, add the long and short pulses together. An index of these codes is provided in Section One.

FALSE OR INVALID CODES

If any codes other than those listed in the Code Index are displayed, recheck to verify the code number is one that is not listed. If a code is displayed that is not in the Code Index, substitute a known-good PCM and recheck for the invalid codes. If the invalid code is gone, replace the original PCM.

RECORD CODES

Record any trouble codes that are displayed by the PCM. This is an important point as the next step is to do a PCM reset (clear the memory). Any codes that are not recorded will be lost. This information could be important if any of the codes were intermittent in nature. Refer to the previous information in this article to distinguish between a hard or soft type of trouble code.

PCM CHART NO. 1 (NO PGM-FI LIGHT)

This repair chart is used to determine why the PGM-FI or check engine light (MIL) was not activated during the first two seconds right after the key was turned on. This bulb check is the first step in the PGM-FI Diagnostics.

CHART NO. 1 TEST PROCEDURE

1)   Turn the key on and observe the oil pressure light. If it is on, go to Step 2. If the oil pressure light is off, check the No. 4 or 5 fuse. If it is blown, locate the cause of the blown fuse and then replace the fuse. If the fuse is okay, locate and repair the open circuit in the wire between the fuse and combination meter.

2)   If the oil pressure light is on in Step 1, turn the key off and connect the Honda Test Harness or BOB to the PCM and wiring harness. Then connect a jumper wire between PCM terminal B6 (A13 on some Accord and Prelude models) and body ground. Turn the key on and observe the PGM-FI or check engine (C/E) light. If the light is on now, go to Step 3. If the light is still off, replace the bulb. If the bulb is okay, locate and repair the open circuit in the PGM-FI or C/E light wire between the PCM and combination meter.

3)   If the PGM-FI or C/E light was on in Step 3, test the PCM ground connections as follows. Turn the key on and use a DVOM to measure the voltage drop between PCM ground connections A2, A4, A16 and A18 (A23, A24 and A26 on later models) and battery negative. If the voltage drop reading is less than 100 mv on all terminals (okay), substitute a known-good PCM and retest the PGM-FI light at KOEO for two seconds. If the PGM-FI or C/E light works okay, replace the original PCM. If any of the readings are over 100 mv, locate and repair the cause of the open or high resistance connection.

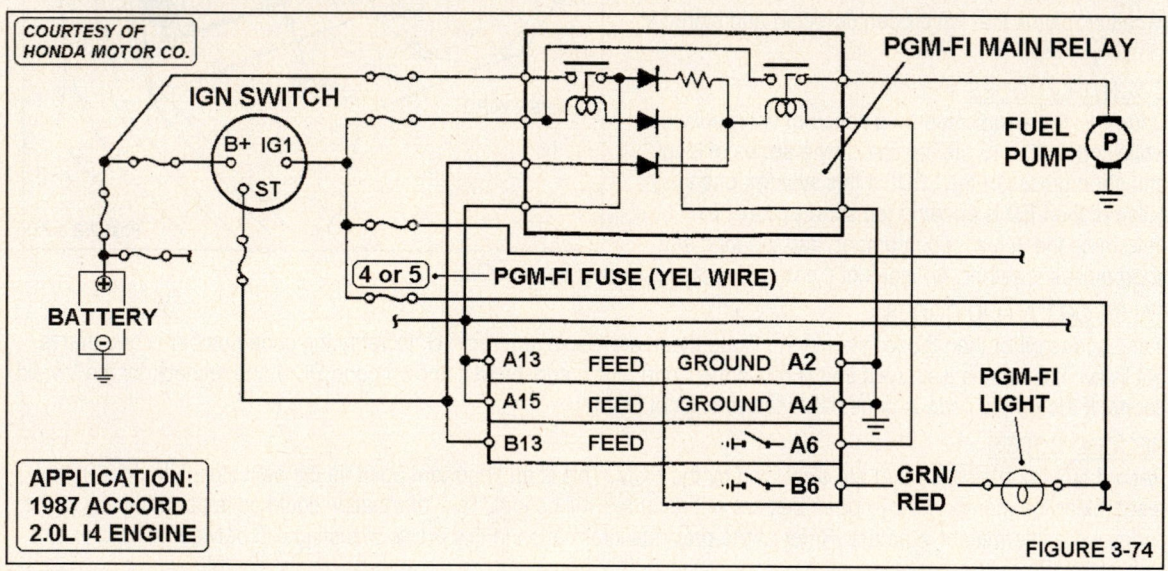

FIGURE 3-74

TIPS: Inspect the connection at the thermostat housing carefully (it may need to be cleaned and reassembled to fix a ground circuit problem).

PCM CHART NO. 2 (PGM-FI LIGHT ON, NO CODES)
Use this chart if the PGM-FI or check engine light remains on for more than two seconds with the key on and no codes are output at the LED on the PCM.

CHART NO. 2 TEST PROCEDURE

1) Attempt to start the engine. If it will not start, go to Step 3. If engine starts and the PGM-FI goes out after two seconds, the fault is gone (it was intermittent). In this case, inspect for any loose wires or connectors at the combination meter and PCM. Retest the system after the inspection. If the light remains on for over two seconds, go to Step 2.

2) If the light remains on for over two seconds in Step 1, check the PCM for any code output on the LED. Is the PCM flashing any codes? If so, refer to the code repair chart. If no codes are output, turn the key off and remove the PCM 'B' connector ('A' connector on later models). Turn the key on and observe the PGM-FI light. If the light is on, locate and repair the short to ground in the light wire between PCM terminal B6 (or A13) and the combination meter. If light is off with the connector removed, substitute a known-good PCM. If the PGM-FI light now works okay, replace the original PCM.

3) If the engine did not start in Step 1, turn the key off. Remove and inspect the fuse that feeds the PGM-FI light in the main fuse box. If the fuse is blown, locate and repair the short, then replace the fuse. If the fuse is okay, remove and inspect the correct fuse from the dash fuse box. If that fuse is blown, locate and repair the short and replace the fuse. If this fuse is okay, turn the key on. Next, remove the connectors to the MAP, TP and EGR Lift sensors (remove one at a time to observe any change). Leave the sensors disconnected as each one is removed. If the PGM-FI light remains on, go to Step 4. If the light goes out after any of these sensors are disconnected, replace the defective sensor and retest the light.

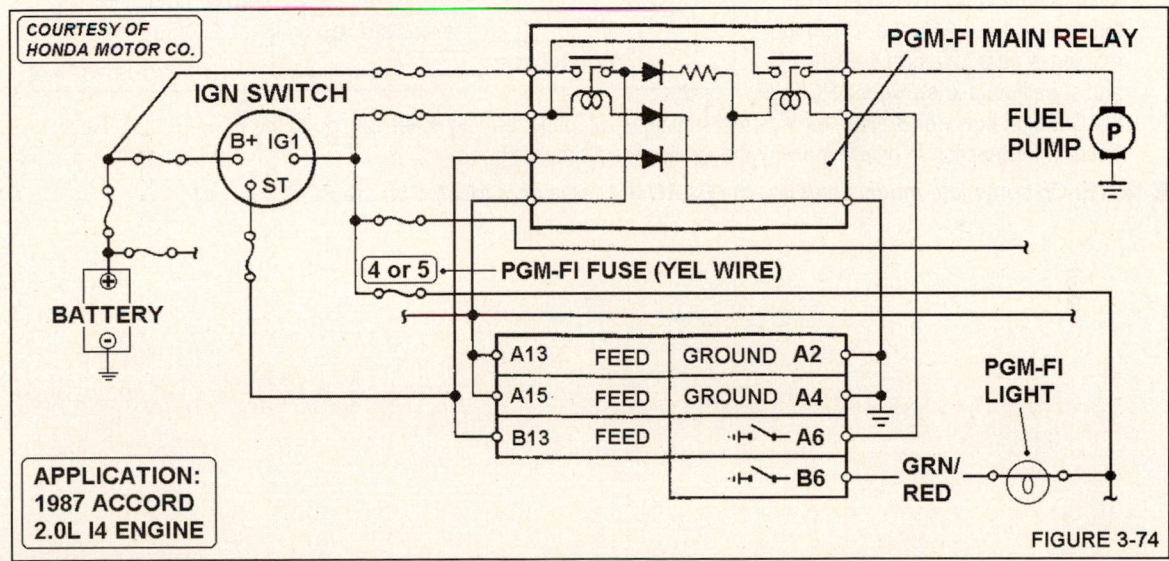

FIGURE 3-74

PCM CHART NO. 2 (PGM-FI LIGHT ON, NO CODES) CONTINUED

Use this chart if the PGM-FI or check engine light remains on for more than two seconds with the key on and no codes are output at the LED on the PCM.

CHART NO. 2 TEST PROCEDURE- CONTINUED

4) If the PGM-FI light remains on in Step 3, remove the BARO or PA sensor connector with the key on (Figure 3-74a). If the PCM red LED begins to flash a Code 13, replace the BARO sensor and retest the light. If Code 13 is not indicated with the BARO sensor connector off, turn the key off and go to Step 5.

5) Next, connect the Honda Test Harness or BOB to the PCM and wiring harness. Leave the PCM 'C' or 'D' connector removed from the PCM, but connected to the wiring harness. Test for continuity between terminals C13, C15 (or D19, D20) and ground at the test harness or BOB. If all of these circuits read open, go to Step 6. If any of them show continuity, locate and repair the particular VREF circuit that is shorted to ground. Retest the operation of the PGM-FI light when repairs are completed.

6) Reconnect the 'C' or 'D' connector to the PCM and all of the sensor connectors. Leave the test harness or BOB connected to the PCM and wiring harness. Turn the key on and measure the voltage drop between terminals A16, A18 (or A23, A24) and battery negative (Figure 3-74a). If the DVOM reads under 100mv (okay), go to Step 7. If the DVOM reads over one 100mv, locate and repair the open circuit or high resistance in the ground connection.

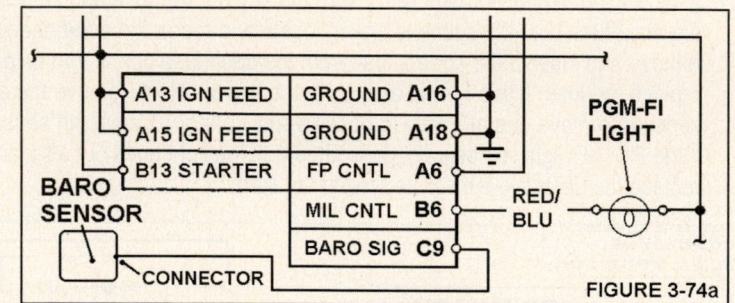

FIGURE 3-74a

7) Measure the voltage between PCM terminals A13 or A15 and A18. If the reading is near 12v, turn the key off and substitute a known-good PCM. If the C/E light now works okay, replace the original PCM. If the reading is not 12v, go to the PGM-FI Main Relay Tests in this section to determine why the ignition feed circuit is open.

■ NOTE: *On some late-model vehicles, the BARO (PA) sensor is located on the PCM.*

PCM RESET (CLEAR CODES) FUNCTION

Turn the key off and remove the correct PCM fuse for 10 seconds (Figure 3-75), then replace the PCM fuse. The codes are cleared once the PCM reset function is finished.

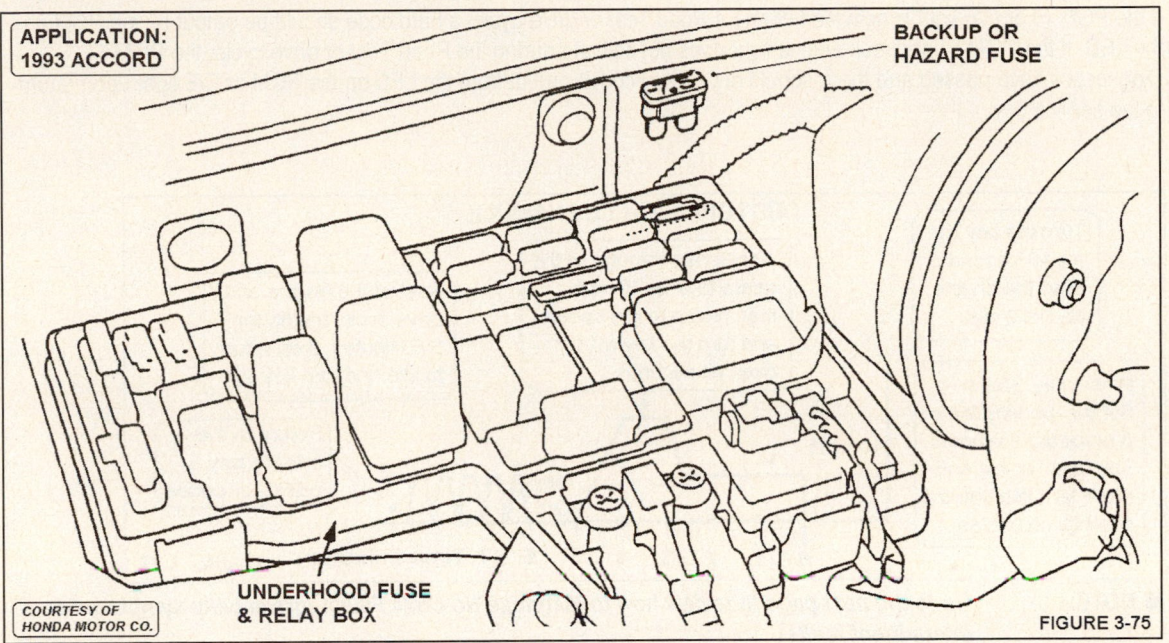

APPLICATION: 1993 ACCORD

BACKUP OR HAZARD FUSE

UNDERHOOD FUSE & RELAY BOX

COURTESY OF HONDA MOTOR CO.

FIGURE 3-75

PCM FAILURE

If the PGM-FI or check engine light remains on for over two seconds and the LED remains on steady without flashing any codes, either the light circuit is grounded or the PCM is faulty. Remove the PCM and turn the key on. If the light remains on, the light circuit is grounded. If the light goes out, replace the PCM and retest the system. If the light does not activate at all, check the PGM-FI light circuit from the ignition switch through the fuse to the light.

PCM RESET FUSE IDENTIFICATION

Accord
1985-89: Fuse No. 11 in underhood fuse/relay box
1990-95: Backup or Backup/ACC fuse in underhood fuse/relay box

Civic
1985-88: Fuse No. 10 in dash fuse box
1989-98: Backup fuse in main fuse box

CR-V
1997-98: Backup fuse in underhood fuse/relay box

Odyssey
1997-98: Backup fuse in underhood fuse/relay box

Passport
1997-98: ECM fuse in underhood fuse block

Prelude
1986-87: Hazard fuse in underhood fuse box
1988-98: Clock or Clock/Radio fuse in underhood fuse box

DRIVE CYCLE OR ROAD TEST

Once it is established that the PGM-FI diagnostics are working and that there are no codes stored in memory, perform a Road Test and attempt to recreate any symptoms or problems listed on the work order.

If the PGM-FI or C/E light is activated during a Road Test or drive cycle, a hard code should be output by the PCM on the LED. If the PGM-FI or check engine light does not activate during the Road Test or drive cycle, the PGM-FI Diagnostics have passed and there should not be any codes output from the LED on the PCM or C/E light upon return to the service bay.

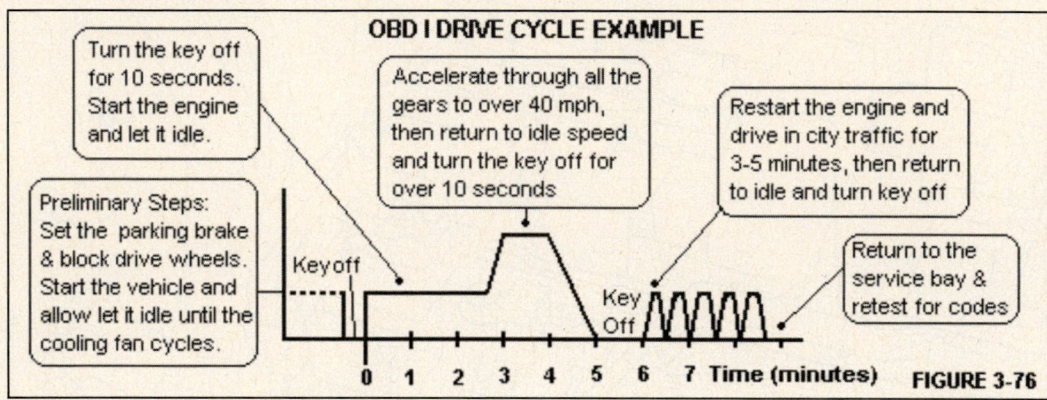

■ **NOTE:**   *Go to the next page to review how to diagnose No Code Faults (driveability symptoms or intermittent faults).*

No Code Fault Diagnosis

<u>DRIVEABILITY SYMPTOMS</u>

If the PGM-FI diagnostics show Pass, attempt to determine if a driveability symptom is present during a Road Test. If a symptom is present, refer to the Driveability Symptom List in Section One to locate the correct symptom for a particular vehicle and problem.

Once the correct symptom is located, work down through the list of system and component checks. At this point, a technician can use their experience and knowledge of testing to determine which symptom test to run first. Choose a particular symptom description (e.g., Idle Concerns) and then refer to that index in the same section to find the driveability symptom. The index is in alphabetical order and the test sequence may require that more than one type of system of component be tested to repair the problem.

<u>INTERMITTENT FAULTS</u>

If no driveability symptom appear during testing, determine from the customer invoice or previous experience during the Road Test if an intermittent fault may exist on the vehicle. If so, go to Intermittent Tests in Section One.

<u>FINAL CHECKOUT PROCEDURE</u>

Once all repairs are completed, do a PCM reset (clear the memory) and Road Test the vehicle one last time. Observe the PGM-FI or check engine light for any signs of a fault being detected or for signs of a driveability symptom that could be present. If a code sets or a symptom is present, repeat the appropriate procedure as outlined in the block chart below.

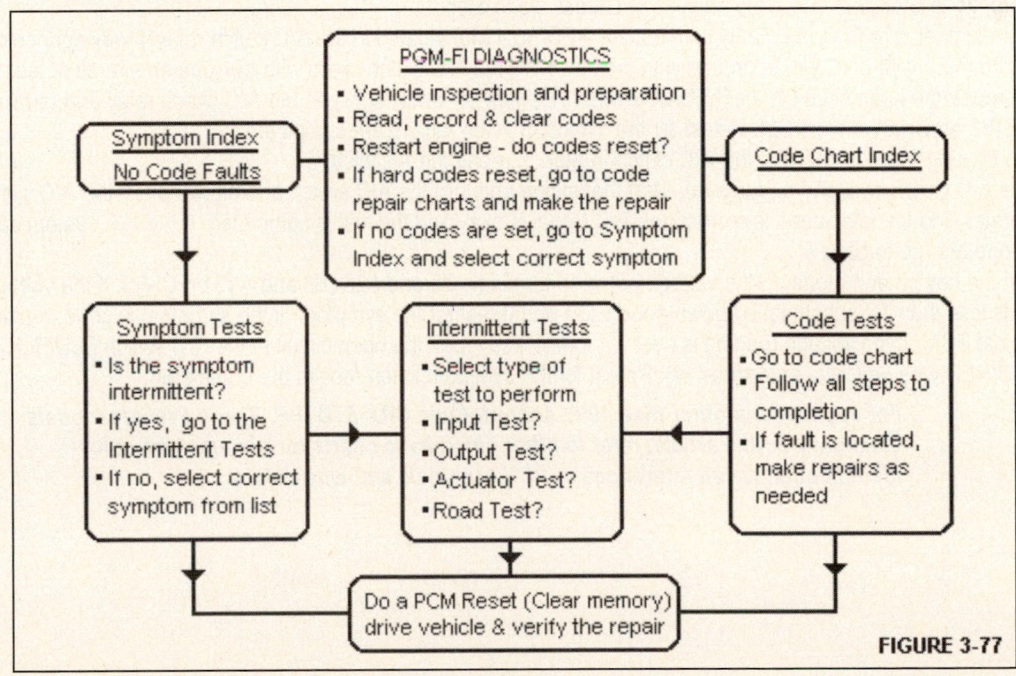

FIGURE 3-77

AIR CONDITIONER SIGNAL (ALL MODELS)

INTRODUCTION

The PCM uses the air conditioning (A/C) signal to determine when the A/C has been engaged. This signal is diagnosed as a symptom, not as a code.

DRIVEABILITY SYMPTOM DIAGNOSIS

A problem with the air conditioning signal can cause the engine idle speed to drop when the A/C is turned on with a warm engine

AIR CONDITIONER SIGNAL TEST

Turn the key off and connect the Honda test harness or BOB to the PCM and wiring harness to allow for circuit testing of components and the wire harness.

1) Leave the 'B' connector removed from the test harness or BOB with the 'A' and 'C' connectors connected to the PCM and test harness or BOB for this step. Turn the key on and measure the voltage between B5 and A26 (B8 and A18 on Civic) at the PCM. If the DVOM reads 5v, go to Step 3.

2) If the DVOM does not read 5v in Step 1, substitute a known-good PCM and retest the voltage. If the DVOM reads 5v after the exchange, replace the original PCM.

3) Turn the key off and reconnect the 'B' connector at the test harness or BOB connector. Turn the key on and momentarily jumper the A/C clutch relay terminals A15 to A26 (B3 to A18 on Civic). An audible clicking noise should be heard from the A/C compressor clutch each time the connection is made. If the clutch clicking noise is heard, go to Step 5. If the clutch noise is not heard, go to Step 4.

4) Use a jumper wire to momentarily connect the RED/BLU wire located in the A/C clutch relay 4-way connector on Accord (YEL wire on Civic) to ground with the key on. If the A/C clutch clicks with the jumper wire as noted, locate and repair the open circuit in the RED/BLU wire (YEL wire on Civic) between the A/C clutch relay and terminal A15 (B3 on Civic) at the PCM. Retest for the symptom when repairs are completed.

5) If no clicking noise is heard at the A/C clutch in Step 3, repair the A/C clutch.

6) If the A/C clutch engaged as required, start the engine and turn the A/C switch and blower on. If the A/C system operates and the idle speed is correct, the A/C signal is okay and the test is completed. If the A/C system does not operate, go to Step 7.

7) Turn the key on and measure the voltage between terminals B5 and A26 (B8 and A18 on Civic). If the voltage reads less than 1v, substitute a known-good PCM and retest for the symptom. If the symptom is gone, replace the original PCM. If the voltage reading is over 1v, locate and repair the open circuit in the A/C switch BLU/BLK wire (BLU/RED wire on Civic) to terminal B8. Retest for the symptom when repairs are completed.

■ **NOTE:**   *For applications other than 1990 Accord, Civic CRX STD, HF, Si and Prelude models discussed in this article, refer to other pin voltage charts for pin and wire color identification for all references to PCM terminals and wire colors.*

ALTERNATOR 'FR' SIGNAL (ALL MODELS)

INTRODUCTION
The PCM uses the alternator 'FR' signal to determine when the alternator is charging. This signal is diagnosed as a symptom (there are no OBD I codes used with this circuit).

DRIVEABILITY SYMPTOM DIAGNOSIS
A problem with the alternator 'FR' signal could cause the engine idle speed to drop when high electrical loads are enabled on the vehicle.

ALTERNATOR 'FR' SIGNAL TEST
Turn the key off and connect the Honda test harness or BOB to the PCM and wiring harness to allow for circuit testing of components and the wire harness.

1) Leave the 'D' connector ('B' connector on Civic) connected to the PCM but disconnected from the test harness or BOB. This allows the 'FR' circuit to be isolated for testing without it being connected to the alternator. Leave all other connectors installed so that the PCM is connected to power and ground.

2) Turn the key on and measure the voltage between terminals D9 and A26 (B14 and A18 on Civic). If the voltage is near 5v, go to Step 3. If the voltage is below 4.75v, turn the key off and substitute a known-good PCM. Retest the voltage. If the reading is 5v, replace the original PCM.

3) If the reading was near 5v in Step 2, turn the key off and remove the 'D' connect ('B' connector on Civic) to the PCM wiring harness. Restart the engine and allow it to warm until the cooling fan cycles. Connect the DVOM to terminals D9 and A26 (B14 and A18 on Civic) and watch the voltage with the headlight switch and rear defogger engaged. If voltage does not decrease as the load devices are engaged, go to Step 4. If the voltage decreases as the load devices are engaged, the alternator 'FR' signal is normal. Turn the key off and remove the alternator "sense" fuse for 10 seconds to reset the PCM.

4) With the key off, remove the 'D' connector ('B' connector on Civic) at the PCM (do not remove it from the wiring harness). Next, disconnect the negative battery cable. Check for continuity between terminal D9 (B14 on Civic) and body ground. If the DVOM reads no continuity, go to Step 5. If it reads continuity, remove the GRN connector at the alternator and recheck for continuity. If continuity exists with the GRN connector removed, locate and repair the short in the WHT/RED wire between the alternator harness and terminal D9 (YEL wire between the alternator harness and terminal B14 on Civic CRX STD or BLU on CRX HF and Si). If the DVOM reads no continuity with the GRN connector off, repair the alternator. Retest for the symptom when repairs are completed.

5) Remove the GRN connector at the alternator. Jumper the WHT/RED wire (Yellow wire on CRX STD or BLU on CRX HF and Si) to battery negative. Repeat the continuity test. If there is no continuity, locate and repair the open circuit in the WHT/RED wire (Yellow wire on CRX STD or BLU on CRX HF and Si) between the alternator connector and terminal D9 (B14 on Civic models). If continuity exists with the WHT/RED wire (YEL wire on Civic CRX STD or BLU on CRX HF and Si) jumped to ground (okay), repair alternator. Retest for the symptom when repairs are completed.

START SIGNAL (ALL MODELS)

INTRODUCTION

The PCM uses the ignition start signal to determine when the engine is in the crank mode. Once this signal is detected, the PCM controls the current signal to the EACV (valve) to provide for easy startup conditions. The start signal is diagnosed as a symptom (there are no OBD I codes).

DRIVEABILITY SYMPTOM DIAGNOSIS

A problem with the start signal could cause a low engine idle speed condition on initial startup with a warm engine.

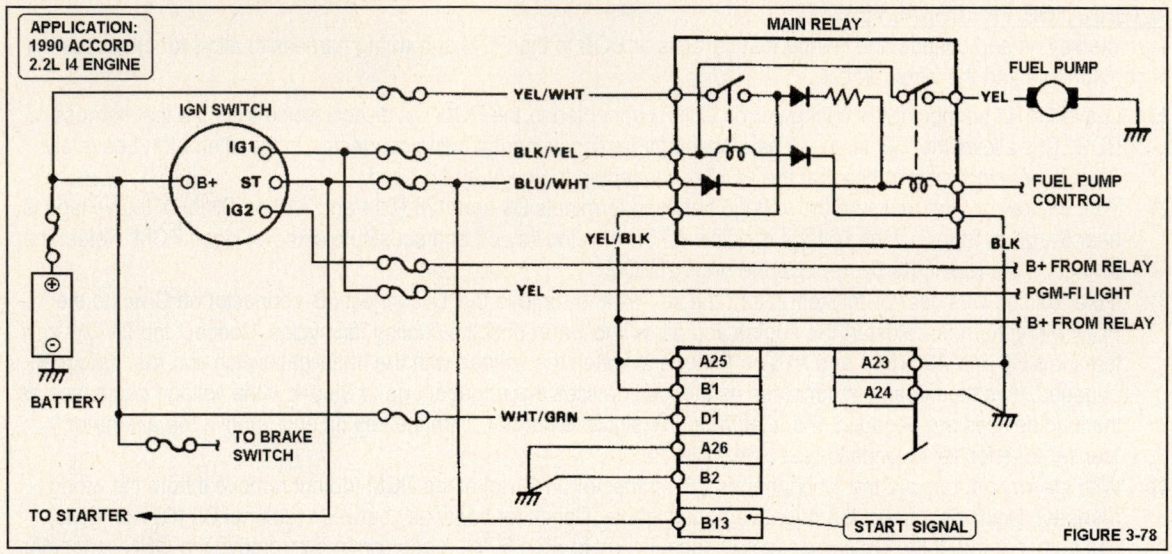

FIGURE 3-78

START SIGNAL TEST

Turn the key off and connect the Honda test harness or BOB to the PCM and wiring harness to allow for circuit testing of components and the wire harness.

1) Connect a DVOM between PCM terminals B13 (B9 on Accord & Prelude 2.1L engines) and A18 in order to test the ignition start signal at startup. Turn the ignition switch quickly to start and watch the DVOM reading. If the DVOM reads from 9-11v until the engine starts, the starter signal is okay and the test is completed.

2) If the voltage reading in Step 1 in not from 9-11v during startup (cranking) conditions, inspect for a blown fuse in the start signal circuit. If the fuse is blown, locate and repair the cause of the blown fuse. Replace the fuse and retest for the condition.

3) If the fuse is okay in Step 2, locate and repair the open circuit in the BLU/WHT wire between the PCM terminal B13 (B9 on Accord & Prelude 2.1L engines) and the fuse or switch. Retest for the condition when repairs are completed.

■ NOTE: *For models other than 1990 Accord and Prelude applications discussed in this article, refer to other pin voltage charts for pin and wire color identification for all references to PCM terminals and wire colors.*

A/T SHIFT POSITION SWITCH (ALL MODELS)

INTRODUCTION

The PCM uses the automatic transaxle (A/T) shift position signal to determine when the transaxle is shifted into park or neutral position. This signal is diagnosed as a symptom.

DRIVEABILITY SYMPTOM DIAGNOSIS

A problem with the A/T shift position could cause the engine idle speed to drop too low when the vehicle is shifted into a drive gear.

A/T SHIFT INDICATOR LAMP TEST

Turn the key off and connect the Honda test harness or BOB to the PCM and wiring harness. Turn the key on and leave the engine off. Watch the A/T shift indicator lights as the gearshift selector is moved between gears. If the lights do not work properly in each position, repair or replace the shift indicator switch. Go to the A/T Shift Signal Test.

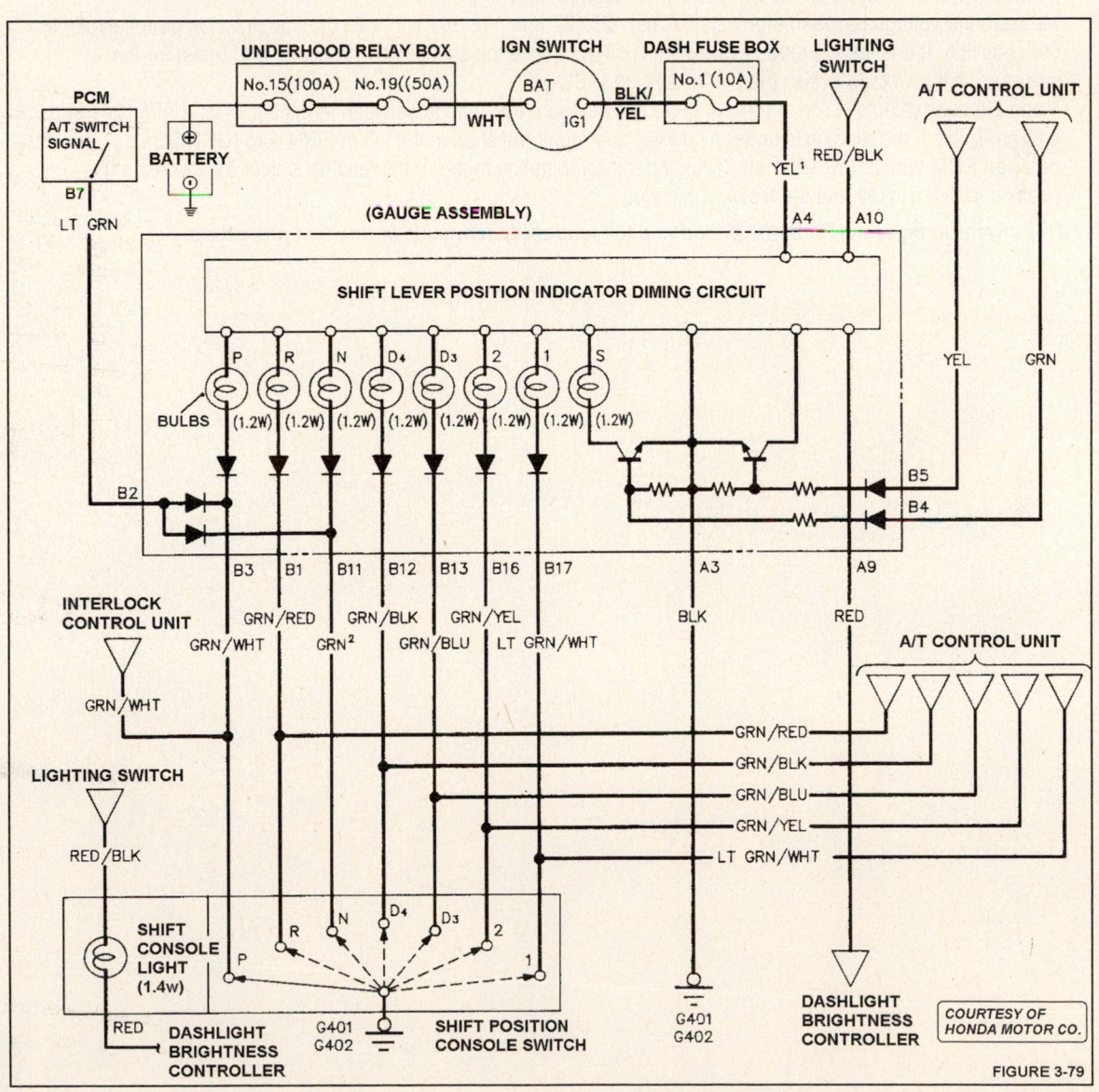

FIGURE 3-79

A/T SHIFT POSITION SWITCH (ALL MODELS)

A/T SHIFT SIGNAL TEST

Turn the key off and connect the Honda test harness or BOB to the PCM and wiring harness to allow for circuit testing of components and the wire harness.

1) Leave the 'B' connector removed from the test harness or BOB to allow the A/T shift signal circuit to be isolated for testing. Leave all other connectors installed so that the PCM is connected to power and ground. Measure the voltage between terminals B7 and A26 (B7 and A18 on Civic). If the voltage is near 5v, go to Step 2. If the voltage is less than 5v, turn the key off. Inspect the B7 terminal at the PCM for a loose connection. If it is okay, substitute a known-good PCM and retest for voltage. If 5 volts is now present, replace the original PCM.

2) Place the transaxle gear selector in neutral position. Reconnect the 'B' connector to the test harness. Measure the voltage between terminals B7 and A26 (B7 and A18 on Civic). If the reading is less than 1v, go to Step 3. If the reading is over 1v, locate and repair the open circuit in the LG or GRN wire between the combination meter and PCM terminal B7. Retest for the symptom when repairs are completed.

3) Measure the voltage between terminals B7 and A26 (B7 and A18 on Civic) with the transmission gear selector in park position. If the voltage reads over 1v (not okay), replace the shift position indicator and retest for the symptom. If the voltage is less than 1v (okay), go to Step 4.

4) Place the gear selector in drive (D4 on CRX) and measure the voltage between terminals B7 and A26 (B11 and A18 on Civic). If the reading is under 5v, locate and repair the short in the LG or GRN wire (GRN/BLK on Civic) between PCM terminal B7 (B11 on Civic) and the combination meter. If the reading is over 5v, the A/T shift position switch is okay and the test is completed.

■ NOTE: *Refer to Figure 3-79 on the previous page to identify terminals in this test procedure.*

BRAKE SWITCH SIGNAL (ACCORD & CIVIC MODELS)

The PCM uses the brake switch signal to determine when the brake pedal is depressed (i.e., a switch signal change from off to on). The brake switch is diagnosed as a symptom as there are no codes set for this circuit on models with OBD I diagnostics.

DRIVEABILITY SYMPTOM DIAGNOSIS

A problem with the brake switch signal could cause the engine to stall or almost stall during deceleration conditions with the brake pedal depressed.

PRELIMINARY STEPS

Turn the key off and connect the Honda test harness or BOB to the PCM and wiring harness to allow for proper circuit testing. Next, check the brake lights at the rear of the vehicle to determine if the lights are on all the time. If they are on all the time, repair the switch or circuit. When repairs are completed (or if the lights are off when they should be off), go to Step 1.

BRAKE SWITCH SIGNAL TEST

1) Depress the brake pedal and verify that the brake lights are on. If none of the brake lights are on, inspect the correct stop or horn fuse in the main fuse box. If the fuse is blown, locate and repair the short and go to Step 2.

2) If the fuse is okay, remove the brake switch 2-wire connector. Use the DVOM to check for continuity across the two switch terminals with the brake pedal depressed. If the meter reads no continuity, replace the faulty brake switch and then go to Step 3.

3) Measure the voltage between PCM terminals D2 and A26 on Accord or C10 and A18 on Civic models (Figure 3-80). If the brake switch is working and properly adjusted, the DVOM will read battery voltage only with the brake pedal depressed. If the brake switch is okay, the Brake Switch Test is completed.

4) If the DVOM does not read battery voltage in this step with the pedal depressed, locate and repair the open signal circuit in the GRN/WHT wire between the brake switch and PCM terminal D2 (C10 on Civic). Retest for the condition when repairs are completed.

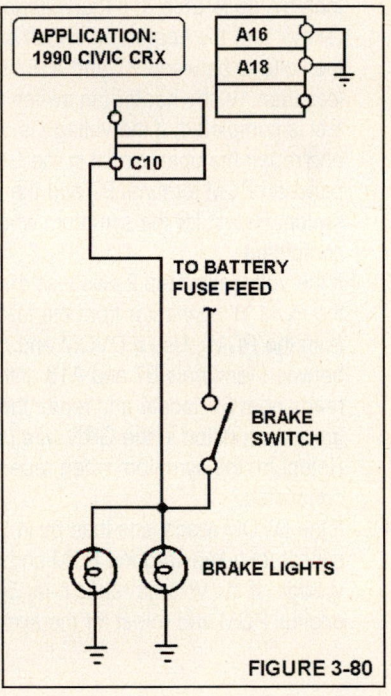

APPLICATION: 1990 CIVIC CRX

A16
A18
C10
TO BATTERY FUSE FEED
BRAKE SWITCH
BRAKE LIGHTS

FIGURE 3-80

HEATER FAN SWITCH (CIVIC CRX HF 1.5L & 1.6L ENGINES)

The PCM uses the heater fan switch signal to determine when the heater fan switch is engaged (i.e., a switch signal change from off to on).

DRIVEABILITY SYMPTOM DIAGNOSIS

A problem with the heater fan switch signal could cause the engine to stall or almost stall during conditions when the heater fan switch is first enabled.

PRELIMINARY STEPS

Turn the key off and connect the Honda test harness or BOB to the PCM and wiring harness to allow for circuit testing of components and the wire harness.

HEATER FAN SWITCH TEST

1) Turn the key to on with the engine off.
2) Use a DVOM to measure between PCM terminals B7 and A18 (Figure 3-81). If the voltage is less than 5v, go to Step 3. If the voltage is near 5v (okay), turn the heater fan switch on and measure the voltage between B7 and A18. If the reading is less than 1v, the heater fan switch is okay and the test is completed. If the voltage is over 1v, locate and repair the open circuit in the GRN wire between PCM terminal B7 and the heater fan switch. Retest for the symptom when repairs are completed.
3) If the voltage in Step 2 was less than 5v, remove the PCM 'B' connector from the test harness (not from the PCM). Use a DVOM and measure between terminals B7 and A18. If the DVOM reads near 5v, locate and repair the short to ground condition in the GRN wire (switch to PCM). Retest for the symptom when repairs are completed.
4) If the DVOM reads less than 5v in Step 3, substitute a known-good PCM and retest the voltage. If the voltage reads near 5v, replace the original PCM and retest for the symptom.

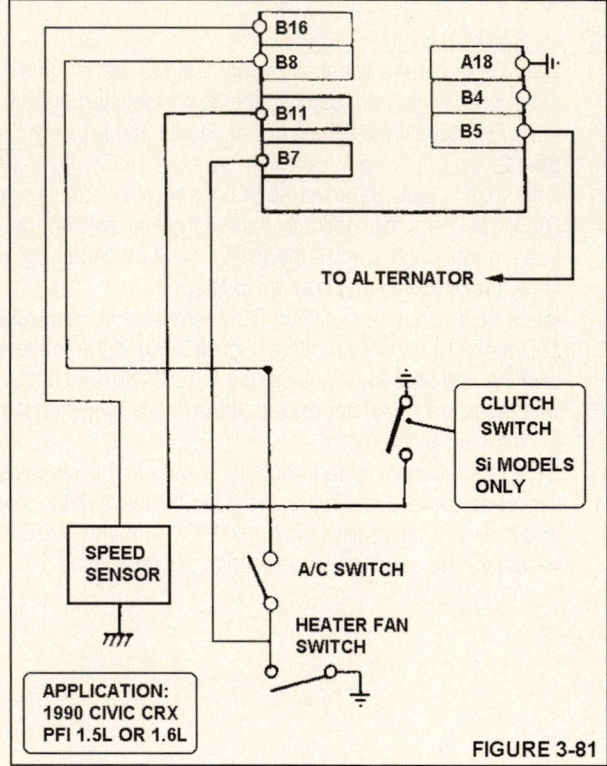

APPLICATION:
1990 CIVIC CRX
PFI 1.5L OR 1.6L

FIGURE 3-81

POWER STEERING SWITCH TEST (ACCORD & PRELUDE MODELS)
The PCM uses the power steering (oil pressure) switch to determine when the power steering load is high (i.e., a switch signal change from off to on).

DRIVEABILITY SYMPTOM DIAGNOSIS
A problem with the power steering pressure switch could cause the engine idle speed to drop too low when the steering wheel is turned.

PRELIMINARY STEPS
Turn the key off and connect the Honda test harness or BOB to the PCM and wiring harness to allow for circuit testing of components and the wire harness.

POWER STEERING SWITCH SIGNAL TEST
Turn the key on and engine off.

1) Measure the voltage between PCM terminals B8 and A18 on Accord (B8 and A26 on Prelude). The reading should be less than 1v. If it is, go to Step 4. If the reading is 12v (open circuit), remove the switch connector (Figure 3-82).

2) Connect a jumper wire across the switch terminals and repeat the test. This is the RED wire to BLK wire on 1990 Accord and the BLU/RED to BLK wire on Prelude. If the reading drops to less than 1v, replace the P/S switch and retest the symptom.

3) If the reading was less than 1v in Step 2, start the engine and turn the wheel while monitoring the voltage reading between terminals B8 and A18 (B8 and A26). If the reading changes to near 12v as the wheel is turned, this change indicates the P/S switch is okay and the test is completed.

4) If the reading in Step 4 was under near 1v as the wheel was turned, remove the PCM 'B' connector from the test harness or BOB (not from the PCM). To verify the PCM feed voltage, measure the voltage between terminals B8 and a ground. If the reading is near 12v, go to Step 5. If the reading is not 12v, substitute a known-good PCM and repeat the test. If the reading is now 12v, replace the original PCM.

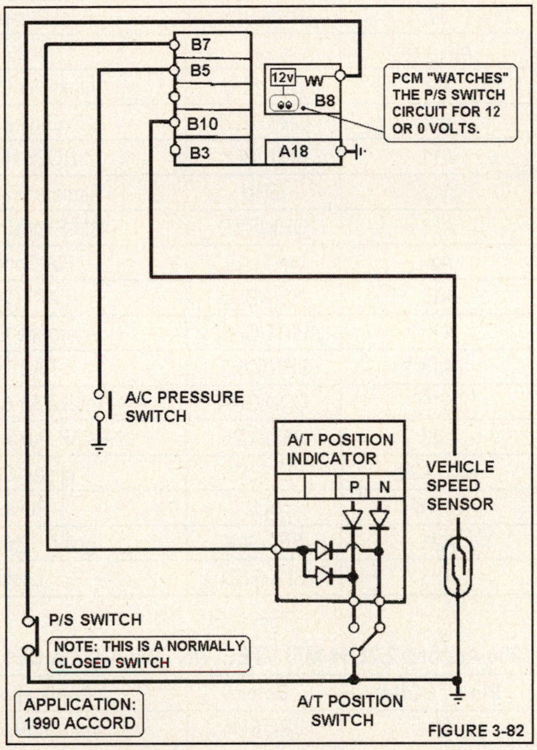

FIGURE 3-82

5) If the reading was near 12v in Step 4, reconnect the 'B' connector to the test harness or BOB. Next, remove the P/S switch 2-wire connector and repeat the voltage test. If the reading is near 12v with the switch connector removed, replace the P/S switch and retest for the symptom. If the reading remains low, locate and repair the short to ground in the RED wire (BLU/RED on Prelude) between PCM terminal B8 and the P/S switch connector. Retest for the symptom when repairs are completed.

■ NOTE: *For models other than the 1990 Accord and Prelude applications discussed in this article, refer to other pin voltage charts for pin and wire color identification for all references to PCM terminals and wire colors.*

REFERENCE INFORMATION
ACCORD PIN CHARTS

1994 Accord 2.2L I4 MFI VTEC VIN CD5, CD7 & CE1 26-P Connector

Pin # / BOB #	Color	Application & Acronym	Value at Idle
A1	BRN	Injector No. 1	At hot idle: 2.0-3.3ms
A2	YEL	Injector No.4	At hot idle: 2.0-3.3ms
A3	RED	Injector No. 2	At hot idle: 2.0-3.3ms
A4	GRN/YEL	VTEC Control Solenoid	At idle: 0v, at high rpm: 12v
A5	BLU	Injector No. 3	At hot idle: 2.0-3.3ms
A6	RED	EGR Control Solenoid	EGR Solenoid On: <1v, Off: 12-14v
A7	GRN/BLK	Fuel Pump Relay Control	Relay On: <1v, Off: 12-14v
A8, A17	---	---	---
A9	BLK/BLU	Intake Air Control Solenoid	Solenoid On: <1v, Off: 12-14v
A10	GRN/WHT	A/T: Engine Mount Solenoid	Solenoid On: <1v, Off: 12-14v
A11	GRN/BLK	HO2S Heater Control	HTRC On: <1v, Off: 12-14v
A12	GRN	Radiator Fan Relay Control	Relay On: <1v, Off: 12-14v
A13	GRN/RED	PGM-FI or Check Engine Light	Light On: <1v, Off: 12-14v
A14	WHT/YEL	FIA Control Solenoid	Solenoid On: <1v, Off: 12-14v
A15	RED/BLU	A/C Clutch Relay	A/C Relay On: <1v, Off: 12-14v
A16	WHT/GRN	Alternator Charging Signal	KOER: 3-5v, w/load: 2-4v
A18	BRN/WHT	FAS TCM Signal	---
A19	ORN/GRN	Intake Air Control Solenoid	Solenoid On: <1v, Off: 12-14v
A20	RED/YEL	EVAP Purge Cutoff Solenoid	Solenoid On: <1v, Off: 12-14v
A21, A22	YEL/GRN	Igniter Control Signal	Idle: 10%, 2500: 60% duty cycle
A23, A24	BLK	Power Ground	<0.1v
A25	YEL/BLK	Ignition Power From Relay	KOEO or KOER: 12-14v
A26	BLK/RED	Logic Ground	<0.1v

1994 Accord 2.2L I4 MFI VTEC VIN CD5, CD7 & CE1 16-P Connector

Pin # / BOB #	Color	Application & Acronym	Value at Idle
B1	YEL/BLK	Ignition Power From Relay	KOEO or KOER: 12-14v
B2	BRN/BLK	Logic Ground	<0.1v
B3	WHT/RED	AFSA TCM Signal	---
B4	GRN	AFSB TCM Signal	---
B5	RED/WHT	A/C Switch Signal	A/C On: 0.1v, Off: 12-14v
B6	---	---	---
B7	LT GRN	A/T: Park/Neutral Switch	In P/N: 0.1v, all others: 11v
B8	GRN	Power Steering Pressure Switch	Wheel straight: 0v, turned: 11v
B9	BLU/RED	Start Signal	KOEC (cranking): 9-11v
B10	ORN	Vehicle Speed Sensor	Moving at 50 mph: 60 Hz
B11	ORN	CYP Sensor	KOER: 0.250mv (AC)
B12	WHT	CYP Sensor	KOER: 0.250mv (AC)
B13	ORN/BLU	TDC Sensor	KOER: 1.00v (AC)
B14	WHT/BLU	TDC Sensor	KOER: 1.00v (AC)
B15	BLU/GRN	CKP Sensor	KOER: 0.900mv (AC)
B16	BLU/YEL	CKP Sensor	KOER: 0.900mv (AC)

■ **NOTE:** *When <1V is shown, this indicates the reading should be less than 1.0v DC.*

1994 Accord 2.2L I4 MFI VTEC VIN CD5, CD7 & CE1 22-P Connector

Pin # / BOB #	Color	Application & Acronym	Value at Idle
D1	WHT/YEL	Keep Alive Power	12-14v
D2	GRN/WHT	Brake Switch Signal	Brake On: 12-14v, Off: 0v
D3, D8, D16	---	---	---
D4	ORN/RED	Service Check Connector	Open: 4.80v, Closed (jumped): 0.1v
D5	BLU/WHT	BARO Signal to TCM	KOEO or KOER: 3v (sea level)
D6	BLU/BLK	VTEC Pressure Switch	At idle: 0v, at high rpm: 12-14v
D7	GRN/RED	Data Link Connector	DLC open: 4.80v, jumped: 2.50 Hz
D9	WHT/RED	Alternator 'FR' Signal	KOER: varies 2.5-3.5v
D10	GRN/RED	Electric Load Detector	KOER: varies 0.5-4.5v
D11	RED/BLK	Throttle Position (TP) Sensor	At hot idle: 0.5v
D12	WHT/BLK	EGR Lift Sensor	At hot idle: 1.0v
D13	RED/WHT	Engine Coolant Temperature	At 180°F: 0.5-0.6v
D14	WHT/RED	HO2S Signal	Hot idle: 0.1v, Accel: 0.5-1.v
D15	RED/YEL	Intake Air Temperature	At 100°F: 2-3v
D17	WHT/YEL	MAP Sensor Signal	In P/N at idle: 0.9v (sea level)
D18	GRN/BLK	VREF out to TCM	4.9-5.1v
D19	YEL/WHT	MAP Sensor VREF	4.9-5.1v
D20	YEL/BLU	Sensor VREF	4.9-5.1v
D21	GRN/WHT	MAP Sensor Ground	<0.1v
D22	GRN/BLU	Sensor Ground	<0.1v

■ **NOTE:** *When <1V is shown, this indicates the reading should be less than 1.0v DC.*

Standard Honda Colors and Abbreviations

Abbreviation	Color	Abbreviation	Color	Abbreviation	Color
BLK	Black	GRN	Green	PNK	Pink
BLU	Blue	LT BLU	Lt. Blue	TAN	Tan
BRN	Brown	LT GRN	Lt. Green	WHT	White
GRY	Gray	ORN	Orange	YEL	Yellow

26-PIN

1	3	5	7	9	11	13	15	17	19	21	23	25
2	4	6	8	10	12	14	16	18	20	22	24	26

16-PIN

1	3	5	7	9	11	13	15
2	4	6	8	10	12	14	16

22-PIN

1	3	5	7	9	11	13	15	17	19	21
2	4	6	8	10	12	14	16	18	20	22

CONNECTOR 'A'   **CONNECTOR 'B'**   **CONNECTOR 'D'**

WIRE SIDE OF HARNESS TERMINALS

Accord Wiring Diagrams - 1994 Accord (Part One)

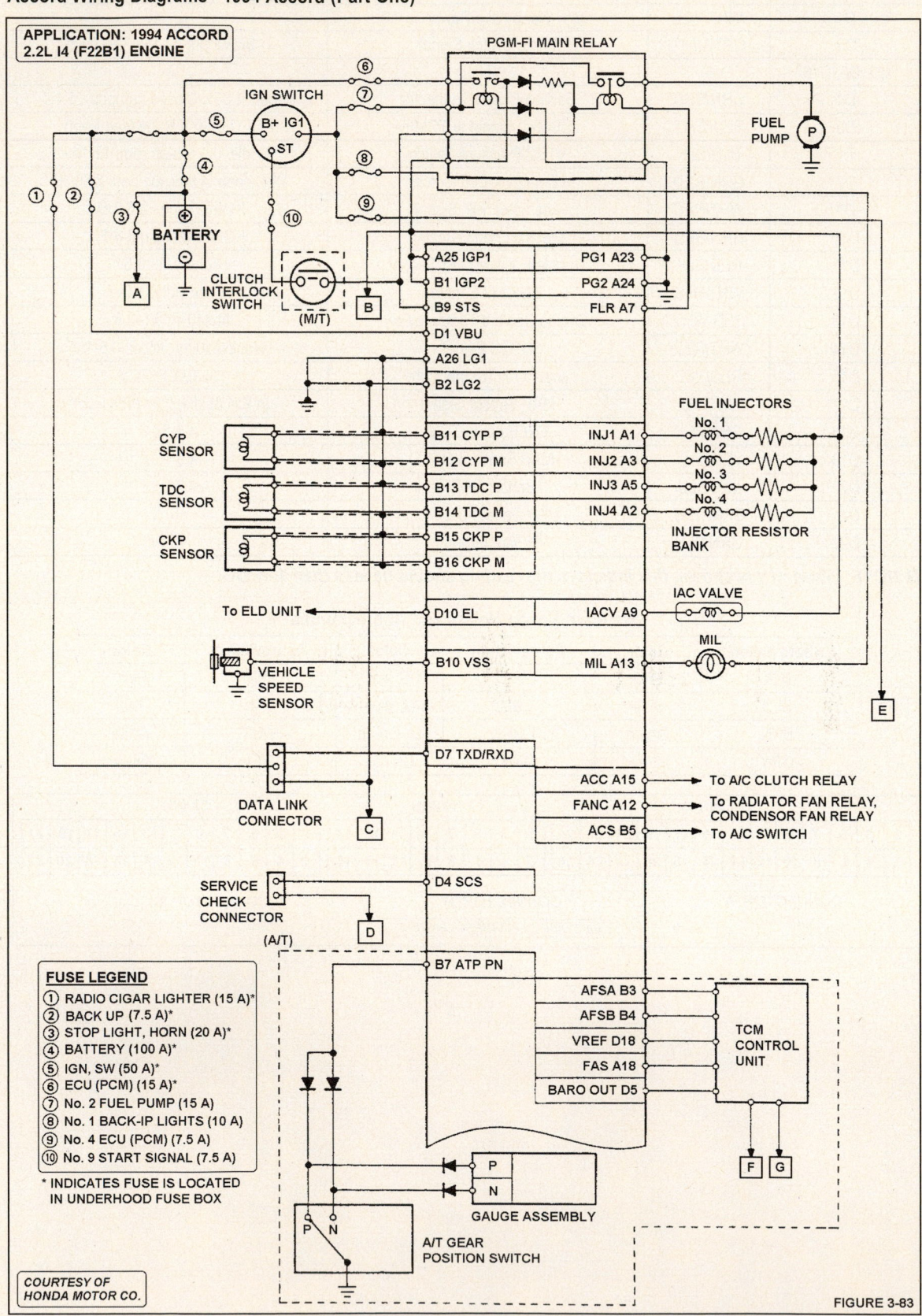

APPLICATION: 1994 ACCORD
2.2L I4 (F22B1) ENGINE

FUSE LEGEND
1. RADIO CIGAR LIGHTER (15 A)*
2. BACK UP (7.5 A)*
3. STOP LIGHT, HORN (20 A)*
4. BATTERY (100 A)*
5. IGN, SW (50 A)*
6. ECU (PCM) (15 A)*
7. No. 2 FUEL PUMP (15 A)
8. No. 1 BACK-IP LIGHTS (10 A)
9. No. 4 ECU (PCM) (7.5 A)
10. No. 9 START SIGNAL (7.5 A)

* INDICATES FUSE IS LOCATED
IN UNDERHOOD FUSE BOX

*COURTESY OF
HONDA MOTOR CO.*

FIGURE 3-83

Accord Wiring Diagrams - 1994 Accord (Part Two)

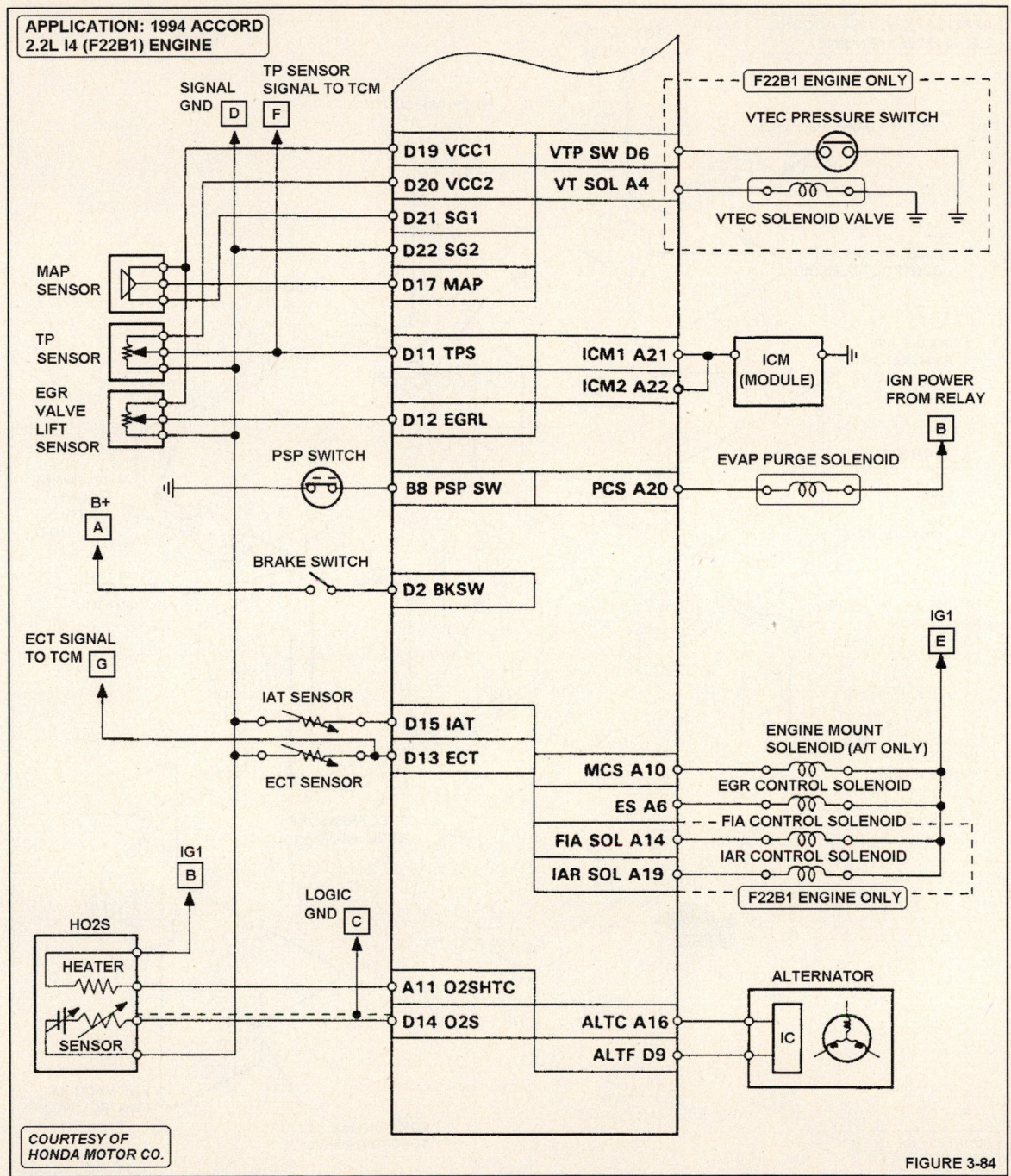

APPLICATION: 1994 ACCORD
2.2L I4 (F22B1) ENGINE

COURTESY OF
HONDA MOTOR CO.

FIGURE 3-84

Accord Vacuum Diagrams - 1994 Accord

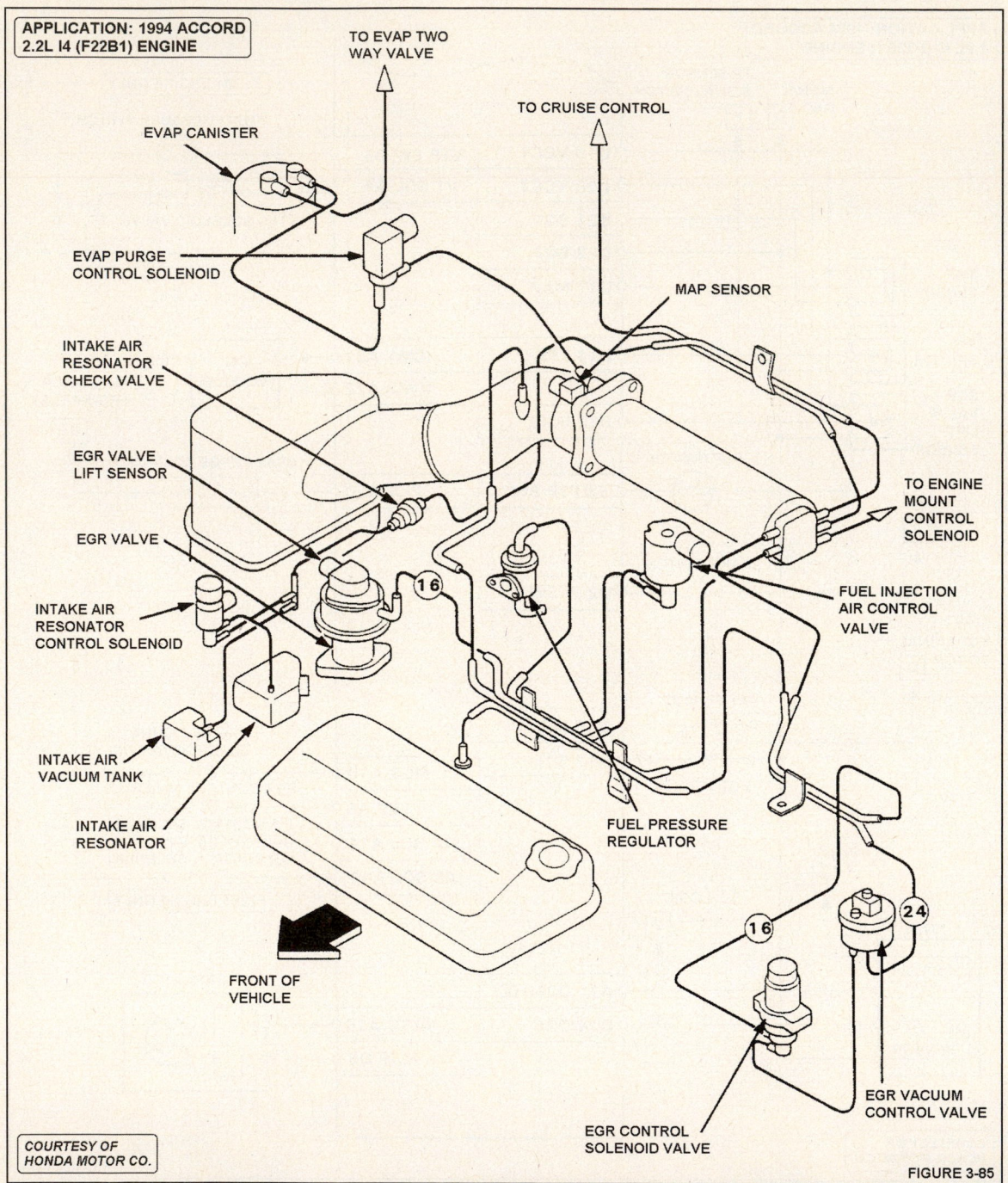

APPLICATION: 1994 ACCORD
2.2L I4 (F22B1) ENGINE

TO EVAP TWO WAY VALVE

TO CRUISE CONTROL

EVAP CANISTER

EVAP PURGE CONTROL SOLENOID

MAP SENSOR

INTAKE AIR RESONATOR CHECK VALVE

EGR VALVE LIFT SENSOR

EGR VALVE

TO ENGINE MOUNT CONTROL SOLENOID

INTAKE AIR RESONATOR CONTROL SOLENOID

FUEL INJECTION AIR CONTROL VALVE

INTAKE AIR VACUUM TANK

INTAKE AIR RESONATOR

FUEL PRESSURE REGULATOR

FRONT OF VEHICLE

EGR CONTROL SOLENOID VALVE

EGR VACUUM CONTROL VALVE

COURTESY OF HONDA MOTOR CO.

FIGURE 3-85

CIVIC Pin Charts

1995 Civic Model VX 1.5L I4 MFI VTEC-E (D15Z1) VIN EH2 26-P Connector

PIN # / BOB #	COLOR	APPLICATION & ACRONYM	VALUE AT IDLE
A1	BRN	Injector No. 1	At hot idle: 2.0-3.3ms
A2	YEL	Injector No. 4	At hot idle: 2.0-3.3ms
A3	RED	Injector No. 2	At hot idle: 2.0-3.3ms
A4	ORN/WHT	VTEC Solenoid Control	At idle: 0v, over 5000 rpm: 12-14v
A5	LT BLUE	Injector No. 3	At hot idle: 2.0-3.3ms
A6	ORN/BLK	HO2S Heater Control	Heater On: <1v, Off: 12-14v
A7	GRN/YEL	Fuel Pump Relay Control	Relay On: <1v, Off: 12-14v
A8	---	---	---
A9	GRN/WHT	Idle Air Control (IAC) Valve	At hot idle: 8-10v
A10	---	---	---
A11	PNK/GRN	EGR Control Solenoid	EGR Solenoid On: <1v, Off: 12-14v
A12	YEL/GRN	Radiator Fan Relay Control	Fan relay On: <1v, Off: 12-14v
A13	GRN/ORN	Check Engine Light	Light On: <1v, Off: 12-14v
A14, A18	---	---	---
A15	BLK/RED	A/C Clutch Relay	A/C Relay On: <1v, Off: 12-14v
A16	WHT/YEL	Alternator Charging Signal	KOER: varies 2.5-3.5v
A17	LGRN	A/T: Lockup Control Solenoid 'A'	Solenoid On: <1v, Off: 12v
A19	YEL	A/T: Lockup Control Solenoid 'B'	Solenoid On: <1v, Off: 12v
A20	RED	EVAP Purge Cutoff Solenoid	Solenoid On: <1v, Off: 12-14v
A21	RED/GRN	Igniter Control Signal	Idle: 10%, 2500: 60% duty cycle
A22	---	---	---
A23, A24	BLK	Power Ground	<0.1v
A25	YEL/BLK	Ignition Power From Relay	KOEO or KOER: 12-14v
A26	BLK/RED	Logic Ground	<0.1v

1995 Civic Model VX 1.5L I4 MFI VTEC-E (D15Z1) VIN EH2 16-P Connector

PIN # / BOB #	COLOR	APPLICATION & ACRONYM	VALUE AT IDLE
B1	YEL/BLK	Ignition Power From Relay	KOEO or KOER: 12-14V
B2	BRN/BLK	Logic Ground	<0.1v
B3	GRN/BLU	ATP D3 Signal	In D3: 0.1v, all others: 11v
B4	GRN/BLK	ATP D4 Signal	In D4: 0.1v, all others: 5v
B5	BLU/RED	A/C Switch Signal	A/C On: 0.1v, Off: 12-14v
B6	---	---	---
B7	PNK/BLK	M/T: Clutch Switch	With clutch in: 0.1v, out: 11v
B7	GRN	A/T: Gear Position Switch	In P/N: 0v, all others: 12v
B8	BRN/RED	Power Steering Pressure Switch	Wheel straight: 0v, turned: 11v
B9	BLU/WHT	Starter (Cranking) Signal	KOEC: 9-11v
B10	YEL/BLU	Vehicle Speed Sensor	Moving at 50 mph: 60 Hz
B11	ORN	CYL Sensor 'P'	KOER: 0.250mv (AC)
B12	WHT	CYL Sensor 'M'	KOER: 0.250mv (AC)
B13	ORN/BLU	TDC Sensor 'P'	KOER: 1.00v (AC)
B14	WHT/BLU	TDC Sensor 'M'	KOER: 1.00v (AC)
B15	BLU/GRN	CKP Sensor Signal 'P'	KOER: 0.900mv (AC)
B16	BLU/YEL	CKP Sensor Signal 'M'	KOER: 0.900mv (AC)

■ NOTE: *When <1V is shown, this indicates the reading should be less than 1.0v DC.*

1995 Civic Model VX 1.5L I4 MFI VTEC-E (D15Z1) VIN EH2 22-P Connector

PIN # / BOB #	COLOR	APPLICATION & ACRONYM	VALUE AT IDLE
D1	WHT/BLU	Keep Alive Power	12-14v
D2	GRN/WHT	Brake Switch Signal	Brake On: 12-14v, off: 0.1v
D3	BLU/YEL	LAF Label (Resistance) Signal	KOER: 0.3-4.9v
D4	BRN	Service Check Connector	Open: 4.80v, Closed (jumped): 0.1v
D5	---	---	---
D6	ORN/BLU	VTEC Pressure Switch	At idle: 0v, over 5000 rpm: 12-14v
D7	LT BLU	Data Link Connector	DLC open: 4.80v, jumped: 2.50 Hz
D8	WHT/BLU	LAF VS+ (drives VS Cell voltage)	At hot idle: near 7v
D9	PNK	Alternator 'FR' Signal	KOER: Varies 2.5-3.5v
D10	GRN/RED	Electric Load Detector	KOER: Varies 0.5-4.5v
D11	PNK/BLK	Throttle Position (TP) Sensor	At hot idle: 0.5v
D12	WHT/BLK	EGR Valve Lift Sensor	At hot idle: 1.2v
D13	RED/WHT	Engine Coolant Temperature	At 180°F: 0.5-0.6v
D14	ORN/BLU	LAF IP+ (HO2S Pump Cell)	KOER: 0.5-5.3v
D15	RED/YEL	Intake Air Temperature	At 100°F: 2-3v
D16	BLU/GRN	LAF Sensor IP- & VS- Ground	KOER: 2.6-2.8v
D17	PNK/WHT	MAP Sensor Signal	In P/N at idle: 0.9v (sea level)
D18	PNK/GRN	MT: Shift-Up Indicator	Shift Light On: <1v, Off: 12-14v
D18	PNK/GRN	A/T: Interlock Control Unit	---
D19	YEL/GRN	MAP Sensor VREF	4.9-5.1v
D20	YEL/WHT	Sensor VREF	4.9-5.1v
D21	GRN/BRN	MAP Sensor Ground	<0.1v
D22	GRN/WHT	Sensor Ground	<0.1v

■ NOTE: *When <1V is shown, this indicates the reading should be less than 1.0v DC.*

Standard Honda Colors and Abbreviations

Abbreviation	Color	Abbreviation	Color	Abbreviation	Color
BLK	Black	GRN	Green	PNK	Pink
BLU	Blue	LT BLU	Lt. Blue	TAN	Tan
BRN	Brown	LGRN	Lt. Green	WHT	White
GRY	Gray	ORN	Orange	YEL	Yellow

26-PIN													16-PIN								22-PIN									
1	3	5	7	9	11	13	15	17	19	21	23	25	1	3	5	7	9	11	13	15	1	3	5	7	9	11	13	15	17	19	21
2	4	6	8	10	12	14	16	18	20	22	24	26	2	4	6	8	10	12	14	16	2	4	6	8	10	12	14	16	18	20	22

CONNECTOR 'A'   CONNECTOR 'B'   CONNECTOR 'D'

WIRE SIDE OF HARNESS TERMINALS

CIVIC WIRING DIAGRAMS - 1995 CIVIC (PART ONE)

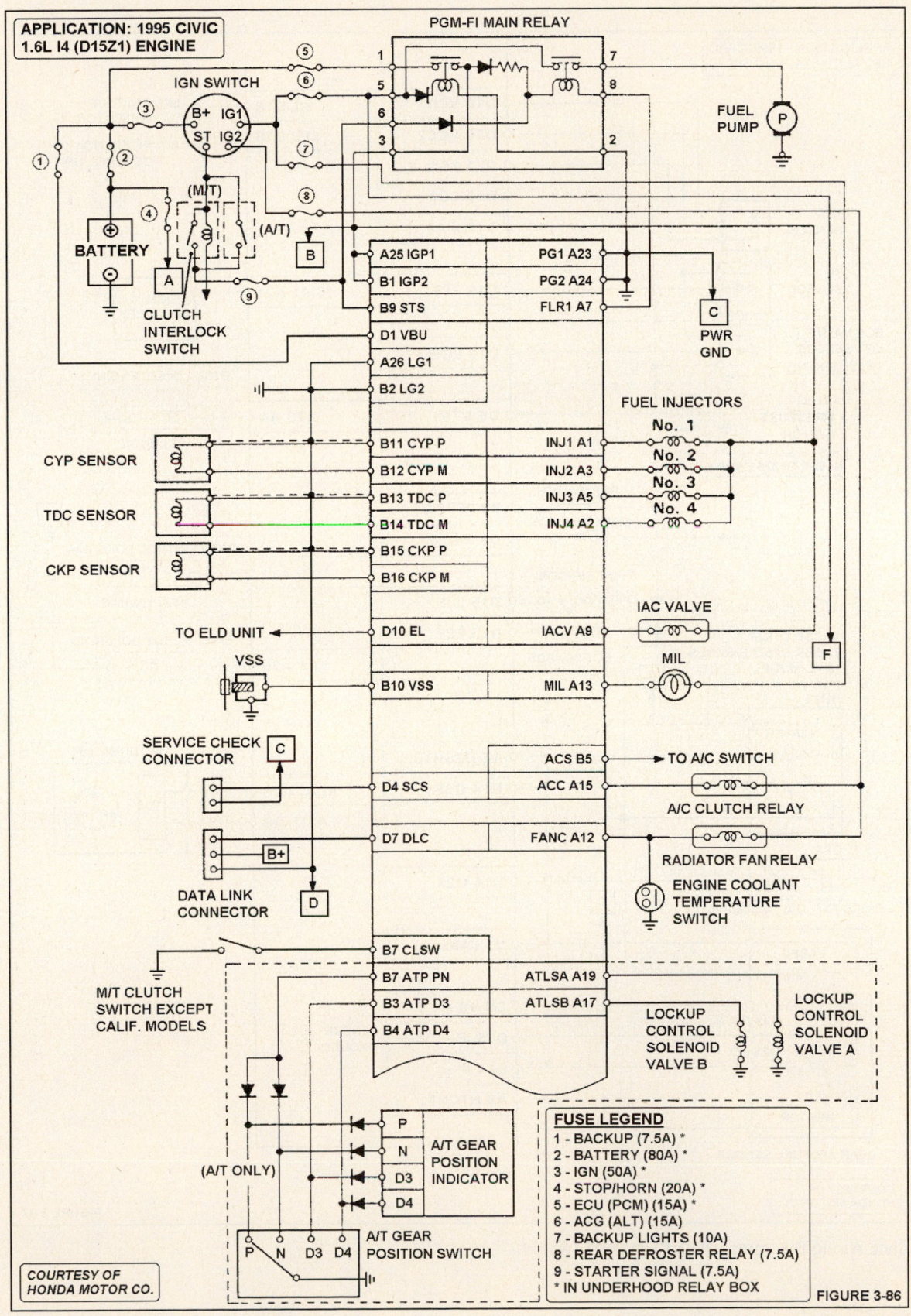

FIGURE 3-86

FUSE LEGEND
1 - BACKUP (7.5A) *
2 - BATTERY (80A) *
3 - IGN (50A) *
4 - STOP/HORN (20A) *
5 - ECU (PCM) (15A) *
6 - ACG (ALT) (15A)
7 - BACKUP LIGHTS (10A)
8 - REAR DEFROSTER RELAY (7.5A)
9 - STARTER SIGNAL (7.5A)
* IN UNDERHOOD RELAY BOX

COURTESY OF HONDA MOTOR CO.

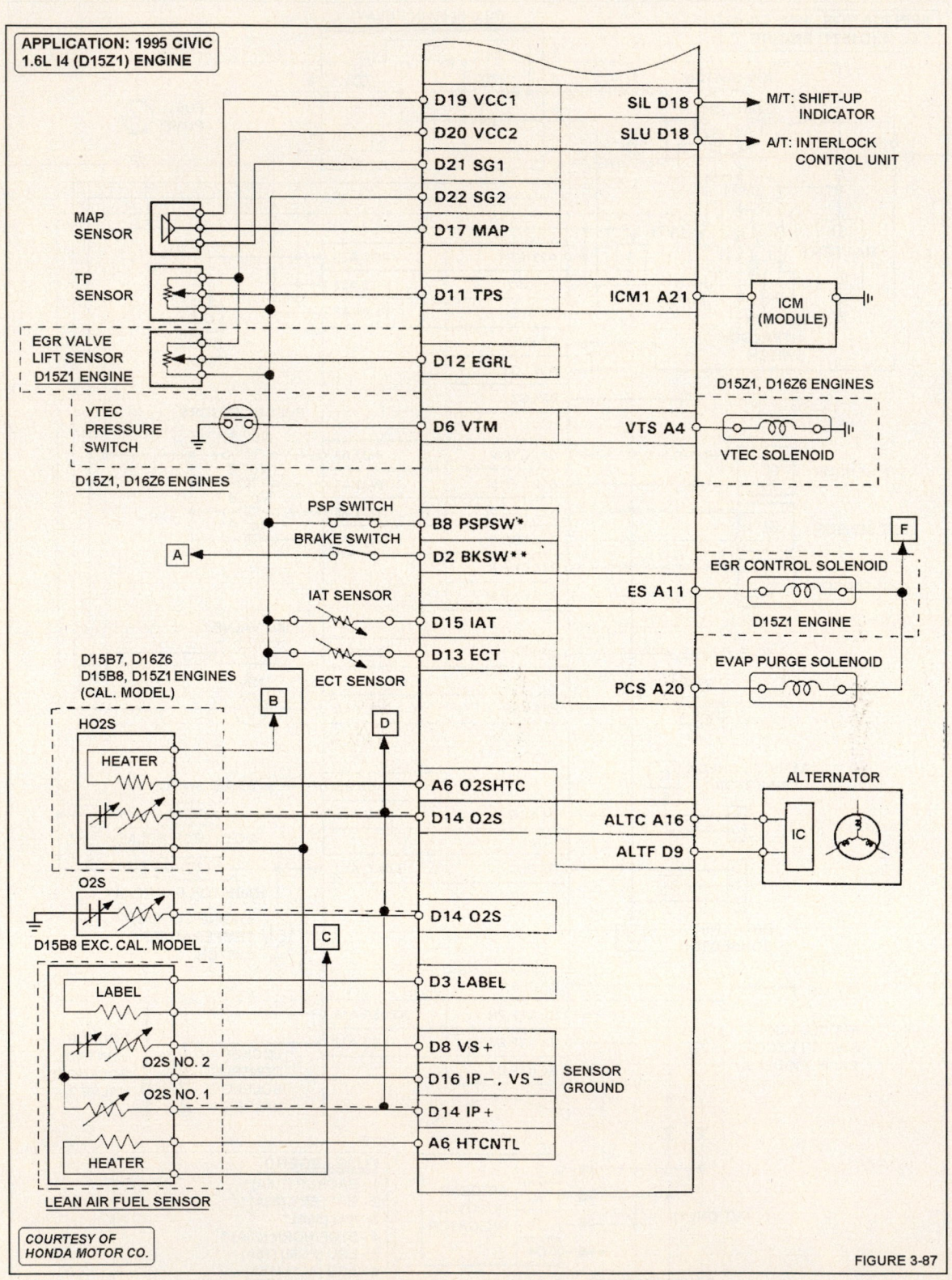

**APPLICATION: 1995 CIVIC
1.6L I4 (D15Z1) ENGINE**

MAP SENSOR — D19 VCC1 — SIL D18 → M/T: SHIFT-UP INDICATOR
D20 VCC2 — SLU D18 → A/T: INTERLOCK CONTROL UNIT
D21 SG1
D22 SG2
D17 MAP

TP SENSOR — D11 TPS — ICM1 A21 — ICM (MODULE)

**EGR VALVE LIFT SENSOR
D15Z1 ENGINE** — D12 EGRL

D15Z1, D16Z6 ENGINES

VTEC PRESSURE SWITCH — D6 VTM — VTS A4 — VTEC SOLENOID

D15Z1, D16Z6 ENGINES

PSP SWITCH — B8 PSPSW*
BRAKE SWITCH — D2 BKSW** — [A]

EGR CONTROL SOLENOID
ES A11 — **D15Z1 ENGINE** — [F]

IAT SENSOR — D15 IAT
D13 ECT
ECT SENSOR

EVAP PURGE SOLENOID
PCS A20

**D15B7, D16Z6
D15B8, D15Z1 ENGINES
(CAL. MODEL)**

HO2S — HEATER — [B] — A6 O2SHTC
D14 O2S — ALTC A16 — ALTERNATOR — IC
ALTF D9
[D]

O2S — D14 O2S

D15B8 EXC. CAL. MODEL

[C]

LABEL — D3 LABEL
O2S NO. 2 — D8 VS+
D16 IP−, VS− — SENSOR GROUND
O2S NO. 1 — D14 IP+
HEATER — A6 HTCNTL

LEAN AIR FUEL SENSOR

*COURTESY OF
HONDA MOTOR CO.*

FIGURE 3-87

Civic Wiring Diagrams - 1995 Civic (Part Two)

CIVIC VACUUM DIAGRAMS

Civic Vacuum Diagrams - 1995 Civic

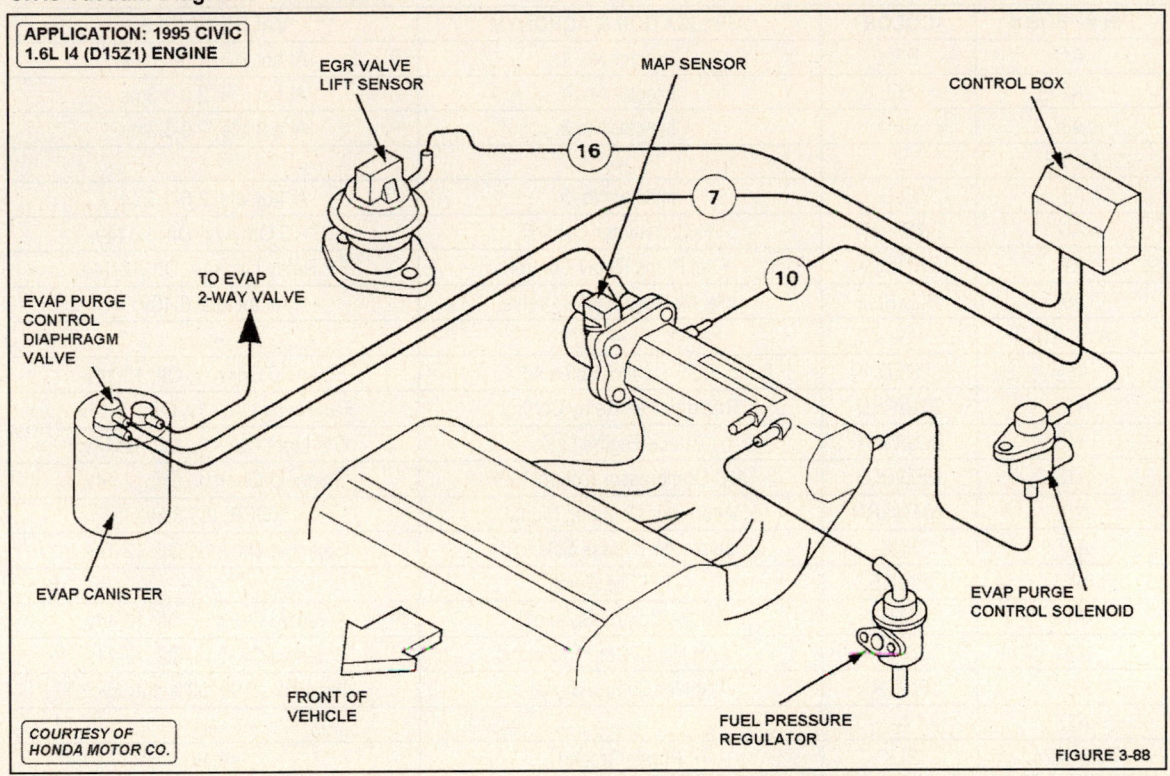

APPLICATION: 1995 CIVIC 1.6L I4 (D15Z1) ENGINE

EGR VALVE LIFT SENSOR

MAP SENSOR

CONTROL BOX

16

7

10

EVAP PURGE CONTROL DIAPHRAGM VALVE

TO EVAP 2-WAY VALVE

EVAP CANISTER

FRONT OF VEHICLE

FUEL PRESSURE REGULATOR

EVAP PURGE CONTROL SOLENOID

COURTESY OF HONDA MOTOR CO.

FIGURE 3-88

PRELUDE PIN CHARTS

1995 Prelude 2.2L I4 MFI VIN BA8 26-Pin Connector

PIN # / BOB #	COLOR	APPLICATION & ACRONYM	VALUE AT IDLE
A1	BRN	Injector No. 1	At hot idle: 2.0-3.3ms
A2	YEL	Injector No. 4	At hot idle: 2.0-3.3ms
A3	RED	Injector No. 2	At hot idle: 2.0-3.3ms
A4, A8	---	---	---
A5	BLU	Injector No. 3	At hot idle: 2.5-3.3ms
A6	ORN/WHT	HO2S Heater Control	HTRC On: <1v, Off: 12-14v
A7	GRN/BLK	Fuel Pump Relay Control	FP Relay On: <1v, Off: 12-14v
A9	BLK/BLU	Idle Air Control (IAC) Valve	At hot idle: 8-10v
A10, A14	---	---	---
A11	RED	EGR Control Solenoid	Solenoid On: <1v, Off: 12-14v
A12	BLU/RED	Radiator Fan Relay Control	Fan Relay On: <1v, Off: 12-14v
A13	BLU/WHT	Check Engine Light	C/E Light On: <1v, Off: 12-14v
A15	RED/BLU	A/C Compressor Control Unit	Solenoid On: <1v, Off: 12-14v
A16	WHT/GRN	Alternator Charging Signal	KOER: 2.5-3.5v
A17	PNK	Intake Air Bypass Solenoid	Solenoid On: <1v, Off: 12-14v
A18	ORN/RED	A/T: TCM Signal	---
A19	WHT	Intake Control Solenoid	Solenoid On: <1v, Off: 12-14v
A20	RED/GRN	EVAP Purge Cutoff Solenoid	Solenoid On: <1v, Off: 12-14v
A21	YEL/GRN	Igniter Control Signal	Idle: 10%, 2500: 60% duty cycle
A22	---	---	---
A23, A24	BLK	Power Ground	<0.1v
A25	YEL/BLK	Ignition Power From Relay	KOEO or KOER: 12-14v
A26	BLK/RED	Logic Ground	<0.1v

1995 Prelude 2.2L I4 MFI VIN BA8 16-Pin Connector

PIN # / BOB #	COLOR	APPLICATION & ACRONYM	VALUE AT IDLE
B1	YEL/BLK	Ignition Power From Relay	KOEO or KOER: 12-14v
B2	BRN/BLK	Logic Ground	<0.1v
B3	ORN	A/T: TCM Signal	---
B4	PNK	A/T: TCM Signal	---
B5	BLU/BLK	A/C Switch Signal	A/C On: 0.1v, Off: 12v
B6	---	---	---
B7	LT GRN	A/T Park/Neutral Switch	In P/N: 0.1v, all others: 11v
B8	RED/GRN	Power Steering Pressure Switch	Wheel straight: 0v, turned: 11v
B9	BLU/RED	Start Signal	KOEC (cranking): 9-11v
B10	ORN	Vehicle Speed Sensor	Moving at 50 mph: 60 Hz
B11	ORN	CYL Sensor	KOER: 0.250mv (AC)
B12	WHT	CYL Sensor	KOER: 0.250mv (AC)
B13	ORN/BLU	TDC Sensor	KOER: 1.00v (AC)
B14	WHT/BLU	TDC Sensor	KOER: 1.00v (AC)
B15	BLU/GRN	CKP Sensor	KOER: 0.900mv (AC)
B16	BLU/YEL	CKP Sensor	KOER: 0.900mv (AC)

■ NOTE: *When <1V is shown, this indicates the reading should be less than 1.0v DC.*

1995 Prelude 2.2L I4 MFI VIN BA8 Engine 22-Pin Connector

PIN #/BOB #	COLOR	APPLICATION & ACRONYM	VALUE AT IDLE
D1	WHT/YEL	Keep Alive Power	12-14v
D2	GRN/WHT	Brake Switch Signal	Brake On: 12-14v, Off: 0v
D3	RED/BLU	Knock Sensor Signal	No knock present: 2.5v
D4	BRN/WHT	Service Check Connector	Open: 4.80v, Closed: 0.1v
D5, D6	---	---	---
D7	LGRN/RED	Data Link Connector	Open: 4.80v, Closed: 0.1v
D8	---	---	---
D9	WHT/RED	Alternator 'FR' Signal	KOER: varies 2.5-3.5v
D10	GRN/BLK	Electrical Load Detector	KOER: varies 0.5-4.5v w/load
D11	RED/BLK	Throttle Position (TP) Sensor	At hot idle: 0.5v
D12	WHT/BLK	EGR Lift Sensor	At idle: 1.0v
D13	YEL/BLU	Engine Coolant Temperature	At 180°F: 0.5-0.6v
D14	WHT	HO2S Signal	Hot idle: 0-1v, Accel: 0.5-1v
D15	RED/YEL	Intake Air Temperature	At 100°F: 2-3v
D16	---	---	---
D17	WHT/BLU	MAP Sensor Signal	In P/N at idle: 0.9v (sea level)
D18	L GRN/BLK	A/T Control Unit Link	---
D19	RED/WHT	Sensor VREF	4.9-5.1v
D20	YEL/WHT	Sensor VREF	4.9-5.1v
D21	BLU/WHT	Sensor Ground	<0.1v
D22	GRN/WHT	Sensor Ground	<0.1v

■ **NOTE:** *When <1V is shown, this indicates the reading should be less than 1.0v DC.*

Standard Honda Colors and Abbreviations

Abbreviation	Color	Abbreviation	Color	Abbreviation	Color
BLK	Black	GRN	Green	PNK	Pink
BLU	Blue	LT BLU	Lt. Blue	TAN	Tan
BRN	Brown	LT GRN	Lt. Green	WHT	White
GRY	Gray	ORN	Orange	YEL	Yellow

26-PIN

1	3	5	7	9	11	13	15	17	19	21	23	25
2	4	6	8	10	12	14	16	18	20	22	24	26

CONNECTOR 'A'

16-PIN

1	3	5	7	9	11	13	15
2	4	6	8	10	12	14	16

CONNECTOR 'B'

22-PIN

1	3	5	7	9	11	13	15	17	19	21
2	4	6	8	10	12	14	16	18	20	22

CONNECTOR 'D'

WIRE SIDE OF HARNESS TERMINALS

Prelude Wiring Diagrams - 1995 Prelude (Part One)

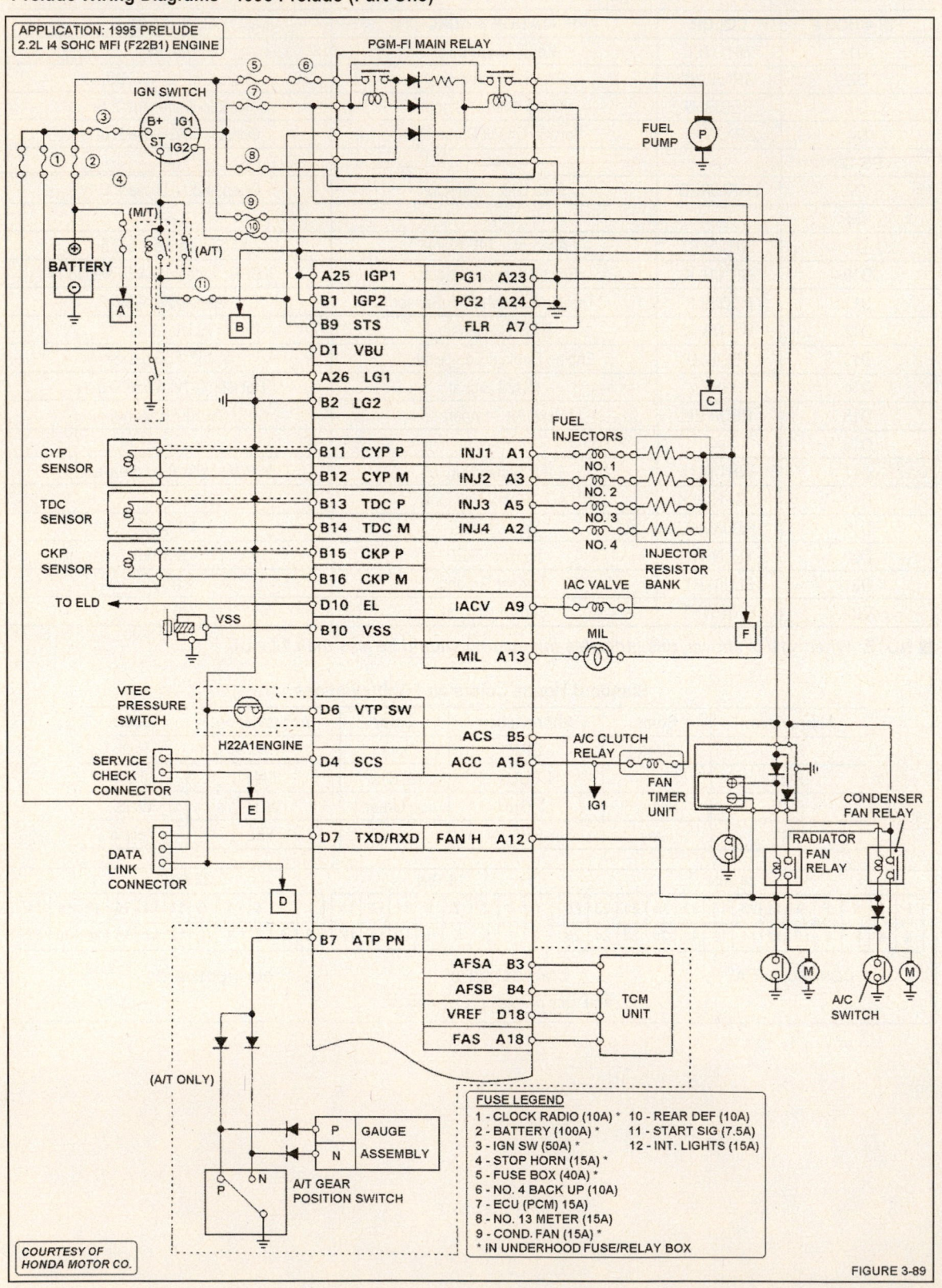

APPLICATION: 1995 PRELUDE
2.2L I4 SOHC MFI (F22B1) ENGINE

PGM-FI MAIN RELAY

IGN SWITCH

FUEL PUMP

BATTERY

CYP SENSOR

TDC SENSOR

CKP SENSOR

TO ELD

VTEC PRESSURE SWITCH

H22A1 ENGINE

SERVICE CHECK CONNECTOR

DATA LINK CONNECTOR

FUEL INJECTORS

INJECTOR RESISTOR BANK

IAC VALVE

MIL

A/C CLUTCH RELAY

FAN TIMER UNIT

CONDENSER FAN RELAY

RADIATOR FAN RELAY

A/C SWITCH

TCM UNIT

(A/T ONLY)

GAUGE ASSEMBLY

A/T GEAR POSITION SWITCH

FUSE LEGEND

1 - CLOCK RADIO (10A) *
2 - BATTERY (100A) *
3 - IGN SW (50A) *
4 - STOP HORN (15A) *
5 - FUSE BOX (40A) *
6 - NO. 4 BACK UP (10A)
7 - ECU (PCM) 15A)
8 - NO. 13 METER (15A)
9 - COND. FAN (15A) *
10 - REAR DEF (10A)
11 - START SIG (7.5A)
12 - INT. LIGHTS (15A)
* IN UNDERHOOD FUSE/RELAY BOX

FIGURE 3-89

Prelude Wiring Diagrams - 1995 Prelude (Part Two)

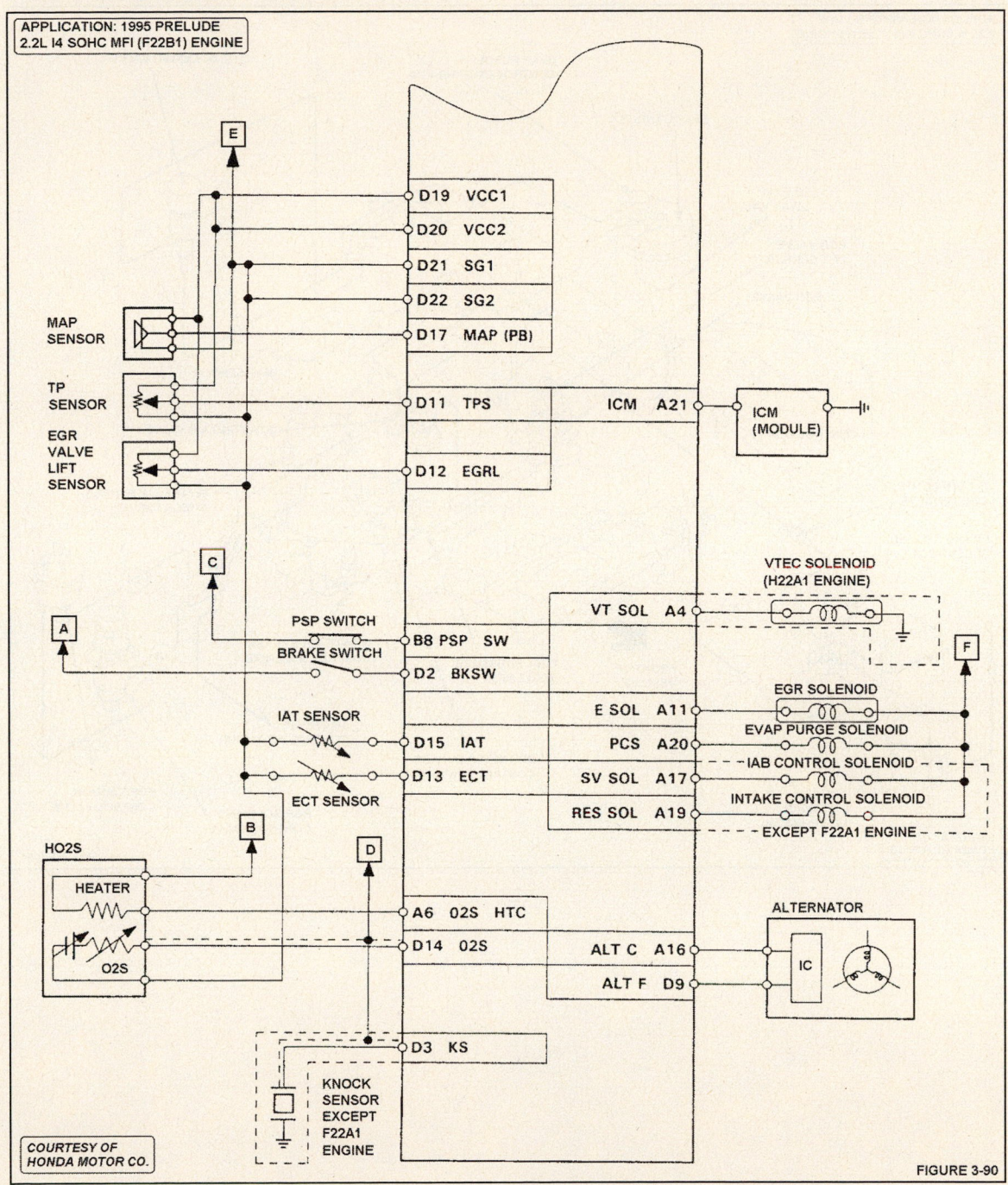

APPLICATION: 1995 PRELUDE
2.2L I4 SOHC MFI (F22B1) ENGINE

MAP SENSOR

TP SENSOR

EGR VALVE LIFT SENSOR

D19 VCC1
D20 VCC2
D21 SG1
D22 SG2
D17 MAP (PB)
D11 TPS — ICM A21 — ICM (MODULE)
D12 EGRL

VTEC SOLENOID (H22A1 ENGINE)
VT SOL A4

PSP SWITCH — B8 PSP SW
BRAKE SWITCH — D2 BKSW

E SOL A11 — EGR SOLENOID
PCS A20 — EVAP PURGE SOLENOID
IAT SENSOR — D15 IAT — IAB CONTROL SOLENOID
ECT SENSOR — D13 ECT — SV SOL A17 — INTAKE CONTROL SOLENOID
RES SOL A19 — EXCEPT F22A1 ENGINE

HO2S
HEATER
O2S

A6 02S HTC
D14 02S — ALT C A16 — ALTERNATOR — IC
ALT F D9

D3 KS
KNOCK SENSOR EXCEPT F22A1 ENGINE

COURTESY OF HONDA MOTOR CO.

FIGURE 3-90

Prelude Vacuum Diagrams - 1995 Prelude

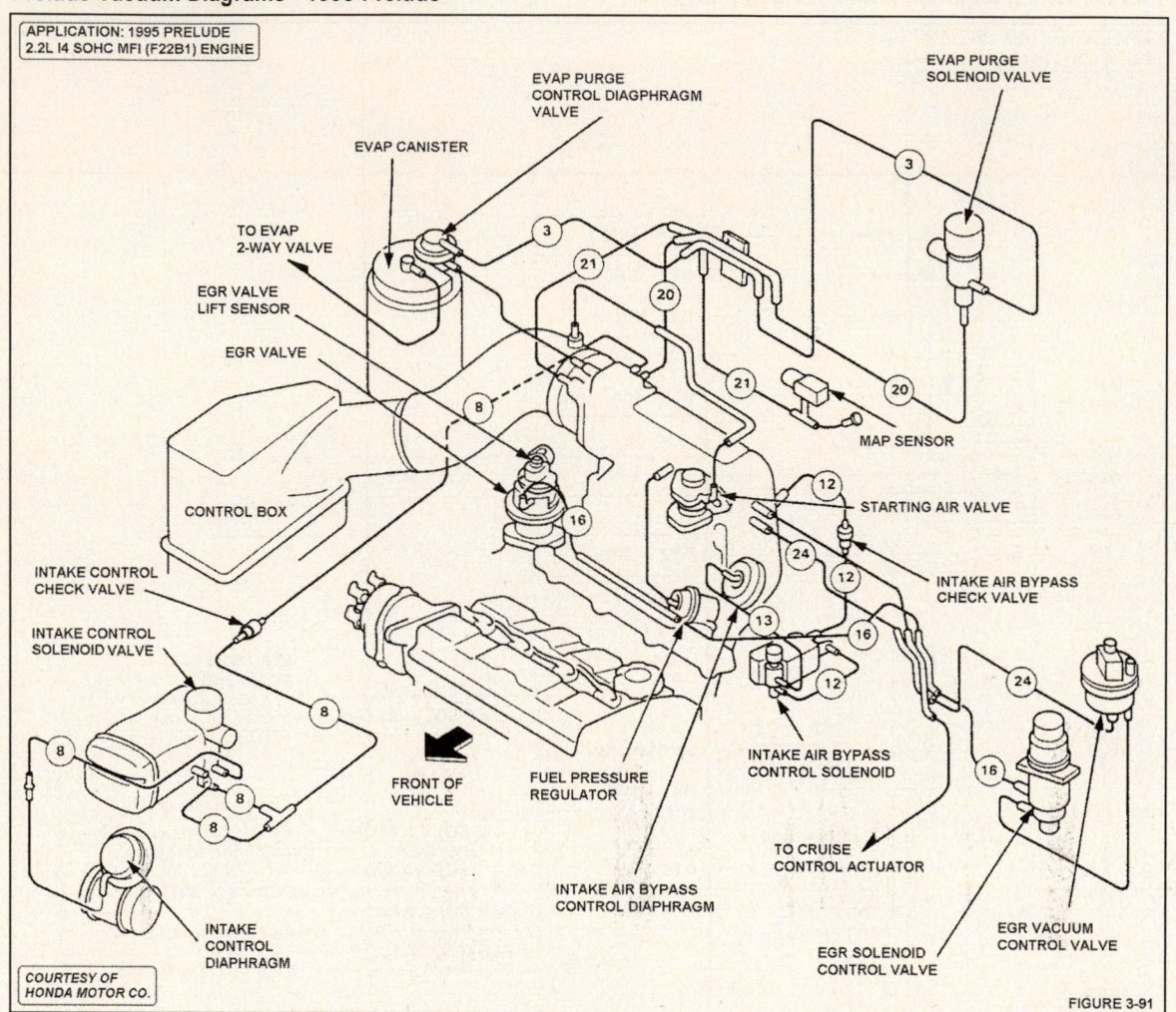

APPLICATION: 1995 PRELUDE
2.2L I4 SOHC MFI (F22B1) ENGINE

EVAP PURGE
CONTROL DIAGPHRAGM
VALVE

EVAP PURGE
SOLENOID VALVE

EVAP CANISTER

TO EVAP
2-WAY VALVE

EGR VALVE
LIFT SENSOR

EGR VALVE

MAP SENSOR

CONTROL BOX

STARTING AIR VALVE

INTAKE CONTROL
CHECK VALVE

INTAKE AIR BYPASS
CHECK VALVE

INTAKE CONTROL
SOLENOID VALVE

INTAKE AIR BYPASS
CONTROL SOLENOID

FRONT OF
VEHICLE

FUEL PRESSURE
REGULATOR

TO CRUISE
CONTROL ACTUATOR

INTAKE
CONTROL
DIAPHRAGM

INTAKE AIR BYPASS
CONTROL DIAPHRAGM

EGR SOLENOID
CONTROL VALVE

EGR VACUUM
CONTROL VALVE

COURTESY OF
HONDA MOTOR CO.

FIGURE 3-91

Technical Service Bulletins

Engine Performance TSB List

TSB Title: *Distributor Repair Procedure*
TSB Number: HSN 89-027, **Date:** 4/17/92
Coverage: 1988-92 Civic, 1990-92 Accord and 1992 Prelude
Description: A distributor sub-assembly is available. The sub-assembly includes the distributor housing, CYL, CRANK and TDC sensors. The use of this sub-assembly allows the replacement of these sensors without replacing the distributor.

TSB Title: *PCM (ECU) Connector Terminal Removal*
TSB Number: HSN 0791-01, **Date:** 7/1/91
Coverage: All Honda Models
Description: Be careful when removing a terminal from the PCM (ECU) connector. If the connector housing is damaged, the wiring harness may have to be replaced. These connectors have a retainer that keeps the terminals from sliding out of the back of the housing. Carefully release the locking tab on each side of the retainer and then swing the retainer out of the way to allow for terminal replacement. Lock the tab when done.

TSB Title: *No Start Condition - Watch the MIL Operation*
TSB Number: HSN 0293-04, **Date:** 2/1/93
Coverage: All Honda with Fuel Injection
Description: The MIL or C/E light should be the first diagnostic tool used on a vehicle with PGM-FI that has a No Start Condition. Turn the key off, wait two seconds, then turn the key on (without starting the engine) and watch the C/E light. It will either come on for two seconds, not come on at all or come on or remain on. Use PGM-FI Chart No. 1 or Chart No. 2 to diagnose the light.

TSB Title: *No Start Condition - Distributor Bearing Problem*
TSB Number: HSN 0293-04, **Date:** 12/11/95
Coverage: 1992 Accord, 1993 Accord Coupe, Sedan and Wagon Models
Description: The distributor bearing may develop excessive clearance. If this happens, the engine may have a No Start Condition. Honda has attempted to contact the owners of the affected vehicles and notify them that the warranty on the affected vehicles will be extended to 6 years or 75,000 miles. The distributor is replaced free on these vehicles.

TSB Title: *Engine Knocking Noise - Related to the EGR Control Solenoid*
TSB Number: HSN 1193-01, **Date:** 11/1/93
Coverage: 1994 Accord - All Models
Description: A knocking or loud tapping noise during acceleration may be heard in the affected models. This noise can be caused by contact between the EGR control solenoid valve and bulkhead body. The noise usually occurs when the EGR solenoid is operating at a rapid rate (in the 1200-2500 rpm range). To fix the noise problem, carefully bend the control box bracket up to supply clearance so that the valve does not contact the body.

Engine Performance TSB List
TSB Title: Noise from Air Cleaner Bracket
TSB Number: HSN 0394-04, **Date:** 3/1/94
Coverage: 1994 Accord - All Models
Description: The front air cleaner bracket may vibrate against the body and make a buzzing noise when accelerating in the 2200-3000 rpm range. To simulate the noise, try tapping on the air cleaner housing cover with your fist. This symptom has been described as a clicking, creaking, rattling, squeaking or ticking noise. It sometimes occurs during deceleration, braking or when more than one person is in the vehicle. If there is a noise coming from the right front corner of the vehicle, check the air cleaner bracket first. To make the repair, first remove the air cleaner housing and the front bracket. The bracket is held on one end by a bolt while the other end is held by a tang. Use a blunt drift and make a slight dimple in the hole in the body that the tang fits into (to put more tension on the bracket). Then put some Molykote M77 lubricant on the tang and reinstall the bracket and air cleaner housing.

TSB Title: Two-Trip Detection Logic Method
TSB Number: HSN 1193-02, **Date:** 11/1/93
Coverage: 1994 Accord & Civic - All Models
Description: A two-trip detection logic method is available to use to diagnose faults in the EGR, HO2S and Fuel Control self-diagnostic functions. With two-trip detection logic, the MIL or C/E light will not come on for Code 1, Code 12 or Code 43 unless the fault occurs on two separate test drives. The second trip must occur after the engine has had time to cool down to a temperature of less than 122°F. The advantage of the two-trip detection logic is that it prevents "false codes" from setting due to any unusual driving occurrences or one of a kind glitches. However, a fault in one of these systems will still set a code and activate the MIL or C/E light if it is present on two trips.

TSB Title: Long Crank Times
TSB Number: HSN 92-043, **Date:** 11/30/93
Coverage: 1992 Civic - All Models
Description: The affected vehicles will exhibit long crank times before the vehicle will start. The problem may be related to a leaking check valve inside the fuel pump. To test for this problem, refer to the Fuel System Tests.

TSB Title: No Start Condition, Radiator Fan or Cruise Control Inoperative
TSB Number: HSN 1085-02, **Date:** 10/1/85
Coverage: Prelude Si Models
Description: The water used during the window tinting process may enter the area below the rear side windows and drip into the ECU or PCM and cause a No Start Condition. It can also disable the Radiator Fan Relay operation and the Cruise Control Operation. If the windows are going to be tinted, this area should be sealed off.

TSB Title: Poor Shift Quality, Torque Converter Clutch "Hunting" Conditions
TSB Number: HSN 0894-05
Date: 8/1/94
Coverage: 1990-94 Accord with A/T, 1988-94 Prelude with A/T
Description: A loose or tight A/T throttle control cable can cause these symptoms. One adjustment method is to make sure the control lever is synchronized with the throttle, but not preloaded (see the Air Induction & Throttle Body articles in this manual). A new A/T throttle cable adjustment method incorporates the use a pressure gauge installed in the throttle 'B' pressure inspection hole in the transmission to check the pressure.

OBD II SYSTEM CONTENTS

Introduction To OBD II Systems

Introduction

REASONS FOR THE OBD II SYSTEM

The OBD II system was developed to accomplish two different objectives. First, it was developed to comply with California and Federal regulations and standards for vehicle emission control monitoring. The initial goal of this system was to detect the degradation or failure of an emission-related component or system that could cause vehicle emissions to rise by 50%.

OBD II CHANGES

However, the program has expanded to include new computer "run" diagnostic "monitors" that are used to verify that an emission control system is operating correctly. This aspect of OBD II has the greatest impact on service technicians as it includes test procedures that require a whole new set of diagnostic procedures. If the operation of the OBD II "monitors" is not understood, a whole new set of driveability problems can appear to the technician who is not trained in this new system. OBD II "monitors" require a particular drive cycle to complete. In some cases, the monitor will not even check a particular system for codes unless a drive cycle is performed.

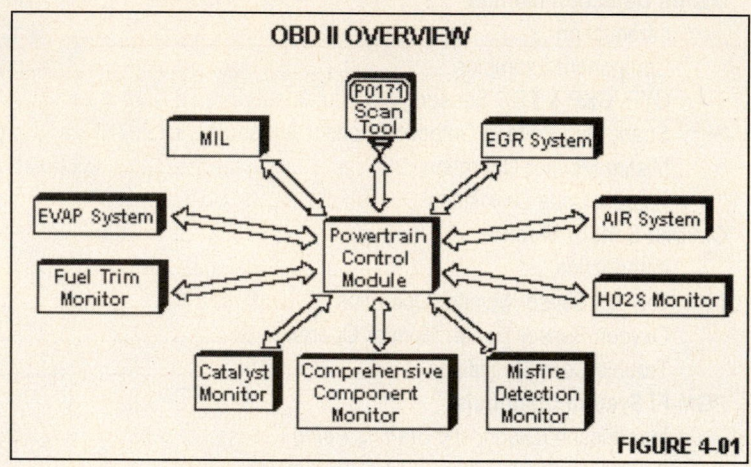

FIGURE 4-01

Second, OBD II introduces changes intended to help with diagnostics. These changes include common terms, connectors and a common diagnostic language for a generic Scan Tool. The changes include:

- Common Diagnostic Connector
- Expanded Malfunction Indicator Light Operation
- Common Codes and Diagnostic Language
- Common Diagnostic Procedures
- New Emissions-Related Procedures, Logic and Sensors
- Expanded Emissions-related Monitoring

IMPORTANT BENEFITS

One important benefit of OBD II is that all vehicles will have a common data output system with a common connector. This allows a generic OBD II certified Scan Tool to read data from any OBD II compliant vehicle and pull codes with a common name and similar descriptions for fault conditions. In the very near future, emissions testing will require the use of an OBD II certifiable Scan Tool to verify that all monitors have run and passed. This important change, shown on the Scan Tool as I/M Readiness Tests, is used to verify that all the readiness tests have run and completed. As of 12/97, Utah is the first state to require that a drive cycle be done and all I/M Readiness tests be completed (with no trouble codes set) to pass their State Emissions Test.

History of OBD Systems

TRANSITION FROM OBD I TO OBD II

To understand the OBD II system, a technician must first understand the OBD I system. It is important to remember that inside every OBD II system is the OBD I diagnostic system with the addition of new emission diagnostic "monitors" that are used to test and verify the emission-related components and systems. Therefore, to understand the OBD II system, a technician must know something about the OBD I system for each vehicle manufacturer, and their particular approach to diagnostics. The articles that follow include a summary of evolution of the OBD I system for each vehicle manufacturer.

HISTORY OF OBD SYSTEMS

Starting in 1978, various manufacturers introduced computer controls of vehicle systems and engine management. These computer-controlled systems involved not only the diagnosis of engine mechanical or component operation, but also the identification of electrical faults and computerized engine control diagnosis. Early attempts at diagnosis involved expensive and specialized diagnostic testers that were connected externally to the computer in series with the wiring connector. These testers were used to monitor the input/output operations of the computer.

By 1980, vehicle manufacturers were designing systems where the computer incorporated internal programs that monitored selected components and stored trouble codes in memory for retrieval at a later time. These codes identified failure conditions that referred a technician to diagnostic repair charts or procedures that helped pin point the problem areas.

While a few vehicle manufacturers began implementation of OBD II mandated programs as early as 1992;most manufacturers did not start their OBD II phase-in period until 1994. All vehicle manufacturers were required to meet OBD II standards by the 1996 model year.

ACURA/HONDA OBD HISTORY

The names and systems that show the history of Acura/Honda On Board Diagnostics are shown in Figure 4-02.

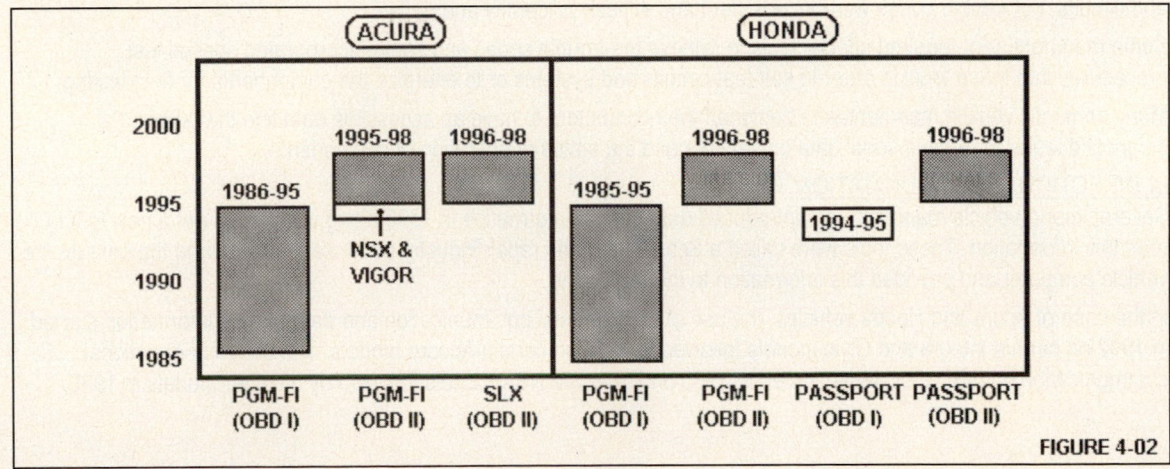

FIGURE 4-02

Computer Diagnostics

OVERVIEW

As vehicle manufacturers began introduction of the engine computer to control emissions, they also developed a new set of diagnostics (tree charts, etc.) that were used to diagnose early emission control systems. General diagnostics of computers fall into two categories: external diagnostics and Internal or On Board Diagnostics. The first level of diagnostics, external diagnostics uses a separate computer diagnostic tool that taps into the computer and runs a series of diagnostic tests. This method of diagnostics was popular in the 1970's and was used in the 1980's on many European vehicles. The second level of diagnostics, On Board Diagnostics, incorporates diagnostic tests into the circuit board of the computer.

In 1980 General Motors incorporated an On Board Computer Program where the "check engine" light came on to inform the vehicle owner that there was a fault in the computer system. The light was turned on when a diagnostic code was set to alert the driver that service was needed on the vehicle.

California formed a government agency, the California Air Resources Board (CARB), in order to monitor the air quality and establish regulations to reduce air pollution. The California Health and Safety Code authorized the Air Resources Board to adopt motor vehicle emissions standards and in-use performance standards which it finds necessary, cost effective and technologically feasible. In 1988, CARB required that all vehicles sold in California incorporate a system with an On-Board Diagnostic program where a "check engine" light would come on to notify the vehicle owner of a potential failure of computer sensors and/or their systems. This system is known as On-Board Diagnostics First Generation and is now referred to as OBD I. OBD II evolved due to problems encountered with OBD I systems. The article that follows contains a summary of the differences between an OBD I system and the OBD II system.

OBD I Versus OBD II

CODE RETRIEVAL

One of the problems with OBD I was that the code retrieval methods were different for each manufacturer and there was no consistency between the engine control systems. Most manufacturers used similar engine computer sensors and circuits, but trouble codes were inconsistent and difficult to identify and define.

Some manufacturers required special tools to retrieve the trouble codes and some incorporated special test procedures into these tools in order to self-test circuits and systems or to energize the components for field testing.

Many domestic vehicle manufacturers designed their computers to have an accessible data line that allowed a diagnostic tester to retrieve serial data on sensors and the status of operation of components.

BI-DIRECTIONAL COMMUNICATION

Several import vehicle manufacturers introduced data stream information in 1990 along with the use of a new tool to read this information. These tools were called a Scan Tool in the repair industry because they scanned the data on the vehicle computer and provided this information to the technician.

In the case of Acura and Honda vehicles, the use of bi-directional communication and data stream information started in 1992 on certain Integra and Civic models followed in 1994 on certain Accord models. The use of bi-directional communication on OBD II systems started in 1995 on the Acura NSX models followed by all other models in 1996.

OBD I Versus OBD II

TEST CONNECTORS

Due to the differences in the connectors, the aftermarket Scan Tool manufacturers had to develop a different cable setup and connector for each vehicle. This is one of the drawbacks to OBD I.

Several examples of the OBD I system connectors from different vehicle manufacturers are shown in Figure 4-03.

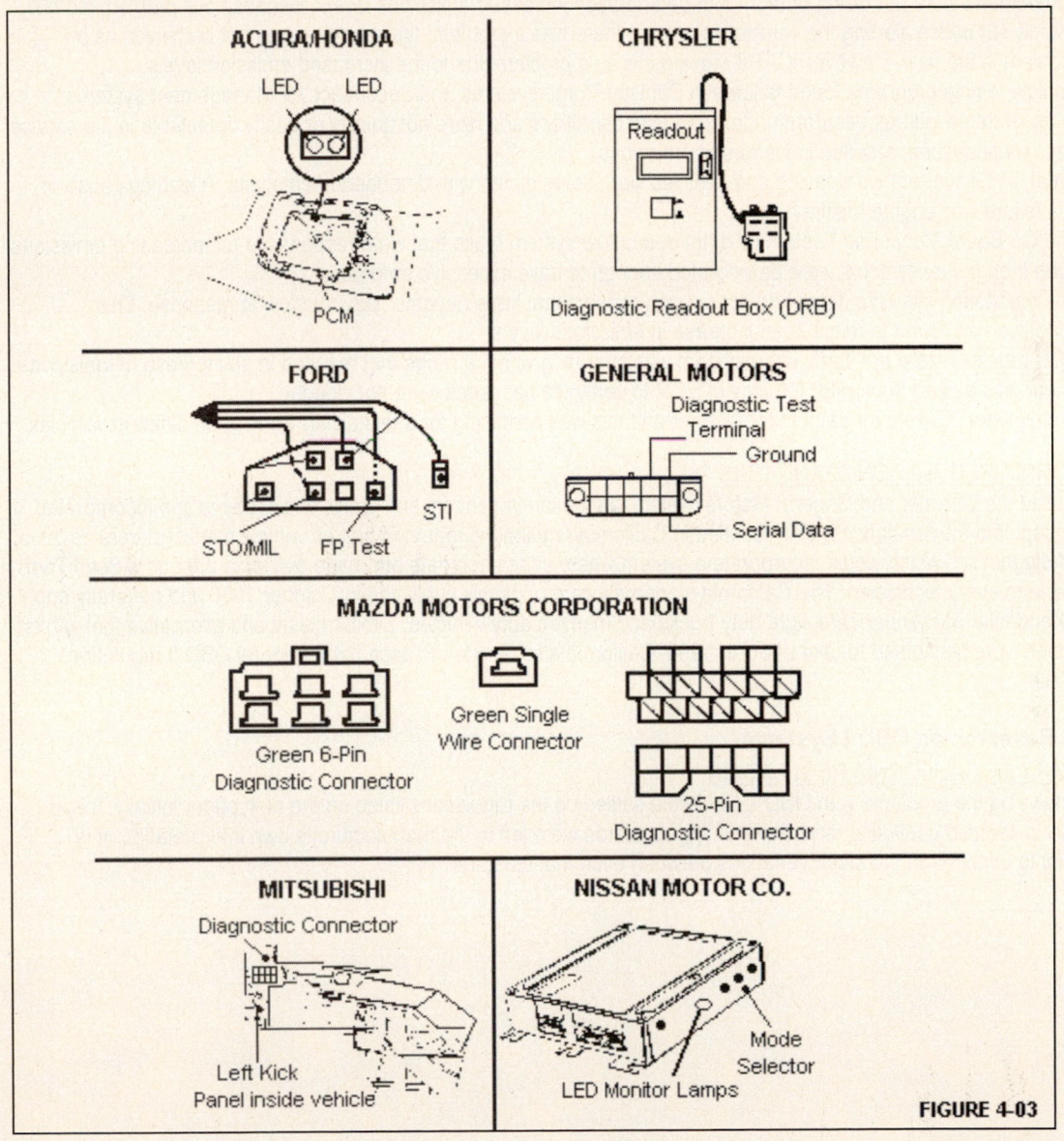

FIGURE 4-03

Government & SAE Regulations

California Regulations

INTRODUCTION

The State of California regulations developed into the OBD II monitors and readiness tests. The California Air Resources Board (CARB) conducted research on vehicle emissions from OBD I vehicles. The results of this study evolved into the regulations that required the On Board monitor "run" tests.

- The study found a significant number of pre-1988 vehicles with degraded emissions components. These components were not failing outright, but deterioration increased emissions levels. Because this problem did not usually set codes alerting the vehicle owner that there was a problem, the condition was not perceived as a problem in the field. However, CARB viewed this as a problem due to the increased emission levels.
- Vehicle testing programs found failures in Canister Purge systems and Secondary Air Management systems. Many of these failures occurred under road load conditions and were not quickly or easily detectable in the service bay. These failures resulted in increased emissions.
- Catalytic Converters were failing and vehicles were being driven with deteriorated catalysts. A leading cause of this failure was engine misfire.
- The On Board Monitoring Systems did not detect fuel system faults that were responsible for increasing emissions even though fuel systems were deteriorated enough to have excessive emissions.
- The monitoring systems did not detect oxygen sensors that have become "lazy" or slow in response. This condition was found to result in an increase in emissions levels.
- EGR monitoring did not verify if the system was operating within a range that resulted in an increase of emissions. There was a need to monitor the flow of EGR to verify the passages were not clogged.
- Codes were different for each manufacturer and this was confusing for a technician working on different vehicles.

CALIFORNIA REGULATION STANDARDS

CARB liked the Chrysler and General Motors system of monitoring sensors and emissions systems and incorporated this concept into the regulations. This resulted in California regulations requiring that all vehicle manufacturers develop a set of diagnostics which would incorporate a system where codes and data are made available through a Scan Tool accessible to every technician. The California standards were originally published in October 1988 and generally apply to 1994 and later passenger cars, light duty trucks and medium duty vehicles. Similar diesel and alternative fuel vehicle regulations were scheduled to take effect in 1996. California also decided to accept the Federal OBD II regulations after 1998.

CARB Research on OBD I Systems

SUMMARY OF CALIFORNIA REGULATION

After analyzing the problems with OBD I, California settled on the regulations listed on the next page. Initially, these regulations seemed confusing, and some of the standards were left to the manufacturer's own interpretation. It is important to understand the slight variations between each manufacturer.

CARB Research On OBD I

SUMMARY OF CALIFORNIA REGULATIONS (CONTINUED)

*   All vehicles should have a Malfunction Indicator Light (MIL) that must remain on if an emission-related fault is detected. The MIL should go out if the fault is not present for two consecutive trips. A fault code can be erased from memory if the fault does not occur for 40 warmup periods.
*   The catalytic converter closest to the engine (if more than one converter is on the car) should be monitored to see if it has deteriorated.
*   The oxygen sensor (HO2S) should be monitored for the output voltage and the response rate. This test should include the HO2S cross counts or switch rate) and should check the sensor transition from rich to lean and lean to rich. The HO2S heater should be monitored for proper operation.
*   The engine should be monitored for misfire and faulty cylinders must be identified. The MIL should flash during the actual misfire and the system should result in a back-up program designed to reduce the misfire.
*   This system should include identification informing the technician that the catalyst is deteriorated, damaged, or has failed an FTP type drive cycle.
*   The fuel delivery system should be monitored for its ability to control fuel. It should determine if the system has a condition that is over the range for optimal fuel control. This system is referred to as fuel trim monitor.
*   The EGR system should be monitored for both high and low flow-rates and EGR solenoid operation should be verified for proper operation.
*   The EVAP System should be monitored for HC flow changes based on fuel tank fill level and the monitor should calculate for these differences.
*   The Secondary Air Management System should be monitored for the control valve operation and for the airflow throughout the system.
*   Any emissions-related component or system not otherwise described that provides input directly or indirectly to the processor should be monitored. This includes all input and output components.

ENVIRONMENTAL PROTECTION AGENCY (EPA)

The Federal EPA also established diagnostic regulations, but in an effort to simplify the development process for vehicle manufacturers, the EPA decided to comply with the California OBD II regulations from 1994 to 1998.

SAE FORMS COMMITTEES TO DEVELOP STANDARDS

The Society of Automotive Engineers (SAE) established committees to help its member engineers coordinate efforts and develop a second generation of diagnostics. Since the new standards required that the vehicle manufacturers provide common terminology and provide information to a generic Scan Tool, the SAE committees were assigned the task of setting new standards to standardize the diagnostic connector and to coordinate the development of the standards between the vehicle manufacturers.

EPA EXPANDS THE SAE STANDARDS

The Federal Government (EPA) reviewed SAE standards and added regulations requiring all manufacturers to meet Federal OBD II standards by 1996. To assist in development of expanded diagnostics, the EPA decided to use the California regulations as federal standards through 1998. Federal and California regulations established a two-year phase-in to take effect from 1994-1996 to allow for design and phase-in of expanded diagnostics.

Industry Regulations
UNDERSTANDING REGULATIONS

The two key government agencies that set the OBD II regulations were the CARB and EPA. These regulations affect various companies who supply the automotive aftermarket (i.e., trade magazines, textbook suppliers and Scan Tool manufacturers) as well as those controlling the actual operation of the vehicle computer and its interaction with a generic Scan Tool. The tables that follow on the next few pages compare the differences between CARB and EPA regulation for OBD II systems. They also define these regulations based on interpretation of the standards by SAE. Some standards were federally mandated and should be on all vehicles. Also, SAE committees recommended additional standards, and these recommended standards may or may not be on a specific vehicle.

INDUSTRY REGULATIONS - SAE RECOMMENDATIONS

The table below contains information mandated by government regulations and the SAE recommended voluntary compliance.

MANDATED BY GOVERNMENT REGULATIONS	SAE RECOMMENDED VOLUNTARY COMPLIANCE
J1930 - Industry Terminology Standardization	J2201 - Additional Guidelines for Generic Scan Tool Interface
J1978 - Standards for Generic OBD II Scan Tool Interface Protocol	J2190 - Enhanced Test Mode Standards
J2205 - Standards for Expanded Scan Tool Interface Protocol	J2008 - Guidelines for Repair Service Information (CARB Standards)
J2008 - Standards for Repair and Service Information (EPA Guidelines)	

INDUSTRY REGULATIONS - CARB AND EPA INVOLVEMENT

The table below contains information that shows the CARB and EPA involvement as far as the government requirements for each item.

CARB	EPA	Government Requirements
	X	1994 Model year - all service information must conform to **J1930**.
X	X	Service manuals must publish a normal range for calculated load values and Mass Airflow Rate at idle and at 2,500 RPM.
	X	The vehicle manufacturer is responsible for ensuring information is available even if the information has to be provided by an intermediary.
	X	The cost of repair information to the independent technician shall not exceed the lowest price that is available to a dealership.
	X	All other information available at a fair and reasonable price, otherwise it is considered not available (a fine of $25K per day could be applied).
	X	Electronic service information must be available by the 1998 model year.
X		Repair procedures must be available which allow effective diagnosis and repairs using a **J1978** Generic Scan Tool and common repair tools.
X	X	**J1978** Scan Tool compatibility - communication protocol. All serial data and enhanced test must be available to a Scan Tool. Scan Tool must inform user which emissions systems are monitored. EPA added requirements that the VIN be accessible off the DLC.
	X	Requires Bi-directional diagnostic control of the computer be available on the Scan Tool meeting **J2205** and **J1979** standards.
	X	1996 model year - all service information must be in the **J2008** format.
	X	Labeling requirements must meet J1877 and J1892 standards

SAE STANDARDS & DOCUMENTATION

Some Scan Tools, service manuals and articles list information referring to the SAE documents. The articles that follow contain a summary of SAE standards and are provided to help explain what is involved in an SAE Standard.

J1930 - COMMON NAMES FOR COMPONENTS

J1930 established common nomenclature for emissions and computer related components and systems. This includes common definitions, abbreviations and acronyms. This standard is designed to provide the technician with a recognizable name for components that apply to all vehicles and is seen as beneficial for technicians who work on multiple lines of vehicles as well as vehicles from different manufacturers.

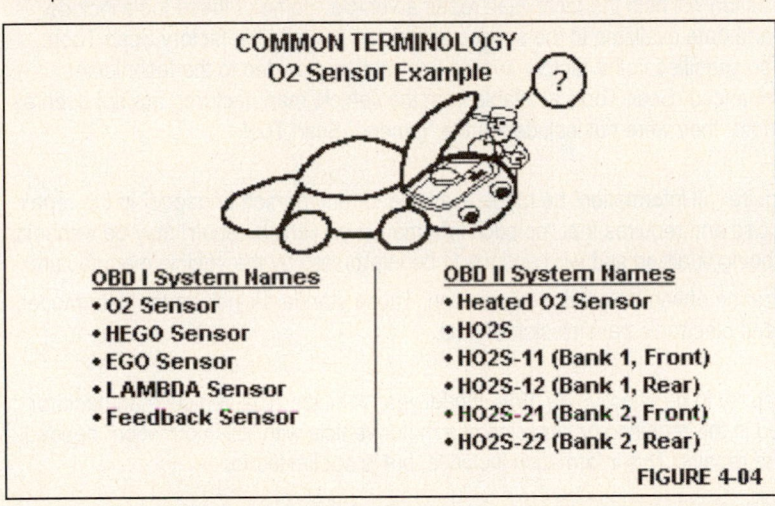

COMMON TERMINOLOGY
O2 Sensor Example

OBD I System Names
* O2 Sensor
* HEGO Sensor
* EGO Sensor
* LAMBDA Sensor
* Feedback Sensor

OBD II System Names
* Heated O2 Sensor
* HO2S
* HO2S-11 (Bank 1, Front)
* HO2S-12 (Bank 1, Rear)
* HO2S-21 (Bank 2, Front)
* HO2S-22 (Bank 2, Rear)

FIGURE 4-04

A few vehicle manufacturers introduced this new terminology as early as 1990. However, by 1994, all manufacturer service publications began to use the J1930 guidelines for terms. Under these guidelines, the engine computer is called the Powertrain Control Module (PCM) and a coolant sensor is called an Engine Coolant Sensor (ECT). Refer to Figure 4-04 for a few examples of this new terminology.

J1978 - GENERIC SCAN TOOL USAGE

J1978 requires that all manufacturers should make readily available to the automotive repair industry data, codes and emissions-related information which can be accessed by a generic microprocessor-based Scan Tool. Fault codes, sensors, component values and freeze frame data stored in the On Board computer must be accessible for download to a Generic Scan Tool.

OBD II GENERIC SCAN TOOL OPERATION

A generic Scan Tool (that connects to the 16-pin connector) should also:

* Clear trouble codes from the vehicle's computer memory
* Display the "current" I/M Readiness Test Status for all on board monitors (see note)
* Retrieve specific data from a vehicle on board computer data stream
* Retrieve "freeze frame" data
* Read the five digit trouble code

TIPS: Acura/Honda OBD II systems are programmed (as are other manufacturers) to run each of the system main monitors when all enable criteria are present on each trip. If a PCM Reset step is done, the status of the I/M Readiness Tests will change from not done to completed the first time a particular monitor test is run. The completed message remains in memory until a test (or component related to a test) fails.

Industry Regulations

EPA EXPANSION OF REGULATIONS

The Federal EPA expanded this regulation to include the capacity to perform bi-directional diagnostic control. To properly define the meaning and function of "bi-directional" testing, vehicle manufacturers require a more specific set of standards. The SAE will incorporate these standards into the J2205 document described next.

J2205 - EXPANDED DIAGNOSTIC PROTOCOL FOR OBD II SCAN TOOLS

OBD II certified Scan Tools incorporate a protocol that enables the technician to access information not specifically required by OBD II standards. This information will help the technician repair a vehicle. Some of these tools include enhanced screens with extra messages and data available to the service technician (usually the factory Scan Tool). The information on these messages will be specified in the factory service information provided to the technician. These advanced programs, offered in "enhanced" Scan Tools available from the vehicle manufacturer, are not seen as important for most vehicle repairs. Therefore, they were not included in the "generic" Scan Tool.

J2008 - SERVICE INFORMATION

The EPA established regulations that require "all information" be made available to "any person engaged" in the repair of the vehicle. The legislation is very specific and requires that "no such information on vehicle repair" may be withheld from any repair technician. J2008 is still being finalized and will continue to be interpreted by the vehicle manufacturer.

J2008 also sets standards for the organization of vehicle service information. These standards include the data model, data type definition, graphics standards and electronic transmission of data.

INFORMATION AVAILABILITY

EPA guidelines require information availability to be effective for 1994 model year vehicles. The vehicle manufacturer must furnish "any person" who is engaged in the repairing or servicing of a motor vehicle with "all information" needed for making emission-related diagnosis and repairs. This information includes, but is not limited to:

- Service manuals
- Technical service bulletins
- Vehicle recalls
- Engine control emissions data stream
- Bi-directional control
- Training information

Any of this information that is provided directly or indirectly to dealers cannot be withheld from the aftermarket technician. The cost of information to repair technicians shall not exceed the lowest price of equivalent information offered to dealerships. All other information shall be made available at a "fair and reasonable price". Otherwise it will be considered as "not available."

J2201 - GENERIC SCAN TOOL TERMINAL DESIGNATION

J2201 sets additional guidelines for generic Scan Tool interface and assigns the designation of terminals for voltage feed, ground and data transmission.

On Board Computer Regulations

J2190 - ENHANCED DIAGNOSTIC TEST MODES

This standard identifies test modes and diagnostics that go beyond the EPA and CARB regulations. These include emissions and non-emissions related systems. Test Modes include an expanded diagnostic routine and may include protocols to establish screens on the Scan Tool with this data:

- Request diagnostic session, freeze frame and diagnostic trouble codes
- Request status of emission monitors, clear diagnostic information
- Request diagnostic data, security access
- Disable or enable normal message transmission
- Request diagnostic data packets, diagnostic routine results
- Enter or exit diagnostic routines, read or write to the EEPROM
- Substitute input sensor values, substitute output control
- Messages from enhanced diagnostic test modes are available on some Scan Tools

J2012 - DIAGNOSTIC TROUBLE CODE (DTC) STANDARDIZATION

Diagnostic Trouble Code (DTC) is a term used to describe an identifier or code that is stored by the On Board Computer when it recognizes a fault in a computer-monitored system. OBD I type codes were stored as one, two, or three digit numbers. SAE J2012 set standards for trouble codes and definitions for emission-related systems. OBD II codes use a five-digit code and codes begin with a letter followed by four numbers.

J1962 - COMMON DIAGNOSTIC CONNECTOR

SAE J1962 establishes standards for the OBD II Diagnostic Link Connector (DLC). It is designated to be a 16-pin connector. The connector should be located beneath the instrument panel and within 12 inches (300 mm) of the vehicle centerline. The DLC is located out of the line of sight of passengers, but easily viewable by a technician from a kneeling position outside the vehicle with the door open. If the connector is located elsewhere, an OEM sticker is added to identify its location.

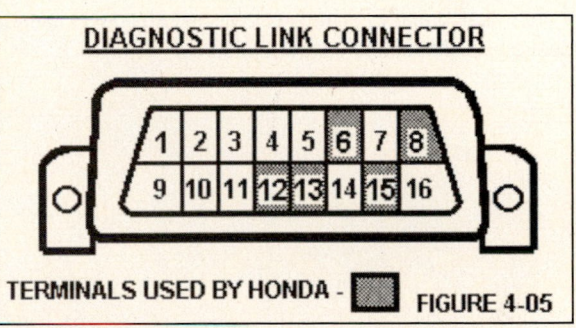

Eight of the DLC pins are assigned SAE labels and 8 are assigned to vehicle manufacturer labels. The 8 pins assigned by SAE include:

- 2 pins for the Serial Data Links
- 2 pins for the ISO- 9141 Serial Data Links
- 1 pin for battery power and one pin each for battery and signal ground

ACURA & HONDA DLC LOCATIONS

Acura Integra - Behind right side of the center console
Acura 2.2CL, 2.3CL, 2.5TL - Under the ashtray
Acura 3.0L CL, 3.2TL, 3.5TL - Under the ashtray
Acura NSX - Behind a removable cover under the glove box
Acura SLX - Under left side of the dash near the kick panel
Honda Accord - Under the ashtray, **Honda CR-V** - Behind right side of center console
Honda Civic - Near left kick panel, **Civic Del Sol** - To right of console under a cover
Honda Odyssey, Prelude- Behind the right side of the center console
Honda Passport - Under the left side of the dash near the kick panel

On Board Computer Regulations

<u>J1979 - DIAGNOSTIC TEST MODE MESSAGES</u>

Defines standard messages used to access fault codes, vehicle data stream information and freeze frame data.

<u>J1850 - SCAN TOOL ACCESS TO EMISSION-RELATED DATA</u>

Regulations from CARB and EPA require that access to emission related Class A data be made available on a standard Diagnostic Data Link (DDL). This link (DDL) is defined in the J1850 standard. Many vehicles include other computers that communicate with serial data. For example, the Climate Control Automatic Air Conditioning System uses signals from the ECT sensor to help determine when to operate the Air Conditioning and Electric Cooling Fan. SAE also developed standards for Vehicle Network and Multiplexing Data Communications (referred to as Class B data). Class B data communications use a system where data is transferred between one or more On Board Computers or modules to eliminate redundant sensors and other system duplication. The modules in this type of system form a multiplex of interactive systems.

Class B data communications have to be able to perform all Class A data functions. However, these two types of communication protocols differ from each other and usually do not communicate in the same format. Scan Tools that communicate with both formats will be available as this standard is defined further and may be made available from the vehicle manufacturer. This means that the Generic Scan Tool (GST) may not be able to access information from computers that control ABS, Air Conditioning, Steering and Suspension, Electronic Transmissions and other related systems.

<u>J2186 - ANTI-TAMPER RELATED STANDARDS</u>

SAE J2186 defines EE Data Link Security standards for computers with Electronically Erasable and re-programmable PROMS. CARB and EPA have established standards intended to eliminate "hot rod" PROMS, which could disable or defeat emissions related control systems.

On Board Computer Regulations

GOVERNMENT REGULATIONS

The table below contains information mandated by government regulations and the SAE recommended voluntary compliance for On-Board Computers.

MANDATED BY GOVERNMENT REGULATIONS	SAE RECOMMENDED VOLUNTARY COMPLIANCE
J2012 - Standards for Diagnostic Trouble Codes (DTCs)	J2186 - CARB approved standards for anti-tamper procedures
J1962 - Standard 16 Pin diagnostic connector	J2178 - Scan Tool message strategy guidelines
J1979 - Standards for Diagnostic Test Modes	J2190 - Enhanced Test Mode Standards
J1850 - Scan tool Communication guidelines for Class B Data Interface	J1724 - Vehicle Electronic Identification Standards
J2186 - EPA mandated Anti-Tamper procedures	

ON BOARD COMPUTER REGULATIONS

The table below contains information that shows the CARB and EPA involvement as far as the government requirements for each item.

CARB	EPA	Government Requirements
X	X	**J2012** Diagnostic Trouble Codes - Uniform identification of fault when detected by the PCM. The PCM must set a code to identify the fault and include conditions where the system reverts to a default mode of operation. Erase fault codes after 40 warm up cycles and MIL not illuminated.
X	X	**J1962** Diagnostic Connector mounted on Instrument Panel driver's side of vehicle with standard pins for serial data, power and ground.
X	X	**J1979** Diagnostic Test Mode Messages - defines standard messages for access to fault codes, vehicle data stream and freeze frame data.
X	X	Scan Tool Interface for Diagnostic Data Link - **SAE J1850** serial data link required for access to emission-related data. Requires three byte headers and CRC and does not allow IBS or checksum.
X	X	Vehicle manufacturers must provide tampering deterrence for PCMs that are programmable (where the PROM is rewritten to change operating parameters). This must include write-protect standards for programmable memory and references **J2186** - Data Link Security for write protection.
	X	Access to vehicle calibration data, odometer and keyless entry codes can be limited, but OEMs must provide "the best means available for providing non-dealer technicians with calibration data necessary to perform repairs".
X	X	Freeze Frame Data Stored with the first fault of any component or system and replace data if there is a subsequent fuel system or misfire fault. EPA added "airflow rate" to required freeze frame data.
X	X	Signal access to the required diagnostic data must be made available through the diagnostic connector using standard messages. Actual values should be identified separately from default or limp home values.
X	X	The vehicle must have one and only one Malfunction Indicator Light (MIL) for emission-related problems. The MIL must remain ON is a malfunction is detected and stays on until three trips indicate the fault is gone. The MIL must blink during a misfire condition. (Note: some manufacturers received an exemption from this standard for some 1994-96 vehicles).

On Board Computer Regulations

REGULATION HIGHLIGHTS

Engine operating parameters should not be changed without the use of manufacturer specialized tools, codes and procedures. CARB specified that any re-programmable computer coded system should include proven write-protection procedures and hardware. The Federal EPA requires that any re-programmable computer codes or operating parameters must be tamper resistant and must conform to SAE J2186 EE Data Link Security standards.

EEPROM UPDATES

The EPA and the vehicle manufacturers have agreed that information to reprogram EEPROMS will be made available to the aftermarket in January of 1998. A generic Scan Tool may include a method to reprogram these devices. This reprogramming step can be accomplished by downloading from the Internet, with a floppy disk or with a CD-ROM.

TAMPER PROTECTION INFORMATION

The possibility that On Board programs may be tampered with or changed from the manufacturer specifications introduces the possibility of additional vehicle problems that would need to be diagnosed. Even before vehicles used computers, it was difficult to diagnose a driveability problem in a modified or performance vehicle. The potential of these changes could make it difficult for committees to balance between allowing individuals to install performance changes and protecting the repair industry from the changes.

There is a need for standards that would allow a PCM to identify a vehicle ID number and current calibration, whether the change is a factory update, or an aftermarket modification. This information should also be made accessible to a repair technician. To help solve this problem, EPA proposed regulations that would force vehicle manufacturers to utilize complicated methods to deter unauthorized reprogramming. These methods would include executable routines that could have copyright protection with encrypted data and mandated electronic access by service facilities to an off-site computer maintained by the vehicle manufacturer. Access to an executable routine would be controlled by the manufacturer and only available to a dealer.

If the vehicle computer method cannot identify any modifications or updates, then the repair technician would need to install the latest factory program. One problem is that these programs were initially available only at a dealer through the use of their specific hardware, software and training information. The SAE and EPA committees are looking into standards for balancing the desire for tamper-proof protection with the need to access information.

■ *NOTE: On late model Acura or Honda vehicles, the PCM is part of the Immobilizer system. If a PCM is replaced or substituted as part of replacing a PCM with a known-good unit, the vehicle may not start due to the fact that the PCM is part of a new Honda security system called the Immobilizer System. In this case, the Immobilizer code for the vehicle must be entered into the new PCM with the Honda PGM-FI Tester. At the time of publication, no aftermarket Scan Tools had this capability.*

J2186 - DIAGNOSTIC DATA LINK SECURITY STANDARDS

Procedures used in tamper protection must discourage tampering yet allow for the service industry to reprogram or service the PCM as deemed necessary by the vehicle manufacturer. Legitimate service of the PROM will be referenced by later EPA regulations. This service is performed if the Scan Tool transmits the proper security codes. However, normal Generic Scan Tool (GST) communications are not affected by this standard.

On Board Computer Monitor Regulations

J2178 - SCAN TOOL MESSAGE STRATEGY GUIDELINES

J2178 sets the standards for how vehicle interface messages are displayed on the Generic Scan Tool. See Figure 1-06 for examples of these messages.

J1724 - VEHICLE ELECTRONIC IDENTIFICATION

SAE is developing a method to provide electronic access to vehicle content information necessary to diagnose, service, test and repair passenger cars and light duty trucks. The SAE committee in charge of this area is currently looking at a wide range of interpretations for this standard.

UNDERSTANDING GOVERNMENT REGULATIONS

The table below contains information that shows how the CARB and EPA were involvement as far as the government requirements for each item related to On Board Computer "Monitor" Regulations.

CARB	EPA	Government Requirements
X	X	Oxygen Sensor must be Monitored for output voltage and response rate for all oxygen sensors once per trip. The results of most recent oxygen sensor evaluation test must be available over the data link connector as serial data. EPA added requirement where the results of the most recent on-board monitoring data and test limits for all systems with specific Monitor evaluation tests be made available.
X	X	Catalyst Monitoring must verify the catalyst is functioning at a steady state of efficiency and has not deteriorated over 1.5X the standard. This test must be monitored once per trip.
X	X	Misfire Monitoring must be continuous under all operating conditions and must identify the cylinder misfiring. Fuel on a Sequential Fuel Injection vehicle must cut off fuel to the misfiring cylinder.
X		EGR System must be Monitored for low and high flow rate once per trip.
X		Evaporative Purge Monitor must check function and leaks once per trip.
X		Secondary AIR management must check for function of the AIR system and switching valves for function and flow once per trip.
X		Fuel system must be monitored for ability to control fuel delivery.
X	X	Comprehensive component monitoring of any emission-related component or system that sends information to or receives information from the On Board computer. CARB also requires a check for out of range values for inputs and functional response for outputs (if applicable). CARB requires continuous monitoring and EPA requires evaluation periodically once per "Drive Cycle".
	X	Monitor for occurrences of deterioration or faults that could cause the exhaust or evaporative emissions to exceed 1.5 times the FTP Standard.
X		Air Conditioning system must be monitored for loss of reactive refrigerant once per trip. Non-reactive refrigerant does not have to be monitored.

OBD II Systems

Malfunction Indicator Lamp

INTRODUCTION

The California Air Resources Board (CARB) and the Federal EPA regulations require that a Malfunction Indicator Lamp (MIL) be illuminated when a fault is detected and that a Diagnostic Trouble Code be stored in the PCM memory.

Many manufacturers already provided "Check Engine Light" diagnostics as required by the 1988 California (OBD I) regulations. OBD II established changes for the operation of the Check Engine Light. A new universal term identifies the light as a Malfunction Indicator Lamp (MIL). However, the lamp in the dash may still be identified with the term "Check Engine" or "Service Engine Soon" for ease of customer understanding.

The OBD II regulations set tight conditions for activating and de-activating the MIL (lamp). This strict set of guidelines has resulted in multiple "levels" of diagnostics with different criteria and conditions for when a fault will cause the MIL to activate and set a code. There are other codes for faults that will not activate the MIL. The guidelines establish how quickly the On Board System must be able to identify a fault, set the DTC in memory and activate the MIL.

OBD II DTC AND MIL REGULATION

There are strict regulations for the conditions to turn off the MIL and to clear codes. In the past, most manufacturers simply removed battery voltage from the computer to clear the codes. These new regulations contain significant changes in how and when the computer turns off the MIL. The vehicle must be driven under specific conditions while the system is monitored. Once a fault is detected, the system that failed must pass three consecutive tests without a fault before the MIL is de-activated. OBD II regulations include:

- Standards to regulate how quickly the computer must identify a fault, activate the MIL and set the DTC
- Standards to regulate the conditions which can turn the MIL Off when the fault is not present
- Standards to establish how long the code remains in memory after a fault is cleared
- Standards to regulate what information must be made available from the vehicle manufacturer for the repair technician to assist in identifying the cause of the fault (Freeze Frame data, code failure conditions, etc.).

FAULTS NOT RELATED TO EMISSION CONTROL SYSTEMS

With OBD II systems, the MIL will not be activated unless the computer has determined that a failure will effect the emissions levels of the vehicle. This means that **only emissions-related faults (and codes)** will cause the PCM to activate the MIL. Be aware that some driveability problems not related to emission control components or systems can cause a code to set without the PCM activating the MIL. However, any trouble codes associated with the fault will still be set in memory for the technician to access.

TIPS: As with OBD I systems, there can be codes with no MIL and symptoms with no codes on systems not related to emission controls. However, on OBD II systems, when diagnosing driveability or emissions related problems, codes are considered "hard" codes. The technician should first read and record the codes - then make the repairs. Once these steps are done, codes and Freeze Frame data can be cleared.

Malfunction Indicator Lamp

UNDERSTANDING MIL CONDITIONS

The three possible MIL conditions are described next.

MIL CONDITION: OFF

This condition indicates that the PCM has not detected any faults in an emissions related component or system, or that the MIL circuit is not working.

MIL CONDITION: ON STEADY

This condition indicates a fault in an emissions related component or system that could affect the vehicle emission levels.

MIL CONDITION: FLASHING

This condition indicates a misfire or fuel system fault that could damage a converter.

■ **NOTE:***In a misfire condition with the MIL On steady, if the driver reaches a vehicle speed and load condition with the engine misfiring at a level that could cause catalyst damage, the MIL would start flashing. It will continue to flash until engine speed and load conditions caused the level of misfire to subside. Then the MIL would go back to the ON steady condition. This could be an intermittent flashing condition.*

ACTIONS OR CONDITIONS TO TURN OFF THE MIL

The PCM will turn off the MIL if any of these actions or conditions occur:

- The codes are cleared with a Generic or Proprietary Scan Tool
- Power to the PCM is removed (at the battery or with the PCM power fuse)
- A vehicle is driven on **three consecutive trips with a warmup cycle** and meets all code set conditions without the PCM detecting any faults

The PCM will set a code if a fault is detected that could cause tailpipe emissions to exceed 1.5 times the FTP Standard. However, the PCM will not de-activate the MIL until the vehicle has been driven on three consecutive trips with vehicle conditions similar to actual conditions present when the fault was detected. This is not merely three vehicle startups and trips. It means three trips where engine operating conditions are met so that the OBD II Monitor that found the fault can run again and pass the diagnostic test.

OBD II WARMUP CYCLE

Once a MIL is off, the code will remain in memory until 40 warmup cycles are completed without the fault reappearing. A warmup cycle is defined as a trip that includes a change in engine temperature of at least 40°F, and where the engine temperature reaches at least 160°F.

SIMILAR CONDITIONS

If a fuel control system or misfire-related code sets, then the vehicle must be driven under conditions similar to when the fault was detected before the PCM will de-activate the MIL. Similar Conditions are:

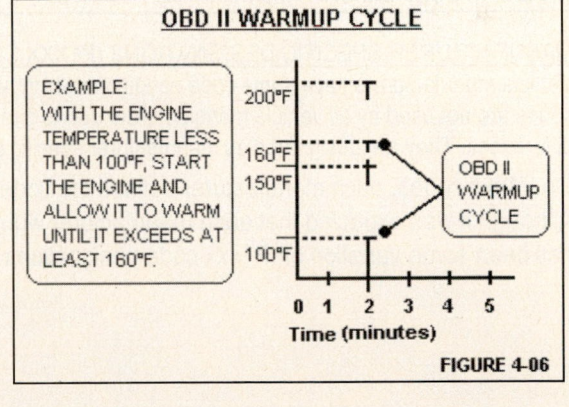

- The vehicle must be driven with engine speed within 375 RPM of the engine speed stored in the Freeze Frame data when the code set
- The vehicle must be driven within engine load ±10% of the engine load value stored in the Freeze Frame data when the code set
- The vehicle must be driven with engine temperature conditions similar to the temperature value stored in Freeze Frame data when the code set

Diagnostic Trouble Codes

<u>INTRODUCTION</u>

One of most helpful features of the OBD II system was an attempt to standardize diagnostic trouble codes. Diagnostic Trouble Code (DTC) is a term used to describe the method applied when the On-Board Computer recognizes and identifies a problem in the computer-monitored systems. On OBD I systems, codes were one, two or three digits displayed as numbers.

<u>SAE J2012 - OBD II DIAGNOSTIC TROUBLE CODES (DTC)</u>

J2012 set standards for trouble codes and definitions for emission-related systems. OBD II codes use a five-digit code and these codes begin with a letter and are followed by four numbers. Since a letter is involved in the numbering, the only way to read codes is with a Scan Tool.

The range of the code designations was designed to allow for future expansion and to allow for manufacturer specific usage for some systems.

Refer to Figure 4-07 for an illustration that shows the OBD II code standardization breakdown.

<u>UNIVERSAL CODE DESIGNATION</u>
<u>EXPLANATION (P0XXX CODES)</u>

The number in the thousandth position indicates if the code is common to all manufacturers (P0xxx codes). This designation, used by most of the vehicle manufacturers, uses a common number

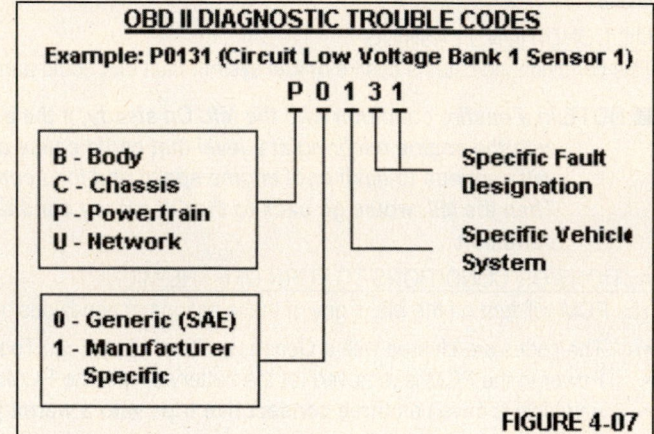

OBD II DIAGNOSTIC TROUBLE CODES

Example: P0131 (Circuit Low Voltage Bank 1 Sensor 1)

P 0 1 3 1

B - Body
C - Chassis
P - Powertrain
U - Network

0 - Generic (SAE)
1 - Manufacturer Specific

Specific Fault Designation

Specific Vehicle System

FIGURE 4-07

and fault message that is assigned to a fault. The fault associated with the code is common to all related vehicle systems and is assigned a universal code designation.

However, code charts are not universal and service procedures vary among different manufacturers. The first letter in the code identifies the system which controls the device that failed (refer to the examples in the table below).

P	Powertrain System	C	Chassis Control System
B	Body Control System	U	UART Data Link, Network Code

<u>MANUFACTURER SPECIFIC DESIGNATIONS (P1XXX CODES)</u>

Vehicle manufacturers have some code conditions which are specific to the design of their individual system. These codes are not used by all vehicle manufacturers due to basic system, diagnostic strategy or implementation differences. They are designated as manufacturer specific codes (P1xxx codes).

On P1xxx codes, each manufacturer defines the code and fault description as their own designation. Although it was expected that each manufacturer would remain consistent across their product line, there has been some variation on P1xxx code designations.

Diagnostic Trouble Codes

DTC NUMBERING EXPLANATION

The number in the hundredth position indicates the specific vehicle system or sub-group that failed. This position should be consistent for P0xxx and P1xxx type codes. The numbers and systems in the list below were established by SAE:

- P0100 - Air Metering and Fuel System fault
- P0200 - Fuel System (fuel injector only) fault
- P0300 - Ignition System or Misfire fault
- P0400 - Emission Control System fault
- P0500 - Idle Speed Control, Vehicle Speed Sensor fault
- P0600 - Computer Output Circuit (relay, solenoid, etc.) fault
- P0700 - Transaxle, Transmission faults

■ **NOTE:** *The tens and ones numbers indicate the part of the system at fault.*

DTC CODE NUMBERING EXAMPLE

DTC P1319 - Acura/Honda Spark Plug Voltage Detection Module Reset Fault:

P - 1st position indicates a Powertrain related code

1 - 2nd (thousandth) position indicates manufacturer specific code

2 - 3rd (hundredth) position indicates ignition system or misfire fault

1 - 4th and 5th position indicates a Spark Plug Voltage Detection Module Reset Fault

SAE DTC CATEGORIES

The two general categories of diagnostic trouble codes are:

- Type A codes - emissions related (PCM activates MIL on the first failure)
- Type B codes - emissions related (PCM activates MIL after the second consecutive trip with a failure)

Diagnostics can be assigned an 'A' or 'B' designation. However, most of the vehicle manufacturer expanded diagnostics seem to be Type 'B' codes that require a minimum of two trips with a fault in order to activate the MIL.

Vehicle manufacturers can add additional DTC categories. For example, some vehicle manufacturers use a 'C' and 'D" category for some vehicles.

COMMON CODE NAMES & DESCRIPTIONS

OBD II guidelines set standards to universalize the Code Name and Description. These universal standards only apply to P0xxx type codes.

■ **NOTE:** *Technicians should be careful because there are many OBD II trouble codes where the same code number will have a different code title.*

Diagnostic Trouble Codes

CONDITIONS TO CLEAR DIAGNOSTIC TROUBLE CODES

Diagnostic trouble codes can be cleared using several different methods (the actual methods vary with different manufacturers). The list below contains a summary of several ways that code can be cleared. The actual conditions for each of the vehicle manufacturers must be determined and followed.

- Regulations allow codes to be cleared by the Diagnostic Program if 40 warm up cycles have occurred since the "last test failed" message cleared and 40 "last test passed" messages occurred. The definition of a warmup cycle varies slightly between vehicle manufacturers.
- The Scan Tool can be used to clear OBD II trouble codes
- On some vehicle systems, when battery power to the PCM is lost, all DTC, Freeze Frame data, Trip, Drive Cycle status and I/M Readiness status information can be lost. Usually the PCM has to be disconnected from power for a period of 5 minutes or longer for this to occur.

■ NOTE:*Do not clear codes unless the diagnostic procedure instructs the technician to do so. Most manufacturers will clear Freeze Frame data that can be used to diagnose the cause of the fault at the same time that codes are cleared (i.e., when codes are cleared, Freeze Frame data is lost).*

Diagnostic Routines

GENERIC OBD I DIAGNOSTIC ROUTINE

The Diagnostic Routine shown in Figure 4-08 can be used to diagnose the cause of Powertrain Control Module faults related to OBD I trouble codes, driveability symptoms or intermittent fault conditions. Follow all steps listed in the chart from top to bottom.

The most important change related to OBD II for repair technicians is that they will have to learn new systems and diagnostic routines. Although the OBD II system is similar to the OBD I system, it has more in-depth control of emissions as well as sensor tests and "monitors" that run thorough emission control system related diagnostics.

While it is true that the OBD II computer diagnostic system can provide the technician with more information, they will still have to quickly and accurately interpret all available information.

OBD II system diagnosis requires skills in reading and interpreting data!

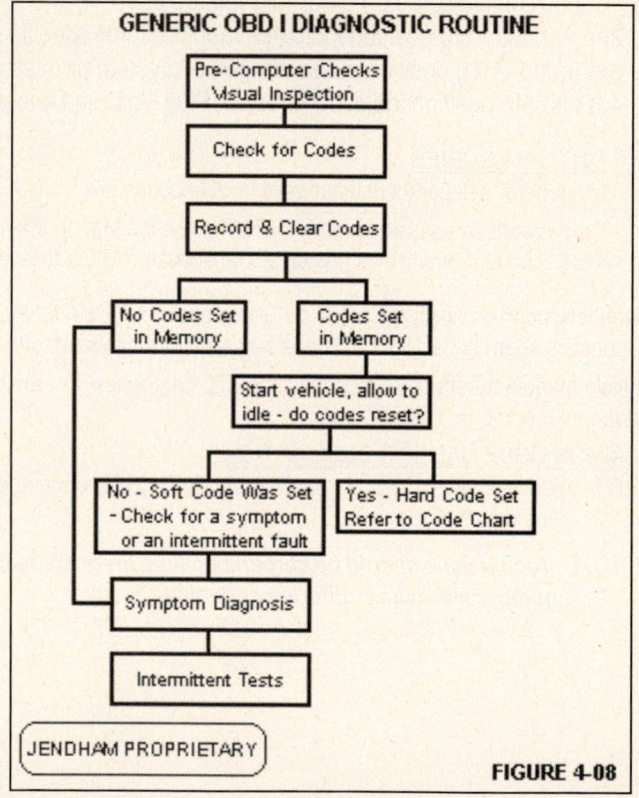

GENERIC OBD I DIAGNOSTIC ROUTINE

- Pre-Computer Checks Visual Inspection
- Check for Codes
- Record & Clear Codes
  - No Codes Set in Memory
  - Codes Set in Memory
    - Start vehicle, allow to idle - do codes reset?
      - No - Soft Code Was Set - Check for a symptom or an intermittent fault
      - Yes - Hard Code Set Refer to Code Chart
- Symptom Diagnosis
- Intermittent Tests

JENDHAM PROPRIETARY

FIGURE 4-08

Diagnostic Routine
GENERIC OBD II DIAGNOSTIC ROUTINE

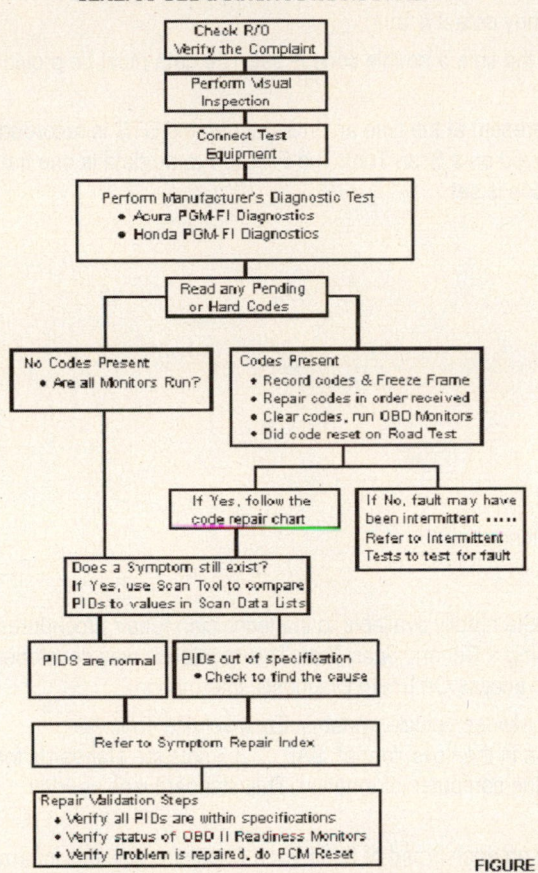

GENERIC OBD II DIAGNOSTIC ROUTINE

FIGURE 4-09

The diagnostic approach used in OBD II systems is more complex than the one used in OBD I systems and can effect how repair technicians approach vehicle diagnostics. The diagnostic routine in Figure 4-09 can be used to diagnose trouble codes, driveability symptoms or intermittent fault conditions on Acura/Honda vehicles with OBD II systems.

KEY POINTS TO CONSIDER
Read the R/O, then verify complaint as written (important for symptoms)

- Do a Visual Inspection, then connect an OBD II Certified Scan Tool
- Turn the key on and read and record all OBD II Pending or Hard Codes
- Record Freeze Frame data and repair the codes in the order received
- If no codes are set, use Scan Tool to check for a No-Code Faults
- Use a Scan Tool to do a PCM Reset and to check the I/M Readiness Status

Freeze Frame Data

INTRODUCTION

CARB and EPA regulations require that the On Board computer store specific Freeze Frame data when the first emission-related fault is detected in the vehicle. Current Freeze Frame data can only be replaced by Freeze Frame data from the Fuel system or Misfire Detection Monitor (if they detect a fault).

Freeze Frame data must contain the data values stored at the time a trouble code is set. The data must be provided in standard units of measurements.

As a result, OBD II systems record in Freeze Frame data present at the time an emission-related DTC is recorded and the MIL is activated. This data can be accessed and displayed on a Scan Tool. The Freeze Frame data is one frame or one instant in time. It records the data present when the code is set.

FREEZE FRAME DATA ITEMS
- Calculated Load Value
- Engine Speed (RPM)
- Short Term and Long Term Fuel Trim Percent
- Fuel System Pressure (on some vehicles)
- Vehicle Speed (MPH)
- Engine Coolant Temperature
- Intake Manifold Pressure
- Closed/Open Loop Status
- Fault Code that triggered the Freeze Frame
- If a Misfire Code is set - identify which cylinder is misfiring

SCAN TOOL GENERAL INFORMATION

CARB regulations require that all vehicle manufacturers make readily available to the technician repair procedures that allow for effective emission-related diagnosis and repair using a Generic Scan Tool. This regulation was developed into SAE J1978 that sets guidelines for a common Scan Tool to access On Board Diagnostic information.

The type of information available on a Scan Tool will vary between vehicle systems. Each service engineer emphasized certain programs and displayed the information in their own format. SAE J2205 sets the standards for how the Scan Tool will interface with the computer and access the computer information. This standard was needed because computers interface in bi-directional formats.

There is a difference in the amount of information available on each brand of Scan Tool. Check for how this information is accessed and displayed on a Scan Tool that is intended for use on OBD II systems.

The Scan Tool gets its power from the vehicle being tested and talks or "interfaces" with the vehicle diagnostic system or diagnostic monitor program.

The following is a list of information that is available on an OBD II Certified Scan Tool (check each individual tool to determine what information is actually available).

CODE INFORMATION
- Current, Pending and History Codes
- Information on whether the MIL has been requested ON (activated)
- Diagnostic test status or I/M Readiness Status (Test Run, Pass or Fail)
- Last test Pass or Fail Message
- Freeze Frame data for the first emission fault condition
- Some manufacturers offer additional "failure records" (up to 5 more records)

The Computer and Driveability Problems

INTRODUCTION

A computer is a collection of components that records and compares values, evaluates data, issues commands, stores information and tests systems. It contains a central processing unit (CPU) that has three major functions:

- Receives input information from sensors
- Processes and evaluates information
- Transmits output command that control components

DRIVEABILITY PROBLEMS

Since driveability problems are difficult for many technicians to diagnose, it is important that they understand the interaction of computer inputs and outputs, devices and signals. The computer contains internal programs that are very specific in what they cover.

As with OBD I systems, vehicles with OBD II systems can have problems that involve symptoms. The computer does not "know" everything. Many engine areas can cause driveability problems without causing a code to set. These engine areas or components must be tested to verify that they are working properly.

Due to improvements in OBD II diagnostics, problems may be detected that would not be detected in an OBD I system.

However, the technician should still check these signals or devices:

- Cam timing to see if it is retarded or advanced
- Canister purge system to see if it is loaded with fuel
- Computer feed and ground circuits for connection or resistance problems
- Cylinder compression: look for readings that are too high, low, or uneven
- EGR valve to determine if it is leaking, sticking, or carbon clogged
- PCM electrical terminals for problems
- Engine for (high) oil consumption, leaking rings, etc.
- Exhaust system for restrictions (is it plugged) and for leaks
- Fuel Injectors clogged, pintle sticking, the wrong part installed
- Fuel system pressure problems
- Ignition Cam or Crank sensor clearance (gap) problems
- PCV system for wrong valve, leaking or damaged valve, system faults
- Secondary Ignition system for low output or leakage problems
- Throttle body min-air-rate adjustment, throttle blade warped or damaged
- Turbocharger component and switching valve for problems
- Engine vacuum problems

Serial Data

Introduction

Serial data is a series of voltage signals pulsed from high voltage to low voltage. These voltage signals can change from a low value (0v) to a higher value (5, 7 or 12 volts). The signals are transmitted through a Serial Data line (Figure 4-10).

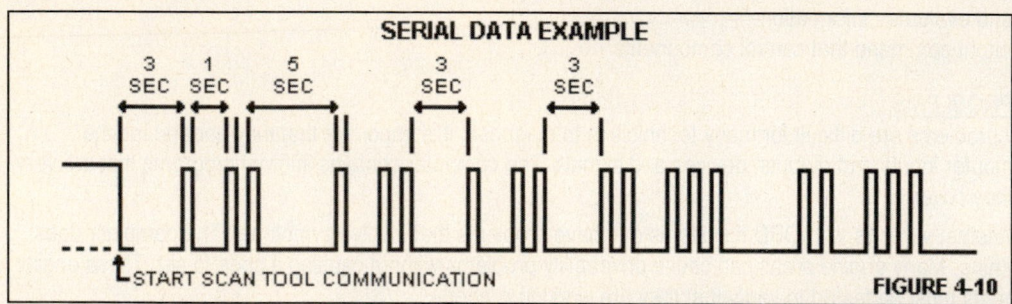

SERIAL DATA EXAMPLE

START SCAN TOOL COMMUNICATION

FIGURE 4-10

Each digital signal is called a "bit" and represents a number. The computer strings together a series of these "bits" in eight-bit segments (Figure 4-11). Eight "bits" or segments make up a word used by the computer. The computer reads these words (in the order they are transmitted) as a series of data entries. This is where the name serial data comes from. A typical serial data stream may have 20 words on it.

Scan Tools read serial data by interfacing with the computer through the Data Link Connector. The Scan Tool contains a CPU that reads the PCM serial data.

BAUD RATE

A serial data stream contains segments of on and off signals that are read as words. The segments represent a message being sent or the segments could be a separation signature between various

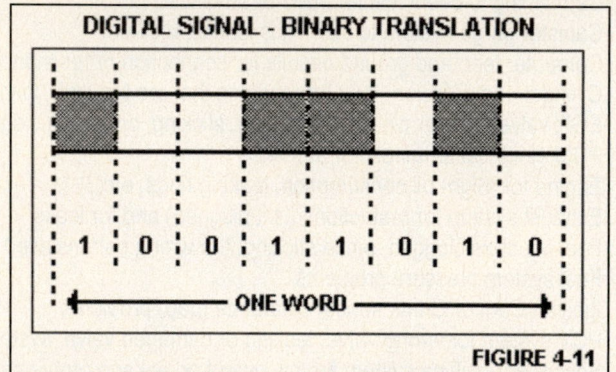

DIGITAL SIGNAL - BINARY TRANSLATION

1 0 0 1 1 0 1 0

ONE WORD

FIGURE 4-11

messages going to and from the computer. These bits are sent at precise time intervals. The speed at which bits are transmitted is known as the computer "Baud Rate."

For example, a PCM baud rate of 8192 indicates that the computer transmitted 8,192 bits of data per second. The baud rates of computers has changed rapidly (e.g., a late model GM computer may have a baud rate of 8192, and this is a much faster rate than that the baud rate of 160 bits per second of some early GM computers).

■ NOTE: *The baud rate speed is important for snapshot features of a Scan Tool.*

UNIDIRECTIONAL AND BI-DIRECTIONAL MICROPROCESSORS

Computers can transmit serial data in one direction only or can receive and transmit data from a Scan Tool (if the data goes two directions, it is referred to as Bi-directional). Chrysler Scan Tool programs have allowed the tool to turn on and turn off solenoids and relays since 1984 and this is a send and receive Bi-directional program. Each vehicle manufacturer decides how much Bi-directional interaction they will allow through a particular Scan Tool.

OBD II Expanded Testing

INTRODUCTION

OBD II regulations focus primarily on establishing requirements that verify that Emission systems are functioning within appropriate standards. All emission faults must set diagnostic codes. Some emissions related faults result in vehicle driveability symptoms while others will not impact vehicle operation. The regulations specify the time and conditions when certain systems have to be monitored and tested. The intent of these diagnostic standards is to validate that the sensors and systems are operating properly and to determine that the systems have not deteriorated beyond a point where tailpipe emissions would exceed 1.5 times the Federal Standard.

On some OBD II systems, vehicle manufacturers may use OBD I type diagnostics to monitor emission-related circuits for electrical faults (i.e., open, shorted and grounded circuits). However, OBD I regulations included exemptions that allowed certain faults in computer controlled devices without having to activate the Check Engine Light and set a trouble code. In effect, the manufacturer could get an exemption. In these situations, an assumption was made that the vehicle owner would bring the vehicle in for repairs due to a driveability symptom and that it was not necessary to activate the light.

OBD II regulations removed this exemption. All emissions faults must set diagnostic codes regardless of the driveability of the vehicle. The regulations require all emissions related systems have to be monitored and their operation validated. To achieve this goal, vehicle manufacturers established tests and "routines" that are run by engine computer. This enhancement has resulted in the development of an entirely new set of diagnostic procedures called monitor run tests.

EXPANDED DIAGNOSTIC TESTING AND MONITORING

On OBD I systems, the engine computer is designed to inform the technician that a component has a general failure. However, long before a component fails, it begins to degrade. The (general) sensor degrading can have an effect on tailpipe emissions and ultimately lead to expensive emissions repairs. The Clean Air Act of 1990 mandated that the vehicle manufacturers monitor the Emission Control systems for proper operation. There are a total of 17 different systems monitored by the on-board computer for faults.

This new type of emission system diagnostic requires that the engine computer continually monitor all of the Emission Control systems in order to determine that all emission control equipment is functioning properly.

Beginning with model year 1996, both the Federal EPA and CARB require all OBD II Monitors must be identified, and that the results of their most recent diagnostic tests be recorded. These test results must be made available to a technician and the information must include the test limits used in comparing the results. This means that the actual test results and the normal limits or parameters must also be available for specific OBD II Monitor run tests. Also, a Generic Scan Tool (GST) could use a standard message content and downloading protocol in order to read and report any faults detected by this system including Freeze Frame data and specific data values.

EMISSION SYSTEMS TESTING

To achieve the goals of the OBD II system (including enhanced tests of emission-related systems), the manufacturers developed specific monitoring programs that are advanced beyond the former OBD I system diagnostics. These programs include expanded Oxygen Sensor, Catalytic Converter, EGR, EVAP system (flow monitoring and diagnosis), Fuel System (fuel trim), Misfire Detection and Secondary AIR Injection function and flow.

DTC Test Conditions

<u>INTRODUCTION</u>

The Comprehensive Component Monitor and Main Monitors are designed to test vehicle emission control components and systems. Some of these monitors run their diagnostic tests only after specific *enable criteria* are met. These individual test conditions must be met in order to run a diagnostic test. Please note that the conditions vary by vehicle and engine configuration. The information related to why a code is set includes *enable criteria* for a test and the actual parameters present when the code is set.

The test conditions in this manual provide a technician with the vehicle manufacturer code criteria used to run a test and to set a code. This allows technicians to duplicate these conditions in order to verify a fault has been repair properly. If the code resets during the next few trips, a hard code is still present. If a pending or history code is stored (the MIL is off), and a hard code does not reset while using the *enable criteria*, the pending code may be due to an intermittent fault or the *enable criteria* may not have been followed correctly.

CODE ID	CODE DESCRIPTION & TEST CONDITIONS
P0131 - Front (1-T)	**HO2S-11 Circuit Low Voltage (Bank 1 Sensor 1)**
	Test Conditions & Results: Engine warmup finished, vehicle running in closed loop with A/T in D2 position (M/T: in 4th gear), **then the test started** and the PCM detected an HO2S-11 voltage fixed below a stored limit.

■ **NOTE:** *It is possible to have one problem set multiple codes. If multiple codes are set, look at the conditions to set the code and find out which conditions are common. Next determine which devices and conditions are common to the codes that set. These clues and steps can help to speed up a diagnosis.*

<u>OBD II DRIVE CYCLE</u>

OBD II regulations also established a vehicle "drive cycle" pattern that allows the CCM and Main Monitors to run and complete their individual tests. The OBD II Monitors that should run during the drive cycle include the CCM, EGR, EVAP, Fuel System, Misfire, Oxygen Sensor and Secondary AIR System. One manufacturer (Ford) has a special code (DTC P1000) that sets if all the Main Monitors have not been run to completion.

<u>OBD II TRIP</u>

Each OBD II Monitor has a specific "trip" or drive pattern that needs to be met before a particular Monitor will operate. This trip may be as simple as turning the key on or as involved as an engine warmup period, followed by a cruise period at highway speeds.

<u>COMPONENTS OR SYSTEMS CONTINUOUSLY MONITORED</u>
- CCM - PCM Input or output devices (both directly or indirectly controlled)
- Misfire Detection - Identify misfires, severity of the misfire, engine cylinder
- Fuel System - Monitor PCM control of the fuel control system (fuel trim)

<u>COMPONENTS OR SYSTEMS MONITORED ONCE PER TRIP</u>
- Catalyst Efficiency
- Oxygen Sensor for activity, response time, and heater operation
- EGR System flow rate, component integrity
- EVAP System flow and leak detection
- Secondary Air System airflow, component integrity

OBD II Diagnostics

Comprehensive Component Monitor

INTRODUCTION

The Comprehensive Component Monitor (CCM) is an on-board strategy designed to monitor a failure in electronic components and circuits (including emission-related and non-emission-related circuits) that provide input or output signals to the PCM. These are systems or devices that are not exclusively monitored by another monitor system. The PCM considers that an input or output signal is inoperative when a failure exists due to an open circuit, out-of-range value or if an on-board rationality or functionality check fails. If an emission-related fault is detected, the PCM will set a code and activate the MIL (requires two consecutive trips).

Tests conducted by the CCM vary depending on the type of hardware, the function of the device and the signal type. Analog signals are checked continuously for opens, shorts and out-of-range values. Some digital signals are checked for both functionality and rationality. These tests require that certain engine conditions be present before the test is performed and that several components be monitored as part of the test. Also, a sensor value can be monitored for change _after_ the PCM sends a command to a device.

Here is a partial list of devices checked by the CCM at key on or with the engine running.

INPUT DEVICE EXAMPLES
- Barometric Pressure Sensor
- Brake Switch
- Camshaft & Crankshaft Sensors
- M/T Clutch Switch
- Cruise Servo Switch
- Engine Coolant Temperature Sensor
- EVAP Purge Sensor
- Fuel Pressure Sensor
- Intake Air Temperature Sensor
- Knock Sensor
- Manifold Absolute Pressure Sensor
- Park Neutral Switch
- Transmission Temperature Sensor
- Transmission Turbine Speed Sensor
- Vehicle Speed Sensor

OUTPUT DEVICE EXAMPLES
- EVAP Canister Purge
- EVAP Purge Vent Solenoid
- Idle Air Control Solenoid
- Ignition Control System
- Transmission Torque Converter Clutch Solenoid
- Transmission Shift Solenoids (Solenoid 'A' or Solenoid 'B')

Comprehensive Component Monitor

CCM AND MIL ACTIVATION

Many PCM sensors and output devices are tested at key on or immediately after engine startup (Figure 4-12). However, some devices (e.g., IAC) are only tested by the CCM after the engine meets certain engine conditions.

The number of times the CCM must detect a fault before it will activate the MIL depends upon the manufacturer. Some faults require that an emission-related circuit fail for two consecutive trips before the PCM set a hard code and activate the MIL.

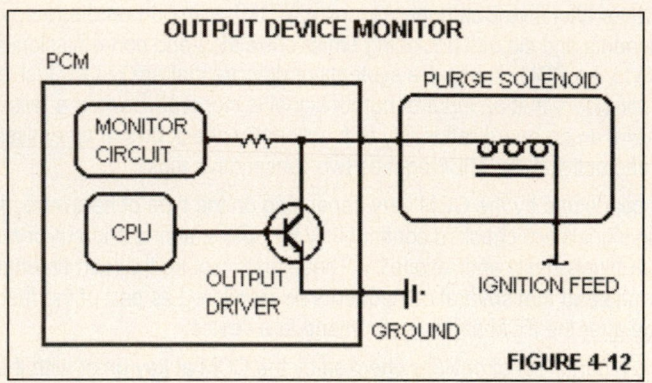

FIGURE 4-12

COMPREHENSIVE COMPONENT MONITOR EXAMPLE

On Acura and Honda vehicles, the CCM can detect a fault and activate the MIL in one or two trips (actual number depends on the device that is tested). In the code P0107 example below, the **(1-T)** indicates that this fault must be detected only on one trip.

CCM Code Description Example

CODE ID	CODE DESCRIPTION & TEST CONDITIONS
P0107 - (1-T)	**MAP Sensor Circuit Low Voltage** *Test Conditions & results:* Key on, then test started and the PCM detected a MAP sensor value of near 0" Hg.

LIST OF DEVICES CHECKED BY THE COMPREHENSIVE COMPONENT MONITOR

The devices tested by Comprehensive Component Monitor are shown below.

Comprehensive Component Monitor

Devices that are checked continuously:

- BARO Sensor - circuit tests
- ECT and IAT Sensor - circuit tests
- Oxygen Sensor - voltage level, activity
- CYP, TDC Sensors - circuit tests, for no signal
- CKP Sensors - circuit tests, for no signal
- EGR, EVAP Solenoids - circuit tests
- Idle Speed Control Motor - circuit tests
- Fuel Injectors - circuit tests
- MAP Sensor - circuit tests, rationality tests
- Oxygen Sensor Heater circuit (shortly after startup)
- TP Sensor - circuit tests, rationality
- Vehicle Speed Sensor - circuit tests
- Some PCM Switches - circuit tests

■ NOTE: *The amount of trips required by the CCM and Main Monitors before a code is set and the MIL is activated <u>varies</u> between vehicle manufacturers.*

Main Monitors

INTRODUCTION

A key difference between first generation On Board Diagnostics (OBD I) and the second generation (OBD II) is the use of several PCM controlled diagnostic monitors contained within the PCM software structure. These monitors are needed to meet OBD II CARB and Federal EPA regulations.

Simply stated, an OBD II Monitor is a diagnostic strategy designed to test the operation of an emissions-related system or component. Some of the OBD II Monitors accomplished this task *directly* by monitoring the action of various input and output devices or sensors connected to the PCM. An example of *direct* monitoring is when the Comprehensive Component Monitor tests the Engine Coolant Temperature or Intake Air Temperature Sensor signals.

Other OBD II Monitors accomplish the task *indirectly* by monitoring the effects of changes to a system or component. The *indirect* method may be accomplished through watching the change or response of a system. This type of test is done by monitoring the input or output signals of a particular device for an "inferred" change.

An example of *indirect* monitoring is when the PCM infers correct or incorrect catalyst action using the Catalyst Monitor to sample signals from the upstream or downstream oxygen sensors. This allows the PCM to determine the oxygen storage efficiency of the converter catalyst.

LIST OF DEVICES CHECKED BY THE CCM AND MAIN MONITORS

The devices or systems tested by Main Monitors are shown in the table below.

Main Monitors
Main Monitors that run continuously:
• Fuel Control System that begins with engine in closed loop
• Misfire Detection test that begins right after startup
Main Monitors run once only per trip:
• Catalyst Efficiency test in closed loop after certain engine temperature, time and VSS requirements are met
• EGR System in closed loop after temperature, time and VSS e requirement is met)
• EVAP System test in closed loop after certain engine temperature, time and VSS requirements are met
• Oxygen Sensor - voltage and response time tests in closed loop after certain engine temperature, time and VSS requirements are met
• Secondary AIR System (in closed loop at off-idle)

MAIN MONITOR EXAMPLE

On Acura and Honda vehicles, the Catalyst Monitor does not set a hard code and activate the MIL until a catalyst fault is detected on three consecutive trips. In the code P0420 example below, the (3-T) indicates that this fault must be detected three times.

Main Monitor Code Description Example

CODE ID	CODE DESCRIPTION & TEST CONDITIONS
P0420 - (3-T)	**Catalyst Efficiency Below Threshold (Bank 1)** *Test Conditions & results:* P0137, P0138, P0141 not set, running in closed loop, VSS at 40-55 mph for 2 minutes, then decelerate to 35 mph with throttle closed, then test started and PCM detected excessive Bank 1 HO2S-12 activity for set time period.

Catalyst Monitor

INTRODUCTION

On OBD II systems, a downstream or post-catalyst Heated Oxygen Sensor (HO2S-2) is used to provide the additional signals needed to monitor the efficiency of the three-way catalyst. On these systems, the PCM compares the signals between the upstream or pre-catalyst and the downstream post-catalyst oxygen sensor *during stable driving conditions* in order to determine the oxygen storage capacity of the catalytic converter.

Catalyst Monitor Operation

To measure catalyst efficiency, the Catalyst Monitor interprets the signals from the pre-catalyst and post-catalyst oxygen sensors. If the three-way catalyst is operating correctly, the post-catalyst signal will have significantly less activity than the pre-catalyst. The activity level from the oxygen sensors in a normal catalyst is shown in Figure 4-13.

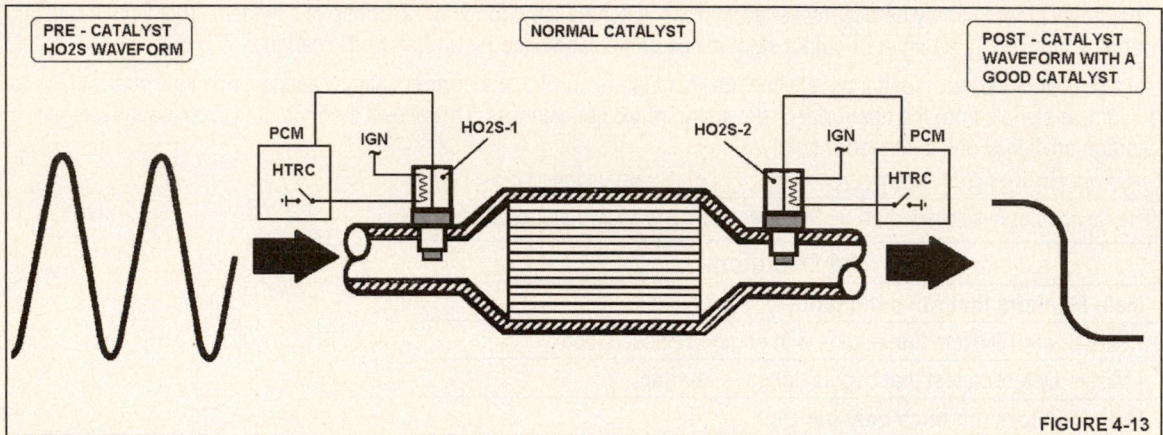

FIGURE 4-13

Once Catalyst Monitor test conditions are met, the diagnostic test begins to monitor the activity level of the heated oxygen sensors. If the post-catalyst activity approaches the activity of the pre-catalyst, the test will fail since the catalyst has degraded (Figure 4-14). If this occurs, the PCM will detect a fault, and set a Catalyst Monitor trouble code.

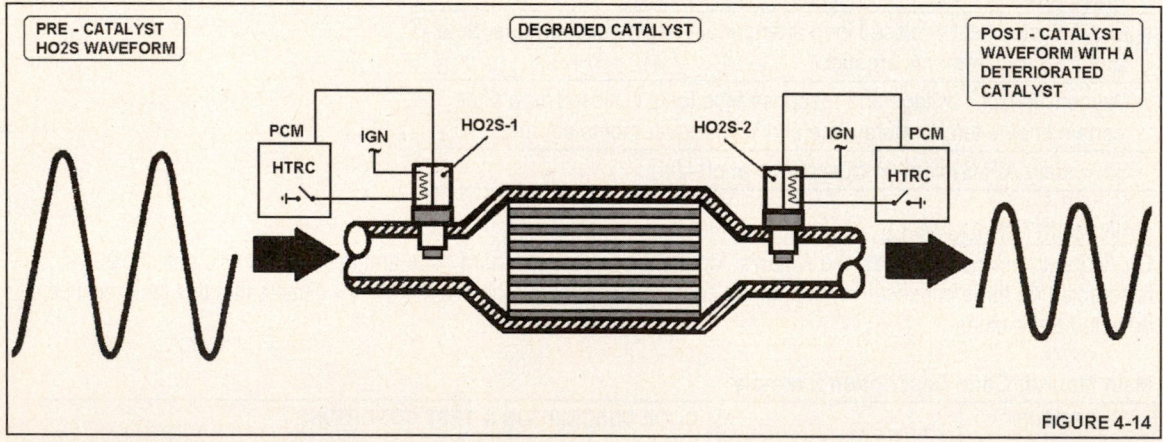

FIGURE 4-14

There are several conditions that can cause the activity level in the rear HO2S to approach the activity level of the front HO2S (Figure 4-14). These conditions include:

- Exhaust leaks present upstream of the rear or post-catalyst
- The rear HO2S-2 contaminated from lead, phosphorus, silica or sulfur
- The use of certain alternate fuels

Catalyst Monitor

COMPONENT EXAMPLES

The three-way catalytic converter (TWC) and warmup three-way catalytic converter (WU-TWC) are used to convert hydrocarbons (HC), carbon monoxide (CO) and oxides of nitrogen (NOx in the exhaust gas into carbon dioxide (CO_2), dinitrogen (N_2) and harmless water vapor (Figure 4-15).

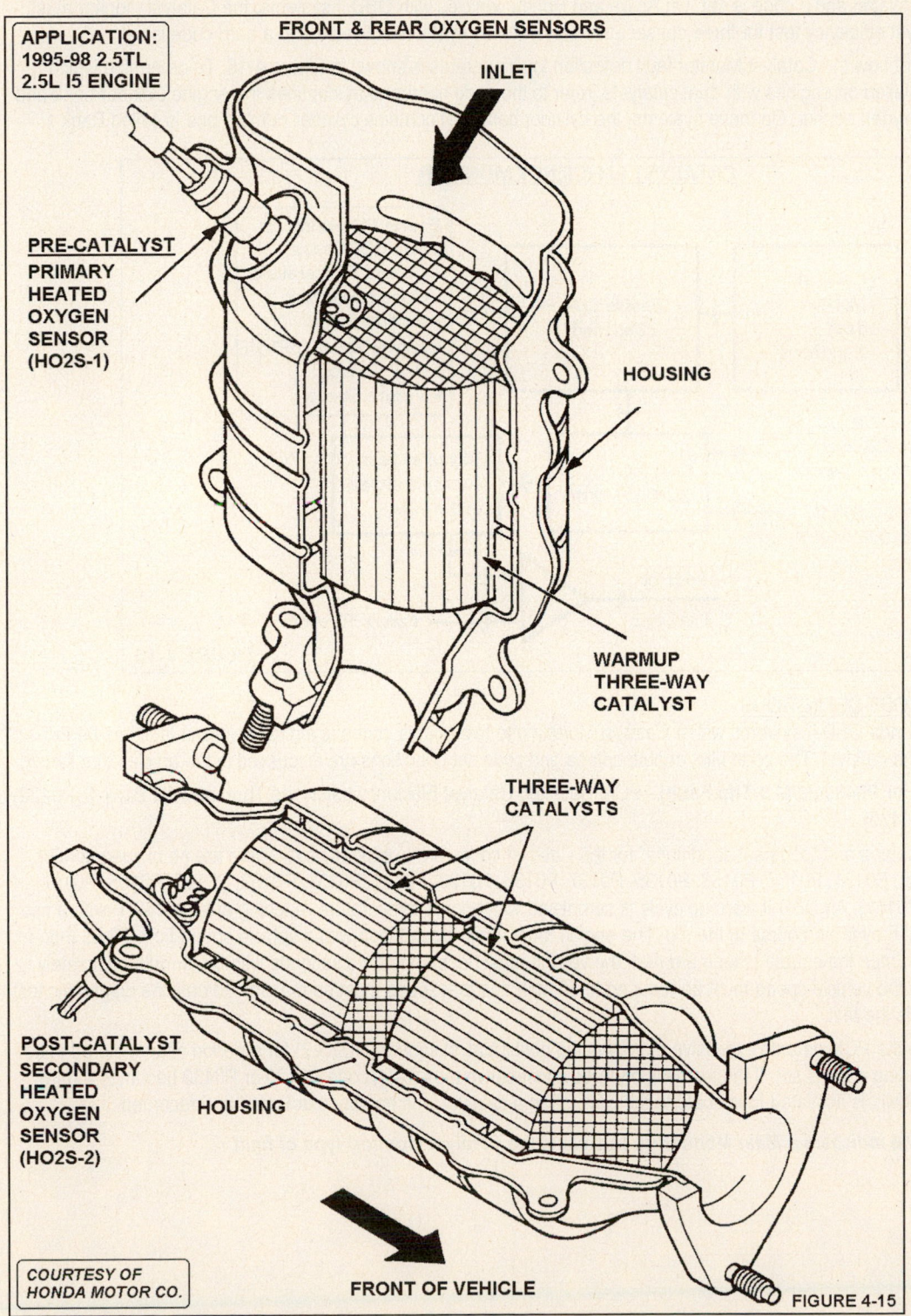

APPLICATION:
1995-98 2.5TL
2.5L I5 ENGINE

FRONT & REAR OXYGEN SENSORS

INLET

PRE-CATALYST
PRIMARY
HEATED
OXYGEN
SENSOR
(HO2S-1)

HOUSING

WARMUP
THREE-WAY
CATALYST

THREE-WAY
CATALYSTS

POST-CATALYST
SECONDARY
HEATED
OXYGEN
SENSOR
(HO2S-2)

HOUSING

COURTESY OF
HONDA MOTOR CO.

FRONT OF VEHICLE

FIGURE 4-15

Catalyst Monitor

CATALYST MONITOR FAULT DETECTION

If a three-way catalyst fault is detected that could cause tailpipe emissions to exceed 1.5 times the Federal Standard, the MIL is activated and a code is set. On Acura and Honda vehicles with OBD II systems, the Catalyst monitor must fail the catalyst efficiency test for three consecutive trips before the MIL is activated and a hard code is set.

An example of how the Catalyst Monitor fault detection logic operates is shown in Figure 4-16. To determine which catalyst that failed on engines with dual catalysts, refer to the code number that identifies the engine cylinder bank and the related oxygen sensor. On these systems, the cylinder bank that contains cylinder number one is called Bank 1.

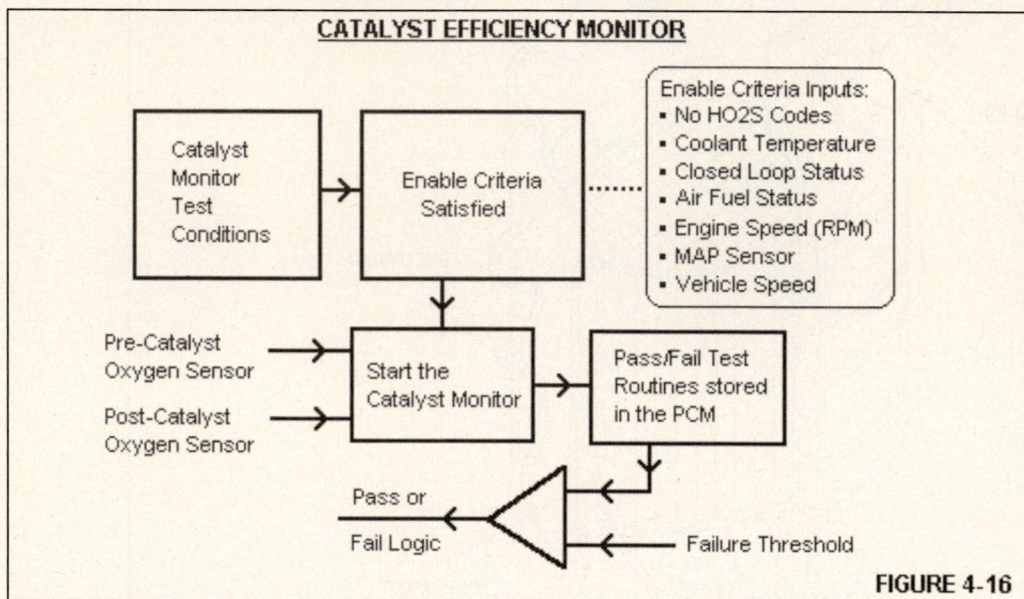

CATALYST EFFICIENCY MONITOR

Enable Criteria Inputs:
- No HO2S Codes
- Coolant Temperature
- Closed Loop Status
- Air Fuel Status
- Engine Speed (RPM)
- MAP Sensor
- Vehicle Speed

Catalyst Monitor Test Conditions → Enable Criteria Satisfied

Pre-Catalyst Oxygen Sensor → Start the Catalyst Monitor → Pass/Fail Test Routines stored in the PCM

Pass or Fail Logic ← Failure Threshold

FIGURE 4-16

TROUBLE CODE DEFINITIONS

Acura and Honda OBD II systems with a Catalyst Monitor use two trouble codes to indicate when a fault has been detected in the catalyst. The code title, enable criteria and code set conditions are discussed in the articles that follow.

Code P0420 or P0430 (M/M 3-Trip Fault) - PCM Detected Catalyst Efficiency below the Threshold for Bank 1 (P0420) or Bank 2 (P0430).

The enable criteria and "code set conditions" for the Catalyst codes are explained next. There are no oxygen sensor codes set (i.e., P0131, P0132, P0133, P0135, P0137, P0138, P0139, P0141, P0151, P0152, P0153, P0155, P0157, P0159 and P0161). An OBD II warmup cycle is completed with engine temperature reaching at least 160°F with a rise of at least 40°F over the course of the trip. The engine must run in closed loop with a vehicle speed from 40-55 mph for two minutes. Once the enable criteria are met, the vehicle must decelerate for 3 seconds with the throttle completely closed. Then the vehicle speed must be reduced to 35 mph and that speed must be maintained until the diagnostic test passes or fails the test.

The first time the PCM detects excessive activity in the post-catalyst sensor (HO2S-2) for a period of time stored in its tables, a pending code is set. If the test fails on three consecutive trips (3-T), code P0420 or P0430 becomes a hard code and the MIL is activated to indicate to the driver than a catalyst fault has been detected and recorded.

■ **NOTE:** *M/M indicates a Main Monitor (in this case, the Catalyst Monitor) type of fault.*

EGR System Monitor

INTRODUCTION

The EGR system is designed to reduce oxides of Nitrogen (NOx) emissions by recirculating exhaust gas through the EGR valve to intake manifold and to the engine.

An EGR valve lift sensor in this system detects the amount of EGR valve lift and sends a signal to the PCM. The PCM contains tables of "ideal" EGR valve lift values for different engine conditions. During certain engine conditions, the PCM compares the sensor "actual" value with an "ideal" value that is determined by signals sent from the other sensors to the PCM. If there is any difference between the two values, the PCM stops current to the EGR control solenoid to reduce vacuum applied to the EGR valve and bring the sensor into the correct operating range.

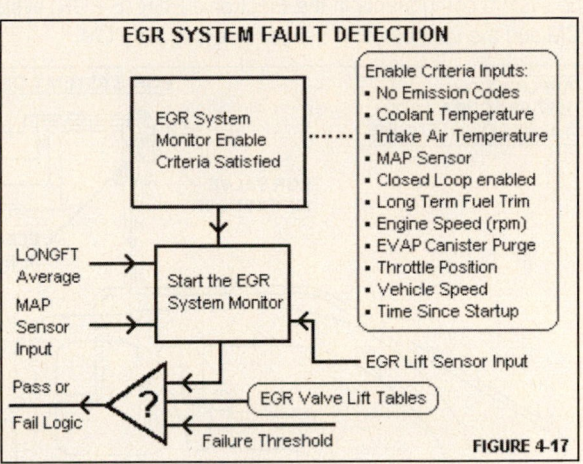

FIGURE 4-17

ELECTRONIC EGR SYSTEM

Certain Acura and Honda models use a new Electronic EGR (E-EGR) system. In this system, the EGR valve is controlled by a duty cycle signal from the PCM to maintain the correct amount of flow through the EGR valve during all engine conditions.

SYSTEM COMPONENTS - 1996 ACCORD 2.2L I4 (F22B1) ENGINE

System components include the EGR valve, EGR valve lift sensor, EGR control solenoid, the EGR vacuum control valve and the various other sensor inputs to the PCM.

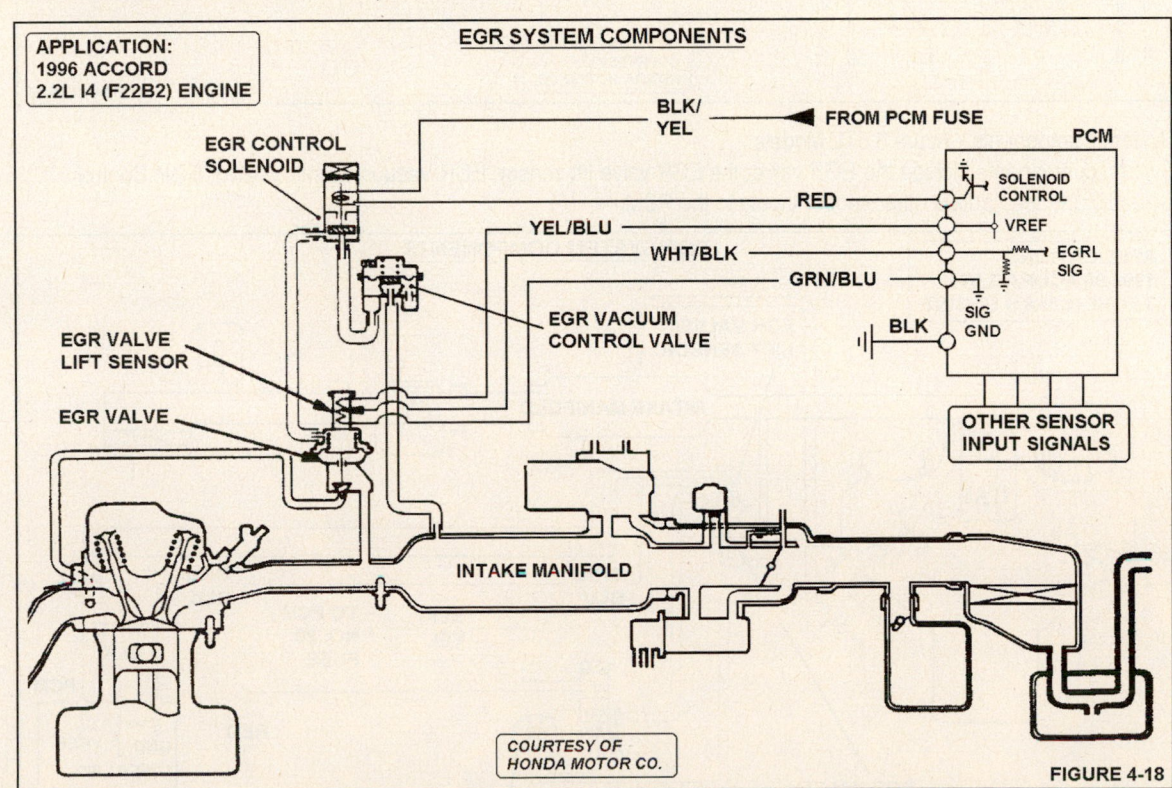

FIGURE 4-18

EGR System Monitor

SYSTEM COMPONENTS - ACURA 3.0CL MODELS

The system components in the Electronic EGR (E-EGR) system include the EGR valve, the EGR valve lift sensor, the PCM and the various other sensor inputs to the PCM.

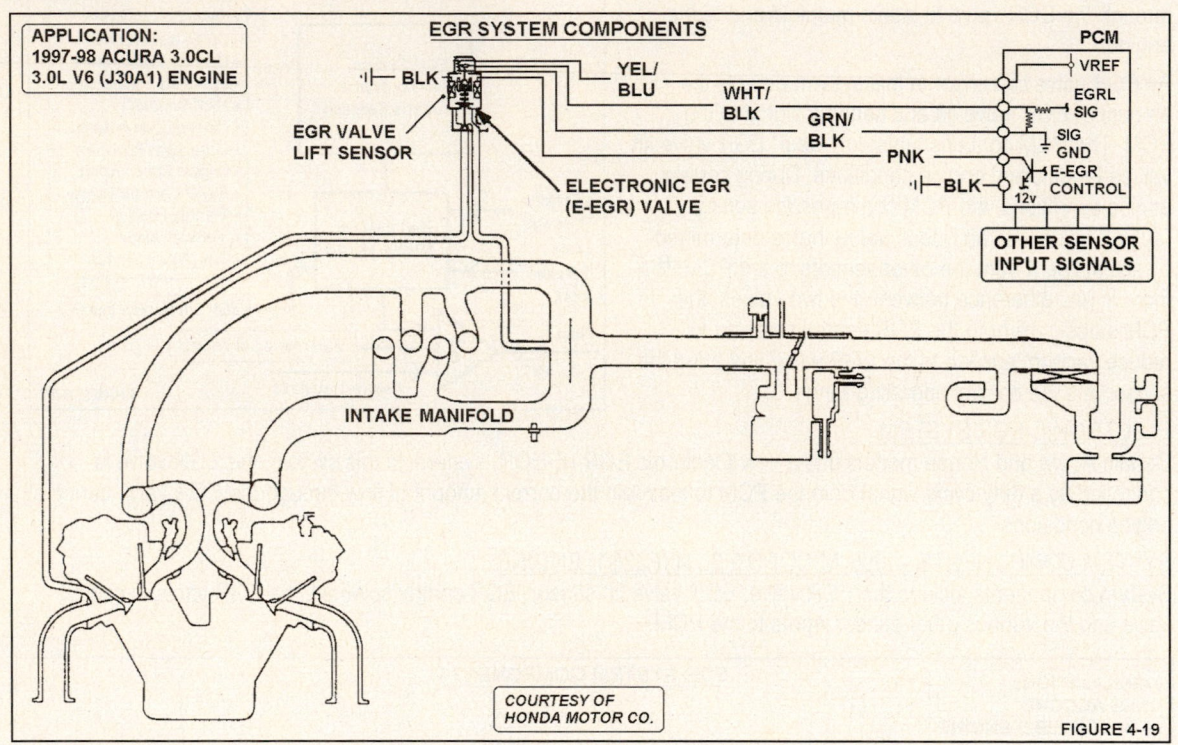

FIGURE 4-19

System Components - Acura 3.5TL Models

System components include the EGR valve, the EGR valve lift sensor, EGR Vacuum Control Valve, EGR Control Solenoid and the various other sensor inputs to the PCM.

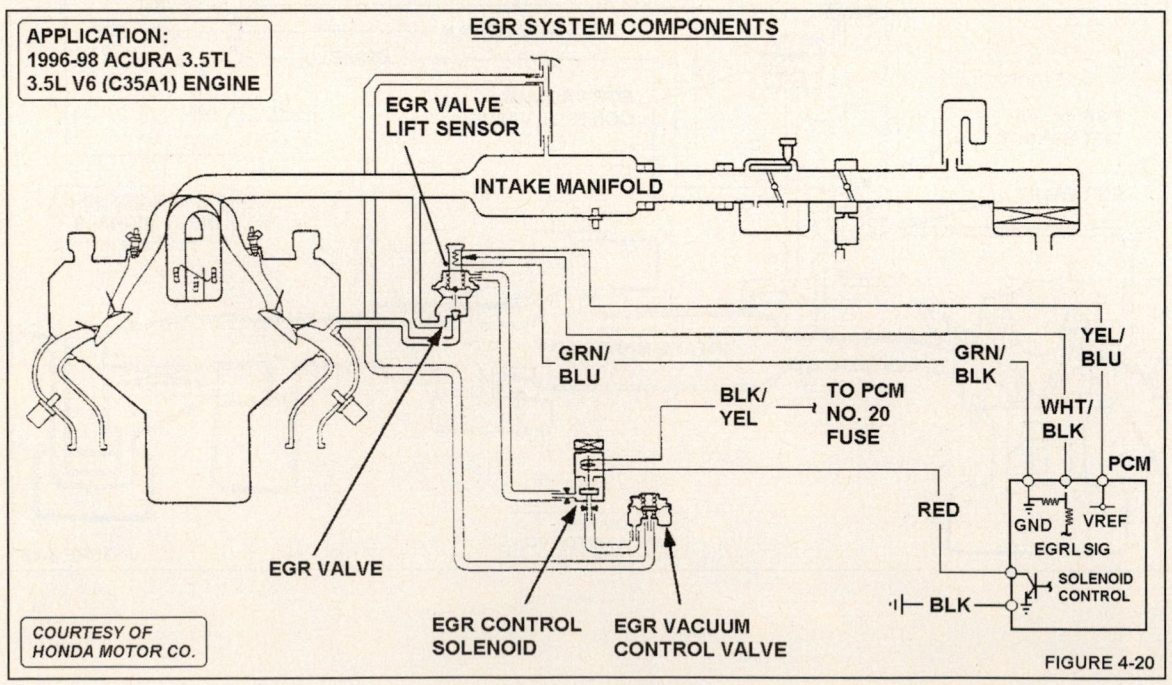

FIGURE 4-20

EGR System Monitor

<u>SYSTEM COMPONENTS - HONDA CIVIC COUPE DX (D16Y5) WITH M/T TRANSMISSION</u>

System components include the EGR valve, EGR valve lift sensor, the PCM and the various other sensor inputs to the PCM.

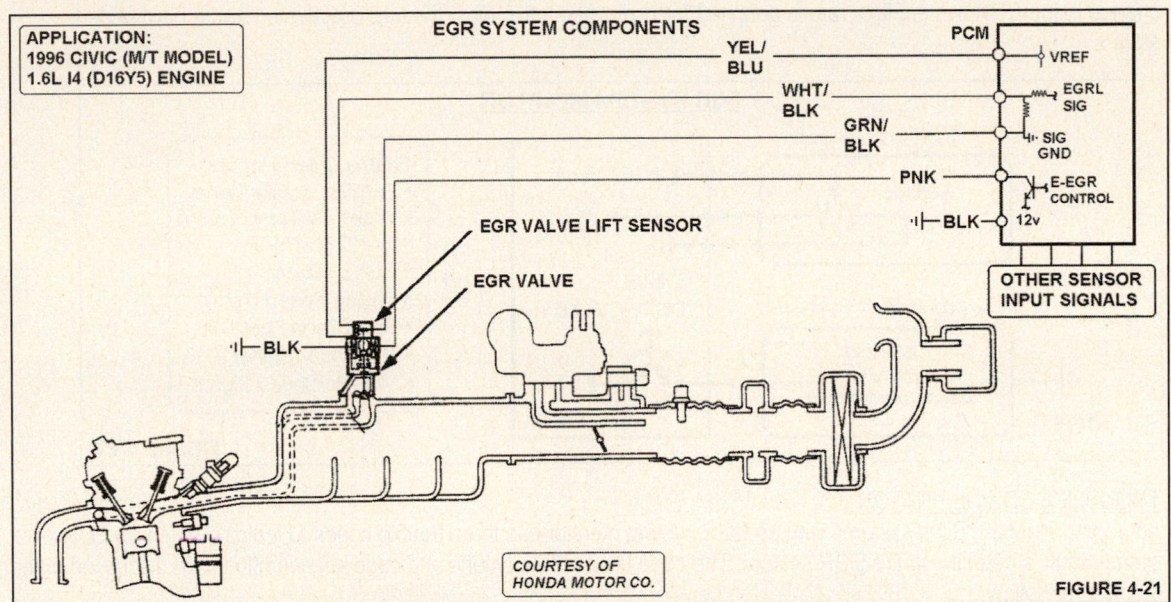

FIGURE 4-21

<u>SYSTEM COMPONENTS - HONDA CIVIC COUPE DX (D16Y5) WITH CVT TRANSMISSION</u>

System components include the EGR valve, EGR valve lift sensor, EGR Control Solenoid Valve, the PCM and the various other sensor inputs to the PCM.

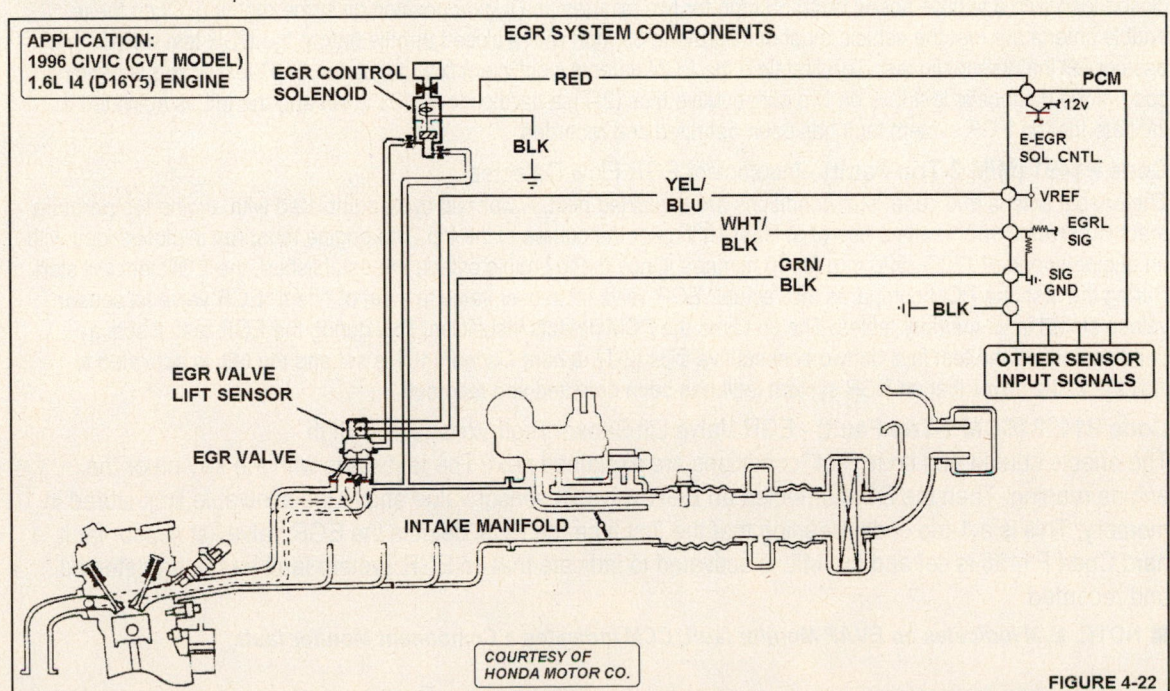

FIGURE 4-22

EGR System Monitor

EGR System Monitor Operation

OBD II regulations require that the EGR System must be monitored for abnormally low or high EGR flow rates that could cause tailpipe emissions to exceed 1.5 times the FTP Standard. If an EGR system component fails, or if a change in the EGR system flow rate is detected for two consecutive trips, the MIL is activated and a hard code is stored.

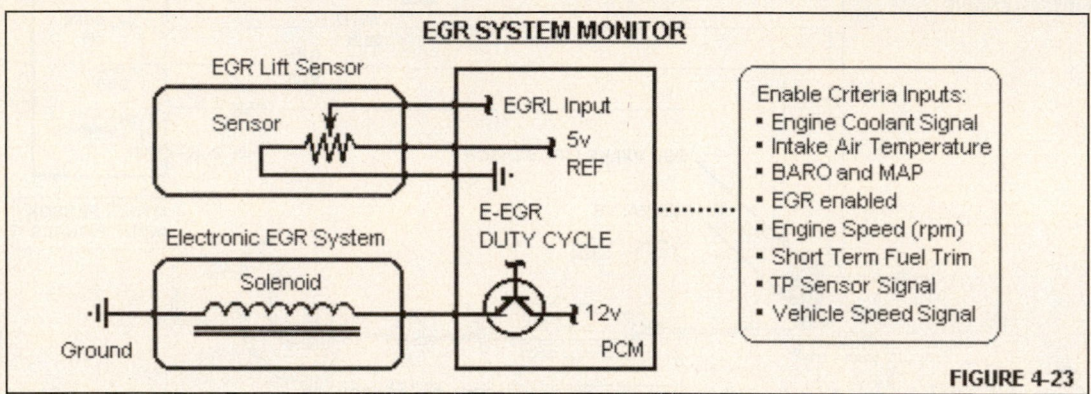

FIGURE 4-23

TROUBLE CODE DEFINITIONS

Acura and Honda OBD II systems with an EGR system monitor use three trouble codes to indicate when an EGR system fault is detected in the EGR system. The code title, enable criteria and code set conditions are discussed in the articles that follow.

Code P0401 (M/M 2-Trip Fault) - Insufficient EGR Flow Detected

The enable criteria and code "set" conditions are explained next.). An OBD II warmup cycle is completed with engine temperature reaching at least 160°F with a rise of at least 40°F over the course of the trip. The engine must run in closed loop with a vehicle speed of 40-55 mph for two minutes (in D_4 gear position on some models). Once these enable criteria are met the vehicle must decelerate to 35 mph with a closed throttle before the EGR test will run and pass or fail the diagnostic test. The first time the PCM detects insufficient flow during the EGR test, it sets a pending code. If the diagnostic test fails on two consecutive trips (2-T), a hard code P0401 is set and the MIL is activated to indicate that an EGR system fault has been detected and recorded.

Code P1491 (M/M 2-Trip Fault) - Insufficient EGR Flow Detected

The enable criteria and code "set" conditions are explained next. A warmup cycle completed with engine temperature reaching at least 160°F with a rise of at least 40°F over the course of the trip. The engine must run in closed loop with an engine speed of 1700-2500 rpm for 10 minutes. Once these enable criteria are established, the EGR test will start. During the test, the PCM compares the "actual" EGR valve lift sensor value to a set of "ideal" EGR valve lift sensor values stored in its memory tables. The first time the PCM detects insufficient flow during the EGR test, it sets a pending code. If the test fails on two consecutive trips (2-T), a hard Code P1491 is set and the MIL is activated to indicate to the driver that an EGR system fault has been detected and recorded.

Code P1498 (CCM 1-Trip Fault) - EGR Valve Lift Sensor High Voltage Too High

The enable criteria and code "set" conditions are explained next. The test starts with the key on or the engine running. Then the PCM detected an EGR valve lift sensor value above the allowable limit stored in memory. This is a 1-trip code meaning that the first time the PCM detects the EGR valve list sensor fault, a hard Code P1498 is set and the MIL is activated to indicate that an EGR system fault has been detected and recorded.

■ NOTE: *M/M indicates an EVAP Monitor fault, CCM indicates a Component Monitor fault.*

EVAP System Monitor

INTRODUCTION

OBD II regulations require that the EVAP system operation be monitored for correct function of the airflow used to purge the EVAP system. Also, the system must be monitored using some form of vacuum or pressure check to verify no leaks are present that could allow fuel vapors stored in the system to escape into the atmosphere. If a leak equal to or greater than 0.040 inch is detected in the EVAP recovery system for two consecutive trips, the MIL is activated and a code is set.

VAPOR PURGE CONTROL SYSTEM DESIGN

On this system, the EVAP canister is purged by drawing fresh air from the EVAP canister and into a port on the intake manifold. The purge vacuum is controlled by the PCM through control over the EVAP purge control solenoid (refer to Figure 4-24).

FUEL TANK VAPOR CONTROL SYSTEM - PURGE FLOW SWITCH DESIGN

On this system, the EVAP canister is purged by drawing fresh air from the EVAP canister and into a port on the intake manifold. The purge vacuum is controlled by the PCM through control over the EVAP purge control solenoid. An EVAP purge flow switch, located in the purge line between the charcoal canister and the EVAP purge control solenoid, is used to determine if proper purge valve operation actually occurs. If the PCM detects that the switch is open or closed under the wrong conditions, a code is set.

SYSTEM COMPONENTS - 1996 ACCORD 2.2L I4 ENGINE

System components include the EVAP canister, EVAP Two-way valve, EVAP purge flow switch, EVAP purge control solenoid, fuel cap, EVAP valve (in the fuel tank), fuel tank, the PCM and various sensor inputs. Note in Figure 4-24 how the PCM monitors the EVAP diagnostic switch circuit to determine if purge occurs under the right conditions.

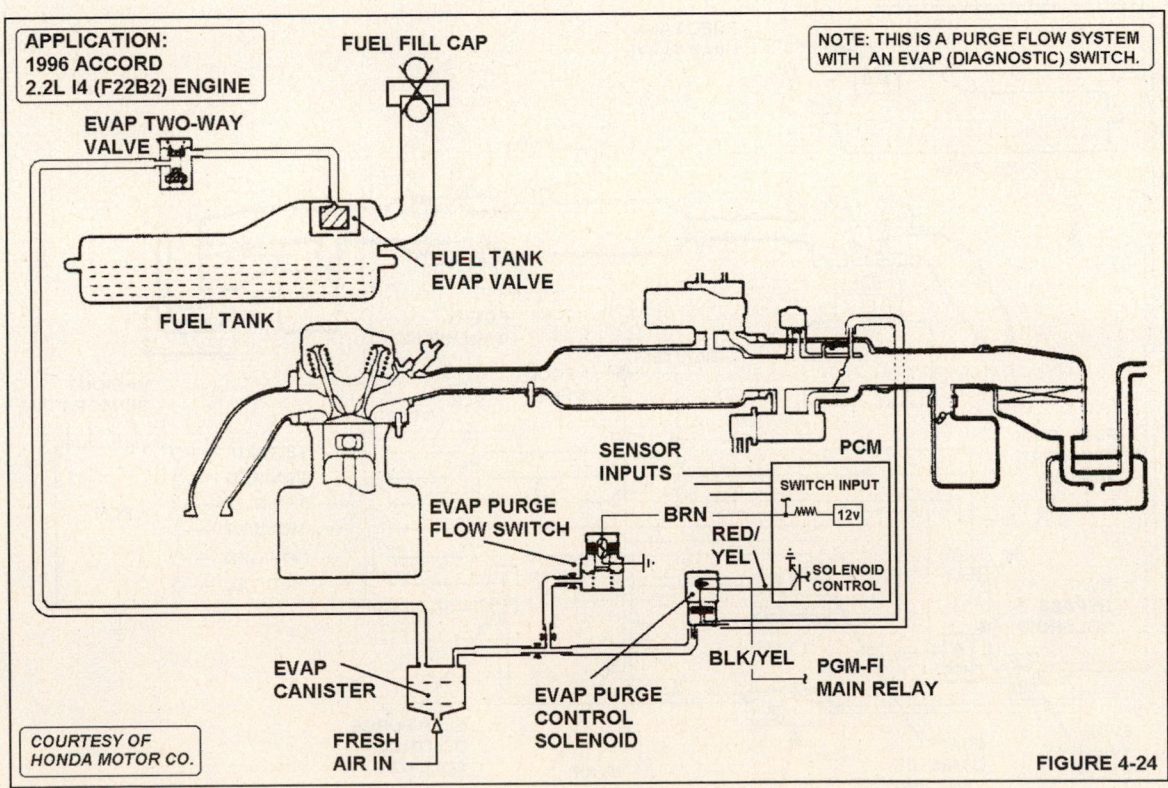

FIGURE 4-24

EVAP System Monitor

<u>FUEL TANK VAPOR CONTROL SYSTEM - PRESSURE SENSOR DESIGN</u>

On vehicles with this system, an EVAP canister is used for the temporary storage of fuel vapor until the vapor can be purged from the canister and burned in the engine.

EVAP canister purging is accomplished by drawing fresh air through the canister and into a port on the throttle body. Once the engine temperature exceeds 167°F, the PCM controls the EVAP purge control solenoid (through a duty cycle signal).

If fuel vapor pressure in the fuel tank reaches a value higher than the set value of the EVAP two-way valve, it opens and regulates the flow of fuel vapor to the canister. This system also includes a fuel tank pressure sensor that is used to verify that no leaks are present that could allow fuel vapors stored in the system to escape into the atmosphere.

On systems with a fuel tank pressure sensor, if a leak equal to or greater than 0.040 inch is detected in the EVAP recovery system for two consecutive trips, the MIL is activated to indicate that an EVAP system leak has been detected and a code is set.

<u>SYSTEM COMPONENTS - 1996 ACCORD 2.2L I4 VTEC ENGINE</u>

System components include the EVAP canister, EVAP two-way valve, EVAP three-way valve, EVAP purge control solenoid, EVAP control canister vent shut valve, EVAP bypass solenoid valve, fuel (filler) cap, fuel tank, fuel tank EVAP valve, fuel tank pressure sensor, the PCM and various sensor inputs. Refer to Figure 4-25 as needed.

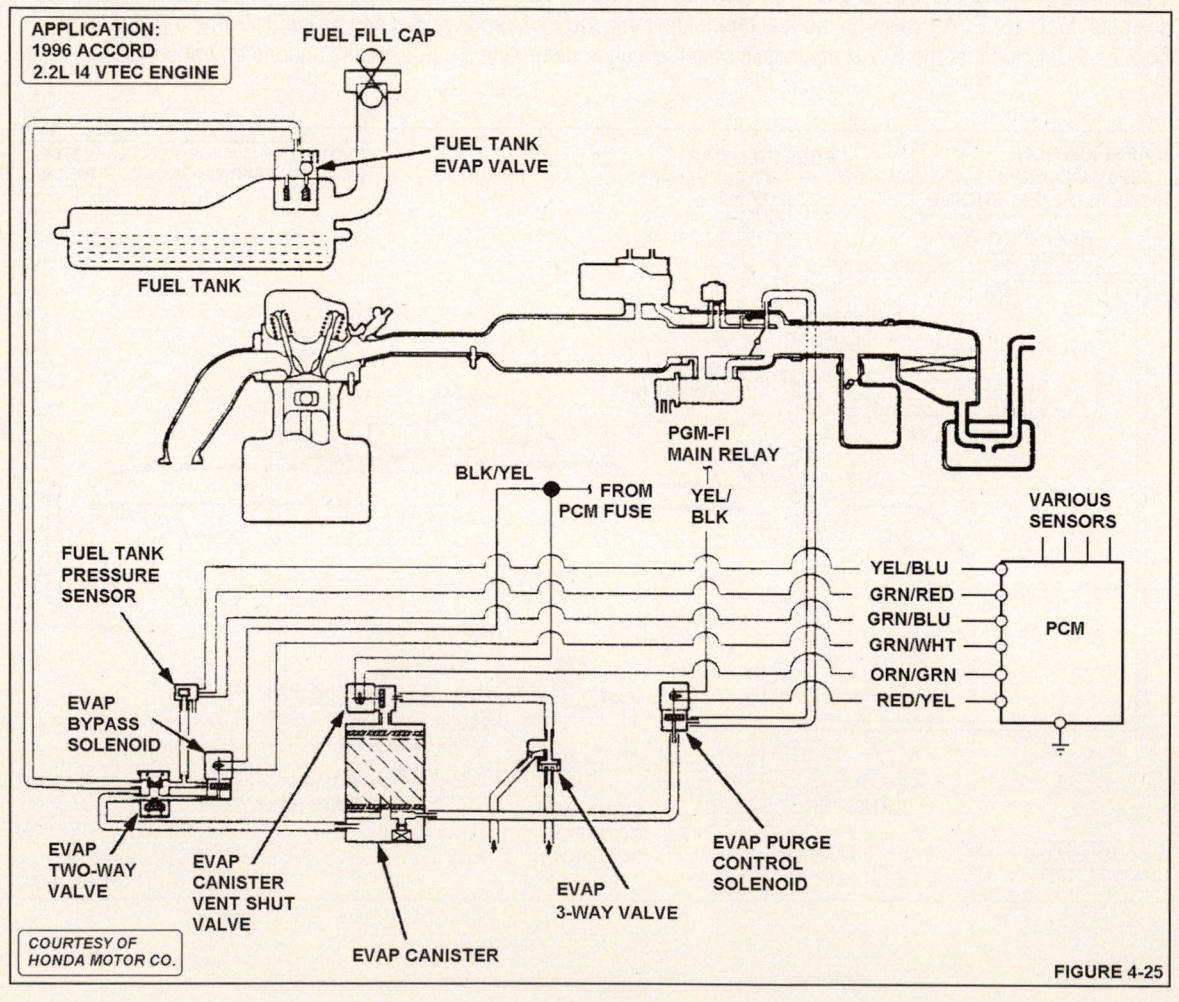

FIGURE 4-25

EVAP System Monitor

FUEL TANK PRESSURE SENSOR

The fuel tank pressure sensor, which is similar in design to a MAP sensor, is used to convert fuel tank absolute pressure into electrical signals. The PCM monitors these signals to determine the fuel tank pressure as part of the EVAP Monitor.

A quick check can be made to determine if the sensor calibration is okay. First, remove the fuel fill cap. Next, turn the key on and use a Scan Tool to monitor the sensor signal voltage. With the cap off, it should read near 2.5v. If it does not, check the sensor circuits (VREF, SIG and GND) for open or shorted conditions.

If a Scan Tool is not available, carefully backprobe the sensor between PCM terminal D15 (on 1996 Civic models) and ground or battery negative to measure the fuel tank pressure sensor voltage with the key on.

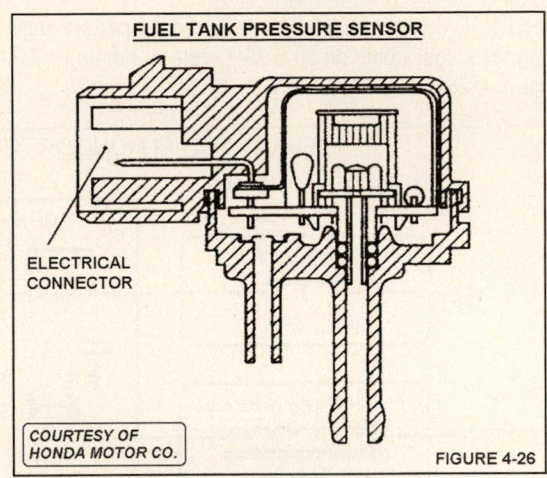

FUEL TANK PRESSURE SENSOR

ELECTRICAL CONNECTOR

COURTESY OF HONDA MOTOR CO.

FIGURE 4-26

Codes P0452, P0453 (CCM 1-Trip Fault) - Fuel Tank Sensor Low or High Voltage

The enable criteria and code "set" conditions are explained next. The test starts with the key on or the engine running. Then the PCM detected a pressure sensor value above or below the allowable limit stored in memory. This is a 1-trip code meaning that the first time the PCM detects a pressure sensor fault, a hard Code P0452 or P0453 is set, the MIL is activated to indicate that an EVAP system fault has been detected and recorded.

COMPONENT EXAMPLES - 1996 CIVIC 1.6L I4 (D16Y5) ENGINE

System components include the EVAP canister, EVAP two-way valve, EVAP purge control solenoid, fuel tank vent valve, fuel fill cap, fuel tank and (fuel tank) EVAP valve.

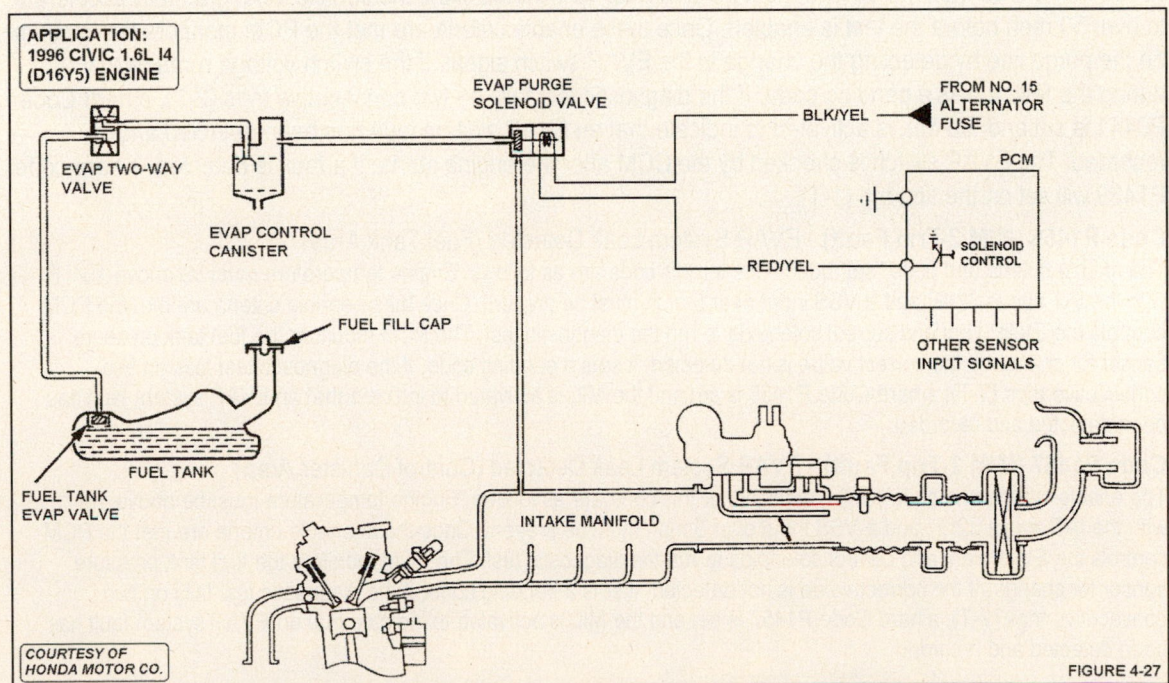

APPLICATION: 1996 CIVIC 1.6L I4 (D16Y5) ENGINE

EVAP TWO-WAY VALVE

EVAP CONTROL CANISTER

FUEL FILL CAP

FUEL TANK

FUEL TANK EVAP VALVE

EVAP PURGE SOLENOID VALVE

BLK/YEL

FROM NO. 15 ALTERNATOR FUSE

PCM

SOLENOID CONTROL

RED/YEL

OTHER SENSOR INPUT SIGNALS

INTAKE MANIFOLD

COURTESY OF HONDA MOTOR CO.

FIGURE 4-27

EVAP System Monitor

EVAP System Monitor Operation

The EVAP System Monitor is enabled once certain engine conditions (enable criteria) are met. An example of how the diagnostic test works on an EVAP system with an EVAP Purge (diagnostic) switch and purge solenoid is shown in Figure 4-28.

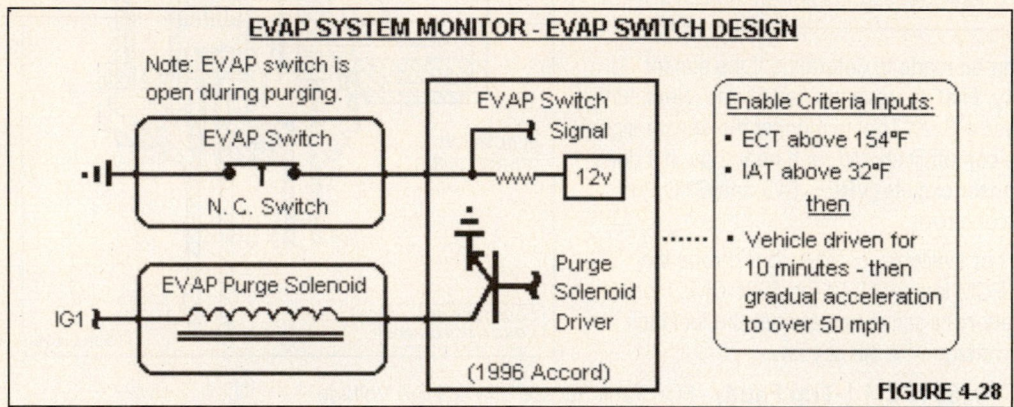

FIGURE 4-28

TROUBLE CODE DEFINITIONS

Acura and Honda OBD II systems with an EVAP System Monitor use three trouble codes to indicate when an EVAP system fault is detected in the EVAP system. The code title, enable criteria and code set conditions are discussed in the articles that follow.

Codes P0441 (M/M 2-Trip Fault) - EVAP System Incorrect Purge Flow (EVAP Switch)

The enable criteria and code "set" conditions for this code are as follows. Engine temperature must be above 154°F and then the vehicle must be driven for 10 minutes. Next the vehicle must gradually accelerate to over 50 mph before the test is enabled. Once these enable criteria are met the PCM monitors the vacuum on the purge line by detecting the change in the EVAP switch signal. If the switch voltage remains low during the test, it sets a pending code. If the diagnostic test fails on two consecutive trips (2-T), a hard Code P0441 is set and the MIL is activated to indicate that an EVAP system fault has been detected and recorded. The EVAP switch is checked by the CCM after the engine starts. If a fault is detected, a hard code **P1459** will set on the first trip (1-T).

Code P1456 (M/M 2-Trip Fault) - EVAP System Leak Detected (Fuel Tank Area)

The enable criteria and code "set" conditions for this code are as follows. Engine temperature must be above 154°F with the IAT above 32°F, and a VSS input over 5 mph must be present. Once these enable criteria are met the PCM controls the EVAP vent and control solenoids to run the diagnostic test. The PCM monitors the fuel tank pressure sensor for change. If the correct value is not detected, it sets a pending code. If the diagnostic test fails on two consecutive trips (2-T), a hard Code P1456 is set and the MIL is activated to indicate that an EVAP system fault has been detected and recorded.

Code P1457 (M/M 2-Trip Fault) - EVAP System Leak Detected (Control Canister Area)

The enable criteria and code "set" conditions for this code are as follows. Engine temperature must be above 154°F with the IAT above 32°F, and a VSS input over 5 mph must be present. Once these enable criteria are met the PCM controls the EVAP vent and control solenoids to run the diagnostic test. The PCM monitors the fuel tank pressure sensor for change. If the correct value is not detected, it sets a pending code. If the diagnostic test fails on two consecutive trips (2-T), a hard Code P1457 is set and the MIL is activated to indicate that an EVAP system fault has been detected and recorded.

■ NOTE: *M/M indicates a Main Monitor fault (in this case, an EVAP Monitor fault).*

Fuel System Monitor

FUEL SYSTEM MONITOR OPERATION

OBD II regulations require that the fuel delivery control system (LONGFT) be tested continuously in order to verify that it can comply with emission standards. If a Fuel system component fails, or if a change in LONGFT is detected that could cause tailpipe emissions to exceed 1.5 times the FTP Standard for two consecutive trips, the MIL is activated and a code is set. When the PCM detects a fault, actual engine operating conditions present at the moment the fault is detected are stored in the Freeze Frame.

Fuel system faults have a higher priority than all other faults except engine misfire. Because of this priority, the Freeze Frame data from a Fuel system fault will overwrite data from faults with a lower priority (i.e., sensors and control device faults).

The Fuel System Monitor is designed to test the adaptive fuel control system. Generally, this task is accomplished using adaptive fuel tables stored in the PCM memory to compensate for variations in Fuel system components due to normal wear and aging. Once closed loop is enabled, the PCM adaptive fuel control strategy "learns" changes needed to correct a fuel system that is biased either rich or lean. These changes are stored in the fuel tables.

Fuel trim correction has two methods of adapting to Fuel system changes: short-term fuel trim (SHRTFT) and long-term fuel trim (LONGFT). The PCM uses SHRTFT and LONGFT together (e.g., if the HO2S signal indicates that the A/F ratio is richer than normal, the PCM can move SHRTFT to a negative range in an attempt to correct the rich condition). If the SHRTFT compensates for this rich condition for too long a period of time, the PCM "learns" this fact and moves LONGFT to a negative range to compensate.

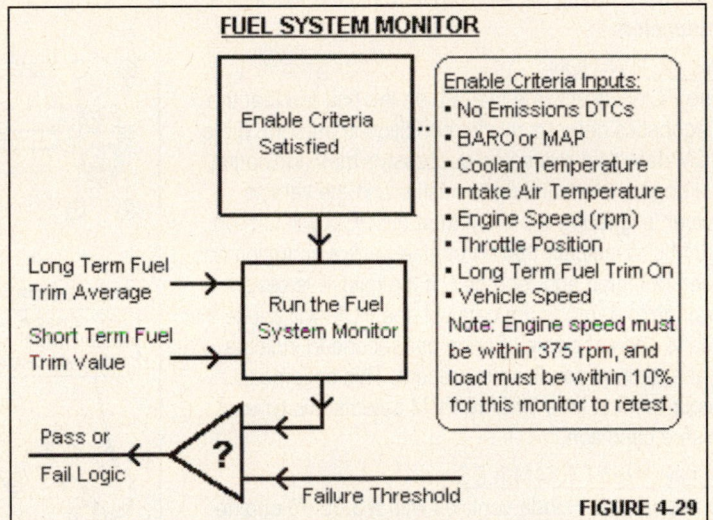

FIGURE 4-29

On a Scan Tool, the SHRTFT and LONGFT parameters are shown as percentages. The SHRTFT range is -30% to +43% and the LONGFT range is -20% to +20%. Refer to Figure 4-29 for an example of how Fuel System Monitor fault detection operates.

TROUBLE CODE DEFINITIONS

Acura and Honda OBD II systems with a Fuel System Monitor use two trouble codes to indicate when a fault in the Fuel system has been detected. The code title, enable criteria and code set conditions are discussed in the articles that follow.

Codes P0171, P0172 (M/M 2-Trip Fault) - Fuel System Too Rich or Too Lean

The enable criteria and code "set" conditions for these codes are as follows. The engine runs until the electric cooling fan cycles. Once the enable criteria are met, the PCM will begins to monitor the amount of change in the fuel trim values based on HO2S signals. If the PCM detects that fuel trim has reached its rich or lean limits for a set period of time, it sets a pending code. If the test fails on two consecutive trips (2-T), a hard code P0171 or P0172 is set and the MIL is activated to indicate a Fuel system fault was detected.

■ NOTE: *M/M indicates a Main Monitor fault (in this case, a Fuel System fault).*

Misfire Detection Monitor

<ins>INTRODUCTION</ins>

The OBD II Misfire Monitor introduced on vehicles with OBD II systems is one of the most important diagnostic changes added with OBD II. Engine misfire can be defined as a condition that allows unburned or partially combusted fuel to flow into the catalytic converter, degrading the converter, shortening its life and increasing emissions levels.

Misfire diagnostics have some special requirements. When the misfire code is set, the computer must record the vehicle speed, engine load and temperature at the time the fault was detected. The Misfire Detection Monitor will not update a Misfire code unless the test conducted during the next trip is within the same parameters.

<ins>MIL ACTIVATION</ins>

The PCM will activate the MIL on the first trip that the diagnostics detect a misfire condition is present. If the PCM detects a misfire on successive trips (indicating the engine is continuing to misfire and the vehicle owner is ignoring the MIL), then the PCM will turn on the MIL to indicate that an engine misfire condition has been detected and recorded. If the misfire is severe enough to damage the catalyst, the PCM will cause the MIL to flash at a rate of once per second to indicate that a serious condition is present. This action can occur the first time that the PCM detects this type of misfire condition.

<ins>COMPONENT EXAMPLES</ins>

On Acura and Honda vehicles with a 3.0L V6 engine with distributor ignition (DI), a dual TDC sensor and CKP sensor are used to provide ignition and fuel control data along with Misfire detection data (Figure 4-30).

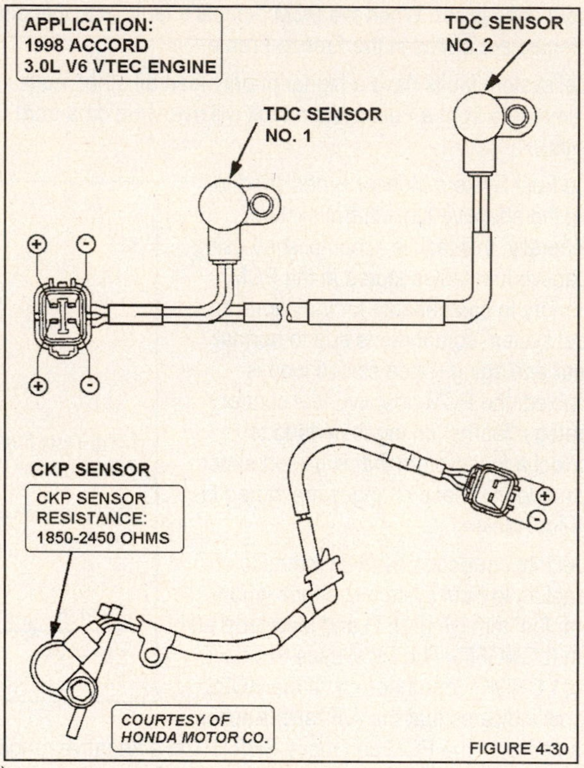

APPLICATION:
1998 ACCORD
3.0L V6 VTEC ENGINE

TDC SENSOR NO. 2

TDC SENSOR NO. 1

CKP SENSOR

CKP SENSOR RESISTANCE: 1850-2450 OHMS

COURTESY OF HONDA MOTOR CO.

FIGURE 4-30

<ins>CKF SENSOR</ins>

Some engines use a Crankshaft Fluctuation Sensor in the distributor (DI models) to detect when misfire conditions are present. These devices are used to read changes in camshaft or crankshaft rotation (Figure 4-31).

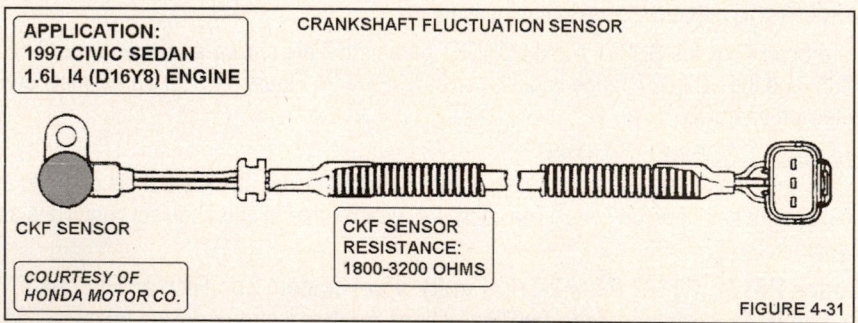

CRANKSHAFT FLUCTUATION SENSOR

APPLICATION:
1997 CIVIC SEDAN
1.6L I4 (D16Y8) ENGINE

CKF SENSOR

CKF SENSOR RESISTANCE: 1800-3200 OHMS

COURTESY OF HONDA MOTOR CO.

FIGURE 4-31

■ NOTE: *The PCM detects engine misfire conditions from minor variance in crankshaft torque. Torque fluctuation data can be influenced from other conditions such as rough roads. On vehicles with ABS, wheel speed sensor data can prevent the PCM and Misfire Monitor from detecting a misfire during rough road conditions.*

Misfire Detection Monitor

CYP SENSORS

1998 Accord 2.3L I4 engines use a Cylinder (CYP) sensor to identify the No. 1 engine cylinder. This signal is used by the PCM as a reference for sequential multiport fuel injection (SMFI) for each cylinder. The variations in the CYP sensor used these engines applications are shown in Figures 4-32 and 4-33.

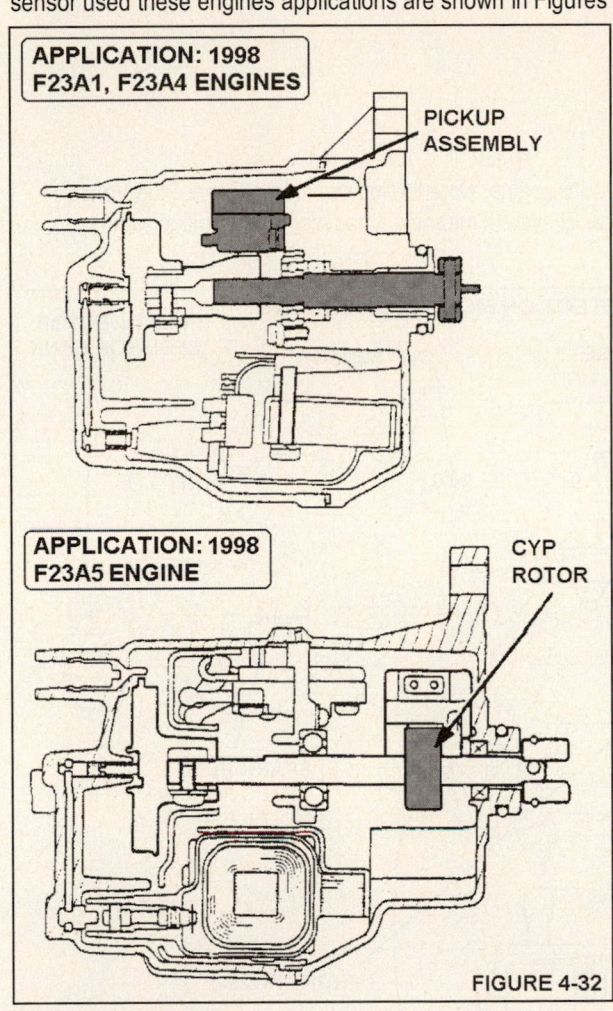

APPLICATION: 1998 F23A1, F23A4 ENGINES

PICKUP ASSEMBLY

APPLICATION: 1998 F23A5 ENGINE

CYP ROTOR

FIGURE 4-32

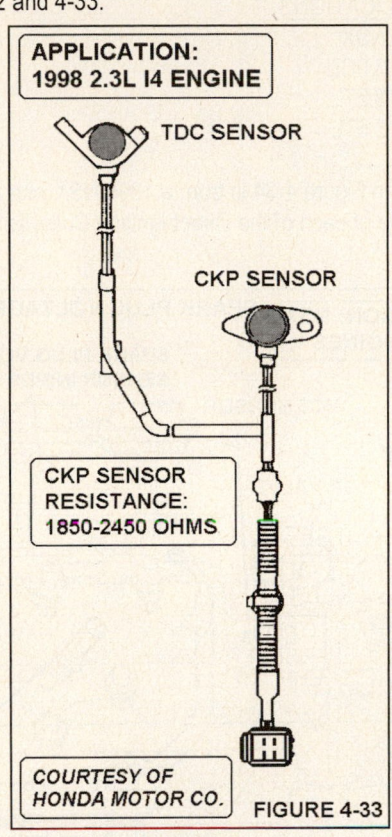

APPLICATION: 1998 2.3L I4 ENGINE

TDC SENSOR

CKP SENSOR

CKP SENSOR RESISTANCE: 1850-2450 OHMS

COURTESY OF HONDA MOTOR CO.

FIGURE 4-33

CKP & TDC SENSORS

The PCM uses the CKP sensor signal to determine the fuel injection timing and ignition timing for each cylinder. The PCM uses the TDC sensor to determine the amount of ignition timing at engine startup.

CKP MISFIRE DETECTION

The CKP sensor signal is also used to help determine if the crankshaft position signal is abnormal. This function is part of the Misfire Detection System.

Misfire Detection Monitor

<u>SPARK PLUG VOLTAGE DETECTION MODULE & SENSOR</u>

Several Acura engines are equipped with a Spark Plug Voltage Sensor (in the coil) and a Spark Plug Detection Module. This system is used to detect when misfire conditions are present (Figure 4-34).

<u>ACURA APPLICATIONS</u>
- 1995-98 NSX
- 1997-98 3.0CL
- 1996-98 3.2TL
- 1996-98 3.5TL

The example in Figure 4-34 is from a 1996 NSX with a 3.0L V6 engine. Note the location of the Spark Plug Voltage Sensors inside of each of the Direct Ignition Coils. This is an EI system meaning it has electronic ignition with no distributor.

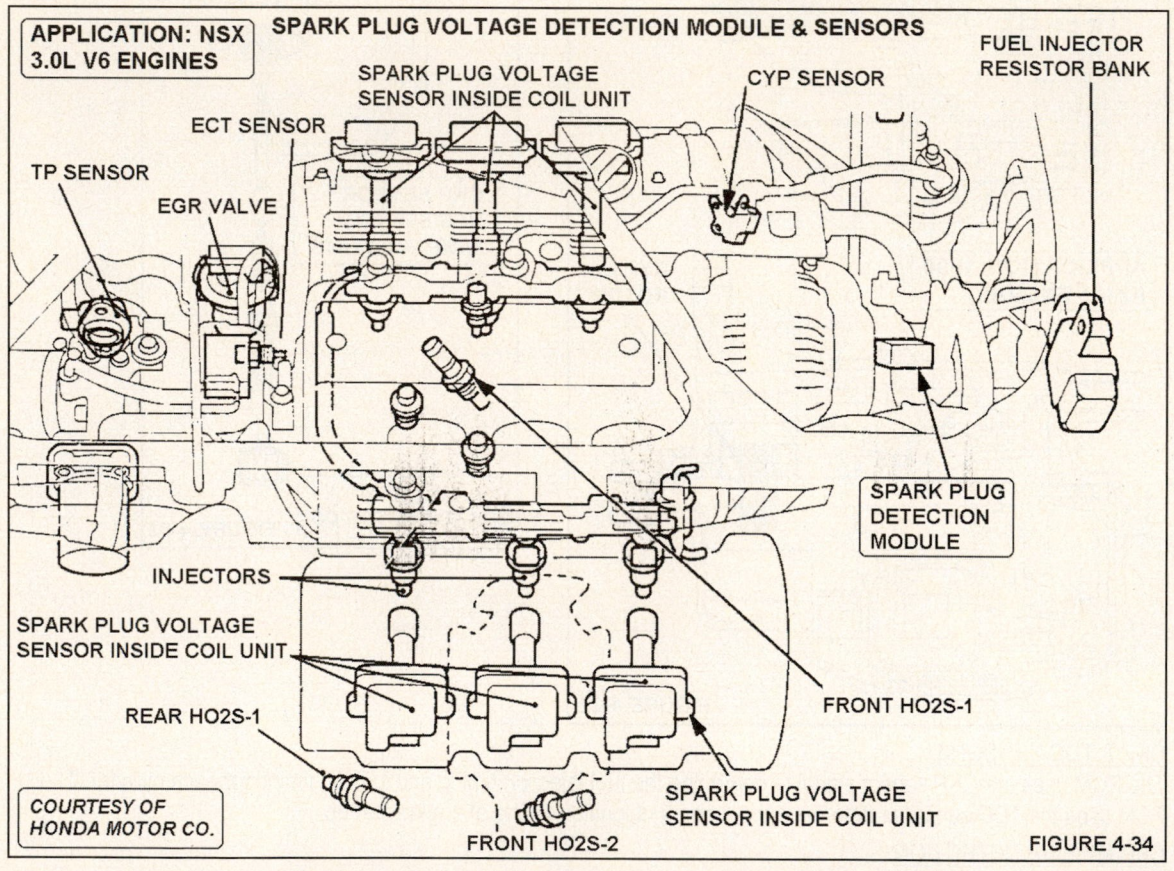

FIGURE 4-34

Misfire Detection Monitor

MISFIRE DETECTION MONITOR OPERATION

OBD II regulations require that the engine be continuously tested for misfires under all engine positive load and speed conditions (defined as accelerating, cruising and idling). The regulations also require that the diagnostics identify which cylinder is misfiring.

Also, the Misfire Detection Monitor must be able to distinguish between a single misfire and multiple misfires. If only a single cylinder is misfiring, then that cylinder must be identified and engine conditions present when the misfire occurred are stored in Freeze Frame. If a misfire is detected that could cause tailpipe emissions to exceed 1.5 times the FTP Standard during a trip, the MIL is activated and a code is set.

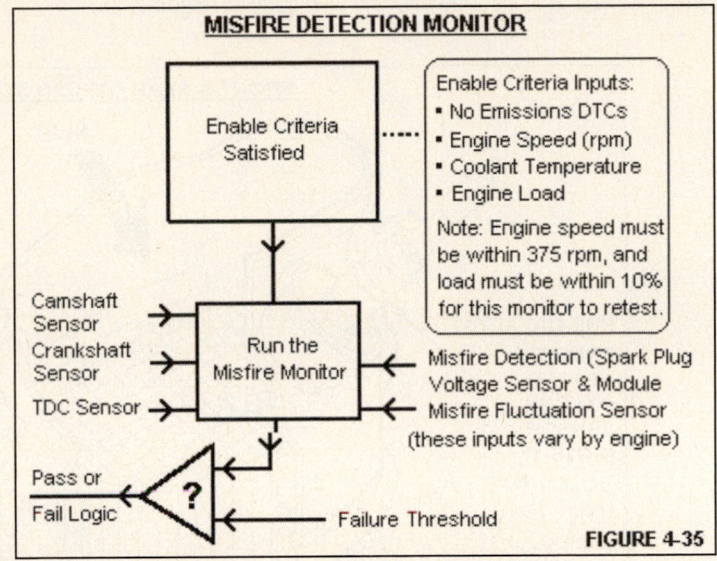

FIGURE 4-35

TYPE 1 MISFIRE

A Type 1 Misfire is an engine misfire condition that could cause tailpipe emissions to exceed 1.5 times the FTP Standard. If a Type 1 Misfire is detected with the engine at similar engine speed, load and temperature conditions during a second trip, the MIL is activated and a code is set. Also, the MIL must be activated if a misfire is detected in two non-consecutive trips that are not more than 80 trips apart under similar conditions.

TYPE 2 MISFIRE

A Type 2 Misfire is an engine misfire so severe that damage to the catalyst can result at the current engine speed, load and temperature. If this fault is detected, the MIL will flash once per second within 200 engine revolutions from the point when the misfire was first detected to warn the driver that continued vehicle operation under similar conditions could cause catalyst damage. The MIL will stop flashing and remain on if the vehicle is no longer operating at engine conditions that could damage the catalyst (Figure 4-35).

TROUBLE CODE DEFINITIONS

Acura and Honda OBD II systems with a Misfire Monitor use several trouble codes to indicate when a misfire condition has been detected. The code title, enable criteria and code set conditions are discussed in the articles that follow.

Codes P0300-P0306, P1300 (M/M 1 or 2-Trip Fault) - Cylinder 1-6 Misfire Detected

Enable criteria and code "set" conditions for these codes are as follows. The PCM begins to monitor for misfire conditions after startup. If a misfire is detected, the PCM sets a pending code. If the test fails on two consecutive trips (2-T), a hard Code P0300-306 is set and the MIL is activated to indicate that a Misfire condition was detected.

Codes P1300 (Monitor 2-Trip Fault) - Random Misfire Detected

Enable criteria and code "set" conditions for this code are as follows. The PCM begins to monitor the engine for random misfire conditions after startup. If a misfire is detected, the PCM sets a pending code. If the test fails on two consecutive trips (2-T), a hard Code P1300 is set and the MIL is activated to indicate a Misfire condition was detected

■ NOTE: *M/M indicates a Main Monitor fault (in this case, a Misfire Monitor fault).*

Oxygen Sensor Monitor

<u>INTRODUCTION</u>

The heated oxygen sensors detect the exhaust gas oxygen content and output a signal to the PCM. During closed loop operation, the PCM receives the HO2S signals and varies the fuel injector pulsewidth in response to these signals. The primary or front HO2S-1 (Bank 1, Sensor 1) is installed in front of the pre-catalyst. The rear or post-catalyst HO2S-2 (Bank 1, Sensor 2) is installed after the post-catalyst (Figure 4-36).

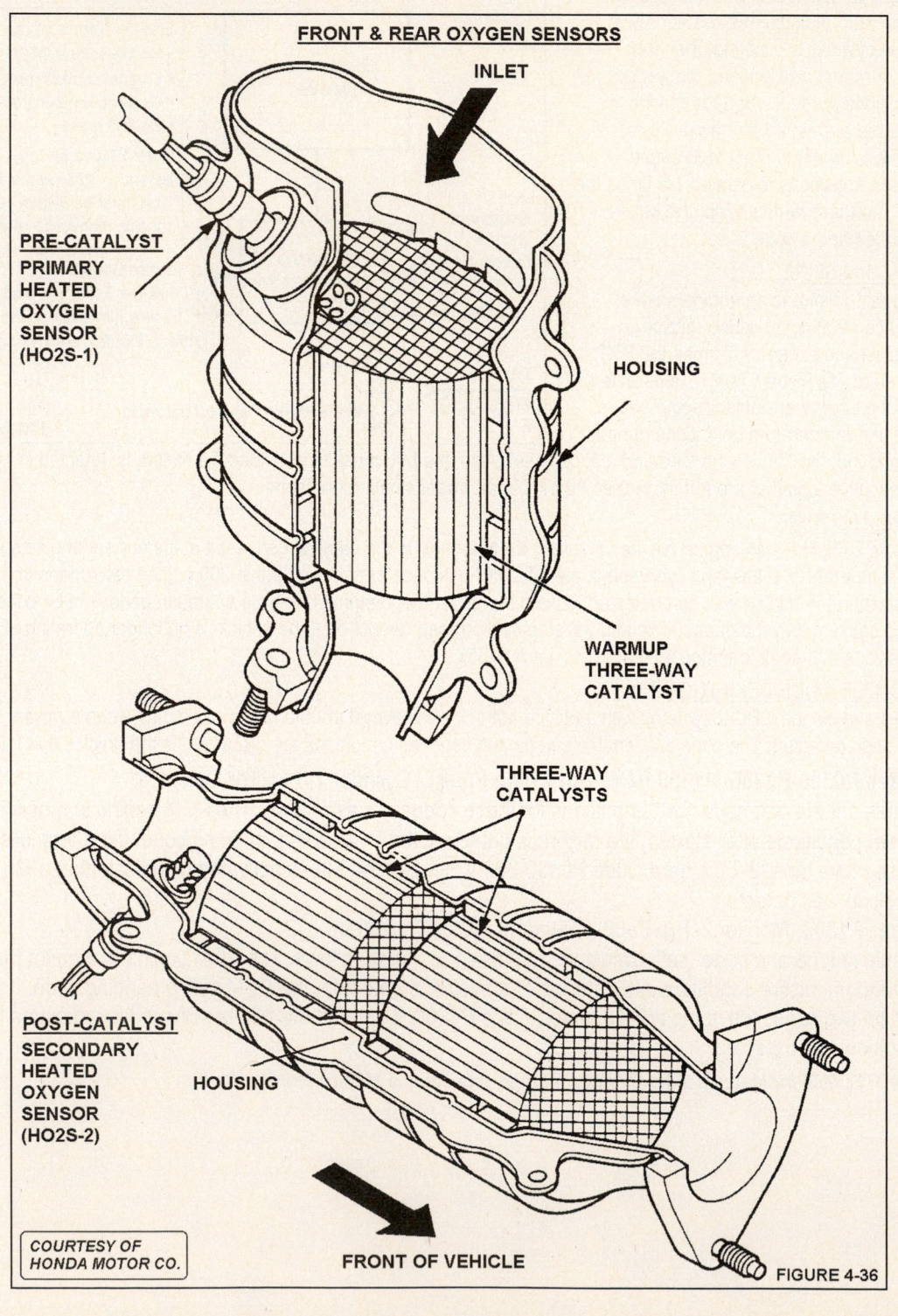

FRONT & REAR OXYGEN SENSORS

INLET

PRE-CATALYST
PRIMARY
HEATED
OXYGEN
SENSOR
(HO2S-1)

HOUSING

WARMUP
THREE-WAY
CATALYST

THREE-WAY
CATALYSTS

POST-CATALYST
SECONDARY
HEATED
OXYGEN
SENSOR
(HO2S-2)

HOUSING

COURTESY OF
HONDA MOTOR CO.

FRONT OF VEHICLE

FIGURE 4-36

Oxygen Sensor Monitor

OXYGEN SENSOR MONITOR OPERATION

Oxygen Sensors are monitored for slow response, lack of switching, heater functions, shifted rich, shifted lean, no activity (open loop), rich or lean indications for too long a period of time, open, grounded and shorted circuit conditions.

OBD II regulations require that the pre-catalyst (HO2S-1) and post-catalyst (HO2S-2) heated Oxygen sensors be monitored for correct response rate and output voltage levels, as well as any other factors that could effect tailpipe emissions. The Oxygen Sensor Monitor diagnostic consists of three separate test areas: run-time to activity, lean-to-rich and rich-to-lean response times and sensor output voltage checks. If the PCM detects an Oxygen sensor fault that could cause tailpipe emissions to exceed 1.5 times the FTP Standard for two consecutive trips, the MIL is activated and a code is set.

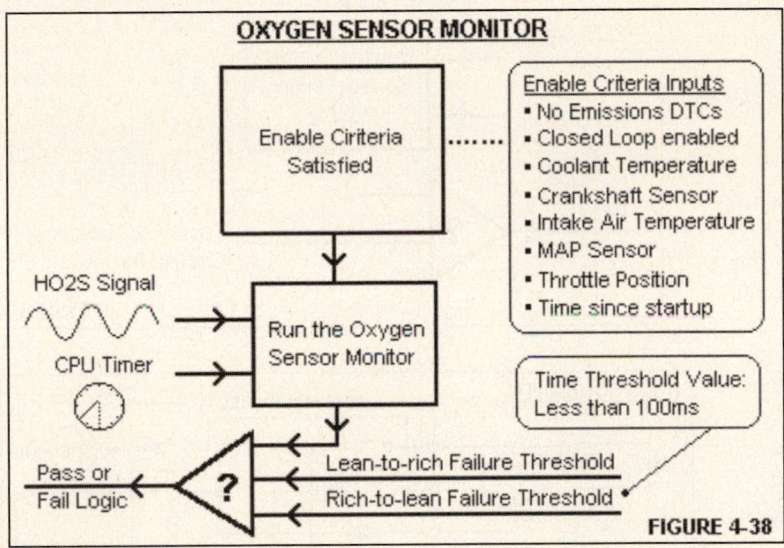

FIGURE 4-38

RESPONSE TIME TEST

Response time is defined as the amount of time it takes for the Oxygen sensor signal to go from rich-to-lean and from lean-to-rich. Once the Oxygen Sensor Monitor is enabled, it compares the average response time to a calibrated value in the PCM memory. If the response time exceeds this value, the PCM determines the sensor is faulty. Refer to Figure 4-38 as needed.

HO2S VOLTAGE TESTS

The Oxygen sensor voltage test is used to monitor the pre-catalyst and post-catalyst sensor signal for no activity (due to an open circuit) or for circuit short-to-voltage or short-to-ground conditions. This part of the Monitor test may also be used to check the pre-catalyst for faults in the fuel control system that may cause the sensor signal to shift in a rich or lean direction.

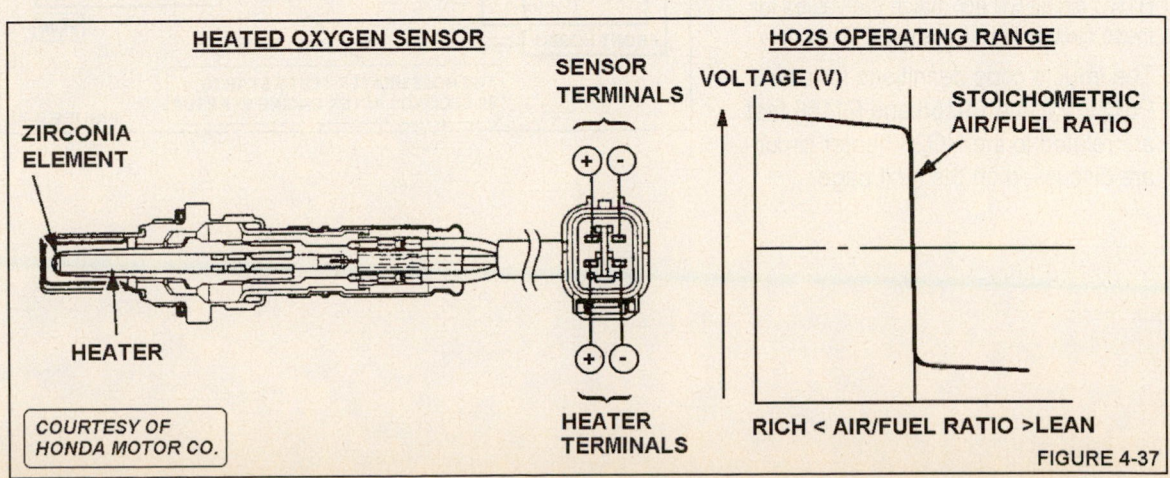

FIGURE 4-37

Oxygen Sensor Heater Monitor

INTRODUCTION

OBD II regulations require that the pre-catalyst and post-catalyst oxygen sensor be monitored for faults in the heater element or circuit. The Oxygen Sensor Heater Monitor test method used on Acura and Honda vehicles includes monitoring the heater circuits for open, grounded or shorted circuit conditions (Figure 4-39).

The first time a HO2S heater circuit fault is detected by the PCM that could cause tailpipe emissions to exceed 1.5 times the FTP Standard (1-T), the MIL is activated and a code is set.

HO2S HEATER CIRCUITS

The HO2S circuits include the signal, sensor ground and heater control (HTRC). On most models, the heater receives power from the main relay through the ignition power circuit (Figure 4-40). However, on the Honda Civic D15Z1 engine with a Lean Air Fuel sensor, the heater circuit receives power from the HTRC relay.

HO2S HEATER CIRCUIT TESTS

During the HO2S heater test, the HO2S Monitor (in this case, the Comprehensive Component Monitor) checks the HO2S heater circuits for fault conditions. This function is depicted in Figure 4-40 by the PCM "eyes" positioned above the HTRC circuits.

The heater circuit tests are accomplished by monitoring the heater circuit for a voltage change 80 seconds after engine startup and once the heater control driver closes the HTRC circuit.

After the HTRC circuit is enabled, if the HO2S heater or its circuits are open, grounded or shorted, the voltage on the HTRC circuit will not match the values for these circuits stored in the PCM memory.

The trouble code definitions for codes P0135, P0141, P0155 and P1166 that are related to the HO2S heater circuit are discussed on the next page.

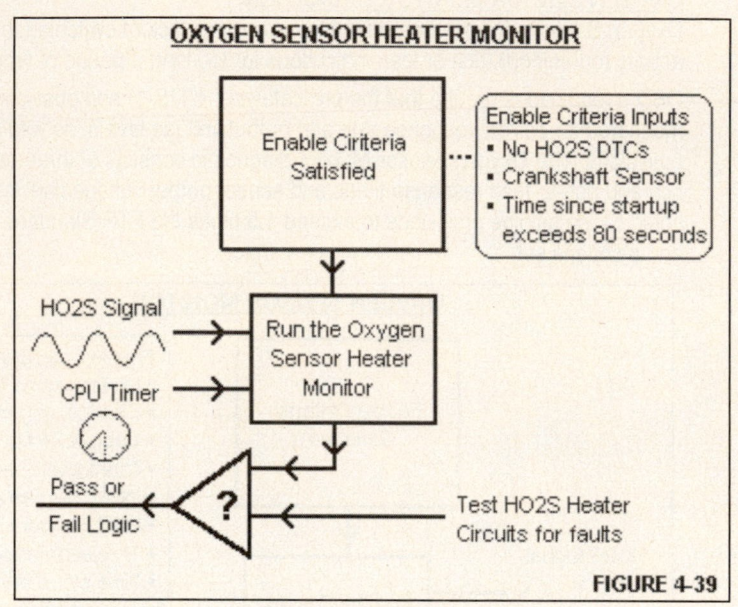

OXYGEN SENSOR HEATER MONITOR

Enable Criteria Satisfied

Enable Criteria Inputs
- No HO2S DTCs
- Crankshaft Sensor
- Time since startup exceeds 80 seconds

HO2S Signal

CPU Timer

Run the Oxygen Sensor Heater Monitor

Pass or Fail Logic

?

Test HO2S Heater Circuits for faults

FIGURE 4-39

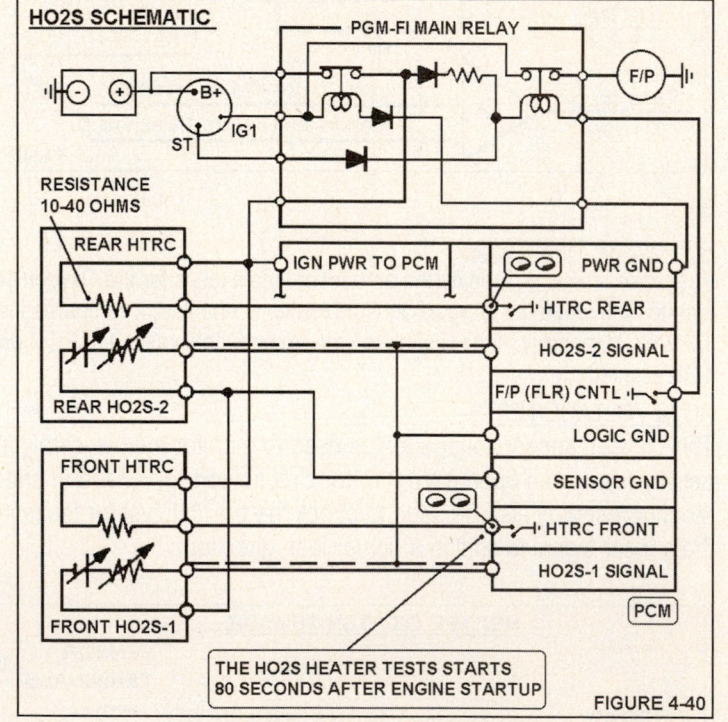

HO2S SCHEMATIC

PGM-FI MAIN RELAY

F/P

B+

ST   IG1

RESISTANCE 10-40 OHMS

REAR HTRC

IGN PWR TO PCM

PWR GND

HTRC REAR

HO2S-2 SIGNAL

REAR HO2S-2

F/P (FLR) CNTL

LOGIC GND

FRONT HTRC

SENSOR GND

HTRC FRONT

HO2S-1 SIGNAL

FRONT HO2S-1

PCM

THE HO2S HEATER TESTS STARTS 80 SECONDS AFTER ENGINE STARTUP

FIGURE 4-40

Oxygen Sensor Monitor

TROUBLE CODE DEFINITIONS

Acura/Honda OBD II systems with an Oxygen Sensor (and Heater) Monitor use several trouble codes to indicate when an oxygen sensor fault has been detected. The code title, enable criteria and code set conditions are discussed in the articles that follow.

Codes P0131, P0137, P0151, P0157, P1162 (CCM 1-Trip Fault) - HO2S Low Voltage

Enable criteria and code "set" conditions for these codes are as follows. An OBD II warmup cycle is completed with engine temperature reaching at least 160°F with a rise of at least 40°F over the course of the trip. The vehicle must be driven with the gear selector in D2 for A/T models (4th gear for M/T models) in closed loop. Once the enable criteria are met, the PCM monitors the HO2S signal. If it detects a signal fixed below 100mv, a hard code is set and the MIL is activated to indicate a fault was detected.

TIPS: *To check the HO2S signal for this type of fault, first run the vehicle at 3000 rpm in P/N until the cooling fan cycles. Prepare the vehicle as described above, then accelerate at WOT while an assistant monitors the HO2S signal. If the signal remains below 100mv, check the fuel pressure. If fuel pressure is okay, check the HO2S circuits. If the circuits are okay, install a new HO2S and repeat the test. If the HO2S signal now reads normal (varies from 100-1000mv), replace the original HO2S.*

Codes P0132, P0138, P0152, P0158, P1164 (CCM 1-Trip Fault) - HO2S High Voltage

Enable criteria and code "set" conditions for these codes are as follows. An OBD II warmup cycle is completed with engine temperature reaching at least 160°F with a rise of at least 40°F over the course of the trip. The vehicle must be driven with the gear selector in D2 for A/T models (4th gear for M/T models) in closed loop. Once the enable criteria are met, the PCM monitors the HO2S signal. If it detects a signal fixed above 900mv, a hard code is set and the MIL is activated to indicate a fault was detected.

TIPS: *To check the HO2S signal for this type of fault, first run the vehicle at 3000 rpm in P/N until the cooling fan cycles. Prepare the vehicle as described above, then decelerate from 40-55 mph (throttle closed) while an assistant monitors the signal during decel. If the signal remains above 900mv, check the HO2S signal for an open circuit fault. If the circuit is okay, install a new HO2S and repeat the test. If the HO2S signal now reads normal (varies from 100-1000mv), replace the original HO2S.*

Codes P0133, P0139, P0153, P0159, P1165 (M/M 2-Trip Fault)- HO2S Slow Response

Enable criteria and code "set" conditions for these codes are as follows. An OBD II warmup cycle is completed with engine temperature reaching at least 160°F with a rise of at least 40°F over the course of the trip. The vehicle must be driven with the gear selector in D2 for A/T models (4th gear for M/T models) in closed loop. Once the enable criteria are met, the PCM begins to monitor the HO2S signal response time. If it detects the average time to switch from rich to lean is more than one second the code is set.

TIPS: *To check the HO2S signal for this type of fault, first run the vehicle at 3000 rpm in P/N until the cooling fan cycles. Prepare the vehicle as described above, then drive the vehicle at 50-65 mph while an assistant monitors the HO2S signal response times on the Scan Tool. If the response times exceed 100 milliseconds, install a new HO2S and repeat the test. If the HO2S response times are now normal (switching from rich to lean in less than one second), replace the original HO2S.*

Codes P0135, P0141, P0155, P0161, P1166 (CCM 1-Trip Fault) - HO2S Heater Fault

Enable criteria and code "set" conditions for these codes are as follows. The engine is started. After 80 seconds, if the PCM detects a fault in the heater circuit, a hard code is set and the MIL is activated to indicate that a heater circuit fault has been detected.

■ NOTE: *M/M indicates an EVAP Monitor fault, CCM indicates a Component Monitor fault.*

PGM-FI DIAGNOSTICS

PGM-FI System Diagnosis

INTRODUCTION (DIAGNOSTIC STARTING POINT)

Refer to the block diagram in Figure 4-41 for instructions that cover the proper steps to follow when testing an Acura/Honda vehicle (1995-98) equipped with OBD II diagnostics.

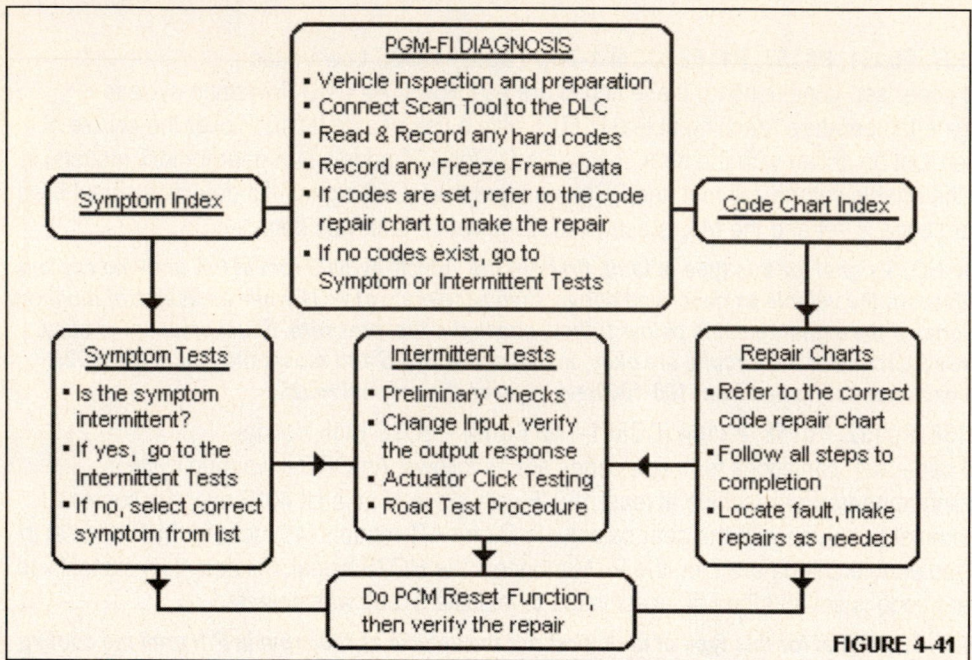

FIGURE 4-41

ACURA/HONDA OBD II DRIVE CYCLE EXAMPLE

After repairing any OBD II trouble codes, a PCM Reset should be done followed by a complete OBD II drive cycle to verify that a repair has been done correctly (Figure 4-42). During the drive cycle, verify that all of the I/M Readiness Monitors run and pass all of their emission-related tests. This is a very important part of the verification process!

If the PGM-FI or C/E light is activated during the drive cycle, the repair has not been completed successfully. Return to the service bay and repeat the Read Codes step. If the light does not activate and no "pending" codes set on the drive cycle, the test has passed and all of the I/M Readiness Monitors should show completed on the Scan Tool.

Once a PCM Reset step is performed, the OBD II Drive Cycle can be used to rebuild the Fuel Control and Idle Speed Control tables stored in the PCM memory. These tables contain "learned" values used to control the injector pulsewidth and IAC motor steps.

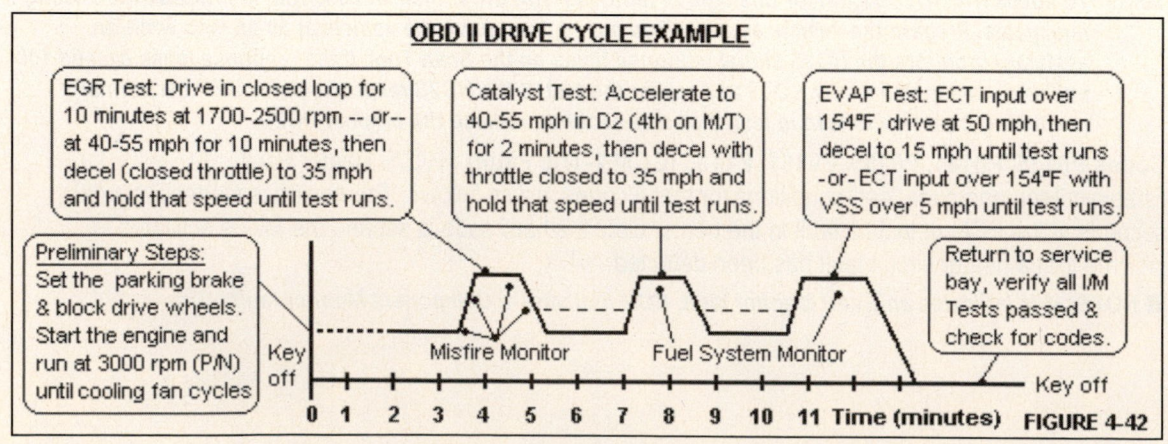

FIGURE 4-42

PGM-FI System Diagnosis

MIL CIRCUIT DIAGNOSIS

If the Malfunction Indicator Lamp (MIL) does not operate correctly, refer to the repair steps listed in the articles that follow to diagnose the MIL operation and the PCM.

MIL CONDITION: LIGHT ON FOR TWO SECONDS, THEN THE LIGHT GOES OFF AT KOEO MODE

This condition indicates the normal operation of the PCM and MIL control circuit. This operation can be used as a bulb check and initial check of the PCM operation.

MIL CONDITION: LIGHT FLASHES
ONCE PER SECOND IN KOER MODE

This condition indicates a fault in an emissions related system that could damage the vehicle catalyst. Use the Scan Tool to read the trouble codes and freeze frame data. Repair the misfire or fuel system fault, do a PCM Reset and a drive cycle.

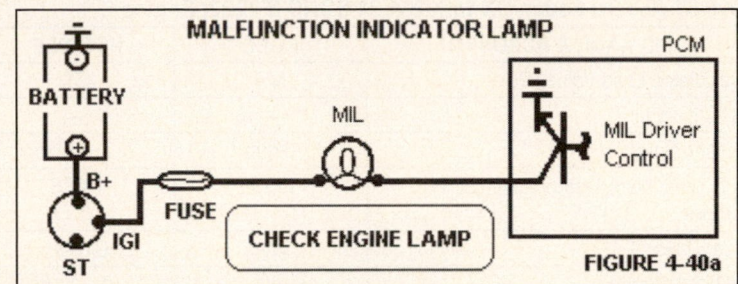

MIL CONDITION: MIL ON WITH NO
CODES STORED IN KOEO OR KOER MODE

This condition can be caused by any of these faults:

* The MIL control wire is shorted to ground between the (MIL) lamp and the PCM.
* The K-Line circuit to the DLC 16-P connector terminal No. 15 is shorted to ground. Turn the key off; remove the PCM connector, check circuit for continuity to ground.
* The PCM is faulty (due to possible shorted MIL control driver circuit)
* An open condition in the PCM direct battery or ignition power circuits
* An open or high resistance condition in the PCM power ground connections

MIL CONDITION: MIL DOES NOT COME ON AT KOEO FOR TWO SECONDS - ENGINE STARTS

This condition can be caused by any of these faults:

* If this problem is intermittent in nature, check the fuse that provides power to the MIL (e.g., the Backup Light fuse - 1996 Accord) for a loose connection or corrosion and the MIL control circuit terminal at the PCM (e.g., terminal A18 - 1996 Accord)
* An open circuit between the MIL (lamp) and PCM control terminal (e.g., terminal 18 on a 1996 Accord) - inspect the bulb connections and the PCM terminals
* The PCM is faulty (due to possible shorted MIL control driver circuit)
* An open condition in the PCM direct battery or ignition power circuits
* An open or high resistance condition in the PCM power ground circuits

MIL CONDITION: MIL DOES NOT COME ON AT KOEO FOR TWO SECONDS - ENGINE WILL NOT START

This condition can be caused by any of these faults:

* Turn the key off and remove the following connectors: the PCM connector that connects to the sensor circuits (e.g., PCM connector C31-P - 1996 Accord), EGR valve lift sensor, fuel tank pressure sensor, MAP sensor and TP sensor. Check for continuity to ground between the MAP sensor VREF and sensor VREF circuits at the PCM (e.g., between terminals C19 (MAP VREF) and C28 (VREF) to ground. If continuity exists, locate the short to ground in the VREF circuit and retest the MIL.
* An open condition in the PCM direct battery or ignition power circuits
* An open or high resistance condition in the PCM power ground connections

Reference Information

PID Chart Example - 1996 Accord Sedan EX

INSTRUCTIONS

Locate the desired input or output parameter and record the known good value. Actual values may vary ± 20%. Read the details in the PID Name box for specific instructions. These PID values were captured using the Scan Tool in OBD II Generic Cartridge.

1996-97 Accord Sedan DX with 2.2L I4 SOHC (F22B2) Engine

PID NAME & ACRONYM	KOEO	HOT IDLE	30 MPH	55 MPH
Calculated Load Value (CLV)	N/A	27%	31%	37%
DTC Number	---	---	---	---
Engine Speed	0 rpm	712 ± 50 rpm	1310 rpm	2210 rpm
Engine Coolant Temperature (ECT) Sensor	185°F	198°F	196°F	196°F
Fuel Status 1	N/A	Closed	Closed	Closed
HO2S-1 (Front)	N/A	0.1-1.1v	0.2-0.8v	0.2-0.8v
HO2S-2 (Rear)	N/A	0.1-1.1v	0.4-0.6v	0.4-0.6v
Intake Air Temperature Sensor	95-140°F	90-102°F	90-102°F	90-102°F
Long Term Fuel Trim (LONGFT)	N/A	± 20%	± 20%	± 20%
Malfunction Indicator Lamp (MIL)	OFF	OFF	OFF	OFF
MAP Sensor	28.8" Hg	10.3" Hg	11" Hg	13.2" Hg
Short Term Fuel Trim (SHRTFT)	N/A	± 20%	± 20%	± 20%
Spark Advance in degrees BTDC	N/A	15° BTDC (± 2°)	33° BTDC	42° BTDC
Throttle Position (TP) Sensor	0%	9.8% (0.51v)	13.5% (0.62v)	16.8% (0.89v)
Vehicle Speed (VSS)	0 mph	0 mph	30 mph	55 mph

PID Chart Example - 1996 Civic Hatchback

INSTRUCTIONS

Locate the desired input or output parameter and record the known good value. Actual values may vary ± 20%. Read the details in the PID Name box for specific instructions. These PIDs were captured with a Mastertech™ with a Honda aftermarket cartridge.

1996-97 Civic Sedan DX with 1.6L I4 SOHC VTEC (D16Y8) Engine

PID NAME & ACRONYM	KOEO	HOT IDLE	30 MPH	55 MPH
A/C Switch	On/Off	On w/ A/C On	Off	Off
A/C Clutch	On/Off	On w/ A/C On	Off	Off
Alternator	N/A	32%	31%	30%
Alternator Control	N/A	14.5v	14.3v	14.1v
BARO Sensor	2.77v	2.77v	2.77v	2.77v
Battery	10.5-12.5v	14.5v	14.3v	14.4v
Calculated Load Value (CLV)	N/A	28%	37%	40%
DTC Number	---	---	---	---
Electronic Load Detector	4.2a	8.2a	10.1a	11.2a
Engine Speed	0 rpm	908 rpm	1410 rpm	2275 rpm
ECT Sensor	152°F	198°F	196°F	196°F
EVAP Purge Duty Cycle	N/A	0%	0-80%	0-90%
Fan Relay	Off	On w/ Fan On	Off	Off
Fuel Status 1	N/A	Closed	Closed	Closed
HO2S-1 (Front)	N/A	0.39v	0.48v	0.76v
HO2S-2 (Rear)	N/A	0.49v	0.55v	0.65v
Oxygen Sensor HTRC	N/A	On	On	On
Idle Air Control (IAC) Motor	N/A	47 counts	---	---
Intake Air Temperature Sensor	3.32v	90-102°F	90-102°F	90-102°F
Knock Advance	N/A	0.5° BTDC	---	---
Long Term Fuel Trim (LONGFT)	N/A	± 20%	± 20%	± 20%
Main Relay (FP Relay)	Off	On	On	On
MAP Sensor	4.8v	1.05v	1.10v	1.21v
MIL On or Off	OFF	OFF	OFF	OFF
MIL Status	OFF	OFF	OFF	OFF
PSP Switch	On	On	Off	Off
SCS (Service Connector)	Open	Open	Open	Open
Shift Lock	N/A	High	High	High
Short Term Fuel Trim (SHRTFT)	N/A	± 20%	± 20%	± 20%
Spark Advance in degrees BTDC	N/A	15.2° BTDC	33° BTDC	42° BTDC
Starter Switch	KOEC: ON	Off	Off	Off
Throttle Position (TP) Sensor	0.55v	0.49v	1.05v	1.12v
VTEC Pressure Switch	N/A	On	On	On
VTEC Solenoid	N/A	Off	Off	Off
Vehicle Speed (VSS)	0 mph	0 mph	30 mph	55 mph

PID Chart Example - 1996 Integra Hatchback

<u>INSTRUCTIONS</u>

Locate the desired input or output parameter and record the known good value. Actual values may vary ± 20%. Read the details in the PID Name box for specific instructions. These PID values were captured using the Scan Tool in OBD II Generic Cartridge.

1996-97 Integra Hatchback with 1.8L I4 SMFI DOHC (B18B1) Engine

PID NAME & ACRONYM	KOEO	HOT IDLE	30 MPH	55 MPH
Calculated Load Value (CLV)	N/A	N/A	N/A	N/A
DTC Number	---	---	---	---
Engine Speed	0 rpm	0 rpm	2365 rpm	2589rpm
Engine Coolant Temperature (ECT) Sensor	N/A	190°F	187°F	189°F
Fuel System 1	Closed	Closed	Closed	Closed
Fuel System 2	N/A	N/A	N/A	N/A
HO2S-1 (Front)	N/A	0.1-0.9v	0.1-0.9v	0.1-0.9v
HO2S-2 (Rear)	N/A	0.4-0.6v	0.4-0.6v	0.4-0.6v
Intake Air Temperature Sensor	N/A	113°F	111°F	106°F
Long Term Fuel Trim (LONGFT)	N/A	2%	3%	2%
Malfunction Indicator Lamp (MIL)	Off	Off	Off	Off
MAP Sensor	N/A	8.2" Hg	22.4" Hg	18" Hg
Short Term Fuel Trim (SHRTFT)	N/A	1-3%	0-3%	0-1%
Spark Advance in degrees BTDC	N/A	16.1° BTDC	35° BTDC	35° BTDC
Throttle Position (TP) Sensor	N/A	9.8%	20%	15-19%
Vehicle Speed (VSS)	0 mph	0 mph	30 mph	55 mph

MAP Sensor Tip - The KOEO MAP PID reading ("Hg) subtracted from the KOER PID reading ("Hg) should equal the vacuum reading from a vacuum gauge connected to manifold vacuum (at the same engine speed).

Spark Timing Tip - The Spark Advance PID value should equal the timing advance observed from a Timing Light when measured at the same engine speed.

Accord Pin Charts - Part One

1996-97 Accord 2.2L I4 MFI VIN CD5, CD7 & CE1 32-Pin 'A' Connector

PIN # / BOB #	COLOR	APPLICATION & ACRONYM	VALUE AT IDLE
A1, A2	YEL, BLU	Injector No. 4, No. 3	At hot idle: 2.0-3.3ms
A3, A4	RED, BRN	Injector No. 2, No. 1	At hot idle: 2.0-3.3ms
A5	ORN/BLU	Rear HO2S-2 Heater Control	HTRC On: <1v, Off: 12-14v
A6	ORN/BLK	Front HO2S-1 Heater Control	HTRC On: <1v, Off: 12-14v
A7	RED	EGR Control Solenoid	Solenoid On: <1v, Off: 12-14v
A9, A22	BRN/BLK	Sensor Ground	<0.1v
A10, A23	BLK	Power Ground	<0.1v
A11, A24	YEL/BLK	Ignition Power From Relay	KOEO or KOER: 12-14v
A12	BLK/BLU	Idle Air Control (IAC) Valve	At hot idle: 8-10v
A13	GRN/WHT	Engine Mount Solenoid	Solenoid On: <1v, Off: 12-14v
A8, 14, 21, 26	---	---	---
A15	RED/YEL	EVAP Purge Cutoff Solenoid	Solenoid On: <1v, Off: 12-14v
A16	GRN/BLK	Fuel Pump Relay Control	Relay On: <1v, Off: 12-14v
A17	RED/BLU	A/C Clutch Relay Control	Relay On: <1v, Off: 12-14v
A18	L/GRN/RED	Check Engine Light	C/E Light On: <1v, Off: 12-14v
A19	WHT/GRN	Alternator Charging Signal	KOER: varies 7-8v
A20	YEL/GRN	Igniter Control Signal	Idle: 10%, 2500: 60% d/cycle
A25	ORN/GRN	Intake Air Resonator Solenoid	Solenoid On: <1v, Off: 12-14v
A27	GRN	Radiator Fan Relay Control	Relay On: <1v, Off: 12-14v
A28	GRN/WHT	EVAP Bypass Solenoid	Solenoid On: <1v, Off: 12-14v
A29	ORN/GRN	EVAP Canister Vent Solenoid	Solenoid On: <1v, Off: 12-14v
A30-A32	---	---	---

1996-97 Accord 2.2L I4 MFI VIN CD5, CD7 & CE1 25-Pin 'B' Connector

PIN # / BOB #	COLOR	APPLICATION & ACRONYM	VALUE AT IDLE
B1-2, B6-7	---	---	---
B3	BLU/YEL	A/T Shift Control Solenoid 'A'	KOER in D2, D3: 12v, D1, D4:<1v
B4	GRN/BLK	A/T: Lockup Control Solenoid 'B'	Lockup On: 12-14v, Off: 0v
B5	YEL	A/T: Lockup Control Solenoid 'A'	Lockup On: 12-14v, Off: 0v
B9-10, 19-21	---	---	---
B8	GRN/BLU	A/T: Gear Position Switch	In D_3: 0v, all others: 12-14v
B11	GRN/WHT	A/T: Shift Control Solenoid 'B'	KOER in D1, D2:12v, D3, D4:<1v
B12	WHT/GRN	Interlock Control Unit Signal	Key on & Brake on: 12-14v
B13	BLU/RED	A/T D4 Indicator Light Switch	In D4 w/light On: 0v, Off: 12-14v
B14	WHT/BLU	Mainshaft Speed Sensor GND	<0.1v
B15	ORN/BLU	Mainshaft Speed Sensor SIG	Engine running: Pulses
B16	GRN/RED	A/T Gear Position Switch	In DR: 0v, all others: 12-14v
B17	GRN/YEL	A/T Gear Position Switch	In D2: 0v, all others: 12-14v
B18	LGRN/WHT	A/T Gear Position Switch	In D1: 0v, all others: 12-14v
B22	BLU/YEL	Sensor Ground	<0.1v
B23	BLU/GRN	Countershaft Speed Sensor	With wheels turning: pulses
B24	LGRN/BLK	A/T Gear Position Switch	In D4: 0v, all others: 12-14v
B25	LT GRN	A/T Gear Position Switch	In P/N: 0v, all others: 12-14v

■ **NOTE:** *When <1V is shown, this indicates the reading should be less than 1.0v DC.*

Accord Pin Charts - Part Two

1996-97 Accord 2.2L I4 MFI VIN CD5, CD7 & CE1 31-Pin 'C' Connector

PIN # / BOB #	COLOR	APPLICATION & ACRONYM	VALUE AT IDLE
C1, C9, C11	---	---	---
C2	BLU	CKP Sensor	KOER: 0.900mv (AC)
C3	GRN	TDC Sensor	KOER: 1.00v (AC)
C4	YEL	CYP Sensor	KOER: 0.250mv (AC)
C5	RED/WHT	A/C Switch Signal	A/C On: <1v, Off: 12-14v
C6	BLU/RED	Start Signal	KOEC (cranking): 9-11v
C7	RED	Service Check Connector	Open: 4.80v, Closed: 0.1v
C8	LT GRN	K-Line Signal	KOEO or KOER: 12-14v
C10	WHT/YEL	Keep Alive Power	12-14v
C12	WHT	CKP Sensor Ground	KOER: 0.900mv (AC)
C13	RED	TDC Sensor Ground	KOER: 1.00v (AC)
C14	BLK	CYP Sensor Ground	KOER: 0.250mv (AC)
C15, 19, 21-31	---	---	---
C16	GRN	Power Steering Press. Switch	Straight: 0v, Turned: 11v
C17	WHT/RED	Alternator 'FR' Signal	KOER: Varies 0.5-4.5v
C18	ORN	Vehicle Speed Sensor	Moving at 50 mph: 60 Hz
C20	BRN	EVAP Purge Flow Switch	Switch On: 12v, Off: 0v

1996-97 Accord 2.2L I4 MFI VIN CD5, CD7 & CE1 16-Pin 'B' Connector

Pin # / BOB #	Color	Application & Acronym	Value at Idle
D1	RED/BLK	Throttle Position (TP) Sensor	At hot idle: 0.5v
D2	RED/WHT	Engine Coolant Temperature	At 180°F: 0.5-0.6v
D3	WHT/YEL	MAP Sensor Signal	In P/N at idle: 0.9v (sea level)
D4	YEL/WHT	MAP Sensor VREF	4.9-5.1v
D5	GRN/WHT	Brake Switch Signal	Brake On: 12-14v, Off: 0.1v
D7	WHT/RED	Front HO2S-1 Signal	Hot idle: 0-1v, Accel: 0.5-1v
D8	RED/YEL	Intake Air Temperature	At 100°F: 2-3v
D9	WHT/BLK	EGR Valve Lift Sensor	At hot idle: 1.2v
D10	YEL/BLU	Sensor VREF	4.9-5.1v
D11	GRN/BLU	Sensor Ground	<0.1v
D12	GRN/WHT	MAP Sensor Ground	<0.1v
D13	RED/WHT	Rear HO2S-2 Sensor Ground	<0.1v
D14	WHT/RED	Rear HO2S-2 Signal	Hot idle: 0-1v, Accel: 0.5-1v
D15	GRN/RED	Fuel Tank Pressure Sensor	Key on, and Fuel Cap off: 2.5v
D16	GRN/RED	Electric Load Detector	KOER: Varies 0.5-4.5v

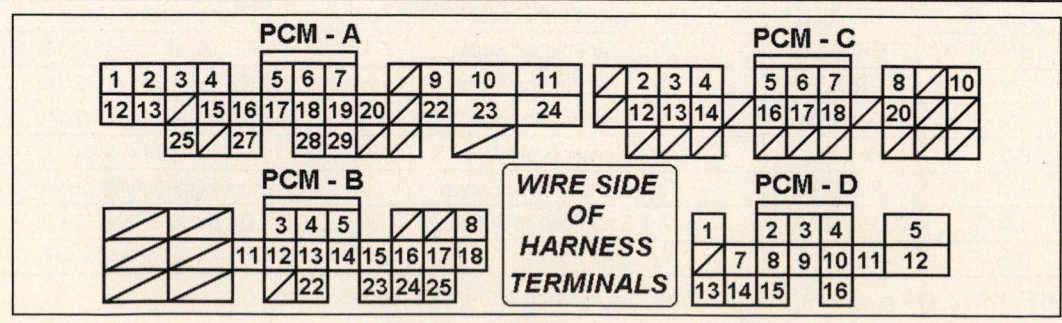

Accord Wiring Diagrams - 1996 Accord (Part One)

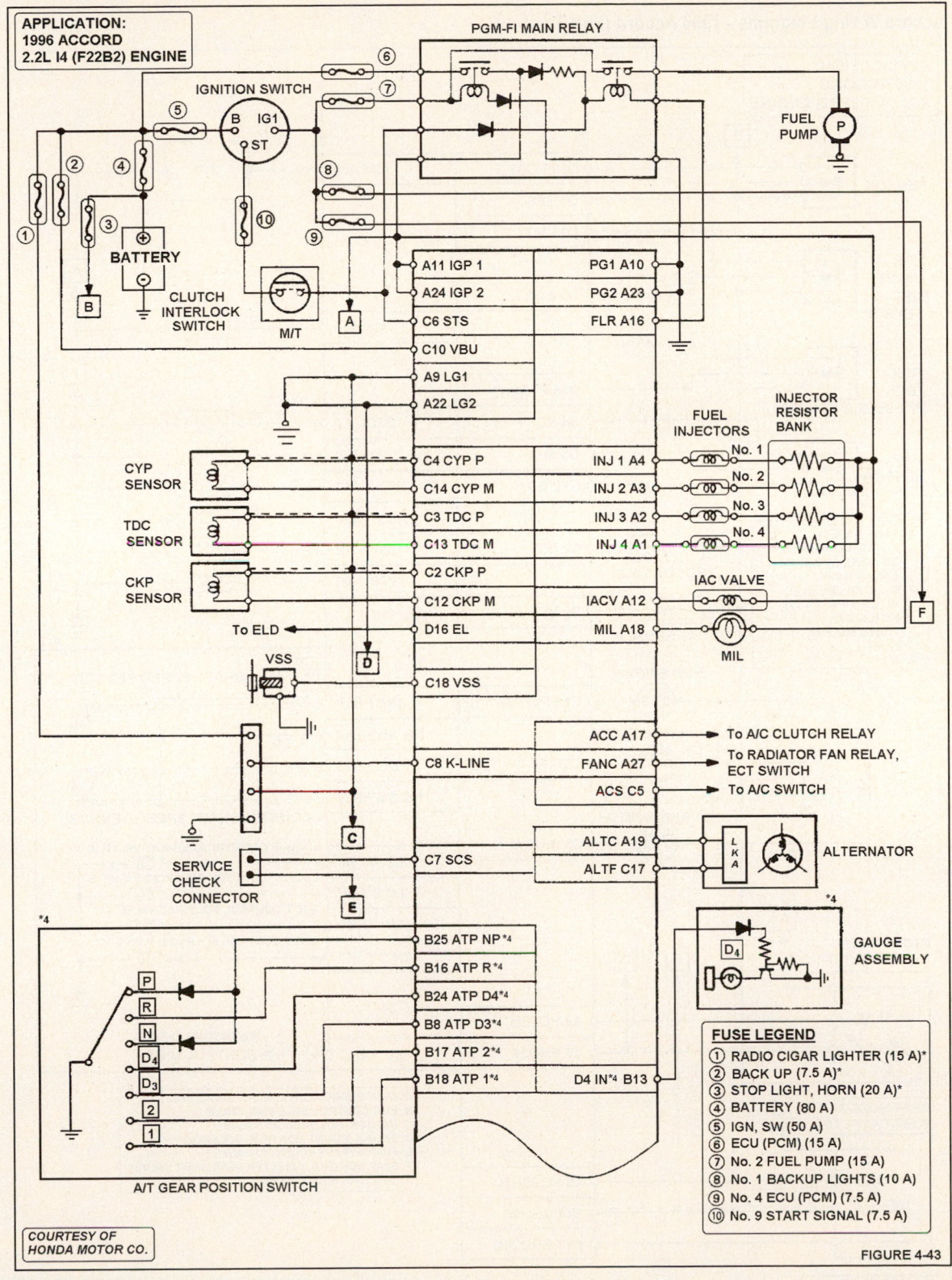

FIGURE 4-43

Accord Wiring Diagrams - 1996 Accord (Part Two)

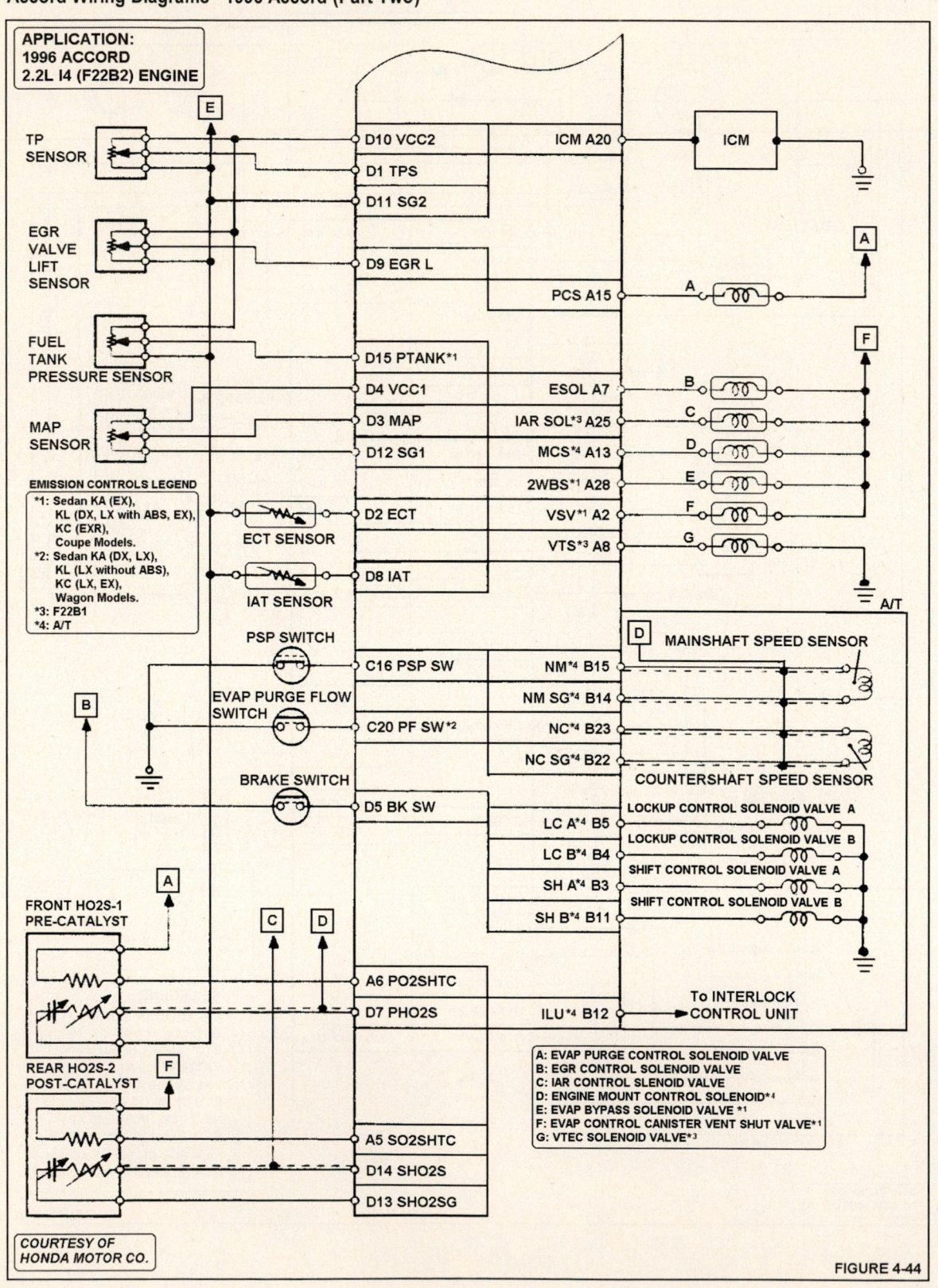

FIGURE 4-44

Accord Vacuum Diagram - 1996 Accord

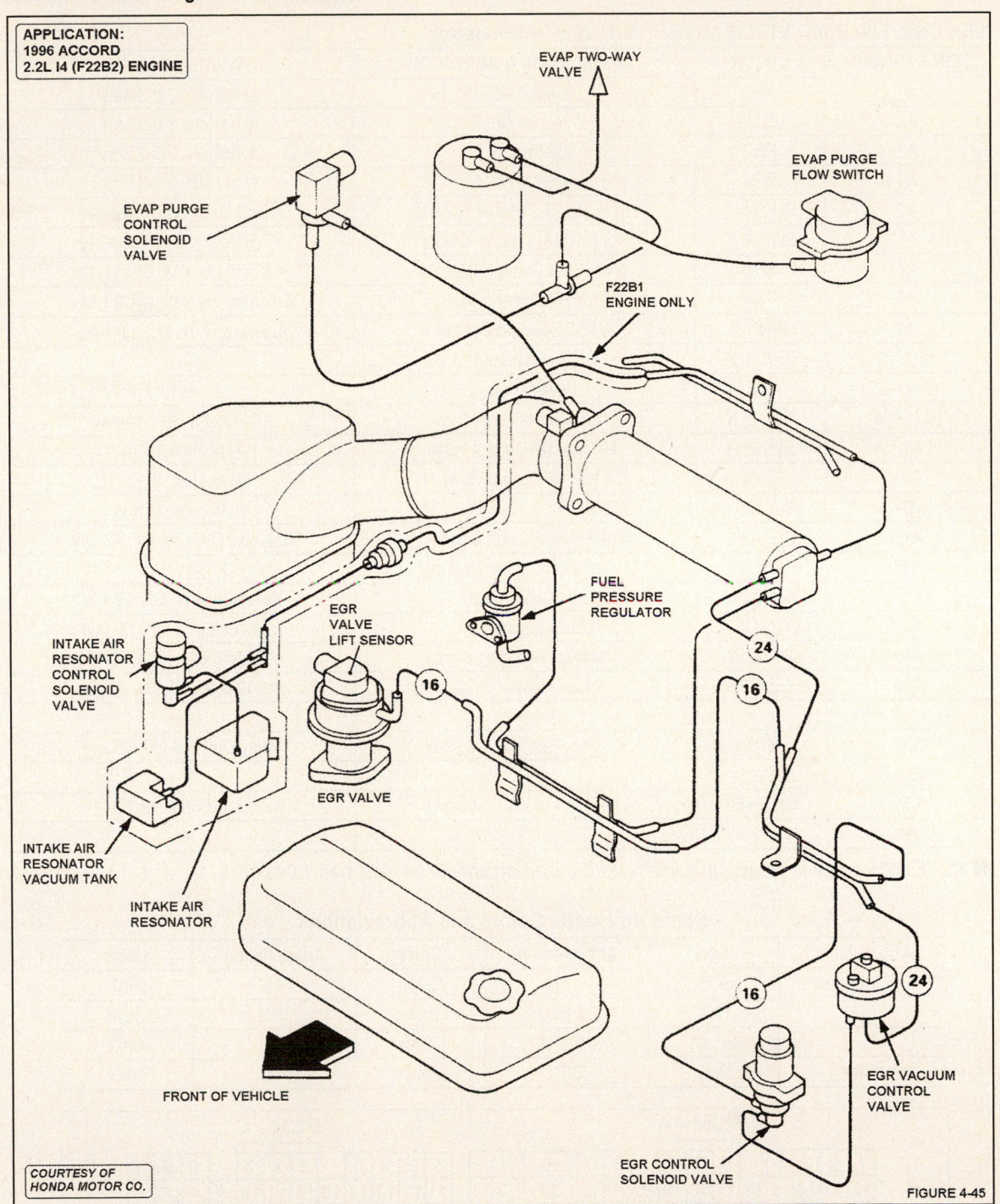

APPLICATION:
1996 ACCORD
2.2L I4 (F22B2) ENGINE

EVAP TWO-WAY VALVE

EVAP PURGE FLOW SWITCH

EVAP PURGE CONTROL SOLENOID VALVE

F22B1 ENGINE ONLY

INTAKE AIR RESONATOR CONTROL SOLENOID VALVE

EGR VALVE LIFT SENSOR

FUEL PRESSURE REGULATOR

INTAKE AIR RESONATOR VACUUM TANK

INTAKE AIR RESONATOR

EGR VALVE

FRONT OF VEHICLE

EGR CONTROL SOLENOID VALVE

EGR VACUUM CONTROL VALVE

COURTESY OF HONDA MOTOR CO.

FIGURE 4-45

Civic Pin Charts - Part One

1996 Civic 1.6L I4 MFI VTEC-E VIN EJ7 (M/T) 32-P 'A' Connector

PIN # / BOB #	COLOR	APPLICATION & ACRONYM	VALUE AT IDLE
A1	YEL	Injector No. 4	At hot idle: 2.0-3.3ms
A2	BLUE	Injector No. 3	At hot idle: 2.0-3.3ms
A3	RED	Injector No. 2	At hot idle: 2.0-3.3ms
A4	BRN	Injector No. 1	At hot idle: 2.0-3.3ms
A5	BLK/WHT	Rear HO2S-2 Heater Control	HTRC On: <1v, Off: 12-14v
A6	BLK/WHT	Front HO2S-1 Heater Control	HTRC On: <1v, Off: 12-14v
A7	RED	CVT: EGR Control Solenoid	Solenoid On: <1v, Off: 12-14v
A7	PNK	M/T: EGR Control Solenoid	Solenoid On: <1v, Off: 12-14v
A8	GRN/YEL	VTEC Solenoid Control	At low rpm: 0v, high rpm: 12v
A9, A22	BRN/BLK	Logic Ground	<0.1v
A10, A23	BLK	Power Ground	<0.1v
A11, A24	YEL/BLK	Ignition Power From Relay	KOEO or KOER: 12-14v
A12	BLK/BLU	M/T Idle Air Control Valve	At hot idle: 8-10v
A13	ORN	A/T Idle Air Control Valve 'N'	At hot idle: 8-10v
A14	BLK/BLU	A/T Idle Air Control Valve 'P'	At hot idle: 8-10v
A15	RED/YEL	EVAP Purge Cutoff Solenoid	Solenoid On: <1v, Off: 12-14v
A16	GRN/YEL	Fuel Pump Relay Control	Relay On: <1v, Off: 12-14v
A17	BLK/RED	A/C Clutch Relay	Relay On: <1v, Off: 12-14v
A18	GRN/ORN	Check Engine Light	Light On: <1v, Off: 12-14v
A19	WHT/GRN	Alternator Charging Signal	KOER: 3-5v, w/load: 2-4v
A20	YEL/GRN	Igniter Control Signal	Idle: 10%, 2500: 60% d/cycle
A21, A25-26	---	---	---
A27	GRN	Radiator Fan Relay Control	Relay On: <1v, Off: 12-14v
A28, A29	---	---	---
A30	WHT/RED	Interlock Control Unit Signal	Key on & Brake on: 12-14v
A31-A32	---	---	---

■ **NOTE:** *When <1V is shown, this indicates the reading should be less than 1.0v DC.*

Standard Honda Colors and Abbreviations

Abbreviation	Color	Abbreviation	Color	Abbreviation	Color
BLK	Black	GRN	Green	PNK	Pink
BLU	Blue	LT BLU	Lt. Blue	TAN	Tan
BRN	Brown	LT GRN	Lt. Green	WHT	White
GRY	Gray	ORN	Orange	YEL	Yellow

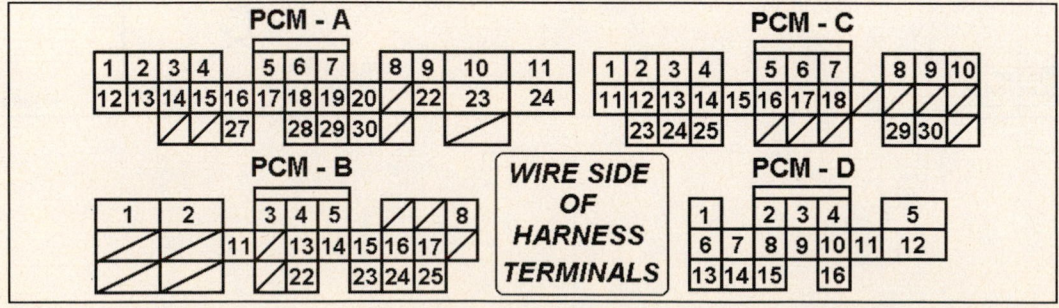

Civic Pin Charts - Part Two

1996 Civic 1.6L I4 MFI VTEC-E VIN EJ7 (M/T) 31-P 'C' Connector

PIN # / BOB #	COLOR	APPLICATION & ACRONYM	VALUE AT IDLE
C1	BLU/RED`	CKF Sensor 'P'	KOER: 0.900mv (AC)
C2	BLU	CKP Sensor 'P'	KOER: 0.900mv (AC)
C3	GRN	TDC Sensor 'P'	KOER: 1.00v (AC)
C4	YEL	CYP Sensor 'P'	KOER: 0.250mv (AC)
C5	BLU/RED	A/C Switch Signal	A/C On: <1v, Off: 12-14v
C6	BLU/ORN	Starter (Cranking) Signal	KOEC: 9-11v
C7	BRN	Service Check Connector	Open: 4.80v, Closed: 0.1v
C8	LT BLU	K-Line Signal	KOEO: 12v, 50% Duty cycle
C9	GRY	TMA Signal	Ignition Switch on: Pulses
C10	WHT/BLU	Keep Alive Power	12-14v
C11	WHT/RED	CKF Sensor Ground	KOER: 0.900mv (AC)
C12	WHT	CKP Sensor Ground	KOER: 0.900mv (AC)
C13	RED	TDC Sensor Ground	KOER: 1.00v (AC)
C14	BLK	CYP Sensor Ground	KOER: 0.250mv (AC)
C15	BLU/BLK	VTEC Pressure Switch	At low rpm: 0v, at high rpm: 12v
C16	GRN	Power Steering Press. Switch	Straight: 0v, Turned: 11v
C17	WHT/RED	Alternator 'FR' Signal	KOER: Varies 0.5-4.5v
C18	BLU/WHT	Vehicle Speed Sensor	Moving over 50 mph: 60 Hz
C19-22, 26-28	---	---	---
C23	BLK	M/T HO2S Pump Cell (IP+)	Ignition Switch on: 0.5-5.3v
C24	RED	M/T HO2S Common IP-& VS-	Hot Engine: 2.6-2.8v
C25	WHT	M/T VS Cell Voltage VS+	Ignition Switch on: 6.5-7.5v
C29	LT GRN	CVT Gear Position Switch	In P/N: 0v, all others: 12v
C29	RED	M/T Clutch Switch Signal	Clutch In: 0v, Clutch Out: 5v
C30	PNK	CVT TMB Signal	Ignition Switch on: Pulses

1996 Civic 1.6L I4 MFI VTEC-E VIN EJ7 (M/T) 16-P 'D' Connector

PIN #/BOB #	COLOR	APPLICATION & ACRONYM	VALUE AT IDLE
D1	RED/BLK	Throttle Position (TP) Sensor	At hot idle: 0.5v
D2	RED/WHT	Engine Coolant Temperature	At 180°F: 0.5-0.6v
D3	RED/GRN	MAP Sensor Signal	In P/N at Idle: 0.9v (sea level)
D4	YEL/RED	MAP Sensor VREF	4.9-5.1v
D5	GRN/WHT	Brake Switch Signal	Brake On: 12-14v, Off: 0.1v
D6	RED/BLU	Knock Sensor	No knock present: 2.5v
D7	WHT	Front HO2S-1 Signal	Hot idle: 0-1v, Accel: 0.5-1v
D7	WHT	M/T Only : Label Signal	Engine running: 0.3-4.9v
D8	RED/YEL	Intake Air Temperature	At 100°F: 2-3v
D9	WHT/BLK	CVT EGR Valve Lift Sensor	At hot idle: 1.2v
D10	YEL/BLU	Sensor VREF	4.9-5.1v
D11	GRN/BLK	Sensor Ground	<0.1v
D12	GRN/WHT	MAP Sensor Ground	<0.1v
D13	GRN/BLK	Rear HO2S-2 Signal GND	Hot idle: 0-1v, Accel: 0.5-1v
D14	WHT/RED	Rear HO2S-2 Signal	Hot idle: 0-1v, Accel: 0.5-1v
D16	GRN/RED	Electric Load Detector	KOER: Varies 2.5-3.5v

■ NOTE: *When <1V is shown, this indicates the reading should be less than 1.0v DC.*

Civic Wiring Diagram - 1996 Civic (Part One)

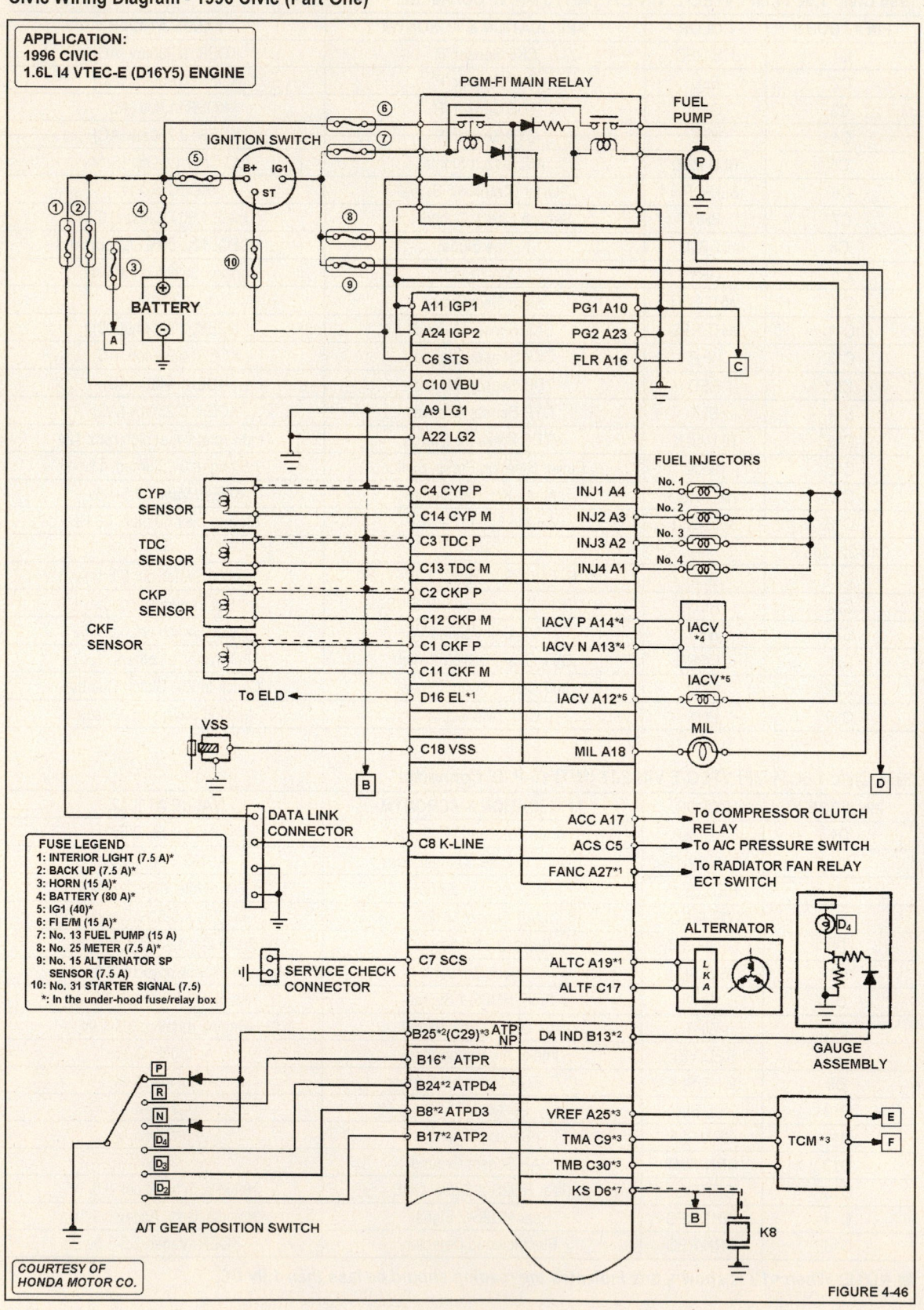

APPLICATION:
1996 CIVIC
1.6L I4 VTEC-E (D16Y5) ENGINE

FUSE LEGEND
1: INTERIOR LIGHT (7.5 A)*
2: BACK UP (7.5 A)*
3: HORN (15 A)*
4: BATTERY (80 A)*
5: IG1 (40)*
6: FI E/M (15 A)*
7: No. 13 FUEL PUMP (15 A)
8: No. 25 METER (7.5 A)*
9: No. 15 ALTERNATOR SP
   SENSOR (7.5 A)
10: No. 31 STARTER SIGNAL (7.5)
   *: In the under-hood fuse/relay box

COURTESY OF
HONDA MOTOR CO.

FIGURE 4-46

Civic Wiring Diagram - 1996 Civic (Part Two)

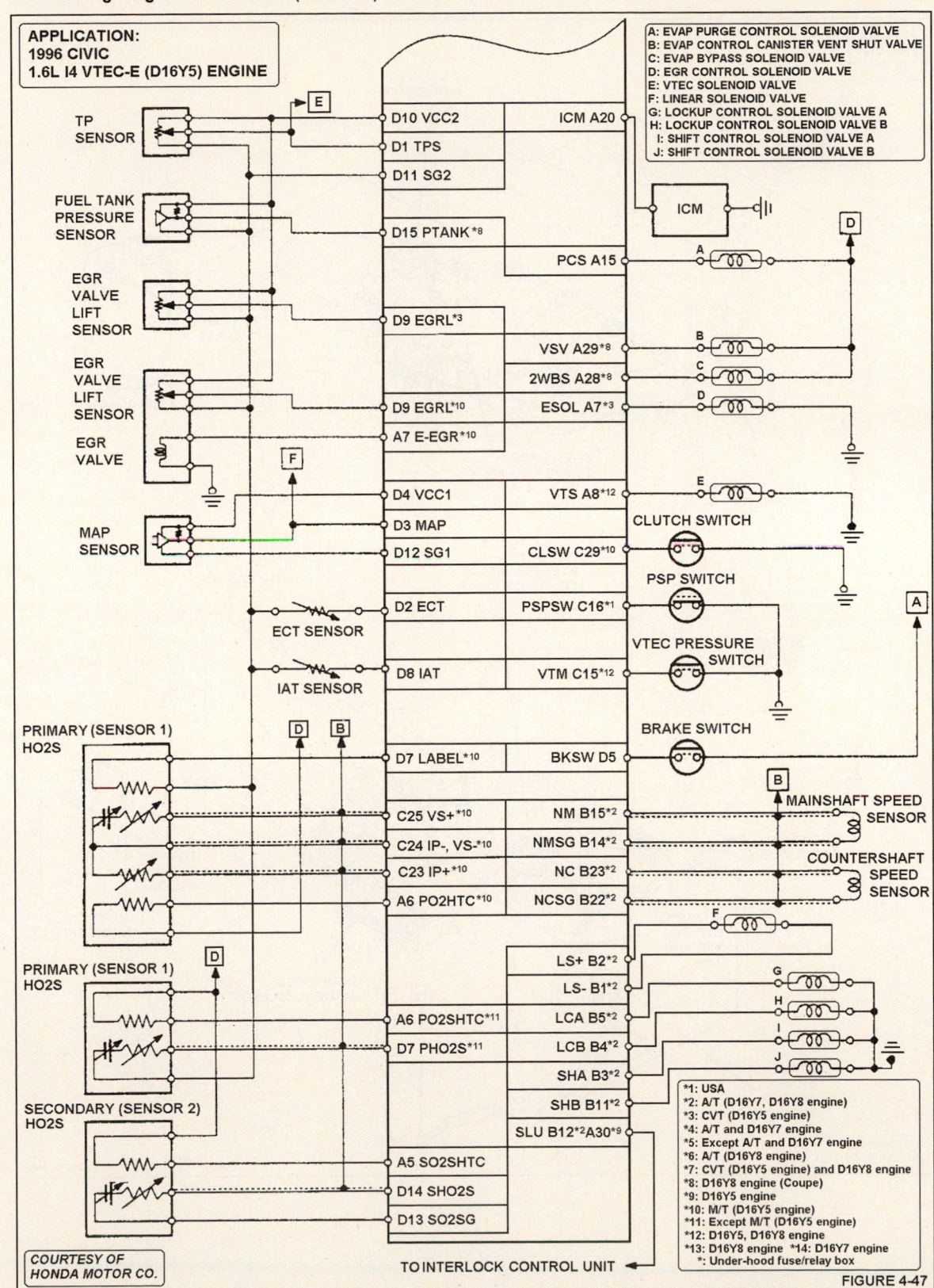

FIGURE 4-47

Civic Vacuum Diagram - 1996 Civic

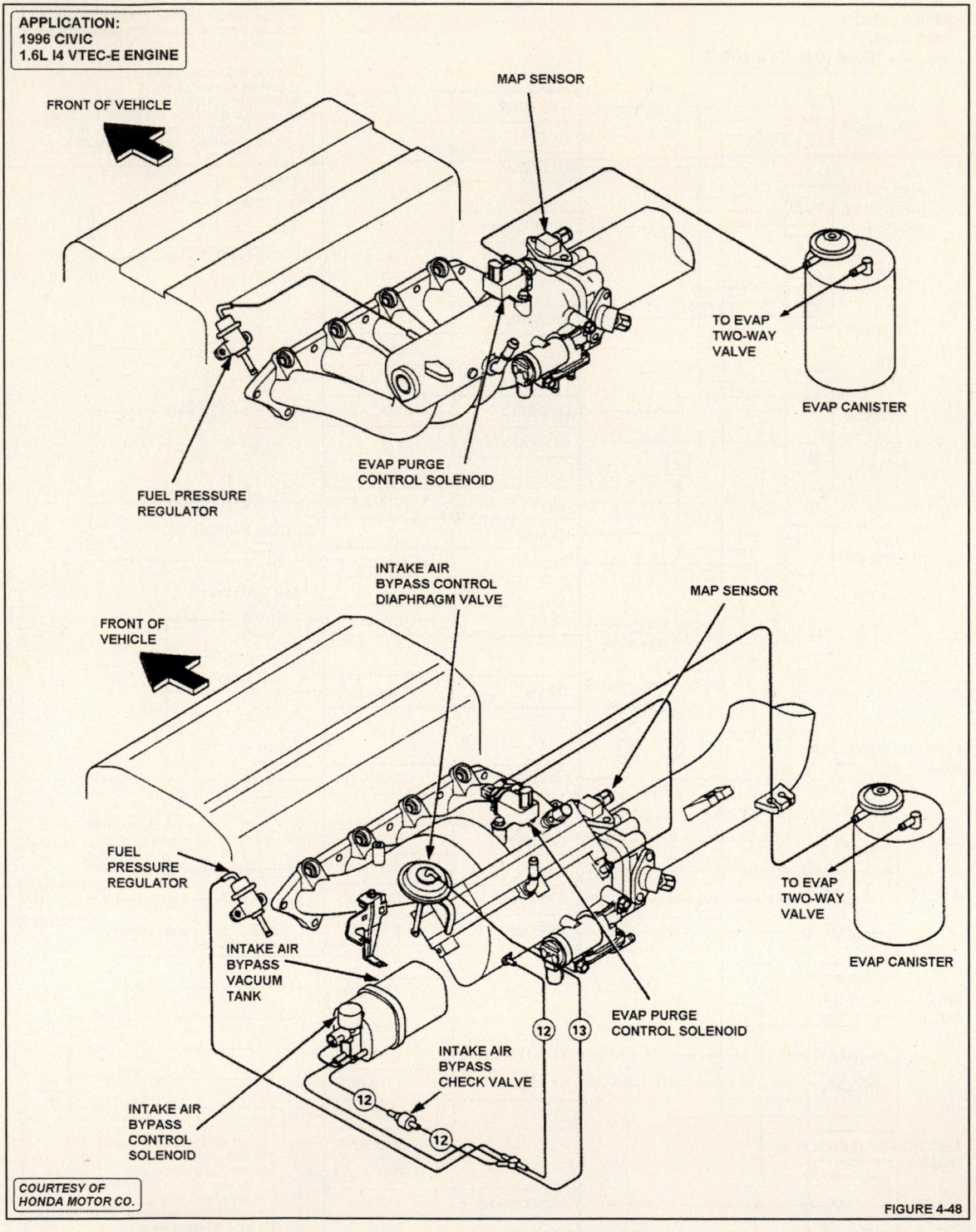

APPLICATION:
1996 CIVIC
1.6L I4 VTEC-E ENGINE

FRONT OF VEHICLE

MAP SENSOR

TO EVAP TWO-WAY VALVE

EVAP CANISTER

EVAP PURGE CONTROL SOLENOID

FUEL PRESSURE REGULATOR

INTAKE AIR BYPASS CONTROL DIAPHRAGM VALVE

MAP SENSOR

FRONT OF VEHICLE

FUEL PRESSURE REGULATOR

TO EVAP TWO-WAY VALVE

EVAP CANISTER

INTAKE AIR BYPASS VACUUM TANK

INTAKE AIR BYPASS CHECK VALVE

EVAP PURGE CONTROL SOLENOID

INTAKE AIR BYPASS CONTROL SOLENOID

*COURTESY OF
HONDA MOTOR CO.*

FIGURE 4-48

Integra Pin Charts - Part One

1996 Integra 1.8L I4 SMFI VIN DC4 32-P 'A' Connector

Pin # / BOB #	Color	Application & Acronym	Value at Idle
A1	YEL	Injector No. 4	At hot idle: 2.0-3.3ms
A2	BLU	Injector No. 3	At hot idle: 2.0-3.3ms
A3	RED	Injector No. 2	At hot idle: 2.0-3.3ms
A4	BRN	Injector No. 1	At hot idle: 2.0-3.3ms
A5	GRN/RED	Rear HO2S-2 Heater Control	HTRC On: <1v, Off: 12-14v
A6	ORN/BLK	Front HO2S-1 Heater Control	HTRC On: <1v, Off: 12-14v
A7-8, A13-14	---	---	---
A9, A22	BRN/BLK	Logic Ground	0.1v
A10, A23	BLK	Power Ground	0.1v
A11, A24	YEL/BLK	Ignition Power From Relay	KOEO or KOER: 12-14v
A12	BLK/BLU	Idle Air Control (IAC) Valve	At hot idle: 8-10v
A15	RED	EVAP Purge Cutoff Solenoid	EVAP solenoid On: <1v, Off: 12-14v
A16	GRN/BLK	Fuel Pump Relay Control	FP Relay On: <1v, Off: 12-14v
A17	BLK/RED	A/C Clutch Relay	A/C Relay On: <1v, Off: 12-14v
A18	GRN/ORN	Check Engine Light	C/E Light On: <1v, Off: 12-14v
A19	WHT/GRN	Alternator Charging Signal	KOER: varies 7-8v
A20	YEL/GRN	Igniter Control Signal	Idle: 10%, 2500: 60% d/cycle
A21, A28-32	---	---	---
A25	WHT/BLK	VREF TCM Signal	Ignition Switch On: 5v, Off; 0v
A26	PNK/BLU	Intake Air Bypass Solenoid	IAB Solenoid On: <1v, Off: 12-14v
A27	GRN	Radiator Fan Relay Control	Fan Relay On: <1v, Off: 12-14v

■ **NOTE:** *When <1V is shown, this indicates the reading should be less than 1.0v DC.*

Standard Honda Colors and Abbreviations

Abbreviation	Color	Abbreviation	Color	Abbreviation	Color
BLK	Black	GRN	Green	PNK	Pink
BLU	Blue	LT BLU	Lt. Blue	TAN	Tan
BRN	Brown	LT GRN	Lt. Green	WHT	White
GRY	Gray	ORN	Orange	YEL	Yellow

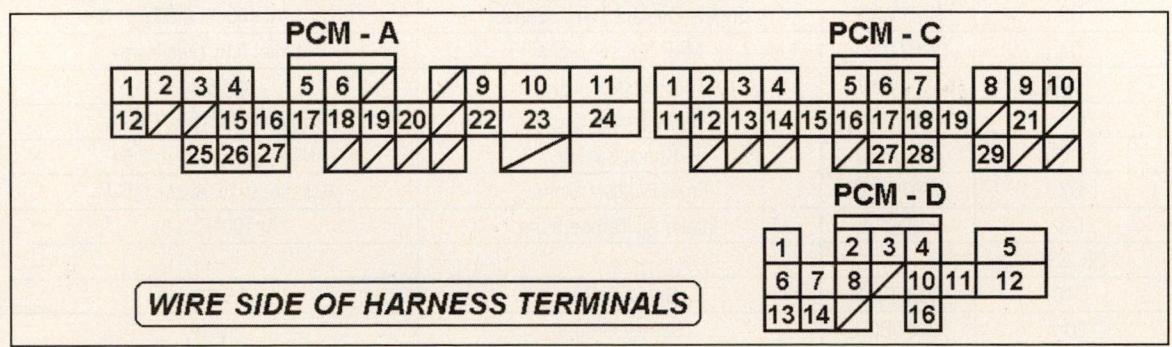

Integra Pin Charts - Part One

1996 Integra 1.8L I4 SMFI VIN DC4 31-P 'C' Connector

Pin # / BOB #	Color	Application & Acronym	Value at Idle
C1	BLU/RED	CKF Sensor 'P'	KOEO: 0.900mv (AC)
C2	BLU/GRN	CKP Sensor 'P'	KOEO: 0.900mv (AC)
C3	ORN/BLU	TDC Sensor 'P'	KOER: 1.00v (AC)
C4	ORN	CYP Sensor 'P'	KOER: 0.250mv (AC)
C5	BLU/RED	A/C Switch Signal	A/C On: <1v, Off: 12-14v
C6	BLU/WHT	Starter Signal	KOEC (Cranking): 9-11V
C7	BRN/WHT	Service Check Connector	Open: 4.80v, jumped: 0.1v
C8	GRN/WHT	K-Line Signal	KOEO: 12-14v
C9	YEL	FAS TCM Signal	At idle: 4.5-5.5v
C10	WHT/BLU	Keep Alive Power	12-14v
C11	WHT/RED	CKF Sensor Ground	KOER: 0.900mv (AC)
C12	BLU/YEL	CKP Sensor Ground	KOER: 0.900mv (AC)
C13	WHT/BLU	TDC Sensor Ground	KOER: 1.00v (AC)
C14	WHT	CYP Sensor Ground	KOER: 0.250mv (AC)
C15	---	---	---
C16	GRN	Power Steering Pressure Switch	Wheel straight: 0v, turned: 11v
C17	WHT/RED	Alternator 'FR' Signal	KOER: Varies 1.5-3.5v
C18	ORN	Vehicle Speed Sensor	Moving at 50 mph: 60 Hz
C19	BLU	A/T: TCM Signal	Ignition Switch on: Pulses
C20	---	---	---
C21	LT GRN	BARO TCM Signal	Ignition Switch on: 2.5-3.5v
C22-26, 30-31	---	---	---
C27	GRY	AFSB TCM Signal	At idle: 4.5-5.5v
C28	GRN/BLU	AFSA TCM Signal	At idle: 4.5-5.5v
C29	LGRN/BLK	A/T Gear Position Switch	In P/N: 0v, all others: 12-14v

1996 Integra 1.8L I4 SMFI VIN DC4 16-P 'D' Connector

Pin # / BOB #	Color	Application & Acronym	Value at Idle
D1	RED/BLK	Throttle Position (TP) Sensor	At hot idle: 0.52v
D2	RED/WHT	Engine Coolant Temp. Sensor	At 180°F: 0.53v
D3	WHT/YEL	MAP Sensor Signal	At Idle: 0.9v (sea level)
D4	YEL/WHT	MAP Sensor VREF	4.9-5.1v
D5	GRN/WHT	Brake Switch Signal	Brake On: 12-14v, Off: 0v
D6	RED/BLU	Knock Sensor	No knock present: 2.5v
D7	WHT/RED	Front HO2S-1 Signal	Hot idle: 0-1v, Accel: 0.5-1v
D8	RED/YEL	Intake Air Temperature	At 100°F: 2-3v
D9, D15	---	---	---
D10	YEL/BLU	Sensor VREF	4.9-5.1v
D11	GRN/BLU	Sensor Ground	<0.1v
D12	GRN/WHT	MAP Sensor Ground	<0.1v
D13	ORN/BLU	Rear HO2S-2 Signal GND	<0.1v
D14	BLU/GRN	Rear HO2S-2 Signal	Hot idle: 0-1v, Accel: 0.5-1v
D16	GRN/RED	Electric Load Detector	KOER: Varies 2.5-3.5v

■ **NOTE:** *When <1V is shown, this indicates the reading should be less than 1.0v DC.*

Integra Wiring Diagram - 1996 Integra (Part One)

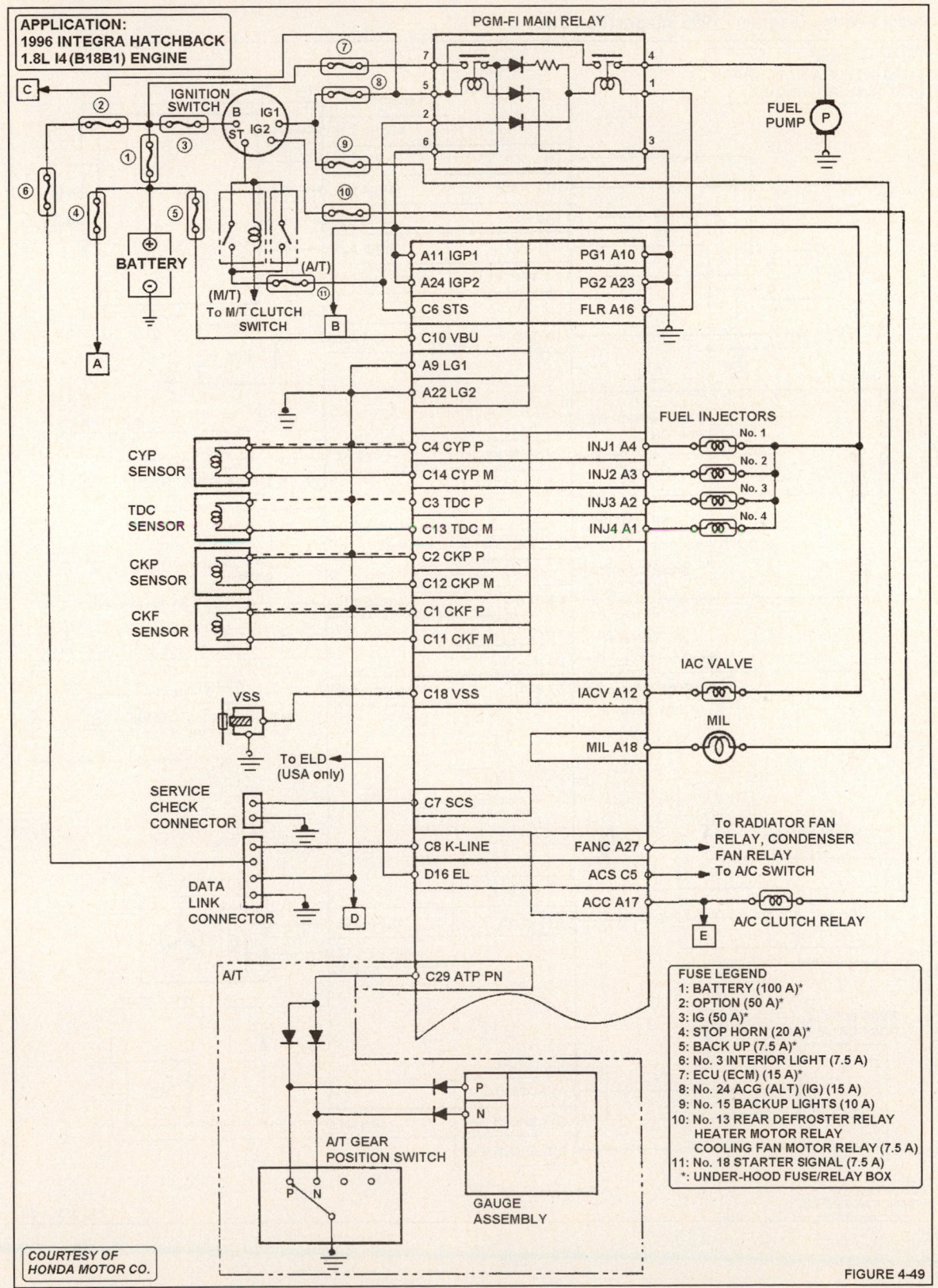

FIGURE 4-49

Integra Wiring Diagram - 1996 Integra (Part Two)

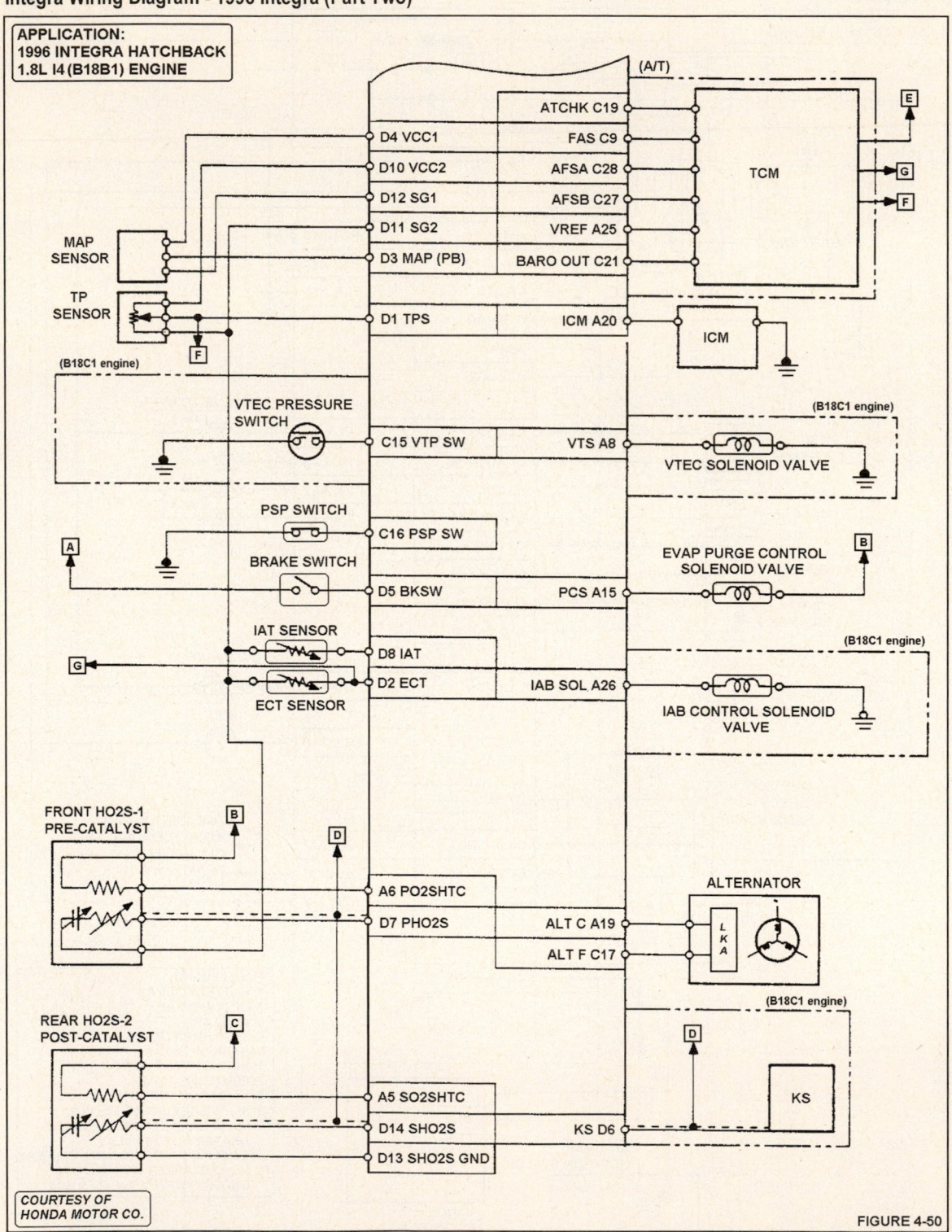

FIGURE 4-50

Integra Vacuum Diagram - 1996 Integra

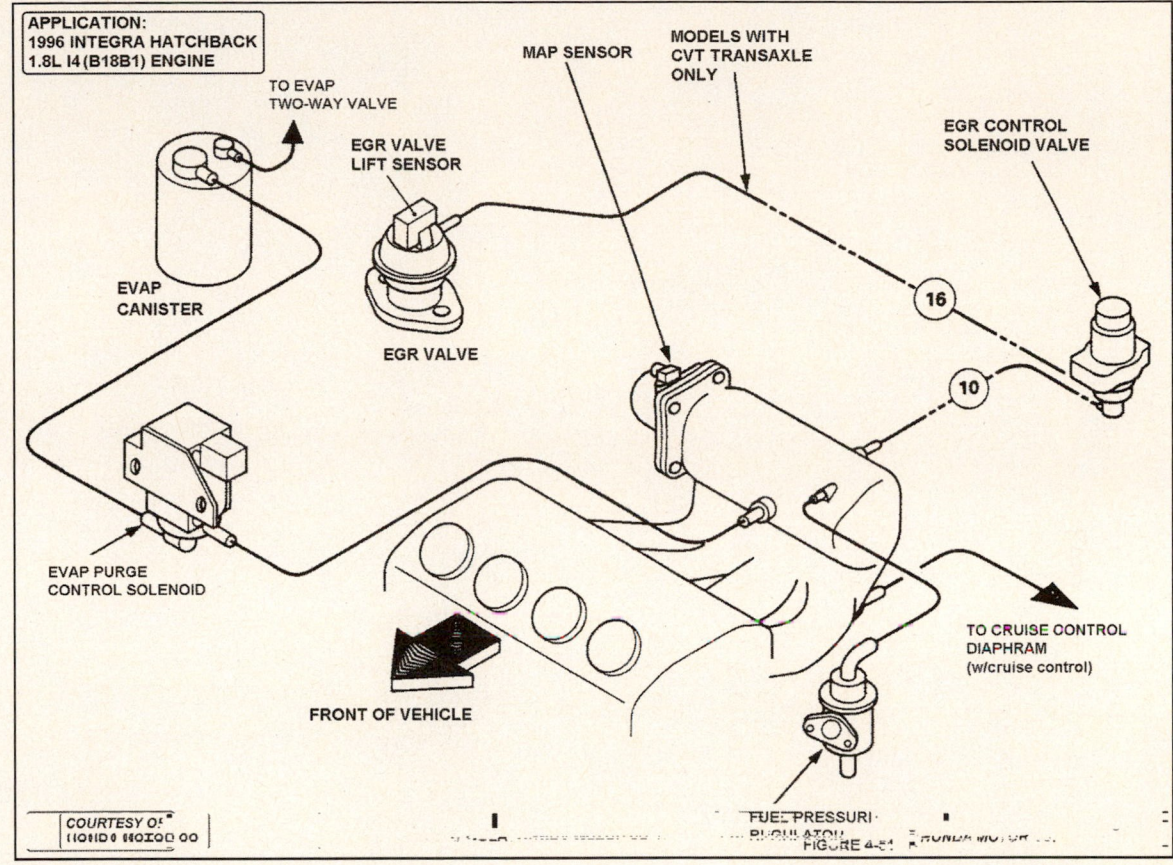

APPLICATION:
1996 INTEGRA HATCHBACK
1.8L I4 (B18B1) ENGINE

TO EVAP
TWO-WAY VALVE

MAP SENSOR

MODELS WITH
CVT TRANSAXLE
ONLY

EGR CONTROL
SOLENOID VALVE

EGR VALVE
LIFT SENSOR

EVAP
CANISTER

EGR VALVE

16

10

EVAP PURGE
CONTROL SOLENOID

TO CRUISE CONTROL
DIAPHRAM
(w/cruise control)

FRONT OF VEHICLE

FUEL PRESSURE
REGULATOR

FIGURE 4-51

COURTESY OF
HONDA MOTOR CO.

HONDA OBD CONTENTS

About This Manual

Introduction

This Acura/Honda Manual contains information on OBD I and OBD II System PID Data, Pin Charts and OBD II Diagnostics. It was developed to assist technicians during repair of problems related to diagnosis of the PCM and its engine subsystems.

Vehicle Coverage

Acura Vehicle coverage includes these models:

- Integra, Legend, NSX, Vigor, 2.2CL, 2.3CL, 2.5TL, 3.0CL, 3.2TL & 3.5RL and SLX (SUV)

Honda Vehicle coverage includes these models:

- Accord, Civic, Civic Del Sol, CR-V, Insight, Odyssey, Prelude and S2000 and Passport (SUV).

Four Acura/Honda Sections

The main features of the Acura/Honda sections are included in:

- Section 5: Introduction to Acura/Honda OBD II Diagnostics
- Section 6: Pin Tables for Acura Vehicles
- Section 7: Pin Tables for Honda Vehicles
- Section 8: Pin Tables for Acura & Honda SUVs

How to Use Each Section

Section 5 - Refer to this section to learn more about how to use PID Data with a Scan Tool or Pin Tables with a Breakout Box and DVOM during diagnosis of OBD I and OBD II Systems.

This section can also be used to learn how to use OBD II System Diagnostics, Trouble Code *enable criteria*, Freeze Frame Data and Example Drive Patterns for OBD II Main Monitors. This section includes example descriptions of common Serial and PID Data items.

Section 6 - Refer to this section to identify PCM descriptions and Pin terminals, wire colors and values for Acura vehicles (1990-2003).

Section 7 - Refer to this section to identify PCM descriptions and Pin terminals, wire colors and values for Honda vehicles (1990-2003).

Section 8 - Refer to this section to identify PCM descriptions and Pin terminals, wire colors and values for Badge vehicles (1994-2003).

Diagnostic Help

The PID Data and Pin Charts contain numerous pieces of diagnostic help found in the "known-good" values throughout this manual.

These values were obtained using a Digital Volt/Ohm Meter (DVOM), Breakout Box (BOB) or an Aftermarket Scan Tool.

What is PID Data?

What is PID Data?

PID is an acronym for Parameter Identification Data used to identify Powertrain Control Module items on both OEM and aftermarket Scan Tools. PID Data or data stream items available for display include:

- PCM analog input signals (ECT, EGRV, IAT, MAP & TP Sensors)
- PCM analog & digital output signals (EGR, IAC & MIL)
- PCM calculated values (LOAD, LONGFT, MISF & SHRTFT)

OBD I Serial Data & OBD II PID Data

PID Data is separated into two types: Onboard Diagnostics Version 1 and Version 2. OBD I PID Data is listed for Acura/Honda vehicles from 1992-95. OBD II PID Data is listed for vehicles from 1995-2003.

How to View PID Data

To view PID Data on a Scan Tool, connect the Scan Tool to the vehicle connector, select either the Generic or Manufacturer setup instructions, and select Parameter ID (data stream) from the menu.

Note: *If a Scan Tool will not power up or read PID Data, the first step is to verify that the power, ground, and Scan Tool cable connections are okay. To test that the Scan Tool is working properly, try it on another vehicle.*

Example of Scan Tool Connection

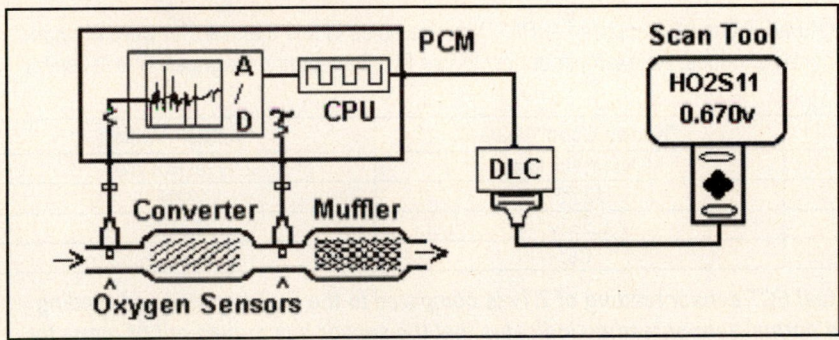

In this example, the Scan Tool is connected to the DLC in order to view "live" HO2S PID Data once it has been converted in the PCM.

How to Use PID Data

Information contained within the PID Data Charts can be used to:

- Validate a previous repair procedure
- Check the operation of a component before or after a repair
- Check the operation of a component or system by viewing "live" data from the vehicle computer data stream

Scan Tool OBD II PID Mode

Vehicles equipped with On Board Diagnostics (OBD) have a unique test mode that can be selected to allow access to vehicle Parameter Identification (PID Data) information. Part of the PID Data is generic in nature and can be accessed by any certified Scan Tool in OBD II Generic mode.

However, on Acura/Honda vehicles, some of the PID Data can only be accessed using a Scan Tool with the Acura/Honda OEM cartridge. This allows the Scan Tool to read the complete list of OEM Data. To learn more about the OEM Data List, refer to the Scan Tool instruction manual.

PID Data Display

An example of PID Data captured with an aftermarket Scan Tool is shown below. This data is from a 1997 Accord LX (2.2L I4 Engine).

PID Acronym	Description	Value at Idle
CLV	Calculated Load	27%
DTC	DTC Number	P0135
FUEL STATUS	Fuel Status 1	CL (Closed Loop)
ECT	ECT Sensor	185ºF
IAT	IAT Sensor	110ºF
LONGFT	Long Term Fuel Trim	+3%
SHRTFT	Short Fuel Trim	-1%

PID Data Comparison

Once actual PID Data has been captured, it can be compared to PID Data examples in this manual that were obtained from vehicles with "known good" operating values. An example of how to use PID Data during diagnosis of a "suspect" ECT sensor is shown below.

PID Acronym	"Known Good" Value	"Actual" Value
ECT	0.6v (at Hot Idle)	2.1v (at Hot Idle)
IAT	125ºF	122ºF
TP	0.532v	0.51v
Fuel Status	CL	CL

Note: *In this example, the actual ECT sensor reading of 2.1v is compared to the ECT known good reading of 0.6v. The higher than normal sensor reading indicates that the sensor has moved out of range for a hot engine.*

Diagnosis with PID Data

An example of how to use PID Data records to diagnose a vehicle with an intermittent stalling problem and no codes is shown below. PID Data recorded at idle speed reveals a defective MAP sensor that caused the MAP sensor signal to suddenly spike very low. When the MAP signal went low, the engine would stumble and die out.

The information in the table (in the Frame View) and picture (in the Graphing View) below shows the MAP Sensor spiked low and caused the engine to stall.

Vehicle: 1994 Accord EX 2.2L VIN CD5 (Frame View)

PID Item	Frame 1	Frame 2	Frame 3	Frame 4	Frame 5	Frame 6
RPM	628	635	640	650	620	625
INJ PW	3.5ms	3.3ms	10.2ms	3.6ms	3.3ms	3.4ms
MAP (Hg)	0.7" Hg	0.75" Hg	0.1" Hg	0.7" Hg	072" Hg	0.75" Hg

Vehicle: 1994 Accord EX 2.2L VIN CD5 (Graphing View)

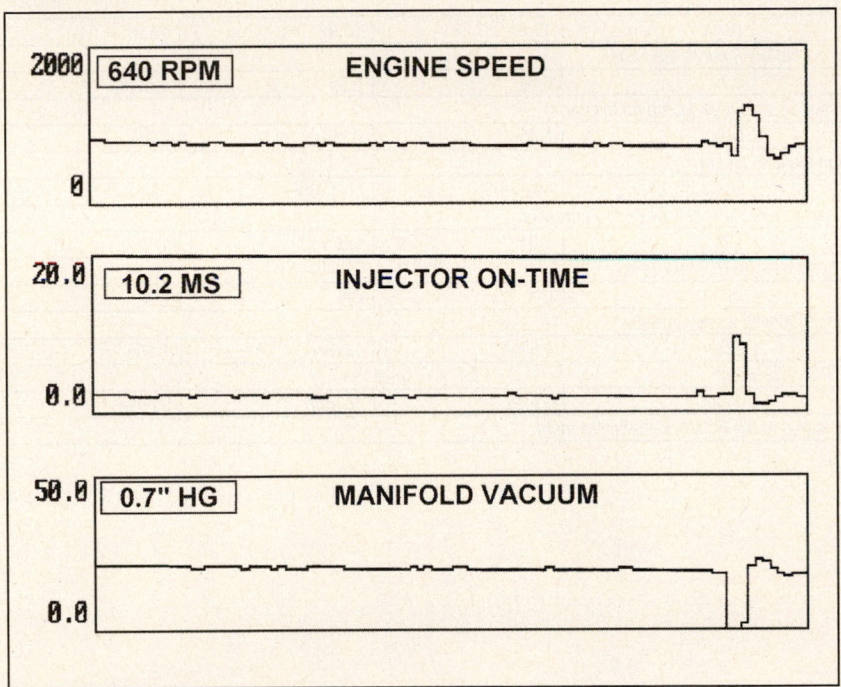

1992 Civic Serial Data Examples

The PID and Serial Data examples in this manual are arranged in table format. Each **table heading** identifies the year, vehicle, engine and the transmission type [All] for both, [A/T] for Automatic or [M/T] for Manual. Tables contain these information fields:

- PCM PID Acronym
- PID - Parameter Identification
- PID value at Key On
- PID value at hot idle
- PID values at 30 & 55 mph

1992 Civic 1.6L I4 MFI VIN EH3 [M/T] - (*This is the Table Heading*)

PCM PID Acronym	Parameter Identification	PID value at Key On	PID value at hot idle	PID value at 30 mph	PID value at 55 mph
BARO	BARO Pressure	2.89v	2.89v	2.89v	2.89v
BARO: Indicates the output from the BARO Sensor (in voltage).					
Battery	Battery Volts	10.7v	13.8v	13.9v	14.0v
BATTERY: Indicates the voltage of the direct battery input to the PCM.					
ECT	ECT Voltage	0.7v	0.5v	0.5v	0.6v
ECTV: Indicates ECT temperature expressed as a voltage (compare to charts)					
HO2S	Heated O2S	N/A	0.65v	0.25v	0.75v
HO2S: Indicates the signal from the heated oxygen sensor.					
IAT °F	Intake Air Temp	50-120°F	50-120°F	50-120°F	50-120°F
IAT °F: Indicates IAT temperature expressed as either (°) Celsius or (°) Fahrenheit					
MAP	MAP Sensor	4.7v	1.01v	1.07v	1.11v
MAP: Indicates the signal from the MAP Sensor to the PCM (in volts).					
PSP	PSP Switch	ON	ON	ON	ON
PSP: Indicates the status of the PSP Switch (Off with wheel turned).					
TP	TP Sensor	0.59v	0.53v	1.00v	1.10v
TP: Throttle Position Sensor signal to the PCM (in volts).					
VSS	Vehicle Speed	0 mph	0 mph	30 mph	55 mph
VSS: Indicates the Vehicle Speed Sensor signal to the PCM (converted to mph).					

PID DATA

1996 Civic PID Data Examples

The PID Data examples in this manual are arranged in table format. Each *table heading* identifies the year, vehicle, engine and the transmission type [All] for both, [A/T] for Automatic or [M/T] for Manual. Tables contain these information fields:

* PCM PID Acronym
* PID - Parameter Identification
* PID values at Key On & at hot idle
* PID values at 30 & 55 mph

1996 Civic 1.6L I4 MFI VIN EJ8 [All] - *(This is the Table Heading)*

PCM PID Acronym	Parameter Identification	PID value at Key On	PID value at hot idle	PID value at 30 mph	PID value at 55 mph
ACS	A/C Switch	OFF	On (AC on)	OFF	OFF
ACS: indicates status of the A/C Switch (the switch is either On or Off).					
ACC	A/C Clutch	OFF	On (AC on)	OFF	OFF
ACC: indicates status of the A/C Clutch (the clutch is either On or Off).					
Alt	Alternator	0v	14.5v	14.3v	14.2v
ALT: Alternator (voltage): indicates the output voltage from the alternator.					
ALTCNTL	Alternator Cntl.	0%	32%	31%	30%
ALTCNTL: indicates PCM alternator control command (in percentage).					
BARO	BARO Pressure	2.77v	2.77v	2.77v	2.77v
BARO: indicates the output from the BARO Sensor (in voltage).					
Battery	Battery Volts	11.8v	14.5v	14.3v	14.4v
Battery: indicates the voltage of the direct battery input to the PCM					
CLV	Engine Load	0%	28%	37%	40%
CLV: indicates the engine load value calculated by the PCM (MAP & RPM).					
DTC	Code Number	---	---	---	---
DTC: indicates the number of a Diagnostic Trouble Code stored in the PCM.					
ECT °F	ECT Degrees	152°F	198°F	196°F	196°F
ECT °F: indicates the ECT temperature expressed as (°) Celsius or (°) Fahrenheit.					
ECTV	ECT Voltage	0.6v	0.6v	0.6v	0.5v
ECTV: indicates the ECT temperature expressed as a voltage (compare to charts).					
ELD	ELD (detector)	4.2a	8.2a	10.1a	11.2a
ELD: indicates amount of amperage detected by the Electronic Load Detector.					
EVAP	Purge Solenoid	0%	0%	0-80%	0-90%
EVAP: indicates the PCM command to the EVAP Purge Solenoid (duty cycle %).					
FAN	Fan Relay Cntl.	OFF	ON	OFF	OFF
Fan: indicates the state of the Cooling Fan Control Relay.					
FUELST1	Fuel Status 1	OL	CL	CL	CL
FuelST1: indicates the PCM control of A/F Mixture (Open or Closed Loop).					
HO2S-1	Front HO2S	0.0v	0.39v	0.48v	0.76v
HO2S-1: indicates the signal from the pre-catalyst heated oxygen sensor.					
HO2S-2	Rear HO2S	0.0v	0.49v	0.55v	0.65v
HO2S-2: indicates the signal from the post-catalyst heated oxygen sensor.					
HTRC	Heater Control	OFF	ON	ON	ON
HTRC: indicates the command state to the heater from the PCM.					

1996 Civic PID Data Examples (Continued)

The PID Data examples in this manual are arranged in table format. Each *table heading* identifies the year, vehicle, engine and the transmission type [All] for both, [A/T] for Automatic or [M/T] for Manual. Tables contain these information fields:

- PCM PID Acronym
- PID - Parameter Identification
- PID values at Key On & at hot idle
- PID values at 30 & 55 mph

1996 Civic 1.6L I4 MFI VIN EJ8 [A/T] - (*This is the Table Heading*)

PCM PID Acronym	Parameter Identification	PID value at Key On	PID value at hot idle	PID value at 30 mph	PID value at 55 mph
IAT °F	Intake Air Temp	50-120 °F	50-120 °F	50-120 °F	50-120 °F
IAT °F: indicates the IAT temperature expressed as (°) Celsius or (°) Fahrenheit.					
IATV	Intake Air Temp	1.7-3.5v	1.7-3.5v	1.7-3.5v	1.7-3.5v
IATV: indicates IAT temperature as a voltage (compare readings to charts).					
IATV	Intake Air Temp	1.7-3.5v	1.7-3.5v	1.7-3.5v	1.7-3.5v
IATV: indicates IAT temperature as a voltage (compare readings to charts).					
IAC	Idle Air Control	N/A	47 counts	Varies	Varies
IAC: indicates the amount of counts commanded to the IAC Valve by the PCM.					
KS ADV	KS Advance	N/A	0.5° BTDC	Varies	Varies
KS ADV: indicates the amount of Knock Sensor Advance commanded by the PCM.					
LONGFT	Long Term FT	N/A	+3%	+2%	+3%
LONGFT: indicates the amount of change commanded to Long Term fuel trim.					
FP	FP Relay	Off	ON	ON	ON
FP: indicates status of the PCM command to the Fuel Pump Relay (On or Off).					
MAP	MAP Sensor	4.8v	1.0v	1.10v	1.21v
MAP: indicates the signal from the MAP Sensor to the PCM (in Volts or kPa).					
LONGFT	Long Term FT	N/A	+3%	+2%	+3%
LONGFT: indicates the amount of change commanded to Long Term fuel trim.					
MIL	MIL On or Off	OFF	ON w/fault	OFF	OFF
MIL: indicates the PCM command to the MIL (On with emission fault detected).					
MILSTAT	MIL Status	OFF	ON w/fault	OFF	OFF
MILSTAT: Indicates the status of the PCM command to the MIL.					
PSP	PSP Switch	OFF	ON (turned)	OFF	OFF
PSP: indicates the status of the PSP Switch (ON with wheels turned).					
SCS	SCS Connector	OPEN	OPEN	OPEN	OPEN
SCS Connector: indicates status of the Service Connector (CLOSED when jumped).					
SHIFT	Shift Lock	N/A	HIGH	HIGH	HIGH
Shift: indicates the status of the Shift Lock Signal.					
SHRTFT	Short Term FT	N/A	+1%	0%	-1%
SHRTFT: indicates the amount of change commanded to Short Term fuel trim.					
SPKADV	Spark Advance	N/A	15° BTDC	33° BTDC	42° BTDC
SPKADV: indicates the Spark Advance command from the PCM in degrees BTDC.					
TP	TP Sensor	0°	0-1°	4°	4-6°
TP: indicates Throttle Position sensor signal to the PCM (in degrees of rotation).					
VSS	Vehicle Speed	0 mph	0 mph	30 mph	55 mph
VSS: indicates the Vehicle Speed Sensor signal to the PCM (converted to mph).					

PIN Tables

What is a Pin Table?

A Pin Table is a term used in this manual to describe a chart or table that contains information on PCM and Breakout Box (BOB) Pins, individual wire colors of PCM circuits, and example values for devices that connect to the PCM. Pin Table information includes:

- Signals from various sensors (ECT, EVPL, IAT, MAP, TP)
- Signals from various switches (A/T Shift, Brake, Heater Fan)
- Signals from oxygen sensors (02S, HO2S-1, HO2S-2)
- Signals from output devices (EGR, EVAP, FP Relay, IAC, INJ)
- Power & ground signals (Direct Battery, Power & Sensor Ground)

Note: Acronyms in these examples are in the Glossary.

OBD I System Pin Table

The OBD I System Pin Charts are separated into three sections. They cover 1990-95 Acura, 1990-95 Honda and 1994-95 "Badge" Vehicles.

OBD II System Pin Charts

OBD II Pin Tables are separated into these three sections: 1995-2003 Acura, 1996-2003 Honda and 1996-2003 "Badge" Vehicles.

How to Connect the BOB & DVOM

To use Pin Chart information with a DVOM, a Breakout Box should be installed. To connect a BOB, first turn the ignition off and then remove the wire harness at the engine computer (PCM). Next, connect the appropriate BOB (with adapters) to the PCM and BOB connectors. This places the BOB between the PCM and wiring so that circuit measurements can be made at the pin connections on the BOB.

PCM to BOB Connection Graphic

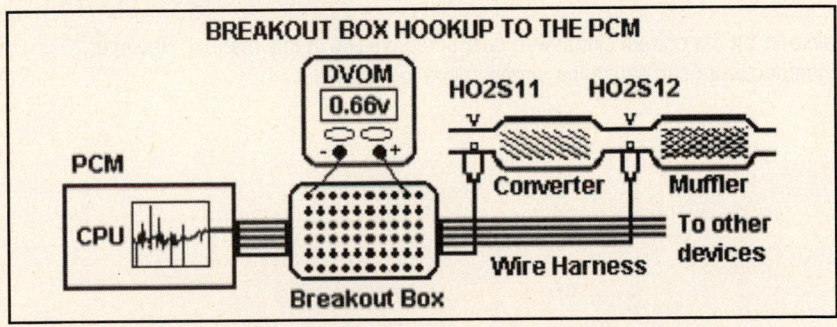

Note: *Read and record all OBD II codes and freeze frame data in the PCM before connecting the BOB as all codes and data are lost if the PCM connector is removed.*

How to use Pin Tables

Information contained within the Pin Tables can be used to:

- Test circuits for open, short to power or short to ground faults
- Check the operation of a component before or after a repair
- Check the operation of a component or system by viewing signals on PCM input/output circuits with a DVOM or Lab Scope

Testing Circuits with a Breakout Box

There are many different Breakout Box designs available for use to test the PCM and its related circuits. However, all of them require that the PCM wire harness is removed and the BOB installed between the PCM and wire harness connector. Several breakout boxes require the use of overlays in order to allow the tool to be used on more than one year or engine type. Always verify that the correct adapter and overlays are used to prevent misdiagnosis.

Power and Ground Circuit Checks

Measurements made at the BOB are accomplished via test leads and probes from the DVOM or a Lab Scope. If any of the terminals on the PCM or BOB are damaged or loose, test measurements made at the Breakout Box can be inaccurate. Be sure to verify that the PCM direct battery, ignition power, power ground and signal ground circuits are okay at the Breakout Box prior to starting a test sequence.

Note: **The voltage drop between battery (+) to KAPWR or VPWR at the BOB should be under 0.1v. The voltage drop between the battery negative (-) post and the ground point at the BOB should be < 0.1v.**

Pin Table Test Example

Once an "actual" PCM reading is recorded, it can be compared to an example from a vehicle with "known-good" values. In the example below from a 1997 Honda Accord 2.2L I4, the "actual" MAP sensor signal is higher than the "known-good" value at idle. In this case, the vehicle would run extremely rich because the MAP value is too high.

PCM Pin #	W/Color	Circuit Tested	Value at Idle
D9	W/BK	EGRV Sensor	0.9-1.0v

Wire Color Changes

Every effort has been made to obtain and list the correct circuit wire colors for vehicles in this manual. However, running changes from the vehicle manufacturers can cause the wrong colors to be listed.

Pin Voltage Table Examples

Pin Tables are arranged in table format in the manual. Each **table heading** identifies the year, vehicle, engine and the transmission type [All] for both, [A/T] for Automatic or [M/T] for Manual. They include:

- PCM Pin Number & Wire Color
- Circuit Description/Acronym for PCM Pin Connector
- Known good operating value at hot idle speed

1995 Civic VX 1.5L I4 VTEC VIN EH2 [All] 26Pin 'A' Connector

PCM Pin #	W/Color	Circuit Description (26-Pin)	Value at Idle
A1, 2	BR/Y	Injector 1, 4	Hot idle: 2.0-3.3ms
A3, 5	R/BL	Injector 2, 3	Hot idle: 2.0-3.3ms
INJ 3: indicates the command from the PCM to Injector 3 (in milliseconds or ms).			
A4	O/W	VTEC Solenoid	Idle:0v, Hi-Speed: 12v
VTEC Solenoid: indicates command from the PCM to the solenoid at idle & hi-rpm.			
A6	O/BK	LAF Sensor Heater Control	HTR on: 1v, off: 12v
HTRC: indicates the command from the PCM to turn the LAF Heater On or Off.			
A7	GN/Y	Fuel Pump Relay Control	Relay on: 1v, off: 12v
FP: indicates the command from the PCM to turn the FP Relay On or Off.			
A8, A10	---	Not Used	---
A9	GN/W	Idle Air Control Valve	Pulse Signals
IAC: indicates the command from the PCM to the IAC Valve (voltage pulse signals).			
A11	PK/GN	EGR Valve Lift Sensor	Hot idle: 1.2v
EGRV: indicates the signal from the EGRV Sensor to the PCM (in volts).			
A12	Y/GN	Radiator Fan Relay	Fan on: 1v, off: 12v
FAN: indicates the command from the PCM to control the Fan Relay.			
A13	GN/O	Check Engine Light	C/E on: 1v, off: 12v
CEL: indicates the command from the PCM to control the C/E Light.			
A14, 18	---	Not Used	---
A15	BK/R	A/C Clutch Relay	A/C on: 1v, off: 12v
A/C Relay: indicates the command from the PCM to the A/C Clutch Relay.			
A16	W/Y	Alternator Charging Signal	Lights off: 12v, on: 0v
ALTC Signal: indicates the command from the PCM to the Alternator.			
A17	GN/BK	A/T: Lockup Solenoid 'A'	LSA on: 12v, off: 0v
A/T Lockup Solenoid: indicates the command from the PCM to control the solenoid.			
A19	YEL	A/T: Lockup Solenoid 'B'	LSB on: 12v, off: 0v
A/T Lockup Solenoid: indicates the command from the PCM to control the solenoid.			
A20	RED	EVAP Purge Solenoid	Sol. on: 1v, off: 12v
EVAP Purge: indicates the command from the PCM to turn the solenoid On or Off.			
A21	RD/GN	Igniter Control Signal	Digital Signals: 0-12-0v
IGNC: indicates the command from the PCM to control the Igniter On/Off function.			
A22	---	Not Used	---
A23, A24	BLK	Power Ground	<0.1v
PWR GND: indicates the voltage level of the Power Ground circuit to the PCM.			
A25	YL/BK	Ignition Power from Relay	12-14v
IGN PWR: indicates the voltage level of the ignition power circuit to the PCM.			
A26	BK/R	Chassis Ground	<0.050v

Pin Voltage Table Examples (Continued)

Pin Tables are arranged in table format in the manual. Each **table heading** identifies the year, vehicle, engine and the transmission type [All] for both, [A/T] for Automatic or [M/T] for Manual. They include:

- PCM Pin Number & Wire Color
- Circuit Description/Acronym for PCM Pin Connector
- Known good operating value at hot idle speed

1995 Civic VS 1.5L I4 VTEC (VIN EH2) [All] 16P 'B' Connector

PCM Pin #	W/Color	Circuit Description (16-Pin)	Value at Idle
B1	W/GN	Ignition Power From Relay	12-14v
IGN PWR: indicates the voltage level of the ignition power circuit to the PCM.			
B2	BR/BK	Logic Ground	<0.050v
B3-4, 6	---	Not Used	---
B5	BL/R	A/C Switch Signal	A/C on: 0v, off: 12v
AC Switch: indicates the value of the A/C Switch circuit with the A/C On or Off.			
B7	GN	M/T: Clutch Switch	Clutch in: 0v, out: 12v
M/T Clutch: indicates the status of the Clutch Switch with the clutch depressed.			
B8	BR/R	PSP Switch	Straight: 0v, Turning: 12v
PSP Switch: indicates the status of the Power Steering Pressure Switch.			
B9	BL/W	Starter Signal	Cranking: 9-11v
B10	Y/BL	Vehicle Speed Sensor	At 50 mph: 60 Hz
B11	O/W	CYP Sensor Signals	0.250mv AC
B12	O/BL	CYP Sensor Ground	<0.050v
B13	O/BL	TDC Sensor	1.00mv AC
B14	W/BL	TDC Sensor	<0.050v
B15	BL/GN	CKP Sensor	0.900mv AC
B16	BL/Y	CKP Sensor Ground	<0.050v

1995 Civic VS 1.5L I4 VTEC VIN EH2 (All) 22P 'D' Connector

PCM Pin	W/Color	Circuit Description (22-Pin)	Value at Idle
D1	W/BL	Keep Alive Power	12-14v
D2	GN/W	Brake Switch Signal	Rake on: 12v, off: 0v
D3	BL/Y	LAF Sensor	Hot idle: 0.5v
D4	BR	Service Check Connector	SCS Open: 5v
D6	O/BL	VTEC Pressure Switch	Idle: 0v, Hi-Speed: 12v
D7	LT BL	Data Link Connector	5v
D8	W/BLU	LAF Sensor VS+ Signal	Hot idle: 0.5v
D9	PK	Alternator 'FR' Signal	Digital Signals: 0-5-0v
D10	PK/BK	Electronic Load Detector	Varies: 0.5-4.5v
D11	PK/BK	TP Sensor	Hot idle: 0.52v
D12	W/BK	EGR Valve Lift Sensor	Hot idle: 1.2v
D13	R/W	ECT Sensor	At 180°F: 0.52v
D14	O/BL	LAF Sensor IP+ Signal	KOER: 0-1v
D15	R/Y	IAT Sensor	At 100°F: 2-3v
D17	W	MAP Sensor Signal	At Idle: 0.8-0.9v
D18	PK/GN	Economy Driving Indicator	Digital Signals
D19	Y/GN	MAP Sensor VREF	4.9-5.1v
D21	GN/W	MAP Sensor Ground	<0.050v

Introduction To OBD II Systems

Reasons for the OBD II System

The OBD II system was developed to accomplish two different objectives. First, it was developed to comply with California and Federal regulations and standards for vehicle emission control monitoring. The initial goal of this system was to detect the degradation or failure of an emission-related component or system that could cause vehicle emissions to rise by 50%.

OBD II CHANGES

However, the program has expanded to include new computer controlled diagnostic tests that are used to verify that an emission control system is operating correctly. This aspect of OBD II has the greatest impact on service technicians as it includes procedures that require a whole new set of diagnostic tests.

If the operation of the OBD II diagnostic "monitors" is not understood, a whole new set of driveability problems can appear to the technician who is not trained in this new system. An OBD II "monitor" requires that a particular drive cycle be completed. In some cases, an OBD II "monitor" will not run a diagnostic test on a particular emission system unless a specific drive cycle is performed.

The second goal of OBD II was to introduce changes intended to help with vehicle diagnostics. These changes include common terms, connectors and a common diagnostic language for a generic Scan Tool. The changes include:

- Common Diagnostic Connector
- Expanded Malfunction Indicator Light Operation
- Common Codes and Diagnostic Language
- Common Diagnostic Procedures
- New Emissions-Related Procedures, Logic and Sensors
- Expanded Emissions-related Monitoring

IMPORTANT BENEFITS

An important benefit of OBD II is that all vehicles have a common data output system and test connector. This allows a generic OBD II certified Scan Tool to read data from any OBD II compliant vehicle and pull codes with a common name and similar descriptions for fault conditions.

Many State I/M Tests now require the use of an OBD II certified Scan Tool to verify that all monitors have run and passed. This information, which appears on the Scan Tool as I/M Readiness Tests, is used to verify that all the I/M Readiness Tests are completed (with no trouble codes set) in order to pass the State Emissions Test.

Transition from OBD I to OBD II

To understand the OBD II system, you should have an understanding of how OBD I system diagnostics operate because inside every OBD II system *is the OBD I diagnostic system* with a series of additional diagnostic tests. These tests or "monitors" are used to verify the operation of all Emission Control related components and systems.

History of OBD II Systems

Starting in 1978, vehicle manufacturers introduced computer control of vehicle systems and engine management in order to identify electrical faults and to aid in the diagnosis of the onboard controllers.

By 1980, vehicle manufacturers had designed systems where the computer incorporated internal programs that monitored selected components and stored trouble codes in memory for retrieval. These codes identified failure conditions that referred the user to diagnostic repair charts or procedures that helped pinpoint the problem areas.

An overview of the evolution of On Board Diagnostics on Asian vehicle applications is shown in the Graphic. Although the California 1990 regulations vary slightly from the CARB OBD II regulations, the EPA adopted the California OBD II for Federal emissions certification, effective with the 1996 model year. In 1998, a new Federal OBD II standard was adopted that effectively eliminated the different status between the California and Federal emissions certification.

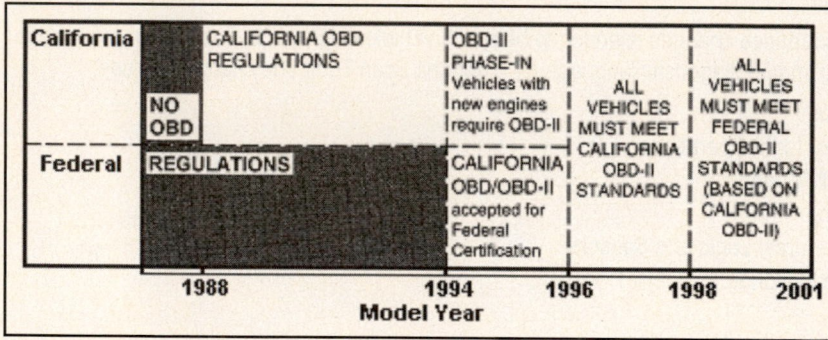

PHASE-IN SYSTEMS

The OBD II system was phased in during the 1995 model year on Acura NSX and Acura 2.5TL models. OBD II systems were expanded to the complete line of Acura and Honda vehicles in model year 1996.

Acura/Honda OBD History

An overview of the evolution of On Board Diagnostics on Acura and Honda vehicle applications is shown in the Graphic below.

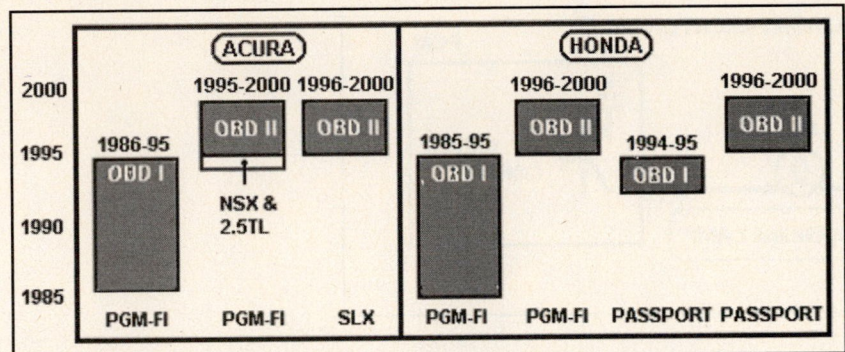

OBD II Certified Scan Tools

The information that should be available on any OBD II certified Scan Tool is contained in the list below:

- Any Current, Pending or History Trouble Codes
- Information on whether a MIL "on" request is present
- I/M Readiness Status (indicates if a monitor ran, passed or failed)
- Last Test Pass or Fail Message
- Freeze Frame Data for the first emission related fault condition

OBD II GENERIC SCAN TOOL OPERATION

A quick check of the Generic side of an OBD II certified Scan Tool should confirm the availability of this type of information:

- Clear trouble codes from the memory in the vehicle computer
- Display the I/M Readiness Test Status for all on board monitors
- Retrieve specific data from the PCM data stream
- Retrieve Freeze Frame Data
- Read any five digit trouble codes stored in memory

Bi-Directional Communication (OBD I)

Acura and Honda introduced the use of bi-directional communication and data stream information during the 1992 model year. A Scan Tool with an Aftermarket Honda cartridge can read serial data on:

- 1994 Accord I4 models and 1995 Accord V6 models.
- 1992 Civic and Integra models.
- 1993-95 Integra models.
- 1994-95 Odyssey models.
- 1994-95 Passport models.

OBD II Terminology

Malfunction Indicator Lamp

OBD II regulations (CARB and EPA) require that the Malfunction Indicator Lamp (MIL) be illuminated when a fault is detected, and that a Diagnostic Trouble Code (DTC) is stored in the PCM memory.

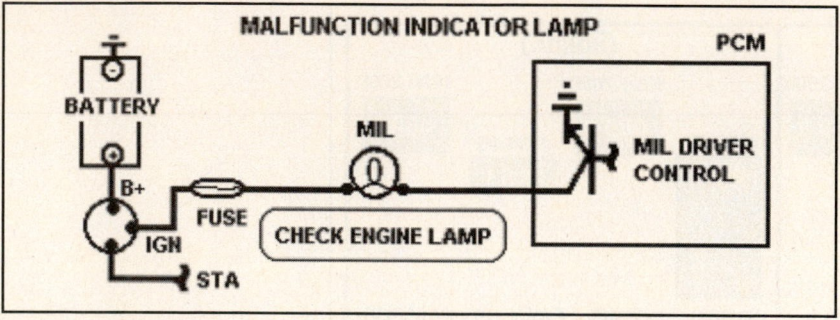

Understanding MIL Conditions

Several of the On/Off conditions for the MIL are described next.

MIL Off all the time - This condition indicates that the PCM has not detected any faults in an emissions related system, or that the MIL circuit is not working.

MIL On all the time - This condition indicates a fault in an emissions related system that could affect the vehicle emission levels.

MIL Flashing - This condition indicates a misfire or fuel system fault that could damage a converter.

Actions or Conditions to Turn off the MIL

The PCM will turn off the MIL if any of these actions occur:

*   The codes are cleared with a Generic or OEM Scan Tool
*   Power to the PCM is removed (at the battery or the PCM fuse)
*   The vehicle is driven on three consecutive trips that include an OBD II Warmup Cycle and meets all code set conditions without detecting any emission-related faults

Similar Conditions

If a fuel control or misfire-related code sets, then the vehicle must be driven under conditions similar to when the fault was detected before the PCM will de-activate the MIL.

Similar Conditions can be defined as described next: The vehicle must be driven within 375 RPM of the engine speed and engine load (±10%) of the engine load value, and with engine temperature conditions similar to the temperature value stored in Freeze Frame data when the code set.

MIL Circuit Diagnosis

If the Malfunction Indicator Lamp (MIL) does not operate correctly, refer to the repair steps listed in the articles that follow to diagnose the MIL operation and the PCM.

MIL CONDITION: LIGHT ON FOR 2 SECONDS, THEN GOES OFF AT KOEO MODE

This is the normal operation of the PCM and MIL control circuit. This step can be used as a bulb check and initial check of PCM operation.

MIL CONDITION: LIGHT FLASHES ONCE PER SECOND IN KOER MODE

This condition indicates a fault in an emissions related system that could damage the vehicle catalyst. Use the Scan Tool to read the trouble codes and freeze frame data. Repair the misfire or fuel system fault. Then perform a PCM Reset step and do a drive cycle.

MIL CONDITION: MIL ON WITH NO CODES STORED IN KOEO OR KOER MODE

This condition can be caused by any of these faults:

- MIL control wire is shorted to ground between the lamp and PCM.
- The K-Line circuit at terminal 15 of the DLC 16P connector is shorted to ground. Turn the key off; remove the PCM connector, check the DLC 16P circuit for continuity to ground.
- PCM is faulty (due to possible shorted MIL control driver circuit).
- An open condition in the PCM battery or ignition power circuits.
- An open or high resistance condition in the PCM ground circuits.

MIL CONDITION: MIL DOES NOT COME ON FOR 2 SECONDS - ENGINE STARTS

This condition can be caused by any of these faults:

- If this problem is intermittent in nature, check the fuse that provides power to the MIL for a loose connection or corrosion. Inspect condition of the MIL control circuit terminal at the PCM.
- An open circuit between the MIL (lamp) and PCM control circuit. Inspect the bulb connections and condition of the PCM terminals.
- PCM is faulty (due to possible shorted MIL control driver circuit).
- An open condition in the PCM battery or ignition power circuits.
- An open or high resistance condition in the PCM ground circuits.

MIL CONDITION: MIL DOES NOT COME ON FOR 2 SECONDS - NO START FAULT

This condition can be caused by any of these faults:

- Turn the key off and remove the following connectors: the PCM connector that connects to the sensor circuits, EGR valve lift sensor, fuel tank pressure sensor, MAP sensor and TP sensor. Check for continuity to ground between the MAP sensor VREF and sensor VREF circuits at the PCM to ground. If continuity exists, locate the short to ground in the VREF circuit and retest.

OBD II Warmup Cycle

Once a MIL is off, the trouble code will remain in memory until 40 warmup cycles are completed without the same fault reoccurring.

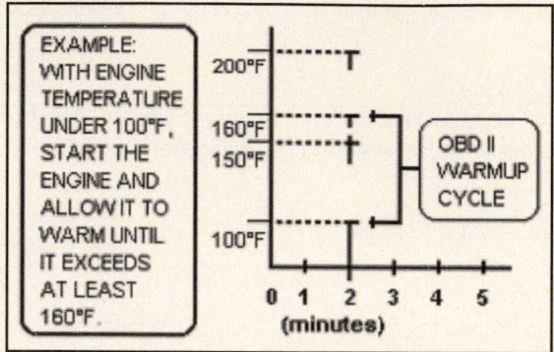

A warmup cycle is defined as a trip that includes a change in engine temperature of at least 40°F, and where the engine temperature reaches at least 160°F.

DTC Numbering Explanation

The number in the hundredth position indicates the specific vehicle system or the sub-group that failed.

This position should be consistent for P0xxx and P1xxx type codes.

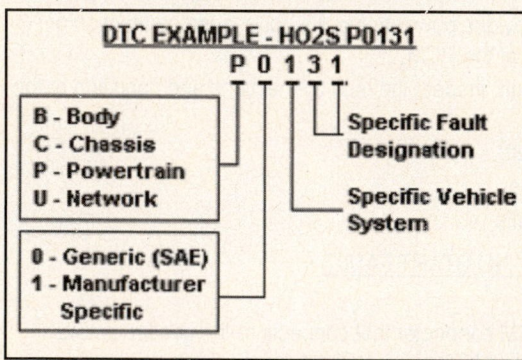

The Society of Automotive Engineers (SAE) established the numbers and the systems shown below.

P0100 - Air Metering and Fuel System fault
P0200 - Fuel System (fuel injector only) fault
P0300 - Ignition System or Misfire fault
P0400 - Emission Control System fault
P0500 - Idle Speed Control, Vehicle Speed Sensor fault
P0600 - Computer Output Circuit (relay, solenoid, etc.) fault
P0700 - Transaxle, Transmission faults

Note: ***The first and tenth digits indicate the type of Emission system that has failed.***

Diagnostic Link Connector

OBD II regulations established standards for use of a unique test connector (the Diagnostic Link Connector or DLC) on all vehicles equipped with the OBD II system.

The 16-pin connector is located beneath the instrument panel and within 12 inches (300 mm) of vehicle centerline. It is located out of the line of sight of passengers, but be easily viewable by a technician from a kneeling position outside the vehicle with the door open.

If it is located elsewhere, an OEM sticker is used to identify its location. Eight pins are assigned SAE labels and eight are assigned vehicle manufacturer labels. The 8 pins assigned by SAE include:

- 2 pins for the Serial Data Links
- 2 pins for the ISO- 9141 Serial Data Links
- 1 pin for battery power, one pin for battery and for signal ground

Data Link Connector Graphic

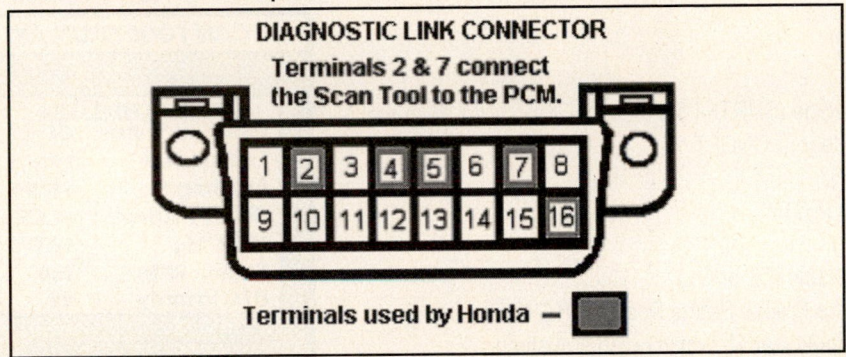

DLC Locations (Acura & Honda)

Year & Model	Location
1996-1997 Accord	Behind the Ash Tray
1998-2003 Accord	Below Dash to the left of the Center Console
1997-2003 Acura CL	Above Shifter (1999) or behind right side of console
1996-2003 Acura NSX	Behind a removable cover under the Glove Box
1996-1998 Acura RL	Behind the right side of Center Console
1999-2001 Acura RL	Behind the Ash Tray (in front of the shifter)
1996-2003 Acura TL	Behind the Ash Tray (1996-98) or behind the Radio
1997-2003 Civic	Near the left side Kick Panel
1997-2001 Civic Del Sol	To the right side of the Center Console (under cover)
1997-2003 CRV	Below the right side of the Center Console
1996-2003 Odyssey	Below the right side of the Center Console
1996 Prelude	Under Cup Holder (behind the shifter)
1997-2003 Prelude	Behind the right side of Center Console

Freeze Frame Data

OBD II Regulations (CARB and EPA) require the onboard computer store specific Freeze Frame Data when an emission-related fault is detected by a computer. The current Freeze Frame data can only be replaced (or overwritten) by Freeze Frame data from the Fuel system or Misfire Detection Monitor (if they detect a fault).

Freeze Frame Data must contain the engine operating conditions (data values) present at the time a trouble code is set. This data must be provided in standard units of measurement.

As a result, OBD II systems record the Freeze Frame at the moment the emission-related trouble code (DTC) is recorded. In the case of a two-trip trouble code, Freeze Frame is the recorded conditions present during the first trip of a two-trip fault (not when the MIL is activated). This data can be accessed and read on an OBD II certified Scan Tool.

Actually, Freeze Frame Data is a recording of one frame or one instant in time. This important information contains the details that describe the engine operating conditions present at the instant a fault is detected and a code is set. The list of these details is shown below.

FREEZE FRAME DATA ITEMS

- Calculated Load Value (CLV)
- Engine Speed (RPM)
- Short Term Fuel Trim Percentage (SHRTFT %)
- Long Term Fuel Trim Percentage (LONGFT %)
- Vehicle Speed (MPH or KPH)
- Engine Coolant Temperature (ECT)
- Manifold AIR Pressure (MAP)
- Closed Loop or Open Loop Status (CL or OL)
- Trouble Code that triggered the Freeze Frame Record (DTC #)
- If a Misfire Code is set – It should identify which cylinder misfired

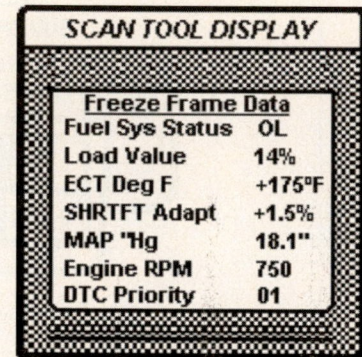

OBD II Trip Definition

An OBD II Trip is vehicle operation (following an engine off period) of such duration and driving modes that all components or systems are monitored at least once by the PCM diagnostics (except the catalyst).

OBD II MONITORS

Comprehensive Component Monitor

The Comprehensive Component Monitor (CCM) is an on-board strategy designed to monitor for failures in emission-related electronic components and circuits that provide input or output signals to the PCM. These are systems or devices that are not exclusively monitored by another monitor system. If the PCM detects that an input or output signal is inoperative due to an out-of-range value, open circuit or if an on-board rationality or functionality check fails, the PCM will set a code in memory and activate the MIL.

Tests conducted by the CCM vary depending on the type of hardware, the function of the device and the signal type. Analog signals are checked continuously for opens, shorts and out-of-range values. Some digital signals are checked for both functionality and rationality. These tests require that certain engine conditions be present before the test is performed and that several components be monitored as part of the test. Also, a sensor value can be monitored for change after the PCM sends a command to a device.

The devices checked by the CCM with the key on or with the engine running include the items in the lists that follow:

INPUT DEVICE EXAMPLES

- Barometric Pressure Sensor
- Brake Switch & Cruise Servo Switch
- Camshaft & Crankshaft Sensors
- M/T Clutch Switch
- Engine Coolant Temperature Sensor
- EVAP Purge Sensor
- Fuel Pressure Sensor
- Intake Air Temperature Sensor
- Knock Sensor
- Manifold Absolute Pressure Sensor
- Park Neutral Switch
- Transmission Temperature Sensor
- Transmission Turbine Speed Sensor
- Vehicle Speed Sensor

OUTPUT DEVICE EXAMPLES

- EVAP Purge and Vent Solenoids
- Idle Air Control Solenoid
- Ignition Control System
- Transmission Torque Converter Clutch Solenoid
- Transmission Shift Solenoids (Solenoid 'A' or Solenoid 'B')

Main Monitors

A key difference between the first version of onboard diagnostics (OBD) and the second version is the use of dedicated diagnostic monitors contained within the PCM software. These monitors are required in order to meet OBD II CARB and U.S. EPA regulations.

Simply stated, an OBD II Monitor is a diagnostic strategy designed to test the operation of an emissions-related system or component. Some of the OBD II Monitors accomplished this task *directly* by monitoring the action of various input and output devices or sensors connected to the PCM. An example of *direct* monitoring is when the Comprehensive Component Monitor tests the Engine Coolant Temperature or Intake Air Temperature Sensor signals.

Other OBD II Monitors accomplish the task *indirectly* by monitoring the effects of changes to a system or component. The *indirect* method may be accomplished through monitoring a change or response in a system. This type of test is done by monitoring the input or output signals of a particular device for an "inferred" change.

An example of *indirect* monitoring is when the PCM infers correct or incorrect catalyst action using the Catalyst Monitor to sample signals from the upstream or downstream oxygen sensors. This allows the PCM to determine the oxygen storage efficiency of the catalyst.

SYSTEMS & DEVICES CHECKED BY THE CCM AND MAIN MONITORS

The systems & devices Tested by these I/M Monitors include:

Main Monitors that run continuously:

*   Fuel Control Monitor (test begins with the engine in closed loop)
*   Misfire Monitor (misfire detection begins right after startup)

Main Monitors that run once only per trip:

*   Catalyst Efficiency Monitor (runs in closed loop after certain engine temperature, time and VSS requirements have been met)
*   EGR System in closed loop (after certain temperature, time and VSS requirements have been met)
*   EVAP System Monitor (runs in closed loop after certain engine temperature, time and VSS requirements have been met)
*   Oxygen Sensor Monitor (Voltage and Response Time Tests (runs in closed loop after certain engine temperature, time and VSS requirements have been met)
*   Secondary AIR System Monitor (runs in closed loop at off-idle)

Note: *Once all of the required enable criteria are met, Acura & Honda OBD II systems are programmed to run all of the OBD II main monitors once each trip.*

Catalyst Monitor

On OBD II systems, a pre-catalyst (front) Heated Oxygen Sensor (HO2S-12) is used to provide the additional signals needed to monitor the efficiency of the three-way catalyst.

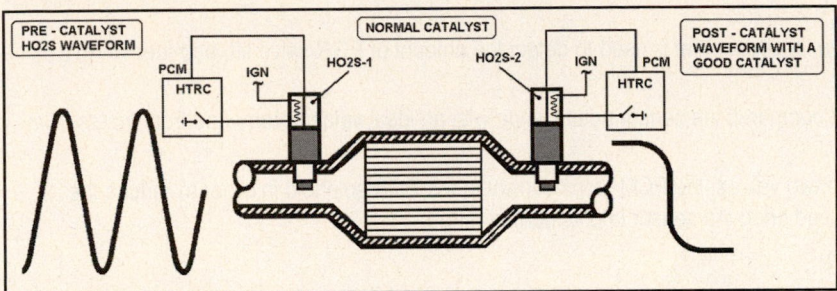

The PCM compares the signals between the pre-catalyst (front) and the post-catalyst (rear) oxygen sensor on these systems *during stable driving conditions* in order to determine the oxygen storage capacity of the catalytic converter.

CATALYST MONITOR OPERATION

The Catalyst Monitor determines the efficiency of the catalyst by monitoring the signals from the pre-catalyst and post-catalyst oxygen sensors. If the three-way catalyst is operating correctly, the post-catalyst signal will have significantly less activity than the pre-catalyst.

CATALYST MONITOR "TRIP" PATTERN

The Monitor "Trip" Pattern shown below can be used to validate an OBD II repair or to "run" the Catalyst Monitor in order to complete the I/M Readiness Test. The example in the Graphic is for DTC P0420.

Catalyst Monitor "Trip" Graphic

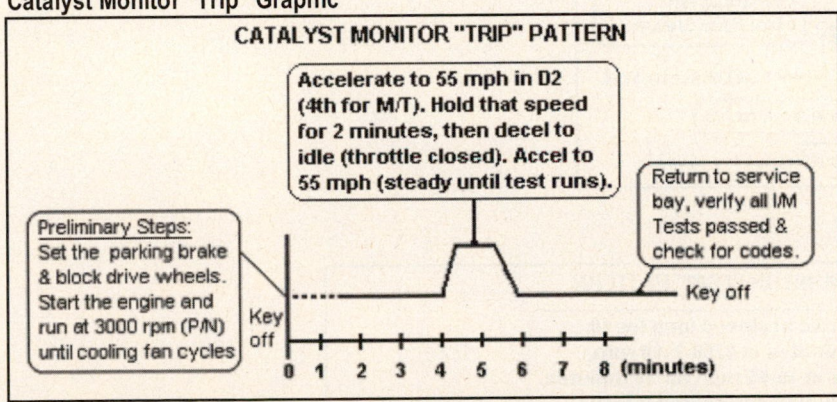

EGR System Monitor

The EGR system is designed to reduce oxides of Nitrogen (NOx) emissions by recirculating exhaust gas through the EGR valve to the intake manifold and to the engine.

EGR VALVE LIFT SENSOR

This system includes an EGR valve lift sensor that is used to detect the amount of EGR valve lift, and then send a signal to the PCM.

During certain conditions the PCM compares the sensor actual value with an ideal value determined from inputs from other engine sensors.

If there is a difference between the two values, the PCM stops current to the EGR solenoid in order to reduce the vacuum applied to the EGR valve and bring the sensor into its normal range.

EGR MONITOR "TRIP" PATTERN

The "Trip" Pattern in the Graphic can be used to validate an OBD II repair or to "run" the EGR System Monitor in order to complete an I/M Readiness Test. The example in the Graphic is for a DTC P0401.

EGR Monitor

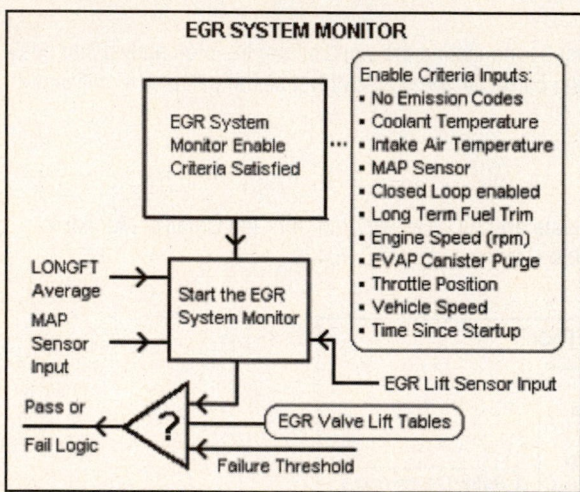

"Trip" Graphic

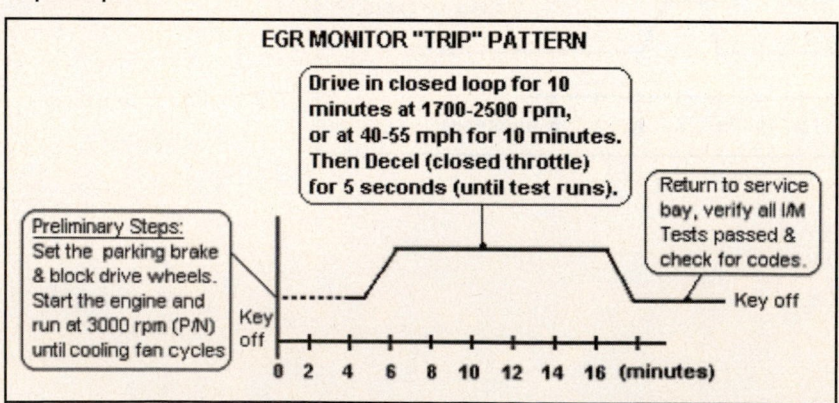

EVAP System Monitor (1996-97)

OBD II regulations require the EVAP system be monitored for correct airflow used to purge the fuel vapors. 1998-2003 systems include a vacuum check to verify there are no leaks that could allow vapors to escape.

If a leak is detected that is larger than 0.040" is detected in the system on two consecutive trips, the PCM will set a trouble code.

FUEL TANK VAPOR CONTROL SYSTEM (EVAP SWITCH DESIGN)

Fuel vapors are purged by drawing fresh air through the canister to a port on the intake manifold. The vacuum to the canister is controlled by an EVAP solenoid. A purge flow switch in the purge line to the canister is monitored by the PCM to detect when vacuum is flowing.

EVAP System Monitor Graphic

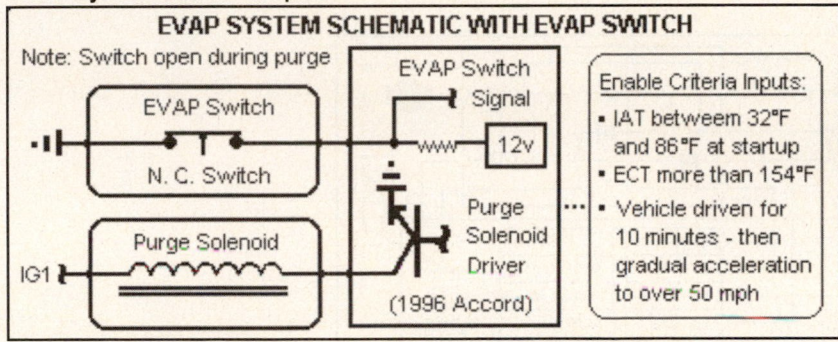

MONITOR "TRIP" PATTERN

The "Trip" Pattern in the Graphic can be used to validate an OBD II repair or to run the EVAP Monitor in order to complete an I/M Readiness Test. This example in the Graphic is for DTC P0441.

EVAP Monitor "Trip" Graphic

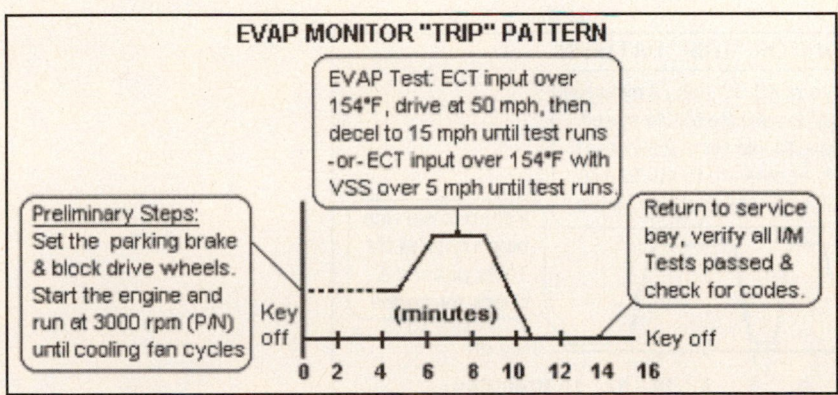

EVAP System Monitor (1998-2003)

The EVAP system is used to minimize the amount of fuel vapor that escapes into the atmosphere. Fuel vapors from the fuel tank are temporarily stored in the EVAP charcoal canister until they can be purged into the engine and then burned.

The EVAP canister is purged by drawing fresh air through it and into a port on the throttle body. The EVAP purge (control) solenoid valve controls the purge vacuum. This valve is opened by the PCM whenever the engine coolant temperature is above 147ºF.

If the vapor pressure in the fuel tank is higher than the set value of the EVAP two-way valve, the valve opens and regulates the flow of fuel vapor into the EVAP canister. The duty cycle signals to the purge valve can be monitored with a Lab Scope as shown in the Graphic.

EVAP System Monitor Graphic

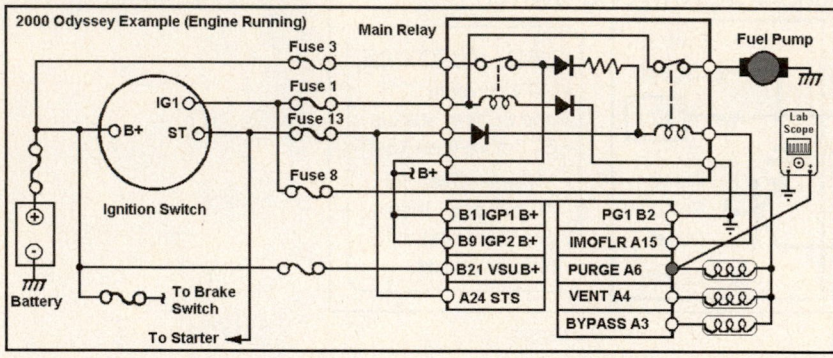

MONITOR "TRIP" PATTERN

The "Trip" Pattern in the Graphic can be used to validate an OBD II repair or to run the EVAP Monitor in order to complete an I/M Readiness Test. This example in the Graphic is for DTC P1456.

EVAP Monitor "Trip" Graphic

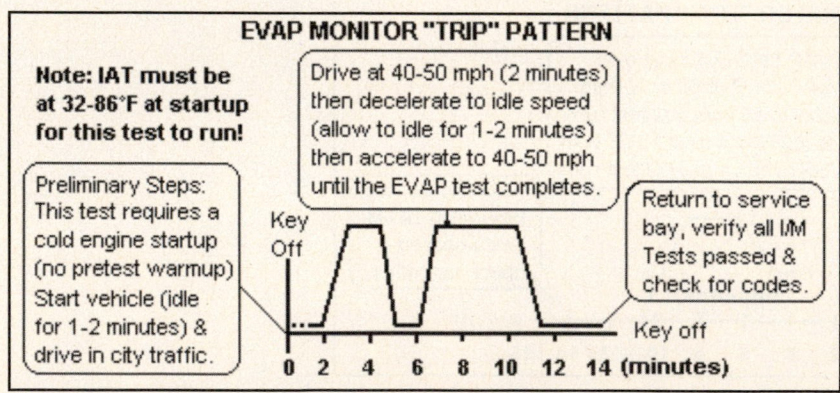

Fuel System Monitor

OBD II regulations require that the fuel delivery system be tested continuously in order to verify it can comply with emission standards.

If a Fuel system component fails, or if a change in long-term fuel trim is detected that could cause tailpipe emissions to exceed 1.5 times the FTP Standard for two consecutive trips, the MIL is activated, a code is set and engine conditions are stored in Freeze Frame. Fuel system faults have a higher priority than all other faults except engine misfire.

Because of this priority, if a Fuel system fault is detected, the PCM will overwrite the Freeze Frame Data from faults with a lower priority.

The Fuel System Monitor is designed to test the adaptive fuel control system. This task is accomplished using adaptive fuel tables stored in the PCM memory to compensate for variations in system components due to normal wear and aging. Once closed loop is enabled, the PCM adaptive fuel control strategy "learns" changes needed to correct a fuel system that is biased either rich or lean.

<u>FUEL TRIM CORRECTIONS</u>

Fuel trim correction has two methods of adapting to Fuel system changes: Short Term fuel trim (SHRTFT) and Long Term fuel trim (LONGFT). If the HO2S signal indicates that the A/F ratio is too rich, the PCM can move SHRTFT to a negative range in an attempt to correct the rich condition. If the SHRTFT compensates for this rich condition for too long a period of time, the PCM "learns" this fact and moves LONGFT to a negative range to compensate. The SHRTFT range is -30% to +43% and LONGFT range is -20% to +20%.

Fuel System Monitor Graphic

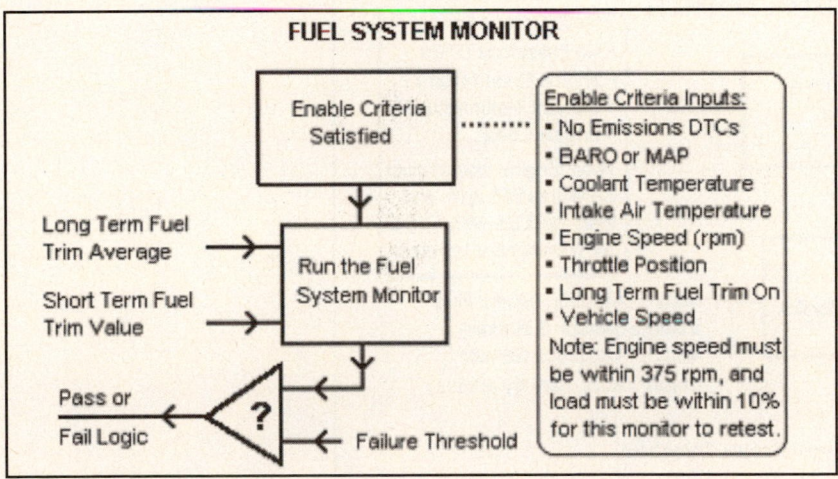

Misfire Monitor

OBD II regulations require that the engine be continuously tested for misfires under all engine positive load and speed conditions. Positive load conditions are defined as accelerating, cruising and idling.

Also, the Misfire Detection Monitor must be able to distinguish between a single misfire and multiple misfires. If a single cylinder is misfiring, then that cylinder must be identified and engine conditions present when the misfire occurred stored in Freeze Frame. If a misfire is detected that could cause tailpipe emissions to exceed 1.5 times the FTP Standard during a trip, the MIL is activated and a code is set.

TYPE 1 MISFIRE

A Type 1 Misfire is an engine misfire condition that could cause tailpipe emissions to exceed 1.5 times the FTP Standard. If a Type 1 Misfire is detected with the engine at similar engine speed, load and temperature conditions during a second trip, the MIL is activated and a code is set. Also, the MIL must be activated if a misfire is detected in two non-consecutive trips that are not more than 80 trips apart.

TYPE 2 MISFIRE

A Type 2 Misfire is an engine misfire so severe that damage to the catalyst can result at the current engine speed, load and temperature. If this type of fault is detected, the MIL will flash once per second within 200 engine revolutions from the point when the misfire was first detected. The MIL will stop flashing and remain on if the vehicle is no longer operating at engine conditions that could damage the catalyst.

Misfire Monitor Graphic

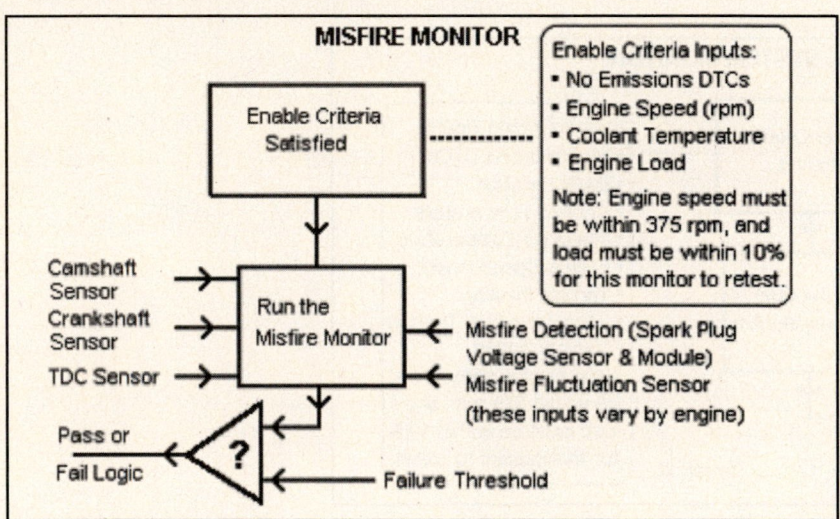

Oxygen Sensor Monitor

OBD II regulations require that the pre-catalyst HO2S-11 (front) and post-catalyst HO2S-12 (rear) heated oxygen sensors be monitored for a correct response rate and output voltage levels, as well as any other factors that could effect tailpipe emissions.

The Oxygen Sensor Monitor consists of three separate tests: run-time to activity, lean-to-rich and rich-to-lean response times, and sensor voltage checks. If the PCM detects an Oxygen sensor fault that could cause tailpipe emissions to exceed 1.5 times the FTP Standard for two consecutive trips, the MIL is activated and a code is set.

RESPONSE TIME TEST

Response time is defined as the amount of time it takes for the Oxygen sensor signal to go from rich-to-lean and from lean-to-rich. Once the Oxygen Sensor Monitor is enabled, it compares the average response time to a calibrated value in the PCM memory. If the response time exceeds this value for two consecutive trips, the PCM sets a code, the MIL is activated and current engine operating conditions are stored in Freeze Frame.

HO2S VOLTAGE TESTS

The Oxygen sensor voltage test is used to monitor the pre-catalyst and post-catalyst sensor signals for no activity (due to an open circuit) or for circuit short-to-voltage or short-to-ground conditions.

This part of the Monitor test is used to check the HO2S-11 signals for faults in the fuel control system that may cause the sensor signal to shift in a rich or lean direction. This test only requires one trip to fail.

Oxygen Sensor Monitor Graphic

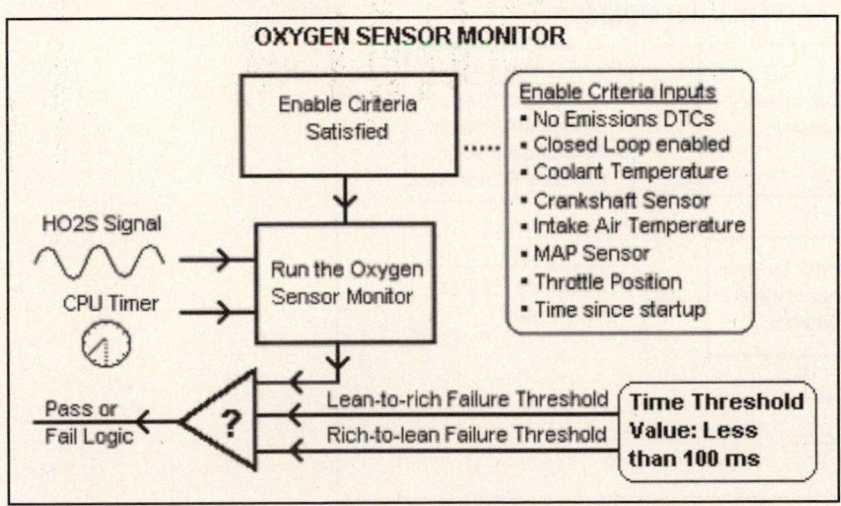

Oxygen Sensor Heater Monitor

OBD II regulations require that the pre-catalyst (HO2S-11 -front) and post-catalyst (HO2S-12 - rear) oxygen sensors be monitored for faults in the heater element or circuit. The Oxygen Sensor Heater Monitor test method used on Acura and Honda vehicles monitors the heater circuits for open, grounded or shorted circuit conditions.

The first time a HO2S heater circuit fault is detected by the PCM that could cause tailpipe emissions to exceed 1.5 times the FTP Standard, the MIL is activated and a code is set. This is a CCM One-Trip code.

HO2S HEATER CIRCUITS

The HO2S circuits include the signal, sensor ground and heater control (HTRC). On most models, the heater receives power from the main relay through the ignition power circuit. However, on the some Honda engines (e.g., D15Z1) with a Lean Air Fuel sensor, the heater circuit receives power from the HTRC relay on a separate circuit.

HO2S HEATER CIRCUIT TESTS

During the heater test, the HO2S Monitor checks the HO2S heater circuits for faults. The PCM controls the heater circuit by toggling the circuit on and off at a fixed rate. The test is accomplished by checking the heater circuit for a voltage change 80 seconds after engine startup (once the heater control driver toggles the HTRC circuit).

After the HTRC circuit is enabled, if the HO2S heater or its circuits are open, grounded or shorted, the voltage on the HTRC circuit will not match the values for these circuits stored in the PCM memory.

Oxygen Sensor Heater Monitor Graphic

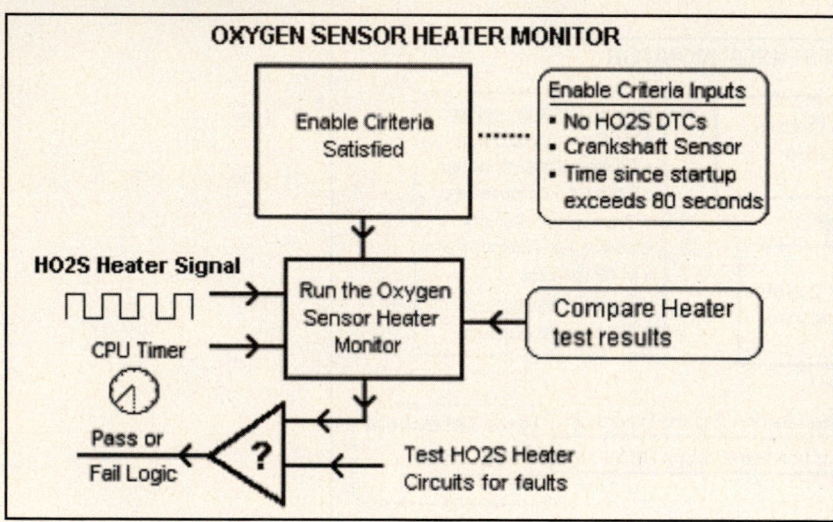

Vehicle Diagnostics

OBD II Diagnostic Routine

The Diagnostic Routine below should be used to diagnose driveability symptoms, intermittent faults and trouble codes on OBD II systems.

OBD II Diagnostic Routine Graphic

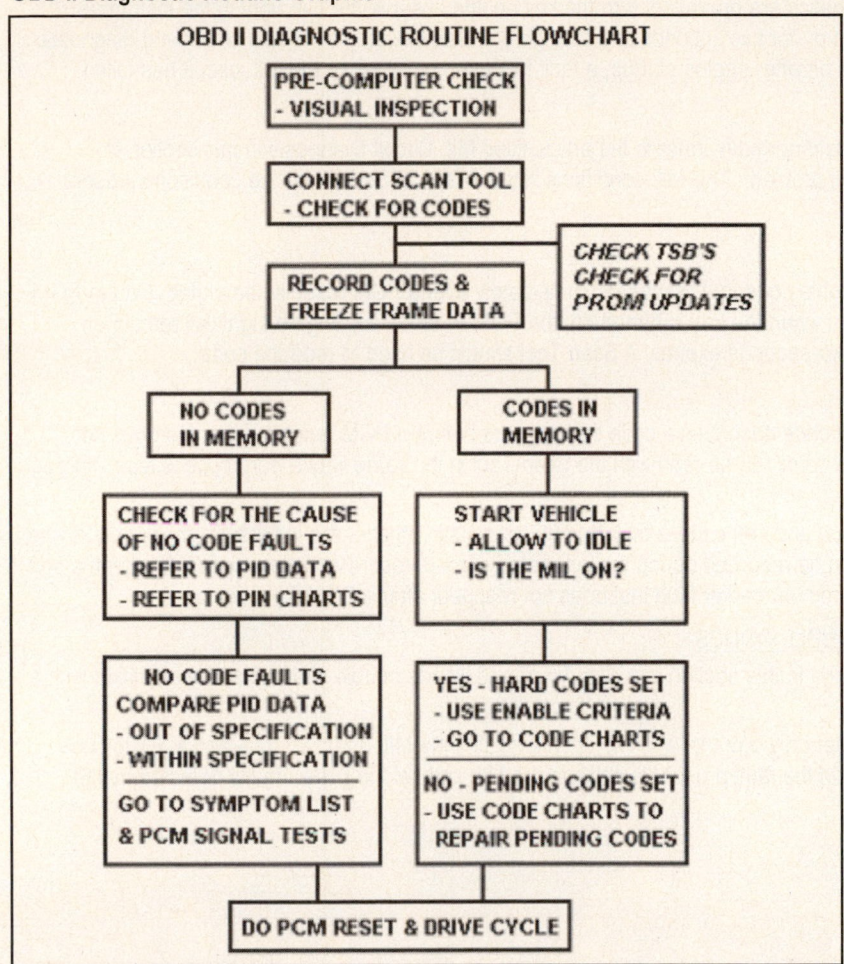

Read, Record & Clear Codes

The Honda PGM-FI system uses the PCM to perform its diagnostic functions. The PCM in the OBD II system is designed to detect failures in emission-related systems or circuits. When a fault is detected, the PCM will set a trouble code in memory, activate the MIL and store current engine operating conditions in Freeze Frame.

DOES THE C/E LIGHT OPERATE PROPERLY? (1995-2003)

To determine if the PGM-FI diagnostics are operating, turn the key on and observe the C/E light in the instrument panel. The C/E light should remain on for two seconds and then go out. Use this step as a bulb check and diagnostic function check. If the light remains on after engine startup, a fault has been detected or the MIL circuit has failed.

MIL CIRCUIT DIAGNOSIS

If the MIL and/or PCM do not operate normally, refer to the article titled MIL Circuit Diagnosis in this section to diagnose the cause of a MIL circuit problem. The MIL conditions listed in this article include the conditions present when the engine will not start.

HARD CODES

The term *hard code* refers to a trouble code that reappears immediately after a PCM reset is completed. On the PGM-FI system, if a *hard code* is present when the key is turned on, the PGM-FI or Check Engine Light will remain on AFTER the bulb check period of two-seconds expires. A Scan Tool should be used to read the code.

SOFT CODES

On PGM-FI systems, the term *soft code* describes a code that appears before a PCM reset is done, but does not reappear after a PCM reset. A *soft code* can be read with the Scan Tool in the same way a *hard code* is read - through the Scan Tool.

The difference between a *hard code* and *soft* code is that the MIL will remain off after the initial bulb check period when a *hard code* exists. A *soft code* can be recorded during the initial step of read, record and clear codes. In effect, a *soft code* is a code stored in memory from an earlier fault that does not reappear after a PCM Reset step.

1-TRIP, 2-TRIP AND 3-TRIP TROUBLE CODES

The Trouble Codes & Enable Criteria in this section includes Acura and Honda trouble codes that are identified with a 1-Trip, 2-Trip or 3-Trip designator.

A 1-Trip code activates the MIL after only one trip, a 2-Trip code activates the MIL after two trips and a 3-Trip code requires three consecutive trips with the fault present to activate the MIL and set the code. Faults detected by the Catalyst Monitor are 3-Trip codes.

PCM Reset Step

Once all repairs to the OBD II system are completed, the PCM should be reset to allow it to relearn certain engine operating information (i.e., fuel trim and idle speed settings).

To perform a PCM reset procedure (clear all codes and Freeze Frame Data from the PCM memory), do the following steps.

Turn the key off and then remove the PCM power fuse from the underhood fuse/relay box for 10 seconds.

This step resets the PCM. If the engine is started after doing this step, any codes that reset are hard codes.

PCM Reset Fuse Locations

- Accord - Backup or AGC fuse in underhood fuse relay box
- Civic, Del Sol, CR-V, Vigor - Backup fuse in main fuse box
- Passport - ECM fuse in underhood fuse block
- Prelude - Clock or Clock/Radio fuse in underhood fuse box
- Integra - Backup or Backup/ACC fuse in underhood fuse box
- Legend - ACG fuse in underdash fuse box
- NSX - Clock Fuse in underhood fuse/relay box
- 2.2CL, 2.3CL, 2.5TL, 3.2TL, 3.5RL - Backup fuse in fuse box

PCM Locations

- 1996-2003 Accord: at bottom front of passenger floor pan
- 1996-2003 Civic: behind the right side kick panel
- 1997-2003 CR-V: behind the right side kick panel (see Graphic)
- 1996-1998 2.3L Odyssey: at bottom front of passenger floor pan
- 1999-2003 3.5L Odyssey: behind at lower center of dash area
- 1996-2003 Passport: behind the center of the dash area
- 1996-2001 Prelude: at bottom front of passenger floor pan

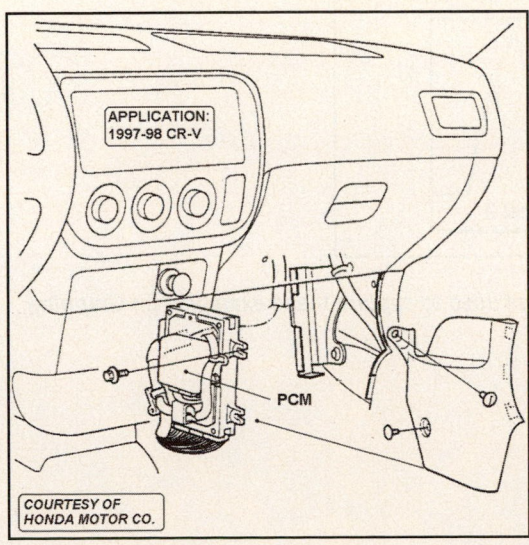

Scan Tool Graphing Mode

A Scan Tool with the Graphing Mode Function can be used to diagnose devices that fail and cause intermittent faults. In the examples below, a Scan Tool was used to collect PID Data and to capture an intermittent fault using the graphic mode test function.

TP Sensor "Static" Test

First turn the key on (engine off) and connect the Scan Tool to read the voltage or percent of throttle opening. Compare the actual reading to the values from a "known-good" vehicle shown below. TP sensor voltage should change smoothly as the throttle is opened and closed.

TP Sensor Chart

Throttle Angle Degrees of Rotation	Voltage
0 Degrees	0.50v
10 Degrees	0.97v
20 Degrees	1.44v
30 Degrees	1.90v
40 Degrees	2.37v
50 Degrees	2.84v
60 Degrees	3.31v
70 Degrees	3.78v
80 Degrees	4.24v

TP Sensor "Dynamic" Test

Warm the engine to normal operating temperature. Then connect a Scan Tool with the "graphing mode" function or a Lab Scope.

Place the gear selector in P/N and block the drive wheels for safety. Observe the waveform for a smooth transition or change as the throttle is momentarily snapped from closed throttle to WOT. Note the extra downward spike in Example No. 2 indicating a faulty TP sensor.

TP Sensor Graphic

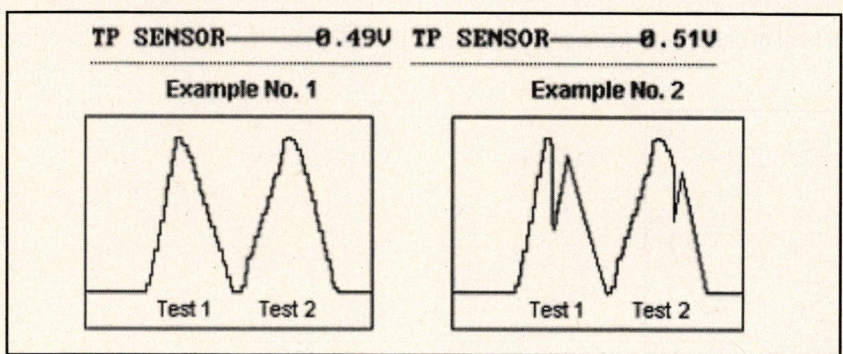

Note: *A Vetronix Mastertech™ with the Honda cartridge was used to capture these examples (in Graphing Mode).*

Trouble Codes and Enable Criteria

Instructions: To use this trouble code list, first read and record all codes in memory along with any Freeze Frame data. This information is used during diagnosis. If a PCM reset function is done, all codes and Freeze Frame data will be lost!

Find the appropriate trouble code with code descriptions and test conditions that indicate the type of fault present when the code was recorded. This information can indicate how to drive the vehicle in order to validate that the original problem is fixed. This is especially true of codes related to the Catalyst, EGR and EVAP Monitors.

P0xxx Trouble Code List (Acura & Honda Models)

Code ID	Code Description & Test Conditions
P0106 CCM 2T All Models	**MAP Sensor Range/Performance** *Test Conditions & Results:* Engine running for one second, then the test started and the PCM detected a MAP sensor value of more than 11.8" Hg.
P0107 CCM 1T All Models	**MAP Sensor Circuit Low Input** *Test Conditions & Results:* Key on or engine running, then the test started and the PCM detected that the MAP sensor value was near 0" Hg.
P0108 CCM 1T All Models	**MAP Sensor Circuit High Input** *Test Conditions & Results:* Engine running, then the test started and the PCM detected that the MAP sensor value was near 29.9" Hg.
P0111 CCM 2T All Models	**IAT Sensor Range/Performance** *Test Conditions & Results:* Engine running, then the test started and the PCM detected the IAT sensor value changed too much in too short an amount of time.
P0112 CCM 1T All Models	**IAT Sensor Circuit Low Input** *Test Conditions & Results:* Key on or engine running, then the test started and the PCM detected that the IAT sensor input was more than 302°F.
P0113 CCM 1T All Models	**IAT Sensor Circuit High Input** *Test Conditions & Results:* Key on or engine running, then the test started and the PCM detected that the IAT sensor value was less than -4°F.
P0116 CCM 1T All Models	**ECT Sensor Range/Performance** *Test Conditions & Results:* Key on or engine running, then the test started and the PCM detected the ECT sensor value changed too much in too short an amount of time.
P0117 CCM 1T All Models	**ECT Sensor Circuit Low Input** *Test Conditions & Results:* Key on or engine running, then the test started and the PCM detected that the ECT sensor input was more than 302°F.

P0xxx Trouble Codes List (Acura & Honda Models)

Code ID	Code Description & Test Conditions
P0118 **CCM 1T** All Models	**ECT Sensor Circuit High Input** *Test Conditions & Results:* Key on, then the test started and the PCM detected an ECT sensor value of under -4°F.
P0122 **CCM 1T** All Models	**TP Sensor Circuit Low Input** *Test Conditions & Results:* Key on or engine running, engine warmed-up, then test started and the PCM detected a closed throttle TP sensor value of less than 10% (0.16v).
P0128 **CCM 2T** All Models	**Thermostat Range/Performance Problem** *Test Conditions & Results:* Vehicle running in closed loop at road load for 10 minutes and the PCM detected that the ECT sensor input did not reach the correct closed loop value.
P0123 **CCM 1T** All Models	**TP Sensor Circuit High Input** *Test Conditions & Results:* Key on or engine running, engine warmup period finished, then test started and the PCM detected a TP sensor value of more than 90% (4.6v).
P0131 **CCM 1T** All Models	**HO2S-11 Circuit Low Input (Front Sensor)** *Test Conditions & Results:* Vehicle running in closed loop in 2nd gear (A/T) or 4th gear (M/T), and the PCM determined that the HO2S-11 signal was fixed below a value in memory.
P0132 **CCM 1T** All Models	**HO2S-11 Circuit High Input (Front Sensor)** *Test Conditions & Results:* Vehicle running in closed loop in 2nd gear (A/T) or 4th gear (M/T), then the test started and the PCM determined that the HO2S-11 signal was fixed above a specified value stored in memory.
P0133 **HO2S 2T** All Models	**HO2S-11 Slow Response (Front Sensor)** *Test Conditions & Results:* Vehicle running in closed loop, then the vehicle was driven at 55 mph in D4 position (A/T) or in 4th gear (M/T), then the test started and then the PCM detected that the HO2S-11 response time from 300-600 mv was "slow", or that the HO2S-11 rich to lean or lean to rich response time was too "slow".
P0135 **CCM 1T** All Models	**HO2S-11 Heater Circuit (Front Sensor)** *Test Conditions & Results:* Engine runtime more than 80 seconds, then test started and the PCM detected a shorted or open circuit condition in the oxygen sensor heater circuit.
P0137 **CCM 1T** All Models	**HO2S-12 Circuit Low Input (Rear Sensor)** *Test Conditions & Results:* Vehicle running in closed loop in 2nd gear, then the test started and the PCM detected that the HO2S-12 voltage was fixed at less than 100 mv.
P0138 **CCM 1T** All Models	**HO2S-12 Circuit High Input (Rear Sensor)** *Test Conditions & Results:* Vehicle running in closed loop in 2nd gear, then the test started and the PCM detected that the HO2S-12 voltage was fixed at more than 600 mv.

P0xxx Trouble Codes List (Acura & Honda Models)

Code ID	Code Description & Test Conditions
P0139 HO2S 2T All Models	**HO2S-12 Slow Response (Rear Sensor)** *Test Conditions & Results:* Vehicle running in closed loop at 55 mph in D4 position (A/T) or in 4th gear (M/T) and the PCM detected that the HO2S-12 response time from 300-600 mv was too slow, or that the rich to lean or lean to rich response time was too "slow".
P0141 CCM 1T All Models	**HO2S-12 Heater Circuit (Bank 1 Sensor 2)** *Test Conditions & Results:* Engine runtime more than 80 seconds, then the PCM detected a shorted or open condition in the HO2S-12 heater circuit.
P0151 CCM 1T All Models	**HO2S-21 Circuit Low Input (Left Front Sensor)** *Test Conditions & Results:* Vehicle running in closed loop in 2nd gear, then the PCM detected that the HO2S-21 voltage was fixed at a value less than 100 mv.
P0152 CCM 1T All Models	**HO2S-21 Circuit High Input (Left Front Sensor)** *Test Conditions & Results:* Vehicle running in closed loop in 2nd gear in closed loop, then test started and the PCM detected that the HO2S-21 voltage was fixed over 900 mv.
P0153 HO2S 2T All Models	**HO2S-21 Slow Response (Left Front Sensor)** *Test Conditions & Results:* Vehicle running in closed loop at 55 mph in D4 position (A/T) or in 4th gear (M/T) and the PCM detected the HO2S-21 response time from 300-600 mv was too slow, or the rich to lean or lean to rich time was slow.
P0155 CCM 1T All Models	**HO2S-21 Heater Circuit (Left Front Sensor)** *Test Conditions & Results:* Engine runtime more than 80 seconds, then the PCM detected a shorted or open circuit condition in the oxygen sensor heater circuit.
P0157 CCM 1T All Models	**HO2S-22 Circuit Low Input (Left Rear Sensor)** *Test Conditions & Results:* Vehicle running in closed loop in 2nd gear, then the PCM detected the HO2S-22 signal was fixed at a value of less than 300 mv.
P0158 CCM 1T All Models	**HO2S-22 Circuit High Input (Left Rear Sensor)** *Test Conditions & Results:* Vehicle running in closed loop in 2nd gear, then the PCM detected the HO2S-22 signal was fixed at more than 600 mv.
P0159 HO2S 2T All Models	**HO2S-22 Slow Response (Left Rear Sensor)** *Test Conditions & Results:* Engine fully warmed up, vehicle running in closed loop at 55 mph with AT in D4 position (4th gear for MT models) and the PCM detected the HO2S-22 response time from 300-600 mv was too slow, or a slow rich to lean or lean to rich response time.
P0161 CCM 1T All Models	**HO2S-22 Heater Circuit (Left Rear Sensor)** *Test Conditions & Results:* Engine runtime more than 80 seconds, then the PCM detected a shorted or open circuit condition in the oxygen sensor heater circuit.

P0xxx Trouble Codes List (Acura & Honda Models)

Code ID	Code Description & Test Conditions
P0171 Fuel 2T All Models	**Fuel System Too Lean (Bank 1)** *Test Conditions & Results:* No PCM codes set, engine running in closed loop, and PCM detected the Long Term fuel trim exceeded the lean limit amount stored in memory.
P0172 Fuel 2T All Models	**Fuel System Too Rich (Bank 1)** *Test Conditions & Results:* No PCM codes set, engine running in closed loop and PCM detected the Long Term fuel trim exceeded the rich limit value stored in memory.
P0174 Fuel 2T All Models	**Fuel System Too Lean (Bank 2)** *Test Conditions & Results:* No PCM codes set, engine running in closed loop, then the PCM detected that the Long Term fuel trim (LONGFT) exceeded the lean limit amount.
P0175 Fuel 2T All Models	**Fuel System Too Rich (Bank 2)** *Test Conditions & Results:* No PCM codes set, engine running in closed loop and the PCM detected that the Long Term fuel trim (LONGFT) exceeded the rich limit value.
P0300 Catalyst 1T Emission 2T All Models	**Multiple Misfire Detected** *Test Conditions & Results:* Engine runtime more than 1 second, then the PCM detected a multiple misfire condition (a misfire in more than one cylinder).
P0301-306 Catalyst 1T Emission 2T All Models	**Cylinder 1, 2, 3, 4, 5 or 6 Misfire Detected** *Test Conditions & Results:* Engine runtime more than 1 second, then the PCM detected a misfire condition in the identified cylinder (i.e., Cylinder 1, 2, 3, 4, 5 or 6).
P0325 CCM 1T All Models	**Knock Sensor Circuit (Rear)** *Test Conditions & Results:* Engine running, then test started and the PCM detected an open or short circuit in the Rear (Left Side) Knock Sensor or its circuit.
P0330 CCM 1T All Models	**Knock Sensor Circuit (Front)** *Test Conditions & Results:* Engine running, then test started and the PCM detected an open or short circuit in the Front (Right Side) Knock Sensor or its circuit.
P0335 CCM 1T All Models	**CKP Sensor 'A' Circuit (No Signal)** *Test Conditions & Results:* Engine cranking or running, then the PCM did not detect any signals from the CKP Sensor 'A'. The engine will crank longer, may buck and jerk, but will start and run without the CKP sensor signal.
P0336 CCM 1T All Models	**CKP Sensor 'A' Circuit Range/Performance** *Test Conditions & Results:* Engine running and the PCM detected an intermittent signal from the CKP Sensor 'A'.
P0401 EGR 2T All Models	**EGR System Insufficient Flow Detected** *Test Conditions & Results:* Vehicle running in closed loop in D4 position at 40-55 mph for 2 minutes, followed by a Decel period to 35 mph with the throttle closed, then the PCM detected low EGR Flow (valve position sensor input).

P0xxx Trouble Codes List (Acura & Honda Models)

Code ID	Code Description & Test Conditions
P0420 **Catalyst 3T** All Models	**Catalyst Efficiency Below Threshold (Bank 1)** *Test Conditions & Results:* DTC P0137, P0138 & P0141 not set, cold engine startup finished, vehicle running at 40-55 mph in closed loop for 2 minutes, followed by a Decel period to 35 mph with the throttle closed, then the PCM detected excessive activity from the HO2S-12 for a set period of time.
P0430 **Catalyst 3T** All Models	**Catalyst Efficiency Below Threshold (Bank 2)** *Test Conditions & Results:* DTC P0137, P0138 & P0141 not set, cold engine startup finished, vehicle running at 40-55 mph in closed loop for 2 minutes, followed by a Decel period to 35 mph with the throttle closed, then the PCM detected excessive activity from the HO2S-22 for a set period of time.
P0441 **EVAP 2T** All Models	**EVAP System Incorrect Purge Flow** *Test Conditions & Results:* Cold engine startup conditions met (IAT input over 14ºF and ECT over 154ºF), engine runtime over 10 minutes, then gradual acceleration to 50-60 rpm, then the PCM detected incorrect purge flow during the EVAP Monitor test period.
P0451 **CCM 1T** All Models	**Fuel Tank Pressure Sensor Range/Performance** *Test Conditions & Results:* Key on or engine running, then the PCM detected the Fuel Tank Pressure (FTP) sensor signal was below the allowable range stored in the memory.
P0452 **CCM 1T** All Models	**Fuel Tank Pressure Sensor Circuit Low Input** *Test Conditions & Results:* Key on or engine running, then the PCM detected the Fuel Tank Pressure (FTP) sensor signal was below the allowable range stored in the memory.
P0453 **CCM 1T** All Models	**Fuel Tank Pressure Sensor Circuit High Input** *Test Conditions & Results:* Key on or engine running, then the PCM detected the Fuel Tank Pressure (FTP) sensor signal was more than the allowable range stored in memory.
P0500 **CCM 1T** All Models	**Vehicle Speed Sensor Circuit Low Input** *Test Conditions & Results:* Vehicle running at road load (cruise speed) and the PCM detected the VSS input was low.
P0501 **CCM 2T** All Models	**Vehicle Speed Sensor Circuit Performance** *Test Conditions & Results:* Vehicle running at road load (cruise) and the PCM detected the VSS input was erratic/low.
P0505 **IAC 2T** All Models	**Idle Speed Control System Fault** *Test Conditions & Results:* Engine running at Cruise speed, followed by a return to idle speed and then the PCM determined that the actual difference between the Actual and Target idle speed was too large an amount.
P0560 **CCM 1T** All Models	**PCM Backup Circuit Low Voltage** *Test Conditions & Results:* Key on or engine running, then the PCM detected a low voltage condition on the PCM Backup Circuit (the Backup/Radio 7.5 amp fuse circuit).

P0xxx Trouble Codes List (Acura & Honda Models)

Code ID	Code Description & Test Conditions
P0700 **TCM 1T** (D4 & MIL Blink) All Models	**Automatic Transaxle Fault** *Test Conditions & Results:* Engine running, then the test started and then the PCM received a TCM signal that an A/T fault occurred (P0700 sets along with other TCM codes).
P0715 **TCM 1T** (D4 & MIL Blink) All Models	**TCM A/T Main Shaft Speed Sensor Circuit** *Test Conditions & Results:* Engine running with VSS inputs, then test started and the PCM detected an open or short condition in the Main Shaft Speed Sensor circuit.
P0720 **TCM 1T** (D4 & MIL Blink) All Models	**TCM A/T Countershaft Speed Sensor Circuit** *Test Conditions & Results:* Engine running, VSS input received, and the PCM detected (from a TCM signal) an open or short fault in the Countershaft Speed Sensor circuit.
P0725 **TCM 1T** (D4 & MIL Blink) All Models	**Automatic Transaxle Fault** *Test Conditions & Results:* Engine running, then the test started and then the PCM received a TCM signal that an A/T fault occurred (P0725 sets along with other TCM codes).
P0730 **TCM 1T** (D4 & MIL Blink) All Models	**TCM A/T Shift Control System Fault** *Test Conditions & Results:* No other TCM codes set, vehicle running at Cruise speed with normal VSS inputs, then the test started and the PCM detected that the lockup clutch did not engage or disengage as commanded.
P0740 **TCM 1T** (D4 & MIL Blink) All Models	**TCM A/T Lockup Clutch System Fault** *Test Conditions & Results:* No other TCM codes set, running at Cruise speed, then the test started and the PCM detected the Lockup Clutch did not engage or disengage.
P0753 **TCM 1T** (D4 & MIL Blink) All Models	**TCM A/T Lockup Solenoid 'A' Circuit** *Test Conditions & Results:* Engine running with normal VSS inputs received, then the test started and the PCM detected that the Solenoid Valve 'A' circuit had an open or shorted circuit condition.
P0758 **TCM 1T** (D4 & MIL Blink) All Models	**TCM A/T Lockup Solenoid 'B' Circuit** *Test Conditions & Results:* Engine running with normal VSS inputs received, then the test started and the PCM detected that the Solenoid Valve 'B' circuit had an open or shorted circuit condition.
P0763 **TCM 1T** Both D4 & MIL Blink (All Models)	**TCM A/T Control Unit or Related Circuit** *Test Conditions & Results:* Engine running with normal VSS inputs received, then the test started and the PCM received a signal from the TCM that indicated a fault in the TCM or one of its related control circuits.
P0780 **TCM 1T** (D4 & MIL Blink) All Models	**Automatic Transaxle Fault Condition** *Test Conditions & Results:* Engine running, then the test started and the PCM received a signal from the TCM that indicated an A/T fault (DTC P0780 sets with other TCM related codes).

P1xxx Trouble Codes List (Acura & Honda Models)

Code ID	Code Description & Test Conditions
P1106 CCM 1T All Models	**BARO Pressure Sensor Range/Performance** *Test Conditions & Results:* Engine running, then test started and the PCM detected that the BARO sensor input did not change enough within a set period of driving time.
P1107 CCM 1T All Models	**BARO Pressure Sensor Circuit Low Input** *Test Conditions & Results:* Key on or engine running, then the test started and the PCM detected that the BARO sensor signal was less than a value stored in memory.
P1108 CCM 1T All Models	**BARO Pressure Sensor Circuit High Input** *Test Conditions & Results:* Key on or engine running, then the test started and the PCM detected that the BARO sensor signal was more than a value stored in memory.
P1121 CCM 1T All Models	**Throttle Position Sensor Input Lower Than Expected** *Test Conditions & Results:* Engine running, then the test started and the PCM detected the TP sensor input was less than the expected value (the CCM rationality test failed).
P1122 CCM 1T All Models	**Throttle Position Sensor Signal Higher Than Expected** *Test Conditions & Results:* Engine running, then the test started and the PCM detected the TP sensor input was more than the expected value (the CCM rationality test failed).
P1128 CCM 1T All Models	**MAP Sensor Value Less Than Expected** *Test Conditions & Results:* Engine running, then the test started and the PCM detected the MAP sensor input was less than the expected value (the CCM rationality test failed).
P1129 CCM 1T All Models	**MAP Sensor Value More Than Expected** *Test Conditions & Results:* Engine running, then the test started and the PCM detected the MAP sensor input was more than the expected value (CCM rationality test failed).
P1149 HO2S 2T All Models	**HO2S-11 Performance (Front or Primary Sensor)** *Test Conditions & Results:* Engine running in closed loop at 55 mph with a steady throttle (in 5th gear on M/T models), then test started and the PCM detected that the HO2S-11 response time in the 300-600 mv range was too slow, or that the rich to lean or lean to rich switch time was too slow.
P1162 CCM 1T All Models	**HO2S-11 Circuit (Front Sensor)** *Test Conditions & Results:* Engine running in closed loop at 1500 rpm (in 3rd gear on M/T) for 1-2 minutes, then the test started and the PCM detected an open or short condition in the HO2S-11 circuit (it failed the CCM circuit test).
P1163 HO2S 2T All Models	**LAF Sensor Slow Response** *Test Conditions & Results:* Engine running in closed loop at 55 mph at throttle steady (in 5th gear on M/T models), then the test started and the PCM detected the LAF sensor response time was too slow, or the R-L or L-R switch time was too slow (it failed the LAF sensor response time test).

P1xxx Trouble Codes List (Acura & Honda Models)

Code ID	Code Description & Test Conditions
P1164 **HO2S 2T** All Models	**LAF Sensor Range/Performance** *Test Conditions & Results:* Vehicle running at over 1500 rpm in closed loop 4th gear, followed by a quick acceleration to WOT, followed by a closed throttle deceleration for 5 seconds (2-3 times), then the PCM detected the LAF sensor response time or the R-L or L-R switch time was too slow.
P1165 **HO2S 2T** All Models	**LAF Sensor Range/Performance** *Test Conditions & Results:* Vehicle running at 55-60 mph at steady throttle in 5th gear for 2-3 minutes, then the PCM detected that the LAF sensor signal was too high or too low.
P1166 **CCM 1T** All Models	**LAF Sensor Heater Circuit** *Test Conditions & Results:* Engine runtime more than 80 seconds, then the test started and the PCM detected an open or short condition in the LAF sensor heater circuit.
P1167 **CCM 1T** All Models	**LAF Sensor Heater VS+ Circuit** *Test Conditions & Results:* Engine runtime more than 80 seconds, then the test started and the PCM detected an open condition in the LAF sensor heater VS+ circuit.
P1168 **HO2S 1T** All Models	**LAF Sensor Label Circuit Low Input** *Test Conditions & Results:* Engine runtime 2 minutes, then the PCM detected that the LAF sensor signal remained below the low threshold limit for too long a period of time.
P1169 **HO2S 1T** All Models	**LAF Sensor Label Circuit High Input** *Test Conditions & Results:* Engine running for 2 minutes, then the PCM detected that the LAF sensor signal remained over the high threshold limit for too long a period of time.
P1201-1206 **Catalyst 1T** **Emission 2T** All Models	**Cylinder 1, 2, 3, 4, 5 or 6 Misfire Detected** *Test Conditions & Results:* Engine runtime more than 10 seconds, then the PCM detected a misfire in the specified cylinder (i.e., in Cylinder 1, 2, 3, 4, 5 or 6).
P1241 **CCM 1T** NSX Models	**Throttle Valve Control Motor 1 Circuit** *Test Conditions & Results:* Engine running and the PCM detected an open or short circuit in the Throttle Valve Control Motor 1 circuit (part of the Drive by Wire system).
P1242 **CCM 1T** NSX Models	**Throttle Valve Control Motor 2 Circuit** *Test Conditions & Results:* Engine running, then the PCM detected an open or short circuit in the Throttle Valve Control Motor 2 circuit (part of the Drive by Wire system).
P1243 **CCM 1T** NSX Models	**Insufficient Throttle Position Detected** *Test Conditions & Results:* Engine running at road load, with at least one wide open throttle event, then test started and PCM detected insufficient throttle position.

P1xxx Trouble Codes List (Acura & Honda Models)

Code ID	Code Description & Test Conditions
P1244 CCM 1T NSX Models	**Insufficient Closed Throttle Position Detected** *Test Conditions & Results:* Engine running at road load, with at least one wide open throttle event, then test started and the PCM detected insufficient closed throttle position.
P1246 CCM 1T NSX Models	**Accelerator Position Sensor 1 Circuit** *Test Conditions & Results:* Key on or engine running, then test started and the PCM detected an open or short condition in the Accelerator Pedal Position Sensor 1 circuit.
P1247 CCM 1T NSX Models	**Accelerator Position Sensor 2 Circuit** *Test Conditions & Results:* Key on or engine running, then test started and the PCM detected an open or short condition in the Accelerator Pedal Position Sensor 2 circuit.
P1248 CCM 1T NSX Models	**Accelerator Pedal Position Sensor Correlation Fault** *Test Conditions & Results:* Key on or engine running, then the PCM detected an incorrect correlation between Accelerator Pedal Position Sensor 1 and 2 signals.
P1253 CCM 1T All Models with VTEC	**VTEC Solenoid Circuit** *Test Conditions & Results:* Engine running, then the PCM detected an open or shorted condition in the Variable Timing Electronic Control (VTEC) solenoid circuit.
P1257 CCM 1T All Models with VTEC	**VTEC System Fault** *Test Conditions & Results:* Engine running at road load, then PCM detected a fault in the operation of the Variable Timing Electronic Control (VTEC) system.
P1258 CCM 1T All Models with VTEC	**VTEC System Fault** *Test Conditions & Results:* Engine running at road load, then the PCM detected a fault in the operation of the VTEC system.
P1259 CCM 1T All Models with VTEC	**VTEC Solenoid or Switch Circuit (Bank 1 - Rear)** *Test Conditions & Results:* Engine running in closed loop, then accelerated in 1st gear to over 6000 rpm for 2 seconds and the PCM detected a fault in the VTEC solenoid or switch.
P1279 CCM 1T All Models with VTEC	**VTEC Solenoid or Switch Circuit (Bank 2 - Front)** *Test Conditions & Results:* Engine running in closed loop, then accelerated in 1st gear to over 6000 rpm for 2 seconds and the PCM detected a fault in the VTEC solenoid or switch.
P1297 CCM 1T All Models	**Electrical Load Detector Circuit Low Input** *Test Conditions & Results:* Engine running with lights on, then the test started and the PCM detected an ELD signal that was less than the allowable range.
P1298 CCM 1T All Models	**Electrical Load Detector Circuit High Input** *Test Conditions & Results:* Engine running with lights on, then the test started and the PCM detected an ELD signal that was more than the allowable range.

P1xxx Trouble Codes List (Acura & Honda Models)

Code ID	Code Description & Test Conditions
P1300 Catalyst 1T Emission 2T All Models	**Random Misfire Detected** *Test Conditions & Results:* Engine runtime more than 1 minute, then the test started and the PCM detected a random misfire in one or more cylinders.
P1301-1306 Catalyst 1T Emission 2T All Models	**Cylinder 1, 2, 3, 4, 5 or 6 Misfire Detected** *Test Conditions & Results:* Engine runtime more than 1 minute, then test started and the PCM detected an engine misfire in a specific cylinder (i.e., Cylinder 1, 2, 3, 4, 5 or 6).
P1316 CCM 1T All Models	**Spark Plug Detection Module Circuit (Bank 2)** *Test Conditions & Results:* Engine running and then the PCM detected an open or short circuit condition in the Spark Detection Module circuit for Cylinder Bank 2.
P1317 CCM 1T All Models	**Spark Plug Detection Module Circuit (Bank 1)** *Test Conditions & Results:* Engine running and then the PCM detected an open or short circuit condition in the Spark Detection Module circuit for Cylinder Bank 1.
P1318 CCM 1T All Models	**Spark Plug Detection Module Reset Fault (Bank 2)** *Test Conditions & Results:* Engine running in closed loop, then test started and the PCM detected an open or shorted condition in the Spark Plug Detection Module Reset circuit for Cylinder Bank 2.
P1319 CCM 1T All Models	**Spark Plug Detection Module Reset Fault (Bank 1)** *Test Conditions & Results:* Engine running in closed loop, then test started and the PCM detected an open or shorted condition in the Spark Plug Detection Module Reset circuit for Cylinder Bank 1.
P1336 CCM 1T All Models	**Crankshaft Speed Fluctuation Sensor Circuit** *Test Conditions & Results:* Engine running, then the test started and the PCM detected an intermittent signal from the Crankshaft Speed Fluctuation (CSF) sensor circuit.
P1336 CCM 1T NSX Models	**Crankshaft Position Sensor 'B' Circuit** *Test Conditions & Results:* Engine running, then the test started and the PCM detected an intermittent signal from Crankshaft Position (CKP) Sensor 'B' circuit.
P1337 CCM 1T All Models	**Crankshaft Speed Fluctuation Sensor Circuit No Signal** *Test Conditions & Results:* Engine running, then test started and the PCM detected that there were no signals from the Crankshaft Speed Fluctuation (CSF) sensor circuit.
P1337 CCM 1T NSX Models	**Crankshaft Position Sensor 'B' Circuit No Signal** *Test Conditions & Results:* Engine running and the PCM did not detect any inputs from the CKP Sensor 'B' circuit.

P1xxx Trouble Codes List (Acura & Honda Models)

Code ID	Code Description & Test Conditions
P1359 **CCM 1T** All Models	**Crankshaft Position Top Dead Center Sensor Circuit** *Test Conditions & Results:* Engine running, then the test started and the PCM detected an open or short condition in the Crankshaft Position Top Dead Center Sensor circuit.
P1361 **CCM 1T** V6 Models	**Top Dead Center 1 Sensor Circuit (Intermittent Signal)** *Test Conditions & Results:* Engine running and the PCM detected an intermittent input from the TDC1 sensor circuit.
P1362 **CCM 1T** V6 Models	**Top Dead Center Sensor 1 Circuit (No Signal)** *Test Conditions & Results:* Engine cranking or running, then the test started and the PCM did not detect any TDC1 sensor inputs. Note: The engine will start and run without any signals from the TDC1 sensor.
P1366 **CCM 1T** V6 Models	**Top Dead Center 2 Sensor Circuit (Intermittent Signal)** *Test Conditions & Results:* Engine running and the PCM detected an intermittent input from the TDC2 sensor circuit.
P1367 **CCM 1T** V6 Models	**Top Dead Center 2 Sensor Circuit (No Signal)** *Test Conditions & Results:* Engine cranking or running, then the test started and the PCM did not detect any input signals from the TDC2 sensor. Note: The engine will start and run without any signals from the TDC2 sensor.
P1381 **CCM 1T** All Models	**Camshaft Position Sensor Circuit (Intermittent Signal)** *Test Conditions & Results:* Engine running and the PCM detected an intermittent signal from the CMP Sensor 'A'.
P1382 **CCM 1T** All Models	**Camshaft Position Sensor Circuit (No Signal)** *Test Conditions & Results:* Engine cranking or running, then the test started and the PCM did not detect any input signals from the CMP sensor. Note: The engine will start and run without any signals from the CMP sensor.
P1386 **CCM 1T** All Models	**Camshaft Position Sensor 'B' Circuit (Intermittent Signal)** *Test Conditions & Results:* Engine running and the PCM detected an intermittent signal from CMP Sensor 'B' Circuit.
P1387 **CCM 1T** All Models	**Camshaft Position Sensor 'B' Circuit (No Signal)** *Test Conditions & Results:* Engine running and then the PCM did not detect any input signals from the CMP Sensor 'B'. Note: The engine will start and run without this input.
P1456 **EVAP 2T** All Models	**EVAP Control System Leak Detected (Fuel Tank Area)** *Test Conditions & Results:* Cold startup finished (IAT input from 32-86° at startup), then with ECT input over 154°F and the VSS at over 5 mph for 2 minutes, the PCM activated the EVAP Control and Vent solenoids to enable the EVAP Leak Test and then it detected an incorrect fuel tank pressure.

P1xxx Trouble Codes List (Acura & Honda Models)

Code ID	Code Description & Test Conditions
P1457 EVAP 2T All Models	**EVAP Control System Leak Detected (Canister Area)** *Test Conditions & Results:* Cold startup finished (IAT input from 32-86° at startup), then with the ECT input over 154°F and the VSS input at over 5 mph for 2 minutes, the PCM enabled the EVAP Control and Vent solenoids to begin the Leak Test and then detected an incorrect fuel tank pressure.
P1459 EVAP 2T All Models	**EVAP Purge Flow Switch Circuit** *Test Conditions & Results:* Engine at idle speed, ECT input more than 154°F, then the test started and the PCM detected that the Purge Flow switch signal was incorrect.
P1486 CCM 2T All Models	**Thermostat Range/Performance Problem** *Test Conditions & Results:* Vehicle running in closed loop at road load for 10 minutes, and PCM detected the ECT sensor input did not reach a normal closed loop value.
P1491 EGR 2T All Models	**EGR Valve Lift Sensor Insufficient Flow Detected** *Test Conditions & Results:* Vehicle running in closed loop at 1700-2500 rpm for 10 minutes, and PCM detected an EGRV sensor signal that indicated insufficient EGR flow.
P1498 CCM 1T All Models	**EGR Valve Lift Sensor High Input** *Test Conditions & Results:* Key on or engine running, then the PCM detected that the EGR Valve Lift sensor input was more than the allowable value stored in PCM memory.
P1508 CCM 1T All Models	**Idle Air Control Valve Circuit** *Test Conditions & Results:* Engine running, then the test started and the PCM detected an open or shorted condition in the IAC valve or its related circuits.
P1509 CCM 1T D16Y7 Engine (A/T)	**Idle Air Control Valve Circuit** *Test Conditions & Results:* Engine running, then the test started and then the PCM detected an open or shorted condition in the Idle Air Control (IAC) valve.
P1519 CCM 1T All Models	**Idle Air Control Valve Circuit** *Test Conditions & Results:* Engine running and the PCM detected an open or short in the Idle Air Control (IAC) valve.
P1607 PCM 1T All Models	**PCM Internal Fault 'A'** *Test Conditions & Results:* Key on or engine running and then the PCM detected an internal fault had occurred.
P1608 (PCM, 1T) All Models	**PCM Internal Fault 'B'** *Test Conditions & Results:* Key on or engine running, then the PCM detected an internal fault had occurred.
P1655 CCM 1T All Models	**TMA or TMB Signal Line Circuit** *Test Conditions & Results:* Engine running, then the PCM detected an open or shorted condition in the TMA or TMB transmission data circuit.

P1xxx Trouble Codes List (Acura & Honda Models)

Code ID	Code Description & Test Conditions
P1655 CCM 1T D4 Blinks All Models	**SEFA or SEAF Signal Line Circuit** *Test Conditions & Results:* Engine running, then the PCM detected an open or shorted condition in the SEFA or SEAF circuit to the TCM.
P1671 CCM 1T All Models	**TCM A/T FI Data Line (No Signal)** *Test Conditions & Results:* Engine runtime over 1 minute, and the PCM did not detect any TCM FI Data Line inputs.
P1672 CCM 1T All Models	**TCM A/T FI Data Line Circuit** *Test Conditions & Results:* Engine runtime over 1 minute, and the PCM detected a fault in the TCM FI Data Line circuit.
P1676 CCM 1T All Models	**TCM A/T FI Data Line Circuit** *Test Conditions & Results:* Engine runtime over 1 minute, and the PCM detected a fault in the TCM FI Data Line circuit.
P1677 CCM 1T All Models	**TCM A/T FI Data Line Circuit** *Test Conditions & Results:* Engine runtime over 1 minute, and the PCM detected a fault in the TCM FI Data Line circuit.
P1678 CCM 1T All Model	**TCM A/T FPTDR Signal Line Circuit** *Test Conditions & Results:* Engine runtime over 1 minute, and PCM detected a fault in the TCM FPTDR Signal circuit.
P1690 CCM 1T All Models	**TCSTB Data Line Circuit** *Test Conditions & Results:* Engine runtime over 1 minute, and the PCM detected a fault in the TCSTB Data Line Circuit.
P1696 CCM 1T All Models	**TCFC Line Low Input** *Test Conditions & Results:* Engine idling for 1 minute, and the PCM detected a low input on the TCFC Data Line Circuit.
P1697 CCM 1T All Models	**TCFC Line High Input** *Test Conditions & Results:* Engine idling for 1 minute, and the PCM detected a high input on the TCFC Data Line circuit.
P1705 CCM 1T D4 Blinks All Models	**TCM A/T Gear Position Switch Circuit Shorted** *Test Conditions & Results:* Engine running, and the PCM detected a shorted condition in the Gear Position switch circuit (the result is no lockup).
P1706 CCM 1T Non-MIL All Models	**TCM A/T Gear Position Switch Circuit Open** *Test Conditions & Results:* Engine running, and then the PCM detected an open condition in the Gear Position switch circuit (the result is no lockup).
P1709 CCM 1T D4 Blinks	**TCM A/T Controller Circuit** *Test Conditions & Results:* Engine running, and the PCM detected an open or short fault in the A/T Controller circuit.
P1738 CCM 1T D4 Blinks All Models	**TCM A/T Controller Circuit** *Test Conditions & Results:* Engine running, and the PCM detected an open or shorted condition in one of the A/T Controller circuits.

P1xxx Trouble Codes List (Acura & Honda Models)

Code ID	Code Description & Test Conditions
P1739 **CCM 1T** **D4 Blinks** All Models	**TCM A/T Controller Circuit** ***Test Conditions & Results:*** Engine running, and the PCM detected a fault in the A/T Controller or one of its circuits.
P1750 **TCM 2T** **D4 Blinks** All Models	**TCM A/T System Fault** ***Test Conditions & Results:*** Engine running at Cruise speed, and the PCM detected a fault in the TCM A/T system.
P1751 **TCM 2T** **D4 Blinks** All Models	**TCM A/T System Fault** ***Test Conditions & Results:*** Engine running at Cruise speed, and then the PCM detected a fault in the TCM A/T system.
P1753 **CCM 1T** **D4 Blinks**	**TCM A/T Lockup Solenoid Valve 'A' Circuit** ***Test Conditions & Results:*** Engine running in gear, then the PCM detected an open or shorted condition in the Lockup Solenoid 'A' circuit.
P1758 **CCM 1T** **D4 Blinks** All Models	**TCM A/T Lockup Solenoid Valve 'B' Circuit** ***Test Conditions & Results:*** Engine running in gear, then the PCM detected an open or shorted condition in the Lockup Solenoid 'B' circuit.
P1768 **CCM 1T** **D4 Blinks** All Models	**TCM A/T Controller Circuit** ***Test Conditions & Results:*** Key on or engine running and then the PCM detected a fault in the A/T Controller circuit.
P1773 **CCM 1T** **D4 Blinks** All Models	**TCM A/T Controller Circuit** ***Test Conditions & Results:*** Key on or engine running and then the PCM detected a fault in A/T Controller or its circuits.
P1785 **CCM 1T** **D4 Blinks** All Models	**TCM A/T Controller Circuit** ***Test Conditions & Results:*** Key on or engine running and then the PCM detected a fault in A/T Controller or its related circuits.
P1790 **CCM 1T** **D4 Blinks** All Models	**TCM A/T TP Sensor Circuit** ***Test Conditions & Results:*** Engine runtime more than 15 seconds, then the PCM detected an incorrect TP sensor input signal from the TCM.
P1791 **CCM 1T** **D4 Blinks** All Models	**TCM A/T Vehicle Speed Sensor Circuit** ***Test Conditions & Results:*** Engine runtime more than 15 seconds, then the PCM detected an incorrect Vehicle Speed sensor input signal from the TCM.
P1792 **CCM 1T** **D4 Blinks** All Models	**TCM A/T ECT Sensor Circuit** ***Test Conditions & Results:*** Engine runtime more than 15 seconds, then the PCM detected an incorrect ECT input signal from the TCM.

Emissions Compliance

PRELIMINARY CHECKS

Prior to starting this symptom test, inspect these underhood items:

* Check that the Charging system voltage is within range (13-15v)
* Check the condition of the PCM main and sensor grounds

Note: *The vehicle fails a state tailpipe emissions test. A rotten egg smell is present (odors do not mean a test will fail).*

Emissions Compliance Repair Chart

Step Number & Action to Take	Yes	No
Step 1 Description: *Diagnostic Check* • Perform a PGM-FI Check of the engine controller diagnostics. • Did the PCM detect any codes?	Repair the codes as necessary.	Go to step 2.
Step 2 Description: *EVAP Failure?* • Did vehicle only fail the EVAP leak test or Purge test (gases okay)?	Go to Step 22.	Go to Step 3.
Step 3 Description: *Analyze Report!* • Identify High or Low Gas readings. • If a drive trace is included, identify the drive mode where the gases failed (e.g., did the gases appear high at first and decrease as the catalyst temperature increased?). • Has I/M test report been analyzed?	Go to Step 4.	If no, carefully repeat all of steps listed under the description: *Analyze the report carefully!*
Step 4 Description: *Create Base Line* • Develop a base line of the tailpipe emissions. Use a calibrated gas analyzer and record the current tailpipe readings. Repeat the test in order to establish true base line. • Watch for related symptoms during the base line test (i.e., exhaust smoke or idle speed concerns). • Has vehicle baseline been done?	Go to Step 5. Note: The vehicle base line report can be a useful tool as the readings can be compared against the final emission readings once all vehicle repairs are completed.	Repeat the base line test step until the vehicle baseline has been completed.
Step 5 Description: *Any Symptoms?* Check for signs of these symptoms: • Rough, low or hunting engine idle. • Signs of exhaust smoke at tailpipe. • Cooling system (normal warmup?). • Did you detect any high emission related symptoms in this step?	Go to the Symptom List in this manual and select the matching symptom.	Go to Step 6.
Step 6 Description: *Initial Checks* Perform these inspections and checks: • Check kinks & moisture in lines • Check for electrical connections for dirt or any backed-out terminals. • Check for "add-on" aftermarket emission controls and components. • Did you find faults or wrong parts?	Repair the fault or service the "add-on" aftermarket component as needed. Go to Step 24 to verify the repairs.	Go to Step 7.

Emissions Compliance Repair Chart - Continued

Step Number & Action to Take	Yes	No
Step 7 Description: *Check CO levels* • Use a calibrated gas analyzer to check the vehicle CO levels. • Were the CO levels excessive?	Note: A high CO level indicates a rich mixture. Go to Step 10.	Go to Step 8.
Step 8 Description: *Check HC levels* • Use a calibrated gas analyzer to check the vehicle HC levels. • Were the HC levels excessive?	Note: A high HC level with normal to low CO indicates a lean condition. Go to Step 16.	Go to Step 9.
Step 9 Description: *Check NOx levels* • Use a calibrated gas analyzer to check the vehicle NOx levels. • Were the NOx levels excessive?	Go to Step 20.	The gas levels are okay. Refer to the Symptom List - look for other symptoms.
Step 10 Description: If the CO levels are too high, check the HC levels • With high CO levels, check the HC levels with calibrated gas analyzer. • Were the HC levels too high?	Go to Step 11 to check for running rich with incomplete combustion.	Go to Step 15 to check for a rich running condition.
Step 11 Description: *Test Ignition Sys* • Inspect ignition secondary devices for leakage (is the rotor shorted?) • Test spark output with spark tester. • Remove and inspect spark plugs. • Was the Ignition system okay?	Go to Step 12.	Make repairs to Ignition primary or secondary system as needed. Go to Step 24 to verify repairs.
Step 12 Description: *Check PCV Sys!* • Inspect PCV components for leaks or broken parts, test the PCV valve • Was the PCV system okay?	Go to Step 13.	Make repairs to the PCV system as needed. Go to Step 24 to verify repairs.
Step 13 Description: *Check Exhaust?* • Check for any leaking components • Test exhaust system for restriction (exhaust backpressure should read less than 1.5 psi at cruise speeds). • Was the Exhaust system okay?	Go to Step 14.	Make repairs to the Exhaust system or related components as needed. Go to Step 24 to verify repairs.
Step 14 Description: *Test the Engine* Check these Base Engine Components: • Check engine compression. • Check for wrong or worn camshaft. • Check valve train components. • Check the timing belt or chain. • Was the Base Engine okay?	Recheck the test results from the previous test steps. If they are okay, the problem is not present at this time.	Make repairs to the Base Engine components as needed. Go to Step 24 to verify repairs.
Step 15 Description: *Rich Condition!* • Test the Fuel system for faults that could cause rich condition (leaking injectors, pressure regulator, etc.). • Check LONGFT & SHRTFT values at idle speed and cruise speed. • Was the Fuel system okay?	Go to Step 18.	Make repairs to the Fuel system as needed. Go to Step 24 to verify repairs.

Emissions Compliance Repair Chart - Continued

Step Number & Action to Take	Yes	No
Step 16 Description: Lean Condition! • Test for leaks or low fuel pressure (use a fuel pressure gauge to test fuel pressure at Cruise and Decel). *Caution: Follow safety precautions!* • Check the LONGFT and SHRTFT readings at Idle and Cruise for signs of a lean running condition. • Check for dirty filter or weak pump. • Was the Fuel system okay?	Go to Step 17.	Make repairs as needed to the Fuel system to repair the lean condition. Go to Step 24 to verify repairs.
Step 17 Description: *Test Ignition Sys* • Inspect ignition secondary devices for leakage (is the rotor shorted?). • Test spark output with spark tester. • Remove and inspect spark plugs. • Use engine analyzer to test ignition • Was the Ignition System okay?	Go to Step 18.	Make repairs to Ignition primary or secondary system as needed. Go to Step 24 to verify repairs.
Step 18 Description: *Check PCV Sys!* • Inspect PCV devices for leaks or broken parts, test valve operation • Was the PCV system okay?	Go to Step 19.	Make repairs to PCV system as needed. Go to Step 24 to verify repairs.
Step 19 Description: *Test the Engine* Check these Base Engine Components: • Check engine compression. • Check for wrong or worn camshaft. • Check valve train and timing belt. • Was the Base Engine okay?	Recheck the test results from the previous test steps. If they are okay, the problem is not present at this time.	Make repairs to the Base Engine components as needed. Go to Step 24 to verify repairs.
Step 20 Description: *Check EGR Sys!* • Perform the EGR Function Test (look for a sticking/leaking valve). • Monitor EGR sensor PID data and compare actual readings to vehicle specific reading in this manual. • Was the EGR system okay?	Go to Step 21.	Make repairs to EGR system as needed. Go to Step 24 to verify repairs.
Step 21 Description: *Other Checks* • Analyze report for when the fault occurs (i.e., did it occur only at idle speed, at cruise speed or both?) • Check Cooling system operation (Look for engine or body changes). • Did you find any faults in this step?	Make repairs to the system or component as needed. Go to Step 24 to verify repairs.	Recheck the test results from the previous test steps. If they are okay, the problem is not present at this time.
Step 22 Description: *EVAP Problem* • When does the EVAP fault occur? • Attempt to verify the EVAP failure. • Check for missing/loose fuel cap. • Check condition of the canister. • Inspect EVAP lines for damage. • Did you find any faults in this step?	Make repairs to the system or component as needed. Go to Step 24 to verify repairs.	Go to Step 23.

Emissions Compliance Repair Chart - Continued

Step Number & Action to Take	Yes	No
Step 23 Description: *Test EVAP Sys* • Perform applicable EVAP system test of components and operation. • Check for missing/loose gas cap. • Was the EVAP system okay?	Verify the test results are okay. The EVAP problem is not present (it may be intermittent)	Make repairs as needed to the EVAP system. Go to Step 24 to verify the repairs.
Step 24 Description: *Repair Verification Test Step (Important!)* • Verify vehicle repairs are done • Key off, remove negative battery cable for 15 minutes to reset the fuel trim tables stored in the KAM • Reconnect negative battery cable. • Run the engine at 2500 rpm in P/N for one minute and then allow it to idle in gear at least 2 minutes. • Perform base line test of emissions with an exhaust gas analyzer. • For I/M Emission Test areas where original gas concentrations are reported in grams per mile: refer to the paragraph below to verify GPM. • All other areas where original gas concentrations are reported in PPM: verify gas levels are in range. • Were all gas levels within range?	Save all of the documentation required by the local or federal emission program laws or statutes.	The gas levels are still too high or one of the gas levels is above the acceptable range. Return to Step 1 of this test procedure and carefully repeat all of the test steps.

Excessive Grams Per Mile Verification Procedure

Follow this procedure to verify excessive grams per mile (GPM) indications using parts per million (PPM) readings. If the vehicle gas readings are excessive, compare the actual GPM readings to the gas cut-point levels needed to pass the required test. Determine how much the actual GPM reading exceeds the cut-point, as this data will provide an indication of how much the PPM reading is over the cut-point. It will also provide an indication of how much the PPM reading needs to be reduced (e.g., if the actual reading is twice the cut-point reading, the baseline reading will have to be cut in half or more).

Example: If the actual HC produced by a vehicle were 1.6 GPM, the cut-point for HC would be 0.8 GPM (the actual reading is twice the cut-point). An HC reading from a test vehicle during the baseline test averages 440 PPM. Before this vehicle will pass an I/M test, the HC reading from the verification test must be 220 PPM (1/2 of baseline).

Summary: This method is meant to give a general idea of how much a PPM reading must be reduced for a vehicle to pass an I/M Test that calculates GPM. Your experience will help determine if the emissions readings were reduced enough for the vehicle to pass an I/M Test.

Driveability Symptom Diagnosis

Cross Reference of PID Data to Symptoms

This article includes a list of Symptoms, PID Descriptions and related Diagnostic Tips. To use this list, locate the symptom that matches the vehicle problem and refer to the PID Description and Diagnostic Tips. Use a Scan Tool to gather PID Data while simulating the condition.

Note: Perform Honda Diagnostics and check for bulletins first!

Symptom: Starting Concerns

PID Description	Diagnostic Tips
Cranking RPM (0-900)	*A steady reading during engine cranking indicates that the PCM received a fuel control reference signal*
Fuel Pump Relay (9-11v during cranking)	*If the reading is within range, the fuel pump relay received ignition power*

Symptom: Idle Speed Concerns

PID Description	Diagnostic Tips
Desired Idle	*Indicates the warm engine desired idle speed commanded by the PCM as it compensates for engine load changes (turn on A/C or blower & check for change).*
Engine Speed - AT, MT	*Indicates engine rpm computed by the PCM from the fuel control reference signal (should be ±80 of Desired Idle)*
Idle Air Control (0-255) Note: The IAC Motor can be tested with an external Noid Light test device.	*Indicates the PCM commanded position of the IAC motor pintle. A reading of 0 counts indicates the pintle is fully extended (no air bypass). If the idle speed is too low with a reading of over 60 counts, the PCM is attempting to correct for the condition. If the idle speed is too high with a reading of 0 counts, the PCM is attempting to correct for the condition.*

Symptom: Stalls, Quits Running

PID Description	Diagnostic Tips
Idle Air Control (0-255) Note: A Code 35 or DTC P0506 may exist if the condition is severe.	*Indicates the PCM commanded position of the IAC motor pintle. If the engine stalls with a reading of over 80 counts, look for other causes of the condition (i.e., Fuel system too rich or too lean, a dirty throttle body bore, restrictions or leaks in the intake manifold, excessive engine loads).*

Symptom: Runs Rough

PID Description	Diagnostic Tips
Long Term Fuel Trim (0-255, -100% to 100%)	*Indicates the amount of fuel correction commanded by the PCM. If the engine runs rough and the LONGFT counts are below 70, check any items that can cause a rich condition. If all okay, check these items: IAC operation; MAP sensor calibration and source vacuum; fuel pressure.*
MAP Sensor (11-105 kPa, 0-5v)	*Monitor MAP sensor with condition present. If readings are out of range, test sensor calibration and vacuum source.*
Throttle Angle (0-100%)	*Monitor TP Angle with the condition present. The reading should hold steady near 0%. If the reading is not steady, check the PCV valve, Ignition system and Base Engine.*

Cross Reference of PID Data to Symptoms

Symptom: Cuts Out, Misses

PID Description	Diagnostic Tips
Engine Speed - AT, MT	Monitor the engine speed with the condition present. If the engine speed decreases suddenly with no or little actual change in rpm, check for EMI on the reference circuit.

Symptom: Surges

PID Description	Diagnostic Tips
O2S or HO2S Volts (0-1132mv)	Warmup the engine and then monitor the Oxygen sensor voltage. The voltage should respond quickly to throttle changes (Snap Accel or Decel operation). If it is sluggish, use a Lab Scope to determine if it should be replaced.
Long Term Fuel Trim (0-255, -100% to 100%)	Monitor LONGFT to determine if it is too far into the negative range (%) or the counts are below 70. If a fault is indicated, check any items that can cause a rich condition.
HO2S Cross Counts (0-20)	Monitor the O2S or HO2S Cross Counts to determine the number of times the sensor switches in a 1-second period. A reading near 0 counts may indicate a lazy O2 Sensor.

Symptom: Hesitation, Sag or Stumble

PID Description	Diagnostic Tips
MAP Sensor (11-105 kPa, 0-5v)	Monitor the MAP Sensor voltage as the throttle is opened. If it does not change smoothly, check the vacuum source.
Throttle Angle (0-100%)	Open and close the throttle while checking the TP Angle for a smooth transition during movement (look for binding).
TP Sensor (0-5v)	Monitor the TP Sensor voltage for a smooth transition.

Symptom: Lack of Power, Sluggish

PID Description	Diagnostic Tips
A/C Clutch (On, Off)	Monitor A/C Clutch signal as the A/C switch is cycled. It should change from Off to On when A/C is selected.
A/C Request (Yes, No)	Monitor A/C Request signal as the A/C switch is cycled. It should change from No to Yes as A/C is selected.
A/C Off for WOT (Yes, No)	Monitor A/C Off for WOT signal as the throttle is moved briefly to WOT - the A/C clutch should disengage at WOT.

Symptom: Emissions Compliance

PID Description	Diagnostic Tips
MAP Sensor (11-105 kPa, 0-5v)	Monitor the MAP Sensor voltage at idle speed and cruise. If the reading is out of range, check the sensor calibration.
O2S or HO2S Volts (0-1132mv)	Warmup the engine and then monitor the O2S voltage. The voltage should vary from 10-1000mv at idle and off-idle steady speeds. Use a Lab Scope to confirm the test.
Long Term Fuel Trim (0-255, -100% to 100%)	Monitor the LONGFT to determine if it is too far into the negative or positive range (in % or counts). If a fault is indicated, check any items that could cause this condition.
MAF Sensor (0-512 g/sec)	Warmup the engine and monitor the MAF Sensor at idle and off-idle speeds. If the sensor is not near 2-10 g/sec at idle, check the Air Intake components and MAF Sensor.

PCM Signal Tests

Air Conditioner Signal Test

INTRODUCTION

The PCM uses the A/C signal to determine when the A/C clutch is engaged. The AC signal is diagnosed as a symptom, not as a code.

DRIVEABILITY SYMPTOM DIAGNOSIS

If the A/C signal circuit is inoperative, this condition can cause the engine idle speed to drop when the A/C is turned on (warm engine).

AIR CONDITIONER SIGNAL TEST

Turn the key off and connect the Honda test harness or BOB to the PCM and wiring harness to set up the vehicle for testing.

1) Leave the 'B' connector off ('A' and 'C' connectors connected to the test harness or BOB for this step). Turn the key on and test the voltage between terminals B5 and A26 (B8 and A18 on Civic).

2) If the DVOM reads 5v, go to Step 3. If the DVOM does not read 5v in Step 1, substitute a known-good PCM and retest. If the DVOM reads 5v after the exchange, replace the original PCM.

3) Turn the key off and reconnect the 'B' connector. Turn the key on and momentarily jumper the A/C clutch relay terminals A15 to A26 (B3 to A18 on Civic). An audible clicking noise should be heard from the A/C compressor clutch each time the connection is made. If the clutch clicking noise is heard, go to Step 5. If the clutch noise is not heard, go to Step 4.

4) Use a jumper wire to momentarily connect the RED/BLU wire located in the A/C clutch relay 4-way connector on Accord (YEL wire on Civic) to ground with the key on. If the A/C clutch clicks under these conditions, locate and repair the open circuit in the RED/BLU wire (YEL wire on Civic) between the A/C clutch relay and terminal A15 (B3 on Civic). Then retest for the condition.

5) If no clicking noise is heard at the A/C clutch in Step 3, repair the A/C clutch.

6) If the A/C clutch engaged as required, start the engine and turn the A/C switch and blower on. If the A/C system operates and the idle speed is correct, the A/C signal is okay and the test is completed. If the A/C system does not operate, go to Step 7.

7) Turn the key on and measure the voltage between terminals B5 and A26 (B8 and A18 on Civic). If the DVOM reads less than 1v, substitute a known-good PCM and retest for the symptom. If the symptom is gone, replace the original PCM. If the DVOM reads over 1v, locate and repair the open circuit in the A/C switch circuit to terminal B8. Retest for the condition when repairs are finished.

Note: **The wire colors and pin numbers in this article cover the 1990 Accord, Civic CRX STD, HF, Prelude and Si models.**

Alternator 'FR' Signal Test

INTRODUCTION

The PCM uses the alternator 'FR' signal to determine when the alternator is charging. The 'FR' signal is diagnosed as a symptom (there are no OBD I codes used with this circuit).

DRIVEABILITY SYMPTOM DIAGNOSIS

A problem with the alternator 'FR' signal could cause the engine idle speed to drop when high electrical loads are enabled on the vehicle.

ALTERNATOR 'FR' SIGNAL TEST (1990 HONDA ACCORD)

Turn the key off and connect the Honda test harness or BOB to the PCM and wiring harness to set up the vehicle for testing.

1) Leave the 'D' connector connected to the PCM but disconnected from the test harness or BOB to allow testing of the 'FR' circuit without it being connected to the alternator. Leave all of the other PCM connectors installed (i.e., the power and ground circuits).

2) Turn the key on and test the voltage between terminals D9 and A26. If the reading is 5v, go to Step 3. If the reading is not 5v, turn the key off. Substitute a known-good PCM and retest. If the reading is 5v, replace the original PCM.

3) If the reading was near 5v in Step 2, turn the key off and remove the 'D' connector to the PCM harness. Restart the engine and allow it to fully warm up. Connect the DVOM to terminals D9 and A26 and watch the reading (headlight switch and rear defogger engaged). If voltage does not decrease under these conditions, go to Step 4. If the voltage decreases as the load devices are engaged, the alternator 'FR' signal is okay. Turn the key off and remove the alternator "sense" fuse for 10 seconds.

4) With the key off, remove the 'D' connector ('B' connector on Civic) at the PCM (leave it attached to the wire harness). Disconnect the negative battery cable. Check for continuity between terminal D9 and body ground. If the DVOM reads no continuity, go to Step 5. If it reads continuity, remove the GRN connector at the alternator and recheck for continuity. If continuity exists with the GRN connector removed, locate and repair the short in the WHT/RED wire between the alternator harness and terminal D9. If the DVOM reads no continuity with the GRN connector off, repair the alternator and then retest for the original condition.

5) Remove the GRN connector at the alternator and jumper the WHT/RED wire to battery negative. Repeat the continuity test. If there is no continuity, repair the open circuit in the WHT/RED wire between the alternator connector and terminal D9. If continuity exists with the WHT/RED wire jumped to ground (okay), repair alternator. Retest for the symptom when repairs are completed.

A/T Shift Position Switch Test

INTRODUCTION

The PCM uses the automatic transaxle (A/T) shift position signal to determine when the transaxle is shifted into park or neutral position. The A/T shift position signal is diagnosed as a symptom.

DRIVEABILITY SYMPTOM DIAGNOSIS

A problem with the A/T shift position could cause the engine idle speed to drop too low when the vehicle is shifted into a drive gear.

A/T SHIFT INDICATOR LAMP TEST

Turn the key off and connect the Honda test harness or BOB to the PCM and wiring harness. Turn the ignition key to (engine off). Watch the A/T shift indicator lights as the shift selector is moved between gears. If the lights do not work properly in each position, repair or replace the shift indicator switch. Go to the A/T Shift Signal Test.

A/T Shift Position Switch Graphic

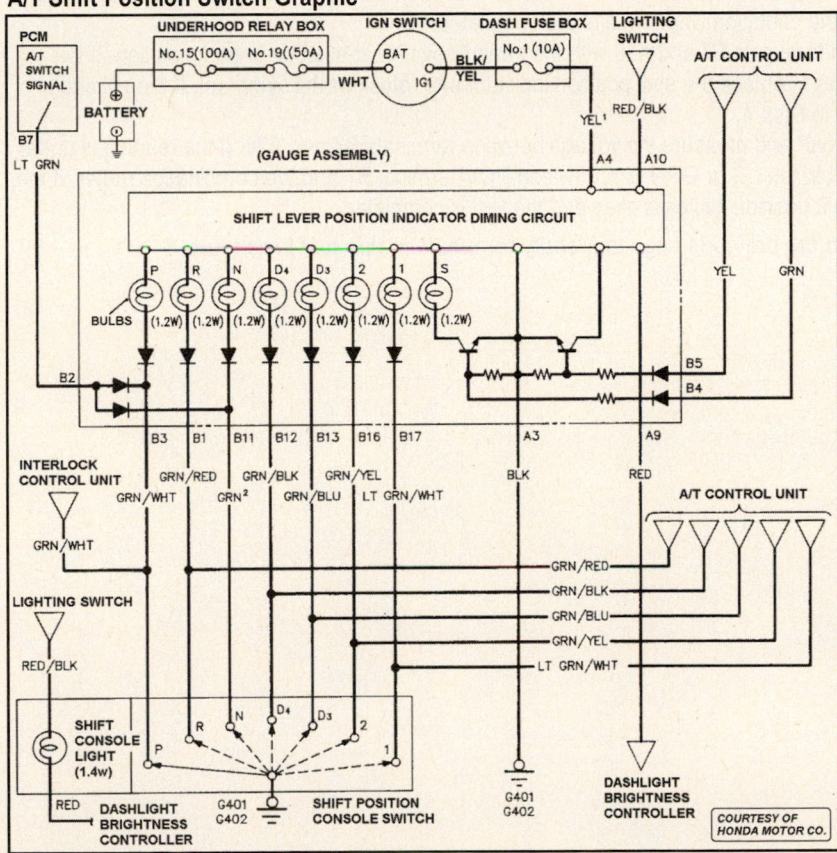

A/T Shift Position Switch Test (Continued)

TEST PROCEDURE

This test procedure was written for a 1990 Honda Accord.

Turn the key off and connect the Honda test harness or BOB to the PCM and wiring harness to set up the vehicle for testing.

1) Leave the 'B' connector removed from the test harness or BOB to allow the A/T shift signal circuit to be isolated for testing. Leave all other connectors installed so that the PCM is connected to power and ground. Measure the voltage between terminals B7 and A26. If the voltage is near 5v, go to Step 2. If the voltage is less than 5v, turn the key off. Inspect the B7 terminal at the PCM for a loose connection. If it is okay, substitute a known-good PCM and retest for voltage. If 5v is now present, replace the original PCM.

2) Place the transaxle gear selector in neutral position. Reconnect the 'B' connector to the test harness. Measure the voltage between terminals B7 and A26 (B7 and A18 on Civic). If the reading is less than 1v, go to Step 3. If the reading is over 1v, locate and repair the open circuit in the LG or GRN wire between the combination meter and PCM terminal B7. Retest for the condition when repairs are completed.

3) Measure the voltage between terminals B7 and A26 with the transmission gear selector in park position. If the voltage reads over 1v (not okay), replace the shift position indicator and retest for the symptom. If the voltage reads less than 1v (okay), go to Step 4.

4) Place the gear selector in "Drive" and measure the voltage between terminals B7 and A26. If the reading is under 5v, locate and repair the short in the LG or GRN wire between PCM terminal B7 and the combination meter. If the reading is over 5v, the A/T shift position switch is okay and the test is completed.

Note: ***Refer to the Graphic on the previous page to identify terminals in this test procedure.***

Brake Switch Signal Test

INTRODUCTION

The PCM uses the brake switch signal to determine when the brake pedal is depressed (i.e., a switch signal change from off to on). The Brake switch signal is diagnosed as a symptom, not as a code as there are no codes for this device on vehicles with OBD I system.

DRIVEABILITY SYMPTOM DIAGNOSIS

A problem with the brake switch signal could cause the engine to stall or almost stall during deceleration with the brake pedal depressed.

PRELIMINARY STEPS

Turn the key off and connect the Honda test harness or BOB to the PCM and wiring harness to allow for proper circuit testing. Next, check the brake lights at the rear of the vehicle to determine if the lights are on all the time. If they are on all the time, refer to the brake light repair section in the correct repair manual or electronic media to make the repair. When repairs are completed (or if the lights are off when they should be on), go to Step 1.

BRAKE SWITCH SIGNAL TEST (1990 CIVIC CRX EXAMPLE)

1) Depress the brake pedal and verify that the brake lights work. If they do not work correctly, inspect the correct stop or horn fuse in the main fuse box. If the fuse is blown, locate and repair the short and go to Step 2.
2) If the fuse is okay, remove the brake switch 2-wire connector. Use the DVOM to check for continuity across the two switch terminals with the brake pedal depressed. If the meter reads no continuity, replace the faulty brake switch. Go to Step 3.
3) Measure the voltage between terminals C10 and A18. If the brake switch is working and properly adjusted, the DVOM will read battery voltage only with the brake pedal depressed. If the brake switch is okay, the Brake Switch Test is completed.
4) If the DVOM does not read battery voltage in this step with the pedal depressed, locate and repair the open signal circuit in the GRN/WHT wire between the brake switch and terminal C10).

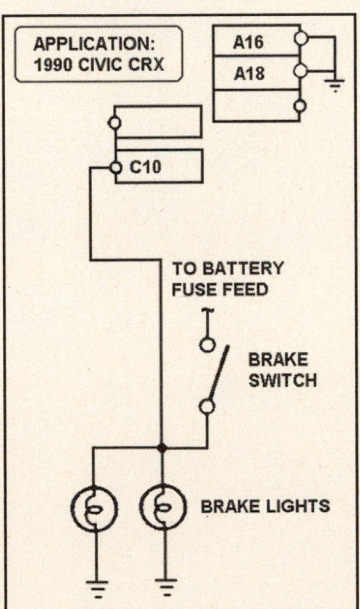

Heater Fan Signal Test

INTRODUCTION

The PCM uses the heater fan switch signal to determine when the heater fan switch is engaged (i.e., the signal changes from off to on).

DRIVEABILITY SYMPTOM DIAGNOSIS

A problem with the heater fan switch signal could cause the engine to stall or almost stall during conditions when the heater fan switch is on.

PRELIMINARY STEPS

Turn the key off and connect the Honda test harness or BOB to the PCM and wiring harness to set up the vehicle for testing.

HEATER FAN SWITCH TEST (1990 CIVIC CRX EXAMPLE)

1) Turn the key to on with the engine off.
2) Use a DVOM to measure between PCM terminals B7 and A18. If the value is less than 5v, go to Step 3. If the value is near 5v (okay), turn the heater fan switch on and measure the voltage between B7 and A18. If the reading is less than 1v, the heater fan switch is okay and the test is completed. If the voltage is over 1v, locate and repair the open circuit in the GRN wire between PCM terminal B7 and the heater fan switch. Retest for the condition when repairs are done.
3) If the value in Step 2 is less than 5v, remove the PCM 'B' connector from the test harness (not from the PCM). Use a DVOM and measure between terminals B7 and A18. If the DVOM reads near 5v, locate and repair the short to ground in the GRN wire (switch to PCM). Retest for the condition when the repairs are completed.
4) If the DVOM reads less than 5v in Step 3, substitute a "known good" PCM and retest the voltage. If the voltage reads near 5v with the "known good" PCM, replace the original PCM and retest for the symptom.

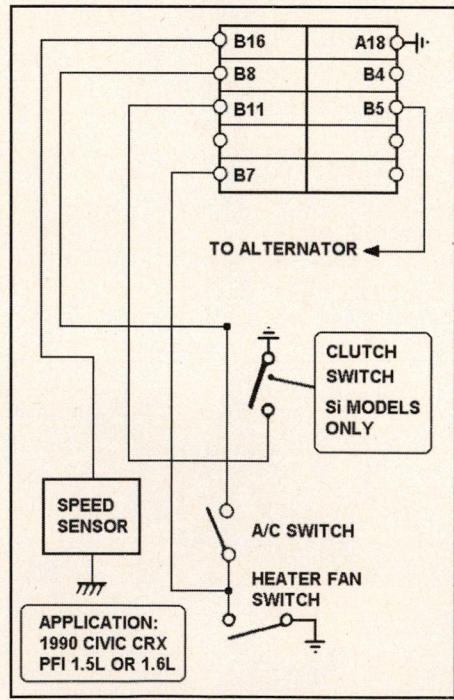

Power Steering Pressure Switch Signal Test

INTRODUCTION

The PCM uses the power steering (oil pressure) switch to determine when the power steering load is high (a signal change from off to on).

DRIVEABILITY SYMPTOM DIAGNOSIS

A problem with the power steering pressure switch could cause the engine idle speed to drop too low when the steering wheel is turned.

PRELIMINARY STEPS (1990 ACCORD EXAMPLE)

Turn the key off and connect the Honda test harness or BOB to the PCM and wiring harness to set up the vehicle for testing.

1) Turn to key on engine off. Measure the voltage between PCM terminals B8 and A18. The reading should be less than 1v. If it is, go to Step 4.

2) If the reading is 12v (open circuit), remove the switch connector. Connect a jumper wire across the switch terminals and repeat the test (jumper the RED to BLK wire). If the reading drops to less than 1v, replace the switch.

3) If the reading was less than 1v in Step 1, start the engine and turn the wheel while watching testing the voltage between terminals B8 and A18. If the reading changes to near 12v as the wheel is turned, the switch is okay.

4) If the reading in Step 3 was near 1v (with the wheel turned), remove the 'B' connector from the test harness or BOB (not at the PCM). Measure the voltage between terminals B8 and ground. If the reading is near 12v, go to Step 5. If it is not, substitute another PCM and retest. If the reading is now 12v, replace the first PCM.

5) If the reading was 12v in Step 3, reconnect the 'B' connector, remove the P/S switch 2-wire connector and retest. If the reading is near 12v with the switch connector removed, replace the P/S switch and retest. If the reading remains low, repair the short to ground in the RED wire between terminal B8 and the P/S switch.

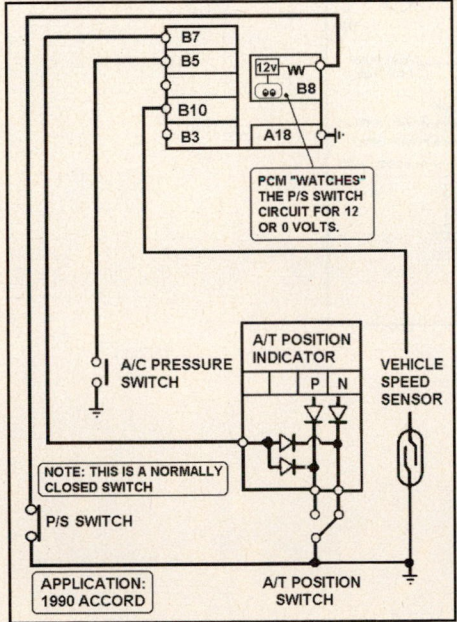

Start Signal Test

INTRODUCTION

The PCM uses the ignition Start signal to determine when the engine is in the crank mode. Once this signal is detected, the PCM controls the current signal to the EACV (valve) to provide for better engine startup. This signal is diagnosed as a symptom (there are no codes).

DRIVEABILITY SYMPTOM DIAGNOSIS

A problem with the start signal could cause a low engine idle speed condition on initial startup with a warm engine.

START SIGNAL TEST (1990 ACCORD EXAMPLE)

Turn the key off and connect the Honda test harness or BOB to the PCM and wiring harness to set up the vehicle for testing.

1) Connect a DVOM between PCM terminals B13 (B9 on Accord & Prelude 2.1L engines) and A18 in order to test the ignition start signal at startup. Turn the ignition switch quickly to start and watch the DVOM reading. If the DVOM reads from 9-11v until the engine starts, the starter signal is okay and the test is completed.

2) If the voltage reading in Step 1 is not from 9-11v during startup (cranking) conditions, inspect for a blown fuse in the start signal circuit. If the fuse is blown, locate and repair the cause of the blown fuse. Replace the fuse and retest for the condition.

3) If the fuse is okay in Step 2, locate and repair the open circuit in the BLU/WHT wire between the PCM terminal B13 (1990 Accord 2.2L engine) and the fuse or switch. Retest for the condition when repairs are completed.

Note: The wire colors and pin numbers in this article cover only the 1990 Accord models.

Start Signal Circuit Graphic

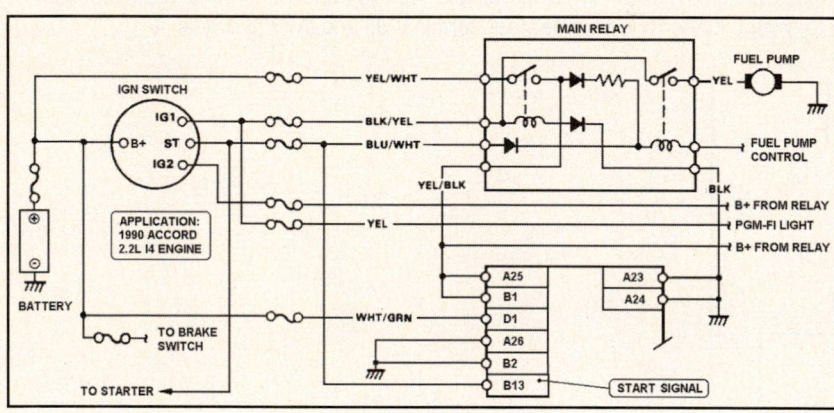

REFERENCE INFORMATION

BARO, ECT, EGRV, IAT, FTP & MAP Sensor Charts

ECT Sensor Conversion Chart

Degrees F	Resistance	Voltage
0°F	14,000 to 20,000 ohms	4.0v
100°F	900-1200 ohms	2-3v
180°F	250-350 ohms	0.5-0.6v

IAT Sensor Conversion Chart

Degrees F	Resistance	Voltage
0°F	14,000 to 20,000 ohms	4.0v
100°F	900-1200 ohms	2-3v
180°F	250-350 ohms	0.5-0.6v

BARO Sensor Chart　　　　MAP Sensor Chart

BARO Pressure (Inches Hg)		Voltage	MAP Pressure (Inches Hg)		Voltage
0" Hg	2.8	3.0v	0" Hg	2.8	3.0v
5" Hg	2.3	2.5v	5" Hg	2.3	2.5v
10" Hg	1.8	2.0v	10" Hg	1.8	2.0v
15" Hg	1.3	1.5v	15" Hg	1.3	1.5v
20" Hg	0.8	1.0v	20" Hg	0.8	1.0v
25" Hg	0.3	0.5v	25" Hg	0.3	0.5v

Fuel Tank Pressure Sensor

Fuel Tank Pressure (Inches Hg)	Voltage
0" Hg	2.8-3.1v
5" Hg	2.3-3.5v
10" Hg	1.8-2.0v
15" Hg	1.3-1.5v
20" Hg	0.8-1.0v
25" Hg	0.3-0.5v

EGR Valve Position Sensor

EGR Valve Position	Voltage
Valve Fully Closed (Idle)	1.0-1.2v
Valve Fully Open (Cruise)	4.5v

Reference Information

1994 Civic PID Data Examples

The PID Examples listed below were captured using a Vetronix Mastertech™ Scan Tool with the Acura/Honda Cartridge.

The list includes example readings at Hot Idle, 30 and 55 MPH. During diagnosis, compare these "known good" readings to the "actual" readings obtained from the vehicle during driving conditions.

1994 Civic Hatchback 1.8L I4 MFI (VIN DB2) [M/T]

PID Acronym	Parameter Description	PID Range	PID Value at Hot Idle	PID Value at 30 mph	PID Value at 55 mph
ACS	A/C Switch	ON or OFF	ON if "on"	OFF	OFF
ACC	A/C Clutch	ON or OFF	ON if "on"	OFF	OFF
ALT	Alternator	0-25.5v	14.2v	14.1v	14.0v
ALTC	Alt. Control	0-100%	43%	11%	16%
BARO	BARO Volts	0-5.1v	2.81v	2.81v	2.82v
Battery	Battery Volts	0-25.5v	14.2v	14.1v	14.0v
Brake	Brake Sw.	ON or OFF	ON if "on"	OFF	OFF
ECT	ECT Sensor	0-5.1v	0.71v	0.59v	0.68v
ELD	ELD Sensor	0-100a	15.9a	16.9a	13.8a
EVAP	EVAP Purge	ON or OFF	OFF	OFF	ON
FAN	Cooling Fan	ON or OFF	ON if "on"	OFF	OFF
FUEL INJ	Fuel Injector	0-99.9 ms	1.82 ms	1.55 ms	1.71ms
HO2S	HO2S Volts	0-1.1v	0.470v	0.700v	0.650v
HTRC	Heater Cntl.	ON or OFF	ON	ON	ON
IAC A	IAC Amps	0-1000ma	247mA	288mA	290mA
IAT	IAT Sensor	0-5.1v	1.26v	1.13v	1.20v
FP Relay	Main Relay	ON or OFF	ON	ON	ON
MAP	MAP Sensor	0-5.1v	0.90v	1.44v	0.96v
MIL	MIL On/Off	ON or OFF	ON if "on"	OFF	OFF
O2 FBST	Fuel Status	CL or OL	CL	CL	CL
RPM	Engine Speed	0-10,000 rpm	673 rpm	2493 rpm	2849 rpm
SCS	Service Connector	OPEN or CLOSED	OPEN	OPEN	OPEN
SP ADV	Spark Advance	-99° to 99°	16° BTDC	43° BTDC	48° BTDC
ST SIG	Starter Signal	ON or OFF	OFF	OFF	OFF
TP	TP Sensor	0-5.1v	0.48v	0.84v	0.92v
VSS	Vehicle Speed	0-155 mph	0 mph	30 mph	55 mph

Note: **The 55 MPH readings in this table were obtained with the vehicle running in fourth gear.**

1996 Accord PID Data Examples

The PID Examples listed below were captured using a Vetronix Mastertech™ Scan Tool with the Acura/Honda Cartridge.

The list includes example readings at Hot Idle, 30 and 55 MPH. During diagnosis, compare these "known good" readings to the "actual" readings obtained from the vehicle during driving conditions.

1996 Accord Sedan 2.2L I4 MFI (VIN CD5) [A/T]

PID Acronym	Parameter Description	PID Range	PID Value at Hot Idle	PID Value at 30 mph	PID Value at 55 mph
ACS	A/C Switch	ON or OFF	ON if "on"	OFF	OFF
ACC	A/C Clutch	ON or OFF	ON if "on"	OFF	OFF
ALT	Alternator	0-25.5v	14.4v	14.3v	14.4v
ALTC	Alt. Control	0-100%	36%	35%	33%
BARO	BARO kPa	10-110	99 kPa	99 kPa	99 kPa
Battery	Battery Volts	0-25.5v	14.4v	14.3v	14.4v
BRAKE	Brake Sw.	ON or OFF	ON if "on"	OFF	OFF
CLV	Engine Load	0-100%	40%	34%	36%
ECT	ECT Sensor	-40 - 304°F	190°F	188°F	191°F
ELD	ELD Sensor	0-100a	15a	12a	14a
EGRV	EGRV signal	0-5.1v	1.2v	2.0v	1.8v
EVAP	EVAP Purge	0-100%	100%	100%	85%
EVAP SW	EVAP Sw.	ON or OFF	ON if "on"	ON	ON
FAN	Cooling Fan	ON or OFF	ON if "on"	OFF	OFF
FSS	Fuel Status	CL or OL	CLOSED	CLOSED	CLOSED
FUEL INJ	Fuel Injector	0-99.9 ms	3.75 ms	3.65 ms	3.60 ms
FUELST1	Fuel Status	CL or OL	CL	CL	CL
HO2S1	HO2S-11 V	0-1100 mv	450 mv	350 mv	650 mv
HTRC1	Heater Cntl.	ON or OFF	ON	ON	ON
HO2S2V	HO2S-12 V	0-1100 mv	650 mv	450 mv	850 mv
HTRC2	Heater Cntl.	ON or OFF	ON	ON	ON
IAC A	IAC Amps	0-1,000ma	474 ma	N/A	N/A
IAC CNT	IAC Counts	0-255	45	N/A	N/A
IAT	IAT Sensor	-40 - 304°F	142°F	138°F	135°F
FP Relay	Main Relay	ON or OFF	ON	ON	ON
MIL	MIL On/Off	ON or OFF	ON if "on"	OFF	OFF
MIL STAT	MIL Status	ON or OFF	OFF	OFF	OFF
MT SOL	Mount Sol.	ON or OFF	ON	ON	ON
PNP	PNP Switch	P-N, Gear	P-N	GEAR	GEAR
RPM	Engine Sp.	0-10,000	770 rpm	1250 rpm	2150 rpm
SCS	Service Plug	CL or OP	OPEN	OPEN	OPEN
SP ADV	Spark ADV	-99 to +99°	15° BTDC	34° BTDC	42° BTDC
ST SIG	Start Signal	ON or OFF	OFF	OFF	OFF
TP	TP Sensor	0-99°	1°	4°	6°
VSS	Vehicle Sp.	0-155 mph	0 mph	30 mph	55 mph

1996 Integra PID Data Examples

The PID Examples listed below were captured using a Vetronix Mastertech™ Scan Tool with the Acura/Honda Cartridge.

The list includes example readings at Hot Idle, 30 and 55 MPH. During diagnosis, compare these "known good" readings to the "actual" readings obtained from the vehicle during driving conditions.

1996 Integra Coupe GS 1.8L I4 MFI (VIN DC2) [M/T]

PID Acronym	Parameter Description	PID Range	PID Value at Hot Idle	PID Value at 30 mph	PID Value at 55 mph
ACS	A/C Switch	ON or OFF	ON if "on"	OFF	OFF
ACC	A/C Clutch	ON or OFF	ON if "on"	OFF	OFF
ALT	Alternator	0-25.5v	14.4v	14.3v	14.4v
ALTC	Alt. Control	0-100%	32%	33%	31%
BARO	BARO kPa	10-110	97 kPa	97 kPa	97 kPa
Battery	Battery Volts	0-25.5v	14.1v	14.2v	14.1v
BRAKE	Brake Switch	ON or OFF	ON if "on"	OFF	OFF
CLV	Engine Load	0-100%	40%	34%	36%
ECT	ECT Sensor	-40 - 304°	172°F	178°F	182°F
ELD	ELD Sensor	0-100a	10a	15a	16a
EGRV	EGRV Sensor	0-5.1v	1.1v	1.6v	1.9v
EVAP	EVAP Purge	0-100%	0%	100%	95%
FAN	Cooling Fan	ON or OFF	ON if "on"	OFF	OFF
FSS	Fuel Status	CL or OL	CLOSED	CLOSED	CLOSED
FUEL INJ	Fuel Injector	0-99.9 ms	3.35 ms	3.30 ms	3.50 ms
FUELST1	Fuel Status	CL or OL	CL	CL	CL
HO2S1V	HO2S-11 V	0-1100 mv	350 mv	450 mv	750 mv
HTRC1	Heater Cntl.	ON or OFF	ON	ON	ON
HO2S2V	HO2S-12 V	0-1100 mv	450 mv	650 mv	750 mv
HTRC2	Heater Cntl.	ON or OFF	ON	ON	ON
IAC A	IAC Amps	0-1000ma	435ma	N/A	N/A
IAC CNT	IAC Counts	0-255	45	N/A	N/A
IAT	IAT Sensor	-40 - 304°	115°F	110°F	112°F
FP Relay	Main Relay	ON or OFF	ON	ON	ON
MIL	MIL On/Off	ON or OFF	ON if "on"	OFF	OFF
MIL STAT	MIL Status	ON or OFF	OFF	OFF	OFF
MT SOL	Mount Sol.	ON or OFF	ON	ON	ON
RPM	Engine Speed	0-10,000	750 rpm	1375 rpm	2550 rpm
SCS	Service Conn.	CL or OP	OPEN	OPEN	OPEN
SP ADV	Spark ADV	-99° to 99°	15° BTDC	35° BTDC	40° BTDC
ST SIG	Starter Signal	ON or OFF	OFF	OFF	OFF
TP	TP Sensor	0-100°	0°	3°	5°
VSS	Vehicle Speed	0-155 mph	0 mph	30 mph	55 mph

1999 Accord PID Data Examples

The PID Examples listed below were captured using a Vetronix Mastertech™ Scan Tool with the Acura/Honda Cartridge.

The list includes example readings at Hot Idle, 30 and 55 MPH. During diagnosis, compare these "known good" readings to the "actual" readings obtained from the vehicle during driving conditions.

1999 Accord Sedan 2.3L I4 VTEC (VIN CG6) [A/T]

PID Acronym	Parameter Description	PID Range	PID Value at Hot Idle	PID Value at 30 mph	PID Value at 55 mph
ACS	A/C Switch	ON or OFF	ON if "on"	OFF	OFF
ACC	A/C Clutch	ON or OFF	ON if "on"	OFF	OFF
ALT	Alternator	0-25.5v	14.4v	14.4v	14.3v
ALTC	Alt. Control	0-100%	35%	34%	34%
BARO	BARO Volts	0-5.1v	2.76v	2.74v	2.73v
Battery	Battery Volts	0-25.5v	14.4v	14.4v	14.3v
BRAKE	Brake Switch	ON or OFF	ON if "on"	OFF	OFF
CLV	Engine Load	0-100%	250%	36%	40%
DTC	DTC Number	0-256	N/A	N/A	N/A
ECT	ECT Sensor	-40 - 304°	190°F	192°F	193°F
ELD	ELD Sensor	0-100a	4.1a	8.1a	10.2a
RPM	Engine Speed	0-10,000	731 rpm	1420 rpm	2265 rpm
EVAP	EVAP Purge	0-100%	0%	0-80%	0-90%
FAN	Fan Relay	ON or OFF	ON if "on"	OFF	OFF
F/STAT	Fuel Status 1	CL or OL	CLOSED	CLOSED	CLOSED
FUEL INJ	Fuel Injector	0-99.9 ms	3.35 ms	3.30 ms	3.50 ms
FUELST1	Fuel Status	CL or OL	CL	CL	CL
HO2S11	HO2S-11 V	0-1100 mv	390 mv	750 mv	240 mv
HO2S12	HO2S-12 V	0-1100 mv	490 mv	560 mv	620 mv
HTRC1	Heater Cntl.	ON or OFF	ON	ON	ON
HTRC2	Heater Cntl.	ON or OFF	ON	ON	ON
IAC CNT	IAC Counts	0-255	50	50	50
IAT	IAT Sensor	-40 - 304°	144°F	139°F	138°F
LONGFT	Long F/T	0-100%	+2%	+1%	+1%
M/R ST	Main Relay	ON or OFF	ON	ON	ON
MAP	MAP Volts	0-5.1v	0.85v	1.11v	1.31v
MIL STAT	MIL Status	ON or OFF	OFF	OFF	OFF
PSP	PSP Switch	ON or OFF	ON if "on"	OFF	OFF
SCS	Service Conn.	CL or OP	OPEN	OPEN	OPEN
SHRTFT	Short F/T	0-100%	-1%	-1%	0%
SP ADV	Spark ADV	-99° to 99°	10° BTDC	15° BTDC	24° BTDC
ST SIG	Starter Signal	ON or OFF	OFF	OFF	OFF
TP	TP Sensor	0-100%	9.8%	11.0%	15%
VSS	Vehicle Sp.	0-155 mph	0 mph	30 mph	55 mph

REFERENCE INFORMATION

1995 Civic 1.5L I4 VTEC (VIN EH2) Wiring Diagrams

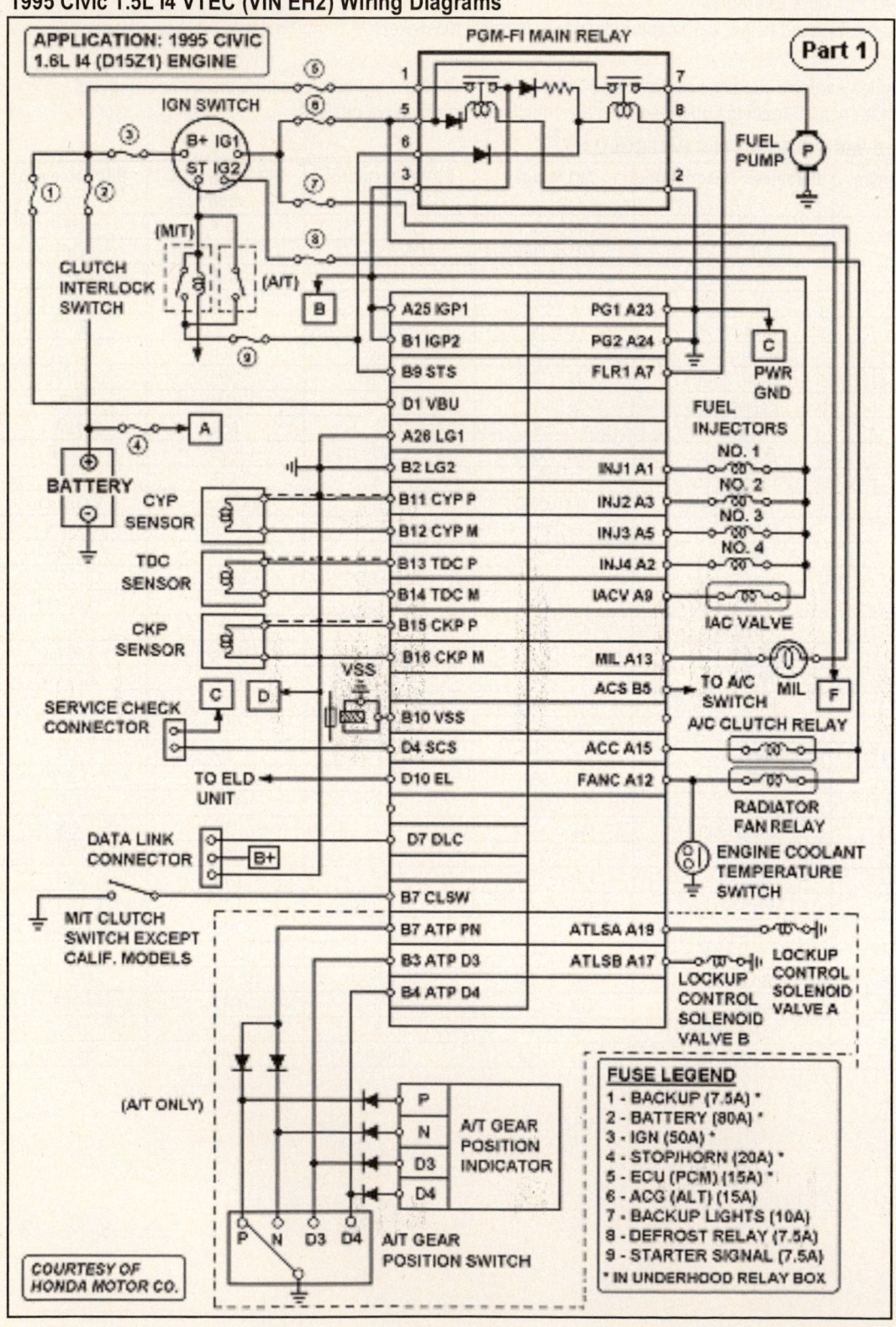

APPLICATION: 1995 CIVIC 1.6L I4 (D15Z1) ENGINE

Part 1

PGM-FI MAIN RELAY

IGN SWITCH
B+ IG1
ST IG2

FUEL PUMP

CLUTCH INTERLOCK SWITCH

(M/T)
(A/T)

BATTERY

CYP SENSOR
TDC SENSOR
CKP SENSOR

SERVICE CHECK CONNECTOR

VSS

DATA LINK CONNECTOR

M/T CLUTCH SWITCH EXCEPT CALIF. MODELS

A25 IGP1
B1 IGP2
B9 STS
D1 VBU
A26 LG1
B2 LG2
B11 CYP P
B12 CYP M
B13 TDC P
B14 TDC M
B15 CKP P
B16 CKP M
B10 VSS
D4 SCS
D10 EL
D7 DLC
B7 CLSW
B7 ATP PN
B3 ATP D3
B4 ATP D4

PG1 A23
PG2 A24
FLR1 A7

PWR GND

FUEL INJECTORS
NO. 1
NO. 2
NO. 3
NO. 4

INJ1 A1
INJ2 A3
INJ3 A5
INJ4 A2
IACV A9

IAC VALVE

MIL A13
ACS B5
ACC A15
FANC A12

TO A/C SWITCH
A/C CLUTCH RELAY
RADIATOR FAN RELAY

ENGINE COOLANT TEMPERATURE SWITCH

TO ELD UNIT

ATLSA A19
ATLSB A17

LOCKUP CONTROL SOLENOID VALVE B
LOCKUP CONTROL SOLENOID VALVE A

(A/T ONLY)

P
N
D3
D4

A/T GEAR POSITION INDICATOR

A/T GEAR POSITION SWITCH

FUSE LEGEND
1 - BACKUP (7.5A) *
2 - BATTERY (80A) *
3 - IGN (50A) *
4 - STOP/HORN (20A) *
5 - ECU (PCM) (15A) *
6 - ACG (ALT) (15A)
7 - BACKUP LIGHTS (10A)
8 - DEFROST RELAY (7.5A)
9 - STARTER SIGNAL (7.5A)
* IN UNDERHOOD RELAY BOX

COURTESY OF HONDA MOTOR CO.

1995 Civic 1.5L I4 VTEC (VIN EH2) Wiring Diagrams

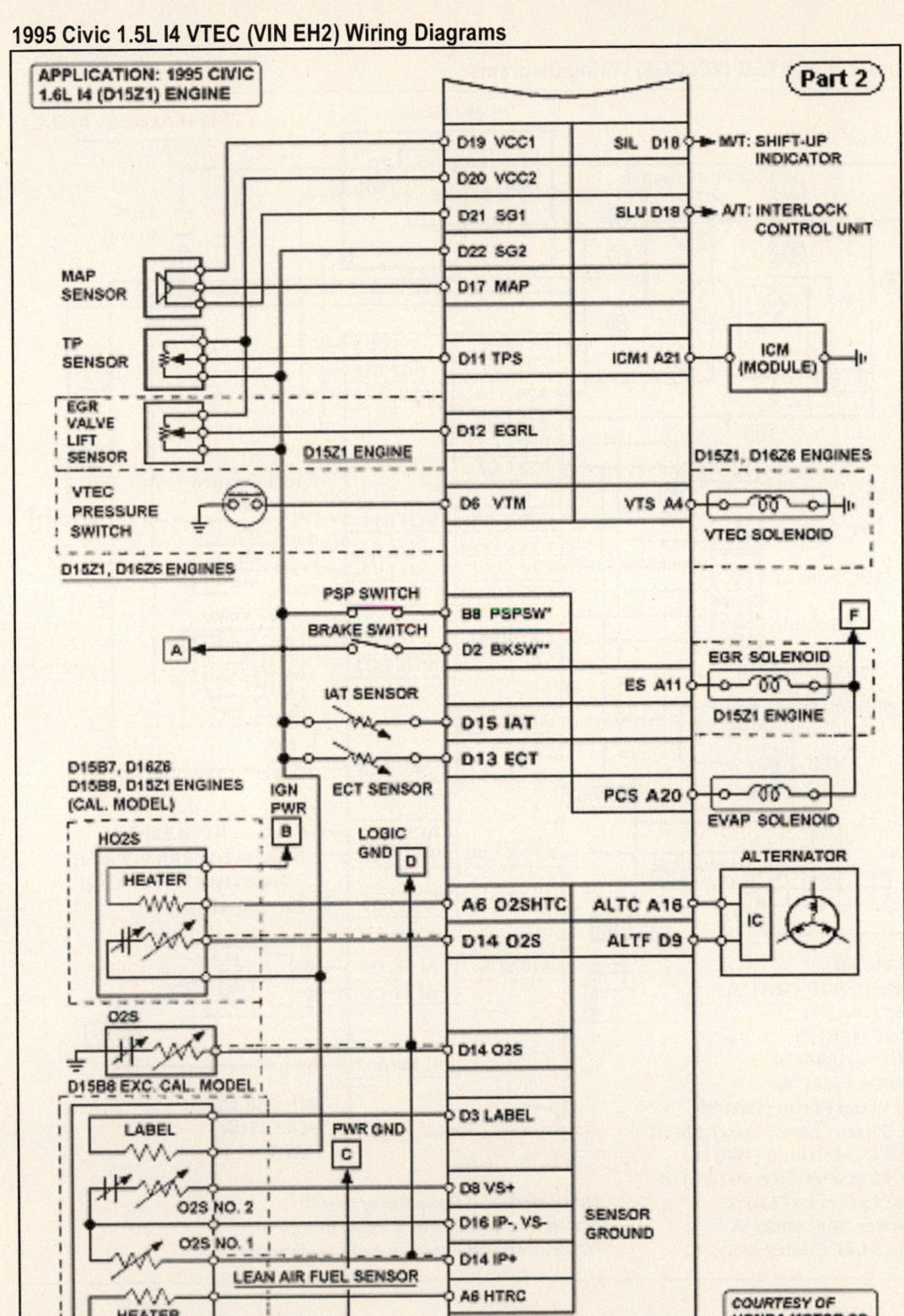

1999 Accord 2.3L I4 VTEC (VIN CG6) Wiring Diagrams

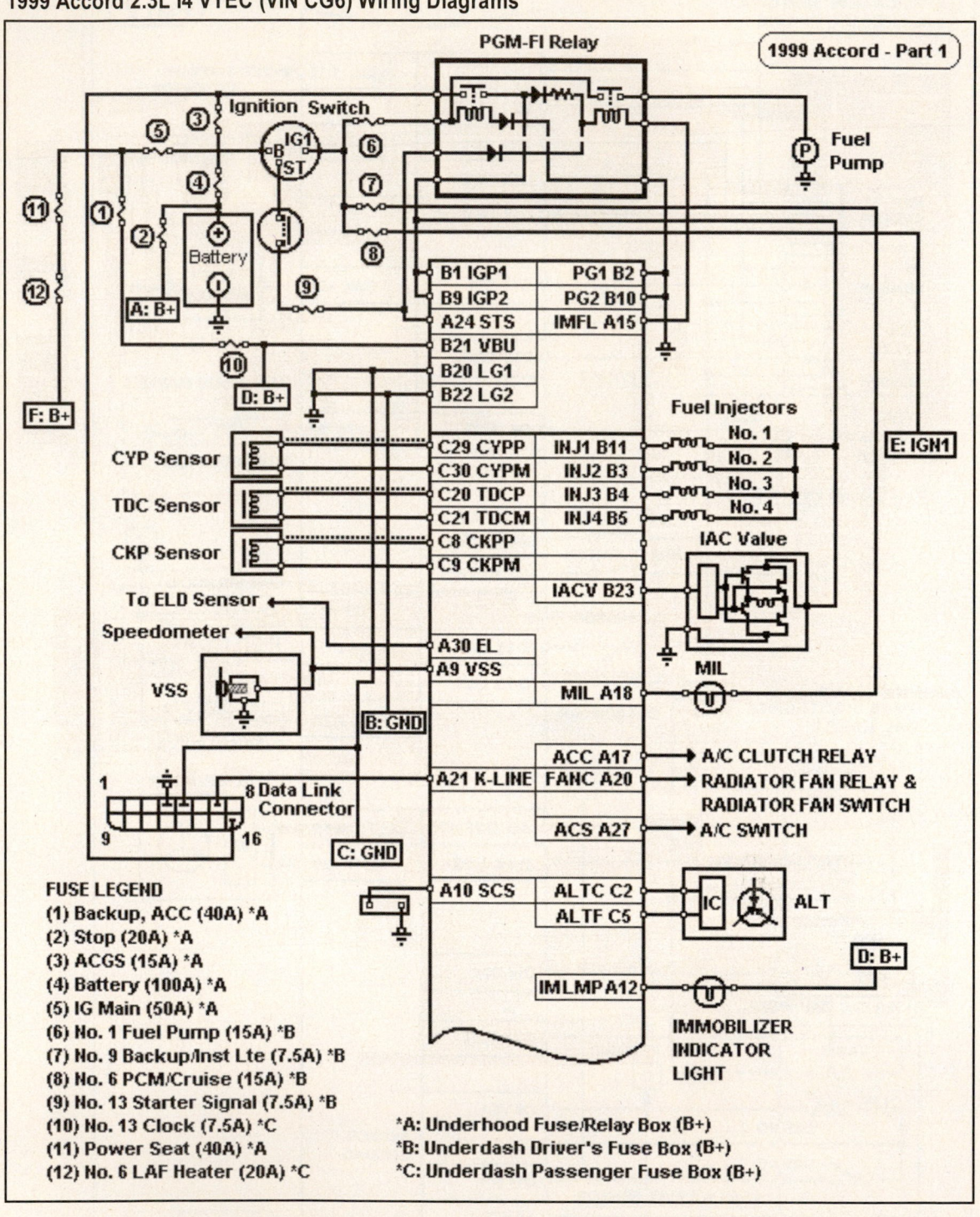

FUSE LEGEND
(1) Backup, ACC (40A) *A
(2) Stop (20A) *A
(3) ACGS (15A) *A
(4) Battery (100A) *A
(5) IG Main (50A) *A
(6) No. 1 Fuel Pump (15A) *B
(7) No. 9 Backup/Inst Lte (7.5A) *B
(8) No. 6 PCM/Cruise (15A) *B
(9) No. 13 Starter Signal (7.5A) *B
(10) No. 13 Clock (7.5A) *C
(11) Power Seat (40A) *A
(12) No. 6 LAF Heater (20A) *C

*A: Underhood Fuse/Relay Box (B+)
*B: Underdash Driver's Fuse Box (B+)
*C: Underdash Passenger Fuse Box (B+)

1999 Accord 2.3L I4 VTEC (VIN CG6) Wiring Diagrams

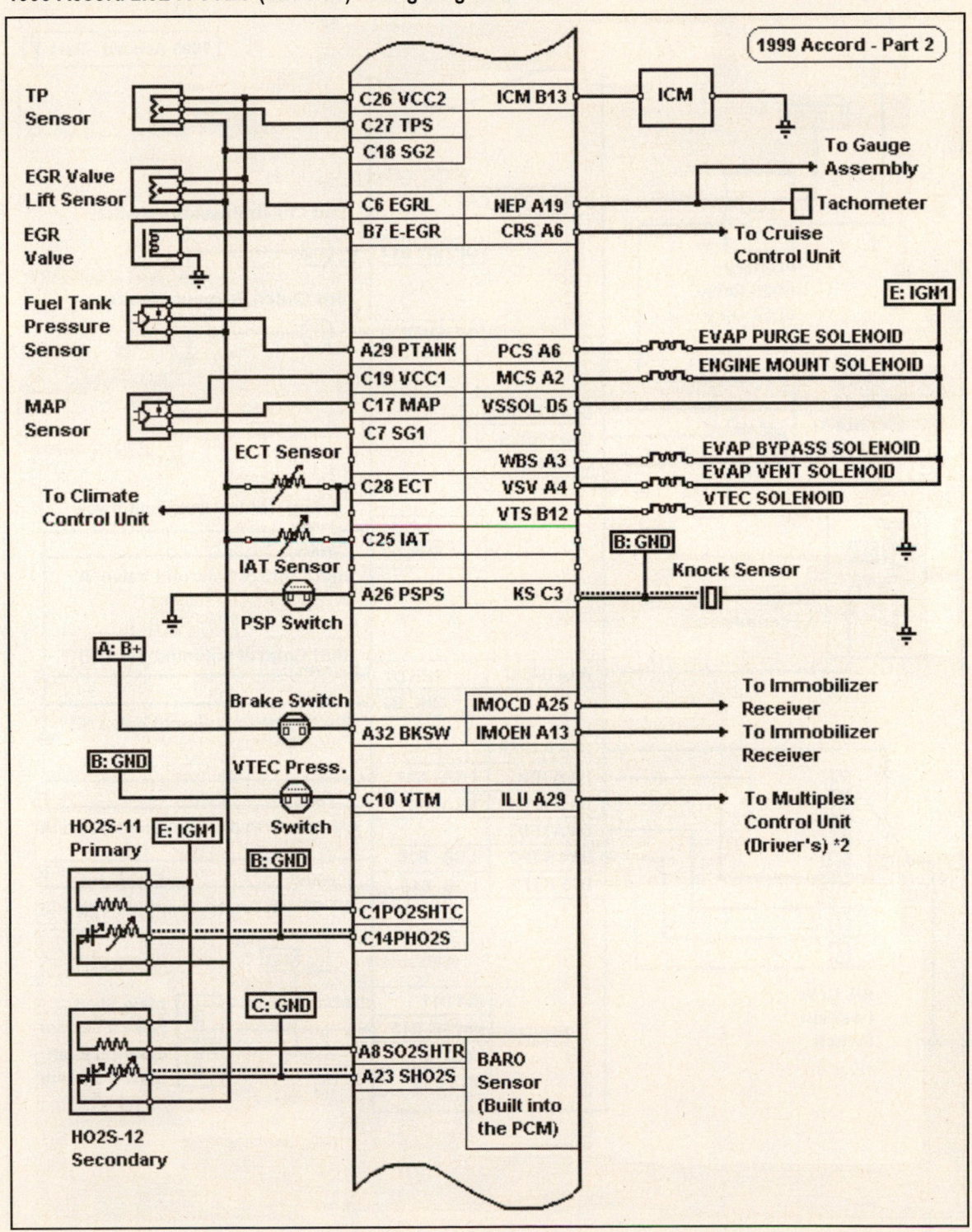

1999 Accord 2.3L I4 VTEC (VIN CG6) Wiring Diagrams

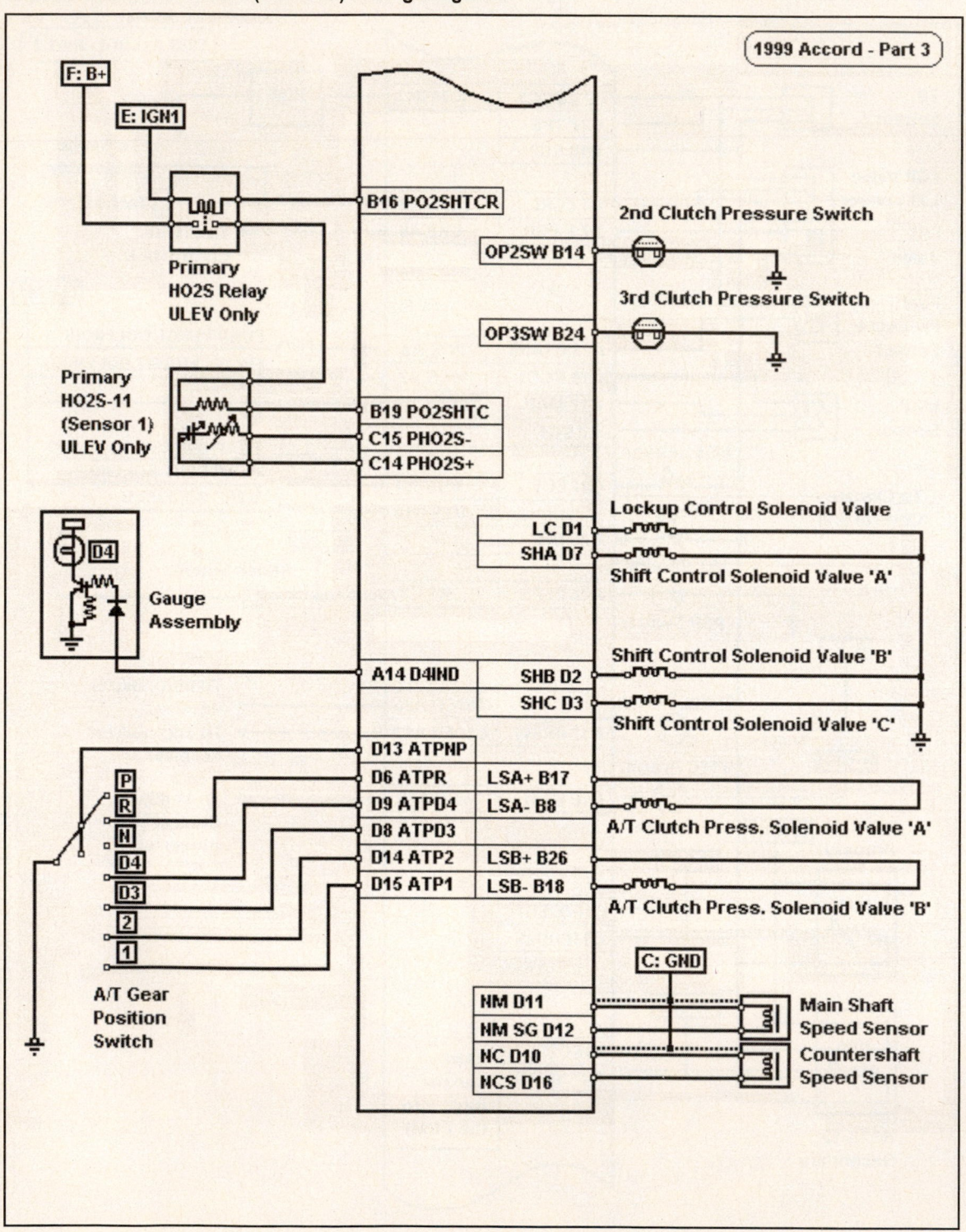

2000 Odyssey 3.5L V6 MFI (VIN RD1) Wiring Diagrams

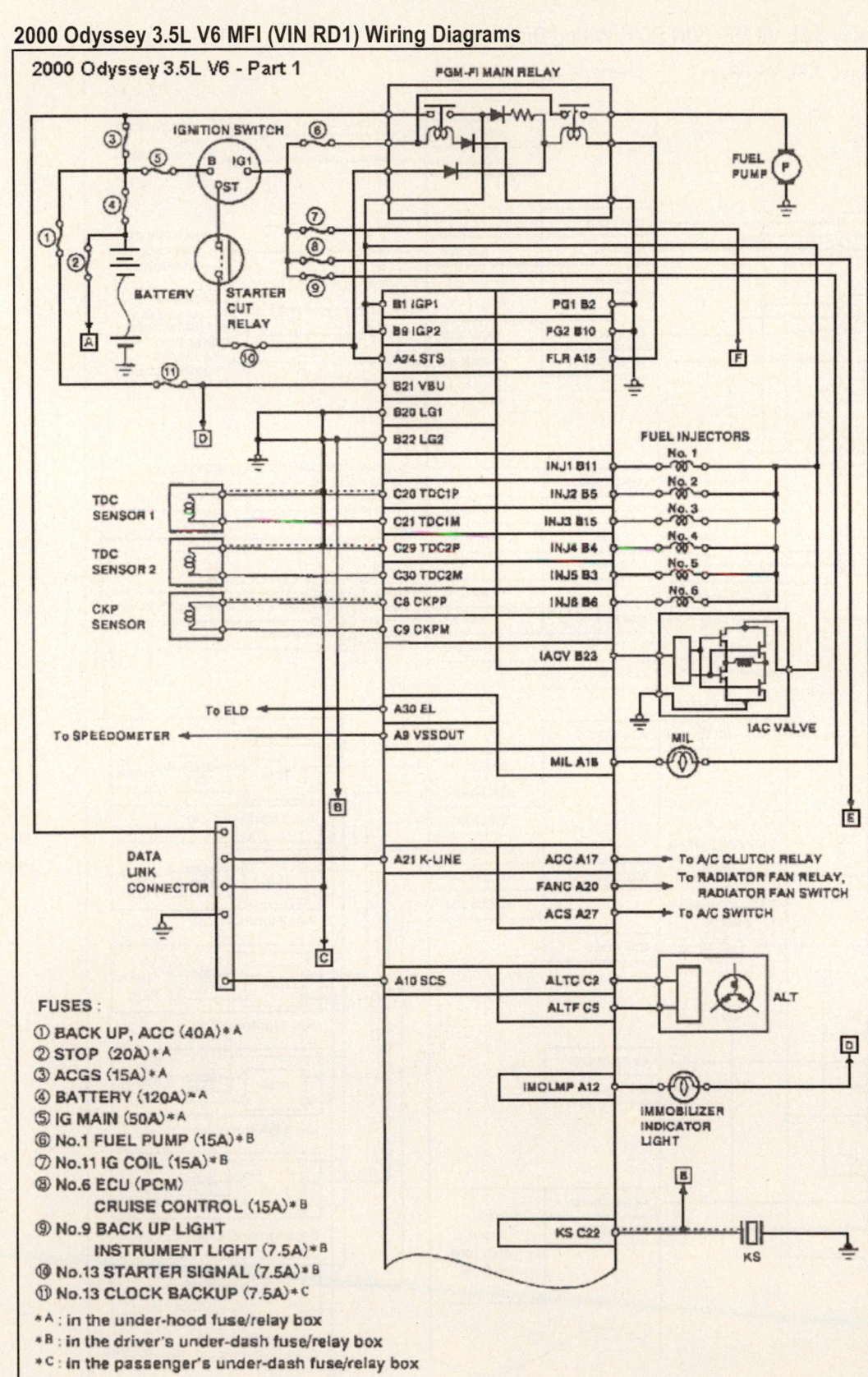

2000 Odyssey 3.5L V6 - Part 1

2000 Odyssey 3.5L V6 MFI (VIN RD1) Wiring Diagrams

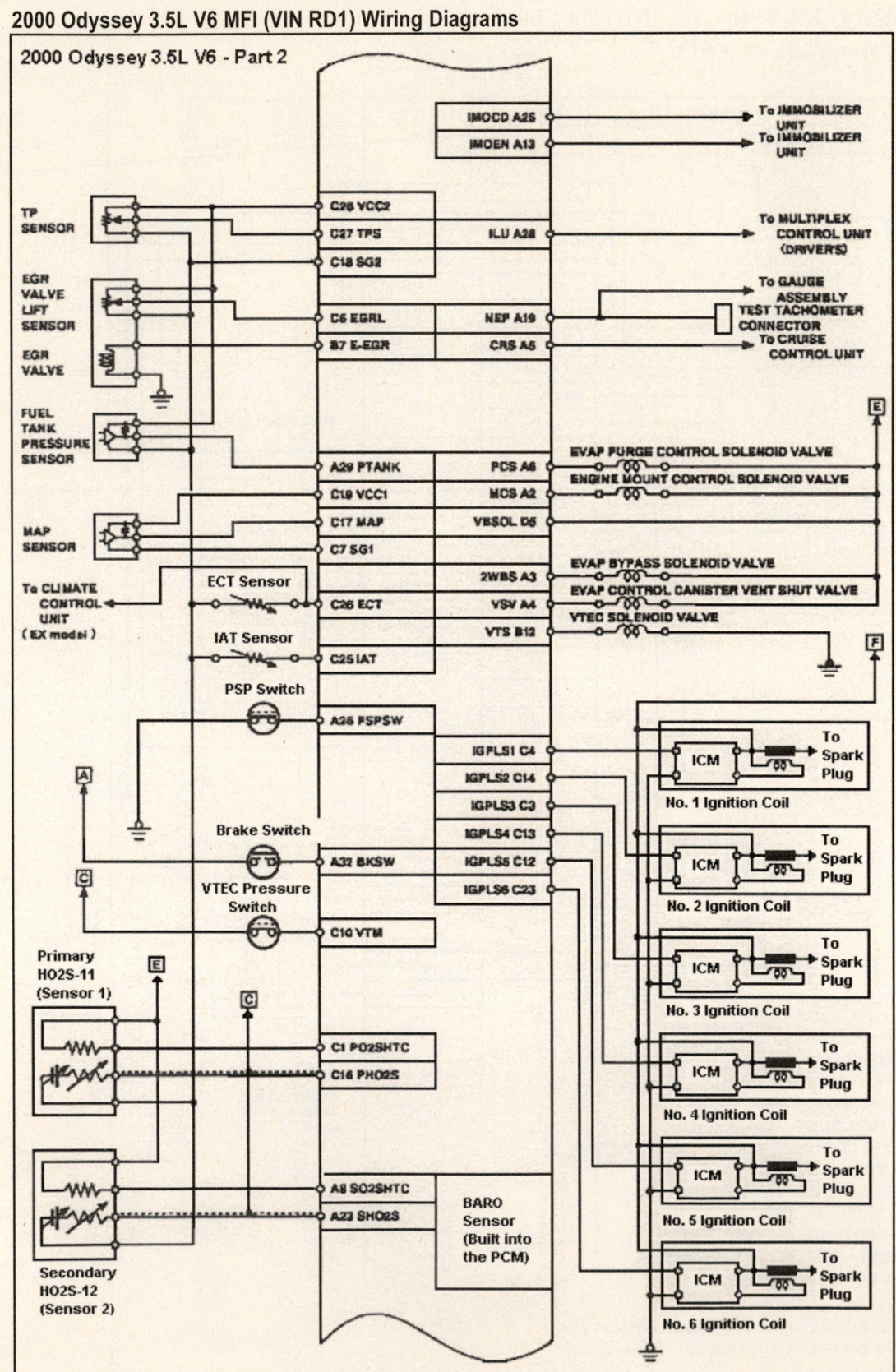

2000 Odyssey 3.5L V6 - Part 2

REFERENCE INFORMATION

Honda OBD Glossary

Honda Glossary of Terms and Acronyms	
(<): Indicates less than the value	(>): Indicates more than the value
ACS: Air Conditioning Switch	ACC: Air Conditioning Clutch
CEL: Check Engine Light	CKP: Crankshaft Position Sensor
CMP: Camshaft Position Sensor	CSS: Countershaft Speed Sensor
DI: Distributor Ignition	DLC: Data Link Connector
EACV: Electronic Idle Air Control	EICV: Electronic Idle Control Valve
EGR: Exhaust Gas Recirculation	EGRV: EGR Valve Lift Sensor
EGR Monitor: OBD II EGR Test	EI: Electronic Ignition
EVAP: Evaporative Emission System	EVAPV: EVAP Purge Vent Solenoid
FAN: Cooling Fan Relay	FP: Fuel Pump Relay
FTP: Fuel Tank Pressure	FWD: Front Wheel Drive
GND: Engine Ground Point	O2S: Oxygen Sensor
HO2S: Heated Oxygen Sensor	HO2S-11: Front Heated O2 Sensor
IAC: Idle Air Control Valve	IAT: Intake Air Temperature
ICM: Ignition Control Module	INJ 1 to INJ 6: Injectors 1 through 6
IMA: Idle Mixture Adjuster	IMM: Immobilizer (Security) System
ITA: Ignition Timing Adjuster	KAM: Keep Alive Memory (B+)
KOEC: Key On, Engine Cranking	KOEO: Key On, Engine Off
KOER: Key On, Engine Running	KS: Knock Sensor
LAMBSE: Short Term Fuel Trim	LCD: Liquid Crystal Display
LSA: Lockup Solenoid 'A' (A/T)	LSB: Lockup Solenoid 'B' (A/T)
LONGFT: Long Term Fuel Trim	LOOP- Engine Operating Loop
LPG: Liquid Petroleum Gas	M/T: Manual Transmission
MAP: Manifold Absolute Pressure	MIL: Malfunction Indicator Lamp
MPH: Miles Per Hours	NC: Normally Closed
NGV: Natural Gas Vehicle	OC: Oxidation Catalyst
OBD I: Onboard Diagnostics Ver. 1	OBD II: Onboard Diagnostics Ver. 2
OHC: Overhead Cam Engine	OSM: Output State Monitor
PCM: Powertrain Control Module	PCS: Purge Control Solenoid
PCV: Positive Crankcase Ventilation	PFI: Port Fuel Injection
PID: Parameter Identification Item	PNP: Park Neutral Position Switch
PWR: Power from Battery or Ignition	PWR GND: Power Ground for PCM
ROM: Read Only Memory	RAM: Random Access Memory
RPM: Revolutions Per Minute	RWD: Rear Wheel Drive
SCP: Standard Corporate Protocol	SSA (A/T): Shift Solenoid 'A'
SSB (A/T): Shift Solenoid 'B'	SSC (A/T): Shift Solenoid 'C'
SHRTFT: Short Term Fuel Trim	SMFI: Sequential Multiport Injection
TA: Throttle Angle	TCS: Traction Control System
TDC: Top Dead Center Position	TOT: Transmission Oil Temperature
TP: Throttle Position	VBU: Voltage Backup (from battery)
VCC: Voltage Control Circuit	VREF: Voltage Reference (PCM)
VSS: Vehicle Speed Sensor	VTEC: Variable Valve Timing with Electronic Control

ACURA CAR CONTENTS

About This Section

Introduction

This section of the manual contains Pin Voltage Tables for Acura vehicles from 1990-2001. It can be used to help you repair Trouble Code and No Code problems related to the PCM.

VEHICLE COVERAGE

The following vehicle applications are covered in this section:

- 1990-2001 Integra Coupe, Hatchback & Sedan
- 1997-2001 CL 2-Door, 4-Door Coupe & Sedan
- 1990-95 Legend Coupe & Sedan
- 1991-2001 NSX Open Top & Coupe Models
- 1996-2001 RL 4-Door Sedan
- 1995-2001 TL 4-Door Sedan
- 1992-1994 Vigor 4-Door Sedan

How to Use This Section

This section can be used to look up the location of a particular pin, a wire color or to find a "known good" value of a circuit. To locate the PCM information for a particular vehicle, find the model, correct engine size (with VIN Code) and finally the year of the vehicle.

For example, to look up the PCM terminals for a 1997 Integra GS-R 1.8L VTEC VIN DC2, go to Contents Page 1 to find this text string.

> **1.8L I4** VTEC VIN DC2, DC4 [All] 79-Pin **(1997-99)** ....

Then turn to Page 6-47 to find the following PCM related information.

The symbol (<) in D11 indicates your reading should be less than the number. The transmission type may be identified in the table title as [A/T] Auto, [M/T] Manual or [All]. Refer to the example directly below.

1997-99 GS-R, Type R 1.8L VTEC VIN DC2 [All] 16P 'D' Connector

PCM Pin #	W/Color	Circuit Description (16-Pin)	Value at Hot Idle
D4	YEL/WHT	MAP Sensor VREF (VCC1)	4.9-5.1v
D5	GRN/WHT	Brake Switch Signal	Brake Off: 12v, On: 0v
D6	RED/BLU	Knock Sensor Signal	No detonation: 18mv AC
D7	WHT/RED	HO2S-11 (B1 S1) Signal	0.1-1.1v
D8	RED/YEL	IAT Sensor Signal	At 100ºF: 2-3v
D11	GRN/BLU	Sensor Ground (SG2)	<0.050v

In this example, the MAP Sensor VREF circuit is connected to Pin D4 of the 16-Pin connector with a YEL/WHT wire. The value at Hot Idle is 4.9-5.1v. This is the reference voltage provided to the MAP sensor from a voltage regulator that is located inside the PCM.

The IAT Sensor circuit is connected to Pin D8 of the same 16-Pin connector with a RED/YEL wire. This signal indicates the voltage from the IAT Sensor with the intake air temperature approximately 100ºF.

INTEGRA Pin Tables

1990-91 Hatchback 1.8L MFI VIN DA9 [All] 18P 'A' Connector

PCM Pin #	W/Color	Circuit Description (18-Pin)	Value at Hot Idle
A1	BRN	Injector 1	2.0-3.3 ms
A2, 4	BLK	Power Ground	<0.1v
A3	RED	Injector 2	2.0-3.3 ms
A5	LT BLU	Injector 3	2.0-3.3 ms
A6	GRN	EVAP Purge Solenoid	Solenoid Off: 12v, On: 1v
A7	YEL	Injector 4	2.0-3.3 ms
A8	RED	EGR Solenoid Control	Solenoid Off: 12v, On: 1v
A9	---	Not Used	---
A10	GRN/YEL	Fuel Pressure Regulator	Solenoid Off: 12v, On: 1v
A11	BLU/YEL	Electronic Air Control Valve	Pulse Signals
A12	GRN/BLK	Fuel Pump Relay Control	Relay Off: 12v, On: 1v
A13	YEL/BLK	Main Relay Power (B+)	12-14v
A14	GRN/BLK	Fuel Pump Relay Control	Relay Off: 12v, On: 1v
A15	YEL/BLK	Main Relay Power (B+)	12-14v
A16	BRN/BLK	Power Ground	<0.1v
A17	---	Not Used	---
A18	BLK/RED	Power Ground	<0.1v

1990-91 Hatchback 1.8L MFI VIN DA9 [All] 20P 'B' Connector

PCM Pin #	W/Color	Circuit Description (20-Pin)	Value at Hot Idle
B1	WHT/GRN	Keep Alive Power (VBU)	12-14v
B2	BLU	A/T: Fast Idle Solenoid	Solenoid Off: 12v, On: 1v
B3	YEL	A/C Clutch Relay Control	Relay Off: 12v, On: 1v
B4	BLU/WHT	A/T: Control Unit Signal	---
B5	---	Not Used	---
B6	GRN/ORN	Malfunction Indicator Lamp	MIL Off: 12v, On: 1v
B7	GRN	A/T: Shift Position Switch	In P/N: 0v, Others: 12v
B8	BLU/RED	A/C Switch Signal	Relay Off: 12v, On: 1v
B9	---	Not Used	---
B10	ORN	CKP Sensor Signal	AC pulse signals
B11	---	Not Used	---
B12	WHT	CKP Sensor Ground	<0.050v
B13	BLU/WHT	Starter Switch Signal	Cranking: 9-11v
B14	BLU	Alternator Charging	Headlights off: 0v, on: 12v
B15	WHT	Igniter Signal	Digital Signals: 0-12-0v
B16	YEL/RED	Vehicle Speed Sensor	Moving: pulse signals
B17	WHT	Igniter Signal	Digital Signals: 0-12-0v
B18	---	Not Used	---
B19	RED	PSP Switch Signal	Straight: 0v, Turning: 12v
B20	BRN	Ignition Timing Adjustment	0.4-4.5v

1990-91 Hatchback 1.8L MFI VIN DA9 [All] 16P 'C' Connector

PCM Pin #	W/Color	Circuit Description (16-Pin)	Value at Hot Idle
C1	BLU/GRN	CYP Sensor Signal	AC pulse signals
C2	BLU/YEL	CYP Sensor Ground	<0.050v
C3	ORN/BLU	TDC Sensor Signal	AC pulse signals
C4	WHT/BLU	TDC Sensor Ground	<0.050v
C5	RED/YEL	IAT Sensor Signal	At 100°F: 2-3v
C6	RED/WHT	ECT Sensor Signal	At 180°F: 0.5-0.6v
C7	RED/BLU	Throttle Angle Sensor	0.5-0.6v
C8	GRN/RED	EGR Valve Lift Sensor	1.1-1.2v
C9	RED/WHT	Atmospheric Press. Sensor	Varies w/alt: 0.5-3.0v
C10	GRN/WHT	Brake Switch Signal	Brake Off: 12v, On: 0v
C11	WHT	MAP Sensor Signal	0.8-0.9v
C12	GRN	Sensor Ground	<0.050v
C13	YEL/WHT	Sensor VREF	4.9-5.1v
C14	GRN/WHT	MAP Sensor Ground	<0.050v
C15	YEL/RED	MAP Sensor VREF	4.9-5.1v
C16	WHT	O2S-11 Signal	0.1-1.1v

Abbreviation	Color	Abbreviation	Color	Abbreviation	Color
BLK	Black	LT BLU	Lt. Blue	TAN	Tan
BLU	Blue	LT GRN	Lt. Green	VIO	Violet
BRN	Brown	ORN	Orange	WHT	White
GRY	Gray	PNK	Pink	YEL	Yellow
GRN	Green	PPL	Purple		

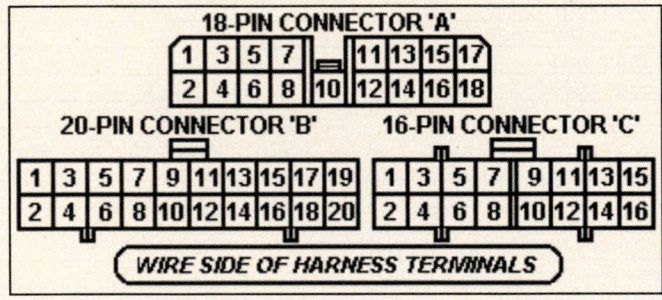

1992-93 Hatchback 1.7L VTEC VIN DB2 [M/T] 26P 'A' Connector

PCM Pin #	W/Color	Circuit Description (26-Pin)	Value at Hot Idle
A1	BRN	Injector 1	2.0-3.3 ms
A2	YEL	Injector 4	2.0-3.3 ms
A3	RED	Injector 2	2.0-3.3 ms
A4	GRN/YEL	VTEC Solenoid Control	0v, Hi-Speed: 12v
A5	LT BLU	Injector 3	2.0-3.3 ms
A6	ORN/BLK	HO2S-11 (B1 S1) Heater	Heater Off: 12v, On: 1v
A7	GRN/BLK	Fuel Pump Relay Control	Relay Off: 12v, On: 1v
A8	GRN/BLK	Fuel Pump Relay Control	Relay Off: 12v, On: 1v
A9	BLU/YEL	Idle Air Control Valve	Pulse Signals
A10-12	---	Not Used	---
A13	GRN/ORN	Malfunction Indicator Lamp	MIL Off: 12v, On: 1v
A14	---	Not Used	---
A15	YEL	A/C Clutch Relay Control	Relay Off: 12v, On: 1v
A16-19	---	Not Used	---
A20	GRN	EVAP Purge Solenoid	Solenoid Off: 12v, On: 1v
A21	YEL/GRN	Igniter Signal	Digital Signals: 0-12-0v
A22	YEL/GRN	Igniter Signal	Digital Signals: 0-12-0v
A23	BLK	Power Ground	<0.1v
A24	BLK	Power Ground	<0.1v
A25	YEL/BLK	Main Relay Power (B+)	12-14v
A26	BLK/RED	Chassis Ground	0.1v

1992-93 Hatchback 1.7L VTEC VIN DB2 [M/T] 16P 'B' Connector

PCM Pin #	W/Color	Circuit Description (16-Pin)	Value at Hot Idle
B1	YEL/BLK	Main Relay Power (B+)	12-14v
B2	BRN/BLK	Chassis Ground	<0.050v
B3-4	---	Not Used	---
B5	BLU/RED	A/C Pressure Switch	Relay Off: 12v, On: 1v
B6-7	---	Not Used	---
B8	RED	PSP Switch Signal	Straight: 0v, Turning: 12v
B9	BLU/WHT	Starter Switch Signal	Cranking: 9-11v
B10	YEL/RED	Vehicle Speed Sensor	Moving: pulse signals
B11	ORN	CYP Sensor Signal	AC pulse signals
B12	WHT	CYP Sensor Ground	<0.050v
B13	ORN/BLU	TDC Sensor Signal	AC pulse signals
B14	WHT/BLU	TDC Sensor Ground	<0.050v
B15	BLU/GRN	CKP Sensor Signal	AC pulse signals
B16	BLU/YEL	CKP Sensor Ground	<0.050v

1992-93 Hatchback 1.7L VTEC VIN DB2 [M/T] 22P 'D' Connector

PCM Pin #	W/Color	Circuit Description (22-Pin)	Value at Hot Idle
D1	WHT/YEL	Keep Alive Power (VBU)	12-14v
D2	GRN/WHT	Brake Switch Signal	Brake Off: 12v, On: 0v
D3	RED/BLU	Knock Sensor Signal	No Detonation: 18mv AC
D4	BRN	Service Check Connector	SCS Open: 4.80v
D5	---	Not Used	---
D6	BLU/BLK	VTEC Pressure Switch	0v, Hi-Speed: 12v
D7	LT BLU	Data Link Connector	No Scan Tool: 5v
D8	---	Not Used	---
D9	BLU	Alternator 'FR' Signal	Digital Signals: 0-5-0v
D10	---	Not Used	---
D11	RED/BLU	TP Sensor Signal	0.5-0.6v
D12	---	Not Used	---
D13	RED/WHT	ECT Sensor Signal	At 180ºF: 0.5-0.6v
D14	WHT	HO2S-11 (B1 S1) Signal	0.1-1.1v
D15	RED/YEL	IAT Sensor Signal	At 100ºF: 2-3v
D18	---	Not Used	---
D17	WHT	MAP Sensor Signal	0.8-0.9v
D18	---	Not Used	---
D19	YEL/RED	MAP Sensor VREF	4.9-5.1v
D20	YEL/WHT	Sensor VREF	4.9-5.1v
D21	GRN/WHT	MAP Sensor Ground	<0.050v
D22	GRN/WHT	Sensor Ground	<0.050v

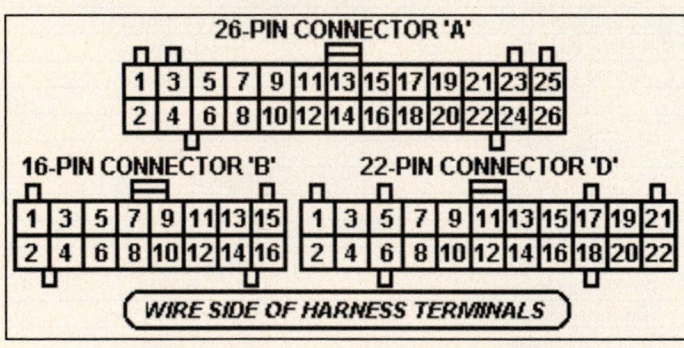

1992-93 Hatchback 1.8L MFI VIN DA9 [All] 26P 'A' Connector

PCM Pin #	W/Color	Circuit Description (26-Pin)	Value at Hot Idle
A1	BRN	Injector 1	2.0-3.3 ms
A2	YEL	Injector 4	2.0-3.3 ms
A3	RED	Injector 2	2.0-3.3 ms
A4	---	Not Used	---
A5	LT BLU	Injector 3	2.0-3.3 ms
A6	ORN/BLK	HO2S-11 (B1 S1) Heater	Heater Off: 12v, On: 1v
A7	GRN/BLK	Fuel Pump Relay Control	Relay Off: 12v, On: 1v
A8	GRN/BLK	Fuel Pump Relay Control	Relay Off: 12v, On: 1v
A9	BLU/YEL	Idle Air Control Valve	Pulse Signals
A10	GRN/YEL	Pressure Regulator Solenoid	Solenoid Off: 12v, On: 1v
A11	RED	EGR Valve Lift Sensor	1.1-1.2v
A12	---	Not Used	---
A13	GRN/ORN	Malfunction Indicator Lamp	MIL Off: 12v, On: 1v
A14	---	---	---
A15	YEL	A/C Clutch Relay Control	Relay Off: 12v, On: 1v
A16-19	---	Not Used	---
A20	GRN	EVAP Purge Solenoid	Solenoid Off: 12v, On: 1v
A21	YEL/GRN	Igniter Signal	Digital Signals: 0-12-0v
A22	YEL/GRN	Igniter Signal	Digital Signals: 0-12-0v
A23	BLK	Power Ground	<0.1v
A24	BLK	Power Ground	<0.1v
A25	YEL/BLK	Main Relay Power (B+)	12-14v
A26	BRN/BLK	Chassis Ground	0.1v

1992-93 Hatchback 1.8L MFI VIN DA9 [All] 16P 'B' Connector

PCM Pin #	W/Color	Circuit Description (16-Pin)	Value at Hot Idle
B1	YEL/BLK	Main Relay Power (B+)	12-14v
B2	BRN/BLK	Chassis Ground	<0.050v
B3-4	---	Not Used	---
B5	BLU/RED	A/C Pressure Switch	Relay Off: 12v, On: 1v
B6	---	Not Used	---
B7	GRN	A/T: Shift Position Switch	In P/N: 0v, Others: 12v
B8	RED	PSP Switch Signal	Straight: 0v, Turning: 12v
B9	BLU/WHT	Starter Switch Signal	Cranking: 9-11v
B10	YEL/RED	Vehicle Speed Sensor	Moving: pulse signals
B11	ORN	CYP Sensor Signal	AC pulse signals
B12	WHT	CYP Sensor Ground	<0.050v
B13	ORN/BLU	TDC Sensor Signal	AC pulse signals
B14	WHT/BLU	TDC Sensor Ground	<0.050v
B15	BLU/GRN	CKP Sensor Signal	AC pulse signals
B16	BLU/YEL	CKP Sensor Ground	<0.050v

1992-93 Hatchback 1.8L MFI VIN DA9 [All] 22P 'D' Connector

PCM Pin #	W/Color	Circuit Description (22-Pin)	Value at Hot Idle
D1	WHT/YEL	Keep Alive Power (VBU)	12-14v
D2	GRN/WHT	Brake Switch Signal	Brake Off: 12v, On: 0v
D3	---	Not Used	---
D4	BRN	Service Check Connector	SCS Open: 4.80v
D5-6	---	Not Used	---
D7	LT BLU	Data Link Connector	No Scan Tool: 5v
D8	RED/WHT	A/T: Control Unit	---
D9	BLU	Alternator 'FR' Signal	Digital Signals: 0-5-0v
D10	---	Not Used	---
D11	RED/BLU	TP Sensor Signal	0.5-0.6v
D12	YEL	EGR Valve Lift Sensor	1.1-1.2v
D13	RED/WHT	ECT Sensor Signal	At 180°F: 0.5-0.6v
D14	WHT	HO2S-11 (B1 S1) Signal	0.1-1.1v
D15	RED/YEL	IAT Sensor Signal	At 100°F: 2-3v
D16	BLU/WHT	A/T: Control Unit	---
D17	WHT	MAP Sensor Signal	0.8-0.9v
D18	---	Not Used	---
D19	YEL/RED	MAP Sensor VREF	4.9-5.1v
D20	YEL/WHT	Sensor VREF	4.9-5.1v
D21	GRN/WHT	MAP Sensor Ground	<0.050v
D22	GRN/WHT	Sensor Ground	<0.050v

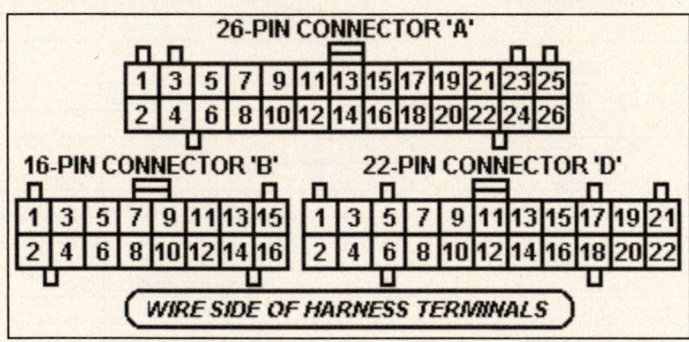

1990-91 GS, LS Sedan 1.8L I4 MFI VIN DB1 [All] 18P A Connector

PCM Pin #	W/Color	Circuit Description (18-Pin)	Value at Hot Idle
A1	BRN	Injector 1	2.0-3.3 ms
A2	BLK	Power Ground	<0.1v
A3	RED	Injector 2	2.0-3.3 ms
A4	BLK	Power Ground	<0.1v
A5	LT BLU	Injector 3	2.0-3.3 ms
A6	GRN	EVAP Purge Solenoid	Solenoid Off: 12v, On: 1v
A7	YEL	Injector 4	2.0-3.3 ms
A8	RED	EGR Solenoid Control	Solenoid Off: 12v, On: 1v
A9	---	Not Used	---
A10	GRN/YEL	Fuel Pressure Regulator	Solenoid Off: 12v, On: 1v
A11	BLU/YEL	Electronic Air Control Valve	Pulse Signals
A12	GRN/BLK	Fuel Pump Relay Control	Relay Off: 12v, On: 1v
A13	YEL/BLK	Main Relay Power (B+)	12-14v
A14	GRN/BLK	Fuel Pump Relay Control	Relay Off: 12v, On: 1v
A15	YEL/BLK	Main Relay Power (B+)	12-14v
A16	BRN/BLK	Power Ground	<0.1v
A17	---	Not Used	---
A18	BLK/RED	Power Ground	<0.1v

1990-91 GS, LS Sedan 1.8L I4 MFI VIN DB1 [All] 20P B Connector

PCM Pin #	W/Color	Circuit Description (20-Pin)	Value at Hot Idle
B1	WHT/GRN	Keep Alive Power (VBU)	12-14v
B2	BLU	A/T: Fast Idle Solenoid	Solenoid Off: 12v, On: 1v
B3	YEL	A/C Clutch Relay Control	Relay Off: 12v, On: 1v
B4	BLU/WHT	A/T: Control Unit Signal	---
B5	---	Not Used	---
B6	GRN/ORN	Malfunction Indicator Lamp	MIL Off: 12v, On: 1v
B7	GRN	A/T: Shift Position Switch	In P/N: 0v, Others: 12v
B8	BLU/RED	A/C Switch Signal	Relay Off: 12v, On: 1v
B9	---	Not Used	---
B10	ORN	CKP Sensor Signal	AC pulse signals
B11	---	Not Used	---
B12	WHT	CKP Sensor Ground	<0.050v
B13	BLU/WHT	Starter Switch Signal	Cranking: 9-11v
B14	BLU	Alternator Charging	Headlights off: 0v, on: 12v
B15	WHT	Igniter Signal	Digital Signals: 0-12-0v
B16	YEL/RED	Vehicle Speed Sensor	Moving: pulse signals
B17	WHT	Igniter Signal	Digital Signals: 0-12-0v
B18	---	Not Used	---
B19	RED	PSP Switch Signal	Straight: 0v, Turning: 12v
B20	BRN	Ignition Timing Adjustment	0.4-4.5v

1990-91 GS, LS Sedan 1.8L I4 MFI VIN DB1 [All] 16P C Connector

PCM Pin #	W/Color	Circuit Description (16-Pin)	Value at Hot Idle
C1	BLU/GRN	CYP Sensor Signal	AC pulse signals
C2	BLU/YEL	CYP Sensor Ground	<0.050v
C3	ORN/BLU	TDC Sensor Signal	AC pulse signals
C4	WHT/BLU	TDC Sensor Ground	<0.050v
C5	RED/YEL	IAT Sensor Signal	At 100ºF: 2-3v
C6	RED/WHT	ECT Sensor Signal	At 180ºF: 0.5-0.6v
C7	RED/BLU	Throttle Angle Sensor	0.5-0.6v
C8	GRN/RED	EGR Valve Lift Sensor	1.1-1.2v
C9	RED/WHT	Atmospheric Press. Sensor	Varies w/alt: 0.5-3.0v
C10	GRN/WHT	Brake Switch Signal	Brake Off: 12v, On: 0v
C11	WHT	MAP Sensor Signal	0.8-0.9v
C12	GRN	Sensor Ground	<0.050v
C13	YEL/WHT	Sensor VREF	4.9-5.1v
C14	GRN/WHT	MAP Sensor Ground	<0.050v
C15	YEL/RED	MAP Sensor VREF	4.9-5.1v
C16	WHT	O2S-11 Signal	0.1-1.1v

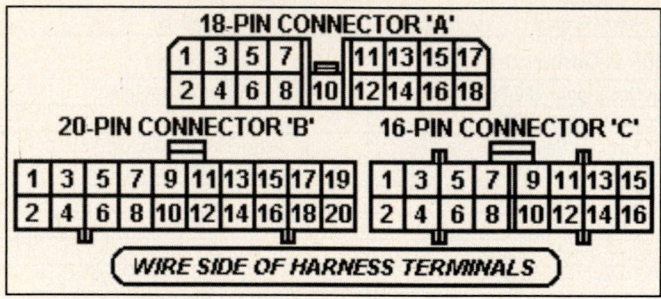

1992-93 GS, RS Sedan 1.8L I4 MFI VIN DB1 [All] 26P A Connector

PCM Pin #	W/Color	Circuit Description (26-Pin)	Value at Hot Idle
A1	BRN	Injector 1	2.0-3.3 ms
A2	YEL	Injector 4	2.0-3.3 ms
A3	RED	Injector 2	2.0-3.3 ms
A4	---	Not Used	---
A5	LT BLU	Injector 3	2.0-3.3 ms
A6	ORN/BLK	HO2S-11 (B1 S1) Heater	Heater Off: 12v, On: 1v
A7	GRN/BLK	Fuel Pump Relay Control	Relay Off: 12v, On: 1v
A8	GRN/BLK	Fuel Pump Relay Control	Relay Off: 12v, On: 1v
A9	BLU/YEL	Idle Air Control Valve	Pulse Signals
A10	GRN/YEL	Pressure Regulator Solenoid	Solenoid Off: 12v, On: 1v
A11	RED	EGR Valve Lift Sensor	1.1-1.2v
A12	---	Not Used	---
A13	GRN/ORN	Malfunction Indicator Lamp	MIL Off: 12v, On: 1v
A14	---	---	---
A15	YEL	A/C Clutch Relay Control	Relay Off: 12v, On: 1v
A16-19	---	Not Used	---
A20	GRN	EVAP Purge Solenoid	Solenoid Off: 12v, On: 1v
A21	YEL/GRN	Igniter Signal	Digital Signals: 0-12-0v
A22	YEL/GRN	Igniter Signal	Digital Signals: 0-12-0v
A23	BLK	Power Ground	<0.1v
A24	BLK	Power Ground	<0.1v
A25	YEL/BLK	Main Relay Power (B+)	12-14v
A26	BRN/BLK	Chassis Ground	0.1v

1992-93 GS, RS Sedan 1.8L I4 MFI VIN DB1 [All] 16P B Connector

PCM Pin #	W/Color	Circuit Description (16-Pin)	Value at Hot Idle
B1	YEL/BLK	Main Relay Power (B+)	12-14v
B2	BRN/BLK	Chassis Ground	<0.050v
B3-4	---	Not Used	---
B5	BLU/RED	A/C Pressure Switch	Relay Off: 12v, On: 1v
B6	---	Not Used	---
B7	GRN	A/T: Shift Position Switch	In P/N: 0v, Others: 12v
B8	RED	PSP Switch Signal	Straight: 0v, Turning: 12v
B9	BLU/WHT	Starter Switch Signal	Cranking: 9-11v
B10	YEL/RED	Vehicle Speed Sensor	Moving: pulse signals
B11	ORN	CYP Sensor Signal	AC pulse signals
B12	WHT	CYP Sensor Ground	<0.050v
B13	ORN/BLU	TDC Sensor Signal	AC pulse signals
B14	WHT/BLU	TDC Sensor Ground	<0.050v
B15	BLU/GRN	CKP Sensor Signal	AC pulse signals
B16	BLU/YEL	CKP Sensor Ground	<0.050v

1992-93 GS, RS Sedan 1.8L I4 MFI VIN DB1 [All] 22P D Connector

PCM Pin #	W/Color	Circuit Description (22-Pin)	Value at Hot Idle
D1	WHT/YEL	Keep Alive Power (VBU)	12-14v
D2	GRN/WHT	Brake Switch Signal	Brake Off: 12v, On: 0v
D3	---	Not Used	---
D4	BRN	Service Check Connector	SCS Open: 4.80v
D5-6	---	Not Used	---
D7	LT BLU	Data Link Connector	No Scan Tool: 5v
D8	RED/WHT	A/T: Control Unit	---
D9	BLU	Alternator 'FR' Signal	Digital Signals: 0-5-0v
D10	---	Not Used	---
D11	RED/BLU	TP Sensor Signal	0.5-0.6v
D12	YEL	EGR Valve Lift Sensor	1.1-1.2v
D13	RED/WHT	ECT Sensor Signal	At 180°F: 0.5-0.6v
D14	WHT	HO2S-11 (B1 S1) Signal	0.1-1.1v
D15	RED/YEL	IAT Sensor Signal	At 100°F: 2-3v
D16	BLU/WHT	A/T: Control Unit	N/A
D17	WHT	MAP Sensor Signal	0.8-0.9v
D18	---	Not Used	---
D19	YEL/RED	MAP Sensor VREF	4.9-5.1v
D20	YEL/WHT	Sensor VREF	4.9-5.1v
D21	GRN/WHT	MAP Sensor Ground	<0.050v
D22	GRN/WHT	Sensor Ground	<0.050v

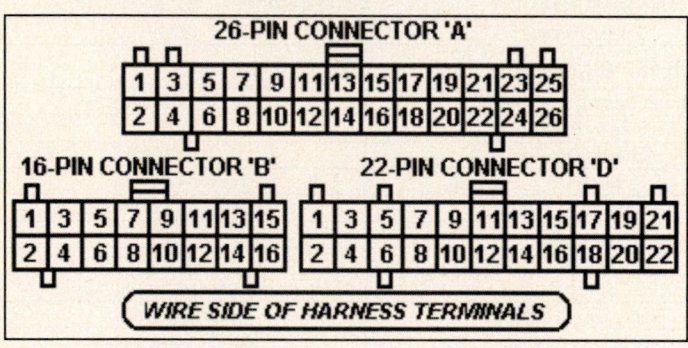

1994 LS, RS Sedan 1.8L I4 MFI VIN DB7 [All] 26P 'A' Connector

PCM Pin #	W/Color	Circuit Description (26-Pin)	Value at Hot Idle
A1	BRN	Injector 1	2.0-3.3 ms
A2	YEL	Injector 4	2.0-3.3 ms
A3	RED	Injector 2	2.0-3.3 ms
A4	---	Not Used	---
A5	LT BLU	Injector 3	2.0-3.3 ms
A6	ORN/BLK	HO2S-11 (B1 S1) Heater	Heater Off: 12v, On: 1v
A7	GRN/BLK	Fuel Pump Relay Control	Relay Off: 12v, On: 1v
A8	---	Not Used	---
A9	BLK/BLU	Idle Air Control Valve	Pulse Signals
A10-11	---	Not Used	---
A12	GRN	Radiator Fan Relay Control	Fan on: 1v, off: 12v
A13	GRN/ORN	Malfunction Indicator Lamp	MIL Off: 12v, On: 1v
A14	---	Not Used	---
A15	BLK/RED	A/C Clutch Relay Control	Relay Off: 12v, On: 1v
A16	WHT/GRN	Alternator Charging Signal	Headlights off: 0v, on: 12v
A17-19	---	Not Used	---
A20	GRN	EVAP Purge Solenoid	Solenoid Off: 12v, On: 1v
A21	YEL/GRN	Igniter Signal	Digital Signals: 0-12-0v
A22	---	Not Used	---
A23	BLK	Power Ground	<0.1v
A24	BLK	Power Ground	<0.1v
A25	YEL/BLK	Main Relay Power (B+)	12-14v
A26	BRN/BLK	Chassis Ground	0.1v

1994 LS, RS Sedan1.8L I4 MFI VIN DB7 [All] 16P 'B' Connector

PCM Pin #	W/Color	Circuit Description (16-Pin)	Value at Hot Idle
B1	YEL/BLK	Main Relay Power (B+)	12-14v
B2	BRN/BLK	Chassis Ground	<0.050v
B3	GRN/BLU	A/T: TCM Signal	N/A
B4	GRY	A/T: TCM Signal	N/A
B5	BLU/RED	A/C Switch Signal	Relay Off: 12v, On: 1v
B6	---	Not Used	---
B7	GRN/BLK	Park/Neutral Signal	In P/N: 0v, Others: 12v
B8	LT GRN	PSP Switch Signal	Straight: 0v, Turning: 12v
B9	BLU/WHT	Starter Switch Signal	Cranking: 9-11v
B10	ORN	Vehicle Speed Sensor	Moving: pulse signals
B11	ORN	CYP Sensor Signal	AC pulse signals
B12	WHT	CYP Sensor Ground	<0.050v
B13	ORN/BLU	TDC Sensor Signal	AC pulse signals
B14	WHT/BLU	TDC Sensor Ground	<0.050v
B15	BLU/GRN	CKP Sensor Signal	AC pulse signals
B16	BLU/YEL	CKP Sensor Ground	<0.050v

1994 LS, RS Sedan 1.8L I4 MFI VIN DB7 [All] 22P 'D' Connector

PCM Pin #	W/Color	Circuit Description (22-Pin)	Value at Hot Idle
D1	WHT/BLU	Keep Alive Power (VBU)	12-14v
D2	GRN/WHT	Brake Switch Signal	Brake Off: 12v, On: 0v
D3	---	Not Used	---
D4	BRN/WHT	Service Check Connector	SCS Open: 4.80v
D5-6	---	Not Used	---
D7	LT BLU	Data Link Connector	No Scan Tool: 5v
D8	LT GRN	A/T: TCM Signal	---
D9	WHT/RED	Alternator 'FR' Signal	Digital Signals: 0-5-0v
D10	GRN/RED	Electronic Load Detector	Varies w/Load: 2.5-3.5v
D11	RED/BLU	TP Sensor Signal	0.5-0.6v
D12	---	Not Used	---
D13	RED/WHT	ECT Sensor Signal	At 180°F: 0.5-0.6v
D14	WHT/RED	HO2S-11 (B1 S1) Signal	0.1-1.1v
D15	RED/YEL	IAT Sensor Signal	At 100°F: 2-3v
D16	WHT/BLK	A/T: TCM Signal	---
D17	WHT/YEL	MAP Sensor Signal	0.8-0.9v
D18	---	Not Used	---
D19	YEL/RED	MAP Sensor VREF	4.9-5.1v
D20	YEL/BLU	Sensor VREF	4.9-5.1v
D21	GRN/WHT	MAP Sensor Ground	<0.050v
D22	GRN/BLU	Sensor Ground	<0.050v

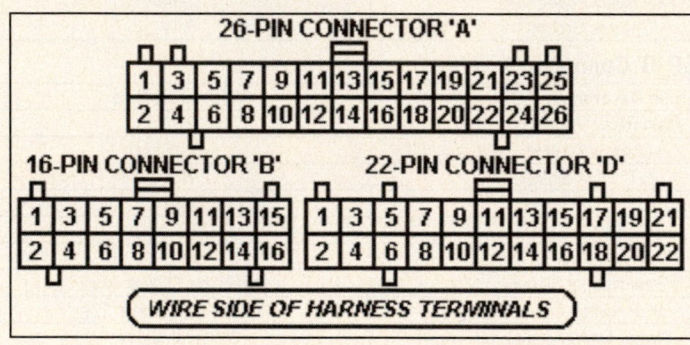

1994 Sedan 1.8L VTEC I4 MFI VIN DB8 [M/T] 26P 'A' Connector

PCM Pin #	W/Color	Circuit Description (26-Pin)	Value at Hot Idle
A1	BRN	Injector 1	2.0-3.3 ms
A2	YEL	Injector 4	2.0-3.3 ms
A3	RED	Injector 2	2.0-3.3 ms
A4	GRN/YEL	VTEC Solenoid Control	0v, Hi-Speed: 12v
A5	LT BLU	Injector 3	2.0-3.3 ms
A6	ORN/BLK	HO2S-11 (B1 S1) Heater	Heater Off: 12v, On: 1v
A7	GRN/BLK	Fuel Pump Relay Control	Relay Off: 12v, On: 1v
A8	---	Not Used	---
A9	BLK/BLU	Idle Air Control Valve	Pulse Signals
A10-11	---	Not Used	---
A12	GRN	Radiator Fan Relay Control	Fan on: 1v, off: 12v
A13	GRN/ORN	Malfunction Indicator Lamp	MIL Off: 12v, On: 1v
A14	---	Not Used	---
A15	BLK/RED	A/C Clutch Relay Control	Relay Off: 12v, On: 1v
A16	WHT/GRN	Alternator Charging Signal	Headlights off: 0v, on: 12v
A17-19	---	Not Used	---
A20	GRN	EVAP Purge Solenoid	Solenoid Off: 12v, On: 1v
A21	YEL/GRN	Igniter Signal	Digital Signals: 0-12-0v
A22	---	Not Used	---
A23	BLK	Power Ground	<0.1v
A24	BLK	Power Ground	<0.1v
A25	YEL/BLK	Main Relay Power (B+)	12-14v
A26	BRN/BLK	Chassis Ground	0.1v

1994 Sedan 1.8L VTEC I4 MFI VIN DB8 [M/T] 16P 'B' Connector

PCM Pin #	W/Color	Circuit Description (16-Pin)	Value at Hot Idle
B1	YEL/BLK	Main Relay Power (B+)	12-14v
B2	BRN/BLK	Chassis Ground	<0.050v
B3-4	---	Not Used	---
B5	BLU/RED	A/C Switch Signal	Relay Off: 12v, On: 1v
B6	---	Not Used	---
B7	---	Not Used	---
B8	LT GRN	PSP Switch Signal	Straight: 0v, Turning: 12v
B9	BLU/WHT	Starter Switch Signal	Cranking: 9-11v
B10	ORN	Vehicle Speed Sensor	Moving: pulse signals
B11	ORN	CYP Sensor Signal	AC pulse signals
B12	WHT	CYP Sensor Ground	<0.050v
B13	ORN/BLU	TDC Sensor Signal	AC pulse signals
B14	WHT/BLU	TDC Sensor Ground	<0.050v
B15	BLU/GRN	CKP Sensor Signal	AC pulse signals
B16	BLU/YEL	CKP Sensor Ground	<0.050v

1994 Sedan 1.8L VTEC I4 MFI VIN DB8 [M/T] 22P 'D' Connector

PCM Pin #	W/Color	Circuit Description (22-Pin)	Value at Hot Idle
D1	WHT/BLU	Keep Alive Power (VBU)	12-14v
D2	GRN/WHT	Brake Switch Signal	Brake Off: 12v, On: 0v
D3	---	Not Used	---
D4	BRN/WHT	Service Check Connector	SCS Open: 4.80v
D5	---	Not Used	---
D6	BLU/BLK	VTEC Pressure Switch	0v, Hi-Speed: 12v
D7	LT BLU	Data Link Connector	No Scan Tool: 5v
D8	---	Not Used	---
D9	WHT/RED	Alternator 'FR' Signal	Digital Signals: 0-5-0v
D10	GRN/RED	Electronic Load Detector	Varies w/Load: 2.5-3.5v
D11	RED/BLU	TP Sensor Signal	0.5-0.6v
D12	---	Not Used	---
D13	RED/WHT	ECT Sensor Signal	At 180ºF: 0.5-0.6v
D14	WHT/RED	HO2S-11 (B1 S1) Signal	0.1-1.1v
D15	RED/YEL	IAT Sensor Signal	At 100ºF: 2-3v
D16	---	Not Used	---
D17	WHT/YEL	MAP Sensor Signal	0.8-0.9v
D18	---	Not Used	---
D19	YEL/RED	MAP Sensor VREF	4.9-5.1v
D20	YEL/BLU	Sensor VREF	4.9-5.1v
D21	GRN/WHT	MAP Sensor Ground	<0.050v
D22	GRN/BLU	Sensor Ground	<0.050v

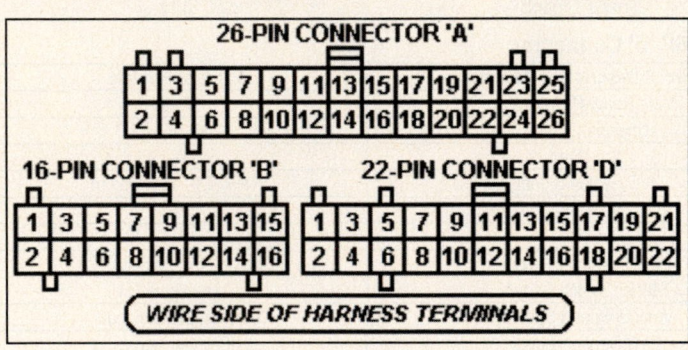

1995 RS, SE Sedan 1.8L I4 MFI VIN DB7 [All] 26P 'A' Connector

PCM Pin #	W/Color	Circuit Description (26-Pin)	Value at Hot Idle
A1	BRN	Injector 1	2.0-3.3 ms
A2	YEL	Injector 4	2.0-3.3 ms
A3	RED	Injector 2	2.0-3.3 ms
A4	---	Not Used	---
A5	LT BLU	Injector 3	2.0-3.3 ms
A6	ORN/BLK	HO2S-11 (B1 S1) Heater	Heater Off: 12v, On: 1v
A7	GRN/BLK	Fuel Pump Relay Control	Relay Off: 12v, On: 1v
A8	---	Not Used	---
A9	BLK/BLU	Idle Air Control Valve	Pulse Signals
A10-11	---	Not Used	---
A12	GRN	Radiator Fan Relay Control	Fan on: 1v, off: 12v
A13	GRN/ORN	Malfunction Indicator Lamp	MIL Off: 12v, On: 1v
A14	---	Not Used	---
A15	BLK/RED	A/C Clutch Relay Control	Relay Off: 12v, On: 1v
A16	WHT/GRN	Alternator Charging Signal	Lights off: 12v, off: 0v
A17-18	---	Not Used	---
A19	YEL/RED	A/T: TCM Signal	N/A
A20	GRN	EVAP Purge Solenoid	Solenoid Off: 12v, On: 1v
A21	YEL/GRN	Igniter Signal	Digital Signals: 0-12-0v
A22	---	Not Used	---
A23	BLK	Power Ground	<0.1v
A24	BLK	Power Ground	<0.1v
A25	YEL/BLK	Main Relay Power (B+)	12-14v
A26	BRN/BLK	Chassis Ground	0.1v

1995 RS, SE Sedan 1.8L I4 MFI VIN DB7 [All] 16P 'B' Connector

PCM Pin #	W/Color	Circuit Description (16-Pin)	Value at Hot Idle
B1	YEL/BLK	Main Relay Power (B+)	12-14v
B2	BRN/BLK	Chassis Ground	<0.050v
B3	GRN/BLU	A/T: TCM Signal	N/A
B4	GRY	A/T: TCM Signal	N/A
B5	BLU/RED	A/C Switch Signal	Relay Off: 12v, On: 1v
B6	---	Not Used	---
B7	GRN/BLK	Park/Neutral Switch	In P/N: 1v, others: 12v
B8	LT GRN	PSP Switch Signal	Straight: 0v, Turning: 12v
B9	BLU/WHT	Starter Switch Signal	Cranking: 9-11v
B10	ORN	Vehicle Speed Sensor	Moving: pulse signals
B11	ORN	CYP Sensor Signal	AC pulse signals
B12	WHT	CYP Sensor Ground	<0.050v
B13	ORN/BLU	TDC Sensor Signal	AC pulse signals
B14	WHT/BLU	TDC Sensor Ground	<0.050v
B15	BLU/GRN	CKP Sensor Signal	AC pulse signals
B16	BLU/YEL	CKP Sensor Ground	<0.050v

1995 RS, SE Sedan 1.8L I4 MFI VIN DB7 [All] 22P 'D' Connector

PCM Pin #	W/Color	Circuit Description (22-Pin)	Value at Hot Idle
D1	WHT/BLU	Keep Alive Power (VBU)	12-14v
D2	GRN/WHT	Brake Switch Signal	Brake Off: 12v, On: 0v
D3	---	Not Used	---
D4	BRN/WHT	Service Check Connector	SCS Open: 4.80v
D5-6	---	Not Used	---
D7	LT BLU	Data Link Connector	Jumped: 2.50v
D8	LT GRN	A/T: TCM Signal	---
D9	WHT/RED	Alternator 'FR' Signal	Digital Signals: 0-5-0v
D10	GRN/RED	Electronic Load Detector	Varies w/Load: 2.5-3.5v
D11	RED/BLU	TP Sensor Signal	0.5-0.6v
D12	---	Not Used	---
D13	RED/WHT	ECT Sensor Signal	At 180ºF: 0.5-0.6v
D14	WHT/RED	HO2S-11 (B1 S1) Signal	0.1-1.1v
D15	RED/YEL	IAT Sensor Signal	At 100ºF: 2-3v
D16	WHT/BLK	A/T: TCM Signal	---
D17	WHT/YEL	MAP Sensor Signal	0.8-0.9v
D18	---	Not Used	---
D19	YEL/RED	MAP Sensor VREF	4.9-5.1v
D20	YEL/BLU	Sensor VREF	4.9-5.1v
D21	GRN/WHT	MAP Sensor Ground	<0.050v
D22	GRN/BLU	Sensor Ground	<0.050v

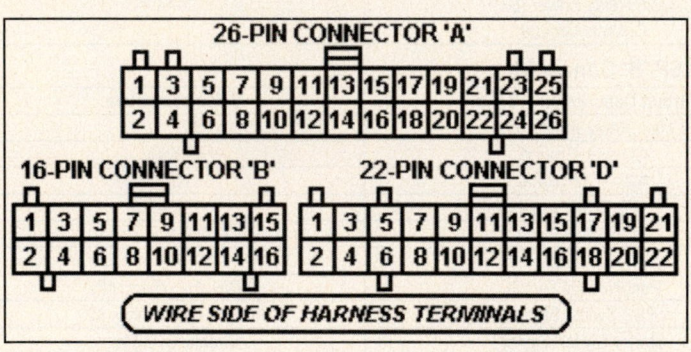

1995 Sedan 1.8L VTEC I4 MFI VIN DB8 [M/T] 26P 'A' Connector

PCM Pin #	W/Color	Circuit Description (26-Pin)	Value at Hot Idle
A1	BRN	Injector 1	2.0-3.3 ms
A2	YEL	Injector 4	2.0-3.3 ms
A3	RED	Injector 2	2.0-3.3 ms
A4	GRN/YEL	VTEC Solenoid Control	0v, Hi-Speed: 12v
A5	LT BLU	Injector 3	2.0-3.3 ms
A6	ORN/BLK	HO2S-11 (B1 S1) Heater	Heater Off: 12v, On: 1v
A7	GRN/BLK	Fuel Pump Relay Control	Relay Off: 12v, On: 1v
A8	---	Not Used	---
A9	BLK/BLU	Idle Air Control Valve	Pulse Signals
A10-11	---	Not Used	---
A12	GRN	Radiator Fan Relay Control	Fan on: 1v, off: 12v
A13	GRN/ORN	Malfunction Indicator Lamp	MIL Off: 12v, On: 1v
A14	---	Not Used	---
A15	BLK/RED	A/C Clutch Relay Control	Relay Off: 12v, On: 1v
A16	WHT/GRN	Alternator Charging Signal	Lights Off: 12v, On: 1v
A17-19	---	Not Used	---
A20	GRN	EVAP Purge Solenoid	Solenoid Off: 12v, On: 1v
A21	YEL/GRN	Igniter Signal	Digital Signals: 0-12-0v
A22	---	Not Used	---
A23	BLK	Power Ground	<0.1v
A24	BLK	Power Ground	<0.1v
A25	YEL/BLK	Main Relay Power (B+)	12-14v
A26	BRN/BLK	Chassis Ground	<0.1v

1995 Sedan 1.8L VTEC I4 MFI VIN DB8 [M/T] 16P 'B' Connector

PCM Pin #	W/Color	Circuit Description (16-Pin)	Value at Hot Idle
B1	YEL/BLK	Main Relay Power (B+)	12-14v
B2	BRN/BLK	Power Ground	<0.1v
B3-4	---	Not Used	---
B5	BLU/RED	A/C Switch Signal	A/C on: 12v
B6-7	---	Not Used	---
B8	LT GRN	PSP Switch Signal	Straight: 0v, Turning: 12v
B9	BLU/WHT	Starter Switch Signal	Cranking: 9-11v
B10	ORN	Vehicle Speed Sensor	Moving: pulse signals
B11	ORN	CYP Sensor Signal	AC pulse signals
B12	WHT	CYP Sensor Ground	<0.050v
B13	ORN/BLU	TDC Sensor Signal	AC pulse signals
B14	WHT/BLU	TDC Sensor Ground	<0.050v
B15	BLU/GRN	CKP Sensor Signal	AC pulse signals
B16	BLU/YEL	CKP Sensor Ground	<0.050v

1995 Sedan 1.8L VTEC I4 MFI VIN DB8 [M/T] 22P 'D' Connector

PCM Pin #	W/Color	Circuit Description (22-Pin)	Value at Hot Idle
D1	WHT/BLU	Keep Alive Power (VBU)	12-14v
D2	GRN/WHT	Brake Switch Signal	Brake Off: 12v, On: 0v
D3	---	Not Used	---
D4	BRN/WHT	Service Check Connector	SCS Open: 4.80v
D5	---	Not Used	---
D6	BLU/BLK	VTEC Pressure Switch	0v, Hi-Speed: 12v
D7	LT BLU	Data Link Connector	Jumped: 2.50v
D8	---	Not Used	---
D9	WHT/RED	Alternator 'FR' Signal	Digital Signals: 0-5-0v
D10	GRN/RED	Electronic Load Detector	Varies w/Load: 2.5-3.5v
D11	RED/BLU	TP Sensor Signal	0.5-0.6v
D12	---	Not Used	---
D13	RED/WHT	ECT Sensor Signal	At 180°F: 0.5-0.6v
D14	WHT/RED	HO2S-11 (B1 S1) Signal	0.1-1.1v
D15	RED/YEL	IAT Sensor Signal	At 100°F: 2-3v
D16	---	Not Used	---
D17	WHT/YEL	MAP Sensor Signal	0.8-0.9v
D18	---	Not Used	---
D19	YEL/RED	MAP Sensor VREF	4.9-5.1v
D20	YEL/BLU	Sensor VREF	4.9-5.1v
D21	GRN/WHT	MAP Sensor Ground	<0.050v
D22	GRN/BLU	Sensor Ground	<0.050v

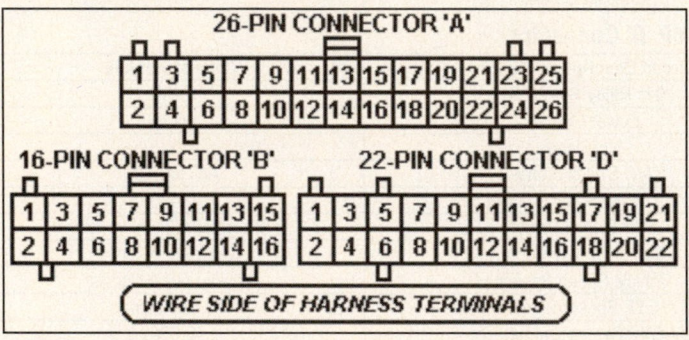

Abbreviation	Color	Abbreviation	Color	Abbreviation	Color
BLK	Black	LT BLU	Lt. Blue	TAN	Tan
BLU	Blue	LT GRN	Lt. Green	VIO	Violet
BRN	Brown	ORN	Orange	WHT	White
GRY	Gray	PNK	Pink	YEL	Yellow
GRN	Green	PPL	Purple		

1996-03 LS, RS Sedan 1.8L I4 MFI VIN DB7 [All] 32P 'A' Connector

PCM Pin #	W/Color	Circuit Description (32-Pin)	Value at Hot Idle
A1	YEL	Injector 4	2.0-3.3 ms
A2	BLU	Injector 3	2.0-3.3 ms
A3	RED	Injector 2	2.0-3.3 ms
A4	BRN	Injector 1	2.0-3.3 ms
A5	GRN/RED	HO2S-12 (B1 S2) Heater	Digital Signals: 0-12-0v
A6	ORN/BLK	HO2S-11 (B1 S1) Heater	Digital Signals: 0-12-0v
A7-8	---	Not Used	---
A9	BRN/BLK	Logic Ground	0.1v
A10	BLK	Power Ground	<0.1v
A11	YEL/BLK	Main Relay Power (B+)	12-14v
A12	BLK/BLU	Idle Air Control Valve	Pulse Signals
A13-14	---	Not Used	---
A15	RED	EVAP Purge Solenoid	Pulse Signals
A16	GRN/BLK	Fuel Pump Relay Control	Relay Off: 12v, On: 1v
A17	BLK/RED	A/C Clutch Relay Control	Relay Off: 12v, On: 1v
A18	GRN/ORN	Malfunction Indicator Lamp	MIL Off: 12v, On: 1v
A19	WHT/GRN	Alternator Charging Signal	Headlights off: 0v, on: 12v
A20	YEL/GRN	Igniter Signal	Digital Signals: 0-12-0v
A21	---	Not Used	---
A22	BRN/BLK	Logic Ground	0.1v
A23	BLK	Power Ground	<0.1v
A24	YEL/BLK	Main Relay Power (B+)	12-14v
A25	WHT/BLK	TCM VREF	4.9-5.1v
A26	---	Not Used	---
A27	GRN	Radiator Fan Relay Control	Fan on: 1v, off: 12v
A28-32	---	Not Used	---

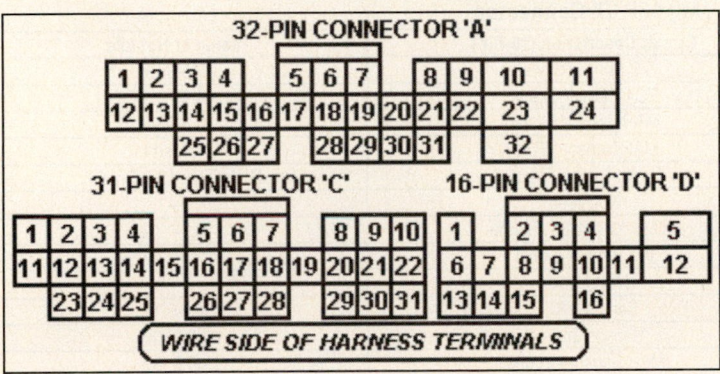

1996-03 LS, RS Sedan 1.8L I4 MFI VIN DB7 [All] 31P 'C' Connector

PCM Pin #	W/Color	Circuit Description (31-Pin)	Value at Hot Idle
C1	BLU/RED	CKF Sensor	AC pulse signals
C2	BLU/GRN	CKP Sensor Signal	AC pulse signals
C3	ORN/BLU	TDC Sensor Signal	AC pulse signals
C4	ORN	CYP Sensor Signal	AC pulse signals
C5	BLU/RED	A/C Switch Signal	A/C Off: 5v, On: 0v
C6	BLU/WHT	Starter Switch Signal	Cranking: 9-11v
C7	BRN/WHT	Service Check Connector	SCS Open: 4.80v
C8	GRN/WHT	K-Line Signal	12v
C9	YEL	FAS TCM Signal	At idle: 4.5-5.5v
C10	WHT/BLU	Keep Alive Power (VBU)	12-14v
C11	WHT/RED	CKF Sensor Ground	<0.050v
C12	BLU/YEL	CKP Sensor Ground	<0.050v
C13	WHT/BLU	TDC Sensor Ground	<0.050v
C14	WHT	CYP Sensor Ground	<0.050v
C15	---	Not Used	---
C16	GRN	PSP Switch Signal	Straight: 0v, Turning: 12v
C17	WHT/RED	Alternator 'FR' Signal	Digital Signals: 0-5-0v
C18	ORN	Vehicle Speed Sensor	Moving: pulse signals
C19	BLU	ATCHK TCM Signal	Pulse Signals
C20	---	Not Used	---
C21	LT GRN	TCM BARO Signal	Varies: 0.5-4.9v
C22-26	---	Not Used	---
C27	GRY	AFSB TCM Signal	At idle: 4.5-5.5v
C28	GRN/BLU	AFSA TCM Signal	At idle: 4.5-5.5v
C29	GRN/BLK	A/T: Gear Position Switch	In P/N: 0v, Others: 12v
C30-31	---	Not Used	---

1996-03 LS, RS Sedan 1.8L I4 MFI VIN DB7 [All] 16P 'D' Connector

PCM Pin #	W/Color	Circuit Description (16-Pin)	Value at Hot Idle
D1	RED/BLK	TP Sensor Signal	0.5-0.6v
D2	RED/WHT	ECT Sensor Signal	At 180°F: 0.5-0.6v
D3	WHT/YEL	MAP Sensor Signal	0.8-0.9v
D4	YEL/WHT	MAP Sensor VREF	4.9-5.1v
D5	GRN/WHT	Brake Switch Signal	Brake Off: 12v, On: 0v
D6	---	Not Used	---
D7	WHT/RED	HO2S-11 (B1 S1) Signal	0.1-1.1v
D8	RED/YEL	IAT Sensor Signal	At 100°F: 2-3v
D9	---	Not Used	---
D10	YEL/BLU	Sensor VREF	4.9-5.1v
D11	GRN/BLU	Sensor Ground	<0.050v
D12	GRN/WHT	MAP Sensor Ground	<0.050v
D13	ORN/BLU	HO2S-12 Ground	<0.050v
D14	BLU/GRN	HO2S-12 (B1 S2) Signal	0.1-1.1v
D15	---	Not Used	---
D16	GRN/RED	Electric Load Detector	Varies: 2.5-3.5v

1996-03 Sedan 1.8L VTEC I4 MFI VIN DB8 [M/T] 32P 'A' Connector

PCM Pin #	W/Color	Circuit Description (32-Pin)	Value at Hot Idle
A1	YEL	Injector 4	2.0-3.3 ms
A2	BLU	Injector 3	2.0-3.3 ms
A3	RED	Injector 2	2.0-3.3 ms
A4	BRN	Injector 1	2.0-3.3 ms
A5	GRN/RED	HO2S-12 (B1 S2) Heater	Digital Signals: 0-12-0v
A6	ORN/BLK	HO2S-11 (B1 S1) Heater	Digital Signals: 0-12-0v
A7	---	Not Used	---
A8	GRN/YEL	VTEC Solenoid Control	0v, Hi-Speed: 12v
A9	BRN/BLK	Logic Ground	0.1v
A10	BLK	Power Ground	<0.1v
A11	YEL/BLK	Main Relay Power (B+)	12-14v
A12	BLK/BLU	Idle Air Control Valve	Pulse Signals
A13-14	---	Not Used	---
A15	RED	EVAP Purge Solenoid	Pulse Signals
A16	GRN/BLK	Fuel Pump Relay Control	Relay Off: 12v, On: 1v
A17	BLK/RED	A/C Clutch Relay Control	Relay Off: 12v, On: 1v
A18	GRN/ORN	Malfunction Indicator Lamp	MIL Off: 12v, On: 1v
A19	WHT/GRN	Alternator Charging Signal	Headlights off: 0v, on: 12v
A20	YEL/GRN	Igniter Signal	Digital Signals: 0-12-0v
A21	---	Not Used	---
A22	BRN/BLK	Logic Ground	0.1v
A23	BLK	Power Ground	<0.1v
A24	YEL/BLK	Main Relay Power (B+)	12-14v
A25	---	Not Used	---
A26	PNK/BLU	Intake Air Bypass Solenoid	Solenoid Off: 12v, On: 1v
A27	GRN	Radiator Fan Relay Control	Fan on: 1v, off: 12v
A28-32	---	Not Used	---

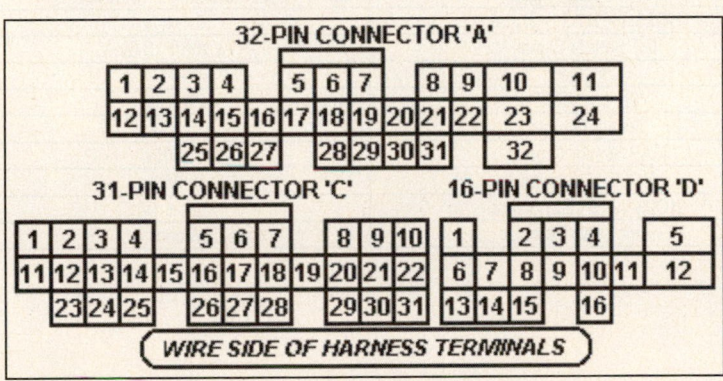

1996-03 Sedan 1.8L VTEC I4 MFI VIN DB8 [M/T] 31P 'C' Connector

PCM Pin #	W/Color	Circuit Description (31-Pin)	Value at Hot Idle
C1	BLU/RED	CKF Sensor	AC pulse signals
C2	BLU/GRN	CKP Sensor Signal	AC pulse signals
C3	ORN/BLU	TDC Sensor Signal	AC pulse signals
C4	ORN	CYP Sensor Signal	AC pulse signals
C5	BLU/RED	A/C Switch Signal	A/C Off: 5v, On: 0v
C6	BLU/WHT	Starter Switch Signal	Cranking: 9-11v
C7	BRN/WHT	Service Check Connector	SCS Open: 4.80v
C8	GRN/WHT	K-Line Signal	12v
C9	---	Not Used	---
C10	WHT/BLU	Keep Alive Power (VBU)	12-14v
C11	WHT/RED	CKF Sensor Ground	<0.050v
C12	BLU/YEL	CKP Sensor Ground	<0.050v
C13	WHT/BLU	TDC Sensor Ground	<0.050v
C14	WHT	CYP Sensor Ground	<0.050v
C15	BLU/BLK	VTEC Pressure Switch	0v, Hi-Speed: 12v
C16	GRN	PSP Switch Signal	Straight: 0v, Turning: 12v
C17	WHT/RED	Alternator 'FR' Signal	Digital Signals: 0-5-0v
C18	ORN	Vehicle Speed Sensor	Moving: pulse signals
C19-31	---	Not Used	---

1996-03 Sedan 1.8L VTEC I4 MFI VIN DB8 [M/T] 16P 'D' Connector

PCM Pin #	W/Color	Circuit Description (16-Pin)	Value at Hot Idle
D1	RED/BLK	TP Sensor Signal	0.5-0.6v
D2	RED/WHT	ECT Sensor Signal	At 180ºF: 0.5-0.6v
D3	WHT/YEL	MAP Sensor Signal	0.8-0.9v
D4	YEL/WHT	MAP Sensor VREF	4.9-5.1v
D5	GRN/WHT	Brake Switch Signal	Brake Off: 12v, On: 0v
D6	RED/BLU	Knock Sensor Signal	No Detonation: 18mv AC
D7	WHT/RED	HO2S-11 (B1 S1) Signal	0.1-1.1v
D8	RED/YEL	IAT Sensor Signal	At 100ºF: 2-3v
D9	---	Not Used	---
D10	YEL/BLU	Sensor VREF	4.9-5.1v
D11	GRN/BLU	Sensor Ground	<0.050v
D12	GRN/WHT	MAP Sensor Ground	<0.050v
D13	ORN/BLU	HO2S-12 Ground	<0.050v
D14	BLU/GRN	HO2S-12 (B1 S2) Signal	0.1-1.1v
D16	GRN/RED	Electric Load Detector	Varies w/Load: 2.5-3.5v

1994-95 RS Coupe 1.8L I4 MFI VIN DC4 [A/T] 26P 'A' Connector

PCM Pin #	W/Color	Circuit Description (26-Pin)	Value at Hot Idle
A1	BRN	Injector 1	2.0-3.3 ms
A2	YEL	Injector 4	2.0-3.3 ms
A3	RED	Injector 2	2.0-3.3 ms
A4	---	Not Used	---
A5	LT BLU	Injector 3	2.0-3.3 ms
A6	ORN/BLK	HO2S-11 (B1 S1) Heater	Heater Off: 12v, On: 1v
A7	GRN/BLK	Fuel Pump Relay Control	Relay Off: 12v, On: 1v
A8	---	Not Used	---
A9	BLK/BLU	Idle Air Control Valve	Pulse Signals
A10-11	---	Not Used	---
A12	GRN	Radiator Fan Relay Control	Fan on: 1v, off: 12v
A13	GRN/ORN	Malfunction Indicator Lamp	MIL Off: 12v, On: 1v
A14	---	Not Used	---
A15	BLK/RED	A/C Clutch Relay Control	Relay Off: 12v, On: 1v
A16	WHT/GRN	Alternator Charging Signal	Headlights off: 0v, on: 12v
A17-18	---	Not Used	---
A19	YEL/RED	A/T: TCM Signal	N/A
A20	GRN	EVAP Purge Solenoid	Solenoid Off: 12v, On: 1v
A21	YEL/GRN	Igniter Signal	Digital Signals: 0-12-0v
A22	---	Not Used	---
A23	BLK	Power Ground	<0.1v
A24	BLK	Power Ground	<0.1v
A25	YEL/BLK	Main Relay Power (B+)	12-14v
A26	BRN/BLK	Chassis Ground	0.1v

1994-95 RS Coupe 1.8L I4 MFI VIN DC4 [A/T] 16P 'B' Connector

PCM Pin #	W/Color	Circuit Description (16-Pin)	Value at Hot Idle
B1	YEL/BLK	Main Relay Power (B+)	12-14v
B2	BRN/BLK	Chassis Ground	<0.050v
B3	GRN/BLU	A/T: TCM Signal	N/A
B4	GRY	A/T: TCM Signal	N/A
B5	BLU/RED	A/C Switch Signal	Relay Off: 12v, On: 1v
B6	---	Not Used	---
B7	GRN/BLK	Park/Neutral Signal	In P/N: 0v, Others: 12v
B8	LT GRN	PSP Switch Signal	Straight: 0v, Turning: 12v
B9	BLU/WHT	Starter Switch Signal	Cranking: 9-11v
B10	ORN	Vehicle Speed Sensor	Moving: pulse signals
B11	ORN	CYP Sensor Signal	AC pulse signals
B12	WHT	CYP Sensor Ground	<0.050v
B13	ORN/BLU	TDC Sensor Signal	AC pulse signals
B14	WHT/BLU	TDC Sensor Ground	<0.050v
B15	BLU/GRN	CKP Sensor Signal	AC pulse signals
B16	BLU/YEL	CKP Sensor Ground	<0.050v

1994-95 RS Coupe 1.8L I4 MFI VIN DC4 [A/T] 22P 'D' Connector

PCM Pin #	W/Color	Circuit Description (22-Pin)	Value at Hot Idle
D1	WHT/BLU	Keep Alive Power (VBU)	12-14v
D2	GRN/WHT	Brake Switch Signal	Brake Off: 12v, On: 0v
D3	---	Not Used	---
D4	BRN/WHT	Service Check Connector	SCS Open: 4.80v
D5-6	---	Not Used	---
D7	LT BLU	Data Link Connector	No Scan Tool: 5v
D8	LT GRN	A/T: TCM Signal	N/A
D9	WHT/RED	Alternator 'FR' Signal	Digital Signals: 0-5-0v
D10	GRN/RED	Electronic Load Detector	Varies w/Load: 2.5-3.5v
D11	RED/BLU	TP Sensor Signal	0.5-0.6v
D12	---	Not Used	---
D13	RED/WHT	ECT Sensor Signal	At 180°F: 0.5-0.6v
D14	WHT/RED	HO2S-11 (B1 S1) Signal	0.1-1.1v
D15	RED/YEL	IAT Sensor Signal	At 100°F: 2-3v
D16	WHT/BLK	A/T: TCM Signal	---
D17	WHT/YEL	MAP Sensor Signal	0.8-0.9v
D18	---	Not Used	---
D19	YEL/RED	MAP Sensor VREF	4.9-5.1v
D20	YEL/BLU	Sensor VREF	4.9-5.1v
D21	GRN/WHT	MAP Sensor Ground	<0.050v
D22	GRN/BLU	Sensor Ground	<0.050v

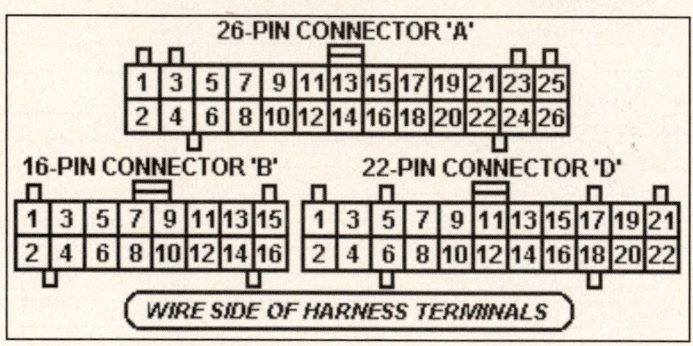

1994-95 RS Coupe 1.8L I4 MFI VIN DC4 [M/T] 26P 'A' Connector

PCM Pin #	W/Color	Circuit Description (26-Pin)	Value at Hot Idle
A1	BRN	Injector 1	2.0-3.3 ms
A2	YEL	Injector 4	2.0-3.3 ms
A3	RED	Injector 2	2.0-3.3 ms
A4	---	Not Used	---
A5	LT BLU	Injector 3	2.0-3.3 ms
A6	ORN/BLK	HO2S-11 (B1 S1) Heater	Heater Off: 12v, On: 1v
A7	GRN/BLK	Fuel Pump Relay Control	Relay Off: 12v, On: 1v
A8	---	Not Used	---
A9	BLK/BLU	Idle Air Control Valve	Pulse Signals
A10-11	---	Not Used	---
A12	GRN	Radiator Fan Relay Control	Fan on: 1v, off: 12v
A13	GRN/ORN	Malfunction Indicator Lamp	MIL Off: 12v, On: 1v
A14	---	Not Used	---
A15	BLK/RED	A/C Clutch Relay Control	Relay Off: 12v, On: 1v
A16	WHT/GRN	Alternator Charging Signal	Headlights off: 0v, on: 12v
A17-19	---	Not Used	---
A20	GRN	EVAP Purge Solenoid	Solenoid Off: 12v, On: 1v
A21	YEL/GRN	Igniter Signal	Digital Signals: 0-12-0v
A22	---	Not Used	---
A23	BLK	Power Ground	<0.1v
A24	BLK	Power Ground	<0.1v
A25	YEL/BLK	Main Relay Power (B+)	12-14v
A26	BRN/BLK	Chassis Ground	0.1v

1994-95 RS Coupe 1.8L I4 MFI VIN DC4 [M/T] 16P 'B' Connector

PCM Pin #	W/Color	Circuit Description (16-Pin)	Value at Hot Idle
B1	YEL/BLK	Main Relay Power (B+)	12-14v
B2	BRN/BLK	Chassis Ground	<0.050v
B3-4	---	Not Used	---
B5	BLU/RED	A/C Switch Signal	Relay Off: 12v, On: 1v
B6	---	Not Used	---
B7	---	Not Used	---
B8	LT GRN	PSP Switch Signal	Straight: 0v, Turning: 12v
B9	BLU/WHT	Starter Switch Signal	Cranking: 9-11v
B10	ORN	Vehicle Speed Sensor	Moving: pulse signals
B11	ORN	CYP Sensor Signal	AC pulse signals
B12	WHT	CYP Sensor Ground	<0.050v
B13	ORN/BLU	TDC Sensor Signal	AC pulse signals
B14	WHT/BLU	TDC Sensor Ground	<0.050v
B15	BLU/GRN	CKP Sensor Signal	AC pulse signals
B16	BLU/YEL	CKP Sensor Ground	<0.050v

1994-95 RS Coupe 1.8L I4 MFI VIN DC4 [M/T] 22P 'D' Connector

PCM Pin #	W/Color	Circuit Description (22-Pin)	Value at Hot Idle
D1	WHT/BLU	Keep Alive Power (VBU)	12-14v
D2	GRN/WHT	Brake Switch Signal	Brake Off: 12v, On: 0v
D3	---	Not Used	---
D4	BRN/WHT	Service Check Connector	SCS open: 4.80v
D5-6	---	Not Used	---
D7	LT BLU	Data Link Connector	No Scan Tool: 5v
D8	---	Not Used	---
D9	WHT/RED	Alternator 'FR' Signal	Digital Signals: 0-5-0v
D10	GRN/RED	Electronic Load Detector	Varies w/Load: 2.5-3.5v
D11	RED/BLU	TP Sensor Signal	0.5-0.6v
D12	---	Not Used	---
D13	RED/WHT	ECT Sensor Signal	At 180ºF: 0.5-0.6v
D14	WHT/RED	HO2S-11 (B1 S1) Signal	0.1-1.1v
D15	RED/YEL	IAT Sensor Signal	At 100ºF: 2-3v
D16	---	Not Used	---
D17	WHT/YEL	MAP Sensor Signal	0.8-0.9v
D18	---	Not Used	---
D19	YEL/RED	MAP Sensor VREF	4.9-5.1v
D20	YEL/BLU	Sensor VREF	4.9-5.1v
D21	GRN/WHT	MAP Sensor Ground	<0.050v
D22	GRN/BLU	Sensor Ground	<0.050v

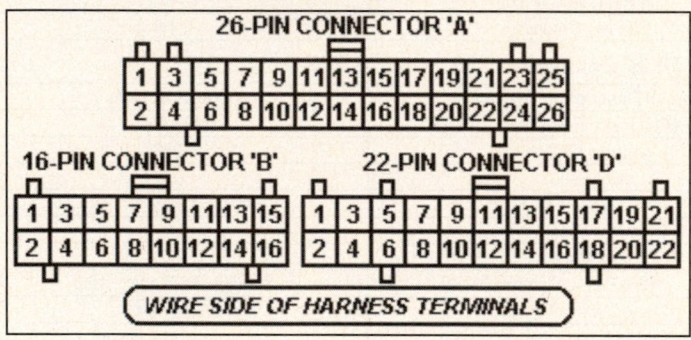

1994-95 GS-R Coupe 1.8L MFI VIN DC2 [M/T] 26P 'A' Connector

PCM Pin #	W/Color	Circuit Description (26-Pin)	Value at Hot Idle
A1	BRN	Injector 1	2.0-3.3 ms
A2	YEL	Injector 4	2.0-3.3 ms
A3	RED	Injector 2	2.0-3.3 ms
A4	GRN/YEL	VTEC Solenoid Control	0v, Hi-Speed: 12v
A5	LT BLU	Injector 3	2.0-3.3 ms
A6	ORN/BLK	HO2S-11 (B1 S1) Heater	Heater Off: 12v, On: 1v
A7	GRN/BLK	Fuel Pump Relay Control	Relay Off: 12v, On: 1v
A8	---	Not Used	---
A9	BLK/BLU	Idle Air Control Valve	Pulse Signals
A10-11	---	Not Used	---
A12	GRN	Radiator Fan Relay Control	Fan on: 1v, off: 12v
A13	GRN/ORN	Malfunction Indicator Lamp	MIL Off: 12v, On: 1v
A14	---	Not Used	
A15	BLK/RED	A/C Clutch Relay Control	Relay Off: 12v, On: 1v
A16	WHT/GRN	Alternator Charging Signal	Headlights off: 0v, on: 12v
A17-19	---	Not Used	---
A20	GRN	EVAP Purge Solenoid	Solenoid Off: 12v, On: 1v
A21	YEL/GRN	Igniter Signal	Digital Signals: 0-12-0v
A22	---	Not Used	---
A23	BLK	Power Ground	<0.1v
A24	BLK	Power Ground	<0.1v
A25	YEL/BLK	Main Relay Power (B+)	12-14v
A26	BRN/BLK	Chassis Ground	0.1v

1994-95 GS-R Coupe 1.8L MFI VIN DC2 [M/T] 16P 'B' Connector

PCM Pin #	W/Color	Circuit Description (16-Pin)	Value at Hot Idle
B1	YEL/BLK	Main Relay Power (B+)	12-14v
B2	BRN/BLK	Chassis Ground	<0.050v
B3-4	---	Not Used	
B5	BLU/RED	A/C Switch Signal	Relay Off: 12v, On: 1v
B6	---	Not Used	---
B7	---	Not Used	---
B8	LT GRN	PSP Switch Signal	Straight: 0v, Turning: 12v
B9	BLU/WHT	Starter Switch Signal	Cranking: 9-11v
B10	ORN	Vehicle Speed Sensor	Moving: pulse signals
B11	ORN	CYP Sensor Signal	AC pulse signals
B12	WHT	CYP Sensor Ground	<0.050v
B13	ORN/BLU	TDC Sensor Signal	AC pulse signals
B14	WHT/BLU	TDC Sensor Ground	<0.050v
B15	BLU/GRN	CKP Sensor Signal	AC pulse signals
B16	BLU/YEL	CKP Sensor Ground	<0.050v

1994-95 GS-R Coupe 1.8L MFI VIN DC2 [M/T] 22P 'D' Connector

PCM Pin #	W/Color	Circuit Description (22-Pin)	Value at Hot Idle
D1	WHT/BLU	Keep Alive Power (VBU)	12-14v
D2	GRN/WHT	Brake Switch Signal	Brake Off: 12v, On: 0v
D3	---	Not Used	---
D4	BRN/WHT	Service Check Connector	SCS Open: 4.80v
D5	---	Not Used	---
D6	BLU/BLK	VTEC Pressure Switch	0v, Hi-Speed: 12v
D7	LT BLU	Data Link Connector	No Scan Tool: 5v
D8	---	Not Used	---
D9	WHT/RED	Alternator 'FR' Signal	Digital Signals: 0-5-0v
D10	GRN/RED	Electronic Load Detector	Varies w/Load: 2.5-3.5v
D11	RED/BLU	TP Sensor Signal	0.5-0.6v
D12	---	Not Used	---
D13	RED/WHT	ECT Sensor Signal	At 180ºF: 0.5-0.6v
D14	WHT/RED	HO2S-11 (B1 S1) Signal	0.1-1.1v
D15	RED/YEL	IAT Sensor Signal	At 100ºF: 2-3v
D16	---	Not Used	---
D17	WHT/YEL	MAP Sensor Signal	0.8-0.9v
D18	---	Not Used	---
D19	YEL/RED	MAP Sensor VREF	4.9-5.1v
D20	YEL/BLU	Sensor VREF	4.9-5.1v
D21	GRN/WHT	MAP Sensor Ground	<0.050v
D22	GRN/BLU	Sensor Ground	<0.050v

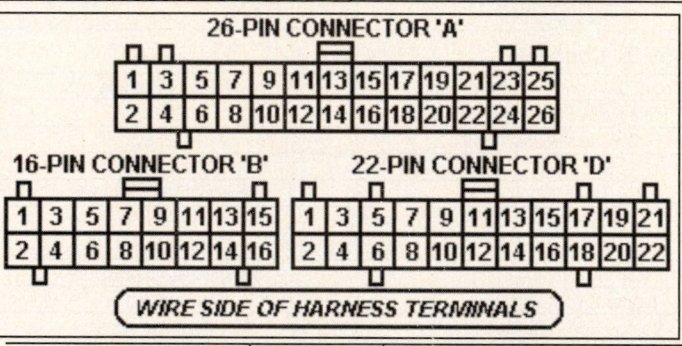

Abbreviation	Color	Abbreviation	Color	Abbreviation	Color
BLK	Black	LT BLU	Lt. Blue	TAN	Tan
BLU	Blue	LT GRN	Lt. Green	VIO	Violet
BRN	Brown	ORN	Orange	WHT	White
GRY	Gray	PNK	Pink	YEL	Yellow
GRN	Green	PPL	Purple		

1996-03 LS-RS-SE Coupe 1.8L MFI VIN DC4 A/T 32P 'A' Connector

PCM Pin #	W/Color	Circuit Description (32-Pin)	Value at Hot Idle
A1	YEL	Injector 4	2.0-3.3 ms
A2	BLU	Injector 3	2.0-3.3 ms
A3	RED	Injector 2	2.0-3.3 ms
A4	BRN	Injector 1	2.0-3.3 ms
A5	GRN/RED	HO2S-12 (B1 S2) Heater	Digital Signals: 0-12-0v
A6	ORN/BLK	HO2S-11 (B1 S1) Heater	Digital Signals: 0-12-0v
A7-8	---	Not Used	---
A9	BRN/BLK	Logic Ground	0.1v
A10	BLK	Power Ground	<0.1v
A11	YEL/BLK	Main Relay Power (B+)	12-14v
A12	BLK/BLU	Idle Air Control Valve	Pulse Signals
A13-14	---	Not Used	---
A15	RED	EVAP Purge Solenoid	Pulse Signals
A16	GRN/BLK	Fuel Pump Relay Control	Relay Off: 12v, On: 1v
A17	BLK/RED	A/C Clutch Relay Control	Relay Off: 12v, On: 1v
A18	GRN/ORN	Malfunction Indicator Lamp	MIL Off: 12v, On: 1v
A19	WHT/GRN	Alternator Charging Signal	Headlights off: 0v, on: 12v
A20	YEL/GRN	Igniter Signal	Digital Signals: 0-12-0v
A21	---	Not Used	---
A22	BRN/BLK	Logic Ground	0.1v
A23	BLK	Power Ground	<0.1v
A24	YEL/BLK	Main Relay Power (B+)	12-14v
A25	WHT/BLK	TCM VREF	4.9-5.1v
A26	---	Not Used	---
A27	GRN	Radiator Fan Relay Control	Fan on: 1v, off: 12v
A28-32	---	Not Used	---

1996-03 LS-RS-SE Coupe 1.8L MFI VIN DC4 A/T 31P 'C' Connector

PCM Pin #	W/Color	Circuit Description (31-Pin)	Value at Hot Idle
C1	BLU/RED	CKF Sensor	AC pulse signals
C2	BLU/GRN	CKP Sensor Signal	AC pulse signals
C3	ORN/BLU	TDC Sensor Signal	AC pulse signals
C4	ORN	CYP Sensor Signal	AC pulse signals
C5	BLU/RED	A/C Switch Signal	A/C Off: 5v, On: 0v
C6	BLU/WHT	Starter Switch Signal	Cranking: 9-11v
C7	BRN/WHT	Service Check Connector	SCS Open: 4.80v
C8	GRN/WHT	K-Line Signal	12v
C9	YEL	FAS TCM Signal	At idle: 4.5-5.5v
C10	WHT/BLU	Keep Alive Power (VBU)	12-14v
C11	WHT/RED	CKF Sensor Ground	<0.050v
C12	BLU/YEL	CKP Sensor Ground	<0.050v
C13	WHT/BLU	TDC Sensor Ground	<0.050v
C14	WHT	CYP Sensor Ground	<0.050v
C15	---	Not Used	---
C16	GRN	PSP Switch Signal	Straight: 0v, Turning: 12v
C17	WHT/RED	Alternator 'FR' Signal	Digital Signals: 0-5-0v
C18	ORN	Vehicle Speed Sensor	Moving: pulse signals
C19	BLU	ATCHK TCM Signal	KOEO on: pulses
C20	---	Not Used	---
C21	LT GRN	TCM BARO Signal	KOEO on: 2.5-3.5v
C22-26	---	Not Used	---
C27	GRY	AFSB TCM Signal	At idle: 4.5-5.5v
C28	GRN/BLU	AFSA TCM Signal	At idle: 4.5-5.5v
C29	GRN/BLK	A/T: Gear Position Switch	In P/N: 0v, Others: 12v
C30-31	---	Not Used	---

1996-03 LS-RS-SE Coupe 1.8L MFI VIN DC4 A/T 16P 'D' Connector

PCM Pin #	W/Color	Circuit Description (16-Pin)	Value at Hot Idle
D1	RED/BLK	TP Sensor Signal	0.5-0.6v
D2	RED/WHT	ECT Sensor Signal	At 180°F: 0.5-0.6v
D3	WHT/YEL	MAP Sensor Signal	0.8-0.9v
D4	YEL/WHT	MAP Sensor VREF	4.9-5.1v
D5	GRN/WHT	Brake Switch Signal	Brake Off: 12v, On: 0v
D6	---	Not Used	---
D7	WHT/RED	HO2S-11 (B1 S1) Signal	0.1-1.1v
D8	RED/YEL	IAT Sensor Signal	At 100°F: 2-3v
D9	---	Not Used	---
D10	YEL/BLU	Sensor VREF	4.9-5.1v
D11	GRN/BLU	Sensor Ground	<0.050v
D12	GRN/WHT	MAP Sensor Ground	<0.050v
D13	ORN/BLU	HO2S-12 Ground	<0.050v
D14	BLU/GRN	HO2S-12 (B1 S2) Signal	0.1-1.1v
D15	---	Not Used	---
D16	GRN/RED	Electric Load Detector	Varies: 2.5-3.5v

1996-03 LS-RS-SE Coupe 1.8L MFI VIN DC4 [M/T] 32P A Connector

PCM Pin #	W/Color	Circuit Description (32-Pin)	Value at Hot Idle
A1	YEL	Injector 4	2.0-3.3 ms
A2	BLU	Injector 3	2.0-3.3 ms
A3	RED	Injector 2	2.0-3.3 ms
A4	BRN	Injector 1	2.0-3.3 ms
A5	GRN/RED	HO2S-12 (B1 S2) Heater	Digital Signals: 0-12-0v
A6	ORN/BLK	HO2S-11 (B1 S1) Heater	Digital Signals: 0-12-0v
A7-8	---	Not Used	---
A9	BRN/BLK	Logic Ground	0.1v
A10	BLK	Power Ground	<0.1v
A11	YEL/BLK	Main Relay Power (B+)	12-14v
A12	BLK/BLU	Idle Air Control Valve	Pulse Signals
A13-14	---	Not Used	---
A15	RED	EVAP Purge Solenoid	Pulse Signals
A16	GRN/BLK	Fuel Pump Relay Control	Relay Off: 12v, On: 1v
A17	BLK/RED	A/C Clutch Relay Control	Relay Off: 12v, On: 1v
A18	GRN/ORN	Malfunction Indicator Lamp	MIL Off: 12v, On: 1v
A19	WHT/GRN	Alternator Charging Signal	Headlights off: 0v, on: 12v
A20	YEL/GRN	Igniter Signal	Digital Signals: 0-12-0v
A21	---	Not Used	---
A22	BRN/BLK	Logic Ground	0.1v
A23	BLK	Power Ground	<0.1v
A24	YEL/BLK	Main Relay Power (B+)	12-14v
A25	---	Not Used	---
A26	---	Not Used	---
A27	GRN	Radiator Fan Relay Control	Fan on: 1v, off: 12v
A28-32	---	Not Used	---

32-PIN CONNECTOR 'A'
31-PIN CONNECTOR 'C'   16-PIN CONNECTOR 'D'
WIRE SIDE OF HARNESS TERMINALS

1996-03 LS-RS-SE Coupe 1.8L MFI VIN DC4 [M/T] 31P C Connector

PCM Pin #	W/Color	Circuit Description (31-Pin)	Value at Hot Idle
C1	BLU/RED	CKF Sensor	AC pulse signals
C2	BLU/GRN	CKP Sensor Signal	AC pulse signals
C3	ORN/BLU	TDC Sensor Signal	AC pulse signals
C4	ORN	CYP Sensor Signal	AC pulse signals
C5	BLU/RED	A/C Switch Signal	A/C Off: 5v, On: 0v
C6	BLU/WHT	Starter Switch Signal	Cranking: 9-11v
C7	BRN/WHT	Service Check Connector	SCS Open: 4.80v
C8	GRN/WHT	K-Line Signal	12v
C9	---	Not Used	---
C10	WHT/BLU	Keep Alive Power (VBU)	12-14v
C11	WHT/RED	CKF Sensor Ground	<0.050v
C12	BLU/YEL	CKP Sensor Ground	<0.050v
C13	WHT/BLU	TDC Sensor Ground	<0.050v
C14	WHT	CYP Sensor Ground	<0.050v
C15	---	Not Used	---
C16	GRN	PSP Switch Signal	Straight: 0v, Turning: 12v
C17	WHT/RED	Alternator 'FR' Signal	Digital Signals: 0-5-0v
C18	ORN	Vehicle Speed Sensor	Moving: pulse signals
C19-31	---	Not Used	---

1996-03 LS-RS-SE Coupe 1.8L MFI VIN DC4 [M/T] 16P D Connector

PCM Pin #	W/Color	Circuit Description (16-Pin)	Value at Hot Idle
D1	RED/BLK	TP Sensor Signal	0.5-0.6v
D2	RED/WHT	ECT Sensor Signal	At 180ºF: 0.5-0.6v
D3	WHT/YEL	MAP Sensor Signal	0.8-0.9v
D4	YEL/WHT	MAP Sensor VREF	4.9-5.1v
D5	GRN/WHT	Brake Switch Signal	Brake Off: 12v, On: 0v
D6	---	Not Used	---
D7	WHT/RED	HO2S-11 (B1 S1) Signal	0.1-1.1v
D8	RED/YEL	IAT Sensor Signal	At 100ºF: 2-3v
D9	---	Not Used	---
D10	YEL/BLU	Sensor VREF	4.9-5.1v
D11	GRN/BLU	Sensor Ground	<0.050v
D12	GRN/WHT	MAP Sensor Ground	<0.050v
D13	ORN/BLU	HO2S-12 Ground	<0.050v
D14	BLU/GRN	HO2S-12 (B1 S2) Signal	0.1-1.1v
D15	---	Not Used	---
D16	GRN/RED	Electric Load Detector	Varies: 2.5-3.5v

1996-03 GS-R Coupe 1.8L I4 MFI VIN DC2 [M/T] 32P 'A' Connector

PCM Pin #	W/Color	Circuit Description (32-Pin)	Value at Hot Idle
A1	YEL	Injector 4	2.0-3.3 ms
A2	BLU	Injector 3	2.0-3.3 ms
A3	RED	Injector 2	2.0-3.3 ms
A4	BRN	Injector 1	2.0-3.3 ms
A5	GRN/RED	HO2S-12 (B1 S2) Heater	Digital Signals: 0-12-0v
A6	ORN/BLK	HO2S-11 (B1 S1) Heater	Digital Signals: 0-12-0v
A7	----	Not Used	---
A8	GRN/YEL	VTEC Solenoid Control	0v, Hi-Speed: 12v
A9	BRN/BLK	Logic Ground	0.1v
A10	BLK	Power Ground	<0.1v
A11	YEL/BLK	Main Relay Power (B+)	12-14v
A12	BLK/BLU	Idle Air Control Valve	Pulse Signals
A13-14	---	Not Used	---
A15	RED	EVAP Purge Solenoid	Pulse Signals
A16	GRN/BLK	Fuel Pump Relay Control	Relay Off: 12v, On: 1v
A17	BLK/RED	A/C Clutch Relay Control	Relay Off: 12v, On: 1v
A18	GRN/ORN	Malfunction Indicator Lamp	MIL Off: 12v, On: 1v
A19	WHT/GRN	Alternator Charging Signal	Headlights off: 0v, on: 12v
A20	YEL/GRN	Igniter Signal	Digital Signals: 0-12-0v
A21	---	Not Used	---
A22	BRN/BLK	Logic Ground	<0.050v
A23	BLK	Power Ground	<0.1v
A24	YEL/BLK	Main Relay Power (B+)	12-14v
A25	---	Not Used	---
A26	PNK/BLU	Intake Air Bypass Solenoid	Solenoid Off: 12v, On: 1v
A27	GRN	Radiator Fan Relay Control	Fan on: 1v, off: 12v
A28-32	---	Not Used	---

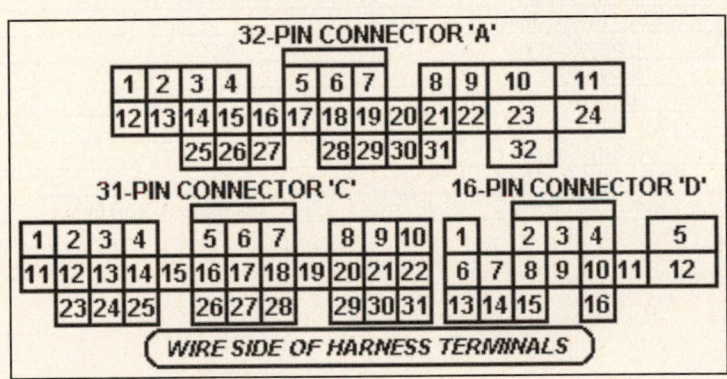

1996-03 GS-R Coupe 1.8L I4 MFI VIN DC2 [M/T] 31P 'C' Connector

PCM Pin #	W/Color	Circuit Description (31-Pin)	Value at Hot Idle
C1	BLU/RED	CKF Sensor	AC pulse signals
C2	BLU/GRN	CKP Sensor Signal	AC pulse signals
C3	ORN/BLU	TDC Sensor Signal	AC pulse signals
C4	ORN	CYP Sensor Signal	AC pulse signals
C5	BLU/RED	A/C Switch Signal	A/C Off: 5v, On: 0v
C6	BLU/WHT	Starter Switch Signal	Cranking: 9-11v
C7	BRN/WHT	Service Check Connector	SCS Open: 4.80v
C8	GRN/WHT	K-Line Signal	12v
C9	---	Not Used	---
C10	WHT/BLU	Keep Alive Power (VBU)	12-14v
C11	WHT/RED	CKF Sensor Ground	<0.050v
C12	BLU/YEL	CKP Sensor Ground	<0.050v
C13	WHT/BLU	TDC Sensor Ground	<0.050v
C14	WHT	CYP Sensor Ground	<0.050v
C15	BLU/BLK	VTEC Pressure Switch	0v, Hi-Speed: 12v
C16	GRN	PSP Switch Signal	Straight: 0v, Turning: 12v
C17	WHT/RED	Alternator 'FR' Signal	Digital Signals: 0-5-0v
C18	ORN	Vehicle Speed Sensor	Moving: pulse signals
C19-31	---	Not Used	---

1996-03 GS-R Coupe 1.8L I4 MFI VIN DC2 [M/T] 16P 'D' Connector

PCM Pin #	W/Color	Circuit Description (16-Pin)	Value at Hot Idle
D1	RED/BLK	TP Sensor Signal	0.5-0.6v
D2	RED/WHT	ECT Sensor Signal	At 180ºF: 0.5-0.6v
D3	WHT/YEL	MAP Sensor Signal	0.8-0.9v
D4	YEL/WHT	MAP Sensor VREF	4.9-5.1v
D5	GRN/WHT	Brake Switch Signal	Brake Off: 12v, On: 0v
D6	RED/BLU	Knock Sensor Signal	No Detonation: 18mv AC
D7	WHT/RED	HO2S-11 (B1 S1) Signal	0.1-1.1v
D8	RED/YEL	IAT Sensor Signal	At 100ºF: 2-3v
D10	YEL/BLU	Sensor VREF	4.9-5.1v
D11	GRN/BLU	Sensor Ground	<0.050v
D12	GRN/WHT	MAP Sensor Ground	<0.050v
D13	ORN/BLU	HO2S-12 Ground	<0.050v
D14	BLU/GRN	HO2S-12 (B1 S2) Signal	0.1-1.1v
D16	GRN/RED	Electric Load Detector	Varies w/Load: 2.5-3.5v

1997 GS & LS 1.8L I4 MFI VIN DB7, DB8 [All] 32P 'A' Connector

PCM Pin #	W/Color	Circuit Description (32-Pin)	Value at Hot Idle
A1	YEL	Injector 4	2.0-3.3 ms
A2	BLU	Injector 3	2.0-3.3 ms
A3	RED	Injector 2	2.0-3.3 ms
A4	BRN	Injector 1	2.0-3.3 ms
A5	GRN/RED	HO2S-12 (B1 S2) Heater	Digital Signals: 0-12-0v
A6	ORN/BLK	HO2S-11 (B1 S1) Heater	Digital Signals: 0-12-0v
A7-8	---	Not Used	---
A9	BRN/BLK	Logic Ground (LG1)	<0.050v
A10	BLK	Power Ground	<0.1v
A11	YEL/BLK	Main Relay Power (B+)	12-14v
A12	BLK/BLU	Idle Air Control Valve	Pulse Signals
A13-14	---	Not Used	---
A15	RED	EVAP Purge Solenoid	Pulse Signals
A16	GRN/BLU	Fuel Pump Relay Control	Relay Off: 12v, On: 1v
A17	BLK/RED	A/C Clutch Relay Control	Relay Off: 12v, On: 1v
A18	GRN/ORN	Malfunction Indicator Lamp	MIL Off: 12v, On: 1v
A19	WHT/GRN	Alternator Charging Signal	Lights on: 0v: off: 12v
A20	YEL/GRN	Igniter Control Signal	Digital Signals: 0-12-0v
A21	---	Not Used	---
A22	BRN/BLK	Logic Ground (LG2)	<0.050v
A23	BLK	Power Ground	<0.1v
A24	YEL/BLK	Main Relay Power (B+)	12-14v
A25	WHT/BLK	A/T: TCM VREF	4.9-5.1v
A26	---	Not Used	---
A27	GRN	Radiator Fan Relay Control	Fan on: 1v, off: 12v
A28-32	---	---	---

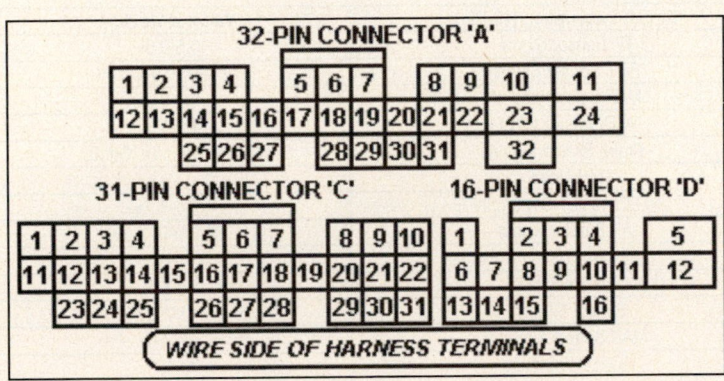

1997 GS & LS 1.8L I4 MFI VIN DB7, DB8 [All] 31P 'C' Connector

PCM Pin #	W/Color	Circuit Description (31-Pin)	Value at Hot Idle
C1	BLU/RED	CKF Sensor Signal	AC pulse signals
C2	BLU	CKP Sensor Signal	AC pulse signals
C3	ORN/BLU	TDC Sensor Signal	AC pulse signals
C4	YEL	CYP Sensor Signal	AC pulse signals
C5	BLU/RED	A/C Switch Signal	A/C Off: 5v, On: 0v
C6	BLU/WHT	Starter Switch Signal	Cranking: 9-11v
C7	BRN/WHT	Service Check Connector	SCS Open: 4.80v
C8	GRN/WHT	K-Line Signal	12v
C9	YEL	FAS TCM Signal	At idle: 4.5-5.5v
C10	WHT/BLU	Keep Alive Power (VBU)	12-14v
C11	WHT/RED	CKF Sensor Ground	<0.050v
C12	BLU/YEL	CKP Sensor Ground	<0.050v
C13	WHT/BLU	TDC Sensor Ground	<0.050v
C14	WHT	CYP Sensor Ground	<0.050v
C15	---	Not Used	---
C16	GRN	PSP Switch Signal	Straight: 0v, Turning: 12v
C17	WHT/RED	Alternator 'FR' Signal	Digital Signals: 0-5-0v
C18	ORN	Vehicle Speed Sensor	Moving: pulse signals
C19	BLU	A/T: BARO to TCM Signal	Varies w/alt: 0-5-4.5v
C20	---	Not Used	---
C21	LT GRN	BARO Signal to TCM	Varies: 0.5-4.9v
C22-26	---	Not Used	---
C27	GRY	A/T: TCM Spark Retard 'A'	At idle: 5v, Shifting: 0v
C28	GRN/BLU	A/T: TCM Spark Retard 'B'	At idle: 5v, Shifting: 0v
C29	GRN/BLK	A/T: Gear Position Switch	In P/N: 0v, Others: 12v
C30-31	---	Not Used	---

1997 GS & LS 1.8L I4 MFI VIN DB7, DB8 [All] 16P 'D' Connector

PCM Pin #	W/Color	Circuit Description (16-Pin)	Value at Hot Idle
D1	RED/BLK	TP Sensor Signal	0.5-0.6v
D2	RED/WHT	ECT Sensor Signal	At 180°F: 0.5-0.6v
D3	WHT/YEL	MAP Sensor Signal	0.8-0.9v
D4	YEL/WHT	MAP Sensor VREF	4.9-5.1v
D5	GRN/WHT	Brake Switch Signal	Brake Off: 12v, On: 0v
D6	---	Not Used	---
D7	WHT	HO2S-11 (B1 S1) Signal	0.1-1.1v
D8	RED/YEL	IAT Sensor Signal	At 100°F: 2-3v
D9	---	Not Used	---
D10	YEL/BLU	Sensor VREF	4.9-5.1v
D11	GRN/BLU	Sensor Ground	<0.050v
D12	GRN/WHT	MAP Sensor Ground	<0.050v
D13	ORN/BLU	HO2S-12 Ground	<0.050v
D14	BLU/RED	HO2S-12 (B1 S2) Signal	0.1-1.1v
D15	---	Not Used	---
D16	GRN/RED	Electric Load Detector	Varies: 2.5-3.5v

1998-99 GS/LS 1.8L I4 MFI VIN DB7, DB8 [All] 32P 'A' Connector

PCM Pin #	W/Color	Circuit Description (32-Pin)	Value at Hot Idle
A1	YEL	Injector 4	2.0-3.3 ms
A2	BLU	Injector 3	2.0-3.3 ms
A3	RED	Injector 2	2.0-3.3 ms
A4	BRN	Injector 1	2.0-3.3 ms
A5	GRN/RED	HO2S-12 (B1 S2) Heater	Digital Signals: 0-12-0v
A6	ORN/BLK	HO2S-11 (B1 S1) Heater	Digital Signals: 0-12-0v
A7-8	---	Not Used	---
A9	BRN/BLK	Logic Ground (LG1)	<0.050v
A10	BLK	Power Ground	<0.1v
A11	YEL/BLK	Main Relay Power (B+)	12-14v
A12	BLK/BLU	Idle Air Control Valve	Pulse Signals
A13-14	---	Not Used	---
A15	RED	EVAP Purge Solenoid	Pulse Signals
A16	GRN/BLU	Fuel Pump Relay Control	Relay Off: 12v, On: 1v
A17	BLK/RED	A/C Clutch Relay Control	Relay Off: 12v, On: 1v
A18	GRN/ORN	Malfunction Indicator Lamp	MIL Off: 12v, On: 1v
A19	WHT/GRN	Alternator Charging Signal	Lights on: 0v: off: 12v
A20	YEL/GRN	Igniter Control Signal	Digital Signals: 0-12-0v
A21	---	Not Used	---
A22	BRN/BLK	Logic Ground (LG2)	<0.050v
A23	BLK	Power Ground	<0.1v
A24	YEL/BLK	Main Relay Power (B+)	12-14v
A25	WHT/BLK	A/T: TCM VREF	4.9-5.1v
A26	---	Not Used	---
A27	GRN	Radiator Fan Relay Control	Fan on: 1v, off: 12v
A28	BLU	EVAP Bypass Solenoid	Solenoid Off: 12v, On: 1v
A29	LT GRN	EVAP Vent Shut Control	Solenoid Off: 12v, On: 1v
A30-32	---	---	---

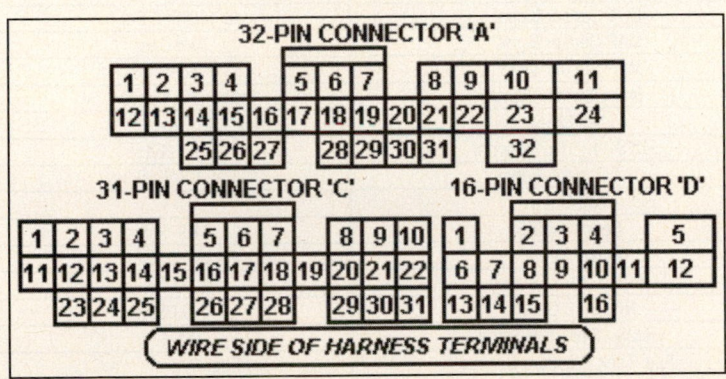

1998-99 GS/LS 1.8L I4 MFI VIN DB7, DB8 [All] 31P 'C' Connector

PCM Pin #	W/Color	Circuit Description (31-Pin)	Value at Hot Idle
C1	BLU/RED	CKF Sensor Signal	AC pulse signals
C2	BLU	CKP Sensor Signal	AC pulse signals
C3	GRN	TDC Sensor Signal	AC pulse signals
C4	YEL	CYP Sensor Signal	AC pulse signals
C5	BLU/RED	A/C Switch Signal	A/C Off: 5v, On: 0v
C6	BLU/WHT	Starter Switch Signal	Cranking: 9-11v
C7	BRN/WHT	Service Check Connector	SCS Open: 4.80v
C8	GRN/WHT	K-Line Signal	12v
C9	YEL	FAS TCM Signal	At idle: 4.5-5.5v
C10	WHT/BLU	Keep Alive Power (VBU)	12-14v
C11	WHT/RED	CKF Sensor Ground	<0.050v
C12	BLU/YEL	CKP Sensor Ground	<0.050v
C13	WHT/BLU	TDC Sensor Ground	<0.050v
C14	WHT	CYP Sensor Ground	<0.050v
C15	---	Not Used	
C16	GRN	PSP Switch Signal	Straight: 0v, Turning: 12v
C17	WHT/RED	Alternator 'FR' Signal	Digital Signals: 0-5-0v
C18	ORN	Vehicle Speed Sensor	Moving: pulse signals
C19	BLU	A/T: BARO to TCM Signal	Varies w/alt: 0-5-4.5v
C20	---	Not Used	---
C21	LT GRN	BARO Signal to TCM	Varies: 0.5-4.9v
C22-26	---	Not Used	---
C27	GRY	A/T: TCM Spark Retard 'A'	At idle: 5v, Shifting: 0v
C28	GRN/BLU	A/T: TCM Spark Retard 'B'	At idle: 5v, Shifting: 0v
C29	GRN/BLK	A/T: Gear Position Switch	In P/N: 0v, Others: 12v
C30-31	---	Not Used	---

1998-99 GS/LS 1.8L I4 MFI VIN DB7, DB8 [All] 16P 'D' Connector

PCM Pin #	W/Color	Circuit Description (16-Pin)	Value at Hot Idle
D1	RED/BLK	TP Sensor Signal	0.5-0.6v
D2	RED/WHT	ECT Sensor Signal	At 180°F: 0.5-0.6v
D3	WHT/YEL	MAP Sensor Signal	0.8-0.9v
D4	YEL/WHT	MAP Sensor VREF	4.9-5.1v
D5	GRN/WHT	Brake Switch Signal	Brake Off: 12v, On: 0v
D6	---	Not Used	---
D7	WHT	HO2S-11 (B1 S1) Signal	0.1-1.1v
D8	RED/YEL	IAT Sensor Signal	At 100°F: 2-3v
D9	---	Not Used	---
D10	YEL/BLU	Sensor VREF	4.9-5.1v
D11	GRN/BLU	Sensor Ground	<0.050v
D12	GRN/WHT	MAP Sensor Ground	<0.050v
D13	ORN/BLU	HO2S-12 Ground	<0.050v
D14	BLU/RED	HO2S-12 (B1 S2) Signal	0.1-1.1v
D15	LT GRN	Fuel Tank Pressure Sensor	Fuel Cap off: 2.5v
D16	GRN/RED	Electric Load Detector	Varies: 2.5-3.5v

2000-003 GS 1.8L MFI VIN DB7, DB8 [All] 32P 'A' Connector

PCM Pin #	W/Color	Circuit Description (32-Pin)	Value at Hot Idle
A1-2	---	Not Used	---
A3	BLU	EVAP Bypass Solenoid	Solenoid Off: 12v, On: 1v
A4	LT GRN	EVAP Vent Shut Control	Solenoid Off: 12v, On: 1v
A5	---	Not Used	---
A6	RED	EVAP Purge Solenoid	Pulse Signals
A7	---	Not Used	---
A8	GRN/RED	HO2S-12 (B1 S2) Heater	Digital Signals: 0-12-0v
A9	---	Not Used	---
A10	BRN/WHT	Service Check Connector	SCS Open: 4.80v
A11	---	Not Used	---
A12	PNK	Immobilizer Indicator Control	Lamp On: 1v, Off: 12v
A13	BLU	IMOEN Immobilizer Signal	Digital Signals
A14	GRN/BLK	A/T: D4 Indicator Light	In D4: 0v, Others: 12v
A15	GRN/YEL	Fuel Pump Relay Control	Relay Off: 12v, On: 1v
A16	---	Not Used	---
A17	BLK/RED	A/C Clutch Relay Control	Relay Off: 12v, On: 1v
A18	GRN/ORN	Malfunction Indicator Light	MIL Off: 12v, On: 1v
A19	BLU	Engine Speed (NEP) Signal	Digital Signals
A20	GRN	Radiator Fan Relay Control	Fan on: 1v, off: 12v
A21	GRN/WHT	K-Line Signal	12v
A22	---	Not Used	---
A23	BLU/RED	HO2S-12 (B1 S2) Signal	0.1-1.1v
A24	BLU/WHT	Starter Switch Signal	Cranking: 9-11v
A25	RED	INOCD Immobilizer Code	Digital Signals
A26	GRN	PSP Switch Signal	Straight: 0v, Turning: 12v
A27	BLU/RED	A/C Switch Signal	A/C Off: 5v, On: 0v
A28	WHT/RED	A/T: Interlock Control Unit	With Brake on: 12v
A29	LT GRN	Fuel Tank Pressure Sensor	Fuel Cap off: 2.5v
A30	GRN/RED	Electric Load Detector	Varies w/Load: 2.5-3.5v
A31	---	Not Used	---
A32	GRN/WHT	Brake Switch Signal	Brake Off: 12v, On: 0v

PCM 104-Pin Connector Graphic

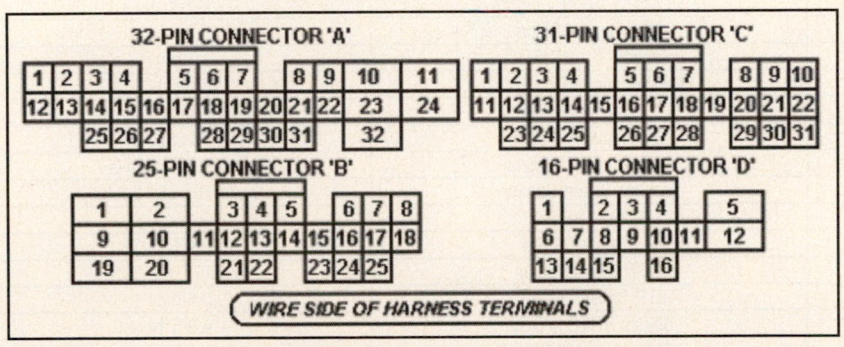

2000-03 GS 1.8L MFI VIN DB7, DB8 [All] 25P 'B' Connector

PCM Pin #	W/Color	Circuit Description (25-Pin)	Value at Hot Idle
B1	YEL/BLK	Main Relay Power (B+)	12-14v
B2	BLK	Power Ground	<0.1v
B3	RED	Injector 2	2.0-3.3 ms
B4	BLU	Injector 3	2.0-3.3 ms
B5	YEL	Injector 4	2.0-3.3 ms
B6-7	---	Not Used	---
B8	WHT/GRN	A/T: Clutch Solenoid LS-	AC Pulse Signals
B9	YEL/BLK	Main Relay Power (B+)	12-14v
B10	BLK	Power Ground	<0.1v
B11	BRN	Injector 1	2.0-3.3 ms
B12	---	Not Used	---
B13	YEL/GRN	Ignition Control Module Signal	Digital Signals: 0-12-0v
B14-15	---	Not Used	---
B16	PNK/BLU	Intake Air Bypass Solenoid	Solenoid Off: 12v, On: 1v
B17	RED/BLU	A/T: Clutch Solenoid LS+	AC Pulse Signals
B18-19	---	Not Used	---
B20	BRN/BLK	Logic Ground (LG1)	<0.050v
B21	WHT/BLU	Keep Alive Power (VBU)	12-14v
B22	BRN/BLK	Logic Ground (LG2)	<0.050v
B23	BLK/BLU	Idle Air Control Valve	DC Pulse Signals

2000-03 GS 1.8L MFI VIN DB7, DB8 [All] 31P 'C' Connector

PCM Pin #	W/Color	Circuit Description (31-Pin)	Value at Hot Idle
C1	BRN/BLK	HO2S-11 (B1 S1) Heater	Digital Signals: 0-12-0v
C2	WHT/GRN	Alternator Charging Signal	Headlights off: 0v, on: 12v
C3	RED/BLU	Knock Sensor Signal	No Detonation: 18mv AC
C4	---	Not Used	---
C5	WHT/RED	Alternator 'FR' Signal	Digital Signals: 0-5-0v
C6	---	Not Used	---
C7	GRN/WHT	MAP Sensor Ground	<0.050v
C8	BLU	CKP Sensor Signal	AC Pulse Signals
C9	BLU/YEL	CKP Sensor Ground	<0.050v
C10-15	---	Not Used	---
C16	WHT	HO2S-11 (B1 S1) Signal	0.1-1.1v
C17	WHT/YEL	MAP Sensor Signal	0.8-0.9v
C18	GRN/BLU	Sensor Ground (SG2)	<0.050v
C19	YEL/RED	MAP Sensor VREF (VCC1)	4.9-5.1v
C20	GRN	TDC Sensor 'P' Signal	A/C Pulse Signals
C21	WHT/BLU	TDC Sensor 'N' Signal	<0.050v
C22	BLU/RED	CKF Sensor 'P' Signal	A/C Pulse Signals
C23	ORN	Vehicle Speed Sensor	Digital Signals: 0-5-0v
C24	---	Not Used	---
C25	RED/YEL	IAT Sensor Signal	At 100°F: 2-3v
C26	RED/WHT	ECT Sensor Signal	At 180°F: 0.5-0.6v
C27	RED/BLK	TP Sensor Signal	0.5-0.6v
C28	YEL/BLU	Sensor VREF (VCC2)	4.9-5.1v
C29	YEL	CYP Sensor 'P' Signal	AC Pulse Signals
C30	WHT	CYP Sensor Ground	<0.050v
C31	WHT/RED	CKF Sensor Ground	<0.050v

2000-03 GS 1.8L MFI VIN DB7, DB8 [All] 16P 'D' Connector

PCM Pin #	W/Color	Circuit Description (16-Pin)	Value at Hot Idle
D1	YEL	A/T: Lockup Solenoid 'A'	LSA Off: 0v, On: 12v
D2	GRN/WHT	A/T: Shift Solenoid 'B'	SSB Off: 0v, On: 12v
D3	GRN/BLK	A/T: Lockup Solenoid 'B'	LSB Off: 0v, On: 12v
D4	---	Not Used	---
D5	BLK/YEL	A/T: Solenoid Feed (B+)	Full: 12v, Partial: Pulses
D6	GRN/RED	A/T: Gear Position Switch	In 'R': 0v, Others: 12v
D7	BLU/YEL	A/T: Shift Solenoid 'A'	SSA Off: 0v, On: 12v
D8	GRN/BLU	A/T: Gear Position Switch	In D3: 0v, Others: 12v
D9	YEL	A/T: Gear Position Switch	In D4: 0v, Others: 12v
D10	BLU	Countershaft Speed Sensor P	Moving: AC Pulse Signals
D11	RED	Mainshaft Speed Sensor 'P'	AC Pulse Signals
D12	WHT	Mainshaft Speed Sensor 'N'	AC Pulse Signals
D13	GRN/BLK	A/T: Gear Position Switch	In P/N: 0v, Others: 12v
D14	GRN/YEL	A/T: Gear Position Switch	In D2: 0v, Others: 12v
D15	GRN/WHT	A/T: Gear Position Switch	In D1: 0v, Others: 12v
D16	GRN	Countershaft Speed Sensor N	Moving: AC Pulse Signals

PCM 104-Pin Connector Graphic

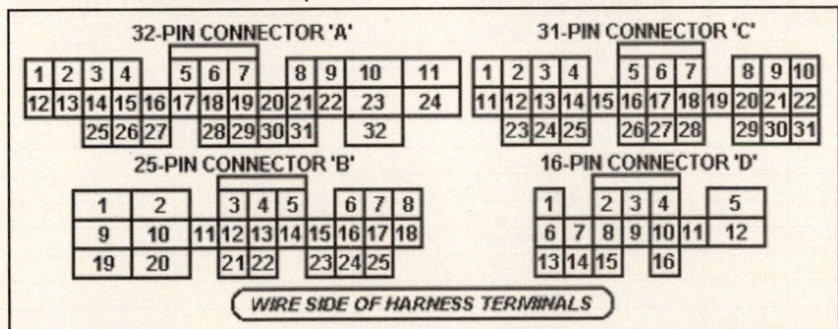

Abbreviation	Color	Abbreviation	Color	Abbreviation	Color
BLK	Black	LT BLU	Lt. Blue	TAN	Tan
BLU	Blue	LT GRN	Lt. Green	VIO	Violet
BRN	Brown	ORN	Orange	WHT	White
GRY	Gray	PNK	Pink	YEL	Yellow
GRN	Green	PPL	Purple		

1997-99 GS-R, Type R 1.8L VTEC VIN DC2 [All] 32P 'A' Connector

PCM Pin #	W/Color	Circuit Description (32-Pin)	Value at Hot Idle
A1	YEL	Injector 4	2.0-3.3 ms
A2	BLU	Injector 3	2.0-3.3 ms
A3	RED	Injector 2	2.0-3.3 ms
A4	BRN	Injector 1	2.0-3.3 ms
A5	GRN/RED	HO2S-12 (B1 S2) Heater	Digital Signals: 0-12-0v
A6	ORN/BLK	HO2S-11 (B1 S1) Heater	Digital Signals: 0-12-0v
A7	---	Not Used	---
A8	GRN/YEL	VTEC Solenoid Control	0v, Hi-Speed: 12v
A9	BRN/BLK	Logic Ground	0.1v
A10	BLK	Power Ground (PG1)	<0.1v
A11	YEL/BLK	Main Relay Power (B+)	12-14v
A12	BLK/BLU	Idle Air Control Valve	Pulse Signals
A13-14	---	Not Used	---
A15	RED	EVAP Purge Solenoid	Pulse Signals
A16	GRN/BLU	Fuel Pump Relay Control	Relay Off: 12v, On: 1v
A17	BLK/RED	A/C Clutch Relay Control	Relay Off: 12v, On: 1v
A18	GRN/ORN	Malfunction Indicator Light	MIL Off: 12v, On: 1v
A19	WHT/GRN	Alternator Charging Signal	Lights on: 0v, off: 12v
A20	YEL/GRN	Igniter Control Signal	Digital Signals: 0-12-0v
A21	---	Not Used	---
A22	BRN/BLK	Logic Ground (LG2)	0.1v
A23	BLK	Power Ground (PG2)	<0.1v
A24	YEL/BLK	Main Relay Power (B+)	12-14v
A25	WHT/BLK	Reference Voltage	4.9-5.1v
A26	PNK/BLU	Intake Air Bypass Solenoid	Solenoid off: 12v, on: 1v
A27	GRN	Radiator Fan Relay Control	Fan on: 1v, off: 12v
A28 ('98-'99)	BLU	EVAP Bypass Solenoid	Solenoid off: 12v, on: 1v
A29 ('98-'99)	LT GRN	EVAP Vent Shut Control	Solenoid off: 12v, on: 1v
A30-32	---	Not Used	---

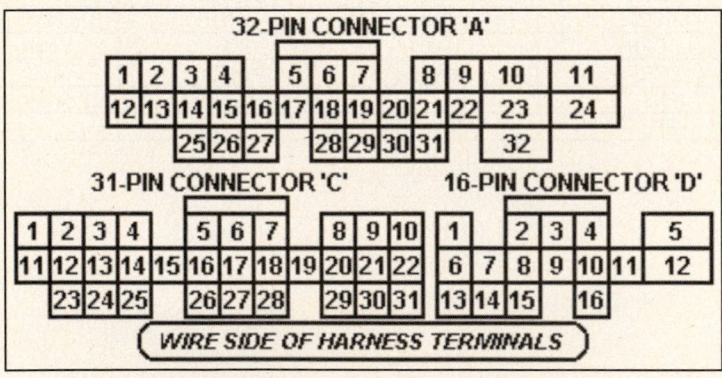

1997-99 GS-R, Type R 1.8L VTEC VIN DC2 [All] 31P 'C' Connector

PCM Pin #	W/Color	Circuit Description (31-Pin)	Value at Hot Idle
C1	BLU/RED	CKF Sensor Signal	AC pulse signals
C2 ('97)	BLU/GRN	CKP Sensor Signal	AC pulse signals
C2 ('98-'99)	BLU	CKP Sensor Signal	AC pulse signals
C3 ('97)	ORN/BLU	TDC Sensor Signal	AC pulse signals
C3 ('98-'99)	GRN	TDC Sensor Signal	AC pulse signals
C4 ('97)	ORN	CYP Sensor Signal	AC pulse signals
C4 ('98-'99)	YEL	CYP Sensor Signal	AC pulse signals
C5	BLU/RED	A/C Switch Signal	A/C Off: 5v, On: 0v
C6	BLU/WHT	Starter Switch Signal	Cranking: 9-11v
C7	BRN/WHT	Service Check Connector	SCS Open: 4.80v
C8	GRN/WHT	K-Line Signal	12v
C9	YEL	A/T: Feedback Signal	At idle: 5v, Shifting: 0v
C10	WHT/BLU	Keep Alive Power (VBU)	12-14v
C11	WHT/RED	CKF Sensor Ground	<0.050v
C12	BLU/YEL	CKP Sensor Ground	<0.050v
C13	WHT/BLU	TDC Sensor Ground	<0.050v
C14	WHT	CYP Sensor Ground	<0.050v
C15	BLU/BLK	VTEC Pressure Switch	0v, Hi-Speed: 12v
C16	GRN	PSP Switch Signal	Straight: 0v, Turned: 12v
C17	WHT/RED	Alternator 'FR' Signal	Digital Signals: 0-5-0v
C18	ORN	Vehicle Speed Sensor	Moving: pulse signals
C19	BLU	A/T: FI Data Line (CHK)	Serial Data
C20	---	Not Used	---
C21	LT GRN	A/T: BARO Signal to TCM	Varies w/alt: 0.5-4.9v
C22-26	---	Not Used	---
C27	GRY	A/T: FI Signal 'B'	At idle: 5v, Shifting: 0v
C28	GRN/BLU	A/T: FI Signal 'A'	At idle: 5v, Shifting: 0v
C29	GRN/BLK	A/T: Gear Position Switch	In P/N: 0v, Others: 12v
C30-31	---	Not Used	---

1997-99 GS-R, Type R 1.8L VTEC VIN DC2 [All] 16P 'D' Connector

PCM Pin #	W/Color	Circuit Description (16-Pin)	Value at Hot Idle
D1	RED/BLK	TP Sensor Signal	0.5-0.6v
D2	RED/WHT	ECT Sensor Signal	At 180°F: 0.5-0.6v
D3	WHT/YEL	MAP Sensor Signal	0.8-0.9v
D4	YEL/WHT	MAP Sensor VREF (VCC1)	4.9-5.1v
D5	GRN/WHT	Brake Switch Signal	Brake Off: 12v, On: 0v
D6	RED/BLU	Knock Sensor Signal	No detonation: 18mv AC
D7	WHT/RED	HO2S-11 (B1 S1) Signal	0.1-1.1v
D8	RED/YEL	IAT Sensor Signal	At 100°F: 2-3v
D9	---	Not Used	---
D10	YEL/BLU	Sensor VREF (VCC2)	4.9-5.1v
D11	GRN/BLU	Sensor Ground (SG2)	<0.050v
D12	GRN/WHT	MAP Sensor Ground (SG1)	<0.050v
D13	ORN/BLU	HO2S-12 Ground	<0.050v
D14	BLU/GRN	HO2S-12 (B1 S2) Signal	0.1-1.1v
D15 ('98-'99)	LT GRN	Fuel Tank Pressure Sensor	Fuel Cap off: 2.5v
D16	GRN/RED	Electric Load Detector	Varies w/Load: 2.5-3.5v

2000-03 GS-R 1.8L VTEC VIN DC2, DC4 [All] 32P 'A' Connector

PCM Pin #	W/Color	Circuit Description (32-Pin)	Value at Hot Idle
A1-2	---	Not Used	---
A3	BLU	EVAP Bypass Solenoid	Solenoid Off: 12v, On: 1v
A4	LT GRN	EVAP Vent Shut Control	Solenoid Off: 12v, On: 1v
A5	---	Not Used	---
A6	RED	EVAP Purge Solenoid	Pulse Signals
A7	---	Not Used	---
A8	GRN/RED	HO2S-12 (B1 S2) Heater	Digital Signals: 0-12-0v
A9	---	Not Used	---
A10	BRN/WHT	Service Check Connector	SCS Open: 4.80v
A11	---	Not Used	---
A12	PNK	Immobilizer Indicator Control	Lamp On: 1v, Off: 12v
A13	BLU	IMOEN Immobilizer Signal	Digital Signals
A14	GRN/BLK	A/T: D4 Indicator Light	In D4: 0v, Others: 12v
A15	GRN/YEL	Fuel Pump Relay Control	Relay Off: 12v, On: 1v
A16	---	Not Used	---
A17	BLK/RED	A/C Clutch Relay Control	Relay Off: 12v, On: 1v
A18	GRN/ORN	Malfunction Indicator Light	MIL Off: 12v, On: 1v
A19	BLU	Engine Speed (NEP) Signal	Digital Signals
A20	GRN	Radiator Fan Relay Control	Fan on: 1v, off: 12v
A21	GRN/WHT	K-Line Signal	12v
A22	---	Not Used	---
A23	BLU/RED	HO2S-12 (B1 S2) Signal	0.1-1.1v
A24	BLU/WHT	Starter Switch Signal	Cranking: 9-11v
A25	RED	INOCD Immobilizer Code	Digital Signals
A26	GRN	PSP Switch Signal	Straight: 0v, Turning: 12v
A27	BLU/RED	A/C Switch Signal	A/C Off: 5v, On: 0v
A28	WHT/RED	A/T: Interlock Control Unit	With Brake on: 12v
A29	LT GRN	Fuel Tank Pressure Sensor	Fuel Cap off: 2.5v
A30	GRN/RED	Electric Load Detector	Varies w/Load: 2.5-3.5v
A31	---	Not Used	---
A32	GRN/WHT	Brake Switch Signal	Brake Off: 12v, On: 0v

PCM 104-Pin Connector Graphic

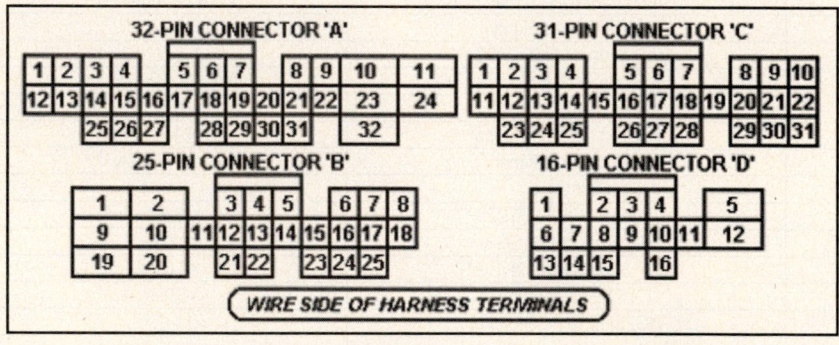

2000-03 GS-R 1.8L VTEC VIN DC2, DC4 [All] 25P 'B' Connector

PCM Pin #	W/Color	Circuit Description (25-Pin)	Value at Hot Idle
B1	YEL/BLK	Main Relay Power (B+)	12-14v
B2	BLK	Power Ground	<0.1v
B3	RED	Injector 2	2.0-3.3 ms
B4	BLU	Injector 3	2.0-3.3 ms
B5	YEL	Injector 4	2.0-3.3 ms
B6-7	---	Not Used	---
B8	WHT/GRN	A/T: Clutch Solenoid LS-	Pulse Signals
B9	YEL/BLK	Main Relay Power (B+)	12-14v
B10	BLK	Power Ground	<0.1v
B11	BRN	Injector 1	2.0-3.3 ms
B12	GRN/YEL	VTEC Solenoid Control	0v, Hi-Speed: 12v
B13	YEL/GRN	Ignition Control Module Signal	Digital Signals: 0-12-0v
B14-15	---	Not Used	---
B16	PNK/BLU	Intake Air Bypass Solenoid	Solenoid Off: 12v, On: 1v
B17	RED/BLU	A/T: Clutch Solenoid LS+	Pulse Signals
B18-19	---	Not Used	---
B20	BRN/BLK	Logic Ground (LG1)	<0.050v
B21	WHT/BLK	Keep Alive Power (VBU)	12-14v
B22	BRN/BLK	Logic Ground (LG2)	<0.050v
B23	BLK/BLU	Idle Air Control Valve	Pulse Signals

2000-03 GS-R 1.8L VTEC VIN DC2, DC4 [All] 31P 'C' Connector

PCM Pin #	W/Color	Circuit Description (31-Pin)	Value at Hot Idle
C1	BRN/BLK	HO2S-11 (B1 S1) Heater	Digital Signals: 0-12-0v
C2	WHT/GRN	Alternator Charging Signal	Headlights off: 0v, on: 12v
C3	RED/BLU	Knock Sensor Signal	No Detonation: 18mv AC
C4	---	Not Used	---
C5	WHT/RED	Alternator 'FR' Signal	Digital Signals: 0-5-0v
C6	---	Not Used	---
C7	GRN/WHT	MAP Sensor Ground	<0.050v
C8	BLU	CKP Sensor Signal	AC Pulse Signals
C9	BLU/YEL	CKP Sensor Ground	<0.050v
C10	BLU/BLK	VTEC Pressure Switch	0v, Hi-Speed: 12v
C11-15	---	Not Used	---
C16	WHT	HO2S-11 (B1 S1) Signal	0.1-1.1v
C17	WHT/YEL	MAP Sensor Signal	0.8-0.9v
C18	GRN/BLU	Sensor Ground (SG2)	<0.050v
C19	YEL/RED	MAP Sensor VREF (VCC1)	4.9-5.1v
C20	GRN	TDC Sensor Signal	AC Pulse Signals
C21	WHT/BLU	TDC Sensor Ground	<0.050v
C22	BLU/RED	CKF Sensor Signal	AC Pulse Signals
C23	ORN	Vehicle Speed Sensor	Moving: pulse signals
C24	---	Not Used	---
C25	RED/YEL	IAT Sensor Signal	At 100ºF: 2-3v
C26	RED/WHT	ECT Sensor Signal	At 180ºF: 0.5-0.6v
C27	RED/BLK	TP Sensor Signal	0.5-0.6v
C28	YEL/BLU	Sensor VREF (VCC2)	4.9-5.1v
C29	YEL	CYP Sensor Signal	AC Pulse Signals
C30	WHT	CYP Sensor Ground	<0.050v
C31	WHT/RED	CKF Sensor Ground	<0.050v

2000-03 GS-R 1.8L VTEC VIN DC2, DC4 [All] 16P 'D' Connector

PCM Pin #	W/Color	Circuit Description (16-Pin)	Value at Hot Idle
D1	YEL	A/T: Lockup Solenoid 'A'	LSA Off: 0v, On: 12v
D2	GRN/WHT	A/T: Shift Solenoid 'B'	SSB Off: 0v, On: 12v
D3	GRN/BLK	A/T: Lockup Solenoid 'B'	LSB Off: 0v, On: 12v
D4	---	Not Used	---
D5	BLK/YEL	A/T: Solenoid Feed (B+)	Full: 12v, Partial: Pulses
D6	GRN/RED	A/T: Gear Position Switch	In 'R': 0v, Others: 12v
D7	BLU/YEL	A/T: Shift Solenoid 'A'	SSA Off: 0v, On: 12v
D8	GRN/BLU	A/T: Gear Position Switch	In D3: 0v, Others: 12v
D9	YEL	A/T: Gear Position Switch	In D4: 0v, Others: 12v
D10	BLU	Countershaft Speed Sensor P	Moving: AC Pulse Signals
D11	RED	Mainshaft Speed Sensor 'P'	AC Pulse Signals
D12	WHT	Mainshaft Speed Sensor 'N'	AC Pulse Signals
D13	GRN/BLK	A/T: Gear Position Switch	In P/N: 0v, Others: 12v
D14	GRN/YEL	A/T: Gear Position Switch	In D2: 0v, Others: 12v
D15	GRN/WHT	A/T: Gear Position Switch	In D1: 0v, Others: 12v
D16	GRN	Countershaft Speed Sensor N	Moving: AC Pulse Signals

PCM 104-Pin Connector Graphic

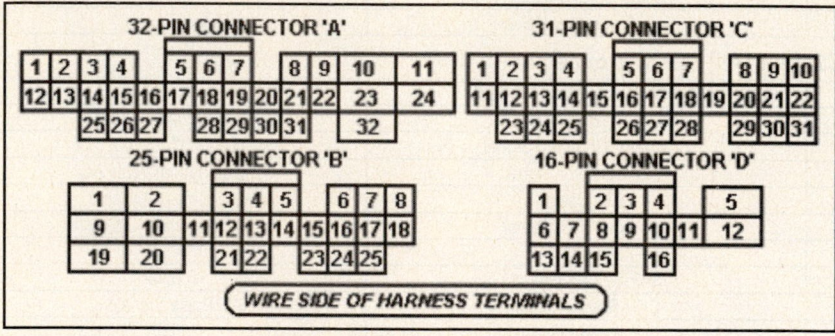

Abbreviation	Color	Abbreviation	Color	Abbreviation	Color
BLK	Black	LT BLU	Lt. Blue	TAN	Tan
BLU	Blue	LT GRN	Lt. Green	VIO	Violet
BRN	Brown	ORN	Orange	WHT	White
GRY	Gray	PNK	Pink	YEL	Yellow
GRN	Green	PPL	Purple		

LEGEND Pin Tables

1990 LS Coupe 2.7L V6 MFI VIN KA3 [All] 18P 'A' Connector

PCM Pin #	W/Color	Circuit Description (18-Pin)	Value at Hot Idle
A1	BRN	Injector 1	2.0-3.3 ms
A2	BLK, BLK	Power Ground	<0.1v
A3	RED	Injector 3	2.0-3.3 ms
A4	BLK, BLK	Power Ground	<0.1v
A5	ORN	Injector 3	2.0-3.3 ms
A6	YEL	Injector 6	2.0-3.3 ms
A7	WHT/BLU	Injector 4	2.0-3.3 ms
A8	BLK/RED	Injector 5	2.0-3.3 ms
A9	---	Not Used	---
A10	WHT	EGR Solenoid Control	Solenoid Off: 12v, On: 1v
A11	BLU/RED	Electronic Air Control Valve	Pulse Signals
A12	GRN/BLK	Fuel Pump Relay Control	Relay Off: 12v, On: 1v
A13	LT GRN	Pressure Regulator Solenoid	Solenoid Off: 12v, On: 1v
A14	RED/BLU	Bypass & Resonator Control	Solenoid Off: 12v, On: 1v
A15	YEL/BLK	Main Relay Power (B+)	12-14v
A16	BRN/WHT	Chassis Ground	<0.050v
A17	YEL/BLU	Keep Alive Power (VBU)	12-14v
A18	BRN/BLK	Chassis Ground	<0.050v

1990 LS Coupe 2.7L V6 MFI VIN KA3 [All] 20P 'B' Connector

PCM Pin #	W/Color	Circuit Description (20-Pin)	Value at Hot Idle
B1	PNK	Air Suction Solenoid	Solenoid Off: 12v, On: 1v
B2	RED/BLU	A/C Pressure Switch	A/C Off: 12v, On: 1v
B3	GRN	Bypass Control Solenoid 'B'	Solenoid Off: 12v, On: 1v
B4	BLU	Cooling Fan Control Unit	N/A
B5	PNK	Information Center Control	N/A
B6	BLU	Malfunction Indicator Lamp	MIL Off: 12v, On: 1v
B7	BLU	M/T: Neutral Switch	In 'N': 0v, Others: 12v
B7	LT GRN	A/T: Shift Position Indicator	In P/N: 0v, Others: 12v
B8	RED/BLU	A/C Clutch Relay Control	Relay Off: 12v, On: 1v
B9	PNK	A/T: Control Unit Signal	---
B9	PNK	M/T: Clutch Switch Signal	Clutch in: 0v, out: 12v
B10	BLU/BLK	Cooling Fan Control Unit	---
B11	YEL/RED	Vehicle Speed Sensor	Moving: pulse signals
B12	RED	PSP Switch Signal	Straight: 0v, Turning: 12v
B13	BLK/WHT	Starter Switch Signal	Cranking: 9-11v
B14	WHT/RED	Alternator Control Signal	Headlights off: 0v, on: 12v
B15	RED/BLU	Igniter Signal	Digital Signals: 0-12-0v
B16	RED/BLU	Igniter Signal	Digital Signals: 0-12-0v
B17	ORN	A/T: Control Unit Signal	---
B18	BRN	Ignition Timing Adjustment	0.4-4.5v
B19	BLU/GRN	CKP Sensor Signal	AC pulse signals
B20	BLU/YEL	CKP Sensor Ground	<0.050v

1990 LS Coupe 2.7L V6 MFI VIN KA3 [All] 16P 'C' Connector

PCM Pin #	W/Color	Circuit Description (16-Pin)	Value at Hot Idle
C1	ORN	CYP Sensor Signal	AC pulse signals
C2	WHT	CYP Sensor Ground	<0.050v
C3	ORN/BLU	TDC Sensor Signal	AC pulse signals
C4	WHT/BLU	TDC Sensor Ground	<0.050v
C5	RED/BLK	IAT Sensor Signal	At 100°F: 2-3v
C6	RED/WHT	ECT Sensor Signal	At 180°: 0.5-0.6v
C7	RED/YEL	Throttle Angle Sensor	0.5-0.6v
C8	WHT/GRN	EGR Valve Lift Sensor	1.1-1.2v
C9	RED	Atmospheric Press. Sensor	2.76-2.96v sea level
C10	RED/BLU	O2S-21 Signal	0.1-1.1v
C11	WHT	MAP Sensor Signal	0.8-0.9v
C12	WHT	O2S-11 Signal	0.1-1.1v
C13	YEL/WHT	Sensor VREF	4.9-5.1v
C14	GRN/WHT	Sensor Ground	<0.050v
C15	YEL/WHT	MAP Sensor VREF	4.9-5.1v
C16	GRN/WHT	MAP Sensor Ground	<0.050v

1990 LS Coupe 2.7L V6 MFI VIN KA3 [All] 5P 'D' Connector

PCM Pin #	W/Color	Circuit Description (5-Pin)	Value at Hot Idle
D1	---	Not Used	---
D2	---	Not Used	---
D3	BLU/GRN	A/T: Control Unit	N/A
D4	BLU/RED	A/T: Control Unit	N/A
D5	BLU/WHT	A/T: Control Unit	N/A

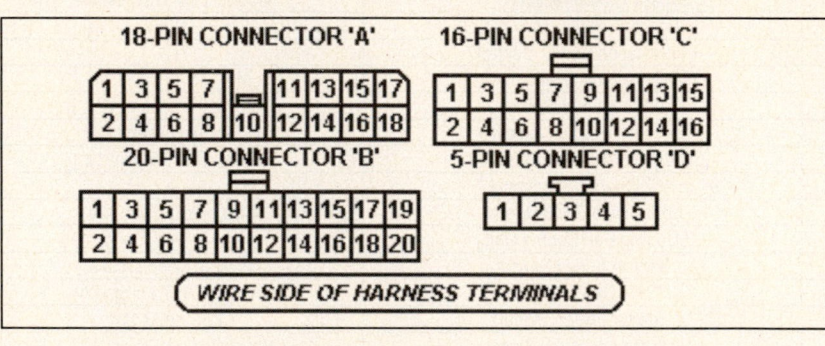

1991-93 L Coupe 3.2L V6 MFI VIN KA8 [All] 26P 'A' Connector

PCM Pin #	W/Color	Circuit Description (26-Pin)	Value at Hot Idle
A1	BRN	Injector 1	2.0-3.3 ms
A3	RED	Injector 3	2.0-3.3 ms
A2	WHT/BLU	Injector 4	2.0-3.3 ms
A4	BLK/RED	Injector 5	2.0-3.3 ms
A5	ORN	Injector 3	2.0-3.3 ms
A6	YEL	Injector 6	2.0-3.3 ms
A7	GRN/BLK	Fuel Pump Relay Control	Relay Off: 12v, On: 1v
A8	GRN	Cooling Fan Control Unit	N/A
A9	BLU/RED	Electronic Air Control Valve	Pulse Signals
A10	GRN/BLU	HO2S-21 (B2 S1) Heater	Heater Off: 12v, On: 1v
A11	WHT	EGR Solenoid Control	Solenoid Off: 12v, On: 1v
A12	GRN/RED	HO2S-11 (B1 S1) Heater	Heater Off: 12v, On: 1v
A13	BLU	Malfunction Indicator Lamp	MIL Off: 12v, On: 1v
A14	RED/BLU	Bypass Low Solenoid	Solenoid Off: 12v, On: 1v
A15	RED/BLU	A/C Clutch Relay Control	Relay Off: 12v, On: 1v
A17	GRY	Air Suction Solenoid	Solenoid Off: 12v, On: 1v
A18	YEL	Bypass High Solenoid	Solenoid Off: 12v, On: 1v
A19	LT GRN	Press. Regulator Solenoid	Solenoid Off: 12v, On: 1v
A20	LT GRN	EVAP Purge Solenoid	Solenoid Off: 12v, On: 1v
A21	PNK	Igniter 1 Signal	Digital Signals: 0-12-0v
A22	BRN	Igniter 2 Signal	Digital Signals: 0-12-0v
A23	BLK	Power Ground	<0.1v
A24	BLK	Power Ground	<0.1v
A25	YEL/BLK	Main Relay Power (B+)	12-14v
A26	BRN/BLK	Chassis Ground	<0.050v

1991-93 L Coupe 3.2L V6 MFI VIN KA8 [All] 16P 'B' Connector

PCM Pin #	W/Color	Circuit Description (16-Pin)	Value at Hot Idle
B1	---	Not Used	---
B2	BRN/BLK	Chassis Ground	<0.050v
B3	RED	Igniter 6 Signal	Digital Signals: 0-12-0v
B4	GRY	Igniter 5 Signal	Digital Signals: 0-12-0v
B5	GRN	PSP Switch Signal	Straight: 0v, Turning: 12v
B6	GRN	Igniter 4 Signal	Digital Signals: 0-12-0v
B7	WHT	M/T: Neutral Switch	In 'N': 0v, Others: 12v
B7	LT GRN	A/T: Shift Position Indicator	In P/N: 0v, Others: 12v
B8	BLU	Igniter 3 Signal	Digital Signals: 0-12-0v
B9	BLU/GRN	CYP Sensor 2 Signal 'P'	AC pulse signals
B10	BLU/YEL	CYP Sensor Ground 2 'N'	<0.050v
B11	ORN/BLU	CYP Sensor 1 Signal 'P'	AC pulse signals
B12	WHT/BLU	CYP Sensor Ground 1 'N'	<0.050v
B13	ORN	CKP Sensor 2 Signal 'P'	AC pulse signals
B14	WHT	CKP Sensor 2 Ground 'N'	<0.050v
B15	ORN/BLU	CKP Sensor 1 Signal 'P'	AC pulse signals
B16	WHT/BLU	CKP Sensor 1 Ground 'N'	<0.050v

1991-93 L Coupe 3.2L V6 MFI VIN KA8 [All] 12P 'C' Connector

PCM Pin #	W/Color	Circuit Description (12-Pin)	Value at Hot Idle
C1	YEL/BLK	Main Relay Power (B+)	12-14v
C2	YEL/RED	Vehicle Speed Sensor	Moving: pulse signals
C3	BLU/BLK	Cooling Fan Control Unit	N/A
C4	BLU	Tachometer Signal	Pulse Signals
C5	RED/BLU	A/C Pressure Switch	A/C Off: 12v, On: 1v
C6	---	Not Used	---
C7	PNK	M/T: Clutch Switch	In 'N': 0v, Others: 12v
C8	---	Not Used	---
C9	PNK	Service Check Connector	SCS open: 4.80v
C10	---	Not Used	---
C11	BLK/WHT	Starter Switch Signal	Cranking: 9-11v

1991-93 L Coupe 3.2L V6 MFI VIN KA8 [All] 22P 'D' Connector

PCM Pin #	W/Color	Circuit Description (22-Pin)	Value at Hot Idle
D1	YEL/BLK	Keep Alive Power (VBU)	12-14v
D2	GRN/WHT	Brake Switch Signal	Brake Off: 12v, On: 0v
D3	WHT	Knock Sensor Signal (Bank 1)	No Detonation: 18mv AC
D4	RED/BLU	Knock Sensor Signal (Bank 2)	No Detonation: 18mv AC
D5-7	---	Not Used	---
D8	BLU/YEL	Ignition Timing Adjustment	0.4-4.5v
D9	WHT/RED	Alternator Control Signal	Headlights off: 0v, on: 12v
D10	---	Not Used	---
D11	RED/BLU	TP Sensor Signal	0.5-0.6v
D12	BLK/WHT	EGR Valve Lift Sensor	1.1-1.2v
D13	RED/WHT	ECT Sensor Signal	At 180°: 0.5-0.6v
D14	WHT	HO21-21 (B2 S1) Signal	0.1-1.1v
D15	RED/YEL	IAT Sensor Signal	At 100°F: 2-3v
D16	RED/BLU	HO2S-11 (B1 S1) Signal	0.1-1.1v
D17	RED	MAP Sensor Signal	0.8-0.9v
D18	---	Not Used	---
D19	YEL/WHT	MAP Sensor VREF	4.9-5.1v
D20	YEL/WHT	Sensor VREF	4.9-5.1v
D21	GRN/WHT	MAP Sensor Ground	<0.050v
D22	GRN/YEL	Sensor Ground	<0.050v

26-PIN CONNECTOR 'A' 12-PIN CONNECTOR 'C'
16-PIN CONNECTOR 'B' 22-PIN CONNECTOR 'D'
WIRE SIDE OF HARNESS TERMINALS

1994-95 L, LS Coupe 3.2L MFI VIN KA8 [A/T] 26P 'A' Connector

PCM Pin #	W/Color	Circuit Description (26-Pin)	Value at Hot Idle
A1	BRN	Injector 1	2.0-3.3 ms
A2	WHT/BLU	Injector 4	2.0-3.3 ms
A3	RED	Injector 2	2.0-3.3 ms
A4	BLK/RED	Injector 5	2.0-3.3 ms
A5	ORN	Injector 3	2.0-3.3 ms
A6	YEL	Injector 6	2.0-3.3 ms
A7	GRN/BLK	Fuel Pump Relay Control	Relay Off: 12v, On: 1v
A8	---	Not Used	---
A9	BLU/RED	Idle Air Control Valve	Pulse Signals
A10	GRN/BLU	HO2S-21 (B2 S1) Heater	Heater Off: 12v, On: 1v
A11	WHT	EGR Solenoid Control	Solenoid Off: 12v, On: 1v
A12	GRN/RED	HO2S-11 (B1 S1) Heater	Heater Off: 12v, On: 1v
A13	BLU	Malfunction Indicator Lamp	MIL Off: 12v, On: 1v
A14	RED/BLU	Bypass Low Solenoid	Solenoid Off: 12v, On: 1v
A15	RED/BLU	A/C Clutch Relay Control	Relay Off: 12v, On: 1v
A17	GRY	Air Suction Solenoid	Solenoid Off: 12v, On: 1v
A18	YEL	Bypass High Solenoid	Solenoid Off: 12v, On: 1v
A19	LT GRN	Press. Regulator Solenoid	Solenoid Off: 12v, On: 1v
A20	LT GRN	EVAP Purge Solenoid	Solenoid Off: 12v, On: 1v
A21	PNK	Igniter 1 Signal	Digital Signals: 0-12-0v
A22	BRN	Igniter 2 Signal	Digital Signals: 0-12-0v
A23	BLK	Power Ground	<0.1v
A24	BLK	Power Ground	<0.1v
A25	YEL/BLK	Main Relay Power (B+)	12-14v
A26	BRN/BLK	Chassis Ground	<0.050v

1994-95 L, LS Coupe 3.2L MFI VIN KA8 [A/T] 16P 'B' Connector

PCM Pin #	W/Color	Circuit Description (16-Pin)	Value at Hot Idle
B1	---	Not Used	---
B2	BRN/BLK	Chassis Ground	<0.050v
B3	RED	Igniter 6 Signal	Digital Signals: 0-12-0v
B4	GRY	Igniter 5 Signal	Digital Signals: 0-12-0v
B5	GRN	PSP Switch Signal	Straight: 0v, Turning: 12v
B6	GRN	Igniter 4 Signal	Digital Signals: 0-12-0v
B7	LT GRN	Park Neutral Switch Signal	In P/N: 0v, Others: 12v
B8	BLU	Igniter 3 Signal	Digital Signals: 0-12-0v
B9	BLU/GRN	CYP Sensor 2 Signal 'P'	AC pulse signals
B10	BLU/YEL	CYP Sensor 2 Ground 'N'	<0.050v
B11	ORN/BLU	CYP Sensor 1 Signal 'P'	AC pulse signals
B12	WHT/BLU	CYP Sensor 1 Ground 'N'	<0.050v
B13	ORN	CKP Sensor 2 Signal 'P'	AC pulse signals
B14	WHT	CKP Sensor 2 Ground 'N'	<0.050v
B15	ORN/BLU	CKP Sensor 1 Signal 'P'	AC pulse signals
B16	WHT/BLU	CKP Sensor 1 Ground 'N'	<0.050v

1994-95 L, LS Coupe 3.2L MFI VIN KA8 [A/T] 12P 'C' Connector

PCM Pin #	W/Color	Circuit Description (12-Pin)	Value at Hot Idle
C1	YEL/BLK	Main Relay Power (B+)	12-14v
C2	YEL/RED	Vehicle Speed Sensor	Moving: pulse signals
C3	BLU/BLK	Cooling Fan Switch	A/C Off: 12v, On: 1v
C4	BLU	Tachometer Signal	Pulse Signals
C5	RED/BLU	A/C Pressure Switch	Relay Off: 12v, On: 1v
C6	GRN/BLK	TCM Signal	N/A
C8	GRY/BLU	TCM Signal	N/A
C9	WHT	Service Check Connector	SCS Open: 4.80v
C11	BLK/WHT	Starter Switch Signal	Cranking: 9-11v
C12	GRY/YEL	TCM Signal	N/A

1994-95 L, LS Coupe 3.2L MFI VIN KA8 [A/T] 22P 'D' Connector

PCM Pin #	W/Color	Circuit Description (22-Pin)	Value at Hot Idle
D1	YEL/BLU	Keep Alive Power (VBU)	12-14v
D2	GRN/WHT	Brake Switch Signal	Brake Off: 12v, On: 0v
D3	WHT	Knock Sensor Signal (Bank 1)	No Detonation: 18mv AC
D4	RED/BLU	Knock Sensor Signal (Bank 2)	No Detonation: 18mv AC
D5	ORN/RED	TCM Signal	N/A
D6-7	---	Not Used	---
D8	BLU/YEL	Ignition Timing Adjustment	0.4-4.5v
D9	WHT/RED	Alternator Control Signal	Headlights off: 0v, on: 12v
D10	---	Not Used	---
D11	RED/BLU	TP Sensor Signal	0.5-0.6v
D12	BLK/WHT	EGR Valve Lift Sensor	1.1-1.2v
D13	RED/WHT	ECT Sensor Signal	At 180°: 0.5-0.6v
D14	WHT	HO2S-11 (B1 S1) Signal	0.1-1.1v
D15	RED/YEL	IAT Sensor Signal	At 100°F: 2-3v
D16	RED/BLU	HO21-21 (B2 S1) Signal	0.1-1.1v
D17	RED	MAP Sensor Signal	0.8-0.9v
D18	BLU	TCM Signal	N/A
D19	YEL/WHT	MAP Sensor VREF	4.9-5.1v
D20	YEL/WHT	Sensor VREF	4.9-5.1v
D21	GRN/WHT	MAP Sensor Ground	<0.050v
D22	GRN/WHT	Sensor Ground	<0.050v

1994-95 L, LS Coupe 3.2L MFI VIN KA8 [M/T] 26P 'A' Connector

PCM Pin #	W/Color	Circuit Description (26-Pin)	Value at Hot Idle
A1	BRN	Injector 1	2.0-3.3 ms
A3	RED	Injector 2	2.0-3.3 ms
A2	WHT/BLU	Injector 4	2.0-3.3 ms
A4	BLK/RED	Injector 5	2.0-3.3 ms
A5	ORN	Injector 3	2.0-3.3 ms
A6	YEL	Injector 6	2.0-3.3 ms
A7	GRN/BLK	Fuel Pump Relay Control	Relay Off: 12v, On: 1v
A8	RED/WHT	M/T: Reverse Lockout Relay	Relay Off: 12v, On: 1v
A9	BLU/RED	Idle Air Control Valve	Pulse Signals
A10	GRN/BLU	HO2S-21 (B2 S1) Heater	Heater Off: 12v, On: 1v
A11	WHT	EGR Solenoid Control	Solenoid Off: 12v, On: 1v
A12	GRN/RED	HO2S-11 (B1 S1) Heater	Heater Off: 12v, On: 1v
A13	BLU	Malfunction Indicator Lamp	MIL Off: 12v, On: 1v
A14	RED/BLU	Bypass Low Solenoid	Solenoid Off: 12v, On: 1v
A15	RED/BLU	A/C Clutch Relay Control	Relay Off: 12v, On: 1v
A16	---	Not Used	---
A17	GRY	Air Suction Solenoid	Solenoid Off: 12v, On: 1v
A18	YEL	Bypass High Solenoid	Solenoid Off: 12v, On: 1v
A19	LT GRN	Press. Regulator Solenoid	Solenoid Off: 12v, On: 1v
A20	LT GRN	EVAP Purge Solenoid	Solenoid Off: 12v, On: 1v
A21	PNK	Igniter 1 Signal	Digital Signals: 0-12-0v
A22	BRN	Igniter 2 Signal	Digital Signals: 0-12-0v
A23	BLK	Power Ground	<0.1v
A24	BLK	Power Ground	<0.1v
A25	YEL/BLK	Main Relay Power (B+)	12-14v
A26	BRN/BLK	Chassis Ground	<0.050v

1994-95 L, LS Coupe 3.2L MFI VIN KA8 [M/T] 16P 'B' Connector

PCM Pin #	W/Color	Circuit Description (16-Pin)	Value at Hot Idle
B1	---	Not Used	---
B2	BRN/BLK	Chassis Ground	<0.050v
B3	RED	Igniter 6 Signal	Digital Signals: 0-12-0v
B4	GRY	Igniter 5 Signal	Digital Signals: 0-12-0v
B5	GRN	PSP Switch Signal	Straight: 0v, Turning: 12v
B6	GRN	Igniter 4 Signal	Digital Signals: 0-12-0v
B7	WHT	M/T: Neutral Switch	In 'N': 0v, Others: 12v
B8	BLU	Igniter 3 Signal	Digital Signals: 0-12-0v
B9	BLU/GRN	CYP Sensor 2 Signal 'P'	AC pulse signals
B10	BLU/YEL	CYP Sensor 2 Ground 'N'	<0.050v
B11	ORN/BLU	CYP Sensor 1 Signal 'P'	AC pulse signals
B12	WHT/BLU	CYP Sensor 1 Ground 'N'	<0.050v
B13	ORN	CKP Sensor 2 Signal 'P'	AC pulse signals
B14	WHT	CKP Sensor 2 Ground 'N'	<0.050v
B15	ORN/BLU	CKP Sensor 1 Signal 'P'	AC pulse signals
B16	WHT/BLU	CKP Sensor 1 Ground 'N'	<0.050v

1994-95 L, LS Coupe 3.2L MFI VIN KA8 [M/T] 12P 'C' Connector

PCM Pin #	W/Color	Circuit Description (12-Pin)	Value at Hot Idle
C1	YEL/BLK	Main Relay Power (B+)	12-14v
C2	YEL/RED	Vehicle Speed Sensor	Moving: pulse signals
C3	BLU/BLK	Cooling Fan Switch	A/C Off: 12v, On: 1v
C4	BLU	Tachometer Signal	Pulse Signals
C5	RED/BLU	A/C Pressure Switch	Relay Off: 12v, On: 1v
C6	---	Not Used	---
C7	PNK	M/T: Clutch Switch	In 'N': 0v, Others: 12v
C8	---	Not Used	---
C9	WHT	Service Check Connector	SCS Open: 4.80v
C10	---	Not Used	---
C11	BLK/WHT	Starter Switch Signal	Cranking: 9-11v
C12	---	Not Used	---

1994-95 L, LS Coupe 3.2L MFI VIN KA8 [M/T] 22P 'D' Connector

PCM Pin #	W/Color	Circuit Description (22-Pin)	Value at Hot Idle
D1	YEL/BLU	Keep Alive Power (VBU)	12-14v
D2	GRN/WHT	Brake Switch Signal	Brake Off: 12v, On: 0v
D3	WHT	Knock Sensor Signal (Bank 1)	No Detonation: 18mv AC
D4	RED/BLU	Knock Sensor Signal (Bank 2)	No Detonation: 18mv AC
D5-7	---	Not Used	---
D8	BLU/YEL	Ignition Timing Adjustment	0.4-4.5v
D9	WHT/RED	Alternator Control Signal	Headlights off: 0v, on: 12v
D10	RED/BLU	Knock Sensor Signal (Bank 2)	No Detonation: 18mv AC
D11	RED/BLU	TP Sensor Signal	0.5-0.6v
D12	BLK/WHT	EGR Valve Lift Sensor	1.1-1.2v
D13	RED/WHT	ECT Sensor Signal	At 180°: 0.5-0.6v
D14	WHT	HO2S-11 (B1 S1) Signal	0.1-1.1v
D15	RED/YEL	IAT Sensor Signal	At 100°F: 2-3v
D16	RED/BLU	HO21-21 (B2 S1) Signal	0.1-1.1v
D17	RED	MAP Sensor Signal	0.8-0.9v
D18	---	Not Used	---
D19	YEL/WHT	MAP Sensor VREF	4.9-5.1v
D20	YEL/WHT	Sensor VREF	4.9-5.1v
D21	GRN/WHT	MAP Sensor Ground	<0.050v
D22	GRN/WHT	Sensor Ground	<0.050v

26-PIN CONNECTOR 'A'
1 3 5 7 9 11 13 15 17 19 21 23 25
2 4 6 8 10 12 14 16 18 20 22 24 26

12-PIN CONNECTOR 'C'
1 3 5 7 9 11
2 4 6 8 10 12

16-PIN CONNECTOR 'B'
1 3 5 7 9 11 13 15
2 4 6 8 10 12 14 16

22-PIN CONNECTOR 'D'
1 3 5 7 9 11 13 15 17 19 21
2 4 6 8 10 12 14 16 18 20 22

WIRE SIDE OF HARNESS TERMINALS

1990 L, LS Sedan 2.7L V6 MFI VIN KA4 [All] 18P 'A' Connector

PCM Pin #	W/Color	Circuit Description (18-Pin)	Value at Hot Idle
A1	BRN	Injector 1	2.0-3.3 ms
A2	BLK	Power Ground	<0.1v
A3	RED	Injector 2	2.0-3.3 ms
A4	BLK	Power Ground	<0.1v
A5	ORN	Injector 3	2.0-3.3 ms
A6	YEL	Injector 6	2.0-3.3 ms
A7	WHT/BLU	Injector 4	2.0-3.3 ms
A8	BLK/RED	Injector 5	2.0-3.3 ms
A10	WHT	EGR Solenoid Control	Solenoid Off: 12v, On: 1v
A11	BLU/RED	Electronic Air Control Valve	Pulse Signals
A12	GRN/BLK	Fuel Pump Relay Control	Relay Off: 12v, On: 1v
A13	LT GRN	Pressure Regulator Solenoid	Solenoid Off: 12v, On: 1v
A14	RED/BLU	Bypass & Resonator Control	Solenoid Off: 12v, On: 1v
A15	YEL/BLK	Main Relay Power (B+)	12-14v
A16	BRN/WHT	Chassis Ground	<0.050v
A17	YEL/BLU	Keep Alive Power (VBU)	12-14v
A18	BRN/WHT	Chassis Ground	<0.050v

1990 L, LS Sedan 2.7L V6 MFI VIN KA4 [All] 20P 'B' Connector

PCM Pin #	W/Color	Circuit Description (20-Pin)	Value at Hot Idle
B1	PNK	Air Suction Solenoid	Solenoid Off: 12v, On: 1v
B2	RED/BLU	A/C Pressure Switch	A/C Off: 12v, On: 1v
B3	GRN	Bypass Control Solenoid 'B'	Solenoid Off: 12v, On: 1v
B4	BLU	Cooling Fan Control Unit	N/A
B5	PNK	Information Center Control	N/A
B6	BLU	Malfunction Indicator Lamp	MIL Off: 12v, On: 1v
B7	BLU	M/T: Neutral Switch	In 'N': 0v, Others: 12v
B7	LT GRN	A/T: Shift Position Indicator	In P/N: 0v, Others: 12v
B8	RED/BLU	A/C Clutch Relay Control	Relay Off: 12v, On: 1v
B9	PNK	M/T: Clutch Switch Signal	Clutch in: 0v, out: 12v
B9	PNK	A/T: Control Unit Signal	N/A
B10	BLU/BLK	Cooling Fan Control Unit	N/A
B11	YEL/RED	Vehicle Speed Sensor	Moving: pulse signals
B12	RED	PSP Switch Signal	Straight: 0v, Turning: 12v
B13	BLK/WHT	Starter Switch Signal	Cranking: 9-11v
B14	WHT/RED	Alternator Control Signal	Lights Off: 12v, On: 1v
B15	RED/BLU	Igniter Signal	Digital Signals: 0-12-0v
B16	RED/BLU	Igniter Signal	Digital Signals: 0-12-0v
B17	ORN	A/T: Control Unit	----
B18	BRN	Ignition Timing Adjustment	0.4-4.5v
B19	BLU/GRN	CKP Sensor Signal	AC pulse signals
B20	BLU/YEL	CKP Sensor Ground	<0.050v

1990 L, LS Sedan 2.7L V6 MFI VIN KA4 [All] 16P 'C' Connector

PCM Pin #	W/Color	Circuit Description (16-Pin)	Value at Hot Idle
C1	ORN	CYP Sensor Signal	AC pulse signals
C2	WHT	CYP Sensor Ground	<0.050v
C3	ORN/BLU	TDC Sensor Signal	AC pulse signals
C4	WHT/BLU	TDC Sensor Ground	<0.050v
C5	RED/BLK	IAT Sensor Signal	At 100°F: 2-3v
C6	RED/WHT	ECT Sensor Signal	At 180°: 0.5-0.6v
C7	RED/YEL	Throttle Angle Sensor	0.5-0.6v
C8	WHT/GRN	EGR Valve Lift Sensor	1.1-1.2v
C9	RED	Atmospheric Press. Sensor	2.76-2.96v sea level
C10	RED/BLU	O2S-21 Signal	0.1-1.1v
C11	WHT	MAP Sensor Signal	0.8-0.9v
C12	WHT	O2S-11 Signal	0.1-1.1v
C13	YEL/WHT	Sensor VREF	4.9-5.1v
C14	GRN/WHT	Sensor Ground	<0.050v
C15	YEL/WHT	MAP Sensor VREF	4.9-5.1v
C16	GRN/WHT	MAP Sensor Ground	<0.050v

1990 L, LS Sedan 2.7L V6 MFI VIN KA4 [All] 5P 'D' Connector

PCM Pin #	W/Color	Circuit Description (5-Pin)	Value at Hot Idle
D1	---	Not Used	---
D2	---	Not Used	---
D3	BLU/GRN	A/T: Control Unit	N/A
D4	BLU/RED	A/T: Control Unit	N/A
D5	BLU/WHT	A/T: Control Unit	N/A

```
18-PIN CONNECTOR 'A'        16-PIN CONNECTOR 'C'
 1  3  5  7  □  11 13 15 17   1  3  5  7  9 11 13 15
 2  4  6  8 10 12 14 16 18    2  4  6  8 10 12 14 16
20-PIN CONNECTOR 'B'        5-PIN CONNECTOR 'D'
 1  3  5  7  9 11 13 15 17 19   1  2  3  4  5
 2  4  6  8 10 12 14 16 18 20

       WIRE SIDE OF HARNESS TERMINALS
```

1991-93 L, LS Sedan 3.2L MFI VIN KA7 [A/T] 26P 'A' Connector

PCM Pin #	W/Color	Circuit Description (26-Pin)	Value at Hot Idle
A1	BRN	Injector 1	2.0-3.3 ms
A2	WHT/BLU	Injector 4	2.0-3.3 ms
A3	RED	Injector 2	2.0-3.3 ms
A4	BLK/RED	Injector 5	2.0-3.3 ms
A5	ORN	Injector 3	2.0-3.3 ms
A6	YEL	Injector 6	2.0-3.3 ms
A7	GRN/BLK	Fuel Pump Relay Control	Relay Off: 12v, On: 1v
A8	GRN	Cooling Fan Control Unit	N/A
A9	BLU/RED	Idle Air Control Valve	Pulse Signals
A10	GRN/BLU	HO2S-21 (B2 S1) Heater	Heater Off: 12v, On: 1v
A11	WHT	EGR Solenoid Control	Solenoid Off: 12v, On: 1v
A12	GRN/RED	HO2S-11 (B1 S1) Heater	Heater Off: 12v, On: 1v
A13	BLU	Malfunction Indicator Lamp	MIL Off: 12v, On: 1v
A14	RED/BLU	Bypass Low Solenoid	Solenoid Off: 12v, On: 1v
A15	RED/BLU	A/C Clutch Relay Control	Relay Off: 12v, On: 1v
A17	GRY	Air Suction Solenoid	Solenoid Off: 12v, On: 1v
A18	YEL	Bypass High Solenoid	Solenoid Off: 12v, On: 1v
A19	LT GRN	Press. Regulator Solenoid	Solenoid Off: 12v, On: 1v
A20	LT GRN	EVAP Purge Solenoid	Solenoid Off: 12v, On: 1v
A21	PNK	Igniter 1 Signal	Digital Signals: 0-12-0v
A22	BRN	Igniter 2 Signal	Digital Signals: 0-12-0v
A23	BLK	Power Ground	<0.1v
A24	BLK	Power Ground	<0.1v
A25	YEL/BLK	Main Relay Power (B+)	12-14v
A26	BRN/BLK	Chassis Ground	<0.050v

1991-93 L, LS Sedan 3.2L MFI VIN KA7 [A/T] 16P 'B' Connector

PCM Pin #	W/Color	Circuit Description (16-Pin)	Value at Hot Idle
B1	---	Not Used	---
B2	BRN/BLK	Chassis Ground	<0.050v
B3	RED	Igniter 6 Signal	Digital Signals: 0-12-0v
B4	GRY	Igniter 5 Signal	Digital Signals: 0-12-0v
B5	GRN	PSP Switch Signal	Straight: 0v, Turning: 12v
B6	GRN	Igniter 4 Signal	Digital Signals: 0-12-0v
B8	BLU	Igniter 3 Signal	Digital Signals: 0-12-0v
B7	LT GRN	A/T: Park Neutral Indicator	In P/N: 0v, Others: 12v
B9	BLU/GRN	CYP Sensor 2 Signal 'P'	AC pulse signals
B10	BLU/YEL	CYP Sensor 2 Ground 'N'	<0.050v
B11	ORN/BLU	CYP Sensor 1 Signal 'P'	AC pulse signals
B12	WHT/BLU	CYP Sensor 1 Ground 'N'	<0.050v
B13	ORN	CKP Sensor 2 Signal 'P'	AC pulse signals
B14	WHT	CKP Sensor 2 Ground 'N'	<0.050v
B15	ORN/BLU	CKP Sensor 1 Signal 'P'	AC pulse signals
B16	WHT/BLU	CKP Sensor 1 Ground 'N'	<0.050v

1991-93 L, LS Sedan 3.2L MFI VIN KA7 [A/T] 12P 'C' Connector

PCM Pin #	W/Color	Circuit Description (12-Pin)	Value at Hot Idle
C1	YEL/BLK	Main Relay Power (B+)	12-14v
C2	YEL/RED	Vehicle Speed Sensor	Moving: pulse signals
C3	BLU/BLK	Cooling Fan Control Unit	---
C4	BLU	Tachometer Signal	Pulse Signals
C5	RED/BLU	A/C Pressure Switch	A/C Off: 12v, On: 1v
C6-8	---	Not Used	---
C9	WHT	Service Check Connector	SCS Open: 4.80v
C10	---	Not Used	---
C11	BLK/WHT	Starter Switch Signal	Cranking: 9-11v
C12	---	Not Used	---

1991-93 L, LS Sedan 3.2L MFI VIN KA7 [A/T] 22P 'D' Connector

PCM Pin #	W/Color	Circuit Description (22-Pin)	Value at Hot Idle
D1	YEL/BLK	Keep Alive Power (VBU)	12-14v
D2	GRN/WHT	Brake Switch Signal	Brake Off: 12v, On: 0v
D3	WHT	Knock Sensor Signal (Bank 1)	No Detonation: 18mv AC
D4	RED/BLU	Knock Sensor Signal (Bank 2)	No Detonation: 18mv AC
D5-7	---	Not Used	---
D8	BLU/YEL	Ignition Timing Adjustment	0.4-4.5v
D9	WHT/RED	Alternator Control Signal	Headlights off: 0v, on: 12v
D10	---	Not Used	---
D11	RED/BLU	TP Sensor Signal	0.5-0.6v
D12	BLK/WHT	EGR Valve Lift Sensor	1.1-1.2v
D13	RED/WHT	ECT Sensor Signal	At 180°: 0.5-0.6v
D14	WHT	HO2S-11 (B1 S1) Signal	0.1-1.1v
D15	RED/YEL	IAT Sensor Signal	At 100°F: 2-3v
D16	RED/BLU	HO21-21 (B2 S1) Signal	0.1-1.1v
D17	RED	MAP Sensor Signal	0.8-0.9v
D18	---	Not Used	---
D19	YEL/WHT	MAP Sensor VREF	4.9-5.1v
D20	YEL/WHT	Sensor VREF	4.9-5.1v
D21	GRN/WHT	MAP Sensor Ground	<0.050v
D22	GRN/YEL	Sensor Ground	<0.050v

```
         26-PIN CONNECTOR 'A'          12-PIN CONNECTOR 'C'
       ┌─────────────────────────┐    ┌─────────────────────┐
       │1│3│5│7│9│11│13│15│17│19│21│23│25│   │1│3│5│7│9│11│
       │2│4│6│8│10│12│14│16│18│20│22│24│26│   │2│4│6│8│10│12│
       └─────────────────────────┘    └─────────────────────┘
  16-PIN CONNECTOR 'B'            22-PIN CONNECTOR 'D'
  ┌───────────────────┐    ┌─────────────────────────────┐
  │1│3│5│7│9│11│13│15│    │1│3│5│7│9│11│13│15│17│19│21│
  │2│4│6│8│10│12│14│16│    │2│4│6│8│10│12│14│16│18│20│22│
  └───────────────────┘    └─────────────────────────────┘
        ( WIRE SIDE OF HARNESS TERMINALS )
```

1991-93 L, LS Sedan 3.2L MFI VIN KA7 [M/T] 26P 'A' Connector

PCM Pin #	W/Color	Circuit Description (26-Pin)	Value at Hot Idle
A1	BRN	Injector 1	2.0-3.3 ms
A2	WHT/BLU	Injector 4	2.0-3.3 ms
A3	RED	Injector 2	2.0-3.3 ms
A4	BLK/RED	Injector 5	2.0-3.3 ms
A5	ORN	Injector 3	2.0-3.3 ms
A6	YEL	Injector 6	2.0-3.3 ms
A7	GRN/BLK	Fuel Pump Relay Control	Relay Off: 12v, On: 1v
A8	GRN	Cooling Fan Control Unit	---
A9	BLU/RED	Electronic Air Control Valve	Pulse Signals
A10	GRN/BLU	HO2S-21 (B2 S1) Heater	Heater Off: 12v, On: 1v
A11	WHT	EGR Solenoid Control	Solenoid Off: 12v, On: 1v
A12	GRN/RED	HO2S-11 (B1 S1) Heater	Heater Off: 12v, On: 1v
A13	BLU	Malfunction Indicator Lamp	MIL Off: 12v, On: 1v
A14	RED/BLU	Bypass Low Solenoid	Solenoid Off: 12v, On: 1v
A15	RED/BLU	A/C Clutch Relay Control	Relay Off: 12v, On: 1v
A17	GRY	Air Suction Solenoid	Solenoid Off: 12v, On: 1v
A18	YEL	Bypass High Solenoid	Solenoid Off: 12v, On: 1v
A19	LT GRN	Pressure Regulator Solenoid	Solenoid Off: 12v, On: 1v
A20	LT GRN	EVAP Purge Solenoid	Solenoid Off: 12v, On: 1v
A21	PNK	Igniter 1 Signal	Digital Signals: 0-12-0v
A22	BLK	Igniter 2 Signal	Digital Signals: 0-12-0v
A23	BLK	Power Ground	<0.1v
A24	BLK	Power Ground	<0.1v
A25	YEL/BLK	Main Relay Power (B+)	12-14v
A26	BRN/BLK	Chassis Ground	<0.050v

1991-93 L, LS Sedan 3.2L MFI VIN KA7 [M/T] 16P 'B' Connector

PCM Pin #	W/Color	Circuit Description (16-Pin)	Value at Hot Idle
B1	---	Not Used	---
B2	BRN/BLK	Chassis Ground	<0.050v
B3	RED	Igniter 6 Signal	Digital Signals: 0-12-0v
B4	GRY	Igniter 5 Signal	Digital Signals: 0-12-0v
B5	GRN	PSP Switch Signal	Straight: 0v, Turning: 12v
B6	GRN	Igniter 4 Signal	Digital Signals: 0-12-0v
B7	WHT	M/T: Neutral Switch	In 'N': 0v, Others: 12v
B8	BLU	Igniter 3 Signal	Digital Signals: 0-12-0v
B9	BLU/GRN	CYP Sensor 2 Signal 'P'	AC pulse signals
B10	BLU/YEL	CYP Sensor Ground 2 'N'	<0.050v
B11	ORN/BLU	CYP Sensor 1 Signal 'P'	AC pulse signals
B12	WHT/BLU	CYP Sensor Ground 1 'N'	<0.050v
B13	ORN	CKP Sensor 2 Signal 'P'	AC pulse signals
B14	WHT	CKP Sensor 2 Ground 'N'	<0.050v
B15	ORN/BLU	CKP Sensor 1 Signal 'P'	AC pulse signals
B16	WHT/BLU	CKP Sensor 1 Ground 'N'	<0.050v

1991-93 L, LS Sedan 3.2L MFI VIN KA7 [M/T] 12P 'C' Connector

PCM Pin #	W/Color	Circuit Description (12-Pin)	Value at Hot Idle
C1	YEL/BLK	Main Relay Power (B+)	12-14v
C2	YEL/RED	Vehicle Speed Sensor	Moving: pulse signals
C3	BLU/BLK	Cooling Fan Control Unit	N/A
C4	BLU	Tachometer Signal	Pulse Signals
C5	RED/BLU	A/C Pressure Switch	A/C Off: 12v, On: 1v
C6	---	Not Used	---
C7	PNK	M/T: Clutch Switch	In 'N': 0v, Others: 12v
C8	---	Not Used	---
C9	PNK	Service Check Connector	SCS Open: 4.80v
C10	---	Not Used	---
C11	BLK/WHT	Starter Switch Signal	Cranking: 9-11v
C12	---	Not Used	---

1991-93 L, LS Sedan 3.2L MFI VIN KA7 [M/T] 22P 'D' Connector

PCM Pin #	W/Color	Circuit Description (22-Pin)	Value at Hot Idle
D1	YEL/BLK	Keep Alive Power (VBU)	12-14v
D2	GRN/WHT	Brake Switch Signal	Brake Off: 12v, On: 0v
D3	WHT	Knock Sensor Signal (Bank 1)	No Detonation: 18mv AC
D4	RED/BLU	Knock Sensor Signal (Bank 2)	No Detonation: 18mv AC
D5-7	---	Not Used	---
D8	BLU/YEL	Ignition Timing Adjustment	0.4-4.5v
D9	WHT/RED	Alternator Control Signal	Headlights off: 0v, on: 12v
D10	---	Not Used	---
D11	RED/BLU	TP Sensor Signal	0.5-0.6v
D12	BLK/WHT	EGR Valve Lift Sensor	1.1-1.2v
D13	RED/WHT	ECT Sensor Signal	At 180°: 0.5-0.6v
D14	WHT	HO2S-11 (B1 S1) Signal	0.1-1.1v
D15	RED/YEL	IAT Sensor Signal	At 100ºF: 2-3v
D16	RED/BLU	HO21-21 (B2 S1) Signal	0.1-1.1v
D17	RED	MAP Sensor Signal	0.8-0.9v
D18	---	Not Used	---
D19	YEL/WHT	MAP Sensor VREF	4.9-5.1v
D20	YEL/WHT	Sensor VREF	4.9-5.1v
D21	GRN/WHT	MAP Sensor Ground	<0.050v
D22	GRN/YEL	Sensor Ground	<0.050v

26-PIN CONNECTOR 'A'   12-PIN CONNECTOR 'C'

16-PIN CONNECTOR 'B'   22-PIN CONNECTOR 'D'

WIRE SIDE OF HARNESS TERMINALS

1994-95 L, LS Sedan 3.2L MFI VIN KA7 [A/T] 26P 'A' Connector

PCM Pin #	W/Color	Circuit Description (26-Pin)	Value at Hot Idle
A1	BRN	Injector 1	2.0-3.3 ms
A2	WHT/BLU	Injector 4	2.0-3.3 ms
A3	RED	Injector 2	2.0-3.3 ms
A4	BLK/RED	Injector 5	2.0-3.3 ms
A5	ORN	Injector 3	2.0-3.3 ms
A6	YEL	Injector 6	2.0-3.3 ms
A7	GRN/BLK	Fuel Pump Relay Control	Relay Off: 12v, On: 1v
A9	BLU/RED	Idle Air Control Valve	Pulse Signals
A10	GRN/BLU	HO2S-21 (B2 S1) Heater	Heater Off: 12v, On: 1v
A11	WHT	EGR Solenoid Control	Solenoid Off: 12v, On: 1v
A12	GRN/RED	HO2S-11 (B1 S1) Heater	Heater Off: 12v, On: 1v
A13	BLU	Malfunction Indicator Lamp	MIL Off: 12v, On: 1v
A14	RED/BLU	Bypass Low Solenoid	Solenoid Off: 12v, On: 1v
A15	RED/BLU	A/C Clutch Relay Control	Relay Off: 12v, On: 1v
A17	GRY	Air Suction Solenoid	Solenoid Off: 12v, On: 1v
A18	YEL	Bypass High Solenoid	Solenoid Off: 12v, On: 1v
A19	LT GRN	Pressure Regulator Solenoid	Solenoid Off: 12v, On: 1v
A20	LT GRN	EVAP Purge Solenoid	Solenoid Off: 12v, On: 1v
A21	PNK	Igniter 1 Signal	Digital Signals: 0-12-0v
A22	BRN	Igniter 2 Signal	Digital Signals: 0-12-0v
A23	BLK	Power Ground	<0.1v
A24	BLK	Power Ground	<0.1v
A25	YEL/BLK	Main Relay Power (B+)	12-14v
A26	BRN/BLK	Chassis Ground	<0.050v

1994-95 L, LS Sedan 3.2L MFI VIN KA7 [A/T] 16P 'B' Connector

PCM Pin #	W/Color	Circuit Description (16-Pin)	Value at Hot Idle
B1	---	Not Used	---
B2	BRN/BLK	Chassis Ground	<0.050v
B3	RED	Igniter 6 Signal	Digital Signals: 0-12-0v
B4	GRY	Igniter 5 Signal	Digital Signals: 0-12-0v
B5	GRN	PSP Switch Signal	Straight: 0v, Turning: 12v
B6	BLU	Igniter 4 Signal	Digital Signals: 0-12-0v
B7	LT GRN	A/T: Neutral Switch	In P/N: 0v, Others: 12v
B8	BLU	Igniter 3 Signal	Digital Signals: 0-12-0v
B9	BLU/GRN	CYP Sensor 2 Signal 'P'	AC pulse signals
B10	BLU/YEL	CYP Sensor 2 Ground 'N'	<0.050v
B11	ORN/BLU	CYP Sensor 1 Signal 'P'	AC pulse signals
B12	WHT/BLU	CYP Sensor 1 Ground 'N'	<0.050v
B13	ORN	CKP Sensor 2 Signal 'P'	AC pulse signals
B14	WHT	CKP Sensor 2 Ground 'N'	<0.050v
B15	ORN/BLU	CKP Sensor 1 Signal 'P'	AC pulse signals
B16	WHT/BLU	CKP Sensor 1 Ground 'N'	<0.050v

1994-95 L, LS Sedan 3.2L MFI VIN KA7 [A/T] 12P 'C' Connector

PCM Pin #	W/Color	Circuit Description (12-Pin)	Value at Hot Idle
C1	YEL/BLK	Main Relay Power (B+)	12-14v
C2	YEL/RED	Vehicle Speed Sensor	Moving: pulse signals
C3	BLU/BLK	Cooling Fan Switch	A/C Off: 12v, On: 1v
C4	BLU	Tachometer Signal	Pulse Signals
C5	RED/BLU	A/C Pressure Switch	Relay Off: 12v, On: 1v
C6	GRN/BLK	TCM Signal	---
C7	---	Not Used	---
C8	GRY/BLU	TCM Signal	---
C9	WHT	Service Check Connector	SCS Open: 4.80v
C10	---	Not Used	---
C11	BLK/WHT	Starter Switch Signal	Cranking: 9-11v
C12	GRY/YEL	TCM Signal	---

1994-95 L, LS Sedan 3.2L MFI VIN KA7 [A/T] 22P 'D' Connector

PCM Pin #	W/Color	Circuit Description (22-Pin)	Value at Hot Idle
D1	YEL/BLU	Keep Alive Power (VBU)	12-14v
D2	GRN/WHT	Brake Switch Signal	Brake Off: 12v, On: 0v
D3	WHT	Knock Sensor Signal (Bank 1)	No Detonation: 18mv AC
D4	RED/BLU	Knock Sensor Signal (Bank 2)	No Detonation: 18mv AC
D5	ORN/RED	TCM Signal	N/A
D6-7	---	Not Used	---
D8	BLU/YEL	Ignition Timing Adjustment	0.4-4.5v
D9	WHT/RED	Alternator Control Signal	Headlights off: 0v, on: 12v
D10	---	Not Used	---
D11	RED/BLU	TP Sensor Signal	0.5-0.6v
D12	BLK/WHT	EGR Valve Lift Sensor	1.1-1.2v
D13	RED/WHT	ECT Sensor Signal	At 180°: 0.5-0.6v
D14	WHT	HO2S-11 (B1 S1) Signal	0.1-1.1v
D15	RED/YEL	IAT Sensor Signal	At 100ºF: 2-3v
D16	RED/BLU	HO21-21 (B2 S1) Signal	0.1-1.1v
D17	RED	MAP Sensor Signal	0.8-0.9v
D18	BLU	TCM Signal	N/A
D19	YEL/WHT	MAP Sensor VREF	4.9-5.1v
D20	YEL/WHT	Sensor VREF	4.9-5.1v
D21	GRN/WHT	MAP Sensor Ground	<0.050v
D22	GRN/WHT	Sensor Ground	<0.050v

26-PIN CONNECTOR 'A'

1	3	5	7	9	11	13	15	17	19	21	23	25
2	4	6	8	10	12	14	16	18	20	22	24	26

12-PIN CONNECTOR 'C'

1	3	5	7	9	11
2	4	6	8	10	12

16-PIN CONNECTOR 'B'

1	3	5	7	9	11	13	15
2	4	6	8	10	12	14	16

22-PIN CONNECTOR 'D'

1	3	5	7	9	11	13	15	17	19	21
2	4	6	8	10	12	14	16	18	20	22

WIRE SIDE OF HARNESS TERMINALS

1994-95 L, LS Sedan 3.2L MFI VIN KA7 [M/T] 26P 'A' Connector

PCM Pin #	W/Color	Circuit Description (26-Pin)	Value at Hot Idle
A1	BRN	Injector 1	2.0-3.3 ms
A2	WHT/BLU	Injector 4	2.0-3.3 ms
A3	RED	Injector 2	2.0-3.3 ms
A4	BLK/RED	Injector 5	2.0-3.3 ms
A5	ORN	Injector 3	2.0-3.3 ms
A6	YEL	Injector 6	2.0-3.3 ms
A7	GRN/BLK	Fuel Pump Relay Control	Relay Off: 12v, On: 1v
A8	RED/WHT	M/T: Reverse Lockout Relay	R/L on: 1v, off: 12v
A9	BLU/RED	Idle Air Control Valve	Pulse Signals
A10	GRN/BLU	HO2S-21 (B2 S1) Heater	Heater Off: 12v, On: 1v
A11	WHT	EGR Solenoid Control	Solenoid Off: 12v, On: 1v
A12	GRN/RED	HO2S-11 (B1 S1) Heater	Heater Off: 12v, On: 1v
A13	BLU	Malfunction Indicator Lamp	MIL Off: 12v, On: 1v
A14	RED/BLU	Bypass Low Solenoid	Solenoid Off: 12v, On: 1v
A15	RED/BLU	A/C Clutch Relay Control	Relay Off: 12v, On: 1v
A17	GRY	Air Suction Solenoid	Solenoid Off: 12v, On: 1v
A18	YEL	Bypass High Solenoid	Solenoid Off: 12v, On: 1v
A19	LT GRN	Pressure Regulator Solenoid	Solenoid Off: 12v, On: 1v
A20	LT GRN	EVAP Purge Solenoid	Solenoid Off: 12v, On: 1v
A21	PNK	Igniter 1 Signal	Digital Signals: 0-12-0v
A22	PNK	Igniter 2 Signal	Digital Signals: 0-12-0v
A23	BLK	Power Ground	<0.1v
A24	BLK	Power Ground	<0.1v
A25	YEL/BLK	Main Relay Power (B+)	12-14v
A26	BRN/BLK	Chassis Ground	<0.050v

1994-95 L, LS Sedan 3.2L MFI VIN KA7 [M/T] 16P 'B' Connector

PCM Pin #	W/Color	Circuit Description (16-Pin)	Value at Hot Idle
B1	---	Not Used	---
B2	BRN/BLK	Chassis Ground	<0.050v
B3	RED	Igniter 6 Signal	Digital Signals: 0-12-0v
B4	GRY	Igniter 5 Signal	Digital Signals: 0-12-0v
B5	GRN	PSP Switch Signal	Straight: 0v, Turning: 12v
B6	GRN	Igniter 4 Signal	Digital Signals: 0-12-0v
B8	BLU	Igniter 3 Signal	Digital Signals: 0-12-0v
B7	LT GRN	M/T Neutral Switch	In 'N': 0v, Others: 12v
B9	BLU/GRN	CYP Sensor 2 Signal 'P'	AC pulse signals
B10	BLU/YEL	CYP Sensor 2 Ground 'N'	<0.050v
B11	ORN/BLU	CYP Sensor 1 Signal 'P'	AC pulse signals
B12	WHT/BLU	CYP Sensor 1 Ground 'N'	<0.050v
B13	ORN	CKP Sensor 2 Signal 'P'	AC pulse signals
B14	WHT	CKP Sensor 2 Ground 'N'	<0.050v
B15	ORN/BLU	CKP Sensor 1 Signal 'P'	AC pulse signals
B16	WHT/BLU	CKP Sensor 1 Ground 'N'	<0.050v

1994-95 L, LS Sedan 3.2L MFI VIN KA7 [M/T] 12P 'C' Connector

PCM Pin #	W/Color	Circuit Description (12-Pin)	Value at Hot Idle
C1	YEL/BLK	Main Relay Power (B+)	12-14v
C2	YEL/RED	Vehicle Speed Sensor	Moving: pulse signals
C3	BLU/BLK	Cooling Fan Switch	A/C Off: 12v, On: 1v
C4	BLU	Tachometer Signal	Pulse Signals
C5	RED/BLU	A/C Pressure Switch	Relay Off: 12v, On: 1v
C6	---	Not Used	---
C7	PNK	M/T Clutch Switch	In 'N': 0v, Others: 12v
C8	---	Not Used	---
C9	WHT	Service Check Connector	SCS Open: 4.80v
C10	---	Not Used	---
C11	BLK/WHT	Starter Switch Signal	Cranking: 9-11v
C12	---	Not Used	---

1994-95 L, LS Sedan 3.2L MFI VIN KA7 [M/T] 22P 'D' Connector

PCM Pin #	W/Color	Circuit Description (22-Pin)	Value at Hot Idle
D1	YEL/BLU	Keep Alive Power (VBU)	12-14v
D2	GRN/WHT	Brake Switch Signal	Brake Off: 12v, On: 0v
D3	WHT	Knock Sensor Signal (Bank 1)	No Detonation: 18mv AC
D4	RED/BLU	Knock Sensor Signal (Bank 2)	No Detonation: 18mv AC
D5-7	---	Not Used	---
D8	BLU/YEL	Ignition Timing Adjustment	0.4-4.5v
D9	WHT/RED	Alternator Control Signal	Headlights off: 0v, on: 12v
D10	---	Not Used	---
D11	RED/BLU	TP Sensor Signal	0.5-0.6v
D12	BLK/WHT	EGR Valve Lift Sensor	1.1-1.2v
D13	RED/WHT	ECT Sensor Signal	At 180°: 0.5-0.6v
D14	WHT	HO2S-11 (B1 S1) Signal	0.1-1.1v
D15	RED/YEL	IAT Sensor Signal	At 100ºF: 2-3v
D16	RED/BLU	HO21-21 (B2 S1) Signal	0.1-1.1v
D17	RED	MAP Sensor Signal	0.8-0.9v
D18	---	Not Used	---
D19	YEL/WHT	MAP Sensor VREF	4.9-5.1v
D20	YEL/WHT	Sensor VREF	4.9-5.1v
D21	GRN/WHT	MAP Sensor Ground	<0.050v
D22	GRN/WHT	Sensor Ground	<0.050v

26-PIN CONNECTOR 'A'   12-PIN CONNECTOR 'C'
16-PIN CONNECTOR 'B'   22-PIN CONNECTOR 'D'
WIRE SIDE OF HARNESS TERMINALS

NSX Pin Tables

1991-94 Coupe 3.0L V6 MFI VIN NA1 [A/T] 26P 'A' Connector

PCM Pin #	W/Color	Circuit Description (26-Pin)	Value at Hot Idle
A1, A3	BRN, RED	Injector 1, Injector 3 Control	2.0-3.3 ms
A2	WHT/BLU	Injector 4 Control	2.0-3.3 ms
A4	BLK/RED	Injector 5 Control	2.0-3.3 ms
A5, A6	ORN, YEL	Injector 3, Injector 6 Control	2.0-3.3 ms
A7, A8	GRN/BLK	Fuel Pump Relay Control	Relay Off: 12v, On: 1v
A9	BLU/RED	Electronic Air Control Valve	Pulse Signals
A10	BLK	HO2S-21 (B2 S1) Heater	Heater Off: 12v, On: 1v
A11	RED	EGR Solenoid Control	EGRS On: <1v
A12	BLK/BLU	HO2S-11 (B1 S1) Heater	Heater Off: 12v, On: 1v
A13	BLU	Malfunction Indicator Lamp	MIL Off: 12v, On: 1v
A14	RED/BLU	Chamber Volume Solenoid	CVS On: <1v
A15	RED/BLU	A/C Clutch Relay Control	Relay Off: 12v, On: 1v
A16	PNK	A/T: Control Unit Signal	Digital Signals
A18	BLU/RED	Radiator Fan High Relay	Fan on: 1v, off: 12v
A20	GRN	EVAP Purge Solenoid	Solenoid On: 1v, off: 12v
A21, A22	WHT	Igniter 1 Signal	At idle: 5 dwell
A23, A24	BLK	Power Ground	<0.1v
A25	YEL/BLK	Main Relay Power (B+)	12-14v
A26	BRN/BLK	Chassis Ground	<0.050v

1991-94 Coupe 3.0L V6 MFI VIN NA1 [A/T] 16P 'B' Connector

PCM Pin #	W/Color	Circuit Description (16-Pin)	Value at Hot Idle
B1	---	Not Used	---
B2	BRN/BLK	Chassis Ground	<0.050v
B3	WHT/RED	Igniter 6 Signal	Digital Signals: 0-12-0v
B4	WHT/YEL	Igniter 5 Signal	Digital Signals: 0-12-0v
B5	GRN	PSP Switch Signal	Straight: 0v, Turning: 12v
B6	WHT/BLU	Igniter 4 Signal	Digital Signals: 0-12-0v
B7	LT GRN	Park Neutral Switch Signal	In P/N: 0v, Others: 12v
B8	WHT/BLK	Igniter 3 Signal	Digital Signals: 0-12-0v
B9	ORN	CYP Sensor 2 Signal	AC pulse signals
B10	WHT	CYP Sensor 2 Ground	<0.050v
B11	BLU/GRN	CYP Sensor 1 Signal	AC pulse signals
B12	BLU/YEL	CYP Sensor 1 Ground	<0.050v
B13	ORN/BLU	CKP Sensor 2 Signal	AC pulse signals
B14	WHT/BLU	CKP Sensor 2 Ground	<0.050v
B15	ORN/BLU	CKP Sensor 1 Signal	AC pulse signals
B16	WHT/BLU	CKP Sensor 1 Ground	<0.050v

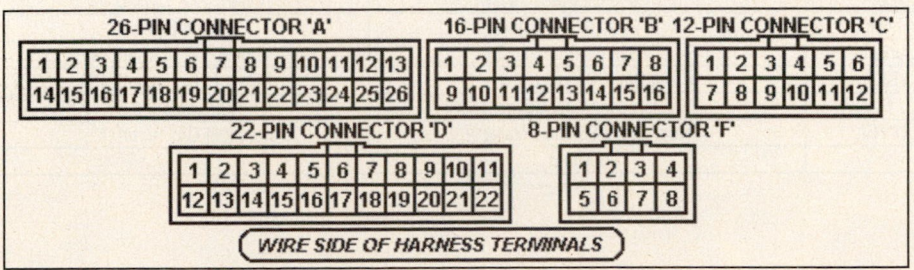

1991-94 Coupe 3.0L V6 MFI VIN NA1 [A/T] 12P 'C' Connector

PCM Pin #	W/Color	Circuit Description (12-Pin)	Value at Hot Idle
C1	YEL/BLK	Main Relay Power (B+)	12-14v
C2	YEL/RED	Vehicle Speed Sensor	Moving: pulse signals
C3	BLU/BLK	Cooling Fan Control Unit	N/A
C4	BLU	Tachometer Signal	Pulse Signals
C5	RED/BLU	A/C Pressure Switch	A/C Off: 12v, On: 1v
C6	WHT/BLK	TCS Signal	Digital Signals
C7	---	Not Used	---
C8	GRN/YEL	TCS Signal	Digital Signals
C9	BLU	Service Check Connector	SCS Open: 4.80v
C10	WHT/RED	TCS Signal	Digital Signals
C11	BLK/WHT	Starter Switch Signal	Cranking: 9-11v
C12	GRY	TCS Signal	Digital Signals

1991-94 Coupe 3.0L V6 MFI VIN NA1 [A/T] 22P 'D' Connector

PCM Pin #	W/Color	Circuit Description (22-Pin)	Value at Hot Idle
D1	WHT/YEL	Keep Alive Power (VBU)	12-14v
D2	---	Not Used	---
D3	ORN	Rear Knock Sensor Signal	No Detonation: 18mv AC
D4	WHT	Front Knock Sensor Signal	No Detonation: 18mv AC
D5	ORN/RED	TCS Signal	Digital Signals
D6	BLK/RED	Fuel Pump Resistor	N/A
D8	BRN	Ignition Timing Adjustment	0.4-4.5v
D9	WHT/RED	Alternator Control Signal	Headlights off: 0v, on: 12v
D10	---	Not Used	---
D11	RED/YEL	TP Sensor Signal	0.5-0.6v
D12	WHT/GRN	EGR Valve Lift Sensor	1.1-1.2v
D13	RED/WHT	ECT Sensor Signal	At 180°: 0.5-0.6v
D14	WHT	HO2S-11 (B1 S1) Signal	0.1-1.1v
D15	RED/BLK	IAT Sensor Signal	At 100°F: 2-3v
D16	RED/BLU	HO21-21 (B2 S1) Signal	0.1-1.1v
D17	WHT	MAP Sensor Signal	0.8-0.9v
D18	BLU	A/T: Control Unit VREF	4.9-5.1v
D19	YEL/WHT	MAP Sensor VREF	4.9-5.1v
D20	YEL/WHT	Sensor VREF	4.9-5.1v
D21	GRN/WHT	MAP Sensor Ground	<0.050v
D22	GRN/WHT	Sensor Ground	<0.050v

1991-94 Coupe 3.0L V6 MFI VIN NA1 [A/T] 8P 'F' Connector

PCM Pin #	W/Color	Circuit Description (8-Pin)	Value at Hot Idle
F1	WHT/GRN	A/T: Control Unit Signal	Digital Signals
F2	BLU	Front Valve Oil Pressure Sw.	Open: 12v, Closed: 1v
F3	WHT/RED	A/T: Control Unit Signal	Digital Signals
F4	BLU/BLK	Rear Valve Oil Pressure Sw.	Open: 12v, Closed: 1v
F5	YEL	Front Spool Solenoid Valve	Solenoid On: 1v, off: 12v
F6	BLU/RED	Acceleration Pedal Sensor 1	ACP closed: 0.5v
F7	BLU/YEL	Rear Spool Solenoid Valve	Solenoid On: 1v, off: 12v
F8	---	Not Used	---

1991-94 Coupe 3.0L V6 MFI VIN NA1 [M/T] 26P 'A' Connector

PCM Pin #	W/Color	Circuit Description (26-Pin)	Value at Hot Idle
A1	BRN	Injector 1	2.0-3.3 ms
A2	WHT/BLU	Injector 4	2.0-3.3 ms
A3	RED	Injector 3	2.0-3.3 ms
A4	BLK/RED	Injector 5	2.0-3.3 ms
A5	ORN	Injector 3	2.0-3.3 ms
A6	YEL	Injector 6	2.0-3.3 ms
A7	GRN/BLK	Fuel Pump Relay Control	Relay Off: 12v, On: 1v
A8	RED	Fuel Pump Relay Control	Relay Off: 12v, On: 1v
A9	BLU/RED	Electronic Air Control Valve	Pulse Signals
A10	BLK	HO2S-21 (B2 S1) Heater	Heater Off: 12v, On: 1v
A11	RED	EGR Solenoid Control	EGRS On: <1v
A12	RED/BLU	HO2S-11 (B1 S1) Heater	Heater Off: 12v, On: 1v
A13	BLU	Malfunction Indicator Lamp	MIL Off: 12v, On: 1v
A14	RED/BLU	Chamber Volume Solenoid	CVS On: <1v
A15	RED/BLU	A/C Clutch Relay Control	Relay Off: 12v, On: 1v
A17	---	Not Used	
A18	BLU/RED	Radiator Fan High Relay	Relay On: <1v
A19	---	Not Used	
A20	GRN	EVAP Purge Solenoid	Solenoid On: 1v, off: 12v
A21	WHT	Igniter 1 Signal	At idle: 5 dwell
A22	WHT/GRN	Igniter 2 Signal	At idle: 5 dwell
A23	BLK	Power Ground	<0.1v
A24	BLK	Power Ground	<0.1v
A25	YEL/BLK	Main Relay Power (B+)	12-14v
A26	BRN/BLK	Chassis Ground	<0.050v

1991-94 Coupe 3.0L V6 MFI VIN NA1 [M/T] 16P 'B' Connector

PCM Pin #	W/Color	Circuit Description (16-Pin)	Value at Hot Idle
B1	---	Not Used	---
B2	BRN/WHT	Chassis Ground	<0.050v
B3	WHT/RED	Igniter 6 Signal	Digital Signals: 0-12-0v
B4	WHT/YEL	Igniter 5 Signal	Digital Signals: 0-12-0v
B5	GRN	PSP Switch Signal	Straight: 0v, Turning: 12v
B6	WHT/BLU	Igniter 4 Signal	Digital Signals: 0-12-0v
B7	LT GRN	M/T: Neutral Position Switch	In N: 0v, others: 12v
B8	WHT/BLK	Igniter 3 Signal	Digital Signals: 0-12-0v
B9	ORN	CYP 2 Signal	AC pulse signals
B10	WHT	CYP 2 Signal	AC pulse signals
B11	BLU/GRN	CYP Sensor 1 Signal	AC pulse signals
B12	BLU/YEL	CYP Sensor 1 Ground	<0.050v
B13	ORN/BLU	CKP Sensor 2 Signal	AC pulse signals
B14	WHT/BLU	CKP Sensor 2 Ground	<0.050v
B15	ORN/BLU	CKP Sensor 1 Signal	AC pulse signals
B16	WHT/BLU	CKP Sensor 1 Ground	<0.050v

1991-94 Coupe 3.0L V6 MFI VIN NA1 [M/T] 12P 'C' Connector

PCM Pin #	W/Color	Circuit Description (12-Pin)	Value at Hot Idle
C1	YEL/BLK	Main Relay Power (B+)	12-14v
C2	YEL/RED	Vehicle Speed Sensor	Moving: pulse signals
C3	BLU/BLK	Cooling Fan Control Unit	N/A
C4	BLU	Tachometer Signal	Pulse Signals
C5	RED/BLU	A/C Pressure Switch	A/C Off: 12v, On: 1v
C6	---	Not Used	---
C7	PNK	Clutch Switch	With Clutch In: 0v
C8-9	---	Not Used	---
C10	BLU	Service Check Connector	SCS Open: 4.80v
C11	BLK/WHT	Starter Switch Signal	Cranking: 9-11v
C12	---	Not Used	---

1991-94 Coupe 3.0L V6 MFI VIN NA1 [M/T] 22P 'D' Connector

PCM Pin #	W/Color	Circuit Description (22-Pin)	Value at Hot Idle
D1	WHT/YEL	Keep Alive Power (VBU)	12-14v
D2	---	Not Used	---
D3	ORN	Rear Knock Sensor Signal	No Detonation: 18mv AC
D4	WHT	Front Knock Sensor Signal	No Detonation: 18mv AC
D5	ORN/RED	TCS Signal	Digital Signals
D6	BLK/RED	Fuel Pump Resistor	N/A
D7	---	Not Used	---
D8	BRN	Ignition Timing Adjustment	0.4-4.5v
D9	WHT/RED	Alternator Control Signal	Headlights off: 0v, on: 12v
D10	---	Not Used	---
D11	RED/YEL	TP Sensor Signal	0.5-0.6v
D12	WHT/GRN	EGR Valve Lift Sensor	1.1-1.2v
D13	RED/WHT	ECT Sensor Signal	At 180°: 0.5-0.6v
D14	WHT	HO2S-11 (B1 S1) Signal	0.1-1.1v
D15	RED/BLK	IAT Sensor Signal	At 100ºF: 2-3v
D16	RED/BLU	HO21-21 (B2 S1) Signal	0.1-1.1v
D17	WHT	MAP Sensor Signal	0.8-0.9v
D18	---	Not Used	---
D19	YEL/WHT	MAP Sensor VREF	4.9-5.1v
D20	YEL/WHT	Sensor VREF	4.9-5.1v
D21	GRN/WHT	MAP Sensor Ground	<0.050v
D22	GRN/WHT	Sensor Ground	<0.050v

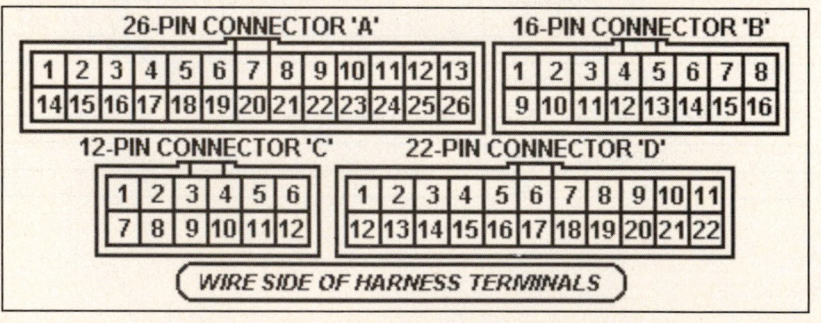

1995-99 Coupe 3.0L V6 VTEC VIN NA1 [A/T] 26P 'A' Connector

PCM Pin #	W/Color	Circuit Description (26-Pin)	Value at Hot Idle
A1	BRN	Injector 1	2.0-3.3 ms
A2	RED	Injector 2	2.0-3.3 ms
A3	BLU	Injector 3	2.0-3.3 ms
A4 ('97)	---	Not Used	---
A4 ('98-'99)	BLK/YEL	EVAP Canister Vent Valve	Solenoid on: 1v, off: 12v
A5	GRN/BLK	Fuel Pump Relay Control 1	Relay Off: 12v, On: 1v
A6	GRN	EGR Solenoid Control	Solenoid Off: 12v, On: 1v
A7	BLU	Malfunction Indicator Lamp	MIL Off: 12v, On: 1v
A8	RED/BLU	A/C Clutch Relay Control	Relay Off: 12v, On: 1v
A9	GRY	HO2S-11 (B1 S1) Heater	Digital Signals: 0-12-0v
A10	LT GRN	HO2S-21 (B2 S1) Heater	Digital Signals: 0-12-0v
A11	PNK	Ignition Coil 1 Control	Digital Signals: 0-12-0v
A12	BLK	Power Ground	<0.1v
A13	YEL/BLK	Main Relay Power (B+)	12-14v
A14	YEL	Injector 4	2.0-3.3 ms
A15	BLK/RED	Injector 5	2.0-3.3 ms
A16	WHT/BLU	Injector 6	2.0-3.3 ms
A17 ('97)	---	Not Used	---
A17 ('98-'99)	BLK/ORN	EVAP Bypass Solenoid Valve	Solenoid on: 1v, off: 12v
A18	BLU/YEL	Front VTEC Solenoid	Idle: 0v, Hi-Speed: 12v
A19	GRN/YEL	Rear VTEC Solenoid	Idle: 0v, Hi-Speed: 12v
A20	PNK/BLU	IAB Solenoid Control	Solenoid off: 12v, on: 1v
A21	RED	EVAP Purge Solenoid	Pulse Signals
A22	GRN/RED	HO2S-12 (B1 S2) Heater	Digital Signals: 0-12-0v
A23	ORN/BLK	HO2S-22 Heater Control	Digital Signals: 0-12-0v
A24	BRN	Ignition Coil 2 Control	Digital Signals: 0-12-0v
A25	BLK	Power Ground	<0.1v
A26	BRN/BLK	Logic Ground	<0.050v

1995-99 Coupe 3.0L V6 VTEC VIN NA1 [A/T] 16P 'B' Connector

PCM Pin #	W/Color	Circuit Description (16-Pin)	Value at Hot Idle
B1	BLU	Front VTEC Pressure Switch	Switch on: 0v, off: 12v
B2	RED	Ignition Coil 6 Control	Digital Signals: 0-12-0v
B3	BLU/BLK	Rear VTEC Pressure Switch	At Low Speed: 0.1v
B4	RED	A/T: Gear Position Switch	P/N: 0v, gear: <5v
B5	ORN	CYP Sensor 2 Signal	AC pulse signals
B6	WHT	CYP Sensor 1 Signal	AC pulse signals
B7	ORN/BLU	CKP Sensor 2 Signal	AC pulse signals
B8	BLU/GRN	CKP Sensor 1 Signal	AC pulse signals
B9	BRN/BLK	Logic Ground	<0.050v
B10	GRY	Ignition Coil 5 Control	Digital Signals: 0-12-0v
B11	GRN	Ignition Coil 4 Control	Digital Signals: 0-12-0v
B12	BLU	Ignition Coil 3 Control	Digital Signals: 0-12-0v
B13	ORN/BLU	CYP Sensor 2 Ground	<0.050v
B14	WHT/BLU	CYP Sensor 1 Ground	<0.050v
B15	WHT/BLU	CKP Sensor 2 Ground	<0.050v
B16	BLU/YEL	CKP Sensor 1 Ground	<0.050v

1995-99 Coupe 3.0L V6 VTEC VIN NA1 [A/T] 12P 'C' Connector

PCM Pin #	W/Color	Circuit Description (12-Pin)	Value at Hot Idle
C1	YEL/BLK	Main Relay Power (B+)	12-14v
C2	BLU/BLK	A/C Switch Signal	A/C Off: 12v, On: 1v
C3	RED/GRN	A/C Pressure Switch Signal	A/C Off: 4-5v, On: 0v
C5	BLU	Service Check Connector	SCS Open: 4.80v
C6	BLK/WHT	Starter Switch Signal	Cranking: 9-11v
C7	ORN	Vehicle Speed Sensor	Moving: pulse signals
C8	GRN	Engine Speed Pulse Signal	Pulses
C9	GRN/YEL	Front Peak Hold Reset	Pulses
C10	GRN/BLK	Rear Peak Hold Reset	Pulses
C11	YEL/RED	Front Misfire Pulse	Pulses
C12	YEL	Rear Misfire Pulse	Pulses

1995-99 Coupe 3.0L V6 VTEC VIN NA1 [A/T] 22P 'D' Connector

PCM Pin #	W/Color	Circuit Description (22-Pin)	Value at Hot Idle
D1	WHT/YEL	Keep Alive Power (VBU)	12-14v
D2	RED/BLU	Knock Sensor 1 Signal	No detonation: 18mv AC
D3 ('97)	BRN	EVAP Purge Flow Switch	Switch on: 0v, off: 12v
D4	YEL/GRN	K-Line Signal	12v
D5	WHT/RED	Alternator 'FR' Signal	Digital Signals: 0-5-0v
D6 ('97)	---	Not Used	---
D6 ('98-'99)	BLU	Fuel Tank Pressure Sensor	Fuel Cap off: 2.5v
D7	RED/WHT	ECT Sensor Signal	At 180ºF: 0.5-0.6v
D8	RED/YEL	IAT Sensor Signal	At 100ºF: 2-3v
D9	WHT/YEL	MAP Sensor Signal	0.8-0.9v
D10	YEL/BLU	Sensor VREF	4.9-5.1v
D11	GRN/BLU	Sensor Ground	<0.050v
D12	RED/BLK	TP Sensor Signal	0.5-0.6v
D13	WHT	Knock Sensor 2 Signal	No detonation: 18mv AC
D14-15	---	Not Used	---
D16	GRN	HO2S-21 (B2 S1) Signal	0.1-1.1v
D17	WHT/RED	HO2S-22 (B2 S2) Signal	0.1-1.1v
D18	BLU/RED	HO2S-11 (B1 S1) Signal	0.1-1.1v
D19	WHT	HO2S-12 (B1 S2) Signal	0.1-1.1v
D20	WHT/BLK	EGR Valve Lift Sensor	1.1-1.2v
D21	BLU/GRN	HO2S-21 (B2 S1) Ground	<0.050v
D22	WHT	HO2S-22 (B2 S2) Ground	<0.050v

Abbreviation	Color	Abbreviation	Color	Abbreviation	Color
BLK	Black	LT BLU	LT Blue	TAN	Tan
BLU	Blue	LT GRN	LT Green	VIO	Violet
BRN	Brown	ORN	Orange	WHT	White
GRY	Gray	PNK	Pink	YEL	Yellow
GRN	Green	PPL	Purple		

1995-99 Coupe 3.0L V6 VTEC VIN NA1 [A/T] 26P 'F' Connector

PCM Pin #	W/Color	Circuit Description (26-Pin)	Value at Hot Idle
F1	YEL/WHT	Sensor VREF	4.9-5.1v
F2	GRN/WHT	Sensor Ground	<0.050v
F3 ('97)	---	Not Used	---
F3 ('98-'99)	BRN	INOCD Immobilizer Code	Digital Signals
F4	RED	Fuel Pump Relay Control	Relay Off: 12v, On: 1v
F5, 17, 19-20	---	Not Used	---
F6	GRY	Brake Switch Signal 2	B/P depressed: 0v
F7	GRN/BLK	Cruise Resume Switch	CR/S released: 0v
F8	GRN/RED	Cruise Control Set Switch	CS/S released: 0v
F9	LT GRN	Cruise Control Main Switch	M/S on: 12v, off: 0v
F10	BLU/ORN	C/C A/T: Gear Position Switch	In 'D', D3, D2: 0v
F11	BLU/BLK	Cruise Indicator Light	Light on: 0v, off: 12v
F12	YEL/RED	Accelerator Position Sensor 2	ACP closed: 0.5v
F13	BLU/RED	Accelerator Position Sensor 1	ACP closed: 0.5v
F14	YEL/WHT	Sensor VREF	4.9-5.1v
F15	GRN/WHT	Sensor Ground	<0.050v
F16	GRN/WHT	Brake Switch Signal 1	Brake Off: 12v, On: 0v
F18	BLU	TCM VREF	4.9-5.1v
F21	GRN/YEL	A/T: FI Data Line 'B'	Pulses
F22	WHT/YEL	A/T: FI Data Line 'A'	Pulses
F24	RED/BLU	TCS Data Line	Pulses

1995-99 Coupe 3.0L V6 VTEC VIN NA1 [A/T] 12P 'G' Connector

PCM Pin #	W/Color	Circuit Description (12-Pin)	Value at Hot Idle
G2, G3	GRN/RED	IGM1, IGM2 Power Source	Key on: 12v, off: 0v
G4	BRN	Motor Phase Out 1	Key on: 0v or pulse
G5	WHT/GRN	Motor Phase Out 2	Key on: 0v or pulse
G7, G8	BLK	Power Ground	<0.1v
G9	ORN	PCM PWR to Motor 1, 3	KOEO: pulse
G10	GRN	PCM PWR to Motor 2, 4	KOEO: pulse
G11	YEL	Motor Phase Out 3	Key on: 0v or pulse
G12	RED	Motor Phase Out 4	Key on: 0v or pulse

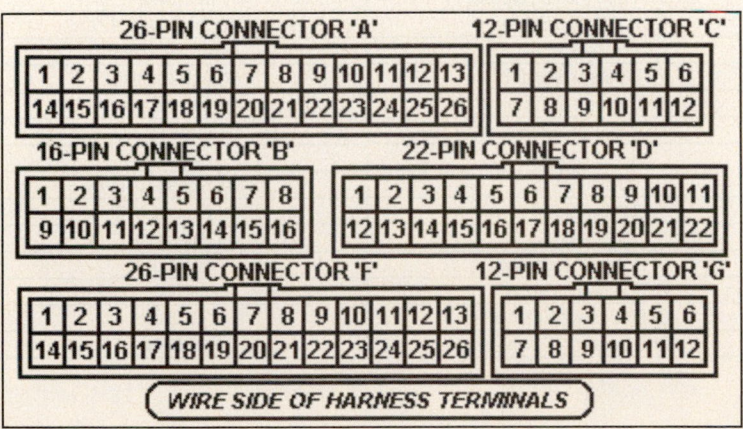

1995-99 Coupe 3.0L V6 VTEC VIN NA1 [M/T] 26P 'A' Connector

PCM Pin #	W/Color	Circuit Description (26-Pin)	Value at Hot Idle
A1	BRN	Injector 1	2.0-3.3 ms
A2	RED	Injector 2	2.0-3.3 ms
A3	BLU	Injector 3	2.0-3.3 ms
A4 ('97)	---	Not Used	---
A4 ('98-'99)	BLK/YEL	EVAP Canister Vent Valve	Solenoid on: 1v, off: 12v
A5	GRN/BLK	Fuel Pump Relay Control 1	Relay Off: 12v, On: 1v
A6	GRN	EGR Solenoid Control	Solenoid Off: 12v, On: 1v
A7	BLU	Malfunction Indicator Lamp	MIL Off: 12v, On: 1v
A8	RED/BLU	A/C Clutch Relay Control	Relay Off: 12v, On: 1v
A9	GRY	HO2S-11 (B1 S1) Heater	Digital Signals: 0-12-0v
A10	LT GRN	HO2S-12 (B1 S2) Heater	Digital Signals: 0-12-0v
A11	PNK	Ignition Coil 1	Digital Signals: 0-12-0v
A12	BLK	Power Ground	<0.1v
A13	YEL/BLK	Main Relay Power (B+)	12-14v
A14	YEL	Injector 4	2.0-3.3 ms
A15	BLK/RED	Injector 5	2.0-3.3 ms
A16	WHT/BLU	Injector 6	2.0-3.3 ms
A17 ('97)	---	Not Used	---
A17 ('98-'99)	BLK/ORN	EVAP Bypass Solenoid Valve	Solenoid on: 1v, off: 12v
A18	BLU/YEL	Front VTEC Solenoid	Idle: 0v, Hi-Speed: 12v
A19	GRN/YEL	Rear VTEC Solenoid	Idle: 0v, Hi-Speed: 12v
A20	PNK/BLU	IAB Solenoid Control	Solenoid Off: 12v, On: 1v
A21	RED/YEL	EVAP Purge Solenoid	Pulse Signals
A22	GRN/RED	HO2S-11 (B1 S1) Heater	Digital Signals: 0-12-0v
A23	ORN/BLK	HO2S-12 (B1 S2) Heater	Digital Signals: 0-12-0v
A24	BRN	Ignition Coil 2	Digital Signals: 0-12-0v
A25	BLK	Power Ground	<0.1v
A26	BRN/BLK	Logic Ground	<0.050v

1995-99 Coupe 3.0L V6 VTEC VIN NA1 [M/T] 16P 'B' Connector

PCM Pin #	W/Color	Circuit Description (16-Pin)	Value at Hot Idle
B1	BLU	Front VTEC Pressure Switch	Switch on: 0v, off: 12v
B2	RED	Ignition Coil 6 Control	Digital Signals: 0-12-0v
B3	BLU/BLK	Rear VTEC Pressure Switch	At Low Speed: 0.1v
B4	LT GRN	Neutral Switch Signal	In 'N': 0v, Others: 12v
B5	ORN	CYP Sensor Signal 2	AC pulse signals
B6	WHT	CYP Sensor Signal 1	AC pulse signals
B7	ORN/BLU	CKP Sensor 2 Signal	AC pulse signals
B8	BLU/GRN	CKP Sensor 1 Signal	AC pulse signals
B9	BRN/BLK	Logic Ground	<0.050v
B10	GRY	Ignition Coil 5	Digital Signals: 0-12-0v
B11	GRN	Ignition Coil 4	Digital Signals: 0-12-0v
B12	BLU	Ignition Coil 3	Digital Signals: 0-12-0v
B13	ORN/BLU	CYP Sensor 2 Ground	<0.050v
B14	WHT/BLU	CYP Sensor 1 Ground	<0.050v
B15	WHT/BLU	CKP Sensor 2 Ground	<0.050v
B16	BLU/YEL	CKP Sensor 1 Ground	<0.050v

1995-99 Coupe 3.0L V6 VTEC VIN NA1 [M/T] 12P 'C' Connector

PCM Pin #	W/Color	Circuit Description (12-Pin)	Value at Hot Idle
C1	YEL/BLK	Main Relay Power (B+)	12-14v
C2	BLU/BLK	A/C Switch Signal	A/C Off: 12v, On: 1v
C3	RED/GRN	A/C Pressure Switch Signal	A/C Off: 4-5v, On: 0v
C4	PNK	Clutch Switch Signal	With Clutch In: 0v
C5	BLU	Service Check Connector	SCS Open: 4.80v
C6	BLK/WHT	Starter Switch Signal	Cranking: 9-11v
C7	ORN	Vehicle Speed Sensor	Moving: pulse signals
C8	GRN	Engine Speed Pulse Signal	Pulses
C9	GRN/YEL	Front Peak Hold Reset	Pulses
C10	GRN/BLK	Rear Peak Hold Reset	Pulses
C11	YEL/RED	Front Misfire Pulse	Pulses
C12	YEL	Rear Misfire Pulse	Pulses

1995-99 Coupe 3.0L V6 VTEC VIN NA1 [M/T] 22P 'D' Connector

PCM Pin #	W/Color	Circuit Description (22-Pin)	Value at Hot Idle
D1	WHT/YEL	Keep Alive Power (VBU)	12-14v
D2	RED/BLU	Knock Sensor 1 Signal	No detonation: 18mv AC
D3 ('97)	BRN	EVAP Purge Flow Switch	Switch on: 0v, off: 12v
D4	YEL/GRN	K-Line Signal	12v
D5	WHT/RED	Alternator 'FR' Signal	Digital Signals: 0-5-0v
D6 ('97)	---	Not Used	---
D6 ('98-'99)	BLU	Fuel Tank Pressure Sensor	Fuel Cap off: 2.5v
D7	RED/WHT	ECT Sensor Signal	At 180°F: 0.5-0.6v
D8	RED/YEL	IAT Sensor Signal	At 100°F: 2-3v
D9	WHT/YEL	MAP Sensor Signal	0.8-0.9v
D10	YEL/BLU	Sensor VREF	4.9-5.1v
D11	GRN/BLU	Sensor Ground	<0.050v
D12	RED/BLK	TP Sensor Signal	0.5-0.6v
D13	WHT	Knock Sensor 2 Signal	No detonation: 18mv AC
D14	---	Not Used	---
D15	---	Not Used	---
D16	GRN	HO2S-21 (B2 S1) Signal	0.1-1.1v
D17	WHT/RED	HO2S-22 (B2 S2) Signal	0.1-1.1v
D18	BLU/RED	HO2S-11 (B1 S1) Signal	0.1-1.1v
D19	WHT	Rear HO2S-11 (B1 S1) Signal	0.1-1.1v
D20	WHT/BLK	EGR Valve Lift Sensor	1.1-1.2v
D21	BLU/GRN	HO2S-21 (B2 S1) Ground	<0.050v
D22	WHT	HO2S-22 (B2 S2) Ground	<0.050v

1995-99 Coupe 3.0L V6 VTEC VIN NA1 [M/T] 26P 'F' Connector

PCM Pin #	W/Color	Circuit Description (26-Pin)	Value at Hot Idle
F1	YEL/WHT	Sensor VREF (VCC1)	4.9-5.1v
F2	GRN/WHT	Sensor Ground	<0.050v
F3 ('97)	---	Not Used	---
F3 ('98-'99)	BRN	INOCD Immobilizer Code	Digital Signals
F4	RED	Fuel Pump Relay Control	Relay Off: 12v, On: 1v
F5	---	Not Used	---
F6	GRY	Brake Switch Signal 2	B/P depressed: 0v
F7	GRN/BLK	Cruise Resume Switch	CR/S released: 0v
F8	GRN/RED	Cruise Control Set Switch	CS/S released: 0v
F9	LT GRN	Cruise Control Main Switch	M/S on: 12v, off: 0v
F10	BLU/ORN	Cruise Control Clutch Switch	Clutch in: 0v, out: 8v
F11	BLU/BLK	Cruise Indicator Light	Light on: 0v, off: 12v
F12	YEL/RED	ACP Sensor 2 Signal (AP2)	ACP closed: 0.5v
F13	BLU/RED	ACP Sensor 1 Signal (AP1)	ACP closed: 0.5v
F14	YEL/WHT	Sensor VREF (VCC3)	4.9-5.1v
F15	GRN/WHT	Sensor Ground (SG3)	<0.050v
F16	GRN/WHT	Brake Switch Signal 1	Brake Off: 12v, On: 0v
F17-26	---	Not Used	---

1995-99 Coupe 3.0L V6 VTEC VIN NA1 [M/T] 12P 'G' Connector

PCM Pin #	W/Color	Circuit Description (12-Pin)	Value at Hot Idle
G1	---	Not Used	---
G2	GRN/RED	IGM1 Power Source	12-14v
G3	GRN/RED	IGM2 Power Source	12-14v
G4	BRN	Motor Phase Out 1	Pulse Signals
G5	WHT/GRN	Motor Phase Out 2	Pulse Signals
G7	BLK	Power Ground	<0.1v
G8	BLK	Power Ground	<0.1v
G9	ORN	PCM PWR to Motor 1, 3	Pulse Signals
G10	GRN	PCM PWR to Motor 2, 4	Pulse Signals
G11	YEL	Motor Phase Out 3	Pulse Signals
G12	RED	Motor Phase Out 4	Pulse Signals

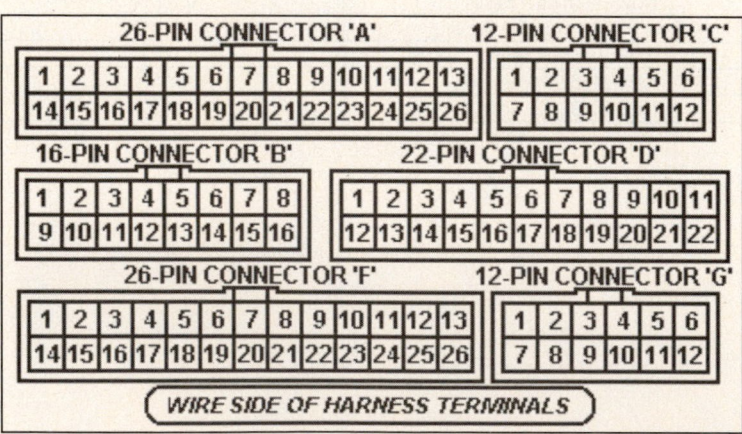

2000-03 Coupe 3.0L V6 VTEC VIN NA1 [All] 26P 'A' Connector

PCM Pin #	W/Color	Circuit Description (26-Pin)	Value at Hot Idle
A1	BRN	Injector 1 Control	2.0-3.3 ms
A2	RED	Injector 2 Control	2.0-3.3 ms
A3	BLU	Injector 3 Control	2.0-3.3 ms
A4	BLK/YEL	EVAP Canister Vent Valve	Solenoid on: 1v, off: 12v
A5	GRN/BLK	Fuel Pump Relay Control 1	Relay Off: 12v, On: 1v
A6	GRN	EGR Solenoid Control	Solenoid Off: 12v, On: 1v
A7	BLU	Malfunction Indicator Lamp	MIL Off: 12v, On: 1v
A8	RED/BLU	A/C Clutch Relay Control	Relay Off: 12v, On: 1v
A9	GRY	HO2S-11 (B1 S1) Heater	Digital Signals: 0-12-0v
A10	LT GRN	HO2S-21 (B2 S1) Heater	Digital Signals: 0-12-0v
A11	PNK	Ignition Coil 1 Control	Digital Signals: 0-12-0v
A12	BLK	Power Ground (PG1)	<0.1v
A13	YEL/BLK	Main Relay Power (B+)	12-14v
A14	YEL	Injector 4 Control	2.0-3.3 ms
A15	BLK/RED	Injector 5 Control	2.0-3.3 ms
A16	WHT/BLU	Injector 6 Control	2.0-3.3 ms
A17	ORN	EVAP Bypass Solenoid Valve	Solenoid on: 1v, off: 12v
A18	YEL/GRN	Front VTEC Solenoid	Idle: 0v, Hi-Speed: 12v
A19	---	Not Used	---
A20	PNK/BLU	IAB Solenoid Control	Solenoid Off: 12v, On: 1v
A21	RED/YEL	EVAP Purge Solenoid	Pulse Signals
A22	GRN/RED	HO2S-12 (B1 S2) Heater	Digital Signals: 0-12-0v
A23	ORN/BLK	HO2S-22 Heater Control	Digital Signals: 0-12-0v
A24	BRN	Ignition Coil 2 Control	Digital Signals: 0-12-0v
A25	BLK	Power Ground (PG2)	<0.1v
A26	BRN/BLK	Logic Ground (LG1)	<0.050v

2000-03 Coupe 3.0L V6 VTEC VIN NA1 [All] 16P 'B' Connector

PCM Pin #	W/Color	Circuit Description (16-Pin)	Value at Hot Idle
B1	BLU	Front VTEC Pressure Switch	Switch on: 0v, off: 12v
B2	RED	Ignition Coil 6 Control	Digital Signals: 0-12-0v
B3	BLU/BLK	Rear VTEC Pressure Switch	At Low Speed: 0.1v
B4	RED	A/T: Gear Position Switch	In P/N: 0v, others: 5v
B4	LT GRN	M/T: Neutral Position Switch	In 'N': 0v, others: 5v
B5	ORN	CYP Sensor 2 Signal	AC pulse signals
B6	WHT	CYP Sensor 1 Signal	AC pulse signals
B7	ORN/BLU	CKP Sensor 2 Signal	AC pulse signals
B8	BLU/GRN	CKP Sensor 1 Signal	AC pulse signals
B9	BRN/BLK	Logic Ground	<0.050v
B10	GRY	Ignition Coil 5 Control	Digital Signals: 0-12-0v
B11	GRN	Ignition Coil 4 Control	Digital Signals: 0-12-0v
B12	BLU	Ignition Coil 3 Control	Digital Signals: 0-12-0v
B13	ORN/BLU	CYP Sensor 2 Ground	<0.050v
B14	WHT/BLU	CYP Sensor 1 Ground	<0.050v
B15	WHT/BLU	CKP Sensor 2 Ground	<0.050v
B16	BLU/YEL	CKP Sensor 1 Ground	<0.050v

2000-03 Coupe 3.0L V6 VTEC VIN NA1 [All] 12P 'C' Connector

PCM Pin #	W/Color	Circuit Description (12-Pin)	Value at Hot Idle
C1	YEL/BLK	Main Relay Power (B+)	12-14v
C2	BLU/BLK	A/C Switch Signal	A/C Off: 12v, On: 1v
C3	RED/GRN	A/C Pressure Switch Signal	A/C Off: 4-5v, On: 0v
C4	PNK	M/T: Clutch Engage Switch	Clutch in: 5v, out: 0v
C5	BLU	Service Check Connector	SCS Open: 4.80v
C6	BLK/WHT	Starter Switch Signal	Cranking: 9-11v
C7	ORN	Vehicle Speed Sensor	Moving: pulse signals
C8	GRN	Engine Speed Pulse Signal	Pulses
C9	GRN/YEL	Front Peak Hold Reset	Pulses
C10	GRN/BLK	Rear Peak Hold Reset	Pulses
C11	YEL/RED	Front Misfire Pulse	Pulses
C12	YEL	Rear Misfire Pulse	Pulses

2000-03 Coupe 3.0L V6 VTEC VIN NA1 [All] 22P 'D' Connector

PCM Pin #	W/Color	Circuit Description (22-Pin)	Value at Hot Idle
D1	WHT/YEL	Keep Alive Power (VBU)	12-14v
D2	RED/BLU	Knock Sensor 1 Signal	No detonation: 18mv AC
D3	WHT/BLK	Air Pump Current Sensor	Pump on: Pulse Signals
D4	YEL/GRN	K-Line Signal	12v
D5	WHT/RED	Alternator 'FR' Signal	Digital Signals: 0-5-0v
D6	BLU	Fuel Tank Pressure Sensor	Fuel Cap off: 2.5v
D7	RED/WHT	ECT Sensor Signal	At 180ºF: 0.5-0.6v
D8	RED/YEL	IAT Sensor Signal	At 100ºF: 2-3v
D9	WHT/YEL	MAP Sensor Signal	0.8-0.9v
D10	YEL/BLU	Sensor VREF (VCC2)	4.9-5.1v
D11	GRN/BLU	Sensor Ground	<0.050v
D12	RED/BLK	TP Sensor Signal	0.5-0.6v
D13	WHT	Knock Sensor 2 Signal	No detonation: 18mv AC
D14-15	---	Not Used	---
D16	GRN	HO2S-21 (B2 S1) Signal	0.1-1.1v
D17	WHT/RED	HO2S-22 (B2 S2) Signal	0.1-1.1v
D18	BLU/RED	HO2S-11 (B1 S1) Signal	0.1-1.1v
D19	WHT	HO2S-12 (B1 S2) Signal	0.1-1.1v
D20	WHT/BLK	EGR Valve Lift Sensor	1.1-1.2v
D21	BLU/GRN	HO2S-21 (B2 S1) Ground	<0.050v
D22	WHT	HO2S-22 (B2 S2) Ground	<0.050v

Abbreviation	Color	Abbreviation	Color	Abbreviation	Color
BLK	Black	LT BLU	LT Blue	TAN	Tan
BLU	Blue	LT GRN	LT Green	VIO	Violet
BRN	Brown	ORN	Orange	WHT	White
GRY	Gray	PNK	Pink	YEL	Yellow
GRN	Green	PPL	Purple		

2000-03 Coupe 3.0L V6 VTEC VIN NA1 [All] 26P 'F' Connector

PCM Pin #	W/Color	Circuit Description (26-Pin)	Value at Hot Idle
F1	YEL/WHT	Sensor VREF (VCC1)	4.9-5.1v
F2, F15	GRN/WHT	Sensor Ground SG1, SG3	<0.050v
F3	BRN	INOCD Immobilizer Code	Digital Signals
F4	RED	Fuel Pump Relay Control	Relay Off: 12v, On: 1v
F6	GRY	Brake Switch Signal 2	B/P depressed: 0v
F7	GRN/BLK	Cruise Resume Switch	CR/S released: 0v
F8	GRN/RED	Cruise Control Set Switch	CS/S released: 0v
F9	LT GRN	Cruise Control Main Switch	M/S on: 12v, off: 0v
F10	BLU/ORN	C/C A/T: Gear Position Switch	In 'D', D3, D2: 0v
F10	BLU/ORN	C/C M/T: Clutch Switch	Clutch in: 0v, out: 8v
F11	BLU/BLK	Cruise Indicator Light	Light on: 0v, off: 12v
F12	YEL/RED	Accelerator Position Sensor 2	ACP closed: 0.5v
F13	BLU/RED	Accelerator Position Sensor 1	ACP closed: 0.5v
F14	YEL/WHT	Sensor VREF (VCC1)	4.9-5.1v
F16	GRN/WHT	Brake Switch Signal 1	Brake Off: 12v, On: 0v
F17	RED/WHT	Reverse Lockout Relay Out	VSS over 30 mph: 0v
F18	BLU	TCM VREF	4.9-5.1v
F19	BLU/YEL	Front VTEC Solenoid Valve	0v, Hi-Speed: 12v
F20	GRY/YEL	Rear VTEC Solenoid Valve	0v, Hi-Speed: 12v
F21	GRN/YEL	A/T: FI Data Line 'B'	Pulse Signals
F22	WHT/YEL	A/T: FI Data Line 'A'	Pulse Signals
F23	BLU/WHT	Air Pump Relay Control	Relay Off: 12v, On: 1v
F24	RED/BLU	TCS Data Line	Pulse Signals
F25	RED	Air Control Solenoid Valve	Pump on: 1v, off: 12v

2000-03 Coupe 3.0L V6 VTEC VIN NA1 [All] 12P 'G' Connector

PCM Pin #	W/Color	Circuit Description (12-Pin)	Value at Hot Idle
G1, G9	---	Not Used	---
G2, G3	GRN/RED	IGM1, IGM2 Power Source	Key on: 12v, off: 0v
G4	BRN	Motor Phase Out 1	Key on: 0v or pulse
G5	WHT/GRN	Motor Phase Out 2	Key on: 0v or pulse
G7, G8	BLK	Power Ground	<0.1v
G9	ORN	PCM PWR to Motor 1, 3	KOEO: pulse
G10	GRN	PCM PWR to Motor 2, 4	KOEO: pulse
G11	YEL	Motor Phase Out 3	Key on: 0v or pulse
G12	RED	Motor Phase Out 4	Key on: 0v or pulse

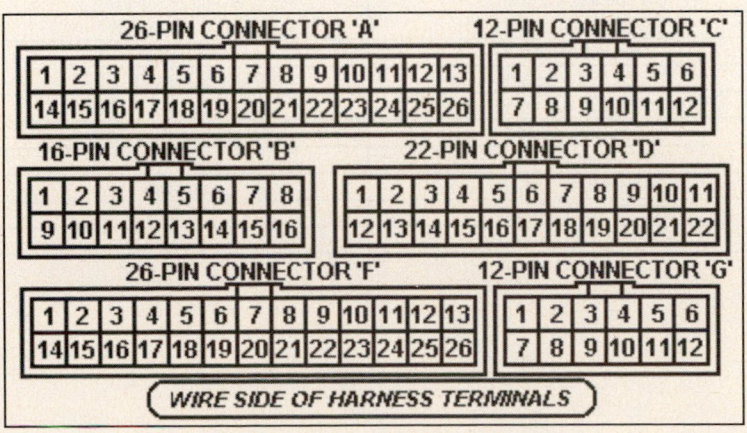

1997-99 Coupe 3.2L V6 VTEC VIN NA2 [A/T] 26P 'A' Connector

PCM Pin #	W/Color	Circuit Description (26-Pin)	Value at Hot Idle
A1	BRN	Injector 1	2.0-3.3 ms
A2	RED	Injector 2	2.0-3.3 ms
A3	BLU	Injector 3	2.0-3.3 ms
A4 ('97)	---	Not Used	---
A4 ('98-'99)	BLK/YEL	EVAP Canister Vent Valve	Solenoid on: 1v, off: 12v
A5	GRN/BLK	Fuel Pump Relay Control 1	Relay Off: 12v, On: 1v
A6	GRN	EGR Solenoid Control	Solenoid Off: 12v, On: 1v
A7	BLU	Malfunction Indicator Lamp	MIL Off: 12v, On: 1v
A8	RED/BLU	A/C Clutch Relay Control	Relay Off: 12v, On: 1v
A9	GRY	HO2S-11 (B1 S1) Heater	Digital Signals: 0-12-0v
A10	LT GRN	HO2S-21 (B2 S1) Heater	Digital Signals: 0-12-0v
A11	PNK	Ignition Coil 1 Control	Digital Signals: 0-12-0v
A12	BLK	Power Ground	<0.1v
A13	YEL/BLK	Main Relay Power (B+)	12-14v
A14	YEL	Injector 4	2.0-3.3 ms
A15	BLK/RED	Injector 5	2.0-3.3 ms
A16	WHT/BLU	Injector 6	2.0-3.3 ms
A17 ('97)	---	Not Used	---
A17 ('98-'99)	BLK/ORN	EVAP Bypass Solenoid Valve	Solenoid on: 1v, off: 12v
A18	BLU/YEL	Front VTEC Solenoid	Idle: 0v, Hi-Speed: 12v
A19	GRN/YEL	Rear VTEC Solenoid	Idle: 0v, Hi-Speed: 12v
A20	PNK/BLU	IAB Solenoid Control	Solenoid Off: 12v, On: 1v
A21	RED	EVAP Purge Solenoid	Pulse Signals
A22	GRN/RED	HO2S-12 (B1 S2) Heater	Digital Signals: 0-12-0v
A23	ORN/BLK	HO2S-22 Heater Control	Digital Signals: 0-12-0v
A24	BRN	Ignition Coil 2 Control	Digital Signals: 0-12-0v
A25	BLK	Power Ground	<0.1v
A26	BRN/BLK	Logic Ground	<0.050v

1997-99 Coupe 3.2L V6 VTEC VIN NA2 [A/T] 16P 'B' Connector

PCM Pin #	W/Color	Circuit Description (16-Pin)	Value at Hot Idle
B1	BLU	Front VTEC Pressure Switch	Switch on: 0v, off: 12v
B2	RED	Ignition Coil 6 Control	Digital Signals: 0-12-0v
B3	BLU/BLK	Rear VTEC Pressure Switch	At Low Speed: 0.1v
B4	RED	A/T: Gear Position Switch	P/N: 0v, gear: <5v
B5	ORN	CYP Sensor 2 Signal	AC pulse signals
B6	WHT	CYP Sensor 1 Signal	AC pulse signals
B7	ORN/BLU	CKP Sensor 2 Signal	AC pulse signals
B8	BLU/GRN	CKP Sensor 1 Signal	AC pulse signals
B9	BRN/BLK	Logic Ground	<0.050v
B10	GRY	Ignition Coil 5 Control	Digital Signals: 0-12-0v
B11	GRN	Ignition Coil 4 Control	Digital Signals: 0-12-0v
B12	BLU	Ignition Coil 3 Control	Digital Signals: 0-12-0v
B13	ORN/BLU	CYP Sensor 2 Ground	<0.050v
B14	WHT/BLU	CYP Sensor 1 Ground	<0.050v
B15	WHT/BLU	CKP Sensor 2 Ground	<0.050v
B16	BLU/YEL	CKP Sensor 1 Ground	<0.050v

1997-99 Coupe 3.2L V6 VTEC VIN NA2 [A/T] 12P 'C' Connector

PCM Pin #	W/Color	Circuit Description (12-Pin)	Value at Hot Idle
C1	YEL/BLK	Main Relay Power (B+)	12-14v
C2	BLU/BLK	A/C Switch Signal	A/C Off: 12v, On: 1v
C3	RED/GRN	A/C Pressure Switch Signal	A/C Off: 4-5v, On: 0v
C5	BLU	Service Check Connector	SCS Open: 4.80v
C6	BLK/WHT	Starter Switch Signal	Cranking: 9-11v
C7	ORN	Vehicle Speed Sensor	Moving: pulse signals
C8	GRN	Engine Speed Pulse Signal	Pulse Signals
C9	GRN/YEL	Front Peak Hold Reset	Pulse Signals
C10	GRN/BLK	Rear Peak Hold Reset	Pulse Signals
C11	YEL/RED	Front Misfire Pulse	Pulse Signals
C12	YEL	Rear Misfire Pulse	Pulse Signals

1997-99 Coupe 3.2L V6 VTEC VIN NA2 [A/T] 22P 'D' Connector

PCM Pin #	W/Color	Circuit Description (22-Pin)	Value at Hot Idle
D1	WHT/YEL	Keep Alive Power (VBU)	12-14v
D2	RED/BLU	Knock Sensor 1 Signal	No detonation: 18mv AC
D3 ('97)	BRN	EVAP Purge Flow Switch	Switch on: 0v, off: 12v
D4	YEL/GRN	K-Line Signal	12v
D5	WHT/RED	Alternator 'FR' Signal	Digital Signals: 0-5-0v
D6 ('97)	---	Not Used	---
D6 ('98-'99)	BLU	Fuel Tank Pressure Sensor	Fuel Cap off: 2.5v
D7	RED/WHT	ECT Sensor Signal	At 180°F: 0.5-0.6v
D8	RED/YEL	IAT Sensor Signal	At 100°F: 2-3v
D9	WHT/YEL	MAP Sensor Signal	0.8-0.9v
D10	YEL/BLU	Sensor VREF	4.9-5.1v
D11	GRN/BLU	Sensor Ground	<0.050v
D12	RED/BLK	TP Sensor Signal	0.5-0.6v
D13	WHT	Knock Sensor 2 Signal	No detonation: 18mv AC
D14-15	---	Not Used	---
D16	GRN	HO2S-21 (B2 S1) Signal	0.1-1.1v
D17	WHT/RED	HO2S-22 (B2 S2) Signal	0.1-1.1v
D18	BLU/RED	HO2S-11 (B1 S1) Signal	0.1-1.1v
D19	WHT	HO2S-12 (B1 S2) Signal	0.1-1.1v
D20	WHT/BLK	EGR Valve Lift Sensor	1.1-1.2v
D21	BLU/GRN	HO2S-21 (B2 S1) Ground	<0.050v
D22	WHT	HO2S-22 (B2 S2) Ground	<0.050v

Abbreviation	Color	Abbreviation	Color	Abbreviation	Color
BLK	Black	LT BLU	LT Blue	TAN	Tan
BLU	Blue	LT GRN	LT Green	VIO	Violet
BRN	Brown	ORN	Orange	WHT	White
GRY	Gray	PNK	Pink	YEL	Yellow
GRN	Green	PPL	Purple		

1997-99 Coupe 3.2L V6 VTEC VIN NA2 [A/T] 26P 'F' Connector

PCM Pin #	W/Color	Circuit Description (26-Pin)	Value at Hot Idle
F1	YEL/WHT	Sensor VREF	4.9-5.1v
F2	GRN/WHT	Sensor Ground	<0.050v
F3 ('97)	---	Not Used	---
F3 ('98-'99)	BRN	INOCD Immobilizer Code	Digital Signals
F4	RED	Fuel Pump Relay Control	Relay Off: 12v, On: 1v
F5, 17, 19-20	---	Not Used	---
F6	GRY	Brake Switch Signal 2	B/P depressed: 0v
F7	GRN/BLK	Cruise Resume Switch	CR/S released: 0v
F8	GRN/RED	Cruise Control Set Switch	CS/S released: 0v
F9	LT GRN	Cruise Control Main Switch	M/S on: 12v, off: 0v
F10	BLU/ORN	C/C A/T: Gear Position Switch	In 'D', D3, D2: 0v
F11	BLU/BLK	Cruise Indicator Light	Light on: 0v, off: 12v
F12	YEL/RED	Accelerator Position Sensor 2	ACP closed: 0.5v
F13	BLU/RED	Accelerator Position Sensor 1	ACP closed: 0.5v
F14	YEL/WHT	Sensor VREF	4.9-5.1v
F15	GRN/WHT	Sensor Ground	<0.050v
F16	GRN/WHT	Brake Switch Signal 1	Brake Off: 12v, On: 0v
F18	BLU	TCM VREF	4.9-5.1v
F21	GRN/YEL	A/T: FI Data Line 'B'	Pulses
F22	WHT/YEL	A/T: FI Data Line 'A'	Pulses
F24	RED/BLU	TCS Data Line	Pulses

1997-99 Coupe 3.2L V6 VTEC VIN NA2 [A/T] 12P 'G' Connector

PCM Pin #	W/Color	Circuit Description (12-Pin)	Value at Hot Idle
G2, G3	GRN/RED	IGM1, IGM2 Power Source	Key on: 12v, off: 0v
G4	BRN	Motor Phase Out 1	Key on: 0v or pulse
G5	WHT/GRN	Motor Phase Out 2	Key on: 0v or pulse
G7, G8	BLK	Power Ground	<0.1v
G9	ORN	PCM PWR to Motor 1, 3	KOEO: pulse
G10	GRN	PCM PWR to Motor 2, 4	KOEO: pulse
G11	YEL	Motor Phase Out 3	Key on: 0v or pulse
G12	RED	Motor Phase Out 4	Key on: 0v or pulse

1997-99 Coupe 3.2L V6 VTEC VIN NA2 [M/T] 26P 'A' Connector

PCM Pin #	W/Color	Circuit Description (26-Pin)	Value at Hot Idle
A1	BRN	Injector 1	2.0-3.3 ms
A2	RED	Injector 2	2.0-3.3 ms
A3	BLU	Injector 3	2.0-3.3 ms
A4 ('97)	---	Not Used	---
A4 ('98-'99)	BLK/YEL	EVAP Canister Vent Valve	Solenoid on: 1v, off: 12v
A5	GRN/BLK	Fuel Pump Relay Control 1	Relay Off: 12v, On: 1v
A6	GRN	EGR Solenoid Control	Solenoid Off: 12v, On: 1v
A7	BLU	Malfunction Indicator Lamp	MIL Off: 12v, On: 1v
A8	RED/BLU	A/C Clutch Relay Control	Relay Off: 12v, On: 1v
A9	GRY	HO2S-11 (B1 S1) Heater	Digital Signals: 0-12-0v
A10	LT GRN	HO2S-12 (B1 S2) Heater	Digital Signals: 0-12-0v
A11	PNK	Ignition Coil 1	Digital Signals: 0-12-0v
A12	BLK	Power Ground	<0.1v
A13	YEL/BLK	Main Relay Power (B+)	12-14v
A14	YEL	Injector 4	2.0-3.3 ms
A15	BLK/RED	Injector 5	2.0-3.3 ms
A16	WHT/BLU	Injector 6	2.0-3.3 ms
A17 ('97)	---	Not Used	---
A17 ('98-'99)	BLK/ORN	EVAP Bypass Solenoid Valve	Solenoid on: 1v, off: 12v
A18	BLU/YEL	Front VTEC Solenoid	Idle: 0v, Hi-Speed: 12v
A19	GRN/YEL	Rear VTEC Solenoid	Idle: 0v, Hi-Speed: 12v
A20	PNK/BLU	IAB Solenoid Control	Solenoid Off: 12v, On: 1v
A21	RED/YEL	EVAP Purge Solenoid	Pulse Signals
A22	GRN/RED	HO2S-11 (B1 S1) Heater	Digital Signals: 0-12-0v
A23	ORN/BLK	HO2S-12 (B1 S2) Heater	Digital Signals: 0-12-0v
A24	BRN	Ignition Coil 2	Digital Signals: 0-12-0v
A25	BLK	Power Ground	<0.1v
A26	BRN/BLK	Logic Ground	<0.050v

1997-99 Coupe 3.2L V6 VTEC VIN NA2 [M/T] 16P 'B' Connector

PCM Pin #	W/Color	Circuit Description (16-Pin)	Value at Hot Idle
B1	BLU	Front VTEC Pressure Switch	Switch on: 0v, off: 12v
B2	RED	Ignition Coil 6 Control	Digital Signals: 0-12-0v
B3	BLU/BLK	Rear VTEC Pressure Switch	At Low Speed: 0.1v
B4	LT GRN	Neutral Switch Signal	In 'N': 0v, Others: 12v
B5	ORN	CYP Sensor Signal 2	AC pulse signals
B6	WHT	CYP Sensor Signal 1	AC pulse signals
B7	ORN/BLU	CKP Sensor 2 Signal	AC pulse signals
B8	BLU/GRN	CKP Sensor 1 Signal	AC pulse signals
B9	BRN/BLK	Logic Ground	<0.050v
B10	GRY	Ignition Coil 5	Digital Signals: 0-12-0v
B11	GRN	Ignition Coil 4	Digital Signals: 0-12-0v
B12	BLU	Ignition Coil 3	Digital Signals: 0-12-0v
B13	ORN/BLU	CYP Sensor 2 Ground	<0.050v
B14	WHT/BLU	CYP Sensor 1 Ground	<0.050v
B15	WHT/BLU	CKP Sensor 2 Ground	<0.050v
B16	BLU/YEL	CKP Sensor 1 Ground	<0.050v

1997-99 Coupe 3.2L V6 VTEC VIN NA2 [M/T] 12P 'C' Connector

PCM Pin #	W/Color	Circuit Description (12-Pin)	Value at Hot Idle
C1	YEL/BLK	Main Relay Power (B+)	12-14v
C2	BLU/BLK	A/C Switch Signal	A/C Off: 12v, On: 1v
C3	RED/GRN	A/C Pressure Switch Signal	A/C Off: 4-5v, On: 0v
C4	PNK	Clutch Switch Signal	With Clutch In: 0v
C5	BLU	Service Check Connector	SCS Open: 4.80v
C6	BLK/WHT	Starter Switch Signal	Cranking: 9-11v
C7	ORN	Vehicle Speed Sensor	Moving: pulse signals
C8	GRN	Engine Speed Pulse Signal	Pulses
C9	GRN/YEL	Front Peak Hold Reset	Pulses
C10	GRN/BLK	Rear Peak Hold Reset	Pulses
C11	YEL/RED	Front Misfire Pulse	Pulses
C12	YEL	Rear Misfire Pulse	Pulses

1997-99 Coupe 3.2L V6 VTEC VIN NA2 [M/T] 22P 'D' Connector

PCM Pin #	W/Color	Circuit Description (22-Pin)	Value at Hot Idle
D1	WHT/YEL	Keep Alive Power (VBU)	12-14v
D2	RED/BLU	Knock Sensor 1 Signal	No detonation: 18mv AC
D3 ('97)	BRN	EVAP Purge Flow Switch	Switch on: 0v, off: 12v
D4	YEL/GRN	K-Line Signal	12v
D5	WHT/RED	Alternator 'FR' Signal	Digital Signals: 0-5-0v
D6 ('97)	---	Not Used	---
D6 ('98-'99)	BLU	Fuel Tank Pressure Sensor	Fuel Cap off: 2.5v
D7	RED/WHT	ECT Sensor Signal	At 180°F: 0.5-0.6v
D8	RED/YEL	IAT Sensor Signal	At 100°F: 2-3v
D9	WHT/YEL	MAP Sensor Signal	0.8-0.9v
D10	YEL/BLU	Sensor VREF	4.9-5.1v
D11	GRN/BLU	Sensor Ground	<0.050v
D12	RED/BLK	TP Sensor Signal	0.5-0.6v
D13	WHT	Knock Sensor 2 Signal	No detonation: 18mv AC
D14	---	Not Used	---
D15	---	Not Used	---
D16	GRN	HO2S-21 (B2 S1) Signal	0.1-1.1v
D17	WHT/RED	HO2S-22 (B2 S2) Signal	0.1-1.1v
D18	BLU/RED	HO2S-11 (B1 S1) Signal	0.1-1.1v
D19	WHT	Rear HO2S-11 (B1 S1) Signal	0.1-1.1v
D20	WHT/BLK	EGR Valve Lift Sensor	1.1-1.2v
D21	BLU/GRN	HO2S-21 (B2 S1) Ground	<0.050v
D22	WHT	HO2S-22 (B2 S2) Ground	<0.050v

1997-99 Coupe 3.2L V6 VTEC VIN NA2 [M/T] 26P 'F' Connector

PCM Pin #	W/Color	Circuit Description (26-Pin)	Value at Hot Idle
F1	YEL/WHT	Sensor VREF (VCC1)	4.9-5.1v
F2	GRN/WHT	Sensor Ground	<0.050v
F3 ('97)	---	Not Used	---
F3 ('98-'99)	BRN	INOCD Immobilizer Code	Digital Signals
F4	RED	Fuel Pump Relay Control	Relay Off: 12v, On: 1v
F5	---	Not Used	---
F6	GRY	Brake Switch Signal 2	B/P depressed: 0v
F7	GRN/BLK	Cruise Resume Switch	CR/S released: 0v
F8	GRN/RED	Cruise Control Set Switch	CS/S released: 0v
F9	LT GRN	Cruise Control Main Switch	M/S on: 12v, off: 0v
F10	BLU/ORN	Cruise Control Clutch Switch	Clutch in: 0v, out: 8v
F11	BLU/BLK	Cruise Indicator Light	Light on: 0v, off: 12v
F12	YEL/RED	ACP Sensor 2 Signal (AP2)	ACP closed: 0.5v
F13	BLU/RED	ACP Sensor 1 Signal (AP1)	ACP closed: 0.5v
F14	YEL/WHT	Sensor VREF (VCC3)	4.9-5.1v
F15	GRN/WHT	Sensor Ground (SG3)	<0.050v
F16	GRN/WHT	Brake Switch Signal 1	Brake Off: 12v, On: 0v
F17-26	---	Not Used	---

1997-99 Coupe 3.2L V6 VTEC VIN NA2 [M/T] 12P 'G' Connector

PCM Pin #	W/Color	Circuit Description (12-Pin)	Value at Hot Idle
G1	---	Not Used	---
G2	GRN/RED	IGM1 Power Source	12-14v
G3	GRN/RED	IGM2 Power Source	12-14v
G4	BRN	Motor Phase Out 1	Pulse Signals
G5	WHT/GRN	Motor Phase Out 2	Pulse Signals
G7	BLK	Power Ground	<0.1v
G8	BLK	Power Ground	<0.1v
G9	ORN	PCM PWR to Motor 1, 3	Pulse Signals
G10	GRN	PCM PWR to Motor 2, 4	Pulse Signals
G11	YEL	Motor Phase Out 3	Pulse Signals
G12	RED	Motor Phase Out 4	Pulse Signals

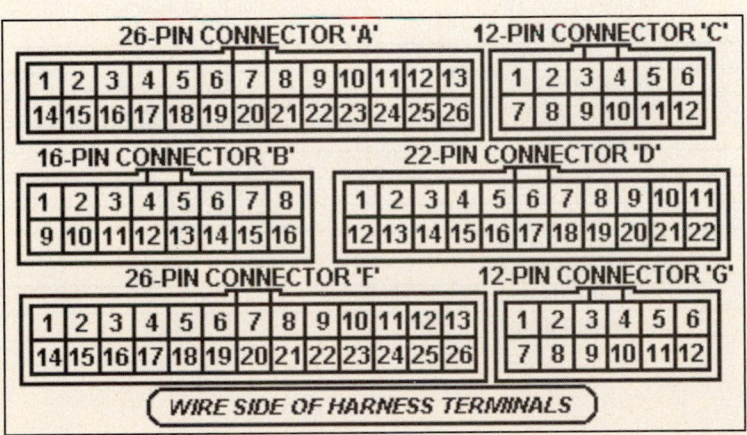

2000-03 Coupe 3.2L V6 VTEC VIN NA2 [All] 26P 'A' Connector

PCM Pin #	W/Color	Circuit Description (26-Pin)	Value at Hot Idle
A1	BRN	Injector 1 Control	2.0-3.3 ms
A2	RED	Injector 2 Control	2.0-3.3 ms
A3	BLU	Injector 3 Control	2.0-3.3 ms
A4	BLK/YEL	EVAP Canister Vent Valve	Solenoid on: 1v, off: 12v
A5	GRN/BLK	Fuel Pump Relay Control 1	Relay Off: 12v, On: 1v
A6	GRN	EGR Solenoid Control	Solenoid Off: 12v, On: 1v
A7	BLU	Malfunction Indicator Lamp	MIL Off: 12v, On: 1v
A8	RED/BLU	A/C Clutch Relay Control	Relay Off: 12v, On: 1v
A9	GRY	HO2S-11 (B1 S1) Heater	Digital Signals: 0-12-0v
A10	LT GRN	HO2S-21 (B2 S1) Heater	Digital Signals: 0-12-0v
A11	PNK	Ignition Coil 1 Control	Digital Signals: 0-12-0v
A12	BLK	Power Ground (PG1)	<0.1v
A13	YEL/BLK	Main Relay Power (B+)	12-14v
A14	YEL	Injector 4 Control	2.0-3.3 ms
A15	BLK/RED	Injector 5 Control	2.0-3.3 ms
A16	WHT/BLU	Injector 6 Control	2.0-3.3 ms
A17	ORN	EVAP Bypass Solenoid Valve	Solenoid on: 1v, off: 12v
A18	YEL/GRN	Front VTEC Solenoid	Idle: 0v, Hi-Speed: 12v
A19	---	Not Used	---
A20	PNK/BLU	IAB Solenoid Control	Solenoid Off: 12v, On: 1v
A21	RED/YEL	EVAP Purge Solenoid	Pulse Signals
A22	GRN/RED	HO2S-12 (B1 S2) Heater	Digital Signals: 0-12-0v
A23	ORN/BLK	HO2S-22 Heater Control	Digital Signals: 0-12-0v
A24	BRN	Ignition Coil 2 Control	Digital Signals: 0-12-0v
A25	BLK	Power Ground (PG2)	<0.1v
A26	BRN/BLK	Logic Ground (LG1)	<0.050v

2000-03 Coupe 3.2L V6 VTEC VIN NA2 [All] 16P 'B' Connector

PCM Pin #	W/Color	Circuit Description (16-Pin)	Value at Hot Idle
B1	BLU	Front VTEC Pressure Switch	Switch on: 0v, off: 12v
B2	RED	Ignition Coil 6 Control	Digital Signals: 0-12-0v
B3	BLU/BLK	Rear VTEC Pressure Switch	At Low Speed: 0.1v
B4	RED	A/T: Gear Position Switch	In P/N: 0v, others: 5v
B4	LT GRN	M/T: Neutral Position Switch	In 'N': 0v, others: 5v
B5	ORN	CYP Sensor 2 Signal	AC pulse signals
B6	WHT	CYP Sensor 1 Signal	AC pulse signals
B7	ORN/BLU	CKP Sensor 2 Signal	AC pulse signals
B8	BLU/GRN	CKP Sensor 1 Signal	AC pulse signals
B9	BRN/BLK	Logic Ground	<0.050v
B10	GRY	Ignition Coil 5 Control	Digital Signals: 0-12-0v
B11	GRN	Ignition Coil 4 Control	Digital Signals: 0-12-0v
B12	BLU	Ignition Coil 3 Control	Digital Signals: 0-12-0v
B13	ORN/BLU	CYP Sensor 2 Ground	<0.050v
B14	WHT/BLU	CYP Sensor 1 Ground	<0.050v
B15	WHT/BLU	CKP Sensor 2 Ground	<0.050v
B16	BLU/YEL	CKP Sensor 1 Ground	<0.050v

2000-03 Coupe 3.2L V6 VTEC VIN NA2 [All] 12P 'C' Connector

PCM Pin #	W/Color	Circuit Description (12-Pin)	Value at Hot Idle
C1	YEL/BLK	Main Relay Power (B+)	12-14v
C2	BLU/BLK	A/C Switch Signal	A/C Off: 12v, On: 1v
C3	RED/GRN	A/C Pressure Switch Signal	A/C Off: 4-5v, On: 0v
C4	PNK	M/T: Clutch Engage Switch	Clutch in: 5v, out: 0v
C5	BLU	Service Check Connector	SCS Open: 4.80v
C6	BLK/WHT	Starter Switch Signal	Cranking: 9-11v
C7	ORN	Vehicle Speed Sensor	Moving: pulse signals
C8	GRN	Engine Speed Pulse Signal	Pulses
C9	GRN/YEL	Front Peak Hold Reset	Pulses
C10	GRN/BLK	Rear Peak Hold Reset	Pulses
C11	YEL/RED	Front Misfire Pulse	Pulses
C12	YEL	Rear Misfire Pulse	Pulses

2000-03 Coupe 3.2L V6 VTEC VIN NA2 [All] 22P 'D' Connector

PCM Pin #	W/Color	Circuit Description (22-Pin)	Value at Hot Idle
D1	WHT/YEL	Keep Alive Power (VBU)	12-14v
D2	RED/BLU	Knock Sensor 1 Signal	No detonation: 18mv AC
D3	WHT/BLK	Air Pump Current Sensor	Pump on: Pulse Signals
D4	YEL/GRN	K-Line Signal	12v
D5	WHT/RED	Alternator 'FR' Signal	Digital Signals: 0-5-0v
D6	BLU	Fuel Tank Pressure Sensor	Fuel Cap off: 2.5v
D7	RED/WHT	ECT Sensor Signal	At 180ºF: 0.5-0.6v
D8	RED/YEL	IAT Sensor Signal	At 100ºF: 2-3v
D9	WHT/YEL	MAP Sensor Signal	0.8-0.9v
D10	YEL/BLU	Sensor VREF (VCC2)	4.9-5.1v
D11	GRN/BLU	Sensor Ground	<0.050v
D12	RED/BLK	TP Sensor Signal	0.5-0.6v
D13	WHT	Knock Sensor 2 Signal	No detonation: 18mv AC
D14-15	---	Not Used	---
D16	GRN	HO2S-21 (B2 S1) Signal	0.1-1.1v
D17	WHT/RED	HO2S-22 (B2 S2) Signal	0.1-1.1v
D18	BLU/RED	HO2S-11 (B1 S1) Signal	0.1-1.1v
D19	WHT	HO2S-12 (B1 S2) Signal	0.1-1.1v
D20	WHT/BLK	EGR Valve Lift Sensor	1.1-1.2v
D21	BLU/GRN	HO2S-21 (B2 S1) Ground	<0.050v
D22	WHT	HO2S-22 (B2 S2) Ground	<0.050v

Abbreviation	Color	Abbreviation	Color	Abbreviation	Color
BLK	Black	LT BLU	LT Blue	TAN	Tan
BLU	Blue	LT GRN	LT Green	VIO	Violet
BRN	Brown	ORN	Orange	WHT	White
GRY	Gray	PNK	Pink	YEL	Yellow
GRN	Green	PPL	Purple		

2000-03 Coupe 3.2L V6 VTEC VIN NA2 [All] 26P 'F' Connector

PCM Pin #	W/Color	Circuit Description (26-Pin)	Value at Hot Idle
F1	YEL/WHT	Sensor VREF (VCC1)	4.9-5.1v
F2, F15	GRN/WHT	Sensor Ground SG1, SG3	<0.050v
F3	BRN	INOCD Immobilizer Code	Digital Signals
F4	RED	Fuel Pump Relay Control	Relay Off: 12v, On: 1v
F6	GRY	Brake Switch Signal 2	B/P depressed: 0v
F7	GRN/BLK	Cruise Resume Switch	CR/S released: 0v
F8	GRN/RED	Cruise Control Set Switch	CS/S released: 0v
F9	LT GRN	Cruise Control Main Switch	M/S on: 12v, off: 0v
F10	BLU/ORN	C/C A/T: Gear Position Switch	In 'D', D3, D2: 0v
F10	BLU/ORN	C/C M/T: Clutch Switch	Clutch in: 0v, out: 8v
F11	BLU/BLK	Cruise Indicator Light	Light on: 0v, off: 12v
F12	YEL/RED	Accelerator Position Sensor 2	ACP closed: 0.5v
F13	BLU/RED	Accelerator Position Sensor 1	ACP closed: 0.5v
F14	YEL/WHT	Sensor VREF (VCC1)	4.9-5.1v
F16	GRN/WHT	Brake Switch Signal 1	Brake Off: 12v, On: 0v
F17	RED/WHT	Reverse Lockout Relay Out	VSS over 30 mph: 0v
F18	BLU	TCM VREF	4.9-5.1v
F19	BLU/YEL	Front VTEC Solenoid Valve	0v, Hi-Speed: 12v
F20	GRY/YEL	Rear VTEC Solenoid Valve	0v, Hi-Speed: 12v
F21	GRN/YEL	A/T: FI Data Line 'B'	Pulse Signals
F22	WHT/YEL	A/T: FI Data Line 'A'	Pulse Signals
F23	BLU/WHT	Air Pump Relay Control	Relay Off: 12v, On: 1v
F24	RED/BLU	TCS Data Line	Pulse Signals
F25	RED	Air Control Solenoid Valve	Pump on: 1v, off: 12v

2000-03 Coupe 3.2L V6 VTEC VIN NA2 [All] 12P 'G' Connector

PCM Pin #	W/Color	Circuit Description (12-Pin)	Value at Hot Idle
G1, G9	---	Not Used	---
G2, G3	GRN/RED	IGM1, IGM2 Power Source	Key on: 12v, off: 0v
G4	BRN	Motor Phase Out 1	Key on: 0v or pulse
G5	WHT/GRN	Motor Phase Out 2	Key on: 0v or pulse
G7, G8	BLK	Power Ground	<0.1v
G9	ORN	PCM PWR to Motor 1, 3	KOEO: pulse
G10	GRN	PCM PWR to Motor 2, 4	KOEO: pulse
G11	YEL	Motor Phase Out 3	Key on: 0v or pulse
G12	RED	Motor Phase Out 4	Key on: 0v or pulse

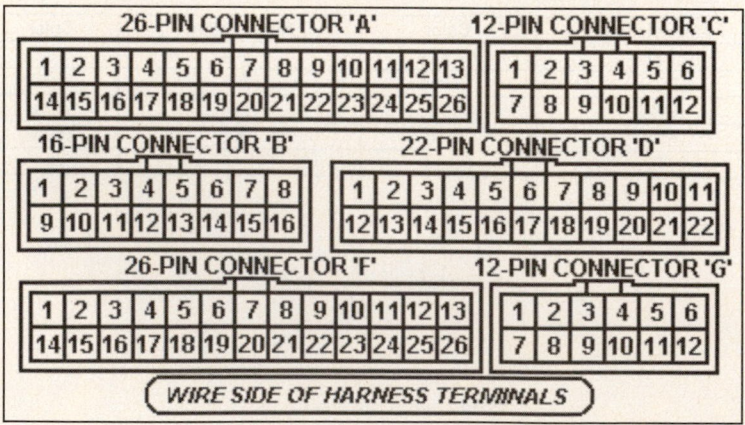

VIGOR Pin Tables

1992 Sedan 2.5L I5 MFI SOHC VIN CC2 [A/T] 26P 'A' Connector

PCM Pin #	W/Color	Circuit Description (26-Pin)	Value at Hot Idle
A1	BRN	Injector 1	2.0-3.3 ms
A2	YEL	Injector 4	2.0-3.3 ms
A3	RED	Injector 2	2.0-3.3 ms
A4	GRN	Injector 5	2.0-3.3 ms
A5	LT BLU	Injector 3	2.0-3.3 ms
A6	PNK/WHT	HO2S-11 (B1 S1) Heater	Heater Off: 12v, On: 1v
A7	GRN/BLK	Fuel Pump Relay Control	Relay Off: 12v, On: 1v
A8	GRN/BLK	Fuel Pump Relay Control	Relay Off: 12v, On: 1v
A9	BLK/BLU	Electronic Air Control Valve	Pulse Signals
A10	ORN	EVAP Purge Solenoid	Solenoid Off: 12v, On: 1v
A11	RED	EGR Solenoid Control	Solenoid Off: 12v, On: 1v
A12	GRN/YEL	Fan Timing Unit Signal	Relay Off: 12v, On: 1v
A13	GRN/RED	Malfunction Indicator Lamp	MIL Off: 12v, On: 1v
A15	RED/BLU	A/C Clutch Relay Control	Relay Off: 12v, On: 1v
A16	WHT/GRN	Alternator Control Signal	Lights Off: 12v, On: 1v
A17	BLK/RED	Bypass Solenoid Signal	Solenoid Off: 12v, On: 1v
A18	PNK	A/T: Shift Acknowledge Signal	Digital Signals
A21	YEL/GRN	Igniter 1 Signal	Digital Signals: 0-12-0v
A22	YEL/GRN	Igniter 2 Signal	Digital Signals: 0-12-0v
A23	BLK	Power Ground	<0.1v
A24	BLK	Power Ground	<0.1v
A25	YEL/BLK	Main Relay Power (B+)	12-14v
A26	BLK/RED	Logic Ground	<0.050v

1992 Sedan 2.5L I5 MFI SOHC VIN CC2 [A/T] 16P 'B' Connector

PCM Pin #	W/Color	Circuit Description (16-Pin)	Value at Hot Idle
B1	YEL/BLK	Main Relay Power (B+)	12-14v
B2	BRN/BLK	Logic Ground	<0.050v
B3	WHT/GRN	A/T: Upshift/Downshift Signal	Digital Signals
B4	WHT/RED	A/T: Upshift/Downshift Signal	Digital Signals
B5	BLU/BLK	A/C Pressure Switch	A/C Off: 12v, On: 1v
B6	---	Not Used	---
B7	YEL/GRN	A/T: Neutral Indicator	In 'N': 0v, Others: 12v
B8	RED	PSP Switch Signal	Straight: 0v, Turning: 12v
B9	BLU/RED	Start (Cranking) Signal	Cranking: 9-11v
B10	ORN	Vehicle Speed Sensor	Moving: pulse signals
B11	ORN	CYP Sensor Signal	AC pulse signals
B12	WHT	CYP Sensor Ground	<0.050v
B13	ORN/BLU	TDC Sensor Signal	AC pulse signals
B14	WHT/BLU	TDC Sensor Ground	<0.050v
B15	YEL/GRN	CKP Sensor Signal	AC pulse signals
B16	BLU/YEL	CKP Sensor Ground	<0.050v

1992 Sedan 2.5L I5 MFI SOHC VIN CC2 [A/T] 12P 'C' Connector

PCM Pin #	W/Color	Circuit Description (12-Pin)	Value at Hot Idle
C1	ORN/WHT	Rear Knock Sensor Signal	No Detonation: 18mv AC
C2	---	Not Used	---
C3	RED/BLU	Front Knock Sensor Signal	No Detonation: 18mv AC
C4-12	---	Not Used	---

1992 Sedan 2.5L I5 MFI SOHC VIN CC2 [A/T] 22P 'D' Connector

PCM Pin #	W/Color	Circuit Description (22-Pin)	Value at Hot Idle
D1	WHT/GRN	Keep Alive Power (VBU)	12-14v
D2	GRN/WHT	Brake Switch Signal	Brake Off: 12v, On: 0v
D3	---	Not Used	---
D4	BRN	Service Check Connector	SCS Open: 4.80v
D5-7	---	Not Used	---
D8	BRN	Ignition Timing Adjustment	0.4-4.5v
D9	WHT/RED	Alternator 'FR' Signal	Digital Signals: 0-5-0v
D10	GRN/RED	Electronic Load Detector	Varies w/Load: 2.5-3.5v
D11	RED/YEL	Throttle Angle Sensor	0.5-0.6v
D12	WHT/BLK	EGR Valve Lift Sensor	1.1-1.2v
D13	YEL/GRN	ECT Sensor Signal	At 180°F: 0.5-0.6v
D14	WHT	HO2S Sensor	0.1-1.1v
D15	WHT/YEL	IAT Sensor Signal	At 100°F: 2-3v
D16	---	Not Used	---
D17	WHT/BLU	MAP Sensor Signal	0.8-0.9v
D18	BLU/WHT	A/T: Control Unit VREF	4.9-5.1v
D19	YEL/WHT	MAP Sensor VREF	4.9-5.1v
D20	YEL/WHT	Sensor VREF	4.9-5.1v
D21	GRN/WHT	MAP Sensor Ground	<0.050v
D22	GRN/WHT	Sensor Ground	<0.050v

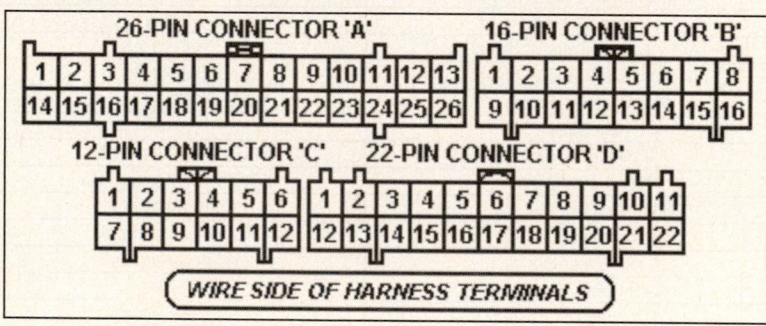

1993 Sedan 2.5L I5 MFI SOHC VIN CC2 [A/T] 26P 'A' Connector

PCM Pin #	W/Color	Circuit Description (26-Pin)	Value at Hot Idle
A1	BRN	Injector 1	2.0-3.3 ms
A2	YEL	Injector 4	2.0-3.3 ms
A3	RED	Injector 2	2.0-3.3 ms
A4	GRN	Injector 5	2.0-3.3 ms
A5	BLU	Injector 3	2.0-3.3 ms
A6	PNK/WHT	HO2S-11 (B1 S1) Heater	Heater Off: 12v, On: 1v
A7	GRN/BLK	Fuel Pump Relay Control	Relay Off: 12v, On: 1v
A8	GRN/BLK	Fuel Pump Relay Control	Relay Off: 12v, On: 1v
A9	BLK/BLU	Idle Air Control Valve	Pulse Signals
A10	ORN	EVAP Purge Solenoid	Solenoid Off: 12v, On: 1v
A11	RED	EGR Solenoid Control	Solenoid Off: 12v, On: 1v
A12	GRN/YEL	Fan Timing Unit Signal	Relay Off: 12v, On: 1v
A13	GRN/RED	Malfunction Indicator Lamp	MIL Off: 12v, On: 1v
A15	RED/BLU	A/C Clutch Relay Control	Relay Off: 12v, On: 1v
A16	WHT/GRN	Alternator Control Signal	Lights Off: 12v, On: 1v
A17	BLU/RED	Bypass Solenoid Signal	Solenoid Off: 12v, On: 1v
A18	PNK	A/T: Shift Acknowledge Signal	Digital Signals
A19-20	BLK/RED	Power Ground	<0.1v
A21	YEL/GRN	Igniter 1 Signal	Digital Signals: 0-12-0v
A22	YEL/GRN	Igniter 2 Signal	Digital Signals: 0-12-0v
A23	BLK	Power Ground	<0.1v
A24	BLK	Power Ground	<0.1v
A25	YEL/BLK	Main Relay Power (B+)	12-14v
A26	BLK/RED	Chassis Ground	<0.050v

1993 Sedan 2.5L I5 MFI SOHC VIN CC2 [A/T] 16P 'B' Connector

PCM Pin #	W/Color	Circuit Description (16-Pin)	Value at Hot Idle
B1	YEL/BLK	Main Relay Power (B+)	12-14v
B2	BRN/BLK	Power Ground	<0.1v
B3	WHT/GRN	A/T: Upshift/Downshift Signal	Digital Signals
B4	WHT/RED	A/T: Upshift/Downshift Signal	Digital Signals
B5	BLU/BLK	A/C Pressure Switch	A/C Off: 12v, On: 1v
B6	---	Not Used	---
B7	YEL/GRN	A/T: Neutral Position Indicator	In P/N: 0v, Others: 12v
B8	RED	PSP Switch Signal	Straight: 0v, Turning: 12v
B9	BLU/RED	Start (Cranking) Signal	Cranking: 9-11v
B10	ORN	Vehicle Speed Sensor	Moving: pulse signals
B11	ORN	CYP Sensor Signal	AC pulse signals
B12	WHT	CYP Sensor Ground	<0.050v
B13	ORN/BLU	TDC Sensor Signal	AC pulse signals
B14	WHT/BLU	TDC Sensor Ground	<0.050v
B15	YEL/GRN	CKP Sensor Signal	AC pulse signals
B16	BLU/YEL	CKP Sensor Ground	<0.050v

1993 Sedan 2.5L I5 MFI SOHC VIN CC2 [A/T] 12P 'C' Connector

PCM Pin #	W/Color	Circuit Description (12-Pin)	Value at Hot Idle
C1	ORN/WHT	Rear Knock Sensor Signal	No Detonation: 18mv AC
C2	---	Not Used	---
C3	RED/BLU	Front Knock Sensor Signal	No Detonation: 18mv AC
C4-12	---	Not Used	---

1993 Sedan 2.5L I5 MFI SOHC VIN CC2 [A/T] 22P 'D' Connector

PCM Pin #	W/Color	Circuit Description (22-Pin)	Value at Hot Idle
D1	WHT/GRN	Keep Alive Power (VBU)	12-14v
D2	GRN/WHT	Brake Switch Signal	Brake Off: 12v, On: 0v
D3	---	Not Used	---
D4	BRN	Service Check Connector	SCS Open: 4.80v
D5-7	---	Not Used	---
D8	BRN	Ignition Timing Adjustment	0.4-4.5v
D9	WHT/RED	Alternator Control Signal	Lights Off: 12v, On: 1v
D10	GRN/RED	Electronic Load Detector	Varies: 2.5-3.5v
D11	RED/YEL	TP Sensor Signal	0.5-0.6v
D12	WHT/BLK	EGR Valve Lift Sensor	1.1-1.2v
D13	YEL/GRN	ECT Sensor Signal	At 180ºF: 0.5-0.6v
D14	WHT	HO2S-11 (B1 S1) Signal	0.1-1.1v
D15	WHT/YEL	IAT Sensor Signal	At 100ºF: 2-3v
D16	---	Not Used	---
D17	WHT/BLU	MAP Sensor Signal	0.8-0.9v
D18	BLU/WHT	A/T: Control Unit VREF	4.9-5.1v
D19	YEL/WHT	MAP Sensor VREF	4.9-5.1v
D20	YEL/WHT	Sensor VREF	4.9-5.1v
D21	GRN/WHT	MAP Sensor Ground	<0.050v
D22	GRN/WHT	Sensor Ground	<0.050v

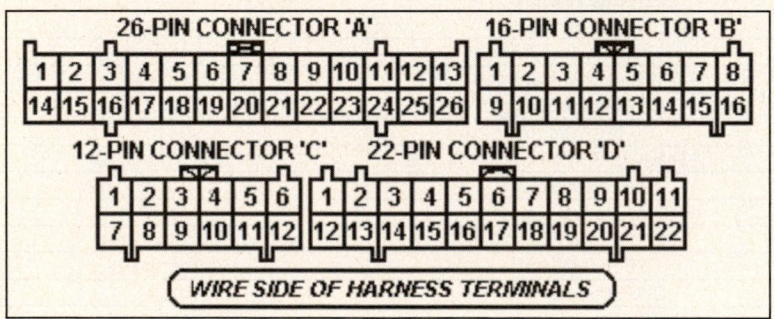

1994 Sedan 2.5L I5 MFI SOHC VIN CC2 [A/T] 26P 'A' Connector

PCM Pin #	W/Color	Circuit Description (26-Pin)	Value at Hot Idle
A1	BRN	Injector 1	2.0-3.3 ms
A2	YEL	Injector 4	2.0-3.3 ms
A3	RED	Injector 2	2.0-3.3 ms
A4	GRN	Injector 5	2.0-3.3 ms
A5	BLU	Injector 3	2.0-3.3 ms
A6	PNK/WHT	HO2S-11 (B1 S1) Heater	Heater Off: 12v, On: 1v
A7	GRN/BLK	Fuel Pump Relay Control	Relay Off: 12v, On: 1v
A8	GRN/BLK	Fuel Pump Relay Control	Relay Off: 12v, On: 1v
A9	BLK/BLU	Electronic Air Control Valve	Pulse Signals
A10	ORN	EVAP Purge Solenoid	Solenoid Off: 12v, On: 1v
A11	RED	EGR Solenoid Control	Solenoid Off: 12v, On: 1v
A12	GRN/YEL	Fan Timing Unit Signal	Relay Off: 12v, On: 1v
A13	GRN/RED	Malfunction Indicator Lamp	MIL Off: 12v, On: 1v
A15	RED/BLU	A/C Clutch Relay Control	Relay Off: 12v, On: 1v
A16	WHT/GRN	Alternator Control Signal	Lights Off: 12v, On: 1v
A17	BLU/RED	Bypass Solenoid Signal	Solenoid Off: 12v, On: 1v
A18	PNK	A/T: Shift Acknowledge Signal	Digital Signals
A20	GRN	EVAP Purge Solenoid	Solenoid Off: 12v, On: 1v
A21	YEL/GRN	Igniter 1 Signal	Digital Signals: 0-12-0v
A22	YEL/GRN	Igniter 2 Signal	Digital Signals: 0-12-0v
A23	BLK	Power Ground	<0.1v
A24	BLK	Power Ground	<0.1v
A25	YEL/BLK	Main Relay Power (B+)	12-14v
A26	BLK/RED	Chassis Ground	<0.050v

1994 Sedan 2.5L I5 MFI SOHC VIN CC2 [A/T] 16P 'B' Connector

PCM Pin #	W/Color	Circuit Description (16-Pin)	Value at Hot Idle
B1	YEL/BLK	Main Relay Power (B+)	12-14v
B2	BRN/BLK	Power Ground	<0.1v
B3	WHT/GRN	A/T: Upshift/Downshift Signal	Digital Signals
B4	WHT/GRN	A/T: Upshift/Downshift Signal	Digital Signals
B5	BLU/BLK	A/C Pressure Switch	A/C Off: 12v, On: 1v
B6	---	Not Used	---
B7	YEL/GRN	A/T: Neutral Position Indicator	In 'N': 0v, Others: 12v
B8	RED	PSP Switch Signal	Straight: 0v, Turning: 12v
B9	BLU/RED	Start (Cranking) Signal	Cranking: 9-11v
B10	ORN	Vehicle Speed Sensor	Moving: pulse signals
B11	ORN	CYP Sensor Signal	AC pulse signals
B12	WHT	CYP Sensor Ground	<0.050v
B13	ORN/BLU	TDC Sensor Signal	AC pulse signals
B14	WHT/BLU	TDC Sensor Ground	<0.050v
B15	BLU/GRN	CKP Sensor Signal	AC pulse signals
B16	BLU/YEL	CKP Sensor Ground	<0.050v

1994 Sedan 2.5L I5 MFI SOHC VIN CC2 [A/T] 12P 'C' Connector

PCM Pin #	W/Color	Circuit Description (12-Pin)	Value at Hot Idle
C1	ORN/WHT	Rear Knock Sensor Signal	No Detonation: 18mv AC
C2	---	Not Used	---
C3	RED/BLU	Front Knock Sensor Signal	No Detonation: 18mv AC
C4-12	---	Not Used	---

1994 Sedan 2.5L I5 MFI SOHC VIN CC2 [A/T] 22P 'D' Connector

PCM Pin #	W/Color	Circuit Description (22-Pin)	Value at Hot Idle
D1	WHT/GRN	Keep Alive Power (VBU)	12-14v
D2	GRN/WHT	Brake Switch Signal	Brake Off: 12v, On: 0v
D3	---	Not Used	---
D4	BRN	Service Check Connector	SCS Open: 4.80v
D5-7	---	Not Used	---
D8	BRN	Ignition Timing Adjustment	0.4-4.5v
D9	WHT/RED	Alternator Control Signal	Lights Off: 12v, On: 1v
D10	GRN/RED	Electronic Load Detector	Varies: 2.5-3.5v
D11	RED/YEL	TP Sensor Signal	0.5-0.6v
D12	WHT/BLK	EGR Valve Lift Sensor	1.1-1.2v
D13	YEL/GRN	ECT Sensor Signal	At 180°F: 0.5-0.6v
D14	WHT	HO2S-11 (B1 S1) Signal	0.1-1.1v
D15	WHT/YEL	IAT Sensor Signal	At 100°F: 2-3v
D16	---	Not Used	---
D17	WHT/BLU	MAP Sensor Signal	0.8-0.9v
D18	---	A/T: Control Unit VREF	4.9-5.1v
D19	YEL/WHT	MAP Sensor VREF	4.9-5.1v
D20	YEL/WHT	Sensor VREF	4.9-5.1v
D21	GRN/WHT	MAP Sensor Ground	<0.050v
D22	GRN/WHT	Sensor Ground	<0.050v

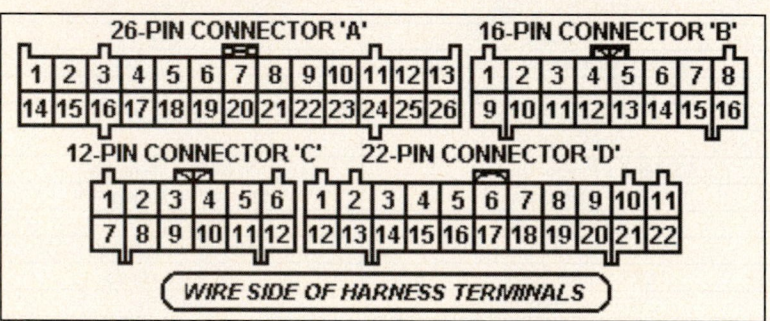

CL Pin Tables

1997 Coupe 2.2L I4 VTEC VIN YA1 [A/T] 32P 'A' Connector

PCM Pin #	W/Color	Circuit Description (32-Pin)	Value at Hot Idle
A1	YEL	Injector 4	2.0-3.3 ms
A3	BLU	Injector 3	2.0-3.3 ms
A2	RED	Injector 2	2.0-3.3 ms
A1	BRN	Injector 1	2.0-3.3 ms
A5	ORN/BLU	HO2S-12 (B1 S2) Heater	Digital Signals: 0-12-0v
A6	ORN/BLK	HO2S-11 (B1 S1) Heater	Digital Signals: 0-12-0v
A7	RED	EGR Solenoid Control	Solenoid Off: 12v, On: 1v
A8	GRN/YEL	VTEC Solenoid Control	Idle; 0v, hi-rpm: 12v
A9, A22	BRN/BLK	Logic Ground	<0.050v
A10, A23	BLK	Power Ground	<0.1v
A11	YEL/BLK	Main Relay Power (B+)	12-14v
A12	BLK/BLU	Idle Air Control Valve	Pulse Signals
A13	GRN/WHT	Engine Mount Solenoid	Solenoid Off: 12v, On: 1v
A15	RED/YEL	EVAP Purge Solenoid	Pulse Signals
A16	GRN/BLK	Fuel Pump Relay Control	Relay Off: 12v, On: 1v
A17	RED/BLU	A/C Clutch Relay Control	Relay Off: 12v, On: 1v
A18	GRN/RED	Malfunction Indicator Lamp	MIL Off: 12v, On: 1v
A19	WHT/GRN	Alternator Charging Signal	Lights Off: 12v, On: 1v
A20	YEL/GRN	Igniter Signal	Digital Signals: 0-12-0v
A24	YEL/BLK	Main Relay Power (B+)	12-14v
A25	ORN/GRN	IAR Solenoid	Solenoid Off: 12v, On: 1v
A27	GRN	Radiator Fan Relay Control	Fan on: 1v, off: 12v

1997 Coupe 2.2L I4 VTEC VIN YA1 [A/T] 16P 'D' Connector

PCM Pin #	W/Color	Circuit Description (16-Pin)	Value at Hot Idle
D1	RED/BLK	TP Sensor Signal	0.5-0.6v
D2	RED/WHT	ECT Sensor Signal	At 180°F: 0.5-0.6v
D3	WHT/YEL	MAP Sensor Signal	0.8-0.9v
D4	YEL/WHT	MAP Sensor VREF	4.9-5.1v
D5	GRN/WHT	Brake Switch Signal	Brake on: 0v, off: 12v
D7	WHT/RED	HO2S-11 (B1 S1) Signal	0.1-1.1v
D8	RED/YEL	IAT Sensor Signal	At 100°F: 2-3v
D9	WHT/BLK	EGR Valve Lift Sensor	1.1-1.2v
D10	YEL/BLU	Sensor VREF	4.9-5.1v
D11	GRN/BLK	Sensor Ground	<0.050v
D12	GRN/WHT	MAP Sensor Ground	<0.050v
D13	RED/WHT	HO2S-22 (B2 S2) Ground	<0.050v
D14	WHT/RED	HO2S-12 (B1 S2) Signal	0.1-1.1v
D16	GRN/RED	Electronic Load Detector	Varies: 2.5-3.5v

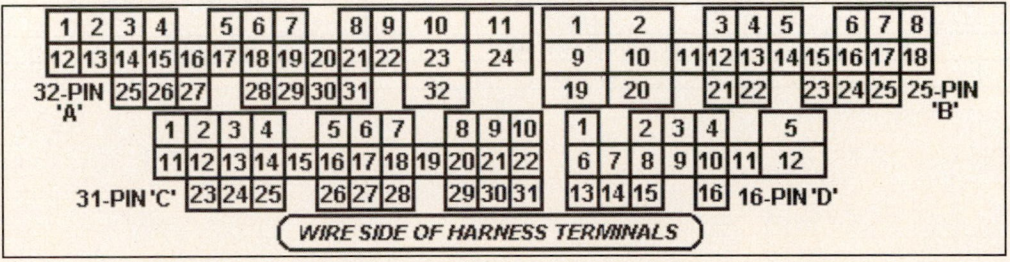

1997 Coupe 2.2L I4 VTEC VIN YA1 [A/T] 25P 'B' Connector

PCM Pin #	W/Color	Circuit Description (25-Pin)	Value at Hot Idle
B1-2	---	Not Used	---
B3	BLU/YEL	A/T: Shift Solenoid 'A'	SSA Off: 12v, On: 1v
B4	GRN/BLK	A/T: Lockup Solenoid 'B'	LSB on: 1v, off: 12v
B5	YEL	A/T: Lockup Solenoid 'A'	LSA on: 1v, off: 12v
B6-7	---	Not Used	---
B8	GRN/BLU	A/T: Gear Position Switch	In D3: 0v, other: 12v
B9-10	---	Not Used	---
B11	GRN/WHT	A/T: Shift Solenoid 'B'	SSB Off: 12v, On: 1v
B12	WHT/GRN	Interlock Control Unit Signal	Key & Brake on: 12v
B13	BLU/RED	A/T: D4 Indicator Light	Light on: 12v, off: 0v
B14	WHT/BLU	Mainshaft Speed Sensor 'N'	<0.050v
B15	ORN/BLU	Mainshaft Speed Sensor 'P'	Pulses
B16	GRN/RED	A/T: Gear Position Switch	In 'D': 0v, others: 12v
B17	GRN/YEL	A/T: Gear Position Switch	In D2: 0v, other: 12v
B18	GRN/WHT	A/T: Gear Position Switch	In D1: 0v, other: 12v
B19-21	---	Not Used	---
B22	BLU/YEL	Countershaft Speed Sensor N	<0.050v
B23	BLU/GRN	Countershaft Speed Sensor P	Wheels turn: pulses
B24	GRN/BLK	A/T: Gear Position Switch	In D4: 0v, other: 12v
B25	LT GRN	A/T: Gear Position Switch	In P/N: 0v, Others: 12v

1997 Coupe 2.2L I4 VTEC VIN YA1 [A/T] 31P 'C' Connector

PCM Pin #	W/Color	Circuit Description (31-Pin)	Value at Hot Idle
C1	---	Not Used	---
C2	BLU	CKP Sensor Signal	AC pulse signals
C3	GRN	TDC Sensor Signal	AC pulse signals
C4	YEL	CYP Sensor Signal	AC pulse signals
C5	RED/WHT	A/C Switch Signal	A/C Off: 12v, On: 1v
C6	BLK/GRN	Starter Switch Signal	Cranking: 9-11v
C7	RED	Service Check Connector	SCS Open: 4.80v
C8	LT GRN	K-Line Signal	12v
C9	---	Not Used	---
C10	WHT/YEL	Keep Alive Power (VBU)	12-14v
C11	---	Not Used	---
C12	WHT	CKP Sensor Ground	<0.050v
C13	RED	TDC Sensor Ground	<0.050v
C14	BLK	CYP Sensor Ground	<0.050v
C15	BLU/BLK	VTEC Pressure Switch	Switch on: 0v, off: 12v
C16	GRN	PSP Switch Signal	Straight: 0v, Turning: 12v
C17	WHT/RED	Alternator 'FR' Signal	Idle: 3-5v, Acc: 2-4v
C18	ORN	Vehicle Speed Sensor	Moving: pulse signals
C19	---	Not Used	---
C20	BRN	EVAP Purge Flow Switch	Switch on: 0v, off: 12v
C21-31	---	Not Used	---

1998-99 Coupe 2.3L I4 VTEC VIN YA3 [A/T] 32P 'A' Connector

PCM Pin #	W/Color	Circuit Description (32-Pin)	Value at Hot Idle
A1, A11	---	Not Used	---
A2	GRN/WHT	Engine Mount Solenoid	0v, Off-Idle: 12v
A3	BLU/BLK	EVAP Bypass Solenoid	Solenoid Off: 12v, On: 1v
A4	ORN/GRN	EVAP Vent Solenoid	Solenoid Off: 12v, On: 1v
A5	BLU/WHT	Cruise Control Signal	Cruise On: pulse signals
A6	RED/YEL	EVAP Purge Solenoid	Pulse Signals
A8	ORN/BLU	HO2S-12 (B1 S2) Heater	Digital Signals: 0-12-0v
A9	ORN	Vehicle Speed Sensor	Moving: pulse signals
A10	RED	Service Check Connector	SCS Open: 4.80v
A12	PNK	Immobilizer Indicator Control	Lamp On: 1v, Off: 12v
A13	ORN/BLU	Immobilizer Enable Signal	Digital Signals
A14	BLU/RED	A/T: D4 Indicator Light	In D4: 0v, Off: 12v
A15	YEL/BLK	Immobilizer Fuel Pump Relay	Relay Off: 12v, On: 1v
A17	RED/BLU	A/C Clutch Relay Control	Relay Off: 12v, On: 1v
A18	GRN/RED	Malfunction Indicator Lamp	MIL Off: 12v, On: 1v
A19	BLU	Engine Speed Pulse	Pulses
A20	GRN	Radiator Fan Relay Control	Relay Off: 12v, On: 1v
A21	LT BLU	K-Line Signal	12v
A16, A22	---	Not Used	---
A23	WHT/RED	HO2S-12 (B1 S2) Signal	0.1-1.1v
A24	BLU/RED	Starter Switch Signal	Cranking: 9-11v
A25	BLU/GRN	INOCD Immobilizer Code	Digital Signals
A26	GRN/YEL	PSP Switch Signal	Straight: 0v, Turning: 12v
A27	RED/WHT	A/C Switch Signal	A/C Off: 12v, On: 1v
A28	WHT/GRN	Interlock Control Unit Signal	Key & Brake on: 12v
A29	LT GRN	Fuel Tank Pressure Sensor	Fuel Cap off: 2.5v
A30	GRN/RED	Electronic Load Detector	Varies: 2.5-3.5v
A31	---	Not Used	---
A32	GRN/WHT	Brake Switch Signal	Brake on: 0v, off: 12v

1998-99 Coupe 2.3L I4 VTEC VIN YA3 [A/T] 16P 'D' Connector

PCM Pin #	W/Color	Circuit Description (16-Pin)	Value at Hot Idle
D1	YEL	A/T: Lockup Solenoid	LUS Off: 12v, On: 1v
D2	GRN/WHT	A/T: Shift Solenoid 'B'	SSB Off: 12v, On: 1v
D3	GRN	A/T Shift Solenoid 'C'	SSA Off: 12v, On: 1v
D4	---	Not Used	---
D5	BLK/YEL	Solenoid Feed (B+)	Solenoid on: 12-14v
D6	GRN/RED	A/T: Gear Position Switch	In 'D': 0v, others: 12v
D7	BLU/YEL	A/T: Shift Solenoid 'A'	SSA Off: 12v, On: 1v
D8	GRN/BLU	A/T: Gear Position Switch	In D3: 0v, Others: 12v
D9	GRN/BLK	A/T: Gear Position Switch	In D4: 0v, Others: 12v
D10	BLU	Countershaft Speed Sensor P	Wheels turning: 12v
D11	RED/BLU	Mainshaft Speed Sensor 'P'	Moving: pulse signals
D12	WHT	Mainshaft Speed, Ground 'N'	<0.050v
D13	LT GRN	A/T: Gear Position Switch	In P/N: 0v, Others: 12v
D14	GRN/YEL	A/T: Gear Position Switch	In D2: 0v, Others: 12v
D15	GRN/WHT	A/T: Gear Position Switch	In D1: 0v, Others: 12v
D16	GRN	Countershaft Speed Sensor N	<0.050v

1998-99 Coupe 2.3L I4 VTEC VIN YA3 [All] 25P 'B' Connector

PCM Pin #	W/Color	Circuit Description (25-Pin)	Value at Hot Idle
B1, B9	YEL/BLK	Main Relay Power (B+)	12-14v
B2, B10	BLK	Power Ground	<0.1v
B3, B4	RED, BLU	Injector 2, Injector 3	2.0-3.3 ms
B5, B11	YEL, BRN	Injector 4, Injector 1	2.0-3.3 ms
B7	PNK	E-EGR Solenoid Control	Digital Signals: 0-12-0v
B8	WHT	A/T: Clutch Solenoid 'A-'	Pulse Signals
B12	GRN/YEL	VTEC Solenoid Control	Idle: 0v, Hi-Speed: 12v
B13	YEL/GRN	Igniter Control Signals	Digital Signals: 0-12-0v
B14	BLU/BLK	A/T: 2nd Clutch Pressure Sw.	Open: 12v, Closed: 1v
B17	RED	A/T: Clutch Solenoid 'A+'	Pulse Signals
B18	GRN	A/T: Clutch Solenoid 'B-'	Pulse Signals
B20, 22	BRN/BLK	Logic Ground (LG1, LG2)	<0.050v
B21	WHT/YEL	Keep Alive Power (VBU)	12-14v
B23	BLK/BLU	Idle Air Control Valve	Pulse Signals
B24	BLU/WHT	3rd Clutch Press. Switch	Open: 12v, Closed: 1v
B25	ORN	A/T" Clutch Solenoid 'B+'	Pulse Signals

1998-99 Coupe 2.3L I4 VTEC VIN YA3 [All] 31P 'C' Connector

PCM Pin #	W/Color	Circuit Description (31-Pin)	Value at Hot Idle
C1	BLK/WHT	HO2S-11 (B1 S1) Heater	Digital Signals: 0-12-0v
C2	WHT/GRN	Alternator Charging Signal	Lights Off: 12v, On: 1v
C3	RED/BLU	Knock Sensor Signal	No Detonation: 18mv AC
C5	WHT/RED	Alternator 'FR' Signal	Digital Signals: 0-5-0v
C6	WHT/BLK	EGR Valve Lift Sensor	1.1-1.2v
C7, C18	GRN/WHT	MAP Sensor, Sensor Ground	<0.050v
C8	BLU	CKP Sensor Signal	AC pulse signals
C9	WHT	CKP Sensor Ground	<0.050v
C10	BLU/BLK	VTEC Pressure Switch	0v, Off-Idle: 12v
C16	WHT	HO2S-11 (B1 S1) Signal	0.1-1.1v
C17	RED/GRN	MAP Sensor Signal	0.8-0.9v
C19	YEL/RED	MAP Sensor VREF	4.9-5.1v
C20	GRN	TDC Sensor Signal	AC pulse signals
C21	RED	TDC Sensor Ground	<0.050v
C23	BLU/WHT	Vehicle Speed Sensor	Moving: pulse signals
C25	RED/YEL	IAT Sensor Signal	At 100ºF: 2-3v
C26	RED/WHT	ECT Sensor Signal	At 180ºF: 0.5-0.6v
C27	RED/BLK	TP Sensor Signal	0.5-0.6v
C28	YEL/BLU	Sensor VREF (VCC2)	4.9-5.1v
C29	YEL	CYP Sensor Signal	AC pulse signals
C30	BLK	CYP Sensor Ground	<0.050v

WIRE SIDE OF HARNESS TERMINALS

1997 Coupe 3.0L V6 VTEC VIN YA2 [A/T] 32P 'A' Connector

PCM Pin #	W/Color	Circuit Description (32-Pin)	Value at Hot Idle
A1	YEL	Injector 4 Control	2.0-3.3 ms
A2	BLU	Injector 3 Control	2.0-3.3 ms
A3	RED	Injector 2 Control	2.0-3.3 ms
A4	BRN	Injector 1 Control	2.0-3.3 ms
A5	BLK/WHT	HO2S-12 (B1 S2) Heater	Digital Signals: 0-12-0v
A6	BLK/WHT	HO2S-11 (B1 S1) Heater	Digital Signals: 0-12-0v
A7	PNK	EGR Solenoid Control	Solenoid Off: 12v, On: 1v
A8	GRN/YEL	VTEC Solenoid Control	Solenoid Off: 12v, On: 1v
A9	BRN/BLK	Logic Ground	<0.050v
A10	BLK	Power Ground	<0.1v
A11	YEL/BLK	Main Relay Power (B+)	12-14v
A12	BLK/BLU	Idle Air Control Valve	Pulse Signals
A13	GRN/WHT	Engine Mount Solenoid	Solenoid Off: 12v, On: 1v
A14	BLK/RED	Injector 5	2.0-3.3 ms
A15	RED/YEL	EVAP Purge Solenoid	Pulse Signals
A16	GRN/BLK	Fuel Pump Relay Control	Relay Off: 12v, On: 1v
A17	RED/BLU	A/C Clutch Relay Control	Relay Off: 12v, On: 1v
A18	GRN/RED	Malfunction Indicator Lamp	MIL Off: 12v, On: 1v
A19	WHT/GRN	Alternator Charging Signal	Lights Off: 12v, On: 1v
A20	YEL/GRN	Igniter Signal	Digital Signals: 0-12-0v
A21	---	Not Used	---
A22	BRN/BLK	Logic Ground	<0.050v
A23	BLK	Power Ground	<0.1v
A24	YEL/BLK	Main Relay Power (B+)	12-14v
A25	WHT/BLU	Injector 6	2.0-3.3 ms
A26	---	Not Used	---
A27	GRN	Radiator Fan Relay Control	Fan on: 1v, off: 12v
A28	BLU	EVAP Bypass Solenoid	Solenoid Off: 12v, On: 1v
A29	GRN/YEL	EVAP Vent Solenoid	Solenoid Off: 12v, On: 1v
A30-32	---	Not Used	---

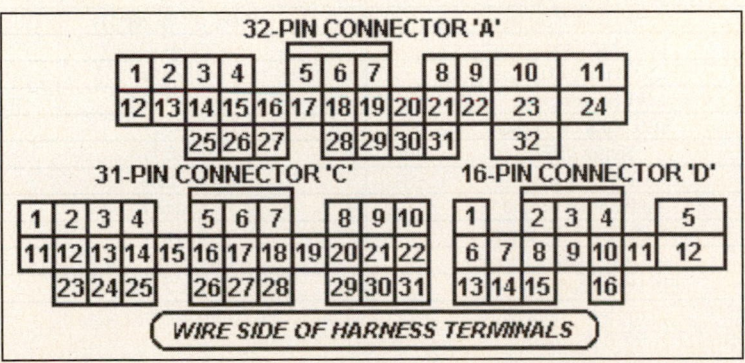

1997 Coupe 3.0L V6 VTEC VIN YA2 [A/T] 31P 'C' Connector

PCM Pin #	W/Color	Circuit Description (31-Pin)	Value at Hot Idle
C1	GRN/BLK	TCM VREF	4.9-5.1v
C2	BLU	CKP Sensor Signal	AC pulse signals
C3	GRN	TDC1 Sensor Signal	AC pulse signals
C4	YEL	TDC2 Sensor Signal	AC pulse signals
C5	RED/WHT	A/C Switch Signal	A/C Off: 12v, On: 1v
C6	BLK/GRN	Starter Switch Signal	Cranking: 9-11v
C7	RED	Service Check Connector	SCS Open: 4.80v
C8	LT GRN	K-Line Signal	12v
C9	---	Not Used	---
C10	WHT/YEL	Keep Alive Power (VBU)	12-14v
C11	---	Not Used	---
C12	WHT	CKP Sensor Ground	<0.050v
C13	RED	TDC1 Sensor Ground	<0.050v
C14	BLK	TDC2 Sensor Ground	<0.050v
C15	BLU/BLK	VTEC Pressure Switch	Switch on: 0v, off: 12v
C16	GRN	PSP Switch Signal	Straight: 0v, Turning: 12v
C17	WHT/RED	Alternator 'FR' Signal	Digital Signals: 0-5-0v
C18	BLU/WHT	Vehicle Speed Sensor	Moving: pulse signals
C19-25	---	Not Used	---
C26	BLU	Engine Speed Pulse Signal	Pulses
C27-28	---	Not Used	---
C29	LT GRN	A/T: Gear Position Switch	In P/N: 0v, Others: 12v
C30	BLU/RED	SEAF ECM Input To TCM	Pulse Signals
C31	RED/BLU	SEAF ECM Output To TCM	Pulse Signals

1997 Coupe 3.0L V6 VTEC VIN YA2 [A/T] 16P 'D' Connector

PCM Pin #	W/Color	Circuit Description (16-Pin)	Value at Hot Idle
D1	RED/BLK	TP Sensor Signal	0.5-0.6v
D2	RED/WHT	ECT Sensor Signal	At 180ºF: 0.5-0.6v
D3	RED/GRN	MAP Sensor Signal	0.8-0.9v
D4	YEL/RED	MAP Sensor VREF	4.9-5.1v
D5	GRN/WHT	Brake Switch Signal	Brake on: 0v, off: 12v
D6	---	Not Used	---
D7	WHT	HO2S-11 (B1 S1) Signal	0.1-1.1v
D8	RED/YEL	IAT Sensor Signal	At 100ºF: 2-3v
D9	WHT/BLK	EGR Valve Lift Sensor	1.1-1.2v
D10	YEL/BLU	Sensor VREF	4.9-5.1v
D11	GRN/BLK	Sensor Ground	<0.050v
D12	GRN/WHT	MAP Sensor Ground	<0.050v
D14	WHT/RED	HO2S-12 (B1 S2) Signal	0.1-1.1v
D15	LT GRN	Fuel Tank Pressure Sensor	Fuel Cap off: 2.5v
D16	GRN/RED	Electronic Load Detector	Varies: 2.5-3.5v

1998-99 Coupe 3.0L V6 VTEC VIN YA2 [A/T] 32P 'A' Connector

PCM Pin #	W/Color	Circuit Description (32-Pin)	Value at Hot Idle
A1	---	Not Used	---
A2	GRN/WHT	Engine Mount Solenoid	Solenoid Off: 12v, On: 1v
A3	BLU/BLK	EVAP Bypass Solenoid	Solenoid Off: 12v, On: 1v
A4	ORN/GRN	EVAP Vent Solenoid	Solenoid Off: 12v, On: 1v
A5	BLU/WHT	Cruise Control Signal	Cruise On: pulse signals
A6	RED/YEL	EVAP Purge Solenoid	Pulse Signals
A8	BLK/WHT	HO2S-12 (B1 S2) Heater	Digital Signals: 0-12-0v
A9	ORN	Vehicle Speed Sensor	Moving: pulse signals
A10	RED	Service Check Connector	SCS Open: 4.80v
A11	---	Not Used	---
A12	PNK	Immobilizer Indicator Control	Lamp On: 1v, Off: 12v
A13	ORN/BLU	Immobilizer Enable Signal	Digital Signals
A14	BLU/RED	A/T: D4 Light Switch	In D4: 0v, Off: 12v
A15	GRN/BLK	IMO Fuel Pump Relay Control	Relay Off: 12v, On: 1v
A17	RED/BLU	A/C Clutch Relay Control	Relay Off: 12v, On: 1v
A18	GRN/RED	Malfunction Indicator Lamp	MIL Off: 12v, On: 1v
A19	BLU	Engine Speed Pulse	Pulses
A20	GRN	Radiator Fan Relay Control	Relay Off: 12v, On: 1v
A21	LT BLU	K-Line Signal	12v
A23	WHT/RED	HO2S-12 (B1 S2) Signal	0.1-1.1v
A24	BLK/GRN	Starter Switch Signal	Cranking: 9-11v
A25	BLU/GRN	INOCD Immobilizer Code	Digital Signals
A26	GRN/YEL	PSP Switch Signal	Straight: 0v, Turning: 12v
A27	RED/WHT	A/C Switch Signal	A/C Off: 12v, On: 1v
A28	WHT/GRN	Interlock Control Unit Signal	Key & Brake on: 12v
A29	LT GRN	Fuel Tank Pressure Sensor	Fuel Cap off: 2.5v
A30	GRN/RED	Electronic Load Detector	Varies: 2.5-3.5v
A31	---	Not Used	---
A32	GRN/WHT	Brake Switch Signal	Brake on: 0v, off: 12v

1998-99 Coupe 3.0L V6 VTEC VIN YA2 [A/T] 16P 'D' Connector

PCM Pin #	W/Color	Circuit Description (16-Pin)	Value at Hot Idle
D1	YEL	A/T: Lockup Solenoid	Solenoid Off: 12v, On: 1v
D2	GRN/WHT	A/T: Shift Solenoid 'B'	Solenoid Off: 12v, On: 1v
D3	GRN	A/T Shift Solenoid 'C'	Solenoid Off: 12v, On: 1v
D5	BLK/YEL	Solenoid Feed (B+)	12-14v
D6	GRN/RED	A/T: Gear Position Switch	In 'D': 0v, others: 12v
D7	BLU/YEL	A/T: Shift Solenoid 'A'	On: <1v, off: 12v
D8	GRN/BLU	A/T: Gear Position Switch	In D3: 0v, other: 12v
D9	GRN/BLK	A/T: Gear Position Switch	In D4: 0v, other: 12v
D10	BLU	Countershaft Speed Sensor P	Moving: AC Pulses
D11	RED/BLU	Mainshaft Speed Sensor 'P'	Pulses
D12	WHT	Mainshaft Speed Sensor 'N'	<0.050v
D13	LT GRN	A/T: Gear Position Switch	In P/N: 0v, Others: 12v
D14	GRN/YEL	A/T: Gear Position Switch	In D2: 0v, other: 12v
D15	GRN/WHT	A/T: Gear Position Switch	In D1: 0v, other: 12v
D16	GRN	Countershaft Speed Sensor N	<0.050v

1998-99 Coupe 3.0L V6 VTEC VIN YA2 [A/T] 25P 'B' Connector

PCM Pin #	W/Color	Circuit Description (25-Pin)	Value at Hot Idle
B1, B9	YEL/BLK	Main Relay Power (B+)	12-14v
B2, B10	BLK	Power Ground	<0.1v
B3, B4	BLK	Injector 5, Injector 4	2.0-3.3 ms
B5, B6	RED, WHT	Injector 2, Injector 6	2.0-3.3 ms
B7	PNK	EGR Solenoid Control	Solenoid Off: 12v, On: 1v
B8	WHT	A/T: Clutch Solenoid 'A-'	Pulse Signals
B11, 15	BRN, BLU	Injector 1, Injector 3	2.0-3.3 ms
B12	GRN/YEL	VTEC Solenoid Control	Idle: 0v, Hi-Speed: 12v
B13	YEL/GRN	Igniter Signal	Digital Signals: 0-12-0v
B14	BLU/BLK	A/T: 2nd Clutch Pressure Sw.	Open: 12v, Closed: 1v
B17	RED	A/T: Clutch Solenoid 'A+'	Pulse Signals
B18	ORN	A/T: Clutch Solenoid 'B-'	Pulse Signals
B20, 22	BRN/BLK	Logic Ground	<0.050v
B21	WHT/YEL	Keep Alive Power (VBU)	12-14v
B23	BLK/BLU	Idle Air Control Valve	Pulse Signals
B24	BLU/WHT	3rd Clutch Press. Switch	Open: 12v, Closed: 1v
B25	ORN	A/T: Clutch Solenoid 'B+'	Pulse Signals

1998-99 Coupe 3.0L V6 VTEC VIN YA2 [A/T] 31P 'C' Connector

PCM Pin #	W/Color	Circuit Description (31-Pin)	Value at Hot Idle
C1	BLK/WHT	HO2S-11 (B1 S1) Heater	Digital Signals: 0-12-0v
C2	WHT/GRN	Alternator Charging Signal	Light on: 0v, off: 12v
C5	WHT/RED	Alternator 'FR' Signal	Digital Signals: 0-5-0v
C6	WHT/BLK	EGR Valve Lift Sensor	1.1-1.2v
C7	GRN/WHT	MAP Sensor Ground	<0.050v
C8	BLU	CKP Sensor Signal	AC pulse signals
C9	WHT	CKP Sensor Ground	<0.050v
C10	BLU/BLK	VTEC Pressure Switch	Switch on: 0v, off: 12v
C16	WHT	HO2S-11 (B1 S1) Signal	0.1-1.1v
C17	RED/GRN	MAP Sensor Signal	0.8-0.9v
C18	GRN/BLK	Sensor Ground	<0.050v
C19	YEL/RED	MAP Sensor VREF	4.9-5.1v
C20	GRN	TDC1 Sensor Signal	AC pulse signals
C21	RED	TDC1 Sensor Ground	<0.050v
C25	RED/YEL	IAT Sensor Signal	At 100ºF: 2-3v
C26	RED/WHT	ECT Sensor Signal	At 180ºF: 0.5-0.6v
C27	RED/BLK	TP Sensor Signal	0.5-0.6v
C28	YEL/BLU	Sensor VREF	4.9-5.1v
C29	YEL	TDC2 Sensor Signal	AC pulse signals
C30	BLK	TDC2 Sensor Ground	<0.050v

WIRE SIDE OF HARNESS TERMINALS

2001-03 Sedan 3.2L V6 VTEC VIN YA4 [A/T] 32P 'A' Connector

PCM Pin #	W/Color	Circuit Description (32-Pin)	Value at Hot Idle
A1	WHT/RED	HO2S-22 (B2 S2) Signal	0.1-1.1v
A2	GRN/WHT	Engine Mount Solenoid	0v, Off-Idle: 12v
A3	BLU	EVAP Bypass Solenoid	Solenoid Off: 12v, On: 1v
A4	GRN/WHT	EVAP Vent Solenoid	Solenoid Off: 12v, On: 1v
A5	BLU/GRN	Cruise Control Signal	Cruise On: pulse signals
A6	RED/YEL	EVAP Purge Solenoid	Pulse Signals
A7	ORN/WHT	Reference Voltage	4.9-5.1v
A8	BLK/WHT	HO2S-12 (B1 S2) Heater	Digital Signals: 0-12-0v
A9	BLU/WHT	Vehicle Speed Sensor	Moving: pulse signals
A10	BRN	Service Check Connector	SCS Open: 4.80v
A11-12	---	Not Used	---
A13	LT BLU	Intake Manifold Runner J32A2	Solenoid Off: 12v, On: 1v
A14	GRN/BLK	D5 Indicator Light	In D5: 0.1v, off: 12v
A15	GRN/YEL	Fuel Pump Relay Control	Relay Off: 12v, On: 1v
A16	---	Not Used	---
A17	RED	A/C Clutch Relay Control	Relay Off: 12v, On: 1v
A18	GRN/ORN	Malfunction Indicator Light	MIL Off: 12v, On: 1v
A19	BLU	Engine Speed Signal	Pulses
A20	GRN	Radiator Fan Relay Control	Relay Off: 12v, On: 1v
A21	GRY	K-Line Signal	12v
A22	RED/GRN	SEAF Serial Data Line J32A2	Digital Signals
A23	---	Not Used	---
A24	BLU/ORN	Starter Switch Signal	Cranking: 9-11v
A25	---	Not Used	---
A26	GRN	PSP Switch Signal	Straight: 0v, Turning: 12v
A27	BLU/RED	A/C Switch Signal	A/C Off: 12v, On: 1v
A28	WHT/RED	Interlock Control Unit Signal	Key & Brake on: 12v
A29	LT GRN	Fuel Tank Pressure Sensor	Fuel Cap off: 2.5v
A30	GRN/RED	Electronic Load Detector	Varies: 2.5-3.5v
A31	YEL/GRN	ECT Sensor Signal to TCM	Digital Signals
A32	WHT/BLK	Brake Switch Signal	Brake on: 0v, off: 12v

2001-03 Sedan 3.2L V6 VTEC VIN YA4 [A/T] 25P 'B' Connector

PCM Pin #	W/Color	Circuit Description (25-Pin)	Value at Hot Idle
B1, B9	YEL/BLK	Main Relay Power (B+)	12-14v
B2, B10	BLK	Power Ground PG1, PG2	<0.1v
B3	BLK/RED	Injector 5 Control	2.0-3.3 ms
B4	YEL	Injector 4 Control	2.0-3.3 ms
B5	RED	Injector 2 Control	2.0-3.3 ms
B6	WHT/BLU	Injector 6 Control	2.0-3.3 ms
B7	BLU/RED	E-EGR Control Solenoid	Digital Signals: 0-12-0v
B8	WHT	A/T: Clutch Solenoid 'A-'	Pulse Signals
B11	BRN	Injector 1 Control	2.0-3.3 ms
B12	GRN/YEL	VTEC Solenoid Control	Idle: 0v, Hi-Speed: 12v
B13	GRN/RED	A/T: Clutch Solenoid 'C+'	Pulse Signals
B14	BLU/WHT	A/T: Gear Position Switch	In P/N: 0v, Others: 12v
B15	BLU	Injector 3 Control	2.0-3.3 ms
B16	---	Not Used	---
B17	RED	A/T: Clutch Solenoid 'A+'	Pulse Signals
B18	GRN	A/T: Clutch Solenoid 'B-'	Pulse Signals
B19	BLU/YEL	A/T: 4th Oil Pressure Switch	12-14v
B20	BRN/YEL	Logic Ground (LG1)	<0.050v
B21	WHT/RED	Keep Alive Power (VBU)	12-14v
B22	BRN/YEL	Logic Ground (LG2)	<0.050v
B23	BLK/RED	Idle Air Control Valve	Pulse Signals
B24	RED/BLU	A/T: Clutch Solenoid 'A-'	Pulse Signals
B25	BRN/BLU	A/T: Clutch Solenoid 'B+'	Pulse Signals

2001-03 Sedan 3.2L V6 VTEC VIN YA4 [A/T] 16P 'D' Connector

PCM Pin #	W/Color	Circuit Description (16-Pin)	Value at Hot Idle
D1	YEL	A/T: Lockup Solenoid	LUS Off: 12v, On: 1v
D2	GRN/WHT	A/T: Shift Solenoid 'B'	SSB Off: 12v, On: 1v
D3	GRN	A/T" Shift Solenoid 'C'	SSA Off: 12v, On: 1v
D4	RED/BLK	A/T: 'N' Gear Position Switch	In 'N': 0v, Others: 12v
D5	BLK/YEL	A/T: Solenoid Feed (B+)	12-14v
D6	WHT	A/T: 'R' Gear Position Switch	In 'R': 0v, Others: 12v
D7	BLU/YEL	A/T: Shift Solenoid 'A'	SSA Off: 12v, On: 1v
D8	YEL	A/T: D4 Gear Position Switch	In D4: 0v, others: 12v
D9	YEL/GRN	A/T: D5 Gear Position Switch	In D5: 0v, Others: 12v
D10	BLU	Countershaft Speed Sensor P	Wheels turning: 12v
D11	RED	Mainshaft Speed Sensor 'P'	Moving: pulse signals
D12	WHT	Mainshaft Speed, Ground 'N'	<0.050v
D13	BLU/WHT	A/T: 3rd Clutch Pressure Sw.	Open: 12v, Closed: 1v
D14	RED	A/T: D3 Gear Position Switch	In D3: 0v, Others: 12v
D15	BLU	A/T: D2 Gear Position Switch	In D2: 0v, Others: 12v
D16	GRN	Countershaft Speed Sensor N	<0.050v

2001-03 Sedan 3.2L V6 VTEC VIN YA4 [A/T] 31P 'C' Connector

PCM Pin #	W/Color	Circuit Description (31-Pin)	Value at Hot Idle
C1	BLK/WHT	HO2S-11 (B1 S1) Heater	Digital Signals: 0-12-0v
C2	WHT/GRN	Alternator Charging Signal	Lights on: 0v, off: 12v
C3	WHT/BLU	Ignition Coil 3 Control	Digital Signals: 0-12-0v
C4	YEL/GRN	Ignition Coil 1 Control	Digital Signals: 0-12-0v
C5	WHT/RED	Alternator 'FR' Signal	Digital Signal: 0-5-0-5v
C6	WHT/BLK	EGR Valve Lift Sensor	1.2v
C7	GRN/WHT	MAP Sensor Ground (SG1)	<0.050v
C8	BLU	CKP Sensor 'P' Signal	AC pulse signals
C9	WHT	CKP Sensor 'N' Signal	<0.050v
C10	BLU/BLK	VTEC Pressure Switch	Idle: 0v, Cruise: 12v
C11	---	Not Used	---
C12	BLK/RED	Ignition Coil 5 Control	Digital Signals: 0-12-0v
C13	BRN	Ignition Coil 4 Control	Digital Signals: 0-12-0v
C14	BLU/RED	Ignition Coil 2 Control	Digital Signals: 0-12-0v
C15	---	Not Used	---
C16	WHT	HO2S-11 (B1 S1) Signal	0.1-1.1v
C17	GRN/RED	MAP Sensor Signal	0.8-0.9v
C18	GRN/YEL	Sensor Ground (SG2)	<0.050v
C19	YEL/RED	MAP Sensor VREF (VCC1)	4.9-5.1v
C20	GRN	TDC1 Sensor 'P' Signal	AC pulse signals
C21	RED	TDC1 Sensor 'N' Signal	<0.050v
C22	RED/BLU	Knock Sensor Signal	No Detonation: 18mv AC
C23	BRN/WHT	Ignition Coil 6 Control	Digital Signals: 0-12-0v
C24	BLU/YEL	A/T: TFT Sensor Signal	0.1-4.2v
C25	RED/YEL	IAT Sensor Signal	Varies w/temp: 0.5-4.9v
C26	RED/WHT	ECT Sensor Signal	At 180ºF: 0.5-0.6v
C27	RED/BLK	TP Sensor Signal	0.5-0.6v
C28	YEL/BLU	Sensor VREF (VCC2)	4.9-5.1v
C29	YEL	TDC2 Sensor 'P' Signal	AC pulse signals
C30	BLK	TDC2 Sensor 'N' Signal	<0.050v

2001-03 Sedan 3.2L V6 VTEC VIN YA4 [A/T] 20P 'E' Connector

PCM Pin #	W/Color	Circuit Description (20-Pin)	Value at Hot Idle
E1	PNK/WHT	Immobilizer Indicator Lamp	Lamp Off: 12v, On: 1v
E2	RED	IMOCD Immobilizer Code	Digital Signals
E3-4	---	Not Used	---
E5	YEL/GRN	TP Sensor Signal to TCM	0.5-0.6v
E6	ORN/GRN	Voltage Reference (ABS)	4.9-5.1v
E7	BLU	PCM to Frame Inhibition	2.5v (normal signal)
E8	BLU/BLK	A/T: Drive Shift Indicator 'C'	2nd, 4th: 0v, 1st, 3rd: 12v
E9	BLU/GRN	A/T: Drive Shift Indicator 'B'	In 4th: 12v, 1st & 3rd: 0v
E10	BLU/RED	A/T: Drive Shift Indicator 'A'	In 4th: 12v, 1st - 3rd: 0v
E11	---	Not Used	---
E12	BLU/WHT	IMOEN Immobilizer Enable	Digital Signals
E13, E15	---	Not Used	---
E14	BRN	A/T: First Hold Down Switch	LH Switch on: 0v, off: 12v
E16	PNK	Frame to PCM Torque Down	TCS On: 5v, Off: 2.5v
E17	LT GRN	A/T: 'P' Gear Position Switch	In 'P': 4v, Others: 0v
E18	ORN	Torque Mgmt. (-) Signal	In Neutral: 12v
E19	YEL	Torque Mgmt. (+) Signal	In Sports Shift: 0v
E20	RED	Torque Mgmt. Mode Signal	In Sports Shift: 0v

RL Pin Voltage Tables

1996-99 Sedan 3.5L V6 MFI VIN KA9 [A/T] 26P 'A' Connector

PCM Pin #	W/Color	Circuit Description (26-Pin)	Value at Hot Idle
A1	BRN	Injector 1	2.0-3.3 ms
A2	RED	Injector 2	2.0-3.3 ms
A3	BLU	Injector 3	2.0-3.3 ms
A4	YEL/GRN	Fuel Pump Relay 1 Control	Relay Off: 12v, On: 1v
A5	BLK/BLU	Idle Air Control Valve	Pulse Signals
A6	RED	EGR Solenoid Control	At off-idle (hot): 0-12-0v
A7	GRN/RED	Malfunction Indicator Lamp	MIL Off: 12v, On: 1v
A8	RED/WHT	A/C Clutch Relay Control	Relay Off: 12v, On: 1v
A9	---	Not Used	---
A10	RED/GRN	Fuel Pressure Regulator	Solenoid Off: 12v, On: 1v
A11	PNK	Ignition Coil 1 Control	Digital Signals: 0-12-0v
A12	BLK	Power Ground	<0.1v
A13	YEL/BLK	Main Relay Power (B+)	12-14v
A14	YEL	Injector 4	2.0-3.3 ms
A15	BLK/RED	Injector 5	2.0-3.3 ms
A16	WHT/BLU	Injector 6	2.0-3.3 ms
A17	LT GRN	Engine Mount Solenoid	Solenoid Off: 12v, On: 1v
A18	GRN/RED	HO2S-11 (B1 S1) Heater	Digital Signals: 0-12-0v
A19	BLK/WHT	HO2S-21 (B2 S1) Heater	Digital Signals: 0-12-0v
A20	RED/BLU	IAB Solenoid Low Control	Under 3200 rpm: 0v
A21	BLK/WHT	HO2S-12 (B1 S2) Heater	Digital Signals: 0-12-0v
A22	YEL	IAB Solenoid High Control	Under 3200 rpm: 0v
A23	RED/YEL	EVAP Purge Solenoid	<90°F: 12v, >90°F: 0v
A24	BRN	Ignition Coil 2 Control	Digital Signals: 0-12-0v
A25	BLK	Power Ground (PG1)	<0.1v
A26	BRN/BLK	Logic Ground (LG1)	<0.050v

1996-99 Sedan 3.5L V6 MFI VIN KA9 [A/T] 16P 'B' Connector

PCM Pin #	W/Color	Circuit Description (16-Pin)	Value at Hot Idle
B1	YEL/BLK	Main Relay Power (B+)	12-14v
B2	RED	Ignition Coil 6 Control	Digital Signals: 0-12-0v
B3	GRN	PSP Switch Signal	Straight: 0v, Turning: 12v
B4	---	Not Used	---
B5	YEL/RED	CYP Sensor 2 Signal	AC pulse signals
B6	YEL	CYP Sensor 1 Signal	AC pulse signals
B7	BLU/RED	CKP Sensor 2 Signal	AC pulse signals
B8	BLU	CKP Sensor 1 Signal	AC pulse signals
B9	BRN/BLK	Logic Ground	<0.050v
B10	GRY	Ignition Coil 5 Control	Digital Signals: 0-12-0v
B11	GRN	Ignition Coil 4 Control	Digital Signals: 0-12-0v
B12	BLU	Ignition Coil 3 Control	Digital Signals: 0-12-0v
B13	BLU/WHT	Vehicle Speed Sensor	Moving: pulse signals
B14	---	Not Used	---
B15	WHT/RED	CKP Sensor 2 Ground	<0.050v
B16	WHT	CKP Sensor 1 Ground	<0.050v

1996-99 Sedan 3.5L V6 MFI VIN KA9 [A/T] 12P 'C' Connector

PCM Pin #	W/Color	Circuit Description (12-Pin)	Value at Hot Idle
C1	---	Not Used	---
C2	PNK	A/C Switch Signal	A/C Off: 12v, On: 1v
C3	GRN/RED	A/C Pressure Switch Signal	A/C Off: 4-5v, On: 0v
C4	GRN	Fuel Pump Relay 2 Control	Relay Off: 12v, On: 1v
C5	RED	Service Check Connector	SCS Open: 4.80v
C6	BLU/RED	Starter Switch Signal	Cranking: 9-11v
C7	---	Not Used	---
C8	BLU	Engine Speed Pulse Signal	Pulses
C9	GRN/RED	Traction Control Inhibit Signal	Pulses
C10	BLK/ORN	T/C Fuel Cut Signal	Pulses
C11	WHT	INOCD Immobilizer Code	Digital Signals
C12	PNK/BLK	T/C Standby Signal	Pulses

1996-99 Sedan 3.5L V6 MFI VIN KA9 [A/T] 22P 'D' Connector

PCM Pin #	W/Color	Circuit Description (22-Pin)	Value at Hot Idle
D1	WHT/YEL	Keep Alive Power (VBU)	12-14v
D2	RED/BLU	Knock Sensor Signal (Bank 1)	No Detonation: 18mv AC
D3	GRN/ORN	Barometric Output Signal	3v (at sea level)
D4	LT GRN	K-Line Signal	12v
D5	WHT/RED	Alternator 'FR' Signal	Digital Signals: 0-5-0v
D6	RED/BLK	TP Sensor Signal	0.5-0.6v
D7	RED/WHT	ECT Sensor Signal	At 180°F: 0.5-0.6v
D8	RED/YEL	IAT Sensor Signal	At 100°F: 2-3v
D9	RED/GRN	MAP Sensor Signal	0.8-0.9v
D10	YEL/RED	MAP Sensor VREF	4.9-5.1v
D11	GRN/WHT	MAP Sensor Ground	<0.050v
D12	BLK	CYP Sensor 1 Ground	<0.050v
D13	YEL	Knock Sensor Signal (Bank 2)	No Detonation: 18mv AC
D14	BLK/RED	CYP Sensor 2 Ground	<0.050v
D15	GRN/YEL	Fuel Tank Pressure Sensor	Fuel Cap off: 2.5v
D16	---	Not Used	---
D17	WHT/BLK	EGR Valve Lift Sensor	1.1-1.2v
D18	BLU/RED	HO21-21 (B2 S1) Signal	0.1-1.1v
D19	WHT	HO2S-11 (B1 S1) Signal	0.1-1.1v
D20	BLU/WHT	TCS VREF	4.9-5.1v
D21	YEL/BLU	Sensor VREF (VCC2)	4.9-5.1v
D22	GRN/BLK	Sensor Ground (SG2)	<0.050v

1996-99 Sedan 3.5L V6 MFI VIN KA9 [A/T] 26P 'E' Connector

PCM Pin #	W/Color	Circuit Description (26-Pin)	Value at Hot Idle
E1	BLK/YEL	Solenoid Feed (B+) (VB SOL)	12-14v
E2	GRN/WHT	Brake Switch Signal	Brake Off: 12v, On: 0v
E3	RED	Linear Solenoid Valve 'P'	Pulses
E4	RED	Mainshaft Speed Sensor	Pulses
E5	BLU	Countershaft Speed Sensor	Wheels turning: 12v
E6	PNK/BLU	TCS Shift Control Signal	Pulses
E7	WHT	A/T: Gear Position Switch	In 'D': 0v, others: 12v
E8	YEL	A/T: Gear Position Switch	In D4: 0v, other: 12v
E9	GRN	A/T: Gear Position Switch	In D3: 0v, other: 12v
E10	BLU	A/T: Gear Position Switch	In D2: 0v, other: 12v
E11	BRN	A/T: Gear Position Switch	In D1: 0v, other: 12v
E12	BLU/YEL	Shift Control Solenoid 'A'	KOER in D2/D3: 12v
E13	YEL	Lockup Control Solenoid 'A'	L/Up on: 12v, off: 0v
E16	WHT	Linear Solenoid Valve 'N'	<0.050v
E17	WHT	Mainshaft Speed Sensor 'N'	<0.050v
E18	GRN	Countershaft Speed Sensor N	<0.050v
E19	WHT/RED	Serial Data Line 'B'	N/A
E20	WHT/GRN	Serial Data Line 'A'	Digital Signals
E22	WHT/GRN	Interlock Control Unit Signal	Key & Brake on: 12v
E23	BRN/YEL	A/T: D4 Indicator Light	D4 lamp on: 0v, off: 12v
E24	LT GRN	A/T: Gear Position Switch	In P/N: 0v, Others: 12v
E25	GRN/WHT	Shift Control Solenoid 'B'	SSB Off: 0v, On: 12v
E26	GRN/BLK	Lockup Control Solenoid 'B'	LSB Off: 0v, On: 12v

1996-99 Sedan 3.5L V6 MFI VIN KA9 [A/T] 8P 'F' Connector

PCM Pin #	W/Color	Circuit Description (8-Pin)	Value at Hot Idle
F1	PNK/BLK	EVAP Bypass Solenoid	Solenoid Off: 12v, On: 1v
F2	WHT/RED	HO2S-12 (B1 S2) Signal	0.1-1.1v
F3	BLU	Right Peak Hold Reset	Digital Signals
F4	YEL	Right Misfire Pulse	Digital Signals
F5	RED/WHT	EVAP Vent Shut Solenoid	Solenoid Off: 12v, On: 1v
F6	GRN/WHT	HO2S Ground	<0.050v
F7	GRN/YEL	Left Peak Hold Reset	Digital Signals
F8	RED	Left Misfire Pulse	Digital Signals

26-PIN CONNECTOR 'A'   16-PIN CONNECTOR 'B'   12-PIN CONNECTOR 'C'
22-PIN CONNECTOR 'D'   26-PIN CONNECTOR 'E'   8-PIN 'F'
WIRE SIDE OF HARNESS TERMINALS

2000-03 Sedan 3.5L V6 MFI VIN KA9 [A/T] 26P 'A' Connector

PCM Pin #	W/Color	Circuit Description (26-Pin)	Value at Hot Idle
A1	BRN	Injector 1 Control	2.0-3.3 ms
A2	RED	Injector 2 Control	2.0-3.3 ms
A3	BLU	Injector 3 Control	2.0-3.3 ms
A4	YEL/GRN	Fuel Pump Relay 1 Control	Relay Off: 12v, On: 1v
A5	BLK/BLU	Idle Air Control Valve	Pulse Signals
A6	RED	Electronic EGR Solenoid Control	At off-idle (hot): 0-12-0v
A7	GRN/RED	Malfunction Indicator Lamp	MIL Off: 12v, On: 1v
A8	RED/WHT	A/C Clutch Relay Control	Relay Off: 12v, On: 1v
A9	---	Not Used	---
A10	RED/GRN	Fuel Pressure Regulator	Solenoid Off: 12v, On: 1v
A11	PNK	Ignition Coil 1 Control	Digital Signals: 0-12-0v
A12	BLK	Power Ground	<0.1v
A13	YEL/BLK	Main Relay Power (B+)	12-14v
A14	YEL	Injector 4	2.0-3.3 ms
A15	BLK/RED	Injector 5	2.0-3.3 ms
A16	WHT/BLU	Injector 6	2.0-3.3 ms
A17	LT GRN	Engine Mount Solenoid	Solenoid Off: 12v, On: 1v
A18	GRN/RED	HO2S-11 (B1 S1) Heater	Digital Signals: 0-12-0v
A19	BLK/WHT	HO2S-21 (B2 S1) Heater	Digital Signals: 0-12-0v
A20	RED/BLU	IAB Solenoid Low Control	Under 3200 rpm: 0v
A21	BLK/WHT	HO2S-12 (B1 S2) Heater	Digital Signals: 0-12-0v
A22	YEL	IAB Solenoid High Control	Under 3200 rpm: 0v
A23	RED/YEL	EVAP Purge Solenoid	<140ºF: 12v, >140ºF: 0v
A24	BRN	Ignition Coil 2 Control	Digital Signals: 0-12-0v
A25	BLK	Power Ground (PG1)	<0.1v
A26	BRN/BLK	Logic Ground (LG1)	<0.050v

2000-03 Sedan 3.5L V6 MFI VIN KA9 [A/T] 16P 'B' Connector

PCM Pin #	W/Color	Circuit Description (16-Pin)	Value at Hot Idle
B1	YEL/BLK	Main Relay Power (B+)	12-14v
B2	RED	Ignition Coil 6 Control	Digital Signals: 0-12-0v
B3	GRN	PSP Switch Signal	Straight: 0v, Turning: 12v
B4	---	Not Used	---
B5	YEL/RED	CYP Sensor 2 'P' Signal	AC pulse signals
B6	YEL	CYP Sensor 1 'P' Signal	AC pulse signals
B7	BLU/RED	CKP Sensor 2 'P' Signal	AC pulse signals
B8	BLU	CKP Sensor 1 'P' Signal	AC pulse signals
B9	BRN/BLK	Logic Ground (LG2)	<0.050v
B10	GRY	Ignition Coil 5 Control	Digital Signals: 0-12-0v
B11	GRN	Ignition Coil 4 Control	Digital Signals: 0-12-0v
B12	BLU	Ignition Coil 3 Control	Digital Signals: 0-12-0v
B13	BLU/WHT	Vehicle Speed Sensor	Moving: pulse signals
B14	---	Not Used	---
B15	WHT/RED	CKP Sensor 2 'N' Signal	<0.050v
B16	WHT	CKP Sensor 1 'N' Signal	<0.050v

2000-03 Sedan 3.5L V6 MFI VIN KA9 [A/T] 12P 'C' Connector

PCM Pin #	W/Color	Circuit Description (12-Pin)	Value at Hot Idle
C1	---	Not Used	---
C2	PNK	A/C Switch Signal	A/C Off: 12v, On: 1v
C3	GRN/RED	A/C Pressure Switch Signal	A/C Off: 4-5v, On: 0v
C4	GRN	Fuel Pump Relay 2 Control	Relay Off: 12v, On: 1v
C5	RED	Service Check Connector	SCS Open: 4.80v
C6	BLU/RED	Starter Switch Signal	Cranking: 9-11v
C7	---	Not Used	---
C8	BLU	Engine Speed Pulse Signal	Pulse Signals
C9	GRN/RED	Traction Control Inhibit Signal	Pulse Signals
C10	BLK/ORN	T/C Fuel Cut Signal	Pulse Signals
C11	WHT	INOCD Immobilizer Code	Digital Signals
C12	PNK/BLK	T/C Standby Signal	Pulse Signals

2000-03 Sedan 3.5L V6 MFI VIN KA9 [A/T] 22P 'D' Connector

PCM Pin #	W/Color	Circuit Description (22-Pin)	Value at Hot Idle
D1	WHT/YEL	Keep Alive Power (VBU)	12-14v
D2	RED/BLU	Knock Sensor Signal (Bank 1)	No Detonation: 18mv AC
D3	GRN/ORN	Barometric Output Signal	3v (at sea level)
D4	LT GRN	K-Line Signal	12v
D5	WHT/RED	Alternator 'FR' Signal	Digital Signals: 0-5-0v
D6	RED/BLK	TP Sensor Signal	0.5-0.6v
D7	RED/WHT	ECT Sensor Signal	At 180°F: 0.5-0.6v
D8	RED/YEL	IAT Sensor Signal	At 100°F: 2-3v
D9	RED/GRN	MAP Sensor Signal	0.8-0.9v
D10	YEL/RED	MAP Sensor VREF	4.9-5.1v
D11	GRN/WHT	MAP Sensor Ground	<0.050v
D12	BLK	CYP Sensor 1 Ground	<0.050v
D13	YEL	Knock Sensor Signal (Bank 2)	No Detonation: 18mv AC
D14	BLK/RED	CYP Sensor 2 Ground	<0.050v
D15	GRN/YEL	Fuel Tank Pressure Sensor	Fuel Cap off: 2.5v
D16	---	Not Used	---
D17	WHT/BLK	EGR Valve Lift Sensor	1.1-1.2v
D18	BLU/RED	HO21-21 (B2 S1) Signal	0.1-1.1v
D19	WHT	HO2S-11 (B1 S1) Signal	0.1-1.1v
D20	BLU/WHT	TCS VREF	4.9-5.1v
D21	YEL/BLU	Sensor VREF (VCC2)	4.9-5.1v
D22	GRN/BLK	Sensor Ground (SG2)	<0.050v

26-PIN CONNECTOR 'A'
1 2 3 4 5 6 7 8 9 10 11 12 13
14 15 16 17 18 19 20 21 22 23 24 25 26

16-PIN CONNECTOR 'B'
1 2 3 4 5 6 7 8
9 10 11 12 13 14 15 16

12-PIN CONNECTOR 'C'
1 2 3 4 5 6
7 8 9 10 11 12

22-PIN CONNECTOR 'D'
1 2 3 4 5 6 7 8 9 10 11
12 13 14 15 16 17 18 19 20 21 22

26-PIN CONNECTOR 'E'
1 2 3 4 5 6 7 8 9 10 11 12 13
14 15 16 17 18 19 20 21 22 23 24 25 26

8-PIN 'F'
1 2 3 4
5 6 7 8

WIRE SIDE OF HARNESS TERMINALS

2000-03 Sedan 3.5L V6 MFI VIN KA9 [A/T] 26P 'E' Connector

PCM Pin #	W/Color	Circuit Description (26-Pin)	Value at Hot Idle
E1	BLK/YEL	Solenoid Feed (B+) (VB SOL)	12-14v
E2	GRN/WHT	Brake Switch Signal	Brake Off: 12v, On: 0v
E3	RED	Linear Solenoid Valve 'P'	Pulses
E4	RED	Mainshaft Speed Sensor	Pulses
E5	BLU	Countershaft Speed Sensor	Wheels turning: 12v
E6	PNK/BLU	TCS Shift Control Signal	Pulses
E7	WHT	A/T: Gear Position Switch	In 'D': 0v, others: 12v
E8	YEL	A/T: Gear Position Switch	In D4: 0v, other: 12v
E9	GRN	A/T: Gear Position Switch	In D3: 0v, other: 12v
E10	BLU	A/T: Gear Position Switch	In D2: 0v, other: 12v
E11	BRN	A/T: Gear Position Switch	In D1: 0v, other: 12v
E12	BLU/YEL	Shift Control Solenoid 'A'	KOER in D2/D3: 12v
E13	YEL	Lockup Control Solenoid 'A'	L/Up on: 12v, off: 0v
E16	WHT	Linear Solenoid Valve 'N'	<0.050v
E17	WHT	Mainshaft Speed Sensor 'N'	<0.050v
E18	GRN	Countershaft Speed Sensor N	<0.050v
E19	WHT/RED	Serial Data Line 'B'	Digital Signals
E20	WHT/GRN	Serial Data Line 'A'	Digital Signals
E22	WHT/GRN	Interlock Control Unit Signal	Key & Brake on: 12v
E23	BRN/YEL	A/T: D4 Indicator Light	D4 lamp on: 0v, off: 12v
E24	LT GRN	A/T: Gear Position Switch	In P/N: 0v, Others: 12v
E25	GRN/WHT	Shift Control Solenoid 'B'	SSB Off: 0v, On: 12v
E26	GRN/BLK	Lockup Control Solenoid 'B'	LSB Off: 0v, On: 12v

2000-03 Sedan 3.5L V6 MFI VIN KA9 [A/T] 8P 'F' Connector

PCM Pin #	W/Color	Circuit Description (8-Pin)	Value at Hot Idle
F1	PNK/BLK	EVAP Bypass Solenoid	Solenoid Off: 12v, On: 1v
F2	WHT/RED	HO2S-12 (B1 S2) Signal	0.1-1.1v
F3	BLU	Right Peak Hold Reset	Digital Signals
F4	YEL	Right Misfire Pulse	Digital Signals
F5	RED/WHT	EVAP Vent Shut Solenoid	Solenoid Off: 12v, On: 1v
F6	GRN/WHT	HO2S-12 Ground	<0.050v
F7	GRN/YEL	Left Peak Hold Reset	Digital Signals
F8	RED	Left Misfire Pulse	Digital Signals

26-PIN CONNECTOR 'A' / 16-PIN CONNECTOR 'B' / 12-PIN CONNECTOR 'C' / 22-PIN CONNECTOR 'D' / 26-PIN CONNECTOR 'E' / 8-PIN 'F' / WIRE SIDE OF HARNESS TERMINALS

TL PIN TABLES

1995-96 Sedan 2.5L I5 MFI VIN UA2 [A/T] 26P 'A' Connector

PCM Pin #	W/Color	Circuit Description (26-Pin)	Value at Hot Idle
A1	BRN	Injector 1	2.0-3.3 ms
A2	RED	Injector 2	2.0-3.3 ms
A3	BLU	Injector 3	2.0-3.3 ms
A4	GRN/BLK	Fuel Pump Relay Control	Relay Off: 12v, On: 1v
A5	BLK/BLU	Idle Air Control Valve	Pulse Signals
A6	RED	EGR Solenoid Control	Solenoid Off: 12v, On: 1v
A7	GRN/RED	Malfunction Indicator Lamp	MIL Off: 12v, On: 1v
A8	RED/BLU	A/C Clutch Relay Control	Relay Off: 12v, On: 1v
A9	PNK/BLU	Intake Air Bypass Solenoid	Solenoid Off: 12v, On: 1v
A10	RED/GRN	Fuel Cutoff Feedback Signal	Pulses
A11	YEL/GRN	Igniter Signal	Digital Signals: 0-12-0v
A12	BLK	Power Ground	<0.1v
A13	YEL/BLK	Main Relay Power (B+)	12-14v
A14	YEL	Injector 4	2.0-3.3 ms
A15	BLK/RED	Injector 5	2.0-3.3 ms
A16	ORN/BLK	HO2S-11 (B1 S1) Heater	Digital Signals: 0-12-0v
A17	GRN/RED	HO2S-12 (B1 S2) Heater	Digital Signals: 0-12-0v
A18	RED	EVAP Purge Solenoid	Pulse Signals
A19	GRN/YEL	Radiator Fan Relay Control	Fan on: 1v: off: 12v
A20	GRN/WHT	Engine Mount Solenoid	Solenoid Off: 12v, On: 1v
A21	WHT/GRN	Alternator Charging Signal	Lights Off: 12v, On: 1v
A22	PNK	TCM A/T Feedback Signal	5v
A25	BLK	Power Ground	<0.1v
A26	BRN/BLK	Logic Ground	<0.050v

1995-96 Sedan 2.5L I5 MFI VIN UA2 [A/T] 12P 'C' Connector

PCM Pin #	W/Color	Circuit Description (12-Pin)	Value at Hot Idle
C1	WHT/GRN	Rear Knock Sensor Signal	No Detonation: 18mv AC
C2	RED/BLU	Front Knock Sensor Signal	No Detonation: 18mv AC
C3	---	Not Used	---
C4	WHT/RED	CKF Sensor	AC pulse signals
C5	BLU/RED	CKF Sensor Ground	<0.050v
C6-12	---	Not Used	---

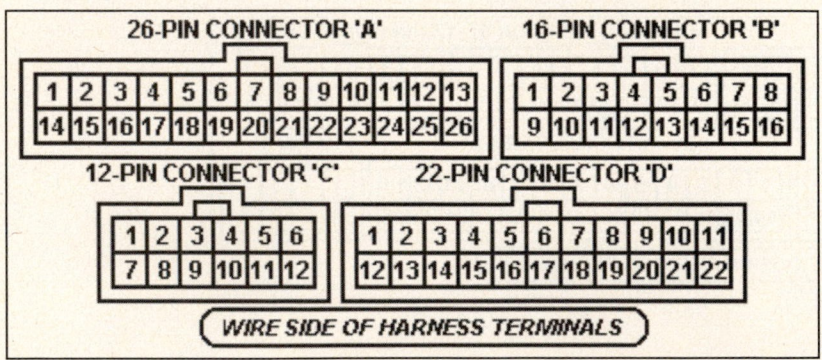

1995-96 Sedan 2.5L I5 MFI VIN UA2 [A/T] 16P 'B' Connector

PCM Pin #	W/Color	Circuit Description (16-Pin)	Value at Hot Idle
B1	YEL/BLK	Main Relay Power (B+)	12-14v
B2	WHT/GRN	TCM A/T FI 'A' Signal	5v
B3	BLU/BLK	A/C Switch Signal	A/C Off: 12v, On: 1v
B4	YEL/GRN	A/T: Gear Position Switch	In P/N: 0v, Others: 12v
B5	BLU/RED	Starter Switch Signal	Cranking: 9-11v
B6	ORN	CYP Sensor Ground	<0.050v
B7	ORN/BLU	TDC Sensor Ground	<0.050v
B8	BLU/GRN	CKP Sensor Ground	<0.050v
B9	BRN/BLK	Logic Ground	<0.050v
B10	WHT/RED	TCM A/T FI ' B' Signal	5v
B11	BRN	EVAP Purge Flow Switch	Switch on: 12v, off: 0.1v
B12	GRN	PSP Switch Signal	Straight: 0v, Turning: 12v
B13	ORN	Vehicle Speed Sensor	Moving: pulse signals
B14	WHT	CYP Sensor Signal	AC pulse signals
B15	WHT/BLU	TDC Sensor Signal	AC pulse signals
B16	BLU/YEL	CKP Sensor Signal	AC pulse signals

1995-96 Sedan 2.5L I5 MFI VIN UA2 [A/T] 22P 'D' Connector

PCM Pin #	W/Color	Circuit Description (22-Pin)	Value at Hot Idle
D1	WHT/GRN	Keep Alive Power (VBU)	12-14v
D2	BLU	TCM A/T FI Data Signal	5v
D3	RED/BLU	Barometric Output Signal	3v (at sea level)
D4	GRN/YEL	K-Line Signal	12v
D5	WHT/RED	Alternator 'FR' Signal	Digital Signals: 0-5-0v
D6	RED/BLK	TP Sensor Signal	0.5-0.6v
D7	RED/WHT	ECT Sensor Signal	At 180°F: 0.5-0.6v
D8	RED/YEL	IAT Sensor Signal	At 100°F: 2-3v
D9	WHT/BLU	MAF Sensor Signal	In P/N: 1.1-1.6v
D10	YEL/WHT	Sensor VREF	4.9-5.1v
D11	GRN/WHT	Sensor Ground	<0.050v
D12	GRN/WHT	Brake Switch Signal	Brake Off: 12v, On: 0v
D13	ORN	Service Check Connector	SCS Open: 4.80v
D14	BLU/RED	HO2S-12 (B1 S2) Signal	0.1-1.1v
D15	---	Not Used	---
D16	GRN/RED	Electric Load Detector	Varies w/Load: 2.5-3.5v
D17	WHT/BLK	EGR Valve Lift Sensor	1.1-1.2v
D18	WHT/RED	HO2S-11 (B1 S1) Signal	0.1-1.1v
D19	BLU/RED	MAF Sensor Ground	<0.050v
D20	BLU/WHT	Solenoid Feed (B+)	4.9-5.1v
D21	YEL/BLU	Sensor VREF	4.9-5.1v
D22	GRN/BLU	Sensor Ground	<0.050v

1997-98 Sedan 2.5L I5 MFI VIN UA2 [A/T] 26P 'A' Connector

PCM Pin #	W/Color	Circuit Description (26-Pin)	Value at Hot Idle
A1	BRN	Injector 1	2.0-3.3 ms
A2	RED	Injector 2	2.0-3.3 ms
A3	BLU	Injector 3	2.0-3.3 ms
A4	GRN/BLK	Fuel Pump Relay Control	Relay Off: 12v, On: 1v
A5	BLK/BLU	Idle Air Control Valve	Pulse Signals
A6	RED	EGR Solenoid Control	Solenoid Off: 12v, On: 1v
A7	GRN/RED	Malfunction Indicator Lamp	MIL Off: 12v, On: 1v
A8	RED/BLU	A/C Clutch Relay Control	Relay Off: 12v, On: 1v
A9	PNK/BLU	Intake Air Bypass Solenoid	Solenoid Off: 12v, On: 1v
A10	RED/GRN	Fuel Cutoff Feedback Signal	Pulses
A11	YEL/GRN	Igniter Signal	Digital Signals: 0-12-0v
A12	BLK	Power Ground	<0.1v
A13	YEL/BLK	Main Relay Power (B+)	12-14v
A14	YEL	Injector 4	2.0-3.3 ms
A15	BLK/RED	Injector 5	2.0-3.3 ms
A16	ORN/BLK	HO2S-11 (B1 S1) Heater	Digital Signals: 0-12-0v
A17	GRN/RED	HO2S-12 (B1 S2) Heater	Digital Signals: 0-12-0v
A18	RED	EVAP Purge Solenoid	Pulse Signals
A19	GRN/YEL	Radiator Fan Relay Control	Fan on: 1v: off: 12v
A20	GRN/WHT	Engine Mount Solenoid	Solenoid Off: 12v, On: 1v
A21	WHT/GRN	Alternator Charging Signal	Lights Off: 12v, On: 1v
A22	PNK	TCM A/T Feedback Signal	5v
A23	YEL/BLU	EVAP Vent Solenoid	Solenoid Off: 12v, On: 1v
A24	WHT/BLU	EVAP Bypass Solenoid	Solenoid Off: 12v, On: 1v
A25	BLK	Power Ground	<0.1v
A26	BRN/BLK	Logic Ground	<0.050v

1997-98 Sedan 2.5L I5 MFI VIN UA2 [A/T] 12P 'C' Connector

PCM Pin #	W/Color	Circuit Description (12-Pin)	Value at Hot Idle
C1	WHT/GRN	Rear Knock Sensor Signal	No Detonation: 18mv AC
C2	RED/BLU	Front Knock Sensor Signal	No Detonation: 18mv AC
C3	---	Not Used	---
C4	WHT/RED	CKF Sensor	AC pulse signals
C5	BLU/RED	CKF Sensor Ground	<0.050v

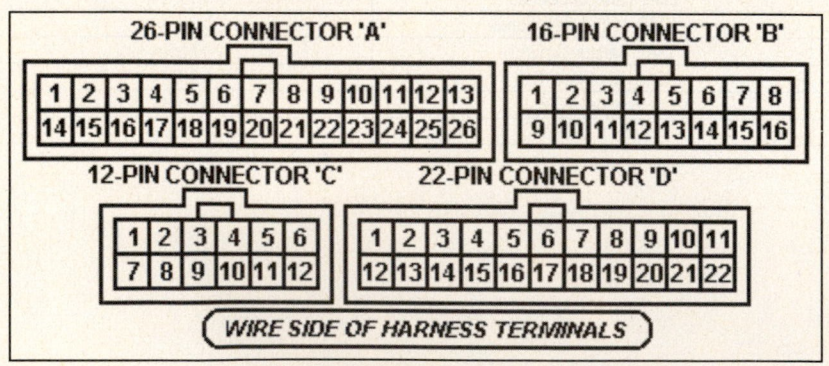

1997-98 Sedan 2.5L I5 MFI VIN UA2 [A/T] 16P 'B' Connector

PCM Pin #	W/Color	Circuit Description (16-Pin)	Value at Hot Idle
B1	YEL/BLK	Main Relay Power (B+)	12-14v
B2	WHT/GRN	TCM A/T FI 'A' Signal	5v
B3	BLU/BLK	A/C Switch Signal	A/C Off: 12v, On: 1v
B4	YEL/GRN	A/T: Gear Position Switch	In P/N: 0v, Others: 12v
B5	BLU/RED	Starter Switch Signal	Cranking: 9-11v
B6	ORN	CYP Sensor Ground	<0.050v
B7	ORN/BLU	TDC Sensor Ground	<0.050v
B8	BLU/GRN	CKP Sensor Ground	<0.050v
B9	BRN/BLK	Logic Ground	<0.050v
B10	WHT/RED	TCM A/T FI ' B' Signal	5v
B11	---	Not Used	---
B12	GRN	PSP Switch Signal	Straight: 0v, Turning: 12v
B13	ORN	Vehicle Speed Sensor	Moving: pulse signals
B14	WHT	CYP Sensor Signal	AC pulse signals
B15	WHT/BLU	TDC Sensor Signal	AC pulse signals
B16	BLU/YEL	CKP Sensor Signal	AC pulse signals

1997-98 Sedan 2.5L I5 MFI VIN UA2 [A/T] 22P 'D' Connector

PCM Pin #	W/Color	Circuit Description (22-Pin)	Value at Hot Idle
D1	WHT/GRN	Keep Alive Power (VBU)	12-14v
D2	BLU	TCM A/T FI Data Signal	5v
D3	RED/BLU	Barometric Output Signal	3v (at sea level)
D4	GRN/YEL	K-Line Signal	12v
D5	WHT/RED	Alternator 'FR' Signal	Digital Signals: 0-5-0v
D6	RED/BLK	TP Sensor Signal	0.5-0.6v
D7	RED/WHT	ECT Sensor Signal	At 180ºF: 0.5-0.6v
D8	RED/YEL	IAT Sensor Signal	At 100ºF: 2-3v
D9	WHT/BLU	MAF Sensor Signal	In P/N: 1.1-1.6v
D10	YEL/WHT	Sensor VREF	4.9-5.1v
D11	GRN/WHT	Sensor Ground	<0.050v
D12	GRN/WHT	Brake Switch Signal	Brake Off: 12v, On: 0v
D13	ORN	Service Check Connector	SCS Open: 4.80v
D14	BLU/RED	HO2S-12 (B1 S2) Signal	0.1-1.1v
D15	PNK	Fuel Tank Pressure Sensor	Fuel Cap off: 2.5v
D16	GRN/RED	Electric Load Detector	Varies w/Load: 2.5-3.5v
D17	WHT/BLK	EGR Valve Lift Sensor	1.1-1.2v
D18	WHT/RED	HO2S-11 (B1 S1) Signal	0.1-1.1v
D19	BLU/RED	MAF Sensor Ground	<0.050v
D20	BLU/WHT	Solenoid Feed (B+)	4.9-5.1v
D21	YEL/BLU	Sensor VREF	4.9-5.1v
D22	GRN/BLU	Sensor Ground	<0.050v

1996 Sedan 3.2L V6 MFI VIN UA3 [A/T] 26P 'A' Connector

PCM Pin #	W/Color	Circuit Description (26-Pin)	Value at Hot Idle
A1	BRN	Injector 1	2.0-3.3 ms
A2	RED	Injector 2	2.0-3.3 ms
A3	BLU	Injector 3	2.0-3.3 ms
A4	GRN/BLK	Fuel Pump Relay Control	Relay Off: 12v, On: 1v
A5	BLK/BLU	Idle Air Control Valve	Pulse Signals
A6	RED	EGR Solenoid Control	Solenoid Off: 12v, On: 1v
A7	GRN/RED	Malfunction Indicator Lamp	MIL Off: 12v, On: 1v
A8	RED/BLU	A/C Clutch Relay Control	Relay Off: 12v, On: 1v
A9	---	Not Used	---
A10	GRN	Pressure Regulator Solenoid	Solenoid Off: 12v, On: 1v
A11	PNK	Ignition Coil 1 Control	Digital Signals: 0-12-0v
A12	BLK	Power Ground	<0.1v
A13	YEL/BLK	Main Relay Power (B+)	12-14v
A14	YEL	Injector 4	2.0-3.3 ms
A15	BLK/RED	Injector 5	2.0-3.3 ms
A16	WHT/BLU	Injector 6	2.0-3.3 ms
A17	---	Not Used	---
A18	GRN/RED	HO2S-21 (B2 S1) Heater	Digital Signals: 0-12-0v
A19	ORN/BLK	HO2S-12 (B1 S2) Heater	Digital Signals: 0-12-0v
A20	RED/BLU	IAB Solenoid 1 Control	Solenoid Off: 12v, On: 1v
A21	ORN/BLK	HO2S-22 (B2 S2) Heater	Digital Signals: 0-12-0v
A22	YEL	IAB Solenoid 2 Control	Solenoid Off: 12v, On: 1v
A23	RED	EVAP Purge Solenoid	Pulse Signals
A24	BRN	Ignition Coil 2 Control	Digital Signals: 0-12-0v
A25	BLK	Power Ground	<0.1v
A26	BRN/BLK	Logic Ground	<0.050v

1996 Sedan 3.2L V6 MFI VIN UA3 [A/T] 16P 'B' Connector

PCM Pin #	W/Color	Circuit Description (16-Pin)	Value at Hot Idle
B1	YEL/BLK	Main Relay Power (B+)	12-14v
B2	RED	Ignition Coil 6 Control	Digital Signals: 0-12-0v
B3	GRN	PSP Switch Signal	Straight: 0v, Turning: 12v
B4	---	Not Used	---
B5	ORN	CYP Sensor 2 Signal	AC pulse signals
B6	WHT	CYP Sensor 1 Signal	AC pulse signals
B7	ORN/BLU	CKP Sensor 2 Signal	AC pulse signals
B8	BLU/GRN	CKP Sensor 1 Signal	AC pulse signals
B9	BRN/BLK	Logic Ground	<0.050v
B10	GRY	Ignition Coil 5 Control	Digital Signals: 0-12-0v
B11	GRN	Ignition Coil 4 Control	Digital Signals: 0-12-0v
B12	BLU	Ignition Coil 3 Control	Digital Signals: 0-12-0v
B13	ORN	Vehicle Speed Sensor	Moving: pulse signals
B14	---	Not Used	---
B15	WHT/BLU	CKP Sensor 2 Ground	<0.050v
B16	BLU/YEL	CKP Sensor 1 Ground	<0.050v

1996 Sedan 3.2L V6 MFI VIN UA3 [A/T] 12P 'C' Connector

PCM Pin #	W/Color	Circuit Description (12-Pin)	Value at Hot Idle
C1	---	Not Used	---
C2	BLU/BLK	A/C Switch Signal	A/C Off: 12v, On: 1v
C3	RED/BLU	A/C Switch Signal 'B'	A/C Off: 4-5v, On: 0v
C4	---	Not Used	---
C5	ORN	Service Check Connector	SCS Open: 4.80v
C6	BLU/RED	Starter Switch Signal	Cranking: 9-11v
C7	---	Not Used	---
C8	BLU	Engine Speed Pulse Sensor	Pulses
C9	GRN/BLK	Traction Control Inhibit Signal	Pulses
C10	GRY/GRN	T/C Fuel Cut Signal	Pulses
C11	---	Not Used	---
C12	LT BLU	T/C Standby Signal	Pulses

1996 Sedan 3.2L V6 MFI VIN UA3 [A/T] 22P 'D' Connector

PCM Pin #	W/Color	Circuit Description (22-Pin)	Value at Hot Idle
D1	WHT/GRN	Keep Alive Power (VBU)	12-14v
D2	RED/BLU	Knock Sensor Signal (Bank 1)	No Detonation: 18mv AC
D3	ORN/RED	Barometric Output Signal	3v (at sea level)
D4	GRN/YEL	K-Line Signal	12v
D5	WHT/RED	Alternator 'FR' Signal	Digital Signals: 0-5-0v
D6	RED/BLK	TP Sensor Signal	0.5-0.6v
D7	RED/WHT	ECT Sensor Signal	At 180°F: 0.5-0.6v
D8	RED/YEL	IAT Sensor Signal	At 100°F: 2-3v
D9	WHT/YEL	MAP Sensor Signal	0.8-0.9v
D10	YEL/WHT	MAP Sensor VREF	4.9-5.1v
D11	GRN/WHT	MAP Sensor Ground	<0.050v
D12	WHT/BLU	CYP Sensor 1 Ground	<0.050v
D13	WHT/GRN	Knock Sensor Signal (Bank 2)	No Detonation: 18mv AC
D14	ORN/BLU	CYP Sensor 2 Ground	<0.050v
D15	BRN	EVAP Purge Flow Switch	Switch on: 0v, off: 12v
D16	---	Not Used	---
D17	WHT/BLK	EGR Valve Lift Sensor	1.1-1.2v
D18	WHT/RED	HO2S-11 (B1 S1) Signal	0.1-1.1v
D19	BLU/RED	HO21-21 (B2 S1) Signal	0.1-1.1v
D20	BLU	TCS VREF	4.9-5.1v
D21	YEL/BLU	Sensor VREF	4.9-5.1v
D22	GRN/BLU	Sensor Ground	<0.050v

1996 Sedan 3.2L V6 MFI VIN UA3 [A/T] 26P 'E' Connector

PCM Pin #	W/Color	Circuit Description (26-Pin)	Value at Hot Idle
E1	BLK/YEL	Solenoid Feed (B+)	12-14v
E2	GRN/WHT	Brake Switch Signal	Brake Off: 12v, On: 0v
E3	RED	Linear Solenoid Valve 'P'	AC Pulse Signals
E4	ORN/BLU	Mainshaft Speed Sensor 'P'	AC Pulse Signals
E5	BLU/GRN	Countershaft Speed Sensor	Wheels turning: 12v
E6	BLU/RED	TCS Shift Control Signal	Pulse Signals
E7	WHT	A/T: Gear Position Switch	In 'D': 0v, others: 12v
E8	YEL	A/T: Gear Position Switch	In D4: 0v, other: 12v
E9	GRN	A/T: Gear Position Switch	In D3: 0v, other: 12v
E10	BLU	A/T: Gear Position Switch	In D2: 0v, other: 12v
E11	BRN	A/T: Gear Position Switch	In D1: 0v, other: 12v
E12	BLU/YEL	Shift Control Solenoid 'A'	In D2/D3: 12v
E13	YEL	Lockup Control Solenoid 'A'	LSA Off: 0v, On: 12v
E14-15	---	Not Used	---
E16	WHT	Linear Solenoid Valve 'N'	Pulses
E17	WHT/BLU	Mainshaft Speed Sensor 'N'	<0.050v
E18	BLU/YEL	Countershaft Speed Sensor N	<0.050v
E19-21	---	Not Used	---
E22	WHT/GRN	Interlock Control Unit Signal	Key & Brake on: 12v
E23	GRN/BLK	A/T: D4 Indicator Light	D4 Light On: 0v, Off: 12v
E24	YEL/GRN	A/T: Gear Position Switch	In P/N: 0v, Others: 12v
E25	GRN/WHT	Shift Control Solenoid 'B'	In D1/D2: 12v
E26	GRN/BLK	Lockup Control Solenoid 'B'	LSB Off: 0v, On: 12v

1996 Sedan 3.2L V6 MFI VIN UA3 [A/T] 8P 'F' Connector

PCM Pin #	W/Color	Circuit Description (8-Pin)	Value at Hot Idle
F1	---	Not Used	---
F2	WHT/RED	HO2S-12 (B1 S2) Signal	0.1-1.1v
F3	GRN/BLK	Right Peak Hold Reset	Digital Signals
F4	YEL	Right Misfire Pulse	Digital Signals
F6	GRN/WHT	HO2S-12 Ground	<0.050v
F7	GRN/YEL	Left Peak Hold Reset	Pulses
F8	YEL/RED	Left Misfire Pulse	Pulses

WIRE SIDE OF HARNESS TERMINALS

1997-98 Sedan 3.2L V6 MFI VIN UA3 [A/T] 26P 'A' Connector

PCM Pin #	W/Color	Circuit Description (26-Pin)	Value at Hot Idle
A1	BRN	Injector 1	2.0-3.3 ms
A2	RED	Injector 2	2.0-3.3 ms
A3	BLU	Injector 3	2.0-3.3 ms
A4	GRN/BLK	Fuel Pump Relay Control	Relay Off: 12v, On: 1v
A5	BLK/BLU	Idle Air Control Valve	Pulse Signals
A6	RED	EGR Solenoid Control	Solenoid Off: 12v, On: 1v
A7	GRN/RED	Malfunction Indicator Lamp	MIL Off: 12v, On: 1v
A8	RED/BLU	A/C Clutch Relay Control	Relay Off: 12v, On: 1v
A9	---	Not Used	---
A10	GRN	Fuel Pressure Regulator	Solenoid Off: 12v, On: 1v
A11	PNK	Ignition Coil 1 Control	Digital Signals: 0-12-0v
A12	BLK	Power Ground	<0.1v
A13	YEL/BLK	Main Relay Power (B+)	12-14v
A14	YEL	Injector 4	2.0-3.3 ms
A15	BLK/RED	Injector 5	2.0-3.3 ms
A16	WHT/BLU	Injector 6	2.0-3.3 ms
A17	---	Not Used	---
A18	GRN/RED	HO2S-21 (B2 S1) Heater	Digital Signals: 0-12-0v
A19	ORN/BLK	HO2S-12 (B1 S2) Heater	Digital Signals: 0-12-0v
A20	RED/BLU	IAB Solenoid 1 Control	Solenoid Off: 12v, On: 1v
A21	ORN/BLK	HO2S-22 (B2 S2) Heater	Digital Signals: 0-12-0v
A22	YEL	IAB Solenoid 2 Control	Solenoid Off: 12v, On: 1v
A23	RED/YEL	EVAP Purge Solenoid	Pulse Signals
A24	BRN	Ignition Coil 2 Control	Digital Signals: 0-12-0v
A25	BLK	Power Ground	<0.1v
A26	BRN/BLK	Logic Ground	<0.050v

1997-98 Sedan 3.2L V6 MFI VIN UA3 [A/T] 16P 'B' Connector

PCM Pin #	W/Color	Circuit Description (16-Pin)	Value at Hot Idle
B1	YEL/BLK	Main Relay Power (B+)	12-14v
B2	RED	Ignition Coil 6 Control	Digital Signals: 0-12-0v
B3	GRN	PSP Switch Signal	Straight: 0v, Turning: 12v
B4	---	Not Used	---
B5	ORN	CYP Sensor 2 Signal	AC pulse signals
B6	WHT	CYP Sensor 1 Signal	AC pulse signals
B7	ORN/BLU	CKP Sensor 2 Signal	AC pulse signals
B8	BLU/GRN	CKP Sensor 1 Signal	AC pulse signals
B9	BRN/BLK	Logic Ground	<0.050v
B10	GRY	Ignition Coil 5 Control	Digital Signals: 0-12-0v
B11	GRN	Ignition Coil 4 Control	Digital Signals: 0-12-0v
B12	BLU	Ignition Coil 3 Control	Digital Signals: 0-12-0v
B13	ORN	Vehicle Speed Sensor	Moving: pulse signals
B14	---	Not Used	---
B15	WHT/BLU	CKP Sensor 2 Ground	<0.050v
B16	BLU/YEL	CKP Sensor 1 Ground	<0.050v

1997-98 Sedan 3.2L V6 MFI VIN UA3 [A/T] 16P 'C' Connector

PCM Pin #	W/Color	Circuit Description (16-Pin)	Value at Hot Idle
C1	---	Not Used	---
C2	BLU/BLK	A/C Switch Signal	A/C Off: 12v, On: 1v
C3	RED/BLU	A/C Switch Signal 'B'	A/C Off: 4-5v, On: 0v
C4	---	Not Used	---
C5	ORN	Service Check Connector	SCS Open: 4.80v
C6	BLU/RED	Starter (Cranking) Signal	Cranking: 9-11v
C7	---	Not Used	---
C8	BLU	Engine Speed Pulse Sensor	Pulses
C9	GRN/BLK	Traction Control Inhibit Signal	Pulses
C10	GRY/GRN	T/C Fuel Cut Signal	Pulses
C11	---	Not Used	---
C12	LT BLU	T/C Standby Signal	Pulses
C13-16	---	Not Used	---

1997-98 Sedan 3.2L V6 MFI VIN UA3 [A/T] 22P 'D' Connector

PCM Pin #	W/Color	Circuit Description (22-Pin)	Value at Hot Idle
D1	WHT/GRN	Keep Alive Power (VBU)	12-14v
D2	RED/BLU	Knock Sensor Signal (Bank 1)	No Detonation: 18mv AC
D3	ORN/RED	Barometric Output Signal	3v (at sea level)
D4	GRN/YEL	K-Line Signal	12v
D5	WHT/RED	Alternator 'FR' Signal	Digital Signals: 0-5-0v
D6	RED/BLK	TP Sensor Signal	0.5-0.6v
D7	RED/WHT	ECT Sensor Signal	At 180°F: 0.5-0.6v
D8	RED/YEL	IAT Sensor Signal	At 100°F: 2-3v
D9	WHT/YEL	MAP Sensor Signal	0.8-0.9v
D10	YEL/WHT	MAP Sensor VREF	4.9-5.1v
D11	GRN/WHT	MAP Sensor Ground	<0.050v
D12	WHT/BLU	CYP Sensor 1 Ground	<0.050v
D13	WHT/GRN	Knock Sensor Signal (Bank 2)	No Detonation: 18mv AC
D14	ORN/BLU	CYP Sensor 2 Ground	<0.050v
D15	PNK	Fuel Tank Pressure Sensor	Fuel Cap off: 2.5v
D17	WHT/BLK	EGR Valve Lift Sensor	1.1-1.2v
D18	WHT/RED	HO2S-11 (B1 S1) Signal	0.1-1.1v
D19	BLU/RED	HO21-21 (B2 S1) Signal	0.1-1.1v
D20	BLU	TCS VREF	4.9-5.1v
D21	YEL/BLU	Sensor VREF	4.9-5.1v
D22	GRN/BLU	Sensor Ground	<0.050v

Abbreviation	Color	Abbreviation	Color	Abbreviation	Color
BLK	Black	LT BLU	Lt. Blue	TAN	Tan
BLU	Blue	LT GRN	Lt. Green	VIO	Violet
BRN	Brown	ORN	Orange	WHT	White
GRY	Gray	PNK	Pink	YEL	Yellow
GRN	Green	PPL	Purple		

1997-98 Sedan 3.2L V6 MFI VIN UA3 [A/T] 26P 'E' Connector

PCM Pin #	W/Color	Circuit Description (26-Pin)	Value at Hot Idle
E1	BLK/YEL	Solenoid Feed (B+)	Sol. On: 12v, off: 0v
E2	GRN/WHT	Brake Switch Signal	Brake Off: 12v, On: 0v
E3	RED	Linear Solenoid Valve 'P'	Pulses
E4	WHT/GRN	Mainshaft Speed Sensor 'P'	Pulses
E5	RED/BLU	Countershaft Speed Sensor P	Moving: AC Pulses
E6	BLU/RED	TCS Shift Control Signal	Pulses
E7	WHT	A/T: Gear Position Switch	In 'D': 0v, others: 12v
E8	YEL	A/T: Gear Position Switch	In D4: 0v, other: 12v
E9	GRN	A/T: Gear Position Switch	In D3: 0v, other: 12v
E10	BLU	A/T: Gear Position Switch	In D2: 0v, other: 12v
E11	BRN	A/T: Gear Position Switch	In D1: 0v, other: 12v
E12	BLU/YEL	Shift Control Solenoid 'A'	SSA Off: 0v, On: 12v
E13	YEL	Lockup Control Solenoid 'A'	L/U on: 12v, off: 0v
E14-15	---	Not Used	---
E16	WHT	Linear Solenoid Valve 'N'	Pulse Signals
E17	WHT/BLU	Mainshaft Speed Sensor 'N'	<0.050v
E18	BLU/YEL	Countershaft Speed Sensor N	<0.050v
E19-21	---	Not Used	---
E22	WHT/GRN	Interlock Control Unit Signal	Key & Brake on: 12v
E23	GRN/BLK	A/T: D4 Indicator Light	D4 Light On: 0v, Off: 12v
E24	YEL/GRN	A/T: Gear Position Switch	In P/N: 0v, Others: 12v
E25	GRN/WHT	Shift Control Solenoid 'B'	SSB Off: 0v, On: 12v
E26	GRN/BLK	Lockup Control Solenoid 'B'	LUS on: 12v, off: 0v

1997-98 Sedan 3.2L V6 MFI VIN UA3 [A/T] 8P 'F' Connector

PCM Pin #	W/Color	Circuit Description (8-Pin)	Value at Hot Idle
F1	WHT/BLU	EVAP Bypass Solenoid	Solenoid Off: 12v, On: 1v
F2	WHT/RED	Rear HO2S-2 Signal	0.1-1.1v
F3	GRN/BLK	Right Peak Hold Reset	Digital Signals
F4	YEL	Right Misfire Pulse	Digital Signals
F5	YEL/BLU	EVAP Vent Shut Solenoid	Solenoid Off: 12v, On: 1v
F6	GRN/WHT	HO2S-12 Ground	<0.050v
F7	GRN/YEL	Left Peak Hold Reset	Digital Signals
F8	YEL/RED	Left Misfire Pulse	Digital Signals

WIRE SIDE OF HARNESS TERMINALS

1999 Sedan 3.2L V6 MFI VIN UA3 [A/T] 32P 'A' Connector

PCM Pin #	W/Color	Circuit Description (32-Pin)	Value at Hot Idle
A1	---	Not Used	---
A2	GRN/WHT	Engine Mount Solenoid	0v, Off-Idle: 12v
A3	BLU	EVAP Bypass Solenoid	Solenoid Off: 12v, On: 1v
A4	GRN/WHT	EVAP Vent Solenoid	Solenoid Off: 12v, On: 1v
A5	BLU/GRN	Cruise Control Signal	Cruise On: pulse signals
A6	RED/YEL	EVAP Purge Solenoid	Pulse Signals
A7	ORN/WHT	Reference Voltage	4.9-5.1v
A8	BLK/WHT	HO2S-12 (B1 S2) Heater	Digital Signals: 0-12-0v
A9	BLU/WHT	Vehicle Speed Sensor	Moving: pulse signals
A10	BRN	Service Check Connector	SCS Open: 4.80v
A11	LT GRN	Gear Position Signal Out	In 'P': 4v, Others: 0v
A12	PNK	Immobilizer Indicator Control	Lamp On: 1v, Off: 12v
A13	BLU	Immobilizer Enable Signal	Digital Signals
A14	GRN/BLK	D4 Indicator Light	In D4: 0v, Off: 12v
A15	GRN/YEL	Immobilizer Fuel Pump Relay	Relay Off: 12v, On: 1v
A17	RED	A/C Clutch Relay Control	Relay Off: 12v, On: 1v
A18	GRN/ORN	Malfunction Indicator Light	MIL Off: 12v, On: 1v
A19	BLU	Engine Speed Signal	Pulses
A20	GRN	Radiator Fan Relay Control	Relay Off: 12v, On: 1v
A21	GRY	K-Line Signal	12v
A22	---	Not Used	---
A23	WHT/RED	HO2S-12 (B1 S2) Signal	0.1-1.1v
A24	BLU/ORN	Starter Switch Signal	Cranking: 9-11v
A25	BLU/GRN	INOCD Immobilizer Code	Digital Signals
A26	GRN	PSP Switch Signal	Straight: 0v, Turning: 12v
A27	BLU/RED	A/C Switch Signal	A/C Off: 12v, On: 1v
A28	WHT/RED	Interlock Control Unit Signal	Key & Brake on: 12v
A29	LT GRN	Fuel Tank Pressure Sensor	Fuel Cap off: 2.5v
A30	GRN/RED	Electronic Load Detector	Varies: 2.5-3.5v
A31	RED/BLU	Throttle Position Sensor Out	0.5v
A32	GRN/WHT	Brake Switch Signal	Brake on: 0v, off: 12v

1999 Sedan 3.2L V6 MFI VIN UA3 [A/T] 25P 'B' Connector

PCM Pin #	W/Color	Circuit Description (25-Pin)	Value at Hot Idle
B1	YEL/BLK	Main Relay Power (B+)	12-14v
B2	BLK	Power Ground (PG1)	<0.1v
B3	BLK/RED	Injector 5 Control	2.0-3.3 ms
B4	YEL	Injector 4 Control	2.0-3.3 ms
B5	RED	Injector 2 Control	2.0-3.3 ms
B6	WHT/BLU	Injector 6 Control	2.0-3.3 ms
B7	PNK	E-EGR Control Solenoid	Digital Signals: 0-12-0v
B8	WHT	A/T: Clutch Solenoid 'A-'	Pulse Signals
B9	YEL/BLK	Main Relay Power (B+)	12-14v
B10	BLK	Power Ground (PG2)	<0.1v
B11	BRN	Injector 1 Control	2.0-3.3 ms
B12	GRN/YEL	VTEC Solenoid Control	Idle: 0v, Hi-Speed: 12v
B13	---	Not Used	---
B14	BLU/WHT	A/T: Gear Position Switch	In P/N: 0v, Others: 12v
B15	BLU	Injector 3 Control	2.0-3.3 ms
B16	---	Not Used	---
B17	RED	A/T: Clutch Solenoid 'A+'	Pulse Signals
B18	GRN	A/T: Clutch Solenoid 'B-'	Pulse Signals
B19	---	Not Used	---
B20	BRN/BLK	Logic Ground (LG1)	<0.050v
B21	WHT/YEL	Keep Alive Power (VBU)	12-14v
B22	BRN/BLK	Logic Ground (LG2)	<0.050v
B23	BLK/BLU	Idle Air Control Valve	Pulse Signals
B24	WHT/RED	A/T: 3rd Clutch Pressure Sw.	Open: 12v, Closed: 1v
B25	GRN	A/T: Clutch Solenoid 'B+'	Pulse Signals

1999 Sedan 3.2L V6 MFI VIN UA3 [A/T] 31P 'C' Connector

PCM Pin #	W/Color	Circuit Description (31-Pin)	Value at Hot Idle
C1	BLK/WHT	HO2S-11 (B1 S1) Heater	Digital Signals: 0-12-0v
C2	WHT/GRN	Alternator Charging Signal	Lights on: 0v, off: 12v
C3	BLU	Ignition Coil 3 Control	Digital Signals: 0-12-0v
C4	YEL/GRN	Ignition Coil 1 Control	Digital Signals: 0-12-0v
C5	WHT/RED	Alternator 'FR' Signal	Digital Signal: 0-5-0-5v
C6	WHT/BLK	EGR Valve Lift Sensor	1.2v
C7	GRN/WHT	MAP Sensor Ground (SG1)	<0.050v
C8	BLU	CKP Sensor Signal	AC pulse signals
C9	WHT	CKP Sensor Ground	<0.050v
C10	BLU/BLK	VTEC Pressure Switch	Idle: 0v, Cruise: 12v
C11	---	Not Used	---
C12	BLK/RED	Ignition Coil 5 Control	Digital Signals: 0-12-0v
C13	YEL	Ignition Coil 4 Control	Digital Signals: 0-12-0v
C14	RED	Ignition Coil 2 Control	Digital Signals: 0-12-0v
C15, C24	---	Not Used	---
C16	WHT	HO2S-11 (B1 S1) Signal	0.1-1.1v
C17	RED/GRN	MAP Sensor Signal	0.8-0.9v
C18	GRN/BLK	Sensor Ground (SG2)	<0.050v
C19	YEL/RED	MAP Sensor VREF (VCC1)	4.9-5.1v
C20	GRN	TDC1 Sensor Signal	AC pulse signals
C21	RED	TDC1 Sensor Ground	<0.050v
C22	RED/BLU	Knock Sensor Signal	No Detonation: 18mv AC
C23	WHT/BLU	Ignition Coil 6 Control	Digital Signals: 0-12-0v
C25	RED/YEL	IAT Sensor Signal	Varies w/temp: 0.5-4.9v
C26	RED/WHT	ECT Sensor Signal	At 180ºF: 0.5-0.6v
C27	RED/BLK	TP Sensor	0.5-0.6v
C28	YEL/BLU	Sensor VREF (VCC2)	4.9-5.1v
C29	YEL	TDC2 Sensor Signal	AC pulse signals
C30	BLK	TDC2 Sensor Ground	<0.050v

1999 Sedan 3.2L V6 MFI VIN UA3 [A/T] 16P 'D' Connector

PCM Pin #	W/Color	Circuit Description (16-Pin)	Value at Hot Idle
D1	YEL	A/T: Lockup Solenoid	LUS Off: 12v, On: 1v
D2	GRN/WHT	A/T: Shift Solenoid 'B'	SSB Off: 12v, On: 1v
D3	GRN	A/T" Shift Solenoid 'C'	SSA Off: 12v, On: 1v
D4	RED/BLK	A/T: Gear Position Switch	In 'N': 0v, Others: 12v
D5	BLK/YEL	Solenoid Feed (B+)	12-14v
D6	WHT	A/T: Gear Position Switch	In 'R': 0v, Others: 12v
D7	BLU/YEL	A/T: Shift Solenoid 'A'	SSA Off: 12v, On: 1v
D8	PNK	A/T: Gear Position Switch	In D3: 0v, Others: 12v
D9	YEL	A/T: Gear Position Switch	In D4: 0v, Others: 12v
D10	BLU	Countershaft Speed Sensor P	Wheels turning: 12v
D11	RED	Mainshaft Speed Sensor 'P'	Moving: pulse signals
D12	WHT	Mainshaft Speed, Ground 'N'	<0.050v
D13	BLU/BLK	A/T: 2nd Clutch Pressure Sw.	Open: 12v, Closed: 1v
D14	BLU	A/T: Gear Position Switch	In D2: 0v, Others: 12v
D15	BRN	A/T: Gear Position Switch	In D1: 0v, Others: 12v
D16	GRN	Countershaft Speed Sensor N	<0.050v

2000-03 Sedan 3.2L V6 MFI VIN UA5 [A/T] 32P 'A' Connector

PCM Pin #	W/Color	Circuit Description (32-Pin)	Value at Hot Idle
A1	WHT/RED	HO2S-22 (B2 S2) Signal	0.1-1.1v
A2	GRN/WHT	Engine Mount Solenoid	0v, Off-Idle: 12v
A3	BLU	EVAP Bypass Solenoid	Solenoid Off: 12v, On: 1v
A4	GRN/WHT	EVAP Vent Solenoid	Solenoid Off: 12v, On: 1v
A5	BLU/GRN	Cruise Control Signal	Cruise On: pulse signals
A6	RED/YEL	EVAP Purge Solenoid	Pulse Signals
A7	ORN/WHT	Reference Voltage	4.9-5.1v
A8	BLK/WHT	HO2S-12 (B1 S2) Heater	Digital Signals: 0-12-0v
A9	BLU/WHT	Vehicle Speed Sensor	Moving: pulse signals
A10	BRN	Service Check Connector	SCS Open: 4.80v
A11-13	---	Not Used	---
A14	GRN/BLK	D4 Indicator Light	In D4: 0v, Off: 12v
A15	GRN/YEL	Fuel Pump Relay Control	Relay Off: 12v, On: 1v
A16	---	Not Used	---
A17	RED	A/C Clutch Relay Control	Relay Off: 12v, On: 1v
A18	GRN/ORN	Malfunction Indicator Light	MIL Off: 12v, On: 1v
A19	BLU	Engine Speed Signal	Pulse Signals
A20	GRN	Radiator Fan Relay Control	Relay Off: 12v, On: 1v
A21	GRY	K-Line Signal	12v
A22-23	---	Not Used	---
A24	BLU/ORN	Starter Switch Signal	Cranking: 9-11v
A25	---	Not Used	---
A26	GRN	PSP Switch Signal	Straight: 0v, Turning: 12v
A27	BLU/RED	A/C Switch Signal	A/C Off: 12v, On: 1v
A28	WHT/RED	Interlock Control Unit Signal	Key & Brake on: 12v
A29	LT GRN	Fuel Tank Pressure Sensor	Fuel Cap off: 2.5v
A30	GRN/RED	Electronic Load Detector	Varies: 2.5-3.5v
A31	YEL/GRN	ECT Sensor Signal to TCM	Digital Signals
A32	WHT/BLK	Brake Switch Signal	Brake on: 0v, off: 12v

2000-03 Sedan 3.2L V6 MFI VIN UA5 [A/T] 25P 'B' Connector

PCM Pin #	W/Color	Circuit Description (25-Pin)	Value at Hot Idle
B1, B9	YEL/BLK	Main Relay Power (B+)	12-14v
B2, B10	BLK	Power Ground (PG1, PG2)	<0.1v
B3	BLK/RED	Injector 5 Control	2.0-3.3 ms
B4	YEL	Injector 4 Control	2.0-3.3 ms
B5	RED	Injector 2 Control	2.0-3.3 ms
B6	WHT/BLU	Injector 6 Control	2.0-3.3 ms
B7	BLU/RED	E-EGR Control Solenoid	Digital Signals: 0-12-0v
B8	WHT	A/T: Clutch Solenoid 'A-'	Pulse Signals
B11	BRN	Injector 1 Control	2.0-3.3 ms
B12	GRN/YEL	VTEC Solenoid Control	Idle: 0v, Hi-Speed: 12v
B13	GRN/RED	A/T: Clutch Solenoid 'C+'	Pulse Signals
B14	BLU/WHT	A/T: Gear Position Switch	In P/N: 0v, Others: 12v
B15	BLU	Injector 3 Control	2.0-3.3 ms
B16	---	Not Used	---
B17	RED	A/T: Clutch Solenoid 'A+'	Pulse Signals
B18	GRN	A/T: Clutch Solenoid 'B-'	Pulse Signals
B19	BLU/YEL	A/T: 4th Oil Pressure Switch	12-14v
B20	BRN/YEL	Logic Ground (LG1)	<0.050v
B21	WHT/RED	Keep Alive Power (VBU)	12-14v
B22	BRN/BLK	Logic Ground (LG2)	<0.050v
B23	BLK/RED	Idle Air Control Valve	Pulse Signals
B24	RED/BLU	A/T: Clutch Solenoid 'A-'	Pulse Signals
B25	BRN/BLU	A/T: Clutch Solenoid 'B+'	Pulse Signals

2000-03 Sedan 3.2L V6 MFI VIN UA5 [A/T] 16P 'D' Connector

PCM Pin #	W/Color	Circuit Description (16-Pin)	Value at Hot Idle
D1	YEL	A/T: Lockup Solenoid	LUS Off: 12v, On: 1v
D2	GRN/WHT	A/T: Shift Solenoid 'B'	SSB Off: 12v, On: 1v
D3	GRN	A/T" Shift Solenoid 'C'	SSA Off: 12v, On: 1v
D4	RED/BLK	A/T: 'N' Gear Position Switch	In 'N': 0v, Others: 12v
D5	BLK/YEL	Solenoid Feed (B+)	12-14v
D6	WHT	A/T: 'R' Gear Position Switch	In 'R': 0v, Others: 12v
D7	BLU/YEL	A/T: Shift Solenoid 'A'	SSA Off: 12v, On: 1v
D8	YEL	A/T: D4 Gear Position Switch	In D4: 0v, others: 12v
D9	YEL/GRN	A/T: D5 Gear Position Switch	In D5: 0v, Others: 12v
D10	BLU	Countershaft Speed Sensor P	Wheels turning: 12v
D11	RED	Mainshaft Speed Sensor 'P'	Moving: pulse signals
D12	WHT	Mainshaft Speed, Ground 'N'	<0.050v
D13	BLU/WHT	A/T: 3rd Clutch Pressure Sw.	Open: 12v, Closed: 1v
D14	RED	A/T: D3 Gear Position Switch	In D3: 0v, Others: 12v
D15	BLU	A/T: D2 Gear Position Switch	In D2: 0v, Others: 12v
D16	GRN	Countershaft Speed Sensor N	<0.050v

2000-03 Sedan 3.2L V6 MFI VIN UA5 [A/T] 31P 'C' Connector

PCM Pin #	W/Color	Circuit Description (31-Pin)	Value at Hot Idle
C1	BLK/WHT	HO2S-11 (B1 S1) Heater	Digital Signals: 0-12-0v
C2	WHT/GRN	Alternator Charging Signal	Lights on: 0v, off: 12v
C3	WHT/BLU	Ignition Coil 3 Control	Digital Signals: 0-12-0v
C4	YEL/GRN	Ignition Coil 1 Control	Digital Signals: 0-12-0v
C5	WHT/RED	Alternator 'FR' Signal	Digital Signal: 0-5-0-5v
C6	WHT/BLK	EGR Valve Lift Sensor	1.2v
C7	GRN/WHT	MAP Sensor Ground (SG1)	<0.050v
C8	BLU	CKP Sensor 'P' Signal	AC pulse signals
C9	WHT	CKP Sensor 'N' Signal	<0.050v
C10	BLU/BLK	VTEC Pressure Switch	Idle: 0v, Cruise: 12v
C11	---	Not Used	---
C12	BLK/RED	Ignition Coil 5 Control	Digital Signals: 0-12-0v
C13	YEL	Ignition Coil 4 Control	Digital Signals: 0-12-0v
C14	RED	Ignition Coil 2 Control	Digital Signals: 0-12-0v
C15	---	Not Used	---
C16	WHT	HO2S-11 (B1 S1) Signal	0.1-1.1v
C17	GRN/RED	MAP Sensor Signal	0.8-0.9v
C18	GRN/YEL	Sensor Ground (SG2)	<0.050v
C19	YEL/RED	MAP Sensor VREF (VCC1)	4.9-5.1v
C20	GRN	TDC1 Sensor 'P' Signal	AC pulse signals
C21	RED	TDC1 Sensor 'N' Signal	<0.050v
C22	RED/BLU	Knock Sensor Signal	No Detonation: 18mv AC
C23	BLU/WHT	Ignition Coil 6 Control	Digital Signals: 0-12-0v
C24	BLU/YEL	A/T: TFT Sensor Signal	0.1-4.2v
C25	RED/YEL	IAT Sensor Signal	Varies w/temp: 0.5-4.9v
C26	RED/WHT	ECT Sensor Signal	At 180°F: 0.5-0.6v
C27	RED/BLK	TP Sensor	0.5-0.6v
C28	YEL/BLU	Sensor VREF (VCC2)	4.9-5.1v
C29	YEL	TDC2 Sensor 'P' Signal	AC pulse signals
C30	BLK	TDC2 Sensor 'N' Signal	<0.050v

2000-03 Sedan 3.2L V6 MFI VIN UA5 [A/T] 20P 'E' Connector

PCM Pin #	W/Color	Circuit Description (20-Pin)	Value at Hot Idle
E1	PNK	Immobilizer Indicator Lamp	Lamp Off: 12v, On: 1v
E2	RED	IMOCD Immobilizer Code	Digital Signals
E3-4	---	Not Used	---
E5	RED/BLU	TP Sensor Signal to TCM	0.5-0.6v
E6	ORN/WHT	Voltage Reference (ABS)	4.9-5.1v
E7	BLU	PCM to Frame Inhibition	2.5v (normal signal)
E8	BLU/BLK	A/T: Drive Shift Indicator 'C'	2nd, 4th: 0v, 1st, 3rd: 12v
E9	BLU/GRN	A/T: Drive Shift Indicator 'B'	In 4th: 12v, 1st & 3rd: 0v
E10	BLU/RED	A/T: Drive Shift Indicator 'A'	In 4th: 12v, 1st - 3rd: 0v
E11	---	Not Used	---
E12	BLU	IMOEN Immobilizer Enable	Digital Signals
E13, E15	---	Not Used	---
E14	BRN	A/T: First Hold Down Switch	LH Switch on: 0v, off: 12v
E16	PNK	Frame to PCM Torque Down	TCS On: 5v, Off: 2.5v
E17	LT GRN	A/T: 'P' Gear Position Switch	In 'P': 4v, Others: 0v
E18	ORN	Torque Mgmt. (-) Signal	In Neutral: 12v
E19	YEL	Torque Mgmt. (+) Signal	In Sports Shift: 0v
E20	RED	Torque Mgmt. Mode Signal	In Sports Shift: 0v

Honda Pin Table Contents

About This Section

Introduction

This section contains Pin Voltage Tables for Honda vehicles from 1990-2001. It can be used to help you repair Trouble Code and No Code problems related to the PCM.

VEHICLE COVERAGE

The following vehicle applications are covered in this section:

- 1990-2001 Accord Coupe, Sedan & Wagon
- 1990-2001 Civic Coupe, Hatchback, Sedan & Wagon
- 1993-97 Civic Del Sol Coupe
- 1997-2001 CR-V Utility Vehicle
- 2000-01 Insight Hatchback
- 1995-2001 Odyssey Van
- 1990-2001 Prelude Coupe
- 2000-01 S2000 Convertible

How to Use This Section

This section can be used to look up the location of a particular pin, a wire color or to find a "known good" value of a circuit. To locate the PCM information for a particular vehicle, find the model, correct engine size (with VIN Code) and finally the year of the vehicle.

For example, to look up the PCM terminals for a 2001 Accord Coupe 3.0L VTEC VIN CG2, go to Contents Page 1 to find this text string.

3.0L V6 VTEC VIN CG2 104-Pin **(2000-01)**......................Page 7-20

Then turn to Page 7-20 to find the following PCM related information.

The symbol (<) in used in some pin voltage readings to indicate that the reading should be less than the number shown in the example.

The transmission type may be identified in the table title as [A/T] Auto, [M/T] Manual or [All]. Refer to the example shown in the table below.

2000-01 Coupe 3.0L V6 VTEC VIN CG2 [All] 32P 'A' Connector

PCM Pin #	W/Color	Circuit Description (32-Pin)	Value at Hot Idle
A16, A25	BLU, RED	Immobilizer Enable, Code	Digital Signals
A17	RED	A/C Clutch Relay Control	Relay Off: 12v, On: 1v
A18	GRN/ORN	Malfunction Indicator Light	MIL Off: 12v, On: 1v
A19	BLU	Engine Speed Pulse (NEP)	AC Pulse Signals
A20	GRN	Radiator Fan Relay Control	Relay Off: 12v, On: 1v
A21	GRY	K-Line Signal	12v

In this example, the Immobilizer Enable signal is connected to Pin A16 of the 32-Pin connector (BLU wire) and the Immobilizer Code signal is connected to Pin A25 (same connector) with a RED wire.

ACCORD Pin Tables

1990-93 Coupe 2.2L I4 MFI VIN CB7 [All] 26P 'A' Connector

PCM Pin #	W/Color	Circuit Description (26-Pin)	Value at Hot Idle
A1	BRN	Injector 1 Control	2.0-3.3 ms
A2	YEL	Injector 4 Control	2.0-3.3 ms
A3	RED	Injector 2 Control	2.0-3.3 ms
A4	---	Not Used	---
A5	BLU	Injector 3 Control	2.0-3.3 ms
A6	ORN	HO2S Heater Control	Relay Off: 12v, On: 1v
A7	GRN/BLK	Fuel Pump Relay	Relay Off: 12v, On: 1v
A8	GRN/BLK	Fuel Pump Relay	Relay Off: 12v, On: 1v
A9	BLK/BLU	Electronic Air Control Valve	Pulse Signals
A10	---	Not Used	---
A11	PNK	EGR Control Solenoid	1.2v
A12	BLU	Radiator Fan Relay	Relay Off: 12v, On: 1v
A13	GRN/RED	Check Engine Light	MIL Off: 12v, On: 1v
A14	---	Not Used	---
A15	RED/BLU	A/C Clutch Relay	Relay Off: 12v, On: 1v
A16	WHT/GRN	Alternator Charging Signal	Lights Off: 12v, On: 0v
A17	---	Not Used	---
A18	BRN/WHT	A/T: Control Unit	---
A19	WHT	Intake Air Regulator Control	Solenoid Off: 12v, On: 1v
A20	RED/GRN	EVAP Purge Solenoid	Solenoid Off: 12v, On: 1v
A21	YEL/GRN	Igniter Control	Hot idle 10% d/cycle
A22	YEL/GRN	Igniter Control	Hot idle 10% d/cycle
A23	BLK	Power Ground	<0.1v
A24	BLK	Power Ground	<0.1v
A25	YEL/BLK	Ignition Power	12v
A26	BLK/RED	Chassis Ground	<0.050v

1990-93 Coupe 2.2L I4 MFI VIN CB7 [All] 16P 'B' Connector

PCM Pin #	W/Color	Circuit Description (16-Pin)	Value at Hot Idle
B1	YEL/BLK	Main Relay Power (B+)	12-14v
B2	BRN/BLK	Power Ground	<0.1v
B3	WHT/RED	A/T: Control Unit	Digital Signals
B4	GRN	A/T: Control Unit	Digital Signals
B5	BLU/BLK	A/C Switch Signal	Switch Off: 12v, On: 1v
B6	---	Not Used	---
B7	LT GRN	A/T: Neutral Position Switch	In N: 0v, Others: 12v
B8	RED	PSP Switch	Straight: 0v, Turning: 11v
B9	BLU/RED	A/T: Start Signal	Cranking: 9-11v
B9	BLK/GRN	M/T: Start Signal	Cranking: 9-11v
B10	ORN	Vehicle Speed Sensor	Moving: pulse signals
B11	ORN	CYP Sensor Signal	AC pulse signals
B12	WHT	CYP Sensor Ground	<0.050v
B13	ORN/BLU	TDC Sensor Signal	AC pulse signals
B14	WHT/BLU	TDC Sensor Ground	<0.050v
B15	BLU/GRN	CKP Sensor Signal	AC pulse signals
B16	BLU/YEL	CKP Sensor Ground	<0.050v

1990-93 Coupe 2.2L I4 MFI VIN CB7 [All] 22P 'D' Connector

PCM Pin #	W/Color	Circuit Description (22-Pin)	Value at Hot Idle
D1	WHT/YEL	Keep Alive Power (VBU)	12-14v
D2	GRN/WHT	Brake Switch Signal	Brake Off: 0v, On: 12v
D3	---	Not Used	---
D4	ORN/RED	Service Check Connector	SCS Open: 4.80v
D5-D8	---	Not Used	---
D9	WHT/RED	Alternator 'FR' Signal	Digital Signals: 0-5-0v
D10	GRN/RED	Electric Load Detector	Varies: 0.5-4.5v
D11	RED/BLK	TP Sensor Signal	0.5-0.6v
D12	WHT/BLK	A/T: EGR Lift Sensor Signal	1.1-1.2v
D13	YEL/GRN	ECT Sensor Signal	At 180°F: 0.52v
D14	WHT	Oxygen Sensor	0.1-1.1v
D15	RED/YEL	IAT Sensor Signal	Varies w/temp. (0.5-4.9v)
D16	---	Not Used	---
D17	WHT/BLU	MAP Sensor Signal	0.8-0.9v
D18	GRN/BLK	A/T: Control Unit	---
D19	RED/WHT	MAP Sensor VREF	4.9-5.1v
D20	YEL/WHT	Sensor VREF	4.9-5.1v
D21	BLU/WHT	Sensor Ground	<0.050v
D22	GRN/WHT	Sensor Ground	<0.050v

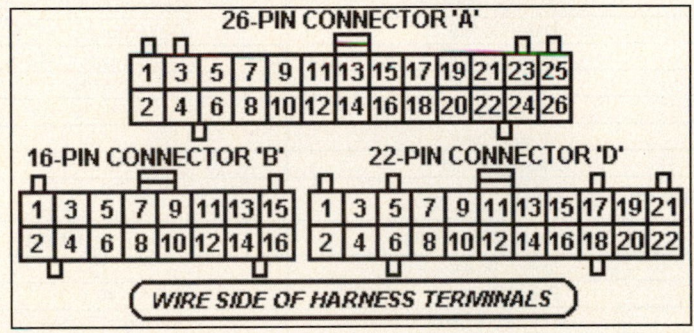

1994-95 Coupe 2.2L I4 VIN CD7 [All] 26P 'A' Connector

PCM Pin #	W/Color	Circuit Description (26-Pin)	Value at Hot Idle
A1	BRN	Injector 1 Control	2.0-3.3 ms
A2	YEL	Injector 4 Control	2.0-3.3 ms
A3	RED	Injector 2 Control	2.0-3.3 ms
A4	---	Not Used	---
A5	BLU	Injector 3 Control	2.0-3.3 ms
A6	ORN/BLK	HO2S Heater Control	Relay Off: 12v, On: 1v
A7	GRN/BLK	Fuel Pump Relay	Relay Off: 12v, On: 1v
A8	GRN/BLK	Fuel Pump Relay	Relay Off: 12v, On: 1v
A9	BLK/BLU	Intake Air Control Solenoid	Solenoid Off: 12v, On: 1v
A10	GRN/WHT	Engine Mount Solenoid	Solenoid Off: 12v, On: 1v
A11	RED	EGR Control Solenoid	Solenoid Off: 12v, On: 1v
A12	GRN	Radiator Fan Relay	Relay Off: 12v, On: 1v
A13	GRN/RED	Check Engine Light	MIL Off: 12v, On: 1v
A14	---	Not Used	---
A15	RED/BLU	A/C Clutch Relay	Relay Off: 12v, On: 1v
A16	WHT/GRN	Alternator Charging Signal	Lights Off: 12v, On: 0v
A17	---	Not Used	---
A18	BRN/WHT	A/T: Control Unit	---
A19	ORN/GRN	Intake Air Regulator Control	Solenoid Off: 12v, On: 1v
A20	RED/YEL	EVAP Purge Solenoid	Solenoid Off: 12v, On: 1v
A21	YEL/GRN	Igniter Control	Digital Signals: 0-12-0v
A22	YEL/GRN	Igniter Control	Digital Signals: 0-12-0v
A23	BLK	Power Ground	<0.1v
A24	BLK	Power Ground	<0.1v
A25	YEL/BLK	Ignition Power	12v
A26	BLK/RED	Chassis Ground	<0.050v

1994-95 Coupe 2.2L I4 VIN CD7 [All] 16P 'B' Connector

PCM Pin #	W/Color	Circuit Description (16-Pin)	Value at Hot Idle
B1	YEL/BLK	Ignition Power	12v
B2	BRN/BLK	Power Ground	<0.1v
B3	WHT/RED	A/T: Control Unit	Digital Signals
B4	GRN	A/T: Control Unit	Digital Signals
B5	RED/WHT	A/C Switch Signal	Switch Off: 12v, On: 1v
B6	---	Not Used	---
B7	LT GRN	A/T: Park/Neutral Switch	In P/N: 0v, Others: 11v
B8	GRN	PSP Switch Signal	Straight: 0v, Turning: 11v
B9	BLU/RED	A/T: Start Signal	Cranking: 9-11v
B9	BLU/RED	M/T: Start Signal	Cranking: 9-11v
B10	ORN	Vehicle Speed Sensor	Moving: pulse signals
B11	ORN	CYP Sensor Signal	AC pulse signals
B12	WHT	CYP Sensor Ground	<0.050v
B13	ORN/BLU	TDC Sensor Signal	AC pulse signals
B14	WHT/BLU	TDC Sensor Ground	<0.050v
B15	BLU/GRN	CKP Sensor Signal	AC pulse signals
B16	BLU/YEL	CKP Sensor Ground	<0.050v

1994-95 Coupe 2.2L I4 VIN CD7 [All] 22P 'D' Connector

PCM Pin #	W/Color	Circuit Description (22-Pin)	Value at Hot Idle
D1	WHT/YEL	Keep Alive Power (VBU)	12-14v
D2	GRN/WHT	Brake Switch Signal	Brake Off: 0v, On: 12v
D3	---	Not Used	---
D4	RED	Service Check Connector	SCS Open: 4.80v
D5	BLU/WHT	A/T: TCM Signal	Digital Signals
D6	---	Not Used	---
D7	GRN/RED	Data Link Connector	5v
D8	---	Not Used	---
D9	WHT/RED	Alternator 'FR' Signal	Digital Signals: 0-5-0v
D10	GRN/RED	Electric Load Detector	Varies: 0.5-4.5v
D11	RED/BLK	TP Sensor Signal	0.5-0.6v
D12	WHT/BLK	A/T: EGR Lift Sensor Signal	1.1-1.2v
D13	RED/WHT	ECT Sensor Signal	At 180°F: 0.52v
D14	WHT/RED	HO2S Signal	0.1-1.1v
D15	RED/YEL	IAT Sensor Signal	Varies w/temp. (0.5-4.9v)
D16	---	Not Used	---
D17	WHT/YEL	MAP Sensor Signal	0.8-0.9v
D18	GRN/BLK	A/T: Control Unit VREF	4.9-5.1v
D19	YEL/WHT	MAP Sensor VREF	4.9-5.1v
D20	YEL/BLU	Sensor VREF	4.9-5.1v
D21	GRN/WHT	MAP Sensor Ground	<0.050v
D22	GRN/BLU	Sensor Ground	<0.050v

Abbreviation	Color	Abbreviation	Color	Abbreviation	Color
BLK	Black	LT BLU	Lt. Blue	TAN	Tan
BLU	Blue	LT GRN	Lt. Green	VIO	Violet
BRN	Brown	ORN	Orange	WHT	White
GRY	Gray	PNK	Pink	YEL	Yellow
GRN	Green	PPL	Purple		

1996-97 Coupe 2.2L I4 VTEC VIN CD7 [All] 32P 'A' Connector

PCM Pin #	W/Color	Circuit Description (32-Pin)	Value at Hot Idle
A1	YEL	Injector 4 Control	2.0-3.3 ms
A2	BLU	Injector 3 Control	2.0-3.3 ms
A3	RED	Injector 2 Control	2.0-3.3 ms
A4	BLU	Injector 1 Control	2.0-3.3 ms
A5	ORN/BLU	HO2S-12 (B1 S2) Heater	Digital Signals: 0-12-0v
A6	ORN/BLK	HO2S-11 (B1 S1) Heater	Digital Signals: 0-12-0v
A7	RED	EGR Solenoid	Solenoid Off: 12v, On: 1v
A8	GRN/YEL	VTEC Solenoid Control	0v, Hi-Speed: 12v
A9	BRN/BLK	Sensor Ground	<0.050v
A10	BLK	Power Ground	<0.1v
A11	YEL/BLK	Main Relay Power (B+)	12-14v
A12	BLK/BLU	Idle Air Control Valve Signal	Pulse Signals
A13	GRN/WHT	Engine Mount Solenoid	Solenoid Off: 12v, On: 1v
A14, 21	---	Not Used	---
A15	RED/YEL	EVAP Purge Solenoid	Solenoid Off: 12v, On: 1v
A16	GRN/BLK	Fuel Pump Relay	Relay Off: 12v, On: 1v
A17	RED/BLU	A/C Clutch Relay	Relay Off: 12v, On: 1v
A18	GRN/RED	Check Engine Light	MIL Off: 12v, On: 1v
A19	WHT/GRN	Alternator Charging Signal	Lights Off: 12v, On: 0v
A20	YEL/GRN	Igniter Control	Digital Signals: 0-12-0v
A22	BRN/BLK	Sensor Ground	<0.050v
A23	BLK	Power Ground	<0.1v
A24	YEL/BLK	Main Relay Power (B+)	12-14v
A25	ORN/GRN	Intake Air Resonator Control	Solenoid Off: 12v, On: 1v
A26	---	Not Used	---
A27	GRN	Radiator Fan Relay	Relay Off: 12v, On: 1v
A28	GRN/WHT	EVAP Bypass Solenoid	Solenoid Off: 12v, On: 1v
A29	ORN/GRN	EVAP Vent Solenoid	Solenoid Off: 12v, On: 1v
A30-32	---	Not Used	---

1996-97 Coupe 2.2L I4 VTEC VIN CD7 [All] 25P 'B' Connector

PCM Pin #	W/Color	Circuit Description (25-Pin)	Value at Hot Idle
B3	BLU/YEL	A/T: Shift Solenoid 'A'	D2-3: 12v, D1-4: 0v
B4	GRN/BLK	A/T: Lockup Solenoid 'B'	LSB On: 12v, Off: 0v
B5	YEL	A/T: Lockup Solenoid 'A'	LSA On: 12v, Off: 0v
B8	GRN/BLU	A/T: Gear Position Switch	In D3: 0v, Others: 12v
B11	GRN/WHT	A/T: Shift Solenoid 'B'	D1-2: 12v, D3-4: 0v
B12	WHT/GRN	Interlock Control Unit Signal	Key & Brake On: 12v
B13	BLU/RED	A/T: D4 Indicator Light Switch	D4 On: 0v, Off: 12v
B14	WHT/BLU	Mainshaft Speed Ground	<0.050v
B15	ORN/BLU	Mainshaft Speed Signal	Moving: AC pulses
B16	GRN/RED	A/T: Gear Position Switch	In 'R': 0v, Others: 12v
B17	GRN/YEL	A/T: Gear Position Switch	In D2: 0v, Others: 12v
B18	GRN/WHT	A/T: Gear Position Switch	In D1: 0v, Others: 12v
B22	BLU/YEL	Sensor Ground	<0.050v
B23	BLU/GRN	Countershaft Speed Sensor	Moving: AC pulses
B24	GRN/BLK	A/T: Gear Position Switch	In D4: 0v, Others: 12v
B25	LT GRN	A/T: Gear Position Switch	In P/N: 0v, Others: 12v

1996-97 Coupe 2.2L I4 VTEC VIN CD7 [All] 31P 'C' Connector

PCM Pin #	W/Color	Circuit Description (31-Pin)	Value at Hot Idle
C2	BLU	CYP Sensor Signal	AC pulse signals
C3	GRN	TDC Sensor Signal	AC pulse signals
C4	YEL	CKP Sensor Signal	AC pulse signals
C5	RED/WHT	A/C Switch Signal	Switch Off: 12v, On: 1v
C6	BLU/RED	Start Signal	Cranking: 9-11v
C7	RED	Service Check Connector	SCS Open: 4.80v
C8	LT GRN	K-Line Signal	12v
C10	WHT/YEL	Keep Alive Power (VBU)	12-14v
C12	WHT	CKP Sensor Ground	<0.050v
C13	RED	TDC Sensor Ground	<0.050v
C14	BLK	CYP Sensor Ground	<0.050v
C16	GRN	PSP Switch	Straight: 0v, Turning: 11v
C17	WHT/RED	Alternator 'FR' Signal	Digital Signals: 0-5-0v
C18	ORN	Vehicle Speed Sensor	Moving: pulse signals
C20	BRN	EVAP Purge Flow Switch	Switch on: 0v, off: 5v
C21-31	---	---	---

1996-97 Coupe 2.2L I4 VTEC VIN CD7 [All] 16P 'D' Connector

PCM Pin #	W/Color	Circuit Description (16-Pin)	Value at Hot Idle
D1	RED/BLK	TP Sensor Signal	0.5-0.6v
D2	RED/WHT	ECT Sensor Signal	At 180°F: 0.52v
D3	WHT/YEL	MAP Sensor Signal	0.8-0.9v
D4	YEL/WHT	MAP Sensor VREF	4.9-5.1v
D5	GRN/WHT	Brake Switch Signal	Brake Off: 0v, On: 12v
D7	WHT/RED	HO2S-11 (B1 S1) Signal	0.1-1.1v
D8	RED/YEL	IAT Sensor Signal	Varies w/temp. (0.5-4.9v)
D9	WHT/BLK	EGR Valve Lift Sensor	1.2v
D10	YEL/BLU	Sensor VREF	4.9-5.1v
D11	GRN/BLU	Sensor Ground	<0.050v
D12	GRN/WHT	MAP Sensor Ground	<0.050v
D13	RED/WHT	HO2S-12 Ground	<0.050v
D14	WHT/RED	HO2S-12 (B1 S2) Signal	0.1-1.1v
D15	GRN/RED	Fuel Tank Pressure Sensor	Fuel Cap off: 2.5v
D16	GRN/RED	Electric Load Detector	Varies: 0.5-4.5v

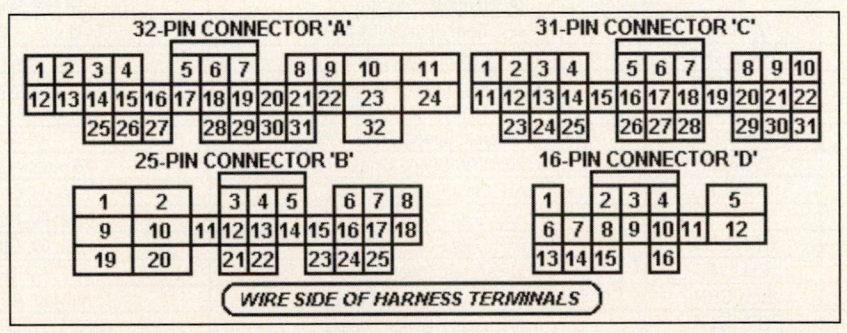

1996-97 Coupe 2.2L I4 MFI VIN CD7 [All] 32P 'A' Connector

PCM Pin #	W/Color	Circuit Description (32-Pin)	Value at Hot Idle
A1	YEL	Injector 4 Control	2.0-3.3 ms
A2	BLU	Injector 3 Control	2.0-3.3 ms
A3	RED	Injector 2 Control	2.0-3.3 ms
A4	BLU	Injector 1 Control	2.0-3.3 ms
A5	ORN/BLU	HO2S-12 (B1 S2) Heater	Digital Signals: 0-12-0v
A6	ORN/BLK	HO2S-11 (B1 S1) Heater	Digital Signals: 0-12-0v
A7	RED	EGR Solenoid Control	Solenoid Off: 12v, On: 1v
A8	---	Not Used	---
A9	BRN/BLK	Sensor Ground	<0.050v
A10	BLK	Power Ground	<0.1v
A11	YEL/BLK	Main Relay Power (B+)	12-14v
A12	BLK/BLU	Idle Air Control Valve Signal	Pulse Signals
A13	GRN/WHT	Engine Mount Solenoid	Solenoid Off: 12v, On: 1v
A14	---	Not Used	---
A15	RED/YEL	EVAP Purge Solenoid	Solenoid Off: 12v, On: 1v
A16	GRN/BLK	Fuel Pump Relay	Relay Off: 12v, On: 1v
A17	RED/BLU	A/C Clutch Relay	Relay Off: 12v, On: 1v
A18	GRN/RED	Check Engine Light	MIL Off: 12v, On: 1v
A19	WHT/GRN	Alternator Charging Signal	Lights Off: 12v, On: 0v
A20	YEL/GRN	Igniter Control	Digital Signals: 0-12-0v
A21, A26	---	Not Used	---
A22	BRN/BLK	Sensor Ground	<0.050v
A23	BLK	Power Ground	<0.1v
A24	YEL/BLK	Main Relay Power (B+)	12-14v
A25	ORN/GRN	Intake Air Resonator Control	Solenoid Off: 12v, On: 1v
A27	GRN	Radiator Fan Relay	Relay Off: 12v, On: 1v
A28	GRN/WHT	EVAP Bypass Solenoid	Solenoid Off: 12v, On: 1v
A29	ORN/GRN	EVAP Vent Solenoid	Solenoid Off: 12v, On: 1v
A30-32	---	Not Used	---

1996-97 Coupe 2.2L I4 MFI VIN CD7 [All] 25P 'B' Connector

PCM Pin #	W/Color	Circuit Description (25-Pin)	Value at Hot Idle
B3	BLU/YEL	A/T: Shift Solenoid 'A'	D2-3: 12v, D1-4: 0v
B4	GRN/BLK	A/T: Lockup Solenoid 'B'	LSB On: 12v, Off: 0v
B5	YEL	A/T: Lockup Solenoid 'A'	LSA On: 12v, Off: 0v
B8	GRN/BLU	A/T: Gear Position Switch	In D3: 0v, Others: 12v
B11	GRN/WHT	A/T: Shift Solenoid 'B'	D1-2: 12v, D3-4: 0v
B12	WHT/GRN	Interlock Control Unit Signal	Key & Brake On: 12v
B13	BLU/RED	A/T: D4 Indicator Light Switch	D4 On: 0v, Off: 12v
B14	WHT/BLU	Mainshaft Speed Ground 'N'	<0.050v
B15	ORN/BLU	Mainshaft Speed Signal 'P'	Moving: AC pulses
B16	GRN/RED	A/T: Gear Position Switch	In 'R': 0v, Others: 12v
B17	GRN/YEL	A/T: Gear Position Switch	In D2: 0v, Others: 12v
B18	GRN/WHT	A/T: Gear Position Switch	In D2: 0v, Others: 12v
B22	BLU/YEL	Countershaft Speed Sensor N	<0.050v
B23	BLU/GRN	Countershaft Speed Sensor	Moving: AC pulses
B24	GRN/BLK	A/T: Gear Position Switch	In D4: 0v, Others: 12v
B25	LT GRN	A/T: Gear Position Switch	In P/N: 0v, Others: 12v

1996-97 Coupe 2.2L I4 MFI VIN CD7 [All] 31P 'C' Connector

PCM Pin #	W/Color	Circuit Description (31-Pin)	Value at Hot Idle
C2	BLU	CYP Sensor Signal	AC pulse signals
C3	GRN	TDC Sensor Signal	AC pulse signals
C4	YEL	CKP Sensor Signal	AC pulse signals
C5	RED/WHT	A/C Switch Signal	Switch Off: 12v, On: 1v
C6	BLU/RED	Start Signal	Cranking: 9-11v
C7	RED	Service Check Connector	SCS Open: 4.80v
C8	LT GRN	K-Line Signal	12v
C10	WHT/YEL	Keep Alive Power (VBU)	12-14v
C12	WHT	CKP Sensor Ground	<0.050v
C13	RED	TDC Sensor Ground	<0.050v
C14	BLK	CYP Sensor Ground	<0.050v
C16	GRN	PSP Switch	Straight: 0v, Turning: 11v
C17	WHT/RED	Alternator 'FR' Signal	Digital Signals: 0-5-0v
C18	ORN	Vehicle Speed Sensor	Moving: pulse signals
C20	BRN	EVAP Purge Flow Switch	Switch on: 0v, off: 5v
C21-31	---	---	---

1996-97 Coupe 2.2L I4 MFI VIN CD7 [All] 16P 'D' Connector

PCM Pin #	W/Color	Circuit Description (16-Pin)	Value at Hot Idle
D1	RED/BLK	TP Sensor Signal	0.5-0.6v
D2	RED/WHT	ECT Sensor Signal	At 180ºF: 0.52v
D3	WHT/YEL	MAP Sensor Signal	0.8-0.9v
D4	YEL/WHT	MAP Sensor VREF	4.9-5.1v
D5	GRN/WHT	Brake Switch Signal	Brake Off: 0v, On: 12v
D7	WHT/RED	HO2S-11 (B1 S1) Signal	0.1-1.1v
D8	RED/YEL	IAT Sensor Signal	Varies w/temp. (0.5-4.9v)
D9	WHT/BLK	EGR Valve Lift Sensor	1.2v
D10	YEL/BLU	Sensor VREF	4.9-5.1v
D11	GRN/BLU	Sensor Ground	<0.050v
D12	GRN/WHT	MAP Sensor Ground	<0.050v
D13	RED/WHT	HO2S-12 Ground	<0.050v
D14	WHT/RED	HO2S-12 (B1 S2) Signal	0.1-1.1v
D15	GRN/RED	Fuel Tank Pressure Sensor	Fuel Cap off: 2.5v
D16	GRN/RED	Electric Load Detector	Varies: 0.5-4.5v

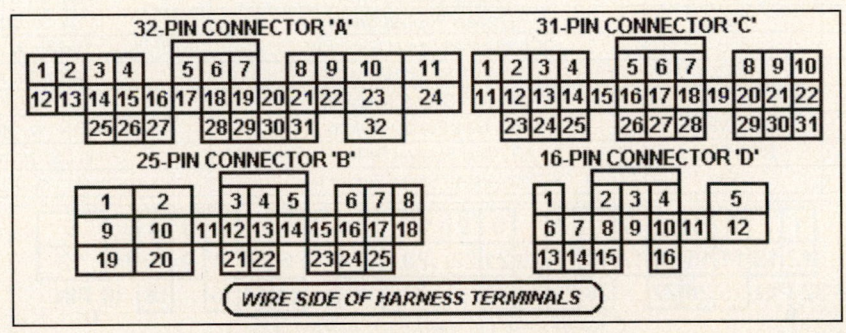

1998-99 Coupe 2.3L I4 VTEC VIN CG3 [All] 32P 'A' Connector

PCM Pin #	W/Color	Circuit Description (32-Pin)	Value at Hot Idle
A2	GRN/WHT	Engine Mount Solenoid	Solenoid Off: 12v, On: 1v
A3	BLU	EVAP Bypass Solenoid	Solenoid Off: 12v, On: 1v
A4	GRN/WHT	EVAP Vent Solenoid	Solenoid Off: 12v, On: 1v
A5	BLU/GRN	Cruise Control Signal	C/C On: pulse signals
A6	RED/YEL	EVAP Purge Solenoid	Solenoid Off: 12v, On: 1v
A8	BLK/WHT	HO2S-12 (B1 S2) Heater	Digital Signals: 0-12-0v
A9	BLU/WHT	A/T: Vehicle Speed Output	Moving: pulse signals
A10	BRN	Service Check Connector	SCS Open: 4.80v
A12	PNK	Immobilizer Indicator Lamp	Lamp Off: 12v, On: 1v
A13, 25	BLU, RED	Immobilizer Enable, Code	Digital Signals
A14	GRN/BLK	A/T: D4 Light Switch	D4 On: 0v, Off: 12v
A15	GRN/YEL	Immobilizer Fuel Pump Relay	Relay Off: 12v, On: 1v
A17	RED	A/C Clutch Relay Control	Relay Off: 12v, On: 1v
A18	GRN/ORN	Malfunction Indicator Light	MIL Off: 12v, On: 1v
A19	BLU	Engine Speed Pulse (NEP)	Digital Signals
A20	GRN	Radiator Fan Relay Control	Relay Off: 12v, On: 1v
A21	GRY	K-Line Signal	12v
A23	WHT/RED	HO2S-12 (B1 S2) Signal	0.1-1.1v
A24	BLU/ORN	Starter Switch Signal	Cranking: 9-11v
A26	GRN	PSP Switch Signal	Straight: 0v, Turning: 11v
A27	BLU/RED	A/C Switch Signal	Switch Off: 12v, On: 1v
A28	WHT/RED	Interlock Control Unit Signal	Key & Brake On: 12v
A29	LT GRN	Fuel Tank Pressure Sensor	Fuel Cap off: 2.5v
A30	GRN/RED	Electric Load Detector	Varies: 0.5-4.5v
A32	WHT/BLK	Brake Switch Signal	Brake Off: 0v, On: 12v

1998-99 Coupe 2.3L I4 VTEC VIN CG3 [All] 16P 'D' Connector

PCM Pin #	W/Color	Circuit Description (16-Pin)	Value at Hot Idle
D1	YEL	A/T: Lockup Control Solenoid	LUS On: 12v, Off: 0v
D2	GRN/WHT	A/T: Shift Solenoid 'B'	SSB Off: 0v, On: 12v
D3	GRN	A/T: Shift Solenoid 'C'	SSC on: 12v, off: 0v
D5	BLK/YEL	A/T: Solenoid Feed (B+)	12-14v
D6	WHT	A/T: Gear Position Switch	In 'R': 0v, Others: 12v
D7	BLU/YEL	A/T: Shift Solenoid 'A'	SSA Off: 0v, On: 12v
D8	PNK	A/T: Gear Position Switch	In D3: 0v, Others: 12v
D9	YEL	A/T: Gear Position Switch	In D4: 0v, Others: 12v
D10	BLU	Countershaft Speed Sensor P	Moving: AC pulses
D11	RED	Mainshaft Speed Sensor 'P'	AC pulse signals
D12	WHT	Mainshaft Speed Sensor 'N'	<0.050v
D13	BLU/WHT	A/T: Gear Position Switch	In P/N: 0v, Others: 12v
D14, 15	BLU, BRN	A/T: Gear Position Switches	In D2: 0v, In D1: 0v
D16	GRN	Countershaft Speed Sensor N	<0.050v

1998-99 Coupe 2.3L I4 VTEC VIN CG3 [All] 25P 'B' Connector

PCM Pin #	W/Color	Circuit Description (25-Pin)	Value at Hot Idle
B1, B9	YEL/BLK	Main Relay Power (B+)	12-14v
B2, B10	BLK	Power Ground	<0.1v
B3, B4	RED, BLU	Injector 2, Injector 3 Control	2.0-3.3 ms
B5, B11	YEL, BRN	Injector 4, Injector 1 Control	2.0-3.3 ms
B7	PNK	E-EGR Control Solenoid	Digital Signals: 0-12-0v
B8, B17	WHT, RED	A/T: Clutch Solenoid 'A-', 'A+'	Pulse Signals
B12	GRN/YEL	VTEC Control Solenoid	0v, Hi-Speed: 12v
B13	YEL/GRN	Ignition Control Signal	Digital Signals: 0-12-0v
B14	BLU/BLK	A/T: 2nd Clutch Pressure Sw.	Open: 12v, Closed: 1v
B16	GRN/RED	HO2S-11 Heater Relay	Relay Off: 12v, On: 1v
B18, B25	GRN, ORN	A/T: Clutch Solenoid 'B-', 'B+'	Pulse Signals
B19	BLK/WHT	HO2S-11 (B1 S1) Heater	Digital Signals: 0-12-0v
B20, B22	BRN/BLK	Logic Ground	<0.050v
B21	WHT/YEL	Keep Alive Power (VBU)	12-14v
B23	BLK/BLU	Idle Air Control Valve	Pulse Signals
B24	BLU/WHT	A/T: 3rd Clutch Pressure Sw.	Open: 12v, Closed: 1v

1998-99 Coupe 2.3L I4 VTEC VIN CG3 [All] 31P 'C' Connector

PCM Pin #	W/Color	Circuit Description (31-Pin)	Value at Hot Idle
C2	WHT/GRN	Alternator Charging Signal	Lights Off: 12v, On: 0v
C3	RED/BLU	Knock Sensor Signal	No Detonation: 18mv AC
C5	WHT/RED	Alternator 'FR' Signal	Digital Signals: 0-5-0v
C6	WHT/BLK	EGR Valve Lift Sensor	1.2v
C7	GRN	MAP Sensor Ground	<0.050v
C8	BLU	CKP Sensor Signal	AC pulse signals
C9	WHT	CKP Sensor Ground	<0.050v
C10	BLU/BLK	VTEC Pressure Switch	0v, Hi-Speed: 12v
C13	WHT	HO2S-11 Heater Relay	Relay Off: 12v, On: 1v
C14	RED	HO2S-11 (B1 S1) Signal	0.1-1.1v
C15	BLU	HO2S-11 Ground	<0.050v
C16	WHT	HO2S-11 (B1 S1) Signal	0.1-1.1v
C17	RED/GRN	MAP Sensor Signal	0.8-0.9v
C18	GRN	Sensor Ground	<0.050v
C19, C28	YEL, YEL	MAP VREF, Sensor VREF	4.9-5.1v
C20	GRN	TDC Sensor Signal	AC pulse signals
C21	RED	TDC Sensor Ground	<0.050v
C23	BLU/WHT	M/T: VSS Input Signal	Moving: 0-5-0v
C24	YEL/GRN	ECT Sensor Signal to TCM	Digital Signals
C25	RED/YEL	IAT Sensor Signal	Varies w/temp (0.5-4.9v)
C26	RED/WHT	ECT Sensor Signal	At 180°F: 0.5-0.6v
C27	RED/BLK	TP Sensor Signal	0.5-0.6v
C29, C30	YEL, BLK	CYP Sensor Signal, Ground	AC pulse signals

1	2		3	4	5		6	7	8		1	2	3	4		5	6	7		8	9	10
9	10		11	12	13	14	15	16	17	18	11	12	13	14	15	16	17	18	19	20	21	22
19	20		21	22		23	24	25				23	24	25		26	27	28		29	30	31

25-PIN 'B'　　(WIRE SIDE OF HARNESS TERMINALS)　　31-PIN 'C'

2000-03 Coupe 2.3L I4 VTEC VIN CG3 [All] 32P 'A' Connector

PCM Pin #	W/Color	Circuit Description (32-Pin)	Value at Hot Idle
A2	GRN/WHT	Engine Mount Solenoid	Solenoid Off: 12v, On: 1v
A3	BLU	EVAP Bypass Solenoid	Solenoid Off: 12v, On: 1v
A4	GRN/WHT	EVAP Vent Solenoid	Solenoid Off: 12v, On: 1v
A5	BLU/GRN	Cruise Control Signal	C/C On: pulse signals
A6	RED/YEL	EVAP Purge Solenoid	Solenoid Off: 12v, On: 1v
A8	BLK/WHT	HO2S-12 (B1 S2) Heater	Digital Signals: 0-12-0v
A9	BLU/WHT	A/T: Vehicle Speed Output	Moving: pulse signals
A10	BRN	Service Check Connector	SCS Open: 4.80v
A12	PNK	Immobilizer Indicator Lamp	Lamp Off: 12v, On: 1v
A13, 25	BLU, RED	Immobilizer Enable, Code	Digital Signals
A14	GRN/BLK	A/T: D4 Light Switch	D4 On: 0v, Off: 12v
A15	GRN/YEL	Immobilizer Fuel Pump Relay	Relay Off: 12v, On: 1v
A17	RED	A/C Clutch Relay Control	Relay Off: 12v, On: 1v
A18	GRN/ORN	Malfunction Indicator Light	MIL Off: 12v, On: 1v
A19	BLU	Engine Speed Pulse (NEP)	Digital Signals
A20	GRN	Radiator Fan Relay Control	Relay Off: 12v, On: 1v
A21	GRY	K-Line Signal	12v
A23	WHT/RED	HO2S-12 (B1 S2) Signal	0.1-1.1v
A24	BLU/ORN	Starter Switch Signal	Cranking: 9-11v
A26	GRN	PSP Switch Signal	Straight: 0v, Turning: 11v
A27	BLU/RED	A/C Switch Signal	Switch Off: 12v, On: 1v
A28	WHT/RED	Interlock Control Unit Signal	Key & Brake On: 12v
A29	LT GRN	Fuel Tank Pressure Sensor	Fuel Cap off: 2.5v
A30	GRN/RED	Electric Load Detector	Varies: 0.5-4.5v
A32	WHT/BLK	Brake Switch Signal	Brake Off: 0v, On: 12v

2000-03 Coupe 2.3L I4 VTEC VIN CG3 [All] 16P 'D' Connector

PCM Pin #	W/Color	Circuit Description (16-Pin)	Value at Hot Idle
D1	YEL	A/T: Lockup Control Solenoid	LUS On: 12v, Off: 0v
D2	GRN/WHT	A/T: Shift Solenoid 'B'	SSB Off: 0v, On: 12v
D3	GRN	A/T: Shift Solenoid 'C'	SSC on: 12v, off: 0v
D5	BLK/YEL	A/T: Solenoid Feed (B+)	12-14v
D6	WHT	A/T: Gear Position Switch	In 'R': 0v, Others: 12v
D7	BLU/YEL	A/T: Shift Solenoid 'A'	SSA Off: 0v, On: 12v
D8	PNK	A/T: Gear Position Switch	In D3: 0v, Others: 12v
D9	YEL	A/T: Gear Position Switch	In D4: 0v, Others: 12v
D10, D16	BLU, GRN	A/T Countershaft Speed P, N	Moving: AC pulses
D11	RED	A/T: Mainshaft Speed 'P'	AC pulse signals
D12	WHT	A/T: Mainshaft Speed 'N'	<0.050v
D13	BLU/WHT	A/T: Gear Position Switch	In P/N: 0v, Others: 12v
D14	BLU	A/T: Gear Position Switch	In D2: 0v, Others: 12v
D15	BRN	A/T: Gear Position Switch	In D1: 0v, Others: 12v

2000-03 Coupe 2.3L I4 VTEC VIN CG3 [All] 25P 'B' Connector

PCM Pin #	W/Color	Circuit Description (25-Pin)	Value at Hot Idle
B1, B9	YEL/BLK	Main Relay Power (B+)	12-14v
B2, B10	BLK	Power Ground	<0.1v
B3, B4	RED, BLU	Injector 2, Injector 3 Control	2.0-3.3 ms
B5, B11	YEL, BRN	Injector 4, Injector 1 Control	2.0-3.3 ms
B7	PNK	E-EGR Control Solenoid	Digital Signals: 0-12-0v
B8, B17	WHT, RED	A/T: Clutch Solenoid 'A-', 'A+'	Pulse Signals
B12	GRN/YEL	VTEC Control Solenoid	0v, Hi-Speed: 12v
B13	YEL/GRN	Ignition Control Signal	Digital Signals: 0-12-0v
B14	BLU/BLK	A/T: 2nd Clutch Pressure Sw.	Open: 12v, Closed: 1v
B16	GRN/RED	HO2S-11 Heater Relay	Relay Off: 12v, On: 1v
B18, B25	GRN, ORN	A/T: Clutch Solenoid 'B-', 'B+'	Pulse Signals
B20, B22	BRN/BLK	Logic Ground	<0.050v
B21	WHT/YEL	Keep Alive Power (VBU)	12-14v
B23	BLK/BLU	Idle Air Control Valve	Pulse Signals
B24	BLU/WHT	A/T: 3rd Clutch Pressure Sw.	Open: 12v, Closed: 1v

2000-03 Coupe 2.3L I4 VTEC VIN CG3 [All] 31P 'C' Connector

PCM Pin #	W/Color	Circuit Description (31-Pin)	Value at Hot Idle
C1	BLK/WHT	HO2S-11 (B1 S1) Heater	Digital Signals: 0-12-0v
C2	WHT/GRN	Alternator Charging Signal	Lights Off: 12v, On: 0v
C3	RED/BLU	Knock Sensor Signal	No Detonation: 18mv AC
C5	WHT/RED	Alternator 'FR' Signal	Digital Signals: 0-5-0v
C6	WHT/BLK	EGR Valve Lift Sensor	1.2v
C7	GRN	MAP Sensor Ground	<0.050v
C8	BLU	CKP Sensor Signal	AC pulse signals
C9	WHT	CKP Sensor Ground	<0.050v
C10	BLU/BLK	VTEC Pressure Switch	0v, Hi-Speed: 12v
C13	WHT	HO2S-11 Heater Relay	Relay Off: 12v, On: 1v
C14	RED	HO2S-11 (B1 S1) Signal	0.1-1.1v
C15	BLU	HO2S-11 Ground	<0.050v
C16	WHT	HO2S-11 (B1 S1) Signal	0.1-1.1v
C17	RED/GRN	MAP Sensor Signal	0.8-0.9v
C18	GRN	Sensor Ground	<0.050v
C19, C28	YEL, YEL	MAP VREF, Sensor VREF	4.9-5.1v
C20	GRN	TDC Sensor Signal	AC pulse signals
C21	RED	TDC Sensor Ground	<0.050v
C23	BLU/WHT	M/T: VSS Input Signal	Moving: 0-5-0v
C24	YEL/GRN	ECT Sensor Signal to TCM	Digital Signals
C25	RED/YEL	IAT Sensor Signal	Varies w/temp (0.5-4.9v)
C26	RED/WHT	ECT Sensor Signal	At 180°F: 0.5-0.6v
C27	RED/BLK	TP Sensor Signal	0.5-0.6v
C29, C30	YEL, BLK	CYP Sensor Signal, Ground	AC pulse signals

WIRE SIDE OF HARNESS TERMINALS

1998-99 Coupe 3.0L V6 VTEC VIN CG2 [All] 32P 'A' Connector

PCM Pin #	W/Color	Circuit Description (32-Pin)	Value at Hot Idle
A2	GRN/WHT	Engine Mount Solenoid	Solenoid Off: 12v, On: 1v
A3	BLU	EVAP Bypass Solenoid	Solenoid Off: 12v, On: 1v
A4	GRN/WHT	EVAP Vent Solenoid	Solenoid Off: 12v, On: 1v
A5	BLU/GRN	Cruise Control Signal	C/C On: pulse signals
A6	RED/YEL	EVAP Purge Solenoid	Solenoid Off: 12v, On: 1v
A8	BLK/WHT	HO2S-12 (B1 S2) Heater	Digital Signals: 0-12-0v
A9	BLU/WHT	A/T: Vehicle Speed Output	Moving: pulse signals
A10	BRN	Service Check Connector	SCS Open: 4.80v
A12	PNK	Immobilizer Indicator Lamp	Lamp Off: 12v, On: 1v
A13, A25	BLU, RED	Immobilizer Enable, Code	Digital Signals
A14	GRN/BLK	A/T: D4 Light Switch	D4 On: 0v, Off: 12v
A15	GRN/YEL	Immobilizer Fuel Pump Relay	Relay Off: 12v, On: 1v
A17	RED	A/C Clutch Relay Control	Relay Off: 12v, On: 1v
A18	GRN/ORN	Malfunction Indicator Light	MIL Off: 12v, On: 1v
A19	BLU	Engine Speed Pulse (NEP)	AC Pulse Signals
A20	GRN	Radiator Fan Relay Control	Relay Off: 12v, On: 1v
A21	GRY	K-Line Signal	12v
A23	WHT/RED	HO2S-12 (B1 S2) Signal	0.1-1.1v
A24	BLU/ORN	Starter Switch Signal	Cranking: 9-11v
A26	GRN	PSP Switch	Straight: 0v, Turning: 11v
A27	BLU/RED	A/C Switch Signal	Switch Off: 12v, On: 1v
A28	WHT/RED	Interlock Control Unit Signal	Key & Brake On: 12v
A29	LT GRN	Fuel Tank Pressure Sensor	Fuel Cap off: 2.5v
A30	GRN/RED	Electric Load Detector	Varies: 0.5-4.5v
A32	WHT/BLK	Brake Switch Signal	Brake Off: 0v, On: 12v

1998-99 Coupe 3.0L V6 VTEC VIN CG2 [All] 16P 'D' Connector

PCM Pin #	W/Color	Circuit Description (16-Pin)	Value at Hot Idle
D1	YEL	Lockup Control Solenoid	LUS On: 12v, Off: 0v
D2	GRN/WHT	A/T: Shift Solenoid 'B'	SSB in 3rd, 4th Gear: 0v
D3	GRN	A/T: Shift Solenoid 'C'	SSC in 2nd, 4th Gear: 0v
D5	BLK/YEL	VB Solenoid Feed (B+)	Sol. on: 12v, off: 0v
D6	WHT	A/T: Gear Position Switch	In 'R': 0v, Others: 12v
D7	BLU/YEL	A/T: Shift Solenoid 'A'	SSA in 1st, 4th Gear: 0v
D8	PNK	A/T: Gear Position Switch	In D3: 0v, Others: 12v
D9	YEL	A/T: Gear Position Switch	In D4: 0v, Others: 12v
D10	BLU	Countershaft Speed Sensor P	Moving: AC pulses
D11	RED	Mainshaft Speed Sensor 'P'	Moving: AC pulses
D12	WHT	Mainshaft Speed Sensor 'N'	Moving: AC pulses
D14	BLU	A/T: Gear Position Switch	In D2: 0v, Others: 12v
D15	BRN	A/T: Gear Position Switch	In D1: 0v, Others: 12v
D16	GRN	Countershaft Speed Sensor N	<0.050v

32-PIN 'A'   16-PIN 'D'   WIRE SIDE OF HARNESS TERMINALS

1998-99 Coupe 3.0L V6 VTEC VIN CG2 [All] 25P 'B' Connector

PCM Pin #	W/Color	Circuit Description (25-Pin)	Value at Hot Idle
B1, B9	YEL/BLK	Main Relay Power (B+)	12-14v
B2, B10	BLK	Power Ground	<0.1v
B3	BLK/RED	Injector 5 Control	2.0-3.3 ms
B4, B5	YEL, RED	Injector 4, Injector 2 Control	2.0-3.3 ms
B6	WHT/BLU	Injector 6 Control	2.0-3.3 ms
B7	PNK	E-EGR Solenoid Control	Digital Signals: 0-12-0v
B8	WHT	A/T: Clutch Solenoid 'A-'	AC Pulse Signals
B11	BRN	Injector 1 Control	2.0-3.3 ms
B12	GRN/YEL	VTEC Control Solenoid	0v, Hi-Speed: 12v
B13	YEL/GRN	Ignition Control Signal	Digital Signals: 0-12-0v
B14	BLU/BLK	A/T: 2nd Clutch Pressure Sw.	Open: 12v, Closed: 1v
B15	BLU	Injector 3 Control	2.0-3.3 ms
B17	RED	A/T: Clutch Solenoid 'A+'	AC Pulse Signals
B18	GRN	A/T: Clutch Solenoid 'B-'	Pulse Signals
B20, B22	BRN/BLK	Logic Ground	<0.050v
B21	WHT/YEL	Keep Alive Power (VBU)	12-14v
B23	BLK/BLU	Idle Air Control Valve	Pulse Signals
B24	BLU/WHT	A/T: 3rd Clutch Pressure Sw.	Open: 12v, Closed: 1v
B25	ORN	A/T: Clutch Solenoid 'B+'	Pulse Signals

1998-99 Coupe 3.0L V6 VTEC VIN CG2 [All] 31P 'C' Connector

PCM Pin #	W/Color	Circuit Description (31-Pin)	Value at Hot Idle
C1	BLK/WHT	HO2S-11 (B1 S1) Heater	Digital Signals: 0-12-0v
C2	WHT/GRN	Alternator Charging Signal	Lights Off: 12v, On: 0v
C3	WHT/BLU	Alternator Charging Signal	Lights Off: 12v, On: 0v
C5	WHT/RED	Alternator 'FR' Signal	Digital Signals: 0-5-0v
C6	WHT/BLK	EGR Valve Lift Sensor	1.2v
C7, C18	GRN/WHT	MAP Sensor, Sensor Ground	<0.050v
C8	BLU	CKP Sensor Signal	AC pulse signals
C9	WHT	CKP Sensor Ground	<0.050v
C10	BLU/BLK	VTEC Pressure Switch	0v, Hi-Speed: 12v
C16	WHT	HO2S-11 (B1 S1) Signal	0.1-1.1v
C17	RED/GRN	MAP Sensor Signal	0.8-0.9v
C19, C28	YEL/RED	MAP VREF, Sensor VREF	4.9-5.1v
C20	GRN	TDC1 Sensor Signal	AC pulse signals
C21	RED	TDC1 Sensor Ground	<0.050v
C25	RED/YEL	IAT Sensor Signal	Varies w/temp. (0.5-4.9v)
C26	RED/WHT	ECT Sensor Signal	At 180°F: 0.5-0.6v
C27	RED/BLK	TP Sensor Signal	0.5-0.6v
C29	YEL	TDC2 Sensor Signal	AC pulse signals
C30	BLK	TDC2 Sensor Ground	<0.050v

1	2		3	4	5		6	7	8		1	2	3	4		5	6	7		8	9	10
9	10	11	12	13	14	15	16	17	18	11	12	13	14	15	16	17	18	19	20	21	22
19	20		21	22		23	24	25			23	24	25		26	27	28		29	30	31

25-PIN 'B'   (WIRE SIDE OF HARNESS TERMINALS)   31-PIN 'C'

2000-03 Coupe 3.0L V6 VTEC VIN CG2 [A/T] 32P 'A' Connector

PCM Pin #	W/Color	Circuit Description (32-Pin)	Value at Hot Idle
A1	YEL/GRN	ECT Signal to TCM	Digital Signals
A2	GRN/WHT	A/T: Engine Mount Solenoid	Solenoid Off: 12v, On: 1v
A3	BLU	EVAP Bypass Solenoid	Solenoid Off: 12v, On: 1v
A4	GRN/WHT	EVAP Vent Solenoid	Solenoid Off: 12v, On: 1v
A5	BLU/GRN	Cruise Control Signal	C/C On: pulse signals
A6	RED/YEL	EVAP Purge Solenoid	Solenoid Off: 12v, On: 1v
A7, 11, 13	WT,GN,BL	ABS-TCS Signals	Digital Signals
A8	BLK/WHT	HO2S-12 (B1 S2) Heater	Digital Signals: 0-12-0v
A9	BLU/WHT	A/T: Vehicle Speed Output	Moving: pulse signals
A10	BRN	Service Check Connector	SCS Open: 4.80v
A12	PNK	Immobilizer Indicator Lamp	Lamp Off: 12v, On: 1v
A14	GRN/BLK	A/T: D4 Light Switch	D4 On: 0v, Off: 12v
A15	GRN/YEL	Immobilizer Fuel Pump Relay	Relay Off: 12v, On: 1v
A16, A25	BLU, RED	Immobilizer Enable, Code	Digital Signals
A17	RED	A/C Clutch Relay Control	Relay Off: 12v, On: 1v
A18	GRN/ORN	Malfunction Indicator Light	MIL Off: 12v, On: 1v
A19	BLU	Engine Speed Pulse (NEP)	AC Pulse Signals
A20	GRN	Radiator Fan Relay Control	Relay Off: 12v, On: 1v
A21	GRY	K-Line Signal	12v
A22, A31	PNK, RED	ABS-TCS Signals	Digital Signals
A23	WHT/RED	HO2S-12 (B1 S2) Signal	0.1-1.1v
A24	BLU/ORN	Starter Switch Signal	Cranking: 9-11v
A26	GRN	PSP Switch	Straight: 0v, Turning: 11v
A27	BLU/RED	A/C Switch Signal	Switch Off: 12v, On: 1v
A28	WHT/RED	Interlock Control Unit Signal	Key & Brake On: 12v
A29	LT GRN	Fuel Tank Pressure Sensor	Fuel Cap off: 2.5v
A30	GRN/RED	Electric Load Detector	Varies: 0.5-4.5v
A32	WHT/BLK	Brake Switch Signal	Brake Off: 0v, On: 12v

2000-03 Coupe 3.0L V6 VTEC VIN CG2 [A/T] 16P 'D' Connector

PCM Pin #	W/Color	Circuit Description (16-Pin)	Value at Hot Idle
D1	YEL	Lockup Control Solenoid	LUS On: 12v, Off: 0v
D2	GRN/WHT	A/T: Shift Solenoid 'B'	SSB in 3rd, 4th Gear: 0v
D3	GRN	A/T: Shift Solenoid 'C'	SSC in 2nd, 4th Gear: 0v
D5	BLK/YEL	VB Solenoid Feed (B+)	Sol. on: 12v, off: 0v
D6	WHT	A/T: Gear Position Switch	In 'R': 0v, Others: 12v
D7	BLU/YEL	A/T: Shift Solenoid 'A'	SSA in 1st, 4th Gear: 0v
D8, D9	PNK, YEL	A/T: Gear Position Switches	In D3: 0v, In D4: 0v
D10, GRN	BLU, GRN	A/T: Countershaft Speed P, N	Moving: AC pulses
D11, D12	RED, WHT	Mainshaft Speed Sensor P, N	Moving: AC pulses
D13	BLU/BLK	A/T: 2nd Clutch Pressure Sw.	Open: 12v, Closed: 1v
D14, 15	BLU, BRN	A/T: Gear Position Switch	D2: 0v, D1: 0v

2000-03 Coupe 3.0L V6 VTEC VIN CG2 [A/T] 25P 'B' Connector

PCM Pin #	W/Color	Circuit Description (25-Pin)	Value at Hot Idle
B1, B9	YEL/BLK	Main Relay Power (B+)	12-14v
B2, B10	BLK	Power Ground	<0.1v
B3	BLK/RED	Injector 5 Control	2.0-3.3 ms
B4, B5	YEL, RED	Injector 4, Injector 2 Control	2.0-3.3 ms
B6	WHT/BLU	Injector 6 Control	2.0-3.3 ms
B7	PNK	E-EGR Solenoid Control	Digital Signals: 0-12-0v
B8	WHT	A/T: Clutch Solenoid 'A-'	AC Pulse Signals
B11	BRN	Injector 1 Control	2.0-3.3 ms
B12	GRN/YEL	VTEC Control Solenoid	0v, Hi-Speed: 12v
B14	BLU/WHT	A/T: Gear Position Switch	In P/N: 0v, Others: 12v
B15	BLU	Injector 3 Control	2.0-3.3 ms
B17	RED	A/T: Clutch Solenoid 'A+'	AC Pulse Signals
B18	GRN	A/T: Clutch Solenoid 'B-'	Pulse Signals
B20, B22	BRN/BLK	Logic Ground (LG1, LG2)	<0.050v
B21	WHT/YEL	Keep Alive Power (VBU)	12-14v
B23	BLK/BLU	Idle Air Control Valve	Pulse Signals
B24	BLU/WHT	A/T: 3rd Clutch Pressure Sw.	Open: 12v, Closed: 1v
B25	ORN	A/T: Clutch Solenoid 'B+'	Pulse Signals

2000-03 Coupe 3.0L V6 VTEC VIN CG2 [A/T] 31P 'C' Connector

PCM Pin #	W/Color	Circuit Description (31-Pin)	Value at Hot Idle
C1	BLK/WHT	HO2S-11 (B1 S1) Heater	Digital Signals: 0-12-0v
C2	WHT/GRN	Alternator Charging Signal	Lights Off: 12v, On: 0v
C3	WHT/BLU	Ignition Coil 3 Control	Digital Signals: 0-12-0v
C4	YEL/GRN	Ignition Coil 1 Control	Digital Signals: 0-12-0v
C5	WHT/RED	Alternator 'FR' Signal	Digital Signals: 0-5-0v
C6	WHT/BLK	EGR Valve Lift Sensor	1.2v
C7, C18	GRN/WHT	MAP Sensor, Sensor Ground	<0.050v
C8, C9	BLU, WHT	CKP Sensor Signal, Ground	AC pulse signals
C10	BLU/BLK	VTEC Pressure Switch	0v, Hi-Speed: 12v
C12	BLK/RED	Ignition Coil 5 Control	Digital Signals: 0-12-0v
C13	BRN	Ignition Coil 4 Control	Digital Signals: 0-12-0v
C14	BLU/RED	Ignition Coil 2 Control	Digital Signals: 0-12-0v
C16	WHT	HO2S-11 (B1 S1) Signal	0.1-1.1v
C17	RED/GRN	MAP Sensor Signal	0.8-0.9v
C19, C28	YEL/RED	MAP VREF, Sensor VREF	4.9-5.1v
C20, C21	GRN, RED	TDC1 Sensor Signal, Ground	AC pulse signals
C23	BRN/WHT	Ignition Coil 6 Control	Digital Signals: 0-12-0v
C25	RED/YEL	IAT Sensor Signal	Varies w/temp. (0.5-4.9v)
C26	RED/WHT	ECT Sensor Signal	At 180°F: 0.5-0.6v
C27	RED/BLK	TP Sensor Signal	0.5-0.6v
C29, C30	YEL, BLK	TDC2 Sensor Signal, Ground	AC pulse signals

1	2		3	4	5		6	7	8		1	2	3	4		5	6	7		8	9	10
9	10	11	12	13	14	15	16	17	18	11	12	13	14	15	16	17	18	19	20	21	22
19	20		21	22		23	24	25			23	24	25		26	27	28		29	30	31

25-PIN 'B'   *WIRE SIDE OF HARNESS TERMINALS*   **31-PIN 'C'**

1990-93 Sedan 2.2L I4 MFI VIN CB7 [All] 26P 'A' Connector

PCM Pin #	W/Color	Circuit Description (26-Pin)	Value at Hot Idle
A1	BRN	Injector 1 Control	2.0-3.3 ms
A2	YEL	Injector 4 Control	2.0-3.3 ms
A3	RED	Injector 2 Control	2.0-3.3 ms
A4	---	Not Used	---
A5	BLU	Injector 3 Control	2.0-3.3 ms
A6	ORN/BLK	HO2S Heater Control	Relay Off: 12v, On: 1v
A7	GRN/BLK	Fuel Pump Relay	Relay Off: 12v, On: 1v
A8	GRN/BLK	Fuel Pump Relay	Relay Off: 12v, On: 1v
A9	BLK/BLU	Electronic Air Control Valve	Pulse Signals
A10	---	Not Used	---
A11	PNK	EGR Control Solenoid	1.2v
A12	BLU	Radiator Fan Relay	Relay Off: 12v, On: 1v
A13	GRN/RED	Check Engine Light	MIL Off: 12v, On: 1v
A14	---	Not Used	---
A15	RED/BLU	A/C Clutch Relay	Relay Off: 12v, On: 1v
A16	WHT/GRN	Alternator Charging Signal	Lights Off: 12v, On: 0v
A17	---	Not Used	---
A18	BRN/WHT	A/T: Control Unit	---
A19	WHT	Intake Air Regulator Control	Solenoid Off: 12v, On: 1v
A20	RED/GRN	EVAP Purge Solenoid	Solenoid Off: 12v, On: 1v
A21	YEL/GRN	Igniter Control	Hot idle 10% d/cycle
A22	YEL/GRN	Igniter Control	Hot idle 10% d/cycle
A23	BLK	Power Ground	<0.1v
A24	BLK	Power Ground	<0.1v
A25	YEL/BLK	Ignition Power	12-14v
A26	BLK/RED	Chassis Ground	<0.050v

1990-93 Sedan 2.2L I4 MFI VIN CB7 [All] 16P 'B' Connector

PCM Pin #	W/Color	Circuit Description (16-Pin)	Value at Hot Idle
B1	YEL/BLK	Main Relay Power (B+)	12-14v
B2	BRN/BLK	Power Ground	<0.1v
B3	WHT/RED	A/T: Control Unit	---
B4	GRN	A/T: Control Unit	---
B5	BLU/BLK	A/C Switch Signal	Switch Off: 12v, On: 1v
B6	---	Not Used	---
B7	LT GRN	A/T: Neutral Position Switch	In N: 0v, Others: 12v
B8	RED	PSP Switch	Straight: 0v, Turning: 11v
B9	BLU/RED	A/T: Start Signal	Cranking: 9-11v
B9	BLK/GRN	M/T: Start Signal	Cranking: 9-11v
B10	ORN	Vehicle Speed Sensor	Moving: pulse signals
B11	ORN	CYP Sensor Signal	AC pulse signals
B12	WHT	CYP Sensor Ground	<0.050v
B13	ORN/BLU	TDC Sensor Signal	AC pulse signals
B14	WHT/BLU	TDC Sensor Ground	<0.050v
B15	BLU/GRN	CKP Sensor Signal	AC pulse signals
B16	BLU/YEL	CKP Sensor Ground	<0.050v

1990-93 Sedan 2.2L I4 MFI VIN CB7 [All] 22P 'D' Connector

PCM Pin #	W/Color	Circuit Description (22-Pin)	Value at Hot Idle
D1	WHT/YEL	Keep Alive Power (VBU)	12-14v
D2	GRN/WHT	Brake Switch Signal	Brake Off: 0v, On: 12v
D3	---	Not Used	---
D4	ORN/RED	Service Check Connector	SCS Open: 4.80v
D5-D8	---	Not Used	---
D9	WHT/RED	Alternator 'FR' Signal	Digital Signals: 0-5-0v
D10	GRN/RED	Electric Load Detector	Varies: 0.5-4.5v
D11	RED/BLK	TP Sensor Signal	0.5-0.6v
D12	WHT/BLK	A/T: EGR Lift Sensor Signal	1.1-1.2v
D13	YEL/GRN	ECT Sensor Signal	At 180°F: 0.52v
D14	WHT	Oxygen Sensor	0.1-1.1v
D15	RED/YEL	IAT Sensor Signal	Varies w/temp. (0.5-4.9v)
D16	---	Not Used	---
D17	WHT/BLU	MAP Sensor Signal	0.8-0.9v
D18	GRN/BLK	A/T: Control Unit VREF	4.9-5.1v
D19	RED/WHT	MAP Sensor VREF	4.9-5.1v
D20	YEL/WHT	Sensor VREF	4.9-5.1v
D21	BLU/WHT	Sensor Ground	<0.050v
D22	GRN/WHT	Sensor Ground	<0.050v

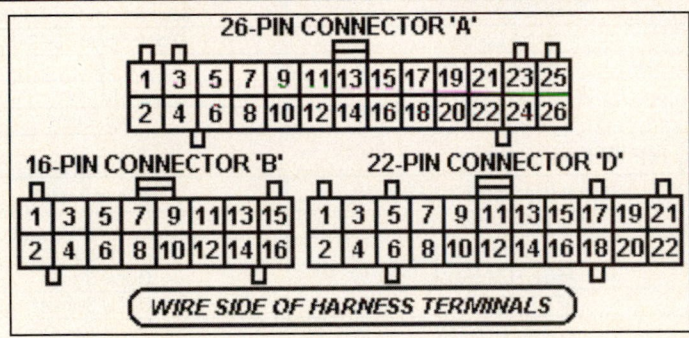

1994 Sedan 2.2L I4 VTEC VIN CD5 [All] 26P 'A' Connector

PCM Pin #	W/Color	Circuit Description (26-Pin)	Value at Hot Idle
A1	BRN	Injector 1 Control	2.0-3.3 ms
A2	YEL	Injector 4 Control	2.0-3.3 ms
A3	RED	Injector 2 Control	2.0-3.3 ms
A4	GRN/YEL	VTEC Control Solenoid	0v, Hi-Speed: 12v
A5	BLU	Injector 3 Control	2.0-3.3 ms
A6	RED	EGR Control Solenoid	Solenoid Off: 12v, On: 1v
A7	GRN/BLK	Fuel Pump Relay	Relay Off: 12v, On: 1v
A8	---	Not Used	---
A9	BLK/BLU	Intake Air Regulator Control	Solenoid Off: 12v, On: 1v
A10	GRN/WHT	Engine Mount Solenoid	Solenoid Off: 12v, On: 1v
A11	GRN/BLK	HO2S Heater Control	Relay Off: 12v, On: 1v
A12	GRN	Radiator Fan Relay Control	Relay Off: 12v, On: 1v
A13	GRN/RED	Check Engine Light	MIL Off: 12v, On: 1v
A14	WHT/YEL	FIA Control Solenoid	Solenoid Off: 12v, On: 1v
A15	RED/BLU	A/C Clutch Relay	Relay Off: 12v, On: 1v
A16	WHT/GRN	Alternator Charging Signal	Lights Off: 12v, On: 0v
A17	---	Not Used	---
A18	BRN/WHT	FAS TCM Signal	Digital Signals
A19	ORN/GRN	Intake Air Regulator Control	Solenoid Off: 12v, On: 1v
A20	RED/YEL	EVAP Purge Solenoid	Solenoid Off: 12v, On: 1v
A21, 22	YEL/GRN	Igniter Control	Digital Signals: 0-12-0v
A23, 24	BLK, BLK	Power Ground	<0.1v
A25	YEL/BLK	Main Relay Power (B+)	12-14v
A26	BLK/RED	Logic Ground	<0.050v

1994 Sedan 2.2L I4 VTEC VIN CD5 [All] 16P 'B' Connector

PCM Pin #	W/Color	Circuit Description (16-Pin)	Value at Hot Idle
B1	YEL/BLK	Main Relay Power (B+)	12-14v
B2	BRN/BLK	Logic Ground	<0.050v
B3	WHT/RED	AFSA TCM Signal	Digital Signals
B4	GRN	AFSB TCM Signal	Digital Signals
B5	RED/WHT	A/C Switch Signal	Switch Off: 12v, On: 1v
B6	---	Not Used	---
B7	LT GRN	A/T: Park/Neutral Switch	In P/N: 0v, Others: 11v
B8	GRN	PSP Switch	Straight: 0v, Turning: 11v
B9	BLU/RED	A/T: Start Signal	Cranking: 9-11v
B9	BLU/RED	M/T: Start Signal	Cranking: 9-11v
B10	ORN	Vehicle Speed Sensor	Moving: pulse signals
B11	ORN	CYP Sensor Signal	AC pulse signals
B12	WHT	CYP Sensor Ground	<0.050v
B13	ORN/BLU	TDC Sensor Signal	AC pulse signals
B14	WHT/BLU	TDC Sensor Ground	<0.050v
B15	BLU/GRN	CKP Sensor Signal	AC pulse signals
B16	BLU/YEL	CKP Sensor Ground	<0.050v

1994 Sedan 2.2L I4 VTEC VIN CD5 [All] 22P 'D' Connector

PCM Pin #	W/Color	Circuit Description (22-Pin)	Value at Hot Idle
D1	WHT/YEL	Keep Alive Power (VBU)	12-14v
D2	GRN/WHT	Brake Switch Signal	Brake Off: 0v, On: 12v
D3	---	Not Used	---
D4	ORN/RED	Service Check Connector	SCS Open: 4.80v
D5	BLU/WHT	A/T: TCM BARO Signal	3v (at sea level)
D6	BLU/BLK	VTEC Pressure Switch	0v, Hi-Speed: 12v
D7	GRN/RED	Data Link Connector	5v
D8	---	Not Used	---
D9	WHT/RED	Alternator 'FR' Signal	Digital Signals: 0-5-0v
D10	GRN/RED	Electric Load Detector	Varies: 0.5-4.5v
D11	RED/BLK	TP Sensor Signal	0.5-0.6v
D12	WHT/BLK	EGR Lift Sensor	1.1-1.2v
D13	RED/WHT	ECT Sensor Signal	At 180°F: 0.51v
D14	WHT/RED	HO2S Signal	0.1-1.1v
D15	RED/YEL	IAT Sensor Signal	Varies w/temp. (0.5-4.9v)
D16	---	Not Used	---
D17	WHT/YEL	MAP Sensor Signal	0.8-0.9v
D18	GRN/BLK	VREF To TCM Signal	4.9-5.1v
D19	YEL/WHT	MAP Sensor VREF	4.9-5.1v
D20	YEL/BLU	Sensor VREF	4.9-5.1v
D21	GRN/WHT	MAP Sensor Ground	<0.050v
D22	GRN/BLU	Sensor Ground	<0.050v

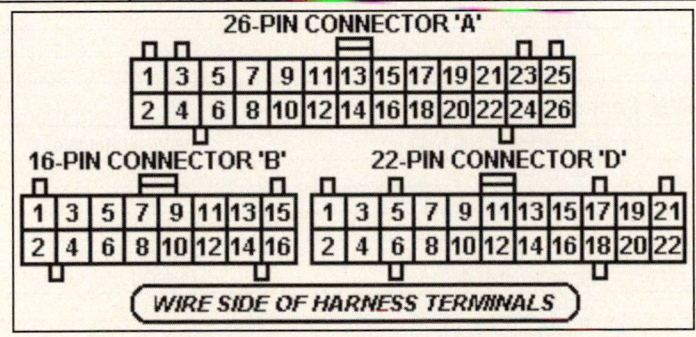

1994 Sedan 2.2L I4 MFI VIN CD5 [All] 26P 'A' Connector

PCM Pin #	W/Color	Circuit Description (26-Pin)	Value at Hot Idle
A1	BRN	Injector 1 Control	2.0-3.3 ms
A2	YEL	Injector 4 Control	2.0-3.3 ms
A3	RED	Injector 2 Control	2.0-3.3 ms
A4	---	Not Used	---
A5	BLU	Injector 3 Control	2.0-3.3 ms
A6	RED	EGR Control Solenoid	Solenoid Off: 12v, On: 1v
A7-8	GRN/BLK	Fuel Pump Relay	Relay Off: 12v, On: 1v
A9	BLK/BLU	Idle Air Control Motor	Pulse Signals
A10	GRN/WHT	Engine Mount Solenoid	Solenoid Off: 12v, On: 1v
A11	GRN/BLK	HO2S Heater Control	Relay Off: 12v, On: 1v
A12	GRN	Radiator Fan Relay	Relay Off: 12v, On: 1v
A13	GRN/RED	Check Engine Light	MIL Off: 12v, On: 1v
A14	---	Not Used	---
A15	RED/BLU	A/C Clutch Relay	Relay Off: 12v, On: 1v
A16	WHT/GRN	Alternator Charging Signal	Lights Off: 12v, On: 0v
A17	---	Not Used	---
A18	BRN/WHT	A/T: Control Unit VREF	5v
A19	ORN/GRN	Intake Air Regulator Control	Solenoid Off: 12v, On: 1v
A20	RED/YEL	EVAP Purge Solenoid	Solenoid Off: 12v, On: 1v
A21	YEL/GRN	Igniter Control	Digital Signals: 0-12-0v
A22	YEL/GRN	Igniter Control	Digital Signals: 0-12-0v
A23	BLK	Power Ground	<0.1v
A24	BLK	Power Ground	<0.1v
A25	YEL/BLK	Ignition Power	12-14v
A26	BLK/RED	Chassis Ground	<0.050v

1994 Sedan 2.2L I4 MFI VIN CD5 [All] 16P 'B' Connector

PCM Pin #	W/Color	Circuit Description (16-Pin)	Value at Hot Idle
B1	YEL/BLK	Ignition Power	12-14v
B2	BRN/BLK	Power Ground	<0.1v
B3	WHT/RED	A/T: Control Unit	Digital Signals
B4	GRN	A/T: Control Unit	Digital Signals
B5	RED/WHT	A/C Switch Signal	Switch Off: 12v, On: 1v
B6	---	Not Used	---
B7	LT GRN	A/T: Park Neutral Switch	In P/N: 0v, Others: 11v
B8	GRN	PSP Switch	Straight: 0v, Turning: 11v
B9	BLU/RED	A/T: Start Signal	Cranking: 9-11v
B9	BLU/RED	M/T: Start Signal	Cranking: 9-11v
B10	ORN	Vehicle Speed Sensor	Moving: pulse signals
B11	ORN	CYP Sensor Signal	AC pulse signals
B12	WHT	CYP Sensor Ground	<0.050v
B13	ORN/BLU	TDC Sensor Signal	AC pulse signals
B14	WHT/BLU	TDC Sensor Ground	<0.050v
B15	BLU/GRN	CKP Sensor Signal	AC pulse signals
B16	BLU/YEL	CKP Sensor Ground	<0.050v

1994 Sedan 2.2L I4 MFI VIN CD5 [All] 22P 'D' Connector

PCM Pin #	W/Color	Circuit Description (22-Pin)	Value at Hot Idle
D1	WHT/YEL	Keep Alive Power (VBU)	12-14v
D2	GRN/WHT	Brake Switch Signal	Brake Off: 0v, On: 12v
D3	---	Not Used	---
D4	ORN/RED	Service Check Connector	SCS Open: 4.80v
D5	BLU/WHT	A/T TCM Signal	Digital Signals
D6	---	Not Used	---
D7	GRN/RED	Data Link Connector	5v
D8	---	Not Used	---
D9	WHT/RED	Alternator 'FR' Signal	Digital Signals: 0-5-0v
D10	GRN/RED	Electric Load Detector	Varies: 0.5-4.5v
D11	RED/BLK	TP Sensor Signal	0.5-0.6v
D12	WHT/BLK	A/T: EGR Lift Sensor Signal	1.1-1.2v
D13	RED/WHT	ECT Sensor Signal	At 180ºF: 0.52v
D14	WHT/RED	HO2S Signal	0.1-1.1v
D15	RED/YEL	IAT Sensor Signal	Varies w/temp. (0.5-4.9v)
D16	---	Not Used	---
D17	WHT/YEL	MAP Sensor Signal	0.8-0.9v
D18	GRN/BLK	A/T: Control Unit VREF	5v
D19	YEL/WHT	MAP Sensor VREF	4.9-5.1v
D20	YEL/BLU	Sensor VREF	4.9-5.1v
D21	GRN/WHT	MAP Sensor Ground	<0.050v
D22	GRN/BLU	Sensor Ground	<0.050v

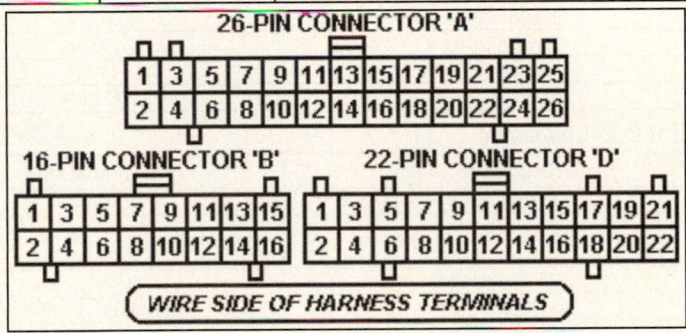

1995 Sedan 2.2L I4 VTEC VIN CD5 [All] 26P 'A' Connector

PCM Pin #	W/Color	Circuit Description (26-Pin)	Value at Hot Idle
A1	BRN	Injector 1 Control	2.0-3.3 ms
A2	YEL	Injector 4 Control	2.0-3.3 ms
A3	RED	Injector 2 Control	2.0-3.3 ms
A4	GRN/YEL	VTEC Solenoid Control	0v, Hi-Speed: 12v
A5	BLU	Injector 3 Control	2.0-3.3 ms
A6	ORN/BLK	HO2S Heater Control	Relay Off: 12v, On: 1v
A7	GRN/BLK	Fuel Pump Relay	Relay Off: 12v, On: 1v
A8	---	Not Used	---
A9	BLK/BLU	Idle Air Control Valve Signal	Pulse Signals
A10	---	Not Used	---
A11	RED	EGR Solenoid	Solenoid Off: 12v, On: 1v
A12	GRN	Radiator Fan Relay	Relay Off: 12v, On: 1v
A13	GRN/RED	Check Engine Light	MIL Off: 12v, On: 1v
A14	---	Not Used	---
A15	RED/BLU	A/C Clutch Relay	Relay Off: 12v, On: 1v
A16	WHT/GRN	Alternator Charging Signal	Lights Off: 12v, On: 0v
A17	---	Not Used	---
A18	BRN/WHT	A/T: Control Unit	Digital Signals
A19	WHT	Intake Air Regulator Control	Solenoid Off: 12v, On: 1v
A20	RED/YEL	EVAP Purge Solenoid	Solenoid Off: 12v, On: 1v
A21	YEL/GRN	Igniter Control	Digital Signals: 0-12-0v
A22	---	Not Used	---
A23	BLK	Power Ground	<0.1v
A24	BLK	Power Ground	<0.1v
A25	YEL/BLK	Ignition Power	12-14v
A26	BLK/RED	Chassis Ground	<0.050v

1995 Sedan 2.2L I4 VTEC VIN CD5 [All] 16P 'B' Connector

PCM Pin #	W/Color	Circuit Description (16-Pin)	Value at Hot Idle
B1	YEL/BLK	Ignition Power	12-14v
B2	BRN/BLK	Power Ground	<0.1v
B3	WHT/RED	A/T: Control Unit	Digital Signals
B4	GRN	A/T: Control Unit	Digital Signals
B5	RED/WHT	A/C Switch Signal	Switch Off: 12v, On: 1v
B6	---	Not Used	---
B7	LT GRN	A/T: Park/Neutral Switch	In P/N: 0v, Others: 11v
B8	RED	PSP Switch	Straight: 0v, Turning: 11v
B9	BLU/RED	A/T: Start Signal	Cranking: 9-11v
B9	BLU/RED	MT: Start Signal	Cranking: 9-11v
B10	ORN	Vehicle Speed Sensor	Moving: pulse signals
B11	ORN	CYP Sensor Signal	AC pulse signals
B12	WHT	CYP Sensor Ground	<0.050v
B13	ORN/BLU	TDC Sensor Signal	AC pulse signals
B14	WHT/BLU	TDC Sensor Ground	<0.050v
B15	BLU/GRN	CKP Sensor Signal	AC pulse signals
B16	BLU/YEL	CKP Sensor Ground	<0.050v

1995 Sedan 2.2L I4 VTEC VIN CD5 [All] 22P 'D' Connector

PCM Pin #	W/Color	Circuit Description (22-Pin)	Value at Hot Idle
D1	WHT/YEL	Keep Alive Power (VBU)	12-14v
D2	GRN/WHT	Brake Switch Signal	Brake Off: 0v, On: 12v
D3	---	Not Used	---
D4	RED	Service Check Connector	SCS Open; 4.80v
D5	BLU/WHT	A/T TCM Signal	---
D6	BLU/BLK	VTEC Pressure Switch	0v, Hi-Speed: 12v
D7	GRN/RED	Data Link Connector	5v
D3	---	Not Used	---
D9	WHT/RED	Alternator 'FR' Signal	Digital Signals: 0-5-0v
D10	GRN/RED	Electric Load Detector	Varies: 0.5-4.5v
D11	RED/BLK	TP Sensor Signal	0.5-0.6v
D12	WHT/BLK	A/T: EGR Lift Sensor Signal	1.1-1.2v
D13	RED/WHT	ECT Sensor Signal	At 180°F: 0.52v
D14	WHT/RED	HO2S Signal	0.1-1.1v
D15	RED/YEL	IAT Sensor Signal	Varies w/temp. (0.5-4.9v)
D16	---	Not Used	---
D17	WHT/YEL	MAP Sensor Signal	0.8-0.9v
D18	GRN/BLK	A/T: Control Unit VREF	5v
D19	YEL/WHT	MAP Sensor VREF	4.9-5.1v
D20	YEL/BLU	Sensor VREF	4.9-5.1v
D21	GRN/WHT	MAP Sensor Ground	<0.050v
D22	GRN/BLU	Sensor Ground	<0.050v

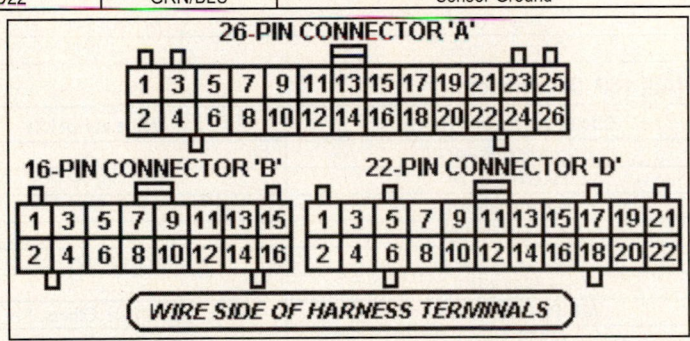

1995 Sedan 2.2L I4 MFI VIN CD5, CD6 [All] 26P 'A' Connector

PCM Pin #	W/Color	Circuit Description (26-Pin)	Value at Hot Idle
A1, 2	BRN, YEL	Injector 1 Control, 4	2.0-3.3 ms
A3, 5	RED, BLU	Injector 2 Control, 3	2.0-3.3 ms
A4	---	Not Used	---
A6	ORN/BLK	HO2S Heater Control	Relay Off: 12v, On: 1v
A7	GRN/BLK	Fuel Pump Relay	Relay Off: 12v, On: 1v
A8	---	Not Used	---
A9	BLK/BLU	Intake Air Control Solenoid	Solenoid Off: 12v, On: 1v
A10	---	Not Used	---
A11	RED	EGR Control Solenoid	Solenoid Off: 12v, On: 1v
A12	GRN	Radiator Fan Relay	Relay Off: 12v, On: 1v
A13	GRN/RED	Check Engine Light	MIL Off: 12v, On: 1v
A14	---	Not Used	---
A15	RED/BLU	A/C Clutch Relay	Relay Off: 12v, On: 1v
A16	WHT/GRN	Alternator Charging Signal	Lights Off: 12v, On: 0v
A17	---	Not Used	---
A18	BRN/WHT	A/T: Control Unit	---
A19	WHT	Intake Air Regulator Control	Solenoid Off: 12v, On: 1v
A20	RED/YEL	EVAP Purge Solenoid	Solenoid Off: 12v, On: 1v
A21	YEL/GRN	Igniter Control	Digital Signals: 0-12-0v
A22	---	Not Used	---
A23	BLK	Power Ground	<0.1v
A24	BLK	Power Ground	<0.1v
A25	YEL/BLK	Ignition Power	12-14v
A26	BLK/RED	Chassis Ground	<0.050v

1995 Sedan 2.2L I4 MFI VIN CD5, CD6 [All] 16P 'B' Connector

PCM Pin #	W/Color	Circuit Description (16-Pin)	Value at Hot Idle
B1	YEL/BLK	Ignition Power	12v
B2	BRN/BLK	Power Ground	<0.1v
B3	WHT/RED	A/T: Control Unit	---
B4	GRN	A/T: Control Unit	---
B5	RED/WHT	A/C Switch Signal	Switch Off: 12v, On: 1v
B6	---	Not Used	---
B7	LT GRN	A/T: Park/Neutral Switch	In P/N: 0v, Others: 11v
B8	RED	PSP Switch	Straight: 0v, Turning: 11v
B9	BLU/RED	A/T: Start Signal	Cranking: 9-11v
B9	BLU/RED	MT: Start Signal	Cranking: 9-11v
B10	ORN	Vehicle Speed Sensor	Moving: pulse signals
B11	ORN	CYP Sensor Signal	AC pulse signals
B12	WHT	CYP Sensor Ground	<0.050v
B13	ORN/BLU	TDC Sensor Signal	AC pulse signals
B14	WHT/BLU	TDC Sensor Ground	<0.050v
B15	BLU/GRN	CKP Sensor Signal	AC pulse signals
B16	BLU/YEL	CKP Sensor Ground	<0.050v

1995 Sedan 2.2L I4 MFI VIN CD5, CD6 [All] 22P 'D' Connector

PCM Pin #	W/Color	Circuit Description (22-Pin)	Value at Hot Idle
D1	WHT/YEL	Keep Alive Power (VBU)	12-14v
D2	GRN/WHT	Brake Switch Signal	Brake Off: 0v, On: 12v
D3	---	Not Used	---
D4	RED	Service Check Connector	SCS Open: 4.80v
D5	BLU/WHT	A/T TCM Signal	Digital Signals
D6	---	Not Used	---
D7	GRN/RED	Data Link Connector	5v
D8	---	Not Used	---
D9	WHT/RED	Alternator 'FR' Signal	Digital Signals: 0-5-0v
D10	GRN/RED	Electric Load Detector	Varies: 0.5-4.5v
D11	RED/BLK	TP Sensor Signal	0.5-0.6v
D12	WHT/BLK	A/T: EGR Lift Sensor Signal	1.1-1.2v
D13	RED/WHT	ECT Sensor Signal	At 180°F: 0.52v
D14	WHT/RED	HO2S Signal	0.1-1.1v
D15	RED/YEL	IAT Sensor Signal	Varies w/temp. (0.5-4.9v)
D16	---	Not Used	---
D17	WHT/YEL	MAP Sensor Signal	0.8-0.9v
D18	GRN/BLK	A/T: Control Unit	---
D19	YEL/WHT	MAP Sensor VREF	4.9-5.1v
D20	YEL/BLU	Sensor VREF	4.9-5.1v
D21	GRN/WHT	MAP Sensor Ground	<0.050v
D22	GRN/BLU	Sensor Ground	<0.050v

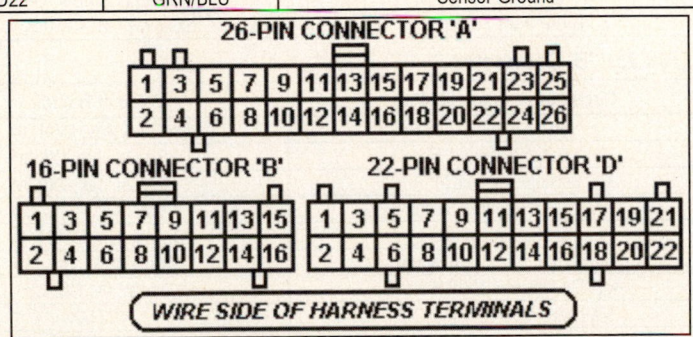

Abbreviation	Color	Abbreviation	Color	Abbreviation	Color
BLK	Black	LT BLU	Light Blue	TAN	Tan
BLU	Blue	LT GRN	Lt. Green	VIO	Violet
BRN	Brown	ORN	Orange	WHT	White
GRY	Gray	PNK	Pink	YEL	Yellow
GRN	Green	PPL	Purple		

1996-97 Sedan 2.2L I4 VTEC VIN CD5 [All] 32P 'A' Connector

PCM Pin #	W/Color	Circuit Description (32-Pin)	Value at Hot Idle
A1	YEL	Injector 4 Control	2.0-3.3 ms
A2	BLU	Injector 3 Control	2.0-3.3 ms
A3	RED	Injector 2 Control	2.0-3.3 ms
A4	BRN	Injector 1 Control	2.0-3.3 ms
A5	ORN/BLU	HO2S-12 (B1 S2) Heater	Digital Signals: 0-12-0v
A6	ORN/BLK	HO2S-11 (B1 S1) Heater	Digital Signals: 0-12-0v
A7	RED	EGR Solenoid	Solenoid Off: 12v, On: 1v
A8	GRN/YEL	VTEC Solenoid	0v, Hi-Speed: 12v
A9, A22	BRN/BLK	Sensor Ground	<0.050v
A10, 23	BLK	Power Ground	<0.1v
A11, A24	YEL/BLK	Main Relay Power (B+)	12-14v
A12	BLK/BLU	Idle Air Control Valve Signal	Pulse Signals
A13	GRN/WHT	Engine Mount Solenoid	Solenoid Off: 12v, On: 1v
A15	RED/YEL	EVAP Purge Solenoid	Solenoid Off: 12v, On: 1v
A16	GRN/BLK	Fuel Pump Relay	Relay Off: 12v, On: 1v
A17	RED/BLU	A/C Clutch Relay	Relay Off: 12v, On: 1v
A18	GRN/RED	Check Engine Light	MIL Off: 12v, On: 1v
A19	WHT/GRN	Alternator Charging Signal	Lights Off: 12v, On: 0v
A20	YEL/GRN	Igniter Control	Digital Signals: 0-12-0v
A25	ORN/GRN	Intake Air Resonator Control	Solenoid Off: 12v, On: 1v
A27	GRN	Radiator Fan Relay	Relay Off: 12v, On: 1v
A28	GRN/WHT	EVAP Bypass Solenoid	Solenoid Off: 12v, On: 1v
A29	ORN/GRN	EVAP Vent Solenoid	Solenoid Off: 12v, On: 1v

1996-97 Sedan 2.2L I4 VTEC VIN CD5 [All] 25P 'B' Connector

PCM Pin #	W/Color	Circuit Description (25-Pin)	Value at Hot Idle
B3	BLU/YEL	Shift Solenoid 'A'	In D2-3: 12v, D1-4: 0v
B4	GRN/BLK	Lockup Solenoid 'B'	LSB On: 12v, Off: 0v
B5	YEL	Lockup Solenoid 'A'	LSA On: 12v, Off: 0v
B8	GRN/BLU	A/T: Gear Position Switch	In D3: 0v, Others: 12v
B11	BLU/YEL	Shift Solenoid 'B'	D2/D3: 12v, D1/D4: 0v
B12	WHT/GRN	Interlock Control Unit Signal	Key & Brake On: 12v
B13	BLU/RED	A/T: D4 Indicator Light Switch	D4 On: 0v, Off: 12v
B14	WHT/BLU	Mainshaft Speed Sensor 'N'	<0.050v
B15	ORN/BLU	Mainshaft Speed Sensor 'P'	Moving: AC pulses
B16	GRN/RED	A/T: Gear Position Switch	In 'R': 0v, Others: 12v
B17	GRN/YEL	A/T: Gear Position Switch	In D2: 0v, Others: 12v
B18	GRN/WHT	A/T: Gear Position Switch	In D1: 0v, Others: 12v
B22	BLU/YEL	Countershaft Speed Sensor N	<0.050v
B23	BLU/GRN	Countershaft Speed Sensor P	Moving: AC pulses
B24	PNK/GRN	A/T: Gear Position Switch	In D4: 0v, Others: 12v
B25	LT GRN	A/T: Gear Position Switch	In P/N: 0v, Others: 12v

1996-97 Sedan 2.2L I4 VTEC VIN CD5 [All] 31P 'C' Connector

PCM Pin #	W/Color	Circuit Description (31-Pin)	Value at Hot Idle
C1	---	Not Used	---
C2	BLU	CKP Sensor Signal	AC pulse signals
C3	GRN	TDC Sensor Signal	AC pulse signals
C4	YEL	CYP Sensor Signal	AC pulse signals
C5	RED/WHT	A/C Switch Signal	Switch Off: 12v, On: 1v
C6	BLU/RED	Start Signal	Cranking: 9-11v
C7	RED	Service Check Connector	SCS Open: 4.80v
C8	LT GRN	K-Line Signal	12v
C10	WHT/YEL	Keep Alive Power (VBU)	12-14v
C12	WHT	CKP Sensor Ground	<0.050v
C13	RED	TDC Sensor Ground	<0.050v
C14	BLK	CYP Sensor Ground	<0.050v
C16	GRN	PSP Switch	Straight: 0v, Turning: 11v
C17	WHT/RED	Alternator 'FR' Signal	Digital Signals: 0-5-0v
C18	ORN	Vehicle Speed Sensor	Moving: pulse signals
C20	BRN	EVAP Purge Flow Switch	Switch on: 0v, off: 5v
C21-31	---	Not Used	---

1996-97 Sedan 2.2L I4 VTEC VIN CD5 [All] 16P 'D' Connector

PCM Pin #	W/Color	Circuit Description (16-Pin)	Value at Hot Idle
D1	RED/BLK	TP Sensor Signal	0.5-0.6v
D2	RED/WHT	ECT Sensor Signal	At 180°F: 0.52v
D3	WHT/YEL	MAP Sensor Signal	0.8-0.9v
D4	YEL/WHT	MAP Sensor VREF	4.9-5.1v
D5	GRN/WHT	Brake Switch Signal	Brake Off: 0v, On: 12v
D7	WHT/RED	HO2S-11 (B1 S1) Signal	0.1-1.1v
D8	RED/YEL	IAT Sensor Signal	Varies w/temp. (0.5-4.9v)
D9	WHT/BLK	EGR Valve Lift Sensor	1.2v
D10	YEL/BLU	Sensor VREF	4.9-5.1v
D11	GRN/BLU	Sensor Ground	<0.050v
D12	GRN/WHT	MAP Sensor Ground	<0.050v
D13	RED/WHT	HO2S-12 Ground	<0.050v
D14	WHT/RED	HO2S-12 (B1 S2) Signal	0.1-1.1v
D15	GRN/RED	Fuel Tank Pressure Sensor	Fuel Cap off: 2.5v
D16	GRN/RED	Electric Load Detector	Varies: 0.5-4.5v

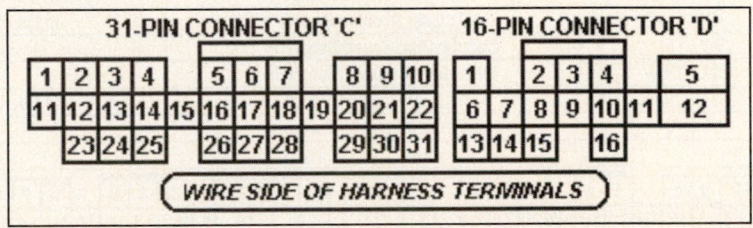

1996-97 Sedan 2.2L I4 MFI VIN CD5 [All] 32P 'A' Connector

PCM Pin #	W/Color	Circuit Description (32-Pin)	Value at Hot Idle
A1	YEL	Injector 4 Control	2.0-3.3 ms
A2	BLU	Injector 3 Control	2.0-3.3 ms
A3	RED	Injector 2 Control	2.0-3.3 ms
A4	BRN	Injector 1 Control	2.0-3.3 ms
A5	ORN/BLU	HO2S-12 (B1 S2) Heater	Digital Signals: 0-12-0v
A6	ORN/BLK	HO2S-11 (B1 S1) Heater	Digital Signals: 0-12-0v
A7	RED	EGR Solenoid	Solenoid Off: 12v, On: 1v
A9, A22	BRN/BLK	Sensor Ground	<0.050v
A10, 23	BLK	Power Ground	<0.1v
A11, 24	YEL/BLK	Main Relay Power (B+)	12-14v
A12	BLK/BLU	Idle Air Control Valve Signal	Pulse Signals
A13	GRN/WHT	Engine Mount Solenoid	Solenoid Off: 12v, On: 1v
A15	RED/YEL	EVAP Purge Solenoid	Solenoid Off: 12v, On: 1v
A16	GRN/BLK	Fuel Pump Relay	Relay Off: 12v, On: 1v
A17	RED/BLU	A/C Clutch Relay	Relay Off: 12v, On: 1v
A18	GRN/RED	Check Engine Light	MIL Off: 12v, On: 1v
A19	WHT/GRN	Alternator Charging Signal	Lights Off: 12v, On: 0v
A20	YEL/GRN	Igniter Control	Digital Signals: 0-12-0v
A25	ORN/GRN	Intake Air Resonator Control	Solenoid Off: 12v, On: 1v
A27	GRN	Radiator Fan Relay	Relay Off: 12v, On: 1v
A28	GRN/WHT	EVAP Bypass Solenoid	Solenoid Off: 12v, On: 1v
A29	ORN/GRN	EVAP Vent Solenoid	Solenoid Off: 12v, On: 1v

1996-97 Sedan 2.2L I4 MFI VIN CD5 [All] 25P 'B' Connector

PCM Pin #	W/Color	Circuit Description (25-Pin)	Value at Hot Idle
B3	BLU/YEL	Shift Solenoid 'A'	In D2-3: 12v, D1-4: 0v
B4	GRN/BLK	Lockup Solenoid 'B'	LSB On: 12v, Off: 0v
B5	YEL	Lockup Solenoid 'A'	LSA On: 12v, Off: 0v
B8	GRN/BLU	A/T: Gear Position Switch	In D3: 0v, Others: 12v
B11	BLU/YEL	Shift Solenoid 'B'	D2/D3: 12v, D1/D4: 0v
B12	WHT/GRN	Interlock Control Unit Signal	Key & Brake On: 12v
B13	BLU/RED	A/T: D4 Indicator Light Switch	D4 On: 0v, Off: 12v
B14	WHT/BLU	Mainshaft Speed Sensor 'N'	<0.050v
B15	ORN/BLU	Mainshaft Speed Sensor 'P'	Moving: AC pulses
B16	GRN/RED	A/T: Gear Position Switch	In 'R': 0v, Others: 12v
B17	GRN/YEL	A/T: Gear Position Switch	In D2: 0v, Others: 12v
B18	GRN/WHT	A/T: Gear Position Switch	In D1: 0v, Others: 12v
B22	BLU/YEL	Countershaft Speed Sensor P	Moving: AC pulses
B23	BLU/GRN	Countershaft Speed Sensor N	<0.050v
B24	PNK/GRN	A/T: Gear Position Switch	In D4: 0v, Others: 12v
B25	LT GRN	A/T: Gear Position Switch	In P/N: 0v, Others: 12v

```
┌──────────────────────────────┐   ┌──────────────────────────────┐
│ 1  2  3  4 │ 5  6  7 │ 8  9  10  11 │   │ 1    │ 2    │ 3 │ 4 │ 5 │ 6 │ 7 │ 8 │
│12 13 14 15 16 17 18 19 20 21 22  23  24│   │ 9    │ 10   │11 12 13 14 15 16 17 18│
│25 26 27 │ 28 29 30 31 │   32   │   │ 19   │ 20   │21 22 │ 23 24 25 │
└──────────────────────────────┘   └──────────────────────────────┘
        32-PIN 'A'      ( WIRE SIDE OF HARNESS TERMINALS )   25-PIN 'B'
```

1996-97 Sedan 2.2L I4 MFI VIN CD5 [All] 31P 'C' Connector

PCM Pin #	W/Color	Circuit Description (31-Pin)	Value at Hot Idle
C1	---	Not Used	---
C2	BLU	CYP Sensor Signal	AC pulse signals
C3	GRN	TDC Sensor Signal	AC pulse signals
C4	YEL	CKP Sensor Signal	AC pulse signals
C5	RED/WHT	A/C Switch Signal	Switch Off: 12v, On: 1v
C6	BLU/RED	Start Signal	Cranking: 9-11v
C7	RED	Service Check Connector	SCS Open: 4.80v
C8	LT GRN	K-Line Signal	12v
C10	WHT/YEL	Keep Alive Power (VBU)	12-14v
C12	WHT	CKP Sensor Ground	<0.050v
C13	RED	TDC Sensor Ground	<0.050v
C14	BLK	CYP Sensor Ground	<0.050v
C16	GRN	PSP Switch	Straight: 0v, Turning: 11v
C17	WHT/RED	Alternator 'FR' Signal	Digital Signals: 0-5-0v
C18	ORN	Vehicle Speed Sensor	Moving: pulse signals
C20	BRN	EVAP Purge Flow Switch	Switch on: 0v, off: 5v

1996-97 Sedan 2.2L I4 MFI VIN CD5 [All] 16P 'D' Connector

PCM Pin #	W/Color	Circuit Description (16-Pin)	Value at Hot Idle
D1	RED/BLK	TP Sensor Signal	0.5-0.6v
D2	RED/WHT	ECT Sensor Signal	At 180°F: 0.52v
D3	WHT/YEL	MAP Sensor Signal	0.8-0.9v
D4	YEL/WHT	MAP Sensor VREF	4.9-5.1v
D5	GRN/WHT	Brake Switch Signal	Brake Off: 0v, On: 12v
D7	WHT/RED	HO2S-11 (B1 S1) Signal	0.1-1.1v
D8	RED/YEL	IAT Sensor Signal	Varies w/temp. (0.5-4.9v)
D9	WHT/BLK	EGR Valve Lift Sensor	1.2v
D10	YEL/BLU	Sensor VREF	4.9-5.1v
D11	GRN/BLU	Sensor Ground	<0.050v
D12	GRN/WHT	MAP Sensor Ground	<0.050v
D13	RED/WHT	HO2S-12 Ground	<0.050v
D14	WHT/RED	HO2S-12 (B1 S2) Signal	0.1-1.1v
D15	GRN/RED	Fuel Tank Pressure Sensor	Fuel Cap off: 2.5v
D16	GRN/RED	Electric Load Detector	Varies: 0.5-4.5v

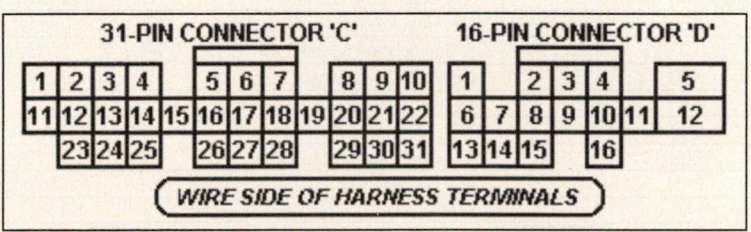

1998-99 Sedan 2.3L I4 MFI VIN CF8 [All] 32P 'A' Connector

PCM Pin #	W/Color	Circuit Description (32-Pin)	Value at Hot Idle
A2	GRN/WHT	A/T: Engine Mount Solenoid	Solenoid Off: 12v, On: 1v
A3	BLU	EVAP Bypass Solenoid	Solenoid Off: 12v, On: 1v
A4	GRN/WHT	EVAP Vent Solenoid	Solenoid Off: 12v, On: 1v
A5	BLU/GRN	Cruise Control Signal	C/C On: pulse signals
A6	RED/YEL	EVAP Purge Solenoid	Solenoid Off: 12v, On: 1v
A8	BLK/WHT	HO2S-12 (B1 S2) Heater	Digital Signals: 0-12-0v
A9	BLU/WHT	A/T: VSS Output Signal	Moving: pulse signals
A10	BRN	Service Check Connector	SCS Open: 4.80v
A12	PNK	Immobilizer Indicator Lamp	Lamp Off: 12v, On: 1v
A13, 25	BLU, RED	Immobilizer Enable, Code	Digital Signals
A14	GRN/BLK	A/T: D4 Light Switch	D4 On: 0v, Off: 12v
A15	GRN/YEL	Immobilizer Fuel Pump Relay	Relay Off: 12v, On: 1v
A17	RED	A/C Clutch Relay Control	Relay Off: 12v, On: 1v
A18	GRN/ORN	Malfunction Indicator Light	MIL Off: 12v, On: 1v
A19	BLU	Engine Speed Pulse (NEP)	Pulse Signals
A20	GRN	Radiator Fan Relay	Relay Off: 12v, On: 1v
A21	GRY	K-Line Signal	12v
A23	WHT/RED	HO2S-12 (B1 S2) Signal	0.1-1.1v
A24	BLU/ORN	Starter Switch Signal	Cranking: 9-11v
A26	GRN	PSP Switch	Straight: 0v, Turning: 11v
A27	BLU/RED	A/C Switch Signal	Switch Off: 12v, On: 1v
A28	WHT/RED	A/T: Interlock Control Unit	Key & Brake On: 12v
A29	LT GRN	Fuel Tank Pressure Sensor	Fuel Cap off: 2.5v
A30	GRN/RED	Electric Load Detector	Varies: 0.5-3.5v
A32	WHT/BLK	Brake Switch Signal	Brake Off: 0v, On: 12v

1998-99 Sedan 2.3L I4 MFI VIN CF8 [All] 16P 'D' Connector

PCM Pin #	W/Color	Circuit Description (16-Pin)	Value at Hot Idle
D1	YEL	A/T: Lockup Control Solenoid	LCS Off: 0v, On: 12v
D2	GRN/WHT	A/T: Shift Solenoid 'B'	SSB Off: 0v, On: 12v
D3	GRN	A/T: Shift Solenoid 'C'	SSC Off: 12v, On: 1v
D5	BLK/YEL	A/T: VB Solenoid Feed (B+)	12-14v
D6	WHT	A/T: Gear Position Switch	In 'R': 0v, Others: 12v
D7	BLU/YEL	Shift Solenoid 'A'	SSA Off: 0v, On: 12v
D8	PNK	A/T: Gear Position Switch	In D3: 0v, Others: 12v
D9	YEL	A/T: Gear Position Switch	In D4: 0v, Others: 12v
D10, D16	BLU, GRN	A/T: Countershaft Speed P, N	Moving: AC pulses
D11	RED	Mainshaft Speed Sensor 'P'	Moving: AC pulses
D12	WHT	Mainshaft Speed Sensor 'N'	<0.050v
D13	BLU/WHT	A/T: Gear Position Switch	In P/N: 0v, Others: 12v
D14	BLU	A/T: Gear Position Switch	In D2: 0v, Others: 12v
D15	BRN	A/T: Gear Position Switch	In D1: 0v, Others: 12v

1998-99 Sedan 2.3L I4 MFI VIN CF8 [All] 25P 'B' Connector

PCM Pin #	W/Color	Circuit Description (25-Pin)	Value at Hot Idle
B1, B9	YEL/BLK	Main Relay Power (B+)	12-14v
B2, B10	BLK	Power Ground	<0.1v
B3	RED	Injector 2 Control	2.0-3.3 ms
B4	BLU	Injector 3 Control	2.0-3.3 ms
B5	YEL	Injector 4 Control	2.0-3.3 ms
B7	PNK	E-EGR Solenoid Control	Digital Signals: 0-12-0v
B8	WHT	A/T: Clutch Solenoid 'A-'	LSA On: pulse signals
B11	BRN	Injector 1 Control	2.0-3.3 ms
B13	YEL/GRN	Ignition Control Signal	Digital Signals: 0-12-0v
B14	BLU/BLK	A/T: 2nd Clutch Pressure Sw.	Open: 12v, Closed: 1v
B16, B19	---	Not Used	---
B17	RED	A/T: Clutch Solenoid 'A+'	LSA On: pulse signals
B18	GRN	A/T: Clutch Solenoid 'B-'	Pulse Signals
B20, 22	BRN/BLK	Logic Ground (LG1, LG2)	<0.050v
B21	WHT/YEL	Keep Alive Power (VBU)	12-14v
B23	BLK/BLU	Idle Air Control Valve	Pulse Signals
B24	BLU/BLK	A/T: 3rd Clutch Pressure Sw.	Open: 12v, Closed: 1v
B25	ORN	A/T: Clutch Solenoid 'B+'	Pulse Signals

1998-99 Sedan 2.3L I4 MFI VIN CF8 [All] 31P 'C' Connector

PCM Pin #	W/Color	Circuit Description (31-Pin)	Value at Hot Idle
C1	BLK/WHT	HO2S-11 (B1 S1) Heater	Digital Signals: 0-12-0v
C2	WHT/GRN	Alternator Charging Signal	Lights Off: 12v, On: 0v
C3	RED/BLU	Knock Sensor Signal	No Detonation: 18mv AC
C5	WHT/RED	Alternator 'FR' Signal	Digital Signals: 0-5-0v
C6	WHT/BLK	EGR Valve Lift Sensor	1.2v
C7	GRN/WHT	MAP Sensor Ground (SG1)	<0.050v
C8	BLU	CKP Sensor Signal	AC pulse signals
C9	WHT	CKP Sensor Ground	<0.050v
C16	WHT	HO2S-11 (B1 S1) Signal	0.1-1.1v
C17	RED/GRN	MAP Sensor Signal	0.8-0.9v
C18	GRN/BLK	Sensor Ground (SG2)	<0.050v
C19	YEL/RED	MAP Sensor VREF (VCC1)	4.9-5.1v
C20	GRN	TDC Sensor Signal	AC pulse signals
C21	RED	TDC Sensor Ground	<0.050v
C23	BLU/WHT	M/T: VSS Input Signal	Moving: 0-5-0v
C25	RED/YEL	IAT Sensor Signal	Varies w/temp. (0.5-4.9v)
C26	RED/WHT	ECT Sensor Signal	At 180°F: 0.5-0.6v
C27	RED/BLK	TP Sensor Signal	0.5-0.6v
C28	YEL/BLU	Sensor VREF (VCC2)	4.9-5.1v
C29	YEL	CYP Sensor Signal	AC pulse signals
C30	BLK	CYP Sensor Ground	<0.050v

1	2	3	4	5	6	7	8	1	2	3	4	5	6	7	8	9	10
9	10	11	12	13	14	15	16	17	18	11	12	13	14	15	16	17	18	19	20	21	22
19	20	21	22	23	24	25	23	24	25	26	27	28	29	30	31

25-PIN 'B'   *WIRE SIDE OF HARNESS TERMINALS*   31-PIN 'C'

2000-03 Sedan 2.3L I4 MFI VIN CF8 [All] 32P 'A' Connector

PCM Pin #	W/Color	Circuit Description (32-Pin)	Value at Hot Idle
A2	GRN/WHT	A/T: Engine Mount Solenoid	Solenoid Off: 12v, On: 1v
A3	BLU	EVAP Bypass Solenoid	Solenoid Off: 12v, On: 1v
A4	GRN/WHT	EVAP Vent Solenoid	Solenoid Off: 12v, On: 1v
A5	BLU/GRN	Cruise Control Signal	C/C On: pulse signals
A6	RED/YEL	EVAP Purge Solenoid	Solenoid Off: 12v, On: 1v
A8	BLK/WHT	HO2S-12 (B1 S2) Heater	Digital Signals: 0-12-0v
A9	BLU/WHT	A/T: VSS Output Signal	Moving: pulse signals
A10	BRN	Service Check Connector	SCS Open: 4.80v
A12	PNK	Immobilizer Indicator Lamp	Lamp Off: 12v, On: 1v
A13, 25	BLU, RED	Immobilizer Enable, Code	Digital Signals
A14	GRN/BLK	A/T: D4 Light Switch	D4 On: 0v, Off: 12v
A15	GRN/YEL	Immobilizer Fuel Pump Relay	Relay Off: 12v, On: 1v
A17	RED	A/C Clutch Relay Control	Relay Off: 12v, On: 1v
A18	GRN/ORN	Malfunction Indicator Light	MIL Off: 12v, On: 1v
A19	BLU	Engine Speed Pulse (NEP)	Pulse Signals
A20	GRN	Radiator Fan Relay	Relay Off: 12v, On: 1v
A21	GRY	K-Line Signal	12v
A23	WHT/RED	HO2S-12 (B1 S2) Signal	0.1-1.1v
A24	BLU/ORN	Starter Switch Signal	Cranking: 9-11v
A26	GRN	PSP Switch	Straight: 0v, Turning: 11v
A27	BLU/RED	A/C Switch Signal	Switch Off: 12v, On: 1v
A28	WHT/RED	A/T: Interlock Control Unit	Key & Brake On: 12v
A29	LT GRN	Fuel Tank Pressure Sensor	Fuel Cap off: 2.5v
A30	GRN/RED	Electric Load Detector	Varies: 0.5-3.5v
A32	WHT/BLK	Brake Switch Signal	Brake Off: 0v, On: 12v

2000-03 Sedan 2.3L I4 MFI VIN CF8 [All] 16P 'D' Connector

PCM Pin #	W/Color	Circuit Description (16-Pin)	Value at Hot Idle
D1	YEL	A/T: Lockup Control Solenoid	LCS Off: 0v, On: 12v
D2	GRN/WHT	A/T: Shift Solenoid 'B'	SSB Off: 0v, On: 12v
D3	GRN	A/T: Shift Solenoid 'C'	SSC Off: 12v, On: 1v
D5	BLK/YEL	A/T: VB Solenoid Feed (B+)	12-14v
D6	WHT	A/T: Gear Position Switch	In 'R': 0v, Others: 12v
D7	BLU/YEL	Shift Solenoid 'A'	SSA Off: 0v, On: 12v
D8	PNK	A/T: Gear Position Switch	In D3: 0v, Others: 12v
D9	YEL	A/T: Gear Position Switch	In D4: 0v, Others: 12v
D10, D16	BLU, GRN	A/T: Countershaft Speed P, N	Moving: AC pulses
D11	RED	Mainshaft Speed Sensor 'P'	Moving: AC pulses
D12	WHT	Mainshaft Speed Sensor 'N'	<0.050v
D13	BLU/WHT	A/T: Gear Position Switch	In P/N: 0v, Others: 12v
D14	BLU	A/T: Gear Position Switch	In D2: 0v, Others: 12v
D15	BRN	A/T: Gear Position Switch	In D1: 0v, Others: 12v

WIRE SIDE OF HARNESS TERMINALS

2000-03 Sedan 2.3L I4 MFI VIN CF8 [All] 25P 'B' Connector

PCM Pin #	W/Color	Circuit Description (25-Pin)	Value at Hot Idle
B1, B9	YEL/BLK	Main Relay Power (B+)	12-14v
B2, B10	BLK	Power Ground	<0.1v
B3	RED	Injector 2 Control	2.0-3.3 ms
B4	BLU	Injector 3 Control	2.0-3.3 ms
B5	YEL	Injector 4 Control	2.0-3.3 ms
B7	PNK	E-EGR Solenoid Control	Digital Signals: 0-12-0v
B8	WHT	A/T: Clutch Solenoid 'A-'	LSA On: pulse signals
B11	BRN	Injector 1 Control	2.0-3.3 ms
B13	YEL/GRN	Ignition Control Signal	Digital Signals: 0-12-0v
B14	BLU/BLK	A/T: 2nd Clutch Pressure Sw.	Open: 12v, Closed: 1v
B16, B19	---	Not Used	---
B17	RED	A/T: Clutch Solenoid 'A+'	LSA On: pulse signals
B18, B25	GRN, ORN	A/T: Clutch Solenoid 'B-', 'B+'	Pulse Signals
B20, 22	BRN/BLK	Logic Ground	<0.050v
B21	WHT/YEL	Keep Alive Power (VBU)	12-14v
B23	BLK/BLU	Idle Air Control Valve	Pulse Signals
B24	BLU/BLK	A/T: 3rd Clutch Pressure Sw.	Open: 12v, Closed: 1v

2000-03 Sedan 2.3L I4 MFI VIN CF8 [All] 31P 'C' Connector

PCM Pin #	W/Color	Circuit Description (31-Pin)	Value at Hot Idle
C1	BLK/WHT	HO2S-11 (B1 S1) Heater	Digital Signals: 0-12-0v
C2	WHT/GRN	Alternator Charging Signal	Lights Off: 12v, On: 0v
C3	RED/BLU	Knock Sensor Signal	No Detonation: 18mv AC
C5	WHT/RED	Alternator 'FR' Signal	Digital Signals: 0-5-0v
C6	WHT/BLK	EGR Valve Lift Sensor	1.2v
C7	GRN/WHT	MAP Sensor Ground (SG1)	<0.050v
C8	BLU	CKP Sensor Signal	AC pulse signals
C9	WHT	CKP Sensor Ground	<0.050v
C16	WHT	HO2S-11 (B1 S1) Signal	0.1-1.1v
C17	RED/GRN	MAP Sensor Signal	0.8-0.9v
C18	GRN/BLK	Sensor Ground (SG2)	<0.050v
C19	YEL/RED	MAP Sensor VREF (VCC1)	4.9-5.1v
C20	GRN	TDC Sensor Signal	AC pulse signals
C21	RED	TDC Sensor Ground	<0.050v
C23	BLU/WHT	M/T: VSS Input Signal	Moving: 0-5-0v
C24	YEL/GRN	ECT Signal to TCM	Digital Signals
C25	RED/YEL	IAT Sensor Signal	Varies w/temp. (0.5-4.9v)
C26	RED/WHT	ECT Sensor Signal	At 180°F: 0.5-0.6v
C27	RED/BLK	TP Sensor Signal	0.5-0.6v
C28	YEL/BLU	Sensor VREF (VCC2)	4.9-5.1v
C29	YEL	CYP Sensor Signal	AC pulse signals
C30	BLK	CYP Sensor Ground	<0.050v

1	2	3	4	5	6	7	8	1	2	3	4	5	6	7	8	9	10
9	10	11	12	13	14	15	16	17	18	11	12	13	14	15	16	17	18	19	20	21	22
19	20	21	22	23	24	25	23	24	25	26	27	28	29	30	31

25-PIN 'B'   *WIRE SIDE OF HARNESS TERMINALS*   **31-PIN 'C'**

1998-99 Sedan 2.3L I4 VTEC VIN CG5, CG6 [All] 32P A Connector

PCM Pin #	W/Color	Circuit Description (32-Pin)	Value at Hot Idle
A2	GRN/WHT	A/T: Engine Mount Solenoid	Solenoid Off: 12v, On: 1v
A3	BLU	EVAP Bypass Solenoid	Solenoid Off: 12v, On: 1v
A4	GRN/WHT	EVAP Vent Solenoid	Solenoid Off: 12v, On: 1v
A5	BLU/GRN	Cruise Control Signal	C/C On: pulse signals
A6	RED/YEL	EVAP Purge Solenoid	Solenoid Off: 12v, On: 1v
A8	BLK/WHT	HO2S-12 (B1 S2) Heater	Digital Signals: 0-12-0v
A9	BLU/WHT	A/T: Vehicle Speed Output	Moving: pulse signals
A10	BRN	Service Check Connector	SCS Open: 4.80v
A12	PNK	Immobilizer Indicator Lamp	Lamp Off: 12v, On: 1v
A13, 25	BLU, RED	Immobilizer Enable, Code	Digital Signals
A14	GRN/BLK	A/T: D4 Light Switch	D4 On: 0v, Off: 12v
A15	GRN/YEL	Immobilizer Fuel Pump Relay	Relay Off: 12v, On: 1v
A17	RED	A/C Clutch Relay Control	Relay Off: 12v, On: 1v
A18	GRN/ORN	Malfunction Indicator Light	MIL Off: 12v, On: 1v
A19	BLU	Engine Speed Pulse (NEP)	Digital Signals
A20	GRN	Radiator Fan Relay Control	Relay Off: 12v, On: 1v
A21	GRY	K-Line Signal	12v
A23	WHT/RED	HO2S-12 (B1 S2) Signal	0.1-1.1v
A24	BLU/ORN	Starter Switch Signal	Cranking: 9-11v
A26	GRN	PSP Switch Signal	Straight: 0v, Turning: 11v
A27	BLU/RED	A/C Switch Signal	Switch Off: 12v, On: 1v
A28	WHT/RED	Interlock Control Unit Signal	Key & Brake On: 12v
A29	LT GRN	Fuel Tank Pressure Sensor	Fuel Cap off: 2.5v
A30	GRN/RED	Electric Load Detector	Varies: 0.5-4.5v
A32	WHT/BLK	Brake Switch Signal	Brake Off: 0v, On: 12v

1998-99 Sedan 2.3L I4 VTEC VIN CG5, CG6 [All] 16P D Connector

PCM Pin #	W/Color	Circuit Description (16-Pin)	Value at Hot Idle
D1	YEL	A/T: Lockup Control Solenoid	LUS On: 12v, Off: 0v
D2	GRN/WHT	A/T: Shift Solenoid 'B'	SSB Off: 0v, On: 12v
D3	GRN	A/T: Shift Solenoid 'C'	SSC on: 12v, off: 0v
D5	BLK/YEL	A/T: Solenoid Feed (B+)	12-14v
D6	WHT	A/T: Gear Position Switch	In 'R': 0v, Others: 12v
D7	BLU/YEL	A/T: Shift Solenoid 'A'	SSA Off: 0v, On: 12v
D8	PNK	A/T: Gear Position Switch	In D3: 0v, Others: 12v
D9	YEL	A/T: Gear Position Switch	In D4: 0v, Others: 12v
Q	BLU, GRN	A/T: Countershaft Speed P, N	Moving: AC pulses
D11	RED	Mainshaft Speed Sensor 'P'	AC pulse signals
D12	WHT	Mainshaft Speed Sensor 'N'	<0.050v
D13	BLU/WHT	A/T: Gear Position Switch	In P/N: 0v, Others: 12v
D14	BLU	A/T: Gear Position Switch	In D2: 0v, Others: 12v
D15	BRN	A/T: Gear Position Switch	In D1: 0v, Others: 12v

1998-99 Sedan 2.3L I4 VTEC VIN CG5, CG6 [All] 25P B Connector

PCM Pin #	W/Color	Circuit Description (25-Pin)	Value at Hot Idle
B1, B9	YEL/BLK	Main Relay Power (B+)	12-14v
B2, B10	BLK	Power Ground	<0.1v
B3, B4	RED, BLU	Injector 2, Injector 3 Control	2.0-3.3 ms
B5, B11	YEL, BRN	Injector 4, Injector 1 Control	2.0-3.3 ms
B7	PNK	E-EGR Control Solenoid	Digital Signals: 0-12-0v
B8, B17	WHT, RED	A/T: Clutch Solenoid 'A-', 'A+'	Pulse Signals
B12	GRN/YEL	VTEC Control Solenoid	0v, Hi-Speed: 12v
B13	YEL/GRN	Ignition Control Signal	Digital Signals: 0-12-0v
B14	BLU/BLK	A/T: 2nd Clutch Pressure Sw.	Open: 12v, Closed: 1v
B16	GRN/RED	HO2S-11 Heater Relay	Relay Off: 12v, On: 1v
B18, B25	GRN, ORN	A/T: Clutch Solenoid 'B-', 'B+'	Pulse Signals
B19	BLK/WHT	HO2S-11 (B1 S1) Heater	Digital Signals: 0-12-0v
B20, B22	BRN/BLK	Logic Ground	<0.050v
B21	WHT/YEL	Keep Alive Power (VBU)	12-14v
B23	BLK/BLU	Idle Air Control Valve	Pulse Signals
B24	BLU/WHT	A/T: 3rd Clutch Pressure Sw.	Open: 12v, Closed: 1v

1998-99 Sedan 2.3L I4 VTEC VIN CG5, CG6 [All] 31P C Connector

PCM Pin #	W/Color	Circuit Description (31-Pin)	Value at Hot Idle
C2	WHT/GRN	Alternator Charging Signal	Lights Off: 12v, On: 0v
C3	RED/BLU	Knock Sensor Signal	No Detonation: 18mv AC
C5	WHT/RED	Alternator 'FR' Signal	Digital Signals: 0-5-0v
C6	WHT/BLK	EGR Valve Lift Sensor	1.2v
C7	GRN	MAP Sensor Ground	<0.050v
C8	BLU	CKP Sensor Signal	AC pulse signals
C9	WHT	CKP Sensor Ground	<0.050v
C10	BLU/BLK	VTEC Pressure Switch	0v, Hi-Speed: 12v
C13	WHT	HO2S-11 Heater Relay	Relay Off: 12v, On: 1v
C14	RED	HO2S-11 (B1 S1) Signal	0.1-1.1v
C15	BLU	HO2S-11 Ground	<0.050v
C16	WHT	HO2S-11 (B1 S1) Signal	0.1-1.1v
C17	RED/GRN	MAP Sensor Signal	0.8-0.9v
C18	GRN	Sensor Ground	<0.050v
C19, C28	YEL, YEL	MAP VREF, Sensor VREF	4.9-5.1v
C20	GRN	TDC Sensor Signal	AC pulse signals
C21	RED	TDC Sensor Ground	<0.050v
C23	BLU/WHT	M/T: VSS Input Signal	Moving: 0-5-0v
C24	YEL/GRN	ECT Sensor Signal to TCM	Digital Signals
C25	RED/YEL	IAT Sensor Signal	Varies w/temp (0.5-4.9v)
C26	RED/WHT	ECT Sensor Signal	At 180°F: 0.5-0.6v
C27	RED/BLK	TP Sensor Signal	0.5-0.6v
C29, C30	YEL, BLK	CYP Sensor Signal, Ground	AC pulse signals

1	2		3	4	5		6	7	8		1	2	3	4		5	6	7		8	9	10
9	10	11	12	13	14	15	16	17	18	11	12	13	14	15	16	17	18	19	20	21	22
19	20	21	22		23	24	25			23	24	25		26	27	28		29	30	31		

25-PIN 'B'   *WIRE SIDE OF HARNESS TERMINALS*   31-PIN 'C'

2000-03 Sedan 2.3L I4 VTEC VIN CG5, CG6 [All] 32P A Connector

PCM Pin #	W/Color	Circuit Description (32-Pin)	Value at Hot Idle
A2	GRN/WHT	A/T: Engine Mount Solenoid	Solenoid Off: 12v, On: 1v
A3	BLU	EVAP Bypass Solenoid	Solenoid Off: 12v, On: 1v
A4	GRN/WHT	EVAP Vent Solenoid	Solenoid Off: 12v, On: 1v
A5	BLU/GRN	Cruise Control Signal	C/C On: pulse signals
A6	RED/YEL	EVAP Purge Solenoid	Solenoid Off: 12v, On: 1v
A8	BLK/WHT	HO2S-12 (B1 S2) Heater	Digital Signals: 0-12-0v
A9	BLU/WHT	A/T: Vehicle Speed Output	Moving: pulse signals
A10	BRN	Service Check Connector	SCS Open: 4.80v
A12	PNK	Immobilizer Indicator Lamp	Lamp Off: 12v, On: 1v
A13, 25	BLU, RED	Immobilizer Enable, Code	Digital Signals
A14	GRN/BLK	A/T: D4 Light Switch	D4 On: 0v, Off: 12v
A15	GRN/YEL	Immobilizer Fuel Pump Relay	Relay Off: 12v, On: 1v
A17	RED	A/C Clutch Relay Control	Relay Off: 12v, On: 1v
A18	GRN/ORN	Malfunction Indicator Light	MIL Off: 12v, On: 1v
A19	BLU	Engine Speed Pulse (NEP)	Digital Signals
A20	GRN	Radiator Fan Relay Control	Relay Off: 12v, On: 1v
A21	GRY	K-Line Signal	12v
A23	WHT/RED	HO2S-12 (B1 S2) Signal	0.1-1.1v
A24	BLU/ORN	Starter Switch Signal	Cranking: 9-11v
A26	GRN	PSP Switch Signal	Straight: 0v, Turning: 11v
A27	BLU/RED	A/C Switch Signal	Switch Off: 12v, On: 1v
A28	WHT/RED	Interlock Control Unit Signal	Key & Brake On: 12v
A29	LT GRN	Fuel Tank Pressure Sensor	Fuel Cap off: 2.5v
A30	GRN/RED	Electric Load Detector	Varies: 0.5-4.5v
A32	WHT/BLK	Brake Switch Signal	Brake Off: 0v, On: 12v

2000-03 Sedan 2.3L I4 VTEC VIN CG5, CG6 [All] 16P D Connector

PCM Pin #	W/Color	Circuit Description (16-Pin)	Value at Hot Idle
D1	YEL	A/T: Lockup Control Solenoid	LUS On: 12v, Off: 0v
D2	GRN/WHT	A/T: Shift Solenoid 'B'	SSB Off: 0v, On: 12v
D3	GRN	A/T: Shift Solenoid 'C'	SSC on: 12v, off: 0v
D5	BLK/YEL	A/T: Solenoid Feed (B+)	12-14v
D6	WHT	A/T: Gear Position Switch	In 'R': 0v, Others: 12v
D7	BLU/YEL	A/T: Shift Solenoid 'A'	SSA Off: 0v, On: 12v
D8	PNK	A/T: Gear Position Switch	In D3: 0v, Others: 12v
D9	YEL	A/T: Gear Position Switch	In D4: 0v, Others: 12v
D10, D16	BLU, GRN	A/T: Countershaft Speed P, N	Moving: AC pulses
D11	RED	Mainshaft Speed Sensor 'P'	AC pulse signals
D12	WHT	Mainshaft Speed Sensor 'N'	<0.050v
D13	BLU/WHT	A/T: Gear Position Switch	In P/N: 0v, Others: 12v
D14	BLU	A/T: Gear Position Switch	In D2: 0v, Others: 12v
D15	BRN	A/T: Gear Position Switch	In D1: 0v, Others: 12v

WIRE SIDE OF HARNESS TERMINALS

2000-03 Sedan 2.3L I4 VTEC VIN CG5, CG6 [All] 25P B Connector

PCM Pin #	W/Color	Circuit Description (25-Pin)	Value at Hot Idle
B1, B9	YEL/BLK	Main Relay Power (B+)	12-14v
B2, B10	BLK	Power Ground	<0.1v
B3, B4	RED, BLU	Injector 2, Injector 3 Control	2.0-3.3 ms
B5, B11	YEL, BRN	Injector 4, Injector 1 Control	2.0-3.3 ms
B7	PNK	E-EGR Control Solenoid	Digital Signals: 0-12-0v
B8, B17	WHT, RED	A/T: Clutch Solenoid 'A-', 'A+'	Pulse Signals
B12	GRN/YEL	VTEC Control Solenoid	0v, Hi-Speed: 12v
B13	YEL/GRN	Ignition Control Signal	Digital Signals: 0-12-0v
B14	BLU/BLK	A/T: 2nd Clutch Pressure Sw.	Open: 12v, Closed: 1v
B16	GRN/RED	HO2S-11 Heater Relay	Relay Off: 12v, On: 1v
B18, B25	GRN, ORN	A/T: Clutch Solenoid 'B-', 'B+'	Pulse Signals
B20, B22	BRN/BLK	Logic Ground	<0.050v
B21	WHT/YEL	Keep Alive Power (VBU)	12-14v
B23	BLK/BLU	Idle Air Control Valve	Pulse Signals
B24	BLU/WHT	A/T: 3rd Clutch Pressure Sw.	Open: 12v, Closed: 1v

2000-03 Sedan 2.3L I4 VTEC VIN CG5, CG6 [All] 31P C Connector

PCM Pin #	W/Color	Circuit Description (31-Pin)	Value at Hot Idle
C1	BLK/WHT	HO2S-11 (B1 S1) Heater	Digital Signals: 0-12-0v
C2	WHT/GRN	Alternator Charging Signal	Lights Off: 12v, On: 0v
C3	RED/BLU	Knock Sensor Signal	No Detonation: 18mv AC
C5	WHT/RED	Alternator 'FR' Signal	Digital Signals: 0-5-0v
C6	WHT/BLK	EGR Valve Lift Sensor	1.2v
C7	GRN	MAP Sensor Ground	<0.050v
C8	BLU	CKP Sensor Signal	AC pulse signals
C9	WHT	CKP Sensor Ground	<0.050v
C10	BLU/BLK	VTEC Pressure Switch	0v, Hi-Speed: 12v
C13	WHT	HO2S-11 Heater Relay	Relay Off: 12v, On: 1v
C14	RED	HO2S-11 (B1 S1) Signal	0.1-1.1v
C15	BLU	HO2S-11 Ground	<0.050v
C16	WHT	HO2S-11 (B1 S1) Signal	0.1-1.1v
C17	RED/GRN	MAP Sensor Signal	0.8-0.9v
C18	GRN	Sensor Ground	<0.050v
C19, C28	YEL, YEL	MAP VREF, Sensor VREF	4.9-5.1v
C20	GRN	TDC Sensor Signal	AC pulse signals
C21	RED	TDC Sensor Ground	<0.050v
C23	BLU/WHT	M/T: VSS Input Signal	Moving: 0-5-0v
C24	YEL/GRN	ECT Sensor Signal to TCM	Digital Signals
C25	RED/YEL	IAT Sensor Signal	Varies w/temp (0.5-4.9v)
C26	RED/WHT	ECT Sensor Signal	At 180ºF: 0.5-0.6v
C27	RED/BLK	TP Sensor Signal	0.5-0.6v
C29, C30	YEL, BLK	CYP Sensor Signal, Ground	AC pulse signals

1	2		3	4	5		6	7	8		1	2	3	4		5	6	7		8	9	10
9	10	11	12	13	14	15	16	17	18	11	12	13	14	15	16	17	18	19	20	21	22
19	20		21	22		23	24	25			23	24	25		26	27	28		29	30	31

25-PIN 'B'   *WIRE SIDE OF HARNESS TERMINALS*   **31-PIN 'C'**

1995 Sedan 2.7L V6 MFI VIN CE6 [A/T] 26P 'A' Connector

PCM Pin #	W/Color	Circuit Description (26-Pin)	Value at Hot Idle
A1	BRN	Injector 1 Control	2.0-3.3 ms
A2	RED	Injector 2 Control	2.0-3.3 ms
A3	BLU	Injector 3 Control	2.0-3.3 ms
A4	GRN/BLK	Fuel Pump Relay	Relay Off: 12v, On: 1v
A5	BLK/BLU	Idle Air Control Valve	Pulse Signals
A6	ORN/BLK	Right HO2S Heater Control	Digital Signals: 0-12-0v
A7	GRN/RED	Check Engine Light	C/E on: <1v, of: 12v
A8	RED/BLU	A/C Clutch Relay	Relay Off: 12v, On: 1v
A9	RED	EGR Control Solenoid	Solenoid Off: 12v, On: 1v
A10	PNK/BLU	Intake Air Bypass Solenoid	Solenoid Off: 12v, On: 1v
A11	YEL/GRN	Igniter Control	Digital Signals: 0-12-0v
A12	BLK	Power Ground	<0.1v
A13	YEL/BLK	Ignition Power	12-14v
A14	YEL	Injector 4 Control	2.0-3.3 ms
A15	BLK/RED	Injector 5 Control	2.0-3.3 ms
A16	WHT/BLU	Injector 6 Control	2.0-3.3 ms
A17	ORN/BLU	Left HO2S Heater Control	Digital Signals: 0-12-0v
A18	GRN/WHT	Engine Mount Solenoid	Solenoid Off: 12v, On: 1v
A19	GRN	Radiator Fan Relay Control	On: <1v, Off: 12v
A21	WHT/GRN	Alternator Charging Signal	Lights on: 12v, off: 0v
A22	BRN/WHT	A/T: TCM Control Unit	N/A
A23	RED/YEL	EVAP Purge Solenoid	Solenoid Off: 12v, On: 1v
A24	BRN	EGR Solenoid	Solenoid Off: 12v, On: 1v
A25	BLK	Power Ground	<0.050v
A26	BRN/BLK	Chassis Ground	<0.050v

1995 Sedan 2.7L V6 MFI VIN CE6 [All] 16P 'B' Connector

PCM Pin #	W/Color	Circuit Description (16-Pin)	Value at Hot Idle
B1	YEL/BLK	Ignition Power	12v
B2	WHT/RED	A/T: TCM Control Unit	N/A
B3	RED/WHT	A/C Switch Signal	Switch Off: 12v, On: 1v
B4	LT GRN	A/T: Park Neutral Position	In P/N: 0v, Others: 11v
B5	BLK/GRN	Starter Switch Signal	Cranking: 9-11v
B6	YEL	CYP Sensor Signal	AC pulse signals
B7	GRN	TDC Sensor Signal	AC pulse signals
B8	BLU	CKP Sensor Signal	AC pulse signals
B9	BRN/BLK	Chassis Ground	<0.050v
B10	GRN	A/T: TCM Control Unit	N/A
B11	---	Not Used	---
B12	GRN	PSP Switch	Straight: 0v, Turning: 11v
B13	ORN	Vehicle Speed Sensor	Moving: pulse signals
B14	BLK	CYP Sensor Ground	<0.050v
B15	RED	TDC Sensor Ground	<0.050v
B16	WHT	CKP Sensor Ground	<0.050v

1995 Sedan 2.7L V6 MFI VIN CE6 [All] 22P 'D' Connector

PCM Pin #	W/Color	Circuit Description (22-Pin)	Value at Hot Idle
D1	WHT/YEL	Keep Alive Power (VBU)	12-14v
D2	RED/WHT	Rear HO2S Ground	<0.050v
D3	BLU/WHT	A/T TCM Signal	Digital Signals
D4	LT GRN	Data Link Connector	5v
D5	WHT/RED	Alternator 'FR' Signal	Digital Signals: 0-5-0v
D6	RED/BLK	TP Sensor Signal	0.5-0.6v
D7	RED/WHT	ECT Sensor Signal	At 180°F: 0.52v
D8	RED/YEL	IAT Sensor Signal	Varies w/temp. (0.5-4.9v)
D9	WHT/YEL	MAP Sensor Signal	0.8-0.9v
D10	YEL/WHT	MAP Sensor VREF	4.9-5.1v
D11	GRN/WHT	MAP Sensor Ground	<0.050v
D12	GRN/WHT	Brake Switch Signal	Brake Off: 0v, On: 12v
D13	RED	Service Check Connector	SCS Open: 4.80v
D14	RED/GRN	A/T TCM Signal	Digital Signals
D15	---	Not Used	---
D16	GRN/RED	Electric Load Detector	Varies: 0.5-4.5v
D17	WHT/BLK	A/T: EGR Lift Sensor Signal	1.2v
D18	WHT/RED	HO2S-11 (B1 S1) Signal	0.1-1.1v
D19	WHT/RED	HO2S-12 (B1 S2) Signal	0.1-1.1v
D20	GRN/BLK	Sensor VREF	4.9-5.1v
D21	YEL/BLU	A/T: EGR Lift Sensor Signal	1.1-1.2v
D22	GRN/BLU	Sensor Ground	<0.050v

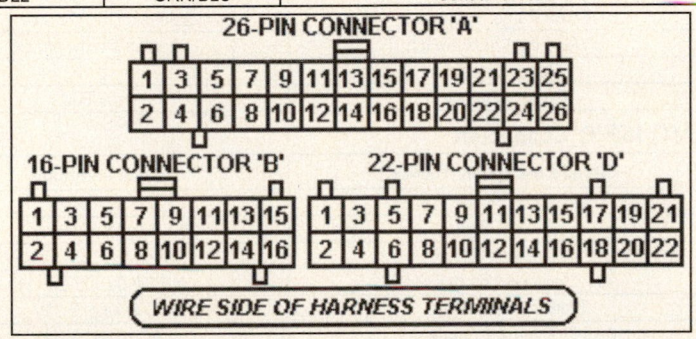

Abbreviation	Color	Abbreviation	Color	Abbreviation	Color
BLK	Black	LT BLU	Lt. Blue	TAN	Tan
BLU	Blue	LT GRN	Lt. Green	VIO	Violet
BRN	Brown	ORN	Orange	WHT	White
GRY	Gray	PNK	Pink	YEL	Yellow
GRN	Green	PPL	Purple		

1996-97 Sedan 2.7L V6 MFI VIN CE6 [A/T] 26P 'A' Connector

PCM Pin #	W/Color	Circuit Description (26-Pin)	Value at Hot Idle
A1	BRN	Injector 1 Control	2.0-3.3 ms
A2	RED	Injector 2 Control	2.0-3.3 ms
A3	BLU	Injector 3 Control	2.0-3.3 ms
A4	GRN/BLK	Fuel Pump Relay	Relay Off: 12v, On: 1v
A5	BLK/BLU	Idle Air Control Valve	Pulse Signals
A6	ORN/BLK	HO2S-11 (B1 S1) Heater	Digital Signals: 0-12-0v
A7	GRN/RED	Check Engine Light	MIL Off: 12v, On: 1v
A8	RED/BLU	A/C Clutch Relay	Relay Off: 12v, On: 1v
A9	RED	EGR Solenoid	Solenoid Off: 12v, On: 1v
A10	PNK/BLU	Intake Air Bypass Solenoid	Solenoid Off: 12v, On: 1v
A11	YEL/GRN	Igniter Control	Digital Signals: 0-12-0v
A12	BLK	Power Ground	<0.1v
A13	YEL/BLK	Main Relay Power (B+)	12-14v
A14	YEL	Injector 4 Control	2.0-3.3 ms
A15	BLK/RED	Injector 5 Control	2.0-3.3 ms
A16	WHT/BLU	Injector 6 Control	2.0-3.3 ms
A17	ORN/BLU	HO2S-12 (B1 S2) Heater	Digital Signals: 0-12-0v
A18	GRN/WHT	Engine Mount Solenoid	Solenoid Off: 12v, On: 1v
A19	GRN	Radiator Fan Relay	Relay Off: 12v, On: 1v
A20	---	Not Used	---
A21	WHT/GRN	Alternator Charging Signal	Lights Off: 12v, On: 0v
A22	BRN/WHT	A/T TCM Signal	Digital Signals
A23	RED/YEL	EVAP Purge Solenoid	Solenoid Off: 12v, On: 1v
A24	BRN	EVAP Purge Flow Switch	Switch on: 0v, off: 5v
A25	BLK	Power Ground	<0.1v
A26	BRN/BLK	Chassis Ground	<0.050v

1996-97 Sedan 2.7L V6 MFI VIN CE6 [A/T] 16P 'B' Connector

PCM Pin #	W/Color	Circuit Description (16-Pin)	Value at Hot Idle
B1	YEL/BLK	Main Relay Power (B+)	12-14v
B2	WHT/RED	A/T TCM Signal	Digital Signals
B3	RED/WHT	A/C Switch Signal	Switch Off: 12v, On: 1v
B4	LT GRN	A/T: Gear Position Indicator	In P/N: 0v, Others: 11v
B5	BLK/GRN	Starter Switch Signal	Cranking: 9-11v
B6	YEL	CYP Sensor Signal	AC pulse signals
B7	GRN	TDC Sensor Signal	AC pulse signals
B8	BLU	CKP Sensor Signal	AC pulse signals
B9	BRN/BLK	Chassis Ground	<0.050v
B10	GRN	A/T: TCM Signal	5v
B11	---	Not Used	---
B12	GRN	PSP Switch	Straight: 0v, Turning: 11v
B13	ORN	Vehicle Speed Sensor	Moving: pulse signals
B14	BLK	CYP Sensor Ground	<0.050v
B15	RED	TDC Sensor Ground	<0.050v
B16	WHT	CKP Sensor Ground	<0.050v

1996-97 Sedan 2.7L V6 MFI VIN CE6 [A/T] 22P 'D' Connector

PCM Pin #	W/Color	Circuit Description (22-Pin)	Value at Hot Idle
D1	WHT/YEL	Keep Alive Power (VBU)	12-14v
D2	RED/WHT	Rear HO2S Sensor Ground	<0.050v
D3	BLU/WHT	A/T: TCM Signal	5v
D4	LT GRN	Data Link Connector	5v
D5	WHT/RED	Alternator 'FR' Signal	Digital Signals: 0-5-0v
D6	RED/BLK	TP Sensor Signal	0.5-0.6v
D7	RED/WHT	ECT Sensor Signal	At 180ºF: 0.52v
D8	RED/YEL	IAT Sensor Signal	Varies w/temp. (0.5-4.9v)
D9	WHT/YEL	MAP Sensor Signal	0.8-0.9v
D10	YEL/WHT	MAP Sensor VREF	4.9-5.1v
D11	GRN/WHT	MAP Sensor Ground	<0.050v
D12	GRN/WHT	Brake Switch Signal	Brake Off: 0v, On: 12v
D13	RED	Service Check Connector	SCS Open: 4.80v
D14	RED/GRN	A/T: TCM Signal	5v
D15	---	Not Used	---
D16	GRN/RED	Electric Load Detector	Varies: 0.5-4.5v
D17	WHT/BLK	A/T: EGR Lift Sensor Signal	1.2v
D18	WHT/RED	HO2S-11 (B1 S1) Signal	0.1-1.1v
D19	WHT/RED	HO2S-12 (B1 S2) Signal	0.1-1.1v
D20	GRN/BLK	A/T: TCM Signal	5v
D21	YEL/BLU	Sensor VREF	4.9-5.1v
D22	GRN/BLU	Sensor Ground	<0.050v

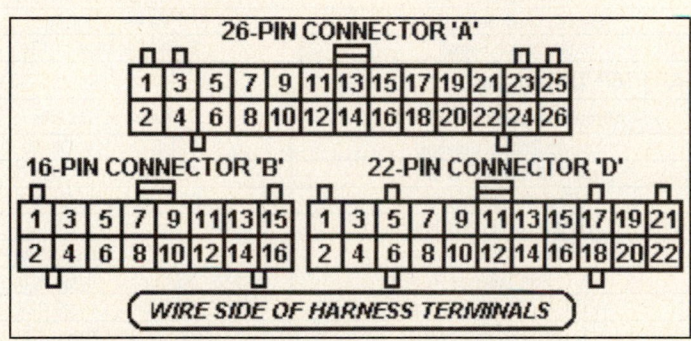

1998-99 Sedan 3.0L V6 VTEC VIN CG1 [A/T] 32P 'A' Connector

PCM Pin #	W/Color	Circuit Description (32-Pin)	Value at Hot Idle
A2	GRN/WHT	Engine Mount Solenoid	Solenoid Off: 12v, On: 1v
A3	BLU	EVAP Bypass Solenoid	Solenoid Off: 12v, On: 1v
A4	GRN/WHT	EVAP Vent Solenoid	Solenoid Off: 12v, On: 1v
A5	BLU/GRN	Cruise Control Signal	C/C On: pulse signals
A6	RED/YEL	EVAP Purge Solenoid	Solenoid Off: 12v, On: 1v
A8	BLK/WHT	HO2S-12 (B1 S2) Heater	Digital Signals: 0-12-0v
A9	BLU/WHT	A/T: Vehicle Speed Output	Moving: pulse signals
A10	BRN	Service Check Connector	SCS Open: 4.80v
A12	PNK	Immobilizer Indicator Lamp	Lamp Off: 12v, On: 1v
A13, A25	BLU, RED	Immobilizer Enable, Code	Digital Signals
A14	GRN/BLK	A/T: D4 Light Switch	D4 On: 0v, Off: 12v
A15	GRN/YEL	Immobilizer Fuel Pump Relay	Relay Off: 12v, On: 1v
A17	RED	A/C Clutch Relay Control	Relay Off: 12v, On: 1v
A18	GRN/ORN	Malfunction Indicator Light	MIL Off: 12v, On: 1v
A19	BLU	Engine Speed Pulse (NEP)	AC Pulse Signals
A20	GRN	Radiator Fan Relay Control	Relay Off: 12v, On: 1v
A21	GRY	K-Line Signal	12v
A23	WHT/RED	HO2S-12 (B1 S2) Signal	0.1-1.1v
A24	BLU/ORN	Starter Switch Signal	Cranking: 9-11v
A26	GRN	PSP Switch	Straight: 0v, Turning: 11v
A27	BLU/RED	A/C Switch Signal	Switch Off: 12v, On: 1v
A28	WHT/RED	Interlock Control Unit Signal	Key & Brake On: 12v
A29	LT GRN	Fuel Tank Pressure Sensor	Fuel Cap off: 2.5v
A30	GRN/RED	Electric Load Detector	Varies: 0.5-4.5v
A32	WHT/BLK	Brake Switch Signal	Brake Off: 0v, On: 12v

1998-99 Sedan 3.0L V6 VTEC VIN CG1 [A/T] 16P 'D' Connector

PCM Pin #	W/Color	Circuit Description (16-Pin)	Value at Hot Idle
D1	YEL	Lockup Control Solenoid	LUS On: 12v, Off: 0v
D2	GRN/WHT	A/T: Shift Solenoid 'B'	SSB in 3rd, 4th Gear: 0v
D3	GRN	A/T: Shift Solenoid 'C'	SSC in 2nd, 4th Gear: 0v
D5	BLK/YEL	VB Solenoid Feed (B+)	Sol. on: 12v, off: 0v
D6	WHT	A/T: Gear Position Switch	In 'R': 0v, Others: 12v
D7	BLU/YEL	A/T: Shift Solenoid 'A'	SSA in 1st, 4th Gear: 0v
D8	PNK	A/T: Gear Position Switch	In D3: 0v, Others: 12v
D9	YEL	A/T: Gear Position Switch	In D4: 0v, Others: 12v
D10	BLU	Countershaft Speed Sensor P	Moving: AC pulses
D11	RED	Mainshaft Speed Sensor 'P'	Moving: AC pulses
D12	WHT	Mainshaft Speed Sensor 'N'	Moving: AC pulses
D14	BLU	A/T: Gear Position Switch	In D2: 0v, Others: 12v
D15	BRN	A/T: Gear Position Switch	In D1: 0v, Others: 12v
D16	GRN	Countershaft Speed Sensor N	<0.050v

Connector pin layout diagram:

```
 1  2  3  4    5  6  7    8  9  10   11    1    2  3  4    5
12 13 14 15 16 17 18 19 20 21 22 23    24    6  7  8  9 10 11   12
32-PIN 25 26 27   28 29 30 31    32      13 14 15   16  16-PIN
  'A'                                               'D'
         WIRE SIDE OF HARNESS TERMINALS
```

1998-99 Sedan 3.0L V6 VTEC VIN CG1 [A/T] 25P 'B' Connector

PCM Pin #	W/Color	Circuit Description (25-Pin)	Value at Hot Idle
B1, B9	YEL/BLK	Main Relay Power (B+)	12-14v
B2, B10	BLK	Power Ground	<0.1v
B3	BLK/RED	Injector 5 Control	2.0-3.3 ms
B4, B5	YEL, RED	Injector 4, Injector 2 Control	2.0-3.3 ms
B6	WHT/BLU	Injector 6 Control	2.0-3.3 ms
B7	PNK	E-EGR Solenoid Control	Digital Signals: 0-12-0v
B8	WHT	A/T: Clutch Solenoid 'A-'	AC Pulse Signals
B11	BRN	Injector 1 Control	2.0-3.3 ms
B12	GRN/YEL	VTEC Control Solenoid	0v, Hi-Speed: 12v
B13	YEL/GRN	Ignition Control Signal	Digital Signals: 0-12-0v
B14	BLU/BLK	A/T: 2nd Clutch Pressure Sw.	Open: 12v, Closed: 1v
B15	BLU	Injector 3 Control	2.0-3.3 ms
B17	RED	A/T: Clutch Solenoid 'A+'	AC Pulse Signals
B18	GRN	A/T: Clutch Solenoid 'B-'	Pulse Signals
B20, B22	BRN/BLK	Logic Ground	<0.050v
B21	WHT/YEL	Keep Alive Power (VBU)	12-14v
B23	BLK/BLU	Idle Air Control Valve	Pulse Signals
B24	BLU/WHT	A/T: 3rd Clutch Pressure Sw.	Open: 12v, Closed: 1v
B25	ORN	A/T: Clutch Solenoid 'B+'	Pulse Signals

1998-99 Sedan 3.0L V6 VTEC VIN CG1 [A/T] 31P 'C' Connector

PCM Pin #	W/Color	Circuit Description (31-Pin)	Value at Hot Idle
C1	BLK/WHT	HO2S-11 (B1 S1) Heater	Digital Signals: 0-12-0v
C2	WHT/GRN	Alternator Charging Signal	Lights Off: 12v, On: 0v
C3	WHT/BLU	Alternator Charging Signal	Lights Off: 12v, On: 0v
C5	WHT/RED	Alternator 'FR' Signal	Digital Signals: 0-5-0v
C6	WHT/BLK	EGR Valve Lift Sensor	1.2v
C7, C18	GRN/WHT	MAP Sensor, Sensor Ground	<0.050v
C8	BLU	CKP Sensor Signal	AC pulse signals
C9	WHT	CKP Sensor Ground	<0.050v
C10	BLU/BLK	VTEC Pressure Switch	0v, Hi-Speed: 12v
C16	WHT	HO2S-11 (B1 S1) Signal	0.1-1.1v
C17	RED/GRN	MAP Sensor Signal	0.8-0.9v
C19, C28	YEL/RED	MAP VREF, Sensor VREF	4.9-5.1v
C20	GRN	TDC1 Sensor Signal	AC pulse signals
C21	RED	TDC1 Sensor Ground	<0.050v
C25	RED/YEL	IAT Sensor Signal	Varies w/temp. (0.5-4.9v)
C26	RED/WHT	ECT Sensor Signal	At 180°F: 0.5-0.6v
C27	RED/BLK	TP Sensor Signal	0.5-0.6v
C29	YEL	TDC2 Sensor Signal	AC pulse signals
C30	BLK	TDC2 Sensor Ground	<0.050v

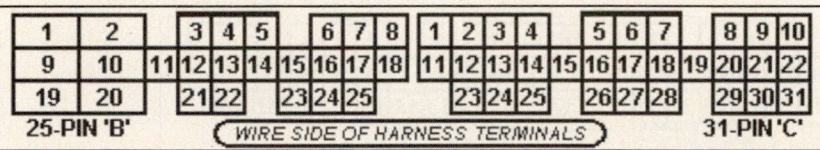

2000-03 Sedan 3.0L V6 VTEC VIN CG1 [A/T] 32P 'A' Connector

PCM Pin #	W/Color	Circuit Description (32-Pin)	Value at Hot Idle
A1	YEL/GRN	ECT Signal to TCM	Digital Signals
A2	GRN/WHT	A/T: Engine Mount Solenoid	Solenoid Off: 12v, On: 1v
A3	BLU	EVAP Bypass Solenoid	Solenoid Off: 12v, On: 1v
A4	GRN/WHT	EVAP Vent Solenoid	Solenoid Off: 12v, On: 1v
A5	BLU/GRN	Cruise Control Signal	C/C On: pulse signals
A6	RED/YEL	EVAP Purge Solenoid	Solenoid Off: 12v, On: 1v
A7, 11, 13	WT,GN,BL	ABS-TCS Signals	Digital Signals
A8	BLK/WHT	HO2S-12 (B1 S2) Heater	Digital Signals: 0-12-0v
A9	BLU/WHT	A/T: Vehicle Speed Output	Moving: pulse signals
A10	BRN	Service Check Connector	SCS Open: 4.80v
A12	PNK	Immobilizer Indicator Lamp	Lamp Off: 12v, On: 1v
A14	GRN/BLK	A/T: D4 Light Switch	D4 On: 0v, Off: 12v
A15	GRN/YEL	Immobilizer Fuel Pump Relay	Relay Off: 12v, On: 1v
A16, A25	BLU, RED	Immobilizer Enable, Code	Digital Signals
A17	RED	A/C Clutch Relay Control	Relay Off: 12v, On: 1v
A18	GRN/ORN	Malfunction Indicator Light	MIL Off: 12v, On: 1v
A19	BLU	Engine Speed Pulse (NEP)	AC Pulse Signals
A20	GRN	Radiator Fan Relay Control	Relay Off: 12v, On: 1v
A21	GRY	K-Line Signal	12v
A22, A31	PNK, RED	ABS-TCS Signals	Digital Signals
A23	WHT/RED	HO2S-12 (B1 S2) Signal	0.1-1.1v
A24	BLU/ORN	Starter Switch Signal	Cranking: 9-11v
A26	GRN	PSP Switch	Straight: 0v, Turning: 11v
A27	BLU/RED	A/C Switch Signal	Switch Off: 12v, On: 1v
A28	WHT/RED	Interlock Control Unit Signal	Key & Brake On: 12v
A29	LT GRN	Fuel Tank Pressure Sensor	Fuel Cap off: 2.5v
A30	GRN/RED	Electric Load Detector	Varies: 0.5-4.5v
A32	WHT/BLK	Brake Switch Signal	Brake Off: 0v, On: 12v

2000-03 Sedan 3.0L V6 VTEC VIN CG1 [A/T] 16P 'D' Connector

PCM Pin #	W/Color	Circuit Description (16-Pin)	Value at Hot Idle
D1	YEL	Lockup Control Solenoid	LUS On: 12v, Off: 0v
D2	GRN/WHT	A/T: Shift Solenoid 'B'	SSB in 3rd, 4th Gear: 0v
D3	GRN	A/T: Shift Solenoid 'C'	SSC in 2nd, 4th Gear: 0v
D5	BLK/YEL	VB Solenoid Feed (B+)	Sol. on: 12v, off: 0v
D6	WHT	A/T: Gear Position Switch	In 'R': 0v, Others: 12v
D7	BLU/YEL	A/T: Shift Solenoid 'A'	SSA in 1st, 4th Gear: 0v
D8, D9	PNK, YEL	A/T: Gear Position Switches	In D3: 0v, In D4: 0v
D10, D16	BLU, GRN	A/T: Countershaft Speed P, N	Moving: AC pulses
D11, D12	RED, WHT	Mainshaft Speed Sensor P, N	Moving: AC pulses
D13	BLU/BLK	A/T: 2nd Clutch Pressure Sw.	Open: 12v, Closed: 1v
D14, 15	BLU, BRN	A/T: Gear Position Switch	D2: 0v, D1: 0v

```
 ┌──────────────────────────────────────────────────────────────┐
 │ │1│2│3│4│ │5│6│7│ │8│9│ 10 │ 11 │  │1│ │2│3│4│ │ 5 │        │
 │ │12│13│14│15│16│17│18│19│20│21│22│ 23 │ 24 │  │6│7│8│9│10│11│ 12 │ │
 │ 32-PIN │25│26│27│ │28│29│30│31│ 32 │        │13│14│15│ 16 │ 16-PIN │
 │  'A'                                                   'D'     │
 │        ⟨  WIRE SIDE OF HARNESS TERMINALS  ⟩                    │
 └──────────────────────────────────────────────────────────────┘
```

2000-03 Sedan 3.0L V6 VTEC VIN CG1 [A/T] 25P 'B' Connector

PCM Pin #	W/Color	Circuit Description (25-Pin)	Value at Hot Idle
B1, B9	YEL/BLK	Main Relay Power (B+)	12-14v
B2, B10	BLK	Power Ground	<0.1v
B3	BLK/RED	Injector 5 Control	2.0-3.3 ms
B4, B5	YEL, RED	Injector 4, Injector 2 Control	2.0-3.3 ms
B6	WHT/BLU	Injector 6 Control	2.0-3.3 ms
B7	PNK	E-EGR Solenoid Control	Digital Signals: 0-12-0v
B8	WHT	A/T: Clutch Solenoid 'A-'	AC Pulse Signals
B11	BRN	Injector 1 Control	2.0-3.3 ms
B12	GRN/YEL	VTEC Control Solenoid	0v, Hi-Speed: 12v
B14	BLU/BLK	A/T: Gear Position Switch	In P/N: 0v, Others: 12v
B15	BLU	Injector 3 Control	2.0-3.3 ms
B17	RED	A/T: Clutch Solenoid 'A+'	AC Pulse Signals
B18	GRN	A/T: Clutch Solenoid 'B-'	Pulse Signals
B20, B22	BRN/BLK	Logic Ground (LG1, LG2)	<0.050v
B21	WHT/YEL	Keep Alive Power (VBU)	12-14v
B23	BLK/BLU	Idle Air Control Valve	Pulse Signals
B24	BLU/WHT	A/T: 3rd Clutch Pressure Sw.	Open: 12v, Closed: 1v
B25	ORN	A/T: Clutch Solenoid 'B+'	Pulse Signals

2000-03 Sedan 3.0L V6 VTEC VIN CG1 [A/T] 31P 'C' Connector

PCM Pin #	W/Color	Circuit Description (31-Pin)	Value at Hot Idle
C1	BLK/WHT	HO2S-11 (B1 S1) Heater	Digital Signals: 0-12-0v
C2	WHT/GRN	Alternator Charging Signal	Lights Off: 12v, On: 0v
C3	WHT/BLU	Ignition Coil 3 Control	Digital Signals: 0-12-0v
C4	YEL/GRN	Ignition Coil 1 Control	Digital Signals: 0-12-0v
C5	WHT/RED	Alternator 'FR' Signal	Digital Signals: 0-5-0v
C6	WHT/BLK	EGR Valve Lift Sensor	1.2v
C7, C18	GRN/WHT	MAP Sensor, Sensor Ground	<0.050v
C8, C9	BLU, WHT	CKP Sensor Signal, Ground	AC pulse signals
C10	BLU/BLK	VTEC Pressure Switch	0v, Hi-Speed: 12v
C12	BLK/RED	Ignition Coil 5 Control	Digital Signals: 0-12-0v
C13	BRN	Ignition Coil 4 Control	Digital Signals: 0-12-0v
C14	BLU/RED	Ignition Coil 2 Control	Digital Signals: 0-12-0v
C16	WHT	HO2S-11 (B1 S1) Signal	0.1-1.1v
C17	RED/GRN	MAP Sensor Signal	0.8-0.9v
C19, C28	YEL/RED	MAP VREF, Sensor VREF	4.9-5.1v
C20, C21	GRN, RED	TDC1 Sensor Signal, Ground	AC pulse signals
C23	BRN/WHT	Ignition Coil 6 Control	Digital Signals: 0-12-0v
C25	RED/YEL	IAT Sensor Signal	Varies w/temp. (0.5-4.9v)
C26	RED/WHT	ECT Sensor Signal	At 180°F: 0.5-0.6v
C27	RED/BLK	TP Sensor Signal	0.5-0.6v
C29, C30	YEL, BLK	TDC2 Sensor Signal, Ground	AC pulse signals

25-PIN 'B' — WIRE SIDE OF HARNESS TERMINALS — 31-PIN 'C'

1991-93 Wagon 2.2L I4 MFI VIN CB9 [All] 26P 'A' Connector

PCM Pin #	W/Color	Circuit Description (26-Pin)	Value at Hot Idle
A1	BRN	Injector 1 Control	2.0-3.3 ms
A2	YEL	Injector 4 Control	2.0-3.3 ms
A3	RED	Injector 2 Control	2.0-3.3 ms
A4	---	Not Used	---
A5	BLU	Injector 3 Control	2.0-3.3 ms
A6	ORN	HO2S Heater Control	Relay Off: 12v, On: 1v
A7	GRN/BLK	Fuel Pump Relay	Relay Off: 12v, On: 1v
A8	GRN/BLK	Fuel Pump Relay	Relay Off: 12v, On: 1v
A9	BLK/BLU	Electronic Air Control Valve	Pulse Signals
A10	---	Not Used	---
A11	PNK	EGR Control Solenoid	1.2v
A12	BLU	Radiator Fan Relay	Relay Off: 12v, On: 1v
A13	GRN/RED	Check Engine Light	MIL Off: 12v, On: 1v
A14, 17	---	Not Used	---
A15	RED/BLU	A/C Clutch Relay	Relay Off: 12v, On: 1v
A16	WHT/GRN	Alternator Charging Signal	Lights Off: 12v, On: 0v
A18	BRN/WHT	A/T: Control Unit	N/A
A19	WHT	Intake Air Regulator Control	Solenoid Off: 12v, On: 1v
A20	RED/GRN	EVAP Purge Solenoid	Solenoid Off: 12v, On: 1v
A21	YEL/GRN	Igniter Control	Hot idle 10% d/cycle
A22	YEL/GRN	Igniter Control	Hot idle 10% d/cycle
A23	BLK	Power Ground	<0.1v
A24	BLK	Power Ground	<0.1v
A25	YEL/BLK	Ignition Power	12-14v
A26	BLK/RED	Chassis Ground	<0.050v

1991-93 Wagon 2.2L I4 MFI VIN CB9 [All] 16P 'B' Connector

PCM Pin #	W/Color	Circuit Description (16-Pin)	Value at Hot Idle
B1	YEL/BLK	Main Relay Power (B+)	12-14v
B2	BRN/BLK	Power Ground	<0.1v
B3	WHT/RED	A/T: Control Unit	---
B4	GRN	A/T: Control Unit	---
B5	BLU/BLK	A/C Switch Signal	Switch Off: 12v, On: 1v
B6	---	Not Used	---
B7	LT GRN	A/T: Neutral Position Switch	In N: 0v, Others: 12v
B8	RED	PSP Switch	Straight: 0v, Turning: 11v
B9	BLU/RED	A/T: Start Signal	Cranking: 9-11v
B9	BLK/GRN	M/T: Start Signal	Cranking: 9-11v
B10	ORN	Vehicle Speed Sensor	Moving: pulse signals
B11	ORN	CYP Sensor Signal	AC pulse signals
B12	WHT	CYP Sensor Ground	<0.050v
B13	ORN/BLU	TDC Sensor Signal	AC pulse signals
B14	WHT/BLU	TDC Sensor Ground	<0.050v
B15	BLU/GRN	CKP Sensor Signal	<AC pulse signals
B16	BLU/YEL	CKP Sensor Ground	<0.050v

1991-93 Wagon 2.2L I4 MFI VIN CB9 [All] 22P 'D' Connector

PCM Pin #	W/Color	Circuit Description (22-Pin)	Value at Hot Idle
D1	WHT/YEL	Keep Alive Power (VBU)	12-14v
D2	GRN/WHT	Brake Switch Signal	Brake Off: 0v, On: 12v
D3	---	Not Used	---
D4	ORN/RED	Service Check Connector	SCS Open: 4.80v
D5-D8	---	Not Used	---
D9	WHT/RED	Alternator 'FR' Signal	Digital Signals: 0-5-0v
D10	GRN/RED	Electric Load Detector	Varies: 0.5-4.5v
D11	RED/BLK	TP Sensor Signal	0.5-0.6v
D12	WHT/BLK	A/T: EGR Lift Sensor Signal	1.1-1.2v
D13	YEL/GRN	ECT Sensor Signal	At 180°F: 0.52v
D14	WHT	Oxygen Sensor	0.1-1.1v
D15	RED/YEL	IAT Sensor Signal	Varies w/temp. (0.5-4.9v)
D16	---	Not Used	---
D17	WHT/BLU	MAP Sensor Signal	0.8-0.9v
D18	GRN/BLK	A/T: Control Unit	---
D19	RED/WHT	MAP Sensor VREF	4.9-5.1v
D20	YEL/WHT	Sensor VREF	4.9-5.1v
D21	BLU/WHT	Sensor Ground	<0.050v
D22	GRN/WHT	Sensor Ground	<0.050v

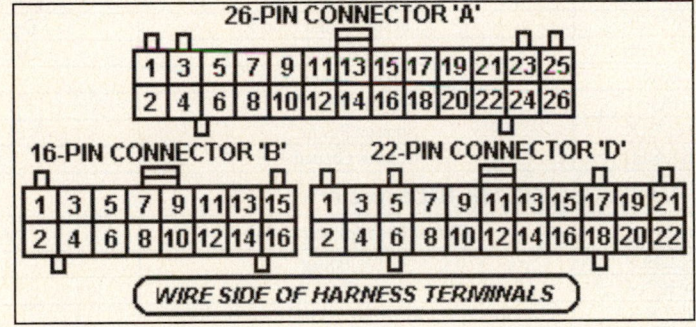

1994-95 Wagon 2.2L I4 MFI VIN CE1 [All] 26P 'A' Connector

PCM Pin #	W/Color	Circuit Description (26-Pin)	Value at Hot Idle
A1	BRN	Injector 1 Control	2.0-3.3 ms
A2	YEL	Injector 4 Control	2.0-3.3 ms
A3	RED	Injector 2 Control	2.0-3.3 ms
A4	---	Not Used	---
A5	BLU	Injector 3 Control	2.0-3.3 ms
A6	RED	EGR Control Solenoid	Solenoid Off: 12v, On: 1v
A7	GRN/BLK	Fuel Pump Relay	Relay Off: 12v, On: 1v
A8	GRN/BLK	Fuel Pump Relay	Relay Off: 12v, On: 1v
A9	BLK/BLU	Idle Air Control Motor	Pulse Signals
A10	GRN/WHT	Engine Mount Solenoid	Solenoid Off: 12v, On: 1v
A11	GRN/BLK	HO2S Heater Control	Relay Off: 12v, On: 1v
A12	GRN	Radiator Fan Relay	Relay Off: 12v, On: 1v
A13	GRN/RED	Check Engine Light	MIL Off: 12v, On: 1v
A14	---	Not Used	---
A15	RED/BLU	A/C Clutch Relay	Relay Off: 12v, On: 1v
A16	WHT/GRN	Alternator Charging Signal	Lights Off: 12v, On: 0v
A17	---	Not Used	---
A18	BRN/WHT	A/T: Control Unit	---
A19	ORN/GRN	Intake Air Regulator Control	Solenoid Off: 12v, On: 1v
A20	RED/YEL	EVAP Purge Solenoid	Solenoid Off: 12v, On: 1v
A21	YEL/GRN	Igniter Control	Digital Signals: 0-12-0v
A22	YEL/GRN	Igniter Control	Digital Signals: 0-12-0v
A23	BLK	Power Ground	<0.1v
A24	BLK	Power Ground	<0.1v
A25	YEL/BLK	Ignition Power	12v
A26	BLK/RED	Chassis Ground	<0.050v

1994-95 Wagon 2.2L I4 MFI VIN CE1 [All] 16P 'B' Connector

PCM Pin #	W/Color	Circuit Description (16-Pin)	Value at Hot Idle
B1	YEL/BLK	Ignition Power	12-14v
B2	BRN/BLK	Power Ground	<0.1v
B3	WHT/RED	A/T: Control Unit	---
B4	GRN	A/T: Control Unit	---
B5	RED/WHT	A/C Switch Signal	Switch Off: 12v, On: 1v
B6	---	Not Used	---
B7	LT GRN	A/T: Park Neutral Switch	In P/N: 0v, Others: 11v
B8	GRN	PSP Switch	Straight: 0v, Turning: 11v
B9	BLU/RED	A/T: Start Signal	Cranking: 9-11v
B9	BLU/RED	M/T: Start Signal	Cranking: 9-11v
B10	ORN	Vehicle Speed Sensor	Moving: pulse signals
B11	ORN	CYP Sensor Signal	AC pulse signals
B12	WHT	CYP Sensor Ground	<0.050v
B13	ORN/BLU	TDC Sensor Signal	AC pulse signals
B14	WHT/BLU	TDC Sensor Ground	<0.050v
B15	BLU/GRN	CKP Sensor Signal	<AC pulse signals
B16	BLU/YEL	CKP Sensor Ground	<0.050v

1994-95 Wagon 2.2L I4 MFI VIN CE1 [All] 22P 'D' Connector

PCM Pin #	W/Color	Circuit Description (22-Pin)	Value at Hot Idle
D1	WHT/YEL	Keep Alive Power (VBU)	12-14v
D2	GRN/WHT	Brake Switch Signal	Brake Off: 0v, On: 12v
D3	---	Not Used	---
D4	ORN/RED	Service Check Connector	SCS Open: 4.80v
D5	BLU/WHT	A/T: TCM Signal	Digital Signals
D6	---	Not Used	---
D7	GRN/RED	Data Link Connector	5v
D8	---	Not Used	---
D9	WHT/RED	Alternator 'FR' Signal	Digital Signals: 0-5-0v
D10	GRN/RED	Electric Load Detector	Varies: 0.5-4.5v
D11	RED/BLK	TP Sensor Signal	0.5-0.6v
D12	WHT/BLK	A/T: EGR Lift Sensor Signal	1.1-1.2v
D13	RED/WHT	ECT Sensor Signal	At 180ºF: 0.52v
D14	WHT/RED	HO2S Signal	0.1-1.1v
D15	RED/YEL	IAT Sensor Signal	Varies w/temp. (0.5-4.9v)
D16	---	Not Used	---
D17	WHT/YEL	MAP Sensor Signal	0.8-0.9v
D18	GRN/BLK	A/T: Control Unit	---
D19	YEL/WHT	MAP Sensor VREF	4.9-5.1v
D20	YEL/BLU	Sensor VREF	4.9-5.1v
D21	GRN/WHT	MAP Sensor Ground	<0.050v
D22	GRN/BLU	Sensor Ground	<0.050v

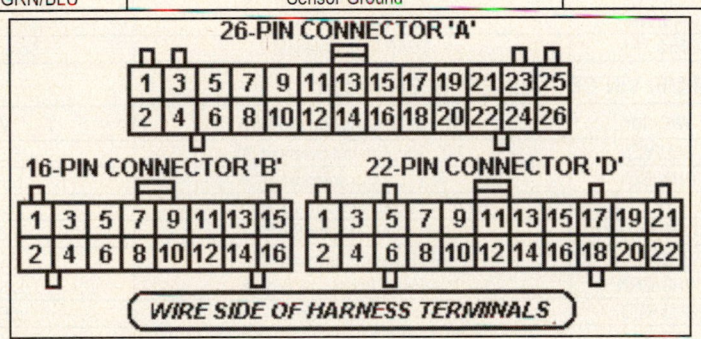

1996-97 Wagon 2.2L I4 MFI VIN CE1 [All] 32P 'A' Connector

PCM Pin #	W/Color	Circuit Description (32-Pin)	Value at Hot Idle
A1	YEL	Injector 4 Control	2.0-3.3 ms
A2	BLU	Injector 3 Control	2.0-3.3 ms
A3	RED	Injector 2 Control	2.0-3.3 ms
A4	BRN	Injector 1 Control	2.0-3.3 ms
A5	ORN/BLU	HO2S-12 (B1 S2) Heater	Digital Signals: 0-12-0v
A6	ORN/BLK	HO2S-11 (B1 S1) Heater	Digital Signals: 0-12-0v
A7	RED	EGR Solenoid	Solenoid on: 1v, off: 12v
A9, 22	BRN/BLK	Sensor Ground	<0.050v
A10	BLK, BLK	Power Ground	<0.1v
A11	YEL/BLK	Main Relay Power (B+)	12-14v
A12	BLK/BLU	Idle Air Control Valve Signal	Pulse Signals
A13	GRN/WHT	Engine Mount Solenoid	Solenoid Off: 12v, On: 1v
A15	RED/YEL	EVAP Purge Solenoid	Solenoid Off: 12v, On: 1v
A16	GRN/BLK	Fuel Pump Relay	Relay Off: 12v, On: 1v
A17	RED/BLU	A/C Clutch Relay Control	Relay Off: 12v, On: 1v
A18	GRN/RED	Check Engine Light	MIL Off: 12v, On: 1v
A19	WHT/GRN	Alternator Charging Signal	Lights Off: 12v, On: 0v
A20	YEL/GRN	Igniter Control	At idle: 5° dwell
A23	BLK	Power Ground	<0.1v
A24	YEL/BLK	Main Relay Power (B+)	12-14v
A25	ORN/GRN	Intake Air Resonator Control	Solenoid Off: 12v, On: 1v
A27	GRN	Radiator Fan Relay Control	Relay Off: 12v, On: 1v
A28	GRN/WHT	EVAP Bypass Solenoid	Solenoid Off: 12v, On: 1v
A29	ORN/GRN	EVAP Vent Solenoid	Solenoid Off: 12v, On: 1v

1996-97 Wagon 2.2L I4 MFI VIN CE1 [All] 25P 'B' Connector

PCM Pin #	W/Color	Circuit Description (25-Pin)	Value at Hot Idle
B3	BLU/YEL	A/T: Shift Control Solenoid 'A'	SSA on: 1v, off: 12v
B4	GRN/BLK	A/T: Lockup solenoid 'B'	LSB On: 12v, Off: 0v
B5	YEL	A/T: Lockup solenoid 'A'	LSA On: 12v, Off: 0v
B8	GRN/BLU	A/T: Gear Position Switch	In D3: 0v, Others: 12v
B11	BLU/YEL	Shift Control Solenoid 'B'	SSB on: 1v, off: 12v
B12	WHT/GRN	Interlock Control Unit Signal	Key & Brake On: 12v
B13	BLU/RED	A/T: D4 Light Switch	In D4: 0v, Others: 12v
B14	WHT/BLU	Mainshaft Speed Ground 'N'	<0.050v
B15	ORN/BLU	Mainshaft Speed Sensor 'P'	Moving: AC pulses
B16	GRN/RED	A/T: Gear Position Switch	In 'R': 0v, Others: 12v
B17, 18	GRN/YEL	A/T: Gear Position Switch	In D2, D1: 0v
B22	BLU/YEL	Countershaft Speed Sensor P	Moving: AC pulses
B23	BLU/YEL	Countershaft Speed Sensor N	<0.050v
B24	GRN/BLU	A/T: Gear Position Switch	In D4: 0v
B25	LT GRN	A/T: Gear Position Switch	In P/N: 0v, Others: 12v

1996-97 Wagon 2.2L I4 MFI VIN CE1 [All] 31P 'C' Connector

PCM Pin #	W/Color	Circuit Description (31-Pin)	Value at Hot Idle
C1	---	Not Used	---
C2	BLU	CYP Sensor	AC pulse signals
C3	GRN	TDC Sensor	AC pulse signals
C4	YEL	CKP Sensor	AC pulse signals
C5	RED/WHT	A/C Switch Signal	Switch Off: 12v, On: 1v
C6	BLU/RED	Starter Switch Signal	Cranking: 9-11v
C7	RED	Service Check Connector	SCS Open: 4.80v
C8	LT GRN	K-Line Signal	12v
C9	---	Not Used	---
C10	WHT/YEL	Keep Alive Power (VBU)	12-14v
C11	---	Not Used	---
C12	WHT	CKP Sensor Ground	<0.050v
C13	RED	TDC Sensor Ground	<0.050v
C14	BLK	CYP Sensor Ground	<0.050v
C15	---	Not Used	---
C16	GRN	PSP Switch	Straight: 0v, Turning: 11v
C17	WHT/RED	Alternator 'FR' Signal	Digital Signals: 0-5-0v
C18	ORN	Vehicle Speed Sensor	Moving: pulse signals
C19	---	Not Used	---
C20	BRN	EVAP Purge Flow Switch	Switch on: 0v, off: 5v
C21-31	---	Not Used	---

1996-97 Wagon 2.2L I4 MFI VIN CE1 [All] 16P 'D' Connector

PCM Pin #	W/Color	Circuit Description (16-Pin)	Value at Hot Idle
D1	RED/BLK	TP Sensor Signal	0.5-0.6v
D2	RED/WHT	ECT Sensor Signal	At 180ºF: 0.52v
D3	WHT/YEL	MAP Sensor Signal	0.8-0.9v
D4	YEL/WHT	MAP Sensor VREF	4.9-5.1v
D5	GRN/WHT	Brake Switch Signal	Brake Off: 0v, On: 12v
D7	WHT/RED	HO2S-11 (B1 S1) Signal	0.1-1.1v
D8	RED/YEL	IAT Sensor Signal	Varies w/temp. (0.5-4.9v)
D9	WHT/BLK	EGR Valve Lift Sensor	1.2v
D10	YEL/BLU	Sensor VREF	4.9-5.1v
D11	GRN/BLU	Sensor Ground	<0.050v
D12	GRN/WHT	MAP Sensor Ground	<0.050v
D13	RED/WHT	HO2S-12 Ground	<0.050v
D14	WHT/RED	HO2S-12 (B1 S2) Signal	0.1-1.1v
D15	GRN/RED	Fuel Tank Pressure Sensor	Fuel Cap off: 2.5v
D16	GRN/RED	Electric Load Detector	Varies: 0.5-4.5v

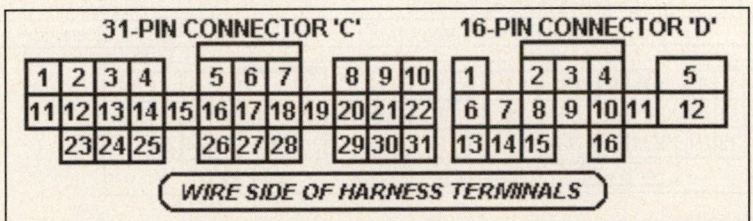

1996-97 Wagon 2.2L I4 VTEC VIN CE1 [All] 32P 'A' Connector

PCM Pin #	W/Color	Circuit Description (32-Pin)	Value at Hot Idle
A1	YEL	Injector 4 Control	2.0-3.3 ms
A2	BLU	Injector 3 Control	2.0-3.3 ms
A3	RED	Injector 2 Control	2.0-3.3 ms
A4	BRN	Injector 1 Control	2.0-3.3 ms
A5	ORN/BLU	HO2S-12 (B1 S2) Heater	Digital Signals: 0-12-0v
A6	ORN/BLK	HO2S-11 (B1 S1) Heater	Digital Signals: 0-12-0v
A7	RED	EGR Solenoid	Solenoid Off: 12v, On: 1v
A8	GRN/YEL	VTEC Control Solenoid	0v, Hi-Speed: 12v
A9, 22	BRN/BLK	Sensor Ground	<0.050v
A10, 23	BLK, BLK	Power Ground	<0.1v
A11, 24	YEL/BLK	Main Relay Power (B+)	12-14v
A12	BLK/BLU	Idle Air Control Valve Signal	Pulse Signals
A13	GRN/WHT	Engine Mount Solenoid	Solenoid Off: 12v, On: 1v
A15	RED/YEL	EVAP Purge Solenoid	Solenoid Off: 12v, On: 1v
A16	GRN/BLK	Fuel Pump Relay	Relay Off: 12v, On: 1v
A17	RED/BLU	A/C Clutch Relay Control	Relay Off: 12v, On: 1v
A18	GRN/RED	Check Engine Light	MIL Off: 12v, On: 1v
A19	WHT/GRN	Alternator Charging Signal	Lights Off: 12v, On: 0v
A20	YEL/GRN	Igniter Control	At idle: 5° dwell
A25	ORN/GRN	Intake Air Resonator Control	Solenoid Off: 12v, On: 1v
A27	GRN	Radiator Fan Relay Control	Relay Off: 12v, On: 1v
A28	GRN/WHT	EVAP Bypass Solenoid	Solenoid Off: 12v, On: 1v
A29	ORN/GRN	EVAP Vent Solenoid	Solenoid Off: 12v, On: 1v

1996-97 Wagon 2.2L I4 VTEC VIN CE1 [All] 25P 'B' Connector

PCM Pin #	W/Color	Circuit Description (25-Pin)	Value at Hot Idle
B3	BLU/YEL	Shift Control Solenoid 'A'	SSA on: 1v, off: 12v
B4	GRN/BLK	Lockup Control Solenoid 'B'	SSB on: 1v, off: 12v
B5	YEL	Lockup Control Solenoid 'A'	LSA on: 1v, off: 12v
B8, 24	GRN/BLU	A/T: Gear Position Switch	In D3: 0v, in D4: 0v
B11	GRN/WHT	Shift Control Solenoid 'B'	Solenoid Off: 12v, On: 1v
B12	WHT/GRN	Interlock Control Unit Signal	Key & Brake On: 12v
B13	BLU/RED	A/T: D4 Light Switch	In D4: 0v, Off: 5v
B14	WHT/BLU	Mainshaft Speed Ground 'N'	<0.050v
B15	ORN/BLU	Mainshaft Speed Signal 'P'	Moving: AC pulses
B16	GRN/RED	A/T: Gear Position Switch	In 'R': 0v, Others: 12v
B17	GRN/YEL	A/T: Gear Position Switch	In D2: 0v, Others: 12v
B18	GRN/YEL	A/T: Gear Position Switch	In D1: 0v, Others: 12v
B22	BLU/YEL	Countershaft Speed Sensor P	Moving: AC pulses
B23	BLU/GRN	Countershaft Speed Sensor N	<0.050v
B25	LT GRN	A/T: Gear Position Switch	In P/N: 0v, Others: 12v

32-PIN 'A'   *WIRE SIDE OF HARNESS TERMINALS*   25-PIN 'B'

1996-97 Wagon 2.2L I4 VTEC VIN CE1 [All] 31P 'C' Connector

PCM Pin #	W/Color	Circuit Description (31-Pin)	Value at Hot Idle
C2	BLU	CYP Sensor Signal	AC pulse signals
C3	GRN	TDC Sensor Signal	AC pulse signals
C4	YEL	CKP Sensor Signal	AC pulse signals
C5	RED/WHT	A/C Switch Signal	Switch Off: 12v, On: 1v
C6	BLU/RED	Starter Switch Signal	Cranking: 9-11v
C7	RED	Service Check Connector	SCS Open: 4.80v
C8	LT GRN	K-Line Signal	12v
C10	WHT/YEL	Keep Alive Power (VBU)	12-14v
C12	WHT	CKP Sensor Ground	<0.050v
C13	RED	TDC Sensor Ground	<0.050v
C14	BLK	CYP Sensor Ground	<0.050v
C16	GRN	PSP Switch	Straight: 0v, Turning: 11v
C17	WHT/RED	Alternator 'FR' Signal	Digital Signals: 0-5-0v
C18	ORN	Vehicle Speed Sensor	Moving: pulse signals
C20	BRN	EVAP Purge Flow Switch	Switch on: 0v, off: 5v

1996-97 Wagon 2.2L I4 VTEC VIN CE1 [All] 16P 'D' Connector

PCM Pin #	W/Color	Circuit Description (16-Pin)	Value at Hot Idle
D1	RED/BLK	TP Sensor Signal	0.5-0.6v
D2	RED/WHT	ECT Sensor Signal	At 180ºF: 0.52v
D3	WHT/YEL	MAP Sensor Signal	0.8-0.9v
D4	YEL/WHT	MAP Sensor VREF	4.9-5.1v
D5	GRN/WHT	Brake Switch Signal	Brake Off: 0v, On: 12v
D7	WHT/RED	HO2S-11 (B1 S1) Signal	0.1-1.1v
D8	RED/YEL	IAT Sensor Signal	Varies w/temp. (0.5-4.9v)
D9	WHT/BLK	EGR Valve Lift Sensor	1.2v
D10	YEL/BLU	Sensor VREF	4.9-5.1v
D11	GRN/BLU	Sensor Ground	<0.050v
D12	GRN/WHT	MAP Sensor Ground	<0.050v
D13	RED/WHT	HO2S-12 Ground	<0.050v
D14	WHT/RED	HO2S-12 (B1 S2) Signal	0.1-1.1v
D15	GRN/RED	Fuel Tank Pressure Sensor	Fuel Cap off: 2.5v
D16	GRN/RED	Electric Load Detector	Varies: 0.5-4.5v

```
      31-PIN CONNECTOR 'C'        16-PIN CONNECTOR 'D'
   ┌──┬──┬──┬──┐ ┌──┬──┬──┐ ┌──┬──┬──┐ ┌──┐ ┌──┬──┬──┐ ┌──┐
   │ 1│ 2│ 3│ 4│ │ 5│ 6│ 7│ │ 8│ 9│10│ │ 1│ │ 2│ 3│ 4│ │ 5│
   ├──┼──┼──┼──┼─┼──┼──┼──┼─┼──┼──┼──┤ ├──┼──┼──┼──┼──┼──┼──┤
   │11│12│13│14│15│16│17│18│19│20│21│22│ │ 6│ 7│ 8│ 9│10│11│12│
   ├──┼──┼──┼─┴┴──┼──┼──┼─┴┴──┼──┼──┤ ├──┼──┼──┼─┴──┤
   │23│24│25│   │26│27│28│   │29│30│31│ │13│14│15│  │16│
   └──┴──┴──┘   └──┴──┴──┘   └──┴──┴──┘ └──┴──┴──┘  └──┘
        WIRE SIDE OF HARNESS TERMINALS
```

Civic Pin Tables

1994-95 Coupe 1.6L I4 VTEC VIN EJ1 26P 'A' Connector

PCM Pin #	W/Color	Circuit Description (26-Pin)	Value at Hot Idle
A1	BRN	Injector 1 Control	2.0-3.3 ms
A2	YEL	Injector 4 Control	2.0-3.3 ms
A3	RED	Injector 2 Control	2.0-3.3 ms
A4	ORN/WHT	VTEC Solenoid	0v, Hi-Speed: 12v
A5	BLU	Injector 3 Control	2.0-3.3 ms
A6	ORN/BLK	HO2S Heater Control	Relay Off: 12v, On: 1v
A7	GRN/YEL	Fuel Pump Relay	Relay Off: 12v, On: 1v
A8	---	Not Used	---
A9	GRN/WHT	Idle Air Control Valve	Pulse Signals
A10-11	---	Not Used	---
A12	YEL/GRN	Radiator Fan Relay	Relay Off: 12v, On: 1v
A13	GRN/ORN	Check Engine Light	MIL Off: 12v, On: 1v
A14	---	Not Used	---
A15	BLK/RED	A/C Clutch Relay	Relay Off: 12v, On: 1v
A16	WHT/YEL	Alternator Charging Signal	Lights Off: 12v, On: 0v
A17	GRN/BLK	A/T: Lockup Control Solenoid	Solenoid Off: 12v, On: 1v
A18	---	Not Used	---
A19	YEL	A/T: Lockup Control Solenoid	Solenoid Off: 12v, On: 1v
A20	RED	EVAP Purge Solenoid	Solenoid Off: 12v, On: 1v
A21	RED/GRN	Igniter Control	Digital Signals: 0-12-0v
A22	---	Not Used	---
A23	BLK	Power Ground	<0.1v
A24	BLK	Power Ground	<0.1v
A25	YEL/BLK	Main Relay Power (B+)	12-14v
A26	BLK/RED	Power Ground	<0.1v

1994-95 Coupe 1.6L I4 VTEC VIN EJ1 16P 'B' Connector

PCM Pin #	W/Color	Circuit Description (16-Pin)	Value at Hot Idle
B1	WHT/GRN	Main Relay Power (B+)	12-14v
B2	BRN/BLK	Logic Ground	<0.050v
B3	GRN/BLU	A/T: Shift Selector Signal	KOEO: 12-14v
B4	GRN/BLK	A/T: Shift Selector Signal	KOEO: 12-14v
B5	BLU/RED	A/C Switch Signal	Switch Off: 12v, On: 1v
B6	---	Not Used	---
B7	GRN	A/T Park/Neutral Switch	In P/N: 0v, Others: 11v
B8	BRN/RED	PSP Switch	Straight: 0v, Turning: 11v
B9	BLU/WHT	Starter Switch Signal	Cranking: 9-11v
B10	YEL/BLU	Vehicle Speed Sensor	Moving: pulse signals
B11	ORN	CYP Sensor Signal	AC pulse signals
B12	WHT	CYP Sensor Ground	<0.050v
B13	ORN/BLU	TDC Sensor Signal	AC pulse signals
B14	WHT/BLU	TDC Sensor Ground	<0.050v
B15	BLU/GRN	CKP Sensor Signal	AC pulse signals
B16	BLU/YEL	CKP Sensor Ground	<0.050v

1994-95 Coupe 1.6L I4 VTEC VIN EJ1 22P 'D' Connector

PCM Pin #	W/Color	Circuit Description (22-Pin)	Value at Hot Idle
D1	WHT/BLU	Keep Alive Power (VBU)	12-14v
D2	GRN/WHT	Brake Switch Signal	Brake Off: 0v, On: 12v
D3	RED/BLU	Knock Sensor Signal	No Detonation: 18mv AC
D4	BRN	Service Check Connector	SCS Open: 4.80v
D5	---	Not Used	---
D6	ORN/BLU	VTEC Pressure Switch	0v, Hi-Speed: 12v
D7	LT BLU	Data Link Connector	5v
D8	---	Not Used	---
D9	PNK	Alternator 'FR' Signal	Digital Signals: 0-5-0v
D10	GRN/RED	Electric Load Detector	Varies: 0.5-4.5v
D11	PNK/BLK	TP Sensor Signal	0.5-0.6v
D12	---	Not Used	---
D13	RED/WHT	ECT Sensor Signal	At 180ºF: 0.51v
D14	WHT	HO2S-11 (B1 S1) Signal	0.1-1.1v
D15	RED/YEL	IAT Sensor Signal	Varies w/temp. (0.5-4.9v)
D16	---	Not Used	---
D17	WHT	MAP Sensor Signal	0.8-0.9v
D18	PNK/GRN	A/T: Interlock Control Unit	Key & Brake On: 12v
D19	YEL/GRN	MAP Sensor VREF	4.9-5.1v
D20	YEL/WHT	Sensor VREF	4.9-5.1v
D21	GRN/BLU	MAP Sensor GND	<0.050v
D22	GRN/WHT	Sensor Ground	<0.050v

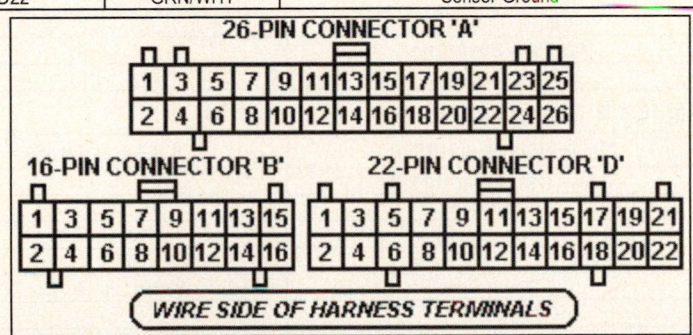

Abbreviation	Color	Abbreviation	Color	Abbreviation	Color
BLK	Black	LT BLU	Lt. Blue	TAN	Tan
BLU	Blue	LT GRN	Lt. Green	VIO	Violet
BRN	Brown	ORN	Orange	WHT	White
GRY	Gray	PNK	Pink	YEL	Yellow
GRN	Green	PPL	Purple		

1994-95 Coupe 1.6L I4 MFI VIN EJ2 [All] 26P 'A' Connector

PCM Pin #	W/Color	Circuit Description (26-Pin)	Value at Hot Idle
A1	BRN	Injector 1 Control	2.0-3.3 ms
A2	YEL	Injector 4 Control	2.0-3.3 ms
A3	RED	Injector 2 Control	2.0-3.3 ms
A4	---	Not Used	---
A5	BLU	Injector 3 Control	2.0-3.3 ms
A6	ORN/BLK	HO2S Heater Control	Relay Off: 12v, On: 1v
A7	GRN/YEL	Fuel Pump Relay	Relay Off: 12v, On: 1v
A8	---	Not Used	---
A9	GRN/WHT	Idle Air Control Valve	Pulse Signals
A10-11	---	Not Used	---
A12	YEL/GRN	Radiator Fan Relay	Relay Off: 12v, On: 1v
A13	GRN/ORN	Check Engine Light	MIL Off: 12v, On: 1v
A14	---	Not Used	---
A15	BLK/RED	A/C Clutch Relay	Relay Off: 12v, On: 1v
A16	WHT/YEL	Alternator Charging Signal	Lights Off: 12v, On: 0v
A17	GRN/BLK	A/T: Lockup Control Solenoid	Solenoid Off: 12v, On: 1v
A18	---	Not Used	---
A19	YEL	A/T: Lockup Control Solenoid	Solenoid Off: 12v, On: 1v
A20	RED	EVAP Purge Solenoid	Solenoid Off: 12v, On: 1v
A21	RED/GRN	Igniter Control	Digital Signals: 0-12-0v
A22	---	Not Used	---
A23	BLK	Power Ground	<0.1v
A24	BLK	Power Ground	<0.1v
A25	YEL/BLK	Main Relay Power (B+)	12-14v
A26	BLK/RED	Power Ground	<0.1v

1994-95 Coupe 1.6L I4 MFI VIN EJ2 [All] 16P 'B' Connector

PCM Pin #	W/Color	Circuit Description (16-Pin)	Value at Hot Idle
B1	WHT/GRN	Main Relay Power (B+)	12-14v
B2	BRN/BLK	Logic Ground	<0.050v
B3	GRN/BLU	A/T: Shift Selector Signal	KOEO: 12-14v
B4	GRN/BLK	A/T: Shift Selector Signal	KOEO: 12-14v
B5	BLU/RED	A/C Switch Signal	Switch Off: 12v, On: 1v
B6	---	Not Used	---
B7	GRN	A/T Park/Neutral Switch	In P/N: 0v, Others: 11v
B8	BRN/RED	PSP Switch	Straight: 0v, Turning: 11v
B9	BLU/WHT	Starter Switch Signal	Cranking: 9-11v
B10	YEL/BLU	Vehicle Speed Sensor	Moving: pulse signals
B11	ORN	CYP Sensor Signal	AC pulse signals
B12	WHT	CYP Sensor Ground	<0.050v
B13	ORN/BLU	TDC Sensor Signal	AC pulse signals
B14	WHT/BLU	TDC Sensor Ground	<0.050v
B15	BLU/GRN	CKP Sensor Signal	AC pulse signals
B16	BLU/YEL	CKP Sensor Ground	<0.050v

1994-95 Coupe 1.6L I4 MFI VIN EJ2 [All] 22P 'D' Connector

PCM Pin #	W/Color	Circuit Description (22-Pin)	Value at Hot Idle
D1	WHT/BLU	Keep Alive Power (VBU)	12-14v
D2	GRN/WHT	Brake Switch Signal	Brake Off: 0v, On: 12v
D3	RED/BLU	Knock Sensor Signal	No Detonation: 18mv AC
D4	BRN	Service Check Connector	SCS Open: 4.80v
D5-6	---	Not Used	---
D7	LT BLU	Data Link Connector	5v
D8	---	Not Used	---
D9	PNK	Alternator 'FR' Signal	Digital Signals: 0-5-0v
D10	GRN/RED	Electric Load Detector	Varies: 0.5-4.5v
D11	PNK/BLK	TP Sensor Signal	0.5-0.6v
D12	---	Not Used	---
D13	RED/WHT	ECT Sensor Signal	At 180°F: 0.51v
D14	WHT	HO2S-11 (B1 S1) Signal	0.1-1.1v
D15	RED/YEL	IAT Sensor Signal	Varies w/temp. (0.5-4.9v)
D16	---	Not Used	---
D17	WHT	MAP Sensor Signal	0.8-0.9v
D18	PNK/GRN	A/T: Interlock Control Unit	Key & Brake On: 12v
D19	YEL/GRN	MAP Sensor VREF	4.9-5.1v
D20	YEL/WHT	Sensor VREF	4.9-5.1v
D21	GRN/BLU	MAP Sensor GND	<0.050v
D22	GRN/WHT	Sensor Ground	<0.050v

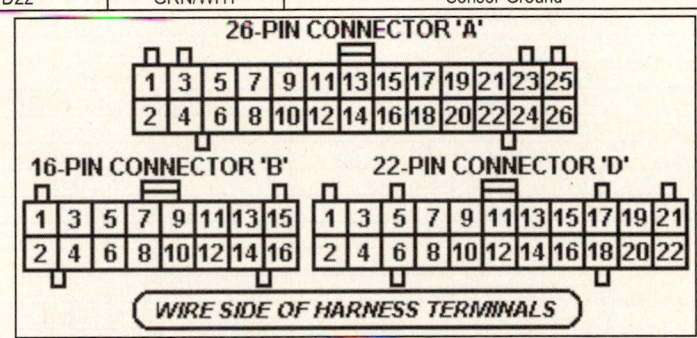

Abbreviation	Color	Abbreviation	Color	Abbreviation	Color
BLK	Black	LT BLU	Lt. Blue	TAN	Tan
BLU	Blue	LT GRN	Lt. Green	VIO	Violet
BRN	Brown	ORN	Orange	WHT	White
GRY	Gray	PNK	Pink	YEL	Yellow
GRN	Green	PPL	Purple		

1996 Coupe 1.6L I4 MFI VIN EJ6 [All] 32P 'A' Connector

PCM Pin #	W/Color	Circuit Description (32-Pin)	Value at Hot Idle
A1	YEL	Injector 4 Control	2.0-3.3 ms
A2	BLU	Injector 4 Control	2.0-3.3 ms
A3	RED	Injector 2 Control	2.0-3.3 ms
A4	BRN	Injector 1 Control	2.0-3.3 ms
A5	BLK/WHT	HO2S-12 (B1 S2) Heater	Digital Signals: 0-12-0v
A6	BLK/WHT	HO2S-11 (B1 S1) Heater	Digital Signals: 0-12-0v
A9, A22	BRN/BLK	Logic Ground	<0.050v
A10, 23	BLK	Power Ground	<0.1v
A11, 24	YEL/BLK	Main Relay Power (B+)	12-14v
A12	BLK/BLU	M/T: Idle Air Control Valve	Pulse Signals
A13	ORN	A/T: Idle Air Control Valve 'N'	Pulse Signals
A14	BLK/BLU	A/T: Idle Air Control Valve 'P'	Pulse Signals
A15	RED/YEL	EVAP Purge Solenoid	Solenoid Off: 12v, On: 1v
A16	GRN/YEL	Fuel Pump Relay	Relay Off: 12v, On: 1v
A17	BLK/RED	A/C Clutch Relay	Relay Off: 12v, On: 1v
A18	GRN/ORN	Check Engine Light	MIL Off: 12v, On: 1v
A19	WHT/GRN	Alternator Charging Signal	Lights Off: 12v, On: 0v
A20	YEL/GRN	Igniter Control	Digital Signals: 0-12-0v
A27	GRN	Radiator Fan Relay	Relay Off: 12v, On: 1v

1996 Coupe 1.6L I4 MFI VIN EJ6 [All] 25P 'B' Connector

PCM Pin #	W/Color	Circuit Description (25-Pin)	Value at Hot Idle
B1	WHT	A/T: Linear Solenoid (-)	Pulse Signals
B2	RED	A/T: Linear Solenoid (+)	Pulse Signals
B3	BLU/YEL	A/T: Shift Solenoid 'A'	SSA on: 1v: off: 12v
B4	GRN/BLK	A/T: Lockup Solenoid 'B'	LSB On: 12v, Off: 0v
B5	YEL	A/T: Lockup Solenoid 'A'	LSA On: 12v, Off: 0v
B8, 17	PNK, BLU	A/T: Gear Position Switch	In D3: 0v, Others: 12v
B11	GRN/WHT	A/T: Shift Solenoid 'B'	SSB on: 1v: off: 12v
B12	WHT/RED	Interlock Control Unit Signal	Key & Brake On: 12v
B13	GRN/BLK	A/T: D4 Indicator Light	In D4 on: 12v, others: 0v
B14	WHT	Mainshaft Speed Sensor 'N'	AC Pulse Signals
B15	RED	Mainshaft Speed Sensor 'P'	AC Pulse Signals
B16	WHT	A/T: Gear Position Switch	In 'R': 0v, Others: 12v
B17	BLU	A/T" Gear Position Switch	In D2: 0v, Others: 12v
B22, 23	GRN	Countershaft Speed Sensor N	Moving: AC pulses
B24	YEL	A/T: Gear Position Switch	In D4: 0v, Others: 12v
B25	LT GRN	A/T: Gear Position Switch	In P/N: 0v, Others: 12v

32-PIN CONNECTOR 'A'   25-PIN CONNECTOR 'B'

WIRE SIDE OF HARNESS TERMINALS

1996 Coupe 1.6L I4 MFI VIN EJ6 [All] 31P 'C' Connector

PCM Pin #	W/Color	Circuit Description (31-Pin)	Value at Hot Idle
C1	BLU/RED	CKF Sensor Signal	AC pulse signals
C2	BLU	CKP Sensor Signal	AC pulse signals
C3	GRN	TDC Sensor Signal	AC pulse signals
C4	YEL	CYP Sensor Signal	AC pulse signals
C5	BLU/RED	A/C Switch Signal	Switch Off: 12v, On: 1v
C6	BLU/ORN	Starter Switch Signal	Cranking: 9-11v
C7	BRN	Service Check Connector	SCS Open: 4.80v
C8	LT BLU	K-Line Signal	12v
C10	WHT/BLU	Keep Alive Power (VBU)	12-14v
C11	WHT/RED	CKF Sensor Ground	<0.050v
C12	WHT	CKP Sensor Ground	<0.050v
C13	RED	TDC Sensor Ground	<0.050v
C14	BLK	CYP Sensor Ground	<0.050v
C16	GRN	PSP Switch	Straight: 0v, Turning: 11v
C17	WHT/RED	Alternator 'FR' Signal	Digital Signals: 0-5-0v
C18	BLU/WHT	Vehicle Speed Sensor	Moving: pulse signals
C30	PNK	A/T: CVT TMB Signal	Pulse Signals

1996 Coupe 1.6L I4 MFI VIN EJ6 [All] 16P 'D' Connector

PCM Pin #	W/Color	Circuit Description (16-Pin)	Value at Hot Idle
D1	RED/BLK	TP Sensor Signal	0.5-0.6v
D2	RED/WHT	ECT Sensor Signal	At 180°F: 0.51v
D3	RED/GRN	MAP Sensor Signal	0.8-0.9v
D4	YEL/RED	MAP Sensor VREF	4.9-5.1v
D5	GRN/WHT	Brake Switch Signal	Brake Off: 0v, On: 12v
D7	WHT	HO2S-11 (B1 S1) Signal	0.1-1.1v
D8	RED/YEL	IAT Sensor Signal	Varies w/temp. (0.5-4.9v)
D9	---	Not Used	---
D10	YEL/BLU	Sensor VREF	4.9-5.1v
D11	GRN/BLK	Sensor Ground	<0.050v
D12	GRN/WHT	MAP Sensor Ground	<0.050v
D13	RED/YEL	HO2S-12 Ground	<0.050v
D14	WHT/RED	HO2S-12 (B1 S2) Signal	0.1-1.1v
D15	---	Not Used	---
D16	GRN/RED	Electric Load Detector	Varies: 0.5-4.5v

31-PIN CONNECTOR 'C' 16-PIN CONNECTOR 'D'

WIRE SIDE OF HARNESS TERMINALS

1997-99 Coupe 1.6L MFI VIN EJ6 [All] 32P 'A' Connector

PCM Pin #	W/Color	Circuit Description (32-Pin)	Value at Hot Idle
A1, A2	YEL, BLU	Injector 4, Injector 3 Control	2.0-3.3 ms
A3, A4	RED, BRN	Injector 2, Injector 1 Control	2.0-3.3 ms
A5	BLK/WHT	HO2S-12 (B1 S2) Heater	Digital Signals: 0-12-0v
A6	BLK/WHT	HO2S-11 (B1 S1) Heater	Digital Signals: 0-12-0v
A7	RED	A/T: EGR Solenoid Control	Solenoid Off: 12v, On: 1v
A7	PNK	M/T: E-EGR Solenoid	Digital Signals: 0-12-0v
A9, A22	BRN/BLK	Logic Ground (LG1, LG2)	<0.050v
A10, 23	BLK	Power Ground (PG1, PG2)	<0.1v
A11, 24	YEL/BLK	Main Relay Power (B+)	12-14v
A12	BLK/BLU	M/T: Idle Air Control Valve	Pulse Signals
A13	ORN	A/T: Idle Air Control Valve 'N'	Pulse Signals
A14	BLK/BLU	A/T: Idle Air Control Valve 'P'	Pulse Signals
A15	RED/YEL	EVAP Purge Solenoid	Solenoid Off: 12v, On: 1v
A16	GRN/YEL	Fuel Pump Relay	Relay Off: 12v, On: 1v
A17	BLK/RED	A/C Clutch Relay	Relay Off: 12v, On: 1v
A18	GRN/ORN	Malfunction Indicator Light	MIL Off: 12v, On: 1v
A19	WHT/GRN	Alternator Charging Signal	Lights Off: 12v, On: 0v
A20	YEL/GRN	Igniter Control	Digital Signals: 0-12-0v
A27	GRN	Radiator Fan Relay Control	Relay Off: 12v, On: 1v
A28	BLU	EVAP Bypass Solenoid	Solenoid Off: 12v, On: 1v
A29	WHT/RED	EVAP Vent Solenoid	Solenoid Off: 12v, On: 1v

1997-99 Coupe 1.6L MFI VIN EJ6 [All] 25P 'B' Connector

PCM Pin #	W/Color	Circuit Description (25-Pin)	Value at Hot Idle
B1	WHT	A/T: Linear Solenoid (-)	Pulse Signals
B2	RED	A/T: Linear Solenoid (+)	Pulse Signals
B3	BLU/YEL	A/T: Shift Solenoid 'A'	SSA On: 12v, Off: 0v
B4	GRN/BLK	A/T: Lockup Solenoid 'B'	LSB On: 12v, Off: 0v
B5	YEL	A/T: Lockup Solenoid 'A'	LSA On: 12v, Off: 0v
B8	PNK	A/T: Gear Position Switch	In D3: 0v, Others: 12v
B11	GRN/WHT	A/T: Shift Solenoid 'B'	SSB Off: 0v, On: 12v
B12	WHT/RED	Interlock Control Unit Signal	Key & Brake On: 12v
B13	GRN/BLK	A/T: D4 Indicator Light	D4 On: 12v, Off: 0v
B14	WHT	Mainshaft Speed Sensor 'P'	AC Pulse Signals
B15	RED	Mainshaft Speed Sensor 'N'	AC Pulse Signals
B16	WHT	A/T: Gear Position Switch	In 'R': 0v, Others: 12v
B17	BLU	A/T: Gear Position Switch	In D2: 0v, Others: 12v
B22	GRN	Countershaft Speed Sensor N	Moving: AC pulses
B23	BLU	Countershaft Speed Sensor P	Moving: AC pulses
B24	YEL	A/T: Gear Position Switch	In D4: 0v, Others: 12v
B25	LT GRN	A/T: Gear Position Switch	In P/N: 0v, Others: 12v

32-Pin 'A'   WIRE SIDE OF HARNESS TERMINALS   25-Pin 'B'

1997-99 Coupe 1.6L MFI VIN EJ6 [All] 31P 'C' Connector

PCM Pin #	W/Color	Circuit Description (31-Pin)	Value at Hot Idle
C1	BLU/RED	CKF Sensor Signal	AC pulse signals
C2	BLU	CKP Sensor Signal	AC pulse signals
C3	GRN	TDC Sensor Signal	AC pulse signals
C4	YEL	CYP Sensor Signal	AC pulse signals
C5	BLU/RED	A/C Switch Signal	Switch Off: 12v, On: 1v
C6	BLU/ORN	Starter Switch Signal	Cranking: 9-11v
C7	BRN	Service Check Connector	SCS Open: 4.80v
C8	LT BLU	K-Line Signal	12v
C10	WHT/BLU	Keep Alive Power (VBU)	12-14v
C11	WHT/RED	CKF Sensor Ground	<0.050v
C12	WHT	CKP Sensor Ground	<0.050v
C13	RED	TDC Sensor Ground	<0.050v
C14	BLK	CYP Sensor Ground	<0.050v
C16	GRN	PSP Switch Signal	Straight: 0v, Turning: 11v
C17	WHT/RED	Alternator 'FR' Signal	Digital Signals: 0-5-0v
C18	BLU/WHT	Vehicle Speed Sensor	Moving: pulse signals
C23	BLK	LAF Pump Cell (IP+)	0.5-3.5v
C24	RED	LAF Common (IP-, VS-)	2.6-2.8v
C25	WHT	LAF VS Cell Voltage (VS+)	7v
C29	LT GRN	A/T: Gear Position Switch	In P/N: 0v, Others: 12v
C29	RED	M/T: Clutch Switch Signal	Clutch In: 0v, Out: 5v
C30	PNK	CVT TMB Signal	Pulse Signals

1997-99 Coupe 1.6L MFI VIN EJ6 [All] 16P 'D' Connector

PCM Pin #	W/Color	Circuit Description (16-Pin)	Value at Hot Idle
D1	RED/BLK	TP Sensor Signal	0.5-0.6v
D2	RED/WHT	ECT Sensor Signal	At 180°F: 0.51v
D3	RED/GRN	MAP Sensor Signal	0.8-0.9v
D4	YEL/RED	MAP Sensor VREF	4.9-5.1v
D5	GRN/WHT	Brake Switch Signal	Brake Off: 0v, On: 12v
D6	RED/BLU	Knock Sensor Signal	No detonation: 18mv AC
D7	WHT	HO2S-11 (B1 S1) Signal	0.1-1.1v
D8	RED/YEL	IAT Sensor Signal	Varies w/temp. (0.5-4.9v)
D9	WHT/BLK	EGR Valve Lift Sensor	1.2v
D10	YEL/RED	Sensor VREF	4.9-5.1v
D11	GRN/BLK	Sensor Ground	<0.050v
D12	GRN/WHT	MAP Sensor Ground	<0.050v
D13	RED/YEL	HO2S-12 Ground	<0.050v
D14	WHT/RED	HO2S-12 (B1 S2) Signal	0.1-1.1v
D15	LT GRN	Fuel Tank Pressure Sensor	Fuel Cap off: 2.5v
D16	GRN/RED	Electric Load Detector	Varies: 0.5-4.5v

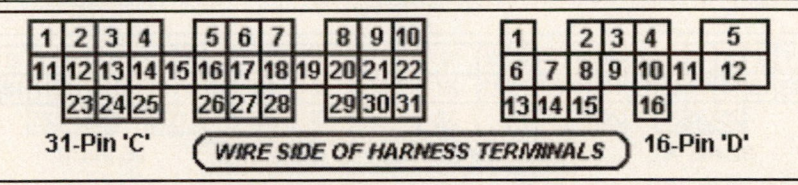

31-Pin 'C'   *WIRE SIDE OF HARNESS TERMINALS*   16-Pin 'D'

2000 Coupe 1.6L MFI VIN EJ6 [All] 32P 'A' Connector

PCM Pin #	W/Color	Circuit Description (32-Pin)	Value at Hot Idle
A1, A2	YEL, BLU	Injector 4, Injector 3 Control	2.0-3.3 ms
A3, A4	RED, BRN	Injector 2, Injector 1 Control	2.0-3.3 ms
A5	BLK/WHT	HO2S-12 (B1 S2) Heater	Digital Signals: 0-12-0v
A6	BLK/WHT	HO2S-11 (B1 S1) Heater	Digital Signals: 0-12-0v
A7	RED	A/T: EGR Solenoid Control	Solenoid Off: 12v, On: 1v
A7	PNK	M/T: E-EGR Solenoid	Digital Signals: 0-12-0v
A9, A22	BRN/BLK	Logic Ground (LG1, LG2)	<0.050v
A10, 23	BLK	Power Ground (PG1, PG2)	<0.1v
A11, 24	YEL/BLK	Main Relay Power (B+)	12-14v
A12	BLK/BLU	M/T: Idle Air Control Valve	Pulse Signals
A13	ORN	A/T: Idle Air Control Valve 'N'	Pulse Signals
A14	BLK/BLU	A/T: Idle Air Control Valve 'P'	Pulse Signals
A15	RED/YEL	EVAP Purge Solenoid	Solenoid Off: 12v, On: 1v
A16	GRN/YEL	Fuel Pump Relay	Relay Off: 12v, On: 1v
A17	BLK/RED	A/C Clutch Relay	Relay Off: 12v, On: 1v
A18	GRN/ORN	Malfunction Indicator Light	MIL Off: 12v, On: 1v
A19	WHT/GRN	Alternator Charging Signal	Lights Off: 12v, On: 0v
A20	YEL/GRN	Igniter Control	Digital Signals: 0-12-0v
A27	GRN	Radiator Fan Relay Control	Relay Off: 12v, On: 1v
A28	BLU	EVAP Bypass Solenoid	Solenoid Off: 12v, On: 1v
A29	WHT/RED	EVAP Vent Solenoid	Solenoid Off: 12v, On: 1v

2000 Coupe 1.6L MFI VIN EJ6 [All] 25P 'B' Connector

PCM Pin #	W/Color	Circuit Description (25-Pin)	Value at Hot Idle
B1	WHT	A/T: Linear Solenoid (-)	Pulse Signals
B2	RED	A/T: Linear Solenoid (+)	Pulse Signals
B3	BLU/YEL	A/T: Shift Solenoid 'A'	SSA On: 12v: Off: 0v
B4	GRN/BLK	A/T: Lockup Solenoid 'B'	LSB On: 12v: Off: 0v
B5	YEL	A/T: Lockup Solenoid 'A'	LSA On: 12v, Off: 0v
B8	PNK	A/T: Gear Position Switch	In D3: 0v, Others: 12v
B11	GRN/WHT	A/T: Shift Solenoid 'B'	SSB Off: 0v, On: 12v
B12	WHT/RED	Interlock Control Unit Signal	Key & Brake On: 12v
B13	GRN/BLK	A/T: D4 Indicator Light	D4 On: 12v, Off: 0v
B14	WHT	Mainshaft Speed Sensor 'P'	AC Pulse Signals
B15	RED	Mainshaft Speed Sensor 'N'	AC Pulse Signals
B16	WHT	A/T: Gear Position Switch	In 'R': 0v, Others: 12v
B17	BLU	A/T: Gear Position Switch	In D2: 0v, Others: 12v
B22	GRN	Countershaft Speed Sensor N	Moving: AC pulses
B23	BLU	Countershaft Speed Sensor P	Moving: AC pulses
B24	YEL	A/T: Gear Position Switch	In D4: 0v, Others: 12v
B25	LT GRN	A/T: Gear Position Switch	In P/N: 0v, Others: 12v

32-Pin 'A'   *WIRE SIDE OF HARNESS TERMINALS*   25-Pin 'B'

2000 Coupe 1.6L MFI VIN EJ6 [All] 31P 'C' Connector

PCM Pin #	W/Color	Circuit Description (31-Pin)	Value at Hot Idle
C1	BLU/RED	CKF Sensor Signal	AC pulse signals
C2	BLU	CKP Sensor Signal	AC pulse signals
C3	GRN	TDC Sensor Signal	AC pulse signals
C4	YEL	CYP Sensor Signal	AC pulse signals
C5	BLU/RED	A/C Switch Signal	Switch Off: 12v, On: 1v
C6	BLU/ORN	Starter Switch Signal	Cranking: 9-11v
C7	BRN	Service Check Connector	SCS Open: 4.80v
C8	LT BLU	K-Line Signal	12v
C10	WHT/BLU	Keep Alive Power (VBU)	12-14v
C11	WHT/RED	CKF Sensor Ground	<0.050v
C12	WHT	CKP Sensor Ground	<0.050v
C13	RED	TDC Sensor Ground	<0.050v
C14	BLK	CYP Sensor Ground	<0.050v
C16	GRN	PSP Switch Signal	Straight: 0v, Turning: 11v
C17	WHT/RED	Alternator 'FR' Signal	Digital Signals: 0-5-0v
C18	BLU/WHT	Vehicle Speed Sensor	Moving: pulse signals
C23	BLK	LAF Pump Cell (IP+)	0.5-3.5v
C24	RED	LAF Common (IP-, VS-)	2.6-2.8v
C25	WHT	LAF VS Cell Voltage (VS+)	7v
C29	LT GRN	A/T: Gear Position Switch	In P/N: 0v, Others: 12v
C29	RED	M/T: Clutch Switch Signal	Clutch In: 0v, Out: 5v
C30	PNK	CVT TMB Signal	Pulse Signals

2000 Coupe 1.6L MFI VIN EJ6 [All] 16P 'D' Connector

PCM Pin #	W/Color	Circuit Description (16-Pin)	Value at Hot Idle
D1	RED/BLK	TP Sensor Signal	0.5-0.6v
D2	RED/WHT	ECT Sensor Signal	At 180°F: 0.51v
D3	RED/GRN	MAP Sensor Signal	0.8-0.9v
D4	YEL/RED	MAP Sensor VREF	4.9-5.1v
D5	GRN/WHT	Brake Switch Signal	Brake Off: 0v, On: 12v
D6	RED/BLU	Knock Sensor Signal	No detonation: 18mv AC
D7	WHT	HO2S-11 (B1 S1) Signal	0.1-1.1v
D8	RED/YEL	IAT Sensor Signal	Varies w/temp. (0.5-4.9v)
D9	WHT/BLK	EGR Valve Lift Sensor	1.2v
D10	YEL/RED	Sensor VREF	4.9-5.1v
D11	GRN/BLK	Sensor Ground	<0.050v
D12	GRN/WHT	MAP Sensor Ground	<0.050v
D13	RED/YEL	HO2S-12 Ground	<0.050v
D14	WHT/RED	HO2S-12 (B1 S2) Signal	0.1-1.1v
D15	LT GRN	Fuel Tank Pressure Sensor	Fuel Cap off: 2.5v
D16	GRN/RED	Electric Load Detector	Varies: 0.5-4.5v

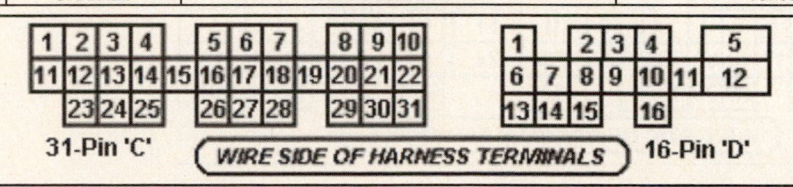

31-Pin 'C'   *WIRE SIDE OF HARNESS TERMINALS*   16-Pin 'D'

1996 Coupe 1.6L I4 VTEC-E VIN EJ7 [All] 32P 'A' Connector

PCM Pin #	W/Color	Circuit Description (32-Pin)	Value at Hot Idle
A1	YEL	Injector 4 Control	2.0-3.3 ms
A2	BLU	Injector 3 Control	2.0-3.3 ms
A3	RED	Injector 2 Control	2.0-3.3 ms
A4	BRN	Injector 1 Control	2.0-3.3 ms
A5	BLK/WHT	HO2S-12 (B1 S2) Heater	Digital Signals: 0-12-0v
A6	BLK/WHT	HO2S-11 (B1 S1) Heater	Digital Signals: 0-12-0v
A7	RED	A/T CVT: EGR Solenoid	Solenoid Off: 12v, On: 1v
A8	GRN/YEL	VTEC Solenoid Control	0v, Hi-Speed: 12v
A9, A22	BRN/BLK	Logic Ground	<0.050v
A10, 23	BLK	Power Ground	<0.1v
A11, 24	YEL/BLK	Main Relay Power (B+)	12-14v
A12	BLK/BLU	M/T: Idle Air Control Valve	Pulse Signals
A13	ORN	Idle Air Control Valve 'N'	Pulse Signals
A14	BLK/BLU	Idle Air Control Valve 'P'	Pulse Signals
A15	RED/YEL	EVAP Purge Solenoid	Solenoid Off: 12v, On: 1v
A16	GRN/YEL	Fuel Pump Relay	Relay Off: 12v, On: 1v
A17	BLK/RED	A/C Clutch Relay	Relay Off: 12v, On: 1v
A18	GRN/ORN	Check Engine Light	MIL Off: 12v, On: 1v
A19	WHT/GRN	Alternator Charging Signal	Lights Off: 12v, On: 0v
A20	YEL/GRN	Igniter Control	Digital Signals: 0-12-0v
A27	GRN	Radiator Fan Relay	Relay Off: 12v, On: 1v
A30	WHT/RED	Interlock Control Unit Signal	Key & Brake On: 12v

1996 Coupe 1.6L I4 VTEC-E VIN EJ7 [All] 16P 'D' Connector

PCM Pin #	W/Color	Circuit Description (16-Pin)	Value at Hot Idle
D1	RED/BLK	TP Sensor Signal	0.5-0.6v
D2	RED/WHT	ECT Sensor Signal	At 180°F: 0.51v
D3	RED/GRN	MAP Sensor Signal	0.8-0.9v
D4	YEL/RED	MAP VREF	4.9-5.1v
D5	GRN/WHT	Brake Switch Signal	Brake Off: 0v, On: 12v
D6	RED/BLU	Knock Sensor Signal	No Detonation: 18mv AC
D7	WHT	HO2S-11 (B1 S1) Signal	0.1-1.1v
D8	RED/YEL	IAT Sensor Signal	Varies w/temp. (0.5-4.9v)
D9	WHT/BLK	EGR Valve Lift Sensor	1.2v
D10	YEL/RED	Sensor VREF	4.9-5.1v
D11, 12	GRN/BLK	MAP Sensor, Sensor Ground	<0.050v
D13	GRN/BLK	HO2S-12 Ground	<0.050v
D14	WHT/RED	HO2S-12 (B1 S2) Signal	0.1-1.1v
D16	GRN/RED	Electric Load Detector	Varies: 0.5-4.5v

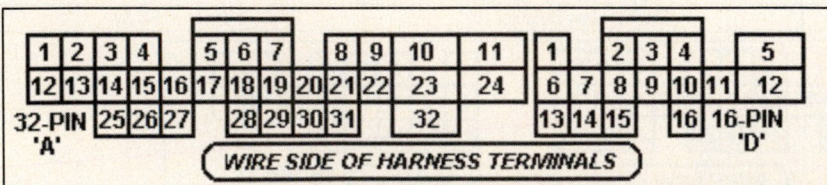

1996 Coupe 1.6L I4 VTEC-E VIN EJ7 [All] 31P 'C' Connector

PCM Pin #	W/Color	Circuit Description (31-Pin)	Value at Hot Idle
C1	BLU/RED	CKF Sensor Signal	AC pulse signals
C2	BLU	CKP Sensor Signal	AC pulse signals
C3	GRN	TDC Sensor Signal	AC pulse signals
C4	YEL	CYP Sensor Signal	AC pulse signals
C5	BLU/RED	A/C Switch Signal	Switch Off: 12v, On: 1v
C6	BLU/ORN	Starter Switch Signal	Cranking: 9-11v
C7	BRN	Service Check Connector	SCS Open: 4.80v
C8	LT BLU	K-Line Signal	12v
C9	GRY	TMA Signal	Pulse Signals
C10	WHT/BLU	Keep Alive Power (VBU)	12-14v
C11	WHT/RED	CKF Sensor Ground	<0.050v
C12	WHT	CKP Sensor Ground	<0.050v
C13	RED	TDC Sensor Ground	<0.050v
C14	BLK	CYP Sensor Ground	<0.050v
C15	BLU/BLK	VTEC Pressure Switch	0v, Hi-Speed: 12v
C16	GRN	PSP Switch	Straight: 0v, Turning: 11v
C17	WHT/RED	Alternator 'FR' Signal	Digital Signals: 0-5-0v
C18	BLU/WHT	Vehicle Speed Sensor	Moving: pulse signals
C23	BLK	M/T: LAF Pump Cell (IP+)	0.5-5.3v
C24	RED	M/T: LAF IP-, VS-	2.6-2.8v
C25	WHT	M/T: LAF VS Cell VS+ Volts	Key on: 6.5-7.5v
C29	LT GRN	A/T CVT Gear Position Switch	In P/N: 0v, Others: 12v
C29	RED	M/T: Clutch Switch Signal	Clutch In: 0v, Out: 5v
C30	PNK	A/T: CVT TMB Signal	Pulse Signals

31-PIN CONNECTOR 'C'

```
 1  2  3  4     5  6  7     8  9 10
11 12 13 14 15 16 17 18 19 20 21 22
   23 24 25    26 27 28    29 30 31
```

WIRE SIDE OF HARNESS TERMINALS

1997-98 Coupe 1.6L VTEC-E VIN EJ7 [All] 32P 'A' Connector

PCM Pin #	W/Color	Circuit Description (32-Pin)	Value at Hot Idle
A1	YEL	Injector 4 Control	2.0-3.3 ms
A2	BLU	Injector 3 Control	2.0-3.3 ms
A3, A4	RED, BRN	Injector 2, Injector 1 Control	2.0-3.3 ms
A5	BLK/WHT	HO2S-12 (B1 S2) Heater	Digital Signals: 0-12-0v
A6	BLK/WHT	HO2S-11 (B1 S1) Heater	Digital Signals: 0-12-0v
A7	RED	A/T: EGR Solenoid	Solenoid on: 1v: off: 12v
A8	GRN/YEL	VTEC Solenoid	0v, Hi-Speed: 12v
A9, A22	BRN/BLK	Logic Ground	<0.050v
A10, 23	BLK	Power Ground	<0.1v
A11, 24	YEL/BLK	Main Relay Power (B+)	12-14v
A13	ORN	Idle Air Control Valve 'N'	Pulse Signals
A14	BLK/BLU	Idle Air Control Valve 'P'	Pulse Signals
A15	RED/YEL	EVAP Purge Solenoid	Solenoid Off: 12v, On: 1v
A16	GRN/YEL	Fuel Pump Relay	Relay Off: 12v, On: 1v
A17	BLK/RED	A/C Clutch Relay	Relay Off: 12v, On: 1v
A18	GRN/ORN	Check Engine Light	MIL Off: 12v, On: 1v
A19	WHT/GRN	Alternator Charging Signal	Lights Off: 12v, On: 0v
A20	YEL/GRN	Igniter Control	Digital Signals: 0-12-0v
A27	GRN	Radiator Fan Relay	Relay Off: 12v, On: 1v
A28	BLU	EVAP Bypass Solenoid	Solenoid Off: 12v, On: 1v
A29	GRN/WHT	EVAP Vent Solenoid	Solenoid Off: 12v, On: 1v
A30	WHT/RED	CVT Interlock Control Unit	Key & Brake On: 12v

1997-98 Coupe 1.6L VTEC-E VIN EJ7 [All] 16P 'D' Connector

PCM Pin #	W/Color	Circuit Description (16-Pin)	Value at Hot Idle
D1	RED/BLK	TP Sensor Signal	0.5-0.6v
D2	RED/WHT	ECT Sensor Signal	At 180ºF: 0.51v
D3	RED/GRN	MAP Sensor Signal	0.8-0.9v
D4	YEL/RED	Sensor VREF	4.9-5.1v
D5	GRN/WHT	Brake Switch Signal	Brake Off: 0v, On: 12v
D6	RED/BLU	Knock Sensor Signal	No Detonation: 18mv AC
D7	WHT	HO2S-11 (B1 S1) Signal	0.1-1.1v
D8	RED/YEL	IAT Sensor Signal	Varies w/temp. (0.5-4.9v)
D9	WHT/BLK	CVT EGR Valve Lift Sensor	1.2v
D10	YEL/BLU	MAP Sensor VREF	4.9-5.1v
D11, D12	GRN/BLK	Sensor Ground, MAP Ground	<0.050v
D13	GRN	HO2S-12 (B1 S2) Signal	0.1-1.1v
D14	BLK	HO2S-12 Ground	<0.050v
D15	LT GRN	Fuel Tank Pressure Sensor	Fuel Cap off: 2.5v
D16	GRN/RED	Electric Load Detector	Varies: 0.5-4.5v

1997-98 Coupe 1.6L VTEC-E VIN EJ7 [All] 31P 'C' Connector

PCM Pin #	W/Color	Circuit Description (31-Pin)	Value at Hot Idle
C1	BLU/RED	CKF Sensor Signal	AC pulse signals
C2	BLU	CKP Sensor Signal	AC pulse signals
C3	GRN	TDC Sensor Signal	AC pulse signals
C4	YEL	CYP Sensor Signal	AC pulse signals
C5	BLU/RED	A/C Switch Signal	Switch Off: 12v, On: 1v
C6	BLU/ORN	Starter Switch Signal	Cranking: 9-11v
C7	BRN	Service Check Connector	SCS Open: 4.80v
C8	LT BLU	K-Line Signal	12v
C9	GRY	TMA Signal	Pulse Signals
C10	WHT/BLU	Keep Alive Power (VBU)	12-14v
C11	WHT/RED	CKF Sensor Ground	<0.050v
C12	WHT	CKP Sensor Ground	<0.050v
C13	RED	TDC Sensor Ground	<0.050v
C14	BLK	CYP Sensor Ground	<0.050v
C15	BLU/BLK	VTEC Pressure Switch	0v, Hi-Speed: 12v
C16	GRN	PSP Switch	Straight: 0v, Turning: 11v
C17	WHT/RED	Alternator 'FR' Signal	Digital Signals: 0-5-0v
C18	BLU/WHT	Vehicle Speed Sensor	Moving: pulse signals
C23	BLK	M/T HO2S Pump Cell (IP+)	Key on: 0.5-5.3v
C24	RED	M/T HO2S IP-, VS-	Hot Engine: 2.6-2.8v
C25	WHT	M/T VS Cell Voltage VS+	Key on: 6.5-7.5v
C29	LT GRN	CVT Gear Position Switch	In P/N: 0v, Others: 12v
C29	RED	M/T Clutch Switch Signal	Clutch In: 0v, Out: 5v
C30	PNK	CVT TMB Signal	Pulse Signals

31-PIN CONNECTOR 'C'

1	2	3	4		5	6	7		8	9	10
11	12	13	14	15	16	17	18	19	20	21	22
	23	24	25		26	27	28		29	30	31

WIRE SIDE OF HARNESS TERMINALS

1999-2000 Coupe 1.6L VTEC-E VIN EJ7 32P 'A' Connector

PCM Pin #	W/Color	Circuit Description (32-Pin)	Value at Hot Idle
A1	YEL	Injector 4 Control	2.0-3.3 ms
A2	BLU	Injector 3 Control	2.0-3.3 ms
A3, A4	RED, BRN	Injector 2, Injector 1 Control	2.0-3.3 ms
A5	BLK/WHT	HO2S-12 (B1 S2) Heater	Digital Signals: 0-12-0v
A6	BLK/WHT	HO2S-11 (B1 S1) Heater	Digital Signals: 0-12-0v
A7	RED	A/T: EGR Solenoid	Digital Signals: 0-12-0v
A7	PNK	M/T: E-EGR Solenoid	Digital Signals: 0-12-0v
A8	GRN/YEL	VTEC Solenoid Control	0v, Hi-Speed: 12v
A9, A22	BRN/BLK	Logic Ground	<0.050v
A10, A23	BLK	Power Ground	<0.1v
A11, A24	YEL/BLK	Main Relay Power (B+)	12-14v
A13	ORN	Idle Air Control Valve 'N'	Pulse Signals
A14	BLK/BLU	Idle Air Control Valve 'P'	Pulse Signals
A15	RED/YEL	EVAP Purge Solenoid	Solenoid Off: 12v, On: 1v
A16	GRN/YEL	Fuel Pump Relay	Relay Off: 12v, On: 1v
A17	BLK/RED	A/C Clutch Relay	Relay Off: 12v, On: 1v
A18	GRN/ORN	Malfunction Indicator Light	MIL Off: 12v, On: 1v
A19	WHT/GRN	Alternator Charging Signal	Lights Off: 12v, On: 0v
A20	YEL/GRN	Igniter Control	Digital Signals: 0-12-0v
A27	GRN	Radiator Fan Relay	Relay Off: 12v, On: 1v
A28	BLU	EVAP Bypass Solenoid	Solenoid Off: 12v, On: 1v
A29	GRN/WHT	EVAP Vent Solenoid	Solenoid Off: 12v, On: 1v
A30	WHT/RED	CVT Interlock Control Unit	Key & Brake On: 12v

1999-2000 Coupe 1.6L VTEC-E VIN EJ7 16P 'D' Connector

PCM Pin #	W/Color	Circuit Description (16-Pin)	Value at Hot Idle
D1	RED/BLK	TP Sensor Signal	0.5-0.6v
D2	RED/WHT	ECT Sensor Signal	At 180°F: 0.51v
D3	RED/GRN	MAP Sensor Signal	0.8-0.9v
D4	YEL/RED	Sensor VREF	4.9-5.1v
D5	GRN/WHT	Brake Switch Signal	Brake Off: 0v, On: 12v
D6	RED/BLU	Knock Sensor Signal	No Detonation: 18mv AC
D7	WHT	LAF Sensor Label Signal	0.3-4.9v
D8	RED/YEL	IAT Sensor Signal	Varies w/temp. (0.5-4.9v)
D9	WHT/BLK	EGR Valve Lift Sensor	1.2v
D10	YEL/BLU	MAP Sensor VREF	4.9-5.1v
D11, D12	GRN/BLK	Sensor Ground, MAP Ground	<0.050v
D13	GRN/BLK	HO2S-12 (B1 S2) Signal	0.1-1.1v
D14	BLK	HO2S-12 Ground	<0.050v
D15	LT GRN	Fuel Tank Pressure Sensor	Fuel Cap off: 2.5v
D16	GRN/RED	Electric Load Detector	Varies: 0.5-4.5v

WIRE SIDE OF HARNESS TERMINALS

1999-2000 Coupe 1.6L VTEC-E VIN EJ7 25P 'B' Connector

PCM Pin #	W/Color	Circuit Description (25-Pin)	Value at Hot Idle
B1	WHT	A/T: Linear Solenoid LS-	Pulse Signals
B2	RED	A/T: Linear Solenoid LS+	Pulse Signals
B3	BLU/YEL	A/T: Shift Solenoid 'A'	SSA Off: 0v, On: 12v
B4	GRN/BLK	A/T: Lockup Solenoid 'B'	LUS On: 12v, Off: 0v
B5	YEL	A/T: Lockup Solenoid 'A'	LUS On: 12v, Off: 0v
B8	PNK	A/T: Gear Position Switch	In D3: 0v, Others: 12v
B11	GRN/WHT	A/T: Shift Solenoid 'B'	SSB Off: 0v, On: 12v
B12	WHT/RED	Interlock Control Unit Signal	Key & Brake On: 12v
B13	GRN/BLK	A/T: D4 Indicator Light Driver	D4 On: 12v, Off: 0v
B14, B15	WHT, RED	Mainshaft Speed Sensor 'N', 'P'	Pulse Signals
B16	WHT	A/T: Gear Position Switch	In 'R': 0v, Others: 12v
B17	BLU	A/T: Gear Position Switch	In D2: 0v, Others: 12v
B22, B23	GRN, BLU	Countershaft speed sensor N, P	<0.050v
B24	YEL	A/T: Gear Position Switch	In D4: 0v, Others: 12v
B25	LT GRN	A/T: Gear Position Switch	In P/N: 0v, Others: 12v

1999-2000 Coupe 1.6L VTEC-E VIN EJ7 31P 'C' Connector

PCM Pin #	W/Color	Circuit Description (31-Pin)	Value at Hot Idle
C1	BLU/RED	CKF Sensor Signal	AC pulse signals
C2	BLU	CKP Sensor Signal	AC pulse signals
C3	GRN	TDC Sensor Signal	AC pulse signals
C4	YEL	CYP Sensor Signal	AC pulse signals
C5	BLU/RED	A/C Switch Signal	Switch Off: 12v, On: 1v
C6	BLU/ORN	Starter Switch Signal	Cranking: 9-11v
C7	BRN	Service Check Connector	SCS Open: 4.80v
C8	LT BLU	K-Line Signal	12v
C9	GRY	ECM Communication to TCM	Digital Signals
C10	WHT/BLU	Keep Alive Power (VBU)	12-14v
C11	WHT/RED	CKF Sensor Ground	<0.050v
C12	WHT	CKP Sensor Ground	<0.050v
C13	RED	TDC Sensor Ground	<0.050v
C14	BLK	CYP Sensor Ground	<0.050v
C15	BLU/BLK	VTEC Pressure Switch	0v, Hi-Speed: 12v
C16	GRN	PSP Switch	Straight: 0v, Turning: 11v
C17	WHT/RED	Alternator 'FR' Signal	Digital Signals: 0-5-0v
C18	BLU/WHT	Vehicle Speed Sensor	Moving: pulse signals
C23	BLK	M/T: LAF Pump Cell (IP+)	0.5-5.3v
C24	RED	M/T: LAF Sensor IP-, VS-	Hot Engine: 2.6-2.8v
C25	WHT	M/T: VS Cell Voltage VS+	Key on: 6.5-7.5v
C29	LT GRN	A/T: Gear Position Switch	In P/N: 0v, Others: 12v
C29	RED	M/T Clutch Switch Signal	Clutch In: 0v, Out: 5v
C30	PNK	ECM Communication to TCM	Digital Signals

1996 Coupe 1.6L I4 VTEC VIN EJ8 [All] 32P 'A' Connector

PCM Pin #	W/Color	Circuit Description (32-Pin)	Value at Hot Idle
A1, A2	YEL, BLU	Injector 4 Control	2.0-3.3 ms
A3, A4	RED, BRN	Injector 2 Control	2.0-3.3 ms
A5	BLK/WHT	HO2S-12 (B1 S2) Heater	Digital Signals: 0-12-0v
A6	BLK/WHT	HO2S-11 (B1 S1) Heater	Digital Signals: 0-12-0v
A8	GRN/YEL	VTEC Solenoid	0v, Hi-Speed: 12v
A9, A22	BRN/BLK	Logic Ground	<0.050v
A10, 23	BLK	Power Ground	<0.1v
A11, 24	YEL/BLK	Main Relay Power (B+)	12-14v
A12	BLK/BLU	M/T: Idle Air Control Valve	Pulse Signals
A13	ORN	A/T: IAC Valve Control 'N'	Pulse Signals
A14	BLK/BLU	A/T: IAC Valve Control 'P'	Pulse Signals
A15	RED/YEL	EVAP Purge Solenoid	Solenoid Off: 12v, On: 1v
A16	GRN/YEL	Fuel Pump Relay	Relay Off: 12v, On: 1v
A17	BLK/RED	A/C Clutch Relay	Relay Off: 12v, On: 1v
A18	GRN/ORN	Check Engine Light	MIL Off: 12v, On: 1v
A19	WHT/GRN	Alternator Charging Signal	Lights Off: 12v, On: 0v
A20	YEL/GRN	Igniter Control	Digital Signals: 0-12-0v
A25	WHT/RED	A/T: TCM VREF Signal	4.9-5.1v
A27	GRN	Radiator Fan Relay	Relay Off: 12v, On: 1v
A28	BLU	EVAP Bypass Solenoid	Solenoid Off: 12v, On: 1v
A29	GRN/WHT	Canister Vent Solenoid	Solenoid Off: 12v, On: 1v

1996 Coupe 1.6L I4 VTEC VIN EJ8 [All] 25P 'B' Connector

PCM Pin #	W/Color	Circuit Description (25-Pin)	Value at Hot Idle
B1	WHT	A/T: Linear Solenoid (-)	Pulse Signals
B2	RED	A/T: Linear Solenoid (+)	Pulse Signals
B3	BLU/YEL	A/T: Shift Solenoid 'A'	SSA Off: 0v, On: 12v
B4	GRN/BLK	Lockup Solenoid 'B'	LSB On: 12v, Off: 0v
B5	YEL	Lockup Solenoid "A"	LSA On: 12v, Off: 0v
B8, 17	PNK, BLU	A/T: Gear Position Switch	In D3: 0v, in D2: 0v
B11	GRN/WHT	A/T: Shift Solenoid 'B'	SSB Off: 0v, On: 12v
B12	WHT/RED	Interlock Control Unit Signal	Key & Brake On: 12v
B13	GRN/BLK	A/T: D4 Indicator Light	In D4: 12v, others: 0v
B14	WHT	Mainshaft Speed Sensor 'N'	<0.050v
B15	RED	Mainshaft Speed Sensor 'P'	AC Pulse Signals
B16, B25	WHT-GRN	A/T: Gear Position Switch	In 'R': 0v, Others: 12v
B22	GRN	Countershaft Speed Sensor N	<0.050v
B23	BLU	Countershaft Speed Sensor P	AC Pulse Signals
B24	YEL	A/T: Gear Position Switch	In D4: 0v, Others: 12v

32-PIN CONNECTOR 'A'   25-PIN CONNECTOR 'B'

WIRE SIDE OF HARNESS TERMINALS

1996 Coupe 1.6L I4 VTEC VIN EJ8 [All] 31P 'D' Connector

PCM Pin #	W/Color	Circuit Description (31-Pin)	Value at Hot Idle
C1	BLU/RED	CKF Sensor Signal	AC pulse signals
C2	BLU, WHT	CKP Sensor Signal	AC pulse signals
C3	GRN	TDC Sensor Signal	AC pulse signals
C4	YEL	CYP Sensor Signal	AC pulse signals
C5	BLU/RED	A/C Switch Signal	Switch Off: 12v, On: 1v
C6	BLU/ORN	Starter Switch Signal	Cranking: 9-11v
C7	BRN	Service Check Connector	SCS Open: 4.80v
C8	LT BLU	K-Line Signal	12v
C9	---	Not Used	---
C10	WHT/BLU	Keep Alive Power (VBU)	12-14v
C11	WHT/RED	CKF Sensor Ground	<0.050v
C12	WHT	CKP Sensor Ground	<0.050v
C13	RED	TDC Sensor Ground	<0.050v
C14	BLK	CYP Sensor Ground	<0.050v
C15	BLU/BLK	VTEC Pressure Switch	0v, Hi-Speed: 12v
C16	GRN	PSP Switch	Straight: 0v, Turning: 11v
C17	WHT/RED	Alternator 'FR' Signal	Digital Signals: 0-5-0v
C18	BLU/WHT	Vehicle Speed Sensor	Moving: pulse signals
C19-31	---	Not Used	---

1996 Coupe 1.6L I4 VTEC VIN EJ8 [All] 16P 'D' Connector

PCM Pin #	W/Color	Circuit Description (16-Pin)	Value at Hot Idle
D1	RED/BLK	TP Sensor Signal	0.5-0.6v
D2	RED/WHT	ECT Sensor Signal	At 180ºF: 0.51v
D3	RED/GRN	MAP Sensor Signal	0.8-0.9v
D4	YEL/RED	MAP Sensor VREF	4.9-5.1v
D5	GRN/WHT	Brake Switch Signal	Brake Off: 0v, On: 12v
D6	RED/BLU	Knock Sensor Signal	No Detonation: 18mv AC
D7	WHT	HO2S-11 (B1 S1) Signal	0.1-1.1v
D8	RED/YEL	IAT Sensor Signal	Varies w/temp. (0.5-4.9v)
D9	---	Not Used	---
D10	YEL/BLU	Sensor VREF	4.9-5.1v
D11	GRN/BLK	Sensor Ground	<0.050v
D12	GRN/WHT	MAP Sensor Ground	<0.050v
D13	GRN/BLK	HO2S-12 Ground	<0.050v
D14	WHT/RED	HO2S-12 (B1 S2) Signal	0.1-1.1v
D15	LT GRN	Fuel Tank Pressure	Fuel Cap off: 2.5v
D16	GRN/RED	Electric Load Detector	Varies: 0.5-4.5v

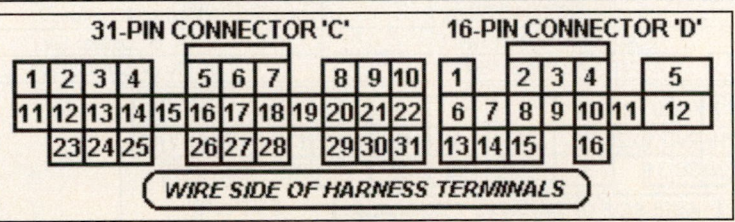

1997-99 Coupe 1.6L VTEC VIN EJ8 [All] 32P 'A' Connector

PCM Pin #	W/Color	Circuit Description (32-Pin)	Value at Hot Idle
A1, A2	YEL, BLU	Injector 4, Injector 3 Control	2.0-3.3 ms
A3, A4	RED, BRN	Injector 2, Injector 1 Control	2.0-3.3 ms
A5	BLK/WHT	HO2S-12 (B1 S2) Heater	Digital Signals: 0-12-0v
A6	BLK/WHT	HO2S-11 (B1 S1) Heater	Digital Signals: 0-12-0v
A8	GRN/YEL	VTEC Solenoid	0v, Hi-Speed: 12v
A9, A22	BRN/BLK	Logic Ground	<0.050v
A10, 23	BLK	Power Ground	<0.1v
A11	YEL/BLK	Main Relay Power (B+)	12-14v
A13	ORN	Idle Air Control Valve 'N'	Pulse Signals
A14	BLK/BLU	Idle Air Control Valve 'P'	Pulse Signals
A15	RED/YEL	EVAP Purge Solenoid	Solenoid Off: 12v, On: 1v
A16	GRN/YEL	Fuel Pump Relay	Relay Off: 12v, On: 1v
A17	BLK/RED	A/C Clutch Relay	Relay Off: 12v, On: 1v
A18	GRN/ORN	Check Engine Light	MIL Off: 12v, On: 1v
A19	WHT/GRN	Alternator Charging Signal	Lights Off: 12v, On: 0v
A20	YEL/GRN	Igniter Control	Digital Signals: 0-12-0v
A24	YEL/BLK	Main Relay Power (B+)	12-14v
A25	WHT/RED	A/T: TCM VREF Signal	4.9-5.1v
A27	GRN	Radiator Fan Relay	Relay Off: 12v, On: 1v
A28	BLU	EVAP Bypass Solenoid	Solenoid Off: 12v, On: 1v
A29	GRN/WHT	EVAP Vent Solenoid	Solenoid Off: 12v, On: 1v

1997-99 Coupe 1.6L I4 VTEC VIN EJ8 [All] 25P 'B' Connector

PCM Pin #	W/Color	Circuit Description (25-Pin)	Value at Hot Idle
B1	WHT	A/T: Linear Solenoid (-)	Pulse Signals
B2	RED	A/T: Linear Solenoid (+)	Pulse Signals
B3	BLU/YEL	A/T: Shift Solenoid 'A'	SSA Off: 0v, On: 12v
B4	GRN/BLK	A/T: Lockup Solenoid 'B'	LSB On: 12v, Off: 0v
B5	YEL	A/T: Lockup Solenoid 'A'	LSA On: 12v, Off: 0v
B8	PNK	A/T: Gear Position Switch	In D3: 0v, Others: 12v
B11	GRN/WHT	A/T: Shift Solenoid 'B'	SSA Off: 0v, On: 12v
B12	WHT/RED	Interlock Control Unit Signal	Key & Brake On: 12v
B13	GRN/BLK	A/T: D4 Indicator Light	D4 On: 12v, Off: 0v
B14	WHT	Mainshaft Speed Sensor 'N'	<0.050v
B15	RED	Mainshaft Speed Sensor 'P'	AC Pulse Signals
B16	WHT	A/T: Gear Position Switch	In 'R': 0v, Others: 12v
B17	BLU	A/T: Gear Position Switch	In D2: 0v, Others: 12v
B22	GRN	Countershaft Speed Sensor N	<0.050v
B23	BLU	Countershaft Speed Sensor P	Moving: AC pulses
B24	YEL	A/T: Gear Position Switch	In D4: 0v, Others: 12v
B25	LT GRN	A/T: Gear Position Switch	In P/N: 0v, Others: 12v

32-Pin 'A'　　WIRE SIDE OF HARNESS TERMINALS　　25-Pin 'B'

1997-99 Coupe 1.6L VTEC VIN EJ8 [All] 31P 'C' Connector

PCM Pin #	W/Color	Circuit Description (31-Pin)	Value at Hot Idle
C1	BLU/RED	CKF Sensor Signal	AC pulse signals
C2	BLU	CKP Sensor Signal	AC pulse signals
C3	GRN	TDC Sensor Signal	AC pulse signals
C4	YEL	CYP Sensor Signal	AC pulse signals
C5	BLU/RED	A/C Switch Signal	Switch Off: 12v, On: 1v
C6	BLU/ORN	Starter Switch Signal	Cranking: 9-11v
C7	BRN	Service Check Connector	SCS Open: 4.80v
C8	LT BLU	K-Line Signal	12v
C9	GRY	ECM Communication to TCM	Digital Signals
C10	WHT/BLU	Keep Alive Power (VBU)	12-14v
C11	WHT/RED	CKF Sensor Ground	<0.050v
C12	WHT	CKP Sensor Ground	<0.050v
C13	RED	TDC Sensor Ground	<0.050v
C14	BLK	CYP Sensor Ground	<0.050v
C15	BLU/BLK	VTEC Pressure Switch	0v, Hi-Speed: 12v
C16	GRN	PSP Switch	Straight: 0v, Turning: 11v
C17	WHT/RED	Alternator 'FR' Signal	Digital Signals: 0-5-0v
C18	BLU/WHT	Vehicle Speed Sensor	Moving: pulse signals
C23	BLK	LAF Pump Cell (IP+)	0.5-3.5v
C24	RED	LAF Common (IP-, VS-)	2.6-2.8v
C25	WHT	LAF VS Cell Voltage (VS+)	7v
C29	LT GRN	A/T: Gear Position Switch	In P/N: 0v, Others: 12v
C29	RED	M/T: Clutch Switch Signal	Clutch In: 0v, Out: 5v
C30	PNK	A/T: CVT TMB Signal	Pulse Signals

1997-99 Coupe 1.6L VTEC VIN EJ8 [All] 16P 'D' Connector

PCM Pin #	W/Color	Circuit Description (16-Pin)	Value at Hot Idle
D1	RED/BLK	TP Sensor Signal	0.5-0.6v
D2	RED/WHT	ECT Sensor Signal	At 180°F: 0.51v
D3	RED/GRN	MAP Sensor Signal	0.8-0.9v
D4	YEL/RED	MAP Sensor VREF	4.9-5.1v
D5	GRN/WHT	Brake Switch Signal	Brake Off: 0v, On: 12v
D6	RED/BLU	Knock Sensor Signal	No Detonation: 18mv AC
D7	WHT	HO2S-11 (B1 S1) Signal	0.1-1.1v
D8	RED/YEL	IAT Sensor Signal	Varies w/temp. (0.5-4.9v)
D10	YEL/RED	Sensor VREF	4.9-5.1v
D11	GRN/BLK	Sensor Ground	<0.050v
D12	GRN/BLK	MAP Sensor Ground	<0.050v
D13	GRN/BLK	HO2S-12 Ground	<0.050v
D14	WHT/RED	HO2S-12 (B1 S2) Signal	0.1-1.1v
D15	LT GRN	Fuel Tank Pressure Sensor	Fuel Cap off: 2.5v
D16	GRN/RED	Electric Load Detector	Varies: 0.5-4.5v

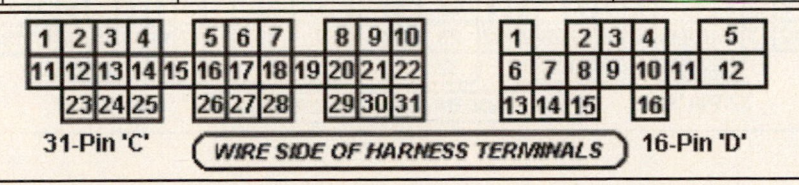

31-Pin 'C'   *WIRE SIDE OF HARNESS TERMINALS*   16-Pin 'D'

2000 Coupe 1.6L VTEC VIN EJ8 [All] 32P 'A' Connector

PCM Pin #	W/Color	Circuit Description (32-Pin)	Value at Hot Idle
A1, A2	YEL, BLU	Injector 4, Injector 3 Control	2.0-3.3 ms
A3, A4	RED, BRN	Injector 2, Injector 1 Control	2.0-3.3 ms
A5	BLK/WHT	HO2S-12 (B1 S2) Heater	Digital Signals: 0-12-0v
A6	BLK/WHT	HO2S-11 (B1 S1) Heater	Digital Signals: 0-12-0v
A8	GRN/YEL	VTEC Solenoid	0v, Hi-Speed: 12v
A9, A22	BRN/BLK	Logic Ground	<0.050v
A10, 23	BLK	Power Ground	<0.1v
A11	YEL/BLK	Main Relay Power (B+)	12-14v
A13	ORN	Idle Air Control Valve 'N'	Pulse Signals
A14	BLK/BLU	Idle Air Control Valve 'P'	Pulse Signals
A15	RED/YEL	EVAP Purge Solenoid	Solenoid Off: 12v, On: 1v
A16	GRN/YEL	Fuel Pump Relay	Relay Off: 12v, On: 1v
A17	BLK/RED	A/C Clutch Relay	Relay Off: 12v, On: 1v
A18	GRN/ORN	Check Engine Light	MIL Off: 12v, On: 1v
A19	WHT/GRN	Alternator Charging Signal	Lights Off: 12v, On: 0v
A20	YEL/GRN	Igniter Control	Digital Signals: 0-12-0v
A24	YEL/BLK	Main Relay Power (B+)	12-14v
A25	WHT/RED	A/T: TCM VREF Signal	4.9-5.1v
A27	GRN	Radiator Fan Relay	Relay Off: 12v, On: 1v
A28	BLU	EVAP Bypass Solenoid	Solenoid Off: 12v, On: 1v
A29	GRN/WHT	EVAP Vent Solenoid	Solenoid Off: 12v, On: 1v

2000 Coupe 1.6L I4 VTEC VIN EJ8 [All] 25P 'B' Connector

PCM Pin #	W/Color	Circuit Description (25-Pin)	Value at Hot Idle
B1	WHT	A/T: Linear Solenoid (-)	Pulse Signals
B2	RED	A/T: Linear Solenoid (+)	Pulse Signals
B3	BLU/YEL	A/T: Shift Solenoid 'A'	SSA Off: 0v, On: 12v
B4	GRN/BLK	A/T: Lockup Solenoid 'B'	LSB On: 12v, Off: 0v
B5	YEL	A/T: Lockup Solenoid 'A'	LSA On: 12v, Off: 0v
B8	PNK	A/T: Gear Position Switch	In D3: 0v, Others: 12v
B11	GRN/WHT	A/T: Shift Solenoid 'B'	SSA Off: 0v, On: 12v
B12	WHT/RED	Interlock Control Unit Signal	Key & Brake On: 12v
B13	GRN/BLK	A/T: D4 Indicator Light	D4 On: 12v, Off: 0v
B14	WHT	Mainshaft Speed Sensor 'N'	<0.050v
B15	RED	Mainshaft Speed Sensor 'P'	AC Pulse Signals
B16	WHT	A/T: Gear Position Switch	In 'R': 0v, Others: 12v
B17	BLU	A/T: Gear Position Switch	In D2: 0v, Others: 12v
B22	GRN	Countershaft Speed Sensor N	<0.050v
B23	BLU	Countershaft Speed Sensor P	Moving: AC pulses
B24	YEL	A/T: Gear Position Switch	In D4: 0v, Others: 12v
B25	LT GRN	A/T: Gear Position Switch	In P/N: 0v, Others: 12v

32-Pin 'A'   WIRE SIDE OF HARNESS TERMINALS   25-Pin 'B'

2000 Coupe 1.6L VTEC VIN EJ8 [All] 31P 'C' Connector

PCM Pin #	W/Color	Circuit Description (31-Pin)	Value at Hot Idle
C1	BLU/RED	CKF Sensor Signal	AC pulse signals
C2	BLU	CKP Sensor Signal	AC pulse signals
C3	GRN	TDC Sensor Signal	AC pulse signals
C4	YEL	CYP Sensor Signal	AC pulse signals
C5	BLU/RED	A/C Switch Signal	Switch Off: 12v, On: 1v
C6	BLU/ORN	Starter Switch Signal	Cranking: 9-11v
C7	BRN	Service Check Connector	SCS Open: 4.80v
C8	LT BLU	K-Line Signal	12v
C9	GRY	ECM Communication to TCM	Digital Signals
C10	WHT/BLU	Keep Alive Power (VBU)	12-14v
C11	WHT/RED	CKF Sensor Ground	<0.050v
C12	WHT	CKP Sensor Ground	<0.050v
C13	RED	TDC Sensor Ground	<0.050v
C14	BLK	CYP Sensor Ground	<0.050v
C15	BLU/BLK	VTEC Pressure Switch	0v, Hi-Speed: 12v
C16	GRN	PSP Switch	Straight: 0v, Turning: 11v
C17	WHT/RED	Alternator 'FR' Signal	Digital Signals: 0-5-0v
C18	BLU/WHT	Vehicle Speed Sensor	Moving: pulse signals
C23	BLK	LAF Pump Cell (IP+)	0.5-3.5v
C24	RED	LAF Common (IP-, VS-)	2.6-2.8v
C25	WHT	LAF VS Cell Voltage (VS+)	7v
C29	LT GRN	A/T: Gear Position Switch	In P/N: 0v, Others: 12v
C29	RED	M/T: Clutch Switch Signal	Clutch In: 0v, Out: 5v
C30	PNK	A/T: CVT TMB Signal	Pulse Signals

2000 Coupe 1.6L VTEC VIN EJ8 [All] 16P 'D' Connector

PCM Pin #	W/Color	Circuit Description (16-Pin)	Value at Hot Idle
D1	RED/BLK	TP Sensor Signal	0.5-0.6v
D2	RED/WHT	ECT Sensor Signal	At 180°F: 0.51v
D3	RED/GRN	MAP Sensor Signal	0.8-0.9v
D4	YEL/RED	MAP Sensor VREF	4.9-5.1v
D5	GRN/WHT	Brake Switch Signal	Brake Off: 0v, On: 12v
D6	RED/BLU	Knock Sensor Signal	No Detonation: 18mv AC
D7	WHT	HO2S-11 (B1 S1) Signal	0.1-1.1v
D8	RED/YEL	IAT Sensor Signal	Varies w/temp. (0.5-4.9v)
D10	YEL/RED	Sensor VREF	4.9-5.1v
D11	GRN/BLK	Sensor Ground	<0.050v
D12	GRN/BLK	MAP Sensor Ground	<0.050v
D13	GRN/BLK	HO2S-12 Ground	<0.050v
D14	WHT/RED	HO2S-12 (B1 S2) Signal	0.1-1.1v
D15	LT GRN	Fuel Tank Pressure Sensor	Fuel Cap off: 2.5v
D16	GRN/RED	Electric Load Detector	Varies: 0.5-4.5v

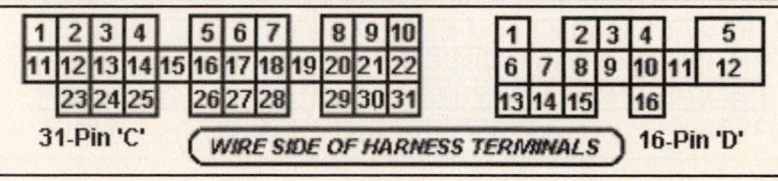

31-Pin 'C'   *WIRE SIDE OF HARNESS TERMINALS*   16-Pin 'D'

2001-03 Coupe 1.7L MFI VIN ES1 [All] 31P Gray 'A' Connector

PCM Pin #	W/Color	Circuit Description (31-Pin)	Value at Hot Idle
A1	BLK/WHT	HO2S-11 (B1 S1) Heater	Digital Signals: 0-12-0v
A2, A3	YEL/BLK	Main Relay Power (B+)	12-14v
A4, A5	BLK	Power Ground (PG2, PG1)	<0.1v
A6	WHT	HO2S-11 (B1 S1) Signal	0.1-1.1v
A7	BLU	CKP Sensor Signal	Digital Signals
A8	YEL	Sensor VREF (VCCR)	4.9-5.1v
A9	RED/BLU	Knock Sensor Signal	No detonation: 18mv AC
A10	GRN/YEL	Sensor Ground (SG2)	<0.050v
A11	GRN/WHT	Sensor Ground (SG1)	<0.050v
A12	BLK/RED	Idle Air Control Valve	Pulse Signals
A13	WHT/BLK	EGR Valve Position Sensor	0.6-1.1v
A14	BLK/WHT	HO2S-12 (B1 S2) Heater	Digital Signals: 0-12-0v
A15	RED/BLK	TP Sensor Signal	0.5-0.6v
A17	BRN/WHT	Injector Mode Signal	Digital Signals
A18	WHT/GRN	Vehicle Speed Signal	Moving: 0-5-0v
A19	GRN/RED	MAP Sensor Signal	0.8-0.9v
A20, A21	YEL/BLU	Sensor VREF (VCC2, VCC1)	4.9-5.1v
A23, A24	BRN/YEL	Logic Ground (LG2, LG1)	<0.1v
A25	WHT/RED	HO2S-12 (B1 S2) Signal	0.1-1.1v
A26	GRN	TDC Sensor Signal	AC pulse signals
A27	BRN	Coil 4 Driver Control	Digital Signals
A28	WHT/BLU	Coil 3 Driver Control	Digital Signals
A29	BLU/RED	Coil 2 Driver Control	Digital Signals
A30	YEL/GRN	Coil 1 Driver Control	Digital Signals

2001-03 Coupe 1.7L MFI VIN ES1 [All] 24P White 'B' Connector

PCM Pin #	W/Color	Circuit Description (24-Pin)	Value at Hot Idle
B2	YEL	Injector 4 Control	2.0-3.3 ms
B3	BLU	Injector 3 Control	2.0-3.3 ms
B4	RED	Injector 2 Control	2.0-3.3 ms
B5	BRN	Injector 1 Control	2.0-3.3 ms
B6	GRN	Fan Relay Control	Relay Off: 12v, On: 1v
B7	RED/BLK	A/T: Linear Solenoid 'A+'	LSA Off: 0v, On: 12v
B8	RED/WHT	ECT Sensor Signal	At 180ºF: 0.51v
B10	WHT/BLU	Alternator Load Signal	Lights Off: 12v, On: 0v
B13	WHT/RED	Alternator 'FR' Signal	Digital Signals: 0-5-0v
B14	BLU/RED	EGR Solenoid Control	Solenoid Off: 12v, On: 1v
B16	BLK/RED	A/T: Lockup Solenoid 'B'	LSB On: 12v: Off: 0v
B17	RED/YEL	IAT Sensor Signal	Varies w/temp. (0.5-4.9v)
B18	WHT/GRN	Alternator Charging Signal	Digital Signals: 0-12-0v
B21	YEL/BLU	EVAP Purge Solenoid	Solenoid Off: 12v, On: 1v

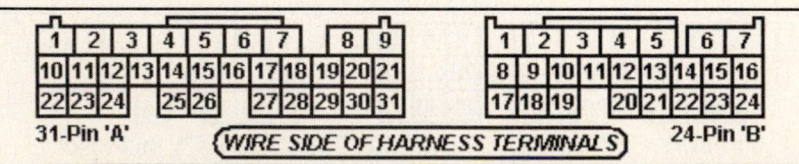

2001-03 Coupe 1.7L MFI VIN ES1 [W/O CVT] 22P C Black Connector

PCM Pin #	W/Color	Circuit Description (22-Pin)	Value at Hot Idle
C1	WHT/BLK	A/T: Linear Solenoid 'A-'	LSA Off: 12v, On: 1v
C2	YEL/BLU	A/T: TCC Solenoid	Solenoid Off: 12v, On: 1v
C4	GRN/WHT	A/T: Shift Solenoid 'B'	SSB Off: 0v, On: 12v
C6	BLU/BLK	A/T: Shift Solenoid 'A'	SSA 12v: Off: 0v
C7	WHT/RED	Mainshaft Speed Sensor 'P'	AC Pulse Signals
C8	BRN/WHT	A/T: Lockup Solenoid 'B-'	LSB On: 12v: Off: 0v
C9	RED	A/T: Gear Position Switch	In D3: 0v, Others: 12v
C10	WHT	A/T: Gear Position Switch	In 'R': 0v, Others: 12v
C11	BLU	A/T: Gear Position Switch	In D2: 0v, Others: 12v
C12	BLU/WHT	A/T: Gear Position Switch	In P/N: 0v, others: 12v
C14	GRN	Countershaft Speed Sensor N	Moving: AC pulses
C15	BLU	Countershaft Speed Sensor P	Moving: AC pulses
C18	BLY/REL	A/T: Gear Position Switch	Forward: 0v, Others: 12v
C20	YEL	A/T: Gear Position Switch	In 'D': 0v, Others: 12v
C21	WHT/GRN	Mainshaft Speed Sensor 'N'	AC Pulse Signals

2001-03 Coupe 1.7L MFI VIN ES1 [All] 31P White 'E' Connector

PCM Pin #	W/Color	Circuit Description (31-Pin)	Value at Hot Idle
E1	GRN/YEL	Immobilizer Fuel Pump Relay	Relay Off: 12v, On: 1v
E3	BRN/YEL	Logic Ground (LG3)	<0.050v
E4	PNK	Sensor Ground (SG3)	<0.050v
E5	YEL/BLU	Sensor VREF (VCC3)	4.9-5.1v
E7	RED/YEL	Main Relay Control	Relay Off: 12v, On: 1v
E9	YEL/BLK	Main Relay Power (B+)	12-14v
E12	BLU/ORN	Cruise Control Unit Signal	Digital Signals
E13	WHT/BLU	Multiplex Control Unit	Digital Signals
E14	LT GRN	Fuel Tank Pressure Sensor	Fuel Cap off: 2.5v
E15	GRN/RED	Electric Load Detector	Varies: 0.5-4.5v
E16	GRN/BLK	PSP Switch Signal	Straight: 0v, Turning: 11v
E18	RED	A/C Clutch Relay	Relay Off: 12v, On: 1v
E20	BLU/RED	EVAP Bypass Solenoid	Solenoid Off: 12v, On: 1v
E21	GRN/RED	EVAP Vent Solenoid	Solenoid Off: 12v, On: 1v
E22	WHT/BLK	Brake Switch Signal	Brake Off: 0v, On: 12v
E23	LT BLU	K-Line Signal	12v
E24	YEL	SEFMJ Signal (Multiplex Unit)	Digital Signals
E25	BLU/WHT	VSS Out Signal	Pulse Signals
E26	BLU	Engine Speed Pulse (NEP)	Digital Signals
E27	RED/BLU	Immobilizer Code Signal	Digital Signals
E29	BRN	Service Check Connector	SCS Open: 4.80v
E30	RED/WHT	WEN Terminal in DLC	0v
E31	GRN/ORN	Malfunction Indicator Lamp	MIL Off: 12v, On: 1v

24-Pin 'C' — WIRE SIDE OF HARNESS TERMINALS — 31-Pin 'E'

2001-03 Coupe 1.7L VTEC VIN ES2 [All] 31P Gray 'A' Connector

PCM Pin #	W/Color	Circuit Description (31-Pin)	Value at Hot Idle
A1	BLK/WHT	HO2S-11 (B1 S1) Heater	Digital Signals: 0-12-0v
A2, A3	YEL/BLK	Main Relay Power (B+)	12-14v
A4, A5	BLK	Power Ground (PG2, PG1)	<0.1v
A6	WHT	HO2S-11 (B1 S1) Signal	0.1-1.1v
A7	BLU	CKP Sensor Signal	Digital Signals
A8	YEL	Sensor VREF (VCCR)	4.9-5.1v
A9	RED/BLU	Knock Sensor Signal	No detonation: 18mv AC
A10	GRN/YEL	Sensor Ground (SG2)	<0.050v
A11	GRN/WHT	Sensor Ground (SG1)	<0.050v
A12	BLK/RED	Idle Air Control Valve	Pulse Signals
A13	WHT/BLK	EGR Valve Position Sensor	0.6-1.1v
A14	BLK/WHT	HO2S-12 (B1 S2) Heater	Digital Signals: 0-12-0v
A15	RED/BLK	TP Sensor Signal	0.5-0.6v
A17	BRN/WHT	Injector Mode Signal	Digital Signals
A18	WHT/GRN	Vehicle Speed Signal	Moving: 0-5-0v
A19	GRN/RED	MAP Sensor Signal	0.8-0.9v
A20, A21	YEL/BLU	Sensor VREF (VCC2, VCC1)	4.9-5.1v
A23, A24	BRN/YEL	Logic Ground (LG2, LG1)	<0.1v
A25	WHT/RED	HO2S-12 (B1 S2) Signal	0.1-1.1v
A26	GRN	TDC Sensor Signal	AC pulse signals
A27	BRN	Coil 4 Driver Control	Digital Signals
A28	WHT/BLU	Coil 3 Driver Control	Digital Signals
A29	BLU/RED	Coil 2 Driver Control	Digital Signals
A30	YEL/GRN	Coil 1 Driver Control	Digital Signals

2001-03 Coupe 1.7L VTEC VIN ES2 [All] 24P White 'B' Connector

PCM Pin #	W/Color	Circuit Description (24-Pin)	Value at Hot Idle
B2, B3	YEL, BLU	Injector 4, Injector 3 Control	2.0-3.3 ms
B4, B5	RED, BRN	Injector 2, Injector 1 Control	2.0-3.3 ms
B6	GRN	Fan Relay Control	Relay Off: 12v, On: 1v
B7	RED/BLK	A/T: Linear Solenoid 'A+'	LSA Off: 0v, On: 12v
B8	RED/WHT	ECT Sensor Signal	At 180°F: 0.51v
B9	BLU/BLK	VTEC Pressure Switch	0v, Hi-Speed: 12v
B10	WHT/BLU	Alternator Load Signal	Lights Off: 12v, On: 0v
B13	WHT/RED	Alternator 'FR' Signal	Digital Signals: 0-5-0v
B14	BLU/RED	EGR Solenoid Control	Solenoid Off: 12v, On: 1v
B15	GRN/YEL	VTEC Solenoid Control	0v, Hi-Speed: 12v
B16	BLK/RED	A/T: Lockup Solenoid 'B'	LSB On: 12v: Off: 0v
B17	RED/YEL	IAT Sensor Signal	Varies w/temp. (0.5-4.9v)
B18	WHT/GRN	Alternator Charging Signal	Digital Signals: 0-12-0v
B21	YEL/BLU	EVAP Purge Solenoid	Solenoid Off: 12v, On: 1v

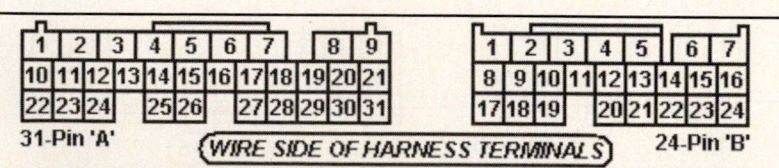

2001-03 Coupe 1.7L VTEC VIN ES2 [W/O CVT] 22P 'C' Connector

PCM Pin #	W/Color	Circuit Description (22-Pin)	Value at Hot Idle
C1	WHT/BLK	A/T: Linear Solenoid 'A-'	LSA Off: 12v, On: 1v
C2	YEL/BLU	A/T: TCC Solenoid	Solenoid Off: 12v, On: 1v
C4	GRN/WHT	A/T: Shift Solenoid 'B'	SSB Off: 0v, On: 12v
C6	BLU/BLK	A/T: Shift Solenoid 'A'	SSA On: 12v, Off: 0v
C7	WHT/RED	Mainshaft Speed Sensor 'P'	AC Pulse Signals
C8	BRN/WHT	A/T: Lockup Solenoid 'B-'	LSB On: 12v, Off: 0v
C9	RED	A/T: Gear Position Switch	In D3: 0v, Others: 12v
C10	WHT	A/T: Gear Position Switch	In 'R': 0v, Others: 12v
C11	BLU	A/T: Gear Position Switch	In D2: 0v, Others: 12v
C12	BLU/WHT	A/T: Gear Position Switch	In P/N: 0v, others: 12v
C14	GRN	Countershaft Speed Sensor N	Moving: AC pulses
C15	BLU	Countershaft Speed Sensor P	Moving: AC pulses
C18	BLY/REL	A/T: Gear Position Switch	Forward: 0v, Others: 12v
C20	YEL	A/T: Gear Position Switch	In 'D': 0v, Others: 12v
C21	WHT/GRN	Mainshaft Speed Sensor 'N'	AC Pulse Signals

2001-03Coupe 1.7L VTEC VIN ES2 [All] 31P White 'E' Connector

PCM Pin #	W/Color	Circuit Description (31-Pin)	Value at Hot Idle
E1	GRN/YEL	Immobilizer Fuel Pump Relay	Relay Off: 12v, On: 1v
E3	BRN/YEL	Logic Ground (LG3)	<0.050v
E4	PNK	Sensor Ground (SG3)	<0.050v
E5	YEL/BLU	Sensor VREF (VCC3)	4.9-5.1v
E7	RED/YEL	Main Relay Control	Relay Off: 12v, On: 1v
E9	YEL/BLK	Main Relay Power (B+)	12-14v
E12	BLU/ORN	Cruise Control Unit Signal	Digital Signals
E13	WHT/BLU	Multiplex Control Unit	Digital Signals
E14	LT GRN	Fuel Tank Pressure Sensor	Fuel Cap off: 2.5v
E15	GRN/RED	Electric Load Detector	Varies: 0.5-4.5v
E16	GRN/BLK	PSP Switch Signal	Straight: 0v, Turning: 11v
E18	RED	A/C Clutch Relay	Relay Off: 12v, On: 1v
E20	BLU/RED	EVAP Bypass Solenoid	Solenoid Off: 12v, On: 0v
E21	GRN/RED	EVAP Vent Solenoid	Solenoid Off: 12v, On: 1v
E22	WHT/BLK	Brake Switch Signal	Brake Off: 0v, On: 12v
E23	LT BLU	K-Line Signal	12v
E24	YEL	SEFMJ Signal (Multiplex Unit)	Digital Signals
E25	BLU/WHT	VSS Out Signal	Pulse Signals
E26	BLU	Engine Speed Pulse (NEP)	Digital Signals
E27	RED/BLU	Immobilizer Code Signal	Digital Signals
E29	BRN	Service Check Connector	SCS Open: 4.80v
E30	RED/WHT	WEN Terminal in DLC	0v
E31	GRN/ORN	Malfunction Indicator Lamp	MIL Off: 12v, On: 1v

```
┌─────────────────────────┐        ┌──────────────────────────────────┐
│ 1  2 │ 3  4  5  6 │ 7   │        │ 1  2  3 │ 4  5  6 │ 7 │ 8  9      │
│ 8  9 10 11 12 13 14 15 │        │ 10 11 12 13 14 15 16 17 18 19 20 21 │
│16 17 18 19 20 │21 22    │        │22 23 24 │25 26│ 27 28 29 30 31     │
└─────────────────────────┘        └──────────────────────────────────┘
   24-Pin 'C'        (WIRE SIDE OF HARNESS TERMINALS)      31-Pin 'E'
```

1990 CRX Coupe 1.5L I4 MFI VIN ED7 [M/T] 18P 'A' Connector

PCM Pin #	W/Color	Circuit Description (18-Pin)	Value at Hot Idle
A1	BRN	Injector 1 Control	2.0-3.3 ms
A2	BLK	Power Ground	<0.1v
A3	RED	Injector 2 Control	2.0-3.3 ms
A4	BLK	Power Ground	<0.1v
A5	BLU	Injector 3 Control	2.0-3.3 ms
A6	GRN	EVAP Purge Solenoid	Solenoid Off: 12v, On: 1v
A7	YEL	Injector 4 Control	2.0-3.3 ms
A8-9	---	Not Used	---
A10	ORN	EGR Solenoid	Solenoid Off: 12v, On: 1v
A11	BLU/YEL	Electronic Air Control Valve	Pulse Signals
A12	GRN/BLK	Fuel Pump Relay	Relay Off: 12v, On: 1v
A13	YEL/BLK	Main Relay Power (B+)	12-14v
A14	GRN/BLK	Fuel Pump Relay	Relay Off: 12v, On: 1v
A15	YEL/BLK	Main Relay Power (B+)	12-14v
A16	BRN/BLK	Power Ground	<0.1v
A17	BLK/YEL	Ignition Power	12-14v
A18	BLK/RED	Power Ground	<0.1v

1990 CRX Coupe 1.5L I4 MFI VIN ED7 [M/T] 20P 'B' Connector

PCM Pin #	W/Color	Circuit Description (20-Pin)	Value at Hot Idle
B1	WHT/GRN	Keep Alive Power (VBU)	12-14v
B2	GRN/YEL	Upshift Indicator Light	Lamp Off: 12v, On: 1v
B3	YEL	A/C Clutch Relay	Relay Off: 12v, On: 1v
B4	YEL/GRN	Radiator Fan Relay	Relay Off: 12v, On: 1v
B5	WHT/YEL	Alternator Charging Signal	Lights Off: 12v, On: 0v
B6	GRN/ORN	Check Engine Light	MIL Off: 12v, On: 1v
B7	GRN	Heater Fan Switch	Fan Off: 0v, On: 12v
B8	BLU/RED	A/C Switch Signal	Switch Off: 12v, On: 1v
B9	GRN/BLK	Reverse Switch Signal	In 'R': 0v, Others: 12v
B10	ORN	CKP Sensor Signal	AC pulse signals
B11	GRN	Clutch Switch Signal	Clutch In: 0v, Out: 12v
B12	WHT	CKP Sensor Ground	<0.050v
B13	BLU/WHT	Start Signal	Cranking: 9-11v
B14	BLU	Alternator 'FR' Signal	Digital Signals: 0-5-0v
B15	WHT	Igniter Control	Digital Signals: 0-12-0v
B16	YEL/RED	Vehicle Speed Sensor	Moving: pulse signals
B17	WHT	Igniter Control	Digital Signals: 0-12-0v
B18	---	Not Used	---
B19	GRN/RED	Electric Load Detector	Varies: 0.5-4.5v
B20	BRN	Ignition Timing Adjuster	0.5-4.5v

1990 CRX Coupe 1.5L I4 MFI VIN ED7 [M/T] 16P 'C' Connector

PCM Pin #	W/Color	Circuit Description (16-Pin)	Value at Hot Idle
C1	BLU/GRN	CYP Sensor Signal	AC pulse signals
C2	BLU/YEL	CYP Sensor Ground	<0.050v
C3	ORN/BLU	TDC Sensor Signal	AC pulse signals
C4	WHT/BLU	TDC Sensor Ground	<0.050v
C5	RED/YEL	IAT Sensor Signal	Varies w/temp. (0.5-4.9v)
C6	RED/WHT	ECT Sensor Signal	At 180°F: 0.51v
C7	RED/BLU	Throttle Angle Sensor Signal	0.5-0.6v
C8	YEL	EGR Valve Lift Sensor	1.2v
C9	RED/WHT	Atmospheric Pressure Sensor	2.76-2.96v at sea level
C10	GRN/WHT	Brake Switch Signal	Brake Off: 0v, On: 12v
C11	WHT	MAP Sensor Signal	0.8-0.9v
C12	GRN/WHT	Sensor Ground	<0.050v
C13	YEL/WHT	Sensor VREF	4.9-5.1v
C14	GRN/WHT	MAP Sensor Ground	<0.050v
C15	YEL/RED	MAP Sensor VREF	4.9-5.1v
C16	WHT	Oxygen Sensor Signal	0.1-1.1v

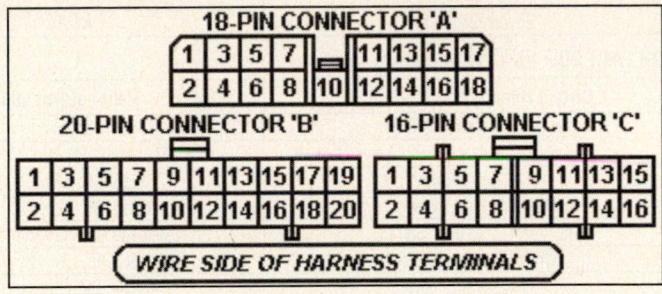

1990-91 CRX Coupe 1.5L I4 MFI VIN ED8 [All] 18P 'A' Connector

PCM Pin #	W/Color	Circuit Description (18-Pin)	Value at Hot Idle
A1	BRN	Injector 1 Control	2.0-3.3 ms
A2	BLK	Power Ground	<0.1v
A3	RED	Injector 2 Control	2.0-3.3 ms
A4	BLK	Power Ground	<0.1v
A5	BLU	Injector 3 Control	2.0-3.3 ms
A6	GRN	EVAP Purge Solenoid	Solenoid Off: 12v, On: 1v
A7	YEL	Injector 4 Control	2.0-3.3 ms
A8	YEL	A/T: Lockup Solenoid 'A'	LSA On: 12v, Off: 0v
A9	---	Not Used	---
A10	RED	A/T: EGR Solenoid (Calif.)	Solenoid Off: 12v, On: 1v
A11	BLU/YEL	Electronic Air Control Valve	Pulse Signals
A12	GRN/BLK	Fuel Pump Relay	Relay Off: 12v, On: 1v
A13	YEL/BLK	Main Relay Power (B+)	12-14v
A14	GRN/BLK	Fuel Pump Relay	Relay Off: 12v, On: 1v
A15	YEL/BLK	Main Relay Power (B+)	12-14v
A16	BRN/BLK	Power Ground	<0.1v
A17	---	Not Used	---
A18	BLK/RED	Power Ground	<0.1v

1990-91 CRX Coupe 1.5L I4 MFI VIN ED8 [All] 20P 'B' Connector

PCM Pin #	W/Color	Circuit Description (20-Pin)	Value at Hot Idle
B1	WHT/GRN	Keep Alive Power (VBU)	12-14v
B2	ORN	Tandem Valve Solenoid	Solenoid Off: 12v, On: 1v
B3	YEL	A/C Clutch Relay	Relay Off: 12v, On: 1v
B4	YEL/GRN	Radiator Fan Relay	Relay Off: 12v, On: 1v
B5	WHT/YEL	Alternator Charging Signal	Lights Off: 12v, On: 0v
B6	GRN/ORN	Check Engine Light	MIL Off: 12v, On: 1v
B7	GRN	A/T: Park/Neutral Switch	In P/N: 0v, Others: 11v
B8	BLU/RED	A/C Switch Signal	Switch Off: 12v, On: 1v
B9	---	Not Used	---
B10	ORN	CYP Sensor Signal	AC pulse signals
B11	GRN/BLK	A/T: D4 Switch Signal	In D4: 0v, Others: 12v
B11	GRN	M/T: Clutch Switch Signal	Clutch In: 0v, Out: 12v
B12	WHT	CYP Sensor Ground	<0.050v
B13	BLU/WHT	Start Signal	Cranking: 9-11v
B14	BLU	Alternator 'FR' Signal	Digital Signals: 0-5-0v
B15	WHT	Igniter Control	Digital Signals: 0-12-0v
B16	YEL/RED	Vehicle Speed Sensor	Moving: pulse signals
B17	WHT	Igniter Control	Digital Signals: 0-12-0v
B18	---	Not Used	---
B19	GRN/RED	Electric Load Detector	Varies: 0.5-4.5v
B20	BRN	Ignition Timing Adjuster	0.5-4.5v

1990-91 CRX Coupe 1.5L I4 MFI VIN ED8 [All] 16P 'C' Connector

PCM Pin #	W/Color	Circuit Description (16-Pin)	Value at Hot Idle
C1	---	Not Used	---
C2	---	Not Used	---
C3	ORN/BLU	TDC Sensor Signal	AC pulse signals
C4	WHT/BLU	TDC Sensor Ground	<0.050v
C5	RED/YEL	IAT Sensor Signal	Varies w/temp. (0.5-4.9v)
C6	RED/WHT	ECT Sensor Signal	At 180ºF: 0.51v
C7	RED/BLU	Throttle Angle Sensor Signal	0.5-0.6v
C8	YEL	EGRV Lift Sensor (California)	1.2v
C9	RED/WHT	Atmospheric Pressure Sensor	2.76-2.96v at sea level
C10	GRN/WHT	Brake Switch Signal	Brake Off: 0v, On: 12v
C11	WHT	MAP Sensor Signal	0.8-0.9v
C12	GRN/WHT	Sensor Ground	<0.050v
C13	YEL/WHT	Sensor VREF	4.9-5.1v
C14	GRN/WHT	MAP Sensor Ground	<0.050v
C15	YEL/RED	MAP Sensor VREF	4.9-5.1v
C16	WHT	Oxygen Sensor Signal	0.1-1.1v

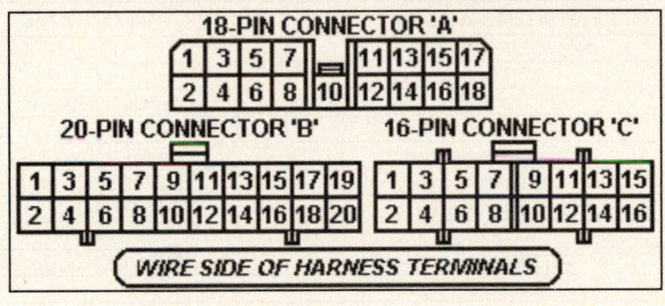

1990-91 CRX Coupe 1.6L I4 MFI VIN ED9 [M/T] 18P 'A' Connector

PCM Pin #	W/Color	Circuit Description (18-Pin)	Value at Hot Idle
A1	BRN	Injector 1 Control	2.0-3.3 ms
A2	BLK	Power Ground	<0.1v
A3	RED	Injector 2 Control	2.0-3.3 ms
A4	BLK	Power Ground	<0.1v
A5	BLU	Injector 3 Control	2.0-3.3 ms
A6	GRN	EVAP Purge Solenoid	Solenoid Off: 12v, On: 1v
A7	YEL	Injector 4 Control	2.0-3.3 ms
A8-9	---	Not Used	
A10	ORN	EGR Solenoid	Solenoid Off: 12v, On: 1v
A11	BLU/YEL	Electronic Air Control Valve	Pulse Signals
A12, 14	GRN/BLK	Fuel Pump Relay	Relay Off: 12v, On: 1v
A13, 15	YEL/BLK	Main Relay Power (B+)	12-14v
A16	BRN/BLK	Power Ground	<0.1v
A17	BLK/YEL	Ignition Power	12-14v
A18	BLK/RED	Power Ground	<0.1v

1990-91 CRX Coupe 1.6L I4 MFI VIN ED9 [M/T] 20P 'B' Connector

PCM Pin #	W/Color	Circuit Description (20-Pin)	Value at Hot Idle
B1	WHT/GRN	Keep Alive Power (VBU)	12-14v
B2	---	Not Used	---
B2	BLU	Fast Idle Control	Solenoid Off: 12v, On: 1v
B3	YEL	A/C Clutch Relay	Solenoid Off: 12v, On: 1v
B4	YEL/GRN	Radiator Fan Relay	Relay Off: 12v, On: 1v
B5	WHT/YEL	Alternator Charging Signal	Lights Off: 12v, On: 0v
B6	GRN/ORN	Check Engine Light	MIL Off: 12v, On: 1v
B7	---	Not Used	---
B8	BLU/RED	A/C Switch Signal	Switch Off: 12v, On: 1v
B9	---	Not Used	---
B10	ORN	CKP Sensor Signal	AC pulse signals
B11	---	Not Used	---
B12	WHT	CKP Sensor Ground	<0.050v
B13	BLU/WHT	Start Signal	Cranking: 9-11v
B14	BLU	Alternator 'FR' Signal	Digital Signals: 0-5-0v
B15	WHT	Igniter Control	Digital Signals: 0-12-0v
B16	YEL/RED	Vehicle Speed Sensor	Moving: pulse signals
B17	WHT	Igniter Control	Digital Signals: 0-12-0v
B18	---	Not Used	---
B19	GRN/RED	Electric Load Detector	Varies: 0.5-4.5v
B20	BRN	Ignition Timing Adjuster	0.5-4.5v

1990-91 CRX Coupe 1.6L I4 MFI VIN ED9 [M/T] 16P 'C' Connector

PCM Pin #	W/Color	Circuit Description (16-Pin)	Value at Hot Idle
C1	BLU/GRN	CYP Sensor Signal	AC pulse signals
C2	BLU/YEL	CYP Sensor Ground	<0.050v
C3	ORN/BLU	TDC Sensor Signal	AC pulse signals
C4	WHT/BLU	TDC Sensor Ground	<0.050v
C5	RED/YEL	IAT Sensor Signal	Varies w/temp. (0.5-4.9v)
C6	RED/WHT	ECT Sensor Signal	At 180°F: 0.51v
C7	RED/BLU	Throttle Angle Sensor Signal	0.5-0.6v
C8	---	Not Used	---
C9	RED/WHT	Atmospheric Pressure Sensor	2.76-2.96v at sea level
C10	GRN/WHT	Brake Switch Signal	Brake Off: 0v, On: 12v
C11	WHT	MAP Sensor Signal	0.8-0.9v
C12	GRN/WHT	Sensor Ground	<0.050v
C13	YEL/WHT	Sensor VREF	4.9-5.1v
C14	GRN/WHT	MAP Sensor Ground	<0.050v
C15	YEL/RED	MAP Sensor VREF	4.9-5.1v
C16	WHT	Oxygen Sensor Signal	0.1-1.1v

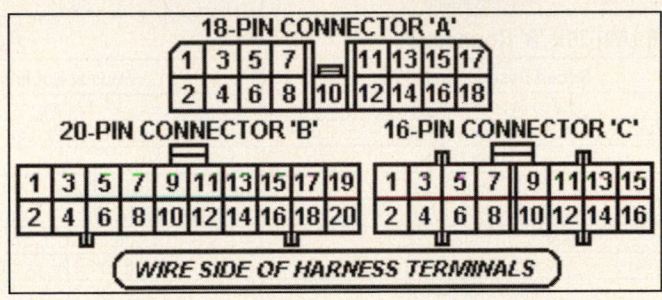

1990-91 Hatchback 1.5L I4 MFI VIN ED6 [All] 18P 'A' Connector

PCM Pin #	W/Color	Circuit Description (18-Pin)	Value at Hot Idle
A1	BRN	Injector 1 Control	2.0-3.3 ms
A2	BLK	Power Ground	<0.1v
A3	RED	Injector 2 Control	2.0-3.3 ms
A4	BLK	Power Ground	<0.1v
A5	BLU	Injector 3 Control	2.0-3.3 ms
A6	GRN	EVAP Purge Solenoid	Solenoid Off: 12v, On: 1v
A7	YEL	Injector 4 Control	2.0-3.3 ms
A8-9	---	Not Used	---
A10	ORN	EGR Solenoid	Solenoid Off: 12v, On: 1v
A11	BLU/YEL	Electronic Air Control Valve	Pulse Signals
A12	GRN/BLK	Fuel Pump Relay	Relay Off: 12v, On: 1v
A13	YEL/BLK	Main Relay Power (B+)	12-14v
A14	GRN/BLK	Fuel Pump Relay	Relay Off: 12v, On: 1v
A15	YEL/BLK	Main Relay Power (B+)	12-14v
A16	BRN/BLK	Power Ground	<0.1v
A17	BLK/YEL	Ignition Power	12-14v
A18	BLK/RED	Power Ground	<0.1v

1990-91 Hatchback 1.5L I4 MFI VIN ED6 [All] 20P 'B' Connector

PCM Pin #	W/Color	Circuit Description (20-Pin)	Value at Hot Idle
B1	WHT/GRN	Keep Alive Power (VBU)	12-14v
B2	GRN/YEL	Upshift Indicator Light	Lamp Off: 12v, On: 1v
B2	---	A/T: Not Used	---
B3	YEL	A/C Clutch Relay	Relay Off: 12v, On: 1v
B4	YEL/GRN	Radiator Fan Relay	Relay Off: 12v, On: 1v
B5	WHT/YEL	Alternator Charging Signal	Lights Off: 12v, On: 0v
B6	GRN/ORN	Check Engine Light	MIL Off: 12v, On: 1v
B7	GRN	Heater Fan Switch	Fan Off: 0v, On: 12v
B8	BLU/RED	A/C Switch Signal	Switch Off: 12v, On: 1v
B9	GRN/BLK	Reverse Switch	In 'R': 0v, Others: 12v
B10	ORN	CKP Sensor Signal	AC pulse signals
B11	GRN	M/T: Clutch Switch	Clutch In: 0v, Out: 5v
B12	WHT	CKP Sensor Ground	<0.050v
B13	BLU/WHT	Start Signal	Cranking: 9-11v
B14	BLU	Alternator 'FR' Signal	Digital Signals: 0-5-0v
B15	WHT	Igniter Control	Digital Signals: 0-12-0v
B16	YEL/RED	Vehicle Speed Sensor	Moving: pulse signals
B17	WHT	Igniter Control	Digital Signals: 0-12-0v
B18	---	Not Used	---
B19	GRN/RED	Electric Load Detector	Varies: 0.5-4.5v
B20	BRN	Ignition Timing Adjuster	0.5-4.5v

1990-91 Hatchback 1.5L I4 MFI VIN ED6 [All] 16P 'C' Connector

PCM Pin #	W/Color	Circuit Description (16-Pin)	Value at Hot Idle
C1	BLU/GRN	CYP Sensor Signal	AC pulse signals
C2	BLU/YEL	CYP Sensor Ground	<0.050v
C3	ORN/BLU	TDC Sensor Signal	AC pulse signals
C4	WHT/BLU	TDC Sensor Ground	<0.050v
C5	RED/YEL	IAT Sensor Signal	Varies w/temp. (0.5-4.9v)
C6	RED/WHT	ECT Sensor Signal	At 180°F: 0.51v
C7	RED/BLU	Throttle Angle Sensor Signal	0.5-0.6v
C8	YEL	EGR Valve Lift Sensor	1.2v
C9	RED/WHT	Atmospheric Pressure Sensor	2.76-2.96v at sea level
C10	GRN/WHT	Brake Switch Signal	Brake Off: 0v, On: 12v
C11	WHT	MAP Sensor Signal	0.8-0.9v
C12	GRN/WHT	Sensor Ground	<0.050v
C13	YEL/WHT	Sensor VREF	4.9-5.1v
C14	GRN/WHT	MAP Sensor Ground	<0.050v
C15	YEL/RED	MAP Sensor VREF	4.9-5.1v
C16	WHT	Oxygen Sensor Signal	0.1-1.1v

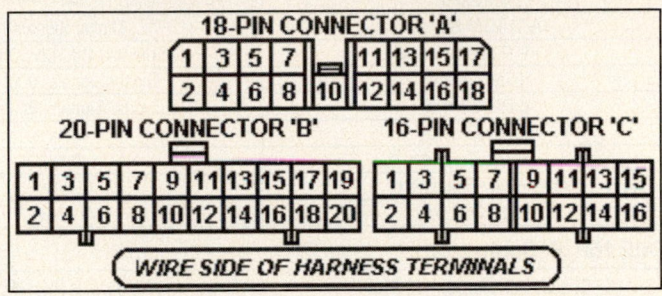

1992-93 Hatchback 1.5L MFI VIN EH2 [All] 26P 'A' Connector

PCM Pin #	W/Color	Circuit Description (26-Pin)	Value at Hot Idle
A1	YEL	Injector 1 Control	2.0-3.3 ms
A2	YEL	Injector 4 Control	2.0-3.3 ms
A3	RED	Injector 2 Control	2.0-3.3 ms
A4	---	Not Used	---
A5	BLU	Injector 3 Control	2.0-3.3 ms
A6	ORN/BLK	HO2S Heater Control	Heater on: 1v, Off: 12v
A7	GRN/YEL	Fuel Pump Relay	Relay Off: 12v, On: 1v
A8	GRN/YEL	Fuel Pump Relay	Relay Off: 12v, On: 1v
A9	GRN/WHT	Electronic Air Control Valve	Pulse Signals
A10-11	---	Not Used	---
A12	YEL/GRN	Radiator Fan Relay	Relay Off: 12v, On: 1v
A13	GRN/ORN	Check Engine Light	MIL Off: 12v, On: 1v
A14	---	Not Used	---
A15	BLK/RED	A/C Clutch Relay	Relay Off: 12v, On: 1v
A16	WHT/YEL	Alternator Charging Signal	Lights Off: 12v, On: 0v
A17	GRN/BLK	A/T: TCM Signal	Digital Signals
A18	---	Not Used	---
A19	YEL	A/T: TCM Signal	Digital Signals
A20	RED	EVAP Purge Solenoid	Solenoid Off: 12v, On: 1v
A21	RED/GRN	Igniter Control	Digital Signals: 0-12-0v
A22	RED/GRN	Igniter Control	Digital Signals: 0-12-0v
A23	BLK	Power Ground	<0.1v
A24	BLK	Power Ground	<0.1v
A25	YEL/BLK	Main Relay Power (B+)	12-14v
A26	BLK/RED	Logic Ground	<0.050v

1992-93 Hatchback 1.5L MFI VIN EH2 [All] 16P 'B' Connector

PCM Pin #	W/Color	Circuit Description (16-Pin)	Value at Hot Idle
B1	WHT/GRN	Main Relay Power (B+)	12-14v
B2	BRN/BLK	Logic Ground	<0.050v
B3	GRN/BLU	A/T: TCM Signal	In D3: 0v, Others: 5v
B4	GRN/BLK	A/T: TCM Signal	In D4: 0v, Others: 5v
B5	BLU/RED	A/C Switch Signal	Switch Off: 12v, On: 1v
B6	---	Not Used	---
B7	GRN	A/T: Park Neutral Switch	In P/N: 0v, Others: 11v
B8	BRN/RED	PSP Switch	Straight: 0v, Turning: 11v
B9	BLU/WHT	Starter Switch Signal	Cranking: 9-11v
B10	YEL/BLU	Vehicle Speed Sensor	Moving: pulse signals
B11	ORN	CYP Sensor Signal	AC pulse signals
B12	WHT	CYP Sensor Ground	<0.050v
B13	ORN/BLU	TDC Sensor Signal	AC pulse signals
B14	WHT/BLU	TDC Sensor Ground	<0.050v
B15	BLU/GRN	CKP Sensor Signal	AC pulse signals
B16	BLU/YEL	CKP Sensor Ground	<0.050v

1992-93 Hatchback 1.5L MFI VIN EH2 [All] 22P 'D' Connector

PCM Pin #	W/Color	Circuit Description (22-Pin)	Value at Hot Idle
D1	WHT/BLU	Keep Alive Power (VBU)	12-14v
D2	GRN/WHT	Brake Switch Signal	Brake Off: 0v, On: 12v
D3	---	Not Used	---
D4	BRN	Service Check Connector	SCS Open: 4.80v
D5-6	---	Not Used	---
D7	LT BLU	Data Link Connector	5v
D8	---	Not Used	---
D9	PNK	Alternator 'FR' Signal	Digital Signals: 0-5-0v
D10	GRN/RED	Electric Load Detector	Varies: 0.5-4.5v
D11	PNK/BLK	TP Sensor Signal	0.5-0.6v
D12	---	Not Used	---
D13	RED/WHT	ECT Sensor Signal	At 180°F: 0.51v
D14	ORN/BLU	HO2S-11 (B1 S1) Signal	0.1-1.1v
D14	WHT	HO2S-11 (B1 S1) Signal (CX)	0.1-1.1v
D15	RED/YEL	IAT Sensor Signal	Varies w/temp. (0.5-4.9v)
D16	---	Not Used	---
D17	WHT	MAP Sensor Signal	0.8-0.9v
D18	PNK/GRN	Economy Driving Indicator	Digital Signals
D19	YEL/GRN	MAP Sensor VREF	4.9-5.1v
D20	YEL/WHT	Sensor VREF	4.9-5.1v
D21	GRN/WHT	MAP Sensor Ground	<0.050v
D21	GRN/BLU	MAP Sensor Ground (CX)	<0.050v
D22	GRN/WHT	Sensor Ground	<0.050v

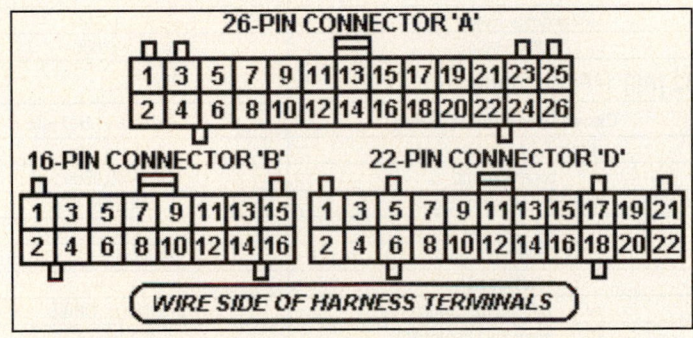

1994-95 Hatchback 1.5L I4 MFI VIN EH2 [All] 26P 'A' Connector

PCM Pin #	W/Color	Circuit Description (26-Pin)	Value at Hot Idle
A1	BRN	Injector 1 Control	2.0-3.3 ms
A2	YEL	Injector 4 Control	2.0-3.3 ms
A3	RED	Injector 2 Control	2.0-3.3 ms
A4	---	Not Used	---
A5	BLU	Injector 3 Control	2.0-3.3 ms
A6	ORN/BLK	HO2S Heater Control	Relay Off: 12v, On: 1v
A7	GRN/YEL	Fuel Pump Relay	Relay Off: 12v, On: 1v
A8	---	Not Used	---
A9	GRN/WHT	Idle Air Control Valve	Pulse Signals
A10	---	Not Used	---
A11	---	Not Used	---
A12	YEL/GRN	Radiator Fan Relay	Relay Off: 12v, On: 1v
A13	GRN/ORN	Check Engine Light	MIL Off: 12v, On: 1v
A14	---	Not Used	---
A15	BLK/RED	A/C Clutch Relay	Relay Off: 12v, On: 1v
A16	WHT/YEL	Alternator Charging Signal	Lights Off: 12v, On: 0v
A17	GRN/BLK	A/T: Lockup Control Solenoid	Solenoid Off: 12v, On: 1v
A18	---	Not Used	---
A19	YEL	A/T: Lockup Control Solenoid	Solenoid Off: 12v, On: 1v
A20	RED	EVAP Purge Solenoid	Solenoid Off: 12v, On: 1v
A21	RED/GRN	Igniter Control	Digital Signals: 0-12-0v
A22	---	Not Used	---
A23	BLK	Power Ground	<0.1v
A24	BLK	Power Ground	<0.1v
A25	YEL/BLK	Main Relay Power (B+)	12-14v
A26	BLK/RED	Logic Ground	<0.050v

1994-95 Hatchback 1.5L I4 MFI VIN EH2 [All] 16P 'B' Connector

PCM Pin #	W/Color	Circuit Description (16-Pin)	Value at Hot Idle
B1	WHT/GRN	Main Relay Power (B+)	12-14v
B2	BRN/BLK	Logic Ground	<0.050v
B3	GRN/BLU	A/T: Shift Selector Signal	In P/N: 12v
B4	GRN/BLK	A/T: Shift Selector Signal	In P/N: 12v
B5	BLU/RED	A/C Switch Signal	Switch Off: 12v, On: 1v
B6	---	Not Used	---
B7	GRN	A/T: Park Neutral Switch	In P/N: 0v, Others: 11v
B7	GRN	M/T: Clutch Switch Signal	Clutch in: 11v
B8	BRN/RED	PSP Switch	Straight: 0v, Turning: 11v
B9	BLU/WHT	Starter Switch Signal	Cranking: 9-11v
B10	YEL/BLU	Vehicle Speed Sensor	Moving: pulse signals
B11	ORN	CYP Sensor Signal	AC pulse signals
B12	WHT	CYP Sensor Ground	<0.050v
B13	ORN/BLU	TDC Sensor Signal	AC pulse signals
B14	WHT/BLU	TDC Sensor Ground	<0.050v
B15	BLU/GRN	CKP Sensor Signal	AC pulse signals
B16	BLU/YEL	CKP Sensor Ground	<0.050v

1994-95 Hatchback 1.5L I4 MFI VIN EH2 [All] 22P 'D' Connector

PCM Pin #	W/Color	Circuit Description (22-Pin)	Value at Hot Idle
D1	WHT/BLU	Keep Alive Power (VBU)	12-14v
D2	GRN/WHT	Brake Switch Signal	Brake Off: 0v, On: 12v
D3	RED/BLU	Knock Sensor Signal	No Detonation: 18mv AC
D4	BRN	Service Check Connector	SCS Open: 4.80v
D5-6	---	Not Used	---
D7	LT BLU	Data Link Connector	5v
D8	---	Not Used	---
D9	PNK	Alternator 'FR' Signal	Digital Signals: 0-5-0v
D10	GRN/RED	Electric Load Detector	Varies: 0.5-4.5v
D11	PNK/BLK	TP Sensor Signal	0.5-0.6v
D12	---	Not Used	---
D13	RED/WHT	ECT Sensor Signal	At 180°F: 0.51v
D14	WHT	HO2S-11 (B1 S1) Signal	0.1-1.1v
D14	ORN/BLU	Oxygen Sensor Signal (CX)	0.1-1.1v
D15	RED/YEL	IAT Sensor Signal	Varies w/temp. (0.5-4.9v)
D16	---	Not Used	---
D17	WHT	MAP Sensor Signal	0.8-0.9v
D18	PNK/GRN	A/T: Interlock Control Unit	Key & Brake On: 12v
D19	YEL/GRN	MAP Sensor VREF	4.9-5.1v
D20	YEL/WHT	Sensor VREF	4.9-5.1v
D21	GRN/BLU	MAP Sensor Ground (CX, EX)	<0.050v
D21	GRN/WHT	MAP Sensor Ground (DC, LX)	<0.050v
D22	GRN/WHT	Sensor Ground	<0.050v

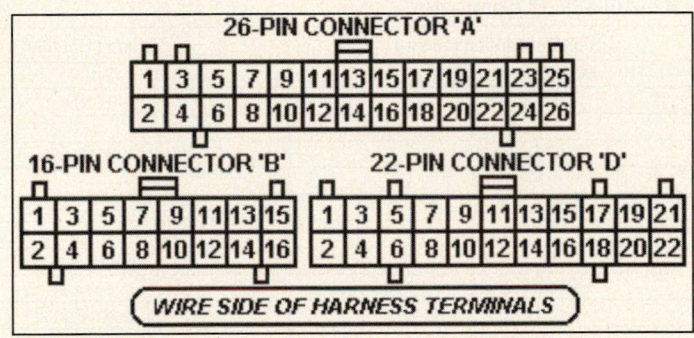

1992-93 Hatchback 1.5L I4 VTEC VIN EH2 [All] 26P 'A' Connector

PCM Pin #	W/Color	Circuit Description (26-Pin)	Value at Hot Idle
A1	BRN	Injector 1 Control	2.0-3.3 ms
A2	YEL	Injector 4 Control	2.0-3.3 ms
A3	RED	Injector 2 Control	2.0-3.3 ms
A4	ORN/WHT	VTEC-E Solenoid	0v, Hi-Speed: 12v
A5	BLU	Injector 3 Control	2.0-3.3 ms
A6	ORN/BLK	HO2S Heater Control	Relay Off: 12v, On: 1v
A7, 8	GRN/YEL	Fuel Pump Relay	Relay Off: 12v, On: 1v
A9	GRN/WHT	Idle Air Control Valve	Pulse Signals
A10	---	Not Used	---
A11	ORN/BLU	EGR Valve Lift Sensor	1.2v
A12	YEL/GRN	Radiator Fan Relay	Relay Off: 12v, On: 1v
A13	GRN/ORN	Check Engine Light	MIL Off: 12v, On: 1v
A14	---	Not Used	---
A15	BLK/RED	A/C Clutch Relay	Relay Off: 12v, On: 1v
A16	WHT/YEL	Alternator Charging Signal	Lights Off: 12v, On: 0v
A17-19	---	Not Used	---
A20	RED	EVAP Purge Solenoid	Solenoid Off: 12v, On: 1v
A21	RED/GRN	Igniter Control	Digital Signals: 0-12-0v
A22	RED/GRN	Igniter Control	Digital Signals: 0-12-0v
A23	BLK	Power Ground	<0.1v
A24	BLK	Power Ground	<0.1v
A25	YEL/BLK	Main Relay Power (B+)	12-14v
A26	BLK/RED	Power Ground	<0.1v

1992-93 Hatchback 1.5L I4 VTEC VIN EH2 [All] 16P 'B' Connector

PCM Pin #	W/Color	Circuit Description (16-Pin)	Value at Hot Idle
B1	WHT/GRN	Main Relay Power (B+)	12-14v
B2	BRN/BLK	Logic Ground	<0.050v
B3-4	---	Not Used	---
B5	BLU/RED	A/C Switch Signal	Switch Off: 12v, On: 1v
B6	---	Not Used	---
B7	GRN	A/T: Park Neutral Switch	In P/N: 0v, Others: 11v
B8	BRN/RED	PSP Switch	Straight: 0v, Turning: 11v
B9	BLU/WHT	Starter Switch Signal	Cranking: 9-11v
B10	YEL/BLU	Vehicle Speed Sensor	Moving: pulse signals
B11	ORN	CYP Sensor Signal	AC pulse signals
B12	WHT	CYP Sensor Ground	<0.050v
B13	ORN/BLU	TDC Sensor Signal	AC pulse signals
B14	WHT/BLU	TDC Sensor Ground	<0.050v
B15	BLU/GRN	CKP Sensor Signal	AC pulse signals
B16	BLU/YEL	CKP Sensor Ground	<0.050v

1992-93 Hatchback 1.5L I4 VTEC VIN EH2 [All] 22P 'D' Connector

PCM Pin #	W/Color	Circuit Description (22-Pin)	Value at Hot Idle
D1	WHT/BLU	Keep Alive Power (VBU)	12-14v
D2	GRN/WHT	Brake Switch Signal	Brake Off: 0v, On: 12v
D3	BLU/YEL	LAF Signal	0.3-4.9v
D4	BRN	Service Check Connector	SCS Open: 4.80v
D6	ORN/BLU	VTEC-E Pressure Switch	0v, Hi-Speed: 12v
D7	LT BLU	Data Link Connector	5v
D8	WHT/BLU	LAF Sensor VS+ Signal	0.5-0.6v
D9	PNK	Alternator 'FR' Signal	Digital Signals: 0-5-0v
D10	GRN/RED	Electric Load Detector	Varies: 0.5-4.5v
D11	LT GRN	Throttle Angle Sensor Signal	0.5-0.6v
D12	WHT/BLK	EGR Valve Lift Sensor	1.2v
D13	RED/WHT	ECT Sensor Signal	At 180°F: 0.51v
D14	ORN/BLU	LAF Sensor IP+ Signal	0.5-5.3v
D15	RED/YEL	IAT Sensor Signal	Varies w/temp. (0.5-4.9v)
D16	BLU/GRN	LAF Sensor IP-, VS- Signal	2.6-2.8v
D17	WHT	MAP Sensor Signal	0.8-0.9v
D18	PNK/GRN	Economy Driving Indicator	---
D19	YEL/GRN	MAP Sensor VREF	4.9-5.1v
D20	YEL/WHT	Sensor VREF	4.9-5.1v
D21	GRN/WHT	MAP Sensor Ground	<0.050v
D22	GRN/WHT	Sensor Ground	<0.050v

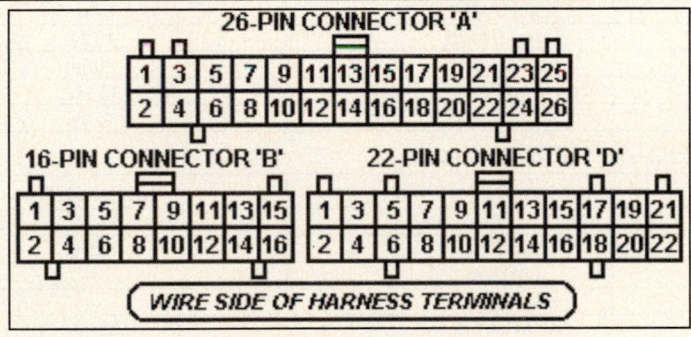

1994-95 Hatchback 1.5L I4 VTEC VIN EH2 [All] 26P 'A' Connector

PCM Pin #	W/Color	Circuit Description (26-Pin)	Value at Hot Idle
A1	BRN	Injector 1 Control	2.0-3.3 ms
A2	YEL	Injector 4 Control	2.0-3.3 ms
A3	RED	Injector 2 Control	2.0-3.3 ms
A4	ORN/WHT	VTEC Solenoid	0v, Hi-Speed: 12v
A5	LT BLU	Injector 3 Control	2.0-3.3 ms
A6	ORN/BLK	HO2S Heater Control	Relay Off: 12v, On: 1v
A7	GRN/YEL	Fuel Pump Relay	Relay Off: 12v, On: 1v
A8	---	Not Used	---
A9	GRN/WHT	Idle Air Control Valve	Pulse Signals
A10	---	Not Used	---
A11	PNK/GRN	EGR Valve Lift Sensor	1.2v
A12	YEL/GRN	Radiator Fan Relay	Relay Off: 12v, On: 1v
A13	GRN/ORN	Check Engine Light	MIL Off: 12v, On: 1v
A14	---	Not Used	---
A15	BLK/RED	A/C Clutch Relay	Relay Off: 12v, On: 1v
A16	WHT/YEL	Alternator Charging Signal	Lights Off: 12v, On: 0v
A17	GRN/BLK	A/T: Lockup Control Solenoid	Solenoid Off: 12v, On: 1v
A18	---	Not Used	---
A19	YEL	A/T: Lockup Control Solenoid	Solenoid Off: 12v, On: 1v
A20	RED	EVAP Purge Solenoid	Solenoid Off: 12v, On: 1v
A21	RED/GRN	Igniter Control	Digital Signals: 0-12-0v
A22	---	Not Used	---
A23	BLK	Power Ground	<0.1v
A24	BLK	Power Ground	<0.1v
A25	YEL/BLK	Main Relay Power (B+)	12-14v
A26	BLK/RED	Power Ground	<0.1v

1994-95 Hatchback 1.5L I4 VTEC VIN EH2 [All] 16P 'B' Connector

PCM Pin #	W/Color	Circuit Description (16-Pin)	Value at Hot Idle
B1	WHT/GRN	Main Relay Power (B+)	12-14v
B2	BRN/BLK	Logic Ground	<0.050v
B3	---	Not Used	---
B4	---	Not Used	---
B5	BLU/RED	A/C Switch Signal	Switch Off: 12v, On: 1v
B6	---	Not Used	---
B7	GRN	M/T: Clutch Switch	Clutch In: 0v, Out: 5v
B8	BRN/RED	PSP Switch	Straight: 0v, Turning: 11v
B9	BLU/WHT	Starter Switch Signal	Cranking: 9-11v
B10	YEL/BLU	Vehicle Speed Sensor	Moving: pulse signals
B11	ORN	CYP Sensor Signal	AC pulse signals
B12	WHT	CYP Sensor Ground	<0.050v
B13	ORN/BLU	TDC Sensor Signal	AC pulse signals
B14	WHT/BLU	TDC Sensor Ground	<0.050v
B15	BLU/GRN	CKP Sensor Signal	AC pulse signals
B16	BLU/YEL	CKP Sensor Ground	<0.050v

1994-95 Hatchback 1.5L I4 VTEC VIN EH2 [All] 22P 'D' Connector

PCM Pin #	W/Color	Circuit Description (22-Pin)	Value at Hot Idle
D1	WHT/BLU	Keep Alive Power (VBU)	12-14v
D2	GRN/WHT	Brake Switch Signal	Brake Off: 0v, On: 12v
D3	BLU/YEL	LAF Sensor Signal	0.3-4.9v
D4	BRN	Service Check Connector	SCS Open: 4.80v
D6	ORN/BLU	VTEC Pressure Switch	0v, Hi-Speed: 12v
D7	LT BLU	Data Link Connector	5v
D8	WHT/BLU	LAF Sensor VS+ Signal	0.5-0.6v
D9	PNK	Alternator 'FR' Signal	Digital Signals: 0-5-0v
D10	GRN/RED	Electric Load Detector	Varies: 0.5-4.5v
D11	PNK/BLK	TP Sensor Signal	0.5-0.6v
D12	WHT/BLK	EGR Valve Lift Sensor	1.2v
D13	RED/WHT	ECT Sensor Signal	At 180°F: 0.51v
D14	ORN/BLU	LAF Sensor IP+ Signal	0.5-5.3v
D15	RED/YEL	IAT Sensor Signal	Varies w/temp. (0.5-4.9v)
D16	BLU/GRN	LAF Sensor IP-, VS- Signal	2.6-2.8v
D17	WHT	MAP Sensor Signal	0.8-0.9v
D18	PNK/GRN	Economy Driving Indicator	Digital Signals
D19	YEL/GRN	MAP Sensor VREF	4.9-5.1v
D20	YEL/WHT	Sensor VREF	4.9-5.1v
D21	GRN/WHT	MAP Sensor Ground	<0.050v
D22	GRN/WHT	Sensor Ground	<0.050v

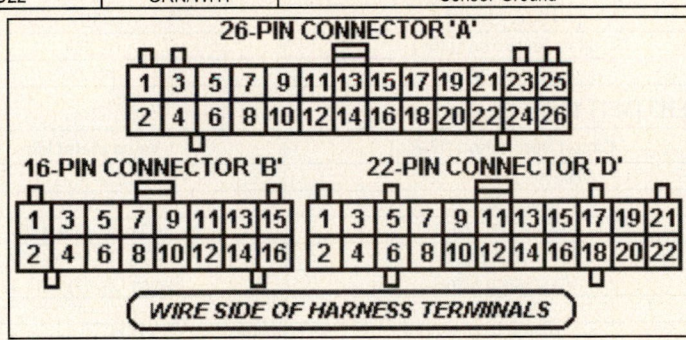

1992-93 Hatchback 1.6L I4 VTEC VIN EH3 [M/T] 26P A Connector

PCM Pin #	W/Color	Circuit Description (26-Pin)	Value at Hot Idle
A1	BRN	Injector 1 Control	2.0-3.3 ms
A2	YEL	Injector 4 Control	2.0-3.3 ms
A3	RED	Injector 2 Control	2.0-3.3 ms
A4	ORN/WHT	VTEC Solenoid	Solenoid Off: 12v, On: 1v
A5	BLU	Injector 3 Control	2.0-3.3 ms
A6	ORN/BLK	HO2S Heater Control	Relay Off: 12v, On: 1v
A7, 8	GRN/YEL	Fuel Pump Relay	Relay Off: 12v, On: 1v
A9	GRN/WHT	Idle Air Control Valve	Pulse Signals
A10-11	---	Not Used	
A12	YEL/GRN	Radiator Fan Relay	Relay Off: 12v, On: 1v
A13	GRN/ORN	Check Engine Light	MIL Off: 12v, On: 1v
A14	---	Not Used	
A15	BLK/RED	A/C Clutch Relay	Relay Off: 12v, On: 1v
A16	WHT/YEL	Alternator Charging Signal	Lights Off: 12v, On: 0v
A17	LT GRN	A/T: TCM Signal	Digital Signals
A18	---	Not Used	---
A19	YEL	A/T: TCM Signal	Digital Signals
A20	RED	EVAP Purge Solenoid	Solenoid Off: 12v, On: 1v
A21	RED/GRN	Igniter Control	Digital Signals: 0-12-0v
A22	RED/GRN	Igniter Control	Digital Signals: 0-12-0v
A23	BLK	Power Ground	<0.1v
A24	BLK	Power Ground	<0.1v
A25	YEL/BLK	Main Relay Power (B+)	12-14v
A26	BLK/RED	Power Ground	<0.1v

1992-93 Hatchback 1.6L I4 VTEC VIN EH3 [M/T] 16P B Connector

PCM Pin #	W/Color	Circuit Description (16-Pin)	Value at Hot Idle
B1	WHT/GRN	Main Relay Power (B+)	12-14v
B2	BRN/BLK	Logic Ground	<0.050v
B3	GRN/BLU	A/T: TCM Signal	In D3: 0v, Others: 5v
B4	GRN/BLK	A/T: TCM Signal	In D4: 0v, Others: 5v
B5	BLU/RED	A/C Switch Signal	Switch Off: 12v, On: 1v
B6	---	Not Used	---
B7	GRN	A/T: Park Neutral Switch	In P/N: 0v, Others: 11v
B8	BRN/RED	PSP Switch	Straight: 0v, Turning: 11v
B9	BLU/WHT	Starter Switch Signal	Cranking: 9-11v
B10	YEL/BLU	Vehicle Speed Sensor	Moving: pulse signals
B11	ORN	CYP Sensor Signal	AC pulse signals
B12	WHT	CYP Sensor Ground	<0.050v
B13	ORN/BLU	TDC Sensor Signal	AC pulse signals
B14	WHT/BLU	TDC Sensor Ground	<0.050v
B15	BLU/GRN	CKP Sensor Signal	AC pulse signals
B16	BLU/YEL	CKP Sensor Ground	<0.050v

1992-93 Hatchback 1.6L I4 VTEC VIN EH3 [M/T] 22P D Connector

PCM Pin #	W/Color	Circuit Description (22-Pin)	Value at Hot Idle
D1	WHT/BLU	Keep Alive Power (VBU)	12-14v
D2	GRN/WHT	Brake Switch Signal	Brake Off: 0v, On: 12v
D4	BRN	Data Link Connector	5v
D6	ORN/BLU	VTEC Pressure Switch	0v, Hi-Speed: 12v
D7	LT BLU	Data Link Connector	5v
D9	PNK	Alternator 'FR' Signal	Digital Signals: 0-5-0v
D10	GRN/RED	Electric Load Detector	Varies: 0.5-4.5v
D11	LT GRN	Throttle Angle Sensor Signal	0.5-0.6v
D13	RED/WHT	ECT Sensor Signal	At 180°F: 0.51v
D14	ORN/BLU	HO2S Signal	0.1-1.1v
D15	RED/YEL	IAT Sensor Signal	Varies w/temp. (0.5-4.9v)
D17	WHT	MAP Sensor Signal	0.8-0.9v
D18	PNK/GRN	Economy Driving Indicator	Digital Signals
D19	YEL/GRN	MAP Sensor VREF	4.9-5.1v
D20	YEL/WHT	Sensor VREF	4.9-5.1v
D21	GRN/BLU	MAP Sensor Ground	<0.050v
D22	GRN/WHT	Sensor Ground	<0.050v

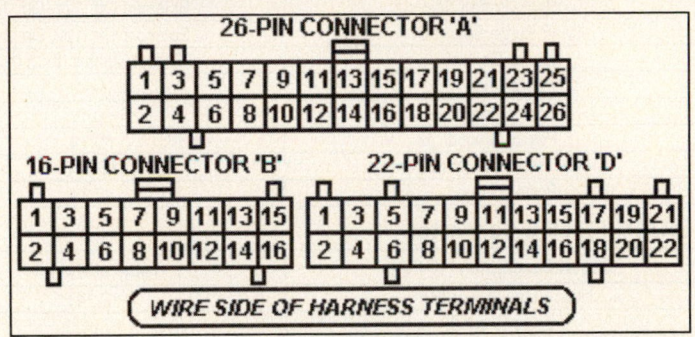

Abbreviation	Color	Abbreviation	Color	Abbreviation	Color
BLK	Black	LT BLU	Lt. Blue	TAN	Tan
BLU	Blue	LT GRN	Lt. Green	VIO	Violet
BRN	Brown	ORN	Orange	WHT	White
GRY	Gray	PNK	Pink	YEL	Yellow
GRN	Green	PPL	Purple		

1994-95 Hatchback 1.6L I4 VTEC VIN EH3 [M/T] 26P A Connector

PCM Pin #	W/Color	Circuit Description (26-Pin)	Value at Hot Idle
A1	BRN	Injector 1 Control	2.0-3.3 ms
A2	YEL	Injector 4 Control	2.0-3.3 ms
A3	RED	Injector 2 Control	2.0-3.3 ms
A4	ORN/WHT	VTEC Solenoid	0v, Hi-Speed: 12v
A5	BLU	Injector 3 Control	2.0-3.3 ms
A6	ORN/BLK	HO2S Heater Control	Relay Off: 12v, On: 1v
A7	GRN/YEL	Fuel Pump Relay	Relay Off: 12v, On: 1v
A8	---	Not Used	---
A9	GRN/WHT	Idle Air Control Valve	Pulse Signals
A10-11	---	Not Used	---
A12	YEL/GRN	Radiator Fan Relay	Relay Off: 12v, On: 1v
A13	GRN/ORN	Check Engine Light	MIL Off: 12v, On: 1v
A14	---	Not Used	---
A15	BLK/RED	A/C Clutch Relay	Relay Off: 12v, On: 1v
A16	WHT/YEL	Alternator Charging Signal	Lights Off: 12v, On: 0v
A17	GRN/BLK	A/T: Lockup Control Solenoid	Solenoid Off: 12v, On: 1v
A18	---	Not Used	---
A19	YEL	A/T: Lockup Control Solenoid	Solenoid Off: 12v, On: 1v
A20	RED	EVAP Purge Solenoid	Solenoid Off: 12v, On: 1v
A21	RED/GRN	Igniter Control	Digital Signals: 0-12-0v
A22	---	Not Used	---
A23	BLK	Power Ground	<0.1v
A24	BLK	Power Ground	<0.1v
A25	YEL/BLK	Main Relay Power (B+)	12-14v
A26	BLK/RED	Power Ground	<0.1v

1994-95 Hatchback 1.6L I4 VTEC VIN EH3 [M/T] 16P B Connector

PCM Pin #	W/Color	Circuit Description (16-Pin)	Value at Hot Idle
B1	WHT/GRN	Main Relay Power (B+)	12-14v
B2	BRN/BLK	Logic Ground	<0.050v
B3	GRN/BLU	A/T: Shift Selector Signal	In D4: 0v, Others: 12v
B4	GRN/BLK	A/T: Shift Selector Signal	In D3: 0v, Others: 12v
B5	BLU/RED	A/C Switch Signal	Switch Off: 12v, On: 1v
B6	---	Not Used	
B7	GRN	A/T: Park Neutral Switch	In P/N: 0v, Others: 12v
B8	BRN/RED	PSP Switch	Straight: 0v, Turning: 11v
B9	BLU/WHT	Starter Switch Signal	Cranking: 9-11v
B10	YEL/BLU	Vehicle Speed Sensor	Moving: pulse signals
B11	ORN	CYP Sensor Signal	AC pulse signals
B12	WHT	CYP Sensor Ground	<0.050v
B13	ORN/BLU	TDC Sensor Signal	AC pulse signals
B14	WHT/BLU	TDC Sensor Ground	<0.050v
B15	BLU/GRN	CKP Sensor Signal	AC pulse signals
B16	BLU/YEL	CKP Sensor Ground	<0.050v

1994-95 Hatchback 1.6L I4 VTEC VIN EH3 [M/T] 22P D Connector

PCM Pin #	W/Color	Circuit Description (22-Pin)	Value at Hot Idle
D1	WHT/BLU	Keep Alive Power (VBU)	12-14v
D2	GRN/WHT	Brake Switch Signal	Brake Off: 0v, On: 12v
D3	RED/BLU	Knock Sensor Signal	No Detonation: 18mv AC
D4	BRN	Service Check Connector	SCS Open: 4.80v
D5	---	Not Used	---
D6	ORN/BLU	VTEC Pressure Switch	0v, Hi-Speed: 12v
D7	LT BLU	Data Link Connector	5v
D8	---	Not Used	---
D9	PNK	Alternator 'FR' Signal	Digital Signals: 0-5-0v
D10	GRN/RED	Electric Load Detector	Varies: 0.5-4.5v
D11	PNK/BLK	TP Sensor Signal	0.5-0.6v
D12	---	Not Used	---
D13	RED/WHT	ECT Sensor Signal	At 180°F: 0.51v
D14	WHT	HO2S Signal	0.1-1.1v
D15	RED/YEL	IAT Sensor Signal	Varies w/temp. (0.5-4.9v)
D16	---	Not Used	---
D17	WHT	MAP Sensor Signal	0.8-0.9v
D18	PNK/GRN	A/T: Interlock Control Unit	Key & Brake On: 12v
D19	YEL/GRN	MAP Sensor VREF	4.9-5.1v
D20	YEL/WHT	Sensor VREF	4.9-5.1v
D21	GRN/BLU	MAP Sensor Ground	<0.050v
D22	GRN/WHT	Sensor Ground	<0.050v

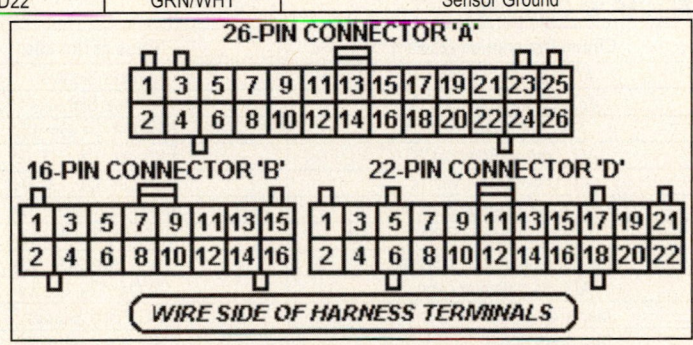

1996 Hatchback 1.6L I4 MFI VIN EJ6 [All] 32P 'A' Connector

PCM Pin #	W/Color	Circuit Description (32-Pin)	Value at Hot Idle
A1	YEL	Injector 4 Control	2.0-3.3 ms
A2	BLU	Injector 3 Control	2.0-3.3 ms
A3	RED	Injector 2 Control	2.0-3.3 ms
A4	BRN	Injector 1 Control	2.0-3.3 ms
A5	BLK/WHT	HO2S-12 (B1 S2) Heater	Digital Signals: 0-12-0v
A6	BLK/WHT	HO2S-11 (B1 S1) Heater	Digital Signals: 0-12-0v
A9	BRN/BLK	Logic Ground	<0.050v
A10, 23	BLK	Power Ground	<0.1v
A11	YEL/BLK	Main Relay Power (B+)	12-14v
A12	BLK/BLU	M/T: Idle Air Control Valve	Pulse Signals
A13	ORN	Idle Air Control Valve 'N'	Pulse Signals
A14	BLK/BLU	Idle Air Control Valve 'P'	Pulse Signals
A15	RED/YEL	EVAP Purge Solenoid	Solenoid Off: 12v, On: 1v
A16	GRN/YEL	Fuel Pump Relay	Relay Off: 12v, On: 1v
A17	BLK/RED	A/C Clutch Relay	Relay Off: 12v, On: 1v
A18	GRN/ORN	Check Engine Light	MIL Off: 12v, On: 1v
A19	WHT/GRN	Alternator Charging Signal	Lights Off: 12v, On: 0v
A20	YEL/GRN	Igniter Control	Digital Signals: 0-12-0v
A22	BRN/BLK	Logic Ground	<0.050v
A23	BLK	Power Ground	<0.1v
A27	GRN	Radiator Fan Relay	Relay Off: 12v, On: 1v

1996 Hatchback 1.6L I4 MFI VIN EJ6 [All] 25P 'B' Connector

PCM Pin #	W/Color	Circuit Description (25-Pin)	Value at Hot Idle
B1	WHT	A/T: Linear Solenoid (-)	Pulse Signals
B2	RED	A/T: Linear Solenoid (+)	Pulse Signals
B3	BLU/YEL	A/T: Shift Solenoid 'A'	SSA Off: 0v, On: 12v
B4	GRN/BLK	A/T: Lockup Solenoid 'B'	LSB On: 12v, Off: 0v
B5	YEL	A/T: Lockup Solenoid 'A'	LSA On: 12v, Off: 0v
B8	PNK	A/T: Gear Position Switch	In D3: 0v, Others: 12v
B11	GRN/WHT	A/T: Shift Solenoid 'B'	SSB Off: 0v, On: 12v
B12	WHT/RED	Interlock Control Unit Signal	Key & Brake On: 12v
B13	GRN/BLK	A/T: D4 Indicator Light	D4 On: 12v, Off: 0v
B14	WHT	Mainshaft Speed Sensor 'N'	AC Pulse Signals
B15	RED	Mainshaft Speed Sensor 'P'	AC Pulse Signals
B16	WHT	A/T: Gear Position Switch	In 'R': 0v, Others: 12v
B17	BLU	A/T: Gear Position Switch	In D2: 0v, Others: 12v
B22	GRN	Countershaft Speed Sensor N	Moving: AC pulses
B23	ORN	Countershaft Speed Sensor P	Moving: AC pulses
B24	YEL	A/T: Gear Position Switch	In D4: 0v, Others: 12v
B25	LT GRN	A/T: Gear Position Switch	In P/N: 0v, Others: 12v

```
            32-PIN CONNECTOR 'A'              25-PIN CONNECTOR 'B'
    ┌──┬──┬──┬──┐  ┌──┬──┬──┐  ┌──┬──┬────┬────┐  ┌──┬──┐  ┌──┬──┬──┐  ┌──┬──┬──┐
    │1 │2 │3 │4 │  │5 │6 │7 │  │8 │9 │ 10 │ 11 │  │1 │2 │  │3 │4 │5 │  │6 │7 │8 │
    ├──┼──┼──┼──┼──┼──┼──┼──┼──┼──┼────┼────┤  ├──┼──┼──┼──┼──┼──┼──┼──┼──┤
    │12│13│14│15│16│17│18│19│20│21│22│ 23 │ 24 │  │9 │ 10 │11│12│13│14│15│16│17│18│
    ├──┼──┼──┤  ├──┼──┼──┼──┤  │        │      │  ├────┼────┤  ├──┼──┤  ├──┬──┬──┐
    │25│26│27│  │28│29│30│31│  │   32   │      │  │ 19 │ 20 │  │21│22│  │23│24│25│
    └──┴──┴──┘  └──┴──┴──┴──┘  └────────┘      │  └────┴────┘  └──┴──┘  └──┴──┴──┘

                    ( WIRE SIDE OF HARNESS TERMINALS )
```

1996 Hatchback 1.6L I4 MFI VIN EJ6 [All] 31P 'C' Connector

PCM Pin #	W/Color	Circuit Description (31-Pin)	Value at Hot Idle
C1	BLU/RED	CKF Sensor Signal	AC pulse signals
C2	BLU	CKP Sensor Signal	AC pulse signals
C3	GRN	TDC Sensor Signal	AC pulse signals
C4	YEL	CYP Sensor Signal	AC pulse signals
C5	BLU/RED	A/C Switch Signal	Switch Off: 12v, On: 1v
C6	BLU/ORN	Starter Switch Signal	Cranking: 9-11v
C7	BRN	Service Check Connector	SCS Open: 4.80v
C8	LT BLU	K-Line Signal	12v
C9	---	Not Used	---
C10	WHT/BLU	Keep Alive Power (VBU)	12-14v
C11	WHT/RED	CKF Sensor Ground	<0.050v
C12	WHT	CKP Sensor Ground	<0.050v
C13	RED	TDC Sensor Ground	<0.050v
C14	BLK	CYP Sensor Ground	<0.050v
C16	GRN	PSP Switch	Straight: 0v, Turning: 11v
C17	WHT/RED	Alternator 'FR' Signal	Digital Signals: 0-5-0v
C18	BLU/WHT	Vehicle Speed Sensor	Moving: pulse signals
C19-29	---	Not Used	---
C30	PNK	A/T: CVT TMB Signal	Pulse Signals
C31	---	Not Used	---

1996 Hatchback 1.6L I4 MFI VIN EJ6 [All] 16P 'D' Connector

PCM Pin #	W/Color	Circuit Description (16-Pin)	Value at Hot Idle
D1	RED/BLK	TP Sensor Signal	0.5-0.6v
D2	RED/WHT	ECT Sensor Signal	At 180°F: 0.51v
D3	RED/GRN	MAP Sensor Signal	0.8-0.9v
D4	YEL/RED	MAP Sensor VREF	4.9-5.1v
D5	GRN/WHT	Brake Switch Signal	Brake Off: 0v, On: 12v
D7	WHT	HO2S-11 (B1 S1) Signal	0.1-1.1v
D8	RED/YEL	IAT Sensor Signal	Varies w/temp. (0.5-4.9v)
D9	---	Not Used	---
D10	YEL/BLU	Sensor VREF	4.9-5.1v
D11	GRN/BLK	Sensor Ground	<0.050v
D12	GRN/WHT	MAP Sensor Ground	<0.050v
D13	RED/YEL	HO2S-12 Ground	0.1-1.1v
D14	WHT/RED	HO2S-12 (B1 S2) Signal	0.1-1.1v
D16	GRN/RED	Electric Load Detector	Varies: 0.5-4.5v

```
 ┌───┬───┬───┬───┐  ┌───┬───┬───┐  ┌───┬───┬──┬──┐  │ 1│ 2│ 3│ 4│ 5│          │ 1│   │ 2│ 3│ 4│      5      │
 │11│12│13│14│15│16│17│18│19│20│21│22│          │ 6│ 7│ 8│ 9│10│11│   12     │
 │  │23│24│25│  │26│27│28│  │29│30│31│          │13│14│15│      16           │
```

31-PIN 'C'   *WIRE SIDE OF HARNESS TERMINALS*   16-PIN 'D'

1997-99 Hatchback 1.6L I4 MFI VIN EJ6 [All] 32P 'A' Connector

PCM Pin #	W/Color	Circuit Description (32-Pin)	Value at Hot Idle
A1, A2	YEL, BLU	Injector 4, Injector 3 Control	2.0-3.3 ms
A3, A4	RED, BRN	Injector 2, Injector 1 Control	2.0-3.3 ms
A5	BLK/WHT	HO2S-12 (B1 S2) Heater	Digital Signals: 0-12-0v
A6	BLK/WHT	HO2S-11 (B1 S1) Heater	Digital Signals: 0-12-0v
A7	RED	A/T: EGR Solenoid Control	Solenoid Off: 12v, On: 1v
A7	PNK	M/T: E-EGR Solenoid	Digital Signals: 0-12-0v
A9, A22	BRN/BLK	Logic Ground (LG1, LG2)	<0.050v
A10, 23	BLK	Power Ground (PG1, PG2)	<0.1v
A11, 24	YEL/BLK	Main Relay Power (B+)	12-14v
A12	BLK/BLU	M/T: Idle Air Control Valve	Pulse Signals
A13	ORN	A/T: Idle Air Control Valve 'N'	Pulse Signals
A14	BLK/BLU	A/T: Idle Air Control Valve 'P'	Pulse Signals
A15	RED/YEL	EVAP Purge Solenoid	Solenoid Off: 12v, On: 1v
A16	GRN/YEL	Fuel Pump Relay	Relay Off: 12v, On: 1v
A17	BLK/RED	A/C Clutch Relay	Relay Off: 12v, On: 1v
A18	GRN/ORN	Malfunction Indicator Light	MIL Off: 12v, On: 1v
A19	WHT/GRN	Alternator Charging Signal	Lights Off: 12v, On: 0v
A20	YEL/GRN	Igniter Control	Digital Signals: 0-12-0v
A27	GRN	Radiator Fan Relay Control	Relay Off: 12v, On: 1v
A28	BLU	EVAP Bypass Solenoid	Solenoid Off: 12v, On: 1v
A29	WHT/RED	EVAP Vent Solenoid	Solenoid Off: 12v, On: 1v

1997-99 Hatchback 1.6L I4 MFI VIN EJ6 [All] 25P 'B' Connector

PCM Pin #	W/Color	Circuit Description (25-Pin)	Value at Hot Idle
B1	WHT	A/T: Linear Solenoid (-)	Pulse Signals
B2	RED	A/T: Linear Solenoid (+)	Pulse Signals
B3	BLU/YEL	A/T: Shift Solenoid 'A'	SSA On: 12v; Off: 0v
B4	GRN/BLK	A/T: Lockup Solenoid 'B'	LSB On: 12v; Off: 0v
B5	YEL	A/T: Lockup Solenoid 'A'	LSA On: 12v; Off: 0v
B8	PNK	A/T: Gear Position Switch	In D3: 0v, Others: 12v
B11	GRN/WHT	A/T: Shift Solenoid 'B'	SSB Off: 0v, On: 12v
B12	WHT/RED	Interlock Control Unit Signal	Key & Brake On: 12v
B13	GRN/BLK	A/T: D4 Indicator Light	D4 On: 12v, Off: 0v
B14	WHT	Mainshaft Speed Sensor 'P'	AC Pulse Signals
B15	RED	Mainshaft Speed Sensor 'N'	AC Pulse Signals
B16	WHT	A/T: Gear Position Switch	In 'R': 0v, Others: 12v
B17	BLU	A/T: Gear Position Switch	In D2: 0v, Others: 12v
B22	GRN	Countershaft Speed Sensor N	Moving: AC pulses
B23	BLU	Countershaft Speed Sensor P	Moving: AC pulses
B24	YEL	A/T: Gear Position Switch	In D4: 0v, Others: 12v
B25	LT GRN	A/T: Gear Position Switch	In P/N: 0v, Others: 12v

32-Pin 'A'   *WIRE SIDE OF HARNESS TERMINALS*   25-Pin 'B'

1997-99 Hatchback 1.6L I4 MFI VIN EJ6 [All] 31P 'C' Connector

PCM Pin #	W/Color	Circuit Description (31-Pin)	Value at Hot Idle
C1	BLU/RED	CKF Sensor Signal	AC pulse signals
C2	BLU	CKP Sensor Signal	AC pulse signals
C3	GRN	TDC Sensor Signal	AC pulse signals
C4	YEL	CYP Sensor Signal	AC pulse signals
C5	BLU/RED	A/C Switch Signal	Switch Off: 12v, On: 1v
C6	BLU/ORN	Starter Switch Signal	Cranking: 9-11v
C7	BRN	Service Check Connector	SCS Open: 4.80v
C8	LT BLU	K-Line Signal	12v
C10	WHT/BLU	Keep Alive Power (VBU)	12-14v
C11	WHT/RED	CKF Sensor Ground	<0.050v
C12	WHT	CKP Sensor Ground	<0.050v
C13	RED	TDC Sensor Ground	<0.050v
C14	BLK	CYP Sensor Ground	<0.050v
C16	GRN	PSP Switch Signal	Straight: 0v, Turning: 11v
C17	WHT/RED	Alternator 'FR' Signal	Digital Signals: 0-5-0v
C18	BLU/WHT	Vehicle Speed Sensor	Moving: pulse signals
C23	BLK	LAF Pump Cell (IP+)	0.5-3.5v
C24	RED	LAF Common (IP-, VS-)	2.6-2.8v
C25	WHT	LAF VS Cell Voltage (VS+)	7v
C29	LT GRN	A/T: Gear Position Switch	In P/N: 0v, Others: 12v
C29	RED	M/T: Clutch Switch Signal	Clutch In: 0v, Out: 5v
C30	PNK	CVT TMB Signal	Pulse Signals

1997-99 Hatchback 1.6L I4 MFI VIN EJ6 16P 'D' Connector

PCM Pin #	W/Color	Circuit Description (16-Pin)	Value at Hot Idle
D1	RED/BLK	TP Sensor Signal	0.5-0.6v
D2	RED/WHT	ECT Sensor Signal	At 180°F: 0.51v
D3	RED/GRN	MAP Sensor Signal	0.8-0.9v
D4	YEL/RED	MAP Sensor VREF	4.9-5.1v
D5	GRN/WHT	Brake Switch Signal	Brake Off: 0v, On: 12v
D6	RED/BLU	Knock Sensor Signal	No detonation: 18mv AC
D7	WHT	HO2S-11 (B1 S1) Signal	0.1-1.1v
D8	RED/YEL	IAT Sensor Signal	Varies w/temp. (0.5-4.9v)
D9	WHT/BLK	EGR Valve Lift Sensor	1.2v
D10	YEL/RED	Sensor VREF	4.9-5.1v
D11	GRN/BLK	Sensor Ground	<0.050v
D12	GRN/WHT	MAP Sensor Ground	<0.050v
D13	RED/YEL	HO2S-12 Ground	<0.050v
D14	WHT/RED	HO2S-12 (B1 S2) Signal	0.1-1.1v
D15	LT GRN	Fuel Tank Pressure Sensor	Fuel Cap off: 2.5v
D16	GRN/RED	Electric Load Detector	Varies: 0.5-4.5v

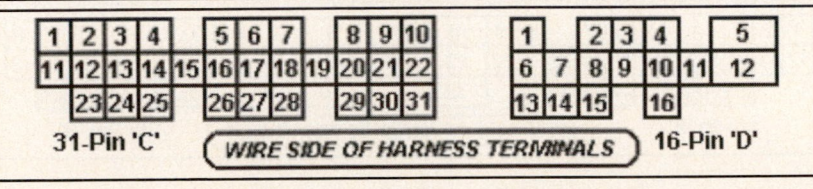

2000 Hatchback 1.6L I4 MFI VIN EJ6 [All] 32P 'A' Connector

PCM Pin #	W/Color	Circuit Description (32-Pin)	Value at Hot Idle
A1, A2	YEL, BLU	Injector 4, Injector 3 Control	2.0-3.3 ms
A3, A4	RED, BRN	Injector 2, Injector 1 Control	2.0-3.3 ms
A5	BLK/WHT	HO2S-12 (B1 S2) Heater	Digital Signals: 0-12-0v
A6	BLK/WHT	HO2S-11 (B1 S1) Heater	Digital Signals: 0-12-0v
A7	RED	A/T: EGR Solenoid Control	Solenoid Off: 12v, On: 1v
A7	PNK	M/T: E-EGR Solenoid	Digital Signals: 0-12-0v
A9, A22	BRN/BLK	Logic Ground (LG1, LG2)	<0.050v
A10, 23	BLK	Power Ground (PG1, PG2)	<0.1v
A11, 24	YEL/BLK	Main Relay Power (B+)	12-14v
A12	BLK/BLU	M/T: Idle Air Control Valve	Pulse Signals
A13	ORN	A/T: Idle Air Control Valve 'N'	Pulse Signals
A14	BLK/BLU	A/T: Idle Air Control Valve 'P'	Pulse Signals
A15	RED/YEL	EVAP Purge Solenoid	Solenoid Off: 12v, On: 1v
A16	GRN/YEL	Fuel Pump Relay	Relay Off: 12v, On: 1v
A17	BLK/RED	A/C Clutch Relay	Relay Off: 12v, On: 1v
A18	GRN/ORN	Malfunction Indicator Light	MIL Off: 12v, On: 1v
A19	WHT/GRN	Alternator Charging Signal	Lights Off: 12v, On: 0v
A20	YEL/GRN	Igniter Control	Digital Signals: 0-12-0v
A27	GRN	Radiator Fan Relay Control	Relay Off: 12v, On: 1v
A28	BLU	EVAP Bypass Solenoid	Solenoid Off: 12v, On: 1v
A29	WHT/RED	EVAP Vent Solenoid	Solenoid Off: 12v, On: 1v

2000 Hatchback 1.6L I4 MFI VIN EJ6 [All] 25P 'B' Connector

PCM Pin #	W/Color	Circuit Description (25-Pin)	Value at Hot Idle
B1	WHT	A/T: Linear Solenoid (-)	Pulse Signals
B2	RED	A/T: Linear Solenoid (+)	Pulse Signals
B3	BLU/YEL	A/T: Shift Solenoid 'A'	SSA On: 12v: Off: 0v
B4	GRN/BLK	A/T: Lockup Solenoid 'B'	LSB On: 12v: Off: 0v
B5	YEL	A/T: Lockup Solenoid 'A'	LSA On: 12v: Off: 0v
B8	PNK	A/T: Gear Position Switch	In D3: 0v, Others: 12v
B11	GRN/WHT	A/T: Shift Solenoid 'B'	SSB Off: 0v, On: 12v
B12	WHT/RED	Interlock Control Unit Signal	Key & Brake On: 12v
B13	GRN/BLK	A/T: D4 Indicator Light	D4 On: 12v, Off: 0v
B14	WHT	Mainshaft Speed Sensor 'P'	AC Pulse Signals
B15	RED	Mainshaft Speed Sensor 'N'	AC Pulse Signals
B16	WHT	A/T: Gear Position Switch	In 'R': 0v, Others: 12v
B17	BLU	A/T: Gear Position Switch	In D2: 0v, Others: 12v
B22	GRN	Countershaft Speed Sensor N	Moving: AC pulses
B23	BLU	Countershaft Speed Sensor P	Moving: AC pulses
B24	YEL	A/T: Gear Position Switch	In D4: 0v, Others: 12v
B25	LT GRN	A/T: Gear Position Switch	In P/N: 0v, Others: 12v

32-Pin 'A'   WIRE SIDE OF HARNESS TERMINALS   25-Pin 'B'

2000 Hatchback 1.6L I4 MFI VIN EJ6 [All] 31P 'C' Connector

PCM Pin #	W/Color	Circuit Description (31-Pin)	Value at Hot Idle
C1	BLU/RED	CKF Sensor Signal	AC pulse signals
C2	BLU	CKP Sensor Signal	AC pulse signals
C3	GRN	TDC Sensor Signal	AC pulse signals
C4	YEL	CYP Sensor Signal	AC pulse signals
C5	BLU/RED	A/C Switch Signal	Switch Off: 12v, On: 1v
C6	BLU/ORN	Starter Switch Signal	Cranking: 9-11v
C7	BRN	Service Check Connector	SCS Open: 4.80v
C8	LT BLU	K-Line Signal	12v
C10	WHT/BLU	Keep Alive Power (VBU)	12-14v
C11	WHT/RED	CKF Sensor Ground	<0.050v
C12	WHT	CKP Sensor Ground	<0.050v
C13	RED	TDC Sensor Ground	<0.050v
C14	BLK	CYP Sensor Ground	<0.050v
C16	GRN	PSP Switch Signal	Straight: 0v, Turning: 11v
C17	WHT/RED	Alternator 'FR' Signal	Digital Signals: 0-5-0v
C18	BLU/WHT	Vehicle Speed Sensor	Moving: pulse signals
C23	BLK	LAF Pump Cell (IP+)	0.5-3.5v
C24	RED	LAF Common (IP-, VS-)	2.6-2.8v
C25	WHT	LAF VS Cell Voltage (VS+)	7v
C29	LT GRN	A/T: Gear Position Switch	In P/N: 0v, Others: 12v
C29	RED	M/T: Clutch Switch Signal	Clutch In: 0v, Out: 5v
C30	PNK	CVT TMB Signal	Pulse Signals

2000 Hatchback 1.6L I4 MFI VIN EJ6 [All] 16P 'D' Connector

PCM Pin #	W/Color	Circuit Description (16-Pin)	Value at Hot Idle
D1	RED/BLK	TP Sensor Signal	0.5-0.6v
D2	RED/WHT	ECT Sensor Signal	At 180°F: 0.51v
D3	RED/GRN	MAP Sensor Signal	0.8-0.9v
D4, D10	YEL/RED	MAP Sensor, Sensor VREF	4.9-5.1v
D5	GRN/WHT	Brake Switch Signal	Brake Off: 0v, On: 12v
D6	RED/BLU	Knock Sensor Signal	No detonation: 18mv AC
D7	WHT	HO2S-11 (B1 S1) Signal	0.1-1.1v
D8	RED/YEL	IAT Sensor Signal	Varies w/temp. (0.5-4.9v)
D9	WHT/BLK	EGR Valve Lift Sensor	1.2v
D11	GRN/BLK	Sensor Ground	<0.050v
D12	GRN/WHT	MAP Sensor Ground	<0.050v
D13	RED/YEL	HO2S-12 Ground	<0.050v
D14	WHT/RED	HO2S-12 (B1 S2) Signal	0.1-1.1v
D15	LT GRN	Fuel Tank Pressure Sensor	Fuel Cap off: 2.5v
D16	GRN/RED	Electric Load Detector	Varies: 0.5-4.5v

```
 1  2  3  4     5  6  7     8  9 10        1     2  3  4      5
11 12 13 14 15 16 17 18 19 20 21 22        6  7  8  9 10 11  12
   23 24 25    26 27 28    29 30 31       13 14 15    16

   31-Pin 'C'     WIRE SIDE OF HARNESS TERMINALS    16-Pin 'D'
```

1990-91 Sedan 1.5L I4 MFI VIN ED3 [All] 18P 'A' Connector

PCM Pin #	W/Color	Circuit Description (18-Pin)	Value at Hot Idle
A1	BRN	Injector 1 Control	2.0-3.3 ms
A2	BLK	Power Ground	<0.1v
A3	RED	Injector 2 Control	2.0-3.3 ms
A4	BLK	Power Ground	<0.1v
A5	BLU	Injector 3 Control	2.0-3.3 ms
A6	GRN	EVAP Purge Solenoid	Solenoid Off: 12v, On: 1v
A7	YEL	Injector 4 Control	2.0-3.3 ms
A8-9	---	Not Used	---
A10	ORN	EGR Solenoid	Solenoid Off: 12v, On: 1v
A11	BLU/YEL	Electronic Air Control Valve	Pulse Signals
A12	GRN/BLK	Fuel Pump Relay	Relay Off: 12v, On: 1v
A13	YEL/BLK	Main Relay Power (B+)	12-14v
A14	GRN/BLK	Fuel Pump Relay	Relay Off: 12v, On: 1v
A15	YEL/BLK	Main Relay Power (B+)	12-14v
A16	BRN/BLK	Power Ground	<0.1v
A17	BLK/YEL	Ignition Power	12-14v
A18	BLK/RED	Power Ground	<0.1v

1990-91 Sedan 1.5L I4 MFI VIN ED3 [All] 20P 'B' Connector

PCM Pin #	W/Color	Circuit Description (20-Pin)	Value at Hot Idle
B1	WHT/GRN	Keep Alive Power (VBU)	12-14v
B2	GRN/YEL	Upshift Indicator Light	Lamp Off: 12v, On: 1v
B3	YEL	A/C Clutch Relay	Relay Off: 12v, On: 1v
B4	YEL/GRN	Radiator Fan Relay	Relay Off: 12v, On: 1v
B5	WHT/YEL	Alternator Charging Signal	Lights Off: 12v, On: 0v
B6	GRN/ORN	Check Engine Light	MIL Off: 12v, On: 1v
B7	GRN	Heater Fan Switch	Fan Off: 0v, On: 12v
B8	BLU/RED	A/C Switch Signal	Switch Off: 12v, On: 1v
B9	GRN/BLK	Reverse Switch	In 'R': 0v, Others: 12v
B10	ORN	CKP Sensor Signal	AC pulse signals
B11	GRN	Clutch Switch	Clutch In: 0v, Out: 5v
B12	WHT	CKP Sensor Ground	<0.050v
B13	BLU/WHT	Start Signal	Cranking: 9-11v
B14	BLU	Alternator 'FR' Signal	Digital Signals: 0-5-0v
B15	WHT	Igniter Control	Digital Signals: 0-12-0v
B16	YEL/RED	Vehicle Speed Sensor	Moving: pulse signals
B17	WHT	Igniter Control	Digital Signals: 0-12-0v
B18	---	Not Used	---
B19	GRN/RED	Electric Load Detector	Varies: 0.5-4.5v
B20	BRN	Ignition Timing Adjuster	0.5-4.5v

1990-91 Sedan 1.5L I4 MFI VIN ED3 [All] 16P 'C' Connector

PCM Pin #	W/Color	Circuit Description (16-Pin)	Value at Hot Idle
C1	BLU/GRN	CYP Sensor Signal	AC pulse signals
C2	BLU/YEL	CYP Sensor Ground	<0.050v
C3	ORN/BLU	TDC Sensor Signal	AC pulse signals
C4	WHT/BLU	TDC Sensor Ground	<0.050v
C5	RED/YEL	IAT Sensor Signal	Varies w/temp. (0.5-4.9v)
C6	RED/WHT	ECT Sensor Signal	At 180°F: 0.51v
C7	RED/BLU	Throttle Angle Sensor Signal	0.5-0.6v
C8	YEL	EGR Valve Lift Sensor	1.2v
C9	RED/WHT	Atmospheric Pressure Sensor	2.76-2.96v at sea level
C10	GRN/WHT	Brake Switch Signal	Brake Off: 0v, On: 12v
C11	WHT	MAP Sensor Signal	0.8-0.9v
C12	GRN/WHT	Sensor Ground	<0.050v
C13	YEL/WHT	Sensor VREF	4.9-5.1v
C14	GRN/WHT	MAP Sensor Ground	<0.050v
C15	YEL/RED	MAP Sensor VREF	4.9-5.1v
C16	WHT	Oxygen Sensor Signal	0.1-1.1v

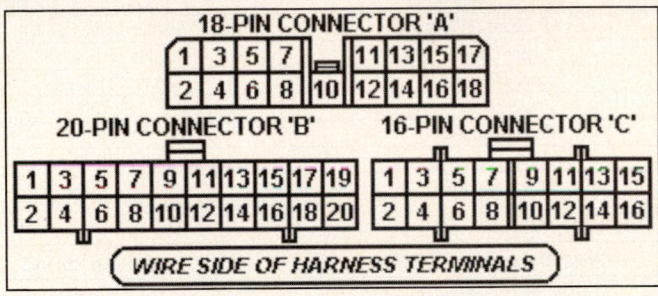

1992-93 Sedan 1.5L I4 MFI VIN EG8 [All] 26P 'A' Connector

PCM Pin #	W/Color	Circuit Description (26-Pin)	Value at Hot Idle
A1	BRN	Injector 1 Control	2.0-3.3 ms
A2	YEL	Injector 4 Control	2.0-3.3 ms
A3	RED	Injector 2 Control	2.0-3.3 ms
A4	BLU	Injector 2 Control	2.0-3.3 ms
A4	---	Not Used	---
A6	ORN/BLK	HO2S-11 (B1 S1) Heater	Heater Off: 12v, On: 1v
A7	GRN/YEL	Fuel Pump Relay	Relay Off: 12v, On: 1v
A8	GRN/YEL	Fuel Pump Relay	Relay Off: 12v, On: 1v
A9	GRN/WHT	Electronic Air Control Valve	Pulse Signals
A10-11	---	Not Used	---
A12	YEL/GRN	Radiator Fan Relay	Relay Off: 12v, On: 1v
A13	GRN/ORN	Check Engine Light	MIL Off: 12v, On: 1v
A14	---	Not Used	---
A15	BLK/RED	A/C Clutch Relay	Relay Off: 12v, On: 1v
A16	WHT/YEL	Alternator Charging Signal	Lights Off: 12v, On: 0v
A17	GRN/BLK	A/T: TCM Signal	In D3: 0v, Others: 5v
A19	YEL	A/T: TCM Signal	In D4: 0v, Others: 5v
A20	RED	EVAP Purge Solenoid	Solenoid Off: 12v, On: 1v
A21	RED/GRN	Igniter Control	Digital Signals: 0-12-0v
A22	RED/GRN	Igniter Control	Digital Signals: 0-12-0v
A23	BLK	Power Ground	<0.1v
A24	BLK	Power Ground	<0.1v
A25	YEL/BLK	Main Relay Power (B+)	12-14v
A26	BLK/RED	Power Ground	<0.1v

1992-93 Sedan 1.5L I4 MFI VIN EG8 [All] 16P 'B' Connector

PCM Pin #	W/Color	Circuit Description (16-Pin)	Value at Hot Idle
B1	WHT/GRN	Main Relay Power (B+)	12-14v
B2	BRN/BLK	Logic Ground	<0.050v
B3	GRN/BLU	A/T: TCM Signal	In D3: 0v, Others: 5v
B4	GRN/BLK	A/T: TCM Signal	In D4: 0v, Others: 5v
B5	BLU/RED	A/C Switch Signal	Switch Off: 12v, On: 1v
B6	---	Not Used	---
B7	GRN	A/T: Park Neutral Switch	In P/N: 0v, Others: 11v
B7	PNK/BLK	M/T: Clutch Pedal Switch	Clutch In: 0v, Out: 5v
B8	BRN/RED	PSP Switch	Straight: 0v, Turning: 11v
B9	BLU/WHT	Starter Switch Signal	Cranking: 9-11v
B10	YEL/BLU	Vehicle Speed Sensor	Moving: pulse signals
B11	ORN	CYP Sensor Signal	AC pulse signals
B12	WHT	CYP Sensor Ground	<0.050v
B13	ORN/BLU	TDC Sensor Signal	AC pulse signals
B14	WHT/BLU	TDC Sensor Ground	<0.050v
B15	BLU/GRN	CKP Sensor Signal	AC pulse signals
B16	BLU/YEL	CKP Sensor Ground	<0.050v

1992-93 Sedan 1.5L I4 MFI VIN EG8 [All] 22P 'D' Connector

PCM Pin #	W/Color	Circuit Description (22-Pin)	Value at Hot Idle
D1	WHT/BLU	Keep Alive Power (VBU)	12-14v
D2	GRN/WHT	Brake Switch Signal	Brake Off: 0v, On: 12v
D3	---	Not Used	---
D4	BRN	Service Check Connector	SCS Open: 4.80v
D5-6	---	Not Used	---
D7	LT BLU	Data Link Connector	5v
D8	---	Not Used	---
D9	PNK	Alternator 'FR' Signal	Digital Signals: 0-5-0v
D10	GRN/RED	Electric Load Detector	Varies: 0.5-4.5v
D11	LT GRN	Throttle Angle Sensor Signal	0.5-0.6v
D12	---	Not Used	---
D13	RED/WHT	ECT Sensor Signal	At 180°F: 0.51v
D14	WHT	O2S Signal (CX)	0.1-1.1v
D14	ORN/BLU	HO2S Signal ([All] others)	0.1-1.1v
D15	RED/YEL	IAT Sensor Signal	Varies w/temp. (0.5-4.9v)
D16	---	Not Used	---
D17	WHT	MAP Sensor Signal	0.8-0.9v
D18	PNK/GRN	Economy Driving Indicator	Digital Signals
D19	YEL/GRN	MAP Sensor VREF	4.9-5.1v
D20	YEL/WHT	Sensor VREF	4.9-5.1v
D21	GRN/WHT	MAP Sensor Ground (DX, LX)	<0.050v
D22	GRN/WHT	Sensor Ground	<0.050v

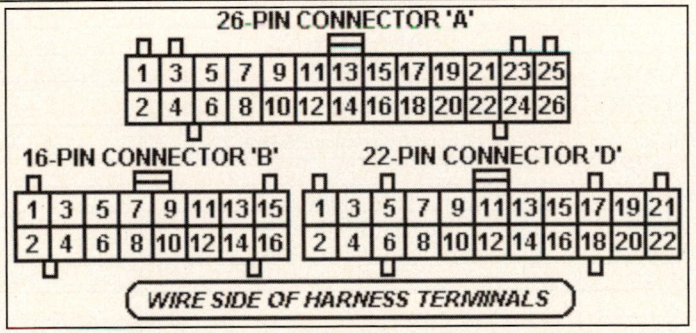

1994-95 Sedan 1.5L I4 MFI VIN EG8 [All] 26P 'A' Connector

PCM Pin #	W/Color	Circuit Description (26-Pin)	Value at Hot Idle
A1	BRN	Injector 1 Control	2.0-3.3 ms
A2	YEL	Injector 4 Control	2.0-3.3 ms
A3	RED	Injector 2 Control	2.0-3.3 ms
A4	---	Not Used	---
A5	BLU	Injector 3 Control	2.0-3.3 ms
A6	ORN/BLK	HO2S Heater Control	Relay Off: 12v, On: 1v
A7	GRN/YEL	Fuel Pump Relay	Relay Off: 12v, On: 1v
A8	---	Not Used	---
A9	GRN/WHT	Idle Air Control Valve	Pulse Signals
A10- 11	---	Not Used	---
A12	YEL/GRN	Radiator Fan Relay	Relay Off: 12v, On: 1v
A13	GRN/ORN	Check Engine Light	MIL Off: 12v, On: 1v
A14	---	Not Used	---
A15	BLK/RED	A/C Clutch Relay	Relay Off: 12v, On: 1v
A16	WHT/YEL	Alternator Charging Signal	Lights Off: 12v, On: 0v
A17	GRN/BLK	A/T: Lockup Control Solenoid	Solenoid Off: 12v, On: 1v
A18	---	Not Used	---
A19	YEL	A/T: Lockup Control Solenoid	Solenoid Off: 12v, On: 1v
A20	RED	EVAP Purge Solenoid	Solenoid Off: 12v, On: 1v
A21	RED/GRN	Igniter Control	Digital Signals: 0-12-0v
A22	---	Not Used	---
A23	BLK	Power Ground	<0.1v
A24	BLK	Power Ground	<0.1v
A25	YEL/BLK	Main Relay Power (B+)	12-14v
A26	BLK/RED	Logic Ground	<0.050v

1994-95 Sedan 1.5L I4 MFI VIN EG8 [All] 16P 'B' Connector

PCM Pin #	W/Color	Circuit Description (16-Pin)	Value at Hot Idle
B1	WHT/GRN	Main Relay Power (B+)	12-14v
B2	BRN/BLK	Logic Ground	<0.050v
B3	GRN/BLU	A/T: Shift Selector Signal	In D3: 0v, Others: 12v
B4	GRN/BLK	A/T: Shift Selector Signal	In D4: 0v, Others: 12v
B5	BLU/RED	A/C Switch Signal	Switch Off: 12v, On: 1v
B6	---	Not Used	---
B7	GRN	A/T: Park Neutral Switch	In P/N: 0v, Others: 11v
B7	GRN	M/T: Clutch Switch Signal	Clutch in: 11v
B8	BRN/RED	PSP Switch	Straight: 0v, Turning: 11v
B9	BLU/WHT	Starter Switch Signal	Cranking: 9-11v
B10	YEL/BLU	Vehicle Speed Sensor	Moving: pulse signals
B11	ORN	CYP Sensor Signal	AC pulse signals
B12	WHT	CYP Sensor Ground	<0.050v
B13	ORN/BLU	TDC Sensor Signal	AC pulse signals
B14	WHT/BLU	TDC Sensor Ground	<0.050v
B15	BLU/GRN	CKP Sensor Signal	AC pulse signals
B16	BLU/YEL	CKP Sensor Ground	<0.050v

1994-95 Sedan 1.5L I4 MFI VIN EG8 [All] 22P 'D' Connector

PCM Pin #	W/Color	Circuit Description (22-Pin)	Value at Hot Idle
D1	WHT/BLU	Keep Alive Power (VBU)	12-14v
D2	GRN/WHT	Brake Switch Signal	Brake Off: 0v, On: 12v
D3	RED/BLU	Knock Sensor Signal	No Detonation: 18mv AC
D4	BRN	Service Check Connector	SCS Open: 4.80v
D5-6	---	Not Used	---
D7	LT BLU	Data Link Connector	5v
D8	---	Not Used	---
D9	PNK	Alternator 'FR' Signal	Digital Signals: 0-5-0v
D10	GRN/RED	Electric Load Detector	Varies: 0.5-4.5v
D11	PNK/BLK	TP Sensor Signal	0.5-0.6v
D12	---	Not Used	---
D13	RED/WHT	ECT Sensor Signal	At 180°F: 0.51v
D14	WHT	HO2S Signal	0.1-1.1v
D15	RED/YEL	IAT Sensor Signal	Varies w/temp. (0.5-4.9v)
D16	---	Not Used	---
D17	WHT	MAP Sensor Signal	0.8-0.9v
D18	PNK/GRN	A/T: Interlock Control Unit	Key & Brake On: 12v
D19	YEL/GRN	MAP Sensor VREF	4.9-5.1v
D20	YEL/WHT	Sensor VREF	4.9-5.1v
D21	GRN/WHT	MAP Sensor GND	<0.050v
D22	GRN/WHT	Sensor Ground	<0.050v

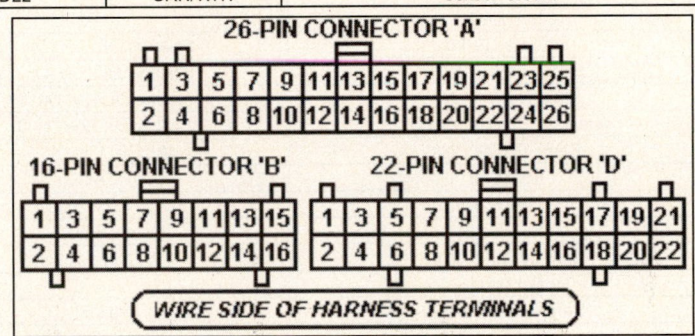

Abbreviation	Color	Abbreviation	Color	Abbreviation	Color
BLK	Black	LT BLU	Lt. Blue	TAN	Tan
BLU	Blue	LT GRN	Lt. Green	VIO	Violet
BRN	Brown	ORN	Orange	WHT	White
GRY	Gray	PNK	Pink	YEL	Yellow
GRN	Green	PPL	Purple		

1990-91 Sedan 1.6L I4 MFI VIN ED4 [All] 18P 'A' Connector

PCM Pin #	W/Color	Circuit Description (18-Pin)	Value at Hot Idle
A1	BRN	Injector 1 Control	2.0-3.3 ms
A2	BLK	Power Ground	<0.1v
A3	RED	Injector 2 Control	2.0-3.3 ms
A4	BLK	Power Ground	<0.1v
A5	BLU	Injector 3 Control	2.0-3.3 ms
A6	GRN	EVAP Purge Solenoid	Solenoid Off: 12v, On: 1v
A7	YEL	Injector 4 Control	2.0-3.3 ms
A8-9	---	Not Used	
A10	ORN	EGR Solenoid Control	Solenoid Off: 12v, On: 1v
A11	BLU/YEL	Electronic Air Control Valve	Pulse Signals
A12	GRN/BLK	Fuel Pump Relay	Relay Off: 12v, On: 1v
A13	YEL/BLK	Main Relay Power (B+)	12-14v
A14	GRN/BLK	Fuel Pump Relay	Relay Off: 12v, On: 1v
A15	YEL/BLK	Main Relay Power (B+)	12-14v
A16	BRN/BLK	Power Ground	<0.1v
A17	BLK/YEL	Ignition Power	12-14v
A18	BLK/RED	Power Ground	<0.1v

1990-91 Sedan 1.6L I4 MFI VIN ED4 [All] 20P 'B' Connector

PCM Pin #	W/Color	Circuit Description (20-Pin)	Value at Hot Idle
B1	WHT/GRN	Keep Alive Power (VBU)	12-14v
B2	---	Not Used	---
B2	BLU	Fast Idle Control	Solenoid Off: 12v, On: 1v
B3	YEL	A/C Clutch Relay	Solenoid Off: 12v, On: 1v
B4	YEL/GRN	Radiator Fan Relay	Relay Off: 12v, On: 1v
B5	WHT/YEL	Alternator Charging Signal	Lights Off: 12v, On: 0v
B6	GRN/ORN	Check Engine Light	MIL Off: 12v, On: 1v
B7	---	Not Used	---
B8	BLU/RED	A/C Switch Signal	Switch Off: 12v, On: 1v
B9	---	Not Used	---
B10	ORN	CKP Sensor Signal	AC pulse signals
B11	---	Not Used	---
B12	WHT	CKP Sensor Ground	<0.050v
B13	BLU/WHT	Start Signal	Cranking: 9-11v
B14	BLU	Alternator 'FR' Signal	Digital Signals: 0-5-0v
B15	WHT	Igniter Control	Digital Signals: 0-12-0v
B16	YEL/RED	Vehicle Speed Sensor	Moving: pulse signals
B17	WHT	Igniter Control	Digital Signals: 0-12-0v
B18	---	Not Used	---
B19	GRN/RED	Electric Load Detector	Varies: 0.5-4.5v
B20	BRN	Ignition Timing Adjuster	0.5-4.5v

1990-91 Sedan 1.6L I4 MFI VIN ED4 [All] 16P 'C' Connector

PCM Pin #	W/Color	Circuit Description (16-Pin)	Value at Hot Idle
C1	BLU/GRN	CYP Sensor Signal	AC pulse signals
C2	BLU/YEL	CYP Sensor Ground	<0.050v
C3	ORN/BLU	TDC Sensor Signal	AC pulse signals
C4	WHT/BLU	TDC Sensor Ground	<0.050v
C5	RED/YEL	IAT Sensor Signal	Varies w/temp. (0.5-4.9v)
C6	RED/WHT	ECT Sensor Signal	At 180°F: 0.51v
C7	RED/BLU	Throttle Angle Sensor Signal	0.5-0.6v
C8	---	Not Used	---
C9	RED/WHT	Atmospheric Pressure Sensor	2.76-2.96v at sea level
C10	GRN/WHT	Brake Switch Signal	Brake Off: 0v, On: 12v
C11	WHT	MAP Sensor Signal	0.8-0.9v
C12	GRN/WHT	Sensor Ground	<0.050v
C13	YEL/WHT	Sensor VREF	4.9-5.1v
C14	GRN/WHT	MAP Sensor Ground	<0.050v
C15	YEL/RED	MAP Sensor VREF	4.9-5.1v
C16	WHT	Oxygen Sensor Signal	0.1-1.1v

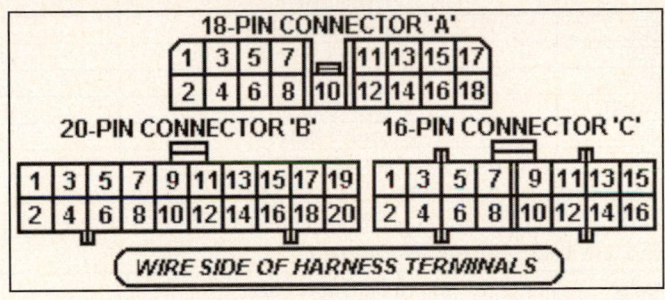

1992-93 Sedan 1.6L I4 VTEC VIN EH9 [All] 26P 'A' Connector

PCM Pin #	W/Color	Circuit Description (26-Pin)	Value at Hot Idle
A1	BRN	Injector 1 Control	2.0-3.3 ms
A2	YEL	Injector 4 Control	2.0-3.3 ms
A3	RED	Injector 2 Control	2.0-3.3 ms
A4	ORN/WHT	VTEC Solenoid	Solenoid Off: 12v, On: 1v
A5	BLU	Injector 3 Control	2.0-3.3 ms
A6	ORN/BLK	HO2S Heater Control	Relay Off: 12v, On: 1v
A7	GRN/YEL	Fuel Pump Relay	Relay Off: 12v, On: 1v
A8	GRN/YEL	Fuel Pump Relay	Relay Off: 12v, On: 1v
A9	GRN/WHT	Electronic Air Control Valve	Pulse Signals
A10-11	---	Not Used	---
A12	YEL/GRN	Radiator Fan Relay	Relay Off: 12v, On: 1v
A13	GRN/ORN	Check Engine Light	MIL Off: 12v, On: 1v
A14	---	Not Used	
A15	BLK/RED	A/C Clutch Relay	Relay Off: 12v, On: 1v
A16	WHT/YEL	Alternator Charging Signal	Lights Off: 12v, On: 0v
A17	GRN/BLK	A/T: TCM Signal	Digital Signals
A18	---	Not Used	
A19	YEL	A/T: TCM Signal	Digital Signals
A20	RED	EVAP Purge Solenoid	Solenoid Off: 12v, On: 1v
A21	RED/GRN	Igniter Control	Digital Signals: 0-12-0v
A22	RED/GRN	Igniter Control	Digital Signals: 0-12-0v
A23	BLK	Power Ground	<0.1v
A24	BLK	Power Ground	<0.1v
A25	YEL/BLK	Main Relay Power (B+)	12-14v
A26	BLK/RED	Power Ground	<0.1v

1992-93 Sedan 1.6L I4 VTEC VIN EH9 [All] 16P 'B' Connector

PCM Pin #	W/Color	Circuit Description (16-Pin)	Value at Hot Idle
B1	WHT/GRN	Main Relay Power (B+)	12-14v
B2	BRN/BLK	Logic Ground	<0.050v
B3	GRN/BLU	A/T: TCM Signal	In D3: 0v, Others: 5v
B4	GRN/BLK	A/T: TCM Signal	In D4: 0v, Others: 5v
B5	BLU/RED	A/C Switch Signal	Switch Off: 12v, On: 1v
B6	---	Not Used	---
B7	GRN	A/T: Park Neutral Switch	In P/N: 0v, Others: 11v
B8	BRN/RED	PSP Switch	Straight: 0v, Turning: 11v
B9	BLU/WHT	Starter Switch Signal	Cranking: 9-11v
B10	YEL/BLU	Vehicle Speed Sensor	Moving: pulse signals
B11	ORN	CYP Sensor Signal	AC pulse signals
B12	WHT	CYP Sensor Ground	<0.050v
B13	ORN/BLU	TDC Sensor Signal	AC pulse signals
B14	WHT/BLU	TDC Sensor Ground	<0.050v
B15	BLU/GRN	CKP Sensor Signal	AC pulse signals
B16	BLU/YEL	CKP Sensor Ground	<0.050v

1992-93 Sedan 1.6L I4 VTEC VIN EH9 [All] 22P 'D' Connector

PCM Pin #	W/Color	Circuit Description (22-Pin)	Value at Hot Idle
D1	WHT/BLU	Keep Alive Power (VBU)	12-14v
D2	GRN/WHT	Brake Switch Signal	Brake Off: 0v, On: 12v
D3	---	Not Used	---
D4	BRN	Service Check Connector	SCS Open: 4.80v
D5	---	Not Used	---
D6	ORN/BLU	VTEC Pressure Switch	0v, Hi-Speed: 12v
D7	LT BLU	Data Link Connector	5v
D8	---	Not Used	---
D9	PNK	Alternator 'FR' Signal	Digital Signals: 0-5-0v
D10	GRN/RED	Electric Load Detector	Varies: 0.5-4.5v
D11	LT GRN	Throttle Angle Sensor Signal	0.5-0.6v
D12	---	Not Used	---
D13	RED/WHT	ECT Sensor Signal	At 180ºF: 0.51v
D14	ORN/BLU	HO2S Signal	0.1-1.1v
D15	RED/YEL	IAT Sensor Signal	Varies w/temp. (0.5-4.9v)
D16	---	Not Used	---
D17	WHT	MAP Sensor Signal	0.8-0.9v
D18	PNK/GRN	Economy Driving Indicator	---
D19	YEL/GRN	MAP Sensor VREF	4.9-5.1v
D20	YEL/WHT	Sensor VREF	4.9-5.1v
D21	GRN/BLU	MAP Sensor GND	<0.050v
D22	GRN/WHT	Sensor Ground	<0.050v

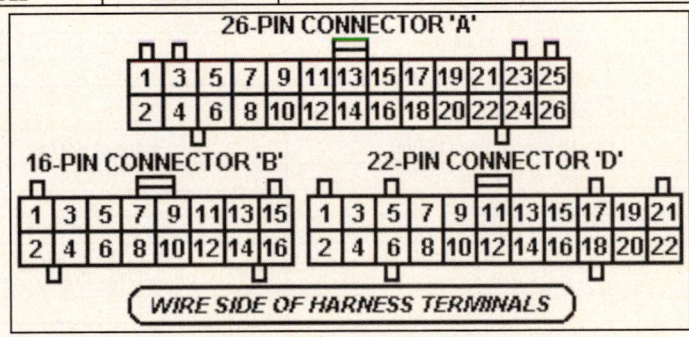

1994-95 Sedan 1.6L I4 VTEC VIN EH9 [All] 26P 'A' Connector

PCM Pin #	W/Color	Circuit Description (26-Pin)	Value at Hot Idle
A1	BRN	Injector 1 Control	2.0-3.3 ms
A2	YEL	Injector 4 Control	2.0-3.3 ms
A3	RED	Injector 2 Control	2.0-3.3 ms
A4	ORN/WHT	VTEC Solenoid	0v, Hi-Speed: 12v
A5	LT BLU	Injector 3 Control	2.0-3.3 ms
A6	ORN/BLK	HO2S Heater Control	Relay Off: 12v, On: 1v
A7	GRN/YEL	Fuel Pump Relay	Relay Off: 12v, On: 1v
A8	---	Not Used	---
A9	GRN/WHT	Intake Air Control Solenoid	Solenoid Off: 12v, On: 1v
A10-11	---	Not Used	---
A12	YEL/GRN	Radiator Fan Relay	Relay Off: 12v, On: 1v
A13	GRN/ORN	Check Engine Light	MIL Off: 12v, On: 1v
A14	---	Not Used	---
A15	BLK/RED	A/C Clutch Relay	Relay Off: 12v, On: 1v
A16	WHT/YEL	Alternator Charging Signal	Lights Off: 12v, On: 0v
A17	GRN/BLK	A/T: Lockup Control Solenoid	Solenoid Off: 12v, On: 1v
A18	---	Not Used	---
A19	YEL	A/T: Lockup Control Solenoid	Solenoid Off: 12v, On: 1v
A20	RED	EVAP Purge Solenoid	Solenoid Off: 12v, On: 1v
A21	RED/GRN	Igniter Control	Digital Signals: 0-12-0v
A22	---	Not Used	---
A23	BLK	Power Ground	<0.1v
A24	BLK	Power Ground	<0.1v
A25	YEL/BLK	Main Relay Power (B+)	12-14v
A26	BLK/RED	Power Ground	<0.1v

1994-95 Sedan 1.6L I4 VTEC VIN EH9 [All] 16P 'B' Connector

PCM Pin #	W/Color	Circuit Description (16-Pin)	Value at Hot Idle
B1	WHT/GRN	Main Relay Power (B+)	12-14v
B2	BRN/BLK	Logic Ground	<0.050v
B3	GRN/BLU	A/T: Shift Selector Signal	In D3: 0v, Others: 5v
B4	GRN/BLK	A/T: Shift Selector Signal	In D4: 0v, Others: 5v
B5	BLU/RED	A/C Switch Signal	Switch Off: 12v, On: 1v
B6	---	Not Used	---
B7	GRN	A/T: Park Neutral Switch	In P/N: 0v, Others: 11v
B8	BRN/RED	PSP Switch	Straight: 0v, Turning: 11v
B9	BLU/WHT	Starter Switch Signal	Cranking: 9-11v
B10	YEL/BLU	Vehicle Speed Sensor	Moving: pulse signals
B11	ORN	CYP Sensor Signal	AC pulse signals
B12	WHT	CYP Sensor Ground	<0.050v
B13	ORN/BLU	TDC Sensor Signal	AC pulse signals
B14	WHT/BLU	TDC Sensor Ground	<0.050v
B15	BLU/GRN	CKP Sensor Signal	AC pulse signals
B16	BLU/YEL	CKP Sensor Ground	<0.050v

1994-95 Sedan 1.6L I4 VTEC VIN EH9 [All] 22P 'D' Connector

PCM Pin #	W/Color	Circuit Description (22-Pin)	Value at Hot Idle
D1	WHT/BLU	Keep Alive Power (VBU)	12-14v
D2	GRN/WHT	Brake Switch Signal	Brake Off: 0v, On: 12v
D3	RED/BLU	Knock Sensor Signal	No Detonation: 18mv AC
D4	BRN	Service Check Connector	SCS Open: 4.80v
D5	---	Not Used	---
D6	ORN/BLU	VTEC Pressure Switch	0v, Hi-Speed: 12v
D7	LT BLU	Data Link Connector	5v
D8	---	Not Used	---
D9	PNK	Alternator 'FR' Signal	Digital Signals: 0-5-0v
D10	GRN/RED	Electric Load Detector	Varies: 0.5-4.5v
D11	PNK/BLK	TP Sensor Signal	0.5-0.6v
D12	---	Not Used	---
D13	RED/WHT	ECT Sensor Signal	At 180ºF: 0.51v
D14	WHT	HO2S Signal	0.1-1.1v
D15	RED/YEL	IAT Sensor Signal	Varies w/temp. (0.5-4.9v)
D16	---	Not Used	---
D17	WHT	MAP Sensor Signal	0.8-0.9v
D18	PNK/GRN	A/T: Interlock Control Unit	Key & Brake On: 12v
D19	YEL/GRN	MAP Sensor VREF	4.9-5.1v
D20	YEL/WHT	Sensor VREF	4.9-5.1v
D21	GRN/BLU	MAP Sensor Ground	<0.050v
D22	GRN/WHT	Sensor Ground	<0.050v

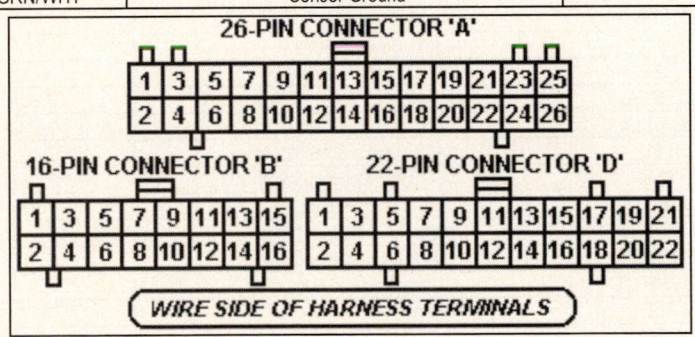

1996 Sedan 1.6L I4 MFI VIN EJ6 [All] 32P 'A' Connector

PCM Pin #	W/Color	Circuit Description (32-Pin)	Value at Hot Idle
A1	YEL	Injector 4 Control	2.0-3.3 ms
A2	BLU	Injector 3 Control	2.0-3.3 ms
A3	RED	Injector 2 Control	2.0-3.3 ms
A4	BRN	Injector 1 Control	2.0-3.3 ms
A5	BLK/WHT	HO2S-12 (B1 S2) Heater	Digital Signals: 0-12-0v
A6	BLK/WHT	HO2S-11 (B1 S1) Heater	Digital Signals: 0-12-0v
A9, A22	BRN/BLK	Logic Ground	<0.050v
A10, A23	BLK	Power Ground	<0.1v
A11	YEL/BLK	Main Relay Power (B+)	12-14v
A12	BLK/BLU	M/T: Idle Air Control Valve	Pulse Signals
A13	ORN	A/T Idle Air Control Valve 'N'	Pulse Signals
A14	BLK/BLU	A/T Idle Air Control Valve 'P'	Pulse Signals
A15	RED/YEL	EVAP Purge Solenoid	Solenoid Off: 12v, On: 1v
A16	GRN/YEL	Fuel Pump Relay	Relay Off: 12v, On: 1v
A17	BLK/RED	A/C Clutch Relay	Relay Off: 12v, On: 1v
A18	GRN/ORN	Check Engine Light	MIL Off: 12v, On: 1v
A19	WHT/GRN	Alternator Charging Signal	Lights Off: 12v, On: 0v
A20	YEL/GRN	Igniter Control	Digital Signals: 0-12-0v
A24	YEL/BLK	Main Relay Power (B+)	12-14v
A27	GRN	Radiator Fan Relay	Relay Off: 12v, On: 1v

1996 Sedan 1.6L I4 MFI VIN EJ6 [All] 25P 'B' Connector

PCM Pin #	W/Color	Circuit Description (25-Pin)	Value at Hot Idle
B1	WHT	A/T: Linear Solenoid (-)	Pulse Signals
B2	RED	A/T: Linear Solenoid (+)	Pulse Signals
B3	BLU/YEL	A/T: Shift Solenoid 'A'	Solenoid Off: 12v, On: 1v
B4	GRN/BLK	A/T: Lockup Solenoid 'B'	LSB On: 12v, Off: 0v
B5	YEL	A/T: Lockup Solenoid 'A'	LSA On: 12v, Off: 0v
B8	PNK	A/T: Gear Position Switch	In D3: 0v, Others: 12v
B11	GRN/WHT	A/T: Shift Solenoid 'B'	Solenoid Off: 12v, On: 1v
B12	WHT/RED	Interlock Control Unit Signal	Key & Brake On: 12v
B13	GRN/BLK	A/T: D4 Indicator Light	D4 On: 12v, Off: 0v
B14	WHT	Mainshaft Speed Sensor 'N'	<0.050v
B15	RED	Mainshaft Speed Sensor 'P'	AC Pulse Signals
B16	WHT	A/T: Gear Position Switch	In 'R': 0v, Others: 12v
B17	BLU	A/T: Gear Position Switch	In D2: 0v, Others: 12v
B22	GRN	Countershaft Speed Sensor N	<0.050v
B23	LT GRN	Countershaft Speed Sensor P	Moving: AC pulses
B24	YEL	A/T: Gear Position Switch	In D4: 0v, Others: 12v
B25	LT GRN	A/T: Gear Position Switch	In P/N: 0v, Others: 12v

1996 Sedan 1.6L I4 MFI VIN EJ6 [All] 31P 'C' Connector

PCM Pin #	W/Color	Circuit Description (31-Pin)	Value at Hot Idle
C1	BLU/RED	CKF Sensor Signal	AC pulse signals
C2	BLU	CKP Sensor Signal	AC pulse signals
C3	GRN	TDC Sensor Signal	AC pulse signals
C4	YEL	CYP Sensor Signal	AC pulse signals
C5	BLU/RED	A/C Switch Signal	Switch Off: 12v, On: 1v
C6	BLU/ORN	Starter Switch Signal	Cranking: 9-11v
C7	BRN	Service Check Connector	SCS Open: 4.80v
C8	LT BLU	K-Line Signal	12v
C9	---	Not Used	---
C10	WHT/BLU	Keep Alive Power (VBU)	12-14v
C11	WHT/RED	CKF Sensor Ground	0.050v
C12	WHT	CKP Sensor Ground	0.050v
C13	RED	TDC Sensor Ground	0.050v
C14	BLK	CYP Sensor Ground	0.050v
C16	GRN	PSP Switch	Straight: 0v, Turning: 11v
C17	WHT/RED	Alternator 'FR' Signal	Digital Signals: 0-5-0v
C18	BLU/WHT	Vehicle Speed Sensor	Moving: pulse signals
C19-29	---	Not Used	---
C30	PNK	A/T: CVT TMB Signal	Pulse Signals
C31	---	Not Used	---

1996 Sedan 1.6L I4 MFI VIN EJ6 [All] 16P 'D' Connector

PCM Pin #	W/Color	Circuit Description (16-Pin)	Value at Hot Idle
D1	RED/BLK	TP Sensor Signal	0.5-0.6v
D2	RED/WHT	ECT Sensor Signal	At 180°F: 0.51v
D3	RED/GRN	MAP Sensor Signal	0.8-0.9v
D4	YEL/RED	MAP Sensor VREF	4.9-5.1v
D5	GRN/WHT	Brake Switch Signal	Brake Off: 0v, On: 12v
D7	WHT	HO2S-11 (B1 S1) Signal	0.1-1.1v
D8	RED/YEL	IAT Sensor Signal	Varies w/temp. (0.5-4.9v)
D9	---	Not Used	---
D10	YEL/BLU	Sensor VREF	4.9-5.1v
D11	GRN/BLK	Sensor Ground	<0.050v
D12	GRN/WHT	MAP Sensor Ground	<0.050v
D13	RED/YEL	HO2S-12 Ground	<0.050v
D14	WHT/RED	HO2S-12 (B1 S2) Signal	0.1-1.1v
D15	---	Not Used	---
D16	GRN/RED	Electric Load Detector	Varies: 0.5-4.5v

31-PIN CONNECTOR 'C'   16-PIN CONNECTOR 'D'

1	2	3	4		5	6	7		8	9	10	1		2	3	4		5
11	12	13	14	15	16	17	18	19	20	21	22	6	7	8	9	10	11	12
	23	24	25		26	27	28		29	30	31	13	14	15		16		

WIRE SIDE OF HARNESS TERMINALS

1997-99 Sedan 1.6L I4 MFI VIN EJ6 [All] 32P 'A' Connector

PCM Pin #	W/Color	Circuit Description (32-Pin)	Value at Hot Idle
A1, A2	YEL, BLU	Injector 4, Injector 3 Control	2.0-3.3 ms
A3, A4	RED, BRN	Injector 2, Injector 1 Control	2.0-3.3 ms
A5	BLK/WHT	HO2S-12 (B1 S2) Heater	Digital Signals: 0-12-0v
A6	BLK/WHT	HO2S-11 (B1 S1) Heater	Digital Signals: 0-12-0v
A7	RED	A/T: EGR Solenoid Control	Solenoid Off: 12v, On: 1v
A7	PNK	M/T: E-EGR Solenoid	Digital Signals: 0-12-0v
A9, A22	BRN/BLK	Logic Ground (LG1, LG2)	<0.050v
A10, 23	BLK	Power Ground (PG1, PG2)	<0.1v
A11, 24	YEL/BLK	Main Relay Power (B+)	12-14v
A12	BLK/BLU	M/T: Idle Air Control Valve	Pulse Signals
A13	ORN	A/T: Idle Air Control Valve 'N'	Pulse Signals
A14	BLK/BLU	A/T: Idle Air Control Valve 'P'	Pulse Signals
A15	RED/YEL	EVAP Purge Solenoid	Solenoid Off: 12v, On: 1v
A16	GRN/YEL	Fuel Pump Relay	Relay Off: 12v, On: 1v
A17	BLK/RED	A/C Clutch Relay	Relay Off: 12v, On: 1v
A18	GRN/ORN	Malfunction Indicator Light	MIL Off: 12v, On: 1v
A19	WHT/GRN	Alternator Charging Signal	Lights Off: 12v, On: 0v
A20	YEL/GRN	Igniter Control	Digital Signals: 0-12-0v
A27	GRN	Radiator Fan Relay Control	Relay Off: 12v, On: 1v
A28	BLU	EVAP Bypass Solenoid	Solenoid Off: 12v, On: 1v
A29	WHT/RED	EVAP Vent Solenoid	Solenoid Off: 12v, On: 1v

1997-99 Sedan 1.6L I4 MFI VIN EJ6 [All] 25P 'B' Connector

PCM Pin #	W/Color	Circuit Description (25-Pin)	Value at Hot Idle
B1	WHT	A/T: Linear Solenoid (-)	Pulse Signals
B2	RED	A/T: Linear Solenoid (+)	Pulse Signals
B3	BLU/YEL	A/T: Shift Solenoid 'A'	SSA On: 12v, Off: 0v
B4	GRN/BLK	A/T: Lockup Solenoid 'B'	LSB On: 12v, Off: 0v
B5	YEL	A/T: Lockup Solenoid 'A'	LSA On: 12v, Off: 0v
B8	PNK	A/T: Gear Position Switch	In D3: 0v, Others: 12v
B11	GRN/WHT	A/T: Shift Solenoid 'B'	SSB Off: 0v, On: 12v
B12	WHT/RED	Interlock Control Unit Signal	Key & Brake On: 12v
B13	GRN/BLK	A/T: D4 Indicator Light	D4 On: 12v, Off: 0v
B14	WHT	Mainshaft Speed Sensor 'P'	AC Pulse Signals
B15	RED	Mainshaft Speed Sensor 'N'	AC Pulse Signals
B16	WHT	A/T: Gear Position Switch	In 'R': 0v, Others: 12v
B17	BLU	A/T: Gear Position Switch	In D2: 0v, Others: 12v
B22	GRN	Countershaft Speed Sensor N	Moving: AC pulses
B23	BLU	Countershaft Speed Sensor P	Moving: AC pulses
B24	YEL	A/T: Gear Position Switch	In D4: 0v, Others: 12v
B25	LT GRN	A/T: Gear Position Switch	In P/N: 0v, Others: 12v

32-Pin 'A' (WIRE SIDE OF HARNESS TERMINALS) 25-Pin 'B'

1997-99 Sedan 1.6L I4 MFI VIN EJ6 [All] 31P 'C' Connector

PCM Pin #	W/Color	Circuit Description (31-Pin)	Value at Hot Idle
C1	BLU/RED	CKF Sensor Signal	AC pulse signals
C2	BLU	CKP Sensor Signal	AC pulse signals
C3	GRN	TDC Sensor Signal	AC pulse signals
C4	YEL	CYP Sensor Signal	AC pulse signals
C5	BLU/RED	A/C Switch Signal	Switch Off: 12v, On: 1v
C6	BLU/ORN	Starter Switch Signal	Cranking: 9-11v
C7	BRN	Service Check Connector	SCS Open: 4.80v
C8	LT BLU	K-Line Signal	12v
C10	WHT/BLU	Keep Alive Power (VBU)	12-14v
C11	WHT/RED	CKF Sensor Ground	<0.050v
C12	WHT	CKP Sensor Ground	<0.050v
C13	RED	TDC Sensor Ground	<0.050v
C14	BLK	CYP Sensor Ground	<0.050v
C16	GRN	PSP Switch Signal	Straight: 0v, Turning: 11v
C17	WHT/RED	Alternator 'FR' Signal	Digital Signals: 0-5-0v
C18	BLU/WHT	Vehicle Speed Sensor	Moving: pulse signals
C23	BLK	LAF Pump Cell (IP+)	0.5-3.5v
C24	RED	LAF Common (IP-, VS-)	2.6-2.8v
C25	WHT	LAF VS Cell Voltage (VS+)	7v
C29	LT GRN	A/T: Gear Position Switch	In P/N: 0v, Others: 12v
C29	RED	M/T: Clutch Switch Signal	Clutch In: 0v, Out: 5v
C30	PNK	CVT TMB Signal	Pulse Signals

1997-99 Sedan 1.6L I4 MFI VIN EJ6 [All] 16P 'D' Connector

PCM Pin #	W/Color	Circuit Description (16-Pin)	Value at Hot Idle
D1	RED/BLK	TP Sensor Signal	0.5-0.6v
D2	RED/WHT	ECT Sensor Signal	At 180°F: 0.51v
D3	RED/GRN	MAP Sensor Signal	0.8-0.9v
D4	YEL/RED	MAP Sensor VREF	4.9-5.1v
D5	GRN/WHT	Brake Switch Signal	Brake Off: 0v, On: 12v
D6	RED/BLU	Knock Sensor Signal	No detonation: 18mv AC
D7	WHT	HO2S-11 (B1 S1) Signal	0.1-1.1v
D8	RED/YEL	IAT Sensor Signal	Varies w/temp. (0.5-4.9v)
D9	WHT/BLK	EGR Valve Lift Sensor	1.2v
D10	YEL/RED	Sensor VREF	4.9-5.1v
D11	GRN/BLK	Sensor Ground	<0.050v
D12	GRN/WHT	MAP Sensor Ground	<0.050v
D13	RED/YEL	HO2S-12 Ground	<0.050v
D14	WHT/RED	HO2S-12 (B1 S2) Signal	0.1-1.1v
D15	LT GRN	Fuel Tank Pressure Sensor	Fuel Cap off: 2.5v
D16	GRN/RED	Electric Load Detector	Varies: 0.5-4.5v

31-Pin 'C' *WIRE SIDE OF HARNESS TERMINALS* 16-Pin 'D'

2000 Sedan 1.6L I4 MFI VIN EJ6 [All] 32P 'A' Connector

PCM Pin #	W/Color	Circuit Description (32-Pin)	Value at Hot Idle
A1, A2	YEL, BLU	Injector 4, Injector 3 Control	2.0-3.3 ms
A3, A4	RED, BRN	Injector 2, Injector 1 Control	2.0-3.3 ms
A5	BLK/WHT	HO2S-12 (B1 S2) Heater	Digital Signals: 0-12-0v
A6	BLK/WHT	HO2S-11 (B1 S1) Heater	Digital Signals: 0-12-0v
A7	RED	A/T: EGR Solenoid Control	Solenoid Off: 12v, On: 1v
A7	PNK	M/T: E-EGR Solenoid	Digital Signals: 0-12-0v
A9, A22	BRN/BLK	Logic Ground (LG1, LG2)	<0.050v
A10, 23	BLK	Power Ground (PG1, PG2)	<0.1v
A11, 24	YEL/BLK	Main Relay Power (B+)	12-14v
A12	BLK/BLU	M/T: Idle Air Control Valve	Pulse Signals
A13	ORN	A/T: Idle Air Control Valve 'N'	Pulse Signals
A14	BLK/BLU	A/T: Idle Air Control Valve 'P'	Pulse Signals
A15	RED/YEL	EVAP Purge Solenoid	Solenoid Off: 12v, On: 1v
A16	GRN/YEL	Fuel Pump Relay	Relay Off: 12v, On: 1v
A17	BLK/RED	A/C Clutch Relay	Relay Off: 12v, On: 1v
A18	GRN/ORN	Malfunction Indicator Light	MIL Off: 12v, On: 1v
A19	WHT/GRN	Alternator Charging Signal	Lights Off: 12v, On: 0v
A20	YEL/GRN	Igniter Control	Digital Signals: 0-12-0v
A27	GRN	Radiator Fan Relay Control	Relay Off: 12v, On: 1v
A28	BLU	EVAP Bypass Solenoid	Solenoid Off: 12v, On: 1v
A29	WHT/RED	EVAP Vent Solenoid	Solenoid Off: 12v, On: 1v

2000 Sedan 1.6L I4 MFI VIN EJ6 [All] 25P 'B' Connector

PCM Pin #	W/Color	Circuit Description (25-Pin)	Value at Hot Idle
B1	WHT	A/T: Linear Solenoid (-)	Pulse Signals
B2	RED	A/T: Linear Solenoid (+)	Pulse Signals
B3	BLU/YEL	A/T: Shift Solenoid 'A'	SSA On: 12v: Off: 0v
B4	GRN/BLK	A/T: Lockup Solenoid 'B'	LSB On: 12v: Off: 0v
B5	YEL	A/T: Lockup Solenoid 'A'	LSA On: 12v: Off: 0v
B8	PNK	A/T: Gear Position Switch	In D3: 0v, Others: 12v
B11	GRN/WHT	A/T: Shift Solenoid 'B'	SSB Off: 0v, On: 12v
B12	WHT/RED	Interlock Control Unit Signal	Key & Brake On: 12v
B13	GRN/BLK	A/T: D4 Indicator Light	D4 On: 12v, Off: 0v
B14	WHT	Mainshaft Speed Sensor 'P'	AC Pulse Signals
B15	RED	Mainshaft Speed Sensor 'N'	AC Pulse Signals
B16	WHT	A/T: Gear Position Switch	In 'R': 0v, Others: 12v
B17	BLU	A/T: Gear Position Switch	In D2: 0v, Others: 12v
B22	GRN	Countershaft Speed Sensor N	Moving: AC pulses
B23	BLU	Countershaft Speed Sensor P	Moving: AC pulses
B24	YEL	A/T: Gear Position Switch	In D4: 0v, Others: 12v
B25	LT GRN	A/T: Gear Position Switch	In P/N: 0v, Others: 12v

```
┌─────────────────────────────────────────────────────────────────────┐
│  1  2  3  4   5  6  7   8  9  10  11    1    2    3  4  5   6  7  8    │
│ 12 13 14 15 16 17 18 19 20 21 22  23   24   9   10  11 12 13 14 15 16 17 18 │
│   25 26 27   28 29 30 31    32        19   20   21 22  23 24 25       │
│                                                                       │
│    32-Pin 'A'      ⟨WIRE SIDE OF HARNESS TERMINALS⟩    25-Pin 'B'     │
└─────────────────────────────────────────────────────────────────────┘
```

2000 Sedan 1.6L I4 MFI VIN EJ6 [All] 31P 'C' Connector

PCM Pin #	W/Color	Circuit Description (31-Pin)	Value at Hot Idle
C1	BLU/RED	CKF Sensor Signal	AC pulse signals
C2	BLU	CKP Sensor Signal	AC pulse signals
C3	GRN	TDC Sensor Signal	AC pulse signals
C4	YEL	CYP Sensor Signal	AC pulse signals
C5	BLU/RED	A/C Switch Signal	Switch Off: 12v, On: 1v
C6	BLU/ORN	Starter Switch Signal	Cranking: 9-11v
C7	BRN	Service Check Connector	SCS Open: 4.80v
C8	LT BLU	K-Line Signal	12v
C10	WHT/BLU	Keep Alive Power (VBU)	12-14v
C11	WHT/RED	CKF Sensor Ground	<0.050v
C12	WHT	CKP Sensor Ground	<0.050v
C13	RED	TDC Sensor Ground	<0.050v
C14	BLK	CYP Sensor Ground	<0.050v
C16	GRN	PSP Switch Signal	Straight: 0v, Turning: 11v
C17	WHT/RED	Alternator 'FR' Signal	Digital Signals: 0-5-0v
C18	BLU/WHT	Vehicle Speed Sensor	Moving: pulse signals
C23	BLK	LAF Pump Cell (IP+)	0.5-3.5v
C24	RED	LAF Common (IP-, VS-)	2.6-2.8v
C25	WHT	LAF VS Cell Voltage (VS+)	7v
C29	LT GRN	A/T: Gear Position Switch	In P/N: 0v, Others: 12v
C29	RED	M/T: Clutch Switch Signal	Clutch In: 0v, Out: 5v
C30	PNK	CVT TMB Signal	Pulse Signals

2000 Sedan 1.6L I4 MFI VIN EJ6 [All] 16P 'D' Connector

PCM Pin #	W/Color	Circuit Description (16-Pin)	Value at Hot Idle
D1	RED/BLK	TP Sensor Signal	0.5-0.6v
D2	RED/WHT	ECT Sensor Signal	At 180°F: 0.51v
D3	RED/GRN	MAP Sensor Signal	0.8-0.9v
D4	YEL/RED	MAP Sensor VREF	4.9-5.1v
D5	GRN/WHT	Brake Switch Signal	Brake Off: 0v, On: 12v
D6	RED/BLU	Knock Sensor Signal	No detonation: 18mv AC
D7	WHT	HO2S-11 (B1 S1) Signal	0.1-1.1v
D8	RED/YEL	IAT Sensor Signal	Varies w/temp. (0.5-4.9v)
D9	WHT/BLK	EGR Valve Lift Sensor	1.2v
D10	YEL/RED	Sensor VREF	4.9-5.1v
D11	GRN/BLK	Sensor Ground	<0.050v
D12	GRN/WHT	MAP Sensor Ground	<0.050v
D13	RED/YEL	HO2S-12 Ground	<0.050v
D14	WHT/RED	HO2S-12 (B1 S2) Signal	0.1-1.1v
D15	LT GRN	Fuel Tank Pressure Sensor	Fuel Cap off: 2.5v
D16	GRN/RED	Electric Load Detector	Varies: 0.5-4.5v

```
 1  2  3  4    5  6  7    8  9 10      1    2  3  4     5
11 12 13 14 15 16 17 18 19 20 21 22    6  7  8  9 10 11  12
23 24 25   26 27 28   29 30 31        13 14 15   16

31-Pin 'C'   WIRE SIDE OF HARNESS TERMINALS   16-Pin 'D'
```

1996 Sedan 1.6L I4 VTEC VIN EJ8 [All] 32P 'A' Connector

PCM Pin #	W/Color	Circuit Description (32-Pin)	Value at Hot Idle
A1, 2	YEL, BLU	Injector 4, Injector 3 Control	2.0-3.3 ms
A3, 4	RED, BRN	Injector 2, Injector 1 Control	2.0-3.3 ms
A5	BLK/WHT	HO2S-12 (B1 S2) Heater	Digital Signals: 0-12-0v
A6	BLK/WHT	HO2S-11 (B1 S1) Heater	Digital Signals: 0-12-0v
A8	GRN/YEL	VTEC Solenoid	0v, Hi-Speed: 12v
A9, A22	BRN/BLK	Logic Ground	<0.050v
A10, 23	BLK	Power Ground	<0.1v
A11, 24	YEL/BLK	Main Relay Power (B+)	12-14v
A12	BLK/BLU	M/T: Idle Air Control Valve	Pulse Signals
A13	ORN	Idle Air Control Valve 'N'	Pulse Signals
A14	BLK/BLU	Idle Air Control Valve 'P'	Pulse Signals
A15	RED/YEL	EVAP Purge Solenoid	Solenoid Off: 12v, On: 1v
A16	GRN/YEL	Fuel Pump Relay	Relay Off: 12v, On: 1v
A17	BLK/RED	A/C Clutch Relay	Relay Off: 12v, On: 1v
A18	GRN/ORN	Check Engine Light	MIL Off: 12v, On: 1v
A19	WHT/GRN	Alternator Charging Signal	Lights Off: 12v, On: 0v
A20	YEL/GRN	Igniter Control	Digital Signals: 0-12-0v
A25	WHT/RED	A/T: TCM VREF Signal	4.9-5.1v
A27	GRN	Radiator Fan Relay	Relay Off: 12v, On: 1v
A28	BLU	EVAP Bypass Solenoid	Solenoid Off: 12v, On: 1v
A29	GRN/WHT	EVAP Vent Solenoid	Solenoid Off: 12v, On: 1v

1996 Sedan 1.6L I4 VTEC VIN EJ8 [All] 25P 'B' Connector

PCM Pin #	W/Color	Circuit Description (25-Pin)	Value at Hot Idle
B1	WHT	A/T: Linear Solenoid (-)	Pulse Signals
B2	RED	A/T: Linear Solenoid (+)	Pulse Signals
B3	BLU/YEL	A/T: Shift Solenoid 'A'	SSA Off: 0v, On: 12v
B4	GRN/BLK	A/T: Lockup Solenoid 'B'	LSB On: 12v, Off: 0v
B5	YEL	A/T: Lockup Solenoid 'A'	LSA On: 12v, Off: 0v
B8	PNK	A/T: Gear Position Switch	In D3: 0v, Others: 12v
B11	GRN/WHT	A/T: Shift Solenoid 'B'	SSB Off: 0v, On: 12v
B12	WHT/RED	Interlock Control Unit Signal	Key & Brake On: 12v
B13	GRN/BLK	A/T: D4 Indicator Light	D4 On: 12v, Off: 0v
B14	WHT	Mainshaft Speed Sensor 'N'	<0.050v
B15	RED	Mainshaft Speed Sensor 'P'	AC Pulse Signals
B16, B25	WHT, YEL	A/T: Gear Position Switch	In 'R': 0v, Others: 12v
B17	BLU	A/T: Gear Position Switch	In D2: 0v, Others: 12v
B22	GRN	Countershaft Speed Sensor N	<0.050v
B23	BLU	Countershaft Speed Sensor P	Pulse Signals
B24	YEL	A/T: Gear Position Switch	In D4: 0v, Others: 12v

32-PIN CONNECTOR 'A' — 25-PIN CONNECTOR 'B'

1 2 3 4 | 5 6 7 | 8 9 10 | 11 | 1 | 2 | 3 4 5 | 6 7 8
12 13 14 15 16 17 18 19 20 21 22 | 23 | 24 | 9 | 10 | 11 12 13 14 15 16 17 18
25 26 27 | 28 29 30 31 | 32 | 19 | 20 | 21 22 | 23 24 25

WIRE SIDE OF HARNESS TERMINALS

1996 Sedan 1.6L I4 VTEC VIN EJ8 [All] 31P 'D' Connector

PCM Pin #	W/Color	Circuit Description (31-Pin)	Value at Hot Idle
C1	BLU/RED	CKF Sensor Signal	AC pulse signals
C2	BLU	CKP Sensor Signal	AC pulse signals
C3	GRN	TDC Sensor Signal	AC pulse signals
C4	YEL	CYP Sensor Ground	AC pulse signals
C5	BLU/RED	A/C Switch Signal	Switch Off: 12v, On: 1v
C6	BLU/ORN	Starter Switch Signal	Cranking: 9-11v
C7	BRN	Service Check Connector	SCS Open: 4.80v
C8	LT BLU	K-Line Signal	12v
C9	---	Not Used	---
C10	WHT/BLU	Keep Alive Power (VBU)	12-14v
C11	WHT/RED	CKF Sensor Ground	<0.050v
C12	WHT	CKP Sensor Ground	<0.050v
C13	RED	TDC Sensor Ground	<0.050v
C14	BLK	CYP Sensor Ground	<0.050v
C15	BLU/BLK	VTEC Pressure Switch	0v, Hi-Speed: 12v
C16	GRN	PSP Switch	Straight: 0v, Turning: 11v
C17	WHT/RED	Alternator 'FR' Signal	Digital Signals: 0-5-0v
C18	BLU/WHT	Vehicle Speed Sensor	Moving: pulse signals
C19-31	---	Not Used	---

1996 Sedan 1.6L I4 VTEC VIN EJ8 [All] 16P 'D' Connector

PCM Pin #	W/Color	Circuit Description (16-Pin)	Value at Hot Idle
D1	RED/BLK	TP Sensor Signal	0.5-0.6v
D2	RED/WHT	ECT Sensor Signal	At 180°F: 0.51v
D3	RED/GRN	MAP Sensor Signal	0.8-0.9v
D4	YEL/RED	MAP Sensor VREF	4.9-5.1v
D5	GRN/WHT	Brake Switch Signal	Brake Off: 0v, On: 12v
D6	RED/BLU	Knock Sensor Signal	No Detonation: 18mv AC
D7	WHT	HO2S-11 (B1 S1) Signal	0.1-1.1v
D8	RED/YEL	IAT Sensor Signal	Varies w/temp. (0.5-4.9v)
D9	---	Not Used	---
D10	YEL/BLU	Sensor VREF	4.9-5.1v
D11	GRN/BLK	Sensor Ground	<0.050v
D12	GRN/WHT	MAP Sensor Ground	<0.050v
D13	GRN/BLK	HO2S-12 Ground	0.1-1.1v
D14	WHT/RED	HO2S-12 (B1 S2) Signal	0.1-1.1v
D15	LT GRN	Fuel Tank Pressure (Coupe)	Fuel Cap off: 2.5v
D16	GRN/RED	Electric Load Detector	Varies: 0.5-4.5v

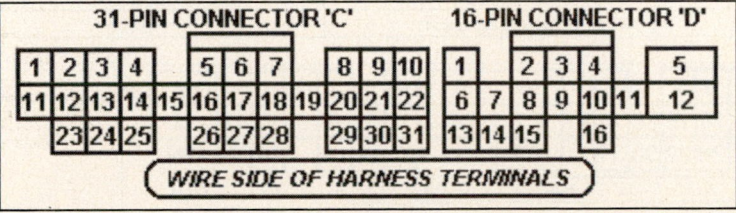

1997-99 Sedan 1.6L I4 VTEC VIN EJ8 [All] 32P 'A' Connector

PCM Pin #	W/Color	Circuit Description (32-Pin)	Value at Hot Idle
A3	BLU	EVAP Bypass Solenoid	Solenoid Off: 12v, On: 1v
A4	GRN/WHT	EVAP Vent Solenoid	Solenoid Off: 12v, On: 1v
A5	BLU/GRN	Cruise Control Signal	C/C On: pulse signals
A6	RED/YEL	EVAP Purge Solenoid	Solenoid Off: 12v, On: 1v
A8	BLK/WHT	HO2S-12 (B1 S2) Heater	Digital Signals: 0-12-0v
A9	LT GRN	A/T: Gear Position Switch	In P/N: 0v, Others: 12v
A10	BRN	Service Check Connector	SCS Open: 4.80v
A14	GRN/BLK	A/T: D4 Light Switch	D4 On: 0v, Off: 12v
A16	GRN/YEL	Fuel Pump Relay Control	Relay Off: 12v, On: 1v
A17	BLK/RED	A/C Clutch Relay Control	Relay Off: 12v, On: 1v
A18	GRN/ORN	Malfunction Indicator Light	MIL Off: 12v, On: 1v
A19	BLU	Engine Speed Pulse (NEP)	Digital Signals
A20	GRN	Radiator Fan Relay Control	Relay Off: 12v, On: 1v
A21	BLU/YEL	K-Line Signal	12v
A22	BLU	A/T: Gear Position Switch	In 'L': 0v, others: 12v
A23	WHT/RED	HO2S-12 (B1 S2) Signal	0.1-1.1v
A24	BLU/WHT	Starter Switch Signal	Cranking: 9-11v
A26	GRN	PSP Switch Signal	Straight: 0v, Turning: 11v
A27	BLU/RED	A/C Switch Signal	Switch Off: 12v, On: 1v
A28	WHT/RED	Interlock Control Unit Signal	Key & Brake On: 12v
A29	LT GRN	Fuel Tank Pressure Sensor	Fuel Cap off: 2.5v
A30	GRN/RED	Electric Load Detector	Varies: 0.5-4.5v
A32	WHT/BLK	Brake Switch Signal	Brake Off: 0v, On: 12v

1997-99 Sedan 1.6L I4 VTEC VIN EJ8 [All] 16P 'D' Connector

PCM Pin #	W/Color	Circuit Description (16-Pin)	Value at Hot Idle
D1	YEL	A/T: Lockup Control Solenoid	LUS On: 12v, Off: 0v
D2	GRN/WHT	A/T: Shift Solenoid 'B'	SSB Off: 0v, On: 12v
D3	GRN	A/T: Shift Solenoid 'C'	SSC On: 12v, off: 0v
D5	BLK/YEL	A/T: Solenoid Feed (B+)	12-14v
D6	WHT	A/T: Gear Position Switch	In 'R': 0v, Others: 12v
D7	BLU/YEL	A/T: Shift Solenoid 'A'	SSA Off: 0v, On: 12v
D8	PNK	A/T: Gear Position Switch	In D3: 0v, Others: 12v
D9	YEL	A/T: Gear Position Switch	In D4: 0v, Others: 12v
D10	BLU	Countershaft Speed Sensor P	Moving: AC pulses
D11	RED	Mainshaft Speed Sensor 'P'	Pulse Signals
D12	WHT	Mainshaft Speed Sensor 'N'	<0.050v
D13	BLU/WHT	A/T: Gear Position Switch	In P/N: 0v, Others: 12v
D14	GRN/BLK	A/T: D4 Indicator Light Driver	D4 on: 5v, off: 0v
D16	GRN	Countershaft Speed Sensor N	<0.050v

1997-99 Sedan 1.6L I4 VTEC VIN EJ8 [All] 25P 'B' Connector

PCM Pin #	W/Color	Circuit Description (25-Pin)	Value at Hot Idle
B1, B9	YEL/BLK	Main Relay Power (B+)	12-14v
B2, B10	BLK	Power Ground	<0.1v
B3, B4	RED, BLU	Injector 2, Injector 3 Control	2.0-3.3 ms
B5, B11	YEL, BRN	Injector 4, Injector 1 Control	2.0-3.3 ms
B7	RED	EGR Control Solenoid	Digital Signals: 0-12-0v
B8	PNK/BLK	A/T: Control Linear Solenoid 'N'	Pulse Signals
B12	GRN/YEL	VTEC Control Solenoid	0v, Hi-Speed: 12v
B13	YEL/GRN	Ignition Control Signal	Digital Signals: 0-12-0v
B15	ORN	Idle Air Control Valve 'N'	Pulse Signals
B17	GRN/WHT	A/T: Control Linear Solenoid 'P'	Pulse Signals
B18	PNK/BLU	A/T: Start Clutch Solenoid 'N'	Pulse Signals
B20, B22	BRN/BLK	Logic Ground (LG1, LG2)	<0.050v
B21	WHT/YEL	Keep Alive Power (VBU)	12-14v
B23	BLK/BLU	Idle Air Control Valve 'P'	Pulse Signals
B24	BLU/WHT	A/T: 3rd Clutch Pressure Sw.	Open: 12v, Closed: 1v
B25	YEL	A/T: Start Clutch Solenoid 'N'	Pulse Signals

1997-99 Sedan 1.6L I4 VTEC VIN EJ8 [All] 31P 'C' Connector

PCM Pin #	W/Color	Circuit Description (31-Pin)	Value at Hot Idle
C1	BLK/WHT	HO2S-11 (B1 S1) Heater	Digital Signals: 0-12-0v
C2	WHT/GRN	Alternator Charging Signal	Lights Off: 12v, On: 0v
C3	RED/BLU	Knock Sensor Signal	No Detonation: 18mv AC
C5	WHT/RED	Alternator 'FR' Signal	Digital Signals: 0-5-0v
C6	WHT/BLK	EGR Valve Lift Sensor	1.2v
C7	GRN/WHT	MAP Sensor Ground (SG1)	<0.050v
C8	BLU	CKP Sensor Signal	AC pulse signals
C9	WHT	CKP Sensor Ground	<0.050v
C10	BLU/BLK	VTEC Pressure Switch	0v, Hi-Speed: 12v
C16	WHT	HO2S-11 (B1 S1) Signal	0.1-1.1v
C17	RED/GRN	MAP Sensor Signal	0.8-0.9v
C18	GRN/BLK	Signal Ground (SG2)	<0.050v
C19, C28	YEL/RED	MAP Sensor, Sensor VREF	4.9-5.1v
C20	GRN	TDC Sensor Signal	AC pulse signals
C21	RED	TDC Sensor Ground	<0.050v
C22	BLU/RED	CKF Sensor Signal	AC pulse signals
C23	BLU/WHT	Vehicle Speed Sensor Signal	Moving: 0-5-0v
C25	RED/YEL	IAT Sensor Signal	Varies w/temp (0.5-4.9v)
C26	RED/WHT	ECT Sensor Signal	At 180°F: 0.5-0.6v
C27	RED/BLK	TP Sensor Signal	0.5-0.6v
C29	YEL	CYP Sensor Signal	AC pulse signals
C30	BLK	CYP Sensor Ground	<0.050v
C31	WHT/RED	CKF Sensor Ground	<0.050v

25-PIN 'B'   WIRE SIDE OF HARNESS TERMINALS   31-PIN 'C'

2000 Sedan 1.6L I4 VTEC VIN EJ8 [All] 32P 'A' Connector

PCM Pin #	W/Color	Circuit Description (32-Pin)	Value at Hot Idle
A3	BLU	EVAP Bypass Solenoid	Solenoid Off: 12v, On: 1v
A4	GRN/WHT	EVAP Vent Solenoid	Solenoid Off: 12v, On: 1v
A5	BLU/GRN	Cruise Control Signal	C/C On: pulse signals
A6	RED/YEL	EVAP Purge Solenoid	Solenoid Off: 12v, On: 1v
A8	BLK/WHT	HO2S-12 (B1 S2) Heater	Digital Signals: 0-12-0v
A9	LT GRN	A/T: Gear Position Switch	In P/N: 0v, Others: 12v
A10	BRN	Service Check Connector	SCS Open: 4.80v
A14	GRN/BLK	A/T: D4 Light Switch	D4 On: 0v, Off: 12v
A16	GRN/YEL	Fuel Pump Relay Control	Relay Off: 12v, On: 1v
A17	BLK/RED	A/C Clutch Relay Control	Relay Off: 12v, On: 1v
A18	GRN/ORN	Malfunction Indicator Light	MIL Off: 12v, On: 1v
A19	BLU	Engine Speed Pulse (NEP)	Digital Signals
A20	GRN	Radiator Fan Relay Control	Relay Off: 12v, On: 1v
A21	BLU/YEL	K-Line Signal	12v
A22	BLU	A/T: Gear Position Switch	In 'L': 0v, others: 12v
A23	WHT/RED	HO2S-12 (B1 S2) Signal	0.1-1.1v
A24	BLU/WHT	Starter Switch Signal	Cranking: 9-11v
A26	GRN	PSP Switch Signal	Straight: 0v, Turning: 11v
A27	BLU/RED	A/C Switch Signal	Switch Off: 12v, On: 1v
A28	WHT/RED	Interlock Control Unit Signal	Key & Brake On: 12v
A29	LT GRN	Fuel Tank Pressure Sensor	Fuel Cap off: 2.5v
A30	GRN/RED	Electric Load Detector	Varies: 0.5-4.5v
A32	WHT/BLK	Brake Switch Signal	Brake Off: 0v, On: 12v

2000 Sedan 1.6L I4 VTEC VIN EJ8 [All] 16P 'D' Connector

PCM Pin #	W/Color	Circuit Description (16-Pin)	Value at Hot Idle
D1	YEL	A/T: Lockup Control Solenoid	LUS On: 12v, Off: 0v
D2	GRN/WHT	A/T: Shift Solenoid 'B'	SSB Off: 0v, On: 12v
D3	GRN	A/T: Shift Solenoid 'C'	SSC on: 12v, off: 0v
D5	BLK/YEL	A/T: Solenoid Feed (B+)	12-14v
D6	WHT	A/T: Gear Position Switch	In 'R': 0v, Others: 12v
D7	BLU/YEL	A/T: Shift Solenoid 'A'	SSA Off: 0v, On: 12v
D8	PNK	A/T: Gear Position Switch	In D3: 0v, Others: 12v
D9	YEL	A/T: Gear Position Switch	In D4: 0v, Others: 12v
D10	BLU	Countershaft Speed Sensor P	Moving: AC pulses
D11	RED	Mainshaft Speed Sensor 'P'	Pulse Signals
D12	WHT	Mainshaft Speed Sensor 'N'	<0.050v
D13	BLU/WHT	A/T: Gear Position Switch	In P/N: 0v, Others: 12v
D14	GRN/BLK	A/T: D4 Indicator Light Driver	D4 on: 5v, off: 0v
D16	GRN	Countershaft Speed Sensor N	<0.050v

WIRE SIDE OF HARNESS TERMINALS

2000 Sedan 1.6L I4 VTEC VIN EJ8 [All] 25P 'B' Connector

PCM Pin #	W/Color	Circuit Description (25-Pin)	Value at Hot Idle
B1, B9	YEL/BLK	Main Relay Power (B+)	12-14v
B2, B10	BLK	Power Ground	<0.1v
B3, B4	RED, BLU	Injector 2, Injector 3 Control	2.0-3.3 ms
B5, B11	YEL, BRN	Injector 4, Injector 1 Control	2.0-3.3 ms
B7	RED	EGR Control Solenoid	Digital Signals: 0-12-0v
B8	PNK/BLK	A/T: Control Linear Solenoid 'N'	Pulse Signals
B12	GRN/YEL	VTEC Control Solenoid	0v, Hi-Speed: 12v
B13	YEL/GRN	Ignition Control Signal	Digital Signals: 0-12-0v
B15	ORN	Idle Air Control Valve 'N'	Pulse Signals
B17	GRN/WHT	A/T: Control Linear Solenoid 'P'	Pulse Signals
B18	PNK/BLU	A/T: Start Clutch Solenoid 'N'	Pulse Signals
B20, B22	BRN/BLK	Logic Ground (LG1, LG2)	<0.050v
B21	WHT/YEL	Keep Alive Power (VBU)	12-14v
B23	BLK/BLU	Idle Air Control Valve 'P'	Pulse Signals
B24	BLU/WHT	A/T: 3rd Clutch Pressure Sw.	Open: 12v, Closed: 1v
B25	YEL	A/T: Start Clutch Solenoid 'N'	Pulse Signals

2000 Sedan 1.6L I4 VTEC VIN EJ8 [All] 31P 'C' Connector

PCM Pin #	W/Color	Circuit Description (31-Pin)	Value at Hot Idle
C1	BLK/WHT	HO2S-11 (B1 S1) Heater	Digital Signals: 0-12-0v
C2	WHT/GRN	Alternator Charging Signal	Lights Off: 12v, On: 0v
C3	RED/BLU	Knock Sensor Signal	No Detonation: 18mv AC
C5	WHT/RED	Alternator 'FR' Signal	Digital Signals: 0-5-0v
C6	WHT/BLK	EGR Valve Lift Sensor	1.2v
C7	GRN/WHT	MAP Sensor Ground (SG1)	<0.050v
C8	BLU	CKP Sensor Signal	AC pulse signals
C9	WHT	CKP Sensor Ground	<0.050v
C10	BLU/BLK	VTEC Pressure Switch	0v, Hi-Speed: 12v
C16	WHT	HO2S-11 (B1 S1) Signal	0.1-1.1v
C17	RED/GRN	MAP Sensor Signal	0.8-0.9v
C18	GRN/BLK	Signal Ground (SG2)	<0.050v
C19, C28	YEL/RED	MAP Sensor, Sensor VREF	4.9-5.1v
C20	GRN	TDC Sensor Signal	AC pulse signals
C21	RED	TDC Sensor Ground	<0.050v
C22	BLU/RED	CKF Sensor Signal	AC pulse signals
C23	BLU/WHT	Vehicle Speed Sensor Signal	Moving: 0-5-0v
C25	RED/YEL	IAT Sensor Signal	Varies w/temp (0.5-4.9v)
C26	RED/WHT	ECT Sensor Signal	At 180ºF: 0.5-0.6v
C27	RED/BLK	TP Sensor Signal	0.5-0.6v
C29	YEL	CYP Sensor Signal	AC pulse signals
C30	BLK	CYP Sensor Ground	<0.050v
C31	WHT/RED	CKF Sensor Ground	<0.050v

1	2		3	4	5		6	7	8		1	2	3	4		5	6	7		8	9	10
9	10	11	12	13	14	15	16	17	18	11	12	13	14	15	16	17	18	19	20	21	22
19	20	21	22		23	24	25			23	24	25		26	27	28		29	30	31

25-PIN 'B'   *WIRE SIDE OF HARNESS TERMINALS*   31-PIN 'C'

2001-03 Sedan 1.7L CNG VIN EN2 [CVT] 31P Gray 'A' Connector

PCM Pin #	W/Color	Circuit Description (31-Pin)	Value at Hot Idle
A1	BLK/WHT	HO2S-11 (B1 S1) Heater	Digital Signals: 0-12-0v
A2, A3	YEL/BLK	Main Relay Power (B+)	12-14v
A4, A5	BLK	Power Ground (PG2, PG1)	<0.1v
A6	WHT	HO2S-11 (B1 S1) Signal	0.1-1.1v
A7	BLU	CKP Sensor Signal	Digital Signals
A10	GRN/YEL	Sensor Ground (SG2)	<0.050v
A11	GRN/WHT	Sensor Ground (SG1)	<0.050v
A12	BLK/RED	Idle Air Control Valve	Pulse Signals
A13	WHT/BLK	EGR Valve Position Sensor	0.6-1.1v
A14	BLK/WHT	HO2S-12 (B1 S2) Heater	Digital Signals: 0-12-0v
A15	RED/BLK	TP Sensor Signal	0.5-0.6v
A17	BRN/WHT	Injector Mode Signal	Digital Signals
A18	WHT/GRN	Vehicle Speed Signal	Moving: 0-5-0v
A19	GRN/RED	MAP Sensor Signal	0.8-0.9v
A20, A21	YEL/BLU	Sensor VREF (VCC2, VCC1)	4.9-5.1v
A23, A24	BRN/YEL	Logic Ground (LG2, LG1)	<0.1v
A25	WHT/RED	HO2S-12 (B1 S2) Signal	0.1-1.1v
A26	GRN	TDC Sensor Signal	AC pulse signals
A27	BRN	Coil 4 Driver Control	Digital Signals
A28	WHT/BLU	Coil 3 Driver Control	Digital Signals
A29	BLU/RED	Coil 2 Driver Control	Digital Signals
A30	YEL/GRN	Coil 1 Driver Control	Digital Signals

2001-03 Sedan 1.7L CNG VIN EN2 [CVT] 24P White 'B' Connector

PCM Pin #	W/Color	Circuit Description (24-Pin)	Value at Hot Idle
B2	YEL	Injector 4 Control	2.0-3.3 ms
B3	BLU	Injector 3 Control	2.0-3.3 ms
B4	RED	Injector 2 Control	2.0-3.3 ms
B5	BRN	Injector 1 Control	2.0-3.3 ms
B6	GRN	Fan Relay Control	Relay Off: 12v, On: 1v
B7	GRN/WHT	CVT: Pulley Control Valve	Pulse Signals
B8	RED/WHT	ECT Sensor Signal	At 180°F: 0.51v
B10	WHT/BLU	Alternator Load Signal	Lights Off: 12v, On: 0v
B13	WHT/RED	Alternator 'FR' Signal	Digital Signals: 0-5-0v
B14	BLU/RED	EGR Solenoid Control	Solenoid Off: 12v, On: 1v

31-Pin 'A'   *WIRE SIDE OF HARNESS TERMINALS*   24-Pin 'B'

B16	BLK/RED	CVT: Pressure Control Valve	Valve On: 12v: Off: 0v
B17	RED/YEL	IAT Sensor Signal	Varies w/temp. (0.5-4.9v)
B18	WHT/GRN	Alternator Charging Signal	Digital Signals: 0-12-0v

2001-03 Sedan 1.7L CNG VIN EN2 [CVT] 22P C Black Connector

PCM Pin #	W/Color	Circuit Description (22-Pin)	Value at Hot Idle
C1	BLK/RED	Pulley Press. Control Valve	Pulse Signals
C2-5	---	Not Used	---
C6	GRN/RED	Inhibitor Solenoid	INH On: 12v; Off: 0v
C7	RED/BLU	Drive Pulley Speed Sensor	Pulse Signals
C8	BLU/RED	Start Press. Control Valve	Pulse Signals
C9	RED	Gear Position Switch	In 2nd: 0v, Others: 12v
C10	WHT	Gear Position Switch	In 'R': 0v, Others: 12v
C11	BLU	Gear Position Switch	In 'L': 0v, Others: 12v
C12	BLU/WHT	Gear Position Switch	In P/N: 0v, others: 12v
C13-14	---	Not Used	---
C15	WHT/GRN	Driven Pulley Speed Sensor	Pulse Signals
C16	GRN/YEL	CVT: Speed Change Control	Pulse Signals
C18	BLU/YEL	Gear Position Switch	In FWD: 0v; Others: 12v
C20	YEL	Gear Position Switch	In 'D': 0v; Others: 12v
C21	WHT/RED	CVT Speed Sensor 1	Pulse Signals

2001-03 Sedan 1.7L CNG VIN EN2 [CVT] 31P White 'E' Connector

PCM Pin #	W/Color	Circuit Description (31-Pin)	Value at Hot Idle
E1	GRN/YEL	Immobilizer Fuel Pump Relay	Relay Off: 12v, On: 1v
E3	BRN/YEL	Logic Ground (LG3)	<0.050v
E4	PNK	Sensor Ground (SG3)	<0.050v
E5	YEL/BLU	Sensor VREF (VCC3)	4.9-5.1v
E7	RED/YEL	Main Relay Control	Relay Off: 12v, On: 1v
E9	YEL/BLK	Main Relay Power (B+)	12-14v
E12	BLU/ORN	Cruise Control Unit Signal	N/A
E13	WHT/BLU	Multiplex Control Unit	N/A
E14	LT GRN	Fuel Tank Pressure Sensor	Fuel Cap off: 2.5v
E15	GRN/RED	Electric Load Detector	Varies: 0.5-4.5v
E16	GRN/BLK	PSP Switch Signal	Straight: 0v, Turning: 11v
E18	RED	A/C Clutch Relay	Relay Off: 12v, On: 1v
E20	GRN	EVAP Bypass Solenoid	Solenoid Off: 12v, On: 1v
E21	GRN/RED	Fuel Tank Temp. Sensor	Varies: 0.1-4.9v
E22	WHT/BLK	Brake Switch Signal	Brake Off: 0v, On: 12v
E23	LT BLU	K-Line Signal	12v
E24	YEL	SEFMJ Signal (Multiplex Unit)	Digital Signals
E25	BLU/WHT	VSS Out Signal	Pulse Signals
E26	BLU	Engine Speed Pulse (NEP)	Digital Signals
E27	RED/BLU	Immobilizer Code Signal	Digital Signals
E29	BRN	Service Check Connector	SCS Open: 4.80v
E30	RED/WHT	WEN Terminal in DLC	0v
E31	GRN/ORN	Malfunction Indicator Lamp	MIL Off: 12v, On: 1v

24-Pin 'C' *(WIRE SIDE OF HARNESS TERMINALS)* 31-Pin 'E'

2001-03 Sedan 1.7L MFI VIN ES1 [All] 31P Gray 'A' Connector

PCM Pin #	W/Color	Circuit Description (31-Pin)	Value at Hot Idle
A1	BLK/WHT	HO2S-11 (B1 S1) Heater	Digital Signals: 0-12-0v
A2, A3	YEL/BLK	Main Relay Power (B+)	12-14v
A4, A5	BLK	Power Ground (PG2, PG1)	<0.1v
A6	WHT	HO2S-11 (B1 S1) Signal	0.1-1.1v
A7	BLU	CKP Sensor Signal	Digital Signals
A8	YEL	Sensor VREF (VCCR)	4.9-5.1v
A9	RED/BLU	Knock Sensor Signal	No detonation: 18mv AC
A10	GRN/YEL	Sensor Ground (SG2)	<0.050v
A11	GRN/WHT	Sensor Ground (SG1)	<0.050v
A12	BLK/RED	Idle Air Control Valve	Pulse Signals
A13	WHT/BLK	EGR Valve Position Sensor	0.6-1.1v
A14	BLK/WHT	HO2S-12 (B1 S2) Heater	Digital Signals: 0-12-0v
A15	RED/BLK	TP Sensor Signal	0.5-0.6v
A17	BRN/WHT	Injector Mode Signal	Digital Signals
A18	WHT/GRN	Vehicle Speed Signal	Moving: 0-5-0v
A19	GRN/RED	MAP Sensor Signal	0.8-0.9v
A20, A21	YEL/BLU	Sensor VREF (VCC2, VCC1)	4.9-5.1v
A23, A24	BRN/YEL	Logic Ground (LG2, LG1)	<0.1v
A25	WHT/RED	HO2S-12 (B1 S2) Signal	0.1-1.1v
A26	GRN	TDC Sensor Signal	AC pulse signals
A27	BRN	Coil 4 Driver Control	Digital Signals
A28	WHT/BLU	Coil 3 Driver Control	Digital Signals
A29	BLU/RED	Coil 2 Driver Control	Digital Signals
A30	YEL/GRN	Coil 1 Driver Control	Digital Signals

2001-03 Sedan 1.7L MFI VIN ES1 [All] 24P White 'B' Connector

PCM Pin #	W/Color	Circuit Description (24-Pin)	Value at Hot Idle
B2	YEL	Injector 4 Control	2.0-3.3 ms
B3	BLU	Injector 3 Control	2.0-3.3 ms
B4	RED	Injector 2 Control	2.0-3.3 ms
B5	BRN	Injector 1 Control	2.0-3.3 ms
B6	GRN	Fan Relay Control	Relay Off: 12v, On: 1v
B7	RED/BLK	A/T: Linear Solenoid 'A+'	LSA Off: 0v, On: 12v
B8	RED/WHT	ECT Sensor Signal	At 180ºF: 0.51v
B10	WHT/BLU	Alternator Load Signal	Lights Off: 12v, On: 0v
B13	WHT/RED	Alternator 'FR' Signal	Digital Signals: 0-5-0v
B14	BLU/RED	EGR Solenoid Control	Solenoid Off: 12v, On: 1v
B16	BLK/RED	A/T: Lockup Solenoid 'B'	LSB On: 12v: Off: 0v
B17	RED/YEL	IAT Sensor Signal	Varies w/temp. (0.5-4.9v)
B18	WHT/GRN	Alternator Charging Signal	Digital Signals: 0-12-0v
B21	YEL/BLU	EVAP Purge Solenoid	Solenoid Off: 12v, On: 1v

31-Pin 'A'   *WIRE SIDE OF HARNESS TERMINALS*   24-Pin 'B'

2001-03 Sedan 1.7L MFI VIN ES1 [W/O CVT] 22P C Black Connector

PCM Pin #	W/Color	Circuit Description (22-Pin)	Value at Hot Idle
C1	WHT/BLK	A/T: Linear Solenoid 'A-'	LSA Off: 12v, On: 1v
C2	YEL/BLU	A/T: TCC Solenoid	Solenoid Off: 12v, On: 1v
C4	GRN/WHT	A/T: Shift Solenoid 'B'	SSB Off: 0v, On: 12v
C6	BLU/BLK	A/T: Shift Solenoid 'A'	SSA On: 12v: Off: 0v
C7	WHT/RED	Mainshaft Speed Sensor 'P'	AC Pulse Signals
C8	BRN/WHT	A/T: Lockup Solenoid 'B-'	LSB On: 12v: Off: 0v
C9	RED	A/T: Gear Position Switch	In D3: 0v, Others: 12v
C10	WHT	A/T: Gear Position Switch	In 'R': 0v, Others: 12v
C11	BLU	A/T: Gear Position Switch	In D2: 0v, Others: 12v
C12	BLU/WHT	A/T: Gear Position Switch	In P/N: 0v, others: 12v
C14	GRN	Countershaft Speed Sensor N	Moving: AC pulses
C15	BLU	Countershaft Speed Sensor P	Moving: AC pulses
C18	BLY/REL	A/T: Gear Position Switch	Forward: 0v, Others: 12v
C20	YEL	A/T: Gear Position Switch	In 'D': 0v, Others: 12v
C21	WHT/GRN	Mainshaft Speed Sensor 'N'	AC Pulse Signals

2001-03 Coupe 1.7L I4 VIN ES1 [All] 31P White 'E' Connector

PCM Pin #	W/Color	Circuit Description (31-Pin)	Value at Hot Idle
E1	GRN/YEL	Immobilizer Fuel Pump Relay	Relay Off: 12v, On: 1v
E3	BRN/YEL	Logic Ground (LG3)	<0.050v
E4	PNK	Sensor Ground (SG3)	<0.050v
E5	YEL/BLU	Sensor VREF (VCC3)	4.9-5.1v
E7	RED/YEL	Main Relay Control	Relay Off: 12v, On: 1v
E9	YEL/BLK	Main Relay Power (B+)	12-14v
E12	BLU/ORN	Cruise Control Unit Signal	Digital Signals
E13	WHT/BLU	Multiplex Control Unit	Digital Signals
E14	LT GRN	Fuel Tank Pressure Sensor	Fuel Cap off: 2.5v
E15	GRN/RED	Electric Load Detector	Varies: 0.5-4.5v
E16	GRN/BLK	PSP Switch Signal	Straight: 0v, Turning: 11v
E18	RED	A/C Clutch Relay	Relay Off: 12v, On: 1v
E20	BLU/RED	EVAP Bypass Solenoid	Solenoid Off: 12v, On: 1v
E21	GRN/RED	EVAP Vent Solenoid	Solenoid Off: 12v, On: 1v
E22	WHT/BLK	Brake Switch Signal	Brake Off: 0v, On: 12v
E23	LT BLU	K-Line Signal	12v
E24	YEL	SEFMJ Signal (Multiplex Unit)	Digital Signals
E25	BLU/WHT	VSS Out Signal	Pulse Signals
E26	BLU	Engine Speed Pulse (NEP)	Digital Signals
E27	RED/BLU	Immobilizer Code Signal	Digital Signals
E29	BRN	Service Check Connector	SCS Open: 4.80v
E30	RED/WHT	WEN Terminal in DLC	0v
E31	GRN/ORN	Malfunction Indicator Lamp	MIL Off: 12v, On: 1v

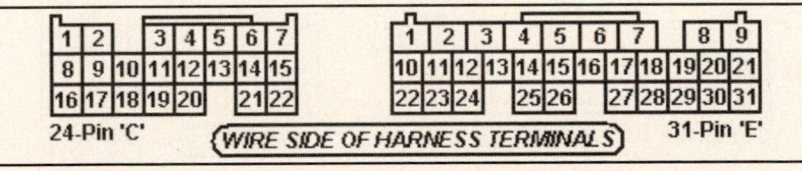

2001-03 Sedan 1.7L VTEC VIN ES1 [All] 31P Gray 'A' Connector

PCM Pin #	W/Color	Circuit Description (31-Pin)	Value at Hot Idle
A1	BLK/WHT	HO2S-11 (B1 S1) Heater	Digital Signals: 0-12-0v
A2, A3	YEL/BLK	Main Relay Power (B+)	12-14v
A4, A5	BLK	Power Ground (PG2, PG1)	<0.1v
A6	WHT	HO2S-11 (B1 S1) Signal	0.1-1.1v
A7	BLU	CKP Sensor Signal	Digital Signals
A8	YEL	Sensor VREF (VCCR)	4.9-5.1v
A9	RED/BLU	Knock Sensor Signal	No detonation: 18mv AC
A10	GRN/YEL	Sensor Ground (SG2)	<0.050v
A11	GRN/WHT	Sensor Ground (SG1)	<0.050v
A12	BLK/RED	Idle Air Control Valve	Pulse Signals
A13	WHT/BLK	EGR Valve Position Sensor	0.6-1.1v
A14	BLK/WHT	HO2S-12 (B1 S2) Heater	Digital Signals: 0-12-0v
A15	RED/BLK	TP Sensor Signal	0.5-0.6v
A17	BRN/WHT	Injector Mode Signal	Digital Signals
A18	WHT/GRN	Vehicle Speed Signal	Moving: 0-5-0v
A19	GRN/RED	MAP Sensor Signal	0.8-0.9v
A20, A21	YEL/BLU	Sensor VREF (VCC2, VCC1)	4.9-5.1v
A23, A24	BRN/YEL	Logic Ground (LG2, LG1)	<0.1v
A25	WHT/RED	HO2S-12 (B1 S2) Signal	0.1-1.1v
A26	GRN	TDC Sensor Signal	AC pulse signals
A27	BRN	Coil 4 Driver Control	Digital Signals
A28	WHT/BLU	Coil 3 Driver Control	Digital Signals
A29	BLU/RED	Coil 2 Driver Control	Digital Signals
A30	YEL/GRN	Coil 1 Driver Control	Digital Signals

2001-03 Sedan 1.7L VTEC VIN ES1 [All] 24P White 'B' Connector

PCM Pin #	W/Color	Circuit Description (24-Pin)	Value at Hot Idle
B2, B3	YEL, BLU	Injector 4, Injector 3 Control	2.0-3.3 ms
B4, B5	RED, BRN	Injector 2, Injector 1 Control	2.0-3.3 ms
B6	GRN	Fan Relay Control	Relay Off: 12v, On: 1v
B7	RED/BLK	A/T: Linear Solenoid 'A+'	LSA Off: 0v, On: 12v
B8	RED/WHT	ECT Sensor Signal	At 180ºF: 0.51v
B9	BLU/BLK	VTEC Pressure Switch	0v, Hi-Speed: 12v
B10	WHT/BLU	Alternator Load Signal	Lights Off: 12v, On: 0v
B13	WHT/RED	Alternator 'FR' Signal	Digital Signals: 0-5-0v
B14	BLU/RED	EGR Solenoid Control	Solenoid Off: 12v, On: 1v
B15	GRN/YEL	VTEC Solenoid Control	0v, Hi-Speed: 12v
B16	BLK/RED	A/T: Lockup Solenoid 'B'	LSB On: 12v: Off: 0v
B17	RED/YEL	IAT Sensor Signal	Varies w/temp. (0.5-4.9v)
B18	WHT/GRN	Alternator Charging Signal	Digital Signals: 0-12-0v
B21	YEL/BLU	EVAP Purge Solenoid	Solenoid Off: 12v, On: 1v

31-Pin 'A'   WIRE SIDE OF HARNESS TERMINALS   24-Pin 'B'

2001-03 Sedan 1.7L VTEC VIN ES1 [NO CVT] 22P C Black Connector

PCM Pin #	W/Color	Circuit Description (22-Pin)	Value at Hot Idle
C1	WHT/BLK	A/T: Linear Solenoid 'A-'	LSA Off: 12v, On: 1v
C2	YEL/BLU	A/T: TCC Solenoid	Solenoid Off: 12v, On: 1v
C4	GRN/WHT	A/T: Shift Solenoid 'B'	SSB Off: 0v, On: 12v
C6	BLU/BLK	A/T: Shift Solenoid 'A'	SSA On: 12v: Off: 0v
C7	WHT/RED	Mainshaft Speed Sensor 'P'	AC Pulse Signals
C8	BRN/WHT	A/T: Lockup Solenoid 'B-'	LSB On: 12v: Off: 0v
C9	RED	A/T: Gear Position Switch	In D3: 0v, Others: 12v
C10	WHT	A/T: Gear Position Switch	In 'R': 0v, Others: 12v
C11	BLU	A/T: Gear Position Switch	In D2: 0v, Others: 12v
C12	BLU/WHT	A/T: Gear Position Switch	In P/N: 0v, others: 12v
C14	GRN	Countershaft Speed Sensor N	Moving: AC pulses
C15	BLU	Countershaft Speed Sensor P	Moving: AC pulses
C18	BLY/REL	A/T: Gear Position Switch	Forward: 0v, Others: 12v
C20	YEL	A/T: Gear Position Switch	In 'D': 0v, Others: 12v
C21	WHT/GRN	Mainshaft Speed Sensor 'N'	AC Pulse Signals

2001-03 Sedan 1.7L VTEC VIN ES1 [All] 31P White 'E' Connector

PCM Pin #	W/Color	Circuit Description (31-Pin)	Value at Hot Idle
E1	GRN/YEL	Immobilizer Fuel Pump Relay	Relay Off: 12v, On: 1v
E3	BRN/YEL	Logic Ground (LG3)	<0.050v
E4	PNK	Sensor Ground (SG3)	<0.050v
E5	YEL/BLU	Sensor VREF (VCC3)	4.9-5.1v
E7	RED/YEL	Main Relay Control	Relay Off: 12v, On: 1v
E9	YEL/BLK	Main Relay Power (B+)	12-14v
E12	BLU/ORN	Cruise Control Unit Signal	N/A
E13	WHT/BLU	Multiplex Control Unit	N/A
E14	LT GRN	Fuel Tank Pressure Sensor	Fuel Cap off: 2.5v
E15	GRN/RED	Electric Load Detector	Varies: 0.5-4.5v
E16	GRN/BLK	PSP Switch Signal	Straight: 0v, Turning: 11v
E18	RED	A/C Clutch Relay	Relay Off: 12v, On: 1v
E20	BLU/RED	EVAP Bypass Solenoid	Solenoid Off: 12v, On: 1v
E21	GRN/RED	EVAP Vent Solenoid	Solenoid Off: 12v, On: 1v
E22	WHT/BLK	Brake Switch Signal	Brake Off: 0v, On: 12v
E23	LT BLU	K-Line Signal	12v
E24	YEL	SEFMJ Signal (Multiplex Unit)	Digital Signals
E25	BLU/WHT	VSS Out Signal	Pulse Signals
E26	BLU	Engine Speed Pulse (NEP)	Digital Signals
E27	RED/BLU	Immobilizer Code Signal	Digital Signals
E29	BRN	Service Check Connector	SCS Open: 4.80v
E30	RED/WHT	WEN Terminal in DLC	0v
E31	GRN/ORN	Malfunction Indicator Lamp	MIL Off: 12v, On: 1v

31-Pin 'E'   *WIRE SIDE OF HARNESS TERMINALS*   24-Pin 'C'

1990-91 Wagon 1.5L I4 MFI VIN EE2 [All] 18P 'A' Connector

PCM Pin #	W/Color	Circuit Description (18-Pin)	Value at Hot Idle
A1	BRN	Injector 1 Control	2.0-3.3 ms
A2	BLK	Power Ground	<0.1v
A3	RED	Injector 2 Control	2.0-3.3 ms
A4	BLK	Power Ground	<0.1v
A5	BLU	Injector 3 Control	2.0-3.3 ms
A6	GRN	EVAP Purge Solenoid	Solenoid Off: 12v, On: 1v
A7	YEL	Injector 4 Control	2.0-3.3 ms
A8	YEL/BLU	A/T: Control Unit Signal	Digital Signals
A9	---	Not Used	---
A10	ORN	EGR Solenoid	Solenoid Off: 12v, On: 1v
A11	BLU/YEL	Electronic Air Control Valve	Pulse Signals
A12	GRN/BLK	Fuel Pump Relay	Relay Off: 12v, On: 1v
A13	YEL/BLK	Main Relay Power (B+)	12-14v
A14	GRN/BLK	Fuel Pump Relay	Relay Off: 12v, On: 1v
A15	YEL/BLK	Main Relay Power (B+)	12-14v
A16	BRN/BLK	Power Ground	<0.1v
A17	---	Not Used	---
A18	BLK/RED	Power Ground	<0.1v

1990-91 Wagon 1.5L I4 MFI VIN EE2 [All] 20P 'B' Connector

PCM Pin #	W/Color	Circuit Description (20-Pin)	Value at Hot Idle
B1	WHT/GRN	Keep Alive Power (VBU)	12-14v
B2	GRN/YEL	Upshift Indicator Light	Lamp Off: 12v, On: 1v
B2	---	Not Used	---
B3	YEL	A/C Clutch Relay	Relay Off: 12v, On: 1v
B4	YEL/GRN	Radiator Fan Relay	Relay Off: 12v, On: 1v
B5	WHT/YEL	Alternator Charging Signal	Lights Off: 12v, On: 0v
B6	GRN/ORN	Check Engine Light	MIL Off: 12v, On: 1v
B7	GRN	A/T: P/N Switch (1989-91)	In P/N: 0v, Others: 11v
B8	BLU/RED	A/C Switch Signal	Switch Off: 12v, On: 1v
B9	GRN/BLK	Reverse Switch	In 'R': 0v, Others: 12v
B10	ORN	CKP Sensor Signal	AC pulse signals
B11	GRN	Clutch Switch	Clutch In: 0v, Out: 5v
B12	WHT	CKP Sensor Ground	0.050v
B13	BLU/WHT	Start Signal	Cranking: 9-11v
B14	BLU	Alternator 'FR' Signal	Digital Signals: 0-5-0v
B15	WHT	Igniter Control	Digital Signals: 0-12-0v
B16	YEL/RED	Vehicle Speed Sensor	Moving: pulse signals
B17	WHT	Igniter Control	Digital Signals: 0-12-0v
B18	---	Not Used	---
B19	GRN/RED	Electric Load Detector	Varies: 0.5-4.5v
B20	BRN	Ignition Timing Adjuster	0.5-4.5v

1990-91 Wagon 1.5L I4 MFI VIN EE2 [All] 16P 'C' Connector

PCM Pin #	W/Color	Circuit Description (16-Pin)	Value at Hot Idle
C1	BLU/GRN	CYP Sensor Signal	AC pulse signals
C2	BLU/YEL	CYP Sensor Ground	<0.050v
C3	ORN/BLU	TDC Sensor Signal	AC pulse signals
C4	WHT/BLU	TDC Sensor Ground	<0.050v
C5	RED/YEL	IAT Sensor Signal	Varies w/temp. (0.5-4.9v)
C6	RED/WHT	ECT Sensor Signal	At 180°F: 0.51v
C7	RED/BLU	Throttle Angle Sensor Signal	0.5-0.6v
C8	YEL	EGR Valve Lift Sensor	1.2v
C9	RED/WHT	Atmospheric Pressure Sensor	2.76-2.96v at sea level
C10	GRN/WHT	Brake Switch Signal	Brake Off: 0v, On: 12v
C11	WHT	MAP Sensor Signal	0.8-0.9v
C12	GRN/WHT	Sensor Ground	<0.050v
C13	YEL/WHT	Sensor VREF	4.9-5.1v
C14	GRN/WHT	MAP Sensor Ground	<0.050v
C15	YEL/RED	MAP Sensor VREF	4.9-5.1v
C16	WHT	Oxygen Sensor Signal	0.1-1.1v

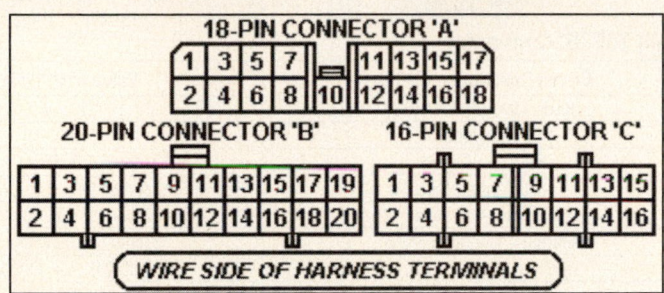

Abbreviation	Color	Abbreviation	Color	Abbreviation	Color
BLK	Black	LT BLU	Lt. Blue	TAN	Tan
BLU	Blue	LT GRN	Lt. Green	VIO	Violet
BRN	Brown	ORN	Orange	WHT	White
GRY	Gray	PNK	Pink	YEL	Yellow
GRN	Green	PPL	Purple		

1990-91 Wagon 1.6L I4 MFI VIN EE4 [All] 18P 'A' Connector

PCM Pin #	W/Color	Circuit Description (18-Pin)	Value at Hot Idle
A1	BRN	Injector 1 Control	2.0-3.3 ms
A2	BLK	Power Ground	<0.1v
A3	RED	Injector 2 Control	2.0-3.3 ms
A4	BLK	Power Ground	<0.1v
A5	BLU	Injector 3 Control	2.0-3.3 ms
A6	GRN	EVAP Purge Solenoid	Solenoid Off: 12v, On: 1v
A7	YEL	Injector 4 Control	2.0-3.3 ms
A8	YEL/BLU	A/T: Control Unit Signal	Digital Signals
A9-10	---	Not Used	---
A11	BLU/YEL	Electronic Air Control Valve	Pulse Signals
A12	GRN/BLK	Fuel Pump Relay	Relay Off: 12v, On: 1v
A13	YEL/BLK	Main Relay Power (B+)	12-14v
A14	GRN/BLK	Fuel Pump Relay	Relay Off: 12v, On: 1v
A15	YEL/BLK	Main Relay Power (B+)	12-14v
A16	BRN/BLK	Power Ground	<0.1v
A17	---	Not Used	---
A18	BLK/RED	Power Ground	<0.1v

1990-91 Wagon 1.6L I4 MFI VIN EE4 [All] 20P 'B' Connector

PCM Pin #	W/Color	Circuit Description (20-Pin)	Value at Hot Idle
B1	WHT/GRN	Keep Alive Power (VBU)	12-14v
B2	---	Not Used	---
B2	BLU	Fast Idle Control	Solenoid Off: 12v, On: 1v
B3	YEL	A/C Clutch Relay	Solenoid Off: 12v, On: 1v
B4	YEL/GRN	Radiator Fan Relay	Relay Off: 12v, On: 1v
B5	WHT/YEL	Alternator Charging Signal	Lights Off: 12v, On: 0v
B6	GRN/ORN	Check Engine Light	MIL Off: 12v, On: 1v
B7	---	Not Used	---
B7	GRN	A/T: P/N Switch	In P/N: 0v, Others: 12v
B8	BLU/RED	A/C Switch Signal	Switch Off: 12v, On: 1v
B9	---	Not Used	---
B10	ORN	CKP Sensor Signal	AC pulse signals
B11	---	Not Used	---
B12	WHT	CKP Sensor Ground	<0.050v
B13	BLU/WHT	Start Signal	Cranking: 9-11v
B14	BLU	Alternator 'FR' Signal	Digital Signals: 0-5-0v
B15	WHT	Igniter Control	Digital Signals: 0-12-0v
B16	YEL/RED	Vehicle Speed Sensor	Moving: pulse signals
B17	WHT	Igniter Control	Digital Signals: 0-12-0v
B18	---	Not Used	---
B19	GRN/RED	Electric Load Detector	Varies: 0.5-4.5v
B20	BRN	Ignition Timing Adjuster	0.5-4.5v

1990-91 Wagon 1.6L I4 MFI VIN EE4 [All] 16P 'C' Connector

PCM Pin #	W/Color	Circuit Description (16-Pin)	Value at Hot Idle
C1	BLU/GRN	CYP Sensor Signal	AC pulse signals
C2	BLU/YEL	CYP Sensor Ground	<0.050v
C3	ORN/BLU	TDC Sensor Signal	AC pulse signals
C4	WHT/BLU	TDC Sensor Ground	<0.050v
C5	RED/YEL	IAT Sensor Signal	Varies w/temp. (0.5-4.9v)
C6	RED/WHT	ECT Sensor Signal	At 180°F: 0.51v
C7	RED/BLU	Throttle Angle Sensor Signal	0.5-0.6v
C8	---	Not Used	---
C8	BLU/WHT	A/T: Control Unit Signal	Digital Signals
C9	RED/WHT	Atmospheric Pressure Sensor	2.76-2.96v at sea level
C10	GRN/WHT	Brake Switch Signal	Brake Off: 0v, On: 12v
C11	WHT	MAP Sensor Signal	0.8-0.9v
C12	GRN/WHT	Sensor Ground	<0.050v
C13	YEL/WHT	Sensor VREF	4.9-5.1v
C14	GRN/WHT	MAP Sensor Ground	<0.050v
C15	YEL/RED	MAP Sensor VREF	4.9-5.1v
C16	WHT	Oxygen Sensor Signal	0.1-1.1v

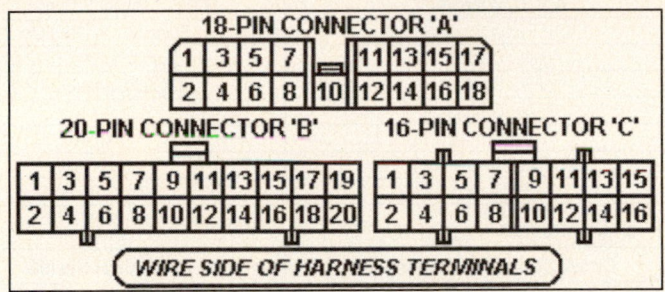

Abbreviation	Color	Abbreviation	Color	Abbreviation	Color
BLK	Black	LT BLU	Lt. Blue	TAN	Tan
BLU	Blue	LT GRN	Lt. Green	VIO	Violet
BRN	Brown	ORN	Orange	WHT	White
GRY	Gray	PNK	Pink	YEL	Yellow
GRN	Green	PPL	Purple		

CIVIC DEL SOL Pin Tables

1993 Coupe 1.5L I4 MFI VIN EG1 [All] 26P 'A' Connector

PCM Pin #	W/Color	Circuit Description (26-Pin)	Value at Hot Idle
A1	BRN	Injector 1 Control	2.0-3.3 ms
A2	YEL	Injector 1 Control	2.0-3.3 ms
A3	RED	Injector 2 Control	2.0-3.3 ms
A4	---	Not Used	---
A5	BLU	Injector 3 Control	2.0-3.3 ms
A6	ORN/BLK	HO2S-11 (B1 S1) Heater	Relay Off: 12v, On: 1v
A7	GRN/YEL	Fuel Pump Relay	Relay Off: 12v, On: 1v
A8	GRN/YEL	Fuel Pump Relay	Relay Off: 12v, On: 1v
A9	GRN/WHT	Idle Air Control Valve	Pulse Signals
A10-11	---	Not Used	---
A12	YEL/GRN	Radiator Fan Relay	Relay Off: 12v, On: 1v
A13	GRN/ORN	Check Engine Light	MIL Off: 12v, On: 1v
A14	---	Not Used	---
A15	BLK/RED	A/C Clutch Relay	Relay Off: 12v, On: 1v
A16	WHT/YEL	Alternator Charging Signal	Lights Off: 12v, On: 0v
A17	LT GRN	A/T: TCM Signal	Digital Signals
A18	---	Not Used	---
A19	YEL	A/T: TCM Signal	Digital Signals
A20	RED	EVAP Purge Solenoid	Solenoid Off: 12v, On: 1v
A21	RED/GRN	Igniter Control	Digital Signals: 0-12-0v
A22	RED/GRN	Igniter Control	Digital Signals: 0-12-0v
A23	BLK	Power Ground	<0.1v
A24	BLK	Power Ground	<0.1v
A25	YEL/BLK	Main Relay Power (B+)	12-14v
A26	BLK/RED	Logic Ground	<0.050v

1993 Coupe 1.5L I4 MFI VIN EG1 [All] 16P 'B' Connector

PCM Pin #	W/Color	Circuit Description (16-Pin)	Value at Hot Idle
B1	WHT/GRN	Main Relay Power (B+)	12-14v
B2	BRN/BLK	Logic Ground	<0.050v
B3	GRN/BLU	A/T: TCM Signal	In D3: 0v, Others: 5v
B4	GRN/BLK	A/T: TCM Signal	In D4: 0v, Others: 5v
B5	BLU/RED	A/C Switch Signal	Relay Off: 12v, On: 1v
B6	---	Not Used	---
B7	GRN	A/T: Park Neutral Switch	In P/N: 0v, Others: 12v
B8	BRN/RED	PSP Switch	Straight: 0v, Turning: 11v
B9	BLU/WHT	Starter Switch Signal	Cranking: 9-11v
B10	YEL/BLU	Vehicle Speed Sensor	Moving: pulse signals
B11	ORN	CYP Sensor Signal	AC pulse signals
B12	WHT	CYP Sensor Ground	<0.050v
B13	ORN/BLU	TDC Sensor Signal	AC pulse signals
B14	WHT/BLU	TDC Sensor Ground	<0.050v
B15	BLU/GRN	CKP Sensor Signal	AC pulse signals
B16	BLU/YEL	CKP Sensor Ground	<0.050v

1993 Coupe 1.5L I4 MFI VIN EG1 [All] 22P 'D' Connector

PCM Pin #	W/Color	Circuit Description (22-Pin)	Value at Hot Idle
D1	WHT/BLU	Keep Alive Power (VBU)	12-14v
D2	GRN/WHT	Brake Switch Signal	Brake Off: 0v, On: 12v
D3	---	Not Used	---
D4	BRN	Service Check Connector	SCS Open: 4.80v
D5	---	Not Used	---
D6	---	Not Used	---
D7	LT BLU	Data Link Connector	5v
D8	---	Not Used	---
D9	PNK	Alternator 'FR' Signal	Varies 1.5-3.5v
D10	GRN/RED	Electric Load Detector	Varies: 0.5-4.5v
D11	LT GRN	TP Sensor Signal	0.5-0.6v
D12	---	Not Used	---
D13	RED/WHT	ECT Sensor Signal	At 180°F: 0.52v
D14	ORN/BLU	HO2S-11 (B1 S1) Signal	0.1-1.1v
D15	RED/YEL	IAT Sensor Signal	Varies w/temp. (0.5-4.9v)
D16	---	Not Used	---
D17	WHT	MAP Sensor Signal	0.8-0.9v
D18	PNK/GRN	Economy Driving Indicator	Digital Signals
D19	YEL/GRN	MAP Sensor VREF	4.9-5.1v
D20	YEL/WHT	Sensor VREF	4.9-5.1v
D21	GRN/WHT	MAP Sensor Ground	<0.050v
D22	GRN/WHT	Sensor Ground	<0.050v

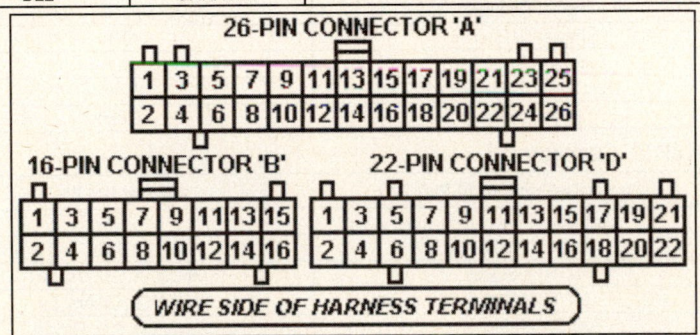

26-PIN CONNECTOR 'A'
1 3 5 7 9 11 13 15 17 19 21 23 25
2 4 6 8 10 12 14 16 18 20 22 24 26

16-PIN CONNECTOR 'B'
1 3 5 7 9 11 13 15
2 4 6 8 10 12 14 16

22-PIN CONNECTOR 'D'
1 3 5 7 9 11 13 15 17 19 21
2 4 6 8 10 12 14 16 18 20 22

WIRE SIDE OF HARNESS TERMINALS

Abbreviation	Color	Abbreviation	Color	Abbreviation	Color
BLK	Black	LT BLU	Lt. Blue	TAN	Tan
BLU	Blue	LT GRN	Lt. Green	VIO	Violet
BRN	Brown	ORN	Orange	WHT	White
GRY	Gray	PNK	Pink	YEL	Yellow
GRN	Green	PPL	Purple		

1993 Coupe 1.6L I4 VTEC VIN EH6 [All] 26P 'A' Connector

PCM Pin #	W/Color	Circuit Description (26-Pin)	Value at Hot Idle
A1	BRN	Injector 1 Control	2.0-3.3 ms
A2	YEL	Injector 4 Control	2.0-3.3 ms
A3	RED	Injector 2 Control	2.0-3.3 ms
A4	ORN/WHT	VTEC Solenoid	0v, Hi-Speed: 12v
A5	BLU	Injector 3 Control	2.0-3.3 ms
A6	ORN/BLK	HO2S-1 Heater Control	Relay Off: 12v, On: 1v
A7	GRN/YEL	Fuel Pump Relay	Relay Off: 12v, On: 1v
A8	GRN/YEL	Fuel Pump Relay	Relay Off: 12v, On: 1v
A9	GRN/WHT	Idle Air Control Valve	Pulse Signals
A10-11	---	Not Used	
A12	YEL/GRN	Radiator Fan Relay	Relay Off: 12v, On: 1v
A13	GRN/ORN	Check Engine Light	MIL Off: 12v, On: 1v
A14	---	Not Used	
A15	BLK/RED	A/C Clutch Relay	Relay Off: 12v, On: 1v
A16	WHT/YEL	Alternator Charging Signal	Lights Off: 12v, On: 0v
A17	LT GRN	A/T TCM Signal	Digital Signal
A18	---	Not Used	---
A19	YEL	A/T TCM Signal	Digital Signals
A20	RED	EVAP Purge Solenoid	Solenoid Off: 12v, On: 1v
A21	RED/GRN	Igniter Control	Digital Signals: 0-12-0v
A22	RED/GRN	Igniter Control	Digital Signals: 0-12-0v
A23	BLK	Power Ground	<0.1v
A24	BLK	Power Ground	<0.1v
A25	YEL/BLK	Main Relay Power (B+)	12-14v
A26	BLK/RED	Logic Ground	<0.050v

1993 Coupe 1.6L I4 VTEC VIN EH6 [All] 16P 'B' Connector

PCM Pin #	W/Color	Circuit Description (16-Pin)	Value at Hot Idle
B1	WHT/GRN	Main Relay Power (B+)	12-14v
B2	BRN/BLK	Logic Ground	<0.050v
B3	GRN/BLU	A/T TCM Signal	In D3: 0v, Others: 5v
B4	GRN/BLK	A/T TCM Signal	In D4: 0v, Others: 5v
B5	BLU/RED	A/C Switch Signal	Relay Off: 12v, On: 1v
B6	---	Not Used	---
B7	GRN	A/T: Park Neutral Switch	In P/N: 0v, Others: 12v
B8	BRN/RED	PSP Switch	Straight: 0v, Turning: 11v
B9	BLU/WHT	Starter Switch Signal	Cranking: 9-11v
B10	YEL/BLU	Vehicle Speed Sensor	Moving: pulse signals
B11	ORN	CYP Sensor Signal	AC pulse signals
B12	WHT	CYP Sensor Ground	<0.050v
B13	ORN/BLU	TDC Sensor Signal	AC pulse signals
B14	WHT/BLU	TDC Sensor Ground	<0.050v
B15	BLU/GRN	CKP Sensor Signal	AC pulse signals
B16	BLU/YEL	CKP Sensor Ground	<0.050v

1993 Coupe 1.6L I4 VTEC VIN EH6 [All] 22P 'D' Connector

PCM Pin #	W/Color	Circuit Description (22-Pin)	Value at Hot Idle
D1	WHT/BLU	Keep Alive Power (VBU)	12-14v
D2	GRN/WHT	Brake Switch Signal	Brake Off: 0v, On: 12v
D4	BRN	Service Check Connector	SCS Open: 4.80v
D6	ORN/BLU	VTEC Pressure Switch	0v, Hi-Speed: 12v
D7	LT BLU	Data Link Connector	5v
D9	PNK	Alternator 'FR' Signal	Varies: 1.5-3.5v
D10	GRN/RED	Electric Load Detector	Varies: 0.5-4.5v
D11	LT GRN	TP Sensor Signal	0.5-0.6v
D13	RED/WHT	ECT Sensor Signal	At 180°F: 0.52v
D14	ORN/BLU	HO2S-11 (B1 S1) Signal	0.1-1.1v
D15	RED/YEL	IAT Sensor Signal	Varies w/temp. (0.5-4.9v)
D17	WHT	MAP Sensor Signal	0.8-0.9v
D18	PNK/GRN	Economy Driving Indicator	N/A
D19	YEL/GRN	MAP Sensor VREF	4.9-5.1v
D20	YEL/WHT	Sensor VREF	4.9-5.1v
D21	GRN/WHT	MAP Sensor Ground	<0.050v
D22	GRN/WHT	Sensor Ground	<0.050v

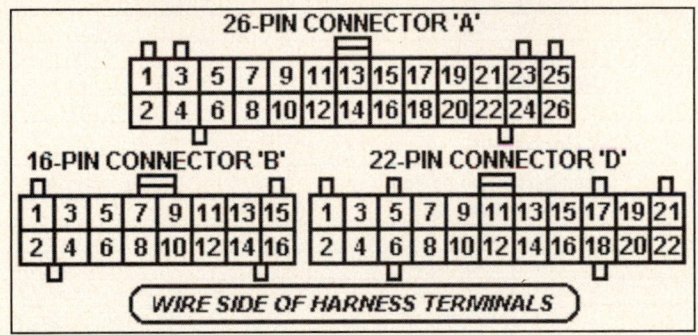

Abbreviation	Color	Abbreviation	Color	Abbreviation	Color
BLK	Black	LT BLU	Lt. Blue	TAN	Tan
BLU	Blue	LT GRN	Lt. Green	VIO	Violet
BRN	Brown	ORN	Orange	WHT	White
GRY	Gray	PNK	Pink	YEL	Yellow
GRN	Green	PPL	Purple		

1994-95 Coupe 1.5L I4 MFI VIN EG1 [All] 26P 'A' Connector

PCM Pin #	W/Color	Circuit Description (26-Pin)	Value at Hot Idle
A1	BRN	Injector 1 Control	2.0-3.3 ms
A2	YEL	Injector 4 Control	2.0-3.3 ms
A3	RED	Injector 2 Control	2.0-3.3 ms
A4	--	Not Used	---
A5	BLU	Injector 3 Control	2.0-3.3 ms
A6	ORN/BLK	HO2S Heater Control	Relay Off: 12v, On: 1v
A7	GRN/YEL	Fuel Pump Relay	Relay Off: 12v, On: 1v
A8	--	Not Used	---
A9	GRN/WHT	Idle Air Control Valve	Pulse Signals
A12	YEL/GRN	Radiator Fan Relay	Relay Off: 12v, On: 1v
A13	GRN/ORN	Check Engine Light	MIL Off: 12v, On: 1v
A15	BLK/RED	A/C Clutch Relay	Relay Off: 12v, On: 1v
A16	WHT/YEL	Alternator Charging Signal	Lights Off: 12v, On: 0v
A17	GRN/BLK	A/T: Lockup Solenoid 'B'	LSB On: 12v, Off: 0v
A19	YEL	A/T: Lockup Solenoid 'A'	LSA On: 12v, Off: 0v
A20	RED	EVAP Purge Solenoid	Solenoid Off: 12v, On: 1v
A21	RED/GRN	Igniter Control	Digital Signals: 0-12-0v
A22	---	Not Used	---
A23	BLK	Power Ground	<0.1v
A24	BLK	Power Ground	<0.1v
A25	YEL/BLK	Main Relay Power (B+)	12-14v
A26	BLK/RED	Logic Ground	<0.050v

1994-95 Coupe 1.5L I4 MFI VIN EG1 [All] 16P 'B' Connector

PCM Pin #	W/Color	Circuit Description (16-Pin)	Value at Hot Idle
B1	WHT/GRN	Main Relay Power (B+)	12-14v
B2	BRN/BLK	Logic Ground	<0.050v
B3	GRN/BLU	A/T: Shift Selector Signal	In D3: 0v, Others: 12v
B4	GRN/BLK	A/T: Shift Selector Signal	In D4: 0v, Others: 12v
B5	BLU/RED	A/C Switch Signal	Relay Off: 12v, On: 1v
B6	---	Not Used	---
B7	GRN	A/T: Park Neutral Switch	In P/N: 0v, Others: 12v
B8	BRN/RED	PSP Switch	Straight: 0v, Turning: 11v
B9	BLU/WHT	Starter Switch Signal	Cranking: 9-11v
B10	YEL/BLU	Vehicle Speed Sensor	Moving: pulse signals
B11	ORN	CYP Sensor Signal	AC pulse signals
B12	WHT	CYP Sensor Ground	<0.050v
B13	ORN/BLU	TDC Sensor Signal	AC pulse signals
B14	WHT/BLU	TDC Sensor Ground	<0.050v
B15	BLU/GRN	CKP Sensor Signal	AC pulse signals
B16	BLU/YEL	CKP Sensor Ground	<0.050v

1994-95 Coupe 1.5L I4 MFI VIN EG1 [All] 22P 'D' Connector

PCM Pin #	W/Color	Circuit Description (22-Pin)	Value at Hot Idle
D1	WHT/BLU	Keep Alive Power (VBU)	12-14v
D2	GRN/WHT	Brake Switch Signal	Brake Off: 0v, On: 12v
D3	RED/BLU	Knock Sensor Signal	No Detonation: 18mv AC
D4	BRN	Service Check Connector	SCS Open: 4.80v
D5-6, 8	---	Not Used	---
D7	LT BLU	Data Link Connector	5v
D9	PNK	Alternator 'FR' Signal	Varies: 1.5-3.5v
D10	GRN/RED	Electric Load Detector	Varies: 0.5-4.5v
D11	LT GRN	TP Sensor Signal	0.5-0.6v
D12	---	Not Used	---
D13	RED/WHT	ECT Sensor Signal	At 180°F: 0.52v
D14	WHT	HO2S-11 (B1 S1) Signal	0.1-1.1v
D15	RED/YEL	IAT Sensor Signal	Varies w/temp. (0.5-4.9v)
D16	---	Not Used	---
D17	WHT	MAP Sensor Signal	0.8-0.9v
D18	PNK/GRN	A/T: Interlock Control Unit	Key & Brake On: 12v
D19	YEL/GRN	MAP Sensor VREF	4.9-5.1v
D20	YEL/WHT	Sensor VREF	4.9-5.1v
D21	GRN/BLU	MAP Sensor Ground	<0.050v
D22	GRN/WHT	Sensor Ground	<0.050v

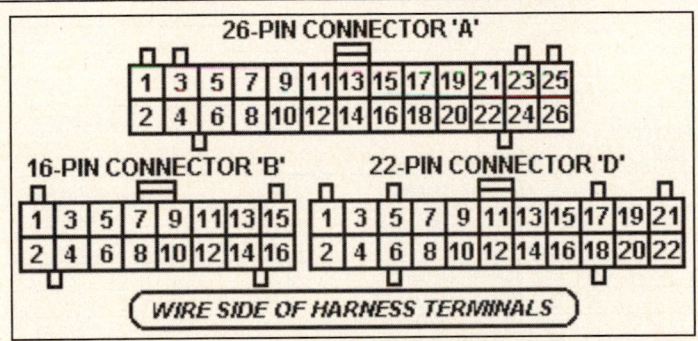

Abbreviation	Color	Abbreviation	Color	Abbreviation	Color
BLK	Black	LT BLU	Lt. Blue	TAN	Tan
BLU	Blue	LT GRN	Lt. Green	VIO	Violet
BRN	Brown	ORN	Orange	WHT	White
GRY	Gray	PNK	Pink	YEL	Yellow
GRN	Green	PPL	Purple		

1994-95 Coupe 1.6L I4 VTEC VIN EH6 [All] 26P 'A' Connector

PCM Pin #	W/Color	Circuit Description (26-Pin)	Value at Hot Idle
A1	BRN	Injector 1 Control	2.0-3.3 ms
A2	YEL	Injector 4 Control	2.0-3.3 ms
A3	RED	Injector 2 Control	2.0-3.3 ms
A4	ORN/WHT	VTEC Solenoid	0v, Hi-Speed: 12v
A5	BLU	Injector 3 Control	2.0-3.3 ms
A6	ORN/BLK	HO2S Heater Control	Relay Off: 12v, On: 1v
A7	GRN/YEL	Fuel Pump Relay	Relay Off: 12v, On: 1v
A8	---	Not Used	---
A9	GRN/WHT	Idle Air Control Valve	Pulse Signals
A10-11	---	Not Used	---
A12	YEL/GRN	Radiator Fan Relay	Relay Off: 12v, On: 1v
A13	GRN/ORN	Check Engine Light	MIL Off: 12v, On: 1v
A14	---	Not Used	---
A15	BLK/RED	A/C Clutch Relay	Relay Off: 12v, On: 1v
A16	WHT/YEL	Alternator Charging Signal	Lights Off: 12v, On: 0v
A17	GRN/BLK	A/T: Lockup Solenoid 'B'	LSB On: 12v, Off: 0v
A18	---	Not Used	---
A19	YEL	A/T: Lockup Solenoid 'A'	LSA On: 12v, Off: 0v
A20	RED	EVAP Purge Solenoid	Solenoid Off: 12v, On: 1v
A21	RED/GRN	Igniter Control	Digital Signals: 0-12-0v
A22	---	Not Used	---
A23	BLK	Power Ground	<0.1v
A24	BLK	Power Ground	<0.1v
A25	YEL/BLK	Main Relay Power (B+)	12-14v
A26	BLK/RED	Logic Ground	<0.050v

1994-95 Coupe 1.6L I4 VTEC VIN EH6 [All] 16P 'B' Connector

PCM Pin #	W/Color	Circuit Description (16-Pin)	Value at Hot Idle
B1	WHT/GRN	Main Relay Power (B+)	12-14v
B2	BRN/BLK	Logic Ground	<0.050v
B3	GRN/BLU	A/T: Shift Selector Signal	In D3: 0v, Others: 12v
B4	GRN/BLK	A/T: Shift Selector Signal	In D4: 0v, Others: 12v
B5	BLU/RED	A/C Switch Signal	Relay Off: 12v, On: 1v
B7	GRN	A/T: Park Neutral Switch	In P/N: 0v, Others: 12v
B8	BRN/RED	PSP Switch	Straight: 0v, Turning: 11v
B9	BLU/WHT	Starter Switch Signal	Cranking: 9-11v
B10	YEL/BLU	Vehicle Speed Sensor	Moving: pulse signals
B11	ORN	CYP Sensor Signal	AC pulse signals
B12	WHT	CYP Sensor Ground	<0.050v
B13	ORN/BLU	TDC Sensor Signal	AC pulse signals
B14	WHT/BLU	TDC Sensor Ground	<0.050v
B15	BLU/GRN	CKP Sensor Signal	AC pulse signals
B16	BLU/YEL	CKP Sensor Ground	<0.050v

1994-95 Coupe 1.6L I4 VTEC VIN EH6 [All] 22P 'D' Connector

PCM Pin #	W/Color	Circuit Description (22-Pin)	Value at Hot Idle
D1	WHT/BLU	Keep Alive Power (VBU)	12-14v
D2	GRN/WHT	Brake Switch Signal	Brake Off: 0v, On: 12v
D3	RED/BLU	Knock Sensor Signal	No Detonation: 18mv AC
D4	BRN	Service Check Connector	SCS Open: 4.80v
D5	---	Not Used	---
D6	ORN/BLU	VTEC Pressure Switch	0v, Hi-Speed: 12v
D7	LT BLU	Data Link Connector	5v
D8	---	Not Used	---
D9	PNK	Alternator 'FR' Signal	Digital Signals: 0-5-0v
D10	GRN/RED	Electric Load Detector	Varies: 0.5-4.5v
D11	LT GRN	TP Sensor Signal	0.5-0.6v
D12	---	Not Used	---
D13	RED/WHT	ECT Sensor Signal	At 180°F: 0.52v
D14	WHT	HO2S Signal	0.1-1.1v
D15	RED/YEL	IAT Sensor Signal	Varies w/temp. (0.5-4.9v)
D16	---	Not Used	---
D17	WHT	MAP Sensor Signal	0.8-0.9v
D18	PNK/GRN	A/T: Interlock Control Unit	Key & Brake On: 12v
D19	YEL/GRN	MAP Sensor VREF	4.9-5.1v
D20	YEL/WHT	Sensor VREF	4.9-5.1v
D21	GRN/WHT	MAP Sensor Ground	<0.050v
D22	GRN/WHT	Sensor Ground	<0.050v

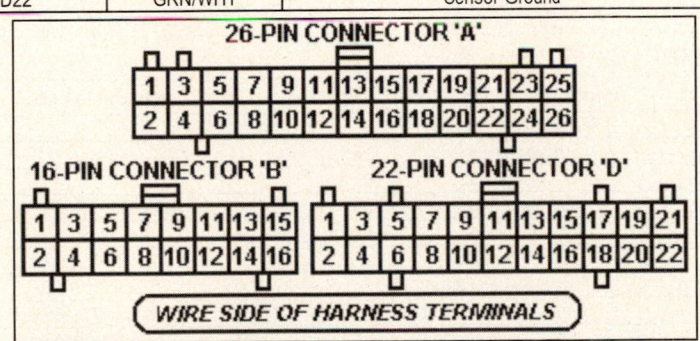

Abbreviation	Color	Abbreviation	Color	Abbreviation	Color
BLK	Black	LT BLU	Lt. Blue	TAN	Tan
BLU	Blue	LT GRN	Lt. Green	VIO	Violet
BRN	Brown	ORN	Orange	WHT	White
GRY	Gray	PNK	Pink	YEL	Yellow
GRN	Green	PPL	Purple		

1994-95 Coupe 1.6L I4 VTEC VIN EG2 [M/T] 26P 'A' Connector

PCM Pin #	W/Color	Circuit Description (26-Pin)	Value at Hot Idle
A1	BRN	Injector 1 Control	2.0-3.3 ms
A2	YEL	Injector 4 Control	2.0-3.3 ms
A3	RED	Injector 2 Control	2.0-3.3 ms
A4	ORN/WHT	VTEC Solenoid	0v, Hi-Speed: 12v
A5	BLU	Injector 3 Control	2.0-3.3 ms
A6	ORN/BLK	HO2S Heater Control	Relay Off: 12v, On: 1v
A7	GRN/YEL	Fuel Pump Relay	Relay Off: 12v, On: 1v
A8	---	Not Used	---
A9	GRN/WHT	Idle Air Control Valve	Pulse Signals
A10-11	---	Not Used	---
A12	YEL/GRN	Radiator Fan Relay	Relay Off: 12v, On: 1v
A13	GRN/ORN	Check Engine Light	MIL Off: 12v, On: 1v
A14	---	Not Used	---
A15	BLK/RED	A/C Clutch Relay	Relay Off: 12v, On: 1v
A16	WHT/YEL	Alternator Charging Signal	Lights Off: 12v, On: 0v
A17-19	---	Not Used	---
A20	RED	EVAP Purge Solenoid	Solenoid Off: 12v, On: 1v
A21	RED/GRN	Igniter Control	Digital Signals: 0-12-0v
A22	---	Not Used	---
A23	BLK	Power Ground	<0.1v
A24	BLK	Power Ground	<0.1v
A25	YEL/BLK	Main Relay Power (B+)	12-14v
A26	BLK/RED	Logic Ground	<0.050v

1994-95 Coupe 1.6L I4 VTEC VIN EG2 [M/T] 16P 'B' Connector

PCM Pin #	W/Color	Circuit Description (16-Pin)	Value at Hot Idle
B1	WHT/GRN	Main Relay Power (B+)	12-14v
B2	BRN/BLK	Logic Ground	<0.050v
B3-4	---	Not Used	---
B5	BLU/RED	A/C Switch Signal	Relay Off: 12v, On: 1v
B6-7	---	Not Used	---
B8	BRN/RED	PSP Switch	Straight: 0v, Turning: 11v
B9	BLU/WHT	Starter Switch Signal	Cranking: 9-11v
B10	YEL/BLU	Vehicle Speed Sensor	Moving: pulse signals
B11	ORN	CYP Sensor Signal	AC pulse signals
B12	WHT	CYP Sensor Ground	<0.050v
B13	ORN/BLU	TDC Sensor Signal	AC pulse signals
B14	WHT/BLU	TDC Sensor Ground	<0.050v
B15	BLU/GRN	CKP Sensor Signal	AC pulse signals
B16	BLU/YEL	CKP Sensor Ground	<0.050v

1994-95 Coupe 1.6L I4 VTEC VIN EG2 [M/T] 22P 'D' Connector

PCM Pin #	W/Color	Circuit Description (22-Pin)	Value at Hot Idle
D1	WHT/BLU	Keep Alive Power (VBU)	12-14v
D2	GRN/WHT	Brake Switch Signal	Brake Off: 0v, On: 12v
D3	RED/BLU	Knock Sensor Signal	No Detonation: 18mv AC
D4	BRN	Service Check Connector	SCS Open: 4.80v
D5	---	Not Used	---
D6	ORN/BLU	VTEC Pressure Switch	0v, Hi-Speed: 12v
D7	LT BLU	Data Link Connector	5v
D8	---	Not Used	---
D9	PNK	Alternator 'FR' Signal	Digital Signals: 0-5-0v
D10	GRN/RED	Electric Load Detector	Varies: 0.5-4.5v
D11	LT GRN	TP Sensor Signal	0.5-0.6v
D12	---	Not Used	---
D13	RED/WHT	ECT Sensor Signal	At 180ºF: 0.52v
D14	WHT	HO2S-11 (B1 S1) Signal	0.1-1.1v
D15	RED/YEL	IAT Sensor Signal	Varies w/temp. (0.5-4.9v)
D16	---	Not Used	---
D17	WHT	MAP Sensor Signal	0.8-0.9v
D18	---	Not Used	---
D19	YEL/GRN	MAP Sensor VREF	4.9-5.1v
D20	YEL/WHT	Sensor VREF	4.9-5.1v
D21	GRN/WHT	MAP Sensor Ground	<0.050v
D22	GRN/WHT	Sensor Ground	<0.050v

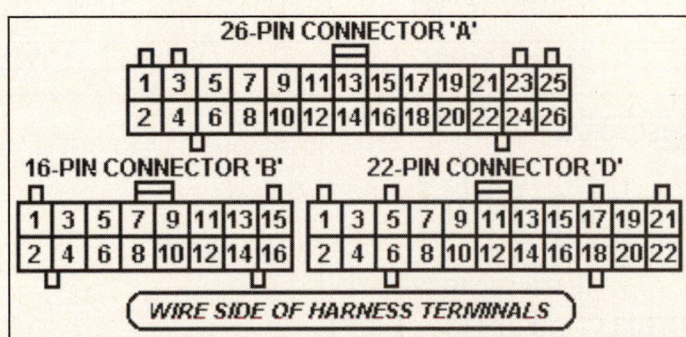

1996-97 Coupe 1.6L I4 VTEC VIN EG2 [M/T] 32P 'A' Connector

PCM Pin #	W/Color	Circuit Description (32-Pin)	Value at Hot Idle
A1	YEL	Injector 4 Control	2.0-3.3 ms
A2	BLU	Injector 3 Control	2.0-3.3 ms
A3	RED	Injector 2 Control	2.0-3.3 ms
A4	BRN	Injector 1 Control	2.0-3.3 ms
A5	BLK/WHT	HO2S-12 (B1 S2) Heater	Digital Signals: 0-12-0v
A6	ORN/BLK	HO2S-11 (B1 S1) Heater	Digital Signals: 0-12-0v
A7	---	Not Used	---
A8	ORN/WHT	VTEC Solenoid	0v, Hi-Speed: 12v
A9	BLK/RED	Logic Ground	<0.050v
A10	BLK	Power Ground	<0.1v
A11	YEL/BLK	Main Relay Power (B+)	12-14v
A12	GRN/WHT	Idle Air Control Valve	Pulse Signals
A13-14	---	Not Used	---
A15	RED	EVAP Purge Solenoid	Solenoid Off: 12v, On: 1v
A16	GRN/YEL	Fuel Pump Relay	Relay Off: 12v, On: 1v
A17	BLK/RED	A/C Clutch Relay	Relay Off: 12v, On: 1v
A18	GRN/ORN	Check Engine Light	MIL Off: 12v, On: 1v
A19	WHT/YEL	Alternator Charging Signal	Lights Off: 12v, On: 0v
A20	RED/GRN	Igniter Control	Digital Signals: 0-12-0v
A21	---	Not Used	---
A22	BRN/BLK	Logic Ground	<0.050v
A23	BLK	Power Ground	<0.1v
A24	YEL/BLK	Main Relay Power (B+)	12-14v
A25-26	---	Not Used	---
A27	YEL/GRN	Radiator Fan Relay	Relay Off: 12v, On: 1v
A28-32	---	Not Used	---

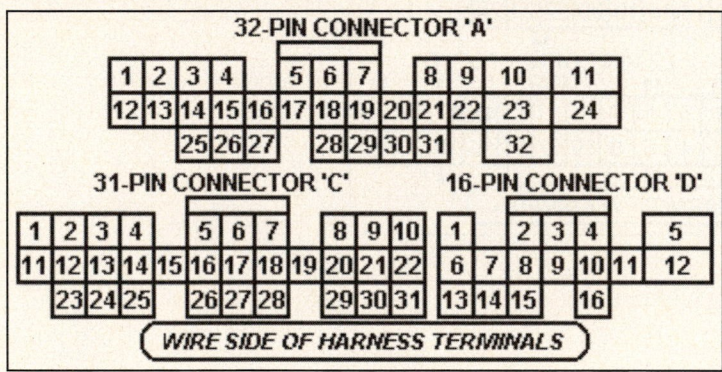

1996-97 Coupe 1.6L I4 VTEC VIN EG2 [M/T] 31P 'C' Connector

PCM Pin #	W/Color	Circuit Description (31-Pin)	Value at Hot Idle
C1	BLU/RED	CKF Sensor Signal	AC pulse signals
C2	BLU/GRN	CKP Sensor Signal	AC pulse signals
C3	ORN/BLU	TDC Sensor Signal	AC pulse signals
C4	ORN	CYP Sensor Signal	AC pulse signals
C5	BLU/RED	A/C Switch Signal	Relay Off: 12v, On: 1v
C6	BLU/WHT	Starter Switch Signal	Cranking: 9-11v
C7	BRN	Service Check Connector	SCS Open: 4.80v
C8	GRN/YEL	K-Line Signal	12v
C9	---	Not Used	---
C10	WHT/BLU	Keep Alive Power (VBU)	12-14v
C11	WHT/RED	CKF Sensor Ground	<0.050v
C12	BLU/YEL	CKP Sensor Ground	<0.050v
C13	WHT/BLU	TDC Sensor Ground	<0.050v
C14	WHT	CYP Sensor Ground	<0.050v
C15	ORN/BLU	VTEC Pressure Switch	0v, Hi-Speed: 12v
C16	BRN/RED	PSP Switch	Straight: 0v, Turning: 11v
C17	PNK	Alternator 'FR' Signal	Varies: 1.5-3.5v
C18	YEL/BLU	Vehicle Speed Sensor	Moving: pulse signals
C19-31	---	Not Used	---

1996-97 Coupe 1.6L I4 VTEC VIN EG2 [M/T] 16P 'D' Connector

PCM Pin #	W/Color	Circuit Description (16-Pin)	Value at Hot Idle
D1	RED/BLU	TP Sensor Signal	0.5-0.6v
D2	RED/WHT	ECT Sensor Signal	At 180°F: 0.52v
D3	WHT	MAP Sensor Signal	0.8-0.9v
D4	YEL/GRN	MAP Sensor VREF	4.9-5.1v
D5	GRN/WHT	Brake Switch Signal	Brake Off: 0v, On: 12v
D6	RED/BLU	Knock Sensor Signal	No Detonation: 18mv AC
D7	WHT	HO2S-11 (B1 S1) Signal	0.1-1.1v
D8	RED/YEL	IAT Sensor Signal	Varies w/temp. (0.5-4.9v)
D9	---	Not Used	---
D10	YEL/WHT	Sensor VREF	4.9-5.1v
D11	GRN/WHT	Sensor Ground	<0.050v
D12	GRN/WHT	MAP Sensor Ground	<0.050v
D13	RED/YEL	HO2S-12 Ground	<0.050v
D14	WHT/RED	HO2S-12 (B1 S2) Signal	0.1-1.1v
D15	---	Not Used	---
D16	GRN/RED	Electric Load Detector	Varies: 0.5-4.5v

1996-97 Coupe 1.6L I4 MFI VIN EH6 [All] 32P 'A' Connector

PCM Pin #	W/Color	Circuit Description (32-Pin)	Value at Hot Idle
A1	YEL	Injector 4 Control	2.0-3.3 ms
A2	BLU	Injector 3 Control	2.0-3.3 ms
A3	RED	Injector 2 Control	2.0-3.3 ms
A4	BRN	Injector 1 Control	2.0-3.3 ms
A5	BLK/WHT	HO2S-12 (B1 S2) Heater	Digital Signals: 0-12-0v
A6	ORN/BLK	HO2S-11 (B1 S1) Heater	Digital Signals: 0-12-0v
A7-8	---	Not Used	---
A9	BLK/RED	Logic Ground	<0.050v
A10	BLK	Power Ground	<0.1v
A11	YEL/BLK	Main Relay Power (B+)	12-14v
A12	---	Not Used	---
A13	ORN	A/T: Idle Air Control Valve 'N'	Pulse Signals
A14	BLK/BLU	A/T: Idle Air Control Valve 'P'	Pulse Signals
A15	RED	EVAP Purge Solenoid	Solenoid Off: 12v, On: 1v
A16	GRN/YEL	Fuel Pump Relay	Relay Off: 12v, On: 1v
A17	BLK/RED	A/C Clutch Relay	Relay Off: 12v, On: 1v
A18	GRN/ORN	Check Engine Light	MIL Off: 12v, On: 1v
A19	WHT/YEL	Alternator Charging Signal	Lights Off: 12v, On: 0v
A20	RED/GRN	Igniter Control	Digital Signals: 0-12-0v
A21	---	Not Used	---
A22	BRN/BLK	Logic Ground	<0.050v
A23	BLK	Power Ground	<0.1v
A24	YEL/BLK	Main Relay Power (B+)	12-14v
A25	GRN/BLK	A/T: Lockup Solenoid 'B'	LSB On: 12v, Off: 0v
A26	YEL	A/T: Lockup Solenoid 'A'	LSA On: 12v, Off: 0v
A27	YEL/GRN	Radiator Fan Relay	Relay Off: 12v, On: 1v
A28-29	---	Not Used	---
A30	WHT/RED	A/T Interlock Control Unit	Key & Brake On: 12v
A31-32	---	Not Used	---

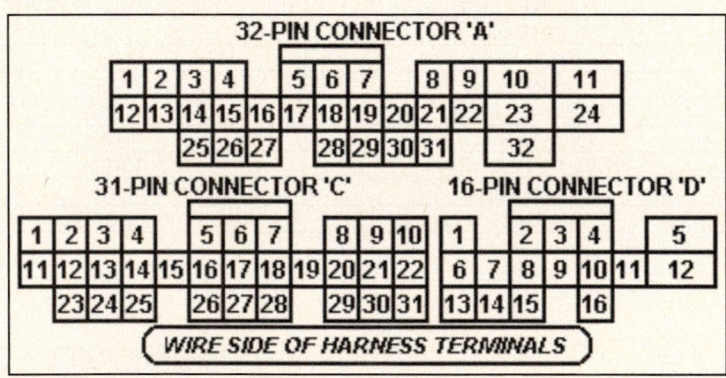

1996-97 Coupe 1.6L I4 MFI VIN EH6 [All] 31P 'C' Connector

PCM Pin #	W/Color	Circuit Description (31-Pin)	Value at Hot Idle
C1	BLU/RED	CKF Sensor Signal	AC pulse signals
C2	BLU/GRN	CKP Sensor Signal	AC pulse signals
C3	ORN/BLU	TDC Sensor Signal	AC pulse signals
C4	ORN	CYP Sensor Signal	AC pulse signals
C5	BLU/RED	A/C Switch Signal	Relay Off: 12v, On: 1v
C6	BLU/WHT	Starter Switch Signal	Cranking: 9-11v
C7	BRN	Service Check Connector	SCS Open: 4.80v
C8	GRN/YEL	K-Line Signal	12v
C9	---	Not Used	---
C10	WHT/BLU	Keep Alive Power (VBU)	12-14v
C11	WHT/RED	CKF Sensor Ground	<0.050v
C12	BLU/YEL	CKP Sensor Ground	<0.050v
C13	WHT/BLU	TDC Sensor Ground	<0.050v
C14	WHT	CYP Sensor Ground	<0.050v
C15	---	Not Used	---
C16	BRN/RED	PSP Switch	Straight: 0v, Turning: 11v
C17	PNK	Alternator 'FR' Signal	Varies: 1.5-3.5v
C18	YEL/BLU	Vehicle Speed Sensor	Moving: pulse signals
C19-26	---	Not Used	---
C27	GRN/BLK	A/T: Gear Position Switch	In D4: 0v, Others: 12v
C28	GRN/BLU	A/T: Gear Position Switch	In D3: 0v, Others: 12v
C29	GRN	A/T: Gear Position Switch	In P/N: 0v, Others: 12v
C30-31	---	Not Used	---

1996-97 Coupe 1.6L I4 MFI VIN EH6 [All] 16P 'D' Connector

PCM Pin #	W/Color	Circuit Description (16-Pin)	Value at Hot Idle
D1	RED/BLU	TP Sensor Signal	0.5-0.6v
D2	RED/WHT	ECT Sensor Signal	At 180°F: 0.52v
D3	WHT	MAP Sensor Signal	0.8-0.9v
D4	YEL/GRN	MAP Sensor VREF	4.9-5.1v
D5	GRN/WHT	Brake Switch Signal	Brake Off: 0v, On: 12v
D6	RED/BLU	Knock Sensor Signal	No Detonation: 18mv AC
D7	WHT	HO2S-11 (B1 S1) Signal	0.1-1.1v
D8	RED/YEL	IAT Sensor Signal	Varies w/temp. (0.5-4.9v)
D9	---	Not Used	---
D10	YEL/WHT	Sensor VREF	4.9-5.1v
D11	GRN/WHT	Sensor Ground	<0.050v
D12	GRN/WHT	MAP Sensor Ground	<0.050v
D13	RED/YEL	HO2S-12 Ground	<0.050v
D14	WHT/RED	HO2S-12 (B1 S2) Signal	0.1-1.1v
D16	GRN/RED	Electric Load Detector	Varies: 0.5-4.5v

1996-97 Coupe 1.6L I4 VTEC VIN EH6 [A/T] 32P 'A' Connector

PCM Pin #	W/Color	Circuit Description (32-Pin)	Value at Hot Idle
A1	YEL	Injector 4 Control	2.0-3.3 ms
A2	BLU	Injector 3 Control	2.0-3.3 ms
A3	RED	Injector 2 Control	2.0-3.3 ms
A4	BRN	Injector 1 Control	2.0-3.3 ms
A5	BLK/WHT	HO2S-12 (B1 S2) Heater	Digital Signals: 0-12-0v
A6	ORN/BLK	HO2S-11 (B1 S1) Heater	Digital Signals: 0-12-0v
A8	ORN/WHT	VTEC Solenoid	0v, Hi-Speed: 12v
A9	BLK/RED	Logic Ground	<0.050v
A10	BLK	Power Ground	<0.1v
A11	YEL/BLK	Main Relay Power (B+)	12-14v
A12	GRN/WHT	M/T: Idle Air Control Valve	Pulse Signals
A13	ORN	A/T: Idle Air Control Valve 'N'	Pulse Signals
A14	BLK/BLU	A/T: Idle Air Control Valve 'P'	Pulse Signals
A15	RED	EVAP Purge Solenoid	Solenoid Off: 12v, On: 1v
A16	GRN/YEL	Fuel Pump Relay	Relay Off: 12v, On: 1v
A17	BLK/RED	A/C Clutch Relay	Relay Off: 12v, On: 1v
A18	GRN/ORN	Check Engine Light	MIL Off: 12v, On: 1v
A19	WHT/YEL	Alternator Charging Signal	Lights Off: 12v, On: 0v
A20	RED/GRN	Igniter Control	Digital Signals: 0-12-0v
A22	BRN/BLK	Logic Ground	<0.050v
A23	BLK	Power Ground	<0.1v
A24	YEL/BLK	Main Relay Power (B+)	12-14v
A25	GRN/BLK	A/T: Lockup Solenoid 'B'	LSB On: 12v, Off: 0v
A26	YEL	A/T: Lockup Solenoid 'A'	LSA On: 12v, Off: 0v
A27	YEL/GRN	Radiator Fan Relay	Relay Off: 12v, On: 1v
A28-29	---	Not Used	---
A30	WHT/RED	A/T Interlock Control Unit	Key & Brake On: 12v
A31-32	---	Not Used	---

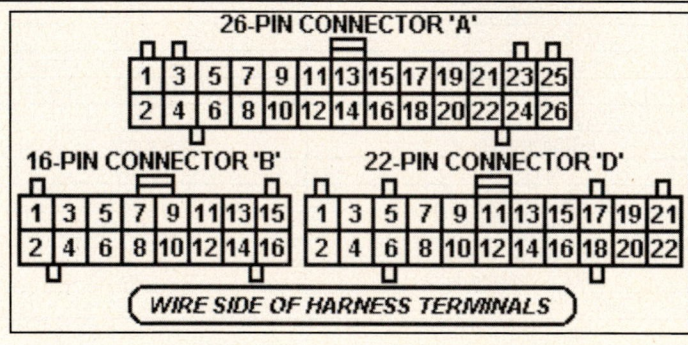

1996-97 Coupe 1.6L I4 VTEC VIN EH6 [A/T] 31P 'C' Connector

PCM Pin #	W/Color	Circuit Description (31-Pin)	Value at Hot Idle
C1	BLU/RED	CKF Sensor Signal	AC pulse signals
C2	BLU/GRN	CKP Sensor Signal	AC pulse signals
C3	ORN/BLU	TDC Sensor Signal	AC pulse signals
C4	ORN	CYP Sensor Signal	AC pulse signals
C5	BLU/RED	A/C Switch Signal	Relay Off: 12v, On: 1v
C6	BLU/WHT	Starter Switch Signal	Cranking: 9-11v
C7	BRN	Service Check Connector	SCS Open: 4.80v
C8	GRN/YEL	K-Line Signal	12v
C9	---	Not Used	---
C10	WHT/BLU	Keep Alive Power (VBU)	12-14v
C11	WHT/RED	CKF Sensor Ground	<0.050v
C12	BLU/YEL	CKP Sensor Ground	<0.050v
C13	WHT/BLU	TDC Sensor Ground	<0.050v
C14	WHT	CYP Sensor Ground	<0.050v
C15	ORN/BLU	VTEC Pressure Switch	0v, Hi-Speed: 12v
C16	BRN/RED	PSP Switch	Straight: 0v, Turning: 11v
C17	PNK	Alternator 'FR' Signal	Varies: 1.5-3.5v
C18	YEL/BLU	Vehicle Speed Sensor	Moving: pulse signals
C27	GRN/BLK	A/T: Gear Position Switch	In D4: 0v, Others: 12v
C28	GRN/BLU	A/T: Gear Position Switch	In D3: 0v, Others: 12v
C29	GRN	A/T: Gear Position Switch	In P/N: 0v, Others: 12v
C30-31	---	Not Used	---

1996-97 Coupe 1.6L I4 VTEC VIN EH6 [A/T] 16P 'D' Connector

PCM Pin #	W/Color	Circuit Description (16-Pin)	Value at Hot Idle
D1	RED/BLU	TP Sensor Signal	0.5-0.6v
D2	RED/WHT	ECT Sensor Signal	At 180°F: 0.52v
D3	WHT	MAP Sensor Signal	0.8-0.9v
D4	YEL/GRN	MAP Sensor VREF	4.9-5.1v
D5	GRN/WHT	Brake Switch Signal	Brake Off: 0v, On: 12v
D6	RED/BLU	Knock Sensor Signal	No Detonation: 18mv AC
D7	WHT	HO2S-11 (B1 S1) Signal	0.1-1.1v
D8	RED/YEL	IAT Sensor Signal	Varies w/temp. (0.5-4.9v)
D9	---	Not Used	---
D10	YEL/WHT	Sensor VREF	4.9-5.1v
D11	GRN/WHT	Sensor Ground	<0.050v
D12	GRN/WHT	MAP Sensor Ground	<0.050v
D13	RED/YEL	HO2S-12 Ground	<0.050v
D14	WHT/RED	HO2S-12 (B1 S2) Signal	0.1-1.1v
D16	GRN/RED	Electric Load Detector	Varies: 0.5-4.5v

1996-97 Coupe 1.6L I4 VTEC VIN EH6 [M/T] 32P 'A' Connector

PCM Pin #	W/Color	Circuit Description (32-Pin)	Value at Hot Idle
A1	YEL	Injector 4 Control	2.0-3.3 ms
A2	BLU	Injector 3 Control	2.0-3.3 ms
A3	RED	Injector 2 Control	2.0-3.3 ms
A4	BRN	Injector 1 Control	2.0-3.3 ms
A5	BLK/WHT	HO2S-12 (B1 S2) Heater	Digital Signals: 0-12-0v
A6	ORN/BLK	HO2S-11 (B1 S1) Heater	Digital Signals: 0-12-0v
A7	---	Not Used	---
A8	ORN/WHT	VTEC Solenoid	0v, Hi-Speed: 12v
A9	BLK/RED	Logic Ground	<0.050v
A10	BLK	Power Ground	<0.1v
A11	YEL/BLK	Main Relay Power (B+)	12-14v
A12	GRN/WHT	Idle Air Control Valve	Pulse Signals
A13-14	---	Not Used	---
A15	RED	EVAP Purge Solenoid	Solenoid Off: 12v, On: 1v
A16	GRN/YEL	Fuel Pump Relay	Relay Off: 12v, On: 1v
A17	BLK/RED	A/C Clutch Relay	Relay Off: 12v, On: 1v
A18	GRN/ORN	Check Engine Light	MIL Off: 12v, On: 1v
A19	WHT/YEL	Alternator Charging Signal	Lights Off: 12v, On: 0v
A20	RED/GRN	Igniter Control	Digital Signals: 0-12-0v
A21	---	Not Used	---
A22	BRN/BLK	Logic Ground	<0.050v
A23	BLK	Power Ground	<0.1v
A24	YEL/BLK	Main Relay Power (B+)	12-14v
A25-26	---	Not Used	---
A27	YEL/GRN	Radiator Fan Relay	Relay Off: 12v, On: 1v
A28-32	---	Not Used	---

32-PIN CONNECTOR 'A'

31-PIN CONNECTOR 'C'   16-PIN CONNECTOR 'D'

WIRE SIDE OF HARNESS TERMINALS

1996-97 Coupe 1.6L I4 VTEC VIN EH6 [M/T] 31P 'C' Connector

PCM Pin #	W/Color	Circuit Description (31-Pin)	Value at Hot Idle
C1	BLU/RED	CKF Sensor Signal	AC pulse signals
C2	BLU/GRN	CKP Sensor Signal	AC pulse signals
C3	ORN/BLU	TDC Sensor Signal	AC pulse signals
C4	ORN	CYP Sensor Signal	AC pulse signals
C5	BLU/RED	A/C Switch Signal	Switch Off: 12v, On: 1v
C6	BLU/WHT	Starter Switch Signal	Cranking: 9-11v
C7	BRN	Service Check Connector	SCS Open: 4.80v
C8	GRN/YEL	K-Line Signal	12v
C9	---	Not Used	---
C10	WHT/BLU	Keep Alive Power (VBU)	12-14v
C11	WHT/RED	CKF Sensor Ground	<0.050v
C12	BLU/YEL	CKP Sensor Ground	<0.050v
C13	WHT/BLU	TDC Sensor Ground	<0.050v
C14	WHT	CYP Sensor Ground	<0.050v
C15	ORN/BLU	VTEC Pressure Switch	0v, Hi-Speed: 12v
C16	BRN/RED	PSP Switch	Straight: 0v, Turning: 11v
C17	PNK	Alternator 'FR' Signal	Varies: 1.5-3.5v
C18	YEL/BLU	Vehicle Speed Sensor	Moving: pulse signals
C19-31	---	Not Used	---

1996-97 Coupe 1.6L I4 VTEC VIN EH6 [M/T] 16P 'D' Connector

PCM Pin #	W/Color	Circuit Description (16-Pin)	Value at Hot Idle
D1	RED/BLU	TP Sensor Signal	0.5-0.6v
D2	RED/WHT	ECT Sensor Signal	At 180°F: 0.52v
D3	WHT	MAP Sensor Signal	0.8-0.9v
D4	YEL/GRN	MAP Sensor VREF	4.9-5.1v
D5	GRN/WHT	Brake Switch Signal	Brake Off: 0v, On: 12v
D6	RED/BLU	Knock Sensor Signal	No Detonation: 18mv AC
D7	WHT	HO2S-11 (B1 S1) Signal	0.1-1.1v
D8	RED/YEL	IAT Sensor Signal	Varies w/temp. (0.5-4.9v)
D9	---	Not Used	---
D10	YEL/WHT	Sensor VREF	4.9-5.1v
D11	GRN/WHT	Sensor Ground	<0.050v
D12	GRN/WHT	MAP Sensor Ground	<0.050v
D13	RED/YEL	HO2S-12 Ground	<0.050v
D14	WHT/RED	HO2S-12 (B1 S2) Signal	0.1-1.1v
D15	---	Not Used	---
D16	GRN/RED	Electric Load Detector	Varies: 0.5-4.5v

CR-V Utility Pin Tables

1997 MPV 2.0L I4 MFI VIN RD1 [A/T] 32P 'A' Connector

PCM Pin #	W/Color	Circuit Description (32-Pin)	Value at Hot Idle
A1	YEL	Injector 4 Control	2.0-3.3 ms
A2	BLU	Injector 3 Control	2.0-3.3 ms
A3	RED	Injector 2 Control	2.0-3.3 ms
A4	BRN	Injector 1 Control	2.0-3.3 ms
A5	BLK/WHT	HO2S-12 (B1 S2) Heater	Digital Signals: 0-12-0v
A6	BLK/WHT	HO2S-11 (B1 S1) Heater	Digital Signals: 0-12-0v
A9, A22	BRN/BLK	Logic Ground	<0.050v
A10, 23	BLK	Power Ground	<0.1v
A11	YEL/BLK	Main Relay Power (B+)	12-14v
A12	BLK/BLU	Idle Air Control Valve	Pulse Signals
A15	RED/YEL	EVAP Purge Solenoid	Solenoid Off: 12v, On: 1v
A16	GRN/YEL	Fuel Pump Relay Control	Relay Off: 12v, On: 1v
A17	BLK/RED	A/C Clutch Relay Control	Relay Off: 12v, On: 1v
A18	GRN/ORN	Check Engine Light	MIL Off: 12v, On: 1v
A19	WHT/GRN	Alternator Charging Signal	Lights Off: 12v, On: 0v
A20	YEL/GRN	Igniter Control	At idle: 5° dwell
A24	YEL/BLK	Main Relay Power (B+)	12-14v
A27	GRN	Radiator Fan Relay Control	R/F on: <1v, off: 12v

1997 MPV 2.0L I4 MFI VIN RD1 [A/T] 25P 'B' Connector

PCM Pin #	W/Color	Circuit Description (25-Pin)	Value at Hot Idle
B1	WHT	A/T: Linear Solenoid (-)	Pulse Signals
B2	RED	A/T: Linear Solenoid (+)	Pulse Signals
B3	BLU/YEL	A/T: Shift Solenoid 'A'	SSA Off: 0v, On: 12v
B4	GRN/BLK	A/T: Lockup Solenoid 'B'	LSB On: 12v, Off: 0v
B5	YEL	A/T: Lockup Solenoid 'A'	LSA On: 12v, Off: 0v
B8	PNK	A/T: Gear Position Switch	In D3: 0v, others: 12v
B11	GRN/WHT	A/T: Shift Solenoid 'B'	SSB Off: 0v, On: 12v
B12	WHT/RED	Interlock Control Unit Signal	Key & Brake On: 12v
B13	GRN/BLK	A/T: D4 Indicator Light Switch	In D4: 12v, others: 0v
B14	WHT	Mainshaft Speed Sensor 'N'	<0.050v
B15	RED	Mainshaft Speed Sensor 'P'	AC Pulse Signals
B16	WHT	A/T: Gear Position Switch	In 'R': 0v, Others: 12v
B17	BLU	A/T: Gear Position Switch	In D2: 0v, Others: 12v
B18	BRN	A/T: Gear Position Switch	In D1: 0v, Others: 12v
B22	GRN	Countershaft Speed Sensor N	<0.050v
B23	BLU	Countershaft Speed Sensor P	Moving: pulse signals
B24	YEL	A/T: Gear Position Switch	In D4: 0v, others: 12v
B25	LT GRN	A/T: Gear Position Switch	In P/N: 0v, Others: 12v

32-PIN CONNECTOR 'A' 25-PIN CONNECTOR 'B'

WIRE SIDE OF HARNESS TERMINALS

1997 MPV 2.0L I4 MFI VIN RD1 [A/T] 31P 'C' Connector

PCM Pin #	W/Color	Circuit Description (31-Pin)	Value at Hot Idle
C1	BLU/RED	CKF Sensor Signal	Idle: 0.900mv (AC)
C2	BLU	CKP Sensor Signal	Idle: 0.900mv (AC)
C3	GRN	TDC Sensor Signal	KOER: 1.00v (AC)
C4	YEL	CYP Sensor Signal	Idle: 0.250mv (AC)
C5	BLU/RED	A/C Switch Signal	Relay Off: 12v, On: 1v
C6	BLU/ORN	Starter Switch Signal	Cranking: 9-11v
C7	BRN	Service Check Connector	SCS Open: 4.80v
C8	LT BLU	K-Line Signal	12v
C9	---	Not Used	---
C10	WHT/BLU	Keep Alive Power (VBU)	12-14v
C11	WHT/RED	CKF Sensor Ground	<0.050v
C12	WHT	CKP Sensor Ground	<0.050v
C13	RED	TDC Sensor Ground	<0.050v
C14	BLK	CYP Sensor Ground	<0.050v
C16	GRN	PSP Switch	Wheel straight: 0v
C17	WHT/RED	Alternator 'FR' Signal	Varies: 1.5-3.5v
C18	BLU/WHT	Vehicle Speed Sensor	Moving: pulse signals
C19-31	---	Not Used	---

1997 MPV 2.0L I4 MFI VIN RD1 [A/T] 16P 'D' Connector

PCM Pin #	W/Color	Circuit Description (16-Pin)	Value at Hot Idle
D1	RED/BLK	TP Sensor Signal	At 0.5-0.6v
D2	RED/WHT	ECT Sensor Signal	At 180°F: 0.5-0.6v
D3	RED/GRN	MAP Sensor Signal	0.8-0.9v
D4	YEL/RED	MAP Sensor VREF	4.9-5.1v
D5	GRN/WHT	Brake Switch Signal	Brake Off: 0v, On: 12v
D7	WHT	HO2S-11 (B1 S1) Signal	0.1-1.1v
D8	RED/YEL	IAT Sensor Signal	Varies w/temp. (0.5-4.9v)
D9	---	Not Used	---
D10	YEL/BLU	Sensor VREF	4.9-5.1v
D11	GRN/BLK	Sensor Ground	<0.050v
D12	GRN/WHT	MAP Sensor Ground	<0.050v
D13	RED/YEL	HO2S-12 Ground	<0.050v
D14	WHT/RED	HO2S-12 (B1 S2) Signal	0.1-1.1v
D16	GRN/RED	Electronic Load Detector	Varies 2.5-3.5v

```
        31-PIN CONNECTOR 'C'        16-PIN CONNECTOR 'D'
     ┌──┬──┬──┬──┐ ┌──┬──┬──┐ ┌──┬──┬──┐ ┌──┐ ┌──┬──┬──┐ ┌──┐
     │1 │2 │3 │4 │ │5 │6 │7 │ │8 │9 │10│ │1 │ │2 │3 │4 │ │5 │
     ├──┼──┼──┼──┼─┼──┼──┼──┼─┼──┼──┼──┼─┼──┼─┼──┼──┼──┼─┼──┤
     │11│12│13│14│15│16│17│18│19│20│21│22│6 │7 │8 │9 │10│11│12│
     ├──┼──┼──┤ ├──┼──┼──┤ ├──┼──┼──┤ ├──┼──┼──┤ ├──┤
     │23│24│25│ │26│27│28│ │29│30│31│ │13│14│15│ │16│
     └──┴──┴──┘ └──┴──┴──┘ └──┴──┴──┘ └──┴──┴──┘ └──┘
        ╭─────────────────────────────────────╮
        │   WIRE SIDE OF HARNESS TERMINALS    │
        ╰─────────────────────────────────────╯
```

1998-03 MPV 2.0L I4 MFI VIN RD1, RD2 [All] 32P 'A' Connector

PCM Pin #	W/Color	Circuit Description (32-Pin)	Value at Hot Idle
A3	BLU	EVAP Bypass Solenoid	Solenoid Off: 12v, On: 1v
A4	GRN/WHT	EVAP Vent Solenoid	Solenoid Off: 12v, On: 1v
A5	BLU	Cruise Control Signal	C/C On: pulse signals
A6	RED/YEL	EVAP Purge Solenoid	Solenoid Off: 12v, On: 1v
A8	BLK/WHT	HO2S-12 (B1 S2) Heater	Digital Signals: 0-12-0v
A10	BRN	Service Check Connector	SCS Open: 4.80v
A14	GRN/BLK	A/T: D4 Light Switch	D4 On: 0v, Off: 12v
A16	GRN/YEL	Fuel Pump Relay Control	Relay Off: 12v, On: 1v
A17	BLK/RED	A/C Clutch Relay Control	Relay Off: 12v, On: 1v
A18	GRN/ORN	Malfunction Indicator Light	MIL Off: 12v, On: 1v
A19	BLU	Engine Speed Pulse (NEP)	Digital Signals
A20	GRN	Radiator Fan Relay Control	Relay Off: 12v, On: 1v
A21	BLU/YEL	K-Line Signal	12v
A23	WHT/RED	HO2S-12 (B1 S2) Signal	0.1-1.1v
A24	BLU/WHT	Starter Switch Signal	Cranking: 9-11v
A26	GRN	PSP Switch Signal	Straight: 0v, Turning: 11v
A27	BLU/RED	A/C Switch Signal	Switch Off: 12v, On: 1v
A28	WHT/RED	Interlock Control Unit Signal	Key & Brake On: 12v
A29	LT GRN	Fuel Tank Pressure Sensor	Fuel Cap off: 2.5v
A30	GRN/RED	Electric Load Detector	Varies: 0.5-4.5v
A32	WHT/BLK	Brake Switch Signal	Brake Off: 0v, On: 12v

1998-03 MPV 2.0L I4 MFI VIN RD1, RD2 [All] 16P 'D' Connector

PCM Pin #	W/Color	Circuit Description (16-Pin)	Value at Hot Idle
D1	YEL	A/T: Lockup Solenoid 'A'	LUS On: 12v, Off: 0v
D2	GRN/WHT	A/T: Shift Solenoid 'B'	SSB Off: 0v, On: 12v
D3	GRN/BLK	A/T: Lockup Solenoid 'B'	LUS On: 12v, Off: 0v
D5	BLK/YEL	A/T: Solenoid Feed (B+)	12-14v
D6	WHT	A/T: Gear Position Switch	In 'R': 0v, Others: 12v
D7	BLU/YEL	A/T: Shift Solenoid 'A'	SSA Off: 0v, On: 12v
D8	PNK	A/T: Gear Position Switch	In D3: 0v, Others: 12v
D8	PNK	M/T: Overdrive Switch	In O/D: 0v, others: 5v
D9	YEL	A/T: Gear Position Switch	In D4: 0v, Others: 12v
D10	BLU	Countershaft Speed Sensor P	Moving: AC pulses
D11	RED	Mainshaft Speed Sensor 'P'	Pulse Signals
D12	WHT	Mainshaft Speed Sensor 'N'	<0.050v
D13	LT GRN	A/T: Gear Position Switch	In P/N: 0v, Others: 12v
D14	BLU	A/T: D4 Indicator Light Driver	In D2: 0v, Off: 12v
D15	BRN	A/T: D4 Indicator Light Driver	In D12: 0v, Off: 12v
D16	GRN	Countershaft Speed Sensor N	<0.050v

1998-03 MPV 2.0L I4 MFI VIN RD1, RD2 [All] 25P 'B' Connector

PCM Pin #	W/Color	Circuit Description (25-Pin)	Value at Hot Idle
B1, B9	YEL/BLK	Main Relay Power (B+)	12-14v
B2, B10	BLK	Power Ground (PG1, PG2)	<0.1v
B3	RED	Injector 2 Control	2.0-3.3 ms
B4	BLU	Injector 3 Control	2.0-3.3 ms
B5	YEL	Injector 4 Control	2.0-3.3 ms
B8	WHT	A/T: Clutch Press. Solenoid 'N'	Pulse Signals
B11	BRN	Injector 1 Control	2.0-3.3 ms
B12	GRN/YEL	VTEC Control Solenoid	0v, Hi-Speed: 12v
B13	YEL/GRN	Ignition Control Signal	Digital Signals: 0-12-0v
B17	RED	A/T: Clutch Press. Solenoid 'P'	Pulse Signals
B18	PNK/BLU	A/T: Start Clutch Solenoid 'N'	Pulse Signals
B19	PNK	A/T: Overdrive Indicator Light	In O/D: 0v, Others: 12v
B20	BRN/BLK	Logic Ground (LG1)	<0.050v
B21	WHT/YEL	Keep Alive Power (VBU)	12-14v
B22	BRN/BLK	Logic Ground (LG2)	<0.050v
B23	BLK/BLU	Idle Air Control Valve	Pulse Signals
B24-25	---	Not Used	---

1998-03 MPV 2.0L I4 MFI VIN RD1, RD2 [All] 31P 'C' Connector

PCM Pin #	W/Color	Circuit Description (31-Pin)	Value at Hot Idle
C1	BLK/WHT	HO2S-11 (B1 S1) Heater	Digital Signals: 0-12-0v
C2	WHT/GRN	Alternator Charging Signal	Lights Off: 12v, On: 0v
C3	RED/BLU	Knock Sensor Signal	No Detonation: 18mv AC
C5	WHT/RED	Alternator 'FR' Signal	Digital Signals: 0-5-0v
C7	GRN/WHT	MAP Sensor Ground (SG1)	<0.050v
C8, C9	BLU, WHT	CKP Sensor Signal, Ground	AC pulse signals
C16	WHT	HO2S-11 (B1 S1) Signal	0.1-1.1v
C17	RED/GRN	MAP Sensor Signal	0.8-0.9v
C18	GRN/BLK	Sensor Ground (SG2)	<0.050v
C19	YEL/RED	Sensor VREF (VCC1)	4.9-5.1v
C20	GRN	TDC Sensor Signal	AC pulse signals
C21	RED	TDC Sensor Ground	<0.050v
C22	BLU/RED	CKF Sensor Signal	AC pulse signals
C23	BLU/WHT	Vehicle Speed Sensor Signal	Moving: 0-5-0v
C25	RED/YEL	IAT Sensor Signal	Varies w/temp (0.5-4.9v)
C26	RED/WHT	ECT Sensor Signal	At 180ºF: 0.5-0.6v
C27	RED/BLK	TP Sensor Signal	0.5-0.6v
C28	YEL/BLU	Sensor VREF (VCC2)	4.9-5.1v
C29	YEL	CYP Sensor Signal	AC pulse signals
C30	BLK	CYP Sensor Ground	<0.050v
C31	WHT/RED	CKF Sensor Ground	<0.050v

25-PIN 'B'   *WIRE SIDE OF HARNESS TERMINALS*   31-PIN 'C'

INSIGHT Pin Tables

2000-03 Hatchback 1.0L I3 VTEC VIN ZE1 [M/T] 32P 'A' Connector

PCM Pin #	W/Color	Circuit Description (32-Pin)	Value at Hot Idle
A1	BLU/WHT	Starter Cut Relay	In Start: 12v
A2	BLK/YEL	Engine Ready Signal	Stop: 12v, No Stop: 0v
A3	BLU	EVAP Bypass Solenoid	Solenoid Off: 12v, On: 1v
A4	GRN/WHT	EVAP Vent Solenoid	Solenoid Off: 12v, On: 1v
A6	RED/YEL	EVAP Purge Solenoid	Solenoid Off: 12v, On: 1v
A7	RED/YEL	DC/DC Converter ECT Signal	Varies: 0.5-4.8v
A9	WHT/GRN	DC/DC Converter Control	Pulse Signals
A10	WHT/RED	Master Power Vacuum Motor	Varies: 1.0-3.0v
A12	PNK	Immobilizer Indicator Lamp	Lamp Off: 12v, On: 1v
A13	BLU/YEL	Immobilizer Enable Signal	Digital Signals
A15	GRN/YEL	Immobilizer Fuel Pump Relay	Relay Off: 12v, On: 1v
A17	RED	A/C Clutch Relay Control	Relay Off: 12v, On: 1v
A18	GRN/ORN	Malfunction Indicator Light	MIL Off: 12v, On: 1v
A19	BLU	Engine Speed Pulse (NEP)	Pulse Signals
A20	BLU/RED	Radiator Fan Relay Control	Relay Off: 12v, On: 1v
A21	GRY	K-Line Signal	12v
A23	BRN/YEL	Heater Standby Signal	Pulse Signals
A24	BLU/ORN	Starter Switch Signal	Cranking: 9-11v
A25	RED	Immobilizer Code Signal	Digital Signals
A27	BLU/BLK	A/C Switch Signal	Switch Off: 12v, On: 1v
A29	LT GRN	Fuel Tank Pressure Sensor	Fuel Cap off: 2.5v
A30	GRN/RED	Electric Load Detector	Varies: 0.5-4.5v
A32	GRN/WHT	Brake Switch Signal	Brake Off: 0v, On: 12v

2000-03 Hatchback 1.0L I3 VTEC VIN ZE1 [M/T] 16P 'D' Connector

PCM Pin #	W/Color	Circuit Description (16-Pin)	Value at Hot Idle
D1	BLU/WHT	Motor Control FSB Signal	Pulse Signals
D2	BLU/RED	Motor Control FSA Signal	Pulse Signals
D3	YEL/RED	Motor Control Standby Signal	Pulse Signals
D4	GRN	HO2S Pump Cell (+) Signal	0.5-3.5v
D5	BLK/WHT	HO2S-11 (B1 S1) Heater	Digital Signals: 0-12-0v
D6	RED/YEL	Motor Control Mode 1 Signal	Pulse Signals
D7	BLU	Engine Torque Signal	Pulse Signals
D8	BLU/BLK	Motor Power Signal	Pulse Signals
D9, D16	---	Not Used	---
D10	RED	HO2S Common IP- (VS-)	2.6-2.8v
D11	BLU	VS Cell Voltage (VS+)	7v
D12	WHT	LAF Label Signal	0.3-4.9v
D13	PNK	'Q' Battery Signal	Pulse Signals
D14	YEL	Motor Torque Signal	Pulse Signals
D15	WHT/RED	Motor Control Mode 2 Signal	Pulse Signals

25-Pin 'B'   *WIRE SIDE OF HARNESS TERMINALS*   31-Pin 'C'

2000-03 Hatchback 1.0L I3 VTEC VIN ZE1 [M/T] 25P 'B' Connector

PCM Pin #	W/Color	Circuit Description (25-Pin)	Value at Hot Idle
B1, B9	YEL/BLK	Main Relay Power (B+)	12-14v
B2, B10	BLK	Power Ground (G101)	<0.1v
B3, B4	RED, BLU	Injector 2, Injector 3 Control	2.0-3.3 ms
B5	GRN	TIM Signal to Gauge Assembly	Digital Signals
B7	PNK	E-EGR Solenoid Control	Digital Signals: 0-12-0v
B11	BRN	Injector 1 Control	2.0-3.3 ms
B12	GRN/YEL	VTEC Control Solenoid	0v, Hi-Speed: 12v
B14	RED	Clutch Switch Signal	Clutch In: 0v, Out: 5v
B15	RED/GRN	ECT Sensor Signal to Gauge	Digital Signals
B16	RED/BLK	Neutral Position Switch Signal	In 'N': 0v, others: 12v
B17	BRN	Service Check Connector	Open: 5v, closed: 0v
B20	BRN/BLK	Power Ground (G101)	<0.1v
B21	WHT/BLU	Keep Alive Power (VBU)	12-14v
B22	BRN/BLK	Power Ground (G101)	<0.1v
B23	BLK/BLU	Idle Air Control Valve	Pulse Signals
B24	GRN/BLK	Reverse Position Switch Signal	In 'R': 12v, Others: 0v

2000-03 Hatchback 1.0L I3 VTEC VIN ZE1 [M/T] 31P 'C' Connector

PCM Pin #	W/Color	Circuit Description (31-Pin)	Value at Hot Idle
C4	WHT	Ignition Coil 1 Control	Digital Signals: 0-12-0v
C5	BLU/WHT	Vehicle Speed Sensor Signal	Moving: 0-5-0v
C6	WHT/BLK	EGR Valve Lift Sensor	1.2v
C7, C18	GRN/WHT	Sensor Ground (SG1, SG2)	<0.050v
C8	BLU	CKP Sensor Signal	AC pulse signals
C9	WHT	CKP Sensor Ground	<0.050v
C10	BLU/BLK	VTEC Pressure Switch	0v, Hi-Speed: 12v
C12	BLK/WHT	HO2S-11 (B1 S1) Heater	Digital Signals: 0-12-0v
C13	WHT/GRN	Ignition Coil 2 Control	Digital Signals: 0-12-0v
C14	WHT/BLK	Ignition Coil 3 Control	Digital Signals: 0-12-0v
C15	WHT/RED	Shift Indicator Lamp Control	Lamp on: 1v, off: 12v
C17	RED/GRN	MAP Sensor Signal	0.8-0.9v
C19	YEL/RED	MAP Sensor VREF (VCC1)	4.9-5.1v
C20	GRN	TDC1 Sensor Signal	AC pulse signals
C21	RED	TDC1 Sensor Ground	<0.050v
C22	BLU/RED	Knock Sensor Signal	No Detonation: 18mv AC
C25	RED/YEL	IAT Sensor Signal	Varies w/temp (0.5-4.9v)
C26	RED/WHT	ECT Sensor Signal	At 180ºF: 0.5-0.6v
C27	RED/BLK	TP Sensor Signal	0.5-0.6v
C28	YEL/BLU	Sensor VREF (VCC2)	4.9-5.1v
C29, C30	YEL, BLK	TDC2 Sensor Signal, Ground	AC pulse signals
C31	WHT/RED	HO2S-12 (B1 S2) Signal	0.1-1.1v

25-Pin 'B' *WIRE SIDE OF HARNESS TERMINALS* 31-Pin 'C'

ODYSSEY Pin Tables

1995 Van 2.2L I4 MFI VIN RA1 [A/T] 26P 'A' Connector

PCM Pin #	W/Color	Circuit Description (26-Pin)	Value at Hot Idle
A1	BRN	Injector 1 Control	2.0-3.3 ms
A2	RED	Injector 2 Control	2.0-3.3 ms
A3	BLU	Injector 3 Control	2.0-3.3 ms
A4	GRN/BLK	Fuel Pump Relay	Relay Off: 12v, On: 1v
A5	BLK/BLU	Idle Air Control Valve	Pulse Signals
A6	ORN/BLK	HO2S-11 (B1 S1) Heater	Digital Signals: 0-12-0v
A7	BLU	Check Engine Light	MIL Off: 12v, On: 1v
A8	RED/BLU	A/C Clutch Relay	Relay Off: 12v, On: 1v
A11	YEL/GRN	Igniter Control	Digital Signals: 0-12-0v
A12	BLK	Power Ground	<0.1v
A13	YEL/BLK	Main Relay Power (B+)	12-14v
A14	YEL	Injector 4 Control	2.0-3.3 ms
A16	RED	EGR Solenoid	Solenoid Off: 12v, On: 1v
A18	GRN/WHT	Engine Mount Solenoid	Solenoid Off: 12v, On: 1v
A19	GRN	Radiator Fan Relay	Relay Off: 12v, On: 1v
A21	WHT/GRN	Alternator Charging Signal	Lights Off: 12v, On: 0v
A22	BRN/WHT	A/T: TCM FAS	N/A
A23	RED/YEL	EVAP Purge Solenoid	Solenoid Off: 12v, On: 1v
A25	BLK	Power Ground	<0.1v
A26	BRN/BLK	Logic Ground	<0.050v

1995 Van 2.2L I4 MFI VIN RA1 [A/T] 16P 'B' Connector

PCM Pin #	W/Color	Circuit Description (16-Pin)	Value at Hot Idle
B1	YEL/BLK	Main Relay Power (B+)	12-14v
B2	WHT/RED	A/T: TCM AFSA	---
B3	RED/WHT	A/C Switch Signal	Relay Off: 12v, On: 1v
B4	LT GRN	A/T: Gear Position Switch	P/N: 0v: others: 12v
B5	BLU/RED	Starter Switch Signal	Cranking: 9-11v
B6	ORN	CYP Sensor Signal	AC pulse signals
B7	ORN/BLU	TDC Sensor Signal	AC pulse signals
B8	BLU/GRN	CKP Sensor Signal	AC pulse signals
B9	BRN/BLK	Logic Ground	<0.050v
B10	GRN	A/T: TCM AFSB	N/A
B12	GRN	PSP Switch	Wheel straight: 0v
B13	ORN	Vehicle Speed Sensor	Moving: pulse signals
B14	WHT	CYP Sensor Ground	<0.050v
B15	WHT/BLU	TDC Sensor Ground	<0.050v
B16	BLU/YEL	CKP Sensor Ground	<0.050v

1995 Van 2.2L I4 MFI VIN RA1 [A/T] 22P 'D' Connector

PCM Pin #	W/Color	Circuit Description (22-Pin)	Value at Hot Idle
D1	WHT/YEL	Keep Alive Power (VBU)	12-14v
D3	BLU/WHT	A/T: TCM BARO Signal	3v (at sea level)
D4	GRN/RED	Data Link Connector	5v
D5	WHT/RED	Alternator 'FR' Signal	Digital Signals: 0-5-0v
D6	RED/BLK	TP Sensor Signal	0.5-0.6v
D7	RED/WHT	ECT Sensor Signal	At 180°F: 0.5-0.6v
D8	RED/YEL	IAT Sensor Signal	Varies w/temp. (0.5-4.9v)
D9	WHT/YEL	MAP Sensor Signal	0.8-0.9v
D10	YEL/WHT	MAP Sensor VREF	4.9-5.1v
D11	GRN/WHT	MAP Sensor Ground	<0.050v
D12	GRN/WHT	Brake Switch Signal	Brake Off: 0v, On: 12v
D13	RED	Service Check Connector	SCS Open: 4.80v
D16	GRN/RED	Electrical Load Detector	Varies: 0.5-4.5v
D17	WHT/BLK	EGR Valve Lift Sensor	1.2v
D18	WHT/RED	HO2S-11 (B1 S1) Signal	0.1-1.1v
D19	---	Not Used	---
D20	GRN/BLK	A/T: TCM VREF	4.9-5.1v
D21	YEL/BLU	Sensor VREF	4.9-5.1v
D22	GRN/BLU	Sensor Ground	<0.050v

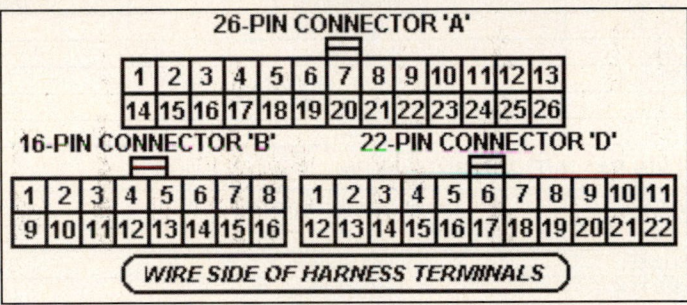

1996-98 Van 2.2L I4 MFI VIN RA1 [A/T] 32P 'A' Connector

PCM Pin #	W/Color	Circuit Description (32-Pin)	Value at Hot Idle
A1	YEL	Injector 4 Control	2.0-3.3 ms
A2	BLU	Injector 3 Control	2.0-3.3 ms
A3	RED	Injector 2 Control	2.0-3.3 ms
A4	BRN	Injector 1 Control	2.0-3.3 ms
A5	ORN/BLK	HO2S-12 (B1 S2) Heater	Digital Signals: 0-12-0v
A6	ORN/BLK	HO2S-11 (B1 S1) Heater	Digital Signals: 0-12-0v
A7	RED	EGR Solenoid	Solenoid Off: 12v, On: 1v
A9	BRN/BLK	Logic Ground	<0.050v
A10	BLK	Power Ground	<0.1v
A11	YEL/BLK	Main Relay Power (B+)	12-14v
A12	BLK/BLU	Idle Air Control Valve	Pulse Signals
A13	GRN/WHT	Engine Mount Solenoid	Solenoid Off: 12v, On: 1v
A15	RED/YEL	EVAP Purge Solenoid	Solenoid Off: 12v, On: 1v
A16	GRN/BLK	Fuel Pump Relay	Relay Off: 12v, On: 1v
A17	RED/BLU	A/C Clutch Relay	Relay Off: 12v, On: 1v
A18	BLU	Check Engine Light	MIL Off: 12v, On: 1v
A19	WHT/GRN	Alternator Charging Signal	Lights Off: 12v, On: 0v
A20	YEL/GRN	Igniter Control	Digital Signals: 0-12-0v
A22	BRN/BLK	Logic Ground	<0.050v
A23	BLK	Power Ground	<0.1v
A24	YEL/BLK	Main Relay Power (B+)	12-14v
A27	GRN	Radiator Fan Relay	Relay Off: 12v, On: 1v
A28-32	---	---	---

1996-98 Van 2.2L I4 MFI VIN RA1 [A/T] 25P 'B' Connector

PCM Pin #	W/Color	Circuit Description (25-Pin)	Value at Hot Idle
B1-2	---	Not Used	---
B3	BLU/YEL	A/T: Shift Solenoid 'A'	SSA Off: 0v, On: 12v
B4	GRN/BLK	A/T: Lockup Solenoid 'B'	LSB On: 12v, Off: 0v
B5	YEL	A/T: Lockup Solenoid 'A'	LSA On: 12v, Off: 0v
B8	GRN	A/T: Gear Position Switch	In D3: 0v, Others: 12v
B11	GRN/WHT	A/T: Shift Solenoid 'B'	SSB Off: 0v, On: 12v
B12	WHT/GRN	Interlock Control Unit Signal	Key & Brake On: 12v
B13	BLU/RED	A/T: D4 Indicator Light Switch	D4 On: 0v, Off: 12v
B14	WHT/BLU	Mainshaft Speed Sensor 'N'	<0.050v
B15	ORN/BLU	Mainshaft Speed Sensor 'P'	AC Pulse Signals
B16	WHT	A/T: Gear Position Switch	In 'R': 0v, Others: 12v
B17	BLU	A/T: Gear Position Switch	In D2: 0v, Others: 12v
B18	BRN	A/T: Gear Position Switch	In D1: 0v, Others: 12v
B22	BLU/YEL	Countershaft Speed Sensor N	<0.050v
B23	BLU/GRN	Countershaft Speed Sensor P	Moving: AC pulses
B24	YEL	A/T: Gear Position Switch	In D4: 0v, Others: 12v
B25	LT GRN	A/T: Gear Position Switch	In P/N: 0v, Others: 12v

1996-98 Van 2.2L I4 MFI VIN RA1 [A/T] 31P 'C' Connector

PCM Pin #	W/Color	Circuit Description (31-Pin)	Value at Hot Idle
C2	BLU	CKP Sensor Signal	AC pulse signals
C3	GRN	TDC Sensor Signal	AC pulse signals
C4	YEL	CYP Sensor Signal	AC pulse signals
C5	RED/WHT	A/C Switch Signal	Switch Off: 12v, On: 1v
C6	BLU/RED	Starter Switch Signal	Cranking: 9-11v
C7	RED	Service Check Connector	SCS Open: 4.80v
C8	LT GRN	K-Line Signal	12v
C9	---	Not Used	---
C10	WHT/YEL	Keep Alive Power (VBU)	12-14v
C12	BLU/YEL	CKP Sensor Ground	<0.050v
C13	WHT/BLU	TDC Sensor Ground	<0.050v
C14	WHT	CYP Sensor Ground	<0.050v
C16	GRN	PSP Switch	Wheels Turned: 11v
C17	WHT/RED	Alternator 'FR' Signal	Digital Signals: 0-5-0v
C18	ORN	Vehicle Speed Sensor	Moving: pulse signals
C19	---	Not Used	---
C20	BRN	EVAP Purge Solenoid	Solenoid Off: 12v, On: 1v
C21-31	---	Not Used	---

1996-98 Van 2.2L I4 MFI VIN RA1 [A/T] 16P 'D' Connector

PCM Pin #	W/Color	Circuit Description (16-Pin)	Value at Hot Idle
D1	RED/BLK	TP Sensor Signal	0.5-0.6v
D2	RED/WHT	ECT Sensor Signal	At 180°F: 0.5-0.6v
D3	WHT/YEL	MAP Sensor Signal	0.8-0.9v
D4	YEL/WHT	MAP Sensor VREF	4.9-5.1v
D5	GRN/WHT	Brake Switch Signal	Brake Off: 0v, On: 12v
D7	WHT/RED	HO2S-11 (B1 S1) Signal	0.1-1.1v
D8	RED/YEL	Intake Air Temperature	Varies w/temp. (0.5-4.9v)
D9	WHT/BLK	EGR Valve Lift Sensor	1.2v
D10	YEL/BLU	Sensor VREF	4.9-5.1v
D11	GRN/BLU	Sensor Ground	<0.050v
D12	GRN/WHT	MAP Sensor Ground	<0.050v
D13	RED/WHT	HO2S-12 Ground	<0.050v
D14	WHT/RED	HO2S-12 (B1 S2) Signal	0.1-1.1v
D16	GRN/RED	Electrical Load Detector	Varies: 0.5-4.5v

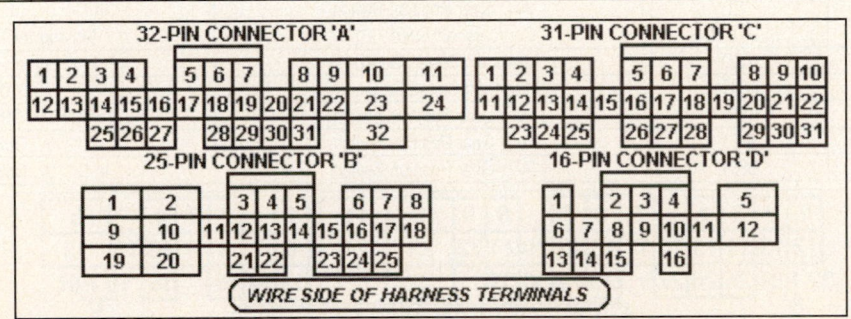

1998 Van 2.3L I4 VTEC SOHC VIN RA3 [A/T] 32P 'A' Connector

PCM Pin #	W/Color	Circuit Description (32-Pin)	Value at Hot Idle
A2	GRN/WHT	Engine Mount Solenoid	Solenoid Off: 12v, On: 1v
A3	GRN/ORN	EVAP Bypass Solenoid	Solenoid Off: 12v, On: 1v
A4	ORN/GRN	EVAP Vent Solenoid	Solenoid Off: 12v, On: 1v
A5	BLU/GRN	Cruise Control Signal	C/C On: Pulse Signals
A6	RED/YEL	EVAP Purge Solenoid	Solenoid Off: 12v, On: 1v
A8	ORN/BLK	HO2S-12 (B1 S2) Heater	Digital Signals: 0-12-0v
A9	BLU/WHT	A/T: VSS Signal from CSS	Moving: pulse signals
A10	RED	Service Check Connector	SCS Open: 4.80v
A12	PNK	Immobilizer Indicator Lamp	Lamp Off: 12v, On: 1v
A13, 25	BLU, RED	Immobilizer Enable, Code	Digital Signals
A14	BLU/RED	A/T: D4 Position Indicator	D4 On: 0v, Off: 12v
A15	ORN/BLK	Immobilizer Fuel Pump Relay	Relay Off: 12v, On: 1v
A17	RED/BLU	A/C Clutch Relay Control	Relay Off: 12v, On: 1v
A18	BLU	Malfunction Indicator Light	MIL Off: 12v, On: 1v
A19	BLU	Engine Speed Pulse (NEP)	Pulse Signals
A20	GRN	Radiator Fan Relay Control	Relay Off: 12v, On: 1v
A21	LT GRN	K-Line Signal	12v
A23	WHT/RED	HO2S-12 (B1 S2) Signal	0.1-1.1v
A24	BLU/RED	Starter Switch Signal	Cranking: 9-11v
A26	GRN	PSP Switch Signal	Straight: 0v, Turning: 11v
A27	RED/WHT	A/C Switch Signal	Switch Off: 5v, On: 0v
A28	WHT/GRN	Interlock Control Unit Signal	Key & Brake On: 12v
A29	WHT/BLU	Fuel Tank Pressure Sensor	Fuel Cap off: 2.5v
A30	ORN/RED	Electric Load Detector	Varies: 0.5-4.5v
A32	GRN/WHT	Brake Switch Signal	Brake Off: 0v, On: 12v

1998 Van 2.3L I4 VTEC SOHC VIN RA3 [A/T] 16P 'D' Connector

PCM Pin #	W/Color	Circuit Description (16-Pin)	Value at Hot Idle
D1	YEL	A/T: Lockup Control Solenoid	LUS On: 12v, Off: 0v
D2	GRN/WHT	A/T: Shift Solenoid 'B'	SSB Off: 0v, On: 12v
D3	GRN	A/T: Shift Solenoid 'C'	SSC on: 12v, off: 0v
D5	BLK/YEL	A/T: Solenoid Feed (B+)	12-14v
D6	WHT	A/T: Gear Position Switch	In 'R': 0v, Others: 12v
D7	BLU/YEL	A/T: Shift Solenoid 'A'	SSA Off: 0v, On: 12v
D8	PNK	A/T: Gear Position Switch	In D3: 0v, Others: 12v
D9	YEL	A/T: Gear Position Switch	In D4: 0v, Others: 12v
D10, D16	BLU, GRN	Countershaft Speed 'P', 'N'	Moving: AC pulses
D11	RED	Mainshaft Speed Sensor 'P'	AC pulse signals
D12	WHT	Mainshaft Speed Sensor 'N'	<0.050v
D13	LT GRN	A/T: Gear Position Switch	In P/N: 0v, Others: 12v
D14	BLU	A/T: Gear Position Switch	In D2: 0v, Others: 12v
D15	BRN	A/T: Gear Position Switch	In D1: 0v, Others: 12v

1998 Van 2.3L I4 VTEC SOHC VIN RA3 [A/T] 25P 'B' Connector

PCM Pin #	W/Color	Circuit Description (25-Pin)	Value at Hot Idle
B1, B9	YEL/BLK	Main Relay Power (B+)	12-14v
B2, B10	BLK	Power Ground	<0.1v
B3, B4	RED, BLU	Injector 2, Injector 3 Control	2.0-3.3 ms
B5, B11	YEL, BRN	Injector 4, Injector 1 Control	2.0-3.3 ms
B7	PNK	E-EGR Control Solenoid	Digital Signals: 0-12-0v
B8	WHT	A/T: Clutch Solenoid 'A-'	Pulse Signals
B12	GRN/YEL	VTEC Control Solenoid	0v, Hi-Speed: 12v
B13	YEL/GRN	Ignition Control Signal	Digital Signals: 0-12-0v
B14	BLU/BLK	A/T: 2nd Clutch Pressure Sw.	Open: 12v, Closed: 1v
B16	GRN/RED	HO2S-11 Heater Relay	Relay Off: 12v, On: 1v
B17	RED	A/T: Clutch Solenoid 'A+'	Pulse Signals
B18	GRN	A/T: Clutch Solenoid 'B-'	Pulse Signals
B19	BLK/WHT	HO2S-11 (B1 S1) Heater	Digital Signals: 0-12-0v
B20, B22	BRN/BLK	Logic Ground (LG1, LG2)	<0.050v
B21	WHT/YEL	Keep Alive Power (VBU)	12-14v
B23	BLK/BLU	Idle Air Control Valve	Pulse Signals
B24	BLU/WHT	A/T: 3rd Clutch Pressure Sw.	Open: 12v, Closed: 1v
B25	ORN	A/T: Clutch Solenoid 'B+'	Pulse Signals

1998 Van 2.3L I4 VTEC SOHC VIN RA3 [A/T] 31P 'C' Connector

PCM Pin #	W/Color	Circuit Description (31-Pin)	Value at Hot Idle
C1	BLK/WHT	HO2S-11 (B1 S1) Heater	Digital Signals: 0-12-0v
C2	WHT/GRN	Alternator Charging Signal	Lights Off: 12v, On: 0v
C3	RED/BLU	Knock Sensor Signal	No Detonation: 18mv AC
C5	WHT/RED	Alternator 'FR' Signal	Digital Signals: 0-5-0v
C6	WHT/BLK	EGR Valve Lift Sensor	1.2v
C7	GRN/WHT	MAP Sensor Ground (SG1)	<0.050v
C8	BLU	CKP Sensor Signal	AC pulse signals
C9	WHT	CKP Sensor Ground	<0.050v
C10	BLU/BLK	VTEC Pressure Switch	0v, Hi-Speed: 12v
C16	WHT	HO2S-11 (B1 S1) Signal	0.1-1.1v
C17	RED/GRN	MAP Sensor Signal	0.8-0.9v
C18	GRN/BLK	Sensor Ground (SG2)	<0.050v
C19	YEL/RED	MAP Sensor VREF (VCC1)	4.9-5.1v
C20	GRN	TDC Sensor Signal	AC pulse signals
C21	RED	TDC Sensor Ground	<0.050v
C25	RED/YEL	IAT Sensor Signal	Varies w/temp (0.5-4.9v)
C26	RED/WHT	ECT Sensor Signal	At 180ºF: 0.5-0.6v
C27	RED/BLK	TP Sensor Signal	0.5-0.6v
C28	YEL/BLU	Sensor VREF (VCC2)	4.9-5.1v
C29	YEL	CYP Sensor Signal	AC pulse signals
C30	BLK	CYP Sensor Ground	<0.050v

25-PIN 'B'   WIRE SIDE OF HARNESS TERMINALS   31-PIN 'C'

1999-2000 Van 3.5L V6 VTEC VIN RL1 [A/T] 32P 'A' Connector

PCM Pin #	W/Color	Circuit Description (32-Pin)	Value at Hot Idle
A2	GRN/WHT	Engine Mount Solenoid	Idle: 0v, off-idle: 12v
A3	BLU	EVAP Bypass Solenoid	Solenoid Off: 12v, On: 1v
A4	GRN/WHT	EVAP Vent Solenoid	Solenoid Off: 12v, On: 1v
A5	BLU/GRN	Cruise Control Signal	C/C On: pulse signals
A6	RED/YEL	EVAP Purge Solenoid	Solenoid Off: 12v, On: 1v
A7	WHT/RED	VREF to TCM (with TCS)	4.9-5.1v
A8	BLK/WHT	HO2S-12 (B1 S2) Heater	Digital Signals: 0-12-0v
A9	BLU/WHT	Vehicle Speed Output	Moving: pulse signals
A10	BRN	Service Check Connector	SCS Open: 4.80v
A11	LT GRN	Gear Position Signal (TCS)	In 'P': 4v
A12	PNK	Immobilizer Indicator Lamp	Lamp Off: 12v, On: 1v
A13, A25	BLU, RED	Immobilizer Enable, Code	Digital Signals
A14	GRN/BLK	D4 Indicator Light	D4 On: 0v, Off: 12v
A15	GRN/YEL	Fuel Pump Relay Control	Relay Off: 12v, On: 1v
A17	RED	A/C Clutch Relay Control	Relay Off: 12v, On: 1v
A18	GRN/ORN	Malfunction Indicator Light	MIL Off: 12v, On: 1v
A19	BLU	Engine Speed Pulse (NEP)	AC Pulse Signals
A20	BLU/RED	Radiator Fan Relay Control	Relay Off: 12v, On: 1v
A21	GRY	K-Line Signal	12v
A23	WHT/RED	HO2S-12 (B1 S2) Signal	0.1-1.1v
A24	BLU/ORN	Starter Switch Signal	Cranking: 9-11v
A26	GRN	PSP Switch Signal	Straight: 0v, Turning: 11v
A27	BLU/RED	A/C Switch Signal	Switch Off: 12v, On: 1v
A28	WHT/RED	Interlock Control Unit Signal	Key & Brake On: 12v
A29	LT GRN	Fuel Tank Pressure Sensor	Fuel Cap off: 2.5v
A30	GRN/RED	Electric Load Detector	Varies: 0.5-4.5v
A31	RED/BLK	TP Sensor Signal to TCS	At idle: 0.5v
A32	WHT/BLK	Brake Switch Signal	Brake Off: 0v, On: 12v

1999-2000 Van 3.5L V6 VTEC VIN RL1 [A/T] 16P 'D' Connector

PCM Pin #	W/Color	Circuit Description (16-Pin)	Value at Hot Idle
D1	YEL	Lockup Control Solenoid	LUS On: 12v, Off: 0v
D2	GRN/WHT	A/T: Shift Solenoid 'B'	SSB in 3rd, 4th Gear: 0v
D3	GRN	A/T: Shift Solenoid 'C'	SSC in 2nd, 4th Gear: 0v
D5	BLK/YEL	VB Solenoid Feed (B+)	12-14v
D6	WHT	A/T: Gear Position Switch	In 'R': 0v, Others: 12v
D7	BLU/YEL	A/T: Shift Solenoid 'A'	SSA in 1st, 4th Gear: 0v
D8, D9	PNK, YEL	A/T: Gear Position Switch	In D3: 0v, In D4: 0v
D10, D16	BLU, GRN	Countershaft Sensor 'P', 'N'	AC Pulses, <0.050v
D11, D12	RED, WHT	Mainshaft Speed Sensor P, N	Moving: AC pulses
D13	BLU/BLK	A/T: 2nd Clutch Pressure Sw.	Open: 12v, Closed: 1v
D14, 15	BLU, BRN	A/T: Gear Position Switch	D2: 0v, D1: 0v

32-Pin 'A'   WIRE SIDE OF HARNESS TERMINALS   16-Pin 'D'

1999-2000 Van 3.5L V6 VTEC VIN RL1 [A/T] 25P 'B' Connector

PCM Pin #	W/Color	Circuit Description (25-Pin)	Value at Hot Idle
B1	YEL/BLK	Main Relay Power (B+)	12-14v
B2	BLK	Power Ground (PG1)	<0.1v
B3	BLK/RED	Injector 5 Control	2.0-3.3 ms
B4	YEL	Injector 2 Control	2.0-3.3 ms
B5	RED	Injector 2 Control	2.0-3.3 ms
B6	WHT/BLU	Injector 6 Control	2.0-3.3 ms
B7	PNK	E-EGR Solenoid Control	Digital Signals: 0-12-0v
B8	WHT	A/T: Clutch Solenoid 'A-'	AC Pulse Signals
B9	YEL/BLK	Main Relay Power (B+)	12-14v
B10	BLK	Power Ground (PG2)	<0.1v
B11	BRN	Injector 1 Control	2.0-3.3 ms
B12	GRN/YEL	VTEC Control Solenoid	0v, Hi-Speed: 12v
B14	BLU/BLK	A/T: Gear Position Switch	In P/N: 0v, Others: 12v
B15	BLU	Injector 3 Control	2.0-3.3 ms
B17	RED	A/T: Clutch Solenoid 'A+'	AC Pulse Signals
B18	GRN	A/T: Clutch Solenoid 'B-'	Pulse Signals
B20	BRN/BLK	Logic Ground (LG1)	<0.050v
B21	WHT/YEL	Keep Alive Power (VBU)	12-14v
B22	BRN/BLK	Logic Ground (LG2)	<0.050v
B23	BLK/BLU	Idle Air Control Valve	Pulse Signals
B24	WHT/RED	A/T: 3rd Clutch Pressure Sw.	Open: 12v, Closed: 1v
B25	ORN	A/T: Clutch Solenoid 'B+'	Pulse Signals

1999-2000 Van 3.5L V6 VTEC VIN RL1 [W/TCS] 12P 'E' Connector

PCM Pin #	W/Color	Circuit Description (12-Pin)	Value at Hot Idle
E1	PNK/BLK	PCM Torque Down Request	Digital Signals: 0-5-0v
E2	---	Not Used	---
E3	GRN/RED	TCS Operation Permission	Digital Signals: 0-5-0v
E4	---	Not Used	---
E5	---	Not Used	---
E6	---	Not Used	---
E7	---	Not Used	---
E8	---	Not Used	---
E9	---	Not Used	---
E10	---	Not Used	---
E11	---	Not Used	---
E12	---	Not Used	---

25-Pin 'B'   12-Pin 'E' Connector   For Vehicles Equipped with ABS / TCS

WIRE SIDE OF HARNESS TERMINALS

1999-2000 Van 3.5L V6 VTEC VIN RL1 [A/T] 31P 'C' Connector

PCM Pin #	W/Color	Circuit Description (31-Pin)	Value at Hot Idle
C1	BLK/WHT	HO2S-11 (B1 S1) Heater	Digital Signals: 0-12-0v
C2	WHT/GRN	Alternator Charging Signal	Lights Off: 12v, On: 0v
C3	BLU	Ignition Coil 3 Control	Digital Signals: 0-12-0v
C4	YEL/GRN	Ignition Coil 1 Control	Digital Signals: 0-12-0v
C5	WHT/RED	Alternator 'FR' Signal	Digital Signals: 0-5-0v
C6	WHT/BLK	EGR Valve Lift Sensor	1.2v
C7	GRN/WHT	MAP Sensor Ground	<0.050v
C8	BLU	CKP Sensor Signal	AC pulse signals
C9	WHT	CKP Sensor Ground	<0.050v
C10	BLU/BLK	VTEC Pressure Switch	0v, Hi-Speed: 12v
C11	---	Not Used	----
C12	BLK/RED	Ignition Coil 5 Control	Digital Signals: 0-12-0v
C13	YEL	Ignition Coil 4 Control	Digital Signals: 0-12-0v
C14	RED	Ignition Coil 2 Control	Digital Signals: 0-12-0v
C15	---	Not Used	----
C16	WHT	HO2S-11 (B1 S1) Signal	0.1-1.1v
C17	RED/GRN	MAP Sensor Signal	0.8-0.9v
C18	GRN/WHT	Sensor Ground	<0.050v
C19	YEL/RED	MAP Sensor VREF (VCC1)	4.9-5.1v
C20	GRN	TDC1 Sensor Signal	AC pulse signals
C21	RED	TDC1 Sensor Ground	<0.050v
C22	RED/BLU	Knock Sensor Signal	No Detonation: 18mv AC
C23	WHT/BLU	Ignition Coil 6 Control	Digital Signals: 0-12-0v
C25	RED/YEL	IAT Sensor Signal	Varies w/temp. (0.5-4.9v)
C26	RED/WHT	ECT Sensor Signal	At 180°F: 0.5-0.6v
C27	RED/BLK	TP Sensor Signal	0.5-0.6v
C28	YEL/RED	Sensor VREF (VCC2)	4.9-5.1v
C29	YEL	TDC2 Sensor Signal	AC pulse signals
C30	BLK	TDC2 Sensor Ground	<0.050v
C31	---	Not Used	----

31-Pin 'C'

1	2	3	4		5	6	7		8	9	10
11	12	13	14	15	16	17	18	19	20	21	22
	23	24	25		26	27	28		29	30	31

WIRE SIDE OF HARNESS TERMINALS

2001-03 Van 3.5L V6 VTEC VIN RL1 [A/T] 31P 'A' Connector

PCM Pin #	W/Color	Circuit Description (31-Pin)	Value at Hot Idle
A1	WHT/YEL	Keep Alive Power (VBU)	12-14v
A2, A3	YEL/BLK	Main Relay Power (B+)	12-14v
A4, A5	BLK	Power Ground (PG2, PG1)	<0.1v
A6	WHT	HO2S-11 (B1 S1) Signal	0.1-1.1v
A9	RED/BLU	Knock Sensor Signal	No detonation: 18mv AC
A10, A22	YEL, BLK	TDC2 Sensor Signal	AC pulse signals
A11, A23	GRN,RED	TDC1 Sensor Signal	AC pulse signals
A12, A24	BLU, WHT	CKP Sensor Signal	AC pulse signals
A13	RED/WHT	ECT Sensor Signal	At 180°F: 0.5-0.6v
A14	RED/YEL	IAT Sensor Signal	Varies w/temp. (0.5-4.9v)
A15	RED/BLK	TP Sensor Signal	0.5-0.6v
A16	WHT/BLK	EGR Valve Position Sensor	0.6-1.1v
A19	WHT/RED	Alternator 'FR' Signal	Digital Signals: 0-5-0v
A20	BLU/BLK	VTEC Pressure Switch	0v, Hi-Speed: 12v
A25, A26	BRN/BLK	Logic Ground (LG2, LG1)	<0.1v
A27	RED/GRN	MAP Sensor Signal	0.8-0.9v
A28, A29	GRN/BLK	Sensor Ground (SG2, SG1)	<0.050v
A30, A31	YEL/BLU	Sensor VREF (VCC2, VCC1)	4.9-5.1v

2001-03 Van 3.5L V6 VTEC VIN RL1 [A/T] 22P 'C' Connector

PCM Pin #	W/Color	Circuit Description (22-Pin)	Value at Hot Idle
C1	WHT	A/T: Pressure Solenoid 'A-'	AC pulse signals
C2	BLU/YEL	A/T: Shift Solenoid 'A'	SSA in 1st, 4th Gear: 0v
C4	PNK	A/T: Gear Position Switch	In D3: 0v, Others: 12v
D5	BLK/YEL	VB Solenoid Feed (B+)	12-14v
D6	WHT	A/T: Gear Position Switch	In 'R': 0v, Others: 12v
C6	WHT	Mainshaft Speed Sensor 'N'	<0.050v
C7	RED	Mainshaft Speed Sensor 'P'	Moving: AC pulses
C8	GRN	A/T: Pressure Solenoid 'B-'	AC pulse signals
C9	GRN/WHT	A/T: Shift Solenoid 'B'	SSB in 3rd, 4th Gear: 0v
C10	YEL	Lockup Control Solenoid	LUS On: 12v, Off: 0v
C11	BLU/BLK	A/T: 2nd Clutch Pressure Sw.	Open: 12v, Closed: 1v
C12	BLU	A/T: Gear Position Switch	In D2: 0v, Others: 12v
C13	YEL	A/T: Gear Position Switch	In D4: 0v, Others: 12v
C14	GRN	Countershaft Sensor 'N'	<0.050v
C15	BLU	Countershaft Sensor 'P'	AC pulse signals
C17	GRN	A/T: Shift Solenoid 'C'	SSC in 2nd, 4th Gear: 0v
C19	WHT/RED	A/T: 3rd Clutch Pressure Sw.	Open: 12v, Closed: 1v

31-Pin 'A' *WIRE SIDE OF HARNESS TERMINALS* 22-Pin 'C'

C20	BRN	A/T: Gear Position Switch	In D1: 0v, Others: 12v
C21	WHT	A/T: Gear Position Switch	In 'R': 0v, Others: 12v
C22	BLU/WHT	A/T: Gear Position Switch	In P/N: 0v, Others: 12v

2001-03 Van 3.5L V6 VTEC VIN RL1 [A/T] 31P 'A' Connector

PCM Pin #	W/Color	Circuit Description (31-Pin)	Value at Hot Idle
B1	---	Not Used	---
B2	GRN/YEL	VTEC Control Solenoid	0v, Hi-Speed: 12v
B3	PNK	E-EGR Solenoid Control	Digital Signals: 0-12-0v
B4	---	Not Used	---
B5	BLK/WHT	HO2S-11 (B1 S1) Heater	Digital Signals: 0-12-0v
B6	WHT/GRN	Alternator Charging Signal	Lights Off: 12v, On: 0v
B7	RED	A/T: Pressure Solenoid 'A+'	AC pulse signals
B8	BLU	Coil 3 Driver Control	Digital Signals: 0-12-0v
B9	RED	Coil 2 Driver Control	Digital Signals: 0-12-0v
B10	YEL/GRN	Coil 1 Driver Control	Digital Signals: 0-12-0v
B11	---	Not Used	---
B12	BLU	Injector 3 Control	2.0-3.3 ms
B13	RED	Injector 2 Control	2.0-3.3 ms
B14	BRN	Injector 1 Control	2.0-3.3 ms
B15	---	Not Used	---
B16	ORN	A/T: Pressure Solenoid 'B+'	AC pulse signals
B17	WHT/BLU	Coil 6 Driver Control	Digital Signals: 0-12-0v
B18	BLK/RED	Coil 5 Driver Control	Digital Signals: 0-12-0v
B19	YEL	Coil 4 Driver Control	Digital Signals: 0-12-0v
B20	WHT/BLU	Injector 6 Control	2.0-3.3 ms
B21	BLK/RED	Injector 5 Control	2.0-3.3 ms
B22	YEL	Injector 4 Control	2.0-3.3 ms
B23	BLK/BLU	Idle Air Control Valve	Pulse Signals
B24	BLK/YEL	VB Solenoid Feed (B+)	12-14v

2001-03 Van 3.5L V6 VTEC VIN RL1 [A/T W/TCS] 17P 'D' Connector

PCM Pin #	W/Color	Circuit Description (17-Pin)	Value at Hot Idle
D1-2	---	Not Used	---
D3	PNK/BLK	TCM Retard Signal to PCM	Digital Signals: 0-5-0v
D4	---	Not Used	---
D5	GRN/RED	TCS Operation Permission	Digital Signals: 0-5-0v
D6-8	---	Not Used	---
D9	WHT/RED	ABS/TCS VREF	4.9-5.1v
D10	---	Not Used	---
D11	LT GRN	A/T Shift Position Signal	In 'P': near 4v
D12-15	---	Not Used	---
D16	RED/BLK	TPS Signal to ABS/TCS Unit	0.5-0.6v
D17	---	Not Used	---

24-Pin 'B'   *WIRE SIDE OF HARNESS TERMINALS*   17-Pin 'D'

2001-03 Van 3.5L V6 VTEC VIN RL1 [A/T] 31P 'E' Connector

PCM Pin #	W/Color	Circuit Description (31-Pin)	Value at Hot Idle
E1	GRN/YEL	Immobilizer Fuel Pump Relay	Relay Off: 12v, On: 1v
E2	BLK/WHT	HO2S-12 (B1 S2) Heater	Digital Signals: 0-12-0v
E3	BLU	Engine Speed (NEP) Signal	Pulse Signals
E4	GRN/OR	Malfunction Indicator Lamp	MIL Off: 12v, On: 1v
E5	GRN/BLK	A/T: Gear Position Switch	In D4: 0v, Others: 12v
E6	YEL/BLU	Sensor VREF (VCC3)	4.9-5.1v
E7	GRN/BLK	Sensor Ground (SG3)	<0.050v
E8	BLU/ORN	Starter Switch Signal	Cranking: 9-11v
E9	WHT/BLK	Brake Switch Signal	Brake Off: 0v, On: 12v
E10	---	Not Used	---
E11	RED/YEL	EVAP Purge Solenoid	Solenoid Off: 12v, On: 1v
E12	RED	A/C Clutch Relay Control	Relay Off: 12v, On: 1v
E13	BLU/RED	Radiator Fan Control Relay	Relay Off: 12v, On: 1v
E14	YEL/GRN	ECT Signal to TCM Unit	At 180°F: 0.5-0.6v
E15	WHT/RED	Illumination (Multiplex Unit)	Digital Signals
E16	BRN	Service Check Connector	SCS Open 5v
E17	BLU/GRN	Cruise Control Signal	C/C On: pulse signals
E18	WHT/RED	HO2S-12 (B1 S2) Signal	0.1-1.1v
E19	GRN/RED	Electric Load Detector	Varies: 0.5-4.5v
E20	GRN	PSP Switch Signal	Straight: 0v, Turning: 11v
E21	BLU/RED	A/C Switch Signal	Switch Off: 12v, On: 1v
E22	GRN/WHT	EVAP Vent Solenoid	Solenoid Off: 12v, On: 1v
E23	BLU	EVAP Bypass Solenoid	Solenoid Off: 12v, On: 1v
E24	GRN/WHT	Engine Mount Solenoid	Idle: 0v, off-idle: 12v
E25	---	Not Used	---
E26	BLU/WHT	VSS Signal to Speedometer	Digital Signals
E27	GRY	K-Line Signal to DCL	12v
E28	LT GRN	Fuel Tank Pressure Sensor	Fuel Cap off: 2.5v
E29	PNK	Immobilizer Indicator Lamp	Lamp Off: 12v, On: 1v
E30	BLU	Immobilizer Enable Signal	Digital Signals
E31	RED	Immobilizer Code Signal	Digital Signals

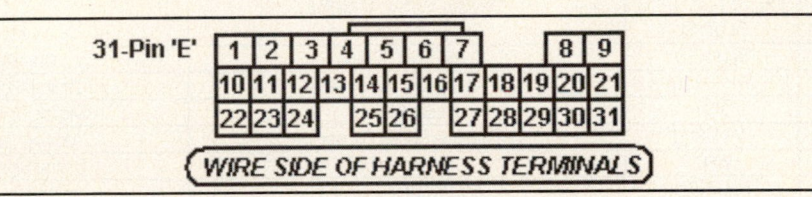

WIRE SIDE OF HARNESS TERMINALS

PRELUDE Pin Tables

1990-91 Coupe 2.0L I4 MFI VIN BA4 [All] 18P 'A' Connector

PCM Pin #	W/Color	Circuit Description (18-Pin)	Value at Hot Idle
A1	BRN	Injector 1 Control	2.0-3.3 ms
A2	BLK	Power Ground	<0.1v
A3	RED	Injector 2 Control	2.0-3.3 ms
A4	BLK	Power Ground	<0.1v
A5	BLU	Injector 3 Control	2.0-3.3 ms
A6	---	Not Used	---
A7	YEL	Injector 4 Control	2.0-3.3 ms
A8	YEL	Bypass Control Solenoid	Solenoid Off: 12v, On: 1v
A9	---	Not Used	---
A10	RED	EGR Solenoid Control	Solenoid Off: 12v, On: 1v
A11	BLU/RED	Electronic Air Control Valve	Pulse Signals
A12	GRN/BLK	Fuel Pump Relay Control	Relay Off: 12v, On: 1v
A13	YEL/BLK	Main Relay Power (B+)	12-14v
A14	GRN/BLK	Fuel Pump Relay Control	Relay Off: 12v, On: 1v
A15	YEL/BLK	Main Relay Power (B+)	12-14v
A16	BRN/BLK	Power Ground	<0.1v
A17	WHT/YEL	Keep Alive Power (VBU)	12-14v
A18	BLK/RED	Power Ground	<0.1v

1990-91 Coupe 2.0L I4 MFI VIN BA4 [All] 20P 'B' Connector

PCM Pin #	W/Color	Circuit Description (20-Pin)	Value at Hot Idle
B1	RED/BLU	A/C Clutch Relay Control	Relay Off: 12v, On: 1v
B2	YEL/BLK	EVAP Purge Solenoid	Solenoid Off: 12v, On: 1v
B3	---	Not Used	---
B4	---	Not Used	---
B5	BRN	Ignition Timing Adjuster	0.4-4.5v
B6	YEL/RED	Check Engine Light	C/E Off: 12v, On: 1v
B7	LT GRN	A/T: Park/Neutral Switch	In P/N: 0v, Others: 12v
B8	RED/GRN	A/C Switch Signal	Switch Off: 12v, On: 1v
B9	---	Not Used	---
B10	BLU/GRN	CKP Sensor Signal	AC pulse signals
B11	RED/BLU	Oxygen Sensor 'B' Signal	0.1-1.1v
B12	BLU/YEL	CKP Sensor Ground	<0.050v
B13	BLU/RED	Starter Switch Signal	Cranking: 9-11v
B14	WHT/RED	Alternator 'FR' Signal	Digital Signals: 0-5-0v
B15	WHT	Igniter Control	Digital Signals: 0-12-0v
B16	WHT/BLU	Vehicle Speed Sensor	Moving: pulse signals
B17	WHT	Igniter Control	Digital Signals: 0-12-0v
B18	BLU/RED	PSP Switch Signal	Straight: 0v, Turning: 11v
B19	YEL/WHT	A/T: Control Unit VREF	4.9-5.1v
B20	---	Not Used	---

1990-91 Coupe 2.0L I4 MFI VIN BA4 [All] 16P 'C' Connector

PCM Pin #	W/Color	Circuit Description (16-Pin)	Value at Hot Idle
C1	ORN	CYP Sensor Signal	AC pulse signals
C2	WHT	CYP Sensor Ground	<0.050v
C3	ORN/BLU	TDC Sensor Signal	AC pulse signals
C4	WHT/BLU	TDC Sensor Ground	<0.050v
C5	WHT/RED	IAT Sensor Signal	Varies w/temp. (0.5-4.9v)
C6	YEL/GRN	ECT Sensor Signal	At 180°F: 0.5-0.6v
C7	RED/YEL	Throttle Angle Sensor Signal	0.5-0.6v
C8	YEL	A/T: EGRV Lift Sensor Signal	1.2v
C9	RED	Atmospheric Pressure Sensor	2.76-2.96v at sea level
C10	---	Not Used	---
C11	WHT/BLU	MAP Sensor Signal	0.8-0.9v
C12	GRN/WHT	Sensor Ground	<0.050v
C13	YEL/WHT	Sensor VREF	4.9-5.1v
C14	BLU/WHT	Sensor Ground	<0.050v
C15	RED/WHT	Sensor VREF	4.9-5.1v
C16	WHT	Oxygen Sensor 'A' Signal	0.1-1.1v

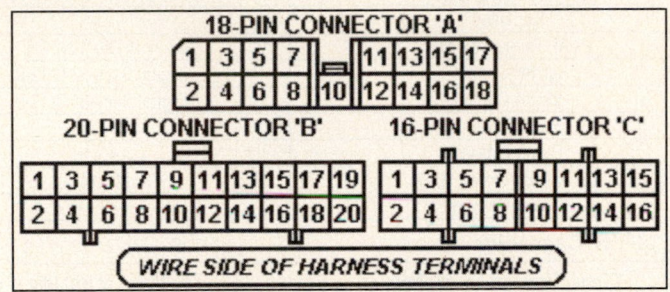

Abbreviation	Color	Abbreviation	Color	Abbreviation	Color
BLK	Black	LT BLU	Lt. Blue	TAN	Tan
BLU	Blue	LT GRN	Lt. Green	VIO	Violet
BRN	Brown	ORN	Orange	WHT	White
GRY	Gray	PNK	Pink	YEL	Yellow
GRN	Green	PPL	Purple		

1990-91 Coupe 2.1L I4 MFI VIN BA4 [All] 26P 'A' Connector

PCM Pin #	W/Color	Circuit Description (26-Pin)	Value at Hot Idle
A1	BRN	Injector 1 Control	2.0-3.3 ms
A2	YEL	Injector 4 Control	2.0-3.3 ms
A3	RED	Injector 2 Control	2.0-3.3 ms
A4	---	Not Used	---
A5	BLU	Injector 3 Control	2.0-3.3 ms
A6	BLK	HO2S-11 (B1 S1) Heater	Relay Off: 12v, On: 1v
A7	GRN/BLK	Fuel Pump Relay Control	Relay Off: 12v, On: 1v
A8	GRN/BLK	Fuel Pump Relay Control	Relay Off: 12v, On: 1v
A9	BLU/RED	Electronic Air Control Valve	Pulse Signals
A10	---	Not Used	---
A11	RED	EGR Solenoid Control	1.2v
A12	---	Not Used	---
A13	YEL/RED	Check Engine Light	C/E Off: 12v, On: 1v
A14	---	Not Used	---
A15	RED/BLU	A/C Clutch Relay Control	Relay Off: 12v, On: 1v
A16-19	---	Not Used	---
A20	YEL/BLK	EVAP Purge Solenoid	Solenoid Off: 12v, On: 1v
A21	WHT	Igniter Control	Digital Signals: 0-12-0v
A22	WHT	Igniter Control	Digital Signals: 0-12-0v
A23	BLK	Power Ground	<0.1v
A24	BLK	Power Ground	<0.1v
A25	YEL/BLK	Main Relay Power (B+)	12-14v
A26	BLK/RED	Power Ground	<0.1v

1990-91 Coupe 2.1L I4 MFI VIN BA4 [All] 16P 'B' Connector

PCM Pin #	W/Color	Circuit Description (16-Pin)	Value at Hot Idle
B1	YEL/BLK	Main Relay Power (B+)	12-14v
B2	BRN/BLK	Power Ground	<0.1v
B3	---	Not Used	---
B4	---	Not Used	---
B5	RED/GRN	A/C Switch Signal	Switch Off: 12v, On: 1v
B6	---	Not Used	---
B7	LT GRN	A/T: Park/Neutral Switch	In P/N: 0v, Others: 12v
B8	BLU/RED	PSP Switch	Straight: 0v, Turning: 11v
B9	BLU/RED	Starter Switch Signal	Cranking: 9-11v
B10	WHT/BLU	Vehicle Speed Sensor	Moving: pulse signals
B11	ORN	CYP Sensor Signal	AC pulse signals
B12	WHT	CYP Sensor Ground	<0.050v
B13	ORN/BLU	TDC Sensor Signal	AC pulse signals
B14	WHT/BLU	TDC Sensor Ground	<0.050v
B15	BLU/GRN	CKP Sensor Signal	AC pulse signals
B16	BLU/YEL	CKP Sensor Ground	<0.050v

1990-91 Coupe 2.1L I4 MFI VIN BA4 [All] 22P 'D' Connector

PCM Pin #	W/Color	Circuit Description (22-Pin)	Value at Hot Idle
D1	WHT/YEL	Keep Alive Power (VBU)	12-14v
D2	---	Not Used	---
D3	---	Not Used	---
D4	BRN	Service Check Connector	SCS Open: 4.80v
D5-8	---	Not Used	---
D9	WHT/RED	Alternator 'FR' Signal	Digital Signals: 0-5-0v
D10	---	Not Used	---
D11	RED/YEL	Throttle Angle Sensor Signal	0.5-0.6v
D12	YEL	A/T: EGRV Lift Sensor Signal	1.2v
D13	YEL/GRN	ECT Sensor Signal	At 180°F: 0.5-0.6v
D14	WHT	HO2S-11 (B1 S1) Signal	0.1-1.1v
D15	WHT/RED	IAT Sensor Signal	Varies w/temp. (0.5-4.9v)
D16	---	Not Used	---
D17	WHT/BLU	MAP Sensor Signal	0.8-0.9v
D18	YEL/WHT	A/T: Control Unit VREF	4.9-5.1v
D19	RED/WHT	MAP Sensor VREF	4.9-5.1v
D20	YEL/WHT	Sensor VREF	4.9-5.1v
D21	BLU/WHT	MAP Sensor Ground	<0.050v
D22	RED/WHT	Sensor VREF	4.9-5.1v

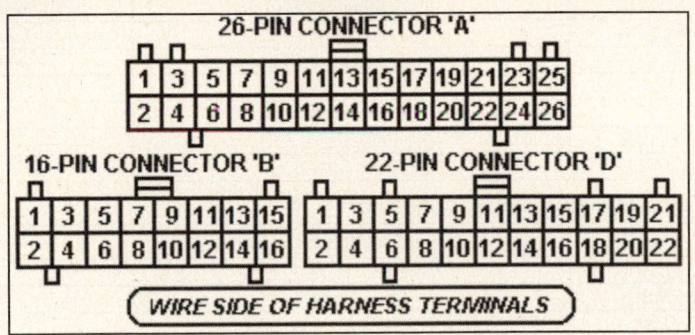

Abbreviation	Color	Abbreviation	Color	Abbreviation	Color
BLK	Black	LT BLU	Lt. Blue	TAN	Tan
BLU	Blue	LT GRN	Lt. Green	VIO	Violet
BRN	Brown	ORN	Orange	WHT	White
GRY	Gray	PNK	Pink	YEL	Yellow
GRN	Green	PPL	Purple		

1992 Coupe 2.2L I4 MFI VIN BA8 [All] 26P 'A' Connector

PCM Pin #	W/Color	Circuit Description (26-Pin)	Value at Hot Idle
A1	BRN	Injector 1 Control	2.0-3.3 ms
A2	YEL	Injector 4 Control	2.0-3.3 ms
A3	RED	Injector 2 Control	2.0-3.3 ms
A4	---	Not Used	
A5	BLU	Injector 3 Control	2.0-3.3 ms
A6	ORN/WHT	HO2S Heater Control	Relay Off: 12v, On: 1v
A7	GRN/BLK	Fuel Pump Relay	Relay Off: 12v, On: 1v
A8	GRN/BLK	Fuel Pump Relay	Relay Off: 12v, On: 1v
A9	BLK/BLU	Electronic Air Control Valve	Pulse Signals
A10	---	Not Used	---
A11	RED	EGR Solenoid	Solenoid Off: 12v, On: 1v
A12	BLU/RED	Radiator Fan Relay	Relay Off: 12v, On: 1v
A13	BLU/WHT	Check Engine Light	MIL Off: 12v, On: 1v
A14	---	Not Used	---
A15	RED/BLU	A/C Compressor Unit	---
A16	WHT/GRN	Alternator Charging Signal	Lights Off: 12v, On: 0v
A17	PNK	Intake Air Bypass Solenoid	Solenoid Off: 12v, On: 1v
A18	ORN/RED	A/T: TCM (shift acknowledge)	N/A
A19	WHT	Intake Air Control Solenoid	Solenoid Off: 12v, On: 1v
A20	RED/GRN	EVAP Purge Solenoid	Solenoid Off: 12v, On: 1v
A21	YEL/GRN	Igniter Control	Digital Signals: 0-12-0v
A22	YEL/GRN	Igniter Control	Digital Signals: 0-12-0v
A23	BLK	Power Ground	<0.1v
A24	BLK	Power Ground	<0.1v
A25	YEL/BLK	Main Relay Power (B+)	12-14v
A26	BLK/RED	Power Ground	<0.1v

1992 Coupe 2.2L I4 MFI VIN BA8 [All] 16P 'B' Connector

PCM Pin #	W/Color	Circuit Description (16-Pin)	Value at Hot Idle
B1	YEL/BLK	Main Relay Power (B+)	12-14v
B2	BRN/BLK	Power Ground	<0.1v
B3	ORN	A/T: TCM Signal (up-shift)	Digital Signals
B4	PNK	A/T: TCM Signal (down-shift)	Digital Signals
B5	BLU/BLK	A/C Switch Signal	Switch Off: 12v, On: 1v
B6	---	Not Used	---
B7	LT GRN	A/T: Park/Neutral Switch	In N: 0v, Others: 12v
B8	RED/GRN	PSP Switch	Straight: 0v, Turning: 11v
B9	BLU/RED	Starter Switch Signal	Cranking: 9-11v
B10	ORN	Vehicle Speed Sensor	Moving: pulse signals
B11	ORN	CYP Sensor Signal	AC pulse signals
B12	WHT	CYP Sensor Ground	<0.050v
B13	ORN/BLU	TDC Sensor Signal	AC pulse signals
B14	WHT/BLU	TDC Sensor Ground	<0.050v
B15	BLU/GRN	CKP Sensor Signal	AC pulse signals
B16	BLU/YEL	CKP Sensor Ground	<0.050v

1992 Coupe 2.2L I4 MFI VIN BA8 [All] 22P 'D' Connector

PCM Pin #	W/Color	Circuit Description (22-Pin)	Value at Hot Idle
D1	WHT/YEL	Keep Alive Power (VBU)	12-14v
D2	GRN/WHT	Brake Switch Signal	Brake Off: 0v, On: 12v
D3	RED/BLU	Knock Sensor Signal Solenoid	Solenoid Off: 12v, On: 1v
D4	BRN/WHT	Service Check Connector	SCS Open: 4.80v
D5-6	---	Not Used	---
D7	GRN/RED	Data Link Connector	5v
D8	---	Not Used	---
D9	WHT/RED	Alternator 'FR' Signal	Digital Signals: 0-5-0v
D10	---	Not Used	---
D11	RED/BLK	Throttle Angle Sensor Signal	0.5-0.6v
D12	WHT/BLK	A/T: EGR Lift Sensor Signal	1.2v
D13	YEL/BLU	ECT Sensor Signal	At 180°F: 0.5-0.6v
D14	WHT	O2S-11 Signal	0.1-1.1v
D15	RED/YEL	IAT Sensor Signal	Varies w/temp. (0.5-4.9v)
D16	---	Not Used	---
D17	WHT/BLU	MAP Sensor Signal	0.8-0.9v
D18	ORN/BLK	A/T: Control Unit VREF	4.9-5.1v
D19	RED/WHT	Sensor VREF	4.9-5.1v
D20	YEL/WHT	Sensor VREF	4.9-5.1v
D21	BLU/WHT	Sensor Ground	<0.050v
D22	GRN/WHT	Sensor Ground	<0.050v

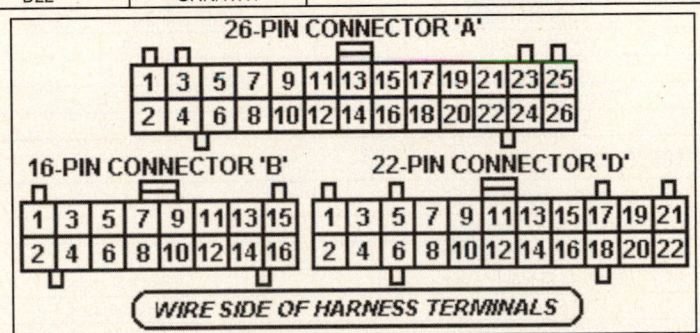

Abbreviation	Color	Abbreviation	Color	Abbreviation	Color
BLK	Black	LT BLU	Lt. Blue	TAN	Tan
BLU	Blue	LT GRN	Lt. Green	VIO	Violet
BRN	Brown	ORN	Orange	WHT	White
GRY	Gray	PNK	Pink	YEL	Yellow
GRN	Green	PPL	Purple		

1993-95 Coupe 2.2L I4 MFI VIN BA8 [All] 26P 'A' Connector

PCM Pin #	W/Color	Circuit Description (26-Pin)	Value at Hot Idle
A1	BRN	Injector 1 Control	2.0-3.3 ms
A2	YEL	Injector 4 Control	2.0-3.3 ms
A3	RED	Injector 2 Control	2.0-3.3 ms
A4	---	Not Used	---
A5	BLU	Injector 3 Control	2.0-3.3 ms
A6	ORN/WHT	HO2S Heater Control	Relay Off: 12v, On: 1v
A7	GRN/BLK	Fuel Pump Relay	Relay Off: 12v, On: 1v
A8	GRN/BLK	Fuel Pump Relay	Relay Off: 12v, On: 1v
A9	BLK/BLU	Idle Air Control Valve	Pulse Signals
A10	---	Not Used	---
A11	RED	EGR Solenoid	Solenoid Off: 12v, On: 1v
A12	BLU/RED	Radiator Fan Relay	Relay Off: 12v, On: 1v
A13	BLU/WHT	Check Engine Light	MIL Off: 12v, On: 1v
A14	---	Not Used	---
A15	RED/BLU	A/C Clutch Relay Control	Relay Off: 12v, On: 1v
A16	WHT/GRN	Alternator Charging Signal	Lights Off: 12v, On: 0v
A17	PNK	Intake Air Bypass Solenoid	Solenoid Off: 12v, On: 1v
A18	ORN/RED	A/T: TCM (shift acknowledge)	N/A
A19	WHT	Intake Control Solenoid	Solenoid Off: 12v, On: 1v
A20	RED/GRN	EVAP Purge Solenoid	Solenoid Off: 12v, On: 1v
A21	YEL/GRN	Igniter Control	Digital Signals: 0-12-0v
A22	YEL/GRN	Igniter Control	Digital Signals: 0-12-0v
A23	BLK	Power Ground	<0.1v
A24	BLK	Power Ground	<0.1v
A25	YEL/BLK	Main Relay Power (B+)	12-14v
A26	BLK/RED	Power Ground	<0.1v

1993-95 Coupe 2.2L I4 MFI VIN BA8 [All] 16P 'B' Connector

PCM Pin #	W/Color	Circuit Description (16-Pin)	Value at Hot Idle
B1	YEL/BLK	Main Relay Power (B+)	12-14v
B2	BRN/BLK	Power Ground	<0.1v
B3	ORN	A/T TCM Signal (Upshift)	Digital Signals
B4	PNK	A/T TCM Signal (Downshift)	Digital Signals
B5	BLU/BLK	A/C Switch Signal	Switch Off: 12v, On: 1v
B6	---	Not Used	---
B7	LT GRN	A/T: Park Neutral Switch	In P/N: 0v, Others: 5v
B8	RED/GRN	PSP Switch	Straight: 0v, Turning: 11v
B9	BLU/RED	Starter Switch Signal	Cranking: 9-11v
B10	ORN	Vehicle Speed Sensor	Moving: pulse signals
B11	ORN	CYP Sensor Signal	AC pulse signals
B12	WHT	CYP Sensor Ground	0.050v
B13	ORN/BLU	TDC Sensor	AC pulse signals
B14	WHT/BLU	TDC Sensor	0.050v
B15	BLU/GRN	CKP Sensor Signal	AC pulse signals
B16	BLU/YEL	CKP Sensor Ground	0.050v

1993-95 Coupe 2.2L I4 MFI VIN BA8 [All] 22P 'D' Connector

PCM Pin #	W/Color	Circuit Description (22-Pin)	Value at Hot Idle
D1	WHT/YEL	Keep Alive Power (VBU)	12-14v
D2	GRN/WHT	Brake Switch Signal	Brake Off: 0v, On: 12v
D3	RED/BLU	Knock Sensor Signal	No Detonation: 18mv AC
D4	BRN/WHT	Service Check Connector	SCS Open: 4.80v
D5	---	Not Used	---
D6	---	Not Used	---
D7	GRN/RED	Data Link Connector	5v
D8	---	Not Used	---
D9	WHT/RED	Alternator 'FR' Signal	Digital Signals: 0-5-0v
D10	GRN/RED	Electrical Load Detector	Varies: 0.5-4.5v
D11	RED/BLK	TP Sensor Signal	0.5-0.6v
D12	WHT/BLK	EGR Valve Lift Sensor	1.2v
D13	YEL/BLU	ECT Sensor Signal	At 180°F: 0.5-0.6v
D14	WHT	HO2S Signal	0.1-1.1v
D15	RED/YEL	IAT Sensor Signal	Varies w/temp. (0.5-4.9v)
D16	---	Not Used	---
D17	WHT/BLU	MAP Sensor Signal	0.8-0.9v
D18	GRN/BLK	A/T: Control Unit VREF	4.9-5.1v
D19	RED/WHT	Sensor VREF	4.9-5.1v
D20	YEL/WHT	Sensor VREF	4.9-5.1v
D21	BLU/WHT	Sensor Ground	<0.050v
D22	GRN/WHT	Sensor Ground	<0.050v

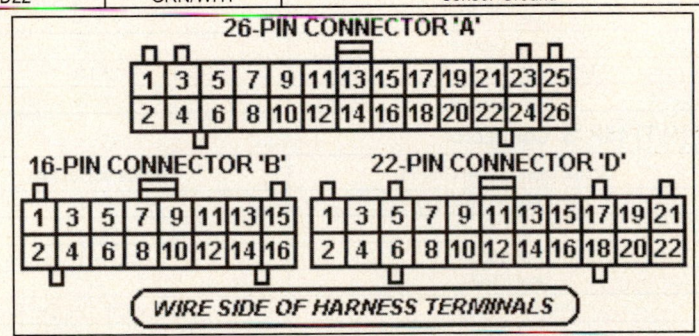

Abbreviation	Color	Abbreviation	Color	Abbreviation	Color
BLK	Black	LT BLU	Lt. Blue	TAN	Tan
BLU	Blue	LT GRN	Lt. Green	VIO	Violet
BRN	Brown	ORN	Orange	WHT	White
GRY	Gray	PNK	Pink	YEL	Yellow
GRN	Green	PPL	Purple		

1993-95 Coupe 2.2L I4 VTEC VIN BB1 [M/T] 26P 'A' Connector

PCM Pin #	W/Color	Circuit Description (26-Pin)	Value at Hot Idle
A1	BRN	Injector 1 Control	2.0-3.3 ms
A2	YEL	Injector 4 Control	2.0-3.3 ms
A3	RED	Injector 2 Control	2.0-3.3 ms
A4	GRN/YEL	VTEC Solenoid	0v, Hi-Speed: 12v
A5	BLU	Injector 3 Control	2.0-3.3 ms
A6	ORN/WHT	HO2S Heater Control	Relay Off: 12v, On: 1v
A7	GRN/BLK	Fuel Pump Relay	Relay Off: 12v, On: 1v
A8	GRN/BLK	Fuel Pump Relay	Relay Off: 12v, On: 1v
A9	BLK/BLU	Idle Air Control Valve	Pulse Signals
A10	---	Not Used	---
A11	RED	EGR Solenoid	Solenoid Off: 12v, On: 1v
A12	BLU/RED	Radiator Fan Relay Control	Relay Off: 12v, On: 1v
A13	BLU/WHT	Check Engine Light	MIL Off: 12v, On: 1v
A14	---	Not Used	---
A15	RED/BLU	A/C Clutch Relay Control	Relay Off: 12v, On: 1v
A16	WHT/GRN	Alternator Charging Signal	Lights Off: 12v, On: 0v
A17	PNK	Intake Air Bypass Solenoid	Solenoid Off: 12v, On: 1v
A18	---	Not Used	---
A19	WHT	Intake Control Solenoid	Solenoid Off: 12v, On: 1v
A20	RED/GRN	EVAP Purge Solenoid	Solenoid Off: 12v, On: 1v
A21	YEL/GRN	Igniter Control	Digital Signals: 0-12-0v
A22	YEL/GRN	Igniter Control	Digital Signals: 0-12-0v
A23	BLK	Power Ground	<0.1v
A24	BLK	Power Ground	<0.1v
A25	YEL/BLK	Main Relay Power (B+)	12-14v
A26	BLK/RED	Power Ground	<0.1v

1993-95 Coupe 2.2L I4 VTEC VIN BB1 [M/T] 16P 'B' Connector

PCM Pin #	W/Color	Circuit Description (16-Pin)	Value at Hot Idle
B1	YEL/BLK	Main Relay Power (B+)	12-14v
B2	BRN/BLK	Power Ground	<0.1v
B3	---	Not Used	---
B4	---	Not Used	---
B5	BLU/BLK	A/C Switch Signal	Switch Off: 12v, On: 1v
B6	---	Not Used	---
B7	---	Not Used	---
B8	RED/GRN	PSP Switch Signal	Straight: 0v, Turning: 11v
B9	BLU/RED	Starter Switch Signal	Cranking: 9-11v
B10	ORN	Vehicle Speed Sensor	Moving: pulse signals
B11	ORN	CYP Sensor Signal	AC pulse signals
B12	WHT	CYP Sensor Ground	<0.050v
B13	ORN/BLU	TDC Sensor Signal	AC pulse signals
B14	WHT/BLU	TDC Sensor Ground	<0.050v
B15	BLU/GRN	CKP Sensor Signal	AC pulse signals
B16	BLU/YEL	CKP Sensor Ground	<0.050v

1993-95 Coupe 2.2L I4 VTEC VIN BB1 [M/T] 22P 'D' Connector

PCM Pin #	W/Color	Circuit Description (26-Pin)	Value at Hot Idle
D1	WHT/YEL	Keep Alive Power (VBU)	12-14v
D2	GRN/WHT	Brake Switch Signal	Brake Off: 0v, On: 12v
D3	RED/BLU	Knock Sensor Signal	No Detonation: 18mv AC
D4	BRN/WHT	Service Check Connector	SCS Open: 4.80v
D5	---	Not Used	---
D6	LT BLU	VTEC Pressure Switch	0v, Hi-Speed: 12v
D7	GRN/RED	DLC TXD, RXD Signals	Digital Signals (0-5-0-5v)
D8	---	Not Used	
D9	WHT/RED	Alternator 'FR' Signal	Digital Signals: 0-5-0v
D10	GRN/RED	Electrical Load Detector	Varies: 0.5-4.5v
D11	RED/BLK	TP Sensor Signal	0.5-0.6v
D12	WHT/BLK	EGR Valve Lift Sensor	1.2v
D13	YEL/BLU	ECT Sensor Signal	At 180°F: 0.5-0.6v
D14	WHT	HO2S Signal	0.1-1.1v
D15	RED/YEL	IAT Sensor Signal	Varies w/temp. (0.5-4.9v)
D16	---	---	---
D17	WHT/BLU	MAP Sensor Signal	0.8-0.9v
D18	---	Not Used	---
D19	RED/WHT	Sensor VREF	4.9-5.1v
D20	YEL/WHT	Sensor VREF	4.9-5.1v
D21	BLU/WHT	Sensor Ground	<0.050v
D22	GRN/WHT	Sensor Ground	<0.050v

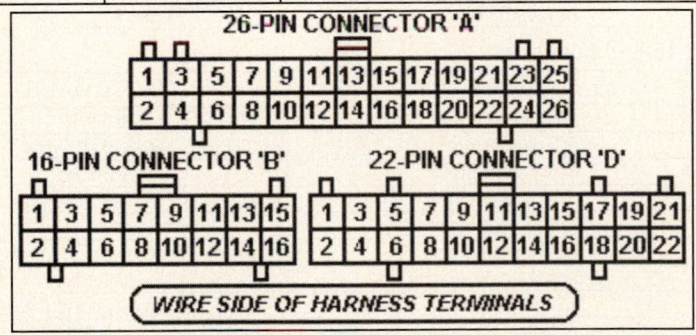

Abbreviation	Color	Abbreviation	Color	Abbreviation	Color
BLK	Black	LT BLU	Lt. Blue	TAN	Tan
BLU	Blue	LT GRN	Lt. Green	VIO	Violet
BRN	Brown	ORN	Orange	WHT	White
GRY	Gray	PNK	Pink	YEL	Yellow
GRN	Green	PPL	Purple		

1996 Coupe 2.2L I4 MFI VIN BA8 [All] 32P 'A' Connector

PCM Pin #	W/Color	Circuit Description (32-Pin)	Value at Hot Idle
A1	YEL	Injector 4 Control	2.0-3.3 ms
A2	BLU	Injector 3 Control	2.0-3.3 ms
A3	RED	Injector 2 Control	2.0-3.3 ms
A4	BRN	Injector 1 Control	2.0-3.3 ms
A5	ORN/BLK	HO2S-12 (B1 S2) Heater	Digital Signals: 0-12-0v
A6	ORN/WHT	HO2S-11 (B1 S1) Heater	Digital Signals: 0-12-0v
A7	RED	EGR Solenoid	Solenoid Off: 12v, On: 1v
A9	BLK/RED	Logic Ground	<0.050v
A10	BLK	Power Ground	<0.1v
A11	YEL/BLK	Main Relay Power (B+)	12-14v
A12	BLK/BLU	Idle Air Control Valve	Pulse Signals
A15	RED/YEL	EVAP Purge Solenoid	Solenoid Off: 12v, On: 1v
A16	GRN/BLK	Fuel Pump Relay	Relay Off: 12v, On: 1v
A17	RED/BLU	A/C Clutch Relay Control	Relay Off: 12v, On: 1v
A18	BLU/WHT	Check Engine Light	MIL Off: 12v, On: 1v
A19	WHT/GRN	Alternator Charging Signal	Lights Off: 12v, On: 0v
A20	YEL/GRN	Igniter Control	Digital Signals: 0-12-0v
A22	BRN/BLK	Logic Ground	<0.050v
A23	BLK	Power Ground	<0.1v
A24	YEL/BLK	Main Relay Power (B+)	12-14v
A25	WHT	Intake Air Control Solenoid	Solenoid Off: 12v, On: 1v
A26	PNK	Intake Air Bypass Solenoid	Solenoid Off: 12v, On: 1v
A27	BLU/RED	Radiator Fan Relay	Relay Off: 12v, On: 1v

1996 Prelude 2.2L I4 MFI VIN BA8 [All] 16P 'D' Connector

PCM Pin #	W/Color	Circuit Description (16-Pin)	Value at Hot Idle
D1	RED/BLK	TP Sensor Signal	0.5-0.6v
D2	YEL/BLU	ECT Sensor Signal	At 180°F: 0.5-0.6v
D3	WHT/YEL	MAP Sensor Signal	0.8-0.9v
D4	YEL/WHT	MAP Sensor VREF	4.9-5.1v
D5	GRN/WHT	Brake Switch Signal	Brake Off: 0v, On: 12v
D6	RED/BLU	Knock Sensor Signal	No Detonation: 18mv AC
D7	WHT	HO2S-11 (B1 S1) Signal	0.1-1.1v
D8	RED/YEL	IAT Sensor Signal	Varies w/temp. (0.5-4.9v)
D9	WHT/BLK	EGR Valve Lift Sensor	1.2v
D10	YEL/WHT	Sensor VREF	4.9-5.1v
D11	GRN/WHT	Sensor Ground	<0.050v
D12	GRN/WHT	MAP Sensor Ground	<0.050v
D13	GRN/BLU	HO2S-12 Ground	<0.050v
D14	WHT/RED	HO2S-12 (B1 S2) Signal	0.1-1.1v
D16	GRN/BLK	Electrical Load Detector	Varies: 0.5-4.5v

Connector pinout diagram: 32-PIN 'A' and 16-PIN 'D'. WIRE SIDE OF HARNESS TERMINALS

1996 Coupe 2.2L I4 MFI VIN BA8 [All] 25P 'B' Connector

PCM Pin #	W/Color	Circuit Description (25-Pin)	Value at Hot Idle
B3	BLU/YEL	A/T: Shift Solenoid 'A'	SSA Off: 0v, On: 12v
B4	WHT/BLK	A/T: Lockup Solenoid 'B'	LSB On: 12v, Off: 0v
B5	RED/WHT	A/T: Lockup Solenoid 'A'	LSA On: 12v, Off: 0v
B8	GRN/BLU	A/T: Gear Position Switch	In D3: 0v, Others: 12v
B11	GRN/YEL	A/T: Shift Solenoid 'B'	SSB Off: 0v, On: 12v
B12	WHT/GRN	Interlock Control Unit Signal	Key & Brake On: 12v
B13	GRN/BLK	A/T: D4 Indicator Light Switch	D4 On: 0v, Off: 12v
B14	WHT/BLU	Mainshaft Speed Sensor 'N'	<0.050v
B15	ORN/BLU	Mainshaft Speed Sensor 'P'	AC Pulse Signals
B16	GRN/RED	A/T: Gear Position Switch	In 'R': 0v, Others: 12v
B17	GRN/YEL	A/T: Gear Position Switch	In D2: 0v, Others: 12v
B18	GRN/WHT	A/T: Gear Position Switch	In D1: 0v, Others: 12v
B22	BLU/YEL	Countershaft Speed Sensor N	<0.050v
B23	BLU/GRN	Countershaft Speed Sensor P	Moving: AC pulses
B24	GRN/BLK	A/T: Gear Position Switch	In D4: 0v, Others: 12v
B25	LT GRN	A/T: Gear Position Switch	In P/N: 0v, Others: 12v

1996 Coupe 2.2L I4 MFI VIN BA8 [All] 31P 'C' Connector

PCM Pin #	W/Color	Circuit Description (31-Pin)	Value at Hot Idle
C1	---	Not Used	---
C2	BLU	CKP Sensor Signal	AC pulse signals
C3	GRN	TDC Sensor Signal	AC pulse signals
C4	YEL	CYP Sensor Signal	AC pulse signals
C5	BLU/BLK	A/C Switch Signal	Switch Off: 12v, On: 1v
C6	BLU/RED	Starter Switch Signal	Cranking: 9-11v
C7	BRN/WHT	Service Check Connector	SCS Open: 4.80v
C8	LT GRN	K-Line Signal	12v
C9	---	Not Used	---
C10	WHT/YEL	Keep Alive Power (VBU)	12-14v
C12	WHT	CKP Sensor Ground	<0.050v
C13	RED	TDC Sensor Ground	<0.050v
C14	BLK	CYP Sensor Ground	<0.050v
C16	RED/GRN	PSP Switch	Straight: 0v, Turning: 11v
C17	WHT/RED	Alternator FR Signal	Digital Signals: 0-5-0v
C18	ORN	Vehicle Speed Sensor	Moving: pulse signals
C19	---	Not Used	---
C20	BRN	EVAP Purge Flow Switch	Switch on: 0v, off: 5v
C21-31	---	Not Used	---

25-PIN 'B'   WIRE SIDE OF HARNESS TERMINALS   31-PIN 'C'

1996 Coupe 2.2L I4 VTEC VIN BB1 [All] 32P 'A' Connector

PCM Pin #	W/Color	Circuit Description (32-Pin)	Value at Hot Idle
A1	YEL	Injector 4 Control	2.0-3.3 ms
A2	BLU	Injector 3 Control	2.0-3.3 ms
A3	RED	Injector 2 Control	2.0-3.3 ms
A4	BRN	Injector 1 Control	2.0-3.3 ms
A5	ORN/BLK	HO2S-12 (B1 S2) Heater	Digital Signals: 0-12-0v
A6	ORN/WHT	HO2S-11 (B1 S1) Heater	Digital Signals: 0-12-0v
A7	RED	EGR Solenoid	Solenoid Off: 12v, On: 1v
A8	GRN/YEL	VTEC Solenoid	0v, Hi-Speed: 12v
A9	BLK/RED	Logic Ground	<0.050v
A10	BLK	Power Ground	<0.1v
A11	YEL/BLK	Main Relay Power (B+)	12-14v
A12	BLK/BLU	Idle Air Control Valve	Pulse Signals
A15	RED/YEL	EVAP Purge Solenoid	Solenoid Off: 12v, On: 1v
A16	GRN/BLK	Fuel Pump Relay	Relay Off: 12v, On: 1v
A17	RED/BLU	A/C Clutch Relay Control	Relay Off: 12v, On: 1v
A18	BLU/WHT	Check Engine Light	MIL Off: 12v, On: 1v
A19	WHT/GRN	Alternator Charging Signal	Lights Off: 12v, On: 0v
A20	YEL/GRN	Igniter Control	Digital Signals: 0-12-0v
A22	BLK/RED	Logic Ground	<0.050v
A23	BLK	Power Ground	<0.1v
A24	YEL/BLK	Main Relay Power (B+)	12-14v
A25	WHT	Intake Air Control Solenoid	Solenoid Off: 12v, On: 1v
A26	PNK	Intake Air Bypass Solenoid	Solenoid Off: 12v, On: 1v
A27	BLU/RED	Radiator Fan Relay Control	Relay Off: 12v, On: 1v

1996 Coupe 2.2L I4 VTEC VIN BB1 [All] 16P 'D' Connector

PCM Pin #	W/Color	Circuit Description (16-Pin)	Value at Hot Idle
D1	RED/BLK	TP Sensor Signal	0.5-0.6v
D2	YEL/BLU	ECT Sensor Signal	At 180ºF: 0.5-0.6v
D3	WHT/YEL	MAP Sensor Signal	0.8-0.9v
D4	YEL/WHT	MAP Sensor VREF	4.9-5.1v
D5	GRN/WHT	Brake Switch Signal	Brake Off: 0v, On: 12v
D6	RED/BLU	Knock Sensor Signal	No Detonation: 18mv AC
D7	WHT	HO2S-11 (B1 S1) Signal	0.1-1.1v
D8	RED/YEL	IAT Sensor Signal	Varies w/temp. (0.5-4.9v)
D9	WHT/BLK	EGR Valve Lift Sensor	1.2v
D10	YEL/WHT	Sensor VREF	4.9-5.1v
D11	GRN/WHT	Sensor Ground	<0.050v
D12	GRN/WHT	MAP Sensor Ground	<0.050v
D13	GRN/BLU	HO2S-12 Ground	<0.050v
D14	GRN/BLU	HO2S-12 (B1 S2) Signal	0.1-1.1v
D16	GRN/BLK	Electrical Load Detector	Varies: 0.5-4.5v

```
+-----------------------------------------------+     +---------------------------+
| 1  2  3  4 | 5  6  7 | 8  9  10 | 11 |         | 1 | 2  3  4 |    5    |
|12 13 14 15 16 17 18 19 20 21 22| 23  | 24      | 6 7 8 | 9 10 11 | 12 |
|32-PIN 25 26 27|28 29 30 31|   32    |          |13 14 15| 16 | 16-PIN |
|  'A'                                 |          |             'D'       |
+-----------------------------------------------+
          WIRE SIDE OF HARNESS TERMINALS
```

1996 Coupe 2.2L I4 VTEC VIN BB1 [All] 25P 'B' Connector

PCM Pin #	W/Color	Circuit Description (25-Pin)	Value at Hot Idle
B1-2	---	Not Used	---
B3	BLU/YEL	A/T: Shift Solenoid 'A'	SSA Off: 0v, On: 12v
B4	WHT/BLK	A/T: Lockup Solenoid 'B'	LSB On: 12v, Off: 0v
B5	RED/WHT	A/T: Lockup Solenoid 'A'	LSA On: 12v, Off: 0v
B6-7	---	Not Used	---
B8	GRN/BLU	A/T: Gear Position Switch	In D3: 0v, Others: 12v
B9-10	---	Not Used	---
B11	GRN/YEL	A/T: Shift Solenoid 'B'	SSB Off: 0v, On: 12v
B12	WHT/GRN	Interlock Control Unit Signal	Key & Brake On: 12v
B13	GRN/BLK	A/T: D4 Indicator Light Switch	D4 On: 0v, Off: 12v
B14	WHT/BLU	Mainshaft Speed Sensor 'N'	<0.050v
B15	ORN/BLU	Mainshaft Speed Sensor 'P'	AC Pulse Signals
B16	GRN/RED	A/T: Gear Position Switch	In 'R': 0v, Others: 12v
B17	GRN/YEL	A/T: Gear Position Switch	In D2: 0v, Others: 12v
B18	GRN/WHT	A/T: Gear Position Switch	In D1: 0v, Others: 12v
B19-21	---	Not Used	---
B22	BLU/YEL	Countershaft Speed Sensor N	<0.050v
B23	BLU/GRN	Countershaft Speed Sensor P	Moving: AC pulses
B24	GRN/BLK	A/T: Gear Position Switch	In D4: 0v, Others: 12v
B25	LT GRN	A/T: Gear Position Switch	In P/N: 0v, Others: 12v

1996 Coupe 2.2L I4 VTEC VIN BB1 [All] 31P 'C' Connector

PCM Pin #	W/Color	Circuit Description (31-Pin)	Value at Hot Idle
C2	BLU	CKP Sensor Signal	AC pulse signals
C3	GRN	TDC Sensor Signal	AC pulse signals
C4	YEL	CYP Sensor Signal	AC pulse signals
C5	BLU/BLK	A/C Switch Signal	Switch Off: 12v, On: 1v
C6	BLU/RED	Starter Switch Signal	Cranking: 9-11v
C7	BRN/WHT	Service Check Connector	SCS Open: 4.80v
C8	LT GRN	K-Line Signal	12v
C10	WHT/YEL	Keep Alive Power (VBU)	12-14v
C12	WHT	CKP Sensor Ground	<0.050v
C13	RED	TDC Sensor Ground	<0.050v
C14	BLK	CYP Sensor Ground	<0.050v
C15	LT BLU	VTEC Pressure Switch	0v, Hi-Speed: 12v
C16	RED/GRN	PSP Switch	Straight: 0v, Turning: 11v
C17	WHT/RED	Alternator 'FR' Signal	Digital Signals: 0-5-0v
C18	ORN	Vehicle Speed Sensor	Moving: pulse signals
C19	---	Not Used	---
C20	BRN	EVAP Purge Flow Switch	Switch on: 0v, off: 5v
C21-31	---	Not Used	---

1	2		3	4	5		6	7	8		1	2	3	4		5	6	7		8	9	10
9	10	11	12	13	14	15	16	17	18		11	12	13	14	15	16	17	18	19	20	21	22
19	20		21	22		23	24	25			23	24	25		26	27	28		29	30	31

25-PIN 'B'   *WIRE SIDE OF HARNESS TERMINALS*   31-PIN 'C'

1997-99 Coupe 2.2L I4 VTEC VIN BB6 [All] 32P 'A' Connector

PCM Pin #	W/Color	Circuit Description (32-Pin)	Value at Hot Idle
A1	YEL	Injector 4 Control	2.0-3.3 ms
A2	BLU	Injector 3 Control	2.0-3.3 ms
A3	RED	Injector 2 Control	2.0-3.3 ms
A4	BRN	Injector 1 Control	2.0-3.3 ms
A5	ORN/BLU	HO2S-12 (B1 S2) Heater	Digital Signals: 0-12-0v
A6	BLK/WHT	HO2S-11 (B1 S1) Heater	Digital Signals: 0-12-0v
A7	ORN	EGR Solenoid Control	Solenoid Off: 12v, On: 1v
A8	GRN/YEL	VTEC Solenoid Control	0v, Hi-Speed: 12v
A9	BRN/BLK	Logic Ground (LG1)	<0.050v
A10	BLK	Power Ground (PG1)	<0.1v
A11	YEL/BLK	Main Relay Power (IGP1)	12-14v
A12	BLK/BLU	Idle Air Control Valve	Pulse Signals
A13	---	Not Used	---
A14	---	Not Used	---
A15	RED/YEL	EVAP Purge Solenoid	Solenoid Off: 12v, On: 1v
A16	GRN/ORN	Fuel Pump Relay	Relay Off: 12v, On: 1v
A17	PNK/BLU	A/C Clutch Relay Control	Relay Off: 12v, On: 1v
A18	GRY/RED	Malfunction Indicator Light	MIL Off: 12v, On: 1v
A19	WHT/GRN	Alternator Charging Signal	Lights Off: 12v, On: 0v
A20	YEL/GRN	Igniter Control	Digital Signals: 0-12-0v
A21	---	Not Used	---
A22	BRN/BLK	Logic Ground (LG2)	<0.050v
A23	BLK	Power Ground (LG1)	<0.1v
A24	YEL/BLK	Main Relay Power (IGP2)	12-14v
A25	WHT	Intake Air Control Solenoid	Solenoid Off: 12v, On: 1v
A26	RED/BLU	Intake Air Bypass Solenoid	Solenoid Off: 12v, On: 1v
A27	GRN	Radiator Fan Relay Control	Relay Off: 12v, On: 1v
A28	GRN/WHT	EVAP Bypass Solenoid	Solenoid Off: 12v, On: 1v
A29	ORN/GRN	EVAP Vent Solenoid	Solenoid Off: 12v, On: 1v
A30-32	---	Not Used	---

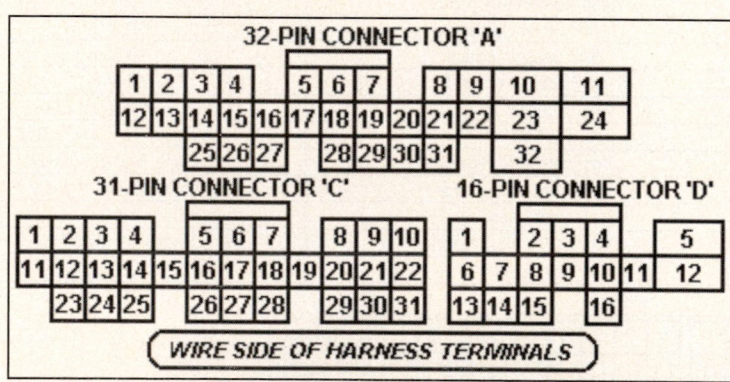

1997-99 Coupe 2.2L I4 VTEC VIN BB6 31P 'C' Connector

PCM Pin #	W/Color	Circuit Description (31-Pin)	Value at Hot Idle
C1	GRN/BLK	VREF Signal to TCM	4.9-5.1v
C2	BLU	CKP Sensor Signal	AC pulse signals
C3	GRN	TDC Sensor Signal	AC pulse signals
C4	YEL	CYP Sensor Signal	AC pulse signals
C5	BLU/ORN	A/C Switch Signal	Switch Off: 12v, On: 1v
C6	BLU/RED	Starter Switch Signal	Cranking: 9-11v
C7	RED/WHT	Service Check Connector	SCS Open: 4.80v
C8	LT GRN	K-Line Signal	12v
C9	---	Not Used	---
C10	WHT/YEL	Keep Alive Power (VBU)	12-14v
C11	---	Not Used	---
C12	WHT	CKP Sensor Ground	<0.050v
C13	RED	TDC Sensor Ground	<0.050v
C14	BLK	CYP Sensor Ground	<0.050v
C15	BLU/BLK	VTEC Pressure Switch	0v, Hi-Speed: 12v
C16	GRN	PSP Switch	Straight: 0v, Turning: 11v
C17	WHT/GRN	Alternator FR Signal	Digital Signals: 0-5-0v
C18	BLU/WHT	Vehicle Speed Sensor	Moving: pulse signals
C19	---	Not Used	---
C20 ('97)	BRN	EVAP Purge Flow Switch	Switch On: 0v, Off: 5v
C22	BRN/YEL	Immobilizer Code Signal	Digital Signals
C23-28	---	Not Used	---
C29	LT GRN	PNP Switch Signal	In P/N: 0v, Others: 12v
C30	GRN/BLU	A/T: TCM SEAF, FITX Signal	Pulse Signals
C31	GRN/YEL	A/T: TCM SEFA, FIRX Signal	Pulse Signals

1997-99 Coupe 2.2L I4 VTEC VIN BB6 16P 'D' Connector

PCM Pin #	W/Color	Circuit Description (16-Pin)	Value at Hot Idle
D1	RED/BLK	TP Sensor Signal	0.5-0.6v
D2	RED/WHT	ECT Sensor Signal	At 180°F: 0.5-0.6v
D3	RED/GRN	MAP Sensor Signal	0.8-0.9v
D4	YEL/RED	MAP Sensor VREF (VCC1)	4.9-5.1v
D5	WHT/BLK	Brake Switch Signal	Brake Off: 0v, On: 12v
D6	RED/BLU	Knock Sensor Signal	No Detonation: 18mv AC
D7	WHT	HO2S-11 (B1 S1) Signal	0.1-1.1v
D8	RED/YEL	IAT Sensor Signal	Varies w/temp. (0.5-4.9v)
D9	WHT/BLK	EGR Valve Lift Sensor	1.2v
D10	YEL/BLU	Sensor VREF	4.9-5.1v
D11	GRN/BLK	Sensor Ground (SG2)	<0.050v
D12	GRN/WHT	MAP Sensor Ground (SG1)	<0.050v
D14	WHT/RED	HO2S-12 (B1 S2) Signal	0.1-1.1v
D15	WHT/BLU	Fuel Tank Pressure Sensor	Fuel Cap off: 2.5v
D16	GRN/RED	Electrical Load Detector	Varies: 0.5-4.5v

2000-01 Coupe 2.2L I4 VTEC VIN BB6 [All] 32P 'A' Connector

PCM Pin #	W/Color	Circuit Description (32-Pin)	Value at Hot Idle
A1	YEL	Injector 4 Control	2.0-3.3 ms
A2	BLU	Injector 3 Control	2.0-3.3 ms
A3	RED	Injector 2 Control	2.0-3.3 ms
A4	BRN	Injector 1 Control	2.0-3.3 ms
A5	ORN/BLU	HO2S-12 (B1 S2) Heater	Digital Signals: 0-12-0v
A6	BLK/WHT	HO2S-11 (B1 S1) Heater	Digital Signals: 0-12-0v
A7	ORN	Electronic EGR Solenoid	Digital Signals: 0-12-0v
A8	GRN/YEL	VTEC Solenoid Control	0v, Hi-Speed: 12v
A9	BRN/BLK	Logic Ground (LG1)	<0.050v
A10	BLK	Power Ground (PG1)	<0.1v
A11	YEL/BLK	Main Relay Power (IGP1)	12-14v
A12	BLK/BLU	Idle Air Control Valve	Pulse Signals
A13	---	Not Used	---
A14	---	Not Used	---
A15	RED/YEL	EVAP Purge Solenoid	Solenoid Off: 12v, On: 1v
A16	GRN/ORN	Fuel Pump Relay	Relay Off: 12v, On: 1v
A17	PNK/BLU	A/C Clutch Relay Control	Relay Off: 12v, On: 1v
A18	GRY/RED	Malfunction Indicator Light	MIL Off: 12v, On: 1v
A19	WHT/GRN	Alternator Charging Signal	Lights Off: 12v, On: 0v
A20	YEL/GRN	Igniter Control	Digital Signals: 0-12-0v
A21	---	Not Used	---
A22	BRN/BLK	Logic Ground (LG2)	<0.050v
A23	BLK	Power Ground (LG1)	<0.1v
A24	YEL/BLK	Main Relay Power (IGP2)	12-14v
A25	WHT	Intake Air Control Solenoid	Solenoid Off: 12v, On: 1v
A26	RED/BLU	Intake Air Bypass Solenoid	Solenoid Off: 12v, On: 1v
A27	GRN	Radiator Fan Relay Control	Relay Off: 12v, On: 1v
A28	GRN/WHT	EVAP Bypass Solenoid	Solenoid Off: 12v, On: 1v
A29	ORN/GRN	EVAP Vent Solenoid	Solenoid Off: 12v, On: 1v
A30-32	---	Not Used	---

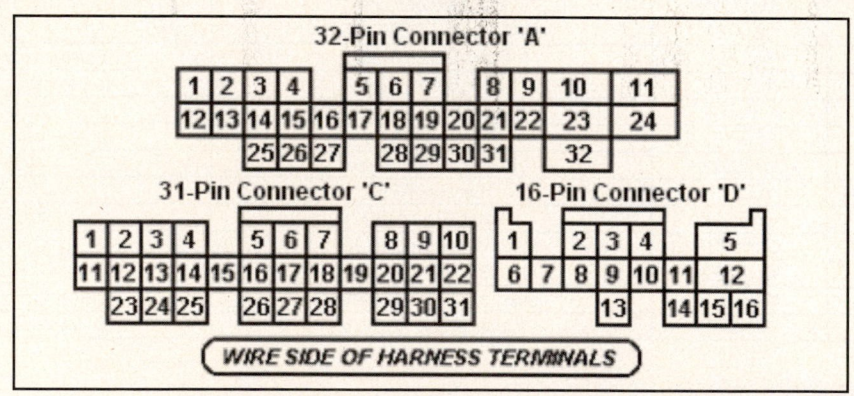

2000-01 Coupe 2.2L I4 VTEC VIN BB6 [All] 31P 'C' Connector

PCM Pin #	W/Color	Circuit Description (31-Pin)	Value at Hot Idle
C1	GRN/BLK	VREF Signal to TCM	4.9-5.1v
C2	BLU	CKP Sensor Signal	AC pulse signals
C3	GRN	TDC Sensor Signal	AC pulse signals
C4	YEL	CYP Sensor Signal	AC pulse signals
C5	BLU/ORN	A/C Switch Signal	Switch Off: 12v, On: 1v
C6	BLU/RED	Starter Switch Signal	Cranking: 9-11v
C7	RED/WHT	Service Check Connector	SCS Open: 4.80v
C8	LT GRN	K-Line Signal	12v
C9	---	Not Used	---
C10	WHT/YEL	Keep Alive Power (VBU)	12-14v
C11	---	Not Used	---
C12	WHT	CKP Sensor Ground	<0.050v
C13	RED	TDC Sensor Ground	<0.050v
C14	BLK	CYP Sensor Ground	<0.050v
C15	BLU/BLK	VTEC Pressure Switch	0v, Hi-Speed: 12v
C16	GRN	PSP Switch	Straight: 0v, Turning: 11v
C17	WHT/GRN	Alternator FR Signal	Digital Signals: 0-5-0v
C18	BLU/WHT	Vehicle Speed Sensor	Moving: pulse signals
C19-20	---	Not Used	---
C22	BRN/YEL	Immobilizer Code Signal	Digital Signals
C23-28	---	Not Used	---
C29	LT GRN	A/T: PNP Switch Signal	In P/N: 0v, Others: 12v
C30	GRN/BLU	A/T: TCM SEAF, FITX Signal	Pulse Signals
C31	GRN/YEL	A/T: TCM SEFA, FIRX Signal	Pulse Signals

2000-01 Coupe 2.2L I4 VTEC VIN BB6 [All] 16P 'D' Connector

PCM Pin #	W/Color	Circuit Description (16-Pin)	Value at Hot Idle
D1	RED/BLK	TP Sensor Signal	0.5-0.6v
D2	RED/WHT	ECT Sensor Signal	At 180°F: 0.5-0.6v
D3	RED/GRN	MAP Sensor Signal	0.8-0.9v
D4	YEL/RED	MAP Sensor VREF (VCC1)	4.9-5.1v
D5	WHT/BLK	Brake Switch Signal	Brake Off: 0v, On: 12v
D6	RED/BLU	Knock Sensor Signal	No Detonation: 18mv AC
D7	WHT	HO2S-11 (B1 S1) Signal	0.1-1.1v
D8	RED/YEL	IAT Sensor Signal	Varies w/temp. (0.5-4.9v)
D9	WHT/BLK	EGR Valve Lift Sensor	1.2v
D10	YEL/BLU	Sensor VREF (VCC2)	4.9-5.1v
D11	GRN/BLK	Sensor Ground (SG2)	<0.050v
D12	GRN/WHT	MAP Sensor Ground (SG1)	<0.050v
D14	WHT/RED	HO2S-12 (B1 S2) Signal	0.1-1.1v
D15	WHT/BLU	Fuel Tank Pressure Sensor	Fuel Cap off: 2.5v
D16	GRN/RED	Electrical Load Detector	Varies: 0.5-4.5v

1992 Coupe 2.3L I4 MFI VIN BB2 [All] 26P 'A' Connector

PCM Pin #	W/Color	Circuit Description (26-Pin)	Value at Hot Idle
A1	BRN	Injector 1 Control	2.0-3.3 ms
A2	YEL	Injector 4 Control	2.0-3.3 ms
A3, 5	RED, BLU	Injector 2 Control, 3	2.0-3.3 ms
A6	ORN/WHT	HO2S Heater Control	Relay Off: 12v, On: 1v
A7, A8	GRN/BLK	Fuel Pump Relay	Relay Off: 12v, On: 1v
A9	BLK/BLU	Electronic Air Control Valve	Pulse Signals
A11	RED	EGR Solenoid	Solenoid Off: 12v, On: 1v
A12	BLU/RED	Radiator Fan Relay Control	Relay Off: 12v, On: 1v
A13	BLU/WHT	Check Engine Light	MIL Off: 12v, On: 1v
A15	RED/BLU	A/C Compressor Unit	---
A16	WHT/GRN	Alternator Charging Signal	Lights Off: 12v, On: 0v
A17	PNK	Intake Air Bypass Solenoid	Solenoid Off: 12v, On: 1v
A18	ORN/RED	A/T: TCM (shift acknowledge)	N/A
A19	WHT	Intake Air Control Solenoid	Solenoid Off: 12v, On: 1v
A20	RED/GRN	EVAP Purge Solenoid	Solenoid Off: 12v, On: 1v
A21, 22	YEL/GRN	Igniter Control	Digital Signals: 0-12-0v
A23, 24	BLK	Power Ground	<0.1v
A25	YEL/BLK	Main Relay Power (B+)	12-14v
A26	BLK/RED	Power Ground	<0.1v

1992 Coupe 2.3L I4 MFI VIN BB2 [All] 16P 'B' Connector

PCM Pin #	W/Color	Circuit Description (16-Pin)	Value at Hot Idle
B1	YEL/BLK	Main Relay Power (B+)	12-14v
B2	BRN/BLK	Power Ground	<0.1v
B3	ORN	A/T: TCM (Upshift compare)	N/A
B4	PNK	A/T: TCM (downshift)	N/A
B5	BLU/BLK	A/C Switch Signal	Switch Off: 12v, On: 1v
B7	LT GRN	A/T: Park/Neutral Switch	In P/N: 0v, Others: 12v
B8	RED/GRN	PSP Switch	Straight: 0v, Turning: 11v
B9	BLU/RED	Starter Switch Signal	Cranking: 9-11v
B10	ORN	Vehicle Speed Sensor	Moving: pulse signals
B11	ORN	CYP Sensor Signal	AC pulse signals
B12	WHT	CYP Sensor Ground	<0.050v
B13	ORN/BLU	TDC Sensor Signal	AC pulse signals
B14	WHT/BLU	TDC Sensor Ground	<0.050v
B15	BLU/GRN	CKP Sensor Signal	AC pulse signals
B16	BLU/YEL	CKP Sensor Ground	<0.050v

Abbreviation	Color	Abbreviation	Color	Abbreviation	Color
BLK	Black	LT BLU	Lt. Blue	TAN	Tan
BLU	Blue	LT GRN	Lt. Green	VIO	Violet
BRN	Brown	ORN	Orange	WHT	White
GRY	Gray	PNK	Pink	YEL	Yellow
GRN	Green	PPL	Purple		

1992 Coupe 2.3L I4 MFI VIN BB2 [All] 22P 'D' Connector

PCM Pin #	W/Color	Circuit Description (22-Pin)	Value at Hot Idle
D1	WHT/YEL	Keep Alive Power (VBU)	12-14v
D2	GRN/WHT	Brake Switch Signal	Brake Off: 0v, On: 12v
D3	RED/BLU	Knock Sensor Signal	No Detonation: 18mv AC
D4	BRN/WHT	Service Check Connector	SCS Open: 4.80v
D5	---	Not Used	---
D6	---	Not Used	---
D7	GRN/RED	Data Link Connector	5v
D8	---	Not Used	---
D9	WHT/RED	Alternator 'FR' Signal	Digital Signals: 0-5-0v
D10	---	Not Used	---
D11	RED/BLK	TP Sensor Signal	0.5-0.6v
D12	WHT/BLK	EGR Valve Lift Sensor	1.2v
D13	YEL/BLU	ECT Sensor Signal	At 180ºF: 0.5-0.6v
D14	WHT	O2S-11 Signal	0.1-1.1v
D15	RED/YEL	IAT Sensor Signal	Varies w/temp. (0.5-4.9v)
D16	---	Not Used	---
D17	WHT/BLU	MAP Sensor Signal	0.8-0.9v
D18	GRN/BLK	A/T: Control Link VREF	4.9-5.1v
D19	RED/WHT	Sensor VREF	4.9-5.1v
D20	YEL/WHT	Sensor VREF	4.9-5.1v
D21	BLU/WHT	Sensor Ground	<0.050v
D22	GRN/WHT	Sensor Ground	<0.050v

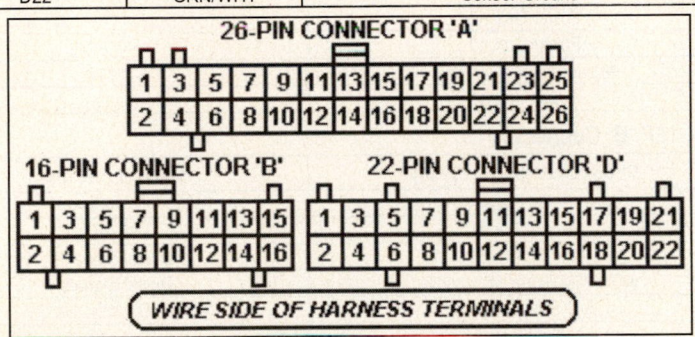

1993-95 Coupe 2.3L I4 MFI VIN BB2 [All] 26P 'A' Connector

PCM Pin #	W/Color	Circuit Description (26-Pin)	Value at Hot Idle
A1	BRN	Injector 1 Control	2.0-3.3 ms
A2	YEL	Injector 4 Control	2.0-3.3 ms
A3	RED	Injector 2 Control	2.0-3.3 ms
A4	---	Not Used	---
A5	BLU	Injector 3 Control	2.0-3.3 ms
A6	ORN/WHT	HO2S Heater Control	Relay Off: 12v, On: 1v
A7	GRN/BLK	Fuel Pump Relay	Relay Off: 12v, On: 1v
A8	GRN/BLK	Fuel Pump Relay	Relay Off: 12v, On: 1v
A9	BLK/BLU	Idle Air Control Valve	Pulse Signals
A10	---	Not Used	---
A11	RED	EGR Solenoid	Solenoid Off: 12v, On: 1v
A12	BLU/RED	Radiator Fan Relay Control	Relay Off: 12v, On: 1v
A13	BLU/WHT	Check Engine Light	MIL Off: 12v, On: 1v
A14	---	Not Used	---
A15	RED/BLU	A/C Compressor Unit	Solenoid Off: 12v, On: 1v
A16	WHT/GRN	Alternator Charging Signal	Lights Off: 12v, On: 0v
A17	PNK	Intake Air Bypass Solenoid	Solenoid Off: 12v, On: 1v
A18	ORN/RED	A/T: TCM Signal	Digital Signals
A19	WHT	Intake Control Solenoid	Solenoid Off: 12v, On: 1v
A20	RED/GRN	EVAP Purge Solenoid	Solenoid Off: 12v, On: 1v
A21	YEL/GRN	Igniter Control	Digital Signals: 0-12-0v
A22	YEL/GRN	Igniter Control	Digital Signals: 0-12-0v
A23	BLK	Power Ground	<0.1v
A24	BLK	Power Ground	<0.1v
A25	YEL/BLK	Main Relay Power (B+)	12-14v
A26	BLK/RED	Power Ground	<0.1v

1993-95 Coupe 2.3L I4 MFI VIN BB2 [All] 16P 'B' Connector

PCM Pin #	W/Color	Circuit Description (16-Pin)	Value at Hot Idle
B1	YEL/BLK	Main Relay Power (B+)	12-14v
B2	BRN/BLK	Power Ground	<0.1v
B3	ORN	A/T: TCM Signal (Upshift)	Digital Signals
B4	PNK	A/T: TCM Signal (downshift)	Digital Signals
B5	BLU/BLK	A/C Switch Signal	Switch Off: 12v, On: 1v
B6	---	Not Used	---
B7	LT GRN	A/T: Park Neutral Switch	P/N: 0v, others: 5v
B8	RED/GRN	PSP Switch	Straight: 0v, Turning: 11v
B9	BLU/RED	Starter Switch Signal	Cranking: 9-11v
B10	ORN	Vehicle Speed Sensor	Moving: pulse signals
B11	ORN	CYP Sensor Signal	AC pulse signals
B12	WHT	CYP Sensor Ground	<0.050v
B13	ORN/BLU	TDC Sensor Signal	AC pulse signals
B14	WHT/BLU	TDC Sensor Ground	<0.050v
B15	BLU/GRN	CKP Sensor Signal	AC pulse signals
B16	BLU/YEL	CKP Sensor Ground	<0.050v

1993-95 Coupe 2.3L I4 MFI VIN BB2 [All] 22P 'D' Connector

PCM Pin #	W/Color	Circuit Description (22-Pin)	Value at Hot Idle
D1	WHT/YEL	Keep Alive Power (VBU)	12-14v
D2	GRN/WHT	Brake Switch Signal	Brake Off: 0v, On: 12v
D3	RED/BLU	Knock Sensor Signal	No Detonation: 18mv AC
D4	BRN/WHT	Service Check Connector	SCS Open: 4.80v
D5-6	---	Not Used	---
D7	GRN/RED	Data Link Connector	5v
D8	---	Not Used	---
D9	WHT/RED	Alternator 'FR' Signal	Digital Signals: 0-5-0v
D10	GRN/RED	Electrical Load Detector	Varies: 0.5-4.5v
D11	RED/BLK	TP Sensor Signal	0.5-0.6v
D12	WHT/BLK	EGR Valve Lift Sensor	1.2v
D13	YEL/BLU	ECT Sensor Signal	At 180°F: 0.5-0.6v
D14	WHT	HO2S Signal	0.1-1.1v
D15	RED/YEL	IAT Sensor Signal	Varies w/temp. (0.5-4.9v)
D16	---	Not Used	---
D17	WHT/BLU	MAP Sensor Signal	0.8-0.9v
D18	GRN/BLK	A/T: Control Unit VRED	4.9-5.1v
D19	RED/WHT	Sensor VREF	4.9-5.1v
D20	YEL/WHT	Sensor VREF	4.9-5.1v
D21	BLU/WHT	Sensor Ground	<0.050v
D22	GRN/WHT	Sensor Ground	<0.050v

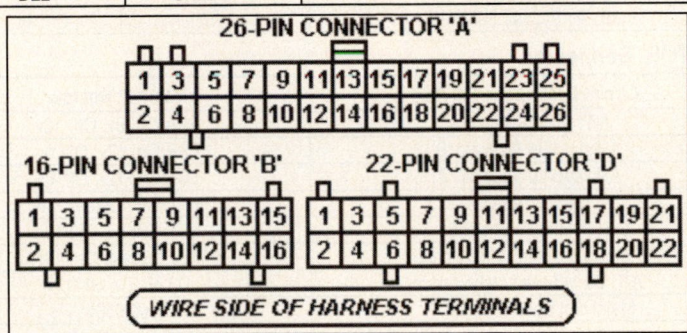

1996 Coupe 2.3L I4 MFI VIN BB2 [All] 32P 'A' Connector

PCM Pin #	W/Color	Circuit Description (32-Pin)	Value at Hot Idle
A1	YEL	Injector 1 Control	2.0-3.3 ms
A2	BLU	Injector 2 Control	2.0-3.3 ms
A3	RED	Injector 3 Control	2.0-3.3 ms
A4	BRN	Injector 4 Control	2.0-3.3 ms
A5	ORN/BLK	HO2S-12 (B1 S2) Heater	Digital Signals: 0-12-0v
A6	ORN/WHT	HO2S-11 (B1 S1) Heater	Digital Signals: 0-12-0v
A7	RED	EGR Solenoid	Solenoid Off: 12v, On: 1v
A9, A22	BLK, BRN	Logic Ground	<0.050v
A10, A23	BLK	Power Ground	<0.1v
A11	YEL/BLK	Main Relay Power (B+)	12-14v
A12	BLK/BLU	Idle Air Control Valve	Pulse Signals
A15	RED/YEL	EVAP Purge Solenoid	Solenoid Off: 12v, On: 1v
A16	GRN/BLK	Fuel Pump Relay	Relay Off: 12v, On: 1v
A17	RED/BLU	A/C Clutch Relay Control	Relay Off: 12v, On: 1v
A18	BLU/WHT	Check Engine Light	MIL Off: 12v, On: 1v
A19	WHT/GRN	Alternator Charging Signal	Lights Off: 12v, On: 0v
A20	YEL/GRN	Igniter Control	Digital Signals: 0-12-0v
A24	YEL/BLK	Main Relay Power (B+)	12-14v
A25	WHT	Intake Air Control Solenoid	Solenoid Off: 12v, On: 1v
A26	PNK	Intake Air Bypass Solenoid	Solenoid Off: 12v, On: 1v
A27	BLU/RED	Radiator Fan Relay Control	Relay Off: 12v, On: 1v
A28	PNK	Idle Air Bypass Solenoid	Solenoid Off: 12v, On: 1v

1996 Coupe 2.3L I4 MFI VIN BB2 [All] 25P 'B' Connector

PCM Pin #	W/Color	Circuit Description (25-Pin)	Value at Hot Idle
B3	BLU/YEL	A/T: Shift Solenoid 'A'	SSA Off: 0v, On: 12v
B4	WHT/BLK	A/T: Lockup Solenoid 'B'	LSB On: 12v, Off: 0v
B5	RED/WHT	A/T: Lockup Solenoid 'A'	LSA On: 12v, Off: 0v
B8	GRN/BLU	A/T: Gear Position Switch	In D3: 0v, Others: 12v
B11	GRN/YEL	A/T: Shift Solenoid 'B'	SSB Off: 0v, On: 12v
B12	WHT/GRN	Interlock Control Unit Signal	Key & Brake On: 12v
B13	GRN/BLK	A/T: D4 Indicator Light Switch	D3 on: 0v, off: 12v
B14	WHT/BLU	Mainshaft Speed Sensor 'N'	<0.050v
B15	ORN/BLU	Mainshaft Speed Sensor 'P'	Moving: AC pulses
B16	GRN/RED	A/T: Gear Position Switch	In 'R': 0v, Others: 12v
B17, 18	GRN/YEL	A/T: Gear Position Switch	In D2, D1: 0v
B22	BLU/YEL	Countershaft Speed Sensor N	<0.050v
B23	BLU/GRN	Countershaft Speed Sensor P	Moving: AC pulses
B24	GRN/BLK	A/T: Gear Position Switch	In D4: 0v, Others: 12v
B25	LT GRN	A/T: Gear Position Switch	In P/N: 0v, Others: 12v

32-PIN CONNECTOR 'A'   **25-PIN CONNECTOR 'B'**

```
┌──┬──┬──┬──┐ ┌──┬──┬──┐ ┌──┬──┬────┬────┐ ┌────┬────┐ ┌──┬──┬──┐ ┌──┬──┬──┐
│ 1│ 2│ 3│ 4│ │ 5│ 6│ 7│ │ 8│ 9│ 10 │ 11 │ │ 1  │ 2  │ │ 3│ 4│ 5│ │ 6│ 7│ 8│
├──┼──┼──┼──┼──┼──┼──┼──┤ ├──┬────┼────┤ ├──┬──┼────┼──┬──┼──┼──┤
│12│13│14│15│16│17│18│19│20│21│22│ 23 │ 24 │ │ 9 │ 10 │11│12│13│14│15│16│17│18│
└──┴──┼──┼──┼──┼──┼──┼──┼──┼──┼────┴────┘ └────┼────┴──┴──┼──┼──┼──┴──┤
      │25│26│27│ │28│29│30│31│ │ 32 │       │19 │ 20 │ │21│22│ │23│24│25│
```

WIRE SIDE OF HARNESS TERMINALS

1996 Coupe 2.3L I4 MFI VIN BB2 [All] 31P 'C' Connector

PCM Pin #	W/Color	Circuit Description (31-Pin)	Value at Hot Idle
C2	BLU	CKP Sensor Signal	AC pulse signals
C3	GRN	TDC Sensor Signal	AC pulse signals
C4	YEL	CYP Sensor Signal	AC pulse signals
C5	BLU/BLK	A/C Switch Signal	Switch Off: 12v, On: 1v
C6	BLU/RED	Starter Switch Signal	Cranking: 9-11v
C7	BRN/WHT	Service Check Connector	SCS Open: 4.80v
C8	LT GRN	K-Line Signal	12v
C10	WHT/YEL	Keep Alive Power (VBU)	12-14v
C12	WHT	CKP Sensor Ground	<0.050v
C13	RED	TDC Sensor Ground	<0.050v
C14	BLK	CYP Sensor Ground	<0.050v
C16	RED/GRN	PSP Switch	Straight: 0v, Turning: 11v
C17	WHT/RED	Alternator 'FR' Signal	Digital Signals: 0-5-0v
C18	ORN	Vehicle Speed Sensor	Moving: pulse signals
C20	BRN	EVAP Purge Flow Switch	Switch on: 0v, off: 5v

1996 Coupe 2.3L I4 MFI VIN BB2 [All] 16P 'D' Connector

PCM Pin #	W/Color	Circuit Description (16-Pin)	Value at Hot Idle
D1	RED/BLK	TP Sensor Signal	0.5-0.6v
D2	YEL/BLU	ECT Sensor Signal	At 180ºF: 0.5-0.6v
D3	WHT/YEL	MAP Sensor Signal	0.8-0.9v
D4	YEL/WHT	MAP Sensor VREF	4.9-5.1v
D5	GRN/WHT	Brake Switch Signal	Brake Off: 0v, On: 12v
D6	RED/BLU	Knock Sensor Signal	No Detonation: 18mv AC
D7	WHT	HO2S-11 (B1 S1) Signal	0.1-1.1v
D8	RED/YEL	IAT Sensor Signal	Varies w/temp. (0.5-4.9v)
D9	WHT/BLK	EGR Valve Lift Sensor	1.2v
D10	YEL/WHT	Sensor VREF	4.9-5.1v
D11	GRN/WHT	Sensor Ground	<0.050v
D12	GRN/WHT	MAP Sensor Ground	<0.050v
D13	GRN/BLU	HO2S-12 Ground	<0.050v
D14	GRN/BLU	HO2S-12 (B1 S2) Signal	Accel: 0.5-1v
D16	GRN/BLK	Electrical Load Detector	Varies: 0.5-4.5v

31-PIN CONNECTOR 'C'   16-PIN CONNECTOR 'D'

1	2	3	4		5	6	7		8	9	10		1		2	3	4		5
11	12	13	14	15	16	17	18	19	20	21	22		6	7	8	9	10	11	12
	23	24	25		26	27	28		29	30	31		13	14	15		16

WIRE SIDE OF HARNESS TERMINALS

S2000 Pin Tables

2000-03 Convertible 2.0L I4 VTEC VIN API [M/T] 32P A Connector

PCM Pin #	W/Color	Circuit Description (32-Pin)	Value at Hot Idle
A1	YEL/GRN	ECT Sensor Signal to Gauge	Digital Signals
A2	RED	Air Control Solenoid Valve	Solenoid Off: 12v, On: 1v
A3	ORN	EVAP Bypass Solenoid	Solenoid Off: 12v, On: 1v
A4	GRN/WHT	EVAP Vent Solenoid	Solenoid Off: 12v, On: 1v
A5	---	Not Used	---
A6	RED/YEL	EVAP Purge Solenoid	Solenoid Off: 12v, On: 1v
A7-8	---	Not Used	---
A9	BLU/WHT	Vehicle Speed Output Signal	Digital Pulses
A10	BRN	Service Check Connector	SCS Open: 4.80v
A11	---	Not Used	---
A12	PNK	Immobilizer Indicator Lamp	Lamp Off: 12v, On: 1v
A13	PNK/BLU	Immobilizer Enable Signal	Digital Signals
A14	---	Not Used	---
A15	GRN/YEL	Immobilizer Fuel Pump Relay	2 sec's after startup: 12v
A16	---	Not Used	---
A17	RED	A/C Clutch Relay Control	Relay Off: 12v, On: 1v
A18	GRN/ORN	Malfunction Indicator Light	MIL Off: 12v, On: 1v
A19	BLU	Engine Speed Pulse (NEP)	Digital Signals
A20	GRN	Radiator Fan Relay Control	Relay Off: 12v, On: 1v
A21	GRY	K-Line Signal	12v
A22-23	---	Not Used	---
A24	BLU/ORN	Starter Switch Signal	Cranking: 9-11v
A25	RED/BLU	Immobilizer Code Signal	Digital Signals
A26	BLU/BLK	PSP Switch Signal	Straight: 0v, Turning: 11v
A27	BLU/RED	A/C Switch Signal	Switch Off: 12v, On: 1v
A28	BLU	Air Pump Relay	Relay Off: 12v, On: 1v
A29	LT GRN	Fuel Tank Pressure Sensor	Fuel Cap off: 2.5v
A30	GRN/RED	Electric Load Detector	Varies: 0.5-4.5v
A31	---	Not Used	---
A32	WHT/BLK	Brake Switch Signal	Brake Off: 0v, On: 12v

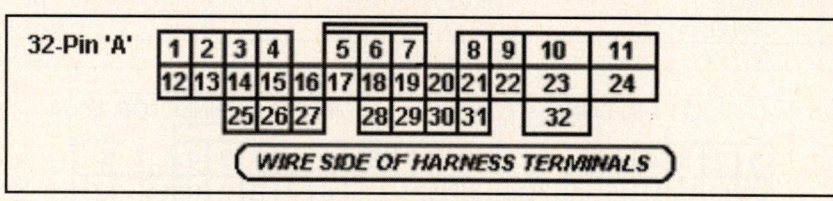

2000-03 Convertible 2.0L I4 VTEC VIN API [M/T] 25P B Connector

PCM Pin #	W/Color	Circuit Description (31-Pin)	Value at Hot Idle
B1	YEL/BLK	Main Relay Power (B+)	12-14v
B2	BLK	Power Ground (PG1)	<0.1v
B3	RED, BLU	Injector 2, Injector 3 Control	2.0-3.3 ms
B4	RED, BLU	Injector 2, Injector 3 Control	2.0-3.3 ms
B5	YEL	Injector 4 Control	2.0-3.3 ms
B7	---	Not Used	---
B8	---	Not Used	---
B9	YEL/BLK	Main Relay Power (B+)	12-14v
B10	BLK	Power Ground (PG1)	<0.1v
B11	BRN	Injector 1 Control	2.0-3.3 ms
B12	GRN/YEL	VTEC Control Solenoid	0v, Hi-Speed: 12v
B13	---	Not Used	---
B14	---	Not Used	---
B15	---	Not Used	---
B16	---	Not Used	---
B17	---	Not Used	---
B18	---	Not Used	---
B19	---	Not Used	---
B20	BRN/YEL	Logic Ground (LG1)	<0.050v
B21	WHT/RED	Keep Alive Power (VBU)	12-14v
B22	BRN/YEL	Logic Ground (LG1)	<0.050v
B23	BLK/RED	Idle Air Control Valve	Pulse Signals
B24	---	Not Used	---
B25	---	Not Used	---

25-Pin 'B'

1	2	3	4	5		6	7	8
9	10	11	12	13	14	15	16	17	18
19	20		21	22		23	24	25

WIRE SIDE OF HARNESS TERMINALS

2000-03 Convertible 2.0L I4 VTEC VIN API [M/T] 31P C Connector

PCM Pin #	W/Color	Circuit Description (31-Pin)	Value at Hot Idle
C1	BLK/WHT	HO2S-11 (B1 S1) Heater	Digital Signals: 0-12-0v
C2	WHT/GRN	Alternator Charging Signal	Lights Off: 12v, On: 0v
C4	WHT	Ignition Coil 1 Control	Digital Signals: 0-12-0v
C5	WHT/RED	Alternator 'FR' Signal	Digital Signals: 0-5-0v
C7	GRN/WHT	MAP Sensor Ground (SG1)	<0.050v
C8	BLU	CKP Sensor Signal	AC pulse signals
C9	WHT	CKP Sensor Ground	<0.050v
C10	BLU/BLK	VTEC Pressure Switch	0v, Hi-Speed: 12v
C11	BLK/WHT	HO2S-12 (B1 S2) Heater	Digital Signals: 0-12-0v
C12	WHT/GRN	Ignition Coil 2 Control	Digital Signals: 0-12-0v
C13	WHT/BLK	Ignition Coil 3 Control	Digital Signals: 0-12-0v
C14	WHT/BLU	Ignition Coil 4 Control	Digital Signals: 0-12-0v
C15	WHT/RED	HO2S-12 (B1 S2) Signal	0.1-1.1v
C16	WHT	HO2S-11 (B1 S1) Signal	0.1-1.1v
C17	GRN/RED	MAP Sensor Signal	0.8-0.9v
C18	GRN/YEL	Sensor Ground (SG2)	<0.050v
C19	YEL/RED	MAP Sensor VREF (VCC1)	4.9-5.1v
C20	GRN	TDC1 Sensor Signal	AC pulse signals
C21	RED	TDC1 Sensor Ground	<0.050v
C22	RED/BLU	Knock Sensor Signal	No Detonation: 18mv AC
C24	WHT/BLK	Air Pump Current Sensor	Pump on: 2-5v
C25	RED/YEL	IAT Sensor Signal	Varies w/temp (0.5-4.9v)
C26	RED/WHT	ECT Sensor Signal	At 180°F: 0.5-0.6v
C27	RED/BLK	TP Sensor Signal	0.3-0.4v
C28	YEL/BLU	Sensor VREF (VCC2)	4.9-5.1v
C29	YEL	TDC2 Sensor Signal	AC pulse signals
C30	BLK	TDC2 Sensor Ground	<0.050v
C31	---	Not Used	---

31-Pin 'C'

1	2	3	4		5	6	7		8	9	10
11	12	13	14	15	16	17	18	19	20	21	22
23	24	25		26	27	28		29	30	31	

WIRE SIDE OF HARNESS TERMINALS

ACURA AND HONDA SUV CONTENTS

About This Section

Introduction

This section contains Pin Voltage Tables for Acura and Honda SUVs from 1990-2003. It can be used to help you repair Trouble Code and No Code problems related to the PCM.

How to Use This Section

This section can be used to look up the location of a particular pin, a wire color or to find a "known good" value of a circuit. To locate the PCM information for a particular vehicle, find the model, correct engine size (with VIN Code) and finally the year of the vehicle.

For example, to look up the PCM terminals for a 1998 Passport 2WD 3.2L MFI VIN W, go to Contents Page 1 to find this text string.

Then turn to Page 8-25 to find the following PCM related information.

The symbol (<) in used in some pin voltage readings to indicate that the reading should be less than the number shown in the example.

The transmission type may be identified in the table title as [A/T] Auto, [M/T] Manual or [All]. Refer to the example shown in the table below.

1998-2001 EX, LX 3.2L VIN W 2WD [A/T] 80P J21 Blue Connector

PCM Pin #	W/Color	Circuit Description (80-Pin)	Value at Hot Idle
12	WHT/RED	CMP Sensor Signal	Digital Signals: 0-5-0v
13	PNK	To be done	---
14	BLK/BLU	MAF Sensor Signal	4.2v
15-16	---	Not Used	---
17	RED	HO2S-22 (B2 S2) Ground	0.1-1.1v

In this example, the CMP Sensor signal is connected to Pin 12 of the PCM 80-Pin connector with a WHT/RED. The MAF Sensor signal is connected to Pin 14 of the same connector with a BLK/BLU wire.

MDX Pin Voltage Tables

2001-03 Utility 3.5L V6 VTEC VIN YD1 [A/T] 32-Pin 'A' Connector

PCM Pin #	W/Color	Circuit Description (32-Pin)	Value at Hot Idle
A1	WHT/RED	HO2S-22 (B2 S2) Signal	0.1-1.1v
A2	RED/GRN	ATF Indicator Light	Lamp Off: 12v, On: 1v
A3	ORN/WHT	EVAP Bypass Solenoid	Solenoid Off: 12v, on: 1v
A4	GRN/WHT	EVAP Vent Solenoid	Solenoid Off: 12v, on: 1v
A5	BLU/GRN	Cruise Control Signal	Cruise On: pulse signals
A6	RED/YEL	EVAP Purge Solenoid	Pulse Signals
A7	BLK	Serial Data Line (VTM-4 Unit)	Digital Signals
A8	BLK/WHT	HO2S-12 (B1 S2) Heater	Digital Signals: 0-12-0v
A9	BLU/WHT	Vehicle Speed Sensor	Digital Signals: 0-5-0v
A10	BRN	Service Check Connector	SCS Open: 4.80v
A11	PNK	FUP Communication Signals	Digital Signals
A12	---	Not Used	---
A13	GRN/WHT	Intake Manifold Runner	Less than 3200 rpm: 5v
A14	GRN/BLK	A/T: D5 Indicator Light	In D5: 0v, Others: 12v
A15	GRN/YEL	Immobilizer Fuel Pump Relay	Relay Off: 12v, On: 1v
A16	---	Not Used	---
A17	RED	A/C Clutch Relay Control	Relay Off: 12v, On: 1v
A18	GRN/ORN	Malfunction Indicator Light	MIL Off: 12v, On: 1v
A19	GRN/BLU	Engine Speed Signal	Pulse Signals
A20	GRN	Radiator Fan Relay Control	Relay Off: 12v, On: 1v
A21	GRY	K-Line Signal	12v
A22	WHT	Serial Data Line (VTM-4 Unit)	Digital Signals
A23	---	Not Used	---
A24	BLU/ORN	Starter Switch Signal	Cranking: 9-11v
A25	---	Not Used	---
A26	LT BLU	PSP Switch Signal	Straight: 0v, Turning: 12v
A27	BLU/RED	A/C Switch Signal	A/C Off: 12v, On: 0v
A28	WHT/GRN	Interlock Control Unit Signal	Key and Brake On: 12v
A29	LT GRN	Fuel Tank Pressure Sensor	With Fuel Cap Off: 2.5v
A30	GRN/RED	Electronic Load Detector	Varies 2.5-3.5v
A31	YEL/GRN	ECT Signal to ECT Gauge	Digital Signals
A32	WHT/BLK	Brake Switch Signal	Brake Off: 12v, On: 0v

2001-03 Utility 3.5L V6 VTEC VIN YD1 [A/T] 25-Pin 'B' Connector

PCM Pin #	W/Color	Circuit Description (25-Pin)	Value at Hot Idle
B1, B9	YEL/BLK	B+ Main Relay IGP1, IGP2	12-14v
B2, B10	BLK	Power Ground PG1, PG2	<0.1v
B3	BLK/RED	Injector 5 Control	2.0-3.3 ms
B4	YEL	Injector 4 Control	2.0-3.3 ms
B5	RED	Injector 2 Control	2.0-3.3 ms
B6	WHT/BLU	Injector 6 Control	2.0-3.3 ms
B7	BLU/RED	E-EGR Control Solenoid	Digital Signals: 0-12-0v
B8	WHT	A/T: Clutch Solenoid 'A-'	Pulse Signals
B11	BRN	Injector 1 Control	2.0-3.3 ms
B12	GRN/YEL	VTEC Solenoid Control	Idle: 0v, Hi-Speed: 12v
B13	GRN/RED	A/T: Clutch Solenoid 'C+'	Pulse Signals
B14	BLU/WHT	A/T: Gear Position Switch	In P/N: 0v, Others: 5v
B15	BLU	Injector 3 Control	2.0-3.3 ms
B16	---	Not Used	---
B17	RED	A/T: Clutch Solenoid 'A+'	Pulse Signals
B18	GRN	A/T: Clutch Solenoid 'B-'	Pulse Signals
B19	BLU/YEL	A/T: 4th Oil Pressure Switch	12-14v
B20, B22	BRN/YEL	Logic Ground (LG1), (LG2)	<0.050v
B21	WHT/RED	Keep Alive Power (VBU)	12-14v
B23	BLK/RED	Idle Air Control Valve	Pulse Signals
B24	RED/BLU	A/T: Clutch Solenoid 'A-'	Pulse Signals
B25	BRN/WHT	A/T: Clutch Solenoid 'B+'	Pulse Signals

2001-03 Utility 3.5L V6 VTEC VIN YD1 [A/T] 20-Pin 'E' Connector

PCM Pin #	W/Color	Circuit Description (20-Pin)	Value at Hot Idle
E1	PNK	Immobilizer Indicator Lamp	Lamp Off: 12v, On: 1v
E2	RED	IMOCD Immobilizer Code	Digital Signals
E3-12	---	Not Used	---
E12	BLU	Immobilizer Enable Signal	Digital Signals
E14	BRN	A/T: First Hold Down Switch	LH Switch on: 0v, off: 12v
E16-20	---	Not Used	---

2001-03 Utility 3.5L V6 MFI VIN YD1 [A/T] 31-Pin 'C' Connector

PCM Pin #	W/Color	Circuit Description (31-Pin)	Value at Hot Idle
C1	BLK/WHT	HO2S-11 (B1 S1) Heater	Digital Signals: 0-12-0v
C2	WHT/GRN	Alternator Charging Signal	Lights on: 0v, off: 12v
C3	WHT/BLU	Ignition Coil 3 Control	Digital Signals: 0-12-0v
C4	YEL/GRN	Ignition Coil 1 Control	Digital Signals: 0-12-0v
C5	WHT/RED	Alternator 'FR' Signal	Digital Signal: 0-5-0-5v
C6	WHT/BLK	EGR Valve Lift Sensor	1.2-2.0v
C7	GRN/WHT	Sensor Ground (SG1)	<0.050v
C8	BLU	CKP Sensor 'P' Signal	0.900v AC
C9	WHT	CKP Sensor 'N' Signal	<0.050v
C10	BLU/BLK	VTEC Pressure Switch	Idle: 0v, Cruise: 12v
C11, 15	---	Not Used	---
C12	BLK/RED	Ignition Coil 5 Control	Digital Signals: 0-12-0v
C13	BRN	Ignition Coil 4 Control	Digital Signals: 0-12-0v
C14	BLU/RED	Ignition Coil 2 Control	Digital Signals: 0-12-0v
C16	WHT	HO2S-11 (B1 S1) Signal	Varies: 0.1-1v
C17	GRN/RED	MAP Sensor Signal	0.8-0.9v
C18	GRN/YEL	Sensor Ground (SG2)	<0.050v
C19	YEL/RED	Sensor VREF (VCC1)	4.9-5.1v
C20	GRN	TDC1 Sensor 'P' Signal	1.00v AC
C21	RED	TDC1 Sensor 'N' Signal	<0.050v
C22	RED/BLU	Knock Sensor Signal	No Detonation: 18mv AC
C23	BRN/WHT	Ignition Coil 6 Control	Digital Signals: 0-12-0v
C24	BLU/YEL	A/T: TFT Sensor Signal	0.1-4.2v
C25	RED/YEL	IAT Sensor Signal	Varies w/temp: 0.5-4.8v
C26	RED/WHT	ECT Sensor Signal	At 180°F: 0.5-0.6v
C27	RED/BLK	TP Sensor	0.5-0.6v
C28	YEL/BLU	Sensor VREF (VCC2)	4.9-5.1v
C29	YEL	TDC2 Sensor 'P' Signal	0.250v AC
C30	BLK	TDC2 Sensor 'N' Signal	<0.050v

2001-03 Utility 3.5L V6 VTEC VIN YD1 [A/T] 16-Pin 'D' Connector

PCM Pin #	W/Color	Circuit Description (16-Pin)	Value at Hot Idle
D1	YEL	A/T: Lockup Solenoid	LUS Off: 12v, On: 1v
D2	GRN/WHT	A/T: Shift Solenoid 'B'	SSB Off: 12v, On: 1v
D3	GRN	A/T" Shift Solenoid 'C'	SSA Off: 12v, On: 1v
D4	RED/BLK	A/T: 'N' Gear Position Switch	In 'N': 0v, Others: 5v
D5	BLK/YEL	Solenoid Feed (B+)	12-14v
D6	WHT	A/T: 'R' Gear Position Switch	In 'R': 0v, Others: 5v
D7	BLU/YEL	A/T: Shift Solenoid 'A'	SSA Off: 12v, On: 1v
D8	YEL	A/T: D4 Gear Position Switch	In D4: 0v, Others: 5v
D9	YEL/GRN	A/T: D5 Gear Position Switch	In D5: 0v, Others: 5v
D10	BLU	Countershaft Speed Sensor P	Wheels turning: pulses
D11	RED	Mainshaft Speed Sensor 'P'	Pulse Signals
D12	WHT	Mainshaft Speed Sensor 'N'	<0.050v
D13	BLU/WHT	A/T: 3rd Oil Pressure Switch	12-14v
D14	RED	A/T: D3 Gear Position Switch	In D3: 0v, Others: 5v
D15	BLU	A/T: D2 Gear Position Switch	In D2: 0v, Others: 5v
D16	GRN	Countershaft Speed Sensor N	<0.050v

SLX Pin Voltage Tables

1996-97 4-Door 3.2L V6 4WD VIN V [A/T] 32P Red 'A' Connector

PCM Pin #	W/Color	Circuit Description (16-Pin)	Value at Hot Idle
A1	RED	Reference Voltage 'A'	4.9-5.1v
A2	YEL	Knock Sensor Signal	No detonation: 18 mv AC
A3	---	Not Used	---
A4	RED/WHT	Keep Alive Power	12-14v
A5	BLU	Idle Air Control 'A' High	Pulse Signals
A6	BLU/WHT	Idle Air Control 'A' Low	Pulse Signals
A7	BLU/BLK	Idle Air Control 'B' Low	Pulse Signals
A8	BLU/RED	Idle Air Control 'B' High	Pulse Signals
A9	PNK	Transmission Fluid Lamp	Lamp Off: 12v, On: 1v
A10	PNK/GRY	Winter Lamp	Lamp On: 1v. off: 12v
A11	GRN/WHT	Power Lamp	Lamp On: 1v. off: 12v
A12	PPL/WHT	AntiLock Brake Light	Lamp Off: 12v, On: 1v
A13	PPL	Malfunction Indicator Lamp	MIL Off: 12v, On: 1v
A14	PNK	Check Transmission Light	Lamp on: 1v, off: 12v
A15	RED/BLU	EVAP Purge Solenoid	Digital Signals: 0-12-0v
A16	BRN/YEL	A/T: Shift Low Band Apply	Sol. on: 1v, off: 12v

1996-97 4-Door 3.2L V6 4WD VIN V [A/T] 32P Red 'B' Connector

PCM Pin #	W/Color	Circuit Description (16-Pin)	Value at Hot Idle
B1	WHT	Reference Voltage 'B'	4.9-5.1v
B2	RED/WHT	Igniter 4 Control	Digital Signals: 0-12-0v
B3	RED/BLU	Igniter 2 Control	Digital Signals: 0-12-0v
B4	RED/GRN	Igniter 6 Control	Digital Signals: 0-12-0v
B5	---	Not Used	---
B6	---	Not Used	---
B7	YEL/RED	Linear EGR Sensor Signal	0.6-0.7v
B8	YEL/GRN	IAT Sensor Signal	Varies w/temp: 0.5-4.8v
B9	---	Not Used	---
B10	GRN/RED	Rough Road Sensor	2.5v
B11	GRN/YEL	PSP Switch Signal	Straight: 12v, Turning: 0v
B12	GRN	Illuminated Switch Signal	Switch on: 0.1v, off: 12v
B13	ORN/GRN	Class 2 Serial Data Link	0v
B14	GRN/RED	A/C Clutch Relay	Relay Off: 12v, On: 1v
B15	---	Not Used	---
B16	---	Not Used	---

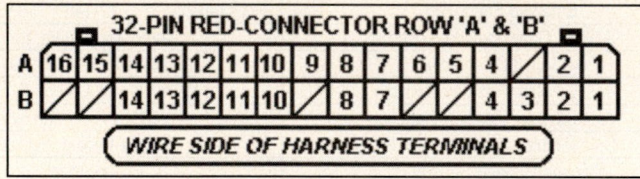

1996-97 4-Door 3.2L 4WD VIN V [A/T] 32P White 'C' Connector

PCM Pin #	W/Color	Circuit Description (16-Pin)	Value at Hot Idle
C1	GRN/RED	Injector 4 Control	2.0-3.3 ms
C2	BRN/BLK	A/T: Shift Solenoid 'B'	SSB Off: 12v, On: 1v
C3	GRN/YEL	Injector 6 Control	2.0-3.3 ms
C4	RED	Igniter 1 Control	Digital Signals: 0-12-0v
C5	YEL	CKP Sensor Signal	Digital Signals: 0-5-0v
C7	BLK/BLU	Power Ground	<0.1v
C8	BLK/PNK	Power Ground	<0.1v
C9	BLK/BLU	Power Ground	<0.1v
C10	BLK/RED	Tachometer Signal	Pulse Signals
C11	---	Not Used	---
C12	WHT/BLU	Alternator Charge Signal	12-14v
C13	YEL/RED	Fuel Gauge PWM Signal	Duty Cycle: 0-100%
C14	PNK	HO2S-21 (B2 S1) Signal	0.1-1.1v
C15	BLU	HO2S-21 (B2 S1) Ground	<0.050v
C16	GRN	HO2S-22 (B2 S2) Signal	0.1-1.1v

1996-97 4-Door 3.2L 4WD VIN V [A/T] 32P White 'D' Connector

PCM Pin #	W/Color	Circuit Description (16-Pin)	Value at Hot Idle
D1	GRN/ORN	Injector 2 Control	2.0-3.3 ms
D2	BRN/BLU	Torque Converter Clutch	TCC on: 1v, off: 12v
D3	GRN/WHT	Injector 1 Control	2.0-3.3 ms
D4	ORN	Serial Data (8192 Baud)	Digital Signals
D5	RED/YEL	Igniter 5 Control	Digital Signals: 0-12-0v
D6	RED/BLU	Igniter 3 Control	Digital Signals: 0-12-0v
D7	WHT/BLK	VSS (4096 pulses per mile)	Moving: 0-5-0-5v
D8	GRN	Sensor 'A' Ground	<0.050v
D9	BLU/YEL	Sensor 'B' Ground	<0.050v
D10	YEL/BLU	MAF Sensor Signal	7-9v
D11	BLU	CMP Sensor Signal	Digital Signals: 0-5-0v
D12	RED	HO2S-12 (B1 S2) Ground	<0.050v
D13	WHT	HO2S-12 (B1 S2) Signal	0.1-1.1v
D14	GRN	HO2S-11 (B1 S1) Ground	<0.050v
D15	RED	HO2S-11 (B1 S1) Signal	0.1-1.1v
D16	BLU	HO2S-22 (B2 S2) Ground	<0.050v

```
  ┌─ 32-PIN WHITE-CONNECTOR ROW 'C' & 'D' ─┐
C │16│15│14│13│12│ / │10│ 9│ 8│ 7│ / │ 5│ 4│ 3│ 2│ 1│
D │16│15│14│13│12│11│10│ 9│ 8│ 7│ 6│ 5│ 4│ 3│ 2│ 1│
  └──────────────────────────────────────────┘
      ( WIRE SIDE OF HARNESS TERMINALS )
```

1996-97 4-Door 3.2L V6 4WD VIN V [A/T] 32P Blue 'E' Connector

PCM Pin #	W/Color	Circuit Description (16-Pin)	Value at Hot Idle
E1	RED	Output Speed Sensor Signal	Moving: AC Pulses
E2	WHT	Output Speed Sensor Ground	<0.050
E3	PPL/RED	A/T: Pressure Control Low	Pulse Signals
E4	PPL/WHT	A/T: Pressure Control High	Pulse Signals
E5	BLK/YEL	Linear EGR Solenoid "High"	Pulse Signals
E6	YEL/PNK	Linear EGR Solenoid "Low"	Pulse Signals
E7	PNK	A/T: Range Signal 'B'	In P/N: 1v, others: 12v
E8	BLU	TP Sensor Signal	0.5-0.6v
E9	BLU/RED	ECT Sensor Signal	At 180°F: 0.54v
E10	---	Not Used	---
E11	YEL/RED	CKP Sensor VREF	4.9-5.1v
E12	PNK/BLU	A/T: Range Signal 'A'	In P/N: 12-14v
E13	RED/WHT	Fuel Pump Relay Control	Relay on: 12v, off: 0v
E14	BRN/WHT	A/T: Shift Solenoid Feed (B+)	12-14v
E15	GRN/ORN	A/C Switch Signal	Switch on: 12v, off: 0v
E16	RED/BLU	B+ from Main Relay	12-14v

1996-97 4-Door 3.2L V6 4WD VIN V [A/T] 32P Blue 'F' Connector

PCM Pin #	W/Color	Circuit Description (16-Pin)	Value at Hot Idle
F1	---	Not Used	---
F2	PNK/YEL	A/T: Range Signal 'C'	In P/N: 1v, others: 12v
F3	PNK/BLK	A/T: Range Signal 'P'	In P/N: 1v, others: 12v
F4	GRN/YEL	Brake Switch Signal	Brake on: 12v, off: 0v
F5	PPL/GRN	Power Switch Signal	Switch on: 0.1v, off: 12v
F6	PPL/GRN	Winter Switch Signal	Switch on: 0.1v, off: 12v
F7	GRN/YEL	A/T: TOT Sensor Signal	Varies w/temp: 0.5-4.8v
F8	RED	MAP Sensor Signal	0.6-1.3v
F9	YEL/BRN	Vacuum Switch Signal	Switch On: 0.1v, off: 12v
F10	GRY/BLU	Cruise Control	Switch On: 0.1v, off: 12v
F11	BLK	A/T: Kickdown Switch	Switch On: 0.1v, off: 12v
F12	---	Not Used	---
F13	GRN	Injector 3 Control	2.0-3.3 ms
F14	BRN/RED	A/T: Shift Solenoid 'A'	SSA Off: 12v, On: 1v
F15	GRN/BLK	Injector 5 Control	2.0-3.3 ms
F16	RED/BLU	B+ from Main Relay	12-14v

```
          ┌─ 32-PIN BLUE-CONNECTOR ROW 'E' & 'F' ─┐
        E │16│15│14│13│12│11│ / │ 9│ 8│ 7│ 6│ 5│ 4│ 3│ 2│ 1│
        F │16│15│14│13│ / │11│10│ 9│ 8│ 7│ 6│ 5│ 4│ 3│ 2│
          └──── WIRE SIDE OF HARNESS TERMINALS ────┘
```

1996-97 4-Door 3.2L V6 4WD VIN V [M/T] 32P Red 'A' Connector

PCM Pin #	W/Color	Circuit Description (16-Pin)	Value at Hot Idle
A1	RED	Reference Voltage 'A'	4.9-5.1v
A2	YEL	Knock Sensor Signal	No detonation: 18 mv AC
A3	---	Not Used	---
A4	RED/WHT	Keep Alive Power	12-14v
A5	BLU	Idle Air Control 'A' High	Pulse Signals
A6	BLU/WHT	Idle Air Control 'A' Low	Pulse Signals
A7	BLU/BLK	Idle Air Control 'B' Low	Pulse Signals
A8	BLU/RED	Idle Air Control 'B' High	Pulse Signals
A9-12	---	Not Used	---
A13	PPL	Malfunction Indicator Lamp	MIL Off: 12v, On: 1v
A14	---	Not Used	---
A15	RED/BLU	EVAP Purge Solenoid	Digital Signals: 0-12-0v
A6	---	Not Used	---

1996-97 4-Door 3.2L V6 4WD VIN V [M/T] 32P Red 'B' Connector

PCM Pin #	W/Color	Circuit Description (16-Pin)	Value at Hot Idle
B1	WHT	Reference Voltage 'B'	4.9-5.1v
B2	RED/WHT	Igniter 4 Control	Digital Signals: 0-12-0v
B3	RED/BLU	Igniter 2 Control	Digital Signals: 0-12-0v
B4	RED/GRN	Igniter 6 Control	Digital Signals: 0-12-0v
B5	---	Not Used	---
B6	---	Not Used	---
B7	YEL/RED	Linear EGR Sensor Signal	0.5-0.6v
B8	YEL/GRN	IAT Sensor Signal	Varies w/temp: 0.5-4.8v
B9	---	Not Used	---
B10	GRN/RED	Rough Road Sensor	2.5v
B11	GRN/YEL	PSP Switch Signal	Straight: 12v, Turning: 0v
B12	GRN	Illuminated Switch Signal	Switch on: 0.1v, off: 12v
B13	ORN/GRN	Class 2 Serial Data Link	0v
B14	GRN/RED	A/C Clutch Relay Control	Relay on: 12v, off: 0v
B15	---	Not Used	---
B16	---	Not Used	---

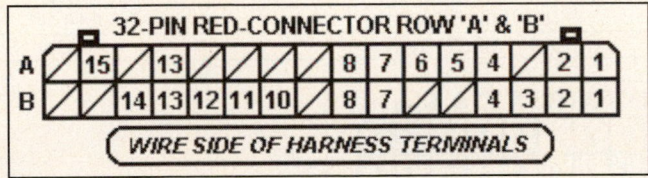

Abbreviation	Color	Abbreviation	Color	Abbreviation	Color
BLK	Black	LT BLU	Lt. Blue	TAN	Tan
BLU	Blue	LT GRN	Lt. Green	VIO	Violet
BRN	Brown	ORN	Orange	WHT	White
GRY	Gray	PNK	Pink	YEL	Yellow
GRN	Green	PPL	Purple		

1996-97 4-Door 3.2L 4WD VIN V [M/T] 32P White 'C' Connector

PCM Pin #	W/Color	Circuit Description (16-Pin)	Value at Hot Idle
C1	GRN/RED	Injector 4 Control	2.0-3.3 ms
C2	---	Not Used	---
C3	GRN/YEL	Injector 6 Control	2.0-3.3 ms
C4	RED	Igniter 1 Control	Digital Signals: 0-12-0v
C5	YEL	CKP Sensor Signal	Digital Signals: 0-5-0v
C6	---	Not Used	---
C7	BLK/BLU	Power Ground	<0.1v
C8	BLK/PNK	Power Ground	<0.1v
C9	BLK/BLU	Power Ground	<0.1v
C10	BLK/RED	Tachometer Signal	Pulse Signals
C11	ORN/BLU	Upshift Lamp Control	Lamp on: 1v, off: 12v
C12	WHT/BLU	Alternator Charge Signal	12-14v
C13	YEL/RED	Fuel Gauge PWM Signal	Duty Cycle: 0-100%
C14	PNK	HO2S-21 (B2 S1) Signal	0.1-1.1v
C15	BLU	HO2S-21 (B2 S1) Ground	<0.050v
C16	GRN	HO2S-22 (B2 S2) Signal	0.1-1.1v

1996-97 4-Door 3.2L 4WD VIN V [M/T] 32P White 'D' Connector

PCM Pin #	W/Color	Circuit Description (16-Pin)	Value at Hot Idle
D1	GRN/ORN	Injector 2 Control	2.0-3.3 ms
D2	---	Not Used	---
D3	GRN/WHT	Injector 1 Control	2.0-3.3 ms
D4	ORN	Serial Data (8192 Baud)	Digital Signals
D5	RED/YEL	Igniter 5 Control	Digital Signals: 0-12-0v
D6	RED/BLU	Igniter 3 Control	Digital Signals: 0-12-0v
D7	WHT/BLK	VSS (4096 pulses per mile)	Moving: 0-5-0-5v
D8	GRN	Sensor 'A' Ground	<0.050v
D9	BLU/YEL	Sensor 'B' Ground	<0.050v
D10	YEL/BLU	MAF Sensor Signal	7-9v
D11	BLU	CMP Sensor Signal	Digital Signals: 0-5-0v
D12	GRN	HO2S-13 (B1 S3) Ground	<0.050v
D13	BLU	HO2S-13 (B1 S3) Signal	0.1-1.1v
D14	GRN	HO2S-11 (B1 S1) Ground	<0.050v
D15	RED	HO2S-11 (B1 S1) Signal	0.1-1.1v
D16	BLU	HO2S-22 (B2 S2) Ground	<0.050v

32-PIN WHITE-CONNECTOR ROW 'C' & 'D'

C | 16 | 15 | 14 | 13 | 12 | 11 | 10 | 9 | 8 | 7 | / | 5 | 4 | 3 | / | 1
D | 16 | 15 | 14 | 13 | 12 | 11 | 10 | 9 | 8 | 7 | 6 | 5 | 4 | 3 | 2 | 1

WIRE SIDE OF HARNESS TERMINALS

1996-97 4-Door 3.2L V6 4WD VIN V [M/T] 32P Blue 'E' Connector

PCM Pin #	W/Color	Circuit Description (16-Pin)	Value at Hot Idle
E1-4	---	Not Used	---
E5	RED/GRN	Linear EGR Solenoid "High"	Pulse Signals
E6	YEL	Linear EGR Solenoid "Low"	Pulse Signals
E7	---	Not Used	---
E8	BLU	TP Sensor Signal	0.5-0.6v
E9	BLU/RED	ECT Sensor Signal	At 180°F: 0.54v
E10	---	Not Used	---
E11	YEL/RED	CKP Sensor VREF	4.9-5.1v
E12	---	Not Used	---
E13	PNK/WHT	Fuel Pump Relay Control	Relay on: 12v, off: 0v
E14	---	Not Used	---
E15	GRN/BLK	A/C Switch Signal	Switch on: 12v, off: 0v
E16	RED/BLU	B+ from Main Relay	12-14v

1996-97 4-Door 3.2L V6 4WD VIN V [M/T] 32P Blue 'F' Connector

PCM Pin #	W/Color	Circuit Description (16-Pin)	Value at Hot Idle
F1-7	---	Not Used	---
F8	RED	MAP Sensor Signal	0.6-1.3v
F9	---	Not Used	---
F10	GRY/BLU	Cruise Control	Switch On: 0.1v, off: 12v
F11	---	Not Used	---
F12	---	Not Used	---
F13	WHT/GRN	Injector 3 Control	2.0-3.3 ms
F14	---	Not Used	---
F15	GRN/BLK	Injector 5 Control	2.0-3.3 ms
F16	RED/BLU	B+ from Main Relay	12-14v

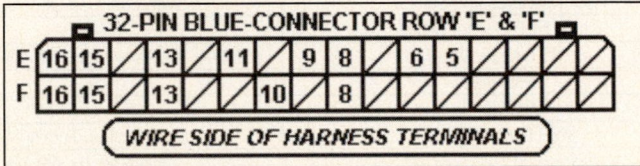

Abbreviation	Color	Abbreviation	Color	Abbreviation	Color
BLK	Black	LT BLU	Lt. Blue	TAN	Tan
BLU	Blue	LT GRN	Lt. Green	VIO	Violet
BRN	Brown	ORN	Orange	WHT	White
GRY	Gray	PNK	Pink	YEL	Yellow
GRN	Green	PPL	Purple		

1998-99 4-Door 3.5L V6 MFI VIN X [All] 32P Red 'A' Connector

PCM Pin #	W/Color	Circuit Description (16-Pin)	Value at Hot Idle
A1	RED	Reference Voltage 'A'	4.9-5.1v
A2	YEL	Knock Sensor Signal	No detonation: 18 mv AC
A3	---	Not Used	---
A4	WHT	Keep Alive Power	12-14v
A5	BLU	Idle Air Control 'A' High	Pulse Signals
A6	BLU/WHT	Idle Air Control 'A' Low	Pulse Signals
A7	BLU/BLK	Idle Air Control 'B' Low	Pulse Signals
A8	BLU/RED	Idle Air Control 'B' High	Pulse Signals
A9	ORN/BLU	A/T: Transmission Fluid Lamp	Lamp on: 1v, off: 12
A10	PNK/GRN	A/T: Winter Lamp	Lamp on: 1v, off: 12
A11	GRY/WHT	A/T: Power Lamp	Lamp on: 1v, off: 12
A12	PPL/WHT	Antilock Brake Light	Lamp Off: 12v, On: 1v
A13	PPL	Malfunction Indicator Lamp	MIL Off: 12v, On: 1v
A14	ORN/BLK	A/T: Check Transmission Light	Lamp on: 1v, off: 12v
A14	ORN/BLU	M/T: Upshift Lamp Relay	Relay on: 1v, off: 12
A15	RED/BLU	EVAP Purge Solenoid	Digital Signals: 0-12-0v
A16	BRN/YEL	A/T: Shift Solenoid Feed (B+)	12-14v
A16	BRN/YEL	M/T: Air Pump Feed (B+)	12-14v

1998-99 4-Door 3.5L V6 MFI VIN X [All] 32P Red 'B' Connector

PCM Pin #	W/Color	Circuit Description (16-Pin)	Value at Hot Idle
B1	RED/YEL	Reference Voltage 'B'	4.9-5.1v
B2	RED/WHT	Igniter 4 Control	Digital Signals: 0-12-0v
B3	RED/BLK	Igniter 2 Control	Digital Signals: 0-12-0v
B4	RED/GRN	Igniter 6 Control	Digital Signals: 0-12-0v
B5	YEL/PPL	Fuel Level Sensor	Tank empty: 1.8v
B6	GRY/RED	Fuel Tank Pressure Sensor	With Fuel Cap Off: 2.5v
B7	YEL/RED	Linear EGR Sensor Signal	0.5-0.6v
B8	YEL/GRN	IAT Sensor Signal	Varies w/temp: 0.5-4.8v
B9	---	Not Used	---
B10	---	Not Used	---
B11	GRN/YEL	PSP Switch Signal	Straight: 12v, Turning: 0v
B12	GRN	Illuminated Switch Signal	Switch on: 0v, off: 12v
B13	ORN/GRN	Class 2 Serial Data Link	0v
B14	GRN/BLK	A/C Clutch Relay Control	Relay Off: 12v, On: 1v
B15	YEL/BRN	Low Fuel Lamp	Lamp on: 1v, off: 12v
B16	YEL/RED	EVAP Vent Solenoid	Tank Empty: 5.7v

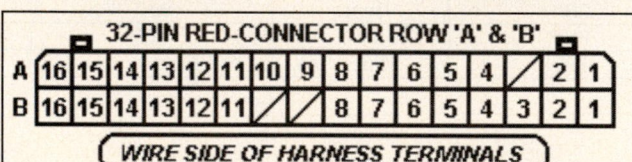

1998-99 4-Door 3.5L V6 MFI VIN X [All] 32P White 'C' Connector

PCM Pin #	W/Color	Circuit Description (16-Pin)	Value at Hot Idle
C1	GRN/RED	Injector 4 Control	2.0-3.3 ms
C2	BRN/BLK	A/T: Shift Solenoid 'B'	SSB Off: 12v, On: 1v
C3	GRN/YEL	Injector 6 Control	2.0-3.3 ms
C4	RED	Igniter 1 Control	Digital Signals: 0-12-0v
C5	YEL	CKP Sensor Signal	Digital Signals: 0-5-0v
C7	BLK/BLU	Power Ground	<0.1v
C8	BLK/PNK	Power Ground	<0.1v
C9	BLK/BLU	Power Ground	<0.1v
C10	BLK/RED	Tachometer Signal	Pulse Signals
C11	YEL/BLK	Variable Intake Manifold Sol.	12-14v (at 3600 rpm)
C12	WHT/BLU	Alternator Charge Signal	12-14v
C13	BLU/PNK	Fuel Gauge PWM Signal	Duty cycle: 0-100%
C14	PNK	HO2S-21 (B2 S1) Signal	0.1-1.1v
C15	BLU	HO2S-21 (B2 S1) Ground	<0.050v
C16	GRN	HO2S-22 (B2 S2) Signal	0.1-1.1v

1998-99 4-Door 3.5L V6 MFI VIN X [All] 32P White 'D' Connector

PCM Pin #	W/Color	Circuit Description (16-Pin)	Value at Hot Idle
D1	GRN/ORN	Injector 2 Control	2.0-3.3 ms
D2	BRN/BLU	A/T: Torque Converter Clutch	TCC on: 1v, off: 12v
D3	GRN/WHT	Injector 1 Control	2.0-3.3 ms
D4	ORN	Serial Data (8192 Baud)	Digital Signals
D5	RED/YEL	Igniter 5 Control	Digital Signals: 0-12-0v
D6	RED/BLU	Igniter 3 Control	Digital Signals: 0-12-0v
D7	GRN/WHT	VSS (4096 pulses per mile)	Moving: 0-5-0-5v
D8	GRN	VREF Sensor 'A' Ground	<0.050v
D9	GRY	VREF Sensor 'B' Ground	<0.050v
D10	YEL	MAF Sensor Signal	7-9v
D11	WHT	CMP Sensor Signal	Digital Signals: 0-5-0v
D12 [A/T]	RED	HO2S-12 (B1 S2) Ground	<0.050v
D12 [M/T]	RED	HO2S-13 (B1 S3) Ground	<0.050v
D13 [A/T]	WHT	HO2S-12 (B1 S2) Signal	0.1-1.1v
D13 [M/T]	WHT	HO2S-13 (B1 S3) Signal	0.1-1.1v
D14	GRN	HO2S-11 (B1 S1) Ground	<0.050v
D15	RED	HO2S-11 (B1 S1) Signal	0.1-1.1v
D16 [A/T]	BLU	HO2S-22 (B2 S2) Ground	<0.050v
D16 [M/T]	BLU	HO2S-12 (B1 S2) Ground	<0.050v

```
 ┌─────────────────────────────────────────────┐
 │      32-PIN WHITE-CONNECTOR ROW 'C' & 'D'     │
 │  C │16│15│14│13│12│ ╱ │10│ 9│ 8│ 7│ ╱ │ 5│ 4│ 3│ 2│ 1│ │
 │  D │16│15│14│13│12│11│10│ 9│ 8│ 7│ 6│ 5│ 4│ 3│ 2│ 1│ │
 │        WIRE SIDE OF HARNESS TERMINALS          │
 └─────────────────────────────────────────────┘
```

1998-99 4-Door 3.5L V6 MFI VIN X [All] 32P Blue 'E' Connector

PCM Pin #	W/Color	Circuit Description (16-Pin)	Value at Hot Idle
E1	RED	Output Shaft Sensor Signal	AC Pulses
E2	WHT	Output Shaft Sensor Ground	<0.050v
E3	PPL/RED	A/T: Pressure Control Low	Pulse Signals
E4	PPL/WHT	A/T: Pressure Control High	Pulse Signals
E5	BLK/YEL	Linear EGR Control High	12-14v
E6	YEL	Linear EGR Control Low	Pulse Signals
E7	BLU/YEL	A/T: Range Switch 'B' Signal	In P/N: 0v, others: 12v
E8	BLU	TP Sensor Signal	0.5-0.8v
E9	BLU/RED	ECT Sensor Signal	At 180°F: 0.54v
E10	---	Not Used	---
E11	YEL/RED	CKP Sensor VREF	4.9-5.1v
E12	BLU/WHT	A/T: Range Switch 'A' Signal	In P/N: 0v, others: 12v
E13	RED/WHT	Fuel Pump Relay Control	Relay on: 12v, off: 0v
E14	BRN/WHT	A/T: Shift Solenoid Feed (B+)	12-14v
E15	GRN/ORN	A/C Switch Signal	Switch on: 12, off: 0v
E16	RED/BLU	B+ from Main Relay	12-14v

1998-99 4-Door 3.5L V6 MFI VIN X [All] 32P Blue 'F' Connector

PCM Pin #	W/Color	Circuit Description (16-Pin)	Value at Hot Idle
F1	---	Not Used	---
F2	BLU/BLK	A/T: Range Signal 'C'	In P/N: 0v, others: 12v
F3	YEL/GRN	A/T: Range Signal 'P'	In P/N: 0v, others: 12v
F4	GRN/YEL	Brake Switch Signal	Switch on: 12v, Off: 0v
F5	PPL/RED	Power Switch Signal	Switch on: 0.1v, off: 12v
F6	PPL/GRN	Winter Switch Signal	Switch on: 0.1v, off: 12v
F7	RED/BLK	A/T: TOT Sensor Signal	Varies w/temp: 0.5-4.8v
F8	GRY/BLK	MAP Sensor Signal	0.6-1.3v
F9	---	Not Used	---
F10	GRY/BLU	Cruise Control	C/C on: 0.1v, off: 12v
F11	LT BLU	A/T: Kickdown Switch	Switch on: 0.1v, off: 12v
F12	ORN/BLU	PCM Diagnostic Enable	Digital Signals
F13	GRN	Injector 3 Control	2.0-3.3 ms
F14	BRN/RED	A/T: Shift Solenoid 'A'	SSA Off: 12v, On: 1v
F15	GRN/BLK	Injector 5 Control	2.0-3.3 ms
F16	RED/BLU	B+ from Main Relay	12-14v

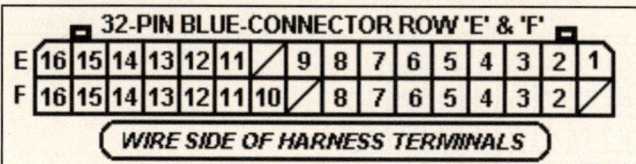

PASSPORT Pin Voltage Tables

1994-95 DX 2.6L I4 MFI VIN E 2WD [M/T] "A28" 26P 'A' Connector

PCM Pin #	W/Color	Circuit Description (26-Pin)	Value at Hot Idle
A28-1	GRN/RED	B+ from Main Relay	12-14v
A28-2	BLK/GRN	Power Ground	<0.1v
A28-3	BLK/YEL	Data Link Connector	4.80v (no Scan Tool)
A28-4	BLU/YEL	Throttle Idle Switch	Idle: 12v, off-idle: 0.1v
A28-5	GRN/BLK	A/C Switch Signal	Switch on: 12v, off: 0v
A28-6	---	Not Used	---
A28-7	GRN/BLU	CKP Sensor Signal	Varies 2.0-3.0v
A28-8	GRN/YEL	Fuel Pump Relay Control	Relay on: 12v, off: 0v
A28-9	BRN/YEL	Air Management Solenoid	Sol. on: 1v. off: 12v
A28-10	LT GRN	EGR Duty Cycle Solenoid	Digital Signals: 0-12-0v
A28-11	---	Not Used	---
A28-12	WHT/BLK	Injector 2 Control	2.0-3.3 ms
Q	PUR/WHT	Injector 1 Control	2.0-3.3 ms
A28-14	GRN/BLK	B+ from Main Relay	12-14v
A28-15	---	Not Used	---
A28-16	---	Not Used	---
A28-17	---	Not Used	---
A28-18	GRN/ORN	Monitor Signals	Digital Signals
A28-19	GRN/YEL	Monitor Signals	Digital Signals
A28-20	WHT	CKP Sensor VREF	4.9-5.1v
A28-21	WHT/GRN	EGR Vacuum Control Valve	Sol on: 1v, off: 12v
A28-22	GRN	Malfunction Indicator Lamp	MIL Off: 12v, On: 1v
A28-23	GRN/WHT	EVAP Purge Solenoid	Digital Signals: 0-12-0v
A28-24	WHT/BLU	Injector 4 Control	2.0-3.3 ms
A28-25	WHT/RED	Injector 3 Control	2.0-3.3 ms
A28-26	BLK/GRN	Logic Ground	<0.050v

A28 26-PIN CONNECTOR

1	2	3	4	5	6	7	8	9	10	11	12	13
14	15	16	17	18	19	20	21	22	23	24	25	26

View is into back of Wire Harness Connector

Abbreviation	Color	Abbreviation	Color	Abbreviation	Color
BLK	Black	LT BLU	Lt. Blue	TAN	Tan
BLU	Blue	LT GRN	Lt. Green	VIO	Violet
BRN	Brown	ORN	Orange	WHT	White
GRY	Gray	PNK	Pink	YEL	Yellow
GRN	Green	PPL	Purple		

1994-95 DX 2.6L I4 MFI VIN E 2WD [M/T] "A27" 22P 'A' Connector

PCM Pin #	W/Color	Circuit Description (22-Pin)	Value at Hot Idle
A27-1	BLU	Engine Speed Sensor	Digital Signals
A27-2	BLK/RED	Power Ground	<0.1v
A27-3	GRN/ORN	MAP Sensor VREF	4.9-5.1v
A27-4	RED/WHT	Keep Alive Power	12-14v
A27-5	WHT/GRN	Vehicle Speed Sensor	Moving: 0-5-0v
A27-6	GRY	HO2S-11 (B1 S1) Ground	<0.050v
A27-7	BLK/WHT	ECT Sensor Ground	<0.050v
A27-8	BRN/WHT	ECT Sensor Signal	At 180°F: 0.52v
A27-9	BLU/WHT	MAP Sensor Signal	0.6-1.3v
A27-10	---	Not Used	---
A27-11	BLK	Power Ground	<0.1v
A27-12	---	Not Used	---
A27-13	BLK/RED	Logic Ground	<0.050v
A27-14	RED	MAF Sensor Signal VREF	12-14v
A27-15	GRN/RED	Wide Open Throttle Switch	At WOT: 12-14v
A27-16	BLK/GRN	Starter Signal	KOEC: 9-11v
A27-17	GRY	MAF Sensor Shield Ground	<0.050v
A27-18	BLU/RED	MAP Sensor Ground	<0.050v
A27-19	BLK	MAF Sensor Ground	<0.050v
A27-20	RED	HO2S-11 (B1 S1) Signal	0.1-1.1v
A27-21	WHT	MAF Sensor Signal	7-9v
A27-22	BLK	Power Ground	<0.1v

A27 22-PIN CONNECTOR

```
 1  2  3  4  5  6  7  8  9 10 11
12 13 14 15 16 17 18 19 20 21 22
```

View is into back of Wire Harness Connector

Abbreviation	Color	Abbreviation	Color	Abbreviation	Color
BLK	Black	LT BLU	Lt. Blue	TAN	Tan
BLU	Blue	LT GRN	Lt. Green	VIO	Violet
BRN	Brown	ORN	Orange	WHT	White
GRY	Gray	PNK	Pink	YEL	Yellow
GRN	Green	PPL	Purple		

1996-97 DX 2.6L I4 MFI VIN E 2WD [M/T] 32P Red 'A' Connector

PCM Pin #	W/Color	Circuit Description (32-Pin)	Value at Hot Idle
A1	GRY/BLU	Reference Voltage 'A'	4.9-5.1v
A2-3	---	Not Used	---
A4	RED/WHT	Keep Alive Power	12-14v
A5	BRN/BLK	Idle Air Control 'A' High	Pulse Signals
A6	BRN/WHT	Idle Air Control 'A' Low	Pulse Signals
A7	BRN/YEL	Idle Air Control 'B' Low	Pulse Signals
A8	GRN/BLU	Idle Air Control 'B' High	Pulse Signals
A9-12	---	Not Used	---
A13	GRN	Malfunction Indicator Lamp	MIL Off: 12v, On: 1v
A14, 16	---	Not Used	---
A15	GRN/WHT	EVAP Purge Solenoid	Purge on, 1v, off: 12-14v

1996-97 DX 2.6L I4 MFI VIN E 2WD [M/T] 32P Red 'B' Connector

PCM Pin #	W/Color	Circuit Description (32-Pin)	Value at Hot Idle
B1	BLU/ORN	Reference Voltage 'B'	4.9-5.1v
B2-6	---	Not Used	---
B7	GRY	Linear EGR Sensor Signal	0.5-0.6v
B8	GRY	IAT Sensor Signal	Varies w/temp: 0.5-4.8v
B9	---	Not Used	---
B10	GRN/WHT	Rough Road Sensor	2.5v
B11	GRN/RED	PSP Switch Signal	Straight: 12v, Turning: 0v
B12	GRN/YEL	Illuminated Switch Signal	Switch on: 0.1v, off: 12v
B13	ORN/BLK	Class 2 Serial Data Link	Varies: 0-7v
B14	GRY/RED	A/C Clutch Relay	Relay on: 12v, off: 0v
B15-16	---	Not Used	---

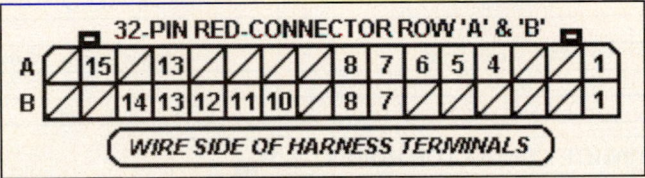

Abbreviation	Color	Abbreviation	Color	Abbreviation	Color
BLK	Black	LT BLU	Lt. Blue	TAN	Tan
BLU	Blue	LT GRN	Lt. Green	VIO	Violet
BRN	Brown	ORN	Orange	WHT	White
GRY	Gray	PNK	Pink	YEL	Yellow
GRN	Green	PPL	Purple		

1996-97 DX 2.6L MFI VIN E 2WD [M/T] 32P White 'C' Connector

PCM Pin #	W/Color	Circuit Description (32-Pin)	Value at Hot Idle
C1	BLU/YEL	Injector 2 Control	2.0-3.3 ms
C2-3	---	Not Used	---
C4	BLU	Igniter Control Signal	Digital Signals: 0-12-0v
C5	RED	CKP Sensor Signal	Digital Signals: 0-5-0v
C6	---	Not Used	---
C7	WHT	Power Ground	<0.1v
C8	BLK/RED	Power Ground	<0.1v
C9	BLK/BLU	Power Ground	<0.1v
C10	---	Not Used	---
C11	YEL/GRN	Upshift Lamp Control	Relay Off: 12v, On: 1v
C12	RED/WHT	Alternator Charge Signal	12-14v
C13	---	Not Used	---
C14	RED	HO2S-11 (B1 S1) Signal	0.1-1.1v
C15	GRN	HO2S-11 (B1 S1) Ground	<0.050v
C16	PNK	HO2S-12 (B1 S2) Signal	0.1-1.1v

1996-97 DX 2.6L MFI VIN E 2WD [M/T] 32P White 'D' Connector

PCM Pin #	W/Color	Circuit Description (32-Pin)	Value at Hot Idle
D1	BLU/RED	Injector 3 Control	2.0-3.3 ms
D2	---	Not Used	---
D3	BLU/BLK	Injector 1 Control	2.0-3.3 ms
D4	RED	Serial Data (8192 Baud)	Digital Signals
D5-6	---	Not Used	---
D7	WHT	Vehicle Speed Sensor	Moving: 0-12v
D8	GRN	Sensor Ground	<0.1v
D9	GRN	Sensor Ground	<0.1v
D10	---	Not Used	---
D11	WHT	CMP Sensor Signal	Digital Signals: 0-5-0v
D12-15	---	Not Used	---
D16	BLU	HO2S-12 (B1 S2) Ground	<0.050v

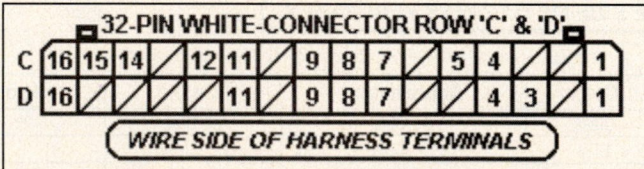

32-PIN WHITE-CONNECTOR ROW 'C' & 'D'

C | 16 | 15 | 14 | / | 12 | 11 | / | 9 | 8 | 7 | / | 5 | 4 | / | / | 1
D | 16 | / | / | / | 11 | / | 9 | 8 | 7 | / | / | 4 | 3 | / | 1

WIRE SIDE OF HARNESS TERMINALS

Abbreviation	Color	Abbreviation	Color	Abbreviation	Color
BLK	Black	LT BLU	Lt. Blue	TAN	Tan
Q	Blue	LT GRN	Lt. Green	VIO	Violet
BRN	Brown	ORN	Orange	WHT	White
GRY	Gray	PNK	Pink	YEL	Yellow
GRN	Green	PPL	Purple		

1996-97 DX 2.6L I4 MFI VIN E 2WD [M/T] 32P Blue 'E' Connector

PCM Pin #	W/Color	Circuit Description (32-Pin)	Value at Hot Idle
E1-4	---	Not Used	---
E5	RED/GRN	Linear EGR Solenoid "High"	Pulse Signals
E6	GRY/RED	Linear EGR Solenoid "Low"	Pulse Signals
E7	---	Not Used	---
E8	BLU	TP Sensor Signal	0.5-0.6v
E9	GRY/BLK	ECT Sensor Signal	At 180°F: 0.54v
E10	---	Not Used	---
E11	WHT/ORN	CKP Sensor VREF	4.9-5.1v
E12	---	Not Used	---
E13	PNK/WHT	Vehicle Speed Sensor	Moving: 0-5-0v
E14	---	Not Used	---
E15	GRN/BLK	A/C Switch Signal	Switch on: 12v, off: 0v
E16	RED/GRN	B+ from Main Relay	12-14v

1996-97 DX 2.6L I4 MFI VIN E 2WD [M/T] 32P Blue 'F' Connector

PCM Pin #	W/Color	Circuit Description (32-Pin)	Value at Hot Idle
F1-7	---	Not Used	---
F8	GRY/RED	MAP Sensor Signal	0.6-1.3v
F9	WHT/GRN	Vacuum Switch Signal	0v
F10-12		Not Used	
F13	BLU/PNK	Injector 4 Control	2.0-3.3 ms
F14-15	---	Not Used	---
F16	RED/GRN	B+ from Main Relay	12-14v

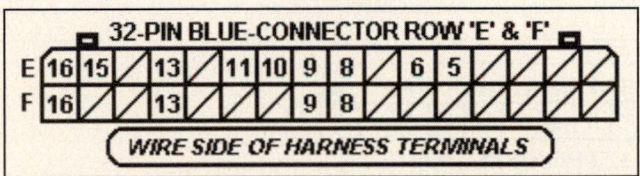

Abbreviation	Color	Abbreviation	Color	Abbreviation	Color
BLK	Black	LT BLU	Lt. Blue	TAN	Tan
BLU	Blue	LT GRN	Lt. Green	VIO	Violet
BRN	Brown	ORN	Orange	WHT	White
GRY	Gray	PNK	Pink	YEL	Yellow
GRN	Green	PPL	Purple		

1994-95 LX 3.2L V6 MFI VIN V 2WD [All] 12P 'A' Connector

PCM Pin #	W/Color	Circuit Description (12-Pin)	Value at Hot Idle
A1	BLU/ORN	TP & MAP Sensor VREF	4.9-5.1v
A2	GRY/RED	A/C Clutch Relay	Relay Off: 12v, On: 1v
A3	---	Not Used	---
A4	BLUPNK	M/T: Shift Indicator Lamp	Lamp on: 1v, off: 12v
A5	GRN	Malfunction Indicator Lamp	MIL Off: 12v, On: 1v
A6	RED/GRN	B+ from Main Relay	12-14v
A7	WHT/RED	EVAP Purge Solenoid	Digital Signals: 0-12-0v
A8	WHT/GRN	EGR Solenoid Control	Sol. on: 1v, off: 12v
A9	YEL/BLU	TCM Signal ECT Input	At 180°F: 0.54v
A10	---	Not Used	---
A11	GRY	Sensor Ground	<0.050v
A12	BLK/ORN	Power Ground	<0.1v

1994-95 LX 3.2L V6 MFI VIN V 2WD [All] 12P 'B' Connector

PCM Pin #	W/Color	Circuit Description (12-Pin)	Value at Hot Idle
B1	RED/WHT	Keep Alive Power	12-14v
B2	BLK/PNK	HO2S-11 (B1 S1) Ground	<0.050v
B3	RED	HO2S-11 (B1 S1) Signal	0.1-1.1v
B4	BRN/BLK	Idle Air Control 'A' High	Pulse Signals
B5	BRN/WHT	Idle Air Control 'A' Low	Pulse Signals
B6	BRN/YEL	Idle Air Control 'B' Low	Pulse Signals
B7	BRN/RED	Idle Air Control 'B' High	Pulse Signals
B8	YEL/VLT	Ignition VREF 'Lo'	<0.050v
B9	YEL/RED	Ignition VREF 'Hi'	Varies: 1.1-2.4v
B10-12	---	Not Used	---

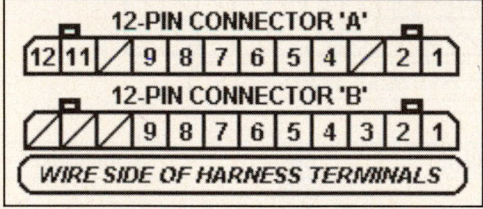

Abbreviation	Color	Abbreviation	Color	Abbreviation	Color
BLK	Black	LT BLU	Lt. Blue	TAN	Tan
BLU	Blue	LT GRN	Lt. Green	VIO	Violet
BRN	Brown	ORN	Orange	WHT	White
GRY	Gray	PNK	Pink	YEL	Yellow
GRN	Green	PPL	Purple		

1994-95 LX 3.2L V6 MFI VIN V 2WD [All] 16P 'C' Connector

PCM Pin #	W/Color	Circuit Description (16-Pin)	Value at Hot Idle
C1	---	Not Used	---
C2	ORN/BLK	Data Link Connector	Open: 4.80v
C3	PNK/WHT	Fuel Pump Relay Control	Relay on: 12v, off: 0v
C4	ORN/YEL	DLC Diagnostic Request	Digital Signals
C5	---	Not Used	---
C6	GRN/YEL	PSP Switch Signal	Straight: 12v, Turning: 0v
C7	ORN	A/T: PNP Switch Signal	In P/N: 0v, others: 12v
C8	---	Not Used	---
C9	YEL/BLK	A/C Switch Signal	Switch on: 12v, off: 0v
C10	---	Not Used	---
C11	BLU/YEL	Injector Pair 1 & 2	2.0-3.3 ms
C12	BLK/ORN	Power Ground	<0.1v
C13	---	Not Used	---
C14	BLU/WHT	Injector Pair 5 & 6	2.0-3.3 ms
C15	BLK/GRN	Power Ground	<0.1v
C16	RED/WHT	Keep Alive Power	12-14v

1994-95 LX 3.2L V6 MFI VIN V 2WD [All] 16P 'D' Connector

PCM Pin #	W/Color	Circuit Description (16-Pin)	Value at Hot Idle
D1	BLK/GRN	Power Ground	<0.1v
D2	BLK/WHT	Sensor Ground	<0.050v
D3-4	---	Not Used	---
D5	YEL	Igniter Control Signal	Digital Signals: 0-12-0v
D6	BLU/RED	TP Sensor Signal	0.5-0.6v
D7	BLU	IAT Sensor Signal	Varies w/temp: 0.5-4.8v
D8	WHT	Vehicle Speed Sensor	Moving: 0-5-0-5v
D9	GRY/RED	MAP Sensor Signal	0.6-1.1v
D10	---	Not Used	---
D11	YEL/GRN	Ignition Control Bypass	Hot idle: 4.9v
D12	GRY/BLK	ECT Sensor Signal	At 180°F: 0.54v
D13-14	---	Not Used	---
D15	BLU/GRN	Injector Pair 3 & 4	2.0-3.3 ms
D16	BLU/GRN	Injector Pair 3 & 4	2.0-3.3 ms

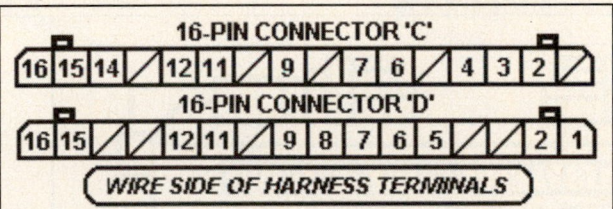

16-PIN CONNECTOR 'C'
16 15 14 / 12 11 / 9 / 7 6 / 4 3 2 /

16-PIN CONNECTOR 'D'
16 15 / / 12 11 / 9 8 7 6 5 / / 2 1

WIRE SIDE OF HARNESS TERMINALS

1996-97 LX 3.2L V6 MFI VIN V 2WD [All] C209 Row 'A' Connector

PCM Pin #	W/Color	Circuit Description (16-Pin)	Value at Hot Idle
A1	RED	Reference Voltage 'A'	4.9-5.1v
A2	YEL	Knock Sensor Signal	No detonation: 18 mv AC
A3, A12	---	Not Used	---
A4	RED/WHT	Keep Alive Power	12-14v
A5	BLU	Idle Air Control 'A' High	Pulse Signals
A6	BLU/WHT	Idle Air Control 'A' Low	Pulse Signals
A7	BLU/BLK	Idle Air Control 'B' Low	Pulse Signals
A8	BLU/RED	Idle Air Control 'B' High	Pulse Signals
A9	ORN/BLK	Transmission Fluid Lamp	Lamp on: 1v. off: 12v
A10	PNK/GRN	Winter Lamp	Lamp on: 1v. off: 12v
A11	PNK/WHT	Power Lamp	Lamp on: 1v. off: 12v
A13	GRN	Malfunction Indicator Lamp	MIL Off: 12v, On: 1v
A14	PPL	Check Transmission Light	Lamp on: 1v, off: 12v
A15	RED/BLU	EVAP Purge Solenoid	Digital Signals: 0-12-0v
A16	BRN/YEL	A/T: Shift Solenoid Feed (B+)	12-14v

1996-97 LX 3.2L V6 MFI VIN V 2WD [All] C209 Row 'B' Connector

PCM Pin #	W/Color	Circuit Description (16-Pin)	Value at Hot Idle
B1	BLU/ORN	Reference Voltage 'B'	4.9-5.1v
B2	RED/WHT	Igniter 4 Control	Digital Signals: 0-12-0v
B3	RED/BLK	Igniter 2 Control	Digital Signals: 0-12-0v
B4	RED/GRN	Igniter 6 Control	Digital Signals: 0-12-0v
B5	ORN/GRN	Fuel Level Sensor	Tank empty: 1.8v
B6	GRY	Fuel Tank Pressure Sensor	With Fuel Cap Off: 2.5v
B7	YEL/RED	Linear EGR Sensor Signal	0.5-0.6v
B8	YEL/GRN	IAT Sensor Signal	Varies w/temp: 0.5-4.8v
B9	---	Not Used	---
B10	GRN/WHT	Rough Road Sensor	At idle: 2.5v
B11	GRN/YEL	PSP Switch Signal	Straight: 12v, Turning: 0v
B12	GRN/YEL	Headlights Switch Signal	Switch on: 0.1v, off: 12v
B13	ORN/BLK	Class 2 Serial Data Link	0v
B14	GRY/RED	A/C Clutch Relay	Relay Off: 12v, On: 1v
B15	PNK	Low Fuel Lamp	Lamp on: 1v, off: 12v
B16	BRN/WHT	EVAP Vent Solenoid	Sol. on: 1v, off: 12v

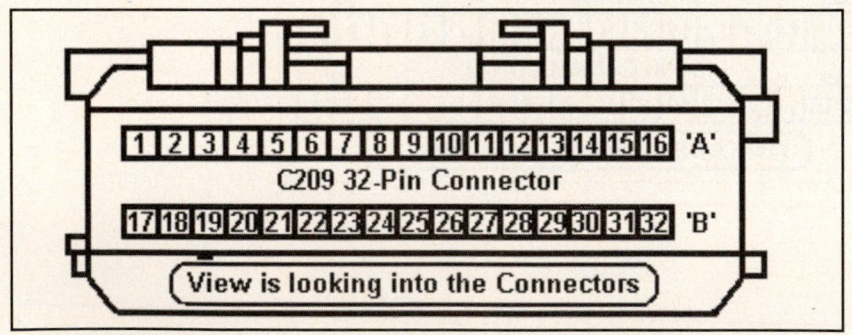

C209 32-Pin Connector

View is looking into the Connectors

1996-97 LX 3.2L MFI VIN V 2WD [All] C211 Row 'C' Connector

PCM Pin #	W/Color	Circuit Description (16-Pin)	Value at Hot Idle
C1	GRN/RED	Injector 4 Control	2.0-3.3 ms
C2	BRN/BLK	A/T: Shift Solenoid 'B'	SSB Off: 12v, On: 1v
C3	GRN/YEL	Injector 6 Control	2.0-3.3 ms
C4	RED	Igniter 1 Control	Digital Signals: 0-12-0v
C5	YEL	CKP Sensor Signal	Digital Signals: 0-5-0v
C7	YEL/BLK	Power Ground	<0.1v
C8	BLK/RED	Power Ground	<0.1v
C9	BLK/BLU	Power Ground	<0.1v
C10	BLK/RED	Tachometer Signal	Pulse Signals
C11	YEL/GRN	M/T: Upshift Lamp Control	Lamp on: 1v, off: 12v
C12	RED/WHT	Alternator Charge Signal	12-14v
C13	YEL/RED	Fuel Gauge PWM Signal	Duty Cycle: 0-100%
C14	WHT	HO2S-21 (B2 S1) Signal	0.1-1.1v
C15	RED	HO2S-21 (B2 S1) Ground	<0.050v
C16	RED	HO2S-22 (B2 S2) Signal	0.1-1.1v

1996-97 LX 3.2L V6 VIN V 2WD [All] C211 Row 'D' Connector

PCM Pin #	W/Color	Circuit Description (16-Pin)	Value at Hot Idle
D1	GRN/ORN	Injector 2 Control	2.0-3.3 ms
D2	RED/YEL	Torque Converter Clutch	TCC on: 1v, off: 12v
D3	GRN/WHT	Injector 1 Control	2.0-3.3 ms
D4	RED	Serial Data (8192 Baud)	Digital Signals
D5	RED/YEL	Igniter 5 Control	Digital Signals: 0-12-0v
D6	RED/BLU	Igniter 3 Control	Digital Signals: 0-12-0v
D7	WHT	Vehicle Speed Sensor	Moving: 0-5-0-5v
D8, D9	GRN	Sensor 'A', 'B' Ground	<0.050v
D10	YEL/BLU	MAF Sensor Signal	7-9v
D11	BLU	CMP Sensor Signal	Digital Signals: 0-5-0v
D12	BLU	HO2S-13 (B1 S3) Ground	<0.050v
D12	BLU	HO2S-12 (B1 S2) Ground	<0.050v
D13	GRN	HO2S-13 (B1 S3) Signal	0.1-1.1v
D13	GRN	HO2S-12 (B1 S2) Signal	0.1-1.1v
D14	BLU	HO2S-11 (B1 S1) Ground	<0.050v
D15	PNK	HO2S-11 (B1 S1) Signal	0.1-1.1v
D16	GRN	HO2S-22 (B2 S2) Ground	<0.050v

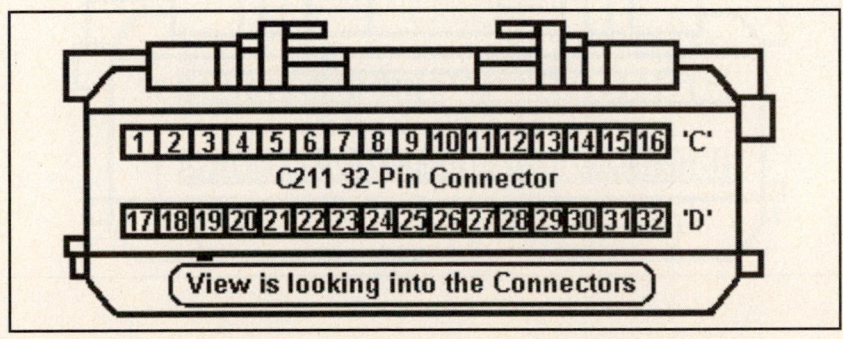

1996-97 LX 3.2L V6 MFI VIN V 2WD [All] C212 Row 'E' Connector

PCM Pin #	W/Color	Circuit Description (32-Pin)	Value at Hot Idle
E1	YEL	VSS "High" Signal	AC Pulse Signals
E2	BRN	VSS "Low" Signal	<0.050v
E3	RED/ORN	A/T: Pressure Control Low	Pulse Signals
E4	RED/BLK	A/T: Pressure Control High	Pulse Signals
E5	RED/GRN	Linear EGR Solenoid "High"	Pulse Signals
E6	YEL	Linear EGR Solenoid "Low"	Pulse Signals
E7	PNK	A/T: Range Signal 'B'	In P/N: 1v, others: 12v
E8	BLU	TP Sensor Signal	0.5-0.6v
E9	BLU/RED	ECT Sensor Signal	At 180°F: 0.54v
E10	---	Not Used	
E11	YEL/RED	CKP Sensor VREF	4.9-5.1v
E12	PNK/BLU	A/T: Range Signal 'A'	In P/N: 1v, others: 12v
E13	PNK/WHT	Fuel Pump Relay Control	Relay on: 12v, off: 0v
E14	BRN/WHT	A/T: Bank Apply Solenoid	Sol. on: 1v, off: 12v
E15	GRN/BLK	A/C Switch Signal	Switch on: 12v, off: 0v
E16	RED/BLU	B+ from Main Relay	12-14v

1996-97 LX 3.2L V6 MFI VIN V 2WD [All] C212 Row 'F' Connector

PCM Pin #	W/Color	Circuit Description (32-Pin)	Value at Hot Idle
F1, 9, 12	---	Not Used	---
F2	BLU/WHT	A/T: Range Signal 'C'	In P/N: 1v, others: 12v
F3	PNK/BLK	A/T: Range Signal 'P'	In P/N: 1v, others: 12v
F4	RED	Brake Switch Signal	On: 12v, Off: 0v
F5	PPL/RED	Power Switch Signal	On: 0.1v, off: 12v
F6	PPL/GRN	Winter Switch Signal	On: 0.1v, off: 12v
F7	GRN/RED	A/T: TOT Sensor Signal	Varies w/temp: 0.5-4.8v
F8	RED	MAP Sensor Signal	0.6-1.3v
F10	GRY/BLU	Cruise Control	On: 0.1v, off: 12v
F11	LT BLU	A/T: Kickdown Switch	On: 0.1v, off: 12v
F13	WHT/GRN	Injector 3 Control	2.0-3.3 ms
F14	YEL/GRN	A/T: Shift Solenoid 'A'	SSA Off: 12v, On: 1v
F15	GRN/BLK	Injector 5 Control	2.0-3.3 ms
F16	RED/BLU	B+ from Main Relay	12-14v

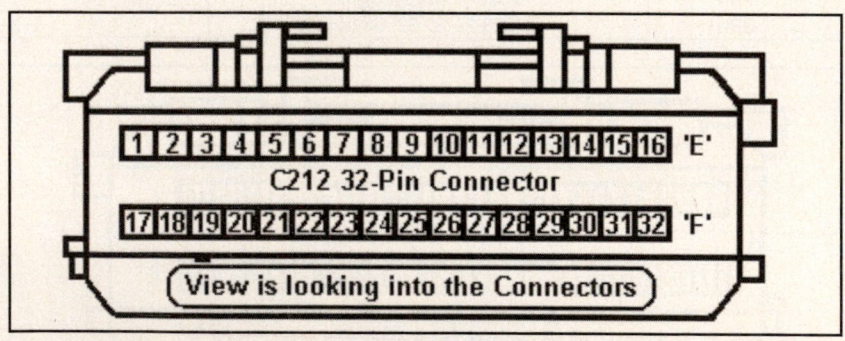

C212 32-Pin Connector

1 2 3 4 5 6 7 8 9 10 11 12 13 14 15 16 'E'

17 18 19 20 21 22 23 24 25 26 27 28 29 30 31 32 'F'

View is looking into the Connectors

1998-03 EX, LX 3.2L VIN W 2WD [A/T] 80P J21 Blue Connector

PCM Pin #	W/Color	Circuit Description (40-Pin)	Value at Hot Idle
1	BLK	Power Ground	<0.1v
2	GRN	Reference Voltage 'A'	4.9-5.1v
3	RED/WHT	Reference Voltage 'B'	4.9-5.1v
4	GRY/RED	A/C Clutch Relay Control	Relay Off: 12v, On: 1v
5	GRN/RED	A/T: Shift Solenoid 'A'	SSA Off: 12v, On: 1v
6	---	Not Used	---
7	BLK/RED	Tachometer Signal	Pulse Signals
8	---	Not Used	---
9	YEL	Knock Sensor 1 Signal	No detonation: 18 mv AC
10	RED	Knock Sensor 2 Signal	No detonation: 18 mv AC
11	---	Not Used	---
12	WHT/RED	CMP Sensor Signal	Digital Signals: 0-5-0v
13	PNK	To be done	---
14	BLK/BLU	MAF Sensor Signal	4.2v
15-16	---	Not Used	---
17	RED	HO2S-22 (B2 S2) Ground	0.1-1.1v
18	---	Not Used	---
19	RED/GRN	Ignition Power (Main Relay)	12-14v
20	RED/WHT	Keep Alive Power	12-14v
21	WHT	HO2S-11 (B1 S1) Ground	<0.050v
22	---	Not Used	---
23	BLU	HO2S-12 (B1 S2) Ground	<0.050v
24	---	Not Used	---
25	RED/BLU	Coil On Plug 3 Control	Digital Signals: 0-12-0v
26	RED/BLK	Coil On Plug 2 Control	Digital Signals: 0-12-0v
27	GRN	IAT Sensor Signal	Varies w/temp: 0.5-4.8v
28	---	Not Used	---
29	RED	MAP Sensor Signal	0.9v
30-36	---	Not Used	---
37	GRN/YEL	PSP Switch Signal	Straight: 12v, Turning: 0v
38	GRN/YEL	Headlights On Signal	Lights Off: 12v, On: 0v
39	RED	Brake Switch Signal	Brake Off: 0v, On: 12v
40	BLK/BLU	Power Ground	<0.1v

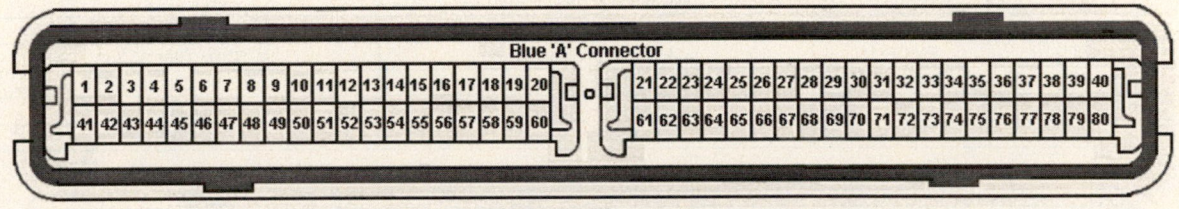

1998-03 EX, LX 3.2L VIN W 2WD [A/T] 80P J21 Blue Connector

PCM Pin #	W/Color	Circuit Description (40-Pin)	Value at Hot Idle
41	GRN	TP Sensor 1 Ground	<0.050v
42	PNK/WHT	Fuel Pump Relay Control	Relay Off: 0v, On: 12v
43	PPL/RED	Pressure Control Solenoid (-)	Pulse Signals
44	ORN/WHT	HO2S-21 (B2 S1) Heater	Digital Signals: 0-12-0v
45	GRN/BLK	A/C Switch Signal	A/C On: 12v, Off: 0v
46	BRN/RED	Brake Switch Signal	Brake Off: 12v, On: 0v
47	PPL/WHT	Pressure Control Solenoid (+)	Pulse Signals
48	BLU	Throttle Motor (-) Signal	Pulse Signals
49	BLK	Knock Sensor 1 Ground	<0.050v
50	ORN	HO2S-22 (B2 S2) Ground	<0.050v
51	---	Not Used	---
52	RED	HO2S-21 (B2 S1) Ground	<0.050v
53	BLK	ECT & EGR Sensor Ground	<0.050v
54	BLK	FTP Sensor Ground	<0.050v
55	---	Not Used	---
56	GRN/BLK	Injector 5 Control	2.0-3.3 ms
57	RED/WHT	Keep Alive Power	12-14v
58	ORN/BLK	Class 2 Serial Data Link	0v
59	---	Not Used	---
60	BLK	Knock Sensor 2 Ground	<0.050v
61	GRN/WHT	TP Sensor 2 Ground	<0.050v
62	GRN/ORN	Injector 2 Control	2.0-3.3 ms
63	RED	A/P Sensor 1 Ground	<0.1v
64	GRN/RED	Injector 4 Control	2.0-3.3 ms
65	BLU/WHT	TP Sensor 2 Signal	0.5-0.6v
66	GRN	Injector 3 Control	2.0-3.3 ms
67	YEL	PCM C.Q. Input (1-3-5)	Digital Signals
68	RED	Knock Sensor Input (2-4-6)	<0.050v
69	GRN/WHT	Injector 1 Control	2.0-3.3 ms
70	---	Not Used	---
71	GRN/WHT	HO2S-22 (B2 S2) Heater	Digital Signals: 0-12-0v
72	WHT/BLU	C/C Resume/Accel Input	Switch Off: 0v, On: 12v
73	WHT/BLU	CKP Sensor Signal	Digital Signals: 0-5-0v
74	BLU/RED	ECT Sensor Signal	At 180°F: 0.54v
75	RED/GRN	Ignition Power (Main Relay)	12-14v
76	PNK/BLK	A/T: Range Signal 'P'	In 'P': 12v, Others: 0v
77	PNK/BLU	A/T: Range Signal 'A'	In 'N': 0v, Others: 12v
78	PNK/YEL	A/T: Range Signal 'C'	In 'C': 0v, Others: 12v
79	PNK	A/T: Range Signal 'B'	In 'B': 0v, Others: 12v
80	BLU	A. P. Sensor 2 Ground	<0.050v

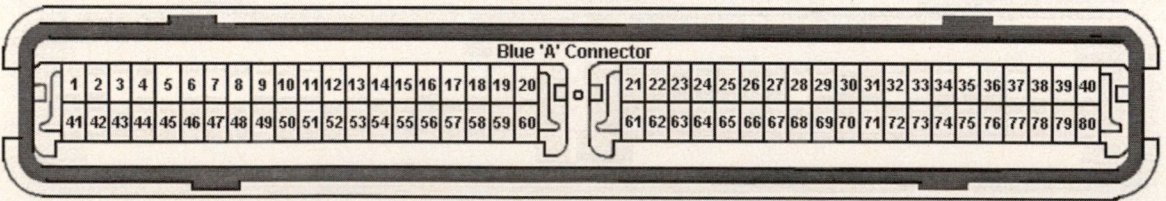

1998-03 EX, LX 3.2L VIN W 2WD [A/T] 80P J22 Red Connector

PCM Pin #	W/Color	Circuit Description (40-Pin)	Value at Hot Idle
1	GRN	Sensor Ground	<0.050v
2	BLK/ORN	HO2S-11 (B1 S1) Heater	Digital Signals: 0-12-0v
3	YEL	A/P Sensor 2 Signal	N/A
4	ORN	Reference Voltage	4.9-5.1v
5	WHT	CKP Sensor Reference	4.9-5.1v
6	GRY/GRN	Cruise Set Switch	Switch Off: 0v, On: 12v
7	RED/BLU	EVAP Purge Solenoid	Valve Off: 12v, On: 1v
8	GRN/WHT	Indicator Control	N/A
9	RED/GRN	Ignition Power (Main Relay)	12-14v
10	---	Not Used	---
11	WHT/GRN	Malfunction Indicator Lamp	MIL Off: 12v, On: 1v
12	GRN	Throttle Motor (+) Signal	Pulse Signals
13	RED/WHT	Coil On Plug 4 Control	Digital Signals: 0-12-0v
14	BLU	HO2S-12 (B1 S2) Ground	<0.050v
15	BLU	HO2S-22 (B2 S2) Ground	<0.050v
16	BLU	HO2S-22 (B2 S2) Ground	<0.050v
17	PNK	HO2S-21 (B1 S1) Ground	<0.050v
18	WHT	HO2S-11 (B1 S1) Ground	<0.050v
19	RED/YEL	Coil On Plug 5 Control	Digital Signals: 0-12-0v
20	YEL/GRN	Sensor Ground	<0.050v
21	BLU/YEL	Sensor Ground	<0.050v
22	BLU/YEL	OSS Sensor 'P' Signal	AC Pulses
24	RED	HO2S-22 (B2 S2) Ground	0.1-1.1v
25	---	Not Used	---
26	BLU/RED	EGR Position Solenoid	Digital Signals: 0-12-0v
27	BLK/YEL	Vehicle Speed Sensor	Moving: 0-5-0-5v
28	GRN/YEL	Injector 6 Control	2.0-3.3 ms
29	---	Not Used	---
30	WHT/GRN	Cruise Engage Switch	Switch Off: 0v, On: 12v
31	GRY/RED	FTP Sensor Reference	4.9-5.1v
32	RED/GRN	Ignition Power (Main Relay)	12-14v
33-34	---	Not Used	---
35	RED/GRN	Coil On Plug 6 Control	Digital Signals: 0-12-0v
36	BLK	A/P Sensor 1 Reference	4.9-5.1v
37	---	Not Used	---
38	RED	TP Sensor 1 Reference	4.9-5.1v
39-40	---	Not Used	---

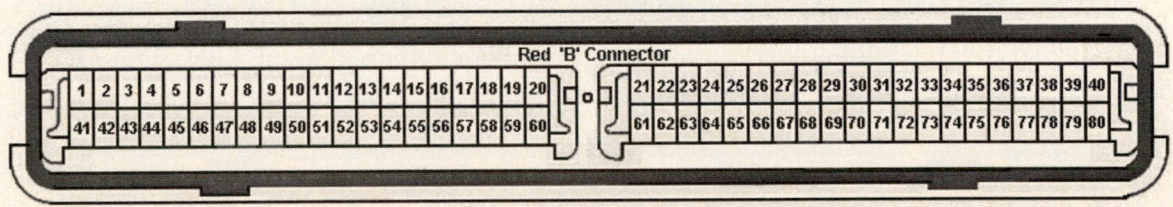

1998-03 EX, LX 3.2L VIN W 2WD [A/T] 80P J22 Red Connector

PCM Pin #	W/Color	Circuit Description (40-Pin)	Value at Hot Idle
41	BLU/ORN	A/P Sensor 3 Ground	<0.050v
42	WHT/RED	HO2S-12 (B1 S2) Heater	Digital Signals: 0-12-0v
43-46	---	Not Used	---
47	WHT/BLK	Water Gauge Signal	0-5v
48	---	Not Used	---
49	RED/WHT	Class 2 Data Line to DLC	0v
50	ORN/GRN	Fuel Level Sensor Signal	Duty Cycle Signal
51	---	Not Used	---
52	GRN	FTP Sensor Signal	With Fuel Cap Off: 2.5v
53	---	Not Used	---
54	BLU	HO2S-12 (B1 S2) Ground	<0.050v
55	BLU/GRN	OSS Sensor 'N' Signal	AC Pulses
56-57	---	Not Used	---
58	ORN/BLU	HO2S-11 (B1 S1) Signal	0.1-1.1v
59	LT GRN	Cruise Set Switch	Switch Off: 0v, On: 12v
60	BLU/YEL	Fuel Level Sensor Reference	12v
61	YEL/GRN	IAT Sensor Signal	Varies w/temp: 0.5-4.8v
62	RED	Coil On Plug 1 Control	Digital Signals: 0-12-0v
63	WHT	HO2S-21 (B2 S1) Signal	0.1-1.1v
64	PNK	HO2S-12 (B1 S2) Signal	0.1-1.1v
65	GRN	HO2S-22 (B2 S2) Signal	0.1-1.1v
66	---	Not Used	---
67	GRY/RED	EGR Position Sensor Signal	1.2-2.0v
68	WHT	A/P Sensor 1 Signal	N/A
69	GRN	Throttle Motor (+) Signal	Pulse Signals
70	---	Not Used	---
71	GRN/WHT	HO2S-22 (B2 S2) Heater	Digital Signals: 0-12-0v
72	RED/GRN	Ignition Power (Main Relay)	12-14v
73	GRN/WHT	Cruise Control Indicator	Lamp Off: 12v, On: 1v
74	YEL/BLK	Intake Air VSV Control	Valve Off: 12v, On: 1v
75	BRN/WHT	EVAP Purge Cut Solenoid	Valve Off: 12v, On: 1v
76	BLU	TP Sensor 1 Signal	0.5-0.8v
77	ORN	A/P Sensor 3 Reference	4.9-5.1v
78	YEL	A/P Sensor 2 Signal	N/A
79	BLU/GRN	A/P Sensor 3 Signal	N/A
80	GRN	MAP Sensor Reference	4.9-5.1v

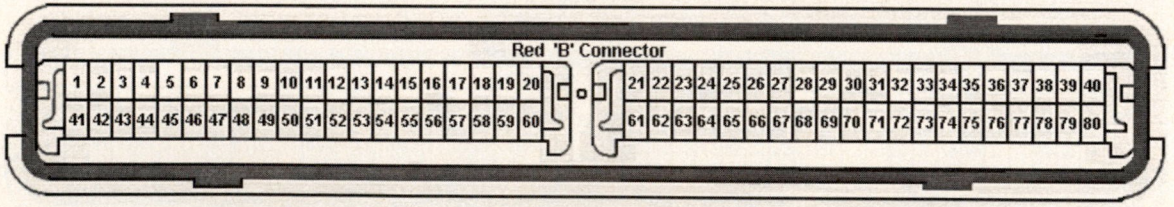

Red 'B' Connector

1998-03 EX, LX 3.2L VIN W 2WD [M/T] 80P J21 Blue Connector

PCM Pin #	W/Color	Circuit Description (40-Pin)	Value at Hot Idle
1	BLK	Power Ground	<0.1v
2	GRN	Reference Voltage 'A'	4.9-5.1v
3	RED/WHT	Reference Voltage 'B'	4.9-5.1v
4	GRY/RED	A/C Clutch Relay Control	Relay Off: 12v, On: 1v
5-6	---	Not Used	---
7	BLK/RED	Tachometer Signal	Pulse Signals
8	---	Not Used	---
9	YEL	Knock Sensor 1 Signal	No detonation: 18 mv AC
10	RED	Knock Sensor 2 Signal	No detonation: 18 mv AC
11	---	Not Used	---
12	WHT/RED	CMP Sensor Signal	Digital Signals: 0-5-0v
13	PNK	To be done	---
14	BLK/BLU	MAF Sensor Signal	4.2v
15-16	---	Not Used	---
17	RED	HO2S-22 (B2 S2) Ground	0.1-1.1v
18	---	Not Used	---
19	RED/GRN	Ignition Power (Main Relay)	12-14v
20	RED/WHT	Keep Alive Power	12-14v
21	WHT	HO2S-11 (B1 S1) Ground	<0.050v
22	---	Not Used	---
23	BLU	HO2S-12 (B1 S2) Ground	<0.050v
24	---	Not Used	---
25	RED/BLU	Coil On Plug 3 Control	Digital Signals: 0-12-0v
26	RED/BLK	Coil On Plug 2 Control	Digital Signals: 0-12-0v
27	GRN	IAT Sensor Signal	Varies w/temp: 0.5-4.8v
28	---	Not Used	---
29	RED	MAP Sensor Signal	0.9v
30-36	---	Not Used	---
37	GRN/YEL	PSP Switch Signal	Straight: 12v, Turning: 0v
38	GRN/YEL	Headlights On Signal	Lights Off: 12v, On: 0v
39	RED	Brake Switch Signal	Brake Off: 0v, On: 12v
40	BLK/BLU	Power Ground	<0.1v

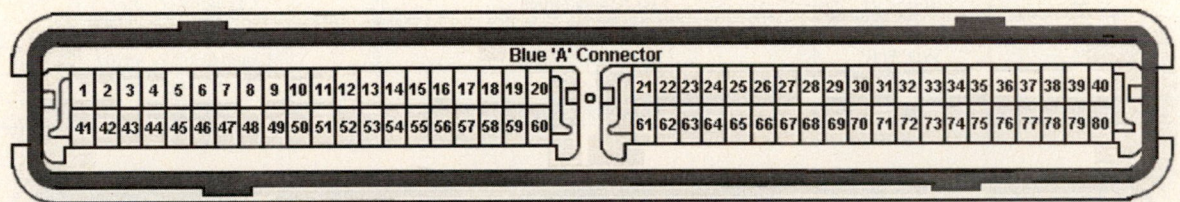

1998-03 EX, LX 3.2L VIN W 2WD [M/T] 80P J21 Blue Connector

PCM Pin #	W/Color	Circuit Description (40-Pin)	Value at Hot Idle
41	GRN	TP Sensor 1 Ground	<0.050v
42	PNK/WHT	Fuel Pump Relay Control	Relay Off: 0v, On: 12v
43	---	Not Used	---
44	ORN/WHT	HO2S-21 (B2 S1) Heater	Digital Signals: 0-12-0v
45	GRN/BLK	A/C Switch Signal	A/C On: 12v, Off: 0v
46	BRN/RED	Brake Switch Signal	Brake Off: 12v, On: 0v
47	---	Not Used	---
48	BLU	Throttle Motor (-) Signal	Pulse Signals
49	BLK	Knock Sensor 1 Ground	<0.050v
50	ORN	HO2S-22 (B2 S2) Ground	<0.050v
51	---	Not Used	---
52	RED	HO2S-21 (B2 S1) Ground	<0.050v
53	BLK	ECT & EGR Sensor Ground	<0.050v
54	BLK	FTP Sensor Ground	<0.050v
55	---	Not Used	---
56	GRN/BLK	Injector 5 Control	2.0-3.3 ms
57	RED/WHT	Keep Alive Power	12-14v
58	ORN/BLK	Class 2 Serial Data Link	0v
59	---	Not Used	---
60	BLK	Knock Sensor 2 Ground	<0.050v
61	GRN/WHT	TP Sensor 2 Ground	<0.050v
62	GRN/ORN	Injector 2 Control	2.0-3.3 ms
63	RED	A/P Sensor 1 Ground	<0.1v
64	GRN/RED	Injector 4 Control	2.0-3.3 ms
65	BLU/WHT	TP Sensor 2 Signal	0.5-0.6v
66	GRN	Injector 3 Control	2.0-3.3 ms
67	YEL	PCM C.Q. Input (1-3-5)	Digital Signals
68	RED	Knock Sensor Input (2-4-6)	<0.050v
69	GRN/WHT	Injector 1 Control	2.0-3.3 ms
70	---	Not Used	---
71	GRN/WHT	HO2S-22 (B2 S2) Heater	Digital Signals: 0-12-0v
72	WHT/BLU	C/C Resume/Accel Input	Switch Off: 0v, On: 12v
73	WHT/BLU	CKP Sensor Signal	Digital Signals: 0-5-0v
74	BLU/RED	ECT Sensor Signal	At 180°F: 0.54v
75	RED/GRN	Ignition Power (Main Relay)	12-14v
76	PNK/BLK	Clutch Switch Signal	Clutch Out: 12v, In: 0v
77-79	---	Not Used	---
80	BLU	A. P. Sensor 2 Ground	<0.050v

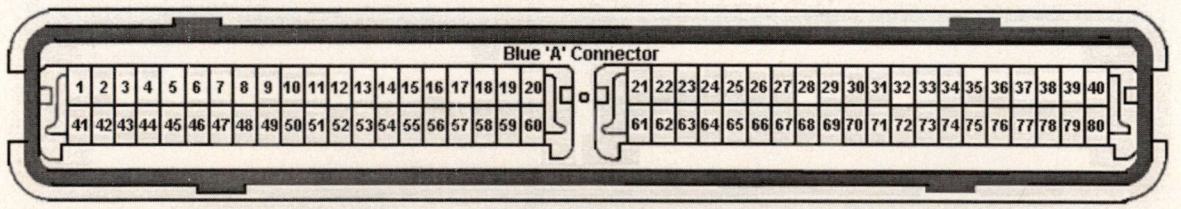

1998-03 EX, LX 3.2L VIN W 2WD [M/T] 80P J22 Red Connector

PCM Pin #	W/Color	Circuit Description (40-Pin)	Value at Hot Idle
1	GRN	Sensor Ground	<0.050v
2	BLK/ORN	HO2S-11 (B1 S1) Heater	Digital Signals: 0-12-0v
3	YEL	A/P Sensor 2 Signal	N/A
4	ORN	Reference Voltage	4.9-5.1v
5	WHT	CKP Sensor Reference	4.9-5.1v
6	GRY/GRN	Cruise Set Switch	Switch Off: 0v, On: 12v
7	RED/BLU	EVAP Purge Solenoid	Valve Off: 12v, On: 1v
8	GRN/WHT	Indicator Control	N/A
9	RED/GRN	Ignition Power (Main Relay)	12-14v
10	---	Not Used	---
11	WHT/GRN	Malfunction Indicator Lamp	MIL Off: 12v, On: 1v
12	GRN	Throttle Motor (+) Signal	Pulse Signals
13	RED/WHT	Coil On Plug 4 Control	Digital Signals: 0-12-0v
14	BLU	HO2S-12 (B1 S2) Ground	<0.050v
15	BLU	HO2S-22 (B2 S2) Ground	<0.050v
16	BLU	HO2S-22 (B2 S2) Ground	<0.050v
17	PNK	HO2S-21 (B1 S1) Ground	<0.050v
18	WHT	HO2S-11 (B1 S1) Ground	<0.050v
19	RED/YEL	Coil On Plug 5 Control	Digital Signals: 0-12-0v
20	YEL/GRN	Sensor Ground	<0.050v
21	BLU/YEL	Sensor Ground	<0.050v
22	---	Not Used	---
24	RED	HO2S-22 (B2 S2) Ground	0.1-1.1v
25	---	Not Used	---
26	BLU/RED	EGR Position Solenoid	Digital Signals: 0-12-0v
27	BLK/YEL	Vehicle Speed Sensor	Moving: 0-5-0-5v
28	GRN/YEL	Injector 6 Control	2.0-3.3 ms
29	---	Not Used	---
30	WHT/GRN	Cruise Engage Switch	Switch Off: 0v, On: 12v
31	GRY/RED	FTP Sensor Reference	4.9-5.1v
32	RED/GRN	Ignition Power (Main Relay)	12-14v
33-34	---	Not Used	---
35	RED/GRN	Coil On Plug 6 Control	Digital Signals: 0-12-0v
36	BLK	A/P Sensor 1 Reference	4.9-5.1v
37	---	Not Used	---
38	RED	TP Sensor 1 Reference	4.9-5.1v
39-40	---	Not Used	---

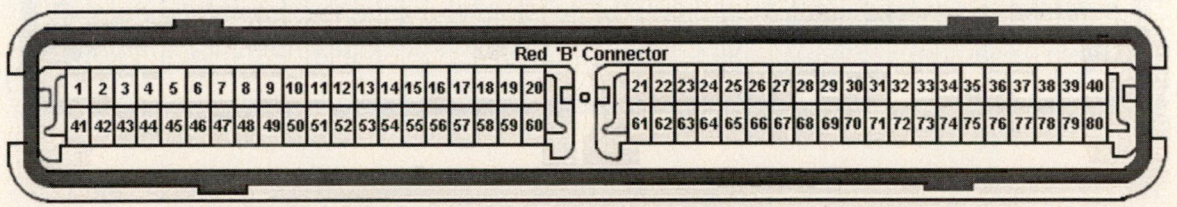

Red 'B' Connector

1998-03 EX, LX 3.2L VIN W 2WD [M/T] 80P J22 Red Connector

PCM Pin #	W/Color	Circuit Description (40-Pin)	Value at Hot Idle
41	BLU/ORN	A/P Sensor 3 Ground	<0.050v
42	WHT/RED	HO2S-12 (B1 S2) Heater	Digital Signals: 0-12-0v
43-46	---	Not Used	---
47	WHT/BLK	Water Gauge Signal	0-5v
48	---	Not Used	---
49	RED/WHT	Class 2 Data Line to DLC	0v
50	ORN/GRN	Fuel Level Sensor Signal	Duty Cycle Signal
51	---	Not Used	---
52	GRN	FTP Sensor Signal	With Fuel Cap Off: 2.5v
53	---	Not Used	---
54	BLU	HO2S-12 (B1 S2) Ground	<0.050v
55-57	---	Not Used	---
58	ORN/BLU	HO2S-11 (B1 S1) Signal	0.1-1.1v
59	LT GRN	Cruise Set Switch	Switch Off: 0v, On: 12v
60	BLU/YEL	Fuel Level Sensor Reference	12v
61	YEL/GRN	IAT Sensor Signal	Varies w/temp: 0.5-4.8v
62	RED	Coil On Plug 1 Control	Digital Signals: 0-12-0v
63	WHT	HO2S-21 (B2 S1) Signal	0.1-1.1v
64	PNK	HO2S-12 (B1 S2) Signal	0.1-1.1v
65	GRN	HO2S-22 (B2 S2) Signal	0.1-1.1v
66	---	Not Used	---
67	GRY/RED	EGR Position Sensor Signal	1.2-2.0v
68	WHT	A/P Sensor 1 Signal	N/A
69	GRN	Throttle Motor (+) Signal	Pulse Signals
70	---	Not Used	---
71	GRN/WHT	HO2S-22 (B2 S2) Heater	Digital Signals: 0-12-0v
72	RED/GRN	Ignition Power (Main Relay)	12-14v
73	GRN/WHT	Cruise Control Indicator	Lamp Off: 12v, On: 1v
74	YEL/BLK	Intake Air VSV Control	Valve Off: 12v, On: 1v
75	BRN/WHT	EVAP Purge Cut Solenoid	Valve Off: 12v, On: 1v
76	BLU	TP Sensor 1 Signal	0.5-0.8v
77	ORN	A/P Sensor 3 Reference	4.9-5.1v
78	YEL	A/P Sensor 2 Signal	N/A
79	BLU/GRN	A/P Sensor 3 Signal	N/A
80	GRN	MAP Sensor Reference	4.9-5.1v

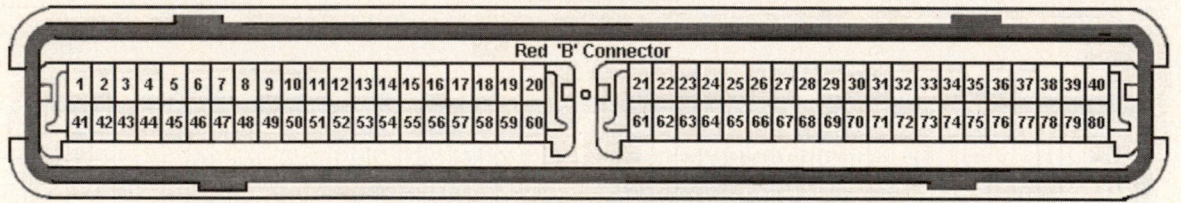

Red 'B' Connector

1994-95 EX, LX 3.2L V6 MFI 4WD VIN V [A/T] 12P 'A' Connector

PCM Pin #	W/Color	Circuit Description (12-Pin)	Value at Hot Idle
A1	BLU/ORN	TP & MAP VREF	4.9-5.1v
A2	GRY/YEL	A/C Clutch Relay	Relay Off: 12v, On: 1v
A3	---	Not Used	---
A4	BLUPNK	M/T: Shift Indicator Lamp	Lamp on: 1v, off: 12v
A5	GRN	Malfunction Indicator Lamp	MIL Off: 12v, On: 1v
A6	RED/GRN	B+ from Main Relay	12-14v
A7	WHT/RED	EVAP Purge Solenoid	Digital Signals: 0-12-0v
A8	WHT/GRN	EGR Solenoid Control	Sol. on: 1v, off: 12
A9	YEL/BLU	ECT Input to TCM	At 180°F: 0.54v
A10	---	Not Used	---
A11	GRY	Sensor Ground	<0.050v
A12	BLK/ORN	Power Ground	<0.1v

1994-95 EX, LX 3.2L V6 MFI 4WD VIN V [A/T] 12P 'B' Connector

PCM Pin #	W/Color	Circuit Description (12-Pin)	Value at Hot Idle
B1	RED/WHT	Keep Alive Power	12-14v
B2	BLK/PNK	HO2S-11 (B1 S1) Ground	<0.050v
B3	RED	HO2S-11 (B1 S1) Signal	0.1-1.1v
B4	BRN/BLK	Idle Air Control 'A' High	Pulse Signals
B5	BRN/WHT	Idle Air Control 'A' Low	Pulse Signals
B6	BRN/YEL	Idle Air Control 'B' Low	Pulse Signals
B7	BRN/RED	Idle Air Control 'B' High	Pulse Signals
B8	YEL/VLT	Electronic Ignition VREF Low	<0.050v
B9	YEL/RED	Electronic Ignition VREF High	Varies: 1.1-2.4v
B10-12	---	Not Used	---

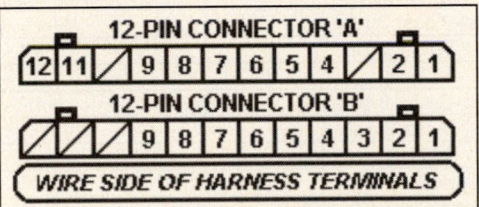

Abbreviation	Color	Abbreviation	Color	Abbreviation	Color
BLK	Black	LT BLU	Light Blue	TAN	Tan
BLU	Blue	LT GRN	Lt. Green	VIO	Violet
BRN	Brown	ORN	Orange	WHT	White
GRY	Gray	PNK	Pink	YEL	Yellow
GRN	Green	PPL	Purple		

1994-95 EX, LX 3.2L V6 MFI 4WD VIN V [A/T] 16P 'C' Connector

PCM Pin #	W/Color	Circuit Description (16-Pin)	Value at Hot Idle
C1	---	Not Used	---
C2	ORN/BLK	Data Link Connector	4.80v (no Scan Tool)
C3	PNK/WHT	Fuel Pump Relay Control	Relay on: 12v, off: 0v
C4	ORN/YEL	DLC Diagnostic Request	Digital Signals
C5	---	Not Used	---
C6	GRN/YEL	PSP Switch Signal	Straight: 12v, Turning: 0v
C7	ORN	PNP Switch Signal	In P/N: 0v, others: 12v
C8	---	Not Used	---
C9	YEL/BLK	A/C Switch Signal	Switch on: 12v, off: 0v
C10	---	Not Used	---
C11	BLU/YEL	Injector Pair 1 & 2	2.0-3.3 ms
C12	BLK/ORN	Power Ground	<0.1v
C13	---	Not Used	---
C14	BLU/WHT	Injector Pair 5 & 6	2.0-3.3 ms
C15	BLK/GRN	Power Ground	<0.1v
C16	RED/WHT	Keep Alive Power	12-14v

1994-95 EX, LX 3.2L V6 MFI 4WD VIN V [A/T] 16P 'D' Connector

PCM Pin #	W/Color	Circuit Description (16-Pin)	Value at Hot Idle
D1	BLK/GRN	Power Ground	<0.1v
D2	BLK/WHT	Sensor Ground	<0.050v
D3-4	---	Not Used	---
D5	YEL	Igniter Control Signal	Digital Signals: 0-12-0v
D6	BLU/RED	TP Sensor Signal	0.5-0.6v
D7	BLU	IAT Sensor Signal	Varies w/temp: 0.5-4.8v
D8	WHT	Vehicle Speed Sensor	Moving: 0-5-0-5v
D9	GRY/RED	MAP Sensor Signal	0.6-1.3v
D10	---	Not Used	---
D11	YEL/GRN	Ignition Control Bypass	Hot idle: 4.9v
D12	GRY/BLK	ECT Sensor Signal	At 180°F: 0.54v
D13-14	---	Not Used	---
D15	BLU/GRN	Injector Pair 3 & 4	2.0-3.3 ms
D16	BLU/GRN	Injector Pair 3 & 4	2.0-3.3 ms

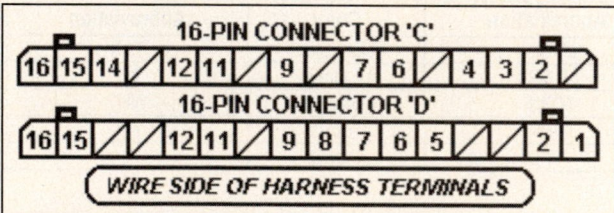

1994-95 EX, LX 3.2L V6 MFI 4WD VIN V [M/T] 12P 'A' Connector

PCM Pin #	W/Color	Circuit Description (12-Pin)	Value at Hot Idle
A1	BLU/ORN	TP & MAP Sensor VREF	4.9-5.1v
A2	GRY/YEL	A/C Clutch Relay	Relay Off: 12v, On: 1v
A3-4	---	Not Used	---
A5	GRN	Malfunction Indicator Lamp	MIL Off: 12v, On: 1v
A6	RED/GRN	B+ from Main Relay	12-14v
A7	WHT/RED	EVAP Purge Solenoid	Digital Signals: 0-12-0v
A8	WHT/GRN	EGR Solenoid Control	Sol. on: 1v, off: 12v
A9-10	---	Not Used	---
A11	GRY	Sensor Ground	<0.050v
A12	BLK/ORN	Power Ground	<0.1v

1994-95 EX, LX 3.2L V6 MFI 4WD VIN V [M/T] 12P 'B' Connector

PCM Pin #	W/Color	Circuit Description (12-Pin)	Value at Hot Idle
B1	RED/WHT	Keep Alive Power	12-14v
B2	BLK/PNK	HO2S-11 (B1 S1) Ground	<0.050v
B3	RED	HO2S-11 (B1 S1) Signal	0.1-1.1v
B4	BRN/BLK	Idle Air Control 'A' High	Pulse Signals
B5	BRN/WHT	Idle Air Control 'A' Low	Pulse Signals
B6	BRN/YEL	Idle Air Control 'B' Low	Pulse Signals
B7	BRN/RED	Idle Air Control 'B' High	Pulse Signals
B8	YEL/VLT	Electronic Ignition VREF Low	<0.050v
B9	YEL/RED	Electronic Ignition VREF High	Digital Pulses: 0-5-0-5v
B10-12	---	Not Used	---

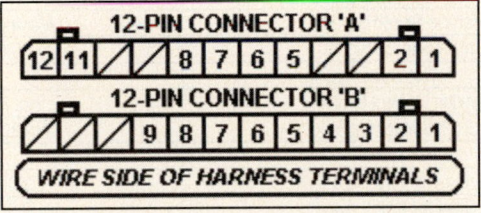

Abbreviation	Color	Abbreviation	Color	Abbreviation	Color
BLK	Black	LT BLU	Light Blue	TAN	Tan
BLU	Blue	LT GRN	Lt. Green	VIO	Violet
BRN	Brown	ORN	Orange	WHT	White
GRY	Gray	PNK	Pink	YEL	Yellow
GRN	Green	PPL	Purple		

1994-95 EX, LX 3.2L V6 MFI 4WD VIN V [M/T] 16P 'C' Connector

PCM Pin #	W/Color	Circuit Description (16-Pin)	Value at Hot Idle
C1	---	Not Used	---
C2	ORN/BLK	Data Link Connector	4.80v (no Scan Tool)
C3	PNK/WHT	Fuel Pump Relay Control	Relay on: 12v, off: 0v
C4	ORN/YEL	DLC Diagnostic Request	Digital Signals
C5	---	Not Used	---
C6	GRN/YEL	PSP Switch Signal	Straight: 12v, Turning: 0v
C7-8	---	Not Used	---
C9	YEL/BLK	A/C Switch Signal	Switch on: 12v, off: 0v
C10	---	Not Used	---
C11	BLU/YEL	Injector Pair 1 & 2	2.0-3.3 ms
C12	BLK/ORN	Power Ground	<0.1v
C13	---	Not Used	---
C14	BLU/WHT	Injector Pair 5 & 6	2.0-3.3 ms
C15	BLK/GRN	Power Ground	<0.1v
C16	RED/WHT	Keep Alive Power	12-14v

1994-95 EX, LX 3.2L V6 MFI 4WD VIN V [M/T] 16P 'D' Connector

PCM Pin #	W/Color	Circuit Description (16-Pin)	Value at Hot Idle
D1	BLK/GRN	Power Ground	<0.1v
D2	BLK/WHT	Sensor Ground	<0.050v
D3-4	---	Not Used	---
D5	YEL	Igniter Control Signal	Idle: 10% d/cycle
D6	BLU/RED	TP Sensor Signal	0.5-0.6v
D7	BLU	IAT Sensor Signal	Varies w/temp: 0.5-4.8v
D8	WHT	Vehicle Speed Sensor	Moving: 0-5-0-5v
D9	GRY/RED	MAP Sensor Signal	0.6-1.3v
D10	---	Not Used	---
D11	YEL/GRN	Ignition Control Bypass	Hot idle: 4.9v
D12	GRY/BLK	ECT Sensor Signal	At 180°F: 0.54v
D13-14	---	Not Used	---
D15	BLU/GRN	Injector Pair 3 & 4	2.0-3.3 ms
D16	BLU/GRN	Injector Pair 3 & 4	2.0-3.3 ms

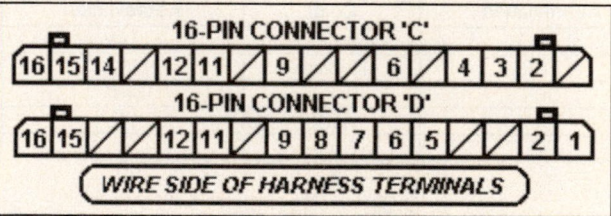

1996-97 EX, LX 3.2L V6 4WD VIN V [A/T] C209 Red 'A' Connector

PCM Pin #	W/Color	Circuit Description (16-Pin)	Value at Hot Idle
A1	RED	Reference Voltage 'A'	4.9-5.1v
A2	YEL	Knock Sensor Signal	No detonation: 18 mv AC
A3, A12	---	Not Used	---
A4	RED/WHT	Keep Alive Power	12-14v
A5	BLU	Idle Air Control 'A' High	Pulse Signals
A6	BLU/WHT	Idle Air Control 'A' Low	Pulse Signals
A7	BLU/BLK	Idle Air Control 'B' Low	Pulse Signals
A8	BLU/RED	Idle Air Control 'B' High	Pulse Signals
A9	ORN/BLK	Transmission Fluid Lamp	Lamp on: 1v, off: 12v
A10	PNK/GRN	Winter Lamp	Lamp on: 1v, off: 12v
A11	PNK/WHT	Power Lamp	Lamp on: 1v, off: 12v
A13	GRN	Malfunction Indicator Lamp	MIL Off: 12v, On: 1v
A14	PPL	Check Transmission Light	Lamp on: 1v, off: 12v
A15	RED/BLU	EVAP Purge Solenoid	Digital Signals: 0-12-0v
A16	BRN/YEL	A/T: Shift Low Band Apply	Sol. on: 1v, off: 12v

1996-97 EX, LX 3.2L V6 4WD VIN V [A/T] C209 Red 'B' Connector

PCM Pin #	W/Color	Circuit Description (16-Pin)	Value at Hot Idle
B1	BLU/ORN	Reference Voltage 'B'	4.9-5.1v
B2	RED/WHT	Igniter 4 Control	Digital Signals: 0-12-0v
B3	RED/BLK	Igniter 2 Control	Digital Signals: 0-12-0v
B4	RED/GRN	Igniter 6 Control	Digital Signals: 0-12-0v
B5	ORN/GRN	Fuel Level Sensor	Tank empty: 1.8v
B6	GRY	Fuel Tank Pressure Sensor	With Fuel Cap Off: 2.5v
B7	YEL/RED	Linear EGR Sensor Signal	0.5-0.6v
B8	YEL/GRN	IAT Sensor Signal	Varies w/temp: 0.5-4.8v
B9	---	Not Used	---
B10	GRN/WHT	Rough Road Sensor	At idle: 2.5v
B11	GRN/YEL	PSP Switch Signal	Straight: 12v, Turning: 0v
B12	GRN/YEL	Headlight Switch Signal	Switch on: 0.1v, off: 12v
B13	ORN/BLK	Class 2 Serial Data Link	0v
B14	GRY/RED	A/C Clutch Relay	Relay Off: 12v, On: 1v
B15	PNK	Low Fuel Lamp	Lamp on: 1v, off: 12v
B16	BRN/WHT	EVAP Vent Solenoid	Sol. on: 1v, off: 12v

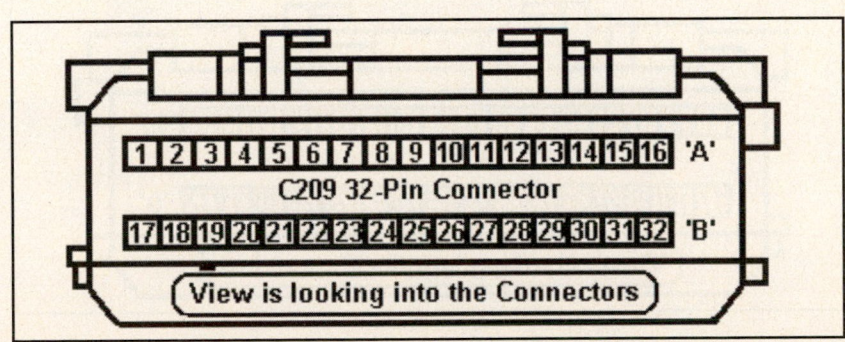

1996-97 EX, LX 3.2L 4WD VIN V [A/T] C211 White 'C' Connector

PCM Pin #	W/Color	Circuit Description (16-Pin)	Value at Hot Idle
C1	GRN/RED	Injector 4 Control	2.0-3.3 ms
C2	BRN/BLK	A/T: Shift Solenoid 'B'	SSB Off: 12v, On: 1v
C3	GRN/YEL	Injector 6 Control	2.0-3.3 ms
C4	RED	Igniter 1 Control	Digital Signals: 0-12-0v
C5	YEL	CKP Sensor Signal	Digital Signals: 0-5-0v
C7	YEL/BLK	Power Ground	<0.1v
C8	BLK/RED	Power Ground	<0.1v
C9	BLK/BLU	Power Ground	<0.1v
C10	BLK/RED	Tachometer Signal	Pulse Signals
C12	RED/WHT	Alternator Charge Signal	12-14v
C13	YEL/RED	Fuel Gauge PWM Signal	Duty cycle: 0-100%
C14	WHT	HO2S-21 (B2 S1) Signal	0.1-1.1v
C15	RED	HO2S-21 (B2 S1) Ground	<0.050v
C16	RED	HO2S-22 (B2 S2) Signal	0.1-1.1v

1996-97 EX, LX 3.2L 4WD VIN V [A/T] C211 White 'D' Connector

PCM Pin #	W/Color	Circuit Description (16-Pin)	Value at Hot Idle
D1	GRN/ORN	Injector 2 Control	2.0-3.3 ms
D2	RED/YEL	Torque Converter Clutch	TCC on: 1v, off: 12v
D3	GRN/WHT	Injector 1 Control	2.0-3.3 ms
D4	RED	Serial Data (8192 Baud)	Digital Signals
D5	RED/YEL	Igniter 5 Control	Digital Signals: 0-12-0v
D6	RED/BLU	Igniter 3 Control	Digital Signals: 0-12-0v
D7	WHT	Vehicle Speed Sensor	Moving: 0-5-0-5v
D8	GRN	Sensor 'A' Ground	<0.050v
D9	GRN	Sensor 'B' Ground	<0.050v
D10	YEL/BLU	MAF Sensor Signal	7-9v
D11	BLU	CMP Sensor Signal	Digital Signals: 0-5-0v
D12	BLU	HO2S-12 (B1 S2) Ground	<0.050v
D13	GRN	HO2S-12 (B1 S2) Signal	0.1-1.1v
D14	BLU	HO2S-11 (B1 S1) Ground	<0.050v
D15	PNK	HO2S-11 (B1 S1) Signal	0.1-1.1v
D16	GRN	HO2S-22 (B2 S2) Ground	<0.050v

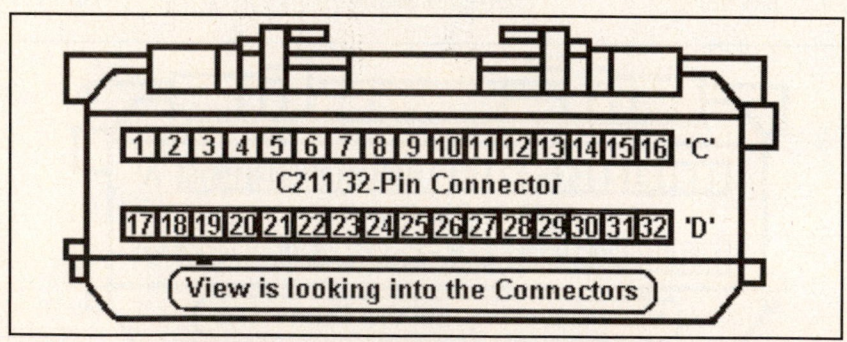

1996-97 EX, LX 3.2L V6 4WD VIN V [A/T] C212 Blue 'E' Connector

PCM Pin #	W/Color	Circuit Description (16-Pin)	Value at Hot Idle
E1	YEL	VSS "High" Signal	AC Pulse Signals
E2	BRN	VSS "Low" Signal	<0.050v
E3	RED/ORN	A/T: Pressure Control 'N'	Pulse Signals
E4	RED/BLK	A/T: Pressure Control 'P'	Pulse Signals
E5	RED/GRN	Linear EGR Solenoid "High"	Pulse Signals
E6	YEL	Linear EGR Solenoid "Low"	Pulse Signals
E7	PNK	A/T: Range Signal 'B'	In P/N: 1v, others: 12v
E8	BLU	TP Sensor Signal	0.5-0.6v
E9	BLU/RED	ECT Sensor Signal	At 180°F: 0.54v
E10	---	Not Used	---
E11	YEL/RED	CKP Sensor VREF	4.9-5.1v
E12	PNK/BLU	A/T: Range Signal 'A'	In P/N: 1v, others: 12v
E13	PNK/WHT	Fuel Pump Relay Control	Relay on: 12v, off: 0v
E14	BRN/WHT	A/T: Shift Solenoid Feed (B+)	12-14v
E15	GRN/BLK	A/C Switch Signal	Switch on: 12v, off: 0v
E16	RED/BLU	B+ from Main Relay	12-14v

1996-97 EX, LX 3.2L V6 4WD VIN V [A/T] C212 Blue 'F' Connector

PCM Pin #	W/Color	Circuit Description (16-Pin)	Value at Hot Idle
F1	---	Not Used	---
F2	BLU/WHT	A/T: Range Signal 'C'	In P/N: 1v, others: 12v
F3	PNK/BLK	A/T: Range Signal 'P'	In P/N: 1v, others: 12v
F4	RED	Brake Switch Signal	Brake on: 12v, Off: 0v
F5	PPL/RED	Power Switch Signal	Switch on: 0.1v, off: 12v
F6	PPL/GRN	Winter Switch Signal	Switch on: 0.1v, off: 12v
F7	GRN/RED	A/T: TOT Sensor Signal	Vaires w/temp. (0.5-4.9v)
F8	RED	MAP Sensor Signal	0.6-1.3v
F9, F12	---	Not Used	---
F10	GRY/BLU	Cruise Control	C/C on: 0.1v, off: 12v
F11	LT BLU	A/T: Kickdown Switch	Switch on: 0.1v, off: 12v
F13	WHT/GRN	Injector 3 Control	2.0-3.3 ms
F14	YEL/GRN	A/T: Shift Solenoid 'A'	Sol. on: 1v, off: 12v
F15	GRN/BLK	Injector 5 Control	2.0-3.3 ms
F16	RED/BLU	B+ from Main Relay	12-14v

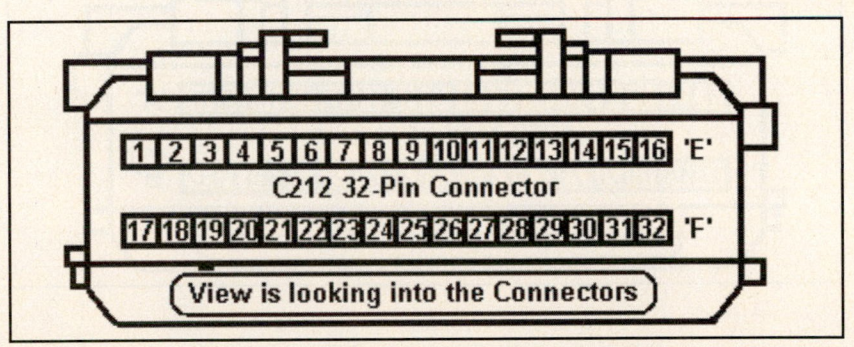

1996-97 EX, LX 3.2L V6 4WD VIN V [M/T] C209 Red 'A' Connector

PCM Pin #	W/Color	Circuit Description (16-Pin)	Value at Hot Idle
A1	RED	Reference Voltage 'A'	4.9-5.1v
A2	YEL	Knock Sensor Signal	No detonation: 18 mv AC
A3	---	Not Used	---
A4	RED/WHT	Keep Alive Power	12-14v
A5	BLU	Idle Air Control 'A' High	Pulse Signals
A6	BLU/WHT	Idle Air Control 'A' Low	Pulse Signals
A7	BLU/BLK	Idle Air Control 'B' Low	Pulse Signals
A8	BLU/RED	Idle Air Control 'B' High	Pulse Signals
A9-12	---	Not Used	---
A13	GRN	Malfunction Indicator Lamp	MIL Off: 12v, On: 1v
A14	PPL	Check Transmission Light	Lamp on: 1v, off: 12v
A15	RED/BLU	EVAP Purge Solenoid	Digital Signals: 0-12-0v
A16	---	Not Used	---

1996-97 EX, LX 3.2L V6 4WD VIN V [M/T] C209 Red 'B' Connector

PCM Pin #	W/Color	Circuit Description (16-Pin)	Value at Hot Idle
B1	BLU/ORN	Reference Voltage 'B'	4.9-5.1v
B2	RED/WHT	Igniter 4 Control	Digital Signals: 0-12-0v
B3	RED/BLK	Igniter 2 Control	Digital Signals: 0-12-0v
B4	RED/GRN	Igniter 6 Control	Digital Signals: 0-12-0v
B5	ORN/GRN	Fuel Level Sensor	Tank empty: 1.8v
B6	GRY	Fuel Tank Pressure Sensor	With Fuel Cap Off: 2.5v
B7	YEL/RED	Linear EGR Sensor Signal	0.5-0.6v
B8	YEL/GRN	IAT Sensor Signal	Varies w/temp: 0.5-4.8v
B9	---	Not Used	---
B10	GRN/WHT	Rough Road Sensor	At idle: 2.5v
B11	GRN/YEL	PSP Switch Signal	Straight: 12v, Turning: 0v
B12	GRN/YEL	Headlight Switch Signal	Switch on: 0.1v, off: 12v
B13	ORN/BLK	Class 2 Serial Data Link	0v
B14	GRY/RED	A/C Clutch Relay	Relay Off: 12v, On: 1v
B15	PNK	Low Fuel Lamp	Lamp on: 1v, off: 12v
B16	BRN/WHT	EVAP Vent Solenoid	Sol. on: 1v, off: 12v

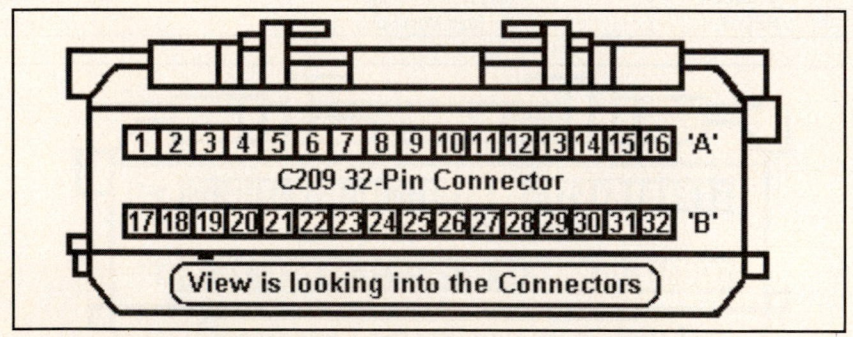

1996-97 EX, LX 3.2L 4WD VIN V [M/T] C211 White 'C' Connector

PCM Pin #	W/Color	Circuit Description (16-Pin)	Value at Hot Idle
C1	GRN/RED	Injector 4 Control	2.0-3.3 ms
C2	---	Not Used	---
C3	GRN/YEL	Injector 6 Control	2.0-3.3 ms
C4	RED	Igniter 1 Control	Digital Signals: 0-12-0v
C5	YEL	CKP Sensor Signal	Digital Signals: 0-5-0v
C7	YEL/BLK	Power Ground	<0.1v
C8	BLK/RED	Power Ground	<0.1v
C9	BLK/BLU	Power Ground	<0.1v
C10	BLK/RED	Tachometer Signal	Pulse Signals
C11	YEL/GRN	Upshift Lamp Control	U/L on: 1v, off: 12
C12	RED/WHT	Alternator Charge Signal	12-14v
C13	YEL/RED	Fuel Gauge PWM Signal	Duty cycle: 0-100%
C14	WHT	HO2S-21 (B2 S1) Signal	0.1-1.1v
C15	RED	HO2S-21 (B2 S1) Ground	<0.050v
C16	RED	HO2S-22 (B2 S2) Signal	0.1-1.1v

1996-97 EX, LX 3.2L 4WD VIN V [M/T] C211 White 'D' Connector

PCM Pin #	W/Color	Circuit Description (16-Pin)	Value at Hot Idle
D1	GRN/ORN	Injector 2 Control	2.0-3.3 ms
D2	---	Not Used	---
D3	GRN/WHT	Injector 1 Control	2.0-3.3 ms
D4	RED	Serial Data (8192 Baud)	Digital Signals
D5	RED/YEL	Igniter 5 Control	Digital Signals: 0-12-0v
D6	RED/BLU	Igniter 3 Control	Digital Signals: 0-12-0v
D7	WHT	Vehicle Speed Sensor	Moving: 0-5-0-5v
D8	GRN	Sensor 'A' Ground	<0.050v
D9	GRN	Sensor 'B' Ground	<0.050v
D10	YEL/BLU	MAF Sensor Signal	7-9v
D11	BLU	CMP Sensor Signal	Digital Signals: 0-5-0v
D12	BLU	HO2S-13 (B1 S3) Ground	<0.050v
D13	GRN	HO2S-13 (B1 S3) Signal	0.1-1.1v
D14	BLU	HO2S-11 (B1 S1) Ground	<0.050v
D15	PNK	HO2S-11 (B1 S1) Signal	0.1-1.1v
D16	GRN	HO2S-22 (B2 S2) Ground	<0.050v

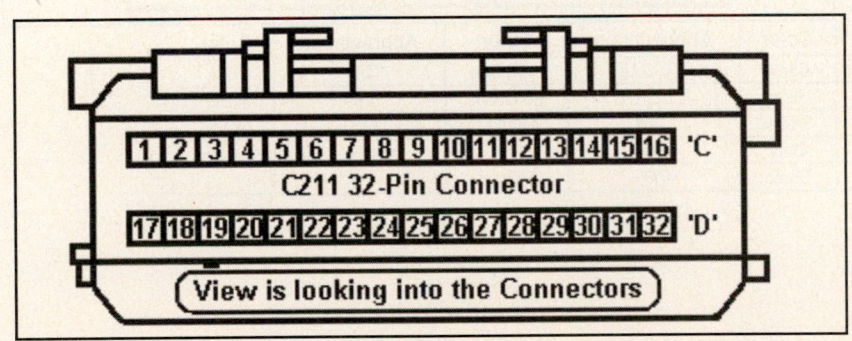

C211 32-Pin Connector

View is looking into the Connectors

1996-97 EX, LX 3.2L V6 4WD VIN V [M/T] C212 Blue 'E' Connector

PCM Pin #	W/Color	Circuit Description (16-Pin)	Value at Hot Idle
E1-4	---	Not Used	---
E5	RED/GRN	Linear EGR Solenoid 'P'	Pulse Signals
E6	YEL	Linear EGR Solenoid 'N'	Pulse Signals
E7	---	Not Used	---
E8	BLU	TP Sensor Signal	0.5-0.6v
E9	BLU/RED	ECT Sensor Signal	At 180°F: 0.54v
E10	---	Not Used	---
E11	YEL/RED	CKP Sensor VREF	4.9-5.1v
E12	---	Not Used	---
E13	PNK/WHT	Fuel Pump Relay Control	Relay on: 12v, off: 0v
E14	---	Not Used	---
E15	GRN/BLK	A/C Switch Signal	Switch on: 12v, off: 0v
E16	RED/BLU	B+ from Main Relay	12-14v

1996-97 EX, LX 3.2L V6 4WD VIN V [M/T] C212 Blue 'F' Connector

PCM Pin #	W/Color	Circuit Description (16-Pin)	Value at Hot Idle
F1-7	---	Not Used	---
F8	RED	MAP Sensor Signal	0.6-1.3v
F9-12	---	Not Used	---
F13	WHT/GRN	Injector 3 Control	2.0-3.3 ms
F14	---	Not Used	---
F15	GRN/BLK	Injector 5 Control	2.0-3.3 ms
F16	RED/BLU	B+ from Main Relay	12-14v

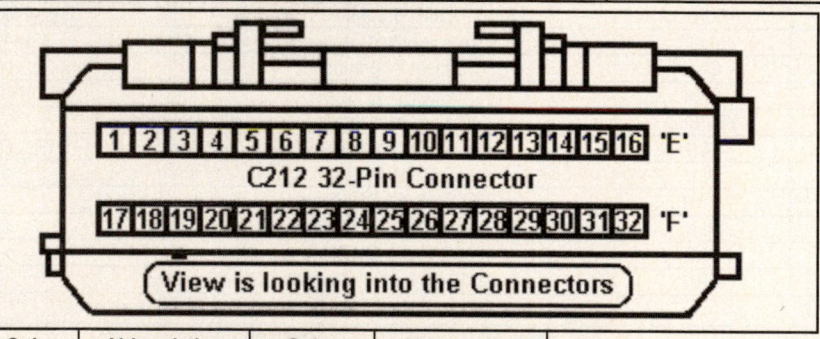

Abbreviation	Color	Abbreviation	Color	Abbreviation	Color
BLK	Black	LT BLU	Light Blue	TAN	Tan
BLU	Blue	LT GRN	Lt. Green	VIO	Violet
BRN	Brown	ORN	Orange	WHT	White
GRY	Gray	PNK	Pink	YEL	Yellow
GRN	Green	PPL	Purple		

1998-03 EX, LX 3.2L VIN W 4WD [A/T] 80P J21 Blue Connector

PCM Pin #	W/Color	Circuit Description (40-Pin)	Value at Hot Idle
1	BLK	Power Ground	<0.1v
2	GRN	Reference Voltage 'A'	4.9-5.1v
3	RED/WHT	Reference Voltage 'B'	4.9-5.1v
4	GRY/RED	A/C Clutch Relay Control	Relay Off: 12v, On: 1v
5	GRN/RED	A/T: Shift Solenoid 'A'	SSA Off: 12v, On: 1v
6	---	Not Used	---
7	BLK/RED	Tachometer Signal	Pulse Signals
8	---	Not Used	---
9	YEL	Knock Sensor 1 Signal	No detonation: 18 mv AC
10	RED	Knock Sensor 2 Signal	No detonation: 18 mv AC
11	---	Not Used	---
12	WHT/RED	CMP Sensor Signal	Digital Signals: 0-5-0v
13	PNK	To be done	---
14	BLK/BLU	MAF Sensor Signal	4.2v
15-16	---	Not Used	---
17	RED	HO2S-22 (B2 S2) Ground	0.1-1.1v
18	---	Not Used	---
19	RED/GRN	Ignition Power (Main Relay)	12-14v
20	RED/WHT	Keep Alive Power	12-14v
21	WHT	HO2S-11 (B1 S1) Ground	<0.050v
22	---	Not Used	---
23	BLU	HO2S-12 (B1 S2) Ground	<0.050v
24	---	Not Used	---
25	RED/BLU	Coil On Plug 3 Control	Digital Signals: 0-12-0v
26	RED/BLK	Coil On Plug 2 Control	Digital Signals: 0-12-0v
27	GRN	IAT Sensor Signal	Varies w/temp: 0.5-4.8v
28	---	Not Used	---
29	RED	MAP Sensor Signal	0.9v
30-36	---	Not Used	---
37	GRN/YEL	PSP Switch Signal	Straight: 12v, Turning: 0v
38	GRN/YEL	Headlights On Signal	Lights Off: 12v, On: 0v
39	RED	Brake Switch Signal	Brake Off: 0v, On: 12v
40	BLK/BLU	Power Ground	<0.1v

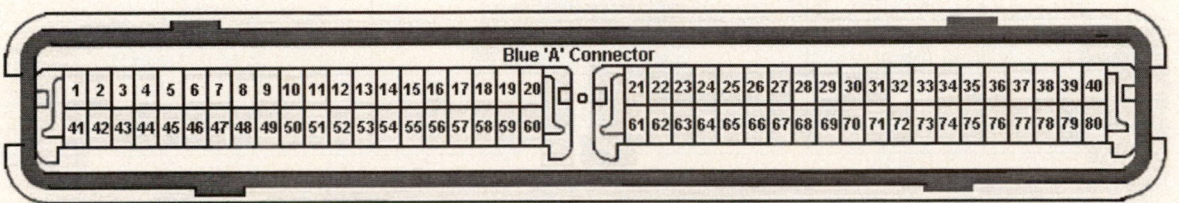

1998-03 EX, LX 3.2L VIN W 4WD [A/T] 80P J21 Blue Connector

PCM Pin #	W/Color	Circuit Description (40-Pin)	Value at Hot Idle
41	GRN	TP Sensor 1 Ground	<0.050v
42	PNK/WHT	Fuel Pump Relay Control	Relay Off: 0v, On: 12v
43	PPL/RED	Pressure Control Solenoid (-)	Pulse Signals
44	ORN/WHT	HO2S-21 (B2 S1) Heater	Digital Signals: 0-12-0v
45	GRN/BLK	A/C Switch Signal	A/C On: 12v, Off: 0v
46	BRN/RED	Brake Switch Signal	Brake Off: 12v, On: 0v
47	PPL/WHT	Pressure Control Solenoid (+)	Pulse Signals
48	BLU	Throttle Motor (-) Signal	Pulse Signals
49	BLK	Knock Sensor 1 Ground	<0.050v
50	ORN	HO2S-22 (B2 S2) Ground	<0.050v
51	----	Not Used	---
52	RED	HO2S-21 (B2 S1) Ground	<0.050v
53	BLK	ECT & EGR Sensor Ground	<0.050v
54	BLK	FTP Sensor Ground	<0.050v
55	----	Not Used	---
56	GRN/BLK	Injector 5 Control	2.0-3.3 ms
57	RED/WHT	Keep Alive Power	12-14v
58	ORN/BLK	Class 2 Serial Data Link	0v
59	----	Not Used	---
60	BLK	Knock Sensor 2 Ground	<0.050v
61	GRN/WHT	TP Sensor 2 Ground	<0.050v
62	GRN/ORN	Injector 2 Control	2.0-3.3 ms
63	RED	A/P Sensor 1 Ground	<0.1v
64	GRN/RED	Injector 4 Control	2.0-3.3 ms
65	BLU/WHT	TP Sensor 2 Signal	0.5-0.6v
66	GRN	Injector 3 Control	2.0-3.3 ms
67	YEL	PCM C.Q. Input (1-3-5)	N/A
68	RED	Knock Sensor Input (2-4-6)	<0.050v
69	GRN/WHT	Injector 1 Control	2.0-3.3 ms
70	----	Not Used	---
71	GRN/WHT	HO2S-22 (B2 S2) Heater	Digital Signals: 0-12-0v
72	WHT/BLU	C/C Resume/Accel Input	Switch Off: 0v, On: 12v
73	WHT/BLU	CKP Sensor Signal	Digital Signals: 0-5-0v
74	BLU/RED	ECT Sensor Signal	At 180°F: 0.54v
75	RED/GRN	Ignition Power (Main Relay)	12-14v
76	PNK/BLK	A/T: Range Signal 'P'	In 'P': 12v, Others: 0v
77	PNK/BLU	A/T: Range Signal 'A'	In 'N': 0v, Others: 12v
78	PNK/YEL	A/T: Range Signal 'C'	In 'C': 0v, Others: 12v
79	PNK	A/T: Range Signal 'B'	In 'B': 0v, Others: 12v
80	BLU	A. P. Sensor 2 Ground	<0.050v

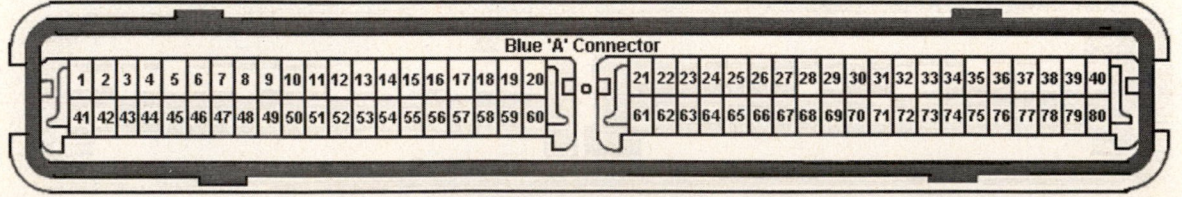

1998-03 EX, LX 3.2L VIN W 4WD [A/T] 80P J22 Red Connector

PCM Pin #	W/Color	Circuit Description (40-Pin)	Value at Hot Idle
1	GRN	Sensor Ground	<0.050v
2	BLK/ORN	HO2S-11 (B1 S1) Heater	Digital Signals: 0-12-0v
3	YEL	A/P Sensor 2 Signal	N/A
4	ORN	Reference Voltage	4.9-5.1v
5	WHT	CKP Sensor Reference	4.9-5.1v
6	GRY/GRN	Cruise Set Switch	Switch Off: 0v, On: 12v
7	RED/BLU	EVAP Purge Solenoid	Valve Off: 12v, On: 1v
8	GRN/WHT	Indicator Control	N/A
9	RED/GRN	Ignition Power (Main Relay)	12-14v
10	---	Not Used	---
11	WHT/GRN	Malfunction Indicator Lamp	MIL Off: 12v, On: 1v
12	GRN	Throttle Motor (+) Signal	Pulse Signals
13	RED/WHT	Coil On Plug 4 Control	Digital Signals: 0-12-0v
14	BLU	HO2S-12 (B1 S2) Ground	<0.050v
15	BLU	HO2S-22 (B2 S2) Ground	<0.050v
16	BLU	HO2S-22 (B2 S2) Ground	<0.050v
17	PNK	HO2S-21 (B1 S1) Ground	<0.050v
18	WHT	HO2S-11 (B1 S1) Ground	<0.050v
19	RED/YEL	Coil On Plug 5 Control	Digital Signals: 0-12-0v
20	YEL/GRN	Sensor Ground	<0.050v
21	BLU/YEL	Sensor Ground	<0.050v
22	BLU/YEL	OSS Sensor 'P' Signal	AC Pulses
24	RED	HO2S-22 (B2 S2) Ground	0.1-1.1v
25	---	Not Used	---
26	BLU/RED	EGR Position Solenoid	Digital Signals: 0-12-0v
27	BLK/YEL	Vehicle Speed Sensor	Moving: 0-5-0-5v
28	GRN/YEL	Injector 6 Control	2.0-3.3 ms
29	---	Not Used	---
30	WHT/GRN	Cruise Engage Switch	Switch Off: 0v, On: 12v
31	GRY/RED	FTP Sensor Reference	4.9-5.1v
32	RED/GRN	Ignition Power (Main Relay)	12-14v
33-34	---	Not Used	---
35	RED/GRN	Coil On Plug 6 Control	Digital Signals: 0-12-0v
36	BLK	A/P Sensor 1 Reference	4.9-5.1v
37	---	Not Used	---
38	RED	TP Sensor 1 Reference	4.9-5.1v
39-40	---	Not Used	---

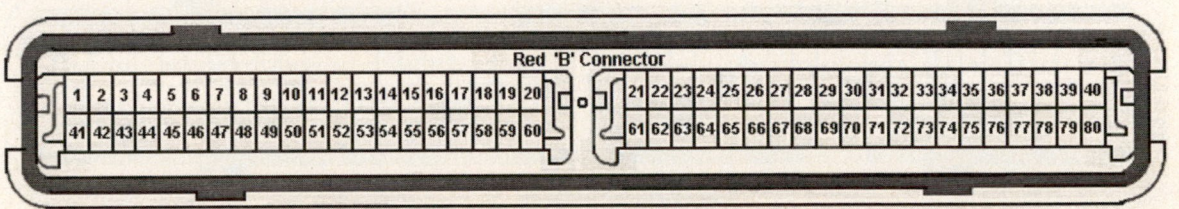

1998-03 EX, LX 3.2L VIN W 4WD [A/T] 80P J22 Red Connector

PCM Pin #	W/Color	Circuit Description (40-Pin)	Value at Hot Idle
41	BLU/ORN	A/P Sensor 3 Ground	<0.050v
42	WHT/RED	HO2S-12 (B1 S2) Heater	Digital Signals: 0-12-0v
43-46	---	Not Used	---
47	WHT/BLK	Water Gauge Signal	0-5v
48	---	Not Used	---
49	RED/WHT	Class 2 Data Line to DLC	0v
50	ORN/GRN	Fuel Level Sensor Signal	Duty Cycle Signal
51	---	Not Used	---
52	GRN	FTP Sensor Signal	With Fuel Cap Off: 2.5v
53	---	Not Used	---
54	BLU	HO2S-12 (B1 S2) Ground	<0.050v
55	BLU/GRN	OSS Sensor 'N' Signal	AC Pulses
56-57	---	Not Used	---
58	ORN/BLU	HO2S-11 (B1 S1) Signal	0.1-1.1v
59	LT GRN	Cruise Set Switch	Switch Off: 0v, On: 12v
60	BLU/YEL	Fuel Level Sensor Reference	12v
61	YEL/GRN	IAT Sensor Signal	Varies w/temp: 0.5-4.8v
62	RED	Coil On Plug 1 Control	Digital Signals: 0-12-0v
63	WHT	HO2S-21 (B2 S1) Signal	0.1-1.1v
64	PNK	HO2S-12 (B1 S2) Signal	0.1-1.1v
65	GRN	HO2S-22 (B2 S2) Signal	0.1-1.1v
66	---	Not Used	---
67	GRY/RED	EGR Position Sensor Signal	1.2-2.0v
68	WHT	A/P Sensor 1 Signal	N/A
69	GRN	Throttle Motor (+) Signal	Pulse Signals
70	---	Not Used	---
71	GRN/WHT	HO2S-22 (B2 S2) Heater	Digital Signals: 0-12-0v
72	RED/GRN	Ignition Power (Main Relay)	12-14v
73	GRN/WHT	Cruise Control Indicator	Lamp Off: 12v, On: 1v
74	YEL/BLK	Intake Air VSV Control	Valve Off: 12v, On: 1v
75	BRN/WHT	EVAP Purge Cut Solenoid	Valve Off: 12v, On: 1v
76	BLU	TP Sensor 1 Signal	0.5-0.8v
77	ORN	A/P Sensor 3 Reference	4.9-5.1v
78	YEL	A/P Sensor 2 Signal	N/A
79	BLU/GRN	A/P Sensor 3 Signal	N/A
80	GRN	MAP Sensor Reference	4.9-5.1v

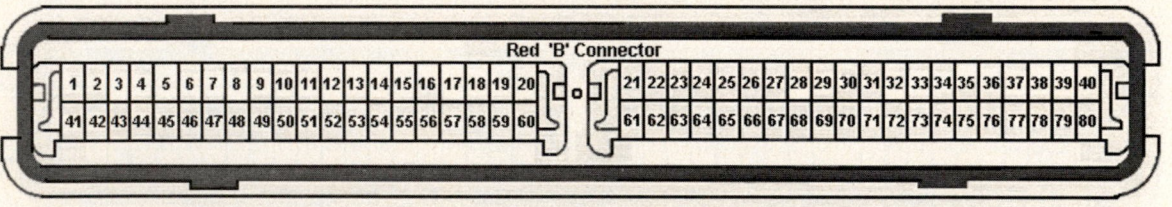

Red 'B' Connector

1998-03 EX, LX 3.2L VIN W 4WD [M/T] 80P J21 Blue Connector

PCM Pin #	W/Color	Circuit Description (40-Pin)	Value at Hot Idle
1	BLK	Power Ground	<0.1v
2	GRN	Reference Voltage 'A'	4.9-5.1v
3	RED/WHT	Reference Voltage 'B'	4.9-5.1v
4	GRY/RED	A/C Clutch Relay Control	Relay Off: 12v, On: 1v
5-6	---	Not Used	---
7	BLK/RED	Tachometer Signal	Pulse Signals
8	---	Not Used	---
9	YEL	Knock Sensor 1 Signal	No detonation: 18 mv AC
10	RED	Knock Sensor 2 Signal	No detonation: 18 mv AC
11	---	Not Used	---
12	WHT/RED	CMP Sensor Signal	Digital Signals: 0-5-0v
13	PNK	To be done	---
14	BLK/BLU	MAF Sensor Signal	4.2v
15-16	---	Not Used	---
17	RED	HO2S-22 (B2 S2) Ground	0.1-1.1v
18	---	Not Used	---
19	RED/GRN	Ignition Power (Main Relay)	12-14v
20	RED/WHT	Keep Alive Power	12-14v
21	WHT	HO2S-11 (B1 S1) Ground	<0.050v
22	---	Not Used	---
23	BLU	HO2S-12 (B1 S2) Ground	<0.050v
24	---	Not Used	---
25	RED/BLU	Coil On Plug 3 Control	Digital Signals: 0-12-0v
26	RED/BLK	Coil On Plug 2 Control	Digital Signals: 0-12-0v
27	GRN	IAT Sensor Signal	Varies w/temp: 0.5-4.8v
28	---	Not Used	---
29	RED	MAP Sensor Signal	0.9v
30-36	---	Not Used	---
37	GRN/YEL	PSP Switch Signal	Straight: 12v, Turning: 0v
38	GRN/YEL	Headlights On Signal	Lights Off: 12v, On: 0v
39	RED	Brake Switch Signal	Brake Off: 0v, On: 12v
40	BLK/BLU	Power Ground	<0.1v

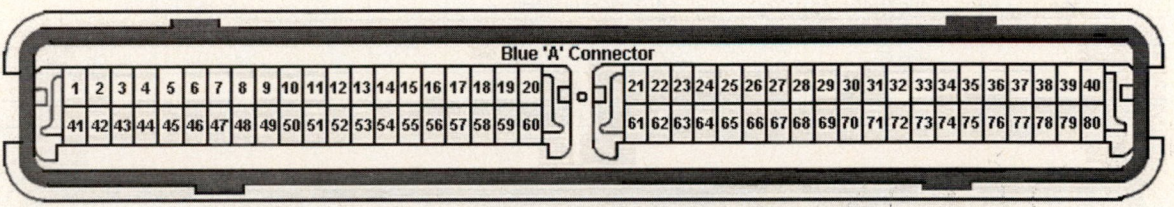

1998-03 EX, LX 3.2L VIN W 4WD [M/T] 80P J21 Blue Connector

PCM Pin #	W/Color	Circuit Description (40-Pin)	Value at Hot Idle
41	GRN	TP Sensor 1 Ground	<0.050v
42	PNK/WHT	Fuel Pump Relay Control	Relay Off: 0v, On: 12v
43	---	Not Used	---
44	ORN/WHT	HO2S-21 (B2 S1) Heater	Digital Signals: 0-12-0v
45	GRN/BLK	A/C Switch Signal	A/C On: 12v, Off: 0v
46	BRN/RED	Brake Switch Signal	Brake Off: 12v, On: 0v
47	---	Not Used	---
48	BLU	Throttle Motor (-) Signal	Pulse Signals
49	BLK	Knock Sensor 1 Ground	<0.050v
50	ORN	HO2S-22 (B2 S2) Ground	<0.050v
51	---	Not Used	---
52	RED	HO2S-21 (B2 S1) Ground	<0.050v
53	BLK	ECT & EGR Sensor Ground	<0.050v
54	BLK	FTP Sensor Ground	<0.050v
55	---	Not Used	---
56	GRN/BLK	Injector 5 Control	2.0-3.3 ms
57	RED/WHT	Keep Alive Power	12-14v
58	ORN/BLK	Class 2 Serial Data Link	0v
59	---	Not Used	---
60	BLK	Knock Sensor 2 Ground	<0.050v
61	GRN/WHT	TP Sensor 2 Ground	<0.050v
62	GRN/ORN	Injector 2 Control	2.0-3.3 ms
63	RED	A/P Sensor 1 Ground	<0.1v
64	GRN/RED	Injector 4 Control	2.0-3.3 ms
65	BLU/WHT	TP Sensor 2 Signal	0.5-0.6v
66	GRN	Injector 3 Control	2.0-3.3 ms
67	YEL	PCM C.Q. Input (1-3-5)	Digital Signals: 0-12-0v
68	RED	Knock Sensor Input (2-4-6)	<0.050v
69	GRN/WHT	Injector 1 Control	2.0-3.3 ms
70	---	Not Used	---
71	GRN/WHT	HO2S-22 (B2 S2) Heater	Digital Signals: 0-12-0v
72	WHT/BLU	C/C Resume/Accel Input	Switch Off: 0v, On: 12v
73	WHT/BLU	CKP Sensor Signal	Digital Signals: 0-5-0v
74	BLU/RED	ECT Sensor Signal	At 180°F: 0.54v
75	RED/GRN	Ignition Power (Main Relay)	12-14v
76	PNK/BLK	Clutch Switch Signal	Clutch Out: 12v, In: 0v
77-79	---	Not Used	---
80	BLU	A. P. Sensor 2 Ground	<0.050v

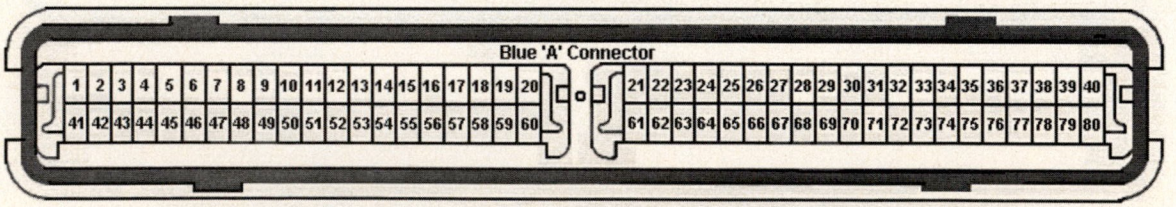

Blue 'A' Connector

1998-03 EX, LX 3.2L VIN W 4WD [M/T] 80P J22 Red Connector

PCM Pin #	W/Color	Circuit Description (40-Pin)	Value at Hot Idle
1	GRN	Sensor Ground	<0.050v
2	BLK/ORN	HO2S-11 (B1 S1) Heater	Digital Signals: 0-12-0v
3	YEL	A/P Sensor 2 Signal	N/A
4	ORN	Reference Voltage	4.9-5.1v
5	WHT	CKP Sensor Reference	4.9-5.1v
6	GRY/GRN	Cruise Set Switch	Switch Off: 0v, On: 12v
7	RED/BLU	EVAP Purge Solenoid	Valve Off: 12v, On: 1v
8	GRN/WHT	Indicator Control	N/A
9	RED/GRN	Ignition Power (Main Relay)	12-14v
10	---	Not Used	---
11	WHT/GRN	Malfunction Indicator Lamp	MIL Off: 12v, On: 1v
12	GRN	Throttle Motor (+) Signal	Pulse Signals
13	RED/WHT	Coil On Plug 4 Control	Digital Signals: 0-12-0v
14	BLU	HO2S-12 (B1 S2) Ground	<0.050v
15	BLU	HO2S-22 (B2 S2) Ground	<0.050v
16	BLU	HO2S-22 (B2 S2) Ground	<0.050v
17	PNK	HO2S-21 (B1 S1) Ground	<0.050v
18	WHT	HO2S-11 (B1 S1) Ground	<0.050v
19	RED/YEL	Coil On Plug 5 Control	Digital Signals: 0-12-0v
20	YEL/GRN	Sensor Ground	<0.050v
21	BLU/YEL	Sensor Ground	<0.050v
22	---	Not Used	---
24	RED	HO2S-22 (B2 S2) Ground	0.1-1.1v
25	---	Not Used	---
26	BLU/RED	EGR Position Solenoid	Digital Signals: 0-12-0v
27	BLK/YEL	Vehicle Speed Sensor	Moving: 0-5-0-5v
28	GRN/YEL	Injector 6 Control	2.0-3.3 ms
29	---	Not Used	---
30	WHT/GRN	Cruise Engage Switch	Switch Off: 0v, On: 12v
31	GRY/RED	FTP Sensor Reference	4.9-5.1v
32	RED/GRN	Ignition Power (Main Relay)	12-14v
33-34	---	Not Used	---
35	RED/GRN	Coil On Plug 6 Control	Digital Signals: 0-12-0v
36	BLK	A/P Sensor 1 Reference	4.9-5.1v
37	---	Not Used	---
38	RED	TP Sensor 1 Reference	4.9-5.1v
39-40	---	Not Used	---

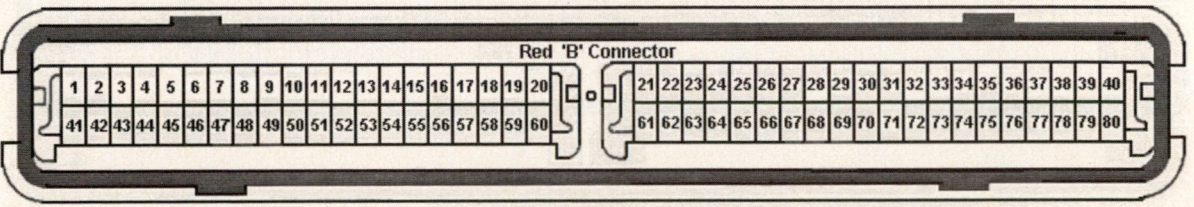

Red 'B' Connector

1998-03 EX, LX 3.2L VIN W 4WD [M/T] 80P J22 Red Connector

PCM Pin #	W/Color	Circuit Description (40-Pin)	Value at Hot Idle
41	BLU/ORN	A/P Sensor 3 Ground	<0.050v
42	WHT/RED	HO2S-12 (B1 S2) Heater	Digital Signals: 0-12-0v
43-46	---	Not Used	---
47	WHT/BLK	Water Gauge Signal	0-5v
48	---	Not Used	---
49	RED/WHT	Class 2 Data Line to DLC	0v
50	ORN/GRN	Fuel Level Sensor Signal	Duty Cycle Signal
51	---	Not Used	---
52	GRN	FTP Sensor Signal	With Fuel Cap Off: 2.5v
53	---	Not Used	---
54	BLU	HO2S-12 (B1 S2) Ground	<0.050v
55-57	---	Not Used	---
58	ORN/BLU	HO2S-11 (B1 S1) Signal	0.1-1.1v
59	LT GRN	Cruise Set Switch	Switch Off: 0v, On: 12v
60	BLU/YEL	Fuel Level Sensor Reference	12v
61	YEL/GRN	IAT Sensor Signal	Varies w/temp: 0.5-4.8v
62	RED	Coil On Plug 1 Control	Digital Signals: 0-12-0v
63	WHT	HO2S-21 (B2 S1) Signal	0.1-1.1v
64	PNK	HO2S-12 (B1 S2) Signal	0.1-1.1v
65	GRN	HO2S-22 (B2 S2) Signal	0.1-1.1v
66	---	Not Used	---
67	GRY/RED	EGR Position Sensor Signal	1.2-2.0v
68	WHT	A/P Sensor 1 Signal	N/A
69	GRN	Throttle Motor (+) Signal	Pulse Signals
70	---	Not Used	---
71	GRN/WHT	HO2S-22 (B2 S2) Heater	Digital Signals: 0-12-0v
72	RED/GRN	Ignition Power (Main Relay)	12-14v
73	GRN/WHT	Cruise Control Indicator	Lamp Off: 12v, On: 1v
74	YEL/BLK	Intake Air VSV Control	Valve Off: 12v, On: 1v
75	BRN/WHT	EVAP Purge Cut Solenoid	Valve Off: 12v, On: 1v
76	BLU	TP Sensor 1 Signal	0.5-0.8v
77	ORN	A/P Sensor 3 Reference	4.9-5.1v
78	YEL	A/P Sensor 2 Signal	N/A
79	BLU/GRN	A/P Sensor 3 Signal	N/A
80	GRN	MAP Sensor Reference	4.9-5.1v

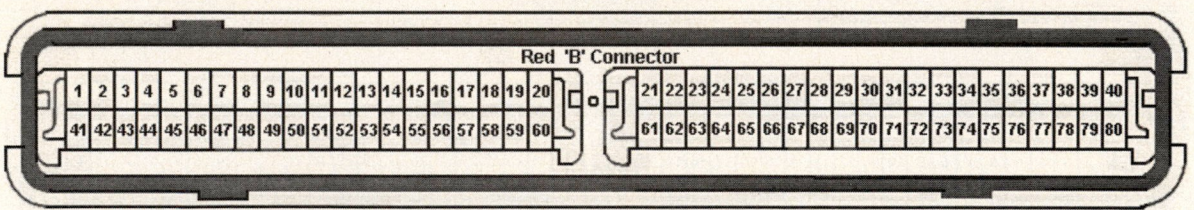

Toyota OBD II Diagnostic Contents

OBD II SYSTEM MONITORS

OBD II SYSTEM

ABOUT THIS SECTION

Introduction

This section contains information on OBD I and OBD II Systems, PID Data and Pin Tables. It assists technicians during repair and diagnosis of problems related to the Powertrain Control Module (PCM) and its subsystems.

Vehicle Coverage

- Car models: Avalon, Camry, Celica, Corolla, Cressida, Echo, MR2, Paseo, Prius, Supra and the Tercel
- SUV models: Land Cruiser, Highlander, RAV4, Sequoia, 4Runner
- Truck models: T-100 Pickup, Tacoma and the Tundra
- Van models: Previa and Sienna Minivans

Five Manual Sections

The main features of this manual are included in 5 sections:

- Section 9: OBD System information for 1996-2003 models
- Section 10: Pin Tables for Cars with fuel injection (1990-2003)
- Section 11: Pin Tables for SUV's with fuel injection (1990-2003)
- Section 12: Pin Tables for Trucks with fuel injection (1990-2003)
- Section 13: Pin Tables for Vans with fuel injection (1990-2003)

How to Use Each Section

Section 1

Refer to this section to learn more about how to use PID Data with a Scan Tool or Pin Tables with a Breakout Box and DVOM during diagnosis of an OBD System, i.e., Version 1 or 2.

This section can also be used to learn how to use OBD II System Tests, Trouble Code *enable criteria*, Freeze Frame Data and Example Drive Patterns for OBD II Main Monitors. This section includes example descriptions of common Serial and PID Data items.

Sections 2, 3, 4 & 5

Refer to these sections to identify PCM descriptions and pin terminals, wire colors and the PCM values for Toyota Car, SUV, Truck and Van applications equipped with fuel injection.

Diagnostic Help

The PID Data, Pin Tables and OBD II System Tests in this section contain numerous pieces of *diagnostic help* in the form of "known-good" values and example specifications.

The values in these examples were obtained using a Digital Volt/Ohm Meter (DVOM), a Breakout Box (BOB) or an Aftermarket Scan Tool.

Vehicle Identification Number

The vehicle identification number (VIN) is a seventeen (17) digit legal identifier of the vehicle. It is located on a plate that is attached to the upper left corner of the instrument panel (viewed from the windshield).

The VIN information includes the country of origin, make, vehicle type, passenger safety equipment, car line, body style, engine, check digit, model year, assembly plant and the vehicle build sequence.

An example of the VIN Code for a 1998 4Runner is included below.

Vehicle Identification Number (VIN) Code Example

1	2	3	4	5	6	7	8	9	10	11	12	13	14	15	16	17
4	T	A	W	N	7	2	N	6	W	Z	0	1	1	8	4	2

VIN Code Decoding Table (1996 4Runner Example)

Position	Interpretation	Code = Description
1	Manufacturing Country	4 = United States
2	Manufacturer	N = New United Motor Mfg (USA)
3	Vehicle Type & Country	A = Truck
4	Body Type	N = 4 x 2 Tacoma Regular Cab P = 4 x 4 Tacoma Regular Cab V = 4 x 4 Tacoma Xtra-Cab W = 4 x 4 Tacoma Xtra-Cab
5	Engine Type	L = 2.4L I4 (2RZ-FE Engine) M = 2.7L I4 (3RZ-FE Engine) N = 3.4L V6 (5VZ-FE Engine)
6	Model	8 = Standard
7	Restraint System	2 = Dual Airbags
8	Line	N = Tacoma
9	Check Digit	2
10	Model Year (Refer to Driver's Door)	L = 1990 M = 1991 N = 1992 P = 1993 R = 1994 S = 1995 T = 1996 V = 1997 W = 1998 X = 1999 Y = 2000 1 = 2001 2 = 2002
11	Assembly Plant	0 - 9: Japan C: Ontario U: Georgetown, United States Z: Fremont, United States
12 to 17	Plant Sequential Number	0-1-1-0-8-4-2

Vehicle Emission Control Information Label

The Vehicle Emission Control Information Label is located under the hood. *This example is for a 2001 Toyota 4Runner.*

TOYOTA   **IMPORTANT VEHICLE INFORMATION**
TOYOTA MOTOR CORPORATION

TEST GROUP : 1TYXTO3 4FFP EVAP FAMILY : 1TYXR013SAKO
SFI, A/F S, TWC(2), HO2S   3.4 LITER

ENGINE TUNEUP SPECIFICATIONS FOR ALL ALTITUDES

VALVE CLEARANCE (ENGINE AT COLD)	INTAKE EXHAUST	0.13-0.23 mm (0.006-0.009 in.) 0.27-0.37 mm (0.011-0.014 in.)

NO OTHER ADJUSTMENTS NEEDED.

THIS VEHICLE CONFORMS TO U.S. EPA NLEV REGULATIONS
APPLICABLE TO GASOLINE-FUELED 2001 MODEL YEAR NEW LEV
LIGHT DUTY TRUCKS AND TO CALIFORNIA REGULATIONS
APPLICABLE TO 2001 MODEL YEAR NEW LEV LIGHT DUTY TRUCKS.

3 4 2 P G F F W
62730   5VZ-FE VZN

CATALYST
OBD II CERTIFIED **PE**
USA & CANADA

SFI, A/F S, TWC (2), HO2S, 3.4 LITER

These designations indicate this vehicle has a three-way catalyst (TWC), an A/F Sensor and an HO2S. The engine is a 3.4L V6.

OBD II Certified

This designator on the VECI label indicates this vehicle was certified for OBD II usage.

50ST (50 States)

If this designator is used, the vehicle conforms to U. S. EPA & State of California regulations for 2001-02 model year new motor vehicles.

CAL (California)

If this designator is used, the vehicle conforms to U.S. EPA and State of California regulations applicable to 2001-02 model year new passenger cars for vehicle introduced into commerce in California.

PID DATA

What is PID Data?

PID is an acronym for Parameter Identification Data used to identify Powertrain Control Module items on both OEM and aftermarket Scan Tools. The complete list of PID Data items for Toyota includes:

- PCM input signals (BARO, ECT, IA/T, MAP & TP Sensors)
- PCM output signals (EGR & EVAP VSV, MIL, S1& S2 Solenoids)
- PCM calculated values (LOAD, LONGFT, MISF & SHRTFT)

OBD I Serial Data & OBD II PID Data

PID Data for Toyota vehicles is separated into On Board Diagnostics Version I (OBD I) and II (OBD II). OBD I Serial Data can be found on 1990-95 vehicles. OBD II PID Data is available for 1994-01 vehicles.

How to View PID Data

To view PID Data on a Scan Tool, connect the Scan Tool to the vehicle connector, select either Generic or Manufacturer OBD II, and then select Parameter ID (PID Data) from the OBD II main menu.

Note: ***If a Scan Tool does not power up or read PID Data, verify that the power, ground, and Scan Tool connections are okay. To verify a Tool operates, try it on another vehicle.***

Example of Scan Tool Connection

In this example, the Scan Tool is connected to the DLC in order to view "live" HO2S PID Data once it has been converted in the PCM.

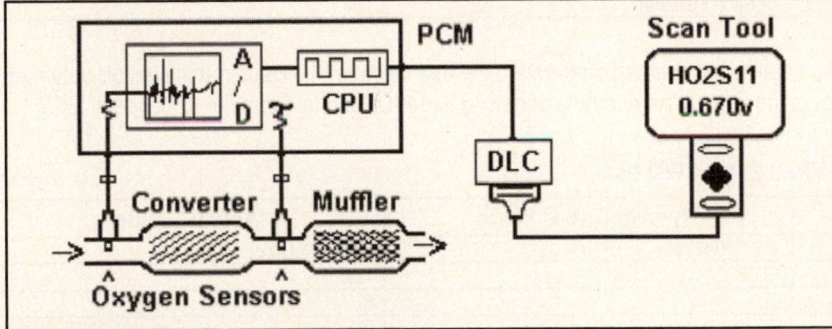

How to Use PID Data

Information contained within the PID Data Charts can be used to:

- Check the operation of a component during diagnosis of a No Code Fault (Driveability Symptom)
- Check the operation of a component or system by viewing current or "live" data from the vehicle computer data stream
- Validate a previous repair procedure

Scan Tool OBD II PID Mode

Vehicles equipped with On Board Diagnostics (OBD) have a unique test mode that can be selected to allow access to vehicle Parameter Identification (PID Data) information. Part of the PID Data is generic in nature and can be viewed by any certified Scan Tool (Generic Mode).

However, on Toyota vehicles, the OEM proprietary PID Data can only be accessed using a Scan Tool with the Toyota OEM cartridge. This feature allows the Scan Tool to read the complete list of OEM Data. To learn more about how to access the Toyota Enhanced Data List, refer to the Scan Tool instructions.

PID Data Display

An example of Generic PID Data captured with an aftermarket Scan Tool is shown in the table below.

1997 Tacoma Pickup 2.4L I4 MFI VIN L Engine PID Data

PID Acronym	Description	Value at Idle
CLV	Calculated Load	22%
ECT	ECT Sensor	195ºF
FUEL SYS #1	Fuel System No. 1	CL (Closed Loop)
IA/T	IA/T Sensor	110ºF
IGN ADV	Ignition Advance	22° BTDC
LONGFT	Long Term Fuel Trim	+3%
MAF	MAF Sensor	3.2 g/sec
O2S B1, S2	O2S Bank1, Sensor 2	0.1-0.9v
RPM	Engine Speed	710
SHRTFT	Short Fuel Trim	-1%

PID Data Comparison

Once the current PID Data has been captured, it can be compared to the examples of PID Data in this section obtained from vehicles with "known good" operating values. An example of how to use PID Data to diagnose a "suspect" ECT sensor is shown in the table below.

1997 Tacoma Pickup 2.4L I4 MFI VIN L Engine PID Data

PID Acronym	"Known Good" Value	"Actual" Value
ECT	195ºF (Hot Idle)	103ºF (Hot Idle)
IA/T	125ºF	122ºF
Fuel SYS #1	CL	CL
TP	8%	9%

Note: *In this example, the actual ECT sensor reading of 103ºF is compared to the ECT known good reading of 195ºF. A lower than normal reading on this sensor indicates that the sensor has moved out of range for a hot engine.*

Diagnosis with PID Data

An example of how to use PID Data records to diagnose a vehicle with an intermittent stalling problem and no codes is shown below. PID Data recorded at idle speed reveals a defective MAP sensor that caused the MAP sensor input to suddenly spike very low. When the MAP signal went low, the engine would stumble and die out.

The information in the table (Frame View) and picture (Graphing View) below shows the MAP Sensor spiked low and this action caused the engine to stall.

Vehicle: 1994 Camry 2.2L 5S-FE VIN S Engine (Frame View)

PID Item	Frame 1	Frame 2	Frame 3	Frame 4	Frame 5	Frame 6
RPM	628	635	640	650	620	625
INJ PW	3.5 ms	3.3 ms	3.2 ms	3.6 ms	3.3 ms	3.4 ms
MAP (Hg)	0.7" Hg	0.75" Hg	0.1" Hg	0.7" Hg	072" Hg	0.75" Hg

Vehicle: 1994 Camry 2.2L 5S-FE VIN S Engine (Graphing View)

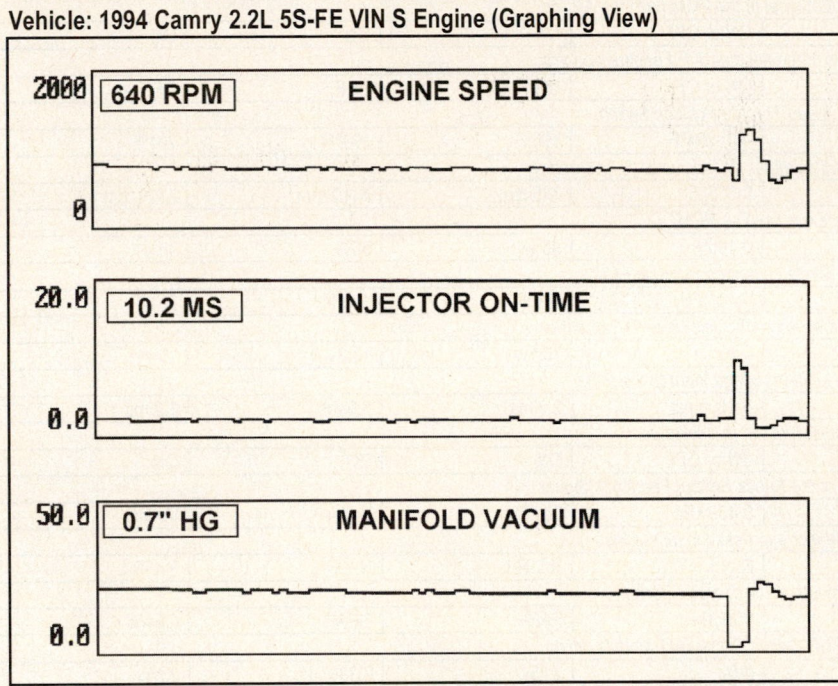

1995 Camry PID Data Example

The PID and Serial Data examples in this section are arranged in table format. Each **table heading** identifies the year, vehicle, engine and transmission (All, A/T - Automatic or M/T - manual). These values were obtained in P/N with A/C Off. Tables contain the following fields:

- PCM PID Acronym
- PID - Parameter Identification
- PID Value Range
- PID Values at Hot Idle, 30 and 55 mph

1995 Camry 2.2L I4 MFI VIN S (All)

PCM PID Acronym	Parameter Identification	PID Value Range	PID Value at Hot Idle	PID Value at 30 mph	PID Value at 55 mph
A/C SIG	A/C Signal	ON or OFF	ON	ON	ON
A/C SIGNAL: Indicates the status of the A/C Signal to the PCM (with the A/C on).					
A/F LEFT	A/F Feedback/L	ON or OFF	ON	ON	ON
A/F FEEDBACK LEFT: Indicates the status of the left side A/F Feedback control.					
CTP SIG	CTP Signal	ON or OFF	ON	OFF	OFF
CTP SIGNAL: Indicates the status of the Closed Throttle Position Switch.					
ECT	ECT Sensor	-40 -284°F	196°F	199°F	201°F
ECT: Indicates ECT temperature expressed as either (°) Celsius or (°) Fahrenheit					
ENGINE	Engine Speed	0-65535	725 rpm	1410 rpm	2300 rpm
Engine Speed: Indicates the engine speed expressed as RPM.					
IAC	IAC Duty Cycle	0-100%	45%	50%	45%
IAC: Indicates the amount of the IAC Duty Cycle command from the PCM.					
IA/T	IA/T Sensor	-4 -212°F	196°F	199°F	201°F
IA/T: Indicates IA/T temperature expressed as either (°) Celsius or (°) Fahrenheit					
IGN	Ignition Voltage	0-25.5v	13.1v	13.2v	13.4v
Ignition Voltage: Indicates the value of the Ignition voltage input to the PCM.					
INJ	Fuel Injector ms	0-1000 ms	2.9 ms	3.0 ms	3.2 ms
INJ: Indicates the amount of fuel injector pulsewidth in milliseconds.					
KNOCK	Knock F/B	ON or OFF	ON	ON	ON
KNOCK FEEDBACK: Indicates the status of the Knock Sensor Feedback Signal.					
MAP	MAP Sensor	0-450 kPa	36 kPa	Varies	Varies
MAP: Indicates the signal from the MAP Sensor expressed in mm of Hg.					
OXL	OXL Signal	R or L	R-L-R-L	R-L-R-L	R-L-R-L
OXL SIGNAL: Indicates the Rich to Lean status of the Oxygen Sensor.					
PNP SIG	PNP Signal	Park/Gear	PARK	GEAR	GEAR
PNP SIGNAL: Indicates the status of the PNP Switch Signal (A/T Models only).					
TAR AFL	Target A/FL	0-5v	2.45v	2.33v	2.55v
TARGET A/FL: Indicates the Target A/F value calculated by the PCM.					
THROT	Throttle position	0-100%	4%	7%	8%
Throttle Position: Indicates the calculated throttle position expressed in percent.					
VSS	Vehicle Speed	0-159	0 mph	30 mph	55 mph
VSS: Indicates the Vehicle Speed Sensor input to the PCM (converted to mph).					

PIN TABLES

Introduction

A Pin Table is a term used in this section to describe a chart or table that contains information on PCM and Breakout Box (BOB) Pins, individual wire colors of PCM circuits, and example values for devices that connect to the PCM. Pin Table information includes:

- Signals from various sensors (MAP, THA, THG, THW & TP)
- Signals from various switches (A/T Shift, Brake, Heater & Fan)
- Signals from oxygen sensors (O2S-11, HO2S-12, HO2S-21)
- Signals from output devices (EGR VSV, EVAP VSV & INJ)
- Power & ground signals (Direct Battery, Power & Sensor Ground)

Note: Acronyms in these examples are in the Glossary.

OBD I System Pin Tables

The OBD I System Pin Tables are separated into four sections that cover 1990-95 Toyota Car, SUV, Truck and Van applications.

OBD II System Pin Tables

The OBD II System Pin Tables are separated into four sections that cover 1994-2002 Toyota Cars, SUV's, Trucks and Vans.

How to Connect a BOB and DVOM

To use Pin Table information with a DVOM, an aftermarket Breakout Box should be installed. To connect a BOB, turn the key off and then remove the wire harness at the engine computer. Next, connect the appropriate BOB (with adapters) to the PCM and BOB connectors.

This location places the BOB between the PCM and the wiring so that circuit measurements can be made at any of the pin connections on the BOB. An example of a breakout box connected to a PCM is shown in the Graphic below.

Breakout Box Hookup Graphic

ν **NOTE:** *Read and record all OBD II codes and freeze frame data in the PCM prior to connecting the BOB as all codes and PID data will be lost if the PCM connector is removed.*

How to Use Pin Tables

Information contained within the Pin Tables can be used to:

- Test circuits for open, short to power or short to ground fault
- Check the operation of a component before or after a repair
- Check the operation of a component or system by viewing signals on the PCM input/output circuits with a DVOM or Lab Scope

Testing Circuits with a Breakout Box

There are several Breakout Box designs available for use to test the PCM and its related circuits. However, all of them require that the PCM wire harness is removed and then the BOB installed between the PCM and wire harness connector. Several breakout boxes require the use of overlays in order to allow the tool to be used on more than one year or engine type. Always verify that the correct adapter and overlays are used to prevent misdiagnosis.

Power and Ground Circuit Checks

Measurements made at the BOB are accomplished via test leads and probes from the DVOM or a Lab Scope. If any of the terminals on the PCM or BOB are damaged or loose, test measurements made at the Breakout Box can be inaccurate. Be sure to verify that the PCM direct battery, ignition power, power ground and signal ground circuits are okay at the Breakout Box prior to starting a test sequence.

ν **NOTE:**   *The voltage drop between battery (+) to battery power or ignition power at the BOB should be less than (<) 0.1v. The voltage drop between the battery negative (post) and to ground (at the BOB) should be less than (<) 0.1v.*

Pin Table Test Example

Once an "actual" PCM reading is recorded, it can be compared to an example from a vehicle with "known-good" values. In the example below from a 1994 Camry 2.2L I4, the "actual" MAP sensor input is higher than the "known-good" value at idle. In this case, the vehicle would run extremely rich because the MAP value is too high.

PCM Pin Table Example

PCM Pin #	Wire Color	PCM Circuit Tested	Known Good Value at Idle	Actual Vehicle Value at idle
2	BLU/YEL	MAP Sensor	1.0-1.5v	2.1v

Wire Color Changes

Every effort has been made to obtain and list the correct wire colors for the PCM and TCM circuits.. However, vehicle manufacturers can make changes to wire colors between model years and this can result in the wrong wire colors being listed.

1996 Camry Pin Table Example (Part 1)

Pin Tables are arranged in table format.. Each *table heading* identifies the year, vehicle, engine and transmission (A/T - Automatic or M/T - manual or All). Tables contain these data fields:

- PCM Pin Number & Wire Color
- Circuit Description/Acronym for PCM Pin Connector
- Known good operating value at hot idle speed

1996 Camry 2.2L I4 MFI VIN G (A/T) 26Pin Connector

PIN No.	Wire Color	Application & Acronym	Value at Idle in P/N
1	BLU/YEL	ECT Solenoid SL	In Lockup: 12-14v
ECT SL: indicates the status of the transmission SL solenoid in lockup position.			
2	VIO	ECT Solenoid S1	3rd or OD: <1v
ECT S1: Indicates the status of the transmission S1 solenoid in 3rd or OD.			
3	WHT/RED	Igniter IGF Signal	Pulses
IGF Signal: Indicates that the IGF signal is shown in pulses.			
4	WHT/RED	CKP Sensor (+) Signal	3-5v AC
CKP (+): Indicates the value of the CKP sensor (+) in AC volts at Hot Idle.			
5	BLK/WHT	G2 (+) Signal	1-3v AC
G2 (+): Indicates the value to the G2 Cam (+) Sensor in AC Volts at Hot Idle.			
6-8	---	Blank	---
9	GRN/RED	IAC RSC Signal	Hot idle: 8-12v
IAC RSC: indicates the command from the PCM to the IAC rotary solenoid close coil.			
10	GRN/YEL	IAC RSO Signal	Hot idle: 8-12v
IAC RSO: indicates the command from the PCM to the IAC rotary solenoid open coil.			
11	YEL	Injector No. 2	2.0-3.3ms
INJ 2: Indicates the amount of injector pulsewidth in milliseconds.			
12	WHT	Injector No. 1	2.0-3.3ms
INJ 1: Indicates the amount of injector pulsewidth in milliseconds.			
13	WHT/BLK	Power Ground	<0.1v
PWR GND: Indicates the voltage for the power ground circuit to the PCM.			
14	BRN	Sensor Ground	<0.050v
SEN GND: Indicates the voltage for the sensor ground circuit to the PCM.			
15, 16, 19	---	Blank	---
17	GRN	CKP Sensor (-) Signal	3-5v AC
CKP (-): Indicates the value of the CKP sensor (-) in AC volts at Hot Idle.			
20	WHT	Igniter IGT Signal	Pulses
IGT Signal: Indicates the value of the IGF signal is shown in pulses.			
22	YEL/BLK	EVAP Purge Solenoid	Idling: 12-14v
EVAP VSV: Indicates the status of the EVAP VSV Solenoid at Hot Idle (off).			
23	GRN	EGR Solenoid	Idling: <1v
EGR VSV: Indicates the status of the EGR VSV Solenoid at Hot Idle (on).			
24	RED/BLK	Injector No. 4	2.0-3.3ms
INJ 4: Indicates the amount of injector pulsewidth in milliseconds.			
25	RED/BLU	Injector No. 3	2.0-3.3ms
INJ 3: Indicates the amount of injector pulsewidth in milliseconds.			
26	WHT/BLK	Power Ground	<0.1v
PWR GND: Indicates the voltage for the power ground circuit to the PCM.			

1996 Camry Pin Table Example (Part 2)

Each Pin Table contains a graphic that shows the position of all of the pin connectors for a particular engine and Powertrain Control Module. This engine has three connectors. Note the empty pins in the graphic.

1996 Camry 2.2L I4 MFI VIN G (A/T) 16Pin Connector

Pin No.	Wire Color	Circuit Description (16-Pin)	Value at Idle in P/N
1	RED	MAP & TP Sensor VREF	4.9-5.1v
SENSOR VREF: Indicates the value of the MAP & TP Sensor VREF Circuit.			
2	BLK/YEL	MAP Sensor input	1.0-1.5v
MAP SIGNAL: Indicates the value of the MAP Sensor at Hot Idle.			
3	BLU/BLK	IA/T Sensor input	At 100ºF: 2.8v
IA/T: Indicates the value of the IA/T Sensor (this value varies with air temperature).			
4	LT GRN	ECT Sensor input	At 180ºF: 0.6v
ECT: Indicates the value of the ECT Sensor (this value varies with coolant temp.).			
5	RED/BLU	O2S12 Signal (Bank 1)	Varies: 0.1-1.0v
O2S12: Indicates the value of the O2S12 (Bank 1, Rear Oxygen Sensor).			
6	WHT	O2S11 Signal (Bank 1)	Varies: 0.1-1.0v
O2S11: Indicates the value of the O2S11 (Bank 1, Front Oxygen Sensor).			
7	BLU/YEL	Vapor Pressure Sensor	2.9-3.7v
V/P SENSOR: Indicates the value of the Vapor Pressure Sensor (varies).			
8	BLK/RED	Vapor Pressure Solenoid	Idling: 12-14v
V/P VSV: Indicates the status of the Vapor Pressure VSV (off at idle).			
9	BRN	Sensor Ground	<0.050v
SEN GND: Indicates the voltage for the sensor ground circuit to the PCM.			
10, 12, 14	---	Blank	---
11	BLK	TP Sensor input	0.3-0.8v
TP SIG: Indicates the value of the TP Sensor at Hot Idle.			
13	WHT	Knock Sensor	No Knock: 2.5v
KS: Indicates the value of the Knock Sensor input (without any knock present).			
15	GRY	Data Link Connector	12-14v
DLC PIN: Indicates the value of the DCL Pin without the Scan Tool connected.			
16	WHT/BLK	Power Ground	<0.1v
PWR GND: Indicates the voltage for the power ground circuit to the PCM.			

Pin Connector Graphic

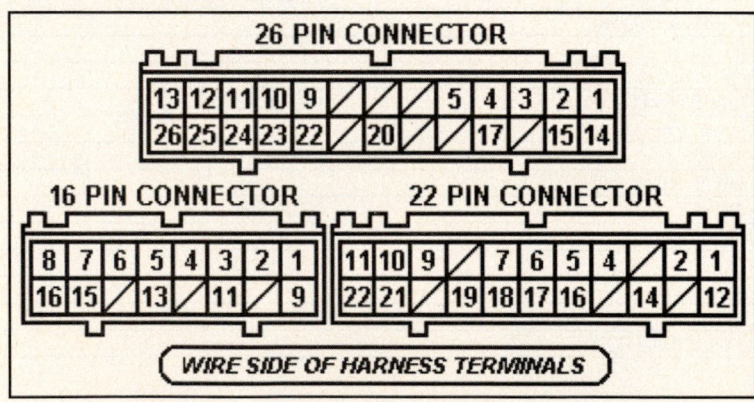

1996 Camry Pin Table Example (Part 3)

Pin Tables are arranged in table format.. Each *table heading* identifies the year, vehicle, engine and transmission (A/T - Automatic or M/T - manual or All). Tables contain these data fields:

- PCM Pin Number & Wire Color
- Circuit Description/Acronym for PCM Pin Connector
- Known good operating value at hot idle speed

1996 Camry 2.2L I4 MFI VIN G (A/T) 22Pin Connector

PIN #	Wire Color	Circuit Description (22-Pin)	Value at Idle in P/N
1	WHT/BLU	Battery Direct	12-14v
BA/TT: Indicates the status of the Battery Direct Circuit to the PCM.			
2	BLK/RED	Defogger/Light Idle up Solenoid	Load On: 12-14v
ELS 1: Indicates the voltage on the Electronic Load Circuit to the PCM (load on).			
3	---	Blank	---
4	GRN/WHT	Stop Light Switch	Brake On: 12-14v
STOP: Indicates the value of the Stop Light Switch Circuit to the PCM (brake on).			
5	GRN/RED	MIL (lamp)	MIL Off: 12-14v
MIL: Indicates the status of the MIL Control Circuit at the PCM (MIL is Off).			
6	BLU/RED	ECT Pattern Select Switch	In NORM: 0.1v
ECT in NORM: Indicates the value of the ECT Select Switch Circuit in NORM mode.			
7	GRN/ORN	Overdrive Main Switch	Switch Off: 12v
OD SWITCH: Indicates the voltage on the Overdrive Main Switch with the switch off.			
8	---	Blank	---
9	VIO/YEL	Vehicle Speed Sensor	At 55 mph: pulses
VSS: Indicates the frequency on the VSS Circuit at 55 mph.			
10	BLK/YEL	A/C Amplifier On Signal	A/C On: 12-14v
A/C ON: Indicates the value of the A/C On Signal from the Amplifier with the A/C on.			
11	BLK/WHT	Starter Switch Signal	KOEC: 9-11v
START: Indicates the value of the Starter Switch Signal to the PCM during cranking.			
12	BLK/ORN	EFI Main Relay B+	12-14v
MR B+: Indicates the value of the EFI Main Relay B+ Circuit to the PCM.			
13, 15, 20	---	Blank	---
14	GRN/RED	Circuit Opening Relay	Idling: 0-1v
FC: Indicates the value of the Circuit Opening Relay Circuit with the throttle closed.			
16	WHT	Data Link Connector	No Scan Tool: 0v
DLC 3: Indicates the value of the OBD II SDL Circuit to the PCM without a Scan Tool.			
17	RED/BLK	A/T: Select Switch - Reverse	In R: 12-14v
A/T R: Indicates the value of the Select Switch Circuit in Reverse Position.			
18	ORN	A/T: Select Switch - 2nd	In 2nd: 14v
A/T 2nd: Indicates the value of the Select Switch Circuit in 2nd Position.			
19	YEL/BLU	A/T: Select Switch - Low	In Low: 12-14v
A/T L: Indicates the value of the Select Switch Circuit in Low Position.			
21	GRN/BLK	A/C Amplifier ACT Signal	A/C On: 4-6v
A/C ACT: Indicates the value of the A/C ACT Signal from the Amplifier with A/C on.			
22	BLK/WHT	Neutral Start Switch	In P or N: 0.1v
NEUT SW: Indicates the value of the Neutral Start Switch Circuit in P or N position.			

OBD II SYSTEMS

Reasons for the OBD II System

The OBD II system was developed to accomplish two different objectives. First, it was developed to comply with California and Federal regulations and standards for vehicle emission control monitoring. The initial goal of this system was to detect the degradation or failure of an emission-related component or system that could cause vehicle emissions to rise by 50%.

OBD II Changes

However, the program has expanded to include new computer controlled diagnostic tests that are used to verify that an emission control system is operating correctly. This aspect of OBD II has the greatest impact on service technicians as it includes procedures that require a whole new set of diagnostic tests.

It is important to understand the operation of the OBD II diagnostic tests (monitors) because these systems operate with a new set of tests and routines. An OBD II "monitor" requires that a particular drive cycle be completed. In some cases, an OBD II "monitor" will not run a diagnostic test on a particular emission system unless a specific drive cycle is performed.

The second goal of OBD II was to introduce changes intended to help with vehicle diagnostics. These changes include common terms, connectors and a common diagnostic language for a generic Scan Tool. The changes include:

- Common Diagnostic Connector
- Expanded Malfunction Indicator Light Operation
- Common Codes and Diagnostic Language
- Common Diagnostic Procedures
- New Emissions-Related Procedures, Logic and Sensors
- Expanded Emissions-related Monitoring

Important Benefits

An important benefit of OBD II is that all vehicles have a common data output system and test connector. This allows a generic OBD II certified Scan Tool to read data from any OBD II compliant vehicle and pull codes with a common name and similar descriptions for fault conditions.

Many State I/M Tests now require the use of an OBD II certified Scan Tool to verify that all monitors have completed their tests and passed. This information, which appears on the Scan Tool as I/M Readiness Tests, is used to verify that all the I/M Readiness Tests are completed (with no trouble codes set) in order to pass the State Emissions Test.

Transition from OBD I to OBD II

To understand the OBD II system, a technician must understand how the OBD I system diagnostics operate. This is important because *OBD I System tests are included in the OBD II System.* OBD II Systems include several additional diagnostic tests. These tests or "monitors" are used to verify the operation of all Emission Control related components and systems.

In effect, in order to work on OBD II Systems, a technician must understand the OBD I System for each vehicle manufacturer, and their particular approach to diagnostics. The articles that follow include a summary of the evolution of the OBD I system.

History of OBD II Systems

Starting in 1978, various manufacturers introduced computer controls of vehicle systems and engine management. These computer-controlled systems involved not only diagnosis of engine mechanical and component operation, but also the identification of electrical faults and computerized engine control diagnosis. Early attempts at diagnosis involved expensive and specialized diagnostic testers that were connected externally to the computer in series with the wiring connector. These testers were used to monitor the input/output operations of the vehicle computer.

By 1980, vehicle manufacturers were designing systems where the computer incorporated internal programs that monitored selected components and stored trouble codes in memory for retrieval at a later time. These codes identified failure conditions that referred repair technicians to diagnostic repair charts or procedures that helped pinpoint the problem areas.

While a few vehicle manufacturers began implementation of OBD II systems as early as 1992, most manufacturers did not begin the phase-in of OBD II systems until 1994. All vehicle manufacturers were required to meet OBD II standards by the 1996 model year.

Phase-In Systems

The OBD II system was introduced on Toyota models in 1994 on the Camry, T-100 and Previa. The use of OBD II was expanded to the complete line of Toyota vehicles in 1996.

Bi-Directional Communication

A Scan Tool with a Toyota cartridge can read OBD I serial data on these vehicles: 1992-94 Camry I4 and 92-93 V6, 1989-92 Cressida, 1992-94 Celica I4, 1994 Celica 1.8L I4, 1993-94 Corolla I4 except the 1.6L, 1993-94 MR2, Supra and T100 V6, 1992-94 4Runner and Truck V6, 1993-94 Supra, 1994 Paseo and the 1994 Tercel.

Toyota OBD History

An overview of the evolution of On Board Diagnostics on Toyota vehicle applications is shown in the Graphic. Although the California 1990 regulations vary slightly from the CARB OBD II regulations, the EPA adopted the California OBD II for Federal emissions certification, effective with the 1996 model year. In 1998, a new Federal OBD II standard was adopted that effectively eliminated the different status between the California and Federal emissions certification.

OBD System Graphic

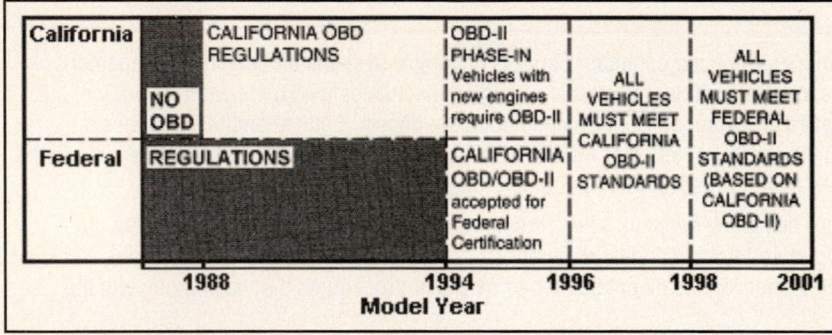

OBD II Certified Scan Tools

A document from the Society of Automotive Engineers (SAE J1978) requires that all vehicle manufacturers make readily available to the automotive repair industry trouble codes and emissions-related information that can be accessed by a Generic Scan Tool (GST). Fault codes, sensor and component values plus any Freeze Frame data stored in the On Board Computer (PCM) must be accessible for download to a generic Scan Tool.

OBD II Generic Scan Tool Functions

A generic Scan Tool should be able to perform these functions:

- Clear trouble codes from the memory in the vehicle computer
- Display the I/M Readiness Test Status for all on board monitors
- Read information from the PCM data stream and Freeze Frame
- Read the five digit trouble codes

Toyota Normal & Check Modes

A Scan Tool with a Toyota Cartridge can be run in both "Normal" and "Check" modes. The Check mode function allows a technician to check for certain codes on one trip. If this function is not available, turn the key off for 10 seconds after the first test. Then restart the vehicle and repeat the test procedure to allow the fault to be detected a second consecutive time in order to enable the two-trip detection.

Malfunction Indicator Lamp

OBD II regulations (CARB and EPA) require that a Malfunction Indicator Lamp (MIL) be illuminated when a fault is detected, and that a Diagnostic Trouble Code (DTC) is stored in the PCM memory.

Malfunction Indicator Lamp Graphic

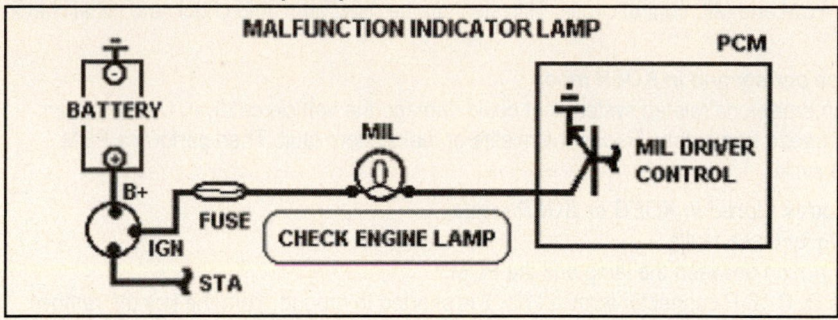

<u>Understanding MIL Conditions</u>

Several of the "on" or "off" conditions for the MIL are described next:

- MIL Off all the time - This indicates the PCM did not detect any faults in the Emission system or the MIL circuit is faulty
- MIL On all the time - This indicates the PCM detected a fault in an Emission system that could affect the emission levels
- MIL Flashing - This indicates the PCM detected a fault in the Fuel system or a Misfire fault that could damage the catalytic converter

Actions or Conditions to turn off the MIL

The PCM will turn off the MIL if any of these actions occur:

- The codes are cleared with a Generic or Proprietary Scan Tool
- Power to the PCM is removed (at the battery or the PCM fuse)
- The vehicle is driven on three consecutive trips that include an OBD II Warmup Cycle and meets all code set conditions without detecting any emission-related faults

Similar Conditions

If a fuel or misfire-related code sets, the vehicle must be driven under conditions similar to when the fault was detected before the PCM will de-activate the MIL. In effect, the vehicle must be driven within 375 RPM of the engine speed and engine load (±10%) of the engine load value, and with engine temperature conditions similar to the temperature value stored in Freeze Frame data when the code set.

Trip Definition

An OBD II Trip is vehicle operation (following an engine off period) of such duration and driving modes that all components or systems are monitored at least once by the PCM diagnostics (except the catalyst).

MIL Circuit Diagnosis

If the Malfunction Indicator Lamp (MIL) does not operate correctly, refer to the repair steps listed in the articles that follow to diagnose the MIL operation and the PCM.

MIL Condition: Light on for 2 seconds, then goes off at KOEO mode

This is the normal operation of the PCM and MIL control circuit. This step can be used as a bulb check and initial check of PCM operation.

MIL Condition: Light flashes once per second in KOER mode

This condition indicates a fault in an emissions related system that could damage the vehicle catalyst. Use the Scan Tool to read the trouble codes and freeze frame data. Repair the misfire or fuel system fault. Then perform a PCM Reset and do the appropriate drive cycle.

MIL Condition: MIL on with no codes stored in KOEO or KOER mode

This condition can be caused by any of these faults:

- MIL control wire is shorted to ground between the lamp and the PCM.
- The J1850 Bus + circuit to the DLC 16-P connector terminal No. 2 is shorted to ground. Turn the key off, remove the PCM connector and check terminal No. 2 in the DLC for continuity to ground.
- PCM is faulty (due to possible shorted MIL control driver circuit).
- An open condition in the PCM battery or ignition power circuits.
- An open or high resistance condition in the PCM ground circuits.

MIL Condition: MIL does not come on for 2 seconds - Engine Starts

This condition can be caused by any of these faults:

- If this problem is intermittent in nature, check the fuse that provides power to the MIL for a loose connection or corrosion. Inspect condition of the MIL control circuit terminal at the PCM.
- An open circuit between the MIL (lamp) and the PCM control circuit. Inspect the bulb connections and condition of the PCM terminals.
- PCM is faulty (due to possible shorted MIL control driver circuit).
- An open condition in the PCM battery or ignition power circuits.
- An open or high resistance condition in the PCM ground circuits.

MIL Condition: MIL does not come on for 2 seconds - No Start fault

This condition can be caused by several conditions - do these steps:

- Turn the key off and remove the following connectors: the PCM connector that connects to these sensors: EGR gas temperature, fuel tank pressure sensor, MAP and Throttle Position.
- Check for continuity to ground between the MAP sensor VREF and the other VREF circuits at the PCM to ground. If continuity exists, locate the short to ground in the VREF circuit and retest.

OBD II Warmup Cycle

Once a MIL is off, the trouble code will remain in memory until 40 warmup cycles are completed without the same fault reoccurring.

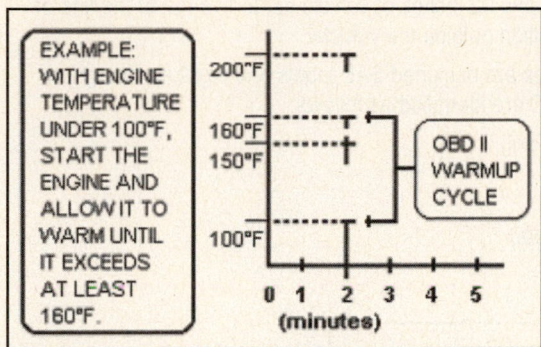

A warmup cycle is defined as a trip that includes a change in engine temperature of at least 40°F, and where the engine temperature reaches at least 160°F.

DTC Numbering Explanation

The number in the hundredth position indicates the specific vehicle system or sub-group in which the failure occurred.

This position should be consistent for P0xxx and P1xxx type codes.

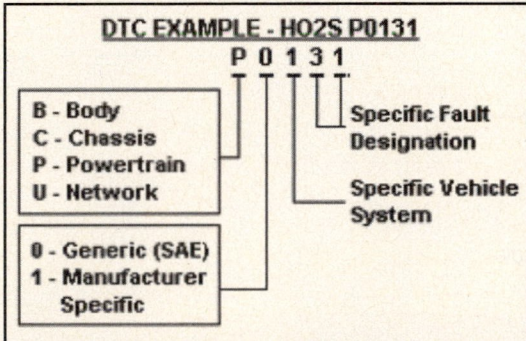

An example of how to interpret an OBD II code is shown in the Graphic to the right.

The Society of Automotive Engineers (SAE) developed the numbers and systems in the list that follows:

P0100 - Air Metering and Fuel System fault
P0200 - Fuel System (fuel injector only) fault
P0300 - Ignition System or Misfire fault
P0400 - Emission Control System fault
P0500 - Idle Speed Control, Vehicle Speed Sensor fault
P0600 - Computer Output Circuit (relay, solenoid, etc.) fault
P0700 - Transaxle, Transmission faults

■ **NOTE:** *The first and tenth digits indicate the type of Emission System that has failed.*

OBD II System Terminology

Diagnostic Link Connector

OBD II regulations established standards for use of a test connector (Data Link Connector or DLC) on vehicles that have an OBD System.

The 16-pin connector is located beneath the instrument panel within 12 inches of vehicle centerline out of the line of sight of passengers, and easily viewable while in a kneeling position outside the vehicle.

An OEM sticker may be added for other locations. Eight DLC pins are assigned SAE labels and eight are assigned vehicle manufacturer labels. The pins used by Toyota in the DLC are identified as follows:

- Terminal No. 2 - SDL (two-way communication with Scan Tool)
- Terminal No. 3 - Chassis ground connection
- Terminal No. 4 - Sensor ground connection
- Terminal No. 7 - SIL (two-way communication with Scan Tool)
- Terminal No. 16 - Battery power connection

Data Link Connector Graphic

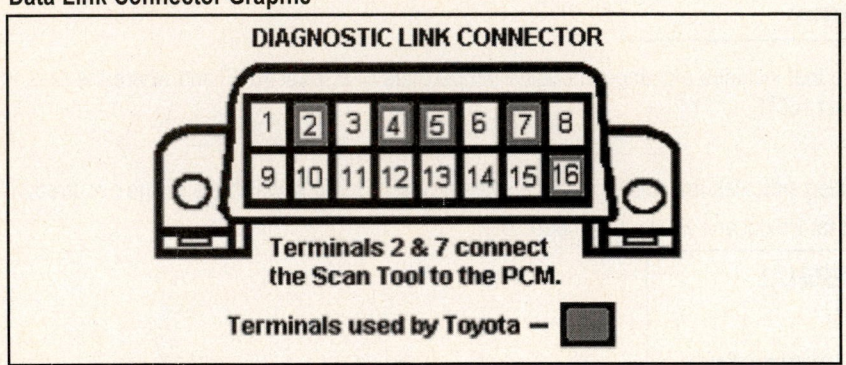

DLC Locations - Lexus
ES 300, GS 300 - Under left side of the dash
LS 400, LX 400, LX 470 - Under the left side of the dash
RX 300 - Behind No. 2 Instrument Panel Cover to right of the console
SC 300, SC 400 - Under the left side of the dash

DLC Locations - Toyota
4Runner - Under center left side of dash to the left of the console
Avalon, Paseo - Below dash to left of the steering column
Camry, Celica, Corolla - Below dash to left of the steering column
Highlander, Land Cruiser - Behind the top RH side of the dash
Previa, RAV4 (1996-2000) - Behind the center of the dash
RAV4 (2001-2002) - Behind the right side kick pad below glove box
Sequoia, Tacoma, Tundra - Behind glove box at top RH side of dash
Supra, T-100 - Behind the left side of the dash
Tercel - At lower left of dash near driver side junction block

Freeze Frame Data

OBD II Regulations (CARB and EPA) require that the vehicle onboard computer store specific Freeze Frame Data when the first emission-related fault is detected. The current readings can only be overwritten by data from the Fuel System or Misfire Monitor.

Freeze Frame Data must contain the engine operating conditions present at the time a code is set. OBD II systems record Freeze Frame Data at the time an emission-related DTC is recorded and the MIL is activated. This data must appear in standard measurements.

Freeze Frame Data can be thought of as a recording of one frame in time. This data contains details that describe the engine operating conditions at the instant a fault is detected and a code is set.

CARB and Toyota Enhanced Signals

Refer to the list of CARB Mandated and Toyota Enhanced Signals.

CARB Mandated Signals	TOYOTA Enhanced Signals
A/FS (A/F) Bank 1 Sensor 2 (Volts)	A/C Signal (On/Off)
A/FS (A/F) Bank 1 Sensor 2 (%)	A/C Cut Signal (On/Off)
Calculated Load (%)	A/C Magnetic Clutch Signal (On/Off)
Coolant Temperature (°C or °F)	Closed Throttle Position Switch (On/Off)
Engine Speed (RPM)	Cylinder 1-6 (Abnormal Variation)
Fuel System Bank 1 (%)	EGR System (On/Off)
Ignition Advance (Degrees)	Electronic Load Signal (On/Off)
Intake Air Temperature (°C or °F)	EVAP Purge VSV (On/Off)
Long Term Fuel Trim (%)	EVAP Vapor Pressure VSV (On/Off)
Manifold Air Pressure Sensor (kPa)	Fuel Cut Idle (On/OFF)
O2S Bank 1 Sensor 1 (Volts)	Fuel Cut TAU - High Load (On/Off)
O2S Bank 1 Sensor 2 (Volts)	Fuel Pump (On/Off)
O2S Bank 1 Sensor 1 (%)	Idle Air Control Duty Ratio (%)
Short Term Fuel Trim (%)	Ignition Counts (0-2000 counts)
Throttle Position (% of opening)	Injector Pulsewidth (ms)
Vehicle Speed (KPH/MPH)	Misfire Engine Speed (RPM)
---	Misfire Load (g/r)
---	O2S L/R Sensor 2 (0-1000ms)
---	O2S R/L Sensor 2 (0-1000ms)
---	Park Neutral Position (On/OFF)
---	Power Steering Oil Press (On/Off)
---	Starter Signal (On during cranking)
---	Stop Light Switch (On/Off)
	Total Fuel Trim Bank 1 (%)

OBD II Certified Scan Tools

An OBD II Certified Scan Tool should be able to display Current, Pending or History Codes, MIL on or off requests, Freeze Frame Data, I/M Readiness Status and Last Test Pass or Fail Messages.

Comprehensive Component Monitor

The Comprehensive Component Monitor (CCM) is an on-board strategy designed to monitor for failures in emission-related electronic components and circuits that provide input or output signals to the PCM. These are systems or devices that are not exclusively monitored by another monitor system. If the PCM detects that an input or output signal is inoperative due to an out-of-range value, open circuit or if an on-board rationality or functionality check fails, the PCM will set a code in memory and activate the MIL.

Tests conducted by the CCM vary depending on the type of hardware, the function of the device and the signal type. Analog signals are checked continuously for opens, shorts and out-of-range values. Some digital signals are checked for both functionality and rationality. These tests require that certain engine conditions be present before the test is performed and that several components are monitored as part of the test. Also, a sensor value can be monitored for change _after_ the PCM sends a command to a device. Here is a list of devices checked by the CCM at key on or engine running.

Input Device Examples
- Barometric Pressure Sensor
- Brake Switch
- Camshaft & Crankshaft Sensors
- Clutch Switch (M/T)
- Cruise Servo Switch
- Engine Coolant Temperature Sensor
- EVAP Pressure Sensor
- Intake Air Temperature Sensor
- Knock Sensor
- Manifold Absolute Pressure Sensor
- Mass Airflow Sensor
- Park Neutral Switch
- Transmission Temperature Sensor
- Transmission Speed Sensor
- Vehicle Speed Sensor

Output Device Examples
- EVAP Purge and Vent Solenoids
- Idle Air Control Solenoid
- Ignition Control System
- Transmission Torque Converter Clutch Solenoid
- Transmission Shift Solenoids (Solenoid S1, S2 and S3)

OBD II Main Monitors

A key difference between the first version of On Board Diagnostics (OBD I) and the second version (OBD II) is the use of several PCM controlled monitors contained within the PCM software structure. These monitors perform diagnostic tests required in order to meet specific California Air Resources Board (CARB) and EPA regulations.

Simply stated, an OBD II Monitor is a diagnostic strategy designed to test the operation of an emissions-related component or system. Some OBD II System Monitors accomplish this task *directly* by monitoring the action of various input and output devices or sensors connected to the PCM. An example of *direct* monitoring is when the Comprehensive Component Monitor monitors the Engine Coolant Temperature or Intake Air Temperature Sensor inputs.

Other OBD II System Monitors accomplish the task *indirectly* by monitoring the effects of changes to a system or component. The *indirect* method may be accomplished through monitoring a change or response in a system. This type of test is done by monitoring the input or output signals of a particular device for an "inferred" change.

An example of *indirect* monitoring is when the PCM infers correct or incorrect catalyst action using the Catalyst Monitor to sample signals from the upstream or downstream oxygen sensors. This allows the PCM to determine the oxygen storage efficiency of the catalyst.

Some of the Main Monitors run continuously while some run only once per trip. The next few articles explain how the Monitors operate.

Continuous Monitors (run all the time)

- Fuel Control System - begins when the engine enters closed loop
- Misfire Detection Test - begins right after startup

Main Monitors that run once only per trip:

- Catalyst Efficiency Test - begins in closed loop after certain engine temperature, time and VSS requirements are met
- EGR System - begins in closed loop after certain engine temperature, time and VSS requirements are met
- EVAP System Test - begins in closed loop after certain engine temperature, time and VSS requirements are met
- O2S Test - voltage and response time tests begin in closed loop after engine temperature, time and VSS requirements are met
- Secondary AIR System Test - begins in closed loop at off-idle

ν **NOTE:** *Once all of the required enable criteria are met, Toyota OBD II systems are programmed to run all the OBD II main monitors once each trip.*

Catalyst Monitor

A downstream (post-catalyst) Heated Oxygen Sensor (HO2S-12) is used to provide the additional signals needed to monitor the efficiency of the three-way catalyst on these systems. The PCM compares the signals between the upstream (pre-catalyst) and downstream oxygen sensor *during stable driving conditions* with the engine warm in order to determine the oxygen storage capacity of the catalytic converter.

Catalyst Monitor Operation

To measure catalyst efficiency, the Catalyst Monitor interprets the signals from the pre-catalyst and post-catalyst oxygen sensors. If the three-way catalyst is operating correctly, the post-catalyst signal will have significantly less activity than the pre-catalyst. Run the engine at 2500 rpm for 3 minutes in P/N and then check the waveforms from both sensors. If the signals are similar, the catalyst may be degraded.

Oxygen Sensor Signal Graphic

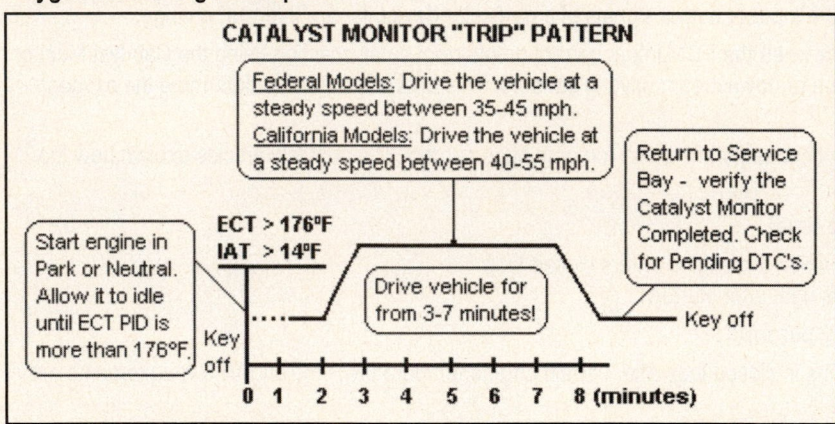

Catalyst Monitor "Trip" Pattern

The Catalyst Monitor "trip" pattern shown below can be used to validate repair of DTC P0420/P0430 or to "run" the Catalyst Monitor to complete the I/M Readiness Test. Federal models require that the vehicle be driven at a different speed than for California models. If the engine is cold on 2002 models, repeat the test (key off after the test).

Catalyst Monitor Trip Graphic

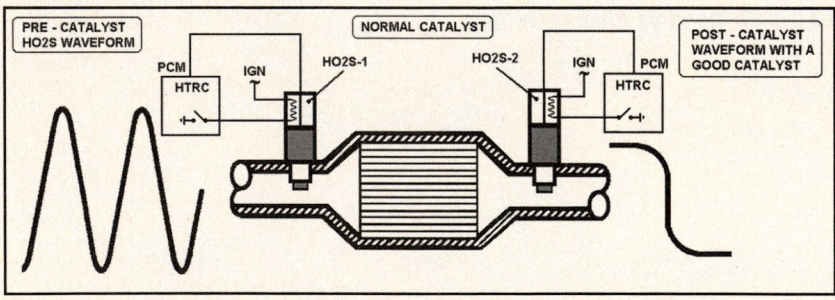

EGR System Monitor

The EGR system is used to reduce the amount of NOx emissions by circulating a portion of the exhaust gas through the EGR valve to the intake manifold. The PCM uses the MAP sensor signal to detect an EGR system fault by determining if the signal is too high or too low.

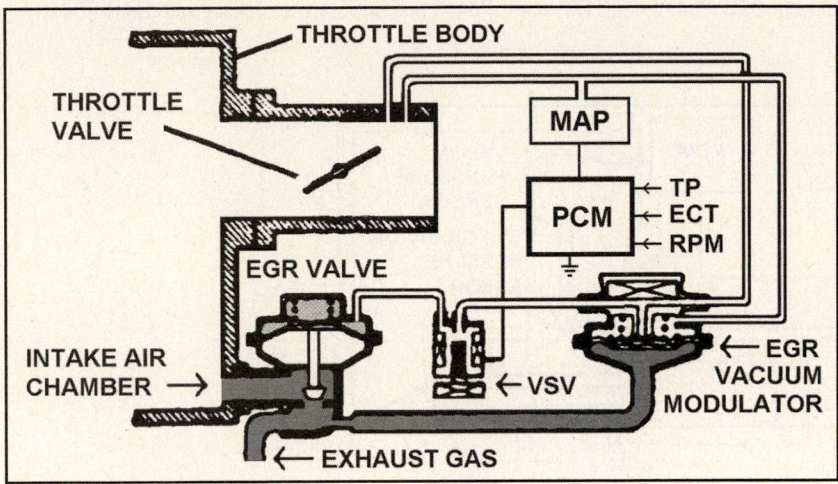

EGR Vacuum Modulator

An EGR vacuum modulator that operates according to engine load controls the amount of EGR gas flow. The PCM turns on the vacuum modulator to allow atmospheric air to "act" on the EGR valve, and to shut off the flow of EGR gas to the intake manifold with a cold engine, during high or low engine load, at high speed, deceleration or at idle.

EGR System Monitor "Trip" Pattern

The "trip" pattern below can be used to validate repair of DTC P0401 or to "run" the EGR Monitor to complete the I/M Readiness Test.

EGR System Trip Graphic

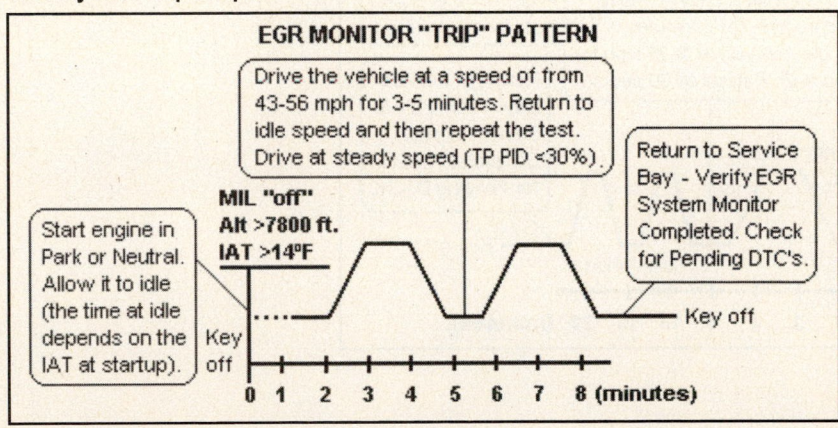

EVAP System Monitor

OBD II regulations require that EVAP system operation be monitored for the correct airflow used to purge the EVAP system. These EVAP systems are monitored using a pressure check to verify that no leaks exist that could allow fuel vapors to escape into the atmosphere. If a leak equal to or greater than 0.040" (0.020" in 2001-02) is detected in the EVAP system for two consecutive trips, a trouble code is set.

The EVAP VSV and Vapor Pressure sensor are used to detect faults in the system. If the PCM detects a fault in one of the devices shown in the Graphic, it sets DTC P0440. If DTC P0441, P0446 or P0450 are set along with DTC P0440, repair these trouble codes first.

EVAP System Graphic

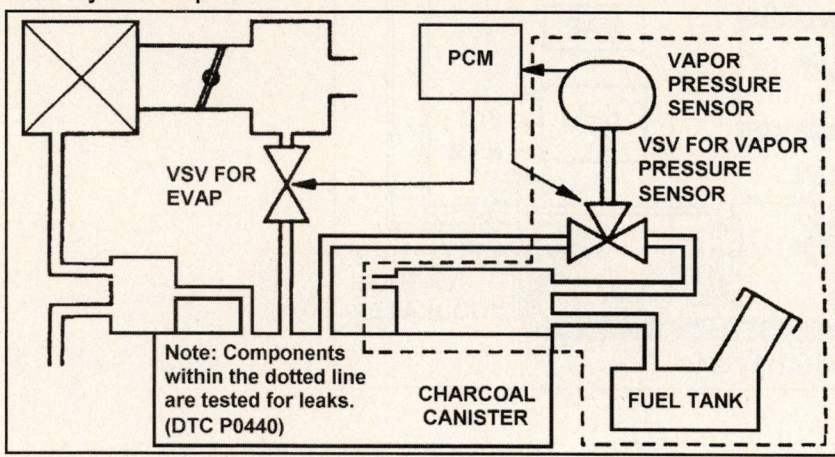

EVAP System Monitor "Trip" Pattern

The "trip" pattern can be used to validate an EVAP DTC (i.e., P0440) or to "run" the EVAP Monitor to complete the Inspection/Maintenance (I/M) Readiness Test.

EVAP Monitor Trip Graphic

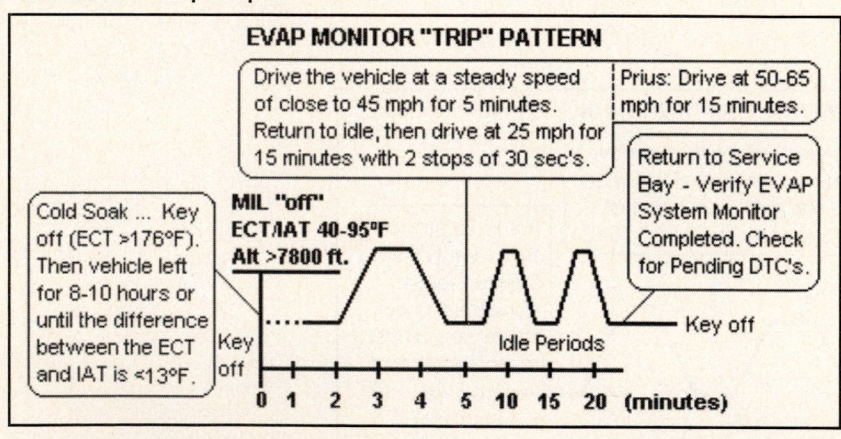

EVAP System Monitor

Diagnosis of DTC P0441 and P0446

The VSV for the Vapor Pressure sensor and the Vapor Pressure sensor are used by the PCM to detect faults in the EVAP system.

The PCM determines if there is a fault in the EVAP system based on the Vapor Pressure signal. If the PCM detects a leak or a fault in any of the components within the dotted line in Graphic below, it will set DTC P0441 or P0446 (depending upon the type of fault that exists).

EVAP System Graphic

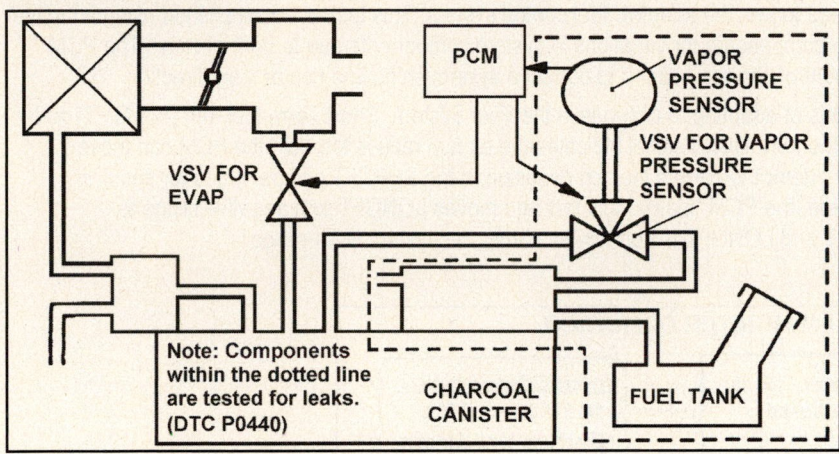

P0440 Code Conditions (2-Trip Detection)

The PCM will set P0440 if it detects the following condition:
- The fuel tank pressure is near atmospheric pressure after the vehicle is driven for 20 minutes (look for a loose or missing cap).

P0441 Code Conditions (2-Trip Detection)

The PCM will set P0441 if it detects one of the following conditions:
- The pressure in the canister does not drop during purge control.
- During purge cutoff, if the pressure in the charcoal canister is very low when compared to the atmospheric pressure value.

P0446 Code Conditions (2-Trip Detection)

The PCM will set P0446 of one of the following conditions is detected:
- With the VSV for the vapor pressure sensor off, the PCM detects that there is no continuity between the sensor and the canister.
- With the VSV for the vapor pressure sensor on, the PCM detects that there is no continuity between the sensor and the fuel tank.
- After purge cutoff is enabled, the pressure in the charcoal canister is maintained at atmospheric pressure.

Fuel System Monitor

OBD II regulations require that the Fuel Delivery Control system be tested continuously in order to verify that it can comply with emission standards. If a Fuel System component fails, or if a change in Long Term fuel trim is detected that could cause tailpipe emissions to exceed 1.5 times the FTP Standard for two consecutive trips, the MIL is activated, a code is set and engine conditions at that time are stored in Freeze Frame.

A Fuel System code is assigned a higher priority than all other codes except for an engine misfire. If a Fuel system fault is detected, the PCM will overwrite Freeze Frame Data for codes with a lower priority.

Fuel Trim Corrections

The Fuel System Monitor is designed to test the adaptive fuel control system. This task is accomplished using adaptive fuel tables stored in the memory to compensate for variations in system components due to normal wear. The PCM adaptive strategy "learns" the amount of change needed to correct a system shifts in a rich or lean direction.

Fuel trim correction has two methods of adapting to changes in the Fuel System: Short Term fuel trim (SHRTFT) and Long Term fuel trim (LONGFT). If the front HO2S signal indicates the air fuel ratio is too rich, the PCM can move SHRTFT to a negative range in an attempt to correct the rich condition. If the SHRTFT compensates for this rich condition for too long a period of time, the PCM "learns" this fact and moves LONGFT to a negative range to compensate. If the total of SHRTFT and LONGFT is from 0% to ±38%, the system is "in-range".

Fuel System Monitor Graphic

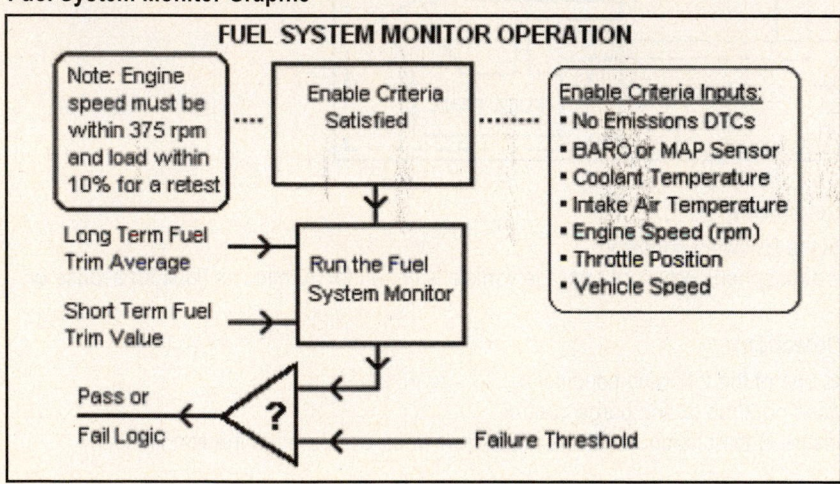

Misfire Monitor

OBD II regulations require that the engine be continuously tested for misfires under all engine positive load and speed conditions. Positive load conditions are defined as accelerating, cruising and idling. In this system, a high frequency CKP sensor is used to detect a misfire.

The PCM monitors crankshaft speed and position through the 'G' and 'NE' inputs. Crankshaft speed normally increases during firing events, and the PCM can detect any change from one firing event to the next in order to monitor for the presence and degree of an engine misfire.

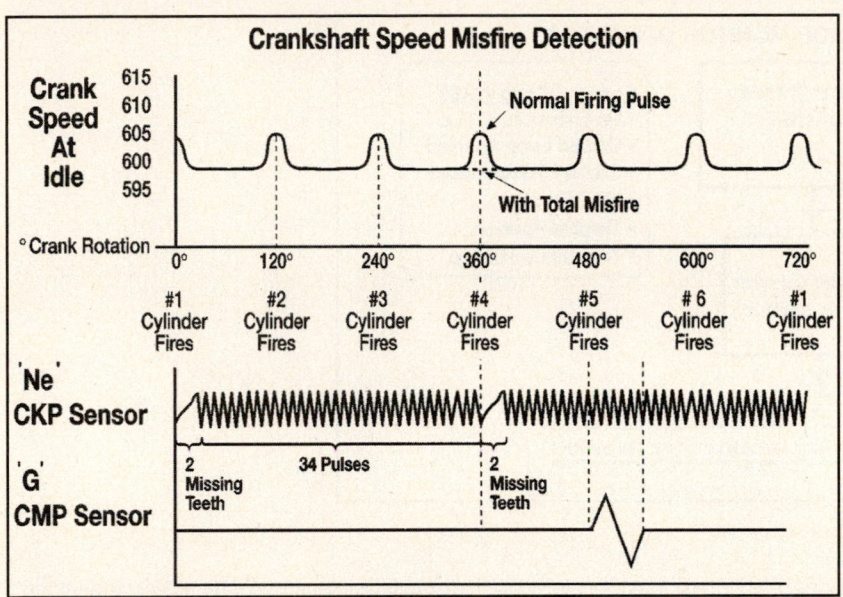

Crankshaft Speed Misfire Detection Graphic

Misfire Conditions

This Monitor can distinguish between a single and a multiple misfire. If only one cylinder is misfiring, then that cylinder must be identified. The conditions present when the misfire occurred are saved in Freeze Frame. If a misfire is detected that could cause tailpipe emissions to exceed 1.5 times the FTP Standard at similar engine speed, load and temperature conditions for two consecutive trips, the MIL is activated and a code is set. Also, the MIL is activated if a misfire is detected in two non-consecutive trips that are not more than 80 trips apart.

If a severe misfire condition is detected that could damage a catalyst, the MIL will flash once per second within 200 engine revolutions from the point when the misfire is first detected. The MIL will stop flashing and remain on if the vehicle stops operating under these conditions.

Oxygen Sensor Monitor

OBD II regulations require that the front heated oxygen sensor (Fuel Control sensor) and the rear heated oxygen sensor (Catalyst sensor) be monitored for correct response rate and output voltage levels, as well as any other factors that could effect tailpipe emissions.

The Oxygen Sensor Monitor consists of three separate tests: run-time to activity, lean-to-rich and rich-to-lean response times, and sensor voltage checks. If the PCM detects an Oxygen sensor fault that could cause tailpipe emissions to exceed 1.5 times the FTP Standard for two consecutive trips, the MIL is activated and a code is set.

Oxygen Sensor Monitor Graphic

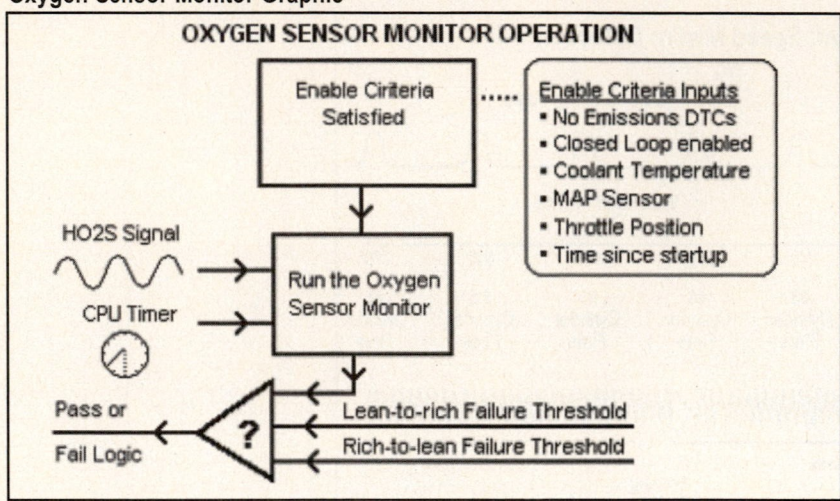

HO2S Voltage Tests

The Oxygen sensor voltage test is designed to test the pre-catalyst and post-catalyst sensors for no activity (due to an open circuit) or for a short-to-voltage or a short-to-ground condition in the circuit.

This test requires that a fault be detected in two consecutive trips to set a code/turn on the MIL.

Voltage Test Graphic

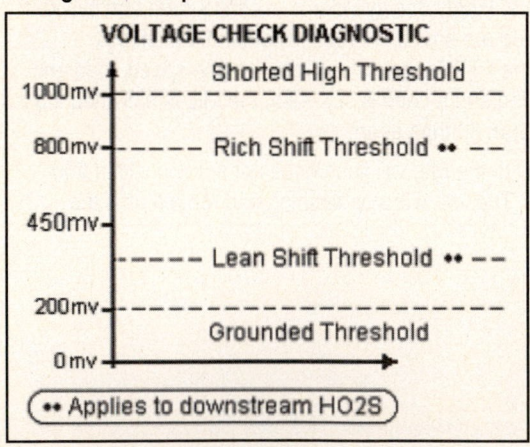

Oxygen Sensor Monitor

Response Time Test

Response time is defined as the amount of time it takes for the Oxygen sensor input to go from rich-to-lean and from lean-to-rich. Once the Oxygen Sensor Monitor is enabled, it compares the average response time to a calibrated value

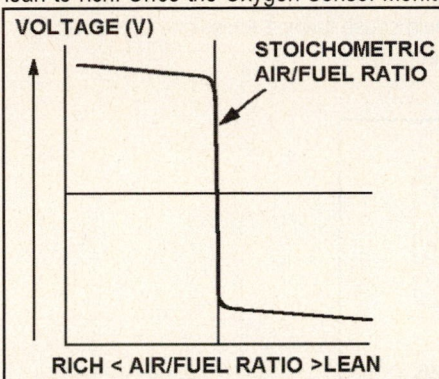

in the PCM memory.

If the response time to change from rich to lean or lean to rich is over one second for two consecutive trips, the MIL is activated, a code is set and current engine conditions are stored in Freeze Frame.

Oxygen Sensor Monitor "Trip" Pattern

The Oxygen Sensor Monitor "Trip" Pattern on this page can be used to validate an OBD II repair or to "run" an Oxygen Sensor Monitor to complete the I/M Readiness Test. The example is for P0130.

HO2S Monitor Trip Graphic

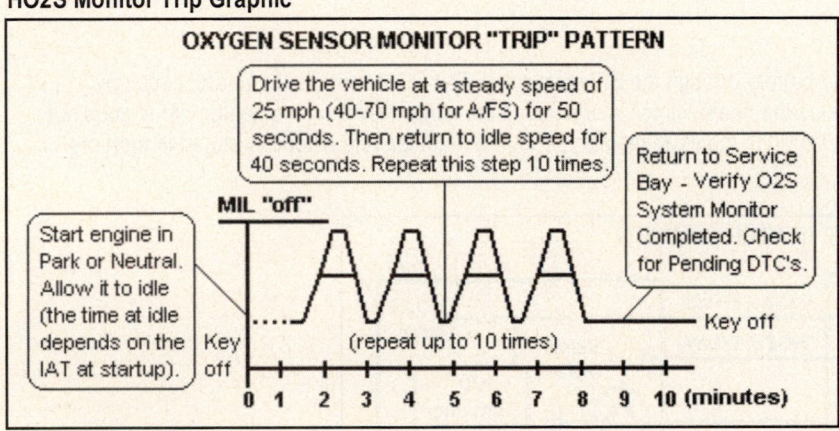

P0130 Code Conditions (2-Trip Detection)

The PCM will set P0130 if it detects the following condition:
- No PCM codes set, engine warmup completed, engine running and the HO2S-11 value remains fixed from >400mv to <550mv.

P1130 Code Conditions - California A/F Sensor (2-Trip Detection)

The PCM will set P1130 if it detects one the following condition:
- No PCM codes set, engine warmup completed, engine running and the A/F sensor value remains fixed from >3.8v to <2.8v.

Oxygen Sensor Heater Monitor

OBD II regulations require that the pre-catalyst and post-catalyst oxygen sensor be monitored for faults in the heater element or circuit. On Toyota vehicles, the PCM tests the operation of the Oxygen Sensor Heater and circuits for open, grounded or shorted circuit conditions by measuring the amount of current to the heater.

The second time a HO2S heater circuit fault is detected by the PCM that could cause tailpipe emissions to exceed 1.5 times the FTP Standard, the MIL is activated and a code is set (two-trip detection).

Oxygen Sensor Heater Monitor Graphic

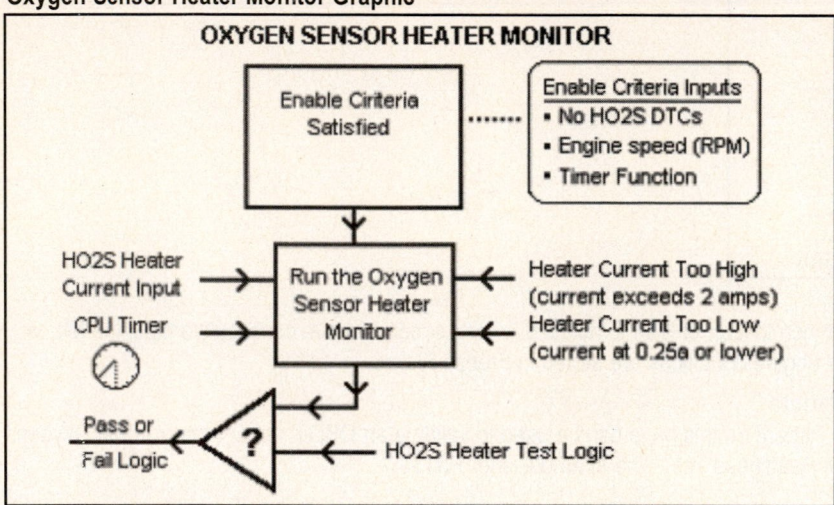

HO2S Heater Circuits

The heater receives power from the battery through the EFI main relay. The PCM controls the heater circuit by switching a transistor to ground. Once the heater circuit is enabled, if the current level on the heater circuit does not match the values for these circuits stored in the PCM memory, a fault is detected and a code is stored in memory.

Heater Circuits Graphic

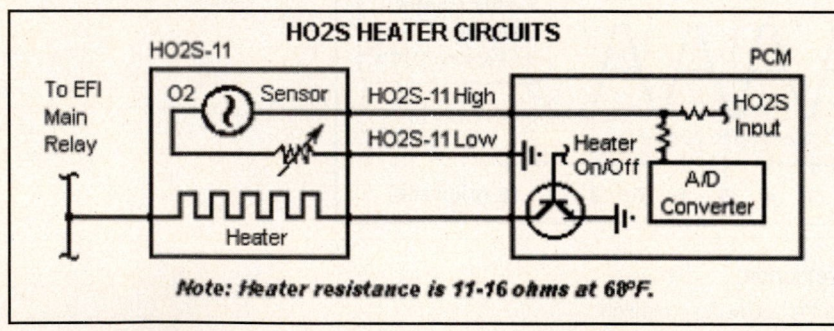

OBD II SYSTEM

Diagnostic Routine

The Diagnostic Routine shown below can be used to diagnose the cause of Code and No Code Faults (Symptoms) on OBD II systems. If a Code Fault is present, refer to the Trouble Code & Enable Criteria List in this section to obtain the details on why a code set.

If the code is related to the Catalyst, EGR, EVAP or Oxygen Sensor Main Monitor, refer to the Drive Pattern Examples in these articles.

No Code Faults

If a No Code Fault is present, refer to the PID Data Examples in this section or the vehicle specific Pin Tables in Sections 2 - 6. If a fault is suspected in a PCM Signal, refer to the test examples in this section.

OBD Flow Chart Graphic

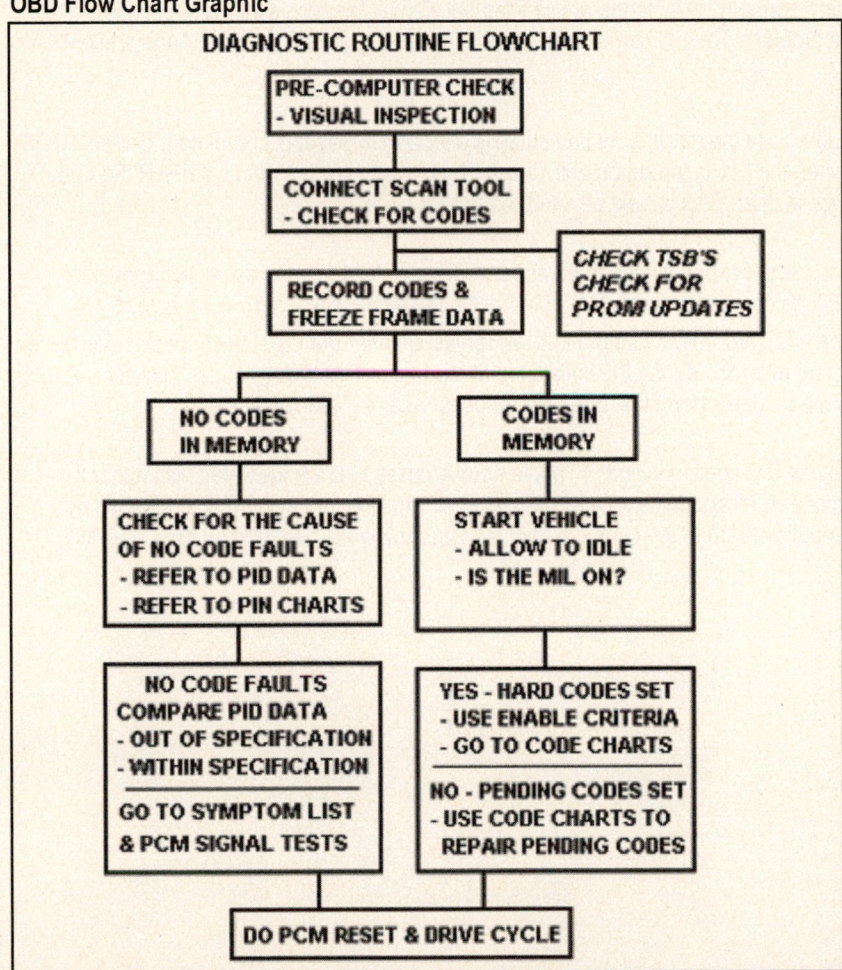

DIAGNOSTIC ROUTINE FLOWCHART

PRE-COMPUTER CHECK
- VISUAL INSPECTION

CONNECT SCAN TOOL
- CHECK FOR CODES

RECORD CODES &
FREEZE FRAME DATA

CHECK TSB'S
CHECK FOR
PROM UPDATES

NO CODES
IN MEMORY

CODES IN
MEMORY

CHECK FOR THE CAUSE
OF NO CODE FAULTS
- REFER TO PID DATA
- REFER TO PIN CHARTS

START VEHICLE
- ALLOW TO IDLE
- IS THE MIL ON?

NO CODE FAULTS
COMPARE PID DATA
- OUT OF SPECIFICATION
- WITHIN SPECIFICATION

GO TO SYMPTOM LIST
& PCM SIGNAL TESTS

YES - HARD CODES SET
- USE ENABLE CRITERIA
- GO TO CODE CHARTS

NO - PENDING CODES SET
- USE CODE CHARTS TO
REPAIR PENDING CODES

DO PCM RESET & DRIVE CYCLE

Read, Record & Clear Codes

The Toyota TCCS system uses the PCM to perform its diagnostic functions. The PCM in the OBD II system is designed to detect failures in emission-related systems or circuits. When a fault is detected, the PCM will set a trouble code in memory, activate the MIL and store current engine operating conditions in Freeze Frame.

Do the Diagnostics Work?

To determine if the PGM-FI diagnostics are operating, turn the key on and observe the PGM-FI light in the instrument panel. The light should remain on for two seconds and then go out. Use this step as a bulb check and diagnostic function check. If the light remains on after the engine is started, the PCM has detected a fault or the MIL circuit has a fault.

MIL Circuit Diagnosis

If the MIL and/or PCM do not operate normally, refer to the article titled MIL Circuit Diagnosis in this section to diagnose the cause of a MIL circuit problem. The MIL conditions listed in this article include the conditions present when the engine will not start.

Hard Codes

The term *hard code* refers to a trouble code that reappears immediately after a PCM reset is completed. On the TCCS system, if a *hard code* is present when the key is turned on, the Check Engine Light will remain on AFTER the bulb check period of two-seconds expires. A Scan Tool should be used to read the code.

Soft Codes

On PGM-FI systems, the term *soft code* describes a code that appears before a PCM reset is done, but does not reappear after a PCM reset. A *soft code* can be read with the Scan Tool in the same way a *hard code* is read.

The difference between a *hard code* and *soft* code is that the MIL will remain off after the initial bulb check period when a *hard code* exists. A *soft code* can be recorded during the initial step of read, record and clear codes. In effect, a *soft code* is a code stored in memory from an earlier fault that does not reappear after a PCM Reset step.

1-Trip and 2-Trip Trouble Codes

The Trouble Codes & Enable Criteria in this section includes Toyota trouble codes that are identified with either a 1-Trip or 2-Trip designator. A 1-Trip code (Fuel and Misfire Only) can set after only one trip, and 2-Trip codes are set after two consecutive trips with the fault present in the component or system. Any faults detected by the Catalyst Monitor are 2-Trip codes.

PCM Reset Step

Once all repairs to the OBD II system are completed, the PCM should be reset to allow it to relearn certain engine operating information (i.e., fuel trim and idle speed settings).

To perform a PCM reset procedure (clear all codes and Freeze Frame Data from the PCM memory), do the following steps. Turn the key off and then remove the PCM power fuse from the Underhood fuse/relay box for 10 seconds. This step resets the PCM. If the engine is started after doing this step, any codes that reset are hard codes.

Component Location Graphic

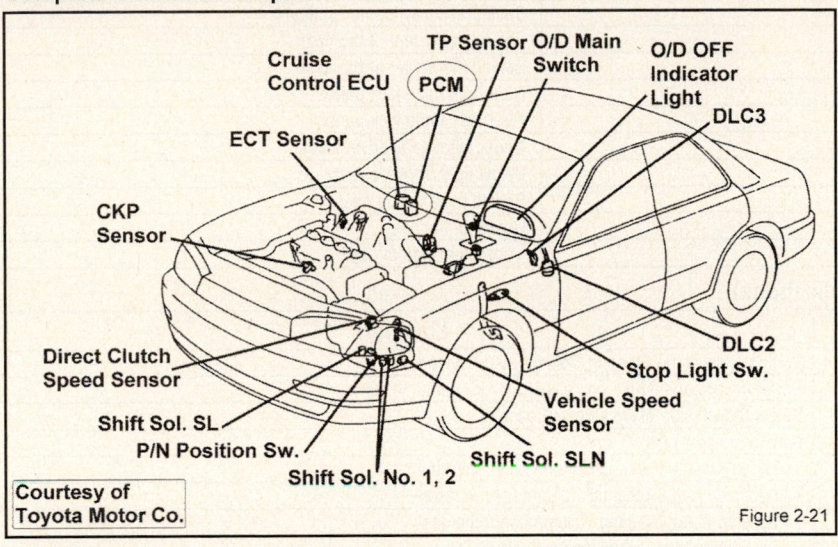

Figure 2-21

PCM Reset Fuse Locations for OBD I/OBD II Applications

- Camry and Celica: EFI 15 amp fuse
- Cressida and Previa: EFI 15 amp fuse
- Sienna: EFI 15 amp fuse
- Corolla, Corolla FX16: STOP 15 amp fuse
- Land Cruiser: EFI 15 amp fuse
- Pickup: EFI 15 amp fuse
- 4Runner, Tacoma: EFI 15 amp fuse
- MR2: AM2 15 amp fuse

DLC & PCM LOCATION TABLES

The tables on this page are provided to help technicians and vehicle owners determine the location of the Powertrain Control Module (PCM) and Diagnostic Link Connector (DLC) on Toyota vehicles.

PCM Location Table (OBD Applications)

Vehicle Application	Location
Avalon, Land Cruiser & Tacoma	Behind the RH side of the dash
Camry & Paseo	Behind the glove box
Celica	Behind ash tray (center of dash)
RAV4	Behind the center of the dash
Corolla	Behind the center of the dash
Pickup	Near right hand kick panel
T100 Pickup	Near right hand kick panel
Previa	Under left side of driver's seat
Supra	Below right front footrest
Tercel	Behind right side of dash
4Runner (I4)	Center of the dash
4Runner (V6)	Under right side of dash

DLC Location Table (OBD II Applications)

Year & Model	Location
1997-2002 4Runner	To right side of the Steering Column near Console
1996 Avalon	Under cover behind fuse box above LH kick panel
1997-2002 Avalon	Below the left side of the Center Console
1996 Camry	Behind the coin box
1997-2002 Camry	To left side of the Steering Column under the dash
1996-2002 Celica	To right side of the Steering Column near Console
1996-2002 Corolla	To left side of the Steering Column under the dash
1996-97 Previa	Under cover at top of right side Instrument Panel
1996-98 Tercel	Behind the fuse box panel above left kick panel
1996-2002 RAV4	Behind the fuse box panel above left kick panel
1006-2002 Sienna	To right side of the Steering Column near Console
1996-99 T-100 Truck	To right side of the Steering Column near Console
2000-2002 Tacoma	To right side of the Steering Column near Console
2001-2002 Tundra	To right side of the Steering Column near Console

Scan Tool Graphing Mode

A Scan Tool with the Graphing Mode Function can be used to diagnose devices that fail and cause intermittent faults. In the examples below, a Scan Tool was used to collect PID Data and to capture an intermittent fault using the graphic mode test function.

TP Sensor "Static" Test

Turn the key on (engine off) and connect a Scan Tool to read the PID voltage or percent of throttle opening. Compare the actual reading to the values from a known-good vehicle shown below. TP sensor voltage should change smoothly as the throttle is opened and closed.

TP Sensor Angle Table

Throttle Angle Degrees of Rotation	Voltage
0 Degrees	0.50v
10 Degrees	0.97v
20 Degrees	1.44v
30 Degrees	1.90v
40 Degrees	2.37v
50 Degrees	2.84v
60 Degrees	3.31v
70 Degrees	3.78v
80 Degrees	4.24v

TP Sensor "Dynamic" Test

Warm the engine to normal operating temperature and then connect a Scan Tool with the *graphing mode* function or connect a Lab Scope.

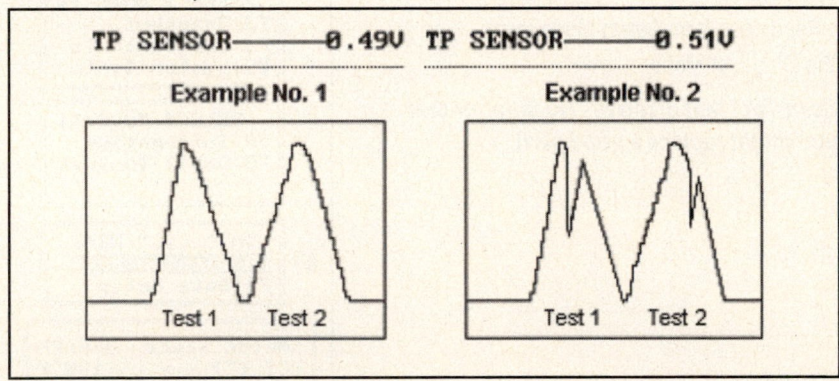

Place the gear selector in P/N and block the drive wheels for safety. Observe the waveform for a smooth transition or change as the throttle is momentarily snapped from closed throttle to WOT. Note the extra downward spike in Example No. 2 indicating a faulty TP sensor.

ν **NOTE:**   *An Aftermarket Scan Tool with graphing capability was used in "graphing mode" to capture these examples.*

How to Access Generic PID Data (OBD II)

The Scan Tool Generic Parameter Identification (PID) Mode allows access to certain data values, analog and digital input and output signals, calculated values and system status.

The list of PID Data that appears in this Mode is available to all OBD II certified Scan Tools.

Scan Tool PID Menus

An example of how to navigate through the Vetronix Scan Tool menus to locate the Generic PID information is shown in the Graphic to the right.

The Graphic contains an example of how to read some of the parameters that are available for this vehicle on an OBD II compatible Scan Tool. The parameters in the right hand column of the example represent known good values for this engine application.

If all of these PID values are within normal range, refer to the Symptom Diagnosis information found this section.

Parameter ID (PID) Information

The proper sequence to follow to obtain a complete Generic PID list for this vehicle is shown in the Graphic.

1) Select F1: Scan Test from the menu.

2) Select Global OBD II from the menu.

3) Select F0: Powertrain from menu.

4) Initializing Communications appears.

5) Select F0: Data List or F1: I/M Readiness from Select Mode menu.

6) Select F0: Display Data from the Data List Menu.

ν **NOTE:**   *Do not use a Scan Tool that displays "No Data" or that displays fault data (as you might replace a good part).*

SCAN TOOL MENUS

(1)
```
        FUNCTION MENU
F1: SCANTEST
F2: DIGITAL METER
F3: OSCILLOCOPE
F4: EMISSION TESTS
F8: TECH TOOLBOX
F9: SETUP
```

(2)
```
   APPLICATIONS ↑↓
-> GLOBAL OBD II
   GM P/T
   GM CHASSIS
```

(3)
```
      MAIN MENU
F0:Powertrain
F1:Replay Data
F9:OBDII Toolpak
```

(4)
```
    Initializing
      OBD II
   Communications
```

(5)
```
   SELECT MODE ↑↓
F0:Datalist
F1:Readiness
F2:DTCs
```

```
   SELECT MODE ↑↓
F3:Snapshot
F4:OBD Controls
F5:System Tests
```

```
   SELECT MODE ↑↓
F8:Information
F9:OBDII Toolpak
```

(6)
```
   DATA LIST MENU
F0:Display Data
F1:Data Setup
```

```
ENGINE SPEED····868RPM
ECT (°)··············180°F
VEHICLE SPEED····0MPH
IGN. TIMING·······12.0°
ENGINE LOAD·······40.0%
MAP (P)·········11.3inHg
TPS (%)···········10.5%
IAT (°)···········122°F
FUEL STAT 1···········0L
ST FT 1··············0.0%
LT FT 1·············-5.4%
O2S B1 S1·······0.045V
```

How to Access Enhanced PID Data (OBD II)

The Toyota Enhanced PID List allows access to certain data values, analog and digital input and output signals, calculated values and system status not available in Generic OBD II Mode.

Toyota has several "data lists" available under the Enhanced PID selection. The PID data available in this mode can be compared to known good readings found here or in other repair manuals.

If all of the PID values are within normal range, refer to the Symptom Diagnosis in this section.

Scan Tool PID Menus

This example shows how to navigate through the Scan Tool (Snap On) menus to locate the OEM PID Information in this section.

Parameter ID (PID) Information

An example of how to obtain a complete OEM PID list is shown in the Graphic.

1) Select F1: Scan Test from the menu.
2) Select Toyota/Lexus from the menu.
3) Select the model year letter ('W').
4) Select the engine size letter ('R').
5) Select ENGINE from the menu.
6) Select A/T WITH 4-SPEED from the Options Menu.
7) Select CODES & DA/TA MENU from the Main Menu.
8) Select CODES & DA/TA from menu.

Examples of the Enhanced PID data can be found in the Reference Information articles in this section.

ν **NOTE:** *Do not use a Scan Tool that displays "No Data" or that displays fault data (as you might replace a good part).*

SCAN TOOL MENUS

HONDA/ACURA (1994-00)
MITSUBISHI (1994-00)
NISSAN/INFINITI (1994-00)
~TOYOTA/LEXUS (1994-00)
GENERIC OBD II

VEHICLE ID SCREENS

10th VIN CHARACTER
VIN: --------W------
VEHICLE: 1983
FOR A/T ID (1994-00):
TURN KEY TO ON

VIN: -- 4--N--R--X--Y
2000 Corolla A/T
ENGINE: 1.8L I4 MPI
PRESS Y TO CONTINUE
PRESS N FOR NEW ID

SELECT SYSTEM:
~ENGINE
4-SPEED TRANS
ANTI-LOCK BRAKES
AIRBAG

SELECT OPTIONS:
M/T WITH A/C
A/T WITH 3-SPEED
~A/T WITH 4-SPEED

MAIN MENU (TOYO ENG)
OTHER SYSTEMS
~CODES & DATA MENU
CUSTOM SETUP
ACTUATOR TESTS

CODES & DATA MENU
~CODES & DATA
1 TRIP CODES

Trouble Codes Lists

Instructions: To use this trouble code list, first read and record all codes in memory along with any Freeze Frame data. This information is used during diagnosis. *If a PCM reset function is done, all codes and Freeze Frame data will be lost!*

Find the appropriate trouble code with code descriptions and test conditions that indicate the type of fault present when the code was recorded. This information can indicate how to drive the vehicle in order to validate that the original problem is fixed. This is especially true of codes related to the Catalyst, EGR and EVAP Monitors.

The information in this trouble code list is not meant to replace the use of a trouble code repair chart found in other manuals or media.

Enable criteria data is important during diagnosis of codes related to Catalyst, EGR and EVAP Monitors. The one **(1T)** and two trip **(2T)** designators indicate the number of trips required to set that code.

OBD II Trouble Code List (P0xxx Codes)

DTC	Trouble Code Title, Conditions & Possible Causes
DTC P2A00 **2T CCM, MIL: Yes** 2003 Avalon, Camry, Highlander, Sienna, Tacoma, Tundra 2.4L VIN D, 2.4L VIN L, 2.7L VIN M, 3.0L VIN F, 3.4L VIN N engines Transmissions: All	**Air Fuel Sensor (Bank 1 Sensor 1) Signal Slow Response Conditions:** Vehicle driven at cruise speed at over 1400 rpm in closed loop at 60 mph, and the PCM detected an unexpected voltage condition on the Bank 1 Air Fuel Sensor 1 (AFS1) circuit. **Possible Causes** • A/F sensor connector is damaged or loose • A/F sensor circuit is open or shorted, or the sensor has failed • A/F sensor heater is damaged or has failed • A/F sensor heater relay circuit is open or the relay has failed • Fuel delivery component has failed (fuel pressure regulator, one or more fuel injectors is leaking or severely restricted) • Induction system problems (air leaks or restricted air filter) • PCM has failed
DTC P2A03 **2T CCM, MIL: Yes** 2003 Avalon, Camry, Highlander, Sienna 3.0L VIN F engine Transmissions: All	**Air Fuel Sensor (Bank 2 Sensor 1) Signal Slow Response Conditions:** Vehicle driven at cruise speed at over 1400 rpm in closed loop at 60 mph, and the PCM detected an unexpected voltage condition on the Bank 2 Air Fuel Sensor 1 (AFS1) circuit. **Possible Causes** • A/F sensor connector is damaged or loose • A/F sensor circuit is open or shorted, or the sensor has failed • A/F sensor heater is damaged or has failed • A/F sensor heater relay circuit is open or the relay has failed • Fuel delivery component has failed (fuel pressure regulator, one or more fuel injectors is leaking or severely restricted) • Induction system problems (air leaks or restricted air filter) • PCM has failed
DTC P0010 **2T CCM, MIL: Yes** 2003 Avalon, Camry, Echo, Highlander, MR2, Sienna 1.5L VIN T, 1.8L VIN R, 3.0L VIN F engines Transmissions: All	**VVT Oil Control Circuit Malfunction (Bank 1) Conditions:** Key on or engine running; and the PCM detected an unexpected voltage condition on the VVT Oil Control Valve Bank 1 circuit. The VVT system controls the intake camshaft in order to provide optimal valve timing during all conditions based signals from the ECT, IAT and TP sensor. The VVT regulates the intake camshaft angle using oil pressure through the Oil Control Valve. This results in the relative position between the camshaft and crankshaft to become optimal. The result is higher torque, better fuel economy and low emissions. **Possible Causes** • OCV assembly connector is damaged or loose • OCV assembly control circuit is open or shorted to ground • OCV assembly is damaged or has failed • PCM has failed
DTC P0011 **2T CCM, MIL: Yes** 2003 Avalon, Camry, Echo, Highlander, MR2, Sienna 1.5L VIN T, 1.8L VIN R, 3.0L VIN F engines Transmissions: All	**Camshaft Position 'A' Over-Advanced Or System Performance (Bank 1) Conditions:** Engine started, ECT sensor more than 158ºF, vehicle driven at an engine speed of 400-4000 rpm, and the PCM detected the valve timing did not change from the "current" valve timing, or the valve timing remain fixed during testing. The VVT system controls the intake camshaft in order to provide optimal valve timing during all conditions based signals from the ECT, IAT and TP sensor. The VVT regulates the intake camshaft angle using oil pressure through the Oil Control Valve. This results in the relative position between the camshaft and crankshaft to become optimal. The result is better engine torque, fuel economy and lower emissions. **Possible Causes** • Engine valve timing malfunction • Camshaft timing oil control valve unit is damaged or has failed • PCM or VVT ECM has failed

OBD II Trouble Code List (P0xxx Codes)

DTC	Trouble Code Title, Conditions & Possible Causes
DTC P0012 **2T CCM, MIL: Yes** 2003 Avalon, Camry, Echo, Highlander, MR2, Sienna 1.5L VIN T, 1.8L VIN R, 3.0L VIN F engines Transmissions: All	**Camshaft Position 'A' Over-Retarded (Bank 1) Conditions:** Engine started, ECT sensor more than 158ºF, vehicle driven at an engine speed of 400-4000 rpm, and the PCM detected the valve timing did not change from the "current" valve timing, or that the valve timing remain fixed during the test period. The VVT system controls the intake camshaft in order to provide optimal valve timing during all conditions. The VVT regulates the intake camshaft angle using oil pressure through the OCV. This causes the relative position between the camshaft and crankshaft to become optimal. The result is improved engine torque, better fuel economy and low emissions. **Possible Causes** • Engine valve timing malfunction • Camshaft timing oil control valve unit is damaged or has failed • PCM or VVT ECM has failed
DTC P0016 **2T CCM, MIL: Yes** 2003 Avalon, Camry, Echo, Highlander, MR2, Sienna 1.5L VIN T, 1.8L VIN R, 3.0L VIN F engines Transmissions: All	**Camshaft Position-Crankshaft Position (Bank 1 Sensor A) Conditions:** Engine started, engine running, and the PCM detected a deviation between the crankshaft position sensor signal and the VVT Sensor 1 signal during the test period. The crankshaft position (NE) sensor consists of a magnet, iron core and pickup coil. The NE sensor signal plate, installed on the crankshaft-timing pulley, has 34 teeth. This sensor generates 34 signals for each engine revolution. The PCM detects the crankshaft angle and engine speed based on the NE signal. It detects the correct cylinder based on signals from the VVT 1 sensor along signals from the crankshaft position sensor. **Possible Causes** • Engine valve timing problem • Engine timing belt mechanical problem (skipped teeth or belt) • PCM has failed
DTC P0018 **2T CCM, MIL: Yes** 2003 Avalon, Camry, Echo, Highlander, MR2, Sienna 3.0L VIN F engine Transmissions: All	**Camshaft Position-Crankshaft Position (Bank 2 Sensor A) Conditions:** Engine started, engine running, and the PCM detected a deviation between the crankshaft position sensor signal and the VVT Sensor 2 signal during the test period. The crankshaft position (NE) sensor consists of a magnet, iron core and pickup coil. The NE sensor signal plate, installed on the crankshaft-timing pulley, has 34 teeth. This sensor generates 34 signals for each engine revolution. The PCM detects the crankshaft angle and engine speed based on the NE signal. It detects the correct cylinder based on signals from the VVT 2 sensor along signals from the crankshaft position sensor. **Possible Causes** • Engine valve timing problem • Engine timing belt mechanical problem (skipped teeth or belt) • PCM has failed
DTC P0020 **2T CCM, MIL: Yes** 2003 Avalon, Camry, Echo, Highlander, MR2, Sienna 3.0L VIN F engine Transmissions: All	**Camshaft Position Sensor Actuator 'A' Circuit (Bank 2) Conditions:** Key on or engine running; and the PCM detected an unexpected voltage condition on the Camshaft Position Sensor 'A' Bank 2 circuit. The VVT system controls the intake camshaft in order to provide optimal valve timing during all conditions based signals from the ECT, IAT and TP sensor. The VVT regulates the intake camshaft angle using oil pressure through the OCV. The result is that relative position between the camshaft and crankshaft becomes optimal. The engine has better torque, fuel economy and lower emissions. **Possible Causes** • OCV assembly connector is damaged or loose • OCV assembly control circuit is open or shorted to ground • OCV assembly is damaged or has failed • PCM has failed
DTC P0021 **2T CCM, MIL: Yes** 2003 Avalon, Camry, Echo, Highlander, MR2, Sienna 3.0L VIN F engine Transmissions: All	**Camshaft Position 'A' Timing Over-Advanced Or System Performance (Bank 2) Conditions:** Engine started, ECT sensor more than 158ºF, vehicle driven at an engine speed of 400-4000 rpm, and the PCM detected the valve timing did not change from the "current" valve timing, or the valve timing remain fixed during testing. The VVT system controls the intake camshaft in order to provide optimal valve timing during all conditions. The VVT regulates the intake camshaft angle using oil pressure through the Oil Control Valve. The result is that relative position between the camshaft and crankshaft becomes optimal. The engine has better torque, fuel economy and lower emissions. **Possible Causes** • Engine valve timing malfunction • Camshaft timing oil control valve unit is damaged or has failed • PCM or VVT ECM has failed

OBD II Trouble Code List (P0xxx Codes)

DTC	Trouble Code Title, Conditions & Possible Causes
DTC P0022 **2T CCM, MIL: Yes** 2003 Avalon, Camry, Echo, Highlander, MR2, Sienna 3.0L VIN F engine Transmissions: All	**Camshaft Position 'A' Timing Over-Retarded (Bank 2) Conditions:** Engine started, ECT sensor over 158°F, engine speed of 400-4000 rpm, and the PCM detected the valve timing did not change from the "current" valve timing, or the valve timing remain fixed. The VVT system controls the intake camshaft in order to provide optimal valve timing during all conditions. The VVT regulates the intake camshaft angle using oil pressure through the OCV. The result is that relative position between the camshaft and crankshaft becomes optimal. The engine has better torque, fuel economy and lower emissions. **Possible Causes** • Engine valve timing malfunction • Camshaft timing oil control valve unit is damaged or has failed • PCM or VVT ECM has failed
DTC P0031 **1T CCM, MIL: Yes** 2003 4Runner, Echo, Land Cruiser, MR2, Sequoia, Tacoma, Tundra All engines Transmissions: All	**Heated Oxygen Sensor (Bank 1 Sensor 1) Heater Circuit Low Code Conditions:** Engine started, and the PCM detected the HO2S-11 heater control circuit indicated less than 0.20 amps during the CCM test period. **Possible Causes** • HO2S-11 heater control circuit is open • HO2S-11 heater assembly is damaged or has failed • PCM has failed
DTC P0032 **1T CCM, MIL: Yes** 2003 4Runner, Echo, Land Cruiser, MR2, Sequoia, Tacoma, Tundra All engines Transmissions: All	**Heated Oxygen Sensor (Bank 1 Sensor 1) Heater Circuit High Input Conditions:** Engine started, and the PCM detected the HO2S-11 heater control circuit indicated more than 2.0 amps during the CCM test period. **Possible Causes** • HO2S-11 heater control circuit is shorted to ground • HO2S-11 heater assembly is damaged or has failed • PCM has failed
DTC P0036 **1T CCM, MIL: Yes** 2003 Highlander, Tacoma 2.4L VIN L, 2.7L VIN M engines Transmissions: All	**Air Fuel Sensor (Bank 1 Sensor 2) Heater Circuit Malfunction Conditions:** Engine started, and the PCM detected the HO2S-12 heater control circuit indicated more than 2.35 amps or less than 0.20 amps. **Possible Causes** • HO2S-12 heater control circuit is open or shorted to ground • HO2S-12 heater assembly is damaged or has failed • HO2S-12 heater power circuit open (test power from EFI relay) • PCM has failed
DTC P0036 **1T CCM, MIL: Yes** 2003 Avalon, Camry, Highlander, Sienna 3.0L VIN F engine Transmissions: All	**Heated Oxygen Sensor (Bank 1 Sensor 2) Heater Circuit Malfunction Conditions:** Engine started, and the PCM detected the HO2S-12 heater control circuit indicated more than 2.35 amps or less than 0.20 amps. **Possible Causes** • HO2S-12 heater control circuit is open or shorted to ground • HO2S-12 heater assembly is damaged or has failed • HO2S-12 heater power circuit open (test power from EFI relay) • PCM has failed
DTC P0037 **2T CCM, MIL: Yes** 2003 4Runner, Echo, Land Cruiser, MR2, Sequoia, Tacoma, Tundra All engines Transmissions: All	**Heated Oxygen Sensor (Bank 1 Sensor 2) Heater Circuit Low Input Conditions:** Engine started, and the PCM detected the HO2S-12 heater control circuit indicated less than 0.20 amps during the CCM test period. **Possible Causes** • HO2S-12 heater control circuit is open • HO2S-12 heater assembly is damaged or has failed • PCM has failed
DTC P0038 **2T CCM, MIL: Yes** 2003 4Runner, Echo, Land Cruiser, MR2, Sequoia, Tacoma, Tundra All engines Transmissions: All	**Heated Oxygen Sensor (Bank 1 Sensor 2) Heater Circuit High Input Conditions:** Engine started, and the PCM detected the HO2S-12 heater control circuit indicated more than 2.0 amps during the CCM test period. **Possible Causes** • HO2S-12 heater control circuit is shorted to ground • HO2S-12 heater assembly is damaged or has failed • PCM has failed
DTC P0051 **2T CCM, MIL: Yes** 2003 4Runner, Land Cruiser, MR2, Sequoia, Tundra 1.8L VIN R, 4.0L VIN U, 4.7L VIN T engines Transmissions: All	**Heated Oxygen Sensor (Bank 2 Sensor 1) Heater Circuit Low Input Conditions:** Engine started, and the PCM detected the HO2S-21 heater control circuit indicated less than 0.20 amps during the CCM test period. **Possible Causes** • HO2S-21 heater control circuit is open • HO2S-21 heater assembly is damaged or has failed • PCM has failed

OBD II Trouble Code List (P0xxx Codes)

DTC	Trouble Code Title, Conditions & Possible Causes
DTC P0052 **2T CCM, MIL: Yes** 2003 4Runner, Land Cruiser, MR2, Sequoia, Tundra 1.8L VIN R, 4.0L VIN U, 4.7L VIN T engines Transmissions: All	**Heated Oxygen Sensor (Bank 2 Sensor 1) Heater Circuit High Input Conditions:** Engine started, and the PCM detected the HO2S-21 heater control circuit indicated more than 2.0 amps during the CCM test period. **Possible Causes** • HO2S-21 heater control circuit is shorted to ground • HO2S-21 heater assembly is damaged or has failed • PCM has failed
DTC P0057 **2T CCM, MIL: Yes** 2003 4Runner, Land Cruiser, Sequoia, Tundra 4.0L VIN U, 4.7L VIN T engines Transmissions: All	**Heated Oxygen Sensor (Bank 2 Sensor 2) Heater Circuit Low Input Conditions:** Engine started, and the PCM detected the HO2S-22 heater control circuit indicated less than 0.20 amps during the CCM test period. **Possible Causes** • HO2S-22 heater control circuit is open • HO2S-22 heater assembly is damaged or has failed • PCM has failed
DTC P0058 **2T CCM, MIL: Yes** 2003 4Runner, Land Cruiser, Sequoia, Tundra 4.0L VIN U, 4.7L VIN T engines Transmissions: All	**Heated Oxygen Sensor (Bank 2 Sensor 2) Heater Circuit High Input Conditions:** Engine started, and the PCM detected the HO2S-22 heater control circuit indicated more than 2.0 amps during the CCM test period. **Possible Causes** • HO2S-22 heater control circuit is shorted to ground • HO2S-22 heater assembly is damaged or has failed • PCM has failed
DTC P0100 **1T CCM, MIL: Yes** 1995 Avalon, Camry, Land Cruiser, Previa, T100, Tacoma, Tercel 1.5L VIN E, 2.4L VIN A, 2.4L VIN R, 2.7L VIN U, 3.0L VIN G, 3.4L VIN V, 4.5L VIN D engines Transmissions: All	**Mass Airflow Sensor Circuit Malfunction Conditions:** Engine running at under 4000 rpm, and the PCM detected an unexpected voltage condition on the MAF sensor circuit. The MAF sensor on this engine includes a hot wire assembly with an air temperature sensor, platinum hot wire and control unit mounted in a plastic housing. This airflow meter works on the principle that the hot wire and temperature sensor located in the intake air bypass of the housing detect any changes in the (incoming) air temperature. **Possible Causes** • MAF sensor signal circuit is open, shorted to ground or power • MAF sensor ground circuit is open between sensor and ground • MAF sensor power circuit is open (check the power to the relay) • MAF sensor has failed, or the PCM has failed
DTC P0100 **1T CCM, MIL: Yes** 1996, 1997, 1998, 1999, 2000, 2001, 2002, 2003 All engines Transmissions: All	**Mass Airflow Sensor Circuit Malfunction Conditions:** Engine started, engine running at under 4000 rpm, and the PCM detected an unexpected low or high voltage condition on the Mass Airflow (MAF) sensor circuit for over 3 seconds during the CCM test. The MAF sensor on this engine includes a hot wire assembly with an air temperature sensor, platinum hot wire and control unit mounted in a plastic housing. This airflow meter works on the principle that the hot wire and temperature sensor located in the intake air bypass of the housing detect any changes in the (incoming) air temperature. **Possible Causes** • MAF sensor signal circuit is open, shorted to ground or power • MAF sensor ground circuit is open between sensor and ground • MAF sensor power circuit is open (check the power to the relay) • MAF sensor has failed, or the PCM has failed
DTC P0101 **2T CCM, MIL: Yes** 1995 Avalon, Camry, Land Cruiser, Previa, T100, Tacoma, Tercel 1.5L VIN E, 2.4L VIN A, 2.4L VIN R, 2.7L VIN U, 3.0L VIN G, 3.4L VIN V, 4.5L VIN D engines Transmissions: All	**Mass Airflow Sensor Signal Range/Performance Conditions:** DTC P0100 not set; engine speed under 900 rpm, throttle valve closed, ECT sensor more than 158°F, and the PCM detected the MAF sensor was more than 2.20v; or with the engine speed over 1500 rpm, the throttle valve closed and the TP sensor above 0.63v, the MAF sensor was below 1.06v. This airflow meter works on the principle where a temperature sensor and hot wire in the intake air bypass area detect any changes in the incoming air. **Possible Causes** • MAF sensor signal circuit is open, shorted to ground or power • MAF sensor is contaminated, damaged or it has failed • PCM has failed
DTC P0101 **2T CCM, MIL: Yes** 1996, 1997, 1998, 1999, 2000, 2001, 2002, 2003 All engines Transmissions: All	**Mass Airflow Sensor Signal Range/Performance Conditions:** DTC P0100 not set, engine speed under 900 rpm, throttle valve closed, ECT sensor over 158°F, and the PCM detected the MAF was above 2.20v; or with the engine speed over 1500 rpm, the throttle valve closed and the TP sensor over 0.63v, the MAF sensor was less than 1.06v. This airflow meter works on the principle that the hot wire and temperature sensor in the intake air bypass of the housing detect any changes in the (incoming) air temperature **Possible Causes** • MAF sensor signal circuit is open, shorted to ground or power • MAF sensor is contaminated, damaged or it has failed • PCM has failed

OBD II Trouble Code List (P0xxx Codes)

DTC	Trouble Code Title, Conditions & Possible Causes
DTC P0102 **2T CCM, MIL: Yes** 2003 4Runner, Echo, Land Cruiser, Sequoia, Tacoma, Tundra 1.5L VIN T, 3.4L VIN N, 4.0L VIN U, 4.7L VIN T engines Transmissions: All	**Mass Airflow Sensor Circuit Low Input Conditions:** DTC P0100 not set, engine started, and the PCM detected an unexpected low voltage condition on the MAF sensor circuit during the CCM test period. This airflow meter works on the principle that the hot wire and temperature sensor located in the intake air bypass of the housing detect any changes in the (incoming) air temperature **Possible Causes** • MAF sensor signal circuit is open or shorted to ground • MAF sensor is contaminated, damaged or it has failed • PCM has failed
DTC P0103 **2T CCM, MIL: Yes** 2003 4Runner, Echo, Land Cruiser, Sequoia, Tacoma, Tundra 1.5L VIN T, 3.4L VIN N, 4.0L VIN U, 4.7L VIN T engines Transmissions: All	**Mass Airflow Sensor Circuit High Input Conditions:** DTC P0100 not set, engine started, and the PCM detected an unexpected high voltage condition on the MAF sensor circuit during the CCM test period. This airflow meter works on the principle that the hot wire and temperature sensor located in the intake air bypass of the housing detect any changes in the (incoming) air temperature **Possible Causes** • MAF sensor signal circuit is shorted to power • MAF sensor is contaminated, damaged or it has failed • PCM has failed
DTC P0105 **1T CCM, MIL: Yes** 1995 Paseo 1.5L VIN E engine Transmissions: All	**Manifold Air Pressure Sensor Circuit Malfunction Conditions:** Key on or engine running; and the PCM detected the MAP sensor signal was less than 0 kPa or more than 130 kPa in the CCM test. **Possible Causes** • MAP sensor signal circuit is open, shorted to ground or power • MAP sensor ground circuit is open between sensor and ground • MAP sensor is damaged or has failed • PCM has failed
DTC P0105 **1T CCM, MIL: Yes** 1996, 1997, 1998, 1999, 2000, 2001, 2002 Paseo, Tercel, Camry, Camry Solara 1.5L VIN E, 2.2L VIN G, 2.4L VIN D engines Transmissions: All	**Manifold Air Pressure Sensor Circuit Malfunction Conditions:** Key on or engine running; and the PCM detected the MAP sensor signal was less than 0 kPa or more than 130 kPa in the CCM test. **Possible Causes** • MAP sensor signal circuit is open, shorted to ground or power • MAP sensor ground circuit is open between sensor and ground • MAP sensor is damaged or has failed • PCM has failed
DTC P0106 **1T CCM, MIL: Yes** 1995 Paseo 1.5L VIN E engine Transmissions: All	**Manifold Air Pressure Sensor Signal Range/Performance Conditions:** DTC P0105 not set, engine started, engine speed at 400-1000 rpm, closed throttle switch indicating "on", ECT sensor input over 158°F, and the PCM detected the MAP sensor input was more than 3.3v for 10 seconds, or it detected it was less than 1.0v at under 2500 rpm with a VTA sensor input of more than 1.82v for 5 seconds. **Possible Causes** • MAP sensor source vacuum line is leaking or restricted • MAP sensor source vacuum line is plugged at intake manifold • MAP sensor is damaged, out-of-calibration or has failed • PCM has failed
DTC P0106 **1T CCM, MIL: Yes** 1996, 1997, 1998, 1999, 2000, 2001, 2002 Paseo, Tercel, Camry, Camry Solara 1.5L VIN E, 2.2L VIN G, 2.4L VIN D engines Transmissions: All	**Manifold Air Pressure Sensor Signal Range/Performance Conditions:** DTC P0105 not set, engine started, engine speed at 400-1000 rpm, closed throttle switch indicating "on", ECT sensor input over 158°F, and the PCM detected the MAP sensor was above 3.3v (10 sec's), or it less below 1.0v at under 2500 rpm (VTA over 1.82v for 5 sec's). **Possible Causes** • MAP sensor source vacuum line is leaking or restricted • MAP sensor source vacuum line is plugged at intake manifold • MAP sensor is damaged, out-of-calibration or has failed • PCM has failed
DTC P0110 **1T CCM, MIL: Yes** 1995 Avalon, Camry, Land Cruiser, Previa, T100, Tacoma, Tercel 1.5L VIN E, 2.4L VIN A, 2.4L VIN R, 2.7L VIN U, 3.0L VIN G, 3.4L VIN V, 4.5L VIN D engines Transmissions: All	**Intake Air Temperature Sensor Circuit Malfunction Conditions:** Key on or engine running; and the PCM detected an unexpected "low" or "high" voltage on the IAT sensor circuit (Scan Tool reads less than -40°F or low voltage or more than 284°F for high voltage). *Note: The IAT sensor is built into the MAF sensor on some engines.* **Possible Causes** • IAT sensor signal circuit is open between the sensor and PCM • IAT sensor signal circuit is shorted to ground or to VREF • IAT sensor ground circuit is open between sensor and ground • IAT sensor is contaminated, damaged or has failed • PCM has failed

OBD II Trouble Code List (P0xxx Codes)

DTC	Trouble Code Title, Conditions & Possible Causes
DTC P0110 **1T CCM, MIL: Yes** 1996, 1997, 1998, 1999, 2000, 2001, 2002, 2003 All engines Transmissions: All	**Intake Air Temperature Sensor Circuit Malfunction Conditions:** Engine started, and the PCM detected an unexpected low or high voltage on the IAT sensor circuit (Scan Tool reads -40ºF or 284ºF). This sensor is located inside the MAF sensor. **Possible Causes** • IAT sensor signal circuit is open, shorted to ground or VREF • IAT sensor ground circuit is open between sensor and ground • IAT sensor is contaminated, damaged or has failed • PCM has failed
DTC P0112 **1T CCM, MIL: Yes** 2003 4Runner, Echo, Land Cruiser, Sequoia, Tacoma, Tundra 1.5L VIN T, 3.4L VIN N, 4.0L VIN U, 4.7L VIN T engines Transmissions: All	**Intake Air Temperature Sensor Circuit Low Input Conditions:** Key on or engine running; and the PCM detected an unexpected low voltage condition on the IAT sensor circuit for over 500 ms. **Possible Causes** • IAT sensor connector is damaged (it may be shorted internally) • IAT sensor ground circuit is shorted to ground • IAT sensor is damaged or has failed • PCM has failed
DTC P0113 **1T CCM, MIL: Yes** 2003 4Runner, Echo, Land Cruiser, Sequoia, Tacoma, Tundra 1.5L VIN T, 3.4L VIN N, 4.0L VIN U, 4.7L VIN T engines Transmissions: All	**Intake Air Temperature Sensor Circuit High Input Conditions:** Key on or engine running; and the PCM detected an unexpected high voltage condition on the IAT sensor circuit for over 500 ms. **Possible Causes** • IAT sensor connector is damaged (it may be open internally) • IAT sensor ground circuit is open • IAT sensor is damaged or has failed • PCM has failed
DTC P0115 **2T CCM, MIL: Yes** 1995 Avalon, Camry, Land Cruiser, Previa, T100, Tacoma, Tercel 1.5L VIN E, 2.4L VIN A, 2.4L VIN R, 2.7L VIN U, 3.0L VIN G, 3.4L VIN V, 4.5L VIN D engines Transmissions: All	**Engine Coolant Temperature Sensor Circuit Malfunction Conditions:** Key on or engine running; and the PCM detected an unexpected "low" or "high" voltage on the ECT sensor circuit (Scan Tool reads less than -40ºF or low voltage or more than 284ºF for high voltage). **Possible Causes** • ECT sensor signal circuit is open, shorted to ground or VREF • ECT sensor ground circuit is open between sensor and ground • ECT sensor is contaminated, damaged or has failed • PCM has failed
DTC P0115 **2T CCM, MIL: Yes** 1996, 1997, 1998, 1999, 2000, 2001, 2002, 2003 All engines Transmissions: All	**Engine Coolant Temperature Sensor Circuit Malfunction Conditions:** Key on or engine running; and the PCM detected an unexpected "low" or "high" voltage on the ECT sensor circuit (Scan Tool reads less than -40ºF or low voltage or more than 284ºF for high voltage). **Possible Causes** • IAT sensor signal circuit is open, shorted to ground or VREF • ECT sensor ground circuit is open between sensor and ground • ECT sensor is contaminated, damaged or has failed • PCM has failed
DTC P0116 **2T CCM, MIL: Yes** 1996, 1997, 1998, 1999, 2000, 2001, 2002 All engines Transmissions: All	**Engine Coolant Temperature Sensor Range/Performance Conditions:** Engine runtime over 20 minutes; and the PCM detected the ECT sensor indicated less than 95ºF during the test. *Note: Check the condition and operation of the Cooling system.* **Possible Causes** • Check for a low coolant level or an incorrect coolant mixture • ECT sensor signal circuit or ground circuit has high resistance • ECT sensor is contaminated, damaged or out-of-calibration • PCM has failed
DTC P0116 **2T CCM, MIL: Yes** 1995 Avalon, Camry, Land Cruiser, Previa, T100, Tacoma, Tercel 1.5L VIN E, 2.4L VIN A, 2.4L VIN R, 2.7L VIN U, 3.0L VIN G, 3.4L VIN V, 4.5L VIN D engines Transmissions: All	**Engine Coolant Temperature Sensor Range/Performance Conditions:** Engine started, engine runtime over 20 minutes, and the PCM detected the ECT sensor indicated less than 95ºF during the test. *Note: Check the condition and operation of the Cooling system.* **Possible Causes** • Check for a low coolant level or an incorrect coolant mixture • ECT sensor signal circuit has high resistance • ECT sensor ground circuit has high resistance • ECT sensor is contaminated, damaged or out-of-calibration • PCM has failed

OBD II Trouble Code List (P0xxx Codes)

DTC	Trouble Code Title, Conditions & Possible Causes
DTC P0116 **2T CCM, MIL: Yes** 2003 All engines Transmissions: All	**Engine Coolant Temperature Sensor Range/Performance Conditions:** Engine started, ECT sensor from 95ºF to 140ºF, IAT sensor more than 19.9ºF, vehicle driven with several changes in the VSS signals, and the PCM detected the ECT sensor signal did not increase more than 37.4ºF after engine was started and the test period completed. **Possible Causes** • Check for problems in the cooling system (i.e., coolant, the fan) • ECT sensor signal circuit or ground circuit has high resistance • ECT sensor is contaminated, damaged or has failed • PCM has failed
DTC P0117 **2T CCM, MIL: Yes** 2003 4Runner, Echo, Land Cruiser, Sequoia, Tacoma, Tundra 1.5L VIN T, 3.4L VIN N, 4.0L VIN U, 4.7L VIN T engines Transmissions: All	**Engine Coolant Temperature Sensor Circuit Low Input Conditions:** Key on or engine running; and the PCM detected an unexpected low voltage condition on the ECT sensor (Scan Tool reads below -40ºF). **Possible Causes** • ECT sensor signal circuit is shorted to ground • ECT sensor is damaged or has failed • PCM has failed
DTC P0117 **2T CCM, MIL: Yes** 2003 4Runner, Echo, Land Cruiser, Sequoia, Tacoma, Tundra 1.5L VIN T, 3.4L VIN N, 4.0L VIN U, 4.7L VIN T engines Transmissions: All	**Engine Coolant Temperature Sensor Circuit High Input Conditions:** Key on or engine running; and the PCM detected an unexpected high voltage condition on the ECT sensor (Scan Tool reads over 284ºF). **Possible Causes** • ECT sensor signal circuit is open • ECT sensor ground circuit is open • ECT sensor is damaged or has failed • PCM has failed
DTC P0120 **2T CCM, MIL: Yes** 1995 Avalon, Camry, Land Cruiser, Previa, T100, Tacoma, Tercel 1.5L VIN E, 2.4L VIN A, 2.4L VIN R, 2.7L VIN U, 3.0L VIN G, 3.4L VIN V, 4.5L VIN D engines Transmissions: All	**TP Sensor or Switch 'A' Circuit Malfunction Conditions:** Key on or engine running; and the PCM detected the TP sensor indicated less than 0.1v with the closed throttle position switch off (in open position), or the TP sensor input indicated 4.9v at any time. **Possible Causes** • TP sensor signal circuit open or shorted to ground • TP sensor ground circuit is open • TP sensor power circuit is open (check VREF circuit at PCM) • TP sensor is damaged or has failed • PCM has failed
DTC P0120 **2T CCM, MIL: Yes** 1996, 1997, 1998, 1999, 2000, 2001, 2002 All engines Transmissions: All	**TP Sensor or Switch 'A' Circuit Malfunction Conditions:** Key on or engine running; and the PCM detected the TP sensor indicated less than 0.1v with the closed throttle position switch off (open position), or the TP sensor was 4.9v at any time. **Possible Causes** • TP sensor signal circuit is open or shorted to ground • TP sensor ground circuit is open, or the power circuit is open • TP sensor is damaged or has failed
DTC P0120 **2T CCM, MIL: Yes** 2003 Avalon, Camry, Camry Solara, Celica, Corolla, Echo, Highlander, Matrix, MR2, Prius, RAV4, Sienna 1.5L VIN B, 1.5L VIN T, 1.8L VIN R, 1.8L VIN Y, 2.0L VIN K, 2.4L VIN D, 3.0L VIN F, 4.7L VIN T Transmissions: All	**TP Sensor or Switch 'A' Circuit Malfunction Conditions:** Engine started, and the PCM detected the TP sensor signal (VTA) was under 0.10v or over 4.90v. The TP sensor, mounted on the throttle body, detects the Throttle Valve opening angle (0.30v-0.70v closed). The PCM uses this signal for A/F ratio correction and power increase changes during all modes of engine operation. **Possible Causes** • VTA sensor signal circuit is open or shorted to ground • VTA sensor ground circuit is open • VTA power circuit (VREF) is open • TP sensor is damaged or has failed • PCM has failed

OBD II Trouble Code List (P0xxx Codes)

DTC	Trouble Code Title, Conditions & Possible Causes
DTC P0120 **2T CCM, MIL: Yes** 2003 4Runner, Avalon, Camry, Highlander, Land Cruiser, Sequoia, Sienna, Tacoma, Tundra 2.4L VIN L, 2.7L VIN M engines, 3.0L VIN F, 4.0L VIN U, 4.7L VIN T Transmissions: All	**Throttle Pedal Position Sensor/Switch 'A' Circuit Malfunction Conditions** Key on or engine running; and the PCM detected the TP sensor 'A' Signal indicated less than 0.10v with the throttle position closed (the switch is open), or the TP sensor signal indicated more than 4.9v at any time. The Electric TP Sensor is mounted on the throttle body. It has two sensors (the electrical throttle system does not use a cable). **Possible Causes** • TP sensor signal circuit open is or shorted to ground • TP sensor ground circuit is open • TP sensor power circuit is open (test VREF circuit at the PCM) • TP sensor is damaged or has failed • PCM has failed
DTC P0121 **2T CCM, MIL: Yes** 1995 Avalon, Camry, Land Cruiser, Previa, T100, Tacoma, Tercel 1.5L VIN E, 2.4L VIN A, 2.4L VIN R, 2.7L VIN U, 3.0L VIN G, 3.4L VIN V, 4.5L VIN D engines Transmissions: All	**TP Sensor or TP Switch 'A' Signal Range/Performance Conditions:** Engine started, VSS input exceeds 19 mph at least once, and the PCM detected that the TP sensor input was out of the applicable range with the VSS reading between 0 and 30 mph. **Possible Causes** • TP sensor signal circuit open or shorted to ground (intermittent) • TP sensor is loose at it mounting or the throttle is binding • TP sensor is damaged or has failed (perform a sweep test) • PCM has failed
DTC P0121 **2T CCM, MIL: Yes** 1996, 1997, 1998, 1999, 2000, 2001, 2002 All engines Transmissions: All	**TP Sensor or TP Switch 'A' Signal Range/Performance Conditions:** Vehicle speed more than 19 mph at least once and the PCM detected the TP sensor input was out of the applicable range with the VSS input reading between 30 mph and 0 mph. **Possible Causes** • TP sensor signal circuit open or shorted to ground (intermittent) • TP sensor is loose at it mounting or the throttle is binding • TP sensor is damaged or has failed (perform a sweep test) • PCM has failed
DTC P0121 **2T CCM, MIL: Yes** 2003 Avalon, Camry, Camry Solara, Celica, Corolla, Echo, Highlander, Matrix, MR2, Prius, Sienna All engines Transmissions: All	**TP Sensor or TP Switch 'A' Signal Range/Performance Conditions:** Vehicle speed more than 19 mph at least once; and the PCM detected the TP sensor input was out of the applicable range with the VSS input reading between 30 mph and 0 mph. **Possible Causes** • TP sensor signal circuit open or shorted to ground (intermittent) • TP sensor is loose at it mounting or the throttle is binding • TP sensor is damaged or has failed (perform a sweep test) • PCM has failed
DTC P0121 **2T CCM, MIL: Yes** 2003 4Runner, Land Cruiser, RAV4, Sequoia, Tacoma, Tundra 2.0L VIN K, 2.4L VIN L, 2.7L VIN M, 3.4L VIN N, 4.0L VIN U, 4.7L VIN T engines Transmissions: All	**Throttle Pedal Position Sensor Switch 'A' Signal Range/Performance Conditions:** Engine started; and the PCM detected the difference between the TP sensor VTA1 and VTA2 signal was out-of-range. The TP sensor, mounted on the throttle body, detects the Throttle Valve opening angle (about 0.70v with the throttle closed). The PCM uses the VTA signal for air/fuel ratio and power increase correction. The Electric TP Sensor is mounted on the throttle body. It has two sensors (the electrical throttle system does not use a cable). **Possible Causes** • TP sensor connector is damaged or it is open • TP sensor is damaged or has failed • PCM has failed
DTC P0122 **2T CCM, MIL: Yes** 2003 4Runner, Land Cruiser, Sequoia, Tacoma, Tundra 3.4L VIN N engine, 4.0L VIN U, 4.7L VIN T Transmissions: All	**Throttle Pedal Position Sensor/Switch 'A' Circuit Low Input Conditions:** Engine started, and the PCM detected an unexpected low voltage (below 0.20v) on the VTA1 signal circuit. The TP sensor detects the Throttle Valve opening angle (0.70v with the throttle closed). The Electric TP Assembly has 2 sensors (this system does not use a throttle cable). **Possible Causes** • Throttle control motor and sensor unit is damaged or failed • TP sensor VC (VREF) circuit is open, or the VTA1 signal circuit is shorted to ground • PCM has failed
DTC P0123 **2T CCM, MIL: Yes** 2003 4Runner, Land Cruiser, Sequoia, Tacoma, Tundra 3.4L VIN N engine, 4.0L VIN U, 4.7L VIN T Transmissions: All	**Throttle Pedal Position Sensor/Switch 'A' Circuit High Input Conditions:** Engine started, and the PCM detected an unexpected high voltage (over 4.80v) on the VTA1 signal circuit. The TP sensor detects the Throttle Valve opening angle (0.70v with the throttle closed). The Electric TP Assembly has 2 sensors (this system does not use a throttle cable). **Possible Causes** • Throttle control motor and sensor unit is damaged or failed • TP sensor VC (VREF) circuit is shorted to the VTA1 circuit • VTA1 ground circuit is open or the VTA1 signal circuit is open • PCM has failed

OBD II Trouble Code List (P0xxx Codes)

DTC	Trouble Code Title, Conditions & Possible Causes
DTC P0125 **1T CCM, MIL: Yes** 1995 Avalon, Camry, Land Cruiser, Previa, T100, Tacoma, Tercel 1.5L VIN E, 2.4L VIN A, 2.4L VIN R, 2.7L VIN U, 3.0L VIN G, 3.4L VIN V, 4.5L VIN D engines Transmissions: All	**Insufficient Coolant Temperature For Closed Loop: Conditions:** DTC P0115 and P0116 not set, engine started, engine running, ECT sensor more than 140ºF, vehicle driven to a speed of 25-62 mph at an engine speed over 1500 rpm, throttle valve not fully closed, and the PCM detected the HO2S signal (internal value) did not change. **Possible Causes** • Check the operation of the thermostat (it may be stuck open) • ECT sensor signal circuit has high resistance • ECT sensor is out-of-calibration, skewed, or it has failed • Inspect for low coolant level or an incorrect coolant mixture
DTC P0125 **1T CCM, MIL: Yes** 1996, 1997, 1998, 1999, 2000, 2001, 2002 All engines Transmissions: All	**Insufficient Coolant Temperature For Closed Loop: Conditions:** **California Models** DTC P0115 and P0116 not set, engine started, ECT sensor more than 140ºF, vehicle driven to a speed of 25-62 mph at an engine speed over 1500 rpm, throttle valve not fully closed, and the PCM detected the A/FS or HO2S signal (internal value) did not change. **Federal Models** Engine started, ECT sensor more than 140ºF, vehicle driven to a speed of 25-62 mph at an engine speed over 1400 rpm, throttle valve not fully closed, and the PCM detected the HO2S-11 input did not exceed 450 mv at least during a test period of 1.5 minutes. **Possible Causes** • Check the operation of the thermostat (it may be stuck open) • ECT sensor signal circuit has high resistance • ECT sensor has failed • Inspect for low coolant level or an incorrect coolant mixture
DTC P0125 **1T CCM, MIL: Yes** 2003 All engines Transmissions: All	**Insufficient Coolant Temperature For Closed Loop Conditions:** **California Models** ECT sensor less than 19.4ºF, engine started, engine runtime over 20 minutes, and the PCM detected the ECT sensor indicated 68ºF or less; or with the ECT sensor from 19.4ºF to 50.0ºF at startup, engine runtime over 5 minutes, the PCM detected the ECT sensor indicated 68ºF or less; or with the ECT sensor more than 19.4ºF at startup, and the engine runtime over 5 minutes, the PCM detected the ECT sensor signal was 68ºF or less; or with the ECT sensor over 50ºF at startup, engine runtime over 2 minutes, the PCM detected the ECT sensor signal did not reach 86ºF during the CCM test period. **Possible Causes** • Check the operation of the thermostat (it may be stuck open) • ECT sensor signal circuit has high resistance • ECT sensor has failed • Inspect for low coolant level for an incorrect coolant mixture
DTC P0128 **2T CCM, MIL: Yes** 2000, 2001, 2002, 2003 All engines Transmissions: All	**Thermostat System Malfunction Conditions:** Engine started, ECT sensor signal below 140ºF at startup, and the PCM detected the ECT sensor did not reach 167ºF after the warmup period expired (engine runtime 5-10 minutes). **Possible Causes** • Check the operation of the thermostat (it may be stuck open) • ECT sensor is out-of-calibration or skewed • Inspect for low coolant level or for an incorrect coolant mixture
DTC P0130 **2T CCM, MIL: Yes** 1995 Avalon, Camry, Land Cruiser, Previa, T100, Tacoma, Tercel 1.5L VIN E, 2.4L VIN A, 2.4L VIN R, 2.7L VIN U, 3.0L VIN G, 3.4L VIN V, 4.5L VIN D engines Transmissions: All	**HO2S-11 (Bank 1 Sensor 1) Circuit Malfunction Conditions:** Engine started, engine warmup completed, engine idling in closed loop, and the PCM detected the HO2S-11 signal was fixed at 400 mv or higher, or that it was fixed at less than 550 mv during the test. **Possible Causes** • HO2S signal circuit is open between the sensor and the PCM • HO2S signal circuit is shorted to sensor or chassis ground • HO2S signal circuit is shorted to VREF or system power (B+) • HO2S is damaged, contaminated or it has failed
DTC P0130 **2T CCM, MIL: Yes** 1996, 1997, 1998, 1999, 2000, 2001, 2002 All engines Transmissions: All	**HO2S-11 (Bank 1 Sensor 1) Circuit Malfunction Conditions:** Engine warmup completed, engine idling in closed loop, and the PCM detected the HO2S-11 signal was fixed at 400 mv or higher, or that it was fixed at less than 550 mv during the test. **Possible Causes** • HO2S signal circuit is open between the sensor and the PCM • HO2S signal circuit is shorted to sensor or chassis ground • HO2S signal circuit is shorted to VREF or system power (B+) • HO2S is damaged, contaminated or it has failed • PCM has failed

OBD II Trouble Code List (P0xxx Codes)

DTC	Trouble Code Title, Conditions & Possible Causes
DTC P0130 **2T CCM, MIL: Yes** 2003 4Runner, Celica, Corolla, Echo, Land Cruiser, Matrix, MR2, Prius, Sequoia, Tundra 1.5L VIN B, 1.5L VIN T, 1.8L VIN R, 1.8L VIN Y, 4.0L VIN U, 4.7L VIN T Transmissions: All	**HO2S-11 (Bank 1 Sensor 1) Circuit Malfunction Conditions:** Engine started, engine warmup completed, engine idling in closed loop, and the PCM detected the HO2S-11 signal was fixed at 400 mv or higher, or that it was fixed at less than 550 mv during the test. **Possible Causes** • HO2S signal circuit is open between the sensor and the PCM • HO2S signal circuit is shorted to sensor ground or to VREF • HO2S is damaged, contaminated or it has failed
DTC P0133 **2T O2S, MIL: Yes** 1995 Avalon, Camry, Land Cruiser, Previa, T100, Tacoma, Tercel 1.5L VIN E, 2.4L VIN A, 2.4L VIN R, 2.7L VIN U, 3.0L VIN G, 3.4L VIN V, 4.5L VIN D engines Transmissions: All	**HO2S-11 (Bank 1 Sensor 1) Slow Response Conditions:** Engine started, engine idling in closed loop, and the PCM detected the HO2S-11 response time from rich-to-lean or from lean-to-rich was 1.1 second or longer during the CCM test. *Note: This fault must be detected at least 3 times at idle speed.* **Possible Causes** • HO2S signal circuit is open or shorted to ground • HO2S element is contaminated, or HO2S heater has failed • Intake air leaks, exhaust manifold leaks or PCV system leaks • MAF sensor out of calibration (it may be dirty or contaminated)
DTC P0133 **2T O2S, MIL: Yes** 1996, 1997, 1998, 1999, 2000, 2001, 2002 All engines Transmissions: All	**HO2S-11 (Bank 1 Sensor 1) Slow Response Conditions:** Engine started, engine idling in closed loop, and the PCM detected the HO2S-11 response time from rich-to-lean or from lean-to-rich was 1.1 second or longer during the CCM test. *Note: This fault must be detected at least 3 times at idle speed.* **Possible Causes** • HO2S signal circuit is open or shorted to ground • HO2S element is contaminated, or HO2S heater has failed • Intake air leaks, exhaust manifold leaks or PCV system leaks • MAF sensor out of calibration (it may be dirty or contaminated)
DTC P0133 **2T O2S, MIL: Yes** 2003 4Runner, Celica, Corolla, Echo, Land Cruiser, Matrix, MR2, Prius, Sequoia, Tundra 1.5L VIN B, 1.5L VIN T, 1.8L VIN R, 1.8L VIN Y, 4.0L VIN U, 4.7L VIN T Transmissions: All	**HO2S-11 (Bank 1 Sensor 1) Slow Response Conditions:** Engine started, engine idling in closed loop, and the PCM detected the HO2S-11 response time from rich-to-lean or from lean-to-rich was 1.1 second or longer during the CCM test. *Note: This fault must be detected at least 3 times at idle speed.* **Possible Causes** • HO2S signal circuit is open or shorted to ground • HO2S element is contaminated, or HO2S heater has failed • Intake air leaks, exhaust manifold leaks or PCV system leaks • MAF sensor out of calibration (it may be dirty or contaminated)
DTC P0134 **2T O2S, MIL: Yes** 2003 Avalon, Camry, Highlander, Sienna, Tacoma 2.4L VIN D, 2.4L VIN L, 2.7L VIN M, 3.0L VIN F Transmissions: All	**Air Fuel Sensor 1 (Bank 1 Sensor 1) Circuit No Activity Detected Conditions:** Engine started, engine runtime over 140 seconds, vehicle driven at a steady speed of 25-81 mph at over 1500 rpm with the throttle valve open, and the PCM detected the A/FS-11 signal did not indicate rich (more than 450 mv) after 65 seconds under these conditions. **Possible Causes** • Air leaks present in the PCV valve, hoses or hose connections • A/FS1 connector is damaged or loose • A/FS1 signal circuit is open or shorted to ground • A/FS1 is damaged, contaminated or it has failed • Fuel Control component faults (sticking injector, low pressure) • Gas leaks in the exhaust system in front of the Oxygen sensor • PCM has failed
DTC P0134 **2T O2S, MIL: Yes** 2003 4Runner, Land Cruiser, MR2, Sequoia, Tacoma, Tundra 1.8L VIN R, 2.7L VIN M, 3.4L VIN N, 4.0L VIN U, 4.7L VIN T engines Transmissions: All	**HO2S-11 (Bank 1 Sensor 1) Circuit No Activity Detected Conditions:** Engine started, engine runtime over 140 seconds, vehicle driven at a steady speed of 25-81 mph at over 1500 rpm with the throttle valve open, and the PCM detected the HO2S-11 signal did not indicate rich (more than 450 mv) after 65 seconds under these conditions. **Possible Causes** • Air leaks present in the PCV valve, hoses or hose connections • Fuel Control component faults (sticking injector, low pressure) • Gas leaks in the exhaust system in front of the Oxygen sensor • HO2S connector is damaged or loose • HO2S signal circuit is open or shorted to ground • HO2S is damaged, contaminated or it has failed • PCM has failed

OBD II Trouble Code List (P0xxx Codes)

DTC	Trouble Code Title, Conditions & Possible Causes
DTC P0135 **2T CCM, MIL: Yes** 1995 Avalon, Camry, Land Cruiser, Previa, T100, Tacoma, Tercel 1.5L VIN E, 2.4L VIN A, 2.4L VIN R, 2.7L VIN U, 3.0L VIN G, 3.4L VIN V, 4.5L VIN D engines Transmissions: All	**HO2S-11 (Bank 1 Sensor 1) Heater Circuit Malfunction Conditions:** Engine started, engine running, and the PCM detected the HO2S-11 heater current exceeded 2 amps, or that it was 0.25 amps or less. **Possible Causes** • HO2S heater control circuit is open or shorted to ground • HO2S heater control circuit is shorted to power • HO2S heater power circuit is open (check power from the relay) • HO2S heater is damaged or has failed • PCM has failed
DTC P0135 **2T CCM, MIL: Yes** 1996, 1997, 1998, 1999, 2000, 2001, 2002 All engines Transmissions: All	**HO2S-11 (Bank 1 Sensor 1) Heater Circuit Malfunction Conditions:** Engine started, engine running, and the PCM detected the HO2S-11 heater current exceeded 2 amps, or that it was 0.25 amps or less. **Possible Causes** • HO2S heater control circuit is open or shorted to ground • HO2S heater control circuit is shorted to power • HO2S heater power circuit is open (check power from the relay) • HO2S heater is damaged or has failed • PCM has failed
DTC P0135 **2T CCM, MIL: Yes** 2003 Celica, Corolla, Matrix, MR2, Prius 1.5L VIN B, 1.8L VIN R, 1.8L VIN Y engines Transmissions: All	**HO2S-11 (Bank 1 Sensor 1) Heater Circuit Malfunction Conditions:** Engine started, engine running, and the PCM detected the HO2S-11 heater current exceeded 2 amps, or that it was 0.25 amps or less. **Possible Causes** • HO2S heater control circuit is open or shorted to ground • HO2S heater control circuit is shorted to power • HO2S heater power circuit is open (check power from the relay) • HO2S heater is damaged or has failed • PCM has failed
DTC P0135 **2T CCM, MIL: Yes** 2003 Avalon, Camry, Highlander, Sienna, Tacoma 2.4L VIN D, 2.4L VIN L, 2.7L VIN M, 3.0L VIN F Transmissions: All	**Air Fuel Sensor 1 (Bank 1 Sensor 1) Heater Circuit Malfunction Conditions:** Engine started, engine running, and the PCM detected the A/FS-11 heater current indicated over 8 amps, or it was less than 0.25 amps. **Possible Causes** • A/FS1 heater control circuit is open or shorted to ground • A/FS1 heater control circuit is shorted to power • A/FS1 heater power circuit is open (check power from relay) • A/FS1 heater is damaged or has failed • PCM has failed
DTC P0136 **2T CCM, MIL: Yes** 1995 Avalon, Camry, Land Cruiser, Previa, T100, Tacoma, Tercel 1.5L VIN E, 2.4L VIN A, 2.4L VIN R, 2.7L VIN U, 3.0L VIN G, 3.4L VIN V, 4.5L VIN D engines Transmissions: All	**HO2S-12 (Bank 1 Sensor 2) Circuit Malfunction Conditions:** Engine started, vehicle driven to a speed of over 25 mph while in closed loop, and the PCM detected the HO2S-12 signal remained fixed at more than 400 mv, or that it was fixed at less than 600 mv. **Possible Causes** • HO2S signal circuit is open or shorted to ground • HO2S signal circuit shorted to VREF or system power (B+) • HO2S is damaged, contaminated or it has failed • PCM has failed
DTC P0136 **2T CCM, MIL: Yes** 1996, 1997, 1998, 1999, 2000, 2001, 2002 All engines Transmissions: All	**HO2S-12 (Bank 1 Sensor 2) Circuit Malfunction Code Conditions:** Vehicle driven to a speed of over 25 mph while in closed loop, and the PCM detected the HO2S-12 signal remained fixed at more than 400 mv, or it was fixed at less than 600 mv. **Possible Causes** • HO2S signal circuit is open or shorted to ground • HO2S signal circuit shorted to VREF or system power (B+) • HO2S is damaged, contaminated or it has failed
DTC P0136 **2T CCM, MIL: Yes** 2003 4Runner, Celica, Corolla, Echo, Land Cruiser, Matrix, MR2, Prius, Sequoia, Tundra 1.5L VIN B, 1.5L VIN T, 1.8L VIN R, 1.8L VIN Y, 4.7L VIN T engines Transmissions: All	**HO2S-12 (Bank 1 Sensor 2) Circuit Malfunction Conditions:** Vehicle driven to a speed of over 25 mph while in closed loop, and the PCM detected the HO2S-12 signal was fixed at more than 400 mv, or that it was fixed at less than 600 mv. **Possible Causes** • HO2S signal circuit is open or shorted to ground • HO2S signal circuit shorted to VREF or system power (B+) • HO2S is damaged, contaminated or it has failed

OBD II Trouble Code List (P0xxx Codes)

DTC	Trouble Code Title, Conditions & Possible Causes
DTC P0136 **2T CCM, MIL: Yes** 2003 Avalon, Camry, Highlander, RAV4, Sienna, Tacoma, Tundra 2.0L VIN K, 2.4L VIN D, 2.4L VIN L, 2.7L VIN M, 3.0L VIN F, 3.4L VIN N Transmissions: All	**HO2S-12 (Bank 1 Sensor 2) Circuit Malfunction Conditions:** Engine started, vehicle driven to a speed of over 31 mph while in closed loop, and the PCM detected the HO2S-12 signal indicated more than 400 mv, or it indicated less than 500 mv during the test. **Possible Causes** • HO2S signal circuit is open or shorted to ground • HO2S signal circuit shorted to VREF or system power (B+) • HO2S is damaged, contaminated or it has failed
DTC P0141 **2T CCM, MIL: Yes** 1995 Avalon, Camry, Land Cruiser, Previa, T100, Tacoma, Tercel 1.5L VIN E, 2.4L VIN A, 2.4L VIN R, 2.7L VIN U, 3.0L VIN G, 3.4L VIN V, 4.5L VIN D engines Transmissions: All	**HO2S-12 (Bank 1 Sensor 2) Heater Circuit Malfunction Conditions:** Engine started, engine running, and the PCM detected the HO2S-12 heater current exceeded 2 amps, or that it was 0.25 amps or less. **Possible Causes** • HO2S heater control circuit is open or shorted to ground • HO2S heater control circuit is shorted to power • HO2S heater power circuit is open (check power from the relay) • HO2S heater is damaged or has failed • PCM has failed
DTC P0141 **2T CCM, MIL: Yes** 1996, 1997, 1998, 1999, 2000, 2001, 2002 All engines Transmissions: All	**HO2S-12 (Bank 1 Sensor 2) Heater Circuit Malfunction Conditions:** Engine started, engine running, and the PCM detected the HO2S-12 heater current exceeded 2 amps, or that it was 0.25 amps or less. **Possible Causes** • HO2S heater control circuit is open, shorted to ground or power • HO2S heater power circuit is open (check power from the relay) • HO2S heater is damaged or has failed • PCM has failed
DTC P0141 **2T CCM, MIL: Yes** 2003 Celica, Corolla, Echo, Matrix, MR2, Prius, RAV4 1.5L VIN B, 1.5L VIN T, 1.8L VIN R, 1.8L VIN Y, 2.0L VIN K engines Transmissions: All	**HO2S-12 (Bank 1 Sensor 2) Heater Circuit Malfunction Conditions:** Engine started, engine running, and the PCM detected the HO2S-12 heater current exceeded 2 amps, or that it was 0.25 amps or less. **Possible Causes** • HO2S heater control circuit is open, shorted to ground or power • HO2S heater power circuit is open (check power from the relay) • HO2S heater is damaged or has failed • PCM has failed
DTC P0150 **2T CCM, MIL: Yes** 1995 Avalon, Camry 3.0L VIN G engines Transmissions: All	**HO2S-21 (Bank 2 Sensor 1) Circuit Malfunction Conditions:** Engine warmup completed, engine idling in closed loop, and the PCM detected the HO2S-21 signal was fixed at 400 mv or higher, or that it was fixed from 400-550 mv during the test. **Possible Causes** • HO2S signal circuit is open between the sensor and the PCM • HO2S signal circuit is shorted to sensor or chassis ground • HO2S is damaged, contaminated or it has failed
DTC P0150 **2T CCM, MIL: Yes** 2003 4Runner, Land Cruiser, MR2, Sequoia, Tundra 1.8L VIN R, 4.0L VIN U, 4.7L VIN T engines Transmissions: All	**HO2S-21 (Bank 2 Sensor 1) Circuit Malfunction Conditions:** Engine warmup completed, engine idling in closed loop, and the PCM detected the HO2S-21 signal was fixed at 400 mv or higher, or that it was fixed at less than 550 mv during the test. **Possible Causes** • HO2S signal circuit is open between the sensor and the PCM • HO2S signal circuit is shorted to sensor ground or to VREF • HO2S is damaged, contaminated or it has failed
DTC P0150 **2T CCM, MIL: Yes** 1996, 1997, 1998, 1999, 2000, 2001, 2002 Camry, Camry Solara, Land Cruiser, Sequoia, Sienna, Supra, Tundra 3.0L VIN D, 3.0L VIN E, 3.0L VIN F, 4.7L VIN T Transmissions: All	**HO2S-21 (Bank 2 Sensor 1) Circuit Malfunction Conditions:** Engine started, engine warmup completed, engine idling in closed loop, and the PCM detected the HO2S-21 signal was fixed at 400 mv or higher, or that it was fixed from 400-550 mv during the test. **Possible Causes** • HO2S signal circuit is open between the sensor and the PCM • HO2S signal circuit is shorted to sensor or chassis ground • HO2S is damaged, contaminated or it has failed • PCM has failed

OBD II Trouble Code List (P0xxx Codes)

DTC	Trouble Code Title, Conditions & Possible Causes
DTC P0153 **2T O2S, MIL: Yes** 1995 Avalon, Camry 3.0L VIN G engine Transmissions: All	**HO2S-21 (B2 S1) Slow Response Conditions:** Engine started, engine idling in closed loop, and the PCM detected the HO2S-21 response time from rich-to-lean or from lean-to-rich was 1.1 second or longer during the CCM test. *Note: This fault must be detected at least 3 times at idle speed.* **Possible Causes** • HO2S signal circuit is open or shorted to ground • HO2S element is contaminated, or HO2S heater has failed • Intake air leaks, exhaust manifold leaks or PCV system leaks • MAF sensor out of calibration (it may be dirty or contaminated)
DTC P0153 **2T O2S, MIL: Yes** 1996, 1997, 1998, 1999, 2000, 2001, 2002 Camry, Camry Solara, Land Cruiser, Sequoia, Sienna, Supra, Tundra 3.0L VIN D, 3.0L VIN E, 3.0L VIN F, 4.7L VIN T	**HO2S-21 (B2 S1) Slow Response Conditions:** Engine started, engine idling in closed loop, and the PCM detected the HO2S-21 response time from rich-to-lean or from lean-to-rich was 1.1 second or longer during the CCM test. *Note: This fault must be detected at least 3 times at idle speed.* **Possible Causes** • HO2S signal circuit is open or shorted to ground • HO2S element is contaminated, or HO2S heater has failed • Intake air leaks, exhaust manifold leaks or PCV system leaks • MAF sensor out of calibration (it may be dirty or contaminated)
DTC P0154 **2T O2S, MIL: Yes** 2003 Avalon, Camry, Highlander, Sienna 3.0L VIN F engine Transmissions: All	**Air Fuel Sensor 2 (Bank 2 Sensor 1) Circuit No Activity Detected Conditions:** Engine started, engine runtime over 140 seconds, vehicle driven at a steady speed of 25-81 mph at over 1500 rpm with the throttle valve open, and the PCM detected the Bank 2 A/FS-21 signal did not indicate rich (more than 450 mv) after 65 seconds during the test. **Possible Causes** • Air leaks present in the PCV valve, hoses or hose connections • A/FS1 connector is damaged or loose • A/FS1 signal circuit is open or shorted to ground • A/FS1 is damaged, contaminated or it has failed • Fuel Control component faults (sticking injector, low pressure) • Gas leaks in the exhaust system in front of the Oxygen sensor
DTC P0154 **2T O2S, MIL: Yes** 2003 4Runner, Land Cruiser, MR2, Sequoia, Tundra 1.8L VIN R, 4.0L VIN U, 4.7L VIN T engines Transmissions: All	**HO2S-21 (Bank 2 Sensor 1) Circuit No Activity Detected Conditions:** Engine runtime over 140 seconds, VSS from 25-81 mph at over 1500 rpm with throttle valve open, and the PCM detected the HO2S-21 signal did not go over 450 mv after 65 seconds. **Possible Causes** • Air leaks present in the PCV valve, hoses or hose connections • Fuel Control component faults (sticking injector, low pressure) • Gas leaks in the exhaust system in front of the Oxygen sensor • HO2S-21 signal circuit is open or shorted to ground • HO2S-21 is damaged, contaminated or it has failed • PCM has failed
DTC P0155 **2T CCM, MIL: Yes** 1995 Avalon, Camry 3.0L VIN G engines Transmissions: All	**HO2S-21 (B2 S1) Heater Circuit Malfunction Conditions:** Engine started, engine running, and the PCM detected the HO2S-21 heater current exceeded 2 amps, or that it was 0.25 amps or less. **Possible Causes** • HO2S heater control circuit is open, shorted to ground or power • HO2S heater power circuit is open (check power from the relay) • HO2S heater is damaged or has failed • PCM has failed
DTC P0155 **2T CCM, MIL: Yes** 1996, 1997, 1998, 1999, 2000, 2001, 2002 Camry, Camry Solara, Land Cruiser, Sequoia, Sienna, Supra, Tundra 3.0L VIN D, 3.0L VIN E, 3.0L VIN F, 4.7L VIN T	**HO2S-21 (B2 S1) Heater Circuit Malfunction Conditions:** Engine started, engine running, and the PCM detected the HO2S-21 heater current exceeded 2 amps, or that it was 0.25 amps or less. **Possible Causes** • HO2S heater control circuit is open or shorted to ground • HO2S heater control circuit is shorted to power • HO2S heater power circuit is open (check power from the relay) • HO2S heater is damaged or has failed • PCM has failed
DTC P0155 **2T CCM, MIL: Yes** 2003 Avalon, Camry, Highlander, Sienna 3.0L VIN F engine Transmissions: All	**Air Fuel Sensor 2 (Bank 2 Sensor 1) Heater Circuit Malfunction Conditions:** Engine started, engine running, and the PCM detected the A/FS-21 heater current indicated over 8 amps, or it was less than 0.25 amps. **Possible Causes** • A/FS1 heater control circuit is open, shorted to ground or power • A/FS1 heater power circuit is open (check power from relay) • A/FS1 heater is damaged or has failed • PCM has failed

OBD II Trouble Code List (P0xxx Codes)

DTC	Trouble Code Title, Conditions & Possible Causes
DTC P0156 **2T CCM, MIL: Yes** 1996, 1997, 1998, 1999, 2000, 2001, 2002, 2003 4Runner, Land Cruiser, RAV4, Sequoia, Tundra 2.0L VIN K, 4.0L VIN U, 4.7L VIN T engines Transmissions: All	**HO2S-22 (Bank 2 Sensor 2) Circuit Malfunction Conditions:** Engine started, engine warmup completed, engine idling in closed loop, and the PCM detected the HO2S-22 signal was fixed at 400 mv or higher, or that it was fixed from 400-550 mv during the test. **Possible Causes** • HO2S signal circuit is open between the sensor and the PCM • HO2S signal circuit is shorted to sensor or chassis ground • HO2S is damaged, contaminated or it has failed
DTC P0161 **2T CCM, MIL: Yes** 1996, 1997, 1998, 1999, 2000, 2001, 2002 Land Cruiser, Sequoia, Tundra 4.7L VIN T engine Transmissions: All	**HO2S-22 (B2 S2) Heater Circuit Malfunction Conditions:** Engine started, engine running, and the PCM detected the HO2S-22 heater current exceeded 2 amps, or that it was 0.25 amps or less. **Possible Causes** • HO2S heater control circuit is open, shorted to ground or power • HO2S heater power circuit is open (check power from the relay) • HO2S heater is damaged or has failed • PCM has failed
DTC P0161 **2T CCM, MIL: Yes** 2003 4Runner, RAV4, Land Cruiser, Sequoia, Tundra 2.0L VIN K, 4.0L VIN U, 4.7L VIN T engines Transmissions: All	**HO2S-22 (B2 S2) Heater Circuit Malfunction Conditions:** Engine started, engine running, and the PCM detected the HO2S-22 heater current exceeded 2 amps, or it was less than 0.25 amps. **Possible Causes** • HO2S heater control circuit is open, shorted to ground or power • HO2S heater power circuit is open (check power from the relay) • HO2S heater is damaged or has failed • PCM has failed
DTC P0170 **2T FUEL, MIL: Yes** 1995 Avalon, Camry, Land Cruiser, Previa, T100, Tacoma, Tercel 1.5L VIN E, 2.4L VIN A, 2.4L VIN R, 2.7L VIN U, 3.0L VIN G, 3.4L VIN V, 4.5L VIN D engines Transmissions: All	**Fuel System Too Rich or Too Lean (Bank 1) Conditions:** DTC P0100, P0101, P0110, P0115, P0120, P0121, P0130, P0133, P0136, P0135, P0136, P0141, P0153, P0155, P0201-206, P0300, P0301-P0306, P0401, P0402 and P0441 not set, engine running in closed loop at a stable engine speed, and the PCM detected the lean or rich fuel trim correction value was more than or less than a calibrated limit in memory. **Possible Causes** • Air leaks present in the exhaust manifold or exhaust pipes • Air being drawn in from leaks in engine gaskets or other seals • Base engine "mechanical" fault affecting one or more cylinders • Fuel control sensor is out of calibration (i.e., ECT, IAT or MAF) • Fuel delivery system supplying too much or too little fuel at idle or cruise (e.g., faulty fuel pump or dirty, restricted fuel filter) • One or more fuel injectors is dirty, leaking or stuck open/closed • HO2S element is contaminated, damaged or it has failed
DTC P0171 **2T FUEL, MIL: Yes** 1996, 1997, 1998, 1999, 2000, 2001, 2002 All engines Transmissions: All	**Fuel System Too Lean (Bank 1) Conditions:** DTC P0100, P0101, P0105, P0110, P0115, P0120, P0121, P0130, P0133, P0136, P0135, P0136, P0141, P0151, P0156, P0161, P0300, P0301-P0306, P0440, P0500 and P0505 not set, ECT sensor more than 158°F, vehicle driven at a constant speed of less than 62 mph with the engine speed over 1500 rpm, and the PCM detected the lean fuel trim correction value was over the limit. **Possible Causes** • A/FS or HO2S is contaminated, deteriorated or it has failed • Air leaks after the MAF sensor, or in the EGR or PCV system • Base engine "mechanical" fault affecting one or more cylinders • Exhaust leaks located in front of the A/FS or HO2S location • Fuel system supplying too little fuel during cruise or idle (faulty fuel pump or fuel filter) • Fuel injector (one or more) dirty or pressure regulator has failed • Vehicle driven low on fuel or until it ran out of fuel
DTC P0171 **2T FUEL, MIL: Yes** 2003 Avalon, Camry, Highlander, Sienna 2.4L VIN D, 3.0L VIN F Transmissions: All Trouble Code ID: **P0171**	**Fuel System Too Lean (Bank 1) Conditions:** DTC P0100, P0101, P0105, P0110, P0115, P0120, P0121, P0130, P0133, P0136, P0135, P0136, P0141, P0151, P0156, P0161, P0300, P0301-P0306, P0440, P0500 and P0505 not set, vehicle speed less than 62 mph with the engine speed over 1500 rpm, ECT sensor more than 158°F, and the PCM detected the lean fuel trim correction value was over the limit. **Possible Causes** • A/FS or HO2S is contaminated, deteriorated or it has failed • Air leaks after the MAF sensor, or in the EGR or PCV system • Base engine "mechanical" fault affecting one or more cylinders • Exhaust leaks located in front of the A/FS or HO2S location • Fuel system supplying too little fuel during cruise or idle (faulty fuel pump or fuel filter) • Fuel injector (one or more) dirty or pressure regulator has failed • Vehicle driven low on fuel or until it ran out of fuel

OBD II Trouble Code List (P0xxx Codes)

DTC	Trouble Code Title, Conditions & Possible Causes
DTC P0171 **2T FUEL, MIL: Yes** 2003 4Runner, Celica, Corolla, Echo, Land Cruiser, Matrix, MR2, Prius, RAV4, Sequoia, Tacoma, Tundra 1.5L VIN B, 1.5L VIN T, 1.8L VIN R, 1.8L VIN Y, 2.0L VIN K, 2.4L VIN L, 2.7L VIN M, 3.4L VIN N, 4.0L VIN U, 4.7L VIN T Transmissions: All	**Fuel System Too Lean (Bank 1) Conditions:** DTC P0100, P0101, P0105, P0110, P0115, P0120, P0121, P0130, P0133, P0136, P0135, P0136, P0141, P0151, P0156, P0161, P0300, P0301-P0306, P0440, P0500 and P0505 not set, ECT sensor more than 158ºF, vehicle driven at a constant speed of less than 62 mph with the engine speed over 1500 rpm, and the PCM detected the lean fuel trim correction value was over the limit. **Possible Causes** • A/FS or HO2S is contaminated, deteriorated or it has failed • Air leaks after the MAF sensor, or in the EGR or PCV system • Base engine "mechanical" fault affecting one or more cylinders • Exhaust leaks located in front of the A/FS or HO2S location • Fuel control sensor is out of calibration (i.e., ECT, IAT or MAP) • Fuel delivery system supplying too little fuel during cruise or idle periods (e.g., faulty fuel pump or dirty, restricted fuel filter) • Fuel injector (one or more) dirty or pressure regulator has failed • Vehicle driven low on fuel or until it ran out of fuel
DTC P0172 **2T FUEL, MIL: Yes** 1996, 1997, 1998, 1999, 2000, 2001, 2002 All engines Transmissions: All	**Fuel System Too Rich (Bank 1) Conditions:** DTC P0100, P0101, P0105, P0110, P0115, P0120, P0121, P0130, P0133, P0136, P0135, P0136, P0141, P0151, P0156, P0161, P0300, P0301-P0306, P0440, P0500 and P0505 not set, ECT sensor more than 158ºF, vehicle driven at a constant speed of less than 62 mph with the engine speed over 1500 rpm, and the PCM detected the rich fuel trim correction value was over the limit. **Possible Causes** • A/FS or HO2S is contaminated, deteriorated or it has failed • Base engine "mechanical" fault affecting one or more cylinders • EVAP system component has failed or canister fuel saturated • Exhaust leaks located in front of the A/FS or HO2S location • Fuel control sensor is out of calibration (i.e., ECT, IAT or MAF) • Fuel delivery system supplying too much fuel during cruise or idle periods (e.g., faulty fuel pump, or faulty pressure regulator) • Fuel injector(s) is leaking or stuck partially open (one or more)
DTC P0172 **2T FUEL, MIL: Yes** 2003 4Runner, Celica, Corolla, Echo, Land Cruiser, Matrix, MR2, Prius, RAV4, Sequoia, Tacoma, Tundra 1.5L VIN B, 1.5L VIN T, 1.8L VIN R, 1.8L VIN Y, 2.0L VIN K, 2.4L VIN L, 2.7L VIN M, 3.4L VIN N, 4.0L VIN U, 4.7L VIN T Transmissions: All	**Fuel System Too Rich (Bank 1) Conditions:** DTC P0100, P0101, P0105, P0110, P0115, P0120, P0121, P0130, P0133, P0136, P0135, P0136, P0141, P0151, P0156, P0161, P0300, P0301-P0306, P0440, P0500 and P0505 not set, ECT sensor more than 158ºF, vehicle driven at a constant speed of less than 62 mph with the engine speed over 1500 rpm, and the PCM detected the rich fuel trim correction value was over the limit. **Possible Causes** • A/FS or HO2S is contaminated, deteriorated or it has failed • Base engine "mechanical" fault affecting one or more cylinders • EVAP system component has failed or canister fuel saturated • Exhaust leaks located in front of the A/FS or HO2S location • Fuel control sensor is out of calibration (i.e., ECT, IAT or MAF) • Fuel delivery system supplying too much fuel during cruise or idle periods (e.g., faulty fuel pump, or faulty pressure regulator) • Fuel injector(s) is leaking or stuck partially open (one or more)
DTC P0172 **2T FUEL, MIL: Yes** 2003 Avalon, Camry, Camry Solara, Highlander, Sienna 2.4L VIN E, 3.0L VIN F engines Transmissions: All	**Fuel System Too Rich (Bank 1) Conditions:** DTC P0100, P0101, P0105, P0110, P0115, P0120, P0121, P0130, P0133, P0136, P0135, P0136, P0141, P0151, P0156, P0161, P0300, P0301-P0306, P0440, P0500 and P0505 not set, ECT sensor more than 158ºF, vehicle driven at a constant speed of less than 62 mph with the engine speed over 1500 rpm, and the PCM detected the rich fuel trim correction value was over the limit. **Possible Causes** • A/FS or HO2S is contaminated, deteriorated or it has failed • Base engine "mechanical" fault affecting one or more cylinders • EVAP system component has failed or canister fuel saturated • Exhaust leaks located in front of the A/FS or HO2S location • Fuel control sensor is out of calibration (i.e., ECT, IAT or MAF) • Fuel delivery system supplying too much fuel during cruise or idle periods (e.g., faulty fuel pump, or faulty pressure regulator) • Fuel injector(s) is leaking or stuck partially open (one or more)

OBD II Trouble Code List (P0xxx Codes)

DTC	Trouble Code Title, Conditions & Possible Causes
DTC P0174 **2T FUEL, MIL: Yes** 2001, 2002, 2003 4Runner, Camry, Camry Solara, Highlander, RAV4, Sequoia, Sienna, Tundra 2.0L VIN K, 3.0L VIN F, 4.0L VIN U, 4.7L VIN T Transmissions: All	**Fuel System Too Lean (Bank 2) Conditions:** DTC P0100, P0101, P0105, P0110, P0115, P0120, P0121, P0136, P0141, P0151, P0156, P0161, P0300, P0301-P0306, P0440, P0500, P0505, P1130, P1133, P1135, 1150, 53 and P1155 not set, ECT sensor more than 158ºF, vehicle driven at a constant speed of less than 62 mph with the engine speed over 1500 rpm, and the PCM detected the lean fuel trim correction value was over the limit. **Possible Causes** • A/FS or HO2S is contaminated, deteriorated or it has failed • Air leaks after the MAF sensor, or in the EGR or PCV system • Base engine "mechanical" fault affecting one or more cylinders • Exhaust leaks located in front of the A/FS or HO2S location • Fuel control sensor is out of calibration (i.e., ECT, IAT or MAF) • Fuel delivery system supplying too little fuel during cruise or idle periods (e.g., faulty fuel pump or dirty, restricted fuel filter) • Fuel injector (one or more) dirty or pressure regulator has failed • MAF sensor is contaminated, out-of-calibration or damaged • Vehicle driven low on fuel or until it ran out of fuel
DTC P0174 **2T FUEL, MIL: Yes** 2003 Avalon, MR2 1.8L VIN R, 3.0L VIN F engines Transmissions: All	**Fuel System Too Lean (Bank 2) Conditions:** DTC P0100, P0101, P0105, P0110, P0115, P0120, P0121, P0136, P0141, P0151, P0156, P0161, P0300, P0301-P0306, P0440, P0500, P0505, P1130, P1133, P1135, 1150, 53 and P1155 not set, ECT sensor more than 158ºF, vehicle driven at a constant speed of less than 62 mph with the engine speed over 1500 rpm, and the PCM detected the lean fuel trim correction value was over the limit. **Possible Causes** • A/FS or HO2S is contaminated, deteriorated or it has failed • Air leaks after the MAF sensor, or in the EGR or PCV system • Base engine "mechanical" fault affecting one or more cylinders • Exhaust leaks located in front of the A/FS or HO2S location • Fuel control sensor is out of calibration (i.e., ECT, IAT or MAF) • Fuel delivery system supplying too little fuel during cruise or idle periods (e.g., faulty fuel pump or dirty, restricted fuel filter) • Fuel injector (one or more) dirty or pressure regulator has failed • MAF sensor is contaminated, out-of-calibration or damaged • Vehicle driven low on fuel or until it ran out of fuel
DTC P0175 **2T FUEL, MIL: Yes** 2001, 2002, 2003 4Runner, Camry, Camry Solara, Highlander, RAV4, Sequoia, Sienna, Tundra 3.0L VIN F, 2.0L VIN K, 4.0L VIN U, 4.7L VIN T Transmissions: All	**Fuel System Too Rich (Bank 2) Conditions:** DTC P0100, P0101, P0105, P0110, P0115, P0120, P0121, P0136, P0141, P0151, P0156, P0161, P0300, P0301-P0306, P0440, P0500, P0505, P1130, P1133, P1135, 1150, 53 and P1155 not set, ECT sensor more than 158ºF, vehicle driven at a constant speed of less than 62 mph with the engine speed over 1500 rpm, and the PCM detected the rich fuel trim correction value was over the limit. **Possible Causes** • A/FS or HO2S is contaminated, deteriorated or it has failed • Base engine "mechanical" fault affecting one or more cylinders • EVAP system component has failed or canister fuel saturated • Exhaust leaks located in front of the A/FS or HO2S location • Fuel control sensor is out of calibration (i.e., ECT, IAT or MAF) • Fuel delivery system supplying too much fuel during cruise or idle periods (e.g., faulty fuel pump, or faulty pressure regulator) • Fuel injector(s) is leaking or stuck partially open (one or more)
DTC P0175 **2T FUEL, MIL: Yes** 2003 Avalon, MR2 1.8L VIN R, 3.0L VIN F engines Transmissions: All	**Fuel System Too Rich (Bank 2) Conditions:** DTC P0100, P0101, P0105, P0110, P0115, P0120, P0121, P0136, P0141, P0151, P0156, P0161, P0300, P0301-P0306, P0440, P0500, P0505, P1130, P1133, P1135, 1150, 53 and P1155 not set, ECT sensor more than 158ºF, vehicle driven at a constant speed of less than 62 mph with the engine speed over 1500 rpm, and the PCM detected the rich fuel trim correction value was over the limit. **Possible Causes** • A/FS or HO2S is contaminated, deteriorated or it has failed • Base engine "mechanical" fault affecting one or more cylinders • EVAP system component has failed or canister fuel saturated • Exhaust leaks located in front of the A/FS or HO2S location • Fuel control sensor is out of calibration (i.e., ECT, IAT or MAF) • Fuel delivery system supplying too much fuel during cruise or idle periods (e.g., faulty fuel pump, or faulty pressure regulator) • Fuel injector(s) is leaking or stuck partially open (one or more)

OBD II Trouble Code List (P0xxx Codes)

DTC	Trouble Code Title, Conditions & Possible Causes
DTC P0201 **2T CCM, MIL: Yes** 1995 Previa 2.4L VIN A engine Transmissions: All	**Fuel Injector 1 Misfire Detected Conditions:** Engine started, engine speed over 500 rpm, and the PCM detected an incorrect voltage condition on the Fuel Injector 1 control circuit. **Possible Causes** • Base engine mechanical fault that affects one or more cylinders • Ignition system problem (coil or plug) that affects one cylinder • Injector control circuit is open or shorted to ground • Injector power circuit is open (check the power from the relay) • Fuel injector is damaged or has failed
DTC P0202 **2T CCM, MIL: Yes** 1995 Previa 2.4L VIN A engine Transmissions: All	**Fuel Injector 2 Misfire Detected Conditions:** Engine started, engine speed over 500 rpm, and the PCM detected an incorrect voltage condition on the Fuel Injector 2 control circuit. **Possible Causes** • Base engine mechanical fault that affects one or more cylinders • Ignition system problem (coil or plug) that affects one cylinder • Injector control circuit is open or shorted to ground • Injector power circuit is open (check the power from the relay) • Fuel injector is damaged or has failed
DTC P0203 **2T CCM, MIL: Yes** 1995 Previa 2.4L VIN A engine Transmissions: All	**Fuel Injector 3 Misfire Detected Conditions:** Engine started, engine speed over 500 rpm, and the PCM detected an incorrect voltage condition on the Fuel Injector 3 control circuit. **Possible Causes** • Base engine mechanical fault that affects one or more cylinders • Ignition system problem (coil or plug) that affects one cylinder • Injector control circuit is open or shorted to ground • Injector power circuit is open (check the power from the relay) • Fuel injector is damaged or has failed
DTC P0204 **2T CCM, MIL: Yes** 1995 Previa 2.4L VIN A engine Transmissions: All	**Fuel Injector 4 Misfire Detected Conditions:** Engine started, engine speed over 500 rpm, and the PCM detected an incorrect voltage condition on the Fuel Injector 4 control circuit. **Possible Causes** • Base engine mechanical fault that affects one or more cylinders • Ignition system problem (coil or plug) that affects one cylinder • Injector control circuit is open or shorted to ground • Injector power circuit is open (check the power from the relay) • Fuel injector is damaged or has failed
DTC P0220 **2T CCM, MIL: Yes** 2003 4Runner, Land Cruiser, Sequoia, Tacoma, Tundra 3.4L VIN N engine, 4.0L VIN U, 4.7L VIN T Transmissions: All	**Throttle Pedal Position Sensor/Switch 'B' Circuit Malfunction Conditions:** Key on or engine running; and the PCM detected the TP sensor 'B' Signal indicated less than 0.50v with the throttle position closed (the switch is open), or the TP sensor signal indicated more than 4.9v at any time. The Electric TP Sensor is mounted on the throttle body. It has two sensors (the electrical throttle system does not use a cable). **Possible Causes** • TP sensor signal circuit open or shorted to ground • TP sensor ground circuit is open • TP sensor power circuit is open (check VREF circuit at PCM) • TP sensor is damaged or failed • PCM has failed
DTC P0220 **2T CCM, MIL: Yes** 2003 4Runner, Land Cruiser, Sequoia, Tacoma, Tundra 3.4L VIN N engine, 4.0L VIN U, 4.7L VIN T Transmissions: All	**Throttle Pedal Position Sensor Switch B Circuit Low Input Conditions:** Key on or engine running; and the PCM detected an unexpected low voltage condition (less than 0.50v) on the VTA2 circuit. The Electric TP Sensor is mounted on the throttle body. It has two sensors (the electrical throttle system does not use a cable). **Possible Causes** • Electric TP sensor connector is damaged or shorted • Electric TP sensor circuit is shorted to ground • Electric TP sensor is damaged or has failed • PCM has failed
DTC P0223 **2T CCM, MIL: Yes** 2003 4Runner, Land Cruiser, Sequoia, Tacoma, Tundra 3.4L VIN N engine, 4.0L VIN U, 4.7L VIN T Transmissions: All	**Throttle Pedal Position Sensor Switch B Circuit High Input Conditions:** Key on or engine running; and the PCM detected an unexpected high voltage condition (more than 4.97v) on the VTA2 circuit. The Electric TP Sensor is mounted on the throttle body. It has two sensors (the electrical throttle system does not use a cable). **Possible Causes** • Electric TP sensor connector is damaged or open • Electric TP sensor circuit is open or shorted to VREF • Electric TP sensor is damaged or has failed • PCM has failed

OBD II Trouble Code List (P0xxx Codes)

DTC	Trouble Code Title, Conditions & Possible Causes
DTC P0230 **2T CCM, MIL: Yes** 2003 4Runner, Land Cruiser, Sequoia, Tundra 4.0L VIN U, 4.7L VIN T engines Transmissions: All	**Fuel Pump Primary Circuit Malfunction Conditions:** Engine started; and the PCM detected an unexpected voltage on the Fuel Pump Primary control circuit (from ST terminal to Starter Relay coil and to the STA terminal of the PCM). **Possible Causes** • Circuit opening relay is damaged or has failed • Fuel pump relay control circuit is open or shorted to ground • Fuel pump relay is damaged or has failed • Fuel pump is damaged or has failed • PCM has failed
DTC P0300 **2T MISFIRE, MIL: Yes** 1995 Avalon, Camry, Land Cruiser, Previa, T100, Tacoma, Tercel 1.5L VIN E, 2.4L VIN A, 2.4L VIN R, 2.7L VIN U, 3.0L VIN G, 3.4L VIN V, 4.5L VIN D engines Transmissions: All	**Multiple Cylinder Misfire Detected Conditions:** DTC P0100, P0101, P0105, P0110, P0115, P0120, P0121, P0125, P0335, P0340, P0500, P0505 and P0510 not set, engine started, vehicle driven to a speed of over 3 mph for 1 minute, and the PCM detected a misfire rate of 1-2% (High Emissions 2T), or a misfire rate of 6-30% (Catalyst Damaging 1T) in two or more cylinders. *Note: If the misfire is severe, the MIL will flash on/off on the 1st trip!* **Possible Causes** • Air leak in the intake manifold, or in the EGR or PCV system • Base engine mechanical fault that affects one or more cylinders • Erratic or interrupted CKP or CMP sensor signals • Fuel delivery component fault that affects one or more cylinders (e.g., a contaminated, dirty or sticking fuel injector) • Ignition system problem (coil or plug) in one or more cylinders
DTC P0301 **2T MISFIRE, MIL: Yes** 1995 Avalon, Camry, Land Cruiser, Previa, T100, Tacoma, Tercel 1.5L VIN E, 2.4L VIN A, 2.4L VIN R, 2.7L VIN U, 3.0L VIN G, 3.4L VIN V, 4.5L VIN D engines Transmissions: All	**Cylinder 1 Misfire Detected Conditions:** DTC P0100, P0101, P0105, P0110, P0115, P0120, P0121, P0335, P0340 and :P0500 not set, engine running, vehicle speed over 3 mph, and the PCM detected a misfire condition in Cylinder 1 in the 200 (Catalyst) or 1000-rpm (High Emissions) revolution range. *Note: If the misfire is severe, the MIL will flash on/off on the 1st trip!* **Possible Causes** • Air leak in the intake manifold, or in the EGR or PCV system • Base engine mechanical fault that affects only one cylinder • Air leak in the intake manifold, or in the EGR or PCV system • Base engine mechanical fault that affects only one cylinder • Fuel component fault that affects only one cylinder (e.g., a dirty or sticking fuel injector) • Ignition system problem (coil or plug) that affects one cylinder
DTC P0302 **2T MISFIRE, MIL: Yes** 1995 Avalon, Camry, Land Cruiser, Previa, T100, Tacoma, Tercel 1.5L VIN E, 2.4L VIN A, 2.4L VIN R, 2.7L VIN U, 3.0L VIN G, 3.4L VIN V, 4.5L VIN D engines Transmissions: All	**Cylinder 2 Misfire Detected Conditions:** DTC P0100, P0101, P0105, P0110, P0115, P0120, P0121, P0335, P0340 and P0500 not set, engine running, vehicle speed over 3 mph, and the PCM detected a misfire condition in Cylinder 2 in the 200 (Catalyst) or 1000-rpm (High Emissions) revolution range. *Note: If the misfire is severe, the MIL will flash on/off on the 1st trip!* **Possible Causes** • Air leak in the intake manifold, or in the EGR or PCV system • Base engine mechanical fault that affects only one cylinder • Fuel component fault that affects only one cylinder (e.g., a dirty or sticking fuel injector) • Ignition system problem (coil or plug) that affects one cylinder
DTC P0303 **2T MISFIRE, MIL: Yes** 1995 Avalon, Camry, Land Cruiser, Previa, T100, Tacoma, Tercel 1.5L VIN E, 2.4L VIN A, 2.4L VIN R, 2.7L VIN U, 3.0L VIN G, 3.4L VIN V, 4.5L VIN D engines Transmissions: All	**Cylinder 3 Misfire Detected Conditions:** DTC P0100, P0101, P0105, P0110, P0115, P0120, P0121, P0335, P0340 and P0500 not set, engine running, vehicle speed over 3 mph, and the PCM detected a misfire condition in Cylinder 3 in the 200 (Catalyst) or 1000-rpm (High Emissions) revolution range. *Note: If the misfire is severe, the MIL will flash on/off on the 1st trip!* **Possible Causes** • Air leak in the intake manifold, or in the EGR or PCV system • Base engine mechanical fault that affects only one cylinder • Fuel component fault that affects only one cylinder (e.g., a dirty or sticking fuel injector) • Ignition system problem (coil or plug) that affects one cylinder
DTC P0304 **2T MISFIRE, MIL: Yes** 1995 Avalon, Camry, Land Cruiser, Previa, T100, Tacoma, Tercel 1.5L VIN E, 2.4L VIN A, 2.4L VIN R, 2.7L VIN U, 3.0L VIN G, 3.4L VIN V, 4.5L VIN D engines Transmissions: All	**Cylinder 4 Misfire Detected Conditions:** DTC P0100, P0101, P0105, P0110, P0115, P0120, P0121, P0335, P0340 and P0500 not set, engine running, vehicle speed over 3 mph, and the PCM detected a misfire condition in Cylinder 4 in the 200 (Catalyst) or 1000-rpm (High Emissions) revolution range. *Note: If the misfire is severe, the MIL will flash on/off on the 1st trip!* **Possible Causes** • Air leak in the intake manifold, or in the EGR or PCV system • Base engine mechanical fault that affects only one cylinder • Fuel component fault that affects only one cylinder (e.g., a dirty or sticking fuel injector) • Ignition system problem (coil or plug) that affects one cylinder

OBD II Trouble Code List (P0xxx Codes)

DTC	Trouble Code Title, Conditions & Possible Causes
DTC P0305 **2T MISFIRE, MIL: Yes** 1995 Avalon, Camry, Land Cruiser, T100, Tacoma 3.0L VIN G, 3.4L VIN V, 4.5L VIN D engines Transmissions: All	**Cylinder 5 Misfire Detected Conditions:** DTC P0100, P0101, P0105, P0110, P0115, P0120, P0121, P0335, P0340 and P0500 not set, engine running, vehicle speed over 3 mph, and the PCM detected a misfire condition in Cylinder 5 in the 200 (Catalyst) or 1000-rpm (High Emissions) revolution range. *Note: If the misfire is severe, the MIL will flash on/off on the 1st trip!* **Possible Causes** • Air leak in the intake manifold, or in the EGR or PCV system • Base engine mechanical fault that affects only one cylinder • Fuel component fault that affects only one cylinder (e.g., a dirty or sticking fuel injector) • Ignition system problem (coil or plug) that affects one cylinder
DTC P0306 **2T MISFIRE, MIL: Yes** 1995 Avalon, Camry, Land Cruiser, T100, Tacoma 3.0L VIN G, 3.4L VIN V, 4.5L VIN D engines Transmissions: All	**Cylinder 6 Misfire Detected Conditions:** DTC P0100, P0101, P0105, P0110, P0115, P0120, P0121, P0335, P0340 and P0500 not set, engine running, vehicle speed over 3 mph, and the PCM detected a misfire condition in Cylinder 6 in the 200 (Catalyst) or 1000-rpm (High Emissions) revolution range. *Note: If the misfire is severe, the MIL will flash on/off on the 1st trip!* **Possible Causes** • Air leak in the intake manifold, or in the EGR or PCV system • Base engine mechanical fault that affects only one cylinder • Fuel component fault that affects only one cylinder (e.g., a dirty or sticking fuel injector) • Ignition system problem (coil or plug) that affects one cylinder
DTC P0300 **2T MISFIRE, MIL: Yes** 1996, 1997, 1998, 1999, 2000, 2001, 2002 All engines Transmissions: All	**Multiple Cylinder Misfire Detected Conditions:** No PCM codes set, engine running, VSS signal over 3 mph, and the PCM detected a misfire rate of 1-2% (High Emissions 2T), or a misfire rate of 6-30% (Catalyst Damaging 1T) in 2 or more cylinders. *Note: If the misfire is severe, the MIL will flash on/off on the 1st trip!* **Possible Causes** • Air leak in the intake manifold, or in the EGR or PCV system • Base engine mechanical fault that affects one or more cylinders • Erratic or interrupted CKP or CMP sensor signals • Fuel delivery component fault that affects one or more cylinders (e.g., a contaminated, dirty or sticking fuel injector) • Ignition system problem (coil or plug) in one or more cylinders • TSB EG011-01 (8/01) contains information related to this code
DTC P0301 **2T MISFIRE, MIL: Yes** 1996, 1997, 1998, 1999, 2000, 2001, 2002 All models All engines Transmissions: All	**Cylinder 1 Misfire Detected Conditions:** DTC P0100, P0101, P0105, P0110, P0115, P0120, P0121, P0335, P0340 and P0500 not set, engine running, vehicle speed over 3 mph, and the PCM detected a misfire condition in Cylinder 1 in the 200 (Catalyst) or 1000-rpm (High Emissions) revolution range. *Note: If the misfire is severe, the MIL will flash on/off on the 1st trip!* **Possible Causes** • Air leak in the intake manifold, or in the EGR or PCV system • Base engine mechanical fault that affects only one cylinder • Fuel delivery component fault that affects only one cylinder (e.g., a contaminated, dirty or sticking fuel injector) • Ignition system problem (coil or plug) that affects one cylinder
DTC P0302 **2T MISFIRE, MIL: Yes** 1996, 1997, 1998, 1999, 2000, 2001, 2002 All models All engines Transmissions: All	**Cylinder 2 Misfire Detected Conditions:** DTC P0100, P0101, P0105, P0110, P0115, P0120, P0121, P0335, P0340 and P0500 not set, engine running, vehicle speed over 3 mph, and the PCM detected a misfire condition in Cylinder 2 in the 200 (Catalyst) or 1000-rpm (High Emissions) revolution range. *Note: If the misfire is severe, the MIL will flash on/off on the 1st trip!* **Possible Causes** • Air leak in the intake manifold, or in the EGR or PCV system • Base engine mechanical fault that affects only one cylinder • Fuel delivery component fault that affects only one cylinder (e.g., a contaminated, dirty or sticking fuel injector) • Ignition system problem (coil or plug) that affects one cylinder
DTC P0303 **2T MISFIRE, MIL: Yes** 1996, 1997, 1998, 1999, 2000, 2001, 2002 All engines Transmissions: All	**Cylinder 3 Misfire Detected Conditions:** DTC P0100, P0101, P0105, P0110, P0115, P0120, P0121, P0335, P0340 and P0500 not set, engine running, vehicle speed over 3 mph, and the PCM detected a misfire condition in Cylinder 3 in the 200 (Catalyst) or 1000-rpm (High Emissions) revolution range. *Note: If the misfire is severe, the MIL will flash on/off on the 1st trip!* **Possible Causes** • Air leak in the intake manifold, or in the EGR or PCV system • Base engine mechanical fault that affects only one cylinder • Fuel delivery component fault that affects only one cylinder (e.g., a contaminated, dirty or sticking fuel injector) • Ignition system problem (coil or plug) that affects one cylinder

OBD II Trouble Code List (P0xxx Codes)

DTC	Trouble Code Title, Conditions & Possible Causes
DTC P0304 **2T MISFIRE, MIL: Yes** 1996, 1997, 1998, 1999, 2000, 2001, 2002 All models All engines Transmissions: All	**Cylinder 4 Misfire Detected Conditions:** DTC P0100, P0101, P0105, P0110, P0115, P0120, P0121, P0335, P0340 and P0500 not set, engine running, vehicle speed over 3 mph, and the PCM detected a misfire condition in Cylinder 4 in the 200 (Catalyst) or 1000-rpm (High Emissions) revolution range. *Note: If the misfire is severe, the MIL will flash on/off on the 1st trip!* **Possible Causes** • Air leak in the intake manifold, or in the EGR or PCV system • Base engine mechanical fault that affects only one cylinder • Fuel delivery component fault that affects only one cylinder (e.g., a contaminated, dirty or sticking fuel injector) • Ignition system problem (coil or plug) that affects one cylinder
DTC P0305 **2T MISFIRE, MIL: Yes** 1996, 1997, 1998, 1999, 2000, 2001, 2002 4Runner, Avalon, Highlander, Land Cruiser, Sequoia, Sienna, Tacoma, Tundra 3.0L VIN D, 3.0L VIN E, 3.0L VIN F, 3.4L VIN N, 4.5L VIN J, 4.7L VIN T Transmissions: All Trouble Code ID: **P0305** Number of Trips to Set Code: **2T** OBD II Monitor Type: **MISFIRE** MIL: YES Schematic: Toyota**P0305**.bmp ; Model clarification string	**Cylinder 5 Misfire Detected Conditions:** DTC P0100, P0101, P0105, P0110, P0115, P0120, P0121, P0335, P0340 and P0500 not set, engine running, vehicle speed over 3 mph, and the PCM detected a misfire condition in Cylinder 5 in the 200 (Catalyst) or 1000-rpm (High Emissions) revolution range. *Note: If the misfire is severe, the MIL will flash on/off on the 1st trip!* **Possible Causes** • Air leak in the intake manifold, or in the EGR or PCV system • Base engine mechanical fault that affects only one cylinder • Fuel delivery component fault that affects only one cylinder (e.g., a contaminated, dirty or sticking fuel injector) • Ignition system problem (coil or plug) that affects one cylinder
DTC P0306 **2T MISFIRE, MIL: Yes** 1996, 1997, 1998, 1999, 2000, 2001, 2002 4Runner, Avalon, Highlander, Land Cruiser, Sequoia, Sienna, Tacoma, Tundra 3.0L VIN D, 3.0L VIN E, 3.0L VIN F, 3.4L VIN N, 4.5L VIN J, 4.7L VIN T Transmissions: All	**Cylinder 6 Misfire Detected Conditions:** DTC P0100, P0101, P0105, P0110, P0115, P0120, P0121, P0335, P0340 and P0500 not set, engine running, vehicle speed over 3 mph, and the PCM detected a misfire condition in Cylinder 6 in the 200 (Catalyst) or 1000-rpm (High Emissions) revolution range. *Note: If the misfire is severe, the MIL will flash on/off on the 1st trip!* **Possible Causes** • Air leak in the intake manifold, or in the EGR or PCV system • Base engine mechanical fault that affects only one cylinder • Fuel delivery component fault that affects only one cylinder (e.g., a contaminated, dirty or sticking fuel injector) • Ignition system problem (coil or plug) that affects one cylinder
DTC P0307 **2T MISFIRE, MIL: Yes** 1998, 1999, 2000, 2001, 2002 Land Cruiser, Sequoia, Tundra Engines: 4.7L VIN T Transmissions: All	**Cylinder 7 Misfire Detected Conditions:** DTC P0100, P0101, P0105, P0110, P0115, P0120, P0121, P0335, P0340 and P0500 not set, engine running, vehicle speed over 3 mph, and the PCM detected a misfire condition in Cylinder 7 in the 200 (Catalyst) or 1000-rpm (High Emissions) revolution range. *Note: If the misfire is severe, the MIL will flash on/off on the 1st trip!* **Possible Causes** • Air leak in the intake manifold, or in the EGR or PCV system • Base engine mechanical fault that affects only one cylinder • Fuel delivery component fault that affects only one cylinder (e.g., a contaminated, dirty or sticking fuel injector) • Ignition system problem (coil or plug) that affects one cylinder
DTC P0308 **2T MISFIRE, MIL: Yes** 1998, 1999, 2000, 2001, 2002 Land Cruiser, Sequoia, Tundra Engines: 4.7L VIN T Transmissions: All; Model clarification string	**Cylinder 8 Misfire Detected Conditions:** DTC P0100, P0101, P0105, P0110, P0115, P0120, P0121, P0335, P0340 and P0500 not set, engine running, vehicle speed over 3 mph, and the PCM detected a misfire condition in Cylinder 8 in the 200 (Catalyst) or 1000-rpm (High Emissions) revolution range. *Note: If the misfire is severe, the MIL will flash on/off on the 1st trip!* **Possible Causes** • Air leak in the intake manifold, or in the EGR or PCV system • Base engine mechanical fault that affects only one cylinder • Fuel delivery component fault that affects only one cylinder (e.g., a contaminated, dirty or sticking fuel injector) • Ignition system problem (coil or plug) that affects one cylinder

OBD II Trouble Code List (P0xxx Codes)

DTC	Trouble Code Title, Conditions & Possible Causes
DTC P0300 **2T MISFIRE, MIL: Yes** 2001, 2002, 2003 Prius All engines Transmissions: All	**Multiple Cylinder Misfire Detected Conditions:** **Trouble Code Conditions** DTC P0100, P0101, P0115, P0116, P0120, P0121, P0335, P0340 and P0500 not set, engine running, VSS signal over 3 mph, and the PCM detected a misfire rate of 1-2% (High Emissions 2T), or a misfire rate of 6-30% (Catalyst Damaging 1T) in two or more cylinder. *Note: If the misfire is severe, the MIL will flash on/off on the 1st trip!* **Possible Causes** • Air leak in the intake manifold, or in the EGR or PCV system • Base engine mechanical fault that affects one or more cylinders • Erratic or interrupted CKP or CMP sensor signals • Fuel delivery component fault that affects one or more cylinders (e.g., a contaminated, dirty or sticking fuel injector) • Ignition system problem (coil or plug) in one or more cylinders • TSB EG006-02 (1/02) contains information related to this code
DTC P0301 **2T MISFIRE, MIL: Yes** 2001, 2002, 2003 Prius All engines Transmissions: All	**Cylinder 1 Misfire Detected Conditions:** DTC P0100, P0101, P0115, P0116, P0120, P0121, P0335, P0340 and P0500 not set, engine running, vehicle speed over 3 mph, and the PCM detected a misfire condition in Cylinder 1 in the 200 (Catalyst) or 1000-rpm (High Emissions) revolution range. *Note: If the misfire is severe, the MIL will flash on/off on the 1st trip!* **Possible Causes** • Air leak in the intake manifold, or in the EGR or PCV system • Base engine mechanical fault that affects only one cylinder • Fuel delivery component fault that affects only one cylinder (e.g., a contaminated, dirty or sticking fuel injector) • Ignition system problem (coil or plug) that affects one cylinder • TSB EG006-02 (1/02) contains information related to this code
DTC P0302 **2T MISFIRE, MIL: Yes** 2001, 2002, 2003 Prius All engines Transmissions: All	**Cylinder 2 Misfire Detected Conditions:** DTC P0100, P0101, P0115, P0116, P0120, P0121, P0335, P0340 and P0500 not set, engine running, vehicle speed over 3 mph, and the PCM detected a misfire condition in Cylinder 2 in the 200 (Catalyst) or 1000-rpm (High Emissions) revolution range. *Note: If the misfire is severe, the MIL will flash on/off on the 1st trip!* **Possible Causes** • Air leak in the intake manifold, or in the EGR or PCV system • Base engine mechanical fault that affects only one cylinder • Fuel delivery component fault that affects only one cylinder (e.g., a contaminated, dirty or sticking fuel injector) • Ignition system problem (coil or plug) that affects one cylinder • TSB EG006-02 (1/02) contains information related to this code
DTC P0303 **2T MISFIRE, MIL: Yes** 2001, 2002, 2003 Prius All engines Transmissions: All	**Cylinder 3 Misfire Detected Conditions:** DTC P0100, P0101, P0115, P0116, P0120, P0121, P0335, P0340 and P0500 not set, engine running, vehicle speed over 3 mph, and the PCM detected a misfire condition in Cylinder 3 in the 200 (Catalyst) or 1000-rpm (High Emissions) revolution range. *Note: If the misfire is severe, the MIL will flash on/off on the 1st trip!* **Possible Causes** • Air leak in the intake manifold, or in the EGR or PCV system • Base engine mechanical fault that affects only one cylinder • Fuel delivery component fault that affects only one cylinder (e.g., a contaminated, dirty or sticking fuel injector) • Ignition system problem (coil or plug) that affects one cylinder • TSB EG006-02 (1/02) contains information related to this code
DTC P0304 **2T MISFIRE, MIL: Yes** 2001, 2002, 2003 Prius All engines Transmissions: All	**Cylinder 4 Misfire Detected Conditions:** DTC P0100, P0101, P0115, P0116, P0120, P0121, P0335, P0340 and P0500 not set, engine running, vehicle speed over 3 mph, and the PCM detected a misfire condition in Cylinder 4 in the 200 (Catalyst) or 1000-rpm (High Emissions) revolution range. *Note: If the misfire is severe, the MIL will flash on/off on the 1st trip!* **Possible Causes** • Air leak in the intake manifold, or in the EGR or PCV system • Base engine mechanical fault that affects only one cylinder • Fuel delivery component fault that affects only one cylinder (e.g., a contaminated, dirty or sticking fuel injector) • Ignition system problem (coil or plug) that affects one cylinder • TSB EG006-02 (1/02) contains information related to this code

OBD II Trouble Code List (P0xxx Codes)

DTC	Trouble Code Title, Conditions & Possible Causes
DTC P0300 **2T MISFIRE, MIL: Yes** 2003 4Runner, Avalon, Camry, Camry Solara, Celica, Corolla, Highlander, Land Cruiser, Matrix, MR2, RAV4, Sequoia, Sienna, Tacoma, Tundra All engines Transmissions: All	**Multiple Cylinder Misfire Detected Conditions:** **Trouble Code Conditions** DTC P0100, P0101, P0102, P0103, P0105, P0110, P0112, P0113, P0115, P0117, P0118, P0120, P0121, P0122, P0123, P0125, P0335, P0340, P0500, P0505 and P0510 not set, engine started, vehicle driven to a speed of over 3 mph for 1 minute, and the PCM detected a misfire rate of 1-2% (High Emissions 2T), or a misfire rate of 6-30% (Catalyst Damaging 1T) in two or more cylinders. *Note: If the misfire is severe, the MIL will flash on/off on the 1st trip! Look at the misfire ratio for all of the cylinders on the Scan Tool. The cylinder with the highest misfire ratio should be checked first!* **Possible Causes** • Air leak in the intake manifold, or in the EGR or PCV system • Base engine mechanical fault that affects two or more cylinders • EGR valve is stuck open or the PCV system has a vacuum leak • Fuel delivery component fault that affects two or more cylinders (e.g., contaminated, dirty or sticking fuel injectors) • Ignition system fault (coil or plug) that affects several cylinders • Mass airflow meter is contaminated, or its signal is out of range
DTC P0301 **2T MISFIRE, MIL: Yes** 2003 4Runner, Avalon, Camry, Camry Solara, Celica, Corolla, Highlander, Land Cruiser, Matrix, MR2, RAV4, Sequoia, Sienna, Tacoma, Tundra All engines Transmissions: All	**Cylinder 1 Misfire Detected Conditions:** **Trouble Code Conditions** DTC P0100, P0101, P0102, P0103, P0105, P0110, P0112, P0113, P0115, P0117, P0118, P0120, P0121, P0122, P0123, P0125, P0335, P0340, P0500, P0505 and P0510 not set, engine started, vehicle driven to a speed of over 3 mph for 1 minute, and the PCM detected a misfire rate of 1-2% (High Emissions 2T), or a misfire rate of 6-30% (Catalyst Damaging 1T) in Cylinder 1. *Note: If the misfire is severe, the MIL will flash on/off on the 1st trip!* **Possible Causes** • Base engine mechanical fault that affects only Cylinder 1 • EGR valve is stuck open or the PCV system has a vacuum leak • Fuel component fault that affects only Cylinder 1 (a contaminated or sticking injector) • Ignition system problem (coil or plug) that affects Cylinder 1
DTC P0302 **2T MISFIRE, MIL: Yes** 2003 4Runner, Avalon, Camry, Camry Solara, Celica, Corolla, Highlander, Land Cruiser, Matrix, MR2, RAV4, Sequoia, Sienna, Tacoma, Tundra All engines Transmissions: All	**Cylinder 2 Misfire Detected Conditions:** **Trouble Code Conditions** DTC P0100, P0101, P0102, P0103, P0105, P0110, P0112, P0113, P0115, P0117, P0118, P0120, P0121, P0122, P0123, P0125, P0335, P0340, P0500, P0505 and P0510 not set, engine started, vehicle driven to a speed of over 3 mph for 1 minute, and the PCM detected a misfire rate of 1-2% (High Emissions 2T), or a misfire rate of 6-30% (Catalyst Damaging 1T) in Cylinder 2. *Note: If the misfire is severe, the MIL will flash on/off on the 1st trip!* **Possible Causes** • Base engine mechanical fault that affects only Cylinder 2 • EGR valve is stuck open or the PCV system has a vacuum leak • Fuel component fault that affects only Cylinder 2 (a contaminated or sticking injector) • Ignition system problem (coil or plug) that affects Cylinder 2
DTC P0303 **2T MISFIRE, MIL: Yes** 2003 4Runner, Avalon, Camry, Camry Solara, Celica, Corolla, Highlander, Land Cruiser, Matrix, MR2, RAV4, Sequoia, Sienna, Tacoma, Tundra All engines Transmissions: All	**Cylinder 3 Misfire Detected Conditions:** **Trouble Code Conditions** DTC P0100, P0101, P0102, P0103, P0105, P0110, P0112, P0113, P0115, P0117, P0118, P0120, P0121, P0122, P0123, P0125, P0335, P0340, P0500, P0505 and P0510 not set, engine started, vehicle driven to a speed of over 3 mph for 1 minute, and the PCM detected a misfire rate of 1-2% (High Emissions 2T), or a misfire rate of 6-30% (Catalyst Damaging 1T) in Cylinder 3. *Note: If the misfire is severe, the MIL will flash on/off on the 1st trip!* **Possible Causes** • Base engine mechanical fault that affects only Cylinder 3 • EGR valve is stuck open or the PCV system has a vacuum leak • Fuel component fault that affects only Cylinder 3 (a contaminated or sticking injector) • Ignition system problem (coil or plug) that affects Cylinder 3
DTC P0304 **2T MISFIRE, MIL: Yes** 2003 4Runner, Avalon, Camry, Camry Solara, Celica, Corolla, Highlander, Land Cruiser, Matrix, MR2, RAV4, Sequoia, Sienna, Tacoma, Tundra All engines Transmissions: All	**Cylinder 4 Misfire Detected Conditions:** **Trouble Code Conditions** DTC P0100, P0101, P0102, P0103, P0105, P0110, P0112, P0113, P0115, P0117, P0118, P0120, P0121, P0122, P0123, P0125, P0335, P0340, P0500, P0505 and P0510 not set, engine started, vehicle driven to a speed of over 3 mph for 1 minute, and the PCM detected a misfire rate of 1-2% (High Emissions 2T), or a misfire rate of 6-30% (Catalyst Damaging 1T) in Cylinder 4. *Note: If the misfire is severe, the MIL will flash on/off on the 1st trip!* **Possible Causes** • Base engine mechanical fault that affects only Cylinder 4 • EGR valve is stuck open or the PCV system has a vacuum leak • Fuel component fault that affects only Cylinder 4 (a contaminated or sticking injector) • Ignition system problem (coil or plug) that affects Cylinder 4

OBD II Trouble Code List (P0xxx Codes)

DTC	Trouble Code Title, Conditions & Possible Causes
DTC P0305 **2T MISFIRE, MIL: Yes** 2003 4Runner, Avalon, Camry, Camry Solara, Highlander, Land Cruiser, Sequoia, Sienna, Tacoma, Tundra 3.0L VIN F, 3.4L VIN N, 4.7L VIN T engines Transmissions: All	**Cylinder 5 Misfire Detected Conditions:** **Trouble Code Conditions** DTC P0100, P0101, P0102, P0103, P0105, P0110, P0112, P0113, P0115, P0117, P0118, P0120, P0121, P0122, P0123, P0125, P0335, P0340, P0500, P0505 and P0510 not set, engine started, vehicle driven to a speed of over 3 mph for 1 minute, and the PCM detected a misfire rate of 1-2% (High Emissions 2T), or a misfire rate of 6-30% (Catalyst Damaging 1T) in Cylinder 5. *Note: If the misfire is severe, the MIL will flash on/off on the 1st trip!* **Possible Causes** • Base engine mechanical fault that affects only Cylinder 5 • EGR valve is stuck open or the PCV system has a vacuum leak • Fuel delivery component fault that affects only Cylinder 5 (e.g., a contaminated, dirty or sticking fuel injector) • Ignition system problem (coil or plug) that affects Cylinder 5
DTC P0306 **2T MISFIRE, MIL: Yes** 2003 4Runner, Avalon, Camry, Camry Solara, Highlander, Land Cruiser, Sequoia, Sienna, Tacoma, Tundra 3.0L VIN F, 3.4L VIN N, 4.7L VIN T engines Transmissions: All	**Cylinder 6 Misfire Detected Conditions:** **Trouble Code Conditions** DTC P0100, P0101, P0102, P0103, P0105, P0110, P0112, P0113, P0115, P0117, P0118, P0120, P0121, P0122, P0123, P0125, P0335, P0340, P0500, P0505 and P0510 not set, engine started, vehicle driven to a speed of over 3 mph for 1 minute, and the PCM detected a misfire rate of 1-2% (High Emissions 2T), or a misfire rate of 6-30% (Catalyst Damaging 1T) in Cylinder 6. *Note: If the misfire is severe, the MIL will flash on/off on the 1st trip!* **Possible Causes** • Base engine mechanical fault that affects only Cylinder 6 • EGR valve is stuck open or the PCV system has a vacuum leak • Fuel delivery component fault that affects only Cylinder 6 (e.g., a contaminated, dirty or sticking fuel injector) • Ignition system problem (coil or plug) that affects Cylinder 6 • Mass airflow meter is contaminated, or its signal is out of range
DTC P0307 **2T MISFIRE, MIL: Yes** 2003 4Runner, Land Cruiser, Sequoia, Tundra Engines: 4.7L VIN T Transmissions: All	**Cylinder 7 Misfire Detected Conditions:** **Trouble Code Conditions** DTC P0100, P0101, P0102, P0103, P0105, P0110, P0112, P0113, P0115, P0117, P0118, P0120, P0121, P0122, P0123, P0125, P0335, P0340, P0500, P0505 and P0510 not set, engine started, vehicle driven to a speed of over 3 mph for 1 minute, and the PCM detected a misfire rate of 1-2% (High Emissions 2T), or a misfire rate of 6-30% (Catalyst Damaging 1T) in Cylinder 7. *Note: If the misfire is severe, the MIL will flash on/off on the 1st trip!* **Possible Causes** • Base engine mechanical fault that affects only Cylinder 7 • EGR valve is stuck open or the PCV system has a vacuum leak • Fuel delivery component fault that affects only Cylinder 7 (e.g., a contaminated, dirty or sticking fuel injector) • Ignition system problem (coil or plug) that affects Cylinder 7 • Mass airflow meter is contaminated, or its signal is out of range
DTC P0308 **2T MISFIRE, MIL: Yes** 2003 4Runner, Land Cruiser, Sequoia, Tundra Engines: 4.7L VIN T Transmissions: All	**Cylinder 8 Misfire Detected Conditions:** **Trouble Code Conditions** DTC P0100, P0101, P0102, P0103, P0105, P0110, P0112, P0113, P0115, P0117, P0118, P0120, P0121, P0122, P0123, P0125, P0335, P0340, P0500, P0505 and P0510 not set, engine started, vehicle driven to a speed of over 3 mph for 1 minute, and the PCM detected a misfire rate of 1-2% (High Emissions 2T), or a misfire rate of 6-30% (Catalyst Damaging 1T) in Cylinder 8. *Note: If the misfire is severe, the MIL will flash on/off on the 1st trip!* **Possible Causes** • Base engine mechanical fault that affects only Cylinder 8 • EGR valve is stuck open or the PCV system has a vacuum leak • Fuel delivery component fault that affects only Cylinder 8 (e.g., a contaminated, dirty or sticking fuel injector) • Ignition system problem (coil or plug) that affects Cylinder 8 • Mass airflow meter is contaminated, or its signal is out of range
DTC P0325 **1T CCM, MIL: Yes** 1995 Previa, T100, Tacoma 2.4L VIN A, 2.4L VIN R, 2.4L VIN U, 2.7L VIN U Transmissions: All	**Knock Sensor 1 Circuit Malfunction Conditions:** Engine started, vehicle driven with the engine speed over 1200 rpm, and the PCM detected an unexpected voltage condition on the Knock Sensor 1 (KS1) circuit during the CCM test. **Possible Causes** • Verify that the Knock Sensor (KS) is tightened to specification • Knock sensor signal circuit is open or shorted to ground • Knock sensor signal circuit is shorted to VREF or system power • Knock sensor is damaged or has failed • PCM has failed

OBD II Trouble Code List (P0xxx Codes)

DTC	Trouble Code Title, Conditions & Possible Causes
DTC P0325 **1T CCM, MIL: Yes** 1995 Avalon, Camry, Land Cruiser, T100, Tacoma 3.0L VIN G, 3.4L VIN V, 4.5L VIN D engines Transmissions: All	**Knock Sensor 1 Circuit Malfunction Conditions:** Engine started, vehicle driven with the engine speed over 2000 rpm, and the PCM detected an unexpected voltage condition on the Knock Sensor 1 (KS1) circuit during the CCM test. **Possible Causes** • Verify that the Knock Sensor (KS) is tightened to specification • Knock sensor signal circuit is open or shorted to ground • Knock sensor signal circuit is shorted to VREF or system power • Knock sensor is damaged or has failed • PCM has failed
DTC P0325 **1T CCM, MIL: Yes** 1996, 1997, 1998, 1999, 2000, 2001, 2002 1.5L VIN C, 1.6L VIN A, 1.6L VIN B, 1.8L VIN B, 1.8L VIN R, 2.0L VIN P, 2.2L VIN G, 2.4L VIN D, 2.4L VIN K, 2.4L VIN L, 2.7L VIN M engines Transmissions: All	**Knock Sensor 1 Circuit Malfunction Conditions:** Engine started, vehicle driven with the engine speed over 1200 rpm, and the PCM detected an unexpected voltage condition on the Knock Sensor 1 (KS1) circuit during the CCM test. **Possible Causes** • Verify that the Knock Sensor (KS) is tightened to specification • Knock sensor signal circuit is open or shorted to ground • Knock sensor signal circuit is shorted to VREF or system power • Knock sensor is damaged or has failed • PCM has failed
DTC P0325 **1T CCM, MIL: Yes** 1996, 1997, 1998, 1999, 2000, 2001, 2002 3.0L VIN D, 3.0L VIN E, 3.0L VIN F, 3.4L VIN N, 4.5L VIN J, 4.7L VIN T Transmissions: All	**Knock Sensor 1 Circuit Malfunction Conditions:** Engine started, vehicle driven with the engine speed over 2000 rpm, and the PCM detected an unexpected voltage condition on the Knock Sensor 1 (KS1) circuit during the CCM test. **Possible Causes** • Verify that the Knock Sensor (KS) is tightened to specification • Knock sensor signal circuit is open or shorted to ground • Knock sensor signal circuit is shorted to VREF or system power • Knock sensor is damaged or has failed • PCM has failed
DTC P0325 **1T CCM, MIL: Yes** 2003 All models All engines Transmissions: All	**Knock Sensor 1 Circuit Malfunction Conditions:** Engine started, vehicle driven with the engine speed over 2000 rpm, and the PCM did not detect a signal on the Knock Sensor 1 circuit. **Possible Causes** • Knock sensor signal circuit is open or shorted to ground • Knock sensor signal circuit is shorted to VREF or system power • Knock sensor is damaged, not tightened properly or has failed • PCM has failed
DTC P0326 **1T CCM, MIL: Yes** 1995 Land Cruiser All engines Transmissions: All	**Knock Sensor 2 Circuit Malfunction Conditions:** Engine started, engine running at 1200 rpm or higher, and the PCM detected an unexpected voltage condition on the Knock Sensor 2 (KS2 circuit during the CCM test. **Possible Causes** • Verify that the Knock Sensor (KS) is tightened to specification • Knock sensor signal circuit is open or shorted to ground • Knock sensor signal circuit is shorted to VREF or system power • Knock sensor is damaged or has failed • PCM has failed
DTC P0327 **1T CCM, MIL: Yes** 2003 Tacoma, Tundra 3.4L VIN N engine Transmissions: All	**Knock Sensor 1 Circuit Low Input (Bank 1) Conditions:** Engine started, engine speed from 1500-5500 rpm, and the PCM detected an unexpected low voltage on the Knock Sensor 1 circuit. **Possible Causes** • Verify that the Knock Sensor (KS) is tightened to specification • Knock sensor signal circuit is shorted to ground • Knock sensor is damaged or has failed • PCM has failed
DTC P0328 **1T CCM, MIL: Yes** 2003 Tacoma, Tundra 3.4L VIN N engine Transmissions: All	**Knock Sensor 1 Circuit High Input (Bank 1) Conditions:** Engine started, engine speed from 1500-5500 rpm, and the PCM detected an unexpected high voltage on the Knock Sensor 1 circuit. **Possible Causes** • Verify that the Knock Sensor (KS) is tightened to specification • Knock sensor signal circuit is open or shorted to power • Knock sensor is damaged or has failed • PCM has failed

OBD II Trouble Code List (P0xxx Codes)

DTC	Trouble Code Title, Conditions & Possible Causes
DTC P0330 **1T CCM, MIL: Yes** 1995 Avalon, Camry, Land Cruiser, Previa, T100, Tacoma, Tercel 3.0L VIN G, 3.4L VIN V, 4.5L VIN D engines Transmissions: All	**Knock Sensor 2 Circuit Malfunction Conditions:** Engine started, vehicle driven with the engine speed over 1200 rpm, and the PCM detected an unexpected voltage condition on the Knock Sensor 1 (KS1) circuit during the CCM test. **Possible Causes** • Verify that the Knock Sensor (KS) is tightened to specification • Knock sensor signal circuit is open or shorted to ground • Knock sensor signal circuit is shorted to VREF or system power • Knock sensor is damaged or has failed • PCM has failed
DTC P0330 **2T CCM, MIL: Yes** 1996, 1997, 1998, 1999, 2000, 2001, 2002 3.0L VIN D, 3.0L VIN E, 3.0L VIN F, 3.4L VIN N, 4.5L VIN J, 4.7L VIN T Transmissions: All	**Knock Sensor 2 Circuit Malfunction Conditions:** Engine started, vehicle driven with the engine speed over 2000 rpm, and the PCM detected an unexpected voltage condition on the Knock Sensor 1 (KS1) circuit during the CCM test. **Possible Causes** • Verify that the Knock Sensor (KS) is tightened to specification • Knock sensor signal circuit is open, shorted to ground or VREF • Knock sensor is damaged or has failed • PCM has failed
DTC P0330 **1T CCM, MIL: Yes** 2003 4Runner, Avalon, Camry, Land Cruiser, Sequoia, Tundra 3.0L VIN F, 4.0L VIN U, 4.7L VIN T engines Transmissions: All	**Knock Sensor 2 Circuit Malfunction Conditions:** Engine started, vehicle driven with the engine speed over 2000 rpm, and the PCM did not detect a signal on the Knock Sensor 2 circuit. **Possible Causes** • Knock sensor signal circuit is open, shorted to ground or VREF • Knock sensor is damaged, not tightened properly or has failed • PCM has failed
DTC P0332 **1T CCM, MIL: Yes** 2003 Tacoma, Tundra 3.4L VIN N engine Transmissions: All	**Knock Sensor 2 Circuit Low Input (Bank 2) Conditions:** Engine started, engine speed from 1500-5500 rpm, and the PCM detected an unexpected low voltage on the Knock Sensor 2 circuit. **Possible Causes** • Verify that the Knock Sensor (KS) is tightened to specification • Knock sensor signal circuit is shorted to ground • Knock sensor had failed, or the PCM has failed
DTC P0333 **1T CCM, MIL: Yes** 2003 Tacoma, Tundra 3.4L VIN N engine Transmissions: All	**Knock Sensor 2 Circuit High Input (Bank 2) Conditions:** Engine speed from 1500-5500 rpm, and the PCM detected an unexpected high voltage on the Knock Sensor 2 circuit. **Possible Causes** • Verify that the Knock Sensor (KS) is tightened to specification • Knock sensor signal circuit is open or shorted to power • Knock sensor has failed, or the PCM has failed
DTC P0335 **2T CCM, MIL: Yes** 1995 Avalon, Camry, Land Cruiser, Previa, T100, Tacoma, Tercel 1.5L VIN E, 2.4L VIN A, 2.4L VIN R, 2.7L VIN U, 3.0L VIN G, 3.4L VIN V, 4.5L VIN D engines Transmissions: All	**Crankshaft Position Sensor 'A' Circuit Malfunction Conditions:** Engine cranking; and the PCM did not detect any CKP Sensor 'A' signals, or with the engine speed over 600 rpm, it did not receive any CKP sensor signals, or the CKP signal was lost. **Possible Causes** • CKP Sensor 'A' signal circuit is open or shorted to ground • CKP Sensor 'A' signal ground circuit is open • CKP Sensor 'A' signal is shorted to VREF or system power • CKP Sensor 'A' is damaged or has failed • PCM has failed
DTC P0335 **1T CCM, MIL: Yes** 1996, 1997, 1998, 1999, 2000, 2001, 2002, 2003 All models All engines Transmissions: All	**Crankshaft Position Sensor 'A' Circuit Malfunction Conditions:** Engine cranking; and the PCM did not detect any CKP Sensor 'A' signals, or with the engine speed over 600 rpm, it did not receive any CKP sensor signals, or the CKP signal was lost. **Possible Causes** • CKP Sensor 'A' signal circuit is open, shorted to ground or shorted to system power • CKP Sensor 'A' signal ground circuit is open • CKP Sensor 'A' is damaged or has failed
DTC P0336 **2T CCM, MIL: Yes** 1996, 1997 4Runner, T100, Tacoma 2.7L VIN M engine Transmissions: All	**Crankshaft Position Sensor 'A' Range/Performance Conditions:** Engine running at idle or cruise speed for one minute, and the PCM detected a variation between the CKP Sensor and the CMP sensor signals. **Possible Causes** • Base engine "mechanical" problem (e.g., valve timing is wrong) • Distributor installation is incorrect • PCM has failed

OBD II Trouble Code List (P0xxx Codes)

DTC	Trouble Code Title, Conditions & Possible Causes
DTC P0339 **2T CCM, MIL: Yes** 2003 4Runner, Camry, Camry Solara, Land Cruiser, MR2, Sequoia, Tacoma, Tundra 1.8L VIN R, 2.4L VIN D, 2.4L VIN L, 3.4L VIN N, 4.0L VIN U, 4.7L VIN T Transmissions: All	**Crankshaft Position Sensor 'A' Circuit Intermittent Conditions:** Engine started, STA signal indicating "off", engine runtime over 3 seconds since STA switched from "on" to "off", engine speed over 1000 rpm, and the PCM did not detect any CKP Sensor 'A' signals for 500 ms. The crankshaft position (NE) sensor consists of a magnet, iron core and pickup coil. The NE sensor signal plate, which has 34 teeth, installed on the crankshaft-timing pulley. This sensor generates 34 signals for each engine revolution. The PCM detects the crankshaft angle and engine speed based on the NE signal. **Possible Causes** • CKP sensor signal circuit is open, shorted to ground or power • CKP Sensor signal ground circuit is open • Crankshaft timing pulley is damaged or out of alignment • CKP Sensor has failed, or the PCM has failed
DTC P0340 **2T CCM, MIL: Yes** 1995 Avalon, Camry, Land Cruiser, Previa, T100, Tacoma, Tercel 1.5L VIN E, 2.4L VIN A, 2.4L VIN R, 2.7L VIN U, 3.0L VIN G, 3.4L VIN V, 4.5L VIN D engines Transmissions: All	**Camshaft Position Sensor Circuit Malfunction Conditions:** Engine cranking; and the PCM did not detect any CMP sensor signals, or with the engine speed over 600 rpm, it did not detect any CMP signals or the CMP signal was interrupted. **Possible Causes** • CMP sensor signal circuit is open or shorted to ground • CMP sensor signal ground circuit is open • CMP sensor signal is shorted to VREF or system power • CMP sensor is damaged or has failed • PCM has failed
DTC P0340 **2T CCM, MIL: Yes** 1996, 1997, 1998, 1999, 2000, 2001, 2002 Other models All other engines Transmissions: All	**Camshaft Position Sensor Circuit Malfunction Conditions:** Engine cranking; and the PCM did not detect any CMP sensor signals, or with the engine speed over 600 rpm, it did not detect any CMP signals, or the CMP signal was interrupted. **Possible Causes** • CMP sensor signal circuit is open or shorted to ground • CMP sensor signal ground circuit is open • CMP sensor signal is shorted to VREF or system power • CMP sensor is damaged or has failed • PCM has failed • TSB EG010-02 (4/02) contains information related to this code
DTC P0340 **2T CCM, MIL: Yes** 2003 All models All engines Transmissions: All	**Camshaft Position Sensor Circuit Malfunction Conditions:** Engine cranking; and the PCM did not detect any CMP sensor signals, or with the engine speed over 600 rpm, it did not detect any CMP signals, or the CMP signal was interrupted. **Possible Causes** • CMP sensor signal circuit is open, shorted to ground or power • CMP sensor signal ground circuit is open • CMP sensor has failed, or the PCM has failed
DTC P0341 **2T CCM, MIL: Yes** 2003 All models All engines Transmissions: All	**Camshaft Position Sensor 'A' Signal Range/Performance Conditions:** Engine cranking; and the PCM detected twelve (12) or more Camshaft Position Sensor 'A' (Bank 1) signals during the test. **Possible Causes** • CMP sensor signal circuit is open, shorted to ground or power • CMP sensor pulley is damaged, or timing belt has jumped teeth • CMP sensor is damaged or has failed • PCM has failed
DTC P0345 **2T CCM, MIL: Yes** 2003 Avalon, Camry, Highlander, Sienna 3.0L VIN F engine Transmissions: All	**Camshaft Position Sensor 'A' Signal Range/Performance Conditions:** Engine cranking; and the PCM detected twelve (12) or more CMP Sensor 'A' (Bank 2) signals during the test. The Left Hand VVT Camshaft Position sensor consists of a magnet, and a circuit board in which a Magnetic Resistive (MR) device is mounted. The VVT signal plate includes three (3) protrusions on its outer surface. **Possible Causes** • VVT sensor signal circuit is open, shorted to ground or power • VVT sensor pulley is damaged, or timing belt has jumped teeth • VVT sensor has failed, or the PCM has failed
DTC P0346 **2T CCM, MIL: Yes** 2003 Avalon, Camry, Highlander, Sienna 3.0L VIN F engine Transmissions: All	**Camshaft Position Sensor 'A' Signal Range/Performance Conditions:** Engine cranking; and the PCM detected twelve (12) or more CMP Sensor 'A' (Bank 2) signals during the test. The Left Hand VVT Camshaft Position sensor consists of a magnet, and a circuit board in which a Magnetic Resistive (MR) device is mounted. The VVT signal plate includes three (3) protrusions on its outer surface. **Possible Causes** • VVT sensor signal circuit is open, shorted to ground or power • VVT sensor pulley is damaged, or timing belt has jumped teeth • VVT sensor had failed, or the PCM has failed

OBD II Trouble Code List (P0xxx Codes)

DTC	Trouble Code Title, Conditions & Possible Causes
DTC P0351 **1T CCM, MIL: Yes** 2003 4Runner, Avalon, Camry, Camry Solara, Echo, Highlander, Land Cruiser, Matrix, MR2, Sienna, Sequoia, Tacoma, Tundra 1.5L VIN T, 1.8L VIN R, 2.4L VIN D, 2.4L VIN L, 2.7L VIN M, 3.0L VIN F, 4.0L VIN U, 4.7L VIN T Transmissions: All	**Ignition Coil No. 1 Primary/Secondary Circuit Malfunction Conditions:** Engine started, and the PCM did not detect a change in the IGF signal on the Ignition Coil No. 1 IGF circuit. This engine uses a Direct Ignition (DI) system where one coil is used to fire one cylinder. The coil high-energy secondary wire is connected to one spark plug. If P0351 to P0356 are all set, check for an open/shorted IGF circuit. **Possible Causes** • IGT1 circuit is open or shorted to ground • Ignition Coil No. 1 is damaged or it has failed • Problem present in the Ignition System • PCM has failed
DTC P0352 **1T CCM, MIL: Yes** 2003 4Runner, Avalon, Camry, Camry Solara, Echo, Highlander, Land Cruiser, Matrix, MR2, Sienna, Sequoia, Tacoma, Tundra 1.5L VIN T, 1.8L VIN R, 2.4L VIN D, 2.4L VIN L, 2.7L VIN M, 3.0L VIN F, 4.0L VIN U, 4.7L VIN T Transmissions: All	**Ignition Coil No. 2 Primary/Secondary Circuit Malfunction Conditions:** Engine started, and the PCM did not detect a change in the IGF signal on the Ignition Coil No. 2 IGF circuit. This engine uses a Direct Ignition (DI) system where one coil is used to fire one cylinder. The coil high-energy secondary wire is connected to one spark plug. If P0351 to P0356 are all set, check for an open/shorted IGF circuit. **Possible Causes** • IGT2 circuit is open or shorted to ground • Ignition Coil No. 2 is damaged or it has failed • Problem present in the Ignition System • PCM has failed
DTC P0353 **1T CCM, MIL: Yes** 2003 4Runner, Avalon, Camry, Camry Solara, Echo, Highlander, Land Cruiser, Matrix, MR2, Sienna, Sequoia, Tacoma, Tundra 1.5L VIN T, 1.8L VIN R, 2.4L VIN D, 2.4L VIN L, 2.7L VIN M, 3.0L VIN F, 4.0L VIN U, 4.7L VIN T Transmissions: All	**Ignition Coil No. 3 Primary/Secondary Circuit Malfunction Conditions:** Engine started, and the PCM did not detect a change in the IGF signal on the Ignition Coil No. 3 IGF circuit. This engine uses a Direct Ignition (DI) system where one coil is used to fire one cylinder. The coil high-energy secondary wire is connected to one spark plug. If P0351 to P0356 are all set, check for an open/shorted IGF circuit. **Possible Causes** • IGT3 circuit is open or shorted to ground • Ignition Coil No. 3 is damaged or it has failed • Problem present in the Ignition System • PCM has failed
DTC P0354 **1T CCM, MIL: Yes** 2003 4Runner, Avalon, Camry, Camry Solara, Echo, Highlander, Land Cruiser, Matrix, MR2, Sienna, Sequoia, Tacoma, Tundra 1.5L VIN T, 1.8L VIN R, 2.4L VIN D, 2.4L VIN L, 2.7L VIN M, 3.0L VIN F, 4.0L VIN U, 4.7L VIN T Transmissions: All	**Ignition Coil No. 4 Primary/Secondary Circuit Malfunction Conditions:** Engine started, and the PCM did not detect a change in the IGF signal on the Ignition Coil No. 4 IGF circuit. This engine uses a Direct Ignition (DI) system where one coil is used to fire one cylinder. The coil high-energy secondary wire is connected to one spark plug. If P0351 to P0356 are all set, check for an open/shorted IGF circuit. **Possible Causes** • IGT4 circuit is open or shorted to ground • Ignition Coil No. 4 is damaged or it has failed • Problem present in the Ignition System • PCM has failed
DTC P0355 **1T CCM, MIL: Yes** 2003 4Runner, Avalon, Camry, Camry Solara, Highlander, Land Cruiser, Sequoia, Sienna, Tundra 3.0L VIN F, 4.0L VIN U, 4.7L VIN T engines Transmissions: All	**Ignition Coil No. 5 Primary/Secondary Circuit Malfunction Conditions:** Engine started, and the PCM did not detect a change in the IGF signal on the Ignition Coil No. 5 IGF circuit. This engine uses a Direct Ignition (DI) system where one coil is used to fire one cylinder. The coil high-energy secondary wire is connected to one spark plug. If P0351 to P0356 are all set, check for an open/shorted IGF circuit. **Possible Causes** • IGT5 circuit is open or shorted to ground • Ignition Coil No. 5 is damaged or it has failed • Problem present in the Ignition System • PCM has failed

OBD II Trouble Code List (P0xxx Codes)

DTC	Trouble Code Title, Conditions & Possible Causes
DTC P0356 **1T CCM, MIL: Yes** 2003 4Runner, Avalon, Camry, Camry Solara, Highlander, Land Cruiser, Sequoia, Sienna, Tundra 3.0L VIN F, 4.0L VIN U, 4.7L VIN T engines Transmissions: All	**Ignition Coil No. 6 Primary/Secondary Circuit Malfunction Conditions:** Engine started, and the PCM did not detect a change in the IGF signal on the Ignition Coil No. 6 IGF circuit. This engine uses a Direct Ignition (DI) system where one coil is used to fire one cylinder. The coil high-energy secondary wire is connected to one spark plug. If P0351 to P0356 are all set, check for an open/shorted IGF circuit. **Possible Causes** • IGT6 circuit is open or shorted to ground • Ignition Coil No. 6 is damaged or it has failed • Problem present in the Ignition System • PCM has failed
DTC P0357 **1T CCM, MIL: Yes** 2003 4Runner, Land Cruiser, Sequoia, Tundra Engines: 4.7L VIN T Transmissions: All	**Ignition Coil No. 7 Primary/Secondary Circuit Malfunction Conditions:** Engine started, and the PCM did not detect a change in the IGF signal on the Ignition Coil No. 7 IGF circuit. This engine uses a Direct Ignition (DI) system where one coil is used to fire one cylinder. The coil high-energy secondary wire is connected to one spark plug. If P0351 to P0358 are all set, check for an open/shorted IGF circuit. **Possible Causes** • IGT7 circuit is open or shorted to ground • Ignition Coil No. 7 is damaged or it has failed • Problem present in the Ignition System • PCM has failed
DTC P0358 **1T CCM, MIL: Yes** 2003 4Runner, Land Cruiser, Sequoia, Tundra Engines: 4.7L VIN T Transmissions: All	**Ignition Coil No. 8 Primary/Secondary Circuit Malfunction Conditions:** Engine started, and the PCM did not detect a change in the IGF signal on the Ignition Coil No. 8 IGF circuit. This engine uses a Direct Ignition (DI) system where one coil is used to fire one cylinder. The coil high-energy secondary wire is connected to one spark plug. If P0351 to P0358 are all set, check for an open/shorted IGF circuit. **Possible Causes** • IGT8 circuit is open or shorted to ground • Ignition Coil No. 8 is damaged or it has failed • Problem present in the Ignition System • PCM has failed
DTC P0385 **2T CCM, MIL: Yes** 1996, 1997 Supra 3.0L VIN D engine Transmissions: All	**Crankshaft Position Sensor 'B' Circuit Malfunction Conditions:** Engine started, engine speed more than 600 rpm, it did not detect any Crankshaft Position (CKP) Sensor 'B' signals during the test. **Possible Causes** • CKP Sensor 'B' signal circuit is open, shorted to ground or shorted to system power • CKP Sensor 'B' signal ground circuit is open • CKP Sensor 'B' has failed, or the PCM has failed
DTC P0401 **2T EGR, MIL: Yes** 1995 Avalon, Camry, Land Cruiser, Previa, T100, Tacoma, Tercel 1.5L VIN E, 2.4L VIN A, 2.4L VIN R, 2.7L VIN U, 3.0L VIN G, 3.4L VIN V, 4.5L VIN D engines Transmissions: All	**EGR System Insufficient Flow Detected Conditions:** Engine started, vehicle driven at a speed of over 50 mph with the engine running in closed loop at a steady throttle for 3-5 minutes, and the PCM detected the EGR gas temperature signal was not 106-140°F higher than the ambient air temperature signal in the test. **Possible Causes** • EGR gas temperature sensor circuit open or shorted to VREF • EGR gas temperature sensor is damaged or has failed • EGR valve is stuck in closed position • EGR valve vacuum hose is disconnected or leaking • PCM has failed
DTC P0401 **2T EGR, MIL: Yes** 1996, 1997, 1998, 1999, 2000, 2001, 2002 All models All engines Transmissions: All	**EGR System Insufficient Flow Detected Conditions:** Engine started, vehicle driven at a speed of over 50 mph with the engine running in closed loop at a steady throttle for 3-5 minutes, and the PCM detected the EGR sensor signal was not 106-140°F higher than the ambient air temperature in the EGR Monitor test. **Possible Causes** • EGR gas temperature sensor circuit open or shorted to VREF • EGR gas temperature sensor is damaged or has failed • EGR valve is stuck in closed position • EGR valve vacuum hose is disconnected or leaking
DTC P0401 **2T EGR, MIL: Yes** 2003 Tacoma 2.4L VIN L, 2.7L VIN M Transmissions: All	**EGR System Insufficient Flow Detected Conditions:** Engine started, vehicle driven at a speed of over 50 mph with the engine running in closed loop at a steady throttle for 3-5 minutes, and the PCM detected the EGR sensor signal was not 106-140°F higher than the ambient air temperature in the EGR Monitor test. **Possible Causes** • EGR gas temperature sensor circuit open or shorted to VREF • EGR gas temperature sensor is damaged or has failed • EGR valve is stuck in closed position • EGR valve vacuum hose is disconnected or leaking

OBD II Trouble Code List (P0xxx Codes)

DTC	Trouble Code Title, Conditions & Possible Causes
DTC P0402 **2T EGR, MIL: Yes** 1995 Avalon, Camry, Land Cruiser, Previa, T100, Tacoma, Tercel 1.5L VIN E, 2.4L VIN A, 2.4L VIN R, 2.7L VIN U, 3.0L VIN G, 3.4L VIN V, 4.5L VIN D engines Transmissions: All	**Excessive EGR Flow Detected Conditions:** Engine started cold (ECT sensor signal less than 86ºF), engine running without load at less than 4000 rpm, and the PCM detected the EGR sensor indicated a high value during the EGR Cutoff Test, or it detected the EGR valve was open during all driving conditions. **Possible Causes** • EGR gas temperature sensor circuit open or shorted to ground • EGR gas temperature sensor is damaged or has failed • EGR Vacuum Switching Valve (VSV) is damaged or has failed • EGR valve stuck partially or EGR valve stuck fully open • EGR VSV control circuit is open or shorted to system power • PCM has failed
DTC P0402 **2T EGR, MIL: Yes** 1996, 1997, 1998, 1999, 2000, 2001, 2002 All other models All engines Transmissions: All	**Excessive EGR Flow Detected Conditions:** Engine started cold (ECT sensor signal less than 86ºF), engine running without load at less than 4000 rpm, and the PCM detected the EGR sensor indicated a high value during the EGR Cutoff test; or it detected the EGR valve was open during all driving conditions. **Possible Causes** • EGR gas temperature sensor circuit open or shorted to ground • EGR gas temperature sensor is damaged or has failed • EGR Vacuum Switching Valve (VSV) is damaged or has failed • EGR valve stuck partially or EGR valve stuck fully open • EGR VSV control circuit is open or shorted to system power • PCM has failed
DTC P0402 **2T EGR, MIL: Yes** 2003 Tacoma 2.4L VIN L, 2.7L VIN M engines Transmissions: All	**Excessive EGR Flow Detected Conditions:** Engine started cold (ECT sensor signal less than 86ºF), engine running without load at less than 4000 rpm, and the PCM detected the EGR sensor indicated a high value during the EGR Cutoff test; or it detected the EGR valve was open during all driving conditions. **Possible Causes** • EGR gas temperature sensor circuit open or shorted to ground • EGR gas temperature sensor is damaged or has failed • EGR Vacuum Switching Valve (VSV) is damaged or has failed • EGR valve stuck partially or EGR valve stuck fully open • EGR VSV control circuit is open or shorted to system power • PCM has failed
DTC P0420 **2T CAT, MIL: Yes** 1995 Avalon, Camry, Land Cruiser, Previa, T100, Tacoma, Tercel 1.5L VIN E, 2.4L VIN A, 2.4L VIN R, 2.7L VIN U, 3.0L VIN G, 3.4L VIN V, 4.5L VIN D engines Transmissions: All	**Catalyst Efficiency Below Normal (Bank 1) Conditions:** DTC P0100, P0101, P0105, P0110, P0115, P0120, P0121, P0335, P0340 and P0500 not set, vehicle driven to a speed of 45-60 mph at 2500-3000 rpm in closed loop for 3-5 minutes, and the PCM detected the voltage amplitudes of the rear and front HO2S were too similar. **Possible Causes** • Air leaks at the exhaust manifold or in the exhaust pipes • Catalytic converter is damaged, contaminated or has failed • Front HO2S or rear HO2S is contaminated with fuel or moisture • Front HO2S or the rear HO2S is loose in its mounting hole • Front HO2S is older (aged) than the rear HO2S (HO2S is lazy)
DTC P0420 **2T CAT, MIL: Yes** 1996, 1997, 1998, 1999, 2000, 2001, 2002 All models All engines Transmissions: Al	**Catalyst Efficiency Below Normal (Bank 1) Conditions:** DTC P0100, P0101, P0105, P0110, P0115, P0120, P0121, P0335, P0340 and P0500 not set, engine started, vehicle driven to a speed of 45-60 mph at 2500-3000 rpm in closed loop for 3-5 minutes, and the PCM detected the voltage amplitudes of the rear HO2S-12 and front HO2S-11 were similar during the Catalyst Monitor test. **Possible Causes** • Air leaks at the exhaust manifold or in the exhaust pipes • Catalytic converter is damaged, contaminated or has failed • Front HO2S or rear HO2S is contaminated with fuel or moisture • Front HO2S or the rear HO2S is loose in its mounting hole • Front HO2S is older (aged) than the rear HO2S (HO2S is lazy)
DTC P0420 **2T CAT, MIL: Yes** 2003 All models All engines Transmissions: All	**Catalyst Efficiency Below Normal (Bank 1) Conditions:** DTC P0100, P0101, P0102, P0103, P0110, P0112, P0113, P0115, P0116, P0117, P0118, P0120, P0121, P0122, P0123, P0335, P0340 and P0500 not set, engine started, vehicle driven to a speed of 45-60 mph at 2500-3000 rpm in closed loop for 3-5 minutes, and the PCM detected too much variation in the voltage amplitudes of the HO2S-12 signal (Bank 1). **Possible Causes** • Catalytic converter is damaged, contaminated or has failed • Front A/FS or rear HO2S is contaminated with fuel or moisture • Front A/FS or the rear HO2S is loose in its mounting hole • Front A/FS is older (aged) than the rear HO2S (HO2S is lazy) • Gas leaks at the exhaust manifold or in the exhaust pipes

OBD II Trouble Code List (P0xxx Codes)

DTC	Trouble Code Title, Conditions & Possible Causes
DTC P0430 **2T CAT, MIL: Yes** 1996, 1997, 1998, 1999, 2000, 2001, 2002 Land Cruiser, Sequoia, Tundra Engines: 4.7L VIN T Transmissions: All	**Catalyst Efficiency Below Normal (Bank 2) Conditions:** DTC P0100, P0101, P0105, P0110, P0115, P0120, P0121, P0335, P0340 and P0500 not set, vehicle speed from 45-60 mph at 2500-3000 rpm in closed loop for 3-5 minutes, and the PCM detected the voltage amplitudes of the rear HO2S-22 and front HO2S-21 were similar. **Possible Causes** • Air leaks at the exhaust manifold or in the exhaust pipes • Catalytic converter is damaged, contaminated or has failed • Front HO2S or rear HO2S is contaminated with fuel or moisture • Front HO2S or the rear HO2S is loose in its mounting hole • Front HO2S is older (aged) than the rear HO2S (HO2S is lazy)
DTC P0430 **2T CAT, MIL: Yes** 2003 4Runner, Avalon, Camry, Land Cruiser, RAV4, Sequoia, Tundra 2.0L VIN K, 3.0L VIN F, 4.0L VIN U, 4.7L VIN T Transmissions: All	**Catalyst Efficiency Below Normal (Bank 2) Conditions:** DTC P0100, P0101, P0102, P0103, P0110, P0112, P0113, P0115, P0116, P0117, P0118, P0120, P0121, P0122, P0123, P0335, P0340 and P0500 not set, engine started, vehicle driven to a speed of 45-60 mph at 2500-3000 rpm in closed loop for 3-5 minutes, and the PCM detected too much variation in the voltage amplitudes of the Rear HO2S-12 for Bank 2 during the Catalyst Monitor test. **Possible Causes** • Catalytic converter is damaged, contaminated or has failed • Front A/FS or rear HO2S is contaminated with fuel or moisture • Front A/FS or the rear HO2S is loose in its mounting hole • Front A/FS is older (aged) than the rear HO2S (HO2S is lazy) • Gas leaks at the exhaust manifold or in the exhaust pipes
DTC P0440 **2T EVAP, MIL: Yes** 1996, 1997, 1998, 1999, 2000, 2001, 2002 All models All engines Transmissions: All	**EVAP Control System Large Leak (0.080") Detected Conditions:** Engine started cold (ECT sensor signal less 86°F), engine runtime over 20 minutes in closed loop, vehicle driven to a speed of 55-60 mph, and the PCM detected the fuel tank pressure indicated the same value as the atmospheric pressure in the EVAP Monitor test. **Possible Causes** • Charcoal canister is clogged, loaded with fuel or with moisture • Fuel filler cap missing, loose (not tightened) or the wrong part • Fuel tank over-fill check valve cracked or damaged • Fuel tank seal leaking, fuel tank cracked or damaged/leaking • Fuel vapor hoses/tubes blocked or restricted, or fuel vapor control valve tube or fuel vapor vent valve assembly blocked • Vacuum hose or tubing cracked, damaged or disconnected • Vapor Pressure sensor incorrectly installed • PCM has failed
DTC P0440 **2T EVAP, MIL: Yes** 2003 Celica, Corolla, Matrix, RAV4 1.8L VIN R, 1.8L VIN Y, 2.0L VIN K engines Transmissions: All	**EVAP Control System Large Leak (0.080") Detected Conditions:** Engine started cold (ECT sensor signal less 86°F), engine runtime over 20 minutes in closed loop, vehicle driven to a speed of 55-60 mph, and the PCM detected the fuel tank pressure indicated the same value as the atmospheric pressure in the EVAP Monitor test. **Possible Causes** • Charcoal canister is clogged, loaded with fuel or with moisture • Fuel filler cap missing, loose (not tightened) or the wrong part • Fuel tank over-fill check valve cracked or damaged • Fuel tank seal leaking, fuel tank cracked or damaged/leaking • Fuel vapor hoses/tubes blocked or restricted, or fuel vapor control valve tube or fuel vapor vent valve assembly blocked • Vacuum hose or tubing cracked, damaged or disconnected • Vapor Pressure sensor incorrectly installed • PCM has failed
DTC P0441 **2T EVAP, MIL: Yes** 1995 Avalon, Camry, Land Cruiser, Previa, T100, Tacoma, Tercel 1.5L VIN E, 2.4L VIN A, 2.4L VIN R, 2.7L VIN U, 3.0L VIN G, 3.4L VIN V, 4.5L VIN D engines Transmissions: All	**EVAP System Incorrect Purge Flow Detected Conditions:** Cold engine startup requirement met (ECT sensor signal less 86°F), engine running at cruise speed for over 3 minutes, VSS input from 55-60 mph, and the PCM detected the canister pressure did not decrease during purge or it remained too low during purge cutoff conditions. **Possible Causes** • Charcoal canister is clogged, loaded with fuel or with moisture • Fuel tank over-fill check valve cracked or damaged • Fuel tank seal leaking, fuel tank cracked or damaged/leaking • Fuel vapor hoses/tubes blocked or restricted, or fuel vapor control valve tube or fuel vapor vent valve assembly blocked • Vacuum hose or tubing cracked, damaged or disconnected • VSV for the canister purge solenoid is open or shorted to ground, or the purge solenoid is damaged or has failed (closed) • VSV for vapor pressure sensor circuit is open, shorted to ground, or sensor has failed • PCM has failed

OBD II Trouble Code List (P0xxx Codes)

DTC	Trouble Code Title, Conditions & Possible Causes
DTC P0441 **2T EVAP, MIL: Yes** 1996, 1997, 1998, 1999, 2000, 2001, 2002, 2003 All models All engines Transmissions: All	**EVAP System Incorrect Purge Flow Detected Conditions:** ECT sensor less than 86ºF at startup, vehicle driven at 55-60 mph for 2-3 minutes, and the PCM detected the EVAP canister pressure did not decrease during purge conditions, or it remained too low during purge cutoff conditions. The vapor pressure sensor, VSV for the canister closed valve (CCV) and the VSV for the vapor-switching valve are used to detect EVAP system faults. The PCM closes the CCV and opens the VSV for vapor switching valve to cause an increase in vacuum in the EVAP system. Once the vacuum reaches a certain point, the PCM closes the VSV to test system operation. **Possible Causes** • Charcoal canister is clogged, loaded with fuel or with moisture • Fuel tank over-fill check valve cracked or damaged • Fuel tank seal leaking, fuel tank cracked or damaged/leaking • Fuel vapor hoses/tubes blocked or restricted, or fuel vapor control valve tube or fuel vapor vent valve assembly blocked • Vacuum hose or tubing cracked, damaged or disconnected • Vapor pressure sensor is damaged or has failed • VSV circuit for the canister purge, VSV for the CCV or the VSV for the pressure switching valve is open or shorted to ground • VSV for the vapor pressure sensor circuit is open or shorted to ground, or the vapor pressure sensor is damaged or has failed
DTC P0442 **2T EVAP, MIL: Yes** 1998, 1999, 2000, 2001, 2002, 2003 All models All engines Transmissions: All	**EVAP System Small Leak (0.040") Detected Conditions:** Engine started; IAT sensor signal from 39-86ºF, fuel tank level from 25-75% for 10 seconds, and the PCM detected the EVAP system was unable to hold a specified vacuum level for a set period of time. After the system is purged, the PCM shuts off the VSV for the purge valve to seal the vacuum in the system, and then monitors the increase in pressure in the system. The pressure should increases slowly. If it increases at too fast a rate, then this code is set. **Possible Causes** • Canister Purge valve is damaged, leaking or has failed • Charcoal canister is loaded with fuel or moisture • Fuel filler cap loose, cross-threaded, incorrect part or damaged • Fuel tank is cracked (leaking), or a leak exists in the 'O' ring • Fuel tank pressure sensor is damaged or has failed • Fuel vapor line(s), fuel pipes or hoses damaged or leaking • PCM has failed
DTC P0442 **2T EVAP, MIL: Yes** 1996, 1997, 1998, 1999, 2000, 2001, 2002, 2003 All models All engines Transmissions: All	**EVAP Vent Control Solenoid Circuit Malfunction Conditions:** Engine started, engine running at cruise speed under light load conditions, VSV for vapor pressure switching valve "on", and the PCM detected a lack of vacuum continuity between the vapor pressure sensor, the charcoal canister and the fuel tank; or with the VSV for the vapor pressure switching valve "off", it detected the pressure in the fuel tank remained at atmospheric pressure; or with the VSV for the CCV "on", it detected the pressure in the charcoal canister and the fuel tank remained near atmospheric pressure. The PCM closes the VSV for the vapor-switching valve, and this action blocks any air from entering the fuel tank side of the system. The pressure rise on the fuel tank under these conditions is minimal. If there was no change in pressure, the PCM determines the VSV for the vapor-switching valve did not close, and it will set this code. **Possible Causes** • Charcoal canister is clogged, loaded with fuel or with moisture • Fuel tank over-fill check valve cracked or damaged • Fuel tank seal leaking, fuel tank cracked, damaged or leaking • Fuel vapor hoses/tubes blocked or restricted, or fuel vapor control valve tube or fuel vapor vent valve assembly blocked • Vacuum hose or tubing cracked, damaged or disconnected • Vapor pressure sensor is damaged or has failed • VSV circuit for the canister purge, VSV for the CCV or the VSV for the pressure switching valve is open or shorted to ground • VSV for the vapor pressure sensor circuit is open or shorted to ground, or the vapor pressure sensor is damaged or has failed

OBD II Trouble Code List (P0xxx Codes)

DTC	Trouble Code Title, Conditions & Possible Causes
DTC P0450 **2T CCM, MIL: Yes** 1996, 1997, 1998, 1999, 2000, 2001, 2002, 2003 All models All engines Transmissions: All	**EVAP Vapor Pressure Sensor Circuit Malfunction Conditions:** Engine started, engine runtime less than 10 seconds since startup, and the PCM detected the Vapor Pressure sensor value was less than -3.5 kPa (-1.0 in. Hg), or the Vapor Pressure sensor was more than or equal to 1.5 kPa (0.4 in. Hg) during testing. The PCM uses the Vapor Pressure Sensor, VSV for the Canister Closed valve and VSV for the Pressure Switching valve to find faults in this system. **Possible Causes** • Vapor pressure sensor signal circuit open or shorted to ground • Vapor pressure sensor ground circuit is open • Vapor pressure sensor power circuit is open • Vapor pressure sensor is damaged or has failed • PCM has failed
DTC P0451 **2T CCM, MIL: Yes** 1996, 1997, 1998, 1999, 2000, 2001, 2002, 2003 All models All engines Transmissions: All	**EVAP Vapor Pressure Sensor Range/Performance Conditions:** Engine started, engine at idle speed with the VSS indicating 0 mph, VSV for Vapor Switching valve "off", and the PCM detected too much change in the pressure sensor value, or the pressure sensor value equaled the opening value of the charcoal canister. The PCM uses the Vapor Pressure Sensor, VSV for the Canister Closed valve and VSV for Pressure Switching valve to find faults in the system. **Possible Causes** • Vapor pressure sensor vacuum hoses loose or damaged • Vapor pressure sensor is damaged or has failed • PCM has failed
DTC P0452 **2T CCM, MIL: Yes** 2003 4Runner, Echo, Land Cruiser, Sequoia, Tacoma, Tundra All engines Transmissions: All	**EVAP Vapor Pressure Sensor Circuit Low Input Conditions:** Engine started; VSV for vapor pressure switching valve "off"; and the PCM detected an unexpected low voltage condition on the vapor pressure sensor circuit during the CCM test. **Possible Causes** • Vapor pressure sensor connector is damaged or open • Vapor pressure sensor circuit is open • Vapor pressure sensor is damaged or has failed • PCM has failed
DTC P0453 **2T CCM, MIL: Yes** 2003 4Runner, Echo, Land Cruiser, Sequoia, Tacoma, Tundra All engines Transmissions: All	**EVAP Vapor Pressure Sensor Circuit High Input Conditions:** Engine started; VSV for vapor pressure switching valve "off"; and the PCM detected an unexpected high voltage condition on the vapor pressure sensor circuit during the CCM test. **Possible Causes** • Vapor pressure sensor connector is damaged or shorted • Vapor pressure sensor circuit is shorted to VREF • Vapor pressure sensor is damaged or has failed • PCM has failed
DTC P0456 **2T EVAP, MIL: Yes** 2003 4Runner, Camry, Camry Solara, Echo, Land Cruiser, Matrix, MR2, RAV4, Sequoia, Sienna, Tacoma, Tundra All engines Transmissions: All	**EVAP System Very Small Leak (0.020") Detected Conditions:** Engine started; IAT sensor signal from 39-86°F, fuel tank level from 25-75% for 10 seconds, and the PCM detected the EVAP system was unable to hold a specified vacuum level for a set period of time. After the system is purged, the PCM shuts off the VSV for the purge valve to seal the vacuum in the system, and then monitors the increase in pressure in the system. The pressure should increase slowly. If it increases at too fast a rate, this code is set. **Possible Causes** • Canister Purge valve is damaged, leaking or has failed • Charcoal canister is loaded with fuel or moisture • Fuel filler cap loose, cross-threaded, incorrect part or damaged • Fuel tank is cracked (leaking), or a leak exists in the 'O' ring • Fuel tank pressure sensor is damaged or has failed • Fuel tank overfill check valve is cracked or is damaged • Fuel vapor line(s), fuel pipes or hoses damaged or leaking • PCM has failed
DTC P0500 **2T EVAP, MIL: Yes** 1995 Avalon, Camry, Land Cruiser, Previa, T100, Tacoma, Tercel 1.5L VIN E, 2.4L VIN A, 2.4L VIN R, 2.7L VIN U, 3.0L VIN G, 3.4L VIN V, 4.5L VIN D engines Transmissions: A/T	**Vehicle Speed Sensor Circuit Malfunction Conditions:** Engine started, ECT sensor more than 158°F, vehicle driven with the engine speed from 1500-5500 rpm, and the PCM did not receive any VSS signals from the Combination Meter during the CCM test. **Possible Causes** • VSS signal circuit is open between the meter and the PCM • VSS signal circuit shorted to ground between meter and PCM • VSS signal circuit shorted to VREF or system power • Combination Meter is damaged or has failed • PCM has failed

OBD II Trouble Code List (P0xxx Codes)

DTC	Trouble Code Title, Conditions & Possible Causes
DTC P0500 **1T CCM, MIL: Yes** 1996 All models All engines Transmissions: A/T	**Vehicle Speed Sensor Circuit Malfunction Conditions:** Engine started, ECT sensor more than 158ºF, P/N switch indicating "off", vehicle driven at an engine speed of 1500-3500 rpm, and the PCM did not receive any VSS signals from the Combination Meter. **Possible Causes** • VSS signal circuit is open between the meter and the PCM • VSS signal circuit shorted to ground between meter and PCM • VSS signal circuit shorted to VREF or system power • Combination Meter is damaged or has failed • PCM has failed
DTC P0500 **1T CCM, MIL: Yes** 1997, 1998, 1999, 2000, 2001, 2002 All models All engines Transmissions: A/T	**Vehicle Speed Sensor Circuit Malfunction Conditions:** Engine started, engine speed more than 2350 rpm, P/N switch indicating "off" for 1 second, throttle angle equal to or less than 13 degrees (º), and the PCM did not detect any VSS signals during the test period (Scan Tool PID ECT = Electronic Controlled Transaxle). **Possible Causes** • VSS signal circuit is open, shorted to ground or to power (B+) • Combination Meter is damaged or has failed • PCM has failed • TSB EG004-02 (2/02) contains information related to this code
DTC P0500 **1T CCM, MIL: Yes** 1996, 1997, 1998, 1999, 2000, 2001, 2002 All All engines Transmissions: M/T	**Vehicle Speed Sensor Circuit Malfunction Conditions:** Engine started, ECT sensor signal over 158ºF, vehicle driven with the engine speed from 1500-5500 rpm, and the PCM did not receive any VSS signals from the Combination Meter. **Possible Causes** • VSS signal circuit is open between the meter and the PCM • VSS signal circuit shorted to ground or to VREF or power (B+) • Combination Meter is damaged or has failed • PCM has failed
DTC P0500 **2T CCM, MIL: Yes** 2003 4Runner, Camry, Camry Solara, Celica, Corolla, Echo, Land Cruiser, Matrix, MR2, RAV4, Sequoia, Tacoma, Tundra 1.5L VIN T, 1.8L VIN R, 1.8L VIN Y, 2.0L VIN K, 2.4L VIN D, 2.4L VIN L, 2.7L VIN M, 3.4L VIN N, 4.0L VIN U, 4.7L VIN T Transmissions: All	**Vehicle Speed Sensor Circuit Malfunction Conditions:** Engine started, vehicle driven with the engine speed from 1500-5500 rpm and back to idle several times, and the PCM did not receive any VSS signals during the test. The VSS (No.1) assembly outputs a 4-pulse signal for every revolution of the rotor shaft, which is generated by the transmission output shaft via the driven gear. **Possible Causes** • VSS signal circuit is open between the meter and the PCM • VSS signal circuit shorted to ground between meter and PCM • VSS No. 1 is damaged or has failed • Combination Meter is damaged or has failed • PCM has failed
DTC P0500 **1T CCM, MIL: Yes** 2003 Avalon, Camry, Highlander, Sienna 3.0L VIN F engine Transmissions: A/T	**Vehicle Speed Sensor Circuit Malfunction Conditions:** No TP sensor codes set, engine runtime 2 seconds with the ECT sensor more than 132ºF and IAT sensor more than 50ºF, P/N switch indicating 'P' or 'N', TP angle less than 13º with the engine speed less than 2350 rpm; or TP angle less than 21º with the engine speed less than 2680 rpm; or TP angle less than 30º with the engine speed less than 2835 rpm; or TP angle less than 30º with the engine speed less than 3250 rpm; and the PCM detected the engine speed was equal or more than the VSS signal speed for 500 ms during testing. **Possible Causes** • ABS speed sensor connector is damaged or open • ABS speed sensor circuit is open or shorted to ground • ABS speed sensor is damaged or has failed • Combination Meter is damaged or has failed • PCM or the ABS controller has failed
DTC P0503 **2T CCM, MIL: Yes** 2003 4Runner, Land Cruiser, MR2, Sequoia, Tundra All engines Transmissions: All	**Vehicle Speed Sensor 'A' Signal Erratic Or Intermittent Conditions:** Engine started, vehicle driven through several transmission shifts and braking events; and the PCM detected an interruption in the VSS signal from the Combination Meter, or the signal was too high. The speed sensor for the ABS detects the wheel speeds and sends the signals to the ABS ECU. The ECU converts these signals into a 4-pulse signal and outputs this signal to the Combination Meter. **Possible Causes** • ABS ECU is damaged or has failed • Combination Meter is damaged or has failed • VSS signal circuit is open between the meter and the PCM • VSS signal circuit shorted to ground between meter and PCM

OBD II Trouble Code List (P0xxx Codes)

DTC	Trouble Code Title, Conditions & Possible Causes
DTC P0504 **2T CCM, MIL: Yes** 2003 4Runner, Land Cruiser, MR2, Sequoia, Tundra All engines Transmissions: All	**Brake Switch 'A' To 'B' Correlation Malfunction Conditions:** Key on or engine running; brake pedal released, and the PCM detected the STP signal indicated "off" while the ST1 signal also indicated "off". The stoplight switch signal is used to prevent the engine from stalling when the brakes are applied suddenly. The stoplight switch uses a duplex system (STP and ST1 signals). **Possible Causes** • Stoplight switch signal circuit is shorted to power • Stoplight switch assembly is damaged or shorted • PCM has failed
DTC P0505 **2T CCM, MIL: Yes** 1995 Avalon, Camry, Land Cruiser, Previa, T100, Tacoma, Tercel 1.5L VIN E, 2.4L VIN A, 2.4L VIN R, 2.7L VIN U, 3.0L VIN G, 3.4L VIN V, 4.5L VIN D engines Transmissions: All	**Idle Control System Malfunction Conditions:** Engine started, engine running at idle speed in closed loop, and the PCM detected the Actual Idle Speed was more than 100-200 rpm above or below the Target Idle Speed. Note: RSO is the acronym for Rotary Solenoid (IAC) Open Coil and RSC is the acronym for the Rotary Solenoid (IAC) Close Coil. **Possible Causes** • RSC or RSO control circuit is open, shorted to ground or power • RSO or RSO power circuit is open (test power from EFI relay) • IAC valve is contaminated, damaged or has failed
DTC P0505 **2T CCM, MIL: Yes** 1996, 1997, 1998, 1999, 2000, 2001, 2002 All models All engines Transmissions: All	**Idle Control System Malfunction Conditions:** Engine started, engine running at idle speed in closed loop, and the PCM detected the Actual Idle Speed was more than 100-200 rpm above or below the Target Idle Speed. Note: RSO is the acronym for Rotary Solenoid (IAC) Open Coil and RSC is the acronym for the Rotary Solenoid (IAC) Close Coil. **Possible Causes** • RSC/RSO connector is damaged, open or shorted • RSC or RSO control circuit is open, shorted to ground or power • RSO or RSO power circuit is open (test power from EFI relay) • IAC valve is contaminated
DTC P0505 **2T CCM, MIL: Yes** 2003 4Runner, Avalon, Camry, Camry, Solara, Celica, Corolla, Echo, Land Cruiser, Matrix, MR2, RAV4, Sequoia, Tacoma, Tundra 1.5L VIN T, 1.8L VIN R, 1.8L VIN Y, 2.0L VIN K, 2.4L VIN D, 2.4L VIN L, 2.7L VIN M, 3.0L VIN F, 4.0L VIN U, 4.7L VIN T Transmissions: All	**Idle Control System Malfunction Conditions:** Engine started, engine running at idle speed n closed loop, and the PCM detected the Actual Idle Speed was more than 100-200 rpm above or below the Target Idle Speed. A Rotary solenoid type of ISC valve is located in front of the air intake chamber and intake air bypassing the throttle valve is directed to the Intake Air Control (IAC) valve via a passage. The PCM controls the idle speed by regulating the amount of intake air volume that bypasses the throttle valve. **Possible Causes** • Air Induction system leaks (check for intake manifold leaks) • Air leaks in the PCV system (at the valve or its related hoses) • Throttle body assembly is damaged or has failed • PCM has failed
DTC P0510 **2T CCM, MIL: Yes** 1995 Avalon, Camry, Land Cruiser, Previa, T100, Tacoma, Tercel 1.5L VIN E, 2.4L VIN A, 2.4L VIN R, 2.7L VIN U, 3.0L VIN G, 3.4L VIN V, 4.5L VIN D engines Transmissions: All	**Closed Throttle Position Switch Circuit Malfunction Conditions:** Engine started, and the PCM detected throttle switch input did not change from Off to On during a normal driving period. Refer to Circuit Test or code repair chart to test code. **Possible Causes** • Closed throttle position switch signal circuit is open or grounded • Closed throttle position switch signal circuit is shorted to power • Closed throttle position switch or TP sensor damaged or failed • PCM has failed
DTC P0510 **2T CCM, MIL: Yes** 1996, 1997 All models All engines Transmissions: All	**Closed Throttle Position Switch Circuit Malfunction Conditions:** Engine started, vehicle driven with VSS signals received, and the PCM did not detect any change in the Closed Throttle Position switch status (from off to on, or from on to off). **Possible Causes** • Closed throttle position switch signal circuit is open or grounded • Closed throttle position switch signal circuit is shorted to power • Closed throttle position switch or TP sensor damaged or failed • PCM has failed

OBD II Trouble Code List (P0xxx Codes)

DTC	Trouble Code Title, Conditions & Possible Causes
DTC P0511 **2T CCM, MIL: Yes** 2003 Avalon, Camry, Echo, Highlander, Matrix, MR2, Sienna, Tacoma 1.5L VIN T, 1.8L VIN R, 2.4L VIN D, 2.4L VIN L, 2.7L VIN M, 3.0L VIN F Transmissions: All	**Idle Control Valve Circuit Malfunction Conditions:** Engine running at idle speed n closed loop, and the PCM detected an unexpected voltage condition on the IAC valve circuit. A Rotary solenoid valve, in front of the air intake chamber, allows intake air to bypass the throttle valve and be directed to the Intake Air Control (IAC) valve through a passage. This configuration allows the PCM to control the engine idle speed by regulating the amount of intake air volume that bypasses the throttle valve. **Possible Causes** • IAC valve connector is damaged or loose • IAC valve control circuit is open, shorted to ground or power • IAC valve is damaged or has failed • PCM has failed
DTC P0513 **2T PCM, MIL: Yes** 2003 Avalon, Camry, Highlander, Sienna 3.0L VIN F engine Transmissions: All	**Unmatched Key Code Detected Conditions:** Key inserted, and the PCM detected that a key with an unregistered key code had been inserted into the ignition lock. **Possible Causes** • An unmatched ignition key was inserted
DTC P0550 **1T CCM, MIL: Yes** 2000, 2001, 2002 Echo 1.5L VIN T engine Transmissions: All	**Power Steering Pressure Sensor Circuit Malfunction Conditions:** Key on or engine running; and the PCM detected the Power Steering Pressure (PSP) sensor was less than 0.26v, or the PCM detected the PSP sensor was over 4.90v during the test. **Possible Causes** • PSP sensor connector is damaged, open or shorted • PSP sensor signal circuit is open, shorted to ground or VREF • PSP sensor ground circuit is open • PSP sensor is damaged or has failed • PCM has failed
DTC P0552 **1T CCM, MIL: Yes** 2003 Echo 1.5L VIN T engine Transmissions: All	**Power Steering Pressure Sensor Circuit Low Input Conditions:** Key on or engine running; and the PCM detected the Power Steering Pressure (PSP) sensor was less than 0.26v during the CCM test. **Possible Causes** • PSP sensor signal circuit is open or shorted to ground • PSP sensor ground circuit is open • PSP sensor is damaged or has failed • PCM has failed
DTC P0553 **1T CCM, MIL: Yes** 2003 Echo 1.5L VIN T engine Transmissions: All	**Power Steering Pressure Sensor Circuit High Input Conditions:** Key on or engine running; and the PCM detected the Power Steering Pressure (PSP) sensor was more than 4.90v during the CCM test. **Possible Causes** • PSP sensor signal circuit is shorted to power • PSP sensor ground circuit is open • PSP sensor is damaged or has failed • PCM has failed
DTC P0560 **2T CCM, MIL: Yes** 2003 All models All engines Transmissions: All	**System Voltage (Backup Power Circuit) Malfunction Conditions:** Key on or engine running; and the PCM detected an unexpected low voltage condition on the Backup Power Circuit during the test. **Possible Causes** • Battery backup circuit is open between battery and the PCM • PCM has failed
DTC P0571 **2T CCM, MIL: Yes** 2003 Tacoma, Tundra 3.4L VIN N engine Transmissions: All	**Brake Switch 'A' Circuit Malfunction Conditions:** Engine started, vehicle driven to cruise speed and then back to idle speed several times, and the PCM did not detect any change in the Brake Switch 'A' circuit status. The signal from this switch is used to determine when the brakes have been applied, and to determine the Fuel Cutoff engine speed during some types of braking operations. **Possible Causes** • Stoplight switch signal circuit is shorted to power • Stoplight switch assembly is damaged or shorted • PCM has failed
DTC P0604 **1T PCM, MIL: Yes** 2003 4Runner, Land Cruiser, MR2, Sequoia, Tacoma, Tundra All engines Transmissions: All	**PCM Internal Control Module Random Access Memory Processing Error Conditions:** Key on, and the PCM detected a processing error in the Internal Control Module Random Access Memory (RAM) function. **Possible Causes** • Clear the codes and retest for this code. If the same code resets, substitute a known good control module and retest. If the trouble code is gone, the original PCM has failed. • TSB TC002-03 (6/03) contains information related to this code

OBD II Trouble Code List (P0xxx Codes)

DTC	Trouble Code Title, Conditions & Possible Causes
DTC P0606 **1T PCM, MIL: Yes** 2003 4Runner, Echo, Land Cruiser, MR2, Sequoia, Tacoma, Tundra All engines Transmissions: All	**ECM/PCM Processing Error Conditions:** Key on, and the PCM detected a processing error occurred. **Possible Causes** • Clear the codes and retest for this code. If the same code resets, substitute a known good control module and retest. If the trouble code is gone, the original PCM has failed. • TSB TC002-03 (6/03) contains information related to this code
DTC P0607 **1T PCM, MIL: Yes** 2003 4Runner, Echo, Land Cruiser, MR2, Sequoia, Tacoma, Tundra All engines Transmissions: All	**Control Module Performance Conditions:** Key on, and the PCM detected a performance problem occurred. **Possible Causes** • Clear the codes and retest for this code. If the same code resets, substitute a known good control module and retest. If the trouble code is gone, the original PCM has failed. • TSB TC002-03 (6/03) contains information related to this code
DTC P0617 **1T CCM, MIL: Yes** 2003 4Runner, Echo, Land Cruiser, MR2, Sequoia, Tacoma, Tundra All engines Transmissions: All	**Starter Relay Circuit High Input Conditions:** Engine started, engine speed over 1000 rpm, system voltage over 10.5v, and the PCM detected the Starter Motor signal indicated high. **Possible Causes** • Park/Neutral switch assembly is damaged or it has failed • Ignition switch is damaged or has failed • PCM has failed
DTC P0657 **1T CCM, MIL: Yes** 2003 4Runner, Land Cruiser, MR2, Sequoia, Tacoma, Tundra 1.8L VIN R, 3.4L VIN N, 4.0L VIN U, 4.7L VIN T Transmissions: All	**PCM Actuator Supply Voltage Circuit Malfunction Conditions:** Key on or engine running; and the PCM detected an unexpected voltage condition on the Actuator Supply Voltage circuit. **Possible Causes** • Actuator supply voltage circuit is open • Clear the codes and retest for this code. If the same code resets, substitute a known good control module and retest. If the trouble code is gone, the original PCM has failed. • TSB TC002-03 (6/03) contains information related to this code
DTC P0705 **2T CCM, MIL: Yes** 2003 Avalon, Camry, Highlander, Sienna, Tacoma, Tundra 2.4L VIN D, 2.4L VIN L, 2.7L VIN M, 3.0L VIN F, 3.4L VIN N engines Transmissions: A/T	**A/T Range Sensor Circuit (PRNDL) Malfunction Conditions:** Key on or engine running; and the PCM detected simultaneous "on" signals (N, 2, L or R) from the Transmission Range sensor circuit. The P/N switch indicates "on" whenever the shift lever is in the 'N' or 'P' position. When it is "on", the NSW circuit to the PCM is grounded to chassis ground through the starter motor relay, and reads 0.00v. When the shift lever is in 'R', 'D' or 'L' position, the switch is "off" and the NSW circuit reads 12.0v. When the shift lever is moved from the 'N' to the 'D' position, the PCM uses this signal to air/fuel ratio correction and idle speed control (estimated control) functions. **Possible Causes** • Park/Neutral switch assembly is shorted • Park/Neutral switch assembly is damaged or has failed • PCM has failed.
DTC P0710 **1T CCM, MIL: Yes** 1996, 1997, 1998, 1999, 2000, 2001, 2002 All models All engines Transmissions: A/T	**Transmission Fluid Temperature Sensor Circuit Malfunction Conditions:** Engine started, and the PCM detected the TFT sensor indicated less than -40°F, or after the engine runtime exceeded 15 minutes, it indicated more than 300°F during the CCM test. **Possible Causes** • TFT sensor signal circuit is open, shorted to ground or shorted to system power • TFT sensor is damaged or has failed • PCM has failed
DTC P0710 **1T CCM, MIL: Yes** 2003 Echo, RAV4, Highlander, Sienna 1.5L VIN T engine, 2.0L VIN K, 3.0L VIN F Transmissions: A/T	**Transmission Fluid Temperature Sensor Circuit Malfunction Conditions:** Engine started, and the PCM detected the TFT sensor indicated less than -40°F, or with the engine runtime over 15 minutes, it detected the TFT sensor indicated more than 300°F. **Possible Causes** • TFT sensor signal circuit is open, shorted to ground or shorted to system power • TFT sensor is damaged or has failed • PCM has failed
DTC P0711 **1T CCM, MIL: Yes** 1996, 1997, 1998, 1999, 2000, 2001, 2002 All models All engines Transmissions: A/T	**Transmission Fluid Temperature Sensor Performance Conditions:** Engine started, engine runtime over 15 seconds, and the PCM detected the ambient air temperature and transmission fluid temperature varied by more than 40°F, or after an engine runtime of 20 minutes and 6.2 miles were traveled, it indicated less than 50°F. **Possible Causes** • TFT sensor signal circuit or ground circuit has high resistance • TFT sensor has failed, or the PCM has failed

<u>OBD II TROUBLE CODE LIST (P0XXX CODES)</u>

DTC	Trouble Code Title, Conditions & Possible Causes
DTC P0711 **1T CCM, MIL: Yes** 2003 Echo, RAV4 1.5L VIN T engine, 2.0L VIN K Transmissions: A/T	**Transmission Fluid Temperature Sensor Performance Conditions:** ECT and IAT sensor signals more than 14ºF, engine runtime over 12 seconds, then after the vehicle was driven for more than 6.2 miles and the engine runtime exceeded 20 minutes, the PCM detected the TFT sensor remained at a value less than 14ºF during the CCM test. **Possible Causes** • TFT sensor signal circuit has high resistance • TFT sensor is damaged or has failed (it may be contaminated) • PCM has failed
DTC P0712 **1T CCM, MIL: Yes** 2003 Echo 1.5L VIN T engine Transmissions: A/T	**Transmission Fluid Temperature Sensor Circuit Low Input Conditions:** Engine started, engine running, and the PCM detected the Transmission Fluid Temperature (TFT) sensor indicated a value of more than 284ºF for 500 ms in the test. **Possible Causes** • TFT sensor signal circuit is shorted • TFT sensor is damaged or has failed • PCM has failed
DTC P0713 **1T CCM, MIL: Yes** 2003 Echo 1.5L VIN T engine Transmissions: A/T	**Transmission Fluid Temperature Sensor Circuit High Input Conditions:** Engine started, engine running, and the PCM detected the TFT sensor indicated a value of less than -40ºF for 500 ms in t he test. **Possible Causes** • TFT sensor signal circuit is open • TFT sensor is damaged or has failed • PCM has failed
DTC P0715 **1T CCM, MIL: Yes** 1998, 1999, 2000, 2001, 2002 Land Cruiser, Sequoia, Tundra All engines Transmissions: A/T	**A/T Input Shaft/Direct Clutch Speed Sensor Circuit Malfunction Conditions:** Engine started, P/N switch indicating "off", vehicle driven in 1st, 2nd or 3rd gear or in Overdrive, the SS1, SS2 and SL (shift valves) and VSS all operating normally, and the PCM detected the ISS signal indicated 100 rpm or less during the CCM Rationality test. **Possible Causes** • O/D direct clutch speed sensor circuit is open • O/D direct clutch speed sensor circuit is shorted to ground • O/D direct clutch speed sensor is damaged or has failed • PCM has failed
DTC P0717 **1T CCM, MIL: Yes** 2003 Avalon, Camry, Highlander, Sienna 3.0L VIN F engine Transmissions: A/T	**A/T Input Shaft Speed Sensor 'A' Circuit No Signal Conditions:** No SS1, SS2, SL shift solenoid or VSS codes set, engine started, P/N switch indicating "off", gear position indicating 1st, 2nd, 3rd gear, or in O/D, no gear change occurring, and the PCM detected the Input Shaft Speed sensor was below 300 rpm, or over 1000 rpm for 4 seconds. **Possible Causes** • Direct clutch connector is damaged (it may be open or shorted) • Direct clutch sensor signal circuit is open or shorted • Direct clutch speed sensor is damaged or has failed • PCM has failed
DTC P0717 **1T CCM, MIL: Yes** 2003 Echo 1.5L VIN T engine Transmissions: A/T	**A/T Turbine Shaft Speed Sensor Circuit No Signal Conditions:** No Shift Solenoid or P/N codes set, engine started, P/N switch indicating "off", gear position indicating 2nd, 3rd gear, or in O/D, no gear change occurring, and the PCM detected the Turbine Shaft Speed (ISS) sensor indicated less than 300 rpm, or more than 1000 rpm for 4 seconds. The PCM detects the rotation speed of the input turbine, and compares the signals from the input turbine speed (NT) sensor to the counter gear speed sensor (NC). The PCM uses this signal to detect the shift time so that it can control the engine torque and hydraulic pressure in response to various driving conditions. **Possible Causes** • Input shaft speed sensor is damaged or loose • Input shaft speed sensor signal (NT) circuit is open or shorted • Input shaft speed sensor is damaged or has failed • PCM has failed
DTC P0724 **2T CCM, MIL: Yes** 2003 Avalon, Camry, Highlander, Sienna, Tacoma, Tundra 2.4L VIN D, 2.4L VIN L, 2.7L VIN M, 3.0L VIN F, 3.4L VIN N engines Transmissions: A/T	**Brake Switch 'B' Circuit High Input Conditions:** Engine started, vehicle driven to cruise speed and then back to idle speed at least 30 times, and the PCM did not detect any change in the Brake Switch 'A' circuit status. The STP 'B' switch signal is used to determine when the brakes have been applied, and to determine the Fuel Cutoff engine speed during periods with the brakes applied. **Possible Causes** • Stoplight switch signal circuit is shorted to power • Stoplight switch assembly is damaged or shorted • PCM has failed

OBD II Trouble Code List (P0xxx Codes)

DTC	Trouble Code Title, Conditions & Possible Causes
DTC P0724 **1T CCM, MIL: Yes** 2003 Echo, MR2 1.5L VIN T, 1.8L VIN R Transmissions: A/T	**Brake Switch 'B' Circuit High Input Conditions:** Engine started, vehicle driven to cruise speed and then back to idle speed at least 30 times, and the PCM did not detect any change in the Brake Switch 'A' circuit status. The STP 'B' switch signal is used to determine when the brakes have been applied, and to determine the Fuel Cutoff engine speed during periods with the brakes applied. **Possible Causes** • Stoplight switch signal circuit is shorted to power • Stoplight switch assembly is damaged or shorted • PCM has failed
DTC P0724 **1T CCM, MIL: Yes** 2003 Avalon, Camry, Echo, Highlander, Sienna 1.5L VIN T, 3.0L VIN F Transmissions: A/T	**A/T Torque Converter Clutch Shift Solenoid Performance Conditions:** Engine started, vehicle driven to over 50 mph and then back to idle speed several times, and the PCM detected that TCC lockup did not occur, or that the TCC remained in lockup position in the "off" range. The PCM uses signals from the CKP, MAF and TP sensors to monitor engagement of the Lockup Clutch solenoid to find a fault. **Possible Causes** • Lockup clutch solenoid is damaged or has failed • Shift solenoid SL is damaged or has failed (mechanical fault) • Shift solenoid SL is stuck in "on" or "off" position • Valve body is blocked or stuck
DTC P0743 **2T CCM, MIL: Yes** 2003 Avalon, Camry, Echo, Highlander, Sienna 1.5L VIN T, 3.0L VIN F Transmissions: A/T	**A/T Torque Converter Clutch Shift Solenoid Circuit Malfunction Conditions:** Vehicle driven to over 50 mph and then back to idle speed several times; and the PCM detected an unexpected voltage on the Shift Solenoid (SL) valve control circuit. The Shift Solenoid valve is turned On/Off" through commands from the PCM to control the hydraulic pressure acting on the lockup relay valve (and it controls the operation of the lockup clutch). **Possible Causes** • Shift solenoid SL connector is damaged or loose • Shift solenoid SL control circuit is open or shorted to ground • Shift solenoid SL is damaged or has failed (electrical fault) • PCM has failed
DTC P0750 **2T CCM, MIL: Yes** 1995 Avalon, Camry, Land Cruiser, Previa, T100, Tacoma, Tercel 1.5L VIN E, 2.4L VIN A, 2.4L VIN R, 2.7L VIN U, 3.0L VIN G, 3.4L VIN V, 4.5L VIN D engines Transmissions: A/T	**A/T Shift Solenoid 1 or 'A' Malfunction (Mechanical) Conditions:** Engine started, vehicle driven under normal driving conditions, and the PCM detected the Actual gear ratio and the Required gear ratio did not match during the CCM Rationality test. **Possible Causes** • A/T component problems (i.e., in clutch, brake or gears) • SS1 or SSA is damaged, stuck "open" or stuck "closed" • Transmission valve body is clogged, dirty or stuck
DTC P0750 **2T CCM, MIL: Yes** 1996, 1997, 1998, 1999, 2000, 2001, 2002 All models All engines Transmissions: A/T	**A/T Shift Solenoid 1 or 'A' Malfunction (Mechanical) Conditions:** Engine started, vehicle driven under normal driving conditions, and the PCM detected the Actual gear ratio and the Required gear ratio did not match during the CCM Rationality test. **Possible Causes** • A/T component problems (i.e., in clutch, brake or gears) • SS1 or SSA is damaged, stuck "open" or stuck "closed" • Transmission valve body is clogged, dirty or stuck
DTC P0750 **2T CCM, MIL: Yes** 2003 Celica, Corolla, Matrix, RAV4 1.8L VIN R, 1.8L VIN Y, 2.0L VIN K engines Transmissions: A/T	**A/T Shift Solenoid 'A' Malfunction (Mechanical) Conditions:** Engine started, vehicle driven under normal driving conditions, and the PCM detected the Actual gear ratio and the Required gear ratio did not match during the CCM Rationality test. **Possible Causes** • A/T component problems (i.e., in clutch, brake or gears) • SSA is damaged, stuck "open" or stuck "closed" • Transmission valve body is clogged, dirty or stuck
DTC P0751 **8T CCM, MIL: Yes** 2003 Avalon, Camry, Echo, Highlander, Sienna 1.5L VIN T, 3.0L VIN F Transmissions: A/T	**A/T Shift Solenoid 'A' Signal Range/Performance Conditions:** Engine started, vehicle driven to a speed over 50 mph, and the PCM detected the Actual gear position did not match the Desired gear position during the CCM test period. The PCM uses inputs from the VSS and Direct Clutch speed sensor to determine the actual gear position (i.e., 1st, 2nd, 3rd or O/D gear). **Possible Causes** • SSA control circuit is open or shorted to ground • SSA control circuit is shorted to system power (B+) • SSA is damaged or has failed (an electrical fault) • PCM has failed

OBD II Trouble Code List (P0xxx Codes)

DTC	Trouble Code Title, Conditions & Possible Causes
DTC P0753 **8T CCM, MIL: Yes** 1995 Avalon, Camry, Land Cruiser, Previa, T100, Tacoma, Tercel 1.5L VIN E, 2.4L VIN A, 2.4L VIN R, 2.7L VIN U, 3.0L VIN G, 3.4L VIN V, 4.5L VIN D engines Transmissions: A/T	**A/T Shift Solenoid 1 or 'A' Circuit Malfunction Conditions:** Engine started, engine running during normal driving conditions; and the PCM detected the Shift Solenoid 1 or 'A' (SS1/SSA) control circuit voltage was "high" with the solenoid "on", or the SS1 or SSA control circuit voltage was "low" with the SS1/SSA commanded "off". **Possible Causes** • SS1 or SSA control circuit is open or shorted to ground • SS1or SSA control circuit is shorted to system power (B+) • SS1 or SSA is damaged or has failed (an electrical fault) • PCM has failed
DTC P0753 **8T CCM, MIL: Yes** 1996, 1997, 1998, 1999, 2000, 2001, 2002 All models All engines Transmissions: A/T	**A/T Shift Solenoid 1 or 'A' Circuit Malfunction Conditions:** Engine started, engine running during normal driving conditions; and the PCM detected the Shift Solenoid 1 or 'A' (SS1/SSA) control circuit voltage was "high" with the solenoid "on", or the SS1 or SSA control circuit voltage was "low" with the SS1/SSA commanded "off". **Possible Causes** • SS1 or SSA control circuit is open, shorted to ground or shorted to system power (B+) • SS1 or SSA has failed, or the PCM has failed
DTC P0753 **8T CCM, MIL: Yes** 2003 Avalon, Camry, Highlander, Sienna 3.0L VIN F engine Transmissions: A/T	**A/T Shift Solenoid 'A' Circuit Malfunction Conditions:** Engine started, engine running during normal driving conditions; and the PCM detected an unexpected voltage condition on the Shift Solenoid 'A' control circuit during the CCM test. **Possible Causes** • SSA control circuit is open, shorted to ground or to power (B+) • SSA is damaged or has failed (an electrical fault) • PCM has failed
DTC P0753 **8T CCM, MIL: Yes** 2003 Celica, Corolla, Matrix, RAV4 1.8L VIN R, 1.8L VIN Y, 2.0L VIN K engines Transmissions: A/T	**A/T Shift Solenoid 'A' Circuit Malfunction Conditions:** Engine running during normal driving conditions; and the PCM detected an unexpected voltage condition on the Shift Solenoid 'A' control circuit during the CCM test. **Possible Causes** • SSA control circuit is open, shorted to ground or to power (B+) • SSA is damaged or has failed (an electrical fault) • PCM has failed
DTC P0755 **2T CCM, MIL: Yes** 1995 Avalon, Camry, Land Cruiser, Previa, T100, Tacoma, Tercel 1.5L VIN E, 2.4L VIN A, 2.4L VIN R, 2.7L VIN U, 3.0L VIN G, 3.4L VIN V, 4.5L VIN D engines Transmissions: A/T	**A/T Shift Solenoid 'B' Malfunction (Mechanical) Conditions:** Engine started, vehicle driven under normal driving conditions, and the PCM detected the Actual gear ratio and the Required gear ratio did not match during the CCM Rationality test. **Possible Causes** • A/T component problems (i.e., in clutch, brake or gears) • SSB is damaged, stuck "open" or stuck "closed" • Transmission valve body is clogged, dirty or stuck
DTC P0755 **2T CCM, MIL: Yes** 1996, 1997, 1998, 1999, 2000, 2001, 2002 All models All engines Transmissions: A/T	**A/T Shift Solenoid 'B' Malfunction (Mechanical) Conditions:** Engine started, vehicle driven under normal driving conditions, and the PCM detected the Actual gear ratio and the Required gear ratio did not match during the CCM Rationality test. **Possible Causes** • A/T component problems (i.e., in clutch, brake or gears) • SSB is damaged, stuck "open" or stuck "closed" • Transmission valve body is clogged, dirty or stuck
DTC P0755 **8T CCM, MIL: Yes** 2003 Celica, Corolla, Matrix 1.8L VIN R, 1.8L VIN Y Transmissions: A/T	**A/T Shift Solenoid 'B' Malfunction (Mechanical) Conditions:** Engine started, vehicle driven under normal driving conditions, and the PCM detected the Actual gear ratio and the Required gear ratio did not match during the CCM Rationality test. **Possible Causes** • A/T component problems (i.e., in clutch, brake or gears) • SSB is damaged, stuck "open" or stuck "closed" • Transmission valve body is clogged, dirty or stuck
DTC P0756 **8T CCM, MIL: Yes** 2003 Avalon, Camry, Echo, Highlander, Sienna 1.5L VIN T, 3.0L VIN F Transmissions: A/T	**A/T Shift Solenoid 'B' Signal Range/Performance Conditions:** Vehicle driven to a speed over 50 mph, and the PCM detected the Actual gear position did not match the Desired gear position. The PCM uses inputs from the VSS and Direct Clutch speed sensor to determine the actual gear position (i.e., 1st, 2nd, 3rd or O/D gear). **Possible Causes** • SSB control circuit is open, shorted to ground or to power (B+) • SSB is damaged or has failed (an electrical fault) • PCM has failed

OBD II Trouble Code List (P0xxx Codes)

DTC	Trouble Code Title, Conditions & Possible Causes
DTC P0758 **8T CCM, MIL: Yes** 1995 Avalon, Camry, Land Cruiser, Previa, T100, Tacoma, Tercel 1.5L VIN E, 2.4L VIN A, 2.4L VIN R, 2.7L VIN U, 3.0L VIN G, 3.4L VIN V, 4.5L VIN D engines Transmissions: A/T	**A/T Shift Solenoid 2 or 'B' Circuit Malfunction Conditions:** Engine started, engine running during normal driving conditions, and the PCM detected the Shift Solenoid 2 or 'B' (SS2/SSB) control circuit voltage was "high" with the solenoid "on", or the SS2 or SSB control circuit voltage was "low" with the SS2/SSB commanded "off". **Possible Causes** • SS2 or SSB control circuit is open or shorted to ground • SS2or SSB control circuit is shorted to system power (B+) • SS2 or SSB is damaged or has failed (an electrical fault) • PCM has failed
DTC P0758 **1T CCM, MIL: Yes** 1996, 1997, 1998, 1999, 2000, 2001, 2002 All models All engines Transmissions: A/T	**A/T Shift Solenoid 2 or 'B' Circuit Malfunction Conditions:** Engine started, engine running during normal driving conditions, and the PCM detected the Shift Solenoid 2 or 'B' (SS2/SSB) control circuit voltage was "high" with the solenoid "on", or the SS2 or SSB control circuit voltage was "low" with the SS2/SSB commanded "off". *Note: This problem must occur eight (8) times during one trip before this code is triggered.* **Possible Causes** • SS2 or SSB control circuit is open or shorted to ground • SS2or SSB control circuit is shorted to system power (B+) • SS2 or SSB is damaged or has failed (an electrical fault) • PCM has failed
DTC P0758 **8T CCM, MIL: Yes** 2003 Celica, Corolla, Matrix, RAV4 1.8L VIN R, 1.8L VIN Y, 2.0L VIN K engines Transmissions: A/T	**A/T Shift Solenoid 'B' Circuit Malfunction Conditions:** Engine started, engine running during normal driving conditions, and the PCM detected an unexpected voltage condition on the Shift Solenoid 'B' control circuit during the CCM test. **Possible Causes** • SSB control circuit is open, shorted to ground or to power (B+) • SSB is damaged or has failed (an electrical fault) • PCM has failed
DTC P0758 **8T CCM, MIL: Yes** 2003 Avalon, Camry, Highlander, Sienna 2.4L VIN D, 3.0L VIN F Transmissions: A/T	**A/T Shift Solenoid 'B' Circuit Malfunction Conditions:** Engine started, engine running during normal driving conditions, and the PCM detected an unexpected voltage condition on the Shift Solenoid 'B' control circuit during the CCM test. **Possible Causes** • SSB control circuit is open, shorted to ground or to power (B+) • SSB is damaged or has failed (an electrical fault) • PCM has failed
DTC P0765 **8T CCM, MIL: Yes** 2003 RAV4 2.0L VIN K engine Transmissions: A/T	**A/T Shift Solenoid 'D' Malfunction (Mechanical) Conditions:** Engine started, engine running during normal driving conditions, and the PCM detected an unexpected voltage condition on the Shift Solenoid 'D' control circuit during the CCM test. **Possible Causes** • SL1 or SL2 is damaged, stuck "open" or stuck "closed" • S4 is damaged, stuck "open" or stuck "closed" • Transmission valve body is clogged, dirty or stuck
DTC P0768 **8T CCM, MIL: Yes** 2003 RAV4 2.0L VIN K engine Transmissions: A/T	**A/T Shift Solenoid 'D' Circuit Malfunction Conditions:** Engine started, engine running during normal driving conditions, and the PCM detected an unexpected voltage condition on the Shift Solenoid 'D' control circuit during the CCM test. **Possible Causes** • SL1, SL2 or S4 circuit is open, shorted to ground or to power • SL1, SL2 or S4 assembly is damaged or has failed • PCM has failed
DTC P0770 **2T CCM, MIL: Yes** 1995 Avalon, Camry, Land Cruiser, Previa, T100, Tacoma, Tercel 1.5L VIN E, 2.4L VIN A, 2.4L VIN R, 2.7L VIN U, 3.0L VIN G, 3.4L VIN V, 4.5L VIN D engines Transmissions: A/T	**A/T Shift Solenoid 'E' (SL) Malfunction (Mechanical) Conditions:** Engine started, vehicle driven at a speed of over 50 mph for 1-3 minutes, and the PCM detected the transmission lockup function did not occur in the "lockup" range, or the "lockup" function was "on" during periods when it should have been "off" during the CCM test. *Note: The A/T clutch or A/T brake clutch will slip with this code set.* **Possible Causes** • A/T component problems (i.e., in clutch, brake or gears) • SL (shift lockup) is damaged, stuck "open" or stuck "closed" • Transmission valve body is clogged, dirty or stuck

OBD II Trouble Code List (P0xxx Codes)

DTC	Trouble Code Title, Conditions & Possible Causes
DTC P0770 **1T CCM, MIL: Yes** 1996, 1997, 1998, 1999, 2000, 2001, 2002 All models All engines Transmissions: A/T	**A/T Shift Solenoid 'E' (SL) Malfunction (Mechanical) Conditions:** Engine started, vehicle driven at a speed of over 50 mph for 1-3 minutes, and the PCM detected the transmission lockup function did not occur in the "lockup" range, or the "lockup" function was "on" during periods when it should have been "off" during the CCM test. *Note: The A/T clutch or A/T brake clutch will slip with this code set.* **Possible Causes** • A/T component problems (i.e., in clutch, brake or gears) • SL (shift lockup) is damaged, stuck "open" or stuck "closed" • Transmission valve body is clogged, dirty or stuck
DTC P0770 **1T CCM, MIL: Yes** 2003 Celica, Corolla, Matrix, RAV4 1.8L VIN R, 1.8L VIN Y, 2.0L VIN K engines Transmissions: A/T	**A/T Shift Solenoid 'E' (SL) Malfunction (Mechanical) Conditions:** Engine started, vehicle driven at a speed of over 50 mph for 1-3 minutes, and the PCM detected the transmission lockup function did not occur in the "lockup" range, or the "lockup" function was "on" during periods when it should have been "off" during the CCM test. *Note: The A/T clutch or A/T brake clutch will slip with this code set.* **Possible Causes** • A/T component problems (i.e., in clutch, brake or gears) • SL (shift lockup) is damaged, stuck "open" or stuck "closed" • Transmission valve body is clogged, dirty or stuck
DTC P0773 **1T CCM, MIL: Yes** 1995 Avalon, Camry, Land Cruiser, Previa, T100, Tacoma, Tercel 1.5L VIN E, 2.4L VIN A, 2.4L VIN R, 2.7L VIN U, 3.0L VIN G, 3.4L VIN V, 4.5L VIN D engines Transmissions: A/T	**A/T Shift Solenoid 'E' (SL) Circuit Malfunction Conditions:** Engine started, engine running during normal driving conditions, and the PCM detected the Shift Lockup (SL) control circuit voltage was "high" with the solenoid "on", or the SL control circuit voltage was "low" with the SL commanded "off" during the CCM Rationality test. **Possible Causes** • SL control circuit is open or shorted to ground • SL control circuit is shorted to system power (B+) • SL is damaged or has failed (an electrical fault) • PCM has failed
DTC P0773 **1T CCM, MIL: Yes** 1996, 1997, 1998, 1999, 2000, 2001 All models All engines Transmissions: A/T	**A/T Shift Solenoid 'E' (SL) Circuit Malfunction Conditions:** Engine started, engine running during normal driving conditions, and the PCM detected the Shift Lockup (SL) control circuit voltage was "high" with the solenoid "on", or the SL control circuit voltage was "low" with the SL commanded "off" during the CCM Rationality test. **Possible Causes** • SL control circuit is open or shorted to ground • SL control circuit is shorted to system power (B+) • SL is damaged or has failed (an electrical fault) • PCM has failed
DTC P0773 **1T CCM, MIL: Yes** 2001 Avalon, Sienna All engines Transmissions: A/T	**A/T Shift Solenoid 'E' (SL) Circuit Malfunction Conditions:** Engine started, engine running during normal driving conditions, and the PCM detected the Shift Lockup (SL) control circuit voltage was "high" with the solenoid "on", or the SL control circuit voltage was "low" with the SL commanded "off" during the CCM Rationality test. **Possible Causes** • SL control circuit is open, shorted to ground or to power (B+) • SL is damaged or has failed (an electrical fault) • PCM has failed • TSB EG16-01 (12/01) contains information related to this code
DTC P0773 **1T CCM, MIL: Yes** 2002 All models All engines Transmissions: A/T	**A/T Shift Solenoid 'E' (SL) Circuit Malfunction Conditions:** Engine started, engine running during normal driving conditions, and the PCM detected the Shift Lockup (SL) control circuit voltage was "high" with the solenoid "on", or the SL control circuit voltage was "low" with the SL commanded "off" during the CCM Rationality test. **Possible Causes** • SL control circuit is open or shorted to ground • SL control circuit is shorted to system power (B+) • SL is damaged or has failed (an electrical fault) • PCM has failed
DTC P0773 **1T CCM, MIL: Yes** 2003 Celica, Corolla, Matrix, RAV4 1.8L VIN R, 1.8L VIN Y, 2.0L VIN K engines Transmissions: A/T	**A/T Shift Solenoid 'E' (SL) Circuit Malfunction Conditions:** Engine started, engine running during normal driving conditions, and the PCM detected the Shift Lockup (SL) control circuit voltage was "high" with the solenoid "on", or the SL control circuit voltage was "low" with the SL commanded "off" during the CCM Rationality test. **Possible Causes** • SL solenoid control circuit is open or shorted to ground • SL solenoid control circuit is shorted to system power (B+) • SL solenoid is damaged or has failed (an electrical fault) • PCM has failed

OBD II Trouble Code List (P0xxx Codes)

DTC	Trouble Code Title, Conditions & Possible Causes
DTC P0787 **1T CCM, MIL: Yes** 2003 Echo 1.5L VIN T engine Transmissions: A/T	**A/T Shift Timing Solenoid (ST) Circuit Low Input Conditions:** Engine started, engine running during normal driving conditions, and the PCM detected an unexpected low voltage condition on the Shift Timing Solenoid (SL) control circuit at least (4) times during testing. **Possible Causes** • ST solenoid connector is damaged or shorted • SL solenoid control circuit is shorted to ground • SL solenoid is damaged or has failed • PCM has failed
DTC P0788 **1T CCM, MIL: Yes** 2003 Echo 1.5L VIN T engine Transmissions: A/T	**A/T Shift Timing Solenoid (ST) Circuit High Input Conditions:** Engine started, engine running during normal driving conditions, and the PCM detected an unexpected high voltage condition on the Shift Timing Solenoid (SL) control circuit at least (4) times during testing. **Possible Causes** • ST solenoid connector is damaged or loose • SL solenoid control circuit is open or shorted to system power • SL solenoid is damaged or has failed • PCM has failed
DTC P0850 **1T CCM, MIL: Yes** 2003 Avalon, Camry, Highlander, Sienna, Tacoma, Tundra 2.4L VIN D, 2.4L VIN L, 2.7L VIN M, 3.0L VIN F, 3.4L VIN N engines Transmissions: A/T	**A/T Park/Neutral Switch Circuit Malfunction Conditions:** Engine started, vehicle at 1500-2500 rpm at a speed over 43 mph for at least 30 seconds, and the PCM detected a continuous "on" "N" signal from the P/N switch. The P/N switch indicates "on" whenever the shift lever is in the 'N' or 'P' position. When it is "on", the NSW circuit to the PCM is grounded to chassis ground through the starter motor relay, and reads 0.00v. When the shift lever is in 'R', 'D' or 'L' position, the switch is "off" and the NSW circuit reads 12.0v. When the shift lever is moved from the 'N' to the 'D' position, the PCM uses this signal to air/fuel ratio correction and idle speed control functions. **Possible Causes** • Park/Neutral switch assembly is shorted • Park/Neutral switch assembly is damaged or has failed • PCM has failed.
DTC P0973 **1T CCM, MIL: Yes** 2003 Echo 1.5L VIN T engine Transmissions: A/T	**A/T Shift Solenoid 1 Circuit Low Input Conditions:** Engine started, engine running during normal driving conditions, and the PCM detected an unexpected low voltage condition on the Shift Solenoid 1 control circuit at least (4) times. **Possible Causes** • SS1 (solenoid) connector is damaged or shorted • SS1 (solenoid) control circuit is shorted to ground • SS1 (solenoid) is damaged or has failed • PCM has failed
DTC P0974 **1T CCM, MIL: Yes** 2003 Echo 1.5L VIN T engine Transmissions: A/T	**A/T Shift Solenoid 1 Circuit High Input Conditions:** Engine started, engine running during normal driving conditions, and the PCM detected an unexpected high voltage condition on the Shift Solenoid 1 control circuit at least (4) times. **Possible Causes** • SS1 (solenoid) connector is damaged or open • SS1 (solenoid) control circuit is open or shorted to power (B+) • SS1 (solenoid) is damaged or has failed • PCM has failed
DTC P0976 **1T CCM, MIL: Yes** 2003 Echo 1.5L VIN T engine Transmissions: A/T	**A/T Shift Solenoid 2 Circuit Low Input Conditions:** Engine started, engine running during normal driving conditions, and the PCM detected an unexpected low voltage condition on the Shift Solenoid 2 control circuit at least (4) times during testing. **Possible Causes** • SS2 (solenoid) connector is damaged or shorted • SS2 (solenoid) control circuit is shorted to ground • SS2 (solenoid) is damaged or has failed • PCM has failed
DTC P0977 **1T CCM, MIL: Yes** 2003 Echo 1.5L VIN T engine Transmissions: A/T	**A/T Shift Solenoid 2 Circuit High Input Conditions:** Engine started, engine running during normal driving conditions, and the PCM detected an unexpected high voltage condition on the Shift Solenoid 2 control circuit at least (4) times during testing. **Possible Causes** • SS2 (solenoid) connector is damaged or open • SS2 (solenoid) control circuit is open or shorted to power (B+) • SS2 (solenoid) is damaged or has failed • PCM has failed

OBD II Trouble Code List (P1xxx Codes)

DTC	Trouble Code Title, Conditions & Possible Causes
DTC P1100 **1T CCM, MIL: Yes** 1995, 1996, 1997 Supra Engines: 3.0L VIN E Transmissions: All	**Barometric Pressure Sensor Circuit Malfunction Conditions:** Key on or engine running; and the PCM detected an unexpected voltage condition on the Barometric Pressure (BARO) sensor circuit during the CCM test. **Possible Causes** • BARO sensor signal circuit is open or shorted to ground • BARO sensor signal circuit shorted to VREF or system power • BARO sensor power circuit is open between sensor and PCM • BARO sensor ground circuit is open between sensor and PCM • BARO sensor is damaged or has failed • PCM has failed
DTC P1120 **1T CCM, MIL: Yes** 1998, 1999, 2000, 2001, 2002 Supra, Land Cruiser, Sequoia, Tundra 3.0L VIN D, 4.7L VIN T Transmissions: All	**Accelerator Pedal Position Sensor Circuit Malfunction Conditions:** Engine started; and the PCM detected the APP sensor VPA reading was equal to or less than 0.2v with a VPA2 reading of equal to or less than 0.5v; or the VPA reading was equal to or more than 4.7v; or the VPA reading indicated from 0.2-1.8v with the VPA2 reading equal to or more than 4.97v; or the VPA reading minus the VPA2 reading was less than 0.02v; or the VPA 2 reading minus the VPA reading was less than 0.02v for 5 seconds. **Possible Causes** • APP sensor circuit is open or shorted to ground • APP sensor circuit is shorted to VREF • APP sensor power circuit is open between sensor and the PCM • APP sensor ground circuit is open between sensor and PCM • APP sensor is damaged or has failed • PCM has failed
DTC P1121 **1T CCM, MIL: Yes** 1998, 1999, 2000, 2001, 2002 Supra, Land Cruiser, Sequoia, Tundra 3.0L VIN D, 4.7L VIN T Transmissions: All	**ETCS Accelerator Pedal Position Sensor Performance Conditions:** Engine started, engine running, and the PCM detected the difference between the Electronic Throttle Control System (ETCS) APP sensor VPA and VPA2 readings was less than 0.7v or 1.7v for more than 2 seconds during the CCM test. **Possible Causes** • APP sensor is damaged or has failed • Throttle assembly or throttle linkage is binding or has failed • PCM has failed
DTC P1125 **1T CCM, MIL: Yes** 1998, 1999, 2000, 2001, 2002 Supra, Land Cruiser, Sequoia, Tundra 3.0L VIN D, 4.7L VIN T Transmissions: All	**ETCS Throttle Control Motor Circuit Malfunction Conditions:** Engine started, engine running, and the PCM detected the motor duty cycle was equal to or more than 80% with a current level less than 0.5A; or the throttle motor current level was more than 16A, or the motor current level was equal to or more than 7A for 600 ms. **Possible Causes** • ETCS throttle motor control (+) circuit is open, shorted to ground or to power (B+) • ETCS throttle motor control (-) circuit is open, shorted to ground or to power (B+) • ETCS throttle motor is damaged or has failed • PCM has failed
DTC P1125 **1T CCM, MIL: Yes** 2001, 2002, 2003 Prius Engines: 1.5L VIN B Transmissions: All	**ETCS Throttle Control Motor Circuit Malfunction Conditions:** Engine started, engine running, and the PCM detected the motor duty cycle was equal to or more than 80% with a current level less than 0.5A; or the throttle motor current level was more than 16A, or the motor current level was equal to or more than 7A for 600 ms. **Possible Causes** • ETCS throttle motor control (+) circuit is open, shorted to ground or to power (B+) • ETCS throttle motor control (-) circuit is open, shorted to ground or to power (B+) • ETCS throttle motor is damaged or has failed • PCM has failed
DTC P1126 **1T CCM, MIL: Yes** 1998, 1999, 2000, 2001, 2002 Supra, Land Cruiser, Sequoia, Tundra 3.0L VIN, 4.7L VIN T Transmissions: All	**Magnetic Clutch Circuit Malfunction Conditions:** Engine started; and the PCM detected the magnetic clutch current was equal to or greater than 1.4 amps, or it was in a range of 0.8 amps to 1.4 amps for 1.5 seconds during the test. **Possible Causes** • ETCS magnetic clutch (+) circuit is open, shorted to ground or to power (B+) • ETCS magnetic clutch (-) circuit is open, shorted to ground or to power (B+) • ETCS magnetic clutch is damaged or has failed • PCM has failed
DTC P1126 **1T CCM, MIL: Yes** 2003 MR2 Engines: 1.8L VIN R Transmissions: All	**Magnetic Clutch Circuit Malfunction Conditions:** Engine started; and the PCM detected the magnetic clutch current was equal to or greater than 1.4 amps, or it was in a range of 0.8 amps to 1.4 amps for 1.5 seconds during the test. **Possible Causes** • ETCS magnetic clutch (+) circuit is open, shorted to ground or to power (B+) • ETCS magnetic clutch (-) circuit is open, shorted to ground or to power (B+) • ETCS magnetic clutch is damaged or has failed • PCM has failed

OBD II Trouble Code List (P1xxx Codes)

DTC	Trouble Code Title, Conditions & Possible Causes
DTC P1127 **1T CCM, MIL: Yes** 1998, 1999, 2000, 2001, 2002 Supra, Land Cruiser, Sequoia, Tundra 3.0L VIN D, 4.7L VIN T Transmissions: All	**ETCS Actuator Power Source Circuit Malfunction Conditions:** Key on and the PCM detected a fault in the Electric Throttle Control system power source. The PCM shuts off power to the throttle motor and magnetic clutch (throttle valve is fully closed by a return spring, so the accelerator pedal can be opened with the throttle valve). **Possible Causes** • Battery connections are dirty or loose • ETCS power source circuit is open (check the ETCS fuse) • PCM has failed
DTC P1127 **1T CCM, MIL: Yes** 2001, 2002, 2003 Prius Engines: 1.5L VIN B Transmissions: All	**ETCS Actuator Power Source Circuit Malfunction Conditions:** Key on, and the PCM detected an unexpected voltage condition on the Electric Throttle Control System (ETCS) power source circuit. *Note: The PCM shuts off power to the throttle motor and magnetic clutch (the accelerator pedal can be opened with the throttle valve).* **Possible Causes** • Battery connections are dirty or loose • ETCS power source circuit is open (check the ETCS fuse) • PCM has failed
DTC P1128 **1T CCM, MIL: Yes** 1998, 1999, 2000, 2001, 2002 Supra, Land Cruiser, Sequoia, Tundra 3.0L VIN D, 4.7L VIN T Transmissions: All	**ETCS Throttle Control Motor Lock Malfunction Conditions:** Engine started, engine running, and the PCM detected the throttle control motor position was "locked" during normal operation of the throttle control motor during the CCM test. **Possible Causes** • ETCS throttle motor is damaged or has failed • Throttle body assembly is damaged or stuck closed • PCM has failed
DTC P1128 **1T CCM, MIL: Yes** 2001, 2002, 2003 Prius Engines: 1.5L VIN B Transmissions: All	**ETCS Throttle Control Motor Lock Malfunction Conditions:** Engine started, engine running, and the PCM detected the throttle control motor position was "locked" during normal operation of the throttle control motor during the CCM test. **Possible Causes** • ETCS throttle motor is damaged or has failed • Throttle body assembly is damaged or stuck • PCM has failed
DTC P1129 **1T CCM, MIL: Yes** 1998, 1999, 2000, 2001, 2002 Supra, Land Cruiser, Sequoia, Tundra 3.0L VIN D, 4.7L VIN T Transmissions: All	**Electric Throttle Control System (ETCS) Malfunction Conditions:** Engine started, engine running, and the PCM detected the Actual throttle opening angle varied too much from the Target throttle opening angle during the CCM Rationality test. **Possible Causes** • ETCS (system) is damaged or has failed • PCM has failed
DTC P1129 **1T CCM, MIL: Yes** 2001, 2002, 2003 Prius Engines: 1.5L VIN B Transmissions: All	**Electric Throttle Control System (ETCS) Malfunction Conditions:** Engine started, engine running, and the PCM detected the Actual throttle opening angle varied too much from the Target throttle opening angle during the CCM Rationality test. **Possible Causes** • ETCS (system) is damaged or has failed • PCM has failed
DTC P1130 **2T O2S, MIL: Yes** 1998, 1999, 2000, 2001, 2002 All models All engines Transmissions: All	**A/F Sensor-11 (Bank 1 Sensor 1) Circuit Malfunction Conditions:** DTC P1135 not set, engine started, engine warmup completed, and the PCM detected the A/F sensor signal was more than 3.8v or that it was less than 2.8v; or the A/F sensor signal was fixed at 3.30v; or the PCM detected an unexpected "high" or "low" voltage condition in the A/F sensor signal circuit during the CCM test. **Possible Causes** • A/FS signal circuit or ground circuit is open (intermittent fault) • A/FS is contaminated, damaged or has failed • PCM has failed
DTC P1130 **2T O2S, MIL: Yes** 2003 RAV4 All engines Transmissions: All	**A/F Sensor-11 (Bank 1 Sensor 1) Circuit Malfunction Conditions:** DTC P1135 not set, engine started, engine warmup completed, and the PCM detected the A/F sensor signal was more than 3.8v or that it was less than 2.8v; or the A/F sensor signal was fixed at 3.30v; or the PCM detected an unexpected "high" or "low" voltage condition in the A/F sensor signal circuit during the CCM test. **Possible Causes** • A/FS signal circuit or ground circuit is open (intermittent fault) • A/FS is contaminated, damaged or has failed • PCM has failed

OBD II Trouble Code List (P1xxx Codes)

DTC	Trouble Code Title, Conditions & Possible Causes
DTC P1133 **1T O2S, MIL: Yes** 1998, 1999, 2000, 2001, 2002 All models All engines Transmissions: All	**A/F Sensor-11 (Bank 1 Sensor 1) Slow Response Conditions:** DTC P1135 not set, vehicle speed over 38 mph at over 1600 rpm for 1-3 minutes, and the PCM detected the A/F sensor response times deteriorated below an acceptable value. **Possible Causes** • A/F sensor signal circuit or ground circuit is open • A/F sensor power circuit is open • A/F sensor is contaminated, damaged or has failed • Intake air leaks, exhaust manifold leaks or PCV system leaks • MAF sensor out of calibration (it may be dirty or contaminated) • PCM has failed
DTC P1133 **2T O2S, MIL: Yes** 2003 RAV4 All engines Transmissions: All	**A/F Sensor-11 (Bank 1 Sensor 1) Slow Response Conditions:** DTC P1135 not set, engine started, vehicle driven to a speed of over 38 mph at over 1600 rpm for 1-3 minutes, and the PCM detected the A/F sensor response times deteriorated below an acceptable value. **Possible Causes** • A/F sensor signal circuit or ground circuit is open • A/F sensor power circuit is open • A/F sensor is contaminated, damaged or has failed • Intake air leaks, exhaust manifold leaks or PCV system leaks • MAF sensor out of calibration (it may be dirty or contaminated) • PCM has failed
DTC P1135 **2T CCM, MIL: Yes** 1998, 1999, 2000, 2001, 2002 All models All engines Transmissions: All	**A/F Sensor-11 (Bank 1 Sensor 1) Heater Circuit Malfunction Conditions:** Engine started, engine running, and the PCM detected the A/F Sensor-11 (A/FS-11) heater current was less than 0.25A, or that it was more than 8.0A at anytime during the CCM test. **Possible Causes** • A/F sensor heater circuit or ground circuit is open • A/F sensor heater power circuit is open • A/F sensor heater is damaged or has failed • PCM has failed
DTC P1135 **2T CCM, MIL: Yes** 2003 RAV4 All engines Transmissions: All	**A/F Sensor-11 (Bank 1 Sensor 1) Heater Circuit Malfunction Conditions:** Engine started; and the PCM detected the A/F Sensor-11 (A/FS-11) heater current was less than 0.25A, or that it was more than 8.0A at anytime during the CCM test. **Possible Causes** • A/F sensor heater circuit or ground circuit is open • A/F sensor heater power circuit is open • A/F sensor heater is damaged or has failed • PCM has failed
DTC P1150 **2T CCM, MIL: Yes** 1998, 1999, 2000, 2001, 2002 All models All engines Transmissions: All	**A/F Sensor-21 (Bank 2 Sensor 1) Circuit Malfunction Conditions:** DTC P1155 not set, engine running in closed loop, and the PCM detected the A/F sensor signal was more than 3.8v or less than 2.8v; or the A/F sensor signal was fixed at 3.30v; or it detected an unexpected "high" or "low" voltage condition on the A/F Sensor signal circuit. **Possible Causes** • A/FS signal circuit or ground circuit is open (intermittent fault) • A/FS power circuit is open (an intermittent fault) • A/FS is contaminated, damaged or has failed • PCM has failed
DTC P1150 **2T CCM, MIL: Yes** 2003 RAV4 All engines Transmissions: All	**A/F Sensor-21 (Bank 2 Sensor 1) Circuit Malfunction Conditions:** DTC P1155 not set; engine running in closed loop, and the PCM detected the A/F sensor signal was more than 3.8v or less than 2.8v; or the A/F sensor signal was fixed at 3.30v; or it detected an unexpected "high" or "low" voltage condition on the A/F Sensor signal circuit. **Possible Causes** • A/FS signal circuit or ground circuit is open (intermittent fault) • A/FS power circuit is open (an intermittent fault) • A/FS is contaminated, damaged or has failed • PCM has failed
DTC P1153 **2T O2S, MIL: Yes** 1998, 1999, 2000, 2001, 2002 All models All engines Transmissions: All	**A/F Sensor-21 (Bank 2 Sensor 1) Slow Response Conditions:** DTC P1155 not set; vehicle speed over 38 mph at over 1600 rpm for 1-3 minutes, and the PCM detected the A/FS-21 response time had deteriorated below an acceptable value. **Possible Causes** • A/F sensor signal circuit or ground circuit is open • A/F sensor power circuit is open • A/F sensor is contaminated, damaged or has failed • Intake air leaks, exhaust manifold leaks or PCV system leaks • MAF sensor out of calibration (it may be dirty or contaminated)

OBD II Trouble Code List (P1xxx Codes)

DTC	Trouble Code Title, Conditions & Possible Causes
DTC P1153 **2T O2S, MIL: Yes** 2003 RAV4 All engines Transmissions: All	**A/F Sensor-21 (Bank 2 Sensor 1) Slow Response Conditions:** DTC P1155 not set; vehicle speed over 38 mph at over 1600 rpm for 1-3 minutes, and the PCM detected the A/FS-21 response time had deteriorated below an acceptable value. **Possible Causes** • A/F sensor signal circuit or ground circuit is open • A/F sensor power circuit is open • A/F sensor is contaminated, damaged or has failed • Intake air leaks, exhaust manifold leaks or PCV system leaks • MAF sensor out of calibration (it may be dirty or contaminated) • PCM has failed
DTC P1155 **2T CCM, MIL: Yes** 1998, 1999, 2000, 2001, 2002 All models All engines Transmissions: All	**A/F Sensor-21 (Bank 2 Sensor 1) Heater Circuit Malfunction Conditions:** Engine started, engine running, and the PCM detected the A/F Sensor-21 (A/FS-21) heater current was less than 0.25A, or that it was more than 8.0A at anytime during the CCM test. **Possible Causes** • A/F sensor heater circuit or ground circuit is open • A/F sensor heater power circuit is open • A/F sensor heater is damaged or has failed • PCM has failed
DTC P1155 **2T CCM, MIL: Yes** 2003 RAV4 All engines Transmissions: All	**A/F Sensor-21 (Bank 2 Sensor 1) Heater Circuit Malfunction Conditions:** Engine started, engine running, and the PCM detected the A/F Sensor-21 (A/FS-21) heater current was less than 0.25A, or that it was more than 8.0A at anytime during the CCM test. **Possible Causes** • A/F sensor heater circuit or ground circuit is open • A/F sensor heater power circuit is open • A/F sensor heater is damaged or has failed • PCM has failed
DTC P1200 **2T CCM, MIL: Yes** 1996, 1997 Supra All engines Transmissions: All	**Fuel Pump Relay/ECU Circuit Malfunction Conditions:** Engine speed less than 1000 rpm; and the PCM detected an unexpected voltage condition on the fuel pump, fuel pump relay ECU input circuit, or on the fuel pump Diagnostic circuit. **Possible Causes** • Fuel Pump ECU control circuit is open or shorted to ground • Fuel Pump ECU control circuit is shorted to system power (B+) • Fuel Pump ECU power circuit is open (check power from relay) • Fuel Pump ECU is damaged or has failed • Fuel Pump is damaged or has failed • PCM has failed
DTC P1300 **1T CCM, MIL: Yes** 1995 Avalon, Camry, Land Cruiser, Previa, T100, Tacoma, Tercel 1.5L VIN E, 2.4L VIN A, 2.4L VIN R, 2.7L VIN U, 3.0L VIN G, 3.4L VIN V, 4.5L VIN D engines Transmissions: All	**Igniter Circuit Malfunction Conditions:** Engine started; and the PCM did not detect any IGF signals after detecting two IGT signals. *Note: The fault must occur for 3 IGF signals after 6 IGT signals in order to set this code.* **Possible Causes** • IGT or IGF signal circuit is open or shorted to ground • Igniter is damaged or has failed • PCM has failed
DTC P1300 **1T CCM, MIL: Yes** 1996, 1997, 1998, 1999 Celica, Camry 1.8L VIN B, 2.2L VIN G Transmissions: All	**Igniter Circuit Malfunction Conditions:** Engine started, engine running, and the PCM did not detect any IGF signals after detecting four IGT1 signals during the CCM test. **Possible Causes** • IGT or IGF signal circuit is open between the igniter and PCM • IGT or IGF signal circuit is shorted to ground • Igniter is damaged or has failed • PCM has failed
DTC P1300 **1T CCM, MIL: Yes** 2000, 2001, 2002 Camry, Camry Solara 2.2L VIN G, 2.2L VIN N Transmissions: All	**Igniter No. 1 Circuit Malfunction Conditions:** Engine started, engine running, and the PCM did not detect any IGF signals after detecting four IGT1 signals during the CCM test. **Possible Causes** • IGT or IGF signal circuit is open between the igniter and PCM • IGT or IGF signal circuit is shorted to ground • Igniter is damaged or has failed • PCM has failed

OBD II Trouble Code List (P1xxx Codes)

DTC	Trouble Code Title, Conditions & Possible Causes
DTC P1300 **1T CCM, MIL: Yes** 1997, 1998, 1999, 2000 4Runner, RAV4, T100, Tacoma 2.0L VIN K, 2.0L VIN P, 2.2L VIN S, 2.4L VIN L, 2.7L VIN M engines Transmissions: All	**Igniter No. 1 Circuit Malfunction Conditions:** Engine started; and the PCM did not detect any IGF signals after detecting four IGT1 signals during the CCM test. **Possible Causes** • IGT or IGF signal circuit is open between the igniter and PCM • IGT or IGF signal circuit is shorted to ground • Igniter is damaged or has failed • PCM has failed
DTC P1300 **1T CCM, MIL: Yes** 1996, 1997, 1998, 1999 Avalon, Land Cruiser, Previa, T100, Tacoma, Tercel All engines Transmissions: All	**Igniter Circuit Malfunction Conditions:** Engine started; and the PCM did not detect any IGF signals after detecting two or more IGT signals during the CCM test. *Note: This problem must occur for 3 IGF signals after 6 IGT signals in order to set this trouble code and activate the MIL.* **Possible Causes** • IGT or IGF signal circuit is open or shorted to ground • Igniter is damaged or has failed • PCM has failed
DTC P1300 **1T CCM, MIL: Yes** 1997, 1998, 1999, 2000, 2001, 2002 Camry, Camry Solara, 4Runner 3.0L VIN F, 3.4L VIN N Transmissions: All	**Igniter Circuit Malfunction Conditions:** Engine started, engine running, and the PCM did not detect any IGF signals after detecting four IGT1 signals during the CCM test. **Possible Causes** • IGT or IGF signal circuit is open between the igniter and PCM • IGT or IGF signal circuit is shorted to ground • Igniter is damaged or has failed • PCM has failed
DTC P1300 **1T CCM, MIL: Yes** 2000, 2001, 2002 Avalon, Echo, MR2, Highlander, Prius, Sienna, Land Cruiser, Sequoia, Tundra All engines Transmissions: All	**Igniter No. 1 Circuit Malfunction Conditions:** Engine started, engine running, and the PCM did not detect any IGF1 signals after detecting two IGT1 signals during the CCM test. *Note: This vehicle uses a Coil-On-Plug design Ignition System.* **Possible Causes** • IGT or IGF signal circuit is open or shorted to ground • Igniter is damaged or has failed • PCM has failed
DTC P1300 **1T CCM, MIL: Yes** 2001, 2002, 2003 RAV4 2.0L VIN K engine Transmissions: All	**Igniter No. 1 Circuit Malfunction Conditions:** Engine started, engine running, and the PCM did not detect any IGF1 signals after detecting two IGT1 signals during the CCM test. **Possible Causes** • IGF circuit is open or shorted to ground • IGT1 circuit is open or shorted to ground • Igniter is damaged or has failed • PCM has failed
DTC P1300 **1T CCM, MIL: Yes** 1998, 1999, 2000 Sienna 3.0L VIN F engine Transmissions: All	**Igniter No. 1 Circuit Malfunction Conditions:** Engine started; and the PCM did not detect any IGF1 signals after detecting 4 IGT1 signals. **Possible Causes** • IGT or IGF signal circuit is open between the igniter and PCM • IGT or IGF signal circuit is shorted to ground • Igniter is damaged or has failed • PCM has failed
DTC P1300 **1T CCM, MIL: Yes** 2000, 2001, 2002 Tacoma Engines: 2.4L VIN L Transmissions: All	**Igniter No. 1 Circuit Malfunction Conditions:** Engine started; and the PCM did not detect any IGF1 signals after detecting 2 IGT1 signals. **Possible Causes** • IGT or IGF signal circuit is open or shorted to ground • Igniter is damaged or has failed • PCM has failed
DTC P1300 **1T CCM, MIL: Yes** 2000, 2001, 2002, 2003 Celica, Corolla, Matrix, Prius All engines Transmissions: All	**Igniter No. 1 Circuit Malfunction Conditions:** Engine started, engine running, and the PCM did not detect any IGF signals after detecting four IGT3 signals. This engine uses a Direct Ignition (DI) system where one coil is used to fire one cylinder. The coil high-energy secondary wire is connected to one spark plug. If P1300 to P1315 are all set, check for an open or shorted IGF circuit. **Possible Causes** • IGT or IGF1 signal circuit is open between the igniter and PCM • IGT or IGF1 signal circuit is shorted to ground • Igniter is damaged or has failed • PCM has failed

OBD II Trouble Code List (P1xxx Codes)

DTC	Trouble Code Title, Conditions & Possible Causes
DTC P1305 **1T CCM, MIL: Yes** 2000, 2001, 2002 Avalon, Echo, MR2, Highlander, Sienna, Land Cruiser, Sequoia, Tundra All engines Transmissions: All	**Igniter No. 2 Circuit Malfunction Conditions:** Engine started; and the PCM did not detect any IGF2 signals after detecting two IGT2 signals. *Note: This vehicle uses a Coil-On-Plug design Ignition System.* **Possible Causes** • IGT or IGF signal circuit is open or shorted to ground • Igniter is damaged or has failed • PCM has failed
DTC P1305 **1T CCM, MIL: Yes** 2000, 2001, 2002, 2003 Celica, Corolla, Matrix, Prius All engines Transmissions: All	**Igniter No. 2 Circuit Malfunction Conditions:** Engine started; and the PCM did not detect any IGF signals after detecting four IGT3 signals. This engine uses a Direct Ignition (DI) system where one coil is used to fire one cylinder. The coil high-energy secondary wire is connected to one spark plug. If P1300 to P1315 are both set, check for an open or shorted condition in the IGF circuit. **Possible Causes** • IGT or IGF2 signal circuit is open between the igniter and PCM • IGT or IGF2 signal circuit is shorted to ground • Igniter is damaged or has failed • PCM has failed
DTC P1305 **1T CCM, MIL: Yes** 2001, 2002, 2003 RAV4 2.0L VIN K engine Transmissions: All	**Igniter No. 2 Circuit Malfunction Conditions:** Engine started; and the PCM did not detect any IGF2 signals after detecting 2 IGT2 signals. **Possible Causes** • IGF circuit is open or shorted to ground • IGT2 circuit is open or shorted to ground • Igniter is damaged or has failed • PCM has failed
DTC P1305 **1T CCM, MIL: Yes** 2000, 2001, 2002 Tacoma Engines: 2.4L VIN L Transmissions: All	**Igniter No. 2 Circuit Malfunction Conditions:** Engine started; and the PCM did not detect any IGF signals after detecting 2 IGT2 signals. **Possible Causes** • IGT or IGF signal circuit is open or shorted to ground • Igniter is damaged or has failed • PCM has failed
DTC P1310 **1T CCM, MIL: Yes** 1997, 1998, 1999, 2000 4Runner, RAV4, T100, Tacoma 2.0L VIN K, 2.0L VIN P, 2.2L VIN S, 2.7L VIN M Transmissions: All	**Igniter No. 2 Circuit Malfunction Conditions:** Engine started; and the PCM did not detect any IGF signals after detecting 4 IGT2 signals. **Possible Causes** • IGT or IGF signal circuit is open between the igniter and PCM • IGT or IGF signal circuit is shorted to ground • Igniter is damaged or has failed • PCM has failed
DTC P1310 **1T CCM, MIL: Yes** 2001, 2002, 2003 RAV4 2.0L VIN K engine Transmissions: All	**Igniter No. 3 Circuit Malfunction Conditions:** Engine started; and the PCM did not detect any IGF3 signals after detecting 2 IGT3 signals. **Possible Causes** • IGF circuit is open or shorted to ground • IGT3 circuit is open or shorted to ground • Igniter is damaged or has failed • PCM has failed
DTC P1310 **1T CCM, MIL: Yes** 1998, 1999, 2000 Sienna 3.0L VIN F engine Transmissions: All	**Igniter No. 2 Circuit Malfunction Conditions:** Engine started; and the PCM did not detect any IGF2 signals after detecting 4 IGT2 signals. **Possible Causes** • IGT or IGF signal circuit is open between the igniter and PCM • IGT or IGF signal circuit is shorted to ground • Igniter has failed, or the PCM has failed
DTC P1310 **1T CCM, MIL: Yes** 2000, 2001, 2002 Tacoma Engines: 2.4L VIN L Transmissions: All	**Igniter No. 3 Circuit Malfunction Conditions:** Engine started; and the PCM did not detect any IGF3 signals after detecting 2 IGT3 signals. **Possible Causes** • IGT or IGF signal circuit is open or shorted to ground • Igniter is damaged or has failed • PCM has failed
DTC P1310 **1T CCM, MIL: Yes** 2000, 2001, 2002 Avalon, Echo, MR2, Highlander, Sienna, Land Cruiser, Sequoia, Tundra All engines and transmissions	**Igniter No. 3 Circuit Malfunction Conditions:** Engine started; and the PCM did not detect any IGF3 signals after detecting two IGT3 signals during the test. *Note: This vehicle uses a Coil-On-Plug design Ignition System.* **Possible Causes** • IGT or IGF signal circuit is open or shorted to ground • Igniter is damaged or has failed

OBD II Trouble Code List (P1xxx Codes)

DTC	Trouble Code Title, Conditions & Possible Causes
DTC P1310 **1T CCM, MIL: Yes** 2000, 2001, 2002, 2003 Celica, Corolla, Matrix, Prius All engines Transmissions: All	**Igniter No. 3 Circuit Malfunction Conditions:** Engine started; and the PCM did not detect any IGF signals after detecting four IGT3 signals. This engine uses a Direct Ignition (DI) system where one coil is used to fire one cylinder. The coil high-energy secondary wire is connected to one spark plug. If P1300 to P1315 are all set, check for an open or shorted IGF circuit. **Possible Causes** • IGT or IGF3 signal circuit is open between the igniter and PCM • IGT or IGF3 signal circuit is shorted to ground • Igniter is damaged or has failed • PCM has failed
DTC P1315 **1T CCM, MIL: Yes** 2000, 2001, 2002 Tacoma Engines: 2.4L VIN L Transmissions: All	**Igniter No. 4 Circuit Malfunction Conditions:** Engine started; and the PCM did not detect any IGF4 signals after detecting 2 IGT4 signals. **Possible Causes** • IGT or IGF signal circuit is open or shorted to ground • Igniter is damaged or has failed • PCM has failed
DTC P1315 **1T CCM, MIL: Yes** 2001, 2002, 2003 RAV4 2.0L VIN K engine Transmissions: All	**Igniter No. 4 Circuit Malfunction Conditions:** Engine started; and the PCM did not detect any IGF4 signals after detecting 2 IGT4 signals. **Possible Causes** • IGF circuit is open or shorted to ground • IGT4 circuit is open or shorted to ground • Igniter is damaged or has failed • PCM has failed
DTC P1315 **1T CCM, MIL: Yes** 2000, 2001, 2002 Avalon, Echo, MR2, Highlander, Sienna, Land Cruiser, Sequoia, Tundra All engines Transmissions: All	**Igniter No. 4 Circuit Malfunction Conditions:** Engine started; and the PCM did not detect any IGF4 signals after detecting 2 IGT4 signals during the CCM test. *Note: This vehicle uses a Coil-On-Plug design Ignition System.* **Possible Causes** • IGT or IGF signal circuit is open or shorted to ground • Igniter is damaged or has failed • PCM has failed
DTC P1315 **1T CCM, MIL: Yes** 2000, 2001, 2002, 2003 Celica, Corolla, Matrix, Prius All engines Transmissions: All	**Igniter No. 4 Circuit Malfunction Conditions:** Engine started; and the PCM did not detect any IGF signals after detecting four IGT4 signals. This engine uses a Direct Ignition (DI) system where one coil is used to fire one cylinder. The coil high-energy secondary wire is connected to one spark plug. If P1300 to P1315 are all set, check for an open or shorted IGF circuit. **Possible Causes** • IGT or IGF4 signal circuit is open between the igniter and PCM • IGT or IGF4 signal circuit is shorted to ground • Igniter is damaged or has failed • PCM has failed
DTC P1320 **1T CCM, MIL: Yes** 2000, 2001, 2002 Avalon, Highlander, Land Cruiser, Sequoia, Sienna, Tundra 3.0L VIN F, 4.7L VIN T Transmissions: All	**Igniter No. 5 Circuit Malfunction Conditions:** Engine started; and the PCM did not detect any IGF5 signals after detecting IGT5 signals during the CCM test. *Note: This vehicle uses a Coil-On-Plug design Ignition System.* **Possible Causes** • IGT or IGF signal circuit is open between the igniter and PCM • IGT or IGF signal circuit is shorted to ground • Igniter is damaged or has failed • PCM has failed
DTC P1325 **1T CCM, MIL: Yes** 2000, 2001, 2002 Avalon, Highlander, Land Cruiser, Sequoia, Sienna, Tundra 3.0L VIN F, 4.7L VIN T Transmissions: All	**Igniter No. 6 Circuit Malfunction Conditions:** Engine started; and the PCM did not detect any IGF6 signals after detecting IGT6 signals during the CCM test. *Note: This vehicle uses a Coil-On-Plug design Ignition System.* **Possible Causes** • IGT or IGF signal circuit is open between the igniter and PCM • IGT or IGF signal circuit is shorted to ground • Igniter is damaged or has failed • PCM has failed
DTC P1330 **1T CCM, MIL: Yes** 1998, 1999, 2000, 2001, 2002 Land Cruiser, Sequoia, Tundra Engines: 4.7L VIN T Transmissions: All	**Igniter No. 7 Circuit Malfunction Conditions:** Engine started; and the PCM did not detect any IGF7 signals after detecting IGT7 signals during the CCM test. *Note: This vehicle uses a Coil-On-Plug design Ignition System.* **Possible Causes** • IGT or IGF signal circuit is open between the igniter and PCM • IGT or IGF signal circuit is shorted to ground • Igniter is damaged or has failed • PCM has failed

OBD II Trouble Code List (P1xxx Codes)

DTC	Trouble Code Title, Conditions & Possible Causes
DTC P1335 **1T CCM, MIL: Yes** 1995 Avalon, Camry, Land Cruiser, Previa, T100, Tacoma, Tercel 1.5L VIN E, 2.4L VIN A, 2.4L VIN R, 2.7L VIN U, 3.0L VIN G, 3.4L VIN V, 4.5L VIN D engines Transmissions: All	**Crankshaft Position Sensor Circuit Malfunction Conditions:** Engine speed over 1000 rpm, and the PCM did not detect any CKP (NE) sensor signals. **Possible Causes** • CKP sensor (+) circuit is open or shorted to ground • CKP sensor (-) circuit is open or shorted to ground • CKP sensor (+), (-) circuit is shorted to VREF or system power • CKP sensor is damaged or has failed • PCM has failed
DTC P1335 **1T CCM, MIL: Yes** 1996, 1997, 1998, 1999, 2000, 2001, 2002, 2003 All models All engines Transmissions: All	**Crankshaft Position Sensor Circuit Malfunction Conditions:** Engine speed over 1000 rpm, and the PCM did not detect any CKP (NE) sensor signals for 500 ms. This code may be set by an intermittent loss of the CKP sensor signal. **Possible Causes** • CKP sensor (+) circuit is open or shorted to ground • CKP sensor (-) circuit is open or shorted to ground • CKP sensor (+), (-) circuit is shorted to VREF or system power • CKP sensor is damaged or has failed • PCM has failed
DTC P1340 **1T CCM, MIL: Yes** 1998, 1999, 2000, 2001, 2002 Land Cruiser, Sequoia, Tundra Engines: 4.7L VIN T Transmissions: All	**Igniter No. 8 Circuit Malfunction Conditions:** Engine started; and the PCM did not detect any IGF8 signals after detecting IGT8 signals during the CCM test. *Note: This vehicle uses a Coil-On-Plug design Ignition System.* **Possible Causes** • IGT or IGF signal circuit is open between the igniter and PCM • IGT or IGF signal circuit is shorted to ground • Igniter is damaged or has failed • PCM has failed
DTC P1345 **1T CCM, MIL: Yes** 2000, 2001, 2002 Echo All engines Transmissions: All	**Variable Valve Timing, CMP Sensor Circuit Malfunction Conditions:** Engine cranking for 4 seconds, and the PCM did not detect a signal from the VVT sensor; or with the engine speed over 600 rpm, the PCM did not detect VVT signal for 5 seconds. **Possible Causes** • VVT sensor (+) signal circuit is open or shorted to ground • VVT sensor (-) signal circuit is open or shorted to ground • VVT sensor is damaged or has failed • PCM has failed
DTC P1346 **1T CCM, MIL: Yes** 2000, 2001, 2002, 2003 Echo, Celica, Corolla, Highlander, Matrix, MR2, Prius, RAV4 All engines Transmissions: All	**Variable Valve Timing, CMP Sensor Range/Performance Conditions:** Engine started; and the PCM detected a deviation between the CKP sensor and the VVT sensor signals due to a mechanical fault in the timing belt or the VVT sensor during testing. **Possible Causes** • Worn timing belt (i.e., the belt may be skipping teeth) • PCM has failed
DTC P1349 **1T CCM, MIL: Yes** 2000, 2001, 2002, 2003 Echo, Celica, Corolla, Matrix, MR2, Prius, RAV4 All engines Transmissions: All	**Variable Valve Timing System (Bank 1) Conditions:** Engine speed from 400-4000 rpm in closed loop, and the PCM detected the valve timing did not change from its initial position, or the current valve timing was fixed during the CCM test. **Possible Causes** • Oil control valve is damaged or has failed • Valve timing is not correct • PCM has failed
DTC P1400 **1T CCM, MIL: Yes** 1995, 1996, 1997, 1998 Supra Engines: 3.0L VIN E Transmissions: All	**Sub-Throttle Position Sensor Circuit Malfunction Conditions:** Engine started; and the PCM detected the Sub-Throttle Position sensor (VTA2) signal was less than 0.25v with the Closed Throttle Position switch "off", or more than 4.90v at any time. **Possible Causes** • VTA2 signal circuit is open or shorted to ground • Sub-Throttle position sensor VREF or ground circuit is open • Sub-Throttle position sensor is damaged or has failed • PCM has failed
DTC P1401 **2T CCM, MIL: Yes** 1995, 1996, 1997, 1998 Supra Engines: 3.0L VIN E Transmissions: All	**Sub-Throttle Position Sensor Range/Performance Conditions:** Vehicle speed over 3 mph, and the PCM detected the difference between the Throttle Position sensor angle and Sub-Throttle Position sensor angle was over 35 degrees. **Possible Causes** • Sub-Throttle position sensor is damaged or has failed • Throttle linkage or throttle body is binding or sticking • PCM has failed

OBD II Trouble Code List (P1xxx Codes)

DTC	Trouble Code Title, Conditions & Possible Causes
DTC P1405 **2T CCM, MIL: Yes** 1995, 1996, 1997, 1998 Supra Engines: 3.0L VIN E Transmissions: All	**Turbo Pressure Sensor Circuit Malfunction Conditions:** Engine started, engine running, and the PCM detected an unexpected voltage condition on the Turbo Boost Pressure sensor circuit during the CCM test. **Possible Causes** • Boost Pressure sensor signal circuit open or shorted to ground • Boost Pressure sensor power or ground circuit is open • Boost Pressure sensor is damaged or has failed • PCM has failed
DTC P1406 **2T CCM, MIL: Yes** 1995, 1996, 1997, 1998 Supra Engines: 3.0L VIN E Transmissions: All	**Turbo Pressure Sensor Circuit Malfunction Conditions:** Engine started, engine running, MAF sensor more than 1.3 g/sec, and the PCM detected the Turbo Pressure sensor signal was less than 1.2v; or with the MAF sensor less than 0.45 g/sec, the Turbo Pressure sensor indicated more than 4.2v during the CCM test. **Possible Causes** • Boost Pressure sensor signal circuit open or shorted to ground • Boost Pressure sensor power or ground circuit is open • Boost Pressure sensor is damaged or has failed • PCM has failed
DTC P1410 **2T CCM, MIL: Yes** 1998, 1999, 2000, 2001, 2002 Avalon, Camry, Camry Solara 3.0L VIN F engine Transmissions: All	**EGR Valve Position Sensor Circuit Malfunction Conditions:** Key on or engine running; and the PCM detected an unexpected voltage condition on the EGR EVP sensor circuit during the test. **Possible Causes** • EVP sensor signal circuit is open between sensor and the PCM • EVP sensor signal circuit is shorted to ground • EVP sensor power (VREF) circuit is open • EVP sensor ground circuit is open • EVP sensor is damaged or has failed • PCM has failed
DTC P1411 **2T CCM, MIL: Yes** 1998, 1999, 2000, 2001, 2002 Avalon, Camry, Camry Solara 3.0L VIN F engine Transmissions: All	**EGR Valve Position Sensor Range/Performance Conditions:** Engine started; ECT sensor signal below 131°F, and the PCM detected the EGR sensor signal was under 0.35v or it indicated a value equal to or greater than 1.65v for 7 seconds. **Possible Causes** • EVP sensor signal circuit is open between sensor and the PCM • EVP sensor signal circuit is shorted to ground • EVP sensor power (VREF) circuit is open • EVP sensor ground circuit is open • EVP sensor is damaged or has failed • PCM has failed
DTC P1430 **1T CCM, MIL: Yes** 2001, 2002, 2003 Prius All engines Transmissions: A/T	**Vacuum Sensor for Absorber & Catalyst System Circuit Malfunction Conditions:** Engine started; and the PCM detected an unexpected voltage condition on the Vacuum Sensor Absorber circuit. *Note: The Scan Tool reads 0 kPa or over 130 kPa if this code sets.* **Possible Causes** • Vacuum sensor signal circuit is open or shorted to ground • Vacuum sensor signal circuit is shorted to VREF • Vacuum sensor power (VREF) circuit is open • Vacuum sensor is damaged or has failed • PCM has failed
DTC P1431 **2T CCM, MIL: Yes** 2001, 2002, 2003 Prius All engines Transmissions: A/T	**Vacuum Sensor for Absorber & Catalyst Performance Conditions:** Key on, engine stopped, ECT sensor more than 32°F, VSV for HC Absorber "off", and the PCM detected the PIM input was less than 1.20v; or engine running at over 1000 rpm, VSV for the HC Absorber "on", ECT sensor more than 32°F, and the PCM detected the PIM input indicated more than 3.96v. *Note: If DTC P0110, P0115, P0120, P0121, P1430 and P1431 are set at the same time, the "common" ground circuit may be open.* **Possible Causes** • Vacuum sensor line (vacuum line) is damaged or disconnected • Vacuum sensor is damaged or has failed • PCM has failed
DTC P1436 **1T CCM, MIL: Yes** 2001, 2002, 2003 Prius All engines Transmissions: A/T	**Variable Valve Malfunction Conditions:** Cold engine startup (ECT sensor from 14-104°F), engine running, ECT sensor more than 113°F, and the PCM detected the Bypass valve operation was not performed correctly during the CCM test. **Possible Causes** • Front exhaust pipe is damaged or leaking • Variable Valve actuator is damaged or has failed • Vacuum line to the Variable Valve actuator is damaged or off

OBD II Trouble Code List (P1xxx Codes)

DTC	Trouble Code Title, Conditions & Possible Causes
DTC P1437 **1T CCM, MIL: Yes** 2001, 2002, 2003 Prius All engines Transmissions: A/T	**Vacuum Line Malfunction Conditions:** ECT sensor signal from 14-104ºF at startup (cold engine); ECT sensor over 113ºF, engine load factor more than 30%, and the PCM detected an unusual negative pressure amount. **Possible Causes** • Check valve is clogged, damaged or has failed • Vacuum sensor line (vacuum line) is damaged or disconnected • Vacuum sensor is damaged or has failed • VSV for the HC Absorber and Catalyst System has failed
DTC P1455 **1T CCM, MIL: Yes** 2001, 2002, 2003 Prius All engines Transmissions: A/T	**Vapor Reducing Fuel Tank Small Leak Detected Conditions:** Engine started, engine running, VSV for the Purge Flow switching valve "on", and the PCM detected the value of the vapor density of the air that flows from the EVAP VSV (vacuum switching valve) to the intake manifold was too high. **Possible Causes** • Fuel system component problems • HO2S is contaminated, damaged or has failed • Hose and/or pipe is damaged or has failed • Ignition system component problems • Mass airflow meter is damaged or has failed • VVT System is damaged or has failed • PCM has failed
DTC P1500 **1T CCM, MIL: Yes** 1995 Avalon, Camry, Land Cruiser, Previa, T100, Tacoma, Tercel 1.5L VIN E, 2.4L VIN A, 2.4L VIN R, 2.7L VIN U, 3.0L VIN G, 3.4L VIN V, 4.5L VIN D engines Transmissions: All	**Starter Signal Circuit Malfunction Conditions:** Engine cranking; and the PCM detected that it did not receive a signal from the starter signal circuit during the test period. **Possible Causes** • Engine starter signal circuit is open or shorted to ground • Engine starter signal circuit is shorted to system power (B+) • Starter has failed • PCM has failed
DTC P1500 **1T CCM, MIL: Yes** 1996, 1997, 1998, 1999, 2000, 2001, 2002 All models All engines Transmissions: All	**Starter Signal Circuit Malfunction Conditions:** Engine cranking and the PCM detected it did not receive a signal on the starter signal circuit. **Possible Causes** • Engine starter signal circuit is open or shorted to ground • Engine starter signal circuit is shorted to system power (B+) • Starter has failed • PCM has failed
DTC P1510 **1T CCM, MIL: Yes** 1996, 1997 Previa All engines Transmissions: A/T	**Boost Pressure Control Circuit Malfunction (SC Enabled) Conditions:** Engine running with the Super magnetic clutch relay "on", and the PCM detected the Intake Air volume flow rate was too high or too low, or with the engine speed at over 2800 rpm, it detected the Intake Air volume flow rate was too low. **Possible Causes** • Supercharger magnetic clutch relay circuit open or shorted • Supercharger magnetic clutch control circuit open or shorted • Supercharger magnetic clutch relay or magnetic clutch is damaged or has failed • Supercharger bypass valve had failed, or the PCM has failed
DTC P1511 **2T CCM, MIL: Yes** 1995, 1996, 1997, 1998 Supra Engines: 3.0L VIN E Transmissions: All	**Turbo Boost Pressure Low Malfunction Conditions:** Engine speed more than 2600 rpm in closed loop, then during a WOT event, the PCM detected +740 mmHg or more of intake pipe pressure during the CCM Rationality test. **Possible Causes** • Actuator for the Waste Gate valve, Intake Air Control valve, Exhaust Gas Control valve is damaged, binding or has failed • Intake Air system is clogged or leaking • VSV control circuit for the Waste Gate valve, Intake Air Control valve, Exhaust Gas Control valve is open • PCM has failed
DTC P1512 **2T CCM, MIL: Yes** 1995, 1996, 1997, 1998 Supra Engines: 3.0L VIN E Transmissions: All	**Turbo Boost Pressure High Malfunction Conditions:** Engine speed less than 3400 rpm in closed loop, then during a WOT event, the PCM detected +150 mmHg or less of intake pipe pressure during the CCM Rationality test. **Possible Causes** • Waste Gate valve, Intake Air Control valve, Exhaust Gas Control valve has failed • Intake Air system is clogged or leaking • VSV control circuit for the Waste Gate valve, Intake Air Control valve, Exhaust Gas Control valve is shorted to ground • PCM has failed

OBD II Trouble Code List (P1xxx Codes)

DTC	Trouble Code Title, Conditions & Possible Causes
DTC P1520 **1T CCM, MIL: Yes** 1998, 1999, 2000, 2001, 2002, 2003 All models All engines Transmissions: All	**Stop Light Switch Circuit Malfunction Conditions:** Vehicle driven to a speed of over 19 mph several times; and the PCM detected the Stop Light signal status did not change at least once under these operating conditions. **Possible Causes** • Stop light switch is shorted to ground • Stop light switch is damaged or has failed • PCM has failed
DTC P1525 **1T CCM, MIL: Yes** 2001, 2002, 2003 Prius All engines Transmissions: A/T	**Resolver Circuit Malfunction Conditions:** Engine started, vehicle driven to a speed of over 12 mph for at least 16 seconds, and the PCM did not detect any vehicle speed signals from the HV ECU SPDO circuit during the CCM test. **Possible Causes** • HV ECU SPDO circuit is open or shorted to ground • HV ECU is damaged or has failed • PCM has failed
DTC P1525 **1T CCM, MIL: Yes** 1999 Land Cruiser All engines Transmissions: All	**Cruise Control Switch Circuit Malfunction Conditions:** Engine started, vehicle driven to a speed of over 35 mph, and the PCM detected an unexpected voltage condition on the Cruise Control switch circuit during the CCM test. **Possible Causes** • Cruise control switch signal circuit is shorted to ground • Cruise control switch is damaged or has failed • PCM has failed
DTC P1566 **1T CCM, MIL: Yes** 1999 Land Cruiser All engines Transmissions: All	**Cruise Control Main Switch Circuit Malfunction Conditions:** Engine started, vehicle driven to a speed of over 35 mph, and the PCM detected an unexpected voltage condition on the Cruise Control Main switch circuit during the CCM test. **Possible Causes** • Cruise control Main switch signal circuit is open • Cruise control Main switch is damaged or has failed • PCM has failed
DTC P1600 **1T CCM, MIL: Yes** 1995 Avalon, Camry, Land Cruiser, Previa, T100, Tacoma, Tercel 1.5L VIN E, 2.4L VIN A, 2.4L VIN R, 2.7L VIN U, 3.0L VIN G, 3.4L VIN V, 4.5L VIN D engines Transmissions: All	**PCM Battery Backup Circuit Malfunction Conditions:** Key on, and the PCM detected an unexpected voltage in the Battery Backup circuit (KAM circuit) during the CCM test. *Note: The PCM will not store any other codes with this code set.* **Possible Causes** • Battery backup circuit is open (check EFI fuse and fuse link) • Battery terminals are corroded or loose • PCM has failed
DTC P1600 **1T CCM, MIL: Yes** 1996, 1997, 1998, 1999, 2000, 2001, 2002, 2003 All models All engines Transmissions: All	**PCM Battery Backup Circuit Malfunction Conditions:** Key on, and the PCM detected an unexpected voltage in the Battery Backup circuit (KAM circuit) during the CCM test. *Note: The PCM will not store any other codes with this code set.* **Possible Causes** • Battery backup circuit is open (check EFI fuse and fuse link) • Battery terminals are corroded or loose • PCM has failed
DTC P1605 **1T CCM, MIL: Yes** 1995, 1996 Avalon, Camry, Land Cruiser, Previa, T100, Tacoma, Tercel 1.5L VIN E, 2.4L VIN A, 2.4L VIN R, 2.7L VIN U, 3.0L VIN G, 3.4L VIN V, 4.5L VIN D engines Transmissions: All	**Knock Control CPU Malfunction Conditions:** Engine started, engine running, and the PCM detected a problem in the Knock Control portion of the controller during the test period. **Possible Causes** • Clear the codes and determine if this code resets. If the same trouble code (P1605 resets), the PCM has failed. • TSB TC002-03 (6/03) contains information related to this code
DTC P1630 **1T CCM, MIL: Yes** 1998 Supra All engines Transmissions: All	**Traction Control System Malfunction Conditions:** Engine runtime over 5 seconds, and the PCM detected an unexpected voltage condition on the Traction Control system circuit, or it received a signal that a TRAC problem existed. **Possible Causes** • ETC+ or ETC signal circuit is open, shorted to ground or power • EFI+ or EFI- signal circuit is open, shorted to ground or power • Throttle Control ECU or the PCM has failed

OBD II Trouble Code List (P1xxx Codes)

DTC	Trouble Code Title, Conditions & Possible Causes
DTC P1633 **1T CCM, MIL: Yes** 1998 Supra All engines Transmissions: All	**Engine Throttle Control System Circuit Malfunction Conditions:** Engine started, engine running, and the PCM detected a problem in the ETCS portion of the circuit located in the engine control module. **Possible Causes** • Clear the codes and determine if this code resets. If the same trouble code (P1633 resets), the PCM has failed. • TSB TC002-03 (6/03) contains information related to this code
DTC P1633 **1T CCM, MIL: Yes** 2001, 2002, 2003 Prius All engines Transmissions: A/T	**ECM ETCS Circuit Malfunction Conditions:** Key on or engine running; and the PCM detected that it lost communication with the ECU ETCS circuit for 1.5 seconds. **Possible Causes** • ETCS signal circuit to the PCM is open • ETCS signal circuit to the PCM is shorted to ground or power • ECU ETCU has failed or the PCM has failed • TSB TC002-03 (6/03) contains information related to this code
DTC P1636 **1T CCM, MIL: Yes** 2001, 2002, 2003 Prius All engines Transmissions: All	**HV ECU Malfunction Conditions:** Key on or engine running; and the PCM detected that it lost communication with the HV ECU module for 1.5 seconds. **Possible Causes** • HV ECU signal (+) or (-) circuit to PCM is open • HV ECU signal (+) or (-) circuit to PCM is shorted to ground • HV ECU has failed or the PCM has failed • TSB TC002-03 (6/03) contains information related to this code
DTC P1637 **1T CCM, MIL: Yes** 2001, 2002, 2003 Prius All engines Transmissions: A/T	**EGSTP Signal Malfunction Conditions:** Key on or engine running; and the PCM did not detect any EGSTP signals from the HV ECU for 2 seconds during the CCM test. **Possible Causes** • EGSTP signal circuit to PCM is open • EGSTP signal circuit to PCM is shorted to ground • EGSTP signal circuit to PCM is shorted to power • HV ECU or the PCM has failed
DTC P1645 **1T CCM, MIL: Yes** 2000, 2001, 2002, 2003 Celica, Corolla, Matrix, MR2 1.8L VIN R, 1.8L VIN Y Transmissions: All	**Body ECU Malfunction Conditions:** Key on or engine running; and the PCM did not detect any signals from the Body or A/C ECU for a period of 3 seconds during the test. **Possible Causes** • A/C ECU had failed (no communication fault) • Body Control ECU has failed (no communication fault) • Communication data bus circuit is open or shorted to ground • Communication data bus circuit is shorted to system power
DTC P1646 **1T CCM, MIL: Yes** 2003 MR2 Engines: 1.8L VIN R Transmissions: All	**Transmission Control ECU Malfunction Conditions:** Key on or engine running; and the PCM did not detect any signals from the Transmission Control ECU for 3 seconds during the test. **Possible Causes** • Communication data bus circuit is open or shorted to ground • Communication data bus circuit is shorted to system power • Transmission Control ECU has failed (no communication fault)
DTC P1656 **1T CCM, MIL: Yes** 2000, 2001, 2002, 2003 Celica, Corolla, Matrix, MR2, Prius, RAV4 All engines Transmissions: All	**Oil Control Valve Circuit Malfunction (Bank 1) Conditions:** Key on or engine running; and the PCM detected an unexpected voltage condition on the Oil Control Valve (OCV) circuit in the test. **Possible Causes** • OCV signal (+) circuit is open or shorted to ground • OCV signal (-) circuit is open or shorted to ground • OCV (valve) is damaged or has failed • PCM has failed
DTC P1663 **1T CCM, MIL: Yes** 2001, 2002 Prius All engines Transmissions: A/T	**Oil Control Valve Circuit Malfunction (Bank 2) Conditions:** Key on or engine running; and the PCM detected an unexpected voltage condition on the Oil Control Valve (OCV) circuit in the test. **Possible Causes** • OCV signal (+) circuit is open or shorted to ground • OCV signal (-) circuit is open or shorted to ground • OCV (valve) is damaged or has failed • PCM has failed

OBD II Trouble Code List (P1xxx Codes)

DTC	Trouble Code Title, Conditions & Possible Causes
DTC P1690 **1T CCM, MIL: Yes** 2000, 2001, 2002, 2003 Celica, Corolla, Matrix All engines Transmissions: All	**Oil Control Valve Circuit Malfunction Conditions:** Key on or engine running; and the PCM detected an unexpected voltage condition on the Oil Control Valve (OCV) circuit in the test. **Possible Causes** • OCV (valve) signal (+) or (-) circuit is open or shorted to ground • OCV (valve) signal circuit is shorted to VREF or system power • OCV (valve) is damaged or has failed • PCM had failed
DTC P1692 **1T CCM, MIL: Yes** 1998, 1999, 2000, 2001, 2002, 2003 Celica, Corolla, Matrix All engines Transmissions: All	**Oil Control Valve Circuit Malfunction (Open) Conditions:** Engine speed less than 6000 rpm, and after the PCM switched the locker arm from low speed to high speed, it detected the Oil Pressure switch signal indicated "on" for 5 seconds. **Possible Causes** • OCV (valve) connector is damaged or open • OCV (valve) control circuit is open • Oil control valve for the VVTL is damaged or has failed • Oil pressure switch for the VVTL is damaged or has failed • PCM has failed
DTC P1693 **1T CCM, MIL: Yes** 1998, 1999, 2000, 2001, 2002, 2003 Celica, Corolla, Matrix All engines Transmissions: All	**Oil Control Valve Circuit Malfunction (Closed) Conditions:** Engine started, ECT sensor more than 140ºF engine speed less than 6000 rpm, then after the PCM switched the locker arm from low speed to high speed, the PCM detected the Oil Pressure switch signal indicated "off" for over 1 second or more during the CCM test. **Possible Causes** • OCV (valve) control circuit is shorted to ground • Oil control valve for the VVTL is damaged or has failed • Oil pressure switch for the VVTL is damaged or has failed • PCM has failed
DTC P1700 **2T CCM, MIL: Yes** 1995, 1996, 1997, 1998 Avalon, Camry, Land Cruiser, Previa, T100, Tacoma, Tercel 1.5L VIN E, 2.4L VIN A, 2.4L VIN R, 2.7L VIN U, 3.0L VIN G, 3.4L VIN V, 4.5L VIN D engines Transmissions: A/T	**Vehicle Speed Sensor '2' Circuit Malfunction Conditions:** Vehicle driven to a speed of over 6 mph, P/N switch indicating "off" (not in Park or Neutral position) for over 4 seconds, TR switch indicating a position other than Neutral, and the PCM did not detect any VSS2 signals after detecting at least 4 VSS1 signals. *Note: This problem must occur at least 500 times (continuously) in order for this trouble code to set.* **Possible Causes** • VSS signal (+) or (-) circuit is open, shorted to ground or power • VSS is damaged or has failed • PCM has failed
DTC P1705 **1T CCM, MIL: Yes** 1995 Avalon, Camry, Land Cruiser, Previa, T100, Tacoma, Tercel 1.5L VIN E, 2.4L VIN A, 2.4L VIN R, 2.7L VIN U, 3.0L VIN G, 3.4L VIN V, 4.5L VIN D engines Transmissions: A/T	**A/T Direct Clutch Speed Sensor Circuit Malfunction Conditions:** Vehicle running, P/N switch at "off" and the PCM detected a Direct Clutch Speed sensor output of 300 rpm or less. **Possible Causes** • Direct clutch speed sensor signal (+) or (-) circuit is open • Direct clutch speed sensor signal (+) or (-) shorted to ground • Direct clutch speed sensor (+) or (-) signal is shorted to power • Direct clutch is damaged or has failed • PCM has failed
DTC P1725 **1T CCM, MIL: Yes** 2000, 2001, 2002, 2003 Echo, RAV4 1.5L VIN T, 2.0L VIN K Transmissions: A/T	**A/T Turbine Shaft Speed Sensor Circuit Malfunction Conditions:** No Shift Solenoid or P/N codes set, engine started, P/N switch indicating "off", gear position indicating 2nd, 3rd gear, or in O/D, no gear change occurring, and the PCM detected the Turbine Shaft Speed (ISS) sensor indicated less than 300 rpm for 4 seconds. **Possible Causes** • Input shaft speed sensor signal (NT) circuit is open or shorted • Input shaft speed sensor is damaged or has failed • PCM has failed
DTC P1730 **1T CCM, MIL: Yes** 2003 RAV4 2.0L VIN K engine Transmissions: A/T	**A/T Revolution Speed Sensor Circuit Malfunction Conditions:** No Shift Solenoid or Neutral codes set, engine started, gear position indicating 2nd, 3rd gear or O/D, no gear change occurring, and the PCM detected the Revolution Speed (NC) sensor indicated less than 300 rpm for 5 seconds. The PCM detects the rotation speed of the counter gear, and compares the signals from the Direct Clutch speed (NT) sensor to the counter gear speed sensor (NC). The PCM uses this signal to detect the shift time so that it can control the engine torque and hydraulic pressure in response to various conditions. **Possible Causes** • Counter gear speed sensor (NC) circuit is open or shorted • Counter gear speed sensor is damaged or has failed • PCM has failed

OBD II Trouble Code List (P1xxx Codes)

DTC	Trouble Code Title, Conditions & Possible Causes
DTC P1760 **1T CCM, MIL: Yes** 2000, 2001, 2002 Echo 1.5L VIN T engine Transmissions: A/T	**A/T Linear Shift Solenoid Circuit Malfunction Conditions:** Engine speed over 500 rpm, P/N switch indicating off, and the PCM detected an unexpected low voltage (under 0.20v) or unexpected high voltage (12.0v) on the SLT solenoid circuit. **Possible Causes** • SLT shift solenoid control circuit is open or shorted to ground • SLT shift solenoid is damaged or has failed • PCM has failed
DTC P1760 **1T CCM, MIL: Yes** 2003 Celica, Corolla, Matrix, RAV4 1.8L VIN R, 2.0L VIN K Transmissions: A/T	**A/T Linear Shift Solenoid Circuit Malfunction Conditions:** Engine started, engine speed over 500 rpm, P/N switch indicating off, and the PCM detected an unexpected low voltage (under 0.20v) or unexpected high voltage (12.0v) on the SLT solenoid circuit. **Possible Causes** • SLT shift solenoid control circuit is open or shorted to ground • SLT shift solenoid is damaged or has failed • PCM has failed
DTC P1765 **1T CCM, MIL: Yes** 1995 Avalon, Camry, Land Cruiser, Previa, T100, Tacoma, Tercel 1.5L VIN E, 2.4L VIN A, 2.4L VIN R, 2.7L VIN U, 3.0L VIN G, 3.4L VIN V, 4.5L VIN D engine w/ A/T	**A/T Linear Shift Solenoid Circuit Malfunction Conditions:** Engine speed over 500 rpm, P/N switch indicating off, and the PCM detected that the Linear Shift Solenoid valve current flow was less than 0.2 amps during the CCM test. **Possible Causes** • SLT shift solenoid control circuit is open or shorted to ground • Linear shift solenoid control circuit is shorted to power • Linear shift solenoid is damaged or has failed • PCM has failed
DTC P1780 **1T CCM, MIL: Yes** 1995 Avalon, Camry, Land Cruiser, Previa, T100, Tacoma, Tercel 1.5L VIN E, 2.4L VIN A, 2.4L VIN R, 2.7L VIN U, 3.0L VIN G, 3.4L VIN V, 4.5L VIN D engines Transmissions: All	**Park/Neutral Position Switch Circuit Malfunction Conditions:** Engine started, engine runtime over 30 seconds, and the PCM detected two or more P/N inputs (Drive, Neutral, 2nd, Low or Reverse) at the same time; or with the engine speed from 1500-2500 rpm with the VSS indicating over 50 mph, it detected the P/N switch indicated "on" for 30 seconds during the CCM test. **Possible Causes** • P/N switch is shorted to ground • P/N switch is out-of-adjustment, damaged or has failed • PCM has failed
DTC P1780 **1T CCM, MIL: Yes** 1996, 1997, 1998, 1999, 2000, 2001, 2002 All models All engines Transmissions: All	**Park/Neutral Position Switch Circuit Malfunction Conditions:** Engine started, engine runtime over 30 seconds, and the PCM detected two or more P/N inputs (Drive, Neutral, 2nd, Low or Reverse) at the same time; or with the engine speed from 1500-2500 rpm with the VSS indicating over 50 mph, it detected the P/N switch indicated "on" for 30 seconds during the CCM test. **Possible Causes** • P/N switch is shorted to ground or wiring harness is shorted • P/N switch is out-of-adjustment, damaged or has failed • PCM has failed
DTC P1780 **1T CCM, MIL: Yes** 2003 Celica, Corolla, Matrix, RAV4 1.8L VIN R, 1.8L VIN Y, 2.0L VIN K engines Transmissions: All	**Park/Neutral Position Switch Circuit Malfunction Conditions:** Engine started, engine runtime over 30 seconds, and the PCM detected two or more P/N inputs (Neutral, 2nd, Low or Reverse) at the same time; or with the engine speed from 1500-5000 rpm with the VSS indicating over 50 mph, MAP sensor at 300 mmHg or more, it detected the P/N switch indicated "on" for 30 seconds in the test. **Possible Causes** • P/N switch is shorted to ground or wiring harness is shorted • P/N switch is out-of-adjustment, damaged or has failed • PCM has failed
DTC P1780 **1T CCM, MIL: Yes** 2000, 2001, 2002 Echo 1.5L VIN T engine Transmissions: All	**A/T Shift Solenoid Valve Circuit Malfunction Conditions:** Engine speed over 500 rpm, P/N switch indicating off, and the PCM detected an unexpected low voltage or unexpected high voltage (12v) on the Shift Solenoid Valve (ST) control circuit. **Possible Causes** • ST shift solenoid connector is damaged or loose • ST shift solenoid control circuit is open or shorted to ground • ST shift solenoid is damaged or has failed • PCM has failed

OBD II Trouble Code List (P2xxx Codes)

DTC	Trouble Code Title, Conditions & Possible Causes
DTC P2102 **1T CCM, MIL: Yes** 2003 4Runner, Land Cruiser, MR2, Sequoia, Tacoma, Tundra 1.8L VIN R, 3.4L VIN N, 4.0L VIN U, 4.7L VIN T Transmissions: All	**Throttle Actuator Control Motor Circuit Low Input Conditions:** Engine started, throttle control motor output duty cycle at 80% or higher, and the PCM detected the throttle motor current was less than 0.5 amps for 2 seconds. The PCM controls the motor position in order to open and close the throttle valve. The opening angle of the throttle valve is sensed by the TP sensor mounted on the throttle body. The PCM uses the TP sensor signal to control the Throttle Valve opening angle (throttle motor) to respond to driving conditions. **Possible Causes** • Throttle motor connector is damaged or open • Throttle control motor circuit is open • Throttle control motor is damaged or has failed • PCM has failed
DTC P2103 **1T CCM, MIL: Yes** 2003 4Runner, Land Cruiser, MR2, Sequoia, Tacoma, Tundra 1.8L VIN R, 3.4L VIN N, 4.0L VIN U, 4.7L VIN T Transmissions: All	**Throttle Actuator Control Motor Circuit High Input Conditions:** Engine started, throttle control motor output duty cycle at 80% or higher, and the PCM detected the throttle motor current was more than 10.0 amps for 600 ms. The PCM controls the motor position in order to open and close the throttle valve. The opening angle of the throttle valve is sensed by the TP sensor mounted on the throttle body. The PCM uses the TP sensor signal to control the Throttle Valve opening angle (throttle motor) to respond to driving conditions. **Possible Causes** • Throttle motor connector is damaged or shorted • Throttle control motor circuit is shorted • Throttle control motor is damaged or has failed • PCM has failed
DTC P2111 **1T CCM, MIL: Yes** 2003 4Runner, Land Cruiser, MR2, Sequoia, Tacoma, Tundra 1.8L VIN R, 3.4L VIN N, 4.0L VIN U, 4.7L VIN T Transmissions: All	**Throttle Actuator Control System Stuck Open Conditions:** Key on or engine started, and the PCM detected the throttle control motor position is stuck open. The PCM controls the motor position in order to open and close the throttle valve. The opening angle of the throttle valve is sensed by the TP sensor mounted on the throttle body. The PCM uses the TP sensor signal to control the Throttle Valve opening angle (throttle motor) to respond to driving conditions. **Possible Causes** • Throttle control motor circuit is open • Throttle control motor is damaged or has failed • Throttle body or throttle valve is damaged or has failed
DTC P2112 **1T CCM, MIL: Yes** 2003 4Runner, Land Cruiser, MR2, Sequoia, Tacoma, Tundra 1.8L VIN R, 3.4L VIN N, 4.0L VIN U, 4.7L VIN T Transmissions: All	**Throttle Actuator Control System Stuck Closed Conditions:** Key on or engine started, and the PCM detected the throttle control motor position is stuck closed. The PCM controls the motor position in order to open and close the throttle valve. The opening angle of the throttle valve is sensed by the TP sensor mounted on the throttle body. The PCM uses the TP sensor signal to control the Throttle Valve opening angle (throttle motor) to respond to driving conditions. **Possible Causes** • Throttle control motor circuit is shorted • Throttle control motor is damaged or has failed • Throttle body or throttle valve is damaged or has failed
DTC P2118 **1T CCM, MIL: Yes** 2003 4Runner, Land Cruiser, MR2, Sequoia, Tacoma, Tundra 1.8L VIN R, 3.4L VIN N, 4.0L VIN U, 4.7L VIN T Transmissions: All	**Throttle Actuator Control Motor Current Performance Conditions:** Key on or engine started, and the PCM detected an unexpected low voltage condition (open circuit) on the ETCS power source circuit. Battery positive voltage is applied to the +BM circuit of the PCM under both Key on and Key off conditions. **Possible Causes** • ETCS power source circuit is open • PCM has failed

OBD II Trouble Code List (P2xxx Codes)

DTC	Trouble Code Title, Conditions & Possible Causes
DTC P2119 **1T CCM, MIL: Yes** 2003 4Runner, Land Cruiser, MR2, Sequoia, Tacoma, Tundra 1.8L VIN R, 3.4L VIN N, 4.0L VIN U, 4.7L VIN T Transmissions: All	**Throttle Actuator Control Throttle Body Performance Conditions:** Engine started, and the PCM detected Actual throttle opening angle continued to vary greatly from the Target opening angel. The idle speed on this vehicle is controlled by the Electronic Throttle Control system (ETCS). This system includes a throttle control motor to operate the throttle valve, a throttle position sensor to detect the accelerator pedal position, and the PCM to control the ETCS and one-valve design of throttle body. The PCM controls this motor in order to control the throttle valve opening to achieve its target speed. **Possible Causes** • ETCS throttle control system • PCM has failed
DTC P2120 **1T CCM, MIL: Yes** 2003 4Runner, Land Cruiser, MR2, Sequoia, Tacoma, Tundra 1.8L VIN R, 3.4L VIN N, 4.0L VIN U, 4.7L VIN T Transmissions: All	**Throttle Pedal Position Sensor/Switch 'D' Circuit Malfunction Conditions:** Engine started, and the PCM detected VPA1 signal indicated less than 0.20v while the VPA2 signal indicated over 0.97 degrees, or the VPA1 signal indicated more than 4.80v for 500 ms. This system (ETCS) does not use a throttle cable. The Accelerator Pedal Position (APP) sensor is mounted on the accelerator pedal bracket. It includes two sensors to detect the accelerator position, and to detect any faults in the APP sensor or its related circuits. **Possible Causes** • APP sensor signal circuit is open or shorted to ground • APP sensor is damaged or has failed • PCM has failed
DTC P2121 **1T CCM, MIL: Yes** 2003 4Runner, Land Cruiser, MR2, Sequoia, Tacoma, Tundra 1.8L VIN R, 3.4L VIN N, 4.0L VIN U, 4.7L VIN T Transmissions: All	**Accelerator Pedal Position Sensor Signal Performance Conditions:** Engine started, IDL signal "off", and the PCM detected the difference between the VPA and VPA2 signal was out-of-range for 2 seconds. This system (ETCS) does not use a throttle cable. The Accelerator Pedal Position (APP) sensor is mounted on the accelerator pedal bracket. It includes two sensors to detect the accelerator position, and to detect any faults in the APP sensor or its related circuits. **Possible Causes** • APP sensor is damaged or has failed • PCM has failed
DTC P2122 **1T CCM, MIL: Yes** 2003 4Runner, Land Cruiser, Sequoia, Tacoma, Tundra 3.4L VIN N, 4.0L VIN U, 4.7L VIN engines Transmissions: All	**Throttle Pedal Position Sensor/Switch 'D' Circuit Low Input Conditions:** Engine started, and the PCM detected VPA1 signal was less than 0.20v while the VPA2 signal indicated over 0.97 degrees for 500 ms. This system (ETCS) does not use a throttle cable. The Accelerator Pedal Position (APP) sensor is mounted on the accelerator pedal bracket. It includes two sensors to detect the accelerator position, and to detect any faults in the APP sensor or its related circuits. **Possible Causes** • APP sensor signal circuit is shorted to ground • APP sensor is damaged or has failed • PCM has failed
DTC P2123 **1T CCM, MIL: Yes** 2003 4Runner, Land Cruiser, Sequoia, Tacoma, Tundra 3.4L VIN N, 4.0L VIN U, 4.7L VIN T engines Transmissions: All	**Throttle Pedal Position Sensor/Switch 'D' Circuit High Input Conditions:** Engine started, and the PCM detected VPA1 signal indicated over 4.80v for 2 seconds. This system (ETCS) does not use a throttle cable. The Accelerator Pedal Position (APP) sensor is mounted on the accelerator pedal bracket. It includes two sensors to detect the accelerator position, and to detect any faults in the APP sensor. **Possible Causes** • APP sensor signal circuit is open • APP sensor is damaged or has failed • PCM has failed
DTC P2125 **1T CCM, MIL: Yes** 2003 4Runner, Land Cruiser, Sequoia, Tacoma, Tundra 3.4L VIN N, 4.0L VIN U, 4.7L VIN T engines Transmissions: All	**Throttle Pedal Position Sensor/Switch 'E' Circuit Malfunction Conditions:** Engine started, and the PCM detected VPA2 signal indicated less than 0.50v while the VPA1 signal indicated over 0.97 degrees, or the VPA1 signal was more than 4.80v or less than 0.20v for 500 ms. This system (ETCS) does not use a throttle cable. The Accelerator Pedal Position (APP) sensor is mounted on the accelerator pedal bracket. It includes two sensors to detect the accelerator position, and to detect any faults in the APP sensor or its circuits. **Possible Causes** • APP sensor signal circuit is open or shorted to ground • APP sensor is damaged or has failed • PCM has failed
DTC P2127 **1T CCM, MIL: Yes** 2003 4Runner, Land Cruiser, Sequoia, Tacoma, Tundra 3.4L VIN N, 4.0L VIN U, 4.7L VIN T engines Transmissions: All	**Throttle Pedal Position Sensor/Switch 'E' Circuit Low Input Conditions:** Engine started, and the PCM detected VPA2 signal was less than 0.20v while the VPA1 signal indicated over 0.97 degrees for 500 ms. The ETCS does not use a throttle cable. The Accelerator Pedal Position sensor is mounted on the accelerator pedal bracket. It includes two sensors to detect the accelerator position or any faults in the APP sensor or its circuits. **Possible Causes** • APP sensor signal circuit is shorted to ground • APP sensor has failed, or the PCM has failed

OBD II Trouble Code List (P2xxx Codes)

DTC	Trouble Code Title, Conditions & Possible Causes
DTC P2128 **1T CCM, MIL: Yes** 2003 4Runner, Land Cruiser, Sequoia, Tacoma, Tundra 3.4L VIN N, 4.0L VIN U, 4.7L VIN T engine Transmissions: All	**Throttle Pedal Position Sensor/Switch 'E' Circuit High Input Conditions:** Engine started, and the PCM detected VPA1 signal was over 4.80v or under 0.20v for 2 seconds. This system (ETCS) does not use a throttle cable. The Accelerator Pedal Position (APP) sensor is mounted on the accelerator pedal bracket. It includes two sensors to detect the accelerator position, and any faults in the APP sensor. **Possible Causes** • APP sensor signal circuit is open • APP sensor is damaged or has failed • PCM has failed
DTC P2135 **1T CCM, MIL: Yes** 2003 4Runner, Land Cruiser, Sequoia, Tundra 4.0L VIN U, 4.7L VIN T Transmissions: All	**Throttle Pedal Position Sensor/Switch 'A'/'B' Voltage Correlation Conditions:** Engine started, and the PCM detected the value of the VPA1 signal less the VPA2 was less than 0.02v, or the VPA1 signal was less than 0.20v with the VPA2 signal less than 0.50v for 400-500 ms. This system (ETCS) does not use a throttle cable. The Accelerator Pedal Position (APP) sensor is mounted on the accelerator pedal bracket. It includes two sensors to detect the accelerator position, and to detect any faults present in the APP sensor. **Possible Causes** • APP sensor signal circuit is open • APP sensor is damaged or has failed • PCM has failed
DTC P2138 **1T CCM, MIL: Yes** 2003 4Runner, Land Cruiser, Sequoia, Tundra 4.0L VIN U, 4.7L VIN T Transmissions: All	**Throttle Pedal Position Sensor/Switch 'D'/'E' Voltage Correlation Conditions:** Engine started, and the PCM detected the value of the VPA1 signal less the VPA2 was less than 0.02v, or the VPA1 signal was less than 0.20v with the VPA2 signal less than 0.50v for 2 seconds. This system (ETCS) does not use a throttle cable. The Accelerator Pedal Position (APP) sensor is mounted on the accelerator pedal bracket. It includes two sensors to detect the accelerator position, and to detect any faults present in the APP sensor. **Possible Causes** • APP sensor signal circuit is open • APP sensor is damaged or has failed • PCM has failed
DTC P2716 **1T CCM, MIL: Yes** 2003 Echo 1.5L VIN T engine Transmissions: A/T	**A/T Pressure Control Solenoid 'D' Circuit Malfunction Conditions:** Engine speed over 500 rpm, P/N switch indicating off, and the PCM detected an unexpected low voltage (under 0.20v) on the Shift Solenoid Valve (SLT) control circuit, or it detected an unexpected high voltage (over 12.0v) on the SLT solenoid control circuit during the CCM test. **Possible Causes** • SLT shift solenoid connector is damaged or loose • SLT shift solenoid control circuit is open or shorted to ground • SLT shift solenoid is damaged or has failed • PCM has failed
DTC P2195 **1T CCM, MIL: Yes** 2003 Avalon, Camry, Echo, Highlander, Sienna, Tacoma 1.5L VIN T, 2.4L VIN D, 2.4L VIN L, 2.7L VIN M, 3.0L VIN F engines Transmissions: All	**Air Fuel Sensor 1 (Bank 1 Sensor 1) Signal Stuck "Lean" Conditions:** Vehicle speed from 25-87 mph at over 1500 rpm with the throttle valve open, and the PCM detected the Air Fuel sensor signal indicated more than 3.80v for 10 seconds. **Possible Causes** • A/FS1 signal circuit is open or shorted to ground • A/FS1 is damaged, contaminated or it has failed • Air induction system is severely restricted • Fuel Control component problems (e.g., low fuel pressure, or one or more severely restricted fuel injectors) • PCM has failed
DTC P2196 **1T CCM, MIL: Yes** 2003 Avalon, Camry, Echo, Highlander, Sienna 1.5L VIN T, 2.4L VIN D, 2.4L VIN L, 2.7L VIN M, 3.0L VIN F engines Transmissions: All	**Air Fuel Sensor 1 (Bank 1 Sensor 1) Signal Stuck "Rich" Conditions:** Vehicle speed from 25-87 mph at over 1500 rpm with the throttle valve open, and the PCM detected the Air Fuel sensor signal indicated more than 2.80v for 10 seconds. **Possible Causes** • A/FS1 signal circuit is open or shorted to ground • A/FS1 is damaged, contaminated or it has failed • Air induction system is leaking (check for PCV system leaks) • Fuel component problem (high fuel pressure, leaking regulator or a leaking injector) • PCM has failed
DTC P2196 **1T CCM, MIL: Yes** 2003 Avalon, Camry, Echo, Highlander, Sienna 1.5L VIN T, 2.4L VIN D, 3.0L VIN F engines Transmissions: All	**Air Fuel Sensor 1 (Bank 2 Sensor 1) Signal Stuck "Lean" Conditions:** Vehicle speed from 25-87 mph at over 1500 rpm with the throttle valve open, and the PCM detected the Air Fuel sensor signal indicated more than 3.80v for 10 seconds. **Possible Causes** • A/FS1 signal circuit is open or shorted to ground • A/FS1 is damaged, contaminated or it has failed • Air induction system is severely restricted • Fuel component problem (e.g., low fuel pressure, or a severely restricted fuel injector) • PCM has failed

OBD II Trouble Code List (P2xxx Codes)

DTC	Trouble Code Title, Conditions & Possible Causes
DTC P2198 **1T CCM, MIL: Yes** 2003 Avalon, Camry, Echo, Highlander, Sienna 1.5L VIN T, 2.4L VIN D, 3.0L VIN F engines Transmissions: All	**Air Fuel Sensor 1 (Bank 2 Sensor 1) Signal Stuck "Rich" Conditions:** Vehicle driven at a steady speed of 25-87 mph at over 1500 rpm with the throttle valve open, and the PCM detected the Air Fuel sensor signal indicated more than 2.80v for 10 seconds. **Possible Causes** • A/FS1 signal circuit is open or shorted to ground • A/FS1 is damaged, contaminated or it has failed • Air induction system is leaking (check for PCV system leaks) • Fuel Control component problems (e.g., high fuel pressure, a leaking pressure regulator, one or more leaking fuel injectors) • PCM has failed
DTC P2237 **1T CCM, MIL: Yes** 2003 Avalon, Camry, Highlander, Sienna, Tacoma 2.4L VIN L, 2.7L VIN M, 3.0L VIN F engines Transmissions: All	**Air Fuel Sensor 1 (Bank 1 Sensor 1) Pumping Current Signal Open Conditions:** Vehicle speed from 25-87 mph at over 1500 rpm with the throttle valve open, and the PCM detected the A/F Sensor AF+ signal was less than 0.50v or more than 4.80v for 5 seconds. **Possible Causes** • A/FS1 signal circuit is open or shorted to ground • A/FS1 is damaged, contaminated or it has failed • A/FS1 heater assembly is damaged or its circuit has failed • A/FS1 heater relay is damaged or has failed • PCM has failed
DTC P2240 **1T CCM, MIL: Yes** 2003 Avalon, Camry, Highlander, Sienna 3.0L VIN F engine Transmissions: All	**Air Fuel Sensor 1 (Bank 2 Sensor 1) Pumping Current Signal Open Conditions:** Engine started, vehicle driven at a steady speed of 25-87 mph at over 1500 rpm with the throttle valve open, and the PCM detected the Air Fuel Sensor AF+ signal indicated less than 0.50v or indicated more than 4.80v for 5 seconds. **Possible Causes** • A/FS1 signal circuit is open or shorted to ground • A/FS1 is damaged, contaminated or it has failed • A/FS1 heater assembly is damaged or its circuit has failed • A/FS1 heater relay is damaged or has failed • PCM has failed
DTC P2725 **1T CCM, MIL: Yes** 2003 Avalon, Camry, Highlander, Sienna 3.0L VIN F engine Transmissions: All	**A/T Pressure Control Solenoid 'E' Circuit Malfunction Conditions:** Engine started, engine warmup period completed, gearshift selector in 'P' or 'N', engine speed 500 rpm or more, and the PCM detected that current flowed to the Shift Solenoid (SLN) control circuit for over 1 second. The Shift Solenoid SLN controls the hydraulic pressure acting on the accumulator control valve when gears are shifted in order to provide smooth gear shifting. **Possible Causes** • Shift solenoid (SLN) connector is damaged or loose • Shift solenoid (SLN) control circuit is open or shorted to ground • Shift solenoid (SLN) is damaged or has failed (electrical fault) • PCM has failed
DTC P3190 **1T CCM, MIL: Yes** 2001, 2002, 2003 Prius All engines Transmissions: A/T	**Poor Engine Power Conditions:** Engine started, engine running at a stable speed 3-5 minutes, HV ECU communicating with the PCM, engine not in start mode, engine target torque at a fixed value or higher, and the PCM detected the ratio of estimated torque to the target torque was less than 20%. **Possible Causes** • Airflow meter is damaged or has failed • Base engine mechanical problems are present • Camshaft position or crankshaft position sensor is damaged • Intake air leaks or restrictions are present • Fuel pressure is too low or too high • Throttle body is damaged or has failed • Water temperature sensor is damaged or has failed • PCM has failed
DTC P3190 **1T CCM, MIL: Yes** 2001, 2002, 2003 Prius All engines Transmissions: A/T	**Engine Does Not Start Conditions:** Engine cranking at a stable speed for a fixed length of time, HV ECU communication with the PCM valid, and the PCM detected the engine remained in engine start mode (it did not start). **Possible Causes** • Airflow meter is damaged or has failed • Base engine mechanical problems are present • Camshaft position or crankshaft position sensor is damaged • Fuel pressure is too low or too high • Throttle body is damaged or has failed • Vehicle is out of fuel • PCM has failed

OBD II Trouble Code List (B2xxx Codes)

DTC	Trouble Code Title, Conditions & Possible Causes
DTC B2785 **1T CCM, MIL: No** 1998, 1999, 2000, 2001, 2002 All models All engines Transmissions: All	**Ignition Switch Signal Always On Conditions:** Key on or engine running; and the PCM detected an unexpected voltage condition on the Ignition Switch signal circuit during the test. **Possible Causes** • Ignition switch signal circuit shorted to system power (B+) • Ignition switch is damaged or has failed (in "on" position) • PCM has failed
DTC B2786 **1T CCM, MIL: No** 1998, 1999, 2000, 2001, 2002 All models All engines Transmissions: All	**Ignition Switch Signal Always Off Conditions:** Key on or engine running; and the PCM detected an unexpected voltage condition on the Ignition Switch signal circuit during the test. **Possible Causes** • EFI main relay is damaged or has failed • Ignition switch signal circuit is open between switch and PCM • Ignition switch is damaged or has failed (in "off" position) • PCM has failed
DTC B2791 **1T CCM, MIL: No** 1998, 1999, 2000, 2001, 2002 All models All engines Transmissions: All	**Key Unlock Warning Switch Off Conditions:** Key on, and the PCM detected an unexpected voltage condition on the Ignition Switch signal circuit during the CCM test. **Possible Causes** • Key unlock warning switch circuit is open or shorted to ground • Key unlock warning switch is damaged or has failed • PCM has failed
DTC B2795 **1T CCM, MIL: No** 1998, 1999, 2000, 2001, 2002 All models All engines Transmissions: All	**Unmatched Key Code Conditions:** Key on, and the PCM received a signal from the Immobilizer system indicating that an "unmatched" key had been inserted. **Possible Causes** • This trouble code (B2795) is set when an unregistered key is inserted. The first step is to clear the code and then insert the key from the customer to detect if B2795 is output. • When a key is found that outputs B2795, register this key. If B2795 is not output, there is a possibility that the unregistered key has been inserted previously (the ECM is operating okay). • Ask the customer how the key was used when the code set.
DTC B2796 **1T CCM, MIL: No** 1998, 1999, 2000, 2001, 2002 All models All engines Transmissions: All	**No Communication With The Immobilizer System Conditions:** Key on, and the PCM detected that it could not communicate with the Immobilizer System. **Possible Causes** • Ignition key is damaged or has failed • Transponder is damaged or has failed • Receiver is damaged or has failed • Wire harness problem in the CODE, RXCH or TXCT circuits. • PCM has failed
DTC B2797 **1T CCM, MIL: No** 1998, 1999, 2000, 2001, 2002 All models All engines Transmissions: All	**Communication Malfunction No. 1 Conditions:** Key on or engine running; and the PCM detected a Communication No. 1 malfunction had occurred with the Immobilizer System. **Possible Causes** • Ignition key is damaged or has failed • Check for signs of add-on equipment (radio, stereos, etc.) • Transponder key amplifier is damaged or has failed • PCM has failed
DTC B2798 **1T CCM, MIL: No** 1998, 1999, 2000, 2001, 2002 All models All engines Transmissions: All	**Communication Malfunction No. 2 Conditions:** Key on or engine running; and the PCM detected a Communication No. 2 malfunction had occurred with the Immobilizer System. **Possible Causes** • Ignition key is damaged or has failed • Check for signs of add-on equipment (radio, stereos, etc.) • Transponder key amplifier is damaged or has failed • PCM has failed
DTC U1001 **1T CCM, MIL: No** 2003 MR2 All engines Transmissions: A/T	**Gearshift Module Communication Circuit Malfunction Conditions:** Key on or engine running; and the PCM detected that it did not receive any communication signals from the Transmission Control ECU (TCM). The TCM and the PCM communicate with each other via the CAN communication line. Note that if U1001 and P1646 set simultaneously, the failsafe function of the PCM will stop the engine. **Possible Causes** • TCM "CAN" communication circuit is open or shorted to ground • Transmission Control ECU is damaged or has failed • PCM has failed

SYMPTOM DIAGNOSIS

Introduction

This section includes a list of Driveability Symptom Descriptions, Checks and Tests. To use this list, locate the symptom that matches the vehicle problem and refer to the Symptom Checks and Tests.

The checks and tests for each symptom do not apply to all engines or vehicle systems. *Toyota Diagnostics should be done prior to this step!*

Driveability Symptom List

Symptom Description	Symptom Checks & Tests
Test 1: Starting Concerns Engine Does Not Crank Hard Start, Long or Erratic Crank Stall After Startup No Start, Normal Crank No Start due to a short to ground on VREF circuit (MIL is Off)	- Check battery, battery circuits to starter - Check for a damaged flywheel, engine compression, base timing and idle speed - Check for a failed fuel pump relay - Check for distributor rotor "punch-through" - Check for faulty ignition module or circuits - Check for short to ground on VREF circuit - Check for lack of starter signal to the PCM
Test 2: Idle Speed Concerns Slow Return to Idle Speed Rolling Idle Speed Fast Idle Speed Low or Slow Idle Speed	- Inspect for intake manifold vacuum leaks - Check PCV valve and for excessive carbon - Check for a restricted exhaust - Check base idle speed and fuel pressure - Check throttle body linkage for binding
Test 3: Stalls, Quits Running At idle Speed During Acceleration During Cruise Speeds During Deceleration	- Inspect for intake manifold vacuum leaks - Check PCV valve and for excessive carbon - Check for a restricted exhaust - Check base idle speed and fuel pressure - Check throttle body linkage for binding - Check for no A/C or electrical load signal
Test 4: Runs Rough At idle Speed During Acceleration During Cruise Speeds	- Inspect for intake manifold vacuum leaks - "Scope" Ignition Secondary components - Check Base timing and idle speed setting - Check fuel pressure and fuel injectors
Test 5: Cuts Out, Misses At idle Speed During Acceleration During Cruise Speeds	- Inspect for intake manifold vacuum leaks - "Scope" Ignition Secondary components - Check fuel pressure and fuel injectors - Check for restricted exhaust or converter
Test 6: Surges During Acceleration During Cruise Speeds During Deceleration	- "Scope" Ignition Secondary components - Check fuel pressure and fuel injectors - Monitor LONGFT value with fault present - Remove/inspect O2S for contamination - Check for restricted exhaust or converter
Test 7: Hesitates, Poor Acceleration During Acceleration During Deceleration	- "Scope" Ignition Secondary components - Check the PCV Valve, related components - Check for sticking or broken engine parts
Test 8: Lack of Power, Sluggish During Acceleration During Cruise Speeds	- "Scope" Ignition Secondary components - Check fuel pressure and fuel injectors - Check for restricted exhaust or converter
Test 9: Emissions Compliance Fails Tailpipe Inspection Fails I/M Readiness Status	- Check Base engine idle speed and timing - Check PCM control of Fuel system (O2S) - OBD II: Do a PCM Reset and Drive Cycle

Test 1: Starting Concerns

Preliminary Checks

Prior to starting this symptom test, carefully perform these steps:

- Verify engine cranks (turns over), verify starter relay operation
- Verify that the EFI Main Relay energizes at key on, engine off
- Check Air Intake system for restrictions (air inlet tubes, dirty filter)
- For No Starts, use Scan Tool to check for a Starter Signal

NOTE: *If the vehicle cranks and will not start, on DI systems, check the rotor for a "punch-through" condition. On EI systems, check the DIS coilpack for faults or leakage in the primary or secondary circuits for the cause of a fault.*

Test 1 Repair Chart

Step Number & Action to Take	YES	NO
Step 1 Description: *No Start Only!* • Check battery cables, state of charge • Does the engine crank normally?	If yes, go to step 2	If no, make repairs to battery, starter or engine mechanical.
Step 2 Description: *Test Ignition Sec.* • Inspect ignition secondary components for damage or leakage (rotor shorted) • Check spark output with spark tester • Use engine analyzer to test secondary • Are Ignition system faults suspected?	If yes, make repairs as needed to the Ignition primary or secondary system. Then retest for the condition.	If no, go to step 3.
Step 3 Description: *Test Fuel System* • Test for leaks or low fuel pressure • Are any fuel system faults suspected?	If yes, make repairs to Fuel system and retest the condition.	If no, go to step 4.
Step 4 Description: *Exhaust restricted?* • Check for any leaking components • Test exhaust system for restriction • Is an exhaust restriction suspected?	If yes, locate the restriction & make repairs. Then retest for the condition.	If no, go to step 5.
Step 5 Description: *Is engine too hot?* • Check for signs of an overheating condition (related to a hard start fault) • Is the engine overheated?	If yes, make repairs to correct the overheating. Then retest the condition.	If no, go to step 6.
Step 6 Description: *Check PCV system* • Inspect PCV system components for broken parts or leaking connections • Test PCV valve operation • Are PCV system faults suspected?	If yes, make repairs as needed to the PCV system. Then retest for the condition.	If no, go to step 7.
Step 7 Description: *Test EVAP system* • Inspect for damaged or disconnected EVAP system components • Are EVAP system faults suspected?	If yes, make repairs as needed to the EVAP system. Then retest the condition.	If no, go to step 8.
Step 8 Description: *Test Base Engine* • Test engine compression • Check the timing chain & valve timing • Check for worn camshaft or valve train • Check for worn or jumped timing chain • Check for manifold gasket leaks • Are any Base Engine faults present?	If yes, make repairs as needed to the Base Engine. Then retest for the condition.	If no, something that caused the fault was missed. Repeat all of the test steps to locate and repair the starting concern.

Test 2: Idle Concerns

Preliminary Checks

Prior to starting this symptom test, inspect these Underhood items:

- All engine related vacuum lines for proper routing and integrity
- All related electrical connectors and wiring harnesses for faults
- Check Air Intake system for restrictions (air inlet tubes, dirty filter)
- Check intake manifold and components for leaks (EGR, IAC, etc.)

NOTE: If a rough idle exists, check base timing and idle speed adjustments along with IAC operation. If okay, check the engine for dirty injectors or excessive carbon buildup.

Test 2 Repair Chart

Step Number & Action to Take	YES	NO
Step 1 Description: *Idle Speed Low?* • Does the warm engine have a low idle speed condition in Park or Neutral?	If yes (low idle or rough idle condition exists), go to step 2.	If no, condition is not present at time. Use Intermittent List
Step 2 Description: *Test IAC operation!* • Disconnect the IAC motor connector • Start the engine at part throttle in P/N • Does the engine start and run smoothly at part throttle in P or N?	If yes, refer to Idle Speed system tests and adjustments in other repair media to set the base idle.	If no, go to step 3.
Step 3 Description: *Compare PID Data* • Connect scan tool, turn off accessories • Start the engine and allow for warmup • Compare specific vehicle known-good Scan Tool readings to actual readings • Use a DVOM to test PCM pin voltages • Are PID & Pin voltages within range?	If yes, meaning all serial data items and Pin Table readings match the known-good values, go to step 4. Look at the IAC readings.	If no (one or more values are out of normal range), refer to Idle Speed tests in other manuals or repair media to make the repairs.
Step 4 Description: *Test Ignition Sec!* • Inspect ignition secondary components for damage or leakage (rotor shorted) • Check spark output with spark tester • Use engine analyzer to test secondary • Are Ignition system faults suspected?	If yes, make repairs as needed to the Ignition primary or secondary system. Then retest for the condition.	If no, go to step 5.
Step 5 Description: *Exhaust restricted?* • Check for any leaking components • Test exhaust system for restriction • Is an exhaust restriction suspected?	If yes, locate the restriction & make repairs. Then retest for the condition.	If no, go to step 6.
Step 6 Description: Test Fuel System! • Test for leaks or low fuel pressure • Are any fuel system faults suspected?	If yes, make repairs to Fuel system and retest the condition.	If no, go to step 7.
Step 7 Description: *Check PCV system* • Inspect PCV system components for broken parts or leaking connections • Test PCV valve operation • Are PCV system faults suspected?	If yes, make repairs as needed to the PCV system. Then retest for the condition.	If no, go to step 8.
Step 8 Description: *Test EVAP system* • Inspect for damaged or disconnected EVAP system components • Are EVAP system faults suspected?	If yes, make repairs as needed to the EVAP system. Then retest the condition.	If no, test Base engine compression and valve timing for reason for condition

Test 3: Stalls, Quits Running

Preliminary Checks

Prior to starting this symptom test, inspect these Underhood items:

- All engine related vacuum lines for proper routing and integrity
- Check Air Intake system for restrictions (air inlet tubes, dirty filter)
- Check intake manifold and components for leaks (EGR, IAC, etc.)

Definition: If the vehicle stalls or quits running and the base timing and idle adjustments are okay, also check the IAC Motor.

Test 3 Repair Chart

Step Number & Action to Take	YES	NO
Step 1 Description: *Does Engine Stall?* • Start warm engine in P/N (allow to idle) • Does the engine stall or almost stall?	If yes, read & repair any codes. With no codes, go to step 4.	If no (meaning the engine does not stall), go to step 2.
Step 2 Description: *Rough idle?* • Does the engine have a warm engine rough idle condition in P or N?	If yes (engine has a warm engine rough idle) go to step 4.	In no faults were detected, condition is not present now.
Step 3 Description: *Compare PID Data* • Connect scan tool, turn off accessories • Start the engine and allow for warmup • Compare the actual vehicle readings to *manual vehicle specific readings*. • Use a DVOM to test PCM pin voltages • Are PID & Pin voltages within range?	If yes, meaning all serial data items and Pin Table readings match the known-good values, go to step 5. Look at the IAC readings.	If no (one or more readings are out of normal range), go the Idle Speed tests in other manuals or repair media to make the repairs.
Step 4 Description: *Recheck for Stall!* • Disconnect the IAC motor connector and recheck for stall or near stall • Start engine at part throttle in P/N. • Does engine stall with IAC removed?	If yes, refer to Idle Speed system tests and adjustments in other repair media to set base idle.	If no, go to step 6.
Step 5 Description: *Test IAC operation* • Start the engine at part throttle in P/N • Disconnect the IAC motor connector • Check for rpm drop on engine stall • Does the engine speed drop or does engine stall with IAC plug removed?	If yes, go to step 3 to read and record PID and Pin Table values. Refer to examples of PID Data in manual.	If no, refer to Idle Speed system tests and adjustments in other repair media to set the base idle.
Step 6 Description: *Test Ignition Sec!* • Inspect ignition secondary components for damage or leakage (rotor shorted) • Use engine analyzer to test secondary • Are Ignition system faults suspected?	If yes, make repairs to Ignition primary or secondary system. Then retest for the condition.	If no, go to step 7
Step 7 Description: *Test Fuel System* • Test for leaks or low fuel pressure • Are any fuel system faults suspected?	If yes, make repairs to Fuel system and retest the condition.	If no, go to step 8.
Step 8 Description: *Check PCV system* • Inspect PCV components for broken parts or leaks, test the valve operation • Are PCV system faults suspected?	If yes, make repairs to the PCV system. Then retest for the condition.	If no, go to step 9.
Step 9 Description: *Test EVAP system* • Inspect for damaged or disconnected EVAP system components • Are EVAP system faults suspected?	If yes, make repairs as needed to the EVAP system. Then retest the condition.	If no, test Base engine compression and valve timing for reason for condition

Test 4: Runs Rough

Preliminary Checks

Prior to starting this symptom test, inspect these Underhood items:

- All engine related vacuum lines for proper routing and integrity
- Check intake manifold and components for leaks (EGR, IAC, etc.)

NOTE: **If the engine runs rough and all engine subsystems are okay, check the engine for excessive carbon buildup.**

Test 4 Repair Chart

Step Number & Action to Take	YES	NO
Step 1 Description: Runs rough? • Start warm engine in P/N (allow to idle) • Does the engine run rough?	If yes, read & repair any codes. With no codes, go to step 3.	If no (the engine does not run rough, go to step 2.
Step 2 Description: Not running rough! • Inspect various Underhood items that could cause a rough running condition (IAC, Ignition system, throttle body). • Were any problems located?	If yes, correct the problem(s). Do a PCM Reset and Idle Relearn if it applies. Verify repair is done	If no obvious faults were detected and repaired during this step, problem is not present at this time.
Step 3 Description: Test IAC operation • Start the engine in P/N (allow it to idle) • Disconnect the IAC motor connector • Check for rough running condition • Next, reconnect the IAC connector • Did the engine run rough with IAC off?	If yes, go to step 4 to read and record PID and Pin Table values. Refer to examples of PID Data in manual.	If no, refer to Idle Speed system tests and adjustments in other repair media to set the base idle.
Step 4 Description: Compare PID Data • Connect scan tool, turn off accessories • Start the engine and allow for warmup • Compare the actual vehicle readings to *manual vehicle specific readings*. • Use a DVOM to test PCM pin voltages • Are PID & Pin voltages within range?	If yes, meaning all serial data items and Pin Table readings match the known-good values, go to step 5. Look at the IAC readings.	If no (one or more readings are out of normal range), go to engine system or component test related to the value that is out of range
Step 5 Description: Test Ignition Sec. • Inspect ignition secondary components for damage or leakage (rotor shorted) • Use engine analyzer to test secondary • Are Ignition system faults suspected?	If yes, make repairs to Ignition primary or secondary system. Then retest for the condition.	If no, go to step 6
Step 6 Description: Test Fuel System • Test for leaks or low fuel pressure • Are any fuel system faults suspected?	If yes, make repairs to Fuel system and retest the condition.	If no, go to step 7.
Step 7 Description: Exhaust restricted? • Check for any leaking components • Test exhaust system for restriction • Is an exhaust restriction suspected?	If yes, locate the restriction & make repairs. Then retest for the condition.	If no, go to step 8.
Step 8 Description: Check PCV system • Inspect PCV components for broken parts or leaks, test the valve operation • Are PCV system faults suspected?	If yes, make repairs to the PCV system. Then retest for the condition.	If no, go to step 9.
Step 9 Description: Test EVAP system • Inspect for damaged or disconnected EVAP system components • Are EVAP system faults suspected?	If yes, make repairs as needed to the EVAP system. Then retest the condition.	If no, test Base engine compression and valve timing for reason for condition

Test 5: Cuts-Out or Misses

Preliminary Checks

Prior to starting this symptom test, inspect these Underhood items:

- All engine related vacuum lines for proper routing and integrity
- Check intake manifold and components for leaks (EGR, IAC, etc.)

Definition: Steady pulsation or jerking that follows engine speed. Usually more pronounced as the engine load increases.

Test 5 Repair Chart

Step Number & Action to Take	YES	NO
Step 1 Description: *Misses at idle?* • Start warm engine in P/N (allow to idle) • Does the engine miss at idle speed?	If yes, read & repair any codes. With no codes, go to step 3.	If no (the engine does not miss at idle, go to step 2.
Step 2 Description: *No miss at idle!* • Inspect various Underhood items that could cause a miss at idle condition (Ignition and Fuel system components) • Were any problems located?	If yes, correct the problem(s). Do a PCM Reset and Idle Relearn if it applies. Verify repair is done	In no obvious faults were detected and repaired during this step, problem is not present at this time.
Step 3 Description: *Compare PID Data* • Connect scan tool, turn off accessories • Start the engine and allow for warmup • Compare the actual vehicle readings to *manual vehicle specific readings*. • Use a DVOM to test PCM pin voltages • Are PID & Pin voltages within range?	If yes, meaning all serial data items and Pin Table readings match the known-good values, go to step 4. OBD II - see misfire history	If no (one or more readings are out of normal range), go to engine system or component test related to the value that is out of range
Step 4 Description: *Test Ignition Sec!* • Inspect ignition secondary components for damage or leakage (rotor shorted) • Use engine analyzer to test secondary • Are Ignition system faults suspected?	If yes, make repairs to Ignition primary or secondary system. Then retest for the condition.	If no, go to step 5
Step 5 Description: *Test Fuel system* • Test for leaks or low fuel pressure • Are any fuel system faults suspected?	If yes, make repairs to Fuel system and retest the condition.	If no, go to step 6.
Step 6 Description: *Exhaust restricted?* • Check for any leaking components • Test exhaust system for restriction • Is an exhaust restriction suspected?	If yes, locate the restriction & make repairs. Then retest for the condition.	If no, go to step 7.
Step 7 Description: *Test EVAP system* • Inspect for damaged or disconnected EVAP system components • Are EVAP system faults suspected?	If yes, make repairs as needed to the EVAP system. Then retest for the fault.	If no, go to step 8
Step 8 Description: *Check PCV system* • Inspect PCV components for broken parts or leaks, test the valve operation • Are PCV system faults suspected?	If yes, make repairs to the PCV system. Then retest for the condition.	If no, test Base engine compression and valve timing for possible causes of the condition. Also check Air Intake system components and the controls to the transaxle or transmission.

Test 6: Surges

Preliminary Checks

Prior to starting this symptom test, inspect these Underhood items:

- Air leaks at the intake manifold mounting surface & throttle body
- All related vacuum lines for leaks, kinks, routing, splits & integrity
- Check charging voltage (13-15v), VSS reading at cruise speeds

Definition: Engine power variation at steady throttle or cruise. It feels like vehicle speeds up/slows down without throttle change.

Test 6 Repair Chart

Step Number & Action to Take	YES	NO
Step 1 Description: *Surge?* • Drive vehicle - verify it bucks or jerks • Does the engine buck or jerk (cruise)?	If yes, read & repair any codes. With no codes, go to step 3.	If no (the engine does not buck or jerk, go to step 2.
Step 2 Description: *No Surge!* • Inspect various Underhood items that could cause an intermittent condition (i.e., fuel, ignition system components) • Were any problems located?	If yes, correct the problem(s). Do a PCM Reset and Road Test to verify the repair is done.	In no obvious faults were detected and repaired during this step, problem is not present at this time.
Step 3 Description: *Compare PID Data* • Connect scan tool, turn off accessories • Drive the vehicle at cruise speeds and compare the actual vehicle readings to *manual vehicle specific readings*. • Use a DVOM to test PCM pin voltages • Are PID & Pin voltages within range?	If yes, meaning all serial data items and Pin Table readings match the known-good values, go to step 4. OBD II - see misfire history	If no (meaning one or more readings are out of range), go to engine system or component test related to the value that is out of range
Step 4 Description: *Check HO2S Input* • Perform a Snap-Accel Test of HO2S at Idle and Cruise speeds. As the throttle is snapped to open in P/N, the signal should quickly go to 0.5-1v. At Decel, the signal should move to 0.0-0.4v. • Did HO2S pass the Accel/Decel Test?	If yes, go to step 5. Note: The tool of choice to test the HO2S operation is the Lab Scope (so that waveforms can be observed).	If no, check for problems at sensor connector (loose or moisture tracking). Inspect sensor tip for white powder & for signs of coolant.
Step 5 Description: *Test Fuel system* • Test for leaks or low fuel pressure (a fuel gauge can be used to test fuel pressure at Cruise and Decel speeds). • Check for a restricted fuel filter. • Monitor LONGFT & SHRTFT at cruise. • Are any fuel system faults suspected?	If yes, make repairs to Fuel system and retest the condition. If SHRTFT reading is near 159, check the PCM ground at the engine grounds.	If no, go to step 6.
Step 6 Description: *Test Ignition Sec!* • Inspect ignition secondary components for damage or leakage (rotor shorted) • Test the spark output with a spark tester. • Remove spark plugs, look for deposits • Use engine analyzer to test secondary • Are Ignition system faults suspected?	If yes, make repairs to Ignition primary or secondary system. Then retest for the original condition.	If no, go to step 7
Step 7 Description: *Exhaust restricted?* • Check for any leaking components • Test exhaust system for restriction • Is an exhaust restriction suspected?	If yes, locate the restriction & make repairs. Then retest for the condition.	In no faults were detected, problem is not present at this time.

Test 7: Hesitation, Sag or Stumble

Preliminary Checks

Prior to starting this symptom test, inspect these Underhood items:

- Air leaks at the intake manifold mounting surface & throttle body
- Check charging voltage (13-15v), PCM main and sensor grounds
- Engine valve timing and compression, and for a worn camshaft

Definition: A momentary lack of response as the accelerator is pushed down that may cause engine to stall if the condition is severe enough.

Test 7 Repair Chart

Step Number & Action to Take	YES	NO
Step 1 Description: *Hesitation or Sag?* • Drive vehicle - verify lack of power. • Does the engine have lack of power?	If yes, read & repair any codes. With no codes, go to step 3.	If no, (engine does not have lack of power) go to step 2.
Step 2 Description: *No Hesitation!* • Inspect various Underhood items that could cause an intermittent condition (i.e., fuel, ignition system components) • Were any problems located?	If yes, correct the problem(s). Do a PCM Reset and Road Test to verify the repair is done.	In no obvious faults were detected and repaired during this step, problem is not present at this time.
Step 3 Description: *Compare PID Data* • Connect scan tool, turn off accessories • Drive the vehicle at part throttle and compare the actual vehicle readings to *manual vehicle specific readings.* • Use a DVOM to test PCM pin voltages • Are PID & Pin voltages within range?	If yes, meaning all serial data items and Pin Table readings match the known-good values, go to step 4. Check the HO2S readings.	If no (meaning one or more readings were out of range), go to engine system or component test related to the value that is out of range
Step 4 Description: *Check HO2S Input* • Perform a Snap-Accel Test of HO2S at Idle and Cruise speeds. As the throttle is snapped to open in P/N, the signal should quickly go to 0.5-1v. At Decel, the signal should move to 0.0-0.4v. • Did HO2S pass the Accel/Decel Test?	If yes, go to step 5. Note: The tool of choice to test the HO2S operation is the Lab Scope (so that waveforms can be observed).	If no, check for problems at sensor connector (loose or moisture tracking). Inspect sensor tip for white powder & for signs of coolant.
Step 5 Description: *Test Fuel system* • Test for leaks or low fuel pressure (a fuel gauge can be used to test fuel pressure at Cruise and Decel speeds). • Check for a restricted fuel filter. • Monitor LONGFT & SHRTFT at cruise. • Are any fuel system faults suspected?	If yes, make repairs to Fuel system and retest the condition. If SHRTFT reading is near 159, check the PCM ground at ICM mounting point.	If no, go to step 6.
Step 6 Description: *Test Ignition Sec!* • Inspect ignition secondary components for damage or leakage (rotor shorted) • Test the spark output with a spark tester. • Remove spark plugs, look for deposits • Use engine analyzer to test secondary • Are Ignition system faults suspected?	If yes, make repairs to Ignition primary or secondary system. Then retest for the original condition.	If no, go to step 7
Step 7 Description: *Exhaust restricted?* • Check for any leaking components • Test exhaust system for restriction • Is an exhaust restriction suspected?	If yes, locate the restriction & make repairs. Then retest for the condition.	In no faults were detected, problem is not present at this time.

Test 8: Lack of Power, Sluggish or Spongy

Preliminary Checks

Prior to starting this symptom test, inspect these Underhood items:

- Air leaks at the intake manifold mounting surface & throttle body
- Check charging voltage (13-15v), PCM main and sensor grounds
- Engine valve timing and compression, and for a worn camshaft

Definition: Engine delivers less than expected power. There is little increase in speed with accelerator pedal pushed part way.

Test 8 Repair Chart

Step Number & Action to Take	YES	NO
Step 1 Description: *Lack of Power?* • Drive vehicle - verify lack of power. • Does the engine have lack of power?	If yes, read & repair any codes. With no codes, go to step 3.	If no, (engine does not have lack of power) go to step 2.
Step 2 Description: *No Lack of Power!* • Inspect various Underhood items that could cause an intermittent condition (i.e., fuel, ignition system components) • Were any problems located?	If yes, correct the problem(s). Do a PCM Reset and Road Test to verify the repair is done.	In no obvious faults were detected and repaired during this step, problem is not present at this time.
Step 3 Description: *Compare PID Data* • Connect scan tool, turn off accessories • Drive the vehicle at part throttle and compare the actual vehicle readings to *manual vehicle specific readings*. • Use a DVOM to test PCM pin voltages • Are PID & Pin voltages within range?	If yes, meaning all serial data items and Pin Table readings match the known-good values, go to step 4. Check the HO2S readings.	If no (meaning one or more readings are out of range), go to engine system or component test related to the value that is out of range
Step 4 Description: *Check HO2S Input* • Perform a Snap-Accel Test of HO2S at Idle and Cruise speeds. As the throttle is snapped to open in P/N, the signal should quickly go to 0.5-1v. At Decel, the signal should move to 0.0-0.4v. • Did HO2S pass the Accel/Decel Test?	If yes, go to step 5. Note: The tool of choice to test the HO2S operation is the Lab Scope (so that waveforms can be observed).	If no, check for problems at sensor connector (loose or moisture tracking). Inspect sensor tip for white powder & for signs of coolant.
Step 5 Description: *Test Fuel system* • Test for leaks or low fuel pressure (a fuel gauge can be used to test fuel pressure at Cruise and Decel speeds). • Check for a restricted fuel filter. • Monitor LONGFT & SHRTFT at cruise. • Are any fuel system faults suspected?	If yes, make repairs to Fuel system and retest the condition. If SHRTFT reading is near 159, check the PCM ground at the engine grounds.	If no, go to step 6.
Step 6 Description: *Test Ignition Sec!* • Inspect ignition secondary components for damage or leakage (rotor shorted) • Test the spark output with a spark tester. • Remove spark plugs, look for deposits • Use engine analyzer to test secondary • Are Ignition system faults suspected?	If yes, make repairs to Ignition primary or secondary system. Then retest for the original condition.	If no, go to step 7
Step 7 Description: *Exhaust restricted?* • Check for any leaking components • Test exhaust system for restriction • Is an exhaust restriction suspected?	If yes, locate the restriction & make repairs. Then retest for the condition.	In no faults were detected, problem is not present at this time.

Test 9: Emissions Compliance

Preliminary Checks

Prior to starting this symptom test, inspect these Underhood items:

- Check that the Charging system voltage is within range (13-15v)
- Check the condition of the PCM main and sensor grounds

Definition: The vehicle fails a state tailpipe emissions test. A rotten egg smell is present (odors do not mean a test will fail).

Test 9 Repair Chart

Step Number & Action to Take	YES	NO
Step 1 Description: *Diagnostic Check!* • Perform an OBD System Check (OBD II) or Diagnostic Circuit Check (OBD I). • Were any codes detected in this step?	If yes, go to other repair manuals or electronic media and repair all codes	If no, go to step 2.
Step 2 Description: *EVAP Failure?* • Did the vehicle only fail the EVAP leak test or Purge Flow test (gases okay)?	If yes, go to step 22.	If no, go to step 3.
Step 3 Description: *Analyze Report!* • Identify any High or Low Gas readings. • If a drive trace is included, identify the drive mode in which the gases failed (e.g., did the gases appear high early in the test and then decrease as the catalyst temperature increased?). • Has the I/M test report been analyzed?	If yes, go to step 4.	If no, carefully repeat all of steps listed under the description: *Analyze the report carefully!*
Step 4 Description: *Develop Base Line* • Develop a base line of the current tailpipe emissions. Use a calibrated gas analyzer and record the current tailpipe readings. Repeat the test more than once to establish a true base line. • Watch for any related symptoms during the base line test (i.e., exhaust smoke or idle speed concerns). • Has the vehicle baseline been done?	If yes, go to step 5. Note: The vehicle base line report can be a useful tool as the readings can be compared against the final emission readings once all vehicle repairs are completed.	If no, repeat the base line test step until the vehicle baseline has been completed.
Step 5 Description: *Any Symptoms?* Check for signs of these symptoms: • Rough, low or "hunting" engine idle. • Signs of exhaust smoke at tailpipe. • Cooling system (did engine warmup?). • Are any of these or any other high emission related symptoms present?	If yes, go to the Symptom List in this manual and select the matching symptom. If it is not listed, refer to other manuals or media.	If no, go to step 6.
Step 6 Description: *Preliminary Checks* Perform these inspections and checks: • Check vacuum lines for kinks, moisture • Check for dirty or loose electrical connections, or backed-out terminals. • Check for the installation or use of aftermarket emission controls and components. • Were any faults or wrong parts found?	If yes, repair the fault or service the aftermarket component as needed. Go to Step 24 to verify the repairs.	If no, go to step 7.

Test 9: Emissions Compliance

Test 9 Repair Chart - Continued

Step Number & Action to Take	YES	NO
Step 7 Description: *Check CO levels* • Use a calibrated gas analyzer to check the vehicle CO levels. • Were the CO levels excessive?	If yes, a high CO level indicates a rich mixture. Go to Step 10.	If no, go to step 8.
Step 8 Description: *Check HC levels* • Use a calibrated gas analyzer to check the vehicle HC levels. • Were the HC levels excessive?	If yes, a high HC level with normal to low CO indicates a lean condition. Go to Step 16.	If no, go to step 9.
Step 9 Description: *Check NOx levels* • Use a calibrated gas analyzer to check the vehicle NOx levels. • Were the NOx levels excessive?	If yes, go to step 20.	If no, the gas levels are okay. Go to the Symptom List - look for other symptoms.
Step 10 Description: *With high CO levels, check the HC levels* • With high CO levels, check the HC levels with a calibrated gas analyzer. • Were the HC levels too high?	If yes, go to step 16 to check for running rich with incomplete combustion.	If no, go to step 15 to check for a rich running condition.
Step 11 Description: *Test Ignition Sys* • Inspect ignition secondary components for damage or leakage (rotor leakage) • Test the spark output with a spark tester. • Remove spark plugs, look for deposits • Use engine analyzer to test secondary • Is the Ignition System okay?	If yes, go to step 12.	If no, make repairs to Ignition primary or secondary system as needed. Go to step 24 to verify repairs.
Step 12 Description: *Check PCV Sys!* • Inspect PCV components for broken parts or leaks, test the valve operation • Is the PCV system okay?	If yes, go to step 13.	If no, make repairs to PCV system as needed. Go to step 24 to verify repairs.
Step 13 Description: *Check Exhaust?* • Check for any leaking components • Test the exhaust system for restriction (the exhaust backpressure reading should be less than 1.5 psi at cruise speeds). • Is the Exhaust system okay?	If yes, go to step 14.	If no, make repairs to the Exhaust system or related components as needed. Go to step 24 to verify repairs.
Step 14 Description: *Test Base Engine* Check these Base Engine Components: • Check engine compression. • Check for wrong or worn camshaft. • Check valve train components. • Check condition of timing belt or chain. • Are Base Engine components okay?	If yes, recheck the test results from the previous test steps. If they are okay, the problem is not present at this time.	If no, make repairs to the Base Engine components as needed. Go to step 24 to verify repairs.
Step 15 Description: *Rich Condition!* • Test the Fuel system for problems that could cause a rich condition (leaking injectors, fuel pressure regulator, etc.). • Monitor the LONGFT & SHRTFT values at idle speed and cruise speed. • Is the Fuel system okay?	If yes, go to step 18.	If no, make repairs to Fuel system as needed. Go to step 24 to verify repairs.

Test 9: Emissions Compliance

Test 9 Repair Chart - Continued

Step Number & Action to Take	YES	NO
Step 16 Description: Lean Condition! • Test for leaks or low fuel pressure (a fuel gauge can be used to test fuel pressure at Cruise and Decel speeds). *Caution: Follow all safety precautions!* • Monitor the LONGFT and SHRTFT readings at engine idle and cruise for signs of a lean running condition. • Check for a weak pump, dirty fuel filter. • Is the Fuel system okay?	If yes, go to step 17.	If no, make repairs as needed to the Fuel system to repair the lean condition. Go to step 24 to verify repairs.
Step 17 Description: *Test Ignition Sys* • Inspect ignition secondary components for damage or leakage (rotor leakage) • Test the spark output with a spark tester. • Remove spark plugs, look for deposits • Use engine analyzer to test secondary • Is the Ignition System okay?	If yes, go to step 18.	If no, make repairs to Ignition primary or secondary system as needed. Go to step 24 to verify repairs.
Step 18 Description: *Check PCV Sys!* • Inspect PCV components for broken parts or leaks, test the valve operation • Is the PCV system okay?	If yes, go to step 19.	If no, make repairs to PCV system as needed. Go to step 24 to verify repairs.
Step 19 Description: *Test Base Engine* Check these Base Engine Components: • Check engine compression. • Check for wrong or worn camshaft. • Check valve train, timing chain or belt. • Are Base Engine components okay?	If yes, recheck the test results from the previous test steps. If they are okay, the problem is not present at this time.	If no, make repairs to the Base Engine components as needed. Go to step 24 to verify repairs.
Step 20 Description: *Check EGR Sys!* • Perform EGR Function Test as needed (check for sticking valve, valve leaks). • Monitor EGR sensor Serial or PID data (compare actual readings to vehicle specific reading in this manual). • Is the EGR system okay?	If yes, go to step 21.	If no, make repairs to EGR system as needed. Go to step 24 to verify repairs.
Step 21 Description: *Additional Checks* Perform the following additional checks: • Analyze report for when the fault occurs (i.e., did it occur only at idle speed, at cruise speed or both?) • Check Cooling system operation (any add-on front Facia or intake changes?) • Are all of these checks okay?	If yes, recheck the test results from the previous test steps. If they are okay, the problem is not present at this time.	If no, make repairs to the system or component as needed. Go to step 24 to verify repairs.
Step 22 Description: *EVAP Problem* • When does the EVAP fault occur? • Attempt to verify the EVAP failure. • Check for loose or damaged fuel cap. • Check condition of the carbon canister. • Inspect EVAP vacuum and/or plastic lines for damage or disconnects. • Are all of these checks okay?	If yes, go to step 23.	If no, make repairs to the system or component as needed. Go to step 24 to verify repairs.

Test 9: Emissions Compliance

Test 9 Repair Chart - Continued

Step Number & Action to Take	YES	NO
Step 23 Description: *Test EVAP Sys* • Perform applicable EVAP system test of components and operation (refer to other manuals or electronic media). • Check for a missing or loose gas cap. • Was the EVAP system okay?	If yes, verify test results. If all okay, the EVAP problem is not present at this time - the problem was intermittent!	If no, make repairs as needed to the EVAP system. Go to step 24 to verify the repairs.
Step 24 Description: *Repair verification* • All vehicle repairs are completed • Key off, remove negative battery cable for over 15 minutes to reset the KAM in the PCM (it contains fuel trim tables) • Reconnect the negative battery cable. • PCM Relearn Steps: run the engine at 2500 rpm in P/N for one minute, then idle the engine in gear for two minutes. • Perform base line test of emissions using exhaust analyzer. • For I/M Emission Test areas where original gas concentrations are reported in grams per mile: refer to the paragraph that follows to verify GPM. • All other areas where original gas concentrations are reported in PPM: verify gas levels are within range. • Are all gas levels within range?	If yes, save any documentation required by the local or federal emission program laws or statutes.	If no, the gas levels are still too high or one of the gas levels is above the acceptable range. Return to step 1 of this test procedure and carefully repeat all of the test steps.

Excessive Grams Per Mile Verification Procedure

Follow this procedure to verify excessive grams per mile (GPM) indications using parts per million (PPM) readings. If the vehicle gas readings are excessive, compare the actual GPM readings to the gas cut-point levels needed to pass the required test. Determine how much the actual GPM reading exceeds the cut-point, as this data will provide an indication of how much the PPM reading is over the cut-point. It will also provide an indication of how much the PPM reading needs to be reduced (e.g., if the actual reading is twice the cut-point reading, the baseline reading will have to be cut in half or more).

Example: If the actual HC produced by a vehicle were 1.6 GPM, the cut-point for HC would be 0.8 GPM (the actual reading is twice the cut-point). An HC reading from a test vehicle during the baseline test averages 440 PPM. Before this vehicle will pass an Inspection/Maintenance (I/M) test, the HC reading from the verification test must be 220 PPM (1/2 of baseline).

Summary: This method is meant to give a general idea of how much a PPM reading must be reduced for a vehicle to pass an Inspection/Maintenance (I/M) Test that calculates GPM. Your experience will help determine if the emissions readings were reduced enough for the vehicle to pass an Inspection/Maintenance (I/M) Test.

Cross Reference of PID Data to Symptoms

This article includes a list of Symptoms, PID Descriptions and related Diagnostic Tips. To use this list, locate the symptom that matches the vehicle problem and refer to the PID Description and Diagnostic Tips. Use a Scan Tool to gather PID Data while simulating the condition.

Note: Perform Toyota Diagnostics and check for bulletins first!

Symptom: Starting Concerns

PID Description	Diagnostic Tips
Cranking RPM (0-900)	*A steady reading during engine cranking indicates that the PCM received a fuel control reference signal*
Fuel Pump Relay (9-11v during cranking)	*If the reading is within range, the fuel pump relay received ignition power*

Symptom: Idle Speed Concerns

PID Description	Diagnostic Tips
Desired Idle	*Indicates the warm engine desired idle speed commanded by the PCM as it compensates for engine load changes (turn on A/C or blower & check for change).*
Engine Speed - A/T, M/T	*Indicates engine rpm computed by the PCM from the fuel control reference signal (should be ±80 of Desired Idle)*
Idle Air Control (0-255) Note: The IAC Motor can be tested with an external Noid Light test device.	*Indicates the PCM commanded position of the IAC motor Pintle. A reading of 0 counts indicates the Pintle is fully extended (no air bypass). If the idle speed is too low with a reading of over 60 counts, the PCM is attempting to correct for the condition. If the idle speed is too high with a reading of 0 counts, the PCM is attempting to correct for the condition.*

Symptom: Stalls, Quits Running

PID Description	Diagnostic Tips
Idle Air Control (0-255) Note: A Code 35 or DTC P0506 may exist if the condition is severe.	*Indicates the PCM commanded position of the IAC motor Pintle. If the engine stalls with a reading of over 80 counts, look for other causes of the condition (i.e., Fuel system too rich or too lean, a dirty throttle body bore, restrictions or leaks in the intake manifold, excessive engine loads).*

Symptom: Runs Rough

PID Description	Diagnostic Tips
Long Term Fuel Trim (0-255, -100% to 100%)	*Indicates the amount of fuel correction commanded by the PCM. If the engine runs rough and the LONGFT counts are below 70, check any items that can cause a rich condition. If all okay, check these items: IAC operation; MAP sensor calibration and source vacuum; fuel pressure.*
MAP Sensor (11-105 kPa, 0-5v)	*Monitor MAP sensor with condition present. If readings are out of range, test sensor calibration and vacuum source.*
Throttle Angle (0-100%)	*Monitor TP Angle with the condition present. The reading should hold steady near 0%. If the reading is not steady, check the PCV valve, Ignition system and Base Engine.*

Cross Reference of PID Data to Symptoms

Symptom: Cuts Out, Misses

PID Description	Diagnostic Tips
Engine Speed - A/T, M/T	*Monitor the engine speed with the condition present. If the engine speed decreases suddenly with no or little actual change in rpm, check for EMI on the reference circuit.*

Symptom: Surges

PID Description	Diagnostic Tips
O2S or HO2S Volts (0-1132mv)	*Warmup the engine and then monitor the Oxygen sensor voltage. The voltage should respond quickly to throttle changes (Snap Accel or Decel operation). If it is sluggish, use a Lab Scope to determine if it should be replaced.*
Long Term Fuel Trim (0-255, -100% to 100%)	*Monitor LONGFT to determine if it is too far into the negative range (%) or the counts are below 70. If a fault is indicated, check any items that can cause a rich condition.*
HO2S Cross Counts (0-20)	*Monitor the O2S or HO2S Cross Counts to determine the number of times the sensor switches in a 1-second period. A reading near 0 counts may indicate a lazy O2 Sensor.*

Symptom: Hesitation, Sag or Stumble

PID Description	Diagnostic Tips
MAP Sensor (11-105 kPa, 0-5v)	*Monitor the MAP Sensor voltage as the throttle is opened. If it does not change smoothly, check the vacuum source.*
Throttle Angle (0-100%)	*Open and close the throttle while checking the TP Angle for a smooth transition during movement (look for binding).*
TP Sensor (0-5v)	*Monitor the TP Sensor voltage for a smooth transition.*

Symptom: Lack of Power, Sluggish

PID Description	Diagnostic Tips
A/C Clutch (On, Off)	*Monitor A/C Clutch signal as the A/C switch is cycled. It should change from Off to On when A/C is selected.*
A/C Request (Yes, No)	*Monitor A/C Request signal as the A/C switch is cycled. It should change from No to Yes as A/C is selected.*
A/C Off for WOT (Yes, No)	*Monitor A/C Off for WOT signal as the throttle is moved briefly to WOT - the A/C clutch should disengage at WOT.*

Symptom: Emissions Compliance

PID Description	Diagnostic Tips
MAP Sensor (11-105 kPa, 0-5v)	*Monitor the MAP Sensor voltage at idle speed and cruise. If the reading is out of range, check the sensor calibration.*
O2S or HO2S Volts (0-1132mv)	*Warmup the engine and then monitor the O2S voltage. The voltage should vary from 10-1000mv at idle and off-idle steady speeds. Use a Lab Scope to confirm the test.*
Long Term Fuel Trim (0-255, -100% to 100%)	*Monitor the LONGFT to determine if it is too far into the negative or positive range (in % or counts). If a fault is indicated, check any items that could cause this condition.*
MAF Sensor (0-512 g/sec)	*Warmup the engine and monitor the MAF Sensor at idle and off-idle speeds. If the sensor is not near 2-10 g/sec at idle, check the Air Intake components and MAF Sensor.*

POWERTRAIN SIGNALS

A/C Compressor Lock Sensor Circuit (1997 Camry 2.2L I4)

Circuit Description

The A/C Compressor Lock Sensor sends one pulse per engine revolution to the PCM.

If the number ratio of the compressor speed divided by the engine speed is smaller than a predetermined value, the PCM turns the compressor off, and the inductor flashes at about one second intervals (Figure 2-25).

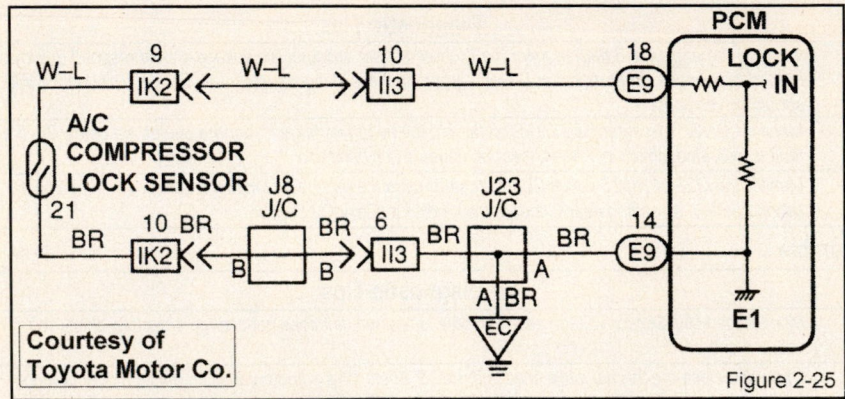

Figure 2-25

A/C Compressor Circuit Graphic

A/C Evaporator Temperature Sensor Circuit (1997 Camry 2.2L I4)

Circuit Description

The PCM uses the signal from the A/C Evaporator Temperature Sensor to detect the temperature inside the cooling unit (Figure 2-26).

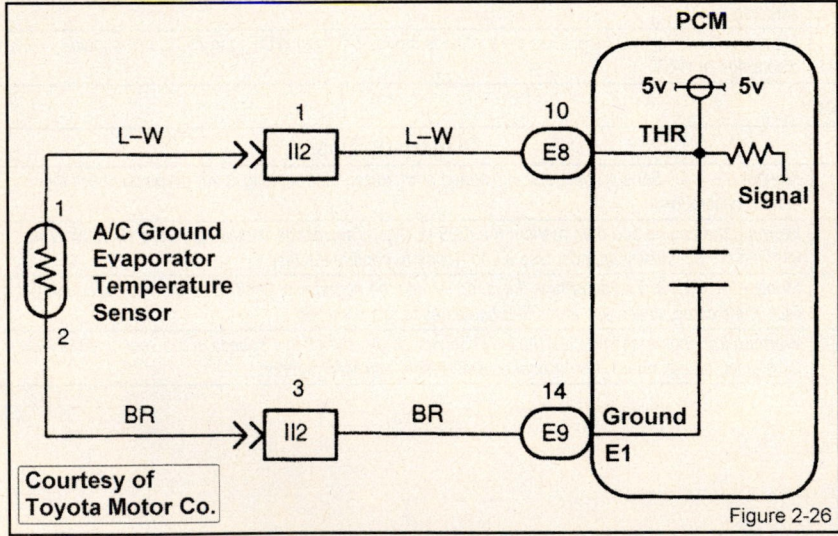

Figure 2-26

POWERTRAIN SIGNALS

Fuel Pump Control Circuit (1997 Camry 2.2L I4)

Circuit Description

The Fuel Pump Control circuit is used to control the operation of the fuel pump relay. The fuel pump is powered through the EFI Main relay, the Circuit Opening relay and the PCM.

Fuel Pump Control Circuit Graphic

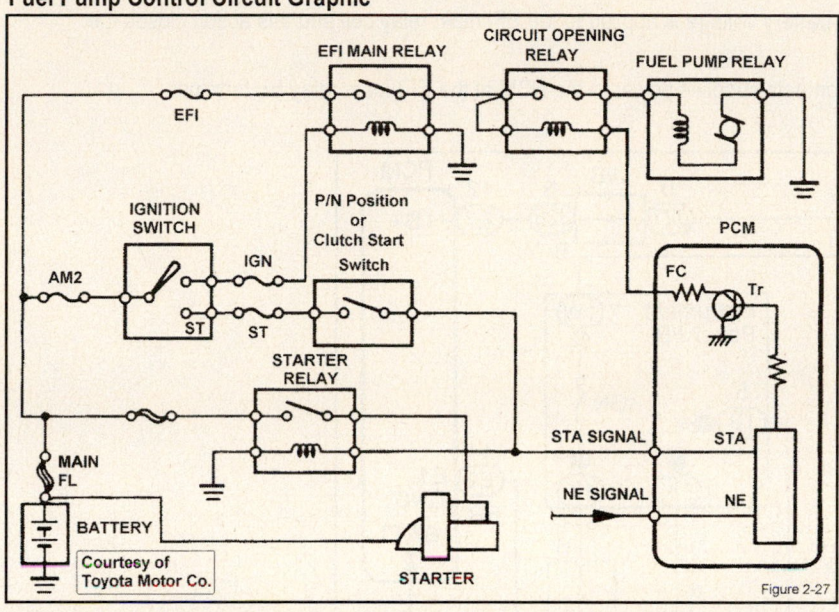

Figure 2-27

Fuel Pump Control Operation

With the engine cranking, current flows from terminal ST of the ignition switch to the Starter relay coil and also to the STA terminal at the PCM.

When the STA signal and NE signal are received by the PCM, the transistor (Tr) is turned on and current flows to the Circuit Opening relay coil. This action causes the relay to switch to on and power is then supplied to the fuel pump so that it can operate.

With the engine running, the NE signal is generated and the PCM keeps the transistor (Tr) on, and this action keeps the Circuit Opening relay energized. This operation enables the fuel pump to run.

Power Source Circuit (1997 Camry 2.2L I4)

Circuit Description

The EFI main relay is used to control the flow of ignition voltage to the PCM. The PCM also is connected to direct battery power through the EFI fuse and to main ground at the E1 connection (Figure 2-28).

Power Source Signal Operation

With the ignition switch turned on, battery voltage is applied to the EFI main relay coil and this action closes the contacts of the relay.

It is the action of closing the coil contacts supplies power to the PCM at the EFI main relay B+ terminal.

Power Source Circuit Graphic

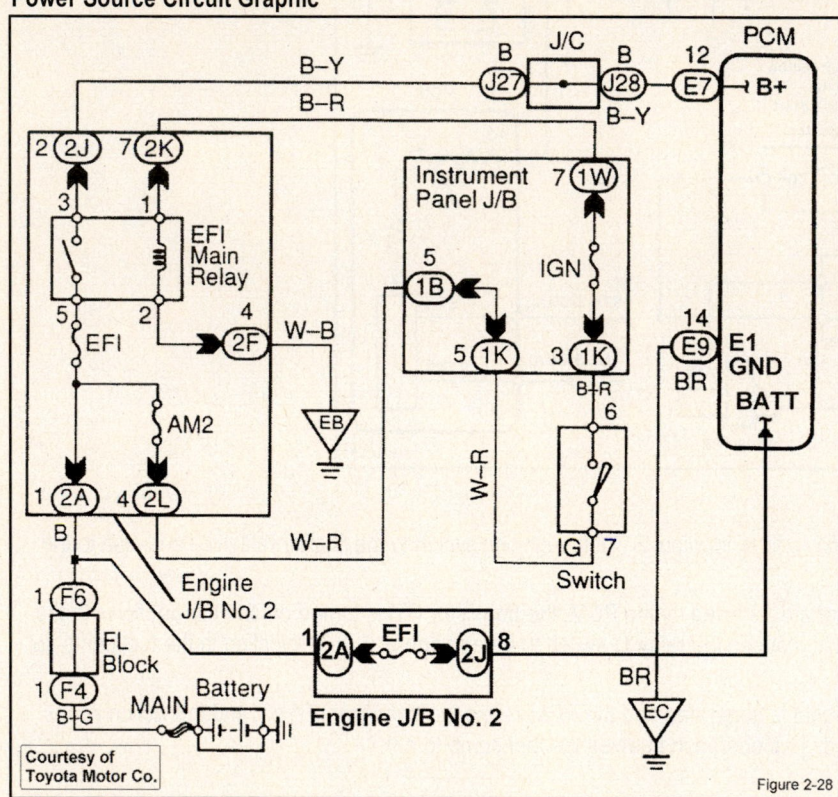

Figure 2-28

Starter Signal Circuit (1997 Camry 2.2L I4)

Circuit Description

During engine cranking, the intake airflow is slow, so fuel vaporization is poor. A rich mixture is required to achieve good engine startup.

Starter Circuit Operation

During cranking, the PCM receives a battery voltage on the STA terminal (Figure 2-29).

The PCM uses this signal to increase fuel injector volume during engine startup and after-startup fuel injection control.

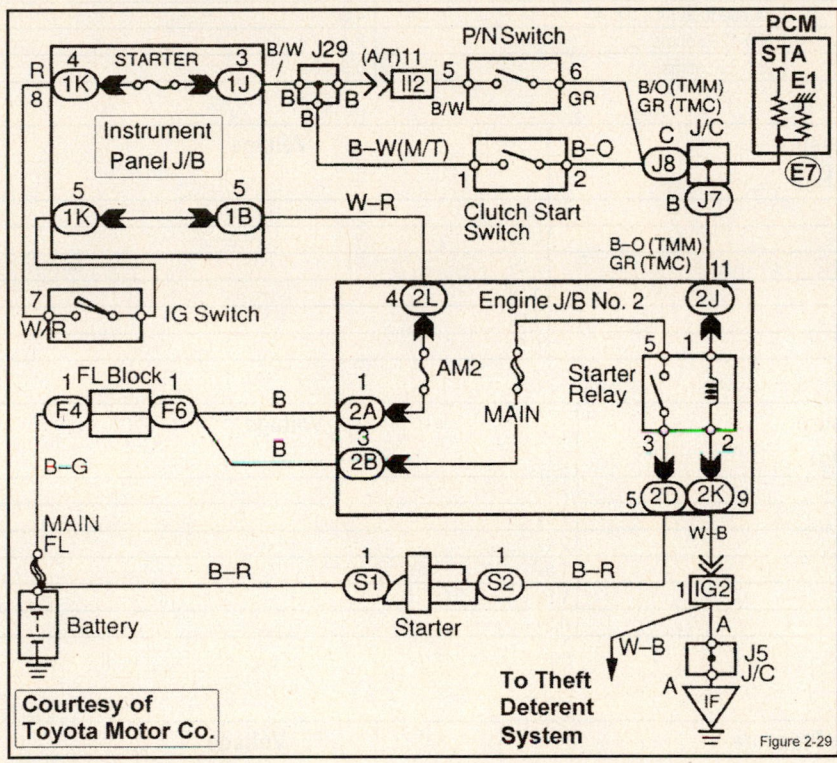

Starter Circuit Graphic

REFERENCE INFORMATION
ECT, IA/T, BARO, MAP Sensor Range Charts

ECT Sensor Conversion Chart

Degrees F	Resistance	Voltage
0°F	14,000 to 20,000 ohms	4.0v
100°F	900-1200 ohms	2-3v
180°F	250-350 ohms	0.5-0.6v

IA/T Sensor Conversion Chart

Degrees F	Resistance	Voltage
0°F	14,000 to 20,000 ohms	4.0v
100°F	900-1200 ohms	2-3v
180°F	250-350 ohms	0.5-0.6v

BARO Sensor Chart

BARO Pressure (Inches Hg)		Voltage
0" Hg	2.8	3.0v
5" Hg	2.3	2.5v
10" Hg	1.8	2.0v
15" Hg	1.3	1.5v
20" Hg	0.8	1.0v
25" Hg	0.3	0.5v

MAP Sensor Chart

MAP Pressure (Inches Hg)		Voltage
0" Hg	2.8	3.0v
5" Hg	2.3	2.5v
10" Hg	1.8	2.0v
15" Hg	1.3	1.5v
20" Hg	0.8	1.0v
25" Hg	0.3	0.5v

Fuel Tank Pressure Sensor

Fuel Tank Pressure (Inches Hg)	Voltage
0" Hg	2.8-3.1v
5" Hg	2.3-3.5v
10" Hg	1.8-2.0v
15" Hg	1.3-1.5v
20" Hg	0.8-1.0v
25" Hg	0.3-0.5v

Solenoid Resistance Chart

Solenoid Resistance	Ohms
ISCO, ISCC Valve	18-22 ohms
RSO, RSC Valve	17-24 ohms
Shift Lock (SL)	22-27 ohms
Solenoid 1 (SS1)	11-15 ohms
Solenoid 2 (SS2)	11-15 ohms
Solenoid 3 (SS3)	11-15 ohms
SLN Solenoid	5.1-5.5 ohms

Wiring Diagram for 1993 Corolla 1.6L I4 (Part 1)

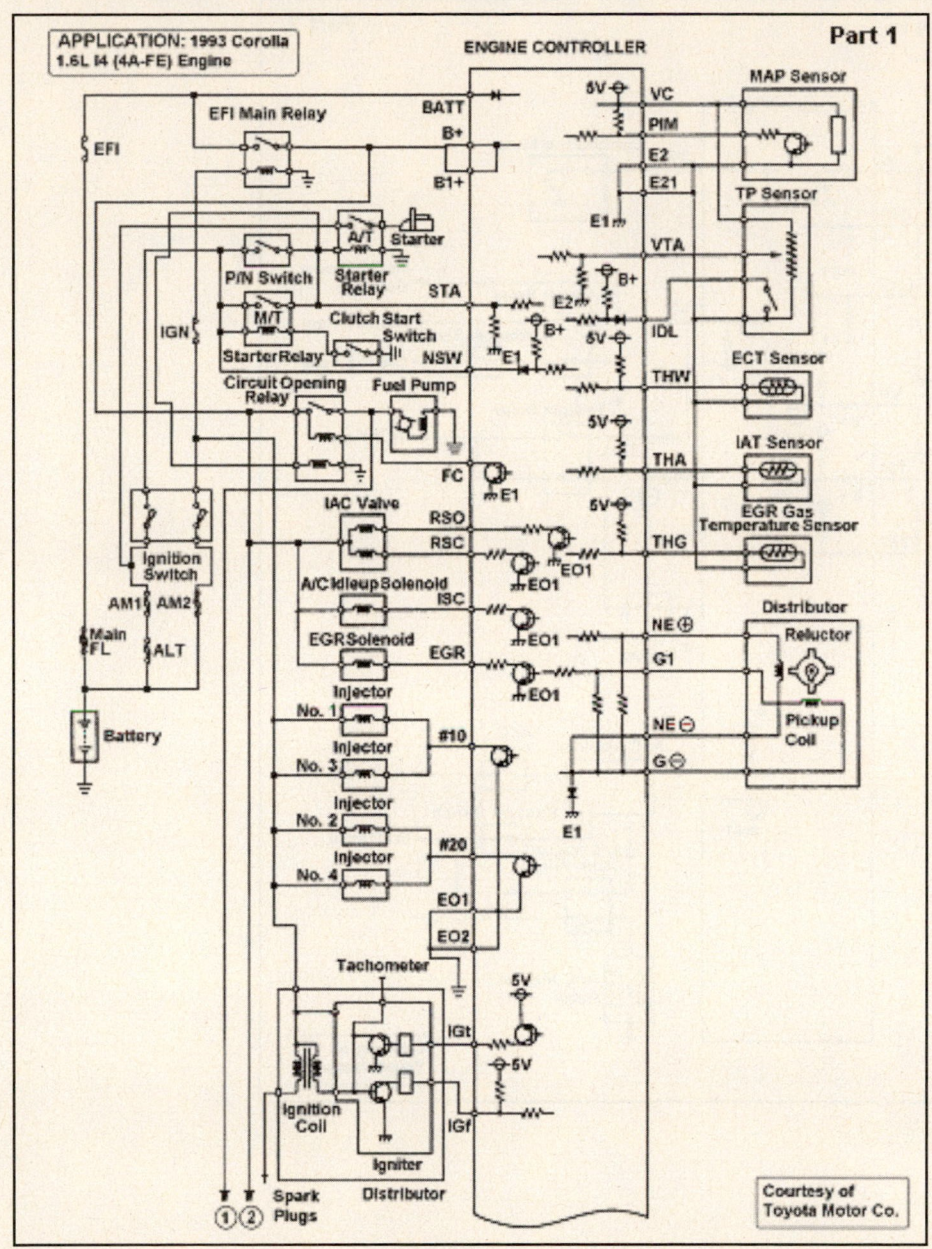

APPLICATION: 1993 Corolla 1.6L I4 (4A-FE) Engine

ENGINE CONTROLLER

Part 1

Courtesy of Toyota Motor Co.

Wiring Diagram for 1993 Corolla 1.6L I4 (Part 2)

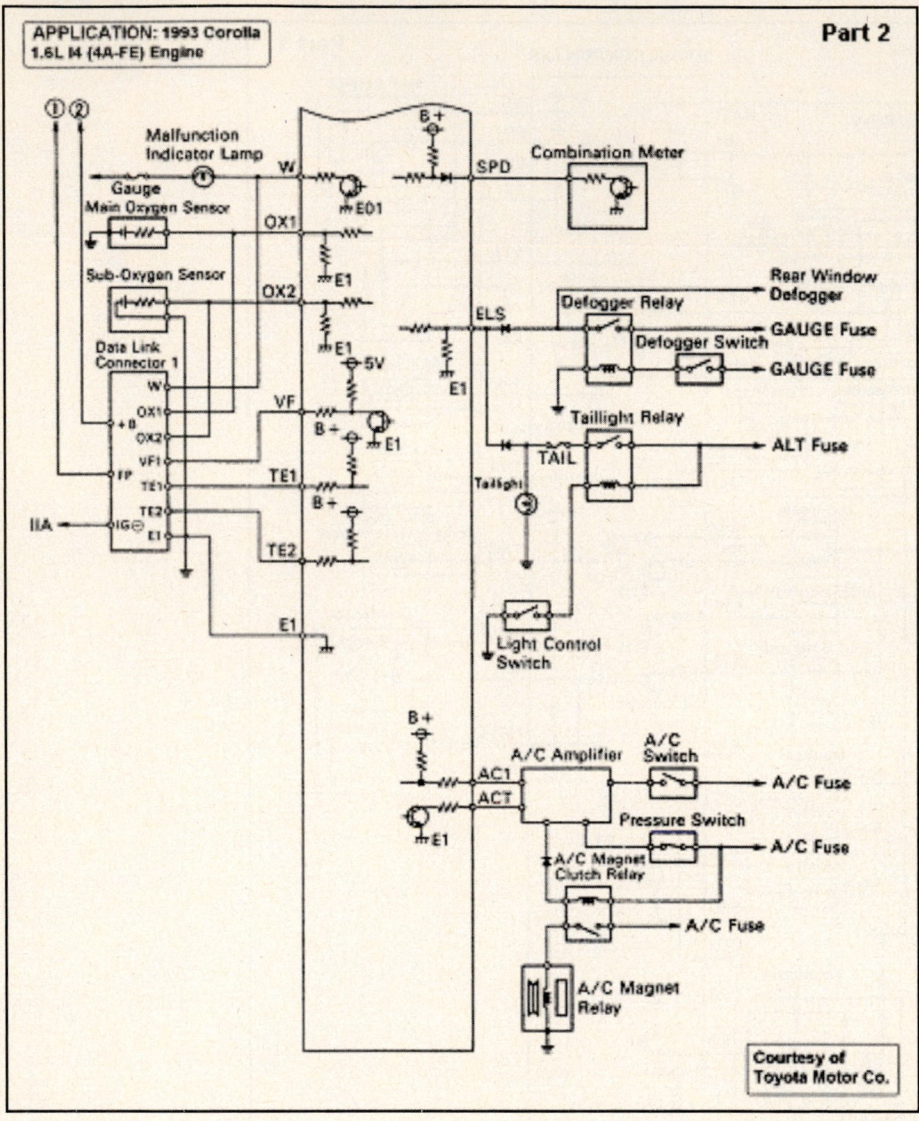

APPLICATION: 1993 Corolla 1.6L I4 (4A-FE) Engine

Part 2

Courtesy of Toyota Motor Co.

Wiring Diagram for 1994 T100 Pickup 3.0L V6 (Part 1)

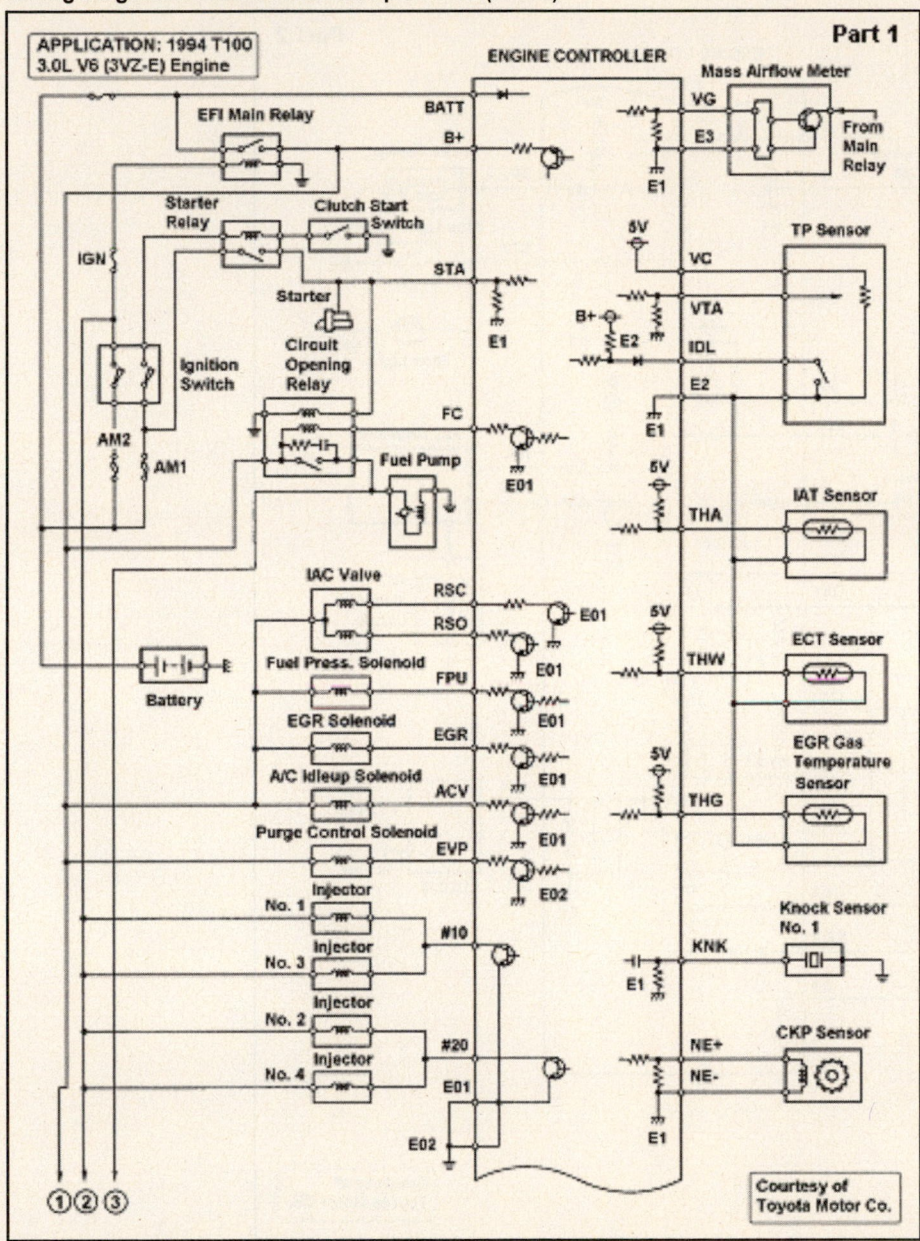

Wiring Diagram for 1994 T100 Pickup 3.0L V6 (Part 2)

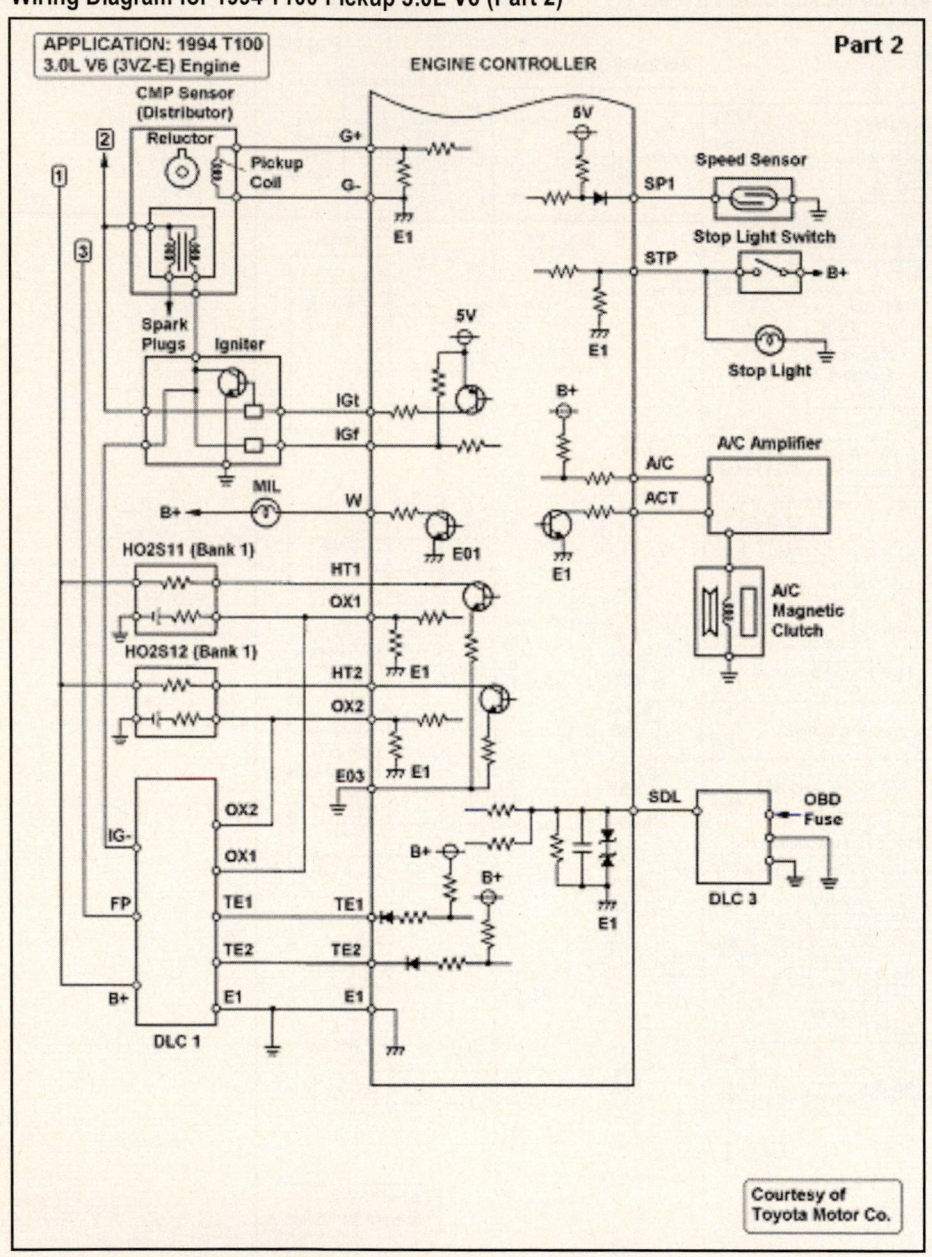

1996 Camry PID Data Examples

The PID Examples listed below were captured using an Aftermarket Scan Tool along with a Toyota Cartridge.

The table includes the PID Value Range and Hot Idle, 30 and 55 mph values. During diagnosis, compare the "known good" readings to the "actual" readings obtained from the vehicle during driving conditions.

1996 Camry Sedan 2.2L I4 VIN G (A/T) - Toyota Enhanced Signals

Parameter Identification	PID Value Range	PID Value at Hot Idle	PID Value at 30 mph	PID Value at 55 mph
A/C Switch	ON or OFF	On if on	OFF	OFF
A/C Clutch	ON or OFF	On if on	OFF	OFF
Alternator	0-25.5v	14.4v	14.3v	14.4v
Alt. Control	0-100%	36%	35%	33%
BARO kPa	0-450	99 kPa	99 kPa	99 kPa
Battery Volts	0-25.5v	14.4v	14.3v	14.4v
Brake Switch	ON or OFF	ON: If on	OFF	OFF
Engine Load	0-100%	40%	34%	36%
ECT Sensor	-40 -284°F	190°F	188°F	191°F
ELD Sensor	0-100	15a	12a	14a
EGRV Sensor	0-5v	1.2v	2.0v	1.8v
EVAP Purge	0-100%	0%	30%	85%
EVAP Switch	ON or OFF	OFF	ON	ON
Cooling Fan	ON or OFF	On if on	OFF	OFF
FP Relay	ON or OFF	ON	ON	ON
Fuel Status	CL or OL	CL	CL	CL
Fuel Injector	0-999	3.45ms	3.65ms	3.95ms
Fuel Status	CL or OL	CL	CL	CL
HO2S1 Volts	0-1000	450mv	350mv	650mv
Heater Control	ON or OFF	ON	ON	ON
HO2S2 Volts	0-1000	650mv	450mv	850mv
Heater Control	ON or OFF	ON	ON	ON
IAC Amps	0-1000	474ma	N/A	N/A
IAC Counts	0-255	45	N/A	N/A
IA/T Sensor	-4 - 212°F	142°F	138°F	135°F
MIL On/Off	ON or OFF	OFF	OFF	OFF
MIL Status	ON or OFF	OFF	OFF	OFF
Engine Mount	ON or OFF	ON	ON	ON
PNP Switch	P-N/GEAR	P-N	GEAR	GEAR
Engine Speed	0-65535	770 rpm	1250 rpm	2150 rpm
Service Conn.	CL/Open	OPEN	OPEN	OPEN
Spark ADV	-20° to 70°	15° BTDC	34° BTDC	42° BTDC
Starter Signal	ON or OFF	OFF	OFF	OFF
TP Sensor	0-100°	1°	4°	6°
Vehicle Speed	0-159	0 mph	30 mph	55 mph

1998 Camry PID Data Examples

The PID Examples listed below were captured using an Aftermarket Scan Tool along with a Toyota Cartridge.

The table includes the PID Value Range and Hot Idle, 30 and 55 mph values. During diagnosis, compare the "known good" readings to the "actual" readings obtained from the vehicle during driving conditions.

1998 Camry 3.0L V6 MFI VIN F (A/T) - Toyota Enhanced Signals

Parameter Identification	PID Value Range	PID Value at Hot Idle	PID Value at 30 mph	PID Value at 55 mph
A/C On Signal	ON or OFF	ON if on	OFF	OFF
A/C Cut Signal	ON or OFF	OFF if on	ON	ON
CTP Signal	ON or OFF	ON	OFF	OFF
CYL Variation	0-100%	0%	0%	0%
EGR System	ON or OFF	OFF	ON	ON
EVAP VSV	0-100%	0%	35%	60%
Press VSV	ON or OFF	OFF	ON	ON
Fuel Cut Idle	ON or OFF	ON	OFF	ON (Decel)
Fuel Cut Light	ON or OFF	OFF	OFF	ON (Decel)
FP Signal	ON or OFF	ON	ON	ON
IAC D/Cycle	0-100%	35%	30%	32%
Revolutions per 1K	0-3000	Varies	Varies	Varies
INJ Pulsewidth	0-999	2.2ms	2.4ms	2.5ms
Intake Control	ON or OFF	OFF	OFF	OFF
Elect. Load	ON or OFF	ON if on	OFF	OFF
Misfire Load	0-999	0 g/r	0 g/r	0 g/r
Misfire RPM	0-65535	0 rpm	0 rpm	0 rpm
O2S1L/R Ratio	0-1000	Varies	Varies	Varies
O2S2L/R Ratio	0-1000	Varies	Varies	Varies
O2S1R/L Ratio	0-1000	Varies	Varies	Varies
O2S2R/L Ratio	0-1000	Varies	Varies	Varies
PNP Switch	ON or OFF	ON in P/N	OFF	OFF
PSP Switch	ON or OFF	ON turned	OFF	OFF
Starter Signal	ON or OFF	OFF	OFF	OFF
Stop Light SW	ON or OFF	ON	OFF	OFF
Total FT BK 1	0-2	0.8-1.2	0.8-1.2	0.8-1.2
Total FT BK 2	0-2	0.8-1.2	0.8-1.2	0.8-1.2

Note: *The O2S L/R (lean/rich) and O2S R/L (rich/lean) Ratio parameters listed in this table are not available on California models with the A/F Ratio Sensor.*

1998 4Runner PID Data Examples

The PID Examples listed below were captured using an Aftermarket Scan Tool along with a Toyota Cartridge.

The table includes the PID Value Range and Hot Idle, 30 and 55 mph values. During diagnosis, compare the "known good" readings to the "actual" readings obtained from the vehicle during driving conditions.

1998 4Runner 3.4L V6 MFI VIN N (A/T) - Toyota Enhanced Signals

Parameter Identification	PID Value Range	PID Value at Hot Idle	PID Value at 30 mph	PID Value at 55 mph
A/C On Signal	ON or OFF	ON if on	OFF	OFF
A/C Cut Signal	ON or OFF	OFF if on	ON	ON
Air Flow Volts	0-1,000 g's	2.3	5.8	24
Cat. Monitor	Done/ND	Not Done	Not Done	Done
CTP Signal	ON or OFF	ON	OFF	OFF
CYL 1-6 Variation	0-100%	0%	0%	0%
ECT Sensor	-40 -284°F	198°F	197°F	196°F
EGR Monitor	Done/ND	Not Done	Done	Done
EVAP Monitor	Done/ND	Not Done	Not Done	Not Done
Fuel Cut IDL	ON or OFF	ON	OFF	ON (Decel)
Fuel Cut Light	ON or OFF	OFF	OFF	ON (Decel)
Fuel System 1	CL or OL	CL	CL	CL
Fuel System 2	CL or OL	CL	CL	CL
F/SYS Monitor	Ready/NR	READY	READY	READY
FP Signal	ON or OFF	ON	ON	ON
IAC D/Cycle	0-100%	35%	30%	32%
IA/T Sensor	-4 -212°F	88°F	90°F	91°F
IGN Advance	0-1000	10° BTDC	22° BTDC	19° BTDC
Calc. Load	0-100%	18%	18%	65%
LONGFT BK 1	0 ±20%	1%	0%	2%
Misfire Load	0-999	0 g/r	0 g/r	0 g/r
Misfire RPM	0-65535	0 rpm	0 rpm	0 rpm
MISF Monitor	Ready/NR	READY	READY	READY
O2S1 Bank 1	0-1000	730mv	250mv	350mv
O2S2 Bank 1	0-1000	650mv	450mv	550mv
O2S Monitor	Ready/NR	READY	READY	READY
O2 Heater Monitor	Ready/NR	READY	READY	READY
PNP Switch	ON or OFF	ON in P/N	OFF	OFF
PSP Switch	ON or OFF	ON turned	OFF	OFF
Engine Speed	0-65535	700 rpm	1010 rpm	1955 rpm
SHRTFT BK 1	0 ±20%	3%	2%	-1%
Starter Signal	ON or OFF	OFF	OFF	OFF
TP Sensor	0-100%	11%	15%	19%
F/T BK1 S1	0-100%	99.2%	99.2%	99.2%
Vehicle Speed	0-159	0 mph	30 mph	55 mph

2000 Corolla PID Data Examples

The PID Examples listed below were captured using an Aftermarket Scan Tool along with a Toyota Cartridge.

The table includes the PID Value Range and Hot Idle, 30 and 55 mph values. During diagnosis, compare the "known good" readings to the "actual" readings obtained from the vehicle during driving conditions.

2000 Corolla 1.8L I4 MFI VIN R (A/T) - Toyota Enhanced Signals

Parameter Identification	PID Value Range	PID Value at Hot Idle	PID Value at 30 mph	PID Value at 55 mph
A/C On Signal	ON or OFF	ON if on	OFF	OFF
A/C Cut Signal	ON or OFF	OFF if on	ON	ON
Calculated Load	0-100%	40	29.4	50.5
CYL 1-4 Variation	0-100%	0	0	0
ECT Sensor	-40 -284°F	180	199	194
EVAP Purge VSV	ON or OFF	OFF	ON	ON
EVAP Vapor VSV	ON or OFF	OFF	OFF	OFF
Engine Speed	0-65535rpm	868	1650	1948
FC IDL (fuel cut)	ON or OFF	OFF	OFF	OFF
FC TAU (fuel cut)	ON or OFF	OFF	OFF	OFF
Fuel Pump Status	ON or OFF	ON	ON	ON
Fuel Status 1	CL or OL	CL	CL	CL
Fuel Status 2	CL or OL	CL	CL	CL
HO2S-11 Bank 1	0-1000v	0.080	0.090	0.065
HO2S-21 Bank 2	0-1000v	0.030	0.080	0.075
HO2S-12 Bank 1	0-1000mv	0.070	0.020	0.490
IAC Duty Cycle	0-100%	33	35	35
IA/T Sensor	-4 -212°F	122	124	115
Ignition: each rev	0-2000	500	700	900
Ignition Timing	-90 to 90°	12.0	20.5	26.0
Injector On-Time	0-999.9 ms	1.5	1.8	2.3
LONGFT Bank 1	0 ±20%	0.0	-4.6	+2.3
Mass Airflow	0-1,000 g's	1.9	5.2	22.5
MIL Control	ON or OFF	OFF	OFF	OFF
Misfire Load (1st)	0-1,000 g's	0	0	0
Misfire RPM (1st)	0-65535	0 rpm	0 rpm	0 rpm
PNP Switch	ON or OFF	ON	OFF	OFF
SHRTFT Bank 1	0 ±20%	0.0	-0.7	-5.8
Starter Signal	ON or OFF	Crank (on)	OFF	OFF
Stop Light Switch	ON or OFF	OFF	OFF	OFF
Stored T/Codes	0-255	0	0	0
Total Fuel Trim	Average 0-5	1.8	1.8	1.9
TP Sensor (%)	0-100%	10.5	13.3	20.3
Vehicle Speed	0-159 mph	0	30	55

OBD I Serial Data Explanations

The Serial Data Explanations in the following table includes Acronym, Data Description, Data Value Range and Serial Data Explanations.

PCM Data Acronym	Serial Data Description	Data Value Range	Serial Data Explanation
A/F FB L & R	Loop Status for left and right cylinder banks	CL or OL	The AFR feedback loop status. OL indicates the PCM ignores the O2S. CL indicates that final injection pulse width is corrected for O2 feedback.
IAC DUTY	Commanded duty cycle ratio of rotary IAC	5-60% 4A, 7A-FE: 15% ±10% 5S-FE: 45% 3SGE: 38%	Displayed value is the commanded duty cycle ratio of the rotary solenoid IAC valve.
IAC STEP #	Commanded position of step type IAC valve	35 ±15 Idle 2JZGE: 22 1UZFE: 35 1FZFE: 40	Displayed value is the commanded (not actual) position. 125 steps = fully open and 0 steps = fully closed.
IDL SIG	Closed Throttle Signal (closed IDL contacts)	ON or OFF	The display is determined by the status of IDL terminal of PCM (low voltage = ON, high voltage = OFF).
INJECTOR	Cyl Pair #10 Injector duration	1.8-5.0ms (idle speed)	Displayed value indicates the commanded injector driver duration for the #10 Injector or group (1 & 3).
KS SIG	Airflow Meter signal indicating volume of air entering engine	35 ±5ms at idle except: '94 LS400: 40ms ±5ms	Displayed value represents the time that the KS signal is low. As the VS signal frequency increases, the time also increases.
MAF SIG	The amount of airflow entering the engine	3.8 ±1.2 g/sec at idle speed	Display is calculated by comparing the VG analog signal to a lookup table stored in ROM in the PCM.
MAP SIG	Absolute Pressure in the intake manifold	Idle speed: 4A, 7A-FE: 9.8 ±1.2 "hg 5S-FE: 9.0 ±1.2 "hg	Displayed value is determined by comparing the analog signal at the PIM terminal of the PCM to a lookup table stored in Read Only Memory (ROM) in the PCM.
PNP SIG	Park/Neutral Switch status	P-N or GEAR	Displayed value is N-P in neutral or park and Gear in other positions.
OX SIG L & R	Main Oxygen Sensor input	Rich / Lean R-L-R-L 8 times in 10 seconds	Displayed value is the signal voltage from the main O2S. A high amount of exhaust oxygen = lean exhaust or <400mv. A low amount of exhaust oxygen = rich exhaust or >600mv.
TARGET A/F L & R	Long Term Fuel Trim for left and right cylinder banks	0-5v Typical: 2.5v ±1.25v	The AFR feedback correction value displayed as correction factor: 2.50v = no correction, <2.50v = rich correction, >2.50v = lean correction.
VS SIG	Airflow Meter signal indicating volume of air entering engine	2.50v at idle except: 22RE: 2.6v 3VZE: 2.8v 1FZF: 1.8v	Displayed value is the same value as the analog voltage signal at the VS terminal of the PCM.

OBD II PID Data Explanations

The PID (Parameter ID) Explanations on this page include the PID Acronym, the PID Description, Hot Idle Range and PID Explanation.

PID Acronym	PID Description	Hot Idle Range	PID Explanation
A/C SIG	Air Conditioning Switch status	ON or OFF (On with clutch "on")	Display reads ON with compressor clutch energized and OFF when compressor off (low temperature).
A/C CUT SIG	Air Conditioning Cut Signal status	ON or OFF	With the A/C On, displayed value should be ON when the PCM pulls ACT terminal low (requesting the A/C compressor cut due to a load).
CALC LOAD	Percent of Max possible engine load	0-100% 7-11% at idle speed	Displayed value calculated mathematically using this formula: actual air volume ÷ maximum possible air volume x 100%.
CTP SW	Closed Throttle Position Signal from IDL switch: contacts closed	ON or OFF (On at idle)	Displayed value determined by comparing VTA analog signal to a lookup table in the PCM. Each 0.5v in amplitude = 10% of opening.
ECT	ECT Sensor	-40 - 284°F (typical value is 176 to 203°F)	The displayed ECT Sensor value is determined by comparing the THW analog voltage to a lookup table stored in ROM in the PCM.
EGRT GAS	Temperature of Exhaust Gas in the recirculation intake port	-40 - 284°F Should be <IA/T but >ECT value	Displayed value determined by comparing THG analog voltage to a lookup table in the PCM. Use Active Test to confirm THG operation.
EGR SYS	EGR System status	ON or OFF	Displayed value should be ON when within specified throttle angle, rpm and load range, and then the PCM de-energizes the EGR VSV (this allows vacuum to pass to the modulator valve and EGR valve).
ELECT LOAD SIG	Read Defogger or Tail Light Circuit status	ON or OFF ON with load on	Displayed value should be ON whenever rear window defogger or tail light relays are energized.
FC IDL	Fuel Cut with CTP switch contacts closed	ON or OFF	Displayed value should be ON when closed throttle Decel is detected above a specified threshold.
FC TAU	Fuel Cut with light load Decel	ON or OFF	Displayed value should be ON when Decel fuel cut is commanded by the PCM with the CTP Switch open.
FUEL SYS 1, 2	Loop Status for Cylinder Bank 1 & 2	CL or OL	Displays the AFR feedback status. OL is forced during Accel, Decel and with a cold engine. CL indicates that the injector duration is corrected for O2S feedback.
IA/T	IA/T Sensor	-4 - 212°F Should be near under-hood temp.	The displayed IA/T Sensor value determined by comparing the THA analog voltage a lookup table stored in ROM in the PCM.

OBD II PID Data Explanations

The PID (Parameter ID) Explanations on this page include the PID Acronym, the PID Description, Hot Idle Range and PID Explanation.

PID Acronym	PID Description	Hot Idle Range	PID Explanation
IGN ADV	Cylinder No. 1 IGN Advance	12° ±5 BTDC	Displayed value is calculated by comparing the relationship between the CKP (NE) and CMP (G) sensor inputs. The gap on the NE sensor (36-2 tooth) identifies No. 1 Cyl TDC and 'G' signal identifies 90° BTDC with No. 1 Cyl on compression.
INJECTOR	Cylinder No. 1 Injector duration	0-1000 ms At idle: 2.2-7.5 ms	Displayed value is determined by monitoring the commanded injector driver duration for Cyl No. 1. *Note: may read normal pulse in fail-safe.*
INT VSV	Intake Control VSV status	ON or OFF	Displayed value should be ON when PCM energizes the VSV (ACIS air valve closes to allow high torque).
LONGFT 1 & 2	Long Term Fuel Trim for Cylinder Bank 1 & 2	0% ±20%	AFR feedback correction value, part of basic injection calculation. Positive value = rich correction for lean condition and lean value = lean correction for rich correction.
MAF	MAF Sensor	0-512 g/sec (Idle speed value is 2.4-4.8 g/sec)	This is the displayed MAF Sensor value calculated by comparing the VG analog voltage a lookup table stored in ROM in the PCM.
O2S B1 S1 O2S B2, S1	Main O2S Signal Voltage for Engine Cylinder Bank 1 or Cylinder Bank 2	0-1000 mv	A high concentration of exhaust oxygen indicates a lean condition and results in a low value. A low concentration of exhaust oxygen indicates a rich condition and results in a high voltage value.
O2 R/L or L/R BK1, S1; B2, S1	Main O2S lean to rich or rich to lean switch time	0-1000 ms	Time in milliseconds for O2S signal to go from >600mv to <400mv (L/R) or from <400mv to >600mv (R/L).
PNP SW	Park/Neutral Switch status	ON or OFF ON in P/N	Displayed value should be ON in P/N and OFF in all other positions.
RPM	Engine Speed	0-65535 (idle speed is 700 rpm)	Engine revolutions calculated by comparing the NE signal with the clock pulses in the PCM.
SHRTFT 1, SHRTFT 2	Short Term Fuel Trim for Cylinder Bank 1 & 2	0% ±20%	AFR feedback correction value applied to injection duration after basic injection calculation. Positive value = rich correction for lean condition and lean value = lean correction for rich correction.
Total FT Bank 1, Total FT Bank 2	Average Total Fuel Trim for Bank 1 & 2	Normal is 1. Typical is 0.8-1.2%.	The total correction including basic and corrected injection duration. A reading of <1.00 is less correction and >1.00 is more correction.

OBD II SYSTEM GLOSSARY

Toyota Glossary of Terms and Acronyms

(<): Indicates less than the value	(>): Indicates more than the value
ABS: Anti-Lock Brake System	ACC: Air Conditioning Clutch
ACIS: Acoustic Control Induction System	ACSD: Automatic Cold Start Device
ACS: Air Conditioning Switch	ACV: A/C Idle-up Solenoid
ACT: A/C Amplifier Signal	A/F: Air/Fuel Ratio
AC1: A/C Amplifier Signal	AS: Air Suction
A/T: Automatic Transmission	BARO: Barometric Pressure
BTDC: Before Top Dead Center	BVSV: Bimetal Vacuum switching valve
B+ or B1+: Battery Positive Voltage	(Cal): California Models
CB: Circuit Breaker	CEL: Check Engine Light
CKP: Crankshaft Position Sensor	CMH: Cold Mixture Heater
CMP: Camshaft Position Sensor	CRS: Child Restraint System
C/C or CCS: Cruise Control System	DI: Distributor Ignition
DLC1: Check Connector 1	DLC2: Diagnosis Communication Link 2
DLC3: OBD II Diagnostic Connector 3	DTC: Diagnostic Trouble Code
ECT: Electronic Control Transmission	ECT: Engine Coolant Temperature
ECM: Engine Control Module	ECU: Electronic Control Unit
EGR: Exhaust Gas Recirculation	EFI: Electronic Fuel Injection
EGR Monitor: OBD II EGR Test	EI: Electronic Ignition
EPROM: Erasable Programmable Read Only Memory	EEPROM: Electrically Erasable Programmable Read Only Memory
ELS: Engine Load Signal	EVAP: Evaporative Emission System
EVP: EGR Valve Position Sensor	ESA: Electronic Spark Advance
E1, E2, E3, E21: Sensor Ground Circuits	EO1, EO2, EO3: Power Ground Circuits
EIS: Engine Immobilizer System	FAN: Cooling Fan Relay
FC: Circuit Opening Relay	(Fed): Federal Models
FP: Fuel Pump Relay	FPU: Fuel Pressure-up Solenoid
FWD: Front Wheel Drive	G-, G+: Engine Revolution Sensor
GVW: Gross Vehicle Weight	HAC: High Altitude Compensation
HAFR: Heated Air Fuel Ratio Sensor	HO2S: Heated Oxygen Sensor
HO2S-11: Pre-catalyst Oxygen Sensor	HT: Oxygen Sensor Heater
IAC: Idle Air Control Valve	IA/T: Intake Air Temperature
ICM: Ignition Control Module	IDL: Closed Throttle Switch
IIGf: Igniter Confirmation Signal	IGt: Igniter Control Signal
INJ 1 to INJ 8: Injectors 1 through 8	ISC: Idle Speed Control
ISC1, ISC2: ISC Valve Control 1 & 2	KAM: Keep Alive Memory (Battery B+)
ISCC: ISC Valve (close) Coil	ISCO: ISC Valve (open) Coil
KOEC: Key On, Engine Cranking	KOEO: Key On, Engine Off
KOER: Key On, Engine Running	KOER: Key On, Engine Running
LCD: Liquid Crystal Display	LED: Light Emitting Diode
LONGFT - Long Term Fuel Trim	LSD - Limited Slip Differential
LSPV: Load Sensing Proportioning Valve	L4: L4 Detect Transfer Switch
MAF: Mass Air Flow	MAP: Manifold Absolute Pressure
MIL: Malfunction Indicator Lamp	MPH: Miles Per Hour
MPX: Multiplex Communication System	M/T: Manual Transmission
MREL: EFI Main Relay	N - A/T: Select Switch Neutral Position

OBD II SYSTEM GLOSSARY

Toyota Glossary of Terms and Acronyms

NC2 -, NC2+: Direct Clutch Speed Sensor	NE -, NE+ -: CKP Sensor
NGV: Natural Gas Vehicle	NSW: Neutral Start Switch
NUMMI: New U/M Manufacturing Inc.	NVRAM: Non-Volatile RAM
OBD I: On Board Diagnostics Version 1	OBD II: On Board Diagnostics Version 2
OEM: Original Equipment Manufacturer	OIL: Oil Temperature Sensor (A/T only)
OILW: Oil Temp. Warning Light (A/T)	OSM: Output State Monitor
OX: Oxygen Sensor	O2S: Oxygen Sensor
PCM: Powertrain Control Module	PCS: Purge Control Solenoid
PCV: Positive Crankcase Ventilation	PFI: Port Fuel Injection
PIM: MAP Sensor	PNP: Park Neutral Position Switch
PPS: Progressive Power Steering	PSP: Power Steering Pressure Switch
PWR: ECT Pattern Select Switch	PWR GND: Power Ground for PCM
RAM: Random Access Memory (PCM)	ROM: Read Only Memory (PCM)
RPM: Revolutions Per Minute	RSC: Rotary Solenoid IAC (close) Coil
RSO: Rotary Solenoid IAC (open) Coil	SCP: Standard Corporate Protocol
SIL: OBD II Data Link Connector	SHRTFT: Short Term Fuel Trim
SIG: Signal	SIL: OBD II Data Link Connector
SL: ECT Select Lockup Solenoid	SOL: Solenoid
SPD: Vehicle Speed Sensor input	ST: ECT Select Timing Solenoid
STA: Starter Relay	STD: Standard
STJ: Cold Start Injector	STP: Stop Light Switch
SP1, SP2: Vehicle Speed Sensor 1 & 2	S1, S2, S3: A/T-ECT Solenoid 1, 2 & 3
S/C, SC: Supercharger	TACO: Tachometer Signal
TCC: Torque Converter Clutch	TCCS: Toyota Computer Control System
TCM: Transmission Control Module	TDC: Top Dead Center
TDCL: Toyota Diagnostic Comm. Link	TEMS: Toyota E-Modulated Suspension
TE1: Data Link Connector	THA: Intake Air Temperature Sensor
THE: A/C Thermistor Signal	THO: Transfer Fluid Temp. Sensor
THW - Engine Coolant Temp. Sensor	TIL: Turbo Indicator Light
TMC: Toyota Motor Corporation (Japan)	TMM: Toyota Motor Manufacturer (USA)
TP: Throttle Position Sensor	TPC: Turbo Pressure Sensor
T-VIS: Toyota Variable Induction System	TSW: Water Temperature Switch
TWC: Three-Way Catalyst	T/C or TC: Turbocharger
VC: Sensor VREF	VCV: Vacuum Control Valve
VG: MAF Sensor input	VREF: PCM reference voltage
VS: Air Flow Meter Signal	VSS: Vehicle Speed Sensor
VSV: Vacuum Switching Valve	VTA: TP Sensor input
W Signal: Malfunction Indicator Light	WOT: Wide Open Throttle Switch
2 - A/T: Select Switch 2nd Gear	2WD: Two Wheel Drive Vehicle (2x2)
4WD: 4WD Indicator Switch	4WD: Four Wheel Drive Vehicle (4x4)
#10: Injector No. 1	#20: Injector No. 2
#30: Injector No. 3	#40: Injector No. 4
#50: Injector No. 5	#60: Injector No. 6
#70: Injector No. 7	#80: Injector No. 8

TOYOTA CAR PIN TABLE CONTENTS

Note: EIS is an abbreviation for Engine Immobilizer System.

Note: ECT is an abbreviation for Electronic Controlled Transmission System.

Note: ECT is an abbreviation for Electronic Controlled Transmission System.

AVALON Pin Tables

1995 Sedan 3.0L V6 VIN G (A/T-ECT) 16 Pin Connector

PCM Pin #	Wire Color	Circuit Description (16 Pin)	Value at Hot Idle
1	WT/BL	A/C Idle-Up Solenoid	A/C Off: 12v, On: 1v
2 (Cal)	BK/RD	EVAP Solenoid Control	12v or 0v
3	GN/RD	MIL (lamp) Control	MIL Off: 12v, On: 1v
5	GY/BK	Check Connector	12-14v
6	RD/YL	Intake Air Solenoid	12v or 0v
7	RD/BK	MAF Sensor Ground	<0.050v
10	PK	HO2S-21 (B2 S1) Heater	1v (Heater On)
11	BL/BK	HO2S-11 (B1 S1) Heater	1v (Heater On)
12	BK/BL	EGR Solenoid Control (VSV)	12v or 0v
14	GN/YL	EGR Gas Temp. Sensor	3.5-4.0v
16	BR	Sensor Ground	<0.050v

1995 Sedan 3.0L V6 VIN G (A/T-ECT) 22 Pin Connector

PCM Pin #	Wire Color	Circuit Description (22 Pin)	Value at Hot Idle
1	BL/RD	Sensor VREF (VC)	4.9-5.1v
4	WT/BL	OD Clutch Speed Sensor (-)	AC pulse signals
5	RD	CKP Sensor Signal (NE+)	AC pulse signals
6	GN	CKP Sensor Signal (NE-)	<0.050v
7	BK/YL	TP Sensor Signal	0.3-0.8v
8	RD	MAF Sensor Signal	1.1-1.5v
9	YL/BL	OD Clutch Speed Sensor (+)	AC pulse signals
13	RD/BL	HO2S-11 (B1 S1) Signal	0.1-1.1v
14	WT	Knock Sensor 1 Signal	<0.075v AC
15	WT	Knock Sensor 2 Signal	<0.075v AC
16	BK/WT	CMP Sensor Signal (G+)	AC pulse signals
17	BL	CMP Sensor Signal (G-)	<0.050v
18	GN/RD	Circuit Opening Relay (FC)	0-3v, off-idle: 12v
19	RD/BL	HO2S-21 (B2 S1) Signal	0.1-1.1v
20	GN/BK	ECT Sensor Signal	At 180°F: 0.51v
21	BL/BK	IAT Sensor Signal	At 100°F: 2.60v
22	BR	Sensor Ground	<0.050v

Pin Connector Graphic

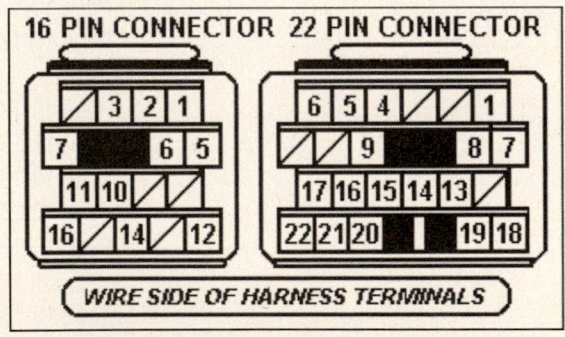

1995 Sedan 3.0L V6 VIN G (A/T-ECT) 28 Pin Connector

PCM Pin #	Wire Color	Circuit Description (28 Pin)	Value at Hot Idle
5	GN/BK	A/C Amplifier Signal (ACT)	Clutch On: 12v, Off: 1.5v
9	RD	Cooling Fan Relay	Relay Off: 12v, On: 1v
12	PK/YL	Vehicle Speed Sensor	At 55 mph: 48 Hz
13	BK	Tachometer Signal (TACO)	Pulse Signals
14	WT/BL	Battery Direct	12-14v
20	BK/YL	A/C Amplifier Signal (AC1)	Clutch On: 1.5v, Off: 12v
21	BK/RD	Defogger/Light Idle Up Signal	Switch On: 12v, Off: 0v
23	BK/OR	EFI Main Relay Power	12-14v
24	GN/WT	Stop Light Switch Signal	Brake Off: 0v, On: 12v
25	PK/BK	HO2S-12 (B1 S2) Heater	1v (Heater On)
26	BK	HO2S-12 (B1 S2) Signal	0.1-1.1v
28	WT	Data Link Connector	12v

1995 Sedan 3.0L V6 VIN G (A/T-ECT) 34 Pin Connector

PCM Pin #	Wire Color	Circuit Description (34 Pin)	Value at Hot Idle
3, 27	YL, BL	A/T-ECT Solenoid SL-, SL+	In Lockup: 12-14v
5	GN	Injector 6 Control	2.0-3.3 ms
6	RD	Injector 5 Control	2.0-3.3 ms
7	BL	Injector 4 Control	2.0-3.3 ms
8	GY	Injector 3 Control	2.0-3.3 ms
9	YL	Injector 2 Control	2.0-3.3 ms
10	WT	Injector 1 Control	2.0-3.3 ms
11	PK	A/T-ECT Solenoid (S1)	In 3rd or OD: 1v
12	WT/RD	Igniter Signal (IGF)	Digital Signal: 0-5-0v
13	BK/WT	Starter Switch Signal	9-11v (cranking)
14	BK/WT	Neutral Start Switch	In P/N: 9-11v (cranking)
15	GY/BK	Igniter Transistor 3 Control	7% duty cycle
16	YL/RD	Igniter Transistor 2 Control	7% duty cycle
17	PK/BK	A/T-ECT Solenoid (S2)	1st or OD: 1v
22, 23	YL, GN	IAC Signals (RSC, RSO)	Pulse Signals
24	WT/GN	Igniter Transistor 1 Control	7% duty cycle
25	BK/RD	Fuel Pressure Up Solenoid	1v (at hot restart)
26	BL/BK	Igniter Transistor 4 Control	7% duty cycle
28, 33-34	WT/BK	Power Ground	<0.1v
29	GN/RD	Igniter Transistor 6 Control	7% duty cycle
30	RD/BK	Igniter Transistor 5 Control	7% duty cycle
32	LB	Closed Throttle Switch	<0.1v
33	WT/BK	Power Ground	<0.1v

Pin Connector Graphic

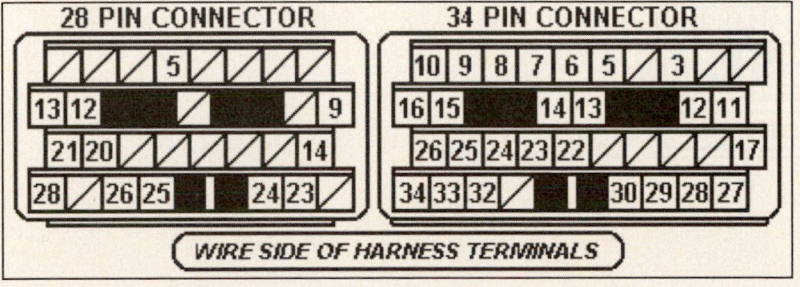

1996 Sedan 3.0L V6 VIN F (A/T-ECT) 16 Pin Connector

PCM Pin #	Wire Color	Circuit Description (16 Pin)	Value at Hot Idle
2 (Cal)	BK/RD	EVAP Purge Solenoid (VSV)	12v or 0v
3	GN/RD	MIL (lamp) Control	MIL Off: 12v, On: 1v
5	GY/BK	Check Connector	12-14v
6	RD/YL	Intake Air Solenoid	12v or 0v
7	RD/BK	MAF Sensor Ground	<0.050v
8	WT/BL	EVAP Vapor Pressure (VSV)	12v or 0v
10	BL/RD	HO2S-21 (B2 S1) Heater	1v (Heater On)
11	BL/BK	HO2S-11 (B1 S1) Heater	1v (Heater On)
12	BK/WT	EGR Solenoid Control (VSV)	12v or 0v
13	PK	EVAP Vapor Pressure Sensor	2.9-3.1v (with hose off)
14	GN/RD	EGR Gas Temp. Sensor	3.5-4.0v
16	BR	Sensor Ground	<0.050v

1996 Sedan 3.0L V6 VIN F (A/T-ECT) 22 Pin Connector

PCM Pin #	Wire Color	Circuit Description (22 Pin)	Value at Hot Idle
1	BL/RD	Sensor VREF (VC)	4.9-5.1v
4	WT/BL	OD Clutch Speed Sensor (-)	AC pulse signals
5	WT	CKP Sensor Signal (NE+)	AC pulse signals
6	OR/BL	CKP Sensor Signal (NE-)	<0.050v
7	BK/YL	TP Sensor Signal	0.3-0.8v
8	RD	MAF Sensor Signal	1.1-1.5v
9	YL/BL	OD Clutch Speed Sensor (+)	AC pulse signals
13	WT	HO2S-11 (B1 S1) Signal	0.1-1.1v
14	WT	Knock Sensor 1 Signal	<0.075v AC
15	WT	Knock Sensor 2 Signal	<0.075v AC
16	WT/BL	CMP Sensor Signal (G+)	AC pulse signals
18	GN	Circuit Opening Relay (FC)	0-3v, off-idle: 12v
19	RD/BL	HO2S-21 (B2 S1) Signal	0.1-1.1v
20	GN/BK	ECT Sensor Signal	At 180ºF: 0.51v
21	BL/BK	IAT Sensor Signal	At 100ºF: 2.60v
22	BR	Sensor Ground	<0.050v

Pin Connector Graphic

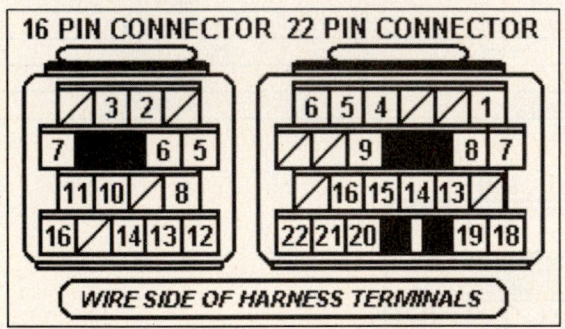

1996 Sedan 3.0L V6 VIN F (A/T-ECT) 28 Pin Connector

PCM Pin #	Wire Color	Circuit Description (28 Pin)	Value at Hot Idle
5	PK	A/C Amplifier Signal (ACT)	Clutch On: 12v, Off: 1.5v
9	RD	Cooling Fan Relay	Relay Off: 12v, On: 1v
12	PK/YL	Vehicle Speed Sensor	At 55 mph: 48 Hz
13	BK	Tachometer Signal (TACO)	Pulse Signals
14	BK/YL	Battery Direct	12-14v
20	BK/YL	A/C Amplifier Signal (AC1)	Clutch On: 1.5v, Off: 12v
21	BK/RD	Defogger Idle Up Signal	Switch On: 12v, Off: 0v
23	BK/OR	EFI Main Relay Power	12-14v
24	GN/WT	Stop Light Switch Signal	Brake Off: 0v, On: 12v
25	PK/BK	HO2S-12 (B1 S2) Heater	1v (Heater On)
26	BK	HO2S-12 (B1 S2) Signal	0.1-1.1v
28	WT	Data Link Connector	12v

1996 Sedan 3.0L V6 VIN F (A/T-ECT) 34 Pin Connector

PCM Pin #	Wire Color	Circuit Description (34 Pin)	Value at Hot Idle
3	YL/GN	A/T-ECT Solenoid (SL-)	In Lockup: 12-14v
5	GN	Injector 6 Control	2.0-3.3 ms
6	RD	Injector 5 Control	2.0-3.3 ms
7	BL	Injector 4 Control	2.0-3.3 ms
8	GY	Injector 3 Control	2.0-3.3 ms
9	YL	Injector 2 Control	2.0-3.3 ms
10	WT	Injector 1 Control	2.0-3.3 ms
11	PK	A/T-ECT Solenoid (S1)	In 3rd or OD: 1v
12	WT/RD	Igniter Signal (IGF)	Digital Signal: 0-5-0v
13	BK/RD	Starter Switch Signal	9-11v (cranking)
14	BK/WT	Neutral Start Switch	In P/N: 9-11v (cranking:
15	GY/BK	Igniter Transistor 3 Control	7% duty cycle
16	YL/RD	Igniter Transistor 2 Control	7% duty cycle
17	PK/BK	A/T-ECT Solenoid (S2)	1st or OD: 1v
22	YL/BK	IAC Signal (RSC)	Pulse Signals
23	GN/BK	IAC Signal (RSO)	Pulse Signals
24	BL/BK	Igniter Transistor 1 Control	7% duty cycle
25	BK/YL	Fuel Pressure Up Solenoid	Hot Restart: 12-14v
27	BL/YL	A/T-ECT Solenoid (SL+)	In Lockup: 12-14v
28, 33-34	WT/BK	Power Ground	<0.1v
32	BL/WT	Closed Throttle Switch	1v, off-idle: 12v

Pin Connector Graphic

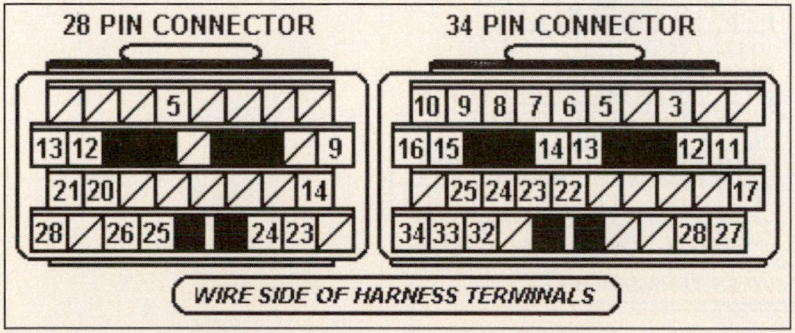

1997 Sedan 3.0L V6 MFI VIN F (A/T-ECT) 16 Pin Connector

PCM Pin #	Wire Color	Circuit Description (16 Pin)	Value at Hot Idle
2	BK/RD	EVAP Purge Solenoid (VSV)	12v or 0v
3	GN/RD	MIL (lamp) Control	MIL Off: 12v, On: 1v
5	BL	Data Link Connector	12-14v
6	BL	Intake Air Solenoid	12v or 0v
7	RD/BK	MAF Sensor Ground	<0.050v
8	WT/BL	EVAP Vapor Pressure (VSV)	12v or 0v
9	BK/YL	Cooling Fan Relay	Relay Off: 12v, On: 1v
10	BL/RD	HO2S-21 (B2 S1) Heater	1v (Heater on)
11	BL/BK	HO2S-11 (B1 S1) Heater	1v (Heater on)
12	BK/WT	EGR Solenoid Control (VSV)	12v or 0v
13	PK	EVAP Vapor Pressure Sensor	2.9-3.1v (with hose off)
14	GN/RD	EGR Gas Temp. Sensor	3.5-4.0v
15	BK/OR	EGR Valve Position Sensor	0.4-1.6v
16	BR	Sensor Ground	<0.050v

1997 Sedan 3.0L V6 MFI VIN F (A/T-ECT) 22 Pin Connector

PCM Pin #	Wire Color	Circuit Description (22 Pin)	Value at Hot Idle
1	BL/RD	Sensor VREF (VC)	4.9-5.1v
2	YL/GN	A/T-ECT Solenoid (SLN-)	During shifting: 12v
4	BL	OD Clutch Speed Sensor (-)	AC pulse signals
5	BK/RD	CKP Sensor Signal (NE+)	AC pulse signals
6	BL	CKP Sensor Signal (NE-)	<0.050v
7	BK/YL	TP Sensor Signal	0.3-0.8v
8	RD	MAF Sensor Signal	1.1-1.5v
9	YL	OD Clutch Speed Sensor (+)	AC pulse signals
13	WT	HO2S-11 (B1 S1) Signal	0.1-1.1v
14	WT	Knock Sensor 1 Signal	<0.075v AC
15	WT	Knock Sensor 2 Signal	<0.075v AC
17	BK/WT	CMP Sensor Signal (G+)	AC pulse signals
18	GN	Circuit Opening Relay (FC)	0-3v, off-idle: 12v
19	BL/BK	HO2S-21 (B2 S1) Signal	0.1-1.1v
20	GN/BK	ECT Sensor Signal	At 180°F: 0.51v
21	BL/BK	IAT Sensor Signal	At 100°F: 2.60v
22	BR	Sensor Ground	<0.050v

Pin Connector Graphic

WIRE SIDE OF HARNESS TERMINALS

1997 Sedan 3.0L V6 MFI VIN F (A/T-ECT) 28 Pin Connector

PCM Pin #	Wire Color	Circuit Description (28 Pin)	Value at Hot Idle
1	YL/BL	A/T Select Switch Low	In Low: 12v, Others: 0v
2	BK/RD	Mirror Heater Switch Signal	Switch On: 12v, Off: 0v
3	GN	Tail Light Switch Signal	Switch On: 12v, Off: 0v
5	PK	A/C Amplifier Signal (ACT)	Clutch On: 12v, Off: 1.5v
8	WT	Data Link Connector	12v
9	RD	Cooling Fan Relay	Relay Off: 12v, On: 1v
10	GN/YL	A/T Select Switch 2nd	In 2nd: 12v, Others: 0v
12	PK	Vehicle Speed Sensor	At 55 mph: 48 Hz
13	BK	Tachometer Signal (TACO)	Pulse Signals
14	BK/YL	Battery Direct	12-14v
15	RD/BK	A/T Select Switch Reverse	In 'R': 12v, Others: 0v
16	BK/YL	A/C Amplifier Signal (AC1)	Clutch On: 1.5v, Off: 12v
17	PK/BK	HO2S-12 (B1 S2) Heater	1v (Heater On)
18	BK	HO2S-12 (B1 S2) Signal	0.1-1.1v
19	BR/WT	ABS/Traction ECU (NEO)	Pulse Signals
21	BK/RD	Defogger Idle Up Signal	Switch On: 12v, Off: 0v
22	BL/RD	A/T Pattern Selector Switch	Norm: 0v, PWR: 12v
23	LG	EFI Main Relay Power	12-14v
24	GN/WT	Stop Light Switch Signal	Brake Off: 0v, On: 12v
25	PK	ABS/Traction ECU (EFI-)	Pulse Signals
26	LG	ABS/Traction ECU (EFI+)	Pulse Signals
27	GN/RD	ABS/Traction ECU (TRC-)	Pulse Signals
28	GN	ABS/Traction ECU (TRC+)	Pulse Signals

1997 Sedan 3.0L V6 MFI VIN F (A/T-ECT) 34 Pin Connector

PCM Pin #	Wire Color	Circuit Description (34 Pin)	Value at Hot Idle
3	GN	A/T-ECT Solenoid (SLN-)	During shifting: 12v
5	GN	Injector 6 Control	2.0-3.3 ms
6	RD	Injector 5 Control	2.0-3.3 ms
7	BL	Injector 4 Control	2.0-3.3 ms
8	BK	Injector 3 Control	2.0-3.3 ms
9	YL	Injector 2 Control	2.0-3.3 ms
10	WT	Injector 1 Control	2.0-3.3 ms
11	PK	A/T-ECT Solenoid (S1)	In 3rd or OD: 1v
12	WT/RD	Igniter Signal (IGF)	Digital Signal: 0-5-0v
13	BK/RD	Starter Switch Signal	9-11v (cranking)
14	BK/WT	Neutral Start Switch	In P/N: 9-11v (cranking)
15	GN/RD	Igniter Transistor 3 Control	7% duty cycle
16	YL/RD	Igniter Transistor 2 Control	7% duty cycle
17	GN/YL	A/T-ECT Solenoid (S2)	1st or OD: 1v
22	YL/BK	IAC Signal (RSC)	Pulse Signals
23	GN/BK	IAC Signal (RSO)	Pulse Signals
24	BL/BK	Igniter Transistor 1 Control	7% duty cycle
27	BL/YL	A/T-ECT Solenoid (SL)	In Lockup: 12-14v
28	WT/BK	Power Ground	<0.1v
31	PK/BK	PSP Switch Signal	Straight: 12v, Turning: 0v
33	WT/BK	Power Ground	<0.1v
34	WT/BK	Power Ground	<0.1v

1998 Sedan 3.0L V6 W/O EIS-TC VIN F Federal 16 Pin Connector

PCM Pin #	Wire Color	Circuit Description (16 Pin)	Value at Hot Idle
2	BK/RD	EVAP Purge Solenoid (VSV)	12v or 0v
3	GN/RD	MIL (lamp) Control	MIL Off: 12v, On: 1v
5	BL	Data Link Connector	12v
6	RD/YL	Intake Air Control Solenoid	12v or 0v
7	RD/BK	MAF Sensor Ground	<0.050v
8	WT/BL	EVAP Vapor Pressure (VSV)	12v or 0v
9	BK/YL	Cooling Fan 1 Relay	Relay Off: 12v, On: 1v
10	BL/RD	HO2S-21 (B2 S1) Heater	1v (Heater on)
11	BL/BK	HO2S-11 (B1 S1) Heater	1v (Heater on)
12	BK/WT	EGR Solenoid Control (VSV)	12v or 0v
13	PK	EVAP Vapor Pressure Sensor	2.9-3.1v (with hose off)
14	GN/RD	EGR Gas Temp. Sensor	3.5-4.0v
15	RD/WT	EGR Valve Position Sensor	0.4-1.6v
16	BR	Sensor Ground	<0.050v

1998 Sedan 3.0L V6 W/O EIS-TC VIN F Federal 22 Pin Connector

PCM Pin #	Wire Color	Circuit Description (22 Pin)	Value at Hot Idle
1	BL/RD	Sensor VREF (VC)	4.9-5.1v
2	YL/GN	A/T-ECT Solenoid (SL-)	In Lockup: 12-14v
4	BL	OD Clutch Speed Signal (-)	AC pulse signals
5	BK/RD	CKP Sensor Signal (NE+)	AC pulse signals
6	BL	CKP Sensor Signal (NE-)	<0.050v
7	BK/YL	TP Sensor Signal	0.3-0.8v
8	RD	MAF Sensor Signal	1.1-1.5v
9	YL	OD Clutch Speed Signal (+)	AC pulse signals
13	WT	HO2S-11 (B1 S1) Signal	0.1-1.1v
14	WT	Knock Sensor 1 Signal	<0.075v AC
15	WT	Knock Sensor 2 Signal	<0.075v AC
17	BK/WT	CMP Sensor Signal (G+)	AC pulse signals
18	GN	Circuit Opening Relay (FC)	0-3v, off-idle: 12v
19	RD/BL	HO2S-21 (B2 S1) Signal	0.1-1.1v
20	GN/BK	ECT Sensor Signal	At 180°F: 0.51v
21	BL/BK	IAT Sensor Signal	At 100°F: 2.60v
22	BR	Sensor Ground	<0.050v

Pin Connector Graphic

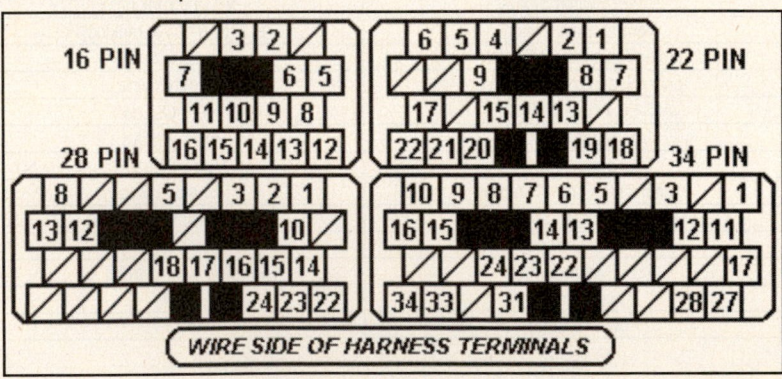

1998 Sedan 3.0L V6 W/O EIS-TC VIN F Federal 28 Pin Connector

PCM Pin #	Wire Color	Circuit Description (28 Pin)	Value at Hot Idle
1	YL	A/T Select Switch Low	In Low: 12v, Others: 0v
2	BK/RD	Mirror Heater Switch Signal	Switch On: 12v, Off: 0v
3	GN	Tail Light Switch Signal	Switch On: 12v, Off: 0v
5	PK	A/C Amplifier Signal (ACT)	Clutch On: 12v, Off: 1.5v
8	WT	Data Link Connector	Activated: Pulses
10	GN/YL	A/T Select Switch 2nd	In 2nd: 12v, Others: 0v
12	PK	Vehicle Speed Sensor	At 55 mph: 48 Hz
13	BK	Tachometer Signal (TACO)	Pulses
14	BK/YL	Battery Direct	12-14v
15	RD/BK	A/T Select Switch Reverse	In 'R': 12v, Others: 0v
16	BK/YL	A/C Amplifier Signal (AC1)	Clutch On: 1.5v, Off: 12v
17	PK/BK	HO2S-12 (B1 S2) Heater	1v (Heater on)
18	BK	HO2S-12 (B1 S2) Signal	0.1-1.1v
22	BL/RD	A/T Pattern Selector Switch	Norm: 0v, PWR: 12v
23	LG	EFI Main Relay Power	12-14v
24	GN/WT	Stop Light Switch Signal	Brake Off: 0v, On: 12v

1998 Sedan 3.0L V6 W/O EIS-TC VIN F Federal 34 Pin Connector

PCM Pin #	Wire Color	Circuit Description (34 Pin)	Value at Hot Idle
1	BR	Sensor Ground	<0.050v
3	GN	A/T-ECT Solenoid (SL-)	In Lockup: 12-14v
5	GN	Injector 6 Control	2.0-3.3 ms
6	RD	Injector 5 Control	2.0-3.3 ms
7	BL	Injector 4 Control	2.0-3.3 ms
8	BK	Injector 3 Control	2.0-3.3 ms
9	YL	Injector 2 Control	2.0-3.3 ms
10	WT	Injector 1 Control	2.0-3.3 ms
11	PK	A/T-ECT Solenoid (S1)	In 3rd or OD: 1v
12	WT/RD	Igniter Signal (IGF)	Digital Signal: 0-5-0v
13	BK/RD	Starter Switch Signal	9-11v (cranking)
14	BK/WT	Neutral Start Switch	In P/N: 9-11v (cranking)
15	GN/RD	Igniter Transistor 3 Control	7% duty cycle
16	YL/RD	Igniter Transistor 2 Control	7% duty cycle
17	GN/YL	A/T-ECT Solenoid (S2)	1st or OD: 1v
22	YL/BK	IAC Signal (RSC)	Pulse Signals
23	GN/BK	IAC Signal (RSO)	Pulse Signals
24	BL/BK	Igniter Transistor 1 Control	7% duty cycle
27	BL/YL	A/T-ECT Solenoid (SL+)	In Lockup: 12-14v
28	WT/BK	Power Ground	<0.1v
31	PK/BK	PSP Switch Signal	Straight: 12v, Turning: 0v
33	WT/BK	Power Ground	<0.1v
34	WT/BK	Power Ground	<0.1v

Pin Connector Graphic

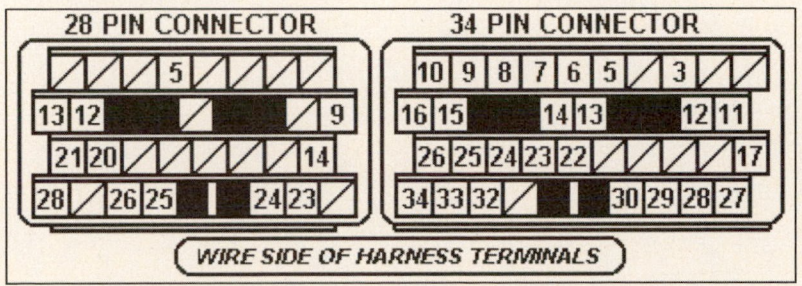

1998-99 Sedan 3.0L V6 W/EIS-TC CAL VIN F 17 Pin Connector

PCM Pin #	Wire Color	Circuit Description (17 Pin)	Value at Hot Idle
4	BK/RD	Transponder Amplifier Code	Inserting key: pulses
5	YL	Transponder Amplifier Signal	Inserting key: pulses
10	GN	Transponder Amplifier Signal	Inserting key: pulses
11	RD/YL	Unlock Warning Switch	No key: 4-5v
13	BR	Sensor Shield Ground	<0.050v
16	PK	Security Indicator Light	Digital Signal

1998-99 Sedan 3.0L V6 W/EIS-TC CAL VIN F 22 Pin Connector

PCM Pin #	Wire Color	Circuit Description (22 Pin)	Value at Hot Idle
1	BK/YL	Battery Direct	12-14v
2	BK/OR	Ignition Switch Power	12-14v
3	GN	Circuit Opening Relay (FC)	0-3v, off-idle: 12v
6	GN/RD	MIL (lamp) Control	MIL Off: 12v, On: 1v
7	BK/RD	Starter Switch Signal	9-11v (cranking)
8	BK/OR	EFI Main Relay Power	12-14v
9	WT/BL	EVAP Vapor Pressure (VSV)	12v or 0v
11	WT	SIL Signal (Scan Tool)	Digital Signal
13	GN	ABS/Traction ECU (TRC+)	Pulses
14	LG	ABS/Traction ECU (EFI+)	Pulses
15	GN/WT	Stop Light Switch Signal	Brake Off: 0v, On: 12v
16	LG	EFI Main Relay Power	12-14v
17	PK	EVAP Vapor Pressure Sensor	2.9-3.1v (with hose off)
18	BK/RD	Mirror Heater Switch Signal	Switch On: 12v, Off: 0v
19	GN	Tail Light Switch Signal	Switch On: 12v, Off: 0v
20	GN/RD	ABS/Traction ECU (TRC-)	Pulses
21	PK	ABS/Traction ECU (EFI-)	Pulses

Pin Connector Graphic

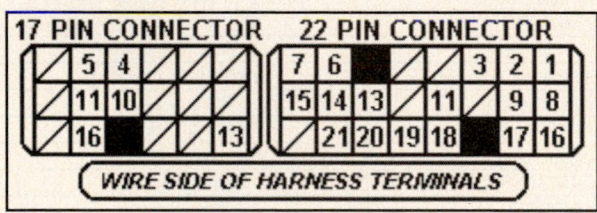

1998-99 Sedan 3.0L V6 W/EIS-TC CAL VIN F 24 Pin Connector

PCM Pin #	Wire Color	Circuit Description (24 Pin)	Value at Hot Idle
1	WT/BK	Power Ground	<0.1v
2	BL/RD	Sensor VREF	4.9-5.1v
3	BL/RD	AFS-11 (B1 S1) Heater	1v (Heater On)
4	BK/RD	AFS-21 (B2 S1) Heater	1v (Heater On)
5	WT	Injector 1 Control	2.0-3.3 ms
6	YL	Injector 2 Control	2.0-3.3 ms
7	BK/RD	EVAP Purge Solenoid (VSV)	12v or 0v
8	WT/BK	Sensor Ground	<0.050v
9	PK/BK	PSP Switch Signal	Straight: 12v, Turning: 0v
10	RD	MAF Sensor Signal	1.1-1.5v
11	BK/WT	AFS-11 (B1 S1) Signal (+)	3.0-3.6v
12	BR	AFS-21 (B2 S1) Signal (+)	3.0-3.6v
13	GN/RD	EGR Gas Temp. Sensor	3.5-4.0v
14	GN/BK	ECT Sensor Signal	At 180ºF: 0.51v
16	BK/RD	CKP Sensor Signal (NE+)	AC pulse signals
17	BR	Shield Ground	<0.050v
18	BR	Sensor Ground	<0.050v
19	RD/BK	MAF Sensor Ground	<0.050v
20	BL	AFS-11 (B1 S1) Signal (AF-)	Fixed at 3v
21	BK/RD	AFS-21 (B2 S2) Signal (-)	Fixed at 3v
22	BL/BK	IAT Sensor Signal	At 100ºF: 2.60v
23	BK/YL	TP Sensor Signal	0.3-0.8v
24	BL	CKP/CMP Sensor Ground (-)	<0.050v

1998-99 Sedan 3.0L V6 W/EIS-TC CAL VIN F 28 Pin Connector

PCM Pin #	Wire Color	Circuit Description (28 Pin)	Value at Hot Idle
8	BK	HO2S-12 (B1 S2) Signal	0.1-1.1v
9	PK/BK	HO2S-12 (B1 S2) Heater	1v (Heater on)
13	PK	A/C Amplifier Signal (ACT)	Clutch On: 12v, Off: 1.5v
14	BL/RD	A/C Amplifier Signal (THWO)	Pulse Signals
16	BR/WT	ABS/Traction ECU (NEO)	Pulse Signals
20	BK/WT	Neutral Start Switch	In P/N: 9-11v (cranking)
22	PK	Vehicle Speed Sensor	At 55 mph: 48 Hz
25	BK/YL	A/C Amplifier Signal (AC1)	Clutch On: 1.5v, Off: 12v
27	BK	Tachometer Signal (TACO)	Pulse Signal

Standard Colors and Abbreviations

Abbreviation	Color	Abbreviation	Color	Abbreviation	Color
BK	Black	GY	Gray	RD	Red
BL	Blue	GN	Green	TN	Tan
BR	Brown	LG	Light Green	VT	Violet
DB	Dark Blue	OR	Orange	WT	White
DG	DK Green	PK	Pink	YL	Yellow

1998-99 Sedan 3.0L V6 W/EIS-TC Calif. VIN F 31 Pin Connector

PCM Pin #	Wire Color	Circuit Description (31 Pin)	Value at Hot Idle
1	BK	Injector 3 Control	2.0-3.3 ms
2	BL	Injector 4 Control	2.0-3.3 ms
3	RD	Injector 5 Control	2.0-3.3 ms
4	GN	Injector 6 Control	2.0-3.3 ms
6	BL	Data Link Connector (TE)	12-14v
10	BK/WT	CMP Sensor Signal (G+)	AC pulse signals
11	BL/BK	Igniter Transistor 1 Control	7% duty cycle
12	YL/RD	Igniter Transistor 2 Control	7% duty cycle
13	GN/RD	Igniter Transistor 3 Control	7% duty cycle
15	YL/BK	IAC Signal (RSC)	Pulse Signals
16	GN/BK	IAC Signal (RSO)	Pulse Signals
17	RD/YL	ACIS Solenoid Control (VSV)	12v or 0v
18	BK/WT	EGR Solenoid Control (VSV)	12v or 0v
21	WT/BK	Power Ground	<0.1v
22	RD/WT	EGR Valve Position Sensor	0.4-1.6v
23 ('98)	BR	Sensor Ground	<0.050v
25	WT/RD	Igniter Signal (IGF)	Digital Signal: 0-5-0v
27	WT	Knock Sensor 1 Signal	<0.075v AC
28	WT	Knock Sensor 2 Signal	<0.075v AC
29	BK/YL	Cooling Fan 1 Relay	Relay Off: 12v, On: 1v
30	WT/BK	Power Ground	<0.1v
31	WT/BK	Power Ground	<0.1v

Pin Connector Graphic

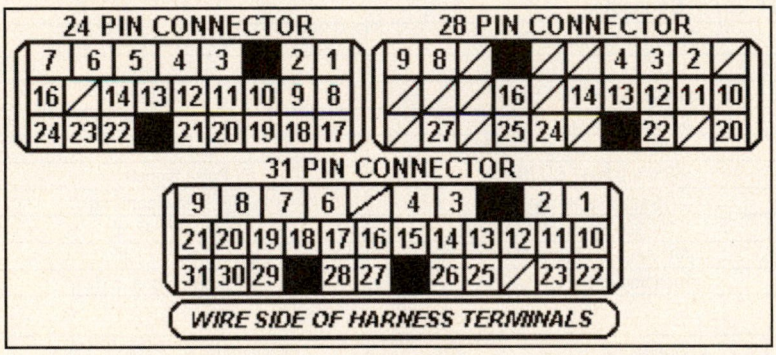

1998-99 Sedan 3.0L V6 W/EIS VIN F Federal 17 Pin Connector

PCM Pin #	Wire Color	Circuit Description (17 Pin)	Value at Hot Idle
4	BK/RD	Transponder Amplifier Code	Inserting key: pulses
5	YL	Transponder Amplifier Signal	Inserting key: pulses
10	GN	Transponder Amplifier Signal	Inserting key: pulses
11	RD/YL	Unlock Warning Switch	No Key: 4-5v
13	BR	Sensor Shield Ground	<0.050v
16	PK	Security Indicator Light	Pulses

1998-99 Sedan 3.0L V6 W/EIS VIN F Federal 22 Pin Connector

PCM Pin #	Wire Color	Circuit Description (22 Pin)	Value at Hot Idle
1	BK/YL	Battery Direct	12-14v
2	BK/OR	Ignition Switch Power	12-14v
3	GN	Circuit Opening Relay (FC)	0-3v, off-idle: 12v
6	GN/RD	MIL (lamp) Control	MIL Off: 12v, On: 1v
7	BK/RD	Starter Switch Signal	9-11v (cranking)
8	BK/OR	EFI Main Relay Power	12-14v
9	WT/BL	EVAP Vapor Pressure (VSV)	12v or 0v
11	WT	SIL Signal (Scan Tool)	Digital Signal
13	GN	ABS/Traction ECU (TRC+)	Pulses
14	LG	ABS/Traction ECU (EFI+)	Pulses
15	GN/WT	Stop Light Switch Signal	Brake Off: 0v, On: 12v
16	LG	EFI Main Relay Power	12-14v
17	PK	EVAP Vapor Pressure Sensor	2.9-3.1v (with hose off)
18	BK/RD	Mirror Heater Switch Signal	Heater On: 12-14v
19	GN	Tail Light Switch Signal	Switch On: 12v, Off: 0v
20	GN/RD	ABS/Traction ECU (TRC-)	Pulses
21	PK	ABS/Traction ECU (EFI-)	Pulses

Pin Connector Graphic

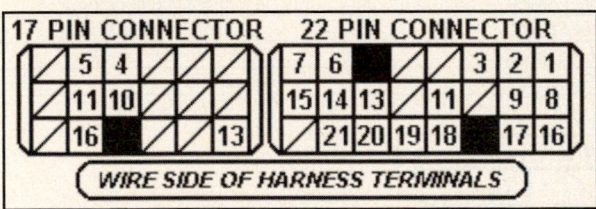

1998-99 Sedan 3.0L V6 W/EIS VIN F Federal 24 Pin Connector

PCM Pin #	Wire Color	Circuit Description (24 Pin)	Value at Hot Idle
2	BL/RD	Sensor VREF	4.9-5.1v
3	BL/BK	HO2S-11 (B1 S1) Heater	1v (Heater on)
4	BL/RD	HO2S-21 (B2 S1) Heater	1v (Heater on)
5	WT	Injector 1 Control	2.0-3.3 ms
6	YL	Injector 2 Control	2.0-3.3 ms
7	BK/RD	EVAP Purge Solenoid (VSV)	12v or 0v
9	PK/BK	PSP Switch Signal	Straight: 12v, Turning: 0v
10	RD	MAF Sensor Signal	1.1-1.5v
11	WT	HO2S-11 (B1 S1) Signal	0.1-1.1v
12	RD/BL	HO2S-21 (B2 S1) Signal	0.1-1.1v
13	GN/RD	EGR Gas Temp. Sensor	3.5-4.0v
14	GN/BK	ECT Sensor Signal	At 180°F: 0.51v
16	BK/RD	CKP Sensor Signal (NE+)	AC pulse signals
17	BR	Shield Ground	<0.050v
18	BR	Sensor Ground	<0.050v
19	RD/BK	MAF Sensor Ground	<0.050v
22	BL/BK	IAT Sensor Signal	At 100°F: 2.60v
23	BK/YL	TP Sensor Signal	0.3-0.8v
24	BL	CKP/CMP Sensor Ground (-)	<0.050v

1998-99 Sedan 3.0L V6 W/EIS VIN F Federal 28 Pin Connector

PCM Pin #	Wire Color	Circuit Description (28 Pin)	Value at Hot Idle
8	BK	HO2S-12 (B1 S2) Signal	0.1-1.1v
9	PK/BK	HO2S-12 (B1 S2) Heater	1v (Heater on)
13	PK	A/C Amplifier Signal (ACT)	Clutch On: 12v, Off: 1.5v
14	BL/RD	A/C Amplifier Signal (THWO)	Pulse Signals
16	BR/WT	ABS/Traction ECU (NEO)	Pulse Signals
20	BK/WT	Neutral Start Switch	In P/N: 9-11v (cranking)
22	PK	Vehicle Speed Sensor	At 55 mph: 48 Hz
25	BK/YL	A/C Amplifier Signal (AC1)	Clutch On: 1.5v, Off: 12v
27	BK	Tachometer Signal (TACO)	Pulse Signals

Pin Connector Graphic

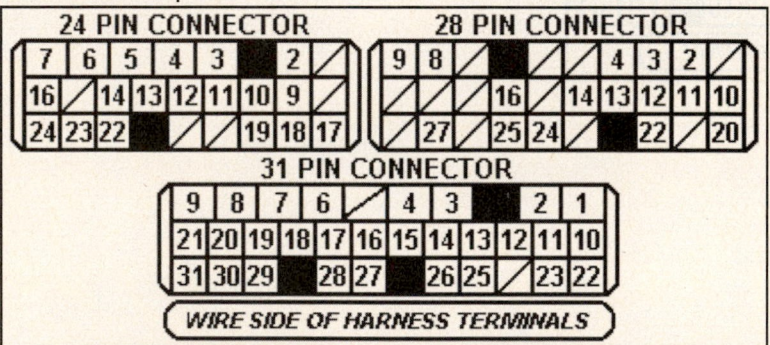

Standard Colors and Abbreviations

Abbreviation	Color	Abbreviation	Color	Abbreviation	Color
BK	Black	GY	Gray	RD	Red
BL	Blue	GN	Green	TN	Tan
BR	Brown	LG	Light Green	VT	Violet
DB	Dark Blue	OR	Orange	WT	White
DG	DK Green	PK	Pink	YL	Yellow

1998-99 Sedan 3.0L V6 W/EIS VIN F Federal 31 Pin Connector

PCM Pin #	Wire Color	Circuit Description (31 Pin)	Value at Hot Idle
1	BK	Injector 3 Control	2.0-3.3 ms
2	BL	Injector 4 Control	2.0-3.3 ms
3	RD	Injector 5 Control	2.0-3.3 ms
4	GN	Injector 6 Control	2.0-3.3 ms
5	---	Not Used	---
6	BL	Data Link Connector (TE)	12v
7-9	---	Not Used	---
10	BK/WT	CMP Sensor Signal (G+)	AC pulse signals
11	BL/BK	Igniter Transistor 1 Control	7% duty cycle
12	YL/RD	Igniter Transistor 2 Control	7% duty cycle
13	GN/RD	Igniter Transistor 3 Control	7% duty cycle
14	---	Not Used	---
15	YL/BK	IAC Signal (RSC)	Pulse Signals
16	GN/BK	IAC Signal (RSO)	Pulse Signals
17	RD/YL	ACIS Solenoid Control (VSV)	12v or 0v
18	BK/WT	EGR Solenoid Control (VSV)	12v or 0v
19-20	---	Not Used	---
21	WT/BK	Power Ground	<0.1v
22	RD/WT	EGR Valve Position Sensor	0.4-1.6v
23 ('98)	BR	Sensor Ground	<0.050v
25	WT/RD	Igniter Signal (IGF)	Digital Signal: 0-5-0v
26	---	Not Used	---
27	WT	Knock Sensor 1 Signal	<0.075v AC
28	WT	Knock Sensor 2 Signal	<0.075v AC
29	BK/YL	Cooling Fan 1 Relay Control	Relay Off: 12v, On: 1v
30	WT/BK	Power Ground	<0.1v
31	WT/BK	Power Ground	<0.1v

Pin Connector Graphic

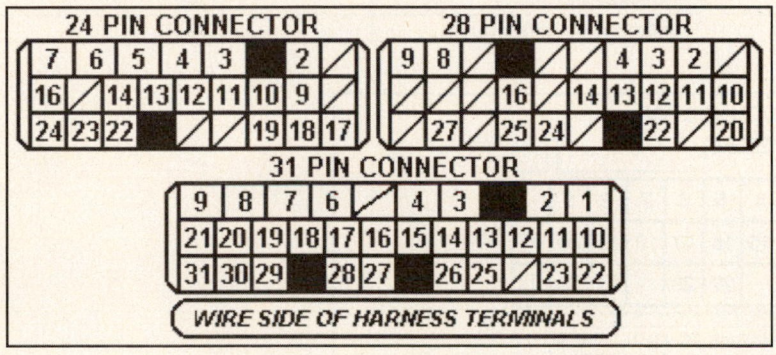

2000-01 Sedan 3.0L V6 DOHC VIN F (A/T-ECT) E4 31P Connector

PCM Pin #	Wire Color	Circuit Description (31 Pin)	Value at Hot Idle
1	YL	Injector 3 Control	1.6-2-9 ms
2	WT	Injector 4 Control	1.6-2-9 ms
3	GN	Injector 5 Control	1.6-2-9 ms
4	GN	Injector 6 Control	1.6-2.9 ms
5	RD	VVT Solenoid Control (OC1-)	12v or 0v
6	RD/BK	VVT Solenoid Control (OC1+)	12v or 0v
7	PK	A/T-ECT Solenoid (S1)	In 3rd or OD: 1v
8-9	---	Not Used	---
10	BK/WT	CMP Sensor Signal (RH+)	AC pulse signals
11	GY	Igniter Transistor 1 Control	6°, at 55 mph: 8° dwell
12	BK/RD	Igniter Transistor 2 Control	6°, at 55 mph: 8° dwell
13	LG/BK	Igniter Transistor 3 Control	6°, at 55 mph: 8° dwell
14	BL/YL	Igniter Transistor 4 Control	6°, at 55 mph: 8° dwell
15	BL	Igniter Transistor 5 Control	6°, at 55 mph: 8° dwell
16	LG	Igniter Transistor 6 Control	6°, at 55 mph: 8° dwell
17	RD/YL	ACIS 1 Control (VSV)	12v or 0v
18	RD/WT	VVT Solenoid Control (OC2-)	12v or 0v
19	YL	A/T-ECT Solenoid (SLN-)	Pulse Signals
20	PK/BL	A/T-ECT Solenoid (SLN+)	Pulse Signals
21	WT/BK	Power Ground (E01)	<0.1v
22	BK/WT	CMP Sensor Signal (LH+)	AC pulse signals
23	GN	O/D Clutch Speed Sensor (-)	AC pulse signals
24	PK/BL	O/D Clutch Speed Sensor (+)	AC pulse signals
25	BK/YL	Igniter Signal (IGF)	Digital Signal: 0-5-0v
26	GN/BK	IAC Signal (RSO)	Pulse Signals
27	WT	Knock Sensor 1 Signal	0.075v AC
28	WT	Knock Sensor 2 Signal	0.075v AC
29	RD/BL	VVT Solenoid Control (OC2+)	12v or 0v
30	WT/BK	Power Ground (E03)	<0.1v
31	WT/BK	Power Ground (E04)	<0.1v

Pin Connector Graphic

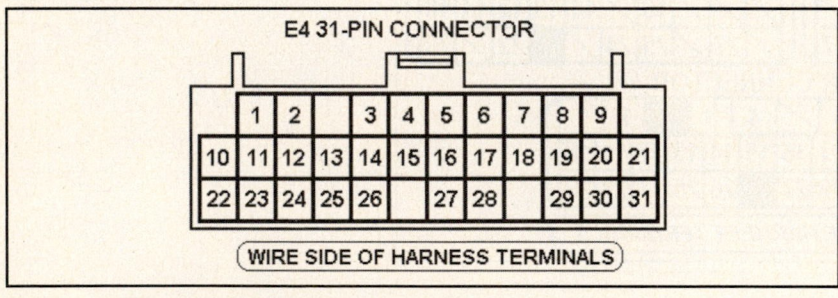

2000-01 Sedan 3.0L V6 DOHC VIN F (A/T-ECT) E5 24P Connector

PCM Pin #	Wire Color	Circuit Description (24 Pin)	Value at Hot Idle
1	WT/BK	Power Ground (E04)	<0.1v
2	BL/RD	Sensor VREF (VC)	4.9-5.1v
3	BK/RD	AFS-11 (B1 S1) Heater	1v (Heater On)
4	BK/RD	AFS-21 (B2 S1) Heater	1v (Heater On)
5	BL	Injector 1 Control	1.6-2-9 ms
6	RD	Injector 2 Control	1.6-2-9 ms
7	BK/RD	EVAP Purge Solenoid (VSV)	12v or 0v
8	WT/BK	Power Ground (E05)	<0.1v
9	BK/RD	PSP Switch Signal	Straight: 12v, Turned: 0v
10	RD	MAF Sensor Signal	1-1.1v
11	BK/RD	AFS-11 (B1 S1) Signal (+)	3.0-3.6v
12	BK/WT	AFS-21 (B2 S1) Signal (+)	3.0-3.6v
13	----	Not Used	---
14	GN/YL	ECT Sensor Signal	At 180ºF: 0.51v
15	WT/RD	VSV ACIS 2 Control	12v or 0v
16	BK/WT	CKP Sensor Signal (NE+)	310-330 Hz
17	BR	Shield Ground (E1)	<0.050v
18	WT	Sensor Ground	<0.050v
19	RD/BK	MAF Sensor Ground (E2G)	<0.050v
20	BR	AFS-11 (B1 S1) Signal (AF-)	3.0-3.6v
21	LB	AFS-21 (B2 S1) Signal (-)	3.0-3.6v
22	BL/BK	IAT Sensor Signal	0.5-3.4v
23	LG	TP Sensor Signal	0.53-1.27v
24	BL	CKP/CMP Sensor Ground (-)	<0.050v

Pin Connector Graphic

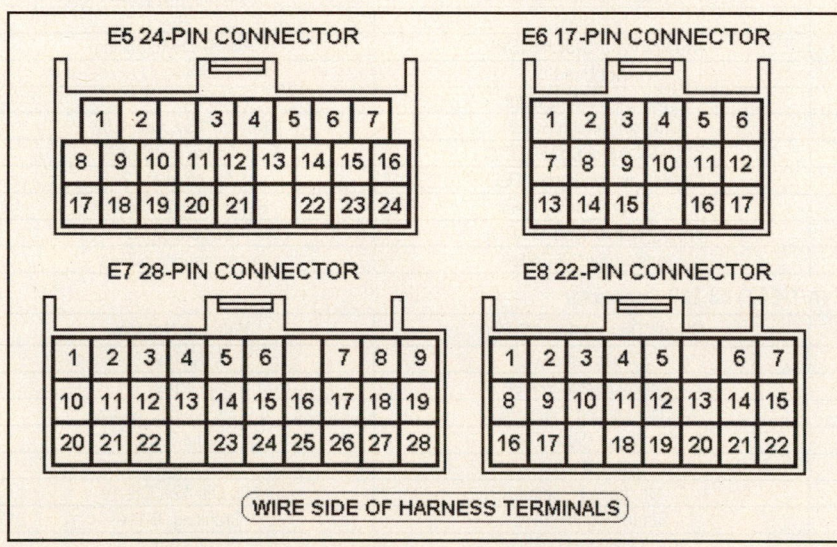

2000-01 Sedan 3.0L V6 DOHC VIN F (A/T-ECT) E6 17P Connector

PCM Pin #	Wire Color	Circuit Description (17 Pin)	Value at Hot Idle
1	GN/YL	A/T-ECT Solenoid (S2)	1st or OD: 1v
2-7	---	Not Used	---
8	RD/BK	A/T Select Switch Reverse	In 'R': 12v, Others: 0v
9-11	---	Not Used	---
12	GN/OR	Overdrive Main Switch	Switch Off: 12v, On: 1v
13	YL	A/T Select Switch Low	In Low: 12v, Others: 0v
14	GN/YL	A/T Select Switch 2nd	In 2nd: 12v, Others: 0v
15	GN/WT	A/T-ECT Solenoid (SL+)	In Lockup: 12-14v

2000-01 Sedan 3.0L V6 DOHC VIN F (A/T-ECT) E7 28P Connector

PCM Pin #	Wire Color	Circuit Description (28 Pin)	Value at Hot Idle
1-2	---	Not Used	---
3	PK/BK	EVAP Vapor Pressure (VSV)	12v or 0v
4	GN	Tail Light Switch Signal	Switch On: 12v, Off: 0v
5	BL	Data Link Connector (TE1)	12-14v
6	LG/RD	A/C Magnetic Clutch (ACMG)	Clutch Off: 0v, On: 12v
7	YL/RD	A/C Amplifier Signal (AC1)	Clutch On: 1.5v, Off: 12v
8	BK	HO2S-12 (B1 S2) Signal	0.1-1.1v
9	LG	HO2S-12 (B1 S2) Heater	1v (Heater On)
10-12	---	Not Used	---
13	BK	Mirror Heater Switch Signal	Switch Off: 1.5v, On: 12v
14	YL/GN	A/C Amplifier Signal (THWO)	Pulse Signals
15	---	Not Used	---
16	YL/BK	ABS/Traction NEO Signal	Pulse Signals
17	---	Not Used	---
18	BL/YL	TXCT Ignition Signal	Inserting key: pulses
19	---	Not Used	---
20	BK/WT	Neutral Start Switch (NSW)	In P/N: 9-11v (cranking)
21	---	Not Used	---
22	VT/WT	Speedometer Indicator	At 55 mph: 48 Hz
23	RD/YL	Unlock Warning Switch	Key In: 1.5v, Out: 4.5v
24	GY	Cruise Control Signal (OD1)	At Cruise in OD: 12v
25	YL	Cruise Control Signal (IDLO)	1.5v, off-idle: 12v
26	GN	Cooling Fan 1 Relay Control	Relay Off: 12v, On: 1v
27	BK	Tachometer Signal (TACO)	Pulse Signals
28	PK/GN	Ignition Switch Code	Inserting key: pulses

2000-01 Sedan 3.0L V6 DOHC VIN F (A/T-ECT) E8 22P Connector

PCM Pin #	Wire Color	Circuit Description (22 Pin)	Value at Hot Idle
1	BK/RD	Direct Battery	12-14v
2	BK/OR	Ignition Switch Power	12-14v
3	GN/BK	Circuit Opening Relay (FC)	0-0.3v, off-idle: 12-14v
4	WT	SIL Signal (Scan Tool)	Digital Signal
5	---	Not Used	---
6	BK/YL	MIL (lamp) Control	MIL Off: 12v, On: 1v
7	BK/RD	Starter Signal (STA)	Cranking: 9-11v
8	BL/OR	EFI Main Relay Power	12-14v
9	GN/OR	Overdrive Lamp Control	At Cruise in OD: 1v
10	GN	EVAP Canister Closed Valve	12v or 0v
11-12	---	Not Used	---
13	BR	ABS/Traction: TRC (+) Signal	DC pulse signals
14	PK	ABS/Traction: ENG (+) Signal	DC pulse signals
15	GN/WT	Brake Switch Signal	Brake Off: 0v, On: 12v
16	BK/WT	EFI Main Relay Power	12-14v
17	PK	EVAP Vapor Pressure Sensor	2.9-3.1v (with hose off)
18-19	---	Not Used	---
20	GN	ABS/Traction: TRC (-) Signal	DC pulse signals
21	GY	ABS/Traction: ENG (-) Signal	DC pulse signals
22	VT/WT	Security Indicator Light	Inserting key: pulses

2002-03 Sedan 3.0L V6 DOHC VIN F (A/T-ECT) E4 31P Connector

PCM Pin #	Wire Color	Circuit Description (31 Pin)	Value at Hot Idle
1	YL	Injector 3 Control	1.6-2-9 ms
2	WT	Injector 4 Control	1.6-2-9 ms
3	BL/RD	Injector 5 Control	1.6-2-9 ms
4	GN	Injector 6 Control	1.6-2.9 ms
5	GR	VVT Solenoid Control (OC1-)	12v or 0v
6	RD/BK	VVT Solenoid Control (OC1+)	12v or 0v
7	PK	A/T-ECT Solenoid (S1)	In 3rd or OD: 1v
8-9	---	Not Used	---
10	BK/WT	CMP Sensor Signal (RH+)	AC pulse signals
11	GY/RD	Igniter Transistor 1 Control	6°, at 55 mph: 8° dwell
12	YL/GY	Igniter Transistor 2 Control	6°, at 55 mph: 8° dwell
13	GY/YL	Igniter Transistor 3 Control	6°, at 55 mph: 8° dwell
14	BL/YL	Igniter Transistor 4 Control	6°, at 55 mph: 8° dwell
15	BL	Igniter Transistor 5 Control	6°, at 55 mph: 8° dwell
16	BK	Igniter Transistor 6 Control	6°, at 55 mph: 8° dwell
17	RD/YL	ACIS 1 Control (VSV)	12v or 0v
18	RD/WT	VVT Solenoid Control (OC2-)	12v or 0v
19	YL	A/T-ECT Solenoid (SLN-)	Pulse Signals
20	PK/BL	A/T-ECT Solenoid (SLN+)	Pulse Signals
21	WT/BK	Power Ground (E01)	<0.1v
22	BK/WT	CMP Sensor Signal (LH+)	AC pulse signals
23	GN	O/D Clutch Speed Sensor (-)	AC pulse signals
24	PK/BL	O/D Clutch Speed Sensor (+)	AC pulse signals
25	BK/YL	Igniter Signal (IGF)	Digital Signal: 0-5-0v
26	GN/BK	IAC Signal (RSO)	Pulse Signals
27	WT	Knock Sensor 1 Signal	0.075v AC
28	WT	Knock Sensor 2 Signal	0.075v AC
29	RD/BL	VVT Solenoid Control (OC2+)	12v or 0v
30	WT/BK	Power Ground (E03)	<0.1v
31	WT/BK	Power Ground (E02)	<0.1v

Pin Connector Graphic

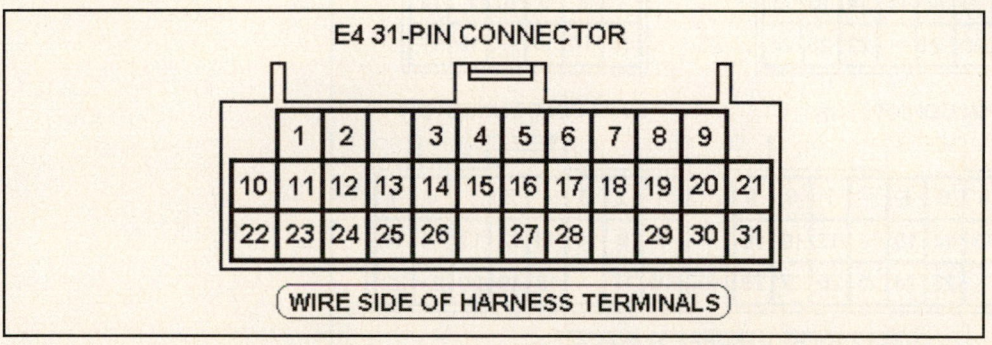

2002-03 Sedan 3.0L V6 DOHC VIN F (A/T-ECT) E5 24P Connector

PCM Pin #	Wire Color	Circuit Description (24 Pin)	Value at Hot Idle
1	WT/BK	Power Ground (E04)	<0.1v
2	BL/RD	Sensor VREF (VC)	4.9-5.1v
3	GN	AFS-11 (B1 S1) Heater	1v (Heater On)
4	BK/RD	AFS-21 (B2 S1) Heater	1v (Heater On)
5	BL	Injector 1 Control	1.6-2-9 ms
6	RD	Injector 2 Control	1.6-2-9 ms
7	BK	EVAP Purge Solenoid (VSV)	12v or 0v
8	WT/BK	Power Ground (E05)	<0.1v
9	GN	PSP Switch Signal	Straight: 12v, Turned: 0v
10	RD	MAF Sensor Signal	1-1.1v
11	GN	AFS-11 (B1 S1) Signal (+)	3.0-3.6v
12	BK/WT	AFS-21 (B2 S1) Signal (+)	3.0-3.6v
13	---	Not Used	---
14	GN/YL	ECT Sensor Signal	At 180ºF: 0.51v
15	WT/RD	VSV ACIS 2 Control	12v or 0v
16	BK/WT	CKP Sensor Signal (NE+)	310-330 Hz
17	BR	Shield Ground (E1)	<0.050v
18	WT	Sensor Ground	<0.050v
19	RD/BK	MAF Sensor Ground (E2)	<0.050v
20	RD	AFS-11 (B1 S1) Signal (AF-)	3.0-3.6v
21	BL	AFS-21 (B2 S1) Signal (-)	3.0-3.6v
22	BL/BK	IAT Sensor Signal	0.5-3.4v
23	LG	TP Sensor Signal	0.53-1.27v
24	BL	CKP/CMP Sensor Ground (-)	<0.050v

Pin Connector Graphic

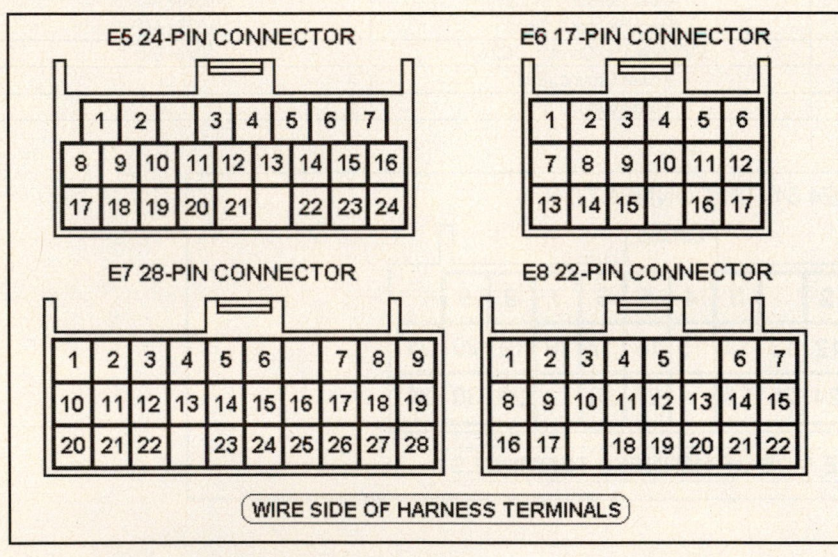

2002-03 Sedan 3.0L V6 DOHC VIN F (A/T-ECT) E6 17P Connector

PCM Pin #	Wire Color	Circuit Description (17 Pin)	Value at Hot Idle
1	GN/YL	A/T-ECT Solenoid (S2)	1st or OD: 1v
2-7	---	Not Used	---
8	RD/BK	A/T Select Switch Reverse	In 'R': 12v, Others: 0v
9-10	---	Not Used	---
11	WT/BL	Active Control Engine Mount (VSV)	12v or 0v
12	GN/OR	Overdrive Main Switch	Switch Off: 12v, On: 1v
13	YL	A/T Select Switch Low	In Low: 12v, Others: 0v
14	GN/YL	A/T Select Switch 2nd	In 2nd: 12v, Others: 0v
15	GN/WT	A/T-ECT Solenoid (SL+)	In Lockup: 12-14v

2002-03 Sedan 3.0L V6 DOHC VIN F (A/T-ECT) E7 28P Connector

PCM Pin #	Wire Color	Circuit Description (28 Pin)	Value at Hot Idle
1	BR	Power Ground (EOM)	<0.1v
2	---	Not Used	---
3	PK/BK	EVAP Vapor Pressure (VSV)	12v or 0v
4	GN	Tail Light Switch Signal	Switch On: 12v, Off: 0v
5	LG/RD	Data Link Connector (TE1)	12-14v
6	LG/BK	A/C Magnetic Clutch (ACMG)	Clutch Off: 0v, On: 12v
7	YL/RD	A/C Amplifier Signal (AC1)	Clutch On: 1.5v, Off: 12v
8	BK	HO2S-12 (B1 S2) Signal	0.1-1.1v
9	LG	HO2S-12 (B1 S2) Heater	1v (Heater On)
10-12	---	Not Used	---
13	BK	Mirror Heater Switch Signal	Switch Off: 1.5v, On: 12v
14	YL/GN	A/C Amplifier Signal (THWO)	Pulse Signals
15	---	Not Used	---
16	YL/BK	ABS/Traction NEO Signal	Pulse Signals
17	---	Not Used	---
18	BL/YL	TXCT Ignition Signal	Inserting key: pulses
19	---	Not Used	---
20	BK/WT	Neutral Start Switch (NSW)	In P/N: 9-11v (cranking)
21	---	Not Used	---
22	VT/WT	Speedometer Indicator	At 55 mph: 48 Hz
23	RD/YL	Unlock Warning Switch	Key In: 1.5v, Out: 4.5v
24	GY	Cruise Control Signal (OD1)	At Cruise in OD: 12v
25	YL	Cruise Control Signal (IDLO)	1.5v, off-idle: 12v
26	GN	Cooling Fan 1 Relay Control	Relay Off: 12v, On: 1v
27	BK	Tachometer Signal (TACO)	Pulse Signals
28	PK/GN	Ignition Switch Code	Inserting key: pulses

2002-03 Sedan 3.0L V6 DOHC VIN F (A/T-ECT) E8 22P Connector

PCM Pin #	Wire Color	Circuit Description (22 Pin)	Value at Hot Idle
1	BK/RD	Direct Battery	12-14v
2	BK/OR	Ignition Switch Power	12-14v
3	GN/BK	Circuit Opening Relay (FC)	0-0.3v, off-idle: 12-14v
4	WT	SIL Signal (Scan Tool)	Digital Signal
6	BK/YL	MIL (lamp) Control	MIL Off: 12v, On: 1v
7	BK/RD	Starter Signal (STA)	Cranking: 9-11v
8	BL/OR	EFI Main Relay Power	12-14v
9	GN/OR	Overdrive Lamp Control	At Cruise in OD: 1v
10	GN	EVAP Canister Closed Valve	12v or 0v
13	BR	ABS/Traction: TRC (+) Signal	DC pulse signals
14	VT	ABS/Traction: ENG (+) Signal	DC pulse signals
15	GN/WT	Brake Switch Signal	Brake Off: 0v, On: 12v
16	BK/WT	EFI Main Relay Power	12-14v
17	PK	EVAP Vapor Pressure Sensor	2.9-3.1v (with hose off)
20	GN/WT	ABS/Traction: TRC (-) Signal	DC pulse signals
21	GY	ABS/Traction: ENG (-) Signal	DC pulse signals
22	VT/WT	Security Indicator Light	Inserting key: pulses

CAMRY Pin Tables

1990-91 Sedan, Wagon 2.0L (A/T) VIN S 10 Pin Connector

PCM Pin #	Wire Color	Circuit Description (10 Pin)	Value at Hot Idle
1	BK/WT	A/T Select Switch Neutral	In 'N': 12v, Others: 0v
2	RD/WT	Check Connector	12-14v
3	BK	A/T: Starter Switch Signal	9-11v (cranking)
3	BK/WT	M/T: Starter Switch Signal	9-11v (cranking)
4	WT	Injector Pair 1 & 3 Control	2.0-3.3 ms
5	WT/BR	Power Ground	<0.1v
7	BR	Sensor Ground	<0.050v
8	WT	Igniter Signal (IGT)	Digital Signal: 0-5-0v
9	YL	Injector Pair 2 & 4 Control	2.0-3.3 ms
10	WT/BK	Power Ground	<0.1v

1990-91 Sedan, Wagon 2.0L (A/T) VIN S 18 Pin Connector

PCM Pin #	Wire Color	Circuit Description (18 Pin)	Value at Hot Idle
1	WT	Distributor Signal (NE+)	AC pulse signals
3	RD	Distributor Signal (G+)	AC pulse signals
4	BK	Distributor Signal (G-)	<0.050v
5	WT/RD	Igniter Signal (IGF)	Digital Signal: 0-5-0v
6	BL	Closed Throttle Switch	<0.1v
7	YL/GN	Check Connector	12-14v
8	GN/RD	MIL (lamp) Control	MIL Off: 12v, On: 1v
9	YL	ISC Motor 1	Pulse Signals
10	GN	ECT Sensor Signal	At 180ºF: 0.51v
11	BK/RD	TP Sensor Signal (VTA)	0.3-0.8v
12 (Cal)	BL/BK	EGR Gas Temp. Sensor	3.5-4.0v
13	WT	Main O2S Signal	0.1-1.1v
14	BR	Sensor Ground	<0.050v
15	BK/WT	A/C Magnetic Clutch (ACMG)	Clutch Off: 0v, On: 12v
16	BL/BK	A/C Amplifier Signal (ACT)	Clutch On: 12v, Off: 1.5v
17 (Cal)	RD/BL	Sub O2S Signal	0.1-1.1v
18	GN/BK	ISC Motor 2	Pulse Signals

1990-91 Sedan, Wagon 2.0L (A/T) VIN S 14-Pin Connector

PCM Pin #	Wire Color	Circuit Description (14-Pin)	Value at Hot Idle
1	WT/RD	EFI Main Relay Power	12-14v
2	WT/BL	Battery Direct	12-14v
3	YL/RD	IAT Sensor Signal	At 100ºF: 2.60v
4	YL/BL	Airflow Meter Signal	2.5v
5	BL/RD	Sensor VREF	4.9-5.1v
6	BL/BK	A/T Signal (L1)	5v, at WOT: 0v
7	YL/RD	A/T Signal (L3)	0v, at WOT: 5v
8	WT/RD	EFI Main Relay Power	12-14v
9	GN	Defogger/Light Idle Up Signal	Switch On: 12v, Off: 0v
10	PK/YL	Vehicle Speed Sensor	At 55 mph: 48 Hz
11	GN/WT	Stop Light Switch Signal	Brake Off: 0v, On: 12v
12	WT/BK	Sensor Ground	<0.050v
13	YL/GN	A/T Signal (L2)	5v, at WOT: 5v
14	YL/BK	A/T Signal (IDL)	0v, at WOT: 12v

Pin Connector Graphic

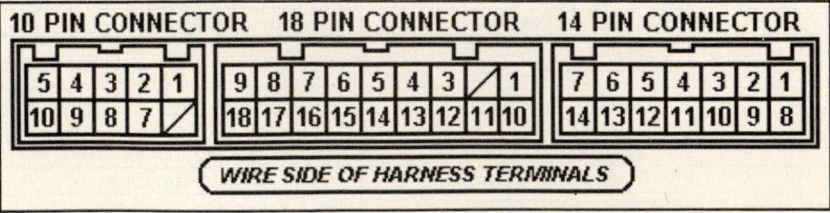

1992-93 Sedan, Wagon 2.2L VIN S (A/T-ECT) 16 Pin Connector

PCM Pin #	Wire Color	Circuit Description (16 Pin)	Value at Hot Idle
1	RD	Sensor VREF	4.9-5.1v
2	BK/YL	MAP Sensor Signal	1-1.5v
3	BL/BK	IAT Sensor Signal	At 100ºF: 2.60v
4	LG	ECT Sensor Signal	At 180ºF: 0.51v
5 (Cal)	RD/BL	Sub O2S Signal	0.1-1.1v
6	WT	Main O2S Signal	0.1-1.1v
7	BR/BK	Check Connector	12-14v
8	RD/WT	Check Connector	12-14v
9	BR	Sensor Ground	<0.050v
10 (Cal)	GY	EGR Gas Temp. Sensor	3.5-4.0v
11	BK	TP Sensor Signal	0.3-0.8v
12	BL	Closed Throttle Switch	1v, at off-idle: 12v
13	WT	Knock Sensor Signal	<0.075v AC
14	GN/WT	Check Connector	12-14v
15	GY	Check Connector	12-14v
16	BR	Sensor Ground	<0.050v

1992-93 Sedan, Wagon 2.2L VIN S (A/T-ECT) 22 Pin Connector

PCM Pin #	Wire Color	Circuit Description (22 Pin)	Value at Hot Idle
1	WT/BL	Battery Direct	12-14v
2	BK/RD	Defogger/Light Idle Up Signal	Switch On: 12v, Off: 0v
4	GN/WT	Stop Light Switch Signal	Brake Off: 0v, On: 12v
5	GN/RD	MIL (lamp) Control	MIL Off: 12v, On: 1v
7	GY/BL	Main Overdrive Switch	Switch Off: 12-14v
8	BK/YL	A/C Amplifier Signal (AC1)	Clutch On: 1.5v, Off: 12v
9	PK/YL	Vehicle Speed Sensor	At 55 mph: 48 Hz
11	BK/WT	Starter Switch Signal	9-11v (cranking)
12	BK/OR	EFI Main Relay Power	12-14v
13	BK/OR	EFI Main Relay Power B1+	12-14v
14	GN/RD	Circuit Opening Relay (FC)	0-3v
20	PK	Cruise Control ECU	At Cruise in OD: 12v
21	GN/BK	A/C Amplifier Signal (ACT)	Clutch On: 12v, Off: 1.5v
22	BK/WT	A/T Neutral Start Switch	In P/N: 9-11v (cranking)

Standard Colors and Abbreviations

Abbreviation	Color	Abbreviation	Color	Abbreviation	Color
BK	Black	GY	Gray	RD	Red
BL	Blue	GN	Green	TN	Tan
BR	Brown	LG	Light Green	VT	Violet
DB	Dark Blue	OR	Orange	WT	White
DG	DK Green	PK	Pink	YL	Yellow

1992-93 Sedan, Wagon 2.2L VIN S (A/T-ECT) 26 Pin Connector

PCM Pin #	Wire Color	Circuit Description (26 Pin)	Value at Hot Idle
1	BL/YL	A/T-ECT Solenoid (SL)	In Lockup: 12-14v
2	PK	A/T-ECT Solenoid (S1)	In 1st: 12-14v
3	WT/RD	Igniter Signal (IGF)	Digital Signal: 0-5-0v
4	RD	Distributor Signal (NE+)	AC pulse signals
5	BL	Distributor Signal (NE-)	<0.050v
6	OR	A/T Select Switch 2nd	In 2nd: 12v, Others: 0v
7	LG	A/C Idle Up Solenoid	A/C Off: 12v, On: 1v
9	GN/RD	ISC Signal (ISCC)	Pulse Signals
10	GN/YL	ISC Signal (ISCO)	Pulse Signals
11	YL	Injector Pair 2 & 4 Control	2.0-3.3 ms
12	WT	Injector Pair 1 & 3 Control	2.0-3.3 ms
13	WT/BK	Power Ground	<0.1v
14	BR	Sensor Ground	<0.050v
15	PK/BL	A/T-ECT Solenoid (S2)	In 2nd: 12v, Others: 0v
16	GN/BK	A/T-ECT Speed Sensor	Moving: varies 0-5v
17	BK	Distributor Signal (G-)	<0.050v
18	YL	Distributor Signal (G+)	AC pulse signals
19	YL/BL	A/T Select Switch Low	In Low: 12v, Others: 0v
20	WT	Igniter Signal (IGT)	Digital Signal: 0-5-0v
22	BL/RD	A/T Pattern Selector Switch	Norm: 0v, PWR: 12v
23	GN	EGR Solenoid Control (VSV)	12v or 0v
26	WT/BK	Power Ground	<0.1v

Pin Connector Graphic

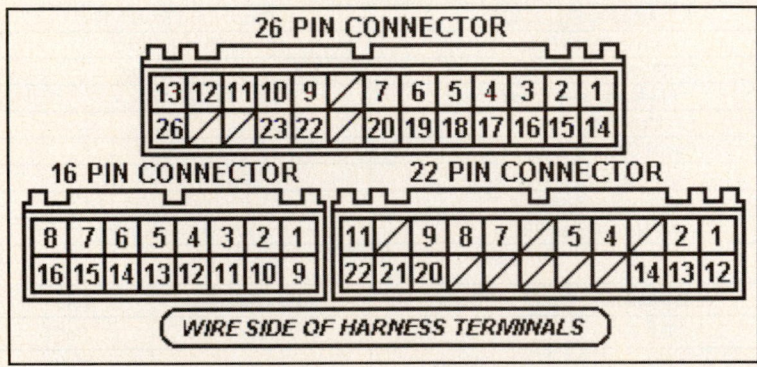

Standard Colors and Abbreviations

Abbreviation	Color	Abbreviation	Color	Abbreviation	Color
BK	Black	GY	Gray	RD	Red
BL	Blue	GN	Green	TN	Tan
BR	Brown	LG	Light Green	VT	Violet
DB	Dark Blue	OR	Orange	WT	White
DG	DK Green	PK	Pink	YL	Yellow

1992-93 Sedan, Wagon 2.2L I4 MFI VIN S (M/T) 12 Pin Connector

PCM Pin #	Wire Color	Circuit Description (12 Pin)	Value at Hot Idle
1	BK/OR	EFI Main Relay Power B1+	12-14v
2	WT/BL	Battery Direct	12-14v
3	BK/YL	A/C Amplifier Signal (AC1)	Clutch On: 1.5v, Off: 12v
4	GN/RD	Circuit Opening Relay (FC)	0-3v, off-idle: 12v
6	GN/BK	A/C Amplifier Signal (ACT)	Clutch On: 12v, Off: 1.5v
7	BK/OR	EFI Main Relay Power	12-14v
8	GN/RD	MIL (lamp) Control	MIL Off: 12v, On: 1v
11	PK/YL	Vehicle Speed Sensor	At 55 mph: 48 Hz
12	BK/RD	Defogger/Light Idle Up Signal	Switch On: 12v, Off: 0v

1992-93 Sedan, Wagon 2.2L I4 MFI VIN S (M/T) 16 Pin Connector

PCM Pin #	Wire Color	Circuit Description (16 Pin)	Value at Hot Idle
1 (Cal)	RD/BL	Sub O2S Signal	0.1-1.1v
2	BK/YL	MAP Sensor Signal	1-1.5v
3	BL/BK	IAT Sensor Signal	At 100ºF: 2.60v
4	LG	ECT Sensor Signal	At 180ºF: 0.51v
5	WT	Knock Sensor Signal	<0.075v AC
6	WT	Main O2S Signal	0.1-1.1v
7	GN/WT	Check Connector	12-14v
8	RD/WT	Check Connector	12-14v
9	BR	Sensor Ground	<0.050v
10	BK	TP Sensor Signal	0.3-0.8v
11	RD	Sensor VREF	4.9-5.1v
12	BL	Closed Throttle Switch	<0.1v
13 (Cal)	GN/RD	EGR Gas Temp. Sensor	3.5-4.0v
15	GN	Check Connector	12-14v
16	BR	Sensor Ground	<0.050v

1992-93 Sedan, Wagon 2.2L I4 MFI VIN S (M/T) 26 Pin Connector

PCM Pin #	Wire Color	Circuit Description (26 Pin)	Value at Hot Idle
1	BK/YL	A/C Idle Up Solenoid	A/C Off: 12v, On: 1v
2	BK/WT	Starter Switch Signal	9-11v (cranking)
3	WT/RD	Igniter Signal (IGF)	Varies: 1-3v
4	RD	Distributor Signal (NE+)	AC pulse signals
5	YL	Distributor Signal (G+)	AC pulse signals
9	GN/RD	ISC Signal (ISCC)	Pulse Signals
10	GN/YL	ISC Signal (ISCO)	Pulse Signals
12	WT	Injector Pair 1 & 3 Control	2.0-3.3 ms
13	WT/BK	Power Ground	<0.1v
17	BL	Distributor Signal (NE-)	<0.050v
18	BK	Distributor Signal (G-)	<0.050v
22	WT	Igniter Signal (IGT)	Digital Signal: 0-5-0v
23	GN	EGR Solenoid Control (VSV)	12v or 0v
24	BR	Sensor Ground	<0.050v
25	YL	Injector Pair 2 & 4 Control	2.0-3.3 ms
26	WT/BK	Power Ground	<0.1v

Pin Connector Graphic

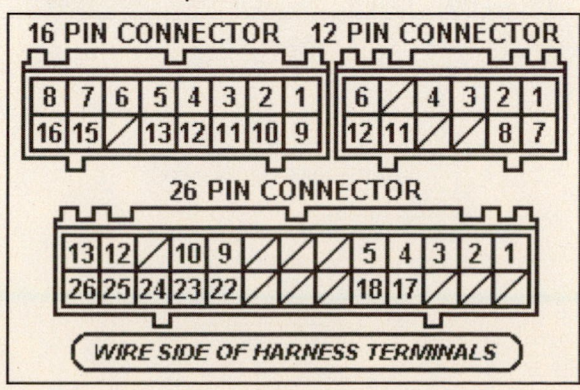

1994-95 Coupe & Sedan, Wagon 2.2L I4 Federal M/T VIN S 26 Pin

PCM Pin #	Wire Color	Circuit Description (26 Pin)	Value at Hot Idle
3	WT/RD	Igniter Signal (IGF)	Digital Signal: 0-5-0v
4 (Cal)	RD	Distributor Signal (NE+)	AC pulse signals
4 (Fed)	RD	Distributor Signal (NE+)	AC pulse signals
5 (Cal)	BL	Distributor Signal (G2+)	AC pulse signals
5 (Fed)	BL	Distributor Signal (NE-)	<0.050v
7	LG	A/C Idle Up Solenoid	12v or 0v
8	BK/RD	Fuel Pressure Up Solenoid	12v or 0v
9	GN/RD	IAC Signal (ISCC)	Pulse Signals
10	GN/YL	IAC Signal (ISCO)	Pulse Signals
11	YL	Injector 2 Control	2.0-3.3 ms
12	WT	Injector 1 Control	2.0 - 5.0ms
13	WT/BK	Power Ground	<0.1v
14	BR	Sensor Ground	<0.050v
17	BK	Distributor Signal (G-)	<0.050v
18 (Cal)	YL	Distributor Signal (G1+)	AC pulse signals
18 (Fed)	YL	Distributor Signal (G+)	AC pulse signals
20	WT	Igniter Signal (IGT)	Digital Signal: 0-5-0v
23	GN	EGR Solenoid Control (VSV)	12v or 0v
24	RD/BK	Injector 4 Control	2.0-3.3 ms
25	RD/BL	Injector 3 Control	2.0 - 5.0ms
26	WT/BK	Power Ground	<0.1v

1994-95 Coupe & Sedan, Wagon 2.2L I4 Federal M/T VIN S 16 Pin

PCM Pin #	Wire Color	Circuit Description (16 Pin)	Value at Hot Idle
1	RD	Sensor VREF (VC)	4.9-5.1v
2	BK/YL	MAP Sensor Signal	1-1.5v
3	BL/BK	IAT Sensor Signal	At 100°F: 2.8v
4	LG	ECT Sensor Signal	At 180°F: 0.6v
5	RD/BL	Sub O2S Signal	0.1-1.1v
6	WT	Main O2S Signal	0.1-1.1v
7	BR/BK	Data Link Connector	12-14v
8	RD/WT	Data Link Connector (TN)	12-14v
9	BR	Sensor Ground	<0.050v
10	GY	EGR Gas Temp. Sensor	3.5-4.0v
11	BK	TP Sensor Signal	0.5v
12	BL	Closed Throttle Switch	<0.1v
13	WT	Knock Sensor	<0.075v AC
14	GN/WT	Data Link Connector (TN)	12-14v
15	GY	Data Link Connector (TN)	12-14v

Standard Colors and Abbreviations

Abbreviation	Color	Abbreviation	Color	Abbreviation	Color
BK	Black	GY	Gray	RD	Red
BL	Blue	GN	Green	TN	Tan
BR	Brown	LG	Light Green	VT	Violet
DB	Dark Blue	OR	Orange	WT	White
DG	DK Green	PK	Pink	YL	Yellow

1994-95 Coupe & Sedan, Wagon 2.2L I4 Federal M/T VIN S 22 Pin

PCM Pin #	Wire Color	Circuit Description (22 Pin)	Value at Hot Idle
1	WT/BL	Battery Direct	12-14v
2	BK/RD	Defogger/Light Idle Up Signal	Switch On: 12v, Off: 0v
4	GN/WT	Stop Light Switch Signal	Brake Off: 0v, On: 12v
5	GN/RD	MIL (lamp) Control	MIL Off: 12v, On: 1v
9	PK/YL	Vehicle Speed Sensor	At 55 mph: 48 Hz
10	BK/YL	A/C Amplifier Signal (AC1)	Clutch On: 1.5v, Off: 12v
11	BK/WT	Starter Switch Signal	9-11v (cranking)
12	BK/OR	EFI Main Relay Power	12-14v
13	BK/OR	EFI Main Relay Power B1+	12-14v
14	GN/RD	Circuit Opening Relay (FC)	0-3v, off-idle: 12v
21	GN/BK	A/C Amplifier Signal (ACT)	Clutch On: 12v, Off: 1.5v
22	BK/WT	A/T Neutral Start Switch	In P/N: 9-11v (cranking)

Pin Connector Graphic

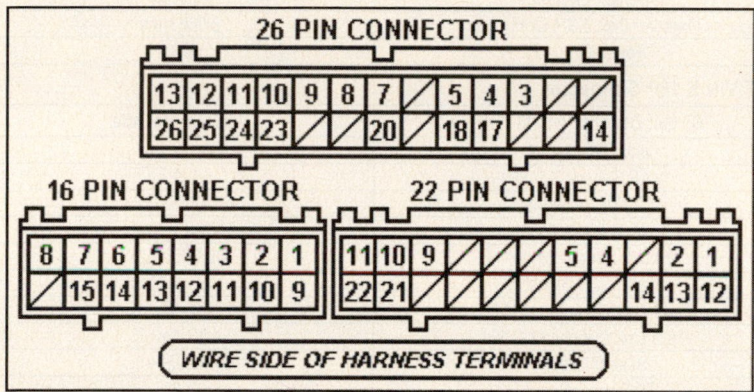

Standard Colors and Abbreviations

Abbreviation	Color	Abbreviation	Color	Abbreviation	Color
BK	Black	GY	Gray	RD	Red
BL	Blue	GN	Green	TN	Tan
BR	Brown	LG	Light Green	VT	Violet
DB	Dark Blue	OR	Orange	WT	White
DG	DK Green	PK	Pink	YL	Yellow

1994-95 Coupe & Sedan, Wagon 2.2L Cal M/T VIN S 26P Connector

PCM Pin #	Wire Color	Circuit Description (26 Pin)	Value at Hot Idle
1	LG	A/C Idle Up Solenoid	A/C On: 1v
2	BK/WT	Starter Switch Signal	9-11v (cranking)
3	WT/RD	Igniter Signal (IGF)	Digital Signal: 0-5-0v
4	RD	Distributor Signal (NE+)	AC pulse signals
5	YL	Distributor Signal (G+)	AC pulse signals
9	GN/RD	IAC Signal (ISCC)	Pulse Signals
10	GN/YL	IAC Signal (ISCO)	Pulse Signals
12	WT	Injector Pair 1 & 3 Control	2.0 - 5.0ms
13	WT/BK	Power Ground	<0.1v
17	BL	Distributor Signal (NE-)	<0.050v
18	BK	Distributor Signal (G-)	<0.050v
22	WT	Igniter Signal (IGT)	Digital Signal: 0-5-0v
23	GN	EGR Solenoid Control (VSV)	12v or 0v
24	BR	Sensor Ground	<0.050v
25	YL	Injector Pair 2 & 4 Control	2.0-3.3 ms
26	WT/BK	Power Ground	<0.1v

1994-95 Coupe & Sedan, Wagon 2.2L Cal M/T VIN S 16P Connector

PCM Pin #	Wire Color	Circuit Description (16 Pin)	Value at Hot Idle
1	RD/BL	Sub O2S Signal	0.1-1.1v
2	BK/YL	MAP Sensor Signal	0.9-1.1v
3	BL/BK	IAT Sensor Signal	At 100ºF: 2.8v
4	LG	ECT Sensor Signal	At 180ºF: 0.6v
5	WT	Knock Sensor Signal	<0.075v AC
6	WT	Main O2S Signal	0.1-1.1v
7	GN/WT	Data Link Connector (TN)	12-14v
8	RD/WT	Data Link Connector (TN)	12-14v
9	BR	Sensor Ground	<0.050v
10	BK	TP Sensor Signal	0.5v
11	RD	Sensor VREF (VC)	4.9-5.1v
12	BL	Closed Throttle Switch	<0.1v
13	GY	EGR Gas Temp. Sensor	3.5-4.0v
15	GY	Data Link Connector	12-14v
16	BR	Sensor Ground	<0.050v

1994-95 Coupe & Sedan, Wagon 2.2L Cal M/T VIN S 12P Connector

PCM Pin #	Wire Color	Circuit Description (12 Pin)	Value at Hot Idle
2	WT/BL	Battery Direct	12-14v
3	BK/YL	A/C Amplifier Signal (AC1)	Clutch On: 1.5v, Off: 12v
4	GN/RD	Circuit Opening Relay (FC)	0-3v, off-idle: 12v
6	GN/BK	A/C Amplifier Signal (ACT)	Clutch On: 12v, Off: 1.5v
7	BK/OR	EFI Main Relay Power	12-14v
8 ('94)	BK/OR	EFI Main Relay Power	12-14v
9	GN/RD	MIL (lamp) Control	MIL Off: 12v, On: 1v
11	PK/YL	Vehicle Speed Sensor	At 55 mph: 48 Hz
12	BK/RD	Defogger/Light Idle Up Signal	Switch On: 12v, Off: 0v

Pin Connector Graphic

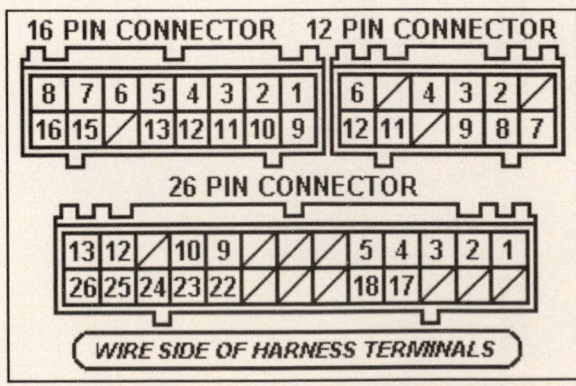

1996 Coupe & Sedan 2.2L I4 MFI VIN G (A/T) 26 Pin Connector

PCM Pin #	Wire Color	Circuit Description (26 Pin)	Value at Hot Idle
1	BL/YL	A/T-ECT Solenoid (SL)	In Lockup: 12-14v
2	PK	A/T-ECT Solenoid (S1)	In 3rd or OD: 1v
3	WT/RD	Igniter Signal (IGF)	Digital Signal: 0-5-0v
4 (TMC)	RD	CKP Sensor Signal (NE+)	AC pulse signals
4 (TMM)	WT/BL	CKP Sensor Signal (NE+)	AC pulse signals
5 (TMC)	BK/WT	Distributor Signal (G2+)	AC pulse signals
5 (TMM)	BL/GN	Distributor Signal (G2+)	AC pulse signals
9	GN/RD	IAC Signal (RSC)	Pulse Signals
10	GN/YL	IAC Signal (RSO)	Pulse Signals
11	YL	Injector 2 Control	2.0-3.3 ms
12	WT	Injector 1 Control	2.0-3.3 ms
13	WT/BK	Power Ground	<0.1v
14	BR	Sensor Ground	<0.050v
17 (TMC)	GN	CKP Sensor Signal (NE-)	<0.050v
17 (TMM)	OR/BL	CKP Sensor Signal (NE-)	<0.050v
20	WT	Igniter Signal (IGT)	Digital Signal: 0-5-0v
22	YL/BK	EVAP Purge Solenoid (VSV)	12v or 0v
23	GN	EGR Solenoid Control (VSV)	12v or 0v
24	RD/BK	Injector 4 Control	2.0-3.3 ms
25	RD/BL	Injector 3 Control	2.0-3.3 ms
26	WT/BK	Power Ground	<0.1v

1996 Coupe & Sedan 2.2L I4 MFI VIN G (A/T) 16 Pin Connector

PCM Pin #	Wire Color	Circuit Description (16 Pin)	Value at Hot Idle
1	RD	Sensor VREF (VC)	4.9-5.1v
2	BK/YL	MAP Sensor Signal	1-1.5v
3	BL/BK	IAT Sensor Signal	At 100°F: 2.8v
4	LG	ECT Sensor Signal	At 180°F: 0.6v
5	RD/BL	O2S-12 (B1 S2) Signal	0.1-1.1v
6	WT	O2S-11 (B1 S1) Signal	0.1-1.1v
7	BL/YL	EVAP Vapor Pressure Sensor	2.9-3.1v (with hose off)
8	BK/RD	EVAP Vapor Pressure (VSV)	12v or 0v
9	BR	Sensor Ground	<0.050v
11	BK	TP Sensor Signal	0.3-0.8v
13	WT	Knock Sensor Signal	<0.075v AC
15	GY	Data Link Connector	12-14v
16 (TMC)	BR	Power Ground	<0.1v
16 (TMM)	WT/BK	Power Ground	<0.1v

Standard Colors and Abbreviations

Abbreviation	Color	Abbreviation	Color	Abbreviation	Color
BK	Black	GY	Gray	RD	Red
BL	Blue	GN	Green	TN	Tan
BR	Brown	LG	Light Green	VT	Violet
DB	Dark Blue	OR	Orange	WT	White
DG	DK Green	PK	Pink	YL	Yellow

1996 Coupe & Sedan 2.2L I4 MFI VIN G (A/T) 22 Pin Connector

PCM Pin #	Wire Color	Circuit Description (22 Pin)	Value at Hot Idle
1	WT/BL	Battery Direct	12-14v
2	BK/RD	Defogger/Light Idle Up Signal	Switch On: 12v, Off: 0v
4	GN/WT	Stop Light Switch Signal	Brake Off: 0v, On: 12v
5	GN/RD	MIL (lamp) Control	MIL Off: 12v, On: 1v
6	BL/RD	A/T Pattern Selector Switch	Norm: 0v, PWR: 12v
7	GN/OR	Overdrive Main Switch	Switch Off: 12v
9	PK/YL	Vehicle Speed Sensor	At 55 mph: 48 Hz
10	BK/YL	A/C Amplifier Signal (AC1)	Clutch On: 1.5v, Off: 12v
11	BK/WT	Starter Switch Signal	9-11v (cranking)
12	BK/OR	EFI Main Relay Power	12-14v
14	GN/RD	Circuit Opening Relay (FC)	0-3v, off-idle: 12v
16	WT	Data Link Connector	12-14v
17	RD/BK	A/T Select Switch Reverse	In 'R': 12v, Others: 0v
18	OR	A/T Select Switch 2nd	In 2nd: 12v, Others: 0v
19	YL/BL	A/T Select Switch Low	In Low: 12v, Others: 0v
21	GN/BK	A/C Amplifier Signal (ACT)	Clutch On: 12v, Off: 1.5v
22	BK/WT	A/T Neutral Start Switch	In P/N: 9-11v (cranking)

Pin Connector Graphic

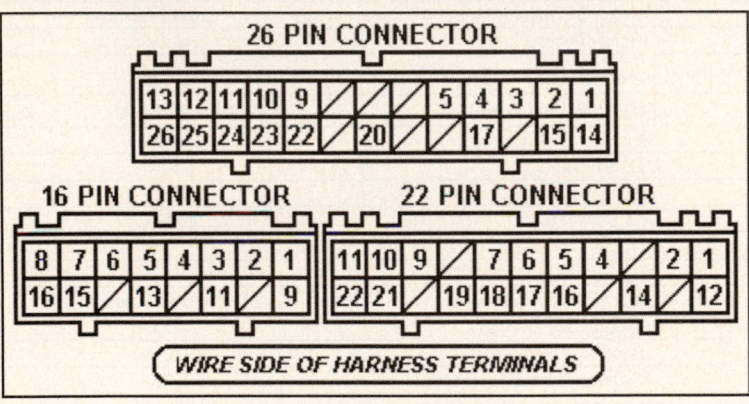

1997 Sedan 2.2L I4 MFI VIN G (A/T) 26 Pin Connector

PCM Pin #	Wire Color	Circuit Description (26 Pin)	Value at Hot Idle
1	PK	A/T-ECT Solenoid (SL)	In Lockup: 12-14v
2 (Cal)	GN	HTAF Heater Control	1v (Heater On)
3	WT/RD	Igniter Signal (IGF)	Digital Signal: 0-5-0v
4	BK/RD	CKP Sensor Signal (NE+)	AC pulse signals
5	BK/WT	CMP Sensor Signal (G+)	AC pulse signals
6	BL/BK	A/T-ECT Solenoid (S2)	1st or OD: 1v
7	PK	A/T-ECT Solenoid (S1)	In 3rd or OD: 1v
8	BL/YL	HO2S-11 (B1 S1) Heater	1v (Heater On)
9	BK/OR	IAC Signal (ISCC)	Pulse Signals
10	WT	IAC Signal (ISCO)	Pulse Signals
11	RD	Injector 2 Control	2.0-3.3 ms
12	BL	Injector 1 Control	2.0-3.3 ms
13	BR	Power Ground	<0.1v
14	BR	Sensor Ground	<0.050v
15 (Cal)	BR	Power Ground	<0.1v
17	BL	CKP Sensor Signal (NE-)	<0.050v
18	WT/BL	A/C System (Lock In)	A/C On: 12v
19	YL/RD	Igniter Transistor 2 Control	Digital Signal: 0-5-0v
20	BK	Igniter Transistor 1 Control	Digital Signal: 0-5-0v
21	PK/BK	HO2S-12 (B1 S2) Heater	1v (Heater On)
22	PK/WT	EVAP Purge Solenoid (VSV)	12v or 0v
23	PK/BK	EGR Solenoid Control (VSV)	12v or 0v
24	WT	Injector 4 Control	2.0-3.3 ms
25	YL	Injector 3 Control	2.0-3.3 ms
26	BR	Power Ground	<0.1v

1997 Sedan 2.2L I4 MFI VIN G (A/T) 16 Pin Connector

PCM Pin #	Wire Color	Circuit Description (16 Pin)	Value at Hot Idle
1	YL	Sensor VREF	4.9-5.1v
2	BK/YL	MAP Sensor Signal	1-1.5v
3	YL/BK	IAT Sensor Signal	At 100°F: 2.8v
4	GN/BK	ECT Sensor Signal	At 180°F: 0.6v
5	BK	HO2S-12 (B1 S2) Signal	0.1-1.1v
6	WT	HO2S-11 (B1 S1) Signal	0.1-1.1v
6 (Cal)	WT	AFS-11 (B1 S1) Signal (AF+)	3.0-3.6v
7	PK	EVAP Vapor Pressure Sensor	2.9-3.1v (with hose off)
8	PK	EVAP Vapor Pressure (VSV)	12v or 0v
9	BR	Sensor Ground	<0.050v
10	BL/WT	A/C Evaporator Temp. Signal	1.4-1.8v (temp. >59°F)
11	LG	TP Sensor Signal	0.3-0.8v
12	BK/BL	PSP Switch Signal	Straight: 12v, Turning: 0v
13	WT	Knock Sensor Signal	<0.075v AC
14 (Cal)	OR	AFS-11 (B1 S1) Signal (AF-)	Fixed at 3v
15	BL/WT	Data Link Connector	12-14v
16	BR	Power Ground	<0.1v

1997 Sedan 2.2L I4 MFI VIN G (A/T) 22 Pin Connector

PCM Pin #	Wire Color	Circuit Description (22 Pin)	Value at Hot Idle
1	BK/YL	Battery Direct	12-14v
2	BK/RD	Defogger/Light Idle Up Signal	Switch On: 12v, Off: 0v
3	BL/RD	Cruise Control ECU	At Cruise in OD: 12v
4	GN/WT	Stop Light Switch Signal	Brake Off: 0v, On: 12v
5	GN/RD	MIL (lamp) Control	MIL Off: 12v, On: 1v
7	GN/OR	Overdrive Main Switch	Switch Off: 12v
8	BK/OR	Tachometer Signal (TACO)	Pulse Signals
9	PK/WT	Vehicle Speed Sensor	At 55 mph: 48 Hz
10	RD/BK	A/C Switch Signal	Clutch On: 1.5v, Off: 12v
11 (TMC)	GY	Starter Switch Signal	9-11v (cranking)
11 (TMM)	BK/WT	Starter Switch Signal	9-11v (cranking)
12	BK/YL	EFI Main Relay Power	12-14v
13	GN	A/C Dual Press Switch	Switch Off: 12v
14	GN/RD	Circuit Opening Relay (FC)	0-3v, off-idle: 12v
15	RD/WT	A/C Lock In Signal	A/C On: 12v
16	WT	Data Link Connector	No Scan Tool: 0v
17	RD/BK	A/T Select Switch Reverse	In 'R': 12v, Others: 0v
18 (TMC)	BL/WT	A/T Select Switch Second	In 2nd: 12v, Others: 0v
18 (TMM)	OR	A/T Select Switch Second	In 2nd: 12v, Others: 0v
19	YL	A/T Select Switch Low	In Low: 12v, Others: 0v
20	YL/BK	Cruise Control ECU	At Cruise in OD: 12v
21	BL/YL	A/C Magnetic Clutch (ACMG)	Clutch Off: 0v, On: 12v
22	BK/WT	A/T Neutral Start Switch	In P/N: 9-11v (cranking)

Pin Connector Graphic

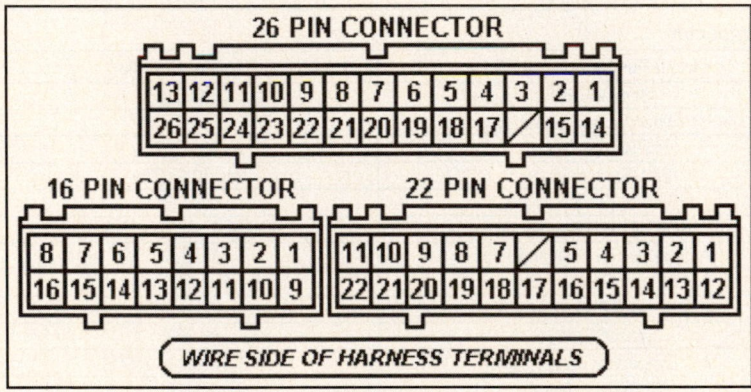

Note: *TMC indicates a vehicle produced by Toyota Motor Corp. (Japan) and TMM indicates a vehicle produced by Toyota Motor Manufacturer (USA).*

1998-99 Sedan W/O EIS 2.2L I4 VIN G (A/T) 26 Pin Connector

PCM Pin #	Wire Color	Circuit Description (26 Pin)	Value at Hot Idle
1	PK	A/T-ECT Solenoid (SL)	In Lockup: 12-14v
2 (Cal)	GN	AFS-11 (B1 S1) Heater	1v (Heater On)
3	WT/RD	Igniter Signal (IGF)	Digital Signal: 0-5-0v
4	BK/RD	CKP Sensor Signal (NE+)	AC pulse signals
5	BK/WT	CMP Sensor Signal (G+)	AC pulse signals
6	BL/BK	A/T-ECT Solenoid (S2)	1st or OD: 1v
7	PK	A/T-ECT Solenoid (S1)	In 3rd or OD: 1v
8	BL/YL	HO2S-11 (B1 S1) Heater	1v (Heater On)
9	BK/OR	IAC Signal (ISCC)	Pulses
10	WT	IAC Signal (ISCO)	Pulses
11	RD	Injector 2 Control	2.0-3.3 ms
12	BL	Injector 1 Control	2.0-3.3 ms
13	BR	Power Ground	<0.1v
14	BR	Power Ground	<0.1v
15 (Cal)	BR	Power Ground	<0.1v
17	BL	CKP/CMP Sensor Ground (-)	<0.050v
18	WT/BL	A/C Lock In Signal	A/C On: 12v
19	YL/RD	Igniter Transistor 2 Control	Digital Signal: 0-5-0v
20	BK	Igniter Transistor 1 Control	Digital Signal: 0-5-0v
21	PK/BK	HO2S-12 (B1 S2) Heater	1v (Heater On)
22	PK/WT	EVAP Purge Solenoid (VSV)	12v or 0v
23	PK/BK	EGR Solenoid Control (VSV)	12v or 0v
24	WT	Injector 4 Control	2.0-3.3 ms
25	YL	Injector 3 Control	2.0-3.3 ms
26	BR	Power Ground	<0.1v

1998-99 Sedan W/O EIS 2.2L I4 VIN G (A/T) 16 Pin Connector

PCM Pin #	Wire Color	Circuit Description (16 Pin)	Value at Hot Idle
1	YL	Sensor VREF (VC)	4.9-5.1v
2	BK/YL	MAP Sensor Signal	1-1.5v
3	YL/BK	IAT Sensor Signal	At 100ºF: 2.8v
4	GN/BK	ECT Sensor Signal	At 180ºF: 0.6v
5	BK	HO2S-12 (B1 S2) Signal	0.1-1.1v
6 (TMC)	BL	AFS-11 (B1 S1) Signal (AF+)	3.0-3.6v
6 (TMM)	WT	AFS-11 (B1 S1) Signal (AF+)	3.0-3.6v
6	WT	HO2S-11 (B1 S1) Signal	0.1-1.1v
7	PK	EVAP Vapor Pressure Sensor	2.9-3.1v (with hose off)
8	PK	EVAP Vapor Pressure (VSV)	12v or 0v
9	BR	Sensor Ground	<0.050v
10	BL/RD	A/C Evaporator Temp. Sensor	EVAP Temp. 59ºF: 2.4v
11	LG	TP Sensor Signal	0.3-0.8v
12	BK/BL	PSP Switch Signal	Straight: 12v, Turning: 0v
13	WT	Knock Sensor Signal	<0.075v AC
14 (TMC)	BK/WT	AFS-11 (B1 S1) Signal (AF-)	Fixed at 3v
14 (TMM)	OR	AFS-11 (B1 S1) Signal (AF-)	Fixed at 3v
15	BL/WT	Data Link Connector (TE)	12-14v
16	BR	Power Ground	<0.1v

1998-99 Sedan W/O EIS 2.2L I4 VIN G (A/T) 22 Pin Connector

PCM Pin #	Wire Color	Circuit Description (22 Pin)	Value at Hot Idle
1	BK/YL	Battery Direct	12-14v
2	BK/RD	Defogger/Light Idle Up Signal	Switch On: 12v, Off: 0v
3	BL/RD	Cruise Control Signal (IDLO)	1.5v, off-idle: 12v
4	GN/WT	Stop Light Switch Signal	Brake Off: 0v, On: 12v
5	GN/RD	MIL (lamp) Control	MIL Off: 12v, On: 1v
7	GN/OR	Overdrive "Off" Indicator	Light Off: 12v
8	BK/OR	Tachometer Signal (TACO)	Pulses
9	PK/WT	Vehicle Speed Sensor	At 55 mph: 48 Hz
10	RD/BK	A/C Switch Signal	Clutch On: 1.5v, Off: 12v
11 (TMC)	GY	Starter Switch Signal	9-11v (cranking)
11 (TMM)	BK/OR	Starter Switch Signal	9-11v (cranking)
12	BK/YL	EFI Main Relay Power	12-14v
13	GN	A/C Dual Press Switch	Switch Off: 12v
14	GN/RD	Circuit Opening Relay (FC)	0-3v
15	RD/WT	A/C Lock In Signal	A/C On: 12v
16	WT	SIL Signal (Scan Tool)	Digital Signal
17	RD/BK	A/T Select Switch Reverse	In 'R': 12v, Others: 0v
18 (TMC)	BL/WT	A/T Select Switch 2nd	In 2nd: 12v, Others: 0v
18 (TMM)	OR	A/T Select Switch 2nd	In 2nd: 12v, Others: 0v
19	YL	A/T Select Switch Low	In Low: 12v, Others: 0v
20	YL/BK	Cruise Control Signal (OD1)	At Cruise in OD: 12v
21	BL/YL	A/C Magnetic Clutch (ACMG)	Clutch Off: 0v, On: 12v
22	BK/WT	A/T Neutral Start Switch	In P/N: 9-11v (cranking)

Pin Connector Graphic

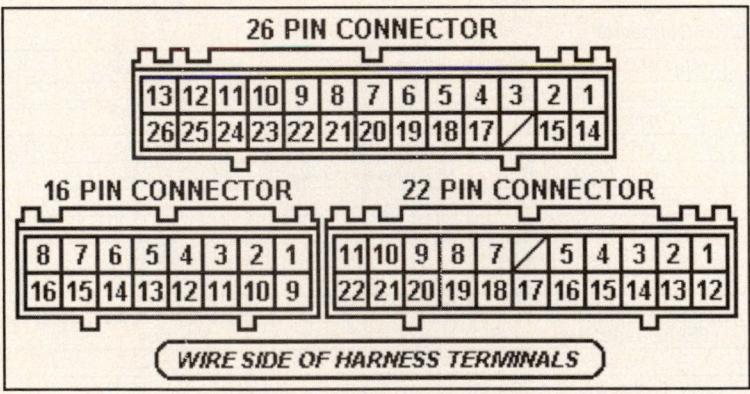

Note: TMC indicates a vehicle produced by Toyota Motor Corp. (Japan) and TMM indicates a vehicle produced by Toyota Motor Manufacturer (USA).

1998-99 Sedan W/EIS 2.2L I4 VIN G (A/T) 26 Pin Connector

PCM Pin #	Wire Color	Circuit Description (26 Pin)	Value at Hot Idle
1 (Fed)	BL/YL	HO2S-11 (B1 S1) Heater	1v (Heater On)
2 (Cal)	GN	AFS-11 (B1 S1) Heater	1v (Heater On)
3	PK/WT	EVAP Purge Solenoid (VSV)	12v or 0v
4	BK/BL	PSP Switch Signal	Straight: 12v, Turning: 0v
6	BK/OR	IAC Signal (ISCC)	Pulse Signals
7	WT	IAC Signal (ISCO)	Pulse Signals
8	PK	A/T-ECT Solenoid (S1)	In 3rd or OD: 1v
9	WT	Injector 4 Control	2.0-3.3 ms
10	YL	Injector 3 Control	2.0-3.3 ms
11	RD	Injector 2 Control	2.0-3.3 ms
12	BL	Injector 1 Control	2.0-3.3 ms
13	BR	Power Ground	<0.1v
14	PK/BK	HO2S-12 (B1 S2) Heater	1v (Heater On)
15 (Cal)	BR	Power Ground	<0.1v
17	WT/RD	Igniter Signal (IGF)	Digital Signal: 0-5-0v
19	WT/BL	A/C Lock In Signal	A/C On: 12v
20	PK	A/T-ECT Solenoid (SL)	In Lockup: 12-14v
21	BL/BK	A/T-ECT Solenoid (S2)	1st or OD: 1v
22	YL/RD	Igniter Transistor 2 Control	Digital Signal: 0-5-0v
23	BK	Igniter Transistor 1 Control	Digital Signal: 0-5-0v
24	BR	Sensor Ground	<0.050v
25	BR	Power Ground	<0.1v
26	BR	Power Ground	<0.1v

1998-99 Sedan W/EIS 2.2L I4 VIN G (A/T) 16 Pin Connector

Q	Wire Color	Circuit Description (16 Pin)	Value at Hot Idle
1	YL	Sensor VREF	4.9-5.1v
2	BK/YL	MAP Sensor Signal	1-1.5v
3	YL/BK	IAT Sensor Signal	At 100ºF: 2.8v
4	GN/BK	ECT Sensor Signal	At 180ºF: 0.6v
5	WT	HO2S-11 (B1 S1) Signal	0.1-1.1v
6 (TMC)	BL	AFS-11 (B1 S1) Signal (AF+)	3.0-3.6v
6 (TMM)	WT	AFS-11 (B1 S1) Signal (AF+)	3.0-3.6v
7	BL/WT	Data Link Connector (TE)	12-14v
8	PK	EVAP Vapor Pressure Sensor	2.9-3.1v (with hose off)
9	BR	Sensor Ground	<0.050v
10	LG	TP Sensor Signal	0.3-0.8v
11	BL/RD	A/C Evaporator Temp. Sensor	EVAP Temp. 59ºF: 2.4v
12	WT	Knock Sensor Signal	<0.075v AC
13	BK	HO2S-12 (B1 S2) Signal	0.1-1.1v
14 (TMC)	BK/WT	AFS-11 (B1 S1) Signal (AF-)	Fixed at 3v
14 (TMM)	OR	AFS-11 (B1 S1) Signal (AF-)	Fixed at 3v
15	PK/BK	EGR Solenoid Control (VSV)	1v (Heater On)
16	PK	EVAP Vapor Pressure (VSV)	12v or 0v

1998-99 Sedan W/EIS 2.2L I4 VIN G (A/T) 22 Pin Connector

PCM Pin #	Wire Color	Circuit Description (22 Pin)	Value at Hot Idle
1	BK/RD	Ignition Switch Power	12-14v
2	BK/YL	Battery Direct	12-14v
3	BL/RD	Cruise Control Signal (IDLO)	1.5v, off-idle: 12v
4	GN/RD	MIL (lamp) Control	MIL Off: 12v, On: 1v
5	GN/OR	Overdrive "Off" Indicator	LED Off: 12v, On: 1v
6	WT	SIL Signal (Scan Tool)	Digital Signal
7	BK/OR	Tachometer Signal (TACO)	Pulses
8	PK/WT	Vehicle Speed Sensor	At 55 mph: 48 Hz
9	GN/WT	Stop Light Switch Signal	Brake Off: 0v, On: 12v
10	RD/BK	A/C Switch Signal	Clutch On: 1.5v, Off: 12v
11 (TMC)	GY	Starter Switch Signal	9-11v (cranking)
11 (TMM)	BK/OR	Starter Switch Signal	9-11v (cranking)
12	BK/YL	EFI Main Relay Power	12-14v
13	BK/RD	Defogger/Light Idle Up Signal	Switch On: 12v, Off: 0v
14	GN/RD	Circuit Opening Relay (FC)	0-3v, off-idle: 12v
15	YL	A/T Select Switch Low	In Low: 12v, Others: 0v
16 (TMC)	BL/WT	A/T Select Switch 2nd	In 2nd: 12v, Others: 0v
16 (TMM)	OR	A/T Select Switch 2nd	In 2nd: 12v, Others: 0v
17	RD/BK	A/T Select Switch Reverse	In 'R': 12v, Others: 0v
18	YL/BK	Cruise Control Signal (OD1)	At Cruise in OD: 12v
19	GN	A/C Dual Pressure Switch	A/C Off: 12v, On: 1v
20	RD/WT	A/C Lock In Signal	A/C On: 12v
21	BL/YL	A/C Magnetic Clutch (ACMG)	Clutch Off: 0v, On: 12v
22	BK/WT	A/T Neutral Start Switch	In P/N: 9-11v (cranking)

Pin Connector Graphic

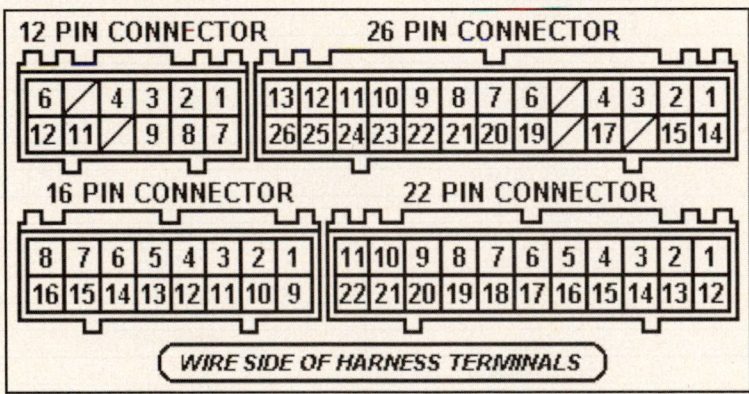

Note: TMC indicates a vehicle produced by Toyota Motor Corp. (Japan) and TMM indicates a vehicle produced by Toyota Motor Manufacturer (USA).

1998-99 Sedan W/EIS 2.2L I4 VIN G (A/T) 12 Pin Connector

PCM Pin #	Wire Color	Circuit Description (12 Pin)	Value at Hot Idle
1	RD/YL	Theft Deterrent ECU	Pulse Signals
2	BR	Sensor Ground	<0.050v
3	RD/BL	Transponder Amplifier Signal	Inserting key: pulses
4	BL/BK	Unlock Warning Switch	No Key: 4-5v
6	BL	CKP/CMP Sensor Ground (-)	<0.050v
7	BK/WT	EFI Main Relay Power	12-14v
8	GN/WT	Transponder Amplifier Code	Inserting key: pulses
9	BL/YL	Transponder Amplifier Signal	Inserting key: pulses
11	BK/WT	CMP Sensor Signal (G+)	AC pulse signals
12	BK/RD	CKP Sensor Signal (NE+)	AC pulse signals

Pin Connector Graphic

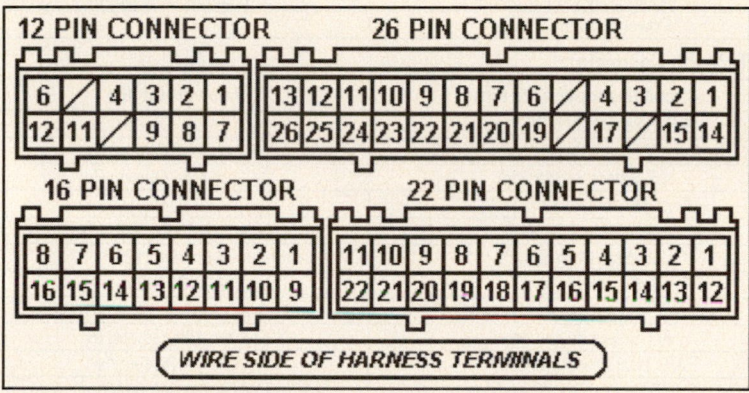

Standard Colors and Abbreviations

Abbreviation	Color	Abbreviation	Color	Abbreviation	Color
BK	Black	GY	Gray	RD	Red
BL	Blue	GN	Green	TN	Tan
BR	Brown	LG	Light Green	VT	Violet
DB	Dark Blue	OR	Orange	WT	White
DG	DK Green	PK	Pink	YL	Yellow

2000-01 Sedan W/O EIS 2.2L I4 VIN G (A/T) E7 22 Pin Connector

PCM Pin #	Wire Color	Circuit Description (22 Pin)	Value at Hot Idle
1	BK/YL	Battery Direct	12-14v
3	GN/RD	Circuit Opening Relay (FC)	0-3v, off-idle: 12v
4	WT	SIL Signal (Scan Tool)	Pulse Signals
5	PK	Overdrive "Off" Indicator	Light Off: 12v
6	GN/RD	MIL (lamp) Control	MIL Off: 12v, On: 1v
7 (TMC)	GY	Starter Switch Signal	9-11v (cranking)
7 (TMM)	BK/OR	Starter Switch Signal	9-11v (cranking)
9	YL/BK	Cruise Control Signal (OD1)	At Cruise in OD: 12v
10	BL/RD	Cruise Control Signal (IDLO)	1.5v, off-idle: 12v
15	GN/WT	Stop Light Switch Signal	Brake Off: 0v, On: 12v
16	BK/YL	EFI Main Relay Power	12-14v
18	BK/RD	Defogger/Light Switch Signal	Switch On: 12v, Off: 0v
22	BR	Power Ground (EOM)	<0.1v

2000-01 Sedan W/O EIS 2.2L I4 VIN G (A/T) E8 28 Pin Connector

PCM Pin #	Wire Color	Circuit Description (28 Pin)	Value at Hot Idle
2	RD/BK	A/T Select Switch Reverse	In 'R': 12v, Others: 0v
3 (TMC)	BL/WT	A/T Select Switch 2nd	In 2nd: 12v, Others: 0v
3 (TMM)	OR	A/T Select Switch 2nd	In 2nd: 12v, Others: 0v
5	BL/WT	Data Link Connector (TE1)	12-14v
6	RD/BK	A/C Switch Signal	Clutch On: 1.5v, Off: 12v
7	GN/OR	Overdrive Main Switch	Switch Off: 12v
8	RD/WT	A/C Lock In Signal	A/C On: 12v
10	PK	EVAP Vapor Pressure Sensor	2.9-3.1v (with hose off)
12	YL	A/T Select Switch Low	In Low: 12v, Others: 0v
17	GN	A/C Dual Pressure Switch	A/C Off: 12v, On: 1v
20	BK/WT	A/T Neutral Start Switch	In P/N: 9-11v (cranking)
22	VT/WT	Speedometer Indicator	At 55 mph: 48 Hz
27	BK/OR	Tachometer Signal (TACO)	Pulse Signals

Pin Connector Graphic

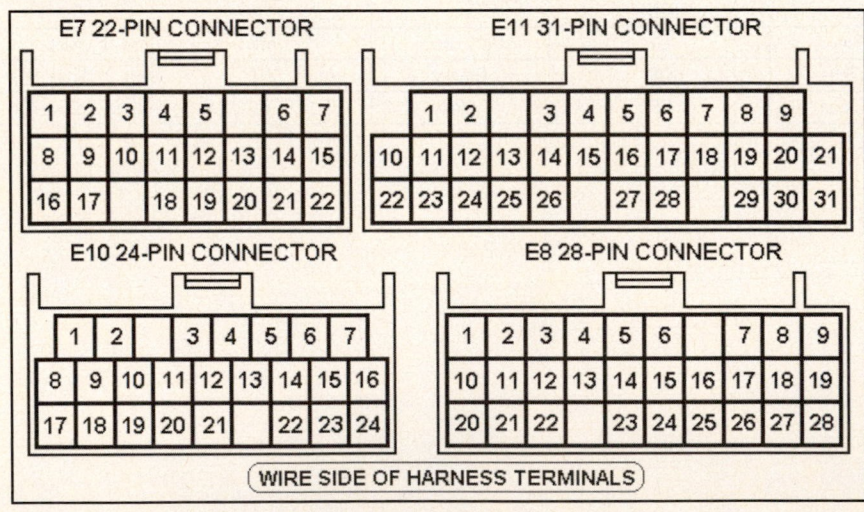

2000-01 Sedan W/O EIS 2.2L I4 VIN G (A/T) E10 24 Pin Connector

PCM Pin #	Wire Color	Circuit Description (24 Pin)	Value at Hot Idle
2	YL	Sensor VREF (VC)	4.9-5.1v
3	PK/BK	HO2S-12 (B1 S2) Heater	1v (Heater On)
4 (Cal)	GN	HAF1A-11 (B1 S1) Heater	1v (Heater On)
4 (Fed)	BL/YL	HO2S-11 (B1 S1) Heater	1v (Heater On)
5	BL/YL	A/C Magnetic Clutch (ACMG)	Clutch Off: 0v, On: 12v
6	PK/BK	EGR Solenoid Control (VSV)	12v or 0v
7	BL/RD	EVAP Canister Closed Valve	12v or 0v
8 (Cal)	BR	Power Ground	<0.1v
9	BK/YL	MAP Sensor Signal	1-1.5v
12	BK/BL	PSP Switch Signal	Straight: 12v, Turning: 0v
13	BK	HO2S-12 (B1 S2) Signal	0.1-1.1v
14 (Cal)	BL	AFS-11 (B1 S1) Signal (AF+)	Fixed at 3.0v
14 (Fed)	WT	HO2S-11 (B1 S1) Signal	0.1-1.1v
16	BK/WT	CMP Sensor Signal (G+)	AC pulse signals
17	BR	Power Ground (E1)	<0.1v
18	BR	Sensor Ground (E2)	<0.050v
22 (Cal)	BK/WT	AFS-11 (B1 S1) Signal (AF-)	Fixed at 3v
23	BL	CKP/CMP Sensor Ground (-)	<0.050v
24	BK/RD	CKP Sensor Signal (NE+)	AC pulse signals

2000-01 Sedan W/O EIS 2.2L I4 VIN G (A/T) E11 31 Pin Connector

PCM Pin #	Wire Color	Circuit Description (31 Pin)	Value at Hot Idle
1	VT	EVAP Vapor Pressure (VSV)	12v or 0v
2	VT/WT	EVAP Purge Solenoid (VSV)	12v or 0v
3	BL	Injector 1 Control	2.0-3.3 ms
4	RD	Injector 2 Control	2.0-3.3 ms
5	YL	Injector 3 Control	2.0-3.3 ms
6	WT	Injector 4 Control	2.0-3.3 ms
7	PK	A/T-ECT Solenoid (SL)	In Lockup: 12-14v
8	PK	A/T-ECT Solenoid (S2)	In 3rd or OD: 1v
9	BL/BK	A/T-ECT Solenoid (S1)	1st or OD: 1v
10	WT/RD	Igniter Signal (IGF)	Digital Signal: 0-5-0v
11	YL/RD	Igniter Transistor 2 Control	Digital Signal: 0-5-0v
12	BK	Igniter Transistor 1 Control	Digital Signal: 0-5-0v
20	WT	IAC Signal (RSD)	Pulses
21	BR	Sensor Ground (E01)	<0.1v
22	YL/BK	IAT Sensor Signal	At 100ºF: 2.8v
23	LG	TP Sensor Signal	0.3-0.8v
24	GN/BK	ECT Sensor Signal	At 180ºF: 0.6v
25	BL/RD	A/C Evaporator Temp. Sensor	EVAP Temp. 59ºF: 2.4v
26	WT/BL	A/C M/C & Lock In Signal	A/C On: 12v, Off: 0v
28	WT	Knock Sensor Signal	<0.075v AC
30	BR	Power Ground (E03)	<0.1v
31	BR	Power Ground (E02)	<0.1v

Note: *TMC indicates a vehicle produced by Toyota Motor Corp. (Japan) and TMM indicates a vehicle produced by Toyota Motor Manufacturer (USA).*

2000-01 Sedan W/EIS 2.2L I4 VIN G (A/T) E7 22 Pin Connector

PCM Pin #	Wire Color	Circuit Description (22 Pin)	Value at Hot Idle
1	BK/YL	Battery Direct	12-14v
2	BK/RD	Ignition Switch Power	12-14v
3	GN/RD	Circuit Opening Relay (FC)	0-3v, off-idle: 12v
4	WT	SIL Signal (Scan Tool)	Digital Signal
5	PK	Overdrive "Off" Indicator	Light Off: 12v
6	GN/RD	MIL (lamp) Control	MIL Off: 12v, On: 1v
7 (TMC)	GY	Starter Switch Signal	9-11v (cranking)
7 (TMM)	BK/OR	Starter Switch Signal	9-11v (cranking)
8	BK/WT	EFI Main Relay Power	12-14v
9	YL/BK	Cruise Control Signal (OD1)	At Cruise in OD: 12v
10	BL/RD	Cruise Control Signal (IDLO)	1.5v, off-idle: 12v
15	GN/WT	Stop Light Switch Signal	Brake Off: 0v, On: 12v
16	BK/YL	EFI Main Relay Power	12-14v
18	BK/RD	Defogger/Light Switch Signal	Switch On: 12v, Off: 0v
22	BR	Power Ground (EOM)	<0.1v

2000-01 Sedan W/EIS 2.2L I4 VIN G (A/T) E8 28 Pin Connector

PCM Pin #	Wire Color	Circuit Description (28 Pin)	Value at Hot Idle
2	RD/BK	A/T Select Switch Reverse	In 'R': 12v, Others: 0v
3 (TMC)	BL/WT	A/T Select Switch 2nd	In 2nd: 12v, Others: 0v
3 (TMM)	OR	A/T Select Switch 2nd	In 2nd: 12v, Others: 0v
4	BK/YL	A/C Amplifier Signal (AC1)	Clutch On: 1.5v, Off: 12v
5	BL/WT	Data Link Connector (TE1)	12-14v
6	RD/BK	A/C Switch Signal	Clutch On: 1.5v, Off: 12v
7	GN/OR	Overdrive Main Switch	Switch Off: 12v
8	RD/WT	A/C Lock In Signal	A/C On: 12v
10	PK	EVAP Vapor Pressure Sensor	2.9-3.1v (with hose off)
12	YL	A/T Select Switch Low	In Low: 12v, Others: 0v
13	LG/BK	A/C Amplifier Signal (ACT)	Clutch On: 12v, Off: 1.5v
14	PK	A/C Amplifier Signal (THWO)	Pulse Signals
17	GN	A/C Dual Pressure Switch	A/C Off: 12v, On: 1v
18	BL/YL	Transponder Amplifier Signal	Inserting key: pulses
19	RD/BL	Transponder Amplifier Signal	Inserting key: pulses
20	BK/WT	A/T Neutral Start Switch	In P/N: 9-11v (cranking)
22	VT/WT	Speedometer Indicator	At 55 mph: 48 Hz
25	BL/BK	Unlock Warning Switch	Inserting key: pulses
26	RD/YL	Theft Deterrent ECU Signal	No key inserted: pulses
27	BK/OR	Tachometer Signal (TACO)	Pulse Signals
28	GN/WT	Transponder Amplifier Code	Inserting key: pulses

Pin Connector Graphic

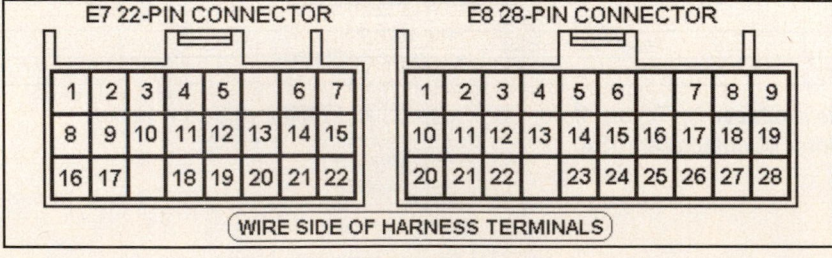

2000-01 Sedan W/EIS 2.2L I4 VIN G (A/T) E10 24 Pin Connector

PCM Pin #	Wire Color	Circuit Description (24 Pin)	Value at Hot Idle
1	PK/BK	EGR Solenoid Control (VSV)	12v or 0v
2	YL	Sensor VREF (VC)	4.9-5.1v
3	BL/RD	EVAP Canister Closed Valve	12v or 0v
4	PK	EVAP Vapor Pressure (VSV)	12v or 0v
5	PK/BK	HO2S-12 (B1 S2) Heater	1v (Heater On)
6 (Cal)	GN	AFS-11 (B1 S1) Heater	1v (Heater On)
6 (Fed)	BL/YL	HO2S-11 (B1 S1) Heater	1v (Heater On)
7	PK/WT	EVAP Purge Solenoid (VSV)	12v or 0v
8	BR	Power Ground (E04)	<0.1v
11	BK/BL	PSP Switch Signal	Straight: 12v, Turning: 0v
13	BK	HO2S-12 (B1 S2) Signal	0.1-1.1v
14 (Cal)	BL	AFS-11 (B1 S1) Signal (AF+)	Fixed at 3.3v
14 (Fed)	WT	HO2S-11 (B1 S1) Signal	0.1-1.1v
15	BK/WT	CMP Sensor Signal (G+)	AC pulse signals
16	BK/RD	CKP Sensor Signal (NE+)	AC pulse signals
17	BR	Power Ground (E1)	<0.1v
18	BR	Sensor Ground (E2)	<0.050v
19	BL/YL	A/C Magnetic Clutch (ACMG)	Clutch Off: 0v, On: 12v
21	BK/YL	MAP Sensor Signal	1-1.5v
23 (Cal)	BK/WT	AFS-11 (B1 S1) Signal (AF-)	Fixed at 3v
24	BL	CKP/CMP Sensor Ground (-)	<0.050v

2000-01 Sedan W/EIS 2.2L I4 VIN G (A/T) E11 31 Pin Connector

PCM Pin #	Wire Color	Circuit Description (31 Pin)	Value at Hot Idle
1	BL	Injector 1 Control	2.0-3.3 ms
2	RD	Injector 2 Control	2.0-3.3 ms
3	YL	Injector 3 Control	2.0-3.3 ms
4	WT	Injector 4 Control	2.0-3.3 ms
6	PK	A/T-ECT Solenoid (SL)	In Lockup: 12-14v
7	PK	A/T-ECT Solenoid (S1)	In 3rd or OD: 1v
8	BL/BK	A/T-ECT Solenoid (S2)	1st or OD: 1v
10	BK	Igniter Transistor 1 Control	Digital Signal: 0-5-0v
11	YL/RD	Igniter Transistor 2 Control	Digital Signal: 0-5-0v
12	WT/RD	Igniter Signal (IGF)	Digital Signal: 0-5-0v
13	BL/RD	A/C Evaporator Temp. Sensor	EVAP Temp. 59°F: 2.4v
18	WT	IAC Signal (RSD)	Pulses
21	BR	Sensor Ground	<0.050v
22	GN/BK	ECT Sensor Signal	At 180°F: 0.6v
23	YL/BK	IAT Sensor Signal	At 100°F: 2.8v
24	LG	TP Sensor Signal	0.3-0.8v
25	WT/BL	A/C M/C & Lock In Signal	A/C On: 12v, Off: 0v
27	WT	Knock Sensor Signal	<0.075v AC
30	BR	Power Ground (E03)	<0.1v
31	BR	Power Ground (E02)	<0.1v

Note: **TMC indicates a vehicle produced by Toyota Motor Corp. (Japan) and TMM indicates a vehicle produced by Toyota Motor Manufacturer (USA).**

2002-03 Sedan 2.4L I4 VIN E W/EIS (A/T) E6 31 Pin Connector

PCM Pin #	Wire Color	Circuit Description (31 Pin)	Value at Hot Idle
1	BK/RD	EFI Main Relay Output (B+)	12-14v
2	BL/RD	Battery Direct (+BM)	12-14v
3	BK/YL	Battery Direct (BATT)	12-14v
4	VT	EVAP Vapor Pressure (VSV)	12v or 0v
5	BK/OR	Tachometer Signal (TACO)	DC pulse signals
6-7, 20, 24, 26	---	Not Used	---
8	BK/WT	EFI Main Relay Control	Relay On: 12v, Off: 0v
9	BK/OR	Ignition Switch Power (B+)	12-14v
10	GN/RD	Circuit Opening Relay (FC)	0-3v, off-idle: 12v
11	WT	SIL Signal (Scan Tool)	Digital Signal
12	GN	Tail Light Switch Signal	Switch Off: 1.5v, On: 12v
13	BK/YL	Mirror Heater Switch Signal	Switch Off: 1.5v, On: 12v
14	PK/BK	Data Link Connector (TC)	12-14v
15	BR	Power Ground (EOM)	<0.1v
18	GN/RD	MIL (lamp) Control	MIL Off: 12v, On: 1v
19	RD	Scan Tool (WFSE)	12v
21	PK	EVAP Vapor Pressure Sensor	2.9-3.1v (with hose off)
22	BL/YL	APP Sensor (VPA) Signal	0.8-1.2v
23	WT/RD	APP Sensor (VPA2) Signal	0.8-1.2v
25	RD	APP Sensor VREF	4.5-5.5v
27	RD	APP Sensor VREF	4.5-5.5
28	LG/BK	APP Sensor Ground (EPA)	<0.050v
29	LG	APP Sensor Ground (EPA2)	<0.050v

2002-03 Sedan 2.4L I4 VIN E W/EIS (A/T) E7 35 Pin Connector

PCM Pin #	Wire Color	Circuit Description (35 Pin)	Value at Hot Idle
1-3, 5-6, 12-13	---	Not Used	---
4	BL	HO2S-12 (B1 S2) Heater	Heater On: <1v
7	OR	Overdrive Indicator Control	LED Off: 12v, On: 0v
8	YL	A/T Select Switch Low Signal	In Low: 12v, Others: 0v
9	BL/WT	A/T Select Switch 2nd Signal	In 2nd: 12v, Others: 0v
10	WT/BL	A/T Select Switch 'D' Signal	In Drive: 12v, Others: 0v
11	RD/BK	A/T Select Switch 'R' Signal	In 'R': 12v, Others: 0v
14	YL/GN	Hot Engine Lamp Control	Lamp Off: 12v, On: 1v
15	GN/WT	Transponder Amplifier Code	Inserting key: pulses
16	VT	Transponder IMLD Signal	Inserting key: pulses
17	VT/WT	Speedometer Indicator	Moving: 0-5-0v
18, 20-21	---	Not Used	---
19	GN/WT	Stop Light Switch Signal	Brake Off: 0v, On: 12v
22	BK	HO2S-12 (B1 S2) Signal	0.1-1.1v
23-25, 28, 32	---	Not Used	---
26	BL/YL	Theft Deterrent ECU Signal	No key inserted: pulses
27	RD/BL	Transponder Amplifier Code	Inserting key: pulses
29	GN/OR	Overdrive Main Switch	Switch Off: 12v, On: 0v
31	PK/BL	A/C Switch Signal	AC On: 9-14v
33	BK	A/C Switch Signal (Auto AC)	AC On: 9-14v
34	BL	Transponder Amplifier Signal	Key Inserted: <1.5v

2002-03 Sedan 2.4L I4 VIN E W/EIS (A/T) E8 32 Pin Connector

PCM Pin #	Wire Color	Circuit Description (24 Pin)	Value at Hot Idle
1	BR	Power Ground (E1)	<0.1v
2	BL/WT	A/C Magnetic Clutch (ACMG)	A/C On: <3.0v
4	WT	Throttle Body Motor (M-)	DC pulse signals
5	BK	Throttle Body Motor (M+)	DC pulse signals
6	WT/BK	Sensor Ground (ME01)	<0.050v
7	WT/BK	Power Ground (E03)	<0.1v
8	WT/BK	Power Ground (E02)	<0.1v
9	GN/WT	Cooling Fan Relay Control	Relay Off: 12v, On: 1v
10	RD/WT	PSP Switch Signal	Straight: 12v, Turning: 0v
11	BK/RD	EVAP Solenoid Control (VSV)	12v or 0v
12	GN	EVAP Canister Closed Valve	12v or 0v
17	BR	Sensor Shield Ground	<0.050v
15	YL	Cam Timing Oil Valve (OCV-)	Pulse signals
16	BK/WT	Cam Timing Oil Valve (OCV+)	Pulse signals
30	YL/BK	Cooling Fan Relay Control	Relay Off: 12v, On: 1v

2002-03 Sedan 2.4L I4 VIN E W/ESI (A/T) E9 35 Pin Connector

PCM Pin #	Wire Color	Circuit Description (35 Pin)	Value at Hot Idle
1	WT	Knock Sensor Signal	<0.075v AC
2-3	---	Not Used	---
4	BK/RD	AFS-11 (B1 S1) Heater	1v (Heater On)
5	---	Not Used	---
6	WT/BK	Power Ground (E05)	<0.1v
7	WT/BK	Power Ground (E04)	<0.1v
8	BK/YL	Neutral Start Switch	In P/N: 9-11v (cranking)
9	BK/WT	Starter Switch Signal (STA)	9-11v (cranking)
10	---	Not Used	---
11	YL	A/T-ECT Solenoid (DSL)	In Lockup: 12-14v
12-15	---	Not Used	---
16	BL/RD	A/T-ECT Solenoid (SL2-)	Moving in 3rd or OD: <1v
17	BL/YL	A/T-ECT Solenoid (SL2+)	Moving in 3rd or OD: <1v
18	PK	A/T-ECT Solenoid (SL1-)	Moving in 1st Gear: <1v
19	RD/BK	A/T-ECT Solenoid (SL1+)	Moving in 1st Gear: <1v
18	WT	IAC Signal (RSD)	Pulses
23	OR	AFS-11 (B1 S1) Signal (AF+)	3.0-3.6v
24	RD	MAF Sensor Signal (VG)	1.1-1.5v
26	RD	Counter Gear Speed (NC+)	AC pulse signals
27	BL	Turbine Speed Sensor (NT+)	AC pulse signals
28-30	---	Not Used	---
31	WT	AFS-11 (B1 S1) Signal (AF-)	Fixed at 3v
32	BL/WT	MAF Sensor Ground (E2G)	<0.050v
33	---	Not Used	---
34	GN	Counter Gear Speed (NC-)	AC pulse signals
35	LG	Turbine Speed Sensor (NT-)	AC pulse signals

2002-03 Sedan 2.4L I4 VIN E W/EIS (A/T) E10 34 Pin Connector

PCM Pin #	Wire Color	Circuit Description (34 Pin)	Value at Hot Idle
1	BL	Injector 1 Control	2.0-3.3 ms
2	RD	Injector 2 Control	2.0-3.3 ms
3	YL	Injector 3 Control	2.0-3.3 ms
4	WT	Injector 4 Control	2.0-3.3 ms
5	---	Not Used	---
6	WT/BK	Power Ground (E02)	<0.1v
7	WT/BK	Power Ground (E01)	<0.1v
8	RD/WT	Igniter Transistor 1 Control	7% duty cycle
9	PK	Igniter Transistor 2 Control	7% duty cycle
10	LG/BK	Igniter Transistor 3 Control	7% duty cycle
11	BL/YL	Igniter Transistor 2 Control	7% duty cycle
12	---	Not Used	---
13	YL	Sensor VREF (VC)	4.9-5.1v
14-15	---	Not Used	---
16	YL/BK	A/T-ECT Solenoid (SLT-)	During shifting: 12v
17	YL/RD	A/T-ECT Solenoid (SLT+)	During shifting: 12v
18	YL	TP Sensor VREF (VC)	4.9-5.1v
19	GN/YL	ECT Sensor Signal	At 180ºF: 0.51v
20	LB	IAT Sensor Signal	At 100ºF: 2.60v
21	LG	TP Sensor Signal	0.4-1.0v
22	---	Not Used	---
23	WT/RD	Igniter Signal (IGF)	Digital Signal: 0-5-0v
24-25	---	Not Used	---
26	BK/WT	CMP Sensor Signal (G22+)	AC pulse signals
27	RD	CKP Sensor Signal (NE+)	AC pulse signals
28	BR	Power Ground (E2)	<0.1v
29	---	Not Used	---
30 (TMC)	GN	A/T ATF Sensor Signal	<1.5v at 239ºF
30 (TMM)	GN/RD	A/T ATF Sensor Signal	<1.5v at 239ºF
31	BK/RD	TP Sensor 2 Signal (VTA2)	2.0-2.9v
32-33	---	Not Used	---
34	GN	CKP/CMP Sensor Ground	<0.050v

1990-91 Sedan, Wagon 2.5L V6 VIN V (A/T-ECT) 26 Pin Connector

PCM Pin #	Wire Color	Circuit Description (26 Pin)	Value at Hot Idle
1	BL	Distributor Signal (NE+)	AC pulse signals
2	YL	Distributor Signal (G2+)	AC pulse signals
3	WT/RD	Igniter Signal (IGF)	Digital Signal: 0-5-0v
4	RD/YL	ISC Motor (ISC4)	Pulses
5	BL/GN	ISC Motor (ISC3)	Pulses
6	GY/YL	ISC Motor (ISC2)	Pulses
7	WT/BL	ISC Motor (ISC1)	Pulses
8	RD/GN	Main O2S Heater	1v (Heater On)
9	BK/RD	Fuel Pressure Up Solenoid	12v or 0v
10	GN/BK	Cold Start Injector Control	1v (at cold startup)
11	YL	Injector Pair 2 & 3 Control	2.0-3.3 ms
12	WT	Injector Pair 1 & 6 Control	2.0-3.3 ms
13	WT/BK	Power Ground	<0.1v
14	BK	Distributor Signal (G-)	<0.050v
15	RD	Distributor Signal (G+)	AC pulse signals
17	BL/YL	A/T-ECT Solenoid (SL)	In Lockup: 12-14v
18	PK/RD	A/T-ECT Solenoid (S2)	1st or OD: 1v
19	PK	A/T-ECT Solenoid (S1)	In 3rd or OD: 1v
20	WT/GN	Igniter Signal (IGT)	Digital Signal: 0-5-0v
22	BL/BK	A/C Acceleration Cut	A/C On: 4.5-5.5v
24	BR	Sensor Ground	<0.050v
25	GY	Injector Pair 4 & 5 Control	2.0-3.3 ms
26	WT/BK	Power Ground	<0.1v

1990-91 Sedan, Wagon 2.5L V6 VIN V (A/T-ECT) 16 Pin Connector

PCM Pin #	Wire Color	Circuit Description (16 Pin)	Value at Hot Idle
1	BL/RD	Airflow Meter Signal	1.5-2.5v
2	YL/RD	Sensor VREF	4.9-5.1v
3	YL	IAT Sensor Signal	At 100ºF: 2.60v
4	GN	ECT Sensor Signal	At 180ºF: 0.51v
5	RD/BL	Sub O2S Signal	0.1-1.1v
6	WT	Main O2S Signal	0.1-1.1v
7	BL/RD	A/T Pattern Selector Switch	Norm: 0v, PWR: 12v
8	GN/RD	Check Connector	12-14v
9	BR	Sensor Ground	<0.050v
10	GY	EGR Gas Temp. Sensor	3.5-4.0v
11	BK	TP Sensor Signal	0.3-0.8v
12	BL	Closed Throttle Switch	0-3v, off-idle: 12v
13	GN/WT	Stop Light Switch Signal	Brake Off: 0v, On: 12v
14	WT	Knock Sensor Signal	<0.075v AC
15	YL/GN	Check Connector	12-14v

1990-91 Sedan, Wagon 2.5L V6 VIN V (A/T-ECT) 22 Pin Connector

PCM Pin #	Wire Color	Circuit Description (22 Pin)	Value at Hot Idle
1	WT/BL	Battery Direct	12-14v
2	BK/OR	Ignition Switch Power	12-14v
3	RD/BK	A/T Select Switch Reverse	In 'R': 12v, Others: 0v
4	BK/YL	EFI Main Relay Power	12-14v
5	GY	MIL (lamp) Control	MIL Off: 12v, On: 1v
6 (Cal)	BL/WT	Sub O2S Signal	0.1-1.1v
7 ('91)	BR/BK	A/T Select Switch Drive	In 'D': 12v, Others: 0v
8 ('91)	GN/BK	A/T-ECT Speed Sensor	Moving: varies 0-5v
9	PK/YL	Vehicle Speed Sensor	At 55 mph: 48 Hz
10	BK/GN	A/C Magnetic Clutch (ACMG)	Clutch Off: 0v, On: 12v
11	BK	A/T: Starter Switch Signal	9-11v (cranking)
11	BK/WT	M/T: Starter Switch Signal	9-11v (cranking)
12	WT/RD	EFI Main Relay Power	12-14v
13	WT/RD	EFI Main Relay Power	12-14v
14	YL/BL	A/T Select Switch Low	In Low: 12v, Others: 0v
15	OR	A/T Select Switch 2nd	In 2nd: 12v, Others: 0v
16	RD	A/T: Select Switch Neutral	In Neutral: 8-14v
20	GN/OR	Main Overdrive Switch	Switch Off: 12-14v
21	YL/BK	Cruise Control ECU	At Cruise in OD: 12v
22	BK/WT	A/T: Neutral Drive Switch	In P/N: 9-11v (cranking)

Pin Connector Graphic

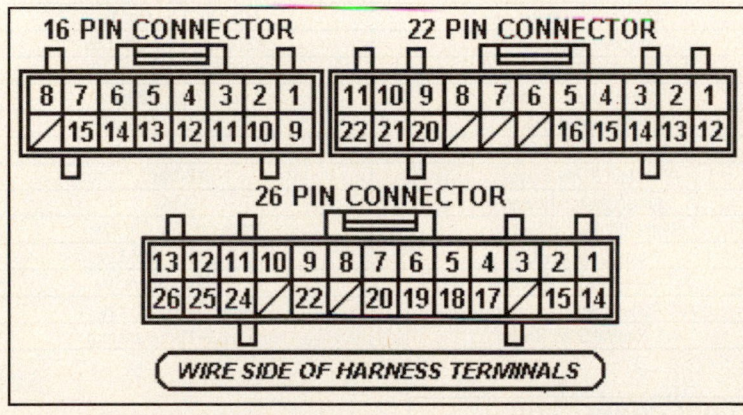

1992-93 Sedan, Wagon 3.0L VIN G (A/T-ECT) 16 Pin Connector

PCM Pin #	Wire Color	Circuit Description (16 Pin)	Value at Hot Idle
1	BL/RD	Sensor VREF (VC)	4.9-5.1v
2	YL/BL	Air Flow Meter Signal	2.5-4.5v
3	BL/BK	IAT Sensor Signal	At 100°F: 2.60v
4	GN	ECT Sensor Signal	At 180°F: 0.51v
5	WT	O2S-11 (B1 S1) Signal	0.1-1.1v
6	WT	Knock Sensor 1 Signal	<0.075v AC
7	GY	Check Connector	12-14v
8	GY/BK	Check Connector	12-14v
9	BR	Sensor Ground	<0.050v
10	BK	TP Sensor Signal	0.3-0.8v
11	BL	Closed Throttle Switch	1v, at off-idle: 12v
12 (Cal)	GY	EGR Gas Temp. Sensor	3.5-4.0v
13	RD/BL	O2S-12 (B1 S2) Signal	0.1-1.1v
14	BK	Knock Sensor 2 Signal	<0.075v AC
15	GN/WT	Check Connector	12-14v
16	GN/BK	A/T-ECT Speed Sensor	Moving: varies 0-5v

1992-93 Sedan, Wagon 3.0L VIN G (A/T-ECT) 22 Pin Connector

PCM Pin #	Wire Color	Circuit Description (22 Pin)	Value at Hot Idle
1	BK/OR	EFI Main Relay Power B1+	12-14v
2	WT/BL	Battery Direct	12-14v
3	RD/YL	EFI Main Relay Power	12-14v
4	GN/RD	MIL (lamp) Control	MIL Off: 12v, On: 1v
6	GN/BK	A/C Amplifier Signal (ACT)	Clutch On: 12v, Off: 1.5v
7	BK/YL	A/C Amplifier Signal (AC1)	Clutch On: 1.5v, Off: 12v
8	PK/YL	Vehicle Speed Sensor	At 55 mph: 48 Hz
9	GY/BL	Main Overdrive Switch	Switch Off: 12-14v
11	BK/WT	A/T: Starter Switch Signal	9-11v (cranking)
12	BK/OR	EFI Main Relay Power	12-14v
13	BK/OR	Ignition Switch Power	12-14v
14	GN/WT	Stop Light Switch Signal	Brake Off: 0v, On: 12v
18	YL/BK	Cruise Control ECU	At Cruise in OD: 12v
19	RD/BK	A/T Select Switch Reverse	In 'R': 12v, Others: 0v
20	BL/RD	A/T Pattern Selector Switch	Norm: 0v, PWR: 12v
22	BK/WT	A/T Neutral Start Switch	In P/N: 9-11v (cranking)

Pin Connector Graphic

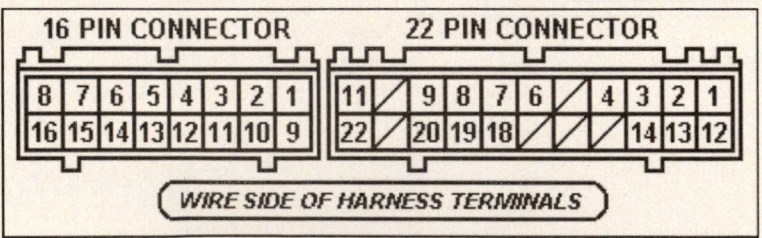

1992-93 Sedan, Wagon 3.0L VIN G (A/T-ECT) 12 Pin Connector

PCM Pin #	Wire Color	Circuit Description (12 Pin)	Value at Hot Idle
2	BK/RD	Fuel Pressure Up Solenoid	12v or 0v
3	LG	Intake Air Control Solenoid	12v or 0v
5 (Cal)	WT	Sub O2S Signal	0.1-1.1v
6	YL	Distributor Signal (G-)	<0.050v
10	BL	Distributor Signal (G2+)	AC pulse signals
11	RD	Distributor Signal (G1+)	AC pulse signals
12	BK	Distributor Signal (NE-)	<0.050v

1992-93 Sedan, Wagon 3.0L VIN G (A/T-ECT) 26 Pin Connector

PCM Pin #	Wire Color	Circuit Description (26 Pin)	Value at Hot Idle
1 (Cal)	PK/BK	HO2S Heater	1v (Heater On)
2	OR	A/T Select Switch 2nd	In 2nd: 12v, Others: 0v
3	YL/BL	A/T Select Switch Low	In Low: 12v, Others: 0v
4	RD/BK	ISC Signal 4	Pulse Signals
5	BL/RD	ISC Signal 3	Pulse Signals
6	GN/WT	ISC Signal 2	Pulse Signals
7	WT/BL	ISC Signal 1	Pulse Signals
8	BL/YL	A/T-ECT Solenoid (SL)	In Lockup: 12-14v
9	PK/BL	A/T-ECT Solenoid (S2)	In 2nd: 12v, Others: 0v
10	PK	A/T-ECT Solenoid (S1)	In 1st: 12-14v
11	YL	Injector 2 Control	2.0-3.3 ms
12	WT	Injector 1 Control	2.0-3.3 ms
13	WT/BK	Power Ground	<0.1v
14	BK	Check Connector	12-14v
15	RD/WT	Check Connector	12-14v
17	WT/RD	Igniter Signal (IGF)	Digital Signal: 0-5-0v
18	WT	Igniter Signal (IGT)	Digital Signal: 0-5-0v
20	GN	Injector 6 Control	2.0-3.3 ms
21	RD	Injector 5 Control	2.0-3.3 ms
22	BL	Injector 4 Control	2.0-3.3 ms
23	GY	Injector 3 Control	2.0-3.3 ms
24	BR	Sensor Ground	<0.050v
25	GN	Cold Start Injector Control	1v (at cold startup)
26	WT/BK	Power Ground	<0.1v

Pin Connector Graphic

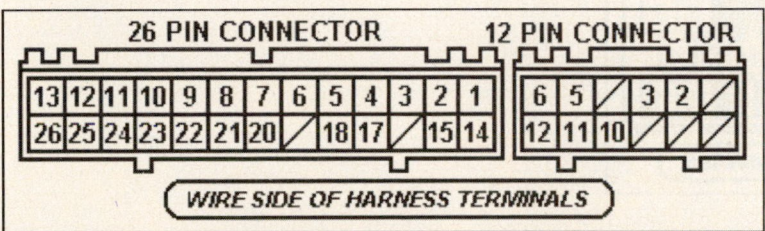

1994-95 Coupe & Sedan 3.0L V6 MFI VIN G (A/T) 16 Pin Connector

PCM Pin #	Wire Color	Circuit Description (16 Pin)	Value at Hot Idle
1	WT/BL	A/C Idle Up Solenoid	12v or 0v
3	GN/RD	MIL (lamp) Control	MIL Off: 12v, On: 1v
5	GY/BK	Check Connector	12-14v
6	RD/YL	Intake Air Control Solenoid	12v or 0v
7	RD/BK	MAF Sensor Ground	<0.050v
10	BL/RD	HO2S-21 (B2 S1) Heater	1v (Heater On)
11	PK/BK	HO2S-11 (B1 S1) Heater	1v (Heater On)
12	BK/BL	EGR Solenoid Control (VSV)	12v or 0v
14 ('94)	GN/YL	EGR Gas Temp. Sensor	3.5-4.0v
14 ('95)	GN/BK	EGR Gas Temp. Sensor	3.5-4.0v
16	BR	Sensor Ground	<0.050v

1994-95 Coupe & Sedan 3.0L V6 MFI VIN G (A/T) 22 Pin Connector

PCM Pin #	Wire Color	Circuit Description (22 Pin)	Value at Hot Idle
1	BL/RD	Sensor VREF (VC)	4.9-5.1v
5	RD	CKP Sensor Signal (NE+)	AC pulse signals
6	GN	CKP Sensor Signal (NE-)	<0.050v
7	BK/YL	TP Sensor Signal	0.3-0.8v
8	RD	MAF Sensor Signal	1.1-1.5v
13	RD/BL	HO2S-11 (B1 S1) Signal	0.1-1.1v
14	WT	Knock Sensor 1 Signal	<0.075v AC
15	WT	Knock Sensor 2 Signal	<0.075v AC
16	BK/WT	CMP Sensor Signal (G+)	AC pulse signals
17	BL	CMP Sensor Signal (G-)	<0.050v
18	GN/RD	Circuit Opening Relay (FC)	0-3v, off-idle: 12v
19	RD/BL	HO2S-21 (B2 S1) Signal	0.1-1.1v
20	GN/BK	ECT Sensor Signal	At 180ºF: 0.51v
21	BL/BK	IAT Sensor Signal	At 100ºF: 2.60v
22	BR	Sensor Ground	<0.050v

Pin Connector Graphic

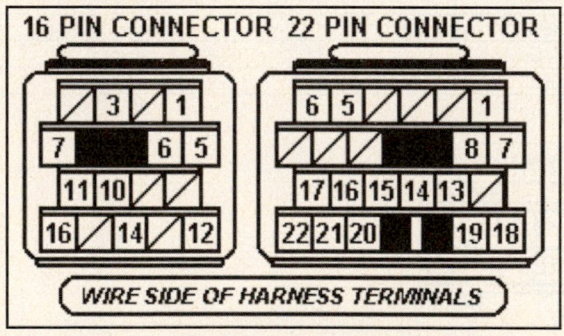

1994-95 Coupe & Sedan 3.0L V6 MFI VIN G (A/T) 28 Pin Connector

PCM Pin #	Wire Color	Circuit Description (28 Pin)	Value at Hot Idle
5	GN/BK	A/C Amplifier Signal (ACT)	Clutch On: 12v, Off: 1.5v
9 ('95)	RD	Cooling Fan Relay	Fan On: <1v
11	GN/WT	Data Link Connector (TN)	12-14v
12	PK/YL	Vehicle Speed Sensor	At 55 mph: 48 Hz
13 ('95)	BK	Tachometer Signal (TACO)	Pulses
14	WT/BL	Battery Direct	12-14v
20	BK/YL	A/C Amplifier Signal (AC1)	Clutch On: 1.5v, Off: 12v
21	BK/RD	Idle-Up Signal	Load On: 12v, Off: 0v
22	BK/OR	EFI Main Relay Power B1+	12-14v
23	BK/OR	EFI Main Relay Power	12-14v
24 ('95)	GN/WT	Stop Light Switch Signal	Brake Off: 0v, On: 12v
25	PK/BK	HO2S-12 (B1 S2) Heater	12v or 0v
26	BK	HO2S-12 (B1 S2) Signal	0.1-1.1v
28	WT	Data Link Connector (TN)	12-14v

Pin Connector Graphic

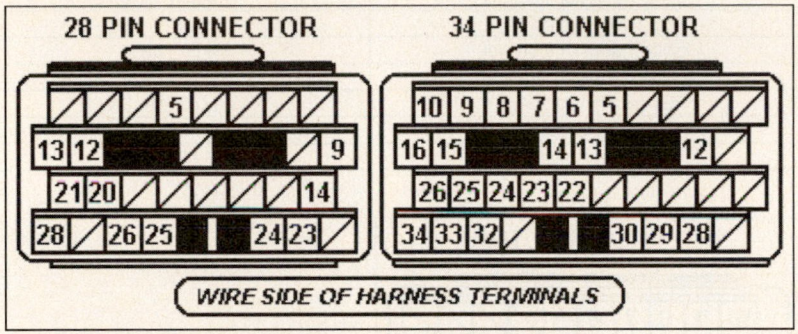

Standard Colors and Abbreviations

Abbreviation	Color	Abbreviation	Color	Abbreviation	Color
BK	Black	GY	Gray	RD	Red
BL	Blue	GN	Green	TN	Tan
BR	Brown	LG	Light Green	VT	Violet
DB	Dark Blue	OR	Orange	WT	White
DG	DK Green	PK	Pink	YL	Yellow

1994-95 Coupe & Sedan 3.0L V6 MFI VIN G (A/T) 34 Pin Connector

PCM Pin #	Wire Color	Circuit Description (34 Pin)	Value at Hot Idle
5	GN	Injector 6 Control	2.0-3.3 ms
6	RD	Injector 5 Control	2.0-3.3 ms
7	BL	Injector 4 Control	2.0-3.3 ms
8	GY	Injector 3 Control	2.0-3.3 ms
9	YL	Injector 2 Control	2.0-3.3 ms
10	WT	Injector 1 Control	2.0-3.3 ms
12	WT/RD	Igniter Signal (IGF)	Digital Signal: 0-5-0v
13	BK/WT	Starter Switch Signal	9-11v (cranking)
14	BK/WT	Neutral Start Switch	In P/N: 9-11v (cranking)
15	GY/BK	Igniter Transistor 3 Control	7% duty cycle
16	GY/BK	Igniter Transistor 2 Control	7% duty cycle
22	YL/BK	IAC Signal (RSC)	Pulse Signals
23	GN/BK	IAC Signal (RSO)	Pulse Signals
24	WT/GN	Igniter Transistor 1 Control	7% duty cycle
25	BK/RD	Fuel Pressure Up Solenoid	1v (at hot restart)
26	BL/BK	Igniter Transistor 4 Control	7% duty cycle
28	WT/BK	Power Ground	<0.1v
29	GN/RD	Igniter Transistor 6 Control	7% duty cycle
30	RD/BK	Igniter Transistor 5 Control	7% duty cycle
32	BL	Closed Throttle Switch	1v, at off-idle: 12v
33	WT/BK	Power Ground	<0.1v
34	WT/BK	Power Ground	<0.1v

Pin Connector Graphic

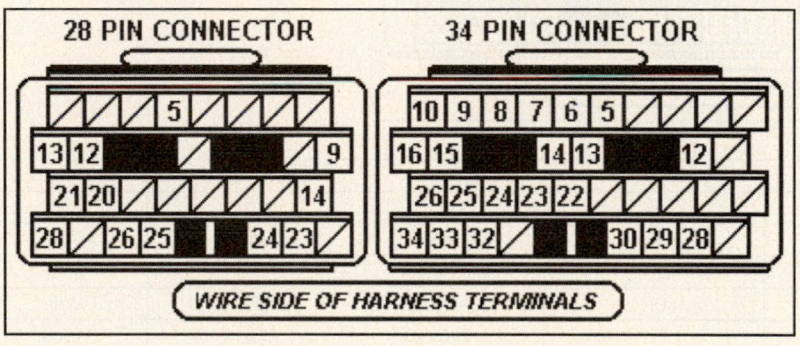

1996 Coupe & Sedan 3.0L V6 VIN F (A/T-ECT) 16 Pin Connector

PCM Pin #	Wire Color	Circuit Description (16 Pin)	Value at Hot Idle
3	GN/RD	MIL (lamp) Control	MIL Off: 12v, On: 1v
5	GY/BK	Check Connector	12-14v
6	RD/YL	Intake Air Control Solenoid	12v or 0v
7	RD/BK	MAF Sensor Ground	<0.050v
10	BL/RD	HO2S-21 (B2 S1) Heater	1v (Heater On)
11	BL/BK	HO2S-11 (B1 S1) Heater	1v (Heater On)
12	BK/WT	EGR Solenoid Control (VSV)	12v or 0v
14	GN/YL	EGR Gas Temp. Sensor	3.5-4.0v
16	BR	Sensor Ground	<0.050v

1996 Coupe & Sedan 3.0L V6 VIN F (A/T-ECT) 22 Pin Connector

PCM Pin #	Wire Color	Circuit Description (22 Pin)	Value at Hot Idle
1	BL/RD	Sensor VREF (VC)	4.9-5.1v
4	WT/BL	OD Clutch Speed Signal (-)	AC pulse signals
5 (TMC)	RD	CKP Sensor Signal (NE+)	AC pulse signals
5 (TMM)	WT	CKP Sensor Signal (NE+)	AC pulse signals
6 (TMC)	GN	CKP Sensor Signal (NE-)	<0.050v
6 (TMM)	OR	CKP Sensor Signal (NE-)	<0.050v
7	BK/YL	TP Sensor Signal	0.3-0.8v
8	RD	MAF Sensor Signal	1.1-1.5v
9	YL/BL	OD Clutch Speed Signal (+)	AC pulse signals
13	WT	HO2S-11 (B1 S1) Signal	0.1-1.1v
14	WT	Knock Sensor 1 Signal	<0.075v AC
15	WT	Knock Sensor 2 Signal	<0.075v AC
16 (TMC)	BK/WT	CMP Sensor Signal (G+)	AC pulse signals
16 (TMM)	WT/BL	CMP Sensor Signal (G+)	AC pulse signals
18	GN/RD	Circuit Opening Relay (FC)	0-3v, off-idle: 12v
19	RD/BL	HO2S-21 (B2 S1) Signal	0.1-1.1v
20	GN/BK	ECT Sensor Signal	At 180ºF: 0.51v
21	BL/BK	IAT Sensor Signal	At 100ºF: 2.60v
22	BR	Sensor Ground	<0.050v

Pin Connector Graphic

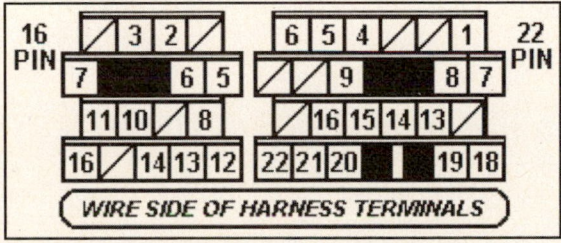

WIRE SIDE OF HARNESS TERMINALS

1996 Coupe & Sedan 3.0L V6 VIN F (A/T-ECT) 28 Pin Connector

PCM Pin #	Wire Color	Circuit Description (28 Pin)	Value at Hot Idle
5	GN/BK	A/C Amplifier Signal (ACT)	Clutch On: 12v, Off: 1.5v
7	YL/BK	Cruise Control ECU	At Cruise in OD: 12v
9	BK/RD	Cooling Fan Relay	Relay Off: 12v, On: 1v
12	PK/YL	Vehicle Speed Sensor	At 55 mph: 48 Hz
13	BK	Tachometer Signal (TACO)	Pulse Signals
14	WT/BL	Battery Direct	12-14v
20	BK/YL	A/C Amplifier Signal (AC1)	Clutch On: 1.5v, Off: 12v
21	BK/RD	Defogger Idle Up Signal	Switch On: 12v, Off: 0v
23	BK/OR	EFI Main Relay Power	12-14v
24	GN/WT	Stop Light Switch Signal	Brake Off: 0v, On: 12v
25	PK/BK	HO2S-12 (B1 S2) Heater	1v (Heater On)
26	BK	HO2S-12 (B1 S2) Signal	0.1-1.1v
28	WT	Data Link Connector	No Scan Tool: 0v

Pin Connector Graphic

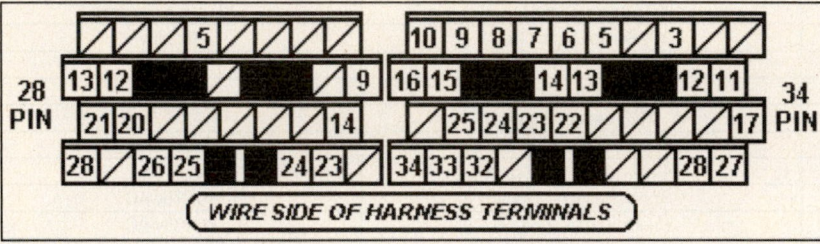

WIRE SIDE OF HARNESS TERMINALS

Standard Colors and Abbreviations

Abbreviation	Color	Abbreviation	Color	Abbreviation	Color
BK	Black	GY	Gray	RD	Red
BL	Blue	GN	Green	TN	Tan
BR	Brown	LG	Light Green	VT	Violet
DB	Dark Blue	OR	Orange	WT	White
DG	DK Green	PK	Pink	YL	Yellow

1996 Coupe & Sedan 3.0L V6 VIN F (A/T-ECT) 34 Pin Connector

PCM Pin #	Wire Color	Circuit Description (34 Pin)	Value at Hot Idle
3	YL/GN	A/T-ECT Solenoid (SLN-)	SLN shifts: 9-14v
5	GN	Injector 6 Control	2.0-3.3 ms
6	RD	Injector 5 Control	2.0-3.3 ms
7	BL	Injector 4 Control	2.0-3.3 ms
8	GY	Injector 3 Control	2.0-3.3 ms
9	YL	Injector 2 Control	2.0-3.3 ms
10	WT	Injector 1 Control	2.0-3.3 ms
11	PK	A/T-ECT Solenoid (S1)	SLN shifts: 9-14v
12	WT/RD	Igniter Signal (IGF)	Digital Signal: 0-5-0v
13	BK/WT	Starter Switch Signal	9-11v (cranking)
14	BK/WT	Neutral Start Switch	In P/N: 9-11v (cranking)
15 (TMC)	GY/BK	Igniter Transistor 3 Control	7% duty cycle
15 (TMM)	GN/RD	Igniter Transistor 3 Control	7% duty cycle
16	YL/RD	Igniter Transistor 2 Control	7% duty cycle
17	PK/BK	A/T-ECT Solenoid (S2)	In Lockup: 12v
22	YL/BK	IAC Signal (RSC)	Pulse Signals
23	GN/BK	IAC Signal (RSO)	Pulse Signals
24 (TMC)	WT/GN	Igniter Transistor 1 Control	7% duty cycle
24 (TMM)	BL/BK	Igniter Transistor 1 Control	7% duty cycle
25	BK/RD	Fuel Pressure Up Solenoid	1v (at hot restart)
27	BK/YL	A/T-ECT Solenoid (SL)	In Lockup: 12v
28	BR	Power Ground	<0.1v
32	BL/WT	Closed Throttle Switch	1v, at off-idle: 12v
33	WT/BK	Power Ground	<0.1v
34	WT/BK	Power Ground	<0.1v

Pin Connector Graphic

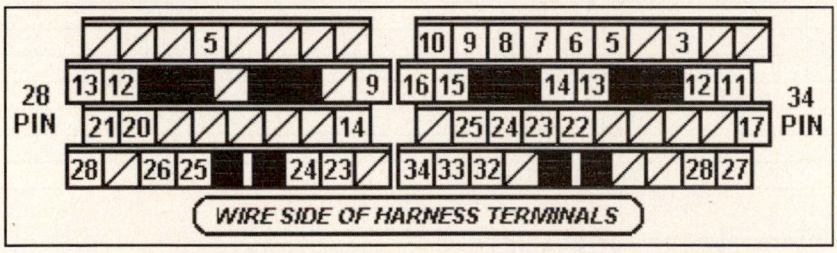

WIRE SIDE OF HARNESS TERMINALS

1997 Sedan 3.0L V6 MFI VIN F W/EIS (A/T-ECT) 16 Pin Connector

PCM Pin #	Wire Color	Circuit Description (16 Pin)	Value at Hot Idle
2	LG	EVAP Purge Solenoid (VSV)	12v or 0v
3	GN/RD	MIL (lamp) Control	MIL Off: 12v, On: 1v
5	BL/WT	Data Link Connector	12-14v
6	RD/YL	Intake Air Control Solenoid	12v or 0v
7	RD/BK	MAF Sensor Ground	<0.050v
8	WT/RD	EVAP Vapor Pressure (VSV)	12v or 0v
9	GN/WT	Cooling Fan 1 Relay	Relay Off: 12v, On: 1v
10	YL/RD	HO2S-21 (B2 S1) Heater	1v (Heater On)
11	BL/BK	HO2S-11 (B1 S1) Heater	1v (Heater On)
12	YL/GN	EGR Solenoid Control (VSV)	12v or 0v
13	BL/RD	EVAP Vapor Pressure Sensor	2.9-3.1v (with hose off)
14	GN/YL	EGR Gas Temp. Sensor	3.5-4.0v
15	WT/GN	EGR Valve Position Sensor	1.1-1.9v
16	BR	Sensor Ground	<0.050v

1997 Sedan 3.0L V6 MFI VIN F W/EIS (A/T-ECT) 22 Pin Connector

PCM Pin #	Wire Color	Circuit Description (22 Pin)	Value at Hot Idle
1	YL	Sensor VREF (VC)	4.9-5.1v
2	BK/YL	A/T-ECT Solenoid (SLN-)	During shifting: 12v
4	WT/BL	OD Clutch Speed Signal (-)	AC pulse signals
5	BK/RD	CKP Sensor Signal (NE+)	AC pulse signals
6	BL	CKP Sensor Signal (NE-)	<0.050v
7	BL	TP Sensor Signal	0.3-0.8v
8	PK	MAF Sensor Signal	1.1-1.5v
9	RD	OD Clutch Speed Signal (+)	AC pulse signals
13	WT	HO2S-11 (B1 S1) Signal	0.1-1.1v
14	WT	Knock Sensor 1 Signal	<0.075v AC
15	WT	Knock Sensor 2 Signal	<0.075v AC
17	BK/WT	CMP Sensor Signal (G+)	AC pulse signals
18	GN/RD	Circuit Opening Relay (FC)	0-3v, off-idle: 12v
19	RD/BL	HO2S-21 (B2 S1) Signal	0.1-1.1v
20	GN/BK	ECT Sensor Signal	At 180°F: 0.51v
21	BL/YL	IAT Sensor Signal	At 100°F: 2.60v
22	BR	Sensor Ground	<0.050v

Pin Connector Graphic

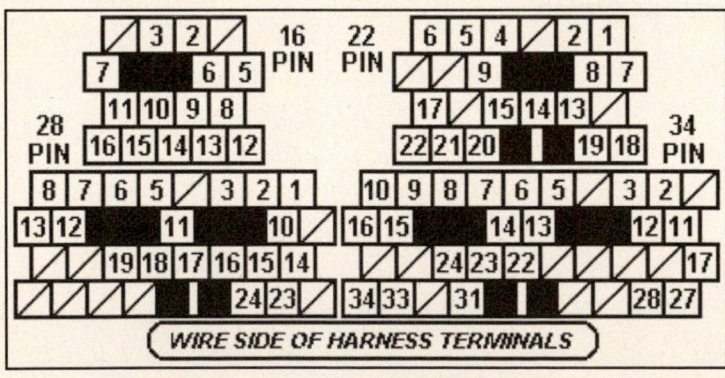

1997 Sedan 3.0L V6 MFI VIN F W/EIS (A/T-ECT) 28 Pin Connector

PCM Pin #	Wire Color	Circuit Description (28 Pin)	Value at Hot Idle
1	YL	A/T Select Switch Low	In Low: 12v, Others: 0v
2	BK/YL	Mirror Heater Switch Signal	Switch On: 12v, Off: 0v
3	GN/OR	Tail Light Switch Signal	Switch On: 12v, Off: 0v
5	GN/BK	A/C Amplifier Signal (ACT)	Clutch On: 12v, Off: 1.5v
6	GN/OR	Overdrive Main Switch	Switch Off: 12v
7	YL/BK	Cruise Control ECU	At Cruise in OD: 12v
8	WT	Data Link Connector	12v
10	BL/WT	A/T Select Switch 2nd	In 2nd: 12v, Others: 0v
11	BL/RD	Cruise Control ECU	At Cruise in OD: 12v
12	PK/WT	Vehicle Speed Sensor	At 55 mph: 48 Hz
13	BK/OR	Tachometer Signal (TACO)	Pulses
14	BK/YL	Battery Direct	12-14v
15	RD/BK	A/T Select Switch Reverse	In 'R': 12v, Others: 0v
16	BK/YL	A/C Amplifier Signal (AC1)	Clutch On: 1.5v, Off: 12v
17	PK/BK	HO2S-12 (B1 S2) Heater	1v (Heater On)
18	BK	HO2S-12 (B1 S2) Signal	0.1-1.1v
19	BR/WT	ABS/Traction ECU (NEO)	Pulse Signals
23	BK/YL	EFI Main Relay Power	12-14v
24	GN/WT	Stop Light Switch Signal	Brake Off: 0v, On: 12v

Pin Connector Graphic

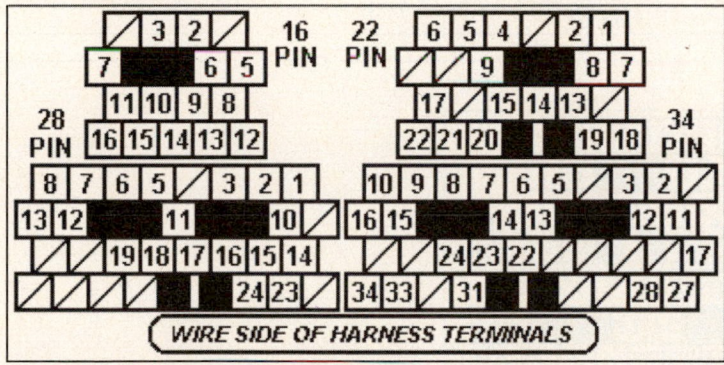

Standard Colors and Abbreviations

Abbreviation	Color	Abbreviation	Color	Abbreviation	Color
BK	Black	GY	Gray	RD	Red
BL	Blue	GN	Green	TN	Tan
BR	Brown	LG	Light Green	VT	Violet
DB	Dark Blue	OR	Orange	WT	White
DG	DK Green	PK	Pink	YL	Yellow

1997 Sedan 3.0L V6 MFI VIN F W/EIS (A/T-ECT) 34 Pin Connector

PCM Pin #	Wire Color	Circuit Description (34 Pin)	Value at Hot Idle
2	BR	Shield Ground	<0.050v
3	WT/BL	A/T-ECT Solenoid (SLN)	During shifting: 12v
5	GN	Injector 6 Control	2.0-3.3 ms
6	RD/BL	Injector 5 Control	2.0-3.3 ms
7	WT	Injector 4 Control	2.0-3.3 ms
8	YL	Injector 3 Control	2.0-3.3 ms
9	RD	Injector 2 Control	2.0-3.3 ms
10	BL	Injector 1 Control	2.0-3.3 ms
11	PK	A/T-ECT Solenoid (S1)	In 3rd or OD: 1v
12	WT/RD	Igniter Signal (IGF)	Digital Signal: 0-5-0v
13	GY	Starter Switch Signal	9-11v (cranking)
14	BK/WT	A/T Neutral Start Switch	In P/N: 9-11v (cranking)
15	GN/BK	Igniter Transistor 3 Control	7% duty cycle
16	BR/YL	Igniter Transistor 2 Control	7% duty cycle
17	BL/BK	A/T-ECT Solenoid (S2)	1st or OD: 1v
22	YL/BK	IAC Signal (RSC)	Pulse Signals
23	RD/WT	IAC Signal (RSO)	Pulse Signals
24	GY	Igniter Transistor 1 Control	7% duty cycle
27	PK/BL	A/T-ECT Solenoid (SL)	In Lockup: 12-14v
28	BR	Power Ground	<0.1v
31	BK/BL	PSP Switch Signal	Straight: 12v, Turning: 0v
33	BR	Power Ground	<0.1v
34	BR	Power Ground	<0.1v

Pin Connector Graphic

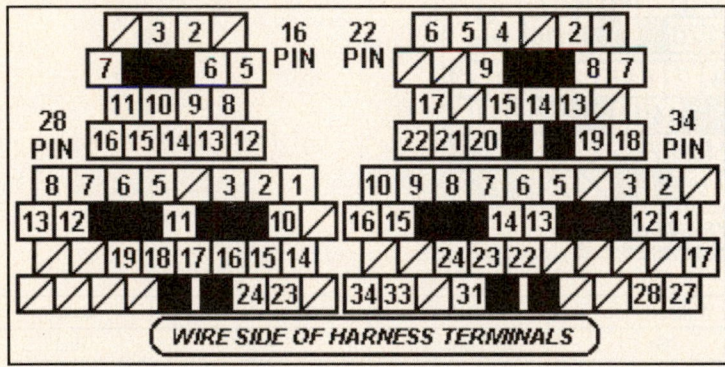

WIRE SIDE OF HARNESS TERMINALS

Standard Colors and Abbreviations

Abbreviation	Color	Abbreviation	Color	Abbreviation	Color
BK	Black	GY	Gray	RD	Red
BL	Blue	GN	Green	TN	Tan
BR	Brown	LG	Light Green	VT	Violet
DB	Dark Blue	OR	Orange	WT	White
DG	DK Green	PK	Pink	YL	Yellow

1998-99 Sedan 3.0L V6 VIN F W/EIS-TC (A/T) 17 Pin Connector

PCM Pin #	Wire Color	Circuit Description (17 Pin)	Value at Hot Idle
4	GN/WT	Transponder Amplifier Code	Inserting key: pulses
5	RD/BL	Transponder Amplifier Signal	Inserting key: pulses
10	BL/YL	Transponder Amplifier Signal	Inserting key: pulses
11	BL/BK	Unlock Warning Switch	No Key: 4-5v
13	BR	Sensor Data Link Ground	<0.050v
16	RD/YL	Theft Deterrent ECU	Pulse Signals

1998-99 Sedan 3.0L V6 VIN F W/EIS-TC (A/T) 22 Pin Connector

PCM Pin #	Wire Color	Circuit Description (22 Pin)	Value at Hot Idle
1	BK/YL	Battery Direct	12-14v
2	BK/RD	Ignition Switch Power	12-14v
3	GN/RD	Circuit Opening Relay (FC)	0-3v, off-idle: 12v
6	GN/RD	MIL (lamp) Control	MIL Off: 12v, On: 1v
7	GY	Starter Switch Signal	9-11v (cranking)
7	BK/OR	Starter Switch Signal	9-11v (cranking)
8	BK/WT	EFI Main Relay Power	12-14v
9	WT/RD	EVAP Vapor Pressure (VSV)	12v or 0v
9	PK	EVAP Vapor Pressure (VSV)	12v or 0v
11	WT	SIL Signal (Scan Tool)	Digital Signal
13	LG	ABS/Traction ECU (TRC+)	Pulse Signals
14	WT	ABS/Traction ECU (EFI+)	Pulse Signals
15	GN/WT	Stop Light Switch Signal	Brake Off: 0v, On: 12v
16	BK/YL	EFI Main Relay Power	12-14v
17 (TMC)	BL/RD	EVAP Vapor Pressure Sensor	2.9-3.1v (with hose off)
17 (TMM)	PK	EVAP Vapor Pressure Sensor	2.9-3.1v (with hose off)
18	BK/YL	Mirror Heater Switch Signal	Switch On: 12v, Off: 0v
19	GN/OR	Tail Light Switch Signal	Switch On: 12v, Off: 0v
20	BL	ABS/Traction ECU (TRC-)	Pulse Signals
21	WT	ABS/Traction ECU (EFI-)	Pulse Signals

1998-99 Sedan 3.0L V6 VIN F W/EIS-TC (A/T) 28 Pin Connector

PCM Pin #	Wire Color	Circuit Description (28 Pin)	Value at Hot Idle
2	RD/BK	A/T Select Switch Reverse	In 'R': 12v, Others: 0v
3 (TMC)	BL/WT	A/T Select Switch 2nd	In 2nd: 12v, Others: 0v
3 (TMM)	OR	A/T Select Switch 2nd	In 2nd: 12v, Others: 0v
4	BL/RD	Cruise Control Signal (IDLO)	1.5v, off-idle: 12v
8	BK	HO2S-12 (B1 S2) Signal	0.1-1.1v
9	PK/BK	HO2S-12 (B1 S2) Heater	1v (Heater On)
10	GN/OR	Overdrive Main Switch	Switch Off: 12v
12	YL	A/T Select Switch Low	In Low: 12v, Others: 0v
13	GN/BK	A/C Amplifier Signal (ACT)	Clutch On: 12v, Off: 1.5v
14	PK	A/C Amplifier Signal (THWO)	Pulse Signals
16	BR/WT	ABS/Traction ECU (NEO)	Pulse Signals
20	BK/WT	Neutral Start Switch	In P/N: 9-11v (cranking)
22	PK/WT	Vehicle Speed Sensor	At 55 mph: 48 Hz
24	YL/BK	Cruise Control Signal (OD1)	At Cruise in OD: 12v
25	BK/YL	A/C Amplifier Signal (AC1)	Clutch On: 1.5v, Off: 12v
27	BK/OR	Tachometer Signal (TACO)	Pulses

1998-99 Sedan 3.0L V6 VIN F W/EIS-TC (A/T) 24 Pin Connector

PCM Pin #	Wire Color	Circuit Description (24 Pin)	Value at Hot Idle
1	BR	Sensor Ground	<0.050v
2	YL	Sensor VREF (VC)	4.9-5.1v
3 (Cal)	BK/RD	AFS-11 (B1 S1) Heater	1v (Heater On)
3 (Fed)	BL/BK	HO2S-11 (B1 S1) Heater	1v (Heater On)
4 (Cal)	BK/WT	AFS-21 (B2 S1) Heater	1v (Heater On)
4 (Fed)	YL/RD	HO2S-21 (B2 S1) Heater	1v (Heater On)
5	BL	Injector 1 Control	2.0-3.3 ms
6	RD	Injector 2 Control	2.0-3.3 ms
7	LG	EVAP Purge Solenoid (VSV)	12v or 0v
8	BR	Power Ground	<0.1v
9	BK/BL	PSP Switch Signal	Straight: 12v, Turning: 0v
10	PK	MAF Sensor Signal	1.1-1.5v
11 (TCM)	GN	AFS-11 (B1 S1) Signal (AF+)	3.0-3.6v
11 (TMM)	BR	AFS-11 (B1 S1) Signal (AF+)	3.0-3.6v
11 (Fed)	WT	HO2S-11 (B1 S1) Signal	0.1-1.1v
12 (TMC)	BK/WT	AFS-21 (B2 S1) Signal (AF+)	3.0-3.6v
12 (TMM)	BL	AFS-21 (B2 S1) Signal (AF+)	3.0-3.6v
12 (Fed)	BK	HO2S-21 (B2 S1) Signal	0.1-1.1v
13	GN/YL	EGR Gas Temp. Sensor	3.5-4.0v
14	GN/BK	ECT Sensor Signal	At 180ºF: 0.51v
16	BK/RD	CKP Sensor Signal (NE+)	AC pulse signals
17	BR	Shield Ground	<0.050v
18	BR	Sensor Ground	<0.050v
19	RD/BK	MAF Sensor Ground	<0.050v
20 (TMC)	BK/RD	AFS-11 (B1 S1) Signal (AF-)	Fixed at 3v
20 (TMM)	RD	AFS-11 (B1 S1) Signal (AF-)	Fixed at 3v
21 (TMC)	BK/WT	AFS-21 (B2 S1) Signal (AF-)	Fixed at 3v
21 (TMM)	BL	AFS-21 (B2 S1) Signal (AF-)	Fixed at 3v
22	BL/YL	IAT Sensor Signal	At 100ºF: 2.60v
23	BL	TP Sensor Signal	0.3-1.1v
24	BL	CKP/CMP Sensor Ground (-)	<0.050v

Pin Connector Graphic

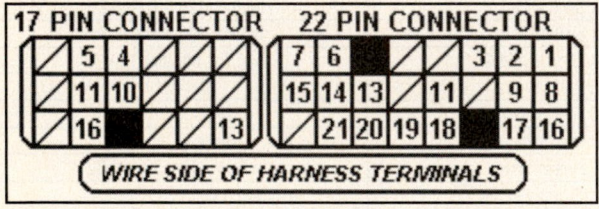

1998-99 Sedan 3.0L V6 VIN F W/EIS-TC (A/T) 31 Pin Connector

PCM Pin #	Wire Color	Circuit Description (31 Pin)	Value at Hot Idle
1	YL	Injector 3 Control	2.0-3.3 ms
2	WT	Injector 4 Control	2.0-3.3 ms
3	RD/BL	Injector 5 Control	2.0-3.3 ms
4	GN	Injector 6 Control	2.0-3.3 ms
6	BL/WT	Data Link Connector (TE)	12-14v
7	PK	A/T-ECT Solenoid (S1)	In 3rd or OD: 1v
8	BL/BK	A/T-ECT Solenoid (S2)	1st or OD: 1v
9	PK/BL	A/T-ECT Solenoid (SL)	In Lockup: 12-14v
10	BK/WT	CMP Sensor Signal (G+)	AC pulse signals
11	GY	Igniter Transistor 1 Control	7% duty cycle
12	BR/YL	Igniter Transistor 2 Control	7% duty cycle
13	GN/BK	Igniter Transistor 3 Control	7% duty cycle
14	RD	OD Clutch Speed Signal (+)	AC pulse signals
15	YL/BK	IAC Signal (RSC)	Pulse Signals
16	RD/WT	IAC Signal (RSO)	Pulse Signals
17	RD/YL	ACIS Solenoid Control (VSV)	12v or 0v
18	YL/GN	EGR Solenoid Control (VSV)	12v or 0v
19	BK/YL	A/T-ECT Solenoid (SLN-)	During shifting: 12v
20	WT/BL	A/T-ECT Solenoid (SLN+)	During shifting: 12v
21	BR	Power Ground	<0.1v
22	WT/GN	EGR Valve Position Sensor	0.4-1.6v
23	BR	A/T: Sensor Ground	<0.050v
24 (Fed)	BR	M/T: Sensor Ground	<0.050v
25	WT/RD	Igniter Signal (IGF)	Digital Signal: 0-5-0v
26	GN	OD Clutch Speed Signal (-)	AC pulse signals
27	WT	Knock Sensor 1 Signal	<0.075v AC
28	WT	Knock Sensor 2 Signal	<0.075v AC
29	GN/WT	Cooling Fan 1 Relay	Relay Off: 12v, On: 1v
30	WT/BK	Power Ground	<0.1v
31	WT/BK	Power Ground	<0.1v

Pin Connector Graphic

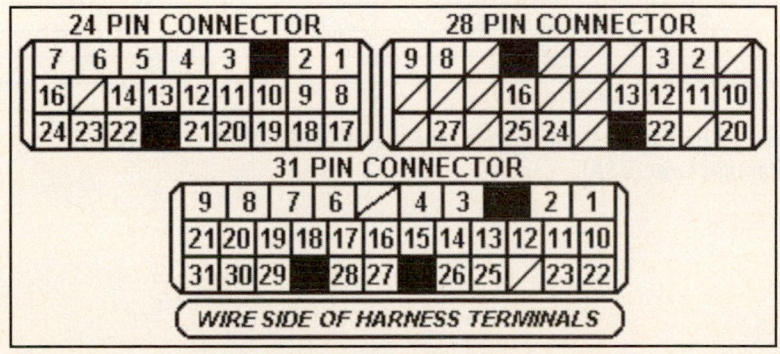

1998 Sedan 3.0L V6 VIN F W/O EIS-TC Federal 28 Pin Connector

PCM Pin #	Wire Color	Circuit Description (28 Pin)	Value at Hot Idle
1	YL	A/T Select Switch Low	In Low: 12v, Others: 0v
2	GN/OR	Tail Light Switch Signal	Switch On: 12v, Off: 0v
3	BK/YL	Mirror Heater Switch Signal	Switch On: 12v, Off: 0v
5	GN/BK	A/C Amplifier Signal (ACT)	Clutch On: 12v, Off: 1.5v
6	GN/OR	Overdrive Main Switch	Switch Off: 12v
7	YL/BK	Cruise Control Signal (OD1)	At Cruise in OD: 12v
8	WT	Data Link Connector	No Scan Tool: 0v
10 (TMC)	BL/WT	A/T Select Switch 2nd	In 2nd: 12v, Others: 0v
10 (TMM)	OR	A/T Select Switch 2nd	In 2nd: 12v, Others: 0v
11	BL/RD	Cruise Control Signal (IDLO)	1.5v, off-idle: 12v
12	PK/WT	Vehicle Speed Sensor	At 55 mph: 48 Hz
13	BK/OR	Tachometer Signal (TACO)	Pulses
14	BK/YL	Battery Direct	12-14v
15	RD/BK	A/T Select Switch Reverse	In 'R': 12v, Others: 0v
16	BK/YL	A/C Amplifier Signal (AC1)	Clutch On: 1.5v, Off: 12v
17	PK/BK	HO2S-12 (B1 S2) Heater	1v (Heater On)
18	BK	HO2S-12 (B1 S2) Signal	0.1-1.1v
23	BK/YL	EFI Main Relay Power	12-14v
24	GN/WT	Stop Light Switch Signal	Brake Off: 0v, On: 12v

1998 Sedan 3.0L V6 VIN F W/O EIS-TC Federal 16 Pin Connector

PCM Pin #	Wire Color	Circuit Description (16 Pin)	Value at Hot Idle
2	LG	EVAP Purge Solenoid (VSV)	12v or 0v
3	GN/RD	MIL (lamp) Control	MIL Off: 12v, On: 1v
5	BL/WT	Data Link Connector	12-14v
6	RD/YL	Intake Air Control Solenoid	12v or 0v
7	RD/BK	MAF Sensor Ground	<0.050v
8	WT/RD	EVAP Vapor Pressure (VSV)	12v or 0v
9	GN/WT	Cooling Fan 1 Relay	Relay Off: 12v, On: 1v
10	YL/RD	HO2S-21 (B2 S1) Heater	1v (Heater On)
11	BL/BK	HO2S-11 (B1 S1) Heater	1v (Heater On)
12	YL/GN	EGR Solenoid Control (VSV)	12v or 0v
13	BL/RD	EVAP Vapor Pressure Sensor	2.9-3.1v (with hose off)
14	GN/YL	EGR Gas Temp. Sensor	3.5-4.0v
15	WT/GN	EGR Valve Position Sensor	1.1-1.9v
16	BR	Sensor Ground	<0.050v

Note: *TMC indicates a vehicle produced by Toyota Motor Corp. (Japan) and TMM indicates a vehicle produced by Toyota Motor Manufacturer (USA).*

1998 Sedan 3.0L V6 VIN F W/O EIS-TC Federal 22 Pin Connector

PCM Pin #	Wire Color	Circuit Description (22 Pin)	Value at Hot Idle
1	YL	Sensor VREF (VC)	4.9-5.1v
2	BK/YL	A/T-ECT Solenoid (SLN-)	During shifting: 12v
4	GN	OD Clutch Speed Signal (-)	AC pulse signals
5	BK/RD	CKP Sensor Signal (NE+)	AC pulse signals
6	BL	CKP Sensor Signal (NE-)	<0.050v
7	BL	TP Sensor Signal	0.3-0.8v
8	PK	MAF Sensor Signal	1.1-1.5v
9	RD	OD Clutch Speed Signal (+)	AC pulse signals
13	WT	HO2S-11 (B1 S1) Signal	0.1-1.1v
14	WT	Knock Sensor 2 Signal	<0.075v AC
15	WT	Knock Sensor 1 Signal	<0.075v AC
17	BK/WT	CMP Sensor Signal (G+)	AC pulse signals
18	GN/RD	Circuit Opening Relay (FC)	0-3v, off-idle: 12v
19	BK	HO2S-21 (B2 S1) Signal	0.1-1.1v
20	GN/BK	ECT Sensor Signal	At 180ºF: 0.51v
21	BL/YL	IAT Sensor Signal	At 100ºF: 2.60v
22	BR	Sensor Ground	<0.050v

Pin Connector Graphic

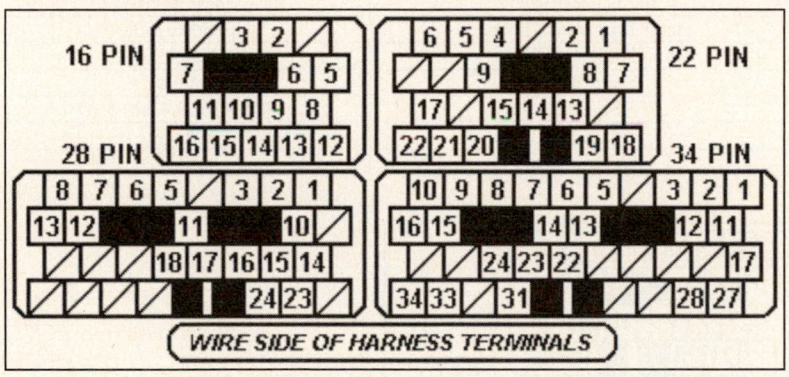

WIRE SIDE OF HARNESS TERMINALS

1998 Sedan 3.0L V6 VIN F W/O EIS-TC Federal 28 Pin Connector

PCM Pin #	Wire Color	Circuit Description (28 Pin)	Value at Hot Idle
1	BR	A/T: Sensor Ground	<0.1v
2	BR	M/T: Sensor Ground	<0.1v
3	WT/BL	A/T-ECT Solenoid (SLN+)	During shifting: 12v
5	GN	Injector 6 Control	2.0-3.3 ms
6	RD/BL	Injector 5 Control	2.0-3.3 ms
7	WT	Injector 4 Control	2.0-3.3 ms
8	YL	Injector 3 Control	2.0-3.3 ms
9	RD	Injector 2 Control	2.0-3.3 ms
10	BL	Injector 1 Control	2.0-3.3 ms
11	PK	A/T-ECT Solenoid (S1)	In 3rd or OD: 1v
12	WT/RD	Igniter Signal (IGF)	Digital Signal: 0-5-0v
13	GY	Starter Switch Signal	9-11v (cranking)
13	BK/OR	Starter Switch Signal	9-11v (cranking)
14	BK/WT	A/T Neutral Start Switch	In P/N: 9-11v (cranking)
15	GN/BK	Igniter Transistor 3 Control	7% duty cycle
16	BR/YL	Igniter Transistor 2 Control	7% duty cycle
17	BL/BK	A/T-ECT Solenoid (S2)	1st or OD: 1v
22	YL/BK	IAC Signal (RSC)	Pulse Signals
23	RD/WT	IAC Signal (RSO)	Pulse Signals
24	GY	Igniter Transistor 1 Control	7% duty cycle
27	PK/BL	A/T-ECT Solenoid (SL)	In Lockup: 12-14v
28	BR	Power Ground	<0.1v
31	BK/BL	PSP Switch Signal	Straight: 12v, Turning: 0v
33	BR	Power Ground	<0.1v
34	BR	Power Ground	<0.1v

Pin Connector Graphic

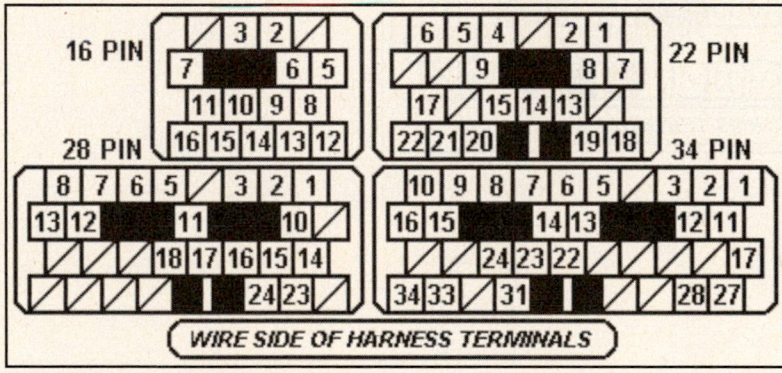

Standard Colors and Abbreviations

Abbreviation	Color	Abbreviation	Color	Abbreviation	Color
BK	Black	GY	Gray	RD	Red
BL	Blue	GN	Green	TN	Tan
BR	Brown	LG	Light Green	VT	Violet
DB	Dark Blue	OR	Orange	WT	White
DG	DK Green	PK	Pink	YL	Yellow

2000-01 Sedan 3.0L V6 VIN F W/EIS-TC (A/T) E11 31P Connector

PCM Pin #	Wire Color	Circuit Description (31 Pin)	Value at Hot Idle
1	YL	Injector 3 Control	1.6-2.9 ms
2	WT	Injector 4 Control	1.6-2.9 ms
3	RD/BL	Injector 5 Control	1.6-2.9 ms
4	GN	Injector 6 Control	1.6-2.9 ms
6	BL/WT	Data Link Connector (TE1)	12-14v
7	VT	A/T-ECT Solenoid (S1)	In 3rd or OD: 1v
8	BL/BK	A/T-ECT Solenoid (S2)	1st or OD: 1v
9	PK/BL	A/T-ECT Solenoid (SL)	In Lockup: 12-14v
10	BK/WT	CMP Sensor Signal (G+)	AC pulse signals
11	GY	Igniter Transistor 1 Control	7% duty cycle
12	BR/YL	Igniter Transistor 2 Control	7% duty cycle
13	GN/BK	Igniter Transistor 3 Control	7% duty cycle
14	RD	OD Clutch Speed Signal (+)	AC pulse signals
15	YL/BK	IAC Signal (RSC)	Pulse Signals
16	RD/WT	IAC Signal (RSO)	Pulse Signals
17	RD/YL	ACIS Solenoid Control (VSV)	12v or 0v
18	YL/GN	EGR Solenoid Control (VSV)	12v or 0v
19	BK/YL	A/T-ECT Solenoid (SLN-)	During shifting: 12v
20	WT/BL	A/T-ECT Solenoid (SLN+)	During shifting: 12v
21	BR	Power Ground (E01)	<0.1v
22	WT/GN	EGR Valve Position Sensor	0.4-1.6v
23	GN/OR	Overdrive (OD) Main Switch	Switch Off: 12v, On: 1v
25	WT/RD	Igniter Signal (IGF)	Digital Signal: 0-5-0v
26	GN	OD Clutch Speed Signal (-)	AC pulse signals
27	WT	Knock Sensor 1 Signal	<0.075v AC
28	WT	Knock Sensor 2 Signal	<0.075v AC
29	GN/WT	Cooling Fan 1 Relay	Relay Off: 12v, On: 1v
30	WT/BK	Power Ground (E03)	<0.1v
31	WT/BK	Power Ground (E02)	<0.1v

Pin Connector Graphic

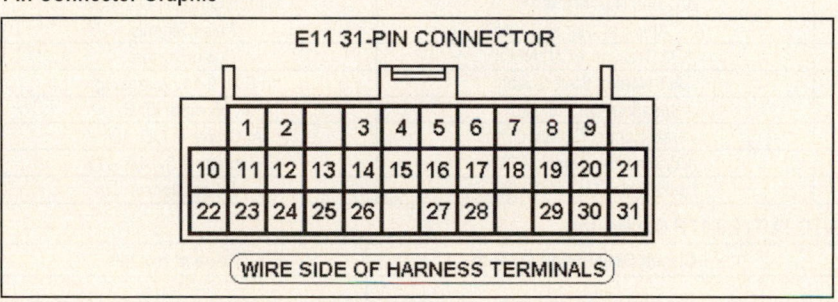

2000-01 Sedan 3.0L V6 VIN F W/EIS-TC (A/T) E7 22P Connector

PCM Pin #	Wire Color	Circuit Description (22 Pin)	Value at Hot Idle
1	BK/YL	Battery Direct	12-14v
2	BK/RD	Ignition Switch Power	12-14v
3	GN/RD	Circuit Opening Relay (FC)	0-3v, off-idle: 12v
6	GN/RD	MIL (lamp) Control	MIL Off: 12v, On: 1v
7 (TMC)	GY	Starter Switch Signal	9-11v (cranking)
7 (TMM)	BK/OR	Starter Switch Signal	9-11v (cranking)
8	BK/RD	EFI Main Relay Power	12-14v
9	PK	Overdrive "Off" Indicator	LED Off: 12v, On: 1v
11	WT	SIL Signal (Scan Tool)	Digital Signal
13	LG	ABS/Traction ECU (TRC+)	Pulse Signals
14	WT	ABS/Traction ECU (EFI+)	Pulse Signals
15	GN/WT	Stop Light Switch Signal	Brake Off: 0v, On: 12v
16	BK/YL	EFI Main Relay Power	12-14v
17	PK	EVAP Vapor Pressure Sensor	2.9-3.1v (with hose off)
18	BK/YL	Mirror Heater Switch Signal	Switch Off: 1.5v, On: 12v
19	GN/OR	Tail Light Switch Signal	Switch Off: 1.5v, On: 12v
20	BL	ABS/Traction ECU (TRC-)	Pulse Signals
21	BK	ABS/Traction ECU (EFI-)	Pulse Signals

2000-01 Sedan 3.0L V6 VIN F W/EIS-TC (A/T) E8 28P Connector

Q	Wire Color	Circuit Description (28 Pin)	Value at Hot Idle
1	WT/RD	EVAP Vapor Pressure (VSV)	12v or 0v
2	RD/BK	A/T Select Switch Reverse	In 'R': 12v, Others: 0v
3 (TMC)	BL/WT	A/T Select Switch 2nd	In 2nd: 12v, Others: 0v
3 (TMM)	OR	A/T Select Switch 2nd	In 2nd: 12v, Others: 0v
4	BL/RD	Cruise Control Signal (IDLO)	1.5v, off-idle: 12v
5	BL/WT	EVAP Canister Closed Valve	12v or 0v
8	BK	HO2S-12 (B1 S2) Signal	0.1-1.1v
9	PK/GN	HO2S-12 (B1 S2) Heater	1v (Heater On)
12	YL	A/T Select Switch Low	In Low: 12v, Others: 0v
13	GN/BK	A/C Amplifier Signal (ACT)	Clutch On: 12v, Off: 1.5v
14	VT	A/C Amplifier Signal (THWO)	Pulse Signals
16	BR/WT	ABS/Traction ECU (NEO)	Pulse Signals
20	BK/WT	A/T Neutral Start Switch	In P/N: 9-11v (cranking)
22	PK/WT	Speedometer Indicator	At 55 mph: 48 Hz
24	YL/BK	Cruise Control Signal (OD1)	At Cruise in OD: 12v
25	BK/YL	A/C Amplifier Signal (AC1)	Clutch On: 1.5v, Off: 12v
27	BK/OR	Tachometer Signal (TACO)	Pulse Signals

2000-01 Sedan 3.0L V6 VIN F W/EIS-TC (A/T) E9 17P Connector

PCM Pin #	Wire Color	Circuit Description (17 Pin)	Value at Hot Idle
1-3, 6-9	---	Not Used	---
4	GN/WT	Transponder Amplifier Code	Inserting key: pulses
5	RD/BL	Transponder Amplifier Signal	Inserting key: pulses
12, 15-16	---	Not Used	---
10	BL/YL	Transponder Amplifier Signal	Inserting key: pulses
11	BL/BK	Unlock Warning Switch	Inserting key: pulses
13	BR	Power Ground (EOM)	<0.1v
16	RD/YL	Theft Deterrent ECU Signal	No key inserted: pulses

2000-01 Sedan 3.0L V6 VIN F W/EIS-TC (A/T) E10 24P Connector

PCM Pin #	Wire Color	Circuit Description (24 Pin)	Value at Hot Idle
1 (TMC)	BR	Sensor Ground (E04)	<0.050v
2	YL	Sensor VREF (VC)	4.9-5.1v
3 (Cal)	BK/RD	AFS-11 (B1 S1) Heater	1v (Heater On)
3 (Fed)	BL/BK	HO2S-11 (B1 S1) Heater	1v (Heater On)
4	YL/RD	HO2S-21 (B2 S1) Heater	1v (Heater On)
5	BL	Injector 1 Control	1.6-2.9 ms
6	RD	Injector 2 Control	1.6-2.9 ms
7	LG	EVAP Purge Solenoid (VSV)	12v or 0v
8	BR	Power Ground (E05)	<0.1v
9	BK/BL	PSP Switch Signal	Straight: 12v, Turning: 0v
10	PK	MAF Sensor Signal	1.1-1.5v
11 (Cal)	BR	AFS-11 (B1 S1) Signal (AF+)	3.0-3.6v
12 (Fed)	BK	HO2S-21 (B2 S1) Signal	0.1-1.1v
13	GN/YL	EGR Gas Temp. Sensor	3.5-4.0v
14	GN/BK	ECT Sensor Signal	At 180ºF: 0.51v
16	BK/RD	CKP Sensor Signal (NE+)	AC pulse signals
17	BR	Power Ground (E1)	<0.050v
18	BR	Sensor Ground (E2)	<0.050v
19	RD/BK	MAF Sensor Ground	<0.050v
20 (Cal)	BK/RD	AFS-11 (B1 S1) Signal (AF-)	Fixed at 3v
22	BL/YL	IAT Sensor Signal	At 100ºF: 2.60v
23	BL	TP Sensor Signal	0.3-1.1v
24	BL	CKP/CMP Sensor Ground (-)	<0.050v

Pin Connector Graphic

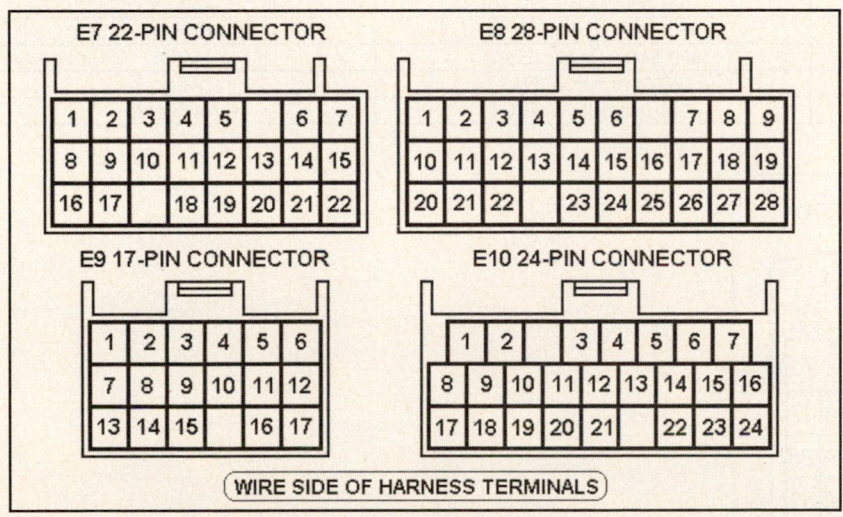

Note: **TMC indicates a vehicle produced by Toyota Motor Corp. (Japan) and TMM indicates a vehicle produced by Toyota Motor Manufacturer (USA).**

2002-03 Sedan 3.0L V6 VIN F W/EIS-TC (A/T) E6 31P Connector

PCM Pin #	Wire Color	Circuit Description (31 Pin)	Value at Hot Idle
1	BK/RD	EFI Main Relay Power	12-14v
2-3	---	Not Used	---
4	VT	EVAP Vapor Pressure (VSV)	12v or 0v
5	BK/OR	Tachometer Signal (TACO)	DC pulse signals
6-7	---	Not Used	---
8	BK/WT	EFI Main Relay Power	12-14v
9	BK/OR	Ignition Switch Power (B+)	12-14v
10	GN/RD	Circuit Opening Relay (FC)	0-3v, off-idle: 12v
11	PK/BK	Data Link Connector (TC)	12-14v
12	GN/RD	MIL (lamp) Control	MIL Off: 12v, On: 1v
13	GN	Tail Light Switch Signal	Switch Off: 0v, On: 12v
14	BK/YL	Mirror Heater Switch Signal	Switch Off: 1.5v, On: 12v
15	BR	Power Ground (EOM)	<0.1v
16	GN/OR	Overdrive (OD) Main Switch	Switch Off: 12v, On: 1v
17	BR/WT	(NEO)	---
18	WT	SIL Signal (Scan Tool)	Digital Signal
19	RD	Scan Tool (WFSE)	12v
20	---	Not Used	---
21	PK	EVAP Vapor Pressure Sensor	2.9-3.1v (with hose off)
22	BL/YL	APP Sensor Signal (VPA)	1.1-1.5v
23	WT/RD	APP Sensor Signal (VPA2)	0.9-2.3v
24	WT	ABS/Traction ECU (ENG+)	Pulse Signals
25	GN	ABS/Traction ECU (TRC+)	Pulse Signals
26	RD	APP Sensor VREF (VCPA)	4.5-5.5v
27	BR/RD	APP Sensor VREF (VCP2)	4.5-5.5v
28	LG/BK	APP Sensor Ground (EPA)	<0.050v
29	LG	APP Sensor Ground (EPA2)	<0.050v
30	BK	ABS/Traction ECU (ENG-)	Pulse Signals
31	BL	ABS/Traction ECU (TRC-)	Pulse Signals

Pin Connector Graphic

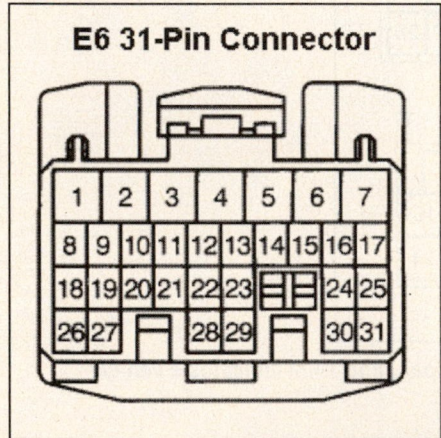

Note: TMC indicates a vehicle produced by Toyota Motor Corp. (Japan) and TMM indicates a vehicle produced by Toyota Motor Manufacturer (USA).

2002-03 Sedan 3.0L V6 VIN F W/EIS-TC (A/T) E7 35P Connector

PCM Pin #	Wire Color	Circuit Description (35 Pin)	Value at Hot Idle
1, 4-5	---	Not Used	---
2	BK/YL	Battery Direct (BATT)	12-14v
3	BL/WT	A/C Magnetic Clutch (ACMG)	A/C On: <3.0v
6	BL/RD	Battery Direct (+BM)	12-14v
7	OR	Overdrive "Off" Indicator	LED Off: 12v, On: 1v
8	YL	A/T Select Switch Low Signal	In Low: 12v, Others: 0v
9	BL/WT	A/T Select Switch 2nd Signal	In 2nd: 12v, Others: 0v
10	WT/BL	A/T Select Switch 'D' Signal	In 'D': 12v, Others: 0v
11	RDBK	A/T Select Switch 'R' Signal	In 'R': 12v, Others: 0v
12-13	---	Not Used	---
14	YL/GN	A/C Amplifier Signal (THWO)	Pulse Signals
15	GN/WT	Transponder Amplifier Code	Inserting key: pulses
16	VT	Transponder IMLD Signal	Inserting key: pulses
17	VT/WT	Speedometer Indicator	Moving: 0-5-0v
18, 20-25	---	Not Used	---
19	GN/WT	Stop Light Switch Signal	Brake Off: 0v, On: 12v
26	BL/YL	Theft Deterrent ECU Signal	No key inserted: pulses
27	RD/BL	Transponder Amplifier Signal	Inserting key: pulses
28-30	---	Not Used	---
31	PK/BL	A/C Switch Signal (Auto AC)	A/C On: 9-14v
31	WT	A/C Switch Signal (Manual AC)	A/C On: 9-14v
32, 35	---	Not Used	---
33	BK	A/C Switch Signal (Auto AC)	A/C On: 9-14v
33	YL/BK	A/C Switch Signal (Manual AC)	A/C On: 9-14v
34	BL	Transponder Amplifier Signal	Key Inserted: <1.5v

2002-03 Sedan 3.0L V6 VIN F W/EIS-TC (A/T) E8 32P Connector

PCM Pin #	Wire Color	Circuit Description (32 Pin)	Value at Hot Idle
1	BR	Power Ground (E1)	<0.050v
2	WT	Throttle Body Motor (M-)	DC pulse signals
3	BK	Throttle Body Motor (M+)	DC pulse signals
4	WT/BL	Power Ground (ME01)	<0.1v
5	YL	HO2S-12 (B1 S2) Heater	1v (Heater On)
6	BL	HO2S-22 (B2 S2) Heater	1v (Heater On)
7	WT/BK	Power Ground (E03)	<0.1v
8	GN/WT	Cooling Fan Relay Control	Relay Off: 12v, On: 1v
9, 13-16	---	Not Used	---
10	RD/WT	PSP Switch Signal	Straight: 12v, Turning: 0v
11	LG	EGR Solenoid Control (VSV)	12v or 0v
12	BL	EVAP Canister Closed Valve	12v or 0v
17	BR	Shield Ground (GE01)	<0.050v
18-26	---	Not Used	---
27	BK/WT	CMP Sensor Signal (G22+)	AC pulse signals
28-32	---	Not Used	---

Pin Connector Graphic

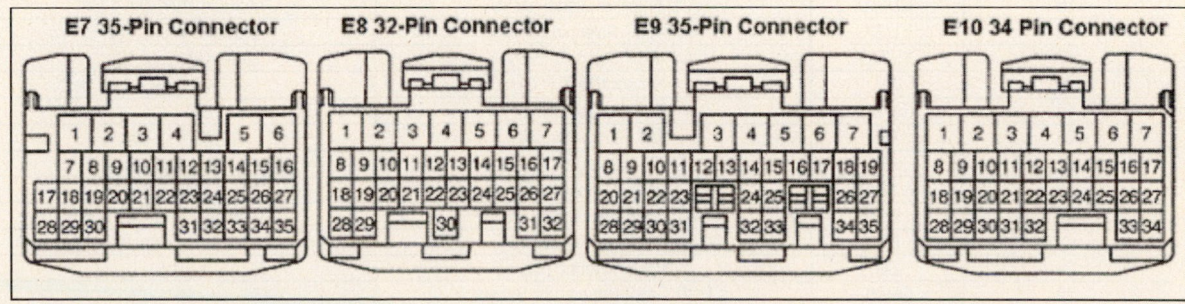

2002-03 Sedan 3.0L V6 VIN F W/EIS-TC (A/T) E9 35P Connector

PCM Pin #	Wire Color	Circuit Description (35 Pin)	Value at Hot Idle
1	BK	Knock Sensor 1 Signal	<0.075v AC
2	WT	Knock Sensor 2 Signal	<0.075v AC
3	BK/RD	AFS-21 (B2 S1) Heater	1v (Heater On)
4	BK/WT	AFS-11 (B1 S1) Heater	1v (Heater On)
5	GN	Injector 6 Control	1.6-2.9 ms
6	WT/BK	Power Ground (E05)	<0.1v
7	WT/BK	Sensor Ground (E04)	<0.050v
8	BK/YL	Neutral Start Switch	In P/N: 9-11v (cranking)
9	BK/WT	Starter Switch Signal (STA)	9-11v (cranking)
10-12	---	Not Used	---
13	BL	A/T-ECT Overdrive Solenoid	Moving in OD: 12-14v
14-15, 25	---	Not Used	---
16	BL/RD	A/T-ECT Solenoid (SL2)	Moving in 3rd or OD: 1v
17	BL/YL	A/T-ECT Solenoid (SL2+)	Moving in 3rd or OD: 1v
18	PK	A/T-ECT Solenoid (SL1-)	Moving in 1st Gear: 1v
19	RD/BK	A/T-ECT Solenoid (SL1+)	Moving in 1st Gear: 1v
20	YL/GN	EGR Solenoid Control (VSV)	12v or 0v
21	WT	HO2S-11 (B1 S1) Signal	0.1-1.1v
22	BR	AFS-11 (B1 S1) Signal (AFR+)	3.0-3.6v
23	OR	AFS-21 (B2 S1) Signal (AFL)	3.0-3.6v
24	RD	MAF Sensor Signal (VG)	1.1-1.5v
26	RD	Clutch Speed Signal (NC+)	AC pulse signals
27	Rd	Clutch Speed Signal (NT+)	AC pulse signals
28	WT/GN	EGR Valve Position Sensor	0.4-1.6v
29	OR	HO2S-22 (B2 S2) Signal	0.1-1.1v
30	BK/RD	AFS-11 (B1 S1) Signal (AF-)	Fixed at 3v
31	WT	AFS-11 (B1 S1) Signal (AF-)	Fixed at 3v
32	BL/WT	MAF Sensor Ground (EG2)	<0.050v
34	GN	Clutch Speed Signal (NC-)	AC pulse signals
35	BL	Clutch Speed Signal (NT-)	AC pulse signals

2002-03 Sedan 3.0L V6 VIN F W/EIS-TC (A/T) E10 34P Connector

PCM Pin #	Wire Color	Circuit Description (34 Pin)	Value at Hot Idle
1	BL	Injector 1 Control	2.0-3.3 ms
2	RD	Injector 2 Control	2.0-3.3 ms
3	YL	Injector 3 Control	2.0-3.3 ms
4	WT	Injector 4 Control	2.0-3.3 ms
5	RD/BL	Injector 5 Control	1.6-2.9 ms
6	WT/BK	Power Ground (E02)	<0.050v
7	WT/BK	Power Ground (E01)	<0.050v
8	RD/WT	Igniter Transistor 1 Control	7% duty cycle
9	PK	Igniter Transistor 2 Control	7% duty cycle
10	LG/BK	Igniter Transistor 3 Control	7% duty cycle
11	BL/YL	Igniter Transistor 4 Control	7% duty cycle
12	GN/RD	Igniter Transistor 5 Control	7% duty cycle
13	BL	Igniter Transistor 6 Control	7% duty cycle
14, 22, 26	---	Not Used	---
15	RD/YL	A/C Amplifier Signal (ACIS)	Clutch On: 12v, Off: 1.5v
16	YL/BK	A/T-ECT Solenoid (STN-)	During Shifting: 12v
17	YL/RD	A/T-ECT Solenoid (STN+)	During shifting: 12v
18	YL	Sensor VREF (VC)	4.9-5.1v
19	GN/BK	ECT Sensor Signal	At 180ºF: 0.51v
20	LB	IAT Sensor Signal	At 100ºF: 2.60v
21	LG	TP Sensor Signal (VTA1)	0.3-1.1v
23	WT/RD	Igniter Signal (IGF)	Digital Signal: 0-5-0v
25	WT	A/C Amplifier Signal (AICV)	Clutch On: 1.5v, Off: 12v
27	RD	CKP Sensor Signal (NE+)	AC pulse signals
28	BR	Power Ground (E2)	<0.1v
29	GN/YL	EGR Gas Temperature Sensor	---
30 (TMC)	GN	A/T ATF Sensor Signal	<1.5v at 239ºF
30 (TMM)	GN/RD	A/T ATF Sensor Signal	<1.5v at 239ºF
31	BK/RD	TP Sensor 2 Signal (VTA2)	2.0-2.9v
32-33	---	Not Used	---
34	GN	CKP/CMP Sensor Ground (-)	<0.050v

CELICA Pin Tables

1990-93 Hatchback Turbo 2.0L MFI VIN S (M/T) 16 Pin Connector

PCM Pin #	Wire Color	Circuit Description (16 Pin)	Value at Hot Idle
1	PK/BL	Sensor VREF	0-5v
2	GY/BK	Airflow Meter Signal	1.5-2.5v
3	GY	IAT Sensor Signal	At 100°F: 2.60v
4	RD	ECT Sensor Signal	At 180°F: 0.51v
5 ('90-'91)	WT	Knock Sensor Signal	<0.075v AC
5 ('92-'93)	RD/WT	Turbo Pressure Sensor	2.5-4.5v
6	WT	O2S Signal	0.1-1.1v
8	PK/YL	Check Connector	12-14v
9	BR	Sensor Ground	<0.050v
10 (Cal)	BL	EGR Gas Temp. Sensor	3.5-4.0v
11	PK/BK	TP Sensor Signal	0.3-0.8v
12	PK	Closed Throttle Switch	1v, at off-idle: 12v
13	RD/WT	Turbo Pressure Sensor	2.5-4.5v
13 ('92-'93)	WT	Knock Sensor Signal	<0.075v AC
14	WT	O2S Check	0.1-1.1v
15	OR	Check Connector	12-14v
16 ('90-'91)	BK	Distributor Signal (G-)	<0.050v

1990-93 Hatchback Turbo 2.0L MFI VIN S (M/T) 22 Pin Connector

PCM Pin #	Wire Color	Circuit Description (22 Pin)	Value at Hot Idle
1	PK	Battery Direct	12-14v
2	OR	Defogger/Light Idle Up Signal	Switch On: 12v, Off: 0v
4	GN/WT	Stop Light Switch Signal	Brake Off: 0v, On: 12v
5	PK	MIL (lamp) Control	MIL Off: 12v, On: 1v
6	GN/RD	Fuel Pump Relay	Relay Off: 12v, On: 1v
9	BL/WT	Vehicle Speed Sensor	At 55 mph: 48 Hz
10	BK/WT	A/C Magnetic Clutch (ACMG)	Clutch Off: 0v, On: 12v
11	BK	Starter Switch Signal	9-11v (cranking)
12	BK/YL	EFI Main Relay Power	12-14v
13	BK/YL	EFI Main Relay Power	12-14v
14 ('92-'93)	GN	Circuit Opening Relay (FC)	0-3v, off-idle: 12v
21	GN/BK	A/C Amplifier Signal (ACT)	Clutch On: 12v, Off: 1.5v

Pin Connector Graphic

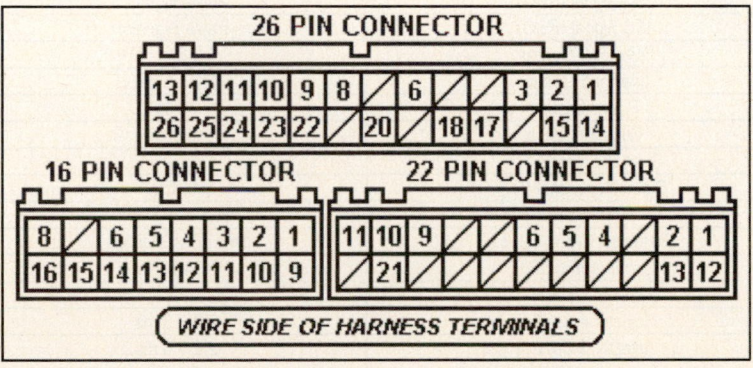

1990-91 Hatchback Turbo 2.0L MFI VIN S (M/T) 26 Pin Connector

PCM Pin #	Wire Color	Circuit Description (26 Pin)	Value at Hot Idle
1	YL	Distributor Signal (NE+)	AC pulse signals
2	BL	Distributor Signal (G2+)	AC pulse signals
3	BK/YL	Distributor Signal (IGF)	Digital Signal: 0-5-0v
4-5, 7	---	Not Used	---
6	BL/RD	Turbo Pressure Solenoid	12v or 0v
8	PK/BK	HO2S Heater	1v (Heater On)
9	GN/WT	ISC Motor (ISC1)	Pulse Signals
10	GN	Cold Start Injector Control	1v (at cold startup)
11	GN	Injector 2 Control	2.0-3.3 ms
12	RD	Injector 1 Control	2.0-3.3 ms
13	WT/BK	Power Ground	<0.1v
14	BR	Sensor Ground	<0.050v
15	GN	Distributor Signal (G1+)	AC pulse signals
16, 19, 21	---	Not Used	---
17	GN	Circuit Opening Relay (FC)	0-3v, off-idle: 12v
18	LG	TN-VIS Solenoid	12v or 0v
20	WT	Igniter Signal (IGT)	0.74-0.76v
22	GN/BK	ISC Motor (ISC2)	Pulse Signals
23	GN/RD	EGR Solenoid Control (VSV)	12v or 0v
24	BL	Injector 4 Control	2.0-3.3 ms
25	YL	Injector 3 Control	2.0-3.3 ms
26	WT/BK	Power Ground	<0.1v

1992-93 Hatchback Turbo 2.0L I4 VIN S (M/T) 26 Pin Connector

PCM Pin #	Wire Color	Circuit Description (26 Pin)	Value at Hot Idle
1	BL/GN	TN-VIS Solenoid	12v or 0v
2	BL/RD	Turbo Pressure Solenoid	12v or 0v
3	BK/YL	Distributor Signal (IGF)	0.72-0.74v
4	YL	Distributor Signal (NE+)	AC pulse signals
5	BL	Distributor Signal (G2+)	AC pulse signals
6	GN/RD	EGR Solenoid Control (VSV)	12v or 0v
7	GN	Cold Start Injector Control	1v (at cold startup)
8	PK/BK	HO2S Heater	1v (Heater On)
9	GN/WT	ISC Signal (RSC)	Pulse Signals
10	GN/BK	ISC Signal (RSO)	Pulse Signals
11	GN	Injector 2 Control	2.0-3.3 ms
12	RD	Injector 1 Control	2.0-3.3 ms
13	WT/BK	Power Ground	<0.1v
14	BR	Sensor Ground	<0.050v
15-16	---	Not Used	---
17	BK	Distributor Signal (G-)	<0.050v
19	---	Not Used	---
18	RD	Distributor Signal (G1+)	AC pulse signals
20	WT	Igniter Signal (IGT)	0.74-0.76v
21-23	---	Not Used	---
24	BL	Injector 4 Control	2.0-3.3 ms
25	YL	Injector 3 Control	2.0-3.3 ms
26	WT/BK	Power Ground	<0.1v

1990-93 Coupe 1.6L I4 MFI VIN A (All) 16 Pin Connector

PCM Pin #	Wire Color	Circuit Description (16 Pin)	Value at Hot Idle
1	BK/YL	EFI Main Relay Power	12-14v
2	PK	Battery Direct	12-14v
4	GN	Circuit Opening Relay (FC)	0-3v, off-idle: 12v
7	GN/BK	A/C Amplifier Signal (ACT)	Clutch On: 12v, Off: 1.5v
8	OR	Data Link Connector	12-14v
9	BK/YL	EFI Main Relay Power	12-14v
10	PK	MIL (lamp) Control	MIL Off: 12v, On: 1v
12	BK/WT	A/C Magnetic Clutch (ACMG)	Clutch Off: 0v, On: 12v
13	BL/WT	Vehicle Speed Sensor	At 55 mph: 48 Hz
14 ('90-'91)	YL/BK	Overdrive Solenoid	Solenoid Off: 12v, On: 1v
15 ('92-'93)	YL/BK	Overdrive Solenoid	Solenoid Off: 12v, On: 1v
16	PK/YL	Data Link Connector	12-14v

1990-93 Coupe 1.6L I4 MFI VIN A (All) 26 Pin Connector

PCM Pin #	Wire Color	Circuit Description (26 Pin)	Value at Hot Idle
1 (Cal)	BL/BK	EGR Solenoid Control (VSV)	12v or 0v
2	BK	A/T: Starter Switch Signal	9-11v (cranking)
3	RD	ECT Sensor Signal	At 180°F: 0.51v
4	GY/BK	MAP Sensor Signal	1-1.5v
5	GY	IAT Sensor Signal	At 100°F: 2.60v
6	BK	Igniter Signal (IGT)	0.74-0.76v
7	BK/YL	Igniter Signal (IGF)	0.72-0.74v
8	RD	Distributor Signal (G1+)	AC pulse signals
9	BK	Distributor Signal (G-)	<0.050v
10 (Cal)	WT	O2S-11 Signal	0.1-1.1v
10 (Fed)	WT	HO2S-11 Signal	0.1-1.1v
11	BK	Starter Switch Signal	9-11v (cranking)
12	WT	Injector 1 & 3 Control	2.0-3.3 ms
13	WT/BK	Power Ground	<0.1v
14	BL	ISC (+) Motor Solenoid	Pulse Signals
15 (Cal)	YL/BK	Overdrive Solenoid	12v or 0v
15 (Fed)	PK/BK	HO2S-11 Heater	1v (Heater On)
16, 22, 24	BR	Sensor Ground	<0.050v
17	PK/BK	Wide Open Throttle Switch	1v, at off-idle: 12v
18	PK/BL	Vacuum Sensor VREF	4.9-5.1v
19	PK	Closed Throttle Switch	1v, at off-idle: 12v
20 (Cal)	BL	EGR Gas Temp. Sensor	3.5-4.0v
21	WT	Distributor Signal (NE+)	AC pulse signals
25	YL	Injector 2 & 4 Control	2.0-3.3 ms
26	WT/BK	Power Ground	<0.1v

Pin Connector Graphic

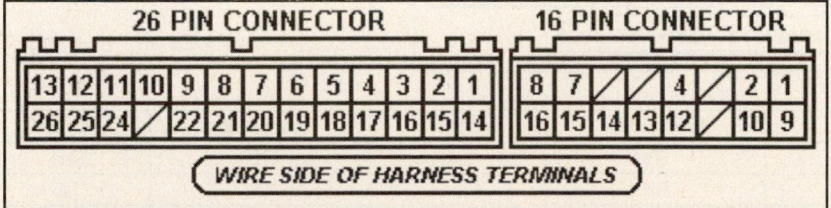

Standard Colors and Abbreviations

Abbreviation	Color	Abbreviation	Color	Abbreviation	Color
BK	Black	GY	Gray	RD	Red
BL	Blue	GN	Green	TN	Tan
BR	Brown	LG	Light Green	VT	Violet
DB	Dark Blue	OR	Orange	WT	White
DG	DK Green	PK	Pink	YL	Yellow

1994-95 Coupe, HB 1.8L I4 MFI VIN A (A/T-ECT) 16 Pin Connector

PCM Pin #	Wire Color	Circuit Description (16 Pin)	Value at Hot Idle
1	RD	Sensor VREF	4.9-5.1v
2	GY/BK	MAP Sensor Signal	1-1.5v
3	BL/RD	IAT Sensor Signal	At 100°F: 2.60v
4	GN	ECT Sensor Signal	At 180°F: 0.51v
5	WT	Sub O2S Signal	0.1-1.1v
6	WT	Main O2S Signal	0.1-1.1v
7	BK	Data Link Connector	12-14v
8	RD/WT	Data Link Connector	12-14v
9	BR	Sensor Ground	<0.050v
10 (Cal)	BK/RD	EGR Gas Temp. Sensor	3.5-4.0v
11	BK/WT	TP Sensor Signal	0.3-0.8v
12	BL/WT	Closed Throttle Switch	1v, at off-idle: 12v
13	WT	Knock Sensor Signal	<0.075v AC
14	LG	Data Link Connector	12-14v
15	YL	Data Link Connector	12-14v

1994-95 Coupe, HB 1.8L I4 MFI VIN A (A/T-ECT) 22 Pin Connector

PCM Pin #	Wire Color	Circuit Description (22 Pin)	Value at Hot Idle
1	PK	Battery Direct	12-14v
2	OR	Defogger/Light Idle Up Signal	Switch On: 12v, Off: 0v
4	GN/WT	Stop Light Switch Signal	Brake Off: 0v, On: 12v
5	RD/GN	MIL (lamp) Control	MIL Off: 12v, On: 1v
9	OR	Vehicle Speed Sensor	At 55 mph: 48 Hz
10	BL/BK	A/C Amplifier Signal (ACT)	Clutch On: 12v, Off: 1.5v
11	BK	Starter Switch Signal	9-11v (cranking)
12	BK/YL	EFI Main Relay Power	12-14v
14	GN/RD	Circuit Opening Relay (FC)	0-3v, off-idle: 12v
21	GN/YL	A/C Amplifier Signal (AC1)	Clutch On: 1.5v, Off: 12v
22	BK/YL	A/T Neutral Start Switch	In P/N: 9-11v (cranking)

Pin Connector Graphic

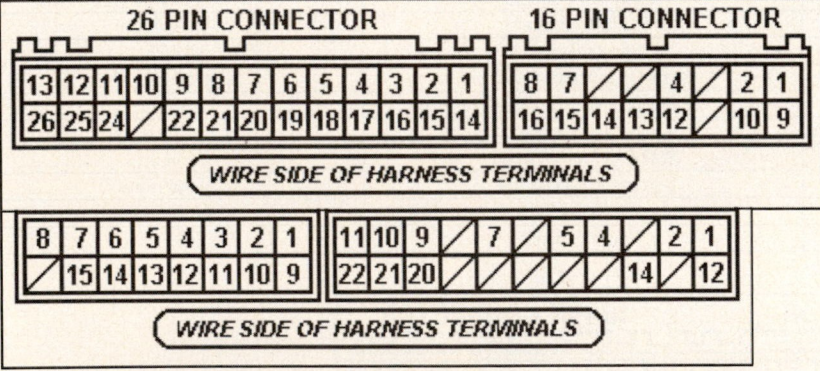

Standard Colors and Abbreviations

Abbreviation	Color	Abbreviation	Color	Abbreviation	Color
BK	Black	GY	Gray	RD	Red
BL	Blue	GN	Green	TN	Tan
BR	Brown	LG	Light Green	VT	Violet
DB	Dark Blue	OR	Orange	WT	White
DG	DK Green	PK	Pink	YL	Yellow

1994-95 Coupe, HB 1.8L I4 MFI VIN A (A/T-ECT) 26 Pin Connector

PCM Pin #	Wire Color	Circuit Description (26 Pin)	Value at Hot Idle
1	BL/YL	A/T-ECT Solenoid (SL)	In Lockup: 12-14v
2	BL/RD	A/T-ECT Solenoid (S1)	In 3rd or OD: 1v
3	BK/YL	Igniter Signal (IGF)	Digital Signal: 0-5-0v
4	YL	Distributor Signal (NE+)	AC pulse signals
5	BK	Distributor Signal (G-)	<0.050v
6	PK/GN	A/T Select Switch 2nd	In 2nd: 12v, Others: 0v
9	GN/WT	ISC Signal (RSC)	Pulse Signals
10	BK/WT	ISC Signal (RSO)	Pulse Signals
11	YL	Injector 2 Control	2.0-3.3 ms
12	WT	Injector 1 Control	2.0-3.3 ms
13	WT/BK	Power Ground	<0.1v
14	BR	Sensor Ground	<0.050v
15	BR/YL	A/T-ECT Solenoid (S2)	1st or OD: 1v
17	BK	Distributor Signal (NE-)	<0.050v
18	RD	Distributor Signal (G+)	AC pulse signals
19	PK/RD	A/T Select Switch Low	In Low: 12v, Others: 0v
20	BK	Igniter Signal (IGT)	Digital Signal: 0-5-0v
22	BL/GN	A/T Select Switch Reverse	In 'R': 12v, Others: 0v
23	BL/BK	EGR Solenoid Control (VSV)	12v or 0v
24	BK/RD	Injector 4 Control	2.0-3.3 ms
25	BK/YL	Injector 3 Control	2.0-3.3 ms
26	WT/BK	Power Ground	<0.1v

Pin Connector Graphic

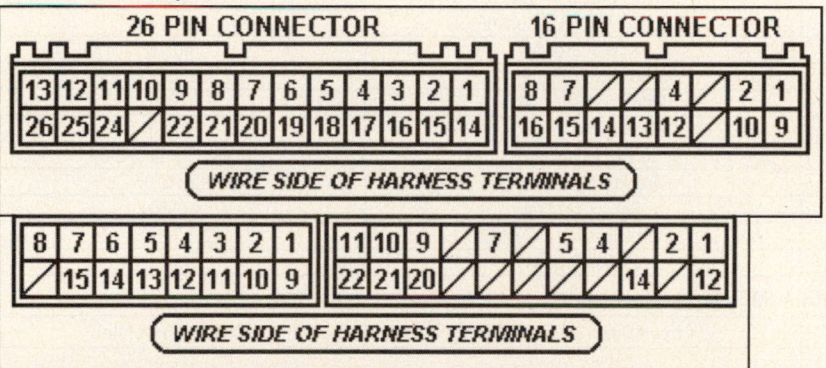

Standard Colors and Abbreviations

Abbreviation	Color	Abbreviation	Color	Abbreviation	Color
BK	Black	GY	Gray	RD	Red
BL	Blue	GN	Green	TN	Tan
BR	Brown	LG	Light Green	VT	Violet
DB	Dark Blue	OR	Orange	WT	White
DG	DK Green	PK	Pink	YL	Yellow

1994-95 Coupe, Hatchback 1.8L I4 VIN A (M/T) 16 Pin Connector

PCM Pin #	Wire Color	Circuit Description (16 Pin)	Value at Hot Idle
2	GY/BK	MAP Sensor Signal	2.5-4.5v
3	BL/RD	IAT Sensor Signal	At 100ºF: 2.60v
4	RD	ECT Sensor Signal	At 180ºF: 0.51v
4	GN	ECT Sensor Signal	At 180ºF: 0.51v
5	WT	Sub O2S Signal	0.1-1.1v
6	WT	Main O2S Signal	0.1-1.1v
7	LG	Data Link Connector	12-14v
8	RD/WT	Data Link Connector	12-14v
9	BR	Sensor Ground	<0.050v
10	BK/WT	TP Sensor Signal	0.3-0.8v
11	RD	Sensor VREF	4.9-5.1v
12	BL/WT	Closed Throttle Switch	1v, at off-idle: 12v
13	BK/RD	EGR Gas Temp. Sensor	3.5-4.0v
14	WT	Knock Sensor Signal	<0.075v AC
15	YL	Data Link Connector	12-14v

1994-95 Coupe, Hatchback 1.8L I4 VIN A (M/T) 26 Pin Connector

PCM Pin #	Wire Color	Circuit Description (26 Pin)	Value at Hot Idle
2	BK	Starter Switch Signal	9-11v (cranking)
3	BK/YL	Igniter Signal (IGF)	Digital Signal: 0-5-0v
4	YL	Distributor Signal (NE+)	AC pulse signals
5	RD	Distributor Signal (G1+)	AC pulse signals
9	GN/WT	ISC Signal (RSC)	Pulses
10	WT/BK	ISC Signal (RSO)	Pulses
12	BK/YL	Injector Pair 1 & 3 Control	2.0-3.3 ms
13	WT/BK	Power Ground	<0.1v
16	BL/BK	EGR Solenoid Control (VSV)	12v or 0v
17	BL	Distributor Signal (NE-)	<0.050v
18	BK	Distributor Signal (G-)	<0.050v
22	WT	Igniter Signal (IGT)	Digital Signal: 0-5-0v
23	RD/YL	A/C Idle Up Solenoid	12v or 0v
24	BR	Sensor Ground	<0.050v
25	BK/RD	Injector Pair 2 & 4 Control	2.0-3.3 ms
26	WT/BK	Power Ground	<0.1v

1994-95 Coupe, Hatchback 1.8L I4 VIN A (M/T) 12 Pin Connector

PCM Pin #	Wire Color	Circuit Description (12 Pin)	Value at Hot Idle
2	PK	Battery Direct	12-14v
4	GN/RD	Circuit Opening Relay (FC)	0-3v, off-idle: 12v
6	GN/YL	A/C Amplifier Signal (ACT)	Clutch On: 12v, Off: 1.5v
7	BK/YL	EFI Main Relay Power	12-14v
8	RD/GN	MIL (lamp) Control	MIL Off: 12v, On: 1v
10	BL/BK	A/C Amplifier Signal (AC1)	Clutch On: 1.5v, Off: 12v
11	OR	Vehicle Speed Sensor	At 55 mph: 48 Hz
12	OR	Mirror Heater Switch Signal	Switch On: 12v, Off: 0v

Pin Connector Graphic

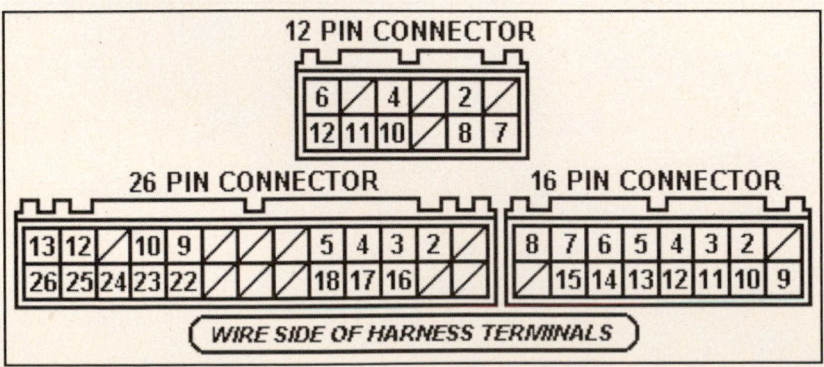

1996-97 Coupe, HB 1.8L I4 MFI VIN B (A/T-ECT) 16 Pin Connector

PCM Pin #	Wire Color	Circuit Description (16 Pin)	Value at Hot Idle
1	RD	Sensor VREF	4.9-5.1v
2	BL	MAP Sensor Signal	2.5-4.5v
3	BL/RD	IAT Sensor Signal	At 100°F: 2.60v
4	GN	ECT Sensor Signal	At 180°F: 0.51v
5	WT	HO2S-12 (B1 S2) Signal	0.1-1.1v
6	WT	HO2S-11 (B1 S1) Signal	0.1-1.1v
9	BR	Sensor Ground	<0.050v
11	BK/WT	TP Sensor Signal	0.3-0.8v
13	WT	Knock Sensor Signal	<0.075v AC
15	YL	Data Link Connector	12-14v
16	BR	Power Ground	<0.1v

1996-97 Coupe, HB 1.8L I4 MFI VIN B (A/T-ECT) 22 Pin Connector

PCM Pin #	Wire Color	Circuit Description (22 Pin)	Value at Hot Idle
1	PK	Battery Direct	12-14v
2	OR	Defogger/Light Idle Up Signal	Switch On: 12v, Off: 0v
4	GN/WT	Stop Light Switch Signal	Brake Off: 0v, On: 12v
5	RD/BK	MIL (lamp) Control	MIL Off: 12v, On: 1v
7	GY/BL	Overdrive Main Switch	Switch Off: 12v
9	OR	Vehicle Speed Sensor	At 55 mph: 48 Hz
10	BL/BK	A/C Amplifier Signal (ACT)	Clutch On: 12v, Off: 1.5v
11	BK	Starter Switch Signal	9-11v (cranking)
12	BK/RD	EFI Main Relay Power	12-14v
14	GN/RD	Circuit Opening Relay (FC)	12-14v
16	WT	Data Link Connector (SDL)	No Scan Tool: 0v
17	RD/WT	A/T Select Switch Reverse	In 'R': 12v, Others: 0v
18	YL/GN	A/T Select Switch 2nd	In 2nd: 12v, Others: 0v
19	YL/RD	A/T Select Switch Low	In Low: 12v, Others: 0v
20	PK	Cruise Control ECU	At Cruise in OD: 12v
21	GN/YL	A/C Amplifier Signal (AC1)	Clutch On: 1.5v, Off: 12v
22	BK/YL	A/T Neutral Start Switch	In P/N: 9-11v (cranking)

Standard Colors and Abbreviations

Abbreviation	Color	Abbreviation	Color	Abbreviation	Color
BK	Black	GY	Gray	RD	Red
BL	Blue	GN	Green	TN	Tan
BR	Brown	LG	Light Green	VT	Violet
DB	Dark Blue	OR	Orange	WT	White
DG	DK Green	PK	Pink	YL	Yellow

1996-97 Coupe, HB 1.8L I4 MFI VIN B (A/T-ECT) 26 Pin Connector

PCM Pin #	Wire Color	Circuit Description (26 Pin)	Value at Hot Idle
1	BL/YL	A/T-ECT Solenoid (SL)	In Lockup: 12-14v
2	BL/RD	A/T-ECT Solenoid (S1)	In 3rd or OD: 1v
3	GN/YL	Igniter Signal (IGF)	Digital Signal: 0-5-0v
4	OR	CKP Sensor Signal (NE+)	AC pulse signals
5	BK	Distributor Signal (G-)	<0.050v
9	GN/WT	ISC Signal (RSC)	Pulse Signals
10	BK/WT	ISC Signal (RSO)	Pulse Signals
11	BK/RD	Injector 2 Control	2.0-3.3 ms
12	BK/WT	Injector 1 Control	2.0-3.3 ms
13	BR	Power Ground	<0.1v
14	BR	Sensor Ground	<0.050v
15	BR/YL	A/T-ECT Solenoid (S2)	1st or OD: 1v
17	WT	CKP Sensor Signal (NE-)	<0.050v
19	PK	HO2S-12 (B1 S2) Heater	1v (Heater On)
20	WT	Igniter Signal (IGT)	Digital Signal: 0-5-0v
23	BK/BK	EGR Solenoid Control (VSV)	12v or 0v
24	BK/RD	Injector 4 Control	2.0-3.3 ms
25	BK/WT	Injector 3 Control	2.0-3.3 ms
26	BR	Power Ground	<0.1v

Pin Connector Graphic

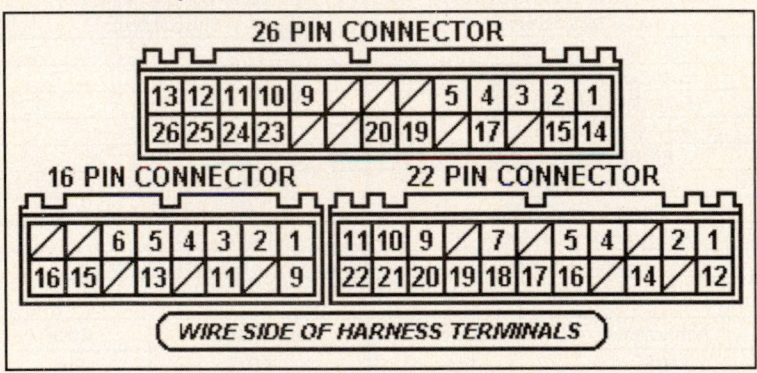

2000-02 GT 1.8L I4 1ZZ-FE DOHC VIN R (All) E2 22P Connector

PCM Pin #	Wire Color	Circuit Description (22 Pin)	Value at Hot Idle
1	WT	Battery Direct	12-14v
2	---	Not Used	---
3	GN/RD	Circuit Opening Relay (FC)	0-3v, off-idle: 12v
4	BL/BK	EVAP Vapor Pressure Sensor	Fuel Cap Off: 2.9-3.7v
5	BR/YL	Overdrive Main Switch	Switch Off: 12v, On: 0v
6	LG/RD	Multiplex Meter Input (MPX2)	Digital Signals
7	WT/BK	Power Ground (E03)	<0.1v
8	BK/OR	Ignition Switch Power	12-14v
9	---	Not Used	---
10	BL/WT	Cruise Control Signal (IDLO)	1.5v, off-idle: 12v
11	WT	SIL Signal (Scan Tool)	Digital Signal
12	YL/BK	A/C Magnetic Clutch (ACMG)	Clutch Off: 0v, On: 12v
13-14	---	Not Used	---
15	RD/BK	MIL (lamp) Control	MIL Off: 12v, On: 1v
16	BK/RD	EFI Main Relay Power	12-14v
17-22	---	Not Used	---

Pin Connector Graphic

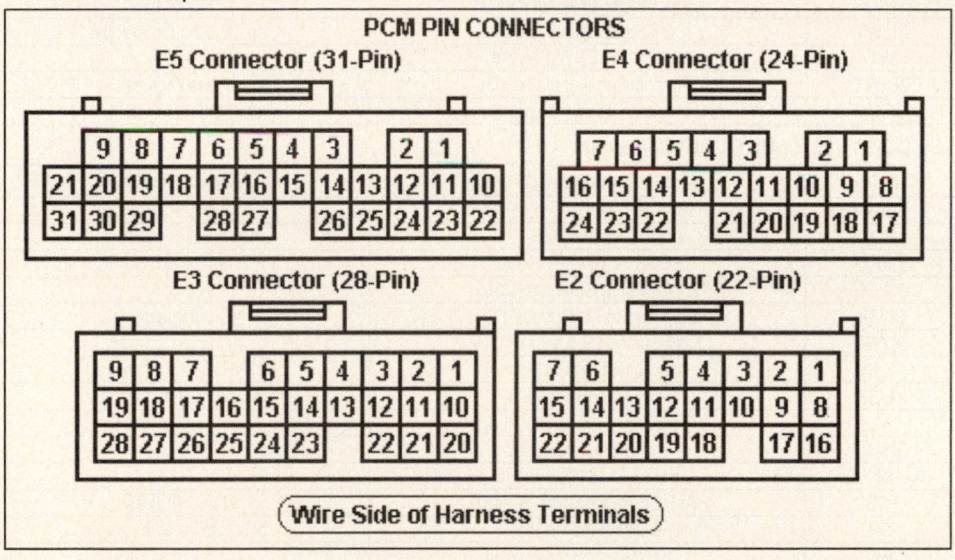

2000-02 GT 1.8L I4 1ZZ-FE DOHC VIN R (All) E3 28P Connector

PCM Pin #	Wire Color	Circuit Description (28 Pin)	Value at Hot Idle
1	BK	Cruise Control Signal (OD1)	At Cruise in OD: 12v
2	RD/BK	A/T Select Switch Reverse	In 'R': 12v, Others: 0v
3	BL/WT	Cooling Fan 3 Relay	Relay Off: 12v, On: 1v
4	BL/YL	Cooling Fan 1 Relay	Relay Off: 12v, On: 1v
5	PK/BK	Data Link Connector (TC)	12-14v
6	GN/WT	Stop Light Switch Signal	Brake Off: 0v, On: 12v
7, 13-15	---	Not Used	---
8	PK	Center Airbag Assembly	Digital Signals
9	RD/BL	ACIS Solenoid Control (VSV)	12v or 0v
10	LG/RD	Multiplex Meter Input (MPX1)	Digital Signals
11	BL	Starter Switch Signal	9-11v (cranking)
12	WT/BK	Power Ground (EC)	<0.1v
16	YL/GN	HO2S-12 (B1 S2) Heater	1v (Heater On)
17, 19-20	---	Not Used	---
18	OR	A/C Dual Pressure Switch	A/C Off: 12v, On: 1v
21	BL/BK	EFI Main Relay Power	12-14v
22	WT/RD	Speedometer Indicator	At 55 mph: 48 Hz
23	GN/OR	EVAP Vapor Pressure (VSV)	12v or 0v
24	BL/WT	A/T Select Switch Drive	In 'D': 12v, Others: 0v
25	WT	HO2S-12 (B1 S2) Signal	0.1-1.1v
27	BR/WT	Tachometer Signal (TACO)	Pulse Signals
28	GN	A/C Magnetic Clutch & Lock Sensor	A/C Off: 12v, On: 1v

2000-02 GT 1.8L I4 1ZZ-FE DOHC VIN R (All) E4 24P Connector

PCM Pin #	Wire Color	Circuit Description (24 Pin)	Value at Hot Idle
1	YL/GN	MAF Sensor Ground (EVG)	<0.050v
2	RD	Sensor VREF	4.9-5.1v
3	YL/RD	HO2S-11 (B1 S1) Heater	1v (Heater On)
4	GN/OR	EVAP Purge Solenoid (VSV)	12v or 0v
6	BL/BK	Cam Timing Oil Control VVT-	AC pulse signals
7	BL/WT	Cam Timing Oil Control VVT+	AC pulse signals
8	PK/BL	A/T: Select Switch Neutral	In P/N: 9-11v (cranking)
9	YL/BK	A/T Select Switch Low	In Low: 12v, Others: 0v
10	YL	Generator Control Signal (L)	12v
11	GN/WT	MAF Sensor Signal	1.1-1.5v
12	BK	HO2S-11 (B1 S1) Signal	0.1-1.1v
13	GY/BL	A/T Oil Temperature Sensor	At 230°F: 0.95v
14	GN	ECT Sensor Signal	1-4v (varies with temp.)
15	BL	CMP Sensor Signal (G2+)	AC pulse signals
16	OR	CKP Sensor Signal (NE+)	AC pulse signals
17	BR	Power Ground (E1)	<0.1v
18	BR	Sensor Ground (E2)	<0.050v
19	BL/YL	A/T Select Switch 2nd	In 2nd: 12v, Others: 0v
20	PK	A/T: Select Switch Park	In P/N: 9-11v (cranking)
21	GN	Oil Pressure Switch Signal	Open: 12v, Closed: 0v
22	BL/RD	IAT Sensor Signal	1-4v (varies with temp.)
23	BK/WT	TP Sensor Signal	0.3-1.0v
24	WT	CKP/CMP Sensor Ground (-)	<0.050v

2000-02 GT 1.8L I4 1ZZ-FE DOHC VIN R (All) E5 31P Connector

PCM Pin #	Wire Color	Circuit Description (31 Pin)	Value at Hot Idle
1	RD	Injector 1 Control	2.0-3.3 ms
2	RD/BL	Injector 2 Control	2.0-3.3 ms
3	RD/WT	Injector 3 Control	2.0-3.3 ms
4	RD/BK	Injector 4 Control	2.0-3.3 ms
5	YL/GN	A/T-ECT Solenoid (SLT-)	Pulse signals
6	YL/RD	A/T-ECT Solenoid (SLT+)	Pulse signals
7	GN	A/T-ECT Solenoid (SL1+)	Pulse signals
8	BR/YL	A/T-ECT Solenoid (S1)	In 3rd or OD: 1v
9	VT	A/T-ECT Solenoid (SL1-)	Pulse Signals
10	RD/BK	Igniter Transistor 1 Control	Digital Signal: 0-5-0v
11	RD/WT	Igniter Transistor 2 Control	Digital Signal: 0-5-0v
12	GN/RD	Igniter Transistor 3 Control	Digital Signal: 0-5-0v
13	RD/YL	Igniter Transistor 4 Control	Digital Signal: 0-5-0v
15	BK	A/T: ECT VSS (-) Signal	Pulse signals
16	WT/BL	A/T: ECT VSS (+) Signal	Pulse signals
17	PK/WT	EVAP Canister Closed Valve	12v or 0v
18	BK/WT	IAC Valve Signal (RSO)	Pulse signals
19	GN/WT	A/T-ECT Solenoid (SL)	In Lockup: 12-14v
20	YL	A/T-ECT Solenoid (S2)	1st or OD: 1v
21	WT/BK	Power Ground (E01)	<0.1v
22	YL/BK	OIL Pressure Switch	Switch On: 12-14v
23	WT	Camshaft Timing Oil Control Valve (-)	Pulse signals
24	GN/OR	Camshaft Timing Oil Control Valve (+)	Pulse signals
25	BK/YL	Igniter Signal (IGF)	Digital Signal: 0-5-0v
27	WT	Knock Sensor Signal	<0.075v AC
28	PK	PSP Switch Signal	Straight: 12v, Turning: 0v
29	BR/WT	A/T-ECT ST Signal	Pulse signals
31	WT/BK	Power Ground (E02)	<0.1v

Pin Connector Graphic

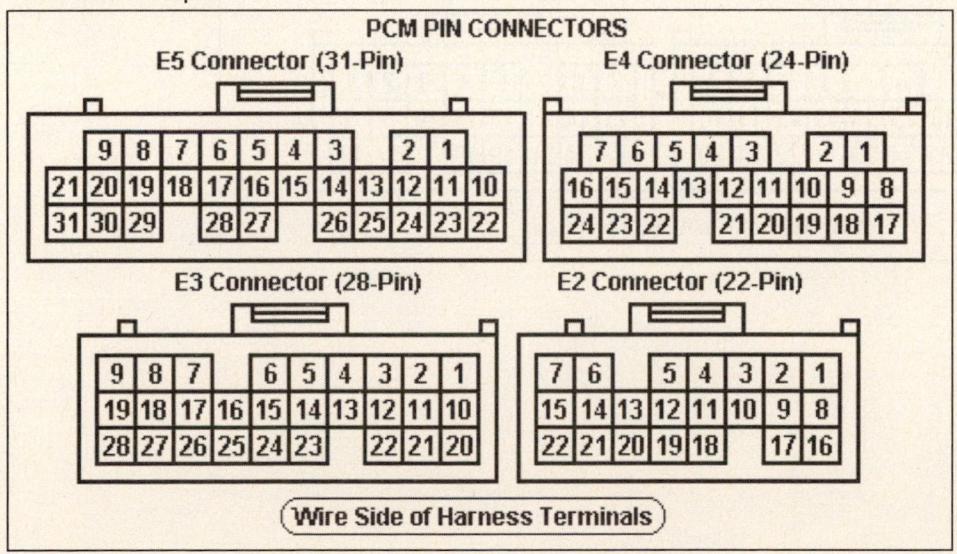

2000-02 GTS 1.8L I4 2ZZ-GE DOHC VIN Y (All) E2 22P Connector

PCM Pin #	Wire Color	Circuit Description (22 Pin)	Value at Hot Idle
1	WT	Battery Direct	12-14v
2	---	Not Used	---
3	GN/RD	Circuit Opening Relay (FC)	0-3v, off-idle: 12v
4	BL/BK	EVAP Vapor Pressure Sensor	Fuel Cap Off: 2.9-3.7v
5	BR/YL	Overdrive Main Switch	Open: 12v, Closed: 0v
6	LG/RD	Multiplex Meter Input (MPX2)	Digital Signals
7	WT/BK	Power Ground (E03)	<0.1v
8	BK/OR	Ignition Switch Power	12-14v
9	BL	A/T: Trans. Shift Switch SFTU	Switch Off: 12v
10	BL/WT	Cruise Control Signal (IDLO)	1.5v, off-idle: 12v
11	WT	SIL Signal (Scan Tool)	Digital Signals
12	YL/BK	A/C Magnetic Clutch (ACMG)	Clutch Off: 0v, On: 12v
13-14	---	Not Used	---
15	RD/BK	MIL (lamp) Control	MIL Off: 12v, On: 1v
16	BK/RD	EFI Main Relay Power	12-14v
17	PK	A/T: Trans. Shift Switch SFTD	Switch Off: 12v
18-22	---	Not Used	---

Pin Connector Graphic

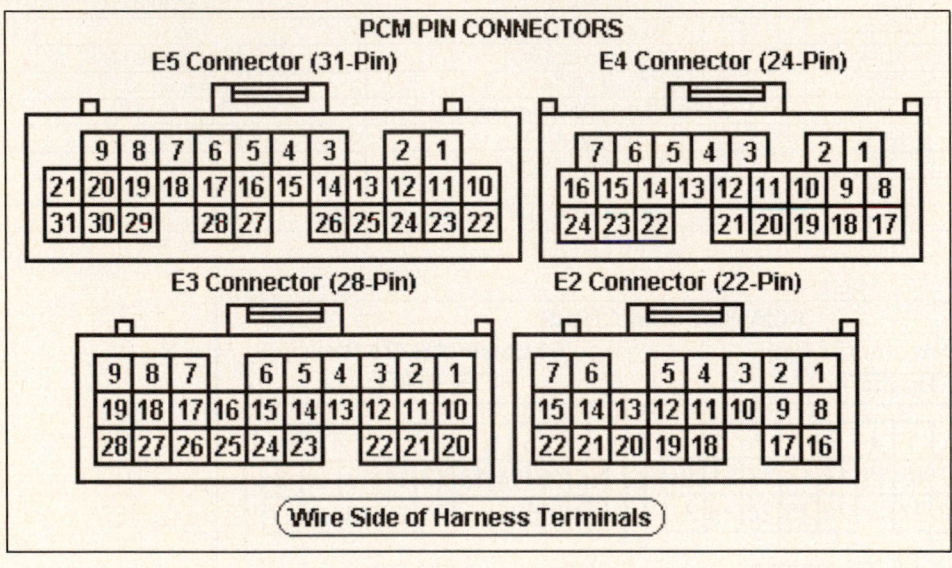

2000-02 GTS 1.8L I4 2ZZ-GE DOHC VIN Y (All) E3 28P Connector

PCM Pin #	Wire Color	Circuit Description (28 Pin)	Value at Hot Idle
1	BK	Cruise Control Signal (OD1)	At Cruise in OD: 12v
2	RD/BK	A/T Select Switch Reverse	In 'R': 12v, Others: 0v
3	BL/WT	Cooling Fan 3 Relay	Relay Off: 12v, On: 1v
4	BL/YL	Cooling Fan 1 Relay	Relay Off: 12v, On: 1v
5	PK/BK	Data Link Connector (TC)	12-14v
6	GN/WT	Stop Light Switch Signal	Brake Off: 0v, On: 12v
7	---	Not Used	---
8	PK	Center Airbag Assembly	Digital Signals
9	RD/BL	ACIS Solenoid Control (VSV)	12v or 0v
10	LG/RD	Multiplex Meter Input (MPX1)	Digital Signals
11	BL	Starter Switch Signal	9-11v (cranking)
12	WT/BK	Power Ground (EC)	<0.1v
13	---	Not Used	---
14	BK/WT	Cruise Control (D) ECU Input	Pulse Signals
15	---	Not Used	---
16	YL/GN	HO2S-12 (B1 S2) Heater	1v (Heater On)
17	---	Not Used	---
18	OR	A/C Dual Pressure Switch	A/C Off: 12v, On: 1v
19-20	---	Not Used	---
21	BL/BK	EFI Main Relay Power	12-14v
22	WT/RD	Speedometer Indicator	At 55 mph: 48 Hz
23	GN/OR	EVAP Vapor Pressure (VSV)	12v or 0v
24	BL/WT	A/T Select Switch Drive	In 'D': 12v, Others: 0v
25	WT	HO2S-12 (B1 S2) Signal	0.1-1.1v
26	---	Not Used	---
27	BR/WT	Tachometer Signal (TACO)	Pulse Signals
28	GN	A/C Magnetic Clutch & Lock Sensor	A/C Off: 12v, On: 1v

2000-02 GTS 1.8L I4 2ZZ-GE DOHC VIN Y (All) E4 24P Connector

PCM Pin #	Wire Color	Circuit Description (24 Pin)	Value at Hot Idle
1	YL/GN	MAF Sensor Ground	<0.050v
2	RD	Sensor VREF	4.9-5.1v
3	YL/RD	HO2S-11 (B1 S1) Heater	1v (Heater On)
4	GN/OR	EVAP Purge Solenoid (VSV)	12v or 0v
6	BL/BK	Cam/Time Oil (OVL-) Valve	DC pulse signals
7	BL/WT	Cam/Time Oil (OVL+) Valve	DC pulse signals
8	PK/BL	A/T: Select Switch Neutral	In P/N: 9-11v (cranking)
9	YL/BK	A/T Select Switch Low	In Low: 12v, Others: 0v
10	YL	Generator Control Signal (L)	12v
11	GN/WT	MAF Sensor Signal	1-3v
12	BK	HO2S-11 (B1 S1) Signal	0.1-1.1v
13	GY/BL	A/T Oil Temperature Sensor	At 230ºF: 0.95v
14	GN	ECT Sensor Signal	1-4v (varies with temp.)
15	BL	CMP Sensor Signal (G2+)	AC pulse signals
16	OR	CKP Sensor Signal (NE+)	AC pulse signals
17	BR	Power Ground (E1)	<0.050v
18	BR	Sensor Ground (E2)	<0.050v
19	BL/YL	A/T Select Switch 2nd	In 2nd: 12v, Others: 0v
20	PK	A/T: Select Switch Park	In P/N: 9-11v (cranking)
21	GY	OIL Level Warning Switch	Open: 12v, Closed: 0v
22	BL/RD	IAT Sensor Signal	1-4v (varies with temp.)
23	BK/WT	TP Sensor Signal	1.1-1.9v
24	WT	CKP/CMP Sensor Ground (-)	<0.050v

2000-02 GTS 1.8L I4 2ZZ-GE DOHC VIN Y (All) E5 31P connector

PCM Pin #	Wire Color	Circuit Description (31 Pin)	Value at Hot Idle
1	RD	Injector 1 Control	2.0-3.3 ms
2	RD/BL	Injector 2 Control	2.0-3.3 ms
3	RD/WT	Injector 3 Control	2.0-3.3 ms
4	RD/BK	Injector 4 Control	2.0-3.3 ms
5	YL/GN	A/T-ECT Solenoid (SLT-)	Pulse signals
6	YL/RD	A/T ECT Solenoid (SLT+)	Pulse signals
7, 9	GN, VT	A/T ECT Solenoid (SL1+, -)	Pulse signals
8, 20	BR, OR	A/T ECT Solenoid (SL2+, -)	Pulse signals
10	RD/BK	Igniter Transistor 1 Control	Digital Signal: 0-5-0v
11	RD/WT	Igniter Transistor 2 Control	Digital Signal: 0-5-0v
12	GN/RD	Igniter Transistor 3 Control	Digital Signal: 0-5-0v
13	RD/YL	Igniter Transistor 4 Control	Digital Signal: 0-5-0v
14	PK/BK	OD Clutch Speed Signal (+)	AC pulse signals
15	BK	A/T ECT VSS (-) Signal	Pulse signals
16	WT/BL	A/T ECT VSS (+) Signal	Pulse signals
17	PK/WT	EVAP Canister Closed Valve	12v or 0v
18	BK/WT	IAC Valve Signal (RSO)	Pulse signals
19	PK/BL	A/T ECT Solenoid (DSL)	In Lockup: 12-14v
21	WT/BK	Power Ground (E01)	<0.1v
22	YL/BK	OIL Pressure Switch	Open: 12v, Closed: 0v
23	WT	Cam Timing Oil Valve (OCV-)	Pulse signals
24	GN/OR	Cam Timing Oil Valve (OCV+)	Pulse signals
25	BK/YL	Igniter Signal (IGF)	Pulse signals
26	GN/WT	OD Clutch Speed Signal (-)	AC pulse signals
27	WT	Knock Sensor Signal	<0.075v AC
28	PK	PSP Switch Signal	Straight: 12v, Turning: 0v
29	YL	A/T-ECT Solenoid (S4)	In Drive: 1v
31	WT/BK	Power Ground (E02)	<0.1v

Pin Connector Graphic (View is into Wire Side of Harness)

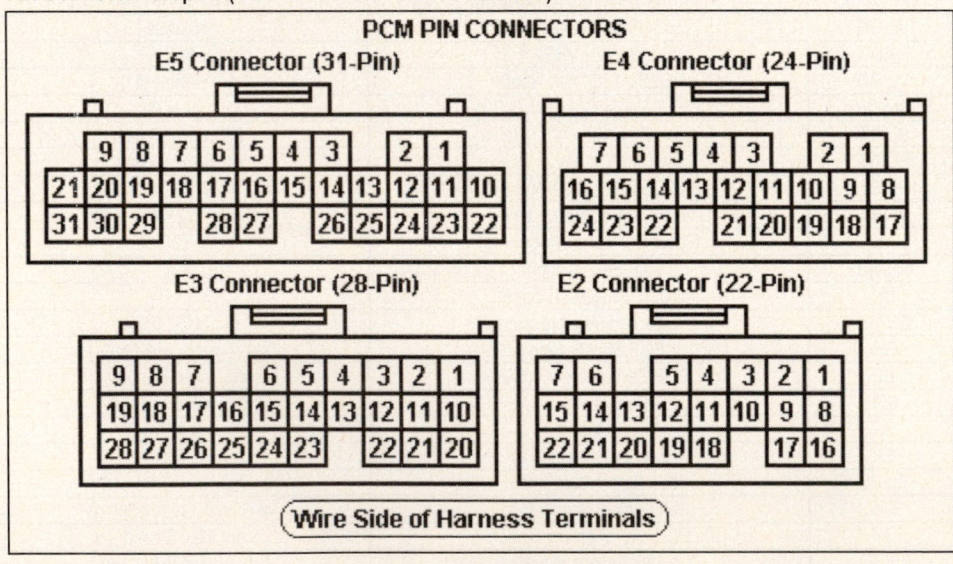

1990-91 Coupe, Hatchback 2.2L VIN S (All) 16 Pin Connector

PCM Pin #	Wire Color	Circuit Description (16 Pin)	Value at Hot Idle
1	PK/BL	Vacuum Sensor VREF	4.9-5.1v
2	GN/BK	Vacuum Sensor Signal	2.5-4.5v
3	GN/RD	IAT Sensor Signal	At 100°F: 2.60v
4	RD	ECT Sensor Signal	At 180°F: 0.51v
6	WT	Main O2S Signal	0.1-1.1v
7	BK	Check Connector	12-14v
8	PK/YL	Check Connector	12-14v
9	BR	Sensor Ground	<0.050v
10 (Cal)	BL	EGR Gas Temp. Sensor	3.5-4.0v
11	PK/BK	TP Sensor Signal	0.3-0.8v
12	PK	Closed Throttle Switch	1v, at off-idle: 12v
14 (Cal)	WT	Sub O2S Signal	0.1-1.1v
15	OR	Check Connector	12-14v
16	BK	Distributor Signal (G-)	<0.050v

1990-91 Coupe, Hatchback 2.2L VIN S (All) 22 Pin Connector

PCM Pin #	Wire Color	Circuit Description (22 Pin)	Value at Hot Idle
1	PK	Battery Direct	12-14v
2	OR	Defogger/Light Idle Up Signal	Switch On: 12v, Off: 0v
4	GN/WT	Stop Light Switch Signal	Brake Off: 0v, On: 12v
5	PK	MIL (lamp) Control	MIL Off: 12v, On: 1v
7	GY/BL	Main Overdrive Switch	Switch Off: 12-14v
8	BK/YL	A/C Amplifier Signal (AC1)	Clutch On: 1.5v, Off: 12v
9	BL/WT	Vehicle Speed Sensor	At 55 mph: 48 Hz
10	BK/WT	A/C Magnetic Clutch (ACMG)	Clutch Off: 0v, On: 12v
11	BK	A/T: Starter Switch Signal	9-11v (cranking)
11	BK/WT	M/T: Starter Switch Signal	9-11v (cranking)
12	BK/YL	EFI Main Relay Power B1+	12-14v
13	BK/YL	EFI Main Relay Power	12-14v
14	GN	Circuit Opening Relay (FC)	0-3v, off-idle: 12v
20	PK	Cruise Control ECU	At Cruise in OD: 12v
21	GN/BK	A/C Amplifier Signal (ACT)	Clutch On: 12v, Off: 1.5v
22	BK/WT	A/T Neutral Start Switch	In P/N: 9-11v (cranking)
22	BK	M/T: Neutral Start Switch	In 'N': 9-11v (cranking)

Standard Colors and Abbreviations

Abbreviation	Color	Abbreviation	Color	Abbreviation	Color
BK	Black	GY	Gray	RD	Red
BL	Blue	GN	Green	TN	Tan
BR	Brown	LG	Light Green	VT	Violet
DB	Dark Blue	OR	Orange	WT	White
DG	DK Green	PK	Pink	YL	Yellow

1990-91 Coupe, Hatchback 2.2L VIN S 26 Pin (All) Connector

PCM Pin #	Wire Color	Circuit Description (26 Pin)	Value at Hot Idle
1	WT	Distributor Signal (NE+)	AC pulse signals
3	BK/YL	Igniter Signal (IGF)	Digital Signal: 0-5-0v
9	GN/WT	ISC Signal (ISCC)	Pulse Signals
10	BK/WT	ISC Signal (ISCO)	Pulse Signals
11	YL	Injector 2 & 4 Control	2.0-3.3 ms
12	WT	Injector 1 & 3 Control	2.0-3.3 ms
13	WT/BK	Power Ground	<0.1v
14	BR	Sensor Ground	<0.050v
15	RD	Distributor Signal (G+)	AC pulse signals
16	BR	Sensor Ground	<0.050v
18	PK/GN	A/T Select Switch 2nd	In 2nd: 12v, Others: 0v
19	PK/RD	A/T Select Switch Low	In Low: 12v, Others: 0v
20	BK	Igniter Signal (IGT)	0.72-0.74v
23	BL/BK	EGR Solenoid Control (VSV)	12v or 0v
26	WT/BK	Power Ground	<0.1v

Pin Connector Graphic

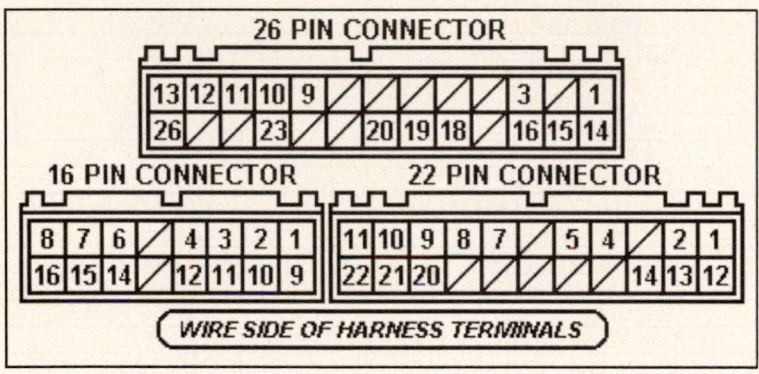

1990-91 Coupe, Hatchback SR 2.2L MFI VIN S 10 Pin Connector

PCM Pin #	Wire Color	Circuit Description (10 Pin)	Value at Hot Idle
1	BK/WT	A/T Neutral Start Switch	In P/N: 9-11v (cranking)
1	BK	M/T: Neutral Start Switch	In 'N': 9-11v (cranking)
3	BK	A/T: Starter Switch Signal	9-11v (cranking)
3	BK/WT	M/T: Starter Switch Signal	9-11v (cranking)
4	WT	Injector 1 & 3 Control	2.0-3.3 ms
5	WT/BK	Power Ground	<0.1v
6	BL/BK	EGR Solenoid Control (VSV)	12v or 0v
7	BR	Sensor Ground	<0.050v
8	WT	Igniter Signal (IGT)	0.72-0.74v
9	YL	Injector 2 & 4 Control	2.0-3.3 ms
10	WT/BK	Power Ground	<0.1v

1990-91 Coupe, Hatchback SR 2.2L MFI VIN S 14-Pin Connector

PCM Pin #	Wire Color	Circuit Description (14-Pin)	Value at Hot Idle
1	BK/YL	EFI Main Relay Power B1+	12-14v
2	PK	Battery Direct	12-14v
3	PK/YL	Check Connector	12-14v
4	GN	Circuit Opening Relay (FC)	0-3v, off-idle: 12v
5	OR	Defogger/Light Idle Up Signal	Switch On: 12v, Off: 0v
6	GN/WT	Stop Light Switch Signal	Brake Off: 0v, On: 12v
7	BK/YL	A/C Amplifier Signal (AC1)	Clutch On: 1.5v, Off: 12v
8	BK/YL	EFI Main Relay Power	12-14v
9	GN	MIL (lamp) Control	MIL Off: 12v, On: 1v
11	BK/WT	A/C Magnetic Clutch (ACMG)	Clutch Off: 0v, On: 12v
12	BL/WT	Vehicle Speed Sensor	At 55 mph: 48 Hz
13	GN/BK	A/C Amplifier Signal (ACT)	Clutch On: 12v, Off: 1.5v
14	BL/GN	Short Pin	Open: 12v, Closed: 0v

1991 Coupe, Hatchback SR 2.2L MFI VIN S (All) 18 Pin Connector

PCM Pin #	Wire Color	Circuit Description (18 Pin)	Value at Hot Idle
1	RD	ECT Sensor Signal	At 180ºF: 0.51v
2	GY/BK	Vacuum Sensor Signal	1-1.5v
3	GN/RD	IAT Sensor Signal	At 100ºF: 2.60v
4	OR	Check Connector	12-14v
5	BK/YL	Igniter Signal (IGF)	Digital Signal: 0-5-0v
6	RD	Distributor Signal (G+)	AC pulse signals
7	BK	Distributor Signal (G-)	<0.050v
8	WT	Main O2S Signal	0.1-1.1v
9	GN/WT	ISC Signal (ISCC)	Pulse Signals
10	BR	Sensor Ground	<0.050v
11	PK/BK	A/T: TP Sensor Signal	0.3-0.8v
11	PK/BK	M/T: PSW Signal	1v, at off-idle: 12v
12	PK/BL	Vacuum Sensor VREF	4.9-5.1v
13	PK	Closed Throttle Switch	1v, at off-idle: 12v
14 (Cal)	BL	EGR Gas Temp. Sensor	3.5-4.0v
15	WT	Distributor Signal (NE+)	AC pulse signals
16	BR	Sensor Ground	<0.050v
17 (Cal)	WT	Sub O2S Signal	0.1-1.1v
18	BK/WT	ISC Signal (ISCO)	Pulse Signals

Pin Connector Graphic

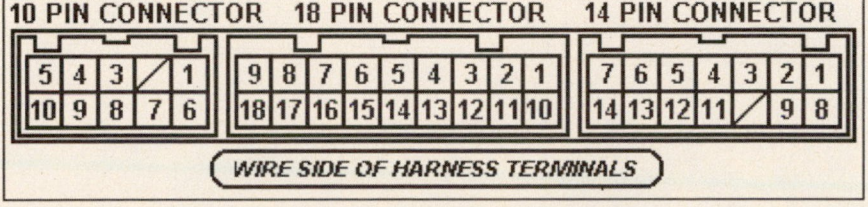

1990-91 Coupe, HB 2.2L MFI VIN S (A/T-ECT) 16 Pin Connector

PCM Pin #	Wire Color	Circuit Description (16 Pin)	Value at Hot Idle
1	PK/BL	Vacuum Sensor VREF	4.9-5.1v
2	GN/BK	Vacuum Sensor Signal	2.5-4.5v
3	GN/RD	IAT Sensor Signal	At 100°F: 2.60v
4	RD	ECT Sensor Signal	At 180°F: 0.51v
6	WT	Main O2S Signal	0.1-1.1v
7	BK	Check Connector	12-14v
8	PK/YL	Check Connector	12-14v
9	BR	Sensor Ground	<0.050v
10 (Cal)	BL	EGR Gas Temp. Sensor	3.5-4.0v
11	PK/BK	TP Sensor Signal	0.3-0.8v
12	PK	Closed Throttle Switch	1v, at off-idle: 12v
14 (Cal)	WT	Sub O2S Signal	0.1-1.1v
15	OR	Check Connector	12-14v
16	BK	Distributor Signal (G-)	<0.050v

1990-91 Coupe, HB 2.2L MFI VIN S (A/T-ECT) 22 Pin Connector

PCM Pin #	Wire Color	Circuit Description (22 Pin)	Value at Hot Idle
1	PK	Battery Direct	12-14v
2	OR	Defogger/Light Idle Up Signal	Switch On: 12v, Off: 0v
4	GN/WT	Stop Light Switch Signal	Brake Off: 0v, On: 12v
5	PK	MIL (lamp) Control	MIL Off: 12v, On: 1v
7	GY/BL	Main Overdrive Switch	Switch Off: 12-14v
8	BK/YL	A/C Amplifier Signal (AC1)	Clutch On: 1.5v, Off: 12v
9	BL/WT	Vehicle Speed Sensor	At 55 mph: 48 Hz
10	BK/WT	A/C Magnet Cutout Relay	A/C Off: 12v, On: 1v
11	BK	A/T: Starter Switch Signal	9-11v (cranking)
12	BK/YL	EFI Main Relay Power B1+	12-14v
13	BK/YL	EFI Main Relay Power	12-14v
14	GN	Circuit Opening Relay (FC)	0-3v, off-idle: 12v
20	PK	Cruise Control ECU	At Cruise in OD: 12v
21	GN/BK	A/C Amplifier Signal (ACT)	Clutch On: 12v, Off: 1.5v
22	BK/WT	A/T Neutral Start Switch	In P/N: 9-11v (cranking)

Standard Colors and Abbreviations

Abbreviation	Color	Abbreviation	Color	Abbreviation	Color
BK	Black	GY	Gray	RD	Red
BL	Blue	GN	Green	TN	Tan
BR	Brown	LG	Light Green	VT	Violet
DB	Dark Blue	OR	Orange	WT	White
DG	DK Green	PK	Pink	YL	Yellow

1990-91 Coupe, HB 2.2L MFI VIN S (A/T-ECT) 26 Pin Connector

PCM Pin #	Wire Color	Circuit Description (26 Pin)	Value at Hot Idle
1	WT	Distributor Signal (NE+)	AC pulse signals
2	BL/GN	A/T Pattern Selector Switch	Norm: 0v, PWR: 12v
3	BK/YL	Igniter Signal (IGF)	Digital Signal: 0-5-0v
4	BL/YL	A/T-ECT Solenoid (SL)	In Lockup: 12-14v
5	BR/YL	A/T-ECT Solenoid (S2)	1st or OD: 1v
6	BL/RD	A/T-ECT Solenoid (S1)	In 3rd or OD: 1v
9	GN/WT	ISC Signal (ISCC)	Pulse Signals
10	BK/WT	ISC Signal (ISCO)	Pulse Signals
11	YL	Injector 2 & 4 Control	2.0-3.3 ms
12	WT	Injector 1 & 3 Control	2.0-3.3 ms
13	WT/BK	Power Ground	<0.1v
14	BR	Sensor Ground	<0.050v
15	RD	Distributor Signal (G+)	AC pulse signals
16	BR	Sensor Ground	<0.050v
17	BL/RD	A/T-ECT Speed Sensor	Moving: 0-5-0v
18	PK/GN	A/T Select Switch 2nd	In 2nd: 12v, Others: 0v
19	PK/RD	A/T Select Switch Low	In Low: 12v, Others: 0v
20	BK	Igniter Signal (IGT)	Digital Signal: 0-5-0v
23	BL/BK	EGR Solenoid Control (VSV)	12v or 0v
26	WT/BK	Power Ground	<0.1v

Pin Connector Graphic

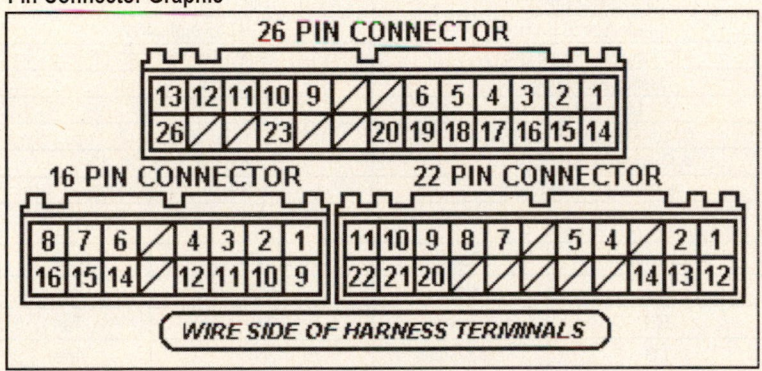

1992-93 Coupe, Hatchback 2.2L I4 VIN S (A/T) 16 Pin Connector

PCM Pin #	Wire Color	Circuit Description (16 Pin)	Value at Hot Idle
1	PK/BL	Sensor VREF	4.9-5.1v
2	GY/BK	MAP Sensor Signal	2.5-4.5v
3	GY	IAT Sensor Signal	At 100ºF: 2.60v
4	RD	ECT Sensor Signal	At 180ºF: 0.51v
5	WT	Sub O2S Signal	0.1-1.1v
6	WT	Main O2S Signal	0.1-1.1v
7	BK	Data Link Connector	12-14v
8	PK/YL	Data Link Connector	12-14v
9	BR	Sensor Ground	<0.050v
10 (Cal)	RD	EGR Gas Temp. Sensor	3.5-4.0v
11	PK/BK	TP Sensor Signal	0.3-0.8v
12	PK	Closed Throttle Switch	1v, at off-idle: 12v
13	WT	Knock Sensor Signal	<0.075v AC
14	LG	Data Link Connector	12-14v
15	YL	Data Link Connector	12-14v

1992-93 Coupe, Hatchback 2.2L I4 VIN S (A/T) 22 Pin Connector

PCM Pin #	Wire Color	Circuit Description (22 Pin)	Value at Hot Idle
1	PK	Battery Direct	12-14v
2	OR	Defogger/Light Idle Up Signal	Switch On: 12v, Off: 0v
4	GN/WT	Stop Light Switch Signal	Brake Off: 0v, On: 12v
5	GN/WT	MIL (lamp) Control	MIL Off: 12v, On: 1v
9	BL/RD	Vehicle Speed Sensor	At 55 mph: 48 Hz
10	BL/BK	A/C Amplifier Signal (ACT)	Clutch On: 12v, Off: 1.5v
11	BK	Starter Switch Signal	9-11v (cranking)
12	BK/YL	EFI Main Relay Power	12-14v
14	GN	Circuit Opening Relay (FC)	0-3v, off-idle: 12v
21	GN/YL	A/C Amplifier Signal (AC1)	Clutch On: 1.5v, Off: 12v
22	BK/WT	A/T Neutral Start Switch	In P/N: 9-11v (cranking)

Standard Colors and Abbreviations

Abbreviation	Color	Abbreviation	Color	Abbreviation	Color
BK	Black	GY	Gray	RD	Red
BL	Blue	GN	Green	TN	Tan
BR	Brown	LG	Light Green	VT	Violet
DB	Dark Blue	OR	Orange	WT	White
DG	DK Green	PK	Pink	YL	Yellow

1992-93 Coupe, Hatchback 2.2L I4 VIN S (A/T) 26 Pin Connector

PCM Pin #	Wire Color	Circuit Description (26 Pin)	Value at Hot Idle
1	BL/YL	A/T-ECT Solenoid (SL)	In Lockup: 12-14v
2	PK	A/T-ECT Solenoid (S1)	In 3rd or OD: 1v
3	BK/YL	Igniter Signal (IGF)	Digital Signal: 0-5-0v
4	YL	Distributor Signal (NE+)	AC pulse signals
5	BK	Distributor Signal (G-)	<0.050v
6	PK/GN	A/T Select Switch 2nd	In 2nd: 12v, Others: 0v
9	GN/WT	ISC Signal (RSC)	Pulse Signals
10	BK/WT	ISC Signal (RSO)	Pulse Signals
11	YL	Injector 2 & 4 Control	2.0-3.3 ms
12	WT	Injector 1 & 3 Control	2.0-3.3 ms
13	WT/BK	Power Ground	<0.1v
14	BR	Sensor Ground	<0.050v
15	BR/YL	A/T-ECT Solenoid (S2)	1st or OD: 1v
17	BK	Distributor Signal (NE-)	<0.050v
18	RD	Distributor Signal (G+)	AC pulse signals
19	PK/RD	A/T Select Switch Low	In Low: 12v, Others: 0v
20	BK	Igniter Signal (IGT)	Digital Signal: 0-5-0v
22	BL/GN	A/T Select Switch Reverse	In 'R': 12v, Others: 0v
23	BK/YL	EGR Solenoid Control (VSV)	12v or 0v
26	WT/BK	Power Ground	<0.1v

Pin Connector Graphic

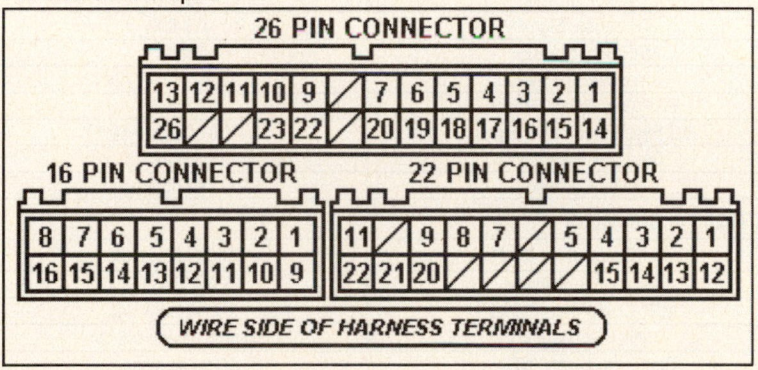

1992-93 Coupe, Hatchback 2.2L I4 VIN S (M/T) 12 Pin Connector

PCM Pin #	Wire Color	Circuit Description (12 Pin)	Value at Hot Idle
1	BK/YL	EFI Main Relay Power B1+	12-14v
2	PK	Battery Direct	12-14v
3	BK/YL	A/C Amplifier Signal (AC1)	Clutch On: 1.5v, Off: 12v
4	GN	Circuit Opening Relay (FC)	0-3v, off-idle: 12v
6	GN/BL	A/C Amplifier Signal (ACT)	Clutch On: 12v, Off: 1.5v
7	BK/YL	EFI Main Relay Power	12-14v
8	PK	MIL (lamp) Control	MIL Off: 12v, On: 1v
10	BK/WT	A/C Magnetic Clutch (ACMG)	Clutch Off: 0v, On: 12v
11	BL/RD	Vehicle Speed Sensor	At 55 mph: 48 Hz
12	OR	Defogger/Light Idle Up Signal	Switch On: 12v, Off: 0v

1992-93 Coupe, Hatchback 2.2L I4 VIN S (M/T) 16 Pin Connector

PCM Pin #	Wire Color	Circuit Description (16 Pin)	Value at Hot Idle
1	WT	Sub O2S Signal	0.1-1.1v
2	GY/BK	MAP Sensor Signal	1-1.5v
3	GY	IAT Sensor Signal	At 100°F: 2.60v
4	RD	ECT Sensor Signal	At 180°F: 0.51v
5	WT	Knock Sensor Signal	<0.075v AC
6	WT	Main O2S Signal	0.1-1.1v
7	BL/GN	Data Link Connector	12-14v
8	PK/YL	Data Link Connector	12-14v
9	BR	Sensor Ground	<0.050v
10	PK/BK	TP Sensor Signal	0.3-0.8v
11	PK/BL	MAP Sensor VREF	4.9-5.1v
12	PK	Closed Throttle Switch	1v, at off-idle: 12v
13 (Cal)	BL	EGR Gas Temp. Sensor	3.5-4.0v
14	YL/RD	A/C Evaporator Temp. Signal	1.4-1.8v (temp. >59°F)
15	BK	Data Link Connector	12-14v
16	BR	Sensor Ground	<0.050v

1992-93 Coupe, Hatchback 2.2L I4 VIN S (M/T) 26 Pin Connector

PCM Pin #	Wire Color	Circuit Description (26 Pin)	Value at Hot Idle
1	YL/BL	A/C Idle Up Solenoid	A/C On: 1v
2	BK	M/T: Clutch Start Signal	9-11v (cranking)
3	BK/YL	Igniter Signal (IGF)	Digital Signal: 0-5-0v
4	YL	Distributor Signal (NE+)	AC pulse signals
5	RD	Distributor Signal (G+)	AC pulse signals
9	GN/WT	ISC Signal (ISCC)	Pulse Signals
10	BK/WT	ISC Signal (ISCO)	Pulse Signals
12	WT	Injector 1 & 3 Control	2.0-3.3 ms
13	WT/BK	Power Ground	<0.1v
17	BL	Distributor Signal (NE-)	<0.050v
18	BK	Distributor Signal (G-)	<0.050v
22	BK	Igniter Signal (IGT)	Digital Signal: 0-5-0v
23	BK/YL	EGR Solenoid Control (VSV)	12v or 0v
24	BR	Sensor Ground	<0.050v
25	YL	Injector 2 & 4 Control	2.0-3.3 ms
26	WT/BK	Power Ground	<0.1v

Pin Connector Graphic

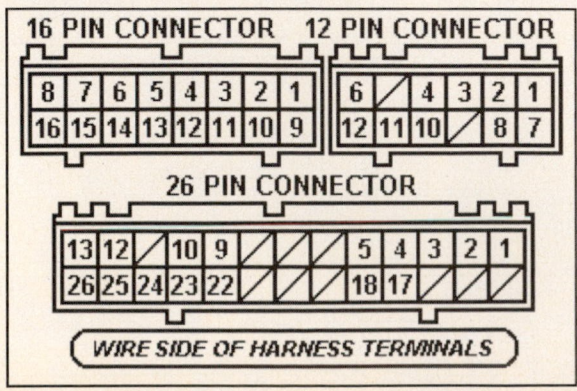

1994 Coupe, HB 2.2L I4 MFI VIN S (All) 16 Pin Connector

PCM Pin #	Wire Color	Circuit Description (16 Pin)	Value at Hot Idle
1	RD	Sensor VREF	4.9-5.1v
2	GY/BK	MAP Sensor Signal	1-1.5v
3	BL/RD	IAT Sensor Signal	At 100ºF: 2.60v
4	GN	ECT Sensor Signal	At 180ºF: 0.51v
5 (Cal)	WT	Sub O2S Signal	0.1-1.1v
6 (Fed)	WT	Main O2S Signal	0.1-1.1v
7	BK	Data Link Connector	12-14v
8	RD/WT	Data Link Connector	12-14v
9	BR	Sensor Ground	<0.050v
10 (Cal)	BK/RD	EGR Gas Temp. Sensor	3.5-4.0v
11	BK/WT	TP Sensor Signal	0.3-0.8v
12	BL/WT	Closed Throttle Switch	1v, at off-idle: 12v
13	WT	Knock Sensor Signal	<0.075v AC
14	LG	Data Link Connector	12-14v
15	YL	Data Link Connector	12-14v

1994 Coupe, HB 2.2L I4 MFI VIN S (All) 22 Pin Connector

PCM Pin #	Wire Color	Circuit Description (22 Pin)	Value at Hot Idle
1	PK	Battery Direct	12-14v
2	OR	Defogger/Light Idle Up Signal	Switch On: 12v, Off: 0v
4	GN/WT	Stop Light Switch Signal	Brake Off: 0v, On: 12v
5	RD/GN	MIL (lamp) Control	MIL Off: 12v, On: 1v
7	GY/BL	Main Overdrive Switch	Switch Off: 12v, On: 0v
8	BL/YL	A/C Amplifier Signal (A/TS)	A/C On: 1.5v
9	BL/RD	Vehicle Speed Sensor	At 55 mph: 48 Hz
10	BL/BK	A/C Amplifier Signal (ACA)	A/C On: 1.5v
11	BK	Starter Switch Signal	9-11v (cranking)
12	BK/YL	EFI Main Relay Power	12-14v
14	GN	Circuit Opening Relay (FC)	0-3v, off-idle: 12v
18	PK/RD	A/T Select Switch 2nd	In 2nd: 12v, Others: 0v
19	BR/YL	A/T Select Switch Low	In Low: 12v, Others: 0v
20	GN/RD	A/T-ECT ECU Signal	Pulse Signals
21	GN/YL	A/C Amplifier Signal (AC1)	Clutch On: 1.5v, Off: 12v
22	BK/WT	Neutral Start Switch	In P/N: 9-11v (cranking)

Standard Colors and Abbreviations

Abbreviation	Color	Abbreviation	Color	Abbreviation	Color
BK	Black	GY	Gray	RD	Red
BL	Blue	GN	Green	TN	Tan
BR	Brown	LG	Light Green	VT	Violet
DB	Dark Blue	OR	Orange	WT	White
DG	DK Green	PK	Pink	YL	Yellow

1994 Coupe, HB 2.2L I4 MFI VIN S (All) 26 Pin Connector

PCM Pin #	Wire Color	Circuit Description (26 Pin)	Value at Hot Idle
1	BL/YL	A/T-ECT Solenoid (SL)	In Lockup: 12-14v
2	PK	A/T-ECT Solenoid (S1)	In 3rd or OD: 1v
3	BK/YL	Igniter Signal (IGF)	Digital Signal: 0-5-0v
4 (TMC)	YL	Distributor Signal (NE+)	AC pulse signals
4 (TMM)	YL	Distributor Signal (NE+)	AC pulse signals
5 (TMC)	BK	Distributor Signal (G2+)	AC pulse signals
5 (TMM)	BK	Distributor Signal (NE-)	<0.050v
7	RD/YL	A/C Idle Up Solenoid	A/C Off: 12v, On: 1v
8	BL/RD	Fuel Pressure Up Solenoid	1v (at hot restart)
9	GN/WT	ISC Signal (ISCC)	Pulse Signals
10	BK/WT	ISC Signal (ISCO)	Pulse Signals
11 (TMC)	BK/WT	Injector 2 Control	2.0-3.3 ms
11 (TMM)	BK/RD	Injector 2 & 4 Control	2.0-3.3 ms
12 (TMC)	BK/YL	Injector 1 Control	2.0-3.3 ms
12 (TMM)	BK/BL	Injector 1 & 3 Control	2.0-3.3 ms
13	WT/BK	Power Ground	<0.1v
14	BR	Sensor Ground	<0.050v
15	BR/YL	A/T-ECT Solenoid (S2)	1st or OD: 1v
17	BK	Distributor Signal (G-)	<0.050v
18 (TMC)	RD	Distributor Signal (G1+)	AC pulse signals
18 (TMM)	RD	Distributor Signal (G+)	AC pulse signals
20	WT	Igniter Signal (IGT)	Digital Signal: 0-5-0v
23	BL/BK	EGR Solenoid Control (VSV)	12v or 0v
24 (TMC)	BK/RD	Injector 4 Control	2.0-3.3 ms
25 (TMM)	BK/RD	Injector 3 Control	2.0-3.3 ms
26	WT/BK	Power Ground	<0.1v

Pin Connector Graphic

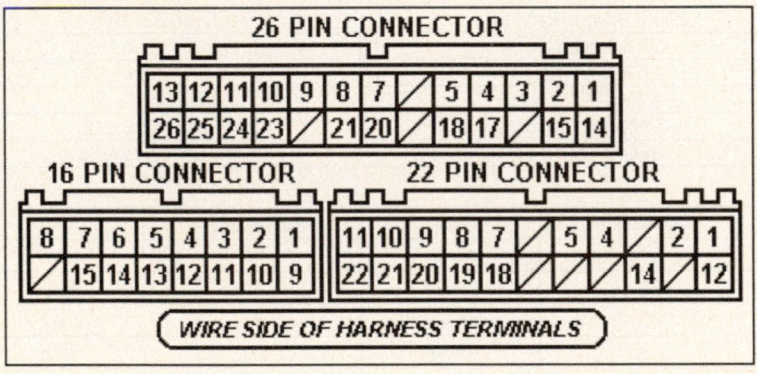

1995 Coupe, HB 2.2L I4 MFI VIN S (All) 16 Pin Connector

PCM Pin #	Wire Color	Circuit Description (16 Pin)	Value at Hot Idle
1	RD	Sensor VREF	4.9-5.1v
2	GY/BK	MAP Sensor Signal	2.5-4.5v
3	BL/RD	IAT Sensor Signal	At 100°F: 2.60v
4	GN	ECT Sensor Signal	At 180°F: 0.51v
5	WT	Sub O2S Signal	0.1-1.1v
6	WT	Main O2S Signal	0.1-1.1v
7	BK	Data Link Connector	12-14v
8	RD/WT	Data Link Connector	12-14v
9	BR	Sensor Ground	<0.050v
10	BK/RD	EGR Gas Temp. Sensor	3.5-4.0v
11	BK/WT	TP Sensor Signal	0.3-0.8v
12	BL/WT	Closed Throttle Switch	1v, at off-idle: 12v
13	WT	Knock Sensor Signal	<0.075v AC
14	LG	Data Link Connector	12-14v
15	YL	Data Link Connector	12-14v

1995 Coupe, HB 2.2L I4 MFI VIN S (All) 22 Pin Connector

PCM Pin #	Wire Color	Circuit Description (22 Pin)	Value at Hot Idle
1	PK	Battery Direct	12-14v
2	OR	Defogger/Light Idle Up Signal	Switch On: 12v, Off: 0v
4	GN/WT	Stop Light Switch Signal	Brake Off: 0v, On: 12v
5	RD/GN	MIL (lamp) Control	MIL Off: 12v, On: 1v
7	GY/BL	Main Overdrive Switch	Switch Off: 12-14v
8	BL/YL	A/C Amplifier A/TS Signal	A/C On: <1.5v
9	OR	Vehicle Speed Sensor	At 55 mph: 48 Hz
10	BL/BK	A/C Amplifier ACA Signal	A/C On: <1.5v
11	BK	Starter Switch Signal	9-11v (cranking)
12	BK/YL	EFI Main Relay Power	12-14v
14	GN/RD	Circuit Opening Relay (FC)	0-3v, off-idle: 12v
18	PK/RD	A/T Select Switch 2nd	In 2nd: 12v, Others: 0v
19	BR/YL	A/T Select Switch Low	In Low: 12v, Others: 0v
20	GN/RD	A/T-ECT ECU Signal	Pulse Signals
21	GN/YL	A/C Amplifier Signal (AC1)	Clutch On: 1.5v, Off: 12v
22	BK/YL	A/T Neutral Start Switch	In P/N: 9-11v (cranking)
22	BK/YL	M/T: Neutral Start Switch	In 'N': 9-11v (cranking)

Standard Colors and Abbreviations

Abbreviation	Color	Abbreviation	Color	Abbreviation	Color
BK	Black	GY	Gray	RD	Red
BL	Blue	GN	Green	TN	Tan
BR	Brown	LG	Light Green	VT	Violet
DB	Dark Blue	OR	Orange	WT	White
DG	DK Green	PK	Pink	YL	Yellow

1995 Coupe, HB 2.2L I4 MFI VIN S (All) 26 Pin Connector

PCM Pin #	Wire Color	Circuit Description (26 Pin)	Value at Hot Idle
1	BL/YL	A/T-ECT Solenoid (SL)	In Lockup: 12-14v
2	PK	A/T-ECT Solenoid (S1)	In 3rd or OD: 1v
3	BK/YL	Igniter Signal (IGF)	Digital Signal: 0-5-0v
4 (TMC)	YL	Distributor Signal (NE+)	AC pulse signals
4 (TMM)	YL	Distributor Signal (NE+)	AC pulse signals
5 (TMC)	BK	Distributor Signal (G2+)	AC pulse signals
5 (TMM)	BK	Distributor Signal (NE-)	<0.050v
7	RD/YL	A/C Idle Up Solenoid	A/C Off: 12v, On: 1v
8	BL/RD	Fuel Pressure Up Solenoid	1v (at hot restart)
9	GN/WT	ISC Signal (ISCC)	Pulse Signals
10	BK/WT	ISC Signal (ISCO)	Pulse Signals
11 (TMC)	BK/WT	Injector 2 Control	2.0-3.3 ms
11 (TMM)	BK/RD	Injector 2 & 4 Control	2.0-3.3 ms
12 (TMC)	BK/BL	Injector 1 Control	2.0-3.3 ms
12 (TMM)	BK/YL	Injector 1 & 3 Control	2.0-3.3 ms
13	WT/BK	Power Ground	<0.1v
14	BR	Sensor Ground	<0.050v
15	BR/YL	A/T-ECT Solenoid (S2)	1st or OD: 1v
17 (TMC)	BK	Distributor G Signal	AC pulse signals
17 (TMM)	BK	Distributor Signal (G-)	<0.050v
18 (TMC)	RD	Distributor Signal (G1+)	AC pulse signals
18 (TMM)	RD	Distributor Signal (G+)	AC pulse signals
20	WT	Igniter Signal (IGT)	Digital Signal: 0-5-0v
21	GN/BK	Water Temperature Switch	Switch On: 0.1v
23	BL/BK	EGR Solenoid Control (VSV)	12v or 0v
24 (TMC)	BK/YL	Injector 4 Control	2.0-3.3 ms
25 (TMM)	BK/RD	Injector 3 Control	2.0-3.3 ms
26	WT/BK	Power Ground	<0.1v

Pin Connector Graphic

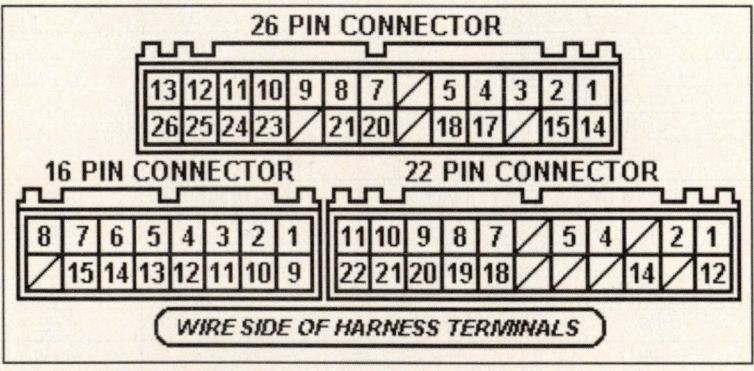

1996-97 Coupe, HB 2.2L I4 VIN G (All) 16 Pin Connector

PCM Pin #	Wire Color	Circuit Description (16 Pin)	Value at Hot Idle
1	RD	Sensor VREF	4.9-5.1v
2	BL	MAP Sensor Signal	1-1.5v
3	BL/RD	IAT Sensor Signal	At 100°F: 2.60v
4	GN	ECT Sensor Signal	At 180°F: 0.51v
5	WT	HO2S-12 (B1 S2) Signal	0.1-1.1v
6	WT	HO2S-11 (B1 S1) Signal	0.1-1.1v
9	BR	Sensor Ground	<0.050v
11	BK/WT	TP Sensor Signal	0.3-0.8v
13	WT	Knock Sensor Signal	<0.075v AC
15	YL	Data Link Connector	12-14v

1996-97 Coupe, HB 2.2L I4 VIN G (All) 22 Pin Connector

PCM Pin #	Wire Color	Circuit Description (22 Pin)	Value at Hot Idle
1	PK	Battery Direct	12-14v
2	OR	Defogger/Light Idle Up Signal	Switch On: 12v, Off: 0v
4	GN/WT	Stop Light Switch Signal	Brake Off: 0v, On: 12v
5	RD/BK	MIL (lamp) Control	MIL Off: 12v, On: 1v
7	GY/BL	Main Overdrive Switch	Switch Off: 12-14v
8	BL/YL	A/C Amplifier Signal (A/TS)	A/C On: 1.5v
9	OR	Vehicle Speed Sensor	At 55 mph: 48 Hz
10	BL/BK	A/C Amplifier Signal (ACA)	A/C On: 1.5v
11	BK	Starter Switch Signal	9-11v (cranking)
12	BK/RD	EFI Main Relay Power	12-14v
14	GN/RD	Circuit Opening Relay (FC)	0-3v, off-idle: 12v
16	WT	SIL Signal (Scan Tool)	Digital Signal
17	RD/WT	A/T Select Switch Reverse	In 'R': 12v, Others: 0v
18	YL/GN	A/T Select Switch 2nd	In 2nd: 12v, Others: 0v
19	YL/RD	A/T Select Switch Low	In Low: 12v, Others: 0v
20	PK	A/T-ECT ECU Signal	Pulse Signals
21	GN/YL	A/C Amplifier Signal (AC1)	Clutch On: 1.5v, Off: 12v
22	BK/YL	A/T Neutral Start Switch	In P/N: 9-11v (cranking)

Standard Colors and Abbreviations

Abbreviation	Color	Abbreviation	Color	Abbreviation	Color
BK	Black	GY	Gray	RD	Red
BL	Blue	GN	Green	TN	Tan
BR	Brown	LG	Light Green	VT	Violet
DB	Dark Blue	OR	Orange	WT	White
DG	DK Green	PK	Pink	YL	Yellow

1996-97 Coupe, HB 2.2L I4 VIN G (All) 26 Pin Connector

PCM Pin #	Wire Color	Circuit Description (26 Pin)	Value at Hot Idle
1	YL	A/T-ECT Solenoid (SL)	In Lockup: 12-14v
2	PK	A/T-ECT Solenoid (S1)	In 3rd or OD: 1v
3	BK/YL	Igniter Signal (IGF)	Digital Signal: 0-5-0v
4	OR	CKP Sensor Signal (NE+)	AC pulse signals
5	BK	Distributor Signal (G2+)	AC pulse signals
9	GN/WT	ISC Signal (ISCC)	Pulse Signals
10	BK/WT	ISC Signal (ISCO)	Pulse Signal
11	BK/WT	Injector 2 Control	2.0-3.3 ms
12	BL	Injector 1 Control	2.0-3.3 ms
13	BR	Power Ground	<0.1v
14	BR	Sensor Ground	<0.050v
15	BR/YL	A/T-ECT Solenoid (S2)	1st or OD: 1v
17	WT	CKP Sensor Signal (NE-)	<0.050v
20	WT	Igniter Signal (IGT)	Digital Signal: 0-5-0v
23	BL/BK	EGR Solenoid Control (VSV)	12v or 0v
24	BK/YL	Injector 4 Control	2.0-3.3 ms
25	BK/RD	Injector 3 Control	2.0-3.3 ms
26	BR	Power Ground	<0.1v

Pin Connector Graphic

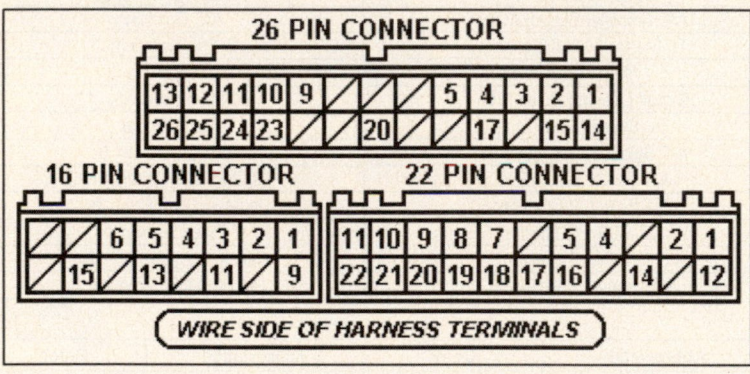

1998-99 Coupe, HB 2.2L I4 VIN G (All) 16 Pin Connector

PCM Pin #	Wire Color	Circuit Description (16 Pin)	Value at Hot Idle
1	RD	Sensor VREF	4.9-5.1v
2	BL	MAP Sensor Signal	1-1.5v
3	BL/RD	IAT Sensor Signal	At 100ºF: 2.60v
4	GN	ECT Sensor Signal	At 180ºF: 0.51v
5	WT	O2S-12 (B1 S2) Signal	0.1-1.1v
6	WT	O2S-11 (B1 S1) Signal	0.1-1.1v
7	RD/WT	EVAP Vapor Pressure Sensor	2.9-3.1v (with hose off)
8	YL/BK	VSV Vapor Press Solenoid	12v or 0v
9	BR	Sensor Ground	<0.050v
11	BK/WT	TP Sensor Signal	0.3-0.8v
13	WT	Knock Sensor Signal	<0.075v AC
15	YL	Data Link Connector (TE)	12-14v
16	BR	Power Ground	<0.1v

1998-99 Coupe, HB 2.2L I4 VIN G (All) 22 Pin Connector

PCM Pin #	Wire Color	Circuit Description (22 Pin)	Value at Hot Idle
1	PK	Battery Direct	12-14v
2	OR	Defogger/Light Idle Up Signal	Switch On: 12v, Off: 0v
4	GN/WT	Stop Light Switch Signal	Brake Off: 0v, On: 12v
5	RD/BK	MIL (lamp) Control	MIL Off: 12v, On: 1v
7	GY/BL	Main Overdrive Switch	Switch Off: 12-14v
8	BL/YL	A/C Amplifier Signal (A/TS)	A/C On: 1.5v
9	OR	Vehicle Speed Sensor	At 55 mph: 48 Hz
10	BL/BK	A/C Amplifier Signal (AC1)	Clutch On: 1.5v, Off: 12v
11	BK	Starter Switch Signal	9-11v (cranking)
12	BK/RD	EFI Main Relay Power	12-14v
14	GN/RD	Circuit Opening Relay (FC)	0-3v, off-idle: 12v
15	GN/BK	A/C Pressure Switch	Switch Closed: 0.1v
16	WT	Data Link Connector (SDL)	0v
17	RD/WT	A/T Select Switch Reverse	In 'R': 12v, Others: 0v
18	YL/GN	A/T Select Switch 2nd	In 2nd: 12v, Others: 0v
19	YL/RD	A/T Select Switch Low	In Low: 12v, Others: 0v
20	PK	A/T-ECT ECU (OD1) Signal	Pulses
21	GN/YL	A/C Amplifier Signal (ACT)	Clutch On: 12v, Off: 1.5v
22	BK/YL	A/T Neutral Start Switch	In P/N: 9-11v (cranking)

Standard Colors and Abbreviations

Abbreviation	Color	Abbreviation	Color	Abbreviation	Color
BK	Black	GY	Gray	RD	Red
BL	Blue	GN	Green	TN	Tan
BR	Brown	LG	Light Green	VT	Violet
DB	Dark Blue	OR	Orange	WT	White
DG	DK Green	PK	Pink	YL	Yellow

1998-99 Coupe, HB 2.2L I4 VIN G (All) 26 Pin Connector

PCM Pin #	Wire Color	Circuit Description (26 Pin)	Value at Hot Idle
1	YL/BK	A/T-ECT Solenoid (SL)	In Lockup: 12-14v
2	PK	A/T-ECT Solenoid (S1)	In 3rd or OD: 1v
3	BK/YL	Igniter Signal (IGF)	Digital Signal: 0-5-0v
4	OR	CKP Sensor Signal (NE+)	AC pulse signals
5	BK	Distributor Signal (G2+)	AC pulse signals
9	GN/WT	ISC Signal (ISCC)	Pulse Signals
10	BK/WT	ISC Signal (ISCO)	Pulse Signals
11	BK/WT	Injector 2 Control	2.0-3.3 ms
12	BL	Injector 1 Control	2.0-3.3 ms
13	BR	Power Ground	<0.1v
14	BR	Sensor Ground	<0.050v
15	BR/YL	A/T-ECT Solenoid (S2)	1st or OD: 1v
17	WT	CKP/CMP Sensor Ground (-)	<0.050v
20	WT	Igniter Signal (IGT)	Digital Signal: 0-5-0v
22	GN/BK	EVAP Purge Solenoid (VSV)	12v or 0v
23	BL/BK	EGR Solenoid Control (VSV)	12v or 0v
24	BK/YL	Injector 4 Control	2.0-3.3 ms
25	BK/RD	Injector 3 Control	2.0-3.3 ms
26	BR	Power Ground	<0.1v

Pin Connector Graphic

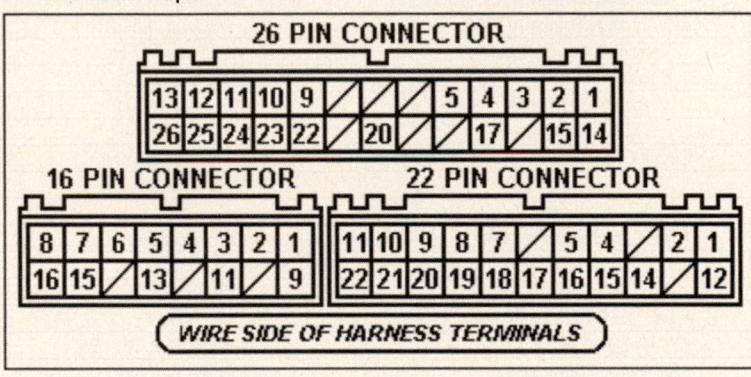

COROLLA Pin Tables

1990-91 Coupe & Sedan, 1.6L I4 4A-FE MFI VIN A 16 Pin Connector

PCM Pin #	Wire Color	Circuit Description (16 Pin)	Value at Hot Idle
1	BK/RD	EFI Main Relay Power	12-14v
2	RD/WT	Battery Direct	12-14v
4	GN/RD	Circuit Opening Relay (FC)	0-3v, off-idle: 12v
5	BK	Defogger/Light Idle Up Signal	Switch On: 12v, Off: 0v
6 ('91)	GN	Cold Start Injector Control	1v (at cold startup)
7	RD/YL	A/C Amplifier Signal (ACT)	Clutch On: 12v, Off: 1.5v
8	BL/WT	Check Connector	12-14v
9	BK/RD	EFI Main Relay Power	12-14v
10	RD/YL	MIL (lamp) Control	MIL Off: 12v, On: 1v
12	BK/WT	A/C Magnetic Clutch (ACMG)	Clutch Off: 0v, On: 12v
13	PK/WT	Vehicle Speed Sensor	At 55 mph: 48 Hz
16	RD/WT	Check Connector	12-14v

1990-91 Coupe & Sedan 1.6L I4 4A-FE MFI VIN A 26 Pin Connector

PCM Pin #	Wire Color	Circuit Description (26 Pin)	Value at Hot Idle
1 (Cal)	BK/YL	EGR Solenoid Control (VSV)	12v or 0v
2	BK/WT	A/T Neutral Start Switch	In P/N: 9-11v (cranking)
3	WT	ECT Sensor Signal	At 180°F: 0.51v
4	GN/RD	Vacuum Sensor Signal	2.5-4.5v
5	YL/BK	IAT Sensor Signal	At 100°F: 2.60v
6	BK	Igniter Signal (IGT)	0.72-0.74v
7	BK/YL	Igniter Signal (IGF)	0.74-0.76v
8	BK	Distributor Signal (G+)	AC pulse signals
9	WT	Distributor Signal (G-)	<0.050v
10	BK	O2S Signal	0.1-1.1v
11	BK	Starter Switch Signal	9-11v (cranking)
12	YL	Injector Pair 1 & 3 Control	2.0-3.3 ms
13	BR	Power Ground	<0.1v
14	BK	ISC Solenoid Control	Pulse Signals
15	RD	Main Overdrive Switch	Switch Off: 12-14v
16	BR	Sensor Ground	<0.050v
17	LG	Wide Open Throttle Switch	1v, at off-idle: 12v
18	YL	Vacuum Sensor VREF	4.9-5.1v
19	BL	Closed Throttle Switch	1v, at off-idle: 12v
20 (Cal)	PK	EGR Gas Temp. Sensor	3.5-4.0v
21	RD	Distributor Signal (NE+)	AC pulse signals
22	BR	Sensor Ground	<0.050v
24	BR	Sensor Ground	<0.050v
25	YL	Injector Pair 2 & 4 Control	2.0-3.3 ms
26	BR	Power Ground	<0.1v

Pin Connector Graphic

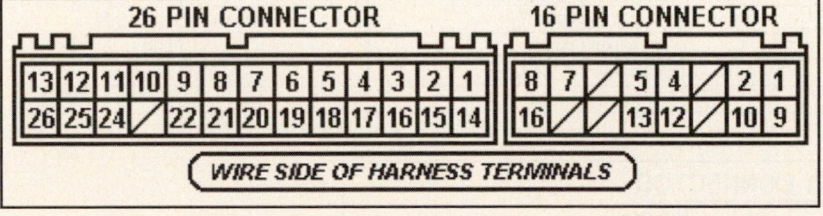

1990-91 Coupe & Sedan 1.6L 4A-GE VIN A (All) 12 Pin Connector

PCM Pin #	Wire Color	Circuit Description (12 Pin)	Value at Hot Idle
1	BK/RD	EFI Main Relay Power B1+	12-14v
2	RD/WT	Battery Direct	12-14v
5	GN/WT	Stop Light Switch Signal	Brake Off: 0v, On: 12v
6	RD/YL	A/C Amplifier Signal (ACT)	Clutch On: 12v, Off: 1.5v
7	BK/RD	EFI Main Relay Power	12-14v
8	RD	MIL (lamp) Control	MIL Off: 12v, On: 1v
10	BK/WT	A/C Magnetic Clutch (ACMG)	Clutch Off: 0v, On: 12v
11	PK/WT	Vehicle Speed Sensor	At 55 mph: 48 Hz

1990-91 Coupe & Sedan 1.6L 4A-GE VIN A (All) 16 Pin Connector

PCM Pin #	Wire Color	Circuit Description (16 Pin)	Value at Hot Idle
1	YL/BK	Airflow Meter VREF	5-9v
2	YL/BL	Airflow Meter Signal	1.5-2.5v
3	BK	IAT Sensor Signal	At 100ºF: 2.60v
4	WT	ECT Sensor Signal	At 180ºF: 0.51v
5 (Cal)	BK	Sub HO2S Signal	0.1-1.1v
6 (Fed)	BK	Main HO2S Signal	0.1-1.1v
7	RD/GN	Check Connector	12-14v
8	RD/WT	Check Connector	12-14v
9	BR	Sensor Ground	<0.050v
10	RD	TP Sensor Signal	0.3-0.8v
11	YL	Sensor VREF (VC)	4.9-5.1v
12	BL	Closed Throttle Switch	1v, at off-idle: 12v
13 (Cal)	RD/GN	EGR Gas Temp. Sensor	3.5-4.0v
14	BK	Knock Sensor Signal	<0.075v AC
16	BR	Sensor Ground	<0.050v

1990-91 Coupe & Sedan 1.6L 4A-GE VIN A (All) 26 Pin Connector

PCM Pin #	Wire Color	Circuit Description (26 Pin)	Value at Hot Idle
1	LG	Fuel Pressure Up Solenoid	1v (at hot restart)
2	BK/WT	Starter Switch Signal	9-11v (cranking)
3	BK/YL	Igniter Signal (IGF)	Digital Signal: 0-5-0v
4	BK/RD	Distributor Signal (NE+)	AC pulse signals
5	BK	Distributor Signal (G+)	AC pulse signals
9	BK/WT	A/C Idle Up Solenoid	A/C On: 1v
11 (Cal)	RD/BK	Sub HO2S Heater	1v (Heater On)
12	GN/RD	Injector Pair 3 & 4 Control	2.0-3.3 ms
13	BR	Power Ground	<0.1v
14 (Fed)	RD/BK	Main HO2S Heater	1v (Heater On)
15	BL/RD	Check Connector	12-14v
16 (Cal)	GN/RD	EGR Solenoid Control (VSV)	12v or 0v
17	WT	Distributor Signal (G-)	<0.050v
22	BK	Igniter Signal (IGT)	Digital Signal: 0-5-0v
24	BR	Sensor Ground	<0.050v
25	GN/YL	Injector Pair 1 & 2 Control	2.0-3.3 ms
26	BR	Power Ground	<0.1v

Pin Connector Graphic

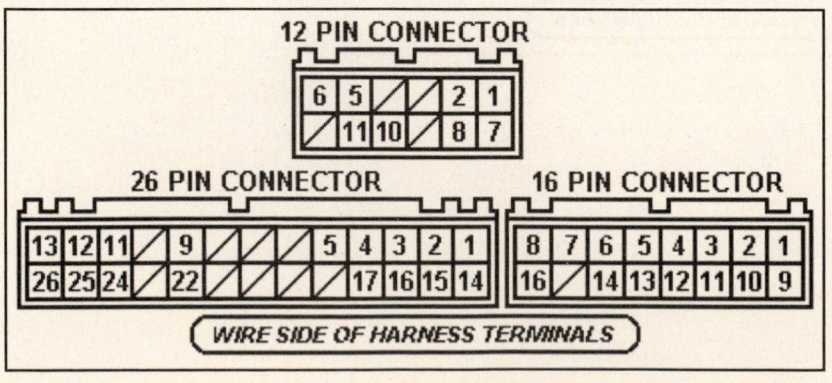

1992 Coupe & Sedan 1.6L I4 MFI VIN A (All) 16 Pin Connector

PCM Pin #	Wire Color	Circuit Description (16 Pin)	Value at Hot Idle
1	BK/RD	EFI Main Relay Power	12-14v
2	RD/WT	Battery Direct	12-14v
4	GN/RD	Circuit Opening Relay (FC)	0-3v, off-idle: 12v
5	BK	Defogger/Light Idle Up Signal	Switch On: 12v, Off: 0v
6	GN	Cold Start Injector Control	1v (at cold startup)
7	RD/YL	A/C Amplifier Signal (ACT)	Clutch On: 12v, Off: 1.5v
8	BL/WT	Check Connector	12-14v
9	BK/RD	EFI Main Relay Power	12-14v
10	RD	MIL (lamp) Control	MIL Off: 12v, On: 1v
12	BK/WT	A/C Magnetic Clutch (ACMG)	Clutch Off: 0v, On: 12v
13	PK/WT	Vehicle Speed Sensor	At 55 mph: 48 Hz
16	RD/WT	Check Connector	12-14v

1992 Coupe & Sedan 1.6L I4 MFI VIN A (All) 26 Pin Connector

PCM Pin #	Wire Color	Circuit Description (26 Pin)	Value at Hot Idle
1 (Cal)	BK/YL	EGR Solenoid Control (VSV)	12v or 0v
2	BK/WT	A/T Neutral Start Switch	In P/N: 9-11v (cranking)
3	WT	ECT Sensor Signal	At 180ºF: 0.51v
4	GN/RD	Vacuum Sensor Signal	2.5-4.5v
5	YL/BK	IAT Sensor Signal	At 100ºF: 2.60v
6	BK	Igniter Signal (IGT)	Digital Signal: 0-5-0v
7	BK/YL	Igniter Signal (IGF)	Digital Signal: 0-5-0v
8	BK	Distributor Signal (G+)	AC pulse signals
9	WT	Distributor Signal (G-)	<0.050v
10	BK	O2S-11 Signal	0.1-1.1v
11	BK	Starter Switch Signal	9-11v (cranking)
12	YL	Injector Pair 1 & 3 Control	2.0-3.3 ms
13	BR	Power Ground	<0.1v
14	BK	ISC Solenoid Control	Pulse Signals
15	RD	Main Overdrive Switch	Switch Off: 12v, On: 0v
16	BR	Sensor Ground	<0.050v
17	LG	Wide Open Throttle Switch	1v, at off-idle: 12v
18	YL	Sensor VREF (VC)	4.9-5.1v
19	BL	Closed Throttle Switch	1v, at off-idle: 12v
20 (Cal)	PK	EGR Gas Temp. Sensor	3.5-4.0v
21	RD	Distributor Signal (NE+)	AC pulse signals
22	BR	Sensor Ground	<0.050v
24	BR	Sensor Ground	<0.050v
25	YL	Injector Pair 2 & 4 Control	2.0-3.3 ms
26	BR	Power Ground	<0.1v

Pin Connector Graphic

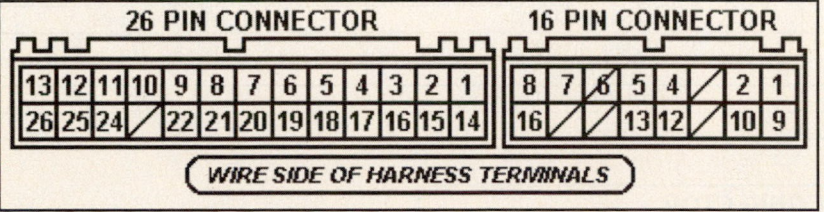

1993-94 Sedan 1.6L I4 MFI VIN A (All) 12 Pin Connector

PCM Pin #	Wire Color	Circuit Description (12 Pin)	Value at Hot Idle
1	BK/RD	EFI Main Relay Power B1+	12-14v
2	RD/WT	Battery Direct	12-14v
4	GN/RD	Circuit Opening Relay (FC)	0-3v, off-idle: 12v
6	RD/YL	A/C Amplifier Signal (ACT)	Clutch On: 12v, Off: 1.5v
7	BK/RD	EFI Main Relay Power	12-14v
8	RD/YL	MIL (lamp) Control	MIL Off: 12v, On: 1v
10	YL/BL	A/C Amplifier Signal (AC1)	Clutch On: 1.5v, Off: 12v
11	PK/WT	Vehicle Speed Sensor	At 55 mph: 48 Hz
12	BK	Defogger/Light Idle Up Signal	Switch On: 12v, Off: 0v

1993-94 Sedan 1.6L I4 MFI VIN A (All) 16 Pin Connector

PCM Pin #	Wire Color	Circuit Description (16 Pin)	Value at Hot Idle
2	GN/RD	MAP Sensor Signal	1-1.5v
3	YL/BK	IAT Sensor Signal	At 100°F: 2.60v
4	WT	ECT Sensor Signal	At 180°F: 0.51v
5 (Cal)	BK	Sub O2S Signal	0.1-1.1v
6 (Fed)	BK	Main O2S Signal	0.1-1.1v
7	GN/WT	Check Connector	12-14v
8	RD/WT	Check Connector	12-14v
9	BR	Sensor Ground	<0.050v
10	LG	TP Sensor Signal	0.3-0.8v
11	YL	Sensor VREF	4.9-5.1v
12	BL	Closed Throttle Switch	1v, at off-idle: 12v
13 (Cal)	WT/BL	EGR Gas Temp. Sensor	3.5-4.0v
15	BL/WT	Check Connector	12-14v
16	BR	Sensor Ground	<0.050v

1993-94 Sedan 1.6L I4 MFI VIN A (All) 26 Pin Connector

PCM Pin #	Wire Color	Circuit Description (26 Pin)	Value at Hot Idle
2	BK	Starter Switch Signal	9-11v (cranking)
3	BK/YL	Igniter Signal (IGF)	Digital Signal: 0-5-0v
4	BK	Distributor Signal (NE+)	AC pulse signals
5	RD	Distributor Signal (G-)	<0.050v
9	BK/WT	ISC Signal (RSC)	Pulse Signals
10	BK/BL	ISC Signal (RSO)	Pulse Signals
12	BK	Injector Pair 1 & 3 Control	2.0-3.3 ms
13	BR	Power Ground	<0.1v
15	BK/WT	A/T Neutral Start Switch	In P/N: 9-11v (cranking)
15	BR	Sensor Ground	<0.050v
16 (Cal)	BK/RD	EGR Solenoid Control (VSV)	12v or 0v
17	WT	Distributor Signal (NE-)	<0.050v
18	GN	Distributor Signal (G+)	AC pulse signals
22	BK	Igniter Signal (IGT)	Digital Signal: 0-5-0v
23	WT/RD	A/C Idle Up Solenoid	A/C On: 1v
24	BR	Sensor Ground	<0.050v
25	BK/RD	Injector Pair 2 & 4 Control	2.0-3.3 ms
26	BR	Power Ground	<0.1v

Pin Connector Graphic

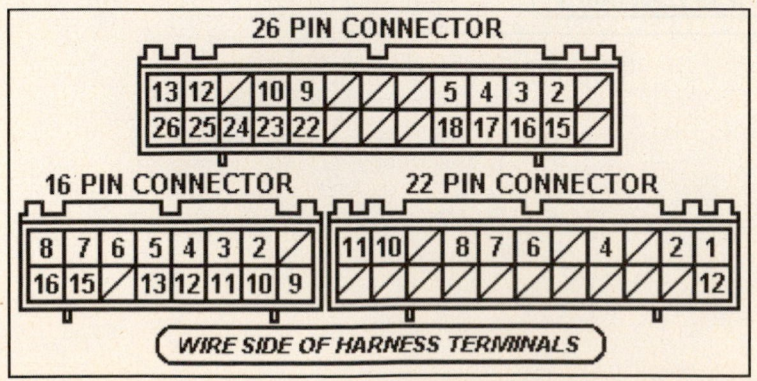

1995 Sedan 1.6L I4 MFI VIN A (All) 16 Pin Connector

PCM Pin #	Wire Color	Circuit Description (16 Pin)	Value at Hot Idle
2	GN/RD	MAP Sensor Signal	1-1.5v
3	YL/BK	IAT Sensor Signal	At 100ºF: 2.60v
4	WT	ECT Sensor Signal	At 180ºF: 0.51v
5 (Cal)	BK	Sub O2S Signal	0.1-1.1v
6 (Fed)	BK	Main O2S Signal	0.1-1.1v
7	GN/WT	Check Connector	12-14v
8	RD/WT	Check Connector	12-14v
9	BR	Sensor Ground	<0.050v
10	LG	TP Sensor Signal	0.3-0.8v
11	YL	Sensor VREF	4.9-5.1v
12	BL	Closed Throttle Switch	1v, at off-idle: 12v
13 (Cal)	WT/BL	EGR Gas Temp. Sensor	3.5-4.0v
15	BL/WT	Check Connector	12-14v

1995 Sedan 1.6L I4 MFI VIN A (All) 22 Pin Connector

PCM Pin #	Wire Color	Circuit Description (22 Pin)	Value at Hot Idle
2	RD/WT	Battery Direct	12-14v
4	GN/RD	Circuit Opening Relay (FC)	0-3v, off-idle: 12v
6	RD/YL	A/C Amplifier Signal (ACT)	Clutch On: 12v, Off: 1.5v
7	BK/RD	EFI Main Relay Power	12-14v
8	RD/YL	MIL (lamp) Control	MIL Off: 12v, On: 1v
10	YL/BL	A/C Amplifier Signal (AC1)	Clutch On: 1.5v, Off: 12v
11	PK/WT	Vehicle Speed Sensor	At 55 mph: 48 Hz
12	BK	Defogger/Light Idle Up Signal	Switch On: 12v, Off: 0v

1995 Sedan 1.6L I4 MFI VIN A (All) 26 Pin Connector

PCM Pin #	Wire Color	Circuit Description (26 Pin)	Value at Hot Idle
2	BK	Starter Switch Signal	9-11v (cranking)
3	BK/YL	Igniter Signal (IGF)	Digital Signal: 0-5-0v
4	BK	Distributor Signal (NE+)	AC pulse signals
5	RD	Distributor Signal (G1+)	AC pulse signals
9	BK/WT	ISC Signal (RSC)	Pulses
10	BK/BL	ISC Signal (RSO)	Pulses
12	BK	Injector Pair 1 & 3 Control	2.0-3.3 ms
13	BR	Power Ground	<0.1v
15	BK/WT	A/T: Neutral Start Switch	In P/N: 9-11v (cranking)
15	BR	M/T: Neutral Start Switch	In 'N': 9-11v (cranking)
16 (Cal)	BK/RD	EGR Solenoid Control (VSV)	12v or 0v
17	WT	Distributor Signal (NE-)	<0.050v
18	GN	Distributor Signal (G-)	<0.050v
22	BK	Igniter Signal (IGT)	Digital Signal: 0-5-0v
23	WT/RD	Idle Up Solenoid Control	A/C On: 1v, Off: 12v
24	BR	Sensor Ground	<0.050v
25	BK/RD	Injector Pair 2 & 4 Control	2.0-3.3 ms
26	BR	Power Ground	<0.1v

Pin Connector Graphic

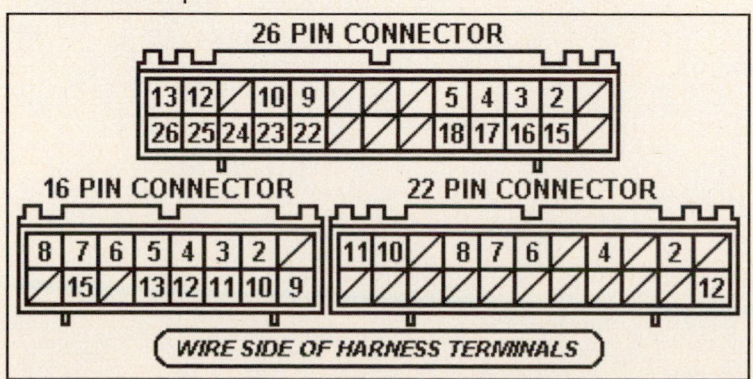

1996-97 Sedan 1.6L I4 MFI VIN A (All) 16 Pin Connector

PCM Pin #	Wire Color	Circuit Description (16 Pin)	Value at Hot Idle
1	YL	Sensor VREF	4.9-5.1v
2	GN/RD	MAP Sensor Signal	1-1.5v
3	YL/BK	IAT Sensor Signal	At 100ºF: 2.60v
4	WT	ECT Sensor Signal	At 180ºF: 0.51v
5	BK	HO2S-12 (B1 S2) Signal	0.1-1.1v
6	BK	O2S-11 (B1 S1) Signal	0.1-1.1v
9	BR	Sensor Ground	<0.050v
11	LG	TP Sensor Signal	0.3-3.1v
13	BK	Knock Sensor Signal	<0.075v AC
15	BL/WT	Data Link Connector	12-14v
16	BR	Power Ground	<0.1v

1996-97 Sedan 1.6L I4 MFI VIN A (All) 22 Pin Connector

PCM Pin #	Wire Color	Circuit Description (22 Pin)	Value at Hot Idle
1	RD/WT	Battery Direct	12-14v
2	GN	Tail Light Switch Signal	Switch On: 12v, Off: 0v
3	BK	Defogger Idle Up Signal	Switch On: 12v, Off: 0v
4	BK/WT	Stop Light Switch Signal	Brake Off: 0v, On: 12v
5	RD/YL	MIL (lamp) Control	MIL Off: 12v, On: 1v
7	LG	Overdrive Main Switch	Switch Off: 12v
9	PK/WT	Vehicle Speed Sensor	At 55 mph: 48 Hz
10 (TMMC)	YL/RD	A/C Amplifier Signal (AC1)	Clutch On: 1.5v, Off: 12v
10 (NUMMI)	YL/BL	A/C On Signal ('97)	Clutch On: 1.5v, Off: 12v
11	BK	Starter Switch Signal	9-11v (cranking)
12 (TMMC)	BK	Main Relay B+	12-14v
12 (NUMMI)	BK/RD	Main Relay B+ ('97)	12-14v
13	BL/BK	A/T Select Switch Drive	In 'D': 12v, Others: 0v
14	GN/RD	Circuit Opening Relay (FC)	0-3v, off-idle: 12v
16	BK	Data Link Connector	12-14v
17	RD/BK	A/T Select Switch Reverse	In 'R': 12v, Others: 0v
18	GN/RD	A/T Select Switch 2nd	In 2nd: 12v, Others: 0v
19	LG	A/T Select Switch Low	In Low: 12v, Others: 0v
20	RD	Cruise Control ECU	At Cruise in OD: 12v
21	RD/YL	A/C Amplifier Signal (ACT)	Clutch On: 12v, Off: 1.5v
22	BK/WT	A/T Neutral Start Switch	In P/N: 9-11v (cranking)

Standard Colors and Abbreviations

Abbreviation	Color	Abbreviation	Color	Abbreviation	Color
BK	Black	GY	Gray	RD	Red
BL	Blue	GN	Green	TN	Tan
BR	Brown	LG	Light Green	VT	Violet
DB	Dark Blue	OR	Orange	WT	White
DG	DK Green	PK	Pink	YL	Yellow

1996-97 Sedan 1.6L I4 MFI VIN A (All) 26 Pin Connector

PCM Pin #	Wire Color	Circuit Description (26 Pin)	Value at Hot Idle
1	BL/YL	A/T-ECT Solenoid (SL)	In Lockup: 12-14v
2	PK	A/T-ECT Solenoid (S1)	In 3rd or OD: 1v
3	BK/YL	Igniter Signal (IGF)	Digital Signal: 0-5-0v
4	BK	CKP Sensor Signal (NE+)	AC pulse signals
5	BK	Distributor Signal (G2+)	AC pulse signals
9	BK/WT	ISC Signal (RSC)	Pulse Signals
10	BK/BL	ISC Signal (RSO)	Pulse Signals
11	BK/RD	Injector 2 Control	2.0-3.3 ms
12	YL	Injector 1 Control	2.0-3.3 ms
13	BR	Power Ground	<0.1v
14	BR	Sensor Ground	<0.050v
15	BR/YL	A/T-ECT Solenoid (S2)	1st or OD: 1v
17	WT	CKP Sensor Signal (NE-)	<0.050v
19	BL/BK	HO2S-11 (B1 S1) Heater	1v (Heater On)
20	YL/GN	Igniter Signal (IGT)	Digital Signal: 0-5-0v
23	BK/WT	EGR Solenoid Control (VSV)	12v or 0v
24	BK	Injector 4 Control	2.0-3.3 ms
25	BK/WT	Injector 3 Control	2.0-3.3 ms
26	BR	Power Ground	<0.1v

Pin Connector Graphic

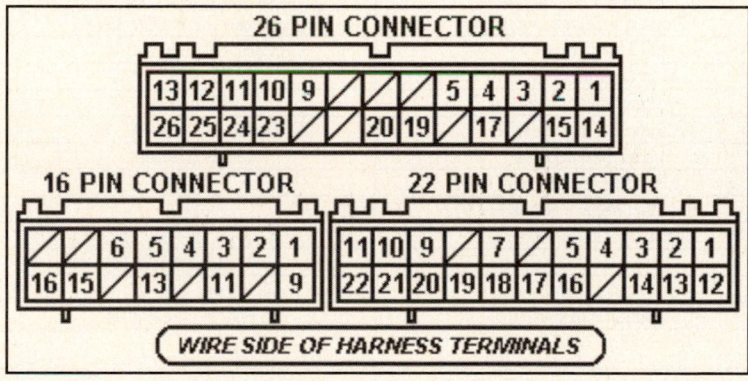

1993-94 Sedan, Wagon 1.8L VIN A (A/T, ECT) 16 Pin Connector

PCM Pin #	Wire Color	Circuit Description (16 Pin)	Value at Hot Idle
1	YL	Sensor VREF	4.9-5.1v
2	GN/RD	MAP Sensor Signal	1-1.5v
3	YL/BK	IAT Sensor Signal	At 100°F: 2.60v
4	WT	ECT Sensor Signal	At 180°F: 0.51v
5 (Cal)	BK	Sub O2S Signal	0.1-1.1v
6 (Fed)	BK	Main O2S Signal	0.1-1.1v
7	BK	Check Connector	12-14v
8	RD/WT	Check Connector	12-14v
9	BR	Sensor Ground	<0.050v
10 (Cal)	WT/BL	EGR Gas Temp. Sensor	3.5-4.0v
11	LG	TP Sensor Signal	0.3-0.8v
12	BL	Closed Throttle Switch	1v, at off-idle: 12v
13	BK	Knock Sensor Signal	<0.075v AC
14	GN/WT	Check Connector	12-14v
15	BL/WT	Check Connector	12-14v
16	BR	Sensor Ground	<0.050v

1993-94 Sedan, Wagon 1.8L VIN A (A/T, ECT) 22 Pin Connector

PCM Pin #	Wire Color	Circuit Description (22 Pin)	Value at Hot Idle
1	RD/WT	Battery Direct	12-14v
2	BK	Defogger/Light Idle Up Signal	Switch On: 12v, Off: 0v
4	GN/WT	Stop Light Switch Signal	Brake Off: 0v, On: 12v
5	RD/YL	MIL (lamp) Control	MIL Off: 12v, On: 1v
7	LG	Main Overdrive Switch	Switch Off: 12-14v
9	PK/WT	Vehicle Speed Sensor	At 55 mph: 48 Hz
10	YL/BL	A/C Amplifier Signal (AC1)	Clutch On: 1.5v, Off: 12v
11	BK	Starter Switch Signal	9-11v (cranking)
12	BK/RD	EFI Main Relay Power	12-14v
13	BK/RD	EFI Main Relay Power B1+	12-14v
14	GN/RD	Circuit Opening Relay (FC)	0-3v, off-idle: 12v
20	RD	Cruise Control ECU	At Cruise in OD: 12v
21	RD/YL	A/C Amplifier Signal (ACT)	Clutch On: 12v, Off: 1.5v
22	BK/WT	Neutral Start Switch	In P/N: 9-11v (cranking)

Standard Colors and Abbreviations

Abbreviation	Color	Abbreviation	Color	Abbreviation	Color
BK	Black	GY	Gray	RD	Red
BL	Blue	GN	Green	TN	Tan
BR	Brown	LG	Light Green	VT	Violet
DB	Dark Blue	OR	Orange	WT	White
DG	DK Green	PK	Pink	YL	Yellow

1993-94 Sedan, Wagon 1.8L VIN A (A/T, ECT) 26 Pin Connector

PCM Pin #	Wire Color	Circuit Description (26 Pin)	Value at Hot Idle
1	BL/YL	A/T-ECT Solenoid (SL)	In Lockup: 12v
2	PK	A/T-ECT Solenoid (S1)	In 3rd or OD: 1v
3	BK/YL	Igniter Signal (IGF)	Digital Signal: 0-5-0v
4	BK	Distributor Signal (NE+)	AC pulse signals
5	GN	Distributor Signal (G-)	<0.050v
6	GN/RD	A/T Select Switch 2nd	In 2nd: 12v, Others: 0v
9	BK/WT	IAC Signal (RSC)	Pulse Signals
10	BK/BL	IAC Signal (RSO)	Pulse Signals
11	BK/RD	Injector Pair 2 & 4 Control	2.0-3.3 ms
12	BK	Injector Pair 1 & 3 Control	2.0-3.3 ms
13	BR	Power Ground	<0.1v
14	BR	Sensor Ground	<0.050v
15	BR/YL	A/T-ECT Solenoid (S2)	1st or OD: 1v
17	WT	Distributor Signal (NE-)	<0.050v
18	RD	Distributor Signal (G1)	AC pulse signals
19	LG	A/T Select Switch Low	In Low: 12v, Others: 0v
20	BK	Igniter Signal (IGT)	Digital Signal: 0-5-0v
21	WT/RD	A/C Idle Up Solenoid	A/C On: 1v, Off: 12v
22	RD/BK	A/T Select Switch Reverse	In 'R': 12v, Others: 0v
23 (Cal)	BK/RD	EGR Solenoid Control (VSV)	12v or 0v
26	BR	Power Ground	<0.1v

Pin Connector Graphic

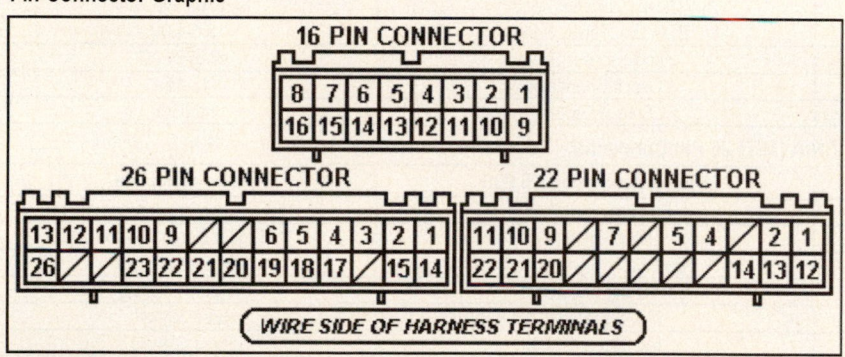

1993-94 Sedan, Wagon 1.8L I4 MFI VIN A (M/T) 12 Pin Connector

PCM Pin #	Wire Color	Circuit Description (12 Pin)	Value at Hot Idle
1	BK/RD	EFI Main Relay Power B1+	12-14v
2	RD/WT	Battery Direct	12-14v
4	GN/RD	Circuit Opening Relay (FC)	0-3v, off-idle: 12v
6	RD/YL	A/C Amplifier Signal (ACT)	Clutch On: 12v, Off: 1.5v
7	BK/RD	EFI Main Relay Power	12-14v
8	RD/YL	MIL (lamp) Control	MIL Off: 12v, On: 1v
10	YL/BL	A/C Amplifier Signal (AC1)	Clutch On: 1.5v, Off: 12v
11	PK/WT	Vehicle Speed Sensor	At 55 mph: 48 Hz
12	BK	Defogger/Light Idle Up Signal	Switch On: 12v, Off: 0v

1993-94 Sedan, Wagon 1.8L I4 MFI VIN A (M/T) 16 Pin Connector

PCM Pin #	Wire Color	Circuit Description (16 Pin)	Value at Hot Idle
2	GN/RD	MAP Sensor Signal	1-1.5v
3	YL/BK	IAT Sensor Signal	At 100°F: 2.60v
4	WT	ECT Sensor Signal	At 180°F: 0.51v
5 (Cal)	BK	Sub O2S Signal	0.1-1.1v
6 (Fed)	BK	Main O2S Signal	0.1-1.1v
7	GN/WT	Check Connector	12-14v
8	RD/WT	Check Connector	12-14v
9	BR	Sensor Ground	<0.050v
10	LG	TP Sensor Signal	0.3-0.8v
11	YL	Sensor VREF	4.9-5.1v
12	BL	Closed Throttle Switch	1v, at off-idle: 12v
13 (Cal)	WT/BL	EGR Gas Temp. Sensor	3.5-4.0v
14	BK	Knock Sensor Signal	<0.075v AC
15	BL/WT	Check Connector	12-14v
16	BR	Sensor Ground	<0.050v

1993-94 Sedan, Wagon 1.8L I4 MFI VIN A (M/T) 26 Pin Connector

PCM Pin #	Wire Color	Circuit Description (26 Pin)	Value at Hot Idle
2	BK	Starter Switch Signal	9-11v (cranking)
3	BK/YL	Igniter Signal (IGF)	Digital Signal: 0-5-0v
4	BK	Distributor Signal (NE+)	AC pulse signals
5	GN	Distributor Signal (G1+)	AC pulse signals
9	BK/WT	IAC Signal (RSC)	Pulse Signals
10	BK/BL	IAC Signal (RSO)	Pulse Signals
12	BK/RD	Injector Pair 1 & 3 Control	2.0-3.3 ms
13	BR	Power Ground	<0.1v
16 (Cal)	BK/RD	EGR Solenoid Control (VSV)	12v or 0v
17	WT	Distributor Signal (NE-)	<0.050v
18	GN	Distributor Signal (G-)	<0.050v
22	BK	Igniter Signal (IGT)	Digital Signal: 0-5-0v
23	WT/RD	ISC Solenoid Control	Pulse Signals
24	BR	Sensor Ground	<0.050v
25	BK/RD	Injector Pair 2 & 4 Control	2.0-3.3 ms
26	BR	Power Ground	<0.1v

Pin Connector Graphic

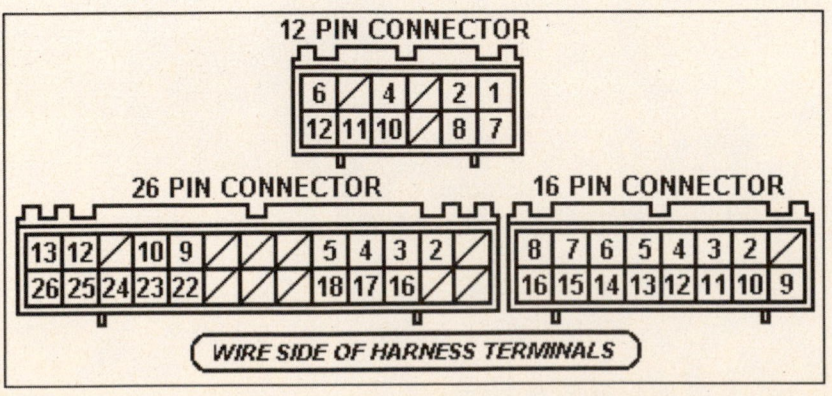

1995 Sedan, Wagon 1.8L I4 VIN A (A/T-ECT) 16 Pin Connector

PCM Pin #	Wire Color	Circuit Description (16 Pin)	Value at Hot Idle
1	YL	Sensor VREF	4.9-5.1v
2	GN/RD	MAP Sensor Signal	1-1.5v
3	YL/BK	IAT Sensor Signal	At 100ºF: 2.60v
4	WT	ECT Sensor Signal	At 180ºF: 0.51v
5 (Cal)	BK	Sub O2S Signal	0.1-1.1v
6 (Fed)	BK	Main O2S Signal	0.1-1.1v
7	BK	Check Connector	12-14v
8	RD/WT	Check Connector	12-14v
9	BR	Sensor Ground	<0.050v
10 (Cal)	WT/BL	EGR Gas Temp. Sensor	3.5-4.0v
11	LG	TP Sensor Signal	0.3-0.8v
12	BL	Closed Throttle Switch	1v, at off-idle: 12v
13	BK	Knock Sensor Signal	<0.075v AC
14	GN/WT	Check Connector	12-14v
15	BL/WT	Check Connector	12-14v

1995 Sedan, Wagon 1.8L I4 VIN A (A/T-ECT) 22 Pin Connector

PCM Pin #	Wire Color	Circuit Description (22 Pin)	Value at Hot Idle
1	RD/WT	Battery Direct	12-14v
2	BK	Tail Light Switch Signal	Switch Off: 12v, On: 1v
3	GN	Defogger Idle Up Signal	Switch On: 12v, Off: 0v
5	RD/YL	MIL (lamp) Control	MIL Off: 12v, On: 1v
7	LG	Main Overdrive Switch	Switch Off: 12v, On: 0v
9	PK/WT	Vehicle Speed Sensor	At 55 mph: 48 Hz
10	YL/BL	A/C Amplifier Signal (AC1)	Clutch On: 1.5v, Off: 12v
11	BK	Starter Switch Signal	9-11v (cranking)
12	BK/RD	EFI Main Relay Power	12-14v
14	GN/RD	Circuit Opening Relay (FC)	0-3v, off-idle: 12v
20	RD	Cruise Control ECU	At Cruise in OD: 12v
21	RD/YL	A/C Amplifier Signal (ACT)	Clutch On: 12v, Off: 1.5v
22	BK/WT	Neutral Start Switch	In PK or N: 12v

Standard Colors and Abbreviations

Abbreviation	Color	Abbreviation	Color	Abbreviation	Color
BK	Black	GY	Gray	RD	Red
BL	Blue	GN	Green	TN	Tan
BR	Brown	LG	Light Green	VT	Violet
DB	Dark Blue	OR	Orange	WT	White
DG	DK Green	PK	Pink	YL	Yellow

1995 Sedan, Wagon 1.8L I4 VIN A (A/T-ECT) 26 Pin Connector

PCM Pin #	Wire Color	Circuit Description (26 Pin)	Value at Hot Idle
1	BL/YL	A/T-ECT Solenoid (SL)	In Lockup: 12-14v
2	PK	A/T-ECT Solenoid (S1)	In 3rd or OD: 1v
3	BK/YL	Igniter Signal (IGF)	Digital Signal: 0-5-0v
4	BK	Distributor Signal (NE+)	AC pulse signals
5	BK	Distributor Signal (G2+)	AC pulse signals
6	GN/RD	A/T Select Switch 2nd	In 2nd: 12v, Others: 0v
7	BL/BK	Sub HO2S Heater	1v (Heater On)
9	BK/WT	ISC Signal (RSC)	Pulse Signals
10	BK/BL	ISC Signal (RSO)	Pulse Signals
11	BK/RD	Injector 2 Control	2.0-3.3 ms
12	BK/YL	Injector 1 Control	2.0-3.3 ms
13	BR	Power Ground	<0.1v
14	BR	Sensor Ground	<0.050v
15	GN/WT	A/T-ECT Solenoid (S2)	1st or OD: 1v
17	WT	Distributor Signal (NE-)	<0.050v
18	RD	Distributor Signal (G1)	AC pulse signals
19	LG	A/T Select Switch Low	In Low: 12v, Others: 0v
20	BK	Igniter Signal (IGT)	Digital Signal: 0-5-0v
21	WT/RD	A/C Idle Up Solenoid	A/C Off: 12v, On: 1v
22	WT/BK	A/T Select Switch Reverse	In 'R': 12v, Others: 0v
23 (Cal)	BK/RD	EGR Solenoid Control (VSV)	12v or 0v
24	BK	Injector 4 Control	2.0-3.3 ms
25	BK/WT	Injector 3 Control	2.0-3.3 ms
26	BR	Power Ground	<0.1v

Pin Connector Graphic

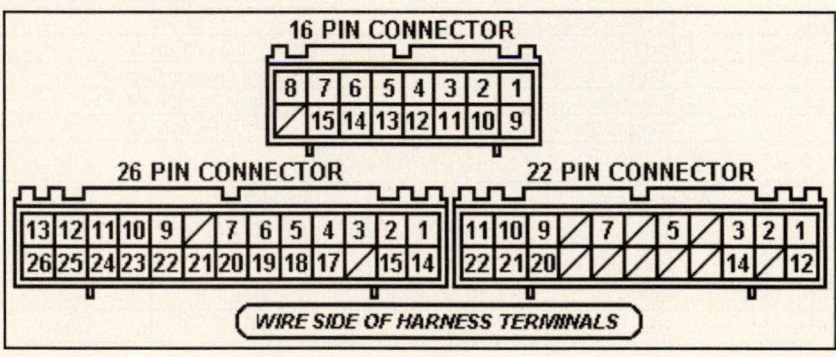

1996-97 Sedan 1.8L I4 MFI VIN B (A/T-ECT) 16 Pin Connector

PCM Pin #	Wire Color	Circuit Description (16 Pin)	Value at Hot Idle
1	YL	Sensor VREF	4.9-5.1v
2	GN/RD	MAP Sensor Signal	1-1.5v
3	YL/BK	IAT Sensor Signal	At 100ºF: 2.60v
4	WT	ECT Sensor Signal	At 180ºF: 0.51v
5	BK	HO2S-12 (B1 S2) Signal	0.1-1.1v
6	BK	O2S-11 (B1 S1) Signal	0.1-1.1v
9	BR	Sensor Ground	<0.050v
11	LG	TP Sensor Signal	0.3-0.8v
13	BK	Knock Sensor Signal	<0.075v AC
15	BL/WT	Data Link Connector	12-14v
16	BR	Power Ground	<0.1v

1996-97 Sedan 1.8L I4 MFI VIN B (A/T-ECT) 22 Pin Connector

PCM Pin #	Wire Color	Circuit Description (22 Pin)	Value at Hot Idle
1	RD/WT	Battery Direct	12-14v
2	GN	Tail Light Switch Signal	Switch On: 12v, Off: 0v
3	BK	Defogger Idle Up Signal	Switch On: 12v, Off: 0v
4	GN/WT	Stop Light Switch Signal	Brake Off: 0v, On: 12v
5	RD/YL	MIL (lamp) Control	MIL Off: 12v, On: 1v
7	LG	Overdrive Main Switch	Switch Off: 12v
9	PK/WT	Vehicle Speed Sensor	At 55 mph: 48 Hz
10 (TMMC)	YL/RD	A/C Cutout Signal	A/C On: 4-6v
10 (NUMMI)	YL/BL	A/C Cutout Signal (1997)	A/C On: 4-6v
11	BK	Starter Switch Signal	9-11v (cranking)
12 (TMMC)	BK	Main Relay B+	12-14v
12 (NUMMI)	BK/RD	Main Relay B+ (1997)	12-14v
13	BL/BK	A/T Select Switch Drive	In 'D': 12v, Others: 0v
14	GN/RD	Circuit Opening Relay (FC)	0-3v, off-idle: 12v
16	BK	Data Link Connector (SDL)	0v
17	RD/BK	A/T Select Switch Reverse	In 'R': 12v, Others: 0v
18	GN/RD	A/T Select Switch 2nd	In 2nd: 12v, Others: 0v
19	LG	A/T Select Switch Low	In Low: 12v, Others: 0v
20	RD	Cruise Control ECU	At Cruise in OD: 12v
21	RD/YL	A/C Amplifier Signal (ACT)	Clutch On: 12v, Off: 1.5v
22	BK/WT	A/T Neutral Start Switch	In PK or N: 12v

Standard Colors and Abbreviations

Abbreviation	Color	Abbreviation	Color	Abbreviation	Color
BK	Black	GY	Gray	RD	Red
BL	Blue	GN	Green	TN	Tan
BR	Brown	LG	Light Green	VT	Violet
DB	Dark Blue	OR	Orange	WT	White
DG	DK Green	PK	Pink	YL	Yellow

1996-97 Sedan 1.8L I4 MFI VIN B (A/T-ECT) 26 Pin Connector

PCM Pin #	Wire Color	Circuit Description (26 Pin)	Value at Hot Idle
1	BL/YL	A/T-ECT Solenoid (SL)	In Lockup: 12-14v
2	PK	A/T-ECT Solenoid (S1)	In 3rd or OD: 1v
3	BL/YL	Igniter Signal (IGF)	Digital Signal: 0-5-0v
4	BK	CKP Sensor Signal (NE+)	AC pulse signals
5	BK	Distributor Signal (G2+)	AC pulse signals
9	BK/WT	ISC Signal (RSC)	Pulse Signals
10	BK/BL	ISC Signal (RSO)	Pulse Signals
11	BK/RD	Injector 2 Control	2.0-3.3 ms
12	YL	Injector 1 Control	2.0-3.3 ms
13	BR	Power Ground	<0.1v
14	BR	Sensor Ground	<0.050v
15	BR/YL	A/T-ECT Solenoid (S2)	1st or OD: 1v
17	WT	CKP Sensor Signal (NE-)	<0.050v
19	BL/BK	HO2S-12 (B1 S2) Heater	1v (Heater On)
20	YL/GN	Igniter Signal (IGT)	Digital Signal: 0-5-0v
23	BK/WT	EGR Solenoid Control (VSV)	12v or 0v
24	BK	Injector 4 Control	2.0-3.3 ms
25	BK/WT	Injector 3 Control	2.0-3.3 ms
26	BR	Power Ground	<0.1v

Pin Connector Graphic

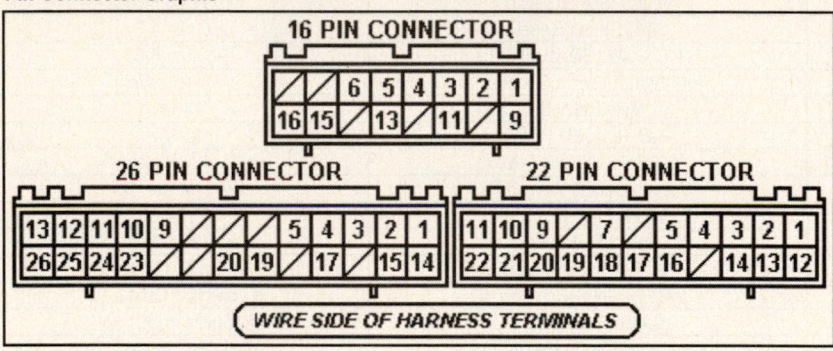

1998-99 Sedan 1.8L I4 MFI VIN R (A/T-ECT) 16 Pin Connector

PCM Pin #	Wire Color	Circuit Description (16 Pin)	Value at Hot Idle
1	YL	Sensor VREF	4.9-5.1v
2	LG/RD	MAP Sensor Signal	1-1.5v
3	YL/BK	IAT Sensor Signal	At 100ºF: 2.60v
4	WT	ECT Sensor Signal	At 180ºF: 0.51v
5	RD	HO2S-12 (B1 S2) Signal	0.1-1.1v
6	WT	HO2S-11 (B1 S1) Signal	0.1-1.1v
7	BL	EVAP Vapor Pressure Sensor	2.9-3.1v (with hose off)
8	PK	HO2S-11 (B1 S1) Heater	1v (Heater On)
9	BR	Sensor Ground	<0.050v
11	LG	TP Sensor Signal	0.3-0.8v
12	RD/WT	Cruise Control Signal (OD1)	At Cruise in OD: 12v
13	BK	Knock Sensor Signal	<0.075v AC
14	PK	HO2S-12 (B1 S2) Heater	1v (Heater On)
15	BL/WT	Data Link Connector	12-14v
16	BR	Power Ground	<0.1v

1998-99 Sedan 1.8L I4 MFI VIN R (A/T-ECT) 22 Pin Connector

PCM Pin #	Wire Color	Circuit Description (22 Pin)	Value at Hot Idle
1	RD/WT	Battery Direct	12-14v
2	BK	Defogger Idle Up Switch	Switch On: 12v, Off: 0v
3	BL/WT	Cruise Control Signal (IDLO)	1.5v, off-idle: 12v
4	GN/WT	Stop Light Switch Signal	Brake Off: 0v, On: 12v
5	RD/YL	MIL (lamp) Control	MIL Off: 12v, On: 1v
8	BK	Tachometer Signal (TACO)	Pulse Signals
9	PK/WT	Vehicle Speed Sensor	At 55 mph: 48 Hz
10	YL/RD	A/C Amplifier Signal (AC1)	Clutch On: 1.5v, Off: 12v
11	BK/WT	Starter Switch Signal	9-11v (cranking)
12	BK	EFI Main Relay Power	12-14v
13	GN	Tail Light Switch Signal	Switch On: 12v, Off: 0v
14	GN/RD	Circuit Opening Relay (FC)	0-3v, off-idle: 12v
16	WT	SIL Signal (Scan Tool)	Digital Signal
17	RD/BK	A/T Select Switch Reverse	In 'R': 12v, Others: 0v
18	GN/RD	A/T Select Switch 2nd	In 2nd: 12v, Others: 0v
19	GN/BK	A/T Select Switch Low	In Low: 12v, Others: 0v
21	RD/BL	A/C Amplifier Signal (ACT)	Clutch On: 12v, Off: 1.5v
22	LG	Overdrive Main Switch	Switch Off: 12v, On: 0v

Standard Colors and Abbreviations

Abbreviation	Color	Abbreviation	Color	Abbreviation	Color
BK	Black	GY	Gray	RD	Red
BL	Blue	GN	Green	TN	Tan
BR	Brown	LG	Light Green	VT	Violet
DB	Dark Blue	OR	Orange	WT	White
DG	DK Green	PK	Pink	YL	Yellow

1998-99 Sedan 1.8L I4 MFI VIN R (A/T-ECT) 26 Pin Connector

PCM Pin #	Wire Color	Circuit Description (26 Pin)	Value at Hot Idle
1	BK/WT	A/T Neutral Start Switch	In P/N: 9-11v (cranking)
4	BK	CKP Sensor Signal (NE+)	AC pulse signals
5	BK	Distributor Signal (G2+)	AC pulse signals
6	BL/RD	PSP Switch Signal	Straight: 12v, Turning: 0v
7	RD	EVAP Vapor Pressure (VSV)	12v or 0v
8	BL/BK	EVAP Purge Solenoid (VSV)	12v or 0v
10	BK/BL	ISC Signal (RSO)	Pulse Signals
11	BK/RD	Injector 2 Control	2.0-3.3 ms
12	YL	Injector 1 Control	2.0-3.3 ms
13	BR	Sensor Ground	<0.050v
14	BR	Power Ground	<0.1v
16	BL/YL	Igniter Signal (IGF)	Digital Signal: 0-5-0v
17	WT	CKP/CMP Sensor Ground (-)	<0.050v
19	RD/BL	Igniter Transistor 2 Control	Digital Signal: 0-5-0v
20	YL/GN	Igniter Transistor 1 Control	Digital Signal: 0-5-0v
21	BL/YL	A/T-ECT Solenoid (SL)	In Lockup: 12-14v
22	BR/YL	A/T-ECT Solenoid (S2)	1st or OD: 1v
23	PK	A/T-ECT Solenoid (S1)	In 3rd or OD: 1v
24	BK	Injector 4 Control	2.0-3.3 ms
25	WT	Injector 3 Control	2.0-3.3 ms
26	BR	Sensor Ground	<0.050v

Pin Connector Graphic

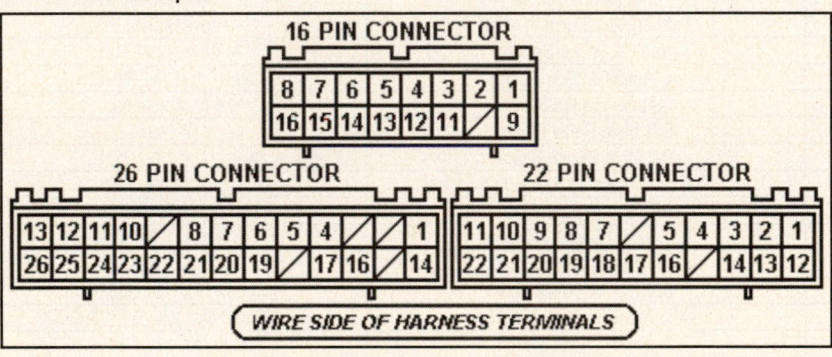

2000-03 Sedan 1.8L I4 MFI VIN R (All) E2 22P Connector

PCM Pin #	Wire Color	Circuit Description (22 Pin)	Value at Hot Idle
1	WT	Battery Direct	12-14v
2	---	Not Used	---
3	GN/RD	Circuit Opening Relay (FC)	0-3v, off-idle: 12v
4	PK	EVAP Vapor Pressure Sensor	2.9-3.1v (with hose off)
5	BR/YL	Overdrive "Off" Indicator	Light Off: 12v, On: 1v
6	GN/OR	Multiplex Meter Input (MPX2)	Digital Signals
7	WT/BK	Power Ground (E03)	<0.1v
8	BK/OR	Ignition Switch Power (B+)	12-14v
9	---	Not Used	---
10	BL/WT	Cruise Control Signal (IDLO)	1.5v, off-idle: 12v
11	WT	SIL Signal (Scan Tool)	Digital Signals
12-14	---	Not Used	---
15	WT	MIL (lamp) Control	MIL Off: 12v, On: 1v
16	BK/RD	EFI No. 1 Fuse Power	12-14v
17-22	---	Not Used	---

2000-03 Sedan 1.8L I4 MFI VIN R (All) E4 28P Connector

PCM Pin #	Wire Color	Circuit Description (28 Pin)	Value at Hot Idle
1	BK	Cruise Control Signal (OD1)	At Cruise in OD: 12v
2	RD/BK	Backup Light Switch Signal	In 'R': 12v, Others: 0v
3	BL/WT	Cooling Fan Relay 3 Control	Relay Off: 12v, On: 1v
4	BL/YL	Cooling Fan Relay 1 Control	Relay Off: 12v, On: 1v
5	PK/BL	Data Link Connector (TC)	12-14v
6	GN/WT	Stop Light Switch Signal (STP)	Brake Off: 0v, On: 12v
7	---	Not Used	---
8	PK	Center Airbag Assembly	Digital Signals
9	RD/BL	ACIS Solenoid (VSV)	12v or 0v
10	LG/RD	Multiplex Meter Input (MPX1)	Digital Signals
11	BL	Starter Switch Signal (STA)	9-11v (cranking)
12	WT/BK	Power Ground (EC)	<0.1v
13-15	---	Not Used	---
16	YL/GN	HO2S-12 (B1 S2) Heater	1v (Heater On)
17	---	Not Used	---
18	OR	A/C Dual Pressure Switch	A/C Off: 12v, On: 1v
19	BR/YL	Overdrive Lamp Control	At Cruise in OD: 1v
20	---	Not Used	---
21	BL/BK	EFI Main Relay Power	12-14v
22	WT/RD	Vehicle Speed Sensor (SPD)	At 55 mph: 48 Hz
23	GN/OR	EVAP Vapor Pressure (VSV)	12v or 0v
24	BK/OR	A/T Select Switch Drive	In 'D': 12v, Others: 0v
25	WT	HO2S-12 (B1 S2) Signal	0.1-1.1v
26	---	Not Used	---
27	BR/WT	Tachometer Signal (TACO)	Pulse Signals
28	GN	A/C Magnetic Clutch (LCK1)	Clutch Off: 0v, On: 12v

Pin Connector Graphic

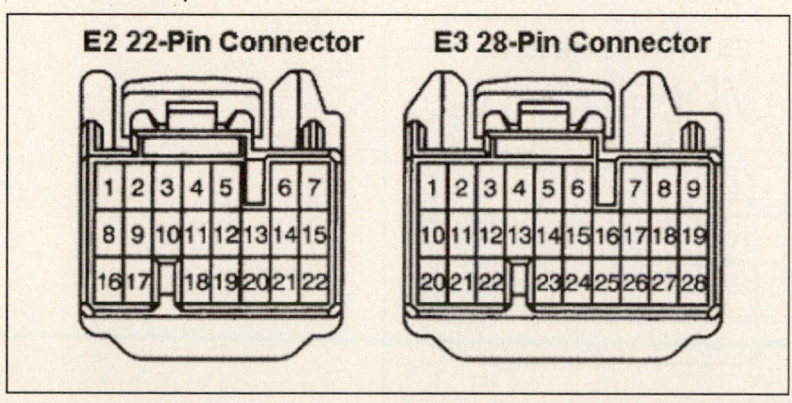

2000-03 Sedan 1.8L I4 MFI VIN R (All) E4 24P Connector

PCM Pin #	Wire Color	Circuit Description (24 Pin)	Value at Hot Idle
1	YL/GN	MAF Sensor Ground (EVG)	<0.050v
2	RD	Sensor VREF (VC)	4.9-5.1v
3	YL/RD	HO2S-11 (B1 S1) Heater	1v (Heater On)
4	GN/OR	EVAP Purge Solenoid (VSV)	12v or 0v
8	PK/BL	A/T Neutral Start Switch	In 'N': 12v, Others: 0v
9	YL/BK	A/T Low Gear Position Switch	In 'LL': 12v, Others: 0v
11	GN/WT	MAF Sensor Signal (VG)	1.0-1.5v
12	BK	HO2S-11 (B1 S1) Signal	0.1-1.1v
13	GN/BL	TFT (fluid) Sensor (THO)	1-4v (varies with temp.)
14	GN	ECT Sensor Signal (THW)	1-4v (varies with temp.)
15	BL	CMP Sensor Signal (G2)	AC pulse signals
16, 24	BK, WT	CKP Sensor Signal (NE+), (NE-)	AC pulse signals
17	BR	Power Ground (E1)	<0.1v
18	BR	Power Ground (E02)	<0.1v
19	BL/YL	A/T 2nd Gear Position Switch	In '2L': 12v, Others: 0v
20	PK	A/T Park Position Switch	In 'P': 12v, Others: 0v
22	BL/RD	IAT Sensor Signal (THA)	1-4v (varies with temp.)
23	BK/WT	TP Sensor Signal (VTA)	1.1-1.9v

2000-03 Sedan 1.8L I4 MFI VIN R (All) E5 31P Connector

PCM Pin #	Wire Color	Circuit Description (31 Pin)	Value at Hot Idle
1	RD	Injector 1 Control	2.0-3.3 ms
2	RD/BL	Injector 2 Control	2.0-3.3 ms
3	RD/WT	Injector 3 Control	2.0-3.3 ms
4	RD/BK	Injector 4 Control	2.0-3.3 ms
5	YL/GN	A/T-ECT Solenoid (SLT-)	Pulse Signals
6	YL/RD	A/T-ECT Solenoid (SLT+)	Pulse Signals
7	WT/BK	Sensor Ground (E03)	<0.050v
8	BR/YL	A/T-ECT Solenoid (S1)	In 3rd or OD: 1v
10	RD/BK	Igniter Transistor 1 Control	Digital Signal: 0-5-0v
11	RD/WT	Igniter Transistor 2 Control	Digital Signal: 0-5-0v
12	GN/RD	Igniter Transistor 3 Control	Digital Signal: 0-5-0v
13	RD/YL	Igniter Transistor 4 Control	Digital Signal: 0-5-0v
15, 16	BK, WT/BL	Turbine Speed Sensor (NT-), (NT+)	AC pulse signals
17	VT/WT	EVAP Canister Closed Valve (VSV)	12v or 0v
18	BK/WT	IAC Valve Signal (RSO)	
19	GN/WT	A/T-ECT Solenoid (SL)	In Lockup: 12-14v
20	YL	A/T-ECT Solenoid (S2)	1st or OD: 1v
21	WT/BK	Power Ground (E01)	<0.1v
22	YL/BK	OIL Pressure Switch (MOPS)	Closed: 1v, Open: 12v
23, 24	WT, GN/OR	Cam/Time Oil Control Valve (-), (+)	Pulse signals
25	BL/YL	Igniter Signal (IGF)	Digital Signal: 0-5-0v
27	WT	Knock Sensor Signal (KNK1)	<0.075v AC
28	PK	PSP Switch Signal	Straight: 12v, Turning: 0v
29	BR/WT	A/T-ECT Solenoid (ST)	Pulse Signals
31	WT/BK	Sensor Ground (E02)	<0.050v

Pin Connector Graphic

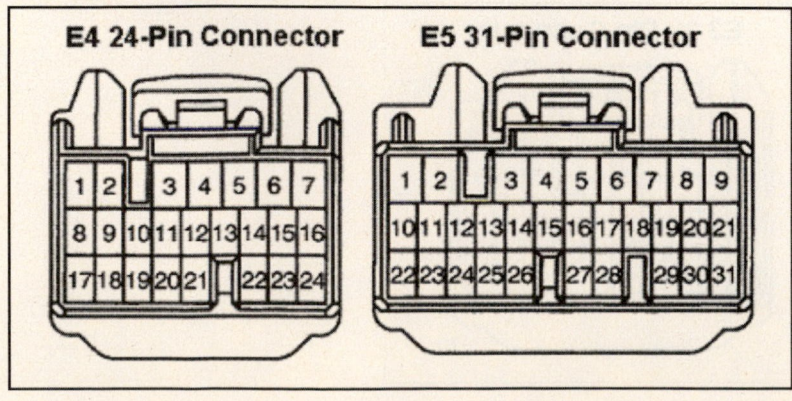

CRESSIDA Pin Tables

1990-92 Sedan 3.0L I6 MFI VIN M (A/T) 16 Pin Connector

PCM Pin #	Wire Color	Circuit Description (16 Pin)	Value at Hot Idle
1	RD/WT	Sensor VREF	KOER: 4.5-5.5v
2	GN/YL	Airflow Meter Signal	1.5-2.5v
3	YL	IAT Sensor Signal	At 100°F: 2.60v
4 ('90)	RD/BL	ECT Sensor Signal	At 180°F: 0.51v
4	GN	ECT Sensor Signal	At 180°F: 0.51v
5 (Cal)	BK	Sub O2S Signal	0.1-1.1v
6	BK	Main O2S Signal	0.1-1.1v
7	GN/RD	Check Connector	12-14v
8	GY	Check Connector	12-14v
9	BR	Sensor Ground	<0.050v
10 (Cal)	GN/WT	EGR Gas Temp. Sensor	3.5-4.0v
11	YL	TP Sensor Signal	0.3-0.8v
12	RD	Closed Throttle Switch	1v, at off-idle: 12v
13	GN/BK	Stop Light Switch Signal	Brake Off: 0v, On: 12v
14	BK	Knock Sensor Signal	<0.075v AC
15	GN/BK	Check Connector	12-14v
16	BK	Distributor Signal (G-)	<0.050v

1990-92 Sedan 3.0L I6 MFI VIN M (A/T) 22 Pin Connector

PCM Pin #	Wire Color	Circuit Description (22 Pin)	Value at Hot Idle
1	WT/RD	Battery Direct	12-14v
2	BK/WT	Ignition Switch Power	12-14v
3	BL/WT	A/T Pattern Selector Switch	Norm: 0v, PWR: 12v
4 ('91)	GY	EFI Main Relay Power	12-14v
4	GN	EFI Main Relay Power	12-14v
5 ('90)	YL/RD	MIL (lamp) Control	MIL Off: 12v, On: 1v
5	RD	MIL (lamp) Control	MIL Off: 12v, On: 1v
6	GN/RD	Fuel Pump Relay	Relay Off: 12v, On: 1v
7	LG	Check Connector	12-14v
8	GN/WT	A/T-ECT Speed Sensor	Moving: 0-5-0v
9	PK/WT	Vehicle Speed Sensor	At 55 mph: 48 Hz
10	BK/WT	A/C Magnetic Clutch (ACMG)	Clutch Off: 0v, On: 12v
11	BK	Starter Switch Signal	9-11v (cranking)
12	BK/RD	EFI Main Relay Power B1+	12-14v
13	BK/RD	EFI Main Relay Power	12-14v
14	OR	A/T Select Switch 2nd	In 2nd: 12v, Others: 0v
15	RD	A/T Pattern Selector Switch	Norm: 0v, PWR: 12v
16	YL/BL	A/T Select Switch Low	In Low: 12v, Others: 0v
20	PK/GN	Main Overdrive Switch	Switch Off: 12v, On: 0v
21	PK/BL	Cruise Control ECU	At Cruise in OD: 12v
22	BK/WT	Neutral Start Switch	In P/N: 9-11v (cranking)

1990-92 Sedan 3.0L I6 MFI VIN M (A/T) 26 Pin Connector

PCM Pin #	Wire Color	Circuit Description (26 Pin)	Value at Hot Idle
1	WT	Distributor Signal (NE+)	AC pulse signals
2	RD	Distributor Signal (G2+)	AC pulse signals
3	BK/YL	Igniter Signal (IGF)	Digital Signal: 0-5-0v
4	GN/OR	ISC Motor (ISC4)	Pulse Signals
5	GN/YL	ISC Motor (ISC3)	Pulse Signals
6	GN/RD	ISC Motor (ISC2)	Pulse Signals
7	GN/WT	ISC Motor (ISC1)	Pulse Signals
8 (Cal)	BK/WT	HO2S Heater	1v (Heater On)
10	GN/RD	EGR Solenoid Control (VSV)	12v or 0v
11	WT	Injector Pair 2 & 6 Control	2.0-3.3 ms
12	YL	Injector Pair 1 & 4 Control	2.0-3.3 ms
13	BR	Power Ground	<0.1v
14	BR	Sensor Ground	<0.050v
15	GN	Distributor Signal (G1+)	AC pulse signals
16	GN/RD	Defogger/Light Idle Up Signal	Switch On: 12v, Off: 0v
17	PK/GN	A/T-ECT Solenoid (SL)	In Lockup: 12-14v
18	BK	A/T-ECT Solenoid (S2)	1st or OD: 1v
19	GY	A/T-ECT Solenoid (S1)	In 3rd or OD: 1v
20	YL/GN	Igniter Signal (IGT)	0.72-0.74v
21	GN/OR	Cruise Control ECU	At Cruise in OD: 12v
24	YL	Cold Start Injector Control	1v (at cold startup)
25	BL	Injector Pair 3 & 5 Control	2.0-3.3 ms
26	BR	Power Ground	<0.1v

Pin Connector Graphic

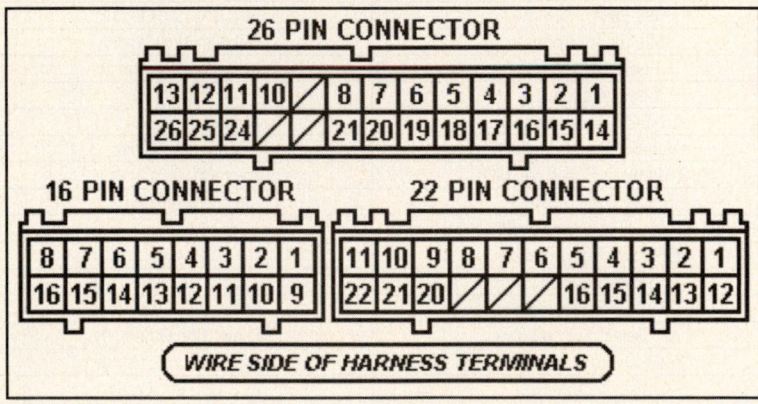

ECHO Pin Tables

2000 Sedan 1.5L I4 DOHC MFI VIN T (All) E4 22 Pin Connector

PCM Pin #	Wire Color	Circuit Description (22 Pin)	Value at Hot Idle
1	BK/YL	Battery Direct	12-14v
2	YL	Water Temp. Cool Indicator	Indicator On: 12, Off: 0v
3	BL	EVAP Canister Closed Valve	12v or 0v
5	YL/RD	MIL (lamp) Control	MIL Off: 12v, On: 1v
6	YL/RD	Water Temp. Hot Indicator	Indicator On: 12v, Off: 0v
8	BK	Tachometer Signal (TACO)	Pulse Signals
9	VT/WT	Speed Signal from C/Meter	At 55 mph: 48 Hz
10	BK/WT	A/C Amplifier Signal (AC1)	Clutch On: 1.5v, Off: 12v
11	BK	Starter Switch Signal	9-11v (cranking)
12	BK/RD	EFI Main Relay Power	12-14v
14	BK	Circuit Opening Relay (FC)	0-3v, off-idle: 12v
16	LG/BK	SIL Signal (Scan Tool)	Digital Signal
17	RD/YL	A/T Select Switch Reverse	In 'R': 12v, Others: 0v
18	GN	A/T Select Switch 2nd	In 2nd: 12v, Others: 0v
19	GN/WT	A/T Select Switch Low	In Low: 12v, Others: 0v
20	BL/WT	Mirror Heater Switch Signal	Switch On: 12-14v
21	BK	A/C Amplifier Signal (ACT)	Clutch On: 12v, Off: 1.5v
22	BK	A/T Neutral Start Switch	Cranking: 9-11v

2000 Sedan 1.5L I4 DOHC MFI VIN T A/T E5 12 Pin Connector

PCM Pin #	Wire Color	Circuit Description (12 Pin)	Value at Hot Idle
1	WT/BL	A/T-ECT Solenoid (SLT+)	Pulse Signals
2	BK/WT	A/T-ECT Solenoid (ST)	Pulse Signals
3	WT/GN	A/T-ECT Solenoid (S1)	In 3rd or OD: 1v
4	GN/OR	A/T: Overdrive "Off" Indicator	Light Off: 12v
5	RD	Turbine Speed Sensor (NT+)	AC pulse signals
6	GN	A/T-ECT Solenoid (SL)	In Lockup: 12-14v
7	WT	A/T-ECT Solenoid (SLT-)	Pulse Signals
9	WT/RD	A/T: ECT Solenoid (S2)	1st or OD: 1v
10	YL	A/T Oil Temperature Sensor	At 230ºF: 0.95v
11	BK	Turbine Speed Sensor (NT-)	AC pulse signals
12	GN/OR	A/T: Overdrive Main Switch	Switch Off: 12v

Pin Connector Graphic

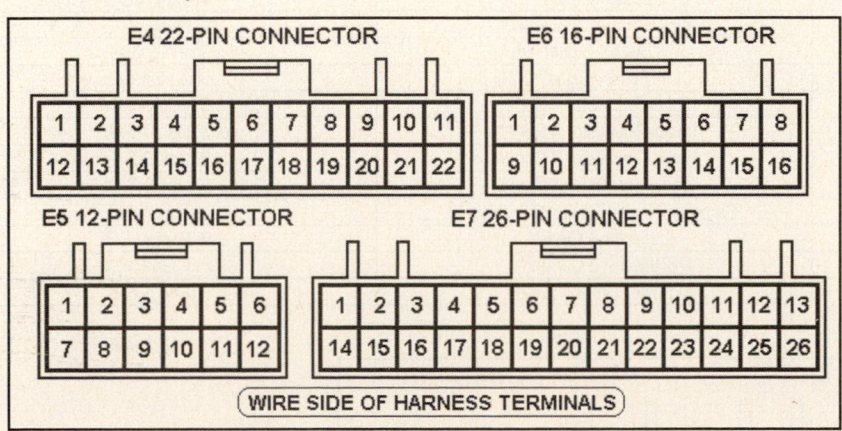

2000 Sedan 1.5L I4 DOHC MFI VIN T (All) E6 16 Pin Connector

PCM Pin #	Wire Color	Circuit Description (16 Pin)	Value at Hot Idle
1	RD/WT	Sensor VREF (VC)	4.9-5.1v
2	PK	MAF Sensor Signal (VG)	1.1-1.5v
3	YL/BK	IAT Sensor Signal	1-4v (varies with temp.)
4	RD/BL	ECT Sensor Signal	1-4v (varies with temp.)
5	BK	HO2S-12 (B1 S2) Signal	0.1-1.1v
6	RD	HO2S-11 (B1 S1) Signal	0.1-1.1v
7	BL/BK	Maximum Hot Switch	Switch On: 12v, Off: 0v
8	BK/OR	HO2S-11 (B1 S1) Heater	1v (Heater On)
9	BR	Sensor Ground (E2)	<0.050v
10	VT	MAF Sensor Ground (EVG)	<0.050v
11	YL/RD	TP Sensor Signal (VTA)	1.1-1.9v
12	YL	PSP Switch Signal	Straight: 12v, Turning: 0v
13	WT	Knock Sensor Signal	<0.075v AC
14	BK	EVAP Vapor Pressure Sensor	Fuel Cap Off: 2.9-3.7v
15	PK/BL	Data Link Connector (TC)	12-14v
16	WT	HO2S-12 (B1 S2) Heater	1v (Heater On)

2000 Sedan 1.5L I4 DOHC MFI VIN T (All) E7 26 Pin Connector

PCM Pin #	Wire Color	Circuit Description (26 Pin)	Value at Hot Idle
1	BR	Power Ground (E03)	<0.1v
2	BK/RD	IAC Signal (RSD)	Pulse Signals
3	YL	Igniter Signal (IGF)	Pulse Signal: 0-5-0v
4	BL	Generator Control Signal (L)	12v
5	RD/YL	A/T Select Switch Drive	In 'D': 12v, Others: 0v
6	GN/WT	Stop Light Switch Signal	Brake Off: 0v, On: 12v
7	---	Not Used	---
8	WT/BL	Water Temperature Switch	Switch Off: 12v
9	WT/GN	EVAP Purge Solenoid (VSV)	12v or 0v
10	RD/WT	Camshaft Timing Oil Control Valve (+)	AC pulse signals
11	BK/YL	Injector 2 Control	2.0-3.3 ms
12	BK/OR	Injector 1 Control	2.0-3.3 ms
13	BR	Power Ground (E1)	<0.1v
14	BR	Power Ground (E01)	<0.1v
15	BR	Case Ground	<0.050v
16	WT	CKP/CMP Sensor Ground (-)	<0.050v
17	OR	CKP Sensor Signal (NE+)	AC pulse signals
18	BK	CMP Sensor Signal (G2+)	AC pulse signals
19	GN/YL	Igniter Transistor 4 Control	Digital Signal: 0-5-0v
20	GN/OR	Igniter Transistor 3 Control	Digital Signal: 0-5-0v
21	GN/BK	Igniter Transistor 2 Control	Digital Signal: 0-5-0v
22	GN/RD	Igniter Transistor 1 Control	Digital Signal: 0-5-0v
23	RD/BK	Camshaft Timing Oil Control Valve (-)	AC pulse signals
24	BK/BL	Injector 4 Control	2.0-3.3 ms
25	BK/WT	Injector 3 Control	2.0-3.3 ms
26	BR	Power Ground (E02)	<0.1v

Pin Connector Graphic

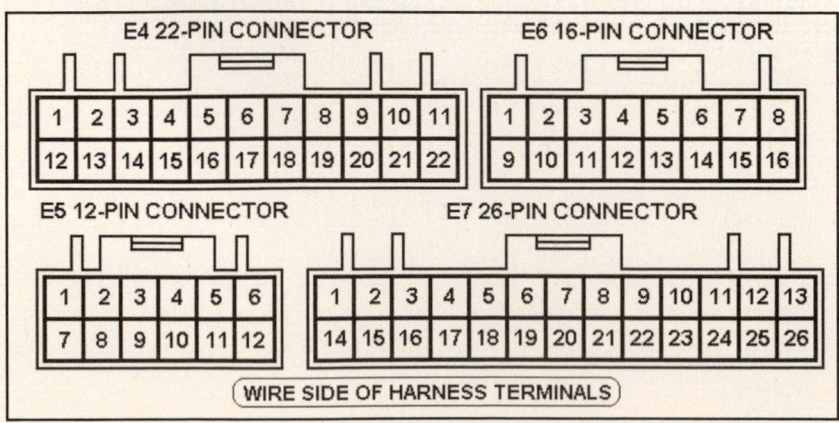

2001-02 Sedan 1.5L I4 DOHC MFI VIN T (All) E4 22 Pin Connector

PCM Pin #	Wire Color	Circuit Description (22 Pin)	Value at Hot Idle
1	BK/YL	Battery Direct	12-14v
2	YL	Water Temp. Cool Indicator	Indicator On: 12, Off: 0v
3	BL	EVAP Canister Closed Valve	12v or 0v
4	---	Not Used	---
5	YL/RD	MIL (lamp) Control	MIL Off: 12v, On: 1v
6	YL/RD	Water Temp. Hot Indicator	Indicator On: 12v, Off: 0v
8	BK	Tachometer Signal (TACO)	Pulse Signals
9	VT/WT	Speed Signal from C/Meter	At 55 mph: 48 Hz
10	BK/WT	A/C Amplifier Signal (AC1)	Clutch On: 1.5v, Off: 12v
11	BK	Starter Switch Signal	9-11v (cranking)
12	BK/RD	EFI Main Relay Power	12-14v
13	---	Not Used	---
14	BK	Circuit Opening Relay (FC)	0-3v, off-idle: 12v
15	---	Not Used	---
16	LG/BK	SIL Signal (Scan Tool)	Digital Signal
17	RD/YL	A/T Select Switch Reverse	In 'R': 12v, Others: 0v
18	GN	A/T Select Switch 2nd	In 2nd: 12v, Others: 0v
19	GN/WT	A/T Select Switch Low	In Low: 12v, Others: 0v
20	BL/WT	Mirror Heater Switch Signal	Switch On: 12-14v
21	BK	A/C Amplifier Signal (ACT)	Clutch On: 12v, Off: 1.5v
22	BK	A/T Neutral Start Switch	Cranking: 9-11v

2001-02 Sedan 1.5L I4 DOHC MFI VIN T A/T E5 12 Pin Connector

PCM Pin #	Wire Color	Circuit Description (12 Pin)	Value at Hot Idle
1	WT/BL	A/T-ECT Solenoid (SLT+)	Pulse Signals
2	BK/WT	A/T-ECT Solenoid (ST)	Pulse Signals
3	WT/GN	A/T-ECT Solenoid (S1)	In 3rd or OD: 1v
4	GN/OR	A/T: Overdrive "Off" Indicator	Light Off: 12v
5	RD	Turbine Speed Sensor (NT+)	AC pulse signals
6	GN	A/T-ECT Solenoid (SL)	In Lockup: 12-14v
7	WT	A/T-ECT Solenoid (SLT-)	Pulse Signals
9	WT/RD	A/T: ECT Solenoid (S2)	1st or OD: 1v
10	YL	A/T Oil Temperature Sensor	At 230ºF: 0.95v
11	BK	Turbine Speed Sensor (NT-)	AC pulse signals
12	GN/OR	A/T: Overdrive Main Switch	Switch Off: 12v

Pin Connector Graphic

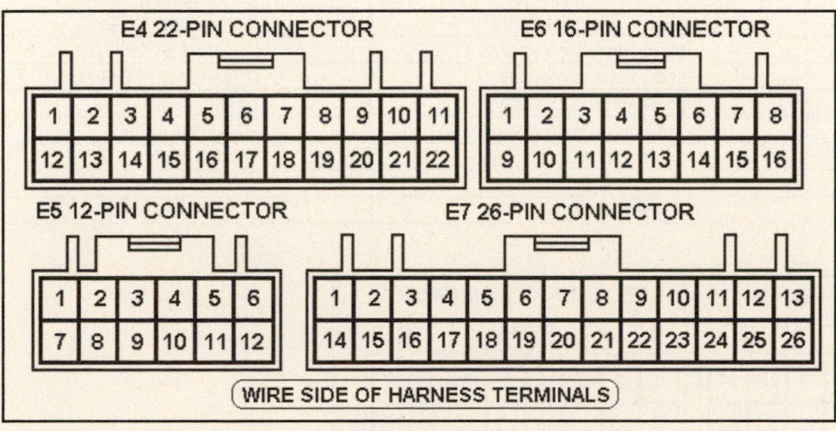

2001-02 Sedan 1.5L I4 DOHC MFI VIN T (All) E6 16 Pin Connector

PCM Pin #	Wire Color	Circuit Description (16 Pin)	Value at Hot Idle
1	RD/WT	Sensor VREF (VC)	4.9-5.1v
2	PK	MAF Sensor Signal (VG)	1.1-1.5v
3	YL/BK	IAT Sensor Signal	1-4v (varies with temp.)
4	RD/BL	ECT Sensor Signal	1-4v (varies with temp.)
5	BK	HO2S-12 (B1 S2) Signal	0.1-1.1v
6	WT	HO2S-11 (B1 S1) Signal	0.1-1.1v
7	BL/BK	Maximum Hot Switch	Switch On: 12v, Off: 0v
8	OR	HO2S-11 (B1 S1) Heater	1v (Heater On)
9	BR	Sensor Ground (E2)	<0.050v
10	VT	MAF Sensor Ground (EVG)	<0.050v
11	YL/RD	TP Sensor Signal (VTA)	1.1-1.9v
12	YL	PSP Switch Signal	Straight: 12v, Turning: 0v
13	WT	Knock Sensor Signal	<0.075v AC
14	BK	EVAP Vapor Pressure Sensor	Fuel Cap Off: 2.9-3.7v
15	PK/BL	Data Link Connector (TC)	12-14v
16	WT	HO2S-12 (B1 S2) Heater	1v (Heater On)

2001-02 Sedan 1.5L I4 DOHC MFI VIN T (All) E7 26 Pin Connector

PCM Pin #	Wire Color	Circuit Description (26 Pin)	Value at Hot Idle
1	BR	Power Ground (E03)	<0.1v
2	BK/RD	IAC Signal (RSD)	Pulse Signals
3	YL	Igniter Signal (IGF)	Pulse Signal: 0-5-0v
4	BL	Generator Control Signal (L)	12v
5	RD/YL	A/T Select Switch Drive	In 'D': 12v, Others: 0v
6	GN/WT	Stop Light Switch Signal	Brake Off: 0v, On: 12v
7	---	Not Used	---
8	WT/BL	Water Temperature Switch	Switch Off: 12v
9	WT/GN	EVAP Purge Solenoid (VSV)	12v or 0v
10	RD/WT	Camshaft Timing Oil Control Valve (+)	AC pulse signals
11	BK/YL	Injector 2 Control	2.0-3.3 ms
12	BK/OR	Injector 1 Control	2.0-3.3 ms
13	BR	Power Ground (E1)	<0.1v
14	BR	Power Ground (E01)	<0.1v
15	BR	Case Ground	<0.050v
16	WT	CKP/CMP Sensor Ground (-)	<0.050v
17	OR	CKP Sensor Signal (NE+)	AC pulse signals
18	BK	CMP Sensor Signal (G2+)	AC pulse signals
19	GN/YL	Igniter Transistor 4 Control	Digital Signal: 0-5-0v
20	GN/OR	Igniter Transistor 3 Control	Digital Signal: 0-5-0v
21	GN/BK	Igniter Transistor 2 Control	Digital Signal: 0-5-0v
22	GN/RD	Igniter Transistor 1 Control	Digital Signal: 0-5-0v
23	RD/BK	Camshaft Timing Oil Control Valve (-)	AC pulse signals
24	BK/BL	Injector 4 Control	2.0-3.3 ms
25	BK/WT	Injector 3 Control	2.0-3.3 ms
26	BR	Power Ground (E02)	<0.1v

Pin Connector Graphic

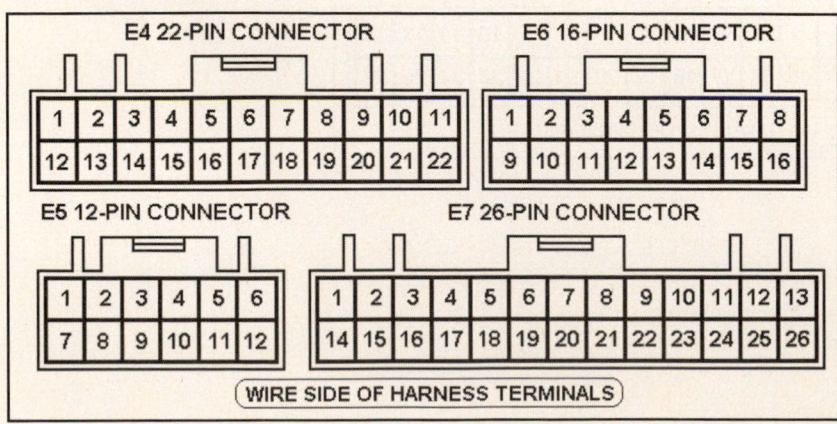

2003 Sedan 1.5L I4 DOHC MFI VIN T (All) E4 31 Pin Connector

PCM Pin #	Wire Color	Circuit Description (31 Pin)	Value at Hot Idle
1	BK/RD	EFI Main Relay Power	12-14v
2	BL/BK	Maximum Hot Switch (MHSW)	Switch Open: 0v, Closed: 12v
3	BK/YL	Battery Direct	12-14v
4	---	Not Used	---
5	BK	Tachometer Signal (TACO)	Pulse Signals
6	YL/RD	Cooling Fan 1 Relay (CF)	Relay Off: 12v, On: 1v
7	WT/BL	A/C Single Pressure Switch (FAN)	Switch Open: 12v, Closed: 1v
8-9	---	Not Used	---
10	BK	Circuit Opening Relay (FC)	0-3v, off-idle: 12v
11	YL/RD	MIL (lamp) Control	MIL Off: 12v, On: 1v
12	BL/WT	Heater Sub1 Relay Control (ELS)	Relay Off: 12v, On: 1v
13	---	Not Used	---
14	YL/BK	Center Airbag Assembly	Digital Signals
15-17	---	Not Used	---
18	LG/BK	Scan Tool Signal (SIL)	Digital Signal
19	RD	Scan Tool (WFSE)	12v
20	PK/BL	Data Link Connector (TC)	12-14v
21	BK	EVAP Vapor Pressure Sensor	Fuel Cap Off: 2.9-3.7v
22-31	---	Not Used	---

2003 Sedan 1.5L I4 DOHC MFI VIN T (All) E5 35 Pin Connector

PCM Pin #	Wire Color	Circuit Description (35 Pin)	Value at Hot Idle
1-2	---	Not Used	---
3	YL/RD	Water Temperature Cool Indicator	Indicator On: 12v, Off: 0v
4	BK/RD	HO2S-12 (B1 S2) Heater	1v (Heater On)
5-6	---	Not Used	---
7	GN/OR	A/T Overdrive "Off" Indicator	Light Off: 12v
8	GN/WT	A/T Select Switch Low	In Low: 12v, Others: 0v
9	GN	A/T Select Switch 2nd	In 2nd: 12v, Others: 0v
10	RD/YL	A/T Select Switch Drive	In 'D': 12v, Others: 0v
11	RD/WT	A/T Select Switch Reverse	In 'R': 12v, Others: 0v
12	---	Not Used	---
13	YL	PSP Switch Signal	Straight: 12v, Turning: 0v
14-16	---	Not Used	---
17	VT/WT	Speed Signal from Combination Meter	At 55 mph: 48 Hz
18	---	Not Used	---
19	GN/WT	Stop Light Switch Signal	Brake Off: 0v, On: 12v
20-27	---	Not Used	---
28	YL/RD	Water Temperature Hot Indicator	Indicator On: 12v, Off: 0v
29	GN/OR	A/T: Overdrive Main Switch	Switch Off: 12v
30	---	Not Used	---
31	BK/WT	A/C Amplifier Signal (AC)	Clutch On: 1.5v, Off: 12v
32	---	Not Used	---
33	BK	A/C Amplifier Signal (ACT)	Clutch On: 12v, Off: 1.5v
34-35	---	Not Used	---

Pin Connector Graphic

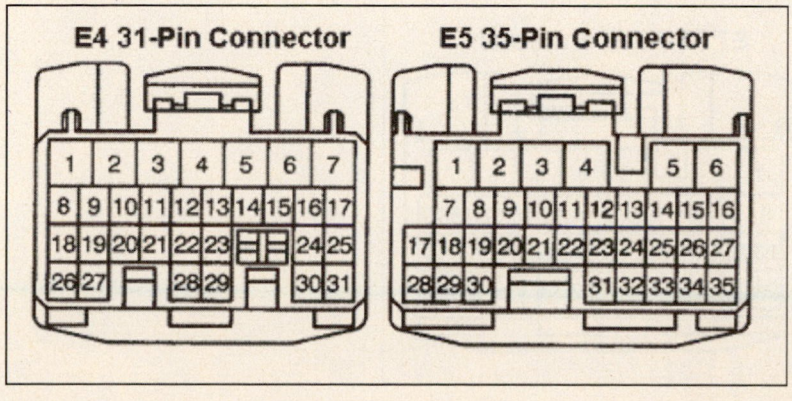

2003 Sedan 1.5L I4 DOHC MFI VIN T (All) E6 34 Pin Connector

PCM Pin #	Wire Color	Circuit Description (34 Pin)	Value at Hot Idle
1	BK/OR	Injector 1 Control	2.0-3.3 ms
2	BK/YL	Injector 2 Control	2.0-3.3 ms
3	BK/WT	Injector 3 Control	2.0-3.3 ms
4	BK/BL	Injector 4 Control	2.0-3.3 ms
5	BK/RD	IAC Signal (RSD)	Pulse Signals
6	BR	Power Ground (E02)	<0.1v
7	BR	Power Ground (E01)	<0.1v
8	GN/RD	Igniter Transistor 1 Control	Digital Signal: 0-5-0v
9	GN/BK	Igniter Transistor 2 Control	Digital Signal: 0-5-0v
10	GN/OR	Igniter Transistor 3 Control	Digital Signal: 0-5-0v
11	GN/YL	Igniter Transistor 4 Control	Digital Signal: 0-5-0v
12	WT/GN	EVAP Purge Solenoid (VSV)	12v or 0v
16	WT	A/T-ECT Solenoid (SLT-)	Pulse Signals
17	WT/BL	A/T-ECT Solenoid (SLT+)	Pulse Signals
18	RD/WT	Sensor VREF (VC)	4.9-5.1v
19	RD/BL	ECT Sensor Signal (THW)	1-4v (varies with temp.)
20	YL/BK	IAT Sensor Signal	1-4v (varies with temp.)
21	YL/RD	TP Sensor Signal (VTA)	1.1-1.9v
23	YL	Igniter Signal (IGF)	Pulse Signal: 0-5-0v
26	BK	CMP Sensor Signal (G2+)	AC pulse signals
27	OR	CKP Sensor Signal (NE+)	AC pulse signals
28	BR	Sensor Ground (E2)	<0.050v
30	YL/GN	A/T Oil Temperature Sensor (OIL)	At 230ºF: 0.95v
34	WT	CKP Sensor Ground (NE-)	<0.050v

2003 Sedan 1.5L I4 DOHC MFI VIN T (All) E7 35 Pin Connector

PCM Pin #	Wire Color	Circuit Description (35 Pin)	Value at Hot Idle
1	WT	Knock Sensor Signal (KNK1)	<0.075v AC
4	BK/RD	HO2S-11 (B1 S1) Heater	1v (Heater On)
5	BR	Power Ground (E03)	<0.1v
7	BL	Generator Control Signal	12-0-12v
8	BK	A/T Neutral Start Switch (NSW)	Cranking: 9-11v
9	BK/YL	Starter Switch Signal (STA)	9-11v (cranking)
12	BK/WT	A/T-ECT Solenoid (ST)	Pulse Signals
13	GN	A/T-ECT Solenoid (SL)	In Lockup: 12-14v
14	WT/RD	A/T: ECT Solenoid (S2)	1st or OD: 1v
15	WT/GN	A/T-ECT Solenoid (S1)	In 3rd or OD: 1v
21	BK	HO2S-12 (B1 S2) Signal	0.1-1.1v
23	WT	HO2S-11 (B1 S1) Signal	0.1-1.1v
24	PK	MAF Sensor Signal (VG)	1.1-1.5v
27	RD	Turbine Speed Sensor (NT+)	AC pulse signals
28	BR	Power Ground (EC)	<0.1v
29	VT	PSP Switch Signal	Straight: 12v, Turning: 0v
32	VT	MAF Sensor Ground (EVG)	<0.050v
35	BK	Turbine Speed Sensor (NT-)	AC pulse signals

Pin Connector Graphic

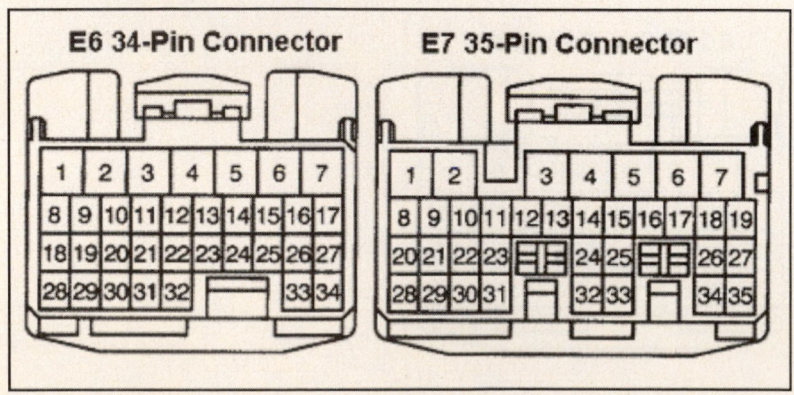

MATRIX Pin Tables

2003 4D Wagon 1.8L DOHC 1ZZ-FE VIN R (All) E3 34P Connector

PCM Pin #	Wire Color	Circuit Description (34 Pin)	Value at Hot Idle
1	YL	Injector 1 Control	2.0-3.3 ms
2	BK	Injector 2 Control	2.0-3.3 ms
3	WT	Injector 3 Control	2.0-3.3 ms
4	BL	Injector 4 Control	2.0-3.3 ms
5	BK/BL	IAC Air Control Valve Signal (RSD)	Pulse Signals
6	WT/BK	Power Ground (E02)	<0.1v
7	WT/BK	Power Ground (E01)	<0.1v
8	RD/BL	Igniter Transistor 1 Control	Digital Signal: 0-5-0v
9	YL/GN	Igniter Transistor 2 Control	Digital Signal: 0-5-0v
10	GY	Igniter Transistor 3 Control	Digital Signal: 0-5-0v
11	WT	Igniter Transistor 4 Control	Digital Signal: 0-5-0v
12	BL/BK	EVAP Vapor Pressure (VSV)	12v or 0v
13	YL	A/C Pressure Switch (HP)	A/C Off: 12v, On: 1v
14	BK/YL	Camshaft Timing Oil Control Valve (-) Left	AC pulse signals
15	YL	Camshaft Timing Oil Control Valve (+) Left	AC pulse signals
16, 17	PK, RD/WT	A/T-ECT Solenoid (SLT-), (SLT+)	Pulse Signals
18	YL	Sensor VREF (VC)	4.9-5.1v
19	WT	ECT Sensor Signal (THW)	1-4v (varies with temp.)
20	YL/BK	IAT Sensor Signal (THA)	1-4v (varies with temp.)
21	LG	TP Sensor Signal (VTA)	0.3-0.9v
23	BL/YL	Igniter Signal (IGF)	Pulse signal: 0-5-0v
24, 25	BL/YL, BK	Camshaft Timing Oil Control Valve (-), (+)	AC pulse signals
26	BK	CMP Sensor Signal (G22+)	AC pulse signals
27, 34	BK, WT	CKP Sensor Signal (NE+), (NE-)	AC pulse signals
28	BR	Power Ground (E2)	<0.1v
30	WT/BL	TFT (fluid) Sensor (THO)	1-4v (varies with temp.)

2003 4D Wagon 1.8L DOHC 1ZZ-FE VIN R (All) E4 35P Connector

PCM Pin #	Wire Color	Circuit Description (35 Pin)	Value at Hot Idle
1	BK, WT	Knock Sensor Signal (+), (-)	<0.075v AC
4	PK	HO2S-11 (B1 S1) Heater	1v (Heater On)
5	WT/BK	Power Ground (E03)	<0.1v
7	BR	Power Ground (E1)	<01v
8	RD	Neutral Start Switch (NSW)	Cranking: 0-3v
9	BK	Starter Signal (STA)	Cranking: 9-11v
11	BK	A/T-ECT Solenoid (DSL)	In Lockup: 12-14v
12	BL/OR	A/T-ECT Solenoid (ST)	In Lockup: 12-14v
13	BL/WT	A/T-ECT Solenoid (SL)	In Lockup: 12-14v
15	RD/YL	A/T-ECT Solenoid (S1)	In 3rd or OD: 1v
16, 17	BL, BL/OR	A/T-ECT Solenoid (SL2-), (SL2+)	Moving in 3rd or OD: <1v
18, 19	BL, RD/YL	A/T-ECT Solenoid (SL1-), (SL1+)	Moving in 1st Gear: <1v
21	WT	HO2S-21 (B2 S1) Signal (OX1B)	0.1-1.1v
23	RD	HO2S-11 (B1 S1) Signal (OX1A)	0.1-1.1v
24	GN	MAF Sensor Signal (VG)	1-3v
28	WT/BK	Power Ground (EC)	<0.1v
29	BL/RD	PSP Switch Signal (PSW)	Straight: 12v, Turning: 0v
32	BL/WT	MAF Sensor Ground (EVG)	<0.050v
33	BK/WT	OIL Pressure Switch Signal (OSW)	Closed: 1v, Open: 12v

Pin Connector Graphic

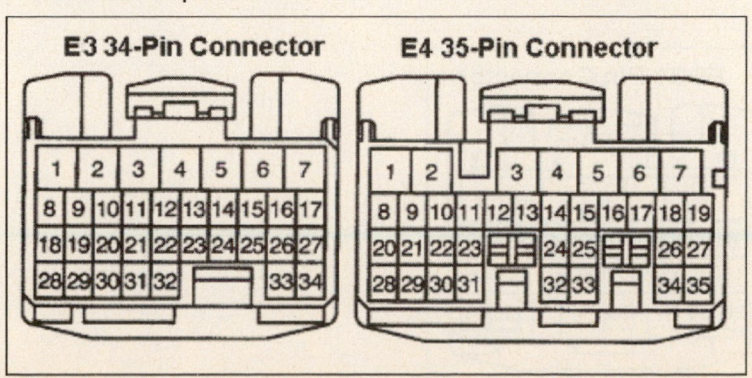

2003 4D Wagon 1.8L DOHC 1ZZ-FE VIN R (All) E5 35P Connector

PCM Pin #	Wire Color	Circuit Description (35 Pin)	Value at Hot Idle
1	BL	EVAP Canister Closed Valve	12v or 0v
2	GN/WT	A/C Magnetic Clutch Relay (ACMG)	Relay Off: 12v, On: 1v
3	---	Not Used	---
4	PK/BK	HO2S-21 (B2 S1) Heater (HT1B)	1v (Heater On)
5	BK	Tachometer Signal (TACO)	Pulse Signals
6	---	Not Used	---
7	LG	Overdrive Indicator Control (ODLP)	Indicator Off: 12v, On: 1v
8	LG/BK	A/T Select Switch Low Signal (L)	In Low: 12v, Others: 0v
9	LG	A/T Select Switch 2nd Signal (2L)	In 2nd: 12v, Others: 0v
10	BL	A/T Select Switch Drive Signal (DL)	In Drive: 12v, Others: 0v
11	RD/BK	A/T Select Switch Reverse (R)	In Reverse: 12v, Others: 0v
12-13	---	Not Used	---
14	YL/RD	Hot Light Indicator Control	Indicator Off: 12v, On: 1v
15-16	---	Not Used	---
17	VT/WT	Speedometer Indicator	At 55 mph: 48 Hz
18	RD/YL	Overdrive Signal to C/C Module (OD1)	0-12-0v
19	GN/WT	Stop Light Switch Signal (STP)	Brake Off: 0v, On: 12v
20-28	---	Not Used	---
29	LG/BK	Overdrive Main Switch Signal (ODMS)	Switch Off: 12v
30	---	Not Used	---
31	YL	A/C Pressure Switch Signal (ACIS)	Switch Open, 12v, Closed: 0v
32	BK/BL	A/C Thermistor Signal	1.4-1.8v (temp. >59ºF)
33	GN/WT	A/C Switch Signal (ACLD)	A/C On: 12v, Off: 0v
34-35	---	Not Used	---

2003 4D Wagon 1.8L DOHC 1ZZ-FE VIN R (All) E6 31P Connector

PCM Pin #	Wire Color	Circuit Description (31 Pin)	Value at Hot Idle
1	BK	EFI Main Relay Power	12-14v
2	---	Not Used	---
3	RD/WT	Battery Direct	12-14v
4	RD	EVAP Pressure Switching (VSV)	12v or 0v
5	---	Not Used	---
6	LG	Cooling Fan 1 Relay (CF)	Relay Off: 12v, On: 1v
7	LG/BK	Cooling Fan 2 Relay (FAN)	Relay Off: 12v, On: 1v
8	BK/WT	EFI Main Relay Control	Relay On: 12v, Off: 1v
9	RD	Ignition Switch Power	12-14v
10	GN/RD	Circuit Opening Relay (FC)	0-3v, off-idle: 12v
11	YL/RD	MIL (lamp) Control	MIL Off: 12v, On: 1v
12	GN	Tail Light Switch Signal	Switch On: 12v, Off: 0v
13	WT	Mirror Heater/Defrost Switch Signal	Heater On: 12-14v
14	YL	Center Airbag Assembly (F/PS)	Digital Signals
15	---	Not Used	---
16	BL/WT	Cruise Control Signal (IDLO)	1.5v, off-idle: 12v
17	---	Not Used	---
18	BL/RD	SIL Signal (Scan Tool)	Digital Signal
19	PK	Scan Tool (WFSE)	12v
20	PK/BK	Data Link Connector (TC)	12-14v
21	BL	EVAP Vapor Pressure Sensor	2.9-3.1v (with hose off)
22-31	---	Not Used	---

Pin Connector Graphic

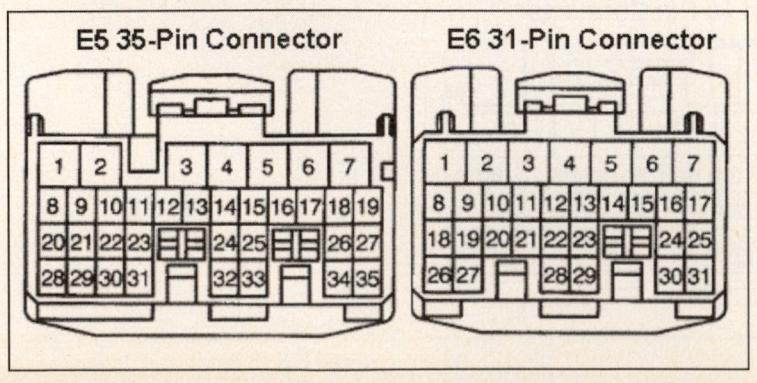

2003 4D Wagon 1.8L DOHC 2ZZ-GE VIN Y (All) E3 34P Connector

PCM Pin #	Wire Color	Circuit Description (34 Pin)	Value at Hot Idle
1	YL	Injector 1 Control	2.0-3.3 ms
2	BK	Injector 2 Control	2.0-3.3 ms
3	WT	Injector 3 Control	2.0-3.3 ms
4	BL	Injector 4 Control	2.0-3.3 ms
5	BK/BL	IAC Air Control Valve Signal (RSD)	Pulse Signals
6	WT/BK	Power Ground (E02)	<0.1v
7	WT/BK	Power Ground (E01)	<0.1v
8	RD/BL	Igniter Transistor 1 Control	Digital Signal: 0-5-0v
9	YL/GN	Igniter Transistor 2 Control	Digital Signal: 0-5-0v
10	GY	Igniter Transistor 3 Control	Digital Signal: 0-5-0v
11	WT	Igniter Transistor 4 Control	Digital Signal: 0-5-0v
12	BL/BK	EVAP Vapor Pressure (VSV)	12v or 0v
13	YL	A/C Pressure Switch (HP)	A/C Off: 12v, On: 1v
14	BK/YL	Camshaft Timing Oil Control Valve (-) Left	AC pulse signals
15	YL	Camshaft Timing Oil Control Valve (+) Left	AC pulse signals
16, 17	PK, RD/WT	A/T-ECT Solenoid (SLT-), (SLT+)	Pulse Signals
18	YL	Sensor VREF (VC)	4.9-5.1v
19	WT	ECT Sensor Signal (THW)	1-4v (varies with temp.)
20	YL/BK	IAT Sensor Signal (THA)	1-4v (varies with temp.)
21	LG	TP Sensor Signal (VTA)	0.3-0.9v
23	BL/YL	Igniter Signal (IGF)	Pulse signal: 0-5-0v
24, 25	BL/YL, BK	Camshaft Timing Oil Control Valve (-), (+)	AC pulse signals
26	BK	CMP Sensor Signal (G22+)	AC pulse signals
27, 34	BK, WT	CKP Sensor Signal (NE+), (NE-)	AC pulse signals
28	BR	Power Ground (E2)	<0.1v
30	WT/BL	TFT (fluid) Sensor (THO)	1-4v (varies with temp.)

2003 4D Wagon 1.8L DOHC 2ZZ-GE VIN Y (All) E4 35P Connector

PCM Pin #	Wire Color	Circuit Description (35 Pin)	Value at Hot Idle
1	BK, WT	Knock Sensor Signal (+), (-)	<0.075v AC
4	PK	HO2S-11 (B1 S1) Heater	1v (Heater On)
5	WT/BK	Power Ground (E03)	<0.1v
7	BR	Power Ground (E1)	<01v
8	RD	Neutral Start Switch (NSW)	Cranking: 0-3v
9	BK	Starter Signal (STA)	Cranking: 9-11v
11	BK	A/T-ECT Solenoid (DSL)	In Lockup: 12-14v
12	BL/OR	A/T-ECT Solenoid (ST)	In Lockup: 12-14v
13	BL/WT	A/T-ECT Solenoid (SL)	In Lockup: 12-14v
15	RD/YL	A/T-ECT Solenoid (S1)	In 3rd or OD: 1v
16, 17	BL, BL/OR	A/T-ECT Solenoid (SL2-), (SL2+)	Moving in 3rd or OD: <1v
18, 19	BL, RD/YL	A/T-ECT Solenoid (SL1-), SL1+)	Moving in 1st Gear: <1v
21	WT	HO2S-21 (B2 S1) Signal (OX1B)	0.1-1.1v
23	RD	HO2S-11 (B1 S1) Signal (OX1A)	0.1-1.1v
24	GN	MAF Sensor Signal (VG)	1-3v
28	WT/BK	Power Ground (EC)	<0.1v
29	BL/RD	PSP Switch Signal (PSW)	Straight: 12v, Turning: 0v
32	BL/WT	MAF Sensor Ground (EVG)	<0.050v
33	BK/WT	OIL Pressure Switch Signal (OSW)	Closed: 1v, Open: 12v

Pin Connector Graphic

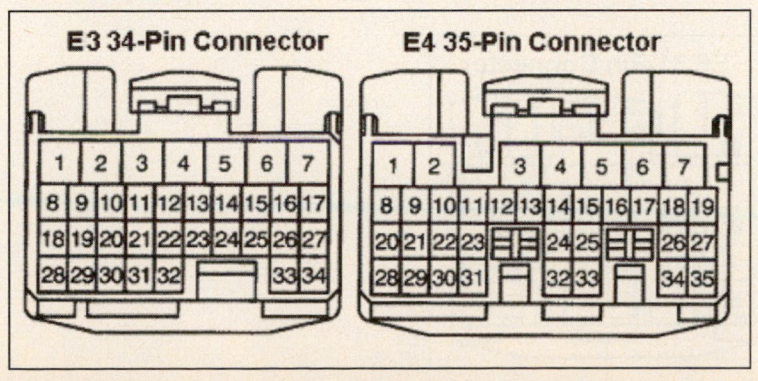

2003 4D Wagon 1.8L DOHC 2ZZ-GE VIN Y (All) E5 35P Connector

PCM Pin #	Wire Color	Circuit Description (35 Pin)	Value at Hot Idle
1	BL	EVAP Canister Closed Valve	12v or 0v
2	GN/WT	A/C Magnetic Clutch Relay (ACMG)	Relay Off: 12v, On: 1v
3	---	Not Used	---
4	PK/BK	HO2S-21 (B2 S1) Heater (HT1B)	1v (Heater On)
5	BK	Tachometer Signal (TACO)	Pulse Signals
6	---	Not Used	---
7	LG	Overdrive Indicator Control (ODLP)	Indicator Off: 12v, On: 1v
8	LG/BK	A/T Select Switch Low Signal (L)	In Low: 12v, Others: 0v
9	LG	A/T Select Switch 2nd Signal (2L)	In 2nd: 12v, Others: 0v
10	BL	A/T Select Switch Drive Signal (DL)	In Drive: 12v, Others: 0v
11	RD/BK	A/T Select Switch Reverse (R)	In Reverse: 12v, Others: 0v
12-13	---	Not Used	---
14	YL/RD	Hot Light Indicator Control	Indicator Off: 12v, On: 1v
15-16	---	Not Used	---
17	VT/WT	Speedometer Indicator	At 55 mph: 48 Hz
18	RD/YL	Overdrive Signal to C/C Module (OD1)	0-12-0v
19	GN/WT	Stop Light Switch Signal (STP)	Brake Off: 0v, On: 12v
20-28	---	Not Used	---
29	LG/BK	Overdrive Main Switch Signal (ODMS)	Switch Off: 12v
30	---	Not Used	---
31	YL	A/C Pressure Switch Signal (ACIS)	Switch Open, 12v, Closed: 0v
32	BK/BL	A/C Thermistor Signal	1.4-1.8v (temp. >59ºF)
33	GN/WT	A/C Switch Signal (ACLD)	A/C On: 12v, Off: 0v
34-35	---	Not Used	---

2003 4D Wagon 1.8L DOHC 2ZZ-GE VIN Y (All) E6 31P Connector

PCM Pin #	Wire Color	Circuit Description (31 Pin)	Value at Hot Idle
1	BK	EFI Main Relay Power	12-14v
2	---	Not Used	---
3	RD/WT	Battery Direct	12-14v
4	RD	EVAP Pressure Switching (VSV)	12v or 0v
5	---	Not Used	---
6	LG	Cooling Fan 1 Relay (CF)	Relay Off: 12v, On: 1v
7	LG/BK	Cooling Fan 2 Relay (FAN)	Relay Off: 12v, On: 1v
8	BK/WT	EFI Main Relay Control	Relay On: 12v, Off: 1v
9	RD	Ignition Switch Power	12-14v
10	GN/RD	Circuit Opening Relay (FC)	0-3v, off-idle: 12v
11	YL/RD	MIL (lamp) Control	MIL Off: 12v, On: 1v
12	GN	Tail Light Switch Signal	Switch On: 12v, Off: 0v
13	WT	Mirror Heater/Defrost Switch Signal	Heater On: 12-14v
14	YL	Center Airbag Assembly (F/PS)	Digital Signals
15	---	Not Used	---
16	BL/WT	Cruise Control Signal (IDLO)	1.5v, off-idle: 12v
17	---	Not Used	---
18	BL/RD	SIL Signal (Scan Tool)	Digital Signal
19	PK	Scan Tool (WFSE)	12v
20	PK/BK	Data Link Connector (TC)	12-14v
21	BL	EVAP Vapor Pressure Sensor	2.9-3.1v (with hose off)
22-31	---	Not Used	---

Pin Connector Graphic

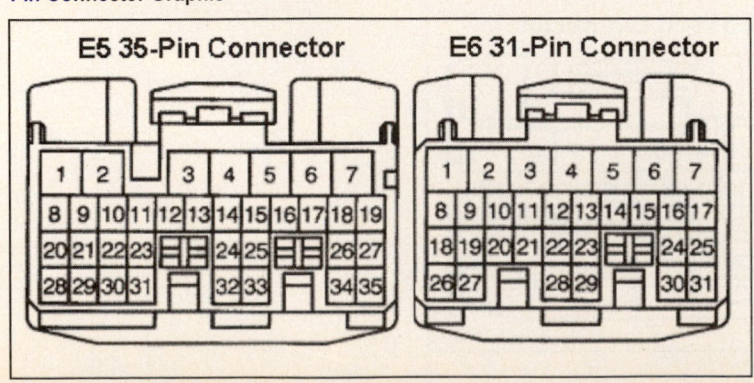

MR2 SPYDER Pin Tables

2000 Coupe 1.8L DOHC 1ZZ-FE VIN R (M/T) E3 28P Connector

PCM Pin #	Wire Color	Circuit Description (28 Pin)	Value at Hot Idle
3	PK/GN	Cooling Fan 3 Relay	Relay Off: 12v, On: 1v
4	BL	Cooling Fan 2 Relay	Relay Off: 12v, On: 1v
5	PK/BK	Data Link Connector (TC)	12-14v
6	GN	Stop Light Switch Signal	Brake Off: 0v, On: 12v
8	BK/WT	Center Airbag Assembly	Digital Signals
10	BL	Multiplex Communication (1)	Digital Signals
11	BK	Starter Switch Signal	9-11v (cranking)
16	RD/YL	ABS ECU Signal	Pulse Signals
18	WT/BL	A/C Dual Pressure Switch	A/C Off: 12v, On: 1v
21	GY	EFI Main Relay Control	Relay Off: 12v, On: 1v
22	VT/WT	Speedometer Indicator	At 55 mph: 48 Hz
23	BL/WT	EVAP Vapor Pressure (VSV)	12v or 0v
27	WT	Tachometer Signal (TACO)	Pulse Signals
28	WT/RD	A/C Magnetic Clutch & Lock Sensor	A/C Off: 12v, On: 1v

2000 Coupe 1.8L DOHC 1ZZ-FE VIN R (M/T) E4 24P Connector

PCM Pin #	Wire Color	Circuit Description (24 Pin)	Value at Hot Idle
1	BK/RD	MAF Sensor Ground (EVG)	<0.050v
2	BL/RD	Sensor VREF (VC)	4.9-5.1v
3	BK/YL	HO2S-11 (B1 S1) Heater	1v (Heater On)
4	WT	EVAP Purge Solenoid (VSV)	12v or 0v
5	BK/WT	HO2S-21 (B2 S1) Heater	1v (Heater On)
6	WT/BK	Power Ground (E03)	<0.1v
8	BL/BK	HO2S-12 (B1 S2) Heater	1v (Heater On)
9	BK	HO2S-12 (B1 S2) Signal	0.1-1.1v
10	YL	Generator Control Signal (L)	12v
11	VT	MAF Sensor Signal	1-3v
12	BK	HO2S-11 (B1 S1) Signal	0.1-1.1v
14	RD/BL	ECT Sensor Signal	1-4v (varies with temp.)
15	WT	CMP Sensor Signal (G2+)	AC pulse signals
16	WT	CKP Sensor Signal (NE+)	AC pulse signals
17	BR	Power Ground (E1)	<01v
18	BR	Sensor Ground (E2)	<0.050v
21	BK	HO2S-21 (B2 S1) Signal	0.1-1.1v
22	YL/BK	IAT Sensor Signal	1-4v (varies with temp.)
23	YL/GN	TP Sensor Signal	1.1-1.9v
24	BK	CKP/CMP Sensor Ground (-)	<0.050v

Pin Connector Graphic

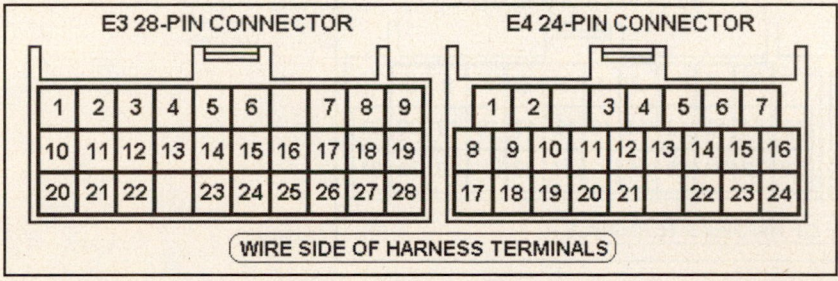

2000 Coupe 1.8L DOHC 1ZZ-FE VIN R (M/T) E2 22P Connector

PCM Pin #	Wire Color	Circuit Description (22 Pin)	Value at Hot Idle
1	BK/YL	Battery Direct	12-14v
2	WT/GN	Immobilizer Security Indicator	LED Off: 0v, On: 12v
3	GN/RD	Circuit Opening Relay (FC)	0-3v, off-idle: 12v
4	GN/BK	EVAP Vapor Pressure Sensor	2.9-3.1v (with hose off)
6	BK	Multiplex Communication (2)	Digital Signals
8	BK/RD	Ignition Switch Power	12-14v
11	WT/GN	SIL Signal (Scan Tool)	Digital Signal
12	BL	A/C Magnetic Clutch (ACMG)	Clutch Off: 0v, On: 12v
15	YL/RD	MIL (lamp) Control	MIL Off: 12v, On: 1v
16	BK	EFI Main Relay Power	12-14v

2000 Coupe 1.8L DOHC 1ZZ-FE VIN R (M/T) E5 31P Connector

PCM Pin #	Wire Color	Circuit Description (31 Pin)	Value at Hot Idle
1	BK/WT	Injector 1 Control	2.0-3.3 ms
2	BK	Injector 2 Control	2.0-3.3 ms
3	BL	Injector 3 Control	2.0-3.3 ms
4	WT	Injector 4 Control	2.0-3.3 ms
6	RD/YL	Engine Throttle Control Power	12-14v
7	GN/RD	Throttle Control Motor (CL-)	Pulse Signals
8, 31	WT/BK	Power Ground (ME01, ME2)	<0.1v
9	GY	Motor Control Shield Ground	<0.050v
10	BK	Igniter Transistor 1 Control	Digital Signal: 0-5-0v
11	BL	Igniter Transistor 2 Control	Digital Signal: 0-5-0v
12	BL/WT	Igniter Transistor 3 Control	Digital Signal: 0-5-0v
13	BL	Igniter Transistor 4 Control	Digital Signal: 0-5-0v
17	BL	EVAP Canister Closed Valve	12v or 0v
18	GN	IAC Valve Signal (RSO)	Pulse Signals
19	RD	Throttle Control Motor (M+)	Pulse Signals
20	YL/RD	Throttle Control Motor (CL+)	Pulse Signals
21	WT/BK	Power Ground (E01)	<0.1v
22	YL/BK	OIL Pressure Switch	Closed: 1v, Open: 12v
23	WT	Camshaft Timing Oil Control Valve (-)	AC pulse signals
24	RD	Camshaft Timing Oil Control Valve (+)	AC pulse signals
25	BK/YL	Igniter Signal (IGF)	Pulse signal: 0-5-0v
27	BK	Knock Sensor Signal	<0.075v AC
29	GN	Throttle Control Motor (M-)	Pulse Signals

Pin Connector Graphic

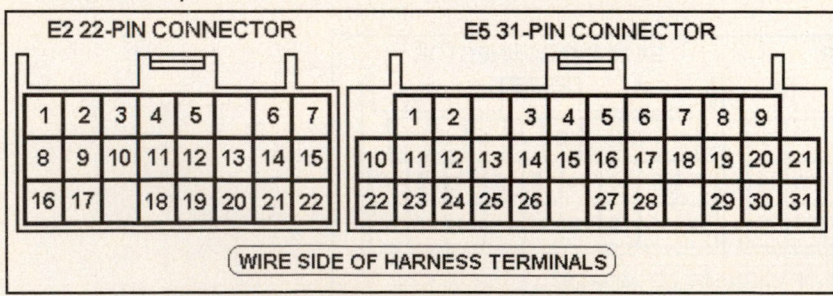

2001-03 Coupe 1.8L DOHC 1ZZ-FE VIN R (M/T) E2 22P Connector

PCM Pin #	Wire Color	Circuit Description (22 Pin)	Value at Hot Idle
1	BK/YL	Battery Direct	12-14v
2	WT/GN	Immobilizer Security Indicator	LED Off: 0v, On: 12v
3	GN/RD	Circuit Opening Relay (FC)	0-3v, off-idle: 12v
4	GN/BK	EVAP Vapor Pressure Sensor	2.9-3.1v (with hose off)
5	BK	Multiplex Meter Input (MPX2)	Digital Signals
6	BK	Multiplex Communication (2)	Digital Signals
7	BL/BK	A/T Neutral Start Switch (NSW)	Cranking: 9-11v
8	BK/RD	Ignition Switch Power	12-14v
9-10	---	Not Used	---
11	WT/GN	SIL Signal (Scan Tool)	Digital Signal
12	BL	A/C Magnetic Clutch (ACMG)	Clutch Off: 0v, On: 12v
13-14	---	Not Used	---
15	YL/RD	MIL (lamp) Control	MIL Off: 12v, On: 1v
16	BK	EFI Main Relay Power	12-14v

2001-03 Coupe 1.8L DOHC 1ZZ-FE VIN R (M/T) E3 28P Connector

PCM Pin #	Wire Color	Circuit Description (28 Pin)	Value at Hot Idle
1	BK/YL	Start Relay Control (STD)	Relay Off: 12v, On: 1v
2	---	Not Used	---
3	YL/GN	Cooling Fan 3 Relay (CF)	Relay Off: 12v, On: 1v
4	BL	Cooling Fan 2 Relay (FAN)	Relay Off: 12v, On: 1v
5	PK/BK	Data Link Connector (TC)	12-14v
6	GN	Stop Light Switch Signal	Brake Off: 0v, On: 12v
7	---	Not Used	---
8	BK/WT	Center Airbag Assembly (F/PS)	Digital Signals
9	---	Not Used	---
10	BL	Multiplex Meter Input (MPX1)	Digital Signals
11	BK	Starter Switch Signal	9-11v (cranking)
12-15	---	Not Used	---
16	RD/YL	ABS ECU Signal	Digital Signals
17	---	Not Used	---
18	WT/BL	A/C Dual Pressure Switch (PRE)	A/C Off: 12v, On: 1v
19	WT/G	Power Steering ECU (PSCT)	Pulse Signals
20	---	Not Used	---
21	GY	EFI Main Relay Control	Relay Off: 12v, On: 1v
22	VT/WT	Speedometer Indicator	At 55 mph: 48 Hz
23	BL/WT	EVAP Vapor Pressure (VSV)	12v or 0v
24	---	Not Used	---
25	BL/OR	Power Steering ECU (PS)	Pulse Signals
26	---	Not Used	---
27	WT	Tachometer Signal (TACO)	Pulse Signals
28	WT/RD	A/C Magnetic Clutch & Lock Sensor	A/C Off: 12v, On: 1v

Pin Connector Graphic

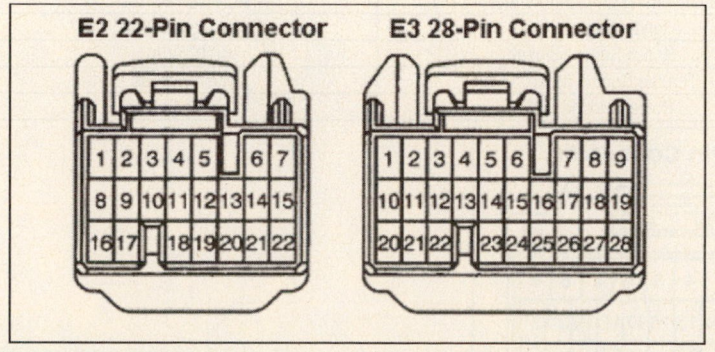

2001-03 Coupe 1.8L DOHC 1ZZ-FE VIN R (M/T) E4 24P Connector

PCM Pin #	Wire Color	Circuit Description (24 Pin)	Value at Hot Idle
1	BK/RD	MAF Sensor Ground (EVG)	<0.050v
2	BL/RD	Sensor VREF (VC)	4.9-5.1v
3	BK/YL	HO2S-11 (B1 S1) Heater	1v (Heater On)
4	WT	EVAP Purge Solenoid (VSV)	12v or 0v
5	BK/WT	HO2S-21 (B2 S1) Heater (HT2A)	1v (Heater On)
6	WT/BK	Power Ground (E03)	<0.1v
7	---	Not Used	---
8	BL/BK	HO2S-12 (B1 S2) Heater	1v (Heater On)
9	BK	HO2S-12 (B1 S2) Signal	0.1-1.1v
10	YL	Generator Control Signal (RL)	12-0-12v
11	VT	MAF Sensor Signal (VG)	1-3v
12	BK	HO2S-11 (B1 S1) Signal	0.1-1.1v
13	YL/BK	TP Sensor Signal (VTA2)	0.3-0.9v
14	RD/BL	ECT Sensor Signal (THW)	1-4v (varies with temp.)
15	WT	CMP Sensor Signal (G2+)	AC pulse signals
16, 24	WT, BK	CKP Sensor Signal (NE+), (NE-)	AC pulse signals
17	BR	Power Ground (E1)	<01v
18	BR	Sensor Ground (E2)	<0.050v
19	BK/BL	Accelerator Pedal Position Sensor (VPA2)	1.1-1.9v
20	BL	Accelerator Pedal Position Sensor (VPA)	1.1-1.9v
21	BK	HO2S-21 (B2 S1) Signal (OX2A)	0.1-1.1v
22	YL/BK	IAT Sensor Signal (THA)	1-4v (varies with temp.)
23	YL/GN	TP Sensor Signal (VTA)	0.3-0.9v

2001-03 Coupe 1.8L DOHC 1ZZ-FE VIN R (M/T) E5 31P Connector

PCM Pin #	Wire Color	Circuit Description (31 Pin)	Value at Hot Idle
1	BK/WT	Injector 1 Control	2.0-3.3 ms
2	BK	Injector 2 Control	2.0-3.3 ms
3	BL	Injector 3 Control	2.0-3.3 ms
4	WT	Injector 4 Control	2.0-3.3 ms
6	RD/YL	Engine Throttle Control Power (+BM)	12-14v
7, 20	GN, YL/RD	Throttle Control Motor (CL-), (CL+)	Pulse Signals
8	WT/BK	Power Ground (ME01, ME2)	<0.1v
9	GY	Motor Control Shield Ground (GE01)	<0.050v
10	BK	Igniter Transistor 1 Control	Digital Signal: 0-5-0v
11	BL	Igniter Transistor 2 Control	Digital Signal: 0-5-0v
12	BL/WT	Igniter Transistor 3 Control	Digital Signal: 0-5-0v
13	BL	Igniter Transistor 4 Control	Digital Signal: 0-5-0v
14, 28	BL, RD/BL	Controller Area Network (+), (-)	0-7-0v
15, 30	PK, BL/OR	Sequential M/T Signal (NEO), (CSMT)	Digital Signals
17	BL	EVAP Canister Closed Valve	12v or 0v
18	GN	IAC Valve Signal (RSO)	Pulse Signals
19	RD	Throttle Control Motor (M+)	Pulse Signals
21	WT/BK	Power Ground (E01)	<0.1v
22	YL/BK	OIL Pressure Switch Signal (MOPS)	Closed: 1v, Open: 12v
23, 24	WT, RD	Camshaft Timing Oil Control Valve (-), (+)	AC pulse signals
25	BK/YL	Igniter Signal (IGF)	Pulse signal: 0-5-0v
27	BK	Knock Sensor Signal	<0.075v AC
29	GN	Throttle Control Motor (M-)	Pulse Signals
31	WT/BK	Power Ground (E02)	<0.1v

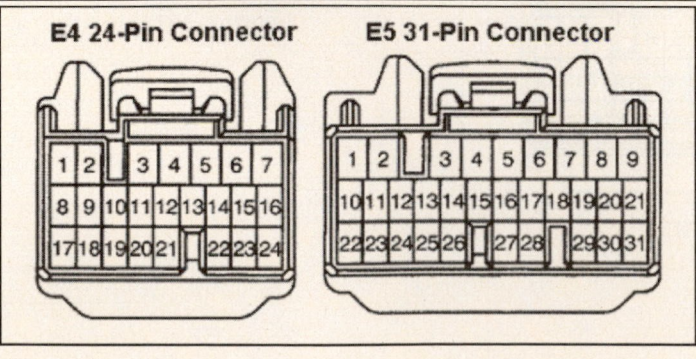

E4 24-Pin Connector   **E5 31-Pin Connector**

MR2 Pin Tables

1991-92 Coupe 2.0L I4 Turbo VIN S (M/T) 16 Pin Connector

PCM Pin #	Wire Color	Circuit Description (16 Pin)	Value at Hot Idle
1	RD/BL	Sensor VREF	4.9-5.1v
2	GY/BK	Airflow Meter Signal	1.5-2.5v
3	YL	IAT Sensor Signal	At 100ºF: 2.60v
4	RD	ECT Sensor Signal	At 180ºF: 0.51v
5	BK	Knock Sensor Signal	<0.075v AC
6	WT	HO2S Signal	0.1-1.1v
8	RD/WT	Check Connector	12-14v
9	BR	Sensor Ground	<0.050v
10 (Cal)	BK/YL	EGR Gas Temp. Sensor	3.5-4.0v
11	WT	TP Sensor Signal	0.3-0.8v
12	PK	Closed Throttle Switch	1v, at off-idle: 12v
13	GN/RD	Turbo Charger Press. Sensor	2.5-4.5v
14	BK	Check Connector	12-14v
15	OR	Check Connector	12-14v
16	RD	Distributor Signal (G-)	<0.050v

1991-92 Coupe 2.0L I4 Turbo VIN S (M/T) 22 Pin Connector

PCM Pin #	Wire Color	Circuit Description (22 Pin)	Value at Hot Idle
1	WT/RD	Battery Direct	12-14v
2	BK	Defogger/Light Idle Up Signal	Switch On: 12v, Off: 0v
4	GN/WT	Stop Light Switch Signal	Brake Off: 0v, On: 12v
5	GN/WT	MIL (lamp) Control	MIL Off: 12v, On: 1v
6	BL/RD	Fuel Pump Relay	Relay Off: 12v, On: 1v
8	GN/YL	ABS ECU Signal	Pulses
9	PK/WT	Vehicle Speed Sensor	At 55 mph: 48 Hz
10	BK/WT	A/C Magnetic Clutch (ACMG)	Clutch Off: 0v, On: 12v
11	RD	Starter Switch Signal	9-11v (cranking)
12	BK/YL	EFI Main Relay Power	12-14v
13	BK/YL	EFI Main Relay Power	12-14v
14	BL/YL	PSP ECU Signal	Pulse Signals
15	BK	PSP ECU Signal	Pulse Signals

Standard Colors and Abbreviations

Abbreviation	Color	Abbreviation	Color	Abbreviation	Color
BK	Black	GY	Gray	RD	Red
BL	Blue	GN	Green	TN	Tan
BR	Brown	LG	Light Green	VT	Violet
DB	Dark Blue	OR	Orange	WT	White
DG	DK Green	PK	Pink	YL	Yellow

1991-92 Coupe 2.0L I4 Turbo VIN S (M/T) 26 Pin Connector

PCM Pin #	Wire Color	Circuit Description (26 Pin)	Value at Hot Idle
1	BK	Distributor Signal (NE+)	AC pulse signals
2	WT	Distributor Signal (G2+)	AC pulse signals
3	WT/RD	Distributor Signal (IGF)	Digital Signal: 0-5-0v
6	BL/RD	Turbo Pressure Solenoid	12v or 0v
8	PK/BK	HO2S-11 (B1 S1) Heater	1v (Heater On)
9	GN/WT	ISC Motor (ISC1)	Pulse Signals
10	GN	Cold Start Injector Control	1v (at cold startup)
11	WT/BL	Injector 2 Control	2.0-3.3 ms
12	RD/WT	Injector 1 Control	2.0-3.3 ms
13	BR	Power Ground	<0.1v
14	BR	Sensor Ground	<0.050v
15	GN	Distributor Signal (G1+)	AC pulse signals
17	GN	Circuit Opening Relay (FC)	0-3v, off-idle: 12v
18	LG	Intake Solenoid Signal (TVIS)	12v or 0v
20	WT	Igniter Signal (IGT)	Digital Signal: 0-5-0v
22	GN/BK	ISC Motor (ISC2)	Pulse Signals
23	GN/RD	EGR Solenoid Control (VSV)	12v or 0v
24	GN/YL	Injector 4 Control	2.0-3.3 ms
25	BK/WT	Injector 3 Control	2.0-3.3 ms
26	BR	Power Ground	<0.1v

Pin Connector Graphic

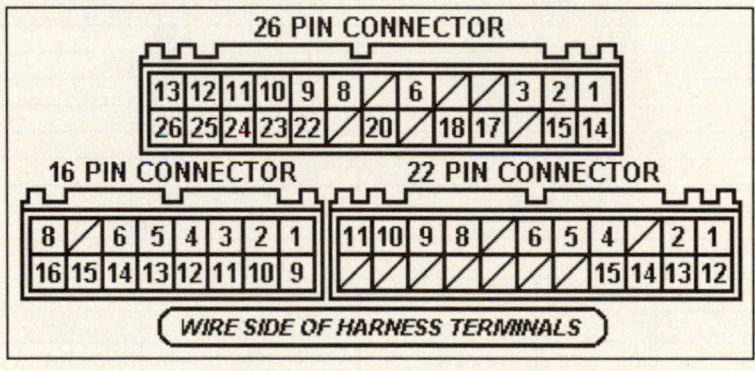

1993 Coupe 2.0L I4 Turbo VIN S (M/T) 16 Pin Connector

PCM Pin #	Wire Color	Circuit Description (16 Pin)	Value at Hot Idle
1	RD/BL	Sensor VREF	4.9-5.1v
2	YL/BL	MAP Sensor Signal	3.5-3.1v
3	BK	IAT Sensor Signal	At 100°F: 2.60v
4	RD	ECT Sensor Signal	At 180°F: 0.51v
5	BL/BK	Turbo Charger Press. Sensor	2.5-4.5v
6	BK	HO2S Signal	0.1-1.1v
8	RD/WT	Check Connector	12-14v
9	BR	Sensor Ground	<0.050v
10 (Cal)	BK/YL	EGR Gas Temp. Sensor	3.5-4.0v
10 (Fed)	BK/YL	Short Pin	Open: 12v, Closed: 0v
11	WT	TP Sensor Signal	0.3-0.8v
12	PK	Closed Throttle Switch	1v, at off-idle: 12v
13	BK	Knock Sensor Signal	<0.075v AC
14	PK/BL	Check Connector	12-14v
15	BK	Check Connector	12-14v

1993 Coupe 2.0L I4 Turbo VIN S (M/T) 22 Pin Connector

PCM Pin #	Wire Color	Circuit Description (22 Pin)	Value at Hot Idle
1	WT/RD	Battery Direct	12-14v
2	BK	Defogger/Light Idle Up Signal	Switch On: 12v, Off: 0v
4	GN/WT	Stop Light Switch Signal	Brake Off: 0v, On: 12v
5	GN/WT	MIL (lamp) Control	MIL Off: 12v, On: 1v
6	BL/RD	Fuel Pump Relay	Relay Off: 12v, On: 1v
7	BL/YL	PSP ECU Signal	Pulse Signals
8	BK	PSP ECU Signal	Pulse Signals
9	PK/WT	Vehicle Speed Sensor	At 55 mph: 48 Hz
10	BK/WT	A/C Magnetic Clutch (ACMG)	Clutch Off: 0v, On: 12v
11	RD	Starter Switch Signal	9-11v (cranking)
12	BK/YL	EFI Main Relay Power	12-14v
13	BK/YL	EFI Main Relay Power B1+	12-14v
14	GN/RD	Circuit Opening Relay (FC)	0-3v, off-idle: 12v
15	GN/YL	ABS ECU Signal	Pulse Signals

Standard Colors and Abbreviations

Abbreviation	Color	Abbreviation	Color	Abbreviation	Color
BK	Black	GY	Gray	RD	Red
BL	Blue	GN	Green	TN	Tan
BR	Brown	LG	Light Green	VT	Violet
DB	Dark Blue	OR	Orange	WT	White
DG	DK Green	PK	Pink	YL	Yellow

1993 Coupe 2.0L I4 Turbo VIN S (M/T) 26 Pin Connector

PCM Pin #	Wire Color	Circuit Description (26 Pin)	Value at Hot Idle
1	GN/BK	Intake Solenoid Signal (TVIS)	12v or 0v
2	BL/OR	Turbo Pressure Solenoid	12v or 0v
3	WT/RD	Distributor Signal (IGF)	Digital Signal: 0-5-0v
4	BK	Distributor Signal (NE+)	AC pulse signals
5	WT	Distributor Signal (G2+)	AC pulse signals
6	BK/RD	EGR Solenoid Control (VSV)	12v or 0v
7	GN	Cold Start Injector Control	1v (at cold startup)
8	RD/WT	HO2S-11 (B1 S1) Heater	1v (Heater On)
9	GN	ISC Signal (RSC)	Pulses
10	RD	ISC Signal (RSO)	Pulses
11	WT/BL	Injector 2 Control	2.0-3.3 ms
12	RD/WT	Injector 1 Control	2.0-3.3 ms
13	BR	Power Ground	<0.1v
14	BR	Sensor Ground	<0.050v
17	RD	Distributor Signal (G-)	<0.050v
20	WT	Igniter Signal (IGT)	Digital Signal: 0-5-0v
24	GN/YL	Injector 4 Control	2.0-3.3 ms
25	BK/WT	Injector 3 Control	2.0-3.3 ms
26	BR	Power Ground	<0.1v

Pin Connector Graphic

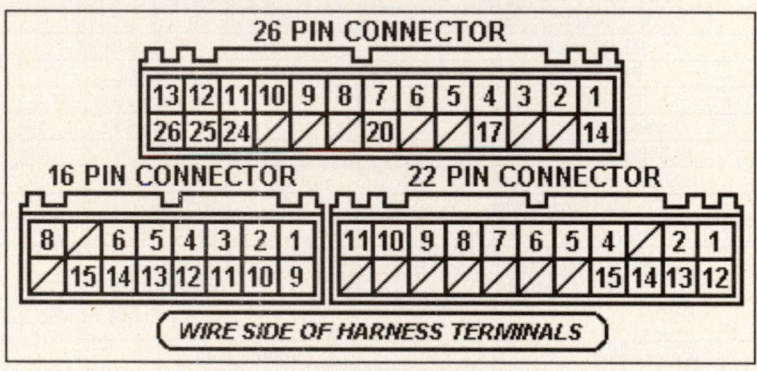

1994-95 Coupe 2.0L I4 Turbo VIN S (M/T) 16 Pin Connector

PCM Pin #	Wire Color	Circuit Description (16 Pin)	Value at Hot Idle
1	RD/BL	Sensor VREF	4.9-5.1v
2	YL/RD	MAP Sensor Signal	1.5-2.5v
3	BK	IAT Sensor Signal	At 100°F: 2.60v
4	RD	ECT Sensor Signal	At 180°F: 0.51v
5	BL/BK	Turbo Pressure Sensor	2.5-4.5v
6	BK	HO2S Signal	0.1-1.1v
8	RD/WT	Data Link Connector	12-14v
9	BR	Sensor Ground	<0.050v
10	BK/YL	EGR Gas Temp. Sensor	3.5-4.0v
11	WT	TP Sensor Signal	0.3-0.8v
12	WT	Closed Throttle Switch	0-3v, off-idle: 12v
13	BK	Knock Sensor Signal	<0.075v AC
14	PK/BL	Data Link Connector	12-14v
15	BK	Data Link Connector	12-14v

1994-95 Coupe 2.0L I4 Turbo VIN S (M/T) 22 Pin Connector

PCM Pin #	Wire Color	Circuit Description (22 Pin)	Value at Hot Idle
1	WT/RD	Battery Direct	12-14v
2	BK	Defogger/Light Idle Up Signal	Switch On: 12v, Off: 0v
4	GN/WT	Stop Light Switch Signal	Brake Off: 0v, On: 12v
5	GN/WT	MIL (lamp) Control	MIL Off: 12v, On: 1v
6	BL/RD	Fuel Pump Relay	Relay Off: 12v, On: 1v
7	BL/YL	PSP ECU Signal	Pulse Signals
8	BK	PSP ECU Signal	Pulse Signals
9	PK/WT	Vehicle Speed Sensor	At 55 mph: 48 Hz
10	BK/WT	A/C Magnetic Clutch (ACMG)	Clutch Off: 0v, On: 12v
11	RD	Starter Switch Signal	9-11v (cranking)
12	BK/YL	EFI Main Relay Power	12-14v
14	GN/RD	Circuit Opening Relay (FC)	0-3v, off-idle: 12v
15	GN/YL	ABS ECU Signal	Pulse Signals

Standard Colors and Abbreviations

Abbreviation	Color	Abbreviation	Color	Abbreviation	Color
BK	Black	GY	Gray	RD	Red
BL	Blue	GN	Green	TN	Tan
BR	Brown	LG	Light Green	VT	Violet
DB	Dark Blue	OR	Orange	WT	White
DG	DK Green	PK	Pink	YL	Yellow

1994-95 Coupe 2.0L I4 Turbo VIN S (M/T) 26 Pin Connector

PCM Pin #	Wire Color	Circuit Description (26 Pin)	Value at Hot Idle
1	GN/BK	TN-VIS Solenoid Control	12v or 0v
2	BL/OR	Turbo Pressure Solenoid	12v or 0v
3	WT/RD	Distributor Signal (IGF)	Digital Signal: 0-5-0v
4	BK	Distributor Signal (NE+)	AC pulse signals
5	WT	Distributor Signal (G2+)	AC pulse signals
6	BK/RD	EGR Solenoid Control (VSV)	12v or 0v
7	GN	Cold Start Injector Control	1v (at cold startup)
8	RD/WT	HO2S-11 (B1 S1) Heater	1v (Heater On)
9	GN	ISC Signal (RSC)	Pulse Signals
10	RD	ISC Signal (RSO)	Pulse Signals
11	WT/BL	Injector 2 Control	2.0-3.3 ms
12	RD/WT	Injector 1 Control	2.0-3.3 ms
13	BR	Power Ground	<0.1v
14	BR	Sensor Ground	<0.050v
17	RD	Distributor Signal (G-)	<0.050v
18	GN	Distributor Signal (G1+)	AC pulse signals
20	WT	Igniter Signal (IGT)	Digital Signal: 0-5-0v
24	GN/YL	Injector 4 Control	2.0-3.3 ms
25	BK/WT	Injector 3 Control	2.0-3.3 ms
26	BR	Power Ground	<0.1v

Pin Connector Graphic

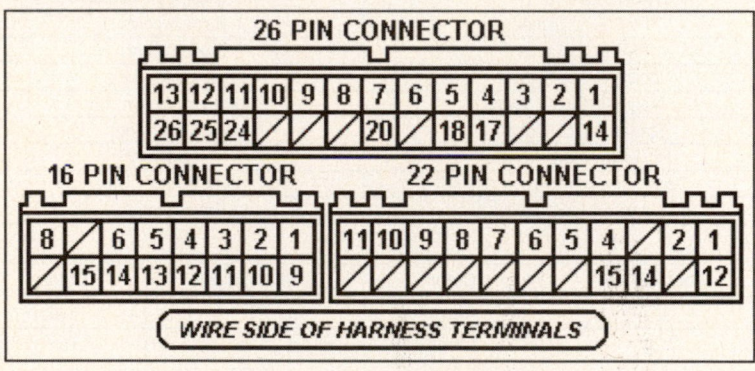

1991-92 Coupe 2.2L I4 MFI VIN S (A/T-ECT) 16 Pin Connector

PCM Pin #	Wire Color	Circuit Description (16 Pin)	Value at Hot Idle
1	RD/BL	Sensor VREF	4.9-5.1v
2	GN/RD	MAP Sensor Signal	1-1.5v
3	YL	IAT Sensor Signal	At 100ºF: 2.60v
4	RD	ECT Sensor Signal	At 180ºF: 0.51v
6	BK	Main O2S Signal	0.1-1.1v
7	BR/BK	Check Connector	12-14v
8	RD/WT	Check Connector	12-14v
9	BR	Sensor Ground	<0.050v
10 (Cal)	BK/YL	EGR Gas Temp. Sensor	3.5-4.0v
11	WT	TP Sensor Signal	0.3-0.8v
12	PK	Closed Throttle Switch	1v, at off-idle: 12v
14 (Cal)	BK	Sub O2S Signal	0.1-1.1v
15	OR	Check Connector	12-14v
16	BK	Distributor Signal (G-)	<0.050v

1991-92 Coupe 2.2L I4 MFI VIN S (A/T-ECT) 22 Pin Connector

PCM Pin #	Wire Color	Circuit Description (22 Pin)	Value at Hot Idle
1	WT/RD	Battery Direct	12-14v
2	BK	Defogger/Light Idle Up Signal	Switch On: 12v, Off: 0v
4	GN/WT	Stop Light Switch Signal	Brake Off: 0v, On: 12v
5	GN/WT	MIL (lamp) Control	MIL Off: 12v, On: 1v
6	BL/YL	PSP ECU Signal (CT)	Pulse Signals
7	LG	Overdrive Main Switch	Switch Off: 12v
8	BK	ABS ECU Signal	Pulse Signals
9	PK/WT	Vehicle Speed Sensor	At 55 mph: 48 Hz
10	BK/WT	A/C Magnetic Clutch (ACMG)	Clutch Off: 0v, On: 12v
11	BK	Starter Switch Signal	9-11v (cranking)
12	BK/YL	EFI Main Relay Power	12-14v
13	BK/YL	EFI Main Relay Power B1+	12-14v
14	GN/RD	Circuit Opening Relay (FC)	0-3v, off-idle: 12v
20	PK/BK	Cruise Control Signal (OD1)	At Cruise in OD: 12v
21	PK/GN	A/C Amplifier Signal (ACT)	Clutch On: 12v, Off: 1.5v
22	BK/WT	A/T Neutral Start Switch	In P/N: 9-11v (cranking)

Standard Colors and Abbreviations

Abbreviation	Color	Abbreviation	Color	Abbreviation	Color
BK	Black	GY	Gray	RD	Red
BL	Blue	GN	Green	TN	Tan
BR	Brown	LG	Light Green	VT	Violet
DB	Dark Blue	OR	Orange	WT	White
DG	DK Green	PK	Pink	YL	Yellow

1991-92 Coupe 2.2L I4 MFI VIN S (A/T-ECT) 26 Pin Connector

PCM Pin #	Wire Color	Circuit Description (26 Pin)	Value at Hot Idle
1	WT	Distributor Signal (NE+)	AC pulse signals
2	BL/GN	A/T Pattern Selector Switch	Norm: 0v, PWR: 12v
3	WT/RD	Igniter Signal (IGF)	0.74-0.76v
4	GY/BK	A/T-ECT Solenoid (SL)	In Lockup: 12-14v
5	BR/WT	A/T-ECT Solenoid (S2)	1st or OD: 1v
6	PK	A/T-ECT Solenoid (S1)	In 3rd or OD: 1v
7	GN	Cold Start Injector Control	1v (at cold startup)
8	GN/BK	Fuel Pressure Up Solenoid	1v (at hot restart)
9	GN	ISC Signal (ISCC)	Pulse Signals
10	RD	ISC Signal (ISCO)	Pulse Signals
11	WT	Injector Pair 2 & 4 Control	2.0-3.3 ms
12	WT	Injector Pair 1 & 3 Control	2.0-3.3 ms
13	WT/BK	Power Ground	<0.1v
14	BR	Sensor Ground	<0.050v
15	RD	Distributor Signal (G+)	AC pulse signals
16	BR	Sensor Ground	<0.050v
17	GN/BK	A/T-ECT Speed Sensor	Moving: 0-5-0v
18	LG	A/T Select Switch 2nd	In 2nd: 12v, Others: 0v
19	GN/RD	A/T Select Switch Low	In Low: 12v, Others: 0v
20	WT	Igniter Signal (IGT)	0.72-0.74v
23	BL/OR	EGR Solenoid Control (VSV)	12v or 0v
26	BR	Power Ground	<0.1v

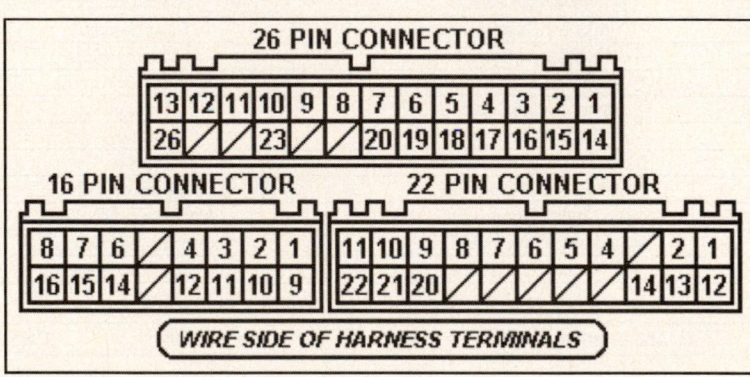

1991-92 Coupe 2.2L I4 MFI VIN S (M/T) 10 Pin Connector

PCM Pin #	Wire Color	Circuit Description (10 Pin)	Value at Hot Idle
1	GN/BK	Fuel Pressure Up Solenoid	1v (at hot restart)
2	GN	Cold Start Injector Control	1v (at cold startup)
3	RD/WT	Starter Switch Signal	9-11v (cranking)
4	WT	Injector Pair 1 & 3 Control	2.0-3.3 ms
5	BR	Power Ground	<0.1v
6	BL/OR	EGR Solenoid Control (VSV)	12v or 0v
7	BR	Sensor Ground	<0.050v
8	WT	Igniter Signal (IGT)	0.72-0.74v
9	WT	Injector Pair 2 & 4 Control	2.0-3.3 ms
10	BR	Power Ground	<0.1v

1991-92 Coupe 2.2L I4 MFI VIN S (M/T) 14-Pin Connector

PCM Pin #	Wire Color	Circuit Description (14-Pin)	Value at Hot Idle
1	BK/YL	EFI Main Relay Power	12-14v
2	WT/RD	Battery Direct	12-14v
3	PK/YL	Check Connector	12-14v
4	GN/RD	Circuit Opening Relay (FC)	0-3v, off-idle: 12v
5	OR	Defogger/Light Idle Up Signal	Switch On: 12v, Off: 0v
7	BL/YL	PSP ECU Signal (CT)	Pulse Signals
8	BK/YL	EFI Main Relay Power	12-14v
9	GN/WT	MIL (lamp) Control	MIL Off: 12v, On: 1v
11	BK/WT	A/C Magnetic Clutch (ACMG)	Clutch Off: 0v, On: 12v
12	PK/WT	Vehicle Speed Sensor	At 55 mph: 48 Hz
14	BK	PSP ECU Signal	Pulse Signals

1991-92 Coupe 2.2L I4 MFI VIN S (M/T) 18 Pin Connector

PCM Pin #	Wire Color	Circuit Description (18 Pin)	Value at Hot Idle
1	RD	ECT Sensor Signal	At 180ºF: 0.51v
2	GN/RD	MAP Sensor Signal	1-1.5v
3	YL	IAT Sensor Signal	At 100ºF: 2.60v
4	OR	Check Connector	12-14v
5	WT/RD	Igniter Signal (IGF)	0.74-0.76v
6	RD	Distributor Signal (G1+)	AC pulse signals
7	BK	Distributor Signal (G-)	<0.050v
8	BK	Main O2S Signal	0.1-1.1v
9	GN	ISC Signal (ISCC)	Pulse Signals
10	BR	Sensor Ground	<0.050v
11	RD/GN	Wide Open Throttle Switch	1v, at off-idle: 12v
12	RD/BL	Sensor VREF	4.9-5.1v
13	PK	Closed Throttle Switch	1v, at off-idle: 12v
14 (Cal)	BK/YL	EGR Gas Temp. Sensor	3.5-4.0v
15	WT	Distributor Signal (NE+)	AC pulse signals
16	BR	Sensor Ground	<0.050v
17 (Cal)	BK	Sub O2S Signal	0.1-1.1v
18	RD	ISC Signal (ISCO)	Pulse Signals

Pin Connector Graphic

1993-95 Coupe 2.2L I4 MFI VIN S (A/T-ECT) 16 Pin Connector

PCM Pin #	Wire Color	Circuit Description (16 Pin)	Value at Hot Idle
1	RD/BL	Sensor VREF	4.9-5.1v
2	GN/RD	MAP Sensor Signal	1-1.5v
3	YL	IAT Sensor Signal	At 100°F: 2.60v
4	RD	ECT Sensor Signal	At 180°F: 0.51v
5 (Cal)	BK	Sub O2S Signal	0.1-1.1v
6	BK	Main O2S Signal	0.1-1.1v
7	BR/BK	Data Link Connector	12-14v
8	RD/WT	Data Link Connector	12-14v
9	BR	Sensor Ground	<0.050v
10 (Cal)	BK/YL	EGR Gas Temp. Sensor	3.5-4.0v
11	WT	TP Sensor Signal	0.3-0.8v
12	PK	Closed Throttle Switch	1v, at off-idle: 12v
13	BK	Knock Sensor Signal	<0.075v AC
14	PK/BL	Data Link Connector	12-14v
15	BK	Data Link Connector	12-14v
16 ('93)	BR	Sensor Ground	<0.050v

1993-95 Coupe 2.2L I4 MFI VIN S (A/T-ECT) 22 Pin Connector

PCM Pin #	Wire Color	Circuit Description (22 Pin)	Value at Hot Idle
1	WT/RD	Battery Direct	12-14v
2	BK	Defogger/Light Idle Up Signal	Switch On: 12v, Off: 0v
3	RD/WT	A/C Evaporator Temperature Signal	1.4-1.8v (temp. >59°F)
4	GN/WT	Stop Light Switch Signal	Brake Off: 0v, On: 12v
5	GN/WT	MIL (lamp) Control	MIL Off: 12v, On: 1v
7	LG	Overdrive Main Switch	Switch Off: 12v
8	BK/WT	A/C Amplifier Signal (AC1)	Clutch On: 1.5v, Off: 12v
9	PK/WT	Vehicle Speed Sensor	At 55 mph: 48 Hz
11	BK	Starter Switch Signal	9-11v (cranking)
12	BK/YL	EFI Main Relay Power	12-14v
13 ('93)	BK/YL	EFI Main Relay Power B1+	12-14v
14	GN/RD	Circuit Opening Relay (FC)	0-3v, off-idle: 12v
15	BL/YL	PSP ECU Signal	Pulse Signals
16	BK	PSP ECU Signal	Pulse Signals
20	PK/BK	Cruise Control Signal (OD1)	At Cruise in OD: 12v
21	PK/GN	A/C Amplifier Signal (ACT)	Clutch On: 12v, Off: 1.5v
22	RD/WT	Neutral Start Switch	In P/N: 9-11v (cranking)

Standard Colors and Abbreviations

Abbreviation	Color	Abbreviation	Color	Abbreviation	Color
BK	Black	GY	Gray	RD	Red
BL	Blue	GN	Green	TN	Tan
BR	Brown	LG	Light Green	VT	Violet
DB	Dark Blue	OR	Orange	WT	White
DG	DK Green	PK	Pink	YL	Yellow

1993-95 Coupe 2.2L I4 MFI VIN S (A/T-ECT) 26 Pin Connector

PCM Pin #	Wire Color	Circuit Description (26 Pin)	Value at Hot Idle
1	GY/BK	A/T-ECT Solenoid (SL)	In Lockup: 12-14v
2	YL/RD	A/T-ECT Solenoid (S1)	In 3rd or OD: 1v
2 ('93)	PK	A/T-ECT Solenoid (S1)	In 3rd or OD: 1v
3	WT/RD	Igniter Signal (IGF)	Digital Signal: 0-5-0v
4	WT	Distributor Signal (NE+)	AC pulse signals
5	GN	Distributor Signal (NE-)	<0.050v
6	LG	A/T Select Switch 2nd	In 2nd: 12v, Others: 0v
7	GN	A/C Idle Up Solenoid	A/C Off: 12v, On: 1v
8	GN/BK	Fuel Pressure Up Solenoid	1v (at hot restart)
9	GN	ISC Signal (ISCC)	Pulse Signals
10	RD	ISC Signal (ISCO)	Pulse Signals
11	WT	Injector Pair 2 & 4 Control	2.0-3.3 ms
12	WT	Injector Pair 1 & 3 Control	2.0-3.3 ms
13	BR	Power Ground	<0.1v
14	BR	Sensor Ground	<0.050v
15	BR/WT	A/T-ECT Solenoid (S2)	1st or OD: 1v
16	GN/BK	A/T-ECT Speed Sensor	Moving: 0-5-0v
17	BK	Distributor Signal (G-)	<0.050v
18	RD	Distributor Signal (G+)	AC pulse signals
19	GN/RD	A/T Select Switch Low	In Low: 12v, Others: 0v
20	WT	Igniter Signal (IGT)	0.72-0.74v
23	BL/OR	EGR Solenoid Control (VSV)	12v or 0v
26	BR	Power Ground	<0.1v

Pin Connector Graphic

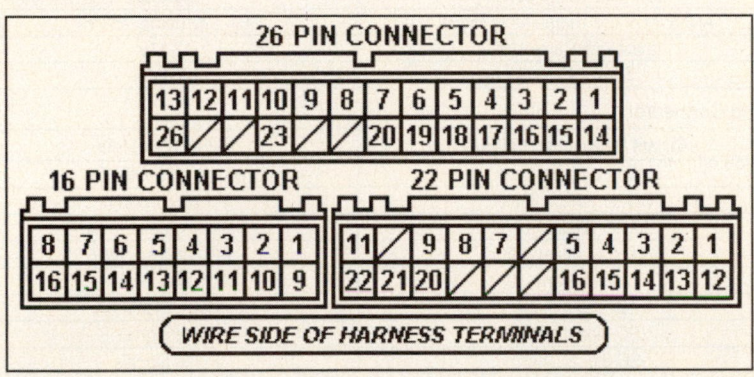

1993-95 Coupe 2.2L I4 MFI VIN S (M/T) 12 Pin Connector

PCM Pin #	Wire Color	Circuit Description (12 Pin)	Value at Hot Idle
1 ('93)	BK/YL	EFI Main Relay Power B1+	12-14v
2	WT/RD	Battery Direct	12-14v
3	BK/WT	A/C Amplifier Signal (AC1)	Clutch On: 1.5v, Off: 12v
4	GN/RD	Circuit Opening Relay (FC)	0-3v, off-idle: 12v
6	PK/GN	A/C Amplifier Signal (ACT)	Clutch On: 12v, Off: 1.5v
7	BK/YL	EFI Main Relay Power	12-14v
8	GN/WT	MIL (lamp) Control	MIL Off: 12v, On: 1v
11	PK/WT	Vehicle Speed Sensor	At 55 mph: 48 Hz
12	BK	Defogger/Light Idle Up Signal	Switch On: 12v, Off: 0v

1993-95 Coupe 2.2L I4 MFI VIN S (M/T) 16 Pin Connector

PCM Pin #	Wire Color	Circuit Description (16 Pin)	Value at Hot Idle
1	BK	Sub O2S Signal	0.1-1.1v
2	GN/RD	MAP Sensor Signal	1-1.5v
3	YL	IAT Sensor Signal	At 100ºF: 2.60v
4	RD	ECT Sensor Signal	At 180ºF: 0.51v
5	BK	Knock Sensor Signal	<0.075v AC
6	BK	Main O2S Signal	0.1-1.1v
7	PK/BL	Data Link Connector	12-14v
8 ('94)	RD/WT	Data Link Connector	12-14v
9	BR	Sensor Ground	<0.050v
10	WT	TP Sensor Signal	0.3-0.8v
11	RD/BL	MAP Sensor VREF	4.9-5.1v
12	PK	Closed Throttle Switch	1v, at off-idle: 12v
13 (Cal)	BK/YL	EGR Gas Temp. Sensor	3.5-4.0v
14	RD/WT	A/C Evaporator Temperature Signal	1.4-1.8v (temp. >59ºF)
15	BK	Data Link Connector	12-14v
16	BR	Sensor Ground	<0.050v

1993-95 Coupe 2.2L I4 MFI VIN S (M/T) 26 Pin Connector

PCM Pin #	Wire Color	Circuit Description (26 Pin)	Value at Hot Idle
1	GN	A/C Idle Up Solenoid	A/C On: 1v, Off: 12v
2	RD/WT	Starter Switch Signal	9-11v (cranking)
3	WT/RD	Igniter Signal (IGF)	Digital Signal: 0-5-0v
4	WT	Distributor Signal (NE+)	AC pulse signals
5	RD	Distributor Signal (G+)	AC pulse signals
7	BK	PSP ECU Signal	Pulse Signals
9	GN	ISC Signal (ISCC)	Pulse Signals
10	RD	ISC Signal (ISCO)	Pulse Signals
12	WT	Injector Pair 1 & 3 Control	2.0-3.3 ms
13	BR	Power Ground	<0.1v
14	BK/YL	Fuel Pressure Up Solenoid	1v (at hot restart)
14 ('93)	GN/BK	Fuel Pressure Up Solenoid	1v (at hot restart)
17	GN	Distributor Signal (NE-)	<0.050v
18	BK	Distributor Signal (G-)	<0.050v
21	BL/YL	PSP ECU Signal	Straight: 12v, Turning: 0v
22	WT	Igniter Signal (IGT)	Digital Signal: 0-5-0v
23	BL/OR	EGR Solenoid Control (VSV)	12v or 0v
24	BR	Sensor Ground	<0.050v
25	WT	Injector Pair 2 & 4 Control	2.0-3.3 ms
26	BR	Power Ground	<0.1v

Pin Connector Graphic

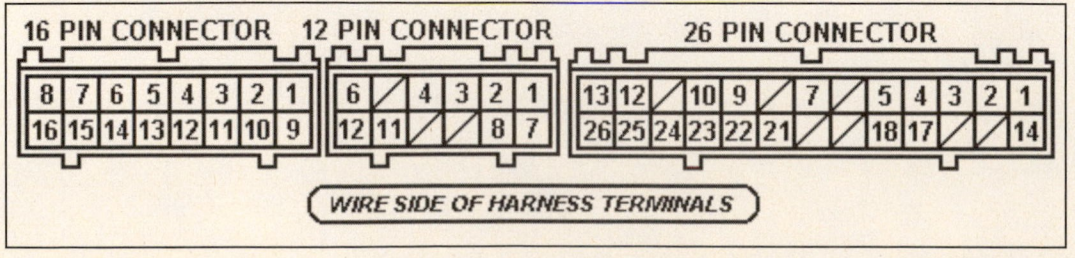

PASEO Pin Tables

1992 Convertible, Coupe 1.5L VIN E (A/T-ECT) 16 Pin Connector

PCM Pin #	Wire Color	Circuit Description (16 Pin)	Value at Hot Idle
1	GN/RD	Sensor VREF	4.9-5.1v
2	PK	MAP Sensor Signal	1-1.5v
3	GN/BK	IAT Sensor Signal	At 100°F: 2.60v
4	GN/BK	ECT Sensor Signal	At 180°F: 0.51v
6	BK	O2S Signal	0.1-1.1v
7	BL	Data Link Connector	12-14v
8	GN/YL	Data Link Connector	12-14v
9	BR	Sensor Ground	<0.050v
10 (Cal)	BL/RD	EGR Gas Temp. Sensor	3.5-4.0v
11	YL/GN	TP Sensor Signal	0.3-0.8v
12	YL/BK	Closed Throttle Switch	1v, at off-idle: 12v
15	PK	Data Link Connector	12-14v
16	BR	Sensor Ground	<0.050v

1992 Convertible, Coupe 1.5L VIN E (A/T-ECT) 22 Pin Connector

PCM Pin #	Wire Color	Circuit Description (22 Pin)	Value at Hot Idle
1	WT/RD	Battery Direct	12-14v
2	GN/RD	A/C High Press. Switch	Switch Off: 12v
4	GN/WT	Stop Light Switch Signal	Brake Off: 0v, On: 12v
5	GN/WT	MIL (lamp) Control	MIL Off: 12v, On: 1v
7	RD/WT	Overdrive Main Switch	Switch Off: 12v, On: 0v
9	YL	Vehicle Speed Sensor	At 55 mph: 48 Hz
10	PK	A/C Magnetic Clutch (ACMG)	Clutch Off: 0v, On: 12v
11	BK	Starter Switch Signal	9-11v (cranking)
12	BK/OR	EFI Main Relay Power	12-14v
13	BK/OR	EFI Main Relay Power B1+	12-14v
14	GN	Circuit Opening Relay (FC)	0-3v, off-idle: 12v
20	RD/GN	Cruise Control ECU	At Cruise in OD: 12v
21	GN/BK	A/C Amplifier Signal (ACT)	Clutch On: 12v, Off: 1.5v
22	BK/BL	Neutral Start Switch	In P/N: 9-11v (cranking)

Standard Colors and Abbreviations

Abbreviation	Color	Abbreviation	Color	Abbreviation	Color
BK	Black	GY	Gray	RD	Red
BL	Blue	GN	Green	TN	Tan
BR	Brown	LG	Light Green	VT	Violet
DB	Dark Blue	OR	Orange	WT	White
DG	DK Green	PK	Pink	YL	Yellow

1992 Convertible, Coupe 1.5L VIN E (A/T-ECT) 26 Pin Connector

PCM Pin #	Wire Color	Circuit Description (26 Pin)	Value at Hot Idle
1	PK/BK	A/T-ECT Solenoid (SL)	In Lockup: 12-14v
2	PK	A/T-ECT Solenoid (S1)	In 3rd or OD: 1v
3	GN/YL	Igniter Signal (IGF)	Digital Signal: 0-5-0v
4	RD	Distributor Signal (NE+)	AC pulse signals
5	GN	Distributor Signal (G+)	AC pulse signals
6	OR	A/T Select Switch 2nd	In 2nd: 12v, Others: 0v
7	BL	Water Temperature Switch	Switch On: 0.1v
8	RD/GN	Throttle Opener Solenoid	12v or 0v
9	BL/WT	Fuel Pressure Up Solenoid	1v (at hot restart)
10	GN/RD	A/C Idle Up Solenoid	A/C On: 1v
11	YL	Injector Pair 2 & 4 Control	2.0-3.3 ms
12	WT	Injector Pair 1 & 3 Control	2.0-3.3 ms
13	BR	Power Ground	<0.1v
14	BR	Sensor Ground	<0.050v
15	PK/GN	A/T-ECT Solenoid (S2)	1st or OD: 1v
16	GN/BK	A/T-ECT Speed Sensor	Moving: 0-5-0v
17	WT	Distributor Signal (NE-)	<0.050v
18	BK	Distributor Signal (G-)	<0.050v
19	YL/BL	A/T Select Switch Low	In Low: 12v, Others: 0v
20	LG	Igniter Signal (IGT)	Digital Signal: 0-5-0v
23	PK/BL	EGR Solenoid Control (VSV)	12v or 0v
26	BR	Power Ground	<0.1v

Pin Connector Graphic

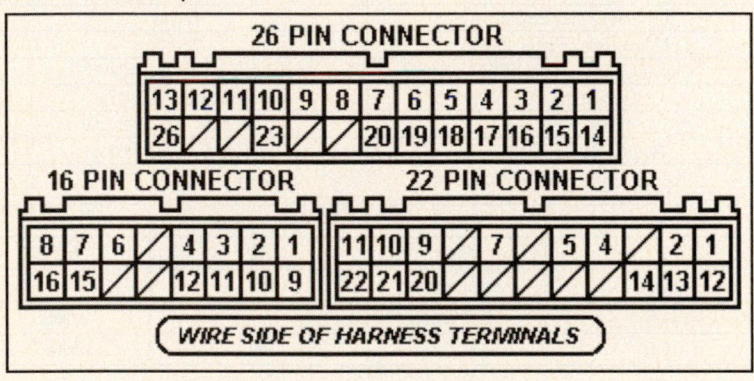

1992 Coupe 1.5L I4 MFI VIN E (M/T) 16 Pin Connector

PCM Pin #	Wire Color	Circuit Description (16 Pin)	Value at Hot Idle
1	BK/OR	EFI Main Relay Power B1+	12-14v
2	WT/RD	Battery Direct	12-14v
4	GN	Circuit Opening Relay (FC)	0-3v, off-idle: 12v
5	GN/RD	Headlight Switch	Lights On: <0.1v
7 (Cal)	RD/GN	Throttle Opener Solenoid	12v or 0v
8	PK	Data Link Connector	12-14v
9	BK/OR	EFI Main Relay Power	12-14v
10	GN/WT	MIL (lamp) Control	MIL Off: 12v, On: 1v
11	PK	A/C Magnetic Clutch (ACMG)	Clutch Off: 0v, On: 12v
13	YL	Vehicle Speed Sensor	At 55 mph: 48 Hz
14	GN/BK	A/C Amplifier Signal (ACT)	Clutch On: 12v, Off: 1.5v
16	GN/YL	Data Link Connector	12-14v

1992 Coupe 1.5L I4 MFI VIN E (M/T) 26 Pin Connector

PCM Pin #	Wire Color	Circuit Description (26 Pin)	Value at Hot Idle
1	BL/WT	Fuel Pressure Up Solenoid	1v (at hot restart)
3	GN/BK	ECT Sensor Signal	At 180°F: 0.51v
4	PK	MAP Sensor Signal	1-1.5v
5	GN/BK	IAT Sensor Signal	At 100°F: 2.60v
6	LG	Igniter Signal (IGT)	Digital Signal: 0-5-0v
7	GN/YL	Igniter Signal (IGF)	Digital Signal: 0-5-0v
8	GN	Distributor Signal (G1+)	AC pulse signals
9	BK	Distributor Signal (G-)	<0.050v
10	BK	O2S Signal	0.1-1.1v
11	BK	Starter Switch Signal	9-11v (cranking)
12	WT	Injector Pair 1 & 3 Control	2.0-3.3 ms
13	BR	Power Ground	<0.1v
14	GN/RD	A/C Idle Up Solenoid	A/C On: 1v
15 (Cal)	PK/BL	EGR Solenoid Control (VSV)	12v or 0v
16	BR	Sensor Ground	<0.050v
15	YL/GN	TP Sensor Signal	0.3-0.8v
18	GN/RD	Sensor VREF	4.9-5.1v
19	YL/BK	Closed Throttle Switch	1v, at off-idle: 12v
20	WT	Distributor Signal (NE-)	<0.050v
21	RD	Distributor Signal (NE+)	AC pulse signals
23 (Cal)	BL/RD	EGR Gas Temp. Sensor	3.5-4.0v
24	BR	Sensor Ground	<0.050v
25	WT	Injector Pair 2 & 4 Control	2.0-3.3 ms
26	BR	Power Ground	<0.1v

Pin Connector Graphic

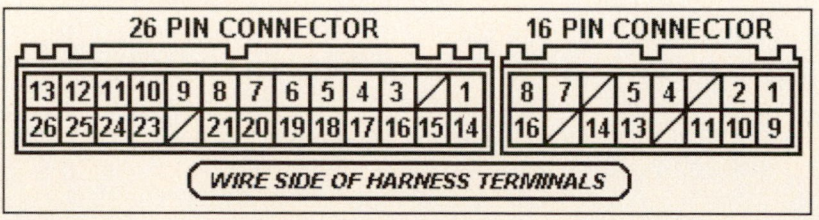

1993-94 Coupe 1.5L I4 MFI VIN E (A/T-ECT) 16 Pin Connector

PCM Pin #	Wire Color	Circuit Description (16 Pin)	Value at Hot Idle
1	GN/RD	Sensor VREF	4.9-5.1v
2	PK	MAP Sensor Signal	1-1.5v
3	GN/BK	IAT Sensor Signal	At 100°F: 2.60v
4	GN/BK	ECT Sensor Signal	At 180°F: 0.51v
6	BK	O2S Signal	0.1-1.1v
7	BL	Data Link Connector	12-14v
8	GN/YL	Data Link Connector	12-14v
9	BR	Sensor Ground	<0.050v
10 (Cal)	BL	EGR Gas Temp. Sensor	3.5-4.0v
11	YL/GN	TP Sensor Signal	0.3-0.8v
12	BL/RD	Closed Throttle Switch	1v, at off-idle: 12v
15	PK	Data Link Connector	12-14v
16	BR	Sensor Ground	<0.050v

1993-94 Coupe 1.5L I4 MFI VIN E (A/T-ECT) 22 Pin Connector

PCM Pin #	Wire Color	Circuit Description (22 Pin)	Value at Hot Idle
1	WT/RD	Battery Direct	12-14v
2	GN/RD	A/C High Pressure Switch	Switch Off: 12v
4	GN/WT	Stop Light Switch Signal	Brake Off: 0v, On: 12v
5	GN/WT	MIL (lamp) Control	MIL Off: 12v, On: 1v
7	RD/WT	Overdrive Main Switch	Switch Off: 12v
9	YL	Vehicle Speed Sensor	At 55 mph: 48 Hz
10	PK/BK	A/C Magnetic Clutch (ACMG)	Clutch Off: 0v, On: 12v
11	BK/RD	Starter Switch Signal	9-11v (cranking)
12	BK/OR	EFI Main Relay Power	12-14v
13	BK/OR	EFI Main Relay Power B1+	12-14v
14	GN	Circuit Opening Relay (FC)	0-3v, off-idle: 12v
20	RD/GN	Cruise Control ECU	At Cruise in OD: 12v
21	GN/BK	A/C Amplifier Signal (ACT)	Clutch On: 12v, Off: 1.5v
22	BK/WT	A/T Neutral Start Switch	In P/N: 9-11v (cranking)

Standard Colors and Abbreviations

Abbreviation	Color	Abbreviation	Color	Abbreviation	Color
BK	Black	GY	Gray	RD	Red
BL	Blue	GN	Green	TN	Tan
BR	Brown	LG	Light Green	VT	Violet
DB	Dark Blue	OR	Orange	WT	White
DG	DK Green	PK	Pink	YL	Yellow

1993-94 Coupe 1.5L I4 MFI VIN E (A/T-ECT) 26 Pin Connector

PCM Pin #	Wire Color	Circuit Description (26 Pin)	Value at Hot Idle
1	PK/BK	A/T-ECT Solenoid (SL)	In Lockup: 12-14v
2	PK	A/T-ECT Solenoid (S1)	In 3rd or OD: 1v
3	GN/YL	Igniter Signal (IGF)	Digital Signal: 0-5-0v
4	RD	Distributor Signal (NE+)	AC pulse signals
5	GN	Distributor Signal (G+)	AC pulse signals
6	OR	A/T Select Switch 2nd	In 2nd: 12v, Others: 0v
7	BL	Water Temperature Switch	Switch On: 0.1v
8	PK/BL	Throttle Opener Solenoid	12v or 0v
9	BL/WT	Fuel Pressure Up Solenoid	1v (at hot restart)
10	GN/RD	A/C Idle Up Solenoid	A/C On: 1v
11	YL	Injector Pair 2 & 4 Control	2.0-3.3 ms
12	WT	Injector Pair 1 & 3 Control	2.0-3.3 ms
13	BR	Power Ground	<0.1v
14	BR	Sensor Ground	<0.050v
15	PK/BK	A/T-ECT Solenoid (S2)	1st or OD: 1v
16	GN/BK	A/T-ECT Speed Sensor	Moving: 0-5-0v
17	WT	Distributor Signal (NE-)	<0.050v
18	BK	Distributor Signal (G-)	<0.050v
19	YL/BL	A/T Select Switch Low	In Low: 12v, Others: 0v
20	LG	Igniter Signal (IGT)	Digital Signal: 0-5-0v
23	PK/BL	EGR Solenoid Control (VSV)	12v or 0v
26	BR	Power Ground	<0.1v

Pin Connector Graphic

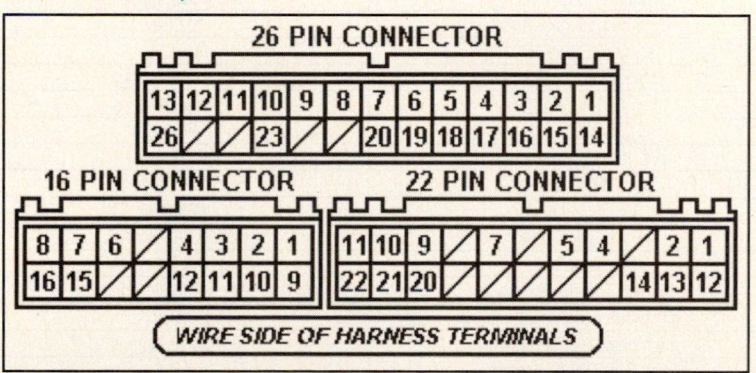

Standard Colors and Abbreviations

Abbreviation	Color	Abbreviation	Color	Abbreviation	Color
BK	Black	GY	Gray	RD	Red
BL	Blue	GN	Green	TN	Tan
BR	Brown	LG	Light Green	VT	Violet
DB	Dark Blue	OR	Orange	WT	White
DG	DK Green	PK	Pink	YL	Yellow

1993-94 Coupe 1.5L I4 MFI VIN E (M/T) 16 Pin Connector

PCM Pin #	Wire Color	Circuit Description (16 Pin)	Value at Hot Idle
1 ('93)	BK/OR	EFI Main Relay Power B1+	12-14v
2	WT/RD	Battery Direct	12-14v
4	GN	Circuit Opening Relay (FC)	0-3v, off-idle: 12v
5	GN/RD	Headlight Switch Signal	Lights On: 0.1v
5	RD/BL	Headlight Switch Signal	Lights On: 0.1v
7 (Cal)	RD/GN	Throttle Opener Solenoid	12v or 0v
8	PK	Data Link Connector	12-14v
9	BK/OR	EFI Main Relay Power	12-14v
10	GN/WT	MIL (lamp) Control	MIL Off: 12v, On: 1v
11	PK/BK	A/C Magnetic Clutch (ACMG)	Clutch Off: 0v, On: 12v
13	YL	Vehicle Speed Sensor	At 55 mph: 48 Hz
14	GN/BK	A/C Amplifier Signal (ACT)	Clutch On: 12v, Off: 1.5v
16	GN/YL	Data Link Connector	12v

1993-94 Coupe 1.5L I4 MFI VIN E (M/T) 26 Pin Connector

PCM Pin #	Wire Color	Circuit Description (26 Pin)	Value at Hot Idle
1	BL/WT	Fuel Pressure Up Solenoid	1v (at hot restart)
2 ('94)	BK/RD	M/T Neutral Start Switch	In P/N: 9-11v (cranking)
3	GN/BK	ECT Sensor Signal	At 180°F: 0.51v
4	PK	MAP Sensor Signal	1-1.5v
5	GN/BK	IAT Sensor Signal	At 100°F: 2.60v
6	LG	Igniter Signal (IGT)	Digital Signal: 0-5-0v
7	GN/YL	Igniter Signal (IGF)	Digital Signal: 0-5-0v
8	GN	Distributor Signal (G+)	AC pulse signals
9	BK	Distributor Signal (G-)	<0.050v
10	BK	O2S Signal	0.1-1.1v
11	BK/RD	Starter Switch Signal	9-11v (cranking)
12	WT	Injector 1 & 3 Control	2.0-3.3 ms
13	BR	Power Ground	<0.1v
14	GN/RD	A/C Idle Up Solenoid	A/C On: 1v
15 (Cal)	PK/BL	EGR Solenoid Control (VSV)	12v or 0v
16	BR	Sensor Ground	<0.050v
17	YL/GN	TP Sensor Signal	0.3-0.8v
18	GN/RD	Sensor VREF	4.9-5.1v
19	BL/RD	Closed Throttle Switch	1v, at off-idle: 12v
20	WT	Distributor Signal (NE-)	<0.050v
21	RD	Distributor Signal (NE+)	AC pulse signals
23 (Cal)	BL	EGR Gas Temp. Sensor	3.5-4.0v
24	BR	Sensor Ground	<0.050v
25	YL	Injector 2 & 4 Control	2.0-3.3 ms
26	BR	Power Ground	<0.1v

Pin Connector Graphic

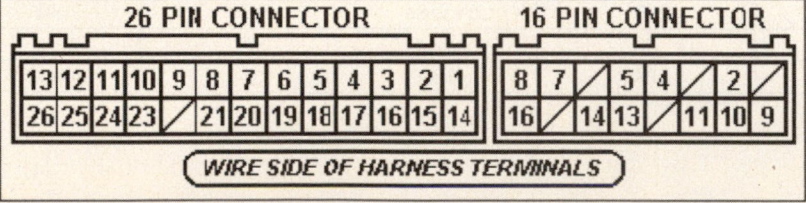

Standard Colors and Abbreviations

Abbreviation	Color	Abbreviation	Color	Abbreviation	Color
BK	Black	GY	Gray	RD	Red
BL	Blue	GN	Green	TN	Tan
BR	Brown	LG	Light Green	VT	Violet
DB	Dark Blue	OR	Orange	WT	White
DG	DK Green	PK	Pink	YL	Yellow

1995 Coupe 1.5L I4 VIN E California (A/T-ECT) 12 Pin Connector

PCM Pin #	Wire Color	Circuit Description (12 Pin)	Value at Hot Idle
1	GN/WT	Stop Light Switch Signal	Brake Off: 0v, On: 12v
2	WT/RD	Battery Direct	12-14v
3	RD/GN	Cruise Control ECU	At Cruise in OD: 12v
4	GN	Circuit Opening Relay (FC)	0-3v, off-idle: 12v
6	GN/BK	A/C Amplifier Signal (ACT)	Clutch On: 12v, Off: 1.5v
7	BK/OR	EFI Main Relay Power	12-14v
8	GN/WT	MIL (lamp) Control	MIL Off: 12v, On: 1v
9	RD/WT	Overdrive Main Switch	Switch Off: 12v
10	PK/BK	A/C Amplifier Signal (AC1)	Clutch On: 1.5v, Off: 12v
11	YL	Vehicle Speed Sensor	At 55 mph: 48 Hz

1995 Coupe 1.5L I4 VIN E California (A/T-ECT) 16 Pin Connector

PCM Pin #	Wire Color	Circuit Description (16 Pin)	Value at Hot Idle
1	BL	Data Link Connector	12-14v
2	PK	MAP Sensor Signal	1-1.5v
3	BL/BK	IAT Sensor Signal	At 100°F: 2.60v
4	GN/BK	ECT Sensor Signal	At 180°F: 0.51v
5	WT	Sub HO2S Signal	0.1-1.1v
6	BK	Main HO2S Signal	0.1-1.1v
7	PK/WT	Data Link Connector	12-14v
8	GN/YL	Data Link Connector	12-14v
9	BR	Sensor Ground	<0.050v
10	YL/GN	TP Sensor Signal	0.3-0.8v
11	GN/RD	Sensor VREF	4.9-5.1v
12	YL/BK	Closed Throttle Switch	1v, at off-idle: 12v
13	BL/RD	EGR Gas Temp. Sensor	3.5-4.0v
14	BK	Knock Sensor Signal	<0.075v AC
15	BL	Data Link Connector	12-14v

Standard Colors and Abbreviations

Abbreviation	Color	Abbreviation	Color	Abbreviation	Color
BK	Black	GY	Gray	RD	Red
BL	Blue	GN	Green	TN	Tan
BR	Brown	LG	Light Green	VT	Violet
DB	Dark Blue	OR	Orange	WT	White
DG	DK Green	PK	Pink	YL	Yellow

1995 Coupe 1.5L I4 VIN E California (A/T-ECT) 26 Pin Connector

PCM Pin #	Wire Color	Circuit Description (26 Pin)	Value at Hot Idle
1	PK/BK	A/T-ECT Solenoid (SL)	In Lockup: 12-14v
2	BK/WT	Starter Switch Signal	9-11v (cranking)
3	GN/RD	Igniter Signal (IGF)	Digital Signal: 0-5-0v
4	BK	CKP Sensor Signal (NE+)	AC pulse signals
7	PK/BK	A/T-ECT Solenoid (S2)	1st or OD: 1v
9	BL/RD	ISC Signal (RSC)	Pulse Signals
10	BL	ISC Signal (RSO)	Pulse Signals
11	WT	Sub HO2S Heater	1v (Heater On)
12	WT	Injector Pair 1 & 3 Control	2.0-3.3 ms
13	BR	Power Ground	<0.1v
14	WT/GN	Main HO2S Heater	1v (Heater On)
15	BK	Neutral Start Switch	In P/N: 9-11v (cranking)
16	BK/WT	EGR Solenoid Control (VSV)	12v or 0v
17	GN	CMP Sensor Signal (G+)	AC pulse signals
18	RD	CMP Sensor Signal (G-)	<0.050v
19	OR	A/T Select Switch 2nd	In 2nd: 12v, Others: 0v
20	YL/BL	A/T Select Switch Low	In Low: 12v, Others: 0v
21	BL/YL	Igniter Transistor 2 Control	Digital Signal: 0-5-0v
22	LG	Igniter Transistor 1 Control	Digital Signal: 0-5-0v
23	GN/BK	A/T-ECT Speed Sensor	Moving: 0-5-0v
24	BR	Sensor Ground	<0.050v
25	YL	Injector Pair 2 & 4 Control	2.0-3.3 ms
26	PK	A/T-ECT Solenoid (S1)	In 3rd or OD: 1v

Pin Connector Graphic

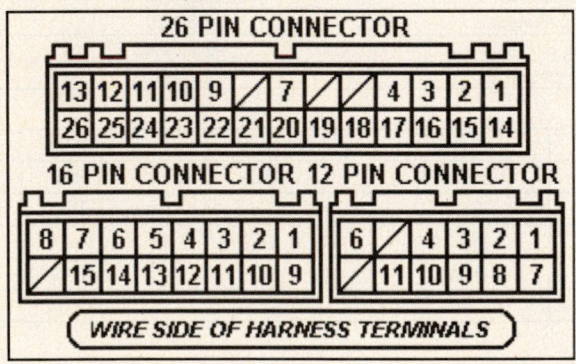

1995 Coupe 1.5L I4 MFI VIN E Fed (A/T-ECT) 16 Pin Connector

PCM Pin #	Wire Color	Circuit Description (16 Pin)	Value at Hot Idle
1	GN/RD	Sensor VREF	4.9-5.1v
2	PK	MAP Sensor Signal	1-1.5v
3	GN/BK	IAT Sensor Signal	At 100ºF: 2.60v
4	GN/BK	ECT Sensor Signal	At 180ºF: 0.51v
6	BK	O2S Signal	0.1-1.1v
7	BL	Data Link Connector	12-14v
8	GN/YL	Data Link Connector	12-14v
9	BR	Sensor Ground	<0.050v
10	BL/RD	EGR Gas Temp. Sensor	3.5-4.0v
11	YL/GN	TP Sensor Signal	0.3-0.8v
12	BL/RD	Closed Throttle Switch	1v, at off-idle: 12v
15	PK	Data Link Connector	12-14v

1995 Coupe 1.5L I4 MFI VIN E Fed (A/T-ECT) 22 Pin Connector

PCM Pin #	Wire Color	Circuit Description (22 Pin)	Value at Hot Idle
1	WT/RD	Battery Direct	12-14v
2	GN/RD	A/C Fan Relay	Relay Off: 12v, On: 1v
4	GN/WT	Stop Light Switch Signal	Brake Off: 0v, On: 12v
5	GN/WT	MIL (lamp) Control	MIL Off: 12v, On: 1v
7	RD/WT	Overdrive Main Switch (OD2)	Switch Off: 12v
9	YL	Vehicle Speed Sensor	At 55 mph: 48 Hz
10	PK/BK	A/C Amplifier Signal (AC1)	Clutch On: 1.5v, Off: 12v
11	BK/WT	Starter Switch Signal	9-11v (cranking)
12	BK/OR	EFI Main Relay Power	12-14v
14	GN	Circuit Opening Relay (FC)	0-3v, off-idle: 12v
20	RD/GN	Overdrive Main Switch (OD1)	Switch Off: 12v
21	GN/BK	A/C Amplifier Signal (ACT)	Clutch On: 12v, Off: 1.5v
22	BK	A/T Neutral Start Switch	In P/N: 9-11v (cranking)

Standard Colors and Abbreviations

Abbreviation	Color	Abbreviation	Color	Abbreviation	Color
BK	Black	GY	Gray	RD	Red
BL	Blue	GN	Green	TN	Tan
BR	Brown	LG	Light Green	VT	Violet
DB	Dark Blue	OR	Orange	WT	White
DG	DK Green	PK	Pink	YL	Yellow

1995 Coupe 1.5L I4 MFI VIN E Fed (A/T-ECT) 26 Pin Connector

PCM Pin #	Wire Color	Circuit Description (26 Pin)	Value at Hot Idle
1	GN/RD	A/T-ECT Solenoid (SL)	In Lockup: 12-14v
2	PK/BL	A/T-ECT Solenoid (S1)	In 3rd or OD: 1v
3	GN/YL	Igniter Signal (IGF)	Digital Signal: 0-5-0v
4	RD	Distributor Signal (NE+)	AC pulse signals
5	GN	Distributor Signal (G+)	AC pulse signals
6	OR	A/T Select Switch 2nd	In 2nd: 12v, Others: 0v
7	BL	Water Temperature Switch	Switch On: 0.1v
8	RD/GN	Throttle Opener Solenoid	12v or 0v
9	BL/WT	Fuel Pressure Up Solenoid	1v (at hot restart)
10	GN/RD	IAC Solenoid (IAC)	Pulse Signals
11	YL	Injector Pair 2 & 4 Control	2.0-3.3 ms
12	WT	Injector Pair 1 & 3 Control	2.0-3.3 ms
13	BR	Power Ground	<0.1v
14	BR	Sensor Ground	<0.050v
15	PK/BK	A/T-ECT Solenoid (S2)	1st or OD: 1v
16	GN/BK	A/T-ECT Speed Sensor	Moving: 0-5-0v
17	WT	Distributor Signal (NE-)	<0.050v
18	BK	Distributor Signal (G-)	<0.050v
19	YL/BL	A/T Select Switch Low	In Low: 12v, Others: 0v
20	LG	Igniter Transistor 1 Control	Digital Signal: 0-5-0v
23	PK/BL	EGR Solenoid Control (VSV)	12v or 0v
26	BR	Power Ground	<0.1v

Pin Connector Graphic

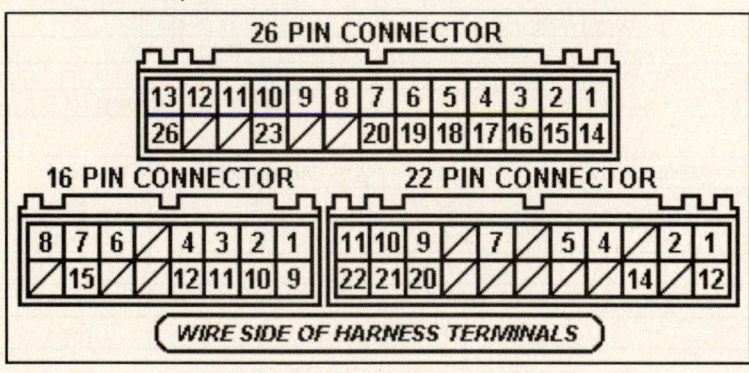

1995 Coupe 1.5L I4 MFI VIN E California (M/T) 16 Pin Connector

PCM Pin #	Wire Color	Circuit Description (16 Pin)	Value at Hot Idle
2	WT/RD	Battery Direct	12-14v
4	GN	Circuit Opening Relay (FC)	0-3v, off-idle: 12v
5	WT/RD	Sub HO2S Heater	1v (Heater On)
6	WT/GN	Main HO2S Heater	1v (Heater On)
7	PK/WT	Data Link Connector	12-14v
8	GN	Data Link Connector	12-14v
9	BK/OR	EFI Main Relay Power	12-14v
10	GN/WT	MIL (lamp) Control	MIL Off: 12v, On: 1v
12	PK/BK	A/C Amplifier Signal (AC1)	Clutch On: 1.5v, Off: 12v
13	YL	Vehicle Speed Sensor	At 55 mph: 48 Hz
14	GN/BK	A/C Amplifier Signal (ACT)	Clutch On: 12v, Off: 1.5v
15	BK/WT	EGR Solenoid Control (VSV)	12v or 0v
16	GN/YL	Data Link Connector	12-14v

1995 Coupe 1.5L I4 MFI VIN E California (M/T) 26 Pin Connector

PCM Pin #	Wire Color	Circuit Description (26 Pin)	Value at Hot Idle
1	BL/RD	ISC Signal (RSC)	Pulses
3	GN/BK	ECT Sensor Signal	At 180ºF: 0.51v
4	PK	MAP Sensor Signal	1-1.5v
5	PK/YL	Igniter Transistor 2 Control	Digital Signal: 0-5-0v
6	GN	Igniter Transistor 1 Control	Digital Signal: 0-5-0v
7	GN/RD	Igniter Signal (IGF)	Digital Signal: 0-5-0v
8	BK	Knock Sensor Signal	<0.075v AC
9	GN	CKP Sensor Signal (NE-)	<0.050v
10	BK	Main HO2S Signal	0.1-1.1v
11	BK/WT	Starter Switch Signal	9-11v (cranking)
12	GN	Injectors 1 & 3 Control	2.0-3.3 ms
13	BR	Power Ground	<0.1v
14	BL/RD	ISC Signal (RSO)	Pulse Signals
15	BL/BK	IAT Sensor Signal	At 100ºF: 2.60v
16	BR	Sensor Ground	<0.050v
17	YL/GN	TP Sensor Signal	0.3-0.8v
18	GN/RD	Sensor VREF	4.9-5.1v
19	YL/BK	Closed Throttle Switch	1v, at off-idle: 12v
20	RD	CMP Sensor Signal (G+)	AC pulse signals
21	BK	CKP Sensor Signal (NE+)	AC pulse signals
22	BL/RD	EGR Gas Temp. Sensor	3.5-4.0v
23	WT	Sub HO2S Signal	0.1-1.1v
24	BR	Power Ground	<0.1v
25	YL	Injectors 2 & 4 Control	2.0-3.3 ms

Pin Connector Graphic

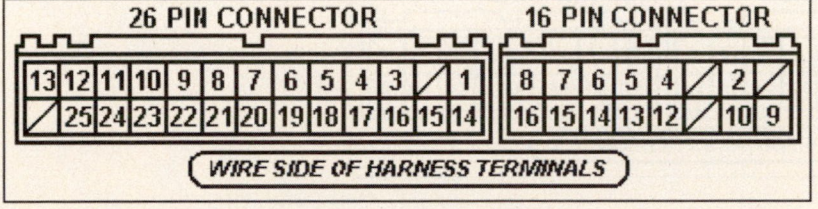

1995 Coupe 1.5L I4 MFI VIN E Federal (M/T) 16 Pin Connector

PCM Pin #	Wire Color	Circuit Description (16 Pin)	Value at Hot Idle
2	WT/RD	Battery Direct	12-14v
4	GN	Circuit Opening Relay (FC)	0-3v, off-idle: 12v
5	RD/GN	Headlight Switch	Switch On: 12v, Off: 0v
7	RD/GN	Throttle Opener Solenoid	12v or 0v
8	PK	Data Link Connector	12-14v
9	BK/OR	EFI Main Relay Power	12-14v
10	GN/WT	MIL (lamp) Control	MIL Off: 12v, On: 1v
11	PK/BK	A/C Amplifier Signal (AC1)	Clutch On: 1.5v, Off: 12v
13	YL	Vehicle Speed Sensor	At 55 mph: 48 Hz
14	GN/BK	A/C Amplifier Signal (ACT)	Clutch On: 12v, Off: 1.5v
16	GN/YL	Data Link Connector	12-14v

1995 Coupe 1.5L I4 MFI VIN E Federal (M/T) 26 Pin Connector

PCM Pin #	Wire Color	Circuit Description (26 Pin)	Value at Hot Idle
1	BL/WT	Fuel Pressure Up Solenoid	12v or 0v
2	BK/RD	Neutral Start Switch	In P/N: 9-11v (cranking)
3	GN/BK	ECT Sensor Signal	At 180ºF: 0.51v
4	PK	MAP Sensor Signal	1-1.5v
5	GN/BK	IAT Sensor Signal	At 100ºF: 2.60v
6	LG	Igniter Signal (IGT)	Digital Signal: 0-5-0v
7	GN/YL	Igniter Signal (IGF)	Digital Signal: 0-5-0v
8	GN	Distributor Signal (G+)	Pulses
9	BK	Distributor Signal (G-)	<0.050v
10	BK	O2S Signal	0.1-1.1v
11	BK/RD	Starter Switch Signal	9-11v (cranking)
12	WT	Injectors 1 & 3 Control	2.0-3.3 ms
13	BR	Power Ground	<0.1v
14	GN/RD	Throttle Opener Solenoid	12v or 0v
15	BK/WT	EGR Solenoid Control (VSV)	12v or 0v
24	BR	Sensor Ground	<0.050v
17	YL/GN	TP Sensor Signal	0.3-0.8v
18	GN/RD	Sensor VREF	4.9-5.1v
19	BL/RD	Closed Throttle Switch	1v, at off-idle: 12v
20	WT	Distributor Signal (NE-)	<0.050v
21	RD	Distributor Signal (NE+)	AC pulse signals
23	BR	EGR Gas Temp. Sensor	3.5-4.0v
24	BR	Sensor Ground	<0.050v
25	YL	Injectors 2 & 4 Control	2.0-3.3 ms
26	BR	Power Ground	<0.1v

Pin Connector Graphic

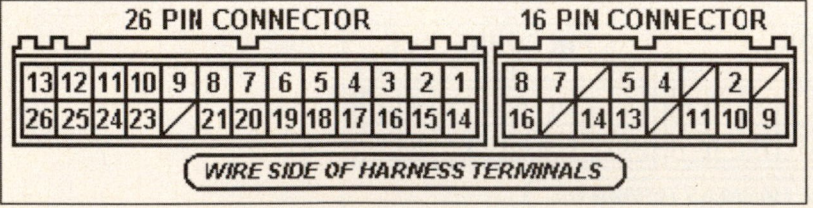

1996 Coupe 1.5L I4 MFI VIN C (A/T-ECT) 16 Pin Connector

PCM Pin #	Wire Color	Circuit Description (16 Pin)	Value at Hot Idle
1	GN/RD	Sensor VREF	4.9-5.1v
2	PK	MAP Sensor Signal	1-1.5v
3	BL/BK	IAT Sensor Signal	At 100ºF: 2.60v
4	GN/BK	ECT Sensor Signal	At 180ºF: 0.51v
5	WT	HO2S-12 (B1 S2) Signal	0.1-1.1v
6	BK	HO2S-11 (B1 S1) Signal	0.1-1.1v
9	BR	Sensor Ground	<0.050v
10	BK	Knock Sensor Signal	<0.075v AC
11	YL/GN	TP Sensor Signal	0.3-0.8v
12	YL/BK	Closed Throttle Switch	1v, at off-idle: 12v
15	GN	Data Link Connector	12-14v
16	BR	Power Ground	<0.1v

1996 Coupe 1.5L I4 MFI VIN C (A/T-ECT) 22 Pin Connector

PCM Pin #	Wire Color	Circuit Description (22 Pin)	Value at Hot Idle
1	WT/RD	Battery Direct	12-14v
4	GN/WT	Stop Light Switch Signal	Brake Off: 0v, On: 12v
5	GY/BL	MIL (lamp) Control	MIL Off: 12v, On: 1v
7	RD/WT	Overdrive Main Switch	Switch Off: 12v, On: 0v
8	RD/BK	A/T Select Switch Reverse	In 'R': 12v, Others: 0v
9	YL	Vehicle Speed Sensor	At 55 mph: 48 Hz
10	GN/BK	A/C Amplifier Signal (AC1)	Clutch On: 1.5v, Off: 12v
11	BK/WT	Starter Switch Signal	9-11v (cranking)
12	BK/RD	EFI Main Relay Power	12-14v
14	GN/BK	Circuit Opening Relay (FC)	0-3v, off-idle: 12v
15	GN/BK	A/T-ECT Speed Sensor	Moving: 0-5-0v
16	WT	Data Link Connector (SDL)	No Scan Tool: 0v
18	GN/RD	A/T Select Switch 2nd	In 2nd: 12v, Others: 0v
19	LG	A/T Select Switch Low	In Low: 12v, Others: 0v
20	YL/RD	Cruise Control ECU	At Cruise in OD: 12v
21	BL	A/C Amplifier Signal (ACT)	Clutch On: 12v, Off: 1.5v
22	BK	A/T Neutral Start Switch	In P/N: 9-11v (cranking)

Standard Colors and Abbreviations

Abbreviation	Color	Abbreviation	Color	Abbreviation	Color
BK	Black	GY	Gray	RD	Red
BL	Blue	GN	Green	TN	Tan
BR	Brown	LG	Light Green	VT	Violet
DB	Dark Blue	OR	Orange	WT	White
DG	DK Green	PK	Pink	YL	Yellow

1996 Coupe 1.5L I4 MFI VIN C (A/T-ECT) 26 Pin Connector

PCM Pin #	Wire Color	Circuit Description (26 Pin)	Value at Hot Idle
1	PK/BK	A/T-ECT Solenoid (SL)	In Lockup: 12-14v
2	PK	A/T-ECT Solenoid (S1)	In 3rd or OD: 1v
3	GN/BK	Igniter Signal (IGF)	Digital Signal: 0-5-0v
4	BK	CKP Sensor Signal (NE+)	AC pulse signals
5	RD	CMP Sensor Signal (G+)	AC pulse signals
6	WT/BL	HO2S-12 (B1 S2) Heater	1v (Heater On)
9	BL/RD	IAC Signal (RSC)	Pulse Signals
10	BL	IAC Signal (RSO)	Pulse Signals
11	YL	Injector Pair 2 & 4 Control	2.0-3.3 ms
12	GN	Injector Pair 1 & 3 Control	2.0-3.3 ms
13	BR	Power Ground	<0.1v
14	BR	Shield Ground	<0.050v
15	PK/GN	A/T-ECT Solenoid (S2)	1st or OD: 1v
17	GN	CKP/CMP Sensor Ground (-)	<0.050v
19	WT/RD	HO2S-11 (B1 S1) Heater	1v (Heater On)
20	LG	Igniter Transistor 1 Control	Digital Signal: 0-5-0v
21	BL/YL	Igniter Transistor 2 Control	Digital Signal: 0-5-0v
23	BK/WT	EGR Solenoid Control (VSV)	12v or 0v

Pin Connector Graphic

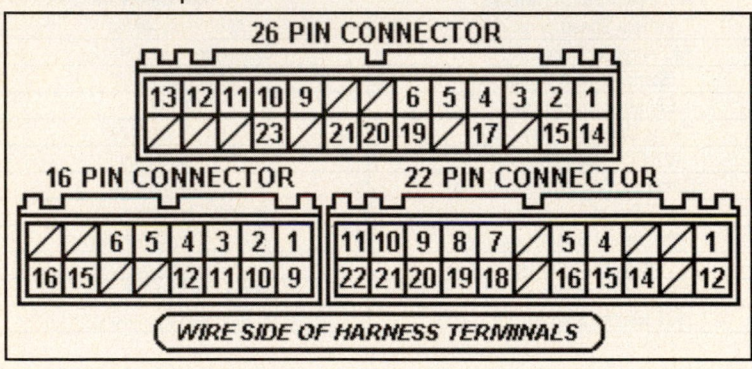

1996 Coupe 1.5L I4 MFI VIN C (M/T) 16 Pin Connector

PCM Pin #	Wire Color	Circuit Description (16 Pin)	Value at Hot Idle
2	WT/RD	Battery Direct	12-14v
4	GN/BK	Circuit Opening Relay (FC)	0-3v, off-idle: 12v
5	WT/RD	HO2S-12 (B1 S2) Heater	1v (Heater On)
6	WT/BL	HO2S-11 (B1 S1) Heater	1v (Heater On)
8	GN	Data Link Connector	12-14v
9	BK/RD	EFI Main Relay Power	12-14v
10	GN/BL	MIL (lamp) Control	MIL Off: 12v, On: 1v
12	GN/BK	A/C Amplifier Signal (AC1)	Clutch On: 1.5v, Off: 12v
13	YL	Vehicle Speed Sensor	At 55 mph: 48 Hz
14	BL	A/C Amplifier Signal (ACT)	Clutch On: 12v, Off: 1.5v
15	BK/WT	EGR Solenoid Control (VSV)	12v or 0v
16	WT	Data Link Connector	12v

1996 Coupe 1.5L I4 MFI VIN C (M/T) 26 Pin Connector

PCM Pin #	Wire Color	Circuit Description (26 Pin)	Value at Hot Idle
1	BL/RD	ISC Signal (RSC)	Pulses
3	GN/BK	ECT Sensor Signal	At 180ºF: 0.51v
4	PK	MAP Sensor Signal	1-1.5v
5	BL/YL	Igniter Transistor 2 Control	Digital Signal: 0-5-0v
6	LG	Igniter Transistor 1 Control	Digital Signal: 0-5-0v
7	GN/BK	Igniter Signal (IGF)	Digital Signal: 0-5-0v
8	BK	Knock Sensor Signal	<0.075v AC
9	GN	CKP Sensor Signal (NE-)	<0.050v
10	BK	HO2S-11 (B1 S1) Signal	0.1-1.1v
11	BK/WT	Starter Switch Signal	9-11v (cranking)
12	GN	Injectors 1 & 3 Control	2.0-3.3 ms
13	BR	Power Ground	<0.1v
14	BL	ISC Signal (RSO)	Pulses
15	BL/BK	IAT Sensor Signal	At 100ºF: 2.60v
16	BR	Sensor Ground	<0.050v
17	YL/GN	TP Sensor Signal	0.3-0.8v
18	GN/RD	Sensor VREF	4.9-5.1v
19	YL/BK	Closed Throttle Switch	0-3v, off-idle: 12v
20	RD	CMP Sensor Signal (G+)	AC pulse signals
21	BK	CKP Sensor Signal (NE+)	AC pulse signals
23	WT	HO2S-12 (B1 S2) Signal	0.1-1.1v
24	BR	Sensor Ground	<0.050v
25	YL	Injectors 2 & 4 Control	2.0-3.3 ms
26	BR	Power Ground	<0.1v

Pin Connector Graphic

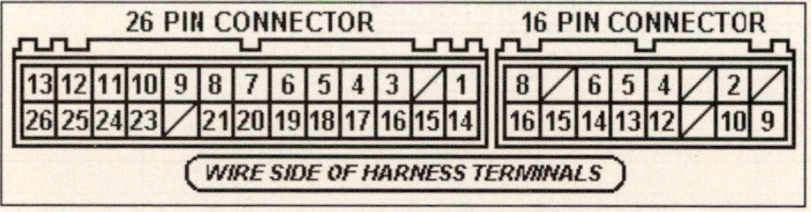

1997-99 Convertible, Coupe 1.5L I4 VIN C (All) 16 Pin Connector

PCM Pin #	Wire Color	Circuit Description (16 Pin)	Value at Hot Idle
1	GN/WT	Stop Light Switch Signal	Brake Off: 0v, On: 12v
2	PK	MAP Sensor Signal	1-1.5v
3	BL/BK	IAT Sensor Signal	At 100°F: 2.60v
4	BK/RD	ECT Sensor Signal	At 180°F: 0.51v
5	WT	HO2S-12 (B1 S2) Signal	0.1-1.1v
6	BK	O2S-11 (B1 S1) Signal	0.1-1.1v
7	BL/WT	EVAP Purge Solenoid (VSV)	12v or 0v
8	GN/YL	EVAP Vapor Pressure (VSV)	12v or 0v
9	BR	Sensor Ground	<0.050v
10	YL/GN	TP Sensor Signal	0.3-0.8v
11	GN/RD	Sensor VREF	4.9-5.1v
12	YL/BK	EVAP Vapor Pressure Sensor	2.5-3.7v
13	WT	SIL Signal (Scan Tool)	Digital Signal
14	BK	Knock Sensor Signal	<0.075v AC
15	GN	Data Link Connector (TE)	12-14v
16	BR	Power Ground	<0.1v

1997-99 Convertible, Coupe 1.5L I4 VIN C (All) 12 Pin Connector

PCM Pin #	Wire Color	Circuit Description (12 Pin)	Value at Hot Idle
1	GY/BL	Cruise Control Signal (IDLO)	1.5v, off-idle: 12v
2	WT/RD	Battery Direct	12-14v
3	YL/RD	Cruise Control Signal (OD1)	At Cruise in OD: 12v
4	GN/BK	Circuit Opening Relay (FC)	0-3v, off-idle: 12v
6	BL	A/C Amplifier Signal (ACT)	Clutch On: 12v, Off: 1.5v
7	BK/RD	EFI Main Relay Power	12-14v
8	GY/BL	MIL (lamp) Control	MIL Off: 12v, On: 1v
9	RD/WT	Overdrive Main Switch	Switch Off: 12v
10	GN/BK	A/C Amplifier Signal (AC1)	Clutch On: 1.5v, Off: 12v
11	YL	Vehicle Speed Sensor	At 55 mph: 48 Hz

Pin Connector Graphic

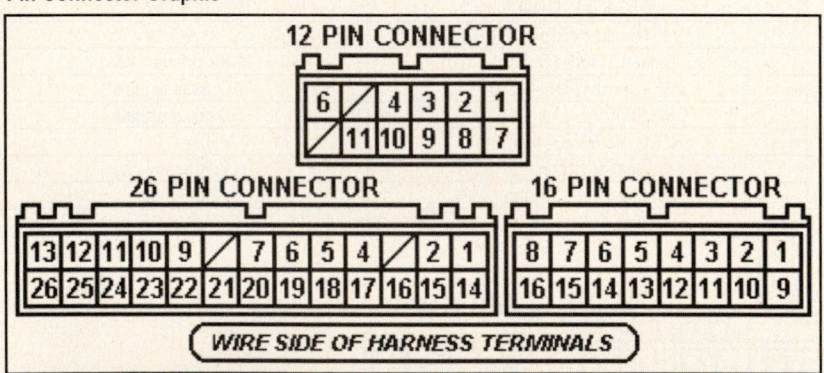

Standard Colors and Abbreviations

Abbreviation	Color	Abbreviation	Color	Abbreviation	Color
BK	Black	GY	Gray	RD	Red
BL	Blue	GN	Green	TN	Tan
BR	Brown	LG	Light Green	VT	Violet
DB	Dark Blue	OR	Orange	WT	White
DG	DK Green	PK	Pink	YL	Yellow

1997-99 Convertible, Coupe 1.5L I4 VIN C (All) 26 Pin Connector

PCM Pin #	Wire Color	Circuit Description (26 Pin)	Value at Hot Idle
1	GY	Igniter Transistor 1 Control	Digital Signal: 0-5-0v
2	BK/WT	Starter Switch Signal	9-11v (cranking)
4	BK	CKP Sensor Signal (NE+)	AC pulse signals
5	LG	A/T Select Switch Low	In Low: 12v, Others: 0v
6	GN/RD	A/T Select Switch 2nd	In 2nd: 12v, Others: 0v
7	RD/BL	Igniter Signal (IGF)	Digital Signal: 0-5-0v
9 ('97)	BL/RD	ISC Signal (RSC)	Pulse Signals
10 ('97)	BL	ISC Signal (RSO)	Pulse Signals
10 ('98-'99)	BL	ISC Signal (RSD)	Pulse Signals
11	RD	Injector 3 Control	2.0-3.3 ms
12	GN	Injector 1 Control	2.0-3.3 ms
13	BR	Sensor Ground	<0.050v
14	BL/YL	Igniter Transistor 2 Control	Digital Signal: 0-5-0v
15	BK	Neutral Start Switch	In P/N: 9-11v (cranking)
15	BK/WT	Neutral Start Switch	In P/N: 9-11v (cranking)
16	GN/BK	A/T-ECT Speed Sensor	Moving: 0-5-0v
17	GN	CKP/CMP Sensor Ground (-)	<0.050v
18	RD	CMP Sensor Signal (G+)	AC pulse signals
19	PK/GN	A/T-ECT Solenoid (S2)	1st or OD: 1v
20	PK	A/T-ECT Solenoid (S1)	In 3rd or OD: 1v
21	WT/RD	HO2S-12 (B1 S2) Heater	1v (Heater On)
22	RD/BK	A/T Select Switch Reverse	In 'R': 12v, Others: 0v
23	PK/BK	A/T-ECT Solenoid (SL)	In Lockup: 12-14v
24	BL	Injector 4 Control	2.0-3.3 ms
25	YL	Injector 2 Control	2.0-3.3 ms
26	BR	Shield Ground	<0.050v

Pin Connector Graphic

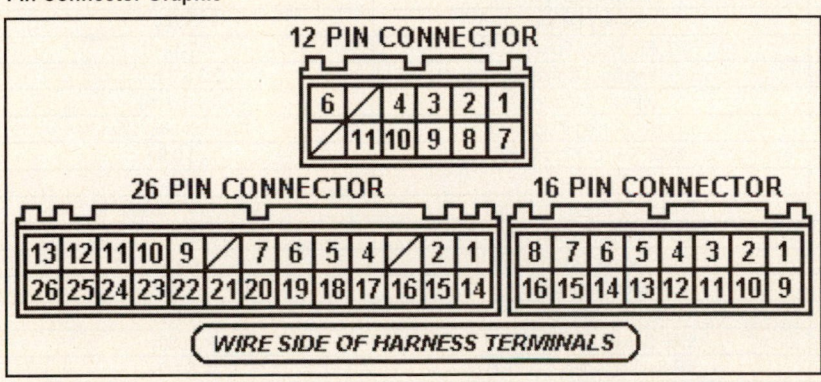

PRIUS Pin Tables

2000-03 Sedan 1.5L I4 Hybrid MFI VIN B (A/T) E7 31P Connector

PCM Pin #	Wire Color	Circuit Description (31 Pin)	Value at Hot Idle
1	BL/WT	Injector 3 Control	2.0-3.3 ms
2	RD/WT	Injector 4 Control	2.0-3.3 ms
3	WT/BK	Power Ground (E03)	<0.1v
4	LG	Fan Relay 1 & 2 Control	Relay Off: 12v, On: 1v
5, 15-20	---	Not Used	---
6	GY	ATC Motor Power Supply	12-14v
7	PK	Throttle Control Motor (M-)	Pulse Signals
8	BL	Throttle Control Motor (M+)	Pulse Signals
9	WT/BK	ETC Motor Ground (ME01)	<0.1v
10	RD	CMP Sensor Signal (G2)	AC pulse signals
11	YL/GN	Igniter Transistor 1 Control	Digital Signal: 0-5-0v
12	WT	Igniter Transistor 2 Control	Digital Signal: 0-5-0v
13	GN	Igniter Transistor 3 Control	Digital Signal: 0-5-0v
14	YL	Igniter Transistor 4 Control	Digital Signal: 0-5-0v
21	WT/BK	Power Ground (E01)	<0.1v
22, 26-27	---	Not Used	---
23	YL/RD	Camshaft Timing Oil Control Valve (+)	AC pulse signals
24	WT/GN	Camshaft Timing Oil Control Valve (-)	AC pulse signals
25	BK/RD	Igniter Signal (IGF)	Pulse Signal: 0-5-0v
28	BK	Knock Sensor Signal (KNK1)	<0.075v AC
29	RD/BL	EVAP Purge Solenoid (VSV)	12v or 0v
30	GY	Motor Shield Ground (GE01)	<0.050v
31	WT/BK	Power Ground (E02)	<0.1v

2000-03 Sedan 1.5L I4 Hybrid MFI VIN B (A/T) E8 24P Connector

PCM Pin #	Wire Color	Circuit Description (24 Pin)	Value at Hot Idle
1, 7-8	---	Not Used	---
2	YL/RD	Sensor VREF (VC)	4.9-5.1v
3	RD/WT	Battery Direct	12-14v
4	BK	EFI Relay Power	12-14v
5	YL	Injector 1 Control	2.0-3.3 ms
6	BK/RD	Injector 2 Control	2.0-3.3 ms
9	GN/RD	Circuit Opening Relay (FC)	0-3v, off-idle: 12v
10	GN	MAF Sensor Signal (VG)	0.3-0.9v
11, 15, 20	---	Not Used	---
12	YL/BK	Engine Oil Pressure Switch	Open: 12v, Closed: 1v
14	WT	ECT Sensor Signal (THW)	1-4v (varies with temp.)
16	RD	CKP Sensor Signal (NE+)	AC pulse signals
17	BR	Power Ground (E1)	<0.1v
18	BR	Sensor Ground (E2)	<0.050v
19	RD	MAF Sensor Ground (EVG)	<0.050v
21	BL	TP Sensor Signal 2 (VTA2)	1.1-1.9v
22	RD/BK	IAT Sensor Signal (THA)	1-4v (varies with temp.)
23	PK	TP Sensor Signal (VTA)	2.0-2.9v
24	GN	CKP/CMP Sensor Ground (-)	<0.050v

Pin Connector Graphic

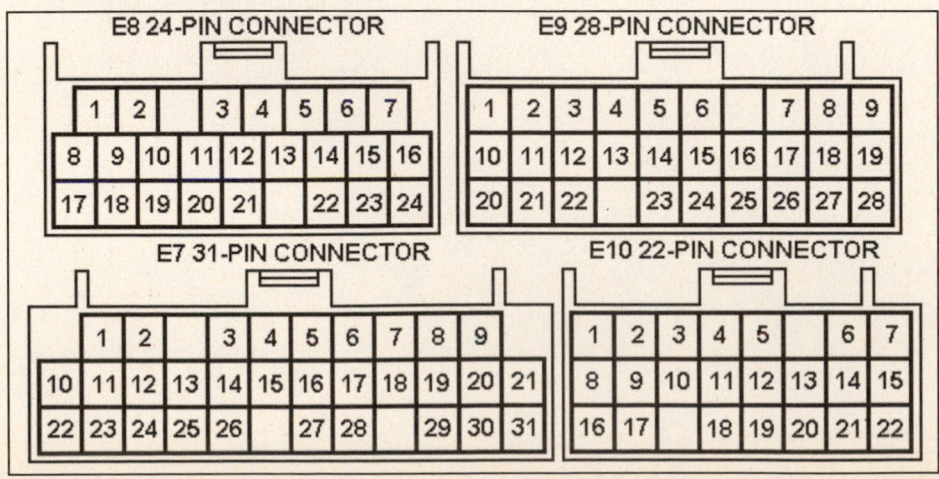

2000-03 Sedan 1.5L I4 Hybrid MFI VIN T (A/T) E9 28 Pin Connector

PCM Pin #	Wire Color	Circuit Description (28 Pin)	Value at Hot Idle
1-5	---	Not Used	---
6	PK/BK	Data Link Connector (TC)	12-14v
7, 9	---	Not Used	---
8	WT/BL	SIL Signal (Scan Tool)	Digital Signal
10	OR	Hybrid ECU Speed Signal	Moving: pulse signals
11	BK/RD	EVAP Purge Switching Valve (VSV)	12v or 0v
12	---	Not Used	---
13	BL	EVAP Canister Closed Valve (VSV)	12v or 0v
14	VT/WT	Hybrid ECU HCLS Signal	1.3-1.9 with vacuum "on"
15, 17	---	Not Used	---
16	RD/YL	Hybrid ECU ESTP Signal	12-14v
18	BR	Hybrid ECU HTE- Signal	Pulse Signals
19	YL	Hybrid ECU HTE+ Signal	Pulse Signals
18	BR	Hybrid ECU THE+ Signal	Pulse Signals
19	YL	Hybrid ECU THE- Signal	Pulse Signals
20	BK/WT	HC Absorber Catalyst Signal	12-14v
21	GN	Hybrid ECU GO Signal	Pulse Signals
22	BL/RD	EVAP Vapor Pressure Sensor (HCC)	2.9-3.1v (with hose off)
23	WT/GN	Outside Air Temperature Sensor (TAM)	Varies: 1-4v
24	---	Not Used	---
25	GN/RD	EFI Main Relay Control	Relay Off: 12v, On: 1v
27	BL	Hybrid ECU ETH- Signal	Pulse Signals
28	PK	Hybrid ECU ETH+ Signal	Pulse Signals

2000-03 Sedan 1.5L I4 Hybrid MFI VIN T (A/T) E10 22 Pin Connector

PCM Pin #	Wire Color	Circuit Description (22 Pin)	Value at Hot Idle
1	PK/BL	HO2S-11 (B1 S1) Heater	1v (Heater On)
2	---	Not Used	---
3	LG	A/C Amplifier Signal (NEO)	Clutch On: 1.5v, Off: 12v
4	---	Not Used	---
5	VT/WT	Speed Signal from C/Meter	At 55 mph: 48 Hz
6	GN//RD	Malfunction Indicator Lamp (MIL) Control	Indicator Off: 12v, On: 1v
7	GN/YL	HO2S-12 (B1 S2) Heater (HT1B)	1v (Heater On)
8	---	Not Used	---
9	BK/WT	Ignition Switch Power	12-14v
10	---	Not Used	---
11	YL	HO2S-12 (B1 S2) Signal (OX1B)	0.1-1.1v
12	WT	HO2S-11 (B1 S1) Signal	0.1-1.1v
13	GY	Multiplex Meter Input (MPX2)	Digital Signals
14	GY/BL	Multiplex Meter Input (MPX1)	Digital Signals
15	---	Not Used	---
16	BR	Oxygen Sensor Ground (E11)	<0.050v
17-21	---	Not Used	---
22	PK/GN	A/C Amplifier Signal (ACT)	Clutch On: 12v, Off: 1.5v

Pin Connector Graphic

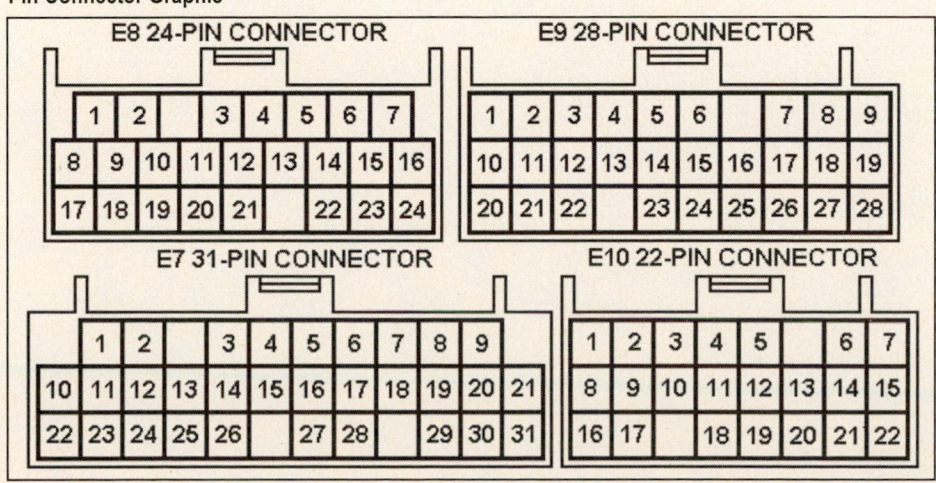

SUPRA Pin Tables

1990-92 Hatchback 3.0L I6 MFI VIN M (All) 10 Pin Connector

PCM Pin #	Wire Color	Circuit Description (10 Pin)	Value at Hot Idle
1	BK/WT	A/T Neutral Start Switch	In P/N: 9-11v (cranking)
1	BK/YL	M/T Clutch Start Switch	In 'N': 9-11v (cranking)
2	GN	Cold Start Injector Control	1v (at cold startup)
3	BK/BL	Starter Switch Signal	9-11v (cranking)
4	WT/BL	Injector Pair 1 & 4 Control	2.0-3.3 ms
5	WT/BK	Power Ground	<0.1v
6 (Cal)	RD/BL	Sub HO2S Heater	1v (Heater On)
7	BR	Sensor Ground	<0.050v
8	WT	Injector Pair 3 & 5 Control	2.0-3.3 ms
9	WT/RD	Injector Pair 2 & 6 Control	2.0-3.3 ms
10	WT/BK	Power Ground	<0.1v

1990-92 Hatchback 3.0L I6 MFI VIN M (All) 18 Pin Connector

PCM Pin #	Wire Color	Circuit Description (18 Pin)	Value at Hot Idle
1	GN	ECT Sensor Signal	At 180°F: 0.51v
2	GN/BK	Igniter Signal (IGF)	0.74-0.76v
3	LG	Igniter Signal (IGT)	0.72-0.74v
4	BL	Distributor Signal (NE+)	AC pulse signals
5	YL	Distributor Signal (G2+)	AC pulse signals
6	RD	Distributor Signal (G1+)	AC pulse signals
7	BK	Distributor Signal (G-)	<0.050v
8	GN/WT	ISC Motor (ISC2)	Pulse Signals
9	WT/YL	ISC Motor (ISC1)	Pulse Signals
10	WT	Knock Sensor Signal	<0.075v AC
11 (Cal)	RD/BL	Sub O2S Signal	0.1-1.1v
12 (Cal)	BK/YL	EGR Gas Temp. Sensor	3.5-4.0v
13	YL/BL	Closed Throttle Switch	0-3v, off-idle: 12v
14	WT/RD	TP Sensor Signal	0.3-0.8v
15	RD/GN	Check Connector	12-14v
16	GY	Check Connector	12-14v
17	RD/BK	ISC Motor (ISC4)	Pulse Signals
18	BL/RD	ISC Motor (ISC3)	Pulse Signals

Standard Colors and Abbreviations

Abbreviation	Color	Abbreviation	Color	Abbreviation	Color
BK	Black	GY	Gray	RD	Red
BL	Blue	GN	Green	TN	Tan
BR	Brown	LG	Light Green	VT	Violet
DB	Dark Blue	OR	Orange	WT	White
DG	DK Green	PK	Pink	YL	Yellow

1990-92 Hatchback 3.0L I6 MFI VIN M (All) 24 Pin Connector

PCM Pin #	Wire Color	Circuit Description (24 Pin)	Value at Hot Idle
1	BK/OR	Ignition Switch Power	12-14v
1	GY	Ignition Switch Power	12-14v
2	BK/YL	Battery Direct	12-14v
3	BL/RD	Sensor VREF	4.9-5.1v
4	GN/RD	Airflow Meter Signal	1.5-2.5v
5	LG	IAT Sensor Signal	At 100°F: 2.60v
6	YL	Fuel Pump Relay	Relay Off: 12v, On: 1v
7	PK	Vehicle Speed Sensor	At 55 mph: 48 Hz
8	RD/BK	EGR Solenoid Control (VSV)	12v or 0v
9	BK/OR	EFI Main Relay Power	12-14v
11	GN/BK	Intake Air Control Solenoid	12v or 0v
12 (Fed)	WT	Main O2S Signal	0.1-1.1v
13	BK/RD	EFI Main Relay Power B1+	12-14v
14	BK/RD	EFI Main Relay Power	12-14v
15 (Cal)	BR	Sensor Ground	<0.050v
16	RD/YL	Headlight Relay Control	Relay On: <1v
17	RD/YL	A/T-ECT ECU Signal	Pulse Signals
17	RD/BL	A/T-ECT ECU Signal	Pulse Signals
18	PK	Defogger/Light Idle Up Signal	Switch On: 12v, Off: 0v
19	GY/GN	MIL (lamp) Control	MIL Off: 12v, On: 1v
20	BL/RD	A/C Magnetic Clutch (ACMG)	Clutch Off: 0v, On: 12v
21	RD/WT	TCM Signal (L3)	0v, at WOT: 5v
22	RD	TCM Signal (L2)	5v, at WOT: 5v
23	BK	TCM Signal (L1)	5v, at WOT: 0v
24	BR	Sensor Ground	<0.050v

Pin Connector Graphic

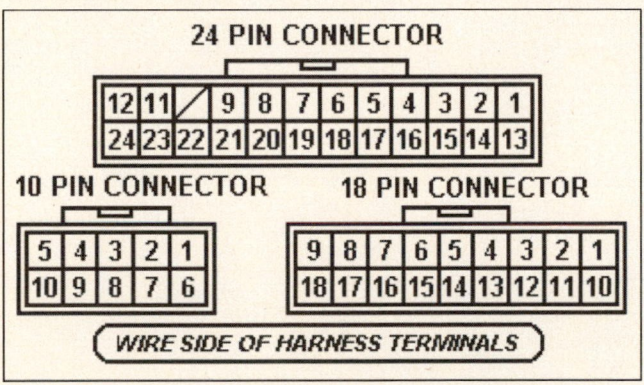

1993-95 Hatchback 3.0L I6 MFI VIN J (All) 40 Pin Connector

PCM Pin #	Wire Color	Circuit Description (40 Pin)	Value at Hot Idle
1	BK/OR	Ignition Switch Power	12-14v
2	PK	Speed Sensor 1 Signal	Moving: 0-5-0v
3	YL	Kickdown Switch	Switch On: <0.1v
4	GN/WT	Stop Light Switch Signal	Brake Off: 0v, On: 12v
6	BL/BK	MIL (lamp) Control	MIL Off: 12v, On: 1v
9	GN/RD	A/T Select Switch 2nd	In 2nd: 12v, Others: 0v
10	GN/BK	A/T Select Switch Low	In Low: 12v, Others: 0v
12	BR/BK	Cruise Control ECU	At Cruise in OD: 12v
15	RD/YL	Defogger Idle Up Signal	Switch On: 12v, Off: 0v
17	GY/RD	Data Link Connector	12-14v
18	GN/YL	A/T Pattern Selector Switch	Norm: 0v, PWR: 12v
19	PK/GN	Data Link Connector	12-14v
20	YL/BL	Data Link Connector	12-14v
21	GN	Fuel Pump ECU	Pulse Signals
22	PK/WT	Fuel Pump ECU	Pulse Signals
23	WT/GN	A/C Magnetic Clutch (ACMG)	Clutch Off: 0v, On: 12v
24	GY	EFI Main Relay Power	12-14v
25	WT/BL	Manual Indicator Light	Light Off: 12-14v
28	PK/GN	Overdrive "Off" Indicator	LED Off: 12v, On: 1v
30 (Cal)	RD/BL	Sub O2S Signal	0.1-1.1v
31	BK/RD	EFI Main Relay Power	12-14v
32	BK/RD	EFI Main Relay Power	12-14v
33	BK/YL	Battery Direct	12-14v
34	BL/RD	A/C Amplifier Signal (AC1)	Clutch On: 1.5v, Off: 12v
35 ('95)	BR/WT	PSP Switch Signal	Straight: 12v, Turning: 0v
36 (Cal)	BR/WT	Sub HO2S Heater	1v (Heater On)

Pin Connector Graphic

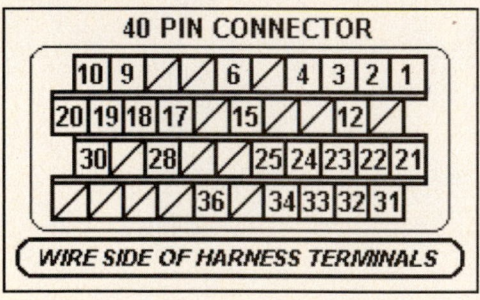

1993-95 Hatchback 3.0L I6 MFI VIN J (All) 80 Pin Connector

PCM Pin #	Wire Color	Circuit Description (80 Pin)	Value at Hot Idle
3	GN	Vehicle Speed 2 Signal (-)	At 55 mph: 48 Hz
4	BR	Sensor Ground	<0.050v
7	GN	Distributor Signal (G-)	<0.050v
8	BK/RD	A/T-ECT Solenoid S3	In Lockup: 12-14v
9	RD/BL	A/T-ECT Solenoid (S2)	1st or OD: 1v
10	WT/RD	A/T-ECT Solenoid (S1)	In 3rd or OD: 1v
15	RD/BK	Injector 6 Control	2.0-3.3 ms
16	RD	Injector 5 Control	2.0-3.3 ms
17	RD/WT	Injector 4 Control	2.0-3.3 ms
18	RD/GN	Injector 3 Control	2.0-3.3 ms
19	RD/YL	Injector 2 Control	2.0-3.3 ms
20	RD/BL	Injector 1 Control	2.0-3.3 ms
23	RD	Vehicle Speed 2 Signal (+)	At 55 mph: 48 Hz
24	OR	A/T Oil Temperature Sensor	Varies: 0.5-4.5v
25	WT	Distributor Signal (G2+)	AC pulse signals
26	RD	Distributor Signal (G1+)	AC pulse signals
27	BK	Distributor Signal (NE+)	AC pulse signals
28 ('94)	BR	Sensor Ground	<0.050v
28	BL/OR	Data Link Connector	12-14v
29	LG	Data Link Connector	12-14v
32	RD/GN	ISC Motor (ISC4)	Pulse Signals
33	GN/OR	ISC Motor (ISC3)	Pulse Signals
34	GN/WT	ISC Motor (ISC2)	Pulse Signals
35	PK/YL	ISC Motor (ISC1)	Pulse Signals
36	WT/BL	Fuel Pressure Up Solenoid	1v (at hot restart)
38 ('94)	GN/RD	Exhaust Bypass Solenoid	12v or 0v
39	GN/YL	Exhaust Control Solenoid	12v or 0v
40 ('94)	GN/BK	Intake Air Control Solenoid	12v or 0v

Standard Colors and Abbreviations

Abbreviation	Color	Abbreviation	Color	Abbreviation	Color
BK	Black	GY	Gray	RD	Red
BL	Blue	GN	Green	TN	Tan
BR	Brown	LG	Light Green	VT	Violet
DB	Dark Blue	OR	Orange	WT	White
DG	DK Green	PK	Pink	YL	Yellow

1993-95 Hatchback 3.0L I6 MFI VIN J (All) 80 Pin Connector

PCM Pin #	Wire Color	Circuit Description (80 Pin)	Value at Hot Idle
41	BL/RD	Sensor VREF	4.9-5.1v
43	YL	TP Sensor Signal	0.3-0.8v
44	BL/YL	ECT Sensor Signal	At 180°F: 0.51v
45	PK/BL	IAT Sensor Signal	At 100°F: 2.60v
46	BR/YL	EGR Gas Temp. Sensor	3.5-4.0v
47	RD/BL	Rear O2S Signal	0.1-1.1v
48	WT	Front O2S Signal	0.1-1.1v
49	WT	Knock Sensor 2 Signal	<0.075v AC
50	WT	Knock Sensor 1 Signal	<0.075v AC
57	RD/WT	Igniter Signal (IGT)	Digital Signal: 0-5-0v
58	RD/YL	Igniter Signal (IGF)	Digital Signal: 0-5-0v
60 ('94)	BL/WT	Waste Gate Solenoid	12v or 0v
64	RD	Closed Throttle Switch	1v, at off-idle: 12v
65	BR/BK	Sensor Ground	<0.050v
66	GN/BK	Airflow Meter Signal	1.5-2.5v
69	BR	Sensor Ground	<0.050v
72	BK/YL	Rear HO2S Heater	1v (Heater On)
73	BK/BL	Front HO2S Heater	1v (Heater On)
74	PK	EVAP Purge Solenoid (VSV)	12v or 0v
75	PK	EGR Solenoid Control (VSV)	12v or 0v
76	BK/WT	Neutral Start Switch	In P/N: 9-11v (cranking)
77	BK	Starter Switch Signal	9-11v (cranking)
79	BR	Power Ground	<0.1v
80	BR	Power Ground	<0.1v

Pin Connector Graphic

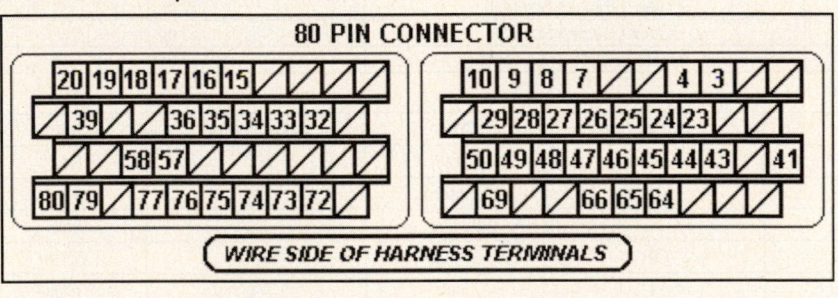

1996 Hatchback 3.0L I6 MFI VIN D (All) 40 Pin Connector

PCM Pin #	Wire Color	Circuit Description (40 Pin)	Value at Hot Idle
1	BK/OR	Ignition Switch Power	12-14v
2	PK	Speed Sensor 1 Signal	Moving: 0-5-0v
3	YL	Kickdown Switch	Open: 12v, Closed: 0v
4	GN/WT	Stop Light Switch Signal	Brake Off: 0v, On: 12v
6	BL/BK	MIL (lamp) Control	MIL Off: 12v, On: 1v
8	GN	Data Link Connector (SDL)	No Scan Tool: 0v
9	GN/RD	A/T Select Switch 2nd	In 2nd: 12v, Others: 0v
10	GN/BK	A/T Select Switch Low	In Low: 12v, Others: 0v
12	BR/BK	Cruise Control ECU	At Cruise in OD: 12v
15	RD/YL	Idle Up Diode Signal	Switch On: 12v, Off: 0v
18	GN/YL	A/T Pattern Selector Switch	Norm: 0v, PWR: 12v
20	YL/BL	Data Link Connector	12-14v
21	GN	Fuel Pump (Control) ECU	Pulse Signals
22	PK/WT	Fuel Pump (DI) ECU	Pulse Signals
23	WT/GN	A/C Magnetic Clutch (ACMG)	Clutch Off: 0v, On: 12v
24	GY	EFI Main Relay Power	12-14v
25	WT/BL	Manual Indicator Light	Light Off: 12-14v
28	PK/GN	Overdrive Main Switch	Switch Off: 12v, On: 0v
30	RD/BL	Sub HO2S Signal	0.1-1.1v
31	BK/RD	EFI Main Relay Power	12-14v
33	BK/YL	Battery Direct	12-14v
34	BL/RD	A/C Amplifier Signal (AC1)	Clutch On: 1.5v, Off: 12v
35	BR/WT	PSP Switch Signal	Straight: 12v, Turning: 0v
36	BR/WT	Sub HO2S Heater	1v (Heater On)

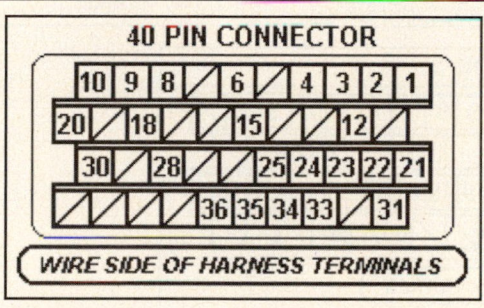

1996 Hatchback 3.0L I6 MFI VIN D (All) 80 Pin Connector

PCM Pin #	Wire Color	Circuit Description (80 Pin)	Value at Hot Idle
3	GN	Vehicle Speed 2 Sensor (-)	Moving: 0-5-0v
5	WT	CKP Sensor Signal (NE2-)	AC pulse signals
6	BK	CKP Sensor Signal (NE2+)	AC pulse signals
7	GN	Distributor Signal (G-)	<0.050v
8	BK/RD	A/T-ECT Solenoid S3	In Lockup: 12-14v
9	RD/BL	A/T-ECT Solenoid (S2)	1st or OD: 1v
10	WT/RD	A/T-ECT Solenoid (S1)	In 3rd or OD: 1v
15	RD/BK	Injector 6 Control	2.0-3.3 ms
16	RD	Injector 5 Control	2.0-3.3 ms
17	RD/WT	Injector 4 Control	2.0-3.3 ms
18	RD/GN	Injector 3 Control	2.0-3.3 ms
19	RD/YL	Injector 2 Control	2.0-3.3 ms
20	RD/BL	Injector 1 Control	2.0-3.3 ms
23	RD	Vehicle Speed 2 Sensor (+)	Moving: 0-5-0v
24	OR	A/T Oil Temperature Sensor	Varies: 0.5-4.5v
25	WT	Distributor Signal (G2+)	AC pulse signals
26	RD	Distributor Signal (G1+)	AC pulse signals
27	BK	Distributor Signal (NE+)	AC pulse signals
30	BR	MAF Sensor Ground	<0.050v
32	RD/GN	ISC Motor (ISC4)	Pulse Signals
33	GN/OR	ISC Motor (ISC3)	Pulse Signals
34	GN/WT	ISC Motor (ISC2)	Pulse Signals
35	PK/YL	ISC Motor (ISC1)	Pulse Signals
36	WT/BL	Fuel Pressure Up Solenoid	1v (at hot restart)
39	GN/BK	Intake Air Control Solenoid	12v or 0v

80 PIN CONNECTOR

```
| 20 | 19 | 18 | 17 | 16 | 15 | / | / | / |     | 10 | 9 | 8 | 7 | 6 | 5 | / | 3 | / |
| / | 39 | / | / | 36 | 35 | 34 | 33 | 32 |     | 30 | / | / | 27 | 26 | 25 | 24 | 23 | / |
| / | / | 58 | 57 | / | / | / | / | / |        | 50 | 49 | 48 | 47 | 46 | 45 | 44 | 43 | / | 41 |
| 80 | 79 | 78 | 77 | 76 | 75 | 74 | 73 | 72 | / |  | / | 69 | / | / | 66 | 65 | 64 | / | / |
```

WIRE SIDE OF HARNESS TERMINALS

1996 Hatchback 3.0L I6 MFI VIN D (All) 80 Pin Connector

PCM Pin #	Wire Color	Circuit Description (80 Pin)	Value at Hot Idle
41	BL/RD	Sensor VREF (VC)	4.9-5.1v
43	YL	TP Sensor Signal	0.3-0.8v
44	BL/YL	ECT Sensor Signal	At 180°F: 0.51v
45	PK/BL	IAT Sensor Signal	At 100°F: 2.60v
46	BR/YL	EGR Gas Temp. Sensor	3.5-4.0v
47	RD/BL	Rear HO2S Signal	0.1-1.1v
48	WT	Front HO2S Signal	0.1-1.1v
49	WT	Knock Sensor 2 Signal	<0.075v AC
50	WT	Knock Sensor 1 Signal	<0.075v AC
57	RD/WT	Igniter Signal (IGT)	Digital Signal: 0-5-0v
58	RD/YL	Igniter Signal (IGF)	Digital Signal: 0-5-0v
64	RD	Closed Throttle Switch	1v, at off-idle: 12v
65	BR/BK	Sensor Ground	<0.050v
66	YL/RD	Mass Airflow Sensor	1.0-1.8v
69	BR	Shield Ground	<0.050v
72	BK/YL	Rear HO2S Heater	1v (Heater On)
73	BK/BL	Front HO2S Heater	1v (Heater On)
74	PK	EVAP Purge Solenoid (VSV)	12v or 0v
75	PK	EGR Solenoid Control (VSV)	1v (Heater On)
76	BK/WT	A/T Neutral Start Switch	In P/N: 9-11v (cranking)
77	BK	Starter Switch Signal	9-11v (cranking)
78	BR	Power Ground	<0.1v
79	BR	Power Ground	<0.1v
80	BR	Power Ground	<0.1v

Standard Colors and Abbreviations

Abbreviation	Color	Abbreviation	Color	Abbreviation	Color
BK	Black	GY	Gray	RD	Red
BL	Blue	GN	Green	TN	Tan
BR	Brown	LG	Light Green	VT	Violet
DB	Dark Blue	OR	Orange	WT	White
DG	DK Green	PK	Pink	YL	Yellow

1997 Hatchback 3.0L I6 MFI VIN D (All) 40 Pin Connector

PCM Pin #	Wire Color	Circuit Description (40 Pin)	Value at Hot Idle
1	BK/OR	Ignition Switch Power	12-14v
2	PK	Speed Sensor 1 Signal	Moving: 0-5-0v
3	YL	Kickdown Switch	Open: 12v, Closed: 0v
4	GN/WT	Stop Light Switch Signal	Brake Off: 0v, On: 12v
6	BL/BK	MIL (lamp) Control	MIL Off: 12v, On: 1v
8	GN	Data Link Connector (SDL)	0v
9	GN/RD	A/T Select Switch 2nd	In 2nd: 12v, Others: 0v
10	GN/BK	A/T Select Switch Low	In Low: 12v, Others: 0v
12	GN/BK	Cruise Control ECU	At Cruise in OD: 12v
15	RD/YL	Idle Up Diode Signal	Switch On: 12v, Off: 0v
18	GN/YL	A/T Pattern Selector Switch	Norm: 0v, PWR: 12v
20	YL/GN	Data Link Connector	12-14v
21	GN	Fuel Pump (Control) ECU	Pulse Signals
22	PK/WT	Fuel Pump (DI) ECU	Pulse Signals
23	WT/GN	A/C Magnetic Clutch (ACMG)	Clutch Off: 0v, On: 12v
24	BK/YL	EFI Main Relay Power	12-14v
25	WT/BL	Manual Indicator Light	Light Off: 12v, On: 1v
28	PK	Overdrive Main Switch	Switch Off: 12v
30	RD/BL	Sub HO2S Signal	0.1-1.1v
31	BK/RD	EFI Main Relay Power	12-14v
32	BR/WT	PSP Switch Signal	Straight: 12v, Turning: 0v
33	BK/WT	Battery Direct	12-14v
34	BL/RD	A/C Amplifier Signal (AC1)	Clutch On: 1.5v, Off: 12v
36	BR/WT	Sub HO2S Heater	1v (Heater On)

Pin Connector Graphic

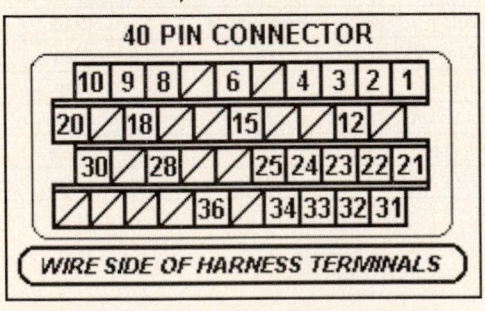

1997 Hatchback 3.0L I6 MFI VIN D (All) 80 Pin Connector

PCM Pin #	Wire Color	Circuit Description (80 Pin)	Value at Hot Idle
3	GN	Vehicle Speed 2 Sensor (-)	Moving: 0-5-0v
4	BK	CKP Sensor Signal (NE2+)	AC pulse signals
5	WT	CKP Sensor Signal (NE2-)	<0.050v
7	GN	Distributor Signal (G-)	<0.050v
8	BK/RD	A/T-ECT Solenoid (SL)	In Lockup: 12-14v
9	RD/BL	A/T-ECT Solenoid (S2)	1st or OD: 1v
10	WT/RD	A/T-ECT Solenoid (S1)	In 3rd or OD: 1v
15	RD/BK	Injector 6 Control	2.0-3.3 ms
16	RD	Injector 5 Control	2.0-3.3 ms
17	RD/WT	Injector 4 Control	2.0-3.3 ms
18	BL	Injector 3 Control	2.0-3.3 ms
19	BL/RD	Injector 2 Control	2.0-3.3 ms
20	RD/BL	Injector 1 Control	2.0-3.3 ms
23	RD	Vehicle Speed 2 Sensor (+)	Moving: 0-5-0v
24	BL/BK	A/T Oil Temperature Sensor	0.5-4.5v
25	WT	Distributor Signal (G2+)	AC pulse signals
26	RD	Distributor Signal (G1+)	AC pulse signals
27	BK	Distributor Signal (NE+)	AC pulse signals
28	BR	MAF Sensor Ground	<0.050v
32	RD/BK	ISC Motor (ISC4)	Pulse Signals
33	BL/BK	ISC Motor (ISC3)	Pulse Signals
34	GN/WT	ISC Motor (ISC2)	Pulse Signals
35	YL/BK	ISC Motor (ISC1)	Pulse Signals
39	GN/YL	Intake Air Control Solenoid	12v or 0v

1997 Hatchback 3.0L I6 MFI VIN D (All) 80 Pin Connector

PCM Pin #	Wire Color	Circuit Description (80 Pin)	Value at Hot Idle
41	BL/RD	Sensor VREF (VC)	4.9-5.1v
43	YL	TP Sensor Signal	0.3-0.8v
44	BL/YL	ECT Sensor Signal	At 180°F: 0.51v
45	GN/BK	IAT Sensor Signal	At 100°F: 2.60v
46	BR/YL	EGR Gas Temp. Sensor	3.5-4.0v
47	RD/BL	HO2S-21 (B2 S1) Signal	0.1-1.1v
48	WT	HO2S-11 (B1 S1) Signal	0.1-1.1v
49	WT	Knock Sensor 2 Signal	<0.075v AC
50	WT	Knock Sensor 1 Signal	<0.075v AC
57	RD/WT	Igniter Signal (IGT)	Digital Signal: 0-5-0v
58	RD/YL	Igniter Signal (IGF)	Digital Signal: 0-5-0v
64	RD/BK	Closed Throttle Switch	1v, at off-idle: 12v
65	WT/BK	Sensor Ground	<0.050v
66	YL/RD	Mass Airflow Sensor	1.0-1.8v
69	BR	Shield Ground	<0.050v
71	GN	HO2S-11 (B1 S1) Heater	1v (Heater On)
72	BK/YL	HO2S-21 (B2 S1) Heater	1v (Heater On)
73	WT/BL	Fuel Pressure Up Solenoid	1v (at hot restart)
74	PK	EVAP Purge Solenoid (VSV)	12v or 0v
75	PK	EGR Solenoid Control (VSV)	1v (Heater On)
76	BK/WT	A/T Neutral Start Switch	In P/N: 9-11v (cranking)
77	BK	Starter Switch Signal	9-11v (cranking)
78	BR	Power Ground	<0.1v
79	BR	Power Ground	<0.1v
80	BR	Power Ground	<0.1v

Pin Connector Graphic

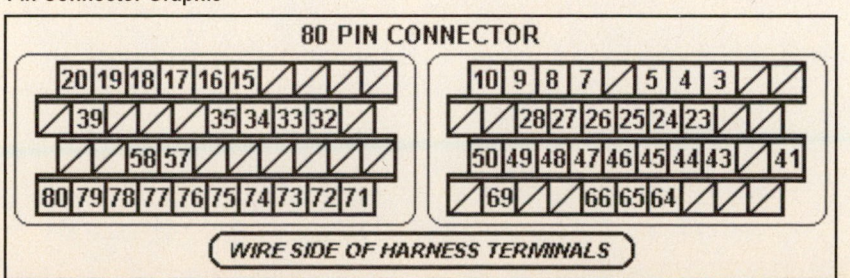

1998 Hatchback 3.0L I6 MFI VIN D (A/T-ECT) 31 Pin Connector

PCM Pin #	Wire Color	Circuit Description (31 Pin)	Value at Hot Idle
1	BL	Injector 3 Control	2.0-3.3 ms
2	RD/WT	Injector 4 Control	2.0-3.3 ms
3	RD	Injector 5 Control	2.0-3.3 ms
4	RD/BK	Injector 6 Control	2.0-3.3 ms
5	GN/YL	Intake Air Control Solenoid	12v or 0v
7	BK	Throttle Control Motor (M-)	Pulse Signals
8	WT	Throttle Control Motor (M+)	Pulse Signals
9	BR	Sensor Ground	<0.050v
10	BL	CMP Sensor Signal (G+)	AC pulse signals
11	RD/WT	Igniter Transistor 1 Control	Digital Signal: 0-5-0v
12	LG	Igniter Transistor 2 Control	Digital Signal: 0-5-0v
13	GN/RD	Igniter Transistor 3 Control	Digital Signal: 0-5-0v
16	GN	PSP Switch Signal	Straight: 12v, Turning: 0v
17	YL/BK	Cam Timing Oil signal (OCV-)	Pulse Signals
18	WT/RD	Cam Timing Oil signal (OCV+)	Pulse Signals
19	YL	Throttle Control Motor (CL-)	Pulse Signals
20	BL	Throttle Control Motor (CL+)	Pulse Signals
21	WT/BR	Power Ground	<0.1v
22	WT	CKP/CMP Sensor Ground (-)	<0.050v
23	BK	CKP Sensor Signal (NE+)	AC pulse signals
24	BK/WT	Neutral Start Switch	In P/N: 9-11v (cranking)
25	RD/YL	Igniter Signal (IGF)	Digital Signal: 0-5-0v
27	WT	Knock Sensor 2 Signal	<0.075v AC
28	WT	Knock Sensor 1 Signal	<0.075v AC
30	BR	Shield Ground	<0.050v
31	WT/BR	Power Ground	<0.1v

1998 Hatchback 3.0L I6 MFI VIN D (A/T-ECT) 22 Pin Connector

PCM Pin #	Wire Color	Circuit Description (22 Pin)	Value at Hot Idle
1	BK/WT	Battery Direct	12-14v
4	GN	Fuel Pump (Control) ECU	Pulse Signals
5	PK/WT	Fuel Pump (DI) ECU	Pulse Signals
6	BL/BK	MIL (lamp) Control	MIL Off: 12v, On: 1v
7	BL/RD	EFI Main Relay Power BM+	12-14v
8	BK/RD	EFI Main Relay Power B2+	12-14v
9	BK/OR	Ignition Switch Power	12-14v
10	BK/YL	EFI Main Relay Power	12-14v
11	BK	SIL Signal (Scan Tool)	Digital Signal
16	BK/RD	EFI Main Relay Power	12-14v
22	WT/BK	ECM/Data Link Ground	<0.050v

1998 Hatchback 3.0L I6 MFI VIN D (A/T-ECT) 28 Pin Connector

PCM Pin #	Wire Color	Circuit Description (28 Pin)	Value at Hot Idle
1	BL/RD	A/C Amplifier Signal (AC1)	Clutch On: 1.5v, Off: 12v
2	BK	Starter Switch Signal	9-11v (cranking)
4	GN/WT	Stop Light Switch Signal	Brake Off: 0v, On: 12v
5	RD	Data Link Connector	12-14v
8	WT	HO2S-12 (B1 S2) Signal	0.1-1.1v
13	WT/GN	A/C Magnetic Clutch (ACMG)	Clutch Off: 0v, On: 12v
16	RD/BK	A/T Select Switch Reverse	In 'R': 12v, Others: 0v
17	GN/RD	A/T Select Switch Drive	In 'D': 12v, Others: 0v
18	BR/YL	EVAP Vapor Pressure Sensor	2.5-3.1v (with hose off)
23	BL	Cruise Control ECU	At Cruise in OD: 12v
24	RD/YL	Cruise Control ECU	At Cruise in OD: 12v
25	WT	Vehicle Speed Sensor	At 55 mph: 48 Hz
26	RD/YL	Idle Up Diode Signal	Switch On: 12v, Off: 0v
28	PK	Overdrive Main Switch	Switch Off: 12v, On: 0v

1998 Hatchback 3.0L I6 MFI VIN D (A/T-ECT) 17 Pin Connector

PCM Pin #	Wire Color	Circuit Description (17 Pin)	Value at Hot Idle
1	WT/RD	A/T-ECT Solenoid (S1)	In 3rd or OD: 1v
2	RD/BL	A/T-ECT Solenoid (S2)	1st or OD: 1v
4	RD	OD Clutch Sensor Signal (+)	Pulse Signals
5	RD/YL	Speed Sensor 2 Signal (+)	Pulse Signals
7	GN/WT	A/T-ECT Solenoid (SLU+)	Pulse Signals
8	YL/GN	A/T-ECT Solenoid (SLN+)	Pulse Signals
9	GN/RD	A/T-ECT Solenoid (SLT+)	Pulse Signals
10	RD	OD Clutch Sensor Signal (-)	Pulse Signals
11	BL/YL	Speed Sensor Signal 2 (-)	Pulse Signals
13	BL/RD	A/T-ECT Solenoid (SLU-)	Pulse Signals
14	PK	A/T-ECT Solenoid (SLN-)	Pulse Signals
15	RD/BK	A/T-ECT Solenoid (SLT-)	Pulse Signals
16	OR	C/C Indicator Light	Light Off: 12v, On: 1v
17	BL/BK	A/T Oil Temperature Sensor	0.5-4.5v

Pin Connector Graphic

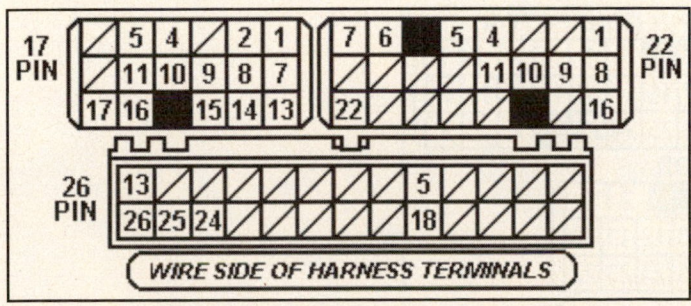

WIRE SIDE OF HARNESS TERMINALS

1998 Hatchback 3.0L I6 MFI VIN D (A/T-ECT) 26 Pin Connector

PCM Pin #	Wire Color	Circuit Description (26 Pin)	Value at Hot Idle
5	GN/YL	A/T Pattern Selector Switch	Norm: 0v, PWR: 12v
13	RD/BK	EVAP Vapor Pressure (VSV)	12v or 0v
18	WT/BL	Manual Indicator Light	Light Off: 12-14v
24	WT	HO2S-22 (B2 S2) Signal	0.1-1.1v
25	GN/YL	HO2S-22 (B2 S2) Heater	1v (Heater On)
26	BL/WT	HO2S-12 (B1 S2) Heater	1v (Heater On)

1998 Hatchback 3.0L I6 MFI VIN D (A/T-ECT) 24 Pin Connector

PCM Pin #	Wire Color	Circuit Description (24 Pin)	Value at Hot Idle
1	WT/BK	Power Ground	<0.1v
2	BL/RD	Sensor VREF	4.9-5.1v
3	BK/YL	HO2S-21 (B2 S1) Heater	1v (Heater On)
4	GN	HO2S-11 (B1 S1) Heater	1v (Heater On)
5	RD/BL	Injector 1 Control	2.0-3.3 ms
6	BL/RD	Injector 2 Control	2.0-3.3 ms
7	YL	EVAP Purge Solenoid (VSV)	12v or 0v
10	YL/RD	MAF Sensor Signal	1.1-1.5v
11	RD/BL	HO2S-21 (B2 S1) Signal	0.1-1.1v
12	WT	HO2S-11 (B1 S1) Signal	0.1-1.1v
13	BL/RD	Mirror Heater Switch Signal	Heater On: 12-14v
14	BL	ECT Sensor Signal	At 180°F: 0.51v
15	GN	Accel Position Sensor (VPA)	0.25-0.9v
16	WT	Accel Position Sensor (VPA2)	1.8-2.7v
17	BR	Sensor Ground	<0.050v
18	BR	Sensor Ground	<0.050v
19	BR	MAF Sensor Ground	<0.050v
20	BL/YL	A/T Select Switch 2nd	In 2nd: 12v, Others: 0v
21	GN/BK	A/T Select Switch Low	In Low: 12v, Others: 0v
22	GN/WT	IAT Sensor Signal	At 100°F: 2.60v
23	YL	TP Sensor (VTA1)	0.4-1.0v
24	RD/BK	TP Sensor (VTA2)	2.0-2.9v

Pin Connector Graphic

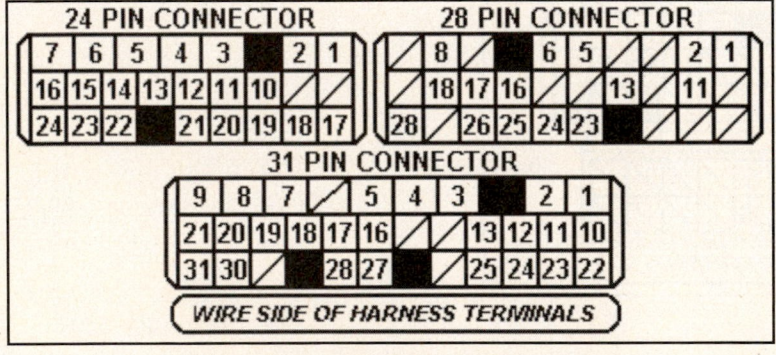

1990-92 Hatchback 3.0L Turbo I6 VIN M (A/T) 16 Pin Connector

PCM Pin #	Wire Color	Circuit Description (16 Pin)	Value at Hot Idle
1	BL/RD	Sensor VREF	4.9-5.1v
2	GN/RD	Airflow Meter Signal	1.5-2.5v
3	LG	IAT Sensor Signal	At 100ºF: 2.60v
4	GN	ECT Sensor Signal	At 180ºF: 0.51v
5	WT	Knock Sensor 1 Signal	<0.075v AC
6	WT	O2S-11 (B1 S1) Signal	0.1-1.1v
7	YL/BK	Oil Pressure Switch	Switch Closed: 0.1v
8	GY	Check Connector	12-14v
9	BR	Sensor Ground	<0.050v
11	WT/RD	TP Sensor Signal	0.3-0.8v
12	YL/BL	Closed Throttle Switch	1v, at off-idle: 12v
13	WT	Knock Sensor 2 Signal	<0.075v AC
15	RD/GN	Check Connector	12-14v
16	BK	Distributor Signal (G-)	<0.050v

1990-92 Hatchback 3.0L Turbo I6 VIN M (A/T) 22 Pin Connector

PCM Pin #	Wire Color	Circuit Description (22 Pin)	Value at Hot Idle
1	BK/YL	Battery Direct	12-14v
2	BK/OR	Ignition Switch Power	12-14v
2 ('90-'91)	GY	Ignition Switch Power	12-14v
4	BK/OR	EFI Main Relay Power	12-14v
5	GY/GN	MIL (lamp) Control	MIL Off: 12v, On: 1v
6	YL	Fuel Pump Relay	Relay Off: 12v, On: 1v
7	GN	Circuit Opening Relay (FC)	0-3v, off-idle: 12v
8	PK	Defogger Diode Signal	Switch On: 12v, Off: 0v
9	PK	Vehicle Speed Sensor	At 55 mph: 48 Hz
10	BL/RD	A/C Magnetic Clutch (ACMG)	Clutch Off: 0v, On: 12v
11	BK/BL	Starter Switch Signal	9-11v (cranking)
12	BK/RD	EFI Main Relay Power	12-14v
13	BK/RD	EFI Main Relay Power B1+	12-14v
17	RD/WT	L3 Signal to TCM	0v, at WOT: 5v
18	RD	L2 Signal to TCM	5v, at WOT: 5v
19	BK	L1 Signal to TCM	5v, at WOT: 0v
20	RD/BL	A/T-ECT ECU Signal	Pulse Signals
21	RD/YL	Headlight Diode Signal	Switch On: 12v, Off: 0v
22	BK/WT	A/T Neutral Start Switch	In P/N: 9-11v (cranking)
22	BK/YL	M/T Clutch Start Switch	In 'N': 9-11v (cranking)

Standard Colors and Abbreviations

Abbreviation	Color	Abbreviation	Color	Abbreviation	Color
BK	Black	GY	Gray	RD	Red
BL	Blue	GN	Green	TN	Tan
BR	Brown	LG	Light Green	VT	Violet
DB	Dark Blue	OR	Orange	WT	White
DG	DK Green	PK	Pink	YL	Yellow

1990-92 Hatchback 3.0L Turbo I6 VIN M (All) 26 Pin Connector

PCM Pin #	Wire Color	Circuit Description (26 Pin)	Value at Hot Idle
1	BL	Distributor Signal (NE+)	AC pulse signals
2	YL	Distributor Signal (G2+)	AC pulse signals
3	GN/BK	Igniter Signal (IGF)	Digital Signal: 0-5-0v
4	RD/BK	ISC Motor (ISC4)	Pulse Signals
5	BL/RD	ISC Motor (ISC3)	Pulse Signals
6	GN/WT	ISC Motor (ISC2)	Pulse Signals
7	WT/YL	ISC Motor (ISC1)	Pulse Signals
8	RD/BK	EGR Solenoid Control (VSV)	1v (Heater On)
9	RD/BL	Fuel Pressure Up Solenoid	1v (at hot restart)
10	RD/BL	HO2S-11 (B1 S1) Heater	1v (Heater On)
11	WT/RD	Injector Pair 2 & 6 Control	2.0-3.3 ms
12	WT/BL	Injector Pair 1 & 4 Control	2.0-3.3 ms
13	WT/BK	Power Ground	<0.1v
14	BR	Sensor Ground	<0.050v
15	RD	Distributor Signal (G1+)	AC pulse signals
16	BK/YL	EGR Gas Temp. Sensor	3.5-4.0v
18	BL/WT	Igniter Signal (IGB)	Digital Signal: 0-5-0v
19	BL/BK	Igniter Signal (IGA)	Digital Signal: 0-5-0v
20	LG	Igniter Signal (IGT)	Digital Signal: 0-5-0v
24	GN	Cold Start Injector Control	1v (at cold startup)
25	WT	Injector Pair 3 & 5 Control	2.0-3.3 ms
26	WT/BK	Power Ground	<0.1v

Pin Connector Graphic

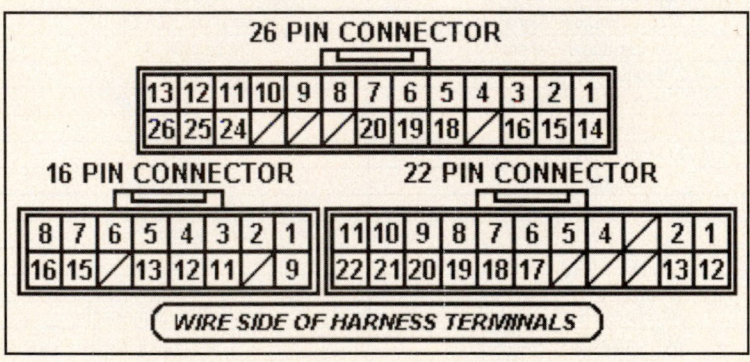

1993 Hatchback 3.0L Twin/Turbo I6 VIN J (All) 40 Pin Connector

PCM Pin #	Wire Color	Circuit Description (40 Pin)	Value at Hot Idle
1	BK/OR	Ignition Switch Power	12-14v
2	PK	Speed Sensor 1 Signal	Moving: 0-5-0v
3	YL	Kickdown Switch	Switch Open: 12v
4	GN/WT	Stop Light Switch Signal	Brake Off: 0v, On: 12v
6	BL/BK	MIL (lamp) Control	MIL Off: 12v, On: 1v
7	RD/BK	A/T Select Switch Reverse	In 'R': 12v, Others: 0v
9	GN/RD	A/T Select Switch 2nd	In 2nd: 12v, Others: 0v
10	GN/BK	A/T Select Switch Low	In Low: 12v, Others: 0v
11	YL/RD	ABS /Traction ECU	Pulse Signals
12	BR/BK	Cruise Control ECU	At Cruise in OD: 12v
13	WT/RD	Traction Control (TRC-)	Pulse Signals
14	OR	Traction Control (TRC+)	Pulse Signals
15	RD/YL	Defogger Diode Signal	Switch On: 12v, Off: 0v
16	BK/WT	Tachometer Signal (TACO)	Pulse Signals
17	GY/RD	Data Link Connector	12-14v
18	GN/YL	A/T Pattern Selector Switch	Norm: 0v, PWR: 12v
19	PK/GN	Data Link Connector	12-14v
20	YL/BL	Data Link Connector	12-14v
21	GN	Fuel Pump ECU	Pulse Signals
22	PK/WT	Fuel Pump ECU	Pulse Signals
23	WT/GN	A/C Magnetic Clutch (ACMG)	Clutch Off: 0v, On: 12v
24	GY	EFI Main Relay Power	12-14v
25	WT/BL	Manual Indicator Light	Light Off: 12-14v
28	PK/GN	Overdrive "Off" Indicator	Switch Off: 12v
31	BK/RD	EFI Main Relay Power	12-14v
32	BK/RD	EFI Main Relay Power	12-14v
33	BK/YL	Battery Direct	12-14v
34	BL/RD	A/C Amplifier Signal (AC1)	Clutch On: 1.5v, Off: 12v
38	PK/BL	Traction Control (NEO)	Pulses

Pin Connector Graphic

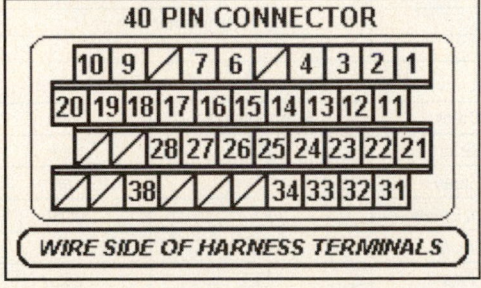

1993 Hatchback 3.0L Twin/Turbo I6 VIN J (All) 80 Pin Connector

PCM Pin #	Wire Color	Circuit Description (80 Pin)	Value at Hot Idle
1	YL	OD Clutch Speed Sensor (-)	Pulses
3	GN	Vehicle Speed Sensor 2 (-)	Pulses
4	BR	Sensor Ground	<0.050v
5	BL	Distributor Signal (G2+)	AC pulse signals
6	OR	Distributor Signal (G1+)	AC pulse signals
7	BK/RD	Distributor Signal (NE+)	AC pulse signals
9	RD/BL	A/T-ECT Solenoid (S2)	1st or OD: 1v
10	WT/RD	A/T-ECT Solenoid (S1)	In 3rd or OD: 1v
12	GN/RD	A/T-ECT Solenoid (SLT-)	Pulse Signals
13	YL/GN	A/T-ECT Solenoid (SLN-)	Pulse Signals
14	GN/BK	A/T-ECT Solenoid (SLU-)	Pulse Signals
15	RD/BK	Injector 6 Control	2.0-3.3 ms
16	RD	Injector 5 Control	2.0-3.3 ms
17	RD/WT	Injector 4 Control	2.0-3.3 ms
18	RD/GN	Injector 3 Control	2.0-3.3 ms
19	RD/YL	Injector 2 Control	2.0-3.3 ms
20	RD/BL	Injector 1 Control	2.0-3.3 ms
21	BL	OD Clutch Speed Sensor (+)	Pulse Signals
23	RD	Vehicle Speed Sensor 2 (+)	Pulse Signals
24	OR	A/T Oil Temperature Sensor	0.5-4.5v
25	BK/WT	Distributor Signal (G2-)	<0.050v
26	WT	Distributor Signal (G1-)	<0.050v
27	BR	Distributor Signal (NE-)	<0.050v
28	BR	MAF Sensor Ground	<0.050v
29	LG	Data Link Connector	12-14v
31	WT/GN	ECT Solenoid (SLT+)	Distributor (NE-) Signal
32	RD/GN	ISC Motor (ISC4)	Pulse Signals
33	GN/OR	ISC Motor (ISC3)	Pulse Signals
34	GN/WT	ISC Motor (ISC2)	Pulse Signals
35	PK/YL	ISC Motor (ISC1)	Pulse Signals
38	GN/RD	Exhaust Bypass Solenoid	12v or 0v
39	GN/YL	EGR Solenoid Control (VSV)	1v (Heater On)
40	GN/BK	Intake Air Control Solenoid	12v or 0v

Standard Colors and Abbreviations

Abbreviation	Color	Abbreviation	Color	Abbreviation	Color
BK	Black	GY	Gray	RD	Red
BL	Blue	GN	Green	TN	Tan
BR	Brown	LG	Light Green	VT	Violet
DB	Dark Blue	OR	Orange	WT	White
DG	DK Green	PK	Pink	YL	Yellow

1993 Hatchback 3.0L Twin/Turbo I6 VIN J (All) 80 Pin Connector

PCM Pin #	Wire Color	Circuit Description (80 Pin)	Value at Hot Idle
41	BL/RD	Sensor VREF (VC)	4,9-5.1v
42	YL/BL	TP Sensor (VTA2)	2.0-2.9v
43	YL	TP Sensor (VTA1)	0.4-1.0v
44	BL/YL	ECT Sensor Signal	At 180°F: 0.51v
45	PK/BL	IAT Sensor Signal	At 100°F: 2.60v
46	BR/YL	EGR Gas Temp. Sensor	3.5-4.0v
47	RD/BL	Rear HO2S Signal	0.1-1.1v
48	WT	Front HO2S Signal	0.1-1.1v
49	WT	Knock Sensor 2 Signal	<0.075v AC
50	WT	Knock Sensor 1 Signal	<0.075v AC
52	BL/OR	Igniter Transistor 6 Control	7% duty cycle
53	PK/BK	Igniter Transistor 5 Control	7% duty cycle
54	GY/GN	Igniter Transistor 4 Control	7% duty cycle
55	GY/BK	Igniter Transistor 3 Control	7% duty cycle
56	BK/OR	Igniter Transistor 2 Control	7% duty cycle
57	RD/WT	Igniter Transistor 1 Control	7% duty cycle
58	RD/YL	Igniter Signal (IGF)	Digital Signal: 0-5-0v
60	BL/WT	Waste Gate Solenoid	12v or 0v
62	BK/YL	Vacuum Sensor Signal	1-1.5v
63	GY/RD	Closed Throttle Switch (IDL2)	1v, at off-idle: 12v
64	RD	Closed Throttle Switch (IDL1)	1v, at off-idle: 12v
65	BR/BK	Sensor Ground	<0.050v
66	GN/BK	MAF Sensor Signal	1.1-1.5v
69	BR	Sensor Ground	<0.050v
71	BK/BL	Front HO2S Heater	1v (Heater On)
72	BR/WT	Rear HO2S Heater	1v (Heater On)
73	WT/BL	Fuel Pressure Up Solenoid	1v (at hot restart)
74	PK	EVAP Purge Solenoid (VSV)	12v or 0v
75	PK	EGR Solenoid Control (VSV)	1v (Heater On)
76	BK/WT	Neutral Start Switch	In P/N: 9-11v (cranking)
77	BK	Starter Switch Signal	9-11v (cranking)
79	BR	Power Ground	<0.1v
80	BR	Power Ground	<0.1v

Pin Connector Graphic

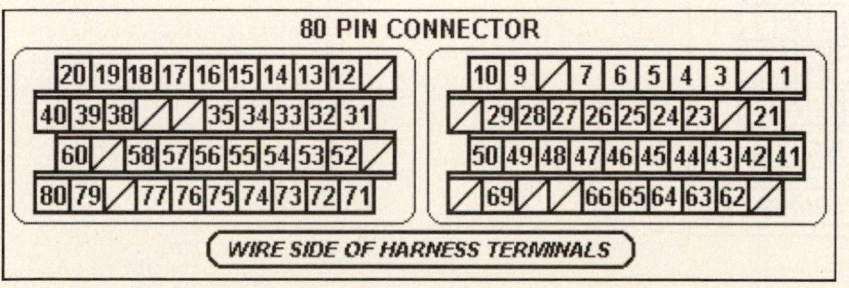

1994-95 Hatchback 3.0L Twin/Turbo I6 VIN J 40 Pin Connector

PCM Pin #	Wire Color	Circuit Description (40 Pin)	Value at Hot Idle
1	BK/OR	Ignition Switch Power	12-14v
2	PK	Speed Sensor 1 Signal	Moving: 0-5-0v
3	YL	Kickdown Switch	Switch Closed: 12v
4	GN/WT	Stop Light Switch Signal	Brake Off: 0v, On: 12v
6	BL/BK	MIL (lamp) Control	MIL Off: 12v, On: 1v
7	RD/BK	A/T Select Switch Reverse	In 'R': 12v, Others: 0v
9	GN/RD	A/T Select Switch 2nd	In 2nd: 12v, Others: 0v
10	GN/BK	A/T Select Switch Low	In Low: 12v, Others: 0v
11	YL/RD	ABS/Traction ECU	Pulses
12	BR/BK	Cruise Control ECU	At Cruise in OD: 12v
13	WT/RD	Traction Control (TRC-)	Pulse Signals
14	OR	Traction Control (TRC+)	Pulse Signals
15	RD/YL	Idle Up Diode Signal	Switch On: 12v, Off: 0v
16	BK/WT	Tachometer Signal (TACO)	Pulse Signals
17	GY/RD	Data Link Connector	12-14v
18	GN/YL	A/T Pattern Selector Switch	Norm: 0v, PWR: 12v
19	PK/GN	Data Link Connector	12-14v
20	YL/BL	Data Link Connector	12-14v
21	GN	Fuel Pump ECU	Pulse Signals
22	PK/WT	Fuel Pump ECU	Pulse Signals
23	WT/GN	A/C Magnetic Clutch (ACMG)	Clutch Off: 0v, On: 12v
24	GY	EFI Main Relay Power	12-14v
25	WT/BL	Manual Indicator Light	Light Off: 12-14v
26	WT	EFI ECU Signal (-)	Pulse Signals
27	BK	EFI ECU Signal (+)	Pulse Signals
28	PK/GN	Overdrive "Off" Indicator	LED Off: 12v, On: 1v
31	BK/RD	EFI Main Relay Power	12-14v
32 ('94)	BK/RD	EFI Main Relay Power B1+	12-14v
33	BK/YL	Battery Direct	12-14v
34	BL/RD	A/C Amplifier Signal (AC1)	Clutch On: 1.5v, Off: 12v
38	PK/BL	Traction Control (NEO)	Pulses

Pin Connector Graphic

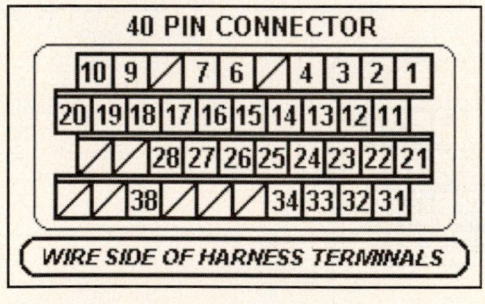

1994-95 Hatchback 3.0L Twin/Turbo I6 VIN J 80 Pin Connector

PCM Pin #	Wire Color	Circuit Description (80 Pin)	Value at Hot Idle
1	YL	OD Clutch Speed Sensor (-)	Pulse Signals
3	GN	Vehicle Speed Sensor 2 (-)	Pulse Signals
4 ('94)	BR	Sensor Ground	<0.050v
5	BL	CMP Sensor Signal (G2-)	<0.050v
6	OR	CMP Sensor Signal (G1-)	<0.050v
7	BR	CKP Sensor Signal (NE-)	<0.050v
7 ('94)	BK/RD	CKP Sensor Signal (NE-)	<0.050v
9	RD/BL	A/T-ECT Solenoid (S2)	1st or OD: 1v
10	WT/RD	A/T-ECT Solenoid (S1)	In 3rd or OD: 1v
12	GN/RD	A/T-ECT Solenoid (SLT-)	Pulse Signals
13	YL/GN	A/T-ECT Solenoid (SLN-)	Pulse Signals
14	GN/BK	A/T-ECT Solenoid (SLU-)	Pulse Signals
15	RD/BK	Injector 6 Control	2.0-3.3 ms
16	RD	Injector 5 Control	2.0-3.3 ms
17	RD/WT	Injector 4 Control	2.0-3.3 ms
18	RD/GN	Injector 3 Control	2.0-3.3 ms
19	RD/YL	Injector 2 Control	2.0-3.3 ms
20	RD/BL	Injector 1 Control	2.0-3.3 ms
21	BL	OD Clutch Speed Sensor (+)	Pulse Signals
23	RD	Vehicle Speed Sensor 2 (+)	Pulse Signals
24	OR	A/T Oil Temperature Sensor	0.5-4.5v
25	BK/WT	CMP Sensor Signal (G2+)	AC pulse signals
26	WT	CMP Sensor Signal (G1+)	AC pulse signals
27	BK/RD	CKP Sensor Signal (NE+)	AC pulse signals
27 ('94)	BR	CKP Sensor Signal (NE+)	AC pulse signals
28	BR	MAF Sensor Ground	<0.050v
29	LG	Data Link Connector	12-14v
31	WT/GN	A/T-ECT Solenoid (SLT+)	Pulses
32	RD/GN	ISC Motor (ISC4)	Pulse Signals
33	GN/OR	ISC Motor (ISC3)	Pulse Signals
34	GN/WT	ISC Motor (ISC2)	Pulse Signals
35	PK/YL	ISC Motor (ISC1)	Pulse Signals
38	GN/RD	Exhaust Bypass Solenoid	12v or 0v
39	GN/YL	EGR Solenoid Control (VSV)	1v (Heater On)
40	GN/BK	Intake Air Control Solenoid	12v or 0v

Standard Colors and Abbreviations

Abbreviation	Color	Abbreviation	Color	Abbreviation	Color
BK	Black	GY	Gray	RD	Red
BL	Blue	GN	Green	TN	Tan
BR	Brown	LG	Light Green	VT	Violet
DB	Dark Blue	OR	Orange	WT	White
DG	DK Green	PK	Pink	YL	Yellow

1994-95 Hatchback 3.0L Twin/Turbo I6 VIN J 80 Pin Connector

PCM Pin #	Wire Color	Circuit Description (80 Pin)	Value at Hot Idle
41	BL/RD	Main Sensor VREF (VC)	4.9-5.1v
42	YL/BL	Sub TP Sensor (VTA2)	2.0-2.9v
43	YL	Main TP Sensor (VTA1)	0.4-1.0v
44	BL/YL	ECT Sensor Signal	At 180ºF: 0.51v
45	PK/BL	IAT Sensor Signal	At 100ºF: 2.60v
46	BR/YL	EGR Gas Temp. Sensor	3.5-4.0v
47	RD/BL	Rear HO2S Signal	0.1-1.1v
48	WT	Front HO2S Signal	0.1-1.1v
49	WT	Knock Sensor 2 Signal	<0.075v AC
50	WT	Knock Sensor 1 Signal	<0.075v AC
52 ('94)	RD/WT	Igniter Transistor 6 Control	Digital Signal: 0-5-0v
52	RD/YL	Igniter Transistor 6 Control	Digital Signal: 0-5-0v
53	BL/OR	Igniter Transistor 5 Control	Digital Signal: 0-5-0v
54	GY/GN	Igniter Transistor 4 Control	Digital Signal: 0-5-0v
55	GY/BK	Igniter Transistor 3 Control	Digital Signal: 0-5-0v
56	BK/OR	Igniter Transistor 2 Control	Digital Signal: 0-5-0v
57	RD/WT	Igniter Transistor 1 Control	Digital Signal: 0-5-0v
58	RD/YL	Igniter Signal (IGF)	Digital Signal: 0-5-0v
60	BL/WT	Waste Gate Solenoid	12v or 0v
62	BK/YL	Turbo Pressure Sensor	2.5-4.5v
63	GY/RD	Closed Throttle Switch (IDL2)	1v, at off-idle: 12v
64	RD	Closed Throttle Switch (IDL1)	1v, at off-idle: 12v
65	BR/BK	Sensor Ground	<0.050v
66	GN/BK	MAF Sensor Signal	1.1-1.5v
69 ('94)	BR	Sensor Ground	<0.050v
71	BK/BL	Front HO2S Heater	1v (Heater On)
72	BR/WT	Rear HO2S Heater	1v (Heater On)
73	WT/BL	Fuel Pressure Up Solenoid	1v (at hot restart)
74	PK	EVAP Purge Solenoid (VSV)	12v or 0v
75	PK	EGR Solenoid Control (VSV)	1v (Heater On)
76	BK/WT	Neutral Start Switch	In P/N: 9-11v (cranking)
77	BK	Starter Switch Signal	9-11v (cranking)
79	BR	Power Ground	<0.1v
80	BR	Power Ground	<0.1v

Pin Connector Graphic

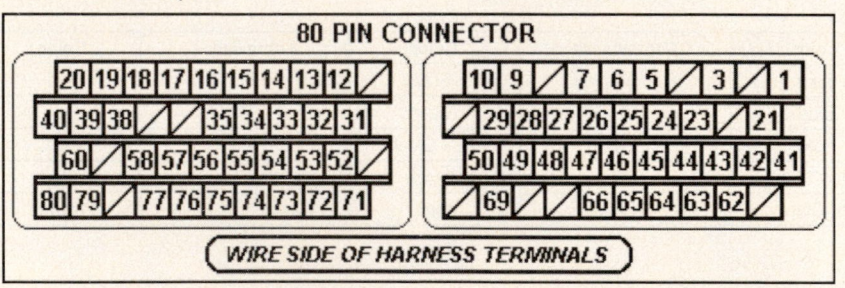

1996 Hatchback 3.0L Twin/Turbo I6 VIN E (All) 40 Pin Connector

PCM Pin #	Wire Color	Circuit Description (40 Pin)	Value at Hot Idle
1	BK/OR	Ignition Switch Power	12-14v
2	PK	Speed Sensor 1 Signal	Moving: 0-5-0v
3	YL	Kickdown Switch	Closed: 12v, Open: 0v
4	GN/WT	Stop Light Switch Signal	Brake Off: 0v, On: 12v
6	BL/BK	MIL (lamp) Control	MIL Off: 12v, On: 1v
7	RD/BK	A/T Select Switch Reverse	In 'R': 12v, Others: 0v
9	GN/RD	A/T Select Switch 2nd	In 2nd: 12v, Others: 0v
10	GN/BK	A/T Select Switch Low	In Low: 12v, Others: 0v
11	YL/RD	ABS /Traction ECU	Pulse Signals
12	BR/BK	Cruise Control ECU	At Cruise in OD: 12v
13	WT/RD	Traction Control (TRC-)	Pulse Signals
14	OR	Traction Control (TRC+)	Pulse Signals
15	RD/YL	Idle Up Diode Signal	Switch On: 12v, Off: 0v
16	BK/WT	Tachometer Signal (TACO)	Pulse Signals
17	GY/RD	Data Link Connector	12-14v
18	GN/YL	A/T Pattern Selector Switch	Norm: 0v, PWR: 12v
19	PK/GN	Data Link Connector	12-14v
20	YL/BL	Data Link Connector	12-14v
21	GN	Fuel Pump ECU	Pulse Signals
22	PK/WT	Fuel Pump ECU	Pulse Signals
23	WT/GN	A/C Magnetic Clutch (ACMG)	Clutch Off: 0v, On: 12v
24	GY	EFI Main Relay Power	12-14v
25	WT/BL	Manual Indicator Light	Light Off: 12-14v
26	WT	Traction Control ECU (EFI-)	Pulse Signals
27	BK	Traction Control ECU (EFI+)	Pulse Signals
28	PK/GN	Overdrive Main Switch	Switch Off: 12v
31	BK/RD	EFI Main Relay Power	12-14v
33	BK/YL	Battery Direct	12-14v
34	BL/RD	A/C Amplifier Signal (AC1)	Clutch On: 1.5v, Off: 12v
38	PK/BL	Traction Control (NEO)	Pulse Signals

Pin Connector Graphic

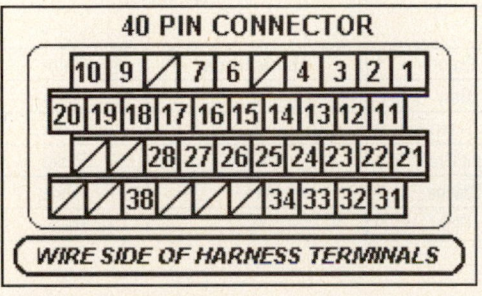

1996 Hatchback 3.0L Twin/Turbo I6 VIN E (All) 80 Pin Connector

PCM Pin #	Wire Color	Circuit Description (80 Pin)	Value at Hot Idle
1	YL	OD Clutch Speed Sensor (-)	Pulses
3	GN	Vehicle Speed Sensor 2 (-)	At 55 mph: 48 Hz
5	BL	CMP Sensor Signal (G2-)	<0.050v
6	OR	CMP Sensor Signal (G1-)	<0.050v
7	BR	CKP Sensor Signal (NE-)	<0.050v
9	RD/BL	A/T-ECT Solenoid (S2)	1st or OD: 1v
10	WT/RD	A/T-ECT Solenoid (S1)	In 3rd or OD: 1v
12	GN/RD	A/T-ECT Solenoid (SLT-)	Pulse Signals
13	YL/GN	A/T-ECT Solenoid (SLN-)	Pulse Signals
14	GN/BK	A/T-ECT Solenoid (SLU-)	Pulse Signals
15	RD/BK	Injector 6 Control	2.0-3.3 ms
16	RD	Injector 5 Control	2.0-3.3 ms
17	RD/WT	Injector 4 Control	2.0-3.3 ms
18	RD/GN	Injector 3 Control	2.0-3.3 ms
19	RD/YL	Injector 2 Control	2.0-3.3 ms
20	RD/BL	Injector 1 Control	2.0-3.3 ms
21	BL	OD Clutch Speed Sensor (+)	Pulse Signals
23	RD	Vehicle Speed Sensor 2 (+)	Pulse Signals
24	OR	A/T Oil Temperature Sensor	0.5-4.5v
25	BK/WT	CMP Sensor Signal (G2+)	AC pulse signals
26	WT	CMP Sensor Signal (G1+)	AC pulse signals
27	BK/RD	CKP Sensor Signal (NE+)	AC pulse signals
28	BR	MAF Sensor Ground	<0.050v
29	LG	Data Link Connector	12-14v
31	WT/GN	A/T-ECT Solenoid (SLT+)	Pulse Signals
32	RD/GN	ISC Motor (ISC4)	Pulse Signals
33	GN/OR	ISC Motor (ISC3)	Pulse Signals
34	GN/WT	ISC Motor (ISC2)	Pulse Signals
35	PK/YL	ISC Motor (ISC1)	Pulse Signals
38	GN/RD	Exhaust Bypass Solenoid	12v or 0v
39	GN/YL	EGR Solenoid Control (VSV)	1v (Heater On)
40	GN/BK	Intake Air Control Solenoid	12v or 0v

Standard Colors and Abbreviations

Abbreviation	Color	Abbreviation	Color	Abbreviation	Color
BK	Black	GY	Gray	RD	Red
BL	Blue	GN	Green	TN	Tan
BR	Brown	LG	Light Green	VT	Violet
DB	Dark Blue	OR	Orange	WT	White
DG	DK Green	PK	Pink	YL	Yellow

1996 Hatchback 3.0L Twin/Turbo I6 VIN E (All) 80 Pin Connector

PCM Pin #	Wire Color	Circuit Description (80 Pin)	Value at Hot Idle
41	BL/RD	Main Sensor VREF (VC)	4.9-5.1v
42	YL/BL	Sub TP Sensor (VTA2)	2.0-2.9v
43	YL	Main TP Sensor (VTA1)	0.4-1.0v
44	BL/YL	ECT Sensor Signal	At 180ºF: 0.51v
45	PK/BL	IAT Sensor Signal	At 100ºF: 2.60v
46	BR/YL	EGR Gas Temp. Sensor	3.5-4.0v
47	RD/BL	HO2S-12 (B1 S2) Signal	0.1-1.1v
48	WT	HO2S-11 (B1 S1) Signal	0.1-1.1v
49	WT	Knock Sensor 2 Signal	<0.075v AC
50	WT	Knock Sensor 1 Signal	<0.075v AC
52	RD/YL	Igniter Transistor 6 Control	7% duty cycle
53	BL/OR	Igniter Transistor 5 Control	7% duty cycle
54	GY/GN	Igniter Transistor 4 Control	7% duty cycle
55	GY/BK	Igniter Transistor 3 Control	7% duty cycle
56	BK/OR	Igniter Transistor 2 Control	7% duty cycle
57	RD/WT	Igniter Transistor 1 Control	7% duty cycle
58	RD/YL	Igniter Signal (IGF)	Digital Signal: 0-5-0v
60	BL/WT	Waste Gate Solenoid	12v or 0v
62	BK/YL	Turbo Pressure Sensor	2.5-4.5v
63	GY/RD	Closed Throttle Switch (IDL2)	1v, at off-idle: 12v
64	RD	Closed Throttle Switch (IDL1)	1v, at off-idle: 12v
65	BR/BK	Sensor Ground	<0.050v
66	GN/BK	Vacuum Sensor Signal	1-1.5v
69	BR/BK	Sensor Ground	<0.050v
71	BK/BL	HO2S-11 (B1 S1) Heater	1v (Heater On)
72	BR/WT	HO2S-12 (B1 S2) Heater	1v (Heater On)
73	WT/BL	Fuel Pressure Up Solenoid	1v (at hot restart)
74	PK	EVAP Purge Solenoid (VSV)	12v or 0v
75	PK	EGR Solenoid Control (VSV)	1v (Heater On)
76	BK/WT	Neutral Start Switch	In P/N: 9-11v (cranking)
77	BK	Starter Switch Signal	9-11v (cranking)
79	BR	Power Ground	<0.1v
80	BR	Power Ground	<0.1v

Pin Connector Graphic

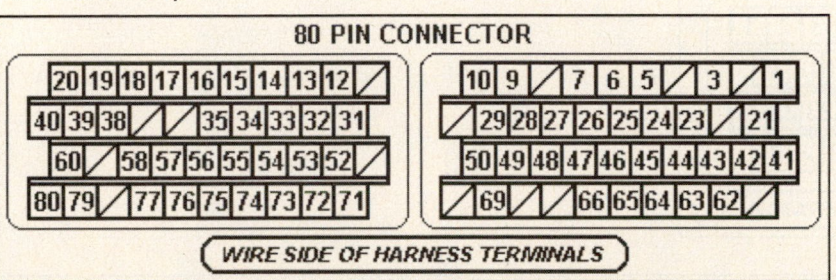

1997-98 Hatchback 3.0L Twin/Turbo I6 MFI VIN E (All) 40 Pin Connector

PCM Pin #	Wire Color	Circuit Description (40 Pin)	Value at Hot Idle
1	BK/OR	Ignition Switch Power	12-14v
2	PK	Speed Sensor 1 Signal	Moving: 0-5-0v
3	YL	Kickdown Switch	Closed: 12v, Closed: 0v
4	GN/WT	Stop Light Switch Signal	Brake Off: 0v, On: 12v
6	BL/BK	MIL (lamp) Control	MIL Off: 12v, On: 1v
7	RD/BK	A/T Select Switch Reverse	In 'R': 12v, Others: 0v
8	GN	Data Link Connector	No Scan Tool: 0v
9	GN/RD	A/T Select Switch 2nd	In 2nd: 12v, Others: 0v
10	GN/BK	A/T Select Switch Low	In Low: 12v, Others: 0v
11	YL/RD	ABS ECU Signal	Pulse Signals
12	BR/BK	Cruise Control ECU	At Cruise in OD: 12v
13	BR	Throttle Control ECU (ETC-)	Pulse Signals
14	YL	Throttle Control ECU (ETC+)	Pulse Signals
15	RD/YL	Idle Up Diode Signal	Load On: 12v
16	BK/WT	Tachometer Signal (TACO)	Pulse Signals
18	GN/YL	A/T Pattern Selector Switch	Norm: 0v, PWR: 12v
20	BL	Data Link Connector	12-14v
21	GN	Fuel Pump ECU	Pulse Signals
22	PK/WT	Fuel Pump ECU	Pulse Signals
23	WT/GN	A/C Magnetic Clutch (ACMG)	Clutch Off: 0v, On: 12v
24	GY	EFI Main Relay Power	12-14v
25	WT/BL	Manual Indicator Light	Light Off: 12v, On: 1v
26	WT	Traction Control ECU (EFI-)	Pulse Signals
27	BK	Traction Control ECU (EFI+)	Pulse Signals
28	PK/GN	Overdrive Main Switch	Switch Off: 12v
31	BK/RD	EFI Main Relay Power	12-14v
33	BK/YL	Battery Direct	12-14v
34	BL/RD	A/C Amplifier Signal (AC1)	Clutch On: 1.5v, Off: 12v
38	PK/BL	Traction Control ECU (NEO)	Pulse Signals
39	GN/OR	Throttle Control ECU (VTO2)	Pulse Signals
40	PK/YL	Throttle Control ECU (VTO1)	Pulse Signals

Pin Connector Graphic

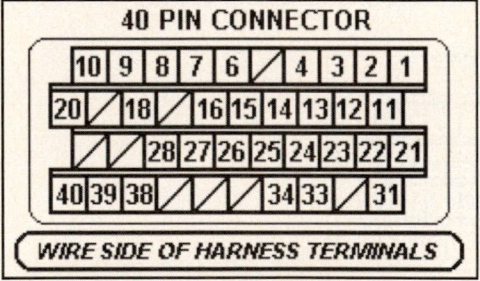

Standard Colors and Abbreviations

Abbreviation	Color	Abbreviation	Color	Abbreviation	Color
BK	Black	GY	Gray	RD	Red
BL	Blue	GN	Green	TN	Tan
BR	Brown	LG	Light Green	VT	Violet
DB	Dark Blue	OR	Orange	WT	White
DG	DK Green	PK	Pink	YL	Yellow

1997-98 Hatchback 3.0L Twin/Turbo I6 MFI VIN E (All) 80 Pin Connector

PCM Pin #	Wire Color	Circuit Description (80 Pin)	Value at Hot Idle
1	YL	OD Clutch Speed Sensor (-)	Pulse Signals
3	GN	Vehicle Speed 2 Sensor (-)	Pulse Signals
6	OR	CMP Sensor Signal (G1-)	<0.050v
7	BR	CKP/CMP Sensor Ground (-)	<0.050v
9	RD/BL	A/T-ECT Solenoid (S2)	1st or OD: 1v
10	WT/RD	A/T-ECT Solenoid (S1)	In 3rd or OD: 1v
12	GN/RD	A/T-ECT Solenoid (SLT-)	Pulse Signals
13	YL/GN	A/T-ECT Solenoid (SLN-)	Pulse Signals
14	GN/BK	A/T-ECT Solenoid (SLU-)	Pulse Signals
15	RD/BK	Injector 6 Control	2.0-3.3 ms
16	RD	Injector 5 Control	2.0-3.3 ms
17	RD/WT	Injector 4 Control	2.0-3.3 ms
18	RD/GN	Injector 3 Control	2.0-3.3 ms
19	RD/YL	Injector 2 Control	2.0-3.3 ms
20	RD/BL	Injector 1 Control	2.0-3.3 ms
21	BL	OD Clutch Speed Sensor (+)	Pulse Signals
23	RD	Vehicle Speed 2 Sensor (+)	Pulse Signals
24	OR	A/T Oil Temperature Sensor	0.5-4.5v
25	BK/WT	CMP Sensor Signal (G2+)	AC pulse signals
26	WT	CMP Sensor Signal (G1+)	AC pulse signals
27	BK/RD	CKP Sensor Signal (NE+)	AC pulse signals
28	BR	MAF Sensor Ground	<0.050v
31	WT/GN	ECT Solenoid (SLT+)	Pulse Signals
32	RD/GN	ISC Motor (ISC4)	Pulse Signals
33	GN/OR	ISC Motor (ISC3)	Pulse Signals
34	GN/WT	ISC Motor (ISC2)	Pulse Signals
35	PK/YL	ISC Motor (ISC1)	Pulse Signals
38	GN/RD	Exhaust Bypass Solenoid	12v or 0v
39	GN/YL	EGR Solenoid Control (VSV)	1v (Heater On)
40	GN/BK	Intake Air Control Solenoid	12v or 0v

Pin Connector Graphic

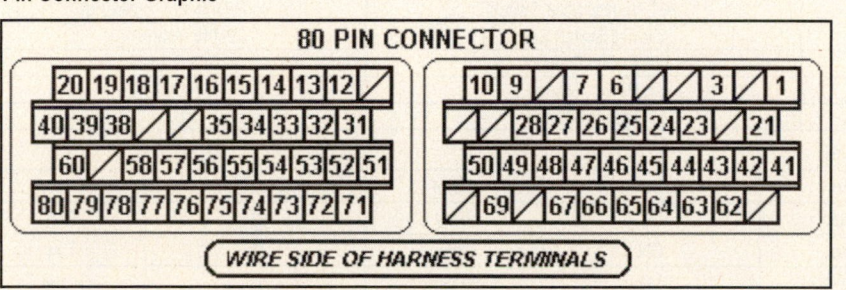

1997-98 Hatchback 3.0L Twin/Turbo I6 MFI VIN E (All) 80 Pin Connector

PCM Pin #	Wire Color	Circuit Description (80 Pin)	Value at Hot Idle
41	BL/RD	Sensor VREF (VC)	4.9-5.1v
42	YL/BL	Sub TP Sensor (VTA2)	2.0-2.9v
43	YL	Main TP Sensor (VTA1)	0.4-1.0v
44	BL/YL	ECT Sensor Signal	At 180°F: 0.51v
45	PK/BL	IAT Sensor Signal	At 100°F: 2.60v
46	BR/YL	EGR Gas Temp. Sensor	3.5-4.0v
47	RD/BL	HO2S-12 (B1 S2) Signal	0.1-1.1v
48	WT	HO2S-11 (B1 S1) Signal	0.1-1.1v
49	WT	Knock Sensor 2 Signal	<0.075v AC
50	WT	Knock Sensor 1 Signal	<0.075v AC
51	WT/BL	Throttle (FAIL) ECU	Pulses
52	RD	Igniter Transistor 6 Control	7% duty cycle
53	BL	Igniter Transistor 5 Control	7% duty cycle
54	BK/RD	Igniter Transistor 4 Control	7% duty cycle
55	LG	Igniter Transistor 3 Control	7% duty cycle
56	WT/RD	Igniter Transistor 2 Control	7% duty cycle
57	RD/WT	Igniter Transistor 1 Control	7% duty cycle
58	RD/YL	Igniter Signal (IGF)	Digital Signal: 0-5-0v
60	BL/WT	Waste Gate Solenoid	12v or 0v
62	BK/YL	Turbo Pressure Sensor	2.5-4.5v
63	GY/RD	Closed Throttle Switch (IDL2)	1v, at off-idle: 12v
64	RD	Closed Throttle Switch (IDL1)	1v, at off-idle: 12v
65	WT/BK	Sensor Ground	<0.050v
66	YL/RD	MAF Sensor Signal	1.1-1.5v
67	PK	Throttle Signal (EFIF)	Pulse Signals
69	BR	Shield Ground	<0.050v
71	BK/BL	HO2S-11 (B1 S1) Heater	1v (Heater On)
72	BR/WT	HO2S-12 (B1 S2) Heater	1v (Heater On)
73	WT/BL	Fuel Pressure Up Solenoid	1v (at hot restart)
74	PK	EVAP Purge Solenoid (VSV)	12v or 0v
75	PK	EGR Solenoid Control (VSV)	1v (Heater On)
76	BK/WT	A/T Neutral Start Switch	In P/N: 9-11v (cranking)
77	BK	Starter Switch Signal	9-11v (cranking)
78	BR	Power Ground	<0.1v
79	BR	Power Ground	<0.1v
80	BR	Power Ground	<0.1v

Standard Colors and Abbreviations

Abbreviation	Color	Abbreviation	Color	Abbreviation	Color
BK	Black	GY	Gray	RD	Red
BL	Blue	GN	Green	TN	Tan
BR	Brown	LG	Light Green	VT	Violet
DB	Dark Blue	OR	Orange	WT	White
DG	DK Green	PK	Pink	YL	Yellow

TERCEL Pin Tables

1990 Coupe, Hatchback 1.5L I4 MFI VIN E (All) 10 Pin Connector

PCM Pin #	Wire Color	Circuit Description (10 Pin)	Value at Hot Idle
1	BK/WT	A/T Neutral Start Switch	In P/N: 9-11v (cranking)
1	BK/RD	M/T Clutch Switch	In 'N': 9-11v (cranking)
3	BK	Starter Switch Signal	9-11v (cranking)
4	WT	Injectors 1 & 3 Control	2.0-3.3 ms
5	BR	Power Ground	<0.1v
6 (Cal)	PK/BL	EGR Solenoid Control (VSV)	1v (Heater On)
7	BR	Sensor Ground	<0.050v
8	LG	Igniter Signal (IGT)	0.72-074v
9	YL	Injectors 2 & 4 Control	2.0-3.3 ms
10	BR	Power Ground	<0.1v

1990 Coupe, Hatchback 1.5L I4 MFI VIN E (All) 14-Pin Connector

PCM Pin #	Wire Color	Circuit Description (14-Pin)	Value at Hot Idle
1	BK/OR	EFI Main Relay Power	12-14v
2	GN/RD	Battery Direct	12-14v
4	GN	Circuit Opening Relay (FC)	0-3v, off-idle: 12v
8	BK/OR	EFI Main Relay Power	12-14v
9	GN/WT	MIL (lamp) Control	MIL Off: 12v, On: 1v
11	BK/RD	A/C Magnetic Clutch (ACMG)	Clutch Off: 0v, On: 12v
12	YL/BL	Vehicle Speed Sensor	At 55 mph: 48 Hz

1990 Coupe, Hatchback 1.5L I4 MFI VIN E (All) 18 Pin Connector

PCM Pin #	Wire Color	Circuit Description (18 Pin)	Value at Hot Idle
1	GN/BK	ECT Sensor Signal	At 180ºF: 0.51v
2	PK	Vacuum Sensor Signal	1-1.5v
3	GN/BK	IAT Sensor Signal	At 100ºF: 2.60v
4	PK	Data Link Connector	12-14v
5	GN/YL	Igniter Signal (IGF)	0.74-0.76v
6	RD	Distributor Signal (G1+)	AC pulse signals
7	BK	Distributor Signal (G-)	<0.050v
8	BK	O2S-11 (B1 S1) Signal	0.1-1.1v
9	GN/YL	Idle Up Solenoid Control	Load On: 12v, Off: 0v
10	BR	Sensor Ground	<0.050v
11	YL/GN	Wide Open Throttle Switch	1v, at off-idle: 12v
12	GN/RD	Vacuum Sensor VREF	4.9-5.1v
13	YL/RD	Closed Throttle Switch	1v, at off-idle: 12v
14 (Cal)	BL/RD	EGR Gas Temp. Sensor	3.5-4.0v
15	WT	Distributor Signal (NE+)	AC pulse signals
16	BR	Sensor Ground	<0.050v
17	GN/YL	Data Link Connector	12-14v
18	BL/WT	Fuel Pressure Up Solenoid	12v or 0v

Pin Connector Graphic

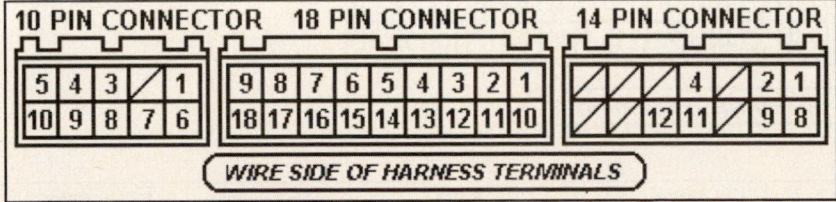

Standard Colors and Abbreviations

Abbreviation	Color	Abbreviation	Color	Abbreviation	Color
BK	Black	GY	Gray	RD	Red
BL	Blue	GN	Green	TN	Tan
BR	Brown	LG	Light Green	VT	Violet
DB	Dark Blue	OR	Orange	WT	White
DG	DK Green	PK	Pink	YL	Yellow

1991-94 Sedan 1.5L I4 MFI VIN E (All) 16 Pin Connector

PCM Pin #	Wire Color	Circuit Description (16 Pin)	Value at Hot Idle
1	BK/OR	EFI Main Relay Power B1+	12-14v
2	WT/RD	Battery Direct	12-14v
4	GN	Circuit Opening Relay (FC)	0-3v, off-idle: 12v
7 (Cal)	RD/GN	Throttle Opener Solenoid	12v or 0v
8	PK	Data Link Connector	12-14v
9	BK/OR	EFI Main Relay Power	12-14v
10	GN/WT	MIL (lamp) Control	MIL Off: 12v, On: 1v
12	PK/BK	A/C Magnetic Clutch (ACMG)	Clutch Off: 0v, On: 12v
13	YL	Vehicle Speed Sensor	At 55 mph: 48 Hz
14	GN/BK	A/C Amplifier Signal (ACT)	Clutch On: 12v, Off: 1.5v
16	GN/YL	Data Link Connector	12v

1991-94 Sedan 1.5L I4 MFI VIN E (All) 26 Pin Connector

PCM Pin #	Wire Color	Circuit Description (26 Pin)	Value at Hot Idle
1	BL/WT	Fuel Pressure Up Solenoid	1v (at hot restart)
2	BK/WT	A/T Neutral Start Switch	In P/N: 9-11v (cranking)
2	BK/RD	M/T: Neutral Start Switch	In 'N': 9-11v (cranking)
3	GN/BK	ECT Sensor Signal	At 180ºF: 0.51v
4	PK	MAP Sensor Signal	1-1.5v
5	GN/BK	IAT Sensor Signal	At 100ºF: 2.60v
6	LG	Igniter Signal (IGT)	Digital Signal: 0-5-0v
7	GN/YL	Igniter Signal (IGF)	Digital Signal: 0-5-0v
9	BK	Distributor Signal (NE-)	<0.050v
10	BK	O2S-11 (B1 S1) Signal	0.1-1.1v
11	BK/RD	Starter Switch Signal	9-11v (cranking)
12	WT	Injectors 1 & 3 Control	2.0-3.3 ms
13	BR	Power Ground	<0.1v
14	GN/RD	Idle-Up Solenoid Control	Load On: 1v
15 (Cal)	PK/BL	EGR Solenoid Control (VSV)	1v (Heater On)
16	BR	Sensor Ground	<0.050v
17	YL/GN	TP Sensor Signal	0.3-0.8v
18	GN/RD	Sensor VREF	4.9-5.1v
19	BL/RD	Closed Throttle Switch	1v, at off-idle: 12v
21	WT	Distributor Signal (NE+)	AC pulse signals
22	BR	Sensor Ground	<0.050v
23 (Cal)	BL/RD	EGR Gas Temp. Sensor	3.5-4.0v
24	BR	Sensor Ground	<0.050v
25	WT	Injectors 2 & 4 Control	2.0-3.3 ms
26	BR	Power Ground	<0.1v

Pin Connector Graphic

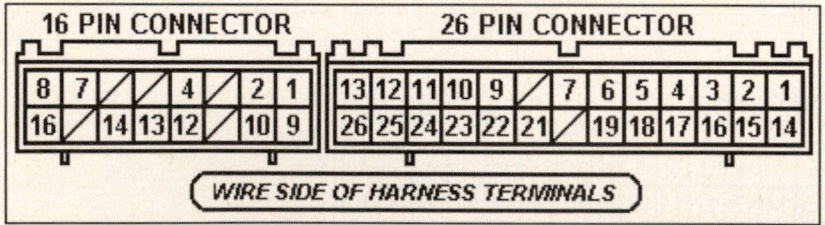

Standard Colors and Abbreviations

Abbreviation	Color	Abbreviation	Color	Abbreviation	Color
BK	Black	GY	Gray	RD	Red
BL	Blue	GN	Green	TN	Tan
BR	Brown	LG	Light Green	VT	Violet
DB	Dark Blue	OR	Orange	WT	White
DG	DK Green	PK	Pink	YL	Yellow

1995-96 Sedan 1.5L I4 MFI VIN C, VIN E (All) 16 Pin Connector

PCM Pin #	Wire Color	Circuit Description (16 Pin)	Value at Hot Idle
2	RD/WT	Battery Direct	12-14v
3	BK	Overdrive Main Switch	Switch Off: 12v, On: 0v
4	GN/BK	Circuit Opening Relay (FC)	0-3v, off-idle: 12v
5	WT/RD	HO2S-12 (B1 S2) Heater	1v (Heater On)
6	WT/BK	HO2S-11 (B1 S1) Heater	1v (Heater On)
8	GN	Data Link Connector	12v
9	BK/RD	EFI Main Relay Power	12-14v
10	GN/BL	MIL (lamp) Control	MIL Off: 12v, On: 1v
11	RD	Overdrive Solenoid	Solenoid Off: 12v, On: 1v
12	BL	A/C Amplifier Signal (AC1)	Clutch On: 1.5v, Off: 12v
13	YL	Vehicle Speed Sensor	At 55 mph: 48 Hz
14	GN/BK	A/C Amplifier Signal (ACT)	Clutch On: 12v, Off: 1.5v
15	BK/WT	EGR Solenoid Control (VSV)	1v (Heater On)
16	WT	Data Link Connector (SDL)	0v

1995-96 Sedan 1.5L I4 MFI VIN C, VIN E (All) 26 Pin Connector

PCM Pin #	Wire Color	Circuit Description (26 Pin)	Value at Hot Idle
1	BL/RD	IAC Signal (RSC)	Pulse Signals
2	BK	Neutral Start Switch	In P/N: 9-11v (cranking)
3	GN/BK	ECT Sensor Signal	At 180ºF: 0.51v
4	PK	MAP Sensor Signal	1-1.5v
5	BL/YL	Igniter Transistor 2 Control	Digital Signal: 0-5-0v
6	LG	Igniter Transistor 1 Control	Digital Signal: 0-5-0v
7	GN/BK	Igniter Signal (IGF)	Digital Signal: 0-5-0v
8	BK	Knock Sensor Signal	<0.075v AC
9	WT	CMP Sensor Signal (G+)	AC pulse signals
10	BK	HO2S-11 (B1 S1) Signal	0.1-1.1v
11	BK/WT	Starter Switch Signal	9-11v (cranking)
12	GN	Injectors 1 & 3 Control	2.0-3.3 ms
13	BR	Power Ground	<0.1v
14	BL	IAC Signal (RSO)	Pulse Signals
15	BL/BK	IAT Sensor Signal	At 100ºF: 2.60v
16	BR	Sensor Ground	<0.050v
17	YL/GN	TP Sensor Signal	0.3-0.8v
18	GN/RD	Sensor VREF	4.9-5.1v
19	YL/BK	Closed Throttle Switch	1v, at off-idle: 12v
20	RD	CMP Sensor Signal (G-)	AC pulse signals
21	BK	CKP Sensor Signal (NE+)	AC pulse signals
23	WT	HO2S-12 (B1 S2) Signal	0.1-1.1v
24	BR	Shield Ground	<0.050v
25	YL	Injectors 2 & 4 Control	2.0-3.3 ms
26	BR	Power Ground	<0.1v

Pin Connector Graphic

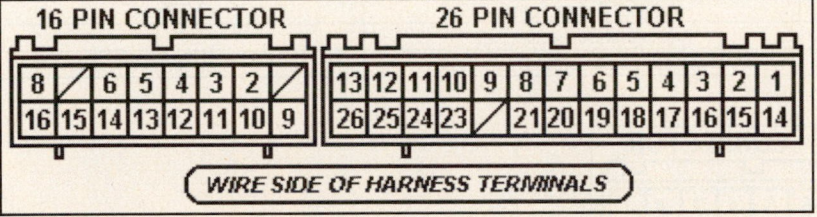

Standard Colors and Abbreviations

Abbreviation	Color	Abbreviation	Color	Abbreviation	Color
BK	Black	GY	Gray	RD	Red
BL	Blue	GN	Green	TN	Tan
BR	Brown	LG	Light Green	VT	Violet
DB	Dark Blue	OR	Orange	WT	White
DG	DK Green	PK	Pink	YL	Yellow

1997-99 Sedan 1.5L I4 MFI VIN C (All) 12 Pin Connector

PCM Pin #	Wire Color	Circuit Description (12 Pin)	Value at Hot Idle
2	WT/RD	Battery Direct	12-14v
4	GN/BK	Circuit Opening Relay (FC)	1v, at off-idle: 12v
6	BL	A/C Amplifier Signal (ACT)	Clutch On: 12v, Off: 1.5v
7	BK/RD	EFI Main Relay Power	12-14v
8	GY/BL	MIL (lamp) Control	MIL Off: 12v, On: 1v
10	GN/BK	A/C Amplifier Signal (AC1)	Clutch On: 1.5v, Off: 12v
11	YL	Vehicle Speed Sensor	At 55 mph: 48 Hz

1997-99 Sedan 1.5L I4 MFI VIN C (All) 26 Pin Connector

PCM Pin #	Wire Color	Circuit Description (26 Pin)	Value at Hot Idle
1	GY	Igniter Transistor 1 Control	Digital Signal: 0-5-0v
2	BK/WT	Starter Switch Signal	9-11v (cranking)
4	BK	CKP Sensor Signal (NE+)	AC pulse signals
7	RD/BL	Igniter Signal (IGF)	Digital Signal: 0-5-0v
9 ('97)	BL/RD	IAC Signal (RSC)	Pulse Signals
10 ('97)	BL	IAC Signal (RSO)	Pulse Signals
10 ('98-'99)	BL	IAC Signal (RSD)	Pulse Signals
11	RD	Injector 3 Control	2.0-3.3 ms
12	GN	Injector 1 Control	2.0-3.3 ms
13	BR	Power Ground	<0.1v
14	BL/YL	Igniter Transistor 2 Control	Digital Signal: 0-5-0v
15	BK	A/T Neutral Start Switch	In P/N: 9-11v (cranking)
15	BK/WT	M/T: Neutral Start Switch	In 'N': 9-11v (cranking)
17	GN	CKP/CMP Sensor Ground (-)	<0.050v
18	RD	CMP Sensor Signal (G+)	AC pulse signals
21	WT/RD	HO2S-12 (B1 S2) Heater	1v (Heater On)
24	BL	Injector 4 Control	2.0-3.3 ms
25	YL	Injector 2 Control	2.0-3.3 ms
26	BR	Shield Ground	<0.050v

1997-99 Sedan 1.5L I4 MFI VIN C (All) 16 Pin Connector

PCM Pin #	Wire Color	Circuit Description (16 Pin)	Value at Hot Idle
2	PK	MAP Sensor Signal	1-1.5v
3	BL/BK	IAT Sensor Signal	At 100ºF: 2.60v
4	BK/RD	ECT Sensor Signal	At 180ºF: 0.51v
5	WT	HO2S-12 (B1 S2) Signal	0.1-1.1v
6	BK	O2S-11 (B1 S1) Signal	0.1-1.1v
7	BL/WT	EVAP Purge Solenoid (VSV)	12v or 0v
8	GN/YL	EVAP Vapor Pressure (VSV)	12v or 0v
9	BR	Sensor Ground	<0.050v
10	YL/GN	TP Sensor Signal	0.3-0.8v
11	GN/RD	Sensor VREF	4.9-5.1v
12	YL/BK	EVAP Vapor Pressure Sensor	12v or 0v
13	WT	SIL Signal (Scan Tool)	Digital Signal
14	BK	Knock Sensor Signal	<0.075v AC
15	GN	Data Link Connector (TE)	12-14v
16	BR	Power Ground	<0.1v

Pin Connector Graphic

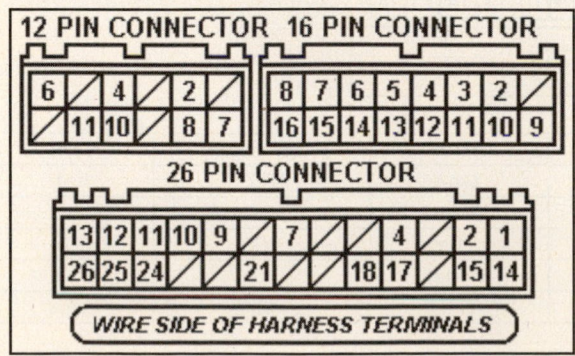

TOYOTA SPORT UTILITY VEHICLE PIN TABLE CONTENTS

Highlander Pin Tables

2001-03 Utility 2.4L I4 VIN D 2AZ-FE (A/T) E5 22 Pin Connector

PCM Pin #	Wire Color	Circuit Description (22 Pin)	Value at Hot Idle
1	BK/WT	Direct Battery	12-14v
2	BK/OR	Ignition Switch Power (IGSW) (IGSW)	12-14v
3	GN/RD	Circuit Opening Relay (FC)	0-3v, at off-idle: 12v
4	WT	SIL Signal (Scan Tool)	Digital Signals
5	BK/RD	O/D OFF Indicator (ODLP)	LED Off: 12v, On: 1v
6	BL/RD	Malfunction Indicator Lamp Control	MIL Off: 12v, On: 1v
7	YL/GN	A/C Magnetic Clutch Relay (ACMG)	Relay Off: 0v, On: 12v
8	PK/BL	EFI Main Relay Control	Relay Off: 0v, On: 12v
9	VT	ABS/Traction ECU (NEO)	Pulse Signals
10-11	---	Not Used	---
12	YL	Body Control ECU Signal (MPX1)	Digital Signals
13	GN/BK	Traction Control Signal (TRC+)	Pulse Signals
14	WT/BL	Traction Control Signal (ENG+)	Pulse Signals
15	GN/WT	Stop Light Switch Signal	Brake Off: 0v, On: 12v
16	WT	EFI Main Relay B+	12-14v
17	WT/BL	Data Link Connector (WFSE)	12v
18	---	Not Used	---
19	BL/WT	Sliding Roof ECU Signal (MPX2)	12v or 0v
20	YL/BK	Traction Control Signal (TRC-)	Pulse Signals
21	WT/GN	Traction Control Signal (ENG-)	Pulse Signals
22	WT/BK	Power Ground (EOM)	<0.1v

2001-03 Utility 2.4L I4 VIN D 2AZ-FE (A/T) E6 28 Pin Connector

PCM Pin #	Wire Color	Circuit Description (28 Pin)	Value at Hot Idle
1	RD/BL	Fan Control Relay 3 Control	Relay Off: 12v, On: 1v
2	RD/BK	A/T Select Switch Reverse	In 'R': 12v, Others: 0v
3-4	---	Not Used	---
5	PK/BK	Data Link Connector 3 Signal (TC)	12v
6	GN/YL	EVAP Vapor Pressure Valve (VSV)	12v or 0v
7	BK/OR	OD Main Switch (ODMS)	Switch Off: 12v, On: 1v
8-9	---	Not Used	---
10	GN/WT	A/T Select Switch Park	In 'P': 12v, Others: 0v
11	RD/WT	A/T Select Switch Neutral	In 'N': 12v, Others: 0v
12	GN/BK	Fan Control Relay 2 Control (PR2)	Relay Off: 12v, On: 1v
13-14	---	Not Used	---
15	RD/YL	Fan Control Relay 1 Control (FAN)	Relay Off: 12v, On: 1v
16-17	---	Not Used	---
18	GY	Transponder Amplifier Signal (RXCK)	Inserting key: pulses
19	PK	Transponder Amplifier Signal (TXCK)	Inserting key: pulses
20	BL/WT	A/T Neutral Start Signal (NSW)	In P/N: 0-3.0v
21	BL/BK	EVAP Vapor Pressure Sensor (PTNK)	2.5-3.1v (hose off)
22	WT/RD	Speedometer Indicator (SPD)	At 55 mph: 48 Hz
23	BL	Unlock Warning Switch	Inserting key: pulses
24	WT	Cruise Control Signal (OD1)	At Cruise in OD: 12v
25	PK/BL	Cruise Control ECU (IDLO)	1.5v, off-idle: 12v
26	LTGN	Theft Deterrent Indicator Control (IMLD)	LED Off: 12v, On: 1v
27	WT/GN	Tachometer Signal (TACO)	DC pulse signals
28	GY/BL	Transponder Amplifier Signal (Code)	Inserting key: pulses

2001-03 Utility 2.4L I4 VIN D 2AZ-FE (A/T) E7 17 Pin Connector

PCM Pin #	Wire Color	Circuit Description (17 Pin)	Value at Hot Idle
1	GN/RD	A/T Select Switch Drive	In 'D': 12v, Others: 0v
2	PK	EVAP Canister Closed Valve (CCV)	12v or 0v
3	---	Not Used	---
4	YL	Generator Control (RL)	12v
5	YL/GN	Engine Oil Pressure Switch (MOPS)	Open: 12v, Closed: 0v
6	RD/BL	A/T-ECT Solenoid DLS	Driving in Lockup: 12v
7	BL/YL	A/T Select Switch Second	In 2nd: 12v, Others: 0v
8	GY/BL	Center Airbag Sensor Assembly (F/PS)	Digital Signals
9-10	---	Not Used	---
11	GN/YL	PSP Switch Signal (PSW)	Straight: 12v, Turning: 0v
12	RD/YL	A/T-ECT Solenoid (S4)	Driving in OD: 12v
13	BL/WT	A/T Select Switch Low	In Low: 12v, Others: 0v
14	BK	HO2S-12 (B1 S2) Heater (HT2B)	Heater Off: 12v, On: 1v
15	WT	HO2S-12 (B1 S2) Signal (OX2B)	0.1-1.1v
16	BL	Starter Switch Signal (STA)	In P/N: 0-3.0v
17	---	Not Used	---

2001-03 Utility 2.4L I4 VIN D 2AZ-FE (A/T) E8 24 Pin Connector

PCM Pin #	Wire Color	Circuit Description (24 Pin)	Value at Hot Idle
1	WT/BK	Power Ground (E04)	<0.1v
2	BL/RD	Sensor VREF (VC)	4.9-5.1v
3	RD/BL	HO2S-22 (B2 S2) Heater (HT1B)	1v (Heater on)
4	BK/RD	AFS-21 (B1 S1) Heater (HAF2A)	1v (Heater on)
5	BK	AFS-21 (B2 S1) Heater (HAF1A)	1v (Heater on)
6	BK/RD	EVAP Purge Solenoid (VSV)	12v or 0v
7	---	Not Used	---
8	WT/BK	Power Ground (E05)	<0.1v
9-10	---	Not Used	---
11	BK	HO2S-22 (B2 S2) Signal (OX1B)	0.1-1.1v
12	YL/GN	MAF Sensor Signal (VG)	1.1-1.8v
13	YL	AFS-21 (B2 S1) Signal (AF2+)	Fixed at 3.3v
14	GN	AFS-11 (B1 S1) Signal (+)	Fixed at 3.3v
15	YL	CMP Sensor Signal (G22+)	AC pulse signals
16	PK	CKP Sensor Signal (NE+)	AC pulse signals
17	BR	Power Ground (E1)	<0.050v
18	WT	Sensor Ground (E2)	<0.050v
19-20	---	Not Used	---
21	RD/BK	MAF Sensor Ground (E2G)	<0.050v
22	BR	AFS-21 (B2 S1) Signal (AF2-)	Fixed at 3.3v
23	RD	AFS-11 (B1 S1) Signal (AF1A-)	Fixed at 3.3v
24	BL	CKP/CMP Sensor Signal (-)	<0.050v

Pin Connector Graphic

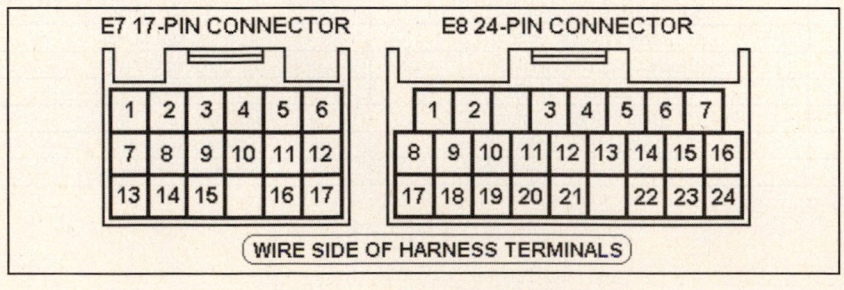

2001-03 Utility 2.4L I4 VIN D 2AZ-FE (A/T) E9 31 Pin Connector

PCM Pin #	Wire Color	Circuit Description (31 Pin)	Value at Hot Idle
1	RD/BL	Injector 1 Control	2.0-3.3 ms
2	YL	Injector 2 Control	2.0-3.3 ms
3	BK/WT	Injector 3 Control	2.0-3.3 ms
4	BL	Injector 4 Control	2.0-3.3 ms
5	BK/RD	A/T-ECT Solenoid (SLT-)	Pulse Signals
6	GN/YL	A/T-ECT Solenoid (SLT+)	Pulse Signals
7	BL/BK	A/T-ECT Solenoid (SL1+)	Pulse Signals
8	RD/BK	A/T-ECT Solenoid (SL2+)	Pulse Signals
9	BL/WT	A/T-ECT Solenoid (SL1-)	Pulse Signals
10	BK/YL	Igniter Transistor 1 Control	6°, 55 mph: 8° dwell
11	YL/GN	Igniter Transistor 2 Control	6°, 55 mph: 8° dwell
12	BL/YL	Igniter Transistor 3 Control	6°, 55 mph: 8° dwell
13	BL/RD	Igniter Transistor 4 Control	6°, 55 mph: 8° dwell
14	WT/RD	Counter Gear Speed (NC+)	AC pulse signals
15	LG	Turbine Speed Sensor (NT-)	AC pulse signals
16	PK	Turbine Speed Sensor (NT+)	AC pulse signals
17	GN/YL	A/T Oil Temperature Signal	At 230°F: <1.5v
18	GN/BK	Idle Air Control Valve (RSD)	Pulse Signals
19	GN/WT	Cam Timing Oil Control (OCV+)	AC pulse signals
20	RD/WT	A/T-ECT Solenoid (SL2-)	At 68°F: 4-5v
21	WT/BK	Power Ground (E01)	<0.1v
22	WT/BL	ECT Sensor Signal (THW)	At 180°F: 0.51v
23	RD/YL	IAT Sensor Signal (THA)	At 100°F: 2.60v
24	BK/WT	TP Sensor Signal (VTA)	0.3-0.8v
25	WT	Igniter Signal (IGF)	Digital Signal: 0-5-0v
26	BL	Counter Gear Speed (NC-)	AC pulse signals
27	WT	Knock Sensor Signal (KNK1)	<0.075v AC
28	---	Not Used	
29	YL/RD	Cam Timing Oil Control (OCV-)	AC pulse signals
30	WT/BK	Power Ground (E03)	<0.1v
31	WT/BK	Power Ground (E02)	<0.1v

Pin Connector Graphic

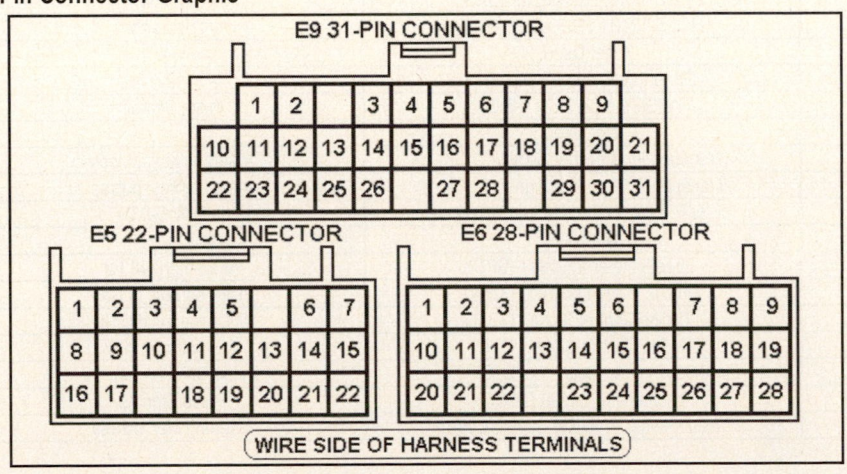

2001-03 Utility 3.0L V6 1MZ-FE VIN F (A/T) E5 22 Pin Connector

PCM Pin #	Wire Color	Circuit Description (22 Pin)	Value at Hot Idle
1	BK/WT	Battery Direct	12-14v
2	BK/OR	Ignition Switch Power (IGSW)	12-14v
3	GN/RD	Circuit Opening Relay (FC)	0-3v, off-idle: 12v
4	WT	SIL Signal (Scan Tool)	Digital Signal
5	GN/YL	EVAP Vapor Pressure Valve (VSV)	12v or 0v
6	BL/RD	Malfunction Indicator Lamp Control	MIL Off: 12v, On: 1v
7	BL	Starter Switch Signal	In P/N: 0-3.0v
8	PK/BL	EFI Main Relay Control	Relay Off: 0v, On: 12v
9	BK/RD	Overdrive "Off" LED Indicator	LED Off: 12v, On: 1v
10	PK	EVAP Canister Closed Valve (CCV)	12v or 0v
11	YL	Multiplex Meter Input (MPX1)	Digital Signals
12	GY/BL	Center Airbag Sensor Assembly (F/PS)	12v or 0v
13	GN/BK	ABS/Traction Signal (TRC+)	Pulse Signals
14	WT/BL	ABS/Traction Signal (ENG+)	Pulse Signals
15	GN/WT	Stop Light Switch Signal	Brake Off: 0v, On: 12v
16	WT	EFI Main Relay Power	12-14v
17	BL/BK	EVAP Vapor Pressure Sensor (PTNK)	2.9-3.7v (hose off)
18	BL/WT	Multiplex Meter Input (MPX2)	Digital Signals
19	---	Not Used	---
20	YL/BK	ABS/Traction Signal (TRC-)	Pulse Signals
21	WT/GN	ABS/Traction Signal (ENG-)	Pulse Signals
22	LG	Theft Deterrent Control (IMLD)	LED Off: 12v, On: 1v

2001-03 Utility 3.0L V6 1MZ-FE VIN F (A/T) E6 28 Pin Connector

PCM Pin #	Wire Color	Circuit Description (28 Pin)	Value at Hot Idle
1	WT/BK	Power Ground (EOM)	<0.1v
2	---	Not Used	---
3	GN/YL	EVAP Vapor Pressure Valve (VSV)	12v or 0v
4	---	Not Used	---
5	PK/BK	Data Link Connector Signal (D3)	12v
6	YL/GN	A/C Magnetic Clutch Relay (ACMG)	Relay Off: 0v, On: 12v
7	---	Not Used	---
8	BK	HO2S-12 (B1 S2) Signal (OXS)	0.1-1.1v
9	BL/RD	HO2S-12 (B1 S2) Heater (HTS)	1v (Heater on)
10-14	---	Not Used	---
15	YL	Generator Control Signal (RL)	12v
16	VT	ABS/Traction ECU (NEO)	Pulse Signals
17	---	Not Used	---
18	GY	Transponder Amplifier Signal (TXCK)	Inserting key: pulses
19	PK	Transponder Amplifier Signal (RXCK)	Inserting key: pulses
20	BL/WT	A/T: Neutral Start Switch (NSW)	In P/N: 0-3.0v
21	---	Not Used	---
22	WT/RD	Speedometer Indicator (SPD)	At 55 mph: 48 Hz
23	BL	Unlock Warning Switch (KSW)	Inserting key: pulses
24	WT	Cruise Control Signal (OD1)	At Cruise in OD: 12v
25	PK/BL	Cruise Control Signal (IDLO)	1.5v, off-idle: 12v
26	RD/BL	Cooling Fan Relay 1 or 2 (CF)	Relay Off: 12v, On: 1v
27	WT/GN	Tachometer Signal (TACO)	Pulse Signals
28	GY/BL	Transponder Amplifier Signal (Code)	Inserting key: pulses

Pin Connector Graphic

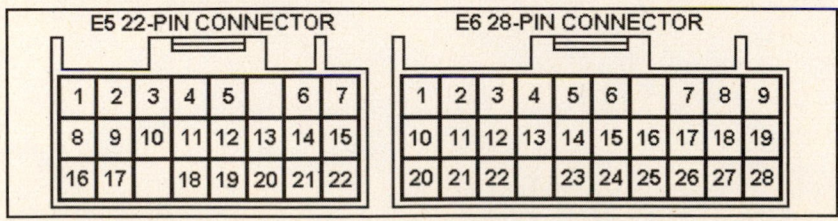

2001-03 Utility 3.0L V6 1MZ-FE VIN F (A/T) E7 17 Pin Connector

PCM Pin #	Wire Color	Circuit Description (17 Pin)	Value at Hot Idle
1	RD/YL	A/T-ECT Solenoid (S4)	Driving in OD: 12v
2	BR/RD	A/T ECT Solenoid (SLT-)	Pulse Signals
3	GN/YL	A/T ECT Solenoid (SLT+)	Pulse Signals
4	GN	Turbine Speed Shaft Signal (NT-)	AC pulse signals
5	PK	Turbine Speed Shaft Signal (NT+)	AC pulse signals
6	---	Not Used	---
7	RD/WT	A/T Select Switch Neutral	In 'N': 12v, Others: 0v
8	RD/BK	A/T Select Switch Reverse	In 'R': 12v, Others: 0v
9	RD/YL	A/T Select Switch Park	In Park: 12v, Others: 0v
10	---	Not Used	---
11	BK	Air Conditioning Signal (VSV)	Pulse Signals
12	---	Not Used	---
13	BL/RD	A/T Select Switch Low	In Low: 12v, Others: 0v
14	BL/YL	A/T Select Switch Second	In 2nd: 12v, Others: 0v
15	---	Not Used	---
16	GN/BK	A/T Select Switch Drive	In 'D': 12v, Others: 0v
17	YL/GN	Engine Oil Pressure Switch (MOPS)	Open: 12v, Closed: 0v

2001-03 Utility 3.0L V6 1MZ-FE VIN F (A/T) E8 24 Pin Connector

PCM Pin #	Wire Color	Circuit Description (24 Pin)	Value at Hot Idle
1	WT/BK	Sensor Ground (E04)	<0.050v
2	BL/RD	Sensor VREF (VC)	4.9-5.1v
3	BK/RD	AFS-11 (B1 S1) Heater (HAFR)	1v (Heater on)
4	BK	AFS-21 (B2 S1) Heater (HAFL)	1v (Heater on)
5	RD/BL	Injector 1 Control	2.0-3.3 ms
6	YL	Injector 2 Control	2.0-3.3 ms
7	BK/RD	EVAP Purge Solenoid (VSV)	12v or 0v
8	WT/BK	Power Ground (E05)	<0.1v
9	GN/YL	PSP Switch Signal (PSW)	Straight: 12v, Turning: 0v
10	YL/GN	MAF Sensor Signal (VG)	1.1-1.5v
11	BL	AFS-11 (B1 S1) signal (AFR+)	3.0-3.6v
12	GN	AFS-21 (B2 S1) Signal (AFL+)	3.0-3.6v
13	GN/RD	ATF Oil Temperature Signal	At 68ºF: 4-5v
14	WT/BL	ECT Sensor Signal (THW)	At 180ºF: 0.51v
15	WT/RD	ACIS Solenoid 2 Control (VSV)	12v or 0v
16	PK	CKP Sensor Signal (NE+)	AC pulse signals
17	BR	Power Ground (E1)	<0.1v
18	WT	Sensor Ground (E2)	<0.050v
19	RD/BK	MAF Sensor Ground (E2G)	<0.050v
20	YL	AFS-11 (B1 S1) Signal (AFR-)	Fixed at 3.3v
21	RD	AFS-21 (B2 S1) Signal (AFL-)	Fixed at 3.3v
22	RD/YL	IAT Sensor Signal (THA)	At 100ºF: 2.60v
23	BK/WT	TP Sensor Signal (VTA)	0.3-1.1v
24	BL	VVT Sensor (-) Ground	<0.050v

Pin Connector Graphic

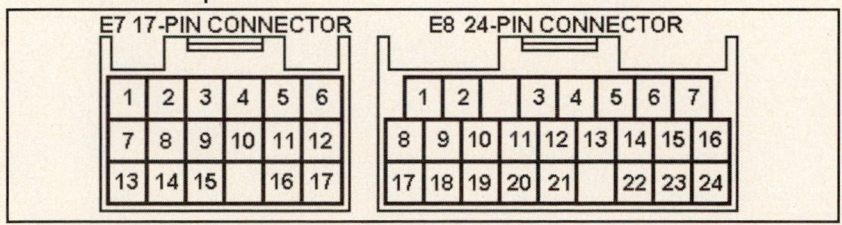

2001-03 Utility 3.0L V6 1MZ-FE VIN F (A/T) E9 31 Pin Connector

PCM Pin #	Wire Color	Circuit Description (31 Pin)	Value at Hot Idle
1	BK/WT	Injector 3 Control	2.0-3.3 ms
2	BL	Injector 4 Control	2.0-3.3 ms
3	BK/RD	Injector 5 Control	2.0-3.3 ms
4	GN	Injector 6 Control	2.0-3.3 ms
5	GN/YL	Cam Timing Oil Valve (OC1-)	Pulse signals
6	WT/GN	Cam Timing Oil Valve (OC1+)	Pulse signals
7	RD/BL	A/T-ECT Solenoid (DSL)	In Lockup: 12-14v
8	RD/WT	A/T ECT Solenoid (SL2-)	During shifting: 12v
9	RD/BK	A/T ECT Solenoid (SL2+)	During shifting: 12v
10	PK	VVT Sensor Left Hand (VV2-)	Pulse signals
11	BK/YL	Igniter Transistor 1 Control	7% duty cycle
12	YL/GN	Igniter Transistor 2 Control	7% duty cycle
13	BL/YL	Igniter Transistor 3 Control	7% duty cycle
14	BL/RD	Igniter Transistor 4 Control	7% duty cycle
15	YL	Igniter Transistor 5 Control	7% duty cycle
16	GN/RD	Igniter Transistor 3 Control	7% duty cycle
17	RD/YL	ACIS 1 Solenoid Control (VSV)	12v or 0v
18	YL/RD	Cam Timing Oil Valve (OC2-)	Pulse signals
19	BL/WT	A/T ECT Solenoid (SL1-)	Pulse Signals
20	BL/BK	A/T ECT Solenoid (SL1+)	Pulse Signals
21	WT/BK	Power Ground (E01)	<0.1v
22	RD	VVT Sensor Left Hand (VV2+)	Pulse signals
23	BL	Counter Gear Speed (NC-)	Pulse signals
24	WT/RD	Counter Gear Speed (NC+)	Pulse signals
25	WT	Igniter Signal (IGF)	Digital Signal: 0-5-0v
26	GN/BK	Idle Air Control Valve (RSO)	Pulse Signals
27	WT	Knock Sensor 1 Signal (KNKR)	<0.075v AC
28	BK	Knock Sensor 2 Signal (KNKL)	<0.075v AC
29	GN/WT	Cam Timing Oil Valve (OC2+)	Pulse signals
30	WT/BK	Power Ground (E03)	<0.1v
31	WT/BK	Power Ground (E02)	<0.1v

Pin Connector Graphic

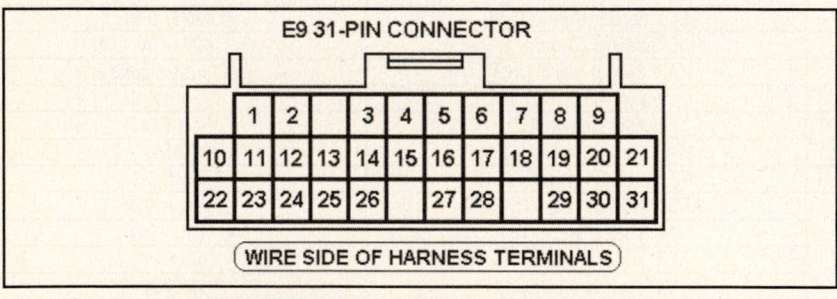

LAND CRUISER Pin Tables

1990 Utility 4.0L I6 MFI VIN F (A/T) 10 Pin Connector

PCM Pin #	Wire Color	Circuit Description (10 Pin)	Value at Hot Idle
1	BL	Check Connector	12-14v
2	BL/RD	Cold Start Injector Control	1v (at cold startup)
3	GY	HO2S-11 (B1 S1) Heater	1v (Heater on)
4	WT/BL	Injectors 1 & 2 & 3 Control	2.0-3.3 ms
5	BR	Power Ground	<0.1v
6	PK	Water Temperature Switch	Open: 12v, Closed: 0v
7	BR	Sensor Ground	<0.050v
8	OR	HO2S-12 (B1 S2) Heater	1v (Heater on)
9	YL	Injectors 4 & 5 & 6 Control	2.0-3.3 ms
10	BR	Power Ground	<0.1v

1990 Utility 4.0L I6 MFI VIN F (A/T) 18 Pin Connector

PCM Pin #	Wire Color	Circuit Description (18 Pin)	Value at Hot Idle
1	RD/WT	ECT Sensor Signal (THW)	At 180°F: 0.51v
2	BL/WT	HO2S-12 (B1 S2) Signal	0.1-1.1v
3 (Cal)	GN/YL	EGR Gas Temperature Sensor	3.5-4.0v
4	GN/WT	Distributor Signal (NE+)	AC pulse signals
6	GN/RD	Distributor Signal (G+)	AC pulse signals
7	GN/BK	Distributor Signal (G-)	AC pulse signals
8	RD/BK	ISC Signal (ISC2)	Pulse Signals
9	BL/BK	ISC Signal (ISC1)	Pulse Signals
10	BR	Sensor Ground	<0.050v
11	WT	HO2S-11 (B1 S1) Signal	0.1-1.1v
12	BR/BK	Sensor Ground	<0.050v
13	GN/WT	Closed Throttle Switch	1v, off-idle: 12v
14	WT/RD	TP Sensor Signal (VTA)	0.3-0.8v
15	BK/WT	Check Connector	12-14v
16	GN	Check Connector	12-14v
17	BL/YL	ISC Signal (ISC4)	Pulse Signals
18	YL	ISC Signal (ISC3)	Pulse Signals

Standard Colors and Abbreviations

Abbreviation	Color	Abbreviation	Color	Abbreviation	Color
BK	Black	GY	Gray	RD	Red
BL	Blue	GN	Green	TN	Tan
BR	Brown	LG	LT Green	VT	Violet
DB	Dark Blue	OR	Orange	WT	White
DG	DK Green	PK	Pink	YL	Yellow

1990 Utility 4.0L I6 MFI VIN F (A/T) 24 Pin Connector

PCM Pin #	Wire Color	Circuit Description (24 Pin)	Value at Hot Idle
1	BK/BL	Ignition Switch Power (IGSW)	12-14v
2	RD/YL	Direct Battery	12-14v
3	BL/RD	Sensor VREF	4.9-5.1v
4	GN/BL	Airflow Meter Signal	2-4v
5	BL/YL	IAT Sensor Signal (THA)	At 100°F: 2.60v
6	BL/RD	Fuel Pressure Up Solenoid	1v (at hot restart)
7	GN/BK	Vehicle Speed Sensor	At 55 mph: 48 Hz
8	BL/WT	EGR Solenoid Control (VSV)	12v or 0v
9	YL/RD	EFI Main Relay	12-14v
11	BK/GN	Igniter Signal (IGF)	Digital Signal: 0-5-0v
12	BK/GN	Igniter Signal (IGF)	Digital Signal: 0-5-0v
13	YL/RD	EFI Main Relay B1+	12-14v
14	YL/RD	EFI Main Relay B+	12-14v
16	BL	Air Injection Solenoid	12v or 0v
17	BK/RD	Starter Switch Signal	9-11v (cranking)
18	BK/WT	Neutral Start Switch	In P/N: 0-3.0v
19	YL/GN	Malfunction Indicator Lamp Control	MIL Off: 12v, On: 1v
20	BK/WT	A/C Magnetic Clutch Relay (ACMG)	Relay Off: 0v, On: 12v
21	GN/WT	Stop Light Switch Signal	Brake Off: 0v, On: 12v
22	YL/BK	4WD Indicator Light	Off: 12-14v, On: 0.1v
24	BR/BK	Sensor Ground	<0.050v

Pin Connector Graphic

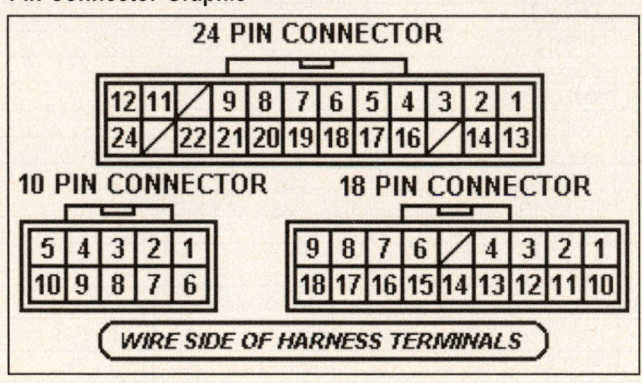

1991-92 Utility 4.0L I6 MFI VIN F (A/T) 12 Pin Connector

PCM Pin #	Wire Color	Circuit Description (12 Pin)	Value at Hot Idle
1	YL/RD	EFI Main Relay	12-14v
2	RD/YL	Direct Battery	12-14v
5	GN/WT	Stop Light Switch Signal	Brake Off: 0v, On: 12v
6	BK/BL	Ignition Switch Power (IGSW)	12-14v
7	YL/RD	EFI Main Relay B+	12-14v
8	YL/WT	Malfunction Indicator Lamp Control	MIL Off: 12v, On: 1v
10	BK/WT	A/C Magnetic Clutch Relay (ACMG)	Relay Off: 0v, On: 12v
11	BK/RD	Vehicle Speed Sensor	At 55 mph: 48 Hz
12	YL/BK	4WD Indicator Light	Off: 12-14v, On: 0.1v

1991-92 Utility 4.0L I6 MFI VIN F (A/T) 16 Pin Connector

PCM Pin #	Wire Color	Circuit Description (16 Pin)	Value at Hot Idle
1	GN/RD	Air Flow Meter VREF	4.9-5.1v
2	GN/BL	Airflow Meter Signal	2-4v
3	RD	EFI Main Relay	12-14v
4	RD/WT	ECT Sensor Signal (THW)	At 180ºF: 0.51v
5 (Cal)	WT	Sub HO2S Signal	0.1-1.1v
6	BL/WT	Main HO2S Signal	0.1-1.1v
7	BK/WT	Check Connector	12-14v
8	GN	Check Connector	12-14v
9	BR	Sensor Ground	<0.050v
10	GN/BK	TP Sensor Signal (VTA)	0.3-0.8v
11	BL/YL	IAT Sensor Signal (THA)	At 100ºF: 2.60v
12	GN/WT	Closed Throttle Switch	1v, off-idle: 12v
13 (Cal)	GN/YL	EGR Gas Temperature Sensor	3.5-4.0v
15	BL	Check Connector	12-14v
16	BR/BK	Sensor Ground	<0.050v

Standard Colors and Abbreviations

Abbreviation	Color	Abbreviation	Color	Abbreviation	Color
BK	Black	GY	Gray	RD	Red
BL	Blue	GN	Green	TN	Tan
BR	Brown	LG	LT Green	VT	Violet
DB	Dark Blue	OR	Orange	WT	White
DG	DK Green	PK	Pink	YL	Yellow

1991-92 Utility 4.0L I6 MFI VIN F (A/T) 26 Pin Connector

PCM Pin #	Wire Color	Circuit Description (26 Pin)	Value at Hot Idle
1	OR	Sub HO2S Heater Control	1v (Heater on)
2	BK/RD	Starter Switch Signal	In P/N: 0-3.0v
3	BK/GN	Igniter Signal (IGF)	Digital Signal: 0-5-0v
4	GN/WT	Distributor Signal (NE+)	AC pulse signals
5	BL/RD	Fuel Pressure Up Solenoid	1v (at hot restart)
7	BL/YL	ISC Signal (ISC4)	Pulse Signals
8	YL	ISC Signal (ISC3)	Pulse Signals
9	RD/BK	ISC Signal (ISC2)	Pulse Signals
10	BL/BK	ISC Signal (ISC1)	Pulse Signals
11	BL/RD	Cold Start Injector Control	1v (at cold startup)
12	WT/BL	Injectors 1 & 2 & 3 Control	2.0-3.3 ms
13	BR	Power Ground	<0.1v
14	GY	Main HO2S Heater Control	1v (Heater on)
15	BK/WT	Neutral Start Switch	In P/N: 0-3.0v
16	BL/WT	EGR Solenoid Control (VSV)	12v or 0v
17	BL	Distributor Signal (G-)	<0.050v
18	GN/RD	Distributor Signal (G+)	AC pulse signals
21	PK	Water Temperature Switch	Switch Closed: 0.1v
22	BK/GN	Igniter Signal (IGF)	Digital Signal: 0-5-0v
23	BL	Air Injection Solenoid	12-14v
24	BR/BK	Sensor Ground	<0.050v
25	YL	Injectors 4 & 5 & 6 Control	2.0-3.3 ms
26	BR	Power Ground	<0.1v

Pin Connector Graphic

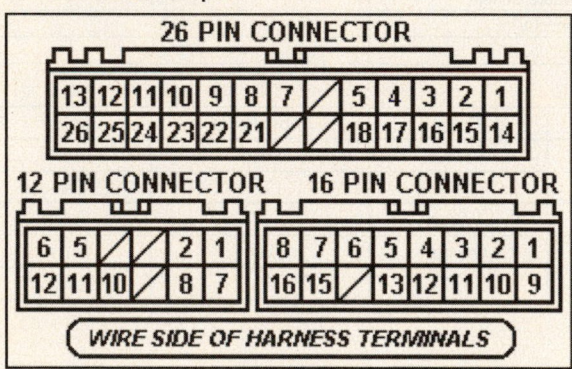

1993-94 Utility 4.5L I6 MFI VIN D (A/T-ECT) 16 Pin Connector

PCM Pin #	Wire Color	Circuit Description (16 Pin)	Value at Hot Idle
1	GN/RD	Air Flow Meter VREF	4.9-5.1v
2	GN/BL	Airflow Meter Signal	2-4v
3	BL/YL	IAT Sensor Signal (THA)	At 100°F: 2.60v
4	RD/WT	ECT Sensor Signal (THW)	At 180°F: 0.51v
5	BL/WT	HO2S-11 (B1 S1) Signal	0.1-1.1v
6	WT	Knock Sensor 1 Signal	<0.075v AC
7	BL	Data Link Connector	12-14v
8 (Cal)	GN/YL	EGR Gas Temperature Sensor	3.5-4.0v
9	BR/BK	Sensor Ground	<0.050v
10	GN/BK	TP Sensor Signal (VTA)	0.3-0.8v
11	GN/WT	Closed Throttle Switch	1v, off-idle: 12v
12	BK/YL	Data Link Connector	12-14v
13	WT	HO2S-12 (B1 S2) Signal	0.1-1.1v
14	BL/WT	Knock Sensor 2 Signal	<0.075v AC
15	BL/YL	Data Link Connector	12-14v
16	BR/BK	Sensor Ground	<0.050v

1993-94 Utility 4.5L I6 MFI VIN D (A/T-ECT) 22 Pin Connector

PCM Pin #	Wire Color	Circuit Description (22 Pin)	Value at Hot Idle
1	YL/RD	EFI Main Relay B1+	12-14v
2	RD/YL	Direct Battery	12-14v
3	RD	EFI Main Relay	12-14v
4	YL/WT	Malfunction Indicator Lamp Control	MIL Off: 12v, On: 1v
7	BK/WT	A/C Magnetic Clutch Relay (ACMG)	Relay Off: 0v, On: 12v
8	BK/RD	Vehicle Speed Sensor	At 55 mph: 48 Hz
9	BL	A/T-ECT TCM (VA)	Pulses
10	BL/YL	A/T-ECT TCM (NE)	Pulses
11	BK/RD	Starter Switch Signal	In P/N: 0-3.0v
12	YL/RD	EFI Main Relay B+	12-14v
13	BK/BL	Ignition Switch Power (IGSW)	12-14v
14	GN/WT	Stop Light Switch Signal	Brake Off: 0v, On: 12v
15	BL/WT	A/T-ECT TCM (ESA3)	4.5-5.5v
16	BL/RD	A/T-ECT TCM (ESA2)	4.5-5.5v
17	BL/GN	A/T-ECT TCM (ESA1)	4.5-5.5v
18	YL/GN	A/T-ECT TCM (ECT)	Pulses
19	BL/BK	A/T-ECT TCM (ECT2)	At 180°F: 0.51v
20	BL/OR	A/T-ECT TCM (ECT1)	KOEO: 12-14v
22	BK/WT	Neutral Start Switch	In P/N: 0-3.0v

Standard Colors and Abbreviations

Abbreviation	Color	Abbreviation	Color	Abbreviation	Color
BK	Black	GY	Gray	RD	Red
BL	Blue	GN	Green	TN	Tan
BR	Brown	LG	LT Green	VT	Violet
DB	Dark Blue	OR	Orange	WT	White
DG	DK Green	PK	Pink	YL	Yellow

1993-94 Utility 4.5L I6 MFI VIN D (A/T-ECT) 12 Pin Connector

PCM Pin #	Wire Color	Circuit Description (12 Pin)	Value at Hot Idle
1	OR	HO2S-11 (B1 S1) Heater	1v (Heater on)
2	GY	HO2S-12 (B1 S2) Heater	1v (Heater on)
6	BL	Distributor Signal (G-)	<0.050v
10	GN	Distributor Signal (G2+)	AC pulse signals
11	RD	Distributor Signal (G1+)	AC pulse signals
12	WT	Distributor Signal (NE+)	AC pulse signals

1993-94 Utility 4.5L I6 MFI VIN D (A/T-ECT) 26 Pin Connector

PCM Pin #	Wire Color	Circuit Description (26 Pin)	Value at Hot Idle
1	YL/BL	Injector 6 Control	2.0-3.3 ms
2	YL/RD	Injector 5 Control	2.0-3.3 ms
3	WT/RD	Fuel Pump Relay Control	Relay Off: 12v, On: 1v
4	BL/YL	ISC Signal (ISC4)	Pulse Signals
5	YL	ISC Signal (ISC3)	Pulse Signals
6	RD/BK	ISC Signal (ISC2)	Pulse Signals
7	BL/BK	ISC Signal (ISC1)	Pulse Signals
8	BL	Air Injection Solenoid	12-14v
19	YL/GN	EVAP Purge Solenoid	12-14v
10	YL	Injector 4 Control	2.0-3.3 ms
11	WT/RD	Injector 2 Control	2.0-3.3 ms
12	WT/BL	Injector 1 Control	2.0-3.3 ms
13	BR	Power Ground	<0.1v
15	OR	Data Link Connector	12-14v
16	BK/WT	Data Link Connector	12-14v
17	BK/GN	Igniter Signal (IGF)	Digital Signal: 0-5-0v
21	BL/RD	Fuel Pressure Up Solenoid	1v (at hot restart)
22	BL/WT	EGR Solenoid Control (VSV)	12v or 0v
23	BK/GN	Igniter Signal (IGF)	Digital Signal: 0-5-0v
24	BR/BK	Sensor Ground	<0.050v
25	WT/GN	Injector 3 Control	2.0-3.3 ms
26	BR	Power Ground	<0.1v

Pin Connector Graphic

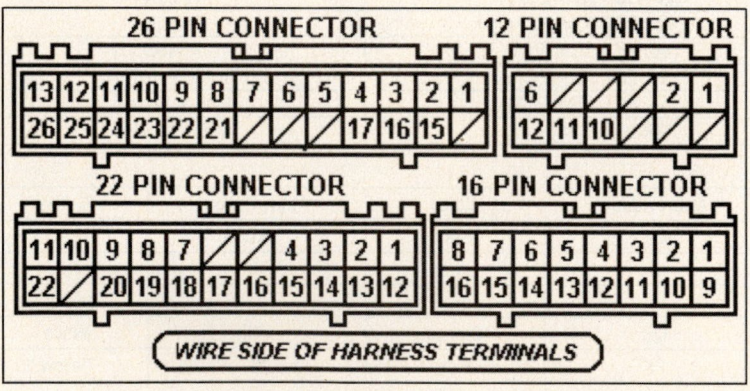

1995-97 Utility 4.5L I6 VIN D, VIN J (A/T-ECT) 12 Pin Connector

PCM Pin #	Wire Color	Circuit Description (12 Pin)	Value at Hot Idle
1	GY/OR	HO2S-11 (B1 S1) Heater	1v (Heater on)
2	PK	A/T Vehicle Speed Sensor (+)	Pulse Signals
4	BL	CKP Sensor Signal (NE+)	AC pulse signals
5	BR	CKP Sensor Signal (NE-)	<0.050v
6	BL	CMP Sensor Signal (G-)	<0.050v
7	GY/OR	HO2S-12 (B1 S2) Heater	1v (Heater on)
8	PK/GN	A/T Vehicle Speed Sensor (-)	Pulse Signals
10	GN	CMP Sensor Signal (G2+)	AC pulse signals
11	RD	CMP Sensor Signal (G1+)	AC pulse signals
12	WT	CMP Sensor Signal (G+)	AC pulse signals

1995-97 Utility 4.5L I6 VIN D, VIN J (A/T-ECT) 16 Pin Connector

PCM Pin #	Wire Color	Circuit Description (16 Pin)	Value at Hot Idle
1	RD/GN	TP Sensor VREF	4.9-5.1v
2	GN	MAF Sensor Signal	1.1-1.5v
3	BL/YL	IAT Sensor Signal (THA)	At 100°F: 2.60v
4	RD/WT	ECT Sensor Signal (THW)	At 180°F: 0.51v
5	PK/BL	HO2S-11 (B1 S1) Signal	0.1-1.1v
6	WT	Knock Sensor 1 Signal	<0.075v AC
7	GN/OR	Data Link Connector	12-14v
9	BR/BK	Sensor Ground	<0.050v
10	GN/BK	TP Sensor Signal (VTA)	0.3-0.8v
11	GN/WT	Closed Throttle Switch	1v, off-idle: 12v
12	RD/WT	A/T Oil Temperature Sensor	At 68°F: 4-5v
13	WT	HO2S-12 (B1 S2) Signal	0.1-1.1v
14	BK	Knock Sensor 2 Signal	<0.075v AC
16	BR/BK	Sensor Ground	<0.050v

Pin Connector Graphic

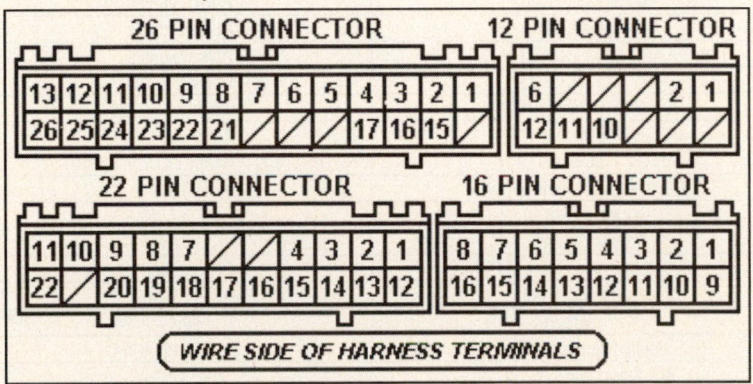

Standard Colors and Abbreviations

Abbreviation	Color	Abbreviation	Color	Abbreviation	Color
BK	Black	GY	Gray	RD	Red
BL	Blue	GN	Green	TN	Tan
BR	Brown	LG	LT Green	VT	Violet
DB	Dark Blue	OR	Orange	WT	White
DG	DK Green	PK	Pink	YL	Yellow

1995-97 Utility 4.5L I6 VIN D, VIN J (A/T-ECT) 22 Pin Connector

PCM Pin #	Wire Color	Circuit Description (22 Pin)	Value at Hot Idle
1	BK/BL	Ignition Switch Power (IGSW)	12-14v
2	RD/YL	Direct Battery	12-14v
3	RD	EFI Main Relay	12-14v
4	YL/RD	Malfunction Indicator Lamp Control	MIL Off: 12v, On: 1v
5	YL/BL	A/T Oil Temperature Lamp	Lamp Off: 12v
6	WT	Data Link Connector (SDL)	0v
7	BK/WT	A/C Magnetic Clutch Relay (ACMG)	Relay Off: 0v, On: 12v
8	BL/WT	Vehicle Speed Sensor	At 55 mph: 48 Hz
9	BK/BL	4WD Detection Transfer (L4)	Switch Closed: 12v
10	BK	Tachometer Signal (TACO)	Pulse Signals
11	BK/RD	Starter Switch Signal	In P/N: 0-3.0v
12	YL	EFI Main Relay B+	12-14v
13	WT/RD	Fuel Pump Relay	Relay Off: 12v, On: 1v
14	GN/WT	Stop Light Switch Signal	Brake Off: 0v, On: 12v
15	RD/BK	A/T Select Switch Reverse	In 'R': 12v, Others: 0v
16	OR	A/T Select Switch 2nd	In 2nd: 12v, Others: 0v
17	GN/WT	A/T Select Switch Low	In Low: 12v, Others: 0v
18	GN/OR	Cruise Control ECU	At Cruise in OD: 12v
19	PK/BL	Overdrive Main Switch	Switch Off: 12v, On: 1v
20	PK/BK	A/T Pattern Select Switch	Norm: 0v, PWR: 12v
21	YL/BL	4WD Detection Transfer (N)	Open: 12v, Closed: 0v
22	BK/WT	Neutral Start Switch	In P/N: 0-3.0v

1995-97 Utility 4.5L I6 VIN D, VIN J (A/T-ECT) 26 Pin Connector

PCM Pin #	Wire Color	Circuit Description (26 Pin)	Value at Hot Idle
1	YL/RD	Injector 5 Control	2.0-3.3 ms
2	YL	Injector 4 Control	2.0-3.3 ms
3	OR	A/T Pattern Select Switch	Norm: 0v, PWR: 12v
4	BL/YL	ISC Signal (ISC4)	Pulse Signals
5	GN/YL	ISC Signal (ISC3)	Pulse Signals
6	RD/BK	ISC Signal (ISC2)	Pulse Signals
7	RD/GN	ISC Signal (ISC1)	Pulse Signals
8	RD/BL	A/T-ECT Solenoid (SL)	In Lockup: 12-14v
9	RD/YL	A/T-ECT Solenoid (S2)	1st or OD: 1v
10	RD	A/T-ECT Solenoid (S1)	3rd or OD: 1v
11	WT/RD	Injector 2 Control	2.0-3.3 ms
12	WT/BL	Injector 1 Control	2.0-3.3 ms
13	BR	Power Ground	<0.1v
14	RD/WT	Circuit Opening Relay (FC)	0-3v, off-idle: 12v
15	YL/BL	Injector 6 Control	2.0-3.3 ms
16	BR	Power Ground	<0.1v
17	BK/YL	Igniter Signal (IGF)	Digital Signal: 0-5-0v
18	RD/WT	A/T Select Switch 2nd	2nd: 12-14v
19	GN/YL	EGR Gas Temperature Sensor	3.5-4.0v
21	BL/RD	Fuel Pressure Up Solenoid	1v (at hot restart)
22	BL/WT	EGR Solenoid Control (VSV)	12v or 0v
23	BK/GN	Igniter Signal (IGF)	Digital Signal: 0-5-0v
24	BR/BK	Sensor Ground	<0.050v
25	WT/GN	Injector 3 Control	2.0-3.3 ms
26	BR	Power Ground	<0.1v

Standard Colors and Abbreviations

Abbreviation	Color	Abbreviation	Color	Abbreviation	Color
BK	Black	GY	Gray	RD	Red
BL	Blue	GN	Green	TN	Tan
BR	Brown	LG	LT Green	VT	Violet
DB	Dark Blue	OR	Orange	WT	White
DG	DK Green	PK	Pink	YL	Yellow

1998 Utility WT/EIS 4.7L V8 VIN T (A/T-ECT) 17 Pin Connector

PCM Pin #	Wire Color	Circuit Description (17 Pin)	Value at Hot Idle
4	BL/RD	Direct Clutch Speed Input (+)	Pulse Signals
5	RD	A/T Vehicle Speed Sensor (+)	Pulse Signals
10	BL/WT	Direct Clutch Speed Input (-)	Pulse Signals
11	GN	A/T Vehicle Speed Sensor (-)	Pulse Signals
17	GN/YL	A/T Oil Temperature Sensor	At 68°F: 4-5v

1998 Utility WT/EIS 4.7L V8 VIN T (A/T-ECT) 28 Pin Connector

PCM Pin #	Wire Color	Circuit Description (28 Pin)	Value at Hot Idle
1	BK	Overdrive Main Switch	Switch Off: 12v, On: 1v
2	RD/BK	A/T Select Switch Reverse	In 'R': 12v, Others: 0v
3	GN	A/T Select Switch 2nd	In 2nd: 12v, Others: 0v
4	GN/BK	A/T Select Switch Low	In Low: 12v, Others: 0v
5	PK/RD	Data Link Connector	12-14v
6	GN/WT	Stop Light Switch Signal	Brake Off: 0v, On: 12v
7	RD/BK	HO2S-22 (B2 S2) Heater	1v (Heater on)
8	BL	HO2S-12 (B1 S2) Heater	1v (Heater on)
9	YL/BK	C/C Indicator Lamp	Lamp On: 0.1v
10	BL	EVAP Vapor Pressure Valve (VSV)	12v or 0v
11	BL/WT	A/T Pattern Select Switch	Norm: 0v, PWR: 12v
12	GN/WT	Idle-Up Signal	Load On: 1v, Off: 12v
13	BL/BK	A/C Amplifier Signal (ACT)	Clutch On: 12v, Off: 1.5v
14	YL/BK	A/C Amplifier Signal (THWO)	Pulse Signals
15	PK	Vehicle Speed Sensor	At 55 mph: 48 Hz
16	BK	Tachometer Signal (TACO)	Pulse Signals
17	BK/RD	Starter Switch Signal	In P/N: 0-3.0v
18	BK	HO2S-12 (B1 S2) Signal	0.1-1.1v
19	GN/YL	A/T Select Switch Drive	In Drive: 12-14v
20	BK/WT	Start Circuit Signal	Cranking: 9-11v
21	WT/BK	Sensor Ground	<0.050v
22	BL/BK	EVAP Vapor Pressure Sensor (PTNK)	2.5-3.1v (with cap off)
25	WT/GN	A/C Amplifier Signal (AC1)	Clutch On: 1.5v, Off: 12v
26	OR	A/T Oil Temperature Lamp	Lamp Off: 12v, On: 1v
27	WT	HO2S-22 (B2 S2) Signal	0.1-1.1v

Pin Connector Graphic

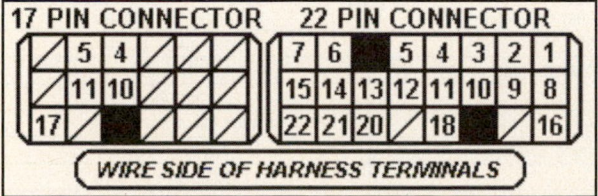

1998 Utility WT/EIS 4.7L V8 VIN T (A/T-ECT) 24 Pin Connector

PCM Pin #	Wire Color	Circuit Description (24 Pin)	Value at Hot Idle
1	WT/BK	Power Ground	<0.1v
2	BL/RD	Sensor VREF	4.9-5.1v
3	YL	HO2S-21 (B2 S1) Heater	1v (Heater on)
4	RD	HO2S-11 (B1 S1) Heater	1v (Heater on)
5	YL	Injector 1 Control	2.0-3.3 ms
6	BK	Injector 2 Control	2.0-3.3 ms
7	BL/BK	EVAP Purge Solenoid (VSV)	12v or 0v
8	RD	ABS ECU Signal	Pulse Signals
9	RD/BK	Accelerator Position Sensor 2	1.8-2.7v
10	BL/YL	MAF Sensor Signal (VG)	1.1-1.5v
11	WT	HO2S-21 (B2 S1) Signal	0.1-1.1v
12	BK	HO2S-11 (B1 S1) Signal	0.1-1.1v
13	RD/YL	TP Sensor Signal (VTA1)	0.4-1.0v
14	GN/BK	ECT Sensor Signal (THW)	At 180ºF: 0.51v
17	BR	Sensor Ground	<0.050v
18	BR/WT	Sensor Ground	<0.050v
19	GN/WT	MAF Sensor Ground (EVG)	<0.050v
20	YL/BK	TP Sensor Signal (VTA2)	2.0-2.9v
21	RD	Accelerator Position Sensor 1	0.25-0.90v
22	YL/BK	IAT Sensor Signal (THA)	At 100ºF: 2.60v

1998 Utility WT/EIS 4.7L V8 VIN T (A/T-ECT) 22 Pin Connector

PCM Pin #	Wire Color	Circuit Description (22 Pin)	Value at Hot Idle
1	RD/YL	Direct Battery	12-14v
2	YL	A/T Pattern Select Switch	2nd: 12-14v
3	BL/RD	A/T Select Switch 2nd	In 2nd: 12v, Others: 0v
4	GN/RD	Fuel Pump Signal (DI)	Pulse Signals
5	GN/WT	Fuel Control Switch Signal	Open: 12v, Closed: 0v
6	WT	Malfunction Indicator Lamp Control	MIL Off: 12v, On: 1v
7	YL/BK	BM (+) Signal	12-14v
8	BK/YL	EFI Main Relay B1+	12-14v
9	BK/RD	Ignition Switch Power (IGSW)	12-14v
10	BK/WT	EFI Main Relay	12-14v
11	PK/WT	SIL Signal (Scan Tool)	12v
12	BL/BK	Transponder Amplifier Signal (Code)	Inserting key: pulses
13	PK/GN	Transponder Amplifier Signal (RXCK)	Inserting key: pulses
14	RD/YL	Transponder Amplifier Signal (RXCK)	Inserting key: pulses
15	YL/GN	A/T Select Switch Park	In P/N: 0-3.0v
16	BK/YL	EFI Main Relay B+	12-14v
18	YL/RD	A/C Amplifier On Signal	Clutch On: 1.5v, Off: 12v
20	RD/BK	Unlock Warning Switch	No Key: 4-5v
21	GN/RD	LED Signal	LED Off: 12v, On: 1v
22	BK/BL	4WD Detection Transfer (L4)	Open: 12v, Closed: 0v

1998 Utility WT/EIS 4.7L V8 VIN T (A/T-ECT) 31 Pin Connector

PCM Pin #	Wire Color	Circuit Description (31 Pin)	Value at Hot Idle
1	BL	Injector 3 Control	2.0-3.3 ms
2	RD	Injector 4 Control	2.0-3.3 ms
3	GN	Injector 5 Control	2.0-3.3 ms
4	RD/BL	Injector 6 Control	2.0-3.3 ms
5	WT	Injector 7 Control	2.0-3.3 ms
6	BK/WT	Injector 8 Control	2.0-3.3 ms
7	WT	Throttle Control Motor (M-)	Pulse Signals
8	RD	Throttle Control Motor (M+)	Pulse Signals
9	WT/BK	Power Ground	<0.1v
10	RD	CMP Sensor Signal (G+)	AC pulse signals
11	BK	Igniter Transistor 1 Control	7% duty cycle
12	RD	Igniter Transistor 2 Control	7% duty cycle
13	BL	Igniter Transistor 3 Control	7% duty cycle
14	GN	Igniter Transistor 4 Control	7% duty cycle
15	YL	Igniter Transistor 5 Control	7% duty cycle
16	BK/YL	Igniter Transistor 6 Control	7% duty cycle
17	WT	Knock Sensor 2 Signal	<0.075v AC
18	BK	Knock Sensor 1 Signal	<0.075v AC
21	WT/BK	Power Ground	<0.1v
22	GN	CMP/CKP Sensor Signal (-)	<0.050v
23	BL	CKP Sensor Signal (NE+)	AC pulse signals
24	BL	Throttle Control Motor (CL-)	Pulse Signals
25	BK/BL	Igniter Transistor 7 Control	7% duty cycle
26	BL/BK	Igniter Transistor 8 Control	7% duty cycle
27	BK/WT	Igniter Signal (IGF1)	Digital Signal: 0-5-0v
28	BK/RD	Igniter Signal (IGF2)	Digital Signal: 0-5-0v
29	GN	Throttle Control Motor (CL+)	Pulse Signals
30	BR	Shield Ground	<0.050v
31	WT/BK	Power Ground	<0.1v

Pin Connector Graphic

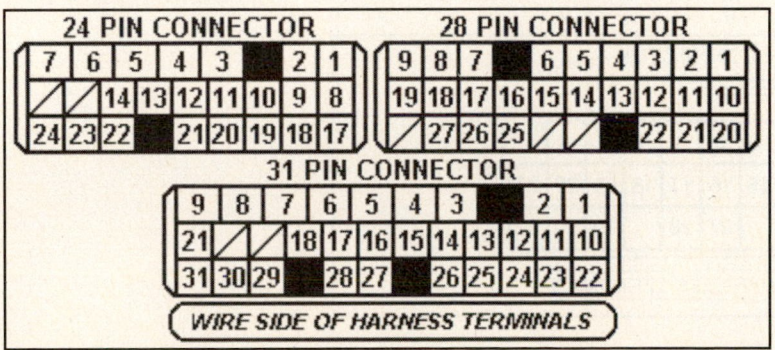

1999-2002 Utility AWD 4.7L 2UZ-FE V8 VIN T E5 31P Connector

PCM Pin #	Wire Color	Circuit Description (31 Pin)	Value at Hot Idle
1	BL	Injector 3 Control	2.0-3.3 ms
2	RD	Injector 4 Control	2.0-3.3 ms
3	GN	Injector 5 Control	2.0-3.3 ms
4	RD/BL	Injector 6 Control	2.0-3.3 ms
5	WT	Injector 7 Control	2.0-3.3 ms
6	BK/WT	Injector 8 Control	2.0-3.3 ms
7	WT	Throttle Control Motor (M-)	Pulse Signals
8	RD	Throttle Control Motor (M+)	Pulse Signals
9	WT/BK	Power Ground (ME01)	<0.1v
10	RD	CMP Sensor Signal (G+)	AC pulse signals
11	BK	Igniter Transistor 1 Control	7% duty cycle
12	RD	Igniter Transistor 2 Control	7% duty cycle
13	BL	Igniter Transistor 3 Control	7% duty cycle
14	GN	Igniter Transistor 4 Control	7% duty cycle
15	YL	Igniter Transistor 5 Control	7% duty cycle
16	BK/YL	Igniter Transistor 6 Control	7% duty cycle
17	WT	Knock Sensor 2 Signal (right)	<0.075v AC
18	BK	Knock Sensor 1 Signal (left)	<0.075v AC
19-20	---	Not Used	---
21	WT/BK	Power Ground (E01)	<0.1v
22	GN	CMP/CKP Sensor Signal (-)	<0.050v
23	BL	CKP Sensor Signal (NE+)	AC pulse signals
24	BL	Throttle Control Motor (CL-)	Pulse Signals
25	BK/BL	Igniter Transistor 7 Control	7% duty cycle
26	BL/BK	Igniter Transistor 8 Control	7% duty cycle
27	BK/WT	Igniter Signal (IGF1)	Digital Signal: 0-5-0v
28	BK/RD	Igniter Signal (IGF2)	Digital Signal: 0-5-0v
29	GN	Throttle Control Motor (CL+)	Pulse Signals
30	BR	Shield Ground (GE01)	<0.050v
31	WT/BK	Power Ground (E02)	<0.1v

Pin Connector Graphic

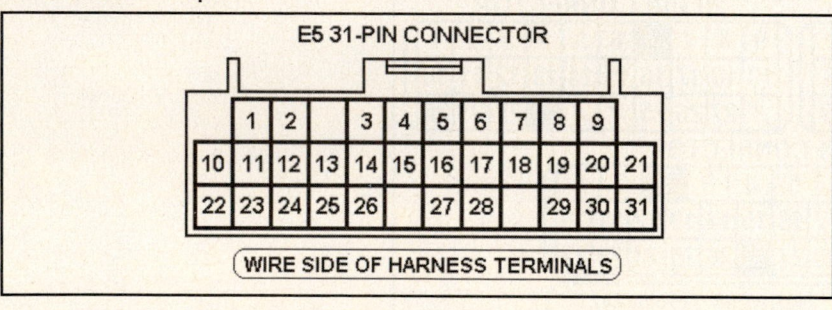

E5 31-PIN CONNECTOR

WIRE SIDE OF HARNESS TERMINALS

1999-2002 Utility AWD 4.7L 2UZ-FE V8 VIN T E6 24P Connector

PCM Pin #	Wire Color	Circuit Description (24 Pin)	Value at Hot Idle
1	WT/BK	Power Ground (E03)	<0.1v
2	BL/RD	Sensor VREF (VC)	4.9-5.1v
3	YL	HO2S-21 (B2 S1) Heater	1v (Heater on)
4	RD	HO2S-11 (B1 S1) Heater	1v (Heater on)
5	YL	Injector 1 Control	2.0-3.3 ms
6	BK	Injector 2 Control	2.0-3.3 ms
7	BL/BK	EVAP Purge Solenoid (VSV)	12v or 0v
8	RD	ABS ECU Signal	Digital Signals
9	RD/BK	Accelerator Position Sensor 2	1.8-2.7v
10	BL/YL	Mass Airflow Sensor (VG)	1.1-1.5v
11	WT	HO2S-21 (B2 S1) Signal	0.1-1.1v
12	BK	HO2S-11 (B1 S1) Signal	0.1-1.1v
13	RD/YL	TP Sensor Signal (VTA)	0.4-1.0v
14	GN/BK	ECT Sensor Signal (THW)	At 180ºF: 0.51v
15-16	---	Not Used	---
17	BR	Power Ground (E1)	<0.1v
18	BR/WT	Sensor Ground (E2)	<0.050v
19	GN/WT	MAF Sensor Ground (EVG)	<0.050v
20	YL/BK	TP Sensor Signal (VTA2)	2.0-2.9v
21	RD	Accelerator Position Sensor 1	0.3-0.90v
22	YL/BK	IAT Sensor Signal (THA)	At 100ºF: 2.60v
23 (01'-02')	RD/BK	HO2S-22 (B2 S2) Heater	1v (Heater on)
24 (01'-02')	BL	HO2S-12 (B1 S2) Heater	1v (Heater on)

1999-2002 Utility AWD 4.7L V8 2UZ-FE VIN T E7 17P Connector

PCM Pin #	Wire Color	Circuit Description (17 Pin)	Value at Hot Idle
1	RD	A/T ECT Solenoid (S1)	In 3rd or OD: 1v
2	WT	A/T ECT Solenoid (S2)	In 1st or OD: 1v
3	GN	A/T ECT Solenoid (SL)	In Lockup: 12-14v
4	BL/RD	Direct Clutch Speed Input (+)	AC pulse signals
5	RD	A/T Vehicle Speed Sensor (+)	AC pulse signals
6-8	---	Not Used	---
9	GN/WT	A/T-ECT Solenoid (SLT+)	Pulse Signals
10	BL/WT	Direct Clutch Speed Input (-)	AC pulse signals
11	GN	A/T Vehicle Speed Sensor (-)	Pulse Signals
12-14	---	Not Used	---
15	GNBK	A/T-ECT Solenoid (SLT-)	Pulse Signals
16	---	Not Used	---
17	GN/YL	A/T Oil Temperature Sensor	At 68ºF: 4-5v

Pin Connector Graphic

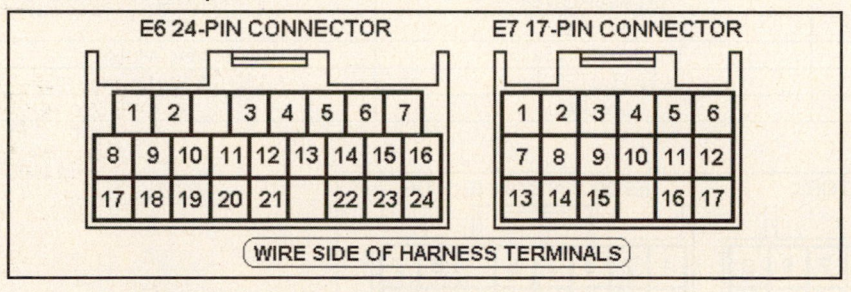

1999-2002 Utility AWD 4.7L 2UZ-FE V8 VIN T E8 28P Connector

PCM Pin #	Wire Color	Circuit Description (28 Pin)	Value at Hot Idle
5	PK/RD	Data Link Connector (TE1)	12-14v
6	GN/WT	Stop Light Switch Signal	Brake Off: 0v, On: 12v
7	RD/BK	HO2S-22 (B2 S2) Heater	1v (Heater on)
8	BL	HO2S-12 (B1 S2) Heater	1v (Heater on)
9	---	Not Used	---
10	BL	EVAP Vapor Pressure Valve (VSV)	12v or 0v
11	---	Not Used	---
12	GN/WT	Defogger & Tail Light Switch	Switch Off: 0v, On: 12v
13	BL/BK	A/C Amplifier Signal (ACT)	Clutch On: 12v, Off: 1.5v
14	YL/BK	A/C Amplifier Signal (THWO)	A/C Off: 12v, On: 1v
15	VT	Vehicle Speed Sensor	At 55 mph: 48 Hz
16	BK	Tachometer Signal (TACO)	Pulse Signals
17	BK/RD	Starter Switch Signal	9-11v (cranking)
18	BK	HO2S-12 (B1 S2) Signal	0.1-1.1v
19	---	Not Used	---
20	BK/WT	Neutral Start Switch (NSW)	In P/N: 0-3.0v
21	WT/BK	Power Ground (EOM)	<0.1v
22	BL/BK	EVAP Vapor Pressure Sensor (PTNK)	2.5-3.1v (with cap off)
23-24	---	Not Used	---
25	WT/GN	A/C Amplifier Signal (AC1)	Clutch On: 1.5v, Off: 12v
26	---	Not Used	---
27	WT	HO2S-22 (B2 S2) Signal	0.1-1.1v
28	---	Not Used	---

1999-2002 Utility AWD 4.7L 2UZ-FE V8 VIN T E9 22P Connector

PCM Pin #	Wire Color	Circuit Description (22 Pin)	Value at Hot Idle
1	RD/YL	Direct Battery	12-14v
2-3	---	Not Used	---
4	GN/RD	Fuel Pump Signal (DI)	12-14v
5	GN/WT	Fuel Control Switch Signal	Closed: 0v, Open: 12v
6	WT	Malfunction Indicator Lamp Control	MIL Off: 12v, On: 1v
7	YL/BK	BM (+) Power	12-14v
8	BK/YL	EFI Main Relay B1+	12-14v
9	BK/RD	Ignition Switch Power (IGSW)	12-14v
10	BK/WT	EFI Main Relay Control	Relay Off: 0v, On: 12v
11	PK/WT	SIL Signal (Scan Tool)	12v
12	BL/BK	Transponder Amplifier Signal (Code)	Inserting key: pulses
13	VT/GN	Transponder Amplifier Signal (RXCK)	Inserting key: pulses
14	RD/YL	Transponder Amplifier Signal (RXCK)	Inserting key: pulses
15	---	Not Used	---
16	BK/YL	EFI Main Relay B+	12-14v
18-19	---	Not Used	---
20	RD/BK	Unlock Warning Switch	Key In: 1.5v, Out: 4-5v
21	GN/RD	Center ECU LED Signal	LED Off: 12v, On: 1v
22	---	Not Used	---

Pin Connector Graphic

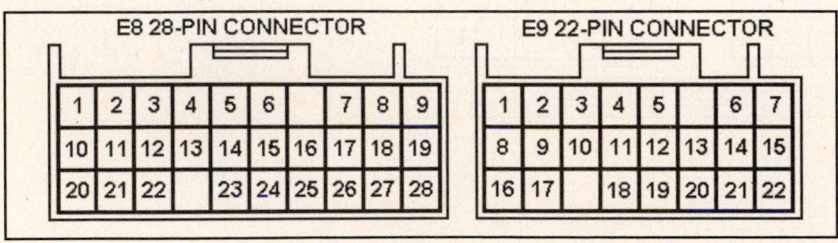

2003 Utility AWD 4.7L 2UZ-FE V8 VIN T E5 34 Pin Connector

PCM Pin #	Wire Color	Circuit Description (34 Pin)	Value at Hot Idle
1	YL	Injector 1 Control	2.0-3.3 ms
2	BK	Injector 2 Control	2.0-3.3 ms
3	BL	Injector 3 Control	2.0-3.3 ms
4	RD	Injector 4 Control	2.0-3.3 ms
5	GN	Injector 5 Control	2.0-3.3 ms
6	WT/BK	Power Ground (E02)	<0.1v
7	WT/BK	Power Ground (E01)	<0.1v
8	RD	Igniter Transistor 2 Control	7% duty cycle
9	BK	Igniter Transistor 1 Control	7% duty cycle
10	BL/BK	Igniter Transistor 8 Control	7% duty cycle
11	GN	Igniter Transistor 4 Control	7% duty cycle
12	YL	Igniter Transistor 5 Control	7% duty cycle
13	BK/BL	Igniter Transistor 7 Control	7% duty cycle
15	RD/GN	A/C Relay Control (ACCR)	Relay Off: 12v, On: 1v
16	BK/WT	Neutral Start Switch (NSW)	In P/N: 0-3.0v
17	BL/RD	Starter Switch Signal (STA)	In P/N: 0-3.0v
18	BL/RD	Sensor VREF (VC)	4.9-5.1v
19	GN/BK	ECT Sensor Signal (THW)	At 180°F: 0.51v
20	YL/BK	IAT Sensor Signal (THA)	At 100°F: 2.60v
21	RD/YL	TP Sensor Signal (VTA1)	0.4-1.0v
22, 32	---	Not Used	---
23	BK/RD	Igniter Signal (IGF2)	Digital Signal: 0-5-0v
24	BK/WT	Igniter Signal (IGF1)	Digital Signal: 0-5-0v
25	BL	Igniter Transistor 3 Control	7% duty cycle
26	BK/YL	Igniter Transistor 6 Control	7% duty cycle
27	BL/RD	EVAP Canister Closed Valve (CCV)	12v or 0v
28	BR/WT	Sensor Ground (E2)	<0.050v
29	GN/WT	MAF Sensor Ground (E2G)	<0.050v
30	BL/YL	Mass Airflow Sensor (VG)	1.1-1.5v
31	YL/BK	TP Sensor Signal (VTA2)	2.0-2.9v
33	GN/WT	Fuel Pump Relay Control (FPR)	Relay Off: 12v, On: 1v
34	BL/BK	EVAP Purge Solenoid (VSV)	12v or 0v

2003 Utility AWD 4.7L 2UZ-FE V8 VIN T E6 35 Pin Connector

PCM Pin #	Wire Color	Circuit Description (35 Pin)	Value at Hot Idle
1	BK	Knock Sensor 1 Signal (KNK1 - left)	<0.075v AC
2	WT	Knock Sensor 2 Signal (KNK2 - right)	<0.075v AC
3	RD/BL	Injector 6 Control	2.0-3.3 ms
4	RD	HO2S-11 (B1 S1) Heater (HT1A)	1v (Heater on)
5	BL	HO2S-12 (B1 S2) Heater (HT1B)	1v (Heater on)
7-8, 20	---	Not Used	---
9	BK/WT	Start Signal (STAR)	In P/N: 0-3.0v
10	WT	ECT Solenoid Control (S2)	12v or 0v
11	RD	ECT Solenoid Control (S1)	12v or 0v
12	GN/BK	ECT Solenoid Control (SLT-)	12v or 0v
13	GN/WT	ECT Solenoid Control (SLT+)	12v or 0v
16	PK/BK	A/T Solenoid Control (SL2-)	Pulse Signals
17	PK/BK	A/T Solenoid Control (SL2+)	Pulse Signals
18	RD/WT	A/T Solenoid Control (SL1-)	Pulse Signals
19	RD/BL	A/T Solenoid Control (SL1+)	Pulse Signals
21	WT	HO2S-22 (B2 S2) Signal (OX2B)	0.1-1.1v
22	WT	HO2S-21 (B2 S1) Signal (OX2A)	0.1-1.1v
23	BK	HO2S-11 (B1 S1) Signal (OX1A)	0.1-1.1v
24	BL	A/T Oil Temperature Sensor 2 (THO2)	At 68°F: 4-5v
25	RD/BK	HO2S-22 (B2 S2) Heater (HT2B)	1v (Heater on)
26	RD	ECT Vehicle Speed Sensor (SP2+)	Pulse Signals
27	BL	ECT Turbine Speed Sensor (NT-)	Pulse Signals
28	---	Not Used	---
29	BK	HO2S-12 (B1 S2) Signal (OX1B)	0.1-1.1v
30-31	---	Not Used	---
32	GN/YL	A/T Oil Temperature Sensor 1 (THO1)	At 68°F: 4-5v
33	YL	HO2S-21 (B2 S1) Heater (HT2A)	1v (Heater on)
34	GN	ECT Vehicle Speed Sensor (SP2-)	Pulse Signals
35	WT	ECT Turbine Speed Sensor (NT+)	Pulse Signals

2003 Utility AWD 4.7L 2UZ-FE V8 VIN T E7 32 Pin Connector

PCM Pin #	Wire Color	Circuit Description (32 Pin)	Value at Hot Idle
1	BR	Power Ground (E1)	<0.1v
2	WT	Throttle Control Motor (M-)	Pulse Signals
3	RD	Throttle Control Motor (M+)	Pulse Signals
4	WT/BK	Power Ground (ME01)	<0.1v
5	BK/WT	Injector 8 Control	2.0-3.3 ms
6	WT	Injector 7 Control	2.0-3.3 ms
7	WT/BK	Power Ground (E03)	<0.1v
11	YL/GN	Neutral Detection Switch (L4)	Switch Open: 0v, Closed: 12v
12	BK/WT	Start Switch Signal (STSW)	Cranking: 9-11v
13-14, 18-20	---	Not Used	---
15	BK	A/T Solenoid Control (SLU-)	Pulse Signals
16	PK/GN	A/T Solenoid Control (SLU+)	Pulse Signals
17	BR	Shield Ground (GE01)	<0.050v
21	BK/OR	Generator Control (RL)	12v
22, 26	---	Not Used	---
23	BL	A/C Lock Sensor (LCK)	12v or 0v
24	GN	CKP Sensor Signal (NE-)	<0.050v
25	BL	CKP Sensor Signal (NE+)	AC pulse signals
27	RD	CMP Sensor Signal (G2+)	AC pulse signals
28-31	---	Not Used	---
32	GN	CMP Sensor Signal (G2-)	AC pulse signals

2003 Utility AWD 4.7L V8 2UZ-FE VIN T E8 35 Pin Connector

PCM Pin #	Wire Color	Circuit Description (35 Pin)	Value at Hot Idle
1	WT/BK	Power Ground (HP)	<0.1v
2, 7-8	---	Not Used	---
3	GN	A/T Select Switch 2nd Signal (2L)	In 2nd: 12v, Others: 0v
4	BK/BL	Low Detection Switch (L4)	Switch Open: 0v, Closed: 12v
5	BL/WT	ECT Pattern Switch 2nd Position (SNW1)	2nd Position: 12v
6	YL/BK	ETCS Power (+BM)	12-14v
9	GN/BK	Shift Lock ECU Control (D)	12v or 0v
10	GN/YL	Shift Lock ECU Control (D)	12v or 0v
11	RD/BK	A/T Select Switch Reverse Signal	In 'R': 1v, Others: 12v
12, 15-16	---	Not Used	---
13	YL/RD	A/C Magnetic Clutch Relay (ACMG)	Relay Off: 0v, On: 12v
14	YL/GN	A/C Amplifier Signal (THWO)	A/C Off: 12v, On: 1v
17	VT	Vehicle Speed Sensor (SPD)	At 55 mph: 48 Hz
18	PK/BK	Body Control ECU Signal (MPX1)	Digital Signals
19	GN/WT	Stop Light Switch Signal	Brake Off: 0v, On: 12v
20-22, 25	---	Not Used	---
23	GN/RD	Shift Lock ECU Control (4)	12v or 0v
26	YL	Transponder Amplifier Signal (IMD)	Inserting key: pulses
27	WT	Transponder Amplifier Signal (IMI)	Inserting key: pulses
28	BL/WT	ECT Pattern Switch Power Signal	Power Position: 12v
29	BK	Body Control ECU Signal (MPX2)	Digital Signals
30	---	Not Used	---
31	BL/BK	A/C Amplifier Signal (ACT)	Relay Off: 12v, On: 1v
32	PK/BK	A/C Amplifier Signal (THE)	A/C Off: 12v, On: 1v
33	GN/BK	A/C Switch Signal (ACLD)	A/C On: 12v, Off: 0v
34-35	---	Not Used	---

Pin Connector Graphic

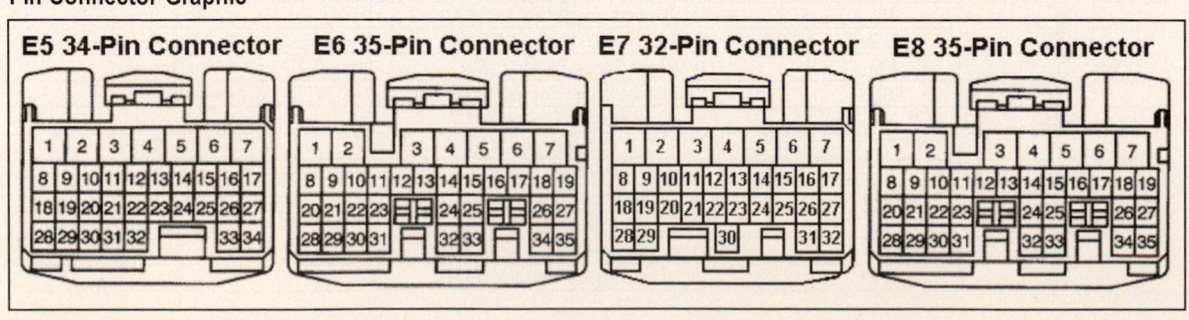

2003 Utility AWD 4.7L 2UZ-FE V8 VIN T E9 31 Pin Connector

PCM Pin #	Wire Color	Circuit Description (31 Pin)	Value at Hot Idle
1	BK/YL	EFI Main Relay Power (EFI Fuse)	12-14v
2	BK/YL	EFI Main Relay Power (EFI Fuse)	12-14v
3	BK/RD	Direct Battery	12-14v
4	BL	EVAP Pressure Switching Valve (VSV)	12v or 0v
5	BK	Tachometer Signal (TACO)	Pulse Signals
6	GN/WT	A/T Select Switch Park Signal (P)	In 'P': 12v, Others: 0v
7	GN/RD	A/T Select Switch Neutral Signal (N)	In 'N': 12v, Others: 0v
8	BK/WT	EFI Main Relay Control	Relay Off: 0v, On: 12v
9	BK/RD	Ignition Switch Power (IGSW)	12-14v
10	BK/WT	Circuit Opening Relay (FC)	0-3v, off-idle: 12v
11	WT	Malfunction Indicator Lamp Control	MIL Off: 12v, On: 1v
12	GN/WT	Body Control ECU	Digital Signals
13	YL/RD	Horn Relay Control	Relay Off: 12v, On: 1v
14	BK	Center Airbag Assembly (F/PS)	Digital Signals
15	WT/BK	Power Ground (EOM)	<0.1v
16	---	Not Used	---
17	WT	Transponder Amplifier Signal (NEO)	Inserting key: pulses
18	VT/WT	SIL Signal (Scan Tool)	Transmitting: pulses
19	WT/RD	Data Link Connector (WFSE)	N/A
20	PK/BK	Data Link Connector (TC)	12v
21	BL/BK	EVAP Vapor Pressure Sensor (PTNK)	2.5-3.1v (with cap off)
22	RD	Accelerator Position Sensor 1 (VPA)	0.3-0.90v
23	RD/BK	Accelerator Position Sensor 2 (VPA2)	1.8-2.7v
24	RD	Traction Control Engine Signal (ENG+)	Pulse Signals
25	YL	Traction Control Signal (TRC+)	Pulse Signals
26	BL/RD	Accelerator Pedal Position Sensor 1 VREF	4.9-5.1v
27	WT	Accelerator Pedal Position Sensor 2 VREF	4.9-5.1v
28	BR/WT	Accelerator Pedal Position Sensor Ground	<0.050v
29	WT/RD	Accelerator Position Sensor 2 (EPA2)	1.8-2.7v
30	GN	Traction Control Engine Signal (ENG-)	Pulse Signals
31	BL	Traction Control Signal (TRC-)	Pulse Signals

Pin Connector Graphic

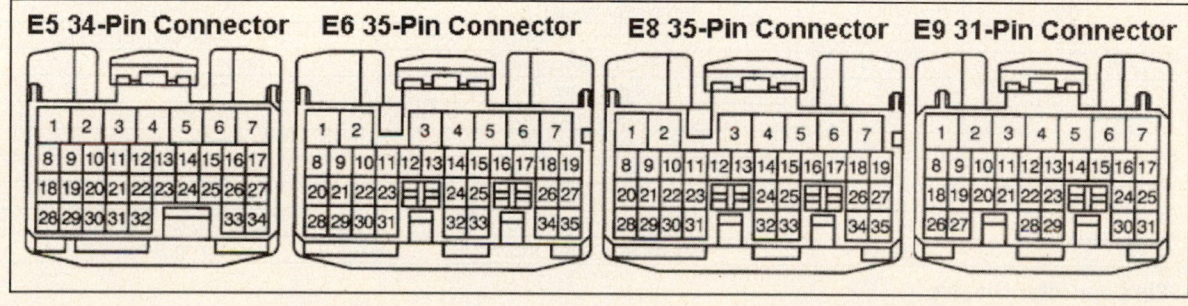

E5 34-Pin Connector   E6 35-Pin Connector   E8 35-Pin Connector   E9 31-Pin Connector

RAV4 Pin Tables

1996-98 Utility 2.0L I4 MFI VIN P (All) 16 Pin Connector

PCM Pin #	Wire Color	Circuit Description (16 Pin)	Value at Hot Idle
1	YL	Sensor VREF (VC)	4.9-5.1v
2	GN/BK	MAP Sensor Signal	Idling: 1-1.5v
3	YL/BK	IAT Sensor Signal (THA)	At 100ºF: 2.60v
4	WT	ECT Sensor Signal (THW)	At 180ºF: 0.51v
5	RD	HO2S-12 (B1 S2) Signal	0.1-1.1v
6	BK	HO2S-11 (B1 S1) Signal	0.1-1.1v
6 (98' - Cal)	WT	AFS-11 (B1 S1) Signal (+)	Fixed at 3.3v
6 (98' - Fed)	WT	HO2S-11 (B1 S1) Signal	0.1-1.1v
7	BL/YL	EVAP Vapor Pressure Sensor (PTNK)	2.5-3.1v (with cap off)
8	RD/WT	EVAP Vapor Pressure Valve (VSV)	12v or 0v
9	BR	Sensor Ground	<0.050v
10	BK	Knock Sensor Signal	<0.075v AC
11	BL/RD	TP Sensor Signal (VTA)	0.3-0.8v
12	RD/BL	4WD Speed Sensor (RR-)	Pulse Signals
13	BL/YL	4WD Speed Sensor (RR+)	Pulse Signals
14 (98' - Cal)	BK	AFS-11 (B1 S1) Signal (-)	Fixed at 3.0v
15	BR	Sensor Ground	<0.050v
16	YL/BK	A/T ECT Solenoid (SL)	In Lockup: 12-14v

1996-98 Utility 2.0L I4 MFI VIN P (All) 22 Pin Connector

PCM Pin #	Wire Color	Circuit Description (22 Pin)	Value at Hot Idle
1	RD/WT	Direct Battery	12-14v
2	BL/RD	4WD ECT Solenoid (SLD-)	Pulse Signals
3	BL/BK	4WD ECT Solenoid (SLD+)	Pulse Signals
4	GN/WT	Stop Light Switch Signal	Brake Off: 0v, On: 12v
5	GN/RD	Malfunction Indicator Lamp Control	MIL Off: 12v, On: 1v
6	BL/WT	Data Link Connector	12-14v
7	LG	Overdrive Main Switch	Switch Off: 12v, On: 1v
8	RD/BK	A/T Select Switch Reverse	In 'R': 12v, Others: 0v
9	PK/WT	Vehicle Speed Sensor	At 55 mph: 48 Hz
10	YL/GN	A/C Amplifier On Signal	Clutch On: 1.5v, Off: 12v
11	BK	Starter Switch Signal	In P/N: 0-3.0v
12	BK/WT	EFI Main Relay B+	12-14v
13	BK	Tachometer Signal (TACO)	Pulse Signals
14	GN/RD	Circuit Opening Relay (FC)	0-3v, off-idle: 12v
15	RD/GN	A/T Pattern Select Switch	Norm: 0v, PWR: 12v
16	WT	SIL Signal (Scan Tool)	0v
17	BK	Integration Relay	Load On: 12-14v
18	GN/BK	A/T Select Switch 2nd	In 2nd: 12v, Others: 0v
19	LG	A/T Select Switch Low	In Low: 12v, Others: 0v
20	YL/BK	Cruise Control ECU	At Cruise in OD: 12v
21	RD/YL	A/C Amplifier Signal (ACT)	Clutch On: 12v, Off: 1.5v
22	BK/WT	Neutral Start Switch (NSW)	Cranking: 9-11v

Pin Connector Graphic

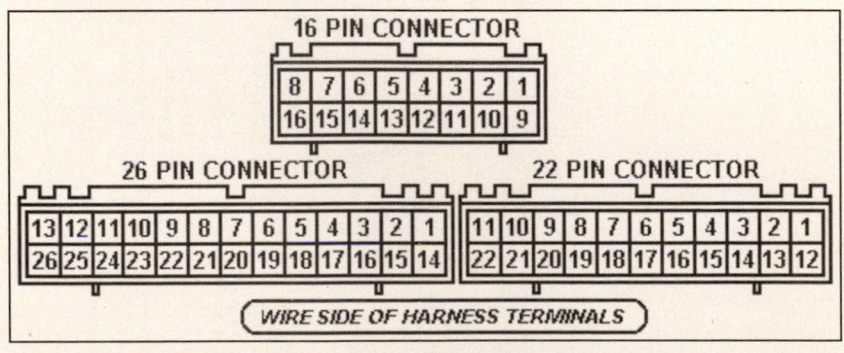

1996-98 Utility 2.0L I4 MFI VIN P (All) 26 Pin Connector

PCM Pin #	Wire Color	Circuit Description (26 Pin)	Value at Hot Idle
1	BR	Sensor Ground	<0.050v
2 (Cal)	RD	AFRS-11 (B1 S1) Heater	1v (Heater on)
2 (Fed)	RD	HO2S-11 (B1 S1) Heater	1v (Heater on)
3	BL/YL	Igniter Signal (IGF)	Digital Signal: 0-5-0v
4	RD	CKP Sensor Signal (NE+)	AC pulse signals
5	BK	CMP Sensor Signal (G+)	AC pulse signals
6	BL	Cruise Control ECU (IDLO)	1.5v, off-idle: 12v
7	WT/RD	4WD Speed Sensor (FR-)	Pulse Signals
8	YL/RD	4WD Speed Sensor (FR+)	Pulse Signals
9	BK/YL	ISC Signal (ISCC)	Pulse Signals
10	BK/BL	ISC Signal (ISCO)	Pulse Signals
11	BK	Injector 2 Control	2.0-3.3 ms
12	BK/RD	Injector 1 Control	2.0-3.3 ms
13	BR	Power Ground	<0.1v
14	BR	Sensor Ground	<0.050v
15	PK	A/T-ECT Solenoid (S2)	1st or OD: 1v
16	LG	A/T-ECT Solenoid (ST)	Pulse Signals
17	PK/BK	A/T-ECT Solenoid (S1)	In 3rd or OD: 1v
18	WT	CKP/CMP Sensor Signal (-)	<0.050v
19	RD/WT	HO2S-12 (B1 S2) Heater	1v (Heater on)
20	BK	Igniter Transistor 1 Control	6°, at 55 mph: 8° dwell
21	BK	Igniter Transistor 2 Control	6°, at 55 mph: 8° dwell
22	PK	EVAP Purge Solenoid (VSV)	12v or 0v
23	BK/WT	EGR Solenoid Control (VSV)	12v or 0v
24	BK/BL	Injector 4 Control	2.0-3.3 ms
25	BK/YL	Injector 3 Control	2.0-3.3 ms
26	BR	Power Ground	<0.1v

Pin Connector Graphic

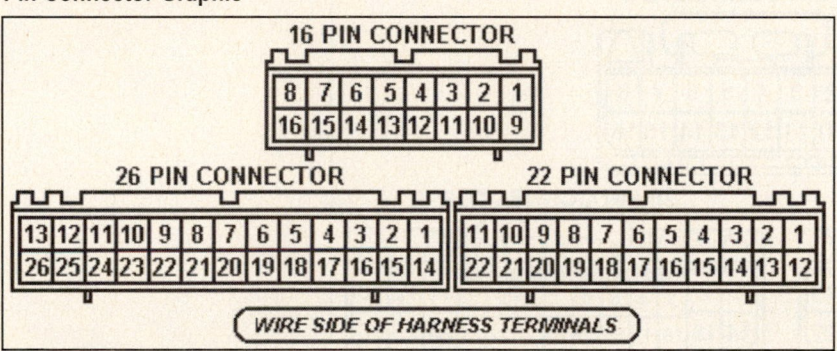

1999-2000 Utility 2.0L I4 MFI VIN P (All) E4 26P Connector

PCM Pin #	Wire Color	Circuit Description (26 Pin)	Value at Hot Idle
1	BR	Sensor Ground	<0.050v
2 (Cal)	RD	AFRS-11 (B1 S1) Heater	1v (Heater on)
2 (Fed)	RD	HO2S-11 (B1 S1) Heater	1v (Heater on)
3	BL/YL	Igniter Signal (IGF)	Digital Signal: 0-5-0v
4	RD	CKP Sensor Signal (NE+)	AC pulse signals
5	BK	CMP Sensor Signal (G+)	AC pulse signals
6	BL	Cruise Control ECU (IDLO)	1.5v, off-idle: 12v
7-8	---	Not Used	---
9	BK/YL	ISC Signal (ISCC)	Pulse Signals
10	BK/BL	ISC Signal (ISCO)	Pulse Signals
11	BK	Injector 2 Control	2.0-3.3 ms
12	BK/RD	Injector 1 Control	2.0-3.3 ms
13	BR	Power Ground	<0.1v
14	BR	Sensor Ground	<0.050v
15-17	---	Not Used	---
18	WT	CKP/CMP Sensor Signal (-)	<0.050v
19	RD/WT	HO2S-12 (B1 S2) Heater	1v (Heater on)
20	BK	Igniter Transistor 1 Control	6°, at 55 mph: 8° dwell
21	BK	Igniter Transistor 2 Control	6°, at 55 mph: 8° dwell
22	PK	EVAP Purge Solenoid (VSV)	12v or 0v
23	BK/WT	EGR Solenoid Control (VSV)	12v or 0v
24	BK/BL	Injector 4 Control	2.0-3.3 ms
25	BK/YL	Injector 3 Control	2.0-3.3 ms
26	BR	Power Ground	<0.1v

Pin Connector Graphic

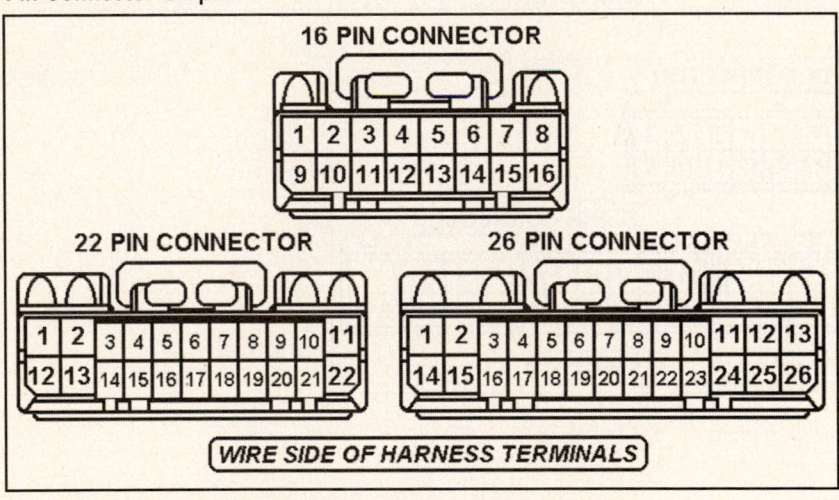

16 PIN CONNECTOR

| 1 | 2 | 3 | 4 | 5 | 6 | 7 | 8 |
| 9 | 10 | 11 | 12 | 13 | 14 | 15 | 16 |

22 PIN CONNECTOR

26 PIN CONNECTOR

WIRE SIDE OF HARNESS TERMINALS

1999-2000 Utility 2.0L I4 MFI VIN P (All) E6 16P Connector

PCM Pin #	Wire Color	Circuit Description (16 Pin)	Value at Hot Idle
1	YL	Sensor VREF	4.9-5.1v
2	GN/BK	MAP Sensor Signal	Idling: 1-1.5v
3	YL/BK	IAT Sensor Signal (THA)	At 100ºF: 2.60v
4	WT	ECT Sensor Signal (THW)	At 180ºF: 0.51v
5	RD	HO2S-12 (B1 S2) Signal	0.1-1.1v
6 (Cal)	WT	AFS-11 (B1 S1) Signal (+)	Fixed at 3.3v
6 (Fed)	WT	HO2S-11 (B1 S1) Signal	0.1-1.1v
7	BL/YL	EVAP Vapor Pressure Sensor (PTNK)	2.5-3.1v (with cap off)
8	RD/WT	EVAP Vapor Pressure Valve (VSV)	12v or 0v
9	BR	Sensor Ground	<0.050v
10	BK	Knock Sensor Signal	<0.075v AC
11	BL/RD	TP Sensor Signal (VTA)	0.3-0.8v
12-13	---	Not Used	---
14 (Cal)	BK	AFS-11 (B1 S1) Signal (-)	Fixed at 3.0v
15	BR	Sensor Ground	<0.050v
16	---	Not Used	---

1999-2000 Utility 2.0L I4 MFI VIN P (All) E6 22P Connector

PCM Pin #	Wire Color	Circuit Description (22 Pin)	Value at Hot Idle
1	RD/WT	Direct Battery	12-14v
2-4	---	Not Used	---
5	GN/RD	Malfunction Indicator Lamp Control	MIL Off: 12v, On: 1v
6	BL/WT	Data Link Connector	12-14v
7-8	---	Not Used	---
9	PK/WT	Vehicle Speed Sensor	At 55 mph: 48 Hz
10	YL/GN	A/C Amplifier Signal (AC1)	Clutch On: 1.5v, Off: 12v
11	BK	Starter Switch Signal	In P/N: 0-3.0v
12	BK/WT	EFI Main Relay B+	12-14v
13	BK	Tachometer Signal (TACO)	Pulse Signals
14	GN/RD	Circuit Opening Relay (FC)	0-3v, off-idle: 12v
15, 18-19	---	Not Used	---
16	WT	SIL Signal (Scan Tool)	12v
17	BK	Integration Relay Signal	Load On: 12-14v
20	YL/BK	Cruise Control Signal (OD1)	At Cruise in OD: 12v
21	RD/YL	A/C Amplifier Signal (ACT)	Clutch On: 12v, Off: 1.5v
22	BK/WT	Start Circuit Signal	Cranking: 9-11v

Pin Connector Graphic

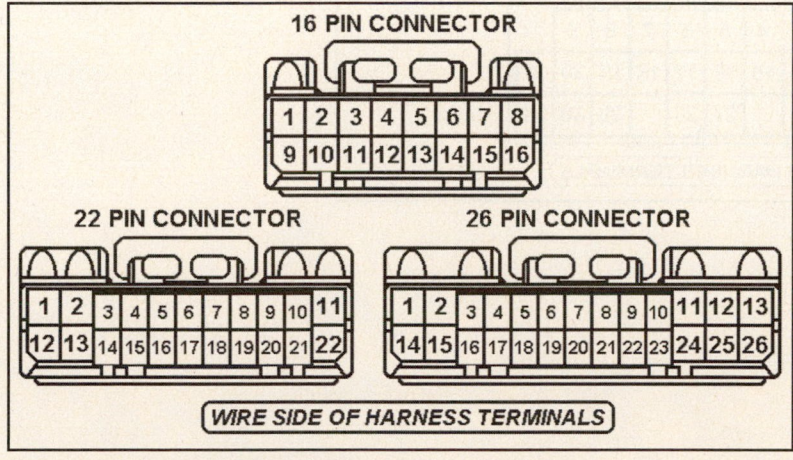

Standard Colors and Abbreviations

Abbreviation	Color	Abbreviation	Color	Abbreviation	Color
BK	Black	GY	Gray	RD	Red
BL	Blue	GN	Green	TN	Tan
BR	Brown	LG	LT Green	VT	Violet
DB	Dark Blue	OR	Orange	WT	White
DG	DK Green	PK	Pink	YL	Yellow

2001-03 Utility 2.0L I4 (All) VIN K E4 31 Pin Connector

PCM Pin #	Wire Color	Circuit Description (31 Pin)	Value at Hot Idle
1	BK/RD	Injector 1 Control	2.0-3.3 ms
2	BK	Injector 2 Control	2.0-3.3 ms
3	WT	Injector 3 Control	2.0-3.3 ms
4	RD	Injector 4 Control	2.0-3.3 ms
5	YL/GN	A/T Solenoid (SLT-)	Pulse Signals
6	YL/RD	A/T Solenoid (SLT+)	Pulse Signals
7	GN/WT	A/T Solenoid (SL1+)	Pulse Signals
8	PK	A/T: Solenoid (SL2+)	Pulse Signals
9	VT	A/T Solenoid (SL1-)	Pulse Signals
10	GN/BK	Igniter Transistor 1 Control	6% duty cycle
11	BL/BK	Igniter Transistor 2 Control	6% duty cycle
12	GN	Igniter Transistor 3 Control	6% duty cycle
13	BK	Igniter Transistor 4 Control	6% duty cycle
14	RD	Counter Gear Speed Input (+)	AC pulse signals
15	B	Turbine Speed Sensor Input -	AC pulse signals
16	WT	Turbine Speed Sensor Input +	AC pulse signals
17	GY	A/T: Trans. Oil Temp. Sensor	At 230ºF: 0.95v
18	BK/BL	Idle Air Control Valve (RSD)	Pulse Signals
19	WT/RD	Cam Timing Oil Valve (OCV+)	Pulse signals
20	YL/BK	A/T Solenoid (SL2-)	Pulse Signals
21	WT/BK	Power Ground (E01)	<0.1v
22	WT	ECT Sensor Signal (THW)	At 180ºF: 0.51v
23	RD/WT	IAT Sensor Signal (THA)	At 100ºF: 2.60v
24	BL/RD	TP Sensor Signal (VTA)	0.3-0.1.0v
25	RD	Igniter Signal (IGF)	Digital Signal: 0-5-0v
26	GN	Counter Gear Speed Input (-)	AC pulse signals
27	WT	Knock Sensor Signal (KNK1)	<0.075v AC
28	---	Not Used	---
29	WT	Cam Timing Oil Valve (OCV-)	Pulse signals
30	WT/BK	Power Ground (E03)	<0.1v
31	WT/BK	Power Ground (E02)	<0.1v

Pin Connector Graphic

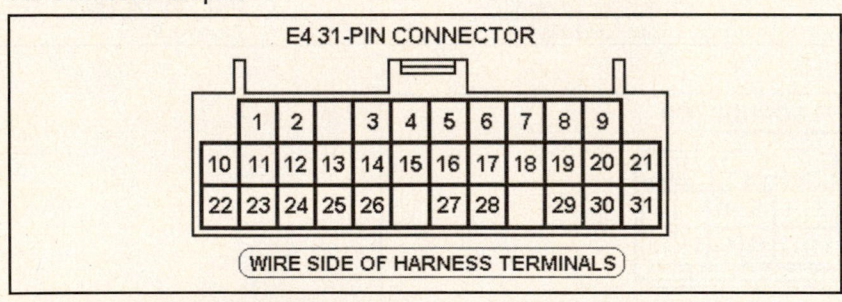

E4 31-PIN CONNECTOR

WIRE SIDE OF HARNESS TERMINALS

2001-03 4D Utility 2.0L I4 VIN H (All) E5 24 Pin Connector

PCM Pin #	Wire Color	Circuit Description (24 Pin)	Value at Hot Idle
1	WT/BK	Power Ground (E04)	<0.1v
2	YL	Sensor VREF (VC)	4.9-5.1v
3	RD/BK	HO2S-11 (B1 S2) Heater (HT1B)	Heater Off: 0v, On: 12v
4	WT/RD	AFS-21 (B2 S1) Heater (HAF2A)	Heater Off: 0v, On: 12v
5	WT/BL	AFS-11 (B1 S1) Heater (HAF1A)	Heater Off: 0v, On: 12v
6	PK	EVAP Purge Solenoid (VSV)	12v or 0v
7	---	Not Used	---
8	WT/BK	Power Ground (E05)	<0.1v
9-10	---	Not Used	---
11	BK	HO2S-11 (B1 S2) Signal (OX1B)	0.1-1.1v
12	BL/WT	MAF Sensor Signal (VG)	1.0-1.5v
13	WT	AFS-21 (B2 S1) Sensor (AF2A+)	Fixed at 3.3v
14	GN	AFS-11 (B1 S1) Sensor (AF1A+)	Fixed at 3.3v
15	YL	CMP Sensor Signal (G22+)	AC pulse signals
16	RD	CKP Sensor Signal (NE+)	AC pulse signals
17	BR	Power Ground (E1)	<0.1v
18	BR	Sensor Ground (E2)	<0.050v
19-20	---	Not Used	---
21	BL	MAF Sensor Ground (E2G)	<0.050v
22	BK	AFS-21 (B2 S1) Sensor (AF2A-)	Fixed at 3.0v
23	RD	AFS-11 (B1 S1) Sensor (AF1A-)	Fixed at 3.0v
24	PK	CKP/CMP Sensor Ground (-)	<0.050v

2001-03 4D Utility 2.0L I4 VIN H (All) E6 17 Pin Connector

PCM Pin #	Wire Color	Circuit Description (17 Pin)	Value at Hot Idle
1	YL	Sensor VREF	4.9-5.1v
2	YL/GN	EVAP Canister Closed Valve (VSV)	12v or 0v
3	GN/OR	EVAP Vapor Pressure Valve (VSV)	12v or 0v
4-5	---	Not Used	---
6	RD/WT	A/T-ECT Solenoid (DSL)	In Lockup: 12-14v
7-10	---	Not Used	---
11	PK	PSP Switch Signal (PSW)	Straight: 12v, Turning: 0v
12	YL	A/T Solenoid (S4)	Pulse Signals
13	---	Not Used	---
14	RD/YL	HO2S-22 (B2 S2) Heater (HT2B)	Heater Off: 12v, On: 1v
15	WT	HO2S-22 (B2 S2) Signal (OX2B)	0.1-1.1v
16	---	Not Used	---
17	LG/BK	Fan Relay Control (FAN)	Relay Off: 12v, On: 1v

Pin Connector Graphic

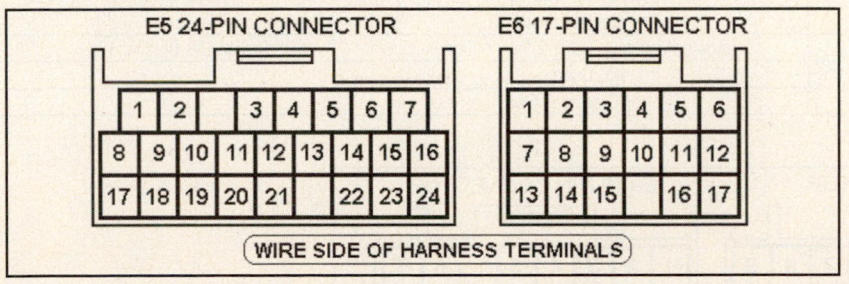

2001-03 4D Utility 2.0L I4 VIN H (All) E7 28 Pin Connector

PCM Pin #	Wire Color	Circuit Description (28 Pin)	Value at Hot Idle
1	BL/BK	A/T Select Switch Drive	In Drive: 12-14v
2	RD/BK	A/T Select Switch Reverse	In 'R': 12v, Others: 0v
3	LG/BK	A/T Select Switch 2nd	In 2nd: 12v, Others: 0v
4	YL/GN	A/C Amplifier (AC1) Signal	Relay Off: 1.5v, On: 12v
5	PK/BK	Data Link Connector (TC)	12v
6	---	Not Used	---
7	YL/BK	Cruise Control Signal (OD1)	At Cruise in OD: 12v
8-11	---	Not Used	---
12	LG	A/T Select Switch Low	In Low: 12v, Others: 0v
13	---	Not Used	---
14	GN	Engine Hot Indicator Control	Indicator Off: 12v, On: 1v
15	RD/YL	A/C Amplifier Signal (ACMG)	Relay Off: 12v, On: 1v
16	PK	Oil Pressure Lamp Control	Lamp Off: 12v, On: 1v
17-19	---	Not Used	---
20	BK/WT	Neutral Start Switch Signal (NSW)	In P/N: 0v, Others: 12v
21	BL/BK	EVAP Vapor Pressure Sensor (PTNK)	2.5-3.1v (with cap off)
22	VT/WT	Vehicle Speed Sensor	At 55 mph: 48 Hz
23-24	---	Not Used	---
25	BL	Cruise Control Signal (IDLO)	1.5v, off-idle: 12v
26	---	Not Used	---
27	BK	Tachometer Signal (TACO)	DC pulse signals
28	---	Not Used	---

2001-03 4D Utility 2.0L I4 VIN H (All) E8 22 Pin Connector

PCM Pin #	Wire Color	Circuit Description (22 Pin)	Value at Hot Idle
1	RD/WT	Direct Battery	12-14v
2	BK/OR	Ignition Switch Power (IGSW)	12-14v
3	GN/RD	Circuit Opening Relay (FC)	Relay Off: 12v, On: 1v
4	WT	SIL Signal (Scan Tool)	12v
5	LG/BK	Overdrive On Lamp Control	Lamp Off: 12v, On: 1v
6	GN/RD	Malfunction Indicator Lamp Control	MIL Off: 12v, On: 1v
7	BK/RD	Starter Switch Signal (STA)	Cranking: 9-11v
8	GN/WT	EFI Main Relay Control	Relay Off: 0v, On: 12v
9	---	Not Used	---
10	LG	Overdrive Main Switch	Switch Off: 12v, On: 1v
11	BL	Center Airbag Assembly (F/PS)	Digital Signals
12-13	---	Not Used	---
14	GY	Combination Meter Hot Light	Light Off: 12v, On: 1v
15	GN/WT	Stop Light Switch Signal	Brake Off: 0v, On: 12v
16	BK/RD	Circuit Opening Relay (FC)	12-14v
17	PK	Data Link Connector (WFSE)	12v
18	PK	Defroster Switch & Taillight Switch Signal	Switch Off: 0v, On: 12v
19-21	---	Not Used	---
22	BR	Power Ground (EOM)	<0.1v

Pin Connector Graphic

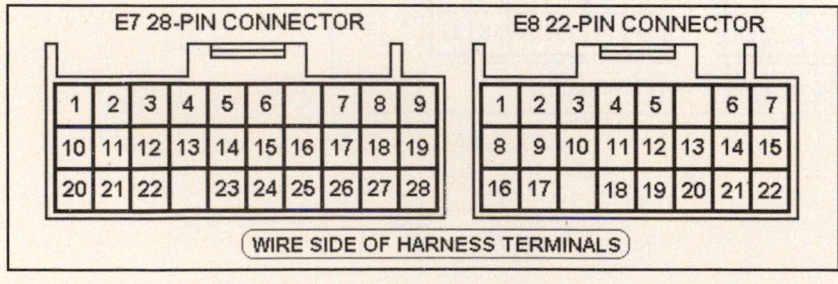

SEQUOIA Pin Tables

2001-02 Utility 4.7L V8 VIN T (All) E4 22 Pin Connector

PCM Pin #	Wire Color	Circuit Description (22 Pin)	Value at Hot Idle
1	BR/RD	Direct Battery	12-14v
2	PK	Neutral Start Switch (NSW)	In P/N: 0-3.0v
3	PK/BK	TC Signal to D6 DLC3	12v
4	VT	Fuel Pump Relay Control	Relay Off: 12v, On: 1v
5	BK/WT	EFI Main Relay Control	Relay Off: 0v, On: 12v
6	GN/OR	Fuel Control Switch Signal	Closed: 0v, Open: 12v
7	WT/GN	BM (+) Power	12-14v
8	BK/YL	EFI Main Relay B1+	12-14v
9	BK/RD	Ignition Switch Power (IGSW)	12-14v
10	OR	Power Ground (EOM)	<0.1v
12	YL	Transponder Amplifier Signal (RXCK)	Inserting key: pulses
13	LG	Transponder Amplifier Signal (RXCK)s	Inserting key: pulses
14	RD/YL	Transponder Amplifier Signal (Code)	Inserting key: pulses
15	GN/YL	Stop Light Switch Signal	Brake Off: 0v, On: 12v
16	BK/YL	EFI Main Relay B1+	12-14v
17	GN/RD	SIL Signal (Scan Tool)	Transmitting: pulses
18	BL	Center Airbag Sensor Assembly (F/PS)	Digital Signals
19	WT/RD	Data Link Connector (WFSE)	N/A
20	YL/BK	A/C Amplifier Signal (THWO)	A/C Off: 12v, On: 1v
21	PK/BL	A/C Amplifier Signal (AC1)	Clutch On: 1.5v, Off: 12v
22	LG/BK	A/C Amplifier Signal (ACT)	Clutch On: 12v, Off: 1.5v

2001-02 Utility 4.7L V8 VIN T (A/T) E5 28 Pin Connector

PCM Pin #	Wire Color	Circuit Description (28 Pin)	Value at Hot Idle
1	LG	A/T Select Switch Low	In 'L': 1v, Others: 12v
2	BL	A/T Select Switch 2nd	In 2nd: 12v, Others: 0v
3	BK/YL	A/T Select Switch Reverse	In 'R': 12v, Others: 0v
4	YL/GN	Tachometer Signal (TACO)	Pulse Signals
5	GN/OR	VSS Signal (Combo Meter)	At 55 mph: 48 Hz
6	GN	Detection Switch Transfer	Open: 12v, Closed: 1v
7	BK	Starter Switch Signal	In P/N: 0-3.0v
8	BL/YL	4WD ECU C/C Signal	12v
9	BL/RD	4WD Detection Transfer (L4)	Open: 12v, Closed: 0v
12	GN/WT	Defogger & Tail Light Switch	Switch Off: 0v, On: 12v
15	BL/WT	OD Main Switch ODMS Input	Open: 12v, Closed: 1v
17	YL	ABS/BA/TRAC/VSC ECU	Digital Signals (NEO)
18	RD	TRAC Engine (ENG-) Signal	Pulse Signals
19	GN	TRAC TRC (TRC-) Signal	Pulse Signals
20	WT/BL	EVAP Canister Closed Valve (CCV)	12v or 0v
21	PK/BL	EVAP Vapor Pressure Valve (VSV)	12v or 0v
22	RD/GN	EVAP Vapor Pressure Sensor (PTNK)	2.5-3.1v (with cap off)
23	VT/WT	Malfunction Indicator Lamp Control	MIL Off: 12v, On: 1v
24	YL/RD	A/T: Oil Temp. Lamp Control	Lamp Off: 12v, On: 1v
25	LG/BK	Security Indictor Light (LED)	LED Off: 12v, On: 1v
26	BL/OR	Overdrive Lamp Control	At Cruise in OD: 1v
27	PK	TRAC Engine (ENG+) Signal	Pulse Signals
28	RD/WT	TRAC TRC (TRC+) Signal	Pulse Signals

Pin Connector Graphic

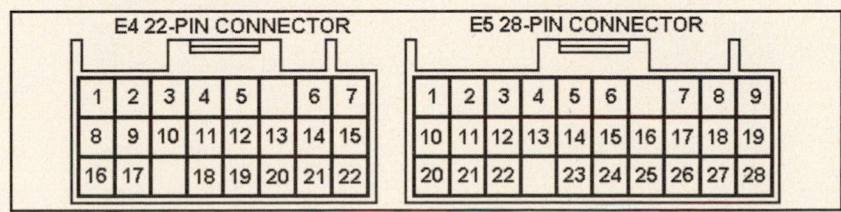

2001-02 Utility 4.7L V8 VIN T (A/T) E6 17 Pin Connector

PCM Pin #	Wire Color	Circuit Description (17 Pin)	Value at Hot Idle
1	RD	A/T ECT Solenoid (S1)	In 3rd or OD: 1v
2	WT/BL	A/T ECT Solenoid (S2)	In 1st or OD: 1v
3	GN/RD	A/T ECT Solenoid (SL)	In Lockup: 12-14v
4	WT	Direct Clutch Speed Input (+)	AC pulse signals
5	YL/RD	A/T: VSS Signal 2 (SP2+)	AC pulse signals
9	BK/RD	A/T-ECT Solenoid (SLT+)	Pulse Signals
10	BK	Direct Clutch Speed Input (-)	AC pulse signals
11	WT/RD	A/T: VSS Signal 2 (SP2-)	AC pulse signals
15	GN/YL	A/T-ECT Solenoid (SLT-)	AC pulse signals
17	RD/YL	A/T Oil Temperature Sensor	At 68°F: 4-5v

2001-02 Utility 4.7L V8 VIN T (A/T) E7 24 Pin Connector

PCM Pin #	Wire Color	Circuit Description (24 Pin)	Value at Hot Idle
1	WT/BK	Power Ground (E03)	<0.1v
2	GN/BK	Sensor VREF (VC)	4.9-5.1v
3	BL/RD	HO2S-21 (B2 S1) Heater	1v (Heater on)
4	BL	HO2S-11 (B1 S1) Heater	1v (Heater on)
5	RD	Injector 1 Control	2.0-3.3 ms
6	WT	Injector 2 Control	2.0-3.3 ms
7	WT/GN	EVAP Purge Solenoid (VSV)	12v or 0v
8	---	Not Used	---
9	BL/YL	Accelerator Position Sensor 2	1.8-2.7v
10	RD/WT	Mass Airflow Sensor (VG)	1.1-1.5v
11	WT	HO2S-21 (B2 S1) Signal	0.1-1.1v
12	BK	HO2S-11 (B1 S1) Signal	0.1-1.1v
13	BK/YL	TP Sensor Signal (VTA)	0.4-1.0v
14	GN/YL	ECT Sensor Signal (THW)	At 180°F: 0.51v
15	RD	HO2S-22 (B2 S2) Signal	0.1-1.1v
16	GN	HO2S-12 (B1 S2) Signal	0.1-1.1v
17	BR	Power Ground (E1)	<0.1v
18	GN/WT	Sensor Ground (E2)	<0.050v
19	BK/WT	MAF Sensor Ground (EVG)	<0.050v
20	PK/BL	TP Sensor Signal (VTA2)	2.0-2.9v
21	GN/RD	Accelerator Position Sensor 1	0.3-0.90v
22	YL/GN	IAT Sensor Signal (THA)	At 100°F: 2.60v
23	YL	HO2S-22 (B2 S2) Heater	1v (Heater on)
24	WT/RD	HO2S-12 (B1 S2) Heater	1v (Heater on)

Pin Connector Graphic

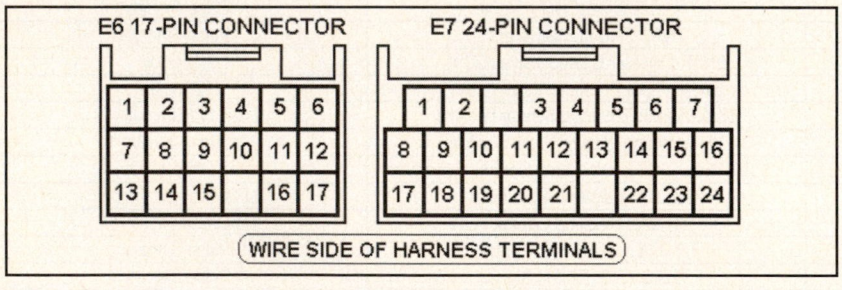

2001-02 Utility 4.7L V8 VIN T (A/T) E8 31 Pin Connector

PCM Pin #	Wire Color	Circuit Description (31 Pin)	Value at Hot Idle
1	GN	Injector 3 Control	2.0-3.3 ms
2	RD/BL	Injector 4 Control	2.0-3.3 ms
3	BL	Injector 5 Control	2.0-3.3 ms
4	YL	Injector 6 Control	2.0-3.3 ms
5	BL/RD	Injector 7 Control	2.0-3.3 ms
6	RD/WT	Injector 8 Control	2.0-3.3 ms
7	PK	Throttle Control Motor (M-)	Pulse Signals
8	VT	Throttle Control Motor (M+)	Pulse Signals
9	WT/BK	Power Ground (ME01)	<0.1v
10	YL	CMP Sensor Signal (G+)	AC pulse signals
11	BK/BL	Igniter Transistor 1 Control	7% duty cycle
12	LG/BK	Igniter Transistor 2 Control	7% duty cycle
13	GN/BK	Igniter Transistor 3 Control	7% duty cycle
14	RD/WT	Igniter Transistor 4 Control	7% duty cycle
15	GN/WT	Igniter Transistor 5 Control	7% duty cycle
16	PK/BL	Igniter Transistor 6 Control	7% duty cycle
17	BK	Knock Sensor 2 Signal (right)	<0.075v AC
18	GY	Knock Sensor 1 Signal (left)	<0.075v AC
19	BL/WT	Throttle Control Motor (CL-)	Pulse Signals
21	WT/BK	Power Ground (E01)	<0.1v
22	RD	CKP Sensor Signal (NE-)	<0.050v
23	GN	CKP Sensor Signal (NE+)	AC pulse signals
24	BL	CMP Sensor Signal (G-)	AC pulse signals
25	GN/RD	Igniter Transistor 7 Control	7% duty cycle
26	LG	Igniter Transistor 8 Control	7% duty cycle
27	BK/RD	Igniter Signal (IGF1)	Digital Signal: 0-5-0v
28	BK/WT	Igniter Signal (IGF2)	Digital Signal: 0-5-0v
29	GN	Throttle Control Motor (CL+)	Pulse Signals
30	BL/RD	Shield Ground (GE01)	<0.050v
31	WT/BK	Power Ground (E02)	<0.1v

Pin Connector Graphic

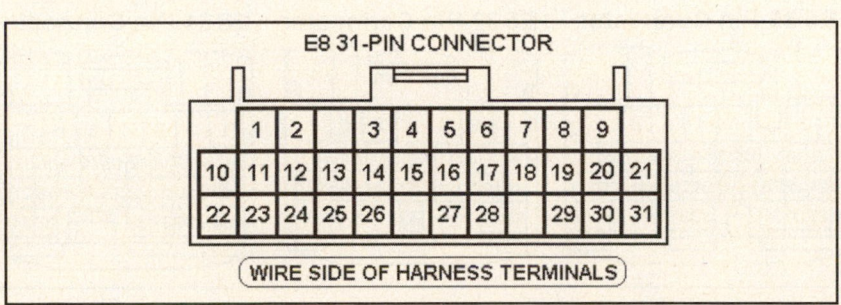

2003 Utility AWD 4.7L V8 VIN T E4 31 Pin Connector

PCM Pin #	Wire Color	Circuit Description (31 Pin)	Value at Hot Idle
1	BK/YL	EFI Main Relay Power (+B)	12-14v
2	BK/YL	EFI Main Relay Power (+B2)	12-14v
3	BK/YL	Direct Battery	12-14v
4	PK/BL	EVAP Pressure Switching Valve (VSV)	12v or 0v
5	YL/GN	Tachometer Signal (TACO)	Pulse Signals
6	GN/WT	A/T Select Switch Park Signal (P)	In 'P': 12v, Others: 0v
7	GN/RD	A/T Select Switch Neutral Signal (N)	In 'N': 12v, Others: 0v
8	BK/WT	EFI Main Relay Control	Relay Off: 0v, On: 12v
9	BK/OR	Ignition Switch Power (IGSW)	12-14v
10	GN/OR	Circuit Opening Relay (FC)	0-3v, off-idle: 12v
11	VT/WT	Malfunction Indicator Lamp Control	MIL Off: 12v, On: 1v
12	GN/YL	Taillight Switch Signal (ELS)	Taillights Off: 0v, On: 12v
13	YL/RD	Horn Relay Control	Relay Off: 12v, On: 1v
14	BL	Center Airbag Assembly (F/PS)	Digital Signals
15	OR	Power Ground (EOM)	<0.1v
16	---	Not Used	---
17	YL	Transponder Amplifier Signal (NEO)	Inserting key: pulses
18	GN/RD	SIL Signal (Scan Tool)	Transmitting: pulses
19	RD	Data Link Connector (WFSE)	N/A
20	PK/BK	Data Link Connector (TC)	12v
21	RD/GN	EVAP Vapor Pressure Sensor (PTNK)	2.5-3.1v (with cap off)
22	GN/RD	Accelerator Position Sensor 1 (VPA)	0.3-0.90v
23	BL/YL	Accelerator Position Sensor 2 (VPA2)	1.8-2.7v
24	PK	Traction Control Engine Signal (ENG+)	Pulse Signals
25	BL	Traction Control Signal (TRC+)	Pulse Signals
26	GN/BK	Accelerator Pedal Position Sensor 1 VREF	4.9-5.1v
27	BL/RD	Accelerator Pedal Position Sensor 2 VREF	4.9-5.1v
28	GN/WT	Accelerator Pedal Position Sensor (EPA)	<0.050v
29	BL/BK	Accelerator Position Sensor 2 (EPA2)	1.8-2.7v
30	RD	Traction Control Engine Signal (ENG-)	Pulse Signals
31	LG	Traction Control Signal (TRC-)	Pulse Signals

Pin Connector Graphic

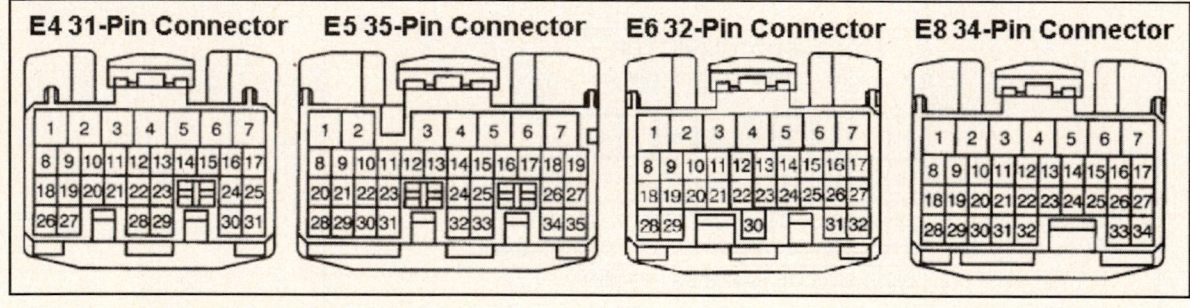

E4 31-Pin Connector   E5 35-Pin Connector   E6 32-Pin Connector   E8 34-Pin Connector

2003 Utility AWD 4.7L V8 VIN T E5 35 Pin Connector

PCM Pin #	Wire Color	Circuit Description (35 Pin)	Value at Hot Idle
1	WT/BK	Power Ground (HP)	<0.1v
2	BL	A/C Magnetic Clutch Relay (ACMG)	Relay Off: 0v, On: 12v
3	GN	A/T Select Switch 2nd Signal (2L)	In 2nd: 12v, Others: 0v
4	BL/RD	Low Detection Switch (L4)	Switch Open: 0v, Closed: 12v
5	BL/WT	ECT Pattern Switch 2nd Position (SNW1)	2nd Position: 12v
6	WT/GN	ETCS Power (+BM)	12-14v
7-8	---	Not Used	---
9	GN/BK	Shift Lock ECU Control (2)	12v or 0v
10	GN/YL	Shift Lock ECU Control (D)	12v or 0v
11	RD/BK	A/T Select Switch Reverse Signal	In 'R': 1v, Others: 12v
12, 15-16	---	Not Used	---
14	WT/RD	A/C Amplifier Signal (THWO)	A/C Off: 12v, On: 1v
17	VT	Vehicle Speed Sensor (SPD)	At 55 mph: 48 Hz
18	PK/BK	Body Control ECU Signal (MPX1)	Digital Signals
19	GN/YL	Stop Light Switch Signal	Brake Off: 0v, On: 12v
20-22, 25	---	Not Used	---
23	GN/RD	Shift Lock ECU Control (4)	12v or 0v
26	BL/YL	Transponder Amplifier Signal (IMD)	Inserting key: pulses
27	PK	Transponder Amplifier Signal (IMI)	Inserting key: pulses
28	BL/WT	ECT Pattern Switch Power Signal	Power Position: 12v
29	BK	Body Control ECU Signal (MPX2)	Digital Signals
30, 34-35	---	Not Used	---
31	GR/BL	A/C Amplifier Signal (A/CS)	Relay Off: 12v, On: 1v
32	PK/BK	A/C Amplifier Signal (THE)	A/C Off: 12v, On: 1v
33	LG/BK	A/C Switch Signal (ACLD)	A/C On: 12v, Off: 0v

2003 Utility AWD 4.7L V8 VIN T E6 32 Pin Connector

PCM Pin #	Wire Color	Circuit Description (32 Pin)	Value at Hot Idle
1	BR	Power Ground (E1)	<0.1v
2	PK	Throttle Control Motor (M-)	Pulse Signals
3	VT	Throttle Control Motor (M+)	Pulse Signals
4	WT/BK	Power Ground (ME01)	<0.1v
5	RD/WT	Injector 8 Control	2.0-3.3 ms
6	BL/RD	Injector 7 Control	2.0-3.3 ms
7	WT/BK	Power Ground (E03)	<0.1v
8-10	---	Not Used	---
11	YL/GN	Neutral Detection Switch (L4)	Switch Open: 0v, Closed: 12v
12	BK	Start Switch Signal (STSW)	Cranking: 9-11v
13-14	---	Not Used	---
15	BK	A/T Solenoid Control (SLU-)	Pulse Signals
16	PK/GN	A/T Solenoid Control (SLU+)	Pulse Signals
17	BL/RD	Shield Ground (GE01)	<0.050v
18-20	---	Not Used	---
21	BK/OR	Generator Control (RL)	12v
22, 26	---	Not Used	---
23	BL	A/C Lock Sensor (LCK)	12v or 0v
24	RD	CKP Sensor Signal (NE-)	<0.050v
25	GN	CKP Sensor Signal (NE+)	AC pulse signals
27	YL	CMP Sensor Signal (G2+)	AC pulse signals
28-31	---	Not Used	---
32	BL	CMP Sensor Signal (G2-)	AC pulse signals

Pin Connector Graphic

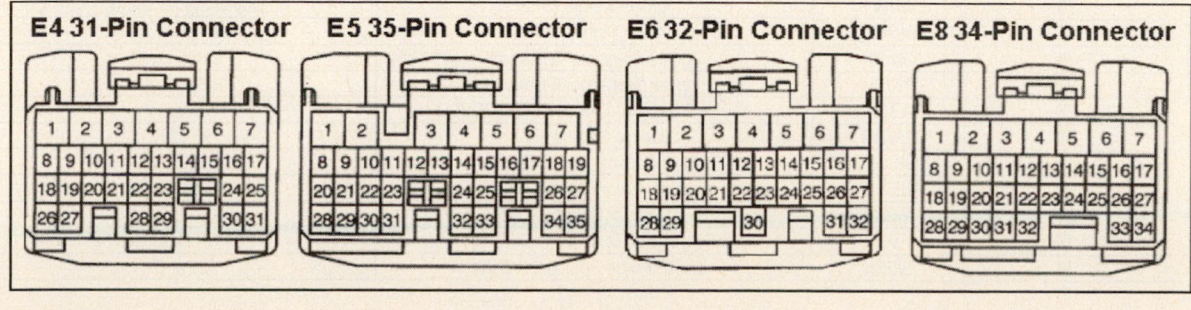

E4 31-Pin Connector   E5 35-Pin Connector   E6 32-Pin Connector   E8 34-Pin Connector

2003 Utility AWD 4.7L V8 VIN T E7 35 Pin Connector

PCM Pin #	Wire Color	Circuit Description (35 Pin)	Value at Hot Idle
1	GY	Knock Sensor 1 Signal (KNK1)	<0.075v AC
2	BK	Knock Sensor 2 Signal (KNK2)	<0.075v AC
3	YL	Injector 6 Control	2.0-3.3 ms
4	BL/RD	HO2S-11 (B1 S1) Heater (HT1A)	1v (Heater on)
5	WT/RD	HO2S-12 (B1 S2) Heater (HT1B)	1v (Heater on)
7-8, 20	---	Not Used	---
9	PK	Start Signal (STAR)	In P/N: 0-3.0v
10	WT	ECT Solenoid Control (S2)	12v or 0v
11	RD	ECT Solenoid Control (S1)	12v or 0v
12	GN/BK	ECT Solenoid Control (SLT-)	12v or 0v
13	GN/WT	ECT Solenoid Control (SLT+)	12v or 0v
16	PK/BK	A/T Solenoid Control (SL2-)	Pulse Signals
17	PK/BK	A/T Solenoid Control (SL2+)	Pulse Signals
18	RD/WT	A/T Solenoid Control (SL1-)	Pulse Signals
19	RD/BL	A/T Solenoid Control (SL1+)	Pulse Signals
21	RD	HO2S-22 (B2 S2) Signal (OX2B)	0.1-1.1v
22	WT	HO2S-21 (B2 S1) Signal (OX2A)	0.1-1.1v
23	BK	HO2S-11 (B1 S1) Signal (OX1A)	0.1-1.1v
24	BL	A/T Oil Temperature Sensor 2 (THO2)	At 68°F: 4-5v
25	YL	HO2S-22 (B2 S2) Heater (HT2B)	1v (Heater on)
26	RD	ECT Vehicle Speed Sensor (SP2+)	Pulse Signals
27	BL	ECT Turbine Speed Sensor (NT-)	Pulse Signals
28	---	Not Used	---
29	GN	HO2S-12 (B1 S2) Signal (OX1B)	0.1-1.1v
30-31	---	Not Used	---
32	GN/YL	A/T Oil Temperature Sensor 1 (THO1)	At 68°F: 4-5v
33	BL	HO2S-21 (B2 S1) Heater (HT2A)	1v (Heater on)
34	GN	ECT Vehicle Speed Sensor (SP2-)	Pulse Signals
35	WT	ECT Turbine Speed Sensor (NT+)	Pulse Signals

2003 Utility AWD 4.7L V8 VIN T E8 34 Pin Connector

PCM Pin #	Wire Color	Circuit Description (34 Pin)	Value at Hot Idle
1	RD	Injector 1 Control	2.0-3.3 ms
2	WT	Injector 2 Control	2.0-3.3 ms
3	GN	Injector 3 Control	2.0-3.3 ms
4	RD/BK	Injector 4 Control	2.0-3.3 ms
5	BL	Injector 5 Control	2.0-3.3 ms
6	WT/BK	Power Ground (E02)	<0.1v
7	WT/BK	Power Ground (E01)	<0.1v
8	LG/BK	Igniter Transistor 2 Control	7% duty cycle
9	BK/BL	Igniter Transistor 1 Control	7% duty cycle
10	LG	Igniter Transistor 8 Control	7% duty cycle
11	RD/WT	Igniter Transistor 4 Control	7% duty cycle
12	GN/WT	Igniter Transistor 5 Control	7% duty cycle
13	GN/RD	Igniter Transistor 7 Control	7% duty cycle
15	BK/OR	A/C Relay Control (ACCR)	Relay Off: 12v, On: 1v
16	PK	Neutral Start Switch (NSW)	In P/N: 0-3.0v
17	BL/RD	Starter Switch Signal (STA)	In P/N: 0-3.0v
18	GN/BK	Sensor VREF (VC)	4.9-5.1v
19	GN/YL	ECT Sensor Signal (THW)	At 180°F: 0.51v
20	YL/GN	IAT Sensor Signal (THA)	At 100°F: 2.60v
21	BK/YL	TP Sensor Signal (VTA1)	0.4-1.0v
22, 32	---	Not Used	---
23	BK/WT	Igniter Signal (IGF2)	Digital Signal: 0-5-0v
24	BK/RD	Igniter Signal (IGF1)	Digital Signal: 0-5-0v
25	GN/BK	Igniter Transistor 3 Control	7% duty cycle
26	PK/BL	Igniter Transistor 6 Control	7% duty cycle
27	WT/BL	EVAP Canister Closed Valve (CCV)	12v or 0v
28	GN/WT	Sensor Ground (E2)	<0.050v
29	BK/WT	MAF Sensor Ground (E2G)	<0.050v
30	RD/WT	Mass Airflow Sensor (VG)	1.1-1.5v
31	PK/BL	TP Sensor Signal (VTA2)	2.0-2.9v
33	VT	Fuel Pump Relay Control (FPR)	Relay Off: 12v, On: 1v
34	BL/BK	EVAP Purge Solenoid (VSV)	12v or 0v

4RUNNER Pin Tables

1990-92 Utility 2.4L I4 MFI VIN R (A/T-ECT) 16 Pin Connector

PCM Pin #	Wire Color	Circuit Description (16 Pin)	Value at Hot Idle
1	GN/YL	TP Sensor VREF	4.9-5.1v
2	GN/BK	Airflow Meter VREF	4.9-5.1v
3	YL/GN	IAT Sensor Signal (THA)	At 100°F: 2.60v
4	GN/BL	ECT Sensor Signal (THW)	At 180°F: 0.51v
5	BK	Knock Sensor Signal	<0.075v AC
6	BK	HO2S-11 (B1 S1) Signal	0.1-1.1v
7	GN/YL	4WD Oil Temperature Sensor	At 230°F: <1.5v
8	YL	Check Connector	12-14v
9	BR/BK	Sensor Ground	<0.050v
10	YL/BL	Airflow Meter Signal	0.5-2.5v
11	YL	TP Sensor Signal (VTA)	0.3-0.8v
12	YL/BL	Closed Throttle Switch	1v, off-idle: 12v
13 (Cal)	GN/WT	EGR Gas Temperature Sensor	3.5-4.0v
15	PK/WT	Check Connector	12-14v

1990-92 Utility 2.4L I4 MFI VIN R (A/T-ECT) 22 Pin Connector

PCM Pin #	Wire Color	Circuit Description (22 Pin)	Value at Hot Idle
1	BK/GN	Direct Battery	12-14v
4	BL/WT	A/T Oil Temperature Indicator	Lamp Off: 12v, On: 1v
5	PK	Malfunction Indicator Lamp Control	MIL Off: 12v, On: 1v
6	GN/WT	Stop Light Switch Signal	Brake Off: 0v, On: 12v
7	RD/BL	A/T Pattern Select Switch	Norm: 0v, PWR: 12v
8	PK/GN	4WD Indicator Switch	Switch Off: 12v, On: 1v
9	GN/BL	Vehicle Speed Sensor	At 55 mph: 48 Hz
11	BK/WT	Starter Switch Signal	9-11v (cranking)
12	WT/RD	EFI Main Relay Power	12-14v
13	WT/RD	EFI Main Relay Power	12-14v
15	BR/BK	Sensor Ground	<0.050v
16	YL/GN	Overdrive Main Switch	Switch Off: 12v, On: 1v
20	PK	Check Connector	12v
21	YL/RD	Cruise Control ECU	At Cruise in OD: 12v

Standard Colors and Abbreviations

Abbreviation	Color	Abbreviation	Color	Abbreviation	Color
BK	Black	GY	Gray	RD	Red
BL	Blue	GN	Green	TN	Tan
BR	Brown	LG	LT Green	VT	Violet
DB	Dark Blue	OR	Orange	WT	White
DG	DK Green	PK	Pink	YL	Yellow

1990-92 Utility 2.4L I4 MFI VIN R (A/T-ECT) 26 Pin Connector

PCM Pin #	Wire Color	Circuit Description (26 Pin)	Value at Hot Idle
1	BK/OR	Distributor Signal (NE+)	AC pulse signals
2	GN	Cold Start Injector Control	1v (at cold startup)
3	BK/YL	Igniter Signal (IGF)	Digital Signal: 0-5-0v
4	RD/YL	A/T-ECT Solenoid (S4)	Pulse Signals
5	RD/GN	A/T-ECT Solenoid (S3)	Pulse Signals
6	RD/WT	A/T-ECT Solenoid (S2)	1st or OD: 1v
7	BL/RD	A/T-ECT Solenoid (S1)	3rd or OD: 1v
10	PK/GN	HO2S-11 (B1 S1) Heater	1v (Heater on)
11	GN	Fuel Pressure Up Solenoid	1v (at hot restart)
12	WT/RD	Injector Pair 1 & 3 Control	2.0-3.3 ms
13	BR	Power Ground	<0.1v
15	BR	Shield Ground	<0.050v
16	BL	A/T Speed Sensor Signal	Moving: 0-5-0-5v
17	PK/WT	A/T Select Switch Low	In Low: 12v, Others: 0v
18	PK/GN	A/T Select Switch 2nd	In 2nd: 12v, Others: 0v
19	PK/RD	A/T Select Switch Neutral	In 'N': 12v, Others: 0v
20	RD	4WD Detection Transfer (L4)	Open: 12v, Closed: 0v
21	BK/BL	Igniter Signal (IGT)	Digital Signal: 0-5-0v
23	GN/RD	Intake Air Solenoid Control	12v or 0v
25	WT	Injector Pair 2 & 4 Control	2.0-3.3 ms
26	BR	Power Ground	<0.1v

Pin Connector Graphic

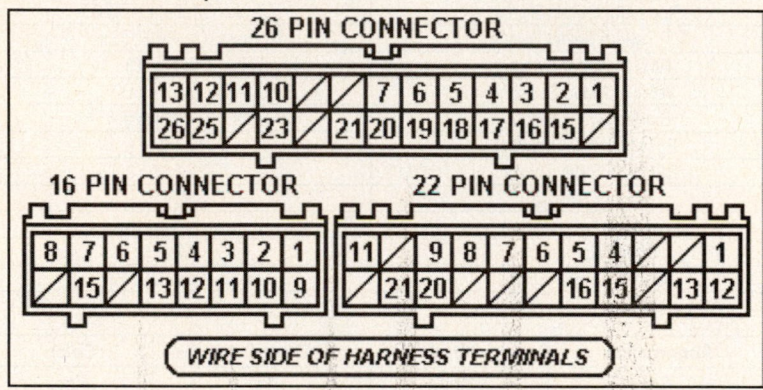

Standard Colors and Abbreviations

Abbreviation	Color	Abbreviation	Color	Abbreviation	Color
BK	Black	GY	Gray	RD	Red
BL	Blue	GN	Green	TN	Tan
BR	Brown	LG	LT Green	VT	Violet
DB	Dark Blue	OR	Orange	WT	White
DG	DK Green	PK	Pink	YL	Yellow

1990-92 Utility 2.4L I4 MFI VIN R (M/T) 10 Pin Connector

PCM Pin #	Wire Color	Circuit Description (10 Pin)	Value at Hot Idle
1	BK/WT	A/T Neutral Start Switch	In P/N: 0-3.0v
1	BK	M/T Clutch Start Switch	9-11v (cranking)
2	GN	Cold Start Injector Control	1v (at cold startup)
3	BK/WT	Starter Switch Signal	9-11v (cranking)
4	WT/RD	Injector Pair 1 & 3 Control	2.0-3.3 ms
5	BR	Power Ground	<0.1v
6	YL	Check Connector	12-14v
7	BR	Sensor Ground	<0.050v
8	BK/BL	Igniter Signal (IGT)	Digital Signal: 0-5-0v
9	WT	Injector Pair 2 & 4 Control	2.0-3.3 ms
10	BR	Power Ground	<0.1v

1990-92 Utility 2.4L I4 MFI VIN R (M/T) 18 Pin Connector

PCM Pin #	Wire Color	Circuit Description (18 Pin)	Value at Hot Idle
1	BK/OR	Distributor Signal (NE+)	AC pulse signals
2	BK	Knock Sensor Signal	<0.075v AC
3 (Cal)	GN/WT	EGR Gas Temperature Sensor	3.5-4.0v
5	BK/YL	Igniter Signal (IGF)	Digital Signal: 0-5-0v
6	YL/BL	Closed Throttle Switch	1v, off-idle: 12v
7	PK/WT	Check Connector	12-14v
8	PK	Malfunction Indicator Lamp Control	MIL Off: 12v, On: 1v
9	GN	Fuel Pressure Up Solenoid	1v (at hot restart)
10	GN/BL	ECT Sensor Signal (THW)	At 180°F: 0.51v
11	YL	TP Sensor Signal (VTA)	0.3-0.8v
12	GN/YL	Sensor VREF	4.9-5.1v
13	BK	HO2S-11 (B1 S1) Signal	0.1-1.1v
14	BR/BK	Sensor Ground	<0.050v
15	PK/GN	HO2S-11 (B1 S1) Heater	1v (Heater on)
17	GN/RD	Intake Air Solenoid Control	12v or 0v

1990-92 Utility 2.4L I4 MFI VIN R (M/T) 14 Pin Connector

PCM Pin #	Wire Color	Circuit Description (14 Pin)	Value at Hot Idle
1	WT/RD	EFI Main Relay Power	12-14v
2	GN/BK	Direct Battery	12-14v
3	YL/GN	IAT Sensor Signal (THA)	At 100°F: 2.60v
4	YL/BL	Airflow Meter Signal	0.5-2.5v
5	GN/BK	Airflow Meter VREF	4.9-5.1v
8	WT/RD	EFI Main Relay Power	12-14v
9	GN/WT	Stop Light Switch Signal	Brake Off: 0v, On: 12v
10	GN/BL	Vehicle Speed Sensor	At 55 mph: 48 Hz
11	GN/YL	4WD Indicator Switch	Switch Off: 12v, On: 1v
12	BR	Sensor Ground	<0.050v
14	YL/RD	Overdrive Relay Control	Relay Off: 12v, On: 1v

Pin Connector Graphic

1993 Utility 2.4L I4 MFI VIN R (A/T-ECT) 26 Pin Connector

PCM Pin #	Wire Color	Circuit Description (26 Pin)	Value at Hot Idle
1	GN	Cold Start Injector Control	1v (at cold startup)
2	PK/GN	Main O2S Heater Control	1v (Heater on)
3	BK/YL	Igniter Signal (IGF)	Digital Signal: 0-5-0v
4	BK/OR	Distributor Signal (NE+)	AC pulse signals
5	RD/GN	A/T-ECT Solenoid (S3)	Pulses
6	RD/WT	A/T-ECT Solenoid (S2)	1st or OD: 1v
7	BL/RD	A/T-ECT Solenoid (S1)	3rd or OD: 1v
8	PK	EGR Solenoid Control (VSV)	12v or 0v
9	GN/RD	Intake Air Solenoid Control	12v or 0v
10	GN	Fuel Pressure Up Solenoid	1v (at hot restart)
11	WT	Injector Pair 2 & 4 Control	2.0-3.3 ms
12	WT/RD	Injector Pair 1 & 3 Control	2.0-3.3 ms
13	BR	Power Ground	<0.1v
14	BR	Sensor Ground	<0.050v
15	GN/RD	Sub O2S Heater Control	1v (Heater on)
16	BL	A/T Speed Sensor Signal	Moving: 0-5-0-5v
17	PK/WT	A/T Select Switch Low	In Low: 12v, Others: 0v
18	PK/GN	A/T Select Switch 2nd	In 2nd: 12v, Others: 0v
19	PK/RD	A/T Select Switch Neutral	In 'N': 12v, Others: 0v
20	RD	4WD Detection Transfer (L4)	Open: 12v, Closed: 0v
22	BK/BL	Igniter Signal (IGT)	Digital Signal: 0-5-0v
26	BR	Power Ground	<0.1v

Pin Connector Graphic

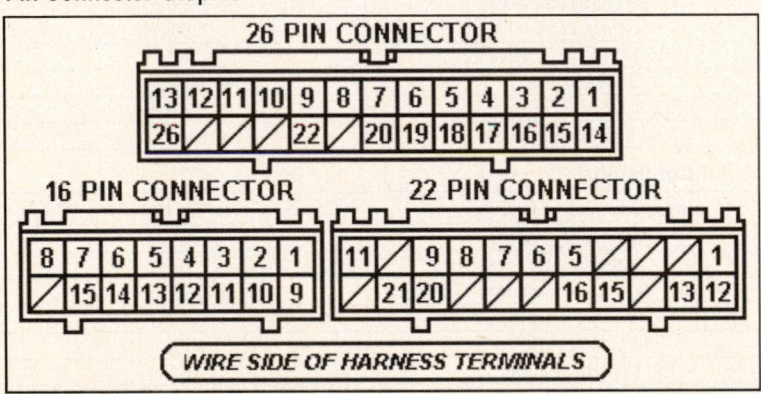

Standard Colors and Abbreviations

Abbreviation	Color	Abbreviation	Color	Abbreviation	Color
BK	Black	GY	Gray	RD	Red
BL	Blue	GN	Green	TN	Tan
BR	Brown	LG	LT Green	VT	Violet
DB	Dark Blue	OR	Orange	WT	White
DG	DK Green	PK	Pink	YL	Yellow

1993 Utility 2.4L I4 MFI VIN R (A/T-ECT) 16 Pin Connector

PCM Pin #	Wire Color	Circuit Description (16 Pin)	Value at Hot Idle
1	GN/YL	TP Sensor VREF	4.9-5.1v
2	YL/BL	Airflow Meter Signal	0.5-2.5v
3	YL/GN	IAT Sensor Signal (THA)	At 100°F: 2.60v
4	GN/BL	ECT Sensor Signal (THW)	At 180°F: 0.51v
5	WT	Sub O2S Signal	0.1-1.1v
6	BK	Main O2S Signal	0.1-1.1v
7	BK	Knock Sensor Signal	<0.075v AC
8	YL	Check Connector	12-14v
9	BR/BK	Sensor Ground	<0.050v
10	YL/BL	Airflow Meter VREF	4.9-5.1v
11	YL	TP Sensor Signal (VTA)	0.3-0.8v
12	YL/BL	Closed Throttle Switch	1v, off-idle: 12v
13 (Cal)	GN/WT	EGR Gas Temperature Sensor	3.5-4.0v
14	PK/GN	Check Connector	12-14v
15	PK/WT	Check Connector	12-14v

1993 Utility 2.4L I4 MFI VIN R (A/T-ECT) 22 Pin Connector

PCM Pin #	Wire Color	Circuit Description (22 Pin)	Value at Hot Idle
1	BK/GN	Direct Battery	12-14v
5	PK	Malfunction Indicator Lamp Control	MIL Off: 12v, On: 1v
6	GN/WT	Stop Light Switch Signal	Brake Off: 0v, On: 12v
7	RD/BL	A/T Pattern Select Switch	Norm: 0v, PWR: 12v
8	PK/GN	4WD Indicator Switch	Switch Off: 12v, On: 1v
9	GN/BL	Vehicle Speed Sensor	At 55 mph: 48 Hz
11	BK/WT	Starter Switch Signal	In P/N: 0-3.0v
12	WT/RD	EFI Main Relay Power	12-14v
13	WT/RD	EFI Main Relay Power	12-14v
15	BR/BK	Sensor Ground	<0.050v
16	YL/GN	Overdrive Main Switch	Switch Off: 12v, On: 1v
20	PK	Check Connector	12-14v
21	YL/RD	Cruise Control ECU	At Cruise in OD: 12v

Standard Colors and Abbreviations

Abbreviation	Color	Abbreviation	Color	Abbreviation	Color
BK	Black	GY	Gray	RD	Red
BL	Blue	GN	Green	TN	Tan
BR	Brown	LG	LT Green	VT	Violet
DB	Dark Blue	OR	Orange	WT	White
DG	DK Green	PK	Pink	YL	Yellow

1993 Utility 2.4L I4 MFI VIN R (M/T) 16 Pin Connector

PCM Pin #	Wire Color	Circuit Description (16 Pin)	Value at Hot Idle
1	PK/GN	Main O2S Heater Control	1v (Heater on)
2	BK/WT	Starter Switch Signal	9-11v (cranking)
3	BK/YL	Igniter Signal (IGF)	Digital Signal: 0-5-0v
4	BK/OR	Distributor Signal (NE+)	AC pulse signals
9	GN/RD	Intake Air Solenoid Control	12v or 0v
10	GN	Fuel Pressure Up Solenoid	1v (at hot restart)
11	GN	Cold Start Injector Control	1v (at cold startup)
12	WT	Injector Pair 2 & 4 Control	2.0-3.3 ms
13	BR	Power Ground	<0.1v
14	RD/GN	Sub O2S Heater Control	1v (Heater on)
15	BK/WT	A/T: Neutral Start Switch	In P/N: 0-3.0v
15	BK	M/T: Clutch Start Switch	9-11v (cranking)

1993 Utility 2.4L I4 MFI VIN R (M/T) 26 Pin Connector

PCM Pin #	Wire Color	Circuit Description (26 Pin)	Value at Hot Idle
1	YL/GN	IAT Sensor Signal (THA)	At 100°F: 2.60v
2	YL/BL	Airflow Meter Signal	0.5-2.5v
3	GN/BK	Airflow Meter VREF	4.9-5.1v
4	GN/BL	ECT Sensor Signal (THW)	At 180°F: 0.51v
5	WT	Sub O2S Signal	0.1-1.1v
6	BK	Main O2S Signal	0.1-1.1v
7	PK/GN	Data Link Connector	12-14v
8	YL	Data Link Connector	12-14v
9	BR/BK	Sensor Ground	<0.050v
10	YL	TP Sensor Signal (VTA)	0.3-0.8v
11	GN/YL	TP Sensor VREF	4.9-5.1v
12	YL/BL	Closed Throttle Switch	1v, off-idle: 12v
13 (Cal)	GN/WT	EGR Gas Temperature Sensor	3.5-4.0v
14	BK	Knock Sensor Signal	<0.075v AC
15	PK/GN	Data Link Connector	12-14v
16	BR/BK	Sensor Ground	<0.050v
22	BK/BL	Igniter Signal (IGT)	Digital Signal: 0-5-0v
23	PK	EGR Solenoid Control (VSV)	12v or 0v
24	BR	Sensor Ground	<0.050v
25	WT	Injector Pair 1 & 3 Control	2.0-3.3 ms
26	BR	Power Ground	<0.1v

1993 Utility 2.4L I4 MFI VIN R (M/T) 12 Pin Connector

PCM Pin #	Wire Color	Circuit Description (12 Pin)	Value at Hot Idle
1	WT/RD	EFI Main Relay Power	12-14v
2	BK/GN	Direct Battery	12-14v
6	BR/YL	4WD Indicator Switch	Switch Off: 12v, On: 1v
7	WT/RD	EFI Main Relay Power	12-14v
8	PK	Malfunction Indicator Lamp Control	MIL Off: 12v, On: 1v
11	GN/BL	Vehicle Speed Sensor	At 55 mph: 48 Hz
12	GN/WT	Stop Light Switch Signal	Brake Off: 0v, On: 12v

Pin Connector Graphic

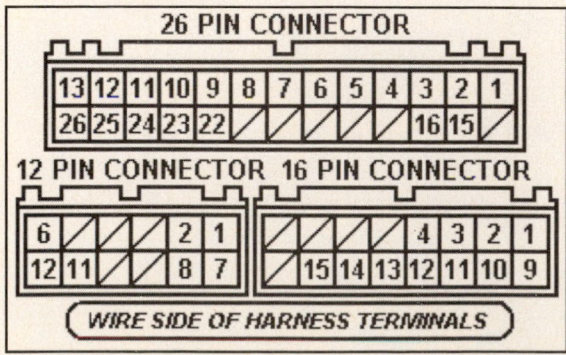

1994-95 Utility 2.4L I4 MFI VIN R (A/T-ECT) 26 Pin Connector

PCM Pin #	Wire Color	Circuit Description (26 Pin)	Value at Hot Idle
1	GN	Cold Start Injector Control	1v (at cold startup)
2	PK/GN	Main O2S Heater Control	1v (Heater on)
3	BK/YL	Igniter Signal (IGF)	Digital Signal: 0-5-0v
4	BK/OR	Distributor Signal (NE+)	AC pulse signals
5	RD/GN	A/T-ECT Solenoid (S3)	Pulse Signals
6	RD/WT	A/T-ECT Solenoid (S2)	1st or OD: 1v
7	BL/RD	A/T-ECT Solenoid (S1)	3rd or OD: 1v
9	GN/RD	Intake Air Solenoid Control	12v or 0v
10	GN	Fuel Pressure Up Solenoid	1v (at hot restart)
11	WT	Injector Pair 2 & 4 Control	2.0-3.3 ms
12	WT/RD	Injector Pair 1 & 3 Control	2.0-3.3 ms
13	BR	Power Ground	<0.1v
14	BR	Sensor Ground	<0.050v
19	BL	A/T Speed Sensor Signal	Moving: 0-5-0-5v
20	BK/BL	Igniter Signal (IGT)	Digital Signal: 0-5-0v
21	PK/WT	A/T Select Switch Low	In Low: 12v, Others: 0v
22	PK/GN	A/T Select Switch 2nd	In 2nd: 12v, Others: 0v
23	PK/RD	A/T Select Switch Neutral	In 'N': 12v, Others: 0v
26	BR	Power Ground	<0.1v

Pin Connector Graphic

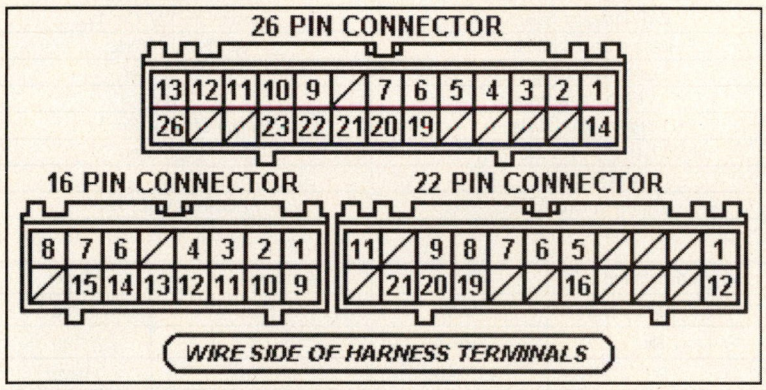

Standard Colors and Abbreviations

Abbreviation	Color	Abbreviation	Color	Abbreviation	Color
BK	Black	GY	Gray	RD	Red
BL	Blue	GN	Green	TN	Tan
BR	Brown	LG	LT Green	VT	Violet
DB	Dark Blue	OR	Orange	WT	White
DG	DK Green	PK	Pink	YL	Yellow

1994-95 Utility 2.4L I4 MFI VIN R (A/T-ECT) 16 Pin Connector

PCM Pin #	Wire Color	Circuit Description (16 Pin)	Value at Hot Idle
1	GN/YL	Sensor VREF	4.9-5.1v
2	YL/BL	Airflow Meter Signal	0.5-2.5v
3	YL/GN	IAT Sensor Signal (THA)	At 100°F: 2.60v
4	GN/BL	ECT Sensor Signal (THW)	At 180°F: 0.51v
6	BK	HO2S-11 (B1 S1) Heater	0.1-1.1v
7	BK	Knock Sensor Signal	<0.075v AC
8	YL	Check Connector	12-14v
9	BR/BK	Sensor Ground	<0.050v
10	GN/BK	Airflow Meter VREF	4.9-5.1v
11	YL	TP Sensor Signal (VTA)	0.3-0.8v
12	YL/BL	Closed Throttle Switch	1v, off-idle: 12v
13 (Cal)	GN/WT	EGR Gas Temperature Sensor	3.5-4.0v
14	PK/GN	Check Connector	12-14v
15	PK/WT	Check Connector	12-14v

1994-95 Utility 2.4L I4 MFI VIN R (A/T-ECT) 22 Pin Connector

PCM Pin #	Wire Color	Circuit Description (22 Pin)	Value at Hot Idle
1	BK/GN	Direct Battery	12-14v
5	PK	Malfunction Indicator Lamp Control	MIL Off: 12v, On: 1v
6	GN/WT	Stop Light Switch Signal	Brake Off: 0v, On: 12v
7	RD/BL	A/T Pattern Select Switch	Norm: 0v, PWR: 12v
8	PK/GN	4WD Indicator Switch	Switch Off: 12v, On: 1v
9	GN/BL	Vehicle Speed Sensor	At 55 mph: 48 Hz
11	BK/YL	Starter Switch Signal	In P/N: 0-3.0v
12	WT/RD	EFI Main Relay Power	12-14v
16	YL/GN	Overdrive Main Switch	Switch Off: 12v, On: 1v
19	RD	4WD Detection Transfer (L4)	Open: 12v, Closed: 0v
20	PK	Check Connector	12-14v
21	YL/RD	Cruise Control ECU	At Cruise in OD: 12v

Standard Colors and Abbreviations

Abbreviation	Color	Abbreviation	Color	Abbreviation	Color
BK	Black	GY	Gray	RD	Red
BL	Blue	GN	Green	TN	Tan
BR	Brown	LG	LT Green	VT	Violet
DB	Dark Blue	OR	Orange	WT	White
DG	DK Green	PK	Pink	YL	Yellow

1994-95 Utility 2.4L I4 MFI VIN R (M/T) 16 Pin Connector

PCM Pin #	Wire Color	Circuit Description (16 Pin)	Value at Hot Idle
1	YL/GN	IAT Sensor Signal (THA)	At 100ºF: 2.60v
2	YL/BL	Airflow Meter Signal	0.5-2.5v
3	GN/BK	Airflow Meter VREF	4.9-5.1v
4	GN/BL	ECT Sensor Signal (THW)	At 180ºF: 0.51v
5 (Cal)	WT	Sub O2S Signal	0.1-1.1v
6	BK	Main O2S Signal	0.1-1.1v
7	PK/GN	Data Link Connector	12-14v
8	YL	Data Link Connector	12-14v
9	BR/BK	Sensor Ground	<0.050v
10	YL	TP Sensor Signal (VTA)	0.3-0.8v
11	GN/YL	TP Sensor VREF	4.9-5.1v
12	YL/BL	Closed Throttle Switch	1v, off-idle: 12v
13 (Cal)	GN/WT	EGR Gas Temperature Sensor	3.5-4.0v
14	BK	Knock Sensor Signal	<0.075v AC
15	PK/WT	Data Link Connector	12-14v

1994-95 Utility 2.4L I4 MFI VIN R (M/T) 26 Pin Connector

PCM Pin #	Wire Color	Circuit Description (26 Pin)	Value at Hot Idle
1	PK/GN	Main O2S Heater Control	1v (Heater on)
2	BK/WT	Starter Switch Signal	In P/N: 0-3.0v
3	BK/YL	Igniter Signal (IGF)	Digital Signal: 0-5-0v
4	BK/OR	Distributor Signal (NE+)	AC pulse signals
9	GN/RD	Intake Air Solenoid Control	12v or 0v
10	GN	Fuel Pressure Up Solenoid	1v (at hot restart)
11	GN	Cold Start Injector Control	1v (at cold startup)
12	WT/RD	Injector Pair 1 & 3 Control	2.0-3.3 ms
13	BR	Power Ground	<0.1v
14 (Cal)	RD/GN	Sub O2S Heater Control	1v (Heater on)
15	BK	Clutch Start Switch	9-11v (cranking)
22	BK/BL	Igniter Signal (IGT)	Digital Signal: 0-5-0v
23	PK	EGR Solenoid Control (VSV)	12v or 0v
24	BR	Sensor Ground	<0.050v
25	WT	Injector Pair 2 & 4 Control	2.0-3.3 ms
26	BR	Power Ground	<0.1v

1994-95 Utility 2.4L I4 MFI VIN R (M/T) 12 Pin Connector

PCM Pin #	Wire Color	Circuit Description (12 Pin)	Value at Hot Idle
2	BK/GN	Direct Battery	12-14v
6	GN/WT	4WD Indicator Switch	Switch Off: 12v, On: 1v
7	WT/RD	EFI Main Relay Power	12-14v
8	PK	Malfunction Indicator Lamp Control	MIL Off: 12v, On: 1v
11	GN/BL	Vehicle Speed Sensor	At 55 mph: 48 Hz
12	GN/WT	Stop Light Switch Signal	Brake Off: 0v, On: 12v

Pin Connector Graphic

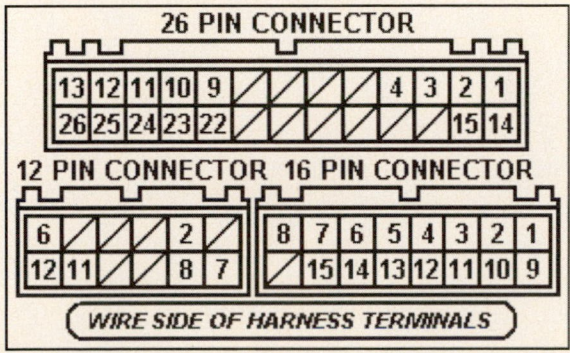

1996 Utility 2.7L I4 MFI VIN M (A/T-ECT) 16 Pin Connector

PCM Pin #	Wire Color	Circuit Description (16 Pin)	Value at Hot Idle
1	GN/YL	TP Sensor VREF	4.9-5.1v
2	BK/RD	MAF Sensor Signal	1.1-1.8v
3	RD/WT	MAF Sensor Ground (E2G)	<0.050v
4	GN/RD	ECT Sensor Signal (THW)	At 180ºF: 0.51v
5	WT	HO2S-11 (B1 S1) Signal	0.1-1.1v
6	GY	Knock Sensor Signal	<0.075v AC
7	PK/WT	Data Link Connector	12-14v
9	BR/BK	Sensor Ground	<0.050v
10	YL	TP Sensor Signal (VTA)	0.3-0.8v
11	YL/BL	Closed Throttle Switch	1v, off-idle: 12v
12	YL/GN	IAT Sensor Signal (THA)	At 100ºF: 2.60v
13	RD	HO2S-12 (B1 S2) Signal	0.1-1.1v
14	PK	EGR Gas Temperature Sensor	3.5-4.0v

1996 Utility 2.7L I4 MFI VIN M (A/T-ECT) 22 Pin Connector

PCM Pin #	Wire Color	Circuit Description (22 Pin)	Value at Hot Idle
2	BK/WT	Direct Battery	12-14v
3	YL	A/T Oil Temperature Indicator	Lamp Off: 12v, On: 1v
4	PK	Malfunction Indicator Lamp Control	MIL Off: 12v, On: 1v
5	BL/OR	Overdrive Main Switch	Switch Off: 12v, On: 1v
6	BL/BK	A/C Amplifier Signal (ACT)	Clutch On: 12v, Off: 1.5v
7	BL/YL	A/C Amplifier On Signal	Clutch On: 1.5v, Off: 12v
8	GN/OR	Vehicle Speed Sensor	At 55 mph: 48 Hz
10	GN/BK	4WD Indicator Switch	Switch Off: 12v, On: 1v
11	BK/WT	Starter Switch Signal	In P/N: 0-3.0v
12	WT/BL	EFI Main Relay Power	12-14v
13	RD/GN	4WD Detection Transfer (N)	Open: 12v, Closed: 0v
14	GN/RD	A/T Pattern Select Switch	Norm: 0v, PWR: 12v
15	LG	A/T Select Switch Low	In Low: 12v, Others: 0v
16	PK/RD	A/T Select Switch 2nd	In 2nd: 12v, Others: 0v
17	RD/YL	A/T Select Switch Reverse	In 'R': 12v, Others: 0v
18	BR/YL	Cruise Control ECU	At Cruise in OD: 12v
19	WT	Data Link Connector (SDL)	0v
21	GN/WT	Stop Light Switch Signal	Brake Off: 0v, On: 12v

Standard Colors and Abbreviations

Abbreviation	Color	Abbreviation	Color	Abbreviation	Color
BK	Black	GY	Gray	RD	Red
BL	Blue	GN	Green	TN	Tan
BR	Brown	LG	LT Green	VT	Violet
DB	Dark Blue	OR	Orange	WT	White
DG	DK Green	PK	Pink	YL	Yellow

1996 Utility 2.7L I4 MFI VIN M (A/T-ECT) 12 Pin Connector

PCM Pin #	Wire Color	Circuit Description (12 Pin)	Value at Hot Idle
1	BK/WT	EVAP Purge Solenoid (VSV)	12v or 0v
2	RD/BK	EVAP Vapor Pressure Valve (VSV)	12v or 0v
4	WT/RD	A/T Vehicle Speed Sensor (-)	Pulse Signals
5	BL	Distributor Signal (G-)	<0.050v
6	GN	CKP Sensor Signal (NE-)	<0.050v
10	YL/RD	A/T Vehicle Speed Sensor (+)	Pulse Signals
11	PK	Distributor Signal (G1+)	AC pulse signals
12	RD	CKP Sensor Signal (NE+)	AC pulse signals

1996 Utility 2.7L I4 MFI VIN M (A/T-ECT) 26-PK Connector

PCM Pin #	Wire Color	Circuit Description (26 Pin)	Value at Hot Idle
3	PK	HO2S-11 (B1 S1) Heater	1v (Heater on)
6	BK	IAC Signal (RSC)	Pulse Signals
7	BK/RD	Idle Air Control Valve (RSO)	Pulse Signals
8	RD/GN	A/T-ECT Solenoid (SL)	In Lockup: 12-14v
9	LG	A/T-ECT Solenoid (S2)	1st or OD: 1v
10	GN/RD	A/T-ECT Solenoid (S1)	3rd or OD: 1v
11	WT	Injectors 2 & 4 Control	2.0-3.3 ms
12	RD	Injectors 1 & 3 Control	2.0-3.3 ms
13	BR	Power Ground	<0.1v
14	GN/YL	Circuit Opening Relay (FC)	0-3v, off-idle: 12v
16	RD/WT	HO2S-12 (B1 S2) Heater	1v (Heater on)
17	BK/YL	Igniter Signal (IGF)	Digital Signal: 0-5-0v
18	GN/BK	EVAP Vapor Pressure Sensor (PTNK)	2.5-3.1v (with cap off)
21	PK/YL	A/T Oil Temperature Sensor	At 230°F: <1.5v
22	PK/BK	EGR Solenoid Control (VSV)	12v or 0v
23	BK/BL	Igniter Signal (IGT)	Digital Signal: 0-5-0v
24	BR	Sensor Ground	<0.050v
25	BR	Power Ground	<0.1v
26	BR	Power Ground	<0.1v

Pin Connector Graphic

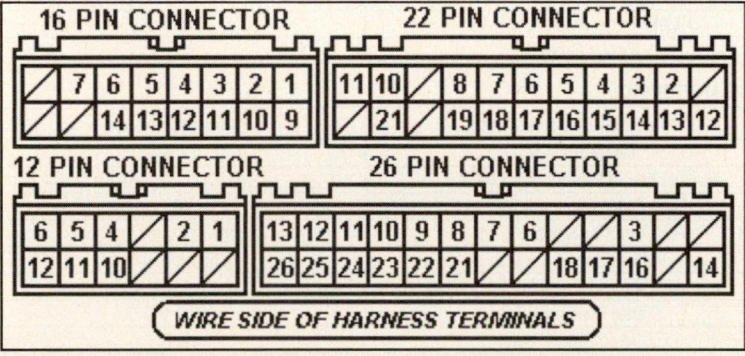

Standard Colors and Abbreviations

Abbreviation	Color	Abbreviation	Color	Abbreviation	Color
BK	Black	GY	Gray	RD	Red
BL	Blue	GN	Green	TN	Tan
BR	Brown	LG	LT Green	VT	Violet
DB	Dark Blue	OR	Orange	WT	White
DG	DK Green	PK	Pink	YL	Yellow

1996 Utility 2.7L I4 MFI VIN M (M/T) 16 Pin Connector

PCM Pin #	Wire Color	Circuit Description (16 Pin)	Value at Hot Idle
1	GN/YL	TP Sensor VREF	4.9-5.1v
2	BK/RD	MAF Sensor Signal	Idling: 1-1.5v
3	PK	EGR Gas Temperature Sensor	3.5-4.0v
4	GN/RD	ECT Sensor Signal (THW)	At 180°F: 0.51v
5	RD	HO2S-12 (B1 S2) Signal	0.1-1.1v
6	WT	HO2S-11 (B1 S1) Signal	0.1-1.1v
7	YL/GN	IAT Sensor Signal (THA)	At 100°F: 2.60v
9	BR/BK	Sensor Ground	<0.050v
10	GN/BK	EVAP Vapor Pressure Sensor (PTNK)	2.5-3.1v (with cap off)
11	YL	TP Sensor Signal (VTA)	0.3-0.8v
12	YL/BL	Closed Throttle Switch	1v, off-idle: 12v
13	GY	Knock Sensor Signal	<0.075v AC
15	PK/WT	Data Link Connector	12v
16	RD/WT	MAF Meter Return	<0.050v

1996 Utility 2.7L I4 MFI VIN M (M/T) 26 Pin Connector

PCM Pin #	Wire Color	Circuit Description (26 Pin)	Value at Hot Idle
2	PK	HO2S-11 (B1 S1) Heater	1v (Heater on)
3	BK/YL	Igniter Signal (IGF)	Digital Signal: 0-5-0v
4	RD	CKP Sensor Signal (NE+)	AC pulse signals
5	PK	Distributor Signal (G1+)	AC pulse signals
6	PK/BK	EGR Solenoid Control (VSV)	12v or 0v
9	BK	IAC Signal (RSC)	Pulse Signals
10	BK/RD	Idle Air Control Valve (RSO)	Pulse Signals
11	WT	Injector Pair 2 & 4 Control	2.0-3.3 ms
12	RD	Injector Pair 1 & 3 Control	2.0-3.3 ms
13, 14	BR	Power Ground	<0.1v
15	RD/WT	HO2S-12 (B1 S2) Heater	1v (Heater on)
17	GN	CKP Sensor Signal (NE-)	AC pulse signals
18	BL	Distributor Signal (G-)	<0.050v
20	BK/BL	Igniter Signal (IGT)	Digital Signal: 0-5-0v
21	RD/BK	EVAP Vapor Pressure Valve (VSV)	12v or 0v
23	BK/WT	EVAP Purge Solenoid (VSV)	12v or 0v
25, 26	BR	Power Ground	<0.1v

1996 Utility 2.7L I4 MFI VIN M (M/T) 22 Pin Connector

PCM Pin #	Wire Color	Circuit Description (22 Pin)	Value at Hot Idle
1	BK/WT	Direct Battery	12-14v
5	PK	Malfunction Indicator Lamp Control	MIL Off: 12v, On: 1v
7	WT	Data Link Connector (SDL)	0v
8	BL/BK	A/C Amplifier Signal (ACT)	Clutch On: 12v, Off: 1.5v
9	GN/OR	Vehicle Speed Sensor	At 55 mph: 48 Hz
10	BL/YL	A/C Amplifier Signal (AC1)	Clutch On: 1.5v, Off: 12v
11	BK/WT	Starter Switch Signal	In P/N: 0-3.0v
12	WT/BL	EFI Main Relay Power	12-14v
14	GN/YL	Circuit Opening Relay (FC)	0-3v, off-idle: 12v
20	GN/WT	Stop Light Switch Signal	Brake Off: 0v, On: 12v
21	GN/BK	4WD Indicator Switch	Switch Off: 12v, On: 1v

Pin Connector Graphic

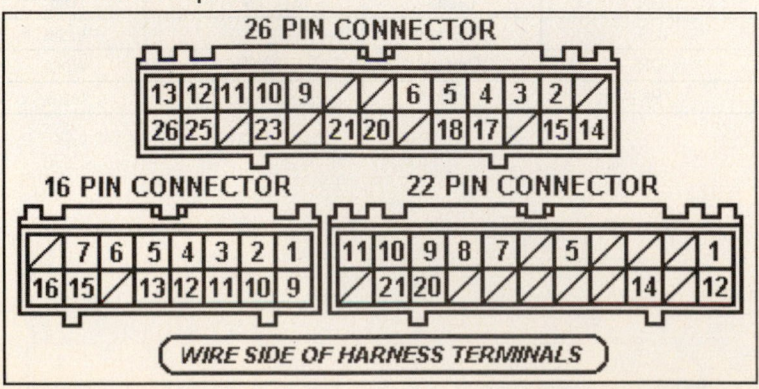

1997 Utility 2.7L I4 MFI VIN M (All) 16 Pin Connector

PCM Pin #	Wire Color	Circuit Description (16 Pin)	Value at Hot Idle
1	GN/YL	Sensor VREF	4.9-5.1
2	BK/RD	MAF Sensor Signal	1.1-1.8v
3	YL/GN	IAT Sensor Signal (THA)	At 100°F: 2.60v
4	GN/RD	ECT Sensor Signal (THW)	At 180°F: 0.51v
5	WT	HO2S-11 (B1 S1) Signal	0.1-1.1v
7	GN/BK	Data Link Connector	12-14v
8	RD/BK	EVAP Vapor Pressure Valve (VSV)	12v or 0v
9	BR/BK	Sensor Ground	<0.050v
10	YL	TP Sensor Signal (VTA)	0.3-0.8v
11	PK	EGR Gas Temperature Sensor	3.5-4.0v
12	GY	Knock Sensor Signal	<0.075v AC
13	RD	HO2S-12 (B1 S2) Signal	0.1-1.1v
15	PK/BK	EGR Solenoid Control (VSV)	12v or 0v

1997 Utility 2.7L I4 MFI VIN M (All) 26 Pin Connector

PCM Pin #	Wire Color	Circuit Description (26 Pin)	Value at Hot Idle
1	RD/WT	HO2S-12 (B1 S2) Heater	1v (Heater on)
2	PK	HO2S-11 (B1 S1) Heater	1v (Heater on)
3	BK/WT	EVAP Purge Solenoid	12-14v
4	YL/BL	Closed Throttle Switch	1v, off-idle: 12v
6	BK	IAC Signal (RSC)	Pulse Signals
7	BK/RD	Idle Air Control Valve (RSO)	Pulse Signals
8	GN/RD	A/T-ECT Solenoid (S1)	3rd or OD: 1v
9	RD/BL	Injector 4 Control	2.0-3.3 ms
10	GN	Injector 3 Control	2.0-3.3 ms
11	WT	Injector 2 Control	2.0-3.3 ms
12	RD	Injector 1 Control	2.0-3.3 ms
13	BR	Power Ground	<0.1v
14	GN/YL	Circuit Opening Relay (FC)	0-3v, off-idle: 12v
15	BK	Tachometer Signal (TACO)	Pulse Signals
17	BK/YL	Igniter Signal (IGF)	Digital Signal: 0-5-0v
20	RD/GN	A/T-ECT Solenoid (SL)	In Lockup: 12-14v
21	LG	A/T-ECT Solenoid (S2)	1st or OD: 1v
22	YL/BK	Igniter Transistor 2 Control	6°, at 55 mph: 8° dwell
23	BK/BL	Igniter Transistor 1 Control	6°, at 55 mph: 8° dwell
24	BR	Sensor Ground	<0.050v
25	BR	Power Ground	<0.1v
26	BR	Power Ground	<0.1v

1997 Utility 2.7L I4 MFI VIN M (All) 12 Pin Connector

PCM Pin #	Wire Color	Circuit Description (12 Pin)	Value at Hot Idle
1	BK/WT	EVAP Purge Solenoid	12-14v
2	OR	A/T: Defogger Idle-Up Signal	Switch On: 12v
3	WT/RD	A/T Vehicle Speed Sensor (-)	AC pulse signals
5	PK	CMP Sensor Signal (G-)	<0.050v
6	OR	CKP Sensor Signal (NE-)	<0.050v
7	RD/WT	MAF Meter Ground	<0.050v
9	YL/RD	A/T Vehicle Speed Sensor (+)	AC pulse signals
10	GN/BK	EVAP Vapor Pressure Sensor (PTNK)	2.5-3.1v
11	PK	CMP Sensor Signal (G+)	AC pulse signals
12	RD	CKP Sensor Signal (NE+)	AC pulse signals

1997 Utility 2.7L I4 MFI VIN M (All) 22 Pin Connector

PCM Pin #	Wire Color	Circuit Description (22 Pin)	Value at Hot Idle
2	BK/WT	Direct Battery	12-14v
3	YL	A/T Oil Temperature Indicator	Lamp Off: 12v, On: 1v
4	PK	Malfunction Indicator Lamp Control	MIL Off: 12v, On: 1v
5	BL/OR	Overdrive Main Switch	Switch Off: 12v, On: 1v
6	BL/BK	A/C Amplifier Signal (ACT)	Clutch On: 12v, Off: 1.5v
7	BL/YL	A/C Amplifier Signal (AC1)	Clutch On: 1.5v, Off: 12v
8	GN/OR	Vehicle Speed Sensor	At 55 mph: 48 Hz
10	GN/BK	4WD Indicator Switch	Switch Off: 12v, On: 1v
11	BK/WT	Starter Switch Signal	In P/N: 0-3.0v
12	WT/BL	EFI Main Relay B+	12-14v
13	RD/GN	4WD Detection Transfer (N)	Open: 12v, Closed: 0v
14	GN/RD	AT ECT Pattern Select Switch	Norm: 0v, PWR: 12v
15	LG	A/T Select Switch Neutral	In 'N': 12v, Others: 0v
16	PK/RD	A/T Select Switch Low	In Low: 12v, Others: 0v
17	RD/YL	A/T Select Switch Reverse	In 'R': 12v, Others: 0v
18	BR/YL	Cruise Control ECU	At Cruise in OD: 12v
19	WT	SIL Signal (Scan Tool)	12v
21	GN/WT	Stop Light Switch Signal	Brake Off: 0v, On: 12v

Pin Connector Graphic

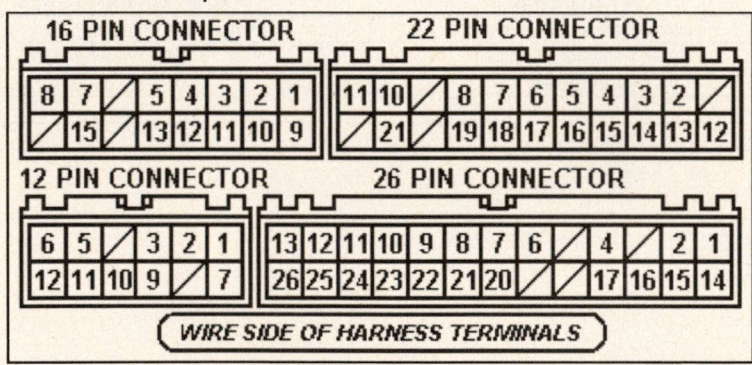

1998-99 Utility 2.7L I4 MFI VIN M (All) 16 Pin Connector

PCM Pin #	Wire Color	Circuit Description (16 Pin)	Value at Hot Idle
1	GN/YL	TP Sensor VREF	4.9-5.1v
2	BK/RD	MAF Sensor Signal	1.1-1.8v
3	YL/GN	IAT Sensor Signal (THA)	At 100°F: 2.60v
4	GN/RD	ECT Sensor Signal (THW)	At 180°F: 0.51v
5	WT	HO2S-11 (B1 S1) Signal	0.1-1.1v
7	GN/BK	Data Link Connector	12-14v
8	RD/BK	EVAP Vapor Pressure Valve (VSV)	12v or 0v
9	BR/BK	Sensor Ground	<0.050v
10	YL	TP Sensor Signal (VTA)	0.3-0.8v
11	PK	EGR Gas Temperature Sensor	3.5-4.0v
12	GY	Knock Sensor Signal	<0.075v AC
13	RD	HO2S-12 (B1 S2) Signal	0.1-1.1v
15	PK/BL	EGR Solenoid Control (VSV)	12v or 0v

1998-99 Utility 2.7L I4 MFI VIN M (All) 26 Pin Connector

PCM Pin #	Wire Color	Circuit Description (26 Pin)	Value at Hot Idle
1	RD/WT	HO2S-12 (B1 S2) Heater	1v (Heater on)
2	PK	HO2S-11 (B1 S1) Heater	1v (Heater on)
3	BK/WT	EVAP Purge Solenoid (VSV)	12v or 0v
4	YL/GN	Closed Throttle Switch	1v, off-idle: 12v
6	PK/BK	IAC Signal (RSC)	Pulse Signals
7	BK/RD	Idle Air Control Valve (RSO)	Pulse Signals
8	GN/RD	AT ECT Solenoid (S1)	3rd or OD: 1v
9	RD/BL	Injector 4 Control	2.0-3.3 ms
10	GN	Injector 3 Control	2.0-3.3 ms
11	WT	Injector 2 Control	2.0-3.3 ms
12	RD	Injector 1 Control	2.0-3.3 ms
13	BR	Power Ground	<0.1v
14	GN/YL	Circuit Opening Relay (FC)	0-3v, off-idle: 12v
15	BK	Tachometer Signal (TACO)	Pulse Signals
16	PK/YL	A/T Oil Temperature Sensor	At 68°F: 4-5v
17	BK/YL	Igniter Signal (IGF)	Digital Signal: 0-5-0v
18	BL/RD	4WD Detection Transfer (L4)	Switch Closed: 0.1v
20	RD/GN	A/T ECT Solenoid (SL)	In Lockup: 12-14v
21	LG	A/T ECT Solenoid (S2)	1st or OD: 1v
22	YL/BK	Igniter Transistor 2 Control	6°, at 55 mph: 8° dwell
23	BK/BL	Igniter Transistor 1 Control	6°, at 55 mph: 8° dwell
24	BR	Sensor Ground	<0.050v
25	BR	Power Ground	<0.1v
26	BR	Power Ground	<0.1v

1998-99 Utility 2.7L I4 MFI VIN M (All) 12 Pin Connector

PCM Pin #	Wire Color	Circuit Description (12 Pin)	Value at Hot Idle
2	OR	A/T: Defogger Idle-up Signal	Load On: 12v
3	WT/RD	A/T ECT Speed Sensor (-)	Pulses
5	PK	CMP Sensor Signal (G-)	AC pulse signals
6	OR	CKP Sensor Signal (NE-)	AC pulse signals
7	RD/WT	MAF Meter Ground	0.050v
9	YL/RD	A/T ECT Speed Sensor (+)	Pulses
10	GN/BK	EVAP Vapor Pressure Sensor (PTNK)	2.5-3.1v
11	PK	CMP Sensor Signal (G+)	AC pulse signals
12	RD	CKP Sensor Signal (NE+)	AC pulse signals

1998-99 Utility 2.7L I4 MFI VIN M (All) 22 Pin Connector

PCM Pin #	Wire Color	Circuit Description (22 Pin)	Value at Hot Idle
2	BL/RD	Direct Battery	12-14v
3	YL	A/T Oil Temperature Indicator	Light Off: 12v, On: 1v
4	PK	Malfunction Indicator Lamp Control	MIL Off: 12v, On: 1v
5	BL/OR	Overdrive Main Switch	Switch Off: 12v, On: 1v
6	BL/BK	A/C Amplifier Signal (ACT)	Clutch On: 12v, Off: 1.5v
7	BL/YL	A/C Amplifier Signal (AC1)	Clutch On: 1.5v, Off: 12v
8	GN/OR	Vehicle Speed Sensor	At 55 mph: 48 Hz
10	GN/BK	4WD Detection Switch	Switch Off: 12v, On: 1v
10 ('99)	BL	4WD Detection Switch	Switch Off: 12v, On: 1v
11	BK/WT	Starter Switch Signal	In P/N: 0-3.0v
12	WT/BL	EFI Main Relay B+	12-14v
13	RD/GN	4WD Detection Transfer (N)	Open: 12v, Closed: 0v
14	GN/RD	AT ECT Pattern Select Switch	Norm: 0v, PWR: 12v
15	LG	A/T Select Switch Low	In Low: 12v, Others: 0v
16	PK/RD	A/T Select Switch 2nd	In 2nd: 12v, Others: 0v
17	RD/YL	A/T Select Switch Reverse	In 'R': 12v, Others: 0v
18	BR/YL	Cruise Control ECU	At Cruise in OD: 12v
19	WT	SIL Signal (Scan Tool)	12v
21	GN/WT	Stop Light Switch Signal	Brake Off: 0v, On: 12v
22	BK	Start Circuit Signal	Cranking: 9-11v

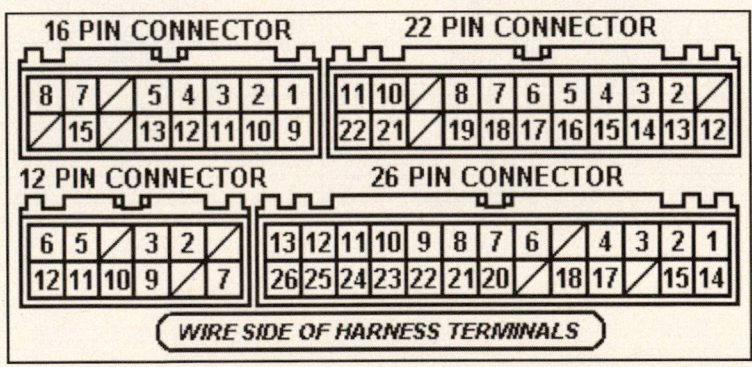

2000 Utility 2.7L I4 MFI VIN M California (All) 31 Pin Connector

PCM Pin #	Wire Color	Circuit Description (31 Pin)	Value at Hot Idle
1	RD	Injector 1 Control	2.0-3.3 ms
2	WT	Injector 2 Control	2.0-3.3 ms
3	GN	Injector 3 Control	2.0-3.3 ms
4	RD/BL	Injector 4 Control	2.0-3.3 ms
5, 6, 21, 31	WT/BK	Power Ground	<0.1v
7	LG/RD	AT ECT Solenoid (S1)	3rd or OD: 1v
8	BL/WT	A/T ECT Solenoid (S2)	1st or OD: 1v
9	GN/RD	A/T ECT Solenoid (SL)	In Lockup: 12-14v
10	BK/YL	Igniter Signal (IGF)	Digital Signal: 0-5-0v
11	BK/BL	Igniter Transistor 1 Control	6°, at 55 mph: 8° dwell
12	YL/BK	Igniter Transistor 2 Control	6°, at 55 mph: 8° dwell
13	BL	Igniter Transistor 3 Control	6°, at 55 mph: 8° dwell
14	BL/YL	Igniter Transistor 4 Control	6°, at 55 mph: 8° dwell
15	BK/RD	Idle Air Control Valve (RSO)	Pulse Signals
26	PK/YL	A/T Oil Temperature Sensor	At 68°F: 4-5v
28	GY	Knock Sensor Signal	<0.075v AC

2000 Utility 2.7L I4 MFI VIN M California (All) 22 Pin Connector

PCM Pin #	Wire Color	Circuit Description (22 Pin)	Value at Hot Idle
1	BL/RD	Direct Battery	12-14v
5	YL	A/T: Oil Temp Lamp Indicator	Lamp Off: 12v, On: 1v
6	PK	Malfunction Indicator Lamp Control	MIL Off: 12v, On: 1v
7	BK/WT	Starter Switch Signal	In P/N: 0-3.0v
9	RD/GN	4WD Detection Transfer (N)	Open: 12v, Closed: 0v
10	BL/OR	O/D OFF (lamp) Indicator	Light Off: 12v, On: 1v
11	BR/YL	A/T Select Switch Drive	In Drive: 12-14v
12	WT	SIL Signal (Scan Tool)	0v
13	BL/YL	A/C Amplifier Signal (AC1)	Clutch On: 1.5v, Off: 12v
14	BL/BK	A/C Amplifier Signal (ACT)	Clutch On: 12v, Off: 1.5v
15	GN/WT	Stop Light Switch Signal	Brake Off: 0v, On: 12v
16	WT/BL	EFI Main Relay B+	12-14v
17	GN/RD	AT ECT Pattern Select Switch	Norm: 0v, PWR: 12v
19	YL/BK	4WD Detection Transfer (L4)	Switch Closed: 0.1v
20	OR	A/T: Defogger Idle-up Signal	Load On: 12v
21	GN/OR	Speedometer Indicator	At 55 mph: 48 Hz
22	BK	Start Circuit Signal	Cranking: 9-11v

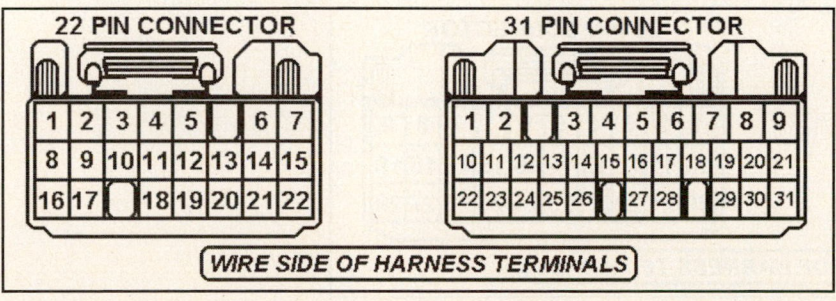

2000 Utility 2.7L I4 MFI VIN M California (All) 28 Pin Connector

PCM Pin #	Wire Color	Circuit Description (28 Pin)	Value at Hot Idle
2	RD/YL	A/T Select Switch Reverse	In 'R': 12v, Others: 0v
3	LG	A/T Select Switch Low	In Low: 12v, Others: 0v
4	PK/RD	A/T Select Switch 2nd	In 2nd: 12v, Others: 0v
5	YL/GN	Closed Throttle Switch	1v, off-idle: 12v
6	GN/YL	Circuit Opening Relay (FC)	0-3v, off-idle: 12v
7	LG/BK	Data Link Connector	12-14v
8	YL	EVAP Vapor Pressure Sensor (PTNK)	2.5-3.1v
10	BL	4WD Detection Switch	Switch Off: 12v, On: 1v
13	BK	Tachometer Signal (TACO)	Pulses
14	YL/RD	A/T ECT Speed Sensor (+)	Pulses
23	WT/RD	A/T ECT Speed Sensor (-)	Pulses
25	BL/WT	Overdrive Main Switch	Switch Off: 12v, On: 1v
28	BL/RD	PSP Switch Signal (PSW)	Straight: 12v, Turned: 0v

2000 Utility 2.7L I4 MFI VIN M California (All) 24 Pin Connector

PCM Pin #	Wire Color	Circuit Description (24 Pin)	Value at Hot Idle
2	GN/YL	TP Sensor VREF	4.9-5.1v
3	RD/WT	HO2S-12 (B1 S2) Heater	1v (Heater on)
4	RD	AFRS-11 (B1 S1) Heater	1v (Heater on)
5	PK/BL	EGR Solenoid Control (VSV)	12v or 0v
7	RD/BK	EVAP Vapor Pressure Valve (VSV)	12v or 0v
8	BK/WT	EVAP Purge Solenoid (VSV)	12v or 0v
9	YL	TP Sensor Signal (VTA)	0.3-0.8v
10	RD	HO2S-12 (B1 S2) Signal	0.1-1.1v
11	BK	AFS-11 (B1 S1) Signal (+)	Fixed at 3.3v
12	GN/RD	ECT Sensor Signal (THW)	At 180ºF: 0.51v
14	BK/RD	MAF Sensor Signal	1.1-1.8v
15	RD	CMP Sensor Signal	AC pulse signals
16	RD	CKP Sensor Signal	AC pulse signals
17	BR	Sensor Ground	<0.050v
18	BL/BK	Sensor Ground	<0.050v
19	PK	EGR Gas Temperature Sensor	3.5-4.0v
20	WT	AFS-11 (B1 S1) Signal (-)	Fixed at 3.3v
21	YL/GN	IAT Sensor Signal (THA)	At 100ºF: 2.60v
22	B/YL	MAF Meter Ground	0.050v
24	GN	CKP/CMP Sensor Signal (G-)	0.050v

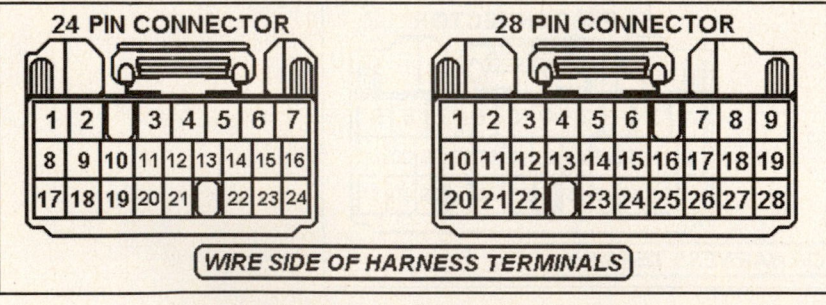

2000 Utility 2.7L I4 MFI VIN M Federal (All) 22 Pin Connector

PCM Pin #	Wire Color	Circuit Description (22 Pin)	Value at Hot Idle
2	BL/RD	Direct Battery	12-14v
3	YL	A/T: Oil Temp. Lamp Indicator	Lamp Off: 12v, On: 1v
4	PK	Malfunction Indicator Lamp Control	MIL Off: 12v, On: 1v
5	BL/OR	O/D OFF (lamp) Indicator	Light Off: 12v, On: 1v
6	BL/BK	A/C Amplifier Signal (ACT)	Clutch On: 12v, Off: 1.5v
7	BL/YL	A/C Amplifier Signal (AC1)	Clutch On: 1.5v, Off: 12v
8	GN/OR	Speedometer Indicator	At 55 mph: 48 Hz
10	BL	4WD Detection Switch	Switch Off: 12v, On: 1v
11	BK/WT	Starter Switch Signal	In P/N: 0-3.0v
12	WT/BL	EFI Main Relay B+	12-14v
13	RD/GN	4WD Detection Transfer (N)	Open: 12v, Closed: 0v
14	GN/RD	AT ECT Pattern Select Switch	Norm: 0v, PWR: 12v
15	LG	A/T Select Switch Low	In Low: 12v, Others: 0v
16	PK/RD	A/T Select Switch 2nd	In 2nd: 12v, Others: 0v
17	RD/YL	A/T Select Switch Reverse	In 'R': 12v, Others: 0v
18	BR/YL	Cruise Control ECU	At Cruise in OD: 12v
19	WT	SIL Signal (Scan Tool)	12v
21	GN/WT	Stop Light Switch Signal	Brake Off: 0v, On: 12v
22	BK	Start Circuit Signal	Cranking: 9-11v

2000 Utility 2.7L I4 MFI VIN M Federal (All) 26 Pin Connector

PCM Pin #	Wire Color	Circuit Description (26 Pin)	Value at Hot Idle
1	RD/WT	HO2S-12 (B1 S2) Heater	1v (Heater on)
2	PK	HO2S-11 (B1 S1) Heater	1v (Heater on)
3	BK/WT	EVAP Purge Solenoid (VSV)	12v or 0v
4	YL/GN	Closed Throttle Switch	1v, off-idle: 12v
6	PK/BL	IAC Signal (RSC)	Pulse Signals
7	BK/RD	Idle Air Control Valve (RSO)	Pulse Signals
8	GN/RD	AT ECT Solenoid (S1)	3rd or OD: 1v
9	RD/BL	Injector 4 Control	2.0-3.3 ms
10	GN	Injector 3 Control	2.0-3.3 ms
11	WT	Injector 2 Control	2.0-3.3 ms
12	RD	Injector 1 Control	2.0-3.3 ms
13	WT/BK	Power Ground	<0.1v
14	GN/YL	Circuit Opening Relay (FC)	0-3v, off-idle: 12v
15	BK	Tachometer Signal (TACO)	Pulses
16	PK/YL	A/T Oil Temperature Sensor	At 68°F: 4-5v
17	BK/YL	Igniter Signal (IGF)	Digital Signal: 0-5-0v
18	BL/RD	4WD Detection Transfer (L4)	Switch Closed: 0.1v
20	RD/GN	A/T ECT Solenoid (SL)	In Lockup: 12-14v
21	LG	A/T ECT Solenoid (S2)	1st or OD: 1v
22	YL/BK	Igniter Transistor 2 Control	6°, at 55 mph: 8° dwell
23	BK/BL	Igniter Transistor 1 Control	6°, at 55 mph: 8° dwell
24	BR	Sensor Ground	<0.050v
25	WT/BK	Power Ground	<0.1v
26	WT/BK	Power Ground	<0.1v

2000 Utility 2.7L I4 MFI VIN M Federal (All) 12 Pin Connector

PCM Pin #	Wire Color	Circuit Description (12 Pin)	Value at Hot Idle
1	BL/WT	Overdrive Main Switch	Switch Off: 12v, On: 1v
2	OR	A/T: Defogger Idle-up Signal	Load On: 12v
3	WT/RD	ECT Speed (-) Sensor	Pulses
6	GN	CKP/CMP Sensor Signal (G-)	0.050v
7	RD/WT	MAF Meter Ground	0.050v
9	YL/RD	A/T ECT Speed Sensor (+)	Pulses
10	LG/BK	EVAP Vapor Pressure Sensor (PTNK)	2.5-3.1v
11	RD	CMP Sensor Signal (G+)	AC pulse signals
12	RD	CKP Sensor Signal (NE+)	AC pulse signals

2000 Utility 2.7L I4 MFI VIN M Federal (All) 16 Pin Connector

PCM Pin #	Wire Color	Circuit Description (16 Pin)	Value at Hot Idle
1	GN/YL	TP Sensor VREF	4.9-5.1v
2	BK/RD	MAF Sensor Signal	1.1-1.8v
3	YL/GN	IAT Sensor Signal (THA)	At 100°F: 2.60v
4	GN/RD	ECT Sensor Signal (THW)	At 180°F: 0.51v
5	WT	HO2S-11 (B1 S1) Signal	0.1-1.1v
7	LG/BK	Data Link Connector	12-14v
8	RD/BK	EVAP Vapor Pressure Valve (VSV)	12v or 0v
9	BL/BK	Sensor Ground	<0.050v
10	YL	TP Sensor Signal (VTA)	0.3-0.8v
11	PK	EGR Gas Temperature Sensor	3.5-4.0v
12	GY	Knock Sensor Signal	<0.075v AC
13	RD	HO2S-12 (B1 S2) Signal	0.1-1.1v
15	PK/BL	EGR Solenoid Control (VSV)	12v or 0v

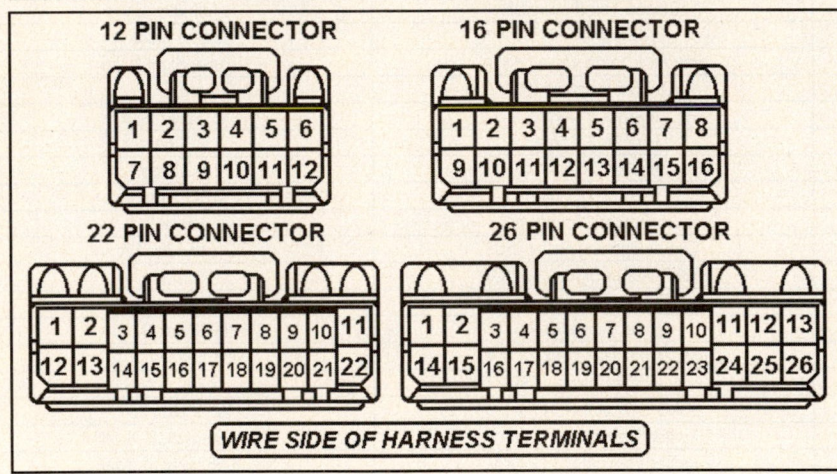

1990 Utility 3.0L V6 MFI VIN V (A/T-ECT) 26 Pin Connector

PCM Pin #	Wire Color	Circuit Description (26 Pin)	Value at Hot Idle
1	WT	Distributor Signal (NE+)	AC pulse signals
2	RD	Distributor Signal (G1+)	AC pulse signals
3	BK/YL	Igniter Signal (IGF)	Digital Signal: 0-5-0v
4	GN/RD	AT ECT Solenoid S4	Pulses
5	YL/BK	AT ECT Solenoid S3	Pulses
6	BK	A/T ECT Solenoid (S2)	1st or OD: 1v
7	WT	AT ECT Solenoid (S1)	3rd or OD: 1v
8	GN	Fuel Pressure Up Solenoid	1v (at hot restart)
9	GN	Cold Start Injector Control	1v (at cold startup)
10	PK/GN	HO2S Heater Control	1v (Heater on)
11	BR	Sensor Ground	<0.050v
12	WT/RD	Injectors 1 & 3 & 5 Control	1.6-2.9 ms
13	BR	Power Ground	<0.1v
14	GN	Distributor Signal (G-)	AC pulse signals
15	BK	Distributor Signal (G2+)	AC pulse signals
16	BR/RD	AT ECT Speed Sensor	Moving: 0-5-0-5v
17	PK/WT	A/T Select Switch Low	In Low: 12v, Others: 0v
18	PK/GN	A/T Select Switch 2nd	In 2nd: 12v, Others: 0v
19	PK/RD	A/T Select Switch Neutral	In 'N': 12v, Others: 0v
20	YL/RD	4WD Detection Transfer (L4)	Switch Closed: 0.1v
21	BK/BL	Igniter Signal (IGT)	Digital Signal: 0-5-0v
22	PK	EGR Solenoid Control (VSV)	12v or 0v
23	GN/RD	Intake Air Solenoid Control	12v or 0v
24	BK/RD	A/C Idle-Up Solenoid	A/C Off: 12v, On: 1v
25	WT	Injectors 2 & 4 & 6 Control	1.6-2.9 ms
26	BR	Power Ground	<0.1v

Pin Connector Graphic

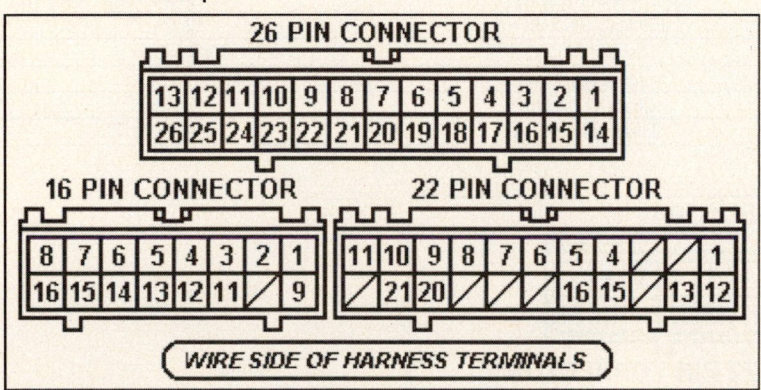

Standard Colors and Abbreviations

Abbreviation	Color	Abbreviation	Color	Abbreviation	Color
BK	Black	GY	Gray	RD	Red
BL	Blue	GN	Green	TN	Tan
BR	Brown	LG	LT Green	VT	Violet
DB	Dark Blue	OR	Orange	WT	White
DG	DK Green	PK	Pink	YL	Yellow

1990 Utility 3.0L V6 MFI VIN V (A/T-ECT) 16 Pin Connector

PCM Pin #	Wire Color	Circuit Description (16 Pin)	Value at Hot Idle
1	GN/BK	Sensor VREF	4.9-5.1v
2	YL/BL	Airflow Meter Signal	0.5-2.5v
3	YL/GN	IAT Sensor Signal (THA)	At 100ºF: 2.60v
4	GN/BL	ECT Sensor Signal (THW)	At 180ºF: 0.51v
5	BK	Knock Sensor Signal	<0.075v AC
6	BK	HO2S-11 (B1 S1) Signal	0.1-1.1v
7	GN/BK	4WD Oil Temperature Sensor	At 68ºF: 4-5v
8	YL/GN	Check Connector	12-14v
9	BR/BK	Sensor Ground	<0.050v
11	YL	TP Sensor Signal (VTA)	0.3-0.8v
12	YL/BL	Closed Throttle Switch	1v, off-idle: 12v
13 (Cal)	GN/WT	EGR Gas Temperature Sensor	3.5-4.0v
14	LG	Transfer Fluid Temp. Sensor	At 230ºF: <1.5v
15	PK/WT	Check Connector	12-14v
16	PK	Coolant Temperature Switch	Switch Closed: 0.1v

1990 Utility 3.0L V6 MFI VIN V (A/T-ECT) 22 Pin Connector

PCM Pin #	Wire Color	Circuit Description (22 Pin)	Value at Hot Idle
1	BK/GN	Direct Battery	12-14v
4	BL/GN	A/T Oil Temperature Indicator	Light Off: 12v, On: 1v
5	PK	Malfunction Indicator Lamp Control	MIL Off: 12v, On: 1v
6	GN/WT	Stop Light Switch Signal	Brake Off: 0v, On: 12v
7	GN/OR	AT ECT Pattern Select Switch	Norm: 0v, PWR: 12v
8	PK/GN	4WD Indicator Switch	Switch Off: 12v, On: 1v
9	GN/BL	Vehicle Speed Sensor	At 55 mph: 48 Hz
10	BK/WT	A/C Amplifier On Signal	Clutch On: 1.5v, Off: 12v
11	BK/WT	Starter Switch Signal	In P/N: 0-3.0v
12	WT/RD	EFI Main Relay B+	12-14v
13	WT/RD	EFI Main Relay B1+	12-14v
15	BR/BK	Sensor Ground	<0.050v
16	YL/GN	Overdrive Main Switch	Switch Off: 12v, On: 1v
20	PK	Check Connector	12-14v
21	YL/RD	Cruise Control ECU	At Cruise in OD: 12v

Pin Connector Graphic

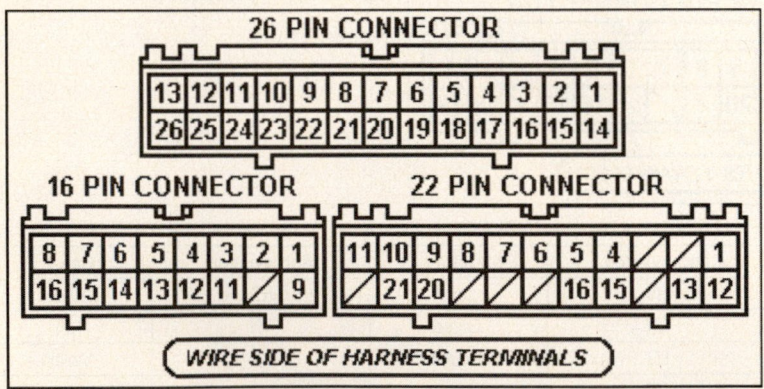

Standard Colors and Abbreviations

Abbreviation	Color	Abbreviation	Color	Abbreviation	Color
BK	Black	GY	Gray	RD	Red
BL	Blue	GN	Green	TN	Tan
BR	Brown	LG	LT Green	VT	Violet
DB	Dark Blue	OR	Orange	WT	White
DG	DK Green	PK	Pink	YL	Yellow

1990 Utility 3.0L V6 MFI VIN V (All) 10 Pin Connector

PCM Pin #	Wire Color	Circuit Description (10 Pin)	Value at Hot Idle
2	GN	Cold Start Injector Control	1v (at cold startup)
3	BK/WT	Starter Switch Signal	In P/N: 0-3.0v
4	WT/RD	Injectors 1 & 3 & 5 Control	1.6-2.9 ms
5	BR	Power Ground	<0.1v
7	BR	Sensor Ground	<0.050v
8	BL/BK	Igniter Signal (IGT)	0.72-0.74v
9	BK/WT	Injectors 2 & 4 & 6 Control	1.6-2.9 ms
10	BR	Power Ground	<0.1v

1990 Utility 3.0L V6 MFI VIN V (All) 18 Pin Connector

PCM Pin #	Wire Color	Circuit Description (18 Pin)	Value at Hot Idle
1	GN/BL	ECT Sensor Signal (THW)	At 180°F: 0.51v
2	BK/YL	Distributor IGF Signal	0.74-0.76v
3	PK/GN	HO2S Heater Control	1v (Heater on)
4	WT	Distributor Signal (NE+)	AC pulse signals
5	BK	Distributor Signal (G2+)	AC pulse signals
6	RD	Distributor Signal (G1+)	AC pulse signals
7	GN	Distributor Signal (G-)	AC pulse signals
8	GN	Fuel Pressure Up Solenoid	1v (at hot restart)
10	BR/BK	Sensor Ground	<0.050v
12	PK	EGR Solenoid Control (VSV)	12v or 0v
13	YL/BL	Closed Throttle Switch	1v, off-idle: 12v
14	YL	TP Sensor Signal (VTA)	0.3-0.8v
15	PK/WT	Check Connector	12-14v
16	YL/GN	Check Connector	12-14v
17	BK/RD	A/C Idle-Up Solenoid	12-14v
18 (Cal)	GN/WT	EGR Gas Temperature Sensor	3.5-4.0v
18 (Fed)	GN/WT	Short Pin	N/A

1990 Utility 3.0L V6 MFI VIN V (All) 24 Pin Connector

PCM Pin #	Wire Color	Circuit Description (24 Pin)	Value at Hot Idle
2	BK/GN	Direct Battery	12-14v
3	GN/BK	Airflow Meter VREF	4.9-5.1v
4	YL/BL	Airflow Meter Signal	0.5-2.5v
5	YL/GN	IAT Sensor Signal (THA)	At 100°F: 2.60v
6	GN/WT	Stop Light Switch Signal	Brake Off: 0v, On: 12v
7	GN/BL	Vehicle Speed Sensor	At 55 mph: 48 Hz
12	BK	HO2S-11 (B1 S1) Signal	0.1-1.1v
13	WT/RD	EFI Main Relay B1+	12-14v
14	WT/RD	EFI Main Relay B+	12-14v
15	PK	Water Temperature Switch	Switch Closed: 0.1v
16	GN/WT	4WD Indicator Switch	Switch Off: 12v, On: 1v
17	BK	Knock Sensor Signal	<0.075v AC
18	GN/RD	Intake Air Solenoid Control	12v or 0v
19	PK	Malfunction Indicator Lamp Control	MIL Off: 12v, On: 1v
20	BK/WT	A/C Magnetic Clutch Relay (ACMG)	Relay Off: 0v, On: 12v
24	BR/BK	Sensor Ground	<0.050v

Pin Connector Graphic

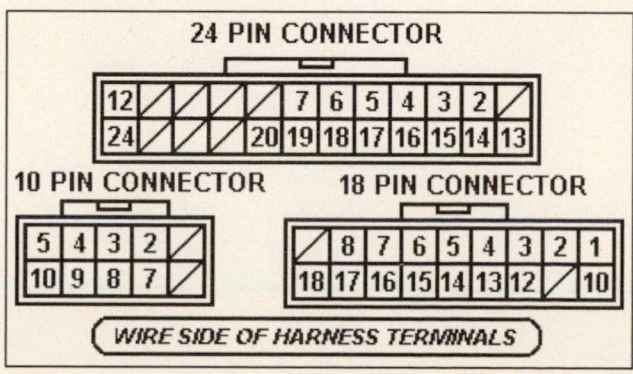

1991-92 Utility 3.0L V6 MFI VIN V (A/T-ECT) 26 Pin Connector

PCM Pin #	Wire Color	Circuit Description (26 Pin)	Value at Hot Idle
1	WT	Distributor Signal (NE+)	AC pulse signals
2	RD	Distributor Signal (G1+)	AC pulse signals
3	BK/YL	Igniter Signal (IGF)	Digital Signal: 0-5-0v
4	GN/RD	AT ECT Solenoid S4	Pulses
5	YL/BK	AT ECT Solenoid S3	Pulses
6	BK	A/T ECT Solenoid (S2)	1st or OD: 1v
7	WT	AT ECT Solenoid (S1)	3rd or OD: 1v
8	GN	Fuel Pressure Up Solenoid	1v (at hot restart)
9	GN	Cold Start Injector Control	1v (at cold startup)
10	PK/GN	HO2S Heater Control	1v (Heater on)
11	BR	Sensor Ground	<0.050v
12	WT/RD	Injectors 1 & 3 & 5 Control	1.6-2.9 ms
13	BR	Power Ground	<0.1v
14	GN	Distributor Signal (G-)	AC pulse signals
15	BK	Distributor Signal (G2+)	AC pulse signals
16	BR/RD	AT ECT Speed Sensor	Moving: 0-5-0-5v
17	PK/WT	A/T Select Switch Low	In Low: 12v, Others: 0v
18	PK/GN	A/T Select Switch 2nd	In 2nd: 12v, Others: 0v
19	PK/RD	A/T Select Switch Neutral	In 'N': 12v, Others: 0v
20	YL/RD	4WD Detection Transfer (L4)	Switch Closed: 0.1v
21	BK/BL	Igniter Signal (IGT)	Digital Signal: 0-5-0v
22	PK	EGR Solenoid Control (VSV)	12v or 0v
23	GN/RD	Intake Air Solenoid Control	12v or 0v
24	BK/RD	A/C Idle-Up Solenoid	A/C Off: 12v, On: 1v
25	WT	Injectors 2 & 4 & 6 Control	1.6-2.9 ms
26	BR	Power Ground	<0.1v

Pin Connector Graphic

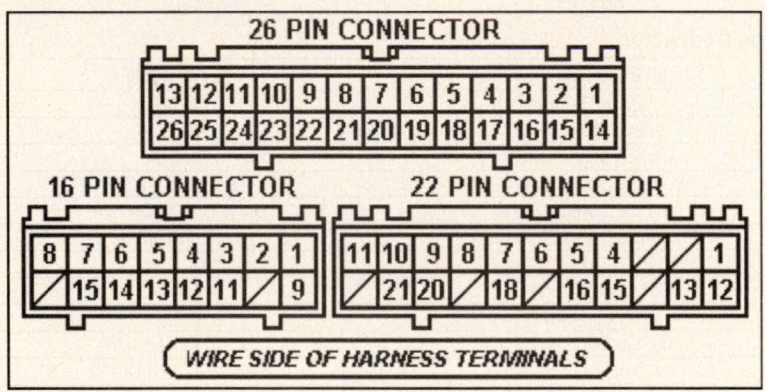

1991-92 Utility 3.0L V6 MFI VIN V (A/T-ECT) 16 Pin Connector

PCM Pin #	Wire Color	Circuit Description (16 Pin)	Value at Hot Idle
1	GN/BK	Airflow Meter VREF	4.9-5.1v
2	RD	Airflow Meter Signal	0.5-2.5v
3	YL/BK	IAT Sensor Signal (THA)	At 100°F: 2.60v
4	GN/BL	ECT Sensor Signal (THW)	At 180°F: 0.51v
5	BK	Knock Sensor Signal	<0.075v AC
6	BK	HO2S-11 (B1 S1) Signal	0.1-1.1v
7	GN/BK	4WD Oil Temperature Sensor	At 68°F: 4-5v
8	YL/GN	Check Connector	12-14v
9	BR/BK	Sensor Ground	<0.050v
11	YL	TP Sensor Signal (VTA)	0.3-0.8v
12	YL/BL	Closed Throttle Switch	1v, off-idle: 12v
13 (Cal)	GN/WT	EGR Gas Temperature Sensor	3.5-4.0v
14	LG	Transfer Fluid Temp. Sensor	At 230°F: <1.5v
15	PK/WT	Check Connector	12-14v

1991-92 Utility 3.0L V6 MFI VIN V (A/T-ECT) 22 Pin Connector

PCM Pin #	Wire Color	Circuit Description (22 Pin)	Value at Hot Idle
1	BK/GN	Direct Battery	12-14v
4	BL/GN	A/T Oil Temperature Indicator	Light Off: 12v, On: 1v
5	PK	Malfunction Indicator Lamp Control	MIL Off: 12v, On: 1v
6	GN/WT	Stop Light Switch Signal	Brake Off: 0v, On: 12v
7	GN/OR	AT ECT Pattern Select Switch	Norm: 0v, PWR: 12v
8	PK/GN	4WD Indicator Switch	Switch Off: 12v, On: 1v
9	GN/BL	Vehicle Speed Sensor	At 55 mph: 48 Hz
10	BK/WT	A/C Amplifier Signal	A/C Off: 12v, On: 1v
11	BK/WT	Starter Switch Signal	In P/N: 0-3.0v
12	WT/RD	EFI Main Relay B+	12-14v
13	WT/RD	EFI Main Relay B1+	12-14v
15	BR/BK	Sensor Ground	<0.050v
16	YL/GN	Overdrive Main Switch	Switch Off: 12v, On: 1v
18	WT/BK	2WD Select (SEL1)	<0.1v
20	PK	Check Connector	12-14v
21	YL/RD	Cruise Control ECU	At Cruise in OD: 12v

Pin Connector Graphic

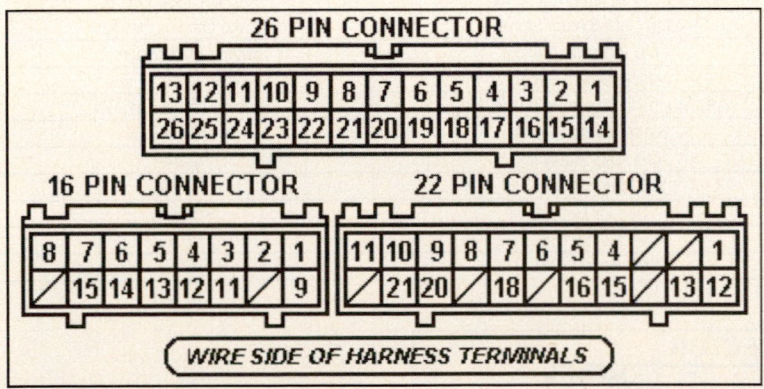

Standard Colors and Abbreviations

Abbreviation	Color	Abbreviation	Color	Abbreviation	Color
BK	Black	GY	Gray	RD	Red
BL	Blue	GN	Green	TN	Tan
BR	Brown	LG	LT Green	VT	Violet
DB	Dark Blue	OR	Orange	WT	White
DG	DK Green	PK	Pink	YL	Yellow

1991-92 Utility 3.0L V6 MFI VIN V (All) 26 Pin Connector

PCM Pin #	Wire Color	Circuit Description (26 Pin)	Value at Hot Idle
1	WT	Distributor Signal (NE+)	AC pulse signals
2	RD	Distributor Signal (G1+)	AC pulse signals
3	BK/YL	Igniter Signal (IGF)	Digital Signal: 0-5-0v
8	GN	Fuel Pressure Up Solenoid	1v (at hot restart)
9	GN	Cold Start Injector Control	1v (at cold startup)
10	PK/GN	HO2S Heater Control	1v (Heater on)
11	BR	Sensor Ground	<0.050v
12	WT/RD	Injectors 1 & 3 & 5 Control	1.6-2.9 ms
13	BR	Power Ground	<0.1v
14	GN	Distributor Signal (G-)	AC pulse signals
15	BK	Distributor Signal (G2+)	AC pulse signals
21	BK/BL	Igniter Signal (IGT)	Digital Signal: 0-5-0v
22	PK	EGR Solenoid Control (VSV)	12v or 0v
23	GN/RD	Intake Air Solenoid Control	12v or 0v
24	BK/RD	A/C Idle-Up Solenoid	A/C Off: 12v, On: 1v
25	WT	Injectors 2 & 4 & 6 Control	1.6-2.9 ms
26	BR	Power Ground	<0.1v

1991-92 Utility 3.0L V6 MFI VIN V (All) 16 Pin Connector

PCM Pin #	Wire Color	Circuit Description (16 Pin)	Value at Hot Idle
1	GN/BK	Airflow Meter VREF	4.9-5.1v
2	YL/RD	Airflow Meter Signal	0.5-2.5v
3	YL/BK	IAT Sensor Signal (THA)	At 100°F: 2.60v
4	GN/BL	ECT Sensor Signal (THW)	At 180°F: 0.51v
5	BK	Knock Sensor Signal	<0.075v AC
6	BK	HO2S-11 (B1 S1) Signal	0.1-1.1v
8	YL/GN	Check Connector	12-14v
9	BR/BK	Sensor Ground	<0.050v
11	YL	TP Sensor Signal (VTA)	0.3-0.8v
12	YL/WT	Closed Throttle Switch	1v, off-idle: 12v
13 (Cal)	GN/WT	EGR Gas Temperature Sensor	3.5-4.0v
15	PK/WT	Check Connector	12-14v

1991-92 Utility 3.0L V6 MFI VIN V (All) 22 Pin Connector

PCM Pin #	Wire Color	Circuit Description (22 Pin)	Value at Hot Idle
1	BK/GN	Direct Battery	12-14v
5	PK	Malfunction Indicator Lamp Control	MIL Off: 12v, On: 1v
6	GN/WT	Stop Light Switch Signal	Brake Off: 0v, On: 12v
8	PK/GN	4WD Indicator Switch	Switch Off: 12v, On: 1v
9	GN/BL	Vehicle Speed Sensor	At 55 mph: 48 Hz
10	BK/WT	A/C Magnetic Clutch Relay (ACMG)	Relay Off: 0v, On: 12v
11	BK/WT	Starter Switch Signal	In P/N: 0-3.0v
12	WT/RD	EFI Main Relay B+	12-14v
13	WT/RD	EFI Main Relay B1+	12-14v
15	BR/BK	Sensor Ground	<0.050v
17	WT/BK	4WD Select (SEL2)	<0.1v
18	WT/BK	2WD Select (SEL1)	<0.1v

Pin Connector Graphic

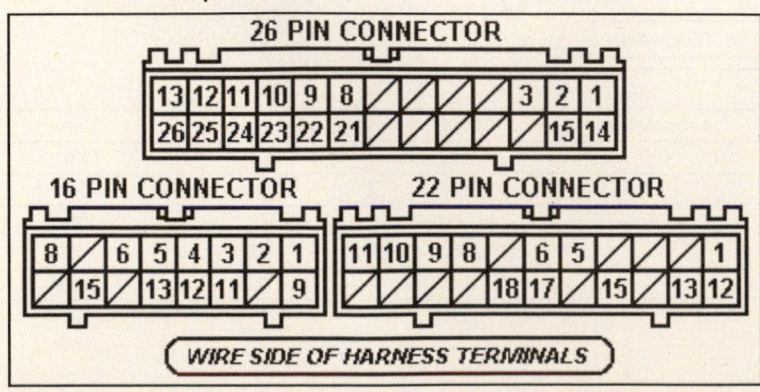

1993-95 Utility 3.0L V6 MFI VIN V (All) 26 Pin Connector

PCM Pin #	Wire Color	Circuit Description (26 Pin)	Value at Hot Idle
1	WT	Distributor Signal (NE+)	AC pulse signals
2	RD	Distributor Signal (G1+)	AC pulse signals
3	BK/YL	Igniter Signal (IGF)	Digital Signal: 0-5-0v
4	GN/RD	AT ECT Solenoid S4	Pulses
5	YL/BK	AT ECT Solenoid S3	Pulses
6	BK	A/T ECT Solenoid (S2)	1st or OD: 1v
7	WT	AT ECT Solenoid (S1)	3rd or OD: 1v
8	GN	Fuel Pressure Up Solenoid	1v (at hot restart)
9	GN	Cold Start Injector Control	1v (at cold startup)
10	PK/GN	Main HO2S Heater Control	1v (Heater on)
11	BR	Sensor Ground	<0.050v
12	WT/RD	Injectors 1 & 3 & 5 Control	1.6-2.9 ms
13	BR	Power Ground	<0.1v
14	GN	Distributor Signal (G-)	AC pulse signals
15	BK	Distributor Signal (G2+)	AC pulse signals
16	BR/RD	AT ECT Speed Sensor	Varies: 0-5-0-5v
17	PK/WT	A/T Select Switch Low	In Low: 12v, Others: 0v
18	PK/GN	A/T Select Switch 2nd	In 2nd: 12v, Others: 0v
19	PK/RD	A/T Select Switch Neutral	In 'N': 12v, Others: 0v
20	YL/RD	4WD Transfer Detect (L4)	Switch Closed: 0.1v
21	BK/BL	Igniter Signal (IGT)	Digital Signal: 0-5-0v
22	PK	EGR Solenoid Control (VSV)	12v or 0v
23	GN/RD	Intake Air Solenoid Control	12v or 0v
24	BK/RD	A/C Idle-Up Solenoid	A/C Off: 12v, On: 1v
25	WT	Injectors 2 & 4 & 6 Control	1.6-2.9 ms
26	BR	Power Ground	<0.1v

Pin Connector Graphic

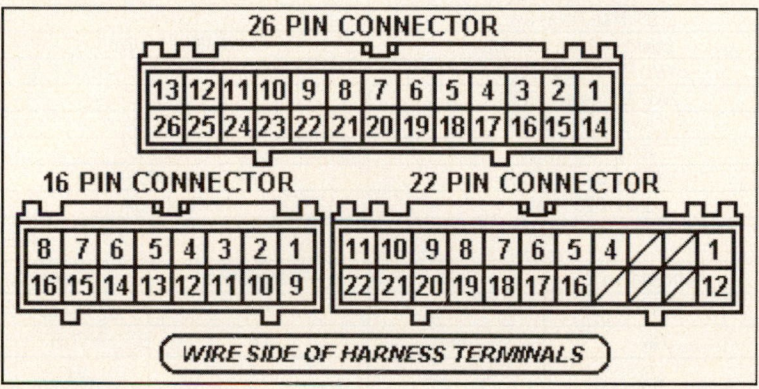

1993-95 Utility 3.0L V6 MFI VIN V (All) 16 Pin Connector

PCM Pin #	Wire Color	Circuit Description (16 Pin)	Value at Hot Idle
1	GN/BK	Airflow Meter VREF	4.9-5.1v
2	YL/RD	Airflow Meter Signal	0.5-2.5v
3	YL/BK	IAT Sensor Signal (THA)	At 100ºF: 2.60v
4	GN/BL	ECT Sensor Signal (THW)	At 180ºF: 0.51v
5	BK	Knock Sensor Signal	<0.075v AC
6	BK	Main HO2S Signal	0.1-1.1v
7	GN/BK	A/T Oil Temperature Sensor	At 68ºF: 4-5v
8	YL/GN	Check Connector	12-14v
9	BR/BK	Sensor Ground	<0.050v
10 (Cal)	WT	Sub HO2S Signal	0.1-1.1v
11	YL	TP Sensor Signal (VTA)	0.3-0.8v
12	YL/WT	Closed Throttle Switch	1v, off-idle: 12v
13 (Cal)	GN/WT	EGR Gas Temperature Sensor	3.5-4.0v
14	LG	Transfer Fluid Temp. Sensor	At 230ºF: <1.5v
15	PK/WT	Check Connector	12-14v

1993-95 Utility 3.0L V6 MFI VIN V (All) 22 Pin Connector

PCM Pin #	Wire Color	Circuit Description (22 Pin)	Value at Hot Idle
1	BK/GN	Direct Battery	12-14v
4	BL/GN	A/T Oil Temperature Indicator	Light Off: 12v, On: 1v
5	PK	Malfunction Indicator Lamp Control	MIL Off: 12v, On: 1v
6	GN/WT	Stop Light Switch Signal	Brake Off: 0v, On: 12v
7	GN/OR	AT ECT Pattern Select Switch	Norm: 0v, PWR: 12v
8	GN/WT	4WD Indicator Switch	Switch Off: 12v, On: 1v
9	GN/BL	Vehicle Speed Sensor	At 55 mph: 48 Hz
10	BK/WT	A/C Magnetic Clutch Relay (ACMG)	Relay Off: 0v, On: 12v
11	BK/WT	MT: Starter Switch Signal	In P/N: 0-3.0v
11	BK/YL	A/T: Starter Switch Signal	In P/N: 0-3.0v
12	WT/RD	EFI Main Relay B+	12-14v
16	YL/GN	Overdrive Main Switch	Switch Off: 12v, On: 1v
17	BR	4WD MT: Select (SEL2)	<0.1v
18	BR	2WD A/T: Select (SEL1)	<0.1v
19	BL/BK	A/C Amplifier Signal (ACT)	Clutch On: 12v, Off: 1.5v
20	PK	Check Connector	12-14v
21	YL/RD	Cruise Control ECU	At Cruise in OD: 12v
22 (Cal)	RD/GN	Sub HO2S Heater	1v (Heater on)

Standard Colors and Abbreviations

Abbreviation	Color	Abbreviation	Color	Abbreviation	Color
BK	Black	GY	Gray	RD	Red
BL	Blue	GN	Green	TN	Tan
BR	Brown	LG	LT Green	VT	Violet
DB	Dark Blue	OR	Orange	WT	White
DG	DK Green	PK	Pink	YL	Yellow

1996-97 Utility 3.4L V6 MFI VIN N (A/T-ECT) 28 Pin Connector

PCM Pin #	Wire Color	Circuit Description (28 Pin)	Value at Hot Idle
1	RD/YL	A/T Select Switch Reverse	In 'R': 12v, Others: 0v
2	PK/RD	A/T Select Switch 2nd	In 2nd: 12v, Others: 0v
3	GN	A/T Select Switch Low	In Low: 12v, Others: 0v
4	YL	A/T Oil Temperature Sensor	At 68°F: 4-5v
5	BL/BK	A/C Amplifier Signal (ACT)	Clutch On: 12v, Off: 1.5v
6	BL/OR	Overdrive Main Switch	Switch Off: 12v, On: 1v
7	BR/YL	Cruise Control ECU	At Cruise in OD: 12v
10	GN/RD	AT ECT Pattern Select Switch	Norm: 0v, PWR: 12v
12	GN/OR	Vehicle Speed Sensor	At 55 mph: 48 Hz
14	BK/WT	Direct Battery	12-14v
17	RD/GN	Transfer Detect Switch (N)	Switch Closed: 0.1v
18	WT	Data Link Connector (SDL)	0v
20	BL/YL	A/C Amplifier Signal (AC1)	Clutch On: 1.5v, Off: 12v
22	WT/BL	EFI Main Relay B+	12-14v
25	GN/WT	Stop Light Switch Signal	Brake Off: 0v, On: 12v
26	GN/BK	4WD Indicator Switch	Switch On: 12v

1996-97 Utility 3.4L V6 MFI VIN N (A/T-ECT) 22 Pin Connector

PCM Pin #	Wire Color	Circuit Description (22 Pin)	Value at Hot Idle
1	GN/BK	Sensor VREF	4.9-5.1v
4	WT/RD	AT ECT Speed (+) Signal	At 55 mph: 48 Hz
5	PK	CKP Sensor Signal (NE+)	AC pulse signals
6	PK	CKP Sensor Signal (NE-)	AC pulse signals
7	BK/YL	TP Sensor Signal (VTA)	0.3-0.8v
8	RD/WT	Mass Airflow Sensor	1.1-1.8v
9	WT/RD	AT ECT Speed (-) Signal	At 55 mph: 48 Hz
10	YL	CMP Sensor Signal (G+)	AC pulse signals
11	BL	CMP Sensor Signal (G-)	AC pulse signals
12	PK/YL	A/T Oil Temperature Sensor	AT 68°F: 4-5v
13	WT	HO2S-11 (B1 S1) Signal	0.1-1.1v
14	YL/GN	IAT Sensor Signal (THA)	At 100°F: 2.60v
15	YL/BK	EVAP Vapor Pressure Sensor (PTNK)	2.5-3.1v
16	GY	Knock Sensor 2 Signal	<0.075v AC
17	BK	Knock Sensor 1 Signal	<0.075v AC
18	BK/WT	Power Ground	<0.1v
19	RD	HO2S-12 (B1 S2) Signal	0.1-1.1v
20	GN/RD	ECT Sensor Signal (THW)	At 180°F: 0.51v
22	BR/BK	Sensor Ground	<0.050v

1996-97 Utility 3.4L V6 MFI VIN N (A/T-ECT) 16 Pin Connector

PCM Pin #	Wire Color	Circuit Description (16 Pin)	Value at Hot Idle
1	YL/GN	Closed Throttle Switch	1v, off-idle: 12v
3	PK	Malfunction Indicator Lamp Control	MIL Off: 12v, On: 1v
4	GN/YL	Circuit Opening Relay (FC)	0-3v, off-idle: 12v
5	PK/WT	Data Link Connector	12-14v
13	PK/BK	EVAP Vapor Pressure Valve (VSV)	12v or 0v
15	WT/GN	EVAP Purge Solenoid (VSV)	12v or 0v
16	BR	Sensor Ground	<0.050v

1996-97 Utility 3.4L V6 MFI VIN N (A/T-ECT) 34-Pin Connector

PCM Pin #	Wire Color	Circuit Description (34-Pin)	Value at Hot Idle
1	BR	Power Ground	<0.1v
5	YL	Injector 6 Control	1.6-2.9 ms
6	BL	Injector 5 Control	1.6-2.9 ms
7	RD/BK	Injector 4 Control	1.6-2.9 ms
8	GN	Injector 3 Control	1.6-2.9 ms
9	WT	Injector 2 Control	1.6-2.9 ms
10	RD	Injector 1 Control	1.6-2.9 ms
11	GN/RD	AT ECT Solenoid (S1)	3rd or OD: 1v
12	BK/YL	Igniter Signal (IGF)	Digital Signal: 0-5-0v
13	BK/WT	Starter Switch Signal	In P/N: 0-3.0v
15	RD/WT	HO2S-12 (B1 S2) Heater	1v (Heater on)
16	PK/GN	HO2S-11 (B1 S1) Heater	1v (Heater on)
17	LG	A/T ECT Solenoid (S2)	1st or OD: 1v
22	BK/RD	IAC Signal (RSC)	Pulse Signals
23	BR/RD	Idle Air Control Valve (RSO)	Pulse Signals
24	BK/BL	Igniter Transistor 1 Control	7% duty cycle
25	BR/YL	Igniter Transistor 2 Control	7% duty cycle
26	BK/WT	Igniter Transistor 3 Control	7% duty cycle
27	RD/GN	A/T ECT Solenoid (SL)	In Lockup: 12-14v
29	GN	4WD Detection Transfer (L4)	Switch Closed: 0.1v
33	BR	Power Ground	<0.1v
34	BR	Power Ground	<0.1v

Pin Connector Graphic

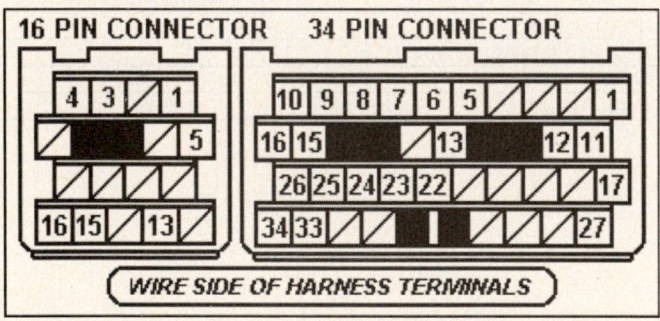

1996-97 Utility 3.4L V6 MFI VIN N (M/T) 16 Pin Connector

PCM Pin #	Wire Color	Circuit Description (16 Pin)	Value at Hot Idle
1 ('96)	GN/BK	Sensor VREF	4.9-5.1v
1 ('97)	GN/YL	Sensor VREF	4.9-5.1v
2 ('96)	RD/WT	MAF Sensor Signal	1.1-1.8v
2 ('97)	BK/RD	MAF Sensor Signal	1.1-1.8v
3	GY	Knock Sensor 2 Signal	<0.075v AC
4	GN/RD	ECT Sensor Signal (THW)	At 180°F: 0.51v
5	WT	HO2S-11 (B1 S1) Signal	0.1-1.1v
6	BK	Knock Sensor 1 Signal	<0.075v AC
7	WT	Data Link Connector	12-14v
8	BK/WT	MAF Sensor Ground (E2G)	<0.050v
9	BR/BK	Sensor Ground	<0.050v
10	BK/YL	TP Sensor Signal (VTA)	0.3-0.8v
12	YL/GN	IAT Sensor Signal (THA)	At 100°F: 2.60v
13	RD	HO2S-12 (B1 S2) Signal	0.1-1.1v

1996-97 Utility 3.4L V6 MFI VIN N (M/T) 26 Pin Connector

PCM Pin #	Wire Color	Circuit Description (26 Pin)	Value at Hot Idle
3	PK/BK	EVAP Vapor Pressure Valve (VSV)	12v or 0v
4	YL/GN	Closed Throttle Switch	1v, off-idle: 12v
5 ('96)	BK/WT	EVAP Purge Solenoid (VSV)	12-14v
5 ('97)	WT/GN	EVAP Purge Solenoid (VSV)	12-14v
6	BK/RD	IAC Signal (RSC)	Pulse Signals
7	BR/RD	Idle Air Control Valve (RSO)	Pulse Signals
8	YL	Injector 6 Control	1.6-2.9 ms
9	BL	Injector 5 Control	1.6-2.9 ms
10	RD/BK	Injector 4 Control	1.6-2.9 ms
11	WT	Injector 2 Control	1.6-2.9 ms
12	RD	Injector 1 Control	1.6-2.9 ms
13	BR	Power Ground	<0.1v
14	GN/YL	Circuit Opening Relay (FC)	0-3v, off-idle: 12v
17	BK/YL	Igniter Signal (IGF)	7% duty cycle
21	BK/WT	Igniter Transistor 3 Control	7% duty cycle
22	BR/YL	Igniter Transistor 2 Control	7% duty cycle
23	BK/BL	Igniter Transistor 1 Control	7% duty cycle
24	BR	Sensor Ground	<0.050v
25	GN	Injector 3 Control	1.6-2.9 ms
26	BR	Power Ground	<0.1v

1996-97 Utility 3.4L V6 MFI VIN N (M/T) 22 Pin Connector

PCM Pin #	Wire Color	Circuit Description (22 Pin)	Value at Hot Idle
2	BK/WT	Direct Battery	12-14v
4	PK	Malfunction Indicator Lamp Control	MIL Off: 12v, On: 1v
6	BL/BK	A/C Amplifier Signal (ACT)	Clutch On: 12v, Off: 1.5v
7	BL/YL	A/C Amplifier Signal (AC1)	Clutch On: 1.5v, Off: 12v
8	GN/OR	Vehicle Speed Sensor	At 55 mph: 48 Hz
9	GN/BK	4WD Detection Switch	Switch On: 12v, Off: 0v
11	BK/WT	Starter Switch Signal	In P/N: 0-3.0v
12	WT/BL	EFI Main Relay Power	12-14v
19	WT	Data Link Connector (SDL)	0v
20	GN/WT	Stop Light Switch Signal	Brake Off: 0v, On: 12v

1996-97 Utility 3.4L V6 MFI VIN N (M/T) 12 Pin Connector

PCM Pin #	Wire Color	Circuit Description (12 Pin)	Value at Hot Idle
3	PK	HO2S-11 (B1 S1) Heater	1v (Heater on)
4	YL/BK	EVAP Vapor Pressure Sensor (PTNK)	2.5-3.1v (with cap off)
5	BL	CMP Sensor Signal (G-)	<0.050v
6	PK	CKP Sensor Signal (NE-)	<0.050v
7	BR	Power Ground	<0.1v
9	RD/WT	HO2S-12 (B1 S2) Heater	1v (Heater on)
11	YL	CMP Sensor Signal (G+)	AC pulse signals
12	PK	CKP Sensor Signal (NE+)	AC pulse signals

Pin Connector Graphic

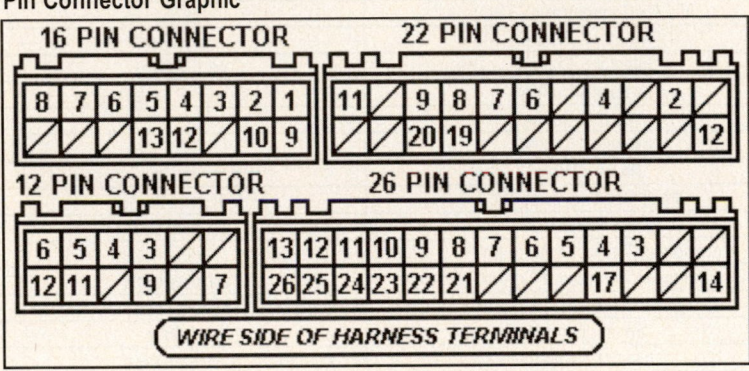

1998 Utility 3.4L V6 MFI VIN N (All) 28 Pin Connector

PCM Pin #	Wire Color	Circuit Description (28 Pin)	Value at Hot Idle
1	LG	A/T Select Switch Low	In Low: 12v, Others: 0v
5	BL/BK	A/C Amplifier Signal (ACT)	Clutch On: 12v, Off: 1.5v
6	BL/OR	Overdrive Main Switch	Switch Off: 12v, On: 1v
7	BR/YL	Cruise Control ECU	At Cruise in OD: 12v
8	WT	SIL Signal (Scan Tool)	12v
10	PK/RD	A/T Select Switch 2nd	In 2nd: 12v, Others: 0v
11	YL/GN	Closed Throttle Switch	1v, off-idle: 12v
12	GN/OR	Vehicle Speed Sensor	At 55 mph: 48 Hz
14	BL/RD	Direct Battery	12-14v
15	RD/YL	A/T Select Switch Reverse	In 'R': 12v, Others: 0v
16	BL/YL	A/C Amplifier Signal (AC1)	Clutch On: 1.5v, Off: 12v
23	WT/BL	EFI Main Relay Power	12-14v
24	GN/WT	Stop Light Switch Signal	Brake Off: 0v, On: 12v

1998 Utility 3.4L V6 MFI VIN N (All) 22 Pin Connector

PCM Pin #	Wire Color	Circuit Description (22 Pin)	Value at Hot Idle
1	GN/BK	Sensor VREF	4.9-5.1v
4	WT/RD	A/T ECT Speed Sensor (-)	Pulse Signals
5	PK	CKP Sensor Signal (NE+)	Pulse Signals
6	PK	CKP Sensor Signal (NE-)	Pulse Signals
7	BK/YL	TP Sensor Signal (VTA)	0.3-0.8v
8	RD/WT	MAF Sensor Signal	1.1-1.8v
9	YL/RD	A/T ECT Speed Sensor (+)	Pulse Signals
12	GN/RD	AT ECT Pattern Select Switch	Norm: 0v, PWR: 12v
13	WT	HO2S-11 (B1 S1) Signal	0.1-1.1v
14	GY	Knock Sensor 2 Signal	<0.075v AC
15	BK	Knock Sensor 1 Signal	<0.075v AC
17	BL	CMP Sensor Signal (G+)	AC pulse signals
18	GN/YL	Circuit Opening Relay (FC)	0-3v, off-idle: 12v
19	RD	HO2S-12 (B1 S2) Signal	0.1-1.1v
20	GN	ECT Sensor Signal (THW)	At 180ºF: 0.51v
21	YL/GN	IAT Sensor Signal (THA)	At 100ºF: 2.60v
22	BR/BK	Sensor Ground	<0.050v

Pin Connector Graphic

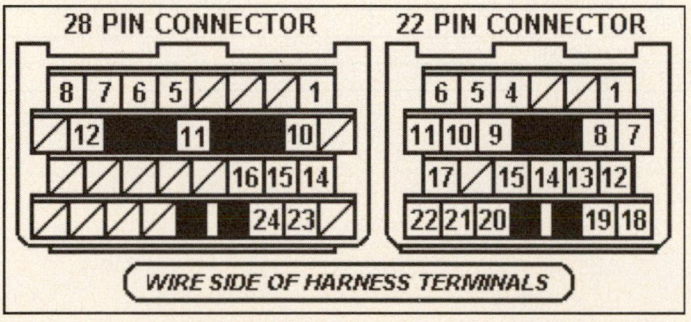

1998 Utility 3.4L V6 MFI VIN N (All) 16 Pin Connector

PCM Pin #	Wire Color	Circuit Description (16 Pin)	Value at Hot Idle
2	WT/GN	EVAP Purge Solenoid (VSV)	12v or 0v
3	PK	Malfunction Indicator Lamp Control	MIL Off: 12v, On: 1v
5	PK/WT	Data Link Connector	12-14v
6	YL	A/T Oil Temperature Sensor	At 68°F: 4-5v
7	BK/WT	MAF Sensor Ground (E2G)	<0.050v
8	PK/BK	EVAP Vapor Pressure Valve (VSV)	12v or 0v
9	RD/GN	4WD Detection Transfer (N)	Open: 12v, Closed: 0v
10	RD/WT	HO2S-12 (B1 S2) Heater	1v (Heater on)
11	PK/BL	HO2S-11 (B1 S1) Heater	1v (Heater on)
13	YL/BK	EVAP Vapor Pressure Sensor (PTNK)	2.5-3.1v (with cap off)
16	BR	Sensor Ground	<0.050v

1998 Utility 3.4L V6 MFI VIN N (All) 34-Pin Connector

PCM Pin #	Wire Color	Circuit Description (34-Pin)	Value at Hot Idle
1	GN/BK	4WD Detection Switch	Switch On: 12v
2	BL/RD	4WD Detection Transfer (L4)	Switch Closed: 0.1v
5	YL	Injector 6 Control	1.6-2.9 ms
6	BL	Injector 5 Control	1.6-2.9 ms
7	RD/BK	Injector 4 Control	1.6-2.9 ms
8	GN	Injector 3 Control	1.6-2.9 ms
9	WT	Injector 2 Control	1.6-2.9 ms
10	RD	Injector 1 Control	1.6-2.9 ms
11	GN/RD	AT ECT Solenoid (S1)	S1: 3rd or OD: 1v
12	BK/YL	Igniter Signal (IGF)	Digital Signal: 0-5-0v
13	BK/WT	Starter Switch Signal	9-11v (cranking)
14	BK	Neutral Start Switch	In P/N: 0-3.0v
15	BK/WT	Igniter Transistor 3 Control	7% duty cycle
16	BK/YL	Igniter Transistor 2 Control	7% duty cycle
17	GN	A/T ECT Solenoid (S2)	S1: 3rd or OD: 1v
22	BK/RD	IAC Signal (RSC)	Pulse Signals
23	BR	Idle Air Control Valve (RSO)	Pulse Signals
24	BK/BL	Igniter Transistor 1 Control	7% duty cycle
27	GN/RD	A/T ECT Solenoid (SL)	In Lockup: 12-14v
28	BR	Power Ground	<0.1v
30	PK/YL	A/T Oil Temperature Sensor	At 68°F: 4-5v
31	PK	PSP Switch Signal (PSW)	Straight: 12v, Turned: 0v
33	BR	Power Ground	<0.1v
34	BR	Power Ground	<0.1v

1999-2000 Utility 3.4L V6 (All) VIN N E9 31 Pin Connector

PCM Pin #	Wire Color	Circuit Description (31 Pin)	Value at Hot Idle
1	GN	Injector 3 Control	1.6-2-9 ms
2	RD/BK	Injector 4 Control	1.6-2-9 ms
3	BL	Injector 5 Control	1.6-2-9 ms
4	YL	Injector 6 Control	1.6-2.9 ms
5	---	Not Used	---
6	PK/WT	Data Link Connector	12-14v
7	LG/RD	AT ECT Solenoid (S1)	3rd or OD: 1v
8	BL/WT	A/T ECT Solenoid (S2)	1st or OD: 1v
9	GN/RD	A/T ECT Solenoid (SL)	In Lockup: 12-14v
10	RD	CMP Sensor Signal (G+)	AC pulse signals
11	BK/BL	Igniter Transistor 1 Control	6°, at 55 mph: 8° dwell
12	BL/YL	Igniter Transistor 2 Control	6°, at 55 mph: 8° dwell
13	GN/WT	Igniter Transistor 3 Control	6°, at 55 mph: 8° dwell
14	YL/RD	A/T ECT Speed Sensor (+)	Pulse Signals
15	BK/RD	IAC Signal (RSC)	Pulse Signals
16	BR/RD	Idle Air Control Valve (RSO)	Pulse Signals
17-20	---	Not Used	---
21	WT/BK	Power Ground (E01)	<0.1v
22-24	---	Not Used	---
25	BK/YL	Igniter Signal (IGF)	Digital Signal: 0-5-0v
26	WT/RD	A/T ECT Speed Sensor (-)	Pulse Signals
27	BK	Knock Sensor 1 Signal	<0.075 VAC
28	GY	Knock Sensor 2 Signal	<0.075 VAC
29	---	Not Used	---
30	WT/BK	Power Ground (E03)	<0.1v
31	WT/BK	Power Ground (E02)	<0.1v

Pin Connector Graphic

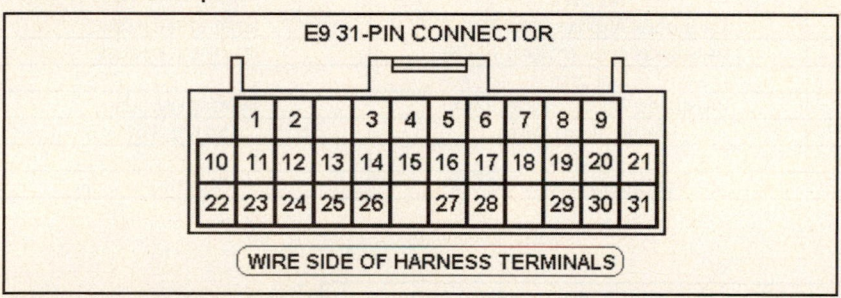

Standard Colors and Abbreviations

Abbreviation	Color	Abbreviation	Color	Abbreviation	Color
BK	Black	GY	Gray	RD	Red
BL	Blue	GN	Green	TN	Tan
BR	Brown	LG	LT Green	VT	Violet
DB	Dark Blue	OR	Orange	WT	White
DG	DK Green	PK	Pink	YL	Yellow

1999-2000 Utility 3.4L V6 VIN N (All) E10 24 Pin Connector

PCM Pin #	Wire Color	Circuit Description (24 Pin)	Value at Hot Idle
1, 3	---	Not Used	---
2	GN/BK	Sensor VREF	4.9-5.1v
4 (Cal)	BL	HAFR-11 (B1 S1) Heater	1v (Heater on)
4 (Fed)	PK/BL	HO2S-11 (B1 S1) Signal	0.1-1.1v
5	RD	Injector 1 Control	1.6-2-9 ms
6	WT	Injector 2 Control	1.6-2-9 ms
7	WT/GN	EVAP Purge Solenoid (VSV)	12v or 0v
8 (Cal)	WT/BK	Power Ground	<0.1v
9	PK	PSP Switch Signal (PSW)	Straight: 12v, Turned: 0v
10	RD/WT	MAF Sensor Signal	1-1.1v
12 (Cal)	WT	AFR-11 (B1 S1) Signal (+)	Fixed at 3.3v
12 (Fed)	WT	HO2S-11 (B1 S1) Heater	1v (Heater on)
13	---	Not Used	---
14	GN	ECT Sensor Signal (THW)	0.5-0.6v
15	PK/YL	A/T Oil Temperature Sensor	At 68°F: 4-5v
16	RD	CKP Sensor Signal (NE+)	AC pulse signals
17	BR	Shield Ground	<0.050v
18	BL/BK	Sensor Ground	<0.050v
19	BK/WT	MAF Sensor Ground (E2G)	<0.050v
20	---	Not Used	---
21 (Cal)	BK	AFR-11 (B1 S1) Signal (-)	Fixed at 3.3v
22	YL/GN	IAT Sensor Signal (THA)	0.5-3.4v
23	BK/YL	TP Sensor Signal (VTA)	0.53-1.27v
24	GN	CMP & CKP Sensor Return	<0.050v

1999-2000 Utility 3.4L V6 VIN N (All) E11 17 Pin Connector

PCM Pin #	Wire Color	Circuit Description (17 Pin)	Value at Hot Idle
1-3	---	Not Used	---
4	GY/RD	Transponder Amplifier Signal (Code)	Inserting key: pulses
5	RD/BK	Transponder Amplifier Signal (RXCK)	Inserting key: pulses
6-9	---	Not Used	---
10	PK/BK	Transponder Amplifier Signal (RXCK)	Inserting key: pulses
11	YL/RD	Unlock Warning Switch	No Key: 4-5v
12-15	---	Not Used	---
16	BL	Theft Security Indicator Light	Digital Signal

Pin Connector Graphic

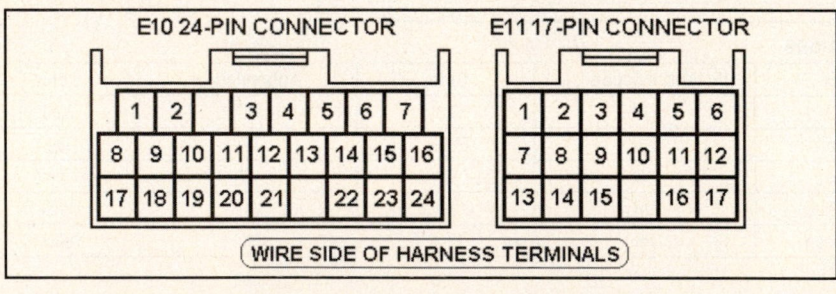

1999-2000 Utility 3.4L V6 (All) VIN N E12 28 Pin Connector

PCM Pin #	Wire Color	Circuit Description (28 Pin)	Value at Hot Idle
2	RD/YL	A/T Select Switch Reverse	In 'R': 12v, Others: 0v
3	PK/RD	A/T Select Switch 2nd	In 2nd: 12v, Others: 0v
4	YL/GN	Cruise Control ECU	At Cruise in OD: 12v
7	YL	A/T Oil Temperature Indicator	Lamp Off: 12v, On: 1v
8	RD	HO2S-12 (B1 S2) Signal	0.1-1.1v
9	RD/WT	HO2S-12 (B1 S2) Heater	1v (Heater on)
10	BL/OR	Overdrive "Off" Indicator	Lamp Off: 12v, On: 1v
11	GN/RD	AT ECT Pattern Select Switch	Norm: 0v, PWR: 12v
12	LG	A/T Select Switch Low	In Low: 12v, Others: 0v
13	BL/BK	A/C Amplifier Signal (ACT)	Clutch On: 12v, Off: 1.5v
17	RD/GN	4WD Detection Transfer (N)	Open: 12v, Closed: 0v
18	BK/BL	4WD ADD Indicator Switch	Open: 12v, Closed: 0v
19	BL/RD	4WD Detection Transfer (L4)	Switch Closed: 0.1v
20	BK	A/T Neutral Start Signal	In P/N: 0-3.0v
22	GN/OR	Speedometer Indicator	At 55 mph: 48 Hz
24	BR/YL	Overdrive Main Switch	Switch Off: 12v, On: 1v
25	BL/YL	A/C Amplifier Signal (AC1)	Clutch On: 1.5v, Off: 12v

1999-2000 Utility 3.4L V6 VIN N (All) E14 22 Pin Connector

PCM Pin #	Wire Color	Circuit Description (22 Pin)	Value at Hot Idle
1	BL/RD	Direct Battery	12-14v
2	BK/BL	Ignition Switch	12-14v
3	GN/YL	Circuit Opening Relay (FC)	Relay On: 1v, Off: 12v
4-5	---	Not Used	
6	PK	Malfunction Indicator Lamp Control	MIL Off: 12v, On: 1v
7	BK/WT	Starter Switch Signal	9-11v (cranking)
8	GY/BK	EFI Main Relay Control	Relay Off: 0v, On: 12v
9	WT/RD	EVAP Vapor Pressure Valve (VSV)	12v or 0v
10	---	Not Used	---
11	WT	SIL Signal (Scan Tool)	0v
12	WT/BK	Power Ground	<0.1v
13-14	---	Not Used	---
15	GN/WT	Brake Switch Signal	Brake Off: 0v, On: 12v
16	WT/BL	EFI Main Relay Power	12-14v
17	YL	EVAP Vapor Pressure Sensor (PTNK)	2.9-3.7v (hose off)
18-22	---	Not Used	---

Pin Connector Graphic

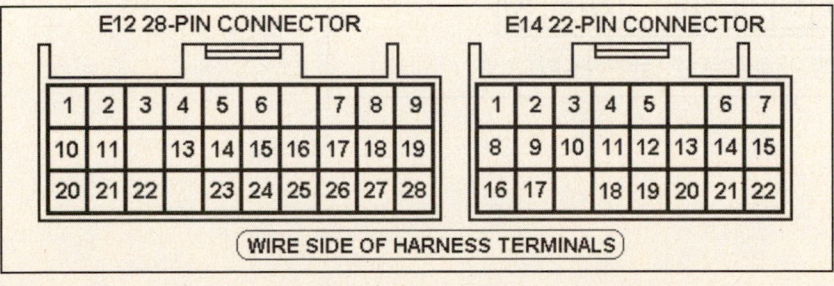

2001-02 Utility 3.4L V6 MFI VIN N (All) E9 31 Pin Connector

PCM Pin #	Wire Color	Circuit Description (31 Pin)	Value at Hot Idle
1	LG	A/T ECT Solenoid (SL)	In Lockup: 12-14v
2	BL/WT	A/T ECT Solenoid (S2)	1st or OD: 1v
3	PK/BL	A/T ECT Solenoid (S3)	3rd or OD: 1v
4	WT/BK	Power Ground (E01)	<0.1v
5	WT/BK	Power Ground (E02)	<0.1v
6	WT/BK	Power Ground (ME01)	<0.1v
7	WT/BK	Power Ground (E03)	<0.1v
8	WT/BK	Power Ground (E04)	<0.1v
9	RD	Throttle Control Motor (M+)	Pulse Signals
10	BK/WT	MAF Sensor Ground (E2G)	<0.050v
11	BL/BK	Sensor Ground (E2)	<0.050v
12	RD/WT	MAF Sensor Signal (VG)	1-1.1v
13	YL/GN	IAT Sensor Signal (THA)	0.5-3.4v
14	WT	AFR-11 (B1 S1) Signal (+)	Fixed at 3.3v
15	BK/YL	TP Sensor Signal (VTA1)	0.53-1.27v
16	---	Not Used	---
17	BR	Shield Ground (GE01)	<0.050v
18	GN	ECT Sensor Signal (THW)	0.5-0.6v
19	BL/RD	ATF Fluid Temp. Input (THO)	At 68°F: 4-5v
20, 28, 30	---	Not Used	---
21	BL	HAFR-11 (B1 S1) Heater	1v (Heater on)
22	GY	Knock Sensor 2 Signal	<0.075 VAC
23	BK	Knock Sensor 1 Signal	<0.075 VAC
24	YL	EVAP Vapor Pressure Sensor (PTNK)	2.9-3.7v (hose off)
25	GN/BK	Sensor VREF	4.9-5.1v
26	BK	AFR-11 (B1 S1) Signal (-)	Fixed at 3.3v
27	BK	HO2S-12 (B1 S2) Signal	0.1-1.1v
29	GN/YL	HO2S-12 (B1 S2) Heater	1v (Heater on)
31	RD	Throttle Control Motor (M-)	Pulse Signals

Pin Connector Graphic

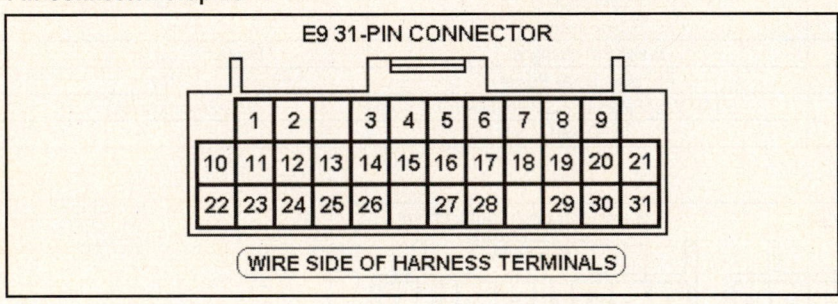

2001-02 Utility 3.4L V6 MFI VIN N (All) E10 24 Pin Connector

PCM Pin #	Wire Color	Circuit Description (24 Pin)	Value at Hot Idle
1	BL	Injector 5 Control	1.6-2-9 ms
2	BK/YL	Igniter Signal (IGF)	Digital Signal: 0-5-0v
3	RD/BK	Injector 4 Control	1.6-2-9 ms
4	GN	Injector 3 Control	1.6-2-9 ms
5	WT	Injector 2 Control	1.6-2-9 ms
6	RD	Injector 1 Control	1.6-2-9 ms
7	GY	Data Link Connector (TC)	12-14v
8	YL	Injector 6 Control	1.6-2.9 ms
9	GN/WT	Igniter Transistor 3 Control	6°, at 55 mph: 8° dwell
10	BL/YL	Igniter Transistor 2 Control	6°, at 55 mph: 8° dwell
11	BK/BL	Igniter Transistor 1 Control	6°, at 55 mph: 8° dwell
12	BL	CKP Sensor Signal (NE+)	AC pulse signals
13	RD	CMP Sensor Signal (G+)	AC pulse signals
14	WT/GN	EVAP Purge Solenoid (VSV)	12v or 0v
15-16	---	Not Used	---
17	BR	Power Ground (E1)	<0.1v
18	PK	PSP Switch Signal (PSW)	Straight: 12v, Turned: 0v
19	BL/BK	EVAP Canister Closed Valve (CCV)	12v or 0v
20	RD/YL	EVAP Vapor Pressure Valve (VSV)	12v or 0v
21	GN	CMP & CKP Sensor Ground	<0.050v
22	WT/RD	A/T ECT Speed Sensor (-)	Pulse Signals
23	YL/RD	A/T ECT Speed Sensor (+)	Pulse Signals
24	---	Not Used	---

2001-02 Utility 3.4L V6 MFI VIN N (All) E11 17 Pin Connector

PCM Pin #	Wire Color	Circuit Description (17 Pin)	Value at Hot Idle
1, 3	---	Not Used	---
2	YL	Throttle Control Motor (CL+)	Pulse Signals
4	GN/YL	TP Sensor 2 (VTA2)	2.0-2.9v
5	RD/YL	A/T-ECT Solenoid (SLT+)	Pulse Signals
6-9	---	Not Used	---
8	BL	Throttle Control Motor (CL-)	Pulse Signals
10	GY	Accel Position Sensor (VPA)	0.25-0.9v
11	YL/BK	A/T-ECT Solenoid (SLT-)	Pulse Signals
12-14	---	Not Used	---
15	BL	Accel Position Sensor (VPA2)	1.8-2.7v
16-17	---	Not Used	---

Pin Connector Graphic

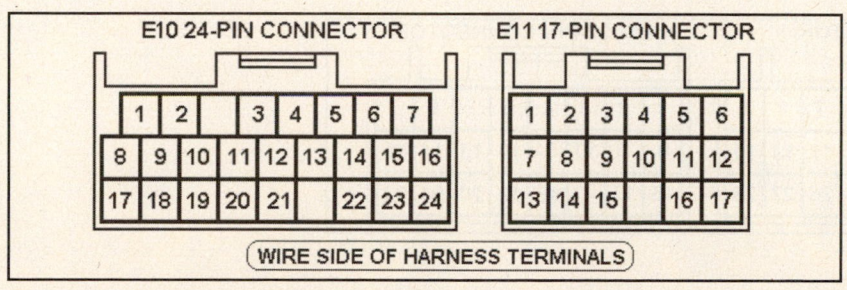

2001-02 UTILITY 3.4L V6 MFI VIN N (ALL) E12 28 PIN CONNECTOR

PCM Pin #	Wire Color	Circuit Description (28 Pin)	Value at Hot Idle
1	GN/RD	A/T: ECT Power Select Signal	Norm: 0v, PWR: 12v
2	BL/RD	4WD Detection Transfer (L4)	Switch Closed: 0.1v
3	BK	Neutral Start Switch Signal	In P/N: 9-12v (cranking)
4	LG	A/T Select Switch Low	In Low: 12v, Others: 0v
5	VT	A/T Select Switch Drive	In Drive: 12-14v
6	WT/BL	A/T Select Switch Reverse	In 'R': 12v, Others: 0v
8	VT/WT	ABS/BA/TRAC/VSC ECU	Digital Signals (NEO)
11	BK/RD	4WD Detection Transfer (N)	Open: 12v, Closed: 0v
13	BK/BL	4WD ADD Indicator Switch	Switch Closed: 0.1v
14	BL/WT	Overdrive Main Switch	Switch Off: 12v, On: 1v
18, 19	PK, BL	TRAC Engine (-), (+) Signals	Pulse Signals
20	GN/YL	O/D OFF (lamp) Indicator	Light Off: 12v, On: 1v
21	YL	A/T: Oil Temp. Lamp Indicator	Lamp Off: 12v, On: 1v
22	GN	Cruise Control Switch Signal	Main Switch On: 1v
25	WT/RD	2WD Detect Transfer Switch	Open: 0v, Closed: 12v
25	GY/BK	4WD Detect Transfer Switch	Open: 0v, Closed: 12v
27, 28	YL, BR	TRAC TRC (-), (+) Signals	Pulse Signals

2001-02 Utility 3.4L V6 MFI VIN N (All) E14 22 Pin Connector

PCM Pin #	Wire Color	Circuit Description (22 Pin)	Value at Hot Idle
1	WT/BL	EFI Main Relay Power	12-14v
2	VT	Malfunction Indicator Lamp Control	MIL Off: 12v, On: 1v
3	BL/BK	A/C Amplifier Signal (ACT)	Clutch On: 12v, Off: 1.5v
4	GY/BK	EFI Main Relay Control	Relay Off: 0v, On: 12v
5	GY/RD	Transponder Amplifier Signal (Code)	Inserting key: pulses
6	GN/OR	Combination Meter SP1 Input	At 55 mph: 48 Hz
7	BK/WT	Starter Switch Signal	In P/N: 0-3.0v
8	GN	ETC Power (B+)	12-14v
9	BL/YL	A/C Amplifier Signal (AC1)	Clutch On: 1.5v, Off: 12v
10	WT/BK	Power Ground (EOM)	<0.1v
11	YL/RD	Unlock Warning Switch	No Key: 4-5v
12	BL	Theft Security Indicator Light	Theft LED On: 12v
13	RD/WT	Data Link Connector (WFSE)	N/A
14	WT	SIL Signal (Scan Tool)	Digital Signal
15	BK/BL	Ignition Switch	12-14v
16	BL/RD	Direct Battery	12-14v
17	GY/RD	Cruise Control Indicator	Lamp On: 1v, Off: 12v
18, 19	RD, PK	Transponder Amplifier Signal (RXCK)s	Inserting key: pulses
20	GN/WT	Stop Light Switch Signal	Brake Off: 0v, On: 12v
22	GN/YL	Circuit Opening Relay (FC)	Relay On: 1v, Off: 12v

Pin Connector Graphic

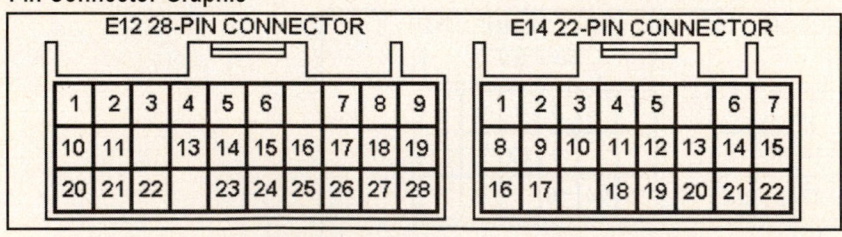

2003 Utility AWD 4.0L V6 VIN U E4 34 Pin Connector

PCM Pin #	Wire Color	Circuit Description (34 Pin)	Value at Hot Idle
1	RD/BL	Injector 1 Control	2.0-3.3 ms
2	BK	Injector 2 Control	2.0-3.3 ms
3	LG	Injector 3 Control	2.0-3.3 ms
4	GN	Injector 4 Control	2.0-3.3 ms
5	YL	Injector 5 Control	2.0-3.3 ms
6, 7	WT/BK	Power Ground (E02), (E01)	<0.1v
8	YL/RD	Igniter Transistor 1 Control	7% duty cycle
9	PK/BL	Igniter Transistor 2 Control	7% duty cycle
10	LG	Igniter Transistor 3 Control	7$ duty cycle
11	LG/BK	Igniter Transistor 4 Control	7% duty cycle
12	GY	Igniter Transistor 5 Control	7% duty cycle
13	BL	Igniter Transistor 6 Control	7% duty cycle
14	WT/GN	A/C Relay Control (ACCR)	Relay Off: 12v, On: 1v
15	WT/BL	ACIS Control (VSV)	12v or 0v
16	PK	Neutral Start Switch (NSW)	In P/N: 0-3.0v
17	BK/YL	Starter Switch Signal (STA)	In P/N: 0-3.0v
18	BL/RD	Sensor VREF (VC)	4.9-5.1v
19	BK/BL	ECT Sensor Signal (THW)	At 180ºF: 0.51v
20	RD/BK	IAT Sensor Signal (THA)	At 100ºF: 2.60v
21	GN/BK	TP Sensor Signal (VTA1)	0.4-1.0v
22-23, 25-26	---	Not Used	---
24	WT/RD	Igniter Signal (IGF1)	Digital Signal: 0-5-0v
27	RD/GN	EVAP Canister Closed Valve (CCV)	12v or 0v
28	BR	Sensor Ground (E2)	<0.050v
29	RD/WT	MAF Sensor Ground (E2G)	<0.050v
30	RD/YL	Mass Airflow Sensor (VG)	1.1-1.5v
31	GN/WT	TP Sensor Signal (VTA2)	2.0-2.9v
33	YL/BK	Fuel Pump Relay Control (FPR)	Relay Off: 12v, On: 1v
34	GN/YL	EVAP Purge Solenoid (VSV)	12v or 0v

2003 Utility AWD 4.0L V6 VIN U E5 35 Pin Connector

PCM Pin #	Wire Color	Circuit Description (35 Pin)	Value at Hot Idle
1	BK	Knock Sensor 1 Signal (KNK1)	<0.075v AC
2	GN	Knock Sensor 2 Signal (KNK2)	<0.075v AC
3	BL	Injector 6 Control	2.0-3.3 ms
4	BK/WT	AFS-11 (B2 S1) Heater (HAFL)	1v (Heater on)
5	RD/BL	AFS-11 (B1 S1) Heater (HAFR)	1v (Heater on)
6, 7	WT/BK	Power Ground (E05), (E04)	<0.050v
8	GN	4WD Switch Signal (4WD)	In 4WD: 12v
9	PK	Start Signal (STAR)	In P/N: 0-3.0v
10	BL/WT	ECT Solenoid Control (S2)	12v or 0v
11	BL/RD	ECT Solenoid Control (S1)	12v or 0v
12	BL/BK	ECT Solenoid Control (SLT-)	12v or 0v
13	GN/YL	ECT Solenoid Control (SLT+)	12v or 0v
15	GN	ECT Solenoid Control (SL)	12v or 0v
16	PK/BL	A/T Solenoid Control (SL2-)	Pulse Signals
17	PK/BK	A/T Solenoid Control (SL2+)	Pulse Signals
18	RD/WT	A/T Solenoid Control (SL1-)	Pulse Signals
19	RD/BL	A/T Solenoid Control (SL1+)	Pulse Signals
20	RD	Knock Sensor 2 Ground	<0.050v
21	WT	HO2S-12 (B1 S2) Signal (OX1B)	0.1-1.1v
22	PK	AFS-11 (B1 S1) Signal (AFR+)	3.0-3.6v
23	YL	AFS-21 (B2 S1) Signal (AFL+)	3.0-3.6v
24	BL	A/T Oil Temperature Sensor 2 (THO2)	At 68ºF: 4-5v
25	GN	HO2S-12 (B1 S2) Heater (HT1B)	1v (Heater on)
26	GN	ECT Vehicle Speed Sensor (SP2+)	Pulse Signals
27	WT/RD	Turbine Speed Sensor (NCO+)	Pulse Signals
28	WT	Knock Sensor 1 Ground	<0.050v
29	BK	HO2S-22 (B2 S2) Signal (OX2B)	0.1-1.1v
30	BL	AFS-11 (B1 S1) Signal (AFR-)	Fixed at 3.3v
31	BR	AFS-21 (B2 S1) Signal (AFL-)	Fixed at 3.3v
32	GN/YL	A/T Oil Temperature Sensor 1 (THO1)	At 68ºF: 4-5v
33	BL	HO2S-22 (B2 S2) Heater (HT2B)	1v (Heater on)
34	RD	ECT Vehicle Speed Sensor (SP2-)	Pulse Signals
35	YL/RD	Turbine Speed Sensor (NCO-)	Pulse Signals

2003 Utility AWD 4.0L V6 VIN U E6 32 Pin Connector

PCM Pin #	Wire Color	Circuit Description (32 Pin)	Value at Hot Idle
1	BR	Power Ground (E1)	<0.1v
2	BL	Throttle Control Motor (M-)	Pulse Signals
3	PK	Throttle Control Motor (M+)	Pulse Signals
4	WT/BK	Power Ground (ME01)	<0.1v
5-6	---	Not Used	---
7	WT/BK	Power Ground (E03)	<0.1v
8-9	---	Not Used	---
10	GN/WT	PSP Switch Signal (PSW)	Straight: 12v, Turning: 0v
11	YL/GN	Neutral Detection Switch (L4)	Switch Open: 0v, Closed: 12v
12	BL/YL	Start Switch Signal (STSW)	Cranking: 9-11v
13	BL/RD	Camshaft Timing Control Valve LH (OC2-)	Pulse Signals
14	BL/WT	Camshaft Timing Control Valve LH (OC2+)	Pulse Signals
15	BL/BK	Camshaft Timing Control Valve RH (OC1-)	Pulse Signals
16	GN/YL	Camshaft Timing Control Valve RH (OC1+)	Pulse Signals
17	BR	Shield Ground (GE01)	<0.050v
18-20	---	Not Used	---
21	BK/OR	Generator Control (RL)	12v
22	---	Not Used	---
23	RD/YL	A/C Lock Sensor (LCK)	12v or 0v
24	WT	CKP Sensor Signal (NE-)	<0.050v
25	BK	CKP Sensor Signal (NE+)	AC pulse signals
26	YL	Variable Valve Timing Sensor LH (VV2+)	AC pulse signals
27	RD	Variable Valve Timing Sensor RH (VV1+)	AC pulse signals
28-31	---	Not Used	---
32	WT	CMP Sensor Signal (G2-)	AC pulse signals

Pin Connector Graphic (1)

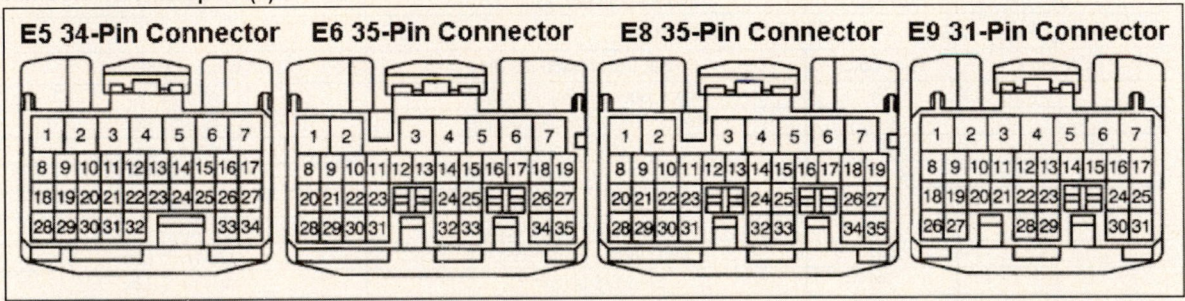

Pin Connector Graphic (2)

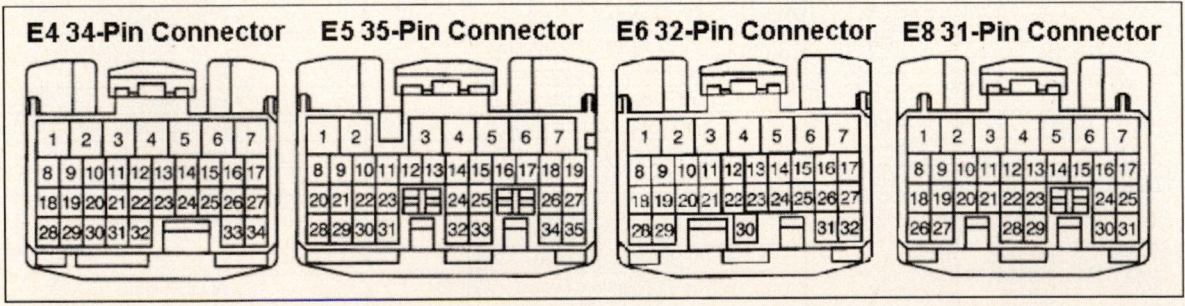

2003 Utility AWD 4.0L V6 VIN U E7 35 Pin Connector

PCM Pin #	Wire Color	Circuit Description (35 Pin)	Value at Hot Idle
1	WT/BK	Power Ground (HP)	<0.1v
2	BK/YL	A/C Magnetic Clutch Relay (ACMG)	Relay Off: 0v, On: 12v
3	GN	Transmission Control Switch Signal (2L)	In 2nd: 12v, Others: 0v
4	WT/BL	Low Detection Switch (L4)	Switch Open: 0v, Closed: 12v
5	BL/WT	ECT Pattern Switch 2nd Position (SNW1)	2nd Position: 12v
6	GN	ETCS Power (+BM)	12-14v
7	---	Not Used	---
8	BL/YL	Shift Lock ECU Control (L)	12v or 0v
9	PK/BL	4WD Low Indicator Control	Indicator Off: 12v, On: 0v
10	GN/YL	Park Neutral Position Switch (D)	In 'D': 12v, Others: 0v
11	RD/YL	A/T Select Switch Reverse (R)	In 'R': 12v, Others: 0v
12	---	Not Used	---
14	BL/BK	A/C Amplifier Signal (THWO)	A/C Off: 12v, On: 1v
15-16	---	Not Used	---
17	VT/RD	Vehicle Speed Sensor (SPD)	At 55 mph: 48 Hz
18	PK/BK	Body Control ECU Signal (MPX1)	Digital Signals
19	GN/YL	Stop Light Switch Signal	Brake Off: 0v, On: 12v
20	BL	Shift Lock ECU Control (3)	12v or 0v
21-22	---	Not Used	---
23	GN/RD	Shift Lock ECU Control (4)	12v or 0v
25	PK	A/T Oil Temperature Indicator	Indicator Off: 12v, On: 1v
26	BL/RD	Transponder Amplifier Signal (IMD)	Inserting key: pulses
27	WT/RD	Transponder Amplifier Signal (IMI)	Inserting key: pulses
28	BL/WT	ECT Pattern Switch Power Signal	Power Position: 12v
29	BK	Body Control ECU Signal (MPX2)	Digital Signals
30	---	Not Used	---
31	GR/BK	A/C Amplifier Signal (A/CS)	Relay Off: 12v, On: 1v
32	GY/GN	A/C Amplifier Signal (THE)	A/C Off: 12v, On: 1v
33	BK/RD	A/C Switch Signal (ACLD)	A/C On: 12v, Off: 0v
34-35	---	Not Used	---

2003 Utility AWD 4.0L V6 VIN U E8 31 Pin Connector

PCM Pin #	Wire Color	Circuit Description (31 Pin)	Value at Hot Idle
1	BK	EFI Main Relay Power (+B)	12-14v
2	BK	EFI Main Relay Power (+B2)	12-14v
3	BL	Direct Battery	12-14v
4	PK/BL	EVAP Pressure Switching Valve (VSV)	12v or 0v
5	BK/WT	Tachometer Signal (TACO)	Pulse Signals
6	GN/WT	A/T Select Switch Park Signal (P)	In 'P': 12v, Others: 0v
7	GN/RD	A/T Select Switch Neutral Signal (N)	In 'N': 12v, Others: 0v
8	WT/GN	EFI Main Relay Control	Relay Off: 0v, On: 12v
9	BK/OR	Ignition Switch Power (IGSW)	12-14v
10	GR/BK	Circuit Opening Relay (FC)	0-3v, off-idle: 12v
11	RD/BK	Malfunction Indicator Lamp Control	MIL Off: 12v, On: 1v
12	YL/GN	Defogger Switch Signal (ELS)	Defogger Off: 0v, On: 12v
13	GN	Taillight Switch Signal (ELS2)	Taillights Off: 12v, On: 1v
14	BL	Center Airbag Assembly (F/PS)	Digital Signals
15	WT/BK	Power Ground (EOM)	<0.1v
16	---	Not Used	---
17	PK	Transponder Amplifier Signal (NEO)	Inserting key: pulses
18	RD/YL	SIL Signal (Scan Tool)	Transmitting: pulses
19	RD/WT	Data Link Connector (WFSE)	12v
20	PK/BL	Data Link Connector (TC)	12v
21	OR	EVAP Vapor Pressure Sensor (PTNK)	2.5-3.1v (with cap off)
22	WT/RD	Accelerator Position Sensor 1 (VPA)	0.3-0.90v
23	RD/BK	Accelerator Position Sensor 2 (VPA2)	1.8-2.7v
24	RD	Traction Control Engine Signal (ENG+)	Pulse Signals
25	BK	Traction Control Signal (TRC+)	Pulse Signals
26	BK/YL	Accelerator Pedal Position Sensor 1 VREF	4.9-5.1v
27	WT/BL	Accelerator Pedal Position Sensor 2 VREF	4.9-5.1v
28	LG/BK	Accelerator Pedal Position Sensor (EPA)	<0.050v
29	VT/WT	Accelerator Position Sensor 2 (EPA2)	1.8-2.7v
30	WT	Traction Control Engine Signal (ENG-)	Pulse Signals
31	YL	Traction Control Signal (TRC-)	Pulse Signals

2003 Utility AWD 4.7L V8 VIN T E4 34 Pin Connector

PCM Pin #	Wire Color	Circuit Description (34 Pin)	Value at Hot Idle
1	RD/BL	Injector 1 Control	2.0-3.3 ms
2	BL	Injector 2 Control	2.0-3.3 ms
3	WT	Injector 3 Control	2.0-3.3 ms
4	VT	Injector 4 Control	2.0-3.3 ms
5	GN	Injector 5 Control	2.0-3.3 ms
6	WT/BK	Power Ground (E02)	<0.1v
7	WT/BK	Power Ground (E01)	<0.1v
8	LG/BK	Igniter Transistor 2 Control	7% duty cycle
9	LG	Igniter Transistor 1 Control	7% duty cycle
10	GN/BK	Igniter Transistor 8 Control	7% duty cycle
11	BL/YL	Igniter Transistor 4 Control	7% duty cycle
12	BK/YL	Igniter Transistor 5 Control	7% duty cycle
13	GN/WT	Igniter Transistor 7 Control	7% duty cycle
14, 22, 32	---	Not Used	---
15	RD/GN	A/C Relay Control (ACCR)	Relay Off: 12v, On: 1v
16	PK	Neutral Start Switch (NSW)	In P/N: 0-3.0v
17	BK/YL	Starter Switch Signal (STA)	In P/N: 0-3.0v
18	BL/RD	Sensor VREF (VC)	4.9-5.1v
19	RD/BL	ECT Sensor Signal (THW)	At 180°F: 0.51v
20	YL/BK	IAT Sensor Signal (THA)	At 100°F: 2.60v
21	YL	TP Sensor Signal (VTA1)	0.4-1.0v
23	RD/WT	Igniter Signal (IGF2)	Digital Signal: 0-5-0v
24	RD/YL	Igniter Signal (IGF1)	Digital Signal: 0-5-0v
25	BL/WT	Igniter Transistor 3 Control	7% duty cycle
26	GN	Igniter Transistor 6 Control	7% duty cycle
27	RD/GN	EVAP Canister Closed Valve (CCV)	12v or 0v
28	BR	Sensor Ground (E2)	<0.050v
29	BK/WT	MAF Sensor Ground (E2G)	<0.050v
30	WT/RD	Mass Airflow Sensor (VG)	1.1-1.5v
31	RD/BK	TP Sensor Signal (VTA2)	2.0-2.9v
33	GN/RD	Fuel Pump Relay Control (FPR)	Relay Off: 12v, On: 1v
34	YL/RD	EVAP Purge Solenoid (VSV)	12v or 0v

2003 Utility AWD 4.7L V8 VIN T E5 35 Pin Connector

PCM Pin #	Wire Color	Circuit Description (35 Pin)	Value at Hot Idle
1	BK	Knock Sensor 1 Signal (KNK1)	<0.075v AC
2	WT	Knock Sensor 2 Signal (KNK2)	<0.075v AC
3	RD	Injector 6 Control	2.0-3.3 ms
4	LG	HO2S-11 (B1 S1) Heater (HT1A)	1v (Heater on)
5	GN/YL	HO2S-12 (B1 S2) Heater (HT1B)	1v (Heater on)
6	BR	4WD Switch Ground	<0.050v
6-8	---	Not Used	---
9	PK	Start Signal (STAR)	In P/N: 0-3.0v
10	WT	ECT Solenoid Control (S2)	12v or 0v
11	RD	ECT Solenoid Control (S1)	12v or 0v
12	GN/BK	ECT Solenoid Control (SLT-)	12v or 0v
13	GN/WT	ECT Solenoid Control (SLT+)	12v or 0v
15	GN	ECT Solenoid Control (SL)	12v or 0v
16	PK/BL	A/T Solenoid Control (SL2-)	Pulse Signals
17	PK/BK	A/T Solenoid Control (SL2+)	Pulse Signals
18	RD/WT	A/T Solenoid Control (SL1-)	Pulse Signals
19	RD/BL	A/T Solenoid Control (SL1+)	Pulse Signals
20, 28, 30-31	---	Not Used	---
21	WT	HO2S-22 (B2 S2) Signal (OX2B)	0.1-1.1v
22	GN	HO2S-21 (B2 S1) Signal (OX2A)	0.1-1.1v
23	RD	HO2S-11 (B1 S1) Signal (OX1A)	0.1-1.1v
24	BL	A/T Oil Temperature Sensor 2 (THO2)	At 68°F: 4-5v
25	PK	HO2S-22 (B2 S2) Heater (HT2B)	1v (Heater on)
26	RD	ECT Vehicle Speed Sensor (SP2+)	Pulse Signals
27	BL	ECT Turbine Speed Sensor (NT+)	Pulse Signals
29	BK	HO2S-12 (B1 S2) Signal (OX1B)	0.1-1.1v
32	GN/YL	A/T Oil Temperature Sensor 1 (THO1)	At 68°F: 4-5v
33	BK/BL	HO2S-21 (B2 S1) Heater (HT2A)	1v (Heater on)
34	GN	ECT Vehicle Speed Sensor (SP2-)	Pulse Signals
35	WT	ECT Turbine Speed Sensor (NT-)	Pulse Signals

2003 Utility AWD 4.7L V8 VIN T E6 32 Pin Connector

PCM Pin #	Wire Color	Circuit Description (32 Pin)	Value at Hot Idle
1	BR	Power Ground (E1)	<0.1v
2	GN	Throttle Control Motor (M-)	Pulse Signals
3	RD	Throttle Control Motor (M+)	Pulse Signals
4	WT/BK	Power Ground (ME01)	<0.1v
5	BR	Injector 8 Control	2.0-3.3 ms
6	WT	Injector 7 Control	2.0-3.3 ms
7	WT/BK	Power Ground (E03)	<0.1v
8-10	---	Not Used	
11	YL/GN	Neutral Detection Switch (L4)	Switch Open: 0v, Closed: 12v
12	BL/YL	Start Switch Signal (STSW)	Cranking: 9-11v
13-14, 18-20	---	Not Used	---
15	BK	A/T Solenoid Control (SLU-)	Pulse Signals
16	PK/GN	A/T Solenoid Control (SLU+)	Pulse Signals
17	BR	Shield Ground (GE01)	<0.050v
21	BK/OR	Generator Control (RL)	12v
22, 26, 28-31	---	Not Used	---
23	RD/YL	A/C Lock Sensor (LCK)	12v or 0v
24	YL	CKP Sensor Signal (NE-)	<0.050v
25	BL	CKP Sensor Signal (NE+)	AC pulse signals
27	BK	CMP Sensor Signal (G2+)	AC pulse signals
32	WT	CMP Sensor Signal (G2-)	AC pulse signals

Pin Connector Graphic (1)

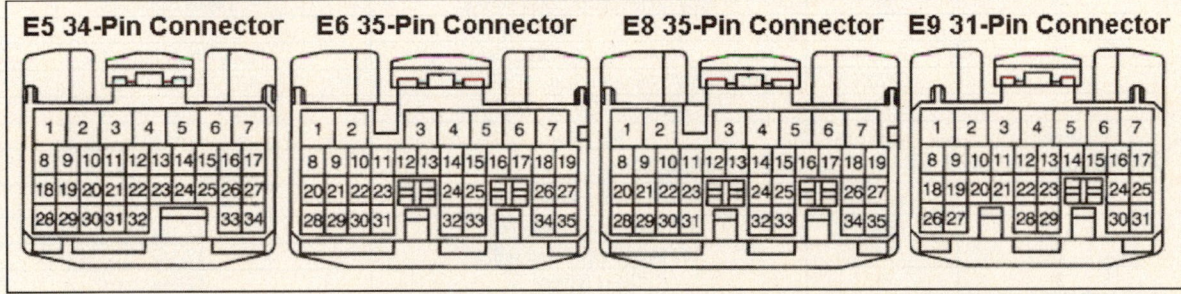

Pin Connector Graphic (2)

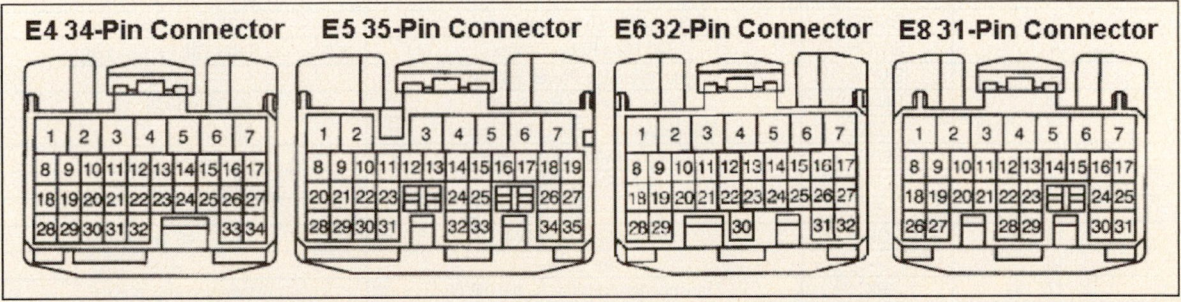

2003 Utility AWD 4.7L V8 VIN T E7 35 Pin Connector

PCM Pin #	Wire Color	Circuit Description (35 Pin)	Value at Hot Idle
1	WT/BK	Power Ground (HP)	<0.1v
2	BL	A/C Magnetic Clutch Relay (ACMG)	Relay Off: 0v, On: 12v
3	GN	A/T Select Switch 2nd Signal (2L)	In 2nd: 12v, Others: 0v
4	WT/BL	Low Detection Switch (L4)	Switch Open: 0v, Closed: 12v
5	BL/WT	ECT Pattern Switch 2nd Position (SNW1)	2nd Position: 12v
6	GN	ETCS Power (+BM)	12-14v
7	---	Not Used	---
8	RD/BL	Shift Lock ECU Control (L)	12v or 0v
9	GN/BK	Shift Lock ECU Control (2)	12v or 0v
10	GN/YL	Shift Lock ECU Control (D)	12v or 0v
11	RD/BK	A/T Select Switch Reverse (R)	In 'R': 1v, Others: 12v
12	---	Not Used	---
14	BL/BK	A/C Amplifier Signal (THWO)	A/C Off: 12v, On: 1v
15-16	---	Not Used	---
17	VT/RD	Vehicle Speed Sensor (SPD)	At 55 mph: 48 Hz
18	PK/BK	Body Control ECU Signal (MPX1)	Digital Signals
19	GN/YL	Stop Light Switch Signal	Brake Off: 0v, On: 12v
20	GN	Shift Lock ECU Control (3)	12v or 0v
21-22	---	Not Used	---
23	GN/RD	Shift Lock ECU Control (4)	12v or 0v
25	---	Not Used	---
26	BL/RD	Transponder Amplifier Signal (IMD)	Inserting key: pulses
27	WT/RD	Transponder Amplifier Signal (IMI)	Inserting key: pulses
28	BL/WT	ECT Pattern Switch Power Signal	Power Position: 12v
29	BK	Body Control ECU Signal (MPX2)	Digital Signals
30	---	Not Used	---
31	BR/BK	A/C Amplifier Signal (A/CS)	Relay Off: 12v, On: 1v
32	GY/GN	A/C Amplifier Signal (THE)	A/C Off: 12v, On: 1v
33	BK/RD	A/C Switch Signal (ACLD)	A/C On: 12v, Off: 0v
34-35	---	Not Used	---

2003 Utility AWD 4.7L V8 VIN T E8 31 Pin Connector

PCM Pin #	Wire Color	Circuit Description (31 Pin)	Value at Hot Idle
1	BK	EFI Main Relay Power (+B)	12-14v
2	BK	EFI Main Relay Power (+B2)	12-14v
3	BL	Direct Battery	12-14v
4	PK/BL	EVAP Pressure Switching Valve (VSV)	12v or 0v
5	BK/WT	Tachometer Signal (TACO)	Pulse Signals
6	GN/WT	A/T Select Switch Park Signal (P)	In 'P': 12v, Others: 0v
7	GN/RD	A/T Select Switch Neutral Signal (N)	In 'N': 12v, Others: 0v
8	WT/GN	EFI Main Relay Control	Relay Off: 0v, On: 12v
9	BK/OR	Ignition Switch Power (IGSW)	12-14v
10	GN/BK	Circuit Opening Relay (FC)	0-3v, off-idle: 12v
11	RD/BK	Malfunction Indicator Lamp Control	MIL Off: 12v, On: 1v
12	BL/RD	Taillight Switch Signal (ELS)	Taillights Off: 0v, On: 12v
13	YL/RD	Horn Relay Control	Relay Off: 12v, On: 1v
14	BL	Center Airbag Assembly (F/PS)	Digital Signals
15	OR	Power Ground (EOM)	<0.1v
16	---	Not Used	---
17	PK	Transponder Amplifier Signal (NEO)	Inserting key: pulses
18	RD/YL	SIL Signal (Scan Tool)	Transmitting: pulses
19	RD/WT	Data Link Connector (WFSE)	12v
20	PK/BL	Data Link Connector (TC)	12v
21	OR	EVAP Vapor Pressure Sensor (PTNK)	2.5-3.1v (with cap off)
22	WT/RD	Accelerator Position Sensor 1 (VPA)	0.3-0.90v
23	RD/BK	Accelerator Position Sensor 2 (VPA2)	1.8-2.7v
24	RD	Traction Control Engine Signal (ENG+)	Pulse Signals
25	BK	Traction Control Signal (TRC+)	Pulse Signals
26	BK/YL	Accelerator Pedal Position Sensor 1 VREF	4.9-5.1v
27	WT/BL	Accelerator Pedal Position Sensor 2 VREF	4.9-5.1v
28	LG/BK	Accelerator Pedal Position Sensor (EPA)	<0.050v
29	VT/WT	Accelerator Position Sensor 2 (EPA2)	1.8-2.7v
30	WT	Traction Control Engine Signal (ENG-)	Pulse Signals
31	YL	Traction Control Signal (TRC-)	Pulse Signals

TOYOTA TRUCK PIN TABLE CONTENTS

PICKUP Pin Tables

1990-92 Pickup 2.4L I4 MFI VIN R (A/T-ECT) 26 Pin Connector

PCM Pin #	Wire Color	Circuit Description (26 Pin)	Value at Hot Idle
1	BK/OR	Distributor Signal (NE+)	AC pulse signals
2	GN	Cold Start Injector	1v (at cold startup)
3	BK/YL	Igniter Signal (IGF)	0.74-0.76v
5	RD/GN	A/T-ECT Solenoid (S3)	In Lockup: 12-14v
6	RD/WT	A/T-ECT Solenoid (S2)	1st or OD: 1v
7	BL/RD	A/T-ECT Solenoid (S1)	3rd or OD: 1v
9 (Cal)	RD/GN	Sub O2S Heater Control	Heater Off: 12v, On: 1v
10	PK/GN	Main O2S Heater Control	Heater Off: 12v, On: 1v
11	GN	Fuel Pressure Up Solenoid	1v (at hot startup)
12	WT/RD	Injector Pair 1 & 3 Control	1.6-2-9 ms
13	BR	Power Ground	<0.1v
15	BR	Sensor Ground	<0.050v
16	BL	A/T Vehicle Speed Sensor	Moving: 0-5-0v
17	PK/WT	A/T Select Switch Low	In Low: 12v, Others: 0v
18	PK/GN	A/T Select Switch 2nd	In 2nd: 12v, Others: 0v
19	VT/RD	A/T Select Switch Neutral	In 'N': 12v, Others: 0v
20	RD	4WD Detection Transfer (L4)	Open: 12v, Closed: 0v
21	BK/BL	Igniter Signal (IGF)	Digital Signal: 0-5-0v
22	PK	EGR Solenoid Control (VSV)	12v or 0v
23	GN/RD	Intake Air Solenoid Control	12v or 0v
24	BK/RD	A/C Idle-Up Solenoid	A/C Off: 12v, On: 1v
25	WT	Injector Pair 2 & 4 Control	1.6-2-9 ms
26	BR	Power Ground	<0.1v

1990-92 Pickup 2.4L I4 MFI VIN R (A/T-ECT) 22 Pin Connector

PCM Pin #	Wire Color	Circuit Description (22 Pin)	Value at Hot Idle
1	BK/GN	Battery Direct	12-14v
4	BL/WT	A/T Oil Temperature Lamp	Lamp Off: 12v, On: 1v
5	PK	MIL (lamp) Control	MIL Off: 12v, On: 1v
6	GN/WT	Stop Light Switch Signal	Brake Off: 0v, On: 12v
7	RD/BL	A/T Pattern Select Switch	Norm: 0v, PWR: 12v
8	PK/GN	4WD Selector Switch	Switch Off: 12v, On: 1v
9	GN/BL	Vehicle Speed Sensor	At 55 mph: 48 Hz
11	BK/WT	Starter Switch Signal	9-11v (cranking)
12	WT/RD	EFI Main Relay Power	12-14v
13	WT/RD	EFI Main Relay Power	12-14v
15	BR/BK	Sensor Ground	<0.050v
16	YL/GN	Overdrive Main Switch	Switch Off: 12v, On: 1v
20	PK	Check Connector	12-14v
21	YL/RD	Cruise Control ECU (OD1)	At Cruise in OD: 12v

1990-92 Pickup 2.4L I4 MFI VIN R (A/T-ECT) 16 Pin Connector

PCM Pin #	Wire Color	Circuit Description (16 Pin)	Value at Hot Idle
1	GN/YL	Sensor VREF (VC)	4.9-5.1v
2	GN/BK	Airflow Meter VREF	4.9-5.1v
3	YL/GN	IAT Sensor Signal (THA)	At 100°F: 2.60v
4	GN/BL	ECT Sensor Signal (THW)	At 180°F: 0.51v
5	BK	Knock Sensor Signal	<0.075v AC
6	BK	Main O2S Signal	0.1-1.1v
7	GN/YL	4WD Oil Temperature Sensor	At 68°F: 4-5v
8	YL	Check Connector	12-14v
9	BR/BK	Sensor Ground	<0.050v
10	YL/BL	Airflow Meter Signal	0.5-2.5v
11	YL	TP Sensor Signal (VTA)	0.3-0.8v
12	YL/BL	Closed Throttle Switch	1v, off-idle: 12v
13	GN/WT	EGR Gas Temperature Sensor (THG)	3.5-4.0v
14 (Cal)	WT	Sub O2S Signal	0.1-1.1v
15	PK/WT	Check Connector	12-14v

Pin Connector Graphic

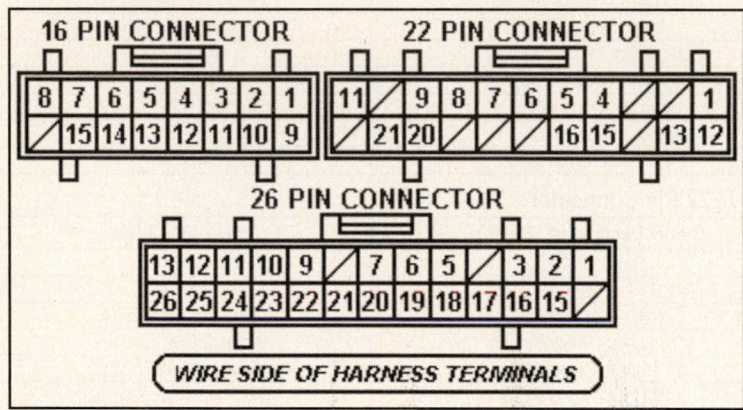

Standard Colors and Abbreviations

Abbreviation	Color	Abbreviation	Color	Abbreviation	Color
BK	Black	GY	Gray	RD	Red
BL	Blue	GN	Green	TN	Tan
BR	Brown	LG	LT Green	VT	Violet
DB	Dark Blue	OR	Orange	WT	White
DG	DK Green	PK	Pink	YL	Yellow

1990 Pickup 2.4L I4 MFI 22R-E VIN R (M/T) 10 Pin Connector

PCM Pin #	Wire Color	Circuit Description (10 Pin)	Value at Hot Idle
1	BK	M/T Clutch Start Switch	9-11v (cranking)
2	GN	Cold Start Injector	1v (at cold startup)
3	BK/WT	Starter Switch Signal	9-11v (cranking)
4	WT/RD	Injector Pair 1 & 3 Control	1.6-2-9 ms
5	BR	Power Ground	<0.1v
6	YL	Check Connector	12-14v
7	BR	Sensor Ground	<0.050v
8	BK/BL	Igniter Signal (IGF)	Digital Signal: 0-5-0v
9	WT	Injector Pair 2 & 4 Control	1.6-2-9 ms
10	BR	Power Ground	<0.1v

1990 Pickup 2.4L I4 MFI 22R-E VIN R (M/T) 14 Pin Connector

PCM Pin #	Wire Color	Circuit Description (14-Pin)	Value at Hot Idle
1	WT/RD	EFI Main Relay Power	12-14v
2	GN/BK	Battery Direct	12-14v
3	YL/GN	IAT Sensor Signal (THA)	At 100ºF: 2.60v
4	YL/BL	Airflow Meter Signal	0.5-2.5v
5	GN/BK	Airflow Meter VREF	4.9-5.1v
7 (Cal)	RD/GN	Sub O2S Heater Control	Heater Off: 12v, On: 1v
8	WT/RD	EFI Main Relay Power	12-14v
9	GN/WT	Stop Light Switch Signal	Brake Off: 0v, On: 12v
10	GN/BL	Vehicle Speed Sensor	At 55 mph: 48 Hz
11	GN/YL	4WD Selector Switch	Switch Off: 12v, On: 1v
12	BR/BK	Sensor Ground	<0.050v
14	YL/RD	Cruise Control ECU (OD1)	At Cruise in OD: 12v

1990 Pickup 2.4L I4 MFI 22R-E VIN R (M/T) 18 Pin Connector

PCM Pin #	Wire Color	Circuit Description (18 Pin)	Value at Hot Idle
1	BK/OR	Distributor Signal (NE+)	AC pulse signals
2	BK	Knock Sensor Signal	<0.075v AC
3	GN/WT	EGR Gas Temperature Sensor (THG)	3.5-4.0v
4 (Cal)	WT	Sub O2 Sensor Signal	0.1-1.1v
5	BK/YL	Igniter Signal (IGF)	Digital Signal: 0-5-0v
6	YL/BL	Closed Throttle Switch	1v, off-idle: 12v
7	PK/WT	Check Connector	12-14v
8	PK	MIL (lamp) Control	MIL Off: 12v, On: 1v
9	GN	Fuel Pressure Up Solenoid	1v (at hot startup)
10	GN/BL	ECT Sensor Signal (THW)	At 180ºF: 0.51v
11	YL	TP Sensor Signal (VTA)	0.3-0.8v
12	GN/YL	Sensor VREF (VC)	4.9-5.1v
13	BK	Main O2S Signal	0.1-1.1v
14	BR/BK	Sensor Ground	<0.050v
15	PK/GN	Main O2S Heater Control	Heater Off: 12v, On: 1v
16	PK	EGR Solenoid Control (VSV)	12v or 0v
17	GN/RD	Intake Air Solenoid Control	12v or 0v
18	BK/RD	A/C Idle-Up Solenoid	A/C Off: 12v, On: 1v

Pin Connector Graphic

1991-92 Pickup 2.4L I4 MFI 22R-E VIN R (M/T) 10 Pin Connector

PCM Pin #	Wire Color	Circuit Description (10 Pin)	Value at Hot Idle
1	BK	M/T Clutch Start Switch	9-11v (cranking)
2	GN	Cold Start Injector	1v (at cold startup)
3	BK/WT	Starter Switch Signal	9-11v (cranking)
4	WT/RD	Injector Pair 1 & 3 Control	1.6-2-9 ms
5	BR	Power Ground	<0.1v
6	YL	Check Connector	12-14v
7	BR	Sensor Ground	<0.050v
8	BK/BL	Igniter Signal (IGF)	Digital Signal: 0-5-0v
9	WT	Injector Pair 2 & 4 Control	1.6-2-9 ms
10	BR	Power Ground	<0.1v

1991-92 Pickup 2.4L I4 MFI 22R-E VIN R (M/T) 14 Pin Connector

PCM Pin #	Wire Color	Circuit Description (14-Pin)	Value at Hot Idle
1	WT/RD	EFI Main Relay B1+	12-14v
2	GN/BK	Battery Direct	12-14v
3	YL/GN	IAT Sensor Signal (THA)	At 100ºF: 2.60v
4	YL/BL	Airflow Meter Signal	0.5-2.5v
5	GN/BK	Airflow Meter VREF	4.9-5.1v
7 (Cal)	RD/GN	Sub O2S Heater Control	Heater Off: 12v, On: 1v
8	WT/RD	EFI Main Relay (B+)	12-14v
9	GN/WT	Stop Light Switch Signal	Brake Off: 0v, On: 12v
10	GN/BL	Vehicle Speed Sensor	At 55 mph: 48 Hz
12	BR/BK	Sensor Ground	<0.050v
14	YL/RD	Cruise Control ECU	At Cruise in OD: 12v

1991-92 Pickup 2.4L I4 MFI 22R-E VIN R (M/T) 18 Pin Connector

PCM Pin #	Wire Color	Circuit Description (18 Pin)	Value at Hot Idle
1	BK/OR	Distributor Signal (NE+)	AC pulse signals
2	BK	Knock Sensor Signal	<0.075v AC
3	GN/WT	EGR Gas Temperature Sensor (THG)	3.5-4.0v
4 (Cal)	WT	Sub O2 Sensor Signal	0.1-1.1v
5	BK/YL	Igniter Signal (IGF)	Digital Signal: 0-5-0v
6	YL/BL	Closed Throttle Switch	1v, off-idle: 12v
7	PK/WT	Check Connector	12-14v
8	PK	MIL (lamp) Control	MIL Off: 12v, On: 1v
9	GN	Fuel Pressure Up Solenoid	1v (at hot startup)
10	GN/BL	ECT Sensor Signal (THW)	At 180ºF: 0.51v
11	YL	TP Sensor Signal (VTA)	0.3-0.8v
12	GN/YL	Sensor VREF (VC)	4.9-5.1v
13	BK	Main O2S Signal	0.1-1.1v
14	BR/BK	Sensor Ground	<0.050v
15	PK/GN	Main O2S Heater Control	Heater Off: 12v, On: 1v
16	PK	EGR Solenoid Control (VSV)	12v or 0v
17	GN/RD	Intake Air Solenoid Control	12v or 0v

Pin Connector Graphic

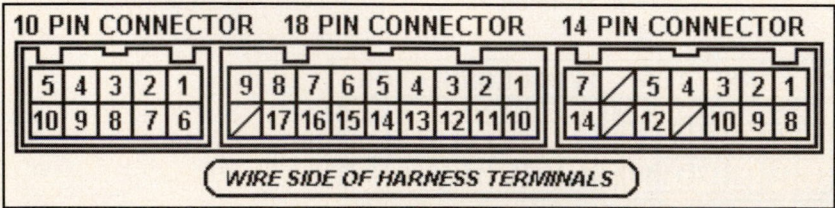

1991-92 Pickup 2WD 2.4L I4 MFI VIN R (M/T) 16 Pin Connector

PCM Pin #	Wire Color	Circuit Description (16 Pin)	Value at Hot Idle
1	WT/RD	EFI Main Relay Power	12-14v
2	BK/GN	Battery Direct	12-14v
6	PK	EGR Solenoid Control (VSV)	12v or 0v
7	GN	Fuel Pressure Up Solenoid	1v (at hot startup)
8	PK/WT	Check Connector	12-14v
9	WT/RD	EFI Main Relay Power	12-14v
10	PK	MIL (lamp) Control	MIL Off: 12v, On: 1v
12	GN/WT	Stop Light Switch Signal	Brake Off: 0v, On: 12v
13	GN/BL	Vehicle Speed Sensor	At 55 mph: 48 Hz
14	YL/GN	Overdrive Main Switch	Switch Off: 12v, On: 1v
15	GN/RD	Intake Air Solenoid Control	12v or 0v
16	YL	Check Connector	12v

1991-92 Pickup 2WD 2.4L I4 MFI VIN R (M/T) 26 Pin Connector

PCM Pin #	Wire Color	Circuit Description (26 Pin)	Value at Hot Idle
1	PK/GN	Main O2S Heater Control	Heater Off: 12v, On: 1v
2	BK/WT	Starter Switch Signal	9-11v (cranking)
3	GN/BL	ECT Sensor Signal (THW)	At 180ºF: 0.51v
4	YL/BL	Airflow Meter Signal	0.5-2.5v
5	YL/GN	IAT Sensor Signal (THA)	At 100ºF: 2.60v
6	BK/BL	Igniter Signal (IGF)	Digital Signal: 0-5-0v
7	BK/YL	Igniter Signal (IGF)	Digital Signal: 0-5-0v
8	GN/WT	EGR Gas Temperature Sensor (THG)	3.5-4.0v
9	GN/BK	Airflow Meter VREF	4.9-5.1v
10	BK	Main O2S Signal	0.1-1.1v
11	GN	Cold Start Injector	1v (at cold startup)
12	WT/RD	Injector 1 & 3 Control	1.6-2-9 ms
13	BR	Power Ground	<0.1v
14 (Cal)	RD/GN	Sub O2S Heater Control	Heater Off: 12v, On: 1v
15	BK	Clutch Start Switch	Clutch Out: 12v, In: 0v
16, 22	BR/BK	Sensor Ground	<0.050v
17	YL	TP Sensor Signal (VTA)	0.3-0.8v
18	GN/YL	Sensor VREF (VC)	4.9-5.1v
19	YL/BL	Closed Throttle Switch	1v, off-idle: 12v
21	BK/OR	Distributor Signal (NE+)	AC pulse signals
23 (Cal)	WT	Sub O2S Signal	0.1-1.1v
24	BR	Sensor Ground	<0.050v
25	WT	Injector 2 & 4 Control	1.6-2-9 ms
26	BR	Power Ground	<0.1v

Pin Connector Graphic

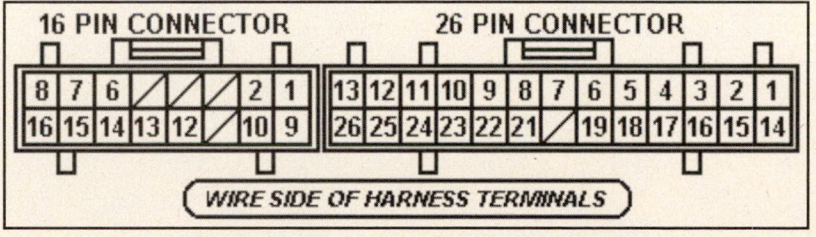

1993-95 Pickup 2.4L I4 MFI VIN R (A/T-ECT) 26 Pin Connector

PCM Pin #	Wire Color	Circuit Description (26 Pin)	Value at Hot Idle
1	GN	Cold Start Injector	1v (at cold startup)
2	PK/GN	Main O2S Heater Control	Heater Off: 12v, On: 1v
3	BK/YL	Igniter Signal (IGF)	Digital Signal: 0-5-0v
4	BK/OR	Distributor Signal (NE+)	AC pulse signals
5	RD/GN	A/T-ECT Solenoid (S3)	In Lockup: 12-14v
6	RD/WT	A/T-ECT Solenoid (S2)	1st or OD: 1v
7	BL/RD	A/T-ECT Solenoid (S1)	3rd or OD: 1v
8	PK	EGR Solenoid Control (VSV)	12v or 0v
9	GN/RD	Intake Air Solenoid Control	12v or 0v
10	GN	Fuel Pressure Up Solenoid	1v (at hot startup)
11	WT	Injector Pair 2 & 4 Control	1.6-2-9 ms
12	WT/RD	Injector Pair 1 & 3 Control	1.6-2-9 ms
13	BR	Power Ground	<0.1v
14	BR	Sensor Ground	<0.050v
15 (Cal)	RD/GN	Sub-HO2S Heater Control	Heater Off: 12v, On: 1v
19	BL	A/T Vehicle Speed Sensor	Moving: 0-5-0v
20	BK/BL	Igniter Signal (IGF)	Digital Signal: 0-5-0v
21	PK/WT	A/T Select Switch Low	In Low: 12v, Others: 0v
22	PK/GN	A/T Select Switch 2nd	In 2nd: 12v, Others: 0v
23	PK/RD	A/T Select Switch Neutral	In 'N': 12v, Others: 0v
26	BR	Power Ground	<0.1v

1993-95 Pickup 2.4L I4 MFI VIN R (A/T-ECT) 22 Pin Connector

PCM Pin #	Wire Color	Circuit Description (22 Pin)	Value at Hot Idle
1	BK/GN	Battery Direct	12-14v
5	PK	MIL (lamp) Control	MIL Off: 12v, On: 1v
6	GN/WT	Stop Light Switch Signal	Brake Off: 0v, On: 12v
7	RD/BL	A/T Pattern Select Switch	Norm: 0v, PWR: 12v
8	PK/GN	4WD Selector Switch	Switch Off: 12v, On: 1v
9	GN/BL	Vehicle Speed Sensor	At 55 mph: 48 Hz
11	BK/WT	Starter Switch Signal	9-11v (cranking)
12	WT/RD	EFI Main Relay (B+)	12-14v
13	WT/RD	EFI Main Relay B1+	12-14v
15	BR/BK	Sensor Ground	<0.050v
16	YL/GN	Overdrive Main Switch	Switch Off: 12v, On: 1v
19	RD	4WD Detection Transfer (L4)	Open: 12v, Closed: 0v
20	PK	Data Link Connector	12-14v
21	YL/RD	Cruise Control ECU	At Cruise in OD: 12v

1993-95 Pickup 2.4L I4 MFI VIN R (A/T-ECT) 16 Pin Connector

PCM Pin #	Wire Color	Circuit Description (16 Pin)	Value at Hot Idle
1	GN/YL	Sensor VREF (VC)	4.9-5.1v
2	YL/BL	Airflow Meter Signal	0.5-2.5v
3	YL/GN	IAT Sensor Signal (THA)	At 100°F: 2.60v
4	GN/BL	ECT Sensor Signal (THW)	At 180°F: 0.51v
5 (Cal)	WT	Sub O2S Signal	0.1-1.1v
6	BK	Main O2S Signal	0.1-1.1v
7	BK	Knock Sensor Signal	<0.075v AC
8	YL	Data Link Connector	12-14v
9	BR/BK	Sensor Ground	<0.050v
10	GN/BK	Airflow Meter VREF	4.9-5.1v
11	YL	TP Sensor Signal (VTA)	0.3-0.8v
12	YL/BL	Closed Throttle Switch	1v, off-idle: 12v
13	GN/WT	EGR Gas Temperature Sensor (THG)	3.5-4.0v
14	PK/GN	Data Link Connector	12-14v
15	PK/WT	Data Link Connector	12-14v

Pin Connector Graphic

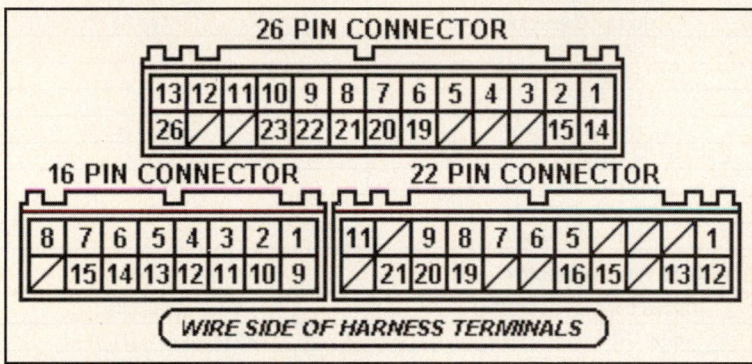

Standard Colors and Abbreviations

Abbreviation	Color	Abbreviation	Color	Abbreviation	Color
BK	Black	GY	Gray	RD	Red
BL	Blue	GN	Green	TN	Tan
BR	Brown	LG	LT Green	VT	Violet
DB	Dark Blue	OR	Orange	WT	White
DG	DK Green	PK	Pink	YL	Yellow

1993-95 Pickup 2.4L I4 MFI 22R-E VIN R (M/T) 12 Pin Connector

PCM Pin #	Wire Color	Circuit Description (12 Pin)	Value at Hot Idle
1	WT/RD	EFI Main Relay B1+	12-14v
2	GN/BK	Battery Direct	12-14v
6	GN/YL	4WD Selector Switch	Switch Off: 12v, On: 1v
7	WT/RD	EFI Main Relay (B+)	12-14v
8	PK	MIL (lamp) Control	MIL Off: 12v, On: 1v
11	GN/BL	Vehicle Speed Sensor	At 55 mph: 48 Hz
12	GN/WT	Stop Light Switch Signal	Brake Off: 0v, On: 12v

1993-95 Pickup 2.4L I4 MFI 22R-E VIN R (M/T) 16 Pin Connector

PCM Pin #	Wire Color	Circuit Description (16 Pin)	Value at Hot Idle
1	YL/GN	IAT Sensor Signal (THA)	At 100°F: 2.60v
2	YL/BL	Airflow Meter Signal	0.5-2.5v
3	GN/BK	Airflow Meter VREF	4.9-5.1v
4	GN/BL	ECT Sensor Signal (THW)	At 180°F: 0.51v
5 (Cal)	WT	Sub O2S Signal	0.1-1.1v
6	BK	Main O2S Signal	0.1-1.1v
7	PK/GN	Data Link Connector	12-14v
8	YL	Data Link Connector	12-14v
9	BR/BK	Sensor Ground	<0.050v
10	YL	TP Sensor Signal (VTA)	0.3-0.8v
11	GN/YL	Sensor VREF (VC)	4.9-5.1v
12	YL/BL	Closed Throttle Switch	1v, off-idle: 12v
13	GN/WT	EGR Gas Temperature Sensor (THG)	3.5-4.0v
14	BK	Knock Sensor Signal	<0.075v AC
15	PK/WT	Data Link Connector	12-14v
16	BR/BK	Data Link Connector	<0.050v

1993-95 Pickup 2.4L I4 MFI 22R-E VIN R (M/T) 26 Pin Connector

PCM Pin #	Wire Color	Circuit Description (26 Pin)	Value at Hot Idle
1	PK/GN	Main O2S Heater Control	Heater Off: 12v, On: 1v
2	BK/WT	Starter Switch Signal	9-11v (cranking)
3	BK/YL	Igniter Signal (IGF)	Digital Signal: 0-5-0v
4	BK/OR	Distributor Signal (NE+)	AC pulse signals
9	GN/RD	Intake Air Solenoid Control	12v or 0v
10	GN	Fuel Pressure Up Solenoid	1v (at hot startup)
11	GN	Cold Start Injector	1v (at cold startup)
12	WT/RD	Injector Pair 1 & 3 Control	1.6-2-9 ms
13	BR	Power Ground	<0.1v
14 (Cal)	RD/GN	Sub O2S Heater Control	Heater Off: 12v, On: 1v
15	BK	M/T Clutch Start Switch	Clutch Out: 12v, In: 0v
22	BK/BL	Igniter Signal (IGF)	Digital Signal: 0-5-0v
23	PK	EGR Solenoid Control (VSV)	12v or 0v
24	BR/BK	Sensor Ground	<0.050v
25	WT	Injector Pair 2 & 4 Control	1.6-2-9 ms
26	BR	Power Ground	<0.1v

Pin Connector Graphic

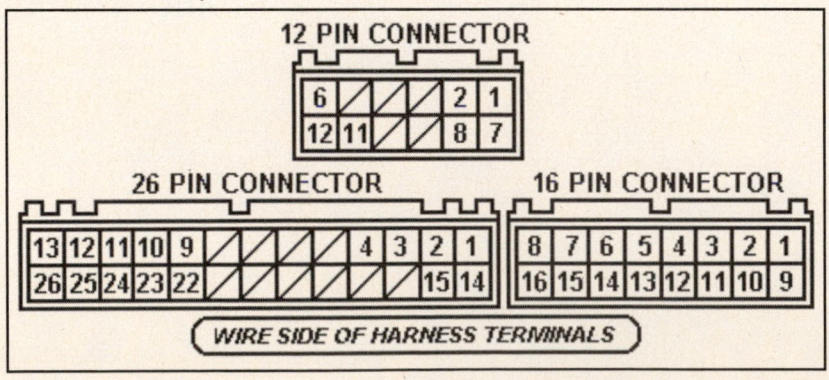

1993-95 Pickup 2WD 2.4L I4 MFI VIN R (M/T) 16 Pin Connector

PCM Pin #	Wire Color	Circuit Description (16 Pin)	Value at Hot Idle
1	WT/RD	EFI Main Relay B1+	12-14v
2	BK/GN	Battery Direct	12-14v
5	PK	EGR Solenoid Control (VSV)	12v or 0v
6	GN	Fuel Pressure Up Solenoid	1v (at hot startup)
7	PK/GN	Data Link Connector	12-14v
8	PK/WT	Data Link Connector	12-14v
9	WT/RD	EFI Main Relay (B+)	12-14v
10	PK	MIL (lamp) Control	MIL Off: 12v, On: 1v
11	BR/BK	Sensor Ground	<0.050v
12	GN/WT	Stop Light Switch Signal	Brake Off: 0v, On: 12v
13	GN/BL	Vehicle Speed Sensor	At 55 mph: 48 Hz
14	YL/GN	Overdrive Main Switch	Switch Off: 12v, On: 1v
15	GN/RD	Intake Air Solenoid Control	12v or 0v
16	YL	Data Link Connector	12-14v

1993-95 Pickup 2WD 2.4L I4 MFI VIN R (M/T) 26 Pin Connector

PCM Pin #	Wire Color	Circuit Description (26 Pin)	Value at Hot Idle
1	PK/GN	Main O2S Heater Control	Heater Off: 12v, On: 1v
2	BK/WT	Neutral Start Switch	In P/N: 9-11v (cranking)
3	GN/BL	ECT Sensor Signal (THW)	At 180ºF: 0.51v
4	YL/BL	Airflow Meter Signal	0.5-2.5v
5	GN/BK	Airflow Meter VREF	4.9-5.1v
6	BK/BL	Igniter Signal (IGF)	Digital Signal: 0-5-0v
7	BK/YL	Igniter Signal (IGF)	Digital Signal: 0-5-0v
10	BK	Main O2S Signal	0.1-1.1v
11	BK/WT	M/T: Starter Switch Signal	9-11v (cranking)
11	BK/YL	A/T: Starter Switch Signal	9-11v (cranking)
12	WT/RD	Injector 1 & 3 Control	1.6-2-9 ms
13	BR	Power Ground	<0.1v
14 (Cal)	RD/GN	Sub O2S Heater Control	Heater Off: 12v, On: 1v
15	GN	Cold Start Injector	1v (at cold startup)
16	BR/BK	Sensor Ground	<0.050v
17	YL	TP Sensor Signal (VTA)	0.3-0.8v
18	GN/YL	Sensor VREF (VC)	4.9-5.1v
19	YL/BL	Closed Throttle Switch	1v, off-idle: 12v
20	YL/GN	IAT Sensor Signal (THA)	At 100ºF: 2.60v
21	BK/OR	Distributor Signal (NE+)	AC pulse signals
22	GN/WT	EGR Gas Temperature Sensor (THG)	3.5-4.0v
23 (Cal)	WT	Sub O2S Signal	0.1-1.1v
24	BR	Sensor Ground	<0.050v
25	WT	Injector 2 & 4 Control	1.6-2-9 ms
26	BR	Power Ground	<0.1v

Pin Connector Graphic

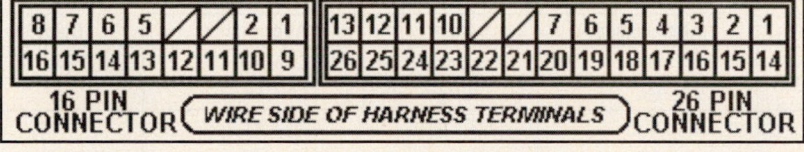

1990 Pickup 3.0L V6 MFI VIN V (A/T-ECT) 26 Pin Connector

PCM Pin #	Wire Color	Circuit Description (26 Pin)	Value at Hot Idle
1	WT	Distributor Signal (NE+)	AC pulse signals
2	RD	Distributor Signal (G1+)	AC pulse signals
3	BK/YL	Igniter Signal (IGF)	0.74-0.76v
4	GN/RD	A/T-ECT Solenoid (S4)	In Lockup: 12-14v
5	YL/BK	A/T-ECT Solenoid (S3)	In Lockup: 12-14v
6	BK	A/T-ECT Solenoid (S2)	1st or OD: 1v
7	PK	A/T-ECT Solenoid (S1)	3rd or OD: 1v
8	GN	Fuel Pressure Up Solenoid	1v (at hot startup)
9	GN	Cold Start Injector	1v (at cold startup)
10	PK/GN	HO2S-11 (B1 S1) Heater	Heater Off: 12v, On: 1v
11	BR	Sensor Ground	<0.050v
12	WT/RD	Injectors 1 & 3 & 5 Control	1.6-2.9 ms
13	BR	Power Ground	<0.1v
14	GN	Distributor Signal (GN-)	<0.050v
15	BK	Distributor Signal (G2+)	AC pulse signals
16	BR/RD	A/T Vehicle Speed Sensor	Moving: 0-5-0v
17	PK/WT	A/T Select Switch Low	In Low: 12v, Others: 0v
18	PK/GN	A/T Select Switch 2nd	In 2nd: 12v, Others: 0v
19	PK/RD	A/T Select Switch Neutral	In 'N': 12v, Others: 0v
20	YL/RD	4WD Detection Transfer (L4)	Open: 12v, Closed: 0v
21	BK/BL	Igniter Signal (IGF)	Digital Signal: 0-5-0v
22	PK	EGR Solenoid Control (VSV)	12v or 0v
23	GN/RD	Intake Air Solenoid Control	12v or 0v
24	BK/RD	A/C Idle-Up Solenoid	A/C Off: 12v, On: 1v
25	WT	Injectors 2 & 4 & 6 Control	1.6-2.9 ms
26	BR	Power Ground	<0.1v

Pin Connector Graphic

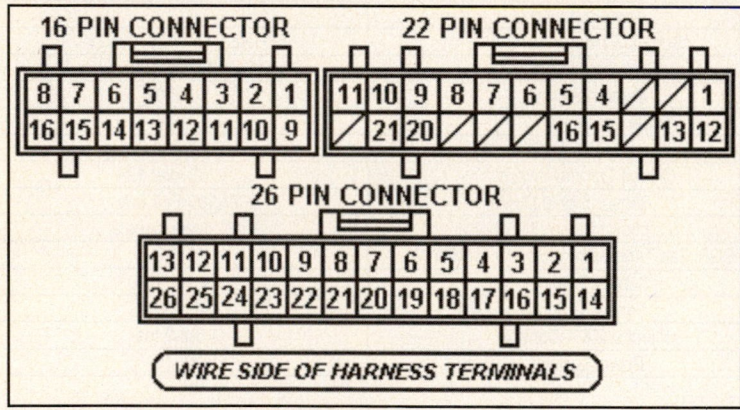

Standard Colors and Abbreviations

Abbreviation	Color	Abbreviation	Color	Abbreviation	Color
BK	Black	GY	Gray	RD	Red
BL	Blue	GN	Green	TN	Tan
BR	Brown	LG	LT Green	VT	Violet
DB	Dark Blue	OR	Orange	WT	White
DG	DK Green	PK	Pink	YL	Yellow

1990 Pickup 3.0L V6 MFI VIN V (A/T-ECT) 16 Pin Connector

PCM Pin #	Wire Color	Circuit Description (16 Pin)	Value at Hot Idle
1	GN/BK	Airflow Meter VREF	4.9-5.1v
2	YL/BL	Airflow Meter Signal	0.5-2.5v
3	YL/GN	IAT Sensor Signal (THA)	At 100°F: 2.60v
4	GN/BL	ECT Sensor Signal (THW)	At 180°F: 0.51v
5	BK	Knock Sensor Signal	<0.075v AC
6	BK	Main O2S Signal	0.1-1.1v
7	GN/BK	4WD Oil Temperature Sensor	At 68°F: 4-5v
8	YL/GN	Check Connector	12-14v
9	BR/BK	Sensor Ground	<0.050v
10 (Cal)	RD/GN	Sub O2S Signal	0.1-1.1v
11	YL	TP Sensor Signal (VTA)	0.3-0.8v
12	YL/BL	Closed Throttle Switch	1v, off-idle: 12v
13	GN/WT	EGR Gas Temperature Sensor (THG)	3.5-4.0v
14	LG	Transfer Oil Temp. Sensor	At 68°F: 4-5v
15	PK/WT	Check Connector	12-14v
16	PK	Coolant Temperature Switch	Open: 12v, Closed: 0v

1990 Pickup 3.0L V6 MFI VIN V (A/T-ECT) 22 Pin Connector

PCM Pin #	Wire Color	Circuit Description (22 Pin)	Value at Hot Idle
1	BK/GN	Battery Direct	12-14v
4	PK/GN	A/T Oil Temperature Lamp	Lamp Off: 12v, On: 1v
5	PK	MIL (lamp) Control	MIL Off: 12v, On: 1v
6	GN/WT	Stop Light Switch Signal	Brake Off: 0v, On: 12v
7	GN/OR	A/T Pattern Select Switch	Norm: 0v, PWR: 12v
8	PK/GN	4WD Selector Switch	Switch Off: 12v, On: 1v
9	GN/BL	Vehicle Speed Sensor	At 55 mph: 48 Hz
10	BK/RD	A/C Magnetic Clutch (ACMG)	Clutch Off: 0v, On: 12v
11	BK/WT	Starter Switch Signal	9-11v (cranking)
12	WT/RD	EFI Main Relay (B+)	12-14v
13	WT/RD	EFI Main Relay B1+	12-14v
15	BR/BK	Sensor Ground	<0.050v
16	YL/GN	Overdrive Main Switch	Switch Off: 12v, On: 1v
20	PK	Check Connector	12-14v
21	YL/RD	Cruise Control ECU	At Cruise in OD: 12v

Standard Colors and Abbreviations

Abbreviation	Color	Abbreviation	Color	Abbreviation	Color
BK	Black	GY	Gray	RD	Red
BL	Blue	GN	Green	TN	Tan
BR	Brown	LG	LT Green	VT	Violet
DB	Dark Blue	OR	Orange	WT	White
DG	DK Green	PK	Pink	YL	Yellow

1990 Pickup 3.0L V6 MFI 3VZ-E VIN V (M/T) 10 Pin Connector

PCM Pin #	Wire Color	Circuit Description (10 Pin)	Value at Hot Idle
1	BK/WT	A/T Neutral Start Switch	In P/N: 9-11v (cranking)
1	BK	M/T Clutch Start Switch	9-11v (cranking)
2	GN	Cold Start Injector	1v (at cold startup)
3	BK/WT	Starter Switch Signal	9-11v (cranking)
4	WT/RD	Injector Pair 1 & 3 Control	1.6-2.9 ms
5	BR	Power Ground	<0.1v
6	YL	Check Connector	12-14v
7	BR	Sensor Ground	<0.050v
8	BK/BL	Igniter Signal (IGF)	Digital Signal: 0-5-0v
9	WT	Injector Pair 2 & 4 Control	1.6-2.9 ms
10	BR	Power Ground	<0.1v

1990 Pickup 3.0L V6 MFI 3VZ-E VIN V (M/T) 14-Pin Connector

PCM Pin #	Wire Color	Circuit Description (14-Pin)	Value at Hot Idle
1	WT/RD	EFI Main Relay B1+	12-14v
2	GN/BK	Battery Direct	12-14v
3	YL/GN	IAT Sensor Signal (THA)	At 100ºF: 2.60v
4	YL/BL	Airflow Meter Signal	0.5-2.5v
5	GN/BK	Airflow Meter VREF	4.9-5.1v
7 (Cal)	RD/GN	Sub O2S Heater Control	Heater Off: 12v, On: 1v
8	WT/RD	EFI Main Relay (B+)	12-14v
9	GN/WT	Stop Light Switch Signal	Brake Off: 0v, On: 12v
10	GN/BL	Vehicle Speed Sensor	At 55 mph: 48 Hz
11	GN/YL	4WD Selector Switch	Switch Off: 12v, On: 1v
12	BR/BK	Sensor Ground	<0.050v
14	YL/RD	Cruise Control ECU	At Cruise in OD: 12v

1990 Pickup 3.0L V6 MFI 3VZ-E VIN V (M/T) 18 Pin Connector

PCM Pin #	Wire Color	Circuit Description (18 Pin)	Value at Hot Idle
1	BK/OR	Distributor Signal (NE+)	AC pulse signals
2	BK	Knock Sensor Signal	<0.075v AC
3	GN/WT	EGR Gas Temperature Sensor (THG)	3.5-4.0v
4 (Cal)	WT	Sub O2S Signal	0.1-1.1v
5	BK/YL	Igniter Signal (IGF)	Digital Signal: 0-5-0v
6	YL/BL	Closed Throttle Switch	1v, off-idle: 12v
7	PK/WT	Check Connector	12-14v
8	PK	MIL (lamp) Control	MIL Off: 12v, On: 1v
9	GN	Fuel Pressure Up Solenoid	1v (at hot startup)
10	GN/BL	ECT Sensor Signal (THW)	At 180ºF: 0.51v
11	YL	TP Sensor Signal (VTA)	0.3-0.8v
12	GN/YL	Sensor VREF (VC)	4.9-5.1v
13	BK	Main O2S Signal	0.1-1.1v
14	BR/BK	Sensor Ground	<0.1v
15	PK/GN	Main O2S Heater Control	Heater Off: 12v, On: 1v
16	PK	EGR Solenoid Control (VSV)	12v or 0v
17	GN/RD	Intake Air Solenoid Control	12v or 0v
18	BK/RD	A/C Idle-Up Solenoid	A/C Off: 12v, On: 1v

Pin Connector Graphic

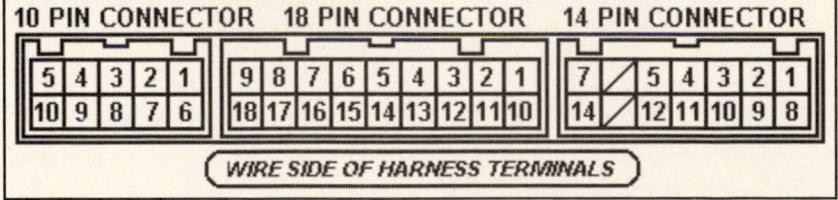

1991-92 Pickup 3.0L V6 MFI VIN V (All) 16 Pin Connector

PCM Pin #	Wire Color	Circuit Description (16 Pin)	Value at Hot Idle
1	GN/BK	Airflow Meter VREF	4.9-5.1v
2	YL/BL	Airflow Meter Signal	0.5-2.5v
3	YL/GN	IAT Sensor Signal (THA)	At 100°F: 2.60v
3 ('92)	YL/BL	IAT Sensor Signal (THA)	At 100°F: 2.60v
4	GN/BL	ECT Sensor Signal (THW)	At 180°F: 0.51v
5	BK	Knock Sensor Signal	<0.075v AC
6	BK	Main O2S Signal	0.1-1.1v
7	GN/BK	A/T Oil Temperature Sensor	At 68°F: 4-5v
8	YL/GN	Check Connector	12-14v
9	BR/BK	Sensor Ground	<0.050v
10 (Cal)	RD/GN	Sub O2S Signal	0.1-1.1v
11	YL	TP Sensor Signal (VTA)	0.3-0.8v
12	YL/BL	Closed Throttle Switch	1v, off-idle: 12v
13	GN/WT	EGR Gas Temperature Sensor (THG)	3.5-4.0v
13 (Fed)	GN/WT	Short Pin	Open: 12v, Closed: 0v
14	LG	Transfer Oil Temp. Sensor	At 68°F: 4-5v
15	PK/WT	Check Connector	12-14v

1991-92 Pickup 3.0L V6 MFI VIN V (All) 22 Pin Connector

PCM Pin #	Wire Color	Circuit Description (22 Pin)	Value at Hot Idle
1	BK/GN	Battery Direct	12-14v
4 ('91)	PK/GN	A/T Oil Temperature Lamp	Lamp Off: 12v, On: 1v
4 ('92)	BL/WT	A/T Oil Temperature Lamp	Lamp Off: 12v, On: 1v
5	PK	MIL (lamp) Control	MIL Off: 12v, On: 1v
6	GN/WT	Stop Light Switch Signal	Brake Off: 0v, On: 12v
7	GN/OR	A/T Pattern Select Switch	Norm: 0v, PWR: 12v
8	PK/GN	A/T 4WD Indicator Switch	Switch Off: 12v, On: 1v
8	GN/WT	M/T 4WD Indicator Switch	Switch Off: 12v, On: 1v
9	GN/BL	Vehicle Speed Sensor	At 55 mph: 48 Hz
10	BK/WT	A/C Amplifier Signal (AC1)	Clutch On: 1.5v, Off: 12v
11	BK/WT	M/T: Starter Switch Signal	9-11v (cranking)
11	BK/RD	A/T: Starter Switch Signal	9-11v (cranking)
12	WT/RD	EFI Main Relay (B+)	12-14v
13	WT/RD	EFI Main Relay B1+	12-14v
15	BR/BK	Sensor Ground	<0.050v
16	YL/GN	Overdrive Main Switch	Switch Off: 12v, On: 1v
17	GY	4WD M/T Select (SEL2)	Open: 12v, Closed: 0v
18	WT/BK	2WD A/T Select (SEL1)	Open: 12v, Closed: 0v
20	YL/WT	Check Connector	12-14v
21	YL/RD	Cruise Control ECU	At Cruise in OD: 12v

1991-92 Pickup 3.0L V6 MFI VIN V (All) 26 Pin Connector

PCM Pin #	Wire Color	Circuit Description (26 Pin)	Value at Hot Idle
1	WT	Distributor Signal (NE+)	AC pulse signals
2	RD	Distributor Signal (G1+)	AC pulse signals
3	BK/YL	Igniter Signal (IGF)	Digital Signal: 0-5-0v
4	GN/RD	A/T-ECT Solenoid (S4)	In Lockup: 12-14v
5	YL/BK	A/T-ECT Solenoid (S3)	In Lockup: 12-14v
6	BK	A/T-ECT Solenoid (S2)	1st or OD: 1v
7	WT	A/T-ECT Solenoid (S1)	3rd or OD: 1v
8	GN	Fuel Pressure Up Solenoid	1v (at hot startup)
9	GN	Cold Start Injector	1v (at cold startup)
10	PK/GN	HO2S Heater Control	Heater Off: 12v, On: 1v
11	BR	Sensor Ground	<0.050v
12	WT/RD	Injectors 1 & 3 & 5 Control	1.6-2.9 ms
13	BR	Power Ground	<0.1v
14	GN	Distributor Signal (GN-)	<0.050v
15	BK	Distributor Signal (G2+)	AC pulse signals
16	BR/RD	A/T Vehicle Speed Sensor	Moving: 0-5-0v
17	PK/WT	A/T Select Switch Low	In Low: 12v, Others: 0v
18	PK/GN	A/T Select Switch 2nd	In 2nd: 12v, Others: 0v
19	PK/RD	A/T Select Switch Neutral	In 'N': 12v, Others: 0v
20	YL/RD	4WD Detection Transfer (L4)	Open: 12v, Closed: 0v
21	BL/BK	Igniter Signal (IGT)	Digital Signal: 0-5-0v
22	PK	EGR Solenoid Control (VSV)	12v or 0v
23	GN/RD	Intake Air Solenoid Control	12v or 0v
24	BK/RD	A/C Amplifier Signal (ACT)	Clutch On: 12v, Off: 1.5v
25	WT	Injectors 2 & 4 & 6 Control	1.6-2.9 ms
26	BR	Power Ground	<0.1v

Pin Connector Graphic

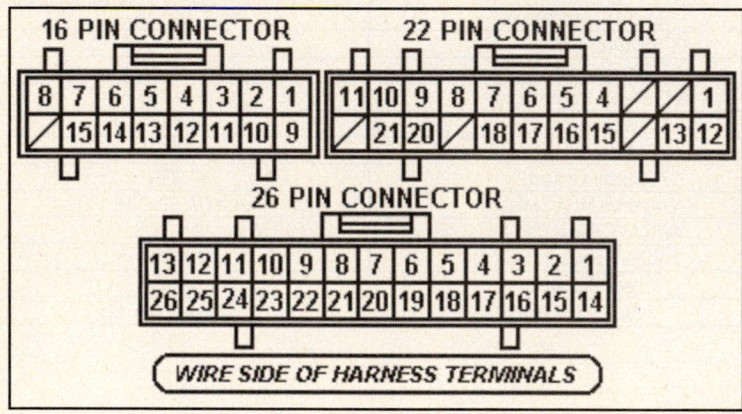

1993 Pickup 3.0L V6 MFI VIN V (All) 16 Pin Connector

PCM Pin #	Wire Color	Circuit Description (16 Pin)	Value at Hot Idle
1	GN/BK	Airflow Meter VREF	4.9-5.1v
2	YL/RD	Airflow Meter Signal	0.5-2.5v
3	YL/BK	IAT Sensor Signal (THA)	At 100°F: 2.60v
4	GN/BL	ECT Sensor Signal (THW)	At 180°F: 0.51v
5	BK	Knock Sensor Signal	<0.075v AC
6	BK	Main O2S Signal	0.1-1.1v
7	GN/BK	A/T Oil Temperature Sensor	At 68°F: 4-6v
8	YL/GN	Data Link Connector	12-14v
9	BR/BK	Sensor Ground	<0.050v
10 (Cal)	RD/GN	Sub O2S Signal	0.1-1.1v
11	YL	TP Sensor Signal (VTA)	0.3-0.8v
12	YL/WT	Closed Throttle Switch	1v, off-idle: 12v
13	GN/WT	EGR Gas Temperature Sensor (THG)	3.5-4.0v
13 (Fed)	GN/WT	Short Pin	Open: 12v, Closed: 0v
14	LG	4WD Oil Temperature Sensor	At 68°F: 4-6v
15	PK/WT	Data Link Connector	12-14v
16	PK/GN	Data Link Connector	12-14v

1993 Pickup 3.0L V6 MFI VIN V (All) 22 Pin Connector

PCM Pin #	Wire Color	Circuit Description (22 Pin)	Value at Hot Idle
1	BK/GN	Battery Direct	12-14v
4	BL/WT	A/T Oil Temperature Lamp	Lamp Off: 12v, On: 1v
5	PK	MIL (lamp) Control	MIL Off: 12v, On: 1v
6	GN/WT	Stop Light Switch Signal	Brake Off: 0v, On: 12v
7	GN/OR	A/T Pattern Select Switch	Norm: 0v, PWR: 12v
8	PK/GN	A/T 4WD Indicator Switch	Switch Off: 12v, On: 1v
8	GN/WT	M/T 4WD Indicator Switch	Switch Off: 12v, On: 1v
9	GN/BL	Vehicle Speed Sensor	At 55 mph: 48 Hz
10	BK/WT	A/C Amplifier Signal (AC1)	Clutch On: 1.5v, Off: 12v
11	BK/RD	A/T: Starter Switch Signal	9-11v (cranking)
11	BK/WT	M/T: Starter Switch Signal	9-11v (cranking)
12	WT/RD	EFI Main Relay (B+)	12-14v
13	WT/RD	EFI Main Relay B1+	12-14v
15	BR/BK	Sensor Ground	<0.050v
16	YL/GN	Overdrive Main Switch	Switch Off: 12v, On: 1v
17	BR	4WD M/T: Select (SEL2)	Open: 12v, Closed: 0v
18	BR	2WD A/T: Select (SEL1)	Open: 12v, Closed: 0v
19	BL/BK	A/C Amplifier Signal (ACT)	Clutch On: 12v, Off: 1.5v
20	YL/WT	Data Link Connector	12-14v
21	YL/RD	Cruise Control ECU	At Cruise in OD: 12v

1993 Pickup 3.0L V6 MFI VIN V (All) 26 Pin Connector

PCM Pin #	Wire Color	Circuit Description (26 Pin)	Value at Hot Idle
1	WT	Distributor Signal (NE+)	AC pulse signals
2	RD	Distributor Signal (G1+)	AC pulse signals
3	BK/YL	Igniter Signal (IGF)	Digital Signal: 0-5-0v
4	GN/RD	A/T-ECT Solenoid (S4)	In Lockup: 12-14v
5	YL/BK	A/T-ECT Solenoid (S3)	In Lockup: 12-14v
6	BK	A/T-ECT Solenoid (S2)	1st or OD: 1v
7	WT	A/T-ECT Solenoid (S1)	3rd or OD: 1v
8	GN	Fuel Pressure Up Solenoid	1v (at hot startup)
9	GN	Cold Start Injector	1v (at cold startup)
10	PK/GN	HO2S-11 (B1 S1) Heater	Heater Off: 12v, On: 1v
11	BR	Sensor Ground	<0.050v
12	WT/RD	Injectors 1 & 3 & 5 Control	1.6-2-9 ms
13	BR	Power Ground	<0.1v
14	GN	Distributor Signal (GN-)	AC pulse signals
15	BK	Distributor Signal (G2+)	AC pulse signals
16	BR/RD	A/T Vehicle Speed Sensor	Moving: 0-5-0v
17	PK/WT	A/T Select Switch Low	In Low: 12v, Others: 0v
18	PK/GN	A/T Select Switch 2nd	In 2nd: 12v, Others: 0v
19	PK/RD	A/T Select Switch Neutral	In 'N': 12v, Others: 0v
20	YL/RD	4WD Detection Transfer (L4)	Open: 12v, Closed: 0v
21	BL/BK	Igniter Signal (IGT)	Digital Signal: 0-5-0v
22	PK	EGR Solenoid Control (VSV)	12v or 0v
23	GN/RD	Intake Air Solenoid Control	12v or 0v
24	BK/RD	A/C Idle-Up Signal	A/C Off: 12v, On: 1v
25	WT	Injectors 2 & 4 & 6 Control	1.6-2-9 ms
26	BR	Power Ground	<0.1v

Pin Connector Graphic

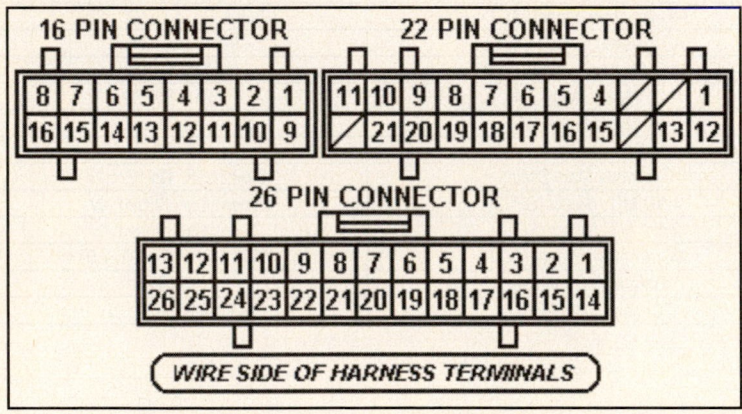

1994-95 Pickup 3.0L V6 MFI VIN V (All) 16 Pin Connector

PCM Pin #	Wire Color	Circuit Description (16 Pin)	Value at Hot Idle
1	GN/BK	Airflow Meter VREF	4.9-5.1v
2	YL/RD	Airflow Meter Signal	0.5-2.5v
3	YL/BK	IAT Sensor Signal (THA)	At 100°F: 2.60v
4	GN/BL	ECT Sensor Signal (THW)	At 180°F: 0.51v
5	BK	Knock Sensor Signal	<0.075v AC
6	BK	HO2S-11 (B1 S1) Signal	0.1-1.1v
7	GN/BK	A/T Oil Temperature Sensor	At 68°F: 4-5v
8	YL/GN	Data Link Connector	12-14v
9	BR/BK	Sensor Ground	<0.050v
10	WT	HO2S-12 (B1 S2) Signal	0.1-1.1v
11	YL	TP Sensor Signal (VTA)	0.3-0.8v
12	YL/WT	Closed Throttle Switch	1v, off-idle: 12v
13	GN/WT	EGR Gas Temperature Sensor (THG)	3.5-4.0v
13 (Fed)	GN/WT	Short Pin	Open: 12v, Closed: 0v
14	LG	Transfer Oil Temp. Sensor	At 68°F: 4-5v
15	PK/WT	Data Link Connector	12-14v
16	PK/GN	Data Link Connector	12-14v

1994-95 Pickup 3.0L V6 MFI VIN V (All) 22 Pin Connector

PCM Pin #	Wire Color	Circuit Description (22 Pin)	Value at Hot Idle
1	BK/GN	Battery Direct	12-14v
4	BL/WT	A/T Oil Temperature Lamp	Lamp Off: 12v, On: 1v
5	PK	MIL (lamp) Control	MIL Off: 12v, On: 1v
6	GN/WT	Stop Light Switch Signal	Brake Off: 0v, On: 12v
7	GN/OR	A/T Pattern Select Switch	Norm: 0v, PWR: 12v
8	PK/GN	A/T 4WD Indicator Switch	Switch Off: 12v, On: 1v
8	GN/WT	M/T 4WD Indicator Switch	Switch Off: 12v, On: 1v
9	GN/BL	Vehicle Speed Sensor	At 55 mph: 48 Hz
10	BK/WT	A/C Amplifier Signal (AC1)	Clutch On: 1.5v, Off: 12v
11	BK/WT	M/T: Starter Switch Signal	9-11v (cranking)
11	BK/RD	A/T: Starter Switch Signal	9-11v (cranking)
12	WT/RD	EFI Main Relay (B+)	12-14v
13	WT/RD	EFI Main Relay B1+	12-14v
15	BR/BK	Sensor Ground	<0.050v
16	YL/GN	Overdrive Main Switch	Switch Off: 12v, On: 1v
17	BR	4WD M/T: Select (SEL2)	Open: 12v, Closed: 0v
18	BR	2WD A/T: Select (SEL1)	Open: 12v, Closed: 0v
19	BL/BK	A/C Amplifier Signal (ACT)	Clutch On: 12v, Off: 1.5v
20	YL/WT	Data Link Connector	12-14v
21	YL/RD	Cruise Control ECU	At Cruise in OD: 12v
22	RD/GN	HO2S-12 (B1 S2) Heater	Heater Off: 12v, On: 1v

1994-95 Pickup 3.0L V6 MFI VIN V (All) 26 Pin Connector

PCM Pin #	Wire Color	Circuit Description (26 Pin)	Value at Hot Idle
1	WT	Distributor Signal (NE+)	AC pulse signals
2	RD	Distributor Signal (G1+)	AC pulse signals
3	BK/YL	Igniter Signal (IGF)	Digital Signal: 0-5-0v
4	GN/RD	A/T-ECT Solenoid (S4)	In Lockup: 12-14v
5	YL/BK	A/T-ECT Solenoid (S3)	In Lockup: 12-14v
6	BK	A/T-ECT Solenoid (S2)	1st or OD: 1v
7	WT	A/T-ECT Solenoid (S1)	3rd or OD: 1v
8	GN	Fuel Pressure Up Solenoid	1v (at hot startup)
9	GN	Cold Start Injector	1v (at cold startup)
10	PK/GN	HO2S-11 (B1 S1) Heater	Heater Off: 12v, On: 1v
11	BR	Sensor Ground	<0.050v
12	WT/RD	Injectors 1 & 3 & 5 Control	1.6-2-9 ms
13	BR	Power Ground	<0.1v
14	GN	Distributor Signal (GN-)	AC pulse signals
15	BK	Distributor Signal (G2+)	AC pulse signals
16	BR/RD	A/T Vehicle Speed Sensor	Moving: 0-5-0v
17	PK/WT	A/T Select Switch Low	In Low: 12v, Others: 0v
18	PK/GN	A/T Select Switch 2nd	In 2nd: 12v, Others: 0v
19	PK/RD	A/T Select Switch Neutral	In 'N': 12v, Others: 0v
20	YL/RD	4WD Detection Transfer (L4)	Open: 12v, Closed: 0v
21	BL/BK	Igniter Signal (IGT)	Digital Signal: 0-5-0v
22	PK	EGR Solenoid Control (VSV)	12v or 0v
23	GN/RD	Intake Air Solenoid Control	12v or 0v
24	BK/RD	A/C Idle-Up Solenoid	A/C Off: 12v, On: 1v
25	WT	Injectors 2 & 4 & 6 Control	1.6-2-9 ms
26	BR	Power Ground	<0.1v

Pin Connector Graphic

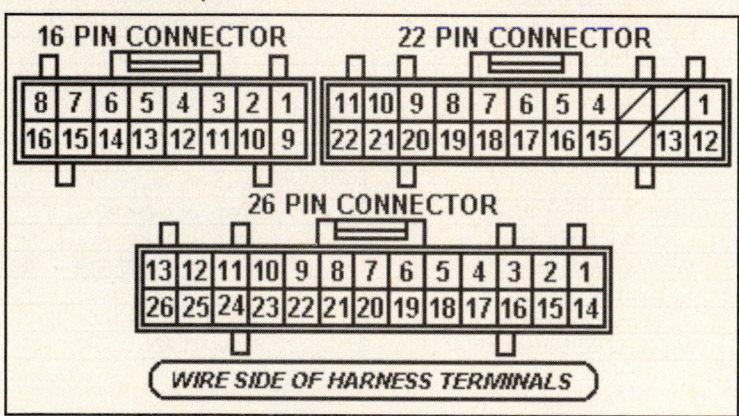

Standard Colors and Abbreviations

Abbreviation	Color	Abbreviation	Color	Abbreviation	Color
BK	Black	GY	Gray	RD	Red
BL	Blue	GN	Green	TN	Tan
BR	Brown	LG	LT Green	VT	Violet
DB	Dark Blue	OR	Orange	WT	White
DG	DK Green	PK	Pink	YL	Yellow

TACOMA Pin Tables

1995 Pickup 2.4L I4 MFI VIN U (All) 22 Pin Connector

PCM Pin #	Wire Color	Circuit Description (22 Pin)	Value at Hot Idle
1	BK/YL	Battery Direct	12-14v
4	GN/OR	A/T Pattern Select Switch	Norm: 0v, PWR: 12v
5	PK	MIL (lamp) Control	MIL Off: 12v, On: 1v
7	WT	Data Link Connector (SDL)	0v
8	BL/BK	A/C Amplifier Signal (ACT)	Clutch On: 12v, Off: 1.5v
9	GN/OR	Vehicle Speed Sensor	At 55 mph: 48 Hz
10	BK/YL	A/C Amplifier Signal (AC1)	Clutch On: 1.5v, Off: 12v
11	BK/WT	Starter Switch Signal	9-11v (cranking)
12	WT/RD	EFI Main Relay (B+)	12-14v
14	GN/YL	Circuit Opening Relay (FC)	0-3v, at off-idle: 12v
17	BL	Overdrive Main Switch	Switch Off: 12v, On: 1v
20	GN/WT	Stop Light Switch Signal	Brake Off: 0v, On: 12v
21	GN/WT	4WD Selector Switch	Switch Off: 12v, On: 1v
22	BK	Neutral Start Switch	In P/N: 9-11v (cranking)

1995 Pickup 2.4L I4 MFI VIN U (All) 16 Pin Connector

PCM Pin #	Wire Color	Circuit Description (16 Pin)	Value at Hot Idle
1	GN/YL	Sensor VREF	4.9-5.1v
2	GY/RD	MAF Sensor Signal (VG)	1.1-1.8v
3	PK	EGR Gas Temperature Sensor (THG)	3.5-4.0v
4	GN/RD	ECT Sensor Signal (THW)	At 180°F: 0.51v
5	BK	HO2S-12 (B1 S2) Signal	0.1-1.1v
6	WT	HO2S-11 (B1 S1) Signal	0.1-1.1v
7	YL/GN	IAT Sensor Signal (THA)	At 100°F: 2.60v
9, 16	BR, BR/WT	Sensor Ground	<0.050v
11	YL	TP Sensor Signal (VTA)	0.3-0.8v
12	YL/BL	Closed Throttle Switch	1v, off-idle: 12v
13	BK	Knock Sensor Signal	<0.075v AC
15	PK/WT	Data Link Connector	12-14v

1995 Pickup 2.4L I4 MFI VIN U (All) 26 Pin Connector

PCM Pin #	Wire Color	Circuit Description (26 Pin)	Value at Hot Idle
1	BK/WT	A/C Idle-Up Solenoid	A/C Off: 12v, On: 1v
2	PK/GN	HO2S-11 (B1 S1) Heater	Heater Off: 12v, On: 1v
3	BK/YL	Igniter Signal (IGF)	Digital Signal: 0-5-0v
4, 17	RD, GN	CKP Sensor Signal (NE+), (NE-)	AC pulse signals
5	YL	Distributor Signal (G2+)	AC pulse signals
6	PK/BK	EGR Solenoid Control (VSV)	12v or 0v
8	GN	Fuel Pressure Up Solenoid	1v (at hot startup)
9	BK	IAC Signal (RSC)	Pulse Signals
10	BK/RD	IAC Signal (RSO)	Pulse Signals
11	WT	Injector Pair 2 & 4 Control	1.6-2-9 ms
12	WT/RD	Injector Pair 1 & 3 Control	1.6-2-9 ms
13, 14	BR	Power Ground	<0.1v
15	RD/WT	HO2S-12 (B1 S2) Heater	Heater Off: 12v, On: 1v
18	BL	Distributor Signal (GN-)	<0.050v
20	BK/BL	Igniter Signal (IGF)	Digital Signal: 0-5-0v
23	WT/GN	EVAP Purge Solenoid (VSV)	12v or 0v
25, 26	BR	Power Ground	<0.1v

Pin Connector Graphic

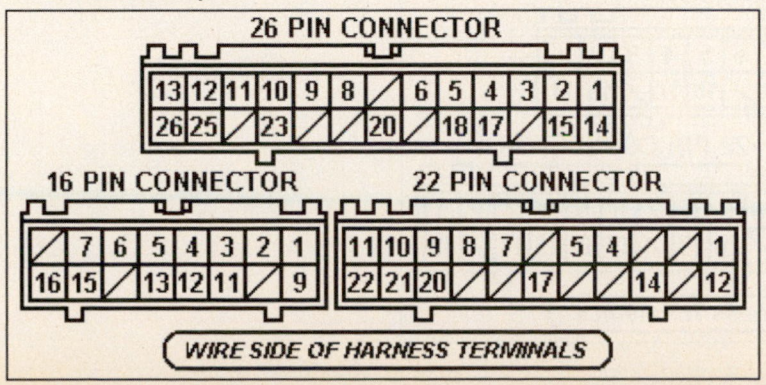

1996-97 Pickup 2.4L I4 MFI VIN L (All) 22 Pin Connector

PCM Pin #	Wire Color	Circuit Description (22 Pin)	Value at Hot Idle
1	BK/YL	Battery Direct	12-14v
5	PK	MIL (lamp) Control	MIL Off: 12v, On: 1v
6	BL	Cruise Control ECU	At Cruise in OD: 12v
7	WT	Data Link Connector (SDL)	0v
8	BL/BK	A/C Amplifier Signal (ACT)	Clutch On: 12v, Off: 1.5v
9	GN/OR	Vehicle Speed Sensor	At 55 mph: 48 Hz
10	BL/YL	A/C Amplifier Signal (AC1)	Clutch On: 1.5v, Off: 12v
11	BK/WT	Starter Switch Signal	9-11v (cranking)
12	WT/RD	EFI Main Relay (B+)	12-14v
14	GN/YL	Circuit Opening Relay (FC)	0-3v, at off-idle: 12v
15	PK/WT	A/T Select Switch Low	In Low: 12v, Others: 0v
16	PK	A/T Select Switch 2nd	In 2nd: 12v, Others: 0v
17	BL	Overdrive Main Switch	Switch Off: 12v, On: 1v
19	GN/RD	Taillight Switch Signal (ELS)	Switch Off: 0v, On: 12v
20	GN/WT	Stop Light Switch Signal	Brake Off: 0v, On: 12v
22	BK	Neutral Start Switch	In P/N: 9-11v (cranking)

1996-97 Pickup 2.4L I4 MFI VIN L (All) 16 Pin Connector

PCM Pin #	Wire Color	Circuit Description (16 Pin)	Value at Hot Idle
1	GN/YL	Sensor VREF	4.9-5.1v
2	GY/RD	MAF Sensor Signal (VG)	1.1-1.8v
3	PK	EGR Gas Temperature Sensor (THG)	3.5-4.0v
4	GN/RD	ECT Sensor Signal (THW)	At 180°F: 0.51v
5	BK	HO2S-12 (B1 S2) Signal	0.1-1.1v
6	WT	HO2S-11 (B1 S1) Signal	0.1-1.1v
7	YL/GN	IAT Sensor Signal (THA)	At 100°F: 2.60v
9	BR/BK	Sensor Ground	<0.050v
11	YL	TP Sensor Signal (VTA)	0.3-0.8v
12	YL/BL or BL	Closed Throttle Switch	1v, off-idle: 12v
13	BK	Knock Sensor Signal	<0.075v AC
15	PK/WT	Data Link Connector	12-14v
16	BR/WT	MAF Sensor Ground	<0.050v

1996-97 Pickup 2.4L I4 MFI VIN L (All) 26 Pin Connector

PCM Pin #	Wire Color	Circuit Description (26 Pin)	Value at Hot Idle
2	PK/GN	HO2S-11 (B1 S1) Heater	Heater Off: 12v, On: 1v
3	BK/YL	Igniter Signal (IGF)	Digital Signal: 0-5-0v
4	RD, GN	CKP Sensor Signal (NE+), (NE-)	AC pulse signals
5, 18	YL, BL	Distributor Signal (G2+), (GN-)	AC pulse signals
6	PK/BK	EGR Solenoid Control (VSV)	12v or 0v
9	BK	IAC Signal (RSC)	Pulse Signals
10	BK/RD	IAC Signal (RSO)	Pulse Signals
11	WT	Injector Pair 2 & 4 Control	1.6-2-9 ms
12	WT/RD	Injector Pair 1 & 3 Control	1.6-2-9 ms
13, 14	BR	Power Ground	<0.1v
15	RD/WT	HO2S-12 (B1 S2) Heater	Heater Off: 12v, On: 1v
20	BK/BL	Igniter Signal (IGF)	Digital Signal: 0-5-0v
23	WT/GN	EVAP Purge Solenoid (VSV)	12v or 0v
25, 26	BR	Power Ground	<0.1v

Pin Connector Graphic

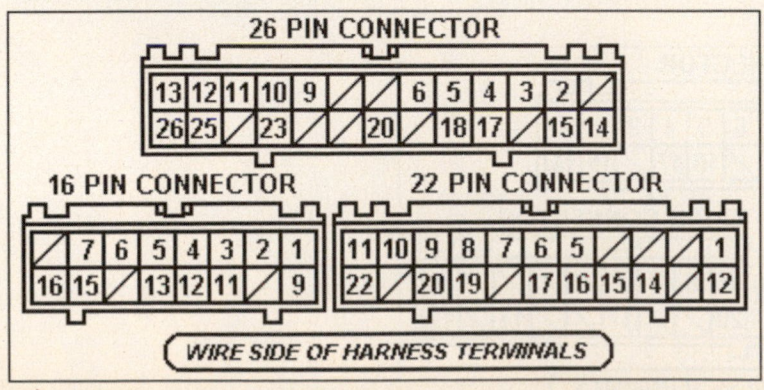

1998-99 Pickup 2.4L I4 MFI VIN L (All) 22 Pin Connector

PCM Pin #	Wire Color	Circuit Description (22 Pin)	Value at Hot Idle
2	BK/YL	Battery Direct	12-14v
4	PK/RD	MIL (lamp) Control	MIL Off: 12v, On: 1v
6	BL/BK	A/C Amplifier Signal (ACT)	Clutch On: 12v, Off: 1.5v
7	BL/YL	A/C Amplifier Signal (AC1)	Clutch On: 1.5v, Off: 12v
8	GN/OR	Speedometer Indicator	At 55 mph: 48 Hz
11	BK/WT	Starter Switch Signal	9-11v (cranking)
12	WT/RD	EFI Main Relay (B+)	12-14v
15	PK/WT	A/T Select Switch Low	In Low: 12v, Others: 0v
16	PK	A/T Select Switch 2nd	In 2nd: 12v, Others: 0v
17	BL	Overdrive Main Switch	Switch Off: 12v, On: 1v
19	WT	SIL (Scan Tool) Signal	12v
21	GN/WT	Stop Light Switch Signal	Brake Off: 0v, On: 12v
22	BK/RD	A/T Neutral Start Switch	In P/N: 9-11v (cranking)

1998-99 Pickup 2.4L I4 MFI VIN L (All) 16 Pin Connector

PCM Pin #	Wire Color	Circuit Description (16 Pin)	Value at Hot Idle
1	GN/YL	Sensor VREF (VC)	4.9-5.1v
2	GY/RD	MAF Sensor Signal (VG)	1.1-1.8v
3	YL/GN	IAT Sensor Signal (THA)	At 100°F: 2.60v
4	GN/RD	ECT Sensor Signal (THW)	At 180°F: 0.51v
5	WT	HO2S-11 (B1 S1) Signal	0.1-1.1v
7	PK/WT	Data Link Connector	12-14v
8	GY/GN	EVAP Vapor Pressure (VSV)	12v or 0v
9	BR/BK	Sensor Ground	<0.050v
10	YL	TP Sensor Signal (VTA)	0.3-0.8v
11	PK	EGR Gas Temperature Sensor (THG)	3.5-4.0v
12	BK	Knock Sensor Signal	<0.075v AC
13	BK	HO2S-12 (B1 S2) Signal	0.1-1.1v
15	PK/BK	EGR Solenoid Control (VSV)	12v or 0v

1998-99 Pickup 2.4L I4 MFI VIN L (All) 12 Pin Connector

PCM Pin #	Wire Color	Circuit Description (12 Pin)	Value at Hot Idle
2	GN/RD	Taillight Switch Signal	Switch Off: 0v, On: 12v
5	BL	CMP Sensor Signal (GN-)	<0.050v
6	GN	CKP Sensor Signal (NE-)	<0.050v
7	BR/WT	MAF Sensor Ground (E2G)	<0.050v
10	RD/YL	EVAP Vapor Pressure Sensor	2.5-3.1v (with fuel cap off)
11	YL	CMP Sensor Signal (G2+)	AC pulse signals
12	RD	CKP Sensor Signal (NE+)	AC pulse signals

Standard Colors and Abbreviations

Abbreviation	Color	Abbreviation	Color	Abbreviation	Color
BK	Black	GY	Gray	RD	Red
BL	Blue	GN	Green	TN	Tan
BR	Brown	LG	LT Green	VT	Violet
DB	Dark Blue	OR	Orange	WT	White
DG	DK Green	PK	Pink	YL	Yellow

1998-99 Pickup 2.4L I4 MFI VIN L (All) 26 Pin Connector

PCM Pin #	Wire Color	Circuit Description (26 Pin)	Value at Hot Idle
1	RD/WT	HO2S-12 (B1 S2) Heater	Heater Off: 12v, On: 1v
2	PK/GN	HO2S-11 (B1 S1) Heater	Heater Off: 12v, On: 1v
2 ('99)	RD/BK	HO2S-11 (B1 S1) Heater	Heater Off: 12v, On: 1v
3	WT/GN	EVAP Purge Solenoid (VSV)	12v or 0v
4	GN/RD	Cruise Control ECU	At Cruise in OD: 12v
6	BK	IAC Signal (RSC)	Pulse Signals
7	BK/RD	IAC Signal (RSO)	Pulse Signals
9	RD/BL	Injector 4 Control	1.6-2-9 ms
10	RD	Injector 3 Control	1.6-2-9 ms
11	WT	Injector 2 Control	1.6-2-9 ms
12	WT/RD	Injector 1 Control	1.6-2-9 ms
13	BR	Power Ground	<0.1v
14	GN/YL	Circuit Opening Relay (FC)	0-3v, at off-idle: 12v
15	BK	Tachometer Signal (TACO)	Pulse Signals
17	BK/YL	Igniter Signal (IGF)	Digital Signal: 0-5-0v
22	BR/YL	Igniter Transistor 2 Control	Digital Signal: 0-5-0v
23	BK/BL	Igniter Transistor 1 Control	Digital Signal: 0-5-0v
24	BR	Shield Ground	<0.050v
25	BR	Power Ground	<0.1v
26	BR	Power Ground	<0.1v

Pin Connector Graphic

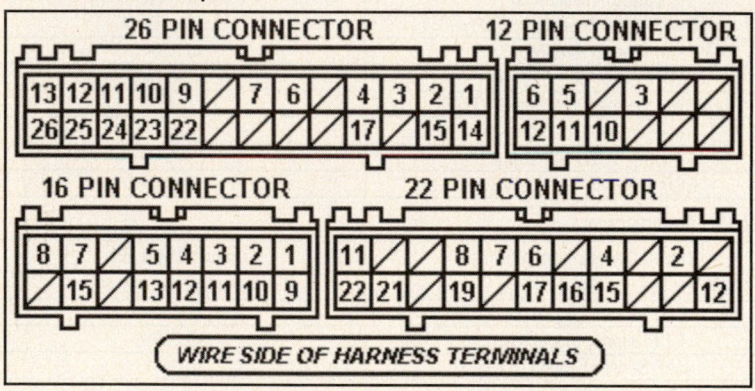

2000 Pickup 2.4L I4 VIN L California (All) E5 22 Pin Connector

PCM Pin #	Wire Color	Circuit Description (22 Pin)	Value at Hot Idle
1	BK/YL	Direct Battery	12-14v
2	RD/BK	A/T Select Switch Reverse	In 'R': 12v, Others: 0v
3	PK/WT	A/T Select Switch Low	In Low: 12v, Others: 0v
4	PK	A/T Select Switch 2nd	In 2nd: 12v, Others: 0v
5	PK	A/T Oil Temperature Indicator	Lamp Off: 12v, On: 1v
6	VT/RD	MIL (lamp) Control	MIL Off: 12v, On: 1v
7	BK/WT	Starter Switch Signal	9-11v (cranking)
8	---	Not Used	
9	BL	A/T Park (lamp) Indicator	Lamp Off: 12v, On: 1v
10	BL/OR	O/D OFF Indicator (ODLP)	Lamp Off: 12v, On: 1v
11	BR/YL	A/T Select Switch Drive	In 'D': 12v, Others: 0v
12	WT	SIL (Scan Tool) Signal	12v
13	BL/YL	A/C Amplifier Signal (AC1)	Clutch On: 1.5v, Off: 12v
14	BL/BK	A/C Amplifier Signal (ACT)	Clutch On: 12v, Off: 1.5v
15	GN/WT	Stop Light Switch Signal	Brake Off: 0v, On: 12v
16	WT/RD	EFI Main Relay (B+)	12-14v
17	GN	A/T Pattern Select Switch	Norm: 0v, PWR: 12v
18-19	---	Not Used	---
20	GN/RD	Defogger Idle-up Signal	Load On: 12v, Off: 0v
21	GN/OR	Speedometer Indicator (SP1)	At 55 mph: 48 Hz
22	BK/RD	A/T: Neutral Start Signal	In P/N: 9-11v (cranking)

2000 Pickup 2.4L I4 VIN L California (All) E6 28 Pin Connector

PCM Pin #	Wire Color	Circuit Description (28 Pin)	Value at Hot Idle
1-4	---	Not Used	---
5	LG/RD	Cruise Control ECU (IDLO)	1.5v, off-idle: 12v
6	GN/YL	Circuit Opening Relay (FC)	0-3v, at off-idle: 12v
7	VT/WT	DLC 7 Signal (TC)	12v
8	RD/YL	EVAP Vapor Pressure Sensor	2.5-3.1v (with fuel cap off)
9-12	---	Not Used	---
13	BK	Tachometer Signal (TACO)	Pulse Signals
14-23	---	Not Used	---
24	---	Not Used	---
25	GN	OD Main Switch (ODMS)	Switch Off: 12v, On: 1v
26-27	---	Not Used	---
28	YL	PSP Switch Signal	Straight: 12v, Turning: 0v

Pin Connector Graphic

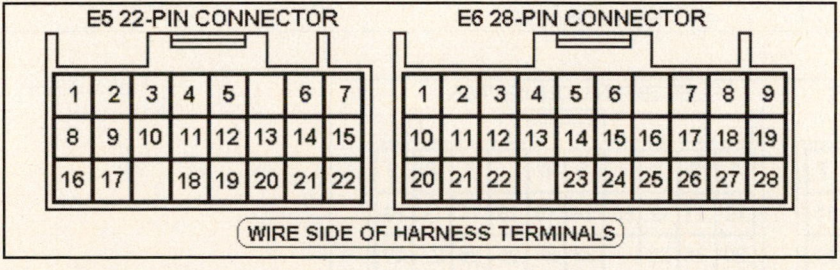

2000 Pickup 2.4L I4 VIN L California (All) E7 24 Pin Connector

PCM Pin #	Wire Color	Circuit Description (24 Pin)	Value at Hot Idle
2	GN/YL	Sensor VREF (VC)	4.9-5.1v
3	RD/WT	HO2S-12 (B1 S2) Heater	Heater Off: 12v, On: 1v
4	WT	AFS-11 (B1 S1) Heater	Heater Off: 12v, On: 1v
6	WT/GN	EVAP Purge Solenoid (VSV)	12v or 0v
7	GY/GN	EVAP Vapor Pressure (VSV)	12v or 0v
9	BK/BL	TP Sensor Signal (VTA)	0.3-0.8v
10	RD	HO2S-12 (B1 S2) Signal	0.1-1.1v
11	WT	AFS-11 (B1 S1) Signal (+)	Fixed at 3.3v
12	GN/RD	ECT Sensor Signal (THW)	At 180ºF: 0.51v
14	GY	MAF Sensor Signal (VG)	1.1-1.8v
15	YL	CMP Sensor Signal	AC pulse signals
16	RD	CKP Sensor Signal	AC pulse signals
17	BR	Shield Ground (E1)	<0.050v
18	BR/BK	Sensor Ground (E2)	<0.050v
20	RD	AFS-11 (B1 S1) Signal (-)	Fixed at 3.3v
21	YL/GN	IAT Sensor Signal (THA)	At 100ºF: 2.60v
22	BK/WT	MAF Sensor Ground (E2G)	<0.050v
24	GN	CKP/CMP Sensor Signal (-)	<0.050v

2000 Pickup 2.4L I4 VIN L California (All) E8 31 Pin Connector

PCM Pin #	Wire Color	Circuit Description (31 Pin)	Value at Hot Idle
1	WT/RD	Injector 1 Control	2.0-3.3 ms
2	WT	Injector 2 Control	2.0-3.3 ms
3	RD	Injector 3 Control	2.0-3.3 ms
4	RD/BL	Injector 4 Control	2.0-3.3 ms
5, 6	WT/BK	Power Ground (E03), (E04)	<0.1v
7	PK/YL	A/T-ECT Solenoid (S1)	3rd or OD: 1v
8	LG	A/T-ECT Solenoid (S2)	1st or OD: 1v
9	RD/GN	A/T-ECT Solenoid (SL)	In Lockup: 12-14v
10	BK/YL	Igniter Signal (IGF)	Digital Signal: 0-5-0v
11	BK/BL	Igniter Transistor 1 Control	6º, 55 mph: 8º dwell
12	BK/WT	Igniter Transistor 2 Control	6º, 55 mph: 8º dwell
13	BL/RD	Igniter Transistor 3 Control	6º, 55 mph: 8º dwell
14	BL/YL	Igniter Transistor 4 Control	6º, 55 mph: 8º dwell
15	BK/RD	IAC Control (RSD)	Pulse Signals
21	WT/BK	Power Ground (E01)	<0.1v
26	YL/RD	A/T Oil Temperature Sensor	At 68ºF: 4-5v
28	GY	Knock Sensor Signal	<0.075v AC
31	WT/BK	Power Ground (E02)	<0.1v

Pin Connector Graphic

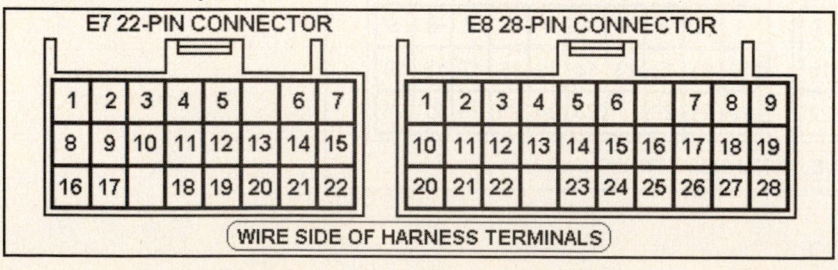

2000 Pickup 2.4L I4 MFI VIN L Federal (All) E5 22 Pin Connector

PCM Pin #	Wire Color	Circuit Description (22 Pin)	Value at Hot Idle
1	---	Not Used	---
2	BK/YL	Direct Battery	12-14v
3	PK	A/T Oil Temperature Indicator	Lamp Off: 12v, On: 1v
4	PK/RD	MIL (lamp) Control	MIL Off: 12v, On: 1v
5	BL/OR	O/D OFF Indicator (ODLP)	Lamp Off: 12v, On: 1v
6	BL/BK	A/C Amplifier Signal (ACT)	Clutch On: 12v, Off: 1.5v
7	BL/YL	A/C Amplifier Signal (AC1)	Clutch On: 1.5v, Off: 12v
8	GN/OR	Speedometer Indicator	At 55 mph: 48 Hz
9-10	---	Not Used	---
11	BK/WT	Starter Switch Signal	9-11v (cranking)
12	WT/RD	EFI Main Relay (B+)	12-14v
13	---	Not Used	---
14	GN	A/T Pattern Select Switch	Norm: 0v, PWR: 12v
15	PK/WT	A/T Select Switch Low	In Low: 12v, Others: 0v
16	PK	A/T Select Switch 2nd	In 2nd: 12v, Others: 0v
17	RD/BK	A/T Select Switch Reverse	In 'R': 12v, Others: 0v
18	RD/YL	Cruise Control ECU (OD1)	At Cruise in OD: 12v
19	WT	SIL (Scan Tool) Signal	Digital Signals
20	---	Not Used	---
21	GN/WT	Stop Light Switch Signal	Brake Off: 0v, On: 12v
22	BK/RD	Start Circuit Signal	Cranking: 9-11v

2000 Pickup 2.4L I4 MFI VIN L Federal (All) E6 12 Pin Connector

PCM Pin #	Wire Color	Circuit Description (12 Pin)	Value at Hot Idle
1	GN	Overdrive Main Switch	Switch Off: 12v, On: 1v
2	GN/RD	Defogger Idle-up Signal	Load On: 12v, Off: 0v
3	WT/RD	A/T Vehicle Speed Sensor (-)	Pulse Signals
5	BL	CMP Sensor Signal (GN-)	<0.050v
6	GN	CKP Sensor Signal (NE-)	<0.050v
7	BR/WT	MAF Meter Ground	<0.050v
8	---	Not Used	---
9	YL/RD	A/T Vehicle Speed Sensor (+)	Pulses
10	RD/YL	EVAP Vapor Pressure Sensor	2.5-3.1v (with fuel cap off)
11	YL	CMP Sensor Signal (G2+)	AC pulse signals
12	RD	CKP Sensor Signal (NE+)	AC pulse signals

Pin Connector Graphic

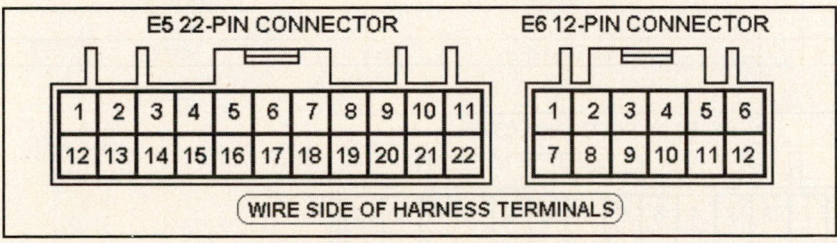

2000 Pickup 2.4L I4 MFI VIN L Federal (All) E7 16 Pin Connector

PCM Pin #	Wire Color	Circuit Description (16 Pin)	Value at Hot Idle
1	GN/YL	Sensor VREF (VC)	4.9-5.1v
2	GY/RD	MAF Sensor Signal (VG)	1.1-1.8v
3	YL/GN	IAT Sensor Signal (THA)	At 100°F: 2.60v
4	GN/RD	ECT Sensor Signal (THW)	At 180°F: 0.51v
5	WT	HO2S-11 (B1 S1) Signal	0.1-1.1v
7	PK/WT	Data Link Connector	12-14v
8	GY/GN	EVAP Vapor Pressure (VSV)	12v or 0v
9	BR/BK	Sensor Ground	<0.050v
10	YL	TP Sensor Signal (VTA)	0.3-0.8v
11	PK	EGR Gas Temperature Sensor (THG)	3.5-4.0v
12	GY	Knock Sensor Signal	<0.075v AC
13	BK	HO2S-12 (B1 S2) Signal	0.1-1.1v
15	PK/BK	EGR Solenoid Control (VSV)	12v or 0v

2000 Pickup 2.4L I4 MFI VIN L Federal (All) E8 26 Pin Connector

PCM Pin #	Wire Color	Circuit Description (26 Pin)	Value at Hot Idle
1	RD/WT	HO2S-12 (B1 S2) Heater	Heater Off: 12v, On: 1v
2	RD/BK	HO2S-11 (B1 S1) Heater	Heater Off: 12v, On: 1v
3	WT/GN	EVAP Purge Solenoid (VSV)	12v or 0v
4	LG/RD	Cruise Control ECU (IDLO)	1.5v, off-idle: 12v
6	BK	IAC Signal (RSC)	Pulse Signals
7	BK/RD	IAC Signal (RSO)	Pulse Signals
8	PK/YL	A/T-ECT Solenoid (S1)	3rd or OD: 1v
9	RD/BL	Injector 4 Control	2.0-3.3 ms
10	RD	Injector 3 Control	2.0-3.3 ms
11	WT	Injector 2 Control	2.0-3.3 ms
12	WT/RD	Injector 1 Control	2.0-3.3 ms
13	BR	Power Ground	<0.1v
14	GN/YL	Circuit Opening Relay (FC)	0-3v, at off-idle: 12v
15	BK	Tachometer Signal (TACO)	Pulse Signals
16	YL/RD	A/T Oil Temperature Sensor	At 68°F: 4-5v
17	BK/YL	Igniter Signal (IGF)	Digital Signal: 0-5-0v
20	RD/GN	A/T-ECT Solenoid (SL)	In Lockup: 12-14v
21	LG	A/T-ECT Solenoid (S2)	1st or OD: 1v
22	BR/YL	Igniter Transistor 2 Control	6°, 55 mph: 8° dwell
23	BK/BL	Igniter Transistor 1 Control	6°, 55 mph: 8° dwell
24	BR	Sensor Ground	<0.050v
25	BR	Power Ground	<0.1v
26	BR	Power Ground	<0.1v

Pin Connector Graphic

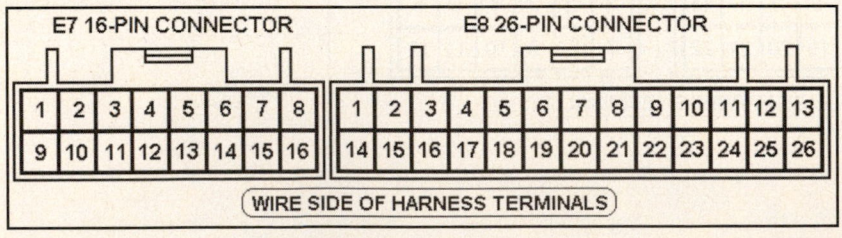

E7 16-PIN CONNECTOR   E8 26-PIN CONNECTOR

WIRE SIDE OF HARNESS TERMINALS

2001 Pickup 2.4L I4 2RZ-FE VIN L (All) E5 22 Pin Connector

PCM Pin #	Wire Color	Circuit Description (22 Pin)	Value at Hot Idle
1	BK/YL	Direct Battery	12-14v
2	RD/BK	A/T Select Switch Reverse	In 'R': 12v, Others: 0v
3	PK/WT	A/T Select Switch Low	In Low: 12v, Others: 0v
4	PK	A/T Select Switch 2nd	In 2nd: 12v, Others: 0v
5	PK	A/T Oil Temperature Indicator	Lamp Off: 12v, On: 1v
6	VT/RD	MIL (lamp) Control	MIL Off: 12v, On: 1v
7	BK/WT	Starter Switch Signal	In P/N: 9-11v (cranking)
8	---	Not Used	---
9	BL	A/T Park (lamp) Indicator	Lamp Off: 12v, On: 1v
10	BL/OR	O/D OFF Indicator (ODLP)	Lamp Off: 12v, On: 1v
11	BR/YL	A/T Select Switch Drive	In 'D': 12v, Others: 0v
12	WT	SIL (Scan Tool) Signal	12v
13	BL/YL	A/C Amplifier Signal (AC1)	Clutch On: 1.5v, Off: 12v
14	BL/BK	A/C Amplifier Signal (ACT)	Clutch On: 12v, Off: 1.5v
15	GN/WT	Stop Light Switch Signal	Brake Off: 0v, On: 12v
16	WT/RD	EFI Main Relay (B+)	12-14v
17	GN	A/T Pattern Select Switch	Norm: 0v, PWR: 12v
18-19	---	Not Used	---
20	GN/RD	Taillight Switch Signal (ELS)	Switch Off: 0v, On: 12v
21	GN/OR	Speedometer Indicator (SP1)	At 55 mph: 48 Hz
22	BK/RD	A/T: Neutral Start Signal	In P/N: 9-11v (cranking)

2001 Pickup 2.4L I4 2RZ-FE VIN L (All) E6 28 Pin Connector

PCM Pin #	Wire Color	Circuit Description (28 Pin)	Value at Hot Idle
1-4	---	Not Used	---
5	LG/RD	Cruise Control ECU (IDLO)	1.5v, off-idle: 12v
6	WT/BL	Circuit Opening Relay (FC)	0-3v, at off-idle: 12v
7	VT/WT	DLC 7 Signal (TC)	12v
8	RD/YL	EVAP Vapor Pressure Sensor	2.5-3.1v (with fuel cap off)
9-12	---	Not Used	---
13	BK	Tachometer Signal (TACO)	Pulse Signals
14-24	---	Not Used	---
25	GN	OD Main Switch (ODMS)	Switch Off: 12v, On: 1v
26-27	---	Not Used	---
28	YL	PSP Switch Signal	Straight: 12v, Turning: 0v

Pin Connector Graphic

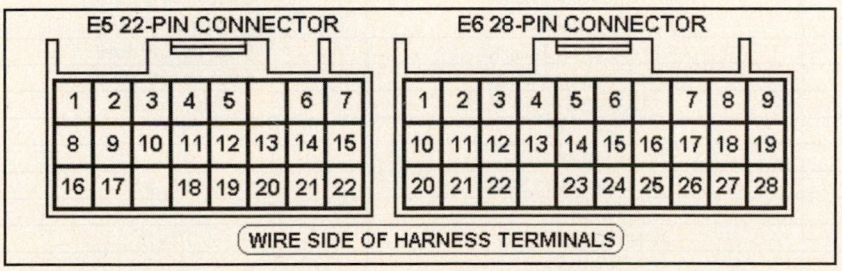

2001 Pickup 2.4L I4 2RZ-FE VIN L (All) E7 24 Pin Connector

PCM Pin #	Wire Color	Circuit Description (24 Pin)	Value at Hot Idle
1	RD/BL	EVAP Closed Canister Vent	12v or 0v
2	GN/YL	Sensor VREF (VC)	4.9-5.1v
3	RD/WT	HO2S-12 (B1 S2) Heater	Heater Off: 12v, On: 1v
4	WT	AFS-11 (B1 S1) Heater	Heater Off: 12v, On: 1v
6	WT/GN	EVAP Purge Solenoid (VSV)	12v or 0v
7	GY/GN	EVAP Vapor Pressure (VSV)	12v or 0v
8	---	Not Used	---
9	YL	TP Sensor Signal (VTA)	0.3-0.8v
10	RD	HO2S-12 (B1 S2) Signal	0.1-1.1v
11	VT	AFS-11 (B1 S1) Signal (+)	Fixed at 3.3v
12	GN/RD	ECT Sensor Signal (THW)	At 180°F: 0.51v
13	---	Not Used	---
14	GY	MAF Sensor Signal (VG)	1.1-1.8v
15	RD	CMP Sensor Signal	AC pulse signals
16	BL	CKP Sensor Signal	AC pulse signals
17	BR	Shield Ground (E1)	<0.050v
18	LG	Sensor Ground (E2)	<0.050v
19	---	Not Used	---
20	PK	AFS-11 (B1 S1) Signal (-)	Fixed at 3.3v
21	YL/GN	IAT Sensor Signal (THA)	At 100°F: 2.60v
22	BK/WT	MAF Sensor Ground (E2G)	<0.050v
23	---	Not Used	---
24	GN	CKP/CMP Sensor Signal (-)	<0.050v

2001 Pickup 2.4L I4 2RZ-FE VIN L (All) E8 31 Pin Connector

PCM Pin #	Wire Color	Circuit Description (31 Pin)	Value at Hot Idle
1	WT/RD	Injector 1 Control	2.0-3.3 ms
2	WT	Injector 2 Control	2.0-3.3 ms
3	RD	Injector 3 Control	2.0-3.3 ms
4	RD/BL	Injector 4 Control	2.0-3.3 ms
5	WT/BK	Power Ground (E03)	<0.1v
6	WT/BK	Power Ground (E04)	<0.1v
7	PK/YL	A/T-ECT Solenoid (S1)	3rd or OD: 1v
8	LG	A/T-ECT Solenoid (S2)	1st or OD: 1v
9	RD/GN	A/T-ECT Solenoid (SL)	In Lockup: 12-14v
10	BK/YL	Igniter Signal (IGF)	Digital Signal: 0-5-0v
11	BK/BL	Igniter Transistor 1 Control	6°, 55 mph: 8° dwell
12	BL	Igniter Transistor 2 Control	6°, 55 mph: 8° dwell
13	BL/RD	Igniter Transistor 3 Control	6°, 55 mph: 8° dwell
14	BL/YL	Igniter Transistor 4 Control	6°, 55 mph: 8° dwell
15	BK/RD	IAC Control (RSD)	Pulse Signals
16-20	---	Not Used	---
21	WT/BK	Power Ground (E01)	<0.1v
26	YL/RD	A/T Oil Temperature Sensor	At 68°F: 4-5v
27	---	Not Used	---
28	BK/YL	Knock Sensor Signal	<0.075v AC
29-30	---	Not Used	---
31	WT/BK	Power Ground (E02)	<0.1v

Pin Connector Graphic

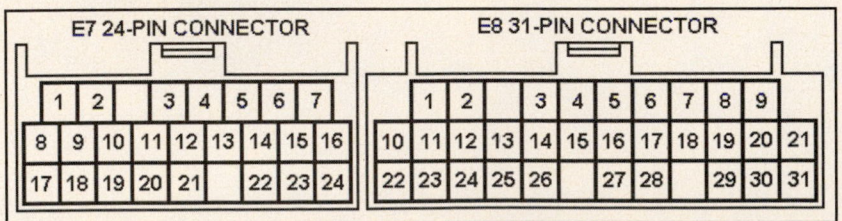

2002-03 Pickup 2.4L I4 2RZ-FE VIN L (All) E5 22 Pin Connector

PCM Pin #	Wire Color	Circuit Description (22 Pin)	Value at Hot Idle
1	BK/YL	Direct Battery	12-14v
2	RD/BK	A/T Select Switch Reverse	In 'R': 12v, Others: 0v
3	PK/WT	A/T Select Switch Low	In Low: 12v, Others: 0v
4	PK	A/T Select Switch 2nd	In 2nd: 12v, Others: 0v
5	PK	A/T Oil Temperature Indicator	Lamp Off: 12v, On: 1v
6	VT/RD	Malfunction Indicator Lamp (MIL) Control	MIL Off: 12v, On: 1v
7	GN	Starter Switch Signal (STA)	In P/N: 9-11v (cranking)
8	---	Not Used	---
9	BL	A/T Park (lamp) Indicator	Lamp Off: 12v, On: 1v
10	BL/OR	O/D OFF Indicator (ODLP)	Lamp Off: 12v, On: 1v
11	BR/YL	A/T Select Switch Drive	In 'D': 12v, Others: 0v
12	WT	SIL (Scan Tool) Signal	12v
13	BL/YL	A/C Amplifier Signal (AC1)	Clutch On: 1.5v, Off: 12v
14	BL/BK	A/C Amplifier Signal (ACT)	Clutch On: 12v, Off: 1.5v
15	GN/WT	Stop Light Switch Signal	Brake Off: 0v, On: 12v
16	WT/RD	EFI Main Relay (B+)	12-14v
17	GN	A/T Pattern Select Switch	Norm: 0v, PWR: 12v
18-19	---	Not Used	---
20	GN/RD	Taillight Switch Signal (ELS)	Switch Off: 0v, On: 12v
21	GN/OR	Speedometer Indicator (SP1)	At 55 mph: 48 Hz
22	BK/RD	A/T: Neutral Start Signal (NSW)	In P/N: 9-11v (cranking)

2002-03 Pickup 2.4L I4 2RZ-FE VIN L (All) E6 28 Pin Connector

PCM Pin #	Wire Color	Circuit Description (28 Pin)	Value at Hot Idle
1-4	---	Not Used	---
5	LG/RD	Cruise Control ECU (IDLO)	1.5v, off-idle: 12v
6	WT/BL	Circuit Opening Relay (FC)	0-3v, at off-idle: 12v
7	YL/BK	DLC 7 Signal (TC)	12v
8	RD/YL	EVAP Vapor Pressure Sensor (PTNK)	2.5-3.1v (with fuel cap off)
9-12	---	Not Used	---
13	LG/BK	Tachometer Signal (TACO)	Pulse Signals
14-24	---	Not Used	---
25	GN	OD Main Switch (ODMS)	Switch Off: 12v, On: 1v
26-27	---	Not Used	---
28	YL	PSP Switch Signal (PSSW)	Straight: 12v, Turning: 0v

Pin Connector Graphic

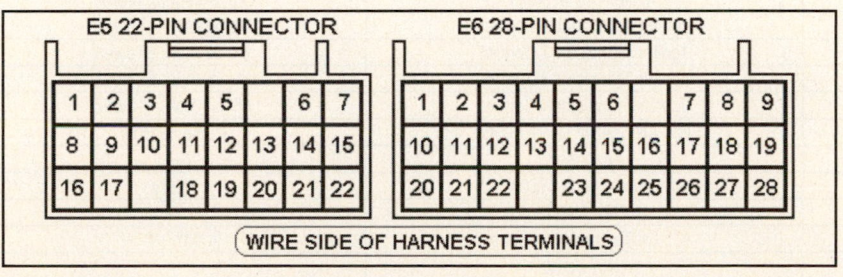

2002-03 Pickup 2.4L I4 2RZ-FE VIN L (All) E7 24 Pin Connector

PCM Pin #	Wire Color	Circuit Description (24 Pin)	Value at Hot Idle
1	RD/BL	EVAP Closed Canister Vent (VSV)	12v or 0v
2	GN/YL	Sensor VREF (VC)	4.9-5.1v
3	RD/WT	HO2S-12 (B1 S2) Heater (HTS)	Heater Off: 0v, On: 12v
4	WT	AFS-11 (B1 S1) Heater (AFHT)	Heater Off: 0v, On: 12v
5	---	Not Used	---
6	WT/GN	EVAP Purge Solenoid (VSV)	12v or 0v
7	GN/BK	EVAP Vapor Pressure (VSV)	12v or 0v
8	---	Not Used	---
9	YL	TP Sensor Signal (VTA)	0.3-0.8v
10	BK	HO2S-12 (B1 S2) Signal (OXS)	0.1-1.1v
11	VT	AFS-11 (B1 S1) Signal (AF+)	Fixed at 3.3v
12	GN/RD	ECT Sensor Signal (THW)	At 180ºF: 0.51v
13	---	Not Used	---
14	GY	MAF Sensor Signal (VG)	1.1-1.8v
15	RD	CMP Sensor Signal (G2+)	AC pulse signals
16	BL	CKP Sensor Signal (NE+)	AC pulse signals
17	BR	Shield Ground (E1)	<0.050v
18	BL/BK	Sensor Ground (E2)	<0.050v
19	---	Not Used	---
20	PK	AFS-11 (B1 S1) Signal (AF-)	Fixed at 3.3v
21	YL/GN	IAT Sensor Signal (THA)	At 100ºF: 2.60v
22	BK/WT	MAF Sensor Ground (E2G)	<0.050v
23	---	Not Used	---
24	GN	CKP/CMP Sensor Signal (NE-)	<0.050v

2002-03 Pickup 2.4L I4 2RZ-FE VIN L (All) E8 31 Pin Connector

PCM Pin #	Wire Color	Circuit Description (31 Pin)	Value at Hot Idle
1	WT/RD	Injector 1 Control	2.0-3.3 ms
2	WT	Injector 2 Control	2.0-3.3 ms
3	RD	Injector 3 Control	2.0-3.3 ms
4	RD/BL	Injector 4 Control	2.0-3.3 ms
5	WT/BK	Power Ground (E03)	<0.1v
6	WT/BK	Power Ground (E04)	<0.1v
7	PK/YL	A/T-ECT Solenoid (S1)	3rd or OD: 1v
8	LG	A/T-ECT Solenoid (S2)	1st or OD: 1v
9	RD/GN	A/T-ECT Solenoid (SL)	In Lockup: 12-14v
10	BK/YL	Igniter Signal (IGF)	Digital Signal: 0-5-0v
11	BK/BL	Igniter Transistor 1 Control	6º, 55 mph: 8º dwell
12	BL	Igniter Transistor 2 Control	6º, 55 mph: 8º dwell
13	BL/RD	Igniter Transistor 3 Control	6º, 55 mph: 8º dwell
14	BL/YL	Igniter Transistor 4 Control	6º, 55 mph: 8º dwell
15	BK/RD	Idle Air Control Valve (RSO)	Pulse Signals
16-20	---	Not Used	---
21	WT/BK	Power Ground (E01)	<0.1v
22-25	---	Not Used	---
26	YL/RD	A/T Oil Temperature Sensor	At 68ºF: 4-5v
27	---	Not Used	---
28	BK	Knock Sensor Signal (KNK)	<0.075v AC
29-30	---	Not Used	---
31	WT/BK	Power Ground (E02)	<0.1v

Pin Connector Graphic

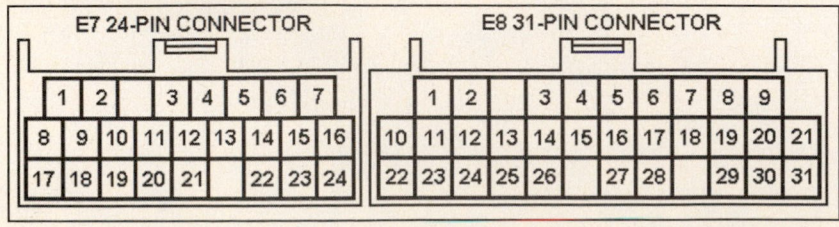

1995 Pickup 2.7L I4 MFI VIN U (A/T-ECT) 22 Pin Connector

PCM Pin #	Wire Color	Circuit Description (22 Pin)	Value at Hot Idle
1	---	Not Used	---
2	BK/YL	Battery Direct	12-14v
3	PK	A/T Oil Temperature Indicator	Lamp Off: 12v, On: 1v
4	PK	MIL (lamp) Control	MIL Off: 12v, On: 1v
5	BL/OR	Overdrive Main Switch	Switch Off: 12v, On: 1v
6	BL/BK	A/C Amplifier Signal (ACT)	Clutch On: 12v, Off: 1.5v
7	BK/YL	A/C Amplifier Signal (AC1)	Clutch On: 1.5v, Off: 12v
8	GN/OR	Vehicle Speed Sensor	At 55 mph: 48 Hz
9	---	Not Used	---
10	GN/WT	4WD Selector Switch	Switch Off: 12v, On: 1v
11	BK/WT	Starter Switch Signal	9-11v (cranking)
12	WT/RD	EFI Main Relay (B+)	12-14v
13	RD/GN	A/T Select Switch Park	In Park: 12v, Others: 0v
14	GN/OR	A/T Pattern Select Switch	Norm: 0v, PWR: 12v
15	PK/WT	A/T Select Switch Low	In Low: 12v, Others: 0v
16	PK	A/T Select Switch 2nd	In 2nd: 12v, Others: 0v
17	---	Not Used	---
18	BL	Cruise Control ECU (OD1)	At Cruise in OD: 12v
19	WT	Data Link Connector (SDL)	0v
20	---	Not Used	---
21	GN/WT	Stop Light Switch Signal	Brake Off: 0v, On: 12v
22	BK	Neutral Start Switch	In P/N: 9-11v (cranking)

1995 Pickup 2.7L I4 MFI VIN U (A/T-ECT) 16 Pin Connector

PCM Pin #	Wire Color	Circuit Description (16 Pin)	Value at Hot Idle
1	GN/YL	Sensor VREF	4.9-5.1v
2	GY/RD	MAF Sensor Signal (VG)	1.1-1.8v
3	BR/WT	Sensor Ground	<0.050v
4	GN/RD	ECT Sensor Signal (THW)	At 180ºF: 0.51v
5	WT	HO2S-11 (B1 S1) Signal	0.1-1.1v
6	BK	Knock Sensor Signal	<0.075v AC
7	PK/WT	Data Link Connector	12-14v
9	BR/BK	Sensor Ground	<0.050v
10	YL	TP Sensor Signal (VTA)	0.3-0.8v
11	YL/BL	Closed Throttle Switch	1v, off-idle: 12v
12	YL/GN	IAT Sensor Signal (THA)	At 100ºF: 2.60v
13	BK	HO2S-12 (B1 S2) Signal	0.1-1.1v
14	PK	EGR Gas Temperature Sensor (THG)	3.5-4.0v

Standard Colors and Abbreviations

Abbreviation	Color	Abbreviation	Color	Abbreviation	Color
BK	Black	GY	Gray	RD	Red
BL	Blue	GN	Green	TN	Tan
BR	Brown	LG	LT Green	VT	Violet
DB	Dark Blue	OR	Orange	WT	White
DG	DK Green	PK	Pink	YL	Yellow

1995 Pickup 2.7L I4 MFI VIN U (A/T-ECT) 26 Pin Connector

PCM Pin #	Wire Color	Circuit Description (26 Pin)	Value at Hot Idle
1	---	Not Used	---
2	BK/WT	A/C Idle-Up Solenoid	A/C Off: 12v, On: 1v
3	PK/GN	HO2S-11 (B1 S1) Heater	Heater Off: 12v, On: 1v
4	GN	Fuel Pressure Up Solenoid	1v (at hot startup)
5	---	Not Used	---
6	BK	IAC Signal (RSC)	Pulse Signals
7	BK/RD	IAC Signal (RSO)	Pulse Signals
8	RD/GN	A/T-ECT Solenoid (SL)	In Lockup: 12-14v
9	LG	A/T-ECT Solenoid (S2)	1st or OD: 1v
10	PK/YL	A/T-ECT Solenoid (S1)	3rd or OD: 1v
11	WT	Injector Pair 2 & 4 Control	1.6-2-9 ms
12	WT/RD	Injector Pair 1 & 3 Control	1.6-2-9 ms
13	BR	Power Ground	<0.1v
14	GN/YL	Circuit Opening Relay (FC)	0-3v, at off-idle: 12v
15	---	Not Used	---
16	RD/WT	HO2S-12 (B1 S2) Heater	Heater Off: 12v, On: 1v
17	BK/YL	Igniter Signal (IGF)	Digital Signal: 0-5-0v
18-19	---	Not Used	---
20	GN	4WD Detection Transfer (L4)	Open: 12v, Closed: 0v
21	YL/RD	A/T Oil Temperature Sensor	At 68°F: 4-5v
22	PK/BK	EGR Solenoid Control (VSV)	12v or 0v
23	BK/BL	Igniter Signal (IGF)	Digital Signal: 0-5-0v
24	BR	Sensor Ground	<0.050v
25	BR	Power Ground	<0.1v
26	BR	Power Ground	<0.1v

1995 Pickup 2.7L I4 MFI VIN U (A/T-ECT) 12 Pin Connector

PCM Pin #	Wire Color	Circuit Description (12 Pin)	Value at Hot Idle
1	WT/GN	EVAP Purge Solenoid (VSV)	12v or 0v
2-3	---	Not Used	---
4	WT/RD	A/T Vehicle Speed Sensor (-)	Pulse Signals
5	BL	Distributor Signal (GN-)	<0.050v
6	GN	CKP Sensor Signal (NE-)	<0.050v
7-9	---	Not Used	---
10	YL/RD	A/T Vehicle Speed Sensor (+)	Pulses
11	YL	Distributor Signal (G2+)	AC pulse signals
12	RD	CKP Sensor Signal (NE+)	AC pulse signals

Pin Connector Graphic

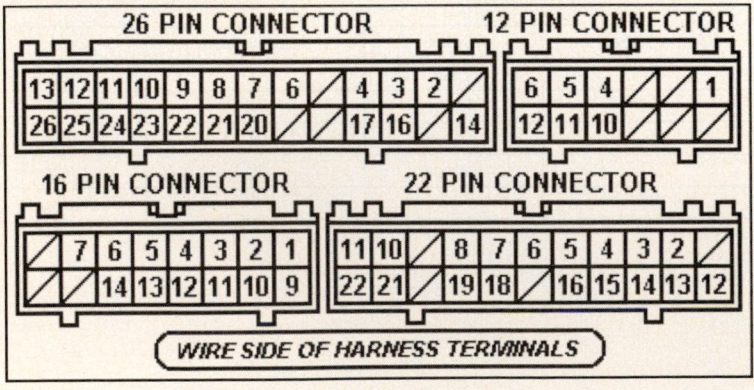

1995 Pickup 2.7L I4 MFI 3RZ-FE VIN U (M/T) 22 Pin Connector

PCM Pin #	Wire Color	Circuit Description (22 Pin)	Value at Hot Idle
1	BK/YL	Battery Direct	12-14v
2-4	---	Not Used	---
5	PK	MIL (lamp) Control	MIL Off: 12v, On: 1v
6	---	Not Used	---
7	WT	Data Link Connector (SDL)	0v
8	BL/BK	A/C Amplifier Signal (ACT)	Clutch On: 12v, Off: 1.5v
9	GN/OR	Vehicle Speed Sensor	At 55 mph: 48 Hz
10	BK/YL	A/C Amplifier Signal (AC1)	Clutch On: 1.5v, Off: 12v
11	BK/WT	Starter Switch Signal	9-11v (cranking)
12	WT/RD	EFI Main Relay (B+)	12-14v
13	---	Not Used	---
14	GN/YL	Circuit Opening Relay (FC)	0-3v, at off-idle: 12v
15-16	---	Not Used	---
17	BL	Overdrive Main Switch	Switch Off: 12v, On: 1v
18-19	---	Not Used	---
20	GN/WT	Stop Light Switch Signal	Brake Off: 0v, On: 12v
21	GN/WT	4WD Selector Switch	Switch Off: 12v, On: 1v
22	BK	Neutral Start Switch	9-11v (cranking)

1995 Pickup 2.7L I4 MFI 3RZ-FE VIN U (M/T) 16 Pin Connector

PCM Pin #	Wire Color	Circuit Description (16 Pin)	Value at Hot Idle
1	GN/YL	Sensor VREF	4.9-5.1v
2	GY/RD	MAF Sensor Signal (VG)	1.1-1.8v
3	PK	EGR Gas Temperature Sensor (THG)	3.5-4.0v
4	GN/RD	ECT Sensor Signal (THW)	At 180ºF: 0.51v
5	BK	HO2S-12 (B1 S2) Signal	0.1-1.1v
6	WT	HO2S-11 (B1 S1) Signal	0.1-1.1v
7	YL/GN	IAT Sensor Signal (THA)	At 100ºF: 2.60v
8	---	Not Used	---
9	BR/BK	Sensor Ground	<0.050v
10	---	Not Used	---
11	YL	TP Sensor Signal (VTA)	0.3-0.8v
12	YL/BL	Closed Throttle Switch	1v, off-idle: 12v
13	BK	Knock Sensor Signal	<0.075v AC
14	---	Not Used	---
15	PK/WT	Data Link Connector	12-14v
16	BR/WT	Sensor Ground	<0.050v

Standard Colors and Abbreviations

Abbreviation	Color	Abbreviation	Color	Abbreviation	Color
BK	Black	GY	Gray	RD	Red
BL	Blue	GN	Green	TN	Tan
BR	Brown	LG	LT Green	VT	Violet
DB	Dark Blue	OR	Orange	WT	White
DG	DK Green	PK	Pink	YL	Yellow

1995 Pickup 2.7L I4 MFI 3RZ-FE VIN U (M/T) 26 Pin Connector

PCM Pin #	Wire Color	Circuit Description (26 Pin)	Value at Hot Idle
1	BK/WT	A/C Idle-Up Solenoid	A/C Off: 12v, On: 1v
2	PK/GN	HO2S-11 (B1 S1) Heater	Heater Off: 12v, On: 1v
3	BK/YL	Igniter Signal (IGF)	Digital Signal: 0-5-0v
4	RD	CKP Sensor Signal	AC pulse signals
5	YL	Distributor Signal (G2+)	AC pulse signals
6	PK/BK	EGR Solenoid Control (VSV)	12v or 0v
8	GN	Fuel Pressure Up Solenoid	1v (at hot startup)
9	BK	IAC Signal (RSC)	Pulse Signals
10	BK/RD	IAC Signal (RSO)	Pulse Signals
11	WT	Injector Pair 2 & 4 Control	1.6-2-9 ms
12	WT/RD	Injector Pair 1 & 3 Control	1.6-2-9 ms
13	BR	Power Ground	<0.1v
14	BR	Sensor Ground	<0.050v
15	RD/WT	HO2S-12 (B1 S2) Heater	Heater Off: 12v, On: 1v
17	GN	CKP Sensor Signal (NE-)	<0.050v
18	BL	Distributor Signal (GN-)	<0.050v
20	BK/BL	Igniter Signal (IGF)	Digital Signal: 0-5-0v
23	WT/GN	EVAP Purge Solenoid (VSV)	12v or 0v
25	BR	Power Ground	<0.1v
26	BR	Power Ground	<0.1v

Pin Connector Graphic

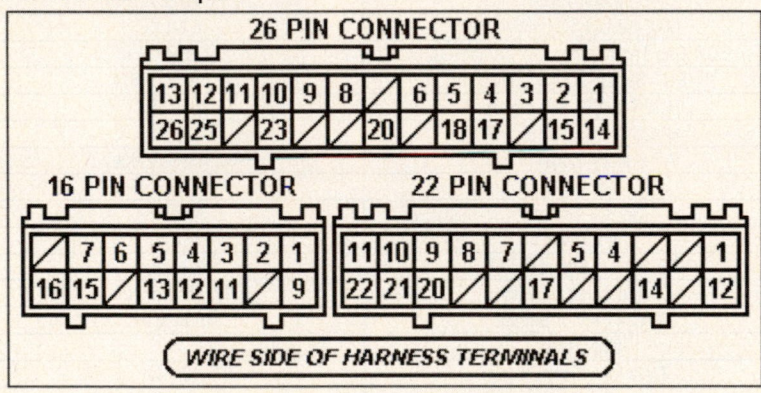

1996 Pickup 2.7L I4 MFI VIN M (A/T-ECT) 22 Pin Connector

PCM Pin #	Wire Color	Circuit Description (22 Pin)	Value at Hot Idle
2	BK/YL	Battery Direct	12-14v
3	PK	A/T Oil Temperature Indicator	Lamp Off: 12v, On: 1v
4	PK	MIL (lamp) Control	MIL Off: 12v, On: 1v
5	BL/OR	Overdrive Main Switch	Switch Off: 12v, On: 1v
6	BL/BK	A/C Amplifier Signal (ACT)	Clutch On: 12v, Off: 1.5v
7	BL/YL	A/C Amplifier Signal (AC1)	Clutch On: 1.5v, Off: 12v
8	GN/OR	Vehicle Speed Sensor	At 55 mph: 48 Hz
10	GN/WT	4WD Selector Switch	Switch Off: 12v, On: 1v
11	BK/WT	Starter Switch Signal	9-11v (cranking)
12	WT/RD	EFI Main Relay (B+)	12-14v
14	GN/OR	A/T Pattern Select Switch	Norm: 0v, PWR: 12v
15	PK/WT	A/T Select Switch Low	In Low: 12v, Others: 0v
16	PK	A/T Select Switch 2nd	In 2nd: 12v, Others: 0v
18	BL	Cruise Control ECU (OD1)	At Cruise in OD: 12v
19	WT	Data Link Connector (SDL)	0v
21	GN/WT	Stop Light Switch Signal	Brake Off: 0v, On: 12v
22	BK	Neutral Start Switch	In P/N: 9-11v (cranking)

1996 Pickup 2.7L I4 MFI VIN M (A/T-ECT) 16 Pin Connector

PCM Pin #	Wire Color	Circuit Description (16 Pin)	Value at Hot Idle
1	GN/YL	Sensor VREF	4.9-5.1v
2	GY/RD	MAF Sensor Signal (VG)	1.1-1.8v
3	BR/WT	Sensor Ground	<0.050v
4	GN/RD	ECT Sensor Signal (THW)	At 180°F: 0.51v
5	WT	HO2S-11 (B1 S1) Signal	0.1-1.1v
6	BK	Knock Sensor Signal	<0.075v AC
7	PK/WT	Data Link Connector	12-14v
9	BR/BK	Sensor Ground	<0.050v
10	YL	TP Sensor Signal (VTA)	0.1-1.0v
11	YL/BL	Closed Throttle Switch	1v, off-idle: 12v
12	YL/GN	IAT Sensor Signal (THA)	At 100°F: 2.60v
13	BK	HO2S-12 (B1 S2) Signal	0.1-1.1v
14	PK	EGR Gas Temperature Sensor (THG)	3.5-4.0v

Standard Colors and Abbreviations

Abbreviation	Color	Abbreviation	Color	Abbreviation	Color
BK	Black	GY	Gray	RD	Red
BL	Blue	GN	Green	TN	Tan
BR	Brown	LG	LT Green	VT	Violet
DB	Dark Blue	OR	Orange	WT	White
DG	DK Green	PK	Pink	YL	Yellow

1996 Pickup 2.7L I4 MFI VIN M (A/T-ECT) 26 Pin Connector

PCM Pin #	Wire Color	Circuit Description (26 Pin)	Value at Hot Idle
3	PK/GN	HO2S-11 (B1 S1) Heater	Heater Off: 12v, On: 1v
6	BK	IAC Signal (RSC)	Pulse Signals
7	BK/RD	IAC Signal (RSO)	Pulse Signals
8	RD/GN	A/T-ECT Solenoid (SL)	In Lockup: 12-14v
9	LG	A/T-ECT Solenoid (S2)	1st or OD: 1v
10	PK/YL	A/T-ECT Solenoid (S1)	3rd or OD: 1v
11	WT	Injector Pair 2 & 4 Control	1.6-2-9 ms
12	WT/RD	Injector Pair 1 & 3 Control	1.6-2-9 ms
13	BR	Power Ground	<0.1v
14	GN/YL	Circuit Opening Relay (FC)	0-3v, at off-idle: 12v
16	RD/WT	HO2S-12 (B1 S2) Heater	Heater Off: 12v, On: 1v
17	BK/YL	Igniter Signal (IGF)	Digital Signal: 0-5-0v
20	GN	4WD Detection Transfer (L4)	Open: 12v, Closed: 0v
21	YL/RD	A/T Oil Temperature Sensor	At 68ºF: 4-5v
22	PK/BK	EGR Solenoid Control (VSV)	12v or 0v
23	BK/BL	Igniter Signal (IGF)	Digital Signal: 0-5-0v
24	BR	Sensor Ground	<0.050v
25	BR	Power Ground	<0.1v
26	BR	Power Ground	<0.1v

1996 Pickup 2.7L I4 MFI VIN M (A/T-ECT) 12 Pin Connector

PCM Pin #	Wire Color	Circuit Description (12 Pin)	Value at Hot Idle
1	WT/GN	EVAP Purge Solenoid (VSV)	12v or 0v
4	WT/RD	A/T Vehicle Speed Sensor (-)	Pulses
5	BL	Distributor Signal (GN-)	<0.050v
6	GN	CKP Sensor Signal (NE-)	<0.050v
10	YL/RD	A/T Vehicle Speed Sensor (+)	Pulse Signal
11	YL	Distributor Signal (G2+)	AC pulse signals
12	RD	CKP Sensor Signal (NE+)	AC pulse signals

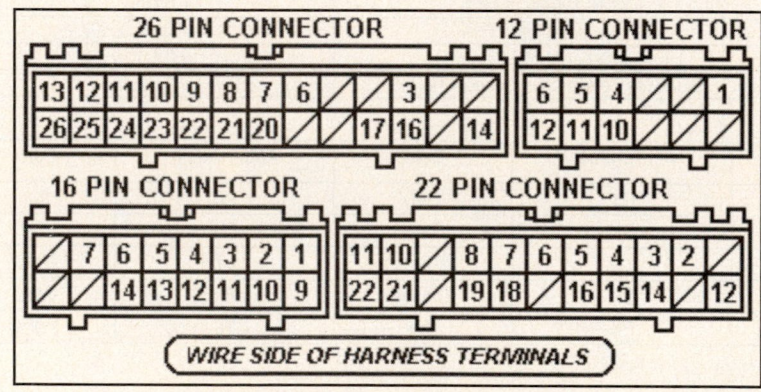

1996 Pickup 2.7L I4 MFI 3RZ-FE VIN M (M/T) 22 Pin Connector

PCM Pin #	Wire Color	Circuit Description (22 Pin)	Value at Hot Idle
1	BK/YL	Battery Direct	12-14v
4	GN/OR	A/T Pattern Select Switch	Norm: 0v, PWR: 12v
5	PK	MIL (lamp) Control	MIL Off: 12v, On: 1v
6	BL	Cruise Control ECU (OD1)	At Cruise in OD: 12v
7	WT	Data Link Connector (SDL)	0v
8	BL/BK	A/C Amplifier Signal (ACT)	Clutch On: 12v, Off: 1.5v
9	GN/OR	Vehicle Speed Sensor	At 55 mph: 48 Hz
10	BK/YL	A/C Amplifier Signal (AC1)	Clutch On: 1.5v, Off: 12v
11	BK/WT	Starter Switch Signal	9-11v (cranking)
12	WT/RD	EFI Main Relay (B+)	12-14v
14	GN/YL	Circuit Opening Relay (FC)	0-3v, at off-idle: 12v
17	BL	Overdrive Main Switch	Switch Off: 12v, On: 1v
20	GN/WT	Stop Light Switch Signal	Brake Off: 0v, On: 12v
21	GN/WT	4WD Selector Switch	Switch Off: 12v, On: 1v
22	BK	Neutral Start Switch	9-11v (cranking)

1996 Pickup 2.7L I4 MFI 3RZ-FE VIN M (M/T) 16 Pin Connector

PCM Pin #	Wire Color	Circuit Description (16 Pin)	Value at Hot Idle
1	GN/YL	Sensor VREF	4.9-5.1v
2	GY/RD	MAF Sensor Signal (VG)	1.1-1.8v
3	PK	EGR Gas Temperature Sensor (THG)	3.5-4.0v
4	GN/RD	ECT Sensor Signal (THW)	At 180ºF: 0.51v
5	BK	HO2S-12 (B1 S2) Signal	0.1-1.1v
6	WT	HO2S-11 (B1 S1) Signal	0.1-1.1v
7	YL/GN	IAT Sensor Signal (THA)	At 100ºF: 2.60v
9	BR/BK	Sensor Ground	<0.050v
11	YL	TP Sensor Signal (VTA)	0.3-0.8v
12	YL/BL	Closed Throttle Switch	1v, off-idle: 12v
13	BK	Knock Sensor Signal	<0.075v AC
15	PK/WT	Data Link Connector	12-14v
16	BR/WT	Sensor Ground	<0.050v

Standard Colors and Abbreviations

Abbreviation	Color	Abbreviation	Color	Abbreviation	Color
BK	Black	GY	Gray	RD	Red
BL	Blue	GN	Green	TN	Tan
BR	Brown	LG	LT Green	VT	Violet
DB	Dark Blue	OR	Orange	WT	White
DG	DK Green	PK	Pink	YL	Yellow

1996 Pickup 2.7L I4 MFI 3RZ-FE VIN M (M/T) 26 Pin Connector

PCM Pin #	Wire Color	Circuit Description (26 Pin)	Value at Hot Idle
2	PK/GN	HO2S-11 (B1 S1) Heater	Heater Off: 12v, On: 1v
3	BK/YL	Igniter Signal (IGF)	Digital Signal: 0-5-0v
4	RD	CKP Sensor Signal (NE+)	AC pulse signals
5	YL	Distributor Signal (G2+)	AC pulse signals
6	PK/BK	EGR Solenoid Control (VSV)	12v or 0v
9	BK	IAC Signal (RSC)	Pulse Signals
10	BK/RD	IAC Signal (RSO)	Pulse Signals
11	WT	Injector Pair 2 & 4 Control	1.6-2-9 ms
12	WT/RD	Injector Pair 1 & 3 Control	1.6-2-9 ms
13	BR	Power Ground	<0.1v
14	BR	Shield Ground	<0.050v
15	RD/WT	HO2S-12 (B1 S2) Heater	Heater Off: 12v, On: 1v
17	GN	CKP Sensor Signal (NE-)	<0.050v
18	BL	Distributor Signal (GN-)	<0.050v
20	BK/BL	Igniter Signal (IGF)	Digital Signal: 0-5-0v
23	WT/GN	EVAP Purge Solenoid (VSV)	12v or 0v
25	BR	Power Ground	<0.1v
26	BR	Power Ground	<0.1v

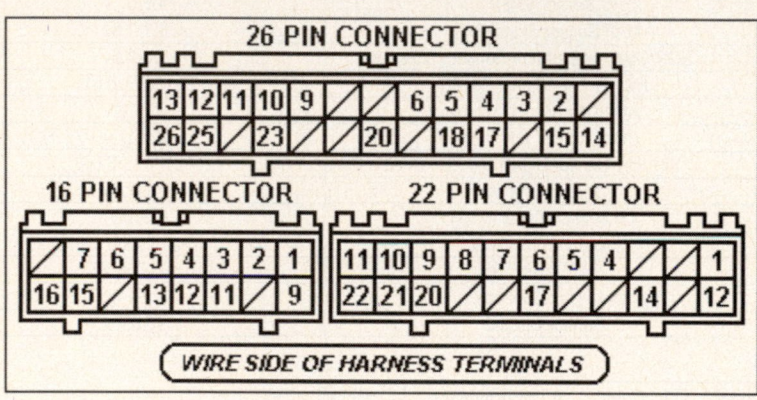

1997 Pickup 2.7L I4 MFI VIN M (All) 22 Pin Connector

PCM Pin #	Wire Color	Circuit Description (22 Pin)	Value at Hot Idle
2	BK/YL	Battery Direct	12-14v
4	PK	MIL (lamp) Control	MIL Off: 12v, On: 1v
5	BL/OR	Overdrive Main Switch	Switch Off: 12v, On: 1v
6	BL/BK	A/C Amplifier Signal (ACT)	Clutch On: 12v, Off: 1.5v
7	BL/YL	A/C Amplifier Signal (AC1)	Clutch On: 1.5v, Off: 12v
8	GN/OR	Vehicle Speed Sensor	At 55 mph: 48 Hz
10	GN/WT	4WD Selector Switch	Switch Off: 12v, On: 1v
11	BK/WT	Starter Switch Signal	9-11v (cranking)
12	WT/RD	EFI Main Relay (B+)	12-14v
13	RD/GN	4WD Detection Transfer (N)	Open: 12v, Closed: 0v
15	PK/WT	A/T Select Switch Low	In Low: 12v, Others: 0v
16	PK	A/T Select Switch 2nd	In 2nd: 12v, Others: 0v
17	RD/BK	A/T Select Switch Reverse	In 'R': 12v, Others: 0v
18	RD/YL	Cruise Control ECU (OD1)	At Cruise in OD: 12v
19	WT	Data Link Connector (SDL)	0v
21	GN/WT	Stop Light Switch Signal	Brake Off: 0v, On: 12v
22	BK	Neutral Start Switch	9-11v (cranking)

1997 Pickup 2.7L I4 MFI VIN M (All) 16 Pin Connector

PCM Pin #	Wire Color	Circuit Description (16 Pin)	Value at Hot Idle
1	GN/YL	Sensor VREF	4.9-5.1v
2	GY/RD	MAF Sensor Signal (VG)	1.1-1.8v
3	YL/GN	IAT Sensor Signal (THA)	At 100°F: 2.60v
4	GN/RD	ECT Sensor Signal (THW)	At 180°F: 0.51v
5	WT	HO2S-11 (B1 S1) Signal	0.1-1.1v
7	PK/WT	Data Link Connector	12-14v
8	GY/GN	EVAP Vapor Pressure (VSV)	12v or 0v
9	BR/BK	Sensor Ground	<0.050v
10	YL	TP Sensor Signal (VTA)	0.3-0.8v
11	PK	EGR Gas Temperature Sensor (THG)	3.5-4.0v
12	BK	Knock Sensor Signal	<0.075v AC
13	BK	HO2S-12 (B1 S2) Signal	0.1-1.1v
15	PK/BK	EGR Solenoid Control (VSV)	12v or 0v

1997 Pickup 2.7L I4 MFI VIN M (All) 12 Pin Connector

PCM Pin #	Wire Color	Circuit Description (12 Pin)	Value at Hot Idle
2	GN/RD	Taillight Switch Signal	Switch Off: 0v, On: 12v
3	WT/RD	AT Vehicle Speed Signal (-)	Pulse Signals
5	BL	CMP Sensor Signal (GN-)	<0.050v
6	GN	CKP Sensor Signal (NE-)	<0.050v
7	BR/WT	MAF Sensor Ground	<0.050v
9	YL/RD	AT Vehicle Speed Signal (+)	Pulse Signals
10	RD/YL	EVAP Vapor Pressure Sensor	2.5-3.1v (with fuel cap off)
11	YL	CMP Sensor Signal (G2+)	AC pulse signals
12	RD	CKP Sensor Signal (NE+)	AC pulse signals

1997 Pickup 2.7L I4 MFI VIN M (All) 26 Pin Connector

PCM Pin #	Wire Color	Circuit Description (26 Pin)	Value at Hot Idle
1	RD/WT	HO2S-12 (B1 S2) Heater	Heater Off: 12v, On: 1v
2	PK/GN	HO2S-11 (B1 S1) Heater	Heater Off: 12v, On: 1v
3	WT/GN	EVAP Purge Solenoid (VSV)	12v or 0v
4	BL	Cruise Control ECU (IDLO)	1.5v, off-idle: 12v
6	BK	IAC Signal (RSC)	Pulse Signals
7	BK/RD	IAC Signal (RSO)	Pulse Signals
8	PK/YL	A/T-ECT Solenoid (S1)	3rd or OD: 1v
9	YL/RD	Injector 4 Control	1.6-2-9 ms
10	WT/GN	Injector 3 Control	1.6-2-9 ms
11	WT	Injector 2 Control	1.6-2-9 ms
12	WT/RD	Injector 1 Control	1.6-2-9 ms
13	BR	Power Ground	<0.1v
14	GN/YL	Circuit Opening Relay (FC)	0-3v, at off-idle: 12v
15	BK	Tachometer Signal (TACO)	Pulse Signals
16	YL/RD	A/T Oil Temperature Sensor	At 68°F: 4-5v
17	BK/YL	Igniter Signal (IGF)	Digital Signal: 0-5-0v
18	GN	4WD Detection Transfer (L4)	Open: 12v, Closed: 0v
22	BR/YL	Igniter Transistor 2 Control	Digital Signal: 0-5-0v
23	BK/BL	Igniter Transistor 1 Control	Digital Signal: 0-5-0v
24	BR	Shield Ground	<0.050v
25	BR	Power Ground	<0.1v
26	BR	Power Ground	<0.1v

Pin Connector Graphic

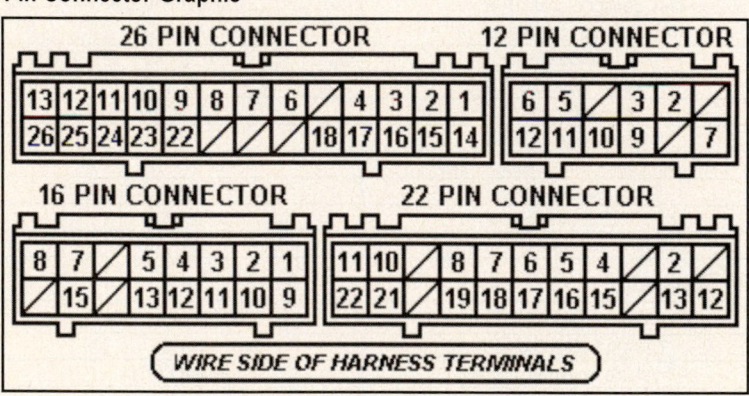

1998-99 Pickup 2.7L I4 MFI VIN M (All) 22 Pin Connector

PCM Pin #	Wire Color	Circuit Description (22 Pin)	Value at Hot Idle
2	BK/YL	Battery Direct	12-14v
3	PK	A/T Oil Temperature Indicator	Light On: 0.1v, Off: 12v
4	PK/RD	MIL (lamp) Control	MIL Off: 12v, On: 1v
5	BL/OR	Overdrive Main Switch	Switch Off: 12v, On: 1v
6	BL/BK	A/C Amplifier Signal (ACT)	Clutch On: 12v, Off: 1.5v
7	BL/YL	A/C Amplifier Signal (AC1)	Clutch On: 1.5v, Off: 12v
8	GN/OR	Speedometer Indicator	At 55 mph: 48 Hz
10	RD/WT	4WD Selector Switch	Switch Off: 12v, On: 1v
11	BK/WT	Starter Switch Signal	9-11v (cranking)
12	WT/RD	EFI Main Relay (B+)	12-14v
13	RD/GN	4WD Detection Transfer (N)	Open: 12v, Closed
14	GN	A/T Pattern Select Switch	Norm: 0v, PWR: 12v
15	PK/WT	A/T Select Switch Low	In Low: 12v, Others: 0v
16	PK	A/T Select Switch 2nd	In 2nd: 12v, Others: 0v
17	RD/BK	A/T Select Switch Reverse	In 'R': 12v, Others: 0v
18	RD/YL	Cruise Control ECU (OD1)	At Cruise in OD: 12v
19	WT	SIL (Scan Tool) Signal	12v
21	GN/WT	Stop Light Switch Signal	Brake Off: 0v, On: 12v
22	BK/RD	Start Circuit Signal	Cranking: 9-11v

1998-99 Pickup 2.7L I4 MFI VIN M (All) 16 Pin Connector

PCM Pin #	Wire Color	Circuit Description (16 Pin)	Value at Hot Idle
1	GN/YL	Sensor VREF	4.9-5.1v
2	GY/RD	MAF Sensor Signal (VG)	1.1-1.8v
3	YL/GN	IAT Sensor Signal (THA)	At 100ºF: 2.60v
4	GN/RD	ECT Sensor Signal (THW)	At 180ºF: 0.51v
5	WT	HO2S-11 (B1 S1) Signal	0.1-1.1v
7	PK/WT	Data Link Connector	12-14v
8	GY/GN	EVAP Vapor Pressure (VSV)	12v or 0v
9	BR/BK	Sensor Ground	<0.050v
10	YL	TP Sensor Signal (VTA)	0.3-0.8v
11	PK	EGR Gas Temperature Sensor (THG)	3.5-4.0v
12	BK	Knock Sensor Signal	<0.075v AC
13	BK	HO2S-12 (B1 S2) Signal	0.1-1.1v
15	PK/BK	EGR Solenoid Control (VSV)	12v or 0v

1998-99 Pickup 2.7L I4 MFI VIN M (All) 12 Pin Connector

PCM Pin #	Wire Color	Circuit Description (12 Pin)	Value at Hot Idle
2 ('98)	GN/RD	Taillight Switch Signal	Lights On: 12v
3	WT/RD	AT Vehicle Speed Signal (-)	Pulse Signals
5	BL	CMP Sensor Signal (GN-)	<0.050v
6	GN	CKP Sensor Signal (NE-)	<0.050v
7	BR/WT	MAF Sensor Ground (E2G)	<0.050v
9	YL/RD	AT Vehicle Speed Signal (+)	Pulse Signals
10	RD/YL	EVAP Vapor Pressure Sensor	2.5-3.1v (with fuel cap off)
11	YL	CMP Sensor Signal (G2+)	AC pulse signals
12	RD	CKP Sensor Signal (NE+)	AC pulse signals

1998-99 Pickup 2.7L I4 MFI VIN M (All) 26 Pin Connector

PCM Pin #	Wire Color	Circuit Description (26 Pin)	Value at Hot Idle
1	RD/WT	HO2S-12 (B1 S2) Heater	Heater Off: 12v, On: 1v
2	PK/GN	HO2S-11 (B1 S1) Heater	Heater Off: 12v, On: 1v
2 ('99)	RD/BK	HO2S-11 (B1 S1) Heater	Heater Off: 12v, On: 1v
3	WT/GN	EVAP Purge Solenoid (VSV)	12v or 0v
4	GN/RD	Cruise Control ECU (IDLO)	1.5v, off-idle: 12v
6	BK	IAC Signal (RSC)	Pulse Signals
7	BK/RD	IAC Signal (RSO)	Pulse Signals
8	PK/YL	A/T-ECT Solenoid (S1)	3rd or OD: 1v
9	RD/BL	Injector 4 Control	1.6-2-9 ms
10	RD	Injector 3 Control	1.6-2-9 ms
11	WT	Injector 2 Control	1.6-2-9 ms
12	WT/RD	Injector 1 Control	1.6-2-9 ms
13	BR	Power Ground	<0.1v
14	GN/YL	Circuit Opening Relay (FC)	0-3v, at off-idle: 12v
15	BK	Tachometer Signal (TACO)	Pulse Signals
16	YL/RD	A/T Oil Temperature Sensor	At 68°F: 4-5v
17	BK/YL	Igniter Signal (IGF)	Digital Signal: 0-5-0v
18	GY	4WD Detection Transfer (L4)	Open: 12v, Closed: 0v
20	RD/GN	A/T-ECT Solenoid (SL)	In Lockup: 12-14v
21	LG	A/T-ECT Solenoid (S1)	3rd or OD: 1v
22	BR/YL	Igniter Transistor 2 Control	Digital Signal: 0-5-0v
23	BK/BL	Igniter Transistor 1 Control	Digital Signal: 0-5-0v
24	BR	Shield Ground	<0.050v
25	BR	Power Ground	<0.1v
26	BR	Power Ground	<0.1v

Pin Connector Graphic

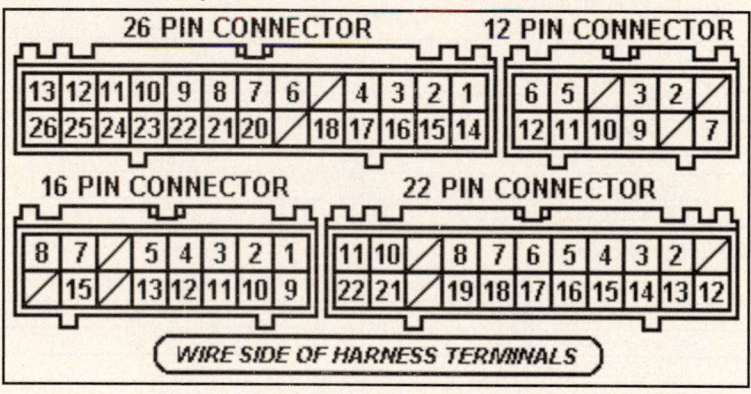

2000 Pickup 2.7L I4 VIN M California (All) E5 22P Connector

PCM Pin #	Wire Color	Circuit Description (22 Pin)	Value at Hot Idle
1	BK/YL	Direct Battery	12-14v
2	RD/BK	A/T Select Switch Reverse	In 'R': 12v, Others: 0v
3	PK/WT	A/T Select Switch Low	In Low: 12v, Others: 0v
4	PK	A/T Select Switch 2nd	In 2nd: 12v, Others: 0v
5	PK	A/T Oil Temperature Indicator	Lamp Off: 12v, On: 1v
6	PK/RD	MIL (lamp) Control	MIL Off: 12v, On: 1v
7	BK/WT	Starter Switch Signal	9-11v (cranking)
9	RD/GN	4WD Detection Transfer (N)	Open: 12v, Closed: 0v
10	BL/OR	O/D OFF Indicator (ODLP)	Lamp Off: 12v, On: 1v
11	RD/YL	Cruise Control ECU (OD1)	At Cruise in OD: 12v
12	WT	SIL (Scan Tool) Signal	12v
13	BL/YL	A/C Amplifier Signal (AC1)	Clutch On: 1.5v, Off: 12v
14	BL/BK	A/C Amplifier Signal (ACT)	Clutch On: 12v, Off: 1.5v
15	GN/WT	Stop Light Switch Signal	Brake Off: 0v, On: 12v
16	WT/RD	EFI Main Relay (B+)	12-14v
17	GN	A/T Pattern Select Switch	Norm: 0v, PWR: 12v
19	GY	4WD Detection Transfer (L4)	Open: 12v, Closed: 0v
21	GN/OR	Speedometer Indicator	At 55 mph: 48 Hz
22	BK/RD	Start Circuit Signal	Cranking: 9-11v

2000 Pickup 2.7L I4 VIN M California (All) E6 28P Connector

PCM Pin #	Wire Color	Circuit Description (28 Pin)	Value at Hot Idle
5	LG/RD	Cruise Control ECU (IDLO)	1.5v, off-idle: 12v
6	GN/YL	Circuit Opening Relay (FC)	0-3v, at off-idle: 12v
7	PK/WT	Data Link Connector	12-14v
8	RD/YL	EVAP Vapor Pressure Sensor	2.5-3.1v (with fuel cap off)
10	RD/WT	4WD Detection Switch	Switch Off: 12v, On: 1v
13	BK	Tachometer Signal (TACO)	Pulse Signals
14	YL/RD	A/T Vehicle Speed Sensor (+)	AC pulse signals
23	WT/RD	A/T Vehicle Speed Sensor (-)	AC pulse signals
25	GN	Overdrive Main Switch	Switch Off: 12v, On: 1v
28	YL	PSP Switch Signal	Straight: 12v, Turning: 0v

Pin Connector Graphic

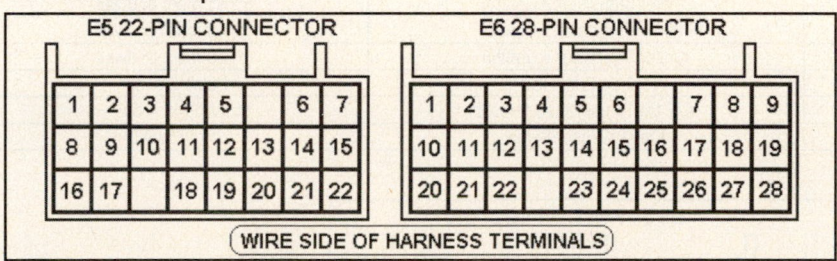

2000 Pickup 2.7L I4 VIN M California (All) E7 24P Connector

PCM Pin #	Wire Color	Circuit Description (24 Pin)	Value at Hot Idle
2	GN/YL	Sensor VREF (VC)	4.9-5.1v
3	RD/WT	HO2S-12 (B1 S2) Heater	Heater Off: 12v, On: 1v
4	WT	AFRS-11 (B1 S1) Heater	Heater Off: 12v, On: 1v
5	PK/BK	EGR Solenoid Control (VSV)	12v or 0v
6	WT/GN	EVAP Purge Solenoid (VSV)	12v or 0v
7	GY/GN	EVAP Vapor Pressure (VSV)	12v or 0v
9	BK/BL	TP Sensor Signal (VTA)	0.3-0.8v
10	BK	HO2S-12 (B1 S2) Signal	0.1-1.1v
11	WT	AFS-11 (B1 S1) Signal (AF+)	Fixed at 3.3v
12	GN/RD	ECT Sensor Signal (THW)	At 180ºF: 0.51v
14	GY/RD	MAF Sensor Signal (VG)	1.1-1.8v
15	YL	CMP Sensor Signal	AC pulse signals
16	RD	CKP Sensor Signal	AC pulse signals
17, 18	BR	Sensor Ground	<0.050v
19	PK	EGR Gas Temperature Sensor (THG)	3.5-4.0v
20	RD	AFS-11 (B1 S1) Signal (AF-)	Fixed at 3.3v
21	YL/GN	IAT Sensor Signal (THA)	At 100ºF: 2.60v
22	BR/WT	MAF Sensor Ground (E2G)	<0.050v
24	GN	CKP/CMP Sensor Signal (-)	<0.050v

2000 Pickup 2.7L I4 VIN M California (All) E8 31P Connector

PCM Pin #	Wire Color	Circuit Description (31 Pin)	Value at Hot Idle
1	WT/RD	Injector 1 Control	2.0-3.3 ms
2	WT	Injector 2 Control	2.0-3.3 ms
3	RD	Injector 3 Control	2.0-3.3 ms
4	RD/BL	Injector 4 Control	2.0-3.3 ms
5, 21, 31	BR	Power Ground	<0.1v
6	WT/BK	Sensor Ground	<0.050v
7	PK/YL	A/T-ECT Solenoid (S1)	3rd or OD: 1v
8	LG	A/T-ECT Solenoid (S2)	1st or OD: 1v
9	RD/GN	A/T-ECT Solenoid (SL)	In Lockup: 12-14v
10	BK/YL	Igniter Signal (IGF)	Digital Signal: 0-5-0v
11	BK/BL	Igniter Transistor 1 Control	6º, 55 mph: 8º dwell
12	BK/WT	Igniter Transistor 2 Control	6º, 55 mph: 8º dwell
13	BL/RD	Igniter Transistor 3 Control	6º, 55 mph: 8º dwell
14	BL/YL	Igniter Transistor 4 Control	6º, 55 mph: 8º dwell
15	BK/RD	IAC Control (RSD)	Pulse Signals
26	YL/RD	A/T Oil Temperature Sensor	At 68ºF: 4-5v
28	GY	Knock Sensor Signal	<0.075v AC

Pin Connector Graphic

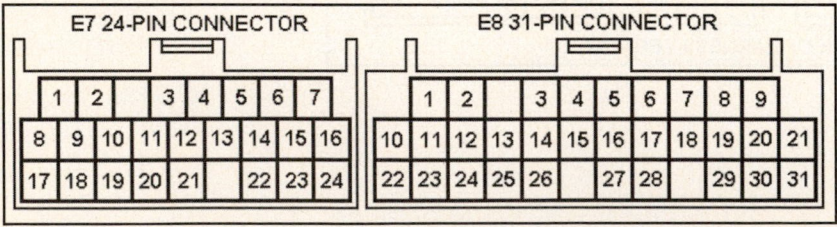

2000 Pickup 2.7L I4 VIN M Federal (All) E5 22 Pin Connector

PCM Pin #	Wire Color	Circuit Description (22 Pin)	Value at Hot Idle
2	BK/YL	Direct Battery	12-14v
3	PK	A/T Oil Temperature Indicator	Lamp Off: 12v, On: 1v
4	PK/RD	MIL (lamp) Control	MIL Off: 12v, On: 1v
5	BL/OR	O/D OFF Indicator (ODLP)	Lamp Off: 12v, On: 1v
6	BL/BK	A/C Amplifier Signal (ACT)	Clutch On: 12v, Off: 1.5v
7	BL/YL	A/C Amplifier Signal (AC1)	Clutch On: 1.5v, Off: 12v
8	GN/OR	Speedometer Indicator	At 55 mph: 48 Hz
10	RD/WT	4WD Detection Switch	Switch Off: 12v, On: 1v
11	BK/WT	Starter Switch Signal	9-11v (cranking)
12	WT/RD	EFI Main Relay Power	12-14v
13	RD/GN	4WD Detection Transfer (N)	Switch Closed: 0v
14	GN	A/T Pattern Select Switch	Norm: 0v, PWR: 12v
15	PK/WT	A/T Select Switch Low	In Low: 12v, Others: 0v
16	PK	A/T Select Switch 2nd	In 2nd: 12v, Others: 0v
17	RD/BK	A/T Select Switch Reverse	In 'R': 12v, Others: 0v
18	BR/YL	Cruise Control ECU (OD1)	At Cruise in OD: 12v
19	WT	SIL (Scan Tool) Signal	12v
21	GN/WT	Stop Light Switch Signal	Brake Off: 0v, On: 12v
22	BK/RD	Start Circuit Signal	Cranking: 9-11v

2000 Pickup 2.7L I4 VIN M Federal (All) E6 12 Pin Connector

PCM Pin #	Wire Color	Circuit Description (12 Pin)	Value at Hot Idle
1	GN	Overdrive Main Switch	Switch Off: 12v, On: 1v
3	WT/RD	A/T Vehicle Speed Sensor (-)	AC pulse signals
5	BL	CMP Sensor Signal (GN-)	<0.050v
6	GN	CKP Sensor Signal (NE-)	<0.050v
7	BR/WT	MAF Sensor Ground (E2G)	<0.050v
9	YL/RD	A/T Vehicle Speed Sensor (+)	AC pulse signals
10	RD/YL	EVAP Vapor Pressure Sensor	2.5-3.1v (with fuel cap off)
11	YL	CMP Sensor Signal (G2+)	AC pulse signals
12	RD	CKP Sensor Signal (NE+)	AC pulse signals

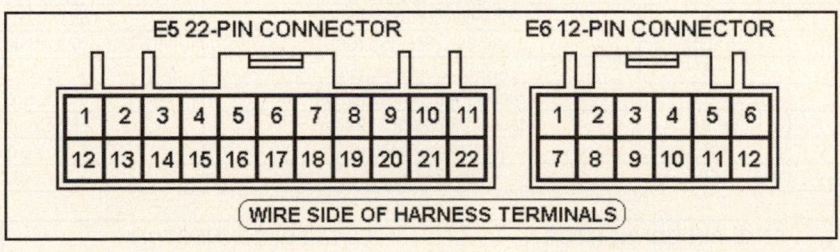

2000 Pickup 2.7L I4 VIN M Federal (All) E7 16 Pin Connector

PCM Pin #	Wire Color	Circuit Description (16 Pin)	Value at Hot Idle
1	GN/YL	Sensor VREF (VC)	4.9-5.1v
2	GY/RD	MAF Sensor Signal (VG)	1.1-1.8v
3	YL/GN	IAT Sensor Signal (THA)	At 100°F: 2.60v
4	GN/RD	ECT Sensor Signal (THW)	At 180°F: 0.51v
5	WT	HO2S-11 (B1 S1) Signal	0.1-1.1v
7	PK/WT	Data Link Connector	12v
8	GY/GN	EVAP Vapor Pressure (VSV)	12v or 0v
9	BR/BK	Sensor Ground	<0.050v
10	YL	TP Sensor Signal (VTA)	0.3-0.8v
11	PK	EGR Gas Temperature Sensor (THG)	3.5-4.0v
12	GY	Knock Sensor Signal	<0.075v AC
13	BK	HO2S-12 (B1 S2) Signal	0.1-1.1v
15	PK/BK	EGR Solenoid Control (VSV)	12v or 0v

2000 Pickup 2.7L I4 VIN M Federal (All) E8 26 Pin Connector

PCM Pin #	Wire Color	Circuit Description (26 Pin)	Value at Hot Idle
1	RD/WT	HO2S-12 (B1 S2) Heater	Heater Off: 12v, On: 1v
2	RD/BK	HO2S-11 (B1 S1) Heater	Heater Off: 12v, On: 1v
3	WT/GN	EVAP Purge Solenoid (VSV)	12v or 0v
4	LG/RD	Cruise Control ECU (IDLO)	1.5v, off-idle: 12v
6	BK	IAC Signal (RSC)	Pulse Signals
7	BK/RD	IAC Signal (RSO)	Pulse Signals
8	PK/YL	A/T-ECT Solenoid (S1)	3rd or OD: 1v
9	RD/BL	Injector 4 Control	2.0-3.3 ms
10	RD	Injector 3 Control	2.0-3.3 ms
11	WT	Injector 2 Control	2.0-3.3 ms
12	WT/RD	Injector 1 Control	2.0-3.3 ms
13	BR	Power Ground	<0.1v
14	GN/YL	Circuit Opening Relay (FC)	0-3v, at off-idle: 12v
15	BK	Tachometer Signal (TACO)	Pulse Signals
16	YL/RD	A/T Oil Temperature Sensor	At 68°F: 4-5v
17	BK/YL	Igniter Signal (IGF)	Digital Signal: 0-5-0v
18	GY	4WD Detection Transfer (L4)	Open: 12v, Closed: 0v
20	RD/GN	A/T-ECT Solenoid (SL)	In Lockup: 12-14v
21	LG	A/T-ECT Solenoid (S2)	1st or OD: 1v
22	BR/YL	Igniter Transistor 2 Control	6°, 55 mph: 8° dwell
23	BK/BL	Igniter Transistor 1 Control	6°, 55 mph: 8° dwell
24	BR	Sensor Ground	<0.050v
25, 26	BR	Power Ground	<0.1v

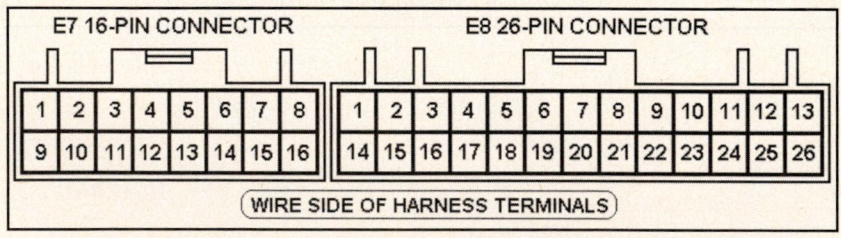

2001-03 Pickup 2.7L I4 3RZ-FE VIN M (All) E5 22 Pin Connector

PCM Pin #	Wire Color	Circuit Description (22 Pin)	Value at Hot Idle
1	BK/YL	Direct Battery	12-14v
2	RD/BK	A/T Select Switch Reverse	In 'R': 12v, Others: 0v
3	RD	A/T Select Switch Low	In Low: 12v, Others: 0v
4	PK	A/T Select Switch 2nd	In 2nd: 12v, Others: 0v
5	OR	A/T Oil Temperature Indicator	Lamp Off: 12v, On: 1v
6	VT/RD	Malfunction Indicator Lamp (MIL) Control	MIL Off: 12v, On: 1v
7	GN	Starter Switch Signal (STA)	9-11v (cranking)
8	---	Not Used	---
9	BL	A/T P/N Position Lamp (TFN)	Lamp Off: 12v, On: 1v
10	BL/OR	O/D OFF Indicator (ODLP)	Lamp Off: 12v, On: 1v
11	BR/YL	Cruise Control Signal (OD1)	At Cruise in OD: 12v
12	WT	SIL (Scan Tool) Signal	12v
13	BL/YL	A/C Amplifier Signal (AC1)	Clutch On: 1.5v, Off: 12v
14	BL/BK	A/C Amplifier Signal (ACT)	Clutch On: 12v, Off: 1.5v
15	GN/WT	Stop Light Switch Signal	Brake Off: 0v, On: 12v
16	WT/RD	EFI Main Relay (B+)	12-14v
17	GN	A/T Pattern Select Switch	Norm: 0v, PWR: 12v
18	---	Not Used	---
19	GY	4WD Detection Transfer (L4)	Open: 12v, Closed: 0v
20	GN/RD	Taillight Switch Signal (ELS)	Switch Off: 0v, On: 12v
21	GN/OR	Speedometer Indicator (SP1)	At 55 mph: 48 Hz
22	BK/YL	A/T Neutral Start Switch Signal (NSW)	9-11v (cranking)

2001-03 Pickup 2.7L I4 3RZ-FE VIN M (All) E6 28 Pin Connector

PCM Pin #	Wire Color	Circuit Description (28 Pin)	Value at Hot Idle
1-4	---	Not Used	---
5	LG/RD	Cruise Control ECU (IDLO)	1.5v, off-idle: 12v
6	WT/BL	Circuit Opening Relay (FC)	0-3v, at off-idle: 12v
7	RD	DLC3 D7 Signal (TC)	12v
8	RD/YL	EVAP Vapor Pressure Sensor (PTNK)	2.5-3.1v (with fuel cap off)
9	---	Not Used	---
10	GNBK	4WD Switch Signal (4WD)	4WD Switch On: 12v
11-12	---	Not Used	---
13	LG/BK	Tachometer Signal (TACO)	Pulse Signals
14	YL/RD	A/T Vehicle Speed Sensor (+)	AC pulse signals
15-17	---	Not Used	---
18	GN/BK	4WD ADD Detection Switch	Open: 12v, Closed: 0v
19	---	Not Used	---
20-22	---	Not Used	---
23	WT/RD	A/T Vehicle Speed Sensor (-)	AC pulse signals
24	---	Not Used	---
25	GN	OD Main Switch (ODMS)	Switch Off: 12v, On: 1v
26-27	---	Not Used	---
28	YL	PSP Switch Signal (PSW)	Straight: 12v, Turning: 0v

Pin Connector Graphic

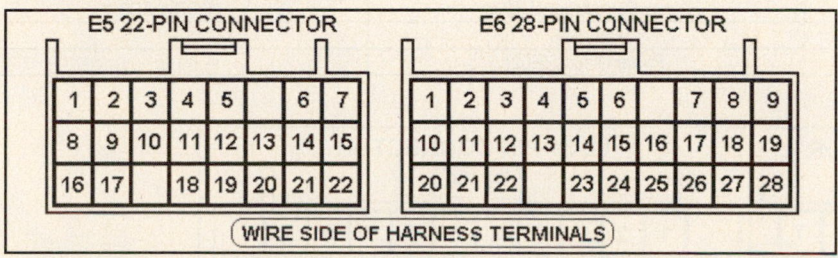

2001-03 Pickup 2.7L I4 3RZ-FE VIN M (All) E7 24 Pin Connector

PCM Pin #	Wire Color	Circuit Description (24 Pin)	Value at Hot Idle
1	RD/BL	EVAP Closed Canister Vent (VSV)	12v or 0v
2	GN/YL	Sensor VREF (VC)	4.9-5.1v
3	RD/WT	HO2S-12 (B1 S2) Heater (HTS)	Heater Off: 12v, On: 1v
4	WT	AFS-11 (B1 S1) Heater (AFHT)	Heater Off: 12v, On: 1v
5	RD/BK	EGR Solenoid Control (VSV)	12v or 0v
6	WT/GN	EVAP Purge Solenoid (VSV)	12v or 0v
7	GN/BK	EVAP Vapor Pressure (VSV)	12v or 0v
8	---	Not Used	---
9	YL	TP Sensor Signal (VTA)	0.3-0.8v
10	BK	HO2S-12 (B1 S2) Signal (OXS)	0.1-1.1v
11	VT	AFS-11 (B1 S1) Signal (AF+)	Fixed at 3.3v
12	GN/RD	ECT Sensor Signal (THW)	At 180ºF: 0.51v
13	---	Not Used	---
14	GY	MAF Sensor Signal (VG)	1.1-1.8v
15	RD	CMP Sensor Signal (G2+)	AC pulse signals
16	BL	CKP Sensor Signal (NE+)	AC pulse signals
17	BR	Power Ground (E1)	<0.050v
18	BL/BK	Sensor Ground (E2)	<0.050v
19	PK/BL	EGR Temperature Sensor (THG)	3.5-4.0v
20	PK	AFS-11 (B1 S1) Signal (AF-)	Fixed at 3.3v
21	YL/GN	IAT Sensor Signal (THA)	At 100ºF: 2.60v
22	BK/WT	MAF Sensor Ground (EVG)	<0.050v
23	---	Not Used	---
24	GN	CKP/CMP Sensor Signal (NE-)	<0.050v

2001-03 Pickup 2.7L I4 3RZ-FE VIN M (All) E8 31 Pin Connector

PCM Pin #	Wire Color	Circuit Description (31 Pin)	Value at Hot Idle
1	WT/RD	Injector 1 Control	2.0-3.3 ms
2	WT	Injector 2 Control	2.0-3.3 ms
3	RD	Injector 3 Control	2.0-3.3 ms
4	RD/BL	Injector 4 Control	2.0-3.3 ms
5	WT/BK	Power Ground (E03)	<0.1v
6	WT/BK	Power Ground (E04)	<0.1v
7	VT	A/T-ECT Solenoid (S1)	3rd or OD: 1v
8	LG	A/T-ECT Solenoid (S2)	1st or OD: 1v
9	RD/WT	A/T-ECT Solenoid (SL)	In Lockup: 12-14v
10	BK/YL	Igniter Signal (IGF)	Digital Signal: 0-5-0v
11	BK/BL	Igniter Transistor 1 Control	6º, 55 mph: 8º dwell
12	BL	Igniter Transistor 2 Control	6º, 55 mph: 8º dwell
13	BL/RD	Igniter Transistor 3 Control	6º, 55 mph: 8º dwell
14	BL/YL	Igniter Transistor 4 Control	6º, 55 mph: 8º dwell
15	BK/RD	Idle Air Control Valve (RSD)	Pulse Signals
16-20	---	Not Used	---
21	WT/BK	Power Ground (E01)	<0.1v
22-25	---	Not Used	---
26	YL/RD	A/T Oil Temperature Sensor	At 68ºF: 4-5v
27	---	Not Used	---
28	GY	Knock Sensor Signal (KNK)	<0.075v AC
29-30	---	Not Used	---
31	WT/BK	Power Ground (E02)	<0.1v

Pin Connector Graphic

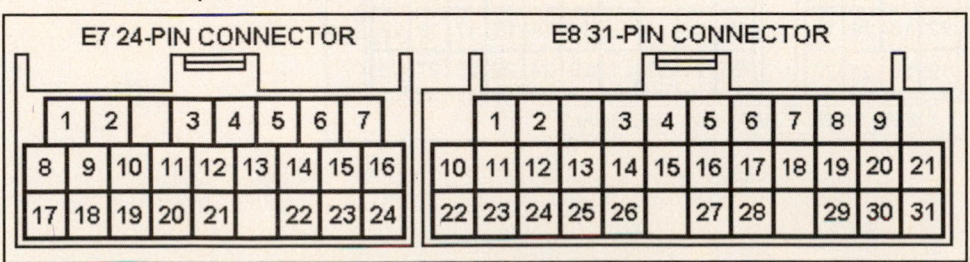

1995 Pickup 3.4L V6 MFI VIN V (A/T-ECT) 28 Pin Connector

PCM Pin #	Wire Color	Circuit Description (28 Pin)	Value at Hot Idle
1	---	Not Used	---
2	PK	A/T Select Switch 2nd	In 2nd: 12v, Others: 0v
3	PK/WT	A/T Select Switch Low	In Low: 12v, Others: 0v
4	PK	A/T Oil Temperature Lamp	Lamp Off: 12v, On: 1v
5	BL/BK	A/C Amplifier Signal (ACT)	Clutch On: 12v, Off: 1.5v
6	BL/OR	Overdrive Main Switch	Switch Off: 12v, On: 1v
7	BL/YL	Cruise Control ECU	At Cruise in OD: 12v
10	GN/OR	A/T Pattern Select Switch	Norm: 0v, PWR: 12v
12	GN/OR	Vehicle Speed Sensor	At 55 mph: 48 Hz
14	BK/YL	Battery Direct	12-14v
17	RD/GN	4WD Detection Transfer (N)	Open: 12v, Closed: 0v
18	WT	Data Link Connector	12-14v
20	BL/BK	A/C Amplifier Signal (AC1)	Clutch On: 1.5v, Off: 12v
22	WT/RD	EFI Main Relay (B+)	12-14v
25	GN/WT	Stop Light Switch Signal	Brake Off: 0v, On: 12v
26	GN/WT	4WD Selector Switch	Open: 12v, Closed: 0v

1995 Pickup 3.4L V6 MFI VIN V (A/T-ECT) 22 Pin Connector

PCM Pin #	Wire Color	Circuit Description (22 Pin)	Value at Hot Idle
1	GN/BK	Sensor VREF	4.9-5.1v
2-3	---	Not Used	---
4	WT/RD	A/T Vehicle Speed Sensor (+)	AC pulse signals
5	RD	CKP Sensor Signal (NE+)	AC pulse signals
6	GN	CKP Sensor Signal (NE-)	<0.050v
7	YL	TP Sensor Signal (VTA)	0.3-0.8v
8	GY/RD	MAF Sensor Signal (VG)	1.1-1.8v
9	WT/RD	A/T Vehicle Speed Sensor (-)	AC pulse signals
10	BK	CMP Sensor Signal (G2+)	AC pulse signals
11	WT	CMP Sensor Signal (GN-)	<0.050v
12	YL/RD	A/T Oil Temperature Sensor	At 68°F: 4-5v
13	WT	HO2S-11 (B1 S1) Signal	0.1-1.1v
14	YL/GN	IAT Sensor Signal (THA)	At 100°F: 2.60v
16	GY	Knock Sensor 2 Signal	<0.075v AC
17	BK	Knock Sensor 1 Signal	<0.075v AC
18, 22	BR/WT	Sensor Ground	<0.050v
19	BK	HO2S-12 (B1 S2) Signal	0.1-1.1v
20	GN/RD	ECT Sensor Signal (THW)	At 180°F: 0.51v
21	PK/GN	EGR Gas Temperature Sensor (THG)	3.5-4.0v

Standard Colors and Abbreviations

Abbreviation	Color	Abbreviation	Color	Abbreviation	Color
BK	Black	GY	Gray	RD	Red
BL	Blue	GN	Green	TN	Tan
BR	Brown	LG	LT Green	VT	Violet
DB	Dark Blue	OR	Orange	WT	White
DG	DK Green	PK	Pink	YL	Yellow

1995 Pickup 3.4L V6 MFI VIN V (A/T-ECT) 16 Pin Connector

PCM Pin #	Wire Color	Circuit Description (16 Pin)	Value at Hot Idle
1-2	---	Not Used	---
3	PK	MIL (lamp) Control	MIL Off: 12v, On: 1v
4	GN/YL	Circuit Opening Relay (FC)	0-3v, at off-idle: 12v
5	PK/WT	Data Link Connector	12-14v
6-7	---	Not Used	---
8	RD/WT	EGR Solenoid Control (VSV)	12v or 0v
9	RD/BK	Fuel Pressure Up Solenoid	1v (at hot startup)
10	BK/RD	A/C Idle-Up Solenoid	A/C Off: 12v, On: 1v
11-14	---	Not Used	---
15	WT/GN	EVAP Purge Solenoid (VSV)	12v or 0v
16	BR	Sensor Ground	<0.050v

1995 Pickup 3.4L V6 MFI VIN V (A/T-ECT) 34-Pin Connector

PCM Pin #	Wire Color	Circuit Description (34-Pin)	Value at Hot Idle
1	BR	Sensor Ground	<0.050v
2-4	---	Not Used	---
5	YL/BK	Injector 6 Control	1.6-2.9 ms
6	WT/BL	Injector 5 Control	1.6-2.9 ms
7	YL/RD	Injector 4 Control	1.6-2.9 ms
8	WT/GN	Injector 3 Control	1.6-2.9 ms
9	WT	Injector 2 Control	1.6-2.9 ms
10	WT/RD	Injector 1 Control	1.6-2.9 ms
11	PK/YL	A/T-ECT Solenoid (S1)	3rd or OD: 1v
12	BK/YL	Igniter Signal (IGF)	Digital Signal: 0-5-0v
13	BK/WT	Starter Switch Signal	9-11v (cranking)
14	BK	Neutral Start Switch	In P/N: 9-11v (cranking)
15	RD/WT	HO2S-12 (B1 S2) Heater	Heater Off: 12v, On: 1v
16	PK/GN	HO2S-11 (B1 S1) Heater	Heater Off: 12v, On: 1v
17	LG	A/T-ECT Solenoid (S2)	1st or OD: 1v
18-21	---	Not Used	---
22	BK/RD	IAC Signal (RSC)	Pulse Signals
23	BR/RD	IAC Signal (RSO)	Pulse Signals
24	BK/BL	Igniter Transistor 1 Control	Digital Signal: 0-5-0v
25	BR/BK	Igniter Transistor 2 Control	Digital Signal: 0-5-0v
26	BK/WT	Igniter Transistor 3 Control	Digital Signal: 0-5-0v
27	RD/GN	A/T-ECT Solenoid (SL)	In Lockup: 12-14v
29	GN	4WD Detection Transfer (L4)	Open: 12v, Closed: 0v
30-31	---	Not Used	---
32	YL/BL	Closed Throttle Switch	1v, off-idle: 12v
33	BR	Power Ground	<0.1v
34	BR	Power Ground	<0.1v

Pin Connector Graphic

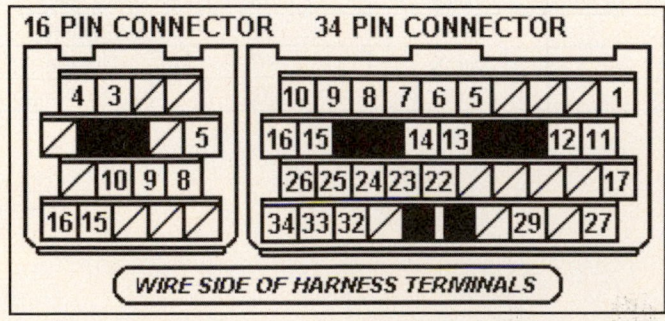

1995 Pickup 3.4L V6 MFI VIN V 5VZ-FE (M/T) E5 22 Pin Connector

PCM Pin #	Wire Color	Circuit Description (22 Pin)	Value at Hot Idle
1	---	Not Used	---
2	BK/YL	Battery Direct	12-14v
3	---	Not Used	---
4	PK	MIL (lamp) Control	MIL Off: 12v, On: 1v
5	---	Not Used	---
6	BL/BK	A/C Amplifier Signal (ACT)	Clutch On: 12v, Off: 1.5v
7	BK/YL	A/C Amplifier Signal (AC1)	Clutch On: 1.5v, Off: 12v
8	GN/OR	Vehicle Speed Sensor	At 55 mph: 48 Hz
9	GN/WT	4WD Selector Switch	Open: 12v, Closed: 0v
10	---	Not Used	---
11	BK/WT	Starter Switch Signal	9-11v (cranking)
12	WT/RD	EFI Main Relay (B+)	12-14v
13-18	---	Not Used	---
19	WT	Data Link Connector (SDL)	0v
20	GN/WT	Stop Light Switch Signal	Brake Off: 0v, On: 12v
21-22	---	Not Used	---

1995 Pickup 3.4L V6 MFI VIN V 5VZ-FE (M/T) E7 16 Pin Connector

PCM Pin #	Wire Color	Circuit Description (16 Pin)	Value at Hot Idle
1	GN/BK	Sensor VREF (VC)	4.9-5.1v
2	GY/RD	MAF Sensor Signal (VG)	1.1-1.8v
3	GY	Knock Sensor 2 Signal	<0.075v AC
4	GN/RD	ECT Sensor Signal (THW)	At 180ºF: 0.51v
5	WT	HO2S-11 (B1 S1) Signal	0.1-1.1v
6	BK	Knock Sensor 1 Signal	<0.075v AC
7	RD	Data Link Connector	12-14v
8	BR/WT	Sensor Ground	<0.050v
9	BR/YL	Sensor Ground	<0.050v
10	YL	TP Sensor Signal (VTA)	0.3-0.8v
11	YL/BL	Closed Throttle Switch	1v, off-idle: 12v
12	YL/GN	IAT Sensor Signal (THA)	At 100ºF: 2.60v
13	BK	HO2S-12 (B1 S2) Signal	0.1-1.1v
14	PK/GN	EGR Gas Temperature Sensor (THG)	3.5-4.0v
15-16	---	Not Used	---

1995 Pickup 3.4L V6 MFI VIN V 5VZ-FE (M/T) E7 12 Pin Connector

PCM Pin #	Wire Color	Circuit Description (12 Pin)	Value at Hot Idle
1-2	---	Not Used	---
3	PK/GN	HO2S-11 (B1 S1) Heater	Heater Off: 12v, On: 1v
4	---	Not Used	---
5	BL	CMP Sensor Signal (GN-)	<0.050v
6	GN	CKP Sensor Signal (NE-)	<0.050v
7	BR	Sensor Ground	<0.050v
8	---	Not Used	---
9	RD/GN	HO2S-12 (B1 S2) Heater	Heater Off: 12v, On: 1v
10	---	Not Used	---
11	YL	CMP Sensor Signal (G2+)	AC pulse signals
12	RD	CKP Sensor Signal (NE+)	AC pulse signals

1995 Pickup 3.4L V6 MFI VIN V 5VZ-FE (M/T) E8 26 Pin Connector

PCM Pin #	Wire Color	Circuit Description (26 Pin)	Value at Hot Idle
2	BK/RD	A/C Idle-Up Solenoid	A/C Off: 12v, On: 1v
5	WT/GN	EVAP Purge Solenoid (VSV)	12v or 0v
6	BK/RD	IAC Signal (RSC)	Pulse Signals
7	BR/RD	IAC Signal (RSO)	Pulse Signals
8	YL/BK	Injector 6 Control	1.6-2.9 ms
9	WT/BL	Injector 5 Control	1.6-2.9 ms
10	YL/RD	Injector 4 Control	1.6-2.9 ms
11	WT	Injector 2 Control	1.6-2.9 ms
12	WT/RD	Injector 1 Control	1.6-2.9 ms
13	BR	Power Ground	<0.1v
14	GN/YL	Circuit Opening Relay (FC)	0-3v, at off-idle: 12v
17	BK/YL	Igniter Signal (IGF)	Digital Signal: 0-5-0v
18	RD/WT	EGR Solenoid Control (VSV)	12v or 0v
19	RD/BK	Fuel Pressure Up Solenoid	1v (at hot startup)
21	BK/WT	Igniter Transistor 3 Control	Digital Signal: 0-5-0v
22	BR/YL	Igniter Transistor 2 Control	Digital Signal: 0-5-0v
23	BK/BL	Igniter Transistor 1 Control	Digital Signal: 0-5-0v
24	BR	Power Ground	<0.1v
25	WT/GN	Injector 3 Control	1.6-2.9 ms
26	BR	Power Ground	<0.1v

Pin Connector Graphic

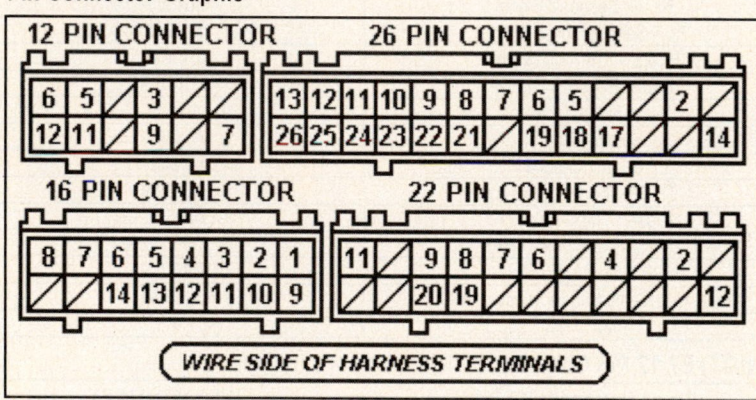

1996 Pickup 3.4L V6 MFI VIN N (A/T-ECT) E5 28 Pin Connector

PCM Pin #	Wire Color	Circuit Description (28 Pin)	Value at Hot Idle
1	RD/BK	A/T Select Switch Reverse	In 'R': 12v, Others: 0v
2	PK	A/T Select Switch 2nd	In 2nd: 12v, Others: 0v
3	PK/WT	A/T Select Switch Low	In Low: 12v, Others: 0v
4	PK	A/T Oil Temperature Lamp	Lamp Off: 12v, On: 1v
5	BL/BK	A/C Amplifier Signal (ACT)	Clutch On: 12v, Off: 1.5v
6	BL/OR	Overdrive Main Switch	Open: 12v, Closed: 0v
7	BL	Cruise Control ECU	At Cruise in OD: 12v
10	GN/OR	A/T Pattern Select Switch	Norm: 0v, PWR: 12v
12	GN/OR	Vehicle Speed Sensor	At 55 mph: 48 Hz
14	BK/YL	Battery Direct	12-14v
17	RD/GN	4WD Detection Transfer (N)	Open: 12v, Closed: 0v
18	WT	Data Link Connector	12-14v
20	BL/YL	A/C Amplifier Signal (AC1)	Clutch On: 1.5v, Off: 12v
22	WT/RD	EFI Main Relay (B+)	12-14v
25	GN/WT	Stop Light Switch Signal	Brake Off: 0v, On: 12v
26	GN/WT	4WD Selector Switch	Clutch On: 1.5v, Off: 12v

1996 Pickup 3.4L V6 MFI VIN N (A/T-ECT) E7 22 Pin Connector

PCM Pin #	Wire Color	Circuit Description (22 Pin)	Value at Hot Idle
1	GN/BK	Sensor VREF	4.9-5.1v
4	WT/RD	AT Vehicle Speed Signal (-)	AC pulse signals
5	RD	CKP Sensor Signal (NE+)	AC pulse signals
6	GN	CKP Sensor Signal (NE-)	<0.050v
7	YL	TP Sensor Signal (VTA)	0.3-0.8v
8	GY/RD	MAF Sensor Signal (VG)	1.1-1.8v
9	YL/RD	AT Vehicle Speed Signal (+)	AC pulse signals
10	BK	CMP Sensor Signal (G2+)	AC pulse signals
11	WT	CMP Sensor Signal (GN-)	<0.050v
12	YL/RD	A/T Oil Temperature Sensor	At 68°F: 4-5v
13	WT	HO2S-11 (B1 S1) Signal	0.1-1.1v
14	YL/GN	IAT Sensor Signal (THA)	At 100°F: 2.60v
15	RD/YL	EVAP Vapor Pressure Sensor	2.5-3.1v (with fuel cap off)
16	GY	Knock Sensor 2 Signal	<0.075v AC
17	BK	Knock Sensor 1 Signal	<0.075v AC
18, 22	BR/WT	Sensor, Power Ground	<0.050v
19	BK	HO2S-12 (B1 S2) Signal	0.1-1.1v
20	GN/RD	ECT Sensor Signal (THW)	At 180°F: 0.51v
21	PK/GN	EGR Gas Temperature Sensor (THG)	3.5-4.0v

Pin Connector Graphic

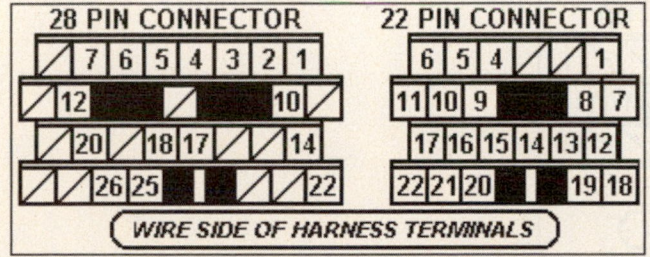

1996 Pickup 3.4L V6 MFI VIN N (A/T-ECT) E6 16 Pin Connector

PCM Pin #	Wire Color	Circuit Description (16 Pin)	Value at Hot Idle
1-2, 6-7	----	Not Used	---
3	PK	MIL (lamp) Control	MIL Off: 12v, On: 1v
4	GN/YL	Circuit Opening Relay (FC)	0-3v, at off-idle: 12v
5	PK/WT	Data Link Connector	12v
8	RD/WT	EGR Solenoid Control (VSV)	12v or 0v
9	RD/BK	Fuel Pressure Up Solenoid	1v (at hot startup)
10	BK/RD	A/C Idle-Up Solenoid	A/C Off: 12v, On: 1v
11-12, 14	---	Not Used	---
13	WT/GN	EVAP Purge Solenoid (VSV)	12v or 0v
15	WT/GN	EVAP Vapor Pressure Sensor	2.5-3.1v (with fuel cap off)
16	BR	Shield Ground	<0.050v

1996 Pickup 3.4L V6 MFI VIN N (A/T-ECT) E8 34-Pin Connector

PCM Pin #	Wire Color	Circuit Description (34-Pin)	Value at Hot Idle
1	BR	Power Ground	<0.1v
5	YL/BK	Injector 6 Control	1.6-2.9 ms
6	WT/BL	Injector 5 Control	1.6-2.9 ms
7	YL/RD	Injector 4 Control	1.6-2.9 ms
8	WT/GN	Injector 3 Control	1.6-2.9 ms
9	WT	Injector 2 Control	1.6-2.9 ms
10	WT/RD	Injector 1 Control	1.6-2.9 ms
11	PK/YL	A/T-ECT Solenoid (S1)	3rd or OD: 1v
12	BK/YL	Igniter Signal (IGF)	Digital Signal: 0-5-0v
13	BK/WT	Starter Switch Signal	9-11v (cranking)
14	BK	Neutral Start Switch	In P/N: 9-11v (cranking)
15	RD/WT	HO2S-12 (B1 S2) Heater	Heater Off: 12v, On: 1v
16	PK/GN	HO2S-11 (B1 S1) Heater	Heater Off: 12v, On: 1v
17	GN	A/T-ECT Solenoid (S2)	1st or OD: 1v
22	BK/RD	IAC Signal (RSC)	Pulse Signals
23	BR/RD	IAC Signal (RSO)	Pulse Signals
24	BK/BL	Igniter Transistor 1 Control	Digital Signal: 0-5-0v
25	BR/YL	Igniter Transistor 2 Control	Digital Signal: 0-5-0v
26	BK/WT	Igniter Transistor 3 Control	Digital Signal: 0-5-0v
27	RD/GN	A/T-ECT Solenoid (SL)	In Lockup: 12-14v
29	GN	4WD Detection Transfer (L4)	Open: 12v, Closed: 0v
32	YL/BL	Closed Throttle Switch	1v, off-idle: 12v
33	BR	Power Ground	<0.1v
34	BR	Power Ground	<0.1v

Pin Connector Graphic

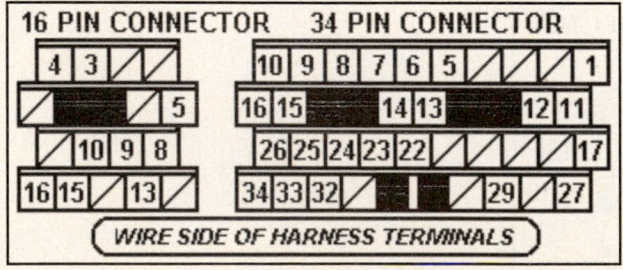

1996 Pickup 3.4L V6 MFI 5VZ-FE VIN N (M/T) E5 22 Pin Connector

PCM Pin #	Wire Color	Circuit Description (22 Pin)	Value at Hot Idle
1, 3, 5	---	Not Used	---
2	BK/YL	Battery Direct	12-14v
4	PK	MIL (lamp) Control	MIL Off: 12v, On: 1v
6	BL/BK	A/C Amplifier Signal (ACT)	Clutch On: 12v, Off: 1.5v
7	BL/YL	A/C Amplifier Signal (AC1)	Clutch On: 1.5v, Off: 12v
8	GN/OR	Vehicle Speed Sensor	At 55 mph: 48 Hz
9	GN/WT	4WD Selector Switch	Open: 12v, Closed: 0v
10, 13-18	---	Not Used	---
11	BK/WT	Starter Switch Signal	9-11v (cranking)
12	WT/RD	EFI Main Relay (B+)	12-14v
19	WT	Data Link Connector (SDL)	0v
20	GN/WT	Stop Light Switch Signal	Brake Off: 0v, On: 12v
21-22	---	Not Used	---

1996 Pickup 3.4L V6 MFI 5VZ-FE VIN N (M/T) E7 16 Pin Connector

PCM Pin #	Wire Color	Circuit Description (16 Pin)	Value at Hot Idle
1	GN/BK	Sensor VREF	4.9-5.1v
2	GY/RD	MAF Sensor Signal (VG)	1.1-1.8v
3	GY	Knock Sensor 2 Signal	<0.075v AC
4	GN/RD	ECT Sensor Signal (THW)	At 180ºF: 0.51v
5	WT	HO2S-11 (B1 S1) Signal	0.1-1.1v
6	BK	Knock Sensor 1 Signal	<0.075v AC
7	RD	Data Link Connector	12v
8	BR/WT	MAF Sensor Ground (EVG)	<0.050v
9	BR/BK	Sensor Ground	<0.050v
10	YL	TP Sensor Signal (VTA)	0.3-0.8v
11	YL/BL	Closed Throttle Switch	1v, off-idle: 12v
12	YL/GN	IAT Sensor Signal (THA)	At 100ºF: 2.60v
13	BK	HO2S-12 (B1 S2) Signal	0.1-1.1v
14	PK/GN	EGR Gas Temperature Sensor (THG)	3.5-4.0v
15-16	---	Not Used	---

1996 Pickup 3.4L V6 MFI 5VZ-FE VIN N (M/T) E6 12 Pin Connector

PCM Pin #	Wire Color	Circuit Description (12 Pin)	Value at Hot Idle
1-2	---	Not Used	---
3	PK/GN	HO2S-11 (B1 S1) Heater	Heater Off: 12v, On: 1v
4	RD/YL	EVAP Vapor Pressure Sensor	2.5-3.1v (with fuel cap off)
5	BL	CMP Sensor Signal (GN-)	<0.050v
6	GN	CKP Sensor Signal (NE-)	<0.050v
7	BR	Power Ground	<0.1v
8	---	Not Used	---
9	RD/GN	HO2S-12 (B1 S2) Heater	Heater Off: 12v, On: 1v
11	YL	CMP Sensor Signal (G2+)	AC pulse signals
12	RD	CKP Sensor Signal (NE+)	AC pulse signals

1996 Pickup 3.4L V6 MFI 5VZ-FE VIN N (M/T) E7 26 Pin Connector

PCM Pin #	Wire Color	Circuit Description (26 Pin)	Value at Hot Idle
2	BK/RD	A/C Idle-Up Solenoid	A/C Off: 12v, On: 1v
3	WT/GN	EVAP Purge Solenoid (VSV)	12v or 0v
5	GY/GN	EVAP Vapor Pressure (VSV)	12v or 0v
6	BK/RD	IAC Signal (RSC)	Pulse Signals
7	BR/RD	IAC Signal (RSO)	Pulse Signals
8	YL/BK	Injector 6 Control	1.6-2.9 ms
9	WT/BL	Injector 5 Control	1.6-2.9 ms
10	YL/RD	Injector 4 Control	1.6-2.9 ms
11	WT	Injector 2 Control	1.6-2.9 ms
12	WT/RD	Injector 1 Control	1.6-2.9 ms
13	BR	Power Ground	<0.1v
14	GN/YL	Circuit Opening Relay (FC)	0-3v, at off-idle: 12v
17	BK/YL	Igniter Signal (IGF)	Digital Signal: 0-5-0v
18	RD/WT	EGR Solenoid Control (VSV)	12v or 0v
19	RD/BK	Fuel Pressure Up Solenoid	1v (at hot startup)
21	BK/WT	Igniter Transistor 3 Control	Digital Signal: 0-5-0v
22	BR/YL	Igniter Transistor 2 Control	Digital Signal: 0-5-0v
23	BK/BL	Igniter Transistor 1 Control	Digital Signal: 0-5-0v
24	BR	Shield Ground	<0.050v
25	WT/GN	Injector 3 Control	1.6-2.9 ms
26	BR	Power Ground	<0.1v

Pin Connector Graphic

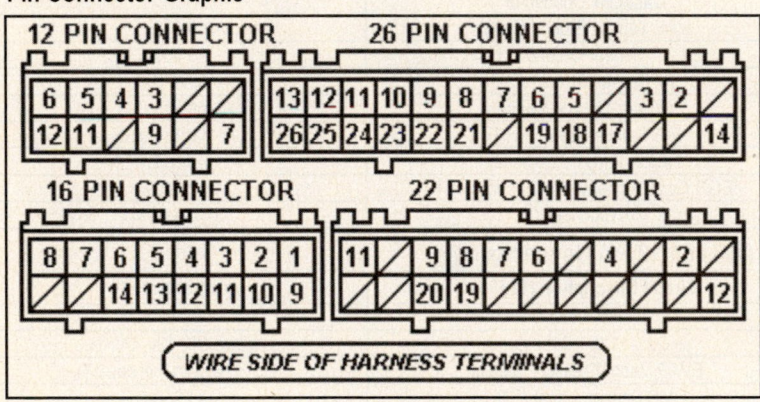

1997 Pickup 3.4L V6 MFI VIN N (A/T-ECT) 28 Pin Connector

PCM Pin #	Wire Color	Circuit Description (28 Pin)	Value at Hot Idle
1	RD/BK	A/T Select Switch Reverse	In 'R': 12v, Others: 0v
2	PK	A/T Select Switch 2nd	In 2nd: 12v, Others: 0v
3	PK/WT	A/T Select Switch Low	In Low: 12v, Others: 0v
4	PK	A/T Oil Temperature Lamp	Lamp Off: 12v, On: 1v
5	BL/BK	A/C Amplifier Signal (ACT)	Clutch On: 12v, Off: 1.5v
6	BL/OR	Overdrive Main Switch	Switch Off: 12v, On: 1v
7	BL	Cruise Control ECU	At Cruise in OD: 12v
10	GN/OR	A/T Pattern Select Switch	Norm: 0v, PWR: 12v
12	GN/OR	Vehicle Speed Sensor	At 55 mph: 48 Hz
14	BK/YL	Battery Direct	12-14v
17	RD/GN	4WD Detection Transfer (N)	Open: 12v, Closed
18	WT	SIL (Scan Tool) Signal	12v
20	BL/YL	A/C Amplifier Signal (AC1)	Clutch On: 1.5v, Off: 12v
22	WT/RD	EFI Main Relay (B+)	12-14v
25	GN/WT	Stop Light Switch Signal	Brake Off: 0v, On: 12v
26	GN/WT	4WD Selector Switch	Switch Off: 12v, On: 1v

1997 Pickup 3.4L V6 MFI VIN N (A/T-ECT) 22 Pin Connector

PCM Pin #	Wire Color	Circuit Description (22 Pin)	Value at Hot Idle
1	GN/BK	Sensor VREF	4.9-5.1v
4	WT/RD	A/T Vehicle Speed Sensor (-)	AC pulse signals
5	RD	CKP Sensor Signal (NE+)	AC pulse signals
6	GN	CKP Sensor Signal (NE-)	<0.050v
7	YL	TP Sensor Signal (VTA)	0.3-0.8v
8	GY/RD	MAF Sensor Signal (VG)	1.1-1.8v
9	YL/RD	A/T Vehicle Speed Sensor (+)	AC pulse signals
10	YL	CMP Sensor Signal (G2+)	AC pulse signals
11	BL	CMP Sensor Signal (GN-)	<0.050v
12	YL/RD	A/T Oil Temperature Sensor	At 68ºF: 4-5v
13	WT	HO2S-11 (B1 S1) Signal	0.1-1.1v
14	YL/GN	IAT Sensor Signal (THA)	At 100ºF: 2.60v
15	RD/YL	EVAP Vapor Pressure Sensor	2.5-3.1v (with fuel cap off)
16	GY	Knock Sensor 2 Signal	<0.075v AC
17	BK	Knock Sensor 1 Signal	<0.075v AC
18, 22	BR/WT	Sensor Ground	<0.050v
19	BK	HO2S-12 (B1 S2) Signal	0.1-1.1v
20	GN/RD	ECT Sensor Signal (THW)	At 180ºF: 0.51v
21	PK/GN	EGR Gas Temperature Sensor (THG)	3.5-4.0v

Pin Connector Graphic

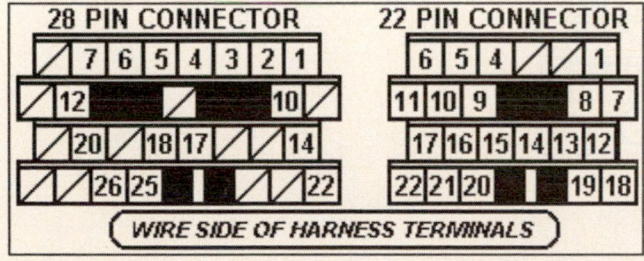

1997 Pickup 3.4L V6 MFI VIN N (A/T-ECT) 16 Pin Connector

PCM Pin #	Wire Color	Circuit Description (16 Pin)	Value at Hot Idle
1	BL	Cruise Control ECU	At Cruise in OD: 12v
3	PK	MIL (lamp) Control	MIL Off: 12v, On: 1v
4	GN/YL	Circuit Opening Relay (FC)	0-3v, at off-idle: 12v
5	PK/WT	Data Link Connector	12-14v
8	RD/WT	EGR Solenoid Control (VSV)	12v or 0v
13	GY/GN	EVAP Vapor Pressure (VSV)	12v or 0v
15	WT/GN	EVAP Purge Solenoid (VSV)	12v or 0v
16	BR	Shield Ground	<0.050v

1997 Pickup 3.4L V6 MFI VIN N (A/T-ECT) 34-Pin Connector

PCM Pin #	Wire Color	Circuit Description (34-Pin)	Value at Hot Idle
1	BR	Power Ground	<0.1v
5	YL/BK	Injector 6 Control	1.6-2.9 ms
6	WT/BL	Injector 5 Control	1.6-2.9 ms
7	YL/RD	Injector 4 Control	1.6-2.9 ms
8	WT/GN	Injector 3 Control	1.6-2.9 ms
9	WT	Injector 2 Control	1.6-2.9 ms
10	WT/RD	Injector 1 Control	1.6-2.9 ms
11	PK/YL	A/T-ECT Solenoid (S1)	3rd or OD: 12-14v
12	BK/YL	Igniter Signal (IGF)	Digital Signal: 0-5-0v
13	BK/WT	Starter Switch Signal	9-11v (cranking)
14	BK	Neutral Start Switch	In P/N: 9-11v (cranking)
15	RD/WT	HO2S-12 (B1 S2) Heater	Heater Off: 12v, On: 1v
16	PK/GN	HO2S-11 (B1 S1) Heater	Heater Off: 12v, On: 1v
17	GN	A/T-ECT Solenoid (S2)	1st or OD: 12-14v
22	BK/RD	IAC Signal (RSC)	Pulse Signals
23	BR/RD	IAC Signal (RSO)	Pulse Signals
24	BK/BL	Igniter Transistor 1 Control	Digital Signal: 0-5-0v
25	BR/YL	Igniter Transistor 2 Control	Digital Signal: 0-5-0v
26	BK/WT	Igniter Transistor 3 Control	Digital Signal: 0-5-0v
27	RD/GN	A/T-ECT Solenoid (SL)	In Lockup: 12-14v
29	GN	4WD Detection Transfer (L4)	Open: 12v, Closed: 0v
33	BR	Power Ground	<0.1v
34	BR	Power Ground	<0.1v

Pin Connector Graphic

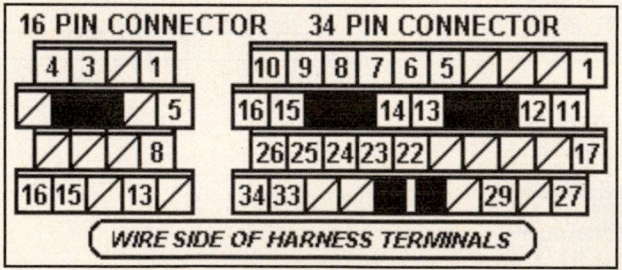

1997 Pickup 3.4L V6 MFI 5VZ-FE VIN N (M/T) 22 Pin Connector

PCM Pin #	Wire Color	Circuit Description (22 Pin)	Value at Hot Idle
2	BK/YL	Battery Direct	12-14v
4	PK	MIL (lamp) Control	MIL Off: 12v, On: 1v
6	BL/BK	A/C Amplifier Signal (ACT)	Clutch On: 12v, Off: 1.5v
7	BL/YL	A/C Amplifier Signal (AC1)	Clutch On: 1.5v, Off: 12v
8	GN/OR	Vehicle Speed Sensor	At 55 mph: 48 Hz
9	GN/WT	4WD Selector Switch	Switch Off: 12v, On: 1v
11	BK/WT	Starter Switch Signal	9-11v (cranking)
12	WT/RD	EFI Main Relay (B+)	12-14v
19	WT	SIL (Scan Tool) Signal	12v
20	GN/WT	Stop Light Switch Signal	Brake Off: 0v, On: 12v

1997 Pickup 3.4L V6 MFI 5VZ-FE VIN N (M/T) 16 Pin Connector

PCM Pin #	Wire Color	Circuit Description (16 Pin)	Value at Hot Idle
1	GN/BK	Sensor VREF	4.9-5.1v
2	GY/RD	MAF Sensor Signal (VG)	1.1-1.8v
3	GY	Knock Sensor 2 Signal	<0.075v AC
4	GN/RD	ECT Sensor Signal (THW)	At 180°F: 0.51v
5	WT	HO2S-11 (B1 S1) Signal	0.1-1.1v
6	BK	Knock Sensor 1 Signal	<0.075v AC
7	RD	Data Link Connector	12-14v
8	BR/WT	MAF Sensor Ground	<0.050v
9	BR/BK	Sensor Ground	<0.050v
10	YL	TP Sensor Signal (VTA)	0.3-0.8v
12	YL/GN	IAT Sensor Signal (THA)	At 100°F: 2.60v
13	BK	HO2S-12 (B1 S2) Signal	0.1-1.1v
14	PK/GN	EGR Gas Temperature Sensor (THG)	3.5-4.0v

1997 Pickup 3.4L V6 MFI 5VZ-FE VIN N (M/T) 12 Pin Connector

PCM Pin #	Wire Color	Circuit Description (12 Pin)	Value at Hot Idle
3	PK/GN	HO2S-11 (B1 S1) Heater	Heater Off: 12v, On: 1v
4	RD/YL	EVAP Vapor Pressure Sensor	2.5-3.1v (with fuel cap off)
5	BL	CMP Sensor Signal (GN-)	<0.050v
6	GN	CKP Sensor Signal (NE-)	<0.050v
7	BR	Power Ground	<0.1v
9	RD/GN	HO2S-12 (B1 S2) Heater	Heater Off: 12v, On: 1v
11	YL	CMP Sensor Signal (G2+)	AC pulse signals
12	RD	CKP Sensor Signal (NE+)	AC pulse signals

Standard Colors and Abbreviations

Abbreviation	Color	Abbreviation	Color	Abbreviation	Color
BK	Black	GY	Gray	RD	Red
BL	Blue	GN	Green	TN	Tan
BR	Brown	LG	LT Green	VT	Violet
DB	Dark Blue	OR	Orange	WT	White
DG	DK Green	PK	Pink	YL	Yellow

1997 Pickup 3.4L V6 MFI 5VZ-FE VIN N (M/T) 26 Pin Connector

PCM Pin #	Wire Color	Circuit Description (26 Pin)	Value at Hot Idle
3	WT/GN	EVAP Vapor Pressure (VSV)	12v or 0v
4	BL	Cruise Control ECU	At Cruise in OD: 12v
5	WT/GN	EVAP Purge Solenoid (VSV)	12v or 0v
6	BK/RD	IAC Signal (RSC)	Pulse Signals
7	BR/RD	IAC Signal (RSO)	Pulse Signals
8	YL/BK	Injector 6 Control	1.6-2.9 ms
9	WT/BL	Injector 5 Control	1.6-2.9 ms
10	YL/RD	Injector 4 Control	1.6-2.9 ms
11	WT	Injector 2 Control	1.6-2.9 ms
12	WT/RD	Injector 1 Control	1.6-2.9 ms
13	BR	Power Ground	<0.1v
14	GN/YL	Circuit Opening Relay (FC)	0-3v, at off-idle: 12v
17	BK/YL	Igniter Signal (IGF)	Digital Signal: 0-5-0v
18	RD/WT	EGR Solenoid Control (VSV)	12v or 0v
21	BK/WT	Igniter Transistor 3 Control	Digital Signal: 0-5-0v
22	BR/YL	Igniter Transistor 2 Control	Digital Signal: 0-5-0v
23	BK/BL	Igniter Transistor 1 Control	Digital Signal: 0-5-0v
24	BR	Shield Ground	<0.050v
25	WT/GN	Injector 3 Control	1.6-2.9 ms
26	BR	Power Ground	<0.1v

Pin Connector Graphic

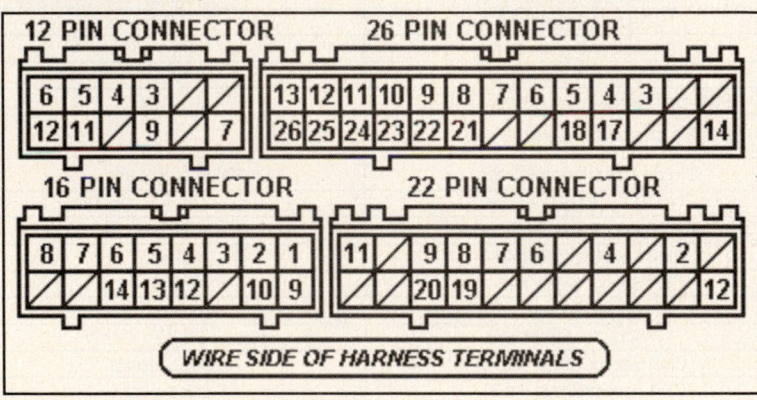

1998-99 Pickup 3.4L V6 MFI VIN N (A/T-ECT) 28 Pin Connector

PCM Pin #	Wire Color	Circuit Description (28 Pin)	Value at Hot Idle
1	PK/WT	A/T Select Switch Low	In Low: 12v, Others: 0v
5	BL/BK	A/C Amplifier Signal (ACT)	Clutch On: 12v, Off: 1.5v
6	BL/OR	Overdrive Main Switch	Switch Off: 12v, On: 1v
7	RD/YL	Cruise Control ECU (OD1)	At Cruise in OD: 12v
8	WT	Data Link Connector	12-14v
10	PK	A/T Select Switch 2nd	In 2nd: 12v, Others: 0v
11	GN/RD	Cruise Control ECU (IDLO)	1.5v, off-idle: 12v
12	GN/OR	Speedometer Indicator	At 55 mph: 48 Hz
14	BK/YL	Battery Direct	12-14v
15	RD/BK	A/T Select Switch Reverse	In 'R': 12v, Others: 0v
16	BL/YL	A/C Amplifier Signal (AC1)	Clutch On: 1.5v, Off: 12v
23	WT/RD	EFI Main Relay (B+)	12-14v
24	GN/WT	Stop Light Switch Signal	Brake Off: 0v, On: 12v

1998-99 Pickup 3.4L V6 MFI VIN N (A/T-ECT) 22 Pin Connector

PCM Pin #	Wire Color	Circuit Description (22 Pin)	Value at Hot Idle
1	GN/BK	Sensor VREF	4.9-5.1v
4	WT/RD	A/T Vehicle Speed Sensor (-)	AC pulse signals
5	RD	CKP Sensor Signal (NE+)	AC pulse signals
6	GN	CKP/CMP Sensor Signal (GN-)	<0.050v
7	YL	TP Sensor Signal (VTA)	0.3-0.8v
8	GY/RD	MAF Sensor Signal (VG)	1.1-1.8v
9	YL/RD	A/T Vehicle Speed Sensor (+)	AC pulse signals
12	GN	A/T Pattern Select Switch	Norm: 0v, PWR: 12v
13	WT	HO2S-11 (B1 S1) Signal	0.1-1.1v
14	GY	Knock Sensor 2 Signal	<0.075v AC
15	BK	Knock Sensor 1 Signal	<0.075v AC
17	YL	CMP Sensor Signal (G2+)	AC pulse signals
18	GN/YL	Circuit Opening Relay (FC)	0-3v, at off-idle: 12v
19	BK	HO2S-12 (B1 S2) Signal	0.1-1.1v
20	GN/RD	ECT Sensor Signal (THW)	At 180°F: 0.51v
21	YL/GN	IAT Sensor Signal (THA)	At 100°F: 2.60v
22	BR/BK	Sensor Ground	<0.050v

Pin Connector Graphic

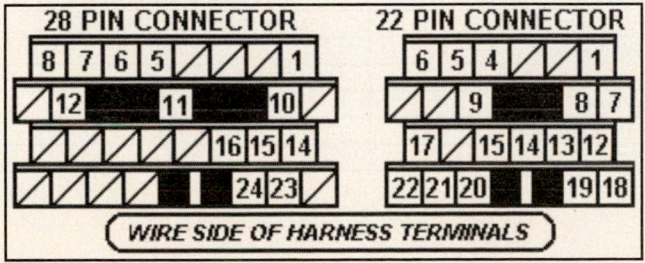

1998-99 Pickup 3.4L V6 MFI VIN N (A/T-ECT) 16 Pin Connector

PCM Pin #	Wire Color	Circuit Description (16 Pin)	Value at Hot Idle
2	WT/GN	EVAP Purge Solenoid (VSV)	12v or 0v
3	PK/RD	MIL (lamp) Control	MIL Off: 12v, On: 1v
5	RD	Data Link Connector	12-14v
6	PK	A/T Oil Temperature Lamp	Lamp Off: 12v, On: 1v
7	BR/WT	MAF Sensor Ground (E2G)	<0.050v
8	GY/GN	EVAP Vapor Pressure (VSV)	12v or 0v
9	RD/GN	4WD Detection Transfer (N)	Open: 12v, Closed
10	RD/WT	HO2S-12 (B1 S2) Heater	Heater Off: 12v, On: 1v
11	PK/GN	HO2S-11 (B1 S1) Heater	Heater Off: 12v, On: 1v
12	RD/WT	EGR Solenoid Control (VSV)	12v or 0v
13	RD/YL	EVAP Vapor Pressure Sensor	2.5-3.1v (with fuel cap off)
14	PK/GN	EGR Gas Temperature Sensor (THG)	3.5-4.0v
16	BR	Shield Ground	<0.050v

1998-99 Pickup 3.4L V6 MFI VIN N (A/T-ECT) 34-Pin Connector

PCM Pin #	Wire Color	Circuit Description (34-Pin)	Value at Hot Idle
1	RD/WT	4WD Selector Switch	Switch Off: 12v, On: 1v
2	GN	4WD Detection Transfer (L4)	Open: 12v, Closed: 0v
2	GY	4WD Detection Transfer (L4)	Open: 12v, Closed: 0v
5	BL	Injector 6 Control	1.6-2.9 ms
6	WT/BL	Injector 5 Control	1.6-2.9 ms
7	BL/RD	Injector 4 Control	1.6-2.9 ms
8	WT/GN	Injector 3 Control	1.6-2.9 ms
9	WT	Injector 2 Control	1.6-2.9 ms
10	WT/RD	Injector 1 Control	1.6-2.9 ms
11	PK/YL	A/T-ECT Solenoid (S1)	3rd or OD: 1v
12	BK/YL	Igniter Signal (IGF)	Digital Signal: 0-5-0v
13	BK/WT	Starter Switch Signal	9-11v (cranking)
14	BK/RD	Neutral Start Switch	In P/N: 9-11v (cranking)
15	BK/WT	Igniter Transistor 3 Control	Digital Signal: 0-5-0v
16	BR/YL	Igniter Transistor 2 Control	Digital Signal: 0-5-0v
17	GN	A/T-ECT Solenoid (S2)	1st or OD: 1v
22	BK/RD	IAC Signal (RSC)	Pulse Signals
23	BR/RD	IAC Signal (RSO)	Pulse Signals
24	BK/BL	Igniter Transistor 1 Control	Digital Signal: 0-5-0v
27	RD/GN	A/T-ECT Solenoid (SL)	In Lockup: 12-14v
28	BR	Power Ground	<0.1v
30	YL/RD	A/T Oil Temperature Sensor	At 68°F: 4-5v
31	YL	PSP Switch Signal	Straight: 12v, Turning: 0v
33	BR	Power Ground	<0.1v
34	BR	Power Ground	<0.1v

2000 Pickup 3.4L V6 DOHC VIN N California E3 22 Pin Connector

PCM Pin #	Wire Color	Circuit Description (22 Pin)	Value at Hot Idle
1	BK/RD	Direct Battery	12-14v
2	BK/OR	Ignition Switch Power	12-14v
3	PK	Circuit Opening Relay (FC)	0-3v, at off-idle: 12v
6	PK/GN	MIL (lamp) Control	MIL Off: 12v, On: 1v
7	BK/WT	Starter Switch Signal	9-11v (cranking)
8	BK/YL	EFI Main Relay	12-14v
9	GY/RD	EVAP Vapor Pressure (VSV)	12v or 0v
11	WT	SIL (Scan Tool) Signal	12v
15	GN/WT	Stop Light Switch Signal	Brake Off: 0v, On: 12v
16	WT/BL	EFI Main Relay (B+)	12-14v
17	RD/GN	EVAP Vapor Pressure Sensor	2.9-3.1v (with fuel cap off)

2000 Pickup 3.4L V6 DOHC VIN N California E6 31 Pin Connector

PCM Pin #	Wire Color	Circuit Description (31 Pin)	Value at Hot Idle
1	GN	Injector 3 Control	1.6-2-9 ms
2	RD/BK	Injector 4 Control	1.6-2-9 ms
3	BL	Injector 5 Control	1.6-2-9 ms
4	YL	Injector 6 Control	1.6-2-9 ms
6	RD	Data Link Connector	12-14v
7	RD	A/T-ECT Solenoid (S1)	3rd or OD: 1v
8	WT/BL	A/T-ECT Solenoid (S2)	1st or OD: 1v
9	GN/RD	A/T-ECT Solenoid (SL)	In Lockup: 12-14v
10	BL	CMP Sensor Signal (G2)	AC pulse signals
11	BK/BL	Igniter Transistor 1 Control	6°, 55 mph: 8° dwell
12	RD/BL	Igniter Transistor 2 Control	6°, 55 mph: 8° dwell
13	LG	Igniter Transistor 3 Control	6°, 55 mph: 8° dwell
14	YL/RD	A/T Vehicle Speed Sensor (+)	AC pulse signals
15	BK/RD	IAC Signal (RSC)	Pulse Signals
16	RD/WT	IAC Signal (RSO)	Pulse Signals
21	BR	Power Ground	<0.1v
25	BK/YL	Igniter Signal (IGF)	Digital Signal: 0-5-0v
26	WT/RD	A/T Vehicle Speed Sensor (-)	AC pulse signals
27	BK	Knock Sensor 1 Signal	<0.075v AC
28	GY	Knock Sensor 2 Signal	<0.075v AC
30	WT/BK	Power Ground (E03)	<0.1v
31	WT/BK	Power Ground (E02)	<0.1v

Pin Connector Graphic

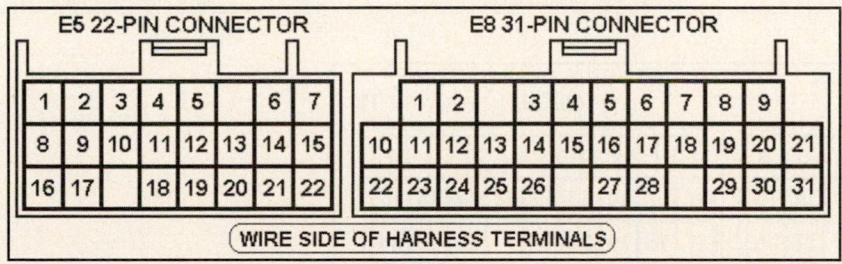

2000 Pickup 3.4L V6 DOHC VIN N California E4 28 Pin Connector

PCM Pin #	Wire Color	Circuit Description (28 Pin)	Value at Hot Idle
2	RD/BK	A/T Select Switch Reverse	In 'R': 12v, Others: 0v
3	BL	A/T Select Switch 2nd	In 2nd: 12v, Others: 0v
4	YL/GN	Cruise Control ECU (IDLO)	1.5v, off-idle: 12v
7	YL/RD	A/T Oil Temperature Indicator	Lamp Off: 12v, On: 1v
8	RD	HO2S-12 (B1 S2) Signal	0.1-1.1v
9	RD/WT	HO2S-12 (B1 S2) Heater	Heater Off: 12v, On: 1v
10	BL/OR	O/D "Off" Indicator (ODLP)	Lamp Off: 12v, On: 1v
12	LG	A/T Select Switch Low	In Low: 12v, Others: 0v
13	BL/BK	A/C Amplifier Signal (ACT)	Clutch On: 12v, Off: 1.5v
17	RD/YL	4WD Detection Transfer (N)	Open: 12v, Closed: 0v
18	GY	4WD ADD Indicator Switch	Open: 12v, Closed: 0v
19	BL/RD	4WD Detection Transfer (L4)	Open: 12v, Closed: 0v
20	PK	Starter Switch Signal	9-11v (cranking)
22	GN/OR	Speedometer Indicator	At 55 mph: 48 Hz
24	BR/YL	Cruise Control ECU (OD1)	At Cruise in OD: 12v
25	BL/YL	A/C Amplifier Signal (AC1)	Clutch On: 1.5v, Off: 12v

2000 Pickup 3.4L V6 DOHC VIN N California E5 24 Pin Connector

PCM Pin #	Wire Color	Circuit Description (24 Pin)	Value at Hot Idle
2	GN/BK	Sensor VREF	4.9-5.1v
4	YL	HAFR-11 (B1 S1) Heater	Heater Off: 12v, On: 1v
5	RD	Injector 1 Control	1.6-2-9 ms
6	WT	Injector 2 Control	1.6-2-9 ms
7	WT/GN	EVAP Purge Solenoid (VSV)	12v or 0v
8	BR	Power Ground	<0.1v
9	YL/RD	PSP Switch Signal	Straight: 12v, Turning: 0v
10	RD/WT	MAF Sensor Signal (VG)	1-1.1v
12	GN	AFS-11 (B1 S1) Signal (AF+)	Fixed at 3.3v
14	GN	ECT Sensor Signal (THW)	0.5-0.6v
15	RD/YL	A/T Oil Temperature Sensor	At 68°F: 4-5v
16	PK	CKP Sensor Signal (NE+)	AC pulse signals
17	BR	Shield Ground (E1)	<0.050v
18	LG	Sensor Ground (E2)	<0.050v
19	BK/WT	MAF Sensor Ground (E2G)	<0.050v
21	RD	AFS-11 (B1 S1) Signal (AF-)	Fixed at 3.3v
22	YL/GN	IAT Sensor Signal (THA)	0.5-3.4v
23	BK/YL	TP Sensor Signal (VTA)	0.53-1.27v
24	PK	CMP & CKP Sensor Return	<0.050v

Pin Connector Graphic

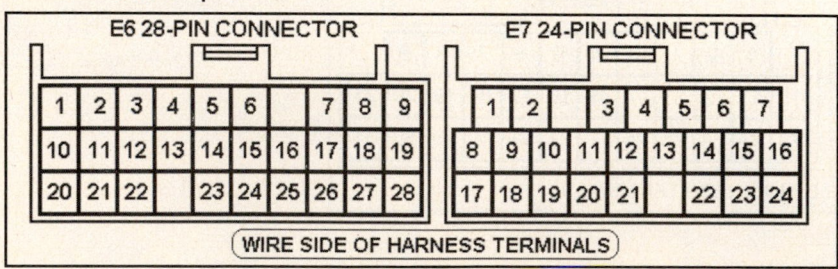

2000 Pickup 3.4L V6 MFI VIN N Federal E3 28 Pin Connector

PCM Pin #	Wire Color	Circuit Description (28 Pin)	Value at Hot Idle
1	LG	A/T Select Switch Low	In Low: 12v, Others: 0v
2	BL/OR	Overdrive Main Switch	Switch Off: 12v, On: 1v
5	BL/BK	A/C Amplifier Signal (ACT)	Clutch On: 12v, Off: 1.5v
7	BR/YL	Cruise Control ECU (OD1)	At Cruise in OD: 12v
8	WT	SIL (Scan Tool) Signal	12v
10	BL	A/T Select Switch 2nd	In 2nd: 12v, Others: 0v
11	YL/GN	Cruise Control ECU (IDLO)	1.5v, off-idle: 12v
12	GN/OR	Speedometer Indicator	At 55 mph: 48 Hz
14	BK/RD	Direct Battery	12-14v
15	RD/BK	A/T Select Switch Reverse	In 'R': 12v, Others: 0v
16	BL/YL	A/C Amplifier Signal (AC1)	Clutch On: 1.5v, Off: 12v
23	WT/BL	EFI Main Relay (B+)	12-14v
24	GN/WT	Stop Light Switch Signal	Brake Off: 0v, On: 12v

2000 Pickup 3.4L V6 MFI VIN N Federal E4 16 Pin Connector

PCM Pin #	Wire Color	Circuit Description (16 Pin)	Value at Hot Idle
2	WT/GN	EVAP Purge Solenoid (VSV)	12v or 0v
3	PK/GN	MIL (lamp) Control	MIL Off: 12v, On: 1v
5	RD	Data Link Connector	12-14v
6	YL/RD	A/T Oil Temperature Sensor	At 68°F: 4-5v
7	BK/WT	MAF Sensor Ground (E2G)	<0.050v
8	GY/RD	EVAP Vapor Pressure (VSV)	12v or 0v
9	RD/YL	4WD Detection Transfer (N)	Switch Closed: 0v
10	RD/WT	HO2S-12 (B1 S2) Heater	Heater Off: 12v, On: 1v
11	PK/BL	HO2S-11 (B1 S1) Heater	Heater Off: 12v, On: 1v
13	RD/GN	EVAP Vapor Pressure Sensor	2.5-3.1v (with fuel cap off)
16	BR	Sensor Ground	<0.050v

Pin Connector Graphic

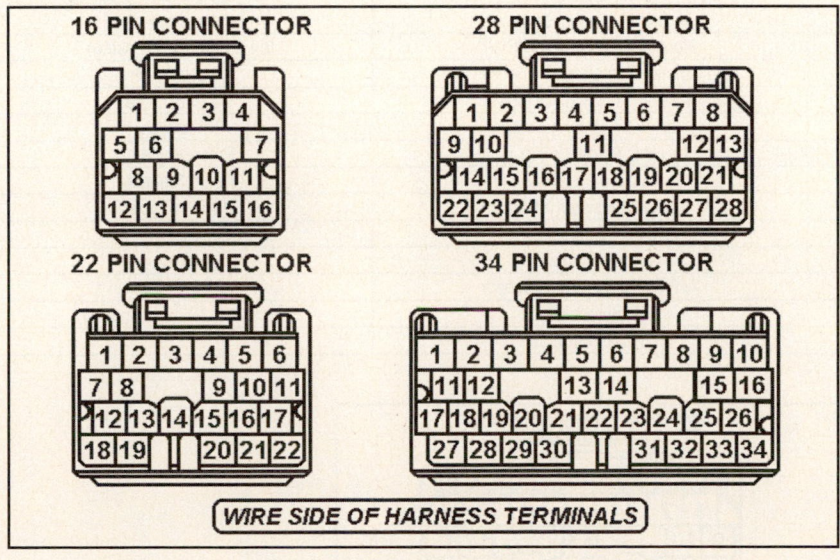

2000 Pickup 3.4L V6 MFI VIN N Federal E5 22 Pin Connector

PCM Pin #	Wire Color	Circuit Description (22 Pin)	Value at Hot Idle
1	GN/BK	Sensor VREF	4.9-5.1v
4	WT/RD	A/T Vehicle Speed Sensor (-)	AC pulse signals
5	PK	CKP Sensor Signal (NE+)	AC pulse signals
6	PK	CKP Sensor Signal (NE-)	<0.050v
7	BK/YL	TP Sensor Signal (VTA)	0.3-0.8v
8	RD/WT	MAF Sensor Signal (VG)	1.1-1.8v
9	YL/RD	A/T Vehicle Speed Sensor (+)	AC pulse signals
13	WT	HO2S-11 (B1 S1) Signal	0.1-1.1v
14	GY	Knock Sensor 2 Signal	<0.075v AC
15	BK	Knock Sensor 1 Signal	<0.075v AC
17	BL	CMP Sensor Signal (G2+)	AC pulse signals
18	YL	Circuit Opening Relay (FC)	0-3v, at off-idle: 12v
19	RD	HO2S-12 (B1 S2) Signal	0.1-1.1v
20	GN	ECT Sensor Signal (THW)	At 180°F: 0.51v
21	YL/GN	IAT Sensor Signal (THA)	At 100°F: 2.60v
22	GN/WT	Sensor Ground (E2)	<0.050v

2000 Pickup 3.4L V6 MFI VIN N Federal E6 34-Pin Connector

PCM Pin #	Wire Color	Circuit Description (34-Pin)	Value at Hot Idle
1	GY	4WD Detection Switch	Open: 12v, Closed: 0v
2	BL/RD	4WD Detection Transfer (L4)	Open: 12v, Closed: 0v
5	YL	Injector 6 Control	1.6-2.9 ms
6	BL	Injector 5 Control	1.6-2.9 ms
7	RD/BK	Injector 4 Control	1.6-2.9 ms
8	GN	Injector 3 Control	1.6-2.9 ms
9	WT	Injector 2 Control	1.6-2.9 ms
10	RD	Injector 1 Control	1.6-2.9 ms
11	RD	A/T-ECT Solenoid (S1)	In 3rd or OD: 1v
12	BK/YL	Igniter Signal (IGF)	Digital Signal: 0-5-0v
13	BK/WT	Starter Switch Signal	9-11v (cranking)
14	PK	Neutral Start Switch (NSW)	In P/N: 9-11v (cranking)
15	LG	Igniter Transistor 3 Control	7% duty cycle
16	RD/BL	Igniter Transistor 2 Control	7% duty cycle
17	WT/BL	A/T-ECT Solenoid (S2)	In 3rd or OD: 1v
22	BK/RD	IAC Signal (RSC)	Pulse Signals
23	RD/WT	IAC Signal (RSO)	Pulse Signals
24	BK/BL	Igniter Transistor 1 Control	7% duty cycle
27	GN/RD	A/T-ECT Solenoid (SL)	In Lockup: 12-14v
28	BR	Power Ground	<0.1v
30	PK/YL	A/T Oil Temperature Sensor	At 68°F: 4-5v
31	YL/RD	PSP Switch Signal	Straight: 12v, Turning: 0v
33, 34	BR	Power Ground	<0.1v

Pin Connector Graphic

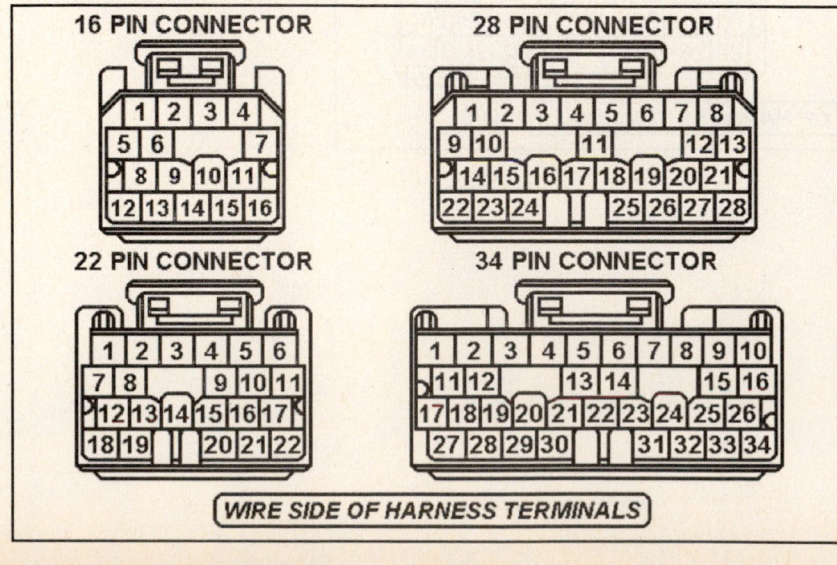

2001 Pickup 3.4L V6 DOHC MFI VIN N (All) E8 31 Pin Connector

PCM Pin #	Wire Color	Circuit Description (31 Pin)	Value at Hot Idle
1	GN	Injector 3 Control	1.6-2-9 ms
2	RD/BK	Injector 4 Control	1.6-2-9 ms
3	BL	Injector 5 Control	1.6-2-9 ms
4	YL	Injector 6 Control	1.6-2.9 ms
6	RD	DLC1 D1 Signal (TE1)	12v
7	VT	A/T-ECT Solenoid (S1)	3rd or OD: 1v
8	LG	A/T-ECT Solenoid (S2)	1st or OD: 1v
9	RD/WT	A/T-ECT Solenoid (SL)	In Lockup: 12-14v
10	RD	CMP Sensor Signal (G2)	AC pulse signals
11	BK/BL	Igniter Transistor 1 Control	6°, 55 mph: 8° dwell
12	GN/BK	Igniter Transistor 2 Control	6°, 55 mph: 8° dwell
13	BK/WT	Igniter Transistor 3 Control	6°, 55 mph: 8° dwell
14	YL/RD	A/T Vehicle Speed Sensor (+)	AC pulse signals
15	BK/RD	IAC Signal (RSC)	Pulse Signals
16	BL/BK	IAC Signal (RSO)	Pulse Signals
17-20	---	Not Used	---
21	WT/BK	Power Ground (E01)	<0.1v
22-24	---	Not Used	---
25	BK/YL	Igniter Signal (IGF)	Digital Signal: 0-5-0v
26	WT/RD	A/T Vehicle Speed Sensor (-)	AC pulse signals
27	BK	Knock Sensor 1 Signal	<0.075v AC
28	GY	Knock Sensor 2 Signal	<0.075v AC
29	---	Not Used	---
30	WT/BK	Power Ground (E03)	<0.1v
31	WT/BK	Power Ground (E02)	<0.1v

2001 Pickup 3.4L V6 DOHC MFI VIN N (All) E5 22 Pin Connector

PCM Pin #	Wire Color	Circuit Description (22 Pin)	Value at Hot Idle
1	BK/YL	Direct Battery	12-14v
2	BK/WT	Ignition Switch Power	12-14v
3	WT/BL	Circuit Opening Relay (FC)	0-3v, at off-idle: 12v
6	PK/GN	MIL (lamp) Control	MIL Off: 12v, On: 1v
7	VT/RD	Starter Switch Signal	9-11v (cranking)
8	BK/OR	EFI Main Relay Control	Relay Off: 0v, On: 12v
9	GN/BK	EVAP Vapor Pressure (VSV)	12v or 0v
10	---	Not Used	---
11	WT	SIL (Scan Tool) Signal	12v
12-14	---	Not Used	---
15	GN/WT	Stop Light Switch Signal	Brake Off: 0v, On: 12v
16	WT/RD	EFI Main Relay (B+)	12-14v
17	RD/YL	EVAP Vapor Pressure Sensor	2.9-3.1v (with fuel cap off)
18-22	---	Not Used	---

Pin Connector Graphic

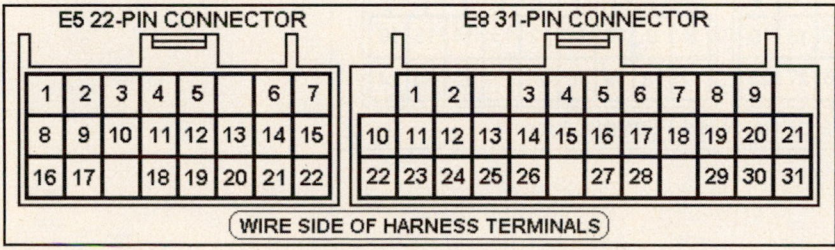

2001 Pickup 3.4L V6 DOHC MFI VIN N (All) E7 24 Pin Connector

PCM Pin #	Wire Color	Circuit Description (24 Pin)	Value at Hot Idle
2	GN/BK	Sensor VREF (VC)	4.9-5.1v
4	WT	AFS-11 (B1 S1) Heater	Heater Off: 12v, On: 1v
5	WT/RD	Injector 1 Control	1.6-2-9 ms
6	BK	Injector 2 Control	1.6-2-9 ms
7	WT/GN	EVAP Purge Solenoid (VSV)	12v or 0v
8	WT/BK	Power Ground	<0.1v
9	BK	PSP Switch Signal	Straight: 12v, Turning: 0v
10	GY	MAF Sensor Signal (VG)	1-1.1v
12	VT	AFS-11 (B1 S1) Signal (AF+)	Fixed at 3.3v
14	GN/RD	ECT Sensor Signal (THW)	0.5-0.6v
15	YL/RD	A/T Oil Temperature Sensor	At 68°F: 4-5v
16	BL	CKP Sensor Signal (NE+)	AC pulse signals
17	BR	Shield Ground (E1)	<0.050v
18	LG	Sensor Ground (E2)	<0.050v
19	BK/WT	MAF Sensor Ground (E2G)	<0.050v
21	PK	AFS-11 (B1 S1) Signal (AF-)	Fixed at 3.3v
22	YL/GN	IAT Sensor Signal (THA)	0.5-3.4v
23	YL	TP Sensor Signal (VTA)	0.53-1.27v
24	GN	CMP/CKP Sensor Ground (-)	<0.050v

2001 Pickup 3.4L V6 DOHC MFI VIN N (All) E6 28 Pin Connector

PCM Pin #	Wire Color	Circuit Description (28 Pin)	Value at Hot Idle
2	RD/BK	A/T Select Switch Reverse	In 'R': 12v, Others: 0v
3	PK	A/T Select Switch 2nd	In 2nd: 12v, Others: 0v
4	LG/RD	Cruise Control ECU (IDLO)	1.5v, off-idle: 12v
7	OR	A/T Oil Temperature Indicator	Lamp Off: 12v, On: 1v
8	BK	HO2S-12 (B1 S2) Signal	0.1-1.1v
9	RD/WT	HO2S-12 (B1 S2) Heater	Heater Off: 12v, On: 1v
10	BL/OR	O/D OFF Indicator (ODLP)	Lamp Off: 12v, On: 1v
12	RD	A/T Select Switch Low	In Low: 12v, Others: 0v
13	BL/BK	A/C Amplifier Signal (ACT)	Clutch On: 12v, Off: 1.5v
17	BL	A/T Park Indicator (TFN)	Lamp Off: 12v, On: 1v
18	GN/BK	4WD ADD Indicator Switch	Open: 12v, Closed: 0v
19	GY	4WD Detection Transfer (L4)	Open: 12v, Closed: 0v
20	BK/YL	Starter Switch Signal (NSW)	9-11v (cranking)
22	GN/OR	Speedometer Indicator (SPI)	At 55 mph: 48 Hz
24	RD/YL	Cruise Control ECU (OD1)	At Cruise in OD: 12v
25	BL/YL	A/C Amplifier Signal (AC1)	Clutch On: 1.5v, Off: 12v

Pin Connector Graphic

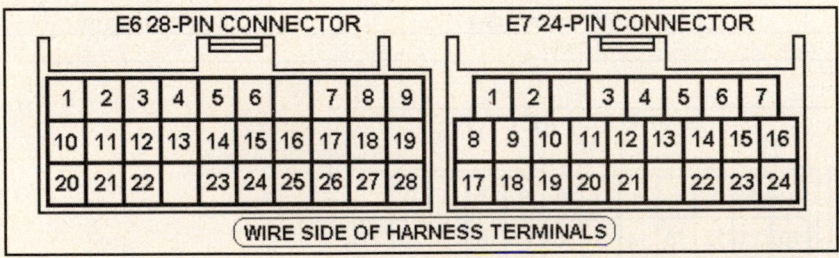

2002-03 Pickup 3.4L V6 DOHC MFI VIN N (All) E5 22 Pin Connector

PCM Pin #	Wire Color	Circuit Description (22 Pin)	Value at Hot Idle
1	BK/RD	EFI Main Relay Control (+B)	Relay Off: 0v, On: 12v
2	GN/OR	MIL (lamp) Control	MIL Off: 12v, On: 1v
3	BL/BK	A/C Amplifier Signal (ACT)	Clutch On: 12v, Off: 1.5v
5	---	Not Used	---
6	GN/OR	Speedometer Indicator (SPI)	At 55 mph: 48 Hz
7	GN	Starter Switch Signal (STA)	In P/N: 0-3v
8	---	Not Used	---
9	BL/YL	A/C Amplifier Signal (AC1)	Clutch On: 1.5v, Off: 12v
10-11	---	Not Used	---
13	YL/RD	Scan Tool (WFSE)	12v
14	WT	SIL (Scan Tool) Signal	12v
15	BK/WT	Ignition Switch (IGSW)	12-14v
16	BK/YL	Direct Battery (BATT)	12-14v
17	LG/RD	Cruise Control ECU (IDLO)	1.5v, off-idle: 12v
18-19	---	Not Used	---
20	GN/WT	Stop Light Switch Signal (STP)	Brake Off: 0v, On: 12v
21	---	Not Used	---
22	WT/BL	Circuit Opening Relay (FC)	0-3v, at off-idle: 12v

2002-03 Pickup 3.4L V6 DOHC MFI VIN N (All) E6 28 Pin Connector

PCM Pin #	Wire Color	Circuit Description (28 Pin)	Value at Hot Idle
1	GN	AT Pattern Select Switch	Norm: 0v, Power: 12v
2	GY	4WD Detection Transfer (L4)	Open: 12v, Closed: 0v
3	YL/GY	A/T Neutral Start Switch (NSW)	In P/N: 9-11v (while cranking)
4	RD	A/T Select Switch Low	In Low: 12v, Others: 0v
5	PK	A/T Select Switch 2nd	In 2nd: 12v, Others: 0v
6	RD/BK	A/T Select Switch Reverse	In 'R': 12v, Others: 0v
7-9	---	Not Used	---
10	RD/YL	Cruise Control ECU (OD1)	At Cruise in OD: 12v
11	BL	A/T Park Indicator (TFN)	Lamp Off: 12v, On: 1v
12	---	Not Used	---
13	GN/BK	4WD Switch Signal (4WD)	Open: 12v, Closed: 0v
14	GN	OD Main Switch (ODMS)	Switch Off: 12v, On: 0v
15-19	---	Not Used	---
20	OR	A/T Oil Temperature Indicator	Lamp Off: 12v, On: 1v
21-29	---	Not Used	---
30	WT/BK	Power Ground (E03)	<0.1v
31	WT/BK	Power Ground (E03)	<0.1v

2002-03 Pickup 3.4L V6 DOHC MFI VIN N (All) E10 17 Pin Connector

PCM Pin #	Wire Color	Circuit Description (17 Pin)	Value at Hot Idle
1	---	Not Used	---
2	YL	Throttle Control Motor (CL+)	Pulses
3	---	Not Used	---
4	VT	TP Sensor Signal (VTA2)	2.0-2.9v
5-7	---	Not Used	---
8	BL	Throttle Control Motor (CL-)	Pulses
9	---	Not Used	---
10	GR	Accelerator Position Sensor (VPA)	0.25-0.90v
15	BL	Accelerator Position Sensor (VPA2)	1.8-2.7v
16-17	---	Not Used	---

Pin Connector Graphic

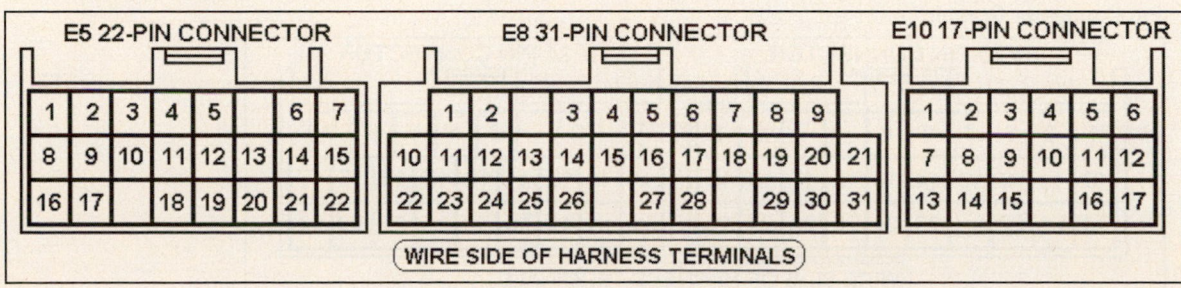

2002-03 Pickup 3.4L V6 DOHC MFI VIN N (All) E7 24 Pin Connector

PCM Pin #	Wire Color	Circuit Description (24 Pin)	Value at Hot Idle
1	WT/BL	Injector 5 Control	1.6-2-9 ms
2	BK/YL	Igniter Signal (IGF)	Digital Signal: 0-5-0v
3	BL/RD	Injector 4 Control	1.6-2-9 ms
4	RD	Injector 3 Control	1.6-2-9 ms
5	BK	Injector 2 Control	1.6-2-9 ms
6	WT/RD	Injector 1 Control	1.6-2-9 ms
7	YL/BK	Data Link Connector (TC)	12v
8	BL	Injector 6 Control	1.6-2-9 ms
9	BK/WT	Igniter Transistor 3 Control	6°, 55 mph: 8° dwell
10	LG/RD	Igniter Transistor 2 Control	6°, 55 mph: 8° dwell
11	BK/BL	Igniter Transistor 1 Control	6°, 55 mph: 8° dwell
12	BL	CKP Sensor Signal (NE+)	AC pulse signals
13	RD	CMP Sensor Signal (G2)	AC pulse signals
14	WT/BK	EVAP Purge Solenoid (VSV)	12v or 0v
16	BL/BK	IAC Signal (RSO)	DC pulse signals
17	BR	Power Ground (E1)	<0.1v
18	BK	PSP Switch Signal (PSW)	Straight: 12v, Turning: 0v
19	PK/BL	EVAP Purge Solenoid (VSV)	12v or 0v
20	GN/BK	EVAP Vapor Pressure Solenoid (VSV)	12v or 0v
21	GN	CMP/CKP Sensor Ground (-)	<0.050v
22	WT/RD	AT Vehicle Speed Sensor (-)	Moving: AC pulse signals
23	YL/RD	AT Vehicle Speed Sensor (+)	Moving: AC pulse signals
24	BK/RD	Idle Speed Control Valve (RSD)	DC pulse signals

2002-03 Pickup 3.4L V6 DOHC MFI VIN N (All) E8 31 Pin Connector

PCM Pin #	Wire Color	Circuit Description (31 Pin)	Value at Hot Idle
1	RD/WT	A/T-ECT Solenoid (SL)	In Lockup: 12-14v
2	LG	A/T-ECT Solenoid (S2)	1st or OD: 1v
3	VT	A/T-ECT Solenoid (S1)	3rd or OD: 1v
4, 5	WT/BK	Power Ground (E01), (E02)	<0.1v
6	WT/BK	Power Ground (ME01)	<0.1v
7, 8	WT/BK	Power Ground (E03), (E04)	<0.1v
9	RD	Throttle Control Motor (M+)	Pulse Signals
10	BK/WT	MAF Sensor Ground (E2G)	<0.050v
11	BL/BK	Sensor Ground (E2)	<0.050v
12	GN	MAF Sensor Signal (VG)	1-1.1v
13	YL/GN	IAT Sensor Signal (THA)	0.5-3.4v
14	VT	AFS-11 (B1 S1) Signal (AF+)	Fixed at 3.3v
15	YL	TP Sensor Signal (VTA)	0.53-1.27v
17	BR	Shield Ground (GE01)	<0.050v
18	GN/RD	ECT Sensor Signal (THW)	0.5-0.6v
19-20	---	Not Used	---
21	WT	AFS-11 (B1 S1) Heater (HAF1)	Heater Off: 12v, On: 1v
22	GY	Knock Sensor 2 Signal (KNK2)	<0.075v AC
23	BK	Knock Sensor 1 Signal (KNK1)	<0.075v AC
24	RD/YL	EVAP Vapor Pressure Sensor (PTNK)	2.9-3.1v (with fuel cap off)
25	GN/YL	Sensor VREF (VC)	4.9-5.1v
26	PK	AFS-11 (B1 S1) Signal (AF-)	Fixed at 3.3v
27	RD	HO2S-12 (B1 S2) Signal (OX2B)	0.1-1.1v
28, 30	---	Not Used	---
29	RD/WT	HO2S-12 (B1 S2) Heater (HT2B)	Heater Off: 12v, On: 1v
31	GN	Throttle Control Motor (M-)	Pulse Signals

Pin Connector Graphic

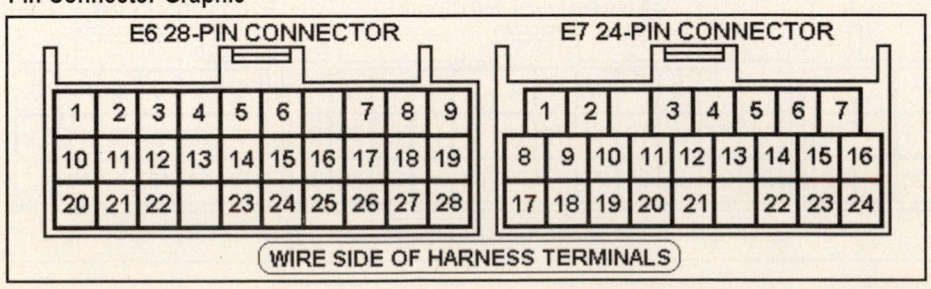

TUNDRA Pin Tables

2000 Pickup 3.4L V6 MFI California VIN N E3 22 Pin Connector

PCM Pin #	Wire Color	Circuit Description (22 Pin)	Value at Hot Idle
1	BK/RD	Direct Battery	12-14v
2	BK/OR	Ignition Switch Power	12-14v
3	YL	Circuit Opening Relay (FC)	0-3v, at off-idle: 12v
6	PK/GN	MIL (lamp) Control	MIL Off: 12v, On: 1v
7	BK/WT	Starter Switch Signal	9-11v (cranking)
8	BK/YL	EFI Main Relay Control	Relay Off: 0v, On: 12v
11	WT	SIL (Scan Tool) Signal	12v
15	GN/WT	Stop Light Switch Signal	Brake Off: 0v, On: 12v
16	WT/BL	EFI Main Relay Power	12-14v
17	RD/GN	EVAP Vapor Pressure Sensor	2.9-3.1v (with fuel cap off)

2000 Pickup 3.4L V6 MFI California VIN N E4 28 Pin Connector

PCM Pin #	Wire Color	Circuit Description (28 Pin)	Value at Hot Idle
2	RD/BK	A/T Select Switch Reverse	In 'R': 12v, Others: 0v
3	BL	A/T Select Switch 2nd	In 2nd: 12v, Others: 0v
4	YL/GN	Cruise Control ECU (IDLO)	1.5v, off-idle: 12v
7	YL/RD	A/T Oil Temperature Indicator	Lamp Off: 12v, On: 1v
8	RD	HO2S-12 (B1 S2) Signal	0.1-1.1v
9	RD/WT	HO2S-12 (B1 S2) Heater	Heater Off: 12v, On: 1v
10	BL/OR	O/D "Off" Indicator (ODLP)	Lamp Off: 12v, On: 1v
12	LG	A/T Select Switch Low	In Low: 12v, Others: 0v
13	BL/BK	A/C Amplifier Signal (ACT)	Clutch On: 12v, Off: 1.5v
17	RD/YL	4WD Detection Transfer (N)	Open: 12v, Closed: 0v
18	GY	4WD ADD Indicator Switch	Open: 12v, Closed: 0v
19	BL/RD	4WD Detection Transfer (L4)	Open: 12v, Closed: 0v
20	PK	Neutral Start Switch (NSW)	In P/N: 9-11v (cranking)
22	GN/OR	Speedometer Indicator	At 55 mph: 48 Hz
24	BR/YL	Cruise Control ECU (OD1)	At Cruise in OD: 12v
25	BL/YL	A/C Amplifier Signal (AC1)	Clutch On: 1.5v, Off: 12v

Pin Connector Graphic

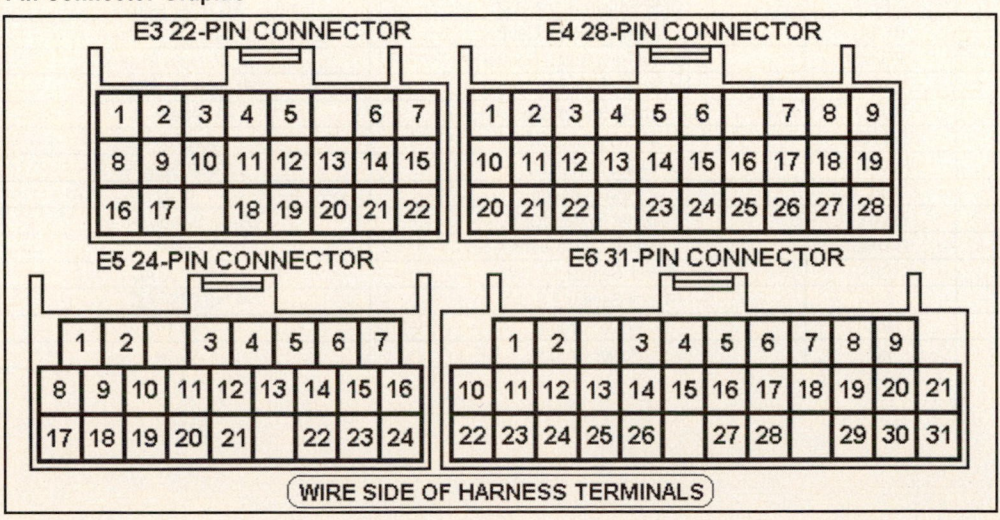

2000 Pickup 3.4L V6 MFI California VIN N E5 24 Pin Connector

PCM Pin #	Wire Color	Circuit Description (24 Pin)	Value at Hot Idle
2	GN/BK	Sensor VREF	4.9-5.1v
4	YL	AFS-11 (B1 S1) Heater	Heater Off: 12v, On: 1v
5	RD	Injector 1 Control	1.6-2-9 ms
6	WT	Injector 2 Control	1.6-2-9 ms
7	WT/GN	EVAP Purge Solenoid (VSV)	12v or 0v
8	BR	Power Ground	<0.1v
9	YL/RD	PSP Switch Signal	Straight: 12v, Turning: 0v
10	RD/WT	MAF Sensor Signal (VG)	1-1.1v
12	GN	AFS-11 (B1 S1) Signal (AF+)	Fixed at 3.3v
14	GN	ECT Sensor Signal (THW)	0.5-0.6v
15	RD/YL	A/T Oil Temperature Sensor	At 68°F: 4-5v
16	PK	CKP Sensor Signal (NE+)	AC pulse signals
17	BR	Power Ground	<0.050v
18	GN/WT	Sensor Ground	<0.050v
19	BK/WT	MAF Sensor Ground (E2G)	<0.050v
21	RD	AFS-11 (B1 S1) Signal (AF-)	Fixed at 3.3v
22	YL/GN	IAT Sensor Signal (THA)	0.5-3.4v
23	BK/YL	TP Sensor Signal (VTA)	0.53-1.27v
24	PK	CMP & CKP Sensor Return	<0.050v

2000 Pickup 3.4L V6 MFI California VIN N E6 31 Pin Connector

PCM Pin #	Wire Color	Circuit Description (31 Pin)	Value at Hot Idle
1	GN	Injector 3 Control	1.6-2-9 ms
2	RD/BK	Injector 4 Control	1.6-2-9 ms
3	BL	Injector 5 Control	1.6-2-9 ms
4	YL	Injector 6 Control	1.6-2-9 ms
6	RD	Data Link Connector	12-14v
7	RD	A/T-ECT Solenoid (S1)	3rd or OD: 1v
8	WT/BL	A/T-ECT Solenoid (S2)	1st or OD: 1v
9	GN/RD	A/T-ECT Solenoid (SL)	In Lockup: 12-14v
10	BL	CMP Sensor Signal (G2)	AC pulse signals
11	BK/BL	Igniter Transistor 1 Control	6°, 55 mph: 8° dwell
12	RD/BL	Igniter Transistor 2 Control	6°, 55 mph: 8° dwell
13	LG	Igniter Transistor 3 Control	6°, 55 mph: 8° dwell
14	YL/RD	A/T Vehicle Speed Sensor (+)	Pulses
15	BK/RD	IAC Signal (RSC)	Pulse Signals
16	RD/WT	IAC Signal (RSO)	Pulse Signals
21	BR	Power Ground	<0.1v
25	BK/YL	Igniter Signal (IGF)	Digital Signal: 0-5-0v
26	WT/RD	A/T Vehicle Speed Sensor (-)	Pulses
27	BK	Knock Sensor 1 Signal	<0.075v AC
28	GY	Knock Sensor 2 Signal	<0.075v AC
30	BR	Power Ground	<0.1v
31	BR	Power Ground	<0.1v

2000 Pickup 3.4L V6 MFI Federal VIN N E3 28 Pin Connector

PCM Pin #	Wire Color	Circuit Description (28 Pin)	Value at Hot Idle
1	LG	A/T Select Switch Low	In Low: 12v, Others: 0v
2	BL/OR	O/D "Off" Indicator (ODLP)	Lamp Off: 12v, On: 1v
5	BL/BK	A/C Amplifier Signal (ACT)	Clutch On: 12v, Off: 1.5v
7	BR/YL	Cruise Control ECU (OD1)	At Cruise in OD: 12v
8	WT	SIL (Scan Tool) Signal	Digital Signals
10	BL	A/T Select Switch 2nd	In 2nd: 12v, Others: 0v
11	YL/GN	Cruise Control ECU (IDLO)	1.5v, off-idle: 12v
12	GN/OR	Speedometer Indicator	At 55 mph: 48 Hz
14	BK/RD	Direct Battery	12-14v
15	RD/BK	A/T Select Switch Reverse	In 'R': 12v, Others: 0v
16	BL/YL	A/C Amplifier Signal (AC1)	Clutch On: 1.5v, Off: 12v
23	WT/BL	EFI Main Relay Power	12-14v
24	GN/WT	Stop Light Switch Signal	Brake Off: 0v, On: 12v

2000 Pickup 3.4L V6 MFI Federal VIN N E4 16 Pin Connector

PCM Pin #	Wire Color	Circuit Description (16 Pin)	Value at Hot Idle
2	WT/GN	EVAP Purge Solenoid (VSV)	12v or 0v
3	PK/GN	MIL (lamp) Control	MIL Off: 12v, On: 1v
5	RD	Data Link Connector	12-14v
6	YL/RD	A/T Oil Temperature Sensor	At 68°F: 4-5v
7	BK/WT	MAF Sensor Ground (E2G)	<0.050v
8	GY/RD	EVAP Vapor Pressure (VSV)	12v or 0v
9	RD/YL	4WD Detection Transfer (N)	Switch Closed: 0v
10	RD/WT	HO2S-12 (B1 S2) Heater	Heater Off: 12v, On: 1v
11	PK/BL	HO2S-11 (B1 S1) Heater	Heater Off: 12v, On: 1v
13	RD/GN	EVAP Vapor Pressure Sensor	2.5-3.1v (with fuel cap off)
16	BR	Sensor Ground	<0.050v

Pin Connector Graphic

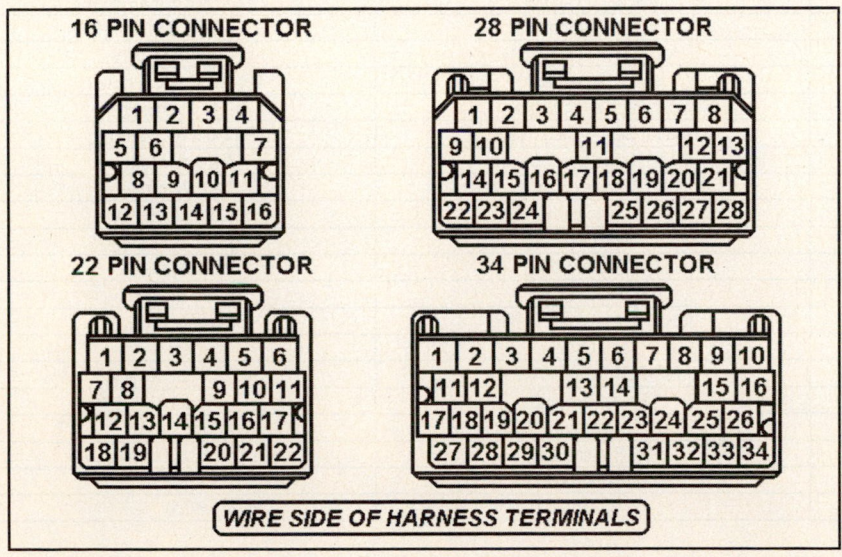

2000 Pickup 3.4L V6 MFI Federal VIN N E5 22 Pin Connector

PCM Pin #	Wire Color	Circuit Description (22 Pin)	Value at Hot Idle
1	GN/BK	Sensor VREF	4.9-5.1v
2-3	---	Not Used	---
4	WT/RD	A/T Vehicle Speed Sensor (-)	AC pulse signals
5	PK	CKP Sensor Signal (NE+)	AC pulse signals
6	PK	CKP Sensor Signal (NE-)	<0.050v
7	BK/YL	TP Sensor Signal (VTA)	0.3-0.8v
8	RD/WT	MAF Sensor Signal (VG)	1.1-1.8v
9	YL/RD	A/T Vehicle Speed Sensor (+)	Pulses
10-12, 16	---	Not Used	---
13	WT	HO2S-11 (B1 S1) Signal	0.1-1.1v
14	GY	Knock Sensor 2 Signal	<0.075v AC
15	BK	Knock Sensor 1 Signal	<0.075v AC
17	BL	CMP Sensor Signal (G2+)	AC pulse signals
18	YL	Circuit Opening Relay (FC)	0-3v, at off-idle: 12v
19	RD	HO2S-12 (B1 S2) Signal	0.1-1.1v
20	GN	ECT Sensor Signal (THW)	At 180ºF: 0.51v
21	YL/GN	IAT Sensor Signal (THA)	At 100ºF: 2.60v
22	GN/WT	Sensor Ground	<0.050v

2000 Pickup 3.4L V6 MFI Federal VIN N E6 34-Pin Connector

PCM Pin #	Wire Color	Circuit Description (34-Pin)	Value at Hot Idle
1	GY	4WD Detection Switch	Open: 12v, Closed: 0v
2	BL/RD	4WD Detection Transfer (L4)	Open: 12v, Closed: 0v
3-4	---	Not Used	---
5	YL	Injector 6 Control	1.6-2.9 ms
6	BL	Injector 5 Control	1.6-2.9 ms
7	RD/BK	Injector 4 Control	1.6-2.9 ms
8	GN	Injector 3 Control	1.6-2.9 ms
9	WT	Injector 2 Control	1.6-2.9 ms
10	RD	Injector 1 Control	1.6-2.9 ms
11	RD	A/T-ECT Solenoid (S1)	S1: 3rd or OD: 1v
12	BK/YL	Igniter Signal (IGF)	Digital Signal: 0-5-0v
13	BK/WT	Starter Switch Signal	9-11v (cranking)
14	PK	Neutral Start Switch (NSW)	In P/N: 9-11v (cranking)
15	LG	Igniter Transistor 3 Control	7% duty cycle
16	RD/BL	Igniter Transistor 2 Control	7% duty cycle
17	WT/BL	A/T-ECT Solenoid (S2)	S2: 3rd or OD: 1v
18-21	---	Not Used	---
22	BK/RD	IAC Signal (RSC)	Pulse Signals
23	RD/WT	IAC Signal (RSO)	Pulse Signals
24	BK/BL	Igniter Transistor 1 Control	7% duty cycle
25-26, 32	---	Not Used	---
27	GN/RD	A/T-ECT Solenoid (SL)	In Lockup: 12-14v
28	BR	Power Ground	<0.1v
30	PK/YL	A/T Oil Temperature Sensor	At 68ºF: 4-5v
31	YL/RD	PSP Switch Signal	Straight: 12v, Turning: 0v
33	BR	Power Ground	<0.1v
34	BR	Power Ground	<0.1v

2001-02 Pickup 3.4L V6 DOHC VIN N (All) E3 22 Pin Connector

PCM Pin #	Wire Color	Circuit Description (22 Pin)	Value at Hot Idle
1	BK/RD	Direct Battery	12-14v
2	BK/OR	Ignition Switch Power	12-14v
3	YL	Circuit Opening Relay (FC)	0-3v, at off-idle: 12v
6	VT/GN	MIL (lamp) Control	MIL Off: 12v, On: 1v
7	BK/WT	Start Switch Signal (STA)	9-11v (cranking)
8	BK/YL	EFI Main Relay Control	Relay Off: 0v, On: 12v
9	GY/RD	EVAP Vapor Pressure (VSV)	12v or 0v
11	WT	SIL (Scan Tool) Signal	12v
15	GN/WT	Stop Light Switch Signal	Brake Off: 0v, On: 12v
16	WT/BL	EFI Main Relay (B+)	12-14v
17	RD/GN	EVAP Vapor Pressure Sensor	2.9-3.1v (with fuel cap off)

2001-02 Pickup 3.4L V6 DOHC MFI VIN N E4 28 Pin Connector

PCM Pin #	Wire Color	Circuit Description (28 Pin)	Value at Hot Idle
2	RD/BK	A/T Select Switch Reverse	In 'R': 12v, Others: 0v
3	BL	A/T Select Switch 2nd	In 2nd: 12v, Others: 0v
4	YL/GN	Cruise Control ECU (IDLO)	1.5v, off-idle: 12v
6	RD	OD Main Switch (ODMS)	Switch Off: 12v, On: 1v
7	YL/RD	A/T Oil Temperature Indicator	Lamp Off: 12v, On: 1v
8	RD	HO2S-12 (B1 S2) Signal	0.1-1.1v
9	RD/WT	HO2S-12 (B1 S2) Heater	Heater Off: 12v, On: 1v
10	BL/OR	O/D OFF Indicator (ODLP)	Lamp Off: 12v, On: 1v
12	LG	A/T Select Switch Low	In Low: 12v, Others: 0v
13	BL/BK	A/C Amplifier Signal (ACT)	Clutch On: 12v, Off: 1.5v
17	RD/YL	4WD Detection Transfer (N)	Open: 12v, Closed: 0v
18	GY	4WD ADD Indicator Switch	Open: 12v, Closed: 0v
19	BL/RD	4WD Detection Transfer (L4)	Open: 12v, Closed: 0v
20	PK	Neutral Start Switch	In P/N: 9-11v (cranking)
22	GN/OR	Speedometer Indicator (SPI)	At 55 mph: 48 Hz
24	BR/YL	Cruise Control ECU (OD1)	At Cruise in OD: 12v
25	BL/YL	A/C Amplifier Signal (AC1)	Clutch On: 1.5v, Off: 12v

Pin Connector Graphic

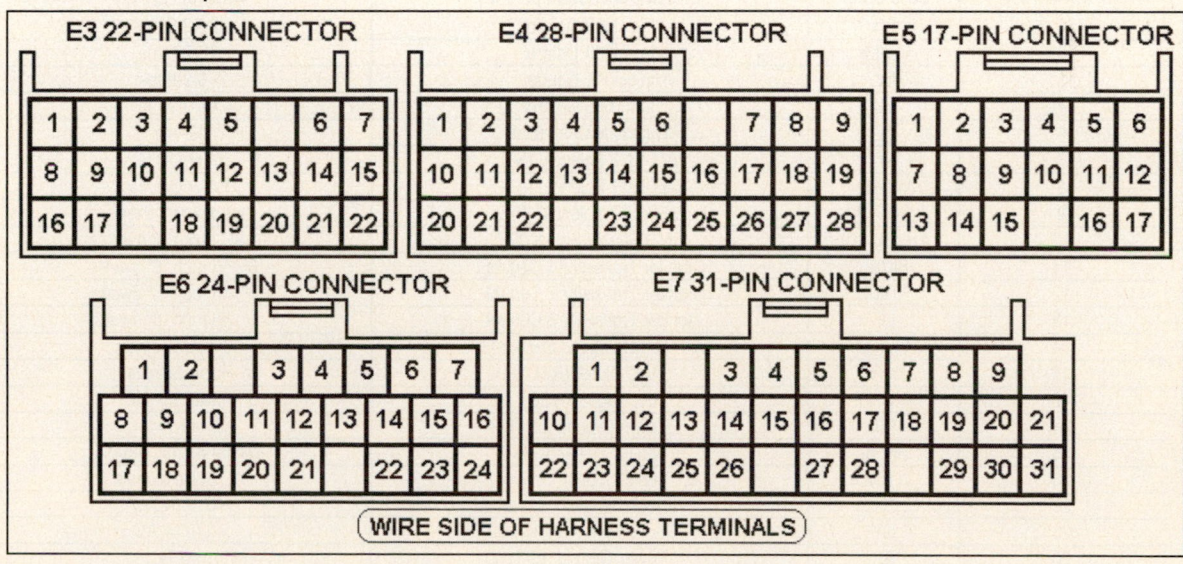

2001-02 Pickup 3.4L V6 DOHC MFI VIN N (All) E5 17 Pin Connector

PCM Pin #	Wire Color	Circuit Description (17 Pin)	Value at Hot Idle
1, 3, 5-7	---	Not Used	---
2	YL	Throttle Control Motor (CL+)	Pulses
4	RD/BK	TP Sensor Signal (VTA2)	2.0-2.9v
8	BL	Throttle Control Motor (CL-)	Pulses
9	---	Not Used	---
10	RD/WT	Accelerator Position Sensor (VPA)	0.25-0.90v
15	BL	Accelerator Position Sensor (VPA2)	1.8-2.7v
16-17	---	Not Used	---

2001-02 Pickup 3.4L V6 DOHC MFI VIN N E6 24 Pin Connector

PCM Pin #	Wire Color	Circuit Description (24 Pin)	Value at Hot Idle
2	GN/BK	Sensor VREF (VC)	4.9-5.1v
4	YL	AFS-11 (B1 S1) Heater	Heater Off: 12v, On: 1v
5	RD	Injector 1 Control	1.6-2-9 ms
6	WT	Injector 2 Control	1.6-2-9 ms
7	WT/GN	EVAP Purge Solenoid (VSV)	12v or 0v
8	WT/BK	Power Ground (E05)	<0.1v
9	YL/RD	PSP Switch Signal	Straight: 12v, Turning: 0v
10	RD/WT	MAF Sensor Signal (VG)	1-1.1v
12	GN	AFS-11 (B1 S1) Signal (AF+)	Fixed at 3.3v
14	GN	ECT Sensor Signal (THW)	0.5-0.6v
15	RD/YL	A/T Oil Temperature Sensor	At 68ºF: 4-5v
16	VT	CKP Sensor Signal (NE+)	AC pulse signals
17	BR	Power Ground	<0.050v
18	GN/WT	Sensor Ground (E2)	<0.050v
19	BK/WT	MAF Sensor Ground (E2G)	<0.050v
21	RD	AFS-11 (B1 S1) Signal (AF-)	Fixed at 3.3v
22	YL/GN	IAT Sensor Signal (THA)	0.5-3.4v
23	BK/YL	TP Sensor Signal (VTA)	0.53-1.27v
24	PK	CMP/CKP Sensor Ground (-)	<0.050v

2001-02 Pickup 3.4L V6 DOHC MFI VIN N E7 31 Pin Connector

PCM Pin #	Wire Color	Circuit Description (31 Pin)	Value at Hot Idle
1	GN	Injector 3 Control	1.6-2-9 ms
2	RD/BK	Injector 4 Control	1.6-2-9 ms
3	BL	Injector 5 Control	1.6-2-9 ms
4	YL	Injector 6 Control	1.6-2-9 ms
6	RD	Data Link Connector	12-14v
7	RD	A/T-ECT Solenoid (S1)	3rd or OD: 1v
8	WT/BL	A/T-ECT Solenoid (S2)	1st or OD: 1v
9	GN/RD	A/T-ECT Solenoid (SL)	In Lockup: 12-14v
10	BL	CMP Sensor Signal (G2)	AC pulse signals
11	BK/BL	Igniter Transistor 1 Control	6º, 55 mph: 8º dwell
12	RD/BL	Igniter Transistor 2 Control	6º, 55 mph: 8º dwell
13	LG	Igniter Transistor 3 Control	6º, 55 mph: 8º dwell
14	YL/RD	A/T Vehicle Speed Sensor (+)	AC pulse signals
15	BK/RD	IAC Signal (RSC)	Pulse Signals
16	RD/WT	IAC Signal (RSO)	Pulse Signals
21	BR	Power Ground (E01)	<0.1v
25	BK/YL	Igniter Signal (IGF)	Digital Signal: 0-5-0v
26	WT/RD	A/T Vehicle Speed Sensor (-)	AC pulse signals
27	BK	Knock Sensor 1 Signal	<0.075v AC
28	GY	Knock Sensor 2 Signal	<0.075v AC
30	BR	Power Ground (E03)	<0.1v
31	WT/BK	Power Ground (E02)	<0.1v

2003 Pickup 3.4L V6 DOHC VIN N (All) E3 22 Pin Connector

PCM Pin #	Wire Color	Circuit Description (22 Pin)	Value at Hot Idle
1	WT/BL	EFI Main Relay (B+)	12-14v
2	VT/GN	MIL (lamp) Control	MIL Off: 12v, On: 1v
3	BL/BK	A/C Amplifier Signal (ACT)	Clutch On: 12v, Off: 1.5v
4	BK/YL	EFI Main Relay Control	Relay Off: 0v, On: 12v
5	---	Not Used	---
6	GN/OR	Speedometer Indicator (SP1)	At 55 mph: 48 Hz
7	BK	Start Switch Signal (STA)	9-11v (cranking)
8	WT/GN	ETCS Power (+BM)	12-14v
9	BL/YL	A/C Amplifier Signal (AC1)	Clutch On: 1.5v, Off: 12v
10-12	---	Not Used	---
13	BL/WT	Data Link Connector (WFSE)	12v
14	WT	SIL (Scan Tool) Signal	12v
15	BK/OR	Ignition Switch Power (IGSW)	12-14v
16	BK/RD	Direct Battery	12-14v
17	YL	Cruise Control Indicator	Indicator Off: 12v, On: 1v
18-19	---	Not Used	---
20	GN/WT	Stop Light Switch Signal	Brake Off: 0v, On: 12v
21	RD/BK	Center Airbag Sensor Assembly (F/PS)	Digital Signals
22	YL	Circuit Opening Relay (FC)	0-3v, at off-idle: 12v

2003 Pickup 3.4L V6 DOHC MFI VIN N E4 28 Pin Connector

PCM Pin #	Wire Color	Circuit Description (28 Pin)	Value at Hot Idle
1, 7	---	Not Used	---
2	BL/RD	4WD Detection Transfer (L4)	Open: 12v, Closed: 0v
3	PK	Neutral Start Switch (NSW)	In 'P' or 'N': 0v
4	LG	A/T Select Switch Signal (L)	In 'L': 12v, Others: 0v
5	BL	A/T Select Switch Signal (2)	In 2nd: 12v, Others: 0v
6	RD/BL	A/T Select Switch Signal (R)	In 'R': 12v, Others: 0v
8	PK/BL	Traction Control ECU Signal (NEO)	Pulse Signals
9	BL/RD	Stop Light Switch (STI-)	Brake Off: 0v, On: 12v
10-12, 16-17	---	Not Used	---
13	GR	4WD Switch Signal (4WD)	4WD Switch On: 12v
14	RD	OD Main Switch (ODMS)	Switch Off: 12v, On: 1v
18	LG	Traction Control ECU Signal (ENG-)	Pulse Signals
19	LG/BK	Traction Control ECU Signal (ENG+)	Pulse Signals
20	BL/OR	O/D OFF Indicator (ODLP)	Lamp Off: 12v, On: 1v
21	YL/RD	A/T Oil Temperature Indicator (OILW)	Lamp Off: 12v, On: 1v
22	GN	Cruise Control Switch, Coast Control Switch	At Cruise in OD: 12v
23-24, 26	---	Not Used	---
25	WT/RD	A/T Select Switch Signal (D)	In 'D': 12v, Others: 0v
27	PK	Traction Control ECU Signal (TRC-)	Pulse Signals
28	VT/GN	Traction Control ECU Signal (TRC+)	Pulse Signals

Pin Connector Graphic

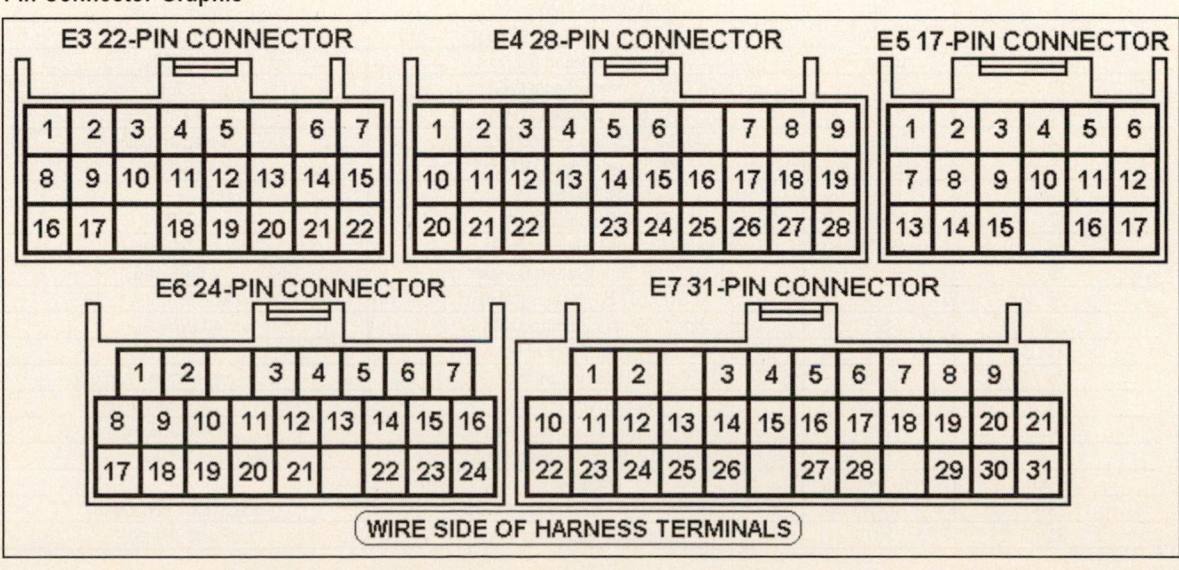

2003 Pickup 3.4L V6 DOHC MFI VIN N (All) E5 17 Pin Connector

PCM Pin #	Wire Color	Circuit Description (17 Pin)	Value at Hot Idle
1, 3, 6-7, 9	---	Not Used	---
2	YL	Throttle Control Motor (CL+)	Pulses
4	RD/BK	TP Sensor Signal (VTA2)	0.3-1.0v
5	BK/RD	A/T-ECT Solenoid (SLT+)	Pulse Signals
8	BL	Throttle Control Motor (CL-)	Pulses
10	RD/WT	Accelerator Position Sensor (VPA)	0.3-1.0v
11	GNWT	A/T-ECT Solenoid (SLT-)	Pulse Signals
12-14, 16-17	---	Not Used	---
15	RD	Accelerator Position Sensor (VPA2)	0.3-1.0v

2003 Pickup 3.4L V6 DOHC MFI VIN N E6 24 Pin Connector

PCM Pin #	Wire Color	Circuit Description (24 Pin)	Value at Hot Idle
1	BL	Injector 5 Control	1.6-2-9 ms
2	BK/YL	Igniter Signal (IGF)	Digital Signal: 0-5-0v
3	RD/BK	Injector 4 Control	1.6-2-9 ms
4	GN	Injector 3 Control	1.6-2-9 ms
5	WT	Injector 2 Control	1.6-2-9 ms
6	RD	Injector 1 Control	1.6-2-9 ms
7	WT/GN	Data Link Connector (TC)	12v
8	YL	Injector 6 Control	1.6-2.9 ms
9	LG	Igniter Transistor 3 Control	6°, 55 mph: 8° dwell
10	RD/BL	Igniter Transistor 2 Control	6°, 55 mph: 8° dwell
11	BK/BL	Igniter Transistor 1 Control	6°, 55 mph: 8° dwell
12	RD	CKP Sensor Signal (NE+)	AC pulse signals
13	BL	CMP Sensor Signal (G2)	AC pulse signals
14	GN/WT	EVAP Purge Solenoid (VSV)	12v or 0v
15-16, 20	---	Not Used	---
17	BR	Power Ground (E01)	<0.1v
18	YL/RD	PSP Switch Signal (PSW)	Straight: 12v, Turning: 0v
19	BL/WT	EVAP Closed Canister Valve (VSV)	12v or 0v
21	GN	CMP/CKP Sensor Ground (NE-)	<0.050v
22	WT/RD	A/T Vehicle Speed Sensor (SP2-)	AC pulse signals
23	YL/RD	A/T Vehicle Speed Sensor (SP2+)	AC pulse signals

2003 Pickup 3.4L V6 DOHC MFI VIN N E7 31 Pin Connector

PCM Pin #	Wire Color	Circuit Description (31 Pin)	Value at Hot Idle
1	GN/RD	A/T-ECT Solenoid (SL)	In Lockup: 12-14v
2	WT/BL	A/T-ECT Solenoid (S2)	1st or OD: 1v
3	RD	A/T-ECT Solenoid (S1)	3rd or OD: 1v
4	WT/BK	Power Ground (E01)	<0.1v
5	WT/BK	Power Ground (E02)	<0.1v
6	BR	Power Ground (ME01)	<0.1v
7	WT/BK	Power Ground (E03)	<0.1v
8	WT/BK	Power Ground (E04)	<0.1v
9	GN	Throttle Control Motor (M+)	Pulse Signals
10	GN/WT	MAF Sensor Ground (E2G)	<0.050v
11	BK/WT	Sensor Ground (E2)	<0.050v
12	RD/YL	MAF Sensor Signal (VG)	1-1.1v
13	YL/GN	IAT Sensor Signal (THA)	0.2-1.0v
14	VT	AFS-11 (B1 S1) Signal (AF1+)	Fixed at 3.3v
15	BK/YL	TP Sensor Signal (VTA)	0.3-1.0v
16, 20	---	Not Used	---
17	BR	Shield Ground (GE01)	<0.050v
18	GN/YL	ECT Sensor Signal (THW)	0.5-0.6v
19	RD/YL	A/T Oil Temperature Sensor	At 68°F: 4-5v
21	YL	AFS-11 (B1 S1) Heater (HTAF1)	Heater Off: 12v, On: 1v
22	GR	Knock Sensor 2 Signal (KNK2)	<0.075v AC
23	BK	Knock Sensor 1 Signal (KNK1)	<0.075v AC
24	RD/BL	EVAP Vapor Pressure Sensor (PTNK)	2.9-3.1v (with fuel cap off)
25	GN/BK	Sensor VREF (VC)	4.9-5.1v
26	PK	AFS-11 (B1 S1) Signal (AF1-)	Fixed at 3.3v
27	RD	HO2S-12 (B1 S2) Signal (OX2B)	0.1-1.1v
28, 30	---	Not Used	---
29	RD/WT	HO2S-12 (B1 S2) Heater (HT2B)	Heater On: 12v, Off: 0v
31	RD	Throttle Control Motor (M-)	Pulse Signals

2000 Pickup 4.7L V8 VIN T 2UZ-FE (All) E3 22 Pin Connector

PCM Pin #	Wire Color	Circuit Description (22 Pin)	Value at Hot Idle
1	BK/RD	Direct Battery	12-14v
4	BK/BL	Fuel Pump Relay	Relay Off: 12v, On: 1v
5	YL	Fuel Pump Switch	Switch On: 1v, Off: 12v
6	PK/GN	MIL (lamp) Control	MIL Off: 12v, On: 1v
7	WT/GN	ETCS Power (+BM)	12-14v
8	WT/BL	EFI Main Relay Power	12-14v
9	BK/OR	Ignition Switch Power	12-14v
10	BK/YL	EFI Main Relay Control	Relay Off: 0v, On: 12v
11	WT	SIL (Scan Tool) Signal	12v
16	WT/BL	EFI Main Relay Power	12-14v

2000 Pickup 4.7L V8 VIN T 2UZ-FE (All) E4 28 Pin Connector

PCM Pin #	Wire Color	Circuit Description (28 Pin)	Value at Hot Idle
1	BL/OR	Overdrive Main Switch	Switch Off: 12v, On: 1v
2	RD/BK	A/T Select Switch Reverse	In 'R': 12v, Others: 0v
3	BL	A/T Select Switch 2nd	In 2nd: 12v, Others: 0v
4	LG	A/T Select Switch Low	In Low: 12v, Others: 0v
5	GN	Data Link Connector	12-14v
6	GN/WT	Stop Light Switch Signal	Brake Off: 0v, On: 12v
7	YL	HO2S-22 (B2 S2) Heater	Heater Off: 12v, On: 1v
8	RD/YL	HO2S-12 (B1 S2) Heater	Heater Off: 12v, On: 1v
9	BL/RD	4WD Detection Transfer (L4)	Open: 12v, Closed: 0v
10	GY/RD	EVAP Vapor Pressure (VSV)	12v or 0v
12	GN/YL	Taillight Switch Signal	Switch Off: 0v, On: 12v
13	BL/BK	A/C Amplifier Signal (ACT)	Clutch On: 12v, Off: 1.5v
15	GN/OR	Speedometer Indicator	At 55 mph: 48 Hz
16	BL/WT	Tachometer Signal (TACO)	Pulse Signals
17	BK/WT	Starter Switch Signal	9-11v (cranking)
18	BK	HO2S-12 (B1 S2) Signal	0.1-1.1v
19	WT/RD	A/T Select Switch Drive	In 'D': 12v, Others: 0v
20	PK	A/T Neutral Start Switch	In P/N: 9-11v (cranking)
22	RD/GN	EVAP Vapor Pressure Sensor	2.5-3.1v (with fuel cap off)
25	BL/YL	A/C Amplifier Signal (AC1)	Clutch On: 1.5v, Off: 12v
26	YL/RD	A/T Oil Temperature Lamp	Lamp Off: 12v
27	WT	HO2S-22 (B2 S2) Signal	0.1-1.1v

Pin Connector Graphic

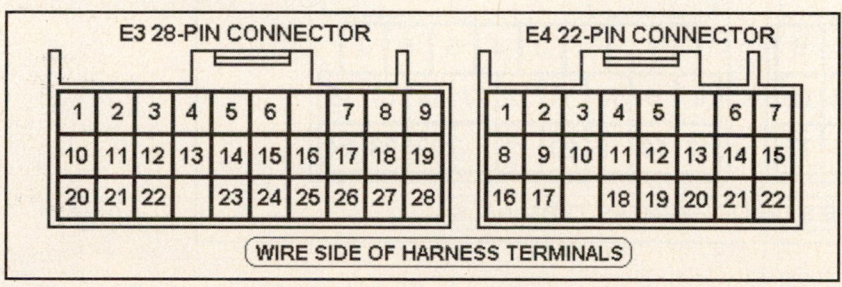

2000 Pickup 4.7L V8 2UZ-FE VIN T (All) E5 17-Pin Connector

PCM Pin #	Wire Color	Circuit Description (17-Pin)	Value at Hot Idle
1	RD	A/T-ECT Solenoid (S1)	S1: 3rd or OD: 1v
2	WT/BL	A/T-ECT Solenoid (S2)	S1: 3rd or OD: 1v
3	GN/RD	A/T-ECT Solenoid (SL)	In Lockup: 12-14v
4	WT	Direct Clutch Speed Input (+)	Pulse Signals
5	YL/RD	A/T Vehicle Speed Sensor (+)	AC pulse signals
9	BK/RD	A/T-ECT Solenoid (SLT+)	Pulse Signals
10	BK	Direct Clutch Speed Input (-)	Pulse Signals
11	WT/RD	A/T Vehicle Speed Sensor (-)	AC pulse signals
15	GN/YL	A/T-ECT Solenoid (SLT-)	Pulse Signals
17	RD/YL	A/T Oil Temperature Sensor	At 68°F: 4-5v

2000 Pickup 4.7L V8 2UZ-FE VIN T (All) E6 24 Pin Connector

PCM Pin #	Wire Color	Circuit Description (24 Pin)	Value at Hot Idle
1	BR	Power Ground	<0.1v
2	GN/BK	Sensor VREF	4.9-5.1v
3	YL	HO2S-21 (B2 S1) Heater	Heater Off: 12v, On: 1v
4	RD	HO2S-11 (B1 S1) Heater	Heater Off: 12v, On: 1v
5	RD	Injector 1 Control	2.0-3.3 ms
6	WT	Injector 2 Control	2.0-3.3 ms
7	WT/GN	EVAP Purge Solenoid (VSV)	12v or 0v
9	RD/BK	Accelerator Position Sensor 2	1.8-2.7v
10	RD/WT	Mass Airflow Sensor	1.1-1.5v
11	WT	HO2S-21 (B2 S1) Signal	0.1-1.1v
12	BK	HO2S-11 (B1 S1) Signal	0.1-1.1v
13	BK/YL	TP Sensor Signal (VTA1)	0.4-1.0v
14	GN	ECT Sensor Signal (THW)	At 180°F: 0.51v
17	BR	Shield Ground	<0.050v
18	GN/WT	Sensor Ground (E2)	<0.050v
19	BK/WT	MAF Sensor Ground (EVG)	<0.050v
20	PK/BL	TP Sensor Signal (VTA2)	2.0-2.9v
21	GN/RD	Accelerator Position Sensor 1	0.25-0.90v
22	YL/GN	IAT Sensor Signal (THA)	At 100°F: 2.60v

Pin Connector Graphic

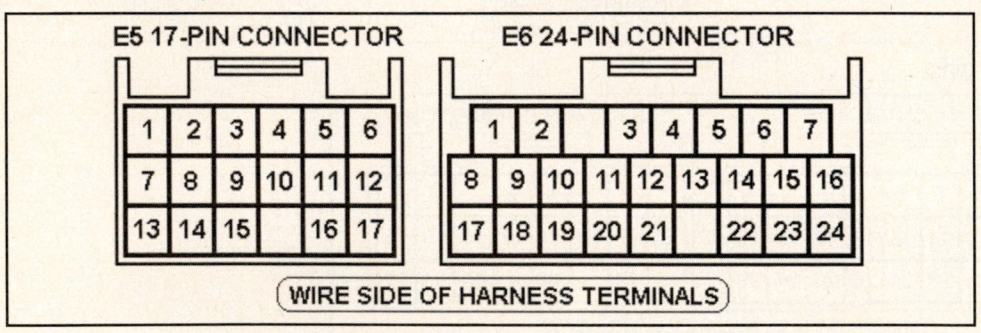

2000 Pickup 4.7L V8 2UZ-FE VIN T (All) E7 31 Pin Connector

PCM Pin #	Wire Color	Circuit Description (31 Pin)	Value at Hot Idle
1	GN	Injector 3 Control	2.0-3.3 ms
2	RD/BK	Injector 4 Control	2.0-3.3 ms
3	BL	Injector 5 Control	2.0-3.3 ms
4	YL	Injector 6 Control	2.0-3.3 ms
5	BL/RD	Injector 7 Control	2.0-3.3 ms
6	GN/YL	Injector 8 Control	2.0-3.3 ms
7	WT	Throttle Control Motor (M-)	Pulse Signals
8	RD	Throttle Control Motor (M+)	Pulse Signals
9	BR	Power Ground	<0.1v
10	YL	CMP Sensor Signal (G2+)	AC pulse signals
11	BK/BL	COP Igniter 1 Control	7% duty cycle
12	LG/BK	COP Igniter 2 Control	7% duty cycle
13	BK/YL	COP Igniter 3 Control	7% duty cycle
14	RD/WT	COP Igniter 4 Control	7% duty cycle
15	GN/WT	COP IGT 5 Control	7% duty cycle
16	PK/BL	COP IGT 6 Control	7% duty cycle
17	BK	Knock Sensor 2 Signal	<0.075v AC
18	GY	Knock Sensor 1 Signal	<0.075v AC
21	BR	Power Ground	<0.1v
22	RD	CMP/CKP Sensor Signal (-)	<0.050v
23	GN	CKP Sensor Signal (NE+)	AC pulse signals
24	BL/WT	Throttle Control Motor (CL-)	Pulses
25	PK	COP IGT 7 Control	7% duty cycle
26	LG	COP IGT 8 Control	7% duty cycle
27	BK/RD	Igniter Signal (IGF1)	Digital Signal: 0-5-0v
28	BK/WT	Igniter Signal (IGF2)	Digital Signal: 0-5-0v
29	BL/BK	Throttle Control Motor (CL+)	Pulse Signals
30	BL/RD	Shield Ground	<0.050v
31	BR	Power Ground	<0.1v

Pin Connector Graphic

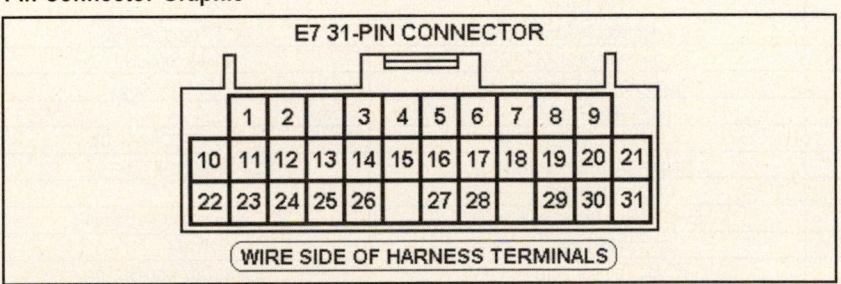

Standard Colors and Abbreviations

Abbreviation	Color	Abbreviation	Color	Abbreviation	Color
BK	Black	GY	Gray	RD	Red
BL	Blue	GN	Green	TN	Tan
BR	Brown	LG	LT Green	VT	Violet
DB	Dark Blue	OR	Orange	WT	White
DG	DK Green	PK	Pink	YL	Yellow

2001-02 Pickup 4.7L V8 VIN T 2UZ-FE (All) E3 22 Pin Connector

PCM Pin #	Wire Color	Circuit Description (22 Pin)	Value at Hot Idle
1	BK/RD	Direct Battery (BATT)	12-14v
2	---	Not Used	---
3	WT/GN	Data Link Connector (TC)	12v
4	BK/BL	Fuel Pump Relay (FPR)	Relay Off: 12v, On: 1v
5	YL	Circuit Opening Relay (FC)	0-3v, at off-idle: 12v
6	VT/GN	MIL (lamp) Control	MIL Off: 12v, On: 1v
7	WT/GN	ETCS Power (+BM)	12-14v
8	WT/BL	EFI Main Relay Power (+B1)	12-14v
9	BK/OR	Ignition Switch Power (IGSW)	12-14v
10	BK/YL	EFI Main Relay Control	Relay Off: 0v, On: 12v
11	WT	SIL (Scan Tool) Signal	Digital Signals
12-15	---	Not Used	---
16	WT/BL	EFI Main Relay Power (+B1)	12-14v
17-18	---	Not Used	---
19	BL/WT	Data Link Connector (WFSE)	12v
20-22	---	Not Used	---

2001-02 Pickup 4.7L V8 VIN T 2UZ-FE (All) E4 28 Pin Connector

PCM Pin #	Wire Color	Circuit Description (28 Pin)	Value at Hot Idle
1	BL/OR	Overdrive Main Switch	Switch Off: 12v, On: 1v
2	RD/BK	A/T Select Switch Reverse	In 'R': 12v, Others: 0v
3	BL	A/T Select Switch Second	In 2nd: 12v, Others: 0v
4	LG	A/T Select Switch Low	In Low: 12v, Others: 0v
5	GN	Data Link Connector	12v
6	GN/WT	Stop Light Switch Signal	Brake Off: 0v, On: 12v
7	YL	HO2S-22 (B2 S2) Heater	Heater Off: 12v, On: 1v
8	RD/YL	HO2S-12 (B1 S2) Heater	Heater Off: 12v, On: 1v
9	BL/RD	4WD Detection Transfer (L4)	Open: 12v, Closed: 0v
10	GY/RD	EVAP Vapor Pressure (VSV)	12v or 0v
11	---	Not Used	---
12	GN/YL	Taillight Switch (ELS)	Switch Off: 0v, On: 12v
13	BL/YL	A/C Amplifier Signal (ACT)	Clutch On: 1.5v, Off: 12v
14	---	Not Used	---
15	GN/OR	Speedometer Indicator (SPD)	At 55 mph: 48 Hz
16	BL/WT	Tachometer Signal (TACO)	Pulse Signals
17	BK/WT	Starter Switch Signal (STA)	9-11v (cranking)
18	---	Not Used	---
19	WT/RD	A/T Select Switch Drive	In 'D': 12v, Others: 0v
20	PK	Neutral Start Switch (NSW)	In 'P' or 'N': 0v
21	---	Not Used	---
22	RD/GN	EVAP Vapor Pressure Sensor	2.5-3.1v (with fuel cap off)
23	---	Not Used	---
24	YL/RD	AT Oil Temperature Lamp	Lamp Off: 12v, On: 1v
25	BL/YL	A/C Amplifier Signal (A/C)	Clutch On: 12v, Off: 1.5v
26	BL/OR	Overdrive Indicator Lamp	Lamp Off: 12v, On: 1v
27-28	---	Not Used	---

Pin Connector Graphic

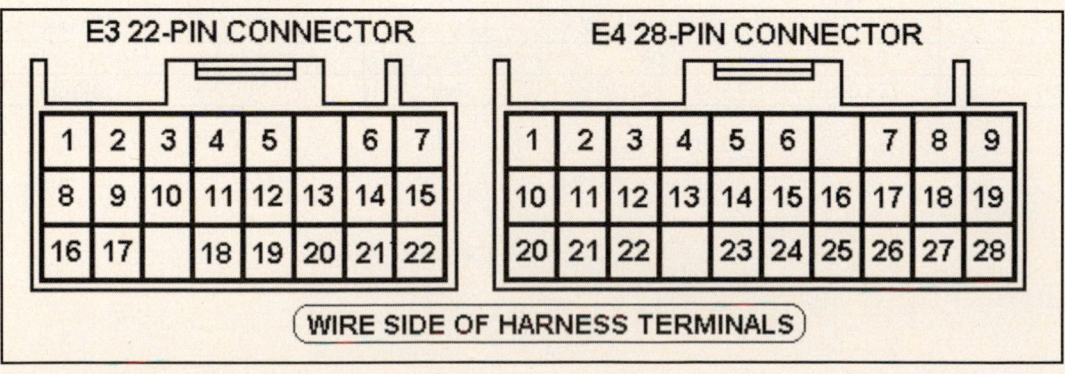

2001-02 Pickup 4.7L V8 2UZ-FE VIN T (All) E5 17-Pin Connector

PCM Pin #	Wire Color	Circuit Description (17-Pin)	Value at Hot Idle
1	RD	A/T-ECT Solenoid (S1)	S1: 3rd or OD: 1v
2	WT/BL	A/T-ECT Solenoid (S2)	S1: 3rd or OD: 1v
3	GN/RD	A/T-ECT Solenoid (SL)	In Lockup: 12-14v
4	WT	Direct Clutch Speed Input (+)	Pulse Signals
5	YL/RD	A/T Vehicle Speed Sensor (+)	AC pulse signals
6-8	---	Not Used	---
9	BK/RD	A/T-ECT Solenoid (SLT+)	Pulse Signals
10	BK	Direct Clutch Speed Input (-)	Pulse Signals
11	WT/RD	A/T Vehicle Speed Sensor (-)	AC pulse signals
12-14	---	Not Used	---
15	GN/YL	A/T-ECT Solenoid (SLT-)	AC pulse signals
16	---	Not Used	---
17	RD/YL	A/T Oil Temperature Sensor	At 68°F: 4-5v

2001-02 Pickup 4.7L V8 2UZ-FE VIN T (All) E6 24 Pin Connector

PCM Pin #	Wire Color	Circuit Description (24 Pin)	Value at Hot Idle
1	BR	Power Ground (E03)	<0.1v
2	GN/BK	Sensor VREF (VC)	4.9-5.1v
3	YL	HO2S-21 (B2 S1) Heater	Heater Off: 12v, On: 1v
4	GN/YL	HO2S-11 (B1 S1) Heater	Heater Off: 12v, On: 1v
5	RD	Injector 1 Control	2.0-3.3 ms
6	WT	Injector 2 Control	2.0-3.3 ms
7	WT/GN	EVAP Purge Solenoid (VSV)	12v or 0v
8	---	Not Used	---
9	BL	Accelerator Position Sensor (VPA2)	1.8-2.7v
10	RD/WT	Mass Airflow Sensor (VG)	1.1-1.5v
11	WT	HO2S-21 (B2 S1) Signal	0.1-1.1v
12	BK	HO2S-11 (B1 S1) Signal	0.1-1.1v
13	BK/YL	TP Sensor Signal (VTA)	0.4-1.0v
14	GN	ECT Sensor Signal (THW)	At 180°F: 0.51v
15	WT	HO2S-22 (B2 S2) Signal	0.1-1.1v
16	BK	HO2S-12 (B1 S2) Signal	0.1-1.1v
17	BR	Shield Ground (E1)	<0.050v
18	GN/WT	Sensor Ground (E2)	<0.050v
19	BK/WT	MAF Sensor Ground (EVG)	<0.050v
20	PK/BL	TP Sensor Signal (VTA2)	2.0-2.9v
21	GN/RD	Accelerator Position Sensor (VPA)	0.25-0.90v
22	YL/GN	IAT Sensor Signal (THA)	At 100°F: 2.60v
23	YL	HO2S-22 (B2 S2) Heater	Heater Off: 12v, On: 1v
24	RD/YL	HO2S-12 (B1 S2) Heater	Heater Off: 12v, On: 1v

Pin Connector Graphic

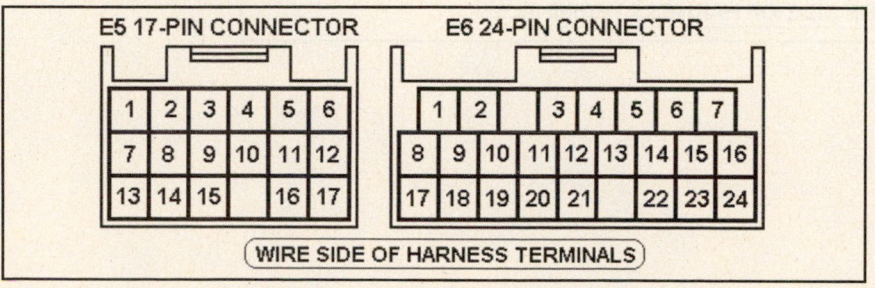

2001-02 Pickup 4.7L V8 2UZ-FE VIN T (All) E7 31 Pin Connector

PCM Pin #	Wire Color	Circuit Description (31 Pin)	Value at Hot Idle
1	GN	Injector 3 Control	2.0-3.3 ms
2	RD/BK	Injector 4 Control	2.0-3.3 ms
3	BL	Injector 5 Control	2.0-3.3 ms
4	YL	Injector 6 Control	2.0-3.3 ms
5	BL/RD	Injector 7 Control	2.0-3.3 ms
6	RD/WT	Injector 8 Control	2.0-3.3 ms
7	WT	Throttle Control Motor (M-)	Pulse Signals
8	RD	Throttle Control Motor (M+)	Pulse Signals
9	BR	Power Ground (ME01)	<0.1v
10	YL	CMP Sensor Signal (G2)	AC pulse signals
11	BK/BL	COP Igniter 1 Control	7% duty cycle
12	LG/BK	COP Igniter 2 Control	7% duty cycle
13	BK/YL	COP Igniter 3 Control	7% duty cycle
14	RD/WT	COP Igniter 4 Control	7% duty cycle
15	GN/WT	COP Igniter 5 Control	7% duty cycle
16	PK/BL	COP Igniter 6 Control	7% duty cycle
17	BK	Knock Sensor 2 Signal	<0.075v AC
18	GY	Knock Sensor 1 Signal	<0.075v AC
19-20	---	Not Used	---
21	BR	Power Ground (E01)	<0.1v
22	RD	CKP Sensor Signal (NE-)	<0.050v
23	WT/GN	CKP Sensor Signal (NE+)	AC pulse signals
24	BL/WT	Throttle Control Motor (CL-)	Pulse Signals
25	PK	COP Igniter 7 Control	7% duty cycle
26	LG	COP Igniter 8 Control	7% duty cycle
27	BK/RD	Igniter Signal (IGF1)	Digital Signal: 0-5-0v
28	BK/WT	Igniter Signal (IGF2)	Digital Signal: 0-5-0v
29	BL/BK	Throttle Control Motor (CL+)	Pulses
30	BL/RD	Shield Ground (GE01)	<0.050v
31	BR	Power Ground (E02)	<0.1v

Pin Connector Graphic

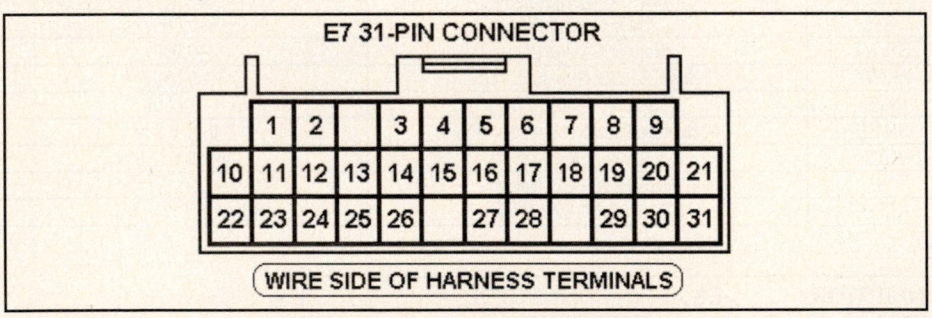

2003 Pickup 4.7L V8 2UZ-FE VIN T (All) E3 31 Pin Connector

PCM Pin #	Wire Color	Circuit Description (31 Pin)	Value at Hot Idle
1	WT/BL	EFI Main Relay Output (+B2)	12-14v
2	WT/BL	EFI Main Relay Output (+B)	12-14v
3	BK/RD	Direct Battery (BATT)	12-14v
4	BK/BL	Fuel Pump Relay (FPR)	Relay Off: 12v, On: 1v
5	BL/WT	Tachometer Signal (TACO)	Pulse Signals
7	WT/GN	ETCS Power (+BM)	12-14v
8	BK/YL	EFI Main Relay Power (+B1)	12-14v
9	BK/OR	Ignition Switch Power (IGSW)	12-14v
10	YL	Circuit Opening Relay (FC)	0-3v, at off-idle: 12v
11	VT/GN	Malfunction Indicator Lamp Control	MIL Off: 12v, On: 1v
12	GN/YL	Taillight Switch (ELS)	Switch Off: 0v, On: 12v
13	---	Not Used	---
14	RD/BK	Center Airbag Sensor Assembly (F/PS)	Digital Signals
15	---	Not Used	---
16	WT/BL	EFI Main Relay Power (+B1)	12-14v
17	---	Not Used	---
18	WT	SIL (Scan Tool) Signal	Digital Signals
19	BL/WT	Data Link Connector (WFSE)	12v
20	WT/GN	Data Link Connector (TC)	12v
21	RD/BL	EVAP Vapor Pressure Sensor (PTNK)	2.5-3.1v (with fuel cap off)
22	RD/GN	Accelerator Position Sensor (VPA)	0.25-0.90v
23	BL	Accelerator Position Sensor (VPA2)	1.8-2.7v
26	PK/GN	Accelerator Pedal Position Sensor (VCPA)	<0.050v
27	GN/RD	Accelerator Pedal Position Sensor (VCP2)	<0.050v
28	GN/WT	Accelerator Pedal Position Sensor (EPA)	<0.050v
29	LG/RD	Accelerator Position Sensor 2 (EPA2)	1.8-2.7v

2003 Pickup 4.7L V8 2UZ-FE VIN T (All) E4 34 Pin Connector

PCM Pin #	Wire Color	Circuit Description (34 Pin)	Value at Hot Idle
1	BL/WT	Power Ground (HP)	<0.1v
2	BL/YL	A/C Magnetic Clutch Relay (ACMG)	Relay Off: 0v, On: 12v
3	GN	Transmission Control Switch Signal (2L)	In 2nd: 12v, Others: 0v
4	BL/RD	Low Detection Switch (L4)	Switch Open: 0v, Closed: 12v
5	BL/WT	ECT Pattern Switch 2nd Position (SNW1)	2nd Position: 12v
6	WT/GN	ETCS Power (+BM)	12-14v
7	---	Not Used	---
8	BL/YL	Shift Lock ECU Control (L)	12v or 0v
9	YL/BK	Park Neutral Switch Signal (STAR)	In P/N: 9-12v
10	GN/YL	Park Neutral Position Switch (D)	In 'D': 12v, Others: 0v
11	RD/YL	A/T Select Switch Reverse (R)	In 'R': 12v, Others: 0v
12	---	Not Used	---
14	BL/BK	A/C Amplifier Signal (THWO)	A/C Off: 12v, On: 1v
15-16	---	Not Used	---
17	VT/RD	Vehicle Speed Sensor (SPD)	At 55 mph: 48 Hz
18	PK/BK	Body Control ECU Signal (MPX1)	Digital Signals
19	GN/WT	Stop Light Switch Signal (STP)	Brake Off: 0v, On: 12v
20	BL	Shift Lock ECU Control (3)	12v or 0v
21-22	---	Not Used	---
23	GN/RD	Shift Lock ECU Control (4)	12v or 0v
25	PK	A/T Oil Temperature Indicator	Indicator Off: 12v, On: 1v
26	BL/RD	Transponder Amplifier Signal (IMD)	Inserting key: pulses
27	WT/RD	Transponder Amplifier Signal (IMI)	Inserting key: pulses
28	BL/WT	ECT Pattern Switch Power Signal	Power Position: 12v
29	BK	Body Control ECU Signal (MPX2)	Digital Signals
30	---	Not Used	---
31	BL	A/C Amplifier Signal (A/CS)	Relay Off: 12v, On: 1v
32	BK/BL	A/C Amplifier Signal (THE)	A/C Off: 12v, On: 1v
33	BL/BK	A/C Switch Signal (ACLD)	A/C On: 12v, Off: 0v
34-35	---	Not Used	---

2003 Utility AWD 4.7L 2UZ-FE V8 VIN T E5 32 Pin Connector

PCM Pin #	Wire Color	Circuit Description (32 Pin)	Value at Hot Idle
1	BR	Power Ground (E1)	<0.1v
2	RD	Throttle Control Motor (M-)	Pulse Signals
3	WT	Throttle Control Motor (M+)	Pulse Signals
4	WT/BK	Power Ground (ME01)	<0.1v
5	RD/WT	Injector 8 Control	2.0-3.3 ms
6	BL/RD	Injector 7 Control	2.0-3.3 ms
7	WT/BK	Power Ground (E03)	<0.1v
11	YL/GN	Neutral Detection Switch (L4)	Switch Open: 0v, Closed: 12v
12	BK	Start Switch Signal (STSW)	Cranking: 9-11v
13-14, 18-20	---	Not Used	---
15	BK	A/T Solenoid Control (SLU-)	Pulse Signals
16	PK/GN	A/T Solenoid Control (SLU+)	Pulse Signals
17	BR	Shield Ground (GE01)	<0.050v
21	BK/OR	Generator Control (RL)	12v
22, 26, 28-31	---	Not Used	---
23	GN/WT	A/C Lock Sensor (LCK)	12v or 0v
24	RD	CKP Sensor Signal (NE-)	<0.050v
25	GN	CKP Sensor Signal (NE+)	AC pulse signals
27	YL	CMP Sensor Signal (G2+)	AC pulse signals
32	BL	CMP Sensor Signal (G2-)	AC pulse signals

2003 Pickup 4.7L V8 2UZ-FE VIN T (All) E6 35 Pin Connector

PCM Pin #	Wire Color	Circuit Description (35 Pin)	Value at Hot Idle
1	BK	Knock Sensor 1 Signal (KNK1 - left)	<0.075v AC
2	WT	Knock Sensor 2 Signal (KNK2 - right)	<0.075v AC
3	YL	Injector 6 Control	2.0-3.3 ms
4	GN/YL	HO2S-11 (B1 S1) Heater (HT1A)	Heater On: 12v, Off: 0v
5	RD/YL	HO2S-12 (B1 S2) Heater (HT1B)	Heater On: 12v, Off: 0v
6	YL/BK	ETCS Power (+BM)	12-14v
7, 12	---	Not Used	---
8	GR	4WD Switch Signal	In 4WD: 12v
9	GN/BK	Shift Lock ECU Control (D)	12v or 0v
10	GN/YL	Shift Lock ECU Control (D)	12v or 0v
11	RD/BK	A/T Select Switch Reverse Signal	In 'R': 1v, Others: 12v
13	YL/RD	A/C Magnetic Clutch Relay (ACMG)	Relay Off: 0v, On: 12v
14	YL/GN	A/C Amplifier Signal (THWO)	A/C Off: 12v, On: 1v
15-16, 20, 30	---	Not Used	---
17	VT	Vehicle Speed Sensor (SPD)	At 55 mph: 48 Hz
18	PK/BK	Body Control ECU Signal (MPX1)	Digital Signals
19	GN/WT	Stop Light Switch Signal	Brake Off: 0v, On: 12v
21	WT	HO2S-22 (B2 S2) Signal (OX2B)	0.1-1.1v
22	WT	HO2S-21 (B2 S1) Signal (OX2A)	0.1-1.1v
23	BK	HO2S-11 (B1 S1) Signal (OX1A)	0.1-1.1v
25	YL/BK	HO2S-22 (B2 S2) Heater (HT2B)	Heater On: 12v, Off: 0v
26	YL	Transponder Amplifier Signal (IMD)	Inserting key: pulses
27	WT	Transponder Amplifier Signal (IMI)	Inserting key: pulses
28	BL/WT	ECT Pattern Switch Power Signal	Power Position: 12v
29	BK	HO2S-12 (B1 S2) Signal (OX1B)	0.1-1.1v
31	BL/BK	A/C Amplifier Signal (ACT)	Relay Off: 12v, On: 1v
32	PK/BK	A/C Amplifier Signal (THE)	A/C Off: 12v, On: 1v
33	YL/GN	HO2S-21 (B2 S1) Heater (HT2A)	Heater On: 12v, Off: 0v
34-35	---	Not Used	---

Pin Connector Graphic

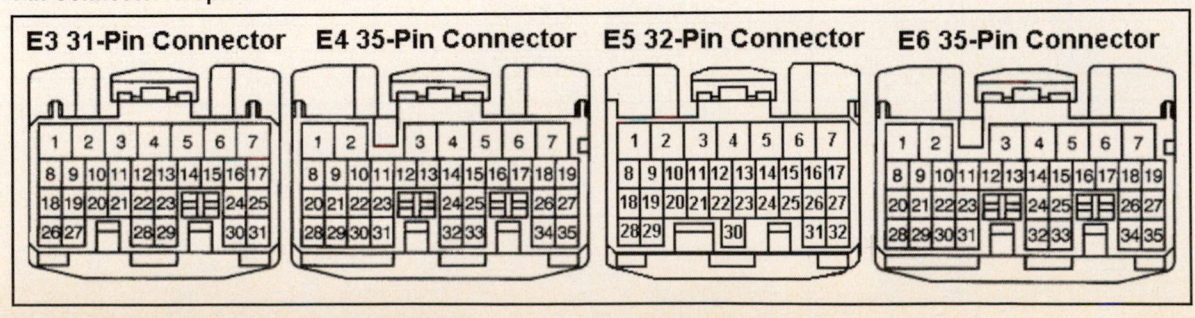

2003 Pickup 4.7L V8 VIN T E7 34 Pin Connector

PCM Pin #	Wire Color	Circuit Description (34 Pin)	Value at Hot Idle
1	RD	Injector 1 Control	2.0-3.3 ms
2	WT	Injector 2 Control	2.0-3.3 ms
3	GN	Injector 3 Control	2.0-3.3 ms
4	RD/BL	Injector 4 Control	2.0-3.3 ms
5	BL	Injector 5 Control	2.0-3.3 ms
6	WT/BK	Power Ground (E02)	<0.1v
7	WT/BK	Power Ground (E01)	<0.1v
8	LG/BK	Igniter Transistor 2 Control	7% duty cycle
9	GN/RD	Igniter Transistor 1 Control	7% duty cycle
10	LG	Igniter Transistor 8 Control	7% duty cycle
11	BK/BL	Igniter Transistor 4 Control	7% duty cycle
12	GN/WT	Igniter Transistor 5 Control	7% duty cycle
13	RD/WT	Igniter Transistor 7 Control	7% duty cycle
14	---	Not Used	---
15	GN/YL	A/C Relay Control (ACCR)	Relay Off: 12v, On: 1v
16	WT/BK	Neutral Start Switch (NSW)	In P/N: 0-3.0v
17	PK	Starter Switch Signal (STA)	In P/N: 0-3.0v
18	GN/BK	Sensor VREF (VC)	4.9-5.1v
19	GN	ECT Sensor Signal (THW)	At 180°F: 0.51v
20	YL/GN	IAT Sensor Signal (THA)	At 100°F: 2.60v
21	BK/YL	TP Sensor Signal (VTA1)	0.4-1.0v
22	---	Not Used	---
23	BL/WT	Igniter Signal (IGF2)	Digital Signal: 0-5-0v
24	BL/BK	Igniter Signal (IGF1)	Digital Signal: 0-5-0v
25	GN/BK	Igniter Transistor 3 Control	7% duty cycle
26	PK/BL	Igniter Transistor 6 Control	7% duty cycle
27	BL/WT	EVAP Canister Closed Valve (VSV)	12v or 0v
28	BK/WT	Sensor Ground (E2)	<0.050v
29	BK/YL	MAF Sensor Ground (E2G)	<0.050v
30	RD/WT	Mass Airflow Sensor (VG)	1.1-1.5v
31	PK/BL	TP Sensor Signal (VTA2)	2.0-2.9v
32	---	Not Used	---
33	GN/RD	Fuel Pump Relay Control (FPR)	Relay Off: 12v, On: 1v
34	YL/RD	EVAP Purge Solenoid (VSV)	12v or 0v

Pin Connector Graphic

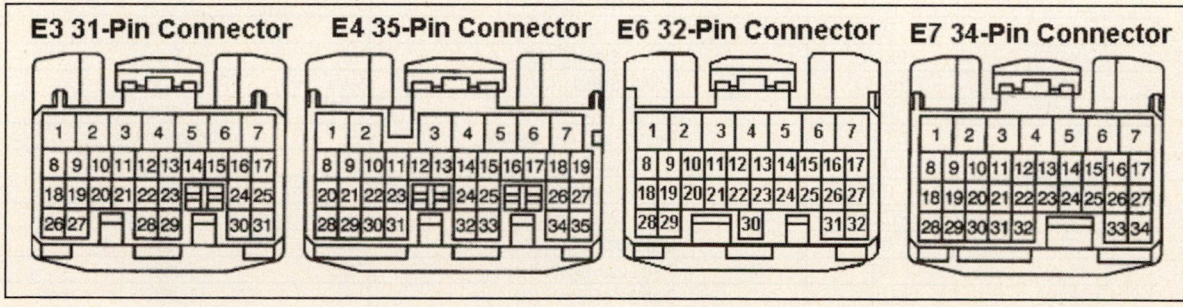

E3 31-Pin Connector E4 35-Pin Connector E6 32-Pin Connector E7 34-Pin Connector

T-100 Pin Tables

1994-95 Pickup 2.7L I4 3RZ-FE VIN U (M/T) 16 Pin Connector

PCM Pin #	Wire Color	Circuit Description (16 Pin)	Value at Hot Idle
1	GN/YL	Sensor VREF (VC)	4.9-5.1v
2	YL/RD	MAF Sensor Signal (VG)	1.1-1.8v
3	GN/WT	EGR Gas Temperature Sensor (THG)	3.5-4.0v
4	GN/BL	ECT Sensor Signal (THW)	At 180ºF: 0.51v
5	WT	HO2S-12 (B1 S2) Signal	0.1-1.1v
6	BK	HO2S-11 (B1 S1) Signal	0.1-1.1v
7	YL/GN	IAT Sensor Signal (THA)	At 100ºF: 2.60v
9	BR/BK	Sensor Ground	<0.050v
11	YL	TP Sensor Signal (VTA)	0.3-0.8v
12	YL/BL	Closed Throttle Switch	1v, off-idle: 12v
13	BK	Knock Sensor Signal	<0.075v AC
14	PK/GN	Data Link Connector	12-14v
15	PK/WT	Data Link Connector	12-14v
16	BR	Sensor Ground	<0.050v

1994-95 Pickup 2.7L I4 3RZ-FE VIN U (M/T) 22 Pin Connector

PCM Pin #	Wire Color	Circuit Description (22 Pin)	Value at Hot Idle
1	BK/GN	Battery Direct	12-14v
5	PK	MIL (lamp) Control	MIL Off: 12v, On: 1v
7	WT	Data Link Connector	12v
8	BL/BK	A/C Amplifier Signal (ACT)	Clutch On: 12v, Off: 1.5v
9	GN	Vehicle Speed Sensor	At 55 mph: 48 Hz
10	BK/RD	A/C Amplifier Signal (AC1)	Clutch On: 1.5v, Off: 12v
11	BK/WT	Starter Switch Signal	9-11v (cranking)
12	WT/RD	EFI Main Relay Power	12-14v
14	GN/YL	Circuit Opening Relay (FC)	0-3v, at off-idle: 12v
20	GN/WT	Stop Light Switch Signal	Brake Off: 0v, On: 12v

1994 Pickup 2.7L I4 3RZ-FE VIN U (M/T) 26 Pin Connector

PCM Pin #	Wire Color	Circuit Description (26 Pin)	Value at Hot Idle
1	BK/RD	A/C Idle-Up Solenoid	A/C Off: 12v, On: 1v
2	BK/GN	HO2S-11 (B1 S1) Heater	Heater Off: 12v, On: 1v
3	BK/YL	Igniter Signal (IGF)	Digital Signal: 0-5-0v
4	WT	Distributor Signal (NE+)	AC pulse signals
5	BK	Distributor Signal (G2+)	AC pulse signals
6 (Cal)	PK	EGR Solenoid Control (VSV)	12v or 0v
7, 16, 19	---	Not Used	---
8	GN	Fuel Pressure Up Solenoid	1v (at hot startup)
9	PK/YL	IAC Signal (RSC)	Pulse Signals
10	PK/RD	IAC Signal (RSO)	Pulse Signals
11	WT	Injector Pair 2 & 4 Control	1.6-2-9 ms
12	WT/RD	Injector Pair 1 & 3 Control	1.6-2-9 ms
13	BR	Power Ground	<0.1v
14	BR	Sensor Ground	<0.050v
15	RD/GN	HO2S-12 (B1 S2) Heater	Heater Off: 12v, On: 1v
17	BK	Distributor Signal (NE-)	<0.050v
18 ('94)	WT	Distributor Signal (GN-)	<0.050v
18 ('95)	GN	Distributor Signal (GN-)	<0.050v
20	BK/BL	Igniter Signal (IGF)	Digital Signal: 0-5-0v
21-22, 24	---	Not Used	---
23	WT/GN	EVAP Purge Solenoid (VSV)	12v or 0v
25	BR	Power Ground	<0.1v
26	BR	Power Ground	<0.1v

Pin Connector Graphic

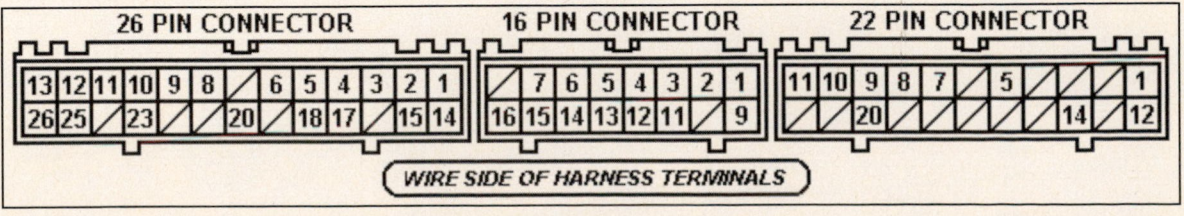

1995 Pickup 2.7L I4 MFI VIN U (A/T-ECT) 12 Pin Connector

PCM Pin #	Wire Color	Circuit Description (12 Pin)	Value at Hot Idle
1	WT/GN	EVAP Purge Solenoid (VSV)	12v or 0v
4	BK/RD	A/T Vehicle Speed Sensor (-)	AC pulse signals
5	GN	Distributor Signal (GN-)	<0.050v
6	BK	CKP Sensor Signal (NE-)	<0.050v
10	BR/RD	A/T Vehicle Speed Sensor (+)	AC pulse signals
11	RD	Distributor Signal (G2+)	AC pulse signals
12	WT	CKP Sensor Signal (NE+)	AC pulse signals

1995 Pickup 2.7L I4 MFI VIN U (A/T-ECT) 22 Pin Connector

PCM Pin #	Wire Color	Circuit Description (22 Pin)	Value at Hot Idle
2	BK/GN	Battery Direct	12-14v
4	PK	MIL (lamp) Control	MIL Off: 12v, On: 1v
5	YL/GN	Overdrive Main Switch	Switch Off: 12v, On: 1v
6	BL/BK	A/C Amplifier Signal (ACT)	Clutch On: 12v, Off: 1.5v
7	BK/RD	A/C Amplifier Signal (AC1)	Clutch On: 1.5v, Off: 12v
8	GN	Vehicle Speed Sensor	At 55 mph: 48 Hz
11	BK/WT	Starter Switch Signal	9-11v (cranking)
12	WT/RD	EFI Main Relay (B+)	12-14v
15	PK/WT	A/T Select Switch Low	In Low: 12v, Others: 0v
16	PK/GN	A/T Select Switch 2nd	In 2nd: 12v, Others: 0v
18	YL/RD	Cruise Control ECU (OD1)	At Cruise in OD: 12v
19	WT	Data Link Connector (SDL)	0v
21	GN/WT	Stop Light Switch Signal	Brake Off: 0v, On: 12v
22	BK/YL	Neutral Start Switch	In P/N: 9-11v (cranking)

Pin Connector Graphic

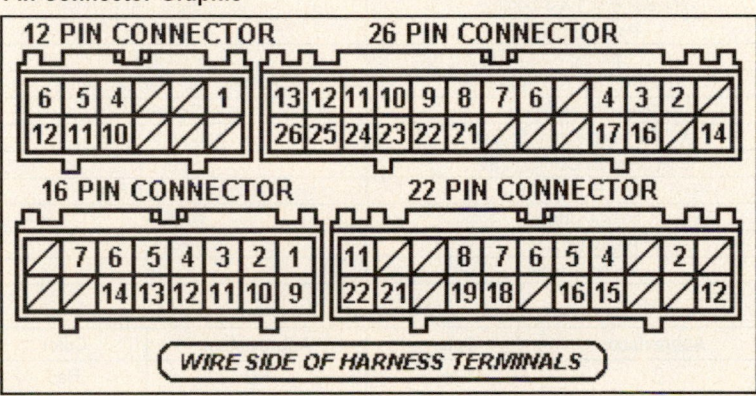

1995 Pickup 2.7L I4 MFI VIN U (A/T-ECT) 16 Pin Connector

PCM Pin #	Wire Color	Circuit Description (16 Pin)	Value at Hot Idle
1	GN/YL	Sensor VREF	4.9-5.1v
2	YL/RD	MAF Sensor Signal (VG)	1.1-1.8v
3	BR	Sensor Ground	<0.050v
4	GN/YL	ECT Sensor Signal (THW)	At 180°F: 0.51v
5	BK	HO2S-11 (B1 S1) Signal	0.1-1.1v
6	BK	Knock Sensor Signal	<0.075v AC
7	PK/WT	Data Link Connector	12-14v
9	BR/BK	Sensor Ground	<0.050v
10	YL	TP Sensor Signal (VTA)	0.3-0.8v
11	YL/BL	Closed Throttle Switch	1v, off-idle: 12v
12	YL/GN	IAT Sensor Signal (THA)	At 100°F: 2.60v
13	WT	HO2S-12 (B1 S2) Signal	0.1-1.1v
14	GN/WT	EGR Gas Temperature Sensor (THG)	3.5-4.0v

1995 Pickup 2.7L I4 MFI VIN U (A/T-ECT) 26 Pin Connector

PCM Pin #	Wire Color	Circuit Description (26 Pin)	Value at Hot Idle
2	BK/RD	A/C Idle-Up Solenoid	A/C Off: 12v, On: 1v
3	PK/GN	HO2S-11 (B1 S1) Heater	Heater Off: 12v, On: 1v
4	GN	Fuel Pressure Up Solenoid	1v (at hot startup)
6	PK/YL	IAC Signal (RSC)	Pulse Signals
7	PK/RD	IAC Signal (RSO)	Pulse Signals
8	YL/BK	A/T-ECT Solenoid (SL)	In Lockup: 12-14v
9	BK/WT	A/T-ECT Solenoid (S2)	1st or OD: 1v
10	WT	A/T-ECT Solenoid (S1)	3rd or OD: 1v
11	WT	Injector Pair 2 & 4 Control	1.6-2-9 ms
12	WT/RD	Injector Pair 1 & 3 Control	1.6-2-9 ms
13	BR	Power Ground	<0.1v
14	GN/YL	Circuit Opening Relay (FC)	0-3v, at off-idle: 12v
16	RD/GN	HO2S-12 (B1 S2) Heater	Heater Off: 12v, On: 1v
17	BK/YL	Igniter Signal (IGF)	Digital Signal: 0-5-0v
21	GN/BK	A/T Oil Temperature Sensor	At 68°F: 4-5v
22 (Cal)	PK	EGR Solenoid Control (VSV)	12v or 0v
23	BK/BL	Igniter Signal (IGF)	Digital Signal: 0-5-0v
24	BR	Sensor Ground	<0.050v
25	BR	Power Ground	<0.1v
26	BR	Power Ground	<0.1v

Standard Colors and Abbreviations

Abbreviation	Color	Abbreviation	Color	Abbreviation	Color
BK	Black	GY	Gray	RD	Red
BL	Blue	GN	Green	TN	Tan
BR	Brown	LG	LT Green	VT	Violet
DB	Dark Blue	OR	Orange	WT	White
DG	DK Green	PK	Pink	YL	Yellow

1996-97 Pickup 2.7L I4 MFI VIN M (A/T-ECT) 12 Pin Connector

PCM Pin #	Wire Color	Circuit Description (12 Pin)	Value at Hot Idle
1	WT/GN	EVAP Purge Solenoid (VSV)	12v or 0v
4	BK/RD	A/T Vehicle Speed Sensor (-)	AC pulse signals
5	GN	Distributor Signal (GN-)	<0.050v
6	BK	CKP Sensor Signal (NE-)	<0.050v
10	BR/RD	A/T Vehicle Speed Sensor (+)	Pulses
11	RD	Distributor Signal (G2+)	AC pulse signals
12	WT	CKP Sensor Signal (NE+)	AC pulse signals

1996-97 Pickup 2.7L I4 MFI VIN M (A/T-ECT) 22 Pin Connector

PCM Pin #	Wire Color	Circuit Description (22 Pin)	Value at Hot Idle
2	BK/GN	Battery Direct	12-14v
4	PK	MIL (lamp) Control	MIL Off: 12v, On: 1v
5	YL/GN	Overdrive Main Switch	Switch Off: 12v, On: 1v
6	BL/BK	A/C Amplifier Signal (ACT)	Clutch On: 12v, Off: 1.5v
7	BK/RD	A/C Amplifier Signal (AC1)	Clutch On: 1.5v, Off: 12v
8	GN	Vehicle Speed Sensor	At 55 mph: 48 Hz
11	BK/WT	Starter Switch Signal	9-11v (cranking)
12	WT/RD	EFI Main Relay Power	12-14v
15	PK/WT	A/T Select Switch Low	In Low: 12v, Others: 0v
16	PK/GN	A/T Select Switch 2nd	In 2nd: 12v, Others: 0v
17	RD/BK	A/T Select Switch Reverse	In 'R': 12v, Others: 0v
19	WT	SIL (Scan Tool) Signal	12v
21	GN/WT	Stop Light Switch Signal	Brake Off: 0v, On: 12v
22	BK/YL	Neutral Start Switch	In P/N: 9-11v (cranking)

Pin Connector Graphic

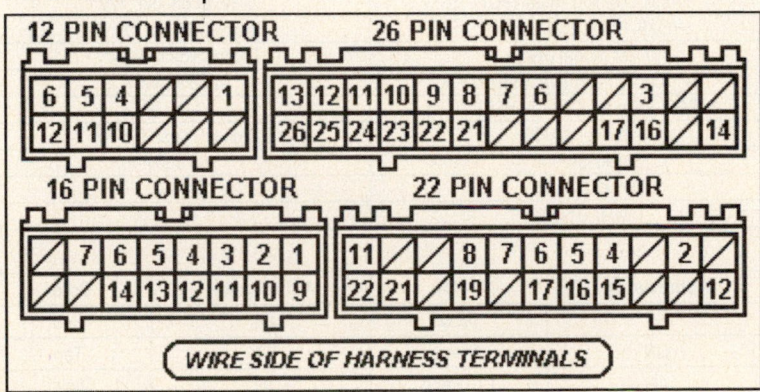

1996-97 Pickup 2.7L I4 MFI VIN M (A/T-ECT) 16 Pin Connector

PCM Pin #	Wire Color	Circuit Description (16 Pin)	Value at Hot Idle
1	GN/YL	Sensor VREF	4.9-5.1v
2	YL/RD	MAF Sensor Signal (VG)	1.1-1.8v
3	BR	Sensor Ground	<0.050v
4	GN/YL	ECT Sensor Signal (THW)	At 180°F: 0.51v
5	BK	HO2S-11 (B1 S1) Signal	0.1-1.1v
6	BK	Knock Sensor Signal	<0.075v AC
7	PK/WT	Data Link Connector	12-14v
9	BR/BK	Sensor Ground	<0.050v
10	YL	TP Sensor Signal (VTA)	0.3-0.8v
11	YL/BL	Closed Throttle Switch	1v, off-idle: 12v
12	YL/GN	IAT Sensor Signal (THA)	At 100°F: 2.60v
13	WT	HO2S-12 (B1 S2) Signal	0.1-1.1v
14	GN/WT	EGR Gas Temperature Sensor (THG)	3.5-4.0v

1996-97 Pickup 2.7L I4 MFI VIN M (A/T-ECT) 26 Pin Connector

PCM Pin #	Wire Color	Circuit Description (26 Pin)	Value at Hot Idle
3	PK/GN	HO2S-11 (B1 S1) Heater	Heater Off: 12v, On: 1v
6	PK/YL	IAC Signal (RSC)	Pulse Signals
7	PK/RD	IAC Signal (RSO)	Pulse Signals
8	YL/BK	A/T-ECT Solenoid (SL)	In Lockup: 12-14v
9	BK/WT	A/T-ECT Solenoid (S2)	1st or OD: 1v
10	WT	A/T-ECT Solenoid (S1)	3rd or OD: 1v
11	WT	Injector Pair 2 & 4 Control	1.6-2-9 ms
12	WT/RD	Injector Pair 1 & 3 Control	1.6-2-9 ms
13	BR	Power Ground	<0.1v
14	GN/YL	Circuit Opening Relay (FC)	0-3v, at off-idle: 12v
16	RD/GN	HO2S-12 (B1 S2) Heater	Heater Off: 12v, On: 1v
17	BK/YL	Igniter Signal (IGF)	Digital Signal: 0-5-0v
21	GN/BK	A/T Oil Temperature Sensor	At 68°F: 4-5v
22	PK	EGR Solenoid Control (VSV)	12v or 0v
23	BK/BL	Igniter Signal (IGF)	Digital Signal: 0-5-0v
24	BR	Shield Ground	<0.050v
25	BR	Power Ground	<0.1v
26	BR	Power Ground	<0.1v

Standard Colors and Abbreviations

Abbreviation	Color	Abbreviation	Color	Abbreviation	Color
BK	Black	GY	Gray	RD	Red
BL	Blue	GN	Green	TN	Tan
BR	Brown	LG	LT Green	VT	Violet
DB	Dark Blue	OR	Orange	WT	White
DG	DK Green	PK	Pink	YL	Yellow

1996-97 Pickup 2.7L I4 MFI 3RZ-FE VIN M (M/T) 22 Pin Connector

PCM Pin #	Wire Color	Circuit Description (22 Pin)	Value at Hot Idle
1	BK/GN	Battery Direct	12-14v
5	PK	MIL (lamp) Control	MIL Off: 12v, On: 1v
7	WT	SIL (Scan Tool) Signal	12v
8	BL/BK	A/C Amplifier Signal (ACT)	Clutch On: 12v, Off: 1.5v
9	GN	Vehicle Speed Sensor	At 55 mph: 48 Hz
10	BK/RD	A/C Amplifier Signal (AC1)	Clutch On: 1.5v, Off: 12v
11	BK/WT	Starter Switch Signal	9-11v (cranking)
12	WT/RD	EFI Main Relay (B+)	12-14v
14	GN/YL	Circuit Opening Relay (FC)	0-3v, at off-idle: 12v
20	GN/WT	Stop Light Switch Signal	Brake Off: 0v, On: 12v

1996-97 Pickup 2.7L I4 MFI 3RZ-FE VIN M (M/T) 16 Pin Connector

PCM Pin #	Wire Color	Circuit Description (16 Pin)	Value at Hot Idle
1	GN/YL	Sensor VREF (VC)	4.9-5.1v
2	YL/RD	MAF Sensor Signal (VG)	1.1-1.8v
3	GN/WT	EGR Gas Temperature Sensor (THG)	3.5-4.0v
4	GN/YL	ECT Sensor Signal (THW)	At 180°F: 0.51v
5	WT	HO2S-12 (B1 S2) Signal	0.1-1.1v
6	BK	HO2S-11 (B1 S1) Signal	0.1-1.1v
7	YL/GN	IAT Sensor Signal (THA)	At 100°F: 2.60v
9	BR/BK	Sensor Ground	<0.050v
11	YL	TP Sensor Signal (VTA)	0.3-0.8v
12	YL/BL	Closed Throttle Switch	1v, off-idle: 12v
13	BK	Knock Sensor Signal	<0.075v AC
15	PK/WT	Data Link Connector	12-14v
16	BR	Sensor Ground	<0.050v

Standard Colors and Abbreviations

Abbreviation	Color	Abbreviation	Color	Abbreviation	Color
BK	Black	GY	Gray	RD	Red
BL	Blue	GN	Green	TN	Tan
BR	Brown	LG	LT Green	VT	Violet
DB	Dark Blue	OR	Orange	WT	White
DG	DK Green	PK	Pink	YL	Yellow

1996-97 Pickup 2.7L I4 MFI 3RZ-FE VIN M (M/T) 26 Pin Connector

PCM Pin #	Wire Color	Circuit Description (26 Pin)	Value at Hot Idle
2	PK/GN	HO2S-11 (B1 S1) Heater	Heater Off: 12v, On: 1v
3	BK/YL	Igniter Signal (IGF)	Digital Signal: 0-5-0v
4	WT	CKP Sensor Signal (NE+)	AC pulse signals
5	RD	Distributor Signal (G2+)	AC pulse signals
6	PK	EGR Solenoid Control (VSV)	12v or 0v
9	PK/YL	IAC Signal (RSC)	Pulse Signals
10	PK/RD	IAC Signal (RSO)	Pulse Signals
11	WT	Injector Pair 2 & 4 Control	1.6-2-9 ms
12	WT/RD	Injector Pair 1 & 3 Control	1.6-2-9 ms
13	BR	Power Ground	<0.1v
14	BR	Shield Ground	<0.050v
15	RD/GN	HO2S-12 (B1 S2) Heater	Heater Off: 12v, On: 1v
17	BK	CKP Sensor Signal (NE-)	<0.050v
18	GN	Distributor Signal (GN-)	<0.050v
20	BK/BL	Igniter Signal (IGF)	Digital Signal: 0-5-0v
23	WT/GN	EVAP Purge Solenoid (VSV)	12v or 0v
25	BR	Power Ground	<0.1v
26	BR	Power Ground	<0.1v

Pin Connector Graphic

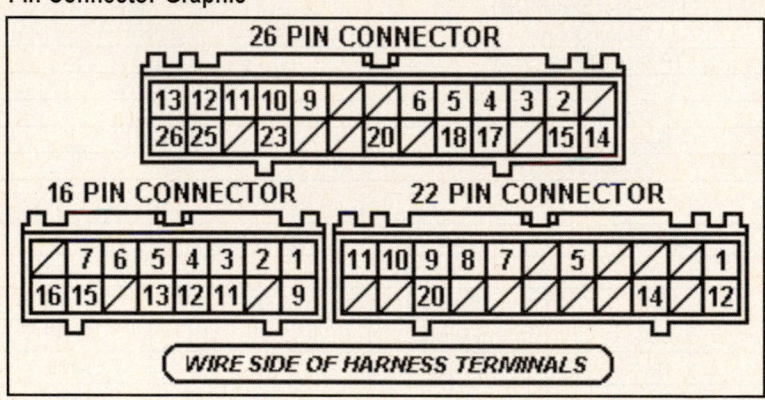

1998 Pickup 2.7L I4 MFI VIN M (All) 12-Pin Connector

PCM Pin #	Wire Color	Circuit Description (12 Pin)	Value at Hot Idle
2	GN/RD	Taillight Switch Signal (ELS)	Lights On: 12v
3	BK/RD	A/T Vehicle Speed Sensor (-)	AC pulse signals
5	GN	CMP Sensor Signal (GN-)	<0.050v
6	BK	CKP Sensor Signal (NE-)	<0.050v
7	BR	MAF Sensor Ground	<0.050v
9	BR/RD	A/T Vehicle Speed Sensor (+)	AC pulse signals
10	RD/BL	EVAP Vapor Pressure Sensor	2.5-3.1v (with fuel cap off)
11	RD	CMP Sensor Signal (G2+)	AC pulse signals
12	WT	CKP Sensor Signal (NE+)	AC pulse signals

1998 Pickup 2.7L I4 MFI VIN M (All) 22 Pin Connector

PCM Pin #	Wire Color	Circuit Description (22 Pin)	Value at Hot Idle
2	BK/GN	Battery Direct	12-14v
4	PK	MIL (lamp) Control	MIL Off: 12v, On: 1v
5	YL/GN	Overdrive Main Switch	Switch Off: 12v, On: 1v
6	BL/BK	A/C Amplifier Signal (ACT)	Clutch On: 12v, Off: 1.5v
7	BK/RD	A/C Amplifier Signal (AC1)	Clutch On: 1.5v, Off: 12v
8	GN	Vehicle Speed Sensor	At 55 mph: 48 Hz
11	BK/WT	Starter Switch Signal	9-11v (cranking)
12	WT/RD	EFI Main Relay Power	12-14v
15	PK/WT	A/T Select Switch Low	In Low: 12v, Others: 0v
16	PK/GN	A/T Select Switch 2nd	In 2nd: 12v, Others: 0v
17	RD/BK	A/T Select Switch Reverse	In 'R': 12v, Others: 0v
19	WT	SIL (Scan Tool) Signal	12v
21	GN/WT	Stop Light Switch Signal	Brake Off: 0v, On: 12v
22	BK/YL	Neutral Start Switch	In P/N: 9-11v (cranking)

Pin Connector Graphic

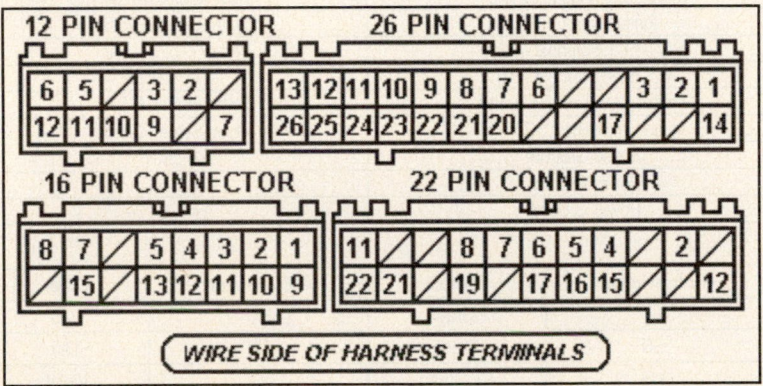

1998 Pickup 2.7L I4 MFI VIN M (All) 16 Pin Connector

PCM Pin #	Wire Color	Circuit Description (16 Pin)	Value at Hot Idle
1	GN/YL	Sensor VREF (VC)	4.9-5.1v
2	YL/RD	MAF Sensor Signal (VG)	1.1-1.8v
3	YL/GN	IAT Sensor Signal (THA)	At 100°F: 2.60v
4	GN/YL	ECT Sensor Signal (THW)	At 180°F: 0.51v
5	BK	HO2S-11 (B1 S1) Signal	0.1-1.1v
7	PK/WT	Data Link Connector	12-14v
8	GN/RD	EVAP Vapor Pressure (VSV)	12v or 0v
9	BR/BK	Sensor Ground	<0.050v
10	YL	TP Sensor Signal (VTA)	0.3-0.8v
11	GN/WT	EGR Gas Temperature Sensor (THG)	3.5-4.0v
12	BK	Knock Sensor Signal	<0.075v AC
13	WT	HO2S-12 (B1 S2) Signal	0.1-1.1v
15	PK	EGR Solenoid Control (VSV)	12v or 0v

1998 Pickup 2.7L I4 MFI VIN M (All) 26 Pin Connector

PCM Pin #	Wire Color	Circuit Description (26 Pin)	Value at Hot Idle
1	RD/GN	HO2S-12 (B1 S2) Heater	Heater Off: 12v, On: 1v
2	PK/GN	HO2S-11 (B1 S1) Heater	Heater Off: 12v, On: 1v
3	WT/GN	EVAP Purge Solenoid (VSV)	12v or 0v
6	PK/YL	IAC Signal (RSC)	Pulse Signals
7	PK/RD	IAC Signal (RSO)	Pulse Signals
8	WT	A/T-ECT Solenoid (S1)	3rd or OD: 1v
9	PK/BK	Injector 4 Control	1.6-2-9 ms
10	BK	Injector 3 Control	1.6-2-9 ms
11	WT	Injector 2 Control	1.6-2-9 ms
12	WT/RD	Injector 1 Control	1.6-2-9 ms
13	BR	Power Ground	<0.1v
14	GN/YL	Circuit Opening Relay (FC)	0-3v, at off-idle: 12v
17	BK/YL	Igniter Signal (IGF)	Digital Signal: 0-5-0v
20	YL/BK	A/T-ECT Solenoid (SL)	In Lockup: 12-14v
21	BK/WT	A/T-ECT Solenoid (S2)	1st or OD: 1v
22	BK/OR	Igniter Transistor 2 Control	Digital Signal: 0-5-0v
23	BK/BL	Igniter Transistor 1 Control	Digital Signal: 0-5-0v
24	BR	Shield Ground	<0.050v
25	BR	Power Ground	<0.1v
26	BR	Power Ground	<0.1v

Standard Colors and Abbreviations

Abbreviation	Color	Abbreviation	Color	Abbreviation	Color
BK	Black	GY	Gray	RD	Red
BL	Blue	GN	Green	TN	Tan
BR	Brown	LG	LT Green	VT	Violet
DB	Dark Blue	OR	Orange	WT	White
DG	DK Green	PK	Pink	YL	Yellow

1993-94 Pickup 3.0L V6 MFI VIN V (All) 16 Pin Connector

PCM Pin #	Wire Color	Circuit Description (16 Pin)	Value at Hot Idle
1	GN/BK	Sensor VREF (VC)	4.9-5.1v
2	BL/YL	MAF Sensor Signal (VG)	1.1-1.8v
3	BL/RD	IAT Sensor Signal (THA)	At 100°F: 2.60v
4	GN/BL	ECT Sensor Signal (THW)	At 180°F: 0.51v
5	BK	Knock Sensor Signal	<0.075v AC
6	BK	HO2S-11 (B1 S1) Signal	0.1-1.1v
7	GN/BL	4WD Oil Temperature Sensor	At 68°F: 4-5v
8	YL/GN	Check Connector	12-14v
9	BR	Sensor Ground	<0.050v
10	WT	HO2S-12 (B1 S2) Signal	0.1-1.1v
11	YL	TP Sensor Signal (VTA)	0.3-0.8v
12	YL/BL	Closed Throttle Switch	1v, off-idle: 12v
13 (Cal)	GN/WT	EGR Gas Temperature Sensor (THG)	3.5-4.0v
15	PK/WT	Data Link Connector	12-14v
16	PK/GN	Data Link Connector	12-14v

1993-94 Pickup 3.0L V6 MFI VIN V (All) 22 Pin Connector

PCM Pin #	Wire Color	Circuit Description (22 Pin)	Value at Hot Idle
1	BK/GN	Battery Direct	12-14v
4	BL/WT	A/T Oil Temperature Lamp	Lamp Off: 12v, On: 1v
5	PK	MIL (lamp) Control	MIL Off: 12v, On: 1v
6	GN/WT	Stop Light Switch Signal	Brake Off: 0v, On: 12v
7	BL/RD	A/T Pattern Select Switch	Norm: 0v, PWR: 12v
8	PK/GN	4WD Selector Switch	Open: 12v, Closed: 0v
9	GN/BL	Vehicle Speed Sensor	At 55 mph: 48 Hz
10	BL/BK	A/C Amplifier Signal (ACT)	Clutch On: 12v, Off: 1.5v
11	BK/WT	Starter Switch Signal	9-11v (cranking)
12	WT/RD	EFI Main Relay Power	12-14v
13	WT/RD	EFI Main Relay Power	12-14v
15	BR/BK	Sensor Ground (E2)	<0.050v
16	YL/GN	Main Overdrive Switch	Switch Off: 12v, On: 1v
17	BR/WT	M/T Ground (SEL2)	<0.050v
17	BR/RD	A/T Short Pin Ground (SEL2)	<0.050v
18	BR/YL	A/T 2WD Ground (SEL1)	<0.050v
19	BK/WT	A/C Amplifier Signal (AC1)	Clutch On: 1.5v, Off: 12v
20	YL/WT	Data Link Connector	12v
21	YL/RD	Cruise Control ECU (OD1)	At Cruise in OD: 12v
22 (Cal)	RD/GN	HO2S-12 (B1 S2) Heater	Heater Off: 12v, On: 1v

1993-94 Pickup 3.0L V6 MFI VIN V (All) 26 Pin Connector

PCM Pin #	Wire Color	Circuit Description (26 Pin)	Value at Hot Idle
1	WT	Distributor Signal (NE+)	AC pulse signals
2	RD	Distributor Signal (G1+)	AC pulse signals
3	BK/YL	Igniter Signal (IGF)	Digital Signal: 0-5-0v
4	WT/GN	EVAP Purge Solenoid (VSV)	12v or 0v
5	YL/BK	A/T-ECT Solenoid (S3)	In Lockup: 12-14v
6	BK	A/T-ECT Solenoid (S2)	1st or OD: 1v
7	WT	A/T-ECT Solenoid (S1)	3rd or OD: 1v
8	GN/YL	Fuel Pressure Up Solenoid	1v (at hot startup)
9	GN	Cold Start Injector	1v (at cold startup)
10	GY/BK	HO2S-11 (B1 S1) Heater	Heater Off: 12v, On: 1v
11	BR	Sensor Ground	<0.050v
12	WT/RD	Injectors 1 & 3 & 5 Control	1.6-2.9 ms
13	BR	Power Ground	<0.1v
14	GN	Distributor Signal (GN-)	<0.050v
15	BK	Distributor Signal (G2+)	AC pulse signals
16	BR/RD	A/T Vehicle Speed Sensor	Moving: 0-5-0v
17	PK/WT	A/T Select Switch Low	In Low: 12v, Others: 0v
18	PK/GN	A/T Select Switch 2nd	In 2nd: 12v, Others: 0v
19	PK/RD	A/T Select Switch Neutral	In 'N': 12v, Others: 0v
20	RD	4WD Detection Transfer (L4)	Open: 12v, Closed: 0v
21	BK/BL	Igniter Signal (IGF)	Digital Signal: 0-5-0v
22 (Cal)	PK	EGR Solenoid Control (VSV)	12v or 0v
23	GN/RD	Intake Air Solenoid Control	12v or 0v
24	BK/RD	A/C Idle-Up Solenoid	A/C Off: 12v, On: 1v
25	WT	Injectors 2 & 4 & 6 Control	1.6-2.9 ms
26	BR	Power Ground	<0.1v

Pin Connector Graphic

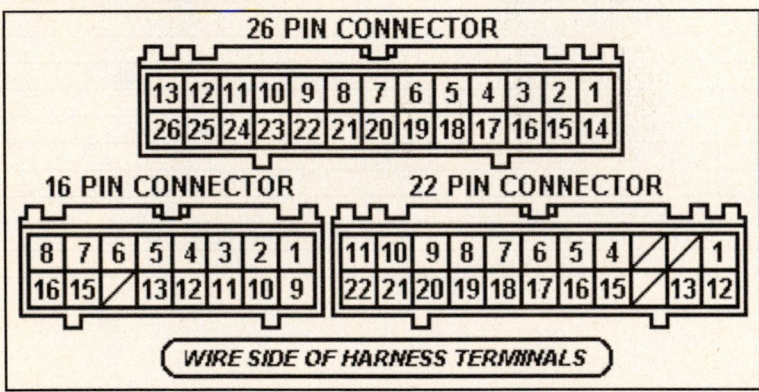

1995 Pickup 3.4L V6 MFI VIN V (A/T-ECT) 28 Pin Connector

PCM Pin #	Wire Color	Circuit Description (28 Pin)	Value at Hot Idle
2	PK/GN	A/T Select Switch 2nd	In 2nd: 12v, Others: 0v
3	PK/WT	A/T Select Switch Low	In Low: 12v, Others: 0v
4	LG	A/T Oil Temperature Lamp	Lamp Off: 12v, On: 1v
5	BL/BK	A/C Amplifier Signal (ACT)	Clutch On: 12v, Off: 1.5v
6	YL/GN	Overdrive Indicator Lamp	Lamp Off: 12-14v
7	YL/RD	Cruise Control ECU (OD1)	At Cruise in OD: 12v
12	GN	Vehicle Speed Sensor	At 55 mph: 48 Hz
14	BK/GN	Battery Direct	12-14v
17	YL	A/T: Parking Indicator Lamp	Lamp Off: 12v, On: 1v
18	WT	SIL (Scan Tool) Signal	12v
20	BK/RD	A/C Amplifier Signal (AC1)	Clutch On: 1.5v, Off: 12v
22	WT/RD	EFI Main Relay Power	12-14v
25	GN/WT	Stop Light Switch Signal	Brake Off: 0v, On: 12v
26	PK/GN	4WD Selector Switch	Switch Off: 12v, On: 1v

1995 Pickup 3.4L V6 MFI VIN V (A/T-ECT) 22 Pin Connector

PCM Pin #	Wire Color	Circuit Description (22 Pin)	Value at Hot Idle
1	GN/BK	Sensor VREF	4.9-5.1v
5	GN	CKP Sensor Signal (NE+)	AC pulse signals
6	BL	CKP Sensor Signal (NE-)	<0.050v
7	YL/BK	TP Sensor Signal (VTA)	0.3-0.8v
8	GY/RD	MAF Sensor Signal (VG)	1.1-1.8v
9	BR/RD	A/T Vehicle Speed Sensor	Moving: 0-5-0v
10	BK	CMP Sensor Signal (G2+)	AC pulse signals
11	WT	CMP Sensor Signal (GN-)	<0.050v
12	GN/BK	A/T Oil Temperature Sensor	At 230ºF: <1.5v
13	WT	HO2S-11 (B1 S1) Signal	0.1-1.1v
14	YL/GN	IAT Sensor Signal (THA)	At 100ºF: 2.60v
16	GY	Knock Sensor 2 Signal	<0.075v AC
17	BK	Knock Sensor 1 Signal	<0.075v AC
18	BR/WT	Sensor Ground	<0.050v
19	RD	HO2S-12 (B1 S2) Signal	0.1-1.1v
20	GN/YL	ECT Sensor Signal (THW)	At 180ºF: 0.51v
21	PK	EGR Gas Temperature Sensor (THG)	3.5-4.0v
22	BR/BK	Sensor Ground	<0.050v

Pin Connector Graphic

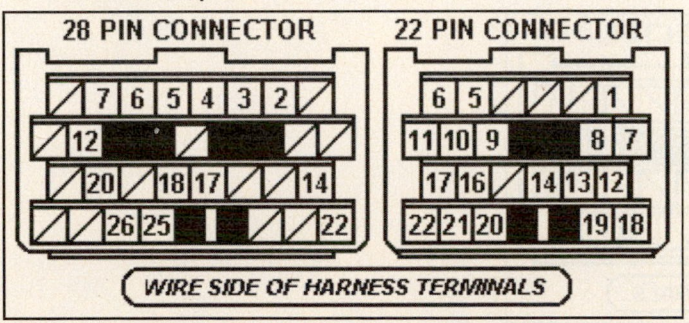

28 PIN CONNECTOR   **22 PIN CONNECTOR**

WIRE SIDE OF HARNESS TERMINALS

1995 Pickup 3.4L V6 MFI VIN V (A/T-ECT) 16 Pin Connector

PCM Pin #	Wire Color	Circuit Description (16 Pin)	Value at Hot Idle
3	PK	MIL (lamp) Control	MIL Off: 12v, On: 1v
4	GN/YL	Circuit Opening Relay (FC)	0-3v, at off-idle: 12v
5	PK/WT	Data Link Connector	12-14v
8	RD/WT	EGR Solenoid Control (VSV)	12v or 0v
9	RD/BK	Fuel Pressure Up Solenoid	1v (at hot startup)
10	BK/RD	A/C Idle-Up Solenoid	A/C Off: 12v, On: 1v
15	WT/GN	EVAP Purge Solenoid (VSV)	12v or 0v
16	BR	Sensor Ground	<0.050v

1995 Pickup 3.4L V6 MFI VIN V (A/T-ECT) 34-Pin Connector

PCM Pin #	Wire Color	Circuit Description (34-Pin)	Value at Hot Idle
1	BR	Power Ground	<0.1v
5	YL/BK	Injector 6 Control	1.6-2.9 ms
6	WT/BL	Injector 5 Control	1.6-2.9 ms
7	YL/RD	Injector 4 Control	1.6-2.9 ms
8	WT/GN	Injector 3 Control	1.6-2.9 ms
9	WT	Injector 2 Control	1.6-2.9 ms
10	WT/RD	Injector 1 Control	1.6-2.9 ms
11	WT	A/T-ECT Solenoid (S1)	3rd or OD: 1v
12	BK/YL	Igniter Signal (IGF)	Digital Signal: 0-5-0v
13	BK/WT	Starter Switch Signal	9-11v (cranking)
14	BK/OR	Neutral Start Switch	In P/N: 9-11v (cranking)
15	RD/GN	HO2S-12 (B1 S2) Heater	Heater Off: 12v, On: 1v
16	PK/GN	HO2S-11 (B1 S1) Heater	Heater Off: 12v, On: 1v
17	BK/WT	A/T-ECT Solenoid (S2)	1st or OD: 1v
22	BK/RD	IAC Signal (RSC)	Pulse Signals
23	BR/RD	IAC Signal (RSO)	Pulse Signals
24	BK/BL	Igniter Transistor 1 Control	Digital Signal: 0-5-0v
25	BR/BK	Igniter Transistor 2 Control	Digital Signal: 0-5-0v
26	BK/WT	Igniter Transistor 3 Control	Digital Signal: 0-5-0v
27	YL/BK	A/T-ECT Solenoid (SL)	In Lockup: 12-14v
29	RD	4WD Detection Transfer (L4)	Open: 12v, Closed: 0v
32	YL/BL	Closed Throttle Switch	1v, off-idle: 12v
33	BR	Power Ground	<0.1v
34	BR	Power Ground	<0.1v

Pin Connector Graphic

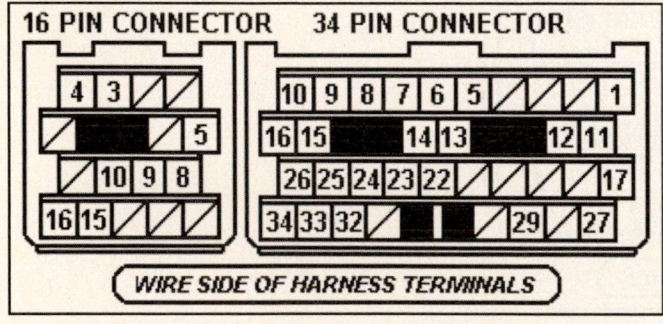

1995 Pickup 3.4L V6 MFI VIN V (M/T) 12 Pin Connector

PCM Pin #	Wire Color	Circuit Description (12 Pin)	Value at Hot Idle
3	PK/GN	HO2S-11 (B1 S1) Heater	Heater Off: 12v, On: 1v
5	WT	CMP Sensor Signal (GN-)	<0.050v
6	BL	CKP Sensor Signal (NE-)	<0.050v
7	BR	Power Ground	<0.1v
9	RD/GN	HO2S-12 (B1 S2) Heater	Heater Off: 12v, On: 1v
11	BK	CMP Sensor Signal (G2+)	AC pulse signals
12	GN	CKP Sensor Signal (NE+)	AC pulse signals

1995 Pickup 3.4L V6 MFI VIN V (M/T) 22 Pin Connector

PCM Pin #	Wire Color	Circuit Description (22 Pin)	Value at Hot Idle
2	BK/GN	Battery Direct	12-14v
4	PK	MIL (lamp) Control	MIL Off: 12v, On: 1v
6	BL/BK	A/C Amplifier Signal (ACT)	Clutch On: 12v, Off: 1.5v
7	BK/RD	A/C Amplifier Signal (AC1)	Clutch On: 1.5v, Off: 12v
8	GN	Vehicle Speed Sensor	At 55 mph: 48 Hz
9	PK/GN	4WD Selector Switch	Switch Off: 12v, On: 1v
11	BK/WT	Starter Switch Signal	9-11v (cranking)
12	WT/RD	EFI Main Relay Power	12-14v
19	WT	SIL (Scan Tool) Signal	12v
20	GN/WT	Stop Light Switch Signal	Brake Off: 0v, On: 12v

Pin Connector Graphic

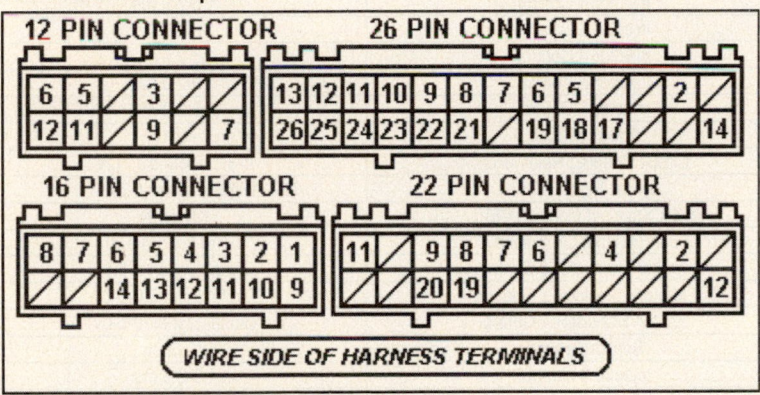

1995 Pickup 3.4L V6 MFI VIN V (M/T) 16 Pin Connector

PCM Pin #	Wire Color	Circuit Description (16 Pin)	Value at Hot Idle
1	GN/YL	Sensor VREF (VC)	4.9-5.1v
2	GY/RD	MAF Sensor Signal (VG)	1.1-1.8v
3	GY	Knock Sensor 2 Signal	<0.075v AC
4	GN/YL	ECT Sensor Signal (THW)	At 180ºF: 0.51v
5	WT	HO2S-11 (B1 S1) Signal	0.1-1.1v
6	BK	Knock Sensor 1 Signal	<0.075v AC
7	PK/WT	Data Link Connector	12-14v
8	BR/WT	Sensor Ground	<0.050v
9	BR	Sensor Ground	<0.050v
10	YL/BK	TP Sensor Signal (VTA)	0.3-0.8v
11	YL/BL	Closed Throttle Switch	1v, off-idle: 12v
12	YL/GN	IAT Sensor Signal (THA)	At 100ºF: 2.60v
13	RD	HO2S-12 (B1 S2) Signal	0.1-1.1v
14	PK	EGR Gas Temperature Sensor (THG)	3.5-4.0v

1995 Pickup 3.4L V6 MFI VIN V (M/T) 26 Pin Connector

PCM Pin #	Wire Color	Circuit Description (26 Pin)	Value at Hot Idle
2	BK/RD	A/C Idle-Up Solenoid	A/C Off: 12v, On: 1v
5	WT/GN	EVAP Purge Solenoid (VSV)	12v or 0v
6	BK/RD	IAC Signal (RSC)	Pulse Signals
7	BR/RD	IAC Signal (RSO)	Pulse Signals
8	YL/BK	Injector 6 Control	1.6-2.9 ms
9	WT/BL	Injector 5 Control	1.6-2.9 ms
10	YL/RD	Injector 4 Control	1.6-2.9 ms
11	WT	Injector 2 Control	1.6-2.9 ms
12	WT/RD	Injector 1 Control	1.6-2.9 ms
13	BR	Power Ground	<0.1v
14	GN/YL	Circuit Opening Relay (FC)	0-3v, at off-idle: 12v
17	BK/YL	Igniter Signal (IGF)	Digital Signal: 0-5-0v
18	RD/WT	EGR Solenoid Control (VSV)	12v or 0v
19	RD/BK	Fuel Pressure Up Solenoid	1v (at hot startup)
21	BK/WT	Igniter Transistor 3 Control	Digital Signal: 0-5-0v
22	BR/BK	Igniter Transistor 2 Control	Digital Signal: 0-5-0v
23	BK/BL	Igniter Transistor 1 Control	Digital Signal: 0-5-0v
24	BR	Sensor Ground	<0.050v
25	WT/GN	Injector 3 Control	1.6-2.9 ms
26	BR	Power Ground	<0.1v

1996 Pickup 3.4L V6 MFI VIN N (A/T-ECT) 28 Pin Connector

PCM Pin #	Wire Color	Circuit Description (28 Pin)	Value at Hot Idle
1	RD/BK	A/T Select Switch Reverse	In 'R': 12v, Others: 0v
2	PK/GN	A/T Select Switch 2nd	In 2nd: 12v, Others: 0v
3	PK/WT	A/T Select Switch Low	In Low: 12v, Others: 0v
4	LG	A/T Oil Temperature Lamp	Lamp Off: 12v, On: 1v
5	BL/BK	A/C Amplifier Signal (ACT)	Clutch On: 12v, Off: 1.5v
6	YL/GN	Overdrive Main Switch	Switch Off: 12v, On: 1v
7	YL/RD	Cruise Control ECU (OD1)	At Cruise in OD: 12v
12	GN	Vehicle Speed Sensor	At 55 mph: 48 Hz
14	BK/GN	Battery Direct	12-14v
17	YL	4WD Detection Transfer (N)	Open: 12v, Closed: 0v
18	WT	SIL (Scan Tool) Signal	12v
20	BK/RD	A/C Amplifier Signal (AC1)	Clutch On: 1.5v, Off: 12v
22	WT/RD	EFI Main Relay Power	12-14v
25	GN/WT	Stop Light Switch Signal	Brake Off: 0v, On: 12v
26	PK/GN	4WD Selector Switch	Switch Off: 12v, On: 1v

1996 Pickup 3.4L V6 MFI VIN N (A/T-ECT) 22 Pin Connector

PCM Pin #	Wire Color	Circuit Description (22 Pin)	Value at Hot Idle
1	GN/BK	Sensor VREF (VC)	4.9-5.1v
5	GN	CKP Sensor Signal (NE+)	AC pulse signals
6	BL	CKP Sensor Signal (NE-)	<0.050v
7	YL/BK	TP Sensor Signal (VTA)	0.3-0.8v
8	GY/RD	MAF Sensor Signal (VG)	1.1-1.8v
9	BR/RD	A/T Vehicle Speed Sensor	Moving: 0-5-0v
10	BK	CMP Sensor Signal (G2+)	AC pulse signals
11	WT	CMP Sensor Signal (GN-)	<0.050v
12	GN/BK	A/T Oil Temperature Sensor	At 230ºF: <1.5v
13	WT	HO2S-11 (B1 S1) Signal	0.1-1.1v
14	YL/GN	IAT Sensor Signal (THA)	At 100ºF: 2.60v
16, 17	GY, BK	Knock Sensor 2 Signal	<0.075v AC
18	BR/WT	Sensor Ground	<0.050v
19 (2WD)	RD	HO2S-21 (B2 S1) Signal	0.1-1.1v
19 (4WD)	RD	HO2S-12 (B1 S2) Signal	0.1-1.1v
20	GN/YL	ECT Sensor Signal (THW)	At 180ºF: 0.51v
21	PK	EGR Gas Temperature Sensor (THG)	3.5-4.0v
22	BR/BK	Sensor Ground	<0.050v

Pin Connector Graphic

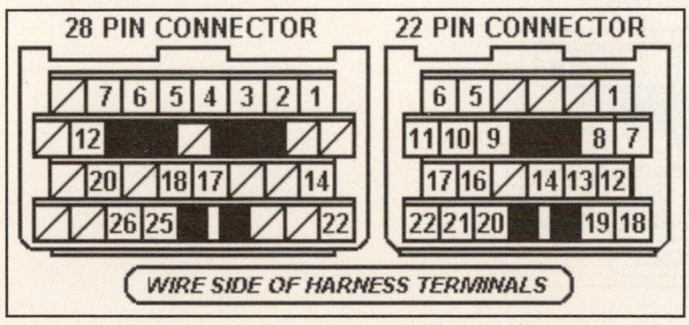

1996 Pickup 3.4L V6 MFI VIN N (A/T-ECT) 16 Pin Connector

PCM Pin #	Wire Color	Circuit Description (16 Pin)	Value at Hot Idle
3	PK	MIL (lamp) Control	MIL Off: 12v, On: 1v
4	GN/YL	Circuit Opening Relay (FC)	0-3v, at off-idle: 12v
5	PK/WT	Data Link Connector	12-14v
8	RD/WT	EGR Solenoid Control (VSV)	12v or 0v
9	RD/BK	Fuel Pressure Up Solenoid	1v (at hot startup)
10	BK/RD	A/C Idle-Up Solenoid	A/C Off: 12v, On: 1v
15	WT/GN	EVAP Purge Solenoid (VSV)	12v or 0v
16	BR	Shield Ground	<0.050v

1996 Pickup 3.4L V6 MFI VIN N (A/T-ECT) 34-Pin Connector

PCM Pin #	Wire Color	Circuit Description (34-Pin)	Value at Hot Idle
1	BR	Power Ground	<0.1v
5	YL/BK	Injector 6 Control	1.6-2.9 ms
6	WT/BL	Injector 5 Control	1.6-2.9 ms
7	YL/RD	Injector 4 Control	1.6-2.9 ms
8	WT/GN	Injector 3 Control	1.6-2.9 ms
9	WT	Injector 2 Control	1.6-2.9 ms
10	WT/RD	Injector 1 Control	1.6-2.9 ms
11	WT	A/T-ECT Solenoid (S1)	3rd or OD: 1v
12	BK/YL	Igniter Signal (IGF)	Digital Signal: 0-5-0v
13	BK/WT	Starter Switch Signal	9-11v (cranking)
14	BK/OR	Neutral Start Switch	In P/N: 9-11v (cranking)
15 (2WD)	RD/GN	HO2S-21 (B2 S1) Heater	Heater Off: 12v, On: 1v
15 (4WD)	RD/GN	HO2S-12 (B1 S2) Heater	Heater Off: 12v, On: 1v
16	PK/GN	HO2S-11 (B1 S1) Heater	Heater Off: 12v, On: 1v
17	BK/WT	A/T-ECT Solenoid (S2)	1st or OD: 1v
22	BK/RD	IAC Signal (RSC)	Pulse Signals
23	BR/RD	IAC Signal (RSO)	Pulse Signals
24	BK/BL	Igniter Transistor 1 Control	Digital Signal: 0-5-0v
25	BR/BK	Igniter Transistor 2 Control	Digital Signal: 0-5-0v
26	BK/WT	Igniter Transistor 3 Control	Digital Signal: 0-5-0v
27	YL/BK	A/T-ECT Solenoid (SL)	In Lockup: 12-14v
29	RD	4WD Detection Transfer (L4)	Open: 12v, Closed: 0v
32	YL/BL	Closed Throttle Switch	1v, off-idle: 12v
33 or 34	BR	Power Ground	<0.1v

Pin Connector Graphic

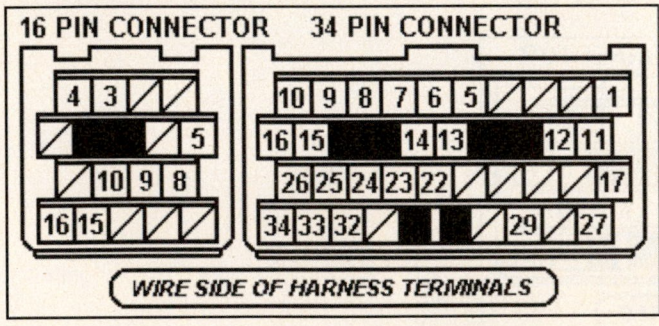

1996 Pickup 3.4L V6 MFI VIN N (M/T) 12 Pin Connector

PCM Pin #	Wire Color	Circuit Description (12 Pin)	Value at Hot Idle
3	PK/GN	HO2S-11 (B1 S1) Heater	Heater Off: 12v, On: 1v
5	WT	CMP Sensor Signal (GN-)	<0.050v
6	BL	CKP Sensor Signal (NE-)	<0.050v
7	BR	Power Ground	<0.1v
9	RD/GN	2WD HO2S-21 (B2 S1) HTR	Heater Off: 12v, On: 1v
9	RD/GN	4WD HO2S-12 (B1 S2) HTR	Heater Off: 12v, On: 1v
11	BK	CMP Sensor Signal (G2+)	AC pulse signals
12	GN	CKP Sensor Signal (NE+)	AC pulse signals

1996 Pickup 3.4L V6 MFI VIN N (M/T) 22 Pin Connector

PCM Pin #	Wire Color	Circuit Description (22 Pin)	Value at Hot Idle
2	BK/GN	Battery Direct	12-14v
4	PK	MIL (lamp) Control	MIL Off: 12v, On: 1v
q	BL/BK	A/C Amplifier Signal (ACT)	Clutch On: 12v, Off: 1.5v
7	BK/RD	A/C Amplifier Signal (AC1)	Clutch On: 1.5v, Off: 12v
8	GN	Vehicle Speed Sensor	At 55 mph: 48 Hz
9	PK/GN	4WD Selector Switch	Switch Off: 12v, On: 1v
11	BK/WT	Starter Switch Signal	9-11v (cranking)
12	WT/RD	EFI Main Relay Power	12-14v
19	WT	SIL (Scan Tool) Signal	12v
20	GN/WT	Stop Light Switch Signal	Brake Off: 0v, On: 12v

Pin Connector Graphic

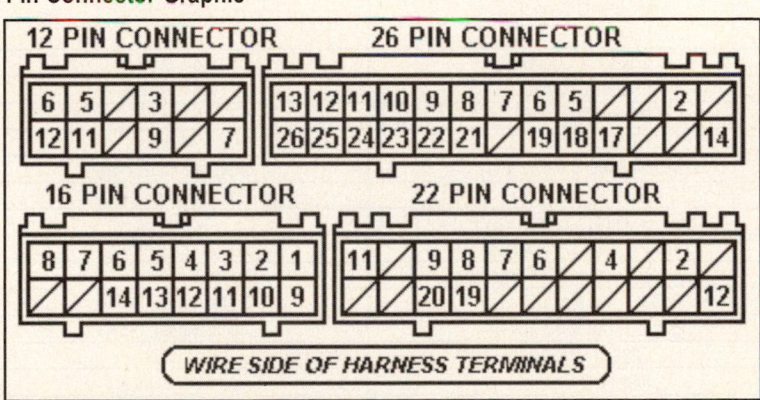

1996 Pickup 3.4L V6 MFI VIN N (M/T) 16 Pin Connector

PCM Pin #	Wire Color	Circuit Description (16 Pin)	Value at Hot Idle
1	GN/BK	Sensor VREF	4.9-5.1v
2	GY/RD	MAF Sensor Signal (VG)	1.1-1.8v
3	GY	Knock Sensor 2 Signal	<0.075v AC
4	GN/YL	ECT Sensor Signal (THW)	At 180ºF: 0.51v
5	WT	HO2S-11 (B1 S1) Signal	0.1-1.1v
6	BK	Knock Sensor 1 Signal	<0.075v AC
7	PK/WT	Data Link Connector	12-14v
8	BR/WT	MAF Sensor Ground	<0.050v
9	BR/BK	Sensor Ground	<0.050v
11	YL/BL	Closed Throttle Switch	1v, off-idle: 12v
10	YL/BK	TP Sensor Signal (VTA)	0.3-0.8v
12	YL/GN	IAT Sensor Signal (THA)	At 100ºF: 2.60v
13 (2WD)	RD	HO2S-21 (B2 S1) Signal	0.1-1.1v
13 (4WD)	RD	HO2S-12 (B1 S2) Signal	0.1-1.1v
14	PK	EGR Gas Temperature Sensor (THG)	3.5-4.0v

1996 Pickup 3.4L V6 MFI VIN N (M/T) 26 Pin Connector

PCM Pin #	Wire Color	Circuit Description (26 Pin)	Value at Hot Idle
2	BK/RD	A/C Idle-Up Solenoid	A/C Off: 12v, On: 1v
5	WT/GN	EVAP Purge Solenoid (VSV)	12v or 0v
6	BK/RD	IAC Signal (RSC)	Pulse Signals
7	BR/RD	IAC Signal (RSO)	Pulse Signals
8	YL/BK	Injector 6 Control	1.6-2.9 ms
9	WT/BL	Injector 5 Control	1.6-2.9 ms
10	YL/RD	Injector 4 Control	1.6-2.9 ms
11	WT	Injector 2 Control	1.6-2.9 ms
12	WT/RD	Injector 1 Control	1.6-2.9 ms
13	BR	Power Ground	<0.1v
14	GN/YL	Circuit Opening Relay (FC)	0-3v, at off-idle: 12v
17	BK/YL	Igniter Signal (IGF)	Digital Signal: 0-5-0v
18	RD/WT	EGR Solenoid Control (VSV)	12v or 0v
19	RD/BK	Fuel Pressure Up Solenoid	1v (at hot startup)
21	BK/WT	Igniter Transistor 3 Control	Digital Signal: 0-5-0v
22	BR/BK	Igniter Transistor 2 Control	Digital Signal: 0-5-0v
23	BK/BL	Igniter Transistor 1 Control	Digital Signal: 0-5-0v
24	BR	Shield Ground	<0.050v
25	WT/GN	Injector 3 Control	1.6-2.9 ms
26	BR	Power Ground	<0.1v

1997 Pickup 3.4L V6 MFI VIN N (A/T-ECT) 28 Pin Connector

PCM Pin #	Wire Color	Circuit Description (28 Pin)	Value at Hot Idle
1	RD/BK	A/T Select Switch Reverse	In 'R': 12v, Others: 0v
2	PK/GN	A/T Select Switch 2nd	In 2nd: 12v, Others: 0v
3	PK/WT	A/T Select Switch Low	In Low: 12v, Others: 0v
4	LG	A/T Oil Temperature Lamp	Lamp Off: 12v, On: 1v
5	BL/BK	A/C Amplifier Signal (ACT)	Clutch On: 12v, Off: 1.5v
6	YL/GN	Overdrive Main Switch	Switch Off: 12v, On: 1v
7	YL/RD	Cruise Control ECU (OD1)	At Cruise in OD: 12v
12	GN	Vehicle Speed Sensor	At 55 mph: 48 Hz
14	BK/GN	Battery Direct	12-14v
17	YL	4WD Detection Transfer (N)	Open: 12v, Closed
18	WT	Data Link Connector (SDL)	0v
20	BK/RD	A/C Amplifier Signal (AC1)	Clutch On: 1.5v, Off: 12v
22	WT/RD	EFI Main Relay Power	12-14v
25	GN/WT	Stop Light Switch Signal	Brake Off: 0v, On: 12v
26	PK/GN	4WD Selector Switch	Switch Off: 12v, On: 1v

1997 Pickup 3.4L V6 MFI VIN N (A/T-ECT) 22 Pin Connector

PCM Pin #	Wire Color	Circuit Description (22 Pin)	Value at Hot Idle
1	GN/BK	Sensor VREF	4.9-5.1v
5	GN	CKP Sensor Signal (NE+)	AC pulse signals
6	BL	CKP Sensor Signal (NE-)	<0.050v
7	YL/BK	TP Sensor Signal (VTA)	0.3-0.8v
8	GY/RD	MAF Sensor Signal (VG)	1.1-1.8v
10	BK	CMP Sensor Signal (G2+)	AC pulse signals
11	WT	CMP Sensor Signal (GN-)	<0.050v
12	GN/BK	A/T Oil Temperature Sensor	At 230ºF: <1.5v
13	WT	HO2S-11 (B1 S1) Signal	0.1-1.1v
14	YL/GN	IAT Sensor Signal (THA)	At 100ºF: 2.60v
15	RD/BL	EVAP Vapor Pressure Sensor	2.5-3.1v (with fuel cap off)
16, 17	GY, BK	Knock Sensor 2, 1 Signal	<0.075v AC
18	BR/WT	Sensor Ground	<0.050v
19 (2WD)	RD	HO2S-21 (B2 S1) Signal	0.1-1.1v
19 (4WD)	RD	HO2S-12 (B1 S2) Signal	0.1-1.1v
20	GN/YL	ECT Sensor Signal (THW)	At 180ºF: 0.51v
21	PK	EGR Gas Temperature Sensor (THG)	3.5-4.0v
22	BR/BK	Sensor Ground	<0.050v

Pin Connector Graphic

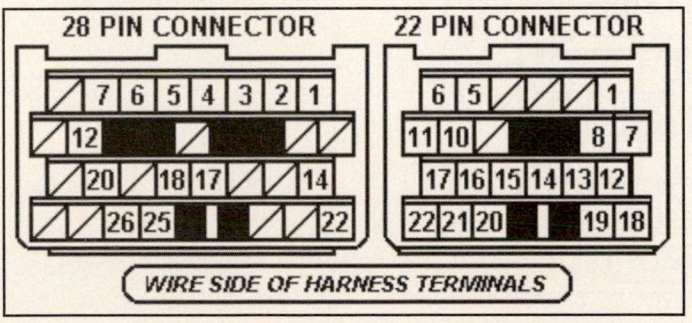

1997 Pickup 3.4L V6 MFI VIN N (A/T-ECT) 16 Pin Connector

PCM Pin #	Wire Color	Circuit Description (16 Pin)	Value at Hot Idle
1	YL/BL	Cruise Control ECU (IDLO)	1.5v, off-idle: 12v
3	PK	MIL (lamp) Control	MIL Off: 12v, On: 1v
4	GN/YL	Circuit Opening Relay (FC)	0-3v, at off-idle: 12v
5	PK/WT	Data Link Connector	12-14v
8	RD/WT	EGR Solenoid Control (VSV)	12v or 0v
13	GN/RD	EVAP Vapor Pressure (VSV)	12v or 0v
15	WT/GN	EVAP Purge Solenoid (VSV)	12v or 0v
16	BR	Shield Ground	<0.050v

1997 Pickup 3.4L V6 MFI VIN N (A/T-ECT) 34-Pin Connector

PCM Pin #	Wire Color	Circuit Description (34-Pin)	Value at Hot Idle
1	BR	Power Ground	<0.1v
5	YL/BK	Injector 6 Control	1.6-2.9 ms
6	WT/BL	Injector 5 Control	1.6-2.9 ms
7	YL/RD	Injector 4 Control	1.6-2.9 ms
8	WT/GN	Injector 3 Control	1.6-2.9 ms
9	WT	Injector 2 Control	1.6-2.9 ms
10	BK	Injector 1 Control	1.6-2.9 ms
11	WT	A/T-ECT Solenoid (S1)	3rd or OD: 1v
12	BK/YL	Igniter Signal (IGF)	Digital Signal: 0-5-0v
13	BK/WT	Starter Switch Signal	9-11v (cranking)
14	BK/OR	Neutral Start Switch	In P/N: 9-11v (cranking)
15	RD/GN	2WD HO2S-21 (B2 S1) HTR	Heater Off: 12v, On: 1v
15	RD/GN	4WD HO2S-12 (B1 S2) HTR	Heater Off: 12v, On: 1v
16	PK/GN	HO2S-11 (B1 S1) Heater	Heater Off: 12v, On: 1v
17	BK/WT	A/T-ECT Solenoid (S2)	1st or OD: 1v
22	BK/RD	IAC Signal (RSC)	Pulse Signals
23	BR/RD	IAC Signal (RSO)	Pulse Signals
24	BK/BL	Igniter Transistor 1 Control	Digital Signal: 0-5-0v
25	BR/BK	Igniter Transistor 2 Control	Digital Signal: 0-5-0v
26	BK/WT	Igniter Transistor 3 Control	Digital Signal: 0-5-0v
27	YL/BK	A/T-ECT Solenoid (SL)	In Lockup: 12-14v
29	RD/BK	4WD Detection Transfer (L4)	Open: 12v, Closed: 0v
33	BR	Power Ground	<0.1v
34	BR	Power Ground	<0.1v

Pin Connector Graphic

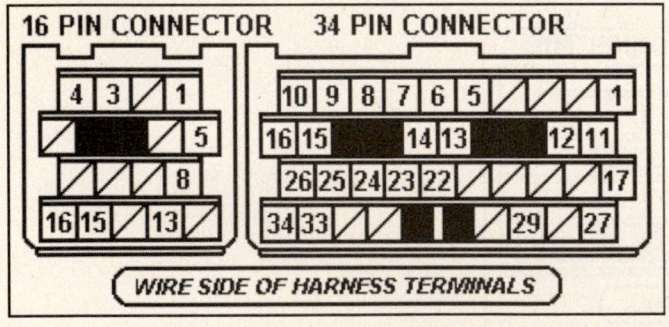

1997 Pickup 3.4L V6 MFI VIN N (M/T) 12 Pin Connector

PCM Pin #	Wire Color	Circuit Description (12 Pin)	Value at Hot Idle
3	PK/GN	HO2S-11 (B1 S1) Heater	Heater Off: 12v, On: 1v
4	RD/BL	EVAP Vapor Pressure Sensor	2.5-3.1v (with fuel cap off)
5	WT	CMP Sensor Signal (GN-)	<0.050v
6	BL	CKP Sensor Signal (NE-)	<0.050v
7	BR	Power Ground	<0.1v
9	RD/GN	HO2S-21 (B2 S1) Heater	Heater Off: 12v, On: 1v
11	BK	CMP Sensor Signal (G2+)	AC pulse signals
12	GN	CKP Sensor Signal (NE+)	AC pulse signals

1997 Pickup 3.4L V6 MFI VIN N (M/T) 22 Pin Connector

PCM Pin #	Wire Color	Circuit Description (22 Pin)	Value at Hot Idle
2	BK/GN	Battery Direct	12-14v
4	PK	MIL (lamp) Control	MIL Off: 12v, On: 1v
6	BL/BK	A/C Amplifier Signal (ACT)	Clutch On: 12v, Off: 1.5v
7	BK/RD	A/C Amplifier Signal (AC1)	Clutch On: 1.5v, Off: 12v
8	GN	Vehicle Speed Sensor	At 55 mph: 48 Hz
9	PK/GN	4WD Selector Switch	Switch Off: 12v, On: 1v
11	BK/WT	Starter Switch Signal	9-11v (cranking)
12	WT/RD	EFI Main Relay Power	12-14v
19	WT	SIL (Scan Tool) Signal	12v
20	GN/WT	Stop Light Switch Signal	Brake Off: 0v, On: 12v

Pin Connector Graphic

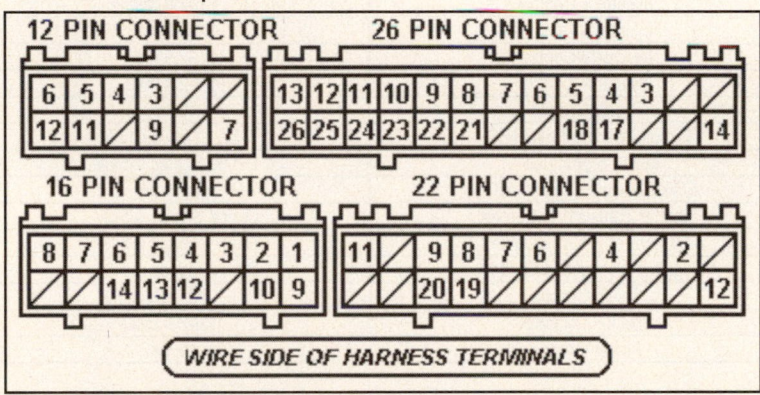

1997 Pickup 3.4L V6 MFI VIN N (M/T) 16 Pin Connector

PCM Pin #	Wire Color	Circuit Description (16 Pin)	Value at Hot Idle
1	GN/BK	Sensor VREF	4.9-5.1v
2	GY/RD	MAF Sensor Signal (VG)	1.1-1.8v
3	GY	Knock Sensor 2 Signal	<0.075v AC
4	GN/YL	ECT Sensor Signal (THW)	At 180°F: 0.51v
5	WT	HO2S-11 (B1 S1) Signal	0.1-1.1v
6	BK	Knock Sensor 1 Signal	<0.075v AC
7	PK/WT	Data Link Connector	12-14v
8	BR/WT	MAF Sensor Ground	<0.050v
9	BR/BK	Sensor Ground	<0.050v
10	YL/BK	TP Sensor Signal (VTA)	0.3-0.8v
12	YL/GN	IAT Sensor Signal (THA)	At 100°F: 2.60v
13	RD	HO2S-21 (B2 S1) Signal	0.1-1.1v
14	PK	EGR Gas Temperature Sensor (THG)	3.5-4.0v

1997 Pickup 3.4L V6 MFI VIN N (M/T) 26 Pin Connector

PCM Pin #	Wire Color	Circuit Description (26 Pin)	Value at Hot Idle
3	GN/RD	EVAP Vapor Pressure (VSV)	12v or 0v
4	YL/BL	Cruise Control ECU (IDLO)	1.5v, off-idle: 12v
5	WT/GN	EVAP Purge Solenoid (VSV)	12v or 0v
6	BK/RD	IAC Signal (RSC)	Pulse Signals
7	BR/RD	IAC Signal (RSO)	Pulse Signals
8	YL/BK	Injector 6 Control	1.6-2.9 ms
9	WT/BL	Injector 5 Control	1.6-2.9 ms
10	YL/RD	Injector 4 Control	1.6-2.9 ms
11	WT	Injector 2 Control	1.6-2.9 ms
12	WT/RD	Injector 1 Control	1.6-2.9 ms
13	BR	Power Ground	<0.1v
14	GN/YL	Circuit Opening Relay (FC)	0-3v, at off-idle: 12v
17	BK/YL	Igniter Signal (IGF)	Digital Signal: 0-5-0v
18	RD/WT	EGR Solenoid Control (VSV)	12v or 0v
21	BK/WT	Igniter Transistor 3 Control	Digital Signal: 0-5-0v
22	BR/BK	Igniter Transistor 2 Control	Digital Signal: 0-5-0v
23	BK/BL	Igniter Transistor 1 Control	Digital Signal: 0-5-0v
24	BR	Shield Ground	<0.050v
25	WT/GN	Injector 3 Control	1.6-2.9 ms
26	BR	Power Ground	<0.1v

1998 Pickup 3.4L V6 MFI VIN N (A/T-ECT) 28 Pin Connector

PCM Pin #	Wire Color	Circuit Description (28 Pin)	Value at Hot Idle
1	PK/WT	A/T Select Switch Low	In Low: 12v, Others: 0v
5	BL/BK	A/C Amplifier Signal (ACT)	Clutch On: 12v, Off: 1.5v
6	YL/GN	Overdrive Main Switch	Switch Off: 12v, On: 1v
7	YL/RD	Cruise Control ECU (OD1)	At Cruise in OD: 12v
8	WT	SIL (Scan Tool) Signal	12v
10	PK/GN	A/T Select Switch 2nd	In 2nd: 12v, Others: 0v
11	YL/BL	Cruise Control ECU (IDLO)	1.5v, off-idle: 12v
12	GN	Vehicle Speed Sensor	At 55 mph: 48 Hz
14	BK/GN	Battery Direct	12-14v
15	RD/BK	A/T Select Switch Reverse	In 'R': 12v, Others: 0v
16	BK/RD	A/C Amplifier Signal (AC1)	Clutch On: 1.5v, Off: 12v
23	WT/RD	EFI Main Relay (B+)	12-14v
24	GN/WT	Stop Light Switch Signal	Brake Off: 0v, On: 12v

1998 Pickup 3.4L V6 MFI VIN N (A/T-ECT) 22 Pin Connector

PCM Pin #	Wire Color	Circuit Description (22 Pin)	Value at Hot Idle
1	GN/BK	Sensor VREF	4.9-5.1v
5	GN	CKP Sensor Signal (NE+)	AC pulse signals
6	BL	CKP/CMP Sensor Signal (-)	<0.050v
7	YL/BK	TP Sensor Signal (VTA)	0.3-0.8v
8	GY/RD	MAF Sensor Signal (VG)	1.1-1.8v
9	BR/RD	A/T Vehicle Speed Sensor	Moving: 0-5-0v
13	WT	HO2S-11 (B1 S1) Signal	0.1-1.1v
14	GY	Knock Sensor 2 Signal	<0.075v AC
15	BK	Knock Sensor 1 Signal	<0.075v AC
17	PK	CMP Sensor Signal (G2+)	AC pulse signals
18	GN/YL	Circuit Opening Relay (FC)	0-3v, at off-idle: 12v
19	RD	HO2S-12 (B1 S2) Signal	0.1-1.1v
20	GN/RD	ECT Sensor Signal (THW)	At 180ºF: 0.51v
21	YL/GN	IAT Sensor Signal (THA)	At 100ºF: 2.60v
22	BR/BK	Sensor Ground	<0.050v

Pin Connector Graphic

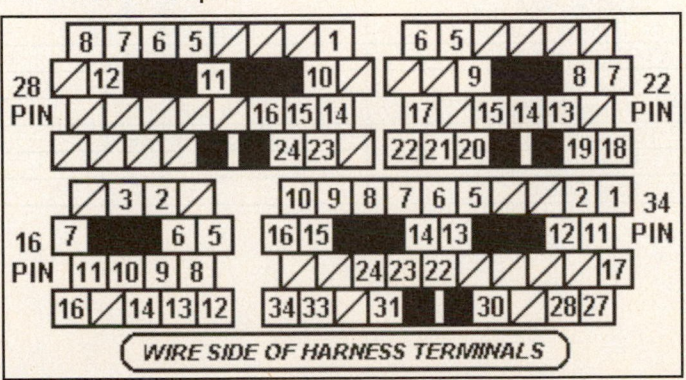

1998 Pickup 3.4L V6 MFI VIN N (A/T-ECT) 16 Pin Connector

PCM Pin #	Wire Color	Circuit Description (16 Pin)	Value at Hot Idle
2	WT/GN	EVAP Purge Solenoid (VSV)	12v or 0v
3	PK	MIL (lamp) Control	MIL Off: 12v, On: 1v
5	PK/WT	Data Link Connector	12c
6	LG	A/T Oil Temperature Indicator	Light Off: 12v, On: 1v
7	BR/WT	MAF Meter Ground	<0.050v
8	GN/RD	EVAP Vapor Pressure (VSV)	12v or 0v
9	YL	4WD Detection Transfer (N)	Open: 12v, Closed
10	RD/GN	HO2S-12 (B1 S2) Heater	Heater Off: 12v, On: 1v
11	PK/GN	HO2S-11 (B1 S1) Heater	Heater Off: 12v, On: 1v
12	RD/WT	EGR Solenoid Control (VSV)	12v or 0v
13	RD/BL	EVAP Vapor Pressure Sensor	2.5-3.1v (with fuel cap off)
14	PK	EGR Gas Temperature Sensor (THG)	3.5-4.0v
16	BR	Shield Ground	<0.050v

1998 Pickup 3.4L V6 MFI VIN N (A/T-ECT) 34-Pin Connector

PCM Pin #	Wire Color	Circuit Description (34-Pin)	Value at Hot Idle
1	PK/GN	4WD Selector Switch	Switch Off: 12v, On: 1v
2	RD/BK	4WD Detection Transfer (L4)	Open: 12v, Closed: 0v
5	YL/GN	Injector 6 Control	1.6-2.9 ms
6	WT/BL	Injector 5 Control	1.6-2.9 ms
7	YL/RD	Injector 4 Control	1.6-2.9 ms
8	WT/GN	Injector 3 Control	1.6-2.9 ms
9	WT	Injector 2 Control	1.6-2.9 ms
10	WT/RD	Injector 1 Control	1.6-2.9 ms
11	WT	A/T-ECT Solenoid (S1)	3rd or OD: 1v
12	BK/YL	Igniter Signal (IGF)	Digital Signal: 0-5-0v
13	BK/WT	Starter Switch Signal	9-11v (cranking)
14	BK/OR	Neutral Start Switch	In P/N: 9-11v (cranking)
15	BK/WT	Igniter Transistor 3 Control	Digital Signal: 0-5-0v
16	BR/BK	Igniter Transistor 2 Control	Digital Signal: 0-5-0v
17	RD/YL	A/T-ECT Solenoid (S2)	1st or OD: 1v
22	BK/RD	IAC Signal (RSC)	Pulse Signals
23	BR/RD	IAC Signal (RSO)	Pulse Signals
24	BK/BL	Igniter Transistor 1 Control	Digital Signal: 0-5-0v
27	YL/BK	A/T-ECT Solenoid (SL)	In Lockup: 12-14v
28	BR	Power Ground	<0.1v
30	GN/BK	A/T Oil Temperature Sensor	At 230ºF: <1.5v
31	PK/BL	PSP Switch Signal	Straight: 12v, Turning: 0v
33	BR	Power Ground	<0.1v
34	BR	Power Ground	<0.1v

TOYOTA VAN PIN TABLE CONTENTS

PREVIA Pin Tables

1991-93 Van Passenger 2.4L I4 VIN A (A/T, M/T) 16 Pin Connector

PCM Pin #	Wire Color	Circuit Description (16 Pin)	Value at Hot Idle
1	BL/BK	Air Flow Meter VREF	4.9-5.1v
2	BL/YL	Air Flow Meter Signal	0.5-2.5v
3	BL/WT	IAT Sensor Signal	At 100°F: 2.60v
4	BL	ECT Sensor Signal	At 180°F: 0.51v
5	BK	O2S-12 (B1 S2) Signal	0.1.1-1.5v
6	BK	HO2S-11 (B1 S1) Signal	0.1.1-1.5v
8	GN	Data Link Connector	12-14v
9	BR/BK	Sensor Ground	<0.050v
10 (Cal)	BR/WT	EGR Gas Temperature Sensor	3.5-4.0v
11	YL/RD	TP Sensor Signal	0.3-0.8v
12	BL/YL	Closed Throttle Switch	0-3v, at off-idle: 12v
14	BK	Knock Sensor Signal	<0.075v AC
16	GN/WT	Data Link Connector	12-14v

1991-93 Van Passenger 2.4L I4 VIN A (A/T, M/T) 22 Pin Connector

PCM Pin #	Wire Color	Circuit Description (22 Pin)	Value at Hot Idle
1	WT/RD	Direct Battery	12-14v
2	BK/RD	Ignition Switch Power	12-14v
4	RD/WT	Engine Oil Level Light	Light On: 1v, Off: 12v
5	GN/RD	MIL (lamp) Control	MIL Off: 12v, On: 1v
6	BK/BL	A/C Amplifier Signal (ACT)	Clutch On: 12v, Off: 1.5v
7	GN	Data Link Connector	12-14v
8	BR/BK	A/T ECT Speed Sensor	Moving: 0-5-0V
9	PK/WT	Vehicle Speed Sensor	At 55 mph: 48 Hz
10	BK/WT	A/C Magnetic Clutch (ACMG)	Clutch Off: 0v, On: 12v
11	BK	Starter Switch Signal	KOEC: 9-11v
12	BK/OR	EFI Main Relay B+	12-14v
13	BK/OR	EFI Main Relay B1+	12-14v
14	BK/BL	A/T Select Switch Low	In Low: 12v, Others: 0v
15	BK/RD	A/T Select Switch 2nd	In 2nd: 12v, Others: 0v
16	BK/YL	A/T Select Switch Neutral	In 'N': 12v, Others: 0v
17	YL/RD	Oil Level 2 Sensor	At 68°F: 4-5v
18	YL	Oil Level 1 Sensor	At 68°F: 4-5v
19	GN/WT	Stop Light Switch	Brake Off: 0v, On: 12v
20	BL/WT	Overdrive Main Switch	Switch Off: 12v, On: 1v
21	PK/YL	Cruise Control ECU	At Cruise in OD: 12v
22	BK/WT	Starter Switch Signal	Cranking: 0-3v
22	BK	Starter Switch Signal	Cranking: 0-3v

Standard Colors and Abbreviations

Abbreviation	Color	Abbreviation	Color	Abbreviation	Color
BK	Black	GY	Gray	RD	Red
BL	Blue	GN	Green	TN	Tan
BR	Brown	LG	LT Green	VT	Violet
DB	Dark Blue	OR	Orange	WT	White
DG	DK Green	PK	Pink	YL	Yellow

1991-93 Van Passenger 2.4L I4 VIN A (A/T, M/T) 26 Pin Connector

PCM Pin #	Wire Color	Circuit Description (26 Pin)	Value at Hot Idle
1	GN/YL	A/T ECT Solenoid (SL)	In Lockup: 12-14v
3	BL/RD	Igniter Signal (IGF)	Digital Signal: 0-5-0v
4	WT	Distributor Signal (NE+)	AC pulse signals
5	GN	Distributor Signal (G2+)	AC pulse signals
6	BK	ISC Signal (ISC1)	Pulse Signals
7	RD	Engine Oil Feeder Motor	Pulse Signals
8	BK/BL	Engine Oil Feeder Relay	Relay Off: 12v, On: 1v
9	BL/BK	Igniter Signal (IGT)	Digital Signal: 0-5-0v
10	GN/BK	HO2S-11 (B1 S1) Heater	1v (Heater on)
11	RD/BK	Cold Start Injector	1v (at cold startup)
12	BK	Injectors 1 & 3 Control	1.6-2.9 ms
13	BR	Power Ground	<0.1v
14	BR/YL	A/T ECT Solenoid (S2)	1st or OD: 1v
15	BR/WT	A/T ECT Solenoid (S1)	3rd or OD: 1v
17	RD	Distributor Signal (G-)	<0.050v
18	BK	Distributor Signal (G1+)	AC pulse signals
19	WT	ISC Signal (ISC2)	Pulse Signals
23	YL	Fuel Pressure Up Solenoid	1v (at hot restart)
24	BR	Sensor Ground	<0.050v
25	BK/BL	Injectors 2 & 4 Control	1.6-2.9 ms
26	BR	Power Ground	<0.1v

Pin Connector Graphic

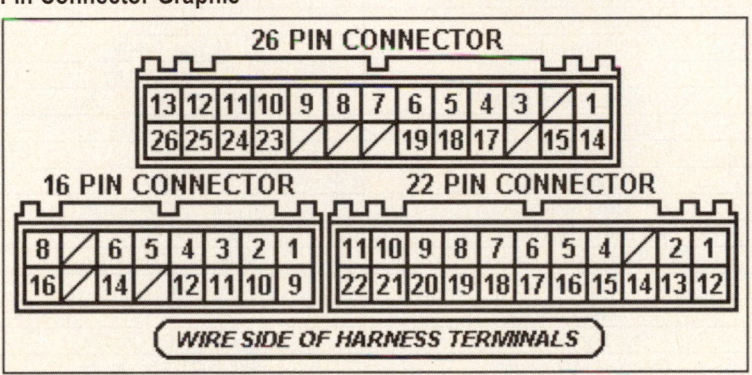

1994 Van Passenger 2.4L I4 VIN A (A/T-ECT) 16 Pin Connector

PCM Pin #	Wire Color	Circuit Description (16 Pin)	Value at Hot Idle
1	BL/BK	Air Flow Meter VREF	4.9-5.1v
2	BL/YL	Air Flow Meter Signal	0.5-2.5v
3	BL/WT	IAT Sensor Signal	At 100°F: 2.60v
4	BL	ECT Sensor Signal	At 180°F: 0.51v
5	BK	O2S-12 (B1 S2) Signal	0.1.1-1.5v
6	BK	HO2S-11 (B1 S1) Signal	0.1.1-1.5v
8	GN	Data Link Connector	12-14v
9	BR/BK	Sensor Ground	<0.050v
10 (Cal)	BR/WT	EGR Gas Temperature Sensor	3.5-4.0v
11	YL/RD	TP Sensor Signal	0.3-0.8v
12	BL/YL	Closed Throttle Switch	0-3v, at off-idle: 12v
14	BK	Knock Sensor Signal	<0.075v AC
16	GN/WT	Data Link Connector	12v

1994 Van Passenger 2.4L I4 VIN A (A/T-ECT) 22 Pin Connector

PCM Pin #	Wire Color	Circuit Description (22 Pin)	Value at Hot Idle
1	WT/RD	Direct Battery	12-14v
2	BK/RD	Ignition Switch Power	12-14v
4	RD/WT	Engine Oil Level Light	Light On: 1v, Off: 12v
5	GN/RD	MIL (lamp) Control	MIL Off: 12v, On: 1v
6	BK/BL	A/C Amplifier Signal (ACT)	Clutch On: 12v, Off: 1.5v
7	GN	Data Link Connector	12-14v
8	BR/BK	A/T ECT Speed Sensor	Moving: 0-5-0V
9	PK/WT	Vehicle Speed Sensor	At 55 mph: 48 Hz
10	BK/WT	A/C Magnetic Clutch (ACMG)	Clutch Off: 0v, On: 12v
11	BK	Starter Switch Signal	KOEC: 9-11v
12	BK/OR	EFI Main Relay B+	12-14v
13	BK/OR	EFI Main Relay B1+	12-14v
14	BK/BL	A/T Select Switch Low	In Low: 12v, Others: 0v
15	BK/RD	A/T Select Switch 2nd	In 2nd: 12v, Others: 0v
16	BK/YL	A/T Select Switch Neutral	In 'N': 12v, Others: 0v
17	YL/RD	Oil Level 2 Sensor	At 230°F: <1.5v
18	YL	Oil Level 1 Sensor	At 230°F: <1.5v
19	GN/WT	Stop Light Switch	Brake Off: 0v, On: 12v
20	BL/WT	Overdrive Main Switch	Switch Off: 12v, On: 1v
21	PK/YL	Cruise Control ECU	At Cruise in OD: 12v
22	BK/WT	Neutral Start Switch	In P/N: Cranking: 0-3v

Standard Colors and Abbreviations

Abbreviation	Color	Abbreviation	Color	Abbreviation	Color
BK	Black	GY	Gray	RD	Red
BL	Blue	GN	Green	TN	Tan
BR	Brown	LG	LT Green	VT	Violet
DB	Dark Blue	OR	Orange	WT	White
DG	DK Green	PK	Pink	YL	Yellow

1994 Van Passenger 2.4L I4 VIN A (A/T-ECT) 26 Pin Connector

PCM Pin #	Wire Color	Circuit Description (26 Pin)	Value at Hot Idle
1	GN/YL	A/T ECT Solenoid (SL)	In Lockup: 12-14v
3	BL/RD	Igniter Signal (IGF)	Digital Signal: 0-5-0v
4	WT	Distributor Signal (NE+)	AC pulse signals
5	GN	Distributor Signal (G2+)	AC pulse signals
6	BK	ISC Signal (ISC1)	Pulse Signals
7	RD	Engine Oil Feeder Motor	Pulse Signals
8	BK/BL	Engine Oil Feeder Relay	Relay Off: 12v, On: 1v
9	BL/BK	Igniter Signal (IGT)	Digital Signal: 0-5-0v
10	GN/BK	HO2S-11 (B1 S1) Heater	1v (Heater on)
11	RD/BK	Cold Start Injector	1v (at cold startup)
12	BK	Injectors 1 & 3 Control	1.6-2.9 ms
13	BR	Power Ground	<0.1v
14	BR/YL	A/T ECT Solenoid (S2)	1st or OD: 1v
15	BR/WT	A/T ECT Solenoid (S1)	3rd or OD: 1v
17	RD	Distributor Signal (G-)	<0.050v
18	BK	Distributor Signal (G1+)	AC pulse signals
19	WT	ISC Signal (ISC2)	Pulse Signals
23	YL	Fuel Pressure Up Solenoid	1v (at hot restart)
24	BR	Sensor Ground	<0.050v
25	BK/BL	Injectors 2 & 4 Control	1.6-2.9 ms
26	BR	Power Ground	<0.1v

Pin Connector Graphic

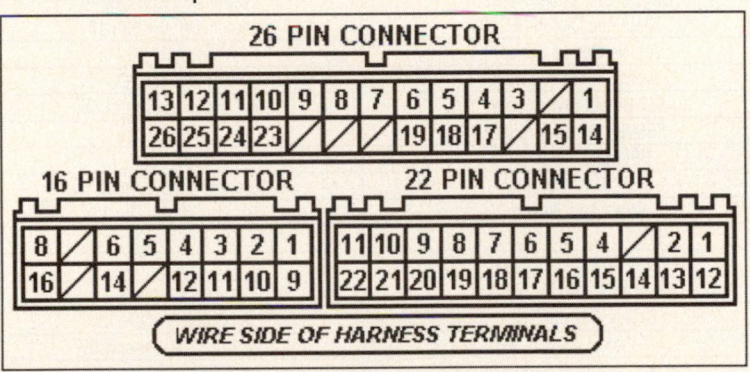

1994 Van Passenger S/C 2.4L VIN A (A/T-ECT) 16 Pin Connector

PCM Pin #	Wire Color	Circuit Description (16 Pin)	Value at Hot Idle
1	BL/BK	Air Flow Meter VREF	4.9-5.1v
2	GN/RD	Air Flow Meter Signal	0.5-2.5v
3	BL/WT	IAT Sensor Signal	At 100°F: 2.60v
4	BL	ECT Sensor Signal	At 180°F: 0.51v
5	BK	HO2S-11 (B1 S1) Signal	0.1.1-1.5v
6	BK	Knock Sensor Signal	<0.075v AC
7	GN/WT	Data Link Connector	12-14v
9	BR/BK	Sensor Ground	<0.050v
10	YL/RD	TP Sensor Signal	0.3-0.8v
11	BL/YL	Closed Throttle Switch	0-3v, at off-idle: 12v
12 (Cal)	BR/WT	EGR Gas Temperature Sensor	3.5-4.0v
13	BK	HO2S-12 (B1 S2) Signal	0.1.1-1.1v
14	GN/WT	Air Flow Meter Ground	<0.050v
15	GN/BK	Data Link Connector	12-14v
16	BR/WT	A/T ECT Speed Sensor	Moving: 0-5-0V

1994 Van Passenger S/C 2.4L VIN A (A/T-ECT) 22 Pin Connector

PCM Pin #	Wire Color	Circuit Description (22 Pin)	Value at Hot Idle
2	WT/RD	Direct Battery	12-14v
3	BL/RD	A/C Amplifier Signal (ACT)	Clutch On: 12v, Off: 1.5v
4	GN/RD	MIL (lamp) Control	MIL Off: 12v, On: 1v
5	RD/GN	S/C Magnetic Clutch Relay	Relay Off: 12v, On: 1v
6	BK/BL	Engine Oil Feeder Relay	Relay Off: 12v, On: 1v
7	BK/WT	A/C Magnetic Clutch (ACMG)	Clutch Off: 0v, On: 12v
8	PK/WT	Vehicle Speed Sensor	At 55 mph: 48 Hz
9	BL/WT	Overdrive Main Switch	Switch Off: 12v, On: 1v
10	BK/BL	A/T Select Switch Low	In Low: 12v, Others: 0v
11	BK	Starter Switch Signal	KOEC: 9-11v
12	BK/OR	EFI Main Relay B+	12-14v
13	GN/OR	PSP Switch Signal	Straight: 12v, Turned: 0v
14	GN/WT	Stop Light Switch	Brake Off: 0v, On: 12v
15	RD/WT	Engine Oil Level Light	Light On: 1v, Off: 12v
16	RD/BK	EGR Solenoid Control	1v (Heater on)
17	RD	Engine Oil Feeder Motor	Pulse Signals
18	PK/YL	Cruise Control ECU	At Cruise in OD: 12v
21	BK/RD	A/T Select Switch 2nd	In 2nd: 12v, Others: 0v
22	BK/WT	Neutral Start Switch	In P/N: Cranking: 0-3v

Standard Colors and Abbreviations

Abbreviation	Color	Abbreviation	Color	Abbreviation	Color
BK	Black	GY	Gray	RD	Red
BL	Blue	GN	Green	TN	Tan
BR	Brown	LG	LT Green	VT	Violet
DB	Dark Blue	OR	Orange	WT	White
DG	DK Green	PK	Pink	YL	Yellow

1994 Van Passenger S/C 2.4L VIN A (A/T-ECT) 12 Pin Connector

PCM Pin #	Wire Color	Circuit Description (12 Pin)	Value at Hot Idle
2	BL/BK	Fuel Pump Relay Control	Relay Off: 12v, On: 1v
3	BK	Data Link Connector	12-14v
4	YL/RD	Oil Level Sensor 2 Signal	At 230ºF: <1.5v
6	WT	Distributor Signal (NE-)	<0.050v
10	YL	Oil Level Sensor 1 Signal	At 230ºF: <1.5v
11	GN	Distributor Signal (G+)	AC pulse signals
12	BK	Distributor Signal (NE+)	AC pulse signals

1994 Van Passenger S/C 2.4L VIN A (A/T-ECT) 26 Pin Connector

PCM Pin #	Wire Color	Circuit Description (26 Pin)	Value at Hot Idle
1	GN/BK	HO2S-11 (B1 S1) Heater	1v (Heater on)
2	GN/WT	HO2S-12 (B1 S2) Heater	1v (Heater on)
3	YL	S/C Bypass Control (AB3)	12v or 0v
4	YL/BK	S/C Bypass Control (AB2)	12v or 0v
5	RD/YL	S/C Bypass Control (AB1)	12v or 0v
6	BL	IAC Signal (ISCC)	Pulse Signals
7	BL/WT	IAC Signal (ISCO)	Pulse Signals
8	GN/YL	A/T ECT Solenoid (SL)	In Lockup: 12-14v
9	BR/YL	A/T ECT Solenoid (S2)	1st or OD: 1v
10	BR/WT	A/T ECT Solenoid (S1)	3rd or OD: 1v
11	BK/BL	Injectors 2 & 4 Control	1.6-2.9 ms
12	BK	Injectors 1 & 3 Control	1.6-2.9 ms
13	BR	Sensor Ground	<0.050v
14	BL/RD	Circuit Opening Relay	0-3v, at off-idle: 12v
16	GN/BK	S/C Bypass Control (AB4)	12v or 0v
17	BL/RD	Igniter Signal (IGF)	Digital Signal: 0-5-0v
18	BL/RD	A/C Amplifier Signal (ACT)	Clutch On: 12v, Off: 1.5v
19	RD/YL	EVAP Purge Solenoid (VSV)	12v or 0v
20	RD/WT	SCB Solenoid Control	12v or 0v
21	RD/BL	A/C Idle-Up Solenoid	A/C Off: 12v, On: 1v
22	BR	Power Ground	<0.1v
23	BL/BK	Igniter Signal (IGT)	Digital Signal: 0-5-0v
24	BR	Sensor Ground	<0.050v
25	BK/YL	A/C Amplifier Signal (AC1)	Clutch On: 1.5v, Off: 12v
26	BR	Power Ground	<0.1v

Pin Connector Graphic

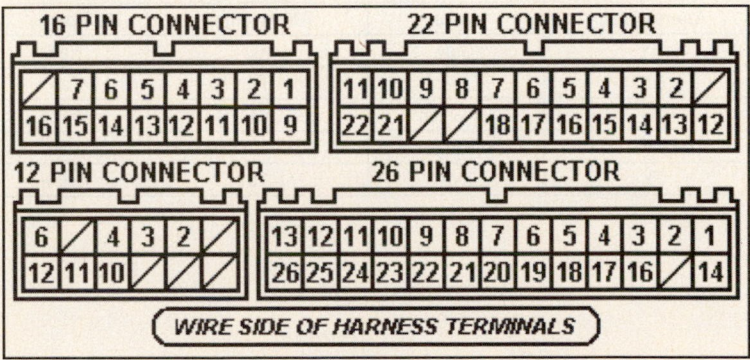

1995 Van Passenger 2.4L I4 VIN A (A/T-ECT) 16 Pin Connector

PCM Pin #	Wire Color	Circuit Description (16 Pin)	Value at Hot Idle
1	BL/BK	Sensor VREF	4.9-5.1v
2	BL/YL	MAF Sensor Signal	1.1-1.8v
3	BL/WT	IAT Sensor Signal	At 100ºF: 2.60v
4	BL	ECT Sensor Signal	At 180ºF: 0.51v
5	BK	O2S-12 (B1 S2) Signal	0.1.1-1.5v
6	BK	HO2S-11 (B1 S1) Signal	0.1.1-1.5v
8	GN	Data Link Connector	12-14v
9	BR/BK	Sensor Ground	<0.050v
10 (Cal)	BR/WT	EGR Gas Temperature Sensor	3.5-4.0v
11	YL/RD	TP Sensor Signal	0.3-0.8v
12	BL/YL	Closed Throttle Switch	0-3v, at off-idle: 12v
14	BK	Knock Sensor Signal	<0.075v AC
16	GN/WT	Data Link Connector	12-14v

1995 Van Passenger 2.4L I4 VIN A (A/T-ECT) 22 Pin Connector

PCM Pin #	Wire Color	Circuit Description (22 Pin)	Value at Hot Idle
1	WT/RD	Direct Battery	12-14v
4	RD/WT	Engine Oil Level Light	Light On: 1v, Off: 12v
5	GN/RD	MIL (lamp) Control	MIL Off: 12v, On: 1v
6	BK/BL	A/C Amplifier Signal (ACT)	Clutch On: 12v, Off: 1.5v
7	GN	Data Link Connector	12-14v
8	BR/WT	A/T ECT Speed Sensor	Moving: 0-5-0v
9	PK/WT	Vehicle Speed Sensor	At 55 mph: 48 Hz
10	BK/WT	A/C Magnetic Clutch (ACMG)	Clutch Off: 0v, On: 12v
11	BK	Starter Switch Signal	Cranking: 0-3v
12	BK/OR	EFI Main Relay Power	12-14v
14	BK/BL	A/T Select Switch Low	In Low: 12v, Others: 0v
15	BK/RD	A/T Select Switch 2nd	In 2nd: 12v, Others: 0v
16	BK/YL	A/T Select Switch Neutral	In 'N': 12v, Others: 0v
17	YL/RD	Oil Level 2 Sensor	At 230ºF: <1.5v
18	YL	Oil Level 1 Sensor	At 230ºF: <1.5v
19	WT/RD	Stop Light Switch	Brake Off: 0v, On: 12v
20	BL/WT	Overdrive Main Switch	Switch Off: 12v, On: 1v
21	PK/YL	Cruise Control ECU	At Cruise in OD: 12v
22	BK/WT	Neutral Start Switch	In P/N: Cranking: 0-3v

Standard Colors and Abbreviations

Abbreviation	Color	Abbreviation	Color	Abbreviation	Color
BK	Black	GY	Gray	RD	Red
BL	Blue	GN	Green	TN	Tan
BR	Brown	LG	LT Green	VT	Violet
DB	Dark Blue	OR	Orange	WT	White
DG	DK Green	PK	Pink	YL	Yellow

1995 Van Passenger 2.4L I4 VIN A (A/T-ECT) 26 Pin Connector

PCM Pin #	Wire Color	Circuit Description (26 Pin)	Value at Hot Idle
1	GN/YL	A/T ECT Solenoid (SL)	In Lockup: 12-14v
3	BL/RD	Igniter Signal (IGF)	Digital Signal: 0-5-0v
4	WT	Distributor Signal (NE+)	AC pulse signals
5	GN	Distributor Signal (G2+)	AC pulse signals
6	BK	ISC Signal (ISCC)	Pulse Signals
7	RD	Engine Oil Feeder Motor	Pulse Signals
8	BK/BL	Engine Oil Feeder Relay	Relay Off: 12v, On: 1v
9	BL/BK	Igniter Signal (IGT)	Digital Signal: 0-5-0v
10	GN/BK	HO2S-11 (B1 S1) Heater	1v (Heater on)
11	RD/BK	Cold Start Injector	1v (at cold startup)
12	BK	Injectors 1 & 3 Control	1.6-2.9 ms
13	BR	Power Ground	<0.1v
14	BR/YL	A/T ECT Solenoid (S2)	1st or OD: 1v
15	BR/WT	A/T ECT Solenoid (S1)	3rd or OD: 1v
17	RD	Distributor Signal (G-)	<0.050v
18	BK	Distributor Signal (G1+)	AC pulse signals
19	WT	ISC Signal (ISCO)	Pulse Signals
23	YL	Fuel Pressure Up Solenoid	1v (at hot restart)
24	BR	Sensor Ground	<0.050v
25	BK/BL	Injectors 2 & 4 Control	1.6-2.9 ms
26	BR	Power Ground	<0.1v

Pin Connector Graphic

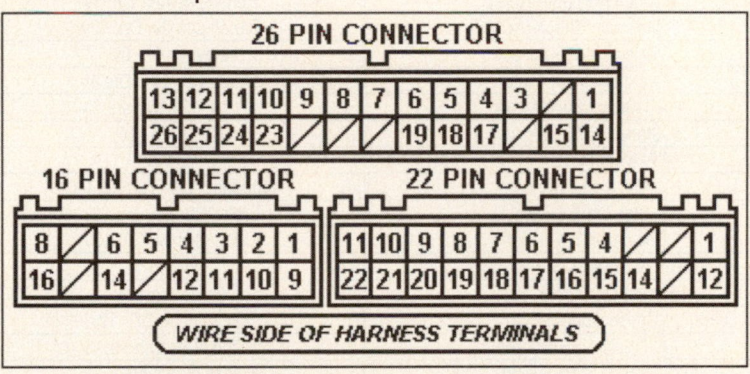

1995 Van Passenger S/C 2.4L VIN A (A/T-ECT) 16 Pin Connector

PCM Pin #	Wire Color	Circuit Description (16 Pin)	Value at Hot Idle
1	BL/BK	Sensor VREF	4.9-5.1v
2	GN/RD	MAF Sensor Signal	1.1-1.8v
3	BL/WT	IAT Sensor Signal	At 100°F: 2.60v
4	BL	ECT Sensor Signal	At 180°F: 0.51v
5	BK	HO2S-11 (B1 S1) Signal	0.1.1-1.5v
6	BK	Knock Sensor Signal	<0.075v AC
7	GN/WT	Data Link Connector	12-14v
9	BR/BK	Sensor Ground	<0.050v
10	YL/RD	TP Sensor Signal	0.3-0.8v
11	BL/YL	Closed Throttle Switch	0-3v, at off-idle: 12v
12 (Cal)	BR/WT	EGR Gas Temperature Sensor	3.5-4.0v
13	BK	HO2S-12 (B1 S2) Signal	0.1.1-1.5v
14	GN/WT	MAF Sensor Ground	<0.050v
15	GN/BK	Data Link Connector	12-14v
16	BR/WT	A/T ECT Speed Sensor	Moving: 0-5-0V

1995 Van Passenger S/C 2.4L VIN A (A/T-ECT) 22 Pin Connector

PCM Pin #	Wire Color	Circuit Description (22 Pin)	Value at Hot Idle
2	WT/RD	Direct Battery	12-14v
3	BL/RD	A/C Amplifier Signal (ACT)	Clutch On: 12v, Off: 1.5v
4	GN/RD	MIL (lamp) Control	MIL Off: 12v, On: 1v
5	RD/GN	S/C Magnetic Clutch Relay	Relay Off: 12v, On: 1v
6	BK/BL	Engine Oil Feeder Relay	Relay Off: 12v, On: 1v
7	BK/WT	A/C Magnetic Clutch (ACMG)	Clutch Off: 0v, On: 12v
8	PK/WT	Vehicle Speed Sensor	Moving: 0-5-0v
9	BL/WT	Overdrive Main Switch	Switch Off: 12v, On: 1v
10	BK/BL	A/T Select Switch Low	In Low: 12v, Others: 0v
11	BK	Starter Switch Signal	KOEC: 9-11v
12	BK/OR	EFI Main Relay Power	12-14v
13	GN/OR	PSP Switch Signal	Straight: 12v, Turned: 0v
14	GN/WT	Stop Light Switch	Brake Off: 0v, On: 12v
15	RD/WT	Engine Oil Level Light	Lamp On: 1v, Off: 12v
16	RD/BK	EGR Solenoid Control (VSV)	12v or 0v
17	RD	Engine Oil Feeder Motor	Pulse Signals
18	PK/YL	Cruise Control ECU	At Cruise in OD: 12v
21	BK/RD	A/T Select Switch 2nd	In 2nd: 12v, Others: 0v
22	BK/WT	Neutral Start Switch	In P/N: Cranking: 0-3v

Standard Colors and Abbreviations

Abbreviation	Color	Abbreviation	Color	Abbreviation	Color
BK	Black	GY	Gray	RD	Red
BL	Blue	GN	Green	TN	Tan
BR	Brown	LG	LT Green	VT	Violet
DB	Dark Blue	OR	Orange	WT	White
DG	DK Green	PK	Pink	YL	Yellow

1995 Van Passenger S/C 2.4L VIN A (A/T-ECT) 12 Pin Connector

PCM Pin #	Wire Color	Circuit Description (12 Pin)	Value at Hot Idle
1	RD	Defogger/Light Idle-Up Signal	Load On: 12v
2	BL/BK	Fuel Pump Relay Control	Relay Off: 12v, On: 1v
3	BK	Data Link Connector	12-14v
4	YL/RD	Oil Level Sensor Signal	At 230ºF: <1.5v
6	WT	CKP Sensor Signal (NE-)	<0.050v
10	YL	Oil Level Sensor Signal	At 230ºF: <1.5v
11	GN	Distributor Signal (G+)	AC pulse signals
12	BK	CKP Sensor Signal (NE+)	AC pulse signals

1995 Van Passenger S/C 2.4L VIN A (A/T-ECT) 26 Pin Connector

PCM Pin #	Wire Color	Circuit Description (26 Pin)	Value at Hot Idle
1	GN/BK	HO2S-11 (B1 S1) Heater	1v (Heater on)
2	GN/WT	HO2S-12 (B1 S2) Heater	1v (Heater on)
3	YL	S/C Bypass Control (AB3)	12v or 0v
4	YL/BK	S/C Bypass Control (AB2)	12v or 0v
5	RD/YL	S/C Bypass Control (AB1)	12v or 0v
6	BL	IAC Signal (RSC)	Pulse Signals
7	BL/WT	IAC Signal (RSO)	Pulse Signals
8	GN/YL	A/T ECT Solenoid (SL)	In Lockup: 12-14v
9	BR/YL	A/T ECT Solenoid (S2)	1st or OD: 1v
10	BR/WT	A/T ECT Solenoid (S1)	3rd or OD: 1v
11	BK/BL	Injectors 2 & 4 Control	1.6-2.9 ms
12	BK	Injectors 1 & 3 Control	1.6-2.9 ms
13, 24	BR	Sensor Ground	<0.050v
14	BL/RD	Circuit Opening Relay	0-3v, at off-idle: 12v
16	GN/BK	S/C Bypass Control (AB4)	12v or 0v
17	BL/RD	Igniter Signal (IGF)	Digital Signal: 0-5-0v
18	BL/RD	A/C Amplifier Signal (ACT)	Clutch On: 12v, Off: 1.5v
19	RD/YL	EVAP Purge Solenoid (VSV)	12v or 0v
20	RD/WT	SCB Solenoid Control	12v or 0v
21	RD/BL	A/C Idle-Up Solenoid	A/C Off: 12v, On: 1v
22	BR	Power Ground	<0.1v
23	BL/BK	Igniter Signal (IGT)	Digital Signal: 0-5-0v
25	BK/YL	A/C Amplifier Signal (AC1)	Clutch On: 1.5v, Off: 12v
26	BR	Power Ground	<0.1v

Pin Connector Graphic

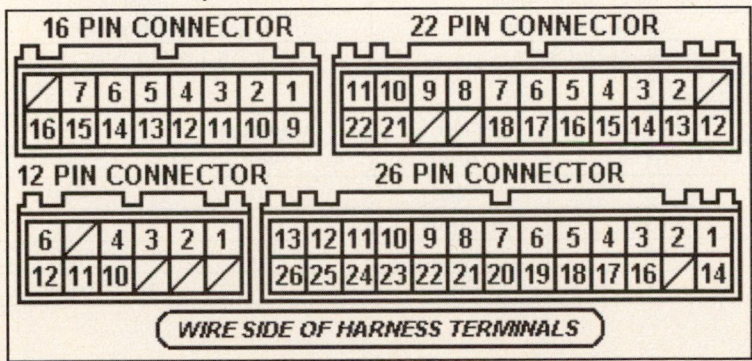

1996-97 Van Passenger 2.4L I4 VIN K (A/T-ECT) 16 Pin Connector

PCM Pin #	Wire Color	Circuit Description (16 Pin)	Value at Hot Idle
1	BL/BK	Sensor VREF	4.9-5.1v
2	GN/RD	MAF Sensor Signal	1.1-1.8v
3	BL/WT	IAT Sensor Signal	At 100°F: 2.60v
4	BL	ECT Sensor Signal	At 180°F: 0.51v
5	BK	HO2S-11 (B1 S1) Signal	0.1.1-1.5v
6	BK	Knock Sensor Signal	<0.075v AC
7	GN/WT	Data Link Connector	12v
9	BR/BK	Sensor Ground	<0.050v
10	YL/RD	TP Sensor Signal	0.3-0.8v
11	BL/YL	Closed Throttle Switch	0-3v, at off-idle: 12v
12	BR/WT	EGR Gas Temperature Sensor	3.5-4.0v
13	BK	HO2S-12 (B1 S2) Signal	0.1.1-1.5v
14	GN/WT	MAF Sensor Ground	<0.050v
15	GN/BK	Data Link Connector	12-14v
16	BR/WT	A/T ECT Speed Sensor	Moving: 0-5-0V

1996-97 Van Passenger 2.4L I4 VIN K (A/T-ECT) 22 Pin Connector

PCM Pin #	Wire Color	Circuit Description (22 Pin)	Value at Hot Idle
2	WT/RD	Direct Battery	12-14v
3	BL/RD	A/C Amplifier Signal (ACT)	Clutch On: 12v, Off: 1.5v
4	GN/RD	MIL (lamp) Control	MIL Off: 12v, On: 1v
5	RD/GN	S/C Magnetic Clutch Relay	Relay On: 12-14v
6	BK/BL	Engine Oil Feeder Relay	Relay Off: 12v, On: 1v
7	BK/WT	A/C Amplifier On Signal	Clutch On: 1.5v, Off: 12v
8	PK/WT	Vehicle Speed Sensor	At 55 mph: 48 Hz
9	BL/WT	Overdrive Main Switch	Switch Off: 12v, On: 1v
10	BK/BL	A/T Select Switch Low	In Low: 12v, Others: 0v
11	BK	Starter Switch Signal	KOEC: 9-11v
12	BK/OR	EFI Main Relay B+	12-14v
13	GN/OR	PSP Switch Signal	Straight: 12v, Turned: 0v
14	GN/WT	Stop Light Switch	Brake Off: 0v, On: 12v
15	RD/WT	Engine Oil Level Light	Light On: 1v, Off: 12v
16	RD/BK	EGR Solenoid Control	1v (Heater on)
17	RD	Engine Oil Feeder Motor	Pulse Signals
18	PK/YL	Cruise Control ECU	At Cruise in OD: 12v
19	BK	A/T Oil Temperature Indicator	Lamp On: 1v, Off: 12v
20	RD/YL	A/T Select Switch Reverse	In 'R': 12v, Others: 0v
21	BK/RD	A/T Select Switch 2nd	In 2nd: 12v, Others: 0v
22	BK/WT	Neutral Start Switch	In P/N: Cranking: 0-3v

Standard Colors and Abbreviations

Abbreviation	Color	Abbreviation	Color	Abbreviation	Color
BK	Black	GY	Gray	RD	Red
BL	Blue	GN	Green	TN	Tan
BR	Brown	LG	LT Green	VT	Violet
DB	Dark Blue	OR	Orange	WT	White
DG	DK Green	PK	Pink	YL	Yellow

1996-97 Van Passenger 2.4L I4 VIN K (A/T-ECT) 12 Pin Connector

PCM Pin #	Wire Color	Circuit Description (12 Pin)	Value at Hot Idle
1	RD	Defogger/Light Idle-Up Signal	Load On: 12-14v
2	BL/BK	Fuel Pump Relay Control	Relay Off: 12v, On: 1v
3	BK	Data Link Connector (SDL)	0v
4	YL/RD	Engine Oil Level Sensor	At 230°F: <1.5v
6	WT	CKP Sensor Signal (NE-)	<0.050v
10	YL	Engine Oil Level Sensor	At 230°F: <1.5v
9	BK/BL	A/T Oil Temperature Sensor	At 230°F: <1.5v
11	GN, BK	Distributor Signal (G+)	AC pulse signals
12	BK	CKP Sensor Signal (NE+)	AC pulse signals

1996-97 Van Passenger 2.4L I4 VIN K (A/T-ECT) 26 Pin Connector

PCM Pin #	Wire Color	Circuit Description (26 Pin)	Value at Hot Idle
1	GN/BK	HO2S-11 (B1 S1) Heater	1v (Heater on)
2	GN/WT	HO2S-12 (B1 S2) Heater	1v (Heater on)
3	YL	S/C Bypass Control (AB3)	12v or 0v
4	YL/BK	S/C Bypass Control (AB2)	12v or 0v
5	RD/YL	S/C Bypass Control (AB1)	12v or 0v
6	BL	IAC Signal (RSC)	Pulse Signals
7	BL/WT	IAC Signal (RSO)	Pulse Signals
8	GN/YL	A/T ECT Solenoid (SL)	In Lockup: 12-14v
9	BR/YL	A/T ECT Solenoid (S2)	S1: 3rd or OD: 1v
10	BR/WT	A/T ECT Solenoid (S1)	S1: 3rd or OD: 1v
11	BK/BL	Injectors 2 & 4 Control	1.6-2.9 ms
12	BK	Injectors 1 & 3 Control	1.6-2.9 ms
13, 24	BR	Sensor Ground	<0.050v
14	BL/RD	Circuit Opening Relay	0-3v, at off-idle: 12v
16	GN/BK	S/C Bypass Control (AB4)	12v or 0v
17	BL/RD	Igniter Signal (IGF)	Digital Signal: 0-5-0v
18	BL/RD	A/C Pressure Switch	Switch Closed: 0.1v
19	RD/YL	EVAP Purge Solenoid (VSV)	12v or 0v
20	RD/WT	SCB Solenoid Control	12v or 0v
21	RD/BL	A/C Idle-Up Solenoid	A/C Off: 12v, On: 1v
22, 26	BR	Power Ground	<0.1v
23	BL/BK	Igniter Signal (IGT)	Digital Signal: 0-5-0v
25	BK/YL	A/C Amplifier Signal (AC1)	Clutch On: 1.5v, Off: 12v

Pin Connector Graphic

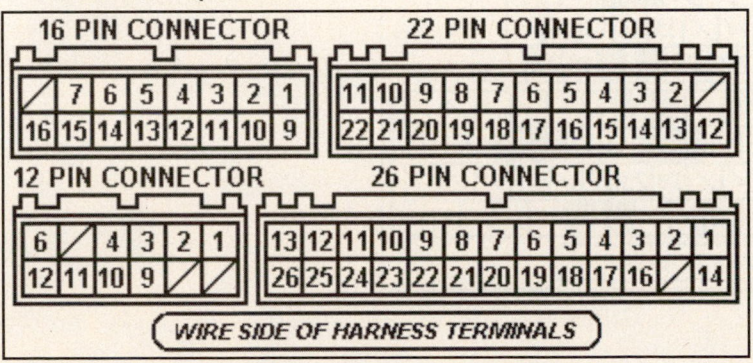

Sienna Pin Tables

1998 Sienna 3.0L V6 MFI VIN F (A/T-ECT) 28 Pin Connector

PCM Pin #	Wire Color	Circuit Description (28 Pin)	Value at Hot Idle
1	YL	A/T Select Switch Low	In Low: 12v, Others: 0v
2	PK	Mirror Switch Signal	Switch Off: 0v, On: 12v
3	GN	Tail Light Signal	Switch Off: 0v, On: 12v
5	GN/BK	A/C Amplifier Signal (ACT)	Clutch On: 12v, Off: 1.5v
6	GN/OR	Overdrive Main Switch	Switch Off: 12v, On: 1v
7	YL/BK	Cruise Control ECU	At Cruise in OD: 12v
8	RD	SIL (Scan Tool Signal)	Digital Signal
10	OR	A/T Select Switch 2nd	In 2nd: 12v, Others: 0v
11	BL	Cruise Control (IDLO) ECU	1.5v, off-idle: 12v
12	PK/YL	Vehicle Speed Sensor	At 55 mph: 48 Hz
13	OR	Tachometer Signal (TACO)	Pulse Signals
14	BK/YL	Direct Battery	12-14v
15	RD/BK	A/T Select Switch Reverse	In 'R': 12v, Others: 0v
16	BK/YL	A/C Amplifier Signal (AC1)	Clutch On: 1.5v, Off: 12v
17	PK/BK	HO2S-12 (B1 S2) Heater	1v (Heater on)
18	WT	HO2S-12 (B1 S2) Signal	0.1.1-1.5v
23	BK/RD	EFI Main Relay B+	12-14v
24	GN/WT	Stop Light Switch	Brake Off: 0v, On: 12v

1998 Sienna 3.0L V6 MFI VIN F (A/T-ECT) 16 Pin Connector

PCM Pin #	Wire Color	Circuit Description (16 Pin)	Value at Hot Idle
2	LG	EVAP Purge Solenoid (VSV)	12v or 0v
3	GN/RD	MIL (lamp) Control	MIL Off: 12v, On: 1v
5	BL/WT	Data Link Connector	12-14v
6	RD/YL	Intake Air Solenoid	12v or 0v
7	RD/BK	MAF Sensor Ground	<0.050v
8	WT/RD	EVAP Vapor Pressure (VSV)	12v or 0v
9	GN/WT	Cooling Fan Relay 1 & 2	Relay Off: 12v, On: 1v
10	YL/RD	HO2S-21 (B2 S1) Heater	1v (Heater on)
11	BL/BK	HO2S-11 (B1 S1) Heater	1v (Heater on)
13	BL/RD	EVAP Vapor Pressure Sensor	2.5-3.7v (with hose off)
16	BR	Shield Ground	<0.050v

Pin Connector Graphic

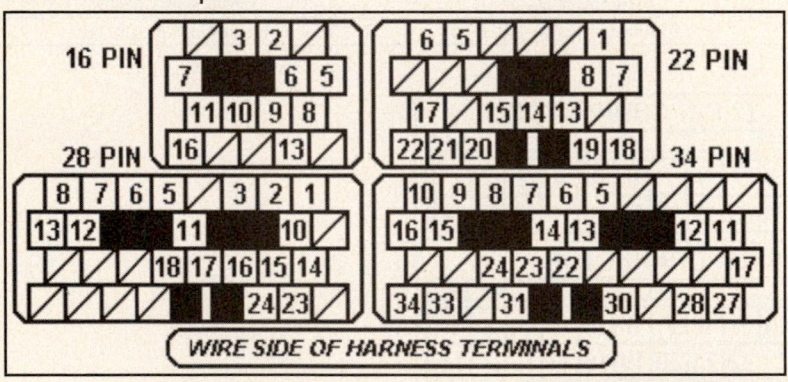

1998 Sienna 3.0L V6 MFI VIN F (A/T-ECT) 22 Pin Connector

PCM Pin #	Wire Color	Circuit Description (22 Pin)	Value at Hot Idle
1	YL	Sensor VREF	4.9-5.1v
5	BK/RD	CKP Sensor Signal (NE+)	AC pulse signals
6	BL	CKP/CMP Sensor Signal (-)	<0.050v
7	BL	TP Sensor Signal	0.3-0.8v
8	PK	MAF Sensor Signal	1.1-1.8v
13	WT	HO2S-11 (B1 S1) Signal	0.1.1-1.5v
14	WT	Knock Sensor Signal 1	<0.075v AC
15	WT	Knock Sensor Signal 2	<0.075v AC
17	BK/WT	CMP Sensor Signal (G+)	AC pulse signals
18	GN/RD	Circuit Opening Relay	0-3v, at off-idle: 12v
19	BK	HO2S-21 (B2 S1) Signal	0.1.1-1.5v
20	GN/BK	ECT Sensor Signal	At 180°F: 0.51v
21	BL/YL	IAT Sensor Signal	At 100°F: 2.60v
22	BR	Sensor Ground	<0.050v

1998 Sienna 3.0L V6 MFI VIN F (A/T-ECT) 34-Pin Connector

PCM Pin #	Wire Color	Circuit Description (34-Pin)	Value at Hot Idle
5	GN	Injector 6 Control	1.6-2.9 ms
6	RD/BL	Injector 5 Control	1.6-2.9 ms
7	WT	Injector 4 Control	1.6-2.9 ms
8	YL	Injector 3 Control	1.6-2.9 ms
9	RD	Injector 2 Control	1.6-2.9 ms
10	BL	Injector 1 Control	1.6-2.9 ms
11	PK	A/T ECT Solenoid (S1)	3rd or OD: 1v
12	WT/RD	Igniter Signal (IGF)	Digital Signal: 0-5-0v
13	GY	Starter Switch Signal	1v (at cold startup)
14	BK/WT	Neutral Start Switch	In P/N: Cranking: 0-3v
15	GN/BK	Igniter Transistor 3 Control	7% duty cycle
16	BR/YL	Igniter Transistor 2 Control	7% duty cycle
17	BL/BK	A/T ECT Solenoid (S2)	1st or OD: 1v
22	YL/BK	IAC Signal (RSC)	Pulse Signals
23	RD/WT	IAC Signal (RSO)	Pulse Signals
24	GY	Igniter Transistor 1 Control	7% duty cycle
27	PK/BL	A/T ECT Solenoid (SL)	In Lockup: 12-14v
28	BR	Power Ground	<0.1v
30	GN/YL	A/T Oil Temperature Sensor	At 230°F: <1.5v
31	BK/BL	PSP Switch Signal	Straight: 12v, Turned: 0v
33	BR	Power Ground	<0.1v
34	BR	Power Ground	<0.1v

Standard Colors and Abbreviations

Abbreviation	Color	Abbreviation	Color	Abbreviation	Color
BK	Black	GY	Gray	RD	Red
BL	Blue	GN	Green	TN	Tan
BR	Brown	LG	LT Green	VT	Violet
DB	Dark Blue	OR	Orange	WT	White
DG	DK Green	PK	Pink	YL	Yellow

1999-2000 Sienna 3.0L 1MZ-FE V6 VIN F E8 22 Pin Connector

PCM Pin #	Wire Color	Circuit Description (22 Pin)	Value at Hot Idle
1	RD/BK	Direct Battery	12-14v
2	BK/OR	Ignition Switch Power	12-14v
3	LG/RD	Circuit Opening Relay (FC)	0-3v, at off-idle: 12v
4-5	---	Not Used	---
6	GN/RD	MIL (lamp) Control	MIL Off: 12v, On: 1v
7 ('00)	GY	Starter Switch Signal	Cranking: 0-3v
8	YL/GN	EFI Main Relay Control	Relay Off: 0v, On: 12v
9	WT/RD	EVAP Vapor Pressure (VSV)	12v or 0v
10	---	Not Used	---
11	RD	SIL (Scan Tool Signal)	Digital Signal
12-14	---	Not Used	---
15	GN/WT	Stop Light Switch Signal	Brake Off: 0v, On: 12v
16	BK/RD	EFI Main Relay Power	12-14v
17	BL/RD	EVAP Vapor Pressure Sensor	2.9-3.1v (with hose off)
18	PK	Heated Mirror Switch Signal	Switch Off: 0v, On: 12v
19	GN	Rear Tail Light Switch Signal	Switch Off: 0v, On: 12v
20-22	---	Not Used	---

1999-2000 Sienna 3.0L 1MZ-FE V6 VIN F E9 28 Pin Connector

PCM Pin #	Wire Color	Circuit Description (28 Pin)	Value at Hot Idle
1	---	Not Used	---
2	RD/BK	A/T Select Switch Reverse	In 'R': 12v, Others: 0v
3	OR	A/T Select Switch 2nd	In 2nd: 12v, Others: 0v
4	BL	Cruise Control ECU (IDLO)	1.5v, off-idle: 12v
5-7	---	Not Used	---
8	WT	HO2S-12 (B1 S2) Signal	0.1.1-1.5v
9	PK/BK	HO2S-12 (B1 S2) Heater	1v (Heater on)
10	---	Not Used	---
11-19	LG/BK	A/C Amplifier Signal (ACT)	Clutch On: 12v, Off: 1.5v
12	YL	A/T Select Switch Low	In Low: 12v, Others: 0v
20	BK/WT	Neutral Start Circuit Signal	In P/N: Cranking: 0-3v
21, 23	---	Not Used	---
22	VT/YL	Vehicle Speed Sensor	At 55 mph: 48 Hz
24	YL/BK	Cruise Control Signal (OD1)	At Cruise in OD: 12v
25	BK/YL	A/C Amplifier Signal (AC1)	Clutch On: 1.5v, Off: 12v
26, 28	---	Not Used	---
27	VT/RD	Tachometer Signal (TACO)	Pulse Signals
28	---	Not Used	---

Pin Connector Graphic

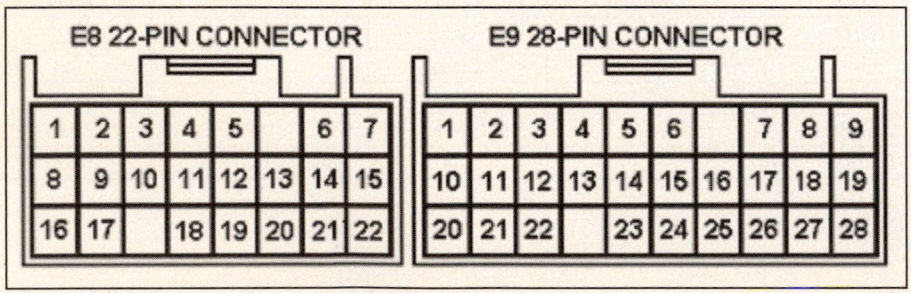

1999-2000 Sienna 3.0L 1MZ-FE V6 VIN F E10 17-Pin Connector

PCM Pin #	Wire Color	Circuit Description (17-Pin)	Value at Hot Idle
1-3, 7-10	---	Not Used	---
4	BR/WT	Transponder Amplifier Code	Inserting key: pulses
5	BR/RD	Transponder Amplifier Signal	Inserting key: pulses
6	BR/YL	Transponder Amplifier Signal	Inserting key: pulses
11	BL/BK	Unlock Warning Switch	No Key: 4-5v
13 ('99)	BK/RD	Shield Ground	<0.050v
16	RD/WT	Theft Security Indicator Light	LED Off: 12v, On: 1v

1999-2000 Sienna 3.0L 1MZ-FE V6 VIN F E11 24-Pin Connector

PCM Pin #	Wire Color	Circuit Description (24-Pin)	Value at Hot Idle
1	WT/BK	Power Ground (E04)	<0.1v
2	YL	Sensor VREF (VC)	4.9-5.1v
3 (Fed)	BL/BK	HO2S-11 (B1 S1) Signal	0.1.1-1.5v
3 (Cal)	BL	AFS-11 (B1 S1) Heater	1v (Heater on)
4 (Fed)	YL/RD	HO2S-21 (B2 S1) Signal	0.1.1-1.5v
4 (Cal)	BL	AFS-21 (B2 S1) Heater	1v (Heater on)
5	BL/RD	Injector 1 Control	1.6-2-9 ms
6	RD	Injector 2 Control	1.6-2-9 ms
7	LG	EVAP Purge Solenoid (VSV)	12v or 0v
8	WT/BK	Power Ground (E05)	<0.1v
9	BK/BL	PSP Switch Signal	Straight: 12v, Turned: 0v
10	PK	MAF Sensor Signal (VG)	1.1-1.5v
11 (Fed)	WT	HO2S-11 (B1 S1) Heater	1v (Heater on)
11 (Cal)	OR	AFS-11 (B1 S1) Signal (AF+)	Fixed at 3.3v
12 (Fed)	BK	HO2S-21 (B2 S1) Heater	1v (Heater on)
12 (Cal)	OR	AFS-21 (B2 S1) Signal (AF+)	Fixed at 3.3v
14	GN/BK	ECT Sensor Signal (THW)	0.5-0.6v
15	GN/YL	A/T: Fluid Temp. Input (THO)	At 68°F: 4-5v
16	BK/RD	CKP Sensor Signal (NE+)	AC pulse signals
17	BR	Shield Ground (E1)	<0.050v
18	BR	Sensor Ground (E2)	<0.050v
19	RD/BK	MAF Sensor Ground (E2G)	<0.050v
20 (Cal)	WT	AFS-11 (B1 S1) Signal (AF-)	Fixed at 3.3v
21 (Cal)	WT	AFS-21 (B2 S1) Signal (AF-)	Fixed at 3.3v
22	BL/YL	IAT Sensor Signal (THA)	0.5-3.4v
23	BL/WT	TP Sensor Signal (VTA1)	0.53-1.27v
24	BL	CMP/CKP Sensor Ground (-)	<0.050v

Pin Connector Graphic

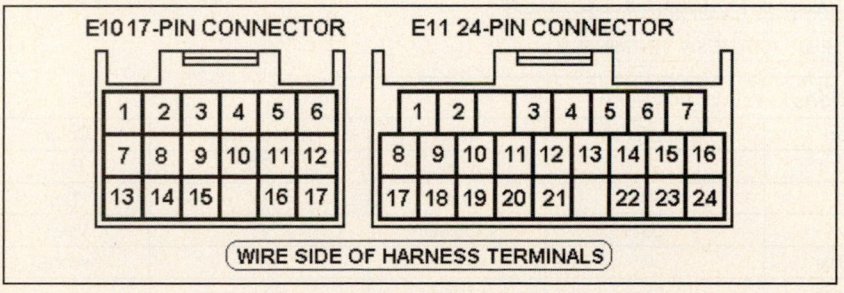

1999-2000 Sienna 3.0L 1MZ-FE V6 VIN F E12 31-Pin Connector

PCM Pin #	Wire Color	Circuit Description (31-Pin)	Value at Hot Idle
1	YL	Injector 3 Control	1.6-2-9 ms
2	WT	Injector 4 Control	1.6-2-9 ms
3	RD/BL	Injector 5 Control	1.6-2-9 ms
4	GN	Injector 6 Control	1.6-2.9 ms
5	---	Not Used	---
6	BL/WT	DLC 1 Signal (TC)	12v
7	VT	A/T-ECT Shift Solenoid (S1)	In 3rd or OD: 1v
8	BL/BK	A/T-ECT Shift Solenoid (S2)	1st or OD: 1v
9	PK/BL	A/T-ECT Shift Solenoid (SL)	In Lockup: 12-14v
10	BK/WT	CMP Sensor Signal (G22+)	AC pulse signals
11	GY	IGT 1 Control	6°, 55 mph: 8° dwell
12	BR/YL	IGT 2 Control	6°, 55 mph: 8° dwell
13	LG/BK	IGT 3 Control	6°, 55 mph: 8° dwell
14	---	Not Used	---
15	YL/BK	IAC Signal (RSC)	Pulse Signals
16	RD/WT	IAC Signal (RSO)	Pulse Signals
17	RD/YL	ACIS Control (VSV)	12v or 0v
18	GN/RD	Overdrive Lamp Control	At Cruise in OD: 1v
19-20	---	Not Used	---
21	WT/BK	Power Ground (E01)	<0.1v
22	---	Not Used	---
23	G/OR	OD Main Sw. Input (ODMS)	Open: 12v, Closed: 1v
25	WT/RD	Igniter Signal (IGF)	Digital Signal: 0-5-0v
26	---	Not Used	---
27	WT	Knock Sensor 1 Signal (right)	<0.075v AC
28	WT	Knock Sensor 2 Signal (left)	<0.075v AC
29	GN/WT	Cooling Fan 1 Relay	Relay Off: 12v, On: 1v
30	WT/BK	Power Ground (E03)	<0.1v
31	WT/BK	Power Ground (E01)	<0.1v

Pin Connector Graphic

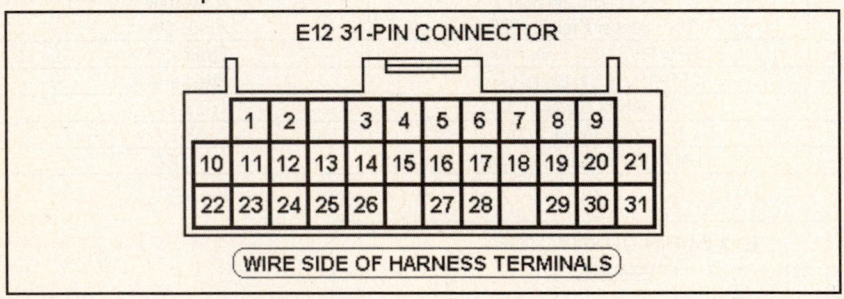

Standard Colors and Abbreviations

Abbreviation	Color	Abbreviation	Color	Abbreviation	Color
BK	Black	GY	Gray	RD	Red
BL	Blue	GN	Green	TN	Tan
BR	Brown	LG	LT Green	VT	Violet
DB	Dark Blue	OR	Orange	WT	White
DG	DK Green	PK	Pink	YL	Yellow

2001-03 Sienna 3.0L 1MZ-FE V6 VIN F (A/T) E8 22 Pin Connector

PCM Pin #	Wire Color	Circuit Description (22 Pin)	Value at Hot Idle
1	RD/BK	Direct Battery	12-14v
2	BK/OR	Ignition Switch Power	12-14v
3	LG/RD	Circuit Opening Relay (FC)	0-3v, at off-idle: 12v
4	RD	SIL (Scan Tool Signal)	12v
5, 11	---	Not Used	---
6	BK/BL	MIL (lamp) Control	MIL Off: 12v, On: 1v
7	GY	Starter Switch Signal	Cranking: 0-3v
8	YL/GN	EFI Main Relay Control	Relay Off: 0v, On: 12v
9	GN/RD	OD Lamp Indicator Control	Lamp Off: 12v, On: 1v
10	BR/RD	EVAP Canister Closed Valve (VSV)	12v or 0v
12	BK	Center Airbag Sensor (F/PS)	Digital Signals
13	OR	TRAC Signal (TRC+)	Pulse Signals
14	GN	TRAC Engine Signal (ENG+)	Pulse Signals
15	GN/WT	Stop Light Switch Signal	Brake Off: 0v, On: 12v
16	BK/RD	EFI Main Relay (B+)	12-14v
17	BL/RD	EVAP Vapor Pressure Sensor (PTNK)	2.9-3.1v (with hose off)
18	PK	Heated Mirror Circuit	Heater On: 12-14v
19	GN	Electric Load Sensor Circuit	Lights On: 12-14v
20	BL/WT	TRAC Signal (TRC-)	Pulse Signals
21	WT	TRAC Engine Signal (ENG-)	Pulse Signals
22	RD/WT	Theft Deterrent Indicator (IMLD)	LED Off: 0v, On:

2001-03 Sienna 3.0L 1MZ-FE V6 VIN F (A/T) E9 28 Pin Connector

PCM Pin #	Wire Color	Circuit Description (28 Pin)	Value at Hot Idle
1	BK/RD	Power Ground (EOM)	<0.1v
2, 10-12	---	Not Used	---
3	WT/RD	EVAP Pressure Switching Valve (VSV)	12v or 0v
4	GN	Tail Light Switch Signal	Switch Off: 0v, On: 12v
5	PK/BK	Data Link Connector (TC)	12v
6	RD/WT	A/C Amplifier Signal (ACMG)	Relay Off: 12v, On: 1v
7	BL/YL	A/C Amplifier Signal (AC)	Clutch On: 12v, Off: 1.5v
8	WT	HO2S-12 (B1 S2) Signal (OXS)	0.1.1-1.5v
9	BL/WT	HO2S-12 (B1 S2) Heater (HTS)	1v (Heater on)
13	VT	Mirror Heater Switch Signal	Switch Off: 0v, On: 12v
14	GY/BK	A/C Amplifier THWO Signal	A/C On: 12-14v
15-17, 21	---	Not Used	---
16	YL/BK	Transponder ECU Signal (NEO)	Inserting key: pulses
18	BR/YL	Transponder Amplifier Signal (TXCT)	Inserting key: pulses
19	BR/RD	Transponder Amplifier Signal (RXCK)	Inserting key: pulses
20	BK/WT	Neutral Start Switch Signal (NSW)	Cranking: 9-11v
22	VT/YL	Vehicle Speed Sensor	At 55 mph: 48 Hz
23	BL/BK	Unlock Warning Switch (KSW)	Switch Open:12v, Closed: 0v
24	BK/BL	Cruise Control Signal (OD1)	At Cruise in OD: 12v
25	BL	Cruise Control ECU (IDLO)	1.5v, off-idle: 12v
26	GN/WT	Cooling Fan Relay 1 (CF)	Relay Off: 12v, On: 1v
27	LG/BK	Tachometer Signal (TACO)	Pulse Signals
28	BR/WT	Transponder Amplifier Signal (Code)	Inserting key: pulses

Pin Connector Graphic

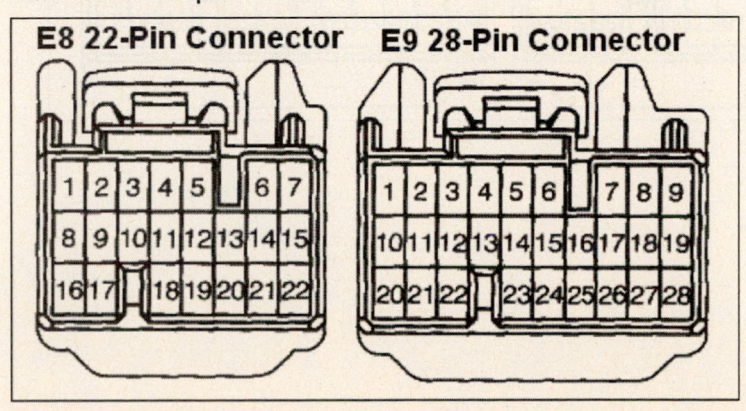

E8 22-Pin Connector   **E9 28-Pin Connector**

2001-03 Sienna 3.0L 1MZ-FE V6 VIN F (A/T) E10 17-Pin Connector

PCM Pin #	Wire Color	Circuit Description (17-Pin)	Value at Hot Idle
1	BL/BK	A/T-ECT Shift Solenoid (S2)	1st or OD: 1v
2-7	---	Not Used	---
8	RD/BK	A/T Select Switch Signal (Reverse)	In 'R': 12v, Others: 0v
9-11	---	Not Used	---
12	GN/OR	Overdrive Main Switch Signal (ODMS)	Open: 12v, Closed: 1v
13	YL	A/T Select Switch Signal (Low)	In Low: 12v, Others: 0v
14	OR	A/T Select Switch Signal (2nd)	In 2nd: 12v, Others: 0v
15	PK/BL	A/T-ECT Shift Solenoid (SL)	In Lockup: 12-14v

2001-03 Sienna 3.0L 1MZ-FE V6 VIN F (A/T) E11 24-Pin Connector

PCM Pin #	Wire Color	Circuit Description (24-Pin)	Value at Hot Idle
1	WT/BK	Power Ground (E04)	<0.1v
2	YL	Sensor VREF (VC)	4.9-5.1v
3	RD	AFS-11 (B1 S1) Heater (HAFR)	1v (Heater on)
4	BL	AFS-21 (B2 S1) Heater (HAFL)	1v (Heater on)
5	WT	Injector 1 Control	1.6-2-9 ms
6	BK	Injector 2 Control	1.6-2-9 ms
7	LG	EVAP Purge Solenoid (VSV)	12v or 0v
8	WT/BK	Power Ground (E05)	<0.1v
9	BK/BL	PSP Switch Signal (PS)	Straight: 12v, Turned: 0v
10	PK	MAF Sensor Signal (VG)	1.1-1.5v
11	BR	AFS-11 (B1 S1) Signal (AFR+)	Fixed at 3.3v
12	BL	AFS-21 (B2 S1) Signal (AFL+)	Fixed at 3.3v
13	GN/YL	Transmission Fluid Temperature Sensor	At 68°F: 4-5v
14	GN/BK	ECT Sensor Signal (THW)	0.5-0.6v
15	BK	Intake Air Control 2 (VSV)	12v or 0v
16	BK/WT	CKP Sensor Signal (NE+)	AC pulse signals
17	BR	DLC Ground (E1)	<0.050v
18	WT	Sensor Ground (E2)	<0.050v
19	RD/BK	MAF Sensor Ground (E2G)	<0.050v
20	BK/RD	AFR-11 (B1 S1) Signal (AFR-)	Fixed at 3.3v
21	BK/WT	AFR-21 (B2 S1) Signal (AFL-)	Fixed at 3.3v
22	BL/YL	IAT Sensor Signal (THA)	0.5-3.4v
23	BL/WT	TP Sensor Signal (VTA1)	0.53-1.27v
24	WT	CMP/CKP Sensor Ground (-)	<0.050v

Pin Connector Graphic

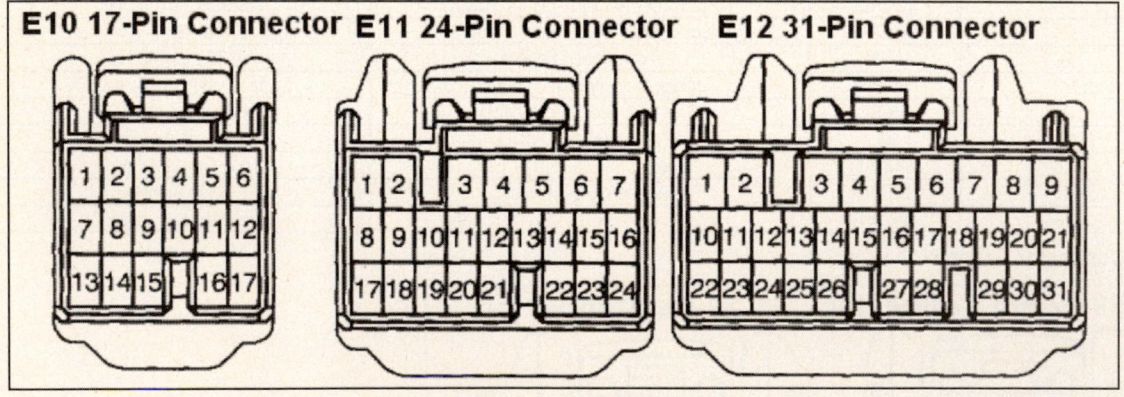

2001-03 Sienna 3.0L 1MZ-FE V6 VIN F (A/T) E12 31-Pin Connector

PCM Pin #	Wire Color	Circuit Description (31-Pin)	Value at Hot Idle
1	BK	Injector 3 Control	1.6-2.9 ms
2	BL	Injector 4 Control	1.6-2.9 ms
3	RD	Injector 5 Control	1.6-2.9 ms
4	GN	Injector 6 Control	1.6-2.9 ms
5	BL/BK	VVT Solenoid RH Control (OC1-)	AC Pulse signals
6	RD/BK	VVT Solenoid RH Control (OC1+)	AC Pulse signals
7	WT/GN	A/T-ECT Solenoid S1	In 3rd or OD: 1v
8-9	---	Not Used	---
10	BK	Cam Timing Oil Control Valve RH (VV1+)	Pulse signals (17-18 Hz)
11	BL/RD	IGT 1 Control	6°, 55 mph: 8° dwell
12	GN/RD	IGT 2 Control	6°, 55 mph: 8° dwell
13	WT/RD	IGT 3 Control	6°, 55 mph: 8° dwell
14	BK/RD	IGT 4 Control	6°, 55 mph: 8° dwell
15	BL/WT	IGT 5 Control	6°, 55 mph: 8° dwell
16	GN/BK	IGT 6 Control	6°, 55 mph: 8° dwell
17	RD/YL	Intake Air Control Valve (ACIS)	12v or 0v
18	BK/WT	VVT Solenoid LH Control (OC2-)	AC Pulse signals
19	GN/WT	A/T-ECT Solenoid (SLN-)	Pulse Signals
20	BK/YL	A/T-ECT Solenoid (SLN+)	Pulse Signals
21	WT/BK	Power Ground (E01)	<0.1v
22	OR	Cam Timing Oil Control Valve LH (VV2+)	Pulse signals (17-18 Hz)
23	YL	Direct Clutch Speed Input (-)	AC Pulse signals
24	BL	Direct Clutch Speed Input (+)	AC pulse signals
25	BK	Igniter Signal (IGF)	Digital Signal: 0-5-0v
26	RD/WT	Idle Air Control Valve (RSO)	Pulse Signals
27	WT	Knock Sensor 1 Signal (KNKR)	<0.075v AC
28	WT	Knock Sensor 2 Signal (KNKL)	<0.075v AC
29	RD	VVT Solenoid LH Control (OC2+)	AC Pulse signals
30	WT/BK	Power Ground (E03)	<0.1v
31	WT/BK	Power Ground (E02)	<0.1v

Pin Connector Graphic

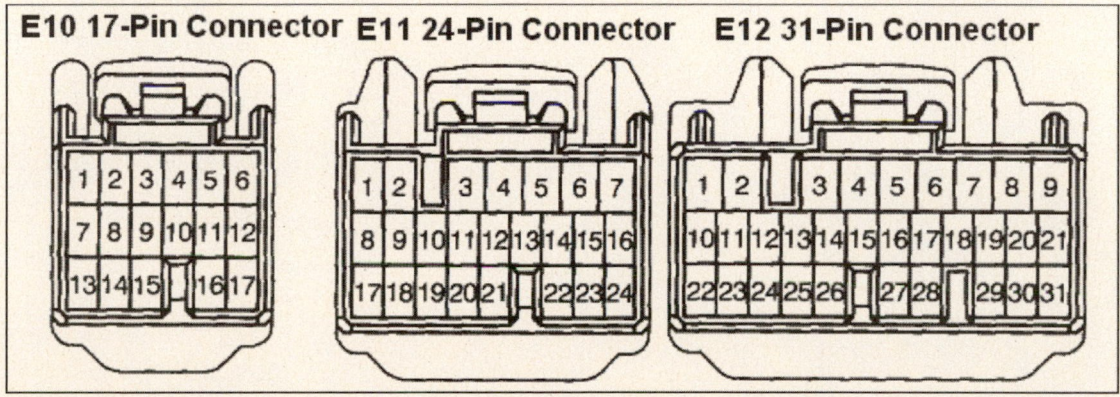

Manual ISBN 1-4018-7412-6/Part No. 27412

With the *Chilton® 2005 Labor Guide*, professional technicians gain access to labor times for vehicle brands and models that conform to current Automotive Aftermarket Industry Association standards. Thousands of labor times for 1981 through 2005 domestic and imported vehicles reflect technicians' use of aftermarket tools and training. Updates based on technical hotline input, Original Equipment Manufacturer (OEM) warranty times, and technical editor evaluation include more diagnostic labor times than ever before. Labor operations have been rewritten to conform to the most recent industry standards. Prior model coverage has been re-evaluated by experts to ensure accuracy. Chilton labor times are accepted by insurance and extended warranty companies.

Labor Guide Manual Benefits:

- 2,500 pages of Chilton labor times
- each OEM is arranged alphabetically by section for easy reference
- improved indexing means easier access to today's repair industry standards

Hardcover manual is 8 7/8" x 11", ©2005

Labor Guide CD-ROM Benefits:

- easy-to-use software to create and print professional-quality estimates and invoices
- three user-defined levels of labor rates correspond to different types of job scenarios, for "real-world" application
- functions as a database of aftermarket labor times for monitoring warranty and insurance claims
- software keeps track of customers and prior estimates for time-saving recall
- customizable application allows service writers to add labor operations and times, and parts companies to add labor times to existing parts ordering systems

CD-ROM ISBN 1-4018-7818-0/Part No. 27818

Previous Year Editions

Chilton 2004 Labor Guide Manual, **ISBN 1-4018-4356-5/Part No. 24356**

Chilton 2004 Labor Guide CD-ROM, **ISBN 1-4018-4357-3/Part No. 24357**

CHILTON MECHANICAL SERVICE MANUALS - ANNUAL EDITIONS

For the most up-to-date service and repair information anywhere, look no further than the newly updated *Chilton® 2005 Mechanical Service Manuals – Annual Editions*! Still the lowest-priced professional repair manuals on the market, this series of manufacturer-based books now features an easier-to-handle, two-volume Asian Manual set. Increased model coverage over the 2004 editions is supported by more illustrations in each section, making fast, accurate repairs and reassembly easier than ever before. With modernized content, it's no wonder that more professionals trust Chilton Professional Manuals for their mechanical service and repair needs.

Mechanical Service Manual Benefits:

- all books are grouped by manufacturer to make accessing information simple
- step-by-step procedures from drive train to chassis and related components help yield fast accurate results
- comprehensive, technically-detailed content is organized by model and system, and is supported by exploded-view illustrations, diagrams, and specification charts for added clarity
- most mechanical systems are included, such as engines, suspensions, steering components, and more
- special tools are described and clearly illustrated so that performing repairs is as easy and quick as possible

Chilton 2005 Ford Mechanical Service Manual
    ISBN 1-4018-6719-7/Part No. 26719
Chilton 2005 General Motors Mechanical Service Manual
    ISBN 1-4018-7146-1/Part No. 27146
Chilton 2005 Chrysler Mechanical Service Manual
    ISBN 1-4018-6718-9/Part No. 26718
Chilton 2005 Asian Mechanical Service Manual (Complete Set of 2 manuals)
    ISBN 1-4018-7180-1/Part No.
Chilton 2005 Asian Mechanical Service Manual, Acura - Mazda
    ISBN 1-4018-6716-2/Part No. 26716
Chilton 2005 Asian Mechanical Service Manual, Mitsubishi - Toyota
    ISBN 1-4018-6717-0/Part No. 26717
Chilton 2005 European Mechanical Service Manual
    ISBN 1-4018-6720-0/Part No. 26720

Manuals are 8 1/2" x 11", ©2005

The *Chilton® Perennial Editions* contain repair and maintenance information for popular mechanical systems that may not be available elsewhere. They offer a wide range of repair information on cars, trucks, vans, and SUVs dating back to the early 1960s, and as current as 2002. Information for 1993 and later model years includes scheduled maintenance interval charts.

Benefits:

- covers the most common vehicle models found in the repair aftermarket today
- gain quick understanding of systems using exploded-view illustrations, diagrams, and charts
- simplify tough jobs with easy-to-follow removal and installation instructions for heater core and other components
- obtain complete coverage of repair procedures from drive train to chassis and associated components

Auto Repair Manual, 1998-2002, 1,426 pages
     ISBN 0-8019-9362-8/Part No. 9362
Auto Repair Manual, 1993-1997, 2,064 pages
     ISBN 0-8019-7919-6/Part No. 7919
Auto Repair Manual, 1988-1992, 1,284 pages
     ISBN 0-8019-7906-4/Part No. 7906
Auto Repair Manual, 1980-1987, 1,344 pages
     ISBN 0-8019-7670-7/Part No. 7670

Import Car Repair Manual, 1998-2002, 1,792 pps
     ISBN 0-8019-9363-6/Part No. 9363
Import Car Repair Manual, 1993-1997, 2,080 pps
     ISBN 0-8019-7920-X/Part No. 7920
Import Car Repair Manual, 1988-1992, 1,632 pages
     ISBN 0-8019-7907-2/Part No. 7907
Import Car Repair Manual, 1980-1987, 1,488 pages
     ISBN 0-8019-7672-3/Part No. 7672

Truck & Van Repair Manual, 1998-2002, 1,408 pages
     ISBN 0-8019-9364-4/Part No. 9364
Truck & Van Repair Manual, 1993-1997, 2,096 pages
     ISBN 0-8019-7921-8/Part No. 7921
Truck & Van Repair Manual, 1991-1995, 1,664 pages
     ISBN 0-8019-7911-0/Part No. 7911
Truck & Van Repair Manual, 1986-1990, 1,536 pages
     ISBN 0-8019-7902-1/Part No. 7902
Truck & Van Repair Manual, 1979-1986, 1,440 pages
     ISBN 0-8019-7655-3/Part No. 7655

SUV Repair Manual, 1998-2002, 1,292 pages
     ISBN 0-8019-9365-2/Part No. 9365

Hardcover manuals are 8 1/2" x 11".

Chilton Collector's Editions - *Reference Manuals for Vintage Vehicles*
Auto Repair Manual, 1964-1971, ISBN 0-8019-5974-8/Part No. 5974,
Truck & Van Repair Manual, 1961-1971, ISBN 0-8019-6198-X/Part No. 6198
Truck & Van Repair Manual, 1971-1978, ISBN 0-8019-7012-1/Part No. 7012

ASE Test Preparation Series

Thomson Delmar Learning

ISBN 1-4018-5182-7
Part No. 25182

(Complete Set: A1-A8, L1, P2 X1, C1)

Thomson Delmar Learning has developed comprehensive ASE Test Preparation Manuals to help automotive technicians increase their success on these certification programs. The material covers the topics one might find during the test process. The booklets include many review questions and answers, as well as detailed descriptions of the repairs involved. Designed to look like the actual test, participants will feel more comfortable with practice, which will translate into greater success in taking the actual tests. The design of the Delmar Learning product also includes helpful test taking hints and student preparation ideas designed to enhance success.

BENEFITS
- The history of the ASE
- Test-taking strategies
- Tasks lists and overview
- Sample test questions
- ASE-style exams
- Explanations to the answers (right and wrong)
- Glossary of terms

(A1) Automotive Engine Repair, 2E

1-4018-2040-9
Part No. 22040

General Engine Diagnosis, Cylinder Head and Valve Train Diagnosis and Repair, Engine Block Diagnosis and Repair, Lubrication and Cooling Systems Diagnosis and Repair, and Fuel, Electrical, Ignition and Exhaust Systems Inspection and Service.

(A2) Automotive Transmissions and Transaxles, 2E

1-4018-2041-7
Part No. 22041

General Transmission/ Transaxle Diagnosis (Mechanical/Hydraulic Systems and Electronic Systems), Transmission/Transaxle Maintenance and Adjustment, In-Vehicle Transmission/Transaxle Repair, Off-Vehicle Transmission/Transaxle Repair.

(A3) Automotive Manual Drive Trains and Axles, 2E

1-4018-2042-5
Part No. 22042

Clutch Diagnosis and Repair, Transmission Diagnosis and Repair, Transaxle Diagnosis and Repair, Drive Shaft/Half Shaft and Universal Joint/Constant Velocity (CV) Joint Diagnosis and Repair (Front and Rear Wheel Drive), Rear Axle Diagnosis and Repair, Four Wheel Drive/All Wheel Drive Component Diagnosis and Repair.

(A4) Automotive Suspension and Steering, 2E

1-4018-2043-3
Part No. 22043

Steering Systems Diagnosis and Repair (Steering Columns and Manual Steering Gears, Power Assisted Steering Units, Steering Linkage), Suspension Systems Diagnosis and Repair (Front Suspensions, Rear Suspensions, Miscellaneous Services), Wheel Alignment Diagnosis, Adjustment and Repair, and Wheel and Tire Diagnosis and Repair.

(A5) Automotive Brakes, 2E

1-4018-2044-1
Part No. 22044

Hydraulic System Diagnosis and Repair, Drum Brake Diagnosis and Repair, Disc Brake Diagnosis and Repair, Power Assist Units Diagnosis and Repair, Miscellaneous Systems Diagnosis and Repair, Antilock Brake Systems (ABS) Diagnosis and Repair.

(A6) Automotive Electrical-Electronic Systems, 2E

1-4018-2045-X
Part No. 22045

General Electrical/Electronic Systems Diagnosis, Battery Diagnosis and Service, Starting Systems Diagnosis and Repair, Charging Systems Diagnosis and Repair, Lighting Systems Diagnosis and Repair, Gauges, Warning Devices and Driver Information Systems Diagnosis and Repair, Horn and Wiper/Washer Diagnosis and Repair.

(A7) Automotive Heating and Air Conditioning, 2E

1-4018-2046-8
Part No. 22046

The manual for A7 includes the following topics: A/C System Diagnosis and Repair, Refrigeration System Component Diagnosis and Repair, Heating and Engine Cooling Systems Diagnosis and Repair, Operating Systems and Related Controls Diagnosis and Repair, Refrigerant Recovery, Recycling, Handling and Retrofit.

(A8) Automotive Engine Performance, 2E

1-4018-2047-6
Part No. 22047

The manual for A8 includes the following topics: General Engine Diagnosis, Ignition System Diagnosis and Repair, Fuel, Air Induction, and Exhaust Systems Diagnosis and Repair, Emissions Control Systems Diagnosis and Repair (Including OBDII), Computerized Engine controls Diagnosis and Repair (Including OBDII), Engine Electrical Systems diagnosis and Repair.

(L1) Automotive Advance Engine Performance, 2E

1-4018-2049-2
Part No. 22049

The manual for L1 includes the following topics: General Powertrain Diagnosis, Computerized Powertrain Controls Diagnosis (Including OBDII), Ignition System Diagnosis, Fuel Systems and Air Induction Systems Diagnosis, Emission Control Systems Diagnosis, I/M Failure Diagnosis.

(P2) Automobile Parts Specialist, 2E

1-4018-2048-4
Part No. 22048

The manual for P2 includes the following topics: General Operations, Customer Relations and Sales Skills, Vehicle Systems Knowledge, Vehicle Identification, Cataloging Skills, Inventory Management, Merchandising.

(X1) Exhaust Systems

1-4018-2050-6
Part No. 22050

Exhaust Systems includes the following topics: Exhaust Systems Inspection and Repair, Emissions Systems Diagnosis, Exhaust System Fabrication, Exhaust System Installation, Exhaust System Repair Regulations.

(C1) Service Consultant

See next page for details

ASE Test Preparation Series in Español!

Thomson Delmar Learning

ISBN 1-4018-1530-8

(Complete Set: A1-A8, L1, P2, X1)

Now available in Español – the first of its kind for Spanish-speaking technicians! This comprehensive package of ASE test preparation booklets are intended for any Spanish-speaking automotive technician who is preparing to take an ASE examination. The series includes questions that relate to each competency required for certification by ASE. In addition to a multitude of questions, the reason why each answer is right or wrong is explained, along with task lists and overview, test-taking strategies, and more.

(A1) Reparación de Motores, 2A Edición
1-4018-1014-4/Part No. 21014

(A2) Transmision Automática/ Eje de Transmision Automática, 2A Edición
1-4018-1015-2/Part No. 21015

(A3) Tren de y Mando Ejes Manuales, 2A Edición
1-4018-1016-0/Part No. 21016

(A4) Suspensión y Dirección, 2A Edición
1-4018-1017-9/Part No. 21017

(A5) Frenos, 2A Edición
1-4018-1018-7/Part No. 21018

(A6) Sistemas Eléctricos/ Electrónicos, 2A Edición
1-4018-1019-5/Part No. 21019

(A7) Calefacción y Aire Acondicionado, 2A Edición
1-4018-1020-9/Part No. 21020

(A8) Funcionamiento de Motores, 2A Edición
1-4018-1021-7/Part No. 21021

(L1) Especialista en el Funciommiato Avansado de Motores, 2A Edición
1-4018-1022-5/Part No. 21022

(P2) Especialista en Partes de Automovil, 2A Edición
1-4018-1023-3/Part No. 21023

(X1) Sistemas de Escape, 2A Edición
1-4018-1024-1/Part No. 21024

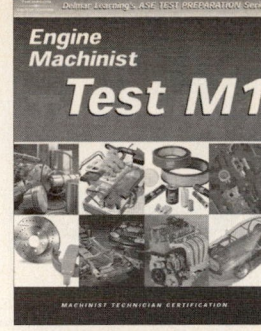

NEW!

ASE Test Preparation Manual - C1 Service Consultant
Thomson Delmar Learning
ISBN 1-4018-2029-8/
Part No. 22029

Prepare to pass the new Service Consultant ASE Exam with help from this new test preparation booklet. The new C1 Exam is designed to measure systems knowledge and people skills of those who come in contact with the customer. It will contain questions on Communications, Product Knowledge, Sales Skills, and Shop Operations.

Service Consultant ASE Test Preparation Manual Benefits:

- the ASE task list is fully up-to-date, while current test prep questions reflect the most recent ASE task changes for the broadest knowledge possible
- hundreds of ASE-style exam questions adequately prepare readers to successfully pass the ASE exam
- readers are given multiple opportunities to check their understanding of critical concepts through sample problems, refresher materials, and competency-specific test questions
- overviews of each task provide a great reference point to help answer difficult ASE questions
- explanations for each answer help the user understand why the response is correct or incorrect

Softcover manual is 8 1/2" x 11", ©2004

ASE Test Preparation Manuals - Engine Machinist
Thomson Delmar Learning
ISBN 0-7668-6283-6/
Part No. 16283
(Complete Set: M1-M3)

With an abundance of up-to-date content, Thomson Delmar Learning's ASE Test Preparation Series contains the most current ASE test preparation material available. Each manual combines refresher materials with an abundance of sample test questions, as well as a wealth of information regarding test-taking strategies and the types of questions found in an ASE exam. In addition to the questions, thorough explanations are provided as to why each answer is correct or incorrect.

Benefits:

- The History section explains why the exams are important to the industry
- test-taking strategies help prepare technicians for the environment they will encounter during the actual exam experience testing first-hand

(M1) Cylinder Head Specialist
0-7668-6280-1/Part No. 16280

(M2) Cylinder Block Specialist
0-7668-6281-X/
Part No. 16281

(M3) Assembly Specialist
0-7668-6282-8/
Part No. 16282

Softcover manuals are 8 1/2" x 11", ©2002

ASE Test Preparation Manuals - Collison Repair
Thomson Delmar Learning
ISBN 1-4018-5120-7/Part No. 25120
(Complete Set: B2-B6)

This fully expanded third edition has been completely updated to provide the most current ASE test preparation material for collision repair and refinishing available anywhere. Each book in the series provides valuable preparation for automotive technicians seeking certification in one or more of the ASE collision repair areas. Readers are afforded scores of opportunities to ascertain their knowledge of critical concepts, through the extensive array of sample problems, ASE-style exams, and competency-specific test questions required for certification by ASE.

Benefits:

- all ASE task lists associated with collision repair and refinishing are fully up-to-date to help sufficiently prepare users for the ASE certification exam
- current, job-related ASE-style exam questions reflecting the most recent ASE task changes test the skills that technicians need to know on the job
- each book contains a general knowledge pretest, a sample test, and additional practice learning that add up to the most real-test practice time available

(B2) Painting and Refinishing, 2E
1-4018-3664-X/Part No. 23664
(B3) Non-Structural Analysis and Damage Repair, 2E
1-4018-3665-8/Part No. 23665
(B4) Structural Analysis and Damage Repair, 2E,
1-4018-3666-6/Part No. 23666
(B5) Mechanical and Electrical Components, 2E,
1-4018-3667-4/Part No. 23667
(B6) Damage Analysis and Estimation, 2E,
1-4018-3668-2/Part No. 23668

Softcover manuals are 8 1/2" x 11", ©2005

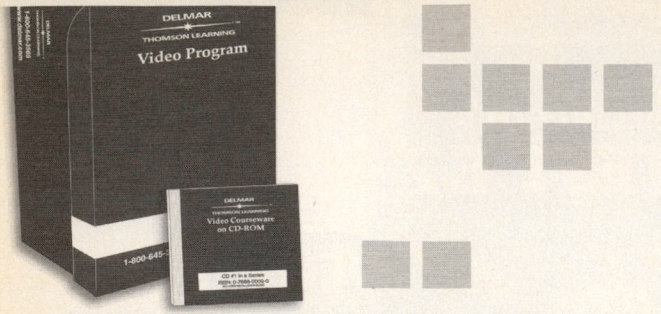

Automotive ASE Preparation Video Series
Thomson Delmar Learning

ISBN 0-7668-3168-X *(Complete Set of 12 Tapes)*
ISBN 0-7668-8042-7 *(Complete Set of 3 CD-ROMs)*

Thomson Delmar Learning's Automotive ASE Test Prep Videos present test takers with a review of the A1-A8, L1, and P2 tests prior to taking the exam. Each tape summarizes key topics and key task areas through live action and animation. Actual technicians, authentic automotive shops, and late-model vehicles are featured for an up-to-date look and feel. Safety is emphasized throughout each tape. An overview tape introduces test takers to the ASE testing style.

BENEFITS OF THE VIDEO SERIES
- lively, easy to follow videos emphasize safety throughout
- covers major task areas and topics for each of the ASE exams
- accompanying Instructor's Guide helps users comprehend and retain information presented

Complete Set of 12 Tapes (with Instructor's Guide), ©2001

Tape 1: Overview of ASE, 0-7668-2484-5
Tape 2: A1 Engine Repair, 0-7668-2485-3
Tape 3: A2 Automatic Transmission, 0-7668-2498-5
Tape 4: A3 Manual Transmission, 0-7668-2499-3
Tape 5: A4 Steering and Suspension, 0-7668-2500-0
Tape 6: A5 Automotive Brakes, 0-7668-2501-9
Tape 7: A6 Electricity/Electronics, 0-7668-2493-4
Tape 8: A7 Air Conditioning, 0-7668-2486-1
Tape 9: A8 Engine Performance, 0-7668-2494-2
Tape 10: P2 Parts Specialist, 0-7668-2487-X
Tape 11: L1 Advanced Engine Performance (Part 1), 0-7668-2491-8
Tape 12: L1 Advanced Engine Performance (Part 2), 0-7668-2492-6

BUNDLES
Bundle 1: Specialty Topics (Set of 4 Tapes) includes Overview of ASE, A1 Engine Repair, A7 Air Conditioning, and P2 Parts Specialist, 0-7668-2483-7
Bundle 2: Engine Performance/Electronics (Set of 4 Tapes) includes L1 Part 1, L1 Part 2, A6 Electricity/ Electronics, and A8 Engine Performance, 0-7668-2490-X
Bundle 3: Undercar (Set of 4 Tapes) includes A2 Automatic Transmissions, A3 Manual Transmissions, A4 Steering and Suspension, and A5 Automotive Brakes, 0-7668-2497-7

CD-ROM COURSEWARE
Based on the ASE Test Prep Series, the CD-ROMs offer the following in addition to the video content:
- Gradebook
- Video Glossary
- Video File Server compatible
- Pre-test/Post-test
- Variety of question types
- Ability to modify
- Remediation

CD-ROM 1: Specialty Topics CD-ROM includes Overview of ASE, A1 Engine Repair, A7 Air Conditioning, and P2 Parts Specialist, 0-7668-2489-6
CD-ROM 2: Engine Performance/Electronics CD-ROM includes L1 Part 1, L1 Part 2, A6 Electricity/ Electronics, and A8 Engine Performance, 0-7668-2496-9
CD-ROM 3: Undercar CD-ROM includes A2 Automatic Transmissions, A3 Manual Transmissions, A4 Steering and Suspension, and A5 Automotive Brakes, 0-7668-2503-5

The ASE "Passing Lane" Package
Thomson Delmar Learning

ISBN 0-7668-4338-6
(Complete Set: A1-A8, L1, P2)

The most comprehensive test preparation for Automotive Tests A1-A8, L1, and P2. Combining the most thorough ASE Test Preparation books with the latest in ASE videos, this package provides a program of self-study for the automotive ASE Tests.

EACH BOOK IN THE SERIES BENEFITS:
- test-taking strategies
- tasks lists and overview
- sample test questions
- ASE-style exams
- explanations to the answers
- glossary of terms

EACH VIDEO IN THE SERIES BENEFITS:
- lively, easy to follow videos emphasize safety throughout
- covers major task areas and topics for each of the ASE exams
- accompanying Activity Sheets help comprehend and retain information

(A1) Automotive Engine Repair Book/Video, 0-7668-4181-2
(A2) Automotive Transmissions and Transaxles Book/Video, 0-7668-4182-0
(A3) Automotive Manual Drive Trains and Axles Book/Video, 0-7668-4183-9
(A4) Automotive Suspension and Steering Book/Video, 0-7668-4184-7
(A5) Automotive Brakes Book/Video, 0-7668-4185-5
(A6) Automotive Electrical-Electronics Systems Book/Video, 0-7668-4186-3
(A7) Automotive Heating and Air Conditioning Book/Video, 0-7668-4187-1
(A8) Automotive Engine Performance Book/Video, 0-7668-4188-X
(L1) Automotive Advanced Engine Performance Book/Video, 0-7668-4189-8
(P2) Automobile Parts Specialist Book/Video, 0-7668-4190-1

Automotive Technician Certification Test Preparation Manual, 2E
Don Knowles

**ISBN 0-7668-1948-5/
Part No. 11948**

The second edition of Certified ASE Master Technician Don Knowles' popular ASE test preparation book adds coverage of the L1 Advanced Engine Performance test to its coverage of automotive tests A1 through A8. All nine tests covered in this book reflect year 2000 task lists, including the updated composite vehicle in the L1 test. This revised edition contains at least one practice question for every ASE task in the tests. Also included is the updated and expanded coverage of electronic automatic transmissions, electronically controlled automatic transmissions, electronically controlled 4 wheel drive and steering, ABS systems, wiring diagrams, and repairing electronic components.

BENEFITS
- a new section has been added on computer-controlled automatic transmissions and transaxles including those used in OBD II vehicles
- new information has been included on electronically-controlled 4WD systems and ABS systems
- the chapter on Electrical/Electronic Systems has been expanded to include information on reading wiring diagrams and inspecting, testing, and repairing electronic components
- a complete chapter has been added to prepare technicians for the Advanced Engine Performance (L1) test

CONTENTS
Engine Repair Automatic Transmission/Transaxle. Manual Drive Train and Axles. Suspension and Steering. Brakes. Electrical/Electronic Systems. Heating, Ventilation, and Air Conditioning Systems. Engine Performance. Advanced Engine Performance.

788 pp, 8½≤ x 11≤, softcover, ©2001

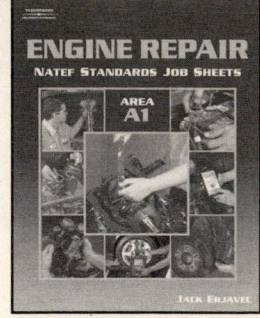

Prepare to Pass the ASE Exam Online

ATCChallenge.com
Thomson Delmar Learning

Updated to the Latest ASE Task Lists!

Thomson Delmar Learning's online ASE test preparation web site has been carefully reviewed and researched by ASE master technicians to include fully updated content on tests A1-A8, L1, P2, X1, and X1. The site offers two different study options so users can choose their study method each time they sign on.

- practice questions provide helpful hints, insight into right and wrong answers, and links back to further reading for each task area
- sample tests prepare users for test day by using ASE-style questions - reflecting the type of questions and task areas on the actual exam - making this the most up-to-date and realistic ASE test preparation study aid available
- one year secure access through any web-enabled computer
- a complete task list including an overview of each task to further enhance study
- automotive dictionary with more than 5,000 terms and Spanish translation
- technical support provided by Thomson Delmar Learning

ATCChallenge.com Plus
Thomson Delmar Learning

ATCChallenge.com Plus is the ideal way to gain the expertise required to pass the *A5* (brakes) and *A7* (heating and air conditioning) ASE exams. Thoroughly reviewed and researched by master automotive technicians, this total online courseware solution may be used effectively in professional training and education courses as well as by individuals preparing for selected ASE exams. While this version includes all the features that *ATCChallenge.com* contains, it also brings with it a variety of new tools for use by students and educators. The biggest addition from the original version is the immediate remediation to the ASE-style questions that brings the user to a file or a video clip that explains the answer to each question.

- combines the *Today's Technician Series*, Erjavec's *Automotive Technology*, and Delmar Learning's *Automotive Video Series* to bring technicians the most comprehensive ASE coverage available in one place
- individuals can track their scores by ASE task, by test, or by question; create personalized testbanks; and generate their own progress reports; making *ATCChallenge.com Plus* an ideal self-study guide and test preparation tool
- a single training director or administrator can easily track average and/or raw scores at the shop-level, by region, or system-wide to gain a measure of learning achievement and program effectiveness
- a complete task list, with an overview of each task to enhances the learning experience, ensures that users are 100% prepared to pass each ASE test

Call Your Thomson Delmar Learning Sales Rep for Part Numbers & Pricing

Visit **ATCChallenge.com** to see the latest modules and a free demo!

ATC Challenge 3.0 CD-ROM
Thomson Delmar Learning

ISBN 0-7668-2982-0

These exciting interactive CD-ROMs have been designed to prepare technicians for successful completion of the Automotive ASE task areas (A1-A8, L1, P2, and F1). This multimedia software assesses strengths and weaknesses by identifying topics needing further study while allowing users to review ASE task areas at their own pace. Explanations, hints, notes, and a glossary aid the user in comprehension, critical thinking and retention. These CD-ROMs offer hundreds of ASE-style questions, a test taking strategy section and LAN compatibility. Not only is *ATC Challenge 3.0* the ultimate in test preparation, but it is also an excellent learning tool!

CD-ROM, ©2001

Site License Available for Multiple Unit Purchases or Multiple Workstations for ATC Challenge 3.0:
User 1: Full Price (List or Net)
Users 2-5:
$80/workstation + Full Price
Users 6-10:
$70/workstation + Full Price
Users 11-20:
$60/workstation + Full Price
Users 21+:
$50/workstation + Full Price

ATC Challenge for P2
Thomson Delmar Learning
ISBN 0-7668-1827-6

This interactive CD-ROM contains material that will help prepare technicians for the Automotive Parts Specialist (P2) certification exam.
CD-ROM, ©2000

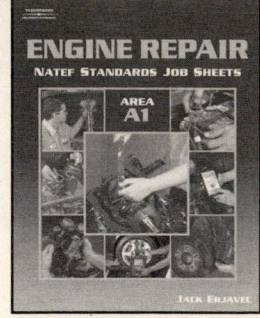

NATEF Standards Job Sheets
Thomson Delmar Learning

ISBN 0-7668-6375-1
(Complete Set: A1-A8)

Each of our eight *NATEF (National Automotive Technicians Education Foundation) Standards Job Sheets* workbooks has been thoughtfully designed to assist users in gaining valuable job preparedness skills and mastering specific technical competencies required for success as a professional automotive technician. The entire series is based on current NATEF standards.

Central to each manual are well-designed and easy-to-read job sheets, each of which contains specific, performance-based objectives, lists of required tools and materials, safety precautions, plus step-by-step procedures to lead users to completion of shop activities.

KEY FEATURES
- easy to use in any automotive education or training program in which NATEF coverage is desired
- completed Job Sheets may be kept as records, providing tangible evidence that instructors are addressing all NATEF tasks while paving the way for program certification

JOB SHEETS AVAILABLE FOR:
(A1) Automotive Engine Repair,
(A2) Automatic Transmissions and Transaxles, 0-7668-6368-9
(A3) Manual Drive Trains and Axles, 0-7668-6369-7
(A4) Automotive Suspension and Steering, 0-7668-6370-0
(A5) Automotive Brakes, 0-7668-6371-9
(A6) Automotive Electrical and 0-7668-6367-0
Electronic Systems, 0-7668-6372-7
(A7) Automotive Heating and Air Conditioning, 0-7668-6373-5
(A8) Automotive Engine Performance, 0-7668-6374-3

All share the following information: 8½≤ x 11≤, softcover, ©2002

AUTOMOTIVE SERVICE MANAGEMENT SERIES

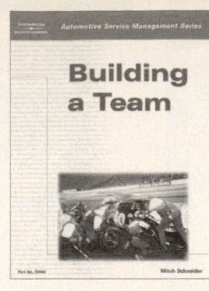

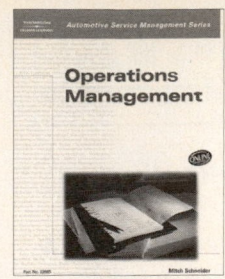

This pioneering eight-book series offers automotive repair shop owners and those wanting to be shop owners the necessary business and customer service skills to run a successful automotive service facility.

The series covers three main topical areas: personnel management, business management, and sales and marketing. Each book provides a framework to help technicians make consistent, high-quality, and productive service a part of every day shop operations. According to the author, "Great performance coupled with increased customer loyalty, trust, and operational excellence will almost always result in increased profits."

Automotive Service Management Series Benefits:

- real-world approach reflects author's experience as a fourth generation technician, a repair & service company owner, and an automotive industry trainer
- all-inclusive coverage spans from designing an automotive repair facility floor plan through financial management techniques, customer/staff relations, and more
- length of each book makes it easy to incorporate this series into workshops, seminars, and training/education courses
- information is available "as is" or for customization

Total Customer Relationship Management
ISBN 1-4018-2657-1/Part No. 22657
From Intent to Implementation
ISBN 1-4018-2658-X/Part No. 22658
Operational Excellence
ISBN 1-4018-2659-8/Part No. 22659
Building a Team
ISBN 1-4018-2660-1/Part No. 22660
The High Performance Shop
ISBN 1-4018-2661-X/Part No. 22661
Safety Communications
ISBN 1-4018-2662-8/Part No. 22662
Managing Dollars with Sense
ISBN 1-4018-2663-6/Part No. 22663
Operations Management
ISBN 1-4018-2665-2/Part No. 22665
Entire Set of 8 Books
ISBN 1-4018-2499-4/Part No. 2499

Softcover manuals are 8 1/2" x 11", ©2003

ABOUT THE AUTHOR

Mitch Schneider is a fourth generation mechanic/technician and is a frequent speaker at major conventions and meetings of automotive industry trade organizations. Schneider is also an award-winning journalist and is a regular contributor and senior contributing editor for *Motor Age* magazine. He provides commentary on the evolving relationship between service dealers, jobbers, warehouse directors and manufacturers.

Schneider has also appeared on the TNN cable show "Truckin' USA" where he hosted the "Tech Tips" segment. In addition to operating the award-winning Schneider's Automotive for 22 years in Simi Valley, CA, he is also the president and founder of Schneider's Future-Tech, a service company specializing in conducting management seminars for automotive service dealers, jobbers, warehouse distribution companies, and manufacturers.

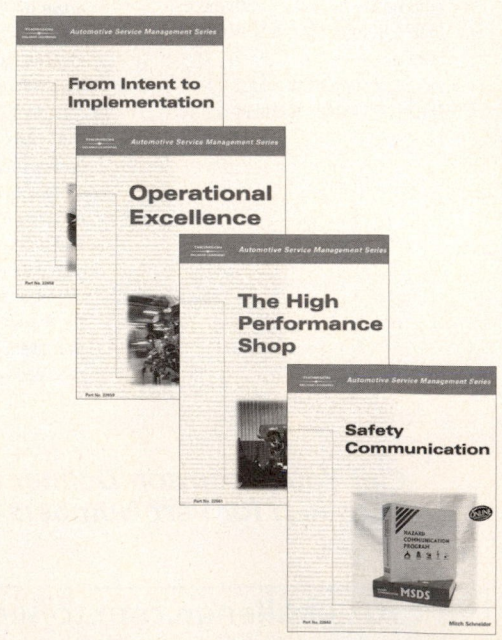

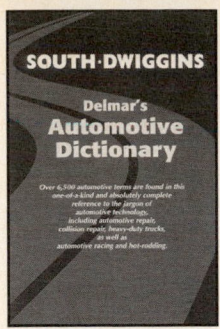

Delmar's Automotive Dictionary
David W. South & Boyce Dwiggins
ISBN 0-8273-7405-4

This handy, ready-reference dictionary provides the automotive engineer, technician, mechanic, student, enthusiast or layperson with a single source for the most up-to-date definitions available of technical, professional and informal terminology used in today's automotive world. It is descriptive and covers the wide scope of terms pertinent to the automotive field. With multiple definitions and aids, and proper pronunciation of terms, this dictionary is a must for all!

BENEFITS

- over 3000 terms comprehensively covering more than 100 subject areas
- enhanced by a list of acronyms and abbreviations
- up-to-date definitions of today's automotive terminology
- aids for proper pronunciation
- each term has multiple definitions

281 pp, 6≤ x 9≤, softcover, ©1997

Practical Problems in Mathematics for Automotive Technicians, 5E
George Morre, Todd Sformo & Larry Sformo
ISBN 0-8273-7944-7

By showing how to apply math solutions to everyday problems, this all-in-one math reference transforms the "remove it and replace it" mechanic into a complete automotive technician. The book builds from math basics to cover more complex topics--not to mention such workplace issues as invoices and scale reading of test meters. Each easy-to-read chapter features step-by-step instructions, diagrams, charts and examples to make the problem-solving process a snap.

256 pp, 7⅛≤ x 9¼≤, softcover, ©1998
Instructor's Manual   **0-8273-7945-5**

Math for the Automotive Trade, 3E
John C. Peterson & William deKryger
ISBN 0-8273-6712-0

Math for Automotive Trades, 3E provides excellent examples and problems that reflect technological requirements of workers in automotive technology. The text has three parts: review of basic mathematics skills, math applications to specific automotive situations, and an examination of measurement aspects beginning with angle and linear measurements and ending with an extensive look at measurement tools used in the automotive trade.

345 pp, 8½≤ x 11≤, softcover, ©1995
Instructor's Manual   **0-8273-6713-9**